MW00592148

Teacher's Edition

Algebra 2

and Trigonometry

TEACHER'S EDITION

ALGEBRA 2

and Trigonometry

John Benson
Evanston Township High School
Evanston, Illinois

Sara Dodge
Evanston Township High School
Evanston, Illinois

Walter Dodge
New Trier High School
Winnetka, Illinois

Charles Hamberg
Illinois Mathematics and
Science Academy
Aurora, Illinois

George Milauskas
Illinois Mathematics and
Science Academy
Aurora, Illinois

Richard Rukin
Evanston Township High School
Evanston, Illinois

McDougal Littell/Houghton Mifflin

Evanston, Illinois

Boston Dallas Phoenix

Authors

John Benson
Teacher, Evanston Township High School
Evanston, Illinois
Recipient, Presidential Award for Excellence in Teaching Mathematics, Illinois, 1987
Teacher, Center for Talent Development, Northwestern University
Head coach, Math Team, Evanston Township High School
Assistant coach, Chicago Area All-Star Math Team
Member, North Central Association Evaluation Teams
Previously taught at Frederick Douglass High School Atlanta, Georgia

Sara Dodge
Teacher, Evanston Township High School
Evanston, Illinois
National Science Foundation Fellow
Teacher, Center for Talent Development, Northwestern University
Delegate, National Council of Teachers of Mathematics, Illinois Council of Teachers of Mathematics
Previously taught at School District 27 Northbrook, Illinois

Walter Dodge
Teacher, New Trier High School
Winnetka, Illinois
Recipient, Presidential Award for Excellence in Teaching Mathematics, Illinois, 1988
Woodrow Wilson Fellow, 1989
Former grader, table leader, exam leader Writing Committee member, Advanced Placement Calculus Test
Past president, Metropolitan Mathematics Club
Football, baseball coach
Computer program developer

Charles Hamberg
Teacher, Illinois Mathematics and Science Academy
Aurora, Illinois
Recipient, Presidential Award for Excellence in Teaching Mathematics, Illinois, 1983
Past president, Council of Presidential Awardees in Mathematics
Recipient, Golden Apple Award, Foundation for Excellence in Teaching
Assistant coach, Chicago Area All-Star Math Team
Previously taught at Adlai E. Stevenson High School
Prairie View, Illinois
Nathan Hale High School
West Allis, Wisconsin
Marina High School
Huntington Beach, California
Mount Morris High School
Mount Morris, Illinois

George Milauskas
Teacher, Illinois Mathematics and Science Academy
Aurora, Illinois
Author, GEOMETRY FOR ENJOYMENT AND CHALLENGE
Editorial Advisory Board member and writer, National Council of Teachers of Mathematics Yearbook on Geometry, 1987
Previously taught at Barrington High School
Barrington, Illinois
New Trier High School
Winnetka, Illinois

Richard Rukin
Teacher, Evanston Township High School
Evanston, Illinois
Midwest chairperson, American Regional Math League
Head coach, Chicago Area All-Star Math Team
Coach, Math Team, Evanston Township High School
Previously taught at Senn High School
Chicago, Illinois

ISBN 0-8123-8801-1

1 2 3 4 5 6 7 8 9 10 – DWO – 99 98 97 96 95

Box 1667, Evanston, Illinois 60204

Printed in the United States of America

CONTENTS

A LETTER TO TEACHERS T vi

Calculators and Computers in Algebra T viii

Classroom Management T x

Writing in the Mathematics Classroom T xii

Cooperative Learning T xiv

Problem Solving T xvi

Using Manipulatives T xvii

A Word on Math Anxiety T xviii

A Guide to Additional Resources T xxiv

STUDENT TEXT

Additional Answers A 1

A Letter to Teachers

When we came together to write a textbook, we tried to implement the ideas that have worked best in our classrooms. We have also tried to write a text flexible enough to enable you to use your own ideas.

It has been our experience that students learn concepts and skills over a period of time. When an idea is first presented and practiced, a student becomes familiar with the concept but does not master it. Mastery comes only when that idea is used over and over, in different contexts and in different applications. For this reason, we do not expect mastery by the end of a section. Instead, students work on a skill or concept, applying the idea in different contexts, throughout many sections. In this way, students come to understand rather than memorize.

Students' recall of algebra skills will vary greatly. We have presented these skills in the context of new material and have included many Algebra 1 problems in the early part of the book.

Our aim, like yours, is to help students move toward more independent learning. We hope you will encourage students to read and use the text and share their ideas. The cooperative learning exercises provide a vehicle for this sharing.

Several strands are interwoven throughout the book. Equations and inequalities, graphing, informal geometry, probability, data analysis, and matrices provide a rich tapestry of mathematical experience.

In our field testing, students found Chapter 1 very interesting. This chapter presents several ways of solving the same problem, reinforcing for students that there can be a variety of ways to solve a problem and that no one particular way is better than another.

The book is clearly organized.

Objectives
are clearly stated so that the student knows exactly what he or she will begin to master.

Part One: Introduction
presents the lesson. Here examples are used to clarify, to teach, and to serve as a reference when students work independently.

Part Two: Sample Problems
illustrates and extends the concepts and skills introduced in Part One. This section represents a second level of development and a second resource for student reference.

There are two subheadings under the heading

Part Three: Exercises and Problems

Warm-up Exercises

can be used as an end-of-class check or can be assigned as skill practice.

Problem Set

is divided into A, B, and C difficulty levels, but only in the Teacher's Edition. Too often we have heard students dismiss a problem with the comment, "I can't do C (or B) problems."

The problem set reflects our experience that students learn by applying a skill or concept in a variety of contexts over time. Students are introduced to ideas slowly, a background is established, and review is reflected in constant application. To help you plan, problems that practice the current topic are labeled P, problems that review and apply previous topics are labeled R, and problems that introduce topics to come are labeled I. Many problems combine two or three purposes. So that students know the structure of the problem sets, it is explained in A LETTER TO STUDENTS, and the problems in Section 2.4 are labeled in the student text.

We have included probability because it is an important topic and is also a marvelous tool for helping students organize their thinking through the systematic examination of options. Matrices are introduced in order to familiarize students with an important vehicle for displaying data. In addition, both topics provide an interesting context for skill practice. Data analysis is included as the topic of the last section in Chapters 4, 5, 6, 8, 9, and 11 and is woven into problem sets throughout subsequent chapters of the book. In our information-laden society, students will need help in order to understand and deal with the daily statistical bombardment.

As we were writing the text, new technology was emerging. We have incorporated technology and hope you and your students will be able to use it.

We hope you and your students are successful and enjoy this book. We would be happy to hear about your experiences.

Good Luck!

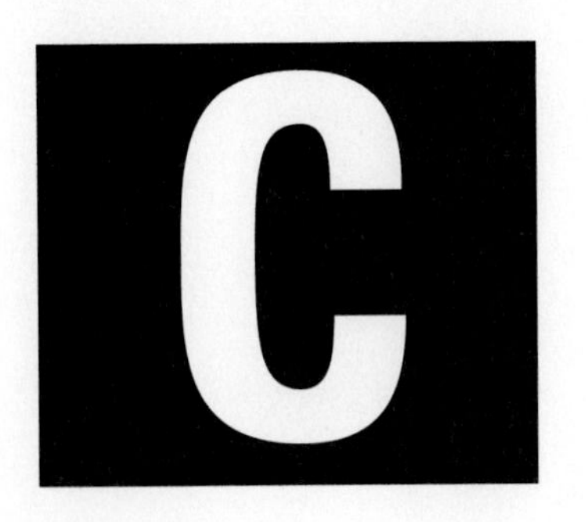

ALCULATORS AND COMPUTERS IN ALGEBRA

Calculators and computers can empower students mathematically. In the past, paper and pencil calculations often became so cumbersome that students missed important mathematical concepts. By including calculators and computers as tools, we can broaden the mathematics we include. Many real-world topics that are not difficult conceptually have traditionally been excluded because they involved calculations that were too troublesome or tedious. With the availability of calculators, computers, and graphing calculators, students can focus on ideas rather than numbers. For example, although students frequently become efficient at graphing a standard parabola, using a graphing calculator or computer graphics software enables them to quickly see the parabola's transformations in shape and/or position as the equation is transformed. Finding reasonable approximations of roots by using a trace function or by zeroing in on the intercepts of the graph clarifies the concept without tedious computation and possible errors.

In our text, we have included use of a scientific calculator, a computer, and a graphing calculator as part of the instruction. The book can, however, be used without hardware. Problems that are cumbersome without a calculator are marked throughout with the symbol . It is also important that all students develop the ability to decide when it is appropriate to use hardware rather than paper and pencil.

We hope that if such hardware is available, teachers will allow students to use graphing calculators and computers as exploration tools. Students can effectively employ graphing calculators for all graphing in one or two dimensions, for evaluating expressions and solving equations, for solving systems, and for matrix operations and data analysis. Computers with spreadsheet and symbolic-manipulation software can be used to further enhance the course.

To facilitate the use of graphing calculators, the text includes a ***Graphing Calculator Handbook,*** beginning on page 745, that provides detailed keystroke instruction for the TI-81, the TI-82, and the Casio fx-7700GB. Throughout the text, where appropriate, the symbol shown below is used to direct students to lessons in the ***Graphing Calculator Handbook*** where they can find assistance.

See p. 745

In the Teacher's Edition, the ***Using Computers*** features provide additional suggestions of ways to integrate computers in classroom work.

LASSROOM MANAGEMENT

As the six of us worked together to write and rewrite this textbook, we talked a great deal about what works best for us in our own classrooms. You are the best judge of what works for you and your class. What we would like to share with you is some of our most successful ideas, which we have included in the Teacher's Edition.

An overview of each chapter on the opening page provides some background you may be able to use as you plan for that chapter.

Class Planning

Time Schedule

Resource References

opens each section to help you plan most efficiently and prepare for each class, especially during the first year you use the book.

Class Opener

is intended as a problem that students work on as soon as they enter the classroom. It can be on an overhead projector, it can be on the board, or a copy can be run off for each student. Copies of Class Openers are included in the *Teacher's Resource Binder* in enlarged type so that they can be used to make transparencies, if your school has the equipment, or to make photocopies. Students work independently or in small groups on these problems. In our classrooms, we also expect students to have their homework out and available for us to check when the bell rings. As we circulate, we record the homework and note where students are having difficulties. In this way, within the first few minutes of class, students are at work, we know the results of last night's assignment, and we know where the learning blocks are.

Class Openers are available in reproducible form in the TEACHER'S RESOURCE BINDER.

Lesson Notes

expand on the student text.

Vocabulary

that is introduced in each section is listed at the bottom of the first lesson page. You may choose to review the words with your students before starting the lesson or after they have gone through the material.

Stumbling Block

Students may shade the wrong portion of the graph. Make sure students test a point to determine which portion of the graph to shade. The origin is a good test point, provided it is not on the boundary line of the inequality.

Stumbling Blocks

are included if there is an error we have seen students make with some consistency.

Checkpoint

provides a few quick quiz questions to help measure student understanding.

Assignment Guide

is based on our experience of teaching the lessons. Since each class is different, you may find it helpful to modify the assignments as you proceed. The assignments as listed here are available as blackline masters in the *Teacher's Resource Binder.*

Problem-Set Notes and Strategies

are comments and/or suggestions about specific problems.

Additional Answers

are those that do not fit next to their respective problems. Other answers can be found in the back of your book, on the tabbed pages.

Many of us have found success with techniques presented by David R. Johnson in his brief booklets *Every Minute Counts* and *Making Every Minute Count Even More* (Palo Alto, Calif.: Dale Seymour, 1982).

We have tried to include material to make you and your class most successful as you work through our book, especially the first year. We hope you and your students have a rewarding time.

RITING IN THE MATHEMATICS CLASSROOM

ALGEBRA 2 provides student writing opportunities in each lesson. Designed to help students internalize mathematical concepts, these brief assignments are based on the lesson material and actively involve the student in the concepts that underlie the problems in the lesson.

Much of language theory is based on the premise that communication is basic to thinking. Vygotsky in *Thought and Language* (1962) writes that putting thoughts into words is the only way we can give form to the myriad images that assault our minds. Britton in *Language and Learning* (1970) theorizes that speaking and writing are "commentary," or a way of making sense out of our random perceptions, and that we must write or speak about an experience to understand it. And Donald Murray in *Write to Learn* (1987) states that writing is the most disciplined way in which we organize our thoughts. As mathematics education continues to emphasize problem solving, we seek ways to translate the symbolic and theoretical language of mathematics into real-world situations. The writing exercises in ALGEBRA 2 give students a solid foundation for using everyday language to record their thoughts about mathematics. With this experience, a student should be better able to deal with everyday situations as well as solve multiple-step word problems.

Communicating Mathematics

Have students write a paragraph that explains the concept of exponential growth.

In addition to facilitating a relationship between mathematics and everyday life, writing empowers the student by giving him or her a sense of ownership about the material. When the writer restates an idea or concept, that idea belongs to the writer and becomes a part of his or her experience. Thus internalized, concepts become more useful and less likely to be forgotten.

As a teacher, you may find completed student writing assignments helpful in making informal assessments. Reading a written explanation or response can be an excellent way of targeting weakness or misunderstanding at the conceptual level. A student's writing may also shed light on why certain kinds of errors are made and what might be done to prevent them.

Communicating Mathematics

Have students prepare a lesson plan on how to solve a system of equations with rational expressions. The lesson should include examples. Choose two or three students to present their lesson to the class.

Metacognition, or the ability to recognize what you do and do not know, is an important thinking skill. Writing can be a valuable tool for arriving at this type of self-knowledge. Confused writing reveals confused thinking, and difficulty in writing about a concept often signals incomplete understanding. Many teachers encourage students to use their writing as a means of discovering where their comprehension is weak or thinking is muddy. You may also suggest that students keep their ALGEBRA 2 writings in a spiral notebook, which can be used for reference and to remind them of their progress throughout the year.

COOPERATIVE LEARNING

Cooperative learning is not new to schools. Recently, cooperative groups have again emerged as an effective way to organize classes for maximum learning. While it is more common to think of elementary school classrooms organized for cooperative learning, recent research shows that high-school students benefit from the interaction of a small group. For those of us who have seen the rewards, small-group activities make sense.

Some students are more willing to participate in a small group than in a large one. It has been our experience that when an activity is open-ended, more students participate in problem solving. The smaller group is often less intimidating to the uncertain student. Through small-group participation, many students gain the confidence to contribute in the large group.

Structuring a class for group work can be achieved in a variety of ways:

- ▼ by designating time to work in class with a suggestion such as "You and Tom have different answers. Why don't you talk about it?"
- ▼ by suggesting that students form groups of two, three, or four informally
- ▼ by assigning groups of two, three, or four with a designated meeting place and time limit

The method you choose will depend upon your class, the physical structure of your room (movable vs. fixed seats), your own experience with small-group activities, and a variety of other factors.

We have included cooperative learning activities with most problem sets.

Cooperative Learning

either is self contained:

or is an extension of a problem in the problem set:

Cooperative Learning

Have students work in groups of two or three and solve the equation $ax^2 + bx + c = 0$ for the general case. Find the sum of the two roots and then find the product. Use these new rules to write a quadratic equation that has two roots whose sum is -2 and whose product is -15. $x^2 + 2x - 15 = 0$

Cooperative Learning

Have students work in small groups to solve problem **36**. Once the region has been identified, have students randomly choose two points from the inside of the region and connect them with a line segment. Repeat this process. Have students draw a conclusion about each line segment and its relationship to the shaded region.

You may find that cooperative work does not come naturally to your students at first. But with practice, cooperative activities can provide opportunities for student involvement, problem solving, and higher-order thinking that may be lacking in a strictly traditional approach. For further reading and information on cooperative learning, the following list of resources may be helpful.

Reading

Aronson, E. *The Jigsaw Classroom.* Beverly Hills, Calif.: Sage, 1978.

Dewey, John. *Experience and Education.* New York: Collier, 1977.

Glasser, William, MD. *Control Theory in the Classroom.* New York: Harper and Row, 1986.

Newman, F.M., and J.A. Thompson. *Effects of Cooperative Learning on Achievement in Secondary Schools.* Madison, Wis.: National Center on Effective Secondary Schools.

Piaget, Jean. *Language and Thought of the Child.* New York: Meridian, 1926.

Sharan, S., and Y. Sharan. *Small-Group Teaching.* Englewood Cliffs, N.J.: Educational Technology Publications, 1976.

Slavin, R.E. "Cooperative Learning." *Review of Educational Research* 50 (2): 315–42.

Tobias, Sheila. *Overcoming Math Anxiety.* New York: W. W. Norton, 1978.

Vygotsky, L.S. *Thought and Language.* Cambridge, Mass.: MIT Press, 1962.

White, Merry. *The Japanese Educational Challenge.* New York: Macmillan, 1987.

Organizations

CENTER FOR RESEARCH ON ELEMENTARY AND MIDDLE SCHOOLS
The Johns Hopkins University
3505 North Charles Street
Baltimore, MD 21218

THE COOPERATIVE LEARNING CENTER
David and Roger Johnson
202 Pattee Hall
University of Minnesota
Minneapolis, MN 55455

INTERNATIONAL ASSOCIATION FOR THE STUDY OF COOPERATION IN EDUCATION
136 Liberty Street
Santa Cruz, CA 95060

Cooperative Learning Activities are available in reproducible form in the TEACHER'S RESOURCE BINDER.

ROBLEM SOLVING

Because the concept of problem solving is understood in so many different ways, we would like to tell you how we have used it in our book.

First, we see a difference between an exercise and a problem. An exercise can be solved by applying a skill and finding an answer. A problem requires more. It requires a student to think, to devise a strategy, to find a method. As teachers, we work with students to figure out what to do with a problem. As Sheila Tobias, author of *Overcoming Math Anxiety,* says, "The difference between a successful and an unsuccessful math student is that the successful student knows what to do when she/he doesn't know what to do."

Students need practice in solving problems just as they must practice skills by working on exercises. Many students come to us convinced that there is one way to find an answer. In Chapter 1 and then throughout our book, we show several methods of solving a single problem. As teachers, we can encourage students to try different methods on their own. We can help students develop a willingness to try, to see that making mistakes can help them learn if they can see what processes do not work and then try again.

In a class period, it is difficult to take the time that students often need, but developing a problem-solving attitude takes time and concentration on process rather than answers. We have found that time spent early in the year has great payoff later as students are willing to try problems, to keep working to find a solution, and to learn.

One of the techniques we have found helpful is to ask questions like "What might be a first step we can try?" or "If we suppose that . . ." or "What if . . . ?" Students are not accustomed to considering problem solving an everyday way of thinking; they usually see it as a process called for in a few pages in their textbook.

If problems as well as exercises are included in tests, you may find that students need more time. It may take some experimenting to know how much can be included in a test timed for a class period. Often, looking at a student's work will show the kind of thinking going on and help us learn how much the student has understood.

In today's changing world, one of the most important skills students need is to be able to learn how to learn. The knowledge they will need will change constantly. If we can help students focus on how to learn and can help them develop the ability to continue to learn on their own, we will have been successful indeed!

USING MANIPULATIVES

In the past, the use of hands-on material has been associated with teaching young children. There is a growing recognition that many older students need concrete experiences in order to be successful with abstract concepts. Thus, greater attention is being paid to the use of manipulative materials and activities in teaching students algebra and geometry.

In *Algebra 2 and Trigonometry*, we have tried to accommodate a variety of learning styles. It has been our experience that although some students are comfortable with abstract concepts, others benefit from a visual representation, and still others need even more concrete experiences. For this reason, we have included visual representations of equations throughout the book in the introduction of concepts and skills and throughout the exercises and problem sets. An example of such a representation is:

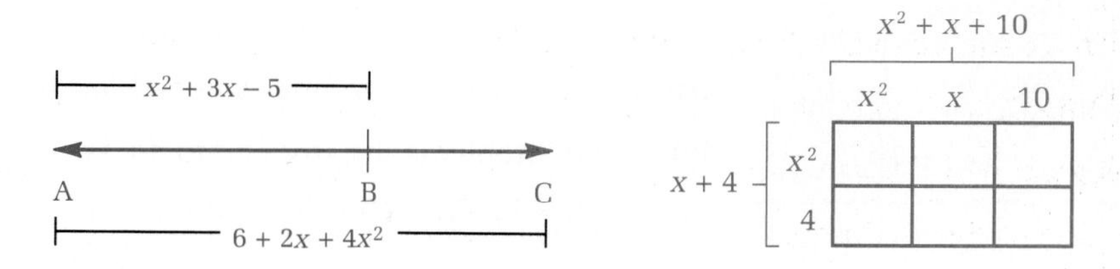

In addition, wherever appropriate, we have included suggestions for using manipulatives in the classroom. These suggestions will be found in the *Teacher's Resource Binder* and referenced in the Teacher's Edition under the heading

Using Manipulatives and Applications

We hope you and your students find this approach helpful.

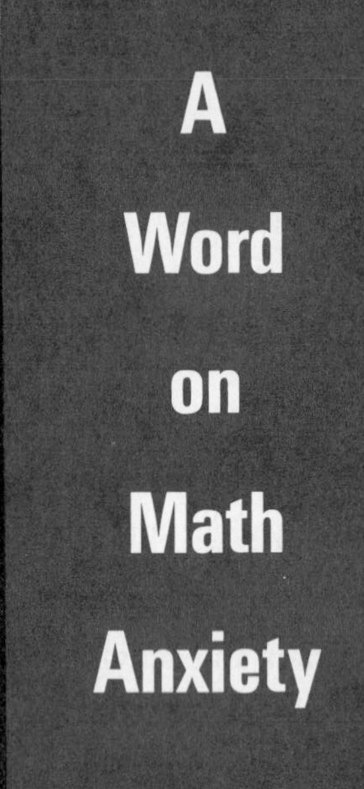

from
Sheila Tobias

▼
▼
▼

Everybody's had it at some time in their lives.

Even students who have gone on to do very well in mathematics and math-based fields of study.

Even math teachers.

"Math anxiety" is not a disease, though it can be prevented and cured.

It's a feeling of helplessness accompanied by real panic, when we have to do something we either don't know or have forgotten how to do.

Most students don't have a full-blown case of math anxiety. They just have some anxious moments when it comes to learning math. In ten years of interviewing students who think they could do better at math if they didn't get the shakes, I have learned much about where math anxiety comes from and what students and their teachers can do to prevent it and overcome it.

Time Pressure

For some students math anxiety starts with timed tests and with the elementary school work sheets that have one hundred problems to add, subtract, multiply, or divide in something like ten minutes. No wonder kids get anxious. Who can think while the clock is ticking away? One of the college students interviewed in a math anxiety clinic said that the only things she ever really learned in elementary school were how to subtract and how to divide. During tests, she said, she kept looking at the clock on her classroom wall, trying to figure out how many minutes she had left for the test and then counting off the number of problems she still had to do. When she realized she had, say, one-half minute for each remaining calculation, she stopped trying altogether and gave up.

Sheila Tobias is the author of *Overcoming Math Anxiety* and *Succeed with Math: Every Student's Guide to Conquering Math Anxiety.*

It's true that in order to solve problems students need to understand the *ideas* on which arithmetic, algebra, geometry, and, later, calculus are based. But it is also true that if students are going to be able to concentrate on the difficult stuff, they are going to have to do the "easy" stuff fast. That's why teachers insist that they practice arithmetic and, later, algebraic skills. You know that the more they practice, the better and faster they can zip through calculations (in arithmetic) and manipulations (in algebra). But students don't know this and become anxious when they have to think under pressure. Teachers need to explain that the better students become at calculations and manipulations, the more they can concentrate on the logic of the step-by-step problem solving that is the focus of this text.

It is mostly at the beginning of every new lesson that students need more time to think. The wise teacher will encourage them to take their time and to develop speed without anxiety.

Time Pressure Exercise

If you think time pressure might be the source of your students' math-anxious moments, try this experiment in class: Move the chairs to face the classroom clock. (If there isn't a clock everyone can see, bring one in.) Then assign the class one-half of their algebra homework under strict time limits. (You might even use an alarm clock that goes off every so many minutes, indicating that time's up.) Then turn the clock to the wall (or cover it with a cloth) and have them do the second half of their homework problems, giving them the feeling of having "all the time in the world" to finish. Watch them work and ask them to compare how they feel under the two artificially constructed situations. Does time pressure make them more nervous? more efficient? Do some of them get more right answers going slowly? fewer right answers? Perhaps, if time pressure is a major problem, you can make some special accommodations—especially during tests—while students get up to speed.

Solving Word Problems

Some students do just fine at calculations and manipulations. Their anxious moments come with word or story problems, or real-world applications of what they've learned in school. What makes word problems difficult? There is a "Great Divide" between a task like "Go to the chalkboard and multiply 14 times 76" and any of the word problems in this text. In the first instance, the student will not know the answer but will surely know what to do. In the multistep word problems featured in this text, students will not only not know the answer, they may not even know where to begin. One thing is certain: Until students learn to cope in this kind of environment of uncertainty, anxiety will develop. The successful student *starts to work the problem* even before he or she figures out what to do. And the courage to do this is the cure for math anxiety.

Another difficulty students have is in *interpretation*. Word problems, particularly in algebra, require the translation of the word sentences and the word questions into mathematical statements and symbols. Since there may be more than one interpretation of word sentences (mathematics statements are more precise than English sentences), there is room for doubt and more uncertainty.

One solution to the problem of interpretation is for students to stay with the English sentences a little longer. They may rewrite the problem in their own words, may even write out the two or three interpretations that suggest themselves. If they estimate an answer (roughly calculated) for each interpretation, they may see which one will produce the most reasonable answer. They can be encouraged to do their homework with a friend or to talk on the telephone with a classmate to see whether their interpretations agree. The important thing to communicate to students of algebra is that differences in interpretation are possible and that the student who reads a problem differently is not dumb.

One of the great strengths of this text are the Warm-up Exercises the authors recommend. Some of the exercises are designed particularly to give students the all-important "windows" into their thinking that successful students in mathematics construct for themselves. The exercises ask students to *explain*, to *describe*, and to *discuss* their *start-up decisions* and to *keep track* of their *decision strategies* as they work a problem. This training in self-monitoring and in writing down what they are thinking is very important in giving students a sense of control over their material and over themselves and can be augmented by the classroom teacher through the following exercise:

> **A**sk students to divide their homework or test page down the middle with a vertical line. On one side of the line, they should write the problems in mathematical symbols and work the solution. On the other side of the line, they should write down one or more plans of action to solve the problem. Writing down their thoughts has two real benefits: As long as their hands keep moving, their minds won't block. And their notes to themselves will keep them from losing their way.

You will probably not want to collect these jottings, because they are your students' personal notes to themselves. But once in a while some class time can be devoted to talking about the left-hand side of the page. Students need to know that their friends—even the ones who are writing all the time—have difficulties, too.

Getting Unstuck

The book you are about to use is the product of a team of teachers who want your students to succeed. The book is arranged so logically that if your students, while doing their homework, forget how you explained something, they don't need to panic, even if it is the last evening before a test. They can simply turn back a few pages and review the previous day's or week's material by themselves.

Still, even the most experienced teachers or the best textbook authors cannot get themselves into every student's mind and figure out exactly what is confusing him or her. Since math anxiety comes from math helplessness and the best cure for math anxiety is getting students back in *control,* it is important for them to learn how to get themselves unstuck. You can help your students by teaching them, systematically, not simply what to do when they remember what to do, but what to do when they forget. One way to stop the end-of-pencil chewing is to teach them the following exercise to do alone or with a friend:

Getting-Unstuck Exercise

Answer, in writing, the following questions:

- What do I know about the problem?
- What are they asking me to find out?
- What is making this problem difficult for me?
- What can I do to make this problem easier for myself?

Working with friends can be very helpful. Educators are finally discovering that though math is not a *spoken* language, it helps when students talk to their friends about their homework. Students struggle together better than they struggle alone, and, in explaining the examples and sample problems to one another, they learn them well.

Who Can Do Math and Why

Some teenagers feel uncomfortable doing algebra and higher math not because they're "dumb" or even "anxious" but because of what sociologists call *role-stereotyping*—that is, because they have the wrong idea that math is for boys and not for girls, for white people and not African Americans and Hispanic Americans. It is true that if you look around, you see fewer women than men and fewer minorities than Caucasians doing higher mathematics, engineering, and science. But that's because of past discrimination and lack of opportunity, not because women and minorities lack some "math gene." Some people disagree, but the evidence is overwhelming: Where opportunities are the same and where there are no prejudices against them, women, African Americans, and Hispanic Americans do very well. Teachers need to be more than color- and gender-blind. So long as these prejudices exist outside the classroom, teachers will need to counter students' sometimes negative assessments of their own potential and skills.

Another related idea that gets in the way of success in math is that students can't be good in both math and English or math and social studies. Certainly, people have their favorites. But really successful people *are* good at both. That's because the skills of good writing and the skills of good problem solving overlap. Writing a logical argument is similar to setting up an equation to solve a problem. Combining like terms in an algebraic statement is good practice for editing out unnecessary words and phrases in an English paper.

As a teacher, you can do much to counter these wrong ideas. You can direct students to biographical information about a more varied group of mathematicians and scientists. Interviews with family members or family friends who work with numbers will quickly dispel the notion that such workers do not need to be good with words.

Another wrong idea and source of math anxiety is that memorizing can get students through. While memorizing has its place—it's appropriate for learning times tables, techniques such as "invert and multiply," and even area and perimeter formulas—algebra is the beginning of the end for math memorizers. Memorizing will cause your students more anxious moments than it will cure.

In the atmosphere of a trusting and noncompetitive classroom, led by a teacher who truly believes his or her students can succeed with math, with ample time for thinking about a problem and with tolerance for error, students can learn to cope with their math-anxious moments and even to learn from their mistakes. But where anxiety persists or seems to be ingrained, where parents are discouraging their daughters or putting too much pressure on their sons, the wise teacher will ask a counselor for help or for referral. Math-anxiety reduction clinics exist in many locales. My books are helpful, too. Not every student will want to major in mathematics or become professionally involved, but as I wrote in *Succeed with Math,* anyone who wants to lead a rich intellectual and professional life needs to know more math.

▼

Two forms of this essay are provided in reproducible form in the TEACHER'S RESOURCE BINDER. Form S is written for students; form P is written for parents.

A Guide to Additional Resources

Teacher's Resource Binder						Additional Resources		
Section	**Practice**	**Enrichment**	**Class Openers**	**Manipulatives**	**Cooperative Learning**	**Transparencies**	**Computers**	**Tests/ Quizzes**
1.1	1		1	1				
1.2	2	1	2			1		
1.3	3		3		1	2		Quiz Forms 1, 2, 3
1.4	4	2	4		2		See Lesson Notes	
1.5	5		5	2	3		See Lesson Notes	
1.6	6		6		4			
1								Test 1
2.1	7		7	3				
2.2	8		8		5		See Lesson Notes	
2.3	9		9					Quiz Forms 1, 2, 3
2.4	10		10	4		3	See Lesson Notes	
2.5	11	3	11		6	4	See Lesson Notes	
2.6	12	4	12		7			
2								Test 2
Exponents and Radicals	13, 14		13	5				

TEACHER'S RESOURCE BINDER						ADDITIONAL RESOURCES		
Section	**Practice**	**Enrichment**	**Class Openers**	**Manipulatives**	**Cooperative Learning**	**Transparencies**	**Computers**	**Tests/ Quizzes**
3.1	15	5	14	6	8		See Lesson Notes	
3.2	16	6	15		9		See Lesson Notes	
3.3	17		16			5	See Lesson Notes	Quiz Forms 1, 2, 3
3.4	18		17		10			
3.5	19		18		11	6	See Lesson Notes	
3.6	20		19				See Lesson Notes	
3								Test 3
4.1	21		20	7				
4.2	22		21		12		See Lesson Notes	
4.3	23	7	22		13		See Lesson Notes	Quiz Forms 1, 2, 3
4.4	24		23					
4.5	25	8	24				See Lesson Notes	
4.6	26		25				See Lesson Notes	
4.7	27		26	8		7.8		
4								Test 4

A complete Assignment Guide will be found on the first page of each problem set. A complete Assignment Guide in reproducible form for all problem sets will be found in the TEACHER'S RESOURCE BINDER.

	TEACHER'S RESOURCE BINDER					ADDITIONAL RESOURCES		
Section	Practice	Enrichment	Class Openers	Manipulatives	Cooperative Learning	Transparencies	Computers	Tests/ Quizzes
5.1	28		27			9		
5.2	29	9	28		14			
5.3	30		29				See Lesson Notes	Quiz Forms 1, 2, 3
5.4	31	10	30			10	See Lesson Notes	
5.5	32		31		15		See Lesson Notes	
5.6	33		32		16		See Lesson Notes	
5.7	34		33	9	17		See Lesson Notes	
5								Test 5
6.1	35	11	34	10		11	See Lesson Notes	
6.2	36		35				See Lesson Notes	
6.3	37	12	36		18	12	See Lesson Notes	Quiz Forms 1, 2, 3
6.4	38		37					
6.5	39		38		19			
6.6	40		39				See Lesson Notes	
6								Test 6
Conics	41, 42		40	11				

TEACHER'S RESOURCE BINDER						ADDITIONAL RESOURCES		
Section	**Practice**	**Enrichment**	**Class Openers**	**Manipulatives**	**Cooperative Learning**	**Transparencies**	**Computers**	**Tests/ Quizzes**
7.1	43	13	41	12	20		See Lesson Notes	
7.2	44		42		21	13	See Lesson Notes	
7.3	45		43		22		See Lesson Notes	Quiz Forms 1, 2, 3
7.4	46		44				See Lesson Notes	
7.5	47	14	45			14	See Lesson Notes	
7.6	48		46					
7								Test 7
Factoring	49, 50		47					
8.1	51		48					
8.2	52		49		23		See Lesson Notes	
8.3	53	15	50		24			Quiz Forms 1, 2, 3
8.4	54	16	51		25	15		
8.5	55		52	13		16	See Lesson Notes	
8								Test 8

A complete Assignment Guide will be found on the first page of each problem set. A complete Assignment Guide in reproducible form for all problem sets will be found in the TEACHER'S RESOURCE BINDER.

TEACHER'S RESOURCE BINDER						ADDITIONAL RESOURCES		
Section	**Practice**	**Enrichment**	**Class Openers**	**Manipulatives**	**Cooperative Learning**	**Transparencies**	**Computers**	**Tests/ Quizzes**
9.1	56	17	53		26	17	See Lesson Notes	
9.2	57		54			18	See Lesson Notes	
9.3	58		55		27		See Lesson Notes	Quiz Forms 1, 2, 3
9.4	59		56		28		See Lesson Notes	
9.5	60	18	57	14			See Lesson Notes	
9.6	61		58				See Lesson Notes	
9								Test 9
10.1	62		59					
10.2	63		60				See Lesson Notes	
10.3	64	19	61		29			Quiz Forms 1, 2, 3
10.4	65	20	62		30	19		
10.5	66		63	15		20		
10								Test 10
Linear Programming	67, 68		64					

Section	Practice	Enrichment	Class Openers	Manipulatives	Cooperative Learning	Transparencies	Computers	Tests/ Quizzes
TEACHER'S RESOURCE BINDER						ADDITIONAL RESOURCES		
11.1	69		65		31	21	See Lesson Notes	
11.2	70		66		32			
11.3	71		67		33			Quiz Forms 1, 2, 3
11.4	72	21	68				See Lesson Notes	
11.5	73	22	69	16		22	See Lesson Notes	
11.6	74		70		34		See Lesson Notes	
11								Test 11
12.1	75	23	71		35	23		
12.2	76		72		36			
12.3	77		73					Quiz Forms 1, 2, 3
12.4	78	24	74		37	24		
12.5	79		75	17	38			
12								Test 12

A complete Assignment Guide will be found on the first page of each problem set. A complete Assignment Guide in reproducible form for all problem sets will be found in the TEACHER'S RESOURCE BINDER.

TEACHER'S RESOURCE BINDER						ADDITIONAL RESOURCES		
Section	**Practice**	**Enrichment**	**Class Openers**	**Manipulatives**	**Cooperative Learning**	**Transparencies**	**Computers**	**Tests/ Quizzes**
13.1	80	25	76		39			
13.2	81		77		40	25	See Lesson Notes	
13.3	82		78				See Lesson Notes	Quiz Forms 1, 2, 3
13.4	83	26	79	18		26		
13								Test 13
Mathematical Induction	84, 85		80					
14.1	86		81		41			
14.2	87		82		42	27	See Lesson Notes	
14.3	88	27	83				See Lesson Notes	Quiz Form 1
14.4	89		84		43			Quiz Form 2
14.5	90	28	85			28		Quiz Form 3
14.6	91		86					
14.7	92		87		44			
14								Test 14

TEACHER'S RESOURCE BINDER						ADDITIONAL RESOURCES		
Section	**Practice**	**Enrichment**	**Class Openers**	**Manipulatives**	**Cooperative Learning**	**Transparencies**	**Computers**	**Tests/ Quizzes**
15.1	93		88		45	29	See Lesson Notes	
15.2	94		89				See Lesson Notes	Quiz Forms 1, 2, 3
15.3	95	29	90			30	See Lesson Notes	
15.4	96	30	91				See Lesson Notes	
15							See Lesson Notes	Test 15
16.1	97		92		46			
16.2	98	31	93					
16.3	99		94			31	See Lesson Notes	Quiz Forms 1, 2, 3
16.4	100		95		47	32	See Lesson Notes	
16.5	101	32	96				See Lesson Notes	
16								Test 16

A complete Assignment Guide will be found on the first page of each problem set. A complete Assignment Guide in reproducible form for all problem sets will be found in the TEACHER'S RESOURCE BINDER.

ALGEBRA 2
and Trigonometry

ALGEBRA 2

and Trigonometry

John Benson
Evanston Township High School
Evanston, Illinois

Sara Dodge
Evanston Township High School
Evanston, Illinois

Walter Dodge
New Trier High School
Winnetka, Illinois

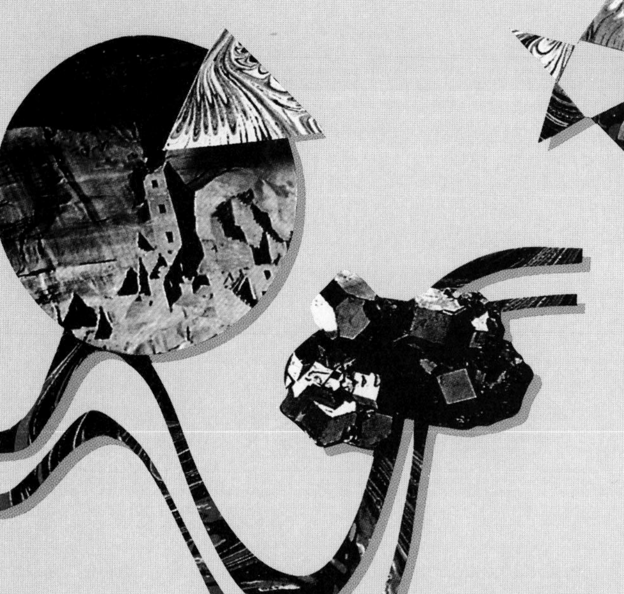

Charles Hamberg
Illinois Mathematics and
Science Academy
Aurora, Illinois

George Milauskas
Illinois Mathematics and
Science Academy
Aurora, Illinois

Richard Rukin
Evanston Township High School
Evanston, Illinois

McDougal Littell / Houghton Mifflin
Evanston, Illinois
Boston Dallas Phoenix

Reviewers

Agnes B. Bailey
Sacramento High School, Retired
Sacramento, California

Judy Bush
Countryside High School
Clearwater, Florida

Dr. Jean Reddy-Clement
State Mathematics Supervisor
Louisiana Department of Education
Baton Rouge, Louisiana

Carol Forman
Catoctin Senior High School
Thurmont, Maryland

Dr. Terry Parks
Director of Mathematics, K–12
Shawnee Mission School District
Shawnee Mission, Kansas

We would like to thank the many teachers and students that field tested this book over an entire school year, 1989–1990. Their suggestions, comments, and criticisms helped to shape a final product we are proud to publish.

ISBN 0-8123-8800-3

Box 1667, Evanston, Illinois 60204

Printed in the United States of America

1 2 3 4 5 6 7 8 9 10 – DWO – 99 98 97 96 95

CONTENTS

A LETTER TO STUDENTS

CHAPTER 1 PROBLEM SOLVING 2

1.1 Introduction to Problem Solving 3
1.2 Graphic Solutions 6
1.3 Geometric Modeling 9
1.4 Finding a Pattern 12
1.5 A Recursive Solution 16
1.6 Demonstrations 19

CHAPTER 2 THE LANGUAGE OF ALGEBRA 22

2.1 The Language of Expressions 23
Career Profile: The Mathematics of Managing Money 29
2.2 Solving Equations 30
2.3 Solving Inequalities 38
2.4 Introduction to Graphing 45
2.5 Matrices 52
2.6 Probability 60
CHAPTER SUMMARY 65
REVIEW PROBLEMS 66
CHAPTER TEST 69

SHORT SUBJECT Exponents and Radicals 70

CHAPTER 3 LINEAR RELATIONSHIPS 78

3.1 Graphing Linear Equations 79
3.2 Systems of Linear Equations 87
Career Profile: Linear Relationships and Tart Cherries 93
3.3 Graphing Inequalities 94
3.4 Functions and Direct Variation 100
3.5 Fitting a Line to Data 107
3.6 Compound Functions 114
Mathematical Excursion: Linear Programming 120
CHAPTER SUMMARY 121
REVIEW PROBLEMS 122
CHAPTER TEST 125

CHAPTER 4 QUADRATIC FUNCTIONS 126

4.1 What Are Quadratics? 127
Historical Snapshot: Algebraic Shorthand 134
4.2 Factoring Quadratics 135
4.3 Graphing Quadratic Functions 142
4.4 Solving Quadratic Equations 149
Career Profile: The Parabola in Baseball 156
4.5 Completing the Square 157
4.6 Applications of Quadratic Functions 164
4.7 Data Displays 171
CHAPTER SUMMARY 177
REVIEW PROBLEMS 178
CHAPTER TEST 181

CHAPTER 5 FUNCTIONS 182
5.1 Relations and Functions 183
5.2 More About Functions 190
5.3 Inverse Functions 198
5.4 Operations on Functions 206
5.5 Discrete Functions 213
Mathematical Excursion: Functions Make Life Simpler 219
5.6 Recursively Defined Functions 220
5.7 Measures of Central Tendency and Percentiles 228
Career Profile: Exponentials and Statistics in Banking 232
CHAPTER SUMMARY 233
REVIEW PROBLEMS 234
CHAPTER TEST 237

CHAPTER 6 GRAPHING 238
6.1 Beyond Plotting Points 239
6.2 Translating Graphs 246
Historical Snapshot: Fun with Graphs 251
6.3 Reflections and Symmetry 252
Career Profile: Inverse Variation in the Desert 259
6.4 Graphing Other Functions 260
6.5 Asymptotes and Holes in Graphs 268
6.6 Measures of Dispersion 275
CHAPTER SUMMARY 279
REVIEW PROBLEMS 280
CHAPTER TEST 283

SHORT SUBJECT Conics 284

CHAPTER 7 SYSTEMS 290

7.1 Solving Problems with Systems of Equations 291

Historical Snapshot: A Surveyor's Dream 297

7.2 Linear Systems with Three Variables 298

Career Profile: A Traffic Jam 305

7.3 Systems of Inequalities 306

7.4 Inverse Variation 312

7.5 Nonlinear Systems 318

7.6 Other Systems of Equations 324

CHAPTER SUMMARY 331

REVIEW PROBLEMS 332

CHAPTER TEST 335

SHORT SUBJECT Factoring to Solve Equations and Inequalities 336

CHAPTER 8 EXTENDING THE REAL-NUMBER SYSTEM 342

8.1 Rational Exponents 343

8.2 Solving Radical Equations 349

Historical Snapshot: When Is a Number Not Real? 356

8.3 Introduction to Complex Numbers 357

8.4 Algebra of Complex Numbers 364

Career Profile: Using a Curve to Trace Disease 371

8.5 The Normal Curve 372

CHAPTER SUMMARY 376

REVIEW PROBLEMS 377

CHAPTER TEST 379

CHAPTER 9 POLYNOMIALS AND POLYNOMIAL FUNCTIONS 380

9.1 The Remainder and Factor Theorems 381
9.2 The Rational-Zeros Theorem 389
Career Profile: Statistical Analysis of Success 395
9.3 Real Zeros of Polynomial Functions 396
9.4 Four Useful Theorems 404
Historical Snapshot: How Much Proof Is Enough? 409
9.5 The Binomial Theorem 410
9.6 Standard Deviation and Chebyshev's Theorem 416
CHAPTER SUMMARY 422
REVIEW PROBLEMS 423
CHAPTER TEST 425

CHAPTER 10 RATIONAL EXPRESSIONS, EQUATIONS, AND FUNCTIONS 426

10.1 Multiplying and Dividing Rational Expressions 427
Career Profile: Teaching Mathematics 433
10.2 Rational Equations and Inequalities 434
Historical Snapshot: Ancient Problems 441
10.3 Adding and Subtracting Rational Expressions with Like Denominators 442
10.4 Adding and Subtracting Rational Expressions with Unlike Denominators 448
10.5 Complex Fractions 454
CHAPTER SUMMARY 461
REVIEW PROBLEMS 462
CHAPTER TEST 465

SHORT SUBJECT Linear Programming 466

CHAPTER 11 EXPONENTIAL AND LOGARITHMIC FUNCTIONS 472

11.1 Definitions and Graphs of the Exponential and Logarithmic Functions 473
11.2 Two Important Exponential and Logarithmic Functions 480
11.3 Properties of Logarithms 487
11.4 Solving Exponential and Logarithmic Equations 494
11.5 Applications of Exponential and Logarithmic Functions 501
Mathematical Excursion: Something Earthshaking 508
11.6 Binomials and Binomial Distributions 509
CHAPTER SUMMARY 514
REVIEW PROBLEMS 515
Career Profile: Logarithmic Intuition 518
CHAPTER TEST 519

CHAPTER 12 COUNTING AND PROBABILITY 520

12.1 Systematic Counting 521
12.2 Sample Spaces, Events, and Probability 528
12.3 Probabilities of Unions 535
12.4 Probabilities of Intersections 542
Historical Snapshot: A Lucky Collaboration 549
12.5 Interpreting Probabilities 550
CHAPTER SUMMARY 556
REVIEW PROBLEMS 557
Career Profile: Statistics and the Public Mind 560
CHAPTER TEST 561

CHAPTER 13 SEQUENCES AND SERIES 562
13.1 Sequences 563
13.2 Series 570
13.3 Arithmetic Sequences and Series 576
Career Profile: Statistics in Predicting Longevity 582
13.4 Geometric Sequences and Series 583
Historical Snapshot: Converging on π 590
CHAPTER SUMMARY 591
REVIEW PROBLEMS 592
CHAPTER TEST 595

SHORT SUBJECT Mathematical Induction 596

CHAPTER 14 INTRODUCTION TO TRIGONOMETRY 602
14.1 The Unit Circle 603
Mathematical Excursion: Systems and Profitability 609
14.2 The Circular Functions 610
14.3 Coordinate Definitions 618
14.4 Inverse Trigonometric Functions 626
14.5 Right Triangles 632
14.6 Radian Measure 640
14.7 Trigonometric Relationships 646
Career Profile: Graphing the Waves of the Heart 652
CHAPTER SUMMARY 653
REVIEW PROBLEMS 654
CHAPTER TEST 657

CHAPTER 15 TRIGONOMETRIC IDENTITIES AND EQUATIONS 658
15.1 Basic Trigonometric Identities 659
Historical Snapshot: The Origin of Trigonometry 665
15.2 The Sum and Difference Identities 666
15.3 The Double- and Half-Angle Identities 673
Career Profile: Trigonometry in Windsurfing 679
15.4 Final Identities and Trigonometric Equations 680
CHAPTER SUMMARY 687
REVIEW PROBLEMS 688
CHAPTER TEST 691

CHAPTER 16 TRIGONOMETRIC GRAPHS AND APPLICATIONS 692
16.1 The Law of Sines 693
16.2 The Law of Cosines 699
Career Profile: Mathematics in Medicine 704
16.3 Graphing Sine Functions 705
16.4 Changing the Period 710
16.5 More Trigonometric Graphs 716
Mathematical Excursion: Soothing Sounds 720
CHAPTER SUMMARY 721
REVIEW PROBLEMS 722
CHAPTER TEST 725

HANDBOOK OF BASIC CONCEPTS 726

TABLES

Normal Distribution Table 732
Table of Squares and Square Roots 733
Common Logarithms of Numbers 734
Natural Logarithms of Numbers 736
Trigonometric Functions 738

GRAPHING CALCULATOR HANDBOOK

Lesson 1 Working with Expressions 745
Lesson 2 Adding and Subtracting Matrices 749
Lesson 3 Scalar Multiplication and Matrix Multiplication 752
Lesson 4 Graphing Inequalities 754
Lesson 5 Graphing Linear Equations 756
Lesson 6 Setting the Viewing Window 759
Lesson 7 Finding Statistical Means 763
Lesson 8 Solving Systems of Linear Equations 766
Lesson 9 Finding a Line of Best Fit 770
Lesson 10 Finding Roots of Quadratic Equations 774
Lesson 11 The Determinant of a Matrix 777
Lesson 12 Permutations and Combinations 779
Lesson 13 Graphing Conic Sections 781
Lesson 14 Working with Absolute Values 784
Lesson 15 Graphing a Function and Its Inverse 787
Lesson 16 Graphing Polynomial Functions 789
Lesson 17 Graphs with Holes and Asymptotes 793
Lesson 18 Finding Standard Deviations 796
Lesson 19 Curve Fitting 798
Lesson 20 Solving Nonlinear Systems Graphically 801
Lesson 21 Applications of Quadratic Functions 805
Lesson 22 Translating Trigonometric Graphs 807
Lesson 23 Stretching and Shrinking Trigonometric Graphs 810
Lesson 24 Compound Transformations of Trigonometric Graphs 812

SELECTED ANSWERS 814

GLOSSARY 865

INDEX 873

A LETTER TO STUDENTS

Welcome to second-year algebra! As high school teachers, we find mathematics enjoyable, challenging, useful, and exciting. We have written this book in the hope that you will share these feelings.

In first-year algebra, you learned a great deal. For many students, a review of these ideas is necessary before new, more challenging topics can be mastered. We have tried to incorporate review into the presentation of new material so that you can review and extend your earlier learning as quickly as possible. You will therefore spend the year learning, reviewing, and retaining what you learn.

In Chapter 1, you will encounter a classic problem and approach it in a variety of ways. Problem solving is a process rather than an answer. We all learn by trying to solve problems. We can solve some problems quickly. Others take time, several tries, and often some frustration. We learn from what we try, and we learn from our mistakes.

We have written this book for you. We want you to be able to read it and use it as a reference. When you work independently, read the text. Look up information. Use the examples and the sample problems. Ask questions. Share questions and ideas with your teacher. Talk to your classmates. Mathematics can help you learn how to learn. We hope your experience this year learning mathematics will be rewarding and exciting.

Let us tell you about some parts of the book.

Part One: Introduction

explains concepts and skills. Examples show these ideas in practice.

Part Two: Sample Problems

applies the information from Part One. The sample problems, along with the examples, are models you can refer to when you solve the Exercises and Problems sections.

Part Three: Exercises and Problems

is where you practice and really learn the concepts and skills of algebra. This is where you learn to be a problem solver, one of the most important tasks you have this year. We have tried to make these exercises and problems worthwhile. In each section you will find problems that

- ▲ *practice the ideas in today's lesson*
- ▲ *review the ideas from yesterday's lesson, last week's lesson, and last month's lesson and apply these ideas to the current topic*
- ▲ *introduce ideas you will learn tomorrow or next week or even later*

Thus, you will learn concepts over days, weeks, and months so that by the end of the year, all these different concepts that seemed puzzling when you first saw them will be firmly integrated in your understanding. To give you an idea of how problem sets combine practicing, reviewing, and introducing concepts, we have labeled the problems in Section 2.4.

At the end of each chapter, you will find these features:

CHAPTER SUMMARY *of ideas, vocabulary, and skills*
REVIEW PROBLEMS
CHAPTER TEST

In addition to the sixteen chapters of the book, there are three components you will find useful:

SHORT SUBJECTS

These topics can be a review for many students, as in the case of Exponents and Factoring, or enrichment, as in the case of Mathematical Induction.

HANDBOOK OF BASIC CONCEPTS

This section is included as a quick reference for topics that are treated in many Algebra 1 courses.

GRAPHING CALCULATOR HANDBOOK

If you have access to a Texas Instruments TI-81 or TI-82 calculator or to a Casio fx-7700GB calculator, this handbook will help you use the calculator to explore the topics treated in your algebra and trigonometry studies.

Good luck. We wish you well and hope that by the end of this year you will share some of our excitement about mathematics.

John Benson Charles L. Hamberg
Sara H. Dodge George Milauskas
Walter Dodge Richard Rukin

STUDENTS MAY BE surprised to spend all of Chapter **1** on one problem. We would like them to begin this year by exploring, making connections, problem solving, and above all, thinking. The handshake problem that is used throughout this chapter is a springboard to solving problems by simulation, graphing, geometric modeling, patterns, and recursion.

The problems in this chapter are open ended to set a tone for discussion and explanation. This should help students overcome the misconception that there is a single approach and one right answer to every problem.

A section on communicating deals with the notion of proof and justification.

The first few days of the course present an ideal opportunity to develop student comfort with large and small group exploration and discussion.

Help the students in your class open the door to the enjoyment of mathematics through making connections and problem solving.

CHAPTER

1 PROBLEM SOLVING

Can you identify the complex patterns in this textile from ancient Peru (C. A.D. 800)?

1.1 INTRODUCTION TO PROBLEM SOLVING

Our society depends more and more on mathematics—in the sciences, in medicine, in business, in government, in the military, in city planning, in transportation. In fact, it is hard to find a field in which mathematical skills and ideas are not important. This book will help you expand your knowledge of mathematics and develop your problem-solving ability.

Five main themes are emphasized throughout this book.

I. Functions
II. Graphing and Modeling
III. Applications
IV. Problem Solving
V. Communicating Mathematics

You will find that a scientific calculator is an important tool for carrying out complex calculations. You will also see some ways in which computers and graphing calculators can make your work more interesting, efficient, and productive.

The Handshake Problem

Solving any problem involves both a *process* and an *answer*. The process—the way you approach the problem and the method you use to find an answer—can be the most interesting and important part of a solution. We encourage you to try solving problems in a variety of ways. By experimenting with different approaches, you can learn a great deal of mathematics.

Now let's look at a problem.

The Handshake Problem

Professor Hill had 25 students in her first-period class. On the first day of class, she asked them to shake hands and introduce themselves to one another. When they were finished, she asked the students, "How many handshakes have just been exchanged?"

In this chapter, your mission will be to explore a number of ways of solving Professor Hill's problem. Along the way, you will apply the techniques you discover to other sorts of problems. You will also discover that it is important to be able to explain why a solution works.

Class Planning

Time Schedule

Average: 1 day
Advanced: 1 day
Honors: $\frac{1}{2}$ day

Resource References

Teacher's Resource Book
Class Opener 1
Practice 1
Manipulatives and Applications 1

Class Opener

A class-opener problem will accompany every lesson. If the problem is available to students when they enter the classroom, the students can develop the habit of beginning to work immediately. Students may or may not always be able to solve each problem completely. Encourage students to try different approaches.

Shake hands with everyone in your class exactly once. How many handshakes occurred? How did you keep track in order to be sure that you shook everyone's hand exactly once?

Answers will vary. Encourage students to share their ideas.

The handshake problem gets students involved in the math as well as providing a way for them to meet each other.

Vocabulary

model

1.1

Lesson Notes

- In this section, we present the classic handshake problem and show a few ways to approach it. Most students have not focused on multiple ways to look at and solve problems in earlier math classes. This chapter sets the stage for looking at a variety of techniques for any problem.

Checkpoint

Checkpoint questions are problems based on the objectives, examples, and sample problems.

1 How many handshakes would be necessary for a group of 10 students in your class? 45

2 John has three shirts and five different ties. How many different shirt and tie combinations are possible? 15

Assignment Guide

Average
2, 3, 4, 6
Advanced
2, 3, 4, 6
Honors
($\frac{1}{2}$ day) 1, 3, 4, 6
($\frac{1}{2}$ day) Section 1.2 1, 2, 5, 6

Additional Answer

1 Answers will vary. Some possible answers are draw a diagram, make a model.

A Way of Solving the Problem

One way to solve Professor Hill's handshake problem is to ask a group of 25 people to shake hands, keeping a record of the number of handshakes. Keeping track of the handshakes, however, can be difficult unless we develop an organized procedure. The following are three of the possible ways of organizing the handshakes.

Method 1

Person 1 can shake hands with each of the others, for a total of 24 handshakes. Then person 2 can shake hands with each of the remaining 23 people, person 3 can shake with each of the remaining 22, and so on. There will be $24 + 23 + 22 + \ldots + 4 + 3 + 2 + 1$, or 300, handshakes in all.

Method 2

Person 2 (P_2) shakes with person 1 (P_1).	1 handshake
P_3 shakes with P_2 and then P_1.	$1 + 2$ more $= 3$ shakes
P_4 shakes with P_3, then P_2, then P_1.	$3 + 3$ more $= 6$ shakes
P_5 shakes with P_4, P_3, P_2, and P_1	$6 + 4$ more $= 10$ shakes
P_6 shakes with P_5, P_4, P_3, P_2, and P_1.	$10 + 5$ more $= 15$ shakes
.	
.	
.	
P_{25} shakes with P_{24} through P_1.	$276 + 24$ more $= 300$ shakes

Again, the total is 300 handshakes.

Method 3

We can choose any convenient kind of symbol and set up a ***model*** of the handshakes. At the right is a model in which the 25 letters A, B, C, . . . , Y are used to represent the 25 students. In the model, each handshake is indicated by a pair of letters. Again, the result is 300.

By using an orderly pattern, we can be sure that we have counted the handshakes without missing any.

AB	BC	. . .	VW	WX	XY
AC	BD	. . .	VX	WY	
AD	BE	. . .	VY		
AE	BF	. . .			
AF	BG	. . .			
AG	BH	. . .			
.	.				
.	.				
.	.				
AX	BY				
AY					

$24 + 23 + \ldots + 3 + 2 + 1$
$= 300$ handshakes in all

Problem Set

1 John must go to court to explain what happened in an accident involving his car and a car driven by Rodney Hott. How might John re-create the events for the judge, showing what led to the accident?

2 A new design for a shopping center was developed by Archie Teck. What could he do to help a possible backer understand his idea? Answers will vary. Possible answer: Build a model.

3 There are eight high schools in the Skyway Athletic Conference. How many games must be scheduled for each school's team to play every other conference team

a One time? 28 **b** Two times? 56 **c** n times? $28n$

4 In a classroom, a row of four seats is to be occupied by four students—Ann, Bob, Carlos, and Dai. Use a model to decide

a In how many ways the four students can be seated in the four seats 24

b In how many of these arrangements Carlos will be in the first seat 6

5 A large square is to be covered with smaller, 1-inch squares in a variety of colors. When the larger square is covered, no row, column, or main diagonal may contain squares of the same color.

a How many distinct arrangements are possible if the large square is 3 by 3 inches and three of the small squares are red, three are white, and three are blue? None

b Find a possible arrangement if the large square is 4 by 4 inches and there are four red, four white, four blue, and four green 1-inch squares.

6 Copy the diagram shown.

a How many dots are in the 10-by-10 graph? 100

b Remove a main diagonal. How many dots are in the remaining diagram? 90

c How many dots are in the lower triangular portion of the new diagram? 45

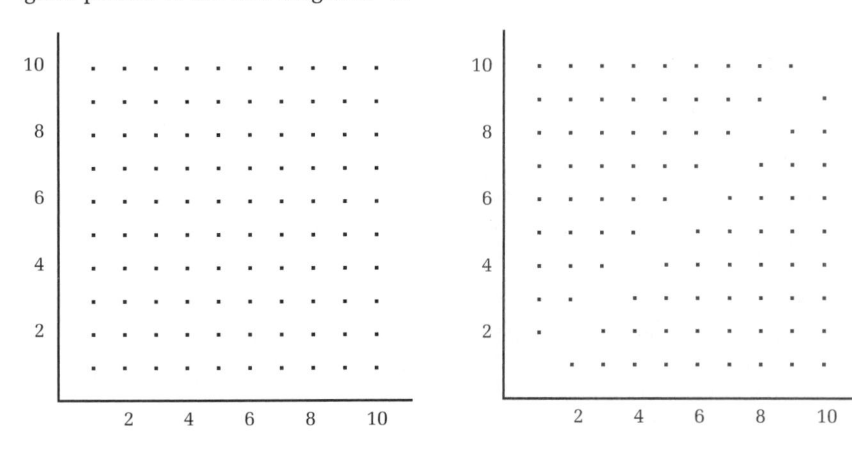

Problem-Set Notes and Strategies

- Encourage your students to draw pictures whenever possible to interpret geometric aspects of a problem. For simple counting problems, list all the possibilities when that is reasonable.
- Each of these problems involves a counting situation which will be explored in more detail throughout the text. Encourage students to try to find a pattern or formula that will work for these problems as well as other similar problems.

Additional Answers

5b Possible answer:

R	W	B	G
G	B	W	R
W	R	G	B
B	G	R	W

Stumbling Block

Stumbling Block is a feature in your Teacher's Edition where we cite errors often made by students. This feature appears wherever appropriate.

Communicating Mathematics

This feature appears in many sections of the Teacher's Edition and is designed to give students the opportunity to write and read about mathematics.

Have students write a paragraph that explains how they feel about the following statement: It is important to read your mathematics textbook.

1.2

Class Planning

Time Schedule

Average: 1 day

Advanced: $\frac{1}{2}$ day

Honors: $\frac{1}{2}$ day

Resource References

Teacher's Resource Book

Class Opener 2

Practice 2

Enrichment 1

Transparency 1

Class Opener

Kai was invited to play a game of darts where the possible scores are 1, 2, 3, 4, 5, or 6. The rule was that Kai would throw two darts. If the second score was greater than the first, Kai wins. Otherwise Kai loses. Is this a "fair" game? Set up a model to see if your answer is correct.

A possible model follows.

11 21 31 41 51 61
12 22 32 42 52 62
13 23 33 43 53 63
14 24 34 44 54 64
15 25 35 45 55 65
16 26 36 46 56 66

Kai wins $\frac{15}{36}$ of the time. It is not "fair".

Lesson Notes

- Section **1.2** introduces two graphical devices to aid in solving problems. Tree diagrams are especially helpful in counting. Branches of the tree may be labeled with the number of possibilities of choosing that branch. Digraphs are useful whenever the direction of exchange is as important as the path itself.

1.2 GRAPHIC SOLUTIONS

Objective

After studying this section, you will be familiar with

- Using a graph to model a problem

Part One: Introduction

Graphing is a powerful method of organizing information. The handshake problem can be modeled by a graph in which each ordered pair (x, y) represents a handshake between person x and person y. For example, the ordered pair (6, 5) represents a handshake between person 6 and person 5.

In Professor Hill's class of 25 students, the value of x and the value of y can range from 1 to 25. There are 625 (that is, 25 · 25) points on the graph.

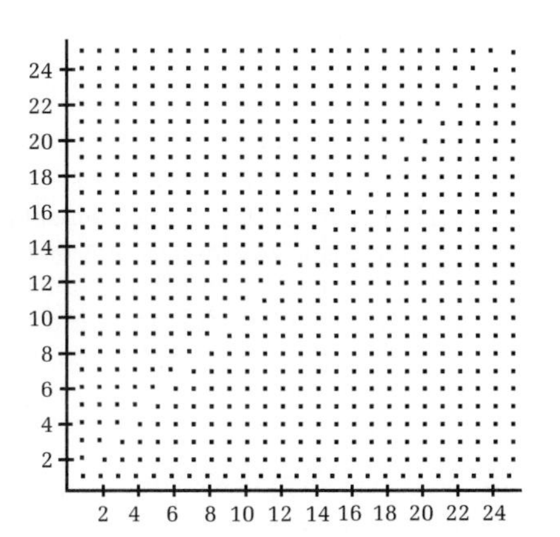

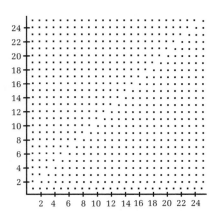

The model is not yet correct. Ordered pairs in which x = y, such as (1, 1), must be eliminated, since no one shakes his or her own hand. When these points are removed, there are 625 − 25, or 600, points left.

Since the ordered pairs (6, 5) and (5, 6) represent the same handshake, we do not need both points. By eliminating all such duplicates, we divide the number of points in half.

There are 300 points left, so there is a total of 300 handshakes. Compare this graph with the model for Method 3 in the preceding section.

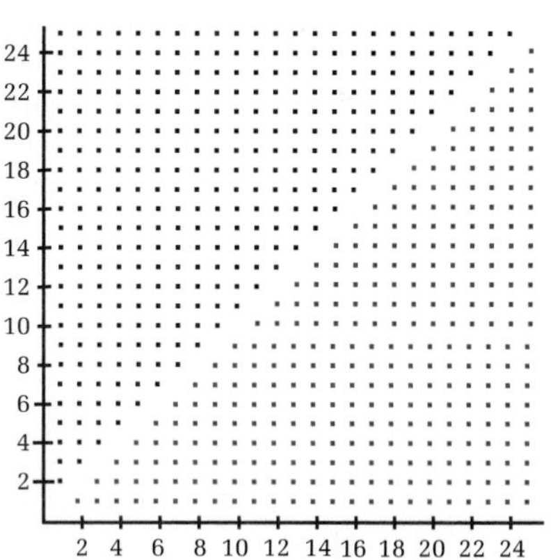

Now consider the following problem.

Paul's Pizza Parlor offers a choice of two styles of pizza—thick and thin crust—and four toppings, one topping per pizza—mushroom, sausage, onions, and green peppers. How many different pizzas can be ordered?

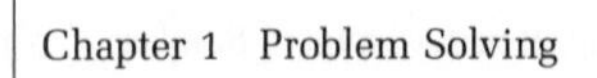

Vocabulary

tree diagram

directed graph

digraph

To solve this problem, we can use a graphing technique known as a ***tree diagram.*** By counting the number of branches at the end of the tree, we can conclude that there are eight choices.

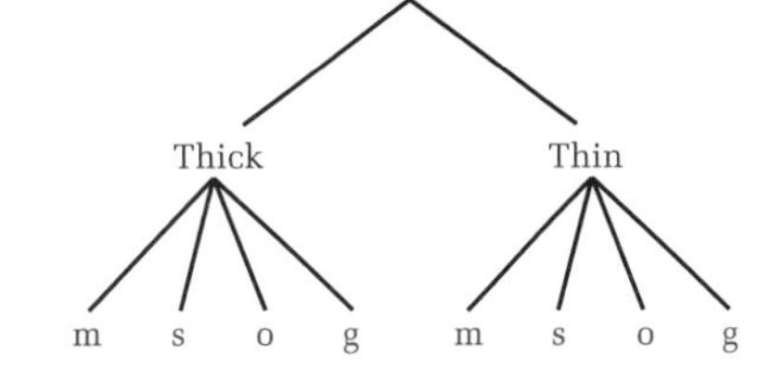

Part Two: Sample Problem

Problem *The path of a rumor can be found by recording who tells the rumor to whom. Suppose that Alice and Greg always tell each other. Bill always tells Doug and Alice. Carol always tells Fran and Greg. Helen and Doug always tell each other. Greg always tells Doug. Helen always tells Carol.*

In order to spread a rumor completely, who should be told first? Who will be the last to know?

Solution In this case we will use a ***directed graph,*** also known as a ***digraph.*** Arrows show the directions in which information flows. Notice that Fran receives but does not pass on information and that Bill passes on information but does not receive it from any other member of the group. Therefore, the only way to ensure that Bill will hear the rumor is to tell him. By following the arrows, we can verify that the rumor will eventually reach all the others in the group. Fran will be the last to know.

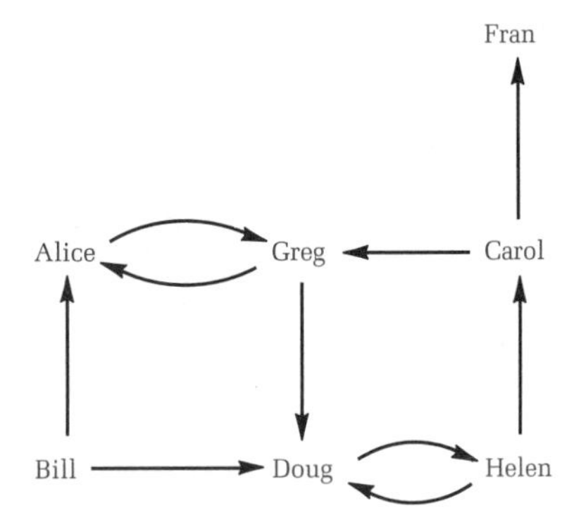

Part Three: Exercises and Problems

1 Paul owes Quincy, Ruth, and Sara $1 each. Ruth owes Quincy and Paul $2 each. Sara owes Ruth and Tom $3 each. Tom owes Ruth $4.

a Draw a digraph and label each arrow with the amount of money owed.

b If each person starts with $10, pays all that he or she owes, and receives all that is owed to him or her, how much money will each have? Paul: $9; Quincy: $13; Ruth: $14; Sara: $5; Tom: $9.

2 By graphing ordered pairs, develop a general formula for the number of handshakes exchanged by any number n of people. $\frac{n(n-1)}{2}$

Checkpoint

1 John has three shirts and five different ties. Draw a tree diagram to illustrate the various choices.

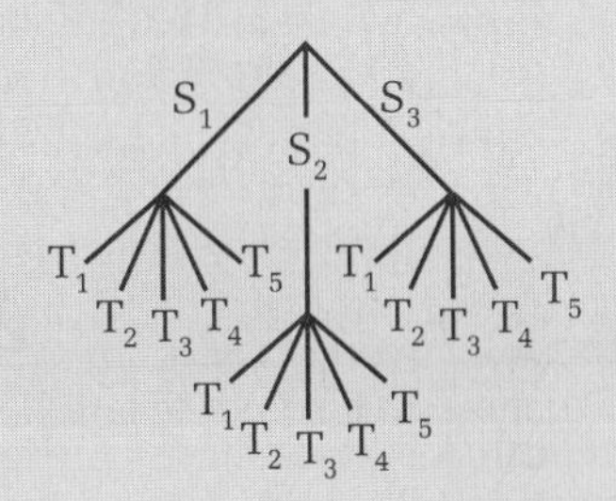

2 How would the tree diagram change if you added the fact that John also has two different colors of socks to wear? There would be 2 more branches at each end of the diagram.

3 Could this problem be solved without a tree diagram? How? Yes. Answers will vary. Any form of list can be used.

Assignment Guide

Average
1, 5
Advanced
($\frac{1}{2}$ day) 1, 5
($\frac{1}{2}$ day) Section 1.3 1, 3
Honors
($\frac{1}{2}$ day) Section 1.1 1, 3, 4, 6
($\frac{1}{2}$ day) 1, 2, 5, 6

Problem-Set Notes and Strategies

- The formula in problem **2** can be related to a story about Gauss in his younger years. It is the formula for finding the sum of the first n integers. The story may be found in "*Men of Mathematics*" by E.T.Bell.

Additional Answer

1a

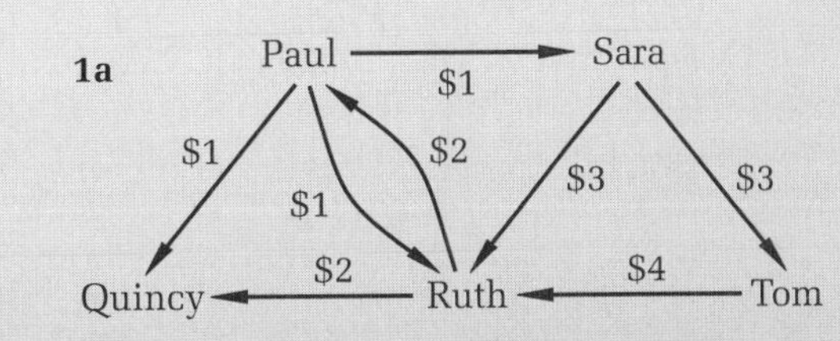

Problem-Set Notes and Strategies, Continued

- Problem **3** can be related to a real world problem such as routing phone calls over long distance lines. How can a call be routed through the shortest path? Are there other paths available if the shortest one is busy?
- Other types of real world problems include finding the shortest route for a traveling salesman or truck driver.
- Problem **6** can be nicely done by modeling the problem with the help of a digraph.

Cooperative Learning

Cooperative Learning activities are provided for most sections of the text. They also appear in reproducable form in the *Teacher's Resource Book*. Answers to all activities can be found in the *Teacher's Resource Book.*

Have students work in groups of three or four to find the 3 by 3 magic square whose rows, columns, and diagonals sum to 15. Use deductive reasoning to find the values. Only the integers from 1 to 9 are necessary.

4 9 2
3 5 7
8 1 6

■

Stumbling Block

When using digraphs, students often neglect the direction of the flow at first. Use an example such as going through a maze with both one- and two-way doors.

■

3 The Travelers planned to travel from their home in Madison, Wisconsin, to Las Vegas, Nevada. Terry Traveler researched several routes, shown in the diagram. The time estimates (in hours) are taken from a road atlas.

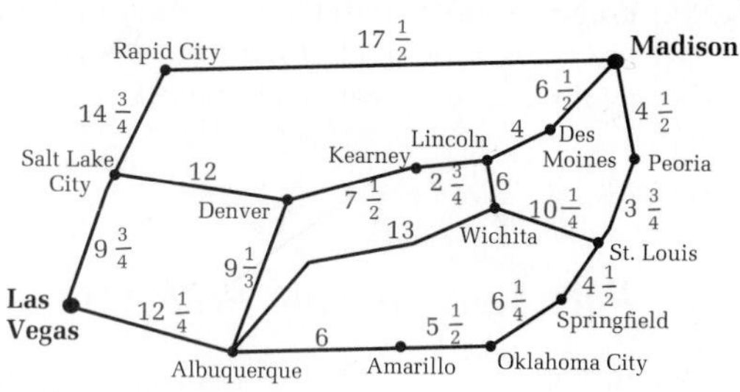

a Which is the fastest route from Madison to Las Vegas?

b What is the estimated travel time for that route? $41\frac{3}{4}$ hr

c Suppose the Travelers decide to visit relatives in Denver on the way. What is the fastest route to Las Vegas via Denver? Via Lincoln, Denver, and Albuquerque

4 Rae Sponsibal is in charge of staffing the refreshment stand for school sporting events. There will be ten such events, and Rae has a staff of five workers. She needs two persons to run the stand at each event, and she wants each person to work the same number of events. Create a working schedule for the five people.

5 A digraph is shown. A path can be identified by listing, in order, the points it passes through—thus, p_4p_3, $p_1p_2p_1$, and $p_1p_2p_5p_3$ are three possible paths.

a Find the shortest path from p_1 to p_4. $p_1p_2p_5p_4$

b Is there a path that passes through all five points? If so, find it. If not, explain why not. Yes; $p_1p_2p_5p_4p_3$

c Is there a path that passes through all five points and then returns to the starting point? If so, what is it? If not, explain why not. No; a path cannot start at p_3, and once a path reaches p_3, it cannot leave.

6 Four jobs need to be done, and four workers are available. The chart shows how long each worker takes to complete each job. If each job is to be assigned to a different worker, how should the jobs be assigned so that the total number of working hours is minimized? Job 1: Roel; Job 2: Noel; Job 3: Joel; Job 4: Zoel.

	Job 1	Job 2	Job 3	Job 4
Joel	3 hr	4 hr	2 hr	3 hr
Noel	4 hr	2 hr	3 hr	2 hr
Zoel	3 hr	3 hr	2 hr	2 hr
Roel	1 hr	3 hr	2 hr	3 hr

Additional Answers

3a via Lincoln and Wichita

Answer for problem **4** is on page **A** 1

1.3 Geometric Modeling

Objective

After studying this section, you will be familiar with
- Using geometric models in the solutions of problems

Part One: Introduction

In order to find the solution to a problem, we often relate the problem to a model. In Section 1.2, we used graphing to model problems. A geometric diagram is another useful type of model.

For Professor Hill's problem, imagine that the 25 people are 25 points on a circle and think of each handshake as a chord of the circle. Twenty-four chords end at each of the 25 points, so the number of chords might seem to be 25(24), or 600. However, since each chord has 2 endpoints, we have counted each chord twice. The number of chords is really $\frac{1}{2}(24 \cdot 25)$, or 300.

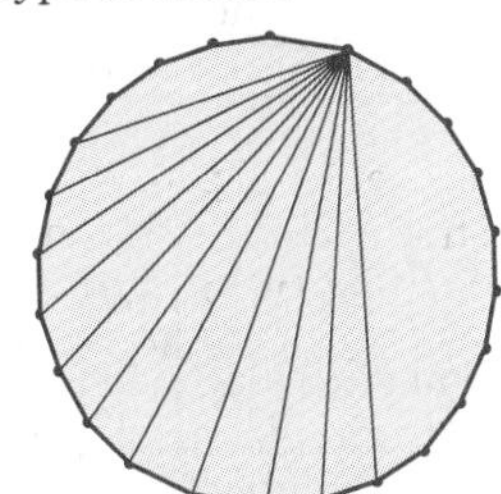

Another geometric model of the handshake problem is a polygon with 25 vertices connected by diagonals. The vertices represent the people. Each of the sides and each diagonal of the polygon represents a handshake.

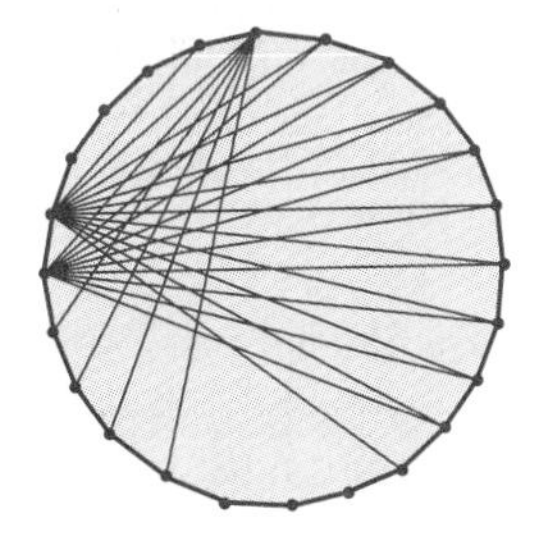

To use this model, we apply a formula from geometry—the number of diagonals in a polygon with n sides is equal to $\frac{n(n-3)}{2}$. Therefore,

$$\text{Total number of segments (handshakes)} = \text{number of diagonals} + \text{number of sides}$$
$$= \frac{n(n-3)}{2} + n$$
$$= \frac{n^2 - 3n + 2n}{2}$$
$$= \frac{n^2 - n}{2}$$

In our problem $n = 25$, so the number of handshakes is $\frac{25^2 - 25}{2}$, or 300.

Vocabulary
network

Class Planning

Time Schedule
Average: $\frac{1}{2}$ day
Advanced: 1 day
Honors: $\frac{1}{2}$ day

Resource References
Teacher's Resource Book
Class Opener 3
Cooperative Learning 1
Practice 3
Transparency 2
Evaluation
Quiz Forms 1, 2, 3

Class Opener

In an 8 team basketball league, each team must play each other team exactly once.

a How many games must be scheduled? The league plays $\frac{8 \cdot 7}{2} = 28$ games.

b How many weeks are needed to complete the schedule if each team plays one game a week? If the league plays 4 games per week, it will take 7 weeks. Here is one possible schedule.

	A	B	C	D	E	F	G	H
A		1	2	3	4	5	6	7
B	1		3	4	5	6	7	2
C	2	3		5	6	7	1	4
D	3	4	5		7	1	2	6
E	4	5	6	7		2	3	1
F	5	6	7	1	2		4	3
G	6	7	1	2	3	4		5
H	7	2	4	6	1	3	5	

By modifying the parameters, schedule making can lead to projects such as students making their own schedules, writing a computer program, reporting on schedules for professional sports, etc.

Lesson Notes

- Geometric models are frequently a key to solving a problem. Encourage students to draw illustrations whenever possible to model the situation. This lesson discusses the interpretation of geometric models and how several different geometric models may be drawn for the same situation.

Checkpoint

1. How many diagonals are there in an octogon (8 sides)? 20
2. An icosagon (20 sides) has how many diagonals? 170
3. If there were 35 students in a class, how many handshakes would be required if everyone were to shake everyone elses hand? 595

Assignment Guide

Average

($\frac{1}{2}$ day) 1, 4, 5

($\frac{1}{2}$ day) Section 1.4 1

Advanced

Day 1 ($\frac{1}{2}$ day) Section 1.2 1, 5
($\frac{1}{2}$ day) 1, 3

Day 2 ($\frac{1}{2}$ day) 4–6
($\frac{1}{2}$ day) Section 1.4 1, 3, 6

Honors

($\frac{1}{2}$ day) 3–5

($\frac{1}{2}$ day) Section 1.4 1, 3, 4–7

Part Two: Sample Problem

Problem *Can a person walk across each of the seven bridges shown in the diagram without crossing over any bridge twice? (This is a famous eighteenth-century problem concerning seven bridges in the city of Königsberg, East Prussia, now Kaliningrad, U.S.S.R.)*

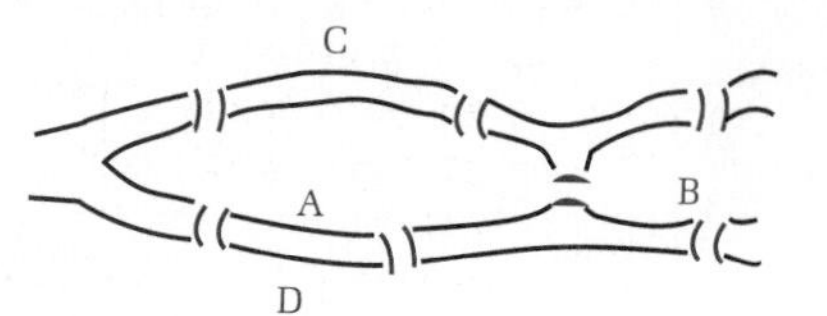

Solution We can create a geometric model known as a ***network.*** Each segment represents a bridge, and each lettered point represents a land region. The problem is to trace the entire network without lifting your pencil from the page and without tracing any segment twice. By starting at each point in turn and trying possible paths, you will find that this is impossible.

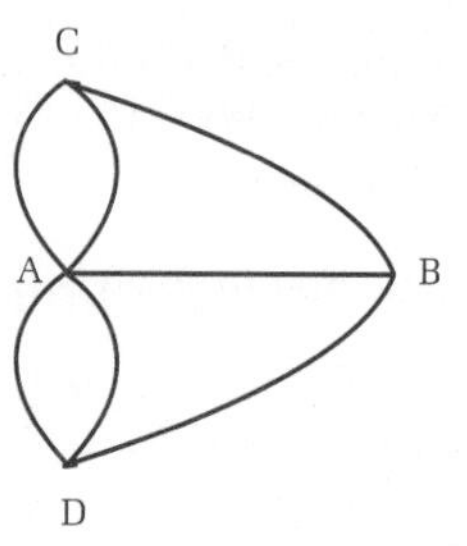

Part Three: Exercises and Problems

1 Find the number of diagonals in a polygon with 25 sides. 275

2 Another general solution to the handshake problem is suggested by the diagrams. (A general solution is one that will work with any number of handshakes.) Use the diagrams to develop a general formula for the number of handshakes. $\frac{n(n-1)}{2}$

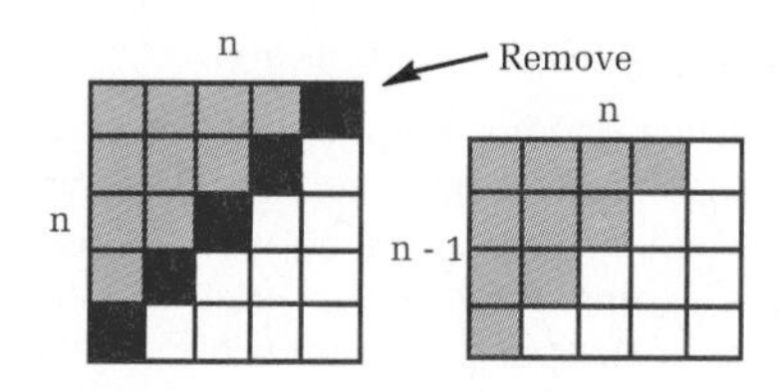

3 Six people are divided into two three-person teams for a singles tennis tournament.

a How many games must be played for each of the people to play each of the others? 15

b Suppose each person on the first team is married to a person on the second team. How many games will be played if spouses may not play each other? 12

c How many games will be played if members of the same team may not play each other? 9

4 Four numbers are placed in squares. The sum of the numbers in the first row is 6. The sum of the numbers in the second row is 12. The sum of the first column is 10, and the sum in the second column is 8. What might the four numbers be? Any numbers of the form a, $6 - a$, $10 - a$, $2 + a$

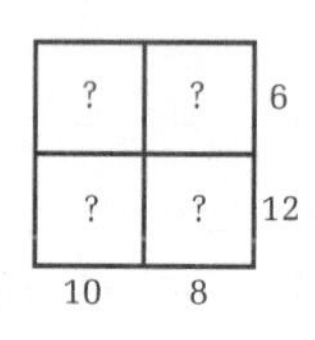

5 See whether you can draw each of the networks shown without lifting your pencil from the paper. You must not go over any of the segments more than once or miss any segment. Try to think of a strategy that you can use to draw such diagrams. Three of these cannot be drawn. Why not? How might you tell just by looking at the network whether it can be completely traced? (Hint: Try counting the segments that meet at each point of intersection. Where must a successful path begin?)

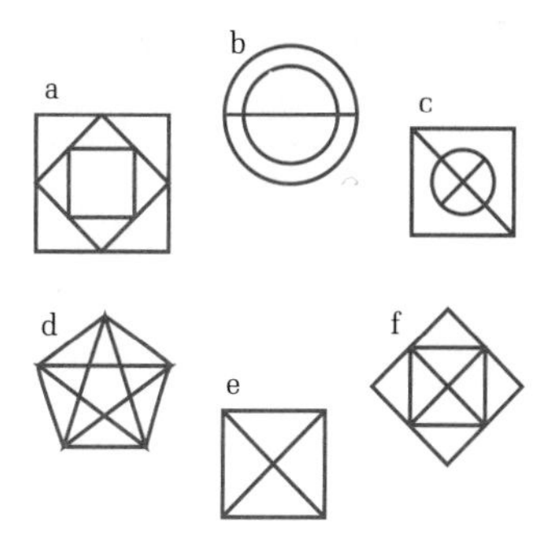

6 Ten baseball teams are in a play-off. How many games will they play if

a Each team plays every other team once? 45
b Each team plays every other team twice? 90

7 In the network shown, arrows indicate the directions of travel from point to point. List the number of ways in which it is possible to go from A to each indicated point.

a B 1 **b** D 1 **c** E 1
d G 1 **e** H 2 **f** I 3
g C 1 **h** F 1 **i** J 3
j K 1 **k** L 10

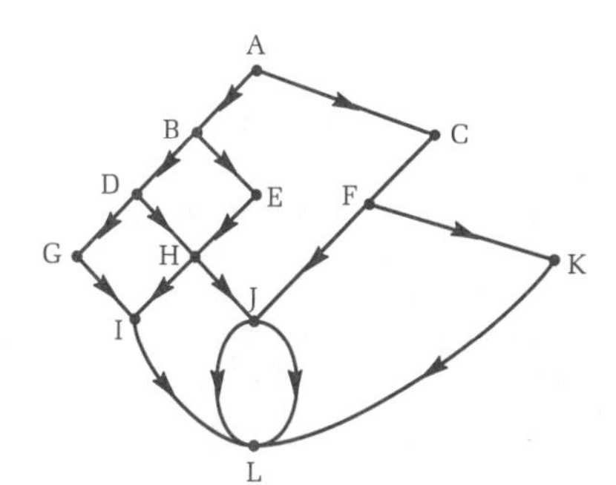

8 a Show how to place 8 checkers on an 8-by-8 checkerboard so that each row, column, and main diagonal contains exactly 1 checker.
b Show how to place 16 checkers on a checkerboard so that exactly 2 checkers will be in each row, column, and main diagonal.

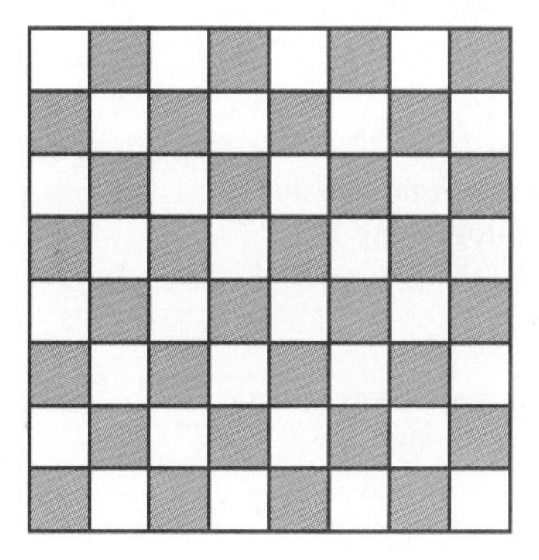

Problem-Set Notes and Strategies

- Once again, encourage your students to draw pictures whenever possible. For counting problems, a listing of all the possibilities is frequently the simplest solution.
- Exercise **5** is a good exercise to assign for students to work on in groups. Encourage students to trace the pictures.
- You might consider providing copies of the checkerboard and checkers or other markers for exercise **8** to encourage trial and error.

Additional Answers

5 Only **a**, **b**, and **d** can be drawn. Any network in which more than two vertices have an odd number of segments leading into them cannot be traced without lifting one's pencil or tracing a segment twice.

Possible answers:

8a

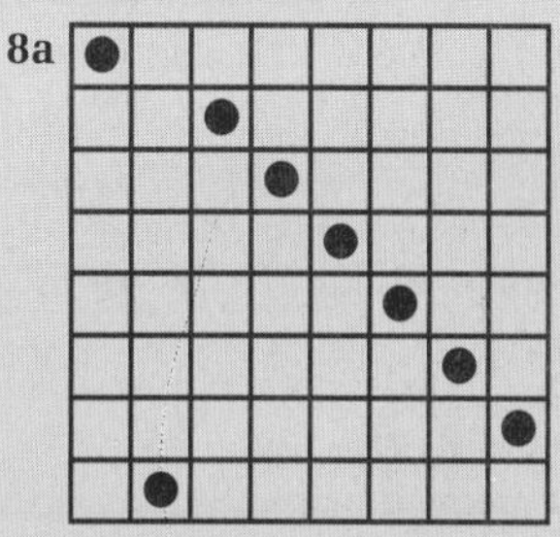

8b

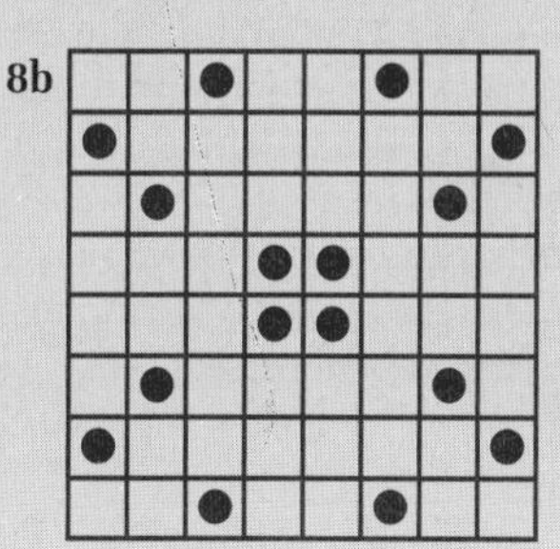

1.4

Class Planning

Time Schedule

Average: $2\frac{1}{2}$ days
Advanced: 1 day
Honors: $\frac{1}{2}$ day

Resource References

Teacher's Resource Book
Class Opener 4
Cooperative Learning 2
Practice 4
Enrichment 2

Class Opener

a How many squares are there on a standard checkerboard? (Hint: the answer is not 64. You might look at simpler cases and look for a pattern.)
Looking at simpler cases...

Dimension	Number of squares
1 x 1	1
2 x 2	4 + 1 = 5
3 x 3	9 + 4 + 1 = 14
M	
8 x 8	64 + 49 + 36 + 25 + 16 + 9 + 4 + 1 = 204

b How many squares in an $n \times n$ array? For an $n \times n$ array, $1^2+2^2+3^2+4^2+\ldots+(n-1)^2+n^2$

Lesson Notes

- This will be the first encounter with sigma (Σ) notation for most of your students. The values above and below the Σ are the starting and stopping values. Summations can be related to loops in programming.

1.4 Finding a Pattern

Objectives

After studying this section, you will be familiar with
- How examining simpler cases can lead to a solution of a problem
- How a computer can be used to simulate a problem

Part One: Introduction

Looking at Simpler Cases

When a problem involves a complicated situation, as the handshake problem does, a common problem-solving strategy is to look at simpler cases, find a pattern, and then apply that pattern to solving the original problem.

Let's look at the handshake problem for two, three, four, or more people.

With two people, there is one handshake.
Adding a third person adds two handshakes.
Adding a fourth person adds three handshakes.
If a fifth person is added, there will be four more handshakes.
If a sixth person is added, there will be five more handshakes.
.
.
.
If an nth person is added, there will be $n - 1$ more handshakes.

Thus, for n people,

$$\text{Total handshakes} = 1 + 2 + 3 + 4 + \ldots + (n - 1)$$

We can write this sum more briefly in ***sigma notation*** (so called because the Greek letter sigma, Σ, is used to stand for "sum"):

$$\sum_{k=2}^{n}(k - 1)$$

This expression means "the sum of all the numbers represented by $k - 1$ when k is each whole number from 2 to n inclusively." It is a general expression for the number of handshakes exchanged by any number of people.

Vocabulary

sigma notation
algorithm

For Professor Hill's 25 students the number of handshakes is

$$\sum_{k=2}^{25}(k-1)$$
$$= (2-1) + (3-1) + (4-1) + (5-1) + \ldots + (25-1)$$
$$= 1 + 2 + 3 + 4 + \ldots + 24$$
$$= 300$$

Computer Simulation

Many of our solutions of Professor Hill's problem have involved repetitious work. This sort of work is just right for a computer. A computer can carry out a large number of operations in a very short time and can generate a large set of data so that patterns can be observed and analyzed.

By itself, a computer does not know how to solve problems. We need to develop a step-by-step process, called an ***algorithm***, that the computer can use to solve the problem we give it. The computer can then repeat the process as many times as we want for whatever input we choose.

The BASIC program that follows uses the system that we called Method 1 (see page 4). If you have access to a computer, enter the program and run it for N = 6, N = 25, and N = 50. (Before running it for the 50-person problem, delete line 70 so that time is not spent waiting for the computer to print every handshake.)

```
 10 REM -----------------THE HANDSHAKE PROBLEM-----------------
 11 REM N = Number of people shaking hands
 12 REM P1 = First person involved in handshake
 13 REM P2 = Second person involved in handshake
 14 REM NS = Total number of handshakes exchanged by all people
 15 REM ------------------------------------------------------
 20 PRINT "Enter the number of people shaking hands==>";
 30 INPUT " ", N
 40 LET NS = 0
 50 FOR P1 =1 to N - 1
 60 FOR P2 = P1 + 1 TO N
 70 PRINT P1; "-- shakes hands with --"; P2
 80 LET NS = NS + 1
 90 NEXT P2
100 NEXT P1
120 PRINT "The total number of handshakes is: "; NS
130 END
```

How is the algorithm used in this program similar to the sigma-notation expression for the general handshake problem?

Lesson Notes, continued

- Algorithms are also introduced in this section. Have students practice the step by step algorithmic process in their thinking skills as if they were explaining it to a person with no mathematical knowledge.

Using Computers

Suggest that your students use the BASIC computer program on page 13 of the text. Discuss the structure of the program and how it counts all possible solutions. If you have students with experience in programming, you might have them help explain how the program works.

Checkpoint

1 Find the value of

$$\sum_{k=3}^{7}(3k+2) \quad 85$$

2 Write the following in sigma notation: 5, 9, 13, 17, 21, 25. One possible solution is

$$\sum_{k=1}^{6}(4k+1)$$

Communicating Mathematics

Have the students write a paragraph describing the importance of being able to recognize and follow patterns in our daily lives. Include a personal example that illustrates how patterns affect our lives.

1.4

Assignment Guide

Average

Day 1 ($\frac{1}{2}$ day) Section 1.3 1, 4, 5
($\frac{1}{2}$ day) 1

Day 2 2, 3, 5, 7

Day 3 4, 6

Advanced

Day 1 ($\frac{1}{2}$ day) Section 1.3 4–6
($\frac{1}{2}$ day) 1, 3, 6

Day 2 ($\frac{1}{2}$ day) 2, 4, 5, 7
($\frac{1}{2}$ day) Section 1.5 1, 2

Honors

($\frac{1}{2}$ day) Section 1.3 3–5
($\frac{1}{2}$ day) 1, 3, 4–7

Problem-Set Notes and Strategies

- The recognition of patterns in problem **1** may be difficult for some students at first. They may find it easy to get the next two numbers, but the concept of finding a general term can be more difficult to carry out.
- Problem **3** utilizes the pigeonhole principle. If you have $n + 1$ pigeons to put into n birdhouses, at least one of the birdhouses will receive more than one pigeon. This principle is used on many counting problems. Exercise **5** may be formulated as a pigeonhole problem.

Part Two: Sample Problem

Problem *Find the value of* $\sum_{k=1}^{100}(2k - 1)$.

Solution Adding 100 numbers would take a long time. Let's look at some simpler cases and try to find a pattern.

$$\sum_{k=1}^{1}(2k - 1) = (2 \cdot 1 - 1) = 1$$

$$\sum_{k=1}^{2}(2k - 1) = (2 \cdot 1 - 1) + (2 \cdot 2 - 1) = 4$$

$$\sum_{k=1}^{3}(2k - 1) = (2 \cdot 1 - 1) + (2 \cdot 2 - 1) + (2 \cdot 3 - 1) = 9$$

$$\sum_{k=1}^{4}(2k - 1) = (2 \cdot 1 - 1) + (2 \cdot 2 - 1) + (2 \cdot 3 - 1) + (2 \cdot 4 - 1) = 16$$

Notice that each value is a perfect square—the first is 1^2, the second is 2^2, and so forth. We might conclude that the answer to the original problem is 100^2, or 10,000. At this point, we are only making a reasonable guess. Later in this book, you will learn how to prove that the guess is in fact correct.

Part Three: Exercises and Problems

1 Find the next two numbers in each sequence and tell what pattern you used to find them.

a 1, 3, 5, 7, 9, 11, . . . 13, 15; consecutive odd numbers
b 1, 4, 7, 10, 13, 16, . . . 19, 22; nth term $= 3n - 2$
c 0, 3, 8, 15, 24, 35, 48, 63, 80, . . . 99, 120; nth term $= n^2 - 1$
d 1, 2, 4, 7, 11, 16, 22, . . . 29, 37; nth term $= \frac{n^2 - n}{2} + 1$
e 1, 2, 4, 8, 16, 32, 64, 128, . . . 256, 512; nth term $= 2^{n-1}$
f 1, 1, 2, 3, 5, 8, 13, 21, 34, 55, . . . 89, 144; each term (after first 2) is sum of 2 preceding terms.

2 Find the first five terms in each sequence described.

a The first term is 7. Each term is 3 less than twice the previous term. 7, 11, 19, 35, 67
b The nth term is equal to $2 \cdot 2^{n+3}$. 32, 64, 128, 256, 512

3 In any group of 367 people you will find at least 2 people with the same birthday. Explain why this is true.

4 Find each sum. **a** $\sum_{k=3}^{11}(2k - 5)$ 81 **b** $\sum_{k=1}^{6}k^2$ 91

5 Big Foot has ten pairs of socks in his drawer. Each pair is a different color. The socks are not paired, and the light bulb in the room is out. How many socks does Mr. Foot need to pull out of the drawer to be sure that at least two of the socks match? 11

Additional Answer

3 Since a year has at most 366 days, in a group of 367 people at least two must have a common birthday.

6 Find the next two rows in the following pattern.

```
          1
        1   1
      1   2   1
    1   3   3   1
  1   4   6   4   1
1   5  10  10   5   1
```

1, 6, 15, 20, 15, 6, 1;
1, 7, 21, 35, 35, 21, 7, 1

7 The set {1, 3, 6, 10, 15, 21, 28, 36, 45, 55, 66, . . .} is called the set of *triangular numbers.*

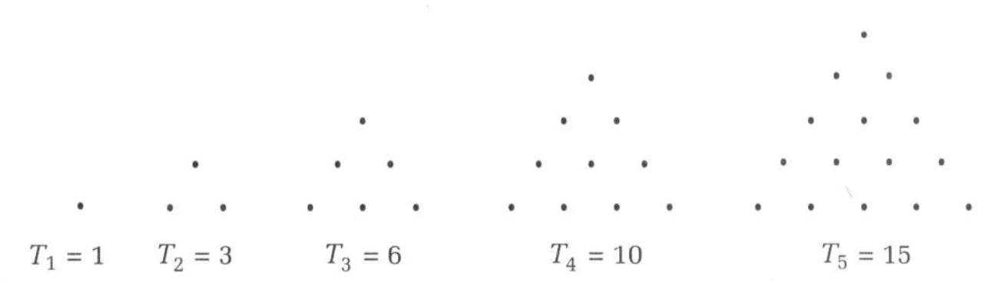

a Find the sums $T_1 + T_2$, $T_2 + T_3$, $T_3 + T_4$, and $T_4 + T_5$ and look for a pattern. 4; 9; 16; 25; the sums are consecutive perfect squares.

b Make a list and a diagram of the first five *square numbers.*

8 In algebra, the equation x = x + 1 has no solutions. The BASIC program in this section has the line

```
80 LET NS = NS + 1
```

What does line 80 do? How is the meaning of this statement different from that of the algebraic equation? It changes the value stored in the variable NS; whereas the equation is a statement of equality, the computer line is a command.

9 A fiddlestick factory opened in January 1987. The graph shows the relationship between monthly profits and the number of sticks produced. Use the graph to answer the following.

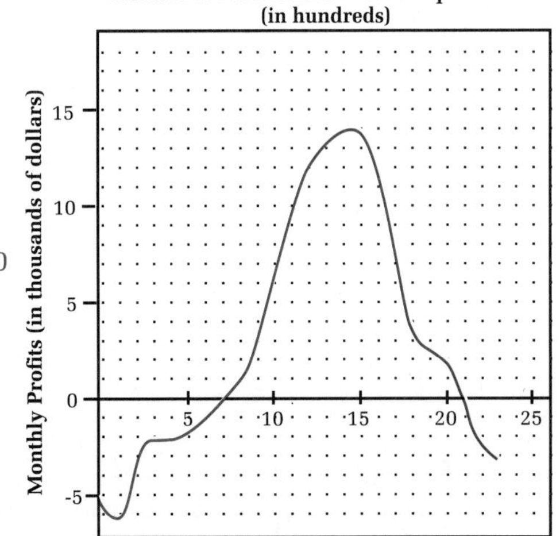

a What are the profits when 900 sticks are produced? $3000

b How many sticks must be produced for the profit to be $12,000? 1200 or 1600

c What are the profits when 400 sticks are produced? What does this result mean and how might you account for it?

d How many sticks must be produced in order for the factory to break even?

e How many must be produced in order for the profits to exceed $6,000?

f Why might the profits begin to decline at some point and become negative when more than 1600 fiddlesticks are produced?

g If you were the manager of the factory, how many sticks would you produce? Why?

Problem-Set Notes and Strategies, continued

- Problem **6** generates the numbers of Pascal's triangle. The next section includes a Pascal program to solve the handshake problem.

Cooperative Learning

Problem **7** is a problem ideally suited for a group project. As well as drawing pictures for part **b**, have the students try to recognize a pattern and find a sigma notation that will generate that pattern. See how many equivalent sigma notations they can find for the same pattern.

Stumbling Block

Students frequently have difficulty relating ordered pairs to real-life situations. Have the students write out the ordered pair in words or use a t-diagram with the real-world labels rather than *x* and *y*.

Additional Answers

9c –$2000 or $2000 is lost. The factory could be just starting and not be efficient.

d 700 or 2100

e More than 10,000.

f Answers will vary. Perhaps the factory is unable to fulfill orders quickly.

g Answers will vary. One answer is 1600.

Additional Answer

7b

1 4 9 16 25

Class Planning

Time Schedule

Average: Optional
Advanced: $\frac{1}{2}$ day
Honors: $\frac{1}{2}$ day

Resource References

Teacher's Resource Book
Class Opener 5
Cooperative Learning 3
Practice 5
Manipulatives and Applications 2

Class Opener

How many rectangles are in a standard checkerboard?
One way to organize a solution is to look at the number of rectangles by rows.

	8 horizontal	
1 x *n*	1 x 1	8
	1 x 2	7
	1 x 3	6
	1 x 4	5
	.	
	.	
	.	
	1 x 8	1
		36

There are 8 rows of 36 = 288.
2 x *n* 7 rows 36
3 x *n* 6 rows of 36
$(8 + 7 + 6 + 5 + 4 + 3 + 2 + 1)36 = 36^2 = 1296$
In general,
$(1 + 2 + \ldots + n)^2$ or $\left(\frac{n(n+1)}{2}\right)^2$.

Lesson Notes

- Recursion is a powerful concept. Most students accept the idea and become comfortable with the technique. It may help to enter the Pascal program provided in the text ahead of class and run it for your students. A

1.5 A Recursive Solution

Objective

After studying this section, you will be familiar with
- Using a recursive pattern to solve a problem

Part One: Introduction

What happens to the total number of handshakes whenever a new person is added to the group?

When a person enters a room containing ten people who have already exchanged handshakes, he or she must shake hands with each of the ten people, increasing the total number of handshakes by ten. Therefore, using H_n to represent the number of handshakes exchanged by n people, we can say that $H_{11} = H_{10} + 10$. In the same way,

$$H_{10} = H_9 + 9$$
$$H_9 = H_8 + 8$$
$$H_8 = H_7 + 7$$
$$H_7 = H_6 + 6$$
$$\vdots$$

We know that for two people there is only one handshake. Using this information and reversing the pattern, we can calculate the number of handshakes exchanged by any larger group of people.

$$H_2 = 1$$
$$H_3 = H_2 + 2 = 3$$
$$H_4 = H_3 + 3 = 6$$
$$H_5 = H_4 + 4 = 10$$
$$\vdots$$
$$H_n = H_{n-1} + (n - 1)$$

A process of this sort, in which each result depends on the preceding result, is called ***recursive.*** We can use a recursive technique to solve a problem whenever we can see a definite relationship between successive results in a series of operations.

The following Pascal program simulates the recursive solution of the handshake problem. If you have access to Pascal, enter the program on a computer and run it for N = 6, N = 25, and N = 50.

Vocabulary

recursive

```
{===================The Handshake Problem====================}

{Recursive solution in Pascal
 The number of handshakes for N people is equal to the
 number of handshakes for the Nth person plus the number
 of handshakes for all the rest of the N - 1 people.}

{===========================================================}

  Program SolveHandshakes;
   Var NumberPeople { Number of people shaking hands }
   TotalShakes: Integer; { Total number of handshakes }

   Function NumberHandshakes (NthPerson: Integer) : Integer;
     Var Person: Integer; { Person shaking hands with Nth person }
   Begin { Function NumberHandshakes }
     If NthPerson = 1 Then NumberHandshakes := 0
     Else begin { Nontrivial Case }
       For Person := 1 To NthPerson - 1 Do
       Writeln (NthPerson:5; ' --shakes hands with--',Person:5);
       NumberHandshakes := (NthPerson - 1) +
                     NumberHandshakes(NthPerson - 1)
     End { Nontrivial Case }
  End; { Function NumberHandshakes }

{-----------------------------------------------------------}

  Begin { Main Program }
   Write ('Enter the number of people ==> ');
   Readln (NumberPeople);
   Writeln;
   TotalShakes := NumberHandshakes(NumberPeople);
   Writeln;
   Writeln ('The total number of handshakes is', TotalShakes : 5)
  End. { Main Program }
```

Compare the algorithm used in this program with Method 2 of organizing handshakes, described on page 4.

Part Two: Sample Problem

Problem *Write a recursive formula that can be used to find the nth term of the sequence 1, 3, 4, 7, 11, 18, . . . , n.*

Solution Notice that, except for the first two numbers, each number in the sequence is the sum of the two preceding numbers. If T_n represents the nth term of the sequence and we know that $T_1 = 1$ and $T_2 = 3$, we can use the formula $T_n = T_{n-1} + T_{n-2}$ to find any subsequent term. (This sequence is known as the Lucas sequence and is related in interesting ways to the famous Fibonacci sequence, 1, 1, 2, 3, 5, 8, 13, . . .)

Lesson Notes, continued

trace option on your computer allows execution line by line and makes it easier to observe what is happening.

Using Computers

Suggest that your students use the Pascal program on page 17 of the text. If any of your students are studying Pascal, you may want to have them explain how the program finds all possible handshakes. This is a meaningful example of how some computer languages such as Pascal support recursive mathematical definitions whereas others such as BASIC do not.

Checkpoint

1. Generate the first five terms of the sequence that has the recursive definition $T_1 = 2$, $T_n = 3T_{n-1}$ 2, 6, 18, 54, 162
2. Find a recursive formula that will generate the sequence 1, 2, 6, 24, 120, 720, ... $T_1=1$, $T_n=n(T_{n-1})$

Assignment Guide

Average

Optional

Advanced

($\frac{1}{2}$ day) Section 1.4 2, 4, 5, 7

($\frac{1}{2}$ day) 1, 2

Honors

($\frac{1}{2}$ day) 1, 2, 4

($\frac{1}{2}$ day) Section 1.6 1, 3, 6, 7, 10, 13

1.5

Problem-Set Notes and Strategies

- Generating recursive formulas is difficult the first time for some students. Practice over time helps clarify the ideas.
- Problem **5** briefly introduces a form of the fundamental counting principle introduced in **12.1**. It also uses the addition rule since there are two possible paths and they are mutually exclusive of one another. Suggest the students model the situation with 10 toys.

Cooperative Learning

Problems **2** and **3** lend themselves to a group working environment. Have students individually write a paragraph for each and then form groups of three or four and compare notes to create an efficient algorithm. Each student should read his algorithm and the benefits of each should be explored to produce a single algorithm from the group.

Communicating Mathematics

If you are familiar with the "*Tower of Hanoi*" puzzle, this is an excellent opportunity for students to discover recursive operations. Use as many discs as people in the group and have the students solve the puzzle with each student assigned one specific size disc to move. Keep track of the number of moves each student makes and ask them to generate a general statement about the best (shortest) solution.

Part Three: Exercises and Problems

1 Draw a coordinate grid at least 36 squares high. Label the lower left-hand corner with the coordinates (0, 0). Put a dot there.
Move right 1 and up 1 and put a dot there.
Move right 1 and up 3 and put a dot there.
Move right 1 and up 5 and put a dot there.
Move right 1 and up 7 and put a dot there.
Move right 1 and up 9 and put a dot there.

a Write the height y of the nth dot in summation notation. $y = \sum_{x=1}^{n}(2x - 1)$

b Label each dot with its coordinates and observe another pattern. Write an equation that describes the relationship between x and y in each ordered pair (x, y). $y = x^2$

c Write a recursive formula that expresses the sum of the coordinates of the nth dot (S_n) in terms of the sum of the preceding dot's coordinates (S_{n-1}). $S_n = S_{n-1} + 2(n - 1)$

2 Explain how you would find all the whole-number divisors of a given positive integer. (You are creating an algorithm.)

3 Create an algorithm to find how many whole-number divisors any given positive integer has.

4 Create an algorithm to find $1 \cdot 2 \cdot 3 \cdot 4 \cdot 5 \cdot \ldots \cdot n$ recursively. (This product, symbolized $n!$, is called n factorial.) $S_1 = 1$, $S_n = n \cdot S_{n-1}$

5 The diagram represents the production lines at a toy factory.

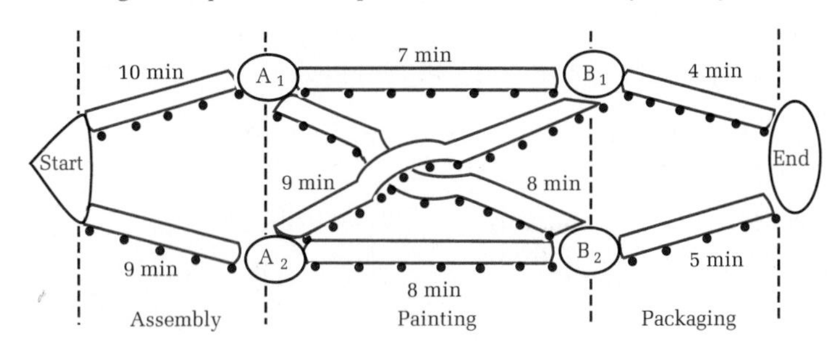

Assembly takes place on two lines. Painting is done on four lines, and packaging is done on two lines. The time it takes for a toy to pass down each line is shown.

a What is the shortest time from start to finish? 21 min

b If line A_1B_1 is shut down, what is the shortest time from start to finish? 22 min

c The parts for 100 toys are fed into the starting position at a rate of 1 toy per second. At each split in the line, half the toys go one direction, and half go the other. How long does it take until the last toy comes out at the end? At least 24 min 38 sec, at most 24 min 40 sec

d Suppose that from the start, 40% of the toys go to A_1, and from A_1, 45% go to B_1. From A_2, 30% go to B_1. What percentage of the toys reach the end by way of B_1? 36%

Additional Answers

2 Answers will vary; e.g. From 1 through the square root of x, check all whole numbers to see if they leave a remainder when divided into x.

3 Answers will vary; e.g. Use the process from problem **2** to build a list of divisors, then count the elements of the list.

1.6 DEMONSTRATIONS

Objective

After studying this section, you will be familiar with

- Some ways in which mathematical ideas are communicated

Part One: Introduction

In mathematics, we must demonstrate the mathematical thinking behind our answers. The type of demonstration depends on the situation, but the clarity of the explanation is always as important as correct mathematics.

Many kinds of formal demonstrations are used in mathematics, including two-column proofs, paragraph proofs, indirect proofs, graphical proofs, and analytic proofs. Most mathematical communication, however, takes place in a more informal way.

In this book we use a variety of informal proofs as well as some formal proofs. We will always give enough information to explain our thinking but will not always include every step. As you proceed in your study of mathematics, you will develop a sense of how much detail is necessary in a given situation. For now, work on communicating your thinking in clear, informal style.

Example *Show that* $n(n + 3) + 2 = (n + 1)(n + 2)$.

To show that the two expressions are equivalent, we can perform the indicated operations and simplify the results.

Left Side	Right Side
$= n(n + 3) + 2$	$= (n + 1)(n + 2)$
$= n^2 + 3n + 2$	$= n^2 + 2n + n + 2$
	$= n^2 + 3n + 2$

Since each side is equal to the same expression, the sides must be equal. For a fuller explanation, we would state what algebraic property was used in each step.

Class Planning

Time Schedule

Average: Optional
Advanced: 1 day
Honors: $1\frac{1}{2}$ days

Resource References

Teacher's Resource Book
Class Opener 6
Cooperative Learning 4
Practice 6
Evaluation
Test Forms 1, 2, 3

Class Opener

The even and the odd intgers can be divided into two sets with no intersection, called disjoint sets. Which set of integers contain the answer to each of the following operations ?

a Add two even integers. Even
b Add two odd integers. Even
c Add an even and an odd integer. Odd
d Multiply two even integers. Even
e Multiply two odd integers. Odd
f Multiply an odd and an even integer. Even

Explain how you arrived at your answer. Prove one of your answers.

Answers will vary. A possible proof follows.

$2n + 2k = 2(n + k)$
$2n + 1 + 2k + 1 = 2(n + k + 1)$
$2n + 1 + 2k = 2(n + k) + 1$
$2n \cdot 2k = 4nk = 2(2nk)$
$(2n + 1)(2k + 1) = 2n \cdot 2k + 2k + 2n + 1 = 2(2nk + k + n) + 1$
$2n(2k + 1) = 2[n(2k + 1)]$

Lesson Notes

- This section emphasizes communicating results to others. Various methods may be used to communicate results. Each has its merits and several methods will be explored in the text. This section focuses on concise, informal methods.

Checkpoint

1. Find all numbers for which three times the number is the same as the number plus 6. 3
2. For any positive integer n, how would you describe all the numbers $4n - 2$? Every other even integer.
3. For any positive integer n, how would you describe all the numbers $6n - 5$? Every third odd integer.

Assignment Guide

Average

Optional

Advanced

1, 2, 3, 6, 7, 10

Honors

Day 1 ($\frac{1}{2}$ day) Section 1.5 1, 2, 4

($\frac{1}{2}$ day) 1, 3, 6, 7, 10, 13

Day 2 4, 5, 8, 11, 14

Problem-Set Notes and Strategies

- In problem **1a**, consider an alternate definition of *even*, different than divisible by two.
- Encourage students to use an algebraic process to solve problem **2** to insure all answers are found.

Part Two: Sample Problems

Problem 1 *Prove that if* $x > 0$, *then* $(1 + x)^2 > 1$.

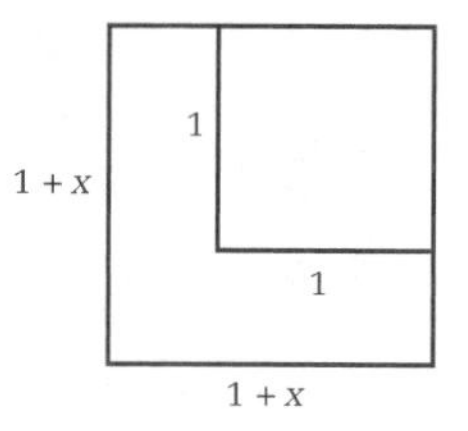

Solution We can use a geometric model. Each side of the large square has a length of $1 + x$. The length of each side of the small square is 1. The large square has an area of $(1 + x)^2$. The small square has an area of 1. The large square is the small square plus more area, so $(1 + x)^2 > 1$.

Problem 2 *Each indicated product of three consecutive integers is divisible by 6. Will this be true for any three consecutive integers?*

$1(2)(3) = 6$ $4(5)(6) = 120$
$2(3)(4) = 24$ $13(14)(15) = 2730$
$3(4)(5) = 60$

Solution We could try calculating all products we can think of, but we couldn't be sure that *every* such product is divisible by 6. Instead, we examine some general characteristics of the numbers in any sequence of three.

Every alternate integer is even, so in every set of three consecutive integers, at least one must be divisible by 2. Every third integer is a multiple of 3, so one of the three consecutive integers must have a factor of 3. Every sixth integer has a factor of 2 and 3. Since the product will always have a factor of 2 and a factor of 3, it will always have a factor of 6. Here is another way of stating the proof:

One of the numbers will be a multiple of 2, so we can call it $2k$.
One of the numbers will be a multiple of 3, so we can call it $3h$.
The third number can be called m.
The product will be $(2k)(3h)(m) = 6(khm)$, a multiple of 6.

Part Three: Exercises and Problems

1 **a** Define *even number*.

b Why would "Every other number is an even number" be an incomplete definition of *even number*?

2 Find all the numbers for which

a The product of the number and itself is 9 −3, 3

b The sum of the number and itself is 9 4.5

c The sum of the number and itself is the same as the product of the number and itself 0, 2

3 For any integer n, how would you describe each number?

a $2n$ Even **b** $2n + 1$ Odd **c** $2n - 1$ Odd **d** $|n|$ Nonnegative integer

Additional Answers

1a A possible answer is: The number has a factor of 2.

b If the first number is odd, then every other number is odd.

4 Explain why the opposite of a negative number is a positive number.

5 Show that when 7 points are labeled on a circle, 21 possible chords are determined.

6 How many three-digit lock combinations can be created if each digit is one of the numbers from 0 through 7? How many of these combinations are possible if no two of the digits can be the same? 512, 336

7 Solve each equation.

a $3^x = 81$ $x = 4$ **b** $\frac{x}{3} = 81$ $x = 243$ **c** $3x = 81$ $x = 27$ **d** $x^2 = 81$ $x = 9$ or $x = -9$

8 Explain in paragraph form why the product of any two even numbers is a multiple of 4. $(2m)(2n) = 4mn$

9 If 50% of the cars entering each fork in the road system take the northern route, what percentage of the cars that enter the system will exit at B? 87.5%

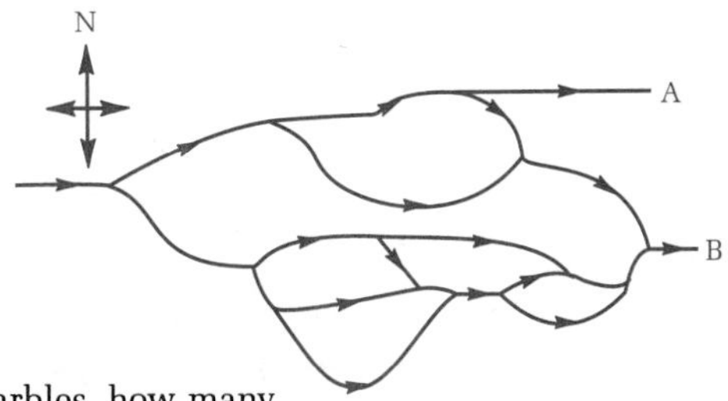

10 If Mimi has 77 marbles and Sam has 31 marbles, how many marbles should Mimi give Sam so that each will have the same number of marbles? Explain how you found the answer. 23; a possible explanation is "by solving $77 - x = 31 + x$."

11 Explain why the sum of any two negative numbers is negative.

12 Identify the pattern to find the missing numbers. $(x, y) \rightarrow |x - y|$

$(3, 5) \rightarrow 2$	$(6, 1) \rightarrow 5$
$(8, 11) \rightarrow 3$	$(8, 12) \rightarrow$? 4
$(2, 7) \rightarrow 5$	$(7, 7) \rightarrow$? 0
$(12, 9) \rightarrow 3$	$(15, 9) \rightarrow$? 6

13 Simplify $\sum_{x=1}^{6}(x + y) - \sum_{x=1}^{6}(x + 3)$. $6y - 18$

14 Suppose that a polygon is formed by connecting *lattice points* (points with integer coordinates) on a graph. Find a formula that expresses the area of the polygon as a function of *I*, the number of lattice points inside the polygon, and S, the number of lattice points on the polygon. Two examples are shown. $A = I - 1 + \frac{S}{2}$

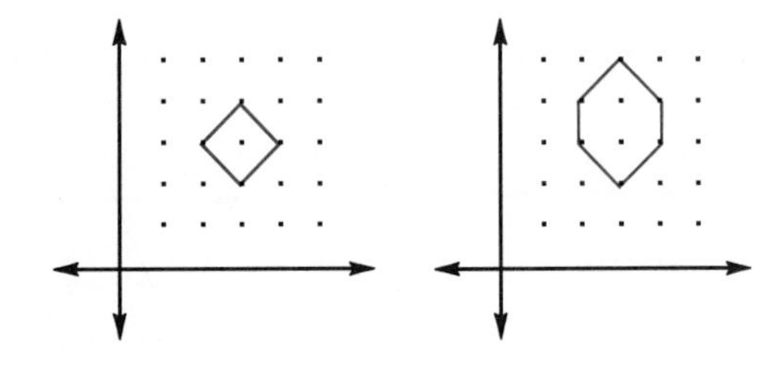

Problem-Set Notes and Strategies, continued

- Problem **7** is a good review of exponential properties. Have students write 81 as 3^4 as an aid to finding the answer.
- Some students will recognize that it is easier to find the cars that exit at **A** in problem **9** and subtract that percentage from 100%.
- In problem **13**, help students to observe and perhaps prove that $\Sigma a - \Sigma b = \Sigma(a - b)$.

Stumbling Block

In **3d**, emphasize that the absolute value is a non-negative value, not a positive value. Have students get in the habit of writing absolute value problems in two parts. This concept will be formalized in the next chapter.

Cooperative Learning

Use problem **8** as an opener for this group lesson. Have students explore the concept of even times odd, even plus even, even plus odd, odd plus odd, even squared and odd squared. This might be a good time to introduce the classic proof that $\sqrt{2}$ is an irrational number. The proof is found in most texts on the history of mathematics.

Additional Answers

4 The number would be the same distance from zero in the opposite direction.

5 The number of chords is the same as the number of shakes between 7 people, $\frac{7(7-1)}{2} = 21$.

11 A possible answer is: Adding a negative value means to add in a negative direction.

SECTION **2.1** REVIEWS some concepts from Algebra 1: The term *expression* is defined, the Distributive Property of Multiplication over Addition is reviewed, a mathematical definition of absolute value is given, and order of operations is reviewed. A common problem is that students think of "the distributive property" as distributing everything over everything. It sometimes helps prevent this error if students get in the habit of stating the full property.

The necessary properties for solving linear and quadratic equations and linear inequalities are presented and graphing on the Cartesian coordinate system is reviewed. Graphs are used to review the midpoint and distance formulas. It is important for students to understand that not every equation has a straight line graph. The availability of function graphers opens exciting possibilities for students to graph functions.

Function notation is introduced in order for students to become comfortable with it and have practice with the concepts of input and output before functions are a focus of work. The concept of function is central to Algebra 2.

Students will be presented with matrix operations and solving matrix equations.

Probability is introduced for the topic itself, as an organizing tool, and to provide a vehicle for practicing topics. Throughout the text, questions are often posed in terms of probability.

CHAPTER

2 THE LANGUAGE OF ALGEBRA

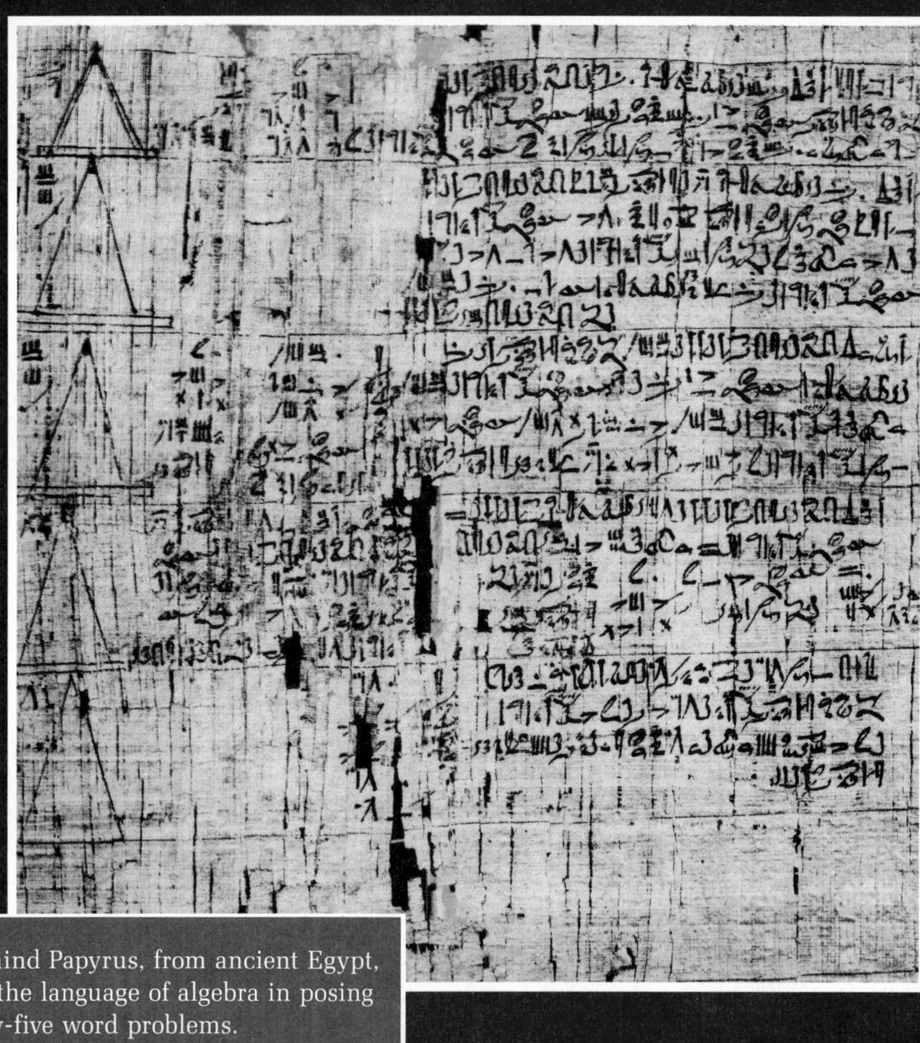

The Rhind Papyrus, from ancient Egypt, used the language of algebra in posing eighty-five word problems.

2.1

THE LANGUAGE OF EXPRESSIONS

Objectives

After studying this section, you will be able to

- Add expressions and recognize equivalent expressions
- Use the Distributive Property of Multiplication over Addition
- Calculate absolute values
- Follow the standard algebraic order of operations

Part One: Introduction

Working with Expressions

An ***expression*** is a mathematical phrase consisting of a number, a variable, or numbers and/or variables connected by operations.

These are expressions:			These are not expressions:
$3x^2 - 2 + 4xy$		x	$9x - 5 = 6$
$\frac{3x + 4}{9 - 2x}$	lwh	$\|4 - 9\|$	$x > 3$

The general rule for adding expressions is that only ***like terms*** may be combined. ***Like terms*** consist of exactly the same variables raised to the same powers.

$$(4x + 54y + 9) + (2y + x^2 + x - 9)$$

To simplify the sum, group and combine the like terms

$$= x^2 + (4x + x) + (54y + 2y) + (9 - 9)$$
$$= x^2 + 5x + 56y$$

The original sum and its simplified form are ***equivalent expressions,*** since they have the same value for all values of x and y.

In the diagram, the area can be represented as the length times the width, $x(x + 4)$, or as the sum of the two smaller rectangles, $x^2 + 4x$. Since the two expressions represent the same area, they must be equivalent.

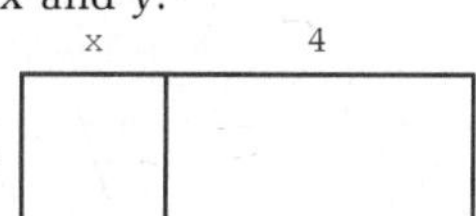

2.1

Class Planning

Time Schedule

All levels: 1 day

Resource References

Teacher's Resource Book
Class Opener 7
Practice 7
Manipulatives and Applications 3

Class Opener

Evaluate $5 + 3x + 5x^2$ if $x = 12$.
$5 + 3 \cdot 12 + 5 \cdot 144 =$
$5 + 36 + 720 = 761$
Student answers will vary, opening a discussion of the need for consistency. Remind students of the conventions we use in algebra.

Lesson Notes

- This chapter reviews the language that will be encountered in Algebra. In this section we define expressions and how to simplify them. Terms and like terms will be defined and students begin the process of understanding the subtle variations of what makes an expression a group of terms and what makes the same expression a factor.

Vocabulary

expression
like terms
equivalent expressions
distribute
opposite
additive inverse
absolute value

Lesson Notes, continued

- The distributive property will be explained both as a method of simplifying expressions and as a method of factoring.

- Absolute values are introduced in an algebraic context as opposed to a geometric approach since the concept of distance only allows specific types of problems in absolute value to be solved. Students should be able to relate the geometric and algebraic definitions to various problems.

The Distributive Property of Multiplication over Addition

The statement $x(x + 4) = x^2 + 4x$ illustrates an important property. To find the product $x(x + 4)$, we ***distribute*** the x over $x + 4$. We are allowed to do so by the Distributive Property of Multiplication over Addition.

The Distributive Property of Multiplication over Addition

If a, b, and c are real numbers, then $a(b + c) = ab + ac$ and $(b + c)a = ba + ca$.

This property can also show how expressions are subtracted. Remember that two numbers with a sum of zero—like 9 and -9, $-\frac{3}{4}$ and $\frac{3}{4}$, and x and $-x$ are known as ***opposites***, or ***additive inverses***, and that subtraction is defined as the inverse of addition.

If a and b are real numbers, $a - b = a + (-b)$.

We subtract an expression by adding its opposite. An opposite can be found by distributing -1 over each term of the expression.

Example *Simplify* $2(6x^2 - 5x + 9) - (3x^2 + 2x - 4)$.

	$2(6x^2 - 5x + 9) - (3x^2 + 2x - 4)$
Distribute and add the opposite	$= (12x^2 - 10x + 18) + (-3x^2 - 2x + 4)$
Group like terms	$= (12x^2 - 3x^2) + (-10x - 2x) + (18 + 4)$
Combine like terms	$= 9x^2 - 12x + 22$

Absolute Value

Mathematicians often need to refer to the numerical value of an expression without regard to its sign. This is the expression's ***absolute value.***

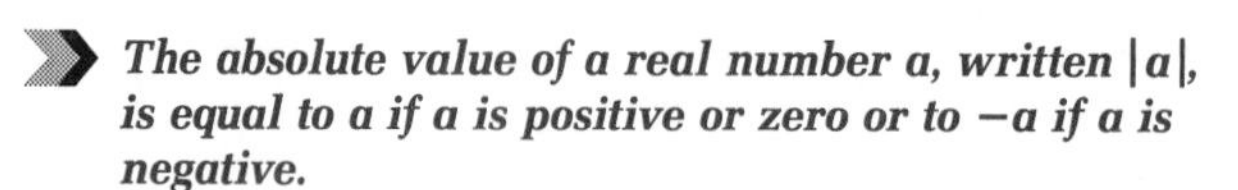

The absolute value of a real number a, written $|a|$, is equal to a if a is positive or zero or to $-a$ if a is negative.

$$|a| = \begin{cases} a \text{ when } a \geq 0 \\ -a \text{ when } a < 0 \end{cases}$$

Since $-a$ is positive when $a < 0$, an absolute value will always be a nonnegative number.

$|6| = 6$ $\quad\left|-\frac{3}{5}\right| = \frac{3}{5}$ $\quad|0| = 0$

Order of Operations

You may recall the standard algebraic order of operations.

Order of Operations in an Expression

1. **Evaluate all expressions within grouping symbols.**
2. **Evaluate all powers and roots.**
3. **Perform all multiplications and divisions from left to right.**
4. **Perform all additions and subtractions from left to right.**

Example *Perform the indicated operations.*
$-3 + 4 \cdot (2 + 7)^2 \div 2 - 5\sqrt{6 + 3}$

	$-3 + 4 \cdot (2 + 7)^2 \div 2 - 5\sqrt{6 + 3}$
Evaluate grouped expressions	$= -3 + 4 \cdot 9^2 \div 2 - 5\sqrt{9}$
Evaluate powers and roots	$= -3 + 4 \cdot 81 \div 2 - 5 \cdot 3$
Multiply and divide	$= -3 + 162 - 15$
Add and subtract	$= 144$

Most scientific calculators use the standard algebraic order of operations. The operations in the preceding example can be performed on a calculator by entering

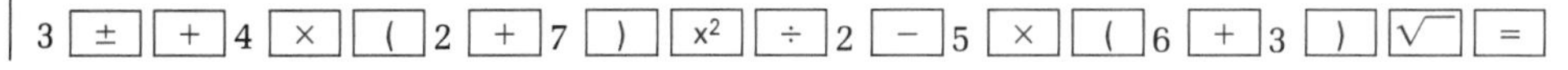

Part Two: Sample Problems

Problem 1 *Express the total area of rectangle* ABCD.

Solution We could use a formula for the total area of the rectangle.

Area = length · width
$= (x + 4)(x + 3)$

We could also express the total area as the sum of the areas of the four smaller rectangles.

Area $= x^2 + 4x + 3x + 12$
$= x^2 + 7x + 12$

Since the expressions $(x + 4)(x + 3)$ and $x^2 + 7x + 12$ express the same area, they are equivalent expressions.

Lesson Notes, continued

- Order of operations are reviewed.

Checkpoint

1. Simplify the expression $(x - 4) - (3x + 2)$. $-2x - 6$
2. Perform the indicated operations. $|3 - 7| + 4^2 - (-5)$ 25
3. Simplify the expression $(3x^2 - 3x + 4) - (x^2 + 5x + 7)$ $2x^2 - 8x - 3$

Assignment Guide

Average
12–15, 17, 18, 20, 21, 23, 25, 28, 33, 35, 36

Advanced
12–15, 17, 22, 28, 30–32, 35, 37, 41

Honors
12–17, 19, 22, 24, 26, 28, 29, 31, 32, 37, 38, 41, 43

Communicating Mathematics

Write a paragraph to explain why an order of operations is necessary and why everyone must agree to use the same order.

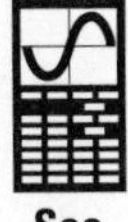

See p. 745

Problem 2 *Let* $y = 4x^2 + (3x)^2$. *Make a table showing the values of* y *when the values of* x *are* $\{-2, -1, 0, 1, 2, 3\}$.

Solution First, we simplify the expression on the right-hand side of the equation.

$$y = 4x^2 + (3x)^2$$
$$= 4x^2 + 9x^2$$
$$= 13x^2$$

Now we can substitute the given values for x in the expression $13x^2$ to find the corresponding values of y.

x	−2	−1	0	1	2	3
y	52	13	0	13	52	117

Problem 3 *Perform the indicated operations.*

$$5 + 2 \cdot 4^2 + |-3 + (-2)| + \frac{\sqrt{9 + 16} - (-3)^2}{2}$$

Solution We follow the standard algebraic order of operations.

$$5 + 2 \cdot 4^2 + |-3 + (-2)| + \frac{\sqrt{9 + 16} - (-3)^2}{2}$$
$$= 5 + 2 \cdot 16 + |-5| + \frac{\sqrt{25} - 9}{2}$$
$$= 5 + 32 + 5 + \frac{-4}{2}$$
$$= 42 - 2$$
$$= 40$$

Part Three: Exercises and Problems

Warm-up Exercises

In problems 1 and 2, simplify each expression.

P

1 $(x + 3) + (x + 5)$ $2x + 8$

2 $(x + 5) - (x + 6)$ -1

P **In problems 3 and 4, use the distributive property to rewrite each expression.**

3 $3(2x + 7)$ $6x + 21$

4 $ax + ay$ $a(x + y)$

P,I **5 a** Write two algebraic expressions to describe the total area of the figure.

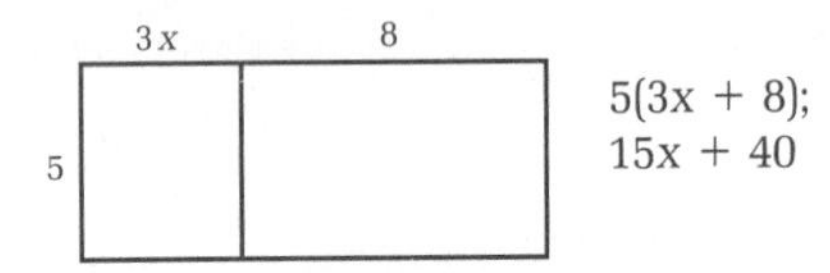

$5(3x + 8)$; $15x + 40$

b Draw an area diagram to illustrate the expression $x(4 + 2x)$. Then find a second algebraic expression that describes the same area. $2x^2 + 4x$

Stumbling Block

In problems such as **2**, students often forget to distribute the negative sign. They correctly subtract x from x, but then add 5 and 6 to get the incorrect solution 11. Suggest that students check their work by substituting 3 for x and simplifying in two ways: substitute before simplifying and after simplifying.

P | **In problems 6–11, perform the indicated operations.**

6 $5 + 3^2 - (4 - 3 \cdot 7)$ 31 **7** $(5 \cdot 2)^2 - 5 \cdot 2^2$ 80 **8** $\sqrt{3^2 + 4^2}$ 5

9 $|8 - 19|$ 11 **10** $|-17 + 9|$ 8 **11** $|-17| + |9|$ 26

Problem Set

A **In problems 12–14, use the Distributive Property of Multiplication over Addition to rewrite each expression. Then combine like terms.**

P | **12** $4(8 - 3x)$ $32 - 12x$ **13** $-3(5 - 7x) - 4x$ $17x - 15$ **14** $8x - (5x + 6)$ $3x - 6$

P,I | **15** Express the total area of the figure in at least two different ways.
$(x + 3)(x + 2)$; $x^2 + 5x + 6$

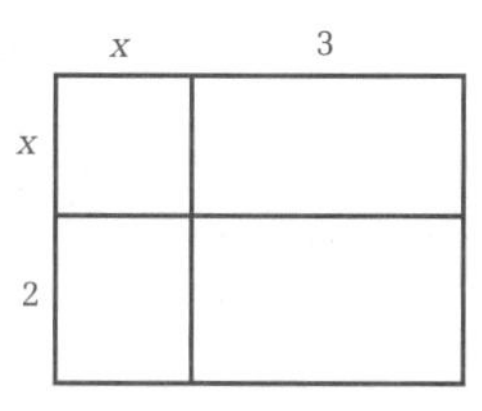

P  | **16** **a** Use a calculator to find the result of the following key sequence.

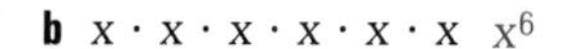
9 [+] 3 [×] 6 [±] [−] 8 [=] −17

b Write an algebraic sentence that represents the calculation in part **a.** $9 + 3(-6) - 8 = -17$

P,I | **17** Let $x = 3$.

a Evaluate $4x + 9$. 21
b Evaluate $7x - 4$. 17
c Evaluate $8x - 3$. 21
d Is $x = 3$ a solution of $4x + 9 = 7x - 4$? No
e Is $x = 3$ a solution of $4x + 9 = 8x - 3$? Yes

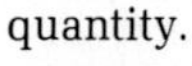

P | **In problems 18–20, simplify each expression.**

18 $5x + 3y + 2(3y - x)$ $3x + 9y$
19 $8x + 5y - (x - y)$ $7x + 6y$
20 $3x \cdot 2x + 5x \cdot x - 4x \cdot 3 + 6 \cdot 2x^2$ $23x^2 - 12x$

P | **21** Write an algebraic expression for each quantity.

a Length AB $27 - 2x$
b Five less than twice a number n $2n - 5$
c Four times the sum of a number x and 7 $4(x + 7)$

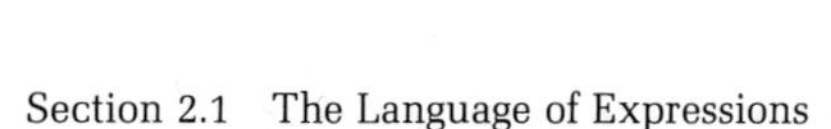

P | **22** Write each expression in simplest form.

a $x + x + x + x + x$ $5x$ **b** $x \cdot x \cdot x \cdot x \cdot x \cdot x$ x^6

Problem-Set Notes and Strategies

- In problems **6** through **11**, emphasize that grouping symbols can take on various forms. In addition to parentheses, we also use brackets, absolute values, radicals, fraction bars and any other indication that one operation must be performed first.

- In **16b**, the answers may not be unique since different calculators work differently. You may have to help individual students with the idiosyncrasies of their particular calculator. Review the use of the ± key as opposed to the +/– on some calculators. Make sure students understand this is a key that changes sign and is different from the subtraction key.

- Problem **17**, some students may not recall the definition of a solution to an equation. Review the terminology *solution, root, zero,* etc.

- Problem **22** is a good review of combining like terms and the properties of exponents.

Problem-Set Notes and Strategies, continued

- Problems **26** and **27** may require some review of sigma notation. Suggest that students write each term and add.

- Using a calculator in problem **31** makes this a good time to discuss rounding errors. Have students multiply $4.45 by 20.3 hours and determine their paycheck to the nearest integer value. Next, have them round the values to the nearest integer value and determine their paycheck. Now discuss the advantages and disadvantages of the proper timing for rounding operations. Students should see the errors introduced by rounding at intermediate steps.

- Problem **35** is a good problem to determine if students remember how to distribute negative signs.
- Students will get practice recording their thinking steps in problems **37** and **38**.

Problem Set, *continued*

B

P **In problems 23–25, perform the indicated operations.**

23 $3|4 - 12| + (-3)^3$ -3 **24** $3(4 - 12) + |-3|^2$ -15 **25** $\left|-21 - |-12|\right|$ 33

P,R **In problems 26 and 27, evaluate each sum.**

26 $\sum_{n=3}^{7}(4n + 5)$ 125 **27** $\sum_{n=1}^{4}2^n$ 30

P **28** Copy and complete the table of values for $y = -5(x - 2) + 1$.

x	**−4**	**−2**	**6**	**3**	**−6**
y	31	21	−19	−4	41

P,I **29** The total area of the figure is equal to $(x + 6)(x + 15)$.

a What are the areas of the small rectangles and the square?

b Write another expression to describe the total area of the figure.

$x^2 + 21x + 90$

	x	15
x	x^2	15x
6	6x	90

P **30** Write an expression for the seating capacity of a school auditorium that contains r rows with 22 seats in each and 12 rows with s seats in each. $22r + 12s$

P,I **31** Evaluate each expression to the nearest hundredth for $x = 0.84$.

a $[(x + 5)x + 6]x + 7$ ≈ 16.16 **b** $x^3 + 5x^2 + 6x + 7$ ≈ 16.16

P **In problems 32–34, evaluate each expression for $(x_1, y_1) = (2, 3)$ and $(x_2, y_2) = (8, 11)$.**

32 $\frac{y_2 - y_1}{x_2 - x_1}$ $\frac{4}{3}$ **33** $\frac{x_1 + x_2}{2}$ 5 **34** $\frac{y_1 + y_2}{2}$ 7

P **In problems 35 and 36, simplify each expression.**

35 $(3x^2 + 9x - 12) - (3x - x^3 + 5x^2)$
$x^3 - 2x^2 + 6x - 12$

36 $(3x^2 + 2y^2 + 4x + 8y) + (x^2 - y^2)$
$4x^2 + y^2 + 4x + 8y$

R **37** Explain, in paragraph form, why the product of any two consecutive natural numbers is even. $n(n + 1)$ is even because either n is even or $n + 1$ is even.

R **38** Is the sum of any two consecutive integers even? Explain your answer. No; $n + (n + 1) = 2n + 1$, which is odd because $2n$ is always even.

P,I **39** Let $x = 7$.

a Evaluate $3x + 9$. 30
b Evaluate $5x - 2$. 33
c Evaluate $6x - 15$. 27
d Is $x = 7$ a solution of $3x + 9 > 5x - 2$? No
e Is $x = 7$ a solution of $3x + 9 \geq 6x - 15$? Yes
f Is $x = 7$ a solution of $6x - 15 < 5x - 2$? Yes

C
P

40 Perform the indicated operations.

$11^2 - |2(-15)| - \sqrt{\frac{4^2}{2}} + 7 \cdot 8 + 3$ 86

P,I

41 Let $x = 9$.

a Evaluate $(x + 4)^2$. 169

b Evaluate $x^2 + 4^2$. 97

c Evaluate $x^2 + 8x + 16$. 169

d Which of the expressions in parts **a–c** are equivalent? **a** and **c**

P,R

42 Evaluate $\sum_{x=3}^{5}(2 + 3x)$. 42

P,I

43 Express the total volume of the box in at least two different ways.
$(x + 2)(x + 3)(x + 4)$; $x^3 + 9x^2 + 26x + 24$

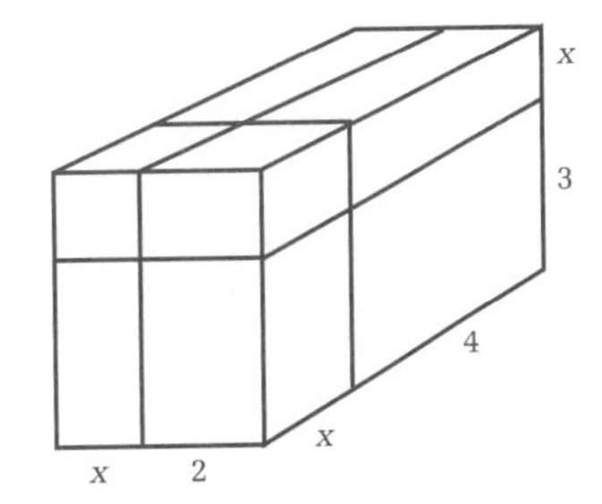

Problem-Set Notes and Strategies, continued

- In problem **41** you might have to review the product of binomials and look at several more examples to convince students that $(a + b)^2 \neq a^2 + b^2$. Suggest that students substitute $a = 3$ and $b = 4$ as a check.
- Students may have difficulty with the hidden box in the three dimensional picture in problem **43**. Have them draw the hidden lines to see each box and calculate the volume of each box separately and add them together.

CAREER PROFILE

THE MATHEMATICS OF MANAGING MONEY

For Rodney Brown, the bottom line is the bottom line

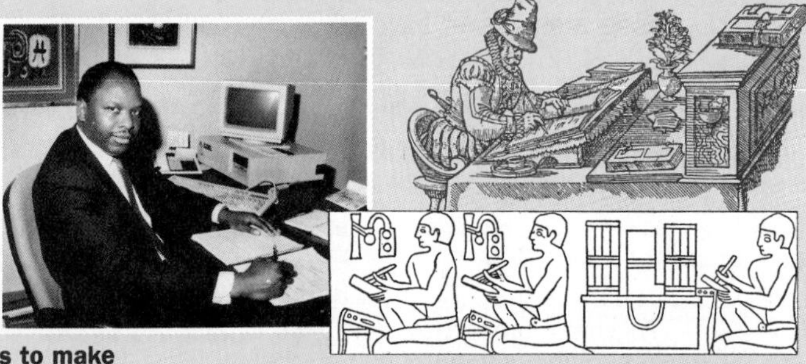

"Mathematics is not only essential to my work, it *is* my work," says Chicago-based Rodney Brown, a Certified Public Accountant. What Mr. Brown does is painstaking and would try most people's patience. Making a long story short, he says: "I use formulas and do a tremendous number of calculations to make sure that a balance sheet adds up correctly." It is essential for businesses or individuals to know how much money they have, how much they have spent and on what, and how much they owe. An accountant goes through all a client's financial transactions and provides that information in a clear and organized manner.

One type of interesting problem Mr. Brown might solve for his clients is that of finding the real value of an investment they have made or are thinking of making. For example, if a client earns 8.9%-per-year interest on an investment of $15,000, the dollar value of the interest is $1335. "They may in fact earn that much money, but the real value of what they receive is considerably less after taxes and inflation," Mr. Brown says. If tax on earnings is 28% and inflation is 6.7%, then the real value of the earnings is (0.089 – 0.067) · $15,000 – (0.28 · $1335). The result is –$43.80!

Mr. Brown graduated with a bachelor's degree in accounting from the University of Illinois at Chicago. "Using mathematics forces you to think logically and clearly," he says.

2.2

Class Planning

Time Schedule

All levels: 1 day

Resource References

Teacher's Resource Book
Class Opener 8
Cooperative Learning 5
Practice 8

Class Opener

Athan needs to rent a car. Dents Rent-a-Car charges $45 plus 12 cents a mile. Trust Us Rent-a-Car charges $60 plus 9 cents a mile. Athan quickly computed his cost and found that the cost would be the same at either company. How far was Athan planning to drive? 500 miles. Some students can reason through this problem without algebra. Seeing the algebra solution should help them see the relationship between the logic they used and the algebraic steps.

Lesson Notes

- Section **2.2** defines the concept of an equation as a statement with two expressions that represent the same value. Five properties of equality are also introduced. The zero product property is very important for further work. Emphasize the point that factoring is helpful in solving equations.

2.2 Solving Equations

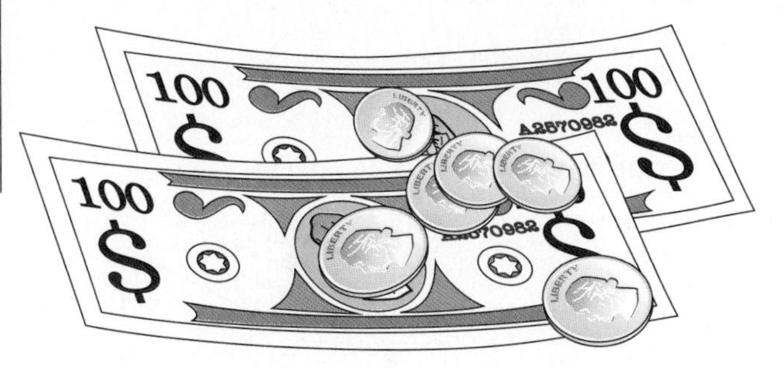

Objectives

After studying this section, you will be able to

- Recognize equivalent equations
- Solve equations by using properties of equality
- Express solutions of equations as ordered pairs

Part One: Introduction

Equivalent Equations

Equations with the same solution(s) are called ***equivalent equations.*** For example, $x + 6 = 17$, $x + 8 = 19$, and $x = 11$ are equivalent equations because all have the same solution, 11.

Some Properties Used in Solving Equations

To solve an equation, we use this general procedure.

> ***Write a series of equivalent equations until you arrive at an equivalent equation of the following form:***
> ***variable = number***

Algebraic properties are the rules that tell us what we can and cannot do. The following properties apply to all real numbers *a*, *b*, and *c*.

Addition and Subtraction Properties of Equality

If $a = b$, then $a + c = b + c$ and $a - c = b - c$.

We can add the same quantity to, or subtract the same quantity from, each side of an equation and the resulting equation is equivalent to the original equation.

Vocabulary

equation
linear equation
ordered pair

Multiplication and Division Properties of Equality

If $a = b$, then $ac = bc$ and (if $c \neq 0$) $\frac{a}{c} = \frac{b}{c}$.

We can multiply or divide each side of an equation by the same numerical quantity (provided that the quantity is not zero) and the resulting equation is equivalent to the original equation.

Zero Product Property

If $a \cdot b = 0$, then $a = 0$ or $b = 0$.
If the product of two or more numbers is zero, at least one of the numbers must be zero.

Solving Linear Equations

To solve an equation means to find all values of the variable that make the equation a true statement. We apply the properties as we work through a solution.

Example *Solve the equation* $3x - 7 = 2x + 4$.

$$\begin{array}{lrcr} & 3x - 7 & = & 2x + 4 \\ \text{Subtract } 2x \text{ from each side} & -2x & & -2x \\ \hline & x - 7 & = & 4 \\ \text{Add 7 to each side} & +7 & & +7 \\ \hline & x & = & 11 \end{array}$$

Check
To check our work, we substitute the value 11 for the variable x in the original equation and see whether the expressions on the left side and the right side have the same value.

Left Side	**Right Side**
$3x - 7$	$= 2x + 4$
$= 3(11) - 7$	$= 2(11) + 4$
$= 26$	$= 26$

When $x = 11$, the sides are equal, so the answer checks.

Ordered Pairs

Although frequently listed in a table, solutions of an equation in two variables, such as x and y, are also often shown as ***ordered pairs*** (x, y), with x as the first number of each pair and y as the second number.

Using Computers

This is an excellent opportunity to introduce symbolic manipulator software such as *Derive*. These programs are best learned in stages and one of their strongest assets is that of solving equations and simplifying algebraic expressions.

Cooperative Learning

Have students work in groups to evaluate the product of $(x-a)(x-b)(x-c) \ldots (x-y)(x-z)$. This activity will help students understand the zero product property and also realize the advantage of thinking through a problem and looking at possibilities before attempting the "obvious" solution . The correct answer is zero since one of the factors is $(x-x)$.

Communicating Mathematics

Write a paragraph to explain the differences and similarities between equations and expressions.

Lesson Notes, continued

- Ordered pairs are introduced. Consider having the student label the table with real-world labels as well as x and y when solving problems. This gives some ownership of the problem to the mathematical model.

2.2

Checkpoint

1. Solve the equation $3x - 2 = 2x + 5$ 7
2. Tom and Jerry have 17 marbles. Tom has three more than Jerry. How many marbles does each have?
   Tom has 10, Jerry has 7.
3. Let $y = 3x - 2$. Find the value of the other variable if $x = 3$, $y = 7$, $x = -5$, $y = -17$.
   7, 3, −17, −5

Assignment Guide

Average
10–13, 15–17, 19, 20, 24, 27, 28, 31, 33, 38, 40, 42, 44
Advanced
10, 11, 13, 14, 17, 22, 23, 25, 27, 29, 32, 37–42
Honors
12–15, 17, 22, 23, 26, 27, 32, 38, 41, 42, 46–49, 54–57

Example *Consider the equation $y = 2(10 - x)$.*

a *Make a table of values for the equation, containing the values $x = -3$, $y = 11$, $x = 0$, $y = 0$, $y = 6$, and $x = \frac{1}{2}$.*

x	−3	4.5	0	10	7	$\frac{1}{2}$
y	26	11	20	0	6	19

b *Write the solutions in the table as ordered pairs.*

$(-3, 26)$, $(4.5, 11)$, $(0, 20)$, $(10, 0)$, $(7, 6)$, $\left(\frac{1}{2}, 19\right)$

c *Is (6, 8) a solution of the equation?*

We substitute 6 and 8 for x and y in the equation.

$$y = 2(10 - x)$$
$$8 = 2(10 - 6)$$
$$8 = 2(4)$$
$$8 = 8$$

Since the equation makes a true statement when $(x, y) = (6, 8)$, this ordered pair is a solution of the equation.

In Section 2.4, you will see that ordered pairs can be used to plot a graph that represents an equation.

Part Two: Sample Problems

Problem 1 *Two runners live 33 miles apart. Jim Schuh runs 8 miles per hour, and Swett Shurt runs 6 miles per hour. If Swett starts running toward Jim's house and Jim starts running toward Swett's house 2 hours later, how long will Jim have been running when they meet?*

Solution Let t = the number of hours Jim runs.
Then $t + 2$ = the number of hours Swett runs.
The distance run by Jim is $8t$, and that run by Swett is $6(t + 2)$.
Since the total distance is 33 miles,

$$8t + 6(t + 2) = 33$$
$$8t + 6t + 12 = 33$$
$$14t + 12 = 33$$
$$14t = 21$$
$$t = 1.5$$

Jim will have been running for 1.5 hours, or 1 hour 30 minutes.

Check
If Jim runs for 1.5 hours at 8 miles per hour, he will run $8 \cdot 1.5$, or 12, miles. If Swett runs for 3.5 hours at 6 miles per hour, he will run $6 \cdot 3.5$, or 21, miles. Since $12 + 21 = 33$, the answer checks.

Problem 2 *The area of rectangle A is equal to the area of rectangle B. Find the dimensions of each rectangle.*

x + 3 | x
x | A
B | 6

Solution The area of rectangle A is $x(x + 3)$ and the area of rectangle B is $6x$.

Since the areas are equal,	$\mathbf{x(x + 3) = 6x}$
use the Distributive Property of Multiplication over Addition	$\mathbf{x^2 + 3x = 6x}$
Regroup	$\mathbf{x^2 - 3x = 0}$
Use the distributive property in reverse	$\mathbf{x(x - 3) = 0}$
Zero Product Property	$\mathbf{x = 0 \text{ or } x - 3 = 0}$
	$\mathbf{x = 0 \text{ or } x = 3}$

The length of a side of a rectangle cannot be zero, so each rectangle measures 3 by 6.

Check
The area of rectangle A is $3 \cdot 6$, or 18, and the area of rectangle B is $6 \cdot 3$, or 18. The areas are equal, so the answer checks.

Part Three: Exercises and Problems

Warm-up Exercises

P — **1** Consider the equation $6(x + 4) = 24$. Which of the steps below would be a useful first step in solving the equation? **a** or **b**
- **a** Distribute the 6 on the left side.
- **b** Divide each side by 6.
- **c** Subtract 4 from each side.

P — **2** Use the Zero Product Property to solve $z(z - 4) = 0$ for z. $z = 0$ or $z = 4$

P,R — **3** Refer to the figure.
- **a** Write a formula for the perimeter P of the figure. $P = 3x + 19$
- **b** Find x if $P = 64$. $x = 15$

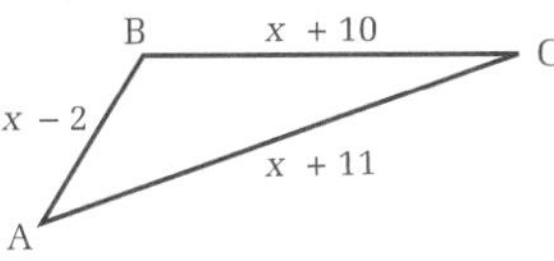
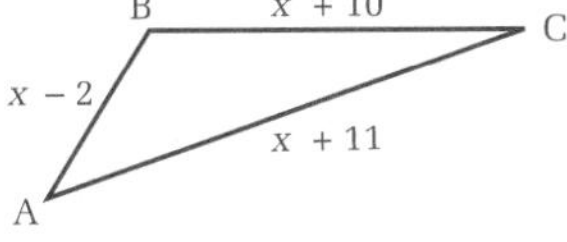

P — **4** Is $z = 36$ a solution of $\sqrt{z} + 30 = z$? Yes

P — **5** Is $(x, y) = (-2, 3)$ a solution of $5y + 6x = 3$? Yes

P — **6** Divide both sides of $5(2x - 7) = 40$ by 5 and solve for x. $x = 7.5$

P,R — **7** Copy and complete the table of values for $x = y - 4$.

x	−4	−7	1	**0**	**5**
y	**0**	**−3**	**5**	4	9

Problem-Set Notes and Strategies

- In problem **3**, you may want to review the concept of perimeter and the fact that it is the sum of the lengths of the sides of the polygon.
- In problems **4** and **5** students are asked to verify that a particular number is a solution. One of the methods of problem solving is the guess and check approach. It is good practice for students to verify that solutions do indeed verify the equation.

Problem-Set Notes and Strategies, continued

- Problem **9** introduces radicals and has students verify that answers are in a particular region. This problem also emphasizes the necessity of estimating results and having an approximation. Certainly $\sqrt{30}$ is between 5 and 6 since 5^2 is 25 and 6^2 is 36.

- The concept of sign change is explored in problem **13**. By multiplying by −1, the absolute value of the number remains the same but the sign changes and the order has reversed. This can be seen here as well as in inequalities when both sides are multiplied by a negative number.

- Although problem **15** has a two dimensional graph, the axes are labeled with real world labels. Emphasize that an ordered pair would still be written as (x, y) but each coordinate has a special interpretation. This concept will be explored further in Chapter **3**.

P,R **8** An empty bushel weighs about 1.2 pounds, and an apple weighs about 0.7 pound.
- **a** What is the approximate weight of 7 apples? 4.9 lb
- **b** Write an expression for the weight of n apples. $0.7n$
- **c** What is the approximate weight of a bushel containing 50 apples? 36.2 lb
- **d** Write a formula for the weight W of a bushel containing n apples. $W = 0.7n + 1.2$

I **9** The numbers between −3 and 5 have been shaded on the number line.

- **a** Is $\sqrt{30}$ shaded? No
- **b** Is $\sqrt{22}$ shaded? Yes
- **c** Is $-\sqrt{22}$ shaded? No
- **d** Is $-\frac{18}{5}$ shaded? No
- **e** Is $\frac{14}{3}$ shaded? Yes
- **f** Is −3 shaded? No

Problem Set

A **In problems 10 and 11, solve each equation.**

P **10** $5(3x + 2) = -50$ $x = -4$ **11** $\frac{3x + 6}{4} = 9$ $x = 10$

P **12** Let $y = 8x + 5$.
- **a** Find x if $y = 2$. $x = -\frac{3}{8}$
- **b** Find x if $y = 2x - 3$. $x = -\frac{4}{3}$
- **c** Find the value of x for which $y = -31$. −4.5

R,I **13** **a** Arrange the numbers 4, −2, 3, and −7 in order from smallest to largest. −7, −2, 3, 4
- **b** Multiply each number in part **a** by −1 and list the results in order from smallest to largest. −4, −3, 2, 7
- **c** What effect did multiplying by −1 have on the order? Reversed order

P **14** Billie's bank contains nickels, dimes, and quarters.
- **a** If Billie has eight nickels, five dimes, and seven quarters, what is the total value of the coins in the bank? $2.65
- **b** Suppose Billie has n nickels, d dimes, and q quarters. Write a formula that can be used to find V, the total value of the coins. $V = 0.05n + 0.1d + 0.25q$

P **15** The graph shows the relationship between a firm's profit P and the number n of widgets it produces.
- **a** Find the profit if 200 widgets are produced. $4000
- **b** How many widgets must be produced for a profit of $6000? 250 or any number between 400 and 600 inclusive
- **c** What happens if no widgets are produced? Loss of $2000

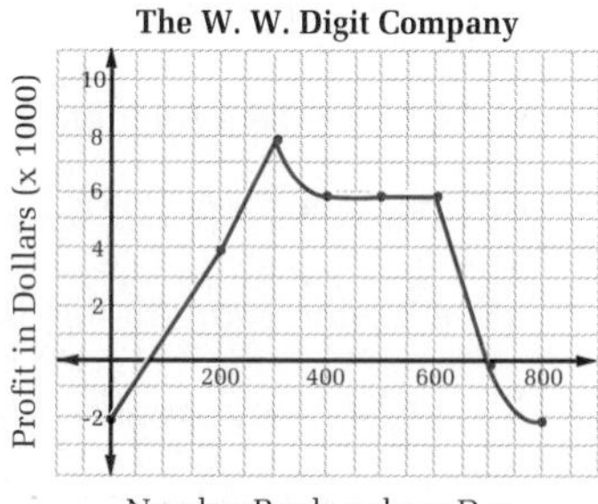

P **16** If $w = 2x + y$, what is

a The value of w for which $(-3, 4)$ is a solution of the equation? -2

b The value of x for which $w = 8$ and $(x, 5)$ is a solution of the equation? $\frac{3}{2}$

P,I **17** Two matrices are equal only if each pair of corresponding entries is equal. In the matrix equation below, for example, w is equal to 7. Find the values of x, y, and z. 4; -8; 64

$$\begin{bmatrix} x+6 & 3y \\ w & 8 \end{bmatrix} = \begin{bmatrix} 10 & -24 \\ 7 & \sqrt{z} \end{bmatrix}$$

P,I **18** Consider the equation $3x - 5y = 48$.

a If $x = 0$, what is y? $-\frac{48}{5}$

b If $y = 0$, what is x? 16

P **19** The perimeter of the rectangle is 62 centimeters. Find the area of the rectangle. (Hint: First find the value of x.) 150 cm^2

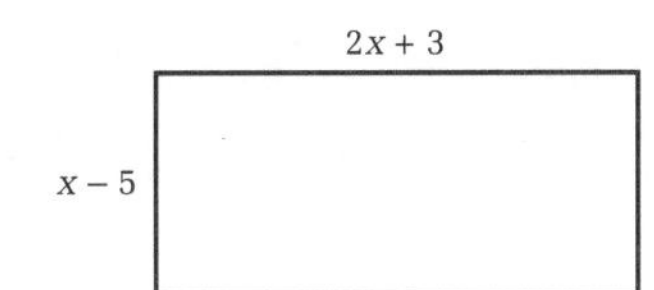

R **20** How many diagonals does a pentagon have? 5

R **21** How many diagonals does a hexagon have? 9

R **22** Write an expression for the sum of five consecutive integers if the smallest of the integers is $x + 1$. $5x + 15$

P,I **23** Two hot-air balloons are flying over a field. The height h, in feet, of the first balloon t seconds from now can be found with the formula $h(t) = 50 + 34t$. The height g, in feet, of the second balloon t seconds from now can be found with the formula $g(t) = 400 - 16t$.

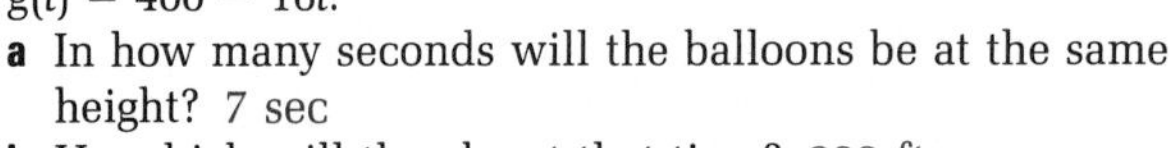

a In how many seconds will the balloons be at the same height? 7 sec

b How high will they be at that time? 288 ft

P **B** **24** Solve $3x + 9y = 12z$ for x. $x = 4z - 3y$

P **25** Solve $2x - 3y = 10y - 6x$ for y. $y = \frac{8x}{13}$

P **26** Solve $x + 2y = 3z$ for y. $y = \frac{3z - x}{2}$

R **27** At what point must you start in order to trace the entire digraph exactly once without lifting your pencil? Explain your reasoning. Point E, since to avoid retracing, a path must begin at a point that is on the E–D loop and that has 2 exit possibilities.

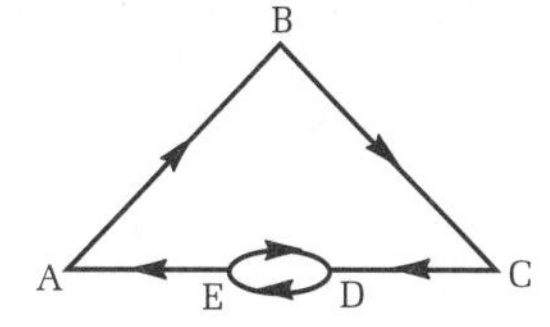

Problem-Set Notes and Strategies, continued

- The matrices in problem **17** will be the first introduction to this topic for some students. Emphasize that each entry in the matrix is distinct and that corresponding entries must be equal for the matrices to be equal. More will be presented in Section **2.5** on matrix operations.
- Intercepts of a graph are explored in problem **18.** This concept will be formalized in Section **2.4** but some students will make the generalization at this point.
- Systems of equations are presented in problem **23** and the student must recognize that both $h(t)$ and $g(t)$ must be equal for the balloons to be the same height. Systems will be discussed in detail in Chapter **3** but many students will recall the process from Algebra 1.
- Problem **27** provides a good review of a digraph from Chapter **1.** Students must examine the number of entrance and exit routes from each point to answer the question.

Problem Set, *continued*

Problem-Set Notes and Strategies, continued

- Problems **28 – 30** illustrate the importance of the zero product property.
- Problem **32** illustrates absolute value problems. Using the definition, every absolute value problem may be solved as two problems.
- Problems **38 – 44** provide an excellent review of and practice with the order of operations and the distributive property.
- Linear equations and graphs are presented in problem **45**. Students should be able to find the intercepts on the graph after working problem **18**. Graphing linear equations is the topic of Section **3.1**

P **In problems 28–30, solve each equation for *y*.**

28 $y(y - 3) = 0$ $y = 0$ or $y = 3$

29 $(y + 3)(2y - 10) = 0$ $y = -3$ or $y = 5$

30 $(y - 4)^2 = 0$ $y = 4$

P,I **31** Suppose that $y = x^2$.

a Find the value of y for which $x = 3$ and that for which $x = -3$. 9; 9

b Find all values of x for which $y = 16$. 4, −4

P,I **32** Find the values of x that satisfy each equation.

a $|x| = 13$ 13, −13 **b** $|x| + 3 = 13$ 10, −10 **c** $|x + 3| = 13$ 10, −16

P **33** Express a in terms of x.
$a = 2x - 11$

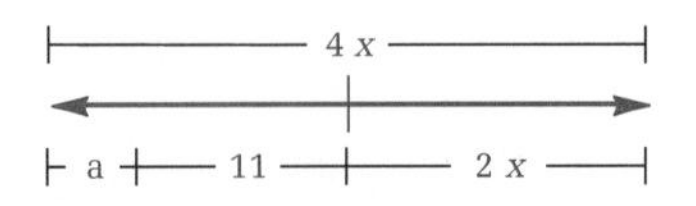

P,R **In problems 34–36, evaluate each expression for (*x*, *y*) = (3, −8).**

34 $3(x - y) + 2x^2$ 51

35 $2|y - x| + 3 + 2x$ 31

36 $(y - x) - |x - y|$ −22

P **37** If $y = 9 - 4x$ and $z = 6x - 11$, for what value of x is $y = z$? 2

R **38** Perform the indicated operations.

a $12 - 4 \cdot 3 + 2 \cdot 3^2$ 18

b $(12 - 4)3 + (2 \cdot 3)^2$ 60

c $12 - 4(3 + 2 \cdot 3)^2$ −312

d $12 - 4|3 + 2| \cdot 3^2$ −168

P,R **In problems 39–44, solve each equation for x and simplify each expression.**

39 $10 - (6 - 3x) = 2(x + 5)$ $x = 6$

40 $x(2x - 4) + 3x = x(5 + 2x) + 6$ $x = -1$

41 $\left[\frac{2}{3}(x - 3) + 7\right] \cdot 6x + 2$ $4x^2 + 30x + 2$

42 $2(3 - x) - 3(4 - x) = 4(5 - x)$ $x = 5.2$

43 $3x = a + 2b$ $x = \frac{a + 2b}{3}$

44 $a(bc + ac - ab) + b(ab + bc - ac)$
$a^2c - a^2b + ab^2 + b^2c$

P,I **45 a** Copy and complete the table of values for the equation $y = \frac{1}{2}x - 3$.

x	−6	−4	0	2	8	12
y	−6	−5	−3	−2	1	3

b Plot the six points on a coordinate system and connect them. What do you see? A straight line

c Is the point (34, 14) on the graph of $y = \frac{1}{2}x - 3$? Yes

P **46** Refer to the figure.

a Write an expression for the surface area of the box. $36x + 144$

b Find x if the surface area is 261.
$x = 3.25$

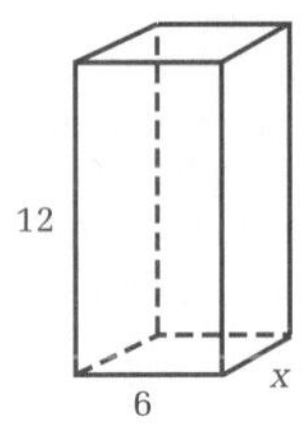

R **C** **47** Explain, in paragraph form, why the product of two odd numbers is an odd number. Let a and b be two odd numbers—i.e., $a = 2k + 1$ and $b = 2n + 1$. The product $(2k + 1)(2n + 1)$ is equivalent to $4kn + 2k + 2n + 1$, which is odd, since $4kn + 2k + 2n$ is even and adding 1 makes it odd.

R,I **48** Refer to the diagram.
- **a** Find the coordinates of point R. (0, −2)
- **b** Find the lengths of $\overline{PR}$ and $\overline{RQ}$. 5; 12
- **c** Find the area of △PQR. 30
- **d** Use the Pythagorean Theorem to find the length of $\overline{PQ}$. 13

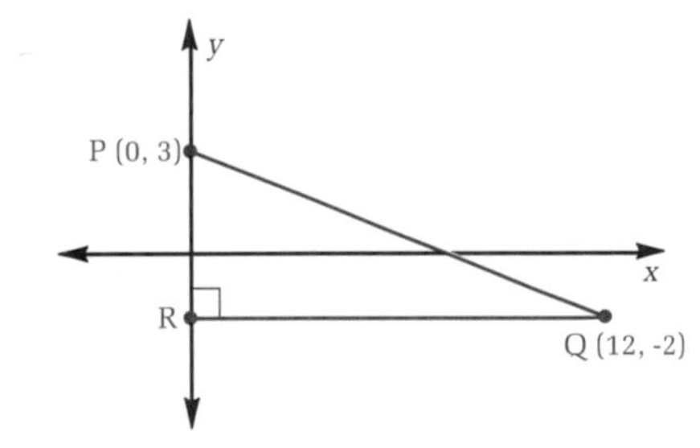

P,R **49** Two sailboats 400 meters apart are sailing toward each other. One is moving at a rate of 10 meters per second, and the other is moving at a rate of 15 meters per second.
- **a** In how many seconds will the boats pass each other? 16 sec
- **b** In how many seconds will the boats be 400 meters apart again? 32 sec

R, I **In problems 50–52, simplify each expression.**

50 $5a \cdot 3a + 5 \cdot 3 \cdot a^2 + 5 \cdot (3a)^2 + 3(5a)^2 + (3 \cdot 5a)^2$ $375a^2$

51 $-2x^2 + (-3)^3x^2 - 4x^2 + x^2$ $-32x^2$

52 $(3x^2 + 2x + 4) - (4 - 2x + 3x^2)$ $4x$

P **53** If $s = 5x - 7$ and $t = x + 10$, for what value of x is $s = 2t - 4$? $\frac{23}{3}$

P,R **54** Make a table of values for the equation $y = 3(x + 2) - 5(x - 3) + 4(x - 4)$, using the integers from −5 to 5 for values of x.

P **55** Find BC if $AB = x^2 + 3x - 5$ and $AC = 6 + 2x + 4x^2$. $3x^2 - x + 11$

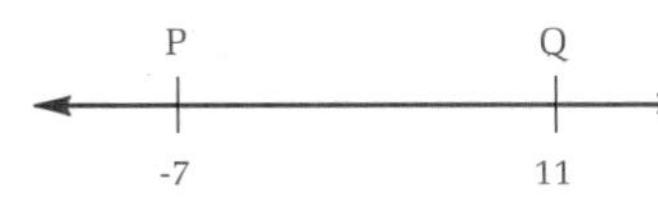

R **56** Refer to the diagram.
- **a** Find the midpoint of $\overline{PQ}$. 2
- **b** Find the two trisection points of $\overline{PQ}$. −1, 5
- **c** Find the three points that divide $\overline{PQ}$ into four equal parts. −2.5, 2, 6.5

P Q
-7 11

R **57** Simplify $52(773p + 851w) - 26(851w + 773p)$. $20{,}098p + 22{,}126w$

Problem-Set Notes and Strategies, continued

- Problem **47** is an excellent problem for critical writing and analytic thinking. Encourage students to mix English prose with their mathematical thoughts. Some may later enter technical fields where they are required to publish results for the lay person to read. These problems provide good practice for that type of writing.
- Problem **48**, reiterate the fact that point R has the same x-coordinate as P and the same y-coordinate as Q.
- Problems **50 – 52** introduce some properties of exponents that should be remembered from Algebra 1. If students are having difficulty, have them write out what $(3a)^2$ means in long form and then simplify the results. This helps solidify their understanding.

Additional Answer

54

x	−5	−4	−3	−2	−1	0	1	2	3	4	5
y	31	29	27	25	23	21	19	17	15	13	11

- Problem **56** is a good review for students to subdivide an interval into equal parts.

2.3

Class Planning

Time Schedule

Average: 2 days
Advanced: 2 days
Honors: $1\frac{1}{2}$ days

Resource References

Teacher's Resource Book
Class Opener 9
Practice 9
Evaluation
Quiz Forms 1, 2, 3

Class Opener

The measure of an obtuse angle is six more than three times the measure of an acute angle. Find the possible range of values for the acute angle's measure.
$28 < x < 58$

Lesson Notes

- Inequalities are the topic of this section. Point out the two main differences between inequalities and equations. First, with equations, generally only a finite number of answers are true, whereas with inequalities there is usually an infinite range of solutions.
- The methods for solving inequalities are basically the same except for the multiplication and division property. When graphing solutions for inequalities, first find those points where equality holds. Second, check the regions of the graph divided by the line to see which are in the solution set.

2.3 Solving Inequalities

Objectives

After studying this section, you will be able to
- Apply properties of inequality
- Solve linear inequalities in one variable
- Graph solutions of inequalities on the number line

Part One: Introduction

Properties of Inequality

In the preceding section, you saw that an equation states that two expressions are equal. If the values of two expressions are not equal, we can write an ***inequality*** using the symbol $\leq$, $\geq$, $\neq$, $>$, or $<$.

The following properties, which apply to all real numbers a, b, and c, are useful in the solution of inequalities.

Addition and Subtraction Properties of Inequality	
If $a > b$, then $a + c > b + c$ and $a - c > b - c$.	We can add the same quantity to, or subtract the same quantity from, each side of an inequality.

Multiplication and Division Properties of Inequality	
If $a > b$ and $c > 0$, then $ac > bc$ and $\frac{a}{c} > \frac{b}{c}$.	We can multiply or divide each side of an inequality by the same positive number.
If $a > b$ and $c < 0$, then $ac < bc$ and $\frac{a}{c} < \frac{b}{c}$.	When we multiply or divide each side of an inequality by a negative number, the order of the inequality is reversed.

Solving an Inequality

To solve an inequality means to find all values of the variable or variables that make the inequality a true statement. The steps we use are similar to those we use in solving equations.

Vocabulary
inequality

Example *Find all values of x for which* $9 - 2x < 17$.

Subtract 9 from each side

$$\begin{array}{r} 9 - 2x < 17 \\ -9 \qquad -9 \\ \hline -2x < 8 \end{array}$$

Divide each side by -2 (remembering that division by a negative number reverses the order)

$$\frac{-2x}{-2} > \frac{8}{-2}$$

$$x > -4$$

To check a solution of an inequality, we follow the same procedure as in checking solutions of equations.

Graphing Solutions on the Number Line

Often, a number line is used to show the solutions of linear inequalities.

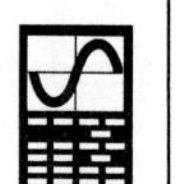

See p. 754

Example *For each inequality, use a number line to represent the possible values of the variable.*

a $x > 3$

Because x is *greater than* 3, the dot at 3 is open to show that 3 is not among the solutions.

b $y \leq 4$

Because y is *less than or equal to* 4, the dot at 4 is solid to show that 4 is among the solutions.

4

c $z \neq -3$

Because z can be *any number except* -3, all the points on the line, except that at -3, represent solutions.

-3

Part Two: Sample Problems

Problem 1 *Solve* $-2 < 3x - 5 < 8$ *for x and graph the solution on a number line.*

Solution The compound inequality $-2 < 3x - 5 < 8$ means

$$-2 < 3x - 5 \textit{ and } 3x - 5 < 8$$

To solve for x, we solve each inequality and combine the solutions in a new compound inequality.

$$\begin{array}{lcl} -2 < 3x - 5 & \text{and} & 3x - 5 < 8 \\ 3 < 3x & & 3x < 13 \\ 1 < x & & x < \frac{13}{3} \end{array}$$

The solution is $1 < x < \frac{13}{3}$. We graph this solution by representing the *intersection* of the graphs of the two parts of the inequality.

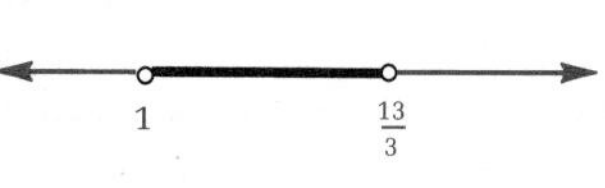

Lesson Notes, continued

This approach will work for higher dimension problems as well. In the one dimensional case, the graphs may seem trivial, but the concept will be extended to two dimensions in Section **3.3**.

- You might want to point out that the inequality ≠ is a combination of the two inequalities < and >. If two numbers are not equal, then either one is larger *or* smaller.

Checkpoint

1. Find all values of x such that $7x - 6 < 15$ $\quad x < 3$
2. Graph $x \neq 7$ on a number line. Graph $x > 7$ or $x < 7$ on a number line.

   7
3. Find all values for x such that $3 - 5x \leq 23$. $\quad x \geq -4$

Stumbling Block

Some students may have a difficult time with the inequalities $x < 4$ and $4 > x$. Point out that the inequalities may be reversed to put the variable on the left if all three parts are reversed, each side and the inequality sign. Suggest that students substitute a value for x as a check.

Communicating Mathematics

Describe in words why the inequality sign must be reversed when multiplying or dividing by a negative number.

Assignment Guide

Average	
Day 1	11–13, 17, 18, 23, 26, 28, 31, 33, 35, 36, 39, 42, 44
Day 2	14, 16, 19, 20, 22, 24, 25, 27, 32, 34, 37, 40, 52, 57
Advanced	
Day 1	11, 12, 15, 17, 18, 22, 23, 26–28, 31–33, 46, 61
Day 2	13, 14, 16, 19–21, 24, 25, 29, 30, 34–38, 40, 48, 51, 52, 57, 64
Honors	
Day 1	13, 14, 18, 20, 21, 25, 30, 31, 33, 36, 37, 39, 40, 49, 52, 56, 57, 65, 66
Day 2	($\frac{1}{2}$ day) 41, 43, 45, 48, 50, 53, 61, 64, 67 ($\frac{1}{2}$ day) Section 2.4 8, 9, 13, 16, 18, 22, 24, 28

Problem-Set Notes and Strategies

- In part **d** of the first problem you may observe students who add $2x$ and 8 to both sides to avoid dividing by a negative number. Although this yields the correct solution, it is not always useful as a technique. This approach does help prove the property of changing the inequality when dividing by a negative number.

Additional Answers

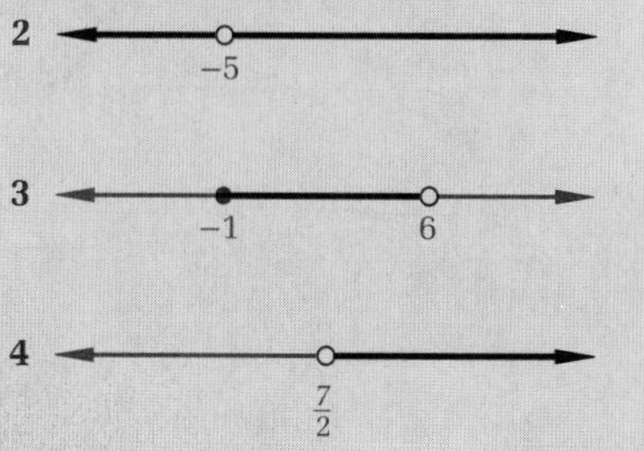

Problem 2 *Aunt Tenna's TV Repair Service charges $45.00 for a visit and $33.75 per hour for labor and parts. Tenna estimates that a given repair job will take at least 2.5 hours but no more than 4.5 hours. What is the possible range of the bill?*

Solution If C = the total charge for a service call and t = the number of hours the call takes, $C = 45 + 33.75t$. In this case, $2.5 \le t \le 4.5$, so

$$C \ge 45 + 33.75(2.5) \quad \text{and} \quad C \le 45 + 33.75(4.5)$$
$$\ge 45 + 84.375 \qquad\qquad \le 45 + 151.875$$
$$\ge 129.375 \qquad\qquad \le 196.875$$

The bill will be at least $129.38 but no more than $196.88.

Problem 3 *Find all possible values of P, the rectangle's perimeter.*

13 − 2x

x − 5

Solution

$$P = 2(x - 5) + 2(13 - 2x)$$
$$= 2x - 10 + 26 - 4x$$
$$= 16 - 2x$$

The length of each side must be positive, so

$$x - 5 > 0 \quad \text{and} \quad 13 - 2x > 0$$
$$x > 5 \qquad\qquad -2x > -13$$
$$x < \frac{13}{2}$$

Since $5 < x < \frac{13}{2}$,

$$P < 16 - 2(5) \quad \text{and} \quad P > 16 - 2\left(\frac{13}{2}\right)$$
$$< 6 \qquad\qquad > 3$$

The perimeter, therefore, is between 3 and 6.

Part Three: Exercises and Problems

Warm-up Exercises

P **1** In which of the following inequalities will the inequality sign be reversed when the inequality is solved for x? **c and d**

a $x - 2 < -8$ **b** $x + 2 < -8$ **c** $-2x < 8$
d $-2x < -8$ **e** $2x < -8$

P **2** Graph $x \ne -5$ on a number line.

P **3** Graph $-1 \le x < 6$ on a number line.

P **4** Solve $14 < 6x - 7$ and graph the solution on a number line. $x > \frac{7}{2}$

P **5** For each verbal description, write a corresponding algebraic statement.

a The sum of y and 10 is less than the product of 3 and y. $y + 10 < 3y$

b The product of 5 and the square of n is not equal to 20. $5n^2 \neq 20$

c One-fifth less than the square root of x is greater than or equal to the square root of 3. $\sqrt{x} - \frac{1}{5} \geq \sqrt{3}$

d The product of 2, π, and r is between 6 and 8. $6 < 2\pi r < 8$

P **6** Describe the result of each operation.

a Multiplying both sides of $-6 < -3$ by 2 $-12 < -6$

b Multiplying both sides of $-6 < -3$ by -2 $12 > 6$

P **7** Describe the result of each operation.

a Dividing both sides of $15 \geq 9$ by -3 $-5 \leq -3$

b Adding -5 to both sides of $15 \geq 9$ $10 \geq 4$

P **8** Write an inequality that describes the shaded area. Solve the inequality for x. $8(3x - 5) < 96$; $x < \frac{17}{3}$

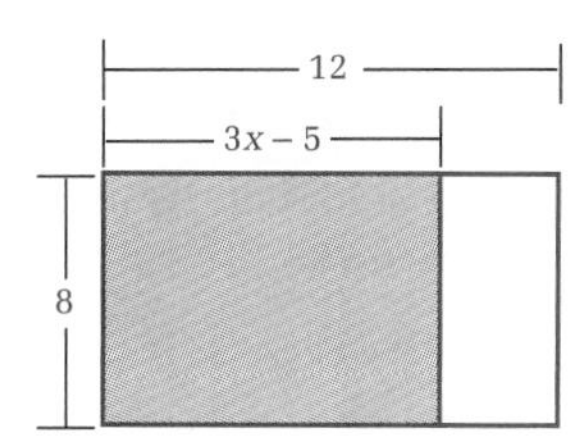

P **9** What must be true of the values of y if 5 times y is greater than 3? $y > \frac{3}{5}$

R **10** Explain why the square of any integral multiple of 3 is divisible by 9. $3k \cdot 3k = 9k^2$, which contains a factor of 9

Problem Set

P **In problems 11–16, solve each inequality and graph the solution on a number line.**

11 $12x + 27 > 3$ $x > -2$ **12** $-7y + 15 \geq 50$ $y \leq -5$ **13** $4(3 - z) < 40$ $z > -7$

14 $-3(2w + 1) \geq 15$ $w \leq -3$ **15** $10 \geq 3x - 11$ $x \leq 7$ **16** $1.5y_2 + 9.2 \leq 6.8$ $y_2 \leq -1.6$

In problems 17–20, express each graph as an inequality.

17 -3 -2 -1 0 1 2 3
$x < -1$

18 -4 -3 -2 -1 0 1 2 3 4
$x \leq 2$ or $x \geq 4$

19 -7 -6 -5 -4 -3 -2 -1 0
$x \neq -5$

20 -7 -6 -5 -4 -3 -2 -1 0
$-3 > x > -5$

P **21** Imagine that a new student asks you how to solve the inequality $9 - 3x \leq 5x + 11$. If the student will have no information except what you give, what will you tell the student? Answers will vary.

Problem-Set Notes and Strategies, continued

- Problem **5** is good practice for students to translate English statements into mathematical expressions.
- Frequently when an inequality has negative numbers on both sides and a student multiplies by a positive number the student feels the inequality should be changed. Problem **6** is intended to show that this is not true.
- Problems **11 – 16** are good practice in the use of the properties of inequality and the distributive property.

Additional Answers

11
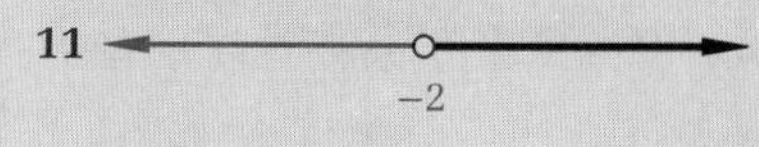

12
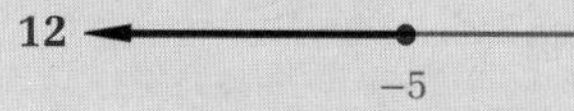

13
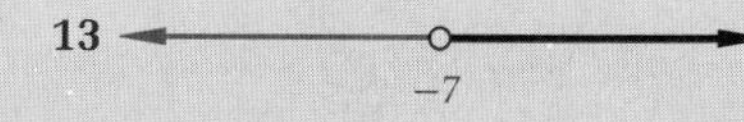

14
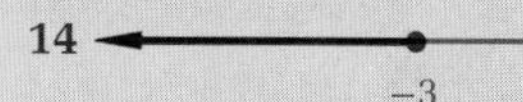

15
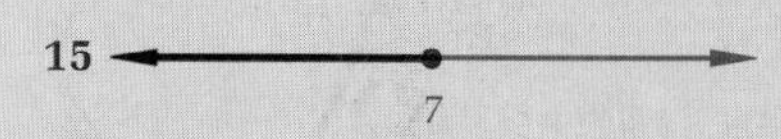

16
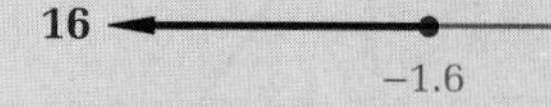

- Students often are confused about the endpoints of an inequality graph where inclusion becomes an issue. Problems **17 – 20** are intended to aid the student in interpreting graphs.

Problem-Set Notes and Strategies, continued

- Linear graphs are introduced in problem **24a**. Make sure that the students graph part **b** on a Cartesian coordinate system as well. The concept of an equation having a horizontal graph is important. It implies that y maintains a constant value which is also what the equation expresses. This topic is explored in detail in the next chapter.

Additional Answer

24a

Possible table:

x	-3	-2	-1	0	1	2
y	-1	1	3	5	7	9

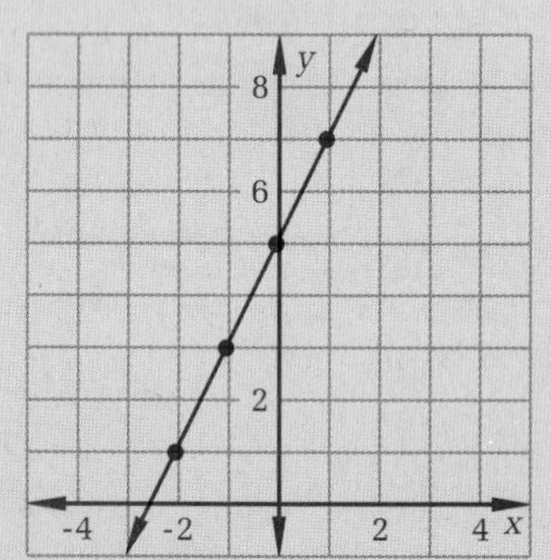

- Problems **28 – 32** review properties of exponents from Algebra I. If necessary, write out the problem using the definition of exponents, i.e. $x^5 = x \cdot x \cdot x \cdot x \cdot x$, to count the number of x's. Use the counting argument to find the answer. Also have the student practice naming the base of the exponential expression. For example, the base of $(4x)^2$ is $4x$ and not just x.
- Problems **35 – 38** are a review of several different properties in one problem. Encourage a step by step solution and care until the students gain confidence when distributing the $-2x^2$ in problem **38**.

Problem Set, *continued*

P **22** Tracey went for a long walk at a rate of 2 miles per hour. Three hours later, her brother Richard rode after her on a tandem bike at a rate of 7 miles per hour.

a How long did it take Richard to meet Tracey? 1.2 hr
b How far had she walked when they met? 8.4 mi
c How long would it take for them to ride back on the bike, traveling at 4 miles per hour? 2 hr 6 min

P **23** Refer to the diagram.

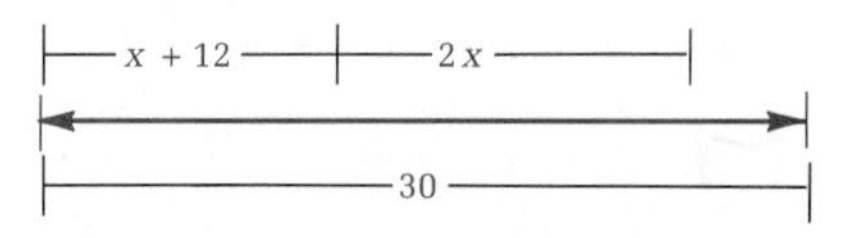

a Write an inequality that represents the diagram. $3x + 12 < 30$
b Solve the inequality for x. $x < 6$

R,I **24** Make a table of values for each equation. Then graph the equation.

a $y = 2x + 5$ **b** $y = -2$

P **25** Find the possible values of the perimeter of the rectangle shown. $14 < P < 28$

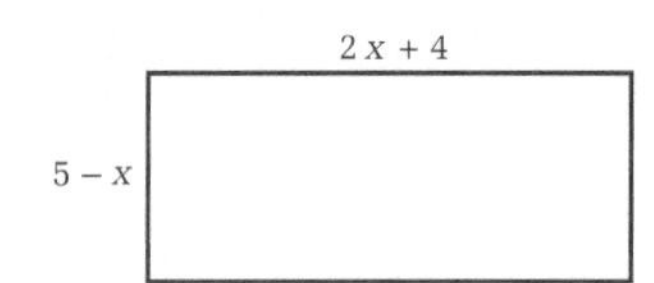

R **26** Determine whether $(x, y) = (5, -2)$ is a solution of each equation.

a $3x + y = 13$ Yes **b** $y = 2x - 10$ No **c** $y = \frac{1}{5}x - 3$ Yes

I **27** Solve the matrix equation for (x, y). $\left(-2, -\frac{1}{2}\right)$

$$4\begin{bmatrix} x \\ y \end{bmatrix} + \begin{bmatrix} 7 \\ 2 \end{bmatrix} = \begin{bmatrix} -1 \\ 0 \end{bmatrix}$$

I **In problems 28–32, find each product.**

28 $x^2 \cdot x^5$ x^7 **29** $3x(4x)^2$ $48x^3$ **30** $5x^5 \cdot 5x^5$ $25x^{10}$

31 $-4x(3x^2 - 7x - 5)$ $-12x^3 + 28x^2 + 20x$ **32** $x^2(3x^5 - 2x)$ $3x^7 - 2x^3$

R **33** The prices at Pavarotti's Pizzas are increasing by 15%. Rewrite the price list to reflect this change.

	Small	Medium	Large
Thin crust	6.50	7.50	8.50
Thick crust	7.50	8.50	9.50

	Small	Medium	Large
Thin crust	7.48	8.63	9.78
Thick crust	8.63	9.78	10.93

R **34** For what value of k is $(5, -3)$ a solution of $4x - 3y = k$? 29

R,I **In problems 35–38, simplify each expression and solve each equation.**

35 $2x(x + 3) + (3x)^2 + (4x)^2$ $27x^2 + 6x$ **36** $0.45x - 8 = 3.2x + 0.5$ $x \approx -3.09$

37 $15x^4y^2 - (3x^3y)(4xy)$ $3x^4y^2$ **38** $5x^3 - 6x^2y - 2x^2(x - 3y)$ $3x^3$

Additional Answer

Answer for problem **24b** is on page A 1.

R | **39** If all moves must be downward, how many different paths are there from A to B? 6

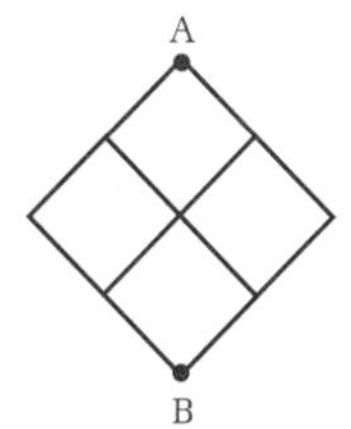

R | **40** Phil A. Telist is thinking of buying 20 stamps for his collection. Some of the stamps he wants cost \$0.90 each; the rest cost \$1.40 each. The total price of all 20 stamps is \$22.00.

a Using x to represent the number of stamps that cost \$0.90, write an expression for the number of stamps that cost \$1.40. $20 - x$

b Write an equation expressing the total price of the stamps as the sum of the prices of each kind. $0.9x + 1.4(20 - x) = 22.00$

c Solve the equation to find how many of each kind of stamp there are. 12 stamps at \$0.90 each; 8 stamps at \$1.40 each

R,I | **41** If $d = \sqrt{x^2 + y^2}$ and $(x, y) = (-5, 12)$, what is the value of d? 13

R | **In problems 42–45, solve each equation.**

42 $10 - x = 2(x + 17)$ $x = -8$

43 $(x + 3)(2x - 9) = 0$ $x = -3$ or $x = \frac{9}{2}$

44 $9(3y + 12) = 0$ $y = -4$

45 $8x_3 - 9 = 3x_3 + 22$ $x_3 = \frac{31}{5}$

B

P | **In problems 46–51, solve each inequality and graph the solution on a number line.**

46 $\frac{4}{3} + 5y > 2$ $y > \frac{2}{15}$

47 $4.3(1.2 - 0.25x) \le -26.66$ $x \ge 29.6$

48 $11 < \frac{2x + 3}{5} \le 13$ $26 < x \le 31$

49 $-\frac{1}{2}(n - 5) < 0$ or $\frac{1}{2}n + 1 < \frac{7}{2}$ $n \ne 5$

P,I | **50** $|z - 5.5| \ge 9$ $z \ge 14.5$ or $z \le -3.5$

P,I | **51** $3 - |y| > 1.6$ $-1.4 < y < 1.4$

P | **52** If the perimeter of the rectangle is at least 62 centimeters, what are the possible values of its area? $A \ge 150$ cm²

$2x + 3$

$x - 5$

P | **In problems 53–55, solve each inequality and graph the solution set on a number line.**

53 $9x + 3 > 15x - 9$ $x < 2$

54 $8x - 7 \le 10x - 15$ $x \ge 4$

55 $-9(x - 3) > -18$ $x < 5$

I | **56** Use the table of values to write an equation expressing the values of y in terms of the values of x. $y = 2^x$

x	0	1	2	3	4	5
y	1	2	4	8	16	32

Problem-Set Notes and Strategies, continued

- By using arrows to indicate a digraph and the flow in problem **39**, the downward paths can be enumerated. Labeling all the vertices will also help in the enumeration process.
- Some students may recall systems of equations from Algebra 1 and prefer to use this process for problem **40**. This is an equivalent method as suggested in the problem. Either $x + y = 20$ or $y = 20 - x$. Both methods yield the same solution and each method has its benefits.
- In problem **41** it might be wise to remind students once again that $\sqrt{x^2 + y^2} \ne x + y$. Urge students to substitute values for x, y as a check.
- Problem **45** involves a subscript x_3. Students may be familiar with the notation from Algebra 1. You might tell students that when working with many variables it is common to subscript them rather than using different letters to keep confusion to a minimum.
- Absolute values in problems **50** and **51** should trigger students to think "two problems" and not just one. Recall the definition of absolute value from Section **2.1**.

Additional Answers

Answers for problems **46 – 51** and **53 – 55** are on page **A** 1.

Problem Set, *continued*

R,I **57** Copy the following matrix and supply the missing data.

	At Bats	Hits	Batting Average (Hits ÷ At Bats)	
Stan	26	8	?	.308
Mickey	40	?	.300	12
Ted	?	12	.364	33

R **58** Find the values of a, b, and c that satisfy the equation
$3ax^2 + (b + 5)x + c - 7 = 12x^2 - 6x - 9$. $a = 4$; $b = -11$; $c = -2$

R **In problems 59–63, solve each equation and simplify each expression.**

59 $(z - 3)(3z + 1)(5z - 8) = 0$

60 $-3|x - y| \div (6y^2 - 18xy) \cdot (2y)^2$

61 $\sqrt{y} + 5 = 6 - 2\sqrt{y}$ $y = \frac{1}{9}$

62 $x - 2(x - 3) = \sqrt{533}$ $x \approx -17.09$

63 $-8^2 - \left(\frac{x}{2}\right)^2 - \frac{1}{4}x - 64 - \frac{x^2 - x}{4}$ $\frac{-x^2}{2} - 128$

C R,I **64** Consider the equation $x = \dfrac{-b + \sqrt{b^2 - 4ac}}{2a}$. What is the value of x if

a $a = 3$, $b = 1$, and $c = 4$? No real value **b** $a = -2$, $b = -5$, and $c = 3$? -3 **c** $a = 2$, $b = 4$, and $c = 1$? ≈ -0.293

I **65 a** Write an equation relating x and y such that the solutions include $(-2, 7)$, $(-1, 4)$, $(0, 1)$, $(1, -2)$, $(2, -5)$, and $(3, -8)$. $y = -3x + 1$

b Draw a graph that represents the equation you wrote in part **a**.

I **66** M is the midpoint of $\overline{AB}$. N is the midpoint of $\overline{BC}$.

a Find the coordinates of points M and N. $\left(3, \frac{7}{2}\right)$; $(5, 1)$

b Find length MN. ≈ 3.2

c Find length AC. ≈ 6.4

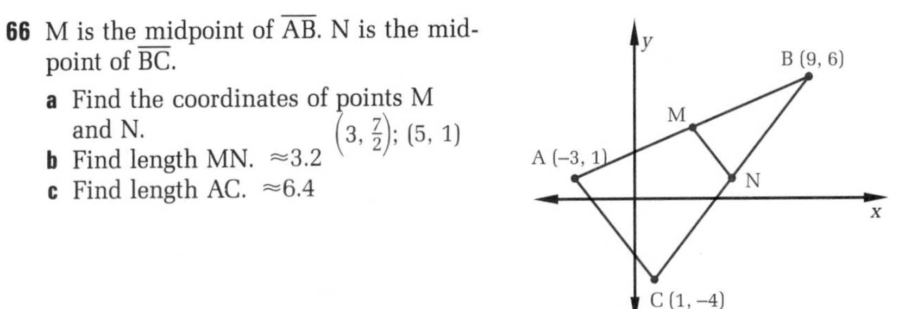

I **67** A cube with edges 3 inches long is painted on the outside, then cut into smaller cubes as shown. If one of these cubes is chosen at random, what is the probability that it

a Has exactly three painted faces? $\frac{8}{27}$

b Has exactly one painted face? $\frac{2}{9}$

c Has exactly four painted faces? 0

d Has no painted faces? $\frac{1}{27}$

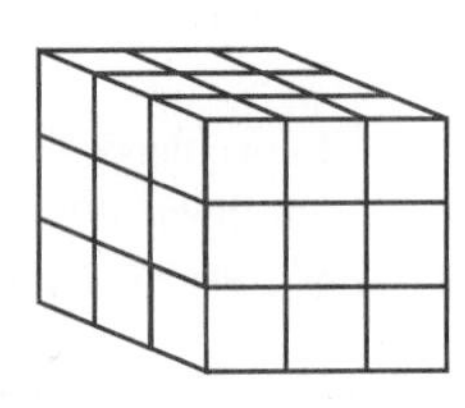

Problem-Set Notes and Strategies, continued

- In problem **58** students need to recall that for polynomials to be equal, the coefficients of the various degree terms must be equal. This sets up a system of equations which may be solved by investigation and substitution.

Additional Answers

59 $z = 3$ or $z = -\frac{1}{3}$ or $z = \frac{8}{5}$

60 $\frac{-2y|x-y|}{y-3x}$ or $\frac{2y|x-y|}{3x-y}$

65b

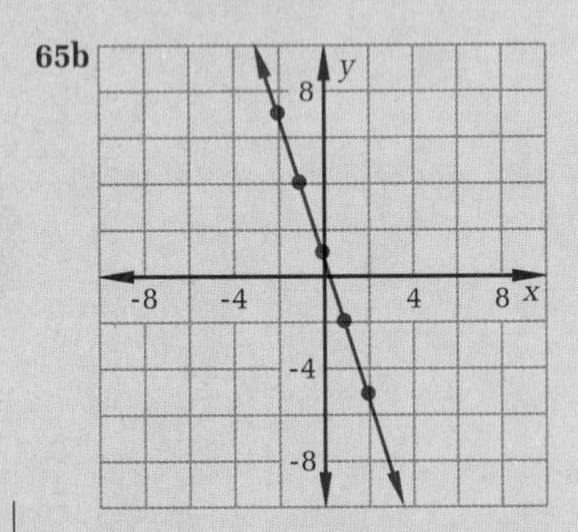

- In problem **64** the signs of a, b and c can cause problems when substituting, especially when b is negative. Students tend to forget that $-b$ is a positive number when b is negative.
- For problem **66** recall that the midpoint is the average of the coordinates of the endpoints.
- Students need to derive the basic counting definition of probability that will be explored in more detail in Section **2.6**. If you have the time, a visual aid of this cube may help to see the various possibilities. The center cube is frequently forgotten.

2.4 Introduction to Graphing

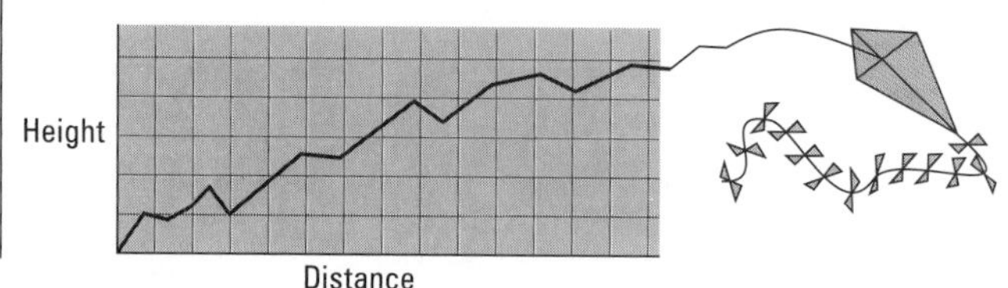

Objectives

After studying this section, you will be able to

- Graph an equation on the Cartesian coordinate system
- Use the midpoint and distance formulas
- Understand function notation

Part One: Introduction

The Graph of an Equation

One of the most important ways to communicate mathematics is to use a graph, since graphs picture what symbols describe.

The graph of $y = \frac{x^2 + 1}{x - 2}$, generated on a graphics calculator, helps us learn about the equation. Where is the y-intercept? How many x-intercepts can we find? What are the possible values of y? What pattern, if any, does the equation describe?

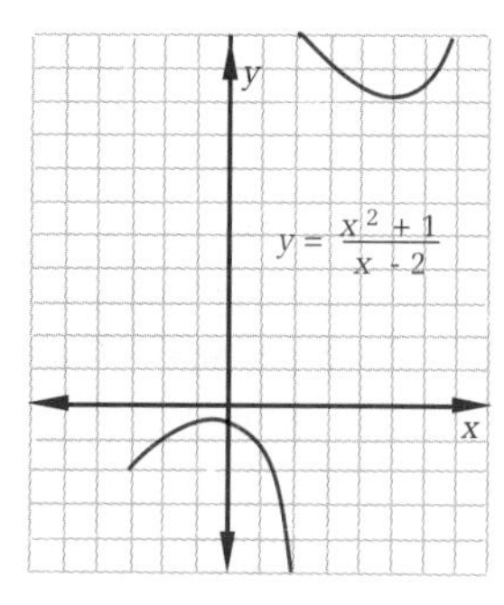

First, let's review some basic terms.

In earlier coursework, you learned about the Cartesian (or rectangular) coordinate system.

The ***x-axis*** and the ***y-axis*** intersect at right angles, dividing the plane into four regions called ***quadrants***, numbered I, II, III, and IV as shown.

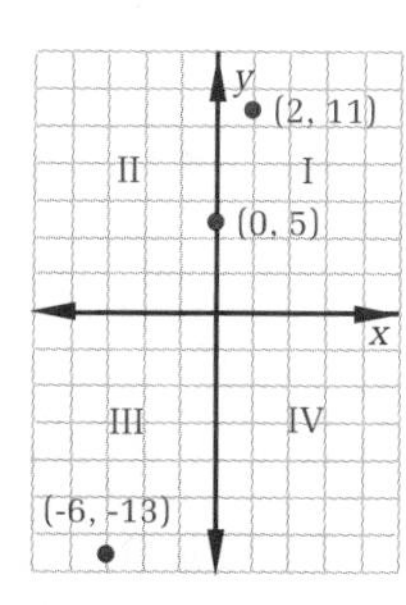

The point of intersection of the axes is called the ***origin***. The coordinates of the origin are (0, 0).

The first coordinate of an ordered pair is called the ***x-coordinate***, or the ***abscissa***. The second coordinate is the ***y-coordinate***, or ***ordinate***.

Class Planning

Time Schedule

Average: 1 day
Advanced: 1 day
Honors: $1\frac{1}{2}$ days

Resource References

Teacher's Resource Book
Class Opener 10
Practice 10
Manipulatives and Applications 4
Transparency 3

Class Opener

Sketch the graph of the equation $y = 2x - 7$ on the coordinate plane. Sketch the graph of $3x + 2y = 42$ on the same axes. Find the coordinates of the point of intersection.

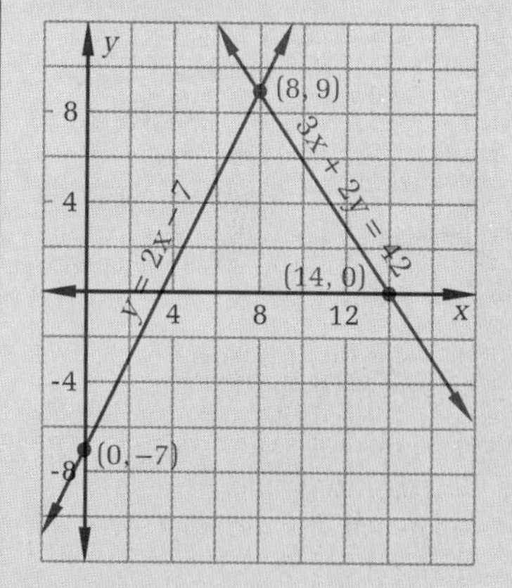

Lesson Notes

- Graphing on the Cartesian plane is presented as a method to communicate mathematics. You might include a little history of René Descartes in this lesson as the "Father of Analytic Geometry" and the impetus behind the discovery of calculus.

Vocabulary

x-axis	*x-coordinate*	*function*
y-axis	*abscissa*	*function notation*
quadrant	*y-coordinate*	*input*
origin	*ordinate*	*output*
	intercept	

Communicating Mathematics

Write a paragraph on the importance of graphing equations.

In the diagram on the preceding page, we marked three points with their coordinates, (2, 11), (−6, −13), and (0, 5). The point (2, 11) is in Quadrant I, (−6, −13) is in Quadrant III, and (0, 5) is on an axis, not in any quadrant.

To graph an equation means to indicate all points that solve the equation.

See p. 756

Example *Graph the equation* $y = 2x + 5$.

We make a table of values.

x	−3	−2	−1	0	1	2	3
y	−1	1	3	5	7	9	11

When we plot these points, they appear to lie on a line. To complete the graph, we connect the points, using arrows to indicate that the line continues in both directions.

The graph of $y = 2x + 5$ crosses the y-axis at (0, 5). The point (0, 5) is called the ***y-intercept*** of the graph.

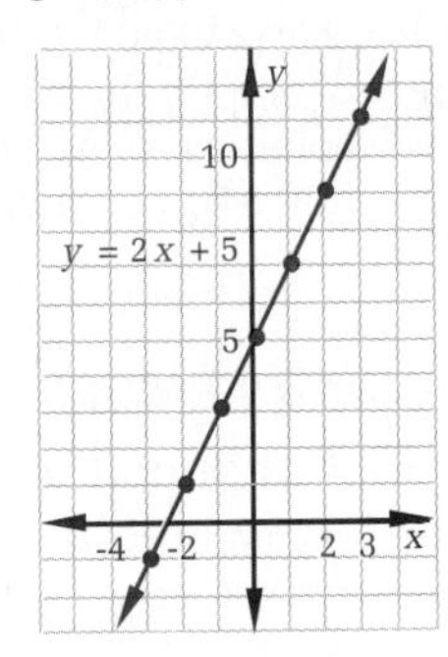

The Midpoint and Distance Formulas

You may remember the formulas used to find the midpoint of a line segment and the length of a line segment.

Lesson Notes, continued

- The midpoint formula can be remembered as the average of the coordinates. Encourage students to remember formulas by association rather than rote memorization.

Example *Find the midpoint of the line segment with endpoints* (−6, 4) *and* (−11, −10).

To find the coordinates of the midpoint of a segment, we find the averages of the corresponding coordinates of the endpoints.

$$x_m = \frac{-6 + (-11)}{2} = -8.5$$

$$y_m = \frac{4 + (-10)}{2} = -3$$

The midpoint is $(x_m, y_m) = (-8.5, -3)$.

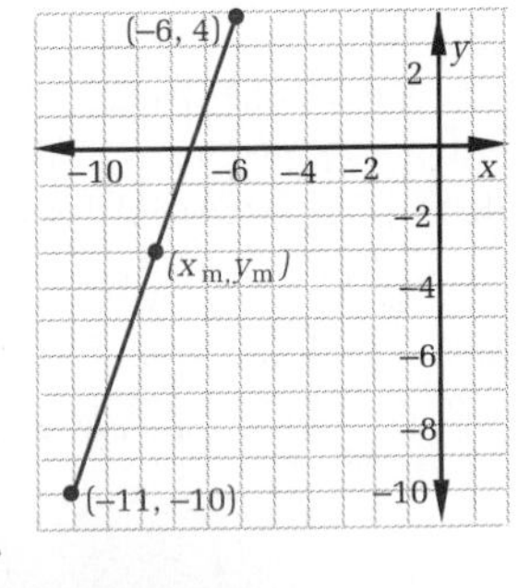

The Midpoint Formula

If a segment has endpoints (x_1, y_1) and (x_2, y_2), then the coordinates of its midpoint (x_m, y_m) are equal to $\left(\frac{x_1 + x_2}{2}, \frac{y_1 + y_2}{2}\right)$.

Example *Find the length of the segment with endpoints (−4, 7) and (2, −1).*

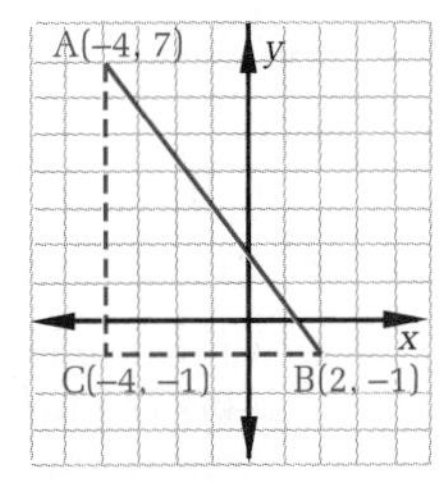

This length is the distance between the two endpoints. We can use the distance formula, which is based on the Pythagorean Theorem. In the figure, when point C is identified at (−4, −1), a right triangle is formed.

The Distance Formula

The distance d between any two points (x_1, y_1) and (x_2, y_2) in the coordinate plane is given by $d = \sqrt{(x_2 - x_1)^2 + (y_2 - y_1)^2}$.

Now we can find d in two ways.

Pythagorean Theorem

$$d = \sqrt{6^2 + 8^2} = \sqrt{36 + 64} = \sqrt{100} = 10$$

Distance Formula

$$d = \sqrt{[2 - (-4)]^2 + (-1 - 7)^2} = \sqrt{6^2 + (-8)^2} = \sqrt{100} = 10$$

Function Notation

The concept of a ***function*** is a key element in the language of algebra and will be fully developed in a later chapter. At this time we will introduce the idea, as well as function notation.

The formula $A = \pi r^2$ expresses the rule for finding the area of a circle A when the radius r is known. To put this in ***function notation*** we write

$A(r) = \pi r^2$

This equation makes three statements: The area A is a function of the radius r.
The radius r is the ***input*** variable.
$A(r)$ is the ***output*** value.

A function is a rule for finding one output value from an input. $A(r) = \pi r^2$ expresses the rule "the function A of r equals π times r squared."

Example *If $f(x) = 3x + 5$, what are $f(4)$ and $f(-7)$?*

$f(4)$ is the output value when the input variable x is 4.

$$f(4) = 3(4) + 5 = 12 + 5 = 17$$

$f(-7)$ is the output value when the input variable x is −7.

$$f(-7) = 3(-7) + 5 = -21 + 5 = -16$$

Lesson Notes, continued

- The distance formula is an application of the Pythagorean Theorem. Have students identify the point necessary to produce a right triangle and find the lengths of the two legs. With this approach, the distance formula seems to derive itself.

- Function notation is introduced in the context of the area of a circle. The statement "The area A is a function of the radius r." can also be interpreted as the area is dependent on the radius. As the radius changes, so does the area. The dependency is the function and should be studied and characterized. Emphasize the uniqueness of the output value for a given input value. This uniqueness is the motivating force behind the definition of a function. More detail will be provided in Chapter **5**.

2.4

Checkpoint

1 In which quadrant is the point (−3, 4)? Quadrant II
2 Find the midpoint of the line segment whose endpoints are (3, −4) and (−7, 12). (−2, 4)
3 Find the distance between the points (1, 5) and (−6, 17). ≈ 13.89

Assignment Guide

Average
7–11, 14, 16, 18–20, 22, 24, 27, 28
Advanced
9, 11, 14, 15, 18–21, 26, 28, 30
Honors
Day 1 ($\frac{1}{2}$ day) Section 2.3 41, 43, 45, 48, 50, 53, 61, 64, 67
($\frac{1}{2}$ day) 8, 9, 13, 16, 18, 22, 24, 28
Day 2 7, 10, 11, 14, 15, 19, 20, 23, 26, 27, 29, 30

Problem-Set Notes and Strategies

- Problems **1 – 6** reflect the presentation and should cause little trouble for the students. You might point out that in problem **3**, $g(0)$ is the y-intercept.

Part Two: Sample Problem

Problem *Refer to the graph of a function g.*
a *Find g(−2).*
b *For what value of x is g(x) = 7?*

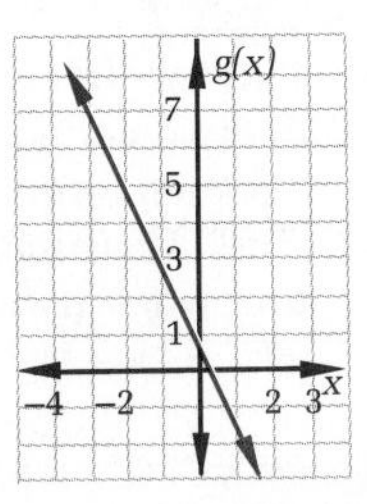

Solution **a** Along the horizontal axis, x represents the input value. Along the vertical axis, g(x) represents the output value. The graph shows that when x is −2, the output g(−2) is 5.
b According to the graph, g(x) = 7 when x = −3.

Part Three: Exercises and Problems

As explained in the "Letter to Students," the problems in this set have been labeled for you.

Warm-up Exercises

Practice **1** Refer to the rectangle.
a Find the coordinates of point C. (11, 5)
b Find the area of RECT. 28

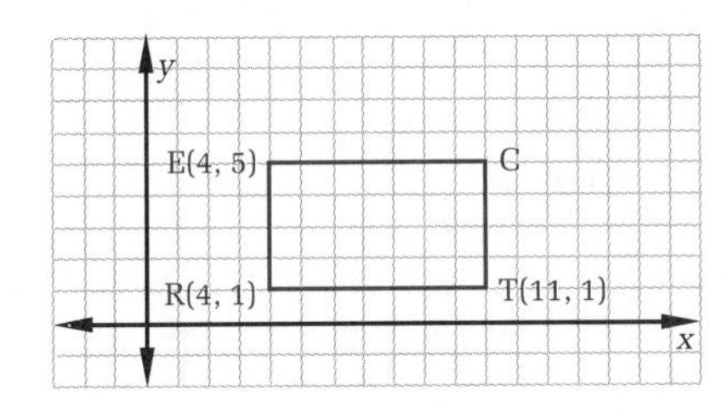

Review **2** Copy and complete the table of values for y = 3x − 4.

x	−2	−1	0	1	2
y	−10	−7	−4	−1	2

Practice **3** Find g(0) if g(x) = 34x − 17. −17

Practice **4** Find the length of the line segment with endpoints (7, 15) and (3, 12). 5

Practice **5** Match each labeled point on the graph with its description.
a x-intercept D
b y-intercept E
c (x, y) A
d x-coordinate C
e y-coordinate B

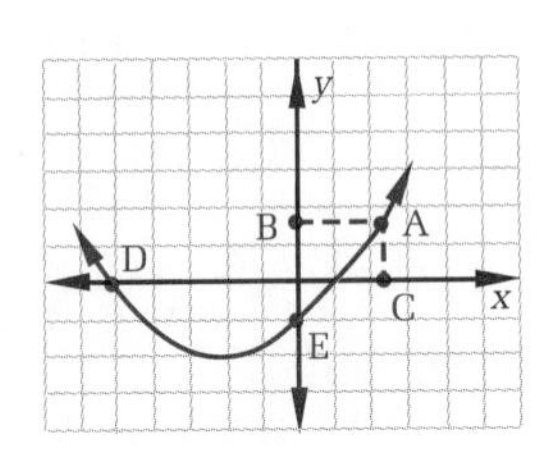

Practice **6** Find the distance between the points (7, 15) and (13, 7). 10

Problem Set

Practice **A**

7 Consider the graph of function Q.
a Find $Q(4)$. -2
b Find $Q(-1)$. 0
c Find $Q(3)$. -2.5
d For what values of x is $Q(x) = 3$? 6 and 8

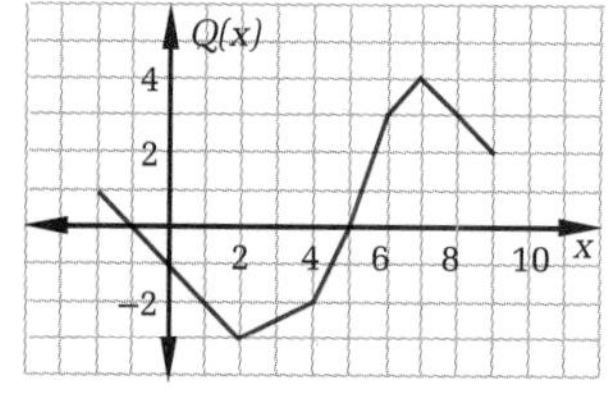

Practice

8 Refer to the diagram.
a Find the length of $\overline{AB}$. 15
b Find the midpoint of $\overline{AB}$. (4, 3.5)

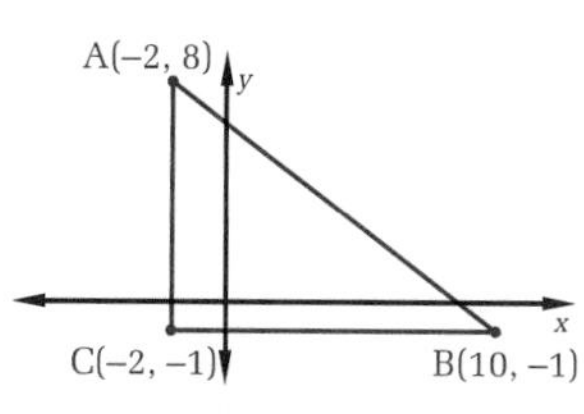

Review

9 If $y = 3x^2 + kx$, for what value of k is $(x, y) = (-2, 16)$? -2

Review

10 If $\angle P$ is acute, what are the possible values of x? $\{x: \frac{7}{4} < x < 13\}$

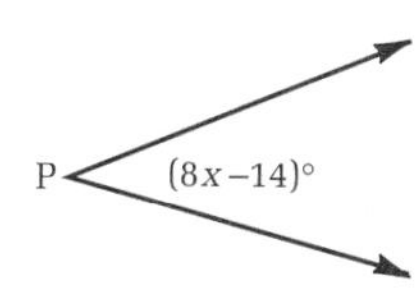

Practice

11 Let $h(x) = 3x - 6$. If the input values x are chosen from $\{-9, 0, 4, 15\}$, what is the set of output values $h(x)$? $\{-33, -6, 6, 39\}$

Review

12 If $x^2 + x > 6x - 5$, determine whether
a $x = 2$ is a solution No
b $x = 3$ is a solution No
c $x = 4$ is a solution Yes
d $x = 5$ is a solution Yes

In problems 13–15, solve for x.

Review

13 $Q + 7x = F$ $x = \frac{F - Q}{7}$

14 $(P + Q)x = R + T$ $x = \frac{R + T}{P + Q}$

15 $ax + bx = c$ $x = \frac{c}{a + b}$

Review

16 The perimeter of the rectangle is 6x.
a Use the information given in the diagram to write an equation for the perimeter. $6x = 10x - 16$
b Solve the equation for x. $x = 4$
c Find the numerical values of the length and width of the rectangle. 4; 8
d Find the numerical value of the perimeter. 24

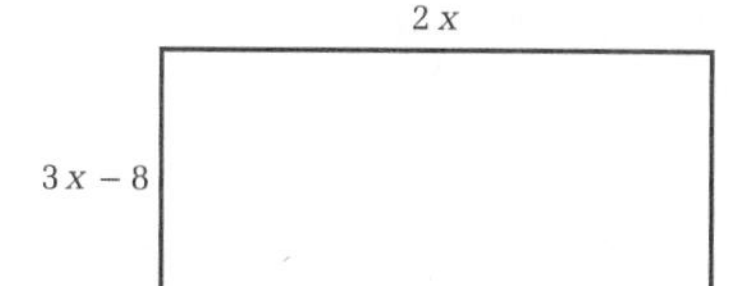

Review

17 What fractional part of AD is BC? $\frac{2}{5}$

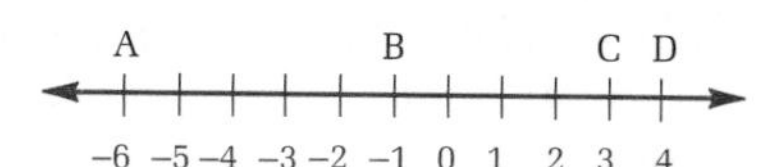

Problem-Set Notes and Strategies, continued

- Problem **7** uses the function notation and students may need to be reminded that $Q(x)$ is just another name for the output or y-coordinate.
- Problem **10** requires that students remember that an acute angle is greater than zero and less than 90°.
- Problems **13 – 15** ask students to solve a general equation with many variables. The importance of factoring as a tool is emphasized in problem **15**.

Problem-Set Notes and Strategies, continued

- The table in problem **18** is a good example of a matrix. It helps the students see numbers in rows and columns and stresses the importance of the position of a number in the matrix.
- Problem **19** is a nice review of the significance of multiplying a number by −1. Part **b** is also a nice review of the significance of the distance between two points and the fact that the distance may be measured in either direction.
- Problem **20** is an exercise in pattern recognition. Some students are instinctive in this ability and others need more time to master the concept. More rules will be developed in the next chapter to recognize linear patterns. Other relationships are discussed throughout the text.

Using Computers

This is an excellent time to bring out the graphics calculator or graphing software and give the students some introductory work. Since several of the problems start with data, it might also be a good time to use a spreadsheet with graphics capability. Problem **20** is an excellent introduction to linear regression using a calculator or a computer.

- A geometric model might help in problem **23**. Similar triangles can be drawn to show the relationship of a segment midpoint and two similar triangles. This problem can also be solved using slopes which students may recall from Algebra 1. Slopes will be discussed in more detail in the next chapter.

Problem Set, *continued*

Review Introduce

18 The following matrix contains statistics of the Great Lakers' five starters in a recent game against Frostbite Falls.

	Field Goals	3-Point Field Goals	Free Throws Made	Free Throws Attempted
Rhea Bounder	1	1	1	3
"Magic" Johansson	10	1	5	8
Leia Upp	6	0	8	10
Zetta Zchott	4	2	4	4
Gardenia Close	2	0	3	5

a What percentage of the starters made at least one 3-point field goal? 60%
b What percentage of the starters made at least half of their free throws? 80%
c How many points were scored by the five starters? 79
d What percentage of those points were scored by the two guards, "Magic" and Zetta? ≈58%

Practice Review

19 Points A and B on a number line have coordinates of −2 and 3 respectively. Find each of the following.
a The coordinate of the reflection of A over the zero point 2
b The coordinate of the reflection of A over B 8

Practice Introduce

20 For each table of values, write an equation that expresses the relationship between the x and y values.

a

y	0	1	2	3	4	5
x	10	9	8	7	6	5

$y = -x + 10$

b

x	2	3	4	$\frac{3}{8}$	$\frac{5}{12}$	$\frac{1}{10}$
y	$\frac{1}{2}$	$\frac{1}{3}$	$\frac{1}{4}$	$\frac{8}{3}$	$\frac{12}{5}$	10

$y = \frac{1}{x}$

Practice **B**

21 Refer to the diagram.
a Find the coordinates of point C. (−1, 2)
b Find AC and BC. 6; 8
c Find AB. 10

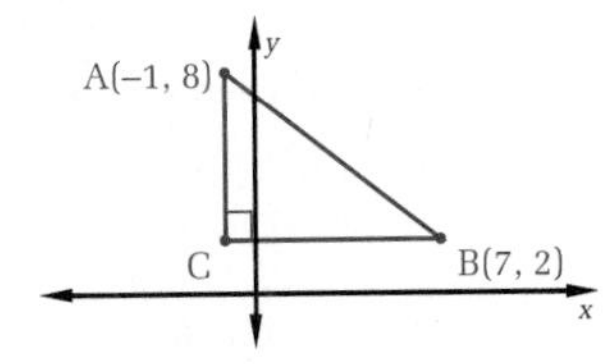

Practice

22 Find the perimeter of a parallelogram PQRS with vertices P(3, 26), Q(10, 2), R(16, 2), and S(9, 26). 62

Practice

23 If one endpoint of a segment is (4, 3) and the segment's midpoint is (9, 11), what is the other endpoint? (14, 19)

Practice Introduce

24 Write the inequalities that describe the shaded region and find the coordinates of the four vertices of the rectangle.
$2 \le x \le 7$ and $-6 \le y \le 4$; (2, 4), (7, 4), (7, −6), (2, −6)

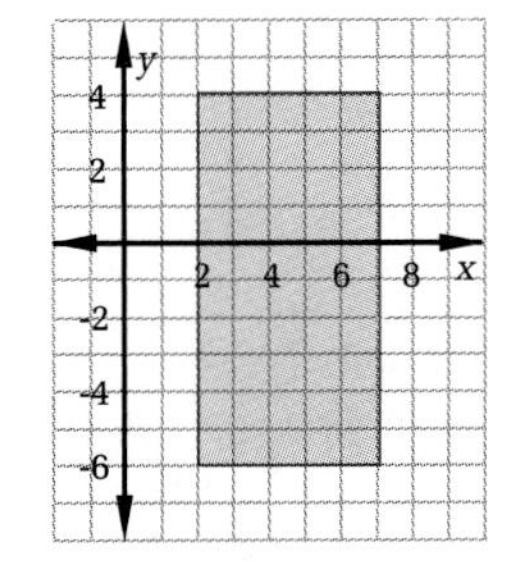

Review

25 Solve $\sum_{n=1}^{3}(x + n) = 36$ for x. $x = 10$

Review

26 A staircase will be built in a stairwell that is 150 inches long and 96 inches high.

a If each step is to be 6 inches high, how many steps will be built? 16

b How deep will each step be? 9.375 in.

c What will be the ratio of each step's height to its depth? 16:25

Practice Introduce

27 Refer to the graph.

a For what values of x is $h(x) = 0$? 2 and 6

b For what values of x is $h(x) < 0$? $\{x: 2 < x < 6\}$

c What is the range of possible values of $h(x)$? $\{h(x): h(x) \ge -3\}$

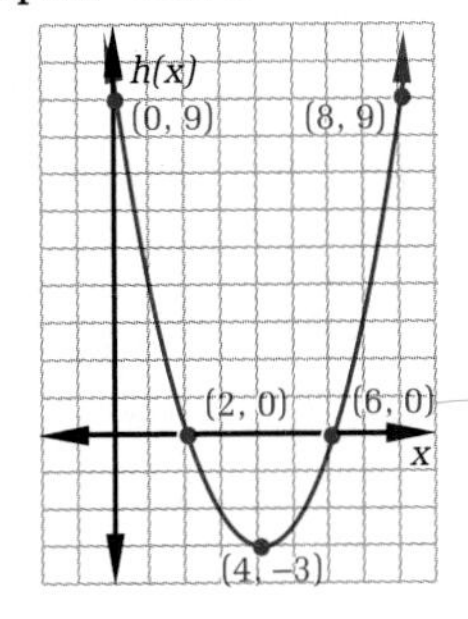

Review

28 Pim owes Sam $5. Sam owes Bom $7. Bom owes Pim $9. What is the minimum amount of money that the people can exchange to discharge all debts? Explain. $4; Sam and Bom each give Pim $2.

Practice

29 Find the area of the rectangle and find the coordinates of the point where the diagonals $\overline{ET}$ and $\overline{RC}$ intersect. 75; (12.5, 7.5)

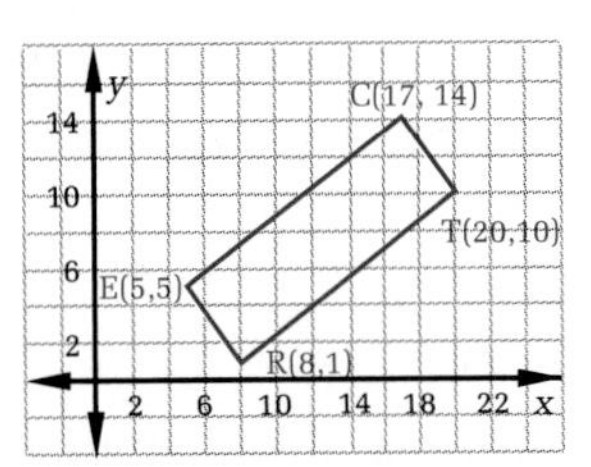

Review

30 Find the units digit of each number by analyzing the pattern of the simpler cases $2^1, 2^2, 2^3, 2^4, 2^5, \ldots$

a 2^{36} 6

b 2^{42} 4

Problem-Set Notes and Strategies, continued

- Linear inequalities are explored in problem **24**. Students must relate their one dimensional understanding to two dimensions. More will be detailed in Chapter **3.3**.
- A graph of the stairwell in problem **26** will help students understand the problem. Locate the bottom of the stairwell at the origin and have it ascend into the first quadrant. Part **c** investigates the slope of the stairwell giving real meaning to the slope as the steepness of the stairwell.
- Remind students that $h(x)$ is another name for the *y*-coordinate in problem **27**.
- Problem **28** may be most easily done by actually using money and observing what happens to the various bills. Then a generalization may begin.
- Knowing that the diagonals of a rectangle bisect each other is the key to solving problem **29**. Then the problem is merely finding the midpoint of either segment.
- Problem **30** involves pattern recognition again. Have students try to derive the rule that the exponent is divided by 4 and the remainder is used to evaluate the units digit using the first four cases.

2.5

Class Planning

Time Schedule

Average: 2 days
Advanced: 2 day
Honors: $2\frac{1}{2}$ days

Resource References

Teacher's Resource Book
Class Opener 11
Cooperative Learning 6
Practice 11
Enrichment 3
Transparency 4

Class Opener

For a school project, Rai had to keep track of the number of commercials on prime nighttime T.V. on ABC, NBC and CBS. His assignment was to see if there was any difference in the number of commercials on different nights of the week on each network. The three students on his team agreed to count the number of commercials every night for a week. Each selected a station. Suggest a method for organizing their data so it can be easily interpreted, compared, and combined with data collected during other weeks of the year.
This leads to the concept of a matrix. You can use the model to illustrate the definition of addition (consider data for several weeks in a row), for scalar multiplication (suppose the stations reduced the total number of commercials by 12%) and matrix multiplication (given the price of a commercial for each network, how much money does each network charge?)
Answers will vary. One possible answer follows.

	M	T	W	Th	F	S	S
CBS							
NBC							
ABC							

2.5 MATRICES

Objectives

After studying this section, you will be able to

- Multiply a matrix by a scalar and solve matrix equations
- Add matrices
- Multiply matrices

Part One: Introduction

Matrix Definitions and Scalar Multiplication

In business, sports, science, and mathematics, it is often convenient to organize data in tables called ***matrices***. A ***matrix*** is a rectangular arrangement of data, usually enclosed in brackets.

$$\begin{array}{cc} & \begin{array}{ccc}\textbf{Column 1} & \textbf{Column 2} & \textbf{Column 3}\end{array} \\ A = & \left[\begin{array}{ccc} 3 & -4 & 18.2 \\ -1 & 11 & -36 \end{array}\right] \begin{array}{l}\textbf{Row 1}\\ \textbf{Row 2}\end{array}\end{array}$$

This matrix contains two horizontal rows and three vertical columns and is called a 2×3 (read "two by three") matrix.

Two matrices are equal if they have the same dimensions and all corresponding terms are equal. For instance, if

$$\begin{bmatrix} 3x & -4y + 2 \\ -15 & 98.6 \end{bmatrix} = \begin{bmatrix} 15 & 38 \\ -z^2 + 1 & t \end{bmatrix}$$

we can conclude that the following four statements are true.

$$3x = 15 \qquad -4y + 2 = 38 \qquad -15 = -z^2 + 1 \qquad 98.6 = t$$

Finally, to multiply a matrix by a number, we multiply each element by that number. This is called ***scalar multiplication.***

Example *Multiply* $-5x\begin{bmatrix} 3 & x \\ -4x & 1 \end{bmatrix}$.

$$-5x\begin{bmatrix} 3 & x \\ -4x & 1 \end{bmatrix} = \begin{bmatrix} -5x(3) & -5x(x) \\ -5x(-4x) & -5x(1) \end{bmatrix} = \begin{bmatrix} -15x & -5x^2 \\ 20x^2 & -5x \end{bmatrix}$$

Vocabulary

matrix
scalar multiplication

Matrix Addition

If two matrices have the same dimensions, we can add them by adding the corresponding terms.

$$\begin{bmatrix} 12 & -3 & 9.3 \\ -6 & 14 & -2 \end{bmatrix} + \begin{bmatrix} 6.5 & 3 & 4.5 \\ -9 & 26 & 2.2 \end{bmatrix} = \begin{bmatrix} 18.5 & 0 & 13.8 \\ -15 & 40 & 0.2 \end{bmatrix}$$

See p. 749

Example *Add* $\begin{bmatrix} 2 & 5 \\ x^2 - 1 & x \end{bmatrix} + \begin{bmatrix} 7 & -4 \\ 2x + 2 & 4 \end{bmatrix}$.

$$\begin{bmatrix} 2 & 5 \\ x^2 - 1 & x \end{bmatrix} + \begin{bmatrix} 7 & -4 \\ 2x + 2 & 4 \end{bmatrix} = \begin{bmatrix} 9 & 1 \\ x^2 + 2x + 1 & x + 4 \end{bmatrix}$$

Matrix Multiplication

To multiply two matrices, we multiply the elements in each *row* of the first matrix by the corresponding elements in each *column* of the second matrix and add the products.

See p. 752

Example *Determine* A · B *for* A = $\begin{bmatrix} 2 & 5 \\ 4 & 3 \end{bmatrix}$ *and* B = $\begin{bmatrix} -7 & 9 & 0 \\ 4 & 6 & -11 \end{bmatrix}$.

Let us proceed one step at a time. We begin by multiplying the first row by the first column.

$$\begin{bmatrix} 2 & 5 \\ & \end{bmatrix}\begin{bmatrix} -7 & & \\ 4 & & \end{bmatrix} = \begin{bmatrix} 2(-7) + 5(4) & \_ & \_ \\ \_ & \_ & \_ \end{bmatrix}$$

$$= \begin{bmatrix} 6 & \_ & \_ \\ \_ & \_ & \_ \end{bmatrix}$$

We multiply the second row by the first column.

$$\begin{bmatrix} & \\ 4 & 3 \end{bmatrix}\begin{bmatrix} -7 & & \\ 4 & & \end{bmatrix} = \begin{bmatrix} 6 & \_ & \_ \\ 4(-7) + 3(4) & \_ & \_ \end{bmatrix}$$

$$= \begin{bmatrix} 6 & \_ & \_ \\ -16 & \_ & \_ \end{bmatrix}$$

We multiply the first row by the second column.

$$\begin{bmatrix} 2 & 5 \\ & \end{bmatrix}\begin{bmatrix} & 9 & \\ & 6 & \end{bmatrix} = \begin{bmatrix} 6 & 2(9) + 5(6) & \_ \\ -16 & \_ & \_ \end{bmatrix}$$

$$= \begin{bmatrix} 6 & 48 & \_ \\ -16 & \_ & \_ \end{bmatrix}$$

We continue in this way until all the elements of the product have been calculated.

$$\begin{bmatrix} 2 & 5 \\ 4 & 3 \end{bmatrix}\begin{bmatrix} -7 & 9 & 0 \\ 4 & 6 & -11 \end{bmatrix} = \begin{bmatrix} 6 & 48 & 2(0) + 5(-11) \\ -16 & 4(9) + 3(6) & 4(0) + 3(-11) \end{bmatrix}$$

$$= \begin{bmatrix} 6 & 48 & -55 \\ -16 & 54 & -33 \end{bmatrix}$$

Lesson Notes

- This section introduces some simple operations of matrices. The term scalar multiplication simply scales a matrix by the constant that it is multiplied by. Relate scaling to the scale found on a map. The fact that 1″ equals 100 miles on the map tells you that all the distances on the map need to be multiplied by a scaling factor to relate to the real world. This is the same idea of scalar multiplication of matrices.
- Matrix addition is intuitive and usually causes no problems. If time permits, you may want to introduce the matrix notation that the entry in the upper left corner is in the 1,1 position as it is in the first row and first column of the matrix. Each entry of a matrix has a unique designation as row position, column position. When multiplying matrices, the entry in the *m,n* position in the solution is found by multiplying the *m*th row of the first matrix by the *n*th column of the second matrix. It should be pointed out that matrix multiplication is definitely not commutative.

Using Computers

This is a good place to use a matrix manipulation package software, such as *Derive*, *Linear Kit*, or the disk included with this text. A calculator with matrix capabilities could also be used effectively. Help the students understand the concepts initially by using a few paper and pencil computations and then using the power of computers and calculators in solving problems involving matrices.

Notice that since we multiply row elements by column elements, the number of elements in each row of the first matrix must equal the number of elements in each column of the second matrix. Because matrices A and B meet this condition, we were able to find the product A · B, but because each of B's rows has three elements and each column of A has only two elements, the product B · A cannot be determined.

In general, the product of an $m \times n$ and an $n \times p$ matrix will be an $m \times p$ matrix.

Part Two: Sample Problems

Problem 1 *If* $A = \begin{bmatrix} 2 & -1 & 3 \\ -5 & 11 & 7.5 \end{bmatrix}$ *and* $B = \begin{bmatrix} 4 & 5 & -3.2 \\ 0 & -9 & 1 \end{bmatrix}$, *what is* $4A + 2B$?

Solution

$$4A + 2B = 4\begin{bmatrix} 2 & -1 & 3 \\ -5 & 11 & 7.5 \end{bmatrix} + 2\begin{bmatrix} 4 & 5 & -3.2 \\ 0 & -9 & 1 \end{bmatrix}$$

$$= \begin{bmatrix} 8 & -4 & 12 \\ -20 & 44 & 30 \end{bmatrix} + \begin{bmatrix} 8 & 10 & -6.4 \\ 0 & -18 & 2 \end{bmatrix}$$

$$= \begin{bmatrix} 16 & 6 & 5.6 \\ -20 & 26 & 32 \end{bmatrix}$$

Problem 2 *The first matrix shows batting statistics of three players for the first six weeks of the season. The second matrix shows the players' statistics for the seventh week.*

$$\begin{array}{l} \\ \\ \textbf{Tom} \\ \textbf{Rick} \\ \textbf{Harry} \end{array} \begin{array}{c} \begin{array}{cc} \textbf{At} & \\ \textbf{Bats} & \textbf{Hits} \end{array} \\ \begin{bmatrix} 140 & 45 \\ 98 & 33 \\ 87 & 18 \end{bmatrix} \end{array} \qquad \begin{array}{l} \\ \\ \textbf{Tom} \\ \textbf{Rick} \\ \textbf{Harry} \end{array} \begin{array}{c} \begin{array}{cc} \textbf{At} & \\ \textbf{Bats} & \textbf{Hits} \end{array} \\ \begin{bmatrix} 15 & 2 \\ 6 & 2 \\ 9 & 1 \end{bmatrix} \end{array}$$

a *What are the three players' statistics for the seven weeks?*

b *What are the players' batting averages after the seven weeks?*

c *If Tom is to be a .300 hitter for a total of 210 at bats during the season, how many of his remaining 55 at bats must be hits?*

Solution **a** To find the total statistics, we add the matrices.

$$\begin{bmatrix} 140 & 45 \\ 98 & 33 \\ 87 & 18 \end{bmatrix} + \begin{bmatrix} 15 & 2 \\ 6 & 2 \\ 9 & 1 \end{bmatrix} = \begin{bmatrix} 140 + 15 & 45 + 2 \\ 98 + 6 & 33 + 2 \\ 87 + 9 & 18 + 1 \end{bmatrix} = \begin{array}{c} \begin{array}{cc} \textbf{At} & \\ \textbf{Bats} & \textbf{Hits} \end{array} \\ \begin{bmatrix} 155 & 47 \\ 104 & 35 \\ 96 & 19 \end{bmatrix} \end{array} \begin{array}{l} \\ \\ \textbf{Tom} \\ \textbf{Rick} \\ \textbf{Harry} \end{array}$$

Checkpoint

1 Find the product of

$$3\begin{bmatrix} 4 & -3 \\ -2 & 5 \end{bmatrix} \quad \begin{bmatrix} 12 & -9 \\ -6 & 15 \end{bmatrix}$$

2 Find the sum of

$$\begin{bmatrix} 12 & -9 \\ -6 & 15 \end{bmatrix} + \begin{bmatrix} -7 & 13 \\ 9 & -11 \end{bmatrix}$$

$$\begin{bmatrix} 5 & 4 \\ 3 & 4 \end{bmatrix}$$

3 Find the product of

$$\begin{bmatrix} 2 & 3 \\ -4 & 1 \end{bmatrix}\begin{bmatrix} -3 & 2 \\ -5 & -1 \end{bmatrix}$$

$$\begin{bmatrix} -21 & 1 \\ 7 & -9 \end{bmatrix}$$

Assignment Guide

Average

Day 1 9, 10, 15, 17, 19–22, 26, 27, 29, 33

Day 2 11–14, 16, 18, 23–25, 30–32, 35, 36

Advanced

Day 1 9, 13, 16, 20, 21, 23, 24, 26, 29, 31, 35

Day 2 10–12, 14, 17–19, 22, 25, 27, 32, 36

Honors

Day 1 10, 12, 13, 16, 17, 20, 21, 24, 25, 27, 31, 35, 37, 41

Day 2 9, 11, 14, 15, 18, 19, 22, 23,26, 29, 30, 32, 36, 38, 40, 42

Day 3 ($\frac{1}{2}$ day) 33, 34, 39, 43
($\frac{1}{2}$ day) Section 2.6 5, 8, 9, 11–14, 17, 20, 21, 23

b A batting average is found by dividing hits by at bats.

Tom's average $= \frac{47}{155} \approx .303$

Rick's average $= \frac{35}{104} \approx .337$

Harry's average $= \frac{19}{96} \approx .198$

c If Tom is to have an average of .300, his total number x of hits divided by his 210 at bats must equal .300.

$$\frac{x}{210} = .300$$
$$x = 63$$

Since Tom already has 47 hits, he needs 63 − 47, or 16, more hits.

Part Three: Exercises and Problems

Warm-up Exercises

In problems 1–3, use the following matrices.

P

$$A = \begin{bmatrix} 3 & 2 \\ -3 & -1 \\ -2 & 1 \end{bmatrix} \qquad B = \begin{bmatrix} -5 & 4 \\ 6 & -3 \\ -6 & 2 \end{bmatrix}$$

1 Find A + B. **2** Find −4B. **3** Find A − B.

4 If $\begin{bmatrix} x_1 & y_1 \\ x_2 & y_2 \end{bmatrix} = \begin{bmatrix} -3 & 2 \\ -4 & -9 \end{bmatrix}$ and $m = \frac{y_2 - y_1}{x_2 - x_1}$, what is the value of m? 11

P **In problems 5 and 6, find each product.**

5 $[3 \quad -7]\begin{bmatrix} 2 \\ -3 \end{bmatrix}$ [27]

6 $\begin{bmatrix} -3 & -1 \\ 5 & -2 \end{bmatrix}\begin{bmatrix} 5 \\ -2 \end{bmatrix}$ $\begin{bmatrix} -13 \\ 29 \end{bmatrix}$

P **7** Consider the 2 × 3 matrix $A = \begin{bmatrix} 2 & 3 & -1 \\ -5 & 0 & 4 \end{bmatrix}$.

a Matrices of what dimensions can be added to A? 2 × 3
b If we can multiply A · B, what must be true of matrix B? B must have three rows.
c If we can multiply C · A, what must be true of matrix C? C must have two columns.

P **8** Multiply the matrices to find George's and Gracie's total pay for the week. George: $326; Gracie: $286

	Regular Hours	Overtime Hours		Hourly Wages	
George	40	12	·	5.75	Regular
Gracie	40	7		8.00	Overtime

Stumbling Block

Students may have a difficult time with matrix multiplication at first. It is not as intuitive as scalar multiplication. Encourage them to write out the process of multiplying the corresponding entries of the first row of the first matrix and the first column of the second matrix to arrive at the answer in the (1, 1) position in the solution matrix.

Problem-Set Notes and Strategies

- Problems **1 – 6** are general problems in the operations of matrices. Problem **4** uses the definition of slope which will be defined in Section **3.1.**
- In problem **7** make sure that the restrictions on the matrices in part **b** and **c** apply only to the number of rows and columns respectively. If matrix **A** is multiplied by matrix **B**, **B** must have three rows, but there is no limitation on the number of columns. Similarly, matrix **C** may have any number of rows.
- Problem **8** is a nice example of matrices in an actual problem. It shows how tables of values may be combined in a natural way.

Additional Answers

1 $\begin{bmatrix} -2 & 6 \\ 3 & -4 \\ -8 & 3 \end{bmatrix}$ **2** $\begin{bmatrix} 20 & -16 \\ -24 & 12 \\ 24 & -8 \end{bmatrix}$ **3** $\begin{bmatrix} 8 & -2 \\ -9 & 2 \\ 4 & -1 \end{bmatrix}$

2.5

Problem-Set Notes and Strategies, continued

- The problems in **9–14** are fairly straightforward. Problem **9** may cause some difficulty combining the scalar multiplication with subtraction. Encourage the student to write out each step of the process to make sure nothing is forgotten.
- Some students may have a calculator that will perform matrix operations. Problem **16** provides the opportunity for students to use their calculators. If students have a computer available and are familiar with programming, it is instructive to write a program that will carry out matrix multiplication for matrices of various sizes.
- Problem **18** introduces the concept of an identity matrix. There is a cooperative learning exercise in this lesson which will carry this concept one step further.

Additional Answers

9 $\begin{bmatrix} 20 & -1 \\ -20 & 4 \\ 7 & 0 \end{bmatrix}$

10 $\begin{bmatrix} 1 & 19 \\ 4 & -8 \\ 8 & -2 \end{bmatrix}$

11 $\begin{bmatrix} 23 & -11 \\ -20 & 26 \\ 19 & -17 \end{bmatrix}$

Answers for problems **17a – c**, and **18a, i – iii** are on page A 1.

Problem Set

A

In problems 9–14, use the following matrices.

P $P = \begin{bmatrix} 4 & 3 \\ -2 & 0 \\ -1 & 2 \end{bmatrix} \quad Q = \begin{bmatrix} -4 & 5 \\ 7 & -2 \\ -5 & 3 \end{bmatrix} \quad R = \begin{bmatrix} -2 & 4 \\ 3 & 1 \end{bmatrix}$

9 Find 3P − 2Q.

10 Find P · R.

11 Find Q · R.

12 Find R · P. Impossible

13 Can the product P · Q be found? Explain. No; the number of columns in P does not equal the number of rows in Q.

14 Can P and R be added? Explain. No; they have different dimensions.

P **15** What are the values of a and b if $\begin{bmatrix} a & 5 \\ -2 & 3b \end{bmatrix} = \begin{bmatrix} -14 & 5 \\ -2 & 22 \end{bmatrix}$? $-14; \frac{22}{3}$

P **16** The following matrices show the numbers of miles driven in a year by each member of a family and the cost per mile of driving each of the family's cars (including gas, insurance, license fees, etc.).

	Miles Driven Car 1	Miles Driven Car 2
Mother	4537	2358
Father	6547	3475
Child 1	1579	1298
Child 2	3543	1579

Cost per Mile (dollars)	
Car 1	0.18
Car 2	0.29

a What is the cost of each person's driving? Mother: \$1500.48; father: \$2186.21; child 1: \$660.64; child 2: \$1095.65

b What is the total cost of the family's driving? \$5442.98

c How much would the family have saved if child 1 had always driven car 1? \$142.78

P **17** Find each product.

a $\begin{bmatrix} 5 & -8 \\ 2 & 6 \end{bmatrix}\begin{bmatrix} -3 & 4 \\ 7 & 9 \end{bmatrix}$ **b** $\begin{bmatrix} 5 & -8 \\ 2 & 6 \end{bmatrix}\begin{bmatrix} 7 & 9 \\ -3 & 4 \end{bmatrix}$ **c** $\begin{bmatrix} 5 & -8 \\ 2 & 6 \end{bmatrix}\begin{bmatrix} 4 & -3 \\ 9 & 7 \end{bmatrix}$

R,I **18** **a** Find each product.

i $\begin{bmatrix} 1 & 0 \\ 0 & 1 \end{bmatrix}\begin{bmatrix} 3 & 2 \\ -1 & 4 \end{bmatrix}$ **ii** $\begin{bmatrix} 1 & 0 \\ 0 & 1 \end{bmatrix}\begin{bmatrix} 7 & 9 \\ -4 & 11 \end{bmatrix}$ **iii** $\begin{bmatrix} 1 & 0 \\ 0 & 1 \end{bmatrix}\begin{bmatrix} a & b \\ c & d \end{bmatrix}$

b The matrix $I = \begin{bmatrix} 1 & 0 \\ 0 & 1 \end{bmatrix}$ is called an *identity matrix*. If B is a 2 × 2 matrix, what is I · B? B

P **19** The matrix shows the wages of several employees. What scalar multiplier should be used to produce a product matrix representing the employees' wages after a 15% pay raise? 1.15

$\begin{bmatrix} 115 & 425 \\ 150 & 175 \\ 135 & 249 \\ 146 & 476 \end{bmatrix}$

R,I **20** **a** Use the left-hand diagram to find an expression equivalent to $(x + 7)^2$. $x^2 + 14x + 49$

b Copy the right-hand diagram and supply the missing measurements. What product does this diagram represent? $(2x + 3)^2$

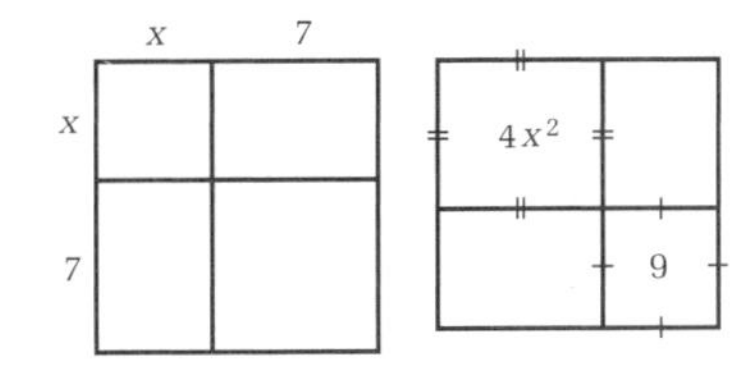

R **21** What are the possible values of the area of a circle if the measure of its radius is between 4 and 10? $\{16\pi < A < 100\pi\}$

R **22** The reflection of the point (−1, 3) over the x-axis is (−1, −3), and its reflection over the y-axis is (1, 3). What is the reflection of the point (5, −7) over the x-axis? (5, 7)

I **23** If you flip a fair coin, what is the probability it will land showing heads? $\frac{1}{2}$

R  **24** Solve each equation. Round your answers to the nearest tenth.

a $\sqrt{x} = 11.36$ $x \approx 129.0$

b $x^2 = 11.36$ $x \approx 3.4$ or $x \approx -3.4$

c $2x = 11.36$ $x \approx 5.7$

d $\frac{x}{2} = 11.36$ $x \approx 22.7$

 **B** P **25** The following matrices show the annual salaries paid to the four categories of employees (A, B, C, and D) at Wally's Widget Company and the number of employees in each category.

Salary (dollars)

	A	B	C	D
3 years experience	26,000	28,000	32,000	39,000
4 years experience	29,500	35,000	37,000	41,000
5 years experience	37,000	40,000	42,000	45,000
6 years experience	42,000	45,000	48,000	49,000

Number of Employees

A	B	C	D
5	3	1	2
0	6	4	3
9	2	1	5
16	3	0	2

a What is the total annual amount paid to the employees by Wally's? $2,390,000

b Suppose that a pay raise of 6% for all employees is put into effect. Write the new salary matrix.

c How much additional money will Wally's pay in salaries each year after the 6% pay raise? $143,400

27,560	29,680	33,920	41,340
31,270	37,100	39,220	43,460
39,220	42,400	44,520	47,700
44,520	47,700	50,880	51,940

R,I **26** Let $(x_1, y_1) = (3, 5)$ and $(x_2, y_2) = (-1, 4)$. Evaluate each expression.

a $\frac{y_2 - y_1}{x_2 - x_1}$ $\frac{1}{4}$

b $\frac{y_1 - y_2}{x_1 - x_2}$ $\frac{1}{4}$

R **27** Evaluate the following for $f(x) = \frac{9x - 63}{x^2 - 9}$.

a $f(7)$ 0

b $f(3)$ Undefined

c $f(-3)$ Undefined

R **28** What fractional part of the graph of $-19 \le x \le 5$ is the segment represented by $-17 \le x \le -2$? $\frac{5}{8}$

Problem-Set Notes and Strategies, continued

- Problem **20** is intended to help students realize that $(x + 7)^2$ is not $x^2 + 7^2$ by using a geometric argument.

Additional Answer

20b

	2x	3
2x	$4x^2$	6x
3	6x	9

- Symmetry is explored in problem **22**. Reflections about an axis or point can help to graph equations and will be explored in more detail in a later chapter.
- In problem **23**, probability is asked on a very intuitive level. You may get answers such as 50-50 or every other time. Explain how these are equivalent answers to $\frac{1}{2}$. The concept will be formally introduced in the next section.
- A rational expression is used in problem **27**. Remind students that division of zero is defined and the answer is zero. Division by zero is not defined and the only correct answer is undefined. Students will sometimes write the answer as ø which is the symbol for the empty set. Explain that this is an answer and not the same as undefined.

Problem-Set Notes and Strategies, continued

- The concept of an inverse is alluded to in problem **32**. A function that undoes the operations of another function may be defined as going backwards to find the input given the output. This will be formalized in Section **5.3.**
- Linear inequalities are explored in problem **34**. More detail is provided on this concept in Chapter **3**.

Cooperative Learning

Suggest that students work in groups of three or four and multiply the 2x2 matrix $\begin{bmatrix} 2 & 4 \\ 5 & 3 \end{bmatrix}$ on both the left and right by the 2x2 matrix $\begin{bmatrix} 1 & 0 \\ 0 & 1 \end{bmatrix}$ and observe the results. Multiply the 3x3 matrix $\begin{bmatrix} 3 & 1 & -2 \\ 5 & 7 & 4 \\ -1 & -3 & 5 \end{bmatrix}$ on both the right and left by the 3x3 matrix $\begin{bmatrix} 0 & 0 & 1 \\ 1 & 0 & 0 \\ 0 & 1 & 0 \end{bmatrix}$ and observe the results. Have the students try to generalize that the first 2x2 matrix they multiplied by was an identity matrix since nothing changed. In the second case, the rows and columns changed places depending on whether the matrix was right multiplied or left multiplied.

Problem Set, *continued*

R **29** Find the perimeter of the trapezoid shown. $19 + \sqrt{41}$

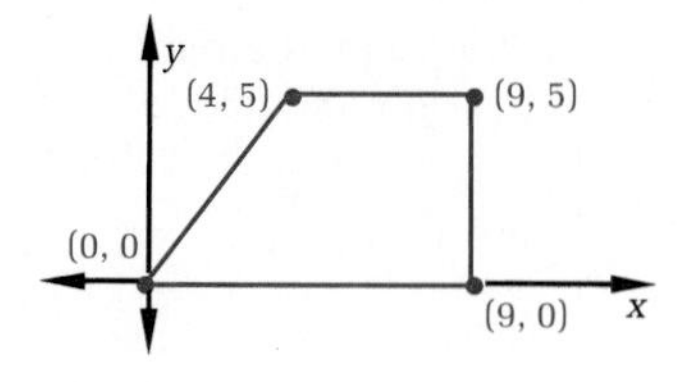

R **30** If $x^2 + 3x - 10 = 0$,

a Is $x = 2$ a solution? Yes
b Is $x = 5$ a solution? No
c Is $x = -2$ a solution? No
d Is $x = -5$ a solution? Yes

P **31** A warehouse has orders to ship refrigerators to stores A, B, C, and D. The refrigerators come in three styles—I, II, and III. The following matrix shows the number of refrigerators to be shipped to each store.

	A	B	C	D
I	1	0	3	8
II	6	2	1	1
III	7	4	2	1

If it costs $40 to ship a refrigerator to store A, $10 to ship one to store B, and $20 to ship one to either store C or store D, what is

a The total cost of shipping for each style of refrigerator? **a** I 260, II 300, III 380
b The total cost of shipping all the refrigerators? $940

R,I **32** Igor takes a number, adds 5, then multiplies the sum by 2. Rogi takes a number, divides by 2, then subtracts 5 from the quotient. If you gave a number to Igor and gave his result to Rogi, what result would Rogi find? The number you gave Igor

R **33** Copy the following matrix, supplying the missing data.

Number of nickels	3	6	7	x	x
Number of dimes	8	2	4	3x	x + 2
Number of quarters	2	6	3	2x	2x
Total value (cents)	145	200	150	85x	65x + 20

I **34** The shaded region represents the intersection of which of the following inequalities? **b** and **e**

a $-4 \le x \le 3$
b $-4 \le y \le 3$
c $x > 7$
d $x \le 7$
e $x < 7$

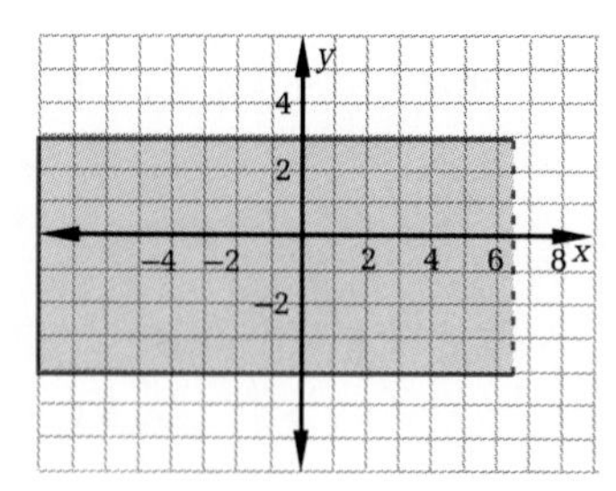

R In problems 35 and 36, find the ordered pair (x, y) that satisfies both equations simultaneously.

35 $\begin{cases} x + y = 32 \\ y = x - 4 \end{cases}$ (18, 14)

36 $\begin{cases} x + y = 13 \\ x - y = 7 \end{cases}$ (10, 3)

C

In problems 37–39, let $y = 10x - 35$ and $z = 1 - 2x$.

R **37** Find all values of x for which $y > 0$. $\{x{:}x > 3.5\}$

38 Find all values of x for which $z \leq 17$. $\{x{:}x \geq -8\}$

39 Find all values of x for which $y \leq z$. $\{x{:}x \leq 3\}$

I **40** Write an equation that relates the values of x and y given in the table. $y = x - 7$

x	3	5	7	9
y	−4	−2	0	2

R,I **41** **a** Plot on a number line the values of x for which $|x - 3| = 2$.
b Plot on a number line the values of x for which $|x - 5| = 1$.
c Look for a pattern in parts **a** and **b.** Then sketch a number line showing the solutions of $|x - 10| = 3$.
d Draw a diagram representing the values of x for which $|x - 7| = k$.

I **42** A pizza parlor offers five toppings that can be added to a basic cheese pizza.
a How many one-topping pizzas are available? 5
b How many two-topping pizzas are available? 10
c How many three-topping pizzas are available? 10
d How many four-topping pizzas are available? 5
e How many five-topping pizzas are available? 1

P **43** Using the following matrices, calculate $C \cdot D$ and $D \cdot C$. Is matrix multiplication commutative?

$C = \begin{bmatrix} 1 & -2 \\ 6 & 5 \end{bmatrix}$ $D = \begin{bmatrix} 3 & 7 \\ 8 & -4 \end{bmatrix}$ $\begin{bmatrix} -13 & 15 \\ 58 & 22 \end{bmatrix}$; $\begin{bmatrix} 45 & 29 \\ -16 & -36 \end{bmatrix}$; no

Problem-Set Notes and Strategies, continued

- Have the students look for recognizable patterns in problem **40**. Linear relationships such as this will be explored in depth in the next chapter and the methods for writing linear equations detailed. For now the relationships are simple enough to recognize with a little perseverance.
- The geometric interpretation of absolute value is explored in problem **41**. It might be worthwhile to work the problem algebraically to verify that the two answers found on the number line are the same as the answers arrived at through the algebraic definition.
- The numbers generated in problem **42** are from Pascal's triangle which was explored in an earlier problem set. These are called combinations and are used in many different applications. Have students notice the symmetry. Permutations and combinations are outlined in detail in chapter **12**. The interested student may want to look ahead.

Communicating Mathematics

Write a paragraph to explain why matrix multiplication is not generally a commutative operation. What conditions must be met in order for the multiplication to be commutative? Find an example of two matrices such that the multiplication is commutative.

Additional Answers

Answers for problems **41a – d** are on page **A** 1.

2.6

Class Planning

Time Schedule

Average: 2 days
Advanced: 2 days
Honors: 1 day

Resource References

Teacher's Resource Book
Class Opener 12
Cooperative Learning 7
Practice 12
Enrichment 4

Class Opener

Given the matrix

$$A = \begin{bmatrix} 12 & 8 & 7 \\ -2 & 5 & -3 \end{bmatrix}$$ and the

matrix $$B = \begin{bmatrix} 7 & 3 \\ 4 & 9 \\ -2 & 1 \end{bmatrix}$$, consider

the product of the two matrices. If an element is selected randomly from the product matrix, what is the probability it will be an even number?

The product matrix is

$$\begin{bmatrix} 102 & 115 \\ 12 & 36 \end{bmatrix}$$

Probability is $\frac{3}{4}$.

Lesson Notes

- This section is one of many exploring concepts in probability and statistics. Probability is valuable both as a topic and for the organization it imposes on solving problems. Probability is explored from a theoretical point of view where each outcome is equally likely.

2.6 Probability

Objective

After studying this section, you will be able to
- Calculate probabilities

Part One: Introduction

In our everyday lives, we cannot always know what will happen when we make decisions. Often, however, the *chances* of events can be determined, thus allowing us to make better-informed decisions. To determine such chances, we use ***probability.***

You will find probability useful in a variety of everyday situations. In addition, by setting up and solving probability problems, you will develop the kind of organized thinking that can be applied in all mathematical studies. Many probability problems are analyzed by means of the following two-part procedure.

Basic Steps for Probability Problems

1. **Find all possible outcomes in an organized manner. This process is called *exhaustive listing.***
2. **Determine which of the possibilities yield the specified result. We will call these winners.**

Probability is calculated according to the formula

$$\text{Probability} = \frac{\text{number of winners}}{\text{number of possibilities}}$$

Vocabulary

probability
exhaustive listing

Part Two: Sample Problems

Problem 1 *If one of the points (x, y) is randomly chosen, what is the probability that its coordinates satisfy the inequality* $3x + 4y > 15$?

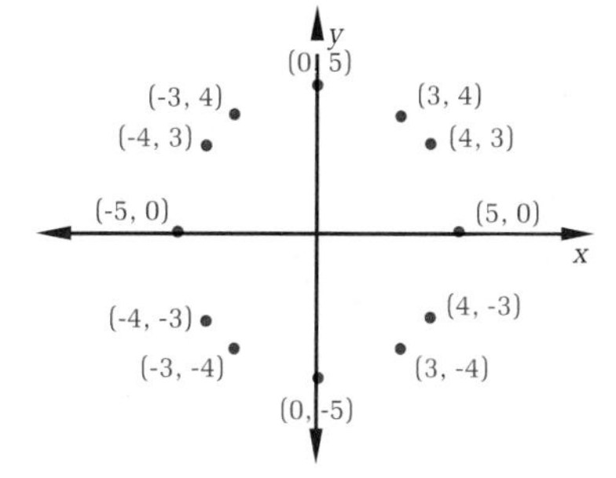

Solution We check each possibility.

(x, y)	**3x + 4y**	**3x + 4y > 15?**
(5, 0)	$3(5) + 4(0) = 15$	No
(4, 3)	$3(4) + 4(3) = 24$	Yes
(3, 4)	$3(3) + 4(4) = 25$	Yes
(0, 5)	$3(0) + 4(5) = 20$	Yes
(−3, 4)	$3(-3) + 4(4) = 7$	No
(−4, 3)	$3(-4) + 4(3) = 0$	No
(−5, 0)	$3(-5) + 4(0) = -15$	No
(−4, −3)	$3(-4) + 4(-3) = -24$	No
(−3, −4)	$3(-3) + 4(-4) = -25$	No
(0, −5)	$3(0) + 4(-5) = -20$	No
(3, −4)	$3(3) + 4(-4) = -7$	No
(4, −3)	$3(4) + 4(-3) = 0$	No

There are 3 winners out of 12 possibilities. The probability is $\frac{3}{12}$, or $\frac{1}{4}$.

Problem 2 *If three of the five figures shown are selected at random, what is the probability that all three will have perimeters greater than 28?*

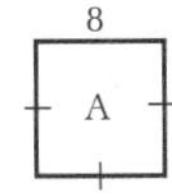

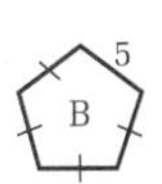

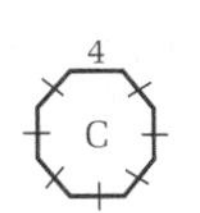

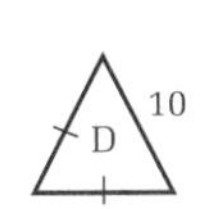

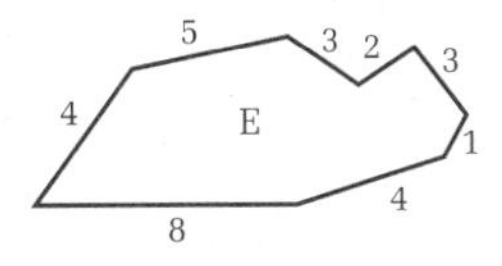

Solution First, we find each perimeter.

Perimeter of figure A = $8 \cdot 4 = 32$
Perimeter of figure B = $5 \cdot 5 = 25$
Perimeter of figure C = $4 \cdot 8 = 32$
Perimeter of figure D = $10 \cdot 3 = 30$
Perimeter of figure E = $8 + 4 + 5 + 3 + 2 + 3 + 1 + 4 = 30$

Lesson Notes, contiinued

You might point out two other common approaches as well. The subjective approach is the likelihood of an event occurring such as a weatherman's statement "There is a 20% chance of rain tomorrow". The empirical approach may also be necessary when the probabilities are not equally likely. For example, suppose a die was loaded so that even values came up twice as often as odd values. The probabilities could be determined by a lengthy experiment of rolling the die many thousand times. As the number of trials increases, the law of large numbers states that the relative frequencies approach the true probability. For this text, only theoretical probability problems will be presented. Encourage listing all possibilities or drawing tree diagrams to view the possibilities.

Checkpoint

1. If a fair coin is flipped twice, what is the probability of it landing heads up both times? $\frac{1}{4}$
2. If a die is tossed once, what is the probability of getting less than a 7? 1
3. What is the probability that you will score 100% on your next Algebra 2 test? Answers may vary. Students should estimate based on past performance.

Assignment Guide

Average

Day 1	7, 9, 10, 13, 16, 17, 19, 20
Day 2	5–8, 11, 12, 14, 16, 18, 21, 22

Advanced

Day 1	8–10, 13, 14, 16, 17, 20, 21
Day 2	5–7, 11, 12, 15, 18, 19, 22, 23

Honors

Day 1	($\frac{1}{2}$ day) Section 2.5 33, 34, 39, 43
	($\frac{1}{2}$ day) 5, 8, 9, 11–14, 17, 20, 21, 23
Day 2	($\frac{1}{2}$ day) 6, 7, 10, 15, 16, 18, 19, 22
	($\frac{1}{2}$ day) Review Problems 1, 5, 7, 9, 11, 15, 18, 19, 21, 23, 25, 30, 35

Problem-Set Notes and Strategies

- If an event is certain not to occur, the probability is 0 and if it is certain to occur, the probability is 1. Problem **1** identifies that all probabilities must be between 0 and 1.
- Problem **3** correlates probability to areas of the rectangle. This concept will be useful when discussing normal curves in Section **8.5**.

Three of the figures are to be selected, so we list all possible threesomes and put a box around each winner—that is, every threesome in which each figure's perimeter is greater than 28. It is important to write the list in an orderly, logical way.

ABC	[ACD]	[ADE]	BCD	BDE	[CDE]
ABD	[ACE]		BCE		
ABE					

The probability is $\frac{4}{10}$, or .4. (Note the similarity between this procedure and some of the approaches we used in solving the handshake problem in Chapter 1.)

Problem 3 *A point within the large rectangle is chosen at random. Find the probability that the point is in the shaded rectangle.*

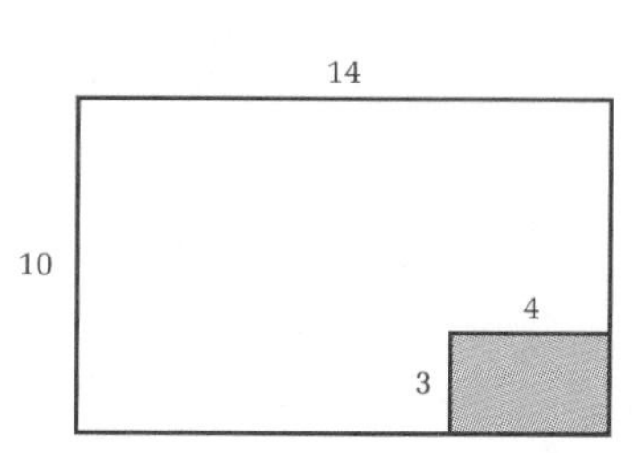

Solution The entire rectangle has an area of 10(14), or 140. The shaded rectangle has an area of 3(4), or 12. The probability is $\frac{12}{140}$, or $\approx .0857$.

Part Three: Exercises and Problems

Warm-up Exercises

1 The probability of a given event's occurring must be between what two numbers? 0 and 1 inclusive

P
P **2** Give an example of a situation in which you would have to estimate the probability of a particular event. Answers will vary.

P **3** A point is chosen at random from within the large rectangle.

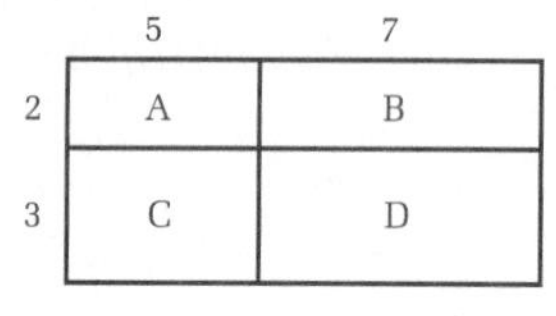

a What is the probability that the point is in region A? In region B? In region C? In region D? $\frac{1}{6}$; $\frac{7}{30}$; $\frac{1}{4}$; $\frac{7}{20}$

b What is the sum of the probabilities you found in part **a**? 1

P **4** If three of the five points shown are chosen at random, what is the probability that they are vertices of a triangle? 1

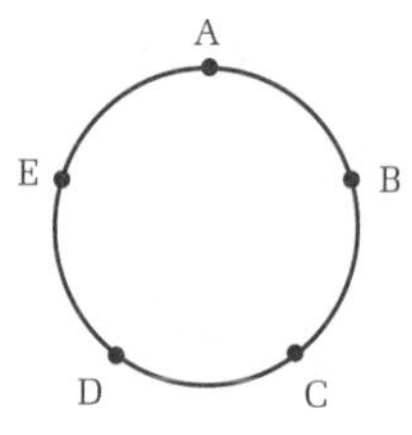

Problem Set

A

P,R **5** For the function $f(x) = \frac{1}{2}x - 5$, an input x is chosen at random from {2, 4, 6, 8, 10, 12, 14, 16}. What is the probability that $f(x) \leq 0$? $\frac{5}{8}$

P,R **6** A point is chosen at random on $\overline{PQ}$.

a If the point's coordinate is n, what is the probability that $0 \leq n \leq 14$? $\frac{7}{12}$

b What is the probability that the point is within 5 units of the point with coordinate 3? $\frac{5}{12}$

P **7** A pancake house has five different syrups. How many different combinations of two different syrups can you use on your pancakes? 10

R,I **8** An engine turns a shaft 120 times per minute. Write a function n to describe the shaft's number of revolutions after t minutes. $n(t) = 120t$

R **9** Determine whether $(x, y) = (2, -5)$ is a solution of each equation.

a $x + y = 7$ No **b** $2x - y = 9$ Yes **c** $3x - y = 11$ Yes

P,R **10** If a number is chosen at random from {−6, −4, −2, 0, 2, 4, 6}, what is the probability that it is a solution of at least one of the following equations? $\frac{5}{7}$

$x^2 + 8x + 12 = 0$ $x^2 = 16$ $-3x + 6 = 0$

R,I **11** Evaluate each expression for $(x_1, y_1) = (3, 5)$, $(x_2, y_2) = (5, -1)$, and $(x_3, y_3) = (6, -4)$.

a $\frac{y_2 - y_1}{x_2 - x_1}$ -3 **b** $\frac{y_3 - y_1}{x_3 - x_1}$ -3 **c** $\frac{y_3 - y_2}{x_3 - x_2}$ -3

B

R **12** Redd and Budd decide to race, with Redd giving Budd a 100-yard head start. Redd runs at 8 yards per second, and Budd runs at 6 yards per second. In how many seconds will Redd catch Budd? 50 sec

R,I **13** Is $(x, y) = (4, 6)$ a solution of both $3x - y = 6$ and $y = 2x - 1$? No

R **In problems 14 and 15, solve for (x, y).**

14 $\begin{bmatrix} 3 & x \\ y & 2 \end{bmatrix}\begin{bmatrix} 5 \\ 6 \end{bmatrix} = \begin{bmatrix} -3 \\ -10 \end{bmatrix}$ $\left(-3, -\frac{22}{5}\right)$

15 $\begin{bmatrix} 4 & 0 \\ 2 & 3 \end{bmatrix}\begin{bmatrix} x \\ y \end{bmatrix} = \begin{bmatrix} -12 \\ 15 \end{bmatrix}$ $(-3, 7)$

Communicating Mathematics

Have the students write a paragraph on the probability of their achieving an "A" in this course. Have them comment on the way they arrived at their answer and its accuracy.

Problem-Set Notes and Strategies, continued

- Problem **7** once again asks for combinations or the numbers from Pascal's triangle. Without the formula, students need to enumerate the solutions.
- Problem **8** evaluates the students understanding of function notation: the number of revolutions is a function of time.
- The formula for slope is used in problem **11** and will be identified in Section **3.1**.
- In problems **14** and **15** make sure that answers are written in parentheses as ordered pairs and not as matrices.

Problem-Set Notes and Strategies, continued

- Problem **16** is asking for observations about intercepts and slopes. This is a good introductory problem for the next chapter and a review of Algebra 1 properties.

Additional Answer

Answer for problem **16** is on page **A** 1.

- In problem **18** the determinant of a matrix is defined. Be prepared to answer questions on the use of a determinant. It can be used to find the inverse of a matrix, to solve systems of equations, to find areas of triangles, to find equations of lines and many other uses.

- Problems **20 – 22** have a small enough sample space to enumerate and observe the number of winning possibilities.

Cooperative Learning

Suggest that students work in groups of three or four and ask each group to roll a single die 100 times and record the results. Ask students to predict the probability of the resulting value, 1 to 6. Each value has equal probability. The observed probability may not be exactly $\frac{1}{6}$, but should be close. Now have the students roll a pair of dice and repeat the experiment using the sum of the numbers on the dice. They should observe that the frequency of observations is not the same for each value from 2 through 11 since each value is not equally likely.

Problem Set, *continued*

R,I **16** Graph each equation on the same coordinate system. What do you observe?

a $y = 1x - 5$ **b** $y = 2x - 5$ **c** $y = 3x - 5$

They all pass through $(0, -5)$, and the lines become more vertical as the coefficient of x increase

R **17** Find matrix A.

$$3A - \begin{bmatrix} -4 & 7 & 0 \\ 2 & -6 & 5 \end{bmatrix} = \begin{bmatrix} 10 & 2 & 15 \\ 19 & 0 & -32 \end{bmatrix}\begin{bmatrix} 2 & 3 & 5 \\ 7 & -2 & -9 \end{bmatrix}$$

I **18** The *determinant* of a 2 × 2 matrix is a number such that

if $M = \begin{bmatrix} a & b \\ c & d \end{bmatrix}$, then $\det M = ad - bc$.

See p. 777

a Find $\det \begin{bmatrix} -4 & 3 \\ 2 & 6 \end{bmatrix}$. -30

b Find $\det \begin{bmatrix} 3 & 2 \\ -4 & 5 \end{bmatrix}$. 23

c Find $\det \begin{bmatrix} 1 & 0 \\ 0 & 1 \end{bmatrix}$. 1

P **19** Suppose that $y = 3x^2 - x$ and $z = x^2 + 4$. If x is chosen at random from $\{-2, -1, 0, 1, 2, 3\}$, what is the probability that $y \le z$? $\frac{1}{2}$

C

P **In problems 20–22, assume that *a* and *b* are chosen at random from {1, 2, 3, 4, 5, 6} and that *a* and *b* can be equal.**

20 What is the probability that $a + b = 7$? $\frac{1}{6}$

21 What is the probability that $a + b$ is even? $\frac{1}{2}$

22 What is the probability that $a \cdot b$ is even? $\frac{3}{4}$

R **23** Consider the matrix equation

$$\begin{bmatrix} 2 & 3 \\ 1 & -2 \end{bmatrix} \cdot A = \begin{bmatrix} -1 \\ 10 \end{bmatrix}.$$

a What are the dimensions of matrix A? 2 × 1

b Find matrix A. $\begin{bmatrix} 4 \\ -3 \end{bmatrix}$

Chapter Summary

Concepts and Procedures

After studying this chapter, you should be able to

- Add expressions and recognize equivalent expressions (2.1)
- Use the Distributive Property of Multiplication over Addition (2.1)
- Calculate absolute values (2.1)
- Follow the standard algebraic order of operations (2.1)
- Recognize equivalent equations (2.2)
- Solve equations by using properties of equality (2.2)
- Express solutions of equations as ordered pairs (2.2)
- Apply properties of inequality (2.3)
- Solve linear inequalities in one variable (2.3)
- Graph solutions of inequalities on the number line (2.3)
- Graph an equation on the Cartesian coordinate system (2.4)
- Use the midpoint and distance formulas (2.4)
- Understand function notation (2.4)
- Multiply a matrix by a scalar and solve matrix equations (2.5)
- Add and multiply matrices (2.5)
- Calculate probabilities (2.6)

Vocabulary

abscissa (2.4)
absolute value (2.1)
additive inverse (2.1)
distribute (2.1)
equation (2.2)
equivalent expressions (2.1)
exhaustive listing (2.6)
expression (2.1)
function (2.4)
function notation (2.4)
inequality (2.3)
input (2.4)
intercept (2.4)
like terms (2.1)
linear equation (2.2)
matrix (2.5)
opposite (2.1)
ordered pair (2.2)
ordinate (2.4)
origin (2.4)
output (2.4)
probability (2.6)
quadrant (2.4)
scalar multiplication (2.5)
x-axis (2.4)
x-coordinate (2.4)
y-axis (2.4)
y-coordinate (2.4)

Properties

Addition and Subtraction Properties of Equality (2.2)
Addition and Subtraction Properties of Inequality (2.3)
Distributive Property of Multiplication over Addition (2.1)
Multiplication and Division Properties of Equality (2.2)
Multiplication and Division Properties of Inequality (2.3)
Zero Product Property (2.2)

2

Class Planning

Time Schedule

Average: 1 day
Advanced: 1 day
Honors: $1\frac{1}{2}$ days

Assignment Guide

Average
1–9, 11, 13–18, 21, 23, 24, 25, 26–29, 33, 34, 36, 38

Advanced
1–10, 12, 14, 15, 17, 19, 21, 22, 25, 26, 28–30, 32, 34

Honors	
Day 1	($\frac{1}{2}$ day) Section 2.6 6, 7, 10, 15, 16, 18, 19, 22
	($\frac{1}{2}$ day) 1, 5, 7, 9, 11, 15, 18, 19, 21, 23, 25, 30, 35
Day 2	2–4, 6, 8, 10, 12, 14, 17, 20, 22, 24, 26–29, 32, 34, 37, 38, 41

Problem-Set Notes and Strategies

- Problem **5**, parts **b** and **c** can lead to the definition of the complement of an event.
- Encourage the student to write out all the equations to be solved in problem **9**.
- In problems **10 – 13** the distributive property and order of operations is essential in finding the correct answer.
- Calculators may be necessary for problem **17**. Students have a hard time with the answer as it seems that the plate makes up more than 40% of the total area.

2 REVIEW PROBLEMS

A

In problems 1–4, simplify each expression and solve each equation.

P **1** $(3x^2 + xy - y^2) + (5x^2 - 6xy - y^2)$ **2** $-3x - 5 = -x - 9$ $x = 2$

3 $\frac{1}{2}x + \frac{1}{3}x = 2$ $x = 2.4$ **4** $x^2 - y^2 + 2x^2 - 3y^2$ $3x^2 - 4y^2$

1 $8x^2 - 5xy - 2y^2$

P,R **5** Refer to the diagram.

a How many triangles can be drawn if {A, B, C, D, E} is the set of possible vertices? 10

b If one of these triangles is chosen at random, what is the probability that point A is one of its vertices? .6

c What is the probability that point A is not a vertex of the triangle chosen in part **b**? .4

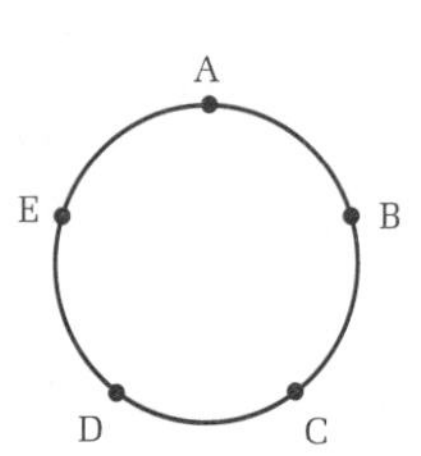

P **In problems 6–8, solve for *x* and graph the solution on a number line.**

6 $19 - 8x = 67$ $x = -6$ **7** $19 - 8x > 67$ $x < -6$ **8** $-3 \le 2x + 7 \le 15$ $-5 \le x \le 4$

9 Solve the matrix equation for x, y, and z.

$$\begin{bmatrix} 3x & x+4 \\ 6 & 8-z \end{bmatrix} = \begin{bmatrix} 2y & 12 \\ 6 & 5y \end{bmatrix}$$ $x = 8$; $y = 12$; $z = -52$

P **In problems 10–13, simplify each expression.**

10 $(x^2 - 9) - x(x + 3) + 3(x - 3)$ -18 **11** $(x + 2)x + (x + 2)3$ $x^2 + 5x + 6$

P,I **12** $x^2(x + 2) + 3x(x + 2) - 5(x + 2)$ $x^3 + 5x^2 + x - 10$ **13** $x^2(x - 3) + 3x(x - 3) + 9(x - 3)$ $x^3 - 27$

P **In problems 14–16, write an equation to represent each verbal description.**

14 The value of *y*, which is eight less than five times a number *n* $y = 5n - 8$

15 The area of a rectangle with a length that is seven units greater than the width $A = w(w + 7)$

16 The length of the hypotenuse *c* of a right triangle having legs with lengths *a* and *b* $c = \sqrt{a^2 + b^2}$

P **17** A penny is dropped into the square box shown. What is the probability that it will land on the circular plate? ≈.40

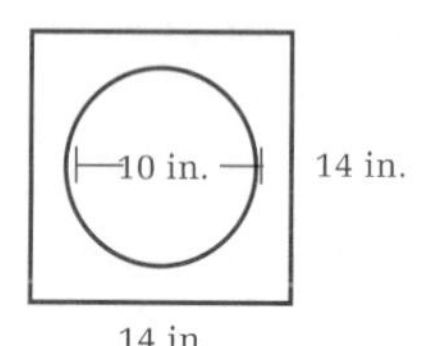

Additional Answers

Answers for problems **6 – 8** are on page **A** 1.

B **18** Evaluate each expression for (x, y, z) = (2, 4, −6).

P **a** $\frac{1}{x} + \frac{1}{y} + \frac{1}{z}$ $\frac{7}{12}$ **b** $\frac{1}{x + y + z}$ Undefined

P **19** A train leaves Transylvania Station at 8:00 A.M. and travels eastward at a rate of 25 miles per hour. Two hours later another train leaves the station, traveling westward at a rate of 35 miles per hour.

a How far apart will the trains be at 1:00 P.M? 230 mi

b At what time will the trains be the same distance from the station? 3:00 P.M.

c At what time will the trains be 370 miles apart? 3:20 P.M.

P **20** Let z = 2x + 4y + 7. Copy the following matrix, supplying the missing values.

$$\begin{matrix} \mathbf{x} \\ \mathbf{y} \\ \mathbf{z} \end{matrix} \begin{bmatrix} 3 & 5 & ? & -1 \\ ? & -4 & 2 & -3 \\ 1 & ? & 15 & ? \end{bmatrix} \quad \begin{matrix} 0 \\ -3 \\ 1; -7 \end{matrix}$$

P  **21** Simplify $-(4.2)^3 - \left(3.7 - \frac{12.4}{6.8 - 3.0}\right)$. ≈ -74.52

P **22** If 18x + 5y = 12x − 18 and y = 4, what is the value of x? $-\frac{19}{3}$

P **23** If 10x − 18 = 3x + 5w and w = −2, what is the value of x? $\frac{8}{7}$

P **24** Find each value for $f(w) = |3w - w^2|$.

a $f(7)$ 28 **b** $f(-7)$ 70 **c** $f(0)$ 0

P **25** Let $y = x^2 + 6x + 48$ and $z = x^2 - 6x$.

a Find the value of x for which y = z. −4

b Write a simplified expression for the sum y + z in terms of x. $2x^2 + 48$

P **26** The base pay for a 40-hour work week is $340. Each hour of overtime work pays $15. Copy the following matrix, supplying the missing elements.

Hours of overtime	6	8	?	?	4; 10
Overtime pay	90	?	60	?	120; 150
Total pay	430	?	?	490	460; 400

P **In problems 27–30, graph the values of x on a number line.**

27 x > 6 or x ≤ −2 **28** $(x - 5)(x + 3) \neq 0$

29 8 − 4x < 2(x − 8) **30** 3 ≤ x ≤ 7

P **In problems 31 and 32, perform the indicated operations.**

31 $2 + [(6^2 - 16^2)3 + 51] + (-20)^2 + 55$ −152 **32** $\sqrt{|-13 - 27|} - 15 + 10 - \frac{5(6)^2}{12}$ 0

Problem-Set Notes and Strategies, continued

- Part **b** of problem **18** is undefined. Make sure the students know the difference between undefined, 0, and ø, the null set.
- A picture will be helpful in solving problem **19**. Label with the rates and times and the equations will be more obvious.
- In problem **24** students may get the same answer for parts **a** and **b**. Remind students about the order of operations and that the absolute value is taken only after $3w - w^2$ is computed. The order does make a difference.
- In problem **27 – 30**, make sure the points of inequality or equality are either open or closed as needed. Students have a tendency to mix up these ideas.
- The order of operations is important in obtaining the correct answer in problems **31** and **32**.

Additional Answers

27 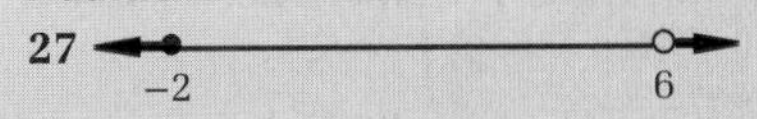

28 

29 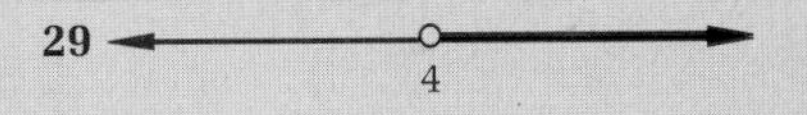

30 

Problem-Set Notes and Strategies, continued

- The chart in problem **33** is called a frequency distribution. Parts **d** and **e** introduce statistics as the science of using data to explore the unknown and make statements about a large population with only a small amount of sample data.

- In problem **34**, the variables are identified by letters and their real world meaning. Suggest to students that it is good practice to identify the variables in a statement such as, "Let B = total bill". It may save some confusion in representing the solution that was asked for in the problem.

- Students are expected to extrapolate the regular salary from the graph in problem **41**. From that, the overtime pay can then be figured out for the balance of the problem. This is a good real world example of the significance of an intercept to a graph.

Review Problems, *continued*

P **33** At Van Buren High School, a group of students were asked about the numbers of TV sets in their homes. The results are shown in the following chart.

Number of TV Sets	Frequency
0	5
1	25
2	50
3 or more	20

a What is the estimated probability that a randomly selected Van Buren student's family has no TV? .05

b What is the estimated probability that a randomly selected Van Buren student's family has at least three TV's? .2

c What is the estimated probability that a randomly selected Van Buren student's family has two or more TV's? .7

d Van Buren has a student population of 900. Predict the number of students in the school whose families do not own a TV. ≈45

e Why might predictions based on this sample be inaccurate? Explain your answer. Answers will vary.

P **34** Mai Tagg, an appliance repairwoman, charges $44 for a house call plus $15 per half-hour for labor. Parts cost extra.

a Write a formula that can be used to find the total bill B for a repair call that takes h hours and requires parts costing p dollars. $B = 44 + 30h + p$

b If a repair call took $3\frac{1}{2}$ hours and the total bill was $196, what was the charge for parts? $47

P **In problems 35–40, solve each equation.**

35 $4.2w + w + 4.3 = 8.1 + 4.2w - 3.8$ $w = 0$

36 $-7(4 - x) = 28$ $x = 8$

37 $7.2y - 37.6 = 98.4 + 4.5y - 60.8$ $y \approx 27.85$

38 $x(2x - 3) = 0$ $x = 0$ or $x = \frac{3}{2}$

39 $4x - 2(3x - 7) = 12$ $x = 1$

P,I **40** $x^2 - 5.3x = 0$ $x = 0$ or $x = 5.3$

C P **41** Refer to the graph.

a Find the pay earned for 3 hours of overtime. $45

b Estimate the total pay for a week in which $4\frac{1}{2}$ hours of overtime were worked. $267.50

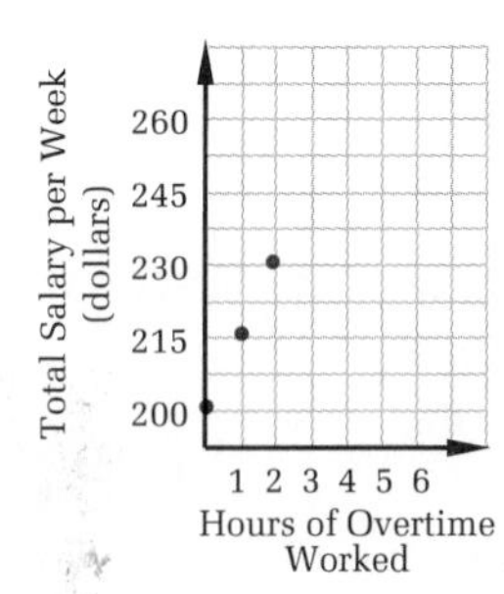

2 Chapter Test

1 Graph each equation on the same coordinate system. What do you observe? The lines are parallel.

a $y = 2x - 3$ b $y = 2x - 1$ c $y = 2x$ d $y = 2x + 2$

2 Find each value for $f(x) = 6x - 2$ and $g(x) = x + 4$.

a $f(3)$ 16 b $g(16)$ 20 c $g(f(3))$ 20

3 The average of 5, x, 10, and 13 is at least 9. Find x. $x \geq 8$

4 What values of x and y satisfy *both* of the following equations.

$$\begin{cases} 3x + 2y = 8 \\ y = 6 - x \end{cases}$$

$x = -4$; $y = 10$

5 Let $3x + 5y = 42$.

a Find y if $x = 6$. 4.8 b Find x if $y = 3$. 9

6 If $5x + 4 \leq 15$ and $3 - x < 4$, what are the possible values of x? $\{x : x \in \mathscr{R}\}$ Graph them on a number line.

In problems 7–9, find the coordinates of A if M is the midpoint of $\overline{AB}$.

7 $M(2, 3)$; $B(7, 4)$ $(-3, 2)$

8 $M(-5, 1)$; $B(3, -1)$ $(-13, 3)$

9 $M(a, b)$; $B(a + m, b - n)$ $(a - m, b + n)$

In problems 10–13, solve each equation or inequality and simplify each expression.

10 $4x(x + 12)(x - 5) = 0$

11 $4x(x + 12) - (x - 5)$ $4x^2 + 47x + 5$

12 $\frac{x - 5}{6} + \frac{1}{3}$ $\frac{x - 3}{6}$

13 $-3(x + 2) > 48$ $x < -18$

10 $x = 0$, $x = -12$, or $x = 5$

14 For what values of x is $|x| > x$ a true statement? $\{x : x < 0\}$

15 Perform the indicated operations.

$$\left| \frac{(-5 + 1.5)(17 - 9)}{(-11 + 9)^2} \right| + 3 \cdot 13 - 5 \div 2 \quad 43.5$$

Resource References

Evaluation

Test Form 1, 2, 3

Additional Answers

1

6

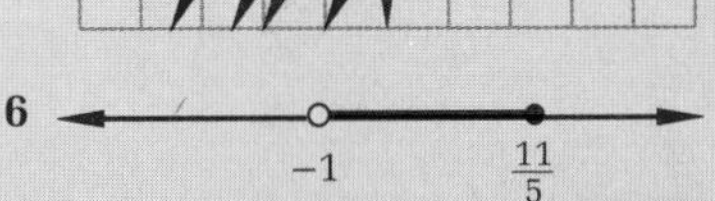

Class Planning

Time Schedule

Average: 4 days
Advanced: 3 days
Honors: 2 days

Resource References

Teacher's Resource Book
Class Opener 13
Practice 13, 14
Manipulatives and Applications 5

Class Opener

For all nonzero values of x and y, tell whether the equation is True or False.

1. $x^3x^4 = x^{12}$ False
2. $x^{3^4} = x^{12}$ True
3. $\frac{x^{12}}{x^4} = x^8$ True
4. $x^2 + x^2 = x^4$ False
5. $x^2 + x^2 = 2x^2$ True
6. $(2x^2)(2x^2) = 4x^2$ False
7. $(x + y)^2 = x^2 + y^2$ False
8. $x^3y^3 = (xy)^3$ True
9. $\sqrt{x^2} = x$ False
10. $(\sqrt{x})^2 = x$ False
11. $\sqrt{x^2 + y^2} = x + y$ False
12. $\sqrt{x^4y^4} = x^2y^2$ True
13. $\frac{x^4}{y^4} = \left(\frac{x}{y}\right)^4$ True
14. $\sqrt{\frac{x}{y}} = \frac{\sqrt{x}}{\sqrt{y}}$ False

SHORT SUBJECT

Exponents and Radicals

Laws of Exponents

In earlier mathematics courses you worked with exponents in a variety of ways. As you review the laws of exponents, remember what exponents represent. In the expression a^n, a is the ***base***, and n is the ***exponent***. The exponent indicates how many times the base is used as a factor, as the following examples show.

$3^4 = 3 \cdot 3 \cdot 3 \cdot 3 \qquad x^5 = x \cdot x \cdot x \cdot x \cdot x \qquad (xy)^3 = (xy)(xy)(xy)$

The Product Rule

If a is a real number and m and n are integers, then $a^m \cdot a^n = a^{m+n}$.

Example

$3^3 \cdot 3^2 = (3 \cdot 3 \cdot 3)(3 \cdot 3) = 3 \cdot 3 \cdot 3 \cdot 3 \cdot 3 = 3^5$, or $3^3 \cdot 3^2 = 3^{3+2} = 3^5$

The Power Rule

If a is a real number and m and n are integers, then $(a^m)^n = a^{mn}$.

Example

$(5^3)^5 = 5^3 \cdot 5^3 \cdot 5^3 \cdot 5^3 \cdot 5^3 = 5^{3+3+3+3+3} = 5^{15}$, or $(5^3)^5 = 5^{3 \cdot 5} = 5^{15}$

The Quotient Rule

If a is a real number and m and n are integers, then $\frac{a^m}{a^n} = a^{m-n}$.

Example

$\frac{3^5}{3^2} = \frac{3 \cdot 3 \cdot 3 \cdot 3 \cdot 3}{3 \cdot 3} = \frac{3 \cdot 3}{3 \cdot 3} \cdot 3 \cdot 3 \cdot 3 = 1 \cdot 3^3 = 3^3$, or $\frac{3^5}{3^2} = 3^{5-2} = 3^3$

You may or may not have worked with negative exponents. The following definition explains what a negative exponent represents.

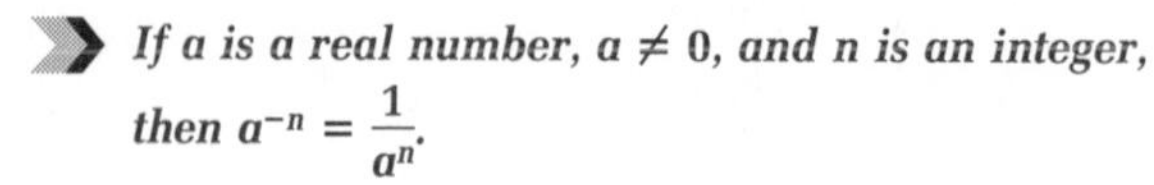

If a is a real number, $a \neq 0$, and n is an integer, then $a^{-n} = \frac{1}{a^n}$.

Vocabulary

base
exponent
standard radical form
conjugates

Example

$$\frac{5^2}{5^3} = \frac{5 \cdot 5}{5 \cdot 5 \cdot 5} = 1 \cdot \frac{1}{5} = \frac{1}{5}, \text{ or } \frac{5^2}{5^3} = 5^{2-3} = 5^{-1}$$

The Product Rule, the Power Rule, and the Quotient Rule apply to expressions with negative exponents as well as those with positive exponents.

The Distributive Property of Exponentiation over Multiplication

If x and y are real numbers and n is an integer, then $x^n \cdot y^n = (x \cdot y)^n$.

Example

$$6^5 \cdot 2^5 = (6 \cdot 2)^5 = 12^5$$

The Distributive Property of Exponentiation over Division

If x and y are real numbers, $y \neq 0$, and n is an integer, then $\frac{x^n}{y^n} = \left(\frac{x}{y}\right)^n$.

Example

$$\frac{6^5}{2^5} = \left(\frac{6}{2}\right)^5 = 3^5$$

There is one more definition we need for our work with exponents.

If a is a real number and $a \neq 0$, then $a^0 = 1$.

You can see why a number raised to the power 0 is equal to 1 by considering that according to the Quotient Rule $\frac{5^3}{5^3} = 5^{3-3} = 5^0$ and that arithmetically $\frac{5^3}{5^3} = \frac{125}{125} = 1$. The definition specifies that $a \neq 0$ because if a were 0, two definitions would be in conflict. On the one hand, we have defined a^0 as 1; on the other, you know that $0^n = 0$. To avoid this inconsistency, we agree that 0^0 is undefined.

Interpreting Radical Expressions

Since the calculator and the computer have become widely used tools, radical expressions are used less frequently by mathematicians. Still, you will see radical symbolism in your studies, so we will look at some properties of radicals. You may recall that if n is a positive integer and $y^n = x$, then $\sqrt[n]{x} = y$ (with the restriction that if n is even, x and y must be positive). This inverse relationship between powers and roots leads us to the following definition.

If x is a real number and n is a positive integer, then $\left(\sqrt[n]{x}\right)^n = x$ whenever $\sqrt[n]{x}$ is defined.

Examples

$$\left(\sqrt[3]{-27}\right)^3 = -27 \qquad \left(\sqrt[4]{16}\right)^4 = 16$$

An expression such as $\left(\sqrt[4]{-16}\right)^4$, however, is undefined in the real-number system, since even roots of negative numbers are not defined.

Lesson Notes

- There are five **Short Subjects** in this book. A short subject is a mini chapter that can be categorized in one of two ways.
- The **Short Subject** can contain material that is review for some of your students. You can decide how much time to spend on these subjects.
- The **Short Subject** can contain enrichment topics that are extensions of concepts presented in a chapter. Again, you have the flexibility to decide how much time to spend with your students on this material.

The value we assign to an expression of the form $\sqrt[n]{x^n}$ depends on whether n is even or odd.

If x is a real number and n is a positive integer, $\sqrt[n]{x^n} = |x|$ if n is even, or $\sqrt[n]{x^n} = x$ if n is odd.

Examples

$\sqrt{(-2)^2} = \sqrt{4} = 2$ $\qquad$ $\sqrt[3]{(-4)^3} = \sqrt[3]{-64} = -4$

Since the radical symbol is most often used in connection with square roots, we will examine some of the properties that allow us to perform operations involving square roots.

If x and y are nonnegative real numbers, then $\sqrt{x}\sqrt{y} = \sqrt{xy}$, $\frac{\sqrt{x}}{\sqrt{y}} = \sqrt{\frac{x}{y}}$ (where $y \neq 0$), and $\sqrt{x}\sqrt{x} = x$.

You might notice that the first two properties are restatements of the distributive properties of exponents in terms of radicals and that the last is a special case of the inverse relationship between powers and roots.

Sample Problems

Problem 1 *Evaluate* $5^4 \cdot \frac{5^3}{5^9}$.

Solution Evaluating the expression will be easier if we simplify it first.

$$5^4 \cdot \frac{5^3}{5^9}$$

$$= 5^{4+3-9}$$

$$= 5^{-2}$$

$$= \frac{1}{5^2}$$

$$= \frac{1}{25}$$

Problem 2 *Simplify* $\frac{3x^5}{(2x^3)^{-1}} \cdot \frac{(x^{-4})^2}{x^2}$.

Solution Notice that after we apply the Power Rule, it makes no difference which of the other rules we apply first.

Method 1

$$\frac{3x^5}{(2x^3)^{-1}} \cdot \frac{(x^{-4})^2}{x^2}$$

Power Rule $= \frac{3x^5}{2^{-1}x^{-3}} \cdot \frac{x^{-8}}{x^2}$

Method 2

$$\frac{3x^5}{(2x^3)^{-1}} \cdot \frac{(x^{-4})^2}{x^2}$$

Power Rule $= \frac{3x^5}{2^{-1}x^{-3}} \cdot \frac{x^{-8}}{x^2}$

Product Rule $= \dfrac{3x^{-3}}{2^{-1}x^{-1}}$ Quotient Rule $= 6x^8x^{-10}$

Quotient Rule $= 6x^{-2}$ Product Rule $= 6x^{-2}$

We could also write the answer in the form $\dfrac{6}{x^2}$.

Problem 3 *Write $\sqrt{252}$ in* ***standard radical form****—that is, the form $a\sqrt{b}$, where b is an integer having no perfect-square factors.*

Solution We need to find perfect-square factors of 252 and use a distributive property. Notice that the order in which we take out perfect-square factors is not important as long as we don't miss any.

Method 1	**Method 2**	**Method 3**
$\sqrt{252}$	$\sqrt{252}$	$\sqrt{252}$
$= \sqrt{9}\sqrt{28}$	$= \sqrt{4}\sqrt{63}$	$= \sqrt{36}\sqrt{7}$
$= 3\sqrt{4}\sqrt{7}$	$= 2\sqrt{9}\sqrt{7}$	$= 6\sqrt{7}$
$= 6\sqrt{7}$	$= 6\sqrt{7}$	

Problem 4 *Simplify $\sqrt{18} + \sqrt{27} + \dfrac{\sqrt{36}}{\sqrt{18}} + \sqrt{50}$.*

Solution

$$\sqrt{18} + \sqrt{27} + \frac{\sqrt{36}}{\sqrt{18}} + \sqrt{50}$$

Distributive properties $= \sqrt{9}\sqrt{2} + \sqrt{9}\sqrt{3} + \sqrt{\dfrac{36}{18}} + \sqrt{25}\sqrt{2}$

Simplify terms $= 3\sqrt{2} + 3\sqrt{3} + \sqrt{2} + 5\sqrt{2}$

Combine like terms $= 9\sqrt{2} + 3\sqrt{3}$

Problem 5 *Multiply $2 + \sqrt{7}$ by $2 - \sqrt{7}$ and simplify your answer.*

Solution

$(2 + \sqrt{7})(2 - \sqrt{7})$

$= 2^2 - 2\sqrt{7} + 2\sqrt{7} - (\sqrt{7})^2$

$= 2^2 - (\sqrt{7})^2$

$= 4 - 7$

$= -3$

Notice that the answer is a rational number. Expressions of the form $a + b\sqrt{c}$ and $a - b\sqrt{c}$ will always have a rational product if a, b, and c are rational. Such expressions are called ***conjugates.***

Assignment Guide

Average	
Day 1	1, 2, 4, 8, 9, 17, 23, 31, 34, 40, 46, 49
Day 2	3, 7, 10, 11, 18, 24, 27, 32, 35, 41, 44, 47, 50, 54
Day 3	5, 6, 12, 15, 19, 21, 25, 28, 36, 39, 42, 45, 51
Day 4	13, 14, 16, 20, 22, 26, 33, 37, 38, 43, 48, 52, 53
Advanced	
Day 1	1–3, 8, 11, 15, 18, 19, 25, 27, 28, 31, 34, 40, 41, 45, 49, 50
Day 2	4, 6, 9, 12, 16, 20, 23, 26, 29, 30, 35, 36, 42, 43, 46, 47
Day 3	5, 7, 10, 14, 17, 21, 22, 24, 32, 33, 37, 38, 44, 48, 53, 54
Honors	
Day 1	1–5, 11, 12, 17, 18, 23, 24, 26–30, 34–36, 40–43, 50, 51, 55, 56
Day 2	6–10, 13–16, 19–22, 25, 31–33, 37–39, 44, 52–54, 57, 58

Problem-Set Notes and Strategies

- In problems **1** and **2** students will frequently forget that the base of an expression such as $(2b^3)^3$ includes the constant 2 as well as the variable.
- In problem **6**, rational exponents are introduced. Some students may know how to do this on a calculator. In this case, the answer works out nicely since 2.5 times 2 equals 5 which is an integer, but some students may try to do the work inside parenthesis first.
- Problem **7** uses the fact that if the bases of an equation are equal, the exponents must also be equal.
- Have students identify the base in problem **9**. Exponents do not distribute over addition as does multiplication. Work out the left side using fraction addition and verify that the answer is not 8.
- In problem **10**, have students substitute some simple numbers to convince themselves that this is truly false.
- Students may have an easier time with problem **11** if they first factor out the greatest common factor of x^5.

Additional Answers

Answers for problems **12a** and **12b** are on page **A** 1.

Problem 6 *Rationalize the numerator of* $\frac{2+\sqrt{5}}{3-\sqrt{6}}$ *(that is, rewrite the expression so that no radical appears in the numerator).*

Solution

Multiply numerator and denominator by conjugate of numerator

$$\frac{2+\sqrt{5}}{3-\sqrt{6}}$$

$$= \frac{2+\sqrt{5}}{3-\sqrt{6}} \cdot \frac{2-\sqrt{5}}{2-\sqrt{5}}$$

$$= \frac{2^2 - 2\sqrt{5} + 2\sqrt{5} - (\sqrt{5})^2}{6 - 3\sqrt{5} - 2\sqrt{6} + \sqrt{30}}$$

$$= \frac{-1}{6 - 3\sqrt{5} - 2\sqrt{6} + \sqrt{30}}$$

A similar process can be used to rationalize a denominator.

Problem Set

A

P **In problems 1 and 2, simplify each expression.**

1 $(2b^3)^3 \cdot 3(b^{-4})^2$ $24b$

2 $\frac{12w^{-4}}{8w^8} \cdot \frac{15w^{12}}{(-6w)^2}$ $\frac{5}{8w^2}$

P **In problems 3–5, indicate whether each statement is True or False (if $x \neq 0$).**

3 $x^{2a} = (x^a)^2$ True

4 $x^a \cdot x^b = x^{ab}$ False

5 $\frac{x^a}{x^a} = x^1$ False

P **6** Evaluate.

a $(5^2)^3$ 15,625 **b** $(5^3)^2$ 15,625 **c** 5^5 3125 P,I **d** $(5^{2.5})^2$ 3125

P,I **7** Find the values of x and y that satisfy both of the following equations.

$$\begin{cases} 2^{2x-y} = 2^3 \\ 3^{x+y} = 3^{12} \end{cases}$$ $x = 5;\ y = 7$

P **In problems 8–10, indicate whether each statement is True or False.**

8 $\left(\frac{3}{5}\right)^{-4} = \left(\frac{5}{3}\right)^4$ True

9 $\left(\frac{1}{3} + \frac{1}{5}\right)^{-1} = 3 + 5$ False

10 $\frac{1}{a+b} = \frac{1}{a} + \frac{1}{b}$ False

P,I **11** How many segments of length x^5 can be made from the segment shown?

$x^{10} + 3x^7 - 2x^5$

$x^5 + 3x^2 - 2$

P,I **12** Make a table of values for each of the following equations, then sketch the equation's graph.

a $y = \sqrt{x}$ **b** $y = 2^x$

P **13** Find an expression for the sum of the volumes of the cube and the box. $2x^6$

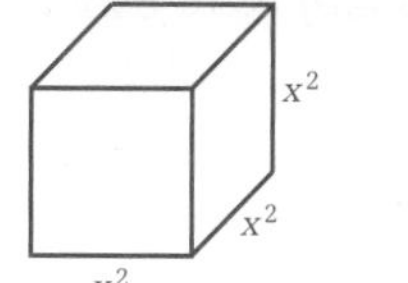

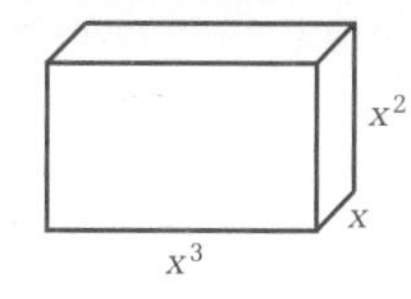

P **14** Evaluate each expression.

a 6^{-2} $\frac{1}{36}$ **b** $(3^{-2} + 1)^{-1}$ $\frac{9}{10}$ **c** $\left(\frac{1}{2} + \frac{1}{3}\right)^{-1}$ $\frac{6}{5}$ **d** $(2^{-1} - 3^{-1})^{-1}$ 6

P **In problems 15 and 16, rewrite each expression without parentheses, using only negative exponents.**

15 $\left(\frac{x^{-3}y^2}{w^{-4}}\right)^5$ $\frac{x^{-15}}{w^{-20}y^{-10}}$ **16** $\left(\frac{x^{-4}y^{-3}}{w^4}\right)^5$ $x^{-20}y^{-15}w^{-20}$

P **17** Write an expression, in simplified form, for the area of the trapezoid. $8x^6y^{14}$

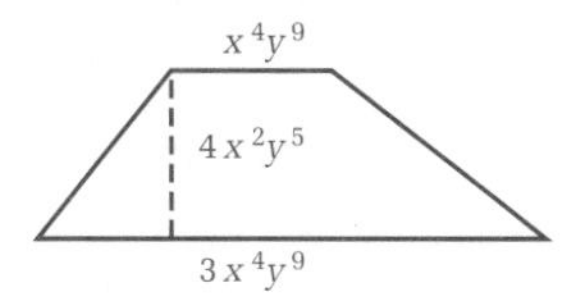

P,I **18** If $x = 9$, $y = 4$, and $n = 3$, what are the values of $(x + y)^n$ and $x^n + y^n$? Is it true that $(x + y)^n = x^n + y^n$? 2197; 793; no

P **19** Simplify $x \cdot x^2 + 2y \cdot 4y^2$. $x^3 + 8y^3$

P **20** Find the ratio of the volume of the small box to that of the large box. $1{:}x^9y^6z^{14}$

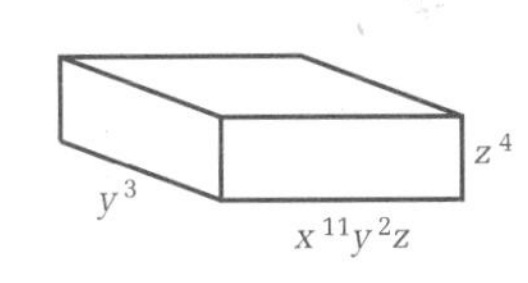

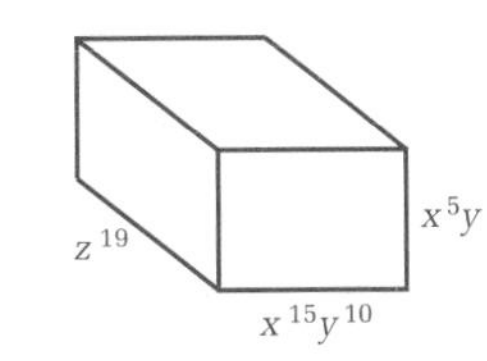

P,I **21** Study the example $16^{\frac{1}{4}} = (2^4)^{\frac{1}{4}} = 2^{\frac{4}{4}} = 2^1 = 2$.
Use this method to evaluate each expression.

a $32^{\frac{1}{5}}$ 2 **b** $64^{\frac{1}{3}}$ 4 **c** $8^{\frac{2}{3}}$ 4 **d** $125^{\frac{4}{3}}$ 625

P,I **22** On a certain planet, there exists a number i such that $i^2 = -1$. Find the value of each expression.

a i^4 1 **b** i^{-2} -1 **c** $i^6 + i^8 + i^{10} + i^{12}$ 0

P,I **23** Use the diagram to find an expression for length s in terms of x.
$s = 1 - 2x + 3x^2$

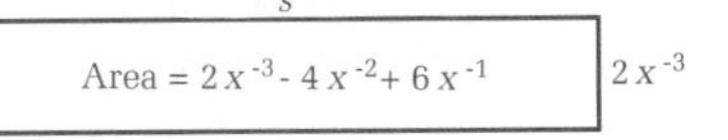

P,I **24** Solve $5 = 5^x = 5^{\frac{1}{3} \cdot y} = (5^z)^{\frac{1}{3}} = w^{\frac{1}{3}}$ for w, x, y, and z. $w = 125$; $x = 1$; $y = 3$; $z = 3$

P **B** **25** Evaluate $\frac{(xy^2)^{18}}{(x^2y)^{19}} \cdot \left(\frac{x^4}{y^4}\right)^5$ for $(x, y) = (5.2, -2)$. $-\frac{1}{8}$

Problem-Set Notes and Strategies, continued

- Remind students that negative exponents, in problem **14,** do not affect the sign of the number, just the placement of the number in the numerator or the denominator.
- You may need to refer students to the area of a trapezoid formula in the Handbook (page 731) for problem **17.**
- In problem **18,** $(x+y)^n \neq x^n + y^n$. It might be interesting to have students find out what values of n will make the inequality an equality.
- Problem **22** introduces the concept of complex numbers. This will be explained in depth in Section **8.3**, but is briefly introduced here as an exercise in exponents.
- In an equation, if the bases are the same, the exponents are the same. Likewise, if the exponents are equal, the bases must also be equal. Both these relationships are necessary to solve problem **24.**
- Encourage students to simplify exponential problems such as **25** before substituting values for the variables. Students with calculators may get carried away with the computations and calculate some very unwieldly numbers in the intermediate steps.

Problem Set, *continued*

I **26** Match each of the following sets of ordered pairs with the corresponding graph.

a (1, 2), (2, 4), (4, 16), (0, 1) **i**

b (0, 0), (1, 1), (−2, 2), (−1, 1) **iv**

c $\left(4, \frac{1}{2}\right), \left(\frac{1}{4}, 2\right)$, (1, 1), $\left(16, \frac{1}{4}\right)$ **iii**

d (0, 0), $\left(\frac{1}{4}, \frac{1}{2}\right)$, (1, 1), (4, 2) **ii**

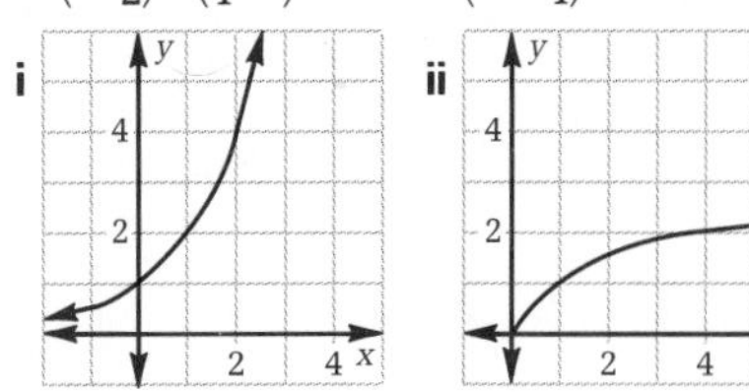

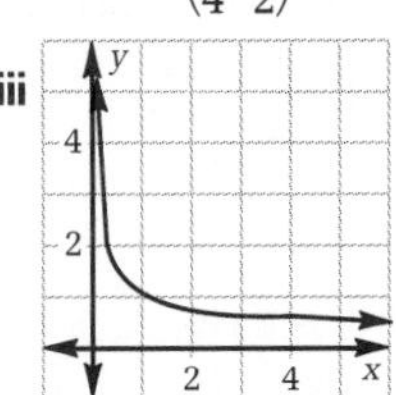

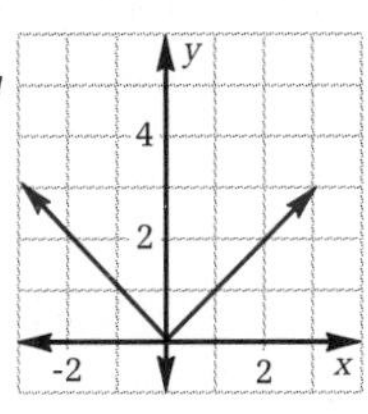

P **In problems 27–29, rewrite each expression without parentheses, using only positive exponents.**

27 $\left(\frac{x^{-6}}{y^{-4}}\right)^5$ $\frac{y^{20}}{x^{30}}$

28 $\left(\frac{x^3}{x^{-2}}\right)^{-5}$ $\frac{1}{x^{25}}$

29 $(x^{-3}y^4)^{-5}$ $\frac{x^{15}}{y^{20}}$

P **30** Rationalize the numerator of $\frac{\sqrt{5} - \sqrt{3}}{2}$. $\frac{1}{\sqrt{5} + \sqrt{3}}$

P **In problems 31–33, simplify each expression.**

31 $\frac{-20xy^3z^2}{24x^4y^2z}$ $-\frac{5yz}{6x^3}$

32 $x^2(x^2 - x + 2) - x^3(x - 1)$ $2x^2$

33 $(-xy)^4(-x^2y^2)^2$ x^8y^8

P,I **34** Refer to the figure.

a What is the area of the figure? $(x + 4)^2$

b What is the sum of the areas of regions I and II? $x^2 + 4^2$

c For what value(s) of x is $(x + 4)^2 = x^2 + 4^2$? 0

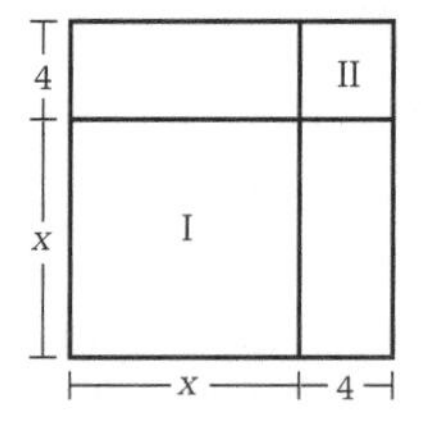

P **In problems 35–38, simplify each expression.**

35 $\frac{xy}{x^2y + xy}$ $\frac{1}{x + 1}$

36 $\frac{10x^3 + 20x^2}{15x^4 + 30x^3}$ $\frac{2}{3x}$

37 $\frac{(3x^2y)^4 \cdot (5x)^3}{(15x^4y^5)^3}$ $\frac{3}{xy^{11}}$

38 $\frac{(xy^5)^{18}}{(xy)^{17}}$ xy^{73}

P,I **39** Find the ordered pair (x, y) that satisfies both of the following equations.

$$\begin{cases} 4^{x+y} = 8^{10} \\ 5^{x-2y} = \frac{1}{5^{12}} \end{cases}$$ (x, y) = (6, 9)

P **In problems 40–43, indicate whether each statement is True or False.**

40 If $x < 0$, then $x^n < 0$. False

41 If $x > 0$, then $x^{-n} > 0$. True

42 $(-2.5)^2 < 2.5^2$ False

43 $\left(\frac{1}{2}\right)^{-5} < \left(\frac{1}{2}\right)^{-4}$ False

P,I **44** Solve each of the following equations for x.

a $\sqrt{x} = 7$ $x = 49$

b $\sqrt{x} = -2$ No real solution

Problem-Set Notes and Strategies, continued

- Conjugates are necessary to simplify problem **30**. Students may be uncomfortable with leaving a radical in the denominator of a fraction. Inform the students that it is sometimes helpful to rationalize numerators to aid in solving problems.
- Problem **34** is another example to illustrate that $(x + 4)^2 \neq x^2 + 4^2$ in general.
- In problem **39**, the base of the first equation must be written as 2 for both sides. Students may miss this. The problem then becomes one where it is necessary to solve a system of equations.
- Problem **44** requires squaring both sides of an equation to find the solution. Remind the student that this sometimes leads to extraneous solutions as in part **b**.

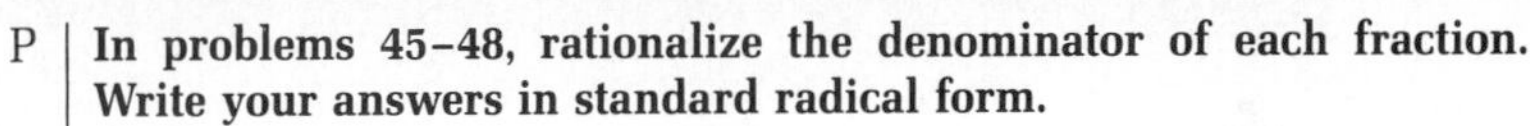

P | **In problems 45–48, rationalize the denominator of each fraction. Write your answers in standard radical form.**

45 $\frac{10}{\sqrt{5}}$ $2\sqrt{5}$ **46** $\frac{\sqrt{30}}{\sqrt{5}}$ $\sqrt{6}$ **47** $\frac{\sqrt{20}}{\sqrt{8}}$ $\frac{1}{2}\sqrt{10}$ **48** $\frac{2}{\sqrt{2}}$ $\sqrt{2}$

P,R | **49** Find the length of the hypotenuse of each triangle.

a

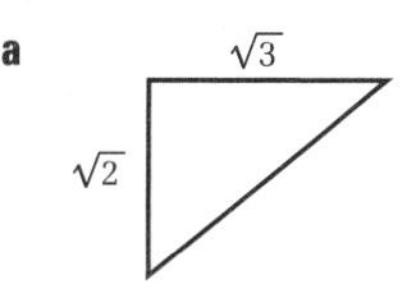

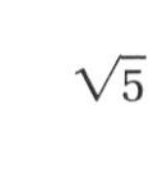

$\sqrt{5}$

b

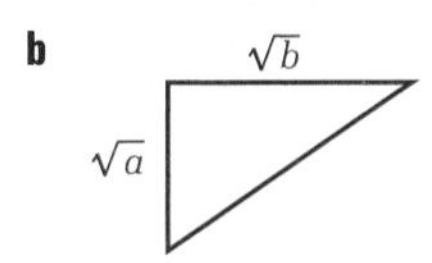

$\sqrt{a + b}$

P | **In problems 50–52, evaluate each expression.**

50 $\sqrt{5}\sqrt{20}$ 10 **51** $\frac{\sqrt{5}}{\sqrt{20}}$ $\frac{1}{2}$ **52** $\frac{\sqrt{2}\sqrt{14}}{\sqrt{7}}$ 2

P | **53** Rationalize the denominator of $\frac{2 + \sqrt{5}}{3 + \sqrt{5}}$, then rationalize the numerator of the resulting expression. $\frac{1 + \sqrt{5}}{4}$; $\frac{-1}{1 - \sqrt{5}}$

I | **54** Match each equation with its graph.

a $y = \frac{1}{\sqrt{x}}$ **iii** **b** $y = \sqrt{x^2}$ **iv** **c** $y = \sqrt{x}$ **ii** **d** $y = 2^x$ **i**

i

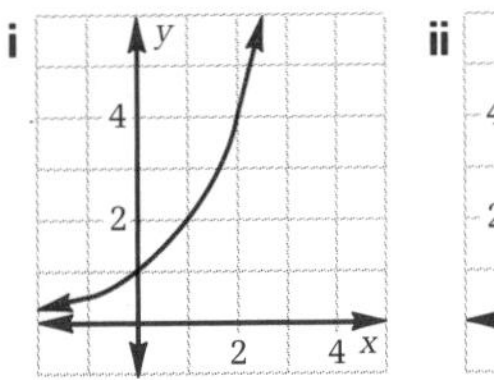

ii

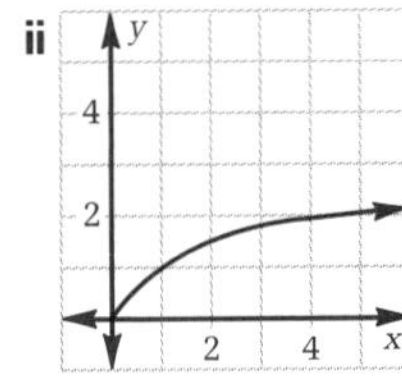

iii

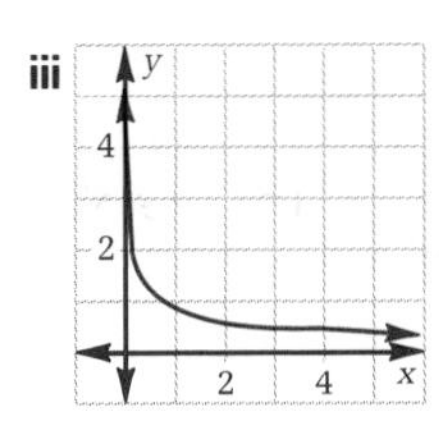

iv

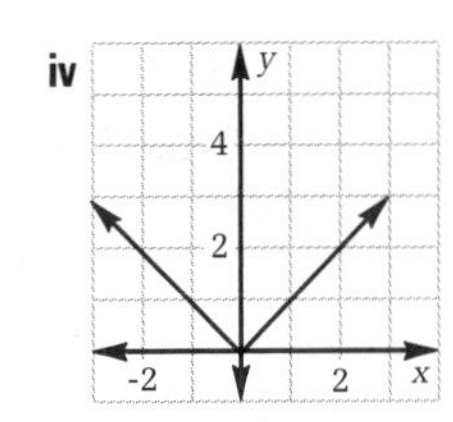

P **C** | **55** Simplify $5\sqrt[3]{16} - \sqrt[3]{54} + 2\sqrt[3]{128}$. $15\sqrt[3]{2}$

I | **56** Find the ordered pair (x, y) that satisfies the following equations simultaneously.

$$\begin{cases} \sqrt{x} + \sqrt{y} = 2 \\ \sqrt{x} - \sqrt{y} = 1 \end{cases} \quad (x, y) = \left(\frac{9}{4}, \frac{1}{4}\right)$$

P,R | **57** Refer to the cube.

a Determine the length of $\overline{AB}$. $2\sqrt{2}$

b Determine the length of $\overline{BC}$. $2\sqrt{3}$

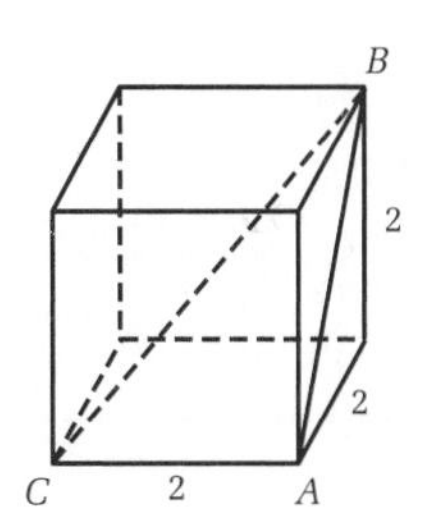

P | **58** Solve $(\sqrt{2n} + 3)(\sqrt{2n} - 3) = 3n - 10$ for n. $n = 1$

Problem-Set Notes and Strategies, continued

- Students may need to plot several points in problem **54** to identify the proper graph. Once they have gotten the answer, have them review the problem again looking for symmetry or generalizations that might aid in the process next time.
- There are several methods of solving problem **56**. One of the easier methods is to first solve for $\sqrt{x}$ and then for $\sqrt{y}$ and then square the results.
- Recognize that in problem **58**, the product of two conjugates is the difference of squares. This leads to a quick solution.

3.1

CHAPTER 3 BEGINS by reviewing additional Algebra 1 and Geometry concepts. *Slope* is defined in several ways and different forms for the equation of a line are reviewed as well as the factors that identify lines as perpendicular or parallel.

Students review both algebraic and graphic techniques for solving systems of equations. Encourage students to identify the algebraic technique that is most useful in solving a particular system. Point out that graphing can be used to approximate or check solutions of a system.

The boundary-value algorithm is presented in this chapter and will be used throughout the text. The concepts of union and intersection with respect to inequalities may be new to some students.

Formal work with data is begun in Section **3.4**. Students will write equations that represent linear functions and identify the restrictions on the domains of these functions. Direct variation is treated as a special linear relationship.

Students will also fit a line to data by estimating the appropriate line and by using the least squares method.

Many students will have their first opportunity to study compound functions and their graphs. Compound functions are used to give another interpretation of the absolute value function.

CHAPTER

3 Linear Relationships

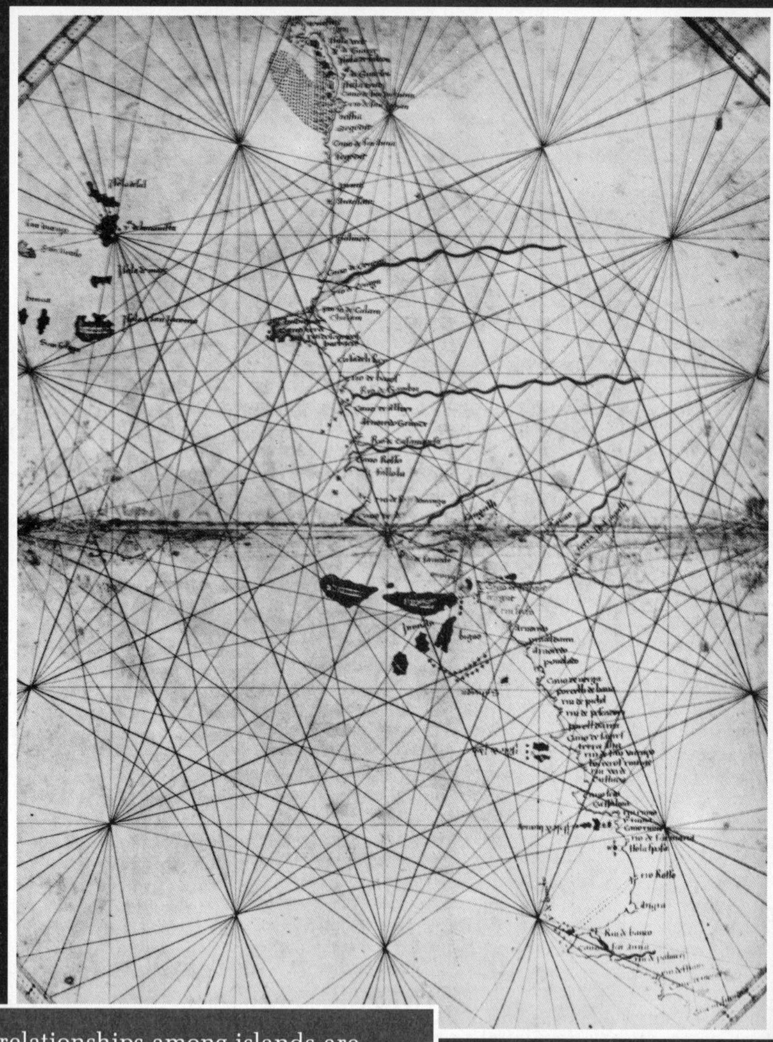

Linear relationships among islands are represented in this antique marine chart showing the West African coast.

3.1 GRAPHING LINEAR EQUATIONS

Objectives

After studying this section, you will be able to

- Determine the slope of a line
- Use the point-slope equation of a line
- Use the slope-intercept equation of a line
- Use the general equation of a line
- Use slope to determine whether two lines are perpendicular or parallel

Part One: Introduction

Slope

Slope, an important concept in graphing equations, is a rate of change from point to point on a line. There are several ways of finding and describing slope. We can choose whatever form is appropriate to a given situation.

1 We can express slope as a *rate of change*.

Slope is the change in y when the change in x is 1.

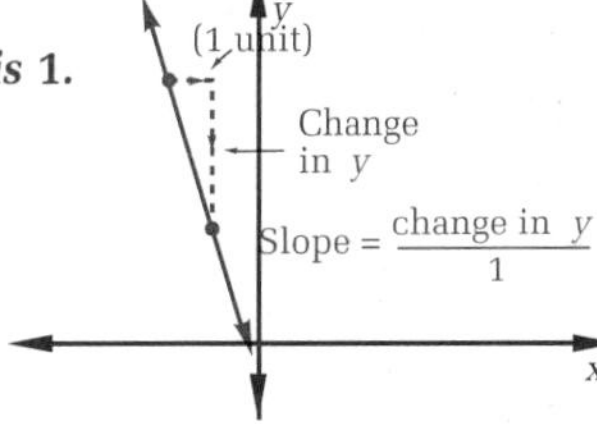

2 We also use the following relationship to find slope.

$$\textbf{Slope} = \frac{\textbf{change in } y}{\textbf{change in } x}$$

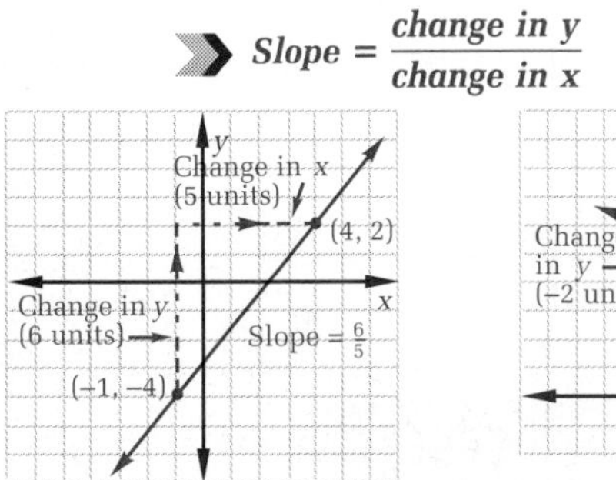

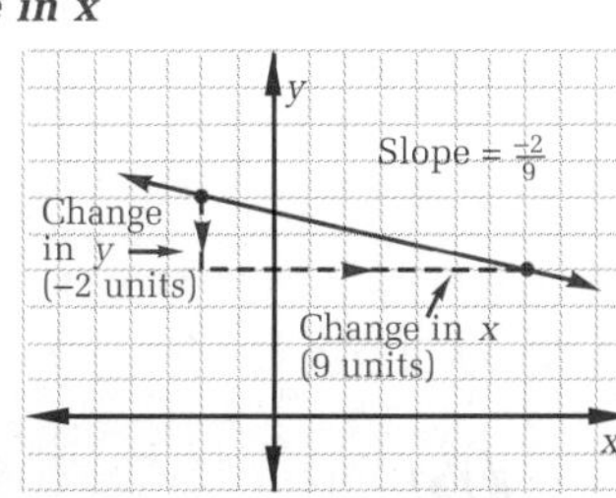

3 If we know the coordinates of two or more points on a line, we can use the slope formula.

Class Planning

Time Schedule

All levels: 1 day

Resource References

Teacher's Resource Book
Class Opener 14
Cooperative Learning 8
Practice 15
Manipulatives and Applications 6
Enrichment 5

Class Opener

Are the points (3, 5), (5, 6), and (8, 13) colinear? No.

Lesson Notes

- This chapter is devoted to linear relationships and how to describe them with both words and mathematical equations. The first concept introduced is the slope of a line related to relative steepness.
- The idea of positive and negative slopes can be related to rising and falling as the line is scanned from left to right. If the line is rising the slope must be positive. If the line is falling, the slope is negative.
- The slope is also the rate of change of one variable with respect to another. Students will relate to miles per hour as they are likely to be driving now or in the near future. The formula for slope utilizes this definition as it calculates the ratio of changes in the two variables.

Vocabulary

slope

The Slope Formula

The slope m of a line containing points (x_1, y_1) and (x_2, y_2) is given by the formula

$$m = \frac{y_2 - y_1}{x_2 - x_1}$$

Example *Find the slope of the line containing points* (4, 7) *and* (−5, 2).

Use slope formula
$$\begin{aligned} m &= \frac{y_2 - y_1}{x_2 - x_1} \\ &= \frac{7 - 2}{4 - (-5)} \\ &= \frac{5}{9} \end{aligned}$$

We can classify lines by their slopes.

Lines that ascend from left to right have *positive slope.*

Lines that descend from left to right have *negative slope.*

Horizontal lines have *zero slope.*

Vertical lines have *undefined slope.*

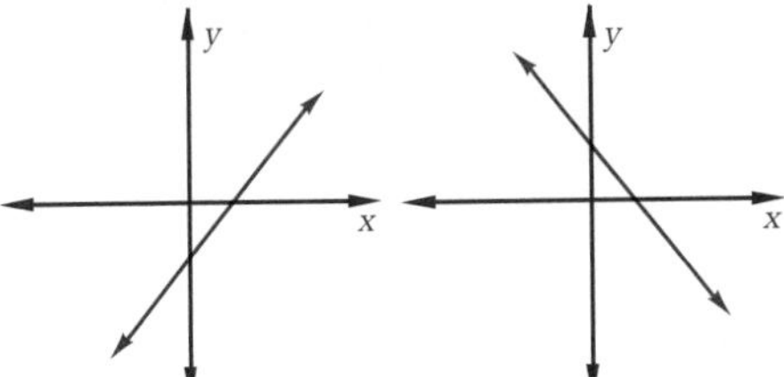
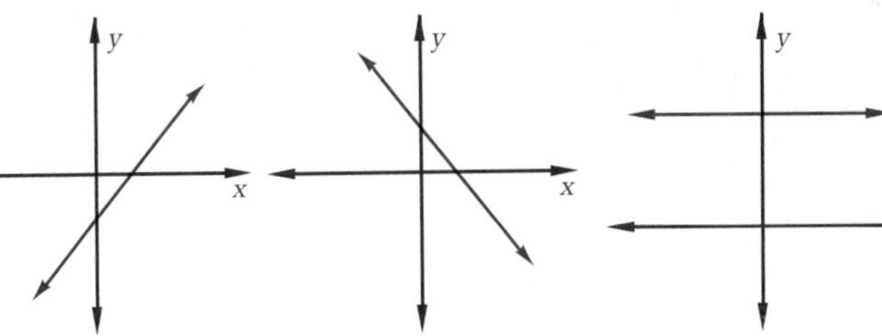

Point-Slope Equation of a Line

If we have a given point, such as (2, 1), and a given slope, such as $\frac{2}{3}$, we can draw a unique line.

In general, if (x_1, y_1) is on a line with slope m, we can write an equation of all points (x, y) on the line by using the slope formula and some algebra. The slope formula tells us that for every $(x, y) \neq (x_1, y_1)$ on the line,

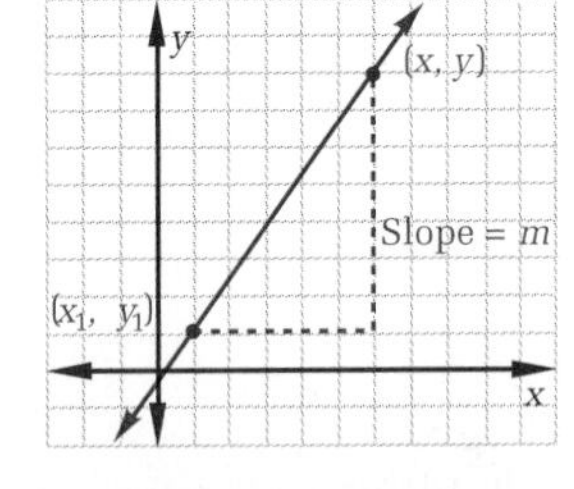

$$\frac{y - y_1}{x - x_1} = m$$

Multiply $\quad y - y_1 = m(x - x_1)$

The Point-Slope Equation of a Line

If a line has slope m and contains point (x_1, y_1), then all points (x, y) on the line satisfy the equation $y - y_1 = m(x - x_1)$.

Lesson Notes, continued

- A horizontal line has no slope as there is no change in the y-coordinates. A vertical line has an undefined slope as the change in the x-coordinate is zero. Many students may have problems with this concept. "Undefined" will have more illustrations when functions are defined in Chapter **5**.
- The point-slope form and slope-intercept forms of a line are introduced and you may point out that the slope-intercept form is a special case of the point-slope form when the y-intercept happens to be known. A point and a slope defines a line since the slope allows you to find any other point.

Example *Write an equation of the line that has a slope of $\frac{15}{17}$ and passes through $\left(\frac{3}{5}, -\frac{7}{11}\right)$.*

We use the point-slope form.

$$y + \frac{7}{11} = \frac{15}{17}\left(x - \frac{3}{5}\right)$$

Slope-Intercept Equation of a Line

The following form of equation is useful with graphing calculators and computer graphing programs. It also allows us to describe a line with function notation, $f(x) = mx + b$.

The Slope-Intercept Equation of a Line

A line with equation $y = mx + b$ has slope m and y-intercept b.

Example *Draw the graph of $f(x) = \frac{2}{3}x - 4$.*

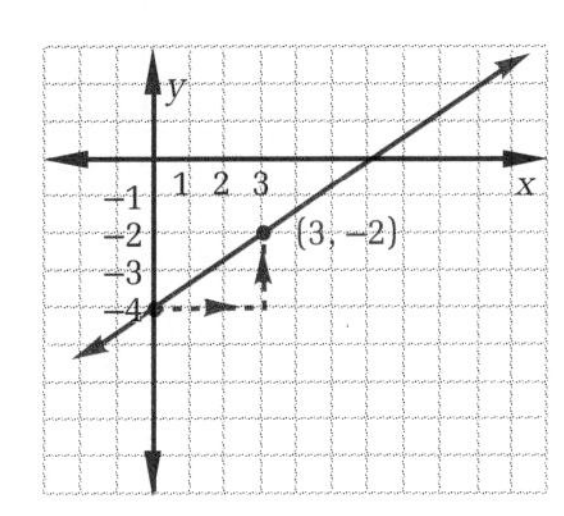

The slope is $\frac{2}{3}$. The y-intercept is -4. From the y-intercept $(0, -4)$, we move three units right and two units up. We repeat this move to find additional points and connect the points with a line.

General Linear Equation of a Line

The form $ax + by = c$ is called the general form of a linear equation. This form can be used to describe any line, including one with undefined slope, such as the graph of $x = 5$.

See p. 756

Example *Graph $2x - 5y = 15$.*

We find the x- and y-intercepts. At the x-intercept, $y = 0$.

$$2x - 5(0) = 15$$
$$2x = 15$$
$$x = \frac{15}{2} = 7.5$$

At the y-intercept, $x = 0$.

$$2(0) - 5y = 15$$
$$-5y = 15$$
$$y = -3$$

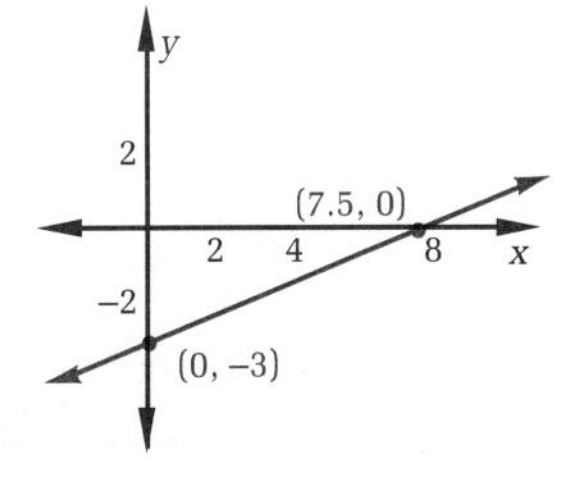

We then connect the two intercepts with a straight line.

Using Computers

This section provides a rich setting for students to explore using computer graphing software or a graphics calculator. Graphics software or a graphing calculator can also be used to solve systems of equations graphically. Using the Zoom feature, students can find results quite accurately. You can also introduce some non-linear systems using the computer or graphics calculator.

Lesson Notes, continued

- The general form of the line is introduced as $Ax + By = C$. Some students may recall that the slope can be found as $-\frac{A}{B}$ and the y-intercept is $\frac{C}{B}$.

Stumbling Block

Students may forget to include the sign of the slope of a line. Have students draw a mental picture of the points or sketch it, and if the line decreases from left to right, the slope must be negative. If it increases from left to right, the slope must be positive. Since we read from left to right, this check on the sign of the slope may be intuitive.

3.1

Lesson Notes, continued

- Parallel and perpendicular lines are also related to slopes.

Cooperative Learning

Have students work in groups of three or four and have them try to prove that for perpendicular lines, the product of their slopes must be −1. Have them draw two perpendicular lines through the origin (not vertical and horizontal) and then find the points on each line with y-coordinates m_1 and m_2 respectively. Then the distance between these points and the origin is $\sqrt{1 + m_1^2}$ and $\sqrt{1 + m_2^2}$ respectively. The distance between the two points is $\sqrt{(m_1 - m_2)^2}$. Use the Pythagorean Theorem to prove that $m_1 m_2 = -1$.

Communicating Mathematics

Write a paragraph describing what slopes indicate, including positive values, negative values, zero slope, and an undefined slope.

Checkpoint

1 Find the slope of the line that passes through the points (1, 3) and (−3, 7). −1

2 Find the intercepts of the line $3x - 4y = 12$. (0, −3) and (4, 0)

3 Write the equation of the line that passes through the point (−3, 4) and has slope = 5. $y - 4 = 5(x + 3)$

Parallel and Perpendicular Lines

Slope can help us determine whether two lines are parallel or perpendicular.

Example *Find the slopes of the parallel lines ℓ and m shown.*

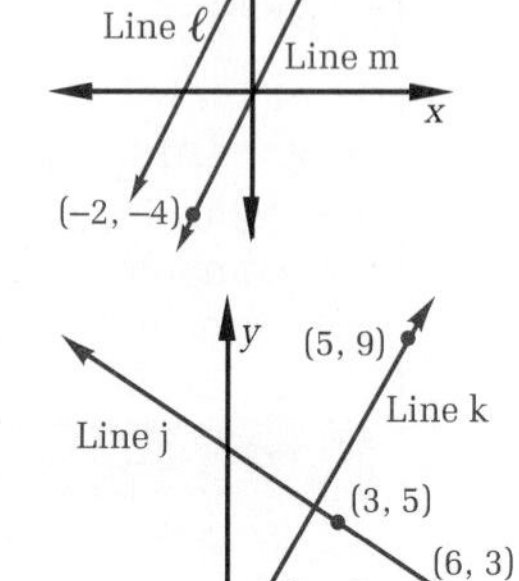

$$\text{Slope of line } \ell = \frac{9 - 5}{2 - 0} = 2$$

$$\text{Slope of line } m = \frac{6 - (-4)}{3 - (-2)} = 2$$

In general, nonvertical lines are parallel only if they have the same slope and different y-intercepts.

Example *Find the slopes of the perpendicular lines j and k shown.*

Line j, Line k, y, x, (5, 9), (3, 5), (6, 3), (1, 3)

$$\text{Slope of line } j = \frac{3 - 5}{6 - 3} = \frac{-2}{3}$$

$$\text{Slope of line } k = \frac{9 - 3}{5 - 1} = \frac{6}{4} = \frac{3}{2}$$

Notice that the product of the slopes, $\frac{-2}{3} \cdot \frac{3}{2}$, is equal to −1.

In general, two nonvertical lines are perpendicular only if the product of their slopes is −1. In other words, the slopes of perpendicular lines are opposite reciprocals.

Example *Lines ℓ and m are perpendicular. Line ℓ has a slope of $-\frac{2}{7}$. Find the slope of line m.*

Lines ℓ and m are perpendicular, so their slopes are opposite reciprocals. Since the slope of line ℓ is $-\frac{2}{7}$, the slope of line m is $\frac{7}{2}$.

Part Two: Sample Problems

Problem 1 *The given line has a slope of −1.4. Find the value of d.*

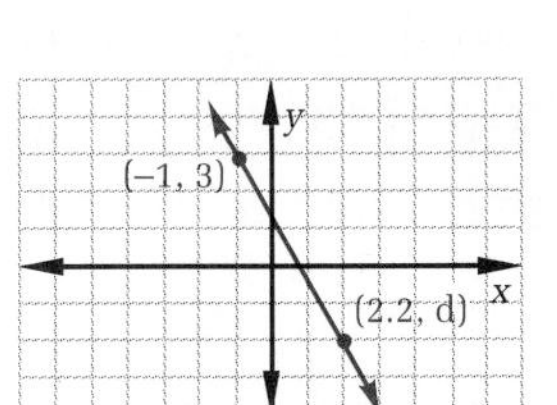

Solution We can use the slope formula or the point-slope form.

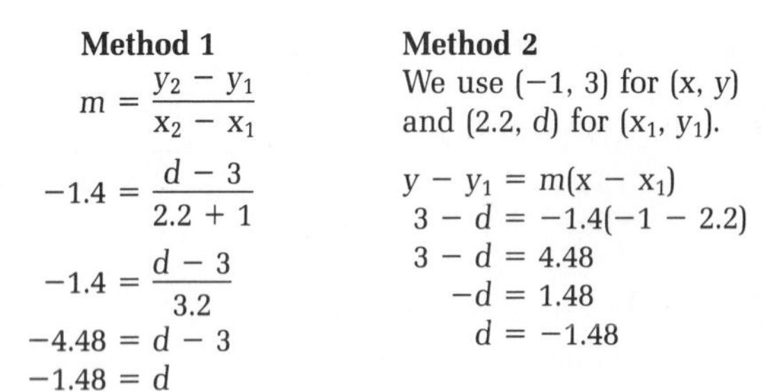

Method 1

$$m = \frac{y_2 - y_1}{x_2 - x_1}$$

$$-1.4 = \frac{d - 3}{2.2 + 1}$$

$$-1.4 = \frac{d - 3}{3.2}$$

$$-4.48 = d - 3$$

$$-1.48 = d$$

Method 2

We use (−1, 3) for (x, y) and (2.2, d) for (x_1, y_1).

$$y - y_1 = m(x - x_1)$$

$$3 - d = -1.4(-1 - 2.2)$$

$$3 - d = 4.48$$

$$-d = 1.48$$

$$d = -1.48$$

Look again at the graph. The y-coordinate of (2.2, d) seems to be close to −1.5, so the answer is reasonable.

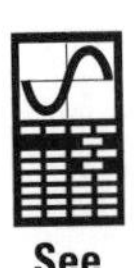

See p. 759

Problem 2 *A company plans to spend $15,000 on new computers. Each JCN PC costs $1100. A Bartlett PC costs $1800 but is faster and easier to use. What combinations of the two brands can the company buy?*

Solution We let x = the number of JCN computers and y = the number of Bartlett computers. The equation 1100x + 1800y = 15,000 indicates how many of each computer can be bought for $15,000. This equation describes a line.

$$\text{x-intercept} = \frac{15{,}000}{1100} \approx 13.6$$

$$\text{y-intercept} = \frac{15{,}000}{1800} \approx 8.3$$

Only whole numbers of computers can be bought, so the company can consider only points with integer coordinates on or below the line. The point (2, 7) is one possibility.

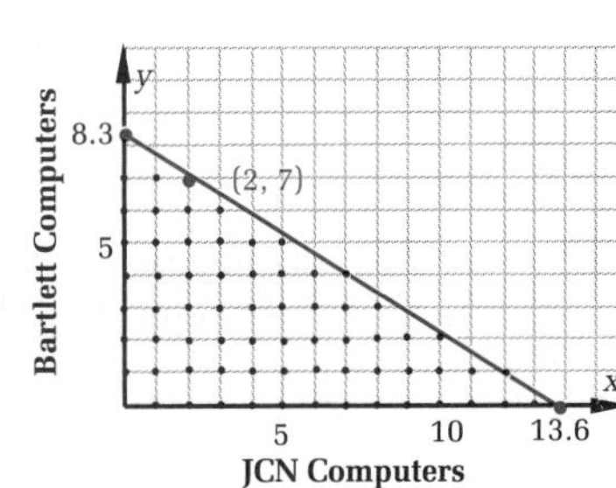

Part Three: Exercises and Problems

Warm-up Exercises

P **1** A line passes through the point (2, 5) and has a slope of $\frac{1}{2}$. Write the coordinates of at least three other points on the line. Answers will vary. Possible answers: (0, 4), (4, 6), (6, 7)

P **2** For what values of k will the graph of y = kx + 3 be a descending line? {k:k < 0}

P **In problems 3 and 4, determine the slope of the line containing the given points.**

3 (5, 2) and (−7, 8) $-\frac{1}{2}$

4 (2, 5) and (8, −7) −2

P **5** The points shown form a line with slope $\frac{2}{3}$. Find a, b, and c. a = 3; b = 9; c = 18

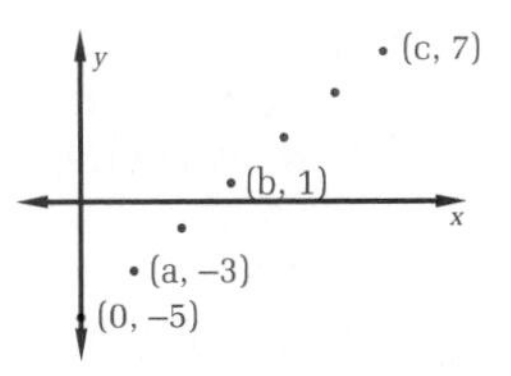

P,R **In problems 6–10, graph each equation.**

6 3x − 5y = 30

7 y = 2

8 x = 5

9 y = 3x − 2

10 y + 2 = −3(x − 3)

Assignment Guide

Average
13–17, 20–23, 26, 27, 29–31
Advanced
13–17, 19, 20, 22–24, 26, 28–30, 32
Honors
13, 14, 16–20, 22–26, 28–30, 32–34

Problem-Set Notes and Strategies

- Point out that in problems **3** and **4** the key is consistency in the subtraction process. It does not matter which point you choose for (x_1, y_1) and (x_2, y_2) as long as you are consistent.
- Problems **7** & **8** may cause problems at first since there is only one variable. Point out that in **7** the y-coordinate must remain constant, but x can vary over all real numbers. This creates a horizontal line. Similarly, problem **8** creates a vertical line.

Additional Answers

6

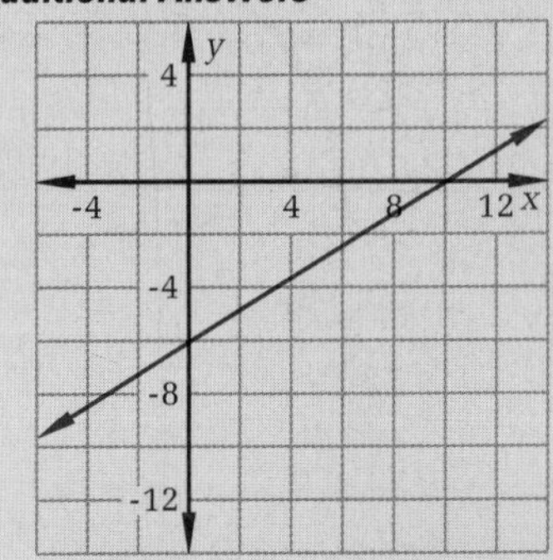

Answers for problems **7** and **8** are on page **A** 1, and the answers for problems **9** and **10** are on page **A** 2.

Problem-Set Notes and Strategies, continued

- In problem **11**, it can be helpful to have students list all possible combinations of four points and identify which will form a rectangle.
- Problem **12** reinforces the need to remember that parallel lines have the same slope and perpendicular lines have slopes that are negative reciprocals.
- Hopefully your students will realize that **16c** is false. Part **d** is rather subtle since using the square root symbol implies that the answer be non-negative.
- Problem **18** is a good exercise to get students into the habit of using functions and function notation.

P,R **11** If four of the six points shown are chosen at random, what is the probability that they are the vertices of a rectangle? $\frac{1}{5}$

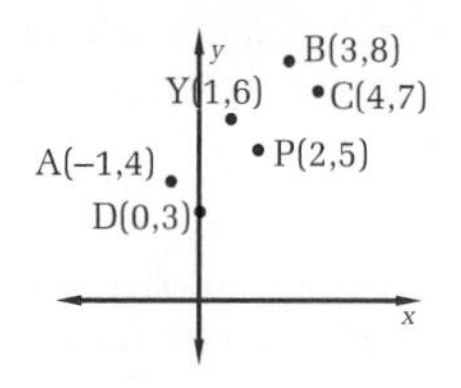

P **12** For what value of k are the graphs of $f(x) = -2x + 6$ and $g(x) = kx + 1$

a Parallel? -2 **b** Perpendicular? $\frac{1}{2}$

Problem Set

A

P **13** List the lines shown in order of the magnitude of their slopes, from smallest to largest. L_4, L_2, L_3, L_1

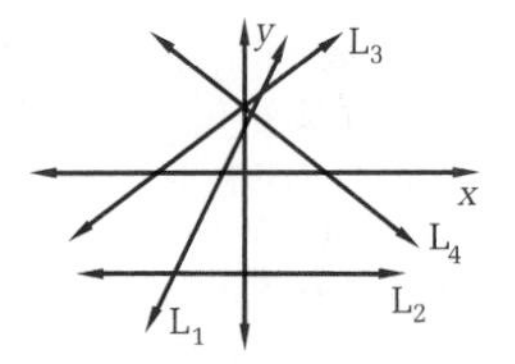

P,R **14** If the midpoint of a line segment is (0, 0) and one endpoint has coordinates (4, 7), what are the coordinates of the other endpoint? $(-4, -7)$

P **15** Draw the line that passes through (3, 6) and (3, 11). Find its slope. Undefined

R **16** Determine whether each of the following algebraic statements is True or False.

a $x^2 \cdot x^5 = x^{10}$ False **b** $x^2 \cdot x^5 = x^7$ True 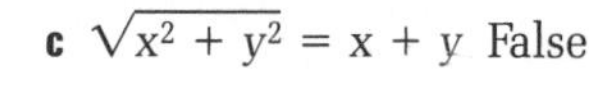 **c** $\sqrt{x^2 + y^2} = x + y$ False

d $\sqrt{(x + y)^2} = x + y$ False **e** $\sqrt{(x + y)^2} = |x + y|$ True

P **17** Find the slope of the line shown. $\frac{4}{3}$

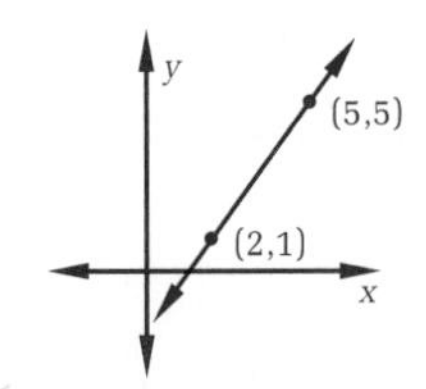

R,I **18** Jerry, an aerobic-exercise teacher, tells Pat to determine his ideal exercise heart rate, R, by subtracting his age, A, from 220 and finding a percentage of the result, corresponding to his fitness level—60% (beginner), 70% (average), or 80% (competitor).

a Write R as a function of A for each fitness level. $R_b = 0.6(220 - A)$; $R_a = 0.7(220 - A)$; $R_c = 0.8(220 - A)$

b If Pat is 23 and in average shape, what is his ideal rate? ≈ 138

c If Lana is a competitive athlete with a rate R of 142, how old is she? $42\frac{1}{2}$

P,R **19** An equation is selected at random from the list below. What is the probability that its graph is an ascending line? $\frac{3}{5}$

$y = 2x - 7$ $y = -4x + 5$ $x - 2y = 6$ $y = 7x + 4$ $y = 3$

P **20** Write an equation to represent each of the following lines. Then draw the line on a coordinate system.

a Slope $-\frac{2}{3}$, y-intercept 4 $y = -\frac{2}{3}x + 4$

b Slope $\frac{3}{5}$, passes through $(-1, 2)$ $y - 2 = \frac{3}{5}(x + 1)$

P,R **21** If $(x, y) = (9, -5)$, which of the following is a true statement? **b**

a $10x + 15y = 12$ **b** $10x + 15y > 12$ **c** $10x + 15y < 12$

R **22** Kathy sells popcorn to raise money for the school French club. A small box sells for $0.30, and a large box sells for $0.50.

a How much money will Kathy collect if she sells 12 small boxes and 15 large boxes? $11.10

b Write an expression for the total amount of money that Kathy would collect if she sold x small boxes of popcorn and 3 more large boxes than small boxes. $0.30x + 0.50(x + 3)$

c If Kathy sold 3 more large boxes than small boxes and collected $25.50, how many boxes of popcorn did Kathy sell all together? 63 boxes

R **23** The first matrix shows the cost of clothing in three shops. The second shows the items that Sue, Sara, and Kai are shopping for.

	Scarf	Gloves	Socks
Le Chic	17.50	15.00	3.50
Le Pen	20.50	13.50	4.75
Bon Ton	22.25	9.75	5.25

	Sue	Sara	Kai
Scarf	2	3	1
Gloves	1	2	3
Socks	4	4	5

a Multiply the cost matrix by the quantity matrix.

b How much would Sue save by shopping at Le Pen rather than Bon Ton?

c Who would spend most at Bon Ton?

a

	Sue	Sara	Kai
Le Chic	64.00	96.50	80.00
Le Pen	73.50	107.50	84.75
Bon Ton	75.25	107.25	77.75

b $1.75 **c** Sara

B

In problems 24 and 25, write an equation for each line described.

P **24** Passes through (3, 1) and (−7, 5) $y - 1 = -\frac{2}{5}(x - 3)$

25 Slope 0, passes through (−5, 0) $y = 0$

P **26** In the diagram, $\overline{AD}$ is perpendicular to $\overline{BC}$. Find the value of k. $k = \frac{23}{4}$

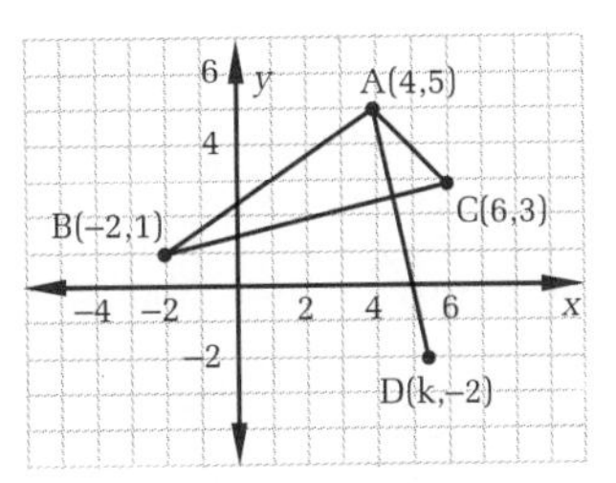

R,I **27** If a car can travel 38 miles on 2.4 gallons of gas, how many miles could the car travel on 6 gallons of gas? 95 mi

Problem-Set Notes and Strategies, continued

- Students need to relate the definition of ascending to a positive slope in problem **19**. Some students may have difficulty with $y = 3$ as it is neither ascending or descending.

Additional Answers

Answers for problems **20a** and **20b** are on page **A** 2.

- Problem **22** illustrates a method for setting up word problems. First, solve an easier problem where you know the answer. Second, set up an expression to illustrate the relationship between expressions. Finally, set up and solve an equation stating that two expressions are equal.

- A nice review of matrix multiplication is provided in problem **23**.

- Problems **24** and **25** provide practice in finding equations of lines given various parts of the line.

- The relationship in problem **27** can be modeled as a proportion problem or a linear equation. Have students point out the significance of the slope in real world terms. It might help to solidify their understanding of slope as a rate of change.

Problem Set, *continued*

R,I **28** Box-seat tickets at Wiggly Field cost \$16. General-admission tickets cost \$10. One night the sales of both types of tickets totaled \$232,672.

a Write an equation for the possible combinations of box-seat tickets B and general-admission tickets G sold. $16B + 10G = 232{,}672$

b Graph the equation you wrote in part **a**.

c Explain why it is not possible that exactly 5015 box seats were sold. Name some other impossible situations. If $B = 5015$, $G = 15243.2$, and only whole numbers of tickets can be sold; answers will vary.

P **29** Consider the points A, B, C, D, E, and F in the diagram. Which of the points satisfy each equation or inequality?

A(3,5), B(7,4), D(1,3), C(7,–1), E(–4,–2), F(0,–4)

a $y = x - 3$ B

b $y > x - 3$ A, D, E

c $y < x - 3$ C, F

P,R,I **30** The table shows the value (in dollars) of a car after x years.

x	**0**	**1**	**2**	**3**	**4**	**5**
Value	**12,600**	**11,100**	**9600**	**8100**	**6600**	**5100**

a What does the slope of the line described by these data mean in terms of the value of the car? The decrease in value (in dollars) per year

b Using the table and common sense, predict the value of the car after ten years. Can't tell. Table predicts −\$2,400, which is impossible. \$300 might be a reasonable guess.

P **31** Let $\frac{x}{7} + \frac{y}{4} = 1$.

a Find the value of y if $x = 0$. 4

b Find the value of x if $y = 0$. 7

C

P,R **32** If one of the following points is chosen at random, what is the probability that it lies on both the graph of $y - 1 = -2(x + 2)$ and the graph of $x - 2y = 2$? $\frac{1}{4}$

(0, 1) (−1, −1) (−0.8, −1.4) (−1.5, 0)

R **33** Let $h(x) = 3x - 10$. Find all the values of a for which $h(a) + h(a + 1) = h(a + 2)$. $\frac{13}{3}$

R,I **34** Find matrix A.

$$A + \begin{bmatrix} 2 & -3 & 7 \\ -5 & 6 & 5 \end{bmatrix} = \begin{bmatrix} 1 & 0 & -9 \\ -4 & 9 & 8 \end{bmatrix}\begin{bmatrix} -1 & 3 & -16 \\ 1 & 3 & 3 \end{bmatrix}$$

P,R,I **35** The graphs of $f(x) = ax + b$ and $g(x) = bx - a$ are perpendicular lines.

a Find $f(0) \cdot g(0)$. 1

b If $\frac{g(1)}{f(0)} = 5$ and $a > 0$, what is the value of a? 2

Problem-Set Notes and Strategies, continued

- A linear equation is used to model problem **28**, yet the domain must be the set of integers since you cannot purchase half a ticket or any portion thereof. When graphing the solution in **b**, the graph should be a set of distinct points at the integer coordinates.

Additional Answer

28b

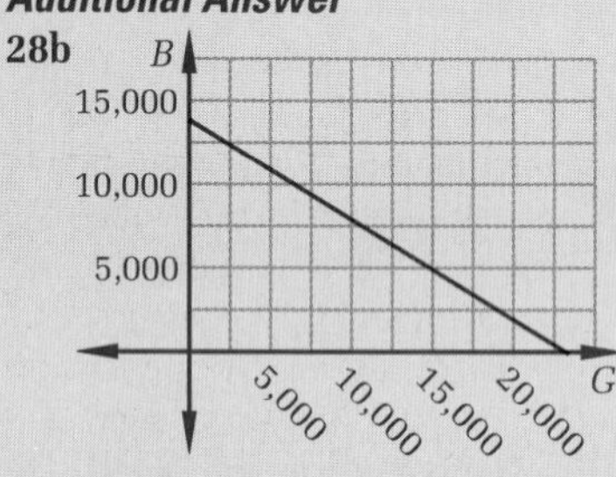

- Problem **30** illustrates that sometimes a graph is only useful for a portion of the domain and common sense must be used to find appropriate values outside this domain.
- Both the x and y intercepts are found in problem **31**. These two points make graphing the equation a quick task. The problem looked difficult with the fractions, but turned out to be rather simple.
- Students may have trouble with the substitution in problem **33**. Assure them that wherever there is an x in the original function, it must be replaced with an a, $a + 1$, or $a + 2$ respectively.
- Problem **35** combines the topics of functions, slopes and simultaneous equations all in one problem.

3.2 SYSTEMS OF LINEAR EQUATIONS

Objectives

After studying this section, you will be able to

- Solve systems of linear equations by graphing
- Solve systems of linear equations algebraically

Part One: Introduction

Systems of Equations

The ordered pairs (6, 0), (0, 4), (−3, 6), and (3, 2) are some of the solutions of the equation $2x + 3y = 12$. The ordered pairs (2.5, 0), (0, −10), (1, −6), (3, 2), and (2, −2) are solutions of the equation $4x - y = 10$. Only one ordered pair satisfies both equations simultaneously. That pair, (3, 2), is the solution of the ***system of linear equations*** that consists of the two equations

$$\begin{cases} 2x + 3y = 12 \\ 4x - y = 10 \end{cases}$$

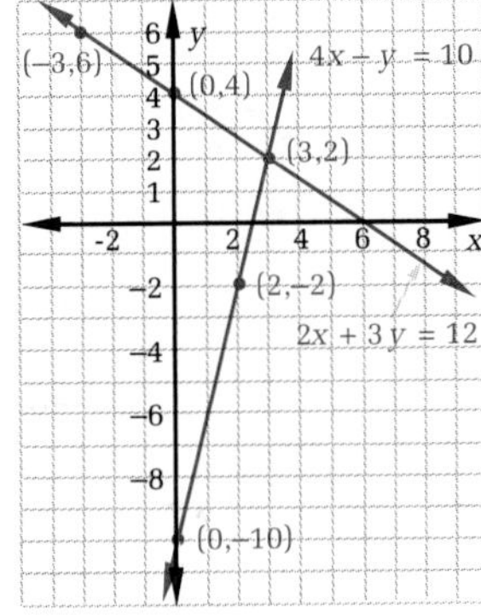

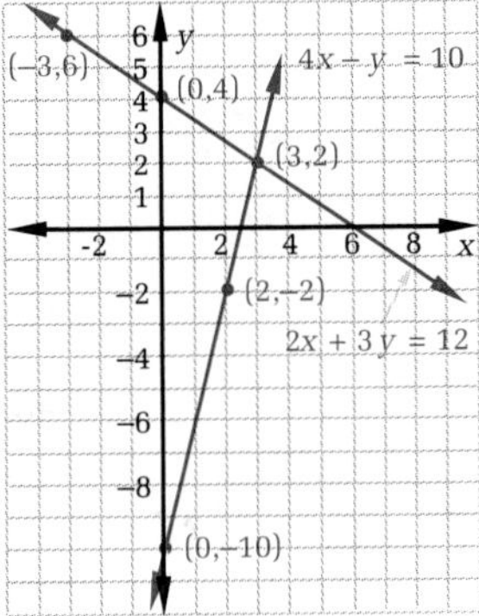

To solve a system means to find all ordered pairs (x, y) that simultaneously make all of the equations in the system true statements. A system may be referred to as a set of ***simultaneous equations*** with simultaneous solutions.

Solving Systems by Graphing

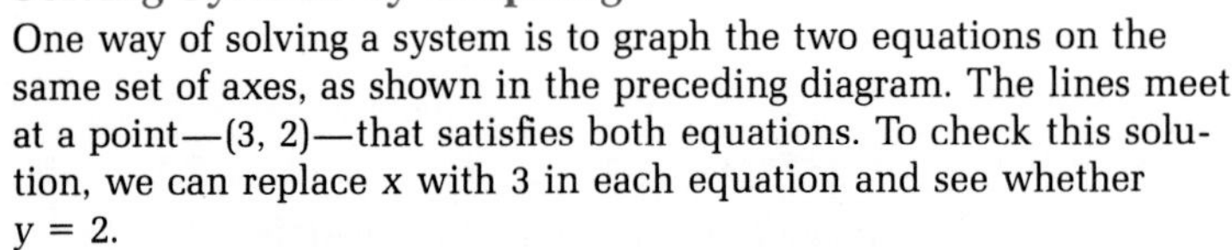

See p. 766

One way of solving a system is to graph the two equations on the same set of axes, as shown in the preceding diagram. The lines meet at a point—(3, 2)—that satisfies both equations. To check this solution, we can replace x with 3 in each equation and see whether $y = 2$.

$$\begin{aligned} 2(3) + 3y &= 12 \\ 6 + 3y &= 12 \\ 3y &= 6 \\ y &= 2 \end{aligned} \qquad \begin{aligned} 4(3) - y &= 10 \\ 12 - y &= 10 \\ 12 - 10 &= y \\ 2 &= y \end{aligned}$$

The pair (3, 2) checks as the solution of the system.

Class Planning

Time Schedule

Average: 2 days
Advanced: 2 days
Honors: $1\frac{1}{2}$ days

Resource References

Teacher's Resource Book
Class Opener 15
Cooperative Learning 9
Practice 16
Enrichment 6

Class Opener

Find the point of intersection of the lines that represent the following equations:
$y = 2x - 8$ and $2x + 3y = 24$
(6, 4)

Lesson Notes

- Three approaches are presented for solving systems of equations. First is the graphing approach which reinforces that we are truly trying to find the point of intersection and that the same point satisfies both equations simultaneously. This is not always the most accurate, but is an excellent method to approximate the solutions. Often an approximation is all that is necessary since the coefficients of the variables are often approximated in a real situation.

Vocabulary

system of linear equations
simultaneous equations
substitution
multiplication-addition method

Solving Systems Algebraically

Graphing is not always the best way to solve a system. Often when we graph a system, we can only estimate the coordinates of the point of intersection. There are two algebraic methods we can use to determine the *exact* solution of a system.

If we can easily represent one variable in terms of the others, ***substitution*** is an appropriate method.

Lesson Notes, continued

- The substitution method works well when one of the equations is easily solved for one of the variables.

Example *Solve the following system.*

$$\begin{cases} 4x + 6y = 26 \\ y = 2x + 1 \end{cases}$$

The second equation is already solved for y. We can use the value $2x + 1$ as a replacement for y in the first equation and solve for x.

$$\begin{aligned} 4x + 6y &= 26 \\ 4x + 6(2x + 1) &= 26 \\ 4x + 12x + 6 &= 26 \\ 16x + 6 &= 26 \\ 16x &= 20 \\ x &= 1.25 \end{aligned}$$

Now we can substitute 1.25 for x in either equation and solve for y. The second equation is the more convenient.

$$\begin{aligned} y &= 2x + 1 \\ &= 2(1.25) + 1 \\ &= 3.5 \end{aligned}$$

The solution is $(x, y) = (1.25, 3.5)$. Since we used the second equation in our solution, we check the answer in the first equation.

Left Side	**Right Side**
$4x + 6y$	$= 26$
$= 4 \cdot 1.25 + 6 \cdot 3.5$	
$= 26$	

The sides are equal, so (1.25, 3.5) checks.

When both equations in a system are in general linear form, it is often easiest to use the ***multiplication-addition method.***

- Finally the multiplication-addition method is introduced as the third method. This method is frequently called the elimination method as it eliminates one of the variables and leaves a single variable equation to solve for by the methods of the last chapter. This method is easiest to generalize to equations with more than two variables and can later be related to elimination methods for matrix solutions.

Example *Solve the following system.*

$$\begin{cases} 4x - y = 17 \\ 2x + 3y = 19 \end{cases}$$

We multiply the first equation by 3 and add the equations.

$$\begin{aligned} \mathbf{12x - 3y} &= \mathbf{51} \\ \mathbf{2x + 3y} &= \mathbf{19} \\ \hline \text{Add} \quad \mathbf{14x} &= \mathbf{70} \\ \mathbf{x} &= \mathbf{5} \end{aligned}$$

Now we substitute 5 for x in one of the equations and solve for y.

$$
\begin{aligned}
2x + 3y &= 19 \\
2(5) + 3y &= 19 \\
10 + 3y &= 19 \\
3y &= 9 \\
y &= 3
\end{aligned}
$$

The solution is $(x, y) = (5, 3)$. To check, substitute 5 for x and 3 for y in the first equation and see whether a true statement results.

Summary

To solve a system of equations, find all ordered pairs (x, y) that satisfy both equations. Both the substitution method and the multiplication-addition method use the same basic strategy.

1. **Eliminate one variable to create a single-variable equation.**
2. **Solve the single-variable equation.**
3. **Substitute the value of the variable in one of the original equations to find the value of the other variable.**
4. **Check your answer in the remaining equation.**

Part Two: Sample Problems

Problem 1 *Find the point where the graphs of* $3x + 4y = -11$ *and* $5x - 6y = 45$ *intersect.*

Solution In this case, the most efficient way to determine the point of intersection is to solve a system by the multiplication-addition method.

Multiply first equation by 3	$\mathbf{9x + 12y = -33}$
Multiply second equation by 2	$\mathbf{10x - 12y = 90}$
Add	$\mathbf{19x = 57}$
	$\mathbf{x = 3}$
Substitute in first equation	$\mathbf{3(3) + 4y = -11}$
	$\mathbf{4y = -20}$
	$\mathbf{y = -5}$

We check the solution $(3, -5)$ in the second equation.

Left Side	**Right Side**
$5x - 6y$	$= 45$
$= 5(3) - 6(-5)$	
$= 45$	

The sides are equal, so the answer checks. The lines intersect at $(3, -5)$.

Using Computers

This is an appropriate time to use a matrix package or system solving software to solve linear systems algebraically. If you have scientific calculators that are capable of matrix inversion, you may be able to introduce your students to solving systems by using matrices at this time. *Linear Kit*, *Derive*, or graphics calculators could be used.

Communicating Mathematics

Write a paragraph describing when the substitution method would be more helpful than the multiplication-addition method and vice versa.

Checkpoint

1 Find the point of intersection of the graphs described by the equations.
$x - y = 12$
$x + y = -6$
$x = 3, y = -9$

2 Find the point of intersection of the graphs described by the equations.
$3x + 5y = 21$
$-4x + 2y = -2$
$(2, 3)$

3 Is the point $(0, 1)$ a solution to the system:
$5x + 7y = 7$
$7x + 5y = 7$
No

Assignment Guide

Average	
Day 1	8, 10, 11, 13, 14, 16, 17, 19, 21, 23, 27
Day 2	9, 12, 15, 18, 20, 22, 24–26, 28, 29, 31
Advanced	
Day 1	8–11, 13–15, 17, 24, 26, 27, 30–32
Day 2	12, 16, 18, 20, 22, 25, 28, 29, 33, 34
Honors	
Day 1	9, 11–13, 15–18, 22, 26,28, 29, 32
Day 2	($\frac{1}{2}$ day) 14, 24, 25, 30, 31, 33, 34
	($\frac{1}{2}$ day) Section 3.3 9–11, 13, 15, 16

Problem-Set Notes and Strategies

- The explanation in problem **1** is important to show whether the students understand the concept of systems of equations.
- Problem **3** will have various answers. Look at the students' answers and ask why they chose a particular method. The text has a specific answer, yet any of the methods may be used.

Problem 2 *P. Nutt wants to sell 35 pounds of mixed nuts for $3.05 per pound. Nutt has cashews selling for $3.25 per pound and peanuts selling for $2.80 per pound. How many pounds of each type of nut should he use in the mixture?*

Solution Let c = the number of pounds of cashews in the mixture and p = the number of pounds of peanuts. The total number of pounds is 35, so $c + p = 35$. The total value of the mixture is $3.05 \cdot 35$, so $3.25c + 2.80p = 3.05(35)$. We therefore have the system

$$\begin{cases} 3.25c + 2.8p = 106.75 \\ c + p = 35 \end{cases}$$

We can rewrite the second equation as $c = 35 - p$ and substitute for c in the first.

$$\begin{aligned} 3.25c + 2.8p &= 106.75 \\ 3.25(35 - p) + 2.8p &= 106.75 \\ 113.75 - 3.25p + 2.8p &= 106.75 \\ -0.45p &= -7 \\ p &\approx 15.6 \end{aligned}$$

Substitute 15.6 for p in second equation.

$$\begin{aligned} c + 15.6 &\approx 35 \\ c &\approx 19.4 \end{aligned}$$

P. Nutt should use about 19.4 pounds of cashews and about 15.6 pounds of peanuts.

Part Three: Exercises and Problems

Warm-up Exercises

P **1** Explain why (4, −5) is or is not a solution of the following system.

$$\begin{cases} y = 3x - 17 \\ 2x + y = 3 \end{cases}$$

It satisfies both equations, so it is a solution.

P **2** Use the diagrams to solve for (x, y).
(x, y) = (12, 4)

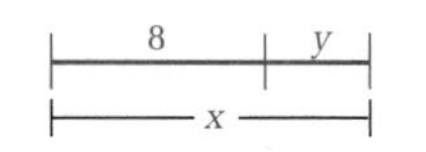

P **3** Of the possible methods—graphing, substitution, guess and check, and multiplication-addition—which would you choose to solve each of the given systems? Explain your choice.

a $\begin{cases} x + y = 18 \\ xy = 81 \end{cases}$ Graphing

b $\begin{cases} y = 2x - 4 \\ 2x + y = 16 \end{cases}$ Substitution or guess and check

c $\begin{cases} w + v = 10 \\ 3.45w + 22.75v = 28.9 \end{cases}$ Multiplication-addition

d $\begin{cases} y = 3x + 1 \\ y - 2 = 4(x + 1) \end{cases}$ Substitution

P **In problems 4–6, find the point of intersection of the graphs of each pair of equations.**

4 $2x - y = 6$ and $y = 2$ (4, 2)
5 $y = 2x + 5$ and $y = -3x + 5$ (0, 5)
6 $3x + y = 12$ and $x + y = 8$ (2, 6)

P 7 Refer to the graphs of $x - 2y = 4$ and $7x + 2y = 8$.

a Approximately where do the lines intersect? (1.5, −1.3)

b Use an algebraic method to find the exact point of intersection. (1.5, −1.25)

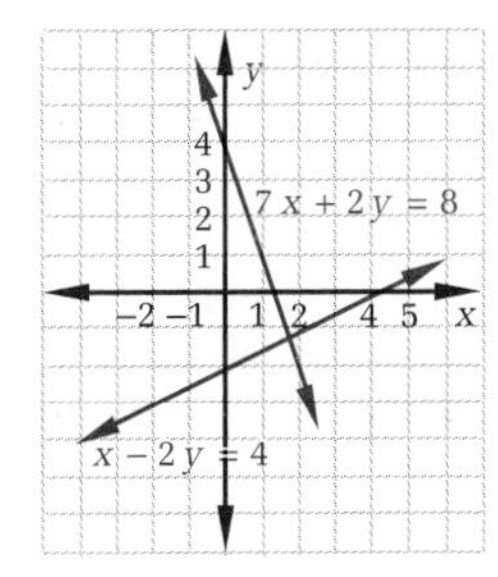

Problem Set

P **A** **8** Use the diagram to solve for x and y.
$x = 8$; $y = 4$

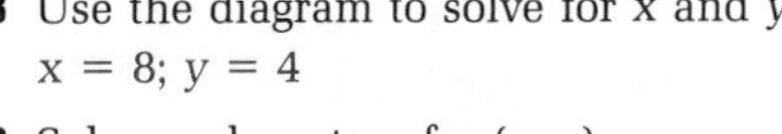

P **9** Solve each system for (x, y).

a $\begin{cases} 3x + y = 22 \\ 2x - y = 13 \end{cases}$ $(x, y) = (7, 1)$

b $\begin{cases} y = 3x - 13 \\ 3x + y = 17 \end{cases}$ $(x, y) = (5, 2)$

c $\begin{cases} 7x - 2y = -6 \\ 5x + 2y = 12 \end{cases}$ $(x, y) = (0.5, 4.75)$

d $\begin{cases} 3x + 2y = 2 \\ 2x + y = 3 \end{cases}$ $(x, y) = (4, -5)$

P **10** A two-egg breakfast with hash browns and toast costs \$1.85. A one-egg breakfast with hash browns and toast is \$1.35. How much is charged for hash browns and toast? 85¢

P **11** Find x and y if the perimeter of rectangle A is 26 and the perimeter of rectangle B is 24. $x \approx 2.14$; $y \approx 2.57$

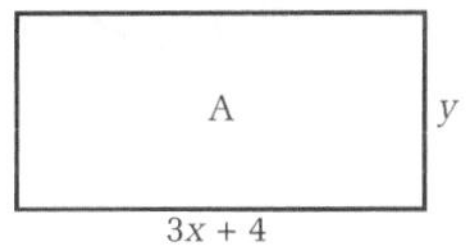

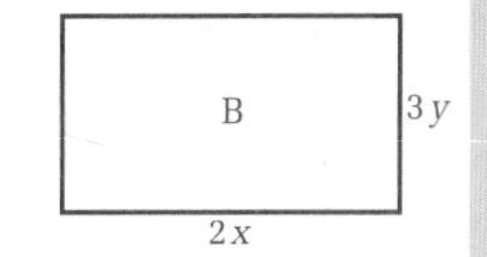

R **12** A committee of three people is to be selected from a group of three women and two men. If the committee will be selected randomly, what is the probability that it will consist of two women and one man? $\frac{3}{5}$

P,R **13** If $f(x) = 3x + 4$ and $g(x) = 3x - 1$, for what value of x is $f(x) = g(x)$? None

R **14** The matrix below shows the values of four stocks on September 1 and October 1. Copy the matrix, supplying the missing information.

Stock	Value per Share, 9/1	Value per Share, 10/1	Change in Value per Share from 9/1 to 10/1	
Flyer Shoes	35.50	38.25	?	2.75
JAMB	67.75	71.50	?	3.75
Cruiser Bikes	81.00	73.75	?	−7.25
Vector Shirt	?	46.50	−3.25	49.75

Problem-Set Notes and Strategies, continued

- Problem **7** illustrates the significance of approximating answers to verify the algebraic solution.
- Different methods may again be applied in problem **9.** Discuss the advantages of one method over another and your reasoning for choosing the method you did. Let the students make their own judgement, but encourage them to give each method fair treatment.
- Problem **12** is a good probability problem and illustrates the hypergeometric probability distribution function. This is a good exercise for group learning and is included in the cooperative learning portion of this section.

Cooperative Learning

Problem **12** lends itself to working in groups. Ask students to find the formula for finding the number of ways to choose two women from three and one man from two. Find the product of these ways and divide by the number of ways of choosing three people for the committee from the five availabe. Next try to find the probability of choosing an 8 person bipartisan subcommittee from a senate committee of 12 Democrats and 8 Republicans.

- Students may work the algebra in problem **13** before they realize the two lines are parallel with different intercepts.

Problem-Set Notes and Strategies, continued

- Problems **19 – 23** review the distributive property and the properties of exponents studied in the short subject at the end of the last chapter.
- In problem **25**, point out that the two regions and the boundary are mutually exclusive and all inclusive of the graph. These terms will be used in the probability portion of Chapter **12** and this is an intuitive approach to their definitions.
- Problem **27** is a proportion problem and if written as a linear equation, the slope or proportionality constant is the price per hot dog. This illustrates the significance of the slope one more time in a real situation.
- When graphing the parts in problem **28**, have students note the transformations and reflections that are taking place. Ask them to point out the symmetry involved and generalize on the effect of adding the 2 inside and outside the absolute value signs.
- Problem **31** requires the student to set up a system of 4 equations in 4 unknowns. These are easily solved by substitution since none of the equations has all four variables in it.

Additional Answer

28

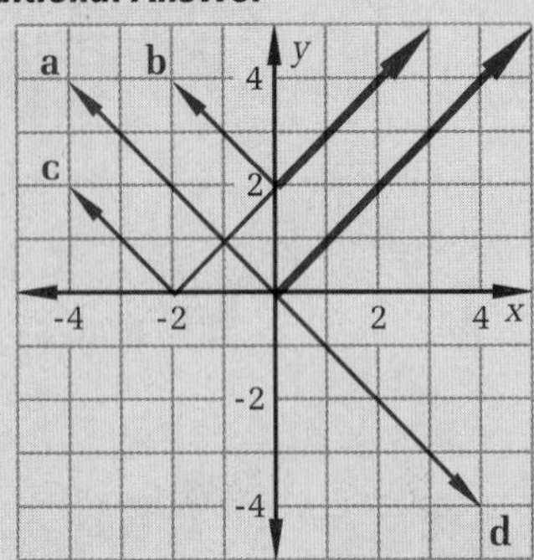

Problem Set, *continued*

R **15** Alex earns \$4.75 per hour plus 1% commission on his total sales.

a Write an equation that describes Alex's wages as a function of number of hours worked h and total sales s. $W = 4.75h + 0.01s$

b If Alex works eight hours a day, how much merchandise must he sell to earn at least \$60 a day? \$2200

R **16** Determine the value of a for which the line passing through $(6, -4)$ and $(2, a)$ has a slope of 12. -52

P **In problems 17 and 18, solve each system for (x, y).**

17 $\begin{cases} y = 4x - 5 \\ y = \frac{1}{2}x + 2 \end{cases}$ $(x, y) = (2, 3)$

18 $\begin{cases} 3x + 2y = 7 \\ 2x - 5y = 12 \end{cases}$ $(x, y) \approx (3.11, -1.16)$

R **In problems 19–23, multiply.**

19 $x^2 \cdot x^5$ x^7

20 $3x(4x)^2$ $48x^3$

21 $5x^5 \cdot 5x^5$ $25x^{10}$

22 $-4x(3x^2 - 7x - 5)$ $-12x^3 + 28x^2 + 20x$

23 $x^2(3x^5 - 2x)$ $3x^7 - 2x^3$

P **In problems 24 and 25, solve each system for (x, y).**

24 $\begin{cases} 3x - 5y = 11 \\ 8x + 10y = 6 \end{cases}$ $(x, y) = (2, -1)$

25 $\begin{cases} x + 3y = 32 \\ y = 2x - 1 \end{cases}$ $(x, y) = (5, 9)$

B

R,I **26** Graph the line that represents the equation $y = 2x - 5$ and name a point *not* on the line. Does the point solve any of the following? Yes—**b** or **c**

a $y = 2x - 5$ **b** $y > 2x - 5$ **c** $y < 2x - 5$

R,I **27** If three hot dogs cost \$7.53, what do you predict to be the cost of eight hot dogs? \$20.08

I **28** Graph the following equations on the same coordinate system.

a $y = |x|$ **b** $y = |x| + 2$
c $y = |x + 2|$ **d** $x = |y|$

P,R **29** If the two equations $x + y = 10$ and $y = mx + 5$ are graphed on the same coordinate plane and if m is randomly chosen from $\left\{-2, -\frac{3}{2}, -1, -\frac{1}{2}, 0, \frac{1}{2}, 1, \frac{3}{2}, 2\right\}$, what is the probability that the two lines will intersect in the first quadrant? $\frac{5}{9}$

P,R **30** Find the value of k for which the slope of the line through $(-2, k)$ and $(5, 3k - 4)$ is 8. 30

P,R **31** Solve the matrix equation for w, x, y, and z.

$$\begin{bmatrix} 2x & x + y \\ z & y - z \end{bmatrix} = \begin{bmatrix} x + 8 & 15 \\ 7 - y & w \end{bmatrix}$$ $w = 7$; $x = 8$; $y = 7$; $z = 0$

P **32** If $x + 2y = 12$ and $3x - 7y = 16$, what is the value of $4x - 5y$? 28

P **33** If two points are selected at random from the following list, what is the probability that the line connecting them will be perpendicular to a line with slope $\frac{2}{3}$? 1

(2, 8) (4, 5) (8, −1) (12, −7)

P,R **34** A Mion car costs $12,895 and gets 32 miles per gallon of gasoline. A Toyord car costs $10,550 and gets 26 miles per gallon of gasoline. Gasoline costs $1.16 per gallon.

a For each car, write an equation describing the cost of buying the car and driving it for x miles. $C_M = 12{,}895 + 1.16\left(\frac{x}{32}\right)$; $C_T = 10{,}550 + 1.16\left(\frac{x}{26}\right)$

b How many miles would owners of the cars need to drive before the cost of buying and driving the cars would be the same? ≈ 280,322 mi

c After how many miles of driving is the difference in the cost of buying and driving the cars $500? ≈220,552 mi and ≈340,092 mi

Problem-Set Notes and Strategies, continued

- All that is asked for in problem **32** is the sum of the two equations.
- All four points in problem **33** are linear and the slope of the line is $-\frac{3}{2}$.
- A graph of the two equations in problem **34** would help to verify the results. By looking at the vertical distance between the lines, the student will be convinced that there are indeed two answers for part **c**.

CAREER PROFILE

LINEAR RELATIONSHIPS AND TART CHERRIES

Amy Iezzoni brings mathematics to the orchard

In order to grow improved varieties of plants, crops, and trees, plant breeders select and combine characteristics of existing varieties. Amy Iezzoni, an associate professor of horticulture at Michigan State University in East Lansing, is working to help breed new types of tart cherries—the kind used in pies. She is seeking alternatives to the 400-year-old Montmorency cherry, the only sour cherry grown in the United States.

"The Montmorency cherry tree is susceptible to numerous diseases and insects. And it would be better to have a series of cherry varieties that ripen at different times and that may bloom later in the spring and hence be less likely to get frosted than Montmorency," explains Dr. Iezzoni.

Dr. Iezzoni has been breeding tart cherries using pollen and seeds that she has collected from all over eastern Europe, the region that she says is the tart cherry's "center of origin and diversity." Her travels have taken her to Yugoslavia, Bulgaria, Hungary, Romania (to the Carpathian Mountains), Poland, and Sweden. "I also got seeds and pollen through the mail from the Soviet Union. Then we grew all these seedlings in East Lansing."

Using matrices, plotting the characteristics of trees bred from different seeds and pollen, and identifying equations that fit her data, Dr. Iezzoni has found that the trees growing across eastern Europe exhibit a range of characteristics that developed as the trees adapted to a range of environments.

Dr. Iezzoni has a bachelor's degree from North Carolina State University and a doctorate from the University of Wisconsin at Madison.

3.3

Class Planning

Time Schedule
Average: 2 days
Advanced: 2 days
Honors: $1\frac{1}{2}$ days

Resource References
Teacher's Resource Book
Class Opener 16
Practice 17
Transparency 5
Evaluation
Quiz Forms 1, 2, 3

Class Opener

Graph $5x + 3y = 15$. Pick five points at random above the line. Substitute each of the pairs into the expression $5x + 3y$ and evaluate.

1 If one of the five points is randomly selected, what is the probability that the value produced will be greater than 15? 1
2 Determine the probability that the value will be less than 15. 0

Lesson Notes

- Graphing linear inequalities is similar to graphing equations in two variables. As with the one dimensional problems in Chapter **2**, the equation is first solved to identify regions of the graph. Each of these regions is then tested to see if it verifies the inequality and finally the boundary is considered to see if it should be included as part of the solution set. Dotted lines indicate that the boundary is not included and solid lines indicate that the boundary is to be included.

3.3 Graphing Inequalities

Objectives
After studying this section, you will be able to
- Graph linear inequalities
- Find the union and the intersection of sets defined by inequalities

Part One: Introduction

Graphing Linear Inequalities

The graph of a linear equation is a set of points that form a straight line. The graph of a linear inequality is a set of points that form a region. The boundary of the region described by a linear inequality is the line found by graphing the inequality as if it were an equation.

We can graph linear inequalities quite accurately by using a ***boundary-value algorithm.***

1 Graph the boundary equation.
2 Test a point in each region.
3 Shade the appropriate region.

Example *Graph* $y > 3x + 5$.

We use a dashed boundary line $y = 3x + 5$ to show that the points on the boundary line, such as (0, 5), are not part of the graph of the inequality.

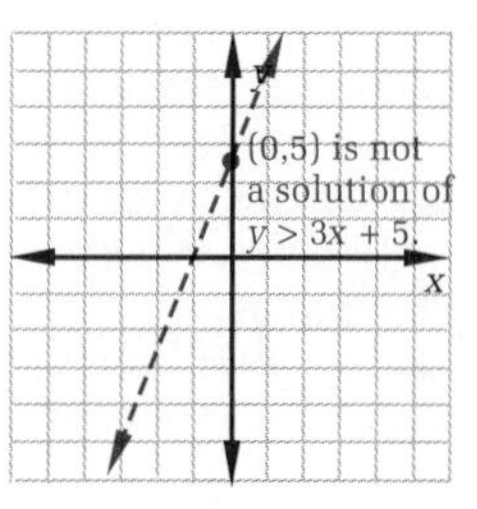

Next we test a point in each of the two regions formed by the boundary line. The values (0, 10) in the left-hand region make the inequality a true statement; the values (0, 0) in the right-hand region do not.

If one point in a region represents a solution of the inequality, all points in the region do. We therefore shade the entire region to the left of the boundary line.

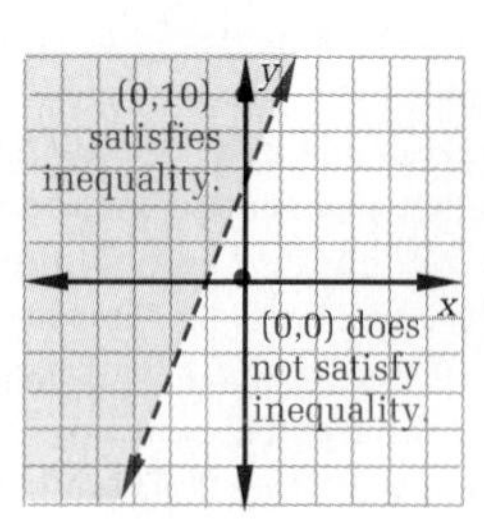

Vocabulary
boundary-value algorithm
union
intersection

Union and Intersection

Since a graph is a set of points, we can consider more than one graph at a time by using the concepts ***union*** (∪) and ***intersection*** (∩) from the language and notation of sets. These operations are defined for sets A and B as follows:

$A \cup B = \{$all elements in set A or in set B or in both sets$\}$
$A \cap B = \{$all elements in set A and in set B$\}$

To define a set in terms of specific conditions, we use a colon to mean "such that." For example, $\{x:x > 0\}$ is read "the set of all numbers x such that x is greater than zero."

See p. 754

Example *Let* $R = \{x:x < 5\}$ *and* $S = \{x:x \geq 3\}$.

a *Find* $R \cup S$.

The single variable tells us to use a one-dimensional graph, a number line. A set that began "$\{(x, y):$" would require a two-dimensional graph.

The union is the set of all real numbers—in set notation, $\{x:x \in \mathfrak{R}\}$, read "the set of all numbers x such that x is an element of the real numbers."

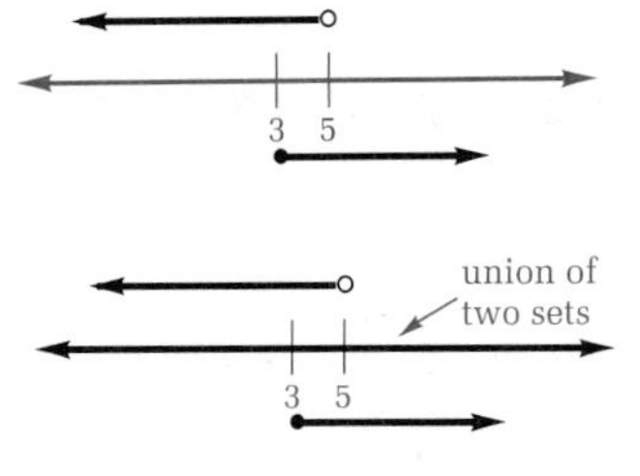

b *Find* $R \cap S$.

The intersection is the set of reals between 3 and 5, including 3 but not 5. In set notation we write $\{x:3 \leq x < 5\}$, read "the set of all numbers x such that x is greater than or equal to 3 and less than 5."

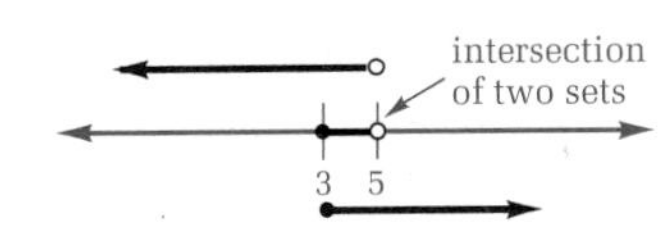

Part Two: Sample Problem

Problem *Let* $A = \{(x, y): y \leq x + 3\}$ *and* $B = \{(x, y): y > -2x - 3\}$. *Graph* $A \cup B$ *and* $A \cap B$.

Solution Since two variables are involved, we must use two-dimensional graphs.

$A \cup B$, the union, consists of all points that belong to at least one of the sets. Notice the boundary lines.

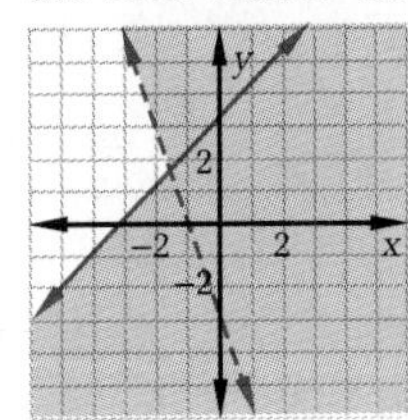

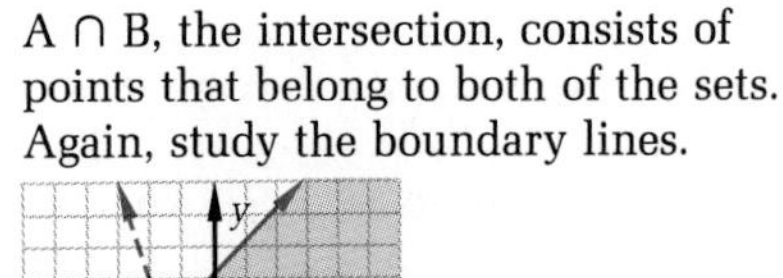

$A \cap B$, the intersection, consists of points that belong to both of the sets. Again, study the boundary lines.

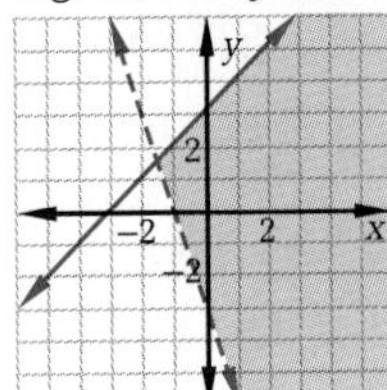

Lesson Notes, continued

- The concepts of intersection (and) and union (or) are introduced to deal with compound inequalities. Emphasize that the key words are *and* and *or*. Intersection implies that the elements have to satisfy *both* of the inequalities. Union implies that either one or the other inequality must be satisfied. The student should be able to relate to both the symbols and the key words *and* and *or*.

Communicating Mathematics

Describe the differences between the terms *union* and *intersection* using a real world example.

Using Computers

Computer software can graph the equality portions of the inequalities in this section and students can check for the correct regions to shade. If you use a symbolic manipulator such as *Derive*, it can check the regions. The *TI81 Graphics Calculator* is capable of graphing inequalities and could be used in conjunction with this section.

■

Stumbling Block

Students frequently look at the inequality sign and shade the lower half of the graph if it is < and the upper half of the graph if it is >. Encourage students to always check a point in one of the regions. If a single point in a region satisfies the inequality, every point in the region will also satisfy it. An obvious choice is the origin if it is not on the boundary.

■

Checkpoint

1 Let $A = \{2, 4, 6, 8, 10\}$ and $B = \{1, 2, 3, 4, 5\}$. Find $A \cap B$ and $A \cup B$.
$A \cap B = \{2, 4\}$
$A \cup B = \{1, 2, 3, 4, 5, 6, 8, 10\}$

2 Graph the inequality $2x - 3y \geq 6$.

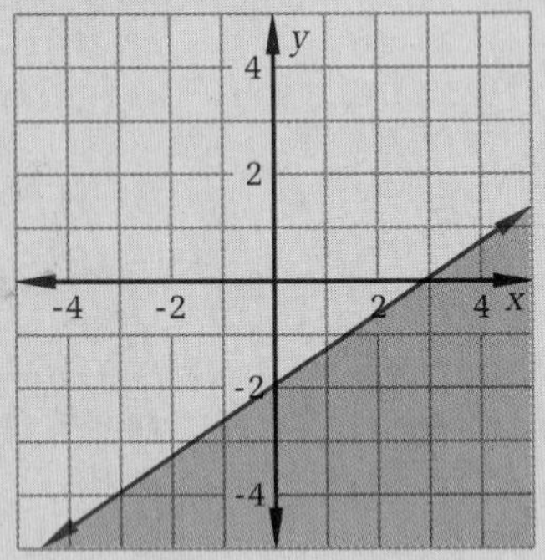

Assignment Guide

Average	
Day 1	9, 11, 13, 14, 16, 17, 19, 20, 22, 23, 27, 29
Day 2	10, 12, 15, 18, 21, 24, 26, 28, 30, 31, 34, 35
Advanced	
Day 1	9–11, 13, 14, 16–19, 21, 23, 24, 28, 29
Day 2	12, 15, 20, 22, 25–27, 30, 31, 34–37
Honors	
Day 1	($\frac{1}{2}$ day) Section 3.2 14, 24, 25, 30, 31, 33, 34 ($\frac{1}{2}$ day) 9–11, 13, 15, 16
Day 2	12, 14, 17–21, 23–26, 29–33, 35–37

Part Three: Exercises and Problems

Warm-up Exercises

1 Graph $y \geq -3x + 4$.

P **In problems 2 and 3, sketch the graph of each set.**

2 $\{(x, y): y < 5\}$ **3** $\{x: x \geq -12\}$

P **4** Sketch the graph of $2x + 3y \geq 12$. Is the point (4, 7) on this graph? Yes

P **5** Let $A = \{2, 4, 6, 8, 13, 14, 15\}$, $B = \{2, 6, 7, 9, 12, 14, 15, 17\}$, and $C = \{4, 7, 8, 9, 12, 13, 17\}$.

a Find $A \cup B$. $\{2, 4, 6, 7, 8, 9, 12, 13, 14, 15, 17\}$
b Find $A \cap B$. $\{2, 6, 14, 15\}$
c Find $A \cap (B \cup C)$. $\{2, 4, 6, 8, 13, 14, 15\}$
d Find $(A \cap B) \cup (A \cap C)$. $\{2, 4, 6, 8, 13, 14, 15\}$

P **6** Consider the positive integers between 0 and 100. Write a simple form of each intersection.

a {multiples of 3} ∩ {multiples of 2} {multiples of 6}
b {multiples of 5} ∩ {multiples of 3} {multiples of 15}
c {multiples of 6} ∩ {multiples of 4} {multiples of 12}

P **7** Graph $\frac{x}{4} + \frac{y}{5} \leq 1$.

P **8** Write an inequality to describe the shaded region. $6x + 5y \geq 30$

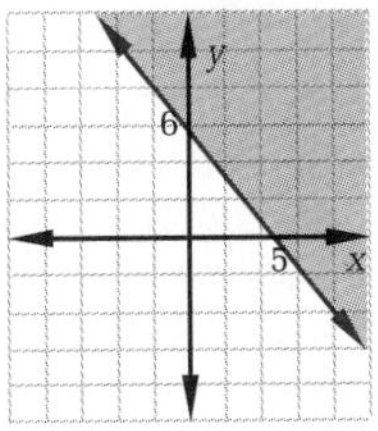

A Problem Set

P **9** Sketch the graph of $\{(x, y): x - 3 < 5\}$.

P **10** Let $A = \{(x, y): y < 2x + 3\}$ and $B = \{(x, y): y < 4x + 3\}$. Sketch the graph of $A \cap B$.

P,R **11** Draw the line that passes through (3, 5) and has a slope of $\frac{2}{3}$.

a Shade the region below the line.
b Write an inequality to describe the graph. $\left\{(x, y): y \leq \frac{2}{3}x + 3\right\}$

R **12** Two perpendicular lines intersect at (0.5, 7.5). Neither line is horizontal. Write equations of the lines. Any equations of the forms $y - 7.5 = m(x - 0.5)$ and $y - 7.5 = -\frac{1}{m}(x - 0.5)$, where $m \neq 0$

Additional Answers

3

Answers for problems **1**, **2**, **7**, and **9 – 11** can be found in the answer pages beginning on **A 1**.

R | **13** Refer to the graph. Estimate the coordinates of the point where the lines intersect; then find the exact coordinates of the point.
Answers will vary; $\left(\frac{11}{8}, \frac{51}{8}\right)$

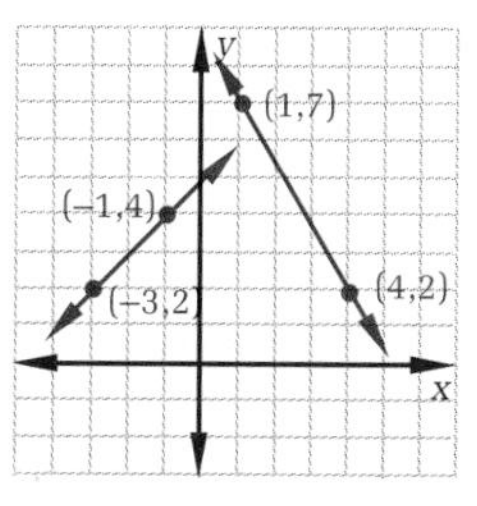

R,I | **14** Find the area of the rectangle shown for each of the following values of x.

a $x = 7$ 176
b $x = -2$ 14
c $x = -6$ -6
d What values of x make sense in this problem? $\{x{:}x > -4\}$

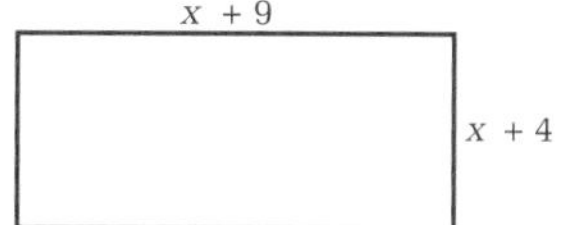

P,R | **15** Let $S = \{(x, y){:}y > x\}$ and $N = \{(x, y){:}y < -x\}$. If a point is selected at random from the following set, what is the probability that it is an element of $S \cap N$? $\frac{1}{8}$

$\{(5, 0), (5, 5), (0, 5), (-5, 5), (-5, 0), (-5, -5), (0, -5), (5, -5)\}$

R | **16** The diagram shows the numbers of All-Star Games won by the American and National Leagues in the years from 1933 through 1989.

a In which decade did the National League win the most games? 1960–1969
b In which decade did the American League win the most games? 1940–1949
c Which league dominated the years 1933–1949? American
d Which league dominated the years 1950–1989? National

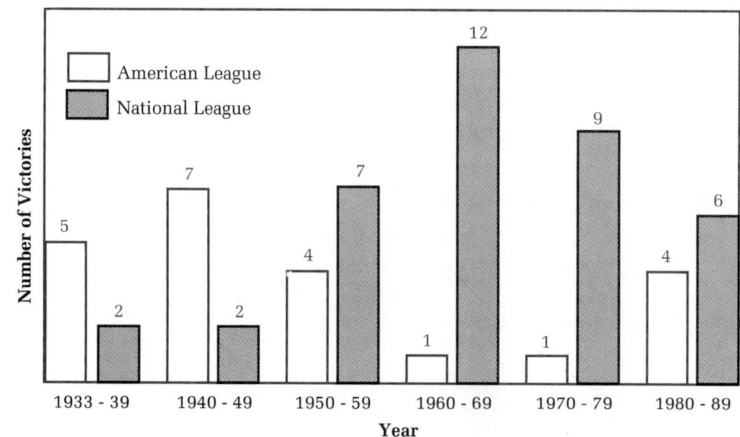

Problem-Set Notes and Strategies

- In problems **1 – 4**, check to make sure a solid line is used for equality and a dotted line when equality is not included. This is one of the most common errors in linear inequalities.
- Emphasize the distinction of the words *and* for intersection and *or* for union.
- Draw both inequalities on the same graph for problem **10** and include the double shaded region. Be careful about the boundary. In this case neither is included.
- There are a multitude of answers for **12**. It is important that the point (0.5, 7.5) satisfy both equations and that the product of the two slopes is –1.
- Have students evaluate their estimate in problem **13**. If they are not close, either their graphing skills need additional work or they need to sharpen their algebra techniques.
- Problem **14** asks students to make a generalization about the domain in part **d**. Emphasize that the domain is not all real numbers. Some may argue that it should be $x > -4$ but that would produce a rectangle with a side of length 0. Some argue that this is a degenerate rectangle, but for this text we will accept $x > -4$.
- It might help to graph the intersection of *S* and *N* in problem **15** to verify the results obtained.

Problem-Set Notes and Strategies, continued

- Encourage students to observe the differences in problem **18** between part **a** and **b** and to try to generalize a result: the sum of a constant times n is the constant times the sum of n. That is, the constant may be factored outside of the summation symbol if it aids solving the problem.
- Students may need help with problem **21** as the concept of intersecting regions of sets related to graphs is still fairly new to most of them.
- Problem **23** is an example of a linear programming problem. This topic will be explored in more detail in a short subject at the end of Chapter **10**.
- Problem **26** is an example of a compound intersection and union problem. Colored chalk, two separate graphs that are combined or working each part on an overhead transparency and overlaying the results are all good techniques.

Additional Answers

19

Answers for problems **20**, **23b**, **24–26**, and **30** can be found in the answer pages beginning on page **A** 1.

Problem Set, *continued*

R **17** Tom mows lawns and whitewashes fences during his summer vacation. He is paid \$12 for each lawn mowed and \$17 for each fence painted. Last month Tom did 105 jobs and made a total of \$1410. How many of each type of job did he do?
75 lawns, 30 fences

R **18** Evaluate each expression.

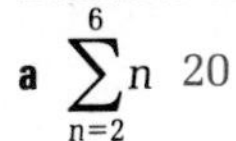

a $\sum_{n=2}^{6} n$ 20 **b** $\sum_{n=2}^{6} 5n$ 100

P,I **19** Sketch the graph of $\{(x, y): x \geq |y|\}$.

P **20** Graph $-y < -x$.

P,R **21** The lines shown are perpendicular. Describe the shaded region algebraically.
$\{(x, y): y \geq -2x + 11\} \cap \left\{(x, y): y \geq \frac{1}{2}x - \frac{3}{2}\right\}$

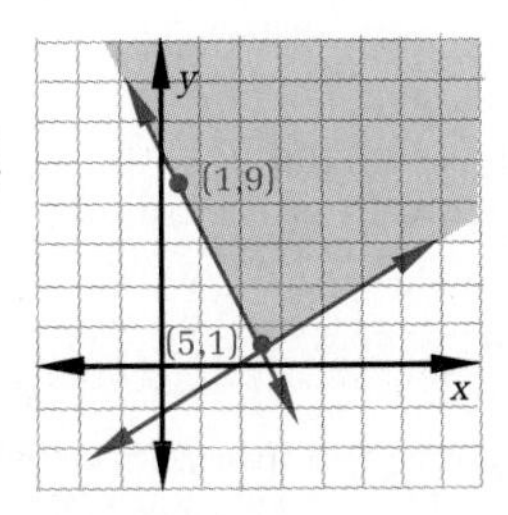

B

R **22** A linear graph contains the points (3, −5) and (7, 12).
a If $(a, 4)$ is also a point on this graph, what is the value of a? $\frac{87}{17}$, or ≈ 5.12
b If $(-7, b)$ is also a point on this graph, what is the value of b? $-\frac{95}{2}$, or -47.5

P,R **23** Lonnie won a contest in which the prize was up to \$85.00 worth of compact discs and audio tapes. Compact discs cost \$12.95, tapes cost \$7.95, and Lonnie must buy at least four items.
a Write two inequalities to describe her options. $12.95c + 7.95t \leq 85$; $c + t \geq 4$
b Graph the inequalities.
c What are the choices Lonnie could make? The numbers of CD's and tapes represented by lattice points in the intersection of the inequalities' graphs
d What combination of CD's and tapes should Lonnie choose if she wants to spend as much as possible of the \$85.00?
1 CD, 9 tapes

P,I **In problems 24–26, draw the graph of each set.**

24 $\{(x, y): x \geq 2y + 1\}$

25 $\{(x, y): y \geq x^2\}$

26 $\{(x, y): y \leq x + 3 \text{ and } 0 \leq x \leq 4\} \cap \{(x, y): y \geq -x + 7 \text{ and } 2 \leq x \leq 6\}$

R **27** Solve the following system for m and b.
$$\begin{cases} 2 = 5m + b \\ 14 = 8m + b \end{cases}$$
$m = 4$; $b = -18$

R **28** Find k_1 and k_2 so that the graphs of $y = 2x + k_1$ and $y = 7x + k_2$ intersect at (4, −3). $k_1 = -11$; $k_2 = -31$

I **29** A number is chosen at random from {−5, −3, −1, 1, 3, 5}. What is the probability that the number cannot be used as a value of x in the function $g(x) = \frac{x + 3}{(x + 5)(x - 1)}$? $\frac{1}{3}$

P **30** Graph $y - 5 \le 2(x + 5)$.

C R **31** In the diagram, △ABC ~ △DEF. Find the values of x and y. 20; 4

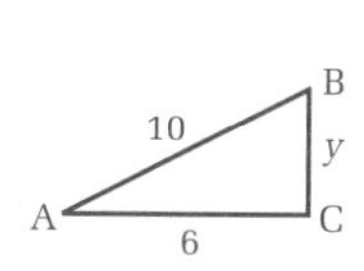

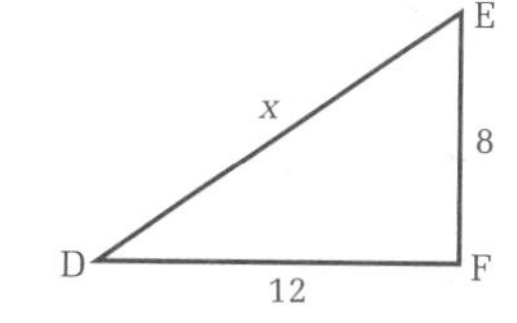

R **32** If (2, −6) and (−4, −12) are points on the graph of $y = ax^2 + b$, what are the values of a and b? $-\frac{1}{2}$; −4

I **33** Graph $|y| + |x| = 4$.

P,I **34** Copy and complete the table of values for the function $f(x) = |x + 10| + |10 - x|$.

x	**−20**	**−10**	**−6**	**0**	**1**	**8**	**10**	**12**	**20**	**50**
f(x)	40	20	20	20	20	20	20	24	40	100

I **35** Consider the sequence 2(1) + 5, 2(2) + 5, 2(3) + 5, . . .

a Find the value of the thirteenth term. 31

b Write an expression that represents the nth term. 2(n) + 5

P,R **36** A point is chosen at random in the square with vertices (0, 0), (0, 15), (15, 15), and (15, 0). What is the probability that the coordinates of the point satisfy the inequality $x + y \le 10$? $\frac{2}{9}$

P **37** According to the triangle-inequality property, the sum of any two sides of a triangle must be greater than the third side of the triangle.

a Use the diagram of △ABC to write three inequalities based on the triangle-inequality property.

b Graph the region that satisfies all three inequalities.

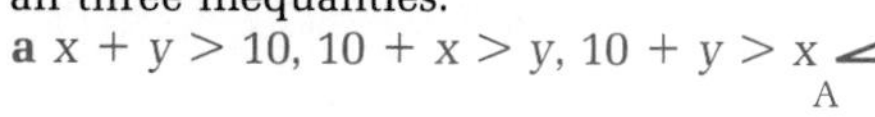

a $x + y > 10$, $10 + x > y$, $10 + y > x$

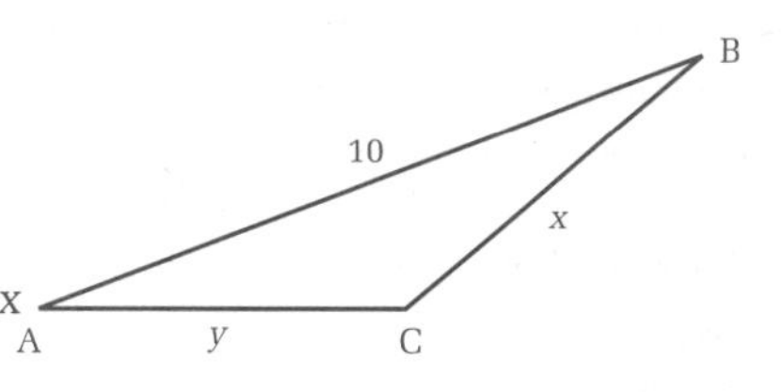

Problem-Set Notes and Strategies, continued

- In problem **29** students are asked to identify domains of a function. Remind them that the denominator of a fraction cannot be equal to zero. Domain and range are explained further in Chapter **5**.
- In problem **32**, a system of equations can be set up for the points in *a* and *b*.
- Problem **33** involves several parts. Remember—both *x* and *y* can be either non-negative or negative. Each possibility sets up a different equation and the answer is the union of all the parts.
- In problem **35**, have students decide how each of the terms differ and link that to a linear equation. This is an intuitive preview of arithmetic sequences in Chapter **13.3**.
- A way to approach problem **36** is to draw a picture and calculate the respective areas. The ratio of the areas is the probability.

Additional Answers

33

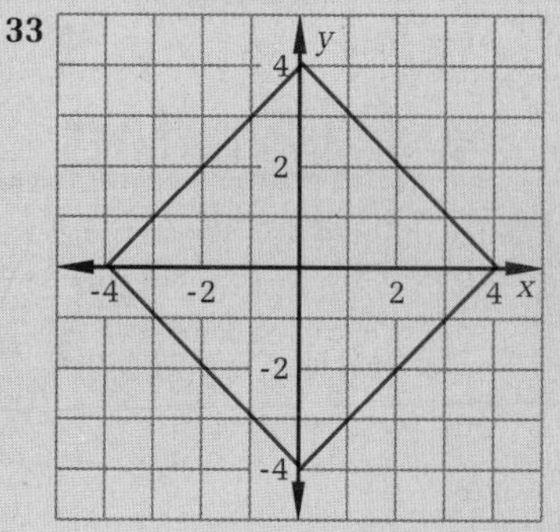

Answer for problems **30** and **37b** can be found in the answer pages beginning on page **A** 1.

3.4

Class Planning

Time Schedule
All levels: 2 days

Resource References
Teacher's Resource Book
Class Opener 17
Cooperative Learning 10
Practice 18

Class Opener

Star wanted to rent some movies. The video store charges a $35 membership fee and a $2 rental fee per movie. List each of the following:

a The cost of renting 3 movies $41
b The cost of renting 4 movies $43
c The cost of renting 5 movies $45
d The cost of 6 movies $47
e The cost of 13 movies $61
f Cost(17)movies $69
g Cost(23) $81
h C(34) $103
i C(45) $125
j C(N) $35 + 2N$

Lesson Notes

- Functions are explored again from a linear context. Any function that has a constant rate of change (i.e. slope) is a linear function.

3.4 Functions and Direct Variation

Objectives
After studying this section, you will be able to
- Write equations that represent linear functions
- Find restrictions on the domain of a function caused by real-world constraints
- Recognize direct variation
- Express direct variation as a proportion

Part One: Introduction

Linear Functions

The following table represents Jon's earnings for a five-week period, where x = his total sales for a week and $f(x)$ = his pay for the week.

x	1000	1250	700	900	1600
$f(x)$	190	200	178	186	214

The quotient [change in $f(x)$] ÷ (change in x) represents the weekly rate of change in earnings compared to sales.

$$\frac{10}{250} = 0.04 \qquad \frac{-22}{-550} = 0.04 \qquad \frac{8}{200} = 0.04 \qquad \frac{28}{700} = 0.04$$

Because this quotient is constant, the data are *linearly related.* It is therefore possible to describe the data with a ***linear function*** of the following form.

$$f(x) = mx + b$$

Remember that slope, m, is a rate of change, so $m = 0.04$. To find the value of b, we substitute any pair of values in the table for x and $f(x)$.

$$f(x) = 0.04x + b$$
$$190 = 0.04(1000) + b$$
$$150 = b$$

Thus a linear function describing Jon's pay, $f(x)$, as a function of his sales, x, is $f(x) = 0.04x + 150$.

Vocabulary
linear function
domain
range
direct variation
directly proportional
constant of variation
constant of proportionality
proportion

Domain

We have not yet discussed what inputs make sense in a function. Mathematically, we can calculate Jon's earnings for $x = -200$ as follows.

$$\begin{aligned} f(-200) &= 0.04(-200) + 150 \\ &= 142 \end{aligned}$$

But what does this result mean? A negative value for Jon's sales does not make sense. To be sure that our results are meaningful, we need to think about what a problem's answer means in terms of the problem. (In this case, since the input is an amount of money, x not only must be nonnegative but must represent a whole number of cents.)

The function f therefore has a restricted ***domain***, or set of inputs, for which its output, $f(x)$, has real-world meaning. The full definition of the function includes its domain.

$$f(x) = 0.04x + 150, \text{ where } x \geq 0 \text{ and } 100x \text{ is a whole number}$$

Domain, and the corresponding set of outputs, ***range***, will be discussed again in Chapter 5.

Direct Variation

There is an important relationship between the perimeter P of a square and the length s of each side. If we double the length of the side, we double the perimeter. If we divide the length of the side by 3, the perimeter is divided by 3. We can represent the relationship between perimeter and side length with the function $P(x) = 4s$ and say that P is ***directly proportional*** to s. In this case, 4 is called the ***constant of variation*** or the ***constant of proportionality***.

As you have seen, a linear relationship between $f(x)$ and x is expressed by a function of the form $f(x) = mx + b$. If $b = 0$, however, the function has the form $f(x) = mx$ and we can say that $f(x)$ varies directly as x or $f(x)$ is directly proportional to x. The slope m is the constant of variation.

Example *The data below are linearly related. Determine whether they represent a direct variation—that is, whether $h(x)$ varies directly as x. If so, find the constant of variation.*

x	4	−7	12
$h(x)$	−24	42	−72

Because $h(x)$ is linearly related to x, $h(x) = mx + b$.

Compute the slope

$$\begin{aligned} m &= \frac{42 - (-24)}{-7 - 4} \\ &= -6 \end{aligned}$$

Substitute to find b

$$\begin{aligned} h(x) &= -6x + b \\ -24 &= -6(4) + b \\ 0 &= b \end{aligned}$$

Since $b = 0$, $h(x)$ varies directly as x. The constant of variation is -6.

Lesson Notes, continued

- The set of inputs for a linear function may not make sense when certain numbers are entered. The set of possible input values that makes sense is referred to as the domain and the corresponding set of output values is the range. These definitions will be stated more formally in Chapter **5**.

- Direct variation problems are presented in this section. Students may relate to the idea that their weekly paycheck depends on the number of hours they work. The more they work, the more they get paid. This represents a direct variation with the constant of proportionality being their hourly wage. A direct variation is a linear equation whose *y*-intercept is zero and the non-zero slope is the constant of proportionality. Direct variations may be easily expressed as ratios or proportions.

Lesson Notes, continued

- Remind students that if $\frac{a}{b} = \frac{c}{d}$, then $ad = bc$. This makes proportion problems easier to solve.

Checkpoint

1. If Johnny B. Goode can do 6 Algebra 2 problems in 10 minutes, how long will it take him to do 15 problems? 25 minutes
2. Are the following ordered pairs linearly related: (1, 1), (2, 4), (3, 6), and (4, 8). No
3. If x and y are directly related and $y = 24$ when $x = 6$, what is the constant of variation? 4

Assignment Guide

Average	
Day 1	9–11, 13, 15, 16, 19, 20, 22, 23, 25, 27–29
Day 2	12, 14, 17, 21, 24, 30, 31, 33, 34, 36, 37, 39, 41
Advanced	
Day 1	9–11, 15, 16, 18, 20, 22, 23, 27–31, 34, 37, 41
Day 2	12, 17, 21, 24, 26, 32, 33, 35, 36, 38–40, 42, 43, 45
Honors	
Day 1	9–12, 16–18, 20, 21, 24, 27, 32–34, 36–38, 41
Day 2	15, 23, 26, 28, 29, 35, 39,

Example *To change a temperature from degrees Celsius C to degrees Fahrenheit F, we use the formula $F = \frac{9}{5}C + 32$. Is this an example of direct variation?*

No. The formula is a function of the form $f(x) = mx + b$, but $b \neq 0$, so F does not vary directly as C.

Proportion

A useful way of expressing direct variation is by means of a ***proportion***. A ***proportion*** is an expression of the form $\frac{a}{b} = \frac{c}{d}$, where a, b, c, and d are real numbers, $b \neq 0$, and $d \neq 0$. If (x_1, y_1) and (x_2, y_2) both satisfy a direct variation $y = mx$, then $y_1 = mx_1$ and $y_2 = mx_2$. Thus, $\frac{y_1}{x_1} = m$ and $\frac{y_2}{x_2} = m$, so $\frac{y_1}{x_1} = \frac{y_2}{x_2}$, which is a proportion.

Example *If x varies directly as y and if x = 14 when y = 15, what is the value of y when x = 7?*

Since x and y vary directly, we can set up a proportion.

$$\frac{\text{y value}}{\text{corresponding x value}} = \frac{\text{y value}}{\text{corresponding x value}}$$

$$\frac{15}{14} = \frac{y}{7}$$

$$7(15) = 14y$$

$$7.5 = y$$

When x = 7, therefore, y = 7.5.

Part Two: Sample Problem

Problem *The number of gallons of gasoline a car uses is directly proportional to the number of miles driven. John's car uses 11.5 gallons to go 200 miles. How many gallons does it need to go 300 miles?*

Solution Let g = the number of gallons used to go 300 miles. We set up a proportion between miles and gallons.

$$\frac{\text{gallons used}}{\text{miles traveled}} = \frac{\text{gallons used}}{\text{miles traveled}}$$

$$\frac{11.5}{200} = \frac{g}{300}$$

$$17.25 = g$$

It takes $17\frac{1}{4}$ gallons to go 300 miles.

Part Three: Exercises and Problems

Warm-up Exercises

P **1** Katy gets paid \$5 to baby-sit for 2 hours. How much should she be paid if she baby-sits for $3\frac{1}{2}$ hours? \$8.75

P **2** Are the values in the table linearly related? Yes

x	3	4	5	6
y	17	21	25	29

P **3** The values in the following table are linearly related. Does g(x) vary directly as x? Yes

x	8	−5	0
g(x)	−32	20	0

P **4** If 18 cans of juice cost \$7.80, how much do 60 cans cost? \$26

P **5** Explain how to determine whether linearly related data represent a direct variation. Answers will vary. Possible answer: If the graph of the linear relationship passes through the origin, the data are directly proportional.

P **In problems 6 and 7, solve for x.**

6 $\frac{x}{3} = \frac{10}{7}$ $x = \frac{30}{7}$

7 $\frac{2x - 4}{7x + 3} = \frac{4}{5}$ $x = -\frac{16}{9}$

P **8** If y varies directly as x and if y = 30 when x = 10, what is the constant of proportionality? 3

Problem Set

A

P **9** Are the values in the table linearly related? Yes

x	−5	−3	0	2	4
y	5	3	0	−2	−4

P **10** If Amy can type $6\frac{1}{4}$ pages in $1\frac{1}{4}$ hours, how many pages can she type in $3\frac{1}{2}$ hours? $17\frac{1}{2}$

R **In problems 11 and 12, solve each system.**

11 $\begin{cases} 5n + 10d = 320 \\ 10n + 5d = 535 \end{cases}$ (n, d) = (50, 7)

12 $\begin{cases} x + \frac{1}{2}y = 12 \\ 5x - \frac{1}{2}y = 18 \end{cases}$ (x, y) = (5, 14)

R **In problems 13 and 14, evaluate each expression for y = 4 and x = −5.**

13 $3 + 2x^2 - \frac{4y}{8}$ 51

14 $x(3 + y) - y(3 + x)$ −27

Problem-Set Notes and Strategies

- Problem **1** can be solved by figuring out that Katy gets \$2.50 per hour and then multiplying by $3\frac{1}{2}$ hours. It may also be solved as a proportion.
- Check for a constant slope in problems **2** and **3**.
- Problem **5** is a good problem to encourage students to use thorough communication skills in their technical writing.
- Problems **11** and **12** review substitution and mult-addition methods from Section **3.2**.
- Here again, remind students that when the −y is distributed in problem **14**, the negative as well as the y is distributed.

3.4

Problem-Set Notes and Strategies, continued

- Problem **15** is an excellent review of the laws of exponents. You might try having students make up their own rectangles and see how elaborate they become.
- Problem **16**, a geometric argument using an isosceles triangle may also be used. Encourage alternate methods.
- Emphasize the concept of problem **17**. This is what makes the definition of absolute value work.
- Problems **18** and **19** ask that a system of equations be solved to determine the slope and intercept of the line passing through two points.
- Problem **23** is a nice application of slope to real problems and shows the significance of a rate of change.
- Remind students to use the value of the coins when working problems with money as in **25**. Have them identify the variables in words and refer to that identification often.

Problem Set, *continued*

R **15** Copy the multiplication chart shown, supplying the missing exponential expressions.

	2^5	2^8	? 2^{11}	
?	?	2^8	?	2^0; 2^5; 2^{11}
2^2	?	?	2^{13}	2^7; 2^{10}

P **16** Explain why line PR must be perpendicular to the line represented by $y = x$.
The slope of $\overleftrightarrow{PR}$ is -1, which is the opposite reciprocal of 1, the slope of $y = x$.

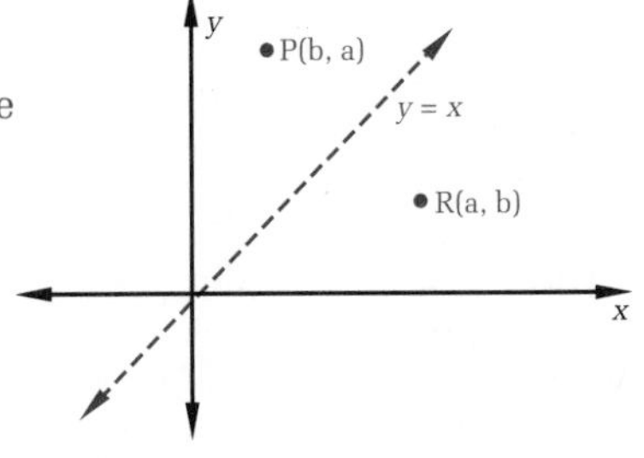

R **17** Does $-x$ always represent a negative value? Explain your answer in paragraph form. No; if x is negative, $-x$ is positive, and if $x = 0$, $-x = 0$.

R,I **In problems 18 and 19, solve each system for *m* and *b*.**

18 $\begin{cases} 22.0 = 31.5m + b \\ 758.6 = 1107.8m + 31.5b \end{cases}$ $m \approx 0.57$; $b \approx 4.1$

19 $\begin{cases} 63 = 9m + b \\ 24 = -4m + b \end{cases}$ $m = 3$; $b = 36$

R **In problems 20–22, determine which elements of $\{-20, -14, -9, -4, 0, 4, 9, 20\}$ are solutions of the equation or inequality.**

20 $|x + 8| = 12$ $-20, 4$

21 $x^2 = 36 - 5x$ $-9, 4$

22 $x + 30 > 0$ $-20, -14, -9, -4, 0, 4, 9, 20$

P **23** The following chart relates the number of miles a compact car is driven to the number of gallons of gas used.

Miles	520	410.3	277.4	471.1
Gallons	18	14.2	9.6	16.3

a Show that the numbers of miles and gallons are directly related.
b Calculate the constant of proportionality. ≈ 28.9
c What does the constant of proportionality represent in this problem? Gas mileage in miles per gallon

P **24** For what values of a and b are the values in the table linearly related? 129; 53

x	65	83	a	111
$h(x)$	30	39	62	b

R **25** Pat has \$3.85 in 25 coins, all of which are quarters and nickels. How many of each type of coin does Pat have? 13 quarters, 12 nickels

B R,I 

26 Graph $x^2 + y^2 = 9$ and $x + y = 3$ on the same coordinate system, then determine the area of the smaller of the two closed regions formed by the graphs. ≈2.569

27 Refer to the diagram. Solve for (x, y).
(x, y) = (−12, 16)

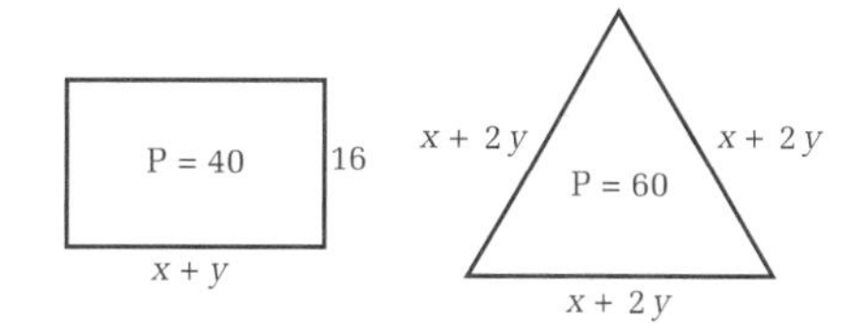

R **In problems 28–31, graph each set.**

28 $\{(x, y): x < 2\}$ **29** $\{x: x < 2\}$ **30** $\{(x, y): y \geq -3\}$ **31** $\{y: y \geq 0\}$

R **32** Let $A = \{(x, y): |x| \leq 10\}$ and $B = \{(x, y): |y| \leq 6\}$.
a Sketch the graph of $A \cap B$.
b Find the area of the graph of $A \cap B$. 240

R **33** If you give Skip an odd number, he triples it and adds 1. If you give him an even number, he divides it by 2.
a What will Skip give back if you give him 17? 52
b What will Skip give back if you give him 52? 26
c Continue the process, giving Skip his own results to work on at least 13 more times. What happens? He begins to repeat the numbers 1, 4, and 2.

P **34** Which of the following are linearly related? **a, c, e**
a Circumference and radius of a circle
b Area and radius of a circle
c Perimeter and side of a square
d Area and side of a square
e Distance traveled and time spent traveling, if speed is constant

P,R **35** In the diagram, △ABE is similar to △CDE.
a Find the length of $\overline{AE}$. 13.5
b Find the length of $\overline{AC}$. 7.5

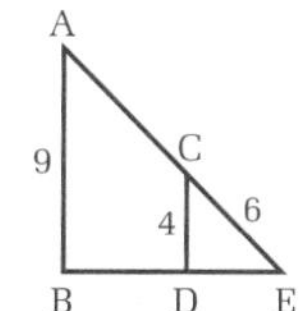

R **36** Find the value of k for which the point (0, 8) lies on the graph of $2x + ky = 24$. 3

R **37** If $f(x) = mx + b$ and if $f(2) = 12$ and $f(-3) = -7$, what are the values of m and b? 3.8; 4.4

R **38** Ma Jones sells Colombian coffee at $3.50 per pound and Bolivian coffee at $4.39 per pound. How much of each type of coffee should Ma put in a blend that will sell for $4 per pound if she wants 200 pounds of blended coffee? Colombian: ≈87.6 lb; Bolivian: ≈112.4 lb

Problem-Set Notes and Strategies, continued

- In problem **26**, students may have to plot some points to graph the circle. Have them take the square root of both sides and think of it as a distance problem.
- Problems **28** and **29** may look identical to many students, but are substantially different in the number of dimensions present in the problem This difference is also in problems **30** and **31.** Help students realize that one variable does not necessarily imply a one dimensional problem.

Cooperative Learning

Ask students to work in groups to use problem **33** as an introduction to pattern making. Try other starting numbers between 1 and 100. Will the process always generate the numbers 1,4 and 2 eventually? Ask the students to generalize this using algebraic expressions.

- Similar triangles are used in problem **35** and students may remember that similar triangles are proportional. Try to get students to generalize that the constant of proportionality is the scaling factor between the triangles analogous to the scale on a map.

Additional Answers

Answers for problems **28 – 31** and **32a** can be found in the answer pages beginning on **A** 1.

Problem-Set Notes and Strategies, continued

- For problems such as number **38**, try setting up an experiment. In this case you could use clear beakers and colored oil and water to show the conservation of fluid.
- Problem **39** is a good review of the student's understanding of function notation and slope. $f(x+4)-f(x)$ is the difference of the y-coordinates when the x-coordinates differ by 4. If the difference of the y-coordinates is 32 as stated, the slope is equal to 8.
- Problem **41** requires a proportion. What is the proportionality constant? How is it different for parts **a** and **b**? Radian measure will be explored further in Chapter **14**.
- Problems **42** and **43** provide practice on function notation.
- Many possible answers are available for problem **44**. Check to make sure that students include the right part of the graph.

Communicating Mathematics

Using problem **45** as a model, describe the differences between fixed costs and variable costs and how these two concepts relate to linear equations and proportion problems.

Problem Set, *continued*

P,R **39** If $f(x)$ is linearly related to x and $f(x + 4) - f(x) = 32$, what is the slope of the graph of f? 8

R,I **40** Use the population curves shown to determine which animal is prey and which animal is predator. Explain your reasoning, using paragraph form. Answers will vary. Possible answer: Animal II seems to be the predator, since its population is smaller and its numbers seem to increase only after the population of Animal I increases.

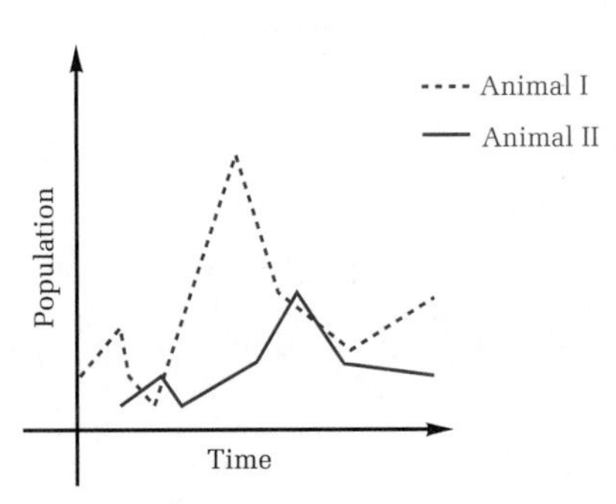

P,I **41** In trigonometry, two different measures of angles are used, degrees and radians, with 180 degrees being equivalent to π radians.

a How many radians are equivalent to 60 degrees? ≈ 1.05
b How many degrees are equivalent to 1.58 radians? ≈ 90.53
c Are radians and degrees directly proportional? Yes

C P,R **42** The outputs of function h are linearly related to the inputs x. If $h(5) = 17$ and $h(8) = 29$, what is $h(x)$? $h(x) = 4x - 3$

P,R **43** Write a linear function f such that $f(4) = 16$ and $f(-2) = 10$. $f(x) = x + 12$

R **44** Write a system of inequalities whose solutions include all of the points shown and whose boundary lines pass through the origin. $\begin{cases} y \le 3x \\ y \ge \frac{1}{4}x \end{cases}$

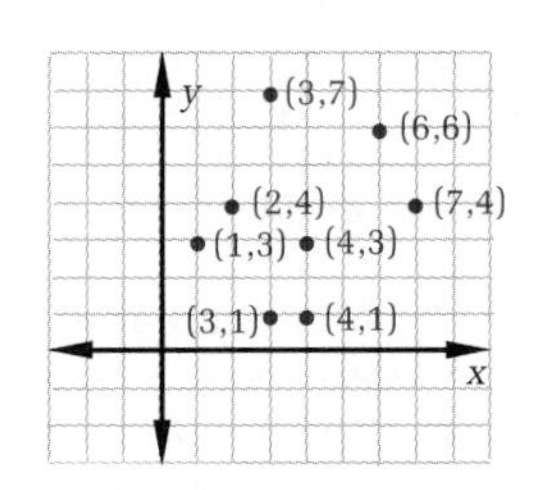

P,R **45** A gas bill is based on the number of therms used plus a fixed monthly charge. The following chart gives gas-bill information for the months of March, April, and May.

Month	Therms Used	Cost of Therms	Fixed Charge
March	457.58	175.71	7.50
April	245.63	94.32	7.50
May	132.47	50.87	7.50

a Find the cost per therm. $0.384
b Write a formula that can be used to find the total bill B for T therms. $B = 7.5 + 0.384T$
c Use this formula to find the bill for using 842.34 therms. $330.96

3.5 FITTING A LINE TO DATA

Objectives

After studying this section, you will be able to

- Fit a line to data by estimation
- Fit a line to data by the least-squares method

Part One: Introduction

Estimating a Fitted Line

In a chemistry experiment, Bea Kerr collected the data shown in the table. When she graphed the corresponding points, the points were not collinear, but they did suggest a linear pattern. Bea used a ruler to draw a line that seemed to pass through the middle of the points.

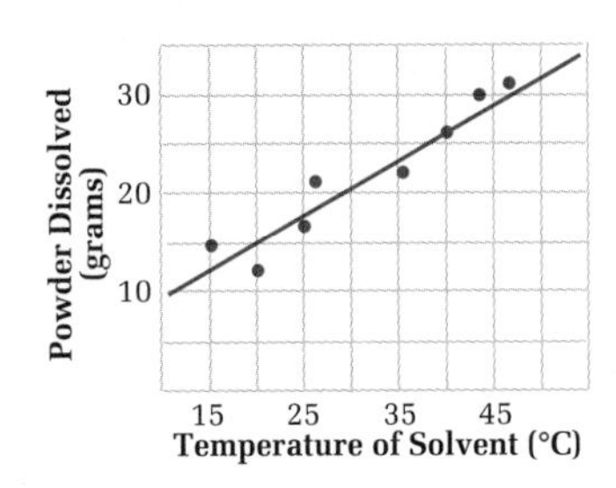

Temperature of solvent (°C)	15	20	25	27	36	40	43	47
Powder dissolved (g)	15	13	17	21	23	26	30	31

The line Bea drew is called a ***fitted line***. Writing an equation that represents a fitted line is easier than writing an equation that accurately describes all the data. The equation of the fitted line can be used to make predictions.

Example *Using the fitted line that Bea drew, predict the amount of powder that would dissolve if the temperature of the solvent were 60°C.*

We use the graph to estimate the coordinates of two points on the line, (15, 13) and (45, 28).

$$\text{Slope} = \frac{28 - 13}{45 - 15} = 0.5$$

Letting T = temperature and P = amount of powder dissolved, we use the point-slope form to write an equation that approximates the relationship between the powder dissolved and the temperature of the solvent.

Class Planning

Time Schedule
All levels: 1 day

Resource References
Teacher's Resource Book
Class Opener 18
Cooperative Learning 11
Practice 19
Transparency 6

Class Opener

Students in a physics class investigated the relationship between the initial velocity of an object projected at a 45° angle and the total distance the object travels. They collected the following data:

v	10	20	30	40	50	60	70	80	90	100
D	7	15	20	30	36	40	48	58	61	71

Their equipment would not let them exceed the initial velocity of 100 feet per second. How far do you think the object would travel if the initial velocity were 140 feet per second? ≈ 95 feet

Lesson Notes

- Fitting an equation to a curve is the second in a series of topics from the statistics curriculum. A linear model is the simplest of models to approximate with an equation. Emphasize to the students that it is important that they first graph the points to establish that there is a linear relationship before going through the formulas for regression. However, a non-linear relationship may exist and the formulas will still fit a line through the points.

Vocabulary
fitted line
method of least squares
regression line
line of best fit

Lesson Notes, continued

- The least squares line will allow predictions to be made for points that are not in the data set. Emphasize to students that these predictions are only valid in the realm of the data points to support the equation. The best predicted values are the closest to the average of the data coordinates, i.e. closest to the point (x average, y average) The further from this point, the less confidence there is in the approximation.
- Note that in order to find the least squares line, a system of linear equations in m and b must be solved. This is the rationale for putting this section in this chapter. Additional statistics topics are intermixed throughout the text.

Checkpoint

1. For the data points (1, 1), (2, 4), (3, 6), and (4, 8), find x average, y average, x^2 average, and xy average.
   2.5, 4.75, 7.5, 14.75
2. For a line of best fit of $y = 3x - 4$, predict the y value for an x of 7. 17

Communicating Mathematics

Since a linear equation is used to model some empirical data, have students describe where the line of best fit provides a reasonable model for the data. Make sure that the students do not use the model for values outside the range of data points.

Set $T = 60$ and solve for P

$$\mathbf{P - 13 = 0.5(T - 15)}$$
$$\mathbf{P - 13 = 0.5(60 - 15)}$$
$$\mathbf{P = 0.5(45) + 13}$$
$$\mathbf{P = 22.5 + 13}$$
$$\mathbf{P = 35.5}$$

Thus, at 60°C the solvent will dissolve about 36 grams.

Fitting Lines by the Method of Least Squares

When there are many data points, graphing the points and fitting a line by estimation is not convenient. Most statisticians use an algebraic method, called the ***method of least squares.*** When people refer to a ***regression line*** or a ***line of best fit,*** they are referring to a line found by the least-squares method.

The line of best fit has an equation of the form $y = mx + b$. The values of m and b are determined by calculating four averages—the x average, the y average, the x^2 average, and the xy average. The constants m and b will satisfy the two equations $y = mx + b$ and $xy = mx^2 + bx$.

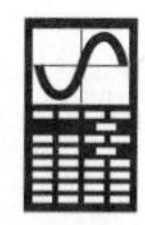

See p. 770

Example *Use the method of least squares to find an equation of the line of best fit for Bea Kerr's data. Use that equation to predict the amount of powder that would dissolve in solvent at 60°C.*

We let x = temperature and y = amount of powder dissolved and set up a table for the four averages.

									Average (to tenths)
x	**15**	**20**	**25**	**27**	**35**	**40**	**43**	**47**	31.5
y	**15**	**13**	**17**	**21**	**23**	**26**	**30**	**31**	22.0
x^2	**225**	**400**	**625**	**729**	**1225**	**1600**	**1849**	**2209**	1107.8
xy	**225**	**260**	**425**	**567**	**805**	**1040**	**1290**	**1457**	758.6

We use the equations $y = mx + b$ and $xy = mx^2 + bx$, replacing x, y, xy, and x^2 with the averages in the table to get a system in m and b.

$$\begin{cases} 22.0 = 31.5m + b \\ 758.6 = 1107.8m + 31.5b \end{cases}$$

When we solve this system with the aid of a calculator or computer, we find that $m \approx 0.57$ and $b \approx 4.1$. An equation of the line of best fit is therefore $y = 0.57x + 4.1$. Using this equation, we set x equal to 60 and solve for y.

$$\begin{aligned} y &= 0.57x + 4.1 \\ &= 0.57(60) + 4.1 \\ &= 38.3 \end{aligned}$$

We predict that at 60°C the solvent will dissolve about 38 grams of powder.

Statisticians usually use a computer or a calculator that has statistics functions to calculate regression lines. To see whether your calculator can determine a line of best fit, check your calculator manual.

Note that the fitted-line method and the least-squares method predict slightly different amounts of dissolved powder at 60°C. Which method is best? In general, the least-squares method will give the line that most accurately represents the known data. Predictions for untested data values, however, are guesses no matter which technique is used.

Part Two: Sample Problem

Problem *Use the data shown in the table to predict the average temperature in September (month 9).*

Month	1	2	3	4	5	6	7	8
Average temperature (°F)	12	22	19	39	51	75	81	83

Solution The data show a fairly consistent increase in temperature each month, but from our experience we know that the weather usually cools off after August. We could still make a guess, but it would be inappropriate to fit a line using either of the methods described in this section.

Part Three: Exercises and Problems

Warm-up Exercises

1 Write an equation of the line containing the points (−4.2, 12) and (9.9, 4). $y - 4 = -0.6(x - 9.9)$

R P **2** In the graph, a line has been drawn to fit the data.

a Choose two points on the line.
b Determine the slope of the line.
c Write an equation of the line.
a Answers will vary. **b** ≈ -0.2
c $t = -0.2d + 16$

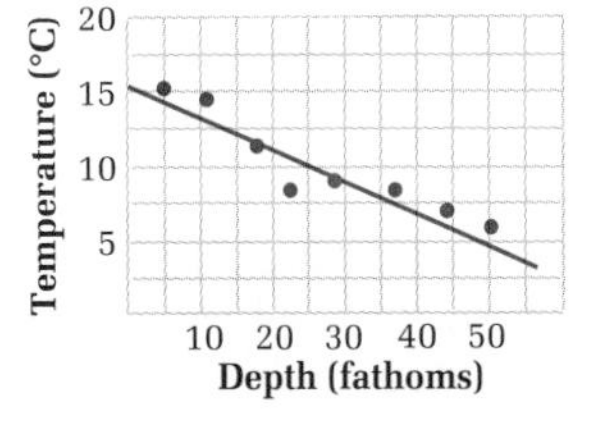

P **3** Use the values in the table to compute
a The x average 25
b The y average ≈ 10.7
c The x^2 average 725
d The xy average 350

x	10	15	20	25	30	35	40
y	−3	5	6	11	14	18	24

Assignment Guide

Average
5, 6, 8–10, 12, 13, 15, 16, 18, 20–22, 25, 28, 31
Advanced
5, 7, 9–13, 15, 16, 18–20, 22, 24, 25, 29, 30, 32–34
Honors
5, 7, 9, 11, 13, 16, 17, 19, 20, 22, 24–26, 30, 33–35

Stumbling Block

A common mistake is to use a least squares line to predict points that are outside the range of the original data. Another is to use current data to try to make predictions about future events or past events. Urge students to use common sense when making predictions about data points. Only values that are within the realm of the existing points should be used to make predictions.

Problem-Set Notes and Strategies

- Problem **1** can be done by using the least squares formulas or by using methods in the beginning of the chapter.

Problem-Set Notes and Strategies, continued

- Even if you have no computers available, it is a good idea to examine the computer program and see the algorithmic steps necessary to find the line of best fit. Data fitting is becoming increasingly more important as the technique is accessible to more individuals.

Using Computers

Many spreadsheet programs such as *Quattro* contain a linear regression component and allow you to graph the scatterplot as well as to determine and graph the fitted line. Symbolic manipulators such as *Derive* and *Mathematica* can do linear regression both algebraically and graphically. Several of the graphics calculators can do linear regression. The program at the beginning of the problem set can also be used in this section.

Additional Answer

5a

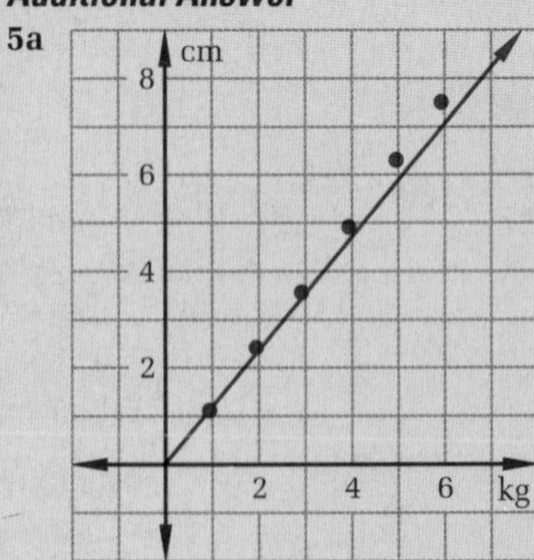

P,R

4 Solve the following system for m and b. Round your answers to the nearest tenth.

$$\begin{cases} 11 = 25m + b \\ 350 = 668m + 25b \end{cases}$$ $m \approx 1.7$; $b \approx -32.6$

Problem Set

The following program will compute a line of best fit by the least-squares method. If you have access to a computer, you may wish to use this program as you work through the problem set.

```
 10 REM This program calculates a best-fit line
 20 REM by the least-squares method
 30 PRINT "Enter the number of data points.==> ";: INPUT N
 40 DIM X(N),Y(N)
 50 FOR I = 1 TO N
 60 PRINT "Enter X(";I;").==> ";: INPUT X(I)
 70 PRINT "Enter Y(";I;").==> ";: INPUT Y(I)
 80 NEXT I
 90 X = 0: Y = 0: X^2 = 0: XY = 0
100 FOR I = 1 TO N
110 X = X + X(I): Y = Y + Y(I): X^2 = X^2 + X(I)^2: XY = XY + X(I) * Y(I)
120 NEXT I
130 X = X/N: Y = Y/N: X^2 = X^2/N: XY = XY/N
140 M = (XY - X * Y)/(X^2 - X * X)
150 B = Y - M * X
160 PRINT "The line of best fit is: ";
170 PRINT "Y = ";M;"X + ";B
180 END
```

A

P

5 In a lab experiment, a spring was stretched by various weights. The masses of the weights and the numbers of centimeters the spring stretched are shown in the following table.

Kilograms	1	2	3	4	5	6
Centimeters	1.3	2.5	3.7	5.0	6.4	7.7

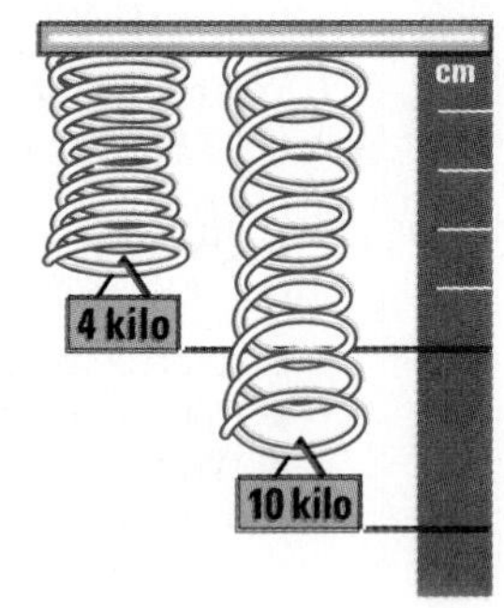

a Graph the data and draw a line that approximately fits the data.

b Use the line to approximate how far the spring will be stretched by a weight having a mass of 10 kilograms. $\approx$12.7 cm

P,R **6** Will earns \$5.50 per hour as a short-order cook.
a Write an equation that describes Will's earnings e for h hours. $e = 5.50h$
b Is e linearly related to h? Yes
c Does e vary directly as h? Yes

P **In problems 7 and 8, find an equation of the line of best fit for each set of data.**

7

x	1	2	3	4	5	6	7	8
y	12,500	11,000	9350	8225	6550	4750	3425	1788

$y = -1527.8x + 14{,}073.6$

8

x	100	450	975	1325	1750	2100
y	2620	12,125	23,590	34,720	46,240	53,975

$y = 25.95x - 99.7$

P **9** Suppose that $f(b)$ is directly proportional to b and $f(7) = 38.4$.
a Find the constant of proportionality. ≈ 5.49
b Find $f(22)$. ≈ 120.7

R,I **10** Which elements of $\left\{(4, 3), (-3, 4), \left(0, \frac{10}{3}\right), (5, 0), (0, -5)\right\}$ are solutions of the following system?

$$\begin{cases} x^2 + y^2 = 25 \\ 2x + 3y = 10 \end{cases}$$ (5, 0)

P **11** Find an equation of the line of best fit for the values in the following table. $y = 3.23x + 17.8$

x	4	12	9	18	35	100	78
y	31	54	49	74	133	340	270

R **12** Sketch the graph of $y < -\frac{1}{2}x$.

P **13** The following table shows temperatures inside a refrigerator during a 10-hour period.

				A.M./P.M.						
Time	8	9	10	11	12	1	2	3	4	5
Temperature (°F)	35	37	39	40	39	36	38	39	37	35

a Why isn't the temperature constant? (Give two reasons.)
b Estimate the slope of the line that fits the data. ≈ 0
c Calculate the line of best fit by the method of least squares. $y = -0.05x + 37.8$
d On a hot day, the refrigerator is left open for ten minutes, and then the temperature is measured. If these data are included in the table, what will be the effect on the line of best fit? Should the data be included? The slope will decrease; answers will vary.

R **14** Refer to the diagram. If B is the midpoint of $\overline{AC}$, what are the values of x and y?
2; 5

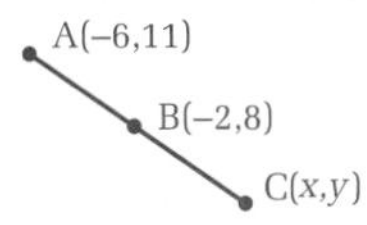

Problem-Set Notes and Strategies, continued

- Problems **7** and **8** require tedious computation and many chances for error if the calculations need to be done by hand. Use computers when available and check a few points to verify the derived equation. Many programs also have x average, y average and the other intermediate terms available since they are used for other statistical terms.
- Problem **9** is a good reminder that $f(x)$ is just another name for y.

Additional Answer

12

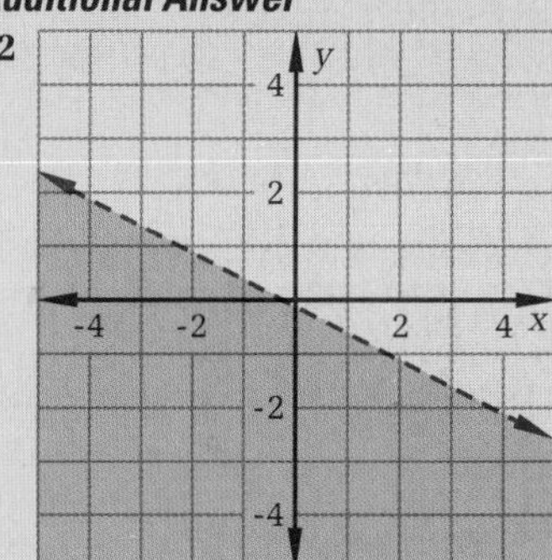

Cooperative Learning

Ask students to work in groups and use the equations $y = mx + b$ and $xy = mx^2 + bx$ to find a general solution for both m and b without having specific data points. Have students check their generalized equations using the data in the example of the text.

Problem-Set Notes and Strategies, continued

- In exercise **16**, all the tables represent linear functions, but only *a* and *b* pass through the origin. Parts **c** and **d** have nonzero *y*-intercepts.
- Problem **18** introduces compound functions defined on split domains. This will be explained in detail in the next section.
- Problem **19** illustrates that once again the obvious solution is not always the correct one. Urge students to examine all possible solutions especially when compound functions are involved.
- Problem **20** explores an alternative and less used method of fitting equations to data. The complexity of the process is reduced as fewer data points are examined at each step. Some accuracy is lost for the ease in complexity and this accounts for this methods limited use. With the advent of personal computers, the complexity issue is fading.

Problem Set, *continued*

R,I **15** The expression $\sqrt{x - 7}$ has a real value when $x - 7$ is positive or equals zero. For what values of x is $x - 7 \geq 0$? $\{x{:}x \geq 7\}$

P,R **16** Which of the following tables represent direct variations? **a, b**

a

x	−2	0	4
y	6	0	−12

b

x	6	$\frac{1}{2}$
y	12	1

c

x	−1	0	1
y	−1	2	5

d

x	1	0
y	9	−3

R **17** The roads in Circleton are all either circles or portions of north-south or east-west radii.

a How long is the shortest path from A to B? ≈43.56 km

b How long is the longest path from A to B if no road is traveled more than once? ≈87.54 km

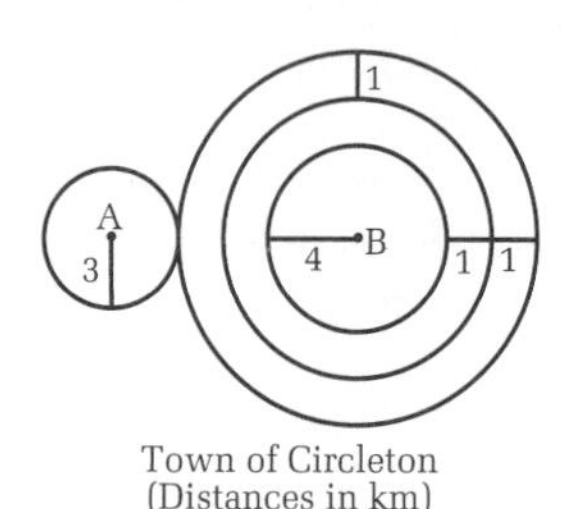

Town of Circleton (Distances in km)

B

I **18** Suppose that $f(x) = 3x$ when $x > 0$ and that $f(x) = x + 1$ when $x \leq 0$. Evaluate each of the following.

a $f(4)$ 12 **b** $f(-2)$ −1 **c** $f(0)$ 1

P,R **19** One bottle of Kook Cola costs \$0.80. A package of six bottles costs \$3.00. What is the least possible cost of ten bottles? \$6 (by buying 2 six-packs and discarding 2 bottles)

P **20** For the data shown,

x	0	2	4	6	8	10
y	3	4	6	8	9	11

a Find x_1, the average of the first three values of x 2

b Find y_1, the average of the first three values of y $\frac{13}{3}$

c Find (x_2, y_2), the averages of the last three values of x and y $\left(8, \frac{28}{3}\right)$

d Write an equation of the line connecting the points (x_1, y_1) and (x_2, y_2) (This method of determining a fitted line is called the *method of group averages.*) $y = \frac{5}{6}x + \frac{8}{3}$

R,I **In problems 21–26, graph each equation.**

21 $y = x$ **22** $y = x^2$ **23** $y = 2x$

24 $y = 2x^2$ **25** $y = \sqrt{x}$ **26** $y = \sqrt{x - 4}$

R,I **In problems 27–30, find the values of *a* that satisfy the equations.**

27 $\frac{a}{4} = \frac{4}{a}$ ± 4

28 $\frac{3a + 2}{7} = \frac{12}{5}$ ≈ 4.93

29 $\frac{3(a - 2) + 5}{6} = \frac{7(a - 1)}{2}$ ≈ 1.1

30 $\frac{3}{a + 2} = \frac{6}{2a + 4}$ All real numbers except -2

R **31** **a** What is the total area of the shaded regions in the diagram? $10x$

b If a point inside the large square is selected at random, what is the probability that it is inside a shaded region? (Express your answer in terms of x.) $\frac{10x}{(x + 5)^2}$

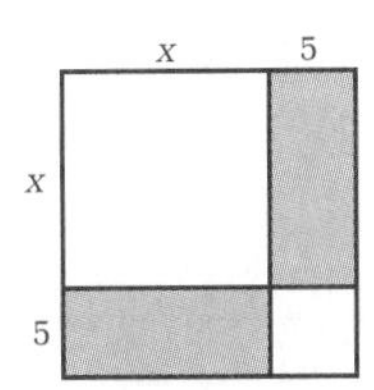

P **32** The distance a car travels in one revolution of the tires is directly proportional to the diameter of the tires. Lee's car travels about 88 inches in one revolution of 28-inch-diameter tires.

a What is the constant of proportionality?

b A truck tire is 48 inches in diameter. How far does the truck travel in one revolution of the tires?

a π **b** ≈ 150.8 in

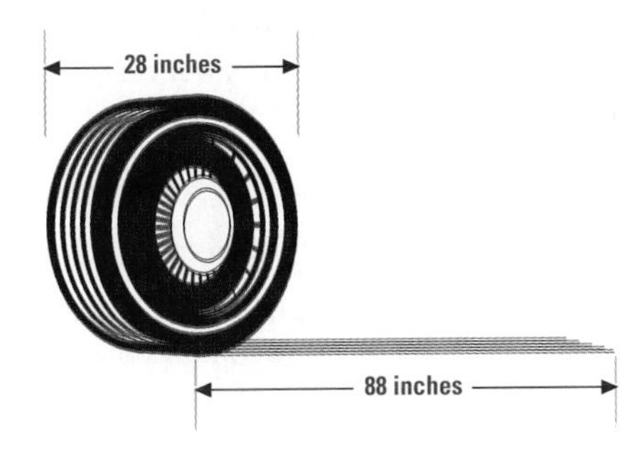

P **33** In function h, h(x) is linearly related to x. Write a formula for function h if $h(5) = 17$ and $h(8) = 29$. $h(x) = 4x - 3$

C R **34** Draw the graph of $-y > 2x + 1$. Then find the area of the closed region formed by the coordinate axes and the graph's boundary line. $\frac{1}{4}$

R **35** **a** The boundary lines of the shaded region shown are parallel. Write a system of inequalities to describe the shaded region.

b Find the area of the portion of the shaded region that lies in Quadrant II.

a $\begin{cases} y \le \frac{3}{4}x + 6 \\ y > \frac{3}{4}x + 2 \end{cases}$ **b** $21\frac{1}{3}$

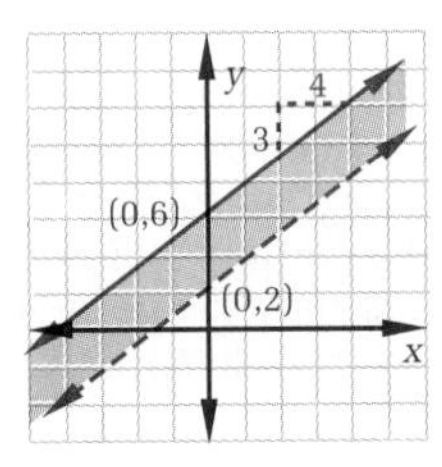

Problem-Set Notes and Strategies, continued

- Problems **21 – 26** may require plotting several points to produce a good graph. Examine the similarities and differences and try to make some generalizations from the graphs. Graphing techniques will be outlined further in Chapter **6**.
- Problems **27 – 30** introduce more complex expressions in proportions. The technique of cross multiplication is still available as long as there is a single fraction on each side of the equation.
- The graph that accompanies problem **31** provides a nice geometric interpretation of the square of the binomial, i.e. $(x + 5)^2$.
- Problem **33** requires writing a linear equation through two known points.
- Problems **34** and **35** review area of triangles and trapezoids.

Additional Answers

21

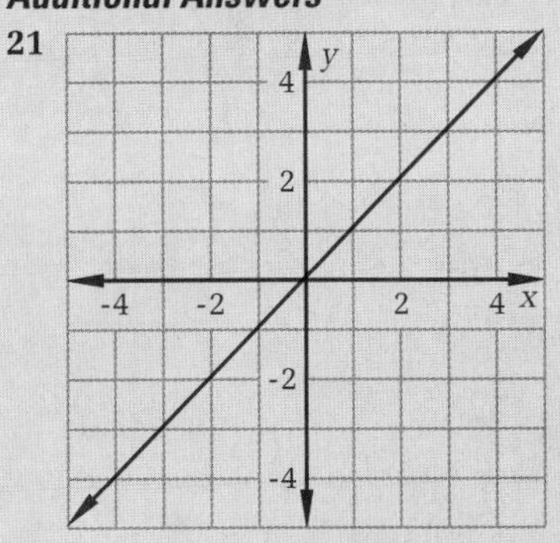

22

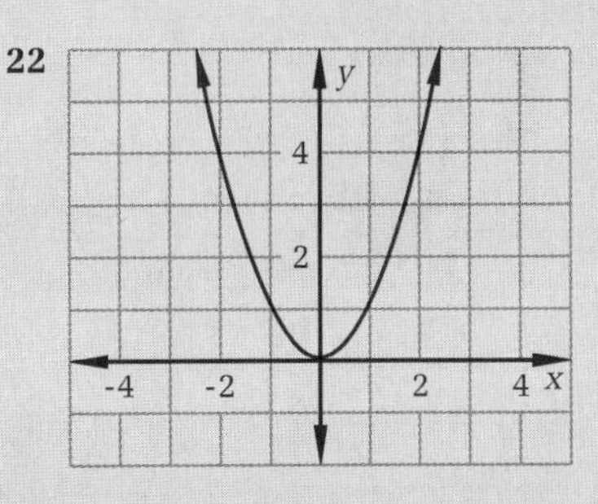

Additional Answers

Answers for problems **23 – 26** can be found in the answer pages beginning on **A 1**.

3.6

3.6 COMPOUND FUNCTIONS

Class Planning

Time Schedule

Average: Optional
Advanced: 2 days
Honors: 2 days

Resource References

Teacher's Resource Book
Class Opener 19
Practice 20

Class Opener

Graph the equation $f(x) = 3x + 6$ for all $x \geq -2$. On the same coordinate system, graph $y = -3x - 6$ for all $x < -2$.

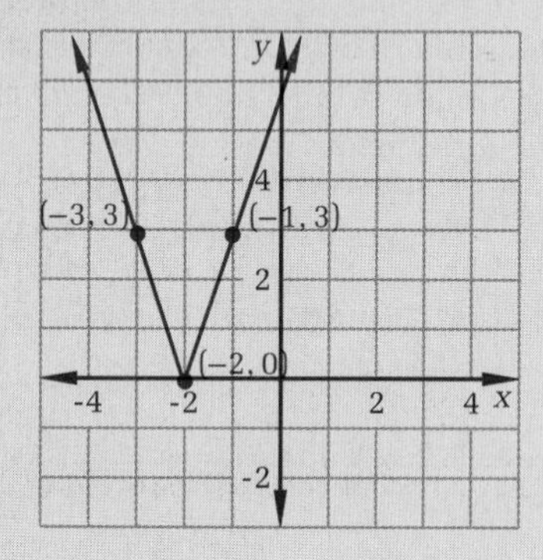

Lesson Notes

- You may decide that this section is optional for some students. This section introduces the topic of compound functions. Many functions are defined differently for different values in the domain. Quantity discounts, flying on a major airline during the week rather than on the weekend, a grade of "B" between 80% and 90%, are each examples of a function defined differently for different values of the domain.

Objectives

After studying this section, you will be able to

- Recognize compound functions
- Graph compound functions
- Recognize the absolute-value function as a compound function

Part One: Introduction

Compound Functions

At Sal's Stationery Store, the cost of a notebook is $1.85. However, if you buy more than five notebooks, the cost of each is $1.75. We can write the following ***compound function*** for the cost $C(n)$ of n notebooks.

$$C(n) = \begin{cases} 1.85n \text{ if } n \leq 5 \\ 1.75n \text{ if } n > 5 \end{cases}$$

If $n \leq 5$, we use the first rule. If $n > 5$, we use the second rule. Thus, the prices of three notebooks and six notebooks can be calculated as follows.

$$C(3) = 1.85(3) = 5.55$$
$$C(6) = 1.75(6) = 10.50$$

Any function with different rules for different inputs is a compound function. Compound functions are said to have ***split domains***.

Graphs of Compound Functions

How do we graph compound functions? Let's look at another example.

$$f(x) = \begin{cases} -x - 4 \text{ if } x < -3 \\ 2x + 5 \text{ if } x \geq -3 \end{cases}$$

To graph $y = f(x)$, we first graph the two equations $y = -x - 4$ and $y = 2x + 5$. Since x is restricted to certain values in each equation, we use only the portions of the graphs in the specified domains.

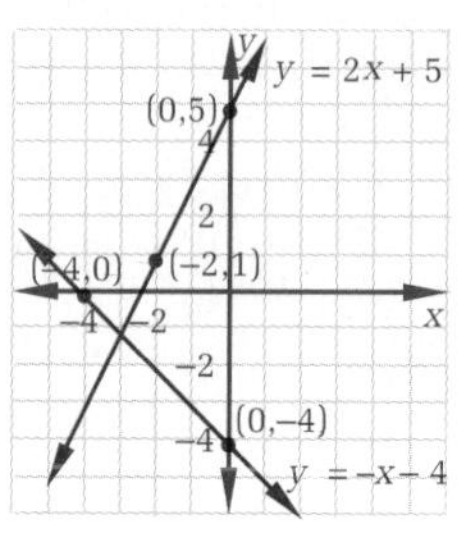

Vocabulary
compound function
split domain

In the graph of $y = -x - 4$, we keep the part of the line to the left of $x = -3$. In the graph of $y = 2x + 5$, we keep the part of the line to the right of and including $x = -3$. The resulting V-shaped graph represents the compound function.

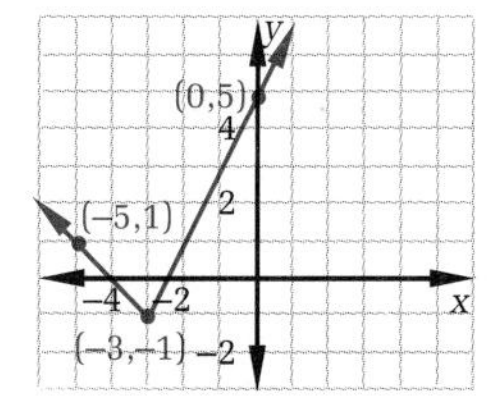

More About Absolute Value

On a number line, the distance from the origin to the point representing a number a is equivalent to the absolute value of a. Formally, absolute value is defined by means of a compound function.

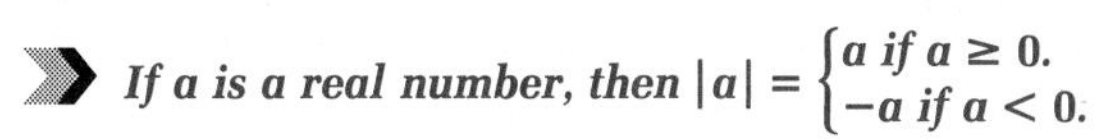

If a is a real number, then |a| = { a if a ≥ 0. / −a if a < 0.

The number line at the right shows that $|-4| = 4$ and $|13.7| = 13.7$. We can also use absolute value to find the distance between any two points on the number line—if the coordinate of one point is a and the coordinate of the other is b, the distance between the points is $|a - b|$.

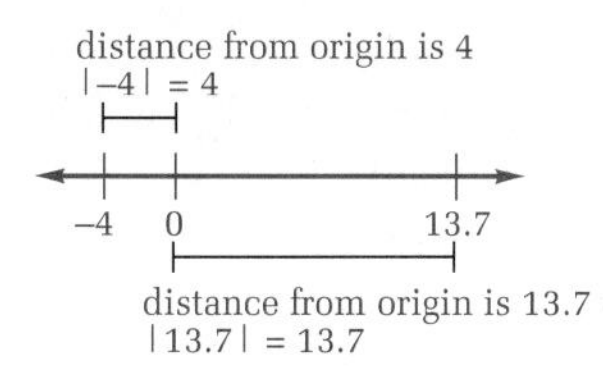

Some important properties of absolute value are given in the following table.

Absolute-Value Properties		
If a and b are real numbers,		
$\lvert a \rvert = \lvert -a \rvert$	$\lvert a-b \rvert = \lvert b-a \rvert$	$\lvert ab \rvert = \lvert a \rvert \lvert b \rvert$
$\lvert a+b \rvert \leq \lvert a \rvert + \lvert b \rvert$	$\lvert a \rvert \geq 0$	$\lvert a \rvert^2 = a^2$

Example *Explain why $|x| = |-x|$.*

Absolute value means distance from 0. The numbers x and $-x$ are on opposite sides of 0 but are the same distance from 0.

See p. 784

Example *Graph $y = |x|$.*

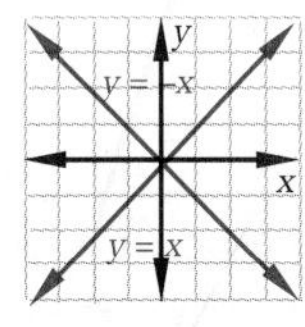

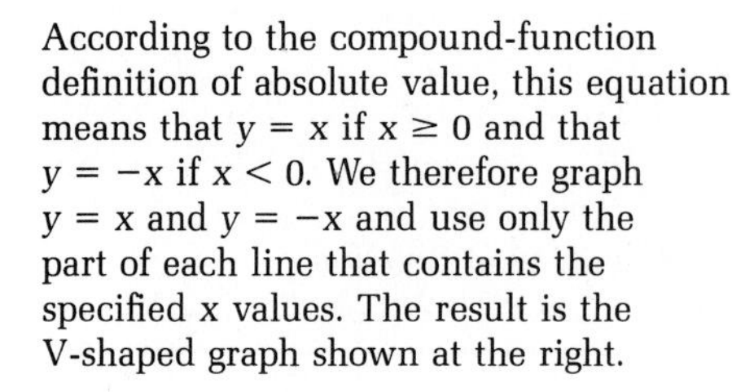

According to the compound-function definition of absolute value, this equation means that $y = x$ if $x \geq 0$ and that $y = -x$ if $x < 0$. We therefore graph $y = x$ and $y = -x$ and use only the part of each line that contains the specified x values. The result is the V-shaped graph shown at the right.

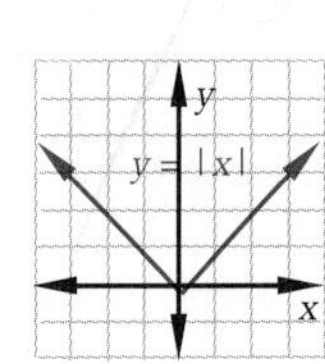

Lesson Notes, continued

The function is split into pieces for various domain elements. Graphs of these functions are drawn by considering only those elements in the domain when drawing that portion of the graph. The absolute value function is an example of a compound function.

Using Computers

Most software will not graph compound functions easily. An article in the fall issue (1990) of the "Consortium-The Newsletter of the Consortium and its Applications" written by Adam Kalai, a high school sophomore describes how to graph compound functions using an ordinary function grapher. Your students could apply this idea to the computer software used at your school.

Checkpoint

1 Solve the equation $|x - 6| = 12$. 18 and −6

2 If $f(x) = \begin{cases} 3x + 4 & \text{if } x > 0 \\ 5x & \text{if } x = 0 \\ -2x - 1 & \text{if } x < 0 \end{cases}$ find $f(0)$, $f(-4)$ and $f(2)$.

0, 7, and 10

3.6

Assignment Guide

Average	
Optional	
Advanced	
Day 1	14–19, 21, 23, 26, 27, 29, 31, 33, 35, 37–39, 42, 45, 48, 49, 54, 56
Day 2	20, 22, 25, 28, 30, 34, 36, 40, 43, 44, 46, 51–53, 55, 57, 58, 60
Honors	
Day 1	14–20, 22, 26, 27, 29, 32, 34, 36, 37, 39, 42, 43, 45, 47, 49, 54, 56, 58, 60
Day 2	25, 28, 30, 31, 35, 38, 40, 44, 46, 48, 51–53, 55, 57, 59, 61, 62

Problem-Set Notes and Strategies

- Graphing the compound function as in problems **1 – 4** helps students see that there really are two parts and that each is significantly different from the other.
- Problem **5** is the signum function used in some mathematics courses and in computer science. It's only purpose is to identify the sign of a number with the flags –1, 0 and 1.

Additional Answers

5

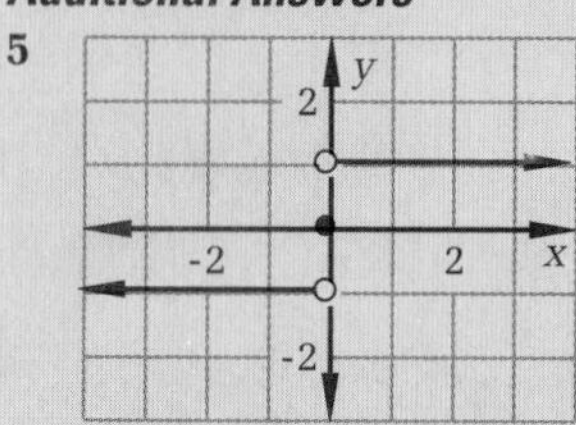

Answer for problem **6a** can be found in the answer pages beginning on **A** 1.

See p. 784

Part Two: Sample Problems

Problem 1 *Solve* $|x - 2| = 6$ *for* x.

Solution The absolute value of an expression is 6 when the expression is equal to either 6 or −6.

$$|x - 2| = 6$$
$$x - 2 = 6 \text{ or } x - 2 = -6$$
$$x = 8 \qquad x = -4$$

Problem 2 *Graph each set.*

a $\{x: |x - 2| = 6\}$ **b** $\{(x, y): |x - 2| = 6\}$

Solution **a** We graph $\{x: |x - 2| = 6\}$ on a number line. From Sample Problem 1, we know that the solution set consists of two points, 8 and −4.

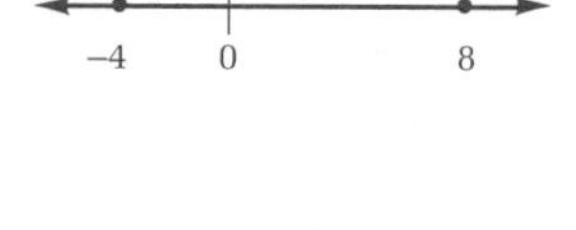

b We graph the set of ordered pairs $\{(x, y): |x - 2| = 6\}$ on a coordinate plane. The graph is the set of all points (x, y) such that x = 8 or x = −4. The graph consists of the two vertical lines shown.

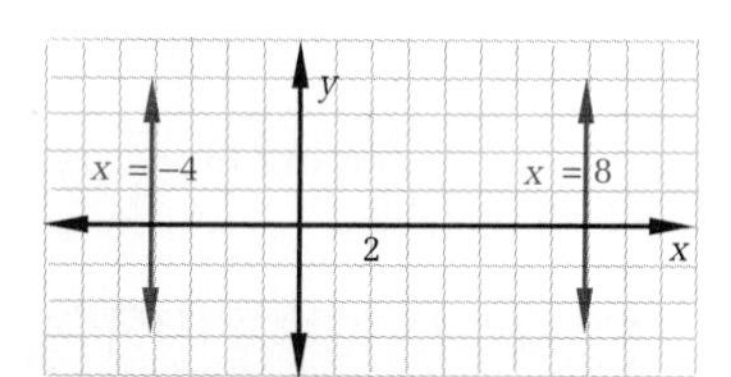

Part Three: Exercises and Problems

Warm-up Exercises

In problems 1–4, use the following compound function.

P

$$g(x) = \begin{cases} -x + 12 \text{ if } x < 8 \\ \frac{1}{2}x - 2 \text{ if } x \ge 8 \end{cases}$$

1 Find g(−2). 14

2 Find g(10). 3

3 Find g(8). 2

4 Which is greater, g(7.9) or g(8.1)? g(7.9)

P

5 Draw the graph of $y = f(x)$ for the following function *f*.

$$f(x) = \begin{cases} 1 \text{ if } x > 0 \\ 0 \text{ if } x = 0 \\ -1 \text{ if } x < 0 \end{cases}$$

P

In problems 6–9, use the following compound function.

$$h(x) = \begin{cases} x + 2 \text{ if } x \le 4 \\ 6 \text{ if } x > 4 \end{cases}$$

6 Draw the graph of $y = h(x)$.

7 Find *h*(0). 2

8 Find *h*(4). 6

9 Find *h*(10). 6

P | **In problems 10–12, solve for w.**

10 $|w + 10| = 17$ $w = -27$ or $w = 7$
11 $|5 - w| = 15$ $w = -10$ or $w = 20$
12 $|w - 5| = 15$ $w = -10$ or $w = 20$

P,R | **13** Tickets for a show cost $2.75 each if fewer than 25 are purchased and $2.40 each if 25 or more are purchased.

a Write a compound function c for the total cost of x tickets. $c(x) = \begin{cases} 2.75x \text{ if } x < 25 \\ 2.4x \text{ if } x \geq 25 \end{cases}$

b Find c(10), c(24), and c(25). 27.5; 66; 60

c What does your answer to part **b** tell you about buying 25 tickets? It's cheaper than buying 24.

Problem Set

A

P | **In problems 14–19, use the following compound function.**

$$f(x) = \begin{cases} 2x \text{ if } 0 < x \leq 10 \\ 20 - \frac{1}{2}x \text{ if } 10 < x < 40 \end{cases}$$

14 Find $f(-4)$. Undefined
15 Find $f(0)$. Undefined
16 Find $f(6)$. 12
17 Find $f(10)$. 20
18 Find $f(30)$. 5
19 Find $f(50)$. Undefined

P,R | **20** Graph each of the following sets.

a $\{x: |x + 3| = 7\}$
b $\{(x, y): |x + 3| = 7\}$

P,R | **21** Graph each of the following equations.

a $y = x$
b $y = -x$
c $y = |x|$

P,R | **22** Graph each equation on the coordinate plane.

a $|x + 5| = 9$
b $y = x + 5$
c $y = |x + 5|$

P | **In problems 23 and 24, solve for x.**

23 $|x + 3| = 9$ $x = -12$ or $x = 6$

24 $4|x - 7| + 2|x - 7| = 12$ $x = 5$ or $x = 9$

R,I | **25** Express the area of the shaded region in terms of x and y. $x^2 - y^2$

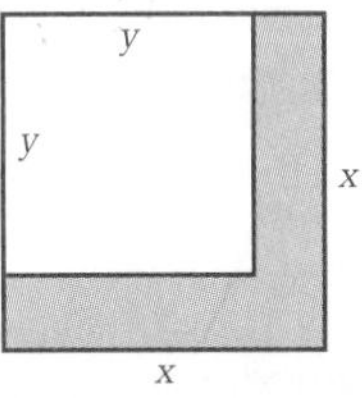

P | **26** Calculators cost $17.95 each if 1–10 are purchased, $15.95 each if 11–200 are purchased, and $14.50 each if more than 200 are purchased.

$c(x) = \begin{cases} 17.95x \text{ if } 1 \leq x \leq 10 \\ 15.95x \text{ if } 11 \leq x \leq 200 \\ 14.5x \text{ if } x > 200 \end{cases}$

a Write a compound function c for the cost of x calculators.

b Find c(7), c(52), and c(345). 125.65; 829.4; 5002.5

c For what purchase ranges would you spend less money by buying more calculators? 10 and 183–200

Problem-Set Notes and Strategies, continued

- Problems **10 – 12** reinforce that absolute value problems are really two problems whether you use the algebraic or geometric definition.
- Some values as in problems **14**, **15** and **19** may not be defined by a compound function.
- Problem **21** is an illustration of the significance of the absolute value operator as a compound function.
- Answers may vary for problem **25**, but should be equivalent.

Additional Answers

20a

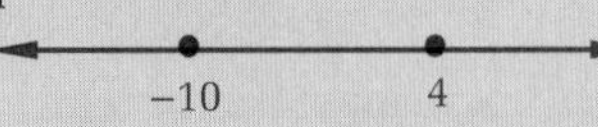

20b

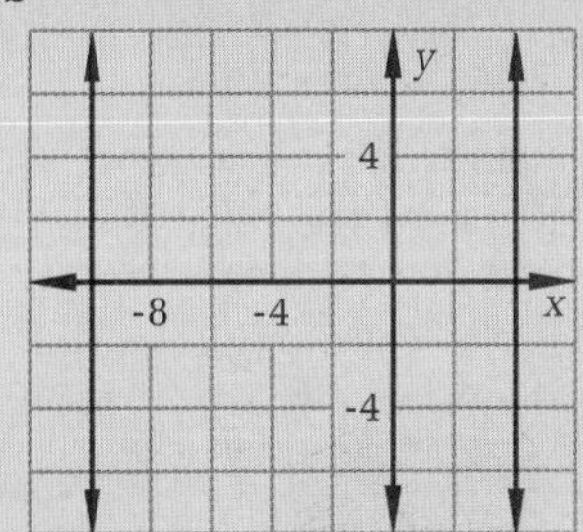

21a

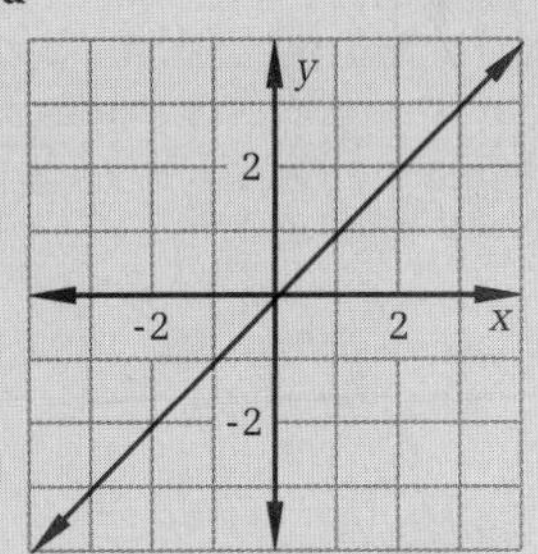

21b

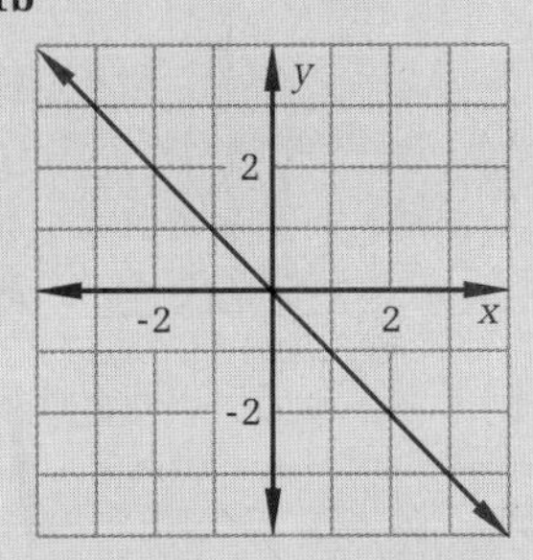

Answers for problems **21c** and **22a – c** can be found in the answer pages beginning on **A** 1.

Problem Set, *continued*

Problem-Set Notes and Strategies, continued

- Have students comment on the estimate in part **d** of problem **27**. Could there be extenuating circumstances that would make this a poor estimate, such as increasing costs of construction since the last part of the tollway was built?

- Factoring formulas for the square of a binomial are verified by direct substitution in problem **31**. Some students will recall these formulas from Algebra 1 and may not have to use the point to verify the results.

- In problem **34**, reiterate that each time there is a variable in an absolute value, there are at least two problems to solve since absolute value is a compound function.

Additional Answer

37

R **27** The following table shows the tolls charged at exits that are various distances from the eastern end of a tollway.

Exit number	1	2	3	4	5	6	7
Miles from east end	25	60	84	130	155	180	210
Toll (dollars)	1.25	3.75	4.00	6.00	7.25	8.50	9.75

a Graph the data, fit a line by estimation, and write an equation of the line. Answers will vary. $t = 0.05d + 0.50$ (where t is toll and d is distance from eastern end)

b Use the method of least squares to determine an equation of the line of best fit. $t = 0.0438d + 0.52$

c A new tollway exit is to be built 100 miles from the east end. What should the toll at that exit be? $4.90

d The tollway is being extended to the west. The toll for the entire trip from the east end to the new west end is to be $10.75. Estimate the length of the extended tollway. $\approx$233.6 mi

R **28** Write a formula for a linear function g such that $g(4) = 24$ and $g(10) = 60$. $g(x) = 6x$

R **29** The graph of a linear function f is perpendicular to the graph of $g(x) = \frac{2}{3}x - 4$. If $f(6) = 14$, what is function f? $f(x) = -\frac{3}{2}x + 23$

R **30** If $f(x)$ varies directly as x and $f(3) = 14$, for what value of x_1 is $f(x_1) = 35$? $\frac{15}{2}$

R,I **31** Evaluate each of the following expressions for $(x, y) = (5, 2)$. Which of the expressions are equivalent? **a** and **c**; **b** and **d**

a $(x - y)^2$ 9 **b** $x^2 - y^2$ 21 **c** $x^2 - 2xy + y^2$ 9 **d** $(x + y)(x - y)$ 21

R **32** In the diagram, x is a number such that $2 \le x \le 5$.

A C B
0 2 x 5 12

a Write a function f for the ratio AC:CB. $f(x) = x:(12 - x)$

b Find $f(4)$. 4:8, or 1:2

R **33** Let $g(x) = \sqrt{x + 5}$. Evaluate each of the following.

a $g(4)$ 3 **b** $g(-5)$ 0 **c** $g(-9)$ No real value

B

In problems 34–36, solve for *p*.

P **34** $|2p + 10| = |-15|$ $p = \frac{5}{2}$ or $p = -\frac{25}{2}$

35 $-5 = 3 - |12 - 3p|$ $p = \frac{4}{3}$ or $p = \frac{20}{3}$

36 $2041 = |16p + 775|$ $p = -176$ or $p = 79.125$

P **37** Graph the following compound function.

$$f(x) = \begin{cases} x + 1 \text{ if } x \ge 2 \\ 3 \text{ if } -2 < x < 2 \\ x + 4 \text{ if } x \le -2 \end{cases}$$

R **38** Refer to the figure.

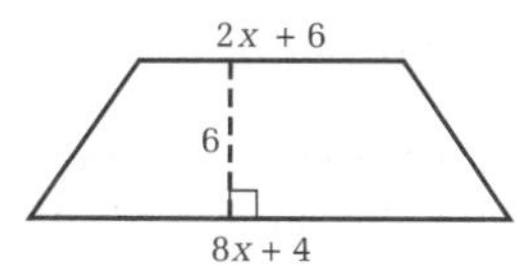

a Write a function to describe the area A of the trapezoid. $A(x) = 30(x + 1)$

b Is the area of the trapezoid linearly related to x? Yes

R **39** If $f(x) = |x - 4|$ and x is randomly chosen from {1, 2, 3, 4, 5, 6, 7, 8}, what is the probability that $f(x) \geq 2$? $\frac{5}{8}$

P,I **40** Sketch the graph of each equation.

a $y \geq |x - 3|$ **b** $y \geq |x + 3|$ **c** $y \geq |x| + 3$ **d** $y \geq |x| - 3$

R,I **41** Let $g(x) = x^2 - x$ and $h(x) = g(x) + 5$.

a Find g(2). 2 **b** Find $h(2)$. 7

R **42** Use a system of inequalities to describe the shaded region. $\begin{cases} y \leq -\frac{4}{7}x + \frac{64}{7} \\ y \geq 4 \\ x \geq 2 \end{cases}$

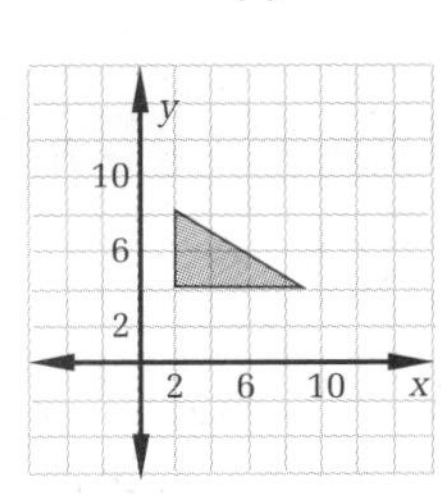

P,R **43** Which of the following statements are true? **a, c**

a $\sqrt{x^2} = |x|$ **b** $\sqrt{x^2} = x$ **c** $\sqrt{x^2} = \begin{cases} x \text{ if } x \geq 0 \\ -x \text{ if } x < 0 \end{cases}$

R **44** Determine which of the given ordered pairs (x, y) satisfy the following matrix equation. **a, b, c, d**

$$\begin{bmatrix} 0 & 1 \\ 1 & 1 \end{bmatrix} \begin{bmatrix} x^2 \\ y^2 \end{bmatrix} = \begin{bmatrix} 16 \\ 25 \end{bmatrix}$$

a (3, 4) **b** (−3, 4) **c** (−3, −4)
d (3, −4) **e** (4, −3) **f** (4, 3)

P,R **In problems 45–53, determine the set of numbers that make each equation or inequality a true statement.**

45 $|x| \geq x$ $\{x{:}x \in \mathcal{R}\}$ **46** $\frac{|x|}{x} = 1$ $\{x{:}x > 0\}$ **47** $|x^2| = x^2$ $\{x{:}x \in \mathcal{R}\}$

48 $|x| \cdot x = x^2$ $\{x{:}x \geq 0\}$ **49** $(\sqrt{x})^2 = |x|$ $\{x{:}x \geq 0\}$ **50** $\sqrt{x^2} = x$ $\{x{:}x \geq 0\}$

51 $|x| - |y| = |x - y|$ $\{(x, y){:}0 \leq y \leq x \text{ or } x \leq y \leq 0\}$ **52** $|x + y| = |x| + |y|$ $\{(x, y){:}xy \geq 0\}$ **53** $(\sqrt{x^2})^2 = x^2$ $\{x{:}x \in \mathcal{R}\}$

P **In problems 54–57, write each absolute-value function as a compound function.**

54 $f(x) = |x + 3|$ $f(x) = \begin{cases} x + 3 \text{ if } x \geq -3 \\ -x - 3 \text{ if } x < -3 \end{cases}$

55 $g(x) = 2|x - 4|$ $g(x) = \begin{cases} 2x - 8 \text{ if } x \geq 4 \\ 8 - 2x \text{ if } x < 4 \end{cases}$

56 $h(x) = |x - 2| + 4$ $h(x) = \begin{cases} x + 2 \text{ if } x \geq 2 \\ 6 - x \text{ if } x < 2 \end{cases}$

57 $R(x) = -|x + 3|$ $R(x) = \begin{cases} -x - 3 \text{ if } x \geq -3 \\ x + 3 \text{ if } x < -3 \end{cases}$

Problem-Set Notes and Strategies, continued

- Problem **41** is the first time a composite function is introduced. These will be explored in Chapter **5** in greater detail. At this point, the answer to part **a** should be used to calculate part **b**.
- In problem **43**, remind students to be careful with the subtle differences between **b** and **c**. Help your students understand the equivalency of parts **a** and **c**.
- Problems **45 – 53** provide a review of positive and negative numbers and the effects of the absolute value function on them.

Additional Answers

40a

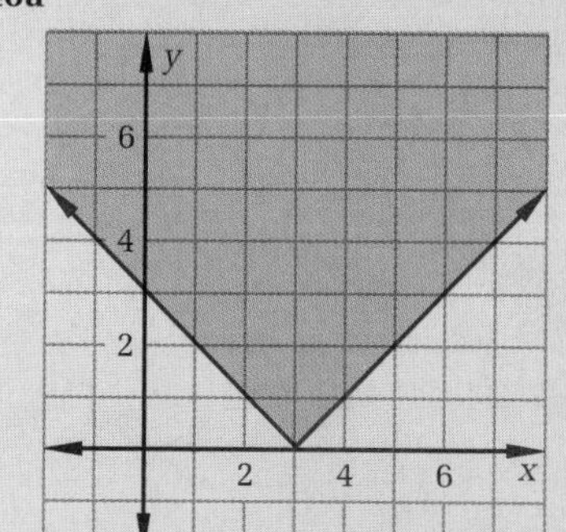

40b

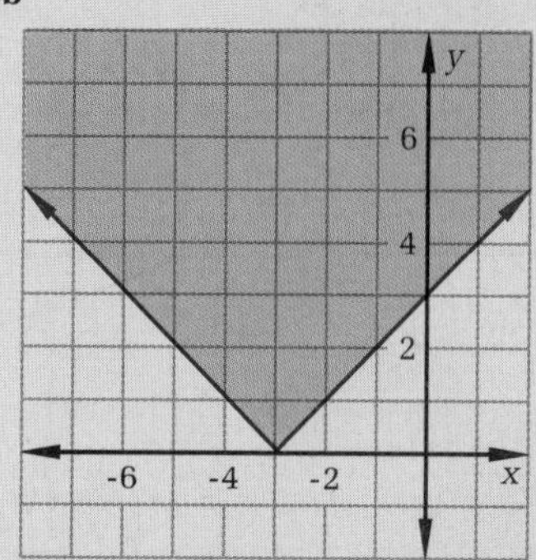

Answers for problems **40c** and **40d** can be found in the answer section beginning on **A** 1.

Problem-Set Notes and Strategies, continued

- You may assign problem **62** to only your better students.

Communicating Mathematics

Using a wholesale business as an example, describe where compound functions may enter into the business model. Examples may include employee wages if overtime is available, sale prices for quantity discounts, and commission percentages for amounts of goods sold by a salesperson.

Problem Set, *continued*

P **In problems 58–61, write each compound function as an absolute-value function.**

58 $f(x) = \begin{cases} x \text{ if } x \geq 0 \\ -x \text{ if } x < 0 \end{cases}$ $f(x) = |x|$

59 $g(x) = \begin{cases} x - 2 \text{ if } x \geq 2 \\ 2 - x \text{ if } x < 2 \end{cases}$ $g(x) = |x - 2|$

60 $L(x) = \begin{cases} 2x + 4 \text{ if } x \geq -2 \\ -2x - 4 \text{ if } x < -2 \end{cases}$ $L(x) = 2|x + 2|$

61 $R(x) = \begin{cases} -x + 3 \text{ if } x \geq 3 \\ x - 3 \text{ if } x < 3 \end{cases}$ $R(x) = -|x - 3|$

P **62** Prove that $-|a| \leq a \leq |a|$ for any real number a. By definition, $|a| = a$ if $a \geq 0$, and $|a| = -a$ if $a < 0$. Therefore, $-|a| = a$ if $a \leq 0$, $-|a| < a$ if $a > 0$, $|a| = a$ if $a \geq 0$, and $|a| > a$ if $a < 0$. Thus, $-|a| \leq a \leq |a|$.

MATHEMATICAL EXCURSION

LINEAR PROGRAMMING

Organizing mathematical decision making

Some of the problems most commonly solved on computers involve linear programming, a method of formulating a problem entirely in terms of linear equations and inequalities.

For example, you could use linear programming in planning a daily diet that would give you the greatest nutritional value with the fewest calories and that would include a wide variety of foods you enjoy. Problems to which linear programming is applied always have many variables subject to defined constraints, with quantities to be minimized or maximized.

Long-distance telephone companies use linear programming in routing calls. Linear programming helps these carriers achieve the most economical use of such equipment as long-distance land lines, repeat amplifiers, and satellite terminals. In fact, a researcher for one long-distance carrier recently developed a method of linear programming that is several times faster than the existing simplex method.

What quantity do you think a long-distance company wants to minimize? What are the variables and constraints in its problem? Think of some other situations in which linear programming could be used to solve a problem.

For more information on linear programming, see the Short Subject that begins on page 466.

3

Chapter Summary

Concepts and Procedures

After studying this chapter, you should be able to

- Determine the slope of a line (3.1)
- Use the point-slope, slope-intercept, and general equations of a line (3.1)
- Use slope to determine whether two lines are perpendicular or parallel (3.1)
- Solve systems of linear equations graphically and algebraically (3.2)
- Graph linear inequalities (3.3)
- Find the union and the intersection of sets defined by inequalities (3.3)
- Write equations that represent linear functions (3.4)
- Find restrictions on the domain of a function caused by real-world constraints (3.4)
- Recognize direct variation (3.4)
- Express direct variation as a proportion (3.4)
- Fit a line to data by estimation and by the least-squares method (3.5)
- Recognize compound functions (3.6)
- Graph compound functions (3.6)
- Recognize the absolute-value function as a compound function (3.6)

Vocabulary

boundary-value algorithm (3.3)
compound function (3.6)
constant of proportionality (3.4)
constant of variation (3.4)
direct variation (3.4)
directly proportional (3.4)
domain (3.4)
fitted line (3.5)
intersection (3.3)
line of best fit (3.5)
linear function (3.4)
method of least squares (3.5)
multiplication-addition method (3.2)
proportion (3.4)
range (3.4)
regression line (3.5)
simultaneous equations (3.2)
slope (3.1)
split domain (3.6)
substitution (3.2)
system of linear equations (3.2)
union (3.3)

3

Class Planning

Time Schedule
All levels: 1 day

Assignment Guide

Average
1, 2, 4, 5, 7–9, 11, 12, 14–17, 19, 20, 22, 24, 27
Advanced
1–3, 5, 7–11, 13, 14, 17, 18, 20, 21, 24–27
Honors
2, 5–7, 9–12, 14, 17–21, 23–27

Problem-Set Notes and Strategies

- Check the slopes in the three parts of problem **2**. By this time students should be more comfortable with the concepts of parallel and perpendicular.
- In problems **3 – 6**, students will likely see absolute values as two part problems.
- Encourage the writing in part **c** of problem **7**.
- Problem **10** is an additional example of mixture problems. These problems come in many styles, but always involve several equations with several unknowns.

3

REVIEW PROBLEMS

A

P **1** Copy and complete the table of values for $y = -3x + 8$, then graph the equation on a coordinate plane.

x	−3	−2	−1	0	1	2	3	4
y	17	14	11	8	5	2	−1	−4

P **2** **a** Write, in point-slope form, an equation of a line. Answers will vary.
b Write, in slope-intercept form, an equation of a line parallel to the first line. Answers will vary.
c Write, in general linear form, an equation of a line perpendicular to the two parallel lines. Answers will vary.

P **In problems 3–6, solve for *x*.**

3 $2.7|x - 4.5| = 13.5$ $x = 9.5$ or $x = -0.5$
4 $-4|x + 7| = 24$ No solution
5 $|x + 5.7| - 2.3 = 14.7$ $x = 11.3$ or $x = -22.7$
6 $\sqrt{2}|3x - 7| = 54.9$ $x \approx 15.27$ or $x \approx -10.61$

P **7** Refer to the data in the following table.

x	5	10	12	15	20	25	30	32	35	40
y	80	85	5	20	45	70	20	95	29	10

a Write an equation of the line of best fit for the data. $y = -0.773x + 63.2$
b Graph the ordered pairs in the table and the equation you wrote in part **a** on the same coordinate system.
c Write a paragraph describing what you found in part **b.** What do you think about the original data? Answers will vary. Students may remark that the data do not seem to be linearly related.

P **8** Let $y - 2 = 3(x - 5)$.
a Find the value of k if the point $(5, k)$ is on the graph of the equation. 2
b Graph the equation.

P **9** Graph $y = x - 3$ and $3x - 5y = 15$ on the same coordinate plane. Approximately where do the graphs intersect? $(0, -3)$

P **10** Bolivian coffee sells for \$4.59 per pound, and Colombian coffee sells for \$5.95 per pound. How many pounds of each coffee should be mixed to produce 4000 pounds of a mixture that will sell for \$5.10 per pound? Bolivian: 2500 lb; Colombian: 1500 lb

Additional Answers
Answers for problems **1**, **7b**, **8b**, and **9** can be found in the answer pages beginning on page **A** 1.

P **11** Sketch each of the described lines.

a Passes through the origin and is parallel to the graph of $y = 3x - 2$

b Passes through $(5, -1)$ and is perpendicular to the graph of $y = \frac{2}{3}x + 4$

c Has an x-intercept that is 4 greater than its y-intercept, with the sum of the intercepts being 14

P **12** The following chart shows toll data for the Indiana Toll Road.

Exit	Mile Marker	Cost
Gary (Westport)	0	—
Portage	9	0.30
Valparaiso	19	0.65
Michigan City	27	0.95
La Porte	37	1.30
Mishawaka	64	2.25
Elkhart	80	2.80
Middlebury	95	3.80
Howe	108	3.80
Fremont	132	4.65
Eastport	157	5.50

a Make a graph showing the relationship between distance and cost, using Gary as a starting point.

b Are these data approximately linearly related? Yes

c What is the cost per mile? $\approx\$0.035$

d Write an equation relating cost to miles driven. $c = 0.035m$

e A new exit is being constructed at mile marker 122. What toll should be charged at that exit? $\approx\$4.27$

P **13** One kite is 200 feet off the ground and is rising at a rate of 6 feet per second. A second kite is 100 feet off the ground and is rising at a rate of 9 feet per second.

a Write expressions that represent the heights of the kites t seconds from now. $200 + 6t$; $100 + 9t$

b How high will each kite be in 24 seconds? 344 ft; 316 ft

c When will the kites be at exactly the same height? In ≈ 33.3 sec

P **14** Refer to the diagram.

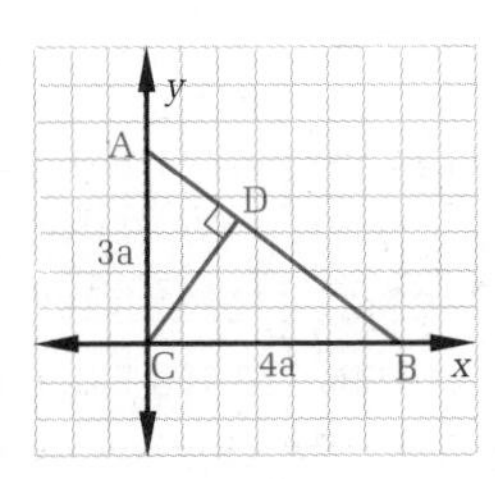

a Find the value of a for which $AB = 15$. 3

b Write the equation of $\overline{AB}$ in general linear form. $3x + 4y = 12a$

c Find the slope of $\overline{CD}$ if $\overline{CD}$ is perpendicular to $\overline{AB}$. $\frac{4}{3}$

P **In problems 15–18, graph each set.**

15 $\{(x, y): y \geq |x| + 2\}$

16 $\{x: -5 < x\}$

17 $\{(x, y): x > -3\}$

18 $\{(x, y): y \geq |x + 2|\}$

Problem-Set Notes and Strategies, continued

- Problem **12** is a good example of using data to build a mathematical model. From the graph it appears the data are linearly related and then an equation is formed to model the data. From this point, the equation can be used to make predictions about points not included in the data set.

Additional Answers

12a

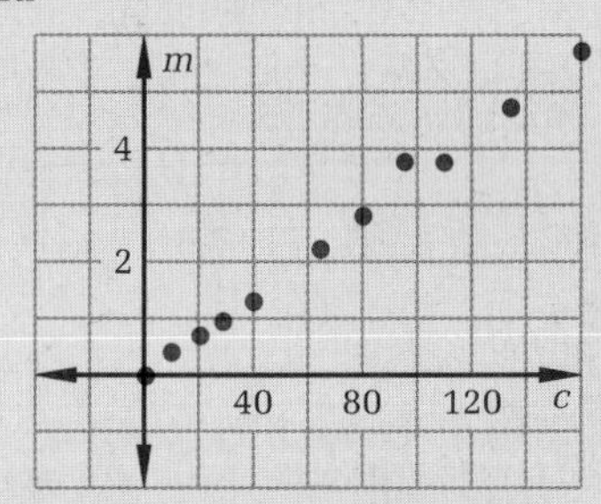

15

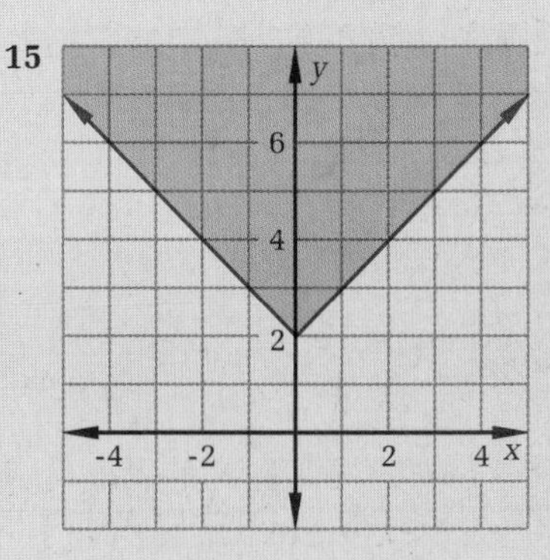

- Areas, distances, systems of equations and geometric formulas are all combined in working problem **19**. Encourage the student to draw a sketch of the graph with the intersection points labeled.

Additional Answers

Answers for problems **11a**, **11b**, **11c**, and **16-18** can be found in the answer pages beginning on page **A** 1.

Problem-Set Notes and Strategies, continued

- In problem **21**, parts **c** and **d** ask you to make predictions outside the realm of the original data points. Have the students comment on what circumstances might make these values incorrect. For instance, maybe the company shuts down during the month of February for inventory and auditing procedures. It is possible that the copier was first purchased in March and it took time for the word to spread that a copy machine was available in the office.

- In problem **23**, reading it a second time and sketching the outcome is a help.

- You might ask in what cases does $|x + y| = |x| + |y|$ in problem **24**. It is not always true, but is sometimes true.

- Problem **26** may trouble students until they figure out that they should graph the entire compound function and eliminate those parts where $f(x)$ is not greater than x.

- In problem **27b**, you might have the students consider factors that could make this prediction incorrect. The state insurance laws may have changed which require greater coverage or significantly less coverage. Values should not be predicted outside the realm of the original data except as a last resort. The accuracy of these values is questionable.

Review Problems, *continued*

P **19** Find the area of the triangle formed by the graphs of $y = 2$, $y = -4x + 18$, and $y = 2x + 18$. 96

R **20** Find the x- and y-intercepts of $Ax + By = C$, where $A \neq 0$ and $B \neq 0$. x-intercept $= \frac{C}{A}$; y-intercept $= \frac{C}{B}$

P **21** The following table shows the cumulative numbers of photocopies that had been made by an office's machine at the end of eight consecutive months, beginning with June 1987.

Month	1	2	3	4	5	6	7	8
Copies	80,452	125,454	172,975	211,426	255,221	302,853	342,598	387,256

a Write an equation of the line of best fit for the data. $y = 43{,}652x + 38{,}345$
b What was the average number of copies produced per month? $\approx 43{,}829$
c A copier needs a major overhaul every million copies. How many months after the beginning of the cumulative record should the office's machine have been overhauled? ≈ 23 mo
d During what month, do you think, was the copier first used? May

B

P,R **22** Suppose that $y = 5x$, $z = 9 - x$, and $t = 12 - 3x$. If $y + 2z = t$, what is the value of x? -1

P,R **23** The intersection of the graphs of $y = 7x + 4$ and $y = 2x - 1$ is the midpoint of a segment with one endpoint at $(-5, 2)$. What is the other endpoint? $(3, -8)$

P **24** Can absolute value be distributed over addition—that is, is $|x + y| = |x| + |y|$? Explain your answer. No; one of the properties of absolute value is that $|a + b| \leq |a| + |b|$, so in some cases $|x + y| \neq |x| + |y|$.

R **25** A sheet of plywood measuring 4 feet by 8 feet is to be cut into equal numbers of 4-foot-by-2-inch strips and 4-foot-by-3-inch strips. What is the maximum number of strips that can be cut from the sheet? 38

C

P **26** Consider the following compound function.

$$f(x) = \begin{cases} x - 5 \text{ if } x \geq 2 \\ -4x + 5 \text{ if } x < 2 \end{cases}$$

Sketch the graph of $\{(x, f(x)):f(x) > x\}$.

P **27** Lia Bility opened an insurance agency in 1982. Her agency's profits for the first six years of operation are shown.

Year	1982	1983	1984	1985	1986	1987
Profit (dollars)	19,600	23,100	27,100	30,400	33,800	37,400

P
a Use the least-squares method to write a linear equation that approximates the data. $y = 3554.3x - 7{,}025{,}000$
b Use your equation to predict the agency's profit in 1990. $\approx \$48{,}057$

Additional Answer

Answer for problem **26** can be found in the answer pages beginning on page **A** 1.

CHAPTER TEST

1 Draw the line represented by each of the following descriptions.

a $y = \frac{3}{2}x - 2$ **b** Slope $-\frac{3}{4}$, contains the point $(1, -1)$

2 Refer to the diagram. Write an equation to represent each of the following.

a A line parallel to $\overline{AB}$

b A line perpendicular to $\overline{BC}$ at the point (2, 1)

a Answers will vary. Possible answer: $y = 2x + 2$

b $y - 1 = \frac{2}{3}(x - 2)$

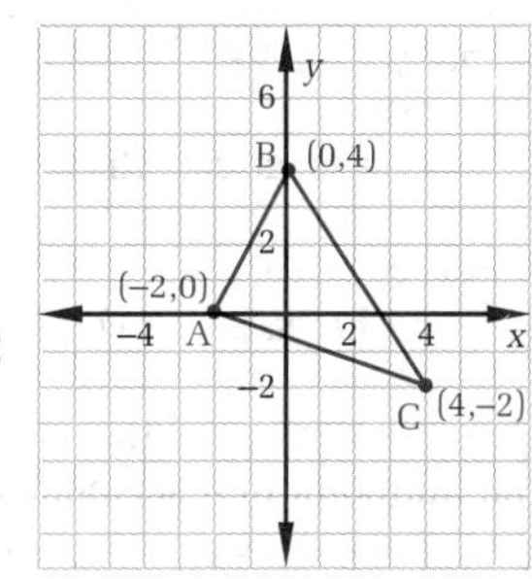

In problems 3 and 4, solve each system.

3 $\begin{cases} 4T + TC = 26 \\ 4T - TC = 14 \end{cases}$ $(T, C) = (5, 1.2)$

4 $\begin{cases} y = 4x - 1 \\ 3x - 2y = 10 \end{cases}$ $(x, y) = (-1.6, -7.4)$

5 Write an equation of the line of best fit for the data in the following table. $y = 0.035x + 0.031$

x	10	35	60	90	145	170	210	250
y	0.35	1.25	2.10	3.25	5.10	6.00	7.40	8.75

6 If S varies directly as T and if S = 45 when T = 18, what is the value of S for each of the following values of T?

a 6 15 **b** 24 60 **c** 42 105 **d** 54 135

7 **a** Graph the region defined by the inequalities $y \leq -\frac{1}{2}x + 8$, $x \geq 0$, and $y \geq 0$.

b Graph the region defined by the inequalities $y \leq -\frac{1}{2}x + 16$, $x \geq 0$, and $y \geq 0$.

c What is the ratio of the area of the region described in part **a** to the area of the region described in part **b**? $\frac{1}{4}$

Resource References

Evaluation

Test Form 1, 2, 3

Additional Answers

1a

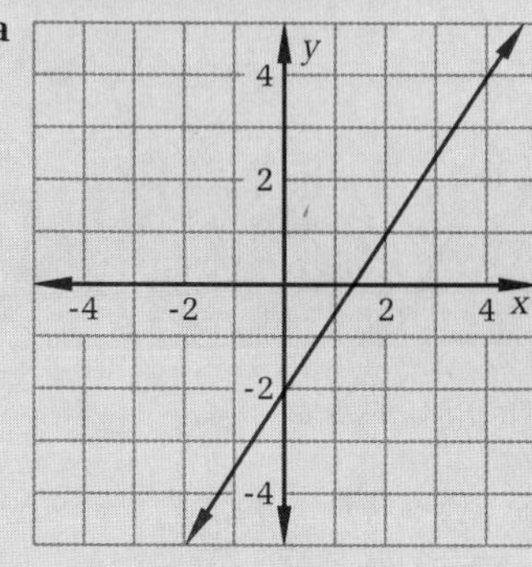

1b

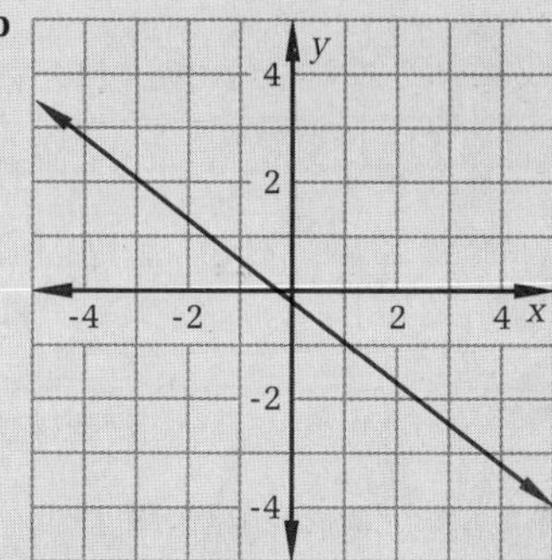

7a

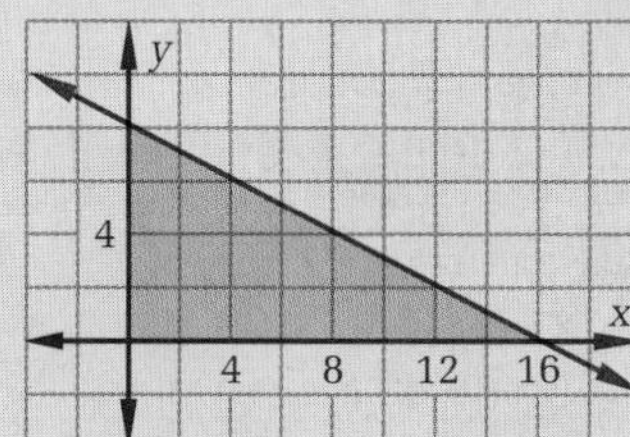

7b

4.1

*T*HIS CHAPTER BEGINS by identifying quadratic equations and expressions and by multiplying binomials.

Students may remember how to factor. The amount of time you spend on factoring will depend on your class.

Graphing quadratic functions (parabolas) is reviewed, with special emphasis on relating the characteristics of the quadratic function to the characteristics of the function's graph. Students will also see the relationship between the real roots of a quadratic equation and the real zeros of a corresponding quadratic function.

Factoring and the quadratic formula are presented as ways to solve quadratic equations, with the discriminant used to determine whether a trinomial is factorable.

A third method for solving quadratic equations, completing the square, is presented in Section **4.5**. This method is also used to identify the minimum and maximum values of a quadratic function and to prove the quadratic formula. The applications of quadratic functions show a connection to real-world applications.

As an extension of the data analysis section, students can gather data, display it using one of the methods suggested, and then write questions appropriate for their display. This activity will depend on your class and on time available.

CHAPTER

4 QUADRATIC FUNCTIONS

The parabola-like forms in *Bush Tucker Dreaming at Wingellina* open outward in the four cardinal directions.

4.1 WHAT ARE QUADRATICS?

Objectives

After studying this section you will be able to
- Recognize quadratic expressions and equations
- Multiply to form quadratics

Part One: Introduction

Setting the Stage

A farmer has 75 feet of fencing with which to enclose a garden. He plans to use an existing wall and wants to have a garden with the greatest possible area. If the plot is as shown, what should the garden's dimensions be?

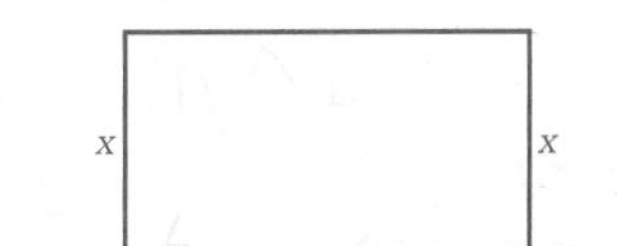

The area of the garden can be found with the following equation.

$$\begin{aligned} A &= lw \\ &= (75 - 2x)x \\ &= 75x - 2x^2 \end{aligned}$$

This equation, $A = 75x - 2x^2$, is a ***quadratic equation.*** To find the maximum possible area, the farmer must solve the equation.

Quadratic Expressions

Quadratic expressions are polynomials of degree two. (If the terms *polynomial* and *degree* are unfamiliar to you, consult the review of these topics in the Handbook.) The following are examples of quadratic expressions.

w^2 $\quad x^2 + y^2$ $\quad bh$ $\quad (x + 2)(x + 3)$ $\quad x^2 + 5x + 6$

The expressions $(x + 2)(x + 3)$ and $x^2 + 5x + 6$ are equivalent. The expression $(x + 2)(x + 3)$ is in *factored* form, and the expression $x^2 + 5x + 6$ is in *expanded* form.

These are not quadratic expressions:

$3x + y$ $\quad \pi d$ $\quad (x + 1)(x + 2)(x + 3)$ $\quad x^3 + 6x^2 + 11x + 6$

The expressions $3x + y$ and πd are linear expressions. The expressions $(x + 1)(x + 2)(x + 3)$ and $x^3 + 6x^2 + 11x + 6$ are equivalent cubic expressions.

Class Planning

Time Schedule

Average: 2 days
Advanced: 1 day
Honors: 1 day

Resource References

Teacher's Resource Book
Class Opener 20
Practice 21
Manipulatives and Applications 7

Class Opener

1 Sketch the graph of $f(x) = x^2 + 4x - 2$.

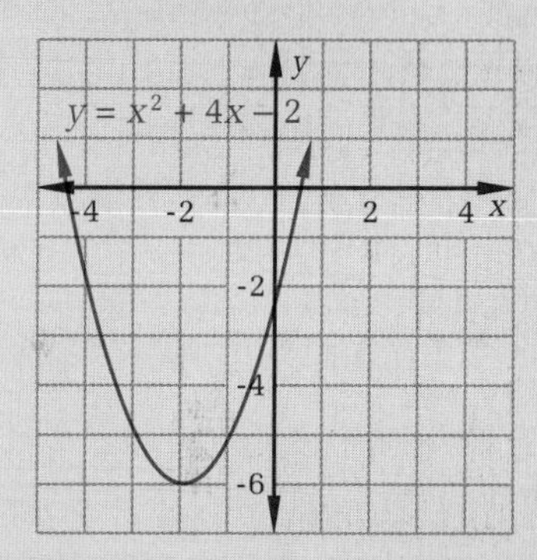

2 Write $(x + 3)(x + 5)$ in expanded form.
$x^2 + 8x + 15$

3 Write $(y - 4)^2$ in expanded form. $y^2 - 8y + 16$

Lesson Notes

- This section sets the stage for the balance of the chapter. Quadratics are explained as second degree polynomials and several examples are given. What is or is not a polynomial should be reviewed for those students who have forgotten.

Vocabulary

quadratic equation
quadratic expression

4.1

Lesson Notes, continued

A polynomial is any expression that can be written in the form $a_n x^n + a_{n-1} x^{n-1} + \ldots + a_1 x^1 + a_0$. It is an expression allowing only non-negative exponents. For the constant, the exponent is not limited to non-negative integers.

- The degree of a polynomial is the largest power of x such that the coefficient is non-zero.

Checkpoint

1 Find an expression for the area of a right triangle whose legs are of length x and $4x - 6$.
$A = x(2x - 3)$

2 Which of the following are quadratic equations:
a $x^2 + 3x - 9 = 0$
b $3x^2y^2 = 0$
c $x^2 - 9 = 3 + x^2 - 12$
Only **a**.

3 Expand $(3x - 5)(3x + 5)$
$9x^2 - 25$

Assignment Guide

Average	
Day 1	7–9, 11, 13, 16, 18, 20, 21, 23, 24, 26, 29, 31
Day 2	10, 12, 14, 15, 17, 19, 22, 25, 28, 30, 32–34
Advanced	
7–12, 15, 16, 18, 19, 21, 23, 25–27, 29, 30, 32, 34–36	
Honors	
9, 12, 14, 15, 19, 21, 23, 24, 26, 27, 29, 30–32, 34–36, 39, 41	

Quadratic Equations

An equation that contains at least one quadratic term and no term of a higher degree is a ***quadratic equation.*** Here are some examples:

$y = x^2 + 5x + 6 \qquad x^2 + y^2 = 36 \qquad xy = 6 \qquad A = 5x + 2x^2$

Quadratic equations are frequently used in solving area, projectile-motion, and locus problems.

Example *Write an equation that represents a circle with its center at the origin and a radius of 5 inches.*

All the points (x, y) of the circle are 5 units from the origin, $(0, 0)$. We can use the distance formula.

$$d = \sqrt{(x_1 - x_2)^2 + (y_1 - y_2)^2}$$
$$5 = \sqrt{(x - 0)^2 + (y - 0)^2}$$
$$5 = \sqrt{x^2 + y^2}$$

Square both sides $\qquad 25 = x^2 + y^2$

An equation of a circle with its center at the origin and a radius of 5 inches is $25 = x^2 + y^2$.

Multiplying to Form Quadratics

Quadratic equations are often used to represent area problems because area involves two dimensions. In the farmer's problem, the area of the garden was expressed as the product of the length and the width of a rectangle, $A = 75x - 2x^2$. Even if the farmer's garden were semicircular, its area would still be represented by a quadratic, $A = \frac{1}{2}\pi r^2$ (see Sample Problem 3). You will frequently encounter quadratics in situations where two linear expressions are multiplied.

Example *Multiply* $(x + 3)(2x - 5)$.

$$(x + 3)(2x - 5) = 2x^2 - 5x + 6x - 15$$
$$= 2x^2 + x - 15$$

Certain products of binomials deserve careful study.

Example *Write each expression in expanded form.*

a $(4x + 9)(4x - 9)$

$$(4x + 9)(4x - 9) = 16x^2 - 36x + 36x - 81$$
$$= 16x^2 - 81$$

This product is the difference of two perfect-square terms; it has no linear term.

b $(5x + 4)^2$

$$(5x + 4)^2 = (5x + 4)(5x + 4)$$
$$= 25x^2 + 20x + 20x + 16$$
$$= 25x^2 + 40x + 16$$

In the expansion of the square of a binomial, the first and last terms are perfect squares, and the middle term is twice the product of the terms of the binomial.

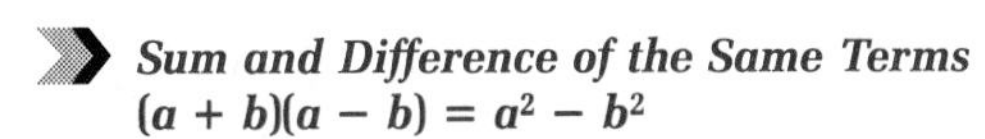

Sum and Difference of the Same Terms
$(a + b)(a - b) = a^2 - b^2$

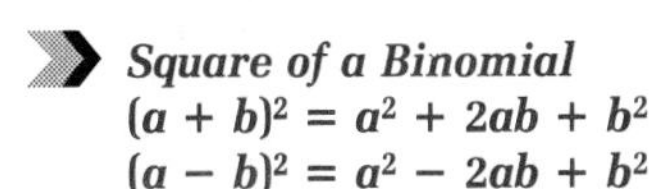

Square of a Binomial
$(a + b)^2 = a^2 + 2ab + b^2$
$(a - b)^2 = a^2 - 2ab + b^2$

Part Two: Sample Problems

Problem 1 *Find an expression for the area of the rectangle shown.*

3x + 7
3x - 7

Solution The area of a rectangle is equal to the product of its length and its width. In this case, $A = (3x - 7)(3x + 7)$, or $A = 9x^2 - 49$.

Problem 2 *Write a quadratic equation stating that the square and the rectangle have the same area.*

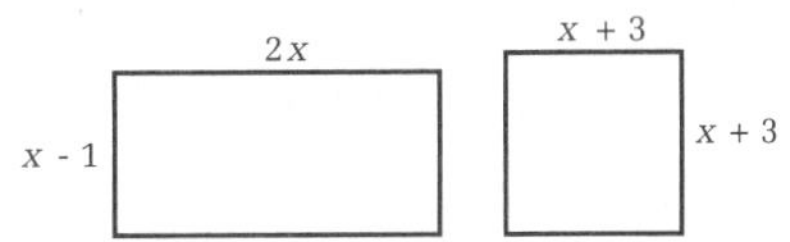

Solution Since the area of the rectangle is equal to the area of the square,

$$lw = (\text{side})^2$$
$$2x(x - 1) = (x + 3)^2$$

Multiply $\quad 2x^2 - 2x = x^2 + 6x + 9$

Combine like terms $\quad x^2 - 8x = 9$

Problem 3 *Find the area of a semicircular garden bounded by 75 feet of fencing and a garden wall.*

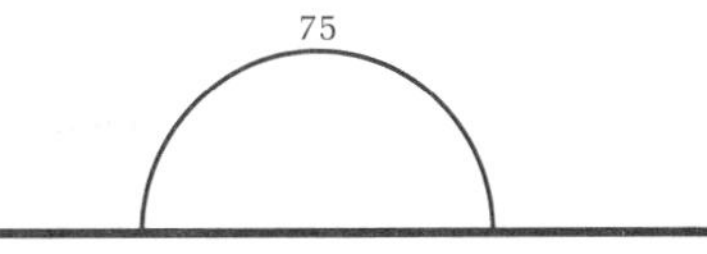

Solution The length of the fence is one-half the circumference of a circle.

$$75 = \frac{1}{2}(2\pi r)$$
$$75 = \pi r$$
$$\frac{75}{\pi} = r$$

Now we can find the area of the garden.

$$A = \frac{1}{2}(\pi r^2)$$
$$= \frac{1}{2}\pi\left(\frac{75}{\pi}\right)^2$$
$$= \frac{75^2}{2\pi} \approx 895 \text{ ft}^2$$

The area of the garden is about 895 square feet.

Communicating Mathematics

Have students write a paragraph explaining why $x^2 + y^2$ is a polynomial of degree 2, but why x^2y^2 is not.

Part Three: Exercises and Problems

Warm-up Exercises

P **1** Match each polynomial with its description.

a $\sqrt{3}x^2 - 2x$ ii
b $10 - 3x$ i
c $2\pi r$ i
d $5x^3 - x^2 + 11x + 2$ iii
e $3x^2 + xy + 3y$ ii

i Linear expression
ii Quadratic expression
iii Cubic expression

P **2** Multiply.

a $(2x + 3)(2x - 3)$ $4x^2 - 9$
b $(2x + 3)(3x - 2)$ $6x^2 + 5x - 6$
c $(3x + 2)(3x - 2)$ $9x^2 - 4$

P **3** Multiply.

a $(3y + 4)^2$ $9y^2 + 24y + 16$
b $(4x_1 - 3)^2$ $16x_1^2 - 24x_1 + 9$

P,R **4** Write an expression for the area of the rectangle. For what values of x will the base have a positive length?
$18x^2 + 49x + 10$; $\left\{x{:}x > -\frac{2}{9}\right\}$

$9x + 2$

$2x + 5$

P,R **5** Each side of a square has a length of $4x + 7$. Write an expression for the area of the square. $16x^2 + 56x + 49$

P,R **6** Find the following products. Then explain the relationship between your answers to parts **a** and **b.**

a $(3x + 1)^2$ $9x^2 + 6x + 1$
b 31^2 961

Because we use a base-10 number system, the digits in part **b** are the same as the coefficients and constant in part **a.**

Problem-Set Notes and Strategies

- Problem **6** is an application of squaring a binomial when the variable is 10. Some students will not recognize this.

Problem Set

A

P **7** Which of the following are quadratic equations? **a, b, e**

a $3 - 5x + x^2 = 0$
b $(6x - 2)(x + 3) = 0$
c $x(x + 1)(x + 2) = 0$
d $5 - 3x + x^2 = x^3$
e $xy + x^2 = 5x + y$

P,R **8** Write an expression for the area of the rectangle. $2x^2 + 11x + 15$

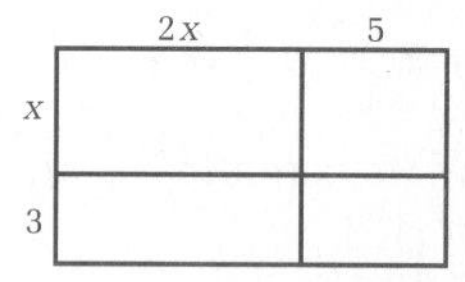

P,R **9** Let $(a, b) = (-2, 5)$.

a Evaluate $(a + b)^2$. 9
b Evaluate $a^2 + b^2$. 29
c Evaluate $a^2 + 2ab + b^2$. 9
d Which two expressions are equivalent? **a** and **c**

P **10** Indicate whether each statement is True or False.

a $(x + 5)(x + 3) = x^2 + 15x + 8$ False
b $(x + 5)(x - 3) = x^2 + 2x - 15$ True
c $(x - 5)(x - 3) = x^2 - 8x - 15$ False
d $(x - 5)(x + 3) = x^2 - 15x - 15$ False

R **In problems 11–14, solve for the variable.**

11 $9y - 2 - 2y + 135 = 0$ $y = -19$
12 $18w + 3 - w - 6 = -12w + 38.2$ $w \approx 1.42$
13 $3(\text{Fred}) + 9 = 12(\text{Fred}) - 11$ $\text{Fred} = \frac{20}{9}$
14 $6.5y_3 = 12y_3 - 25 + 4.1y_3$ $y_3 = \frac{125}{48} \approx 2.6$

P **15** Expand each expression.

a $(5x - 3)(5x + 3)$ $25x^2 - 9$
b $(6x + 5)(6x + 5)$ $36x^2 + 60x + 25$
c $(4y^3 + 6x)(4y^3 - 6x)$ $16y^6 - 36x^2$
d $(9a - b)^2$ $81a^2 - 18ab + b^2$
e $(x - 1)(x + 1)(x^2 + 1)$ $x^4 - 1$

R **16** Simplify each expression.

a $\frac{x^4 \cdot x^5}{x^{12}}$ $\frac{1}{x^3}$
b $\frac{x^{-3}y^5z^{12}}{x^4y^{-2}z^{11}}$ $\frac{y^7z}{x^7}$

P **17** Write an equation for a circle with its center at (0, 0) and a radius of 4. $x^2 + y^2 = 16$

P,R **18** The area of the semicircular region is 16π. Find the figure's perimeter. ≈ 29.09

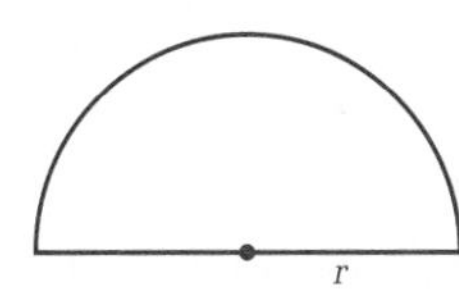

R **19** A mail-order firm's shipping charge is based on the weight of the item to be shipped and the distance it is to be shipped.

Distance (miles)	1–100	101–400	401–900	901–1500	>1500
Cost per pound	\$0.43	\$0.65	\$0.85	\$1.10	\$1.50

a Find the charge for shipping a 12-pound package 750 miles. \$10.20
b Find the charge for shipping a 7.5-pound package 340 miles. \$4.88
c Find the charge for shipping a 15-pound package 1240 miles. \$16.50
d If the shipping charge for a package is \$12, what are the possible values of its weight and the shipping distance?
27.9 lb, 1–100 mi; 18.5 lb, 101–400 mi; 14.1 lb, 401–900 mi; 10.9 lb, 901–1500 mi; 8 lb, > 1500 mi

R **20** Solve each system for x and y.

a $\begin{cases} 2x + 3y = -4 \\ 4x - 5y = 8 \end{cases}$ **a** $(x, y) = \left(\frac{2}{11}, \frac{-16}{11}\right)$ **b** $(x, y) = \left(\frac{3}{7}, \frac{23}{7}\right)$

b

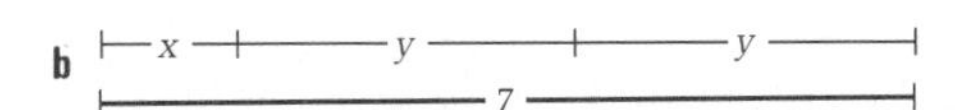

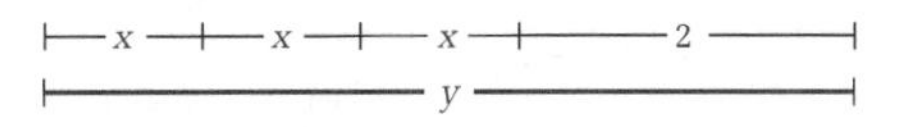

Problem-Set Notes and Strategies, continued

- As another example of conjugate problems in **15**, you might have students find the product of $(x + 7)(-x + 7)$ as conjugates. This is also the difference of squares in a slightly different context.
- Students may forget to include the diameter as part of the perimeter in problem **18**.
- Problem **19** is a review and application of compound functions. Part **d** points out that answers may not be unique for a given output, in other words this function is not one-to-one.

Problem-Set Notes and Strategies, continued

- In problem **21**, students use the zero product property to solve equations. Part **d** emphasizes that it is impossible to factor the sum of squares in the real numbers.
- In problem **23**, students discover that it is impossible to take square roots of negative numbers in the real number system. Since, some students may have calculators that can do complex arithmetic, there may be some unusual calculator results.
- It can be interesting to see the various ways that students divide the shaded area in problem **28** to compute its value. If you have the time, it is interesting to prove that they are all equivalent.

Additional Answers

25a, 25b

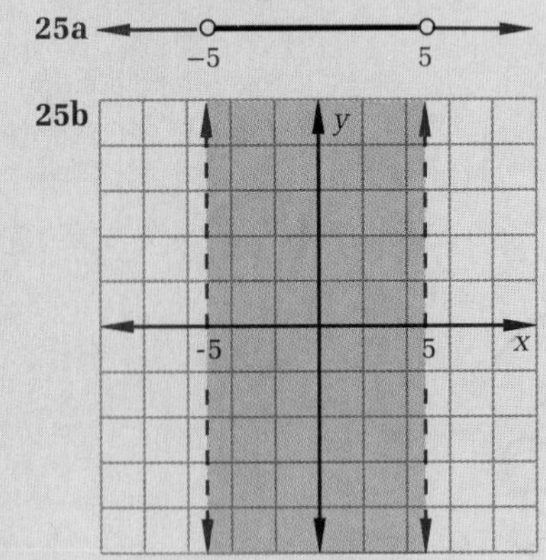

Problem Set, *continued*

B

R,I **21** Solve each equation for x.

a $(3x - 9)(x + 4) = 0$ $x = 3$ or $x = -4$

b $6x^3(15x - 5) = 0$ $x = 0$ or $x = \frac{1}{3}$

c $9(x^2 - 9)(x + 5) = 0$ $x = 3$, $x = -3$, or $x = -5$

d $(x^2 + 4)(x - 3) = 0$ $x = 3$

R,I **22** If $(a + 4)x^2 + bx + 10 = 5x^2 + (2b - 7)x + c$, what are the values of a, b, and c? 1; 7; 10

I **23** Use a calculator to evaluate $\sqrt{x - 7}$, substituting the values {9, 8, 7, 6, 5} for x. Explain your results. ≈1.414, 1, 0, error, error; the calculator works only with real numbers.

R **24** Write an equation expressing the distance d traveled by a car moving at 60 miles per hour as a function of the time t the car travels. Is this function linear? $d = 60t$; yes

R **25** **a** Graph $|x| < 5$ on a number line.

b Graph $|x| < 5$ on a rectangular coordinate system.

R **26** Students in a chemistry class at Tech High determined the number N of grams of a certain salt that would dissolve in 150 grams of water at various Celsius temperatures C. Use their data to estimate the value of N at 60°C. 111 g

C	10	18	30	34	44	52
N	30	41	64	68	83	99

P **27** Write an equation for a circle with its center at (−2, 5) and a radius of 6. $(x + 2)^2 + (y - 5)^2 = 36$

P,R **28** Write an expression for the area of the shaded region. $x^2 + x$

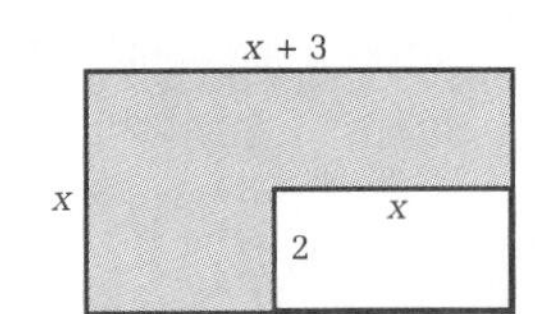

R **29** Find the values of b and b + 3 for which the ordered pairs are linearly related. 3.5; 6.5

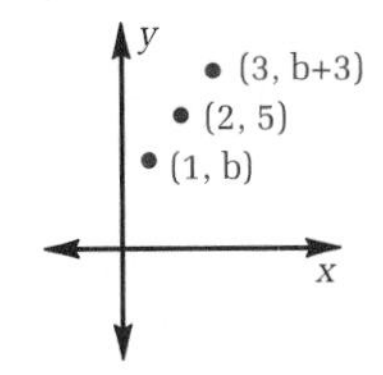

R **30** Solve for w_1.

a $|w_1 - 4| = 5$ $w_1 = 9$ or $w_1 = -1$

b $|-3w_1|\, w_1 = 12$ $w_1 = 2$

R **31** **a** For what values of a and b is $|a + b| < |a| + |b|$? $\{(a, b): ab < 0\}$

b For what values of a and b is $|a + b| = |a| + |b|$? $\{(a, b): ab \geq 0\}$

I **32** Use a computer or a graphing calculator to graph the following equations.

a $y = x^2$ **b** $y = x^2 + 3$ **c** $y = (x + 3)^2$

R **33** Graph the line segment represented by $y = 5 - 2x$ and $1 \le y \le 7$.

P,R **34** Write an equation stating that the rectangle and the square have

a Equal areas $10x^2 + 2x = 16x^2$

b Equal perimeters $16x = 14x + 2$

R **35** The sum of the first n natural numbers is equal to $\frac{n(n + 1)}{2}$.

a If the sum of the first n natural numbers is 5050, what is the value of n? 100

b Find n if the average of the first n natural numbers is 2500. 4999

R **36** The accounting department at Widgets, Inc., reported the following data on the cost c (in dollars) of producing x gidgets.

x	0	100	200	300	400
c	700	2600	4500	6500	8350

a Use the method of least squares to find an equation of the line of best fit. $c = 19.2x + 690$

b Use your answer to part **a** to estimate the cost of producing 800 gidgets. $16,050

c It is not reasonable to assume that your formula will be an accurate predictor of the production cost for any number of gidgets. Why not? Answers will vary. More machinery and workers might be needed, so that the relationship would no longer be linear.

C R **37** Find the missing value. 12 (for $y = 2$) or 204 (for $y = -14$)

$$\begin{bmatrix} x & 4 \\ 2 & y \end{bmatrix}^2 = \begin{bmatrix} 72 & -24 \\ -12 & ? \end{bmatrix}$$

R **38** Find the value of x for which 60% of the area of the square is in the shaded region. ≈ 3.68

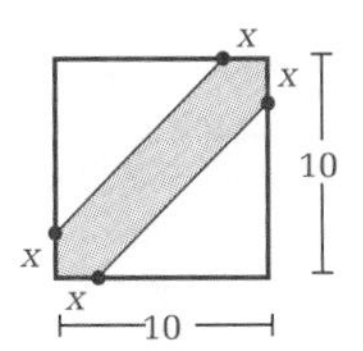

R,I **39** Refer to the graph.

a Write an expression for $d(x)$, the distance from (0, 0) to $(x, f(x))$. $\sqrt{10x^2 + 18x + 9}$

b For what value of x is $d(x)$ a minimum? -0.9

c What is the minimum value of $d(x)$? $\frac{3\sqrt{10}}{10}$

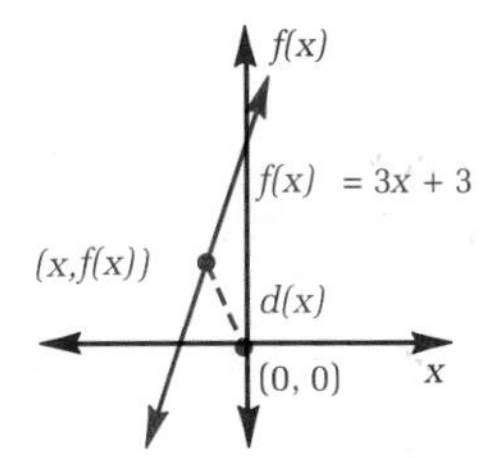

Problem-Set Notes and Strategies, continued

- Problem **32** is a preview of graphing parabolas which is discussed in Section **4.3**. Have students plot points and connect them with a smooth curve.
- In problem **36** spend some time going over the different answers that arise from part **c**. Explain that a model is only as good as the data upon which it is based.
- Problem **39** may be solved as a quadratic or by using slopes. If the distance formula is used, let students know that if you minimize the square of the distance you have also minimized the distance. Some students will remember from geometry that the shortest distance is that which is perpendicular to the line $3x + 3$ and use slopes to find the point.

Additional Answers

33

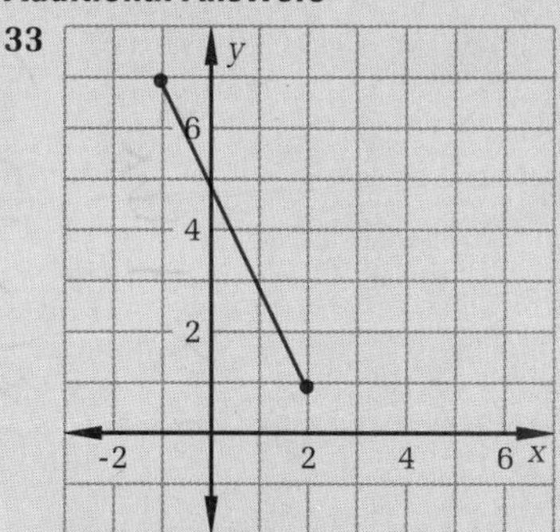

Answers for problems **32a – c** and **40** can be found in the answer pages beginning on **A 1**.

Problem-Set Notes and Strategies, continued

- Problem **41** provides a good analysis of compound functions as it might relate to prospective business students.

Additional Answer

41b

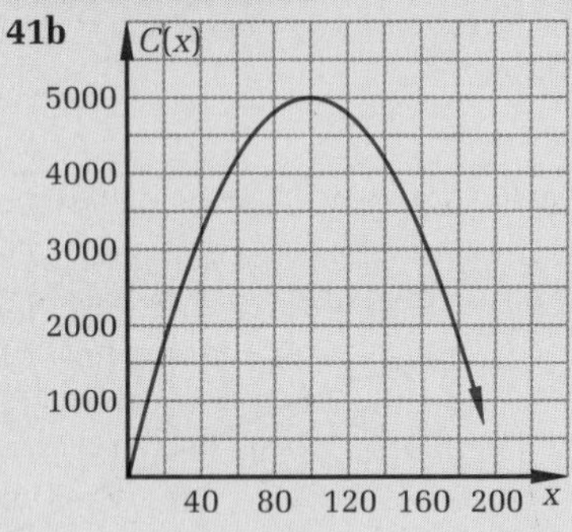

I **40** Graph the following functions on the same set of axes.

a $f(x) = 2x(2x)$
b $g(x) = (2x + 3)(2x - 3)$

R **41** If ten or fewer pieces of software are purchased, the cost of each piece of software is $95. If more than ten pieces of software are purchased, the cost per piece is reduced by $0.50 for each piece of software over ten.

a Write a compound function C for the cost of x pieces of software. $C(x) = \begin{cases} 95x \text{ if } x \le 10 \\ [95 - 0.50(x - 10)] \text{ if } x > 10 \end{cases}$
b Draw the graph of the cost function.
c For what number of pieces of software is the total cost the greatest? 100
d Is it cheaper to buy 100 pieces of software or 125 pieces of software? 125 pieces of software

HISTORICAL SNAPSHOT

ALGEBRAIC SHORTHAND

Syncopated algebra: the beginning of a standard notation system

Can you imagine algebra problems presented in words rather than symbols? That was originally how algebra problems were written and solved.

Diophantus (*c.* A.D. 255), considered by some to be the father of algebra, first proposed a system of algebraic notation in his treatise *Arithmetica*. His system of "syncopated algebra" was derived by dropping letters from words to form mathematical abbreviations.

Diophantus used abbreviations for powers of numbers, relationships, and operations. He replaced unknown numbers with letters and used notations for some exponents. A sample of his notation system, with English equivalents, is shown in the chart above.

Here is the English equivalent of an expression Diophantus wrote using his notation:

PP2 C3 x 5 M P4 U6

Idea Represented	Diophantus	English Equivalent
Unknown Number	ς	x
Square (Power)	Δ^{Υ}	P
Cube	κ^{Υ}	C
Fourth (Square, Square)	$\Delta^{\Upsilon}\Delta$	PP
Fifth (Square, Cube)	$\Delta\kappa^{\Upsilon}$	PC
Sixth (Cube, Cube)	$\kappa^{\Upsilon}\kappa$	CC
Minus	⋔	M
Unit	μ	U

Written in the symbolic notation we use today, the equation reads

$$2x^4 + 3x^3 + 5x - 4x^2 - 6, \text{ or}$$
$$2x^4 + 3x^3 - 4x^2 + 5x - 6$$

Use syncopated algebra to write equations for your classmates to solve. Remember that for Diophantus, this form of notation was state of the art!

4.2

FACTORING QUADRATICS

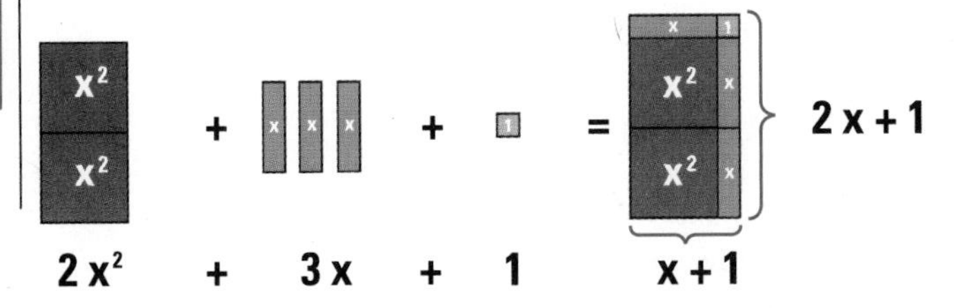

Objectives

After studying this section, you will be able to

- Remove common factors from expressions
- Factor trinomial products of linear binomials
- Factor special products

Part One: Introduction

Common Factors

In Section 4.1, you saw how two linear factors can be multiplied to form a quadratic expression. Mathematicians frequently reverse this process, starting with a product and trying to find the factors. ***Factoring*** is the inverse of multiplying.

Example *Find factors of* $9x^3 + 45x^2 + 54x$.

The terms have a ***common factor*** of 9x, which we can factor out by applying the Distributive Property of Multiplication over Addition in reverse.

$$\begin{aligned} 9x^3 + 45x^2 + 54x &= 9x(x^2) + 9x(5x) + 9x(6) \\ &= 9x(x^2 + 5x + 6) \end{aligned}$$

Factoring Trinomials

In the preceding example, we found two factors of $9x^3 + 45x^2 + 54x$. We may not, however, have *completely* factored the expression. Let's see whether the trinomial factor can itself be factored.

Example *Factor, if possible,* $x^2 + 5x + 6$.

The terms have no common factor. If the expression can be factored, it must be a product of binomials, $(x + \triangle)(x + \square)$, where $\triangle$ and $\square$ are numbers having a product of 6. The only possibilities are 1 and 6 or 2 and 3. (We could consider -1 and -6 or -2 and -3, but for these pairs, the coefficient of the linear term would be negative.) Since $6 + 1 = 7$ and $2 + 3 = 5$ and since the coefficient of the linear term is 5, we use 2 and 3.

$$x^2 + 5x + 6 = (x + 2)(x + 3)$$

4.2

Class Planning

Time Schedule

All levels: 2 days

Resource References

Teacher's Resource Book
Class Opener 21
Cooperative Learning 12
Practice 22

Class Opener

Carlos was doing his homework. In his book, the numbers for problem **23** had been replaced with a # sign. He looked in the back of the book to see the answer and figured out what the original problem must have been. Can you duplicate Carlos' feat? The problem said:

23 Multiply $(x + \#\#)(x + \#\#)$.

The answer in the back was

23 $x^2 + 12x + 32$.

$(x + 8)(x + 4)$

Lesson Notes

- Factoring is a helpful method, when possible, of solving equations. This section introduces factoring some trinomials. The first step is to factor out the greatest common factor. This isn't always necessary, but usually makes the problem simpler.
- If what remains is a trinomial it may be possible to factor into a product of binomials. This is generally a trial and error process. The way to succeed at factoring is to practice.

Vocabulary

factoring
common factor

Lesson Notes, continued

- If the polynomial that remains is a binomial, check to see if it is the difference of squares, a special case described in the previous section.
- Students frequently get discouraged when factoring trinomials because there is no one method that always works well. Remind students that practice will improve their skills.

Using Computers

Factoring Quadratics, a program in the *Algebra Solver Kit* disk, can be used to show students that the algorithm used to factor trinomials can be programmed into a computer. Other sources of factoring programs are the *Mathematics Exploration Toolkit* or *Derive*.

Stumbling Block

Students will frequently forget to factor out the greatest common factor; for example, they will factor $4x^2 - 64$ as the product $(2x-8)(2x+8)$ and not take out the 4 first. This can lead to errors in solving equations.

Communicating Mathematics

Have students explain the significance of factoring out the greatest common factor when factoring. Why is it necessary to factor "completely" when factoring polynomials?

An area diagram is another way of representing the multiplication of two binomials. In the figure shown, the area of the large rectangle is equivalent to the sum of the areas of the four smaller rectangles.

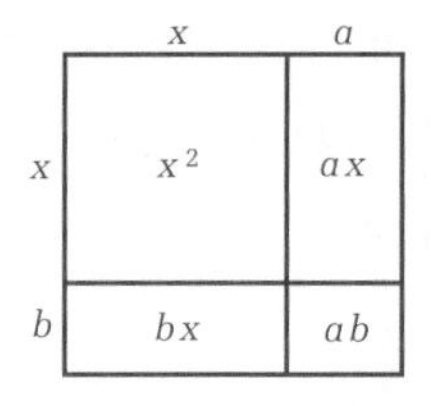

$$(x + b)(x + a) = x^2 + ax + bx + ab$$
$$= x^2 + (a + b)x + ab$$

In the preceding example, $a + b = 5$ and $ab = 6$, so the numbers 2 and 3 are the correct constant terms in the binomial factors.

Special Quadratics

The special forms of multiplication presented in Section 4.1 suggest special forms of factoring.

Example *Factor the quadratic* $81x^2 - 16$.

This is a difference of two squares.

$$81x^2 - 16 = (9x)^2 - 4^2$$
$$= (9x + 4)(9x - 4)$$

It is useful to know that the factors of a difference of two squares are always the sum and the difference of the square roots of its terms.

The pattern that you have used to square a binomial,

$$(a + b)^2 = a^2 + 2ab + b^2$$

can be used in reverse to factor perfect-square trinomials.

Example *Factor the quadratic* $4x^2 + 12x + 9$.

We can use an area diagram to illustrate our solution.

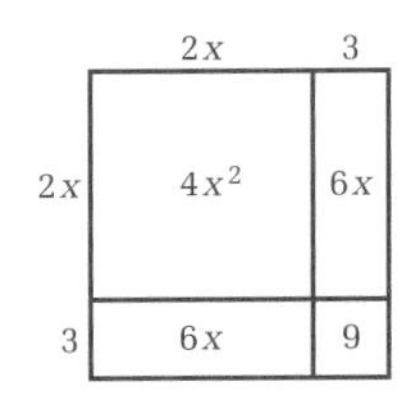

$$4x^2 + 12x + 9 = (2x)^2 + 2(2x)(3) + 3^2$$
$$= (2x + 3)^2$$

Part Two: Sample Problems

Problem 1 *Factor the quadratic* $3x^2 - 48$.

Solution Since the terms have a common factor of 3, we will remove this factor first.

$$3x^2 - 48 = 3(x^2 - 16)$$

Now $x^2 - 16$ can be factored as a difference of two squares.

$$3(x^2 - 16) = 3(x + 4)(x - 4)$$

Problem 2 *Factor the quadratic* $x^2 - 10x - 24$.

Solution Since the constant term, -24, is negative, we need to express it as the product of one positive and one negative factor. Let's look at possible factors of 24 and their differences.

Factors	Difference
24 and 1	23
12 and 2	10
8 and 3	5
6 and 4	2

The coefficient of the linear term is -10, so we use 2 and -12.

$$x^2 - 10x - 24 = (x - 12)(x + 2)$$

Note If the middle term of the trinomial had been $+10x$, the signs in the binomial factors would have been reversed.

$$x^2 + 10x - 24 = (x + 12)(x - 2)$$

Problem 3 *Factor the quadratic* $2x^2 - 5x - 12$.

Solution There is no common factor. The coefficient of the quadratic term is 2, so the binomial factors will be of the form

$$(2x + \triangle)(x + \square)$$

The pair of factors of -12 might be -1 and 12, 1 and -12, -2 and 6, 2 and -6, -3 and 4, or 3 and -4. We could try each pair, but we can use logic and what we know about binomial factors to find the correct one. We need numbers that will give us a sum of -5 (the coefficient of the linear term) when one of them is added to the product of the other and 2 (the coefficient of the quadratic term).

$$2 \cdot 3 + (-4) = 2 \quad \text{No}$$
$$2 \cdot (-4) + 3 = -5 \quad \text{Yes}$$

Therefore, the factored form of $2x^2 - 5x - 12$ is $(2x + 3)(x - 4)$. We can check this result by multiplying.

$$(2x + 3)(x - 4) = 2x^2 - 8x + 3x - 12$$
$$= 2x^2 - 5x - 12$$

Part Three: Exercises and Problems

Warm-up Exercises

1 Match each quadratic expression with the method or methods you would use to factor it completely.

a $4x^2 + 64$ i
b $4x^2 - 64$ i, iii
c $x^2 - 26x + 25$ ii
d $4x^2 - 81$ iii
e $16x^2 - 20x - 6$ i and ii
f $x^2 - 10x + 25$ iv

i Removing a common factor
ii Trinomial factoring
iii Difference-of-squares factoring
iv Perfect-square-trinomial factoring

P

Checkpoint

1 Factor $4x + 28$, if possible. $4(x + 7)$
2 Factor $x^2 + 7x - 3$, if possible. Prime.
3 Factor $4x^2 - 4x + 1$, if possible. $(2x - 1)^2$

Assignment Guide

Average
Day 1 10–13, 15, 17, 20, 21, 24, 26, 28, 29, 33, 35–37, 39
Day 2 14, 16, 18, 19, 22, 23, 25, 30, 31, 34, 38, 40a, b, 42, 43

Advanced
Day 1 10–14, 16–20, 24, 25, 27, 29, 31, 33, 35, 36, 38, 39
Day 2 21–23, 26, 28, 30, 32, 34, 37, 40a, b, 41, 43, 44, 46, 48

Honors
Day 1 14, 17–20, 26, 28, 30–34, 36, 38, 40, 43, 44
Day 2 13, 25, 29, 35, 37, 39, 41, 44–51

Problem-Set Notes and Strategies

- Remind students that the first step in factoring is to take out the greatest common factor. This is the only way to factor part **a** of problem **1** and makes part **b** easier to factor.

Problem-Set Notes and Strategies, continued

- In problem **8**, the difference of squares is imbedded in another difference of squares problem.

- An interesting extension for problem **16** is to prove that 0.9999.... = 1. An alternative approach to the proof is included as a cooperative learning assignment.

Cooperative Learning

Problem **16** in this section leads to an interesting lesson in infinity. Have students generalize that since $\frac{7}{9} = 0.7777...$ and $\frac{8}{9} = 0.8888...$, then $\frac{9}{9}$ should $= 0.9999...$ This is interesting because $\frac{9}{9} = 1$ and now equals 0.9999... Have students prove this with the common algorithm to change repeating decimals to fractions. As an alternative, have them work out the decimal expansion of $\frac{1}{3}$ and $\frac{2}{3}$. Add the two fractions together to get 1 and add the decimal expansions together to get 0.9999... This will be proven again in Section **13.4** using a geometric series.

P **In problems 2–4, factor each expression.**

2 $x^2 + 7x + 12$ $(x + 3)(x + 4)$ **3** $x^2 + 14x + 49$ $(x + 7)^2$ **4** $8x^2 - 10x - 3$ $(4x + 1)(2x - 3)$

P,R **5** Write an expression for the length of a side of the square shown. $3x + 2$

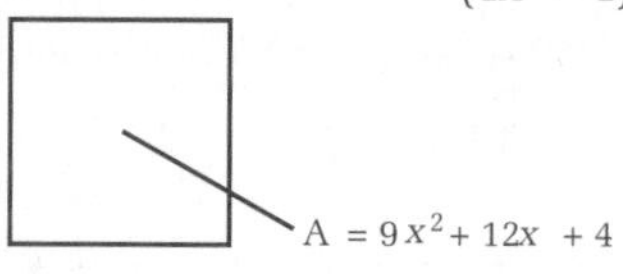

P **In problems 6–8, factor each expression completely.**

6 $5x^2 - 30x + 45$ $5(x - 3)^2$ **7** $2x^2 + 4x - 6$ $2(x + 3)(x - 1)$ **8** $x^4 - y^4$ $(x^2 + y^2)(x + y)(x - y)$

P,R **9** Express the area of the shaded region as a product of linear factors. $\pi(a + b)(a - b)$

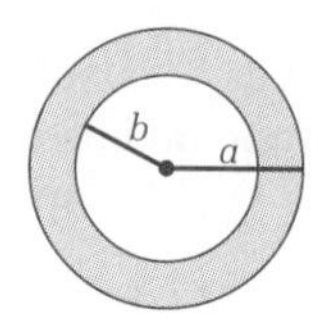

Problem Set

A

In problems 10–12, factor out the common factor.

P **10** $12y^3 + 16y^2 + 8y$ $4y(3y^2 + 4y + 2)$ **11** $21y^2 + 28$ $7(3y^2 + 4)$ **12** $4x^3 + 7x$ $x(4x^2 + 7)$

P **13 a** Copy the area diagram shown, supplying the missing terms.

b Use the diagram to find two expressions for the area of the large rectangle. $(2x + 5)(3x + 4)$; $6x^2 + 23x + 20$

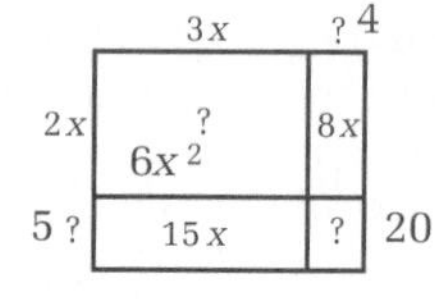

P,R **14** Indicate whether each statement is True or False.

a $(3x + 5)(2x - 4) = 6x^2 - 14x - 20$ False

b $(3x + 5)(2x + 4) = 6x^2 + 14x + 20$ False

R **15** The price of a shirt is \$11.93, and the sales-tax rate is 6.5%. How much sales tax must a purchaser of the shirt pay? What will the total cost of the shirt be? \$0.78; \$12.71

R 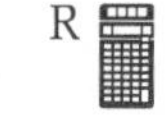 **16 a** Use a calculator to find the decimal equivalents of $\frac{1}{9}$, $\frac{2}{9}$, $\frac{3}{9}$, and $\frac{4}{9}$. $0.\overline{1}$, $0.\overline{2}$, $0.\overline{3}$, $0.\overline{4}$

b Predict the decimal equivalents of $\frac{7}{9}$, $\frac{8}{9}$, and $\frac{9}{9}$. $0.\overline{7}$, $0.\overline{8}$, 1

P,R **In problems 17–19, find the largest perfect-square factor of each term.**

17 144 144 **18** $200x^3$ $100x^2$ **19** $32x^6$ $16x^6$

P,R **20** Let $f(x) = (2x - 3)(2x + 3)$. What are the constant, linear, and quadratic terms of function f? -9; 0; $4x^2$

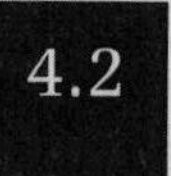

R **In problems 21–23, simplify each expression.**

21 $(3x)^2 - 3x^2$ $6x^2$ **22** $5x \cdot 3x - 15x^2$ 0 **23** $8x - (5x + 6)$ $3x - 6$

R **24** Expand each product.

a $(a + b)(c + d)$ $ac + ad + bc + bd$ **b** $(x + 2)(y - 3)$ $xy + 2y - 3x - 6$

P **25 a** Write an equation that expresses the perimeter P of the rectangle as a function of x and y. $P = 6x + 4y - 8$

b What is the value of P if $(x, y) = (5, 2)$? 30

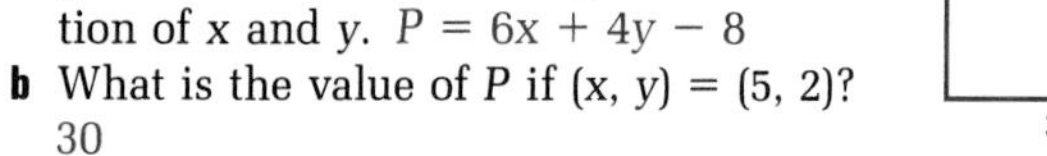

B R,I **26 a** Copy and complete the table of values for $y = 6x^2 + x - 3$.

x	−4	−3	−2	−1	0	1	2	3	4
y	89	48	19	2	−3	4	23	54	97

b Sketch the graph of $y = 6x^2 + x - 3$.

R **27** Solve the following matrix equation for w, x, y, and z.

$$\begin{bmatrix} |x - 3| & 14 \\ -12z & |w| + 4 \end{bmatrix} = \begin{bmatrix} 7 & 2|y + 5| \\ -|2| & 7 \end{bmatrix}$$

$w = \pm 3$; $x = 10$ or $x = -4$; $y = 2$ or $y = -12$; $z = \frac{1}{6}$

P **In problems 28–30, factor each expression completely.**

28 $49w^2 - 196$ $49(w + 2)(w - 2)$ **29** $196x^2 + 49w^2$ $49(4x^2 + w^2)$ **30** $196x^2 + 42xw + 49w^2$ $7(28x^2 + 6xw + 7w^2)$

I **31** Use a computer or a graphing calculator to graph each equation.

a $y = x^2$ **b** $y = -x^2$ **c** $y = 3x^2$ **d** $y = -3x^2$

P,R **32 a** Expand $(a - b)^3$. $a^3 - 3a^2b + 3ab^2 - b^3$

b Expand $(a - b)(a^2 + 2ab + b^2)$. $a^3 + a^2b - ab^2 - b^3$

c Expand $(a - b)(a^2 - 2ab + b^2)$. $a^3 - 3a^2b + 3ab^2 - b^3$

d Which two expressions are equivalent? Explain. **a** and **c**; **c** is a partially expanded form of **a**.

R,I **33 a** Find the x-intercepts of the graph shown. −5, −1, 5

b Solve $f(x) = 0$ for x. $x = -5$, $x = -1$, or $x = 5$

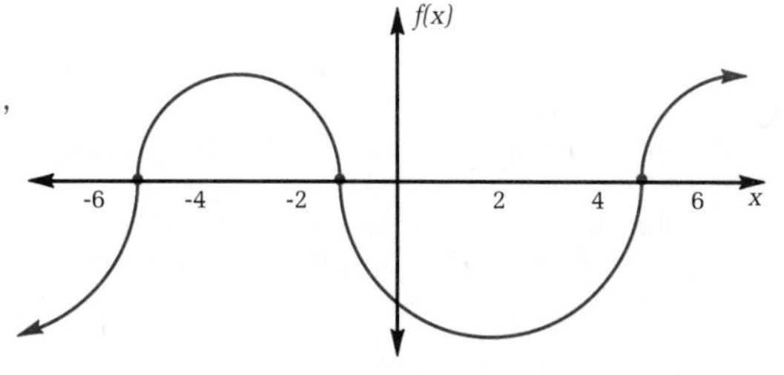

P **34** Factor each expression.

a $[f(x)]^2 - [g(x)]^2$ $[f(x) + g(x)][f(x) - g(x)]$ **b** $(x + 3)^2 - 16$ $(x + 7)(x - 1)$ **c** $x^2 + 6x - 7$ $(x + 7)(x - 1)$

Problem-Set Notes and Strategies, continued

- Students may forget that the 3 is part of the base in $(3x)^2$ in problem **21**, leading to an incorrect answer. Have the students identify base and exponent.
- In problem **31**, try to have students identify both the reflections and dilations that occur in the four parts.
- If students can do part **a** of problem **34**, it is easier to convince them of the validity of part **b** using the difference of squares. If not, square the binomial and then factor the simplified trinomial.

Additional Answer

26b

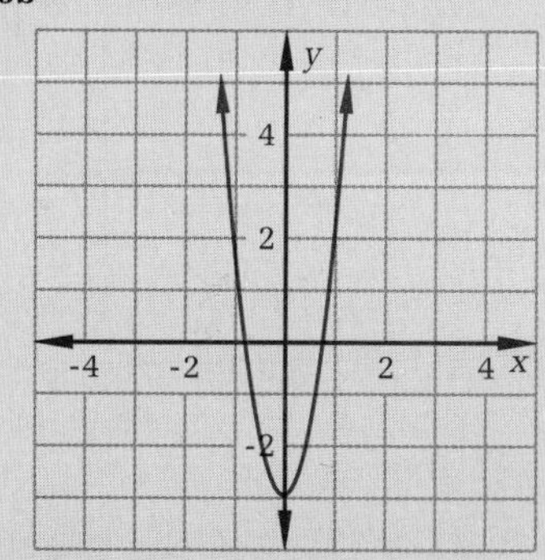

Answers to problems **31a**, **b**, **c**, and **d** can be found in the answer pages beginning on **A** 1.

Problem-Set Notes and Strategies, continued

- Problem **37** attempts to connect solving equations and factoring. In part **a** you could calculate the square root of both sides or subtract 9 from each side and factor the result. Both methods achieve the same answers.

Additional Answers

40a

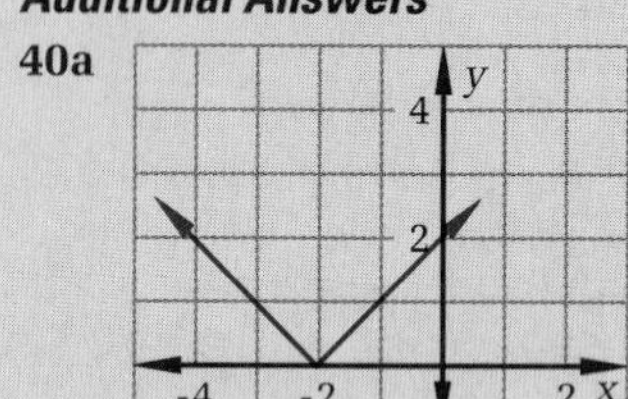

40b

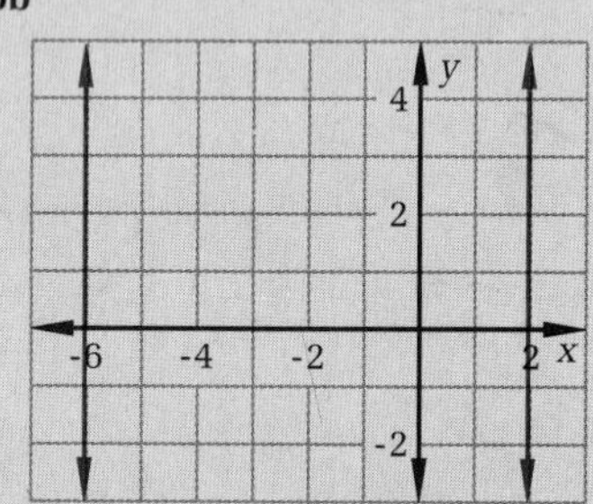

40c

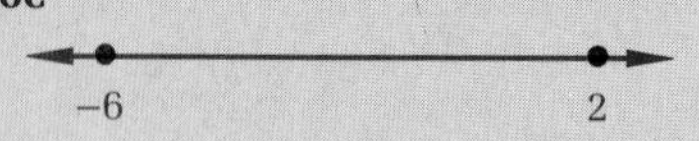

40d

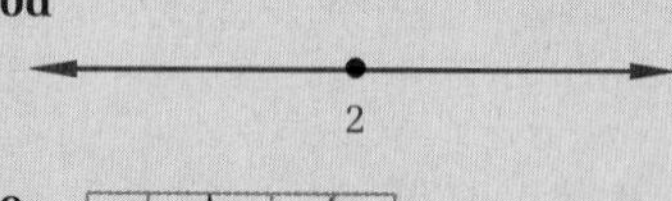

40e

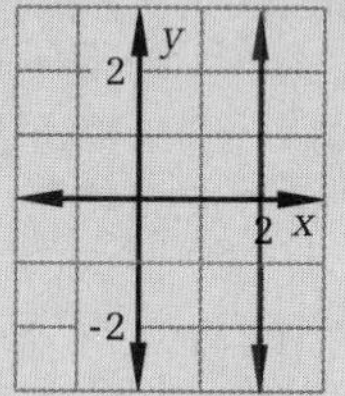

Problem Set, *continued*

R **35** **a** Explain why any even number can be written in the form $2k$, where k is an integer. Evens are divisible by 2.
b Explain why the sum of any two even numbers is even. $2a + 2b = 2(a + b)$

R **36** A point in square ABCD is selected at random. Find the probability that the point is in the shaded region. $\frac{8}{9}$

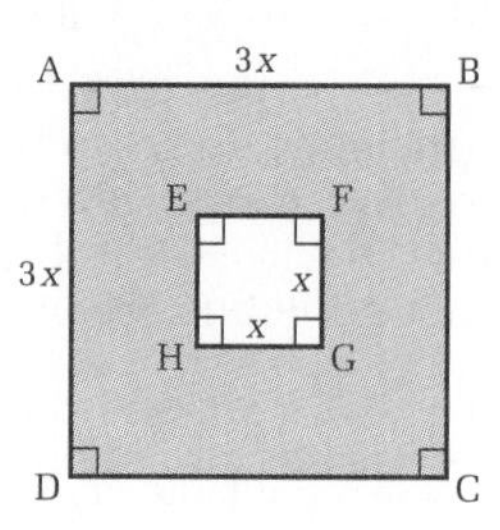

P **37** **a** Solve $x^2 = 9$ for x. $x = 3$ or $x = -3$
b Solve $y^2 = 9$ for y. $y = 3$ or $y = -3$
c Solve $(x - 3)^2 = 9$ for $(x - 3)$. $x - 3 = 3$ or $x - 3 = -3$
d Solve $|x| = 3$ for x. $x = 3$ or $x = -3$

I **38** The equation of the axis of symmetry of the graph shown is $x = 66.7$. Is $f(66)$ or $f(67)$ greater in value? Explain your answer. $f(67)$; since (67, 0) is closer to the axis of symmetry, $f(67)$ is closer to the vertex.

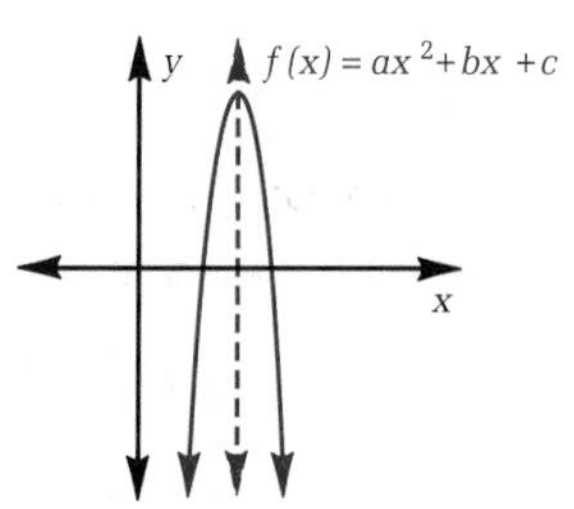

P,R **39** A rectangular rose garden's length is twice its width. The garden is surrounded by a 3-foot-wide walkway having a total area of 99 square feet. Find the dimensions of the rose garden. 3.5 ft by 7 ft

R **40** Graph each set.
a $\{(x, y): y = |x + 2|\}$
b $\{(x, y): 4 = |x + 2|\}$
c $\{x: 4 = |x + 2|\}$
d $\{x: 4 = x + 2\}$
e $\{(x, y): 4 = x + 2\}$

P **41** Factor each expression completely.
a $x^4 - x^2$ $x^2(x + 1)(x - 1)$
b $x^4 - y^2$ $(x^2 + y)(x^2 - y)$
c $x^4y^4 - y^4$ $y^4(x^2 + 1)(x + 1)(x - 1)$

R,I **42** If $g(x) = x^2 + 2x + 1$ and $g(x) - f(x) = x^2$, what is $f(x)$? $f(x) = 2x + 1$

R **43** Both Orville's plane and Wilbur's plane fly 140 miles per hour when there is no wind. Orville flew from Kitty to Hawk in 8 hours with a 10-mile-per-hour head wind. Wilbur flew the same route in 8 hours 40 minutes with a 20-mile-per-hour head wind. What is the distance from Kitty to Hawk? 1040 mi

R,I **44** Refer to the following table of values.

x	0	2	4	6	8	10
y	3	4	6	8	9	11

a Find x_1, the average of the first three values of x. 2
b Find y_1, the average of the first three values of y. $\frac{13}{3}$
c Find (x_2, y_2), the averages of the last three values of x and y. $\left(8, \frac{28}{3}\right)$
d Write an equation of the line connecting the points (x_1, y_1) and (x_2, y_2). (This method of finding a fitted line is known as the *method of group averages.*) $y = \frac{5}{6}x + \frac{8}{3}$

R **45** Factor $x^4 + 6x^2 + 9 - y^2$. $(x^2 + 3 - y)(x^2 + 3 + y)$

P,R **46** **a** Write a function A to represent the area of rectangle ABCD.
b What is the domain of function A?
c For what value of x is the area of ABCD half the area of the original square?
a $A(x) = 100 - x^2$ b $\{x{:}0 < x < 10\}$ c $5\sqrt{2}$

P,R **47** Solve the following equation for x.
$(x - 1)8x^2 + (x - 1)(-2x) + (-3)(x - 1) = 0$ $x = 1$, $x = -\frac{1}{2}$, or $x = \frac{3}{4}$

C R **48** Solve the following system for (x, y). $(x, y) = (1, 2)$
$$\begin{cases} 3^{x+y} = 27 \\ 3^{x-y} = \frac{1}{3} \end{cases}$$

R **49** Consider the following compound function.
$$f(x) = \begin{cases} 1 - x \text{ if } 1 \le x < 2 \text{ or } 3 \le x \le 4 \\ x \text{ if } 0 \le x < 1 \text{ or } 2 \le x < 3 \end{cases}$$
a Find $f\left(\frac{1}{2}\right)$, $f(2.5)$, $f(3.5)$, and $f(4.5)$. $\frac{1}{2}$; 2.5; -2.5; undefined
b Graph $y = f(x)$.

R **50** If the ratio of the area of rectangle ABCF to the area of rectangle CDEF is 3:1, what is the value of x? $\frac{1}{2}$

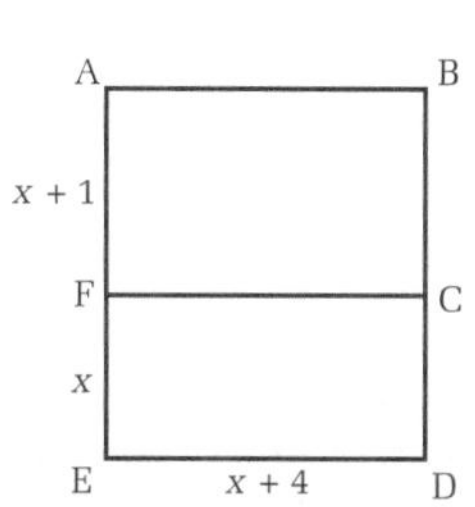

P,R **51** Sketch a graph showing all the values of (x, y) for which $(x - y)^2 > (x + y)^2$.

Problem-Set Notes and Strategies, continued

- Problem **44** illustrates a less frequently used method of fitting lines to data. You might want to compare these two methods to find the accuracy of the second method. Because so many calculators and computers have regression available, it is easy to see why this group averages method is not used much anymore.
- In problem **45**, the four term polynomial requires the student to recognize that if they group terms, the expression can be factored. First, factor the trinomial into a binomial squared and then into the difference of squares.
- Factoring the GCF is illustrated in problem **47**. If this were multiplied out it would yield a cubic polynomial and be more difficult to factor.
- Problem **48** requires that each part of the equations be written with the same base and then the exponents set equal.

Additional Answer

49b

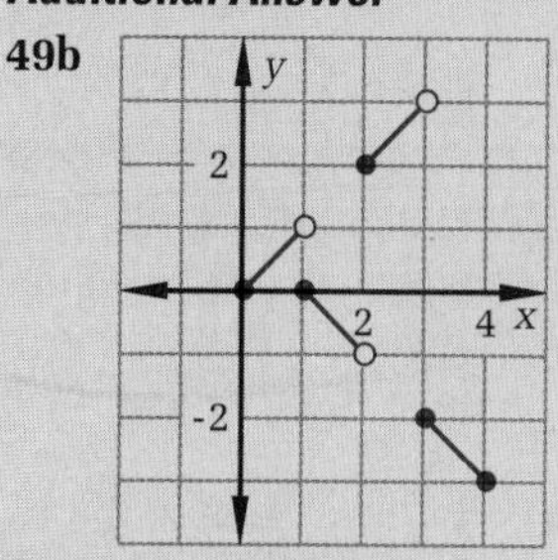

Answer to problem **51** can be found in the answer pages beginning on **A** 1.

4.3

Class Planning

Time Schedule

All levels: 2 days

Resource References

Teacher's Resource Book
- Class Opener 22
- Cooperative Learning 13
- Practice 23
- Enrichment 7

Evaluation
- Quiz Forms 1, 2, 3

Class Opener

1. Factor $f(x) = x^2 - 5x - 24$.
   $(x - 8)(x + 3)$
2. Use the factored form to make a table of values and sketch the graph of the quadratic function in part **1**.

x	y
−3	0
−2	−10
−1	−18
0	−24
1	−28
2	−30
3	−30
4	−28
5	−24
6	−18
7	−10
8	0

Lesson Notes

- Many of the properties of quadratic graphs are introduced in this chapter. The vertex is the maximum or minimum point on a parabola depending on the sign of the coefficient for the quadratic term. When defining a linear function, it was noted that the slope remained constant.

4.3 Graphing Quadratic Functions

Objectives

After studying this section, you will be able to
- Recognize quadratic functions and their graphs
- Relate characteristics of a quadratic function to characteristics of its graph
- Find the real zeros of quadratic functions

Part One: Introduction

Graphing Quadratic Functions

You are familiar with the graphs of functions of the form $f(x) = mx + b$. Such functions are called linear functions because their graphs are lines. In this section, you will be introduced to ***quadratic functions*** and their graphs.

A ***quadratic function*** is a function that can be written in the form $f(x) = ax^2 + bx + c$, where a, b, and c are real numbers and $a \neq 0$. Let's construct a table of values for the simple quadratic function $f(x) = x^2$ and then plot the corresponding points.

x	−5	−4	−3	−2	−1	0	1	2	3	4
$f(x)$	25	16	9	4	1	0	1	4	9	16

The graph of this function is a U-shaped curve called a ***parabola***.

Notice that the parabola opens upward and that the point (0, 0) is lower than any other point on the graph. This point is the parabola's ***vertex.*** It represents the minimum value of the function. When a parabola opens downward, its vertex represents the maximum value of the function.

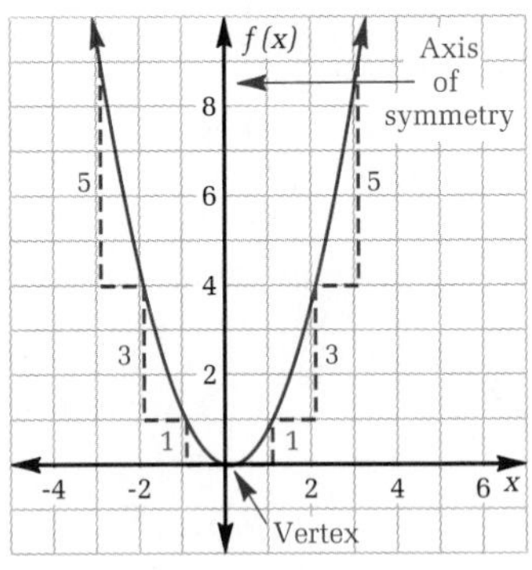

Let's look at the pattern of the values in the table. Notice that as the x values increase and decrease by 1 to the right and the left of the vertex, the values of $f(x)$ change in the pattern 1, 3, 5, 7, . . .

The line that divides the parabola into two halves that are mirror images of each other is called the graph's ***axis of symmetry.*** The axis of symmetry of this parabola has the equation $x = 0$.

Vocabulary
quadratic function
parabola
vertex
axis of symmetry
vertex form
real zero
root

More About Parabolas

As you have seen, determining the slope and the y-intercept of a linear function allows us to make a quick sketch of the function's graph. In the same way, when we want to sketch the graph of a quadratic function, we can rewrite the function in a form that gives us important information about its graph.

Example *Graph the quadratic function* $f(x) = x^2 + 6x + 9$.

We begin by writing the function in factored form.

$$\begin{aligned} f(x) &= x^2 + 6x + 9 \\ &= x^2 + 2(x)(3) + 3^2 \\ &= (x + 3)^2 \end{aligned}$$

By examining the squared binomial, we see that $x + 3 = 0$ when $x = -3$, so we choose values of x to the left and the right of -3.

x	−6	−5	−4	−3	−2	−1	0
f(*x*)	9	4	1	0	1	4	9

The point $(-3, 0)$ is the vertex of the parabola, and the axis of symmetry is $x = -3$. The values of $f(x)$ increase in the pattern 1, 3, 5, 7, . . .

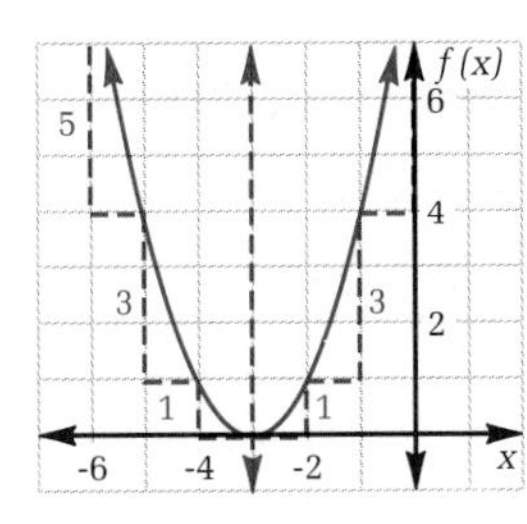

Example *Graph* $f(x) = x^2 + 10x + 29$.

We need to rewrite the quadratic expression so that it contains a perfect-square trinomial.

$$\begin{aligned} f(x) &= x^2 + 10x + 29 \\ &= x^2 + 10x + 25 + 4 \\ &= (x + 5)^2 + 4 \end{aligned}$$

Since $x + 5 = 0$ when $x = -5$, we choose values to the left and the right of -5.

x	−8	−7	−6	−5	−4	−3	−2
f(*x*)	13	8	5	4	5	8	13

When we plot the points, we see that the parabola's vertex is $(-5, 4)$ and its axis of symmetry is $x = -5$. Once again, the values of $f(x)$ change in the pattern 1, 3, 5, 7, . . .

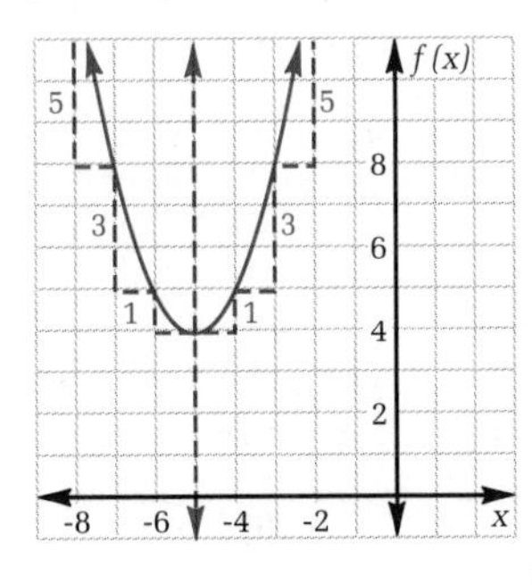

Lesson Notes, continued

When looking at a parabola, the slopes between points on the curve are changing, not constant. There is a pattern to the slopes which will be explored throughout the chapter.

- Symmetry is also noted in the case of a parabola. A cooperative learning lesson is included in this section to expand on the symmetry concept for other graphs as well.

Using Computers

Graphing quadratic functions lends itself to a discovery lesson using the *Omnifarious Plotter*, *Mathematics Exploration Toolkit*, *Derive*, the *Function Analyzer*, or a graphing calculator. Let students vary the parameters of the quadratic function and discover the effects on the graph. Use trace and zoom to determine the roots of the extremes of quadratic functions. Using graphing software or a graphics calculator to solve quadratic equations allows the student to concentrate on setting up the problem. Technology can do the actual solving. In addition, *Derive* or a graphics calculator can help students focus on finding approximate solutions.

4.3

Communicating Mathematics

Parabolas exhibit a certain symmetry through the vertex. Write a paragraph explaining why recognizing symmetry before graphing an equation can make the task easier.

A quadratic function written in the form $f(x) = a(x - h)^2 + k$ is said to be written in ***vertex form,*** since the ordered pair (h, k) identifies the vertex of the function's graph. When a is positive, the parabola opens upward; when a is negative, the parabola opens downward. As the values of x change by 1 to the left and the right of the vertex, the values of $f(x)$ change in the pattern $1a$, $3a$, $5a$, $7a$, . . .

Zeros of Quadratic Functions

Now let's study the x-intercepts of the graph of a quadratic function and see how they are related to a corresponding quadratic equation.

Let's find the x-intercepts of $f(x) = 3\left(x - \frac{1}{3}\right)^2 - \frac{4}{3}$. We can see that the parabola opens upward and $\left(\frac{1}{3}, -\frac{4}{3}\right)$ is the vertex (the minimum point). As the values of x change by 1 to the left and the right of the vertex, the values of $f(x)$ will increase in the pattern 3(1), 3(3), 3(5), . . .

Examining the graph, we find that the x-intercepts are (1, 0) and $\left(-\frac{1}{3}, 0\right)$—that is, the function is equal to zero when $x = 1$ and $x = -\frac{1}{3}$. These values of x are called the ***real zeros*** of the function and correspond to the real ***roots*** (solutions) of the equation $f(x) = 0$.

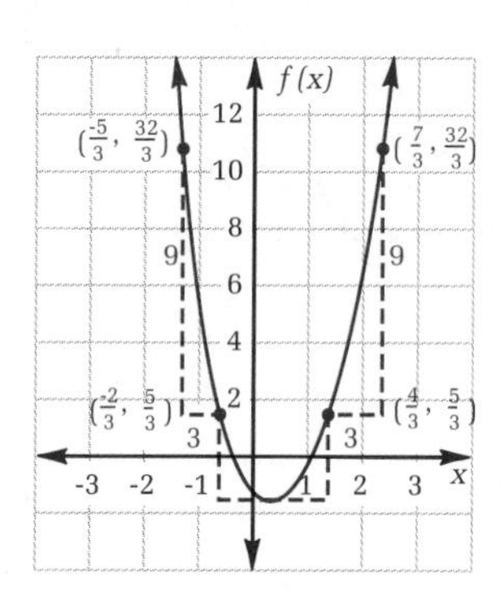

Part Two: Sample Problems

Problem 1 *Find the real zeros of the function* $h(x) = x^2 - 10x + 25$.

Solution We first rewrite the function in vertex form so that we can identify the vertex and the pattern of change in the values of $h(x)$.

$$h(x) = x^2 - 10x + 25$$
$$= (x - 5)^2$$

The vertex of this parabola is (5, 0), and the axis of symmetry is $x = 5$. The values of $h(x)$ change in the pattern 1, 3, 5, . . . The graph intersects the x-axis at only one point, (5, 0). Therefore, 5 is the only zero of the function.

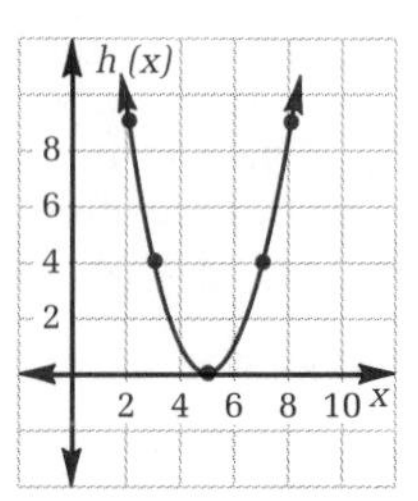

Problem 2 *Graph* $f(x) = -\frac{1}{2}x^2$.

Solution We can rewrite the function in vertex form as $f(x) = -\frac{1}{2}(x + 0)^2 + 0$. The vertex is (0, 0), and the axis of symmetry is $x = 0$. The values of $f(x)$ will change in the pattern $-\frac{1}{2}(1)$, $-\frac{1}{2}(3)$, $-\frac{1}{2}(5)$, . . . to the left and the right of the vertex. The parabola opens downward. Using all this information, we can sketch a graph without constructing a table of values.

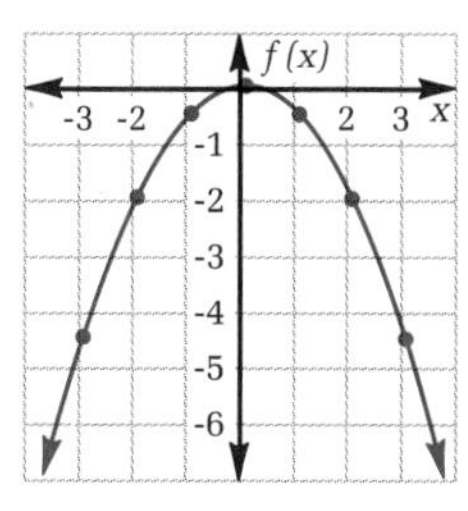

Problem 3 *Solve the garden problem presented at the beginning of Section* 4.1.

Solution We want to find the dimensions that will produce a garden with the maximum area. The function expressing the area of the garden is $f(x) = 75x - 2x^2$.

Let's solve the problem by determining the characteristics of the function's graph. First, let's find the function's zeros by letting $f(x) = 0$ and using the Zero Product Property.

$$f(x) = 75x - 2x^2$$
$$0 = x(75 - 2x)$$
$$x = 0 \text{ or } 75 - 2x = 0$$
$$x = 37.5$$

Thus, the x-intercepts of the function's graph are 0 and 37.5.

Since parabolas are symmetrical, the x-coordinate of the parabola's vertex will be midway between the two intercepts. The x-coordinate of the vertex is therefore $\frac{0 + 37.5}{2}$, or 18.75. The vertex represents the maximum value of the function, so the garden's area will be greatest when $x = 18.75$—that is, when the garden is 18.75 feet wide and $75 - 2(18.75)$, or 37.5, feet long. The maximum area is 18.75(37.5), or 703.125, square feet.

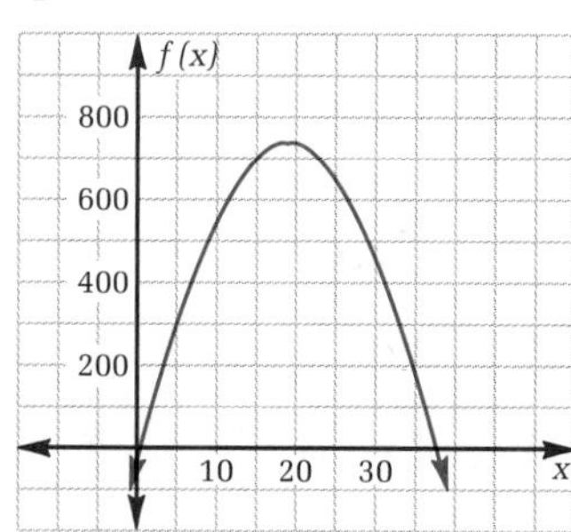

Checkpoint

1 Find the real zeros of $f(x) = 4x^2 - 4x + 1$. $\frac{1}{2}$

2 Graph the function $g(x) = x^2 - 1$.

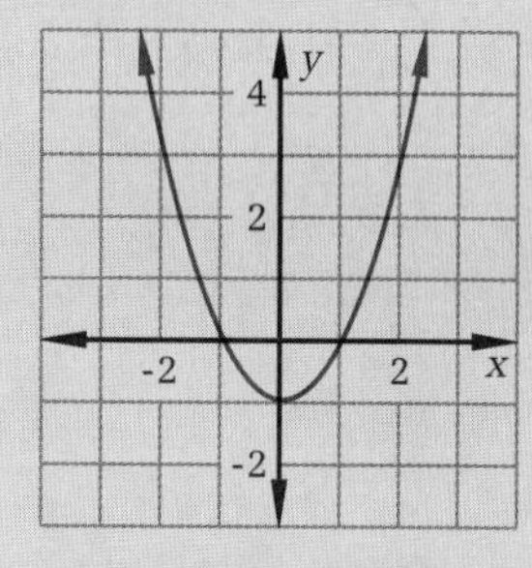

3 Graph the function $h(x) = (x - 1)^2$.

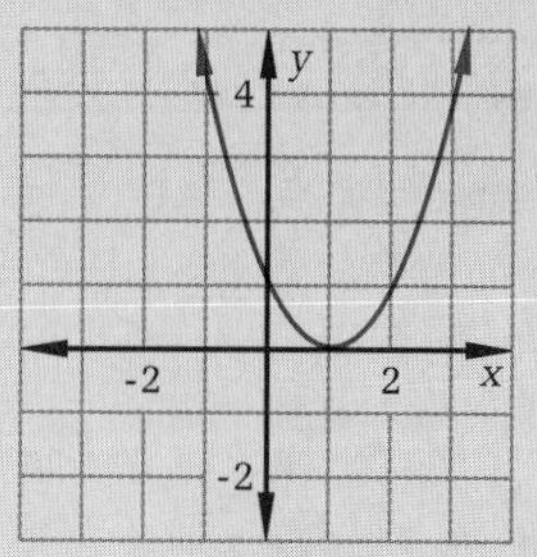

Assignment Guide

Average	
Day 1	4, 6–8, 13, 15, 18, 20–22, 25, 26, 29, 30, 32
Day 2	5, 11, 12, 14, 17, 19, 23, 24, 27, 28, 31, 33, 34a, c, 36
Advanced	
Day 1	4, 6–8, 10, 12, 13, 15, 17, 19, 22, 25, 26, 29, 30, 32
Day 2	5, 9, 11, 21, 23, 24, 27, 28, 31, 33, 34a, c, 35–38
Honors	
Day 1	4, 6, 7, 9, 11, 12, 19, 22, 24, 29, 30, 32, 34, 35
Day 2	8, 10, 21, 23, 25–28, 31, 33, 36–40

Part Three: Exercises and Problems

Warm-up Exercises

P

1 For what values of k does the equation $x^2 + kx + 9 = 0$ have exactly one real root? -6 or 6

Problem-Set Notes and Strategies

- In problem **4**, students may begin to understand the translations, reflections and dilations that are taking place. Once the first graph is drawn, the other graphs are transformations.
- The solution 6, 3 is the one that is usually more obvious in problem **5**. The second solution of −6 and −3 may be less obvious to students.
- Problems **8**, **9** and **10** use the zero product property and factoring. More will be discussed about these methods in the next section.

Additional Answers

11

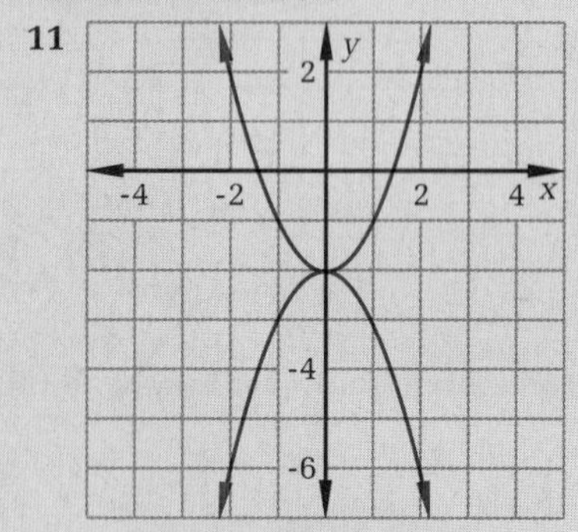

12

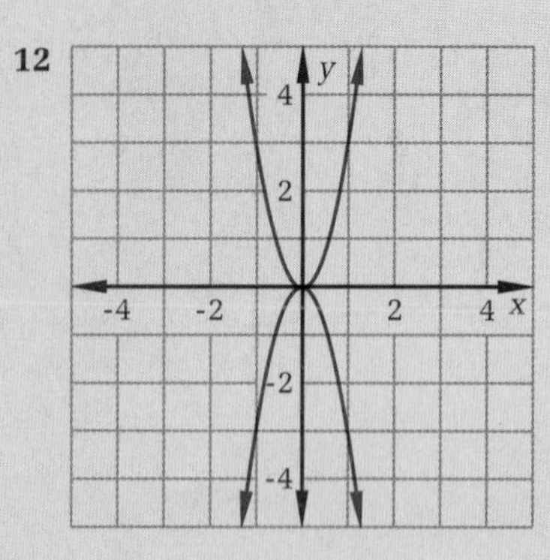

Answers to problems **4a – d** can be found in the answer pages beginning on **A** 1.

P **In problems 2 and 3, write an equation in the form $y = a(x - h)^2 + k$ (vertex form) for each parabola.**

2

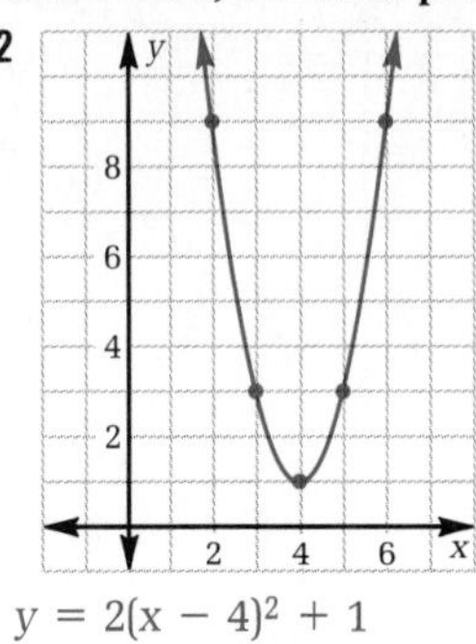

$y = 2(x - 4)^2 + 1$

3

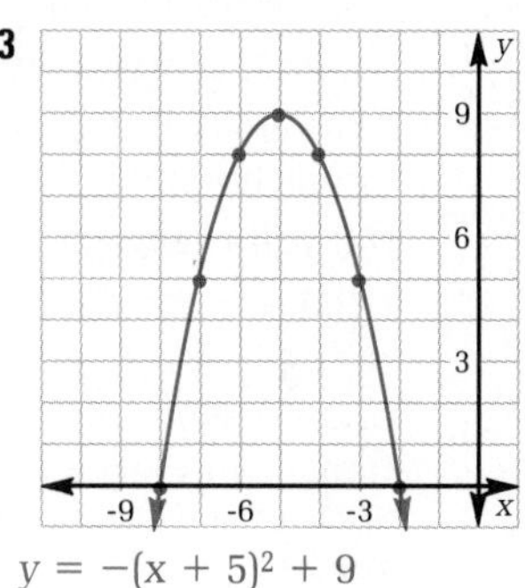

$y = -(x + 5)^2 + 9$

Problem Set

A

P **4** Sketch the graph of each equation.
a $y = x^2$ **b** $y = x^2 + 4$ **c** $y = (x - 4)^2$ **d** $y = 4x^2$

R,I **5** For what values of a and b is $x^2 + ax + 9 = (x + b)^2$? $a = 6$ and $b = 3$ or $a = -6$ and $b = -3$

P **6** Match each of the following equations with its graph.
a $y = 2(x + 3)^2$ iv **b** $y = (x - 2)^2$ iii **c** $y = -x^2$ ii **d** $y = x^2 - 2$ i

i

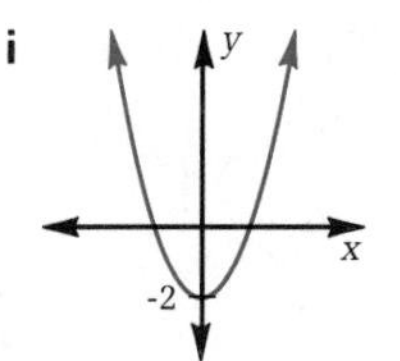

ii

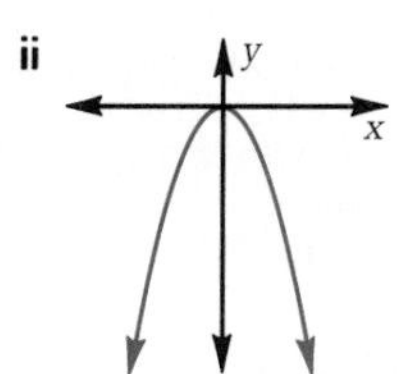

iii

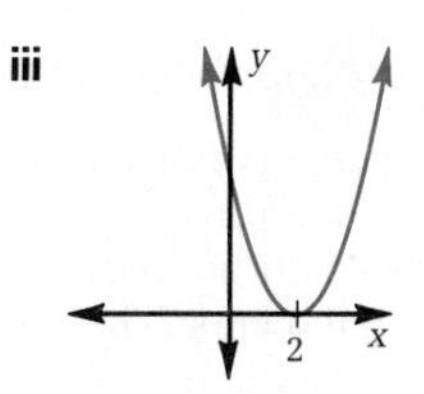

iv

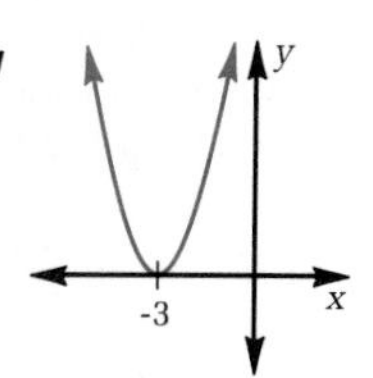

P **7** Find the real zeros of the function $g(x) = x^2 + 14x + 49$. -7

In problems 8–10, solve for x.

R,I **8** $(2x - 6)(x + 3) = 0$
$x = 3$ or $x = -3$

9 $(x + 1)(x^2 - 4) = 0$
$x = -1$, $x = 2$, or $x = -2$

10 $(3x - 5)(3x + 5) = 0$
$x = \frac{5}{3}$ or $x = -\frac{5}{3}$

P **In problems 11 and 12, graph each pair of equations on the same coordinate system.**

11 $y = x^2 - 2$ and $y = -x^2 - 2$

12 $y = 3x^2$ and $y = -3x^2$

R **In problems 13–18, simplify each expression and solve each equation.**

13 $3x - 6x = 12 - 18$ $x = 2$

14 $-2y + 3x - 4(2x - 3y)$ $10y - 5x$

15 $4.5w - 3.2y + 6w - 9.3y$ $10.5w - 12.5y$

16 $8x - (3x + 7)$ $5x - 7$

17 $2.7x - 3.2x = 14.6 - 3$ $x = -23.2$

18 $3(x + 5) + 2(4x - 3)$ $11x + 9$

P | **19** Refer to the graph. For what value(s) of x is each statement true?

a $f(x) = 0$ −1 and 5
b $f(x) < 0$ $\{x: -1 < x < 5\}$
c $f(x)$ is a minimum. 2

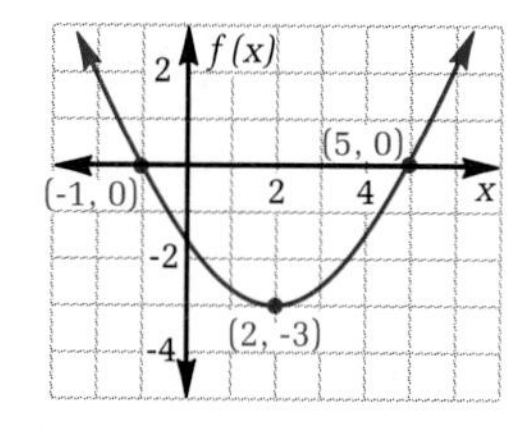

R | **20** Gem knows that she has used at least 65 paper clips from her box of 500 but cannot remember the exact number she has used. Write an algebraic sentence describing the number of paper clips left in the box. $n \le 435$

I | **21** Let $a = 2$, $b = 7$, and $c = -15$.

a Evaluate $\dfrac{-b + \sqrt{b^2 - 4ac}}{2a}$. $\frac{3}{2}$ **b** Evaluate $\dfrac{-b - \sqrt{b^2 - 4ac}}{2a}$. −5

c Substitute your answers to parts **a** and **b** for x in $2x^2 + 7x - 15 = 0$. Are these values solutions of the equation? Yes

R | **22** Express the area of the large parallelogram in at least two different ways.
$(x + 2)(x + 4)$; $x^2 + 6x + 8$

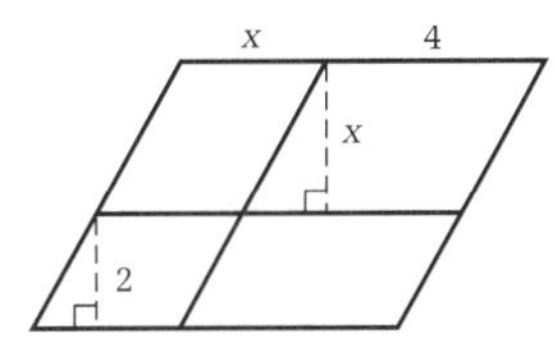

P | **23** Graph $y = 2(x + 5)^2 - 3$.

P **B** | **24** Find the real zeros of the function $h(x) = x^2 + x - 6$. −3, 2

R | **In problems 25–28, identify each statement as True or False.**

25 $(2x)^2 - (3y)^2 = (2x + 3y)(2x - 3y)$ True

26 $x - y = (\sqrt{x} + \sqrt{y})(\sqrt{x} - \sqrt{y})$, where $x > 0$ and $y > 0$ True

27 $x^2 - y^2 = (x - y)(x - y)$ False

28 $x^2 - 2xy + y^2 = (x - y)(x - y)$ True

R,I | **29** Solve each equation for x.

a $(x - 4)(x + 7) = 0$ $x = 4$ or $x = -7$ **b** $x^2 + 3x - 28 = 0$ $x = 4$ or $x = -7$

R,I | **30** If one of the numbers in $\{-4, -3, -2.5, -2, -1, 0, 1, 2, 2.5, 3, 4\}$ is selected at random, what is the probability that it is a solution of the equation $(y - 3)(y + 4)(2y - 5) = 0$? $\frac{3}{11}$

P,R,I | **31** The quadratic function $h(t) = 4t^2 - 24t + 36$ gives the height $h(t)$, in feet above the ground, of a sparrow at time t, in seconds.

a Factor $4t^2 - 24t + 36$. $4(t - 3)^2$
b How far above the ground is the sparrow initially (at $t = 0$)? 36 ft
c When is the sparrow on the ground? After 3 sec
d When is the sparrow again at its initial height? After 6 sec

Problem-Set Notes and Strategies, continued

- In problem **19**, the vertical axis is labeled as $f(x)$. Remind students that this is another name for the y-coordinate. It is more intuitive to ask where y is 0, or negative or at a minimum on the graph. Students will eventually make the generalization to $f(x)$.
- Problem **21** introduces the quadratic formula which will be explained in detail in the next section. Most students will have seen this in Algebra 1. Some may remember it and when it is applicable. Encourage others to generalize where the a, b and c came from.

Additional Answer

23

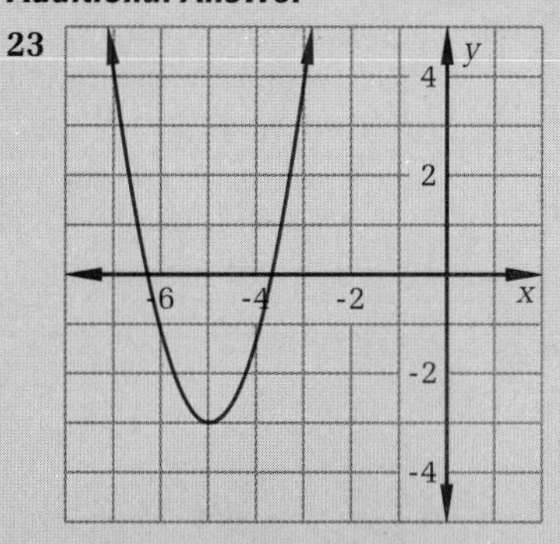

- Applications of a quadratic function are used in problem **31**. As well as solving the problem, it might be worthwhile to graph the parabola and identify those values of t that make sense in the problem.

Problem-Set Notes and Strategies, continued

- Problem **33** is explored further in a cooperative learning exercise.

Cooperative Learning

Problem **33** has students graph the equation $xy = 0$. The students should generate the graph that includes the x- and y-axis since at least one of the factors must be zero. This generates two lines. Can these two lines be written as a system of linear equations? if so, how? What if the equation were changed to $xy = 1$? Sketch a new graph to include points that satisfy this equation. In what quadrant(s) does the graph lie? Why do you suppose it is not in the other quadrant(s)? Can this graph be written as a system of linear equations? Try other constants such as –1, 2, –2 or any other you think is interesting. Will any other number generate a system of linear equations? What does this say about the zero product property?

- Problem **36** introduces the student to the formula for the sum of cubes. There is a similar formula for the difference of cubes which is $(x - a)(x^2 + xa + a^2)$. Some students will generalize this formula and observe the cancelling effects.

- Problem **39** revisits proportions with quadratic terms. In these cases, it is no longer feasible to refer to the constant of proportion as the slope of the line.

Problem Set, *continued*

R **32** **a** Write an expression for the length of a side of square ABCD if the area of the shaded region is $4x^2 - 4xy$ and $x > y$. $2x - y$
b Find AE. $2(x - y)$

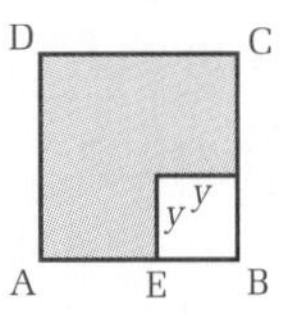

R,I **33** **a** Sketch a graph showing the ordered pairs (x, y) for which $xy = 0$.
b Explain your result in part **a.** If the product of two numbers is zero, at least one of the numbers must be zero, so the graph of $xy = 0$ is the x- and y-axes.

R **34** Solve each equation for b.
a $13 = |5 - 3b|$ $b = -\frac{8}{3}$ or $b = 6$
b $|3 - b| = |b - 3|$ $\{b : b \in \mathcal{R}\}$
c $|b + 3| = |b - 3|$ $b = 0$

R **35** A point (x, y) inside the rectangle is chosen at random. What is the probability that $x + y < 10$? $\frac{1}{3}$

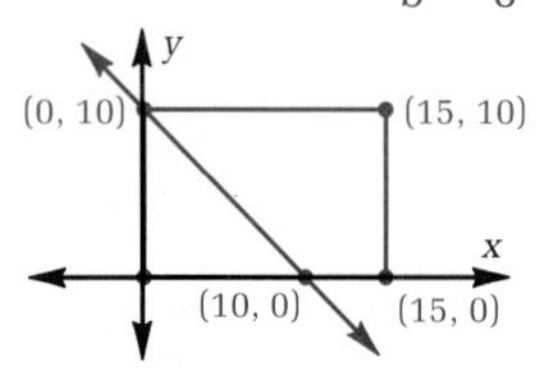

R,I **36** **a** Expand $(x + a)(x^2 - ax + a^2)$. $x^3 + a^3$ **b** Factor $x^3 + 27$. $(x + 3)(x^2 - 3x + 9)$

R **In problems 37 and 38, factor each expression completely.**

37 $2x^2 + 13x + 15$ $(2x + 3)(x + 5)$ **38** $3y(x^2 - 5x) + 6(x^2 - 5x)$ $3x(x - 5)(y + 2)$

C R,I **39** The surface area of a cube varies as the square of the edge of the cube.
a Write an equation expressing the surface area A as a function of the edge e if $A = 96$ when $e = 4$. $A = 6e^2$
b For what value of e will the numerical value of the area be 18 times the numerical value of the length of the edge? 3

R **40** Consider two numbers a and b such that $a = b$.

	$a = b$
Multiply both sides by a	$a^2 = ab$
Subtract b^2 from both sides	$a^2 - b^2 = ab - b^2$
Factor	$(a + b)(a - b) = b(a - b)$
Divide both sides by $a - b$	$a + b = b$
Substitute b for a	$b + b = b$
Add	$2b = b$
Divide both sides by b	$2 = 1$

What is wrong with this "proof" that $2 = 1$? $a - b = 0$, so division by $a - b$ is undefined.

Additional Answer

Answer to problem **33a** can be found in the answer pages beginning on **A** 1.

4.4 SOLVING QUADRATIC EQUATIONS

Objectives

After studying this section, you will be able to

- Solve quadratic equations by factoring
- Solve quadratic equations by using the quadratic formula
- Recognize when factoring can be used to solve a quadratic equation

Part One: Introduction

Solving Quadratics by Factoring

In this section, we will find the zeros of quadratic functions algebraically, by setting the quadratic function equal to zero, factoring, and then using the Zero Product Property.

Example *Find the zeros of* $f(x) = 2x^2 - 5x - 12$.

$$f(x) = 2x^2 - 5x - 12$$

Set $f(x) = 0$ $\quad 0 = 2x^2 - 5x - 12$

Factor $\quad 0 = (2x + 3)(x - 4)$

Zero Product Property $\quad 2x + 3 = 0 \quad \text{or } x - 4 = 0$

$$2x = -3 \qquad x = 4$$

$$x = -1.5$$

Remember, the real zeros of a quadratic function f correspond to the x-intercepts of its graph and the real roots of $f(x) = 0$.

Example *For what values of x are the areas of the rectangle and the square equal?*

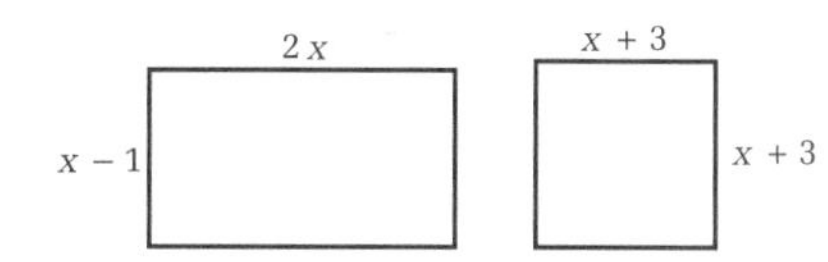

In Sample Problem 2 of Section 4.1, we found that the values of x must satisfy the equation $x^2 - 8x = 9$.

$$x^2 - 8x = 9$$

Rewrite the equation so that one side is 0 $\quad x^2 - 8x - 9 = 0$

Factor $\quad (x - 9)(x + 1) = 0$

Zero Product Property $\quad x - 9 = 0 \text{ or } x + 1 = 0$

$$x = 9 \qquad x = -1$$

Vocabulary
quadratic formula
discriminant

Class Planning

Time Schedule
All levels: 2 days

Resource References
Teacher's Resource Book
Class Opener 23
Practice 24

Class Opener

Find the zeros of
$f(x) = 12x^2 - 7x - 10$

$x = \frac{5}{4}$ or $x = -\frac{2}{3}$

Lesson Notes

- Factoring as a method of solving quadratic equations is explored in this section. Factoring can be a quick method of solving polynomial equations. Once a polynomial has been factored, the zero product property may be employed to find the roots of the equation. Sometimes, however, the polynomial is difficult to factor.

Lesson Notes, continued

- The Quadratic Formula is available to solve all quadratic equations. Insist that the student memorize the entire quadratic formula including the hypothesis, i.e. "If $ax^2 + bx + c = 0$ and $a \neq 0$, then ...." The proof of the formula is left to the next section. Emphasize that the Quadratic Formula always works to find out the roots of a quadratic equation.
- The portion of the formula under the radical is the discriminant because it discriminates between what type of roots are available for the equation. If the discriminant is positive and a perfect square, the roots are rational. If the discriminant is positive but not a perfect square, the roots are irrational. If the discriminant is negative, the roots do not exist in the real number system but do exist in the field of complex numbers which will be explored in Chapter **8**.

Stumbling Block

When solving linear equations such as $2x = 8$, it is admissible to divide both sides by 2. When solving a quadratic equation such as $2x^2 = 8x$ students may divide by the greatest common factor of $2x$ and still get 4 as the answer. Remind them that dividing by a variable may lead to losing an answer. In this example 0 is also a solution but was lost by dividing instead of factoring.

These values of x satisfy the equation, but do they both satisfy the conditions of the problem? If $x = 9$, the rectangle measures 18 by 8 and has an area of 144, and the square measures 12 by 12 and has an area of 144. If $x = -1$, however, the rectangle becomes an impossible figure with sides having negative lengths. Therefore, $x = 9$ is the only solution.

Solving Quadratics with the Quadratic Formula

If the quadratic polynomial in an equation of the form $ax^2 + bx + c = 0$ cannot be factored, we can solve the equation by using the ***quadratic formula.***

The Quadratic Formula

If $ax^2 + bx + c = 0$ and $a \neq 0$, then

$$x = \frac{-b \pm \sqrt{b^2 - 4ac}}{2a}$$

A proof of the quadratic formula is presented in the next section. For now, let's see how it can be used.

See p. 774

Example *Solve* $2x^2 - 6x - 13 = 0$.

The expression $2x^2 - 6x - 13$ is not factorable over the integers, so we will use the quadratic formula. The equation is already in the form $ax^2 + bx + c = 0$, with $a = 2$, $b = -6$, and $c = -13$.

$$x = \frac{-b \pm \sqrt{b^2 - 4ac}}{2a}$$

$$= \frac{-(-6) \pm \sqrt{(-6)^2 - 4(2)(-13)}}{2(2)}$$

$$= \frac{6 \pm \sqrt{36 + 104}}{4}$$

$$= \frac{6 \pm \sqrt{140}}{4}$$

Thus, $x \approx 4.46$ or $x \approx -1.46$.

When to Use Factoring

Factoring is often the quickest way to solve a quadratic equation, but it doesn't always work. How can you tell when to use the quadratic formula to solve such equations and when to factor?

The way to see whether a quadratic trinomial can be factored is to evaluate the ***discriminant***—the expression, $b^2 - 4ac$, that is under the radical sign in the quadratic formula. Let's calculate the discriminant for some equations in which the quadratic trinomial can be factored.

Equation	Discriminant
$2x^2 - 5x - 12 = 0$	$(-5)^2 - 4(2)(-12) = 121$, or 11^2
$4x^2 - 12x + 9 = 0$	$(-12)^2 - 4(4)(9) = 0$
$-16t^2 + 96t = 0$	$96^2 - 4(-16)(0) = 96^2$

Now let's look at some equations containing expressions that cannot be factored.

Equation	Discriminant
$2x^2 - 6x - 13 = 0$	$(-6)^2 - 4(2)(-13) = 140$
$x^2 - 6x + 13 = 0$	$(-6)^2 - 4(1)(13) = -16$
$x^2 + 7x + 14 = 0$	$7^2 - 4(1)(14) = -7$

Whenever the value of the discriminant is a perfect square, the equation can be solved by factoring. If it is not a perfect square, you have already evaluated part of the quadratic formula and can quickly finish the solution. (When the value of the discriminant is negative, as in the last two instances above, the equation has no real solutions.)

Part Two: Sample Problems

Problem 1 *Which of the following equations can be solved by factoring?*

a $2x^2 - 5x - 14 = 0$ **b** $4x^2 + 36x + 81 = 0$

c $3y^2 - 8y - 16 = 0$ **d** $2w^2 + 5w + 4 = 0$

Solution

Equation	Discriminant	Factorable?
a $2x^2 - 5x - 14 = 0$	$(-5)^2 - 4(2)(-14) = 137$	No
b $4x^2 + 36x + 81 = 0$	$36^2 - 4(4)(81) = 0$	Yes
c $3y^2 - 8y - 16 = 0$	$(-8)^2 - 4(3)(-16) = 256$, or 16^2	Yes
d $2w^2 + 5w + 4 = 0$	$5^2 - 4(2)(4) = -7$	No

Equations **b** and **c** can be solved by factoring.

Problem 2 *Solve* $x^2 = 16$ *for* x.

Solution

Method 1

$$\mathbf{x^2 = 16}$$

$$\mathbf{x^2 - 16 = 0}$$

Factor $\mathbf{(x + 4)(x - 4) = 0}$

$$\mathbf{x + 4 = 0 \text{ or } x - 4 = 0}$$

$$\mathbf{x = -4 \qquad x = 4}$$

Method 2

$$\mathbf{x^2 = 16}$$

Take the square root of each side $\mathbf{\sqrt{x^2} = \sqrt{16}}$

Remember, $\sqrt{x^2} = |x|$ $\mathbf{|x| = 4}$

$$\mathbf{x = 4 \text{ or } x = -4}$$

Checkpoint

1 Let $a = 5$, $b = -4$, and $c = -1$, and evaluate the formula $\frac{-b \pm \sqrt{b^2 - 4ac}}{2a}$
1 and $-\frac{1}{5}$

2 Solve the equation $x^2 - 3x - 4 = 0$ for x.
4 and -1

3 Is 5 a solution of the equation $x^2 - 9x + 45 = 0$?
No.

Assignment Guide

Average

Day 1 4, 5, 8, 9, 11, 15, 16, 18–20, 22, 24, 26, 28, 29, 31

Day 2 6, 7, 10, 12–14, 17, 21, 23, 25, 27, 30, 33, 35, 36, 40, 41

Advanced

Day 1 4–7, 10, 11, 15, 16, 18–21, 24, 26, 28, 29, 31, 33, 36, 38

Day 2 8, 9, 12–14, 22, 23, 25, 27, 30, 32, 34, 35, 37, 40, 41, 45, 47

Honors

Day 1 6–8, 10, 11, 14, 16, 18, 22, 23, 26, 28, 29, 31, 35, 36, 38, 44

Day 2 9, 15, 19, 25, 27, 30, 32–34, 37, 39–42, 45–49

Communicating Mathematics

Write a paragraph describing the effect the discriminant has on the type of solutions to a quadratic equation. See if you can find a quick method for determining the sign of the discriminant.

Problem 3 *A rectangle's length is 4 inches greater than its width. Its area is 360 square inches. Find the dimensions of the rectangle.*

Solution If we let w represent the width of the rectangle, then $w + 4$ represents the length of the rectangle.

$w + 4$

w $A = 360$

$$w(w + 4) = 360$$
$$w^2 + 4w = 360$$
$$w^2 + 4w - 360 = 0$$

The value of the discriminant is $4^2 - 4(1)(-360)$, or 1456, so the quadratic trinomial cannot be factored.

$$w = \frac{-4 + \sqrt{1456}}{2} \text{ or } w = \frac{-4 - \sqrt{1456}}{2}$$

Since the value of w must be positive, we reject the second solution.

$$w = \frac{-4 + \sqrt{1456}}{2} \approx \frac{-4 + 38.16}{2} \approx 17.08$$

$$w + 4 \approx 21.08$$

The rectangle measures about 17.08 inches by 21.08 inches.

Part Three: Exercises and Problems

Warm-up Exercises

1 Let $(a, b, c) = (3, 5, -2)$.

P **a** Evaluate $\frac{-b + \sqrt{b^2 - 4ac}}{2a}$. $\frac{1}{3}$ **b** Evaluate $\frac{-b - \sqrt{b^2 - 4ac}}{2a}$. -2

P **2** Solve each equation.

a $2x^2 - 6x = 0$ $x = 0$ or $x = 3$ **b** $9y^2 - 36 = 0$ $y = -2$ or $y = 2$ **c** $4z^2 + 16z = 0$ $z = 0$ or $z = -4$

P **3** Solve each equation.

a $x^2 - 9x + 20 = 0$ $x = 4$ or $x = 5$ **b** $y^2 + 6y - 16 = 0$ $y = -8$ or $y = 2$ **c** $2z^2 + 5z - 3 = 0$ $z = \frac{1}{2}$ or $z = -3$

Problem Set

P **A** **In problems 4–6, solve for x.**

4 $x^2 - 6x + 5 = 0$ $x = 5$ or $x = 1$ **5** $4x^2 - 5x - 6 = 0$ $x = 2$ or $x = -\frac{3}{4}$ **6** $6x^2 + 11x + 3 = 0$ $x = -\frac{1}{3}$ or $x = -\frac{3}{2}$

R **7** For what values of x are ℓ_1 and ℓ_2 parallel? -8 and 4

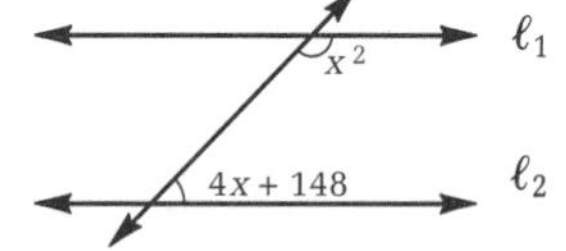

R **8** Find the real zeros of $g(x) = x^2 - 5x - 6$. -1 and 6

Problem-Set Notes and Strategies

- You may need to remind students that dividing by a variable is not an acceptable method of solving quadratic equations for problem **2a** and **c**.

- Students may need a reminder that same side interior angles of parallel lines cut by a transversal are supplementary in order to solve problem **7**.

R **9** Graph the following equations on the same coordinate system.

a $y = (x - 3)^2 + 4$ **b** $x = (y - 3)^2 + 4$

R **10** Graph $f(x) = x^2 - 9x + 18$ and find the x-intercepts of the function. (3, 0) and (6, 0)

P **11** Solve for y.

a $(2y - 3)(y + 1)(3y) = 0$ $y = \frac{3}{2}$, $y = -1$, or $y = 0$ **b** $(y - 2)(2y - 2)(3y - 2) = 0$ $y = 2$, $y = 1$, or $y = \frac{2}{3}$ **c** $(2y + 1)(2y - 1)(3y) = 0$ $y = -\frac{1}{2}$, $y = \frac{1}{2}$, or $y = 0$

P **In problems 12–14, solve for x.**

12 $2x - 6 = 0$ $x = 3$ **13** $2x^2 - 5x - 3 = 0$ $x = -\frac{1}{2}$ or $x = 3$ **14** $x^2 + 3x + 4 = 0$ No real solutions

P **15** If $y = x^2 - 3x - 6$, what is the value of x when $y = -2$? -1 or 4

B

P,R **16** The square of an integer is equal to the product of the integer and 7. Find the integer. 0 or 7

R **17** The length of $\overline{AB}$ is less than 60. What are the possible values of x? $\{x{:}3 < x < 18\}$

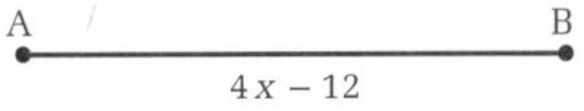

R **18** The slopes of eight lines are $\frac{1}{2}$, $-\frac{2}{5}$, 2, -2, -0.4, 0.5, 2.5, and $-\frac{1}{2}$. If two of the lines are selected at random, what is the probability

a That the lines are parallel? $\frac{1}{14}$

b That the lines are perpendicular? $\frac{5}{28}$

P **19** Consider the equation $(x + 7)^2 - 5(x + 7) - 14 = 0$.

a Substitute z for $x + 7$ and solve for z. $z = 7$ or $z = -2$

b Use your answer to part **a** to solve the original equation for x. $x = 0$ or $x = -9$

R **In problems 20 and 21, factor each expression completely.**

20 $x^2 - 9y^2$ $(x + 3y)(x - 3y)$ **21** $x^2 - 6xy + 9y^2$ $(x - 3y)^2$

R **22** If $\triangle AEF \sim \triangle BCD$, what is the length of $\overline{AE}$? 15

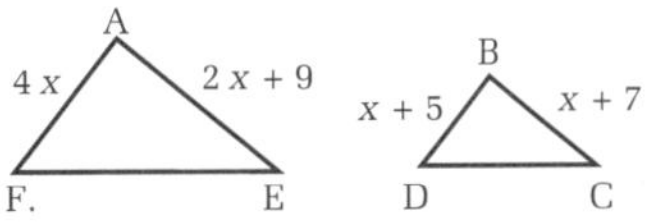

P **23** Solve $2\sqrt{5}x^2 + 3\sqrt{5}x = 0$ for x. $x = 0$ or $x = -\frac{3}{2}$

R **24** When there is no wind, Amelia flies her plane from Oklahoma City to San Francisco in exactly 4 hours, at an average speed of 325 miles per hour. If she were flying into a 25-mile-per-hour head wind, how long would the trip take? 4 hr 20 min

I **25** Write a quadratic equation whose solutions are $x = -3$ and $x = 7.5$. $2x^2 - 9x - 45 = 0$

R **26** Suppose that $y \geq 7$. Solve $4x + 11 \geq 3y$ for x and graph the solution on a number line. $x \geq \frac{5}{2}$

Problem-Set Notes and Strategies, continued

- Problem **9** illustrates that either variable can be the squared term.
- Problem **10** previews that the x-intercepts and the zeros or roots of a function are synonymous.
- Problem **14** has no real solutions, but after Chapter **8**, we will show that it does have two complex solutions.
- In problem **18**, the number of combinations must be determined. Some students may have figured out a formula by this time. Others may still be listing all the possibilities. The combination formula will be developed in Chapter **12**.
- Problem **19** illustrates that substitution is a viable method for solving many types of equations, not just systems of linear equations.
- Problem **25** asks students to reverse the process of solving equations. If $x = -3$ is a solution, that indicates that $x + 3 = 0$ and if the equation were set equal to zero then $x + 3$ is a factor. By the same logic, $x - 7.5$ or equivalently $2x - 15$ is also a factor. Multiply the factors together to arrive at the equation. Answers will vary but should be a scalar multiple of this equation.

Additional Answers

26

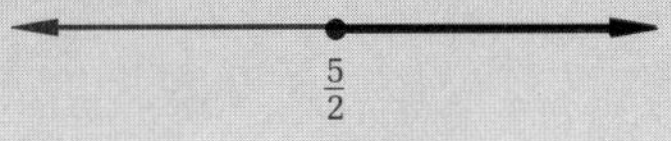

Answer to problem **9** can be found in the answer pages beginning on **A** 1.

Problem-Set Notes and Strategies, continued

- Problem **29** is a good physical application of quadratics. Students in physics will relate well to this problem and we will see this type of problem again in Section **4.6**.

Additional Answer

29a

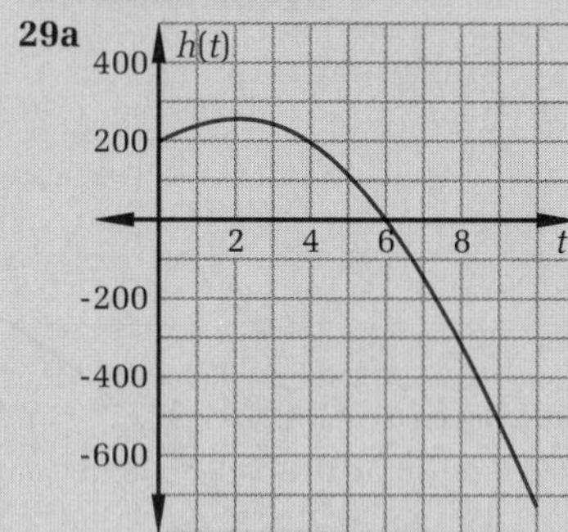

- The Pythagorean Theorem may be used to solve for x in problem **30**. Once x is known, the area and length of the sides can easily be computed.

- Have the students recognize that the data in problem **34** does look very linear from the graph. Without this observation, the regression line that was computed would be meaningless.

- Problems **35** and **36** can be solved by factoring or by multiplying the expressions and then factoring. The first method saves time and is less troublesome.

Problem Set, *continued*

R **27** Hazel bought eight pounds of cashews and four pounds of macadamia nuts for $72.48. Filbert bought three pounds of cashews and five pounds of macadamia nuts for $60.29. What is the price per pound of each type of nut? Cashews: $4.33; macadamias: $9.46

P **28** Solve $[f(x)]^2 + 5[f(x)] - 14 = 0$ for $f(x)$. $f(x) = -7$ or $f(x) = 2$

P,I **29** The height $h(t)$, in feet above the ground, of a ball thrown into the air from the top of a building is given by the function $h(t) = -16(t - 2)^2 + 264$, where t is the number of seconds the ball has been in the air.

a Sketch a graph of function h for $0 \le t \le 10$.
b From what height was the ball thrown? 200 ft
c What height does the ball reach? 264 ft
d When does the ball strike the ground? After $\approx$ 6.06 sec
e For what values of t does the function have meaning? $\{t: 0 \le t \le \approx 6.06\}$

P,R **30** Solve for x to find
a The length of $\overline{AC}$ 20
b The area of the rectangle 192

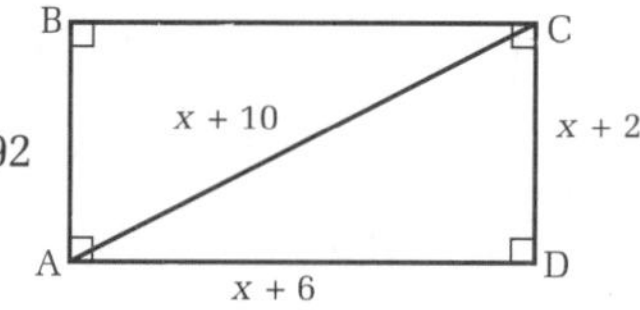

I **31** Copy and complete the table of values for $f(x) = \sqrt{3 - x}$.

x	−1	0	1	2	3	4	5
y	2	$\sqrt{3}$	$\sqrt{2}$	1	0	Undef.	Undef.

P,R **32** The width and the length of a rectangle are in the ratio 3:4. If both the length and the width were decreased by 15 centimeters, the rectangle's area would be 252 square centimeters. Find the dimensions of the rectangle. 27 cm by 36 cm

R **33** If $f(2) = 7$ and $f(5) = 16$, is $f(x)$ directly proportional to x? No

R **34** **a** Copy the graph shown and draw a line that approximately fits the data.
b Use the method of least squares to determine an equation of the line of best fit. **a** Answers will vary.
b $y = 938x - 1{,}863{,}085$

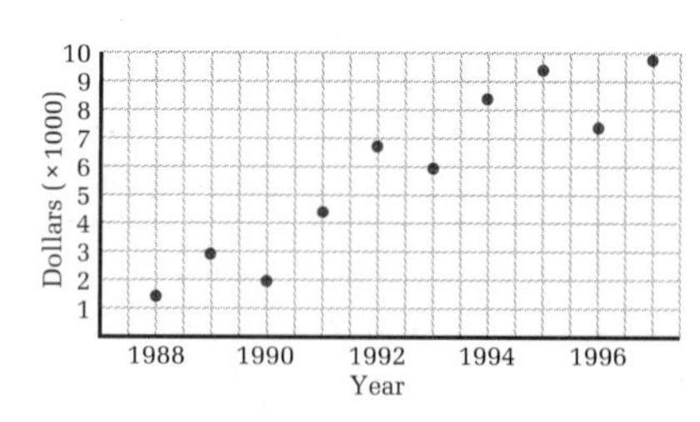

P **In problems 35 and 36, solve each equation for x.**

35 $2(x - 3)^2 - 9(x - 3) - 5 = 0$
$x = 8$ or $x = \frac{5}{2}$

36 $2(3x + 4) - 4x(3x + 4) = 0$
$x = -\frac{4}{3}$ or $x = \frac{1}{2}$

P,R **37** Solve each equation for x.

a $|x|^2 - 5|x| = 0$ $x = -5, x = 0, \text{ or } x = 5$

b $|x|^2 - 7|x| - 8 = 0$ $x = -8 \text{ or } x = 8$

R **38** Refer to the diagram. Draw a graph showing the possible values of (x, y).

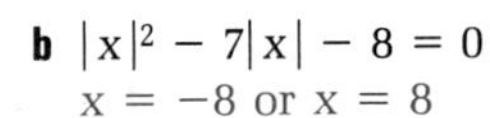
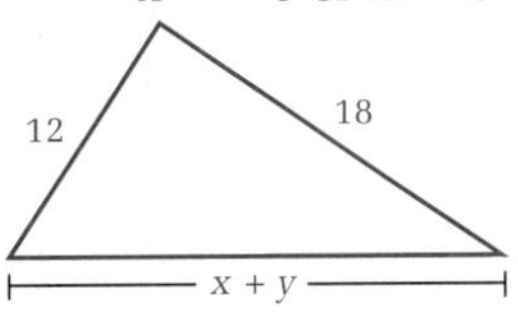

In problems 39–42, factor each expression completely.

R **39** $9x^2 - 27xy$ $9x(x - 3y)$

40 $9x^2 + 24xy + 16y^2$ $(3x + 4y)^2$

41 $wx + wy + zx + zy$ $(w + z)(x + y)$

42 $xy + 8x - 2y - 16$ $(x - 2)(y + 8)$

R **43** Sky's airplane is at an altitude of 5000 feet and rising at a rate of 475 feet per minute. Penny's airplane is at 30,000 feet and descending at a rate of 650 feet per minute. When will the planes be at the same altitude? In ≈ 22.2 min

P **44** **a** Copy and complete the following steps to solve for x.

$x^2 + (a + b)x + ab = 0$ $(x + b)(x + a) = 0$

$x^2 + ax + bx + ab = 0$ $x + b = 0$ or $x + a = 0$

$x(x + a) + b(x + a) = 0$ $x = -b$ $x = -a$

b Solve $x^2 + (2c + 3d)x + 6cd = 0$ for x. $x = -2c$ or $x = -3d$

c Solve $x^2 + (3 + a)x + 3a = 0$ for x. $x = -3$ or $x = -a$

R **45** Suppose that $y = 12$ when $x = 6$ and that y varies directly as x.

a Write an equation expressing y as a function of x. $y = 2x$

b Graph the function you wrote in part **a** for $\{y:y \leq 12\}$.

P,R **46** Routes 30 and 66 are perpendicular. On Route 30, ten miles from the intersection of the roads, is a car traveling toward the intersection at 35 miles per hour. At the same moment, a car traveling on Route 66 is passing through the intersection at 45 miles per hour.

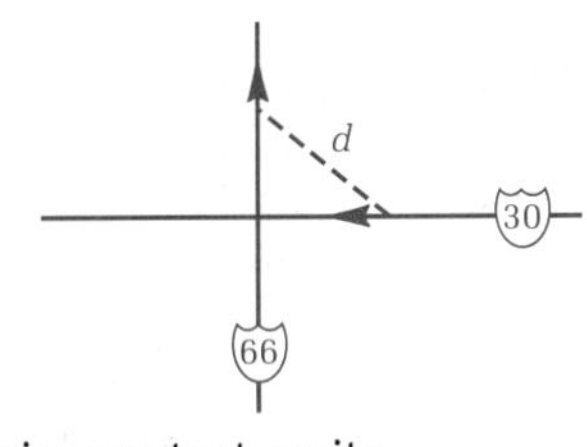

a Assuming that the speeds of the cars remain constant, write an equation expressing the square of the distance d, in miles, between the cars as a function of time t, in hours. $d^2 = 3250t^2 - 700t + 100$

b What is the minimum value of the square of the distance between the cars? ≈62.3

c When is this minimum value reached? After ≈0.108 hr, or ≈6.5 min

d What is the closest to each other the cars get? ≈7.89 mi

I **47** If $x > 0$, what happens to the value of $\frac{1}{x}$ as the value of x

a Increases? It approaches zero.

b Decreases? It increases without limit.

Additional Answer

38

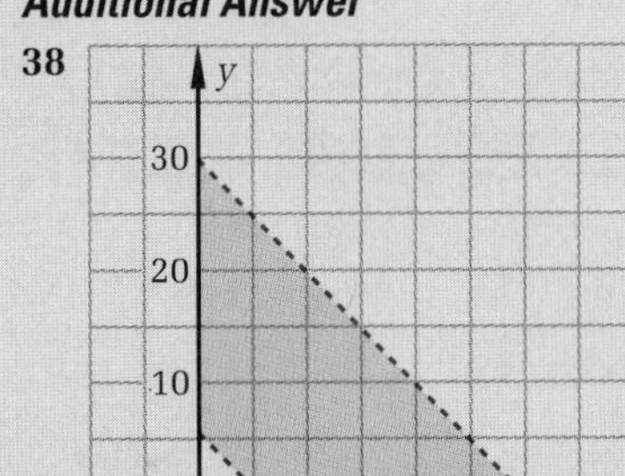

Answer for problem **45b** can be found in the answer pages beginning on **A** 1.

Problem-Set Notes and Strategies, continued

- Problems **41**, **42** and **44** illustrate factoring by grouping.

- Problem **47** asks students to consider limits. What is happening as x gets large? The value of $\frac{1}{x}$ must get smaller. Likewise as x approaches zero, the value of $\frac{1}{x}$ gets larger. Try this with some actual values to convince the doubting student.

Problem-Set Notes and Strategies, continued

- Problem **48** could be used to generalize the distance formula to three dimensions or the Pythagorean Theorem to three dimensions.

Problem Set, *continued*

C P,R **48** The length of a diagonal of a rectangular box is 10. The box's total surface area is 224. Find the sum of the lengths of all the edges. (Hint: What is $[x + y + z]^2$?) 72

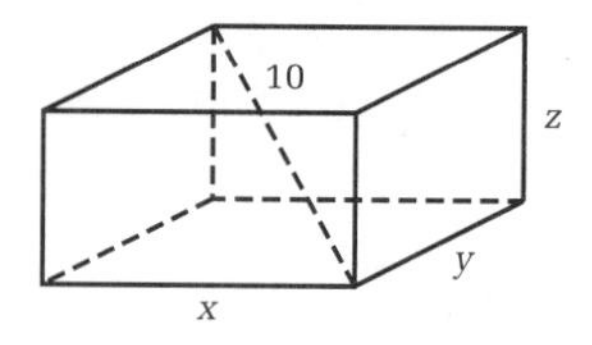

R **49** Given that $f(y) = y^2$, rewrite $\frac{f(x + h) - f(x)}{h}$ in simplest form. $2x + h$

R **50** Find the measure of minor arc CP of the circle shown. 144.5

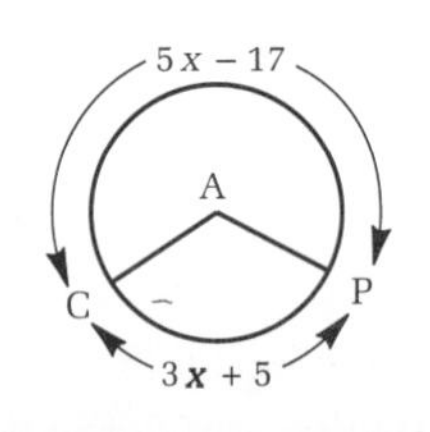

CAREER PROFILE

The Parabola in Baseball

Robert K. Adair demystifies the game

Robert K. Adair, a professor of physics at Yale University, provides a shining example for anyone who wants to combine a career with a hobby. Dr. Adair is the author of two books on physics for readers who are not necessarily physicists: *The Great Design: Particles, Fields, and Creation*, and the most recent, *The Physics of Baseball*.

"In baseball," Dr. Adair explains, "there are essentially two areas where physics comes into play: the flight of the ball—whether thrown or batted—and the impact of the bat hitting the ball." Mathematics helps determine the distance, speed, and direction of the ball's flight.

One example of a "baseball mathematics" problem is that of calculating the distance a given home run would have traveled in an open field. A batted baseball follows the path of a parabola, which can be plotted based on the angle at which the ball leaves the bat, the height at which the ball hits the stands, and how far the ball is from home plate when it hits. To find the total distance in open space, we would find the second x-intercept of the parabola whose first intercept is at the origin, (0, 0).

Dr. Adair has bachelor's and doctoral degrees in physics, both from the University of Wisconsin in Madison.

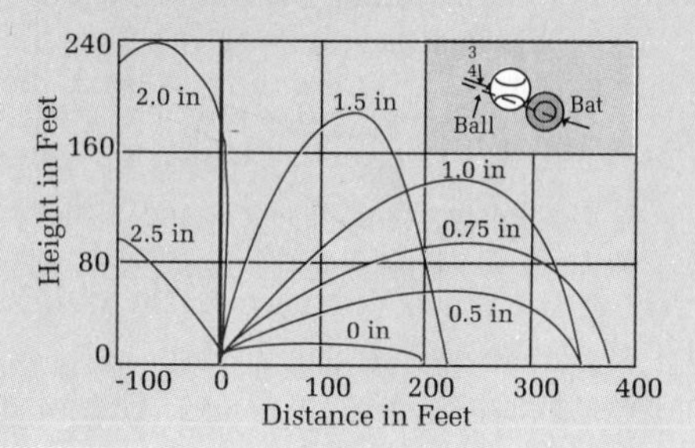

4.5 Completing the Square

Objectives

After studying this section, you will be able to

- Use the method of completing the square to determine the maximum or minimum value of a quadratic function
- Solve quadratic equations by completing the square
- Prove the quadratic formula

Part One: Introduction

Rewriting a Function in Vertex Form

As you have seen, the maximum or minimum value of a quadratic function corresponds to the y-coordinate of the vertex of the function's graph.

Consider the function $f(x) = x^2 + 12x + 23$. What is the minimum value of f? It is not easy to identify the minimum or maximum of a function in this form. If, however, the equation were in vertex form, $f(x) = a(x - h)^2 + k$, we would know that the minimum is k.

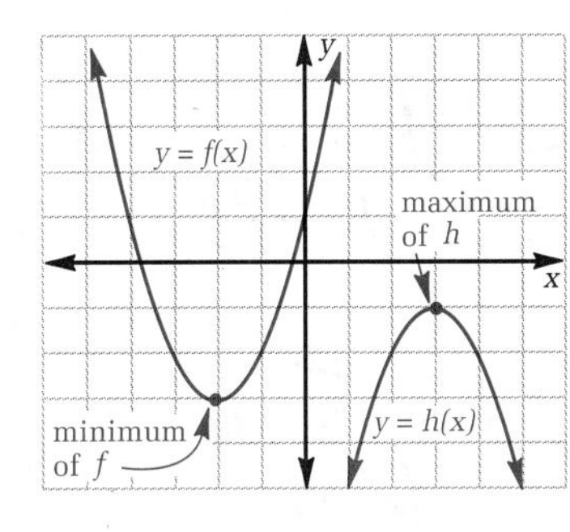

By using a process called ***completing the square,*** we can rewrite a quadratic function of the form $f(x) = ax^2 + bx + c$ in the form $f(x) = a(x - h)^2 + k$.

Example *Write $f(x) = x^2 + 12x + 23$ in vertex form, and find the minimum value of the function.*

We want to rewrite the quadratic expression so that it contains a trinomial of the form $x^2 + 2ax + a^2$—the square of a binomial $x + a$. To do so, we add to the first two terms the square of half the linear coefficient. Then, so that we do not change the value of the expression, we must subtract the same value.

$$f(x) = x^2 + 12x + 23$$

Rewrite, leaving space to add and subtract a quantity $\quad = (x^2 + 12x + \_\_) - \_\_ + 23$

Complete the square by adding and subtracting $\left(\frac{1}{2} \cdot 12\right)^2$ $\quad = (x^2 + 12x + 36) - 36 + 23$

Factor the perfect-square trinomial and add the constants $\quad = (x + 6)^2 - 13$

The minimum value of the function can now be identified—it is -13.

Class Planning

Time Schedule

All levels: 2 days

Resource References

Teacher's Resource Book
Class Opener 24
Practice 25
Enrichment 8

Class Opener

Carlos was doing another homework problem and again discovered that some numbers were missing from his book. This time part of the answer was missing also. Carlos is pretty clever, so he figured out what all the missing numbers must have been. Try it.

13 Factor $x^2 + 18x +$ ##.
14 Factor $x^2 -$ ##$x + 64$.

The answers are:
13 $(x + 9)^2$
14 $(x - 8)^2$

The missing number for problem **13** is 81 and for problem **14** is –16.

Lesson Notes

- Completing the square is a useful process in writing quadratic equations in vertex form. It will be used again in the short subject on conics and elsewhere throughout the text. The process is important in maximum and minimum problems in Algebra.

Vocabulary

completing the square

4.5

Using Computers

Completing the Square on *The Algebra Solver Kit* can show students the possibilities of computer programs. Another option is to have students graph the quadratic function both before and after completing the square. If the graphs coincide, the student has completed the square correctly. If they do not coincide, the graph can show where the error has been made.

Communicating Mathematics

Completing the square allows a trinomial to be written as an expression containing a square of a binomial. Justify that the step of adding and subtracting a specific value in the expression produces an equivalent expression.

Solving Equations by Completing the Square

We can also use the process of completing the square to solve a quadratic equation.

Example *Solve the equation* $x^2 - 14x - 5 = 0$ *by completing the square.*

$$x^2 - 14x - 5 = 0$$

Isolate the constant term $\quad x^2 - 14x = 5$

Add $\left[\frac{1}{2}(-14)\right]^2$ to each side (completing the square on the left) $\quad x^2 - 14x + 49 = 5 + 49$

Factor and add $\quad (x - 7)^2 = 54$

Take the square root of each side $\quad |x - 7| = \sqrt{54}$

$$|x - 7| = 3\sqrt{6}$$

$$x - 7 = 3\sqrt{6} \quad \text{or } x - 7 = -3\sqrt{6}$$

$$x = 7 + 3\sqrt{6} \qquad x = 7 - 3\sqrt{6}$$

Note In solving an equation, you add the same quantity to both sides to complete a square. When completing the square in a quadratic function, you must add and subtract the same quantity. The two methods are related but not identical.

Deriving the Quadratic Formula

In the preceding section, we presented the quadratic formula and asked you to use it without proof. Now we will show you how the quadratic formula can be derived.

$$ax^2 + bx + c = 0$$

Divide each side by a $\quad x^2 + \frac{b}{a}x + \frac{c}{a} = 0$

Add $-\frac{c}{a}$ to each side $\quad x^2 + \frac{b}{a}x = -\frac{c}{a}$

Add $\left(\frac{b}{2a}\right)^2$ to each side $\quad x^2 + \frac{b}{a}x + \left(\frac{b}{2a}\right)^2 = -\frac{c}{a} + \left(\frac{b}{2a}\right)^2$

Factor the trinomial $\quad \left(x + \frac{b}{2a}\right)^2 = -\frac{c}{a} + \frac{b^2}{4a^2}$

On the right side, find a common denominator and add $\quad \left(x + \frac{b}{2a}\right)^2 = \frac{b^2 - 4ac}{4a^2}$

Take the square root of each side $\quad x + \frac{b}{2a} = \pm\frac{\sqrt{b^2 - 4ac}}{2a}$

Add $-\frac{b}{2a}$ to each side $\quad x = \frac{-b \pm \sqrt{b^2 - 4ac}}{2a}$

As you study this proof of the quadratic formula, you will recognize that the formula is nothing more than the solution of the equation $ax^2 + bx + c = 0$ by completing the square.

Part Two: Sample Problems

Problem 1 *Sketch the graph of* $g(x) = 2x^2 - 16x + 21$.

Solution Identifying the characteristics of the graph will be easier if we rewrite the function in vertex form.

$$g(x) = 2x^2 - 16x + 21$$

Factor 2 out of the first two terms, leaving room to complete the square $= 2(x^2 - 8x + \_\_) - \_\_ + 21$

Complete the square (Note that here we are adding and subtracting 32, not 16.) $= 2(x^2 - 8x + 16) - 32 + 21$

$= 2(x - 4)^2 - 11$

The vertex is $(4, -11)$. The graph opens upward, and the values of $g(x)$ change in the pattern $2(1)$, $2(3)$, $2(5)$, . . .

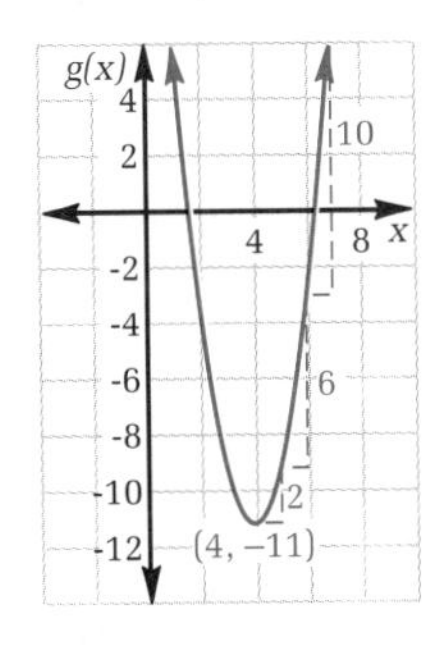

Problem 2 *Find the maximum rectangular area that can be enclosed by an existing wall and 75 feet of fencing. What will the dimensions of the enclosure be?*

Solution This is the farmer's problem with which we began this chapter and which we solved in Section 4.3, Sample Problem 2. This time, we will solve it by completing the square.

$$A = x(75 - 2x)$$
$$= -2x^2 + 75x$$

Factor out -2 $\quad = -2\left(x^2 - \frac{75}{2}x\right)$

Complete the square $\quad = -2\left[x^2 - \frac{75}{2}x + \left(-\frac{75}{4}\right)^2\right] - (-2)\left(-\frac{75}{4}\right)^2$

$$= -2\left(x - \frac{75}{4}\right)^2 + \frac{5625}{8}$$

The maximum area is $\frac{5625}{8}$, or 703.125, square feet. This area is obtained when the enclosure's width (x) is $\frac{75}{4}$, or 18.75, feet and its length is $75 - 2\left(\frac{75}{4}\right)$, or 37.5, feet.

Stumbling Block

The coefficient of the squared term must be 1 in order to utilize the complete-the-square process. If students neglect to factor out the leading coefficient, they may forget to multiply the number they added, thus creating an expression that is not equivalent. Caution students on these two errors.

Checkpoint

1 What number should be added to $x^2 - 8x$ to complete the square?
16

2 What is the first step in completing the square of $5x^2 + 35x$?
Factor out the common factor, 5.

3 Find the minimum value of the function $f(x) = x^2 - 16x$.
−64

4.5

Assignment Guide

Average	
Day 1	7, 9, 11, 12, 14, 16, 18–20, 23, 25, 27, 29, 31, 32, 34
Day 2	8, 10, 13, 15, 17, 21, 22, 24, 26, 28, 33, 35–38
Advanced	
Day 1	7, 9, 11, 12, 14, 16, 18–20, 22, 23, 25–27, 29, 31, 32, 34
Day 2	8, 10, 13, 15, 17, 21, 24, 28, 30, 33, 35–41, 44
Honors	
Day 1	7, 9, 10, 13, 15, 16, 18–21, 23, 25–27, 29, 31, 32, 35, 39
Day 2	14, 17, 24, 28, 30, 32–34, 36, 38, 40–44, 46

Problem-Set Notes and Strategies

- Problems **1**, **2** and **3** provide practice with the process of completing the square. Have students work these problems at the end of class and observe if anyone is having real difficulty with them.
- In problem **6**, students begin the process of finding the value of x that minimizes the functions.

Additional Answers

Answers for problems **12 – 15** can be found in the answer pages beginning on **A 1**.

Part Three: Exercises and Problems

Warm-up Exercises

P **1** What constant term is needed to complete each perfect-square trinomial?

a $x^2 + 8x + \underline{\ ?\ }$ 16 **b** $x^2 - 6x + \underline{\ ?\ }$ 9 **c** $x^2 + 9x + \underline{\ ?\ }$ $\frac{81}{4}$

d $4x^2 + 8x + \underline{\ ?\ }$ 4 **e** $4x^2 + 9x + \underline{\ ?\ }$ $\frac{81}{16}$

P **2** What linear term is needed to complete each perfect-square trinomial?

a $x^2 + \underline{\ ?\ } + 9$ $6x$ **b** $y^2 - \underline{\ ?\ } + 64$ $16y$

c $y^4 + \underline{\ ?\ } + 16x^2$ $8y^2x$ **d** $x^2 + \underline{\ ?\ } + 30.25$ $11x$

P **3** The expression $x^2 + 8x$ can be represented geometrically as shown. Find the area of the region that would complete the square. 16

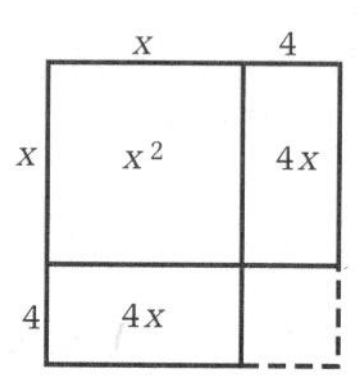

P **4** Find the coordinates of the vertex of each equation's graph.

a $y = (x - 3)^2 + 4$ (3, 4) **b** $y = (2x + 1)^2 + 5$ $\left(-\frac{1}{2}, 5\right)$

c $y = x^2 + 6x + 15$ (−3, 6) **d** $y = x^2 - 10x + 5$ (5, −20)

P **5** Solve each equation by completing the square.

a $x^2 + 10x = 11$ $x = 1$ or $x = -11$ **b** $x^2 - 14x = 15$ $x = 15$ or $x = -1$

c $x^2 + 6x = 5$ $x = -3 \pm \sqrt{14}$ **d** $x^2 - 50x = -625$ $x = 25$

P **6** Find the minimum value of each function.

a $f(x) = x^2 + 12x + 50$ 14 **b** $g(x) = x^2 - 6x + 3$ −6

Problem Set

P **A**

In problems 7–11, rewrite each polynomial in the form $a(x - h)^2 + k$.

7 $x^2 - 8x + 45$ $(x - 4)^2 + 29$ **8** $x^2 + 10x + 11$ $(x + 5)^2 - 14$

9 $2x^2 + 100x + 50$ $2(x + 25)^2 - 1200$ **10** $x^2 + 7x + 4$ $\left(x + \frac{7}{2}\right)^2 - \frac{33}{4}$

11 $x^2 - 50x - 22$ $(x + 25)^2 - 647$

P **In problems 12–15, sketch the graph of each equation.**

12 $y = (x - 3)^2 + 4$ **13** $y = (2x - 1)^2 - 3$

14 $y = x^2 + 6x + 12$ **15** $y = x^2 - 14x + 45$

R **16** Solve each equation for y.

a $y^2 = 4y - 4$ $y = 2$ **b** $y^2 = 9y + 36$ $y = 12$ or $y = -3$

R **17** Given that $\ell_1 \parallel \ell_2$, solve for x and find $m\angle 1$. $x = 19$; $m\angle 1 = 91$

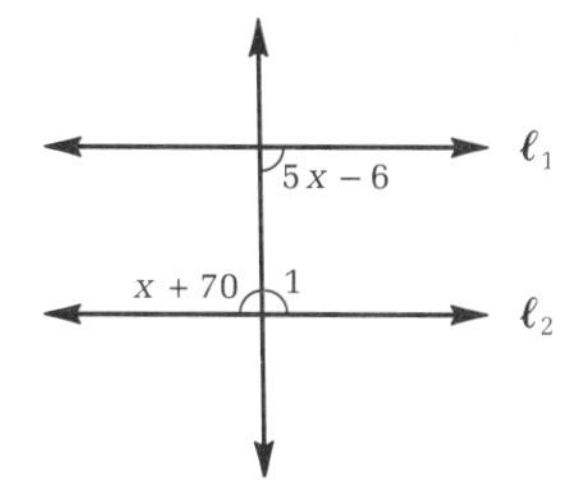

R **18** Write a quadratic equation whose roots are 2 and −5. $x^2 + 3x - 10 = 0$

R,I **19 a** Write a function for $A(x)$, the area of rectangle A. $A(x) = 3x^2 + 6x$

b Write a function for $B(x)$, the area of rectangle B. $B(x) = 2x + 4$

c Find the value(s) of x for which the areas are equal. $\frac{2}{3}$

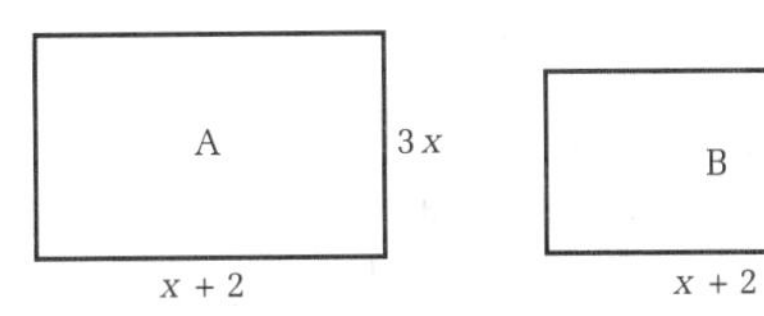

P **In problems 20 and 21, find the values of *k* for which the expressions are perfect-square trinomials.**

20 $x^2 + 8x + k$ 16

21 $x^2 + kx + 25$ ± 10

R **22** Use the diagram to solve for (x, y, z). $(x, y, z) = (6, 1, 161)$

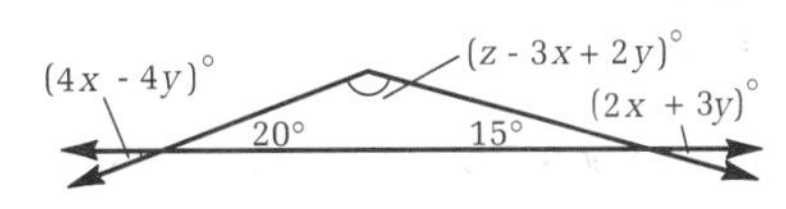

R,I **23** Given the two functions $f(x) = x + 5$ and $g(x) = x - 5$, evaluate each of the following.

a $f(7)$ 12 **b** $g[f(7)]$ 7 **c** $g(a)$ $a - 5$ **d** $f[g(a)]$ a

R,I **24** The perimeter of the figure shown is 20. For what values of x and y will the figure have the greatest possible area? $x = 5$; $y = 5$

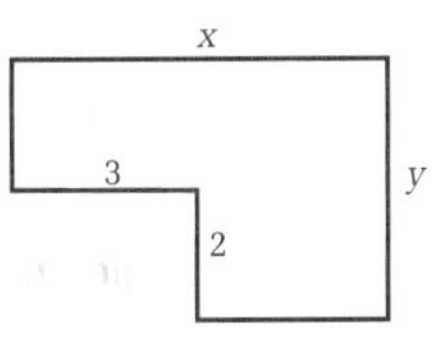

R,I **B** **25** Solve $2x^2 - (x + 1)(x + 2) = 8$ for x. $x = 5$ or $x = -2$

R,I **26** Write an equation of the parabola in the form $y = ax^2 + bx + c$. $y = x^2 - 2x - 3$

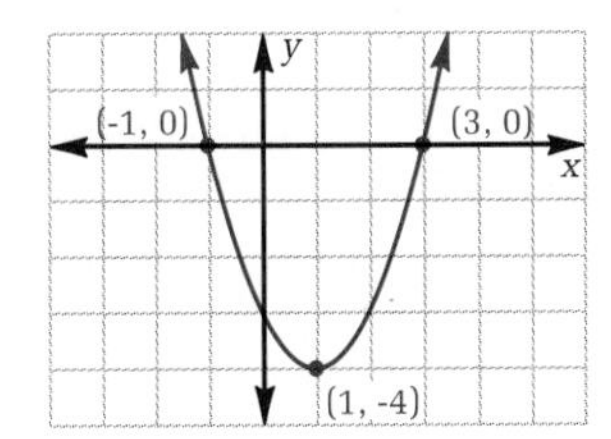

Problem-Set Notes and Strategies, continued

- In problem **17**, students may need to be reminded that opposite interior angles of parallel lines cut by a transversal are congruent.
- Problem **19** is an introduction to applications of quadratic equations that will be examined in more detail in Section **4.6.**
- Problem **22** is a geometric problem which can be expressed as an algebraic problem of three equations in three variables.
- Composite functions are used in problem **23**. Suggest that students follow the order of operations and operate on the inside function first. Then use the answer in the remaining function.
- Problem **26** may be done by putting points in vertex form and finding *a*, *h* and *k* or by substituting the points into the equation $ax^2 + bx + c = 0$ and solving the system of three equations in *a*, *b* and *c*. A parabola is uniquely determined by any three points using the latter method.

Problem Set, *continued*

P,I **27** For what value(s) of k does $f(x) = x^2 + kx + 16$ have exactly one real zero? ± 8

P **28** Write an equation to describe the graph shown. Any equation of the form $y = a(x - 4)^2 - 1$

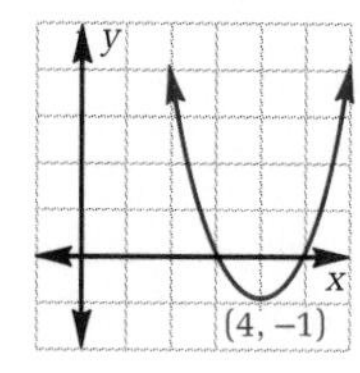

R,I **29** Consider the equation $(12x)^2 - 6(12x) - 7 = 0$.

a Substitute z for $12x$ and solve for z. $z = 7$ or $z = -1$

b Use your answer to part **a** to solve the original equation for x. $x = \frac{7}{12}$ or $x = -\frac{1}{12}$

P,R,I **30** **a** Write a function for $d(x)$, the vertical distance between the two curves.

b For what value of x is $d(x)$ a minimum? 1

c What is the minimum value of $d(x)$? 2

a $d(x) = 2x^2 - 4x + 4$

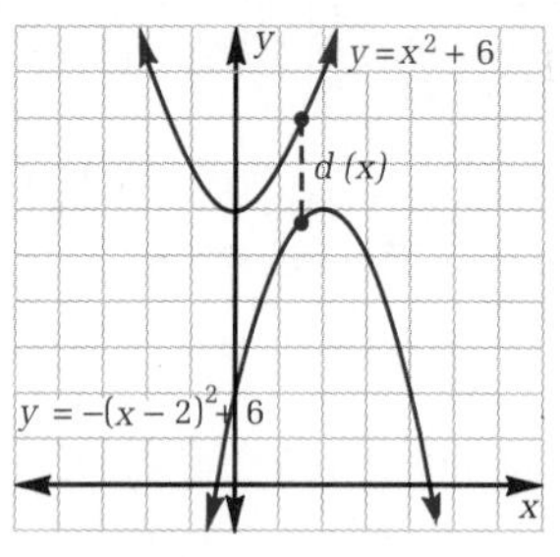

P **In problems 31 and 32, use the quadratic formula to find the roots of each equation.**

31 $x^2 - 3x + 1 = 0$ $x = \frac{3 \pm \sqrt{5}}{2}$

32 $3x^2 + 1 = 8x$ $x = \frac{4 \pm \sqrt{13}}{3}$

P,I **33** An archer shoots an arrow. Its heights (in inches) at three times (in seconds) are given in the table.

Time (t)	3	5	8
Distance above ground (s)	326	350	146

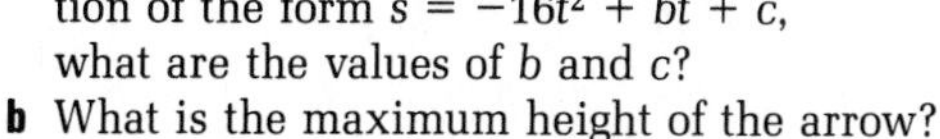

a If the arrow's path satisfies an equation of the form $s = -16t^2 + bt + c$, what are the values of b and c?

b What is the maximum height of the arrow?

c How far above the ground was the arrow when it was released (at $t = 0$)?

a 140; 50 b 356.25 in. c 50 in.

R,I **34** Write, in the form $ax^2 + bx + c = 0$, a quadratic equation whose roots are

a -5 and 7 $x^2 - 2x - 35 = 0$

b $\frac{1}{2}$ and 4 $2x^2 - 9x + 4 = 0$

Problem-Set Notes and Strategies, continued

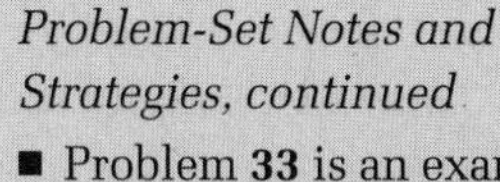

- Problem **33** is an example of a projectile motion problem and these problems will be explored in detail in the next section. For now have the students substitute points and generate a system of linear equations to find b and c.
- Answers may vary for problem **34**, but should differ by no more than a constant.

R **35** Solve the following system for (x, y). (x, y) = (3, −1.5)

$$\begin{cases} [4\ \ -2]\begin{bmatrix} x \\ y \end{bmatrix} = [15] \\ [x\ \ y]\begin{bmatrix} 3 \\ 4 \end{bmatrix} = [3] \end{cases}$$

R **36** Solve each equation for y.

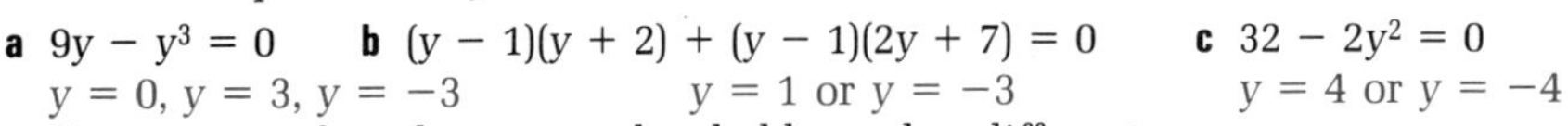

a $9y - y^3 = 0$ $y = 0, y = 3, y = -3$ **b** $(y - 1)(y + 2) + (y - 1)(2y + 7) = 0$ $y = 1$ or $y = -3$ **c** $32 - 2y^2 = 0$ $y = 4$ or $y = -4$

R **37** Different sizes of staples are used to hold together different numbers of sheets of paper. A $\frac{1}{4}$-inch staple can hold about 30 sheets, a $\frac{3}{8}$-inch staple can hold about 70 sheets, and a $\frac{1}{2}$-inch staple can hold about 100 sheets. Estimate the number of sheets a $\frac{5}{8}$-inch staple might hold? Between 130 and 140 sheets.

R **38** If the area of the trapezoid is 30, what is the value of x? 4

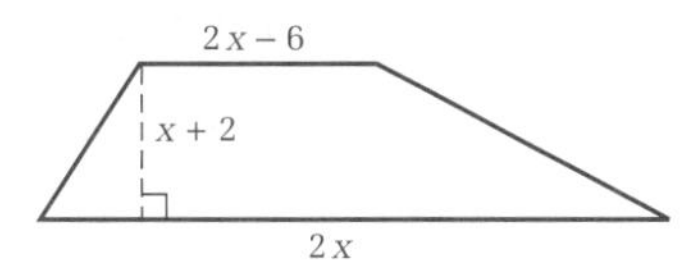

R,I **C** **In problems 39 and 40, solve for x.**

39 $2^{x^2-5x} = 2^6$ $x = 6$ or $x = -1$

40 $2^{x^2+5x} = \frac{1}{16}$ $x = -4$ or $x = -1$

R,I **41** If $f(x) = \frac{9x + 4}{x^2 + 2x - 5}$, what are the restrictions on the value of x? $x \neq -1 \pm \sqrt{6}$

R **42** Find the coordinate of x if the sum of its distance from 6 and the square of its distance from 3 is a minimum. What is the minimum sum? $\frac{7}{2}; \frac{11}{4}$

P,I **43** **a** Find the coordinates of the vertex of the graph of $y = 4x^2 + 17x - 12$. $\left(-\frac{17}{8}, -\frac{481}{16}\right)$

b Find the coordinates of the vertex of the graph of $y = ax^2 + bx + c$. $\left(-\frac{b}{2a}, -\frac{b^2 - 4ac}{4a}\right)$

P,R **44** Find the value(s) of k for which the graph of $y = kx$ intersects the graph of $y = x^2 + 25$ in only one point. ±10

R **45** Consider the set $A_x = \{n_1, n_2, \ldots, n_x\}$. Write a function for $f(x)$, the number of subsets of A_x. $f(x) = 2^x$

P,R **46** The difference between two numbers is 12, and the product of the numbers is 50. Find the sum of the squares of the numbers. (Hint: Start by finding $[x - y]^2$.) 244

Problem-Set Notes and Strategies, continued

- In problems **39** and **40**, as long as the base is the same, the exponents may be set equal and the problem solved.
- Students should recall that a denominator of a fraction may not equal zero for problem **41**.
- Problem **43b** produces a general formula for the vertex of a parabola.
- Problem **45** may take some explanation. Since an element is either in a subset or not in a subset, there are 2 possibilities. Since there are x elements in the original set and each has 2 possibilities there must be 2^x different possible subsets.
- In problem **46**, since $(x - y)^2 = x^2 - 2xy + y^2$, we know from the statement of the problem that $(12)^2 = x^2 - 2xy + y^2$. Using algebra, $144 + 2xy = x^2 + y^2$. Since the product of the numbers is 50, $144 + 2(50) = x^2 + y^2 = 244$.

4.6

Class Planning

Time Schedule
All levels: 1 day

Resource References
Teacher's Resource Book
Class Opener 25
Practice 26

Class Opener

Cat L. Brander needs to build a corral for his livestock. One side of the corral can be determined by the barn, but he needs to fence in the other three sides. He decided on a rectangular shape and has enough material to build 300 feet of fence. What dimentions should Cat use to make the corral as large as possible?
75 by 150

Lesson Notes

- This section explores applications of algebraic techniques to word problems. Emphasize the importance of the quadratic as a model for maximum and minimum problems. A business tries to maximize profits and minimize costs. An engineer tries to maximize the efficiency of a new engine design. A chemist tries to maximize the efficiency of a new drug while minimizing any side effects that may occur.

4.6 APPLICATIONS OF QUADRATIC FUNCTIONS

Objectives

After studying this section, you will be able to

- Analyze and solve problems involving maximum and minimum values
- Analyze and solve projectile-motion problems

Part One: Introduction

Maximums and Minimums

Problems in business and in the physical and social sciences can often be solved by mathematical modeling. Whenever a problem can be modeled by a quadratic function, for example, we can find the maximum or minimum value of the function, and thus the best or worst outcome of the problem situation.

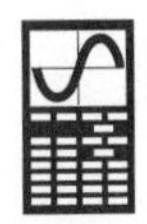

See p. 805

Example *Andy wants the greatest possible harvest from his apple orchard. The orchard contains 220 dwarf trees per acre now, and Andy harvests, on the average, 1300 apples per tree. Because his trees are planted in rows of 10, he adds or removes trees only in groups of 10. Andy's research has shown that for every additional 10 trees he plants per acre, the average yield per tree in the orchard will decrease by 50 apples. How many trees per acre should the orchard contain to produce the maximum harvest?*

Since Andy plants trees in multiples of 10, we will let x represent the number of groups of 10 trees he adds to each acre of the orchard. The number of trees on an acre will therefore be $220 + 10x$, and the average yield per tree will be $1300 - 50x$ apples.

Total yield per acre = (number of trees)(apples per tree)

$$= (220 + 10x)(1300 - 50x)$$
$$= 286{,}000 + 2000x - 500x^2$$
$$= -500(x^2 - 4x) + 286{,}000$$

Complete the square
$$= -500(x^2 - 4x + 4) + 286{,}000 + 2000$$
$$= -500(x - 2)^2 + 288{,}000$$

The maximum harvest, 288,000 apples per acre, occurs when $x = 2$. Andy should therefore plant 20 more trees per acre, so that the orchard contains 240 trees per acre. (If the value of x at the vertex had been negative, Andy would have needed to remove trees to obtain the maximum harvest.)

Projectile Motion

The term *projectile motion* refers to the motion of an object acted on only by the force of gravity. This is one of the many physical situations that can be modeled by quadratic functions. In this case, the height $h(t)$ of a projectile at any time t can be found with the quadratic function

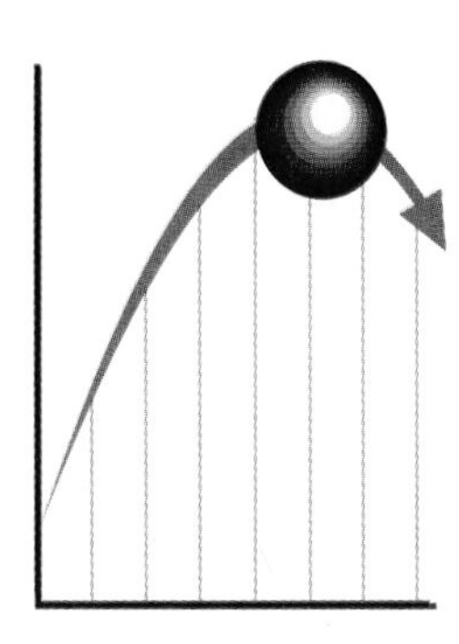

$$h(t) = -\frac{1}{2}gt^2 + v_0t + h_0$$

where g is the acceleration due to gravity, v_0 is the initial upward velocity of the object (the velocity at time zero), and h_0 is the initial height of the object. On earth, the value of g is approximately 9.8 meters per second per second (9.8 m/sec²), or 32 feet per second per second (32 ft/sec²).

See p. 805

Example *A soccer ball resting on the ground is kicked with an initial upward velocity of 48 feet per second.*

a *Write a function for $h(t)$, the height in feet of the ball at time t.*

We know that the problem can be modeled by a function of the form $h(t) = -\frac{1}{2}gt^2 + v_0t + h_0$. The initial height, h_0, is 0, and the initial velocity, v_0, is 48 feet per second. Since this velocity is stated in feet per second, we use 32 feet per second per second for g.

$$\begin{aligned} h(t) &= -\frac{1}{2}gt^2 + v_0t + h_0 \\ &= -\frac{1}{2}(32)t^2 + 48t + 0 \\ &= -16t^2 + 48t \end{aligned}$$

b *How many seconds after being kicked will the ball hit the ground?*

When the ball hits the ground, its height will be 0, so we set $h(t)$ equal to 0 and solve for t.

$$\begin{aligned} 0 &= -16t^2 + 48t \\ 0 &= -16t(t - 3) \\ t &= 0 \text{ or } t = 3 \end{aligned}$$

The ball is on the ground at time zero (when it is kicked) and will return to the ground 3 seconds later.

c *What is the greatest height the ball will reach?*

The maximum height of the ball is the value of the function at the vertex of its graph. We can use either of two methods to determine this value.

Lesson Notes, continued

- If students are currently taking physics, relate quadratic equations to problems in projectile motion. The first major equation in Newtonian Mechanics is $s = -\frac{1}{2}at^2 + v_0t + s_0$ which describes the position of a particle in time. This is a similar equation for projectile motion which is utilized in this section.
- You may want to point out to students that a branch of mathematics called Operations Research is devoted entirely to solving maximum and minimum problems.

Communicating Mathematics

Discover and write several examples of how maximizing and minimizing certain functions are carried out in your daily life.

Method 1: Completing the Square

$$\begin{aligned} h(t) &= -16t^2 + 48t \\ &= -16(t^2 - 3t) \\ &= -16\left[t^2 - 3t + \left(\frac{3}{2}\right)^2\right] - (-16)\left(\frac{3}{2}\right)^2 \\ &= -16\left(t - \frac{3}{2}\right)^2 + 36 \end{aligned}$$

The ball will reach a maximum height of 36 feet, 1.5 seconds after being kicked.

Method 2: Symmetry of the Graph

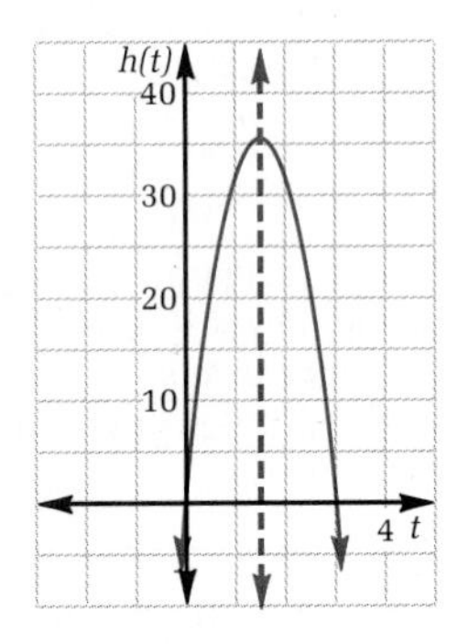

A parabolic graph of the function can also be used to find the maximum height. In part **b,** we found that the solution of the quadratic equation $0 = -16t^2 + 48t$ is $t = 0$ or $t = 3$. Thus, the t intercepts of the parabola are 0 and 3. We know that a parabola is symmetrical about its vertex, so the function's value will be at a maximum when t is midway between 0 and 3—that is, when $t = 1.5$.

$$\begin{aligned} h(t) &= -16t^2 + 48t \\ &= -16(1.5)^2 + 48(1.5) \\ &= -16(2.25) + 72 \\ &= 36 \end{aligned}$$

Again, the maximum height is 36 feet.

Part Two: Sample Problems

Problem 1 *The graphs of $y = x^2 + 8$ and $y = 2x$ are shown. Let $d(x)$ be the vertical distance between the graphs.*

a *Write a function for $d(x)$.*

b *For what values of x is the vertical distance 10?*

c *For what value of x is the vertical distance a minimum?*

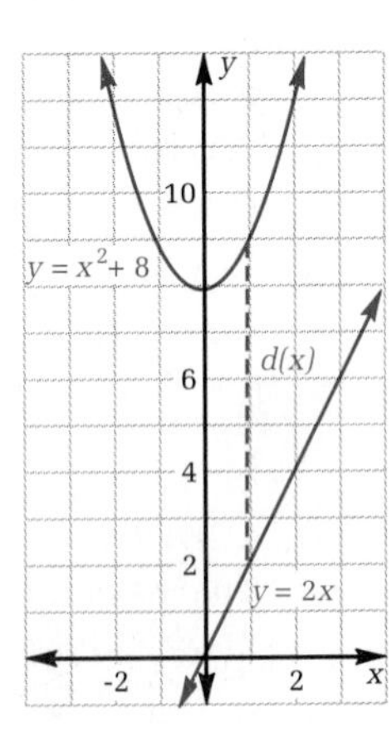

Solution

a For any value of x, the vertical distance between the graphs will be the difference between the corresponding y values.

$$d(x) = (x^2 + 8) - 2x$$
$$= x^2 - 2x + 8$$

b To find the value of x for which $d(x)$ is 10, we substitute 10 for $d(x)$ and solve the equation.

$$d(x) = x^2 - 2x + 8$$
$$10 = x^2 - 2x + 8$$
$$0 = x^2 - 2x - 2$$

By the quadratic formula,

$$x = \frac{2 \pm \sqrt{4 + 8}}{2}$$
$$= \frac{2 \pm 2\sqrt{3}}{2}$$

The distance between the graphs is 10 when x is approximately 2.73 or −0.73.

c Since $d(x)$ is a quadratic function, the distance will be a minimum at the vertex of its graph.

$$\mathbf{d(x) = x^2 - 2x + 8}$$
Complete the square $$\mathbf{= (x^2 - 2x + 1) + 8 - 1}$$
$$\mathbf{= (x - 1)^2 + 7}$$

The vertex is at (1, 7). The minimum distance of 7 is reached when $x = 1$.

Problem 2 *In 1981, there were 1000 fish in Ekal Lake. The fish population of the lake peaked at 8000 in 1986. If the number of fish varied according to a quadratic function of the form $p(x) = a(x - h)^2 + k$, where x is the number of years after 1981, what are the values of a, h, and k? When were there 6000 fish in the lake?*

Solution The maximum population of 8000 was reached 5 years after 1981. The vertex of the parabola is therefore (5, 8000), so $h = 5$ and $k = 8000$.

$$p(x) = a(x - h)^2 + k$$
$$= a(x - 5)^2 + 8000$$

In 1981, when the value of x was 0, there were 1000 fish, so we can substitute 1000 for $p(x)$ and 0 for x and solve for a.

$$p(x) = a(x - 5)^2 + 8000$$
$$1000 = a(-5)^2 + 8000$$
$$-7000 = 25a$$
$$-280 = a$$

Thus, our function is $p(x) = -280(x - 5)^2 + 8000$. To find out when there were 6000 fish, we let $p(x) = 6000$ and solve for x.

Using Computers

Here is another excellent area to use graphing software, graphing calculators, or symbolic manipulators. The technology enables students to concentrate on the meaning and mathematization of the problem and lets the computer/calculator worry about the solving algorithm. This is one of the important insights students can gain and this section is an excellent place for this to happen.

Checkpoint

1. In the equation $h(x) = -16t^2 + 24t + 125$, what is the significance of the −16?
   It is half of the gravitational force of 32 feet per second squared.
2. In the equation $h(x) = -16t^2 + 24t + 125$, what is the significance of the 24?
   It is the initial velocity of the projectile.
3. In the equation $h(x) = -16t^2 + 24t + 125$, what is the significance of the 125?
   It is the initial height of the projectile.

4.6

Assignment Guide

Average
7–13, 15–17, 20–22
Advanced
8–14, 16, 17, 20–22, 24, 25
Honors
8, 9, 11–14, 17, 21, 22, 24–27

Problem-Set Notes and Strategies

- Problem **6** is a business problem. If the variable represents the number of 10 cent increases and not the value of the increases, the computation is simpler.
- Problem **7** represents a parabola with the vertex at a minimum and not a maximum. Care should be taken to also evaluate the endpoints of the domain to find the maximum.

$$6000 = -280(x - 5)^2 + 8000$$
$$\frac{-2000}{-280} = (x - 5)^2$$
$$(x - 5)^2 \approx 7.1429$$
$$|x - 5| \approx 2.6726$$
$$x \approx 7.6726 \text{ or } x \approx 2.3274$$

There were 6000 fish after about $2\frac{1}{3}$ years and again after about $7\frac{2}{3}$ years—that is, sometime in 1983 and sometime in 1988.

Part Three: Exercises and Problems

Warm-up Exercises

In problems 1–4, use the projectile-motion function to determine the initial upward velocity (v_0) and the initial height (h_0) of the object.

P **1** $h(t) = -\frac{1}{2}(32)t^2 + 45t$ $v_0 = 45$ ft/sec; $h_0 = 0$ ft

2 $h(t) = -\frac{1}{2}(9.8)t^2 + 32t + 2$ $v_0 = 32$ m/sec; $h_0 = 2$ m

3 $h(t) = -4.9t^2 + 25t + 0.5$
$v_0 = 25$ m/sec; $h_0 = 0.5$ m

4 $h(t) = -16t^2 + 31t$
$v_0 = 31$ ft/sec; $h_0 = 0$ ft

P **5** Each of the following functions describes the motion of an object acted on by the force of gravity. Find the greatest height reached by each object. (Hint: Complete the square.)

a $h(t) = -16t^2 + 64t + 100$ 164 ft

b $h(t) = -4.9t^2 + 19.11t + 5$ $\approx$23.6 m

P **6** A Star-Go-Stop store charges $2 for a one-day rental of a video cassette. On the average, 200 cassettes are rented from the store each day. If a survey indicates that the store's rentals will decrease by an average of 5 per day for each 10-cent increase in rental charge, what should the store charge to maximize its income? $3 per cassette

Problem Set

A

P **7** The Epic Toothpick Company can produce up to 700 cases of toothpicks per day. The company's profit P, in thousands of dollars, is given by the formula $P = (x - 3)^2 + 5$, where x represents the number of hundreds of cases produced. What is the maximum profit the company can make in a single day? $21,000

P,R **8** **a** Write a function for $A(x)$, the area of the rectangle shown. $A(x) = x^2 + 6x - 16$

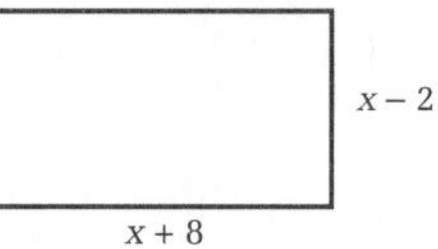

b For what values of x does this function have meaning? $\{x: x > 2\}$

c Find the area of the rectangle when $x = 10$. 144

d For what values of x is the area greater than 75? $\{x: x > 7\}$

P **9** Jacob is standing at the edge of a 3000-foot-deep canyon. He kicks a ball into the air with an initial upward velocity of 32 feet per second.

a When will the ball return to the height from which it was kicked? In 2 sec
b What is the greatest height above the canyon's edge the ball will reach? 16 ft
c When will the ball hit the floor of the canyon? In ≈14.7 sec

R **10** What is the set of two-digit numbers in which the sum of the units digit and the tens digit is less than 4? {10, 11, 12, 20, 21, 30}

P **11** A rectangular plot ABCD is to be created by dividing and fencing a garden as shown. If 100 feet of fencing is used, what is the maximum area of ABCD? $833.\overline{3}$ ft²

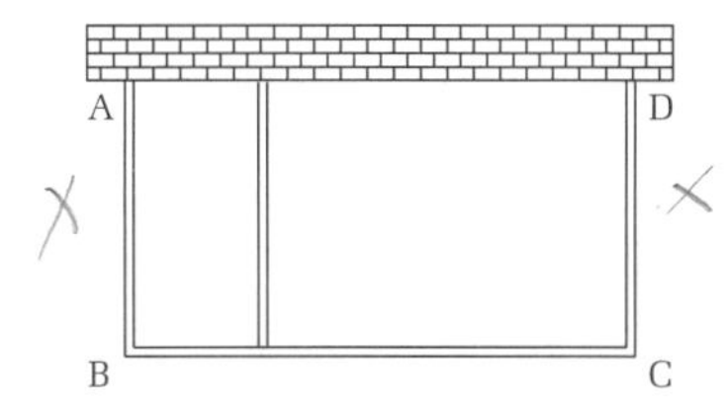

R **12** **a** Find the x-intercepts of the graph of the function $f(x) = x^2 - 7x + 10$. 2 and 5
b For what value of x will the value of $f(x)$ be a minimum? $\frac{7}{2}$

P **13** Isis sells lemon ices at her lemonade stand. She charges $1.20 per ice and sells an average of 50 per day. If she could sell 1 more ice each day for each 2 cents she decreased the price, at what price should Isis sell her ices to maximize her income from them? $1.10

P **14** Jesse dives into a pool from the 3-meter springboard, with an initial upward velocity of 2.94 meters per second.

a Write a function that describes Jesse's vertical motion. $h(t) = -\frac{1}{2}(9.8)t^2 + 2.94t + 3$
b What is Jesse's maximum height above the water? ≈3.44 m
c After how many seconds does Jesse hit the water? ≈1.14 sec

R   **15** Solve $0.043x + 0.172x - 11.342 = 0.623x + 7.52$ for x. $x \approx -46.23$

R,I **16** Solve each of the following equations for x.

a $|x| = 9$ $x = \pm 9$ **b** $x^2 = 81$ $x = \pm 9$

P,I **17** Refer to the diagram of triangle ABC.

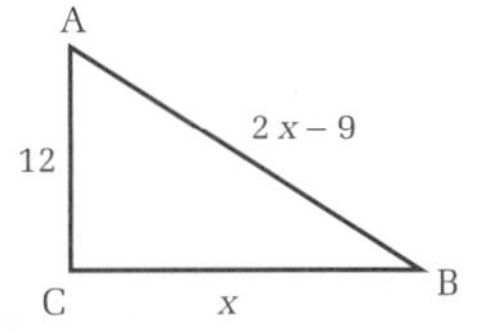

a Solve for x, leaving your answer in simplified radical form. $x = 6 + \sqrt{57}$
b Between what two consecutive integers is the radical term of your answer to part **a**? $\sqrt{57}$ is between 7 and 8.
c Use your answer to part **b** to estimate the length of $\overline{CB}$. Between 13 and 14

Problem-Set Notes and Strategies, continued

- Problem **9** is a vertical motion problem with an initial height of 3000. This problem could also be done with the cliff at height zero and the canyon bottom at −3000 feet. Each will give the same answers.
- Labeling the picture may help establish the formula in problem **11**. Two equations could be generated and substitution used.

Problem Set, *continued*

R,I **In problems 18–20, solve for the variable.**

18 $2x(x - 5) = 0$
$x = 0$ or $x = 5$

19 $2b(b - 3) + 5(b - 3) = 0$
$b = -\frac{5}{2}$ or $b = 3$

20 $4w^2 - 3w = 1$
$w = -\frac{1}{4}$ or $w = 1$

P **21** Find the value(s) of x for which ∠PQR is a right angle. $x = 15$ or $x = -6$

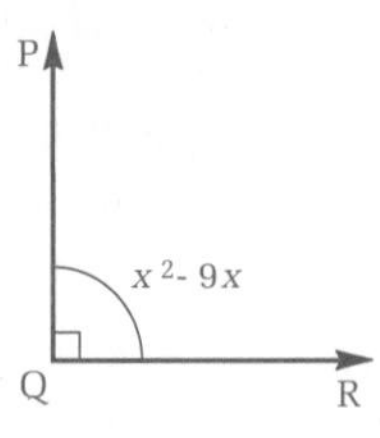

B P **22** A water-skier jumps from a 6-foot-high ski jump with an initial upward velocity of 18 feet per second.

a When will the skier reach her greatest height above the water? In ≈0.56 sec

b When will the skier reach the height of the jump again? In ≈1.13 sec

c When will the skier land on the water? In ≈1.39 sec

R **23** A number increased by 20% of itself exceeds 80. Find the numbers. $\{n : n > 66\frac{2}{3}\}$

R,I **24** Graph $y = f(x) = x^2 - 6x - 16$ and $y = g(x) = f(2x)$ on the same coordinate plane. What are the x-intercepts of the graphs of f and g?
$-2, 8; -1, 4$

R,I **25** Find the value(s) of x for which AC is the geometric mean of CB and AB. (You may recall that the geometric mean of two numbers a and b is $\pm\sqrt{ab}$.)
$x = \frac{-1 + \sqrt{5}}{2}$

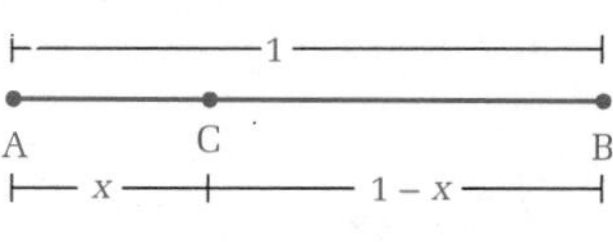

C P **26** A model rocket is fired into the air. After 0.83 seconds, when its fuel is completely consumed, the rocket is 123 meters above the ground, traveling at a rate of 110.25 meters per second.

a When is the rocket 500 meters above the ground? $t \approx 5.04$ sec and $t \approx 19.12$ sec

b What is the greatest height reached by the rocket? ≈743.16 m

c How much time elapses between the rocket's launch and its return to earth? ≈24.40 sec

R,I **27** An isosceles trapezoid has sides as shown.

a Write an equation expressing the area A of the trapezoid as a function of x.

b Use a graphing calculator or a computer spreadsheet program to find the approximate value of A_m, the maximum area of the trapezoid.

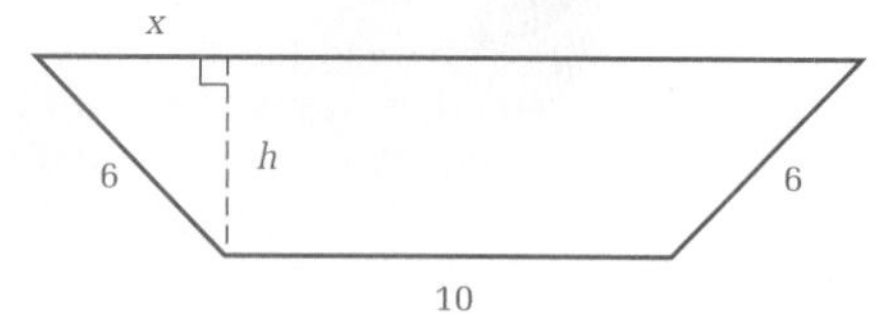

a $A = (10 + x)\sqrt{36 - x^2}$ **b** ≈68.19

Problem-Set Notes and Strategies, continued

- In number **23**, remind students that a 20% increase means that the value is 1.20 times its original value or the original value plus .20 times the original value.
- In problem **25**, the sum of *x* and 1 is the golden ratio. This is a famous number related to the golden rectangle and information may be found in most math history texts. There is also a book called *The Divine Proportion* published by Dover Press which is devoted to this number. An interesting fact is that its reciprocal is equal to itself minus 1.
- Be careful in problem **26**. The projectile motion model takes effect only after the fuel is consumed. This needs to be taken into consideration when solving the problem. Before that time, there are other forces acting in addition to gravity. This is a good problem for prospective and current physics students.

4.7 DATA DISPLAYS

Objectives

After studying this section, you will be able to

- Read and interpret typical data displays
- Construct and interpret stem-and-leaf plots
- Construct and interpret box-and-whisker plots

Common Data Displays

The United States is rapidly becoming an "information society." In the future, much of the information on which you will base important decisions will consist of numerical data. It will be essential for you to be able to read, interpret, and work with this information.

Numerical data may be displayed in many ways. The displays that follow are typical of the types seen in newspapers, magazines, and computer-produced reports. Study each until you understand the information it contains.

Horizontal Bar Graph

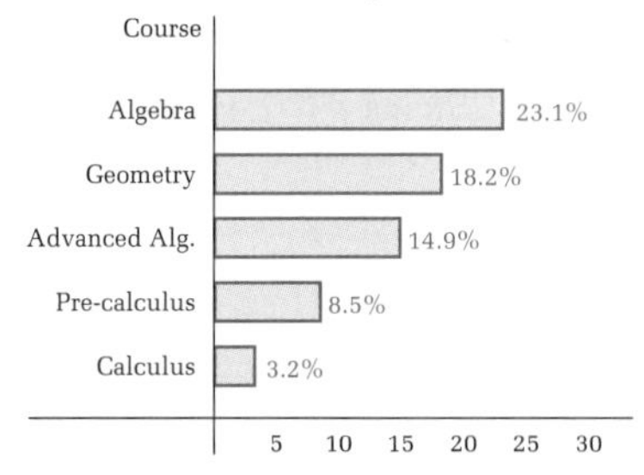

Percent of students in Apollo High School enrolled in college-preparatory math classes–1989

Vertical Bar Graph

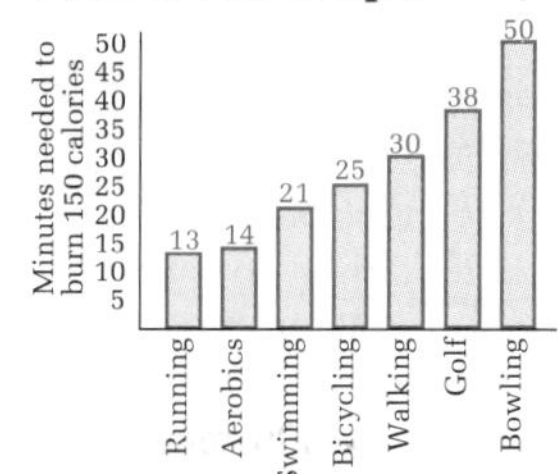

Frequency Polygon

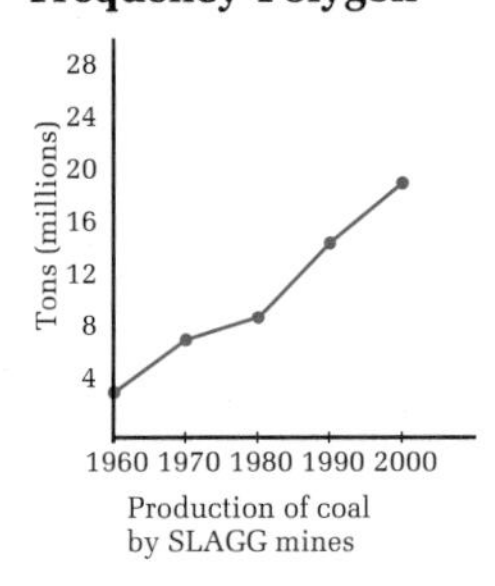

Production of coal by SLAGG mines

Circle (Pie) Graph

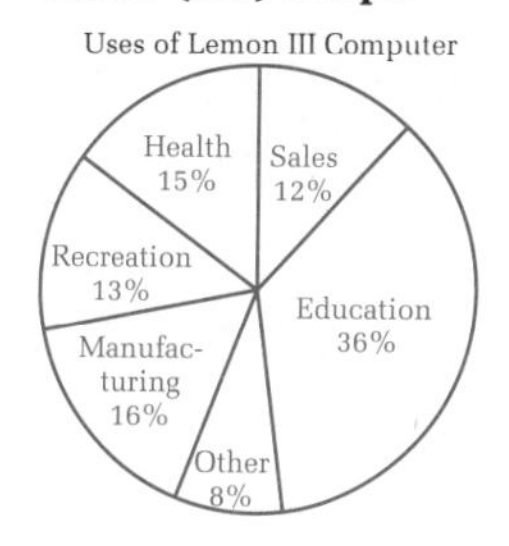

Class Planning

Time Schedule
All levels: 1 day

Resource References
Teacher's Resource Book
Class Opener 26
Practice 27
Manipulatives and Applications 8
Transparency 7, 8

Class Opener

The following lists the number of major volcano errruptions and landslides and the number of resulting deaths in ten year periods from 1900 to 1980.

Years	Erruptions and Landslides
1900-1909	6
1910-1919	1
1920-1929	0
1930-1939	0
1940-1949	0
1950-1959	6
1960-1969	6
1970-1979	8

Estimated Deaths
34,500
5,000
5,029
7,288
2,866

What would be an efficient way to communicate this information? Answers will vary. Students may have difficulty displaying all the information. This can lead to a discussion of the problems of accurately communicating data as well as different styles of graphs — stem and leaf, box and whisker, etc — and when each is most helpful.

Vocabulary
horizontal bar graph
vertical bar graph
frequency polygon
circle graph, chart
pictograph
artistic graph
stem-and-leaf plot
stem, leaf
legend
outlier scores
median
box-and-whisker plot
quartile

4.7

Lesson Notes

- This section introduces the graphical representations of data that statisticians frequently use. Students will be familiar with several of these types of graphs through reading newspapers and magazines. Stem-and-leaf plots are used to allow the data itself to give a picture of the distribution of values. A quick glance can tell if the data are skewed or symmetric. The middle of the data can be estimated easily and the actual data values are available if any computation is necessary. This plot may be changed to a bar chart for visual effects, but the data values are lost. A box-and-whisker plot allows the user to determine how the data is dispersed about the middle. The middle, in this case the median, is the center line and the data in the box comprise the middle 50%. A visual interpretation of the range from lowest to highest as well as how the points are dispersed is shown by the size of the box and length of the whiskers. These can be related to the standard deviation in Chapter **6**.

Chart

Production cost of a service manual for a 1990 Special Motors engine

Number of copies	Number of pages 48	64	80	96
10,000	2.60	3.00	3.30	3.20
8,000	2.80	3.15	3.40	3.35
6,000	3.05	3.40	3.65	3.75
4,000	3.35	3.75	3.95	4.15

Pictograph

Number of garages built by Slapmup Construction Co.

1986

1987

1988

1989

= 10 garages

More elaborate data displays can be generated by using computer graphics. We call such displays ***artistic graphs.***

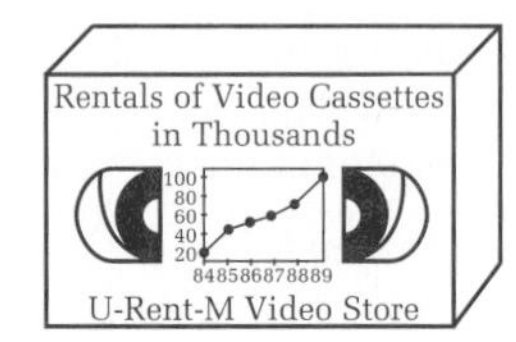

Stem-and-Leaf Plots

The following data represent the daily numbers of patients treated at the Mercy Hospital emergency room during August 1989.

16, 25, 31, 18, 16, 14, 37, 15, 26, 22, 30, 45, 34, 56, 27, 31, 19, 27, 46, 25, 33, 42, 19, 29, 34, 44, 28, 34, 26, 31, 20

We can organize these data in a ***stem-and-leaf plot.***

To draw a stem-and-leaf plot, we identify the low number of patients treated, 14, and the high number of patients treated, 56. We use the tens digits as the ***stems*** of the plot. The stems are listed vertically with a line drawn to the right.

Stems
1
2
3
4
5

We now enter each data value. The first number, 16, has a tens digit of 1. Its units digit, 6, becomes a ***leaf.*** We put the leaf 6 to the right of the stem 1. The second number, 25, has a tens digit of 2 and a units digit of 5. We put the leaf 5 to the right of the stem 2. We continue through all the data, producing the plot shown below.

Stems	Leaves
1	6
2	5
3	
4	
5	

Stems	Leaves									
1	6	8	6	4	5	9	9			
2	5	6	2	7	7	5	9	8	6	0
3	1	7	0	4	1	3	4	4	1	
4	5	6	2	4						
5	6									

We now rearrange each row of leaves in order from smallest to largest. Beneath the plot, we give a ***legend*** to explain the meaning of the data.

Stems	Leaves									
1	4	5	6	6	8	9	9			
2	0	2	5	5	6	6	7	7	8	9
3	0	1	1	1	3	4	4	4	7	
4	2	4	5	6						
5	6									

4 | 2 indicates that 42 patients were treated one day in August 1989

A stem-and-leaf plot can easily be converted into a bar graph.

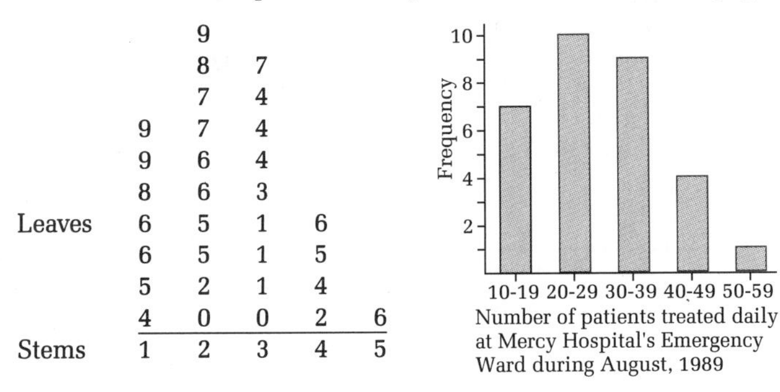

Box-and-Whisker Plots

The following data are the scores of 19 golfers finishing 18 holes in the Metro Open, arranged in ascending order.

63, 64, 66, 67, 67, 68, 68, 68, 69, 70, 71, 71, 71, 72, 73, 75, 76, 76, 78

The low score is 63, and the high score is 78. These are called the ***outlier scores*** of the data. The score 70 is called the ***median*** of the data, since it is the middle score of the set of data.

Now consider the nine scores below 70. Their median (the fifth score) is 67. Of the nine scores above 70, the median score is 73. We will use the two outliers (63 and 78), the median of the entire data (70), and the medians of the two halves (67 and 73) to draw a ***box-and-whisker plot*** of the golf scores.

We draw a number line, labeling the endpoints 63 and 78 and indicating the points that correspond to the values 67, 70, and 73.

The values 67, 70, and 73 are used as boundaries of the plot's boxes. The segments to the left and right of the boxes are the whiskers. We observe that the box-and-whisker plot is divided into four sections (two whiskers and two boxes). Included within each section is one-fourth (25%) of the data. Each of the numbers 67, 70, and 73, which divide the data into the four sections, is called a ***quartile***. The number 67 is designated Q_1, the number 70 (the median) is designated Q_2, and the number 73 is designated Q_3.

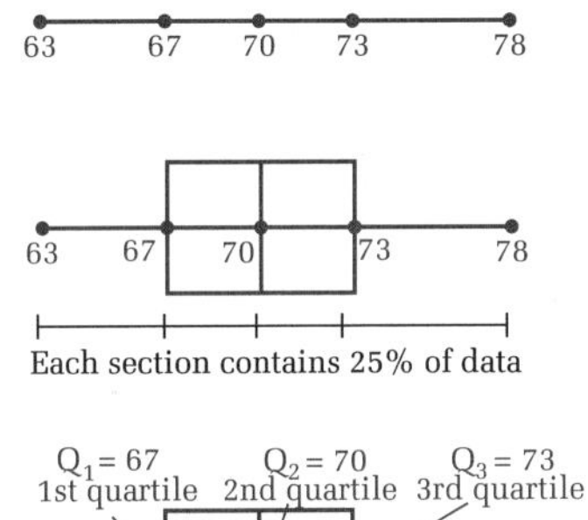

Sample Problem

Problem *The following data represent the numbers of words typed per minute on a typing test by 10 candidates for an executive secretarial job. Draw a box-and-whisker plot of the data.*

52, 64, 77, 81, 90, 94, 99, 100, 102, 124

Using Computers

A spreadsheet program such as *Quattro* or *Excel* with an excellent graphical component allows students to experiment with displaying data and creating impressive charts of data. The *Algebra Solver Kit* can be used for box and whisker and stem and leaf diagrams.

The work on composition of functions can be enhanced using a symbol manipulator such as *Derive.* A discovery laboratory using this software can help students see the relationships involved in forging the composition of two functions.

Checkpoint

1. With two-digit data, the leaf unit of a stem-and-leaf diagram will be which part?
   The units digit of the number.
2. The second quartile goes by another name. What is it?
   The median.
3. The right whisker of a box-and-whisker plot represents what portion of the data?
   The upper quartile or 25%.

4.7

Assignment Guide

Average
1–7, 9, 11
Advanced
1–9, 12, 13
Honors
1, 2, 5–10, 12–14

Problem-Set Notes and Strategies

- There are an even number of data points in problem **1** which may raise some questions. Point out to the students that the median is the average of the two middle terms and then they should be able to find the first and third quartile.
- Help students model their graph for problem **2** similar to the frequency polygon shown in the example. Suggest that students label each of the axes as that is the key to reading graphs. It might help to draw the plots on graph paper.

Solution The smallest and largest values are 52 and 124. These are the outlier values. The median in this case is the arithmetic mean of the two middle values, 90 and 94, so the median (Q_2) is 92. The median of the values below 92 (Q_1) is 77. The median of the values above 92 (Q_3) is 100.

Using the outliers and Q_1, Q_2, and Q_3, we draw the plot.

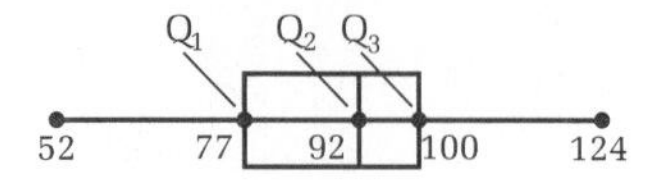

Problem Set

A

P **1** Jordan Michaels scored the following numbers of points in the first 30 games of the 1989–1990 basketball season.

37, 33, 41, 21, 29, 31, 39, 30, 22, 30, 18, 53, 32, 36, 32,
29, 23, 41, 34, 37, 40, 27, 33, 35, 21, 19, 20, 54, 29, 35

a Display this data in a stem-and-leaf plot.
b Summarize the data by means of a box-and-whisker plot.

P **2** The table below shows the records of the Apollo High School baseball team from 1980 through 1989.

	Wins	Losses
1980	16	13
1981	17	12
1982	23	8
1983	14	15
1984	19	10
1985	17	12
1986	22	7
1987	15	14
1988	13	16
1989	18	13

a On the same set of axes, draw frequency polygons to represent the number of wins per year and the number of losses per year.
b Find the average number of wins per year. 17.4
c In what year did the team win the greatest percentage of the games played? 1986
d By how much does the team's average number of wins per year exceed their average number of losses per year? 5.4

R **3** Determine whether each of the following statements is True or False.

a $\sqrt{36} = \sqrt{4}\sqrt{9}$ True
b $\sqrt{36} = \sqrt{16} + \sqrt{20}$ False
c $\sqrt{10} = \sqrt{5}\sqrt{2}$ True
d $\sqrt{10} = \sqrt{4} + \sqrt{6}$ False

Additional Answers

2a

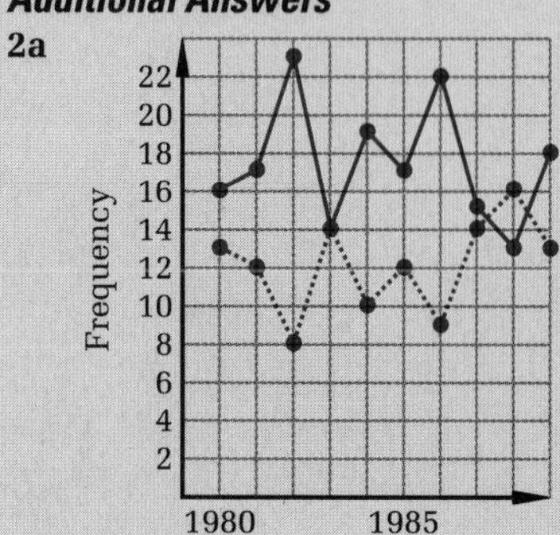

1a

5	3 4
4	0 1 1
3	0 0 1 2 2 3 3 4 5 5 6 7 7 9
2	0 1 1 2 3 7 9 9 9
1	8 9

2 | 1 means 21 points scored during a game

1b

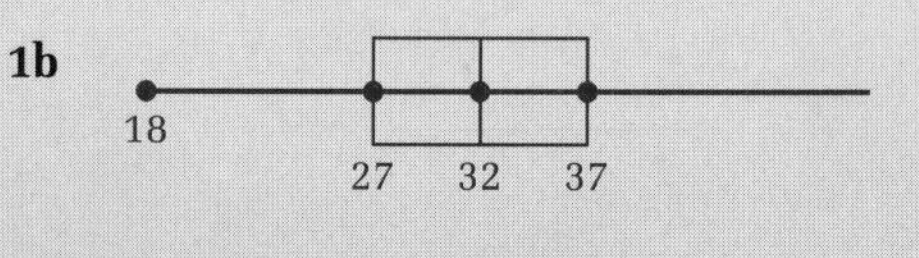

P **4** The stem-and-leaf plot below represents the numbers of radar-detector units sold per week at TV Hut during a 15-week period.

Stem	Leaves					
3	1	1	7			
2	0	2	3	5	6	9
1	3	5	7	7		
0	8	9				

2 | 3 indicates 23 radar-detector units sold during a week

a What was the greatest number of units sold during a week? 37
b What was the least number of units sold during a week? 8
c How many units were sold during the ninth week? Cannot be determined
d In how many weeks were more than 20 units sold? 8

I **5** There are 17 elements in set A and 10 elements in set B. If A ∪ B contains 21 elements, what is the value of x? 1

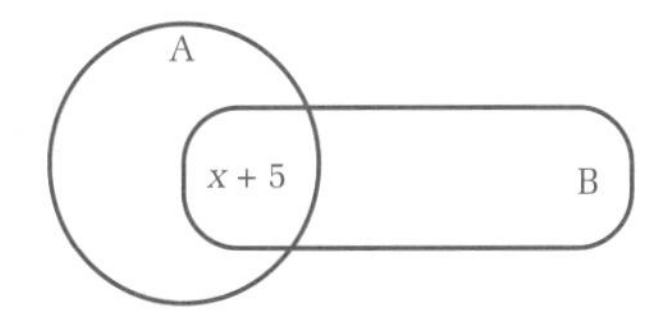

B

P **6** The box-and-whisker plot represents the scores of 85 students on a 90-item multiple-choice history exam.

a What were the high and low scores on the exam? 16, 88
b By how much does the highest score exceed the lowest score? 72
c The middle 50% of the scores are between what two numbers? 62 and 81
d What scores are below Q_3? Those below 81
e What is the average of the students' scores on the exam? Cannot be determined

R **7** Solve the following system for (x, y). (x, y) = (2, 5)

$$\begin{cases} 2^{x+3y} = 2^{17} \\ 2^{9x-y} = 2^{13} \end{cases}$$

P **8** Analyze a newspaper or magazine advertisement in which numerical data are used to convince the reader to buy the advertised product. Discuss whether the advertisement offers a convincing argument. Answers will vary.

P **9** The numbers of acres of corn planted in Sweetcorn County during a six-year period is shown by the frequency polygon.

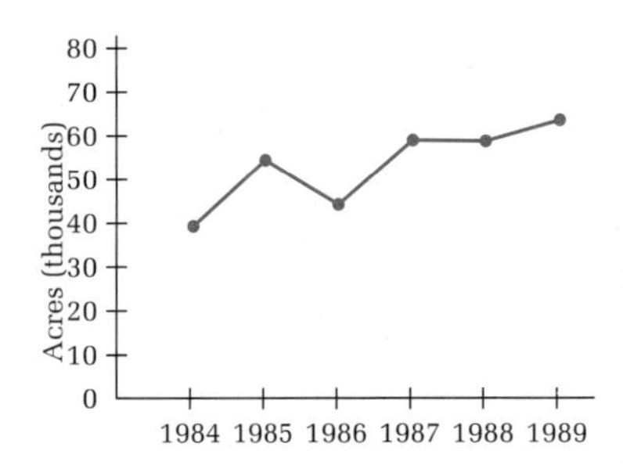

a By what percentage did the acreage planted in corn increase from 1984 to 1989? ≈62.5%
b By what percentage did the acreage planted in corn decrease from 1985 to 1986? ≈18.2%
c The 1988 crop produced about 90 bushels of corn per acre. How much corn was produced in Sweetcorn County that year? ≈5,400,000 bu

Problem-Set Notes and Strategies, continued

- Students need to read a box-and-whisker plot in problem **6**. Make sure students realize that even though the median is different from the average it is still a measure of the center of the data. With extreme values as in this data set, the median is usually a better measure of the middle.

- Problem **8** is an exercise in deciding on the value of a statistic. Have students read their examples and comment on their arguments. This is an excellent question for group discussion.

Additional Answer

10

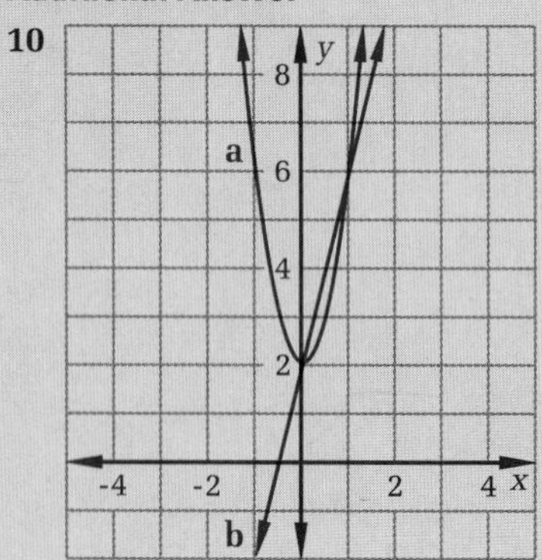

■ Problem **12** is an additional exercise in reading a stem-and-leaf diagram. The mode is introduced in part **c** and is another alternative measure of the center of the data.

Communicating Mathematics

A bar graph and a circle (pie) chart may be used to display the exact same data. Comment on when each might be used most effectively.

■ In problem **14**, you will probably have to remind students that the formula for the surface area of a sphere is $4\pi r^2$.

Problem Set, *continued*

R **10** Graph the following two equations on the same coordinate system.

a $y = 4x^2 + 2$ **b** $y = 4x + 2$

R **11** **a** Write an expression for the area of the shaded region. $x^2 + 6x$

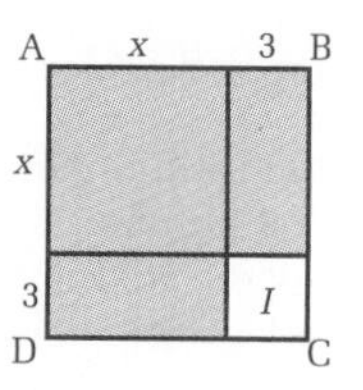

b What is the area of region I? 9

c Write two equivalent expressions for the area of ABCD. $(x + 3)^2$; $x^2 + 6x + 9$

C P **12** Mr. Matt Hematics made a stem-and-leaf plot of his pupils' scores on a test.

Stem	Leaves
9	1 4 7 7 8
8	3 5 8 8 8 9
7	0 1 2 5 6 6 6 6 6 8
6	2 5
5	0 2 9 9
4	
3	7
2	3

8 | 3 means a score of 83 on the test

a What were the highest and lowest scores? 98, 23

b A grade of C was given for a score in the 68–76 range. How many students received a C? 9

c What score occurred the greatest number of times? (This value is called the *mode* of the data.) 76

d Jeff scored 65 points. How many students scored fewer? 7

e How many students took the test? 29

f Who received the highest score? Cannot be determined

R **13** **a** What is the probability that a randomly selected two-digit number is a multiple of 7? $\frac{13}{90}$

b What is the probability that a randomly selected two-digit number is even and a multiple of 3? $\frac{1}{6}$

R **14** The radius of a large spherical balloon is 4 feet and is increasing at a rate of 2.5 feet per minute.

a When will the balloon's surface area be 400 square feet? In ≈0.66 min

b When the balloon is fully inflated, its radius is 14.8 feet. What is the surface area of the fully inflated balloon? ≈2752.5 ft^2

4

Chapter Summary

Concepts and Procedures

After studying this chapter, you should be able to

- Recognize quadratic expressions and equations (4.1)
- Multiply to form quadratics (4.1)
- Remove common factors from expressions (4.2)
- Factor trinomial products of linear binomials (4.2)
- Factor special products (4.2)
- Recognize quadratic functions and their graphs (4.3)
- Relate characteristics of a quadratic function to characteristics of its graph (4.3)
- Find the real zeros of quadratic functions (4.3)
- Solve quadratic equations by factoring (4.4)
- Solve quadratic equations by using the quadratic formula (4.4)
- Recognize when factoring can be used to solve a quadratic equation (4.4)
- Use the method of completing the square to determine the maximum or minimum value of a quadratic function (4.5)
- Solve quadratic equations by completing the square (4.5)
- Prove the quadratic formula (4.5)
- Analyze and solve problems involving maximum and minimum values (4.6)
- Analyze and solve projectile-motion problems (4.6)
- Read and interpret typical data displays (4.7)
- Construct and interpret stem-and-leaf plots (4.7)
- Construct and interpret box-and-whisker plots (4.7)

Vocabulary

artistic graph (4.7)
axis of symmetry (4.3)
box-and-whisker plot (4.7)
chart (4.7)
circle graph (4.7)
common factor (4.2)
completing the square (4.5)
discriminant (4.4)
factoring (4.2)
frequency polygon (4.7)
horizontal bar graph (4.7)
leaf (4.7)
legend (4.7)
median (4.7)
outlier score (4.7)
parabola (4.3)
pictograph (4.7)
quadratic equation (4.1)
quadratic expression (4.1)
quadratic formula (4.4)
quadratic function (4.3)
quartile (4.7)
real zero (4.3)
root (4.3)
stem (4.7)
stem-and-leaf plot (4.7)
vertex (4.3)
vertex form (4.3)
vertical bar graph (4.7)

4

Class Planning

Time Schedule
All levels: 1 day

Assignment Guide

Average
1, 3, 5, 7–10, 12, 13, 15, 16, 18, 19, 22–24, 26, 29, 31, 32, 33, 36, 38

Advanced
2, 3, 6, 8, 10, 13, 14, 16, 17, 19, 22–24, 26, 29, 31–34, 36–38

Honors
6, 9, 10, 11, 14, 17–19, 21, 23–25, 27–29, 31, 32, 34–36, 39–42

Problem-Set Notes and Strategies

- In problems **7 – 9**, make sure students are aware that when taking square roots, there are two answers, both a plus and a minus.
- Problem **16** may be solved by a little logic and deductive reasoning.

4 REVIEW PROBLEMS

P **A** **In problems 1–6, expand each expression.**

1 $3x(2x^2 - 4x)$ $6x^3 - 12x^2$

2 $(2x - y)(4x + 3y)$ $8x^2 + 2xy - 3y^2$

3 $(6y + z)(6y - z)$ $36y^2 - z^2$

4 $(6x - 4)(x + 5)$ $6x^2 + 26x - 20$

5 $(3x - 5)^2$ $9x^2 - 30x + 25$

6 $(x + 1)(x - 1)(x - 1)$ $x^3 - x^2 - x + 1$

P **In problems 7–9, solve for the variable by taking the square root of each side.**

7 $x^2 = 100$
$x = \pm 10$

8 $(y - 2)^2 = 49$
$y = 9$ or $y = -5$

9 $(2z - 1)^2 = 25$
$z = 3$ or $z = -2$

P **10** Write an equation of the parabola shown. $y = \frac{1}{2}(x - 2)^2 - 5$

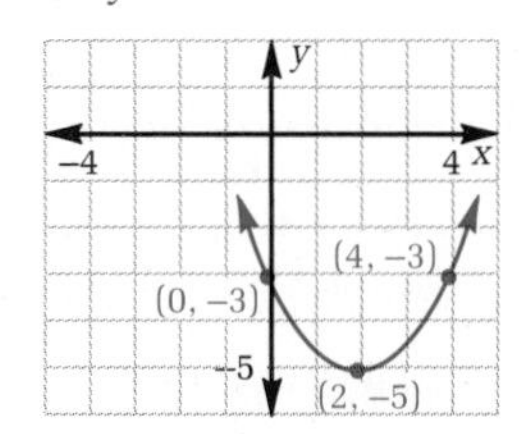

R **11** Graph the inequality $y \geq -4x - 10$. If a point is selected at random from $\{(-3, -2), (-1, -4), (2, 3), (5, -30), (0, -8), (2, -30), (3, -30), (8, -30)\}$, what is the probability that it represents a solution of the inequality? $\frac{5}{8}$

P **In problems 12–14, use the Zero Product Property to solve each equation.**

12 $(x + 1)(x - 3) = 0$
$x = -1$ or $x = 3$

13 $(x + 4)(x - 4) = 0$
$x = \pm 4$

14 $(2x + 1)(x + 3) = 0$
$x = -\frac{1}{2}$ or $x = -3$

R **15** **a** Write an equation expressing the area A of the triangle as a function of x. $A = 2x^2 - 6x$

b What is the domain of the function you wrote in part **a**? $\{x : x > 3\}$

c Find the area of the triangle if the base is three times the height. 216

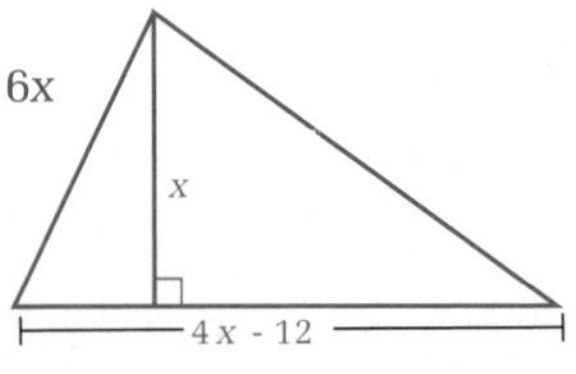

P,R **16** In a certain cubic polynomial the constant term is 5, the coefficient in the linear term is 7, and the coefficient in the quadratic term is twice the exponent in that term. The coefficient in the cubic term is the sum of the other two coefficients and the constant. Find the polynomial. $16x^3 + 4x^2 + 7x + 5$

P **In problems 17–22, factor each expression completely.**

17 $6y^2 - 5y + 1$ $(3y - 1)(2y - 1)$

18 $3x^2 - 5x$ $x(3x - 5)$

19 $6x^2 + 13x + 6$ $(3x + 2)(2x + 3)$

20 $4y^3 - 12y^2$ $4y^2(y - 3)$

21 $z^2 - 17z + 72$ $(z - 9)(z - 8)$

22 $6x^3 - 12x^2 + 6x$ $6x(x - 1)^2$

P **23** In 1988, there were 427 fires in Dawson City. The causes of the fires, as determined by the city's fire department, are shown in the graph.

a How many fires were caused by arson? 47

b How many fires were due to heating equipment? 56

c How many fires were due to electrical problems? 77

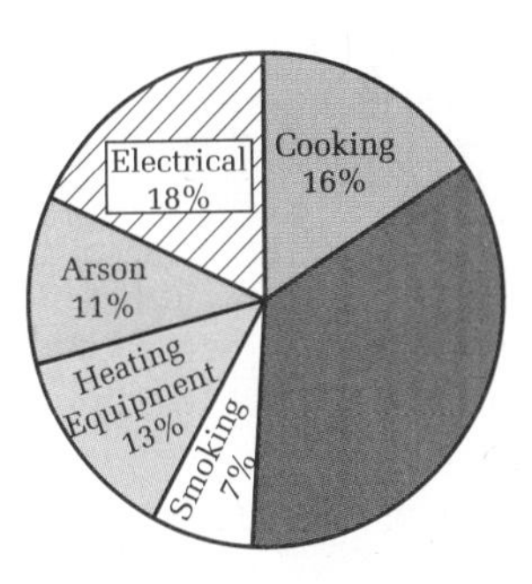

B **In problems 24–27, solve for x.**

P **24** $x^2 + 10x = 24$ $x = 2$ or $x = -12$

25 $x^2 - 8x = 5$ $x = 4 \pm \sqrt{21}$

26 $2x^2 - 10x - 31 = 0$ $x = \frac{5 \pm \sqrt{87}}{2}$

27 $3x^2 + 18x - 2 = 0$ $x = \frac{-9 \pm \sqrt{87}}{3}$

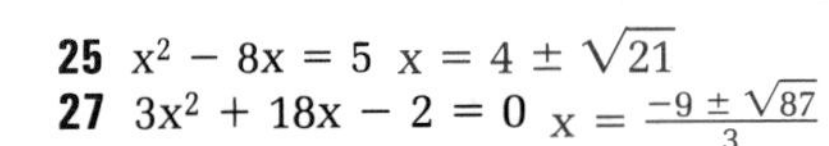

R **28** The diagrams show the numbers of angles smaller than 180° formed by lines that intersect at a single point. When n lines intersect at a point, they form $2n(n - 1)$ such angles.

a How many angles smaller than 180° are formed by five lines intersecting at a point? By six lines? By seven lines? 40; 60; 84

b How many lines intersect to form 3444 angles smaller than 180°? 42

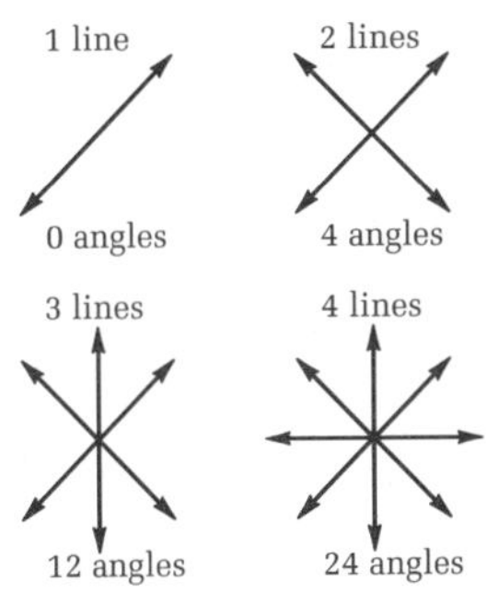

P **29** A dolphin's vertical movement is described by the function $h(t) = -4(t - 3)^2$, where $h(t)$ is the dolphin's height, in feet, relative to the surface of the water at time t, in seconds.

a How far below the surface is the dolphin when $t = 0$? 36 ft

b When does the dolphin reach the surface? At $t = 3$

c When is the dolphin at a depth of 100 feet? At $t = -2$ and $t = 8$

d Graph the dolphin's height as a function of time.

R **30** If $y > 13$, for what values of x is $3x - 2 \geq y$? Graph your solution on a number line. $\{x{:}x > 5\}$

R,I **31** Expand $\left(x + \frac{1}{x}\right)^3$. $x^3 + 3x + \frac{3}{x} + \frac{1}{x^3}$

Problem-Set Notes and Strategies, continued

- In problems **17 – 22**, remind students to factor out the GCF when factoring completely. Otherwise the directions haven't been followed and an incorrect answer may be found.

- Although pie charts are usually expressed in percentages as in problem **23**, point out to students that they could have just as easily been labeled with real values.

Additional Answers

29d

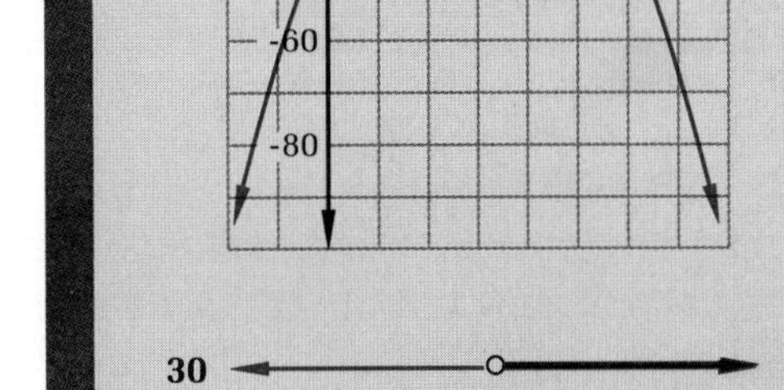

30

5

- Problem **31** introduces the binomial theorem. At this point, the product needs to be calculated using a two step multiplication. The Binomial Theorem is introduced in Section **9.5**.

Problem-Set Notes and Strategies, continued

- In problem **32**, students are asked to interpret a characteristic bar graph. This problem also is an application of union and intersection.

Additional Answer

36 No such points.

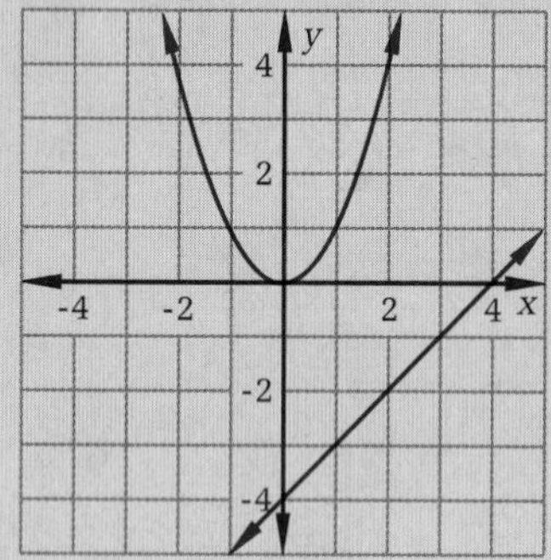

- Problem **37** is similar to experiments students may encounter in a physics or physical science class.

- Problem **38** is another preview of maximum.

- Problem **41** is a problem that arose in a discussion with the owner of a cherry orchard in southern Michigan.

Review Problems, *continued*

P **32** According to a newspaper graphic, the six most popular high-school sports among girls in 1987–1988 had the following numbers of participants (in thousands).

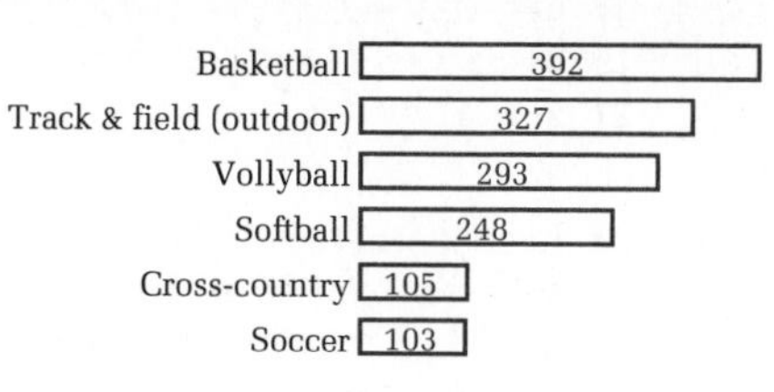

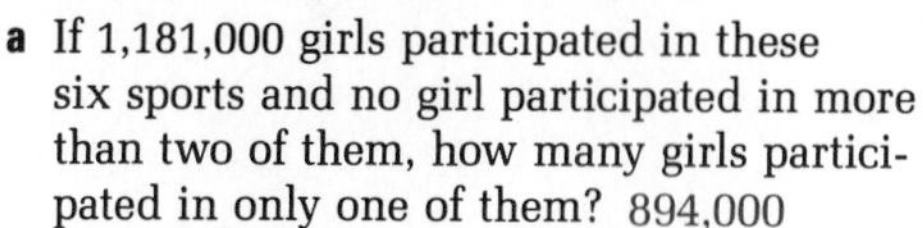

a If 1,181,000 girls participated in these six sports and no girl participated in more than two of them, how many girls participated in only one of them? 894,000

b If 1,181,000 girls participated in these sports, what percentage of the girls participated in basketball? In softball? In cross-country? ≈33.2%; ≈21%; ≈8.9%

P **In problems 33–35, solve for x.**

33 $(x + 2)(x - 3) = x^2 - 11$ $x = 5$

34 $x^4 - 13x^2 + 36 = 0$ $x = \pm 3$ or $x = \pm 2$

35 $(x + 1)(2x) + (x^2 + 3) = (2x + 3)(2x - 3) + 9$ $x = 3$ or $x = -1$

P **36** Graph the equations $y = x^2$ and $y = x - 4$ on the same coordinate system. List the real points of intersection. No such points

R **37** The tank shown is filled with water. A solid glass cube with edge x is then placed in the tank.

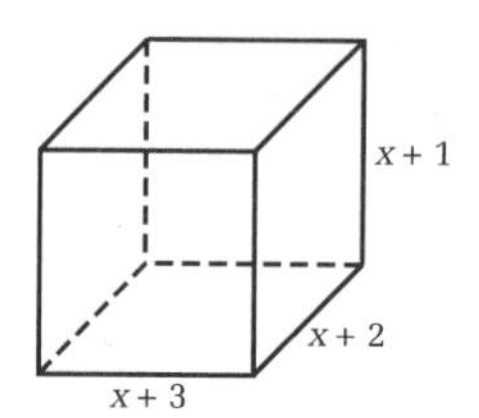

a Write an equation expressing the volume V of the water remaining in the tank as a function of x. $V = 6x^2 + 11x + 6$

b What is the domain of the function you wrote in part **a**? $\{x:x > 0\}$

P **38** **a** Write a function for $A(x)$, the area of the rectangle shown. $A(x) = -2x^2 + 12x + 144$

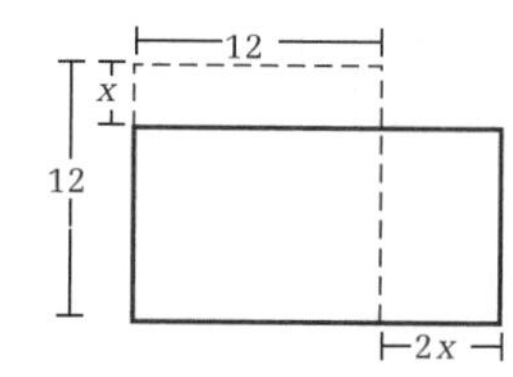

b For what value of x is $A(x)$ a maximum? 3

c What is the maximum area of the rectangle? 162

P **In problems 39 and 40, factor each expression.**

39 $(x + 1)^2 - y^2$ $(x + 1 + y)(x + 1 - y)$

40 $x^4 - 2x^2 + 1$ $(x - 1)^2(x + 1)^2$

C P **41** The trees in a cherry orchard bear an average of 204 pounds of fruit each when there are 90 trees per acre. For each additional tree planted per acre, the average yield per tree decreases by 2 pounds. What number of trees per acre would produce the maximum harvest? 96 trees per acre

R **42** Rewrite $\sum_{n=1}^{4} nx^{n-1}$ as a polynomial expression. $4x^3 + 3x^2 + 2x + 1$

4 CHAPTER TEST

1 Determine whether each statement is True or False.

a $x^2 + 9x + 18 = (x + 3)(x + 6)$ True
b $x^2 - 9x - 18 = (x - 3)(x + 6)$ False
c $16x^2 - 9y^2 = (4x + 3y)(4x - 3y)$ True
d $x^2 + y^2 = (x + y)(x + y)$ False

2 Write an expression for the area of the largest rectangle in the diagram. $x^2 + 5x$

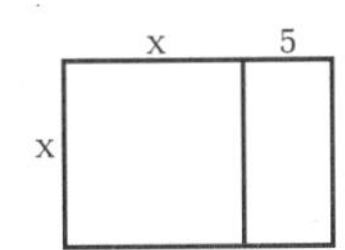

3 Solve each equation for x.

a $3x^2 + 10x - 8 = 0$
$x = \frac{2}{3}$ or $x = -4$

b $2x^2 - 3x - 4 = 0$
$x = \frac{3 \pm \sqrt{41}}{4}$

4 If $f(x) = x^2 + 8x + 5$, what is the minimum value of $f(x)$? -11

5 Find the real zeros of the function $g(x) = 2x^2 - 5x - 3$. $-\frac{1}{2}, 3$

6 Write an expression for the length of the side of the square shown. $4x + 3$

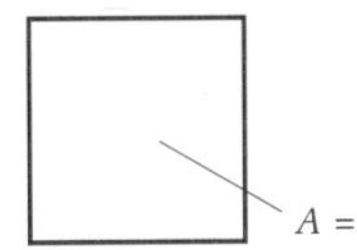

7 Graph each of the following equations on the same coordinate system.

a $y = x^2$
b $y = (x + 3)^2$
c $y = (x + 3)^2 + 5$

8 For what value(s) of x are the perimeters of the figures equal? -5 and 5

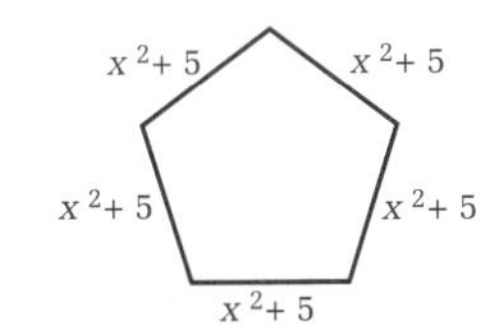

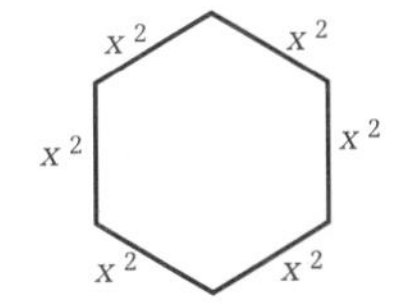

9 The geyser Old Faithful spouts water into the air up to 185 feet. If it takes 0.5 seconds for the water to reach a height of 185 feet, what is the upward velocity of the water? 378 ft/sec

10 Solve by completing the square.

a $x^2 - 8x = 20$
$x = 10$ or $x = -2$

b $y^2 + 5y = \frac{75}{4}$
$y = \frac{5}{2}$ or $y = -\frac{15}{2}$

11 Simplify each expression and solve each equation.

a $6z^2 - 4z - 5 = 2(3z^2 - z) - (5z - 3)$ $z = \frac{8}{3}$
b $(6z^2 - 4z - 5) + 2(3z^2 - z) - (5z - 3)$ $12z^2 - 11z - 2$
c $(6z^2 - 4z - 5) - 2(3z^2 - z) = 5z - 3$ $z = -\frac{2}{7}$

4

Resource References
Evaluation
Test Form 1, 2, 3

Additional Answer

7

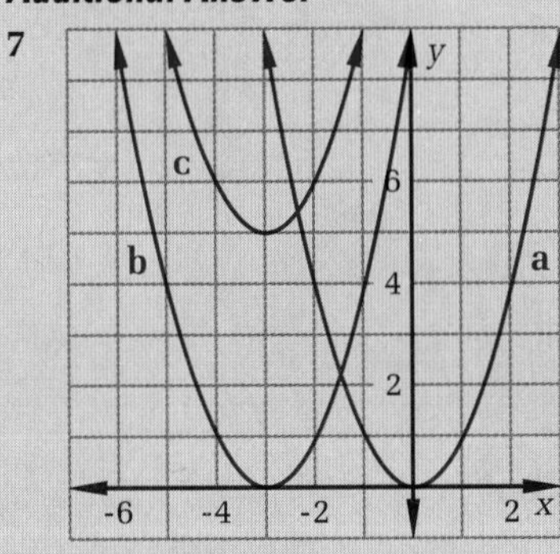

5.1

TO A LARGE degree, the function concept separates Algebra 1 and Algebra 2. It is central to the study of trigonometry, calculus, and computer science.

Six different ways of representing a relation and a function are given so that the concrete learner, the visual learner, and the abstract learner will all have something to relate to.

The definitions of domain and range are formally presented. The concept of function depends on which variable is the input and which is the output. It is very important for students to understand that *y*, as well as *x*, can be an input. For this reason, we do not use the vertical line test.

Restrictions on the domain and range of functions are explained. Composite functions lead to the concept of inverse functions. Here we present the inverse of a function and the relationship between the graphs of a function and its inverse. Students will see that operations on the input of a function yield results different from those produced by operations on the output of the function.

The factorial, combination, and permutation functions are the discrete functions studied in this chapter. Students will see how to interpret recursive definitions of functions and relate them to equivalent closed-form definitions.

CHAPTER

5 FUNCTIONS

In this photograph, taken directly under the St. Louis Gateway Arch, a dramatic increase in height is shown as a function of decreasing width.

5.1 RELATIONS AND FUNCTIONS

Objectives

After studying this section, you will be able to

- Represent relations
- Recognize functions

Part One: Introduction

Relations

Peter has written a computer program that lists the factors of positive integers. When he inputs 1, the program outputs 1. When he inputs 2, the program outputs 1 and 2. When he inputs 8, the program outputs 1, 2, 4, and 8. In short, for each number Peter inputs, the program produces a set of outputs *related* to the input. This ***relation*** can be expressed as a set of ordered pairs of the form (input, output).

 ***A relation is a set of ordered pairs.***

Relations can be represented in a variety of ways. Here are some of the ways of representing Peter's relation for the input values 1, 2, and 8.

Method 1: Ordered Pairs

(1, 1), (2, 1), (2, 2), (8, 1)
(8, 2), (8, 4), (8, 8)

Method 2: Table

Number	1	2	2	8	8	8	8
Factor	1	1	2	1	2	4	8

Method 3: Mapping Diagram

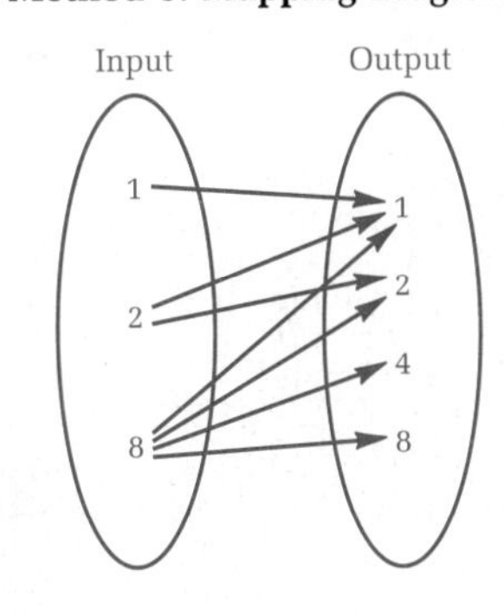

Method 4: Graph

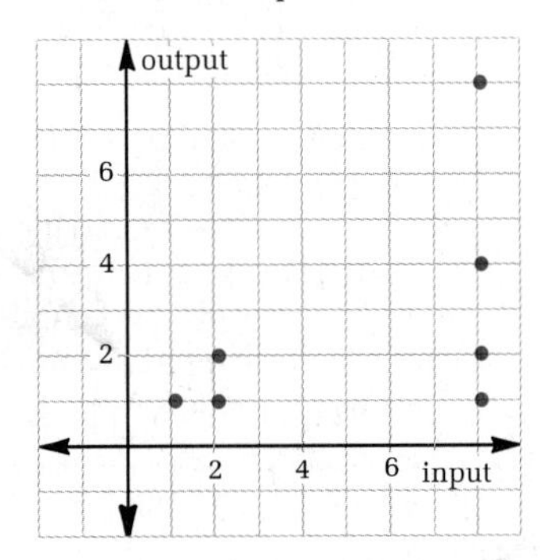

Class Planning

Time Schedule

Average: 2 days
Advanced: $1\frac{1}{2}$ day
Honors: 1 day

Resource References

Teacher's Resource Book
Class Opener 27
Practice 28
Transparency 9

Class Opener

Copy and complete the table for each number given. List each of the following: the factors, the greatest prime factor (GPF), the least prime factor (LPF), and the number of distinct primes.

	Factors	GPF	LPF	Distinct Primes
24	1,24; 2,12; 3,8; 4,6; 2^3,3	3	2	2
18	1,18; 2,9; 3,6; 2,3^2	3	2	2
31	1,31	31	31	2
64	1,64; 2,32; 4,16; 8,8; 2^6	2	2	1
72	1,72; 2,36; 3,24; 4,18; 6,12; 8,9; 3^2,2^3	3	2	2
121	1,121; 11,11; 11^2	11	11	1
101	1,101	101	101	2

Lesson Notes

- The first section of this chapter formalizes the definition of a relation and a function. The idea of a relation seems so simple that students may get confused. A relation is simply a set of ordered pairs, a set of points in the Cartesian plane.

Vocabulary

relation
function
domain
range

Lesson Notes, continued

There need be no pattern to these points whatsoever. A single point is a relation. The empty set is a relation. The entire plane is a relation. Any set of points that you can possibly think of is a relation.

- A function is a very special set of points. For each first coordinate, there is *exactly* one second coordinate that corresponds to it. Every *x* has a *unique y*. For a given input, there is only one output. This uniqueness is what distinguishes a function from a relation.
- The set of possible inputs that make sense for a function is called the domain of the function. The unique set of outputs that correspond to those inputs is the range of the function.
- The idea of a function, i.e. an input, a rule and a unique output will be used throughout this class and the remainder of the student's mathematical career.

Stumbling Block

Students frequently think that the definition of a function implies a one-to-one correspondence. In a function, each input must have a unique output. No one says that each output must come from a unique input. Have students draw a horizontal line on a graph. Choose several different *x*-coordinates and have students identify all *y* values on the line that correspond to that *x*. Ask them if it is unique for each *x*. Now ask them about the *y* values. Are they unique? Does the definition of a function say anything about that? No! Each *x* has a unique *y*, not vice versa!

Method 5: Input-Output Diagram

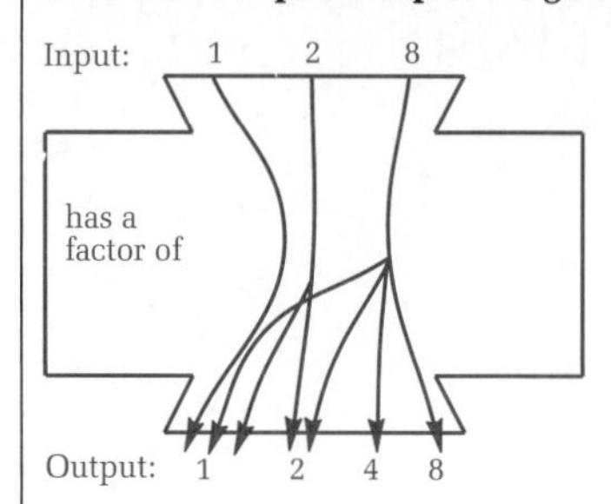

Method 6: Formula

In this case, we cannot write a formula, but we can express the rule governing the relation in set notation:

$$\{(x, y): x \in \{1, 2, 8\} \text{ and } y \text{ is a factor of } x\}$$

Notice that in a relation an input may be paired with more than one output.

Functions

A function is a special type of relation. To get a notion of how functions differ from other relations, consider a familiar function machine—a scientific calculator. Each time you use a calculator to find the square root of 13, for example, you get the same result. It is this consistency of output that distinguishes functions from relations that are not functions.

 ***A function is a relation that pairs each input with exactly one output.***

The ***domain*** of a function is the set of all inputs for which the function produces a meaningful output. The ***range*** of a function is the set of all the function's meaningful outputs.

Thus, a function can be said to have three parts—an input, a rule, and an output. The rule pairs an input with its unique output. Each input is an element of the function's domain. The outputs that the rule generates make up the function's range.

Example *Does the following input set and rule define a function?*
Input set: $\{-2, -1, 0, 1, 2\}$
Rule: "Has an absolute value of"

The mapping diagram shows the result of applying the rule "has an absolute value of" to the elements of $\{-2, -1, 0, 1, 2\}$. Since each input is mapped to exactly one output, the relation defined by the given information is a function.

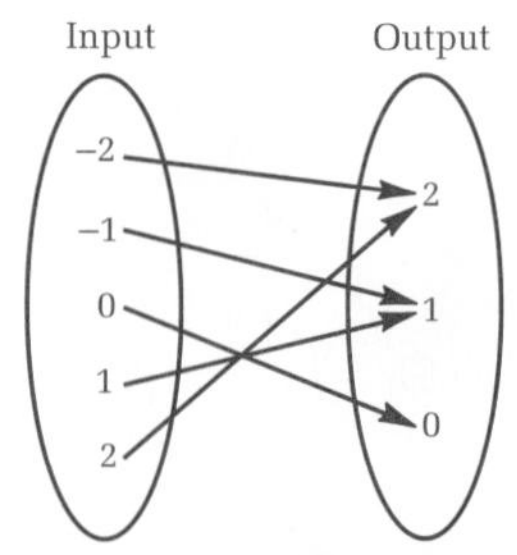

Part Two: Sample Problems

Problem 1 *Consider the relation defined by the equation $x^2 + y^2 = 25$. Is y a function of x? Is x a function of y?*

Solution The graph of $x^2 + y^2 = 25$ is a circle with its center at the origin and a radius of 5. When $x = 3$, the value of y is 4 or −4. Since a single input value (x) produces two different output values (y), y is not a function of x. Similarly, when $y = 3$, the value of x is 4 or −4. Since a single input value (y) produces two different output values (x), x is not a function of y.

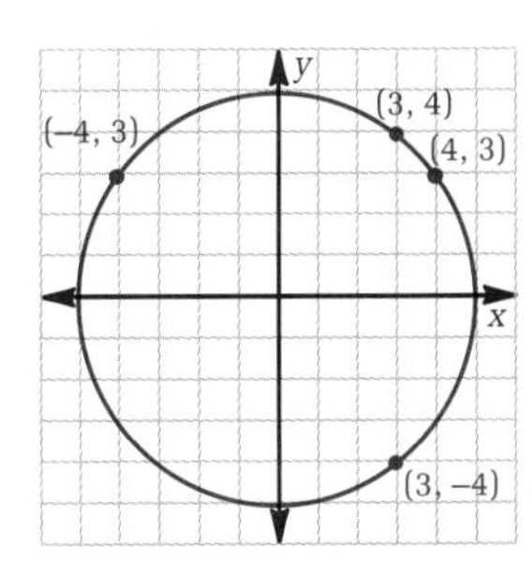

Problem 2 *Represent the function $f(x) = x^2$ for the inputs {−2, −1, 0, 1, 2}.*

Solution The methods used to represent functions are the same as those used to represent other relations.

Method 1: Ordered Pairs

(−2, 4), (−1, 1), (0, 0), (1, 1), (2, 4)

Method 2: Table

x	−2	−1	0	1	2
$f(x)$	4	1	0	1	4

Method 3: Mapping Diagram

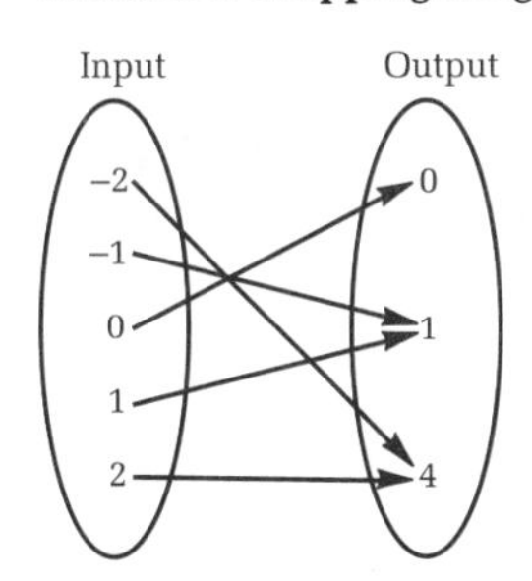

Method 4: Graph

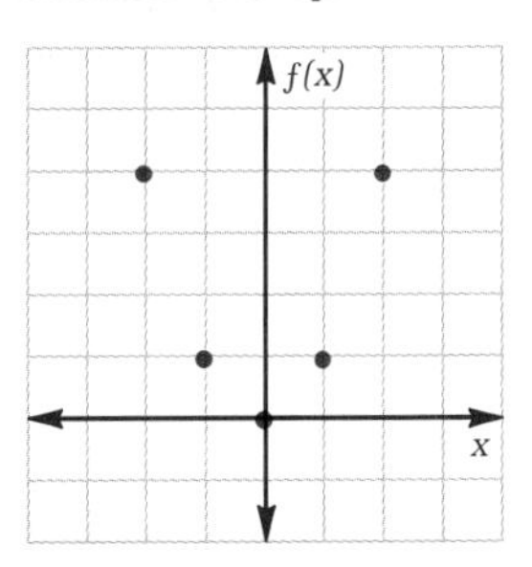

Method 5: Input-Output Diagram

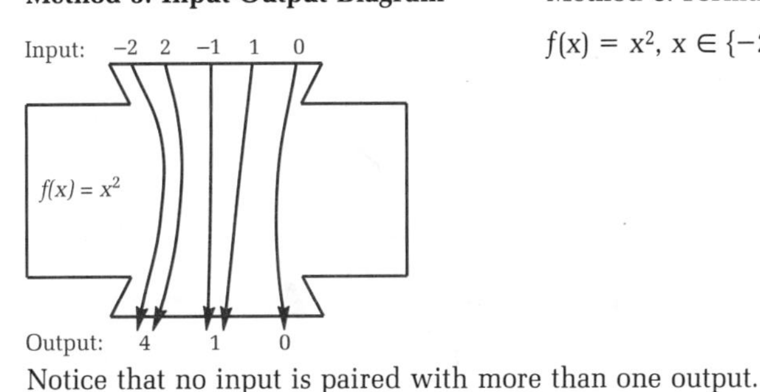

Method 6: Formula

$f(x) = x^2, x \in \{-2, -1, 0, 1, 2\}$

Notice that no input is paired with more than one output.

Checkpoint

1. Does the following set of ordered pairs represent a function?
{(1, 1), (2, 2), (3, 3), (4, 4), (5, 5)}
Yes
2. Does the following set of ordered pairs represent a function?
{(1, 1), (2, 1), (3, 1), (4, 1), (5, 1)}
Yes
3. Does the following set of ordered pairs represent a function?
{(1, 1), (1, 2), (1, 3), (1, 4), (1, 5)}
No
4. Represent the function three different ways.
$f(x) = 3x - 9$, for the inputs {1, 3, 5, 7, 9}
Answers will vary as to the methods chosen. The function can be expressed as a set of ordered pairs, as a mapping diagram, in table format, as a graph, or as an input-output diagram.

Assignment Guide

Average

Day 1 7, 9, 12, 14, 16, 18, 20–22, 26

Day 2 8, 10, 11, 13, 15, 17, 19, 23, 24, 27, 32

Advanced

Day 1 8, 11, 14, 19, 20, 22, 26, 27, 30, 32

Day 2 ($\frac{1}{2}$ day) 7, 9, 10, 15, 16, 21, 23, 24, 28, 29, 31
($\frac{1}{2}$ day) Section 5.2 7, 8, 11, 13–16, 18–21, 23, 24, 26, 27

Honors

7, 11, 14, 15, 17, 20–24, 27, 28, 30–32

Problem-Set Notes and Strategies

- In problem **3**, b is not a function of a because for a value of $a = 1$, b could be either 1 or -1.
- Problem **4** uses a pattern that was introduced back in Chapter **1**. It is interesting to note that functions may be defined using sigma notation. How the rule is defined is unimportant. What is important is that each input gets paired with only one output.

Additional Answers

1a $\{(-2,5), (-1,2), (0,5), (1,2)\ (2,0)\}$

b

Input	−2	−1	0	1	2
Output	5	2	5	2	0

c

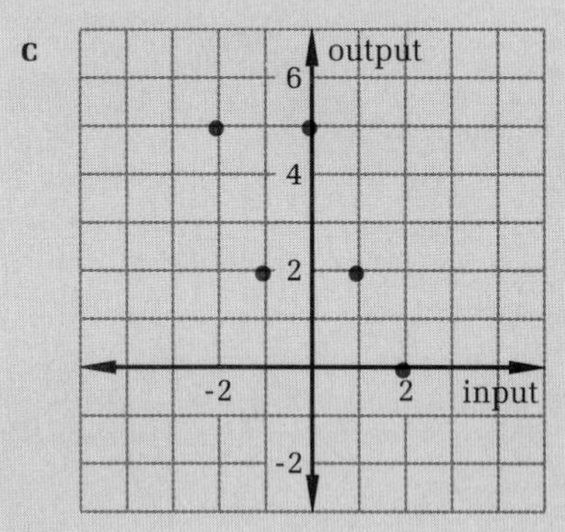

d

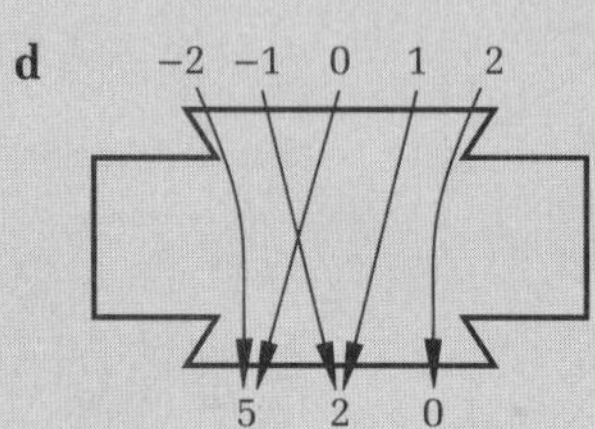

Part Three: Exercises and Problems

Warm-up Exercises

P **1** Represent the relation shown in the mapping diagram in each of the following ways.

- **a** A set of ordered pairs
- **b** A table
- **c** A graph
- **d** An input-output diagram

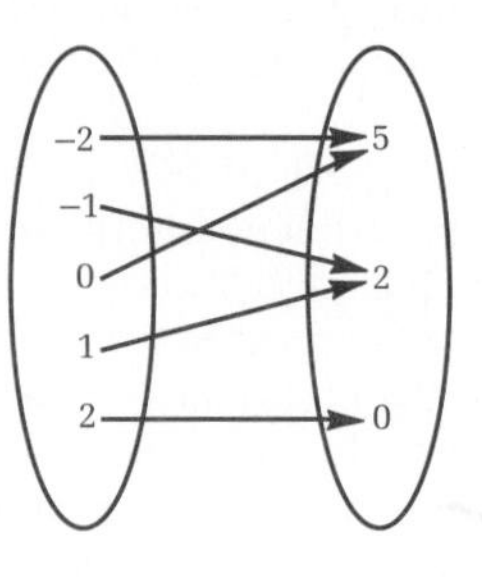

P **2** Which of the following sets of ordered pairs represent functions? **b, c, d**

- **a** $\{(-2, 3), (-3, 4), (-2, 5)\}$
- **b** $\{(3, -2), (4, -2), (-2, 5)\}$
- **c** $\{(1, -1), (0, 0), (-1, 1)\}$
- **d** $\{(-1, 1), (0, 0), (1, 1)\}$
- **e** $\{(1, -1), (0, 0), (1, 1)\}$

P **3** If $a = |b|$, is a a function of b? Is b a function of a? Yes; no

P,R,I **4** Examine the mapping diagram of a function f.

- **a** Predict the output for an input of 5. 15
- **b** What formula did you use to make your prediction? $f(n) = \sum_{k=1}^{n} k$

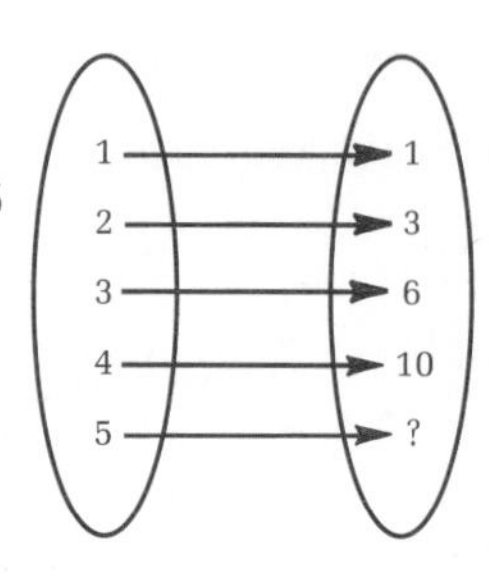

P,R **5** The graph shown represents a relation in which y is a function of x.

- **a** What is the function's domain?
- **b** What is the function's range?
- **c** Represent the function as a set of ordered pairs.

a $\{-4, -2, 0, 2, 5\}$
b $\{-4, -2, 0, 5\}$
c $\{(-4, -2), (-2, 0), (0, 5), (2, -4), (5, -4)\}$

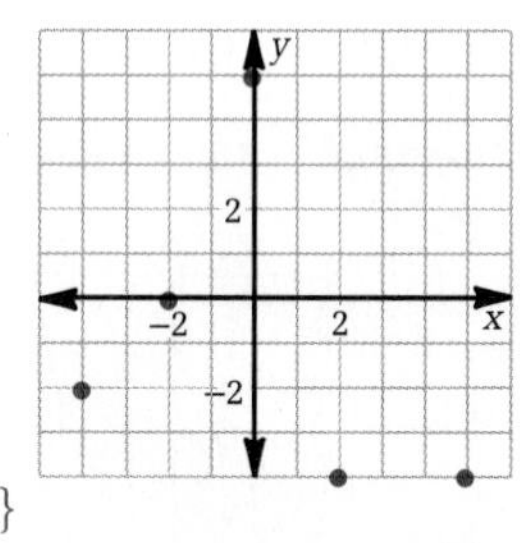

P,R,I **6** Let $f(x) = x^2$ and $g(x) = \sqrt{x}$.

- **a** Find $f(4)$. 16
- **b** Find $g(16)$. 4
- **c** Find $g(9)$. 3
- **d** Find $f(3)$. 9

Problem Set

A P

In problems 7 and 8, indicate whether the mapping diagram represents a relation, a function, or both.

7

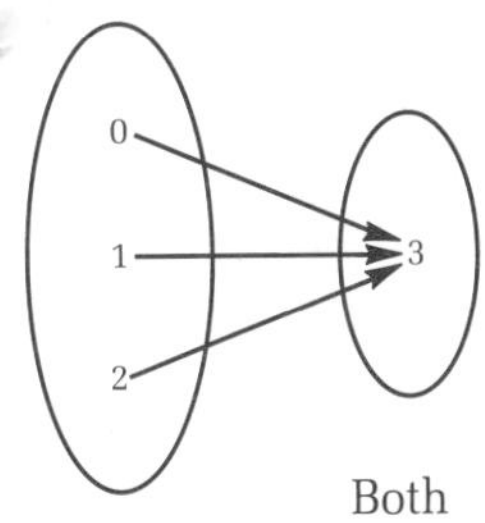

Both

8

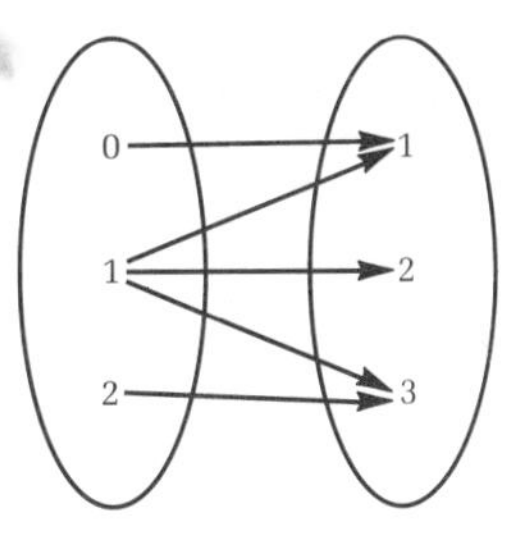

Relation

P **9** Represent the function $f(x) = |x - 2|$ for the inputs $\{-2, 0, 2, 4, 6\}$ in each of the following forms.

a A set of ordered pairs
b A table
c A mapping diagram
d A graph
e An input-output diagram

P,R **10** The graph represents the equation $y = g(x)$, where $g(x)$ is a function.

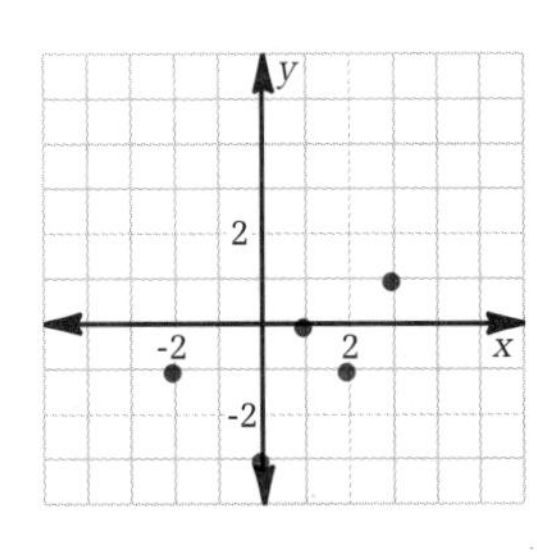

a Find $g(1)$. 0
b Find $g(2)$. -1
c Solve $g(x) = 1$ for x. $x = 3$
d Solve $g(x) = -1$ for x. $x = -2$ or $x = 2$
e What is the domain of function g?
f What is the range of function g?

e $\{-2, 0, 1, 2, 3\}$ f $\{-3, -1, 0, 1\}$

P **11** Is it possible for a function to have more elements in its range than in its domain? Explain your answer. No; since each input can have only one output, no. of outputs $\leq$ no. of inputs.

R,I **12** **a** Find the area of the rectangle if $x = 5$.
b Find the area of the rectangle if $x = 0$.
c What are the possible values of x?

a 60 b Impossible figure c $\{x:x > 2\}$

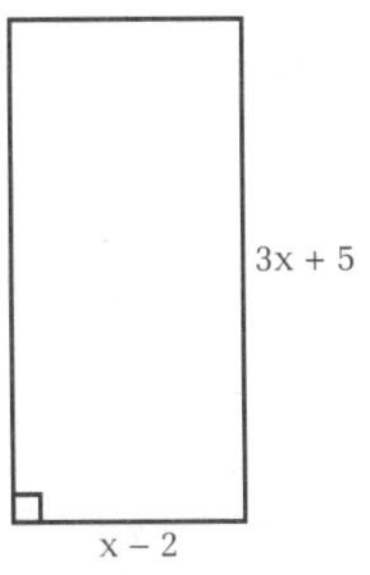

R **13** Solve $9(2x - 3) < 3x + 3$ for x and graph the solution set on a number line. $x < 2$

Problem-Set Notes and Strategies, continued

- Problem **9** illustrates how a function can be represented in various ways. The given situation should determine the most appropriate way to represent the function. There is no single method that is best for all situations.
- For problem **10** the only trouble students usually have is the listing of the domain and range.

Additional Answers

9a $\{(-2,4), (0,2), (2,0), (4,2), (6,4)\}$

9b

x	-2	0	2	4	6
f(x)	4	2	0	2	4

9c

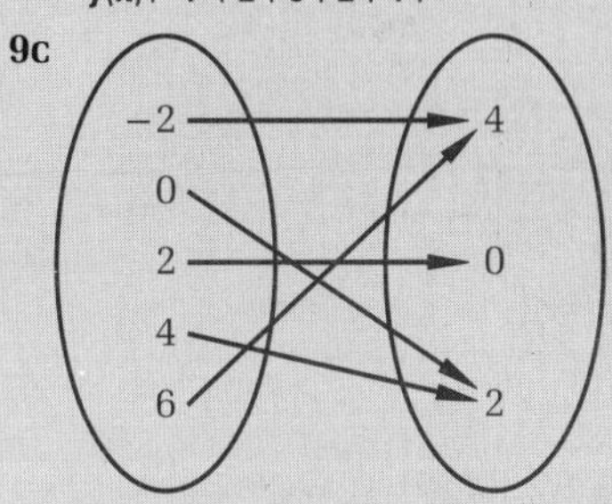

9d

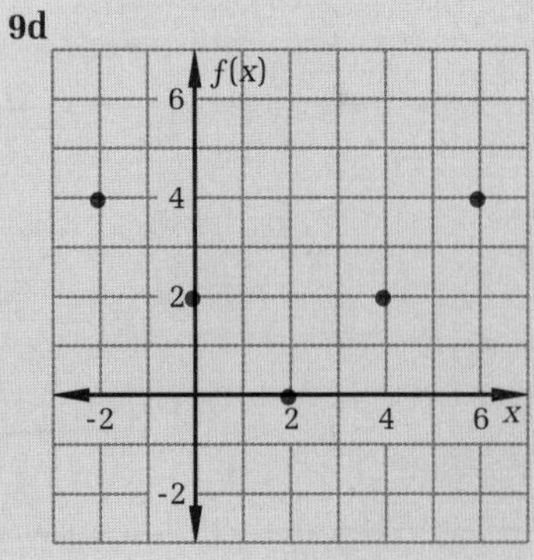

9e

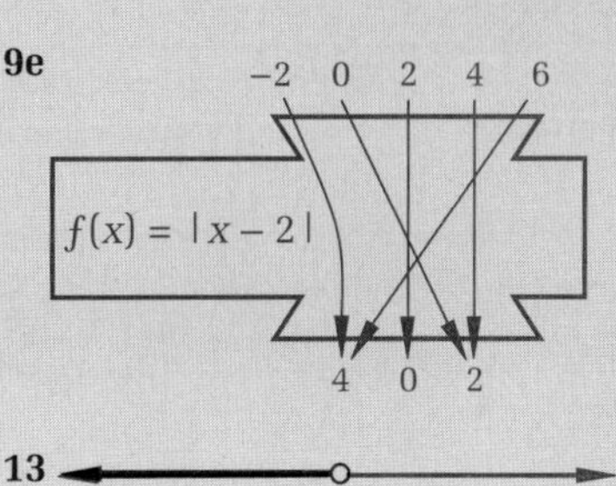

13

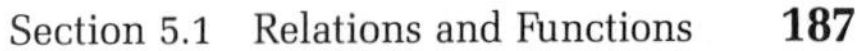

Problem-Set Notes and Strategies, continued

- Problem **15** is an example of the binomial distribution function which is the subject of Section **11.6**.
- Problem **19** is an example of a compound function. The domain is split using three different rules. The mapping can still be a function as long as the different rules do not overlap and cause multiple range elements.

Additional Answer

19

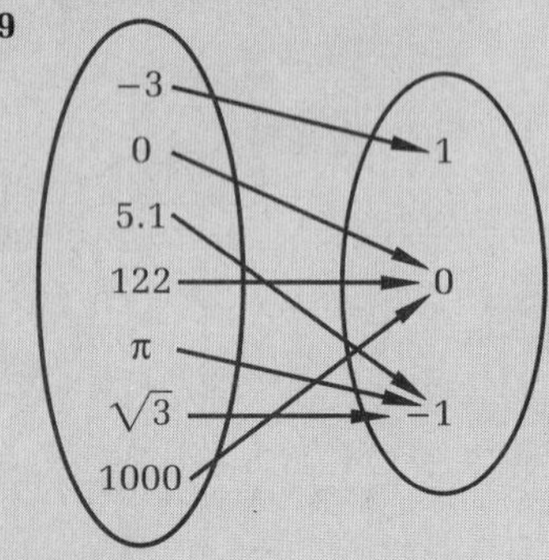

Problem Set, *continued*

P,I **14** Find the factorial key, [x!], on a scientific calculator. Try applying the factorial function to a variety of inputs x. What is the greatest input value for which the calculator can evaluate a factorial? Answers will vary. For many scientific calculators, x must be ≤ 69.

R **15** An algebra class experimented with flipping a set of five coins to see how many of the five would come up heads. The graph shows the results of a number of trials of the experiment.

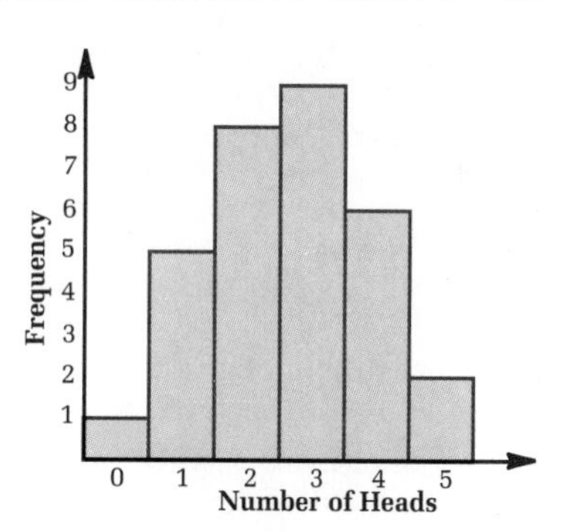

a How many times did three coins come up heads? 9

b How many times did all five coins come up heads? 2

c What is the total number of trials represented by the graph? 31

P,R **16** A certain photocopier can make 80 copies per minute.

a Write a function C to represent the number of copies produced in q minutes. $C(q) = 80q$

b Find $C(3.5)$. 280

R **17** Indicate whether each statement is True or False.

a $\sqrt{8} = \sqrt{4}\sqrt{2}$ True **b** $\sqrt{16} = \sqrt{-4}\sqrt{-4}$ False

R **18** The formula used to convert degrees Fahrenheit (F) to degrees Celsius (C) is $C = \frac{5}{9}(F - 32)$. For what values of F is $C < 20$? $\{F{:}F < 68\}$

B

P **19** Draw a mapping diagram that represents the function described by the following four statements.

i The domain is $\{-3, 0, 5.1, 122, \pi, \sqrt{3}, 1000\}$.
ii If the input is even, the output is 0.
iii If the input is odd, the output is 1.
iv If the input is not an integer, the output is −1.

P,I **20 a** Write a function A to describe the area of the shaded region. $A(x) = 5x + 36$

b For what values of x does function A have meaning? $\{x{:}x > -3\}$

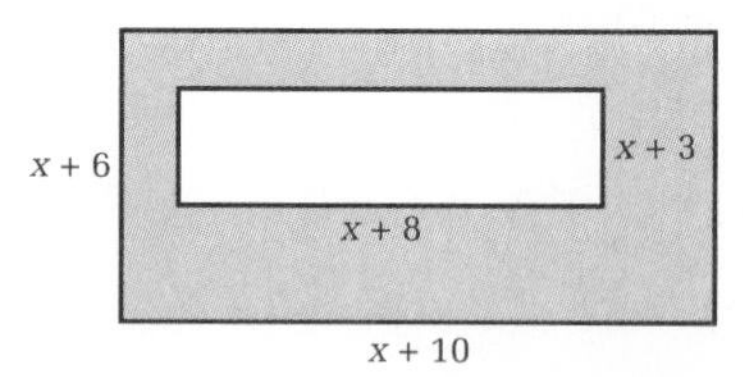

P **21** Consider a function p such that $p(\text{Reagan}) = 1980$, $p(\text{Bush}) = 1988$, $p(\text{Lincoln}) = 1860$, and $p(\text{Truman}) = 1948$.

a What do you think the rule for function p is?

b Find $p(\text{Kennedy})$ and $p(\text{Cleveland})$. 1960; 1884

a $p(x)$ = year in which x was first elected President of the U.S.

P In problems 22–24, each table represents a relation. For each relation, indicate whether y is a function of x and whether x is a function of y.

22

x	−3	−1	1	3	5
y	27	−1	1	27	125

Yes; no

23

x	9	4	1	0	1	4	9
y	−3	−2	−1	0	1	2	3

No; yes

24

x	−2	−1	0	1	2	3
y	2	1	0	−1	−2	−3

Yes; yes

R,I **25** If the graph of $y = x$ is the perpendicular bisector of $\overline{AB}$, what are the coordinates of point B? (6, 1)

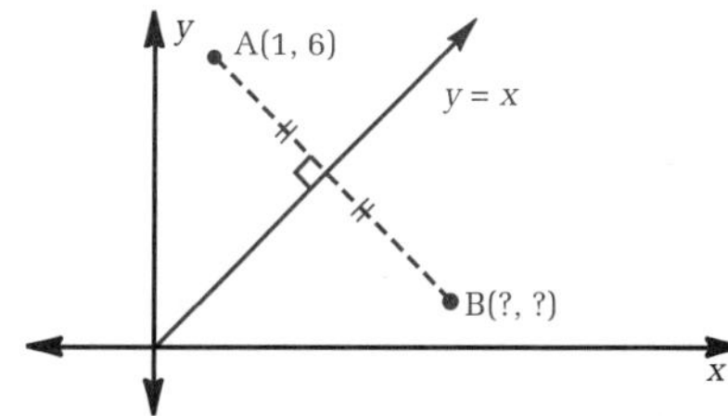

R **In problems 26–29, solve for x.**

26 $x^2 - 8x + 15 = 0$ $x = 3$ or $x = 5$

27 $2x^2 - 3x - 12 = 0$ $x = \frac{3 \pm \sqrt{105}}{4}$

28 $(2^x)^2 - 12(2^x) + 32 = 0$ $x = 2$ or $x = 3$

29 $x^3 - (x - 3)^3 = 27$ $x = 0$ or $x = 3$

C P,R **30** **a** Write a function M to describe the arithmetic mean (the average) of 15, 20, 30, 55, and x. $M(x) = \frac{1}{5}(x + 120)$
b Graph the equation $y = M(x)$.
c For what value of x is the arithmetic mean 0? −120
d For what value of x is the arithmetic mean 30? 30

P,R,I **31** Consider the graphs of $y = 2x^2 + 5$ and $y = x^2$.
a Write a function d to describe the vertical distance between the two curves.
b For what value(s) of x is $d(x) = 30$?
a $d(x) = x^2 + 5$ **b** −5 and 5

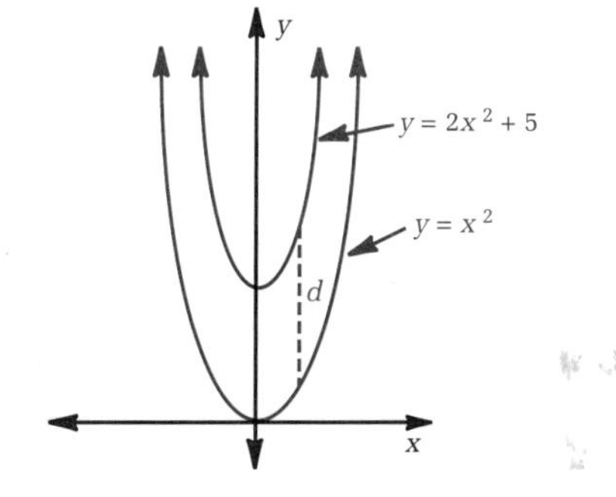

P,I **32** Find the key for the tangent function, tan, on a calculator. Try entering 20 tan to find tan 20, then "undo" the function by pressing INV tan
a Find tan 45. 1
b Find tan 75. ≈3.73
c If tan $x = 2$, what is the value of x? ≈63.43

Problem-Set Notes and Strategies, continued

- The symmetry in problem **25** allows us to solve this several different ways. Geometric methods could be used, algebraic formulas for slope and midpoint could be used, and a relationship about symmetry with respect to the line $y = x$ (which will be explored in Section **5.3**) could be used.

Additional Answer

30

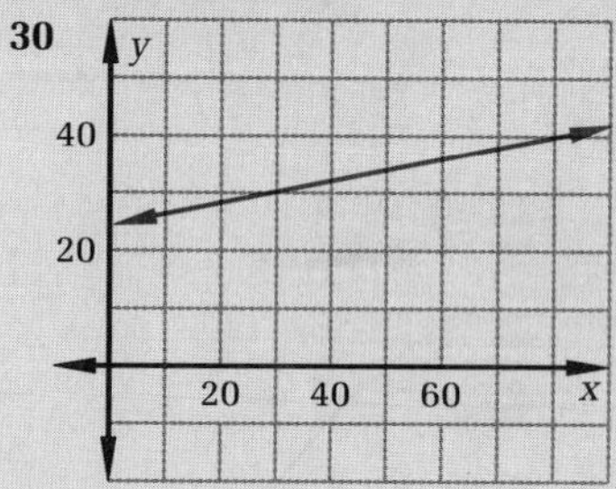

Communicating Mathematics

Have students write a paragraph that identifies three different situations in their life that represent functions. That is, for a given input value, there is exactly one output value.

5.2 More About Functions

Class Planning

Time Schedule
Average: 2 days
Advanced: 1 day
Honors: 2 days

Resource References
Teacher's Resource Book
Class Opener 28
Cooperative Learning 14
Practice 29
Enrichment 9

Class Opener

1 Find the set of all outputs for the functions $f(x) = \frac{24}{\sqrt{x+2}}$ if x is selected from the set $\{7, 4, 2, 0, -2, -6\}$
$7 \rightarrow 8, 4 \rightarrow 9.8, 2 \rightarrow 12, 0 \rightarrow 16.97, -2 \rightarrow$ undefined, $-6 \rightarrow$ undefined.

2 Given $f(x) = \frac{x^2 - 1}{x - 1}$ and $g(x) = \sqrt{x^2 + 4x + 4}$
- a Find $f(7)$. 8
- b Use the output of $f(7)$ as input for g. Find $g(f(7))$ 10
- c Find $g(7)$. 9
- d Use the output of $g(7)$ as input for f. Find $f(g(7))$ 10
- e Does $g(f(7)) = f(g(7))$? Yes

Lesson Notes

- The domain of a function is the set of values for which the function makes sense. In most problems that entails leaving out two problem areas. Fractions may not be divided by zero, so eliminate all values that will cause a zero to appear in the denominator. Similarly, the even roots of a negative number are not defined.

Objectives

After studying this section, you will be able to
- Identify restrictions on the domain of a function
- Interpret composite functions
- Recognize inverse functions

Part One: Introduction

Restrictions on the Domain of a Function

Consider the function $f(x) = \frac{3}{x}$. Try to evaluate $f(0)$ with a calculator. You should get an error display, since division by zero is undefined. Because $f(0)$ is undefined, zero is not an element of the domain of function f.

Since the domain of a function includes only the input values for which the function produces a meaningful output, we frequently need to specify that certain values are excluded from the domain. The following chart shows some functions, along with the domain and the range of each.

Function	Restrictions	Domain	Range
$h(y) = 4(y + 3)$	None	$\{y : y \in \mathcal{R}\}$	$\{h(y) : h(y) \in \mathcal{R}\}$
$f(x) = \frac{16}{x - 2}$	Input cannot be 2	$\{x : x \neq 2\}$	$\{f(x) : f(x) \neq 0\}$
$g(z) = \sqrt{z}$	Input must be nonnegative	$\{z : z \geq 0\}$	$\{g(z) : g(z) \geq 0\}$

In some cases, the domain of a function is restricted because of the nature of the problem we are using it to solve. For example, if a function's inputs represent numbers of passengers on various DC-10 airliners, the domain must be limited to nonnegative integers less than or equal to the maximum seating capacity of the plane.

Composition of Functions

If we evaluate the function $g(x) = \sqrt{9x}$ for the input value 4, we find that $g(4) = \sqrt{9(4)} = \sqrt{36} = 6$. There is, however, another way of thinking about this process. We can think of function g as a ***composite*** of two functions, $f(x) = 9x$ and $h(x) = \sqrt{x}$. We can evaluate $g(4)$ by using the output of $f(4)$ as the input of function h.

Vocabulary
composite
composition of functions
inverse functions

$f(x) = 9x$ $\qquad$ $h(x) = \sqrt{x}$

$f(4) = 9(4)$ $\qquad$ $h(36) = \sqrt{36}$

$= 36$ $\qquad$ $= 6$

Whenever we use the output of one function as the input of another function, we are creating a ***composition of functions.*** In this case, we say that $g(x)$ is equivalent to $h[f(x)]$ or $h \circ f(x)$. Both of these expressions are read "h of f of x."

Example *Let $f(x) = x + 1$ and $g(x) = 2x + 3$. Find $f[g(-4)]$.*

Method 1
We write a simplified expression for $f[g(x)]$, then evaluate the expression for $x = -4$.

$$f[g(x)] = f(2x + 3)$$
$$= (2x + 3) + 1$$
$$= 2x + 4$$
$$f[g(-4)] = 2(-4) + 4$$
$$= -4$$

Method 2
We evaluate $g(-4)$, then use the result as the input of f.

$$g(-4) = 2(-4) + 3$$
$$= -8 + 3$$
$$= -5$$
$$f(-5) = -5 + 1$$
$$= -4$$

Inverses

If $f(x) = 5x$ and $g(x) = \frac{x}{5}$, we can find $f[g(x)]$ and $g[f(x)]$ as follows.

$$f[g(x)] = f\left(\frac{x}{5}\right) = 5\left(\frac{x}{5}\right) = x$$

$$g[f(x)] = g(5x) = \frac{5x}{5} = x$$

The output of each composite is the same as the input for any input value that is in the domains of both functions. Because of this relationship, functions f and g are said to be ***inverses*** of each other.

 ***If $f[g(x)] = g[f(x)] = x$ for all values of x in the domains of f and g, then f and g are inverse functions.***

A notation for the inverse of a function f is INVf. In the case of $f(x) = 5x$ and $g(x) = \frac{x}{5}$, we can say that INV$f(x) = \frac{x}{5}$ and that INV$g(x) = 5x$.

Lesson Notes, continued

In other words, the square root of −9 doesn't make sense in the real numbers so it must be left out of the domain. Once the even roots of a negative number and the values that produce a zero in the denominator have been eliminated, the function is in pretty good shape. At this point in a students career, that is all they need to be concerned about.

- As we progress into the transcendental functions, other problems will surface. Values may be left out due to common sense if the problem is a word problem. For example if we were counting students in the class, only non-negative integers should be allowed.
- Order of operations is the key to solving composite functions. Always start from the inside parenthesis and work your way out. Composite functions are functions inside of functions. Composites lead to the discussion of inverse.
- An inverse function undoes what the function does. For example, subtracting 3 from a number is the inverse of adding 3 to the number. One operation undoes what the other is doing. The inverse is discussed in more detail in the next section.

Stumbling Block

Composite functions look deceivingly like multiplication problems. Students have to learn to look carefully at the problem and identify the input variable. Each of the other letters is the name of a function. In other words, $f(g(x))$ means f of g of x and not f times g of x. Operations on functions such as multiplication will be discussed in Section **5.4**.

Communicating Mathematics

Have students write a paragraph that describes ways in which their life is a composite of functions. In other words, identify examples where the ouptut of some part of life has become the input of another part of life.

Checkpoint

1 Find the restriction(s) on the domain of the function $f(x) = \frac{3x+4}{x-7}$. $x \neq 7$

2 Let $f(x) = 3x + 4$ and $g(x) = x - 7$. Find $f(g(10))$ and $g(f(10))$. $f(g(10)) = 13$ and $g(f(10)) = 27$

3 Let $f(x) = 3x + 4$ and $g(x) = \frac{x-4}{3}$. Are f and g inverse functions? Yes

Part Two: Sample Problems

Problem 1 *Find the domain of each function.*

a $s(n) = \sqrt{2n + 3}$

b $A(x, y) = (3x - 5)(y + 10)$, *the formula for the area of the rectangle shown*

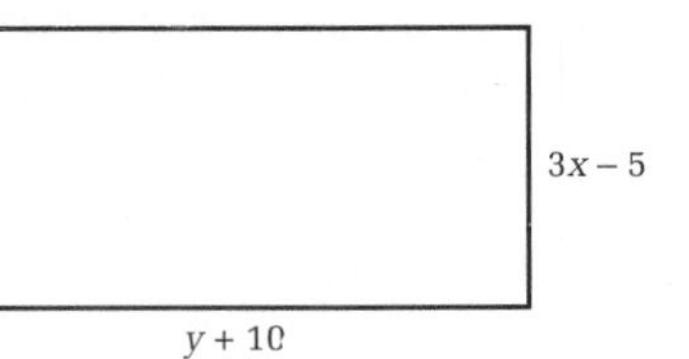

Solution **a** In the real-number system, the operation of taking a square root is defined only for nonnegative inputs.

$$2n + 3 \geq 0$$
$$2n \geq -3$$
$$n \geq -\frac{3}{2}$$

The domain is $\left\{n{:}n \geq -\frac{3}{2}\right\}$.

b The length and the width of a rectangle must be positive.

$$3x - 5 > 0 \qquad y + 10 > 0$$
$$3x > 5 \qquad y > -10$$
$$x > \frac{5}{3}$$

The domain is $\left\{(x, y){:}x > \frac{5}{3} \text{ and } y > -10\right\}$.

Problem 2 *Let* $f(x) = x + 2$ *and* $g(x) = x^2$. *Find the value(s) of x for which* $f[g(x)] = g[f(x)]$.

Solution First we write simplified expressions for $f[g(x)]$ and $g[f(x)]$.

$$f[g(x)] = f(x^2) = x^2 + 2$$

$$g[f(x)] = g(x + 2) = (x + 2)^2 = x^2 + 4x + 4$$

Now we set the composites equal to each other and solve for x.

$$f[g(x)] = g[f(x)]$$
$$x^2 + 2 = x^2 + 4x + 4$$
$$-2 = 4x$$
$$-\frac{1}{2} = x$$

The composites $f[g(x)]$ and $g[f(x)]$ are equal only when $x = -\frac{1}{2}$.

Problem 3 Let $h(x) = \sqrt{x + 7}$ and $g(x) = x^2 - 7$. Determine whether h and g are inverse functions.

Solution If h and g are inverse functions, $h[g(x)]$ will equal x and $g[h(x)]$ will equal x for all the values of x for which h and g are defined.

$$h[g(x)] = h(x^2 - 7) = \sqrt{(x^2 - 7) + 7} = \sqrt{x^2} = |x|$$

$$g[h(x)] = g(\sqrt{x + 7}) = (\sqrt{x + 7})^2 - 7 = (x + 7) - 7 = x$$

Since $h[g(x)] \neq x$ when $x < 0$, h and g are not inverse functions. (In the next section, you will discover that we *can* call h and g inverse functions if we specify a restriction in the domain of g.)

Part Three: Exercises and Problems

Warm-up Exercises

P **1** Determine the domain and the range of each of the following functions.

a $f(x) = 3 + 5\sqrt{x}$ Domain: $\{x:x \geq 0\}$; range: $\{f(x):f(x) \geq 3\}$

b $g(x) = \frac{x - 3}{5 + x}$ Domain: $\{x:x \neq -5\}$; range: $\{g(x):g(x) \neq 1\}$

c $S(n) = 180(n - 2)$, the formula for the sum of the measures of the angles of an n-sided polygon

d $h(y) = \frac{2y^2}{y^2 + 5y + 6}$ Domain: $\{y:y \neq -2 \text{ and } y \neq -3\}$; range: $\{h(y):h(y) \geq 0 \text{ or } h(y) \leq -48\}$;

P **2** Refer to the input-output diagram of the composite $g \circ f$.

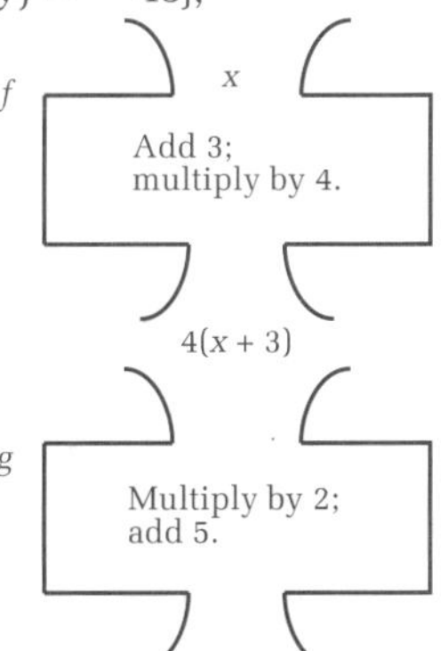

a What is the output for an input of −2? 13

b Write a simplified expression for $g \circ f(x)$. $8x + 29$

P **3** Let $h(x) = 3x + 2$ and $j(x) = x^2$.

a Find $h[j(4)]$. 50

b Find $j[h(4)]$. 196

c Is composition of functions commutative? No

Assignment Guide

Average	
Day 1	8, 10, 11, 13, 14, 19, 21, 23, 25, 28
Day 2	7, 9, 12, 15–18, 20, 22, 24, 27, 29, 30
Advanced	
Day 1	($\frac{1}{2}$ day) Section 5.1 7, 9, 10, 15, 16, 21, 23, 24, 28, 29, 31
	($\frac{1}{2}$ day) 7, 8, 11, 13–16, 18–21, 23, 24, 26, 27
Day 2	($\frac{1}{2}$ day) 9, 10, 12, 17, 22, 25, 28–30
	($\frac{1}{2}$ day) Section 5.3 10–12
Honors	
Day 1	7, 8, 13, 14, 17, 18, 20, 22, 27, 31
Day 2	9, 11, 12, 19, 21, 24, 26, 28–30, 32, 33

Problem-Set Notes and Strategies

- It is easier to find the domain than the range in problem **1**. The range is more difficult as you have to consider each domain element and observe what type of element it will produce. In the case of **1a**, the square root of a number is always nonnegative and so the smallest $f(x)$ will ever be is 3. Others need to be reasoned similarly. If a graph is available, the y-coordinates can be checked and observed to determine what values they assume.
- In problem **2** when composite functions are carried out in succession, it is helpful to combine the functions into a single function. Caution your students that

Additional Answer

1c Domain: {integers greater than 2}
Range: {positive multiples of 180}

Problem-Set Notes and Strategies, continued

the domain must be carefully studied because there may be some output of the first function which is not in the domain of the second function. The corresponding domain of this output must be eliminated from the domain of the first function.

- Problem **4** illustrates a property common to all inverse functions. As a matter of fact, it is the definition of inverse functions.

- As the composition of functions is reversed, the domain of the composition may change drastically. Problem **11** illustrates this fact with a square and square root function.

P,I **4** If function h is the inverse of function g, what is the value of each composite?

a $h \circ g(7)$ 7 **b** $g \circ h(7)$ 7 **c** $h[g(9)]$ 9
d $g[h(9)]$ 9 **e** $h[g(x)]$ x

P,I **5** Explain why $f(x) = 3x + 2$ and $g(x) = \frac{1}{3}x - \frac{2}{3}$ are inverse functions. $g[f(x)] = f[g(x)] = x$ for all x

P **6** Let $f(x) = 2x + 1$ and $g(x) = x^2 - 2$.

a Evaluate $f(3)$ and $g(3)$. 7; 7
b Compute $f(3) \cdot g(3)$ and $g(3) \cdot f(3)$. 49; 49
c Compute $f \circ g(3)$ and $g \circ f(3)$. 15; 47
d Explain why the expressions in part **b** are equivalent but those in part **c** are not. Multiplication is commutative, but composition of functions is not.

A Problem Set

P **7** Let $f(x) = x^2 + 4$ and $g(x) = \sqrt{x - 4}$.

a What are the restrictions on the domain of f? None
b What are the restrictions on the domain of g? $x \geq 4$
c Write a simplified expression for $f[g(x)]$. What is the domain of $f \circ g$? x; $\{x{:}x \geq 4\}$
d Write a simplified expression for $g[f(x)]$. What is the domain of $g \circ f$? $|x|$; $\{x{:}x \in \mathcal{R}\}$
e Find $f[g(5)]$ and $g[f(5)]$. 5; 5
f Find $f[g(-4)]$ and $g[f(-4)]$. Undefined; 4

P **8** Refer to the mapping diagrams.

a Find $g[f(3)]$. 11
b What is the domain of $g[f(x)]$? {3, 7}
c What is the range of $g[f(x)]$? {11, 13}

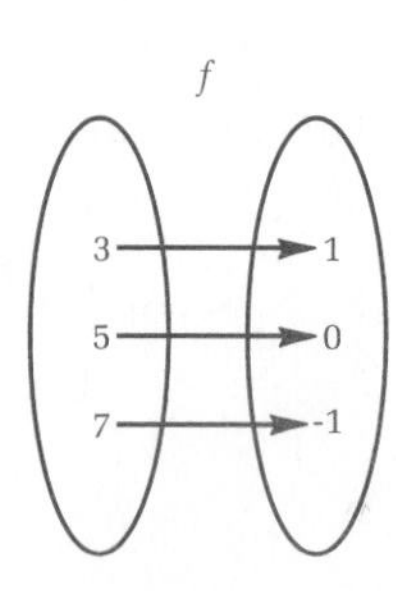

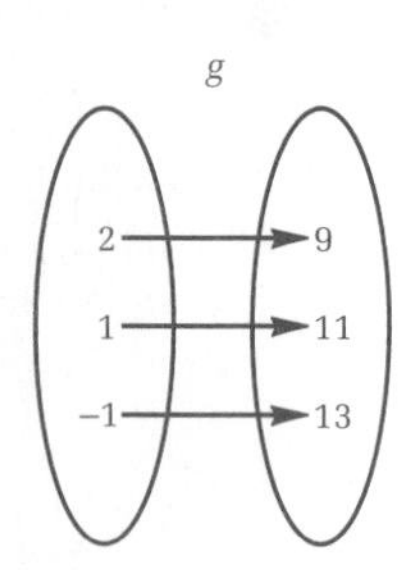

P **9** If $g(x) = 3x + 2$ and $r(x) = 2 - 5x$, what are $g \circ r(y)$ and $r \circ g(y)$? $8 - 15y$; $-8 - 15y$

P **10** For what real values of x is $\sqrt{2x + 10}$ not defined over the real numbers? $\{x{:}x < -5\}$

P,I **11** Let $s(x) = x^2$ and $t(x) = \sqrt{x}$.

a Find $s[t(4)]$ and $t[s(4)]$. 4; 4
b Find $t[s(-9)]$ and $s[t(-9)]$. 9; undefined
c Are s and t inverse functions? No (unless the domain of s is restricted)

R | **12** The area of the rectangle shown is 36. Find the value of x. 1 or −6

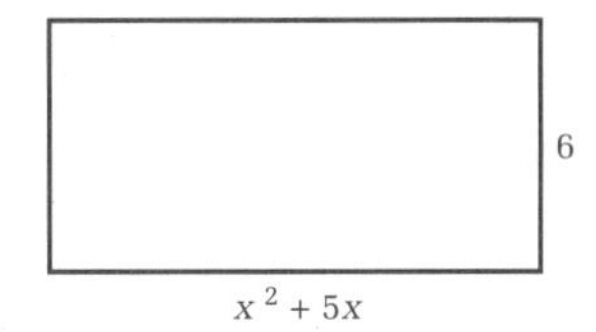

P,I | **13** If $f(x) = 2x + 5$ and $g(x) = 2x - 5$, are f and g inverses? Explain your answer. No; $f[g(x)] = 4x - 5$ and $g[f(x)] = 4x + 5$, so $f[g(x)] \neq g[f(x)] \neq x$.

P | **14** If $f(x) = 5x + 3$ and $g(x) = x^2 - 3$, for what value(s) of x is $f(x) = g(x)$? −1 and 6

R | **15** Is the relation represented by the mapping diagram a function?

a Yes

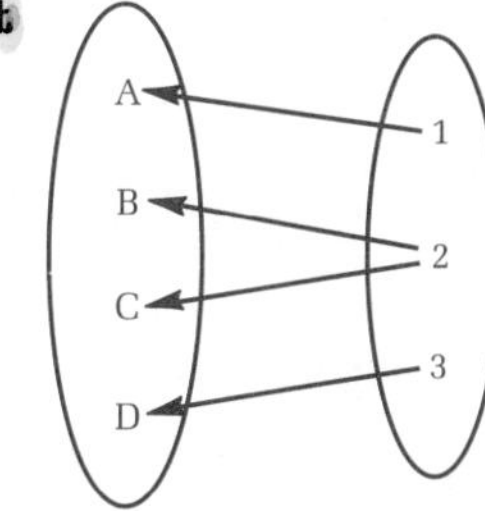

b No

A, B, C, D ← 1, 2, 3

P | **16** If $A(x) = x^2 + 5$ and $B(x) = x - 2$, for what value(s) of x is $A[B(x)] = B[A(x)]$? $\frac{3}{2}$

I | **17** The "greatest-integer" function of x, symbolized [x], outputs the greatest integer that is less than or equal to x—for example, [5.2] = 5, [−1.3] = −2, and [7] = 7. Evaluate each of the following expressions.

a [11.4] 11 **b** [0] 0 **c** [−8.3] −9
d [−5.9] −6 **e** [π] 3 **f** [$-\sqrt{2}$] −2
g $\left[\frac{77}{3}\right]$ 25 **h** [x + 1] − [x] 1

P | **18** If $p(x) = x^2 + 8$ and $q(x) = 2x$, for what value(s) of x is $p[q(x)] = q[p(x)]$? −2 and 2

P,R,I | **19** **a** Write a function m describing the measure of ∠ABC in terms of x.
b What are the restrictions on the domain of function m?
c How does the measure of ∠ABC change as the value of x increases from 2 to 5?
d How does the measure of ∠DBC change as the value of x increases from 2 to 5?

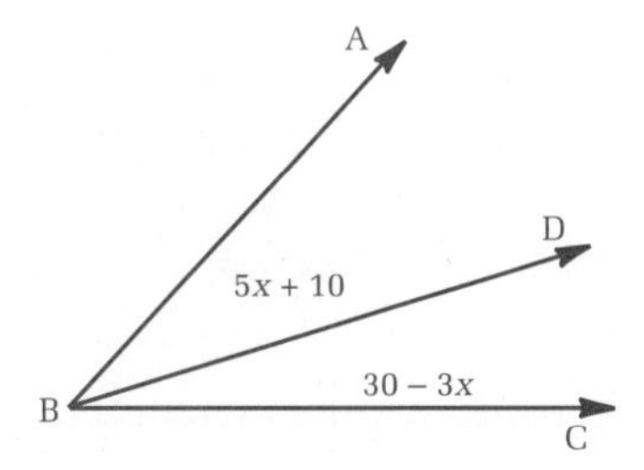

Problem-Set Notes and Strategies, continued

- Problem **13** illustrates that even though the operations are the inverse of one another, the functions may not be inverse functions because of the difference in domain. In the next section, how to insure that a function has an inverse and functions with restricted domains will be studied.
- In problem **17**, students may have difficulty applying the greatest integer function to negative numbers. Since they are to find "...the greatest integer less than or equal to x", in the case of negatives, the greatest integer is to the left of the value if the value is not itself an integer.

Additional Answers

19a $m(x) = 2x + 40$
b $-2 < x < 10$
c It increases from 44 to 50.
d It decreases from 24 to 15.

Problem-Set Notes and Strategies, continued

- Problem **20** is an interesting problem. Students may think that if a function is not written with f, it must be a different function. The name and variable is unimportant; it's the rule and the corresponding domain and range that determine the function.

- It is not obvious by looking at problem **22** that f and g are inverse functions. Doing the calculations will verify their relationship. The inverse of the function $\frac{1}{x}$ will be explored in a cooperative learning lesson in this section.

- Problem **25** illustrates a real-world situation that students will be able to relate to. Many of them have jobs and are learning to handle their own finances and pay income taxes.

- Problem **27** is an example of a function whose inverse is itself. This concept is explored further in a cooperative learning lesson.

Problem Set, *continued*

R **20** Explain why $g(x) = x^2$ and $h(y) = y^2$ represent the same function. Same domain, same rule, same range

P,I **21** Let $f(x) = x^2 + 1$ and $g(x) = \sqrt{x - 1}$.

a Find $f[g(8)]$. 8 **b** Find $g[f(8)]$. 8

c Are f and g inverse functions? No (unless the domain of f is restricted)

B

P,I **22** Let $f(x) = \frac{1}{x - 2}$, where $x \neq 2$, and $g(x) = \frac{1 + 2x}{x}$, where $x \neq 0$.

a Find $f[g(4)]$ and $g[f(4)]$. 4; 4

b Are f and g inverse functions? Yes

P,I **23** Suppose that function s is the inverse of function r and that $r(5) = 11$.

a Find $r \circ s(5)$. 5 **b** Find $s \circ r(5)$. 5 **c** Find $s(11)$. 5

P,R **24** Suppose that $F(A, B) = A + B$ and $G(A, B) = A \cdot B$ are functions for which the inputs A and B are matrices.

a Find $F\left(\begin{bmatrix} 3 & 7 \\ 5 & 9 \end{bmatrix}, \begin{bmatrix} 8 & 11 \\ 20 & 17 \end{bmatrix}\right)$. $\begin{bmatrix} 11 & 18 \\ 25 & 26 \end{bmatrix}$

b Find $G\left(\begin{bmatrix} 2 & 3 \\ 1 & -4 \end{bmatrix}, \begin{bmatrix} 5 & -6 \\ -3 & 8 \end{bmatrix}\right)$. $\begin{bmatrix} 1 & 12 \\ 17 & -38 \end{bmatrix}$

c What is the domain of F? {(A, B):dimensions of A = dimensions of B}

d What is the domain of G? {(A, B):no. of columns in A = no. of rows in B}

P **25** The total cost of x hamburgers priced at $1.39 each is given by the formula $C(x) = 1.39x$. The sales tax charged on a purchase of y dollars, at a tax rate of 7 percent, is given by the formula $T(y) = 0.07y$.

a If $h(x) = T(C(x))$, what is the formula that represents function h? $h(x) = 0.0973x$

b What does the output of h represent? The sales tax charged on a purchase of x hamburgers

P **26** Let $f(x) = \frac{x}{x^2 - 9}$ and $g(x) = \frac{1}{x}$.

a Write a simplified expression for $g \circ f(x)$. What is the domain of $g \circ f$? $\frac{x^2 - 9}{x}$; $\{x{:}x \neq -3, x \neq 0, \text{ and } x \neq 3\}$

b Write a simplified expression for $f \circ g(x)$. What is the domain of $f \circ g$? $\frac{x}{1 - 9x^2}$; $\left\{x{:}x \neq -\frac{1}{3}, x \neq 0, \text{ and } x \neq \frac{1}{3}\right\}$

P,R,I **27** Let $f(x) = \frac{x}{x - 1}$.

a Find $f[f(2)]$. 2 **b** Find $f[f(3)]$. 3

c Find $f[f(-5)]$. −5 **d** Find $f[f(1)]$. Undefined

e What is INVf? INV$f(x) = f(x)$ for $\{x{:}x \neq 1\}$

P,R,I **28** Suppose that $f(n) = 4 + f(n - 1)$ and that $f(1) = 12$.

a Find $f(2)$. 16 **b** Find $f(3)$. 20 **c** Find $f(4)$. 24

P,R **29** If P(x) represents the perimeter of the triangle shown, what are the domain and the range of function P?
Domain: $\{x{:}5 < x < 14\}$;
range: $\{P(x){:}27 < P(x) < 54\}$

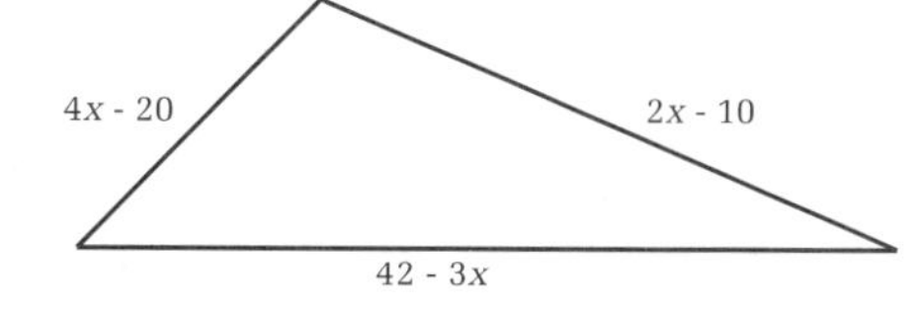

P,R,I **30** If two of the functions listed below are selected at random, what is the probability that they are inverses? $\frac{4}{21}$

$f(x) = 2x$ $g(x) = x + x$ $h(x) = \frac{1}{2}x$ $k(x) = x - 2$

$r(x) = \frac{x}{2}$ $s(x) = x - x$ $t(x) = \frac{1}{2x}$

C P **31** Suppose that the inputs of a function represent the numbers of sides of various polygons and that the outputs of the function represent the numbers of the polygons' diagonals.

a Write a formula for the function. $D(n) = \frac{n(n-3)}{2}$
b What is the function's domain? {integers ≥ 3}
c What is the function's range? $\{0, 2, 5, 9, 14, \ldots\}$

P,I **32** Below is part of a chart used to calculate sales tax in Yoknapatawpha County.

Amount	Tax
1–19¢	1¢
20–39¢	2¢
40–59¢	3¢
60–79¢	4¢
80–99¢	5¢

Write a function T that can be used to find the amount of sales tax charged on any purchase price p, in cents. (Hint: Use the greatest-integer notation described in problem 18.) $T(p) = \left[\frac{p}{20}\right] + 1$

R **33** A rocket is fired upward. After 3 seconds, its fuel runs out. At that instant, it is 1550 feet above the ground and has an upward velocity of 2800 feet per second.

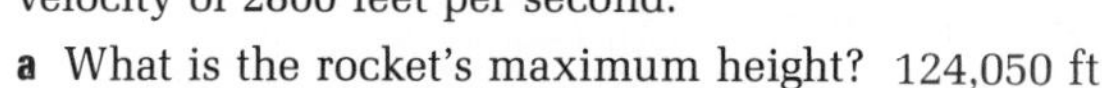

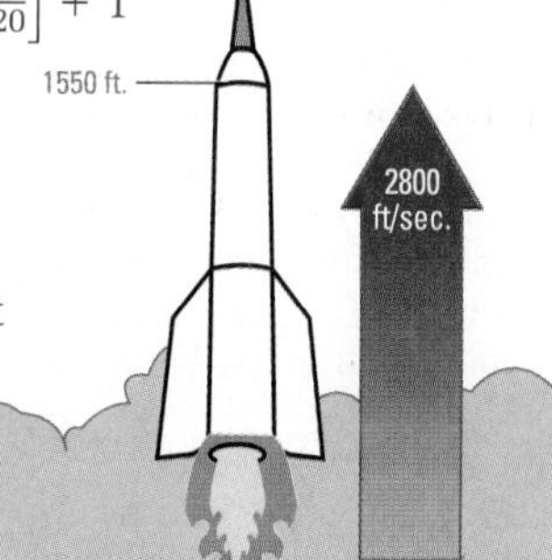

a What is the rocket's maximum height? 124,050 ft
b How many seconds (from the time of its launch) does it take for the rocket to reach its maximum height? 90.5 sec
c What is the total time from the rocket's launch until it strikes the ground? $\approx$178.55 sec

Cooperative Learning

Problems **22** and **27** suggest that the inverse of some functions is not obvious. Have students work in small groups to consider the function $f(x) = \frac{1}{x}$.

1. First find the domain of f.
2. Make a table of values using both positive and negative numbers as inputs for the table.
3. What happens to the function as the values of x become very large or very small?
4. What happens to the values of f as x gets closer to zero?
5. Plot several of these points and graph the function being careful to remember the domain.
6. Evaluate $f(f(x))$. What do you notice?
7. Plot the line $y = x$ on top of your existing graph. What do you notice about the relationship between f and this line?
8. What is the relationship between $f(x)$ and $f(\frac{1}{x})$?
9. Try to make some generalizations about this relationship and the function f and the line $y = x$.

Problem-Set Notes and Strategies, continued

- Some method of counting the possibilities needs to be found in problem **30**. After the possibilities are counted, each pair must be checked for inverses and the winning possibilities counted.

5.3

Class Planning

Time Schedule

Average: 2 days
Advanced: $1\frac{1}{2}$ days
Honors: 2 days

Resource References

Teacher's Resource Book
Class Opener 29
Practice 30
Evaluation
Quiz Forms 1, 2, 3

Class Opener

Given $f(x) = 5x + 7$. If the range of outputs for f is $\{17, 12, 8, 3, -2, -7, -12, -21\}$, find all ordered pairs of the function.

$(2, 17), (1, 12), (\frac{1}{5}, 8), (-\frac{4}{5}, 3), (-\frac{9}{5}, -2), (-\frac{14}{5}, -7), (-\frac{19}{5}, -12), (-\frac{24}{5}, -17)$

Lesson Notes

- To find the inverse of a function, first, interchange x and y in the equation. Second, solve the new equation for y. This new value of y is the output value of the function. The first step of interchanging x and y will help students remember why the point (b, a) is on the inverse of a function if the point (a, b) is on the function. It also provides some measure of sense for the symmetry about the line $y = x$. Since we are interchanging x and y it is reasonable to assume that $x = y$ or equivalently $y = x$.
- As noted in the text, the inverse of a function is not always a function itself.

Vocabulary
reflection

5.3 INVERSE FUNCTIONS

Objectives

After studying this section, you will be able to

- Find the inverse of a given function
- Recognize the relationship between the graphs of a function and its inverse

Part One: Introduction

Finding Inverses

In the preceding section, you learned to recognize when two functions are inverses of each other. The next step is to be able to find the inverse of a given function.

One way of finding a function's inverse is to think of the function as a process involving a sequence of steps and then to reverse this process. The function $f(x) = 3x + 5$, for example, tells us to multiply the input by 3, then add 5; so INVf must consist of the inverses of these operations performed in reverse order—that is, subtracting 5, then dividing by 3.

$$f(x) = 3x + 5 \qquad \text{INV}f(x) = \frac{x - 5}{3}$$

Now consider the input-output diagram for two inverse functions, f and INVf. If an input a in function f produces an output b, then inputting b in INVf must result in an output of a. In other words, if the point with coordinates (a, b) is on the graph of f, then the point (b, a) must be on the graph of INVf.

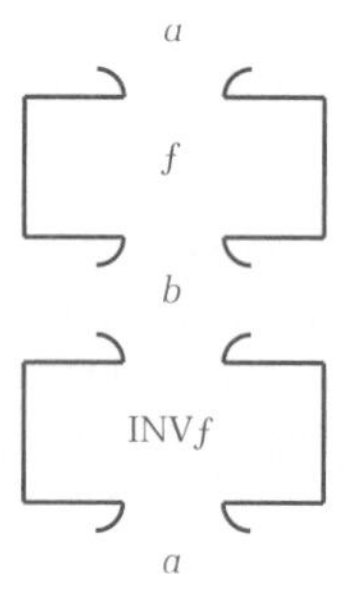

This relationship suggests another way of finding the inverse of a function; we can interchange the input and output variables in a corresponding equation and then solve the new equation for the output variable.

Example *If* $f(x) = 9 - 2x$, *what is INVf?*

We rewrite the function as the equation $y = 9 - 2x$. Interchanging x and y in the equation produces $x = 9 - 2y$, which we solve for y.

$$x = 9 - 2y$$

$$2y = 9 - x$$

$$y = \frac{9 - x}{2}$$

$$\text{Thus, INV}f(x) = \frac{9 - x}{2}.$$

In a number of cases, a given function does not have an inverse function unless we restrict its domain. Suppose, for example, that $g(x) = x^2$. If we interchange the variables in the equation $y = x^2$ and solve for y, we obtain $y = \pm\sqrt{x}$, in which y is not a function of x. If, however, we restrict the domain of g to $\{x : x \geq 0\}$, we can say that $g(x) = x^2$ and $\text{INV}g(x) = \sqrt{x}$.

Graphing Inverse Functions

Let's plot some points (a, b) that are on the graph of $y = 2x + 1$ and then plot the corresponding points of the form (b, a).

(a, b)	**(b, a)**
(−3, −5)	(−5, −3)
(−2, −3)	(−3, −2)
(−1, −1)	(−1, −1)
(0, 1)	(1, 0)
(1, 3)	(3, 1)
(2, 5)	(5, 2)

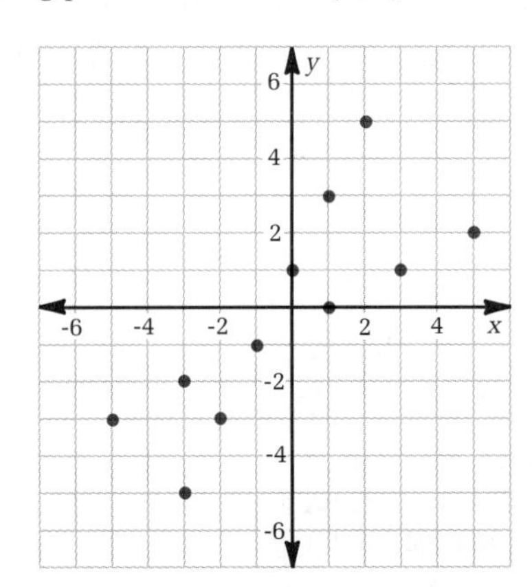

Notice that the pairs of points form a symmetrical arrangement. Now let's look at the graphs of several functions and the corresponding inverse functions.

Lesson Notes, continued

- An additional notation for the inverse of a function that is frequently used is $f^{-1}(x)$. The exponent is an inverse notation, not a power. For the student who understands function notation this is not a problem as f is not a number or variable but a name of a function.

Using Computers

Students can gain insights into the relationship of a function and its inverse with a program capable of graphing such as the *Omnifarious Plotter, Derive, The Function Analyzer,* and the *Mathematics Exploration Toolkit.* A graphics calculator could also be used effectively. A symbol manipulator such as *Derive* can be used to show that the composition of a function and its inverse is the identity.

Checkpoint

1. Let $f(x) = x + 7$ and $g(x) = x - 7$. Are the functions f and g inverse functions? Yes
2. Let $f(x) = 5x + 3$, find the inverse of the function f. $\text{INV}f = \frac{(x-3)}{5}$
3. Let the function f be defined as the following set $\{(1, 3), (2, 4), (3, 5), (4, 6), (5, 7)\}$. Find the function INV$f$. $\{(3, 1), (4, 2), (5, 3), (6, 4), (7, 5)\}$
4. Is the inverse of the function represented by $\{(1, 1), (2, 2), (3, 1), (4, 4), (5, 1)\}$ also a function? No

$y = f(x) = 5x$ and $y = \text{INV}f(x) = \frac{x}{5}$

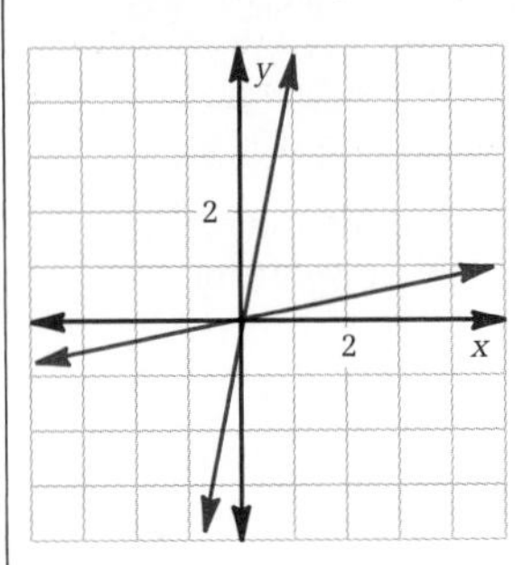

$y = f(x) = \sqrt{x}$ and $y = \text{INV}f(x) = x^2$ for domain $\{x: x \geq 0\}$

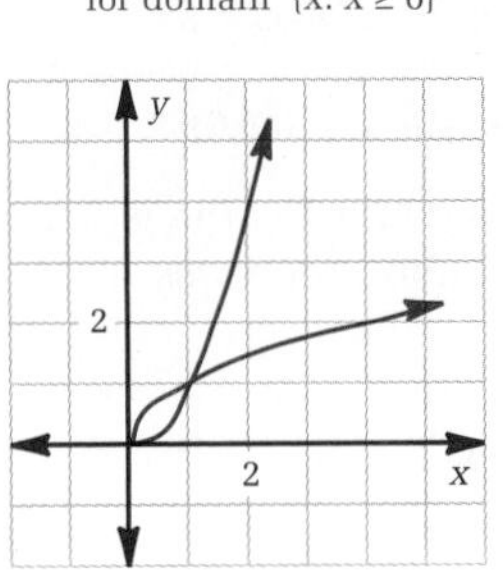

$y = f(x) = 2x + 5$ and $y = \text{INV}f(x) = \frac{x-5}{2}$

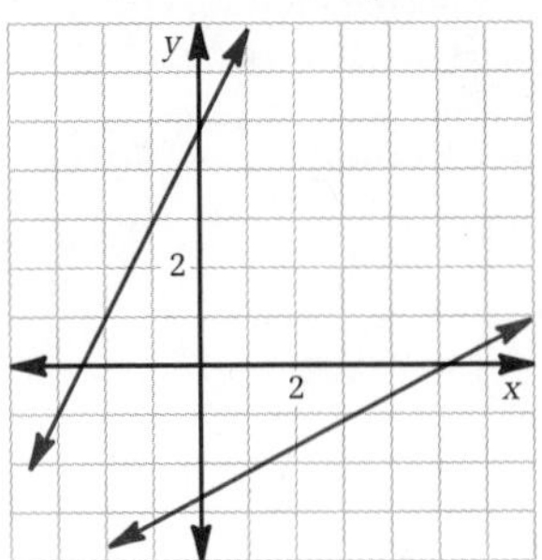

It appears that in each case the graphs of the function and its inverse are symmetric about the line $y = x$ (the line that diagonally bisects Quadrants I and III). Since this line divides the coordinate plane in such a way that the graphs of a function and its inverse are "mirror images" of each other, we say that the two graphs are ***reflections*** of each other over the line $y = x$.

Part Two: Sample Problems

See p. 787

Problem 1 *Let $g(x) = x^2 + 3$, with a domain of $\{x:x \geq 0\}$, and $h(x) = \sqrt{x - 3}$, with a domain of $\{x:x \geq 3\}$.*

a *Are g and h inverse functions? Why or why not?*
b *What is the range of g? What is the range of h? What is significant about the domains and ranges of these functions?*
c *Graph g, h, and h ∘ g on the same coordinate system. What is significant about the graphs?*

Solution

a Functions g and h are inverses, since $g[h(x)] = h[g(x)] = x$ for all values of x in the functions' domains.
b The range of g is $\{g(x):g(x) \geq 3\}$, and the range of h is $\{h(x):h(x) \geq 0\}$. Because the functions are inverses, the values in the range of g form the domain of h, and the values in the range of h form the domain of g.
c The graphs of $y = g(x)$ and $y = h(x)$ are reflections of each other over the graph of $y = h[g(x)]$, which is equivalent to $y = x$ for the values in the restricted domain.

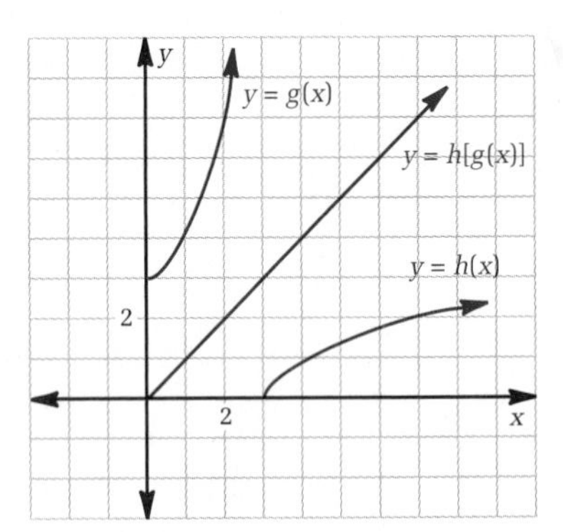

Problem 2 *Let $f(x) = 2x + 4$. Find INVf and graph both functions on the same coordinate system.*

Solution

Method 1
Since f involves multiplying by 2, then adding 4, INVf will involve subtracting 4, then dividing by 2.

$$\text{INV}f(x) = \frac{x - 4}{2}$$

Method 2
Exchanging x and y in the equation $y = 2x + 4$ produces $x = 2y + 4$, which we solve for y.

$$x = 2y + 4$$
$$x - 4 = 2y$$
$$\frac{x - 4}{2} = y$$
$$\text{INV}f(x) = \frac{x - 4}{2}$$

Since f and INVf are linear functions, we can find a few points on the graph of $y = f(x)$, find the reflections of these points over the line $y = x$, and connect each set of points with a straight line.

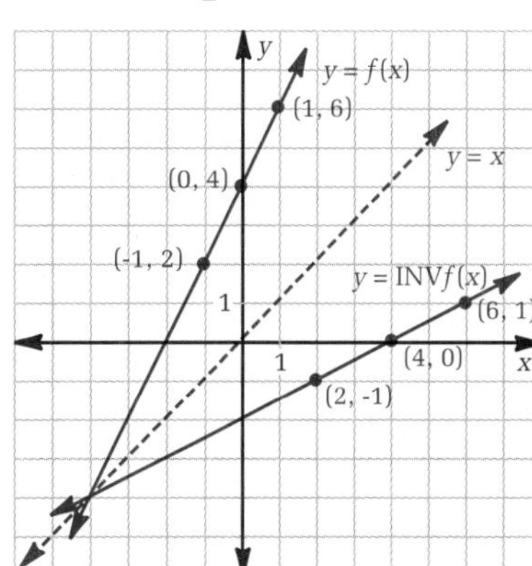

Assignment Guide

Average	
Day 1	10–12, 14, 17, 18, 21, 22, 24, 25, 28–30, 34
Day 2	13, 15, 16, 19, 20, 23, 27, 31–33, 35, 37
Advanced	
Day 1	($\frac{1}{2}$ day) Section 5.2 9, 10, 12, 17, 22, 25, 28–30 ($\frac{1}{2}$ day) 10–12
Day 2	13, 14, 16, 17–19, 21, 23, 26, 27, 29, 30, 32, 33
Honors	
Day 1	11, 14, 16–18, 20, 21, 27, 29–34, 37
Day 2	13, 15, 19, 22, 23, 25, 26, 28, 35, 36, 38

Part Three: Exercises and Problems

Warm-up Exercises

1 The points (0, 1), (2, 4), (4, 16), and (5, 25) are on the graph of a function.

P **a** List four points on the graph of the function's inverse. (1, 0), (4, 2), (16, 4), (25, 5)

b Graph all eight points, along with the equation $y = x$, on a coordinate system.

c How do the two sets of points relate to the graph of $y = x$?
They are reflections of each other over the graph of $y = x$.

P **In problems 2–7, find the inverse of each function.**

2 $f(x) = 10x$ INV$f(x) = \frac{x}{10}$

3 $g(x) = x - 6$ INV$g(x) = x + 6$

4 $h(x) = 6x + 3$ INV$h(x) = \frac{x - 3}{6}$

5 $p(x) = x^2 - 1$, where $x \geq 0$

6 $q(x) = \sqrt{x - 1}$, where $x \geq 1$
INV$q(x) = x^2 + 1$, where $x \geq 0$

7 $r(x) = \frac{1}{2x}$, *where* $x \neq 0$
INV$r(x) = \frac{1}{2x}$

Problem-Set Notes and Strategies

- A considerable amount of work with inverse functions rests on the symmetry about the line $y = x$. Make sure that students make this connection in problem **1**.
- In problem **6**, make sure that students restrict the domain of the inverse function. In this case the inverse is defined everywhere, but the function is not. The domain of the inverse must be restricted appropriately so that the range of the inverse is the same as the domain of the function.

Additional Answer

5 INV $p(x) = \sqrt{x + 1}$ where $x \geq -1$

Answer to problem **1b** can be found in the answer pages beginning on **A1**.

Problem-Set Notes and Strategies, continued

- Help students make the connection between the domain and range of G and INVG in problem **8**.
- Problem **9** is a good problem to identify a restricted domain.

Additional Answers

8a Domain: {−4, −1, 0, 6, 8}
Range: {−2, −1, 1, 3, 4}
b (−2,−4),(−1,1),(3,0),(1,6),(4,8)
c Domain: {−2, −1, 1, 3, 4}
Range: {−4, −1, 0, 6, 8}

9a

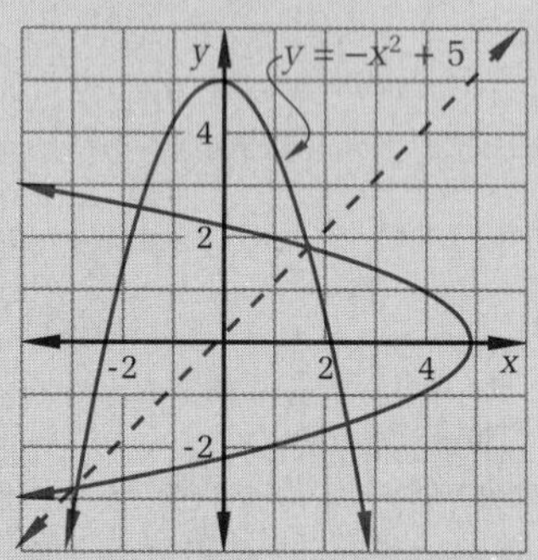

Stumbling Block

Not every function has an inverse that is also a function. For instance, $y = x^2$ is a function and the mechanical steps of the inverse process will yield:

$$x = y^2$$
$$\sqrt{x} = \sqrt{y^2}$$
$$\sqrt{x} = |y|$$
$$\sqrt{x} = y \text{ or } -\sqrt{x} = y$$

So $y = \pm\sqrt{x}$.
This is not a function, since if $x = 4$, $y = 2$ or $y = -2$. Students frequently ignore the ± sign and write $y = \sqrt{x}$ and claim this is the inverse function.

P **8** The graph represents a function G, with inputs x and outputs y.
a What are the domain and the range of G?
b Find the coordinates of the points representing INVG.
c What are the domain and the range of INVG?

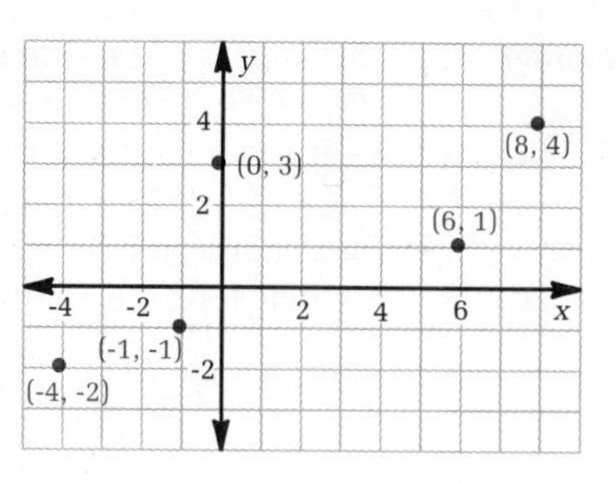

P **9 a** Graph $y = g(x) = -x^2 + 5$. On the same coordinate system, sketch the reflection of the graph over the line $y = x$.
b Does the reflection represent a function? No
c Is there a function INVg? If so, what is it? Yes, if the domain of g is suitably restricted—for instance, to $\{x{:}x \geq 0\}$, in which case INV$g(x) = \sqrt{-x + 5}$

Problem Set

A
P,R **10** Consider the input-output diagram.
a If the output is 17, what was the input? 6
b If the output is 133, what was the input? 64
c If the output is n, what was the input? $\frac{n-5}{2}$

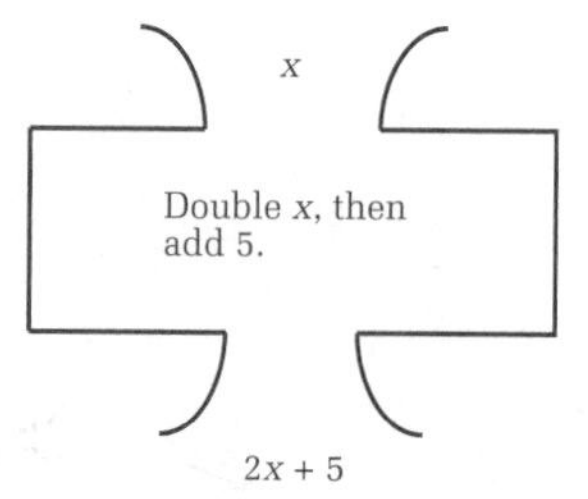

P **11** If a function h can be represented by {(3, 7), (4, 4), (5, 1), (6, −2), (7, 0)}, how can INVh be represented?
{(−2, 6), (0, 7), (1, 5), (4, 4), (7, 3)}

P **12** Let $f(x) = -2x + 8$ and $g(x) = -\frac{1}{2}x + 4$.
a Graph the two functions and the line $y = x$ on the same coordinate system.
b Find $f[g(6)]$ and $g[f(-10)]$. 6; −10
c Show that $f[g(x)] = x$. $f[g(x)] = -2\left(-\frac{1}{2}x + 4\right) + 8 = (x - 8) + 8 = x$
d Are f and g inverses? Yes

P **13** Copy and complete the table of values for two inverse functions, f and g.

x	0	1	2	3	4	5	8
$f(x)$	5	3	0	2	1	4	8
$g(x)$	2	4	3	1	5	0	8

Additional Answer
Answer for problem **12a** can be found in the answer pages beginning on **A1**.

P | **14** If $\mathrm{INV}f(x) = 12 - \frac{1}{2}x$, what is function f? $f(x) = 24 - 2x$

P,R | **15** The graphs of five functions are shown. If two of the graphs are selected at random, what is the probability that they represent inverse functions? $\frac{1}{10}$

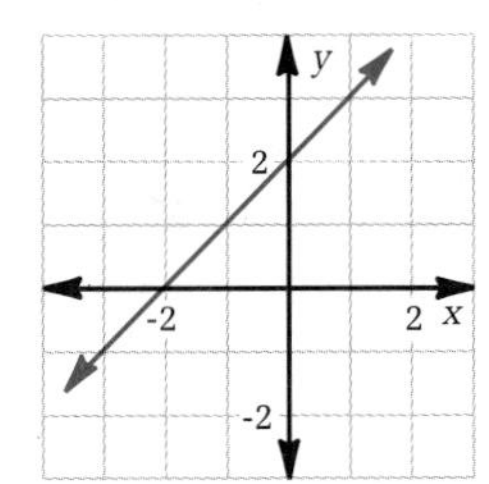

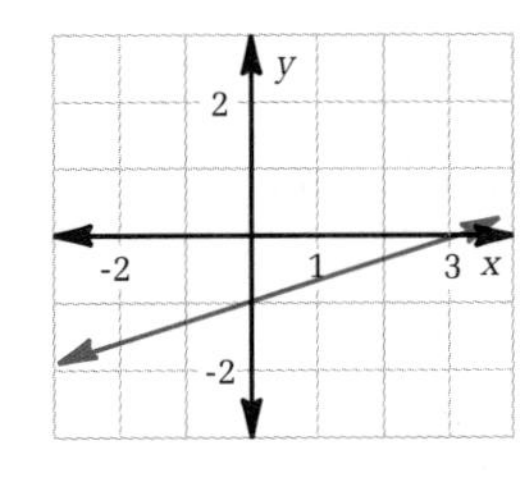

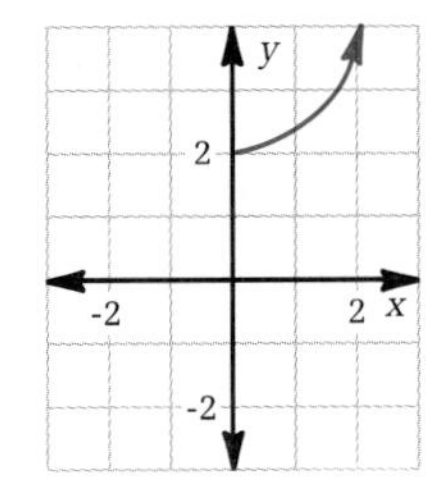

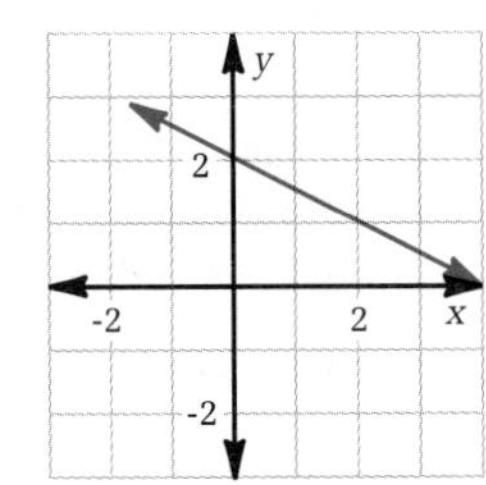

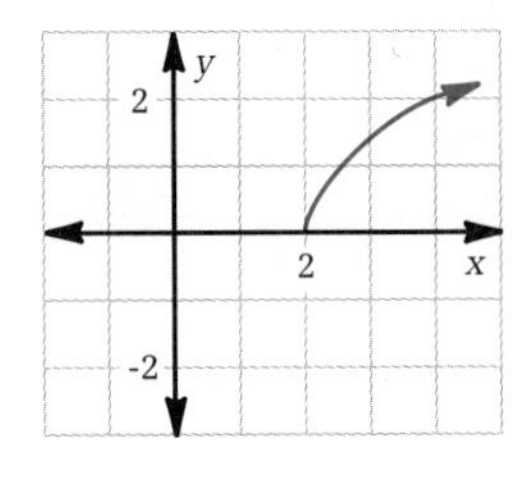

R | **16** If $f(5) = 14$ and $f(2 \cdot 5) = 3[f(5)]$, what is the value of $f(10)$? 42

P,R | **17** Let $f(x) = x^2$, with a domain of $\{x{:}1 < x < 10\}$.

a What is the range of f? $\{f(x){:}1 < f(x) < 100\}$

b If $g(x) = \mathrm{INV}f(x)$, what is function g? $g(x) = \sqrt{x}$

c What are the domain and the range of g?

d Could you have identified the domain and the range of the inverse of f without actually finding g? If so, how? Yes; domain of INVf = range of f and range of INVf = domain of f.

c Domain: $\{x{:}1 < x < 100\}$; range: $\{g(x){:}1 < g(x) < 10\}$

P,I | **18** The equation $S = 180(n - 2)$ represents a function that outputs the sum of the angles of a polygon when the number of the polygon's sides is input.

a What would the inverse of the function do?

b Write an equation corresponding to the inverse function. $n = \frac{S}{180} + 2$

c What is the domain of the inverse function? {positive multiples of 180}

I | **19** The letters a and b can be arranged in only two ways, as ab or as ba. In how many ways can the three letters a, b, and c be arranged? 6

P | **20** If $f(x) = x^2$ and the domain of f is restricted to $\{x{:}x \le 0\}$, what is INVf? $\mathrm{INV}f(x) = -\sqrt{x}$

Problem-Set Notes and Strategies, continued

- Problem **14** illustrates the characteristic that INV(INVf) = f.
- In problem **15**, have students identify how many combinations are possible first and then examine for inverse relations.
- Problem **16** illustrates an operation on functions that will be introduced in the next section. Students may try to distribute the 3 and ask what is $f(15)$.
- Problem **17** illustrates the relationship between domain and range of functions and their inverses. At times it may be easier to find the range of a function by finding the inverse and finding the domain of the inverse. These relationships should be emphasized.
- Permutations are hinted at in problem **19**. These are formally defined in Section **5.5**.

Additional Answer

18a Output the number of sides of a polygon when the sum of its angles is input.

Problem Set, *continued*

P,R,I **21** Sketch the reflection of each of the following graphs over the line $y = x$ and indicate whether the reflection represents a function.

a
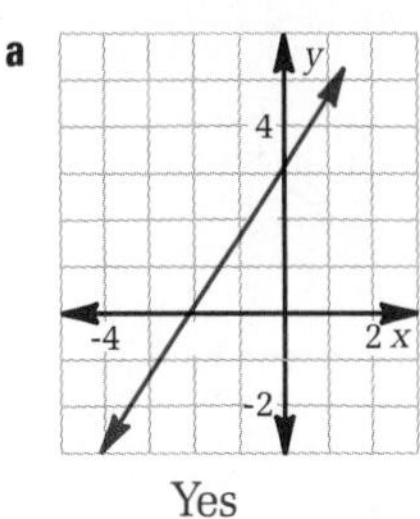

Yes

b Yes

c Yes

d No

P,I **22** **a** Graph $y = h(x) = (x - 3)^2 + 4$. On the same coordinate system, sketch the reflection of the graph over the line $y = x$.

b Does the reflection represent a function? No

c Is there a function INV*h*? If so, what is it? Yes, if the domain of *h* is suitably restricted—for instance, to $\{x{:}x \geq 3\}$, in which case $\text{INV}h(x) = \sqrt{x - 4} + 3$

B

P **23** For each function *f*, find INV*f*. Specify restrictions on the domain of *f* if necessary.

a $f(x) = 3x + 4$ $\text{INV}f(x) = \frac{x - 4}{3}$

b $f(x) = \dfrac{9}{x}$ $\text{INV}f(x) = \frac{9}{x}$, where $x \neq 0$

c $f(x) = \dfrac{3}{4} - \dfrac{1}{2}x$ $\text{INV}f(x) = \frac{3}{2} - 2x$

d $f(x) = (x + 3)^2$ Answers will vary. Possible answer: If $x \geq -3$, $\text{INV}f(x) = \sqrt{x} - 3$

R **24** The area of the rectangle is 104 square centimeters. Find the lengths of its sides. ≈15.65 cm and ≈6.65 cm

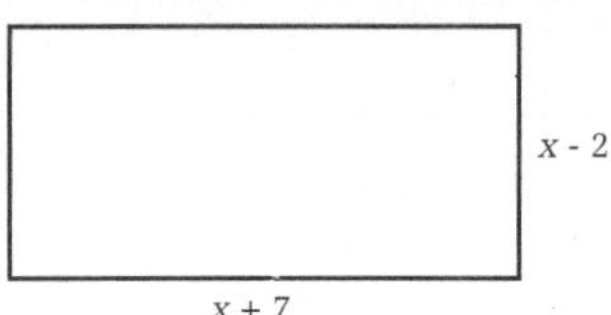

P **25** If $g(x) = -\frac{2}{3}(x + 5)$, what is INV*g*? $\text{INV}g(x) = -\frac{3}{2}x - 5$

P **26** The function $A(x) = x^2$ outputs the area of a square with side length *x*. The function $D(x)$ doubles its input.

a Is $A \circ D$ equal to $D \circ A$? No

b What is the geometric meaning of $A[D(x)]$? Area of square with side length 2x

c What is the geometric meaning of $D[A(x)]$? Total area of two squares, each with side length x

P **27** Let $R(x) = \frac{1}{x}$ and $O(x) = -x$.

a Is *R* its own inverse? Yes

b Is *O* its own inverse? Yes

c Find some other functions that are their own inverses. Answers will vary.

P,R,I **28** Graph each of the following functions, then reflect the graph over the line $y = x$ and indicate whether the result represents a function.

a $f(x) = 2x - 6$ Yes

b $g(x) = 2x^2 - 3$ No

c $h(x) = |x - 2|$ No

d $j(x) = 2^x$ Yes

Additional Answers

21a

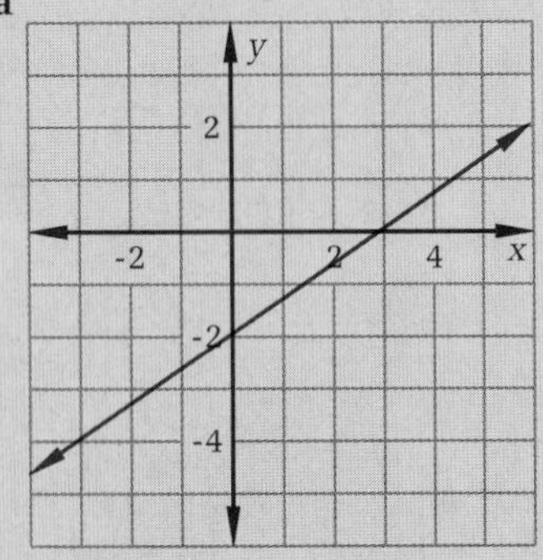

21b

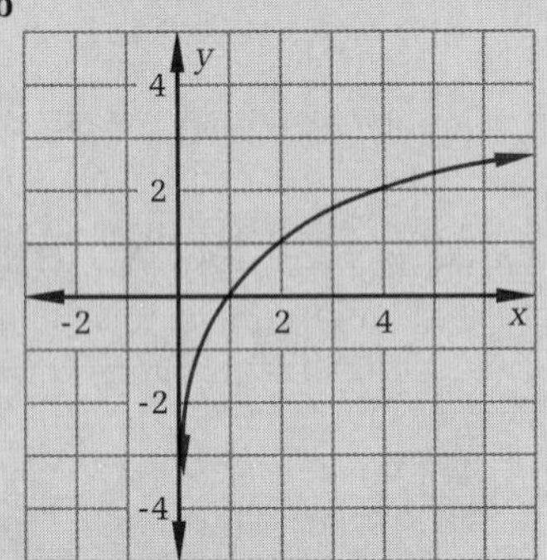

Problem-Set Notes and Strategies, continued

- Make sure zero is eliminated from the domain in problem **23b**. Students have a tendency to look at the inverse and forget that a function may not be defined for a particular number, but the function's inverse may be.
- A geometric argument is given in problem **26** to illustrate that function composition is not commutative.
- Part **d** of problem **28** may cause some graphing problems. Point out that the range can never be less than or equal to zero. Have the students plot points to illustrate the curve. Exponential curves will be explored in Section **6.4**.

Additional Answers

Answers for problems **21c**, **d**, **22a**, **28a**, **b**, **c**, and **d** can be found in the answer pages beginning on **A** 1.

P,I **In problems 29–31, find the equation of each graph's axis of symmetry.**

29

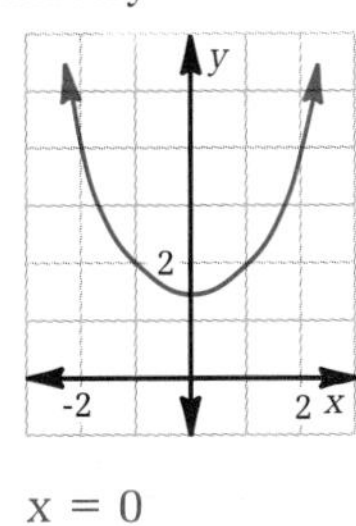

$x = 0$

30

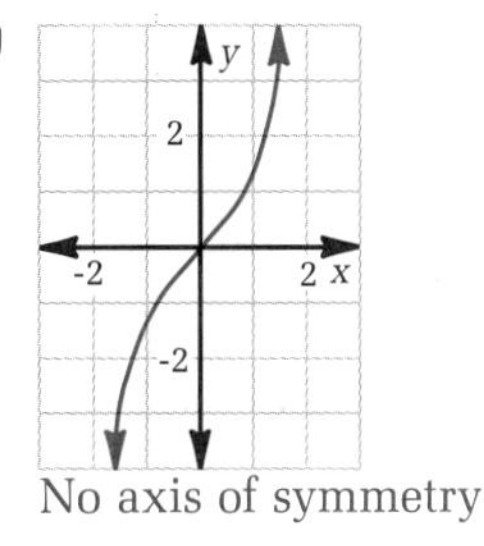

No axis of symmetry

31

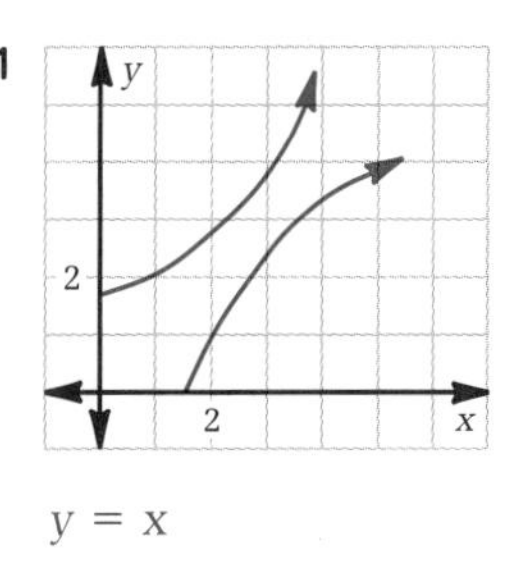

$y = x$

I **32** Ralph's Pretty Good Pizza offers five pizza toppings and lets customers choose any two toppings for the price of one. How many different pairs of toppings can customers select? 10

I **33** On a calculator with a [10ˣ] key, enter 2.5 [10ˣ]. With the result still displayed, press the [log] key. Repeat this sequence for the inputs 9.2, 4.96, and −10.5. What can you conclude about the 10^x and log functions? They are inverses.

P **34** Suppose that the sales tax, $T(x)$, charged on a purchase of x dollars is given by the function $T(x) = 0.07x$.

a What is the inverse of T? $\text{INV}T(x) = \frac{x}{0.07}$

b What does the output of INVT represent? Purchase price on which a sales tax of x dollars is charged

R **35** Determine the equation of the line of best fit for the data points (2, 7), (5, 10), (7, 14), and (10, 16). $y = 1.18x + 4.69$

R **36** Two 2 × 2 matrices are called *inverse matrices* if their product is $\begin{bmatrix} 1 & 0 \\ 0 & 1 \end{bmatrix}$. If two of the following matrices are selected at random, what is the probability that they are inverses? $\frac{1}{5}$

$$\begin{bmatrix} 2 & 0 \\ 0 & 2 \end{bmatrix} \quad \begin{bmatrix} 1 & 2 \\ 2 & 5 \end{bmatrix} \quad \begin{bmatrix} 0.5 & 0 \\ 0 & 0.5 \end{bmatrix} \quad \begin{bmatrix} 1 & 2 \\ -2 & 5 \end{bmatrix} \quad \begin{bmatrix} 5 & -2 \\ -2 & 1 \end{bmatrix}$$

C

P **37** Write a formula that represents the inverse of a quadratic function of the form $f(x) = a(x - h)^2 + k$, where the domain of f is restricted to $\{x{:}x \geq h\}$.

$\text{INV}f(x) = \sqrt{\frac{x-k}{a}} + h$

P,R **38** Let $f\left(\begin{bmatrix} a & b \\ c & d \end{bmatrix}\right) = 2\begin{bmatrix} a & b \\ c & d \end{bmatrix} + 4\begin{bmatrix} 1 & 0 \\ 0 & 1 \end{bmatrix}$.

a Find $f\left(\begin{bmatrix} 3 & 5 \\ 2 & -1 \end{bmatrix}\right)$. $\begin{bmatrix} 10 & 10 \\ 4 & 2 \end{bmatrix}$

b Find INVf.

b $\text{INV}f\left(\begin{bmatrix} a & b \\ c & d \end{bmatrix}\right) = \frac{1}{2}\begin{bmatrix} a-4 & b \\ c & d-4 \end{bmatrix}$

c Show that $f \circ \text{INV}f\left(\begin{bmatrix} a & b \\ c & d \end{bmatrix}\right) = \begin{bmatrix} a & b \\ c & d \end{bmatrix}$.

Problem-Set Notes and Strategies, continued

- Problem **30** has no axis of symmetry, but still looks symmetric. Have students think about folding the graph twice, once over the *x*-axis and again over the *y*-axis. Other types of symmetry are explored in Section **6.3**.
- The formula for combinations for problem **32** will finally be introduced in Section **5.5**. Some students may have figured it out by now or are relating them to Pascal's triangle.
- Care must be taken in problem **33** so that students don't see this as a proof of the fact that 10^x and $\log x$ are inverse functions. We have only shown it is true for a finite set of numbers. A first step in the proof process is observation and conjecture, but this is still just a conjecture. This will be formalized in Chapter **11**.
- Problem **38b** may require some additional thought. Students are not yet familiar with matrices. The identity matrix has not been formally taught but included in an exercise.

Communicating Mathematics

Have students write a paragraph that explains the relationship between the domain and range of a function and the domain and range of the corresponding inverse function.

Additional Answers

Answers for problems **37** and **38c** can be found in the answer pages beginning on **A1**.

5.4

Class Planning

Time Schedule
All levels: 2 days

Resource References
Teacher's Resource Book
Class Opener 30
Practice 31
Enrichment 10
Transparency 10

Class Opener

Given $f(x) = x^2 - 3x + 7$

1 Write an expression for $3f(x)$
$3x^2 - 9x + 21$
2 Write an expression for $f(3x)$
$9x^2 - 9x + 7$
3 Evaluate $f(2)$ 5
4 Evaluate $3f(2)$ 15
5 Evaluate $f(3 \cdot 2)$ 25

5.4 Operations on Functions

Objectives

After studying this section, you will be able to

- Perform operations on the outputs of functions
- Perform operations on the inputs of functions

Part One: Introduction

Operations on Outputs

The solution to a problem frequently involves more than merely evaluating a function. Sometimes we must perform a mathematical operation on the input before evaluating the function; in other cases, we must perform an operation on the function's output.

In this section, we will explore operations on functions. The functions we will deal with will have numerical inputs and outputs so that we can see patterns more easily. Because these functions represent numbers, we will be able to apply the rules of numbers to the outputs—for example, $3[f(x)] + 5[f(x)] = 8[f(x)]$, and $2[f(x)] \cdot 4[f(x)] = 8[f(x)]^2$. Here are some more examples of operations on outputs.

Example *Let $f(x) = x^2 + 2x + 1$. Write an expression for $3[f(x)]$.*

$$3[f(x)] = 3(x^2 + 2x + 1)$$
$$= 3x^2 + 6x + 3$$

Example *If $5[h(x)] = x + h(x)$, what is $h(x)$?*

Remember that we consider $h(x)$ to represent a number.

$$\mathbf{5[h(x)] = x + h(x)}$$

Subtract $h(x)$ from each side $\quad \mathbf{5[h(x)] - h(x) = x}$

$$\mathbf{4[h(x)] = x}$$

Divide each side by 4 $\quad \mathbf{h(x) = \frac{x}{4}}$

Example *Let* $g(x) = x^2 - x$. *Evaluate* $5[g(3)] - [g(3)]^2$.

We first evaluate $g(3)$.

$g(x) = x^2 - x$

$g(3) = 3^2 - 3$

$= 6$

Now we substitute 6 for $g(3)$ in $5[g(3)] - [g(3)]^2$.

$5[g(3)] - [g(3)]^2$

$= 5(6) - 6^2$

$= -6$

Operations on Inputs

There are some important differences between operating on an input of a function and operating on an output. In most cases, when operating on an input, we must know what the function represents before we do any work. Here are some examples.

Example *Let* $f(x) = x^2 + 2x + 1$. *Write an expression for* $f(3x)$.

$f(3x) = (3x)^2 + 2(3x) + 1$

$= 9x^2 + 6x + 1$

Notice that $f(3x)$ indicates a different operation from $3[f(x)]$, which was found in a previous example. The expression $f(3x)$ indicates that the input of function f is tripled, whereas $3[f(x)]$ represents a tripling of the function's output.

Example *If* $g(x) = x^2$, *what is* $g(w - 5) - g(w^3)$?

$g(w - 5) - g(w^3) = (w - 5)^2 - (w^3)^2$

$= w^2 - 10w + 25 - w^6$

Example *Let* $f(x) = 3x + 1$. *Is* $f(a + b)$ *equal to* $f(a) + f(b)$?

Let's try evaluating $f(a + b)$ and $f(a) + f(b)$ for $a = 4$ and $b = 5$.

$f(x) = 3x + 1$

$f(a + b) = 3(4 + 5) + 1$

$= 3(9) + 1$

$= 28$

$f(x) = 3x + 1$

$f(a) + f(b) = f(4) + f(5)$

$= [3(4) + 1] + [3(5) + 1]$

$= 13 + 16$

$= 29$

In this case, $f(a + b) \neq f(a) + f(b)$. This example shows that, as a rule, a function cannot be distributed over addition.

Lesson Notes

- Functions can be thought of as numbers. They may be added, subtracted, multiplied, divided, etc. These operations may be carried out on the input to the function as well as the output.
- Care must be taken to see that a distinction is made between $3f(x)$ and $f(3x)$ and the difference is when the multiplication by 3 is performed.
- Order of operations is still valid in function operations.

Using Computers

The *Function Analyzer*, a program specifically designed to permit changing inputs and outputs, constants, and operations can be used effectively here. In addition, this software enables us to see results in data and in graphs. *Derive*, the *Mathematics Exploration Toolkit*, and a graphics calculator could also be used.

Checkpoint

1 Let $f(x) = x + 5$. Evaluate $5f(x)$ and $f(5x)$. Are they equal? $5x + 25, 5x + 5$, No

2 If $f(x) = x + 5$, find $f(x)^2 - 3f(x)$. $x^2 + 7x + 10$

3 Let $f(x) = 2x + 3$. Evaluate $f(f(x)) - f(x)$. Is this the same as $f(x)$? $2x + 6$, No

Part Two: Sample Problems

Problem 1 *Let $f(x) = x^3 + x^2 - 6$. If $x = 5$, what are the values of $4[f(x)]$ and $f(4x)$?*

Solution The expression $4[f(x)]$ indicates an operation on output, and the expression $f(4x)$ indicates an operation on input.

$$\begin{aligned} 4[f(5)] &= 4(5^3 + 5^2 - 6) \\ &= 4(125 + 25 - 6) \\ &= 4 \cdot 144 \\ &= 576 \end{aligned}$$

$$\begin{aligned} f(4 \cdot 5) &= f(20) \\ &= 20^3 + 20^2 - 6 \\ &= 8000 + 400 - 6 \\ &= 8394 \end{aligned}$$

Notice that $4[f(x)] \neq f(4x)$ if $x = 5$.

Problem 2 *If $f(x) = x^2 + 2x + 3$, what are the values of x for which $f(2x) = 2[f(x)]$?*

Solution

$$\begin{aligned} f(2x) &= 2[f(x)] \\ (2x)^2 + 2(2x) + 3 &= 2(x^2 + 2x + 3) \\ 4x^2 + 4x + 3 &= 2x^2 + 4x + 6 \\ 2x^2 &= 3 \\ x^2 &= \frac{3}{2} \\ x &= \pm\sqrt{\frac{3}{2}} \\ &\approx \pm 1.225 \end{aligned}$$

Problem 3 *A function f has the property that for all real values of x, $f(3x) = 3[f(x)]$. Find $f(0)$.*

Solution We are told that $f(3x) = 3[f(x)]$ for all real inputs, so we can write the equation $f(0) = 3[f(0)]$ and solve it for $f(0)$. (Remember that $f(0)$ represents a number.)

$$\mathbf{f(0) = 3[f(0)]}$$

Subtract $f(0)$ from each side $\quad \mathbf{0 = 3[f(0)] - f(0)}$

$$\mathbf{0 = 2[f(0)]}$$

Divide each side by 2 $\quad \mathbf{0 = f(0)}$

Part Three: Exercises and Problems

Warm-up Exercises

1 Solve $3 \cdot f(x) + 2 = x + 2 \cdot f(x)$ for $f(x)$. $f(x) = x - 2$

P **2** Solve $8 \cdot g(x) = 6x - 7 \cdot g(x)$ for $g(x)$. $g(x) = \frac{2}{5}x$

P,R **3** Let $f(x) = x - 7$ and $g(x) = \sqrt{x}$. Evaluate each of the following expressions.

a $f(3) + g(3)$ $-4 + \sqrt{3}$ **b** $f(3) \cdot g(3)$ $-4\sqrt{3}$ **c** $\frac{f(3)}{g(3)}$ $-\frac{4}{\sqrt{3}}$
d $f[g(3)]$ $\sqrt{3} - 7$ **e** $4[f(3)] + 7[g(3)]^2$ 5

P,R **4** **a** Write a function for $A_r(x)$, the area of the rectangle.
b Write a formula for $A_t(x)$, the area of the triangle.
c Let $A(x) = A_r(x) + A_t(x)$. What is the significance of function A?
d Find $A(20)$. $A(20) = \approx 1106.41$

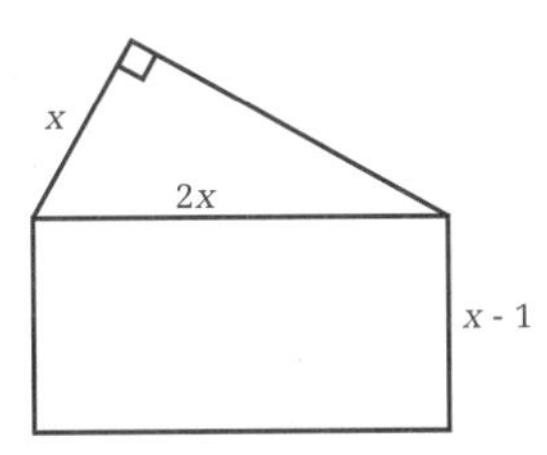

P **5** Let $f(x) = 3x - 4$. Simplify each expression.

a $3 \cdot f(x) + 6$ $9x - 6$ **b** $[f(x)]^2 - 7$ $9x^2 - 24x + 9$
c $[f(x) - 2][3 \cdot f(x) + 4]$ $27x^2 - 78x + 48$ **d** $f[f(x)]$ $9x - 16$
e $4[f(x)]^2 - 10[f(x)]$ $36x^2 - 126x + 104$

P **6** If $f(x) = 2x - 8$, what is the value of w for which $f(w - 3) = 0$? 7

P,I **7** The diagram shows the graphs of $y = f(x)$ and $y = g(x)$. Sketch the graph of $y = f(x) + g(x)$.

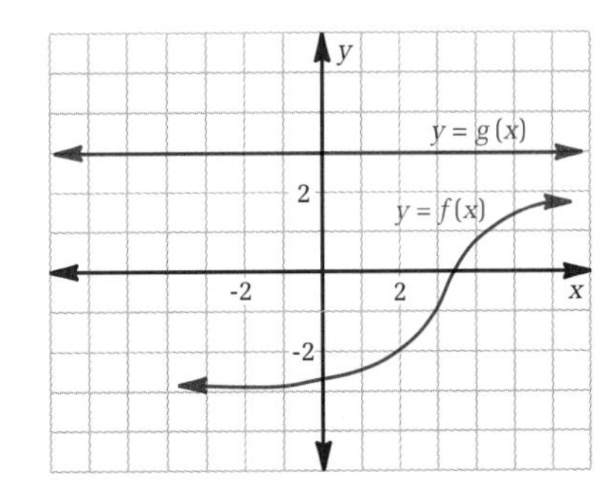

P **8** If $h(x) = x^2$, is $3[h(x)]$ or $h(3x)$ greater? $h(3x)$ if $x \neq 0$

Problem Set

A

P **9** Let $f(x) = \sqrt{x - 5}$. Simplify each expression.

a $f(3x)$ $\sqrt{3x - 5}$ **b** $f(x + 5)$ $\sqrt{x}$ **c** $f(x^2)$ $\sqrt{x^2 - 5}$
d $f(2x - 1)$ $\sqrt{2x - 6}$ **e** $f[f(x)]$ $\sqrt{\sqrt{x - 5} - 5}$

Assignment Guide

Average	
Day 1	9, 11, 12, 14, 16, 19, 20, 26, 27
Day 2	10, 13, 15, 17, 18, 21, 23, 24, 28
Advanced	
Day 1	9, 10, 12, 14, 15, 17, 20, 21, 26, 28
Day 2	11, 13, 16, 18, 19, 22–24, 27, 29
Honors	
Day 1	9–12, 17, 19–21, 25, 26, 28–30
Day 2	13–16, 18, 22–24, 27, 31, 32

Problem-Set Notes and Strategies

- Problems **1** and **2** are a good illustration of how functions may be treated as variables. If students have difficulty, let $f(x) = y$ and solve for y.
- Addition of ordinates may be used to graph problem **7**. This addition of the y-coordinates will be very useful when it is time to graph trig equations.
- Problem **9** illustrates how composition of functions can be written as an operation on the function f. It might be interesting to see if students can write each of these operations as a composition of two or more functions using the original function f.

Additional Answers

4a $A_r(x) = 2x^2 - 2x$
b $A^t(x) = \frac{x^2\sqrt{3}}{2}$
c It represents the total area of the figure.

Answer for problem **7** can be found in the answer pages beginning on **A** 1.

Problem Set, *continued*

P **10** If $f(x) + 3 \cdot f(x) = 8[x + f(x)]$, what is $f(x)$? $f(x) = -2x$

P,R **11** The function $A(x) = x^2$ represents the area of the square shown. Draw an area diagram to represent each of the following expressions.

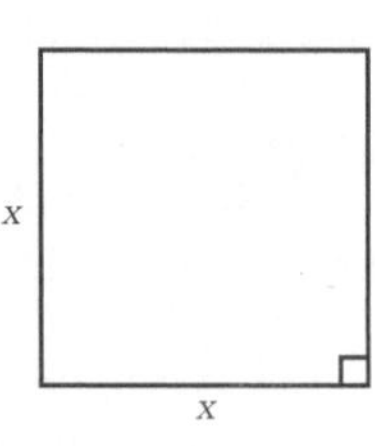

a $3[A(x)]$
b $A(2x)$
c $A(x + 2)$

P,R **12 a** Copy and complete the table for $f(x) = x^2 + 2x + 1$.

x	**−4**	**−3**	**−2**	**−1**	**0**	**1**	**2**	**3**	**4**
***f*(*x*)**	9	4	1	0	1	4	9	16	25
***f*(*x* − 1)**	16	9	4	1	0	1	4	9	16
***f*(*x* + 1)**	4	1	0	1	4	9	16	25	36

b Use the table you generated in part **a** to graph $y = f(x)$, $y = f(x - 1)$, and $y = f(x + 1)$ on the same coordinate system.

R **13** If $f(x) = x^2 - 7$ and $f[g(x)] = x^6 - 7$, what is $g(x)$? $g(x) = x^3$

R **14** Let $f(x) = x + 2$ and $g(x) = -x + 2$. Graph f, g, and $f + g$ on the same coordinate system.

R **15** For what values of x are the y-coordinates of the graph shown less than zero? $\{x{:}x < -4 \text{ or } 4 < x < 8\}$

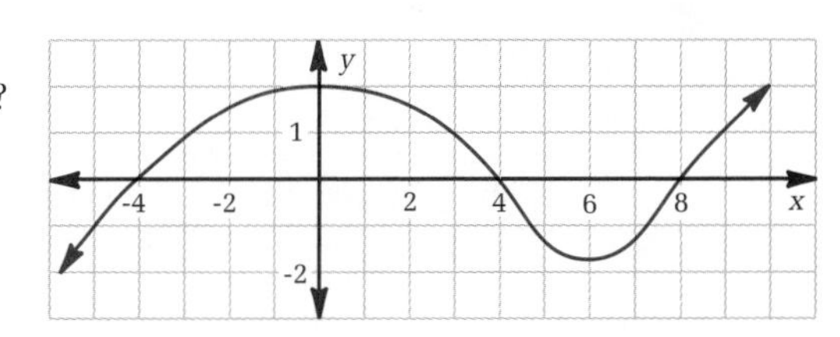

R,I **16** Basso the Clown has four different-sized tires on his miniature car. How many different ways are there for him to put the tires on the car? 24

P,R **17** The function $A(x, y) = xy$ outputs the area of a rectangle with side lengths x and y. Draw an area diagram to represent each of the following expressions.

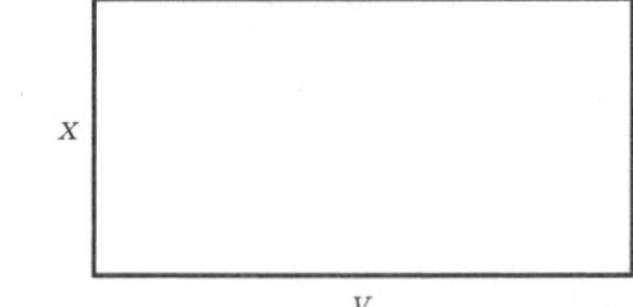

a $A(3x, y)$ **b** $A(x, 3y)$
c $A(3x, 3y)$ **d** $3[A(x, y)]$

P,I **18** Let $f(x) = x + 1$ and $g(x) = 3$. Graph each of the following on the same coordinate system.

a $y = f(x)$ **b** $y = g(x)$ **c** $y = f(x) + g(x)$ **d** $y = f(x) - g(x)$

Problem-Set Notes and Strategies, continued

- Have the students observe the translations in problem **12** and try to make some generalizations. These concepts will be formalized in Section **6.2**.
- Problem **13** will be solved by inspection by some students. Others will need to be convinced by finding INVf and showing that INV$f(f(g(x))) = g(x)$.
- Permutations are introduced informally in problem **16**. These are formally presented in Section **5.5**.

Stumbling Block

Operations on functions do not follow many of the properties we have discussed. In particular, the Distributive Property of Multiplication over Addition or Subtraction does not apply to operations on functions. Does $2f(a + b) = f(2a + 2b)$? Try this with the function $f(x) = 3x - 5$ and use the numbers $a = 2$ and $b = 7$. $2f(a + b) = 44$ and $f(2a + 2b) = 49$. Reiterate that operations on functions are not the same as operations on real numbers.

Additional Answers

Answers for problems **11a – c**, **12b**, **14**, **17a – d**, and **18** can be found in the answer pages beginning on **A 1**.

I **19** Which of the following statements describes the sequence, 1, 3, 5, 7, 9, 11, . . . ? c

a $f(1) = 1$ and $f(2) = 3$ **b** $f(n) = f(n - 1) + 2$

c $f(1) = 1$ and $f(n + 1) = f(n) + 2$

P,I **20** The diagram shows the graphs of $y = g(x)$ and $y = h(x)$. Sketch the graph of $y = g(x) + h(x)$.

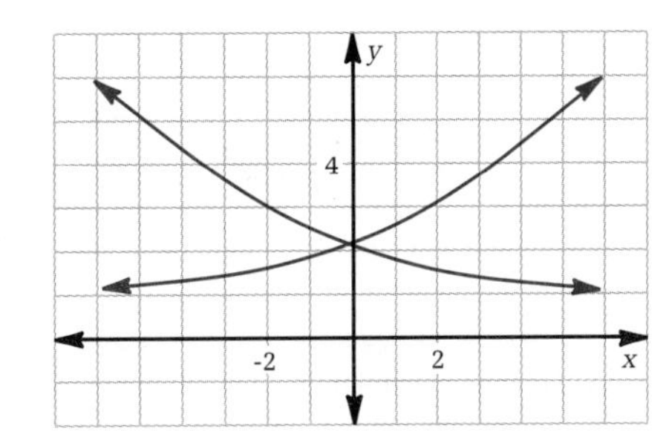

B

P **21** Let $h(x) = x^2 - 4$ and $g(x) = x - 7$.

a If $3[Q(x)] = h(x) - g(x)$, what is $Q(x)$? $Q(x) = \frac{x^2 - x + 3}{3}$

b If $h(x) \cdot R(x) = h[g(x)]$, what is $R(x)$? $R(x) = \frac{x^2 - 14x + 45}{x^2 - 4}$

c Find $h[g(x)]$. $x^2 - 14x + 45$

d Find $g[h(x)]$. $x^2 - 11$

P,I **22** The diagram shows the graphs of $y = f(x)$ and $y = g(x)$, where $f(x) = x^2$.

a Write a formula for function g. $g(x) = 2$

b Sketch the graph of $y = f(x) + g(x)$.

c Write an expression for $f(x) + g(x)$. $x^2 + 2$

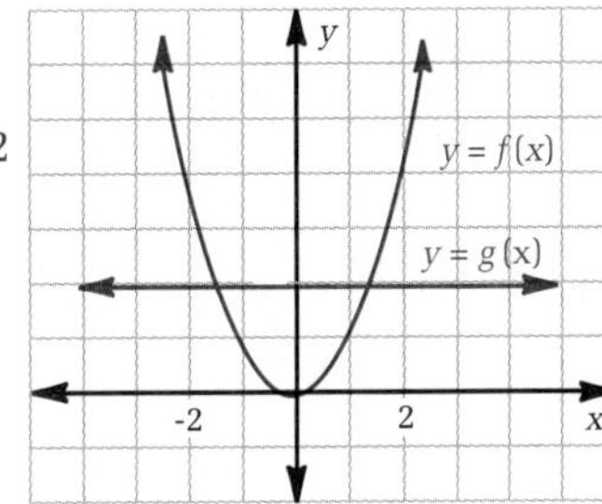

P,R **23** The function $V(x) = x^3$ outputs the volume of a cube with edge x. Draw a diagram to represent each of the following expressions.

a $3[V(x)]$

b $V(3x)$

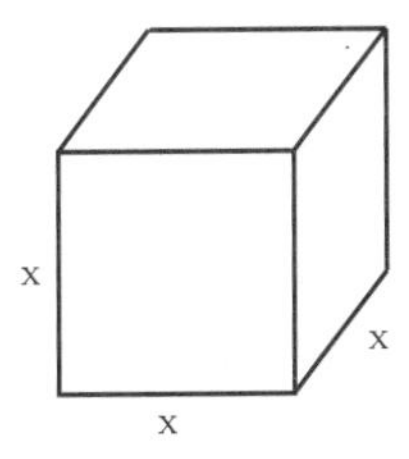

I **24** Suppose that $h(x + 1) = h(x) + h(x - 1)$ and that $h(0) = 3$ and $h(1) = 12$.

a Find $h(2)$. 15 **b** Find $h(3)$. 27 **c** Find $h(4)$. 42

P **25** If $f(x) = \sqrt{x + 4}$ and $f[g(x)] = |x - 2|$, what is $g(x)$? $g(x) = x^2 - 4x$

P,R **26** Let $w(x) = 3x - 1$ and $x = 5$. Find $w(x^2) - [w(x)]^2$. -122

Problem-Set Notes and Strategies, continued

- Recursive functions are used in problem **19**. Students will need to recall a recursive definition first introduced in Section **1.5**. Recursive functions will be presented formally in Section **5.6**.
- In problem **20** encourage students to just add the *y*-coordinates of the respective functions to arrive at the sum.

Additional Answers

20

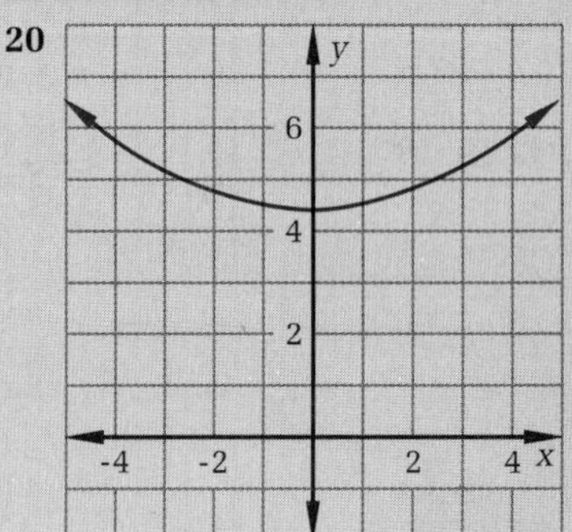

Answers for problems **22b**, **23a**, and **23b** can be found in the answer pages beginning on **A** 1.

- The function $f(x) + g(x)$ in problem **22** is a translation of the graph. Translations will be discussed in detail in Section **6.2**.
- Problem **24** generates a recursive sequence similar to the Fibonacci sequence. It uses the last two terms to generate the next.

Problem-Set Notes and Strategies, continued

- Problem **27** is interesting in that it introduces the log function as f, and one of the properties of logs. Logarithms are studied in Chapter **11**.
- Refer to problem **30**. You might ask where the domain could be restricted to make the reflection a function of x. The answers could vary, but what would be the most reasonable restriction? Answer: $-\{x: -90° < x < 90°\}$

Additional Answers

30

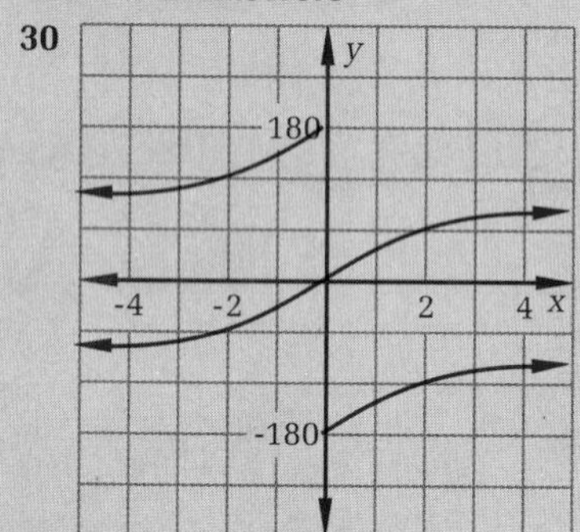

31b

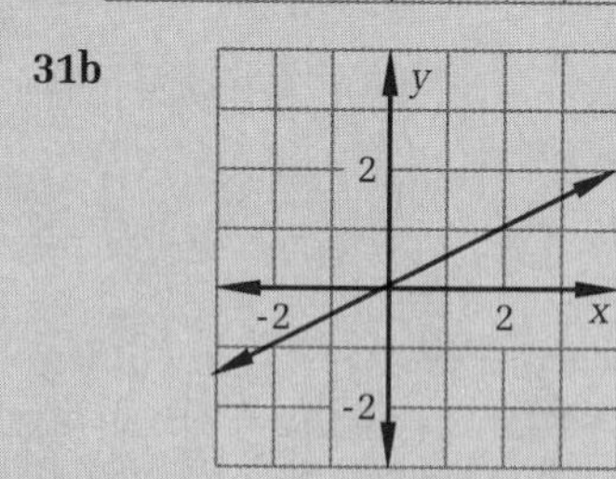

32a $f(g(x)) = \begin{cases} \sqrt{10} \text{ if } x < 4 \\ 0 \text{ if } x = 4 \\ 9 \text{ if } x > 4 \end{cases}$

Answer for problem **32b** can be found in the answer pages beginning on **A** 1.

Communicating Mathematics

Suppose $f(x) = 3x$ and $g(x) = x^2$. Have students discuss the difference between $f(g(x))$ and $g(f(x))$ in terms of operations on functions and inputs and outputs.

Problem Set, *continued*

P,R,I **27** Suppose that $f(xy) = f(x) + f(y)$ and that $f(2) = 0.3010$ and $f(3) = 0.4771$.

a Find $f(6)$. 0.7781 **b** Find $f(12)$. 1.0791 **c** Find $f(9)$. 0.9542 **d** Find $f(18)$. 1.2552

P **28** If $g(x) = x^3$, is $g(2x)$ or $2 \cdot g(x)$ greater? $g(2x)$ if $x > 0$, $2 \cdot g(x)$ if $x < 0$, neither if $x = 0$

C I **29** Alfredo's calculator was run over by a truck, so that only the keys for the digits 1, 3, 5, and 7 work now.

a How many four-digit numbers can Alfredo enter on his calculator if he does not repeat digits? 24

b How many four-digit numbers can he enter if digits can be repeated? 256

c How many four-digit numbers can he enter in which at least one digit is repeated? 232

R,I **30** The diagram shows the graph of $y = \tan x$ for the domain $\{x: -180 \le x \le 180\}$. Sketch the reflection of this graph over the line $y = x$. Does the reflection represent a function of x? No

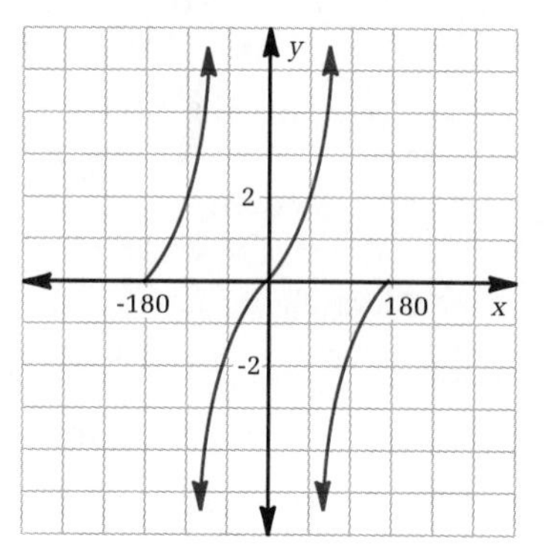

R **31** Function h has the following three properties.

i $5[h(x)] = h(x) + \dfrac{x^2}{h(x)}$

ii $h(x) > 0$

iii The domain of function h is $\{x: x > 0\}$.

a Find $h(x)$. $h(x) = \frac{x}{2}$ **b** Graph $y = h(x)$.

R **32** Consider the following compound functions.

$$f(x) = \begin{cases} x^2 \text{ if } -4 < x < 1 \\ 5 \text{ if } x = 1 \\ \sqrt{x} \text{ if } x > 1 \end{cases} \qquad g(x) = \begin{cases} 10 \text{ if } x < 4 \\ 0 \text{ if } x = 4 \\ -3 \text{ if } x > 4 \end{cases}$$

a Find $f[g(x)]$.

b Sketch the graph of $f \circ g$.

c What are the domain and the range of $f \circ g$? Domain: $\{x: x \in \mathcal{R}\}$; range: $\{0, \sqrt{10}, 9\}$

5.5 DISCRETE FUNCTIONS

Objectives

After studying this section, you will be able to

- Use the factorial function
- Use the permutation function
- Use the combination function

Part One: Introduction

The Factorial Function

Any function whose domain is a set of numbers that can be listed, such as the set of integers, is called a ***discrete function.*** In this section, we will look at three discrete functions.

First, let's look at the ***factorial*** function.

 ***The factorial function F is defined by the formula***

$$F(n) = n(n-1)(n-2)(n-3)\ldots(2)(1)$$

where n is a nonnegative integer.

An exclamation point (!), is used to symbolize the word *factorial*, so another way to express $F(n)$ is $n!$, read "n factorial." By definition, $0! = 1$ and $1! = 1$.

Example *Compute 6!.*

$$6! = 6 \cdot 5 \cdot 4 \cdot 3 \cdot 2 \cdot 1$$
$$= 720$$

Example *The rock group B4 recorded 8 songs to be released on a compact disc. In how many different ways can the songs be arranged on the disc?*

There are 8 choices for the first song. For each possible first song, there are 7 choices left for the second song. For each of the $8 \cdot 7$ ways to choose the first 2 songs, there are 6 ways to choose the third, and so on. The number of ways all 8 songs can be arranged on the disc is $8 \cdot 7 \cdot 6 \cdot 5 \cdot 4 \cdot 3 \cdot 2 \cdot 1 = 8! = 40{,}320$.

In general, $n!$ represents the number of ways n items can be arranged. Becoming familiar with the factorial key on a scientific calculator will save you effort and time.

5.5

Class Planning

Time Schedule

All levels: 2 days

Resource References

Teacher's Resource Book

Class Opener 31

Cooperative Learning 15

Practice 32

Class Opener

1. Find a formula for the number of diagonals of a polygon. (Hint: try a simple case and look for a pattern). $\frac{n(n-3)}{2}$
2. Find a formula for the number of games necessary in a singles elimination tournament with n players entered. $n(n-1)$, since every player except the winner will lose exactly once.

Vocabulary

discrete function
factorial
permutation
combination

5.5

Lesson Notes

- Simple counting arguments are introduced in this section. The ideas of permutation and combination are discovered with their factorial definitions.
- You might point out that the numbers in Pascal's Triangle are combinations and convince students of the symmetry of combinations.
- Many calculators have permutation and combination buttons now which allow easy computation of the two.
- Emphasize that the only difference between the two terms is deciding whether order is important. If yes, use a permutation. If not, use a combination.
- These formulas will be used for some statistical arguments in later sections and in the Binomial Theorem in Chapter **9**.

Using Computers

A scientific calculator with the ability to work with permutations and combinations and factorials can be used effectively here. A spreadsheet can be used to generate values of discrete functions.

The Permutation Function

Suppose that B4's recording company recorded 19 of the group's songs during a concert tour but decided to put only 8 of them on a compact disc. How many different arrangements of 8 songs can be made from the 19 songs that were recorded?

There are 19 choices for the first song, 18 for the next, and so on until 8 songs have been selected. The number of possible arrangements is therefore $19 \cdot 18 \cdot 17 \cdot 16 \cdot \ldots \cdot 12$, or 3,047,466,240. Note that

$$\begin{aligned} & 19 \cdot 18 \cdot 17 \cdot 16 \cdot 15 \cdot 14 \cdot 13 \cdot 12 \\ &= \frac{19 \cdot 18 \cdot 17 \cdot 16 \cdot 15 \cdot 14 \cdot 13 \cdot 12 \cdot 11 \cdot 10 \cdot 9 \cdot 8 \cdot 7 \cdot 6 \cdot 5 \cdot 4 \cdot 3 \cdot 2 \cdot 1}{11 \cdot 10 \cdot 9 \cdot 8 \cdot 7 \cdot 6 \cdot 5 \cdot 4 \cdot 3 \cdot 2 \cdot 1} \\ &= \frac{19!}{11!} \\ &= \frac{19!}{(19-8)!} \end{aligned}$$

This suggests a formula for the ***permutation*** function.

The permutation function P is defined by the formula

$$P(n, r) = \frac{n!}{(n-r)!}$$

In general, $P(n, r)$ represents the number of ways that n objects can be *permuted*, or arranged, into groups of r objects. The domain of the permutation function is the set of *pairs* of nonnegative integers (n, r) such that $n \geq r$.

The Combination Function

B4's manager insisted that the group be allowed to select which 8 songs would be on the compact disc, though the recording company could decide how they would be arranged. In how many ways can the group choose 8 songs from the 19 without regard to their order?

This is an application of the ***combination*** function C, where $C(n, r)$ is the number of sets of r objects that can be chosen from n objects. To compute $C(19, 8)$, let's take a closer look at the permutation function P.

$$\begin{aligned} P(19, 8) &= \text{total number of ways 19 songs can be arranged into groups of 8} \\ &= \begin{pmatrix}\text{number of ways 8 songs can} \\ \text{be chosen from a group of 19}\end{pmatrix}\begin{pmatrix}\text{number of ways 8 songs} \\ \text{can be arranged}\end{pmatrix} \end{aligned}$$

The first factor is C(19, 8), the number we want. The second factor equals 8!. Therefore,

$$P(19, 8) = C(19, 8) \cdot 8!$$

$$C(19, 8) = \frac{P(19, 8)}{8!}$$

$$= \frac{19!}{(19-8)!8!} = 75{,}582$$

There are 75,582 ways in which the group can choose 8 songs from the 19.

 ***The combination function C is defined by the formula***

$$C(n, r) = \frac{n!}{(n-r)!r!}$$

Part Two: Sample Problems

Problem 1 *Ms. Paskal's algebra class has 29 students and 29 desks. For the sake of variety, students change the seating arrangement each day. How many days must pass before they need to repeat a seating arrangement?*

Solution The problem asks for the number of ways in which 29 students can be arranged, so we use the factorial function. Since 29! days is $\approx 8.84 \times 10^{30}$ days, or about 2.42×10^{28} years, they will never repeat a seating arrangement.

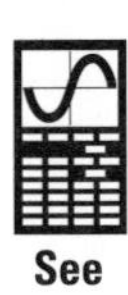

See p. 779

Problem 2 *Ms. Paskal gives 200 extra points to the 4 students who arrive first to class. How many days might she have to give out extra points if*

a *The same 4 students cannot receive extra points twice?*
b *The same 4 students can receive extra points more than once as long as they enter in a different order each time?*

Solution **a** The problem is to determine how many sets of 4 students can be chosen, in any order, from a class of 29. We use the combination function.

$$C(29, 4) = \frac{29!}{25!4!} = 23{,}751$$

Ms. Paskal might have to give extra points for 23,751 days.

b Since order is important, we use the permutation function.

$$P(29, 4) = \frac{29!}{25!} = 29 \cdot 28 \cdot 27 \cdot 26 = 570{,}024$$

Ms. Paskal might have to give extra points for 570,024 days.

In each case, Ms. Paskal could have to give out extra points for the rest of her career.

Checkpoint

1. Evaluate 6!. 720
2. Find the number of ways to arrange 10 different colored shirts in your closet. 3,628,800
3. You have three tickets to the concert Saturday night and want to take two of your six friends. How many different pairs could you choose from? 15
4. The band is scheduled to play 4 of their 12 top ten hits during the night. How many different arrangements of these hits could there be? 11,880

Assignment Guide

Average	
Day 1	11, 18, 19, 23, 24, 26, 29, 30, 33, 38
Day 2	10, 12–17, 20–22, 25, 27, 28, 32, 35
Advanced	
Day 1	12–17, 20, 23, 28–30, 34, 38
Day 2	10, 11, 19, 21, 22, 25–27, 32, 33, 36
Honors	
Day 1	12, 18, 20–23, 26, 28, 33–35, 37, 40
Day 2	10, 11, 13–17, 19, 24, 25, 27, 29–31, 36, 38, 39, 41

Communicating Mathematics

Have students write a paragraph stating the difference between a permutation and a combination. They should include at least two examples to illustrate their point.

5.5

Problem-Set Notes and Strategies

- The question of order is addressed in problem **6**. In part **a** the only concern was the number of groups of winners, not which was first, second, or third, so order was not important and a combination was used. In part **b**, the number of arrangements was asked for and (a, b, c) is a different arrangement than (a, c, b) so permutations must be used.
- Part **e** of problem **7** is an interesting conjecture. Examine the formula and see if students can figure out why it is always so.
- Problem **10** outlines some properties of the factorial function that might be helpful when combinations or permutations of numbers greater than 69 appear. The largest factorial most calculators will calculate is 69!

Stumbling Block

Students may have difficulty identifying a given situation as involving a combination or a permutation. The only difference is order. Have students ask themselves the question "Is the pair (a, b) the same as the pair (b, a)?" If the answer is yes, it's a combination; if not, a permutation.

- The symmetry of the combination function is illustrated in problem **12**. $C(7, 3) = C(7, 4)$. Order is not important so combinations were used.

Part Three: Exercises and Problems

Warm-up Exercises

In problems 1–4, evaluate each expression.

P **1** 5! 120 **2** 10! 3,628,800 **3** 15! $\approx 1.3 \times 10^{12}$ **4** 45! $\approx 1.2 \times 10^{56}$

P **5** Harry Temmert, the morning DJ on XWTR radio, has 12 songs to play between 7 and 8 A.M. In how many different orders can Harry play these songs? 479,001,600

P **6** There are 107 competitors in a 10-kilometer race. A first prize, a second prize, and a third prize are awarded.

a How many different groups of three people can win prizes? 198,485

b How many different arrangements of winners of the first, second, and third prizes are there? 1,190,910

P **7** Let P be the permutation function and C be the combination function.

a Evaluate $P(7, 2)$. 42 **b** Evaluate $C(7, 2)$. 21

c Evaluate $P(15, 6)$. 3,603,600 **d** Evaluate $C(15, 6)$. 5005

e What conjecture might you make on the basis of your results in parts **a–d**? $P(a, b) \geq C(a, b)$

P **8** How many different license plates are possible if each plate contains a group of six letters and/or digits? 36^6, or $\approx 2.18 \times 10^9$

P **9** If a license plate must have two letters followed by four digits, how many different license plates are possible? $26^2 \cdot 10^4$, or 6.76×10^6

Problem Set

A

P **10** Evaluate each expression.

a $\frac{9!}{8!}$ 9 **b** $\frac{97!}{96!}$ 97 **c** $\frac{(n+1)!}{n!}$ $n + 1$

P **11** Martino's Pizza offers a special pizza with 4 toppings. There are 11 toppings available to choose from. Guido decides to order a different 4-topping pizza each day.

a Does order matter when selecting toppings for a pizza? No

b How many days can Guido go without repeating? 330 days

P,R **12** Seven points are selected on a circle.

a How many triangles have three of these points as vertices? 35

b How many quadrilaterals have four of these points as vertices? 35

P **In problems 13–16, evaluate the combination function for each pair of inputs.**

13 $C(5, 4)$ 5 **14** $C(8, 7)$ 8 **15** $C(12, 11)$ 12 **16** $C(51, 50)$ 51

P **17** What pattern do you see in problems 13–16? $C(n, n - 1) = n$

R,I **18** Find the smallest value of x for which $P(10, x) > 1000$. 4

R,I **19** Find the mean (average) of the elements of $\{-5, -3, -1, 11, 17, 29\}$. 8

R,I **In problems 20–22, let $P_2(x)$ represent the probability of a roll of x when two dice are rolled. Find each probability.**

20 $P_2(7)$ $\frac{1}{6}$

21 $P_2(12)$ $\frac{1}{36}$

22 $P_2(13)$ 0

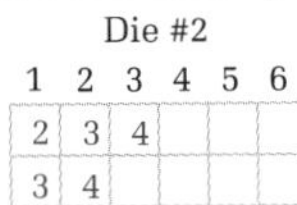

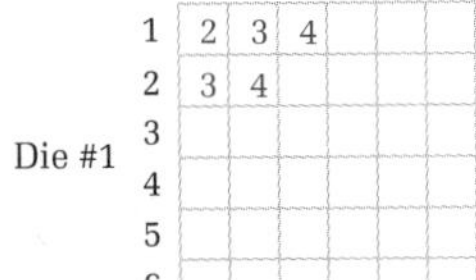

P **23** **a** Evaluate $C(6, 2)$ and $C(6, 4)$. 15; 15
b Evaluate $C(6, 1)$ and $C(6, 5)$. 6; 6
c Evaluate $C(8, 2)$ and $C(8, 6)$. 28; 28
d Evaluate $C(8, 3)$ and $C(8, 5)$. 56; 56
e Describe any patterns you see. $C(n, r) = C(n, n - r)$

R **24** Give the range of $f(x) = 3x^2 - 2$ for each domain.
a $\{-2, -1, 0, 1, 2\}$
$\{-2, 1, 10\}$
b $\{x: -2 \le x \le 2\}$
$\{y: -2 \le y \le 10\}$

R **25** If $f(x) = x + 5$ and $g(y) = y - 5$, what is $3[f(4)] - 4[g(3)]$? 35

B

R **26** **a** If $f(x) = \frac{2}{3}x - 6$, what is INVf? INV$f(x) = \frac{3}{2}(x + 6)$
b Graph $y = f(x)$ and $y =$ INV$f(x)$ on the same coordinate system.
c What is the point of intersection of the two graphs? $(-18, -18)$

R **27** Copy and complete the pattern of factorials. If the pattern were to continue, what would 0! be? What would (−1)! be?
1; $\frac{1}{0}$, which is undefined

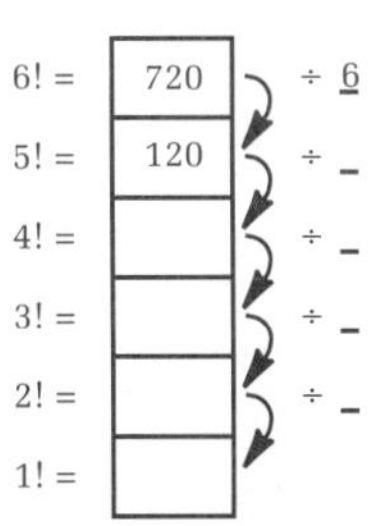

P **28** Find the smallest positive integer n for which $\frac{7!}{n}$ is not an integer. 11

Problem-Set Notes and Strategies, continued

- Problems **13 – 17** can be related to Pascal's Triangle which was used in the first section of the text. Other patterns are also available and will be explored in a cooperative-learning lesson.
- A new statistical term, the mean, is introduced in problem **19**. Other measures of central tendency will be examined in the last section of this chapter.
- Problems **20 – 22** require the student to evaluate the number of combinations when rolling dice and the number of combinations of getting winners. Chapter **12** deals with counting and probability arguments in more detail.
- An interesting pattern of combinations is observed in problem **23**. More detail on these patterns are available in a cooperative learning segment in this section.

Additional Answers

26b

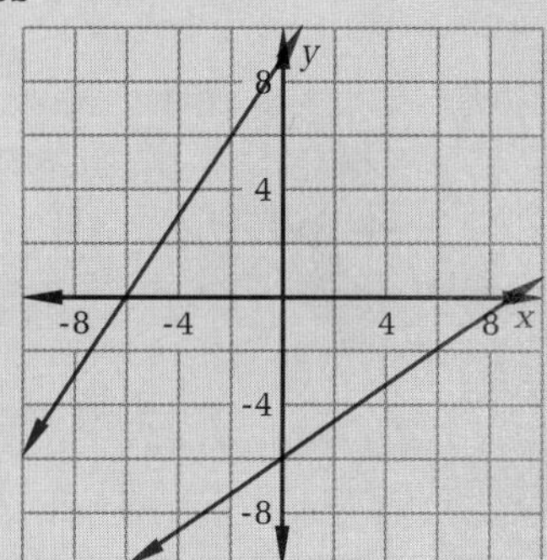

Answer for problem **27** can be found in the answer pages beginning on **A** 1.

- Problem **28** is a nice illustration of why 0! is defined to be 1 and why factorials are not defined for the negative numbers.

Problem-Set Notes and Strategies, continued

- Problem **29** is not at all obvious. Students should realize that they were not asked to find the inverse function, just to state if one exists.

- Use the Distributive Property of Multiplication over Addition in problem **33** and notice that $29 + 13 = 42$ which is $7 \cdot 6$ and when multiplied by 5! yields 7!
- Problem **35** part **c** is difficult to count. For the number of ways of choosing 2 diagonals, there are C(1175,2) = 689,725 different ways. To make sure they have a common endpoint, first pick a diagonal from the 1175 possible. Once it is chosen, there are 46 other diagonals from that point for a total of 54,050 possible winners. The probability is the ratio of the two.

Additional Answer

38 Possible answer:

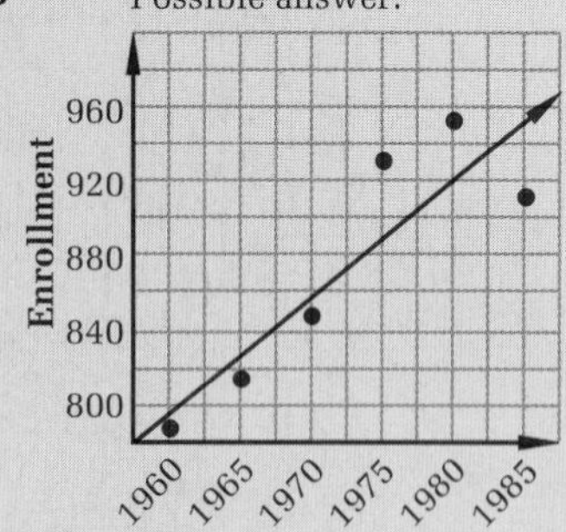

- Problem **39** is another example of a recursive function which will be explored in the next section.

Problem Set, *continued*

P,R **29** If $f(x) = x!$, does f have an inverse function? Yes

P,R **30** Use the factorial, permutation, or combination function to solve the handshake problem presented in Section 1.1. Then explain why your choice of function is appropriate. C(25, 2) = 300; we need to determine the number of sets of two people that can be formed from a set of 25, and order does not matter.

P **31** Find the greatest positive integer n for which $\frac{10!}{2^n}$ is an integer. 8

P,R **32** If C(9, 3) = C(9, x) and $x \neq 3$, what is the value of x? 6

P,R **33** Solve $29 \cdot 5! + 13 \cdot 5! = k!$ for k. $k = 7$

R **34** Solve $[f(x)]^2 - 12[f(x)] + 32 = 0$ for $f(x)$. $f(x) = 8$ or $f(x) = 4$

R **35** Consider a 50-sided polygon.

a How many vertices does it have? 50

b How many diagonals does it have? 1175

c If two diagonals are drawn at random, what is the probability that they have a common endpoint? $\frac{46}{587}$

R **36** The graph shows postage costs for first-class mailings in Mathamerica.

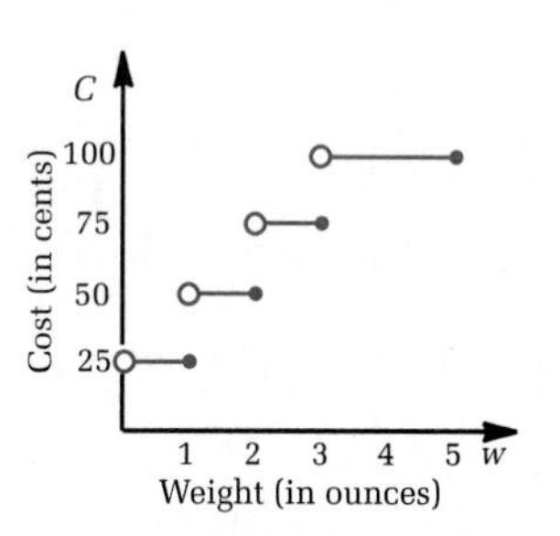

a What is the cost of mailing a 2.5-ounce package? \$0.75

b What is the cost of mailing a 1-ounce package? \$0.25

c What input values w, in ounces, give an output of \$1.00? $\{w: 3 < w \leq 5\}$

d Find the possible outputs if $0.8 < w < 2.3$. 25, 50, 75

P,R **37** Let $f(x) = x!$ and $g(x) = \frac{f(x)}{x}$, where $x > 0$.

a Find g(3), g(4), g(5), and g(6). 2; 6; 24; 120

b Use factorial notation to write a simple expression for g(x). $(x - 1)!$

R **38** The following table shows enrollments in mathematics classes at Parker High School.

Year	1960	1965	1970	1975	1980	1985
Enrollment	787	815	849	930	952	911

a Graph the data and draw a fitted line. Answers will vary.

b Use the method of least squares to write an equation of the line of best fit. $y = 6.35x - 11{,}660$

I **39** Function f is defined by two rules: $f(1) = 5$ and $f(n) = 2 \cdot f(n - 1)$.

a Evaluate $f(2)$. 10

b Evaluate $f(4)$. 40

C
I

40 $S(a)$ is a function whose input is an angle measure, as shown in the diagram. The output is the y-coordinate of point P, where one side of the angle intersects the square. Find each of the following.

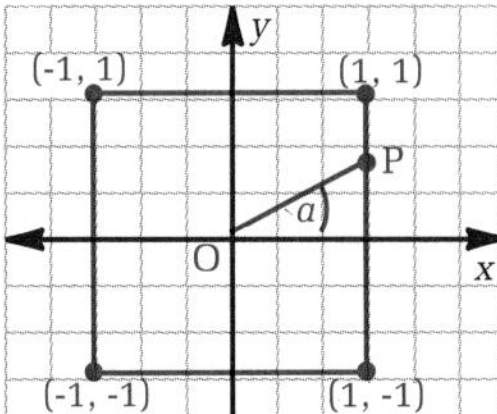

a $S(45)$ 1
b $S(0)$ 0
c $S(30)$ $\frac{1}{\sqrt{3}}$, or ≈ 0.5774
d $S(60)$ 1
e $S(90)$ 1

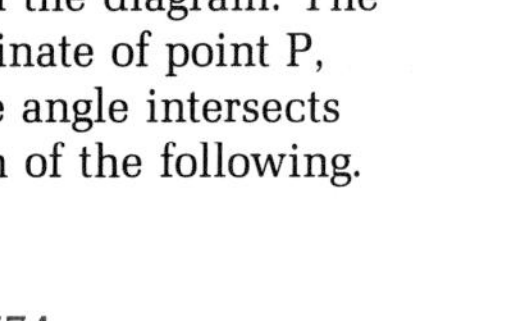

P,R **41** Show that $C(n - 1, k - 1) + C(n - 1, k) = C(n, k)$.

MATHEMATICAL EXCURSION

FUNCTIONS MAKE LIFE SIMPLER

Predictable outcomes in common experiences

If you have ever paid to take a taxi, ride on public transportation, or take a bus, train, or plane trip, you have used a function. If you can imagine having to negotiate your fare for a given trip, then you may be able to appreciate how functions simplify life.

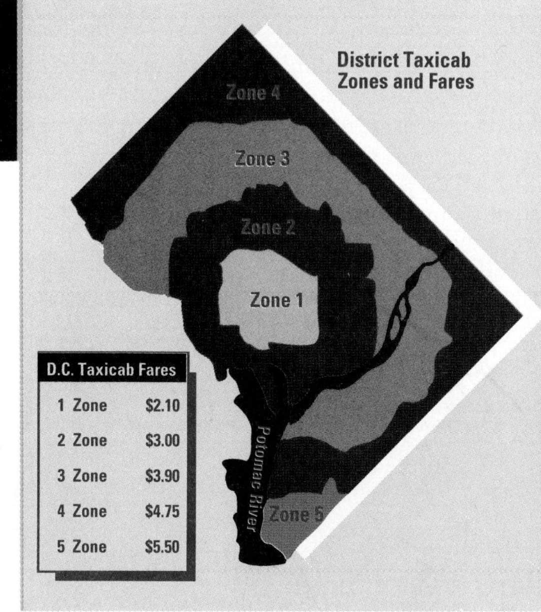

D.C. Taxicab Fares	
1 Zone	$2.10
2 Zone	$3.00
3 Zone	$3.90
4 Zone	$4.75
5 Zone	$5.50

In certain cities, for example, the distance you ride on a bus or subway determines how much you pay for the trip. There is usually a table or chart of zones showing the fares for travel between zones. That there is a given fare for a given distance seems obvious: what arrangement could be simpler? It *is* simple because there is a "fare function"; that is, fare is a function of distance.

Suppose the fare structure were not a function. Then there might be more than one fare for some or all distances. How could that be? In one scenario, the driver or ticket agent might bargain with each passenger for the fare that passenger would pay. In that case, there might be two or more different values for the same trip.

Another example of how we use relations that are functions occurs whenever we practice at a game of skill. A basketball player uses the knowledge that the height of the ball at a given point after a shot is taken is a function of force and direction. Only because the ball's height is a function can a player, through practice, strive for perfection. If the ball's path were not a function, it would never be possible to predict the effect of taking a shot, no matter how perfectly it was executed.

Cooperative Learning

Problem **23** of this section had you observe the pattern that $C(n, r) = C(n, n - r)$. Have students work in small groups to prove this pattern using the formula for combinations. Another interesting formula was proven in problem **41**. Have students work in small groups to discover a pattern and show a proof for the following formulas:

$C(1, 0) + C(1, 1) = ?$

$C(2, 0) + C(2, 1) + C(2, 2) = ?$

$C(3, 0) + C(3, 1) + C(3, 2) + C(3, 3) = ?$

$C(4, 0) + C(4, 1) + C(4, 2) + C(4, 3) + C(4, 4) = ?$

$C(5, 0) + C(5, 1) + C(5, 2) + C(5, 3) + C(5, 4) + C(5, 5) = ?$

In general, what is $\sum_{i=0}^{n} C(n, i) = ?$ 2^n

Additional Answer

41 $C(n-1,k-1) + C(n-1,k)$

$$= \frac{(n-1)!}{[n-1-(k-1)]!\,(k-1)} + \frac{(n-1)!}{(n-1-k)!\,k!}$$

$$= \frac{k \cdot (n-1)!}{(n-k)\,k!} + \frac{(n-k) \cdot (n-1)!}{(n-k)!\,k!}$$

$$= \frac{[k + (n-k)] \cdot (n-1)!}{(n-k)!\,k!}$$

$$= \frac{n \cdot (n-1)!}{(n-k)!\,k!} = \frac{n!}{(n-k)!\,k!} = C(n,k)$$

Class Planning

Time Schedule
All levels: 2 days

Resource References
Teacher's Resource Book
Class Opener 32
Cooperative Learning 16
Practice 33

Class Opener

Given $f(1) = 3$ and $f(n + 1) = f(n) + 2$ for every positive integer n. Find $f(1990)$. (Hint: use simple cases and look for a pattern).
$f(1) = 3, f(2) = 5, f(3) = 7, \ldots$
$f(n) = 2n + 1$
$f(1990) = 3981$

Lesson Notes

- If any of your students are going to study computers or are currently doing any programming, suggest that they pay special attention as recursive programming can be a problem-solving technique and is difficult to master. A recursive function uses past output from the function as input. In programming it is a program that calls itself as subprogram.

5.6 Recursively Defined Functions

Objectives
After studying this section, you will be able to
- Interpret recursive definitions of functions
- Relate recursive definitions to equivalent closed-form definitions

Part One: Introduction

Recursive Definitions
Count the rays and angles in each diagram below.

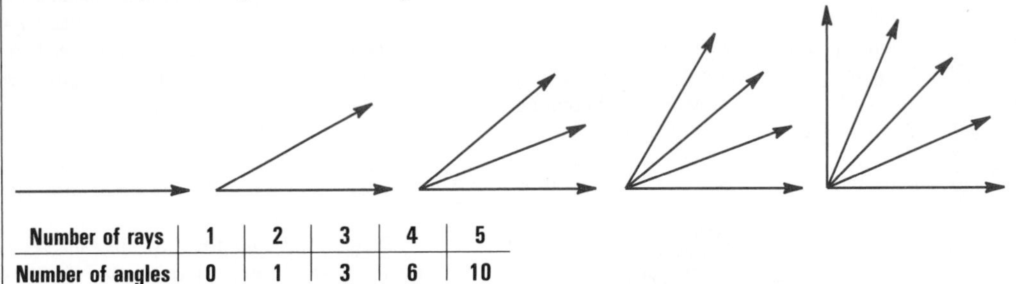

Number of rays	1	2	3	4	5
Number of angles	0	1	3	6	10

We can identify a pattern in the sequence of the inputs (the numbers of rays) and the outputs (the numbers of angles). Each time an input n is increased by 1, the corresponding output increases by n. If we express the relationship between inputs and outputs in function notation, we can represent the pattern in the following way.

$$f(1) = 0$$
$$f(2) = f(1) + 1 = 0 + 1 = 1$$
$$f(3) = f(2) + 2 = 1 + 2 = 3$$
$$f(4) = f(3) + 3 = 3 + 3 = 6$$
$$f(5) = f(4) + 4 = 6 + 4 = 10$$
$$\vdots$$
$$f(n + 1) = f(n) + n$$

Vocabulary
recursive definition
closed form
Fibonacci numbers
first differences
second differences

A more concise way of expressing the pattern is

$f(1) = 0$

$f(n + 1) = f(n) + n$ for all positive integers $n \geq 1$

This two-part statement is called a ***recursive definition*** of function f. Every recursive definition specifies a beginning value—in this case, $f(1) = 0$—and a way to calculate each successive output in terms of the preceding output or outputs—in this case, $f(n + 1) = f(n) + n$.

Most of the functions you have worked with up to now have been represented in ***closed form***. A closed-form formula, such as $f(x) = x^2 + 1$, $g(x) = 2x + 3$, or $h(x) = 8$, shows the relationship between each input and its corresponding output. A recursive definition, on the other hand, shows how each output value is related to one or more previous output values. For this reason, it may be difficult to evaluate a recursively defined function for a particular input value; but representing a function recursively is often the best way of clarifying a numerical pattern. Many functions with discrete domains can, in fact, be represented in both closed and recursive forms.

Lesson Notes, continued

- Mathematically, intermediate values are difficult to compute as every intermediate step must be performed along the way and there is no direct route for a recursively defined function. This illustrates the necessity to also have closed-form functions that are equivalent.

Exploring Recursively Defined Functions

We can often find interesting relationships between recursively defined functions and functions that can be expressed in closed form.

Example *Express the following recursively defined function in closed form.*

$f(0) = 4$

$f(n + 1) = f(n) + 3$

Let's evaluate the function for successive inputs and graph the corresponding ordered pairs.

$f(0) = 4$

$f(1) = f(0) + 3 = 4 + 3 = 7$

$f(2) = f(1) + 3 = 7 + 3 = 10$

$f(3) = f(2) + 3 = 10 + 3 = 13$

$f(4) = f(3) + 3 = 13 + 3 = 16$

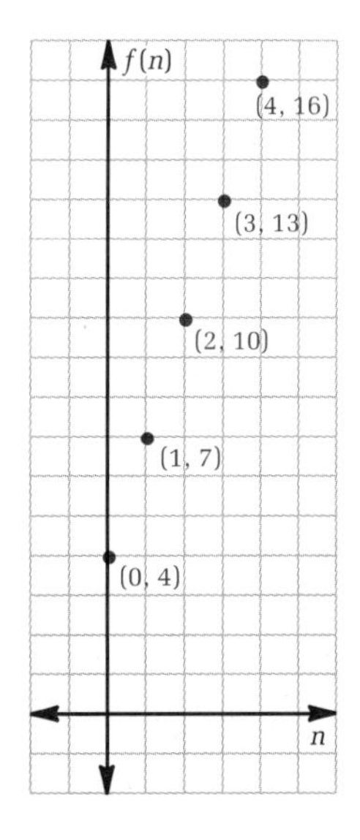

It seems that the points all lie on a line having a slope of 3—that is, a line representing a function of the form $f(x) = 3x + b$. Since we know that $f(0) = 4$, we can write the equation $4 = 3(0) + b$, or $b = 4$. We can therefore express f by means of the closed-form formula

$f(x) = 3x + 4$, where x is a nonnegative integer

Using Computers

A spreadsheet is ideal here also since it mimics the thought process involved in a recursively defined function. Spreadsheets can also be used to create finite difference tables quickly. The Towers of Hanoi problem is an elegant program when done recursively in Pascal and would make an excellent research project for some of your students.

Communicating Mathematics

Have students describe a situation where the output of the first part becomes the input of the second and the output of the second part is the input of the third part and so on. Try to come up with an example from everyday experiences.

Example *Express the following recursively defined function in closed form.*

$$g(0) = 4$$
$$g(n + 1) = 3[g(n)]$$

Again, let's generate a few outputs.

$$g(0) = 4$$
$$g(1) = 3[g(0)] = 3(4) = 12$$
$$g(2) = 3[g(1)] = 3(3)(4) = 3(12) = 36$$
$$g(3) = 3[g(2)] = 3(3^2)(4) = 3(36) = 108$$

This pattern seems to involve multiplying successive powers of 3 by 4, so that a corresponding closed-form formula would be $g(x) = 4(3^x)$, where x is a nonnegative integer. We can test whether this formula in fact represents the function by using both it and the recursive definition to calculate g(4).

Closed Form	Recursive Form
$g(x) = 4(3^x)$	$g(n + 1) = 3[g(n)]$
$g(4) = 4(3^4)$	$g(4) = 3[g(3)]$
$= 4(81)$	$= 3(108)$
$= 324$	$= 324$

The closed-form formula does seem to represent the function.

Part Two: Sample Problems

Problem 1 *Find the first eight outputs of the following recursively defined function.*

$$f(1) = 1$$
$$f(2) = 1$$
$$f(n + 1) = f(n) + f(n - 1), \text{ where } n \geq 2$$

Solution

$$f(1) = 1$$
$$f(2) = 1$$
$$f(3) = 1 + 1 = 2$$
$$f(4) = 2 + 1 = 3$$
$$f(5) = 3 + 2 = 5$$
$$f(6) = 5 + 3 = 8$$
$$f(7) = 8 + 5 = 13$$
$$f(8) = 13 + 8 = 21$$

The values generated by this function are the ***Fibonacci numbers*** a famous sequence that has many interesting properties. You will surely encounter them again.

Problem 2 *Identify the pattern of outputs of the following function.*

$$f(0) = 1$$

$$f(n + 1) = \frac{1}{1 + f(n)}$$

Solution

$$f(0) = 1$$

$$f(1) = \frac{1}{1 + f(0)} = \frac{1}{1 + 1} = \frac{1}{2}$$

$$f(2) = \frac{1}{1 + f(1)} = \frac{1}{1 + \frac{1}{2}} = \frac{1}{\frac{3}{2}} = \frac{2}{3}$$

$$f(3) = \frac{1}{1 + f(2)} = \frac{1}{1 + \frac{2}{3}} = \frac{1}{\frac{5}{3}} = \frac{3}{5}$$

$$f(4) = \frac{1}{1 + f(3)} = \frac{1}{1 + \frac{3}{5}} = \frac{1}{\frac{8}{5}} = \frac{5}{8}$$

This pattern is not easy to identify, nor can we easily write a corresponding closed-form formula. The pattern is, however, a very interesting one. The numerators and the denominators of the successive outputs are successive Fibonacci numbers. Fibonacci numbers occur in surprising places.

Problem 3 *Is the closed-form formula corresponding to the following function linear or quadratic?*

$$f(0) = 3$$

$$f(n + 1) = f(n) + 2n$$

Solution One way to identify the degree of a function is to examine the pattern of the differences between successive outputs. If these differences, called ***first differences,*** are constant, the function is linear; if the differences between the first differences (the ***second differences***) are constant, the function is quadratic.

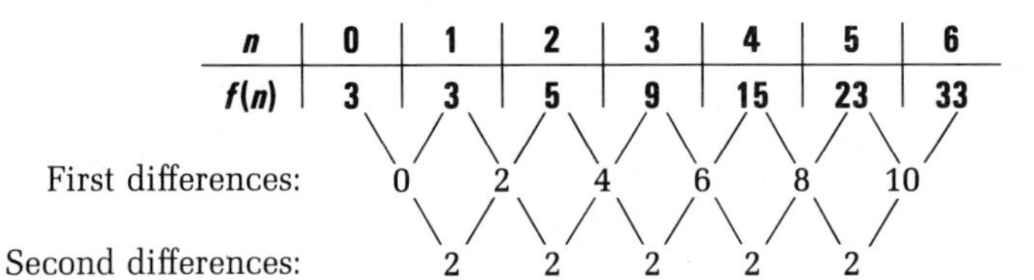

n	0	1	2	3	4	5	6
$f(n)$	3	3	5	9	15	23	33

First differences: 0 2 4 6 8 10

Second differences: 2 2 2 2 2

Since the second differences are the same, this recursively defined function has a quadratic closed-form formula. (In this case, the formula is $f(x) = x^2 - x + 3$, where x is a nonnegative integer.)

Checkpoint

1 Let $f(0) = 5$ and $f(n + 1) = f(n) + 2$, find the first five terms of the recursively defined function.
5, 7, 9, 11, 13

2 Find a closed form function for the recursively defined function in 1 above.
$g(x) = 2x + 5$

3 Find a recursive definition for the function $g(x) = x!$
$f(0) = 1, f(n + 1) = (n + 1) \cdot f(n)$

4 Let $f(1) = 13$ and $f(n + 1) = f(n) - n + 3$. Is the closed form formula corresponding to f linear or quadratic? Quadratic

Part Three: Exercises and Problems

Warm-up Exercises

1 Let $f(1) = 10$ and $f(n + 1) = 4 \cdot f(n)$. Find $f(2)$, $f(3)$, and $f(4)$. 40; 160; 640

P **2** Copy and complete the table for the quadratic function $f(x) = x^2 + 3x + 1$ and find the first and second differences.

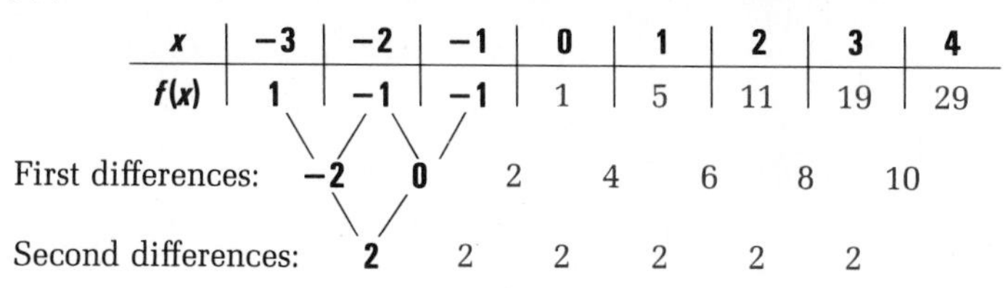

x	−3	−2	−1	0	1	2	3	4
$f(x)$	1	−1	−1	1	5	11	19	29

First differences: −2, 0, 2, 4, 6, 8, 10

Second differences: 2, 2, 2, 2, 2, 2

P **3** Consider the following recursively defined function.

$f(0) = 5$
$f(n) = f(n - 1) + 3$

Make a table of the first six values of $f(n)$. Is the closed-form formula corresponding to f linear or quadratic or neither? Linear

P **In problems 4 and 5, define f by $f(1) = 1$ and $f(n + 1) = (n + 1) \cdot f(n)$.**

4 Find $f(2)$, $f(3)$, $f(4)$, and $f(5)$. 2; 6; 24; 120

5 Express f in closed form. $f(x) = x!$, where x is a positive integer

A Problem Set

P **6** If $f(1) = 12$ and $f(n) = 2n + f(n - 1)$, what are $f(2)$, $f(3)$, $f(4)$, and $f(10)$? 16; 22; 30; 120

P **7** Copy and complete the table for a quadratic function f.

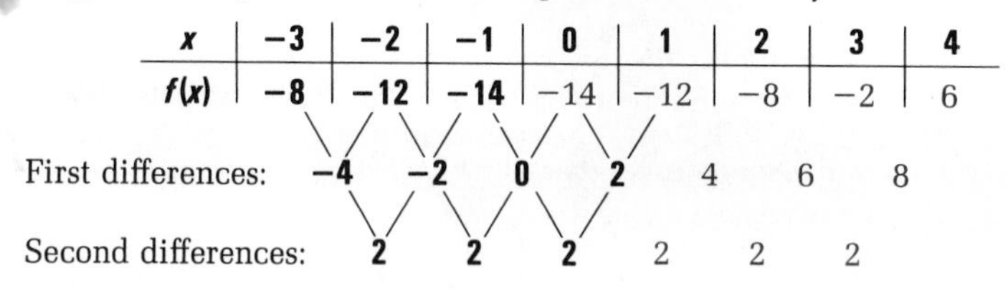

x	−3	−2	−1	0	1	2	3	4
$f(x)$	−8	−12	−14	−14	−12	−8	−2	6

First differences: −4, −2, 0, 2, 4, 6, 8

Second differences: 2, 2, 2, 2, 2, 2

R **8** Refer to the six points A through F.

a How many segments can be drawn having two of these points as endpoints? 15

b How many triangles can be drawn having three of these points as vertices? 20

A B C D E F

Assignment Guide

Average	
Day 1	6, 9–11, 13, 16, 17, 20, 21, 24, 26
Day 2	7, 8, 12, 14, 15, 18, 19, 22, 23, 25
Advanced	
Day 1	7, 9, 10, 12, 13, 16, 17, 20, 22
Day 2	6, 8, 11, 14, 15, 18, 19, 21, 23–26
Honors	
Day 1	8, 9, 11, 13–15, 18, 21, 24, 25, 27
Day 2	6, 7, 10, 12, 16, 17, 19, 20, 22, 23, 26, 28, 29

Problem-Set Notes and Strategies

- Problems **4** and **5** illustrate the factorial function which is neither linear or quadratic. You might have students find both the first and second differences for this function to verify that.
- Problem **8** is an example of combinations from the last section. Note that order is not important in either case.
- You might comment on the fact that the first differences are all zero, yet the closed form definition of the function does not fit the model of a linear function discussed in Chapter **3**.

P | **9** Let $F(1) = -1$ and $F(n + 1) = -1 \cdot F(n)$.

a Find $F(2)$, $F(3)$, and $F(4)$. 1; −1; 1

b Find $F(k)$ if k is odd. −1

c Find $F(k)$ if k is even. 1

d Write a closed-form formula for F. $F(x) = (-1)^x$, where x is a positive integer

R | **10** Refer to the graph of function f.

a What is the domain of f?

b What is the range of f?

a $\{x : x \in \mathcal{R}\}$

b $\{f(x) : f(x) \geq -3\}$

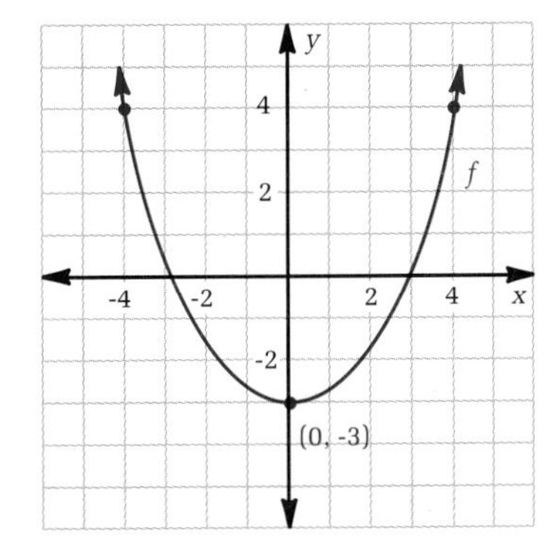

P | **11** If $f(1) = 0$ and $f(n + 1) = \frac{f(n) + 36}{2}$, what is $f(8)$? 35.71875

R | **12** If $\frac{n!}{(n-2)!} = 132$, what is the value of n? 12

R | **13** Let $3 \cdot f(x) + 8x = 5 \cdot f(x) - 2$. Find an expression for $f(x)$. $4x + 1$

P,R | **14** Consider the following recursively defined function.

$$h(1) = \begin{bmatrix} -1 & 0 \\ 0 & 1 \end{bmatrix}$$

$$h(n + 1) = h(n) \cdot \begin{bmatrix} 1 & 0 \\ 0 & -1 \end{bmatrix}$$

Find $h(2)$, $h(3)$, $h(4)$, $h(5)$, and $h(6)$. $\begin{bmatrix} -1 & 0 \\ 0 & -1 \end{bmatrix}$; $\begin{bmatrix} -1 & 0 \\ 0 & 1 \end{bmatrix}$; $\begin{bmatrix} -1 & 0 \\ 0 & -1 \end{bmatrix}$; $\begin{bmatrix} -1 & 0 \\ 0 & 1 \end{bmatrix}$; $\begin{bmatrix} -1 & 0 \\ 0 & -1 \end{bmatrix}$

R | **15** There are 64 squares on a chessboard.

a In how many ways can 3 pawns occupy different squares on the chessboard? 41,664

b In how many ways can a king, a queen, and a rook occupy different squares on the chessboard? 249,984

P | **16** Let $F(1) = 6$ and $F(n + 1) = F(n) + n + 1$.

a Find $F(2)$, $F(3)$, $F(4)$, and $F(5)$. 8; 11; 15; 20

b Is the closed-form formula corresponding to F linear or quadratic? Quadratic

Problem-Set Notes and Strategies, continued

- Problem **12** is a good exercise in manipulating factorials. Hopefully the student has a good understanding of the definition of factorial and that $n! = n(n-1)(n-2)!$.
- See if students can recognize a pattern in problem **14** and predict the value of $h(26)$ without carrying out all the intermediate matrices.
- In problem **15b**, the order implied by three different pieces necessitates the use of a permutation.

Problem-Set Notes and Strategies, continued

- Problem **19** is a good problem in estimation skills and approximation. In real situations, many times the coefficients and constants in the problems are themselves approximations and so a good estimate for the answer is all that is necessary.

Additional Answers

21a

n	0	1	2	3	4	5
$F(n)$	−4	2	8	14	20	26

23

n	1	2	3	4	5	6
$F(n)$	4	6	22	94	492	3046

- Problem **25** is done with a guess and check method. Observation and pattern recognition is needed. Some students will have trouble with this as they don't immediately see a pattern and tend to give up easily.

Problem Set, *continued*

R **In problems 17 and 18, solve each system.**

17 $\begin{cases} 4x - 2y = -13 \\ 2x + 2y = -5 \end{cases}$ $(x, y) = \left(-3, \frac{1}{2}\right)$

18 $\begin{cases} y = -\frac{5}{3}x + 6 \\ 4y + x = 4 \end{cases}$ $(x, y) = \left(\frac{60}{17}, \frac{2}{17}\right)$

R **19** Sketch the graphs of $y = x^2 - x - 2$ and $y = x - 1$ on the same set of axes.

a Approximately where do the graphs intersect? $(\approx-0.5, \approx-1.5)$, $(\approx2.5, \approx1.5)$

b Substitute $x - 1$ for y in the first equation and solve for x. $x \approx -0.41$ or $x \approx 2.41$

B

20 Let $f(0) = -9$ and $f(n) = f(n - 1) + 4$.

P **a** Find the slope and the y-intercept of the linear function corresponding to f. Slope: 4; y-intercept: −9

b Write a recursive definition of $g(x) = 3x + 10$, where x is a nonnegative integer. $g(0) = 10$, $g(n) = g(n - 1) + 3$

P **21** Let $F(0) = -4$ and $F(n) = 6 + F(n - 1)$.

a Construct a table of the first six values of $F(n)$.

b Find the first differences of $F(n)$. 6

c Write a closed-form formula for F. $F(x) = 6x - 4$, where x is a nonnegative integer

R **22** How many six-digit license-plate numbers can be created from the digits 1, 2, 3, 4, 5, 6, 7, 8, and 9 if

a Digits can be repeated? 531,441

b No digit can be repeated? 60,480

P **23** Consider the following recursively defined function.

$F(1) = 4$
$F(2) = 6$
$F(n) = n \cdot F(n - 1) + F(n - 2)$

Construct a table of the first six values of $F(n)$.

P **24** Little Bo Peep counted her sheep to go to sleep. Each night she had to count 12 more sheep than the previous night before she could sleep. If 22 sheep made Peep sleep the first night, how many sheep did she count in a week? 406

P **25** Use the table to write a recursive definition of F. $F(1) = 12$, $F(n + 1) = 3[F(n)] + 4$

n	1	2	3	4	5	6
$F(n)$	12	40	124	376	1132	3400

P **26** Suppose that

$F(1) = 4$
$F(n) = 4 + g(n)$, where $n \geq 1$
$g(n) = 2 \cdot g(n - 1)$

Copy and complete the following table.

n	1	2	3	4	5	6
F(*n*)	4	4	4	4	4	4
g(*n*)	0	0	0	0	0	0

C
P **27** Consider the following compound function.

$$F(n) = \begin{cases} 1 \text{ if } n = 1 \\ F(n-1) + n \text{ if } n \text{ is odd and } n > 1 \\ F(n-1) - n \text{ if } n \text{ is even and } n > 1 \end{cases}$$

a Construct a table of the first eight values of $F(n)$.
b Find $F(497)$. 249
c Find $F(5234)$. -2617

P **28** In the game Tower of Hanoi, the goal is to move the tower of disks, one disk at a time, from one peg to another peg without ever putting a larger disk on top of a smaller disk.

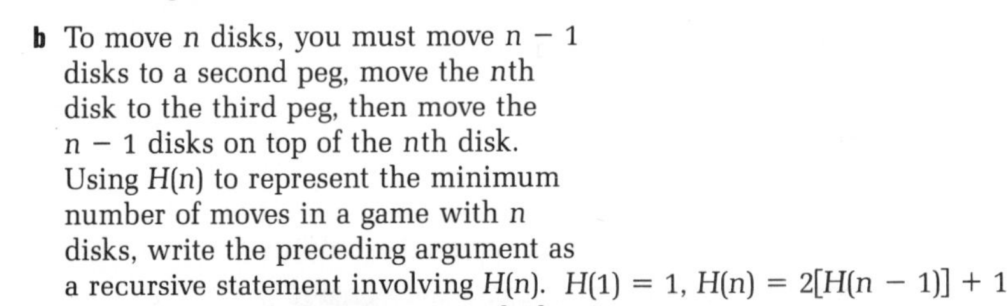

a Copy and complete the following table.

Number of disks	1	2	3	4
Minimum number of moves needed to finish game	1	3	7	15

b To move n disks, you must move $n - 1$ disks to a second peg, move the nth disk to the third peg, then move the $n - 1$ disks on top of the nth disk. Using $H(n)$ to represent the minimum number of moves in a game with n disks, write the preceding argument as a recursive statement involving $H(n)$. $H(1) = 1$, $H(n) = 2[H(n - 1)] + 1$
c Test your recursive statement with the data from part **a**.
$H(1) = 1$; $H(2) = 2(1) + 1 = 3$; $H(3) = 2(3) + 1 = 7$; $H(4) = 2(7) + 1 = 15$

P **29** A binary number consists of a string of 0's and 1's. Let $F(n)$ be the number of different n-digit binary numbers in which the first digit is a 1 and no two consecutive digits are 1's. (For example, 1010 qualifies but 1011 does not.)

a Find $F(1)$, $F(2)$, $F(3)$, $F(4)$, $F(5)$, and $F(6)$. 1; 1; 2; 3; 5; 8
b What do you think might be a recursive definition of F?
$F(1) = 1$; $F(2) = 1$; $F(n + 1) = F(n) + F(n - 1)$, where $n \geq 2$

Problem-Set Notes and Strategies, continued

- Problem **27** generates a familiar sequence of the squares of the integers, but it lists each square twice. Try to have the students generate a closed form definition for this.
- Problem **29** has another example of how the Fibonacci sequence crops up in the oddest places.

Cooperative Learning

The Tower of Hanoi problem in problem **28** of the text is excellent for a group project. Several programming languages support recursive programs. If you have one available, group the students accordingly so that you have someone in each group with programming experience. Have them try to write a subroutine that will calculate the steps necessary for solving the Tower of Hanoi puzzle. Most texts on Pascal programming will cite this as an example when doing recursive programming since it is such an elegant example. Also have the students find a closed form definition for the least number of moves to finish the puzzle. You might share the history of the problem with your students. The story goes that there are three towers in Hanoi with 64 disks which the Buddhist monks must move from one tower to another. When all disks have been moved, the world will come to an end. Calculate approximately how many years it will take to move all the disks if it takes one second to transfer a disk from one tower to the next.
Approximately 6×10^{11} years

Additional Answer

27a

n	1	2	3	4	5	6	7	8
F(*n*)	1	−1	2	−2	3	−3	4	−4

5.7

Class Planning

Time Schedule
All levels: 1 day

Resource References
Teacher's Resource Book
Class Opener 33
Cooperative Learning 17
Practice 34
Manipulatives and Applications 9

Class Opener

The following is a list of the number of nations that have competed in the olympics since 1900.

1900	22	**1904**	12	**1908**	23
1912	28	**1920**	29	**1924**	44
1928	46	**1932**	37	**1936**	49
1948	59	**1952**	69	**1960**	83
1964	93	**1968**	112	**1972**	122
1976	92	**1980**	81	**1984**	141
1988	160				

What is the average number of teams that have competed?
Mean = 68.5, Median = 59.
What does that average mean?
The average is not meaningful.
How can this information be communicated?
Answers will vary but data should be grouped such as pre-WWII and post-WWII.

5.7 Measures of Central Tendency and Percentiles

Objectives

After studying this section, you will be able to

- Determine the mean, the mode, and the median of a set of data
- Interpret percentiles and percentile ranks

Mean, Mode, and Median

A fishing derby was held to see who could catch the greatest number of bass weighing over 2 pounds. Fifteen contestants entered, catching the following numbers of eligible bass.

2, 2, 3, 3, 3, 3, 3, 4, 5, 5, 7, 11, 12, 12, 31

A newspaper reporter writing a story about the derby asked the contest manager, "What was the average number of bass weighing over 2 pounds caught by the contestants?" The contest manager responded, "Which average do you mean?"

The manager first added the numbers of eligible fish the contestants caught and divided the sum by the number of contestants, 15.

$$\frac{\text{Total number of bass}}{\text{Number of contestants}} = \frac{106}{15} = 7\tfrac{1}{15}$$

He told the reporter, "The ***arithmetic mean*** of the numbers of fish caught is $7\frac{1}{15}$." An arithmetic mean is also simply called a ***mean.***

"On the other hand," the contest manager said, "three is the most common number of fish caught." The number that occurs the greatest number of times in a set of data is another type of average, called the ***mode*** of the data.

"Then again," he continued, "the number 4 is at the middle of the set of data. Seven contestants caught more than 4 fish, and 7 caught fewer." The middle number of an ordered set of data is the median.

The mean, the mode, and the median are called ***measures of central tendency*** of a set of data.

Vocabulary

arithmetic mean	*percentile rank*
mean	*lower quartile*
mode	*upper quartile*
measure of central tendency	*bimodal*

Percentiles and Percentile Ranks

After a medical examination, Vera received this report.

	Triglycerides	Cholesterol
Result	113	248
Percentile rank	79	61

The triglyceride level of 113 with a **percentile rank** of 79 means that 79 percent of the people tested had a result less than 113. What does the cholesterol percentile rank of 61 mean?

There are several percentiles commonly used in data reporting. The 25th percentile of a distribution corresponds to Q_1 and is called the ***lower quartile.*** The 75th percentile corresponds to Q_3 and is called the ***upper quartile.*** The 50th percentile is the median (Q_2). A box-and-whisker plot (see Section 4.7) identifies the 25th, 50th, and 75th percentiles of a set of data.

Problem 1 *The following data represent Victor Vector's scores on his first nine quizzes in physics class.*

16, 13, 19, 20, 20, 18, 11, 19, 17

Arranging the data from smallest to largest gives

11, 13, 16, 17, 18, 19, 19, 20, 20

a *Determine the mean, the mode, and the median of the quiz scores.*
b *Victor earned a score of 19 on the next quiz. Determine the new mean, mode, and median.*

Solution **a** The sum of Victor's scores is 153.

$$\text{Mean} = \frac{\text{sum of quiz scores}}{\text{number of quiz scores}} = \frac{153}{9} = 17$$

The scores 19 and 20 appear most often and the same number of times. The data is said to be ***bimodal,*** with the two modes being 19 and 20.

The median (middle) score is 18.

b The data for the 10 quiz scores are

11, 13, 16, 17, 18, 19, 19, 19, 20, 20

The sum of these scores is 172.

$$\text{Mean} = \frac{\text{sum of quiz scores}}{\text{number of quiz scores}} = \frac{172}{10} = 17.2$$

The mode is 19, since this score occurs the greatest number of times.

Since there is an even number of data, we have two middle scores, 18 and 19. The median is the mean of the two middle scores, $\frac{18 + 19}{2}$, or 18.5.

Lesson Notes

- Finding the middle of a data set is not always as easy as it sounds. Three different methods of finding the middle are presented in this section and each has its uses. A cooperative learning lesson is included to illustrate the pitfalls of one method over the other.
- This section continues with the introduction of statistics into the curriculum.
- Percentiles are also included and will be familiar to many of your students. Most of them have recently taken or will take the SAT test for college entrance and percentile rankings as well as raw scores are reported. Be prepared for the question of why no one ever gets into the 100th percentile. Review the definition of percentile with them – it is the point where that percentage scored *less* than that point. If a student were in the 100th percentile, that means that 100% of the people scored less than that point. Where does that leave the student?

Using Computers

The *Algebra Solver Kit* calculates mean, median, and mode and does stem-and-leaf plots. You might also introduce box-and-whisker plots in this section. A computer software program takes the calculation drudgery out of this section as well as strengthening the concepts.

■

Stumbling Block

Students may have difficulty remembering the difference between the three measures of central tendency. Stress the definitions of these terms.

■

Checkpoint

1 Consider the data set {1, 1, 2, 2, 3, 4, 4, 4, 5}. Find the mean, median, and mode.
$2\frac{8}{9}$, 3, 4

2 For the same data set, the 5 is what percentile?
89th percentile

3 Find the upper and lower quartile and the median for the following data set.
{2, 3, 5, 5, 6, 7, 7, 8, 10, 14, 14}
Lower quartile = 5
Median = 7
Upper quartile = 10

Assignment Guide

Average
1–5, 9, 11–13
Advanced
1, 3–6, 8, 9, 11–13
Honors
1–3, 5–7, 9, 10, 13, 14

Problem-Set Notes and Strategies

- For problem **3** students may have to review the techniques for drawing a stem-and-leaf plot.
- In problem **4**, the most appropriate measure should be the median since the data are skewed. Have the students draw a stem-and-leaf plot of the data and examine the shape of the data.

Additional Answers

4b Answers may vary. The mean is probably best, since no student actually works the mean number of hours, and the mode doesn't indicate that a majority of students work more than 10 hours a week.

Problem 2 *In a recent candy drive Dan Teen sold 130 packages of sugarless gum. Dan wanted to know how his sales compared with those of others in his club. The data below represent the numbers of packages of sugarless gum sold by all members in his club. Find the percentile rank of Dan's sales.*

22, 35, 67, 71, 88, 92, 100, 112, 130, 145, 155, 171, 188

Solution The percentile rank of Dan's total sales of 130 is determined by dividing 8 by 13, because his sales exceeded those of 8 of the 13 club members. Since $\frac{8}{13} \approx 0.615$, Dan's sales exceeded those of 61 percent of the club members. We can therefore say that his percentile rank is 61, although percentile ranks are more meaningful for large sets of data.

Problem Set

A

P **1** Ashvin scored 550 on a standardized exam. He was told that his score exceeded the scores of 85,000 of the 113,000 who took the exam. What was his percentile rank? 75

P,R **2** Nachum Oer had the following scores in his first eight games in a recent bowling tournament.

194, 223, 246, 235, 267, 211, 229, 271

Nachum estimates that he needs a mean of 243 for 12 games in order to win some of the prize money. What must he average in his last four games to win some money? 260

R **3** Make a stem-and-leaf plot and a box-and-whisker plot of the following data, which represent the heights, in inches, of students in an eleventh-grade chemistry class.

61, 80, 63, 72, 79, 63, 72, 76, 69, 67, 65, 66,
71, 65, 73, 72, 65, 70, 64, 73, 66, 68, 70, 71, 72

P **4** In a survey of an advanced algebra class at Apollo High School, the 21 class members reported the numbers of hours per week that they worked at jobs outside of school. The reported hours were as follows:

10, 16, 15, 12, 0, 6, 19, 14, 15, 6,
0, 0, 10, 20, 18, 24, 7, 0, 12, 10, 15

a Determine the mean, the mode, and the median of the data. ≈10.9; 0; 12

b In your opinion, which of the three measures of central tendency best describes the data? Explain your reasoning.

c Bob transferred into the class after the survey was taken. He works 19 hours per week. How much does his presence affect the mean, the median, and the mode? Mean increases to ≈11.3; mode is unchanged; median is unchanged.

3a

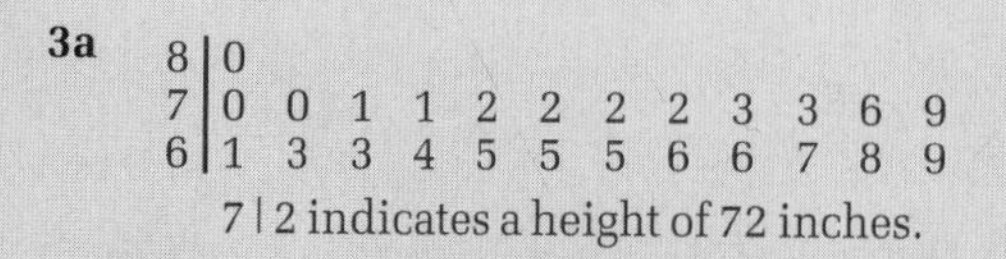

7 | 2 indicates a height of 72 inches.

3b

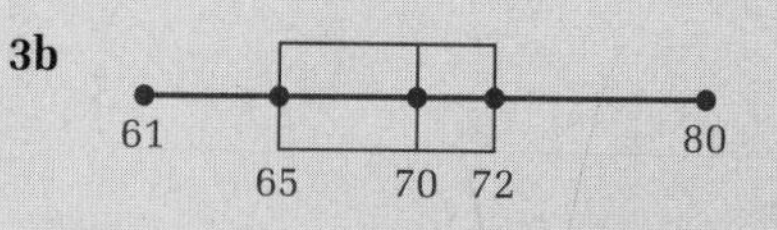

R **5** Explain why you can't make a stem-and-leaf plot showing the data represented by a given box-and-whisker plot. A box-and-whisker plot does not show individual scores, as a stem-and-leaf plot does.

B

R **6** So far this basketball season, Devan has made 36% of the field goals he has attempted. If he has attempted 84 field goals, how many shots in a row must he make to raise his percentage to 40%? 6

P,R **7** What means can the scores 126, 160, 90, x, and 120 have if x is a whole number and x is the median of the data? 123.2, 123.4, 123.6, 123.8, 124, 124.2, 124.4

P **8** What modes are possible for the data {12, 16, x, 9, x, 15, x} if x is a natural number and the mean is less than 14? 9, 12, 15

P **9** Find a table of numerical data in a newspaper or magazine and determine the mean, the mode, and the median of the data. Answers will vary.

I **10** In the frequency graph shown, M_o represents the mode, M_e represents the median, and $\bar{x}$ represents the mean. Copy each of the following frequency graphs and indicate where M_o, M_e, and $\bar{x}$ might be.

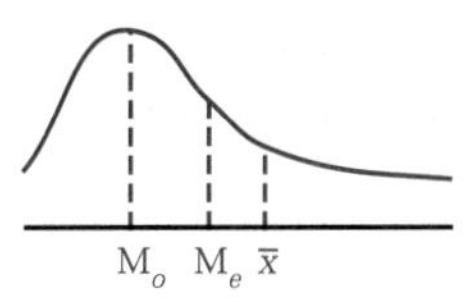

a

b

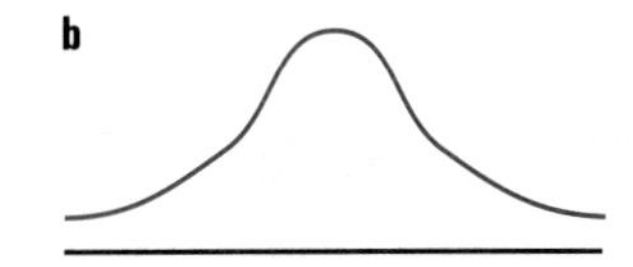

c

In problems 11 and 12, refer to the chart, which shows the numbers of people scoring various numbers of correct responses on a computer-aptitude exam.

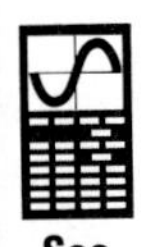

See p. 763

P **11** Find the mean, the mode, and the median number of correct answers. ≈8.1; 8; 8

P **12** Choose four scores and find the percentile rank of each. Answers will vary.

Number of Correct Responses	Number of People with This Score
15	62
14	320
13	602
12	890
11	1123
10	2365
9	4698
8	7642
7	3537
6	2135
5	1601
4	753
3	351
2	162
1	74
0	19

Problem-Set Notes and Strategies, continued

- Problems **7** and **8** require students to consider a range of possibilities making sure that x is a natural number.
- The graph in **10b** is the normal curve and will be discussed in more detail in Section **8.5**.

Communicating Mathematics

Discuss situations where the percentile ranking of the data has more significance than the raw data itself. Now describe an additional situation where the data itself is more important than the percentile rankings.

- The chart in problems **11** and **12** is called a frequency distribution. Students need to realize that the numbers in the second column must be added to get the total number of responses. The first value of 62 means that there were 62 people who each scored 15 correct responses. These values are also necessary to find the mean, median and mode.
- It might be interesting to consolidate all the data from the class and draw a stem-and-leaf diagram or a bar chart for the data used in problems **11** and **12**. It should be almost a normal curve and students may be surprised at the shape of the graph.

10a

$\bar{x}$ M_e M_o

10b

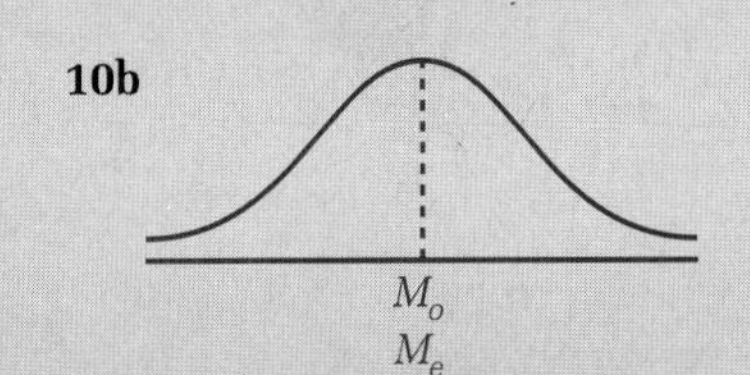

10c

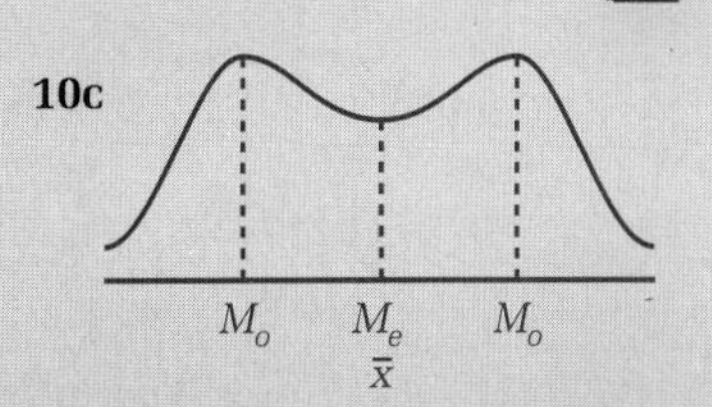

Cooperative Learning

Have students work in small groups to consider the following situation. You have just finished school and are looking for work. You are currently interviewing with a filming company named Code Act. Mr Poler Ride is the manager of the company and seems to like your resume. He has offered you a job and indicated that the average employee in the company makes $37,500 annually. You like this figure and agree to work. After the first month you discover on your pay check that you are paid at the rate of $15,000 annually and are unhappy. The data is as follows.

Mr Ride	$100,000
Assistant	50,000
Worker 1	30,000
Worker 2	30,000
Worker 3	30,000
Worker 4	30,000
Worker asst	15,000
Worker asst	15,000

Should you have accepted the average salary as an estimate for your wage? Would the median salary have been any better? How about the mode? Are any of these statistics useful in this situation?

P,R **13** Roll 3 dice, noting the total of the numbers on the upturned faces. Do this 25 times and record the 25 totals.

a Determine the mean, the mode, and the median of the rolls. Answers will vary.
b Draw a box-and-whisker plot to display the data. Answers will vary.

C P,R **14** The values x, 17, y, y, and 23, where x and y are whole numbers, have a mean less than 9. What are the possible values of (x, y)?
(4, 0), (3, 0), (2, 0), (2, 1), (1, 1), (1, 0), (0, 2), (0, 1), (0, 0)

CAREER PROFILE

EXPONENTIALS AND STATISTICS IN BANKING

Jessalyn Wilscam evaluates businesses' prospects

"The bank I work for offers training classes in accounting and special classes in statistics and finance," says Jessalyn Wilscam, a vice president with a bank in New York City. Why? "I approve loans to companies, and I need to understand the information they are providing to justify the loan. The customers I work with are largely newspapers, radio, and cable TV. If the company is a newspaper, for example, it may say that its sales will go up 8% in the next year."

Looking at statistical data, Ms. Wilscam must decide whether she agrees with what the company projects. Based on her decision, she works out how much to lend the company.

"I also have to figure out how much money the loan will bring in compared with how much it's costing us, because the money we use to make the loan is itself borrowed."

This involves straightforward calculation using exponentials. Ms. Wilscam uses a computer with a spreadsheet program to do the calculation for her. Her own role is to decide whether the numbers make sense.

What does Ms. Wilscam like most about her job? "I travel, I look at different companies and industries, I meet people. It's an analytical job, which I like."

Ms. Wilscam grew up in Omaha and has a political science degree from Newcome College of Tulane University. Besides the statistics class offered by her bank, she had four years of mathematics in high school. She says she wishes she had taken calculus, since she uses so much mathematics in her job.

This graph shows the money the bank pays out and takes in on $1,000,000 over a twelve-month period. Considering that the bank cannot loan the interest it is paying its depositor or investor, what will happen to the bank's profit margin if the loan is not repaid when it is due?

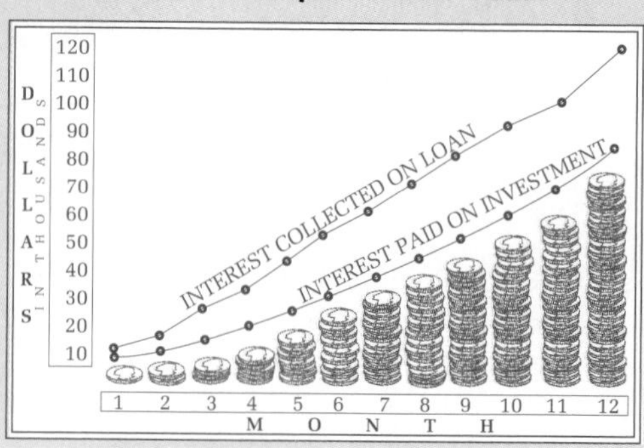

5

CHAPTER SUMMARY

CONCEPTS AND PROCEDURES

After studying this chapter, you should be able to

- Represent relations (5.1)
- Recognize functions (5.1)
- Identify restrictions on the domain of a function (5.2)
- Interpret composite functions (5.2)
- Recognize inverse functions (5.2)
- Find the inverse of a given function (5.3)
- Recognize the relationship between the graphs of a function and its inverse (5.3)
- Perform operations on the outputs of functions (5.4)
- Perform operations on the inputs of functions (5.4)
- Use the factorial function (5.5)
- Use the permutation function (5.5)
- Use the combination function (5.5)
- Interpret recursive definitions of functions (5.6)
- Relate recursive definitions to equivalent closed-form definitions (5.6)
- Determine the mean, the mode, and the median of a set of data (5.7)
- Interpret percentiles and percentile ranks (5.7)

VOCABULARY

arithmetic mean (5.7)
bimodal (5.7)
closed form (5.6)
combination (5.5)
composite (5.2)
composition of functions (5.2)
discrete function (5.5)
domain (5.1)
factorial (5.5)
Fibonacci numbers (5.6)
first differences (5.6)
function (5.1)
inverse functions (5.2)
lower quartile (5.7)
mean (5.7)
measure of central tendency (5.7)
mode (5.7)
percentile rank (5.7)
permutation (5.5)
range (5.1)
recursive definition (5.6)
reflection (5.3)
relation (5.1)
second differences (5.6)
upper quartile (5.7)

Class Planning

Time Schedule

Average: 2 days
Advanced: 2 days
Honors: 1 day

Assignment Guide

Average	
Day 1	1–4, 11, 12, 17–21, 27, 31, 32, 34
Day 2	5–10, 13, 16, 24–26, 33, 35
Advanced	
Day 1	1–4, 11, 12, 17–21, 27, 29, 31, 32, 34
Day 2	5–10, 13, 15, 16, 22, 24–26, 33, 35
Honors	
	2, 4, 5, 8–13, 16–21, 25, 30, 33–36

Problem-Set Notes and Strategies

- In problem **5**, you might suggest that students think of discrete as something that can be counted as opposed to measured.
- Make sure the students have a restricted domain where necessary in problem **8**. It is easy to forget this and just go through the mechanics of finding the inverse. Without this restricted domain, the inverse is useless as we don't know where it may be used.

5 REVIEW PROBLEMS

A

P **1** Is the relation represented by the following table a function? Yes

Input	−8	−4	0	1	4	5.77
Output	0	0	0	1	0	−1

P **2** Let $f(x) = 3x - 5$. Write a simplified expression for each of the following.

a $f(6x)$ $18x - 5$ **b** $f(x + 3)$ $3x + 4$ **c** $f(4y + 3)$ $12y + 4$

P **3** Identify the domain and the range of each of the following functions.

a $f(x) = \frac{1}{x^2}$ $\{x{:}x \neq 0\}$; $\{f(x){:}f(x) > 0\}$

b $g(x) = 2x - 2$ $\{x{:}x \in \mathcal{R}\}$; $\{g(x){:}g(x) \in \mathcal{R}\}$

c $h(x) = |x + 6|$ $\{x{:}x \in \mathcal{R}\}$; $\{h(x){:}h(x) \geq 0\}$

P **4** Let $r(x) = 5x + 1$ and $s(x) = x^2 - 5$.

a Find $r[s(2)]$. -4 **b** Find $s[r(2)]$. 116

P **5** Which of the following functions are discrete? **a** and **c**

	Input	Output
a	Annual income	Federal income tax paid
b	Volume of a tree	Number of sheets of paper made from the tree
c	Number of tickets sold	Total revenue from a baseball game
d	Gallons of gasoline	Miles driven
e	Number of minutes spent exercising	Pulse rate

P **6** If $g(x) = \frac{x}{3} - 2$, what is INV(g)? $\text{INV}g(x) = 3(x + 2)$

P **7** If C is the combination function and $C(10, 3) = n!$, what is the value of n? 5

P **8** Find $\text{INV}f \circ f(x)$ for each of the following functions.

a $f(x) = 2x - 5$ x **b** $g(z) = -3z$ z

c $p(x) = \sqrt{x}$ x, where $x \geq 0$ **d** $r(y) = -\frac{1}{y}$ y, where $y \neq 0$

B

P **9** Let $f(0) = 3$ and $f(n) = n[f(n - 1)]$.

a Find $f(1)$, $f(2)$, $f(3)$, and $f(4)$. 3; 6; 18; 72

b Write a closed-form formula for function f. $f(x) = 3x!$ where x is a nonnegative integer

P,R **10** Bill Yard participated in the Nine Ball Pool Tournament. The graph shows the numbers of balls he pocketed in the games he played.

a How many games did Bill play? How many balls did he pocket all together?
b Find the mean, the mode, and the median number of balls pocketed by Bill.
a 38; 168 **b** ≈4.42; 8; 4

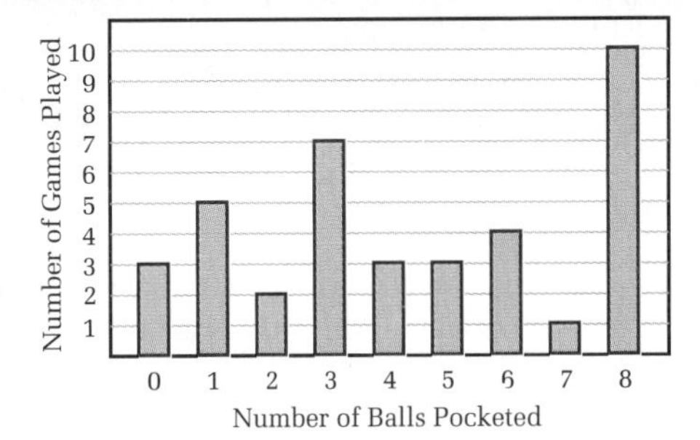

P **In problems 11–15, identify the domain of each function.**

11 $g(x) = \dfrac{4x}{x^2 - 4}$ **12** $h(x) = \sqrt{6 - 2x}$ **13** $F(x) = x!$

14 $t(x) = \dfrac{6}{|x|^5 - 32}$ $\{x{:}x \neq -2 \text{ and } x \neq 2\}$

15 $p(1) = 2,\ p(n + 1) = \dfrac{1}{[p(n)]^2}$ {positive integers}

P,R **16** Find the following matrix product for $f(x) = x^2$ and $g(x) = 2x + 3$.
$\begin{bmatrix} f(2) & f(3) \\ f(-1) & f(-6) \end{bmatrix}\begin{bmatrix} g(4) \\ g(-6) \end{bmatrix}$ $\begin{bmatrix} -37 \\ -313 \end{bmatrix}$

P **17** Which of the following graphs represent y as a function of x? Which represent x as a function of y? **a** and **c**; **b** and **c**

a
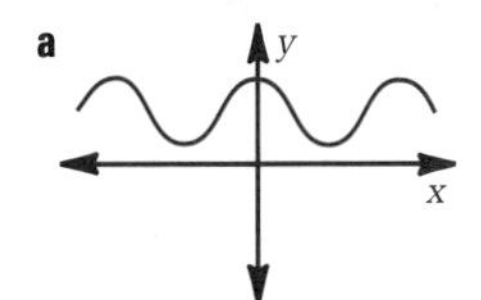

b
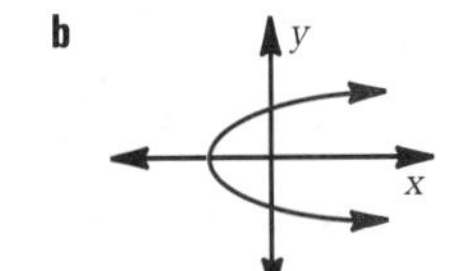

c
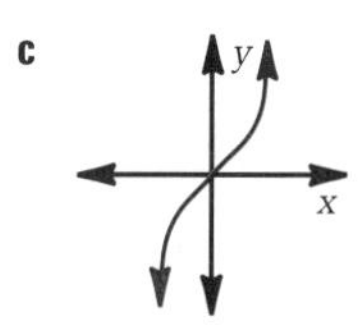

P **18** A certain compact-disc player can be programmed to play the songs on a disc in any order.

a In how many possible orders can the songs on a nine-song disc be played if each is played only once? 362,880
b If it takes 43 minutes to play the whole disc, how long would it take to play the songs on the disc in every possible way? 15,603,840 min, or ≈30 yr, if played nonstop

P **19** When playing the board game Klu, a player can be any of eight different characters. If three people are to play the game, in how many ways can they choose their characters? 336

P **In problems 20–23, graph each function, then sketch the reflection of the graph over the line $y = x$ and indicate whether the reflection represents a function.**

20 $f(x) = 4x + 1$ Yes
21 $g(x) = (x - 2)^2 + 3$ No
22 $h(x) = \dfrac{1}{x + 2}$ Yes
23 $j(x) = |x| + 2$ No

Problem-Set Notes and Strategies, continued

- Problem **10** is a good exercise in interpreting data from graphs and calculating statistics from those same graphs. Usually we have the data and construct the graph from the data.
- Problem **16** may get confusing for the student as it combines function notation and matrix products together. Encourage the student to do a step at a time.
- Order is important in problem **18** so a permutation is used.
- Problem **19** has no implied order, therefore a combination may be used.

Additional Answers

11 $\{x{:}\ x \neq 2 \text{ and } x \neq -2\}$
12 $\{x{:}\ x \leq 3\}$
13 {nonnegative integers}

20
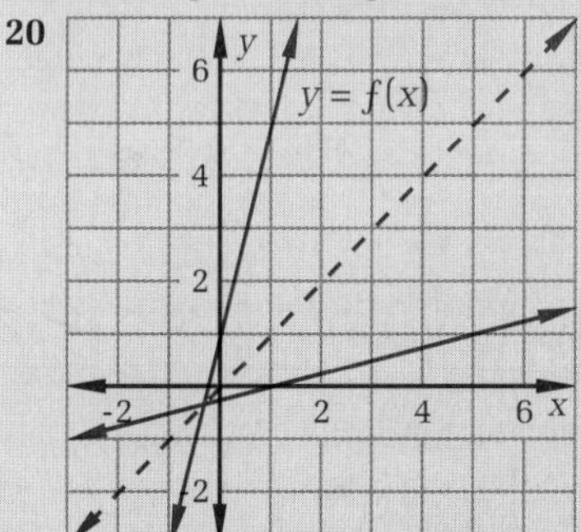

21
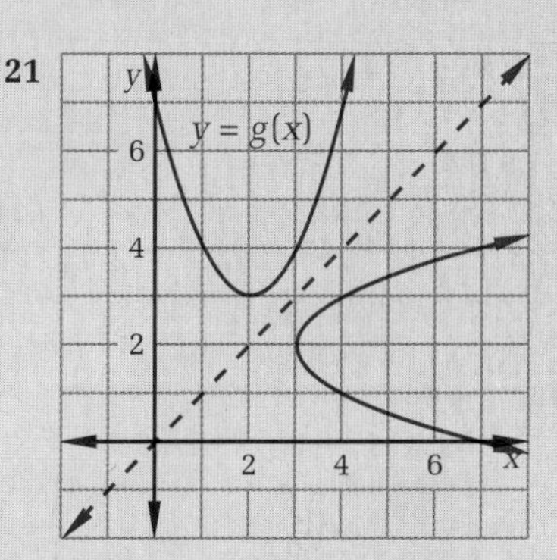

Answers for problems **22** and **23** can be found in the answer pages beginning on **A** 1.

Problem-Set Notes and Strategies, continued

- Problem **25** will require some recall about isosceles right triangles.

Additional Answers

26 $INVp(x) = \frac{x+3}{6}$

27 If $x \geq 0$, $INVr(x) = \sqrt{\frac{x+3}{2}}$

32 $f(n) = -\frac{n}{2} - 1$, where $n \geq 0$

33 $g(n) = 4^{\frac{n+1}{2}}$, where $n \in$ {odd positive integers}

- Composite functions are reviewed in problem **30**. Remind students that $q(p(x))$ is not a multiplication, but a composition of functions.

- Problems **32** and **33** are slightly different from normal. In **32**, the recursion is going backward since the $n-1$ term is defined in terms of the n term. This should cause no problem once the student recognizes the difference. In problem **33**, only the odd numbered terms may by calculated from the recursive definition since $n+1$ and $n-1$ are two apart. Again, this should cause no problem in finding the closed form definition once it is recognized.

- It is difficult to see recursive formulas in closed form definitions. Problem **35** may take some time for students to answer and may need additional explanation.

Review Problems, *continued*

P **24** Function f has a domain of {0, 1, 2, 3, 4, 5} and a range of $\left\{\frac{1}{3}, \frac{1}{4}, \frac{1}{5}, \frac{1}{6}, \frac{1}{7}, \frac{1}{8}\right\}$. Each element of the range is paired with exactly one element of the domain.

a What is the domain of INVf? $\left\{\frac{1}{3}, \frac{1}{4}, \frac{1}{5}, \frac{1}{6}, \frac{1}{7}, \frac{1}{8}\right\}$
b What is the range of INVf? {0, 1, 2, 3, 4, 5}

R **25** In the figure shown, each smaller square's vertices are the midpoints of the sides of the next larger square. The outer square has sides 4 centimeters long. If a point within the figure is picked at random, what is the probability that the point lies in the shaded region? $\frac{1}{32}$

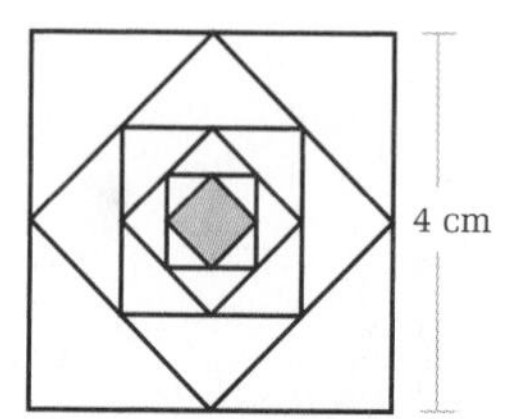

P **In problems 26–29, find the inverse of each function. Specify restrictions on the domain of the function if necessary.**

26 $p(x) = 6x - 3$ | **27** $r(x) = 2x^2 - 3$

28 $s(x) = \frac{1}{\sqrt{x}}$ $INVs(x) = \frac{1}{x^2}$, where $x > 0$ | **29** $t(x) = \frac{1}{2}x + \frac{5}{8}$ $INVt(x) = 2x - \frac{5}{4}$

P **30** If $p(x) = \frac{x^2 + 3x}{x^2 - 2x - 15}$ and $q(x) = \frac{1}{x}$, what are $q[p(x)]$ and $p[q(x)]$? $\frac{x-5}{x}$; $\frac{1}{1-5x}$

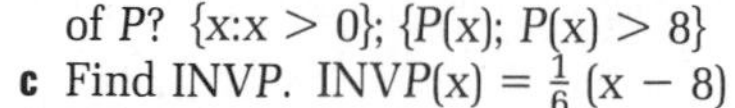

P **31** **a** Write a function P that represents the perimeter of the rectangle. $P(x) = 6x + 8$
b What are the domain and the range of P? $\{x: x > 0\}$; $\{P(x); P(x) > 8\}$
c Find INVP. $INVP(x) = \frac{1}{6}(x - 8)$
d What does the output of INVP represent? The value of x corresponding to the given perimeter

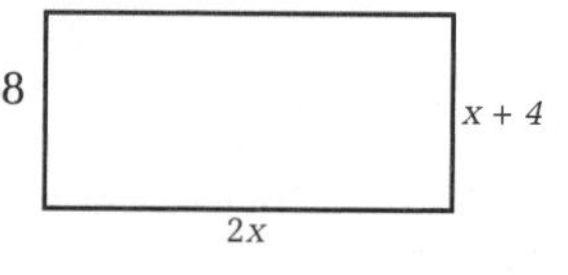

P **In problems 32 and 33, find a closed-form formula for each function.**

32 $f(0) = -1$
$f(n-1) = \frac{1}{2} + f(n)$

33 $g(1) = 4$
$g(n+1) = 4[g(n-1)]$

P **34** Let $g(x) = x^2 - 3$ and $h(x) = 1 - x$. Write a simplified expression for each of the following.
a $g \circ h(x)$ $x^2 - 2x - 2$ **b** $h \circ g(x)$ $4 - x^2$ **c** $h(x + 6)$ $-x - 5$
d $g(x + 6)$ $x^2 + 12x + 33$ **e** $g(6x)$ $36x^2 - 3$ **f** $h(6x)$ $1 - 6x$

C P **35** Write a recursive definition of the function $h(k) = k^2 + 3k + 1$, where k is a positive integer.
$h(1) = 5$, $h(n + 1) = h(n) + 2(n + 1)$

P **36** If $p(x) = 3x^2 - 5$, is $-5[p(x)]$ or $p(-5x)$ greater?
$-5[p(x)]$ if $-\frac{\sqrt{3}}{3} < x < \frac{\sqrt{3}}{3}$; neither if $x = \pm\frac{\sqrt{3}}{3}$; $p(-5x)$ if $x < -\frac{\sqrt{3}}{3}$ or $x > \frac{\sqrt{3}}{3}$

Chapter Test

1 The input values x for the following relation are the set of positive integers.

$$r = \begin{cases} 2 \text{ if } x \text{ is a multiple of } 2 \\ 3 \text{ if } x \text{ is a multiple of } 3 \\ 1 \text{ if } x \text{ is a multiple of neither 2 nor 3} \end{cases}$$

Is r a function of x? No

2 Suppose that $h(y) = \sqrt{y - 5}$ and that $g(x) = x + 2$ for the domain $\{x{:}1 \le x \le 5\}$.

a What is the domain of function h? $\{y{:}y \ge 5\}$
b If $R(x) = h[g(x)]$, what is the domain of R? $\{x{:}3 \le x \le 5\}$
c What is the range of R? $\{R(x){:}0 \le R(x) \le \sqrt{2}\}$

3 Refer to the diagram.

a Write a closed-form formula for F.
b Write a recursive definition of F.

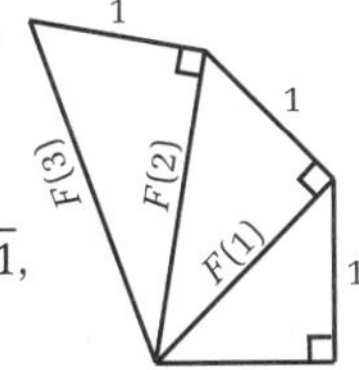

a $F(x) = \sqrt{x + 1}$
b $F(1) = \sqrt{2}$, $F(n + 1) = \sqrt{[F(n)]^2 + 1}$, where $x \in \{1, 2\}$

4 If $f(x) = x^3 + 1$ and $g(x) = |x|$, what are $g \circ f(-3)$ and $f \circ g(-3)$? 26; 28

5 If two of the following functions are selected at random, what is the probability that they are inverse functions? $\frac{1}{5}$

$f(x) = 2x + 10$ $\quad g(x) = \sqrt{x}$ $\quad h(x) = x^2, x \ge 0$

$m(x) = \dfrac{1}{x + 2}$ $\quad n(x) = \frac{1}{2}x - 5$ $\quad p(x) = \dfrac{1}{x} - 2$

6 A congressional committee is to consist of 3 of the 435 members of the House of Representatives. In how many possible ways can the committee be formed? 13,624,345

7 The data below represent the speeds, in miles per hour, of drivers ticketed on Interstate 314 by Officer "Radar" Gunn.

72, 78, 67, 70, 73, 69, 83, 68, 71, 75, 80, 73, 76

a Determine the mean, the mode, and the median of the data. ≈73.5 mph; 73 mph; 73 mph
b Draw a box-and-whisker plot to represent the data.
c Use the plot to determine the speeds that have percentile ranks less than 75. Speeds less than 77 mph

Resource References
Evaluation
Test Form 1, 2, 3

Additional Answer

7

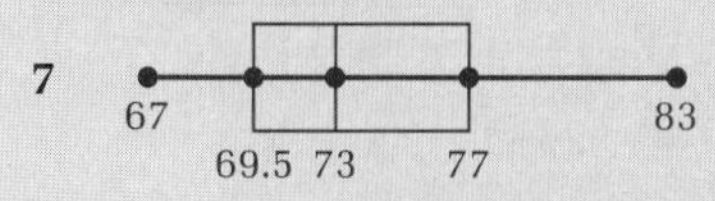

6.1

CHAPTER

6 GRAPHING

This photo of the distorted reflections of buildings in a set of coordinate axes brings to mind reflections of graphs.

FOR SOME STUDENTS, graphing parabolas and absolute value functions may be a review. The concept of a basic parabola, $y = x^2$, is then related to changes in the standard equation and the resulting stretching, shrinking, reflecting, and shifting of the graph vertically and horizontally.

Using tracings of a function on an overhead may help students visualize the reflection of an object on the coordinate plane or recognize symmetry with respect to a line or a point. Another approach is to have students cut up transparencies and manipulate them. There is also a nice manipulative exercise in the *Teacher's Resource Book*.

In Section **6.4**, students see how to graph cubic, quartic, reciprocal, and exponential functions. A graphing device allows students to quickly graph these functions and make connections to their equations. If technology is not available, it is sometimes helpful to prepare transparencies for the overhead so that students can see several different graphs.

Students will see how asymptotes can be used to sketch the graph of a function and how to identify restrictions on a function that lead to holes in the function's graph.

The data-analysis section is an extension of Section **5.7**. It interprets the meanings of scores dispersed within a distribution of data. The measures of dispersion along with the measures of central tendency provide a more complete interpretation of data.

6.1 BEYOND PLOTTING POINTS

Objectives

After studying this section, you will be able to

- Graph parabolas
- Graph absolute-value functions
- Stretch, shrink, and reflect graphs

Part One: Introduction

Graphs of Parabolas

In Section 4.3, we saw that the graph of $f(x) = x^2$ has a vertex at the origin and is ∪-shaped. Some of the ordered pairs on the graph of this function are $(-2, 4)$, $(0, 0)$, $(1, 1)$, and $(3, 9)$. In general, the graph of a function of the form $f(x) = a(x - h)^2 + k$, where a, h, and k are constants, is called a parabola and has a characteristic ∪ shape.

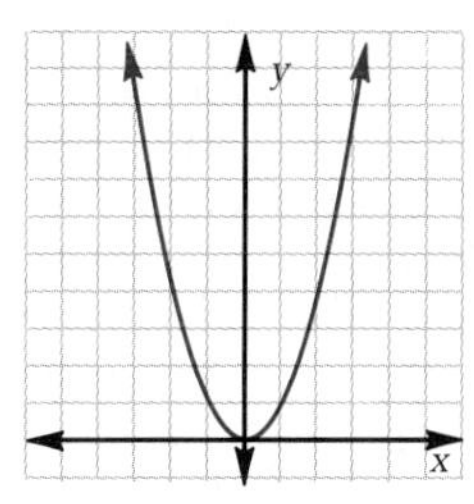

Absolute-Value Graphs

The graph of the absolute-value function $f(x) = |x|$ is similar to the graph of $f(x) = x^2$. The graph has a vertex and a line of symmetry, and $f(x) = f(-x)$. The characteristic shape is a V. Some points on the graph are $(-2, 2)$, $(-1, 1)$, $(0, 0)$, $(1, 1)$, and $(2, 2)$.

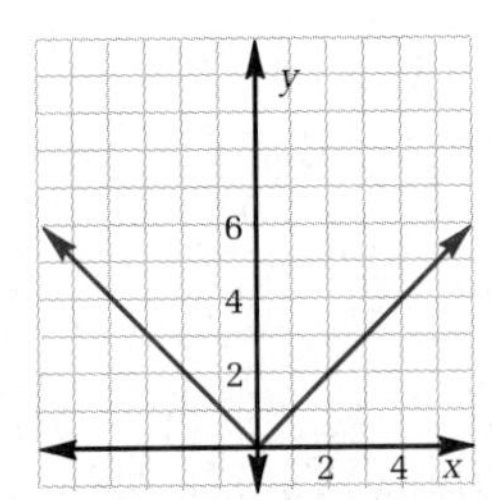

Class Planning

Time Schedule

Average: 2 days
Advanced: 2 days
Honors: 1 day

Resource References

Teacher's Resource Book
- Class Opener 34
- Practice 35
- Manipulatives and Applications 10
- Enrichment 11

Transparency 11

Class Opener

Sketch the graph of $f(x) = x^2$, $f(x) = 4x^2$, and $f(x) = \frac{1}{4}x^2$ on the same coordinate axes.

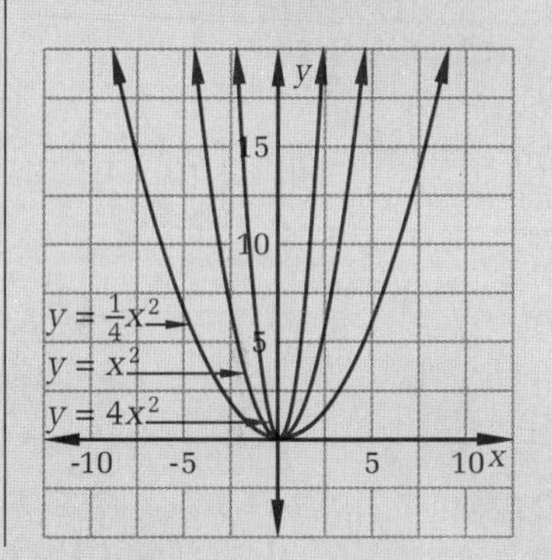

Lesson Notes

- This section uses the standard parabola, $y = x^2$ and the standard absolute value equation $y = |x|$ and organizes some of the related parabolas as transformations of stretching, shrinking, and reflecting.

Vocabulary

stretch
shrink
reflect

6.1

Stretching, Shrinking, and Reflecting

Let's look at the graphs of some functions of the form $y = ax^2$, where $a \neq 1$. How do these graphs differ from the graph of the standard parabola, $f(x) = x^2$?

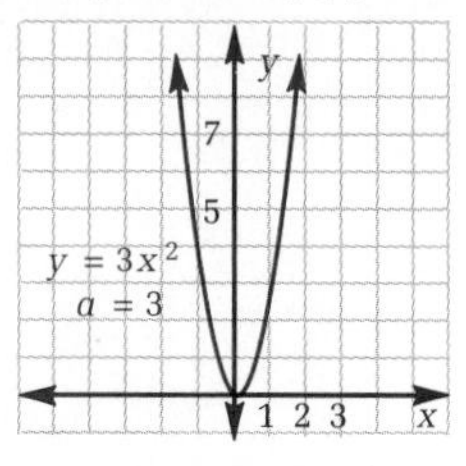

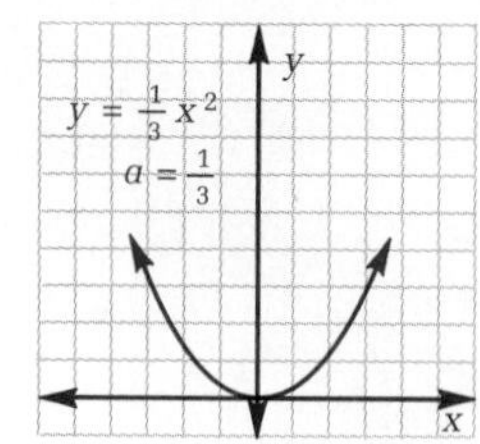

These graphs suggest that if the output of $f(x) = x^2$ is multiplied by a positive constant a, $a > 1$, the graph will appear to be ***stretched*** along the y-axis. If $0 < a < 1$, the graph is ***shrunk*** along the y-axis.

What happens if the coefficient a is negative?

These graphs appear to be similar to the earlier graphs with positive coefficients, except that these are ***reflected*** over the x-axis.

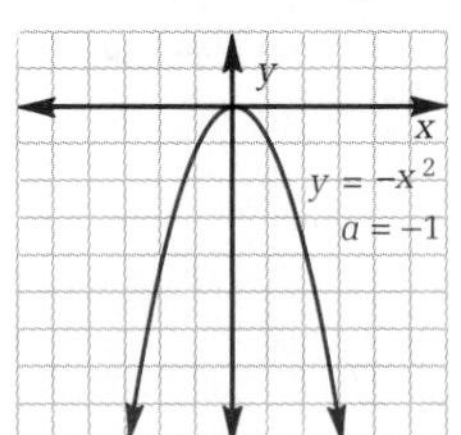

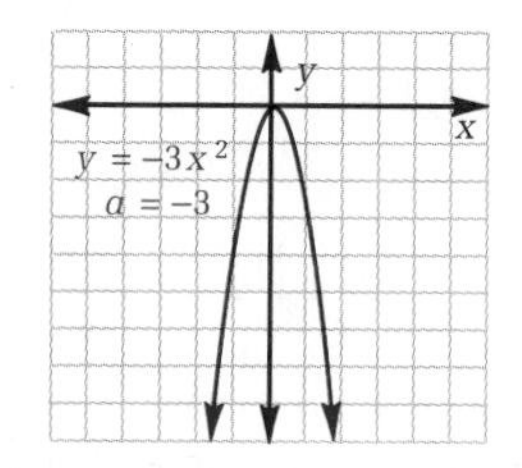

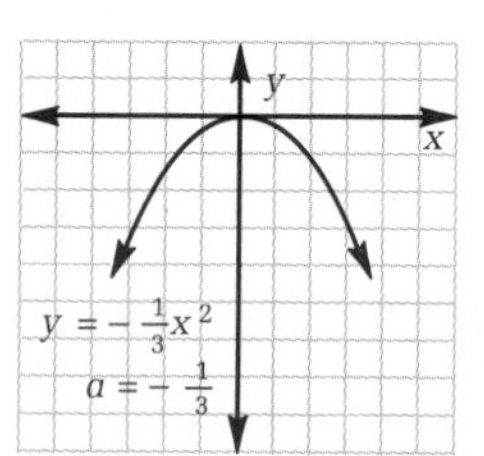

Example *Sketch the graph of* $g(x) = -0.5x^2$.

We can accurately draw a graph by using a plotter, or we can sketch the graph without plotting points. Since the coefficient is negative, the graph of g is a transformation of the graph of $f(x) = x^2$. The graph is reflected over the x-axis, opening downward. Because $|a| < 1$, the graph will be shrunk vertically by a factor of 0.5.

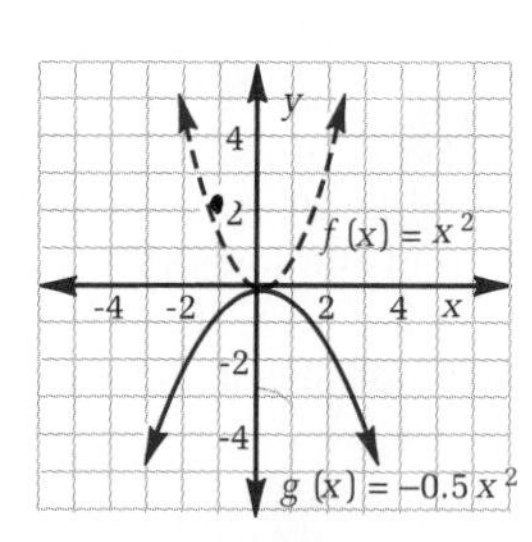

If the output of the standard absolute-value function, $f(x) = |x|$, is multiplied by a constant a, the graph of f will also be transformed.

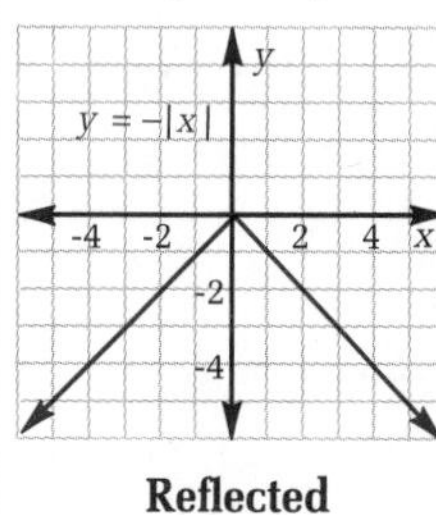

Reflected

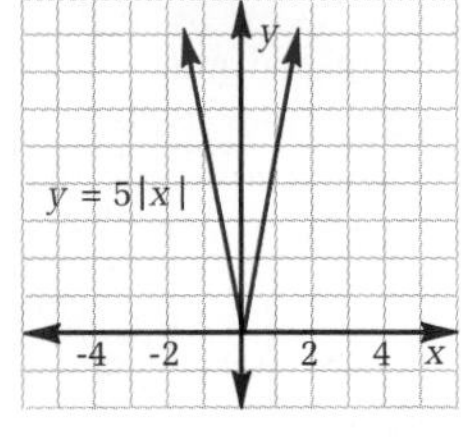

Stretched Vertically

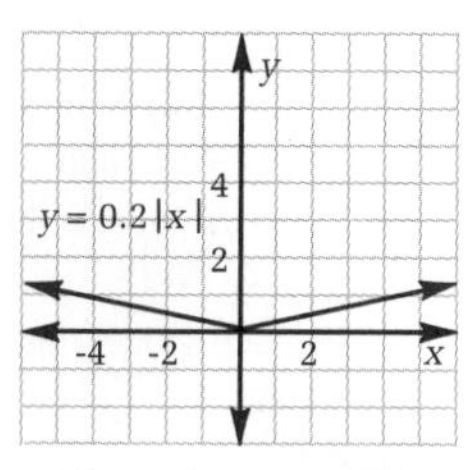

Shrunk Vertically

Checkpoint

1 If a function f has a point (6, 3) on the graph, name a point on the graph of $f(\frac{x}{3})$ and $f(-2x)$

(18, 3) and (−3, 3)

2 The graph of f follows.

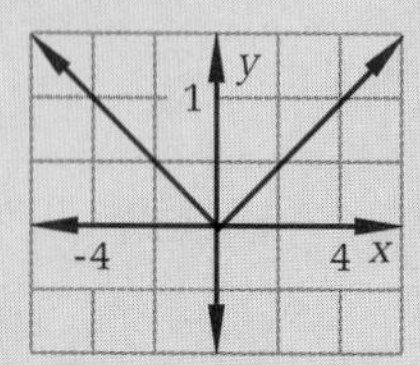

Use this graph to find $f(4x)$ and $4f(x)$.

Graph of $f(4x)$:

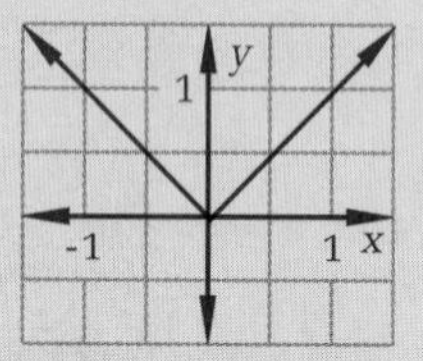

Graph of $4f(x)$:

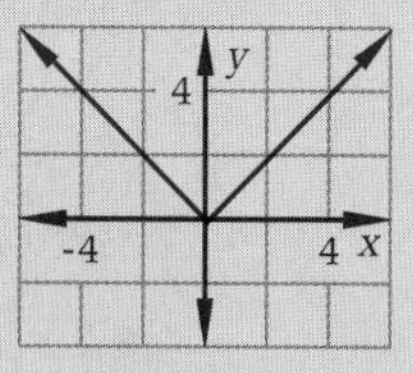

3 Using the same graph as in problem **2**, draw a picture of $f(x)$ and $-f(x)$ on the same graph.

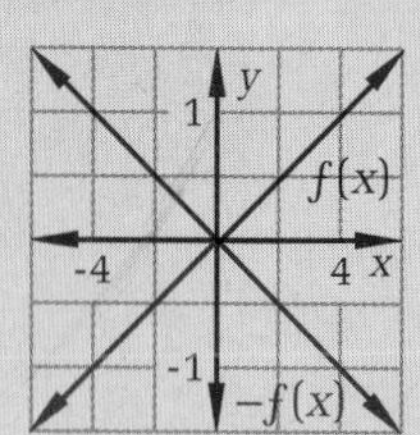

Part Two: Sample Problems

Problem 1 *Suppose (5, 1) is a point on the graph of $y = h(x)$. For each of the following, name a point on the graph.*

a $y = h(2x)$
b $y = -1.5[h(x)]$

Solution When the input to h is 5, the output is 1.

a In $y = h(2x)$, the input is 5. Then, $2x = 5$ and $x = \frac{5}{2}$, or 2.5. One point on the graph is (2.5, 1).
b Since $h(5) = 1$, $-1.5[h(5)] = -1.5(1)$, or -1.5.

Thus, (5, −1.5) is a point on the graph of $y = -1.5[h(x)]$. In **a,** the input is changed. In **b,** the output is changed.

Problem 2 *Use the graph of f to graph g and h.*

a $g(x) = 2[f(x)]$
b $h(x) = f(2x)$

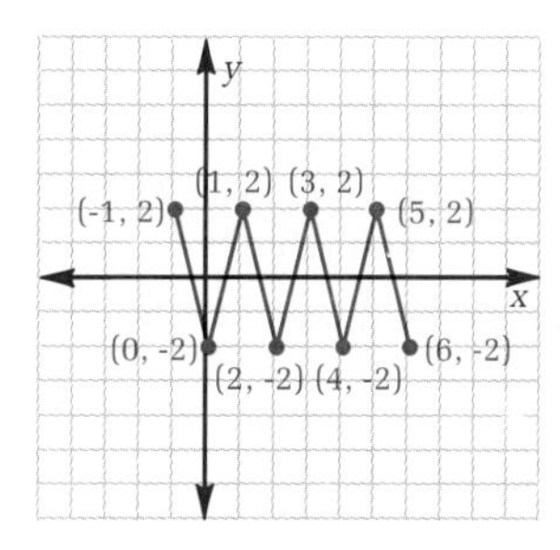

Solution

a For each input, the *output* of g is twice the output of f, so the graph of g is stretched vertically by a factor of 2.

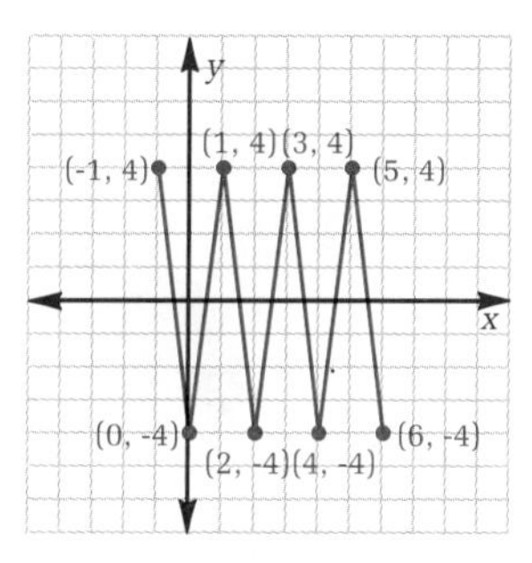

b For any given output, the *input* of h is one-half the input of f, so the graph of h is shrunk horizontally by a factor of $\frac{1}{2}$.

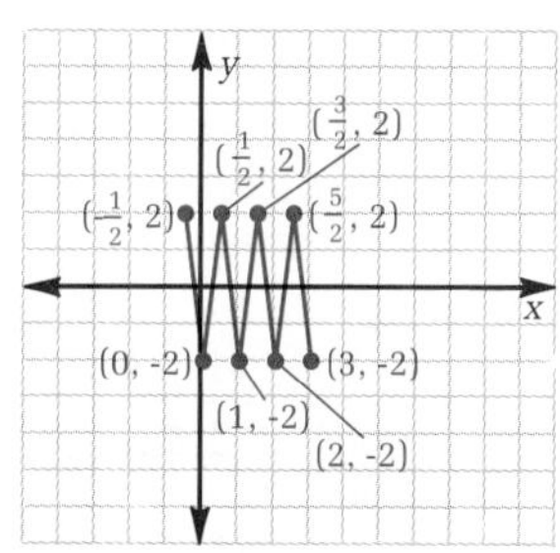

Notice that when the input is multiplied by a constant, the transformation is horizontal. When the output is multiplied by a constant, the transformation is vertical.

Stumbling Block

When multiplying by a constant, students frequently just enlarge the graph by that amount. This is fine if the output is multiplied by that amount. However, the input being multiplied is counterintuitive. If it is multiplied by a constant, the graph is multiplied by the reciprocal of that constant. There is no easy way to remember this except to always check one or two points on the new graph.

Using Computers

Computer graphing software or a graphics calculator is excellent for discovering, introducing and reinforcing the stretching, shrinking, and reflecting of graphs. The computer and calculator permit students to actually see the changes. The *Omnifarious Plotter* works well here.

Assignment Guide

Average

Day 1 11, 12, 14, 16, 18, 19, 22, 24, 26, 28, 36

Day 2 13, 15, 17, 20, 23, 25, 27, 29, 30, 31, 39

Advanced

Day 1 11–14, 17–19, 22, 24, 26, 28, 29, 36

Day 2 15, 16, 20, 23, 25, 27, 30–32, 38, 39

Honors

12, 13, 15, 16, 18, 19, 21, 22, 26, 27, 30, 31, 35, 36, 39

Problem-Set Notes and Strategies

- Students can work through problems **1 – 7** checking each point as they go until they build their confidence. This is a slow process to internalize the dilations of graphs.

Communicating Mathematics

Write a paragraph describing what happens to a graph when the input and output are multiplied by a constant respectively. Include both positive and negative constants.

Part Three: Exercises and Problems

Warm-up Exercises

In problems 1–4, copy the graph of *f*, and on the same axes sketch the graph described.

P

1 The graph of f stretched vertically by a factor of 2

2 The graph of f stretched horizontally by a factor of 3

3 The graph of f shrunk horizontally by a factor of $\frac{1}{3}$

4 The graph of f shrunk vertically by a factor of $\frac{1}{2}$

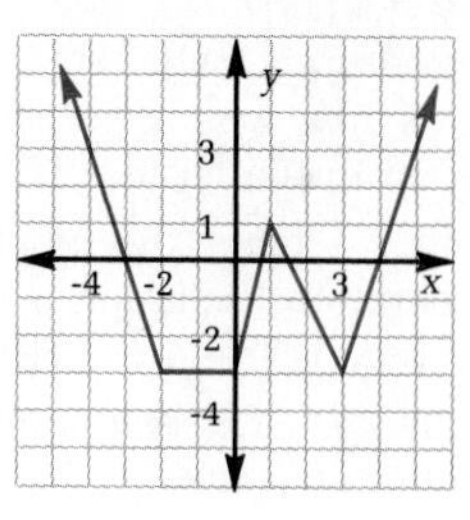

P **In problems 5–7, refer to the table of values for $f(x)$, then copy and complete each table.**

x	−2.0	−1.5	−1.0	−0.5	0	0.5	1.0	1.5	2.0	2.5	3.0
$f(x)$	7	2	−1	1	2	$-\frac{1}{2}$	−3	−8	−15	−22	−40

5

x	−1	0	1	1.5
$f(2x)$	7	2	−15	−40

6

x	−1	0	1	2
$\frac{1}{2}[f(x)]$	$-\frac{1}{2}$	1	$-\frac{3}{2}$	$-\frac{15}{2}$

7

x	−2	−1	0	1	2
$f\left(\frac{1}{2}x\right)$	−1	1	2	$-\frac{1}{2}$	−3

P **8** Copy each diagram and fill in the blanks.

a $f(x) = x^2$

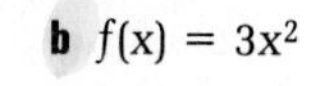

b $f(x) = 3x^2$

c $f(x) = \frac{1}{3}x^2$

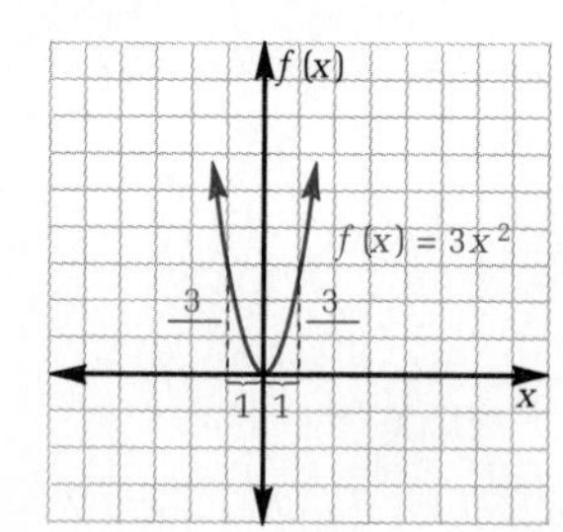

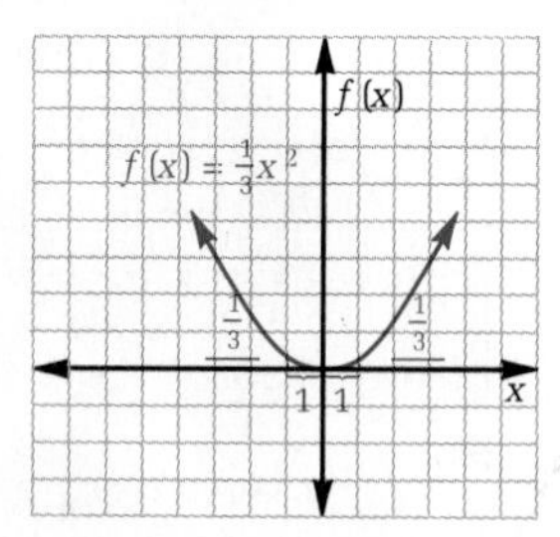

Additional Answers

Answers for problems **1 – 4** can be found in the answer pages beginning on **A 1**.

P **In problems 9 and 10, describe how the graph of *g* or *h* compares with the graph of *f*, where $f(x) = x^2$.**

9 $g(x) = f\left(\frac{1}{2}x\right)$
Stretched horizontally by a factor of 2

10 $h(x) = \frac{1}{2}[f(x)]$
Shrunk vertically by a factor of $\frac{1}{2}$

Problem-Set Notes and Strategies, continued

- In problems **9** and **10**, encourage students to try a few points to get the impact of the dilation.

A Problem Set

P **In problems 11–13, determine the value of *a*, given the graph of $f(x) = a|x|$.**

11

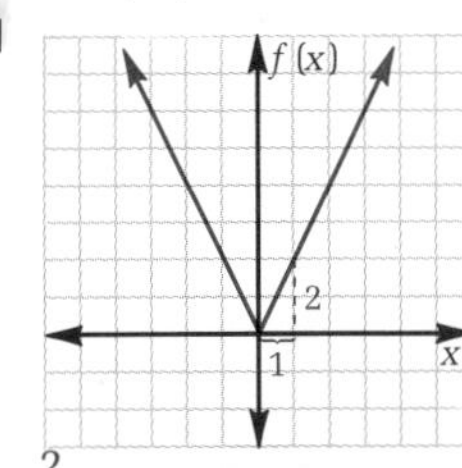

2

12

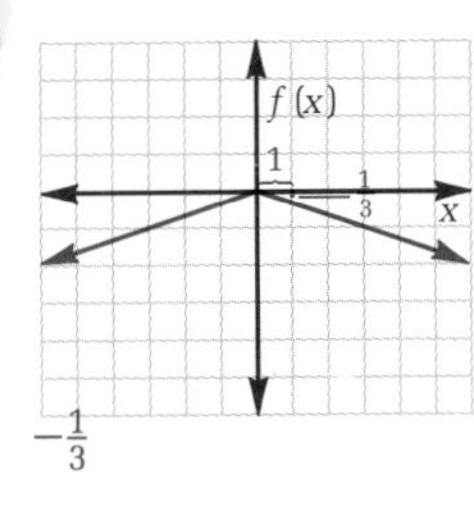

$-\frac{1}{3}$

13

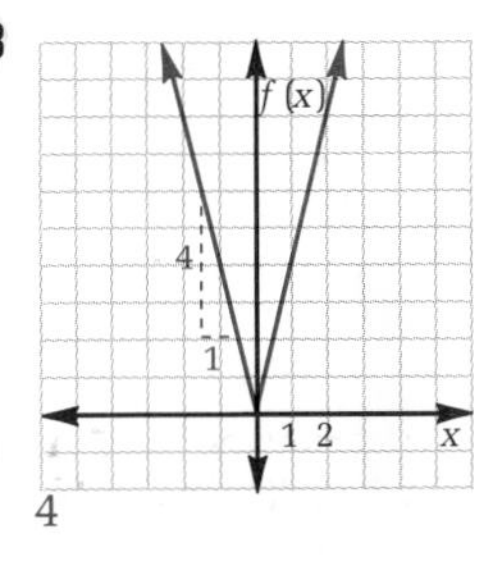

4

P **In problems 14 and 15, describe how the graph of each function differs from the graph of $y = x^2$.**

14 $f(x) = 2.5x^2$ Stretched vertically by a factor of 2.5

15 $g(x) = -2.5x^2$ Stretched vertically by a factor of 2.5 and reflected over the x-axis

- Notice that in problem **15**, the reflection is over the *x*-axis. If the equation had been $(-x)^2$, the reflection would be over the *y*-axis.

R **16** Refer to the diagram.

a Find (x, y). (10, −2)

b Find AB. $\sqrt{89}$

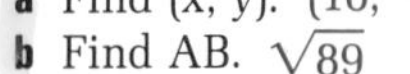

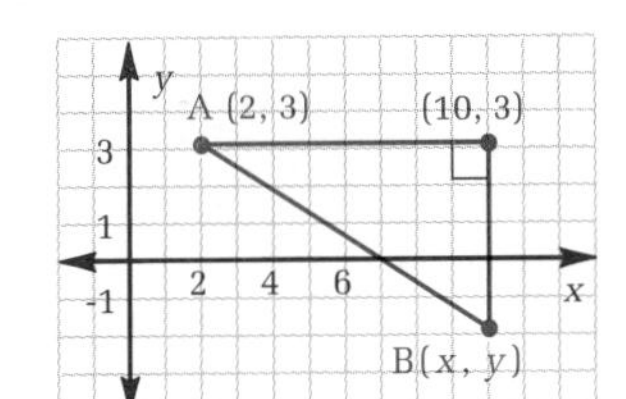

- You may get some interesting answers for **17**. Try to get the student thinking in terms of distance both from the mirror and behind the mirror.

I **17** Joe saw his image in a mirror. Where do you think the mirror is?
Halfway between his face and image

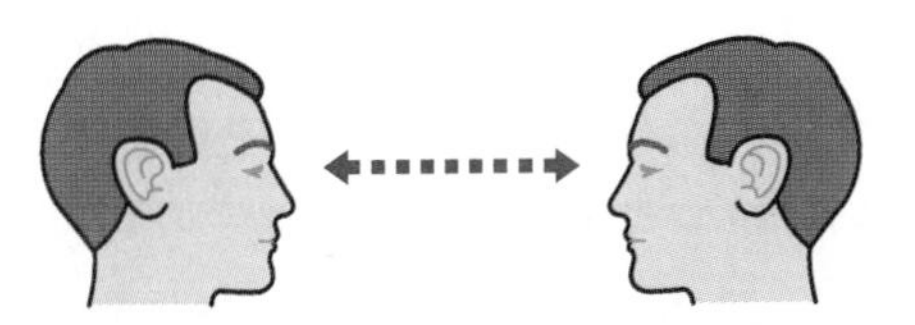

P,R **18** Let the domain of $f(x) = |x|$ be {−3, −2, −1, 0, 1, 2, 3}.

a Find the range of $y = 2[f(x)]$. {0, 2, 4, 6}

b Graph $y = f(x)$ and $y = -2[f(x)]$ on the same axes.

Additional Answer

18b

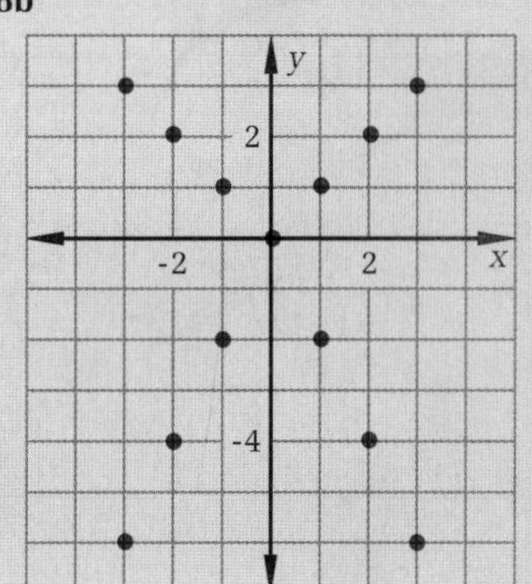

P **In problems 19–21, name a point on the graph of each equation if (2, 6) is a point on the graph of $y = f(x)$.**

19 $y = 4[f(x)]$ (2, 24)

20 $y = -5[f(x)]$ (2, −30)

21 $y = \frac{2}{7}[f(x)]$ $\left(2, \frac{12}{7}\right)$

Problem-Set Notes and Strategies, continued

- If students simplify the expression first in problem **23**, their work will be simpler.
- In problem **29**, students may work out all the possibilities. Some students will be able to generalize the solution of each.
- Problem **30** involves a system of nonlinear equations. This is solved by substitution as in the linear case. This previews Section **7.5**.

Problem Set, *continued*

R **22** For which of the following is y a function of x? **a, b, c**

a

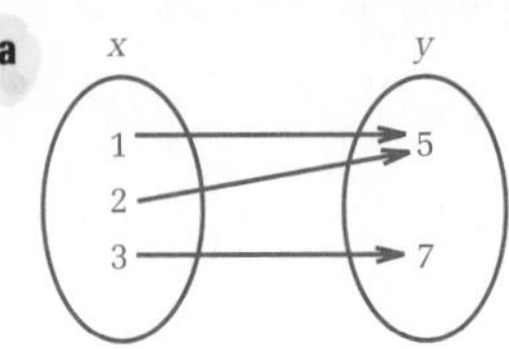

b

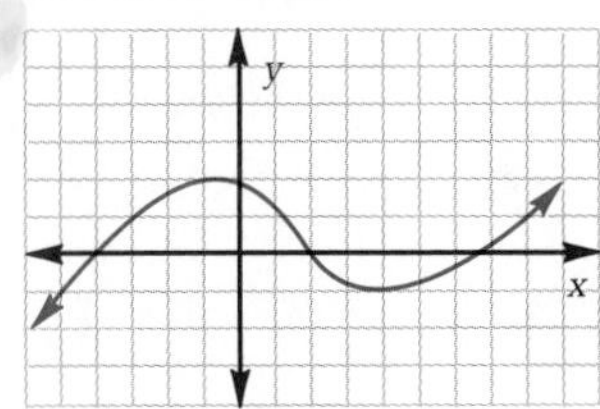

c

x	−6	2	8	10
y	−4	−4	0	1

R,I **23** Evaluate $\frac{x^5y^{-17}}{x^{-3}y^5} \cdot \frac{x^2y^{18}}{x^9y^{-3}}$ if $x = 9.6$ and $y = -3.2$. -3

R **24** The height of a ball above the ground (in meters) at time t seconds is given by $h(t) = -4.9t^2 + 39.2t$.

a How high is the ball at $t = 0$ seconds? 0 m

b When will the ball again be on the ground? 8 sec

c When is the ball 50 meters above the ground? $\approx$1.6 sec and $\approx$6.4 sec

B

P **25** Given $\{(x, f(x)):(x, f(x)) = (-2, 0), (3, 4), (7, 12), (9, -13)\}$, find the set of ordered pairs $(x, 4[f(x)])$.
$\{(x, 4[f(x)]:(x, 4[f(x)]) = (-2, 0), (3, 16), (7, 48), (9, -52)\}$

P,R **26** Determine the value of a for each function.

a $f(x) = ax^2$, with a graph passing through $(1, 9)$ 9

b $g(x) = a|x|$, with a graph passing through $(9, 4)$ $\frac{4}{9}$

P **27** Let $f(x) = 3 + 6x$ and $g(x) = x^2 + 4x$. Evaluate $f(g(1))$ and $g(f(1))$. $f(g(1)) = 33$; $g(f(1)) = 117$

P **28** Refer to the table below. If $g(x) = k[f(x)]$, what is the value of k? 12

x	−3	−2	−1	0	1	2	3
f(x)	1.5	1	0.5	0	0.5	1	1.5
g(x)	18	12	6	0	6	12	18

P **29** Suppose $f(x) = x^2$ and b is a constant greater than 0. List the following range elements from smallest to largest: $f(b)$, $2[f(b)]$, $f(2b)$, $-f(b)$, $-2[f(b)]$, $f(-3b)$.
$-2[f(b)]$, $-f(b)$, $f(b)$, $2[f(b)]$, $f(2b)$, $f(-3b)$

P,I **30** Let $f(x) = k^2x^2$ and $g(x) = -3kx + 10$.

a If $k = 1$, where do the graphs of f and g intersect? $(-5, 25)$, $(2, 4)$

b Where do the graphs intersect in terms of k? $\left(\frac{-5}{k}, 25\right)$, $\left(\frac{2}{k}, 4\right)$

P **31** Let $f(x) = |x|$.

a Copy and complete the following table, and graph $y = f(x) + 2$.

x	−3	−2	−1	0	1	2	3
$f(x) + 2$	5	4	3	2	3	4	5

b Copy and complete the following table, and graph $y = f(x) - 2$.

x	−3	−2	−1	0	1	2	3
$f(x) - 2$	1	0	−1	−2	−1	0	1

c Describe how the graphs in parts **a** and **b** differ from the graph of f.

R **32** Let $f(x) = -\frac{1}{3}x^2$ and $g(x) = x - 4$. Find all the values of x for which $f(x) = g(x)$. $x \approx -5.3$ or $x \approx 2.3$

C

In problems 33 and 34, solve for x.

R,I **33** $\frac{x-2}{4} = \frac{16}{x-2}$ $x = 10$ or $x = -6$ **34** $\frac{x+3}{5} = \frac{11}{x-3}$ ± 8

R **35** The total number of children and adults who attended a recent football game was 30,783. Children's tickets cost \$4.50 each, and adult tickets cost \$6.75 each. The total revenue from the sale of tickets was \$180,054. How many of each type of ticket were sold? Adult: 18,458; children's: 12,325

P,I **36** If $f\left(\begin{bmatrix} a & b \\ c & d \end{bmatrix}\right) = ad - bc$, then what is each of the following?

a $f\left(2\begin{bmatrix} a & b \\ c & d \end{bmatrix}\right)$ $4ad - 4bc$ **b** $2\left[f\left(\begin{bmatrix} a & b \\ c & d \end{bmatrix}\right)\right]$ $2ad - 2bc$

c How do your answers to parts **a** and **b** compare?
Part **a** is double part **b**.

R **37** On January 1, an employee was given two options. Which option should he choose? Explain your answer.

a Double his current income for the coming year.

b Receive 6% increase in salary on the first day of each month, beginning on February 1.
Over the entire year, he gains the most money with option **a**.

R,I **38** **a** What is the domain of $f(x) = \frac{3x(x-2)}{(x-2)}$? $\{x : x \neq 2\}$

b If the expression $(x - 2)$ is canceled, the result is the function $g(x) = 3x$. What is the domain of g? $\{x : x \in \mathcal{R}\}$

R **39** Find the number of atoms in a copper medallion if its mass is 3.1 grams and each copper atom has a mass of 1.07×10^{-22} grams. Express your answer in scientific notation. $\approx 2.9 \times 10^{22}$ atoms

Problem-Set Notes and Strategies, continued

- Problem **31** asks students to generalize about translations of graphs. This is the topic of the next section and will be explained in more detail.
- Problem **32** requires the use of the quadratic formula.
- In both **33** and **34**, students may cross multiply and solve the resulting quadratic equation.
- Problem **35** is another form of a mixture problem. A system of equations in two variables can use one equation for adults and one for children.
- Problem **36** presents a property of the determinant. Students have had a brief introduction to determinants in a problem earlier.
- Problem **38** previews holes in graphs discussed in Section **6.5**.

Additional Answers

31a

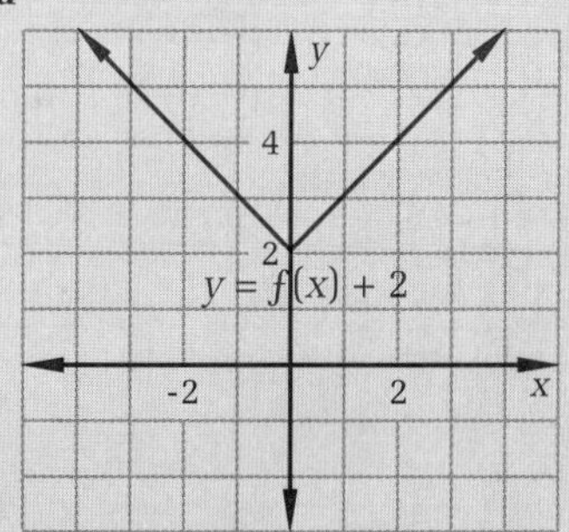

31b

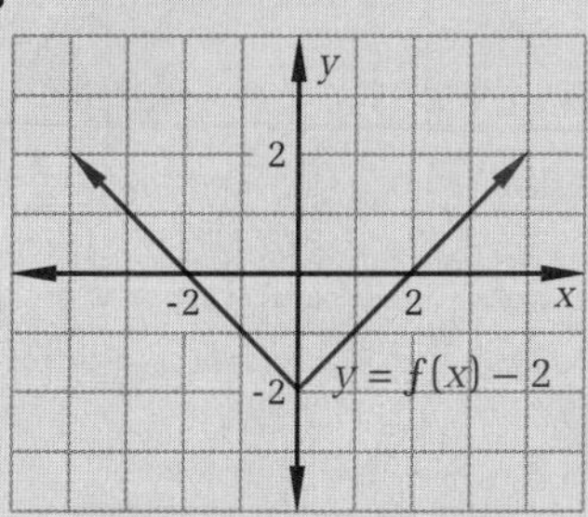

31c The graph of $y = f(x) + 2$ is moved up two units, and the graph of $y = f(x) - 2$ is moved down two units.

6.2

Class Planning

Time Schedule

Average: 3 days
Advanced: 3 days
Honors: 2 days

Resource References

Teacher's Resource Book
Class Opener 35
Practice 36

Class Opener

1 Sketch the graphs of $f(x) = x^2 + 6$ and $f(x) = x^2 - 4$ on the same axes.

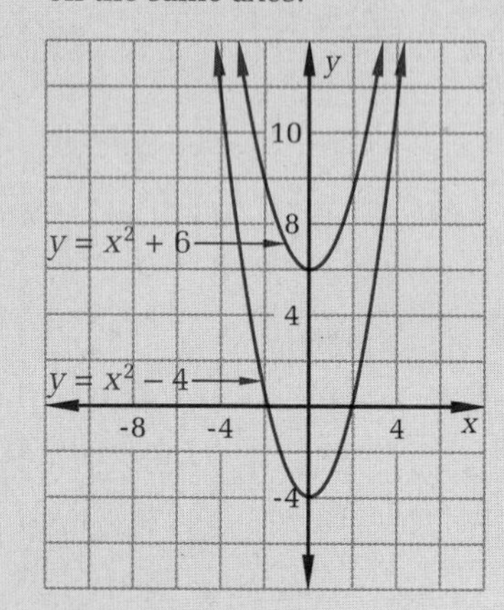

2 Write a sentence describing the relationship between the graphs.
Answers will vary. One possible answer is the graph of $y = x^2 + 6$ is the graph of $y = x^2$ translated up 6 units. The graph of $y = x^2 - 4$ is the graph of $y = x^2$ translated down 4 units.

Communicating Mathematics

Prepare a lesson on how you would explain the concept of slope and y-intercept to someone who never heard of these topics, using only the knowledge of translations and stretching and shrinking presented in the last section.

6.2 TRANSLATING GRAPHS

Objectives

After studying this section, you will be able to

- Shift graphs vertically
- Shift graphs horizontally

Part One: Introduction

Shifting Graphs Vertically

If we add a value to the output of a function, how is the graph affected? What happens if we subtract a value from the output?

Let $f(x) = |x|$, and define the functions g and h as follows.

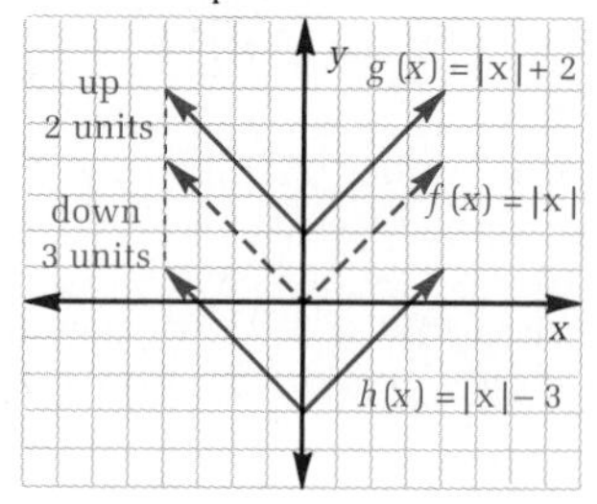

$g(x) = f(x) + 2 \qquad h(x) = f(x) - 3$

$= |x| + 2 \qquad = |x| - 3$

In general, when we add a value to the output of a function, the graph is shifted *upward*. When we subtract a value from the output of a function, the graph is shifted *downward*.

Shifting Graphs Horizontally

How does the graph of a function change if the addition or subtraction is performed on the input?

Again, let $f(x) = |x|$, and define the functions p and r as follows.

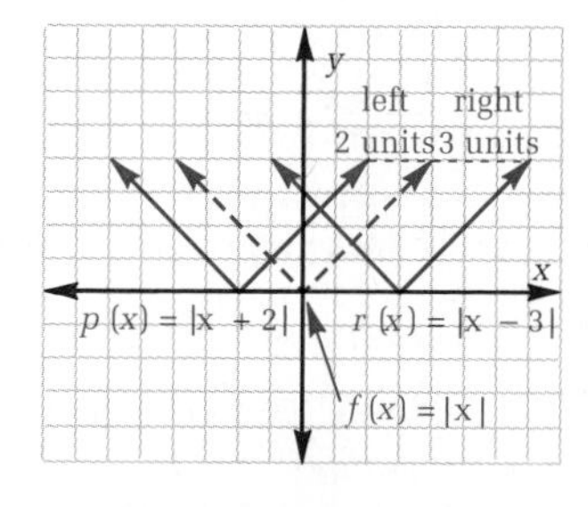

$p(x) = f(x + 2) \qquad r(x) = f(x - 3)$

$= |x + 2| \qquad = |x - 3|$

In general, adding to the input of a function shifts the graph *left*, and subtracting from the input of a function shifts the graph *right*.

Now we can combine the vertical and horizontal shifts with the stretches and shrinks, discussed in Section 6.1.

If $g(x) = a \cdot f(x - h) + k$, where f is a function, then, compared with the graph of f,

1 ***The graph of g is stretched vertically by a factor of $|a|$***
2 ***The graph of g is shifted right h units and up k units***

Part Two: Sample Problems

Problem 1 *Suppose (3, 3) is a point on the graph of $y = f(x)$. Name a point on the graph of $y = f(x - 3) - 7$.*

Solution Compared with the graph of $y = f(x)$, the graph of $y = f(x - 3) - 7$ is shifted three units to the right and seven units down. Thus the point (3, 3) becomes $(3 + 3, 3 - 7)$, or $(6, -4)$.

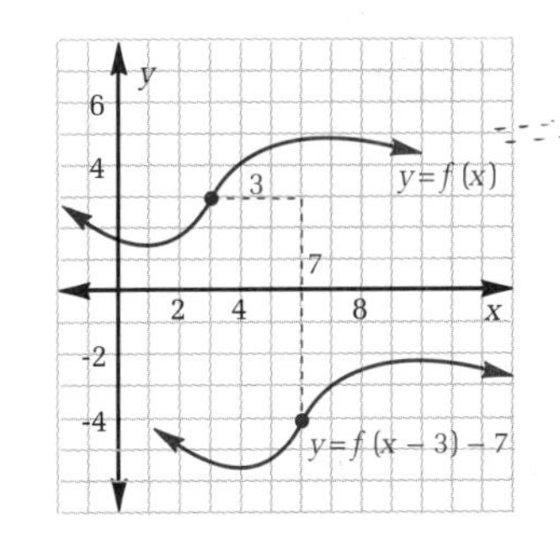

Problem 2 *Sketch the graph of $g(x) = x^2 - 16x + 35$.*

Solution If the function were in the form $g(x) = (x - h)^2 + k$, we could describe its graph as the graph of $f(x) = x^2$ translated h units horizontally and k units vertically. To change g into the form $g(x) = (x - h)^2 + k$, we complete the square. (There is a review of this process in Section 4.5.)

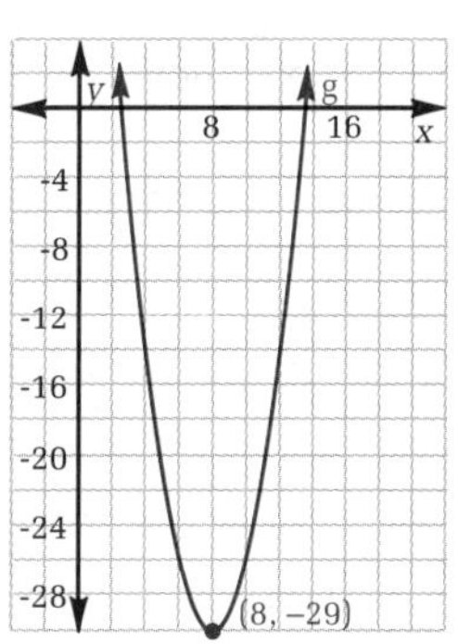

$$\begin{aligned} g(x) &= x^2 - 16x + 35 \\ &= x^2 - 16x + \underline{\quad} - \underline{\quad} + 35 \\ &= x^2 - 16x + \underline{64} - \underline{64} + 35 \\ &= (x - 8)^2 - 29 \end{aligned}$$

Remember that when we add a number and its opposite to a polynomial, we do not change the value of the polynomial.

Part Three: Exercises and Problems

Warm-up Exercises

In problems 1–4, indicate whether the graph of the equation is a translation up, down, to the left, or to the right of the graph of $y = f(x)$.

P

1 $y = f(x - 17)$ Right
2 $y = f(x) - 17$ Down
3 $y = f(x) + 1$ Up
4 $y = f(x + 3.2)$ Left

P **In problems 5–8, name a point on the graph of each equation if (20, 5) is a point on the graph of $y = g(x)$.**

5 $y = g(x) - 12$ (20, −7)
6 $y = g(x) + 1$ (20, 6)
7 $y = g(x + 16)$ (4, 5)
8 $y = g(x - 1)$ (21, 5)

Class Opener, continued

3 Sketch the graphs of $f(x) = (x - 2)^2$ and $f(x) = (x + 4)^2$ on the same axes.

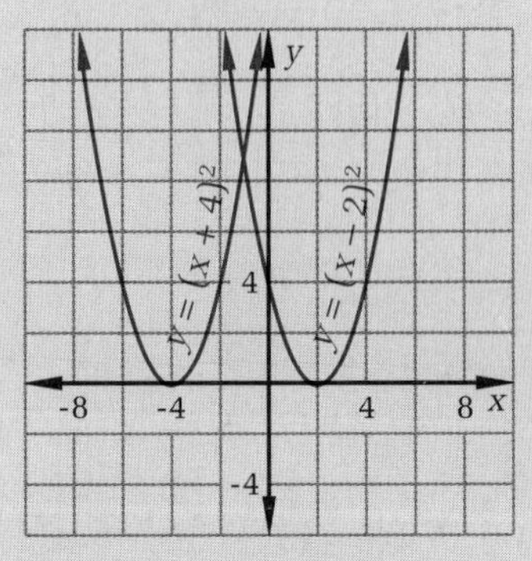

4 Write a sentence describing the relationship between the two graphs.
Answers will vary. One possible answer is the graph of $y = (x - 2)^2$ is the graph of $y = x^2$ translated 2 units to the right. The graph of $y = (x + 4)^2$ is the graph of $y = x^2$ translated 4 units to the left.

Lesson Notes

- Section 2 extends the concept of transformation of a parabola to shifting graphs vertically and horizontally. Once students become familiar with the concept and characteristics of transforming the parabola, they can learn a great deal about the graph without actually plotting a set of points.

Problem-Set Notes and Strategies

- Problems **1 – 12** are translations of graphs.

Problem-Set Notes and Strategies, continued

- In problems **13 – 16** it might help to write the parabola in vertex form. Students will eventually become familiar with both notations and be able to shift between the two.
- Students will have to graph each parabola separately in problems **17 – 22** or do a mental shift of the function. Problem **22** is the only one that may cause confusion as students are not accustomed to problems that do not have answers.
- Problem **23** relates the method of finding the translation for a quadratic polynomial to completing the square done in Chapter **4**.

Checkpoint

1 Suppose the graph of the function $f(x)$ passes through the origin. Name a point on the graph of $y = f(x - 7) + 3$ (7, 3)

2 The equation of the unit circle is given by $x^2 + y^2 = 1$. Sketch the graph of $(x - 2)^2 + (y + 1)^2 = 1$.

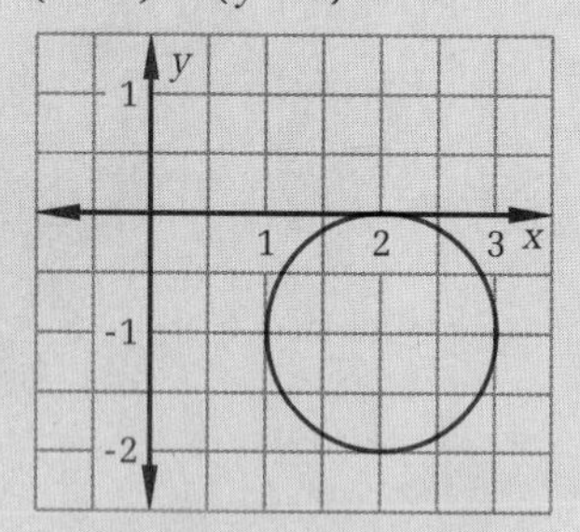

3 Indicate whether the graph is translated up, down, right, or left for the following functions:

$y = f(x - 3)$ right
$y = f(x) - 3$ down
$y = f(x) + 3$ up
$y = f(x + 3)$ left

P **In problems 9–12, write the equation of each graph, assuming that each is a translation of the graph of $y = x^2$.**

9

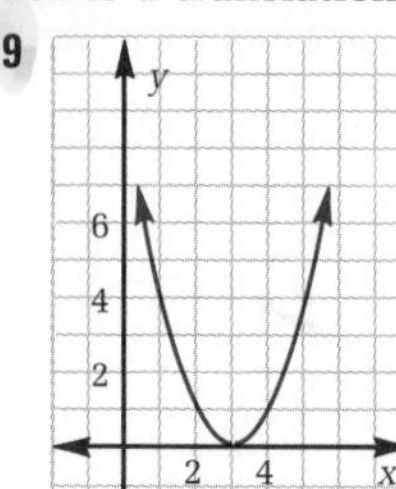

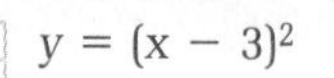

$y = (x - 3)^2$

10

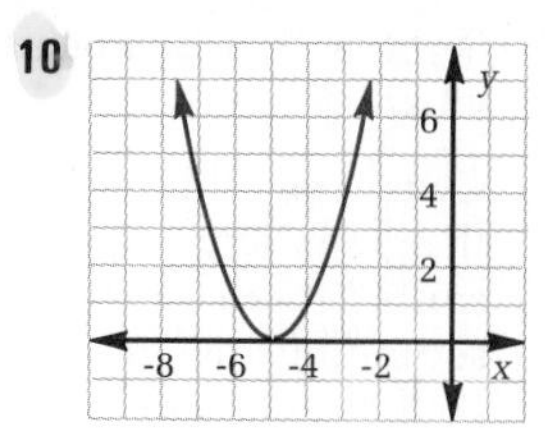

$y = (x + 5)^2$

11

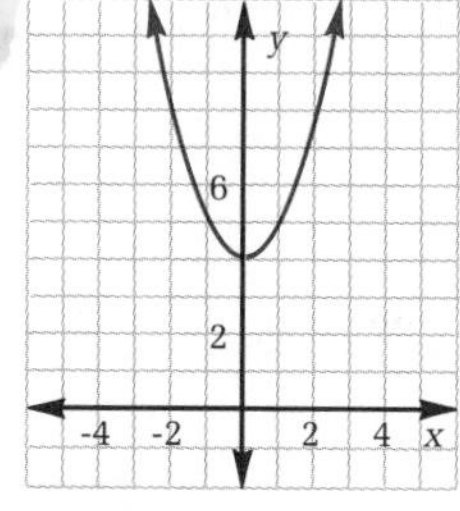

$y = x^2 + 4$

12

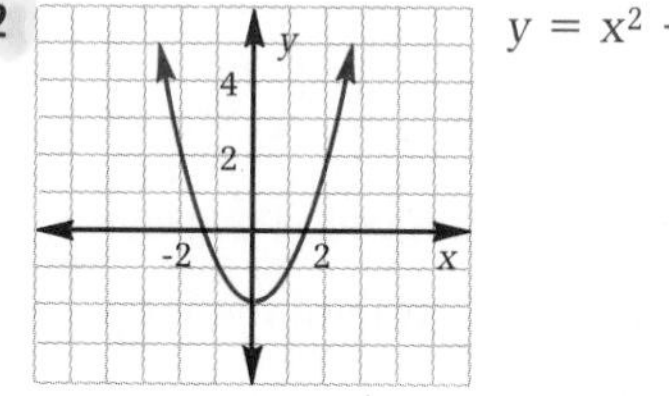

$y = x^2 - 2$

Problem Set

A

P **In problems 13–16, let $f(x) = x^2$. Determine the vertex of each equation's graph and state whether it is the highest or lowest point on the graph.**

13 $2[f(x - 4)] = y$ (4, 0); lowest
14 $-5[f(x - 2)] + 3 = y$ (2, 3); highest
15 $-3[f(x + 2)] = y$ (−2, 0); highest
16 $4[f(x + 7)] + 3 = y$ (−7, 3); lowest

P **In problems 17–22, find the values of x that satisfy each equation, using the graph of $f(x) = y$ shown.**

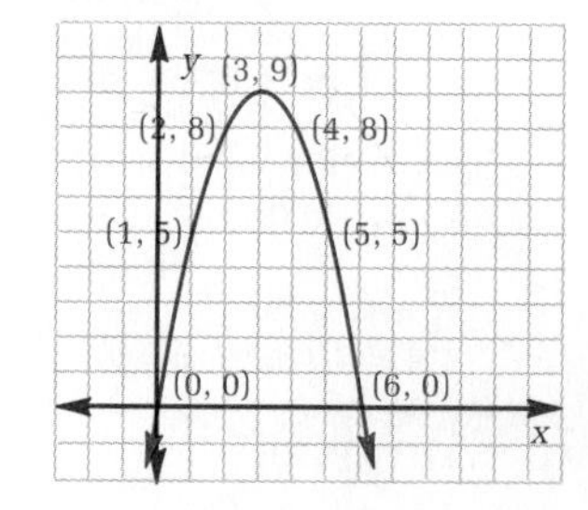

17 $f(x) = 0$ $x = 0$ or $x = 6$
18 $f(x + 4) = 0$ $x = -4$ or $x = 2$
19 $f(x - 2) = 0$ $x = 2$ or $x = 8$
20 $f(x - 9) = 0$ $x = 9$ or $x = 15$
21 $f(x) - 5 = 0$ $x = 1$ or $x = 5$
22 $f(x) - 10 = 0$ No such values

P,R **23** **a** Complete the square in $g(x) = x^2 - 8x + 12$ by filling in the blanks. 16; −16

$$g(x) = (x^2 - 8x + \_\_\_\_) - \_\_\_\_ + 12$$

b If $g(x) = (x - h)^2 + k$, what are the values of h and k? 4; −4

R **24** Simplify $\sqrt{8} + \sqrt{12} + \sqrt{18} + \sqrt{27} + \sqrt{50}$. $10\sqrt{2} + 5\sqrt{3}$

R,I | **25** **a** Draw the graph of the reflection of $y = x + 2$ over the line $y = x$.
b What is the equation of the reflected graph? $y = x - 2$
c What is the equation of the reflection of the graph $y = x + b$ over the line $y = x$? $y = x - b$

P | **26** Find k.
-3

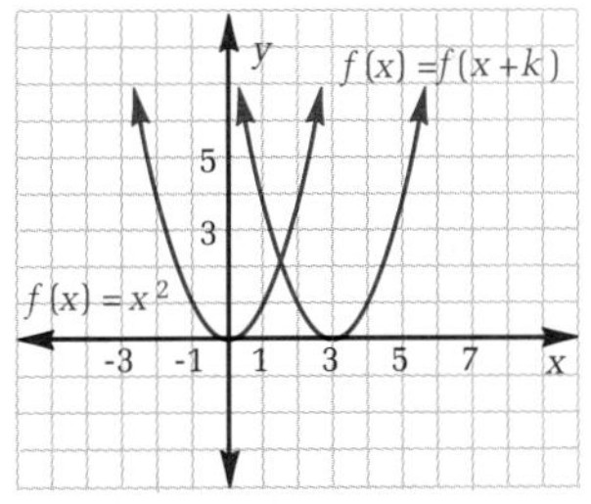

P | **27** Graph each function on the same set of axes.
a $f(x) = -2$ **b** $g(x) = |x|$ **c** $h(x) = f(x) + g(x)$

P,R | **In problems 28–31, find the y-intercept of the graph of each equation if $f(x) = \frac{1}{2}x - 4$.**

28 $f(x + 4) = y$ -2
29 $f(x) - 7 = y$ -11
30 $f(2x) = y$ -4
31 $f(x - 2) = y$ -5

R | **32** How many arrangements of the given letters are possible?
a A, B, C 6 **b** A, B, C, D 24 **c** A, B, C, D, E 120

P,R | **33** Let $f(x) = x^2 + 3$ and $g(x) = -x^2 - 3$.
a Graph $y = f(x)$ and $y = g(x)$ on the same set of axes.
b Graph $y = f(x) + g(x)$.

R | **34** Solve $\frac{5^{2n+7}}{5^3} = 5^{14}$ for n. $n = 5$

R,I | **In problems 35 and 36, refer to the diagram.**

35 Find the coordinates of the point symmetric to P with respect to the x-axis.
36 Find the coordinates of the point symmetric to P with respect to the point (0, 0).

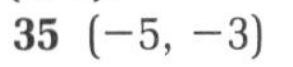

35 $(-5, -3)$ **36** $(5, -3)$

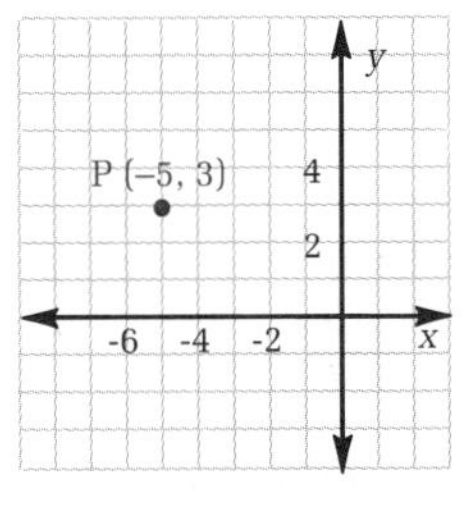

P,R **B** | **37** Find k so that $x^2 + kx + 25$ will be a perfect square. $k = \pm 10$

P,R | **38** Complete the square to find the vertex of the parabola described by $f(x) = x^2 + 6x + 12$. $(-3, 3)$

Problem-Set Notes and Strategies, continued

- In problem **25**, since the graph of $y = x + 2$ is parallel to $y = x$, the reflection of the first through the second is a translation of the line below $y = x$.
- In problems **28 – 31**, students may need to be reminded that the y-intercept is the point of the graph where $x = 0$.
- In problem **36**, symmetry with respect to the origin is the same as symmetry with respect to the x-axis and y-axis at the same time. Reflections and symmetry are discussed more fully in the next section.
- Some students will forget that there are two possibilities for problem **37** and only find the obvious one. In problem **44**, the point (2, 5) gives h and k directly. The second point (4, −3) is necessary to compute a, the stretch factor.

Assignment Guide

Average	
Day 1	13, 15, 17, 19, 21, 23–25, 28, 30, 34–36
Day 2	14, 16, 18, 20, 22, 26, 29, 31, 32, 37, 38
Day 3	27, 33, 39, 41, 43, 44, 48, 49, 52
Advanced	
Day 1	13–16, 23–25, 27, 28, 30, 32, 34–36
Day 2	17–22, 26, 29, 31, 33, 37, 38, 43
Day 3	39–42, 44–53
Honors	
Day 1	13, 15, 17–22, 25, 26, 28, 30, 33, 35, 36, 38, 39
Day 2	27, 34, 41, 42, 44, 45–47, 49, 52, 53, 55–57

Additional Answers

Answers to problems **25**, **27**, and **33** can be found in the answer pages beginning on **A1**.

6.2

Problem Set, *continued*

P,R **In problems 39–42, graph each equation without making a table.**

39 $f(x) = \frac{1}{2}|x - 3|$

40 $h(x) = \frac{1}{2}|x + 3|$

41 $g(x) = \frac{1}{2}|x| + 3$

42 $j(x) = \frac{1}{2}|x| - 3$

R **43** Solve for (x, y). (x, y) = (4, −2)

$$\begin{cases} 5x + 4y = 12 \\ 3x - 5y = 22 \end{cases}$$

P,R **44** Let $f(x) = a(x - h)^2 + k$. If the vertex of the graph of f is (2, −5) and the graph passes through (4, −3), what is f?
$f(x) = \frac{1}{2}(x - 2)^2 - 5$

P **In problems 45–47, use the graph of $y = f(x)$ to graph each equation.**

45 $y = f(x + 3)$
46 $y = f(x) - 2$
47 $y = f(x + 3) - 2$

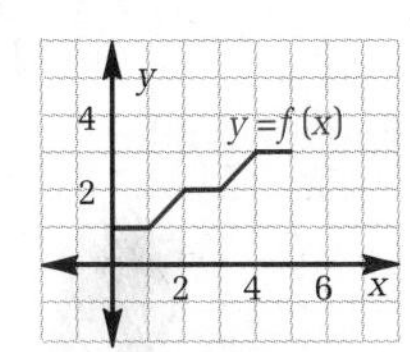

R **48** Mr. Ted trots down a different three of seven possible lanes each day. How many days will pass before he must repeat his choice of three lanes? 35 days

R **In problems 49 and 50, rewrite each expression as a power of x.**

49 $\frac{(x^{a+2})^5}{(x^{2a+1})^4}$ x^{6-3a}

50 $\left(\frac{x^{a+1}}{x^{b-1}}\right)^5$ $x^{5a-5b+10}$

R **51** Graph $3x - 2y < 19$.

P,R **52** Let $f(x) = x + 6$ and $g(x) = 2x - 3$.

a Find the point of intersection of the graphs of $y = f(x)$ and $y = g(x)$. (9, 15)

b Find the point of intersection of the graphs of $y = f(x + 4)$ and $y = g(x + 4)$. (5, 15)

P,R **53** If $f(x) = a(x - 3)^2 + k$ and the graph of f passes through (3, 10) and (6, −26), what are a and k? $a = -4$; $k = 10$

R **54** Let $f(x) = 2x - 3$ and $g(x) = f(-x)$. Graph f and g on the same set of axes.

Stumbling Block

As in the last section, adding values to the input of a function produces a translation that is counterintuitive. Adding to the input produces a left shift and subtracting produces a right shift. Have students verify their shifts with several points on the graph.

Problem-Set Notes and Strategies, continued

- Problems **49 – 50** are both algebraic manipulations of exponents.

Additional Answers

39

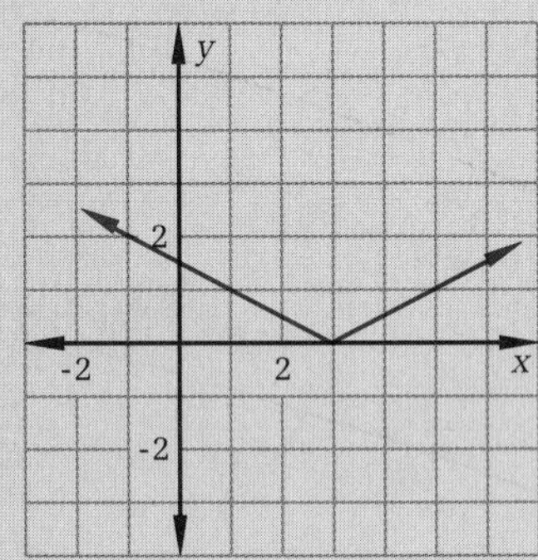

40

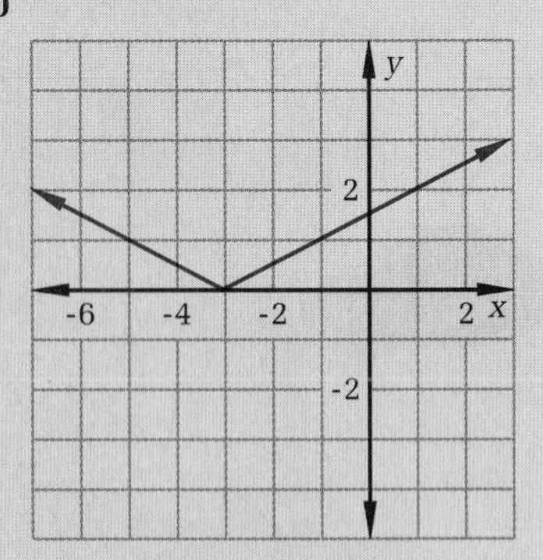

Answers to problems **41**, **42**, **45 – 47**, **51** and **54** can be found in the answer pages beginning on **A** 1.

C R **55** Let $F(1) = 1$ and $F(n + 1) = 2 \cdot F(n)$, where n is a positive integer.

a Copy and complete the table.

n	1	2	3	4	5	6	7	8
$F(n)$	1	2	4	8	16	32	64	128

b Write a closed-form formula for F. $F(n) = 2^{n-1}$

R **56** In the diagram, ABCD ~ BCFE.

a Write a proportion relating the sides. $\frac{x+1}{x} = \frac{x}{1}$

b Solve your proportion for x. $x = \frac{1+\sqrt{5}}{2} \approx 1.62$

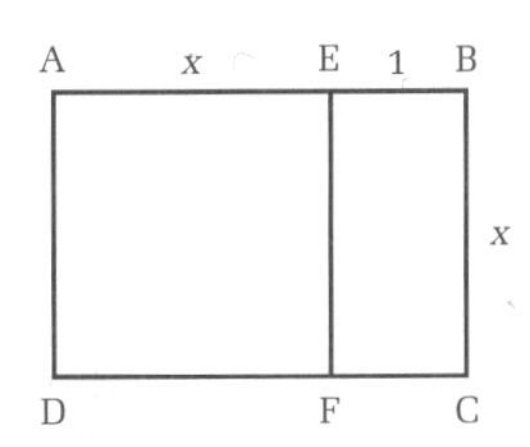

R **57** The formula $h(x) = -0.0035(x - 160)^2 + 90$ describes the height in feet of a baseball at a horizontal distance of x feet from home plate.

a How high does the ball go? 90 ft

b How far does the ball go? ≈320 ft

c When is the ball 50 feet above the ground? When $x \approx 53.1$ ft and $x \approx 266.9$ ft

Problem-Set Notes and Strategies, continued

- Problem **56** generates the golden ratio once again.

Using Computers

This is another section in which computer graphing software or a graphics calculator enables students to see the translations of graphs. This provides excellent reinforcement of concepts. The *Omnifarious Plotter* is very useful here.

HISTORICAL SNAPSHOT

Fun with Graphs

Who was Descartes when he was at home?

Have you ever wondered what fun is to a mathematician? Mathematical fun often means solving problems that have no real applications. Such problems are usually extremely puzzling, and their solutions are often amusing. Two such problems, which you can learn more about on your own, are the seven bridges of Königsburg (see Section 1.3) and Penrose tilings. For now, we will concentrate on playful graphing.

René Descartes (1596–1650) discovered that an equation can be written for every geometric shape and that there is a geometric representation for every equation. The first set of examples shows the graphs of various quadratics, with which you are now somewhat familiar.

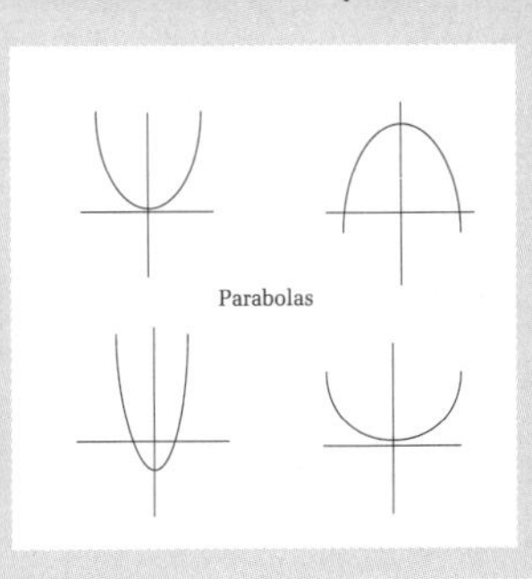

Parabolas

Having made his discovery, Descartes applied himself to writing equations that would yield interesting graphs. Other mathematicians, including Johann Bernoulli (1667–1748), Blaise Pascal (1623–1662), and the eighteenth century Guido Grandi, joined the fun, and the results include the four graphs below.

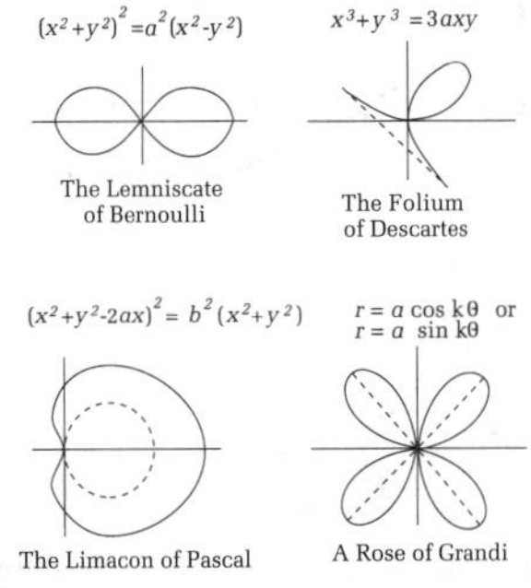

The Lemniscate of Bernoulli

The Folium of Descartes

The Limacon of Pascal

A Rose of Grandi

6.3

Class Planning

Time Schedule
All levels: 3 days

Resource References
Teacher's Resource Book
Class Opener 36
Cooperative Learning 18
Practice 37
Enrichment 12
Transparency 12
Evaluation
Quiz Forms 1, 2, 3

Class Opener

1 Sketch the graph of $f(x) = (x-2)^2$ and $f(x) = -(x-2)^2$ on the same axes. Write a sentence describing the relationship between the two graphs.

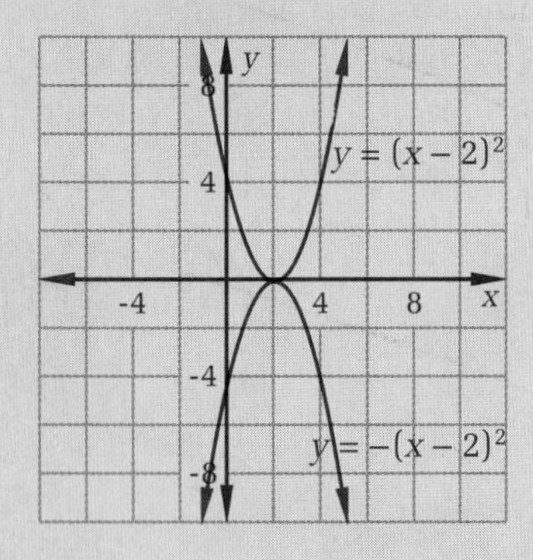

Answers will vary. One possible answer is: the graph of $y = (x-2)^2$ reflected over the x-axis.

2 Write an equation of the graph that is symmetric to the graph of $f(x) = (x-2)^2$ about the y-axis.
$y = (-x-2)^2$

6.3 REFLECTIONS AND SYMMETRY

Objectives

After studying this section, you will be able to

- Reflect objects graphed on a coordinate system
- Recognize symmetry with respect to a line or a point

Part One: Introduction

Reflections

We are familiar with a ***reflection*** as a mirror image. In mathematics, an object can be reflected over a line or a point. For the reflections of triangle ABC, only the vertices are labeled, but every point of the triangle is reflected. In the diagram we see that when an object is reflected over the y-axis, each point (a, b) is reflected to $(-a, b)$. When an object is reflected over the x-axis, each point (a, b) is reflected to $(a, -b)$.

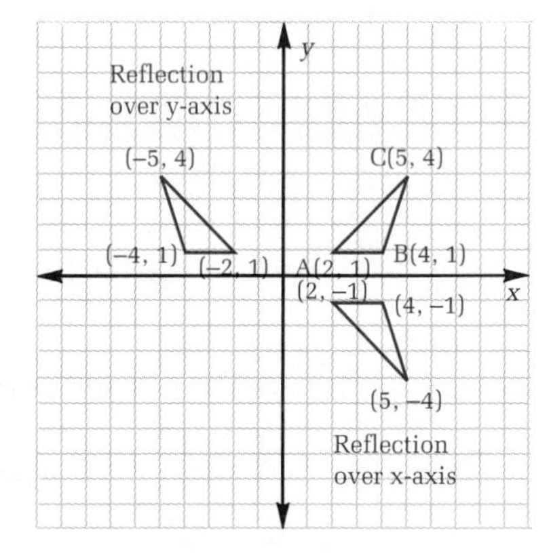

Now look at the following reflections.

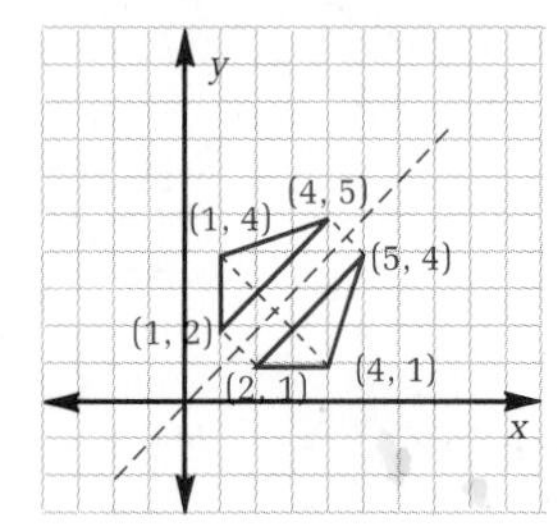

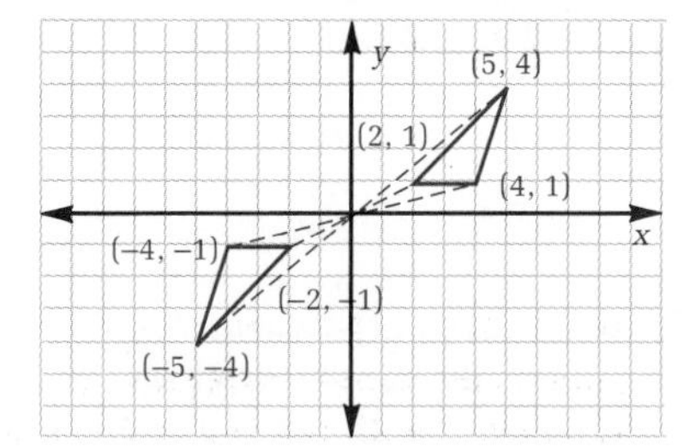

Vocabulary
reflection
symmetry

When an object is reflected over the line $y = x$, each point (a, b) is reflected to (b, a). When an object is reflected over the origin, each point (a, b) is reflected to $(-a, -b)$.

Example *Write an equation for the reflection over the x-axis of the graph of* $y = (x - 3)^2 + 4$.

The reflection over the x-axis of any point (x, y) is $(x, -y)$. To find the equation of the graph's reflection, we substitute $-y$ for y in the original equation. The value of x is unchanged.

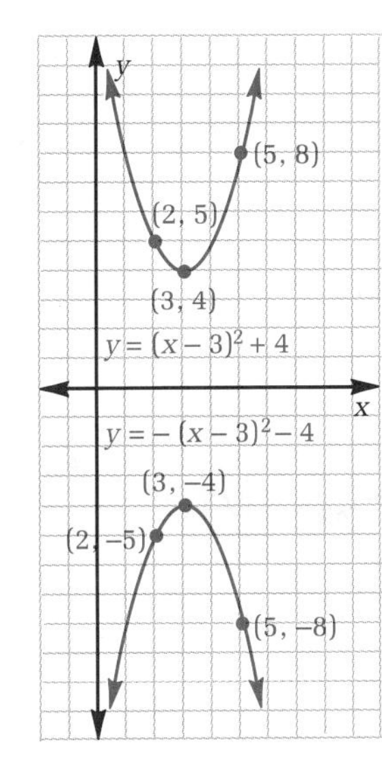

$$-y = (x - 3)^2 + 4$$
$$y = -[(x - 3)^2 + 4]$$
$$y = -(x - 3)^2 - 4$$

To check that $y = -(x - 3)^2 - 4$ is the equation of the reflection, we verify that this graph has the vertex $(3, -4)$ and opens downward.

Symmetry

Some graphs appear unchanged when they are reflected over a line or a point.

Example *Find the reflection over the y-axis of the graph of* $f(x) = x^2 + 4$.

The reflection of a point (x, y) over the y-axis is $(-x, y)$. Let's graph f and reflect several points. In function f, if we replace x with $-x$, the output is unchanged, since $x^2 = (-x)^2$. Therefore, the reflection is the same as the original graph.

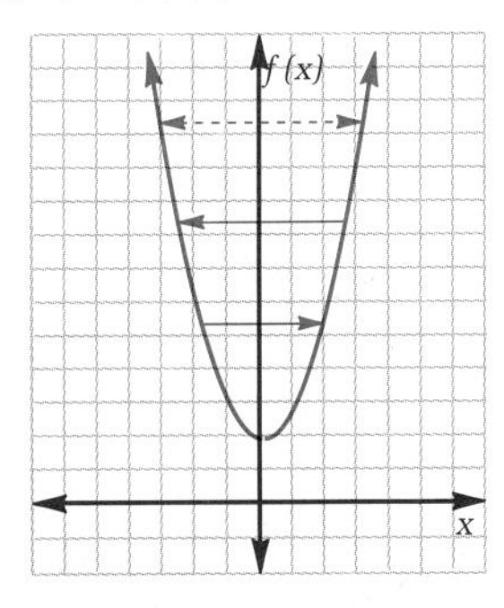

When the result of reflecting a graph over a line or point is the same graph, we say that the graph is symmetric with respect to that line or point.

Lesson Notes

- Symmetry is discussed in this section of this chapter. It is explored both geometrically and algebraically. If students can recognize symmetry in a graph, their work can be simpler. Remind students that symmetry can be with respect to a line or a point.

Communicating Mathematics

Write a paragraph describing the significance of recognizing symmetry in graphing equations. If you knew a graph was symmetric with respect to the *y*-axis, how would you graph it?

Using Computers

As in earlier sections of this chapter, graphing software and a graphics calculator provide a way for students to see graphs reflect over a line. For many students, this visualization reinforces understanding.

6.3

Cooperative Learning

A graphing calculator or computer graphing program will be helpful in this exercise.

1 Graph the function $y = 2$. Does the graph have any symmetry?
2 Graph the function $y = x$ and again note the symmetry.
3 Graph the functions $y = x^2$, x^3, x^4, and x^5 and note any symmetry these graphs might have.
4 What generalizations can you make about the graphs and the exponents?
5 Graph $y = x^5 + x^3 + x$ and note the symmetry.
6 Graph $y = x^4 + x^2 + 2$ and note the symmetry.

Try to generalize any symmetry about polynomials and their exponents. You may graph other combinations if it helps.

Students should conclude that if all exponents are odd the function is symmetric to the origin and if all exponents are even, the graph is symmetric to the y-axis.

Stumbling Block

For students, symmetry about the origin is often one of the more difficult of the symmetries. It is found by replacing x with $-x$ and y with $-y$ and reevaluating the function. Suggest that students think of folding the graph once on the x-axis and the again on the y-axis. The location of the point will be symmetric with respect to the origin.

Part Two: Sample Problems

Problem 1 *Which of the following graphs are symmetric with respect to the x-axis, the y-axis, the line y = x, and the origin?*

a

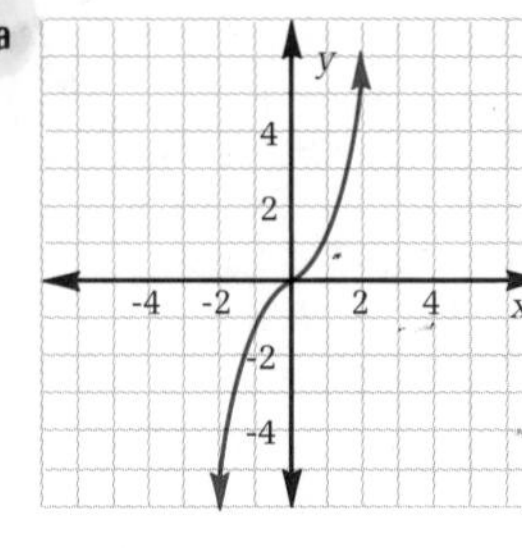

b

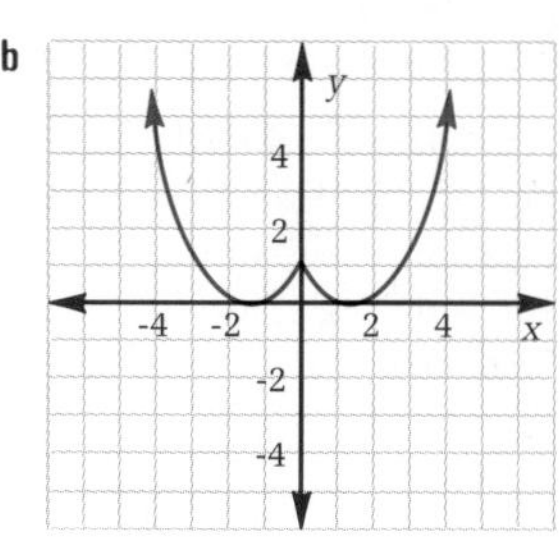

c

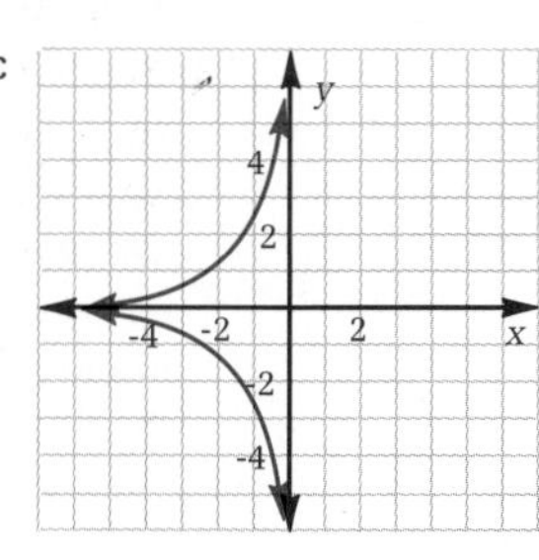

d

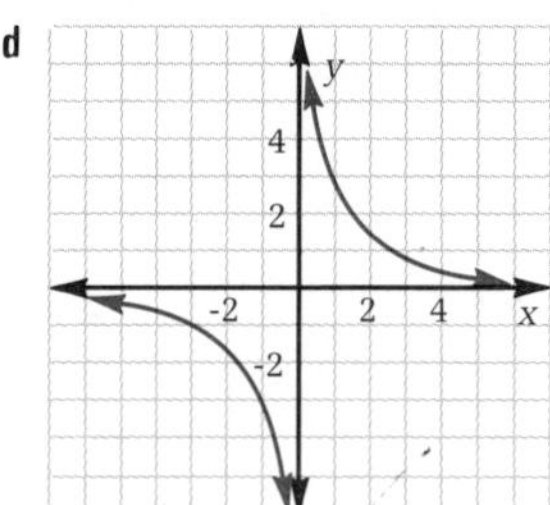

e

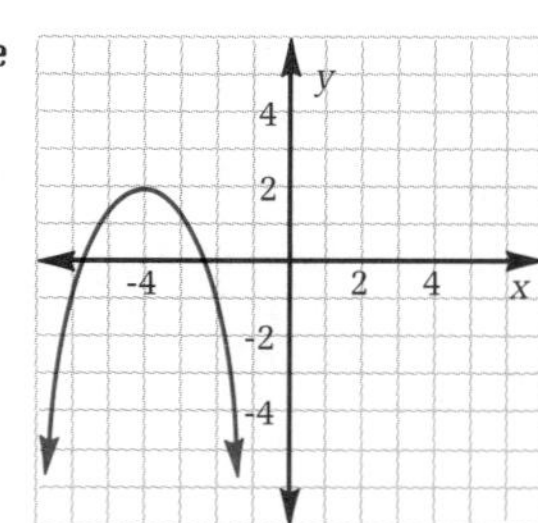

f

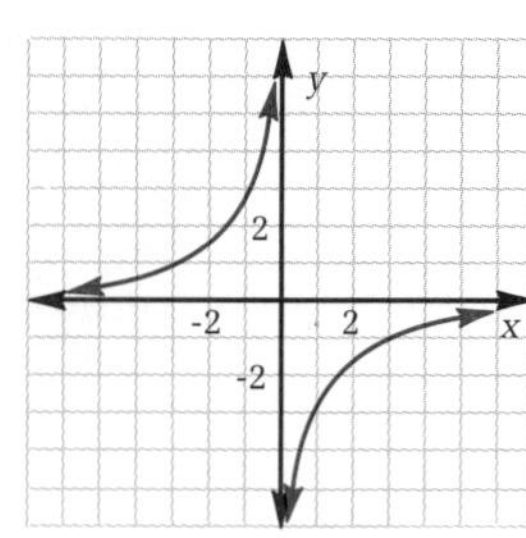

Solution Graph **c** is symmetric with respect to the x-axis. Graph **b** is symmetric with respect to the y-axis. Graphs **d** and **f** are symmetric with respect to y = x. Graphs **a, d,** and **f** are symmetric with respect to the origin.

Although graph **e** is not symmetric with respect to the y-axis, the x-axis, y = x, or the origin, it is symmetric with respect to the line x = −4. Recall from Section 4.3 that this line is called the axis of symmetry of the parabola.

Problem 2 *Write the equation of function g, which represents the reflection of the graph of* $f(x) = \frac{1}{4}(x - 5)^2 + 7$ *over the origin.*

Solution We can find g graphically or algebraically.

Method 1
Reflect the vertex and several other points of the graph of f over the origin. We can see that the reflected parabola has a vertex at $(-5, -7)$, opens downward, and is stretched vertically by the same factor as the graph of f. Therefore, the equation of g is

$$g(x) = -\frac{1}{4}(x + 5)^2 - 7.$$

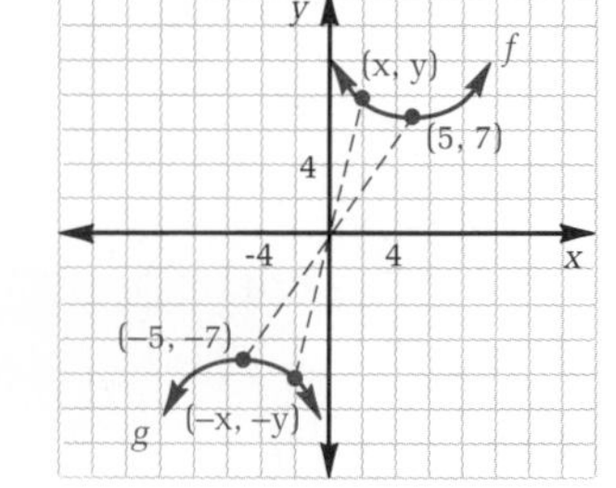

Study the graphs of f and g. A reflection over a point is also called a half-turn. Can you see why?

Method 2

When the graph of $y = \frac{1}{4}(x - 5)^2 + 7$ is reflected over the origin, each point (x, y) is reflected to (−x, −y). To find the equation of the reflected graph, we replace x with −x and y with −y, then solve for y.

$$-y = \frac{1}{4}(-x - 5)^2 + 7$$

$$-y = \frac{1}{4}[-1(x + 5)]^2 + 7$$

$$-y = \frac{1}{4}(x + 5)^2 + 7$$

$$y = -\frac{1}{4}(x + 5)^2 - 7$$

Thus, $g(x) = -\frac{1}{4}(x + 5)^2 - 7$.

Problem 3 *Graph $y = 3^x$, its reflection over the y-axis, and its reflection over the line $y = x$.*

Solution To graph $y = 3^x$, use a graphing device or make a table of values and plot points.

To find the equation of the graph's reflection over the y-axis, replace x with −x.

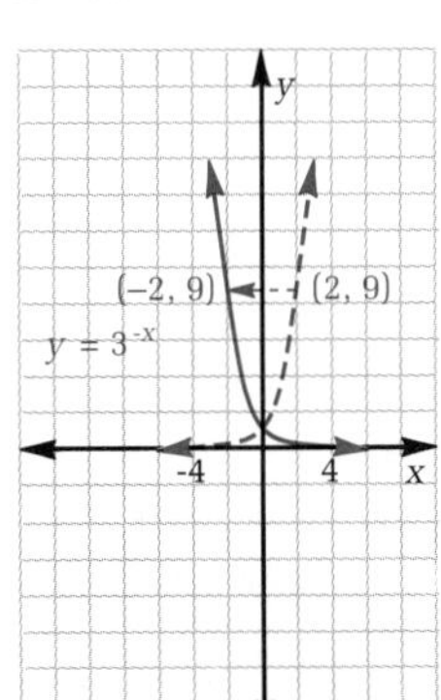

To find the equation of the graph's reflection over the line y = x, exchange y and x.

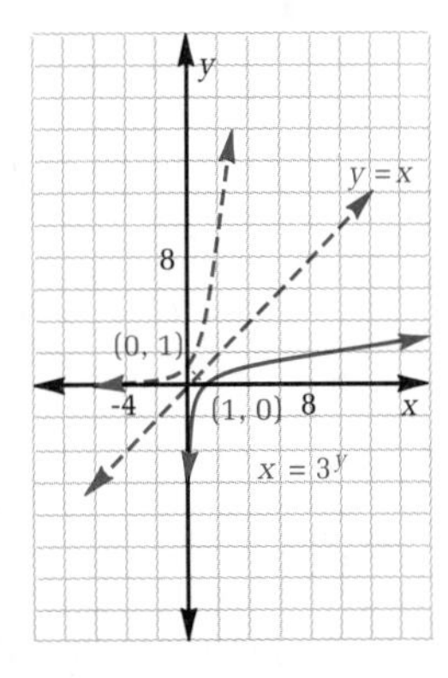

Remember that the reflection of a function's graph over the line $y = x$ is the inverse of the function. The equation of the inverse of $y = 3^x$ is $x = 3^y$, which is graphed at the right above.

Checkpoint

1 The following graph is symmetric with respect to what line(s)?

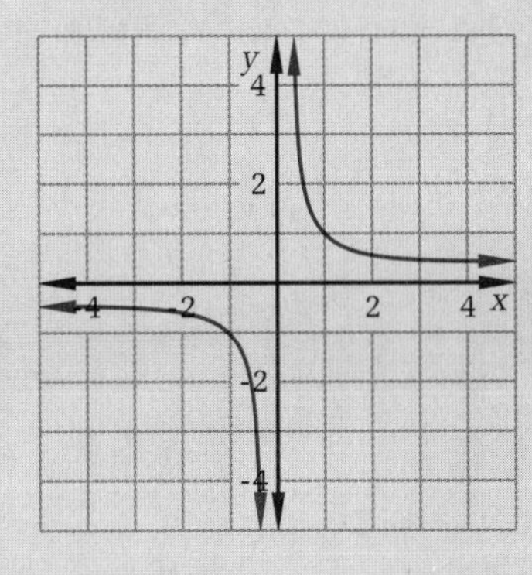

The line $y = x$ and the line $y = -x$.

2 The parabola $y = (x - 4)^2$ has an axis of symmetry about which line? $x = 4$

3 For the given function, find the graph that is symmetric with relation to the origin. $y = 3x - 7$ $y = 3x + 7$

4 Reflect the following graph over the x-axis.

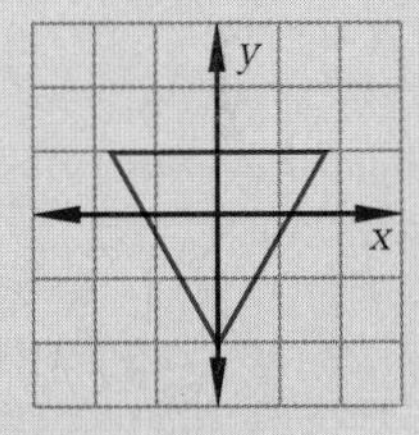

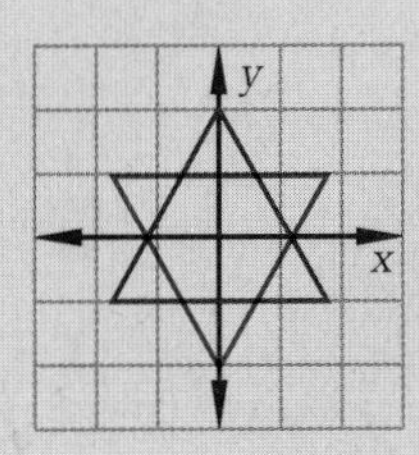

6.3

Assignment Guide

Average	
Day 1	13, 15, 20, 22, 23, 27, 28, 31, 32
Day 2	14, 18, 19, 26, 29, 34, 36, 38, 42
Day 3	16, 17, 21, 24, 25, 30, 33, 35, 37, 39, 40
Advanced	
Day 1	13, 15, 17, 19, 22, 23, 26, 27, 29 – 31
Day 2	14, 16, 18, 20, 21, 25, 28, 32, 34, 35, 38
Day 3	24, 33, 36, 37, 39 – 43
Honors	
Day 1	13–18, 20, 22–27, 30
Day 2	19, 21, 29, 31, 35, 37, 38, 40, 42
Day 3	32–34, 36, 39, 41, 43, 44

Problem-Set Notes and Strategies

- In problem **10**, the coordinates (x, y) are replaced with $(x, -y)$ and the equation solved for y.
- In problem **11**, the coordinates (x, y) are replaced with $(-x, y)$ and the equation solved for y. Answers may vary as the text answer is equivalent to $(-x-6)^2+4$.
- In problem **12**, the coordinates (x, y) are replaced with $(-x, -y)$ and the equation solved for y.

Part Three: Exercises and Problems

Warm-up Exercises

In problems 1–4, give the coordinates of each point.

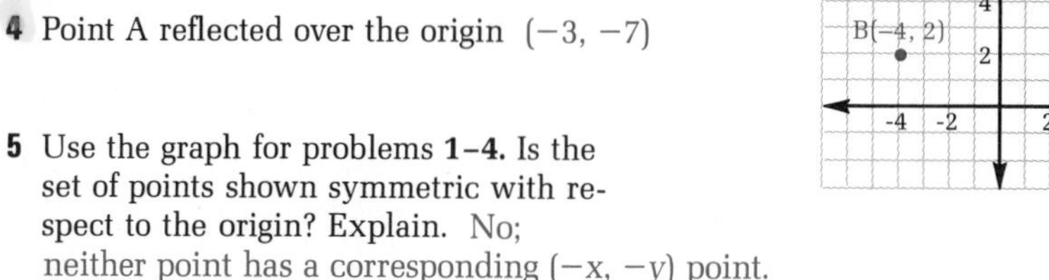

1 Point A reflected over the x-axis $(3, -7)$

P **2** Point B reflected over the line $y = x$ $(2, -4)$

3 Point B reflected over the y-axis $(4, 2)$

4 Point A reflected over the origin $(-3, -7)$

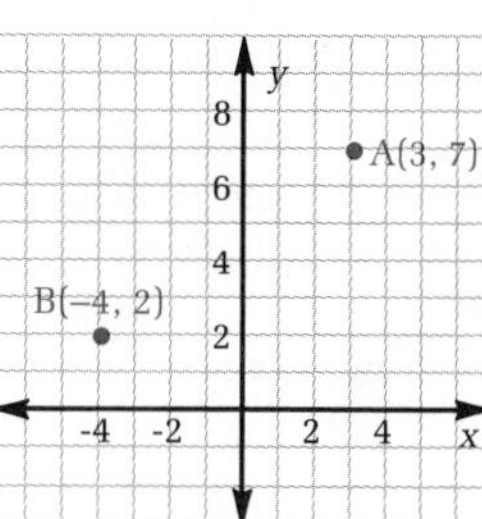

5 Use the graph for problems **1–4.** Is the set of points shown symmetric with respect to the origin? Explain. No; neither point has a corresponding $(-x, -y)$ point.

P **In problems 6–9, determine the equation of the axis of symmetry of each graph.**

6

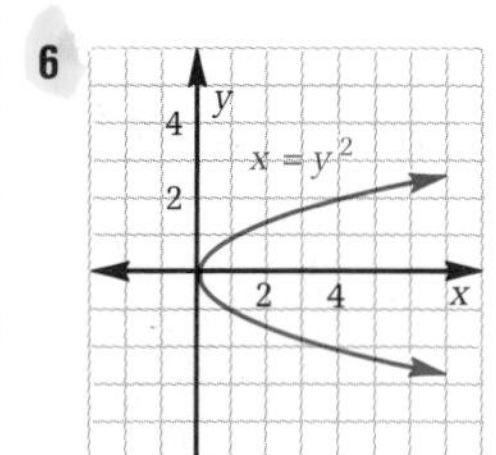

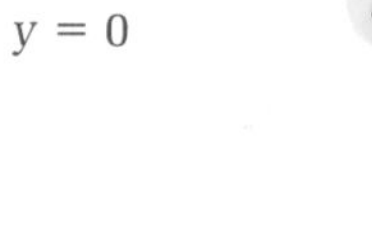

$y = 0$

7

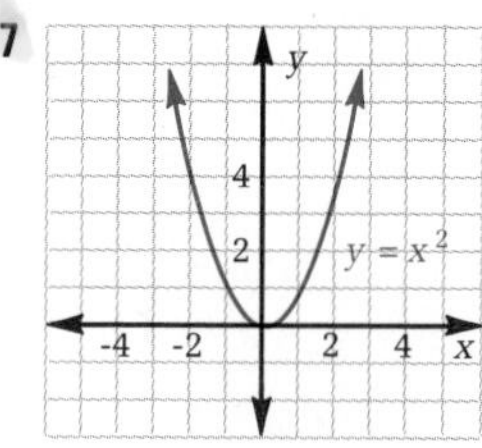

$x = 0$

8

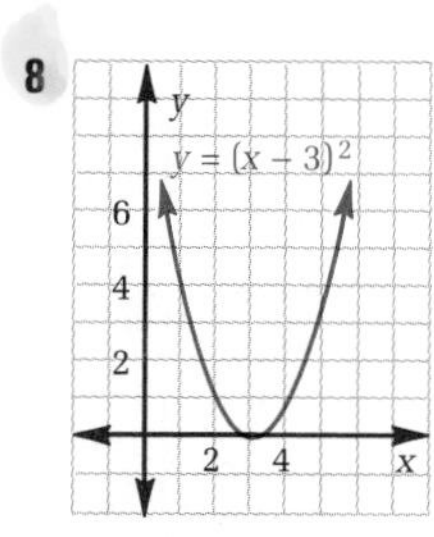

$x = 3$

9

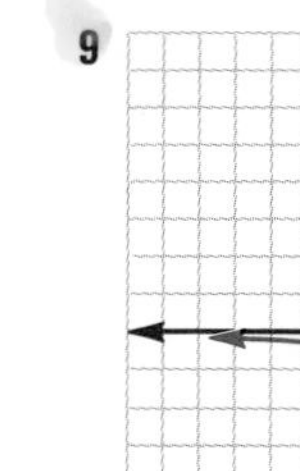

$y = x$ or $y = -x$

P **In problems 10–12, write the equation of the reflection of the graph of $y = (x - 6)^2 + 4$ over each line or point.**

10 The x-axis $y = -(x - 6)^2 - 4$

11 The y-axis $y = (x + 6)^2 + 4$

12 The origin $y = -(x + 6)^2 - 4$

Problem Set

A

P **In problems 13–18, give the coordinates of the reflection of the point (−2, 5) over each line or point.**

13 The x-axis (−2, −5)

14 The line $y = -x$ (−5, 2)

15 The y-axis (2, 5)

16 The line $y = 4$ (−2, 3)

17 The line $y = x$ (5, −2)

18 The origin (2, −5)

P **In problems 19–21, determine the equation of the axis of symmetry of each graph.**

19

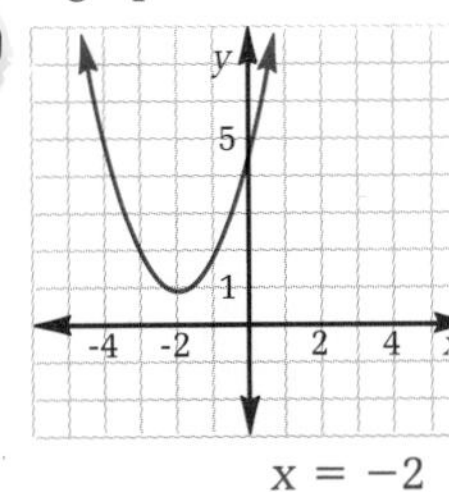

$x = -2$

20

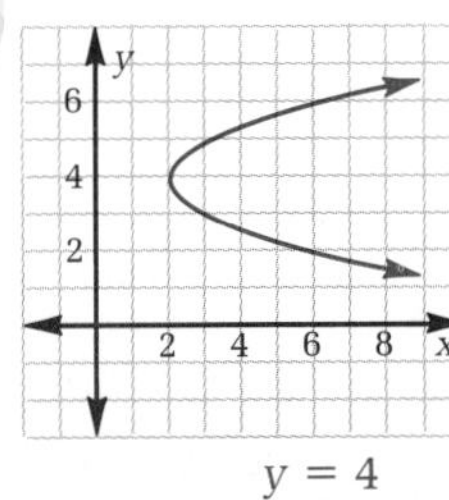

$y = 4$

21

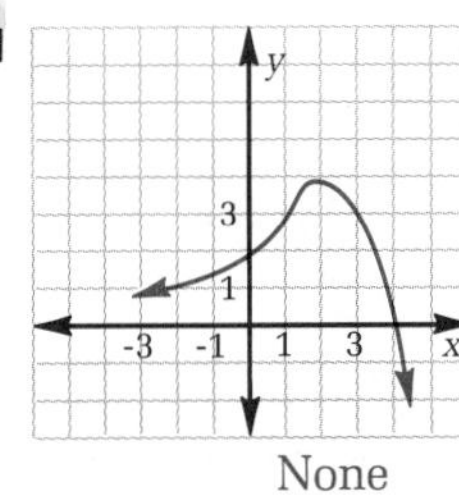

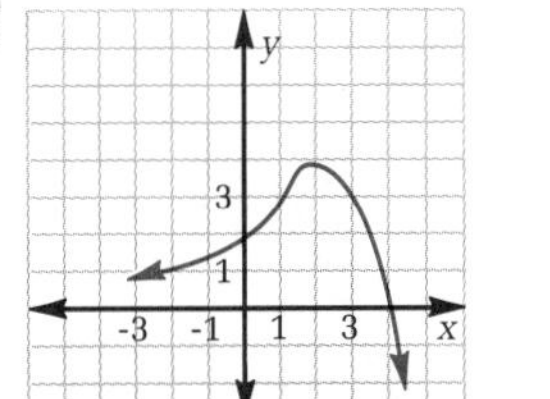

None

R **22** Write a linear function in the form $f(x) = ax + b$ to represent the table below. $f(x) = 3x + 5$

x	−4	−3	−2	−1	0	1	2	3	4
$f(x)$	−7	−4	−1	2	5	8	11	14	17

R **In problems 23–25, graph each set on a number line.**

23 $\{x{:}-2 < x < 5\}$

24 $\{x{:}x \geq 2\}$

25 $\{x{:}x \neq 4\}$

P **26** The coordinates of the vertices of a triangle are (−2, 5), (3, 7), and (6, 1). What are the coordinates of the vertices of the triangle's reflection over

a The x-axis? (−2, −5), (3, −7), (6, −1)

b The origin? (2, −5), (−3, −7), (−6, −1)

B

P **27** Let $f(5) = 7$.

a If g is the reflection of f over the x-axis, what is g(5)? −7

b If h is the reflection of f over the y-axis, what is h(−5)? 7

c If j is the reflection of f over the line $y = x$, what is j(7)? 5

d If k is the reflection of f over the origin, what is k(−5)? −7

R **28** Dick mixed 6 liters of a solvent with 24 liters of water. What percentage of the solution was solvent? 20%

P **29** **a** Graph $y = 2^x$.

b Graph its reflection over the y-axis.

c Graph its reflection over the line $y = x$.

Problem-Set Notes and Strategies, continued

- Students frequently answer problem **28** as 25% since 6 is one fourth of 24. But the problem asks what percent of the solution was solvent and there are 30 liters of solution.

Additional Answers

23 **24** **25**

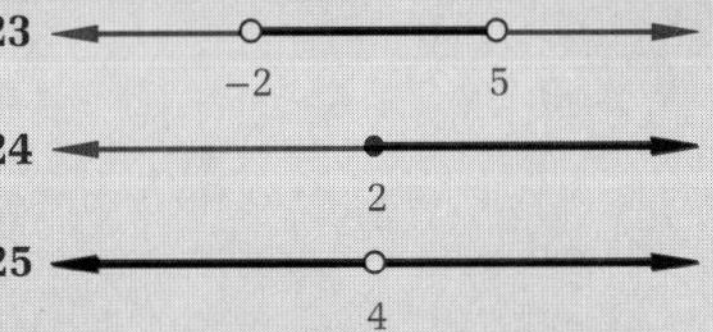

29a

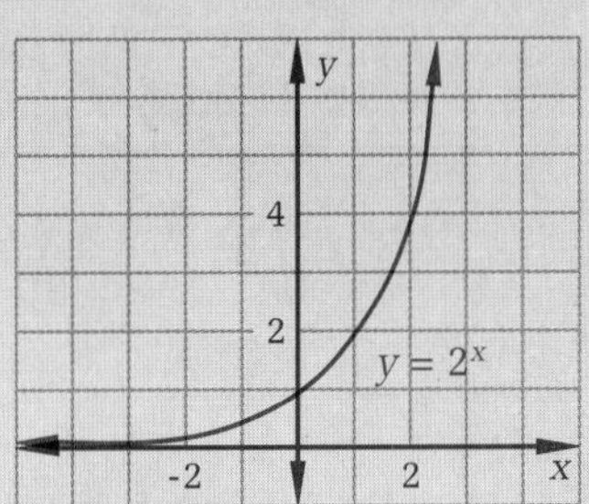

29b

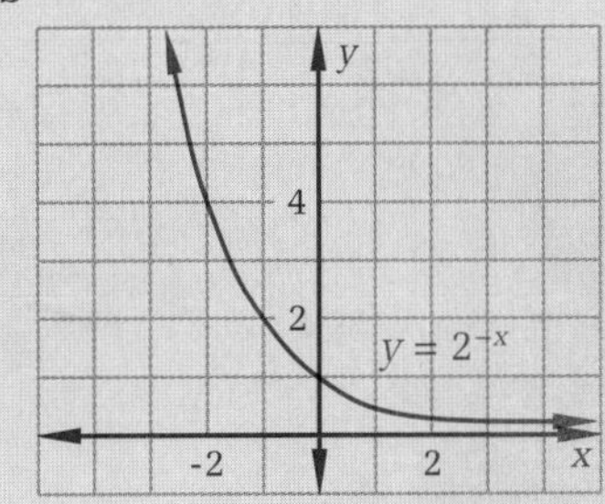

29c

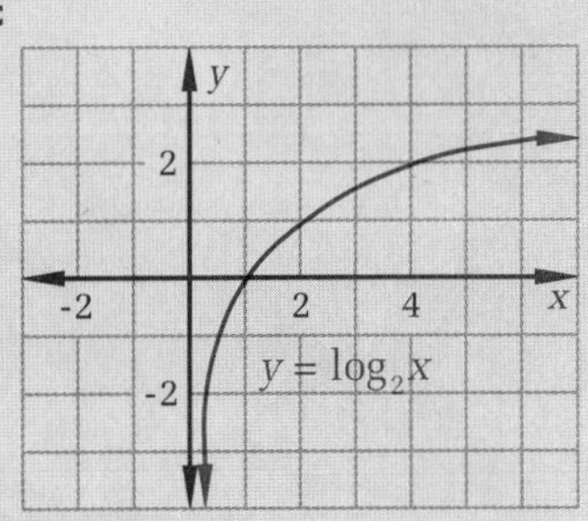

Problem-Set Notes and Strategies, continued

- Problem **33** will yield many geometric interpretations of an isosceles triangle.
- Problem **35** is a nice exercise in multiple transformations. If the student ends up at the correct point, he or she has demonstrated a good working knowledge of symmetry.

Additional Answer

34

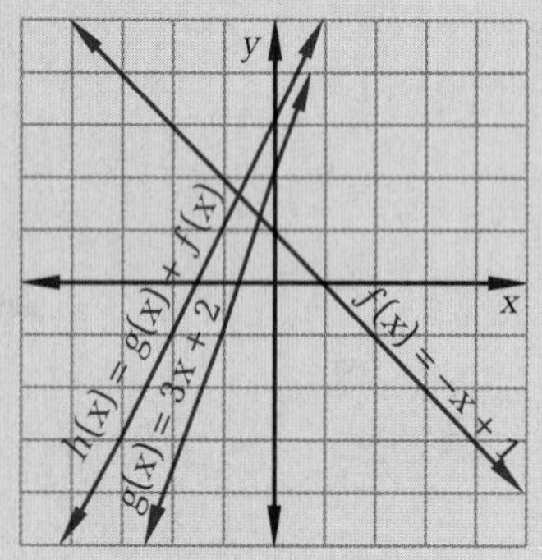

Problem Set, *continued*

R **30** The point $(-4, 3)$ is on the graph of $y = f(x)$. Find the coordinates of a point on the graph of $y = f(x - 2) + 5$. $(-2, 8)$

P **31** Let $f(x) = x^2$.

a Find the equation of the reflection of the graph of $y = f(x)$ over the y-axis. $y = x^2$

b Find the equation of the reflection of the graph of $y = f(x + 2)$ over the y-axis. $y = (x - 2)^2$

R **32** Paula charges a fee of \$55 plus \$20 per hour to paint rooms in a house.

a Write a formula to describe the cost if the job takes h hours. $C(h) = 20h + 55$

b How many hours must Paula work in order to earn at least \$150? At least 4.75 hr

P **33** Identify at least four characteristics of the figure shown. Answers will vary. Possible answer: A and A′ are reflections over $y = x$; $AB = A'B$; $\overline{AA'} \perp$ line $y = x$; M is the midpoint of $\overline{AA'}$.

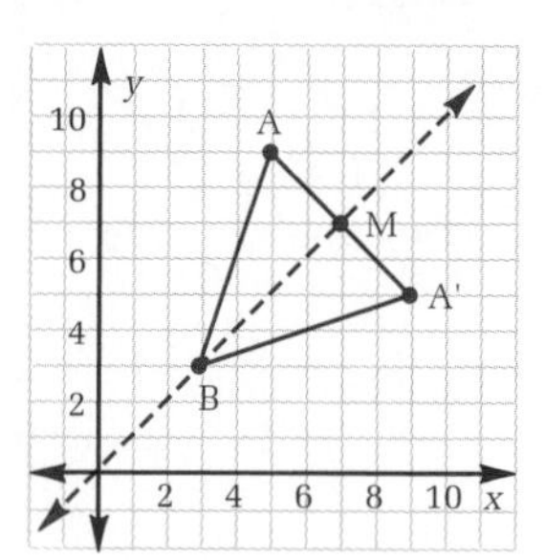

R,I **34** Graph the functions f, g, and h on the same axes, where $f(x) = -x + 1$, $g(x) = 3x + 2$, and $h(x) = f(x) + g(x)$.

P,I **35** If the point $(-3, 4)$ is reflected in succession about the x-axis, the line $y = x$, and the origin, what will be the coordinates of the final point? $(4, 3)$

R **36** If $f(A) = 3A$, where A is a matrix, what is the value of $f(A)$ for the input shown?

$$A = \begin{bmatrix} 1.4 & -4 \\ 6 & \frac{2}{3} \end{bmatrix} \quad f(A) = \begin{bmatrix} 4.2 & -12 \\ 18 & 2 \end{bmatrix}$$

P **37** Half of the parabola $y = f(x)$ is shown, along with the parabola's axis of symmetry. One of the values of x for which $f(x) = 0$ is $4 + \sqrt{3}$. Find the other value of x for which $f(x) = 0$.
$x = 4 - \sqrt{3}$

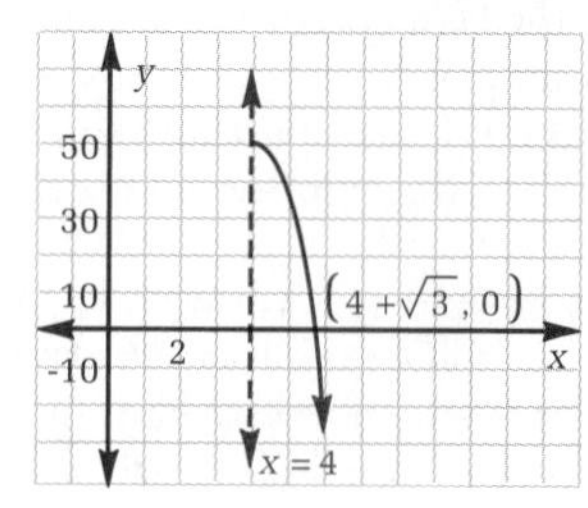

P In problems 38–41, identify the line(s) of symmetry of the graph of each function.

38 $f(x) = x^2 + 4$ $x = 0$

39 $f(x) = \frac{1}{x}$ $y = x$ or $y = -x$

40 $f(x) = |x + 5| - 3$ $x = -5$

41 $f(x) = 2x + 3$ $y = -\frac{1}{2}x + k, k \in \mathcal{R}$

R **42** Let $a = 13$ and $b = 5$.

a Compute $|a - b|$. 8

b Compute $|b - a|$. 8

c Is it true that for any x and y, $|x - y| = |y - x|$? Why or why not?
Yes; both are the distance between x and y on a number line.

C P,R **43 a** Find the area of $\triangle ABC$. 20

b Let A′ be the reflection of A over the y-axis. Find the area of $\triangle A'BC$. 20

c Let A″ be the reflection of A over the x-axis. Find the area of $\triangle A''BC$. 25

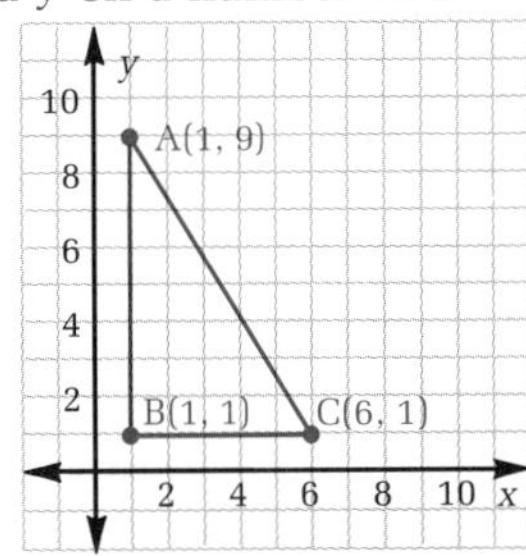

R **44** Let $N(x) = 4x - 3$. Find $\frac{N(a + b) - N(a)}{b}$ for any real a and b such that $b \neq 0$. 4

Problem-Set Notes and Strategies, continued

- A student may also list the line of symmetry as $y = 2x + 3$ for problem **41**. Every line is symmetric with respect to itself.
- Problem **43** yields the slope of the linear equation and is part of the definition of the derivative in elementary calculus.

CAREER PROFILE

INVERSE VARIATION IN THE DESERT

Joel Brown applies mathematics to ecology

Joel Brown, an associate professor of biological sciences at the University of Illinois at Chicago, has studied extensively desert ecology, focusing on how long desert rodents feed in a certain area. Dr. Brown has written and published articles on biology, ecology, biomathematics, and applied mathematics. He uses calculus, vectors, and sequences and series.

The graph shows the fitness of a forager—a kangaroo rat, gerbil, fox squirrel, or other animal that eats primarily seeds and grains—as the animal spends time in a particular feeding area, or *patch*. The x-axis represents energy gained in the patch by eating. The y-axis represents the probability of surviving predation—the probability that the animal will *not* be caught by, say, an owl. The energy the animal gains by harvesting the patch is inversely related to the animal's safety there.

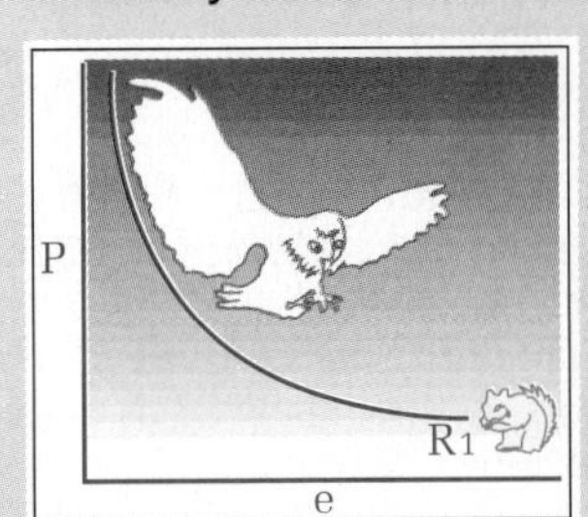

Dr. Brown holds a bachelor's degree in zoology from Pomona College in Claremont, California, and a doctorate in ecology and evolutionary biology, with a minor in economics, from the University of Arizona in Tucson.

6.4

Class Planning

Time Schedule
Average: Optional
Advanced: 2 days
Honors: 2 days

Resource References
Teacher's Resource Book
Class Opener 37
Practice 38

Class Opener

Graph $y = \sqrt{x}$.

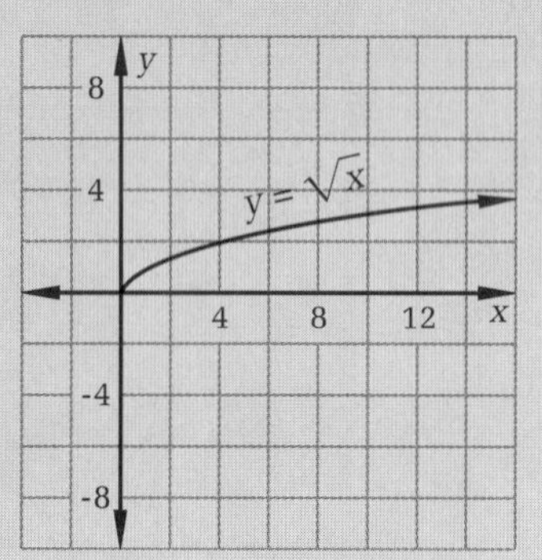

Lesson Notes

- Section **6.4** explores higher degree polynomials and reciprocal and exponential functions. Properties of polynomials such as symmetry and the possible roots are examined. Reciprocal graphs also have symmetry and a restricted domain since there are variables in a denominator. Exponential functions are introduced with real numbers for a domain and positive numbers for a range.

6.4 GRAPHING OTHER FUNCTIONS

Objectives

After studying this section, you will be able to

- Graph cubic and quartic polynomial functions
- Graph reciprocal functions
- Graph exponential functions

Part One: Introduction

Cubic and Quartic Polynomial Functions

The graph of the cubic function $f(x) = x^3$ is shown. A ***cubic function*** is a polynomial function of degree three.

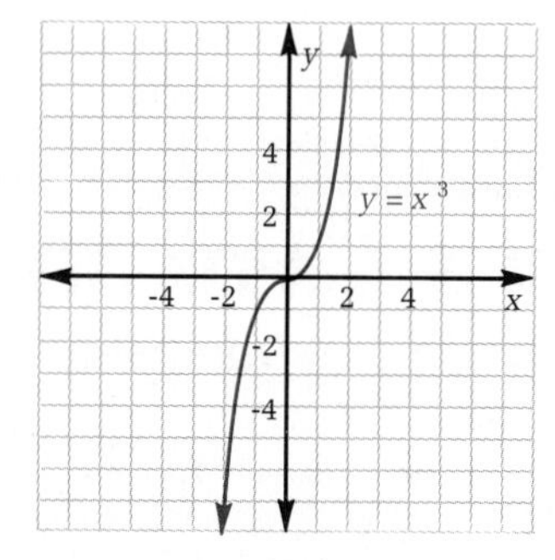

Using what you have seen about stretching, shrinking, reflecting, translating, and symmetry, predict the graphs of
$g(x) = f(x - 4) = (x - 4)^3$,
$h(x) = -5[f(x)] = -5x^3$, and
$j(x) = f(x) + 3 = x^3 + 3$. Check your predictions by graphing these functions.

Study this graph of the cubic function $p(x) = x^3 - 6.5x^2 + 0.5x + 35$.

The graph of p has three x-intercepts. As you learned in Chapter 4, these intercepts correspond to the real zeros of p and the real roots of $p(x) = 0$. From a graph, we can estimate the zeros of p to be -2, 3.5, and 5. A cubic function can have at most three real zeros. How many zeros does $f(x) = x^3$ have?

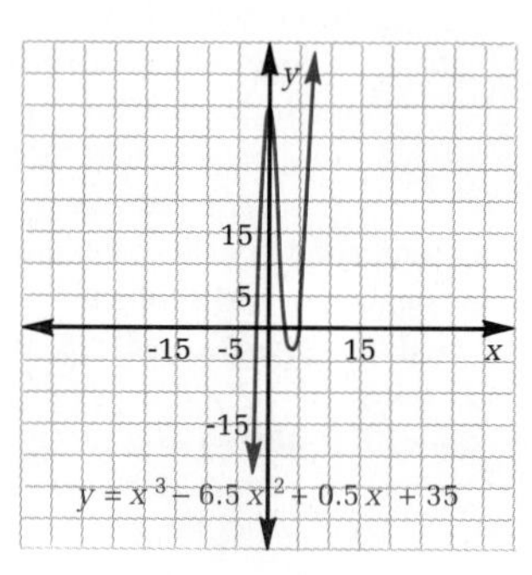

Vocabulary
cubic function
quartic function
reciprocal function
exponential function

Below are some graphs of polynomial functions of degree four, called ***quartic functions***

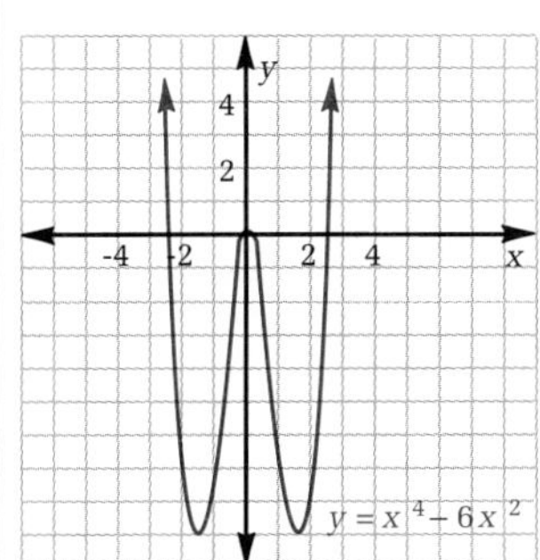

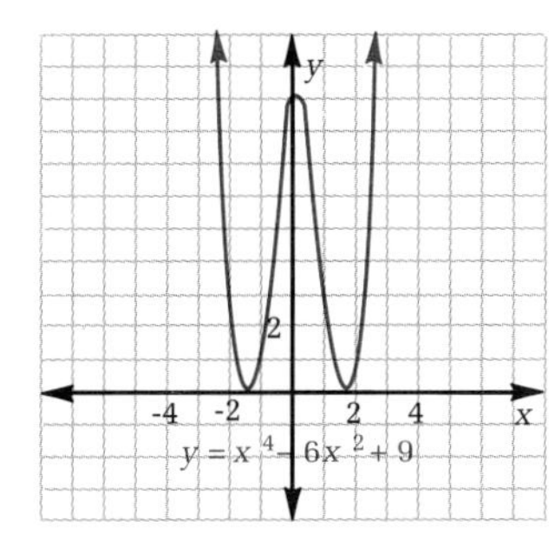

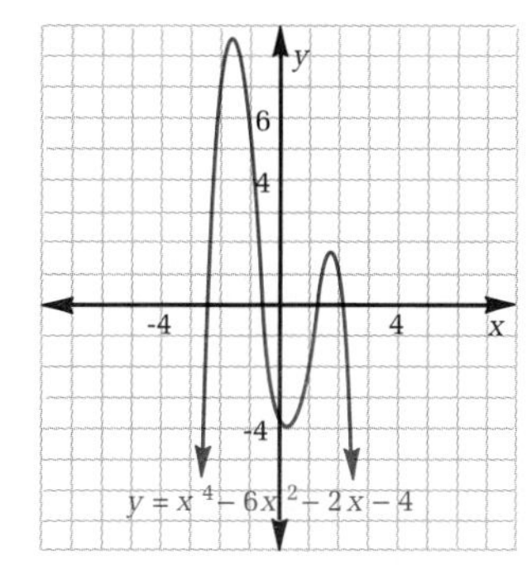

The graphs of many quartic functions have a characteristic W shape and at most four x-intercepts.

Reciprocal Functions

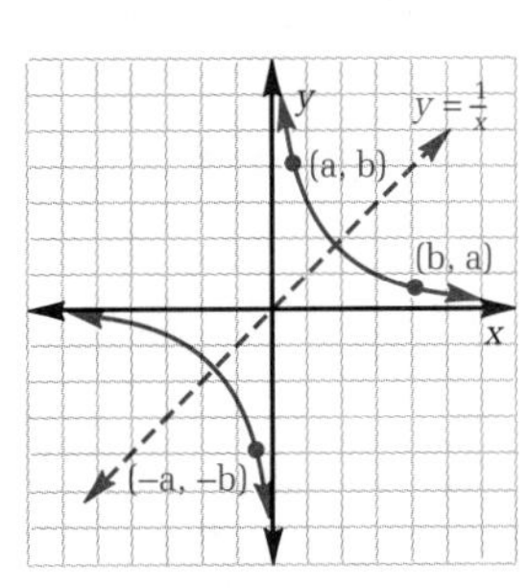

Study the graph of $y = \frac{1}{x}$, called a ***reciprocal function***, since y is the reciprocal of x. Because this function is undefined for $x = 0$, its domain is all real numbers except 0. Some ordered pairs on the graph are (1, 1), (−1, −1), $\left(2, \frac{1}{2}\right)$, $\left(\frac{1}{2}, 2\right)$, $\left(-2, -\frac{1}{2}\right)$, and $\left(-\frac{1}{2}, -2\right)$. If the values of x and y are interchanged, another point on the graph is always produced, so the graph is symmetric with respect to the line $y = x$.

If (x, y) is on the graph, the point (−x, −y) is also on the graph. Therefore, the function is also symmetric with respect to the origin. From the graph we see that the range of the function is all real numbers except 0.

Example *Let $f(x) = \frac{1}{x}$. Graph the function h, where $h(x) = f(x - 3)$.*

First let's solve for $h(x)$. $h(x) = f(x - 3)$

$$= \frac{1}{x - 3}$$

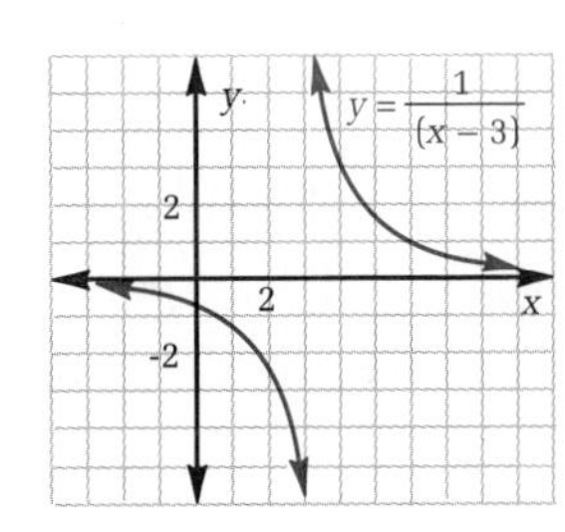

Since $h(x)$ is undefined when $x = 3$, the domain of h is all real numbers except 3. The graph of h is the graph of f shifted three units to the right. The graph is symmetric with respect to the line $y = x - 3$ and the point (3, 0). From the graph we see that the range of h is all real numbers except 0.

6.4

Communicating Mathematics

Find an application of an exponential function and come to class prepared to present it. Bring a graph of the function and explain the real world significance of the points.

Exponential Functions

Graph the ***exponential functions*** $f(x) = 2^x$, $g(x) = 10^x$, and $h(x) = 0.5^x$. If you do not have a graphing calculator or a computer with a graphing program, generate tables of values, perhaps using the x^y key on a scientific calculator. Be sure to include negative values, 0, and positive values for x.

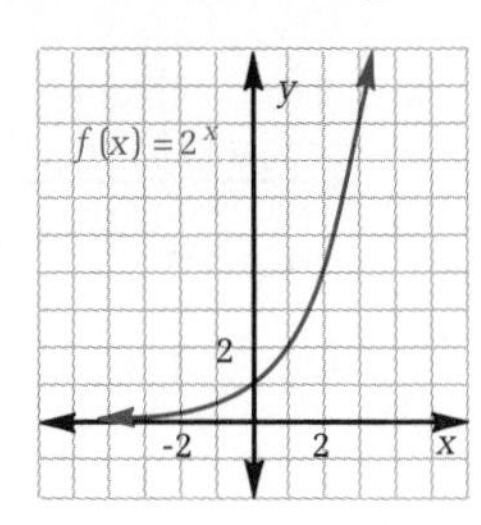

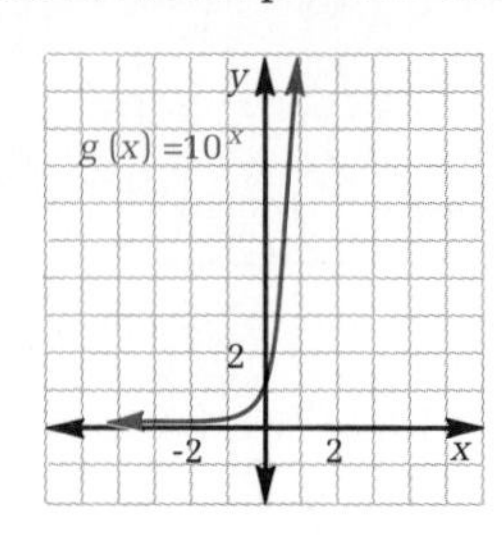

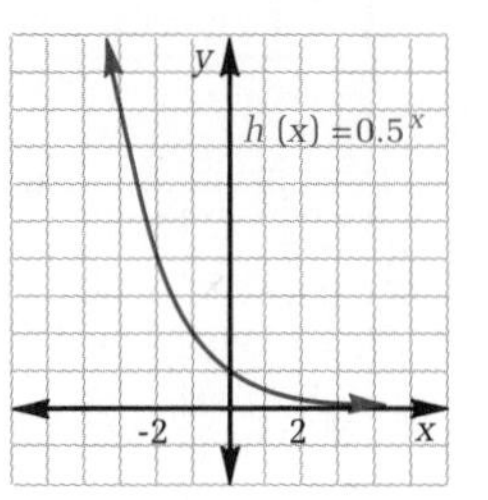

The graphs of f, g, and h all pass through the point (0, 1). The domain of each function is the real numbers. From the graphs, we see that the range of each function is the set of all positive real numbers.

Exponential functions have many applications in the study of natural phenomena. These functions are studied in a later chapter.

Part Two: Sample Problems

Problem 1 *Five sets of experimental data are graphed below. Match each graph with the type of function that best describes the data: (i) exponential, (ii) reciprocal, (iii) quadratic, (iv) linear, or (v) cubic.*

a

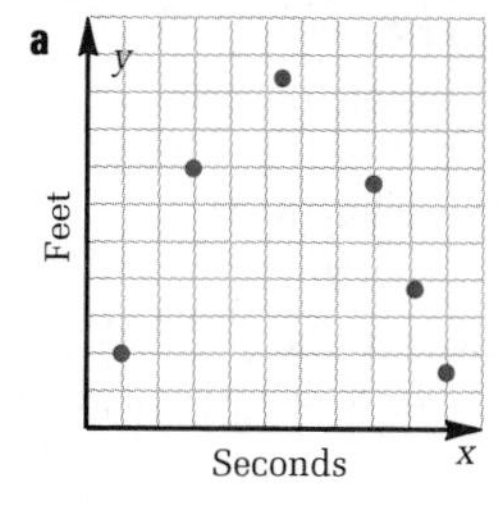

b

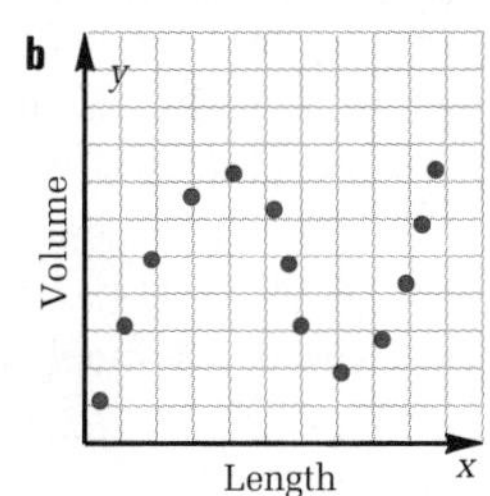

c

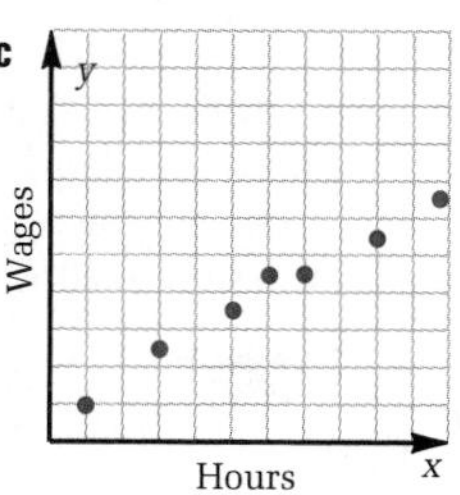

d

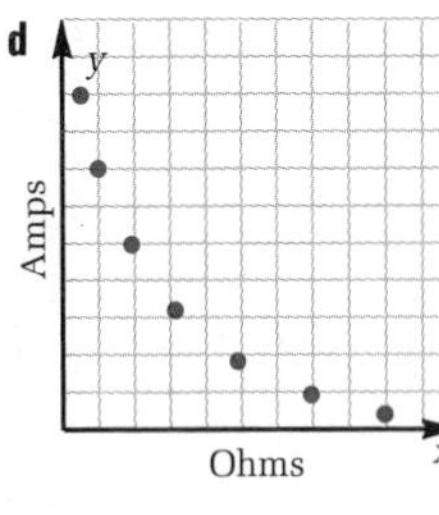

e

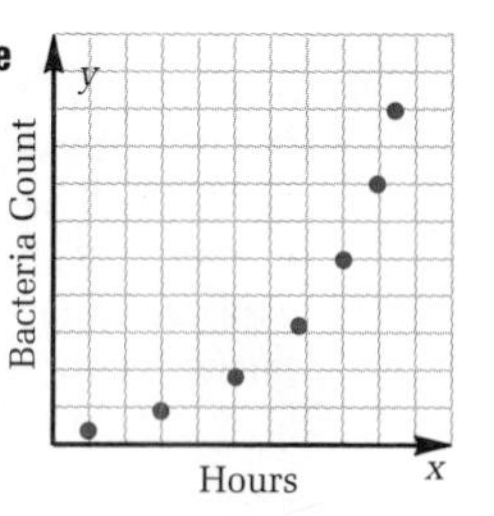

Solution **a** iii **b** v **c** iv **d** ii **e** i

See p. 789

Problem 2 *A piece of corrugated cardboard 10 inches square can be made into an open box with a volume of 48 cubic inches, provided that a square is cut from each corner before the cardboard is folded. What are the possible dimensions of the resulting box?*

Solution Let's start by sketching the original piece of cardboard and the resulting open box.

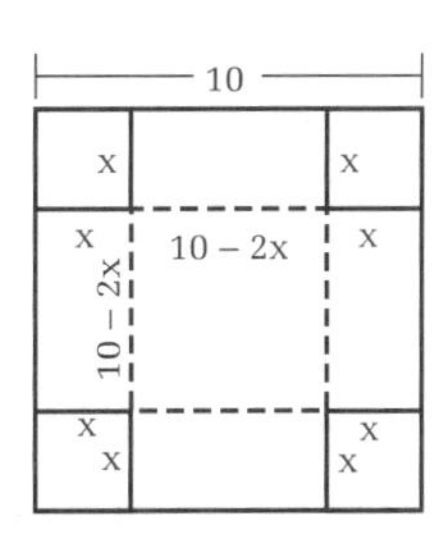

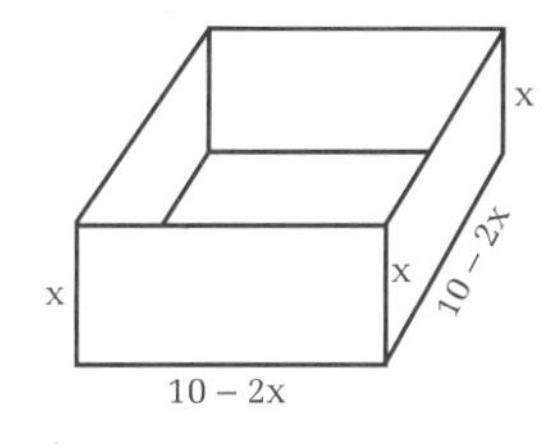

The volume of a rectangular box is $V = lwh$. In this problem, $V = 48$, $l = 10 - 2x$, $w = 10 - 2x$, and $h = x$. Substituting these values,

$$48 = (10 - 2x)(10 - 2x)x, \text{ or}$$

$$0 = x(10 - 2x)^2 - 48$$

Let $p(x) = x(10 - 2x)^2 - 48$. The real zeros of this cubic polynomial function are the values of x for which $p(x) = 0$. These are the values of x we are looking for.

We can graph p and approximate the zeros as 3, 0.63, and 6.3. Let's use these values to make a table.

x	*h*	*l*	*w*	*V*	
3	3	4	4	48	**Exactly!**
0.63	0.63	8.74	8.74	48.12	**Very close!**
6.3	6.3	−2.6			**Impossible!**

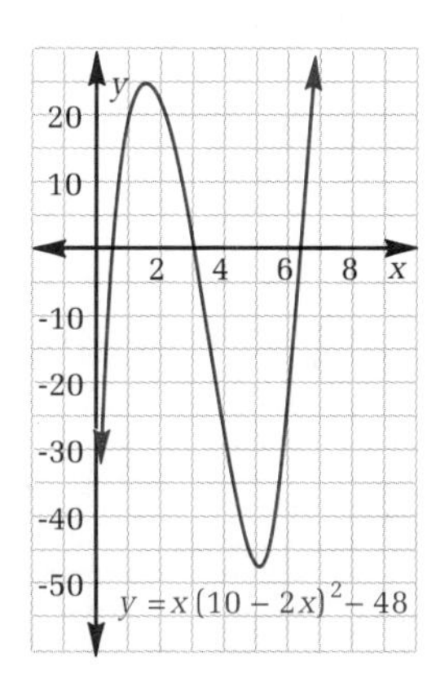

There are two sets of possible dimensions, 4 inches by 4 inches by 3 inches and 8.74 inches by 8.74 inches by 0.63 inch.

Checkpoint

1 Graph the function $f(x) = x^4 - 3$.

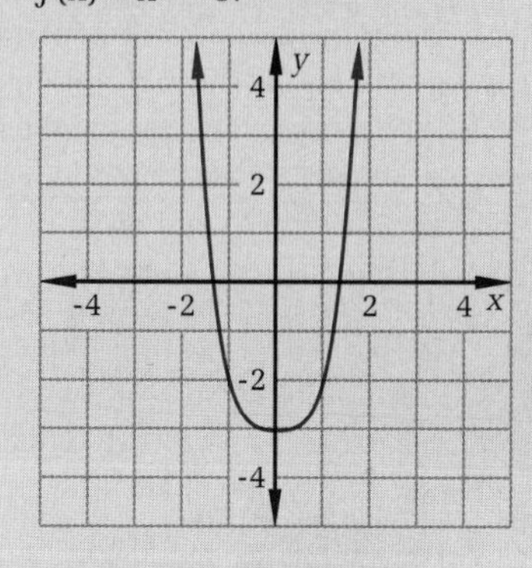

2 Identify the following graphs as linear, quadratic, cubic, quartic or exponential:
$y = x^3 + 3x^4$ quartic
$y = 3 + 4^x$ exponential
$y = 3x + 4$ linear
$y = 3x^3 + 3x^2$ cubic

3 How many distinct zeros does the graph in exercise **1** have?
2

Assignment Guide

Average
Optional
Advanced
Day 1 4, 5, 8, 11, 13, 15, 17, 19, 21, 24, 25
Day 2 6, 7, 9, 12, 14, 16, 18, 22, 23, 26, 28
Honors
Day 1 5, 8–10, 15–18, 20, 21, 24
Day 2 7, 11–13, 19, 22, 23, 25–29

Problem-Set Notes and Strategies

- Note the similarities in the symmetries of graphs **a**, **b** and **c** of problem **2**. Each is not defined at zero and approaches zero as x gets very large in absolute value.
- Note the different ways of asking the same question in parts **a** and **b** of problem **3**. Students should be familiar with each interpretation of these concepts.

Additional Answers

1a

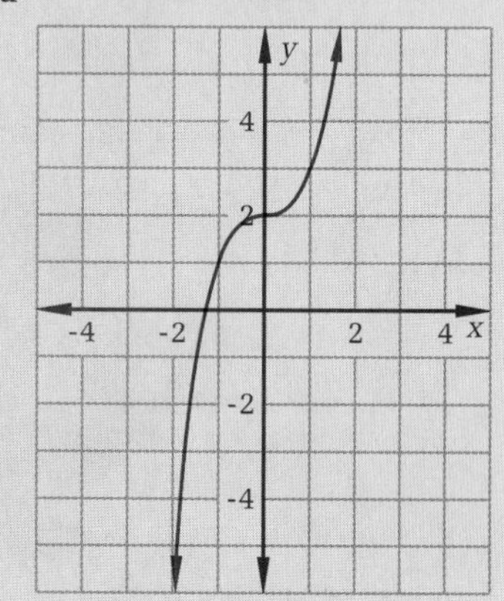

1b

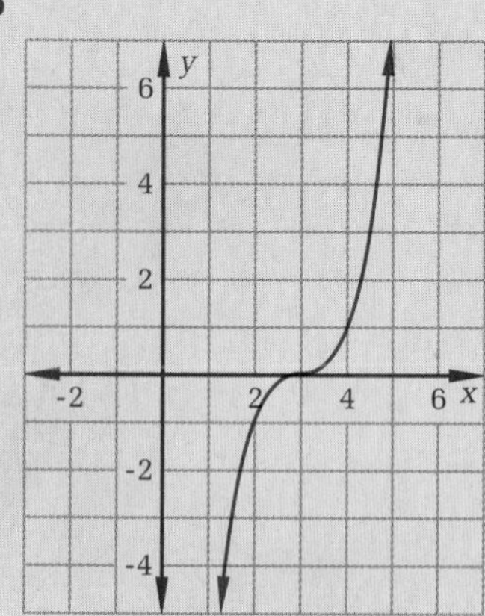

Part Three: Exercises and Problems

Warm-up Exercises

P **1** Let $f(x) = x^3$.

a Graph $g(x) = f(x) + 2$. **b** Graph $h(x) = f(x - 3)$.

P **2** From the equations listed, choose one to describe each graph shown.

i $y = \frac{1}{x}$ **ii** $y = \frac{1}{x^2}$ **iii** $y = -\frac{1}{x}$

a

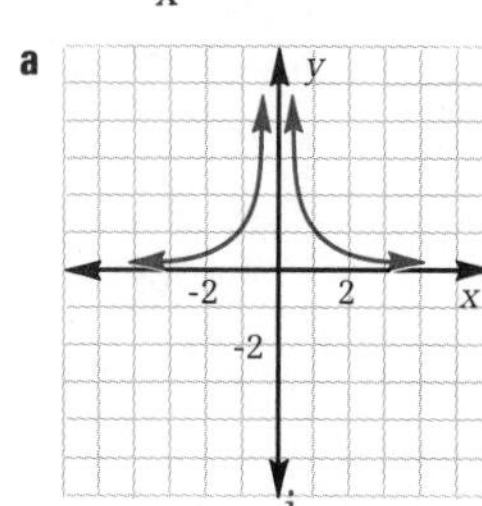

i

b

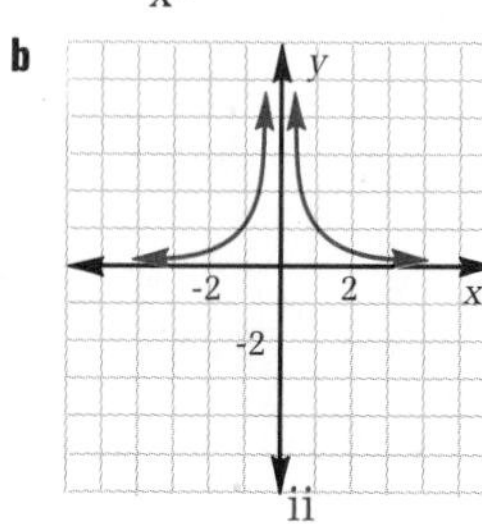

ii

c

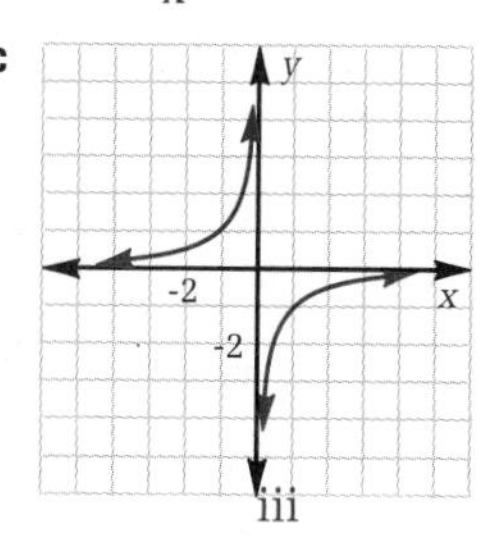

iii

P **3** The graph of f is shown.

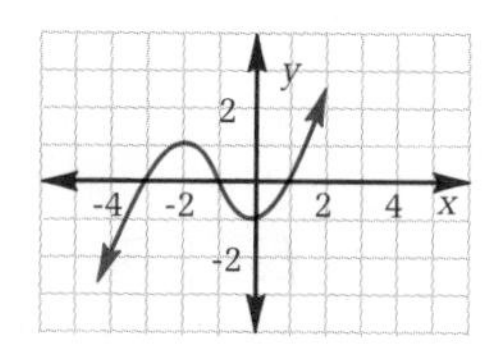

a Find the real zeros of f. $-3, -1, 1$

b For what values of x does $f(x) = 0$?

c Find the real zeros of $f(x + 2)$.

$-5, -3, -1$

b $x = -3$, $x = -1$, and $x = 1$

A Problem Set

P **4** Identify each function as (i) linear, (ii) quadratic, (iii) exponential, or (iv) other.

a $f(x) = 5 + 6x - 2x^2$ ii

b $h(x) = 9x - 3$ i

c $F(x) = 4^x$ iii

d $g(x) = x^3 - 8x$ iv

e $P(x) = 6 + 4x$ i

f $h(x) = x^4 + 5$ iv

g $C(x) = 4$ i

h $f(x) = (1 + x^2)^2$ iv

P **5** How many distinct real zeros does each function have?

a

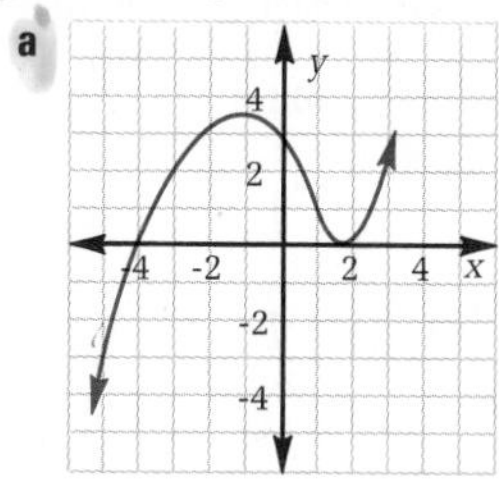

2

b

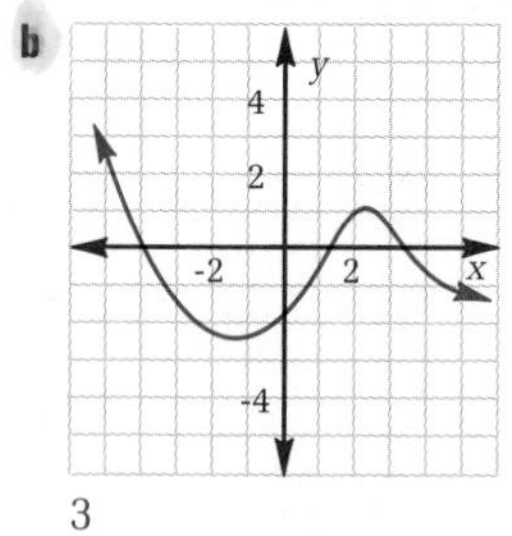

3

c

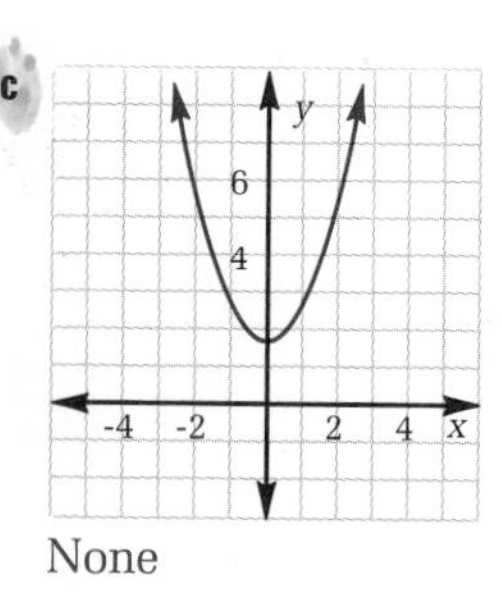

None

P | **6** Tell whether each statement is True or False. Explain your answer.

a The graphs of $y = \frac{1}{x}$ and $x = \frac{1}{y}$ are identical. True; both equations are equivalent to $xy = 1$.

b The graphs of $y = x$ and $y = \frac{x^2}{x}$ are not identical. True; 0 is in the domain of $y = x$ but not $y = \frac{x^2}{x}$.

R | **7** The graph of $y = f(x)$ contains the point (3, 2). The graph of $y = g(x)$ contains the point (3, 5). What point must lie on the graph of each of the following functions?

a $y = f(x) + g(x)$ (3, 7) **b** $y = f(x) \cdot g(x)$ (3, 10) **c** $y = \frac{f(x)}{g(x)}$ $\left(3, \frac{2}{5}\right)$

R | **8** Use the quadratic formula to find x.

a $2x^2 - 3x - 4 = 0$ $x = \frac{3 \pm \sqrt{41}}{4}$, or $x \approx 2.35$ or $x \approx -0.85$ **b** $4x^2 - 20x + 25 = 0$ $\frac{5}{2}$

P,R | **9** Let $f(x) = \frac{1}{x - 3}$.

a Where is f undefined? At $x = 3$

b If $h(x) = f(2x)$, where is h undefined? At $x = \frac{3}{2}$

R | **10** The current in a stream runs at 5 miles per hour. If a boat can go 15 miles per hour on still water, how fast can the boat go downstream? How fast upstream? 20 mph; 10 mph

P | **11** Use the graph of $f(x) = x^4 - 10x^2 + 9$ to find the number of real zeros of each of the following functions.

a f 4
b $h(x) = f(x) - 9$ 3
c $g(x) = f(x - 4)$ 4
d $l(x) = f(x) + 16$ 2
e $m(x) = f(x) + 20$ 0
f $n(x) = f(x) - 12$ 2

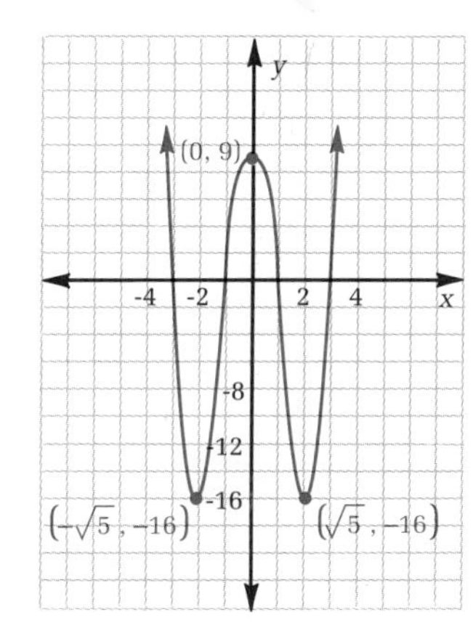

P | **12** Use the graph of $f(x) = \sqrt{x}$ to sketch the graph of each of the following functions.

a $h(x) = \sqrt{x - 3}$ **b** $g(x) = \sqrt{x + 5}$
c $l(x) = \sqrt{x + 7}$ **d** $m(x) = \sqrt{x - 3} - 4$

I | **13** The relation $y = \frac{k}{x}$, where k is a constant, is called an inverse variation—we say that y is inversely proportional to x. If y is inversely proportional to x and y = 10 when x = 4, what is the value of k? 40

Problem-Set Notes and Strategies, continued

- Problem **6** explores equivalency in functions. Some functions may take on the same values at all places in the domains, but have slightly different domains. This is true in part **b**. Concepts such as this are interesting to examine on the graph and will be explored in the next section.
- Problem **11** is a good review of translations of a graph. If a student masters this problem without drawing several graphs, he/she has a good grasp of the translation process.
- As in direct proportions, the first step in solving an inverse variation is to find the constant of proportionality. These are discussed in greater detail in Section **7.4**.

Additional Answers

12a

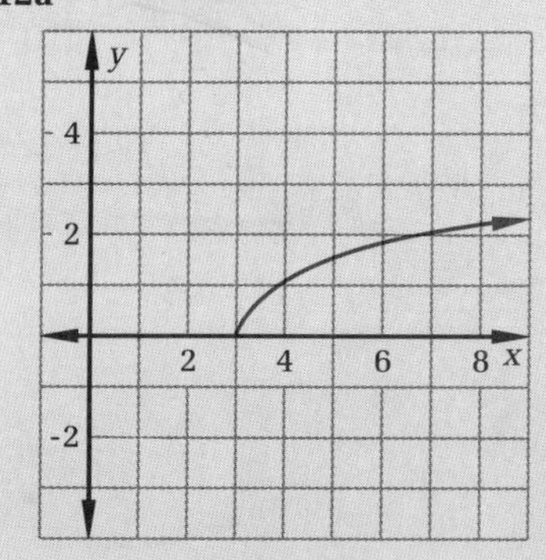

12b

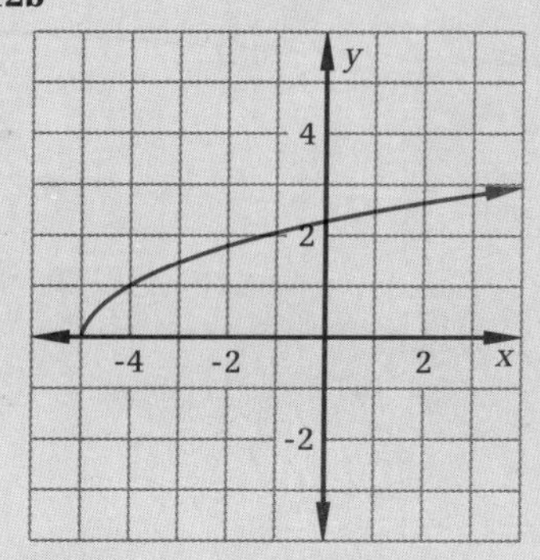

Answers for problems **12c** and **12d** can be found in the answer pages beginning on **A 1**.

Problem Set, *continued*

P **14** Which of the following is the graph of $y = \frac{1}{x^4 + 2}$? a

a

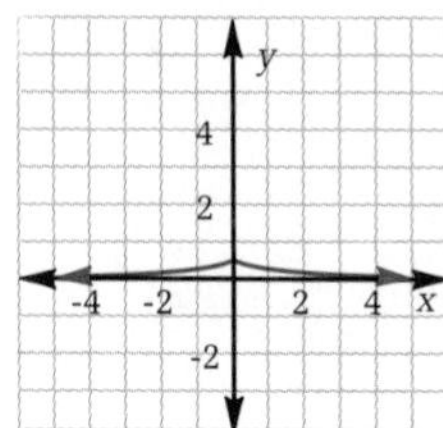

b

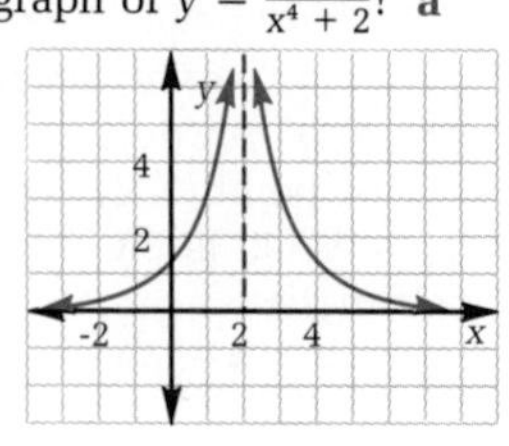

c

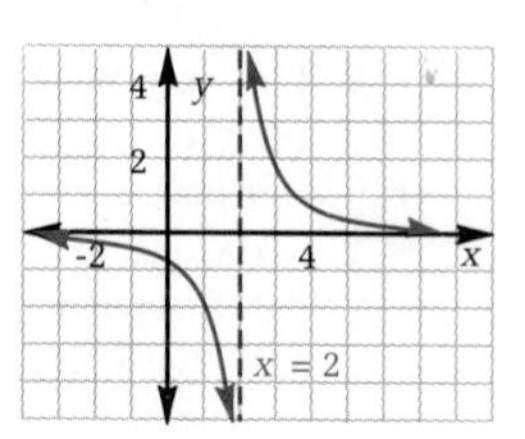

P **In problems 15–18, identify the domain of each function.**

15 $f(x) = 2^x$ $\{x:x \in \mathcal{R}\}$

16 $g(x) = \frac{1}{x}$ $\{x:x \neq 0\}$

17 $h(x) = \frac{1}{x - 4}$ $\{x:x \neq 4\}$

18 $l(x) = \sqrt{x - 4}$ $\{x:x \geq 4\}$

P **19** Let $f(x) = x^3 - 6x^2 + 5$.

a Draw a graph of f.

b Estimate the real zeros of f. 1, ≈5.85, ≈−0.85

P **20** An open box is formed by taking an 8-inch-by-15-inch piece of cardboard and cutting a square of side x from each corner. The volume of the box is 88 cubic inches. Find all possible values of x. $x = 2$ or $x \approx 1.35$

P,I **21** Copy and complete the table for $f(x) = 4^x$ and describe the pattern of the differences between successive values of $f(x)$.

x	−4	−3	−2	−1	0	1	2	3	4
$f(x)$	≈0.0039	≈0.0156	0.0625	0.25	1	4	16	64	256

B

P **22** Use the graph of $f(x) = x^3 - 4x$ to find the real zeros of each of the following functions.

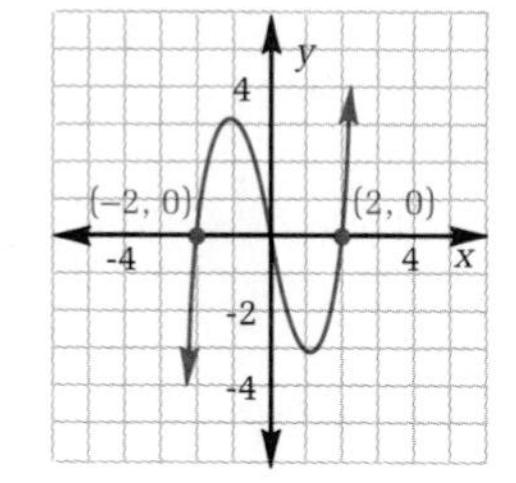

a f −2, 0, 2

b $g(x) = f(x - 4)$ 2, 4, 6

c $h(x) = f(x + 7)$ −9, −7, −5

d $n(x) = f(2x)$ −1, 0, 1

e $m(x) = f\left(\frac{1}{4}x\right)$ −8, 0, 8

I **23** Find the distance between (4, −2, 3) and (6, 5, −9) if the distance between two points (x_1, y_1, z_1) and (x_2, y_2, z_2) in three-dimensional space is given by the formula

$$d = \sqrt{(x_2 - x_1)^2 + (y_2 - y_1)^2 + (z_2 - z_1)^2}.$$ $\sqrt{197}$, or ≈14.04

Problem-Set Notes and Strategies, continued

- The function in problem **14** has a domain of all real numbers since $x^4 + 2$ will never equal zero. This eliminates the graphs in both **b** and **c**. By process of elimination, the correct answer must be **a**.
- For problems **15 – 18** the only numbers that must be eliminated are those that cause division by zero and negative numbers under even radicals.
- It might be interesting to discover which box in problem **20** also yields the largest volume. This turns out to be max/min problem which occupies a great deal of computer time in the real world.
- The formula in problem **23** is an extension of the distance formula for two dimensions. Ask your students if they could generalize a formula for the distance between two points in four dimensions. Now ask them to visualize that distance. This may precipitate a discussion on the way a mathematician approaches a problem.

Additional Answers

19a

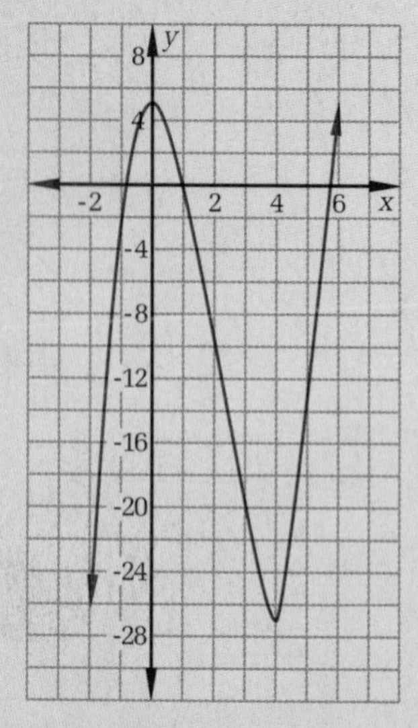

21 Successive differences increase by a factor of 4.

P **24** Solve each of the following for x if $f(x) = 2^x$.

a $f(x) = 8$ $x = 3$ **b** $f(x) = 16$ $x = 4$ **c** $f(x - 1) = 1$ $x = 1$

P **25** Use the graph of $f(x) = \frac{1}{x}$ to sketch the graph of each of the following functions.

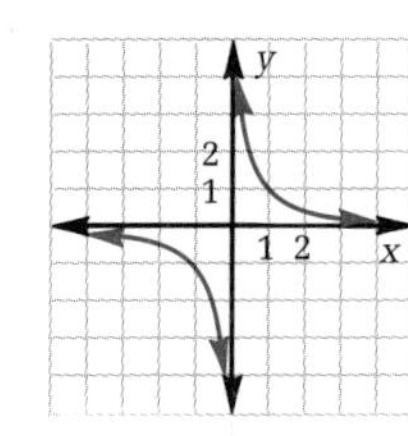

a $g(x) = f(x + 5)$
b $h(x) = f(x) + 2$
c $l(x) = f(x + 3) - 2$

C P **26** Use the graph of $f(x) = \frac{1}{x^2}$ to find the real zeros of each of the following functions.

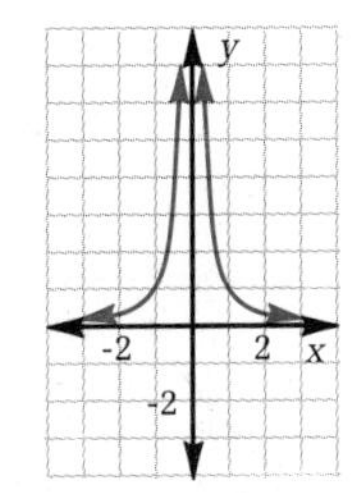

a f None

b $g(x) = f(x + 3)$ None
c $h(x) = f(x) - 4$ $\frac{1}{2}$ and $-\frac{1}{2}$
d $l(x) = -f(x) + 4$ $\frac{1}{2}$ and $-\frac{1}{2}$

R **27** Let $f(x) = \frac{1}{x}$ and $g(x) = x$.

a Find $f \circ g(x)$. $\frac{1}{x}$ **b** Find $g \circ f(x)$. $\frac{1}{x}$

c Explain your answers to parts **a** and **b.**
g maps x to itself, so compound f, g functions are preserved.

I **28** Refer to problem **13.** The intensity of light I is inversely proportional to the square of the distance d from the light source.

a If the intensity is 12 lumens when the distance is 4 feet, what is the intensity when the distance is 20 feet? 0.48 lumen
b Graph the function relating intensity and distance for distances greater than zero.

P,R,I **29** Let $f(x) = x^3$ and $g(x) = x$. The graphs of these functions intersect at (0, 0), (1, 1), and (−1, −1) as shown. Where do the graphs of the following pairs of functions intersect?

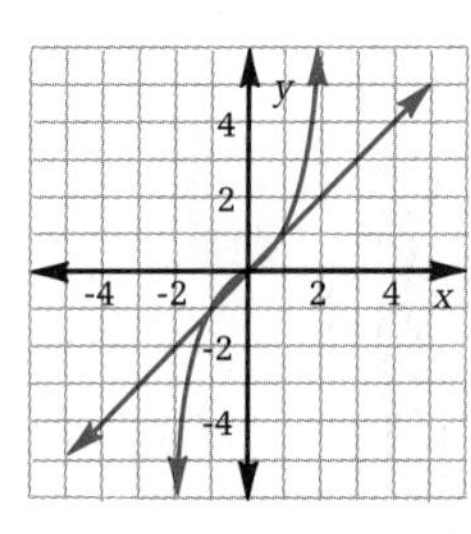

a $\begin{cases} m(x) = f(x) + 3 \\ k(x) = g(x) + 3 \end{cases}$ (0, 3), (1, 4), and (−1, 2)

b $\begin{cases} l(x) = f(x - 4) \\ m(x) = g(x - 4) \end{cases}$ (4, 0), (5, 1), and (3, −1)

c $\begin{cases} r(x) = f(x + 3) - 2 \\ s(x) = g(x + 3) - 2 \end{cases}$ (−3, −2), (−2, −1), and (−4, −3)

Problem-Set Notes and Strategies, continued

- The function g in problem **27** maps x to itself. This is a special function and has a special name. It is called the identity function. Your students are already familiar with identities for addition, multiplication and matrix multiplication. Identities may be generalized to other operations such as functions.
- Problem **28** expands on the concept of inverse variation. As one variable gets larger, the other gets smaller.
- Problem **29** is another review of function translations. It would be helpful to graph these on transparencies and slide them around to see how the points of intersection change.

Additional Answers

25a

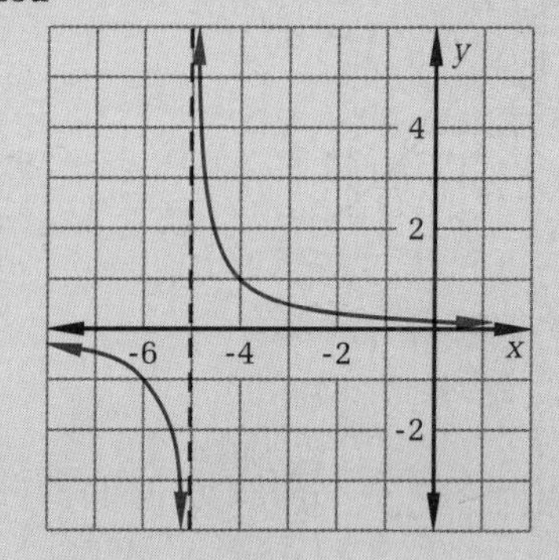

25b

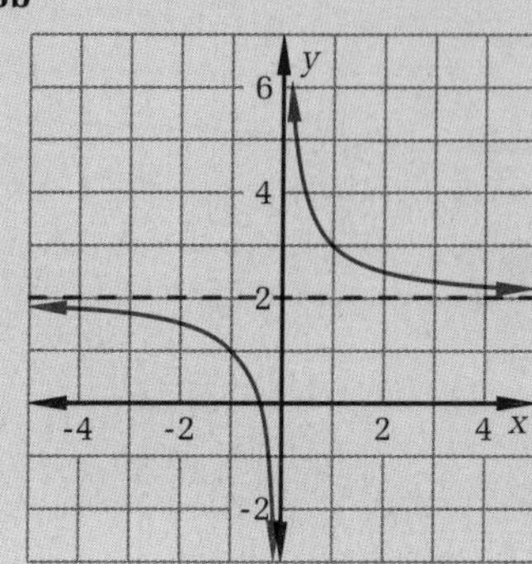

Answers for problems **25c** and **28b** can be found in the answer pages beginning on **A** 1.

6.5

Class Planning

Time Schedule
Average: Optional
Advanced: 2 days
Honors: 2 days

Resource References
Teacher's Resource Book
Class Opener 38
Cooperative Learning 19
Practice 39

Class Opener

Graph $f(x) = \frac{1}{x}$ and $g(x) = \frac{x}{x}$ on the same coordinate system.

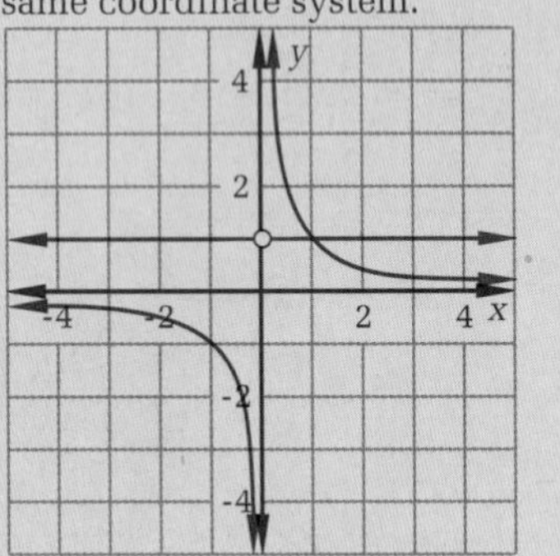

Lesson Notes

- Asymptotes and holes are examined in this section on graphing. You may want to emphasize to students that if a fraction has a zero in the denominator, it is undefined. If a value must be restricted for the variable because the denominator would equal zero, the value is a candidate for a vertical asymptote. If the same factor or value results in a zero in the numerator, the value is frequently cancelled out of the function.

6.5 Asymptotes and Holes in Graphs

Objectives
After studying this section, you will be able to
- Recognize asymptotes of graphs
- Recognize holes in graphs

Part One: Introduction

Exploring Graphs
Study the following graphs and the equations they represent, looking for patterns.

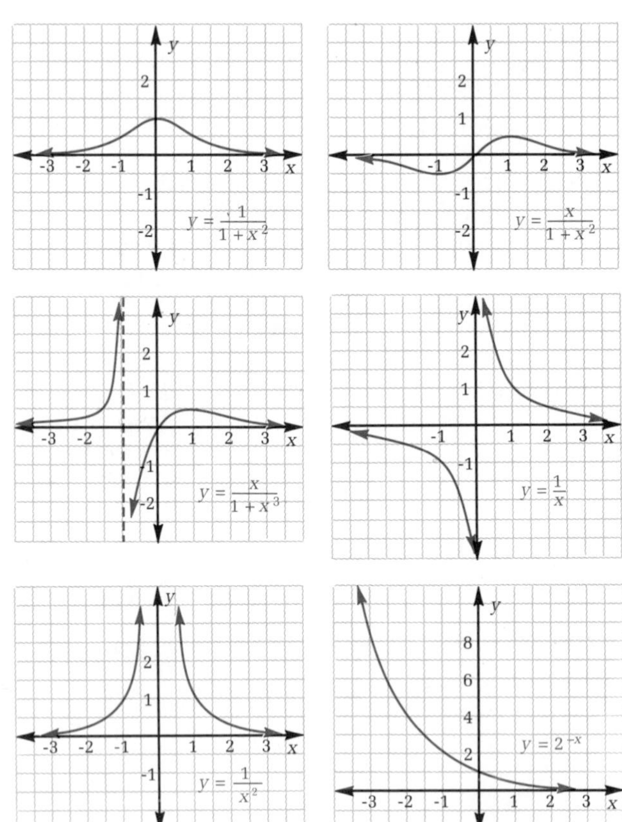

Vocabulary
horizontal asymptote
vertical asymptote
hole

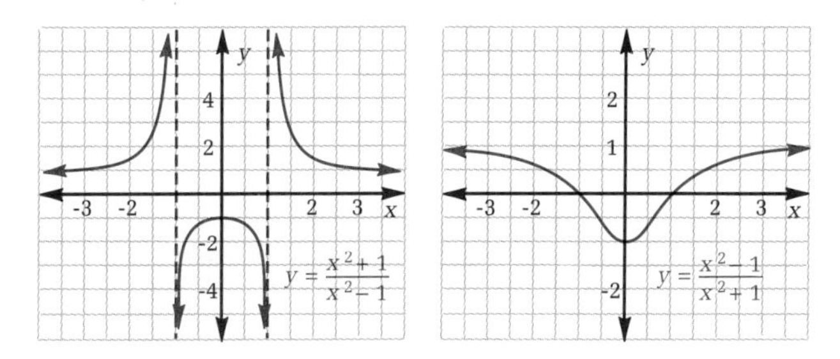

You may have noticed that some of the graphs appear to be symmetrical. What kinds of symmetry do these graphs exhibit? Could you predict whether an equation's graph is symmetrical and, if it is, what type of symmetry it has? Some of the graphs are discontinuous—what characteristics of the equations might account for the gaps?

One of the benefits of graphing calculators and computer graphing programs is that they allow us to quickly generate graphs of complicated functions that would be difficult to graph by hand. If you have access to a function grapher, use it to examine the graphs of a variety of functions. Note how changes in a function affect its graph, and try to draw some conclusions about the relationships between functions and their graphs.

Asymptotes

In some of the preceding diagrams, the graph approaches a horizontal line, called a ***horizontal asymptote,*** or a vertical line, called a ***vertical asymptote.*** A full understanding of asymptotes and their relation to functions requires knowledge that you will gain in future mathematics classes. For now, however, there are some observations we can make and some tentative conclusions we can draw.

The graph of $y = \frac{1}{1 + x^2}$, for example, approaches the line represented by $y = 0$ (the x-axis), since the value of $\frac{1}{1 + x^2}$ becomes closer and closer to 0 as the absolute value of x increases. The line $y = 0$ is therefore a horizontal asymptote of this graph. The easiest way to predict whether a function's graph has a horizontal asymptote is to evaluate the function for several great and small input values and determine whether the outputs approach some specific value.

If a function has a restricted domain, the restriction will be reflected in the function's graph. Frequently, it appears as a vertical asymptote.

For example, the value of $f(x) = \frac{(x - 5)(x + 15)}{x + 5}$ is undefined when $x = -5$, since the denominator of $f(x)$ is equal to 0 for that input. Also, as the x values approach -5, $|x + 5|$ becomes smaller and smaller, so $|f(x)|$ becomes correspondingly greater and greater. The line $x = -5$ is therefore a vertical asymptote of the function's graph.

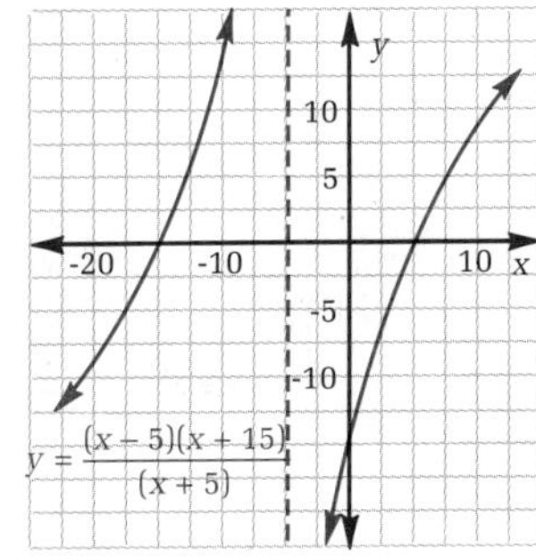

Lesson Notes, continued

It still must be excluded from the domain and therefore causes a hole in the graph. If it cannot be cancelled, it causes a vertical asymptote at that point. Horizontal asymptotes occur when we examine the function as *x* grows very large or small. These are typically examples left for a calculus class but can be explored at an elementary level. See the cooperative learning exercise for a presentation of oblique or slant asymptotes.

Checkpoint

1 Find both vertical and horizontal asymptotes for the function $y = \frac{x^2}{(x^2 - 4)}$.

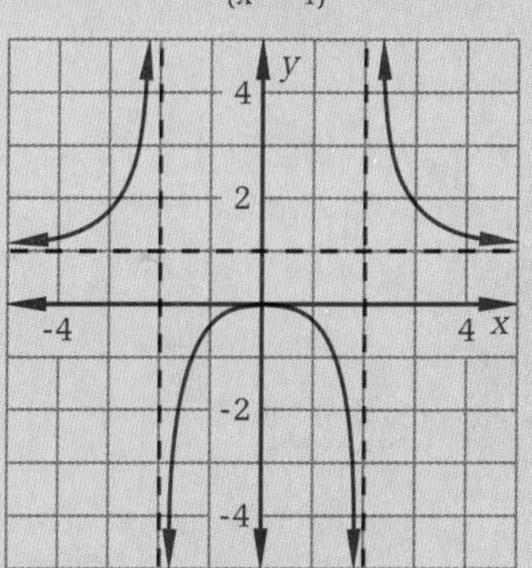

Vertical asymptotes at $x = 2$ and $x = -2$, horizontal asymptotes at $y = 1$.

6.5

Checkpoint, continued

2 Find any holes or asymptotes in the graph of $\frac{x+2}{(x^2-4)}$.

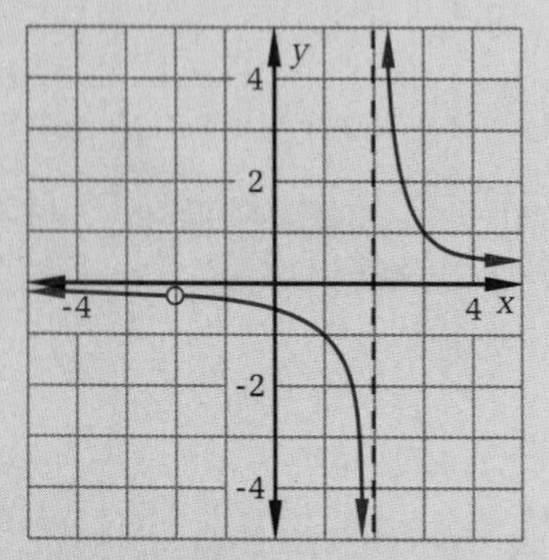

Vertical asymptote at $x = 2$, horizontal asymptote at $y = 0$ and hole at $x = -2$.

3 What are the values where the function $\frac{x}{(x-3)(x^2-16)}$ has vertical asymptotes?
$x = -4$, 3, and 4

Assignment Guide

Average	
Optional	
Advanced	
Day 1	6, 8, 9, 11, 13–15, 18, 19, 21, 24, 25
Day 2	7, 10, 16, 17, 20, 22, 23, 27–29
Honors	
Day 1	6, 8, 9, 11, 12, 14, 18, 19, 23–25, 30
Day 2	7, 10, 16, 20, 21, 27–29, 31, 32

Holes in Graphs

See p. 793

Another way in which a restriction can be reflected in a function's graph is as a ***hole.***

Example *Graph the function* $g(x) = \frac{(x+5)(x+15)}{x+5}$.

Since $\frac{x+5}{x+5} = 1$ for almost every value of x, the graph is basically the same as the graph of the linear function $p(x) = x + 15$. When $x = -5$, however, the value of g(x) is undefined, so the graph has a hole at $x = -5$. We indicate this by an open dot.

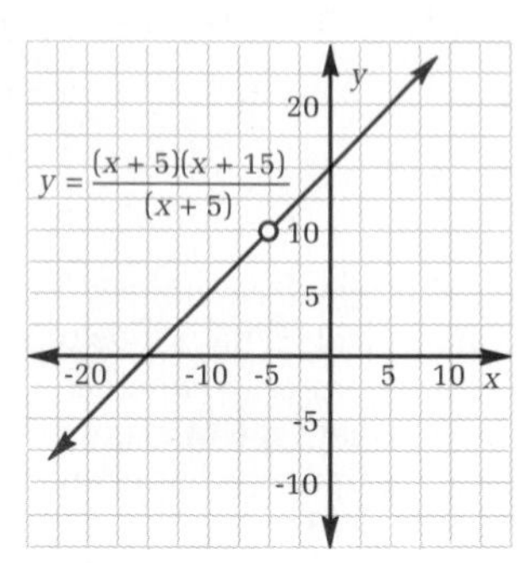

Whenever a function contains a quotient whose numerator and denominator have one or more common factors involving the variable, its graph is likely to have one or more holes. Different graphing calculators and programs react differently to the presence of a hole —many show a continuous graph, ignoring the hole; some start to generate the graph and then stop when they reach the hole; some show a break too small to detect. For this reason, it is useful to be able to identify holes by simply examining the equations of functions.

Part Two: Sample Problem

See p. 793

Problem *The graph of* $y = \frac{x}{x-2}$ *has one vertical and one horizontal asymptote. Write the equation of each asymptote.*

Solution When we use a computer program to generate the graph, it appears that the graph approaches the line $y = 1$ at the left and right sides of the display. By substituting great and small values for x in the equation, we confirm that y approaches 1 as $|x|$ increases, and thus that $y = 1$ is the equation of the horizontal asymptote.

Since the denominator of $\frac{x}{x-2}$ is 0 when $x = 2$, and the numerator is not equal to 0, the equation of the vertical asymptote is $x = 2$.

Part Three: Exercises and Problems

Warm-up Exercises

P

1 The graph of $y = \frac{1}{(x + 2)(x - 3)}$ has two vertical asymptotes. Write their equations. $x = -2$ and $x = 3$

P

2 Let $f(x) = \frac{x(x - 3)(x + 2)}{x(x - 3)(x + 5)}$.

a What are the equations of the vertical asymptotes of the graph of f? $x = -5$

b What are the x values of the holes in the graph? 0 and 3

P

3 Write the equations of the horizontal and vertical asymptotes of the following graphs.

a

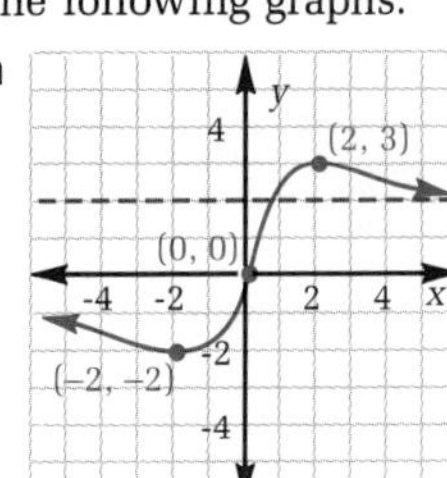

$y = 2$; $y = -1$

b

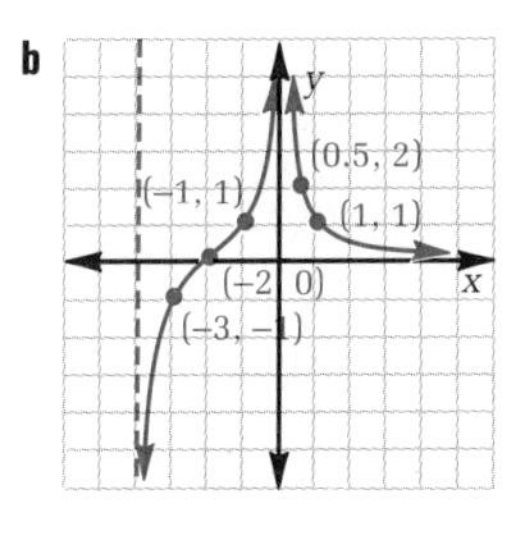

$x = -4$, $x = 0$; $y = 0$

P

4 The graph of $y = \frac{2x}{x + 1}$ has one horizontal asymptote and one vertical asymptote. Write their equations. $y = 2$; $x = -1$

P

5 What is the value of y if $y = \frac{7 - \frac{3}{x}}{1 + \frac{2}{x}}$ and x has the given value?

a 1000 ≈ 6.983

b 10,000 ≈ 6.9983

c 100,000 ≈ 6.99983

d 1,000,000 ≈ 6.999983

Problem Set

A

P

6 The graph of $y = \frac{1}{x^2 + 4x - 21}$ has two vertical asymptotes. Write their equations. $x = -7$ and $x = 3$

P

7 Write the equations of the horizontal and vertical asymptotes of the following graphs.

a

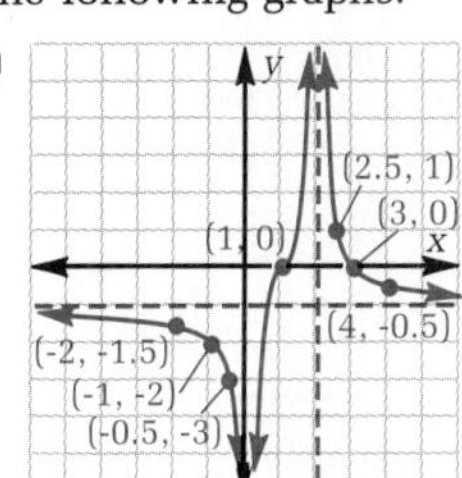

$y = -1$; $x = 0$, $x = 2$

b

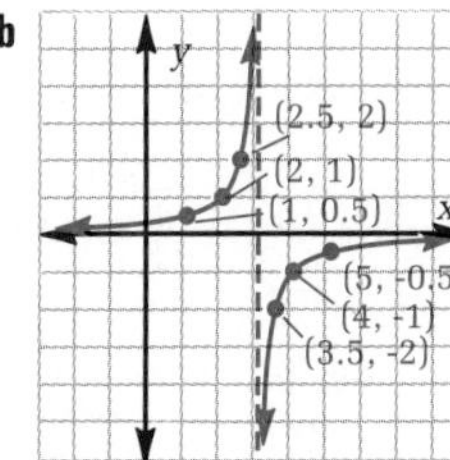
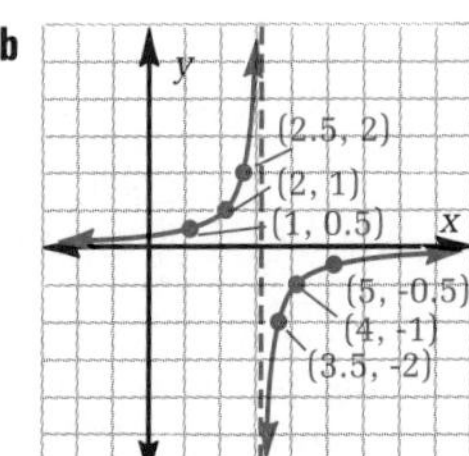

$y = 0$; $x = 3$

P

8 Graph $y = \frac{x(x + 1)}{(x + 1)}$.

Problem-Set Notes and Strategies

- In problem **4**, if the numerator and denominator have the same degree, the horizontal asymptote is the ratio of the leading coefficients since all the other terms are overpowered by the highest power in the fraction.
- Problem **5**, as x gets larger, both $\frac{2}{x}$ and $\frac{3}{x}$ approach zero and are therefore negligible.

Stumbling Block

Students can be careless about holes in the graph. Too often when some factor is cancelled it is just forgotten. Have the students get in the habit of writing down the domain of the function before they begin to evaluate or simplify. This will eliminate forgetting about any trouble spots in the function.

Additional Answer

8

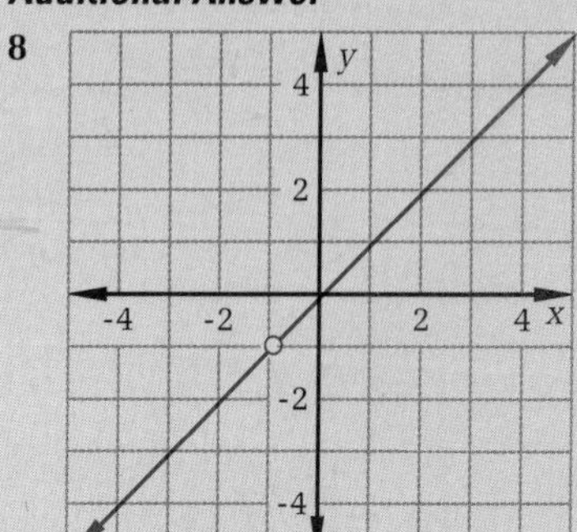

Problem Set, *continued*

P,R **9** The graph of $y = f(x)$ is shown. Sketch the graph of $y = f(x - 2) + 5$.

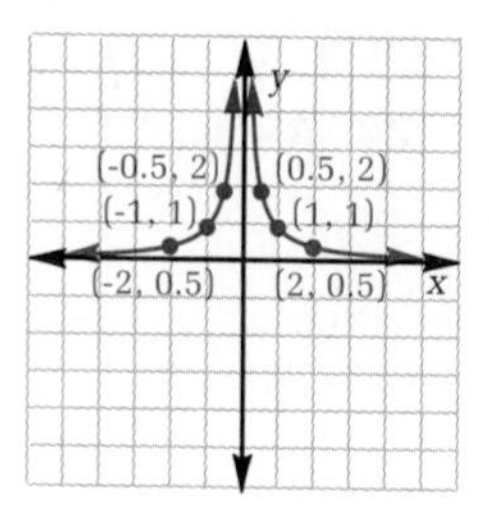

P **10** Sketch the graph of each function.

a $f(x) = \dfrac{1}{x + 2}$ **b** $g(x) = \dfrac{x - 2}{x^2 - 4}$

P **11** Suppose the graph of a function f has vertical asymptotes with equations $x = 2$ and $x = 5$.

a Find the vertical asymptotes of the graph of $f(x + 3)$. $x = -1, x = 2$

b Find the vertical asymptotes of the graph of $f\left(\frac{1}{2}x\right)$. $x = 4, x = 10$

P,R **12** The graph of $y = f(x)$ is shown.

a Sketch the graph of $y = f(x + 1)$.

b Write the equations of the vertical asymptotes of the graph of $y = f(x + 1)$. $x = -4, x = 2$

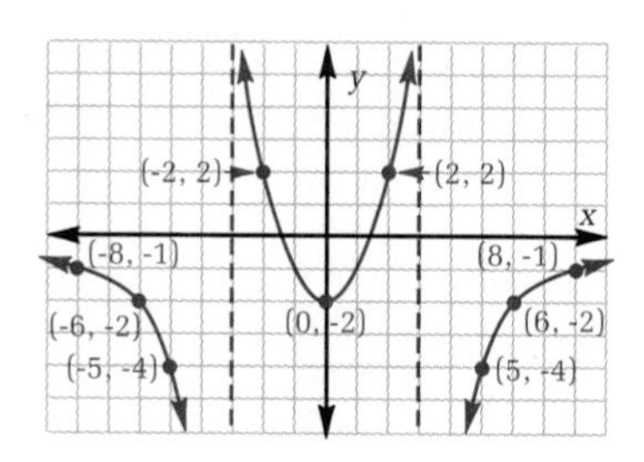

R **In problems 13–15, solve for x.**

13 $2^x = 8$ $x = 3$ **14** $3^x = \dfrac{1}{81}$ $x = -4$ **15** $2^x = \sqrt{2}$ $x = \frac{1}{2}$

B

P **16** Sketch a graph of a function with vertical asymptotes at $x = 2$ and $x = 4$ and a horizontal asymptote at $y = 1$.
Answers will vary. A possible response is shown.

R,I **17** Solve the following system for (x, y). $(x, y) = (-2, -3)$

$$\begin{cases} 3x - 8y = 18 \\ x + 2y = -8 \end{cases}$$

R **18** In how many ways can nine baseball players be arranged in a batting order? 362,880

P **19** Let $f(x) = \frac{4x^2 - 8x}{x^2 + 3x - 10}$.

a Factor completely the numerator and the denominator of $f(x)$. $\frac{4x(x - 2)}{(x + 5)(x - 2)}$

b What are the equations of the asymptotes of f? $x = -5$; $y = 4$

c At what value of x is there a hole in the graph? 2

Problem-Set Notes and Strategies, continued

- In problems **13 – 15**, all terms need to be written as an exponential expression with the same base.

Cooperative Learning

In addition to vertical and horizontal asymptotes, a graph may have a slant asymptote. Graph the function $y = \frac{x^2 + x - 7}{x}$ for all real values except zero. As x gets closer to zero the numerator approaches −7 and the denominator gets smaller and smaller. The fraction is growing without bound in both positive and negative directions.
What happens as x get increasingly large in both the positive and negative directions? Try to simplify the function and think about the values that the function is approaching. Can you find a generalization for all functions that fit a certain classification?
If the numerator is one degree larger than the denominator, the function has a slant asymptote. Rewrite the function as $y = x + 1 - \frac{7}{x}$. As x gets large, the $\frac{7}{x}$ gets so close to zero that it may be disregarded. The graph approaches $y = x + 1$ which is a straight line. The division may get complicated, but there is always a slant asymptote.

Additional Answers

Answers for problems **9, 10, 12,** and **16** can be found in the answer pages beginning on **A1**.

R | **20** Write the equation of the axis of symmetry of the graph shown. $x = 6$

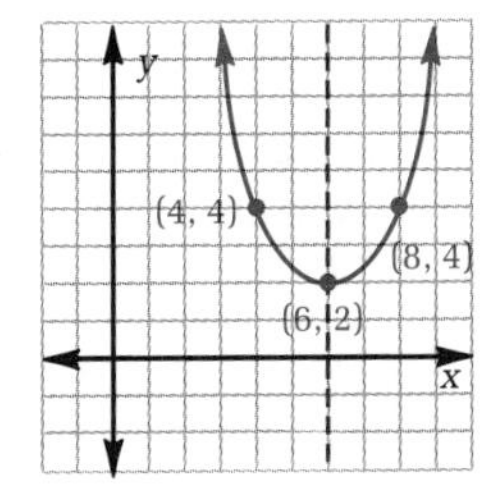

R | **21** Rationalize the numerator of $\frac{\sqrt{5} - \sqrt{3}}{2}$. $\frac{1}{\sqrt{5} + \sqrt{3}}$

R,I | **22** Richard and Mary can paddle a canoe in still water at a rate of 5 miles per hour. A stream has a current of 2.5 miles per hour.

- **a** How fast can they paddle downstream? 7.5 mph
- **b** How fast can they paddle upstream? 2.5 mph
- **c** If Richard and Mary paddled for 3 hours downstream and then 3 hours upstream, how far were they from where they started? 15 mi downstream

P | **23** Write an equation of the function with the given graph. $y = 2\frac{(x + 2)(x - 4)}{(x + 2)(x - 4)}$

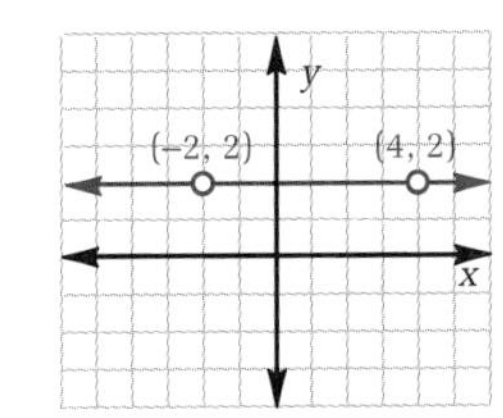

R | **24** Write the equations of the axes of symmetry of the curve shown.
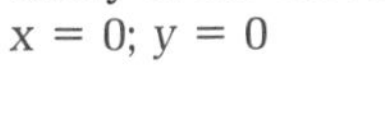
$x = 0$; $y = 0$

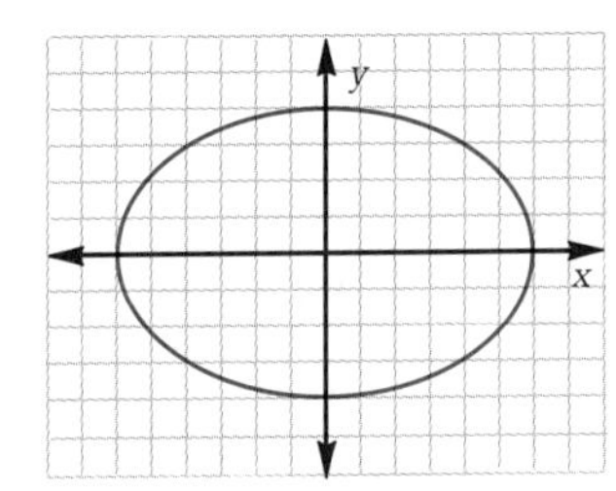

R | **25** Let f be a linear function such that $f(4) = 12$ and $f(5) = 15$.

- **a** Find $f(6)$. 18
- **b** Find $f(3)$. 9
- **c** Write a formula for $f(x)$. $f(x) = 3x$

R,I | **26** Solve the system for (x, y). $(x, y) = (-2, 8)$ or $(x, y) = (2, 8)$

$$\begin{cases} y + x^2 = 12 \\ y - x^2 = 4 \end{cases}$$

R,I | **27** The perimeter of the figure shown becomes 248 if the value of y is increased by 50% and the value of x is increased by 25%. Find x and y. $x = 12$; $y = 12$

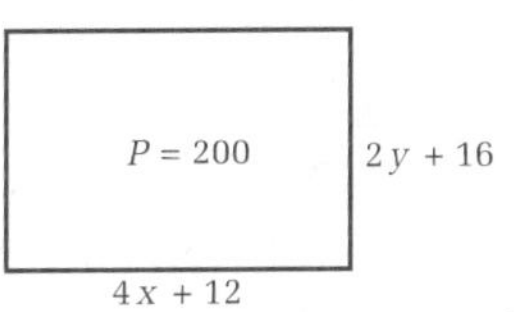

Problem-Set Notes and Strategies, continued

- Problem **22** introduces distance word problems with related rates which are presented in a later chapter. Have students draw a picture and label everything they know about the graph. The basic equation used is $d = rt$.
- Some students will notice that problem **24** is also symmetric with respect to the origin.
- Problem **26** is a system of nonlinear equations but may be solved using the same methods as systems of linear equations. This is a system of two parabolas. You might ask your students what is the maximum number of points in which two distinct parabolas can intersect.

Communicating Mathematics

Describe the difference between a vertical and horizontal asymptote and how each is likely to be found. Also describe the process of finding any holes in the graph.

Problem-Set Notes and Strategies, continued

- Problem **28** introduces a three dimensional coordinate system. In the last section we discovered a distance formula for three dimensions. It might be helpful to use that to find the distance of point *D* from the origin.

- Tangent lines and tangents of angles will be explored in depth in the trig chapters. This is an opportunity to point out the relationship between tangent lines and slopes in problem **32**.

Problem Set, *continued*

R,I **28** In three-dimensional graphing, three mutually perpendicular number lines are used as axes. A point is identified by an ordered triplet of the form (x, y, z). Find the coordinates of A, B, C, and D.
A(3, 4, 0), B(0, 4, 5), C(3, 0, 5), D(3, 4, 5)

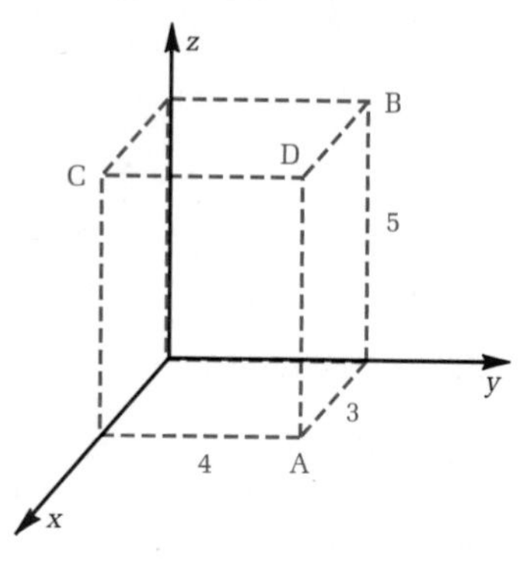

P **29** Write an equation of the function with the given graph. $y = \frac{(x + 1)(x - 3)}{(x - 3)}$

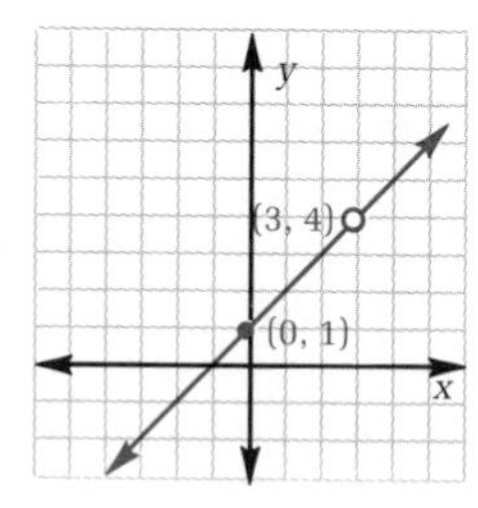

R **30** The graph of $f(x) = x^3$ is shown. Find the x- and y-intercepts of the graphs of each of the following functions.

a f (0, 0); (0, 0)
b $h(x) = f(x - 3)$ (3, 0); (0, −27)
c $g(x) = f(x + 2)$ (−2, 0); (0, 8)
d $l(x) = f(x) + 8$ (−2, 0); (0, 8)

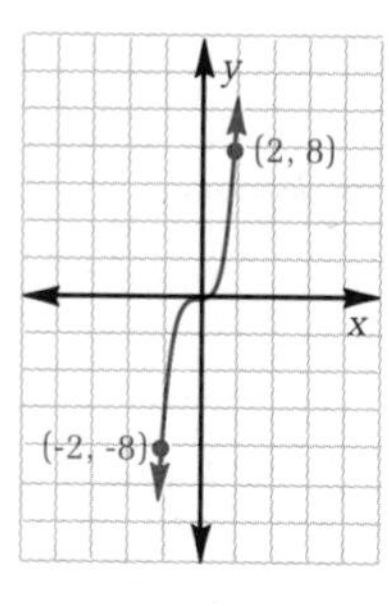

P **C** **31** Determine the vertical asymptotes of $f(x) = \frac{x^3 - x - 5}{x^2 - 5x - 6}$ x = 6; x = −1

R **32** Lines ℓ, m, and n are tangent to the circle. Estimate the slopes of these lines.
Slope ℓ: 1; slope m: undefined; slope n: 0

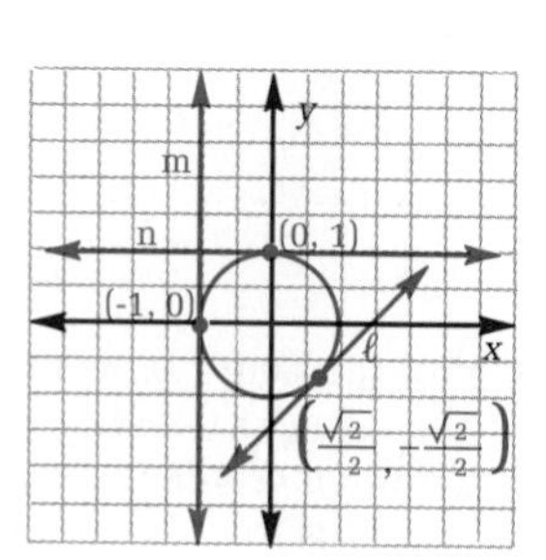

6.6 Measures of Dispersion

Objectives

After studying this section, you will be able to

- Compute the range of a set of data
- Compute the deviation scores and the mean deviation of a set of data

Range

Measures of central tendency provide information only about the "averages" or "centers" of a distribution of data. To find out more, we need to analyze variations of data within the distribution. These variations are analyzed using ***measures of dispersion***, which include ***range, deviation score***, and ***mean deviation.*** (You will learn about another important measure of dispersion, called *standard deviation*, in Section 8.5.)

The data below represent the numbers of miles Mary Thon ran during 7 days of preparation for the Mathem City Marathon.

15, 10, 15, 8, 20, 8, 15

The low and high numbers of miles run are 8 and 20. The difference between the highest number and the lowest number of miles run is 20 − 8, or 12, miles. So we say that the ***range*** of the data is 12.

 ***Range = greatest number − smallest number***

Example *Find the range of each set of data.*

a

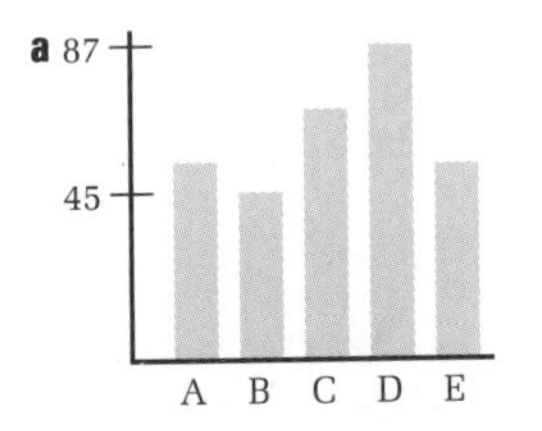

High number = 87
Low number = 45
Range = 87 − 45
= 42

b

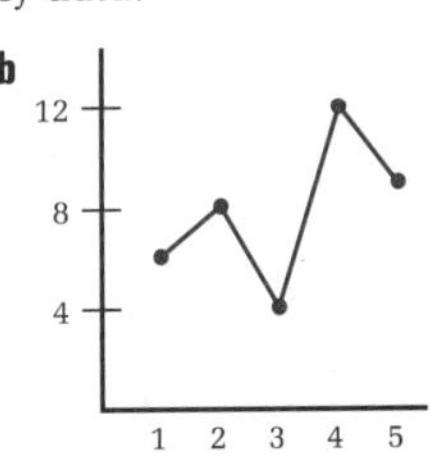

High number = 12
Low number = 4
Range = 12 − 4
= 8

c

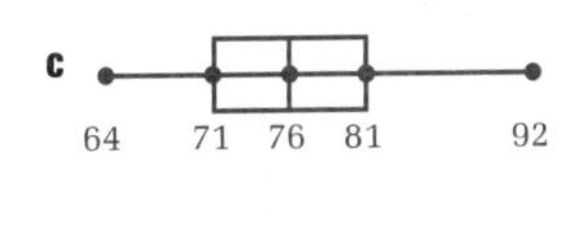

High number = 92
Low number = 64
Range = 92 − 64
= 28

Class Planning

Time Schedule

All levels: 1 day

Resource References

Teacher's Resource Book
Class Opener 39
Practice 40

Class Opener

Each year the Cy Young award is awarded to the best pitcher in baseball. During the years 1956-1966 and 1978-1989, each winner won the following number of games.

Overall

Yr	Name	W
1956	Newcombe	27
1957	Spahn	21
1958	Turley	21
1959	Wynn	22
1960	Law	20
1961	Ford	25
1962	Drysdale	25
1963	Koufax	25
1964	Chance	20
1965	Koufax	26
1966	Koufax	27

American League

Yr	Name	W
1978	Guidry	25
1979	Flanagan	23
1980	Stone	25
1981	Fingers	6
1982	Vuckovich	18
1983	Hoyt	24
1984	Hernandez	9
1985	Saberhagen	20
1986	Clemens	24
1987	Clemens	20
1988	Viola	24
1989	Saberhagen	23

Vocabulary

measure of dispersion
range
deviation score
mean deviation

6.6

Class Opener, continued

Compute the average wins for each period.

Average 1956-1966 23.55

Average 1978-1989 20.08

How do they compare?

The 1984 and 1981 years show a large deviation.

Explain the difference.

Baseball fans will know it was caused by the winners being relief pitchers.

Lesson Notes

- Three different measures of dispersion are discussed in this section, the range, deviation score and mean deviation. Measures of dispersion are important in finding a total picture of the data set. These three methods are fairly simple to compute.

Checkpoint

1. Your first 5 test scores in this class are 95, 82, 90, 74, and 89. Find the range, deviation score and mean deviation for these values.
   range = 21
   deviation scores are 9, –4, 4, –12, and 3 respectively.
   mean deviation is $\frac{32}{5} = 6.4$
2. What does a range of 35 tell you for a list of 37 test scores?
   The difference from highest score to lowest score was 35.
3. If the mean deviation of 20 trout caught in Lake Merced was 6 inches, how long was the longest trout caught that day?
   Can't tell; mean deviation tells us nothing about the specific data points, just how they are dispersed.

Deviation Scores and Mean Deviation

At the William Tell Open, the 8 members of the Antelope Valley Archery Club listed their numbers of bull's-eyes from least to greatest.

41, 43, 44, 47, 48, 48, 49, 50

Hawkeye hit 50 of 50 bull's-eyes, whereas Diana hit 43 of 50 bull's-eyes. Hawkeye and Diana wanted to know how their totals compared with the mean number of bull's-eyes recorded.

The mean number of bull's-eyes recorded was

$$\frac{41 + 43 + 44 + 47 + 48 + 48 + 49 + 50}{8} = 46.25$$

To compare any number in a set of data with the mean, we find its ***deviation score***.

Deviation score = score − mean

We will determine the deviation score of both archers in order to make comparisons with the mean for all the club members.

Hawkeye's deviation score = 50 − 46.25 = 3.75

Diana's deviation score = 43 − 46.25 = −3.25

Hawkeye's performance was 3.75 bull's-eyes above the mean. Diana's performance was 3.25 bull's-eyes below the mean.

Aunt Chovie owns 2 pizza parlors in Mathem City. The data below represent the numbers of pizzas sold at the 2 parlors each Saturday night for the past 6 weeks.

Parlor A 16, 12, 22, 21, 43, 54

Parlor B 31, 34, 29, 40, 30, 28

For purposes of scheduling employees, Aunt Chovie needs information about which pizza parlor has been more consistent in its Saturday sales. To do this, she uses the ***mean deviation*** of the data for each parlor.

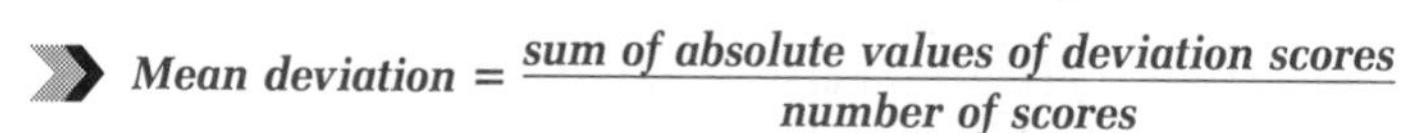

$$\textit{Mean deviation} = \frac{\textit{sum of absolute values of deviation scores}}{\textit{number of scores}}$$

The mean number of pizzas sold at Parlor A was $\frac{168}{6}$, or 28.

Score	Deviation Score (Score − Mean)	Absolute Value of Deviation Score
16	16 − 28 = −12	|−12| = 12
12	12 − 28 = −16	|−16| = 16
22	22 − 28 = −6	|−6| = 6
21	21 − 28 = −7	|−7| = 7
43	43 − 28 = 15	|15| = 15
54	54 − 28 = 26	|26| = 26
Sum = 168	Sum = 0	Sum = 82

$$\text{The mean deviation} = \frac{12 + 16 + 6 + 7 + 15 + 26}{6}$$
$$= \frac{82}{6} = 13\frac{2}{3}$$

The mean number of pizzas sold at Parlor B was $\frac{192}{6}$, or 32.

Score	Deviation Score (Score − Mean)	Absolute Value of Deviation Score
31	$31 - 32 = -1$	$\lvert -1 \rvert = 1$
34	$34 - 32 = 2$	$\lvert 2 \rvert = 2$
29	$29 - 32 = -3$	$\lvert -3 \rvert = 3$
40	$40 - 32 = 8$	$\lvert 8 \rvert = 8$
30	$30 - 32 = -2$	$\lvert -2 \rvert = 2$
28	$28 - 32 = -4$	$\lvert -4 \rvert = 4$
Sum = 192	Sum = 0	Sum = 20

$$\text{The mean deviation} = \frac{1 + 2 + 3 + 8 + 2 + 4}{6}$$
$$= \frac{20}{6} = 3\frac{1}{3}$$

Since the mean deviation for Parlor B is less than the mean deviation for Parlor A, the Saturday pizza sales of Parlor B tend to be closer to the mean than those of Parlor A. Because Parlor B is more consistent than Parlor A in pizza sales, Aunt Chovie knows that Parlor B will need a more consistent number of employees on Saturday.

Note that for each pizza parlor the sum of the deviation scores was zero. Can you explain why?

Problem Set

P,R **1** The graph shows the number of goals scored by Flash Falett during his recent seven-game streak of scoring at least one goal per game.

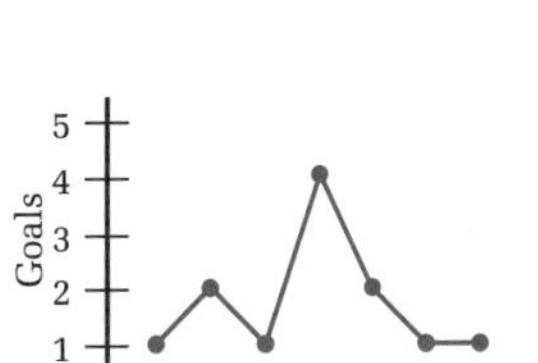
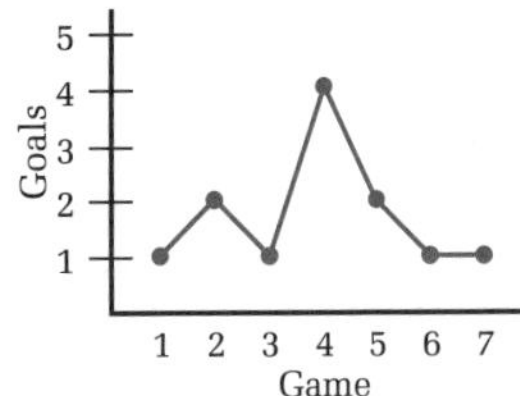

a Find the total number of goals Flash scored during the streak. 12

b Determine the range of goals scored. 3

c Find the mean, the mode, and the median of the data. Mean: ≈1.71; mode: 1; median: 1

P **2** The mean deviation of the data 10, 12, 17, 25, and x is y.

a Find x if the mean is 18. 26

b Find y if the mean is 18. 6

Assignment Guide

Average
1–4, 7
Advanced
1–4, 6, 7
Honors
1–7

Using Computers

Use the *Algebra Solver Kit* with this section. Program for stem-and-leaf plots and measures of dispersion can enhance this section and provide a nice verification for students.

Problem-Set Notes and Strategies

- Problem **1** is a review of reading data from graphs and the different measures of central tendency.
- You might discuss which would be the better measure of central tendency in problem **2**. Since the data are skewed, the median is the better measure. It seems reasonable that the mean deviation would also be the better measure of dispersion.

Problem-Set Notes and Strategies, continued

- Problem **6** is a problem in working with the formulas for mean and mean deviation.
- Problem **7** shows another advantage of the stem-and-leaf diagram. Not only does it give a picture of the data, it also makes all the data points available for computation.

Communicating Mathematics

Write a paragraph about the given situations and how the measures of dispersion allow you to make better judgements.

1 You have just received the results of your fourth test in this class and find out that for the 32 students who took the test, the average was 80%. Your score was 78%. How do you feel?

2 You have now found out that in addition to the fact that the average score is 80%, 30 students scored 80%, 1 scored 82% and 1 scored 78%. Verify that the average is 80% and describe your feelings after you heard this information.

Problem Set, *continued*

P **3** Average monthly temperatures for Twosine and Mathem City are given in the following table. Which city has the more consistent average temperature during these months? Twosine

Month	Twosine	Mathem City
May	75	45
June	85	95
July	90	85
August	70	85
September	80	90

P **4 a** Copy and complete the following table.

Scores	Deviation Score	Absolute Value of Deviation Score
23	−23	23
32	−14	14
50	4	4
54	8	8
71	25	25

b Find the mean deviation score of the data. 14.8

P **5** In most skewed distributions, the difference between the arithmetic mean and the median is about one-third the difference between the arithmetic mean and the mode. Is this true for the data shown? No

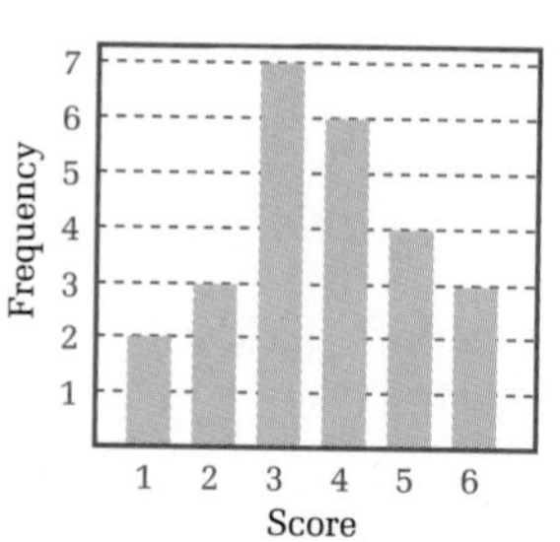

P **6** The mean deviation is found by averaging deviations about the mean. The median deviation is found by averaging deviations about the median. Find the mean and median deviations for the following data. Mean deviation: ≈19.6; median deviation: $15.\overline{5}$

41, 42, 44, 45, 46, 47, 63, 91, 111

P **7** Mr. Matt Hematics's test scores are shown.

a Find the range of the scores. 65

b Find the mean, median, and mode.

c Calculate the mean deviation. ≈12.5

b mean ≈ 70.5; median = 70; mode = 66

Stem	Leaves
9	3 6 8 9
8	4 5 9
7	0 1 3 4 4 7 8 8
6	3 4 4 6 6 6 8 9
5	3 4 7
4	6
3	4 6

5 | 7 means a score of 57 points

6

Chapter Summary

Concepts and Procedures

After studying this chapter, you should be able to

- Graph parabolas (6.1)
- Graph absolute-value functions (6.1)
- Stretch, shrink, and reflect graphs (6.1)
- Shift graphs vertically (6.2)
- Shift graphs horizontally (6.2)
- Reflect objects graphed on a coordinate system (6.3)
- Recognize symmetry with respect to a line or a point (6.3)
- Graph cubic and quartic polynomial functions (6.4)
- Graph reciprocal functions (6.4)
- Graph exponential functions (6.4)
- Recognize asymptotes of graphs (6.5)
- Recognize holes in graphs (6.5)
- Compute the range of a set of data (6.6)
- Compute the deviation scores and the mean deviation of a set of data (6.6)

Vocabulary

cubic function (6.4)
deviation score (6.6)
exponential function (6.4)
hole (6.5)
horizontal asymptote (6.5)
mean deviation (6.6)
measures of dispersion (6.6)
quartic function (6.4)
range (6.6)
reciprocal function (6.4)
reflect (6.1, 6.3)
shift (6.2)
shrink (6.1)
stretch (6.1)
symmetry (6.3)
vertical asymptote (6.5)

Class Planning

Time Schedule

All levels: 1 day

Assignment Guide

Average
1, 2, 4, 5, 8–10, 13–17, 19
Advanced
2–10, 12, 13, 15, 17, 18a–c, 19
Honors
2, 3, 6–10, 12–15, 17–20

Additional Answer

1

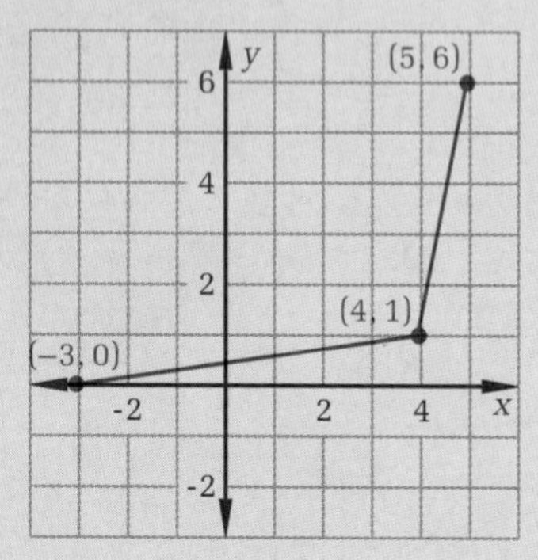

6 Review Problems

A P,R **1** Graph the reflection of $y = f(x)$ over the line $y = x$.

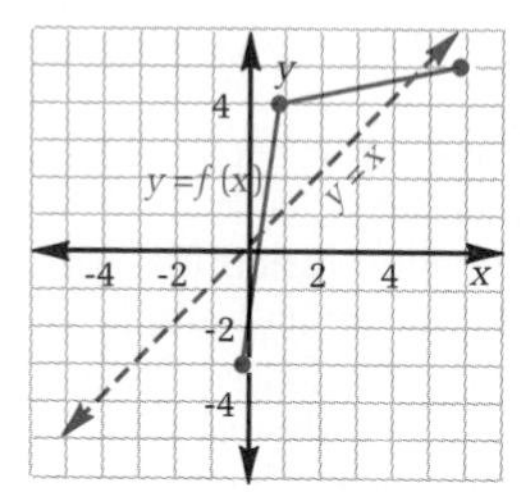

P **2 a** Find the range of the data shown. 7

b Find the mean number of free throws made per game. 3.9

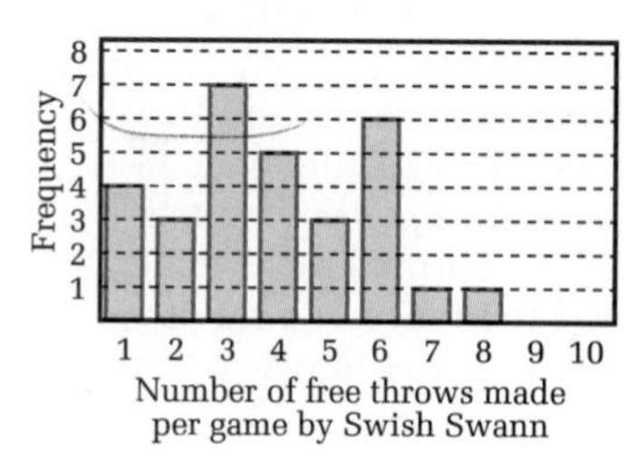

P **3** Let $f(x) = \frac{x^2 + 3x - 28}{x^2 + 8x - 48}$.

a Evaluate the numerator if $x = 4$ and $x = -12$. 0; 80

b Evaluate the denominator if $x = 4$ and $x = -12$. 0; 0

c Describe the behavior of the graph of f at $x = 4$ and $x = -12$. Has hole at $x = 4$ and a vertical asymptote at $x = -12$

P,R **4** Find the range of the data shown. 58

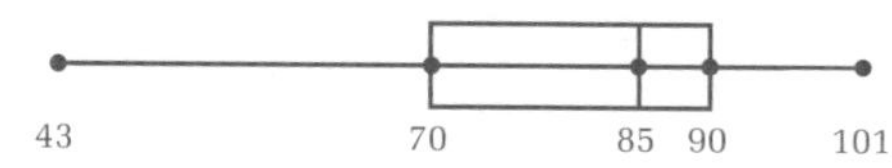

R **5** An airplane that flies at 425 miles per hour in still air is flying into a head wind of 75 miles per hour. How far will the plane fly in 5.5 hours? 1925 mi

P **6** The graph of $y = f(x)$ contains the point (3, 2). What point must be on the graph of each of the following functions?

a $g(x) = \frac{1}{f(x)}$ $\left(3, \frac{1}{2}\right)$

b $h(x) = [f(x)]^2$ (3, 4)

P **7** Match each function with its graph.

a $f(x) = 2^x$ v **b** $g(x) = \frac{1}{x}$ ii **c** $h(x) = \sqrt{x}$ iii

d $l(x) = x^3 - x$ i **e** $m(x) = x^4 - 5x^2 + 4$ iv

i

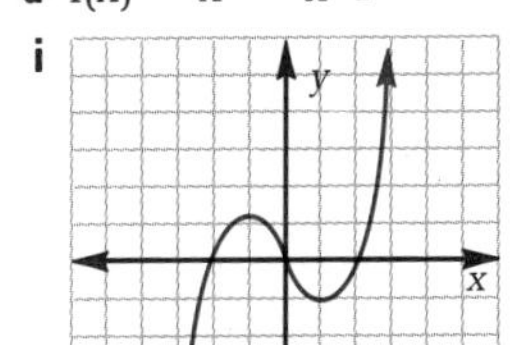

ii

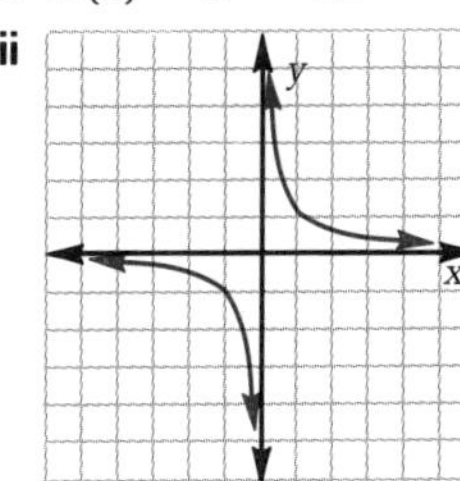

iii

iv

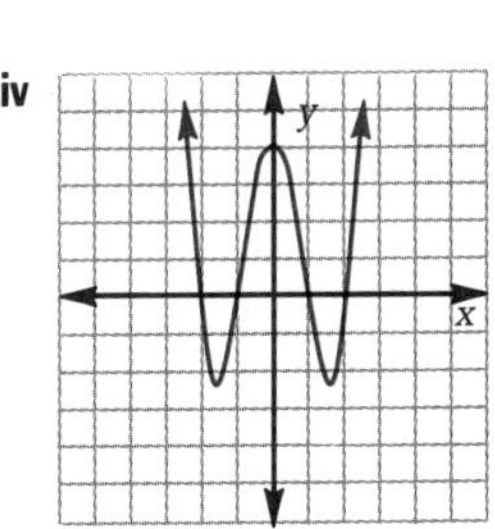

v

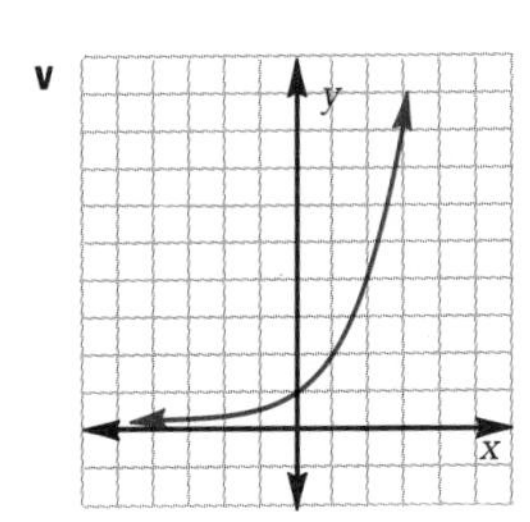

P **8** **a** Find the deviation score of each of the following data. -8; -4; -3; 4; 11

16, 20, 21, 28, 35

b Find the mean deviation score of the data. 6

P **9** If $g(x) = 6[f(x)]$, what are the coordinates of point A?
(2, 54)

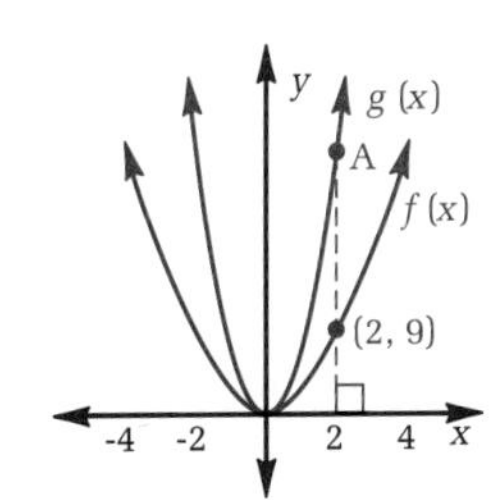

P,R **10** Let the domain of $f(x) = 3x - 5$ be $\{-3, -2, -1, 0, 1, 2, 3\}$. Find the range of each function.

a $y = f(x)$ **b** $y = 4[f(x)]$ **c** $y = -f(x)$
$\{-14, -11, -8, -5, -2, 1, 4\}$ $\{-56, -44, -32, -20, -8, 4, 16\}$ $\{-4, -1, 2, 5, 8, 11, 14\}$

R **11** Rex made 48 ounces of a mixture consisting of 60% saline solution and 40% active ingredients.

a How many ounces of saline were in the mixture? How many ounces of active ingredients? 28.8 oz; 19.2 oz

b If Rex added 10 ounces of saline solution to the mixture, what percentage of the new mixture was saline? $\approx$66.9%

Problem-Set Notes and Strategies

- Problem **11** is a mixture problem that might be encountered in a chemistry class. Students may have trouble keeping the quantities straight. There are two things happening in this problem. First, the amount of liquid is changing. Second, the percentage of saline is also changing. These two must be kept separate in different equations and then solved simultaneously.

Additional Answers

13a

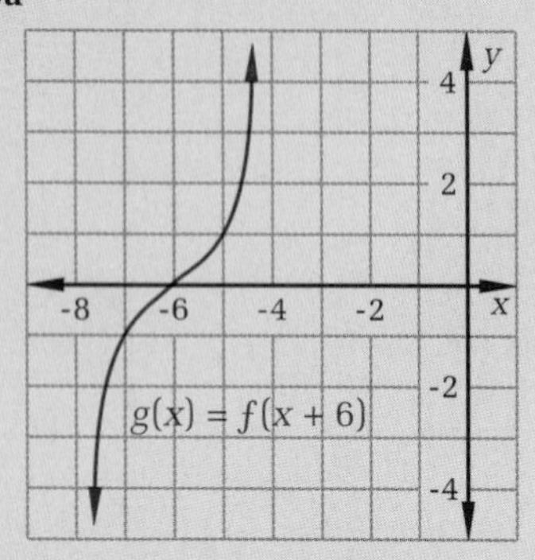

13b

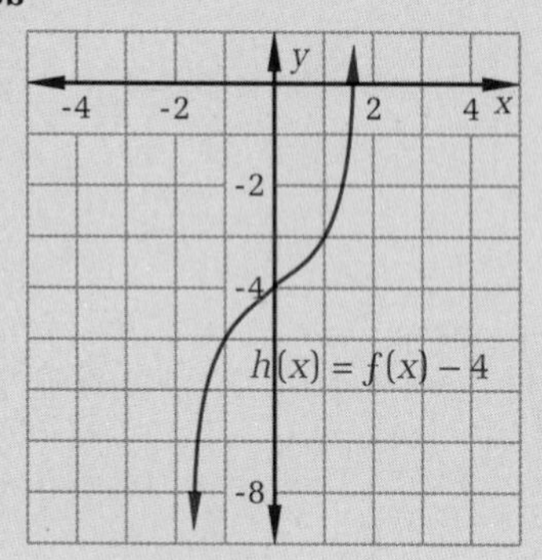

13c

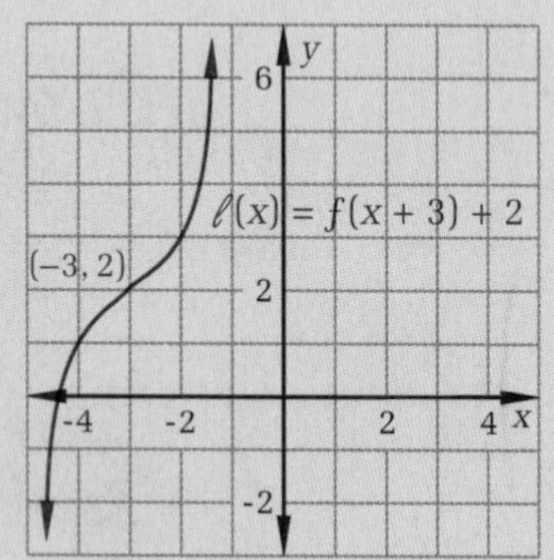

Answer for problem **19** can be found in the answer pages beginning on **A** 1.

Review Problems, *continued*

P **12** Write an equation of a function whose graph has vertical asymptotes at $x = 3$ and $x = 2$ and a hole at $x = -2$. Answers will vary. Possible answer: $f(x) = \frac{x + 2}{(x + 2)(x - 3)(x - 2)}$

P **13** Use the graph of $f(x) = x^3$ to sketch the graph of each function.

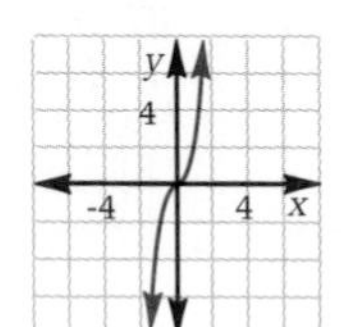

a $g(x) = f(x + 6)$
b $h(x) = f(x) - 4$
c $l(x) = f(x + 3) + 2$

P **14** The point (3, 9) is on the graph of $f(x) = x^2$. Find the corresponding point on the graph of each of the following functions.

a $y = f(x + 3)$ (0, 9)
b $y = f(x - 7)$ (10, 9)
c $y = f(x) + 4$ (3, 13)
d $y = f(x) - 9$ (3, 0)
e $y = f(2x)$ (1.5, 9)

P **15** Supply the missing information for each graph.

a

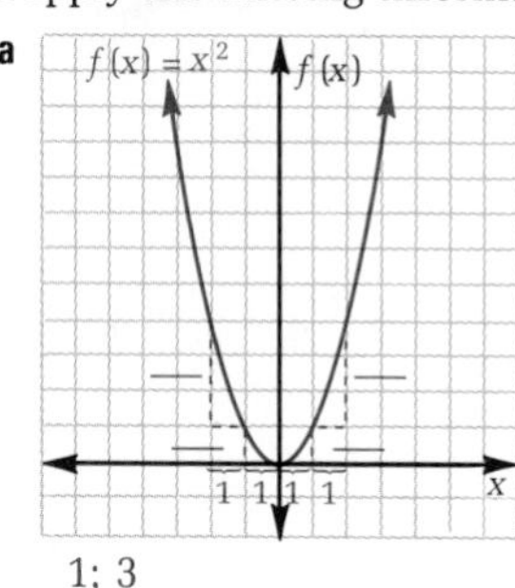

1; 3

b

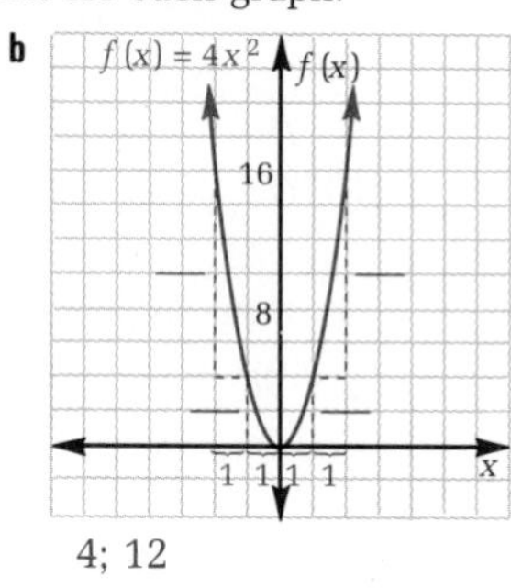

4; 12

c

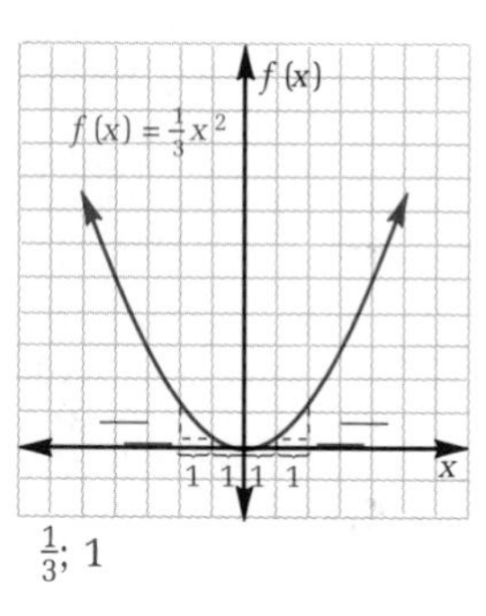

$\frac{1}{3}$; 1

R **B** **16** Find the values of x for which $y = 2x^2 - 3x - 2$ and $y = 0$. $x = -\frac{1}{2}$ or $x = 2$

P,R **17** Let $f(x) = 2(x - 3)^2 + 6$.

a Find the vertex of the graph of f. (3, 6)
b Find the axis of symmetry of the graph of f. $x = 3$
c If the graph of $y = f(x)$ is reflected over the x-axis, what is the equation of the reflected graph? $y = -2(x - 3)^2 - 6$

P,R **18** Let $f(x) = x^2$. Find the vertex of the graph of each of the following equations.

a $y = f(x + 2)$ (−2, 0)
b $y = f(x - 6)$ (6, 0)
c $y = f(x) + 5$ (0, 5)
d $y = f(x) - 7$ (0, −7)
e $y = f(x + 2) - 3$ (−2, −3)
f $y = f(x - 7) + 5$ (7, 5)

P,R **19** Graph $y = x^2 - 2$ and its reflection over the line $y = x$. What is an equation of the reflection? $y = \pm\sqrt{x + 2}$

C R **20** Let $f(x) = x + 2$ and $g(x) = |x|$. Find all ordered pairs (x, y) such that $y = f(x) = g(x)$. (−1, 1)

6

CHAPTER TEST

In problems 1–3, describe how the graph of each function differs from the graph of $f(x) = x^2$.

1 $f(x) = \frac{x^2}{4}$ Shrunk vertically

2 $f(x) = \frac{3}{2}x^2$ Stretched vertically

3 $f(x) = -\frac{2}{3}x^2$ Shrunk vertically and reflected over the x-axis

4 Let $f(x) = x^2$.

a Copy and complete the table and graph $y = f(x + 2)$.

x	−4	−3	−2	−1	0
$x + 2$	−2	−1	0	1	2
$f(x + 2)$	4	1	0	1	4

b Describe how the graph you drew in part **a** differs from the graph of $y = x^2$.
It is shifted two units to the left.

5 Name a point on the graph of each equation if (3, 17) is a point on the graph of $y = g(x)$.

a $y = g(x) - 10$ (3, 7)

b $y = g(x - 10)$ (13, 17)

c $y = g(x + 3) + 5$ (0, 22)

6 Use the graph of $f(x) = \left(\frac{1}{2}\right)^x$ to find the y-intercept of the graph of each of the following functions.

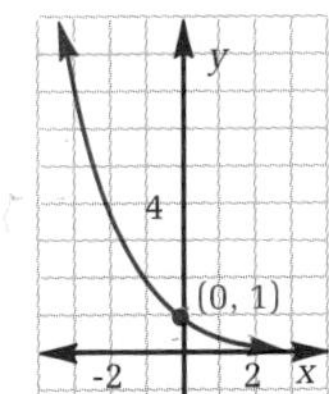

a f (0, 1)

b $g(x) = f(x) - 4$ (0, −3)

c $h(x) = f(x) + 9$ (0, 10)

d $m(x) = f(x - 3)$ (0, 8)

e $n(x) = f(x + 5)$ $\left(0, \frac{1}{32}\right)$

7 Write the equation of the reflection of the graph of $y = 3(x - 2)^2 + 9$ over each line.

a The x-axis $y = -3(x - 2)^2 - 9$

b The y-axis $y = 3(x + 2)^2 + 9$

c The line $x = 2$ $y = 3(x - 2)^2 + 9$

8 Let $f(x) = \frac{x^2 - x - 6}{x^2 - 5x + 6}$.

a What is the equation of the vertical asymptote of the graph of f? $x = 2$

b What is the x value of the hole in the graph of f? $x = 3$

c What are the real zeros of f? $x = -2$

9 The data 5, x, 37, 41, and y are arranged from smallest to largest. If the range is 64 and the mean is 35, what is (x, y)? (23, 69)

6.

Resource References
Evaluation
Test Form 1, 2, 3

Additional Answer
4

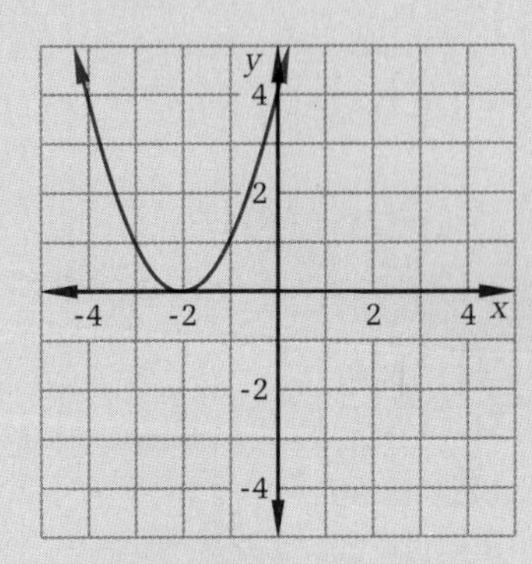

Class Planning

Time Schedule

Average: 2 days
Advanced: 3 days
Honors: 3 days

Resource References

Teacher's Resource Book
Class Opener 40
Practice 41, 42
Manipulatives
and Applications 11

Class Opener

Sketch the graph of $x^2 + y^2 = 36$.

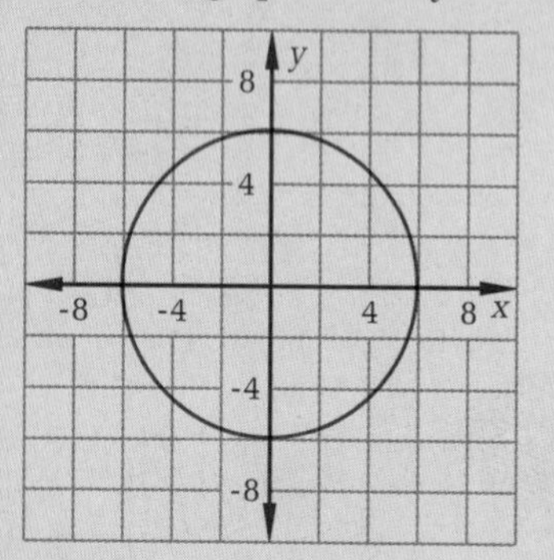

To extend the introduction, ask a few "what if?" questions.
What are the intercepts?
(0, 6), (0, −6), (6, 0), (−6, 0)
If we multiply x^2 by 9 and y^2 by 4, what are the intercepts?
(2, 0), (−2, 0), (0, 3), (0, −3)

Lesson Notes

- A **circle** is the set of all points in a plane each of which is equidistant from a fixed point called the center. The fixed distance is called the radius.

SHORT SUBJECT

Conics

Conics with Centers at the Origin

When a plane intersects a double cone, interesting and useful mathematical curves called conic sections are determined. There are four conic sections that we use frequently.

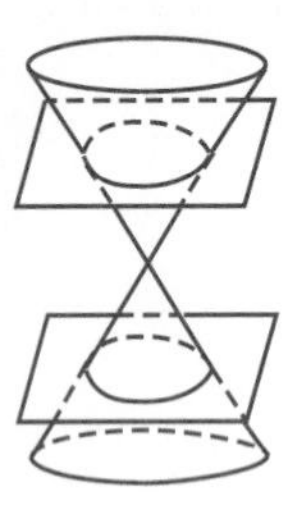

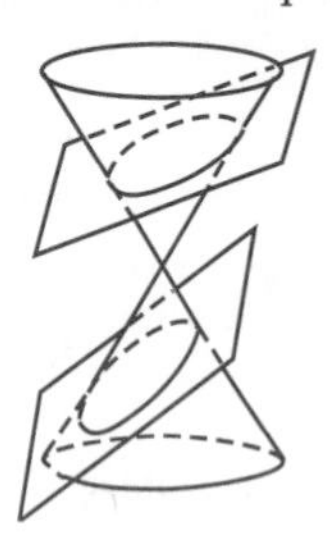

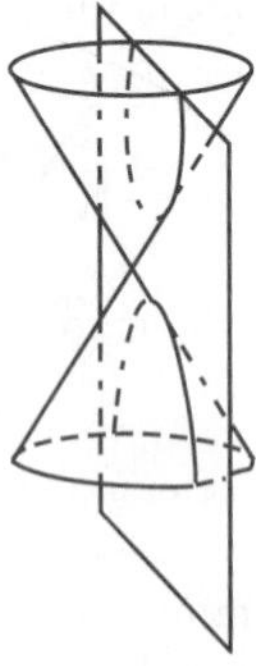

Circle **Ellipse** **Parabola** **Hyperbola**

You are probably familiar with the equations of circles and parabolas.

See p. 781

Example *Describe the graph of the equation* $x^2 + y^2 = 25$.

The graph is a circle with its center at the origin and a radius of 5.
x-intercepts: −5 and 5
y-intercepts: −5 and 5
Range of x values: $-5 \le x \le 5$
Range of y values: $-5 \le y \le 5$
Radius, r: 5
Symmetry: The graph is symmetric about the x-axis, the y-axis, and the origin.

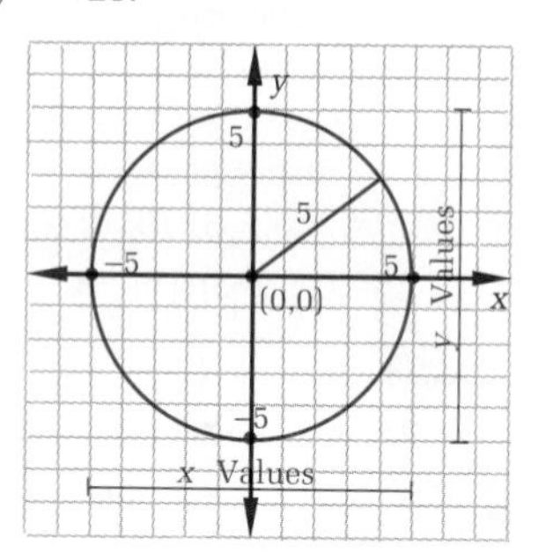

If we divide both sides of the original equation by 25, the equivalent equation is in a form that resembles the intercept form of a line: $\frac{x^2}{25} + \frac{y^2}{25} = 1$.

What happens if we change one of the denominators in this equation?

Vocabulary

ellipse
major axis
minor axis
hyperbola
asymptote

Example *Describe the graph of the equation* $\frac{x^2}{9} + \frac{y^2}{25} = 1$.

x-intercepts: ± 3
y-intercepts: ± 5
Range of x values: $-3 \le x \le 3$
Range of y values: $-5 \le y \le 5$
Symmetry: x-axis, y-axis, origin
The graph of $\frac{x^2}{9} + \frac{y^2}{25} = 1$ is an ellipse.

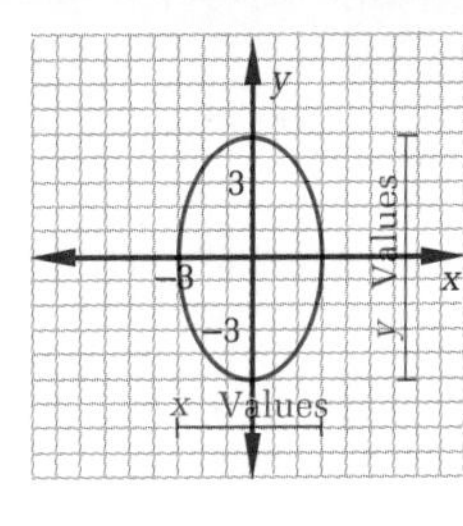

An ellipse is a circle that has been stretched along a line passing through its center. An ellipse with a center at (0, 0), symmetric to the x- and y-axes, has an equation of the form $\frac{x^2}{a^2} + \frac{y^2}{b^2} = 1$.

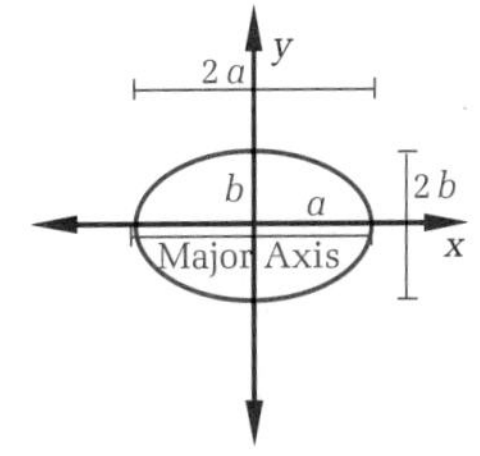

The longest diagonal of the ellipse is called the ***major axis***; the shortest diagonal is called the ***minor axis.*** For the ellipse $\frac{x^2}{9} + \frac{y^2}{25} = 1$, the y-axis is the major axis, and the x-axis is the minor axis.

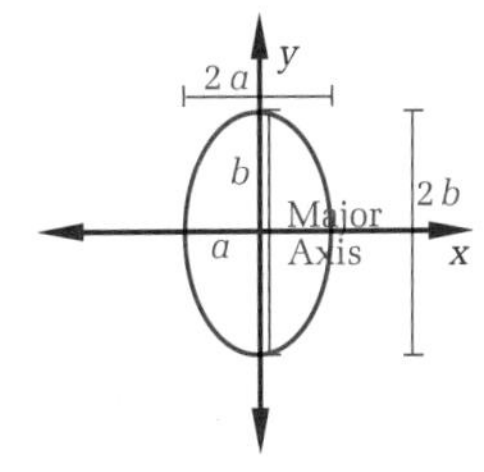

What happens if the x and y terms are subtracted instead of added?

Example *Describe the graph of the equation* $\frac{x^2}{9} - \frac{y^2}{25} = 1$.

x-intercepts: ± 3
y-intercepts: If $x = 0$, $\frac{y^2}{25} = -1$, but no real value of y satisfies this equation, so there are no y-intercepts. In fact, there is no real value of y for the interval $-3 < x < 3$. Thus, the graph will have values only for the range of $x \ge |3|$. To see the graph, we can plot some points.

x	± 3	$\approx \pm 4.2$	± 4
y	0	± 5	$\approx \pm 4.4$

For this equation, x^2 and y^2 have the same value for any set of four ordered pairs (x, y), (−x, −y), (x, −y), and (−x, y), and the graph is symmetric about the x-axis, the y-axis, and the origin.

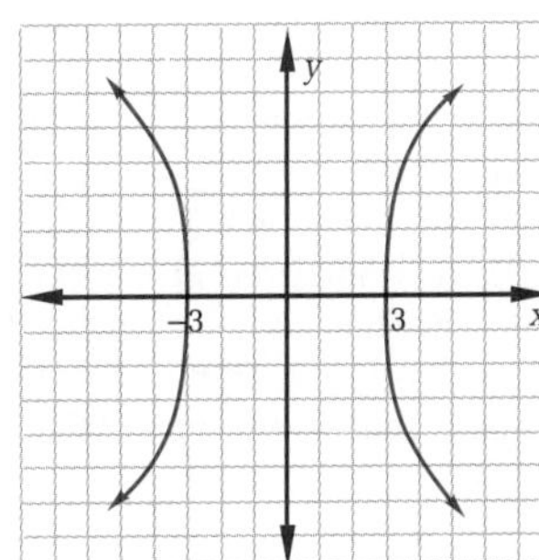

Lesson Notes, continued

- An **ellipse** is the set of all points in a plane such that the sum of the distance from each point to two fixed points, called the foci, is a constant.
- A **parabola** is the set of all points in a plane each of which is the same distance from a fixed line, called the directrix, as it is from a fixed point, called the focus.
- A **hyperbola** is the set of all points in a plane such that for each point the absolute value of the difference of its distances from two fixed points, called the foci, is a constant.
- This short subject deals with the conic sections, the first of many analytic geometry topics. The concepts of stretching, shrinking and translations which were explored in the last chapter will be utilized throughout this section. Symmetry will be explored for all the models. The hyperbola is the only one of the conics to have asymptotes which are used to draw its graph.

This conic is a ***hyperbola***, with an equation of the form $\frac{x^2}{a^2} - \frac{y^2}{b^2} = 1$. As x gets very large, the graph approaches the lines $y = \frac{5}{3}x$ and $y = -\frac{5}{3}x$, which are called ***asymptotes.***

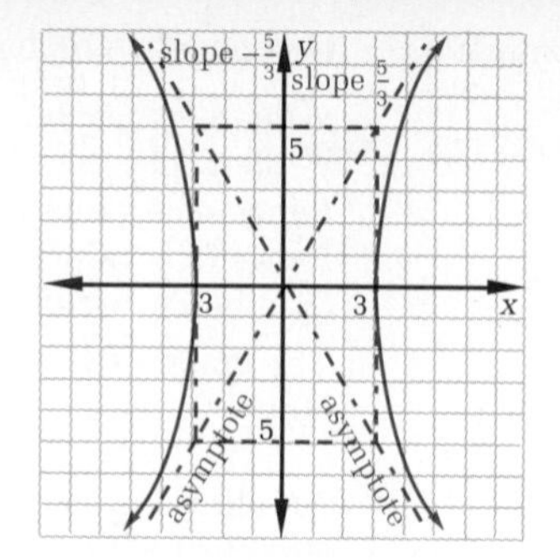

If we interchange x and y in the equation, then the hyperbola is rotated 90°, and the resulting equation is $\frac{y^2}{9} - \frac{x^2}{25} = 1$.

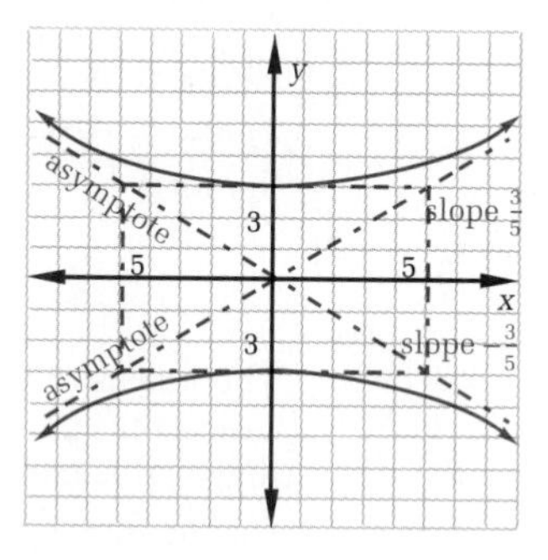

We have already studied the conic with an equation of the form $y = ax^2$: a parabola with a vertex at the origin and a vertical axis of symmetry. The graph of the equation $x = ay^2$ is also a parabola, but one with a horizontal axis of symmetry.

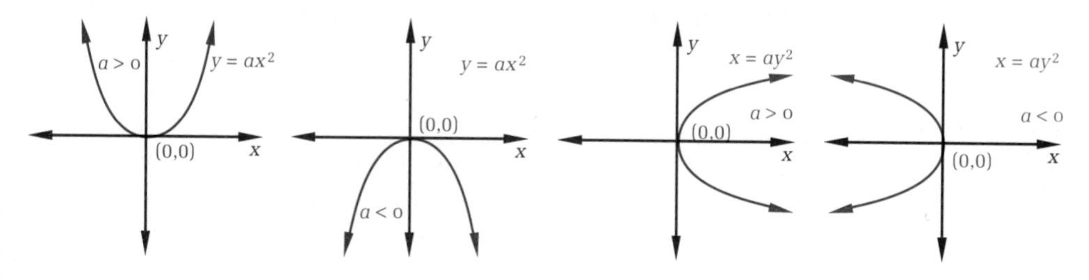

Conics with Centers Not at the Origin

Once we recognize the equation of a conic with a center at the origin, we extend our study to the general equation of a conic.

Remember from our earlier work with parabolas that an equation in the form $y = (x - 4)^2 + 7$, or $y - 7 = (x - 4)^2$, is a parabola that opens upward and has a vertex at (4, 7).

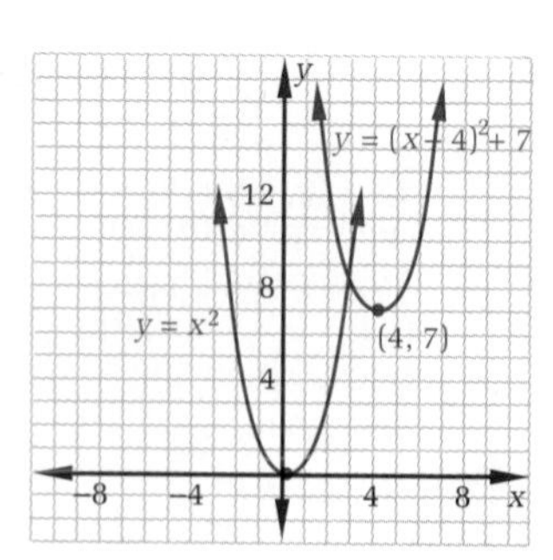

A circle with the equation $(x - h)^2 + (y - k)^2 = r^2$ has a center at (h, k) and radius r.

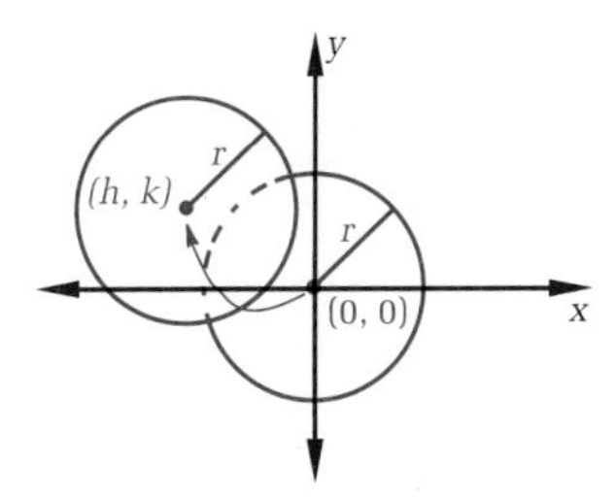

The ellipse $\frac{(x - h)^2}{a^2} + \frac{(y - k)^2}{b^2} = 1$ has a center at (h, k). The vertices are a units right and left of (h, k) and b units up and down from (h, k).

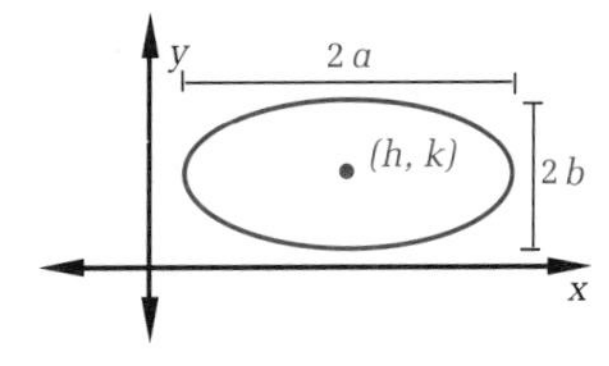

The hyperbola $\frac{(x - h)^2}{a^2} - \frac{(y - k)^2}{b^2} = 1$ has a center at (h, k), has vertices $(h + a, k)$ and $(h - a, k)$, and opens left and right.

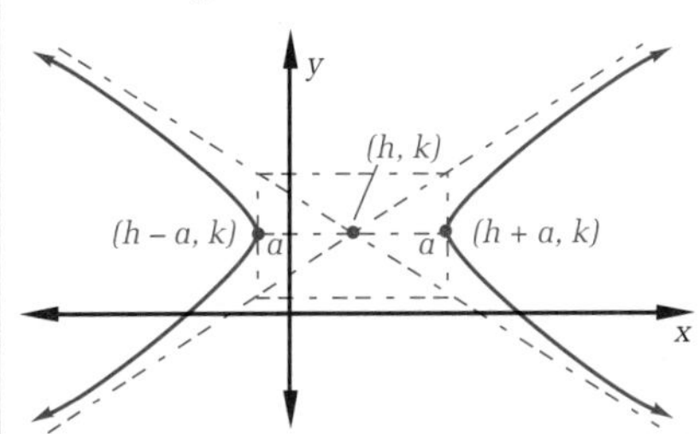

The hyperbola $\frac{(y - k)^2}{b^2} - \frac{(x - h)^2}{a^2} = 1$ has a center at (h, k), has vertices $(h, k + b)$ and $(h, k - b)$, and opens up and down.

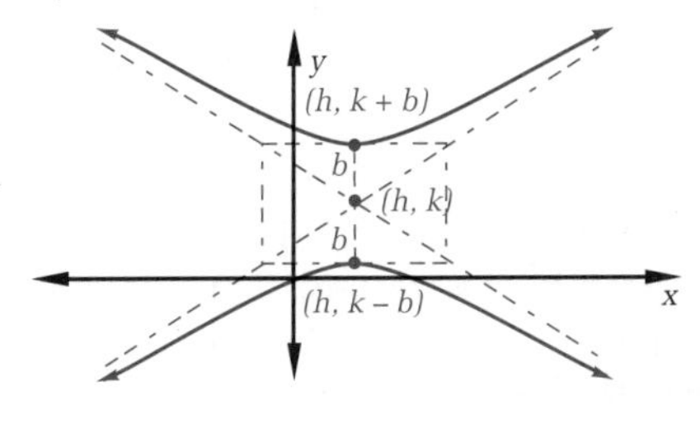

The graph of the hyperbola approaches asymptotes that pass through a center (h, k) and have slopes $\frac{b}{a}$ and $-\frac{b}{a}$.

Sample Problems

See p. 781

Problem 1 *Graph the ellipse* $4x^2 + 25y^2 = 400$.

Solution Change the equation to the form $\frac{x^2}{a^2} + \frac{y^2}{b^2} = 1$.

$$4x^2 + 25y^2 = 400$$

$$\frac{4x^2}{400} + \frac{25y^2}{400} = 1$$

$$\frac{x^2}{100} + \frac{y^2}{16} = 1$$ This is the equation of an ellipse with center $(0, 0)$.

If $y = 0$, $x = 10$ or -10. We move ten units to the left and right of the center. If $x = 0$, $y = 4$ or -4. We move four units up and down from the center. Use these points to sketch the curve. The major axis is horizontal.

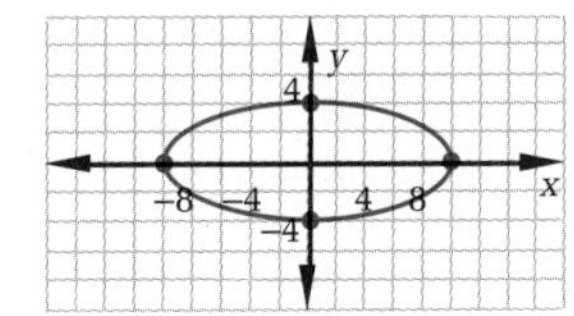

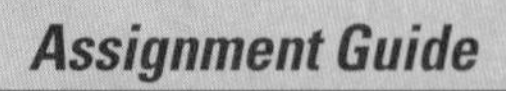

Assignment Guide

Average	
Day 1	1–7, 9, 13, 14, 16, 17, 22
Day 2	8, 15, 18, 19, 20, 21, 24
Advanced	
Day 1	1–9, 13, 14, 20, 22
Day 2	11, 12, 15–17, 21, 23
Day 3	10, 18, 19, 24–26, 28
Honors	
Day 1	1–9, 13, 14, 20, 22
Day 2	11, 12, 15–17, 21, 23
Day 3	10, 18, 19, 24–28

Problem-Set Notes and Strategies

- Students should be able to identify the conics in problems **1 – 6** quickly from the equation by examining the degree of each variable, its coefficient and the sign of the quadratic term.
- Students will use many techniques they learned in the past to draw the graphs in problems **8 – 13**. Pay special attention to problem **12**. It is a degenerate circle because the distance to all the points is zero. If the completing-the-square method was not studied in depth, you may need to spend some time reviewing that concept to do problems **10** and **11**.

Additional Answers

1. Hyperbola; center: (0, 0); vertices: $(0, \pm\frac{2\sqrt{2}}{3})$; asymptotes: $y = \pm\frac{2\sqrt{2}}{3}x$
2. Parabola; vertex (−8, 8)
3. Ellipse; center (0, 0); vertices (±3, 0) and $(0, \pm 2\sqrt{2})$
4. Line; y-intercept (0, 8), x-intercept (−9, 0)
5. Circle; center (0, 0); radius $2\sqrt{2}$
6. Parabola; vertex (0, 8)

Problem 2 *Graph the hyperbola* $25x^2 - 4y^2 = 400$.

Solution Change the equation to the form $\frac{x^2}{a^2} - \frac{y^2}{b^2} = 1$ by dividing by 400. The resulting equation, $\frac{x^2}{16} - \frac{y^2}{100} = 1$, is a hyperbola. It opens left and right and has center (0, 0) and vertices (4, 0) and (−4, 0). This hyperbola looks a bit like a parabola but the curves are not the same. The graph of a hyperbola "straightens out" and approaches two asymptotes, determined by the diagonals of the rectangle. The lengths of the sides of the rectangle are 2a and 2b. The slopes of the asymptotes can be seen in the rectangle to be $\pm\frac{10}{4}$, or ±2.5. The equations of the asymptotes are $y = 2.5x$ and $y = -2.5x$.

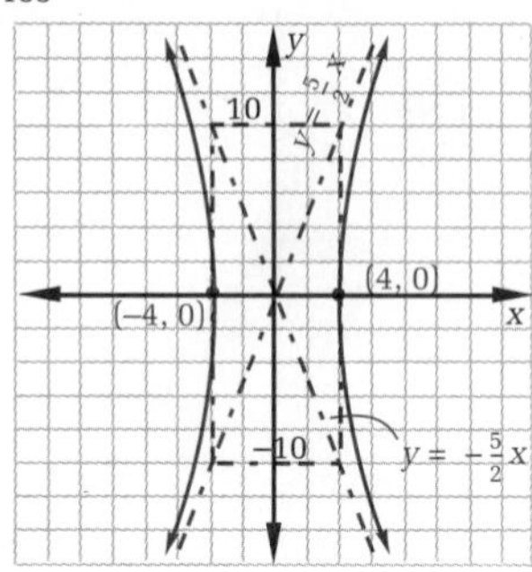

Problem Set

In problems 1–6, identify each equation as representing a line, a circle, an ellipse, a parabola, or a hyperbola. Give any information you can about the figure.

1 $9y^2 - 8x^2 = 72$
2 $9y = (x + 8)^2 + 72$
3 $9y^2 + 8x^2 = 72$
4 $9y - 8x = 72$
5 $9y^2 + 9x^2 = 72$
6 $9y - 8x^2 = 72$

P **7** Find an equation for each conic below.

a

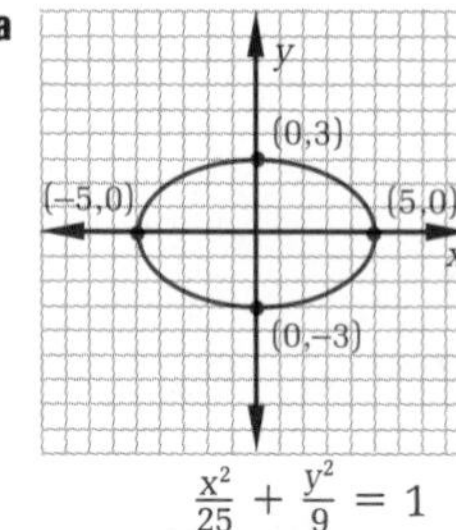

$\frac{x^2}{25} + \frac{y^2}{9} = 1$

b

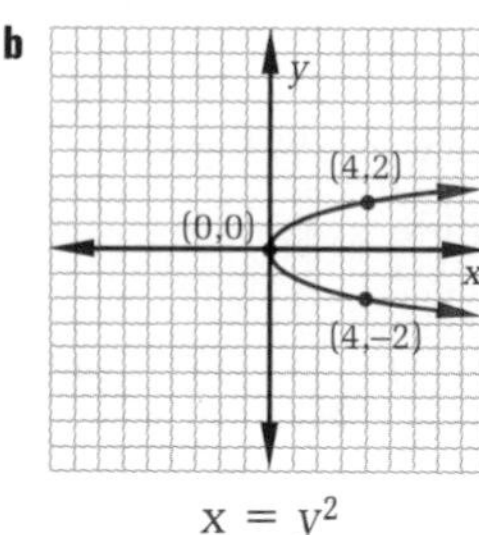

$x = y^2$

c

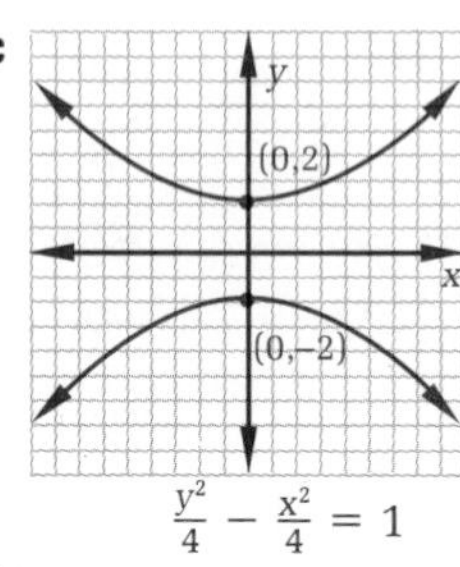

$\frac{y^2}{4} - \frac{x^2}{4} = 1$

P **In problems 8–13, determine the conic represented by each equation. List any important features, such as center, radius, and vertex. Then graph each conic.**

8 $4(x - 6)^2 + (y + 2)^2 = 36$
9 $\frac{(y + 5)^2}{49} - \frac{(x + 2)^2}{25} = 1$
10 $x^2 - y^2 - 6y + 7 = 0$
11 $x^2 + 10x + y^2 - 8y = 0$
12 $x^2 + y^2 = 0$ (Be careful!)
13 $x - 4 = 2(y + 3)^2$

P **14** Graph $x^2 + y^2 = 81$ and identify the curve.

Answers for problems **8 – 14** can be found in the answer pages beninning on A1.

P **15** Which of the following equations have graphs that are conics? Name the conics. All except **e**

a $x^2 + y^2 = 41$ Circle
b $3x^2 + 75y^2 = 675$ Ellipse
c $4y^2 - 9x^2 = 36$ Hyperbola
d $y = \frac{1}{2}(x + 3)^2 + 10$ Parabola
e $y = 2x^3 + 4x^2 + 7x + 4$
f $x = 3(y - 2)^2 - 5$ Parabola

B

In problems 16–19, graph each equation.

P **16** $(x - 4)^2 + (y + 2)^2 = 36$

17 $\frac{(y + 4)^2}{25} + \frac{(x - 3)^2}{16} = 1$

18 $\frac{(y - 4)^2}{16} - \frac{(x + 2)^2}{49} = 1$

19 $\frac{(x - 2)^2}{36} + \frac{(y + 3)^2}{9} = 1$

P,R **20** If one of the following equations is selected at random, what is the probability that its graph is an ellipse? $\frac{1}{3}$

a $\frac{x^2}{4} = 1 + \frac{y^2}{9}$
b $\frac{y^2}{16} + \frac{x^2}{25} = 1$
c $\frac{x^2}{16} - \frac{y^2}{9} = 1$
d $\frac{x^2}{49} + \frac{1}{25} = \frac{y^2}{4}$
e $\frac{x}{16} + \frac{y}{25} = 1$
f $\frac{y^2}{5} + \frac{x^2}{4} = 1$

P **21** Write an equation of the hyperbola shown. $\frac{x^2}{9} - \frac{y^2}{4} = 1$

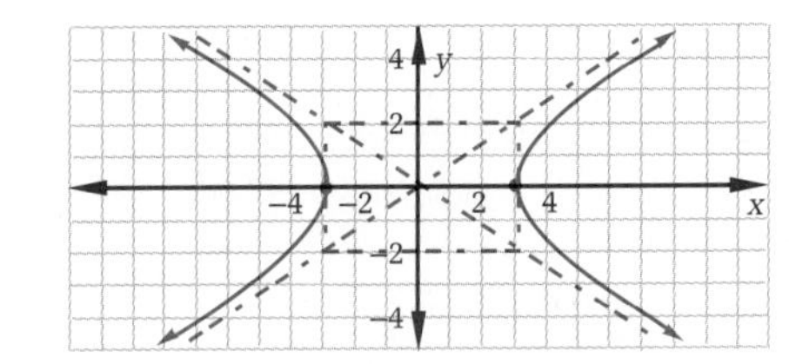

P **22** Write an equation of the circle with center (6, 1) and containing the point (2, 1). $(x - 6)^2 + (y - 1)^2 = 16$

P **23** Graph $\frac{(x + 6)^2}{16} - \frac{(y - 2)^2}{9} = 1$.

P **24** Find an equation of the circle centered at (−3, 7) and containing the point (0, 11). $(x + 3)^2 + (y - 7)^2 = 25$

C

P **25** **a** Find the x-intercepts of $g(x) = -2x^2 + 8$. ± 2
b Find the x-intercepts of $y = g(x + 3)$. $-1, -5$

P **26** Graph the following conics on the same set of axes.

a $\frac{x^2}{9} - \frac{y^2}{9} = 1$
b $\frac{x^2}{4} - \frac{y^2}{4} = 1$
c $\frac{x^2}{1} - \frac{y^2}{1} = 1$
d $x^2 - y^2 = 0$

P **27** Find all x- and y-intercepts of $\frac{(x - 3)^2}{4} - \frac{y^2}{9} = 1$. x-intercepts: 1, 5; y-intercepts: $\pm\frac{3\sqrt{5}}{2}$

P **28** **a** Change $f(x) = x^2 + 3x - 12$ into the form $f(x) = a(x - h)^2 + k$. $f(x) = \left(x + \frac{3}{2}\right)^2 - \frac{57}{4}$
b Identify the curve. Parabola
c Find the vertex. $\left(-\frac{3}{2}, -\frac{57}{4}\right)$

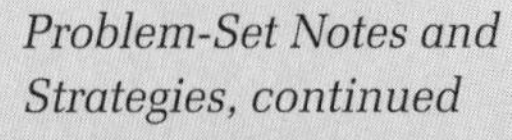

Problem-Set Notes and Strategies, continued

- In problems like **18**, **21** and **23** where graphing a hyperbola is required, students may draw the hyperbola opening in the wrong direction. When the equation is in standard form, only one of the squared variables is positive and that is always the direction in which the hyperbola opens. If the students try to prove that fact, they will find it easier to remember. It is not difficult to prove and the proof is fairly intuitive.
- Problem **26** leads the student to the case of a degenerate hyperbola. Point out to students what is happening to the graph in terms of the curves, especially the curvature of each half of the asymptote.

Additional Answers

16

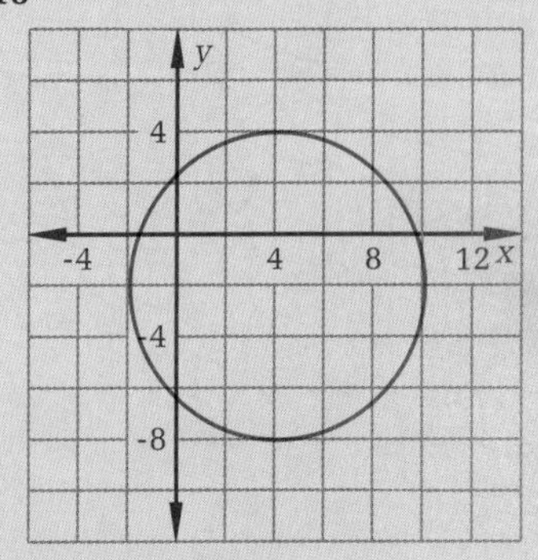

17

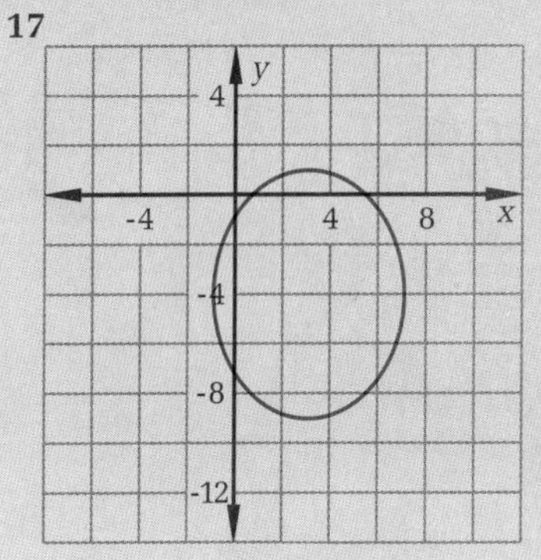

Answers for problems **18**, **19**, **23**, and **26** can be found in the answer pages baeinning on **A** 1.

7.1

CHAPTER 7 SYSTEMS

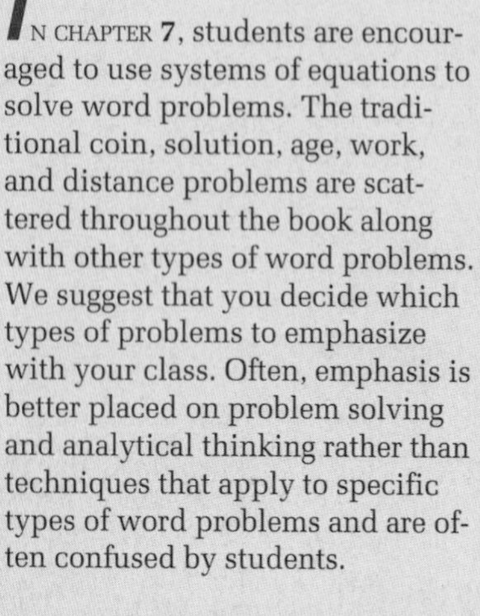

IN CHAPTER **7**, students are encouraged to use systems of equations to solve word problems. The traditional coin, solution, age, work, and distance problems are scattered throughout the book along with other types of word problems. We suggest that you decide which types of problems to emphasize with your class. Often, emphasis is better placed on problem solving and analytical thinking rather than techniques that apply to specific types of word problems and are often confused by students.

The study of systems of equations is extended to include equations in three or more variables. We suggest that you take the opportunity to expose all your students to three-dimensional graphing.

Systems of inequalities prepare students to deal with advanced topics, such as linear programming. Students will also study inverse variation, which has applications in science, economics, and art.

Studying nonlinear and mixed systems of equations will help prepare students for further studies in mathematics that involve quadratic and higher-degree systems.

By describing this scene as a system of equations, you could calculate the conditions under which the paths of two given cars would cross.

7.1 Solving Problems with Systems of Equations

Objective

After studying this section, you will be able to

- Use systems of equations to solve problems

Part One: Introduction

You have already studied how to solve systems of linear equations by making a graph and by using two algebraic methods. Look carefully at the solution to the following word problem.

Example *Paul E. Bank has \$2.25 in quarters and dimes. He has 15 coins. Find the number of quarters and the number of dimes he has.*

Let d and q be the number of dimes and quarters respectively.

The total value of the coins is \$2.25.

Value of dimes + value of quarters = total value

$10d + 25q = 225$ (cents)

The total number of coins is 15.

$d + q = 15$

We have a system of two equations in two unknowns. The system's graph, shown in the diagram, indicates that $(d, q) = (10, 5)$.

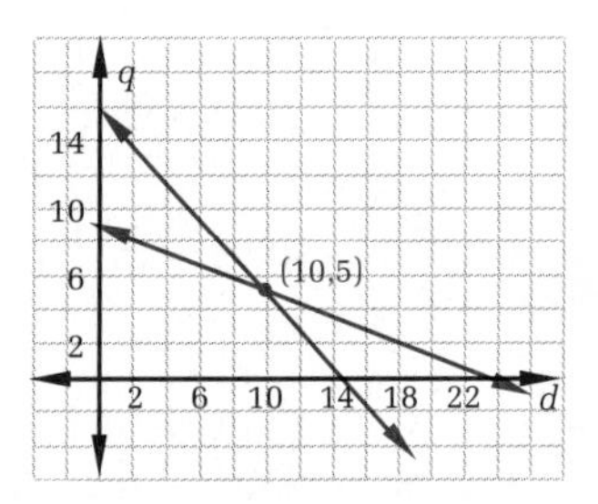

We could also solve the system algebraically, using substitution. Since $d = 15 - q$,

$$10(15 - q) + 25q = 225$$
$$150 - 10q + 25q = 225$$
$$15q = 75$$
$$q = 5, \text{ so } d = 15 - 5 = 10$$

There are 10 dimes and 5 quarters.

7.1

Class Planning

Time Schedule

All levels: 2 days

Resource References

Teacher's Resource Book

Class Opener 41
Cooperative Learning 20
Practice 43
Manipulatives and Applications 12
Enrichment 13

Class Opener

Sketch the graph of $3x + 5y = 30$ and the graph of $y = -2x - 1$ on the same coordinate system. What is the significance of the point of intersection?

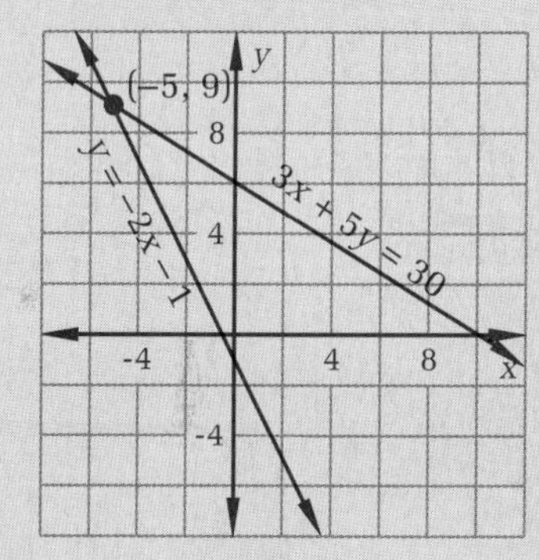

The point of intersection is a solution to both equations.

7.1

Lesson Notes

- Encourage the student to be organized. Write down the given information in an orderly fashion. Suggest that they draw pictures when possible. Remind students that calculations done on paper, rather than mentally, can be referred to at a later time.

- Encourage students to set up variables for each of the unknowns in the problem and write statements identifying each of these variables. Remind students to make sure the question is answered at the end of the problem. Frequently, finding the value of a variable is an intermediate step to answering the question posed in the problem. The equation $x = 25$ may not be sufficient; an uninformed reader would have to search to find what x represents.

Using Computers

This is an excellent time to use graphing software or a graphing calculator. Students can concentrate on the equation, the word problem, or the meaning of the solution and let the computer or calculator find the points of intersection. For linear systems you also can use the linear system solver provided in the *Algebra Solver Kit*.

If we change the total number of coins in the preceding example, the graph of the equation for the total number of coins will change.

If there were 12 coins, $d + q$ would equal 12. (Keep in mind that both d and q must be whole numbers.) The graph shows that $d = 5$ and $q = 7$.

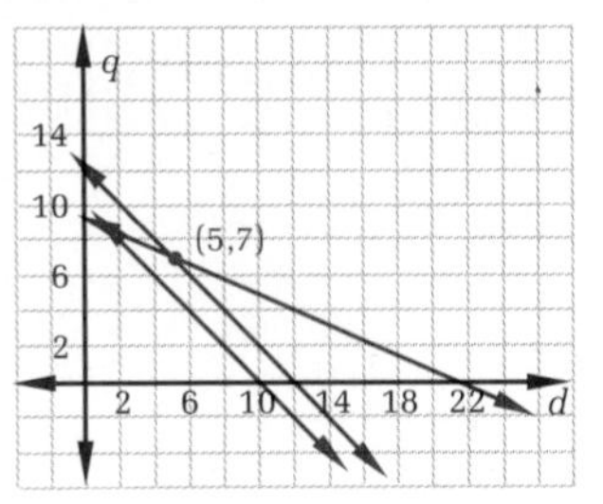

What would happen if there were 10 coins? The intersection of the graph would be $\left(1\frac{2}{3}, 8\frac{1}{3}\right)$. These coordinates are not whole numbers. Even though the system has a solution, the word problem does not. If 10 coins have a value of \$2.25, they cannot be all dimes and quarters.

You will find that some systems have no solutions and that others have infinitely many.

Example *Solve the following system.*

$$\begin{cases} y = 2x - 4 \\ y = 2x + 1 \end{cases}$$

Let's substitute $2x - 4$ for y in the second equation.

$$\begin{aligned} y &= 2x + 1 \\ 2x - 4 &= 2x + 1 \\ -2x \quad & \quad -2x \\ \hline -4 &= 1 \end{aligned}$$

The statement $-4 = 1$ is false. Therefore, the system has no solution. We get the same result by graphing. The slopes are equal but the y-intercepts are different. The graphs are parallel.

Example *Solve the following system.*

$$\begin{cases} y = 5x - 3 \\ 2y = 10x - 6 \end{cases}$$

We substitute $5x - 3$ for y in the second equation.

$$\begin{aligned} 2(5x - 3) &= 10x - 6 \\ 10x - 6 &= 10x - 6 \end{aligned}$$

This statement is an identity. All points that are solutions of $y = 5x - 3$ are also solutions of $2y = 10x - 6$, since one equation is a multiple of the other. This system has an infinite number of solutions, represented by $\{(x, y): y = 5x - 3\}$.

Part Two: Sample Problems

Problem 1 *Clem Chem needs to make 60 milliliters of a 9% solution of sulfuric acid for an experiment. He has available a 4% and a 12% solution. How many milliliters of each solution should he use?*

Solution Let x represent the number of milliliters of the 4% solution, and let y represent the number of milliliters of the 12% solution.

Amount of acid	$0.04x + 0.12y = 0.09(60) = 5.4$
Amount of solution	$x + y = 60$

$$\begin{cases} 4x + 12y = 540 \\ x + y = 60 \end{cases}$$

We can substitute $60 - x$ for y in the first equation.

$$4x + 12(60 - x) = 540$$
$$4x + 720 - 12x = 540$$
$$-8x = -180$$
$$x = 22.5, \text{ so } y = 60 - 22.5 = 37.5$$

Clem needs 22.5 milliliters of the 4% solution and 37.5 milliliters of the 12% solution.

Problem 2 *A 1400-mile flight from Arconia to Citron took 4 hours with a tail wind. On the return flight, with a head wind of the same speed, the flight took 5 hours. What was the wind speed? How long would the flight have taken if there had been no wind?*

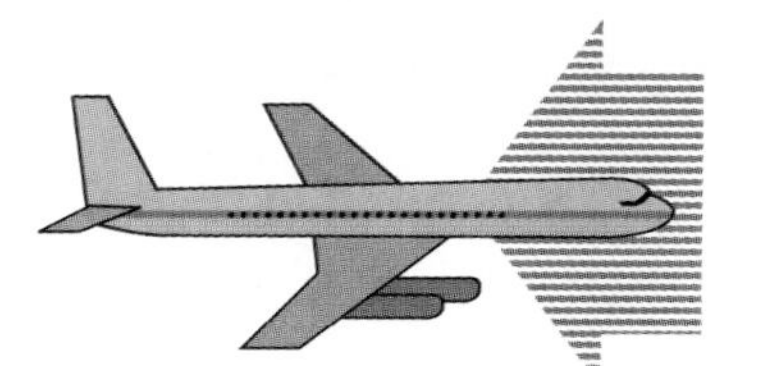

Solution First let's organize the information.

Let p = the speed of the plane in still air,
w = the speed of the wind,
$p + w$ = the speed of the plane with the tail wind,
$p - w$ = the speed of the plane against the head wind.

	Time	Rate	Distance
With the wind	4 hours	$p + w$	1400 miles
Against the wind	5 hours	$p - w$	1400 miles

Remember, time · rate = distance.

$$\begin{cases} 4(p + w) = 1400 \\ 5(p - w) = 1400 \end{cases}$$

We first simplify each equation by dividing out the numerical factors. Then we add the resulting equations.

$$p + w = 350$$
$$\underline{p - w = 280}$$
$$2p = 630$$
$$p = 315, \text{ so } w = 35$$

The wind speed is 35 miles per hour. The plane can travel 315 miles per hour with no wind. The flight would have taken $1400 \div 315$, or ≈ 4.44, hours if there had been no wind.

Checkpoint

1. The sum of two numbers is 16. If the difference of the same two numbers is 4, find both of the numbers. 6 and 10
2. You and Mary work at the same record shop. Mary is paid \$4.50 per hour and you are paid \$4.00 per hour. If together you worked for 30 hours and were paid \$126.50, how many hours did you work? 17
3. Banana splits cost \$3.25 each and milk shakes are \$1.95 each at the new ice cream shop in the mall. If you want to take your 5 friends for a treat but only have \$15.00, how many banana splits can you afford to buy if everyone has some treat? 2

7.1

Assignment Guide

Average	
Day 1	10–14, 18, 19, 21, 24–26
Day 2	15–17, 20, 22, 23, 27, 28, 31, 33, 34, 36, 37
Advanced	
Day 1	10–15, 18, 19, 21, 24–27, 38
Day2	16, 17, 20, 22, 23, 28, 31–34, 37, 40
Honors	
Day 1	10–15, 17–19, 21, 22, 24, 25
Day 2	20, 23, 26–28, 30–36, 38–41

Problem-Set Notes and Strategies

- The diagram in problem **4** should convince students to draw diagrams to aid in problem solving. This is the identical problem posed in the first checkpoint, but expressed in pictures instead of words. Students should try to communicate ideas by both methods.
- Problem **9** is a system of equations in three variables, but students should notice that by adding the equations both *y* and *z* are eliminated.

Communicating Mathematics

Have students write a paragraph that explains how they would solve problem **9**. Have them justify the process they used. Can this process be used to solve every system of equations?

Part Three: Exercises and Problems

Warm-up Exercises

P,R **1** Is (x, y) = (4, −3) a solution of the following system? Yes

$$\begin{cases} 2x + y = 5 \\ 4x + 2y = 10 \end{cases}$$

P,R **2** Is (x, y, z) = (−2, 4, 5) a solution of the following system? Yes

$$\begin{cases} x + y + z = 7 \\ 2x - 2y + 3z = 3 \\ x + y - z = -3 \end{cases}$$

P **3** A plane flies 350 miles per hour with a head wind and 428 miles per hour in the opposite direction when the wind is a tail wind.

a Find the plane's speed in still air. 389 mph

b Find the wind speed. 39 mph

P **4** Use the diagrams to solve for (r, c). (r, c) = (10, 6)

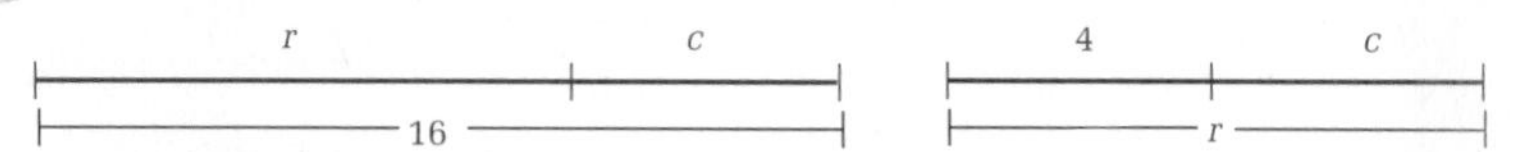

P,R **In problems 5–7, determine whether each system of equations has no solutions, one solution, or many solutions.**

5 $\begin{cases} y = 2x + 5 \\ y = 3x - 7 \end{cases}$ One solution

6 $\begin{cases} y = 3x - 5 \\ y = 3x + 5 \end{cases}$ No solution

7 $\begin{cases} y = 3x - 5 \\ 5 = 3x - y \end{cases}$ Many solutions

P **8** The sum of the ages of Zelda Zee and her sister Zuider is 45. Zelda's age is three more than twice Zuider's age. How old is each girl? Zelda is 31; Zuider is 14.

P,R **9** What is the value of x in the following system? 13

$$\begin{cases} 2x + 3y + 2z = 17 \\ 3x - 3y - 2z = 48 \end{cases}$$

Problem Set

A

In problems 10–12, determine whether each system of equations has no solutions, one solution, or many solutions.

P,R **10** $\begin{cases} y = 2x + 1 \\ y = 2x + 2 \end{cases}$ No solution

11 $\begin{cases} 3x + y = 4 \\ 9x + 3y = 12 \end{cases}$ Many solutions

12 $\begin{cases} y = 3x - 5 \\ y = 25x - 4 \end{cases}$ One solution

P **13** Use the diagrams to solve for (x, y). (x, y) = (6, 4)

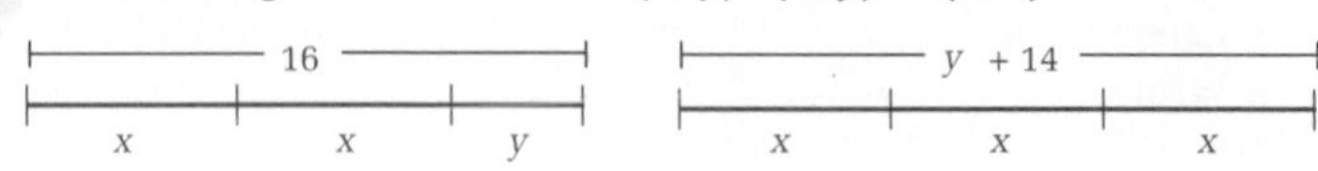

P,R **In problems 14–16, solve each system.**

14 $\begin{cases} 2x + 3y = 28 \\ 3x - y = 9 \end{cases}$ $(x, y) = (5, 6)$

15 $\begin{cases} x - 3y = 7 \\ -2x + 6y = 13 \end{cases}$ No solution

16 $\begin{cases} 2x + 3y = 5 \\ 4x + 6y = 5 \end{cases}$ No solution

P **17** Sam is three times as old as Mary. In six years, Sam will be twice as old as Mary. Find each person's present age. Mary is 6; Sam is 18.

P,R **18** Solve the following system for (x, y, z). $(x, y, z) = \left(\frac{13}{2}, \frac{13}{4}, \frac{91}{4}\right)$

$$\begin{cases} x = 2y \\ z = 7y \\ 3x + 5y - z = 13 \end{cases}$$

P,R **19** Dents Rent-a-Car charges $22.50 per day plus $0.18 per mile. Save-Us Rental charges $28.95 per day plus $0.15 per mile.

a Write an equation for the total cost C_d (in dollars) of a one-day trip of m miles in a car rented from Dents. $C_d = 22.5 + 0.18m$

b Write an equation for the total cost C_s (in dollars) of traveling m miles in one day in a car rented from Save-Us. $C_s = 28.95 + 0.15m$

c How many miles would one need to drive in one day in order for the two rental costs to be the same? 215 mi

B

P **20** The perimeter of each rectangle is 22. Find the values of x and y. x = 4; y = 2

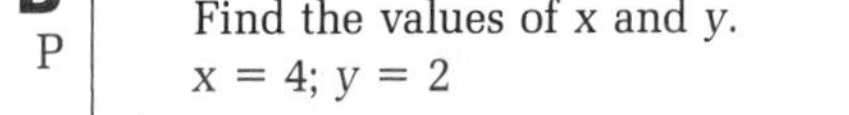

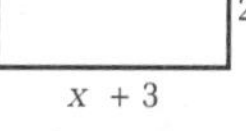

R,I **21** The graph of $y = mx + b$ has a slope of 3 and contains the point (4, −8). Find the value of b. −20

R **22** During the first 10 games of the soccer season, Carlos had a total of 47 shots on goal.

a How many shots on goal did he average per game? 4.7

b How many shots on goal must he have in the next 4 games to increase his average to 6 shots on goal per game? 37

P **23** Port and Starboard are paddling canoes. They travel 20 miles downstream in 4 hours. They return upstream in 6 hours.

a How fast is the current? ≈0.83 mph

b How fast can they paddle in still water? ≈4.17 mph

P,R **24** Solve the following equation for (x, y), then check your solution by means of matrix multiplication. $(x, y) = (-2, 5)$

$$\begin{bmatrix} 3 & -4 \\ 2 & 5 \end{bmatrix} \begin{bmatrix} x \\ y \end{bmatrix} = \begin{bmatrix} -26 \\ 21 \end{bmatrix}$$

R **25** Let $f(x) = (x - 3)(x + 4)(x - 9)$.

a Find the x-intercepts of the graph of $y = f(x)$. 3, −4, and 9

b Find the x-intercepts of the graph of $y = f(2x)$. $\frac{3}{2}$, −2, and $\frac{9}{2}$

R **26** If $f(x) = 2x^2$, for what value(s) of x is $f(x) + 2 = 9$? $\approx \pm 1.87$

Problem-Set Notes and Strategies, continued

- A chart might be helpful in problem **17** to identify Sam and Mary's age both now and six years from now. Once the chart is done, writing the equation becomes an easy task.
- Problem **19** is an application of a noncontrived word problem. Some students may have parents that travel. They might ask if these comparisons are taken into consideration when they rent cars.
- If a picture and chart is prepared for problem **23**, the equations are easy to write. Encourage the student to write and draw all the information that is available in some organized fashion. There may be superfluous information, but all necessary information will also be available.
- Problem **24** is a good illustration of how matrices may be used to write systems of equations. It is leading to the topics many will study later in their education in linear algebra.
- Make sure students realize that there are two answers to problem **26**.

Problem-Set Notes and Strategies, continued

- For problem **27**, borrow a graduated cylinder from the chemistry department and use water and oil to demonstrate this experiment. (Use ml instead of liters.) Students will understand mixture problems easier with a visual model.
- Problem **33** involves both systems of equations and proportions.
- You might relate problem **34** to problem **19** and let students see that to a mathematician, these problems are identical. The model for both is a system of 2 linear equations with two variables.

Additional Answers

34a

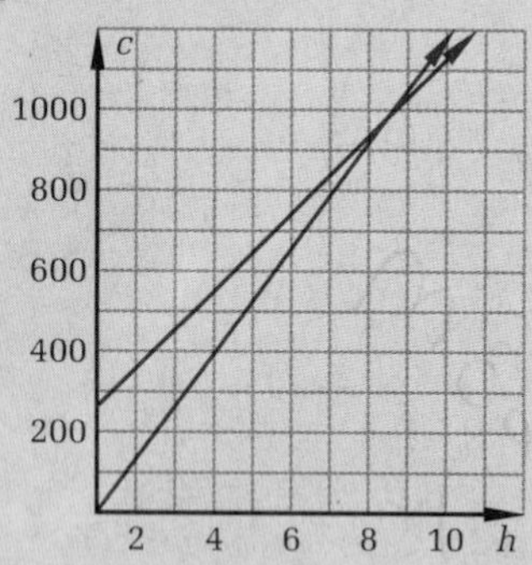

- Problem **36** has an interesting consequence which most students will not notice. If the equation were written as $ax^2 + bx + c = 0$, then the sum of the roots is $-\frac{b}{a}$ and the product of the roots is $\frac{c}{a}$. You might challenge the students to see if that is always the case. See the cooperative learning exercise for this section.

Problem Set, *continued*

P **27** A 140-liter batch of a mixture contains 25 liters of iodine.

- **a** If 60 liters of iodine is added to the mixture, what percentage of the mixture will be iodine? 42.5%
- **b** Find the least number of liters of iodine that must be added to the original mixture to create a mixture that is at least 70% iodine. ≈243.4 L

P,R **28** Find (m, b) if (2, 5) and (7, 11) are points on the graph of $y = mx + b$. $(m, b) = \left(\frac{6}{5}, \frac{13}{5}\right)$

P,R **In problems 29 and 30, solve the system.**

29 $\begin{cases} y = -3x + 4 \\ 12x + 4y = 2 \end{cases}$
No solution

30 $\begin{cases} 3x - 4y = 12 \\ y = \frac{3}{4}x - 3 \end{cases}$
$\{(x, y): 3x - 4y = 12\}$

P **31** The perimeters of the square and the triangle are equal. Find (x, y) if three times the perimeter of the square exceeds twice the perimeter of the triangle by 24. $(x, y) = (14, 6)$

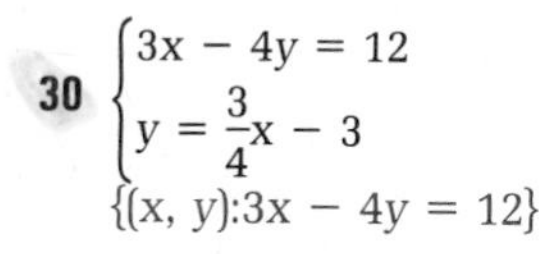
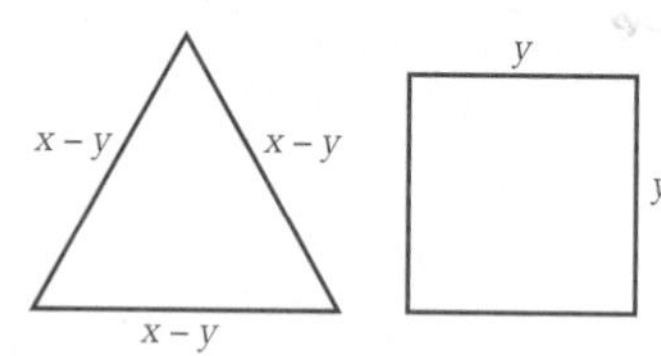

P,R **32** Two of the following equations are selected at random to form a system. Find the probability that the system has one solution. $\frac{4}{5}$

- **a** $y = 3x + 7$
- **b** $y = -5x + 4$
- **c** $5x + y = 20$
- **d** $3x - y = 14$
- **e** $y = -15x + 2$

R **33** If $x = 24$ when $y = 6$ and y varies directly as x, what are the values of x and y when the sum of x and twice y is 36?
$x = 24$; $y = 6$

P **34** Carmony Computer Consultants charges a $255 fee plus $85 per hour for its work. Holliday Computer Works charges a flat rate of $115 per hour.

- **a** Write a cost function for each company, then graph the two functions on the same coordinate system. $C_c = 255 + 85h$; $C_h = 115h$
- **b** For jobs of how many hours is it cheaper to hire Carmony? Jobs of more than 8.5 hr

R **35** Let $y = -5x^2 - 20x + 6$. Find q if $y = -5(x^2 + qx) + 6$. $q = 4$

I,R **36** Let $3x^2 + 5x - 2 = 0$.

- **a** Solve the equation for x. $x = \frac{1}{3}$ or $x = -2$
- **b** Find the sum of the equation's roots. $-\frac{5}{3}$
- **c** Find the product of the equation's roots. $-\frac{2}{3}$

P **37** Solve the following system for (x, y). $(x, y) = (3, 4)$

$$\begin{cases} 2^{x+y} = 128 \\ 5^{2x-y} = 25 \end{cases}$$

P,R **38** Refer to the triangle.

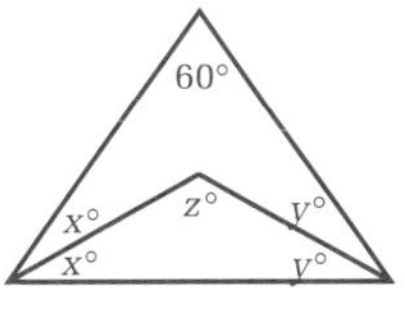

a Find the value of z. 120°

b Can you find the values of x and y? If the triangles are isosceles, $x = 30°$, $y = 30°$. If the triangles are not isosceles we know only that $x + y = 60°$.

C P **39** Mr. Fairwrench's radiator contains seven quarts of a mixture that is 35% antifreeze and 65% water. He needs a mixture that is 50% antifreeze and 50% water. How many quarts should he drain from his radiator and replace with pure antifreeze? ≈1.62 qt

P **In problems 40 and 41, solve each system for (x, y).**

40 $\begin{cases} x^2 = 2y^2 \\ 3x = y^2 \end{cases}$ $(x, y) = (0, 0)$, $(x, y) = (6, 3\sqrt{2})$, or $(x, y) = (6, -3\sqrt{2})$

41 $\begin{cases} y = |x - 2| + 5 \\ y = \frac{1}{2}|x| \end{cases}$ No solution

HISTORICAL SNAPSHOT

A SURVEYOR'S DREAM

Planning a capital city

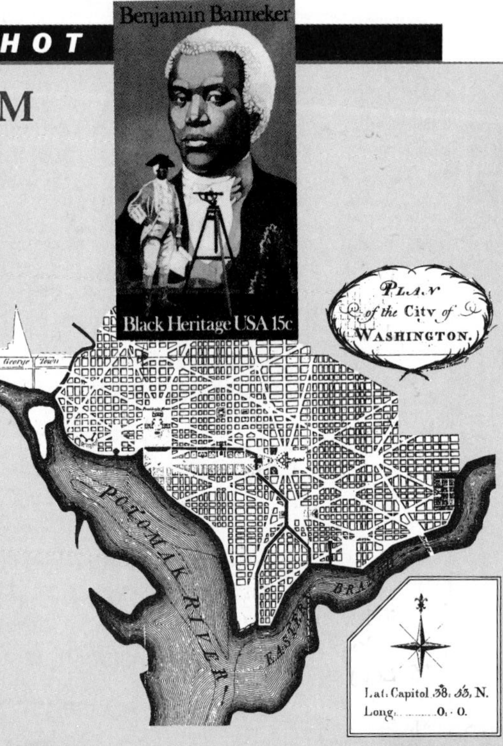

In general, large cities have grown spontaneously around an original settlement. As the city grew outward, streets and bridges were built along the paths that best served the needs of the community.

However certain capital cities have been planned from the very beginning. Brasília, the capital of Brazil, is one example. Washington, D.C., is another.

The first capital of the United States was New York City. It was President George Washington who wanted the capital moved to its present location. He hired a French architect named Pierre Charles L'Enfant to create a plan for the new city. Then Washington put together a six-person team to carry out L'Enfant's plan.

Benjamin Banneker, a self-taught mathematician, became the first black presidential appointee when President Washington chose him as the surveyor for this team. Banneker was instrumental in selecting the sites for the White House, the Capitol, and other buildings still standing today. When L'Enfant resigned and left for France with his plans, Banneker reconstructed the plans from memory. It is largely because of Banneker that we are able to enjoy the beauty of Washington, D.C., today.

Describe situations in which you think linear systems helped Banneker as he faced the tremendous task of helping to design, survey, and lay out the nation's new capital.

Problem-Set Notes and Strategies, continued

- Problem **39** is a mixture problem in reverse. In this case something is being taken away and replaced, not just added. Have students draw pictures of the beginning, intermediate and final steps of the process and try to reason what the equations should be. The constant in this problem is the amount of fluid both drained and then replaced. The percentage of antifreeze in the two cases is different, but the amount is constant.

- Students frequently lose 0 as one of the solutions for x in problem **40** because they divide both sides of the equation by x. Make them aware that when they divide by variables, they frequently are discarding solutions of the original problems. Have them identify the conics and graph them to determine how many solutions are available.

Cooperative Learning

Have students work in groups of two or three and solve the equation $ax^2 + bx + c = 0$ for the general case. It may take a moment or two before they realize that the quadratic formula will give them the two roots. Find the sum of the two roots and then find the product. Use these two new rules to write a quadratic equation that has two roots whose sum is −2 and whose product is −15.
$x^2 + 2x - 15$

7.2

Class Planning

Time Schedule
All levels: 2 days

Resource References
Teacher's Resource Book
Class Opener 42
Cooperative Learning 21
Practice 44
Transparency 13

Class Opener

Solve the following system of equations for x, y, and z.

$$\begin{cases} 3x + 2y = 8 \\ 2x - 3z = 31 \\ 3y + 2z = 14 \end{cases}$$

$x = \frac{-136}{19}$ or ≈ -7.16,
$y = \frac{280}{19}$ or ≈ 14.74,
$z = \frac{-287}{19}$ or ≈ -15.11

Lesson Notes

- This section extends the work of solving systems of linear equations to three dimensions. Systems of equations with three variables are solved by reducing the system to two equations with two variables. This is an iterative process because we redo the same procedure each time with one less variable and one less equation. The method is extendable to larger systems with more variables.

7.2 Linear Systems with Three Variables

Objectives

After studying this section, you will be able to
- Solve systems of equations in three unknowns
- Graph equations that contain three variables

Part One: Introduction

The strategy for solving systems of three equations in three unknowns involves combining pairs of equations to eliminate one of the variables, leaving a system of two equations in two unknowns.

Example *Solve the following system.*

$$\begin{cases} x + 2y + 4z = 8 & (1) \\ x - 2y + 5z = -3 & (2) \\ 4x + 4y - 3z = 1 & (3) \end{cases}$$

In this problem, it is easiest to eliminate y. Notice that we can obtain an equation that does not contain y by adding equations (1) and (2).

$$\begin{array}{lrcl} (1) & x + 2y + 4z &=& 8 \\ (2) & x - 2y + 5z &=& -3 \\ \hline & 2x \quad\quad + 9z &=& 5 \end{array}$$

Next, we multiply (2) by 2 and add the result to (3).

$$\begin{array}{rcl} 2x - 4y + 10z &=& -6 \\ 4x + 4y - 3z &=& 1 \\ \hline 6x \quad\quad + 7z &=& -5 \end{array}$$

Now we have two equations in x and z.

$$\begin{cases} 2x + 9z = 5 \\ 6x + 7z = -5 \end{cases}$$

When we solve this system (by the multiplication-addition method), we find that $z = 1$ and $x = -2$. To find y, we substitute these values in equation (1).

Vocabulary
trace

$$x + 2y + 4z = 8$$
$$-2 + 2y + 4(1) = 8$$
$$2y + 2 = 8$$
$$y = 3$$

The solution of the system is therefore $(x, y, z) = (-2, 3, 1)$. To check this answer, substitute -2 for x, 3 for y, and 1 for z in each equation.

We can also use this technique to find the equation of a parabola.

Example *Write an equation that represents a parabola containing the three points* $(-3, -2)$, $(-1, 3)$, *and* $(5, 6)$.

Since each point lies on the parabola, each ordered pair must be a solution of an equation of the form $y = ax^2 + bx + c$.

Substitute $(-3, -2)$ $\quad \mathbf{-2 = a(-3)^2 + b(-3) + c}$
Substitute $(-1, 3)$ $\quad \mathbf{3 = a(-1)^2 + b(-1) + c}$
Substitute $(5, 6)$ $\quad \mathbf{6 = a(5)^2 + b(5) + c}$

These equations form a system in a, b, and c.

$$\begin{matrix} (1) \\ (2) \\ (3) \end{matrix} \begin{cases} 9a - 3b + c = -2 \\ a - b + c = 3 \\ 25a + 5b + c = 6 \end{cases}$$

Eliminating c is easiest. We multiply equation (2) by -1 and add the result to (1) and (3).

$$\begin{array}{rl} (1) & 9a - 3b + c = -2 \\ -1(2) & \underline{-a + b - c = -3} \\ & 8a - 2b \quad = -5 \end{array} \qquad \begin{array}{rl} -1(2) & -a + b - c = -3 \\ (3) & \underline{25a + 5b + c = 6} \\ & 24a + 6b \quad = 3 \\ & 8a + 2b \quad = 1 \end{array}$$

Now we add the two new equations.

$$\begin{array}{l} 8a - 2b = -5 \\ \underline{8a + 2b = 1} \\ 16a \quad = -4 \\ a = -\frac{1}{4} \end{array}$$

We substitute $-\frac{1}{4}$ for a in one of the equations to find b.

$$8a - 2b = -5$$
$$8\left(-\frac{1}{4}\right) - 2b = -5$$
$$-2 - 2b = -5$$
$$-2b = -3$$
$$b = \frac{3}{2}$$

Now we substitute the values of a and b for these variables in (2) to find c.

Lesson Notes, continued

- Using this process, a parabola may be found through any three nonlinear points on the graph. This process is known as polynomial interpolation and may be extended to any number of points. For example, if you had 7 points and wanted to find an equation, the general equation $ax^6 + bx^5 + cx^4 + dx^3 + ex^2 + fx + g = y$ could be used with the seven points to generate 7 equations in a, b, c, d, e, f and g, and then solved for the coefficients. The process is identical.

- Graphs of these systems occur in three dimensions. These graphs are difficult to draw and may be worthwhile to draw ahead of time on an overhead transparency to save time. Students may have a hard time seeing the results at first, but should pick it up with time.

Using Computers

Use computer software or a calculator that can solve matricies, such as the *TI81*, the *Sharp Graphing Calculator*, or the *HP* calculators. The linear system solver in the *Algebra Solver Kit* uses matrix notation – something students should become familiar with. If you want to introduce Gauss Jordan row reduction form at this point, the software *Linear Kit* (Wiley & Sons) is ideal. You can do row reduction step by step with the *TI81 Graphics calculator.*

$$a - b + c = 3$$
$$\left(-\frac{1}{4}\right) - \frac{3}{2} + c = 3$$
$$-\frac{7}{4} + c = 3$$
$$c = \frac{19}{4}$$

An equation that represents the parabola is $y = -\frac{1}{4}x^2 + \frac{3}{2}x + \frac{19}{4}$.

Graphing Equations with Three Variables

If a linear equation contains three variables, its graph is a plane in three-dimensional space. To graph such an equation, we use a three-dimensional coordinate system consisting of three axes, each of which is perpendicular to the other two. Taken two at a time, these axes determine three coordinate planes—an xy plane, a yz plane, and an xz plane.

Example *Graph* $x + 2y + 4z = 8$.

Since three points determine a plane, we can locate the graph of this equation by identifying the points at which it intersects the three axes. To find each intercept, we substitute 0 for two of the variables and solve for the third. In this case, the x-intercept is 8, the y-intercept is 4, and the z-intercept is 2.

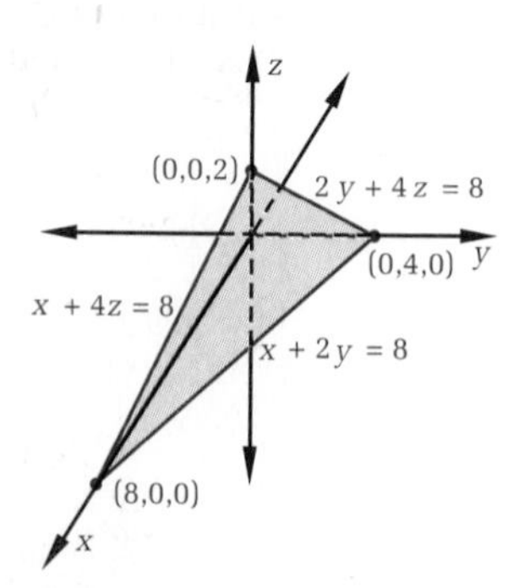

Each of the intersections of this plane with the coordinate planes can be found by substituting 0 for one of the variables and graphing the resulting line. If we let $z = 0$, we can see that the graph intersects the xy plane along the line $x + 2y = 8$. In the same way, we can find that it intersects the other coordinate planes along the lines $2y + 4z = 8$ and $x + 4z = 8$. These lines are called ***traces.***

The solution of a system of three equations in three variables is the intersection of three planes. This intersection may be a point, a line, or the empty set.

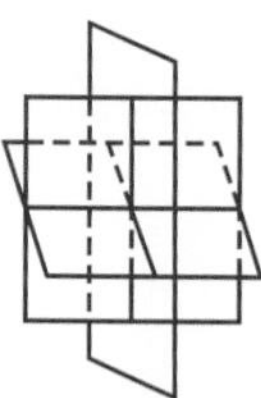

Planes have a point in common–one solution

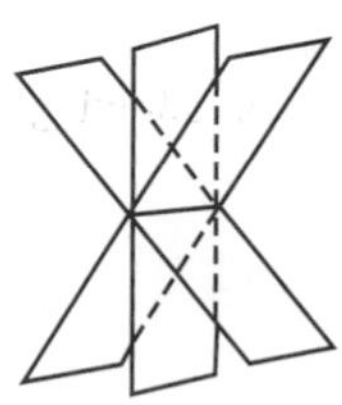

Planes have a line in common–infinite solutions

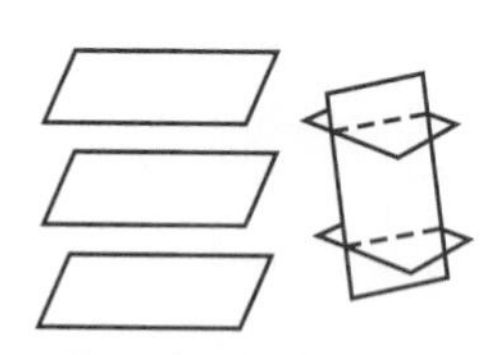

Planes have no points in common–no solutions

There is a final possibility. If two equations of the system are multiples of the third, the solution is a plane.

Part Two: Sample Problem

Problem *Tickets for all 1150 seats for the Thespis High School play were sold. Student tickets sold for \$1.50 and adult tickets sold for \$2.50, with senior citizens getting a \$0.50 discount off the adult price. A total of \$2450 was taken in. If student and adult tickets had cost \$0.30 more, the senior citizens could have got in free, and the total income would have been the same. How many senior-citizen tickets were sold?*

Solution Let x represent the number of student tickets,
y represent the number of adult tickets,
z represent the number of senior-citizen tickets.

Total number of tickets	(1)	$x + y + z = 1150$
Actual income	(2)	$1.5x + 2.5y + 2z = 2450$
Income if senior-citizen tickets had been free	(3)	$1.8x + 2.8y + 0z = 2450$

We can easily eliminate z, since it does not appear in the third equation.

Multiply (1) by −2	$-2x - 2y - 2z = -2300$
Add (2)	$1.5x + 2.5y + 2z = 2450$
Multiply by 2 to	$-0.5x + 0.5y = 150$
clear the decimals	$-x + y = 300$

Now we form a system with this equation and (3).

$$\begin{cases} -x + y = 300 \\ 1.8x + 2.8y = 2450 \end{cases}$$

When we solve this system, we find that x = 350 and y = 650. By substituting these values in (1), we find that z = 150, so 150 senior-citizen tickets were sold.

Part Three: Exercises and Problems

Warm-up Exercises

P **1** Is (w, x, y, z) = (3, −4, 2, 11) a solution of the following system? No

$$\begin{cases} w + x + y = 1 \\ x + y + z = 9 \\ wxy = 24 \end{cases}$$

P **2** Find the value of b in the following system. 23

$$\begin{cases} a + b + c + d + e = 82 \\ a + c + d + e = 59 \end{cases}$$

Checkpoint

1 Shera has two sisters, Sheila and Shauna. If the sum of the ages of the sisters is 21 and Shera is twice as old Shiela, how old are the three girls if Shauna's age is half the difference of Shera's and Shiela's age? 12, 6, 3

2 Find a quadratic equation that goes through the points (0, 4), (2, 2) and (5, 44). $y = 3x^2 - 7x + 4$

3 The sum of the three numbers is 6. If the first number is half the second number and the third number is three times as the first, what are the numbers? 1, 2, 3

Assignment Guide

Average	
Day 1	8, 10–12, 14, 16, 17, 21, 23
Day 2	9, 15, 18, 19, 22, 25, 26a, 28, 31
Advanced	
Day 1	8, 10, 12, 15, 18–20, 25, 31
Day 2	9, 11, 13, 14, 16, 17, 21–23, 27, 28
Honors	
Day 1	8–11, 13, 16–22, 25, 26
Day 2	14, 15, 23, 24, 27, 28, 30–32

Problem-Set Notes and Strategies

- Problem **4** may be solved using either the substitution method or the multiplication-addition methods or a combination of both. Problem **5** is best solved using the mult-addition method.

Communicating Mathematics

Problem **10** is a geometric interpretation of a three-variable system of equations. Have students contrive a word problem that could be represented by this diagram.

Additional Answer

11d

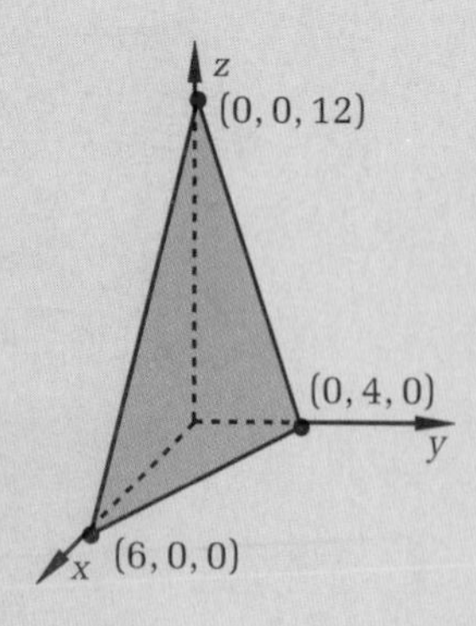

P **3** Find the coordinates of points A, B, and C.
$(4, 0, 0)$; $(4, 5, 0)$; $(4, 5, -6)$

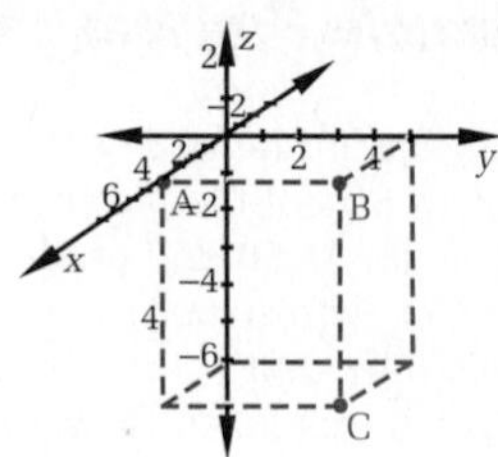

P **In problems 4 and 5, solve each system for (x, y, z).**

4 $\begin{cases} x + y = 8 \\ y + z = 6 \\ x + z = 24 \end{cases}$
$(x, y, z) = (13, -5, 11)$

5 $\begin{cases} x + y + z = 15 \\ x - y - z = -5 \\ -x - y + z = -1 \end{cases}$
$(x, y, z) = (5, 3, 7)$

P,R **6** A parabola crosses the x-axis at $(5, 0)$ and $(-3, 0)$, and its vertex is at $(1, 10)$. Write an equation that represents the parabola.
$y = -\frac{5}{8}x^2 + \frac{5}{4}x + \frac{75}{8}$

P **7** Solve the following system for (x, y, z). $(x, y, z) = (1, 2, 4)$
$\begin{cases} x = 7 - y - z \\ x = 1 - 2y + z \\ x = 1 + 4y - 2z \end{cases}$

Problem Set

A

In problems 8 and 9, solve for (x, y, z).

P **8** $\begin{cases} 3x - 4y + 6z = 39 \\ 4x + 2y + z = 2 \\ 2x - 2y + 3z = -4 \end{cases}$
$(x, y, z) = (-47, 60, 70)$

9 $\begin{cases} y = 2x + 1 \\ z = 3x \\ z + 2x - y = 15 \end{cases}$
$(x, y, z) = \left(\frac{16}{3}, \frac{35}{3}, 16\right)$

P **10** Use the diagram to find (x, y, z). $(x, y, z) = \left(9\frac{1}{2}, 5\frac{1}{2}, 14\frac{1}{2}\right)$

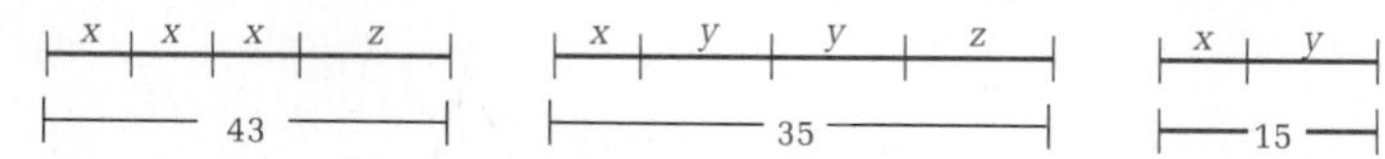

P **11** Consider the plane represented by $2x + 3y + z = 12$.
a Find the plane's x-intercept. 6
b Find the plane's y-intercept. 4
c Find the plane's z-intercept. 12
d Use your answers to parts **a**, **b**, and **c** to graph the plane.

P,R **12** The sum of Bill's age and his father's age is 64. In 5 years, Bill's father's age will be 10 less than 3 times Bill's age then. How old is each person now? Bill is 16; his father is 48.

R **13** Write an equation of the parabola that contains the points $(3, -1)$, $(0, -7)$, and $(2, 9)$. $y = -6x^2 + 20x - 7$

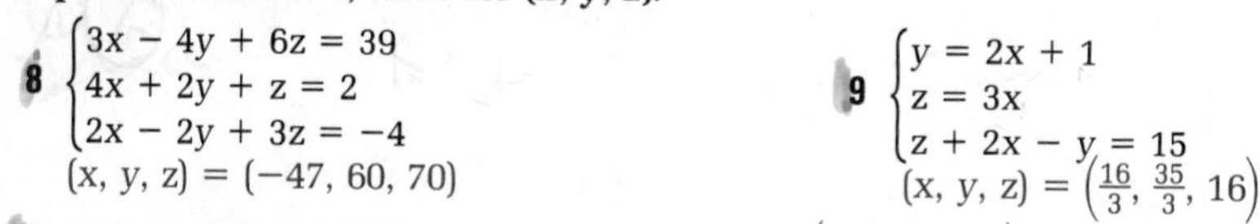

P **In problems 14 and 15, solve each system.**

14 $\begin{cases} x + 2y - 3z = 3 \\ 3x - 4y + 6z = 4 \\ y - 2z = 2 \end{cases}$ $(x, y, z) = (2, -4, -3)$

15 $\begin{cases} 9x + 3y + z = 5 \\ 4x + 2y + z = 7 \\ x + y + z = 7 \end{cases}$ $(x, y, z) = (-1, 3, 5)$

P **16** Consider the equations $3x + 4y + 6z = 12$ and $3x + 4y + 6z = 24$.

a Graph the equations on one set of axes.
b What do you notice about the planes? They are parallel.

P **17** A box of 40 coins contains nickels, dimes, and quarters. The total value of the coins is $4.00. If there were 3 times as many nickels, half as many dimes, and 6 more quarters, the value of the coins would be $5.50. How many quarters are in the box? 4

B

P,R **18** Show that $\begin{bmatrix} x \\ y \\ z \end{bmatrix} = \begin{bmatrix} 3 \\ -5 \\ 6 \end{bmatrix}$ is a solution of $\begin{bmatrix} 1 & 2 & 3 \\ -2 & 5 & 1 \\ 4 & -1 & 1 \end{bmatrix}\begin{bmatrix} x \\ y \\ z \end{bmatrix} = \begin{bmatrix} 11 \\ -25 \\ 23 \end{bmatrix}$

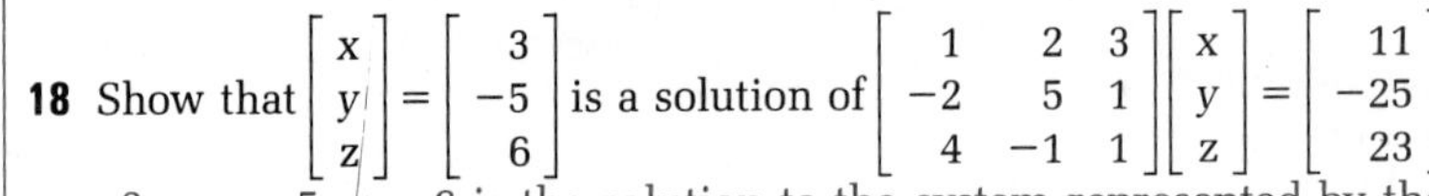

$x = 3$, $y = -5$, $z = 6$ is the solution to the system represented by the matrix.

P,R **19** Write the following system as a matrix equation.

$\begin{cases} 3x + 4y + z = 11 \\ 2x - y + 5z = 9 \\ -4x + 7y + 2z = -4 \end{cases}$ $\begin{bmatrix} 3 & 4 & 1 \\ 2 & -1 & 5 \\ -4 & 7 & 2 \end{bmatrix}\begin{bmatrix} x \\ y \\ z \end{bmatrix} = \begin{bmatrix} 11 \\ 9 \\ -4 \end{bmatrix}$

R,I **20** How many different 5-card poker hands can be dealt from a standard deck of 52 cards? 2,598,960

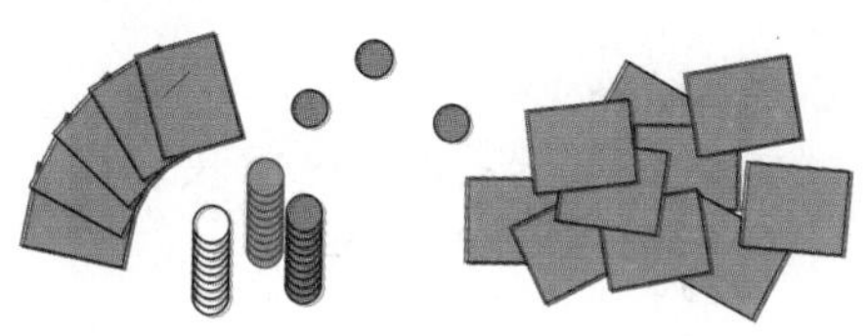

R **21** Solve the following system for (x, y). $\{(x, y): x + y = 5\}$

$\begin{cases} 2^x 2^y = 32 \\ 5^x 5^y = 3125 \end{cases}$

R **22** Refer to the diagrams.

x + 2   6   x + 2

a Write a function S for the area of the square. $S(x) = (x + 2)^2$
b Write a function T for the area of the triangle. $T(x) = 3(x + 2)$
c For what value of x is $T(x) = S(x)$? $x = 1$

R **23** Let A, B, and C be three points such that the coordinates of A are (1, 5), the coordinates of B are (6, 3), and C is the intersection of $\overline{AB}$ and the line $y = x$.

a Find the coordinates of point D, the reflection of B over the line $y = x$. (3, 6)
b Find the distance from A to B. $\sqrt{29}$, or ≈ 5.39
c Find the distance from A to C to D. $\sqrt{29}$, or ≈ 5.39

Problem-Set Notes and Strategies, continued

- Problem **16** illustrates the fact that two equations in three dimensions do not have to intersect. Parallel planes in three dimensions are comparable to parallel lines in two dimensions.

Additional Answer

16a

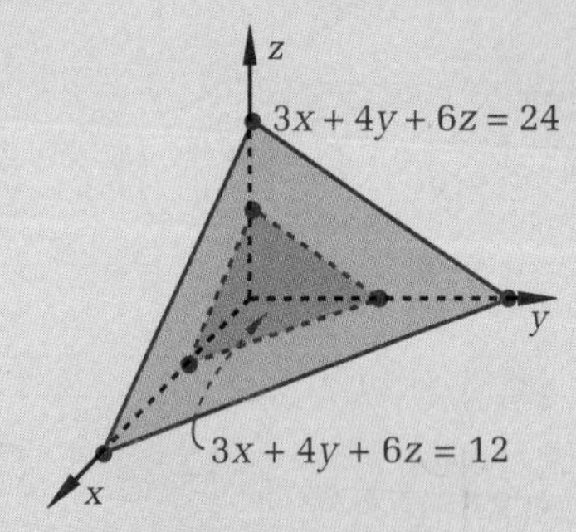

- The variables in problem **17** will identify the number of nickels, dimes, and quarters. A common mistake is to forget that each coin also has a value assigned to it and this must be considered when the total value is $4.00.

- Problem **20** is a good review of combinations. In this case, order is not important so the formula used was for a combination.

Problem-Set Notes and Strategies, continued

- Problem **26** is quadratic by the constant second differences. Finding the closed form formula will seem difficult for some students.

Additional Answer

26a Yes, $f(n+1) - f(n)$ increases uniformly as n increases.

- Problem **29** is similar to finding a parabola given three points. Here six points are given in the pictures and any six may be chosen to solve the problem. Have students compare the points they used and verify that the answer remains the same.

Problem Set, *continued*

P **24** Tickets at a ballpark sell for \$4.50, \$7.00, and \$12.50. One day a total of 24,202 tickets were sold for a total of \$151,924. The sum of the numbers of \$7.00 tickets and \$12.50 tickets sold was 702 less than the number of \$4.50 tickets sold. How many of each kind of ticket were sold?
\$4.50 tickets: 12,452; \$7.00 tickets: 9270; \$12.50 tickets: 2480

P **25** Swenson's Bakery sells three kinds of cakes. The amounts of flour, sugar, and eggs used for each cake are shown below, along with the bakery's daily consumption of these ingredients.

	Cake A	Cake B	Cake C		Amount Used Daily
Flour (cups)	5	5	4	**Flour (cups)**	90
Sugar (cups)	3	2	2	**Sugar (cups)**	46
Eggs	1	2	3	**Eggs**	35

How many of each kind of cake are baked each day?
8 of A, 6 of B, 5 of C

R **26** Consider the following table of values for a function f.

n	3	4	5	6	7	8
$f(n)$	0	2	5	9	14	20

a Does f appear to be a quadratic function? Why or why not?
b Write a closed-form formula for f. $f(n) = \frac{1}{2}n^2 - \frac{3}{2}n$

P,R **27** A truck radiator contains 14 quarts of a mixture that is 75% antifreeze and 25% water. How much should be drained and replaced with pure water for the radiator to contain 14 quarts of a mixture that is 60% antifreeze and 40% water? 2.8 qt

P **28** The average of three numbers is 18. If the first number is increased by five, the second number is doubled, and the third number is tripled, the average becomes 34. If the first number is decreased by five, the second number is tripled, and the third number is doubled, the average becomes 48. Find the largest of the three numbers. 49

C P **29** If the number y of angles determined by n rays with a common endpoint is given by $y = an^2 + bn + c$, what is (a, b, c)? $(a, b, c) = \left(\frac{1}{2}, -\frac{1}{2}, 0\right)$

$y = 0$ $y = 1$ $y = 3$ $y = 6$ $y = 10$ $y = 15$

P **30** The sum of the digits of a three-digit number is 10. The sum of the tens digit and the units digit equals the hundreds digit. The number represented if the hundreds digit is omitted is one less than three times the hundreds digit. Find the three-digit number. 514

P **31** Refer to the diagram.

a Find the coordinates of the midpoints of $\overline{BC}$, $\overline{AB}$, and $\overline{AC}$. (0, 3, 5), (4, 3, 0), (4, 0, 5)

b Write an equation that represents the plane that passes through the midpoints found in part **a.**
$15x + 20y + 12z = 120$

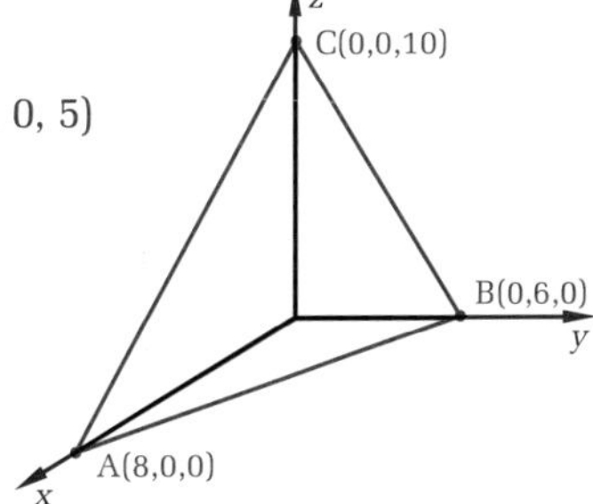

P **32** Solve the following system for (a, b, c, d, e). $(a, b, c, d, e) = (9, 2, 1, 5, 7)$

$$\begin{cases} a + b + c + d = 17 \\ a + b + c + e = 19 \\ a + b + d + e = 23 \\ a + c + d + e = 22 \\ b + c + d + e = 15 \end{cases}$$

Problem-Set Notes and Strategies, continued

- For Problem **31b**, the three traces may be found and from these the equation of the plane generated.
- Problem **32** generalizes the concept of systems of equations into five variables and five equations. The concept is the same as with three variables, just longer.

Cooperative Learning

Problem **32** would be a good problem for groups of two or three students.

CAREER PROFILE

A TRAFFIC JAM

Ken Kohl plots "green time"

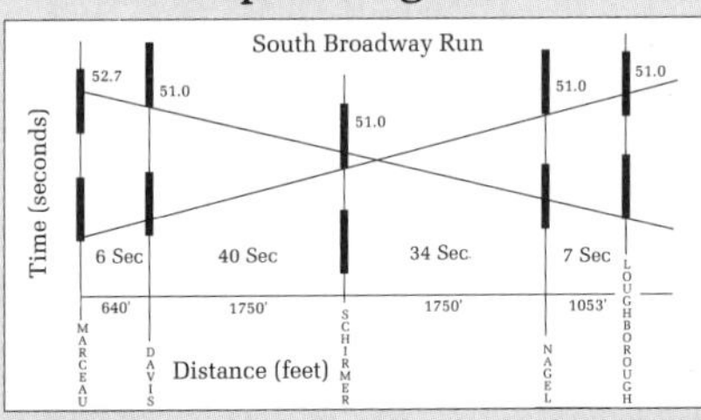

The word *gridlock* describes a phenomenon in which no cars on an entire metropolitan area's streets and freeways can move. The most desirable solution to gridlock is to reduce the number of cars on the road in a city. But even given a manageable amount of traffic, traffic engineering is still necessary for smooth traffic flow. And traffic engineering involves mathematics.

"We use a lot of basic mathematics and calculus," says Ken Kohl, a traffic engineer in St. Louis, Missouri. "I work with graphs to coordinate city traffic." One particular application Mr. Kohl uses is a time-space diagram to coordinate "green times"—the lengths of time traffic lights remain green in one cycle—at major intersections.

The space-time diagram shown is a graph of distance along a given street (South Broadway) versus time. The green times at five successive intersections are represented by solid bars. The slope of each line represents the speed of a car. The two lines in this graph represent cars traveling in opposite directions on South Broadway, each at 30 miles per hour. "Using this graph helps me time the lights so as to establish a smooth traffic flow in *both* directions," Mr. Kohl says.

Mr. Kohl holds an engineering degree from the University of Missouri-Rolla.

7.3

Class Planning

Time Schedule
All levels: 2 days

Resource References
Teacher's Resource Book
Class Opener 43
Cooperative Learning 22
Practice 45
Evaluation
Quiz Forms 1, 2, 3

Class Opener

Graph $y \le 2x + 9$ and $y < -\frac{2}{3}x - 1$ on the same coordinate plane.

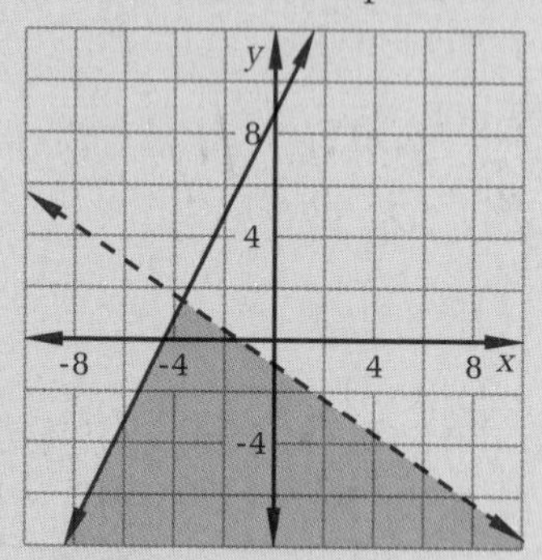

Lesson Notes

- Systems of linear inequalities are done exactly as single linear inequalities except there is more than one. The only concern is identifying the proper intersection of regions as the answer. Be careful to note whether the points of intersection of the boundaries are included or excluded from the solution set.

7.3 Systems of Inequalities

Objective

After studying this section, you will be able to
- Graph systems of inequalities

Part One: Introduction

In Section 3.3 you learned methods of graphing linear inequalities and their intersections. We will extend these methods to the graphing of systems of inequalities.

Example *Graph the following system.*

$$\begin{cases} -4 \le x \le 12 \\ y \ge \dfrac{3}{4}x - 2 \end{cases}$$

We first graph the vertical lines representing $x = -4$ and $x = 12$. The graph of $y = \frac{3}{4}x - 2$ has a slope of $\frac{3}{4}$ and a y-intercept of -2. Because the point (0, 0) satisfies $y \ge \frac{3}{4}x - 2$, we shade the region above this line and between the two vertical lines.

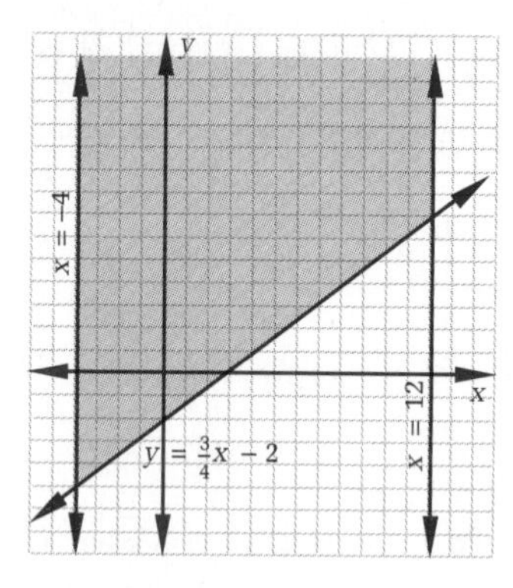

Example *Graph the following system.*

$$\begin{cases} -3 \le y \le 7 \\ y \ge 2x + 3 \\ x \ge -5 \end{cases}$$

The boundaries of the shaded region are the graphs of $y = -3$, $y = 7$, $y = 2x + 3$, and $x = -5$.

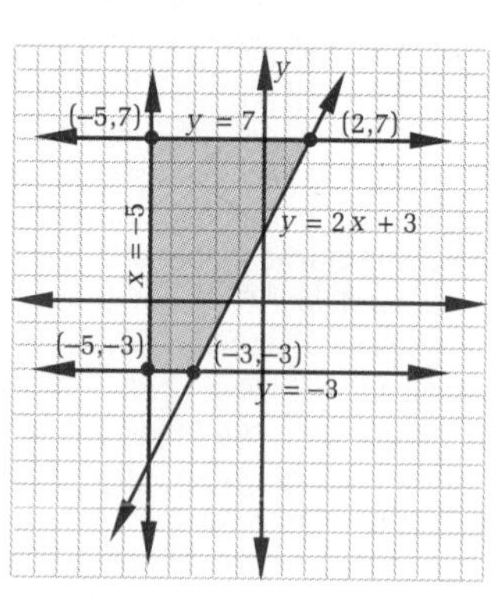

Part Two: Sample Problem

Problem *Graph the following system.*

$$\begin{cases} y > x + 2 \\ y < 4 - x^2 \end{cases}$$

Solution The boundaries are the line representing $y = x + 2$ and the parabola representing $y = -x^2 + 4$. We graph these with dashes (since they are not parts of the solution set) and test some convenient points to find the solution shown in the graph.

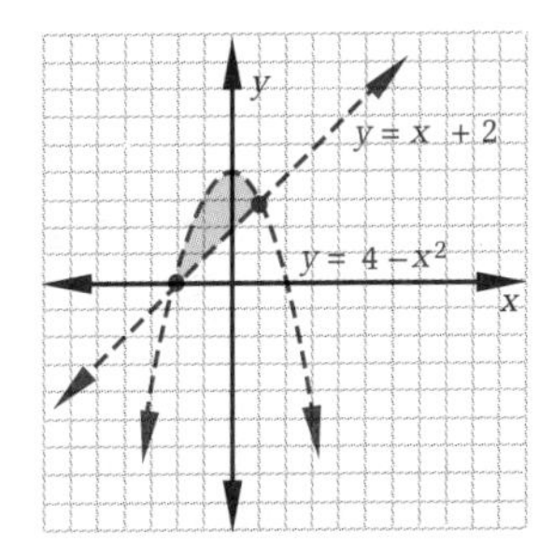

Part Three: Exercises and Problems

Warm-up Exercises

In problems 1 and 2, solve each system by graphing.

P **1** $\begin{cases} 1 > y > -3 \\ -2 < x < 4 \end{cases}$ **2** $\begin{cases} x + y > 8 \\ x + y < 2 \end{cases}$

P,R **3** Write an inequality to describe the graph shown.
$y \le -\frac{3}{4}x + 6$

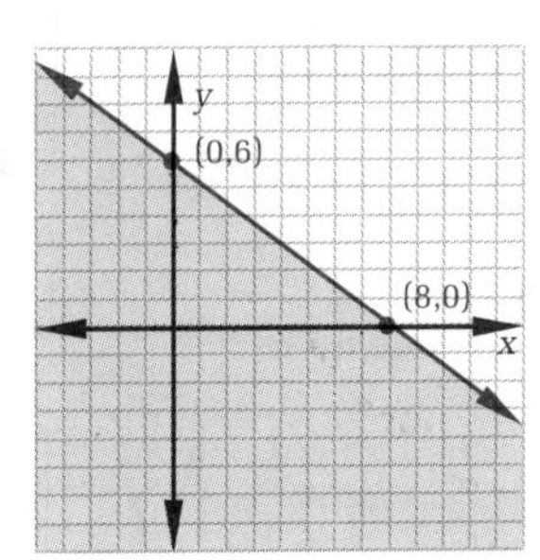

P **4** Graph the solution set of the following system.

$$\begin{cases} x + 2y \le 10 \\ y \ge 2 \\ x \ge -2 \end{cases}$$

P,R **5** Graph the following system and identify the vertices of the region it defines. (−9, −2), (−9, 0), (4, −2), (4, 0)

$$\begin{cases} -9 \le x \le 4 \\ -2 \le y \le 0 \end{cases}$$

R **6** Is (0, 0) a solution of all of the following inequalities? No

$2x + 3y \le 5$ $x < 2$ $y < -3$ $y \le \frac{1}{2}x + 2$

Using Computers

Although most graphing software graphs only equalities, you can use the software to graph the boundaries of the regions to be shaded. If your graphing software offers printing of graphs, you can have students do the shading by hand on the printout. The *TI81 Graphics Calculator* allows you to do some graphing of inequalities, especially with restricted domains and shading between two curves. The *Omnifarious Plotter, Mathematics Exploration Kit,* and *Derive* are appropriate for use here.

Checkpoint

1 Graph the solution set of
$\begin{cases} y > x \\ y \ge -x \end{cases}$

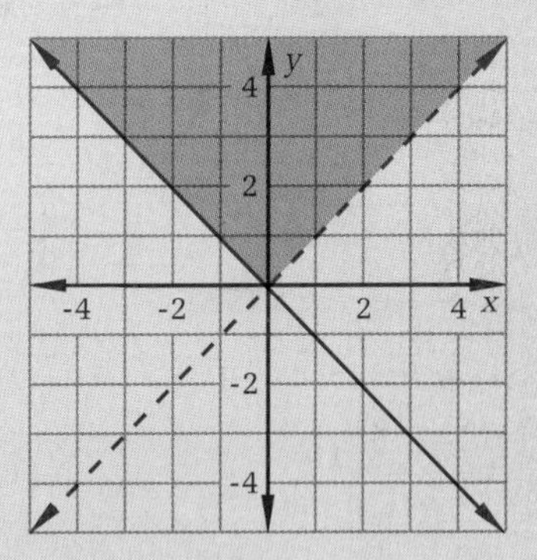

2 Is the point (3, −4) a solution of the following system of inequalities.
$\begin{cases} y \le x - 7 \\ 3x + 2y > 2 \end{cases}$
No

Additional Answers

Answers for problems **1**, **2**, **4**, and **5** can be found in the answer pages beginning on page **A 1**.

Checkpoint, continued

3 Graph the solution of the following system.

$$\begin{cases} y \geq x^2 \\ y < x \\ 2y < -2x + 10 \end{cases}$$

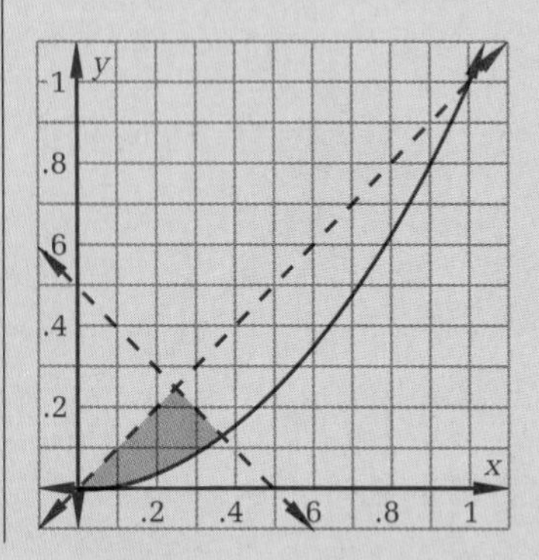

Assignment Guide

Average	
Day 1	8, 10, 12, 13, 16–18, 22, 26
Day 2	9, 11, 14, 19, 20, 23–25, 28, 33
Advanced	
Day 1	8–12, 17, 18, 30
Day 2	13–15, 19, 20, 23, 25, 27, 28, 32, 33
Honors	
Day 1	9, 10, 12, 14–16, 19–21, 25, 27–29
Day 2	18, 23, 26, 30–33, 35, 37

Problem-Set Notes and Strategies

- Students need to recognize that the graphs in problem **2** never intersect and therefore there is no solution to that system. Often, they just leave the two shaded regions and don't even think about the intersection.

P,R **7** An investment firm can invest at most \$80,000 in two companies, BIMCO and Amalgamated Snarfs. BIMCO stock sells for \$55 per share, and Amalgamated Snarfs stock sells for \$87 per share. Translate this situation into an inequality. (Let x equal the number of shares of BIMCO purchased, and let y equal the number of shares of Amalgamated Snarfs purchased.) $55x + 87y \leq 80{,}000$

Problem Set

A **In problems 8 and 9, graph the solution of each system.**

P **8** $\begin{cases} 3x + 2y \leq 6 \\ 3x + 2y \leq 12 \end{cases}$

9 $\begin{cases} -2 < x < 6 \\ -4 < y + 2x < 8 \end{cases}$

R **10** What are the coordinates of the vertices of the shaded region in the diagram shown?
(2, 1), (6, 0), (9, 6), (2, 6)

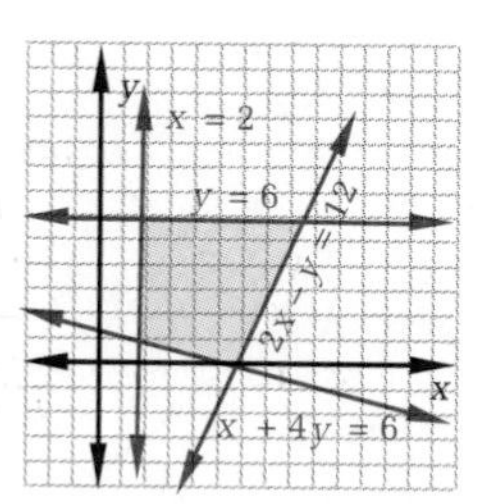

P **11** Graph the solution of the following system.

$$\begin{cases} y \geq x \\ 6x + 5y \leq 30 \end{cases}$$

In problems 12 and 13, solve for (x, y, z).

R **12** $\begin{cases} x + y = 11 \\ x + z = 17 \\ y + z = 42 \end{cases}$
$(x, y, z) = (-7, 18, 24)$

13 $\begin{cases} x + 2y - z = 9 \\ 3y + z = 26 \\ z = 5 \end{cases}$
$(x, y, z) = (0, 7, 5)$

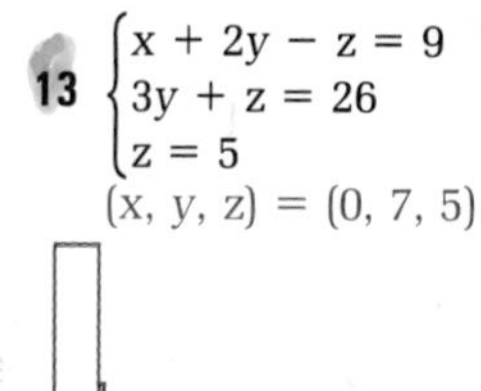

R,I **14** A toy-manufacturing company has at most 200 pounds of plastic available for producing 2 kinds of toys. Three pounds of plastic are needed to produce a set of toy soldiers, and 7 pounds of plastic are required to produce a set of dolls. Write an inequality to describe this situation. (Let x equal the number of toy soldiers produced and y equal the number of dolls produced.)
$3x + 7y \leq 200$

P **15** Graph $2 \leq x + y \leq 10$.

R **16** Solve $3y^2 + 2y - 1 = 0$ for y. $y = \frac{1}{3}$ or $y = -1$

R **17** Graph the following equations on the same set of axes.

a $y = x^2$ **b** $y = (x - 2)^2$ **c** $y = (x - 2)^2 + 5$

Additional Answers

Answers for problems **8**, **9**, **11**, **15**, and **17** can be found in the answer pages beginning on page **A** 1.

R **18** Refer to the diagram.

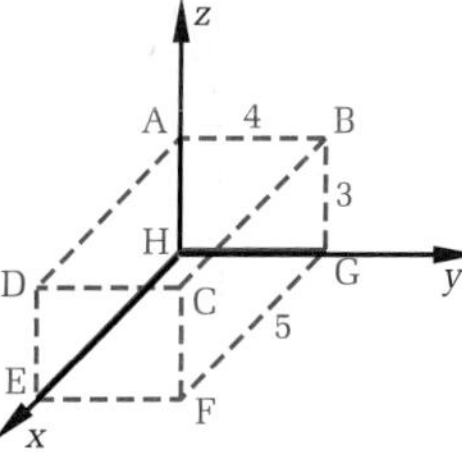
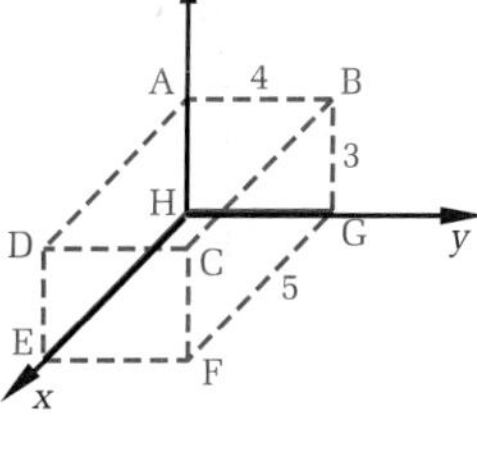

a Find the coordinates of points A, B, C, D, E, F, G, and H.

b Write equations that represent the planes ABCD, EFGH, DCFE, and BCFG. ABCD: $z = 3$; EFGH: $z = 0$; DCFE: $x = 5$; BCFG: $y = 4$

a (0,0,3); (0,4,3); (5,4,3); (5,0,3); (5,0,0); (5,4,0); (0,4,0); (0,0,0)

B

P,I **19** Consider the following system.

$$\begin{cases} x \geq 0 \\ y \geq 0 \\ y \leq -\frac{1}{2}x + 5 \\ y \geq x - 3 \end{cases}$$

a Find the coordinates of the vertices of the region defined by the system. $(0, 0)$, $(3, 0)$, $\left(\frac{16}{3}, \frac{7}{3}\right)$, $(0, 5)$

b Evaluate $f(x, y) = 2x + 3y$ at each vertex. $f(0, 0) = 0$; $f(3, 0) = 6$; $f\left(\frac{16}{3}, \frac{7}{3}\right) = \frac{53}{3}$; $f(0, 5) = 15$

P **20** Graph the solution set of the following system.

$$\begin{cases} y \leq 2x + 3 \\ 4x - 2y \leq 5 \\ -3 \leq x \leq 5 \end{cases}$$

P **21** Write a system of inequalities that represents the shaded region in the diagram shown.

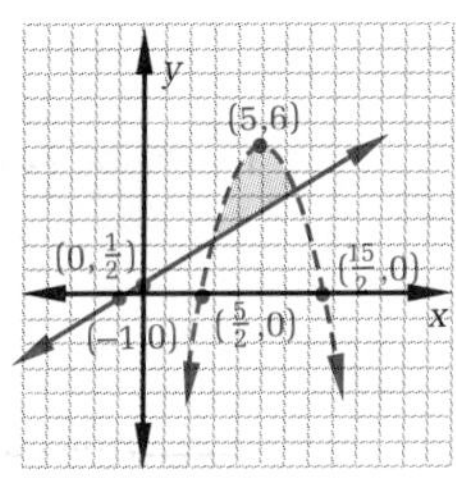

$$\begin{cases} 2y - x \geq 1 \\ y \leq = -\frac{24}{25}x^2 + \frac{48}{5}x - 18 \end{cases}$$

R **22** Kem E. Cal heated 500 milliliters of an 8% saline solution until all but 400 milliliters of the solution had evaporated. What percentage of the remaining solution was salt? 10%

P **23** A point is randomly selected from {(0, 0), (1, 3), (2, 2), (−2, −2)}. What is the probability that the chosen point lies in the region defined by the following system? $\frac{1}{2}$

$$\begin{cases} y < 2x + 6 \\ y > 2x - 6 \\ y < -2x + 6 \\ y > -2x - 6 \end{cases}$$

R **24** Snoopy flies his Sopwith Camel from his base to his destination and back, a total of 240 miles. If he had flown 20 miles per hour faster, the trip would have taken an hour less. How fast did Snoopy fly? 60 mph

Problem-Set Notes and Strategies, continued

- Problems **5**, **10** and **19** are identifying convex regions of a graph that will be the feasible regions for linear programming. This topic is covered in a short subject after Chapter **10**.
- In problem **21** students should use the vertex form of a parabola and solve for *a*.
- The percentages in problem **22** are not too difficult if you realize that the salt has not evaporated with the water. That is the constant in this equation. All but 400 ml evaporating is another way of saying that 100 ml had evaporated and 400 ml were left. Students will realize that there are many ways of wording a unique mathematical statement in the English language.
- Problem **24** may be solved with one equation or with a system of 2 equations. In either case the equation $d = rt$ is used.

Additional Answers

20

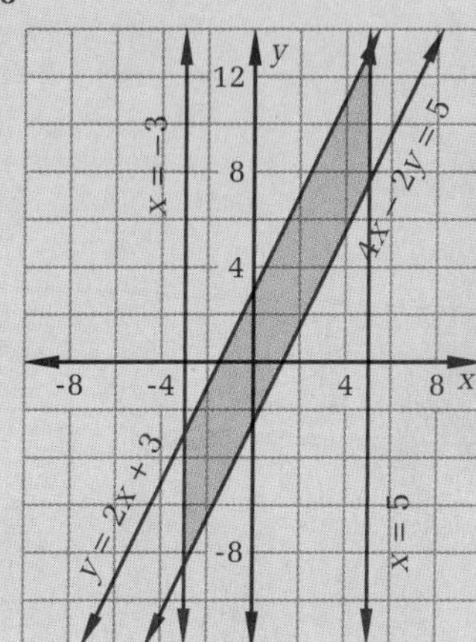

Problem-Set Notes and Strategies, continued

- Slopes are interpreted as rates of change again in problem **25**. In this case it was the ratio of sides, but interpreted that as x changes 5 times, y changes only 3 and so the rate of change is $\frac{3}{5}$.
- Problem **26** requires some geometric concepts that students may have forgotten about 45°-45°-90° triangles. Make sure they remember this triangle because it will be needed again for the trigonometry chapters.

Problem Set, *continued*

P **25** Refer to the diagram.

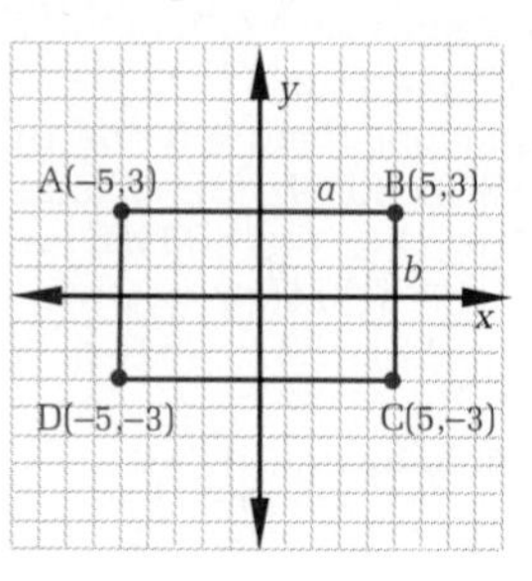

a Find the ratio of length b to length a. $\frac{3}{5}$

b Write equations that represent the lines containing the diagonals of the rectangle. $y = \frac{3}{5}x$; $y = -\frac{3}{5}x$

c How could you use your answer to part **b** to answer part **a**? Since the slope of $\overleftrightarrow{BD}$ is $\frac{3}{5}$, the ratio of the change in y to the change in x between any two points on the line is $\frac{3}{5}$.

P **26** Write a system of inequalities that describes a square region whose center is at the origin, whose vertices are on the axes, and whose area is 50 square units.
$\begin{cases} y \le x + 5 \\ y \le -x + 5 \\ y \ge -x - 5 \\ y \ge x - 5 \end{cases}$

P,I **27** Machine A produces 200 widgets per hour. Machine B produces 375 widgets per hour. Machine A runs at least 2 hours per day, and machine B runs at least 3 hours per day. The 2 machines together run no more than 10 hours per day. Write at least 4 inequalities to describe this situation. (Let x equal the number of hours machine A runs and y equal the number of hours machine B runs.)
$x \ge 2$; $y \ge 3$; $x + y \le 10$; $200x + 375y \ge 1525$

R,I **28** If $f(x, y) = 4x + 3y$, what is the minimum value of f for the domain $\{(-3, 1), (4, 7), (5, -7)\}$? $f(-3, 1) = -9$

P **29** Find the area of the region defined by the following system. 30
$$\begin{cases} 2 \le y \le 7 \\ 4y < 5x + 3 \\ 5x + 2y < 54 \end{cases}$$

R **30** Solve the following system for (x, y, z). $(x, y, z) = (-2, 6, -2)$
$$\begin{cases} 3x + y - 2z = 4 \\ -6x + 4y + 4z = 28 \\ -7x + 2y = 26 \end{cases}$$

R **31** Plane A leaves New York for London at 8:50 A.M. Plane B leaves New York for London at 9:40 A.M. When plane B leaves, the planes are 600 miles apart. One hour later the planes are 550 miles apart. Find the speed of each plane.
A: 720 mph; B: 770 mph

P **32** Find the area of the region defined by the following system. 78
$$\begin{cases} 0 \le x \le 12 \\ y \ge 0 \\ 2y - x \le 10 \\ 2y + x \le 22 \end{cases}$$

P **33** Write a system of inequalities that represents the shaded region in the diagram.

$$\begin{cases} y < 8 \\ x < 0 \\ y > x - 5 \\ y > -3x - 7 \end{cases}$$

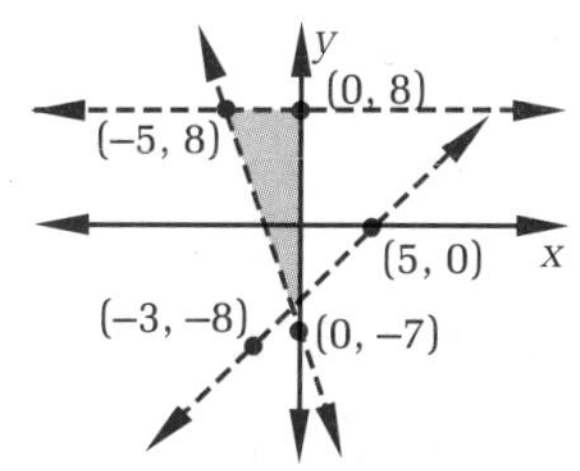

P **34** Graph the region defined by the following system and find its area.

$$\begin{cases} y \le \frac{1}{2}x + 6 \\ y \le -2x + 6 \\ y \ge 2x - 6 \\ y \ge -\frac{1}{2}x - 6 \end{cases}$$

C
R,I **35** A matrix can be used to represent a directed graph.

For example, $\begin{bmatrix} 0 & 2 & 1 \\ 0 & 1 & 1 \\ 3 & 0 & 0 \end{bmatrix}$ represents

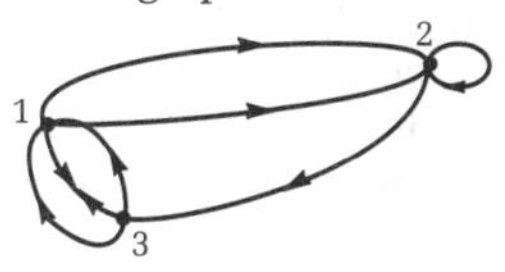

The number in row a, column b, indicates the number of direct paths from a to b.

a Construct a matrix to represent the digraph shown.

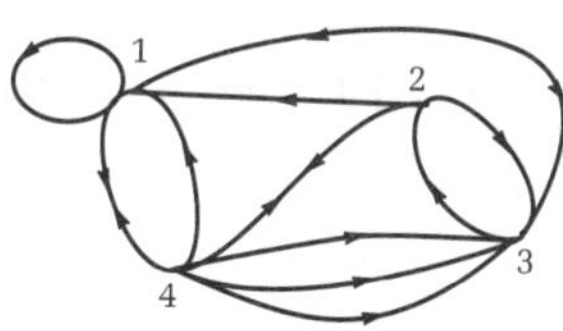

$\begin{bmatrix} 1 & 0 & 1 & 1 \\ 1 & 0 & 1 & 1 \\ 1 & 1 & 0 & 0 \\ 2 & 1 & 3 & 0 \end{bmatrix}$

b Draw the digraph represented by the following matrix.

$$\begin{bmatrix} 0 & 1 & 2 & 3 \\ 1 & 0 & 1 & 0 \\ 1 & 1 & 0 & 0 \\ 2 & 1 & 0 & 1 \end{bmatrix}$$

P **In problems 36 and 37, graph the solution set of each system.**

36 $\begin{cases} y \le -x + 3 \\ y \le x + 3 \\ y \ge -x - 3 \\ y \ge x - 3 \\ x \le 2 \\ x \ge -2 \\ y \le 2 \\ y \ge -2 \end{cases}$

37 $\begin{cases} y < x + 2 \\ y < 2 \\ y < -x + 4 \\ y > x - 4 \\ y > -2 \\ y > -x - 2 \end{cases}$

Stumbling Block

Students may shade the wrong portion of the graph. Make sure students always test a point to determine which portion of the graph to shade. The origin is a good test point provided it is not on the boundary line of the inequality.

- The matrices in problem **35** are sometimes called adjacency matrices and are useful in solving many different problems.

Cooperative Learning

Have students work in small groups to solve problem **36.** Once the region has been identified, have students randomly choose two points from the inside of the region and connect them with a line segment. Now choose two different points and connect them. Repeat this process again. Have students draw a conclusion about each line segment and its relationship to the shaded region. This type of region is called a convex region and is useful in a multitude of applied problems.

Communicating Mathematics

Suppose your best friend had been absent today from class. Write a paragraph to him/her that explains how to solve a system of inequalities.

Additional Answers

Answers for problems **34**, **35b**, **36**, and **37** can be found in the answer pages beginning on page **A** 1.

7.4

Class Planning

Time Schedule
All levels: 2 days

Resource References
Teacher's Resource Book
Class Opener 44
Practice 46

Class Opener

Sketch the graph of all points (x, y) so that $xy = 12$.

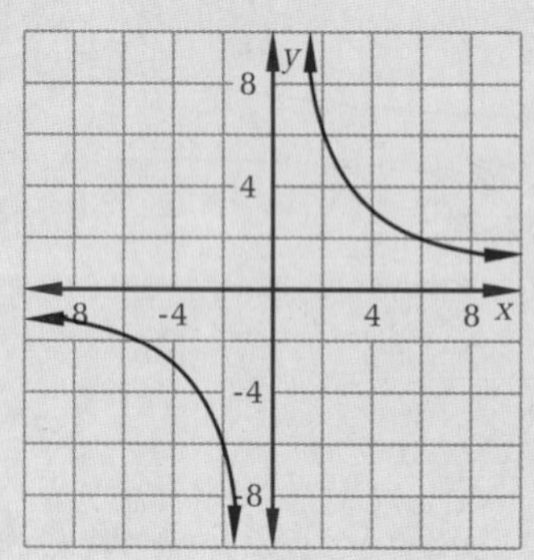

Lesson Notes

- An inverse variation is a relation where one quantity increases, as the other decreases. This relationship holds when both variables are the same sign.
- To solve an inverse-variation problem, a constant of proportion must be given or solved for.

Using Computers

Any graphing software or graphing calculator can be used to advantage here. You can see the graphs generated by equations of the form $xy = k$ and can solve many of the word problems using a graphing calculator.

7.4 Inverse Variation

Objectives

After studying this section, you will be able to
- Solve inverse-variation problems
- Graph equations of the form $xy = k$

Part One: Introduction

The Graph of $xy = k$

The graph of an equation of the form $xy = k$ or $y = \frac{k}{x}$, where k is a nonzero constant, can be represented by a hyperbola with its center at the origin and with the x-axis and y-axis as asymptotes. If $k > 0$, the hyperbola lies in the first and third quadrants; if $k < 0$, the hyperbola lies in the second and fourth quadrants.

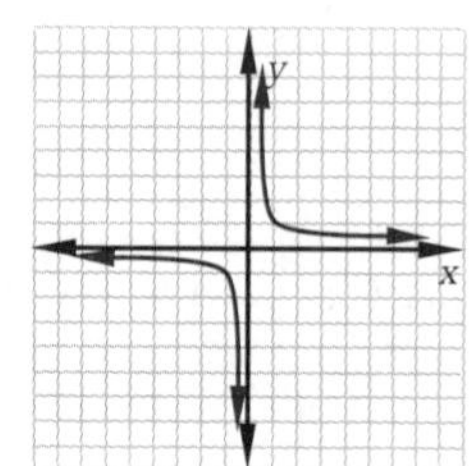

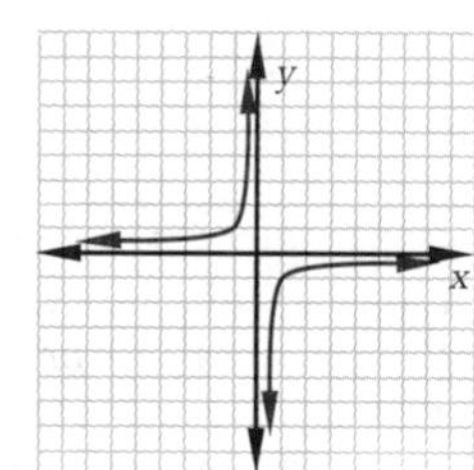

Example *Graph* $xy = -9$.

The graph is a hyperbola lying in the second and fourth quadrants. By plotting a few points and applying a knowledge of symmetry and asymptotes as discussed in Chapter 6, we graph the equation as shown.

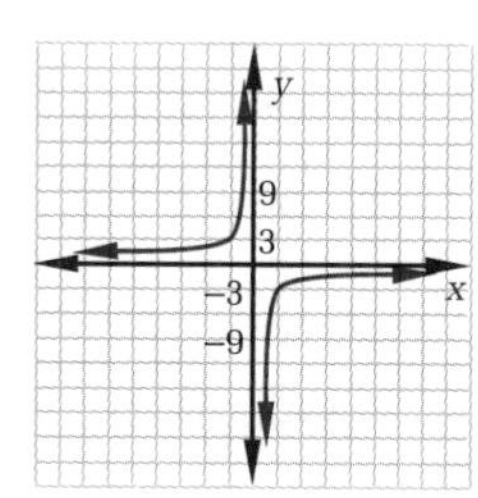

Inverse-Variation Problems

In Section 3.4, you learned that two quantities x and y are said to vary directly if they are related by an equation of the form $y = mx$, where m is the constant of variation. A characteristic of direct variation with a positive constant of variation is that as the value of one variable increases, the value of the other variable also increases proportionally.

Vocabulary
inverse variation

Many real-world relationships, however, are different—the faster you travel, the *less* time it takes to reach your destination; the smaller the volume into which a gas is forced, the *greater* the gas's pressure. Such relationships are examples of ***inverse variation*** and are represented by equations of the form $xy = k$, or $y = \frac{k}{x}$, or $y = k \cdot \frac{1}{x}$, where k is the constant of variation. When two quantities vary inversely, the absolute value of one quantity decreases proportionally as the absolute value of the other increases.

Example *If x varies inversely as y, and if y = 6 when x = 12, what is the value of x when y = 9?*

To find the value of k, the constant of variation, we substitute

$$xy = k$$
$$12 \cdot 6 = k$$
$$72 = k$$

Substitute again

$$xy = 72$$
$$x \cdot 9 = 72$$
$$x = 8$$

Part Two: Sample Problems

Problem 1 *Bankers use the "rule of 72" to approximate the number of years it will take for an investment to double in value. According to this rule, the number of years is inversely proportional to the percentage value of the annual interest rate, with a constant of variation of approximately 72. If you invest money at 9%, approximately how many years will it take for the money to double?*

Solution $\text{Years} = \frac{72}{\text{interest rate}} = \frac{72}{9} = 8$

The money will double in about eight years.

Problem 2 *Graph xy = 12 and identify the vertices of the hyperbola.*

Solution In this inverse variation, the constant of variation is 12. Since the constant is positive, the graph will lie in the first and third quadrants.

We plot several points (x, y) such that $xy = 12$ and sketch the graph.

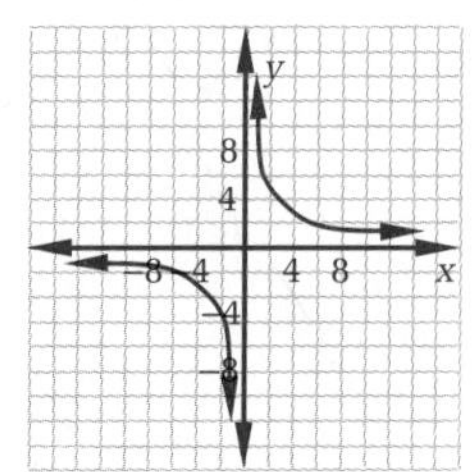

Because the vertices must lie on the diagonal line $y = x$, we can substitute x for y in $xy = 12$, obtaining $x^2 = 12$. Thus, $x = \pm\sqrt{12}$. The vertices are $(\sqrt{12}, \sqrt{12})$ and $(-\sqrt{12}, -\sqrt{12})$.

Checkpoint

1. If f varies inversely as s, and f is 48 when s is 16, find the constant of variation. 768
2. If y varies inversely as x and $y = 12$ when $x = 3$, find y when $x = 5$. $7\frac{1}{5}$ or $\frac{36}{5}$
3. The force of gravity is inversely proportional to the square of the distance from the center of the earth. If you weigh 120 lbs on the surface of the earth, how much would you weigh 250 miles above the surface? (Earth's radius ≈4000 miles.) Approximately 106.3 lbs

Stumbling Block

A common error with respect to inverse variation is to think that if $y = 3x$ is a direct variation with a constant of 3, then $y = \frac{x}{3}$ must be the inverse variation since it is the inverse of the function. While this is true, the terminology *inverse* means something entirely different when applied to inverse variation. *Inverse function* refers to the relationship between two functions, while *inverse relation* refers to two variables and their relationship to one another in the equation.

Communicating Mathematics

In inverse variations, the sign of the constant determines which quadrant(s) the graph will lie in. Discuss with your class how the graph of an inverse variation changes as the absolute values of the constant grow larger and smaller.

Assignment Guide

Average

Day 1 9, 11, 13, 16, 18, 19, 21–23, 27, 28

Day 2 10, 12, 14, 15, 20, 25, 30, 32, 34, 38, 41

Advanced

Day 1 10, 12–14, 16, 18, 20, 21, 28, 29, 32

Day 2 15, 17, 20, 25, 26, 30, 32, 34, 38, 41, 47

Honors

Day 1 10, 12–15, 17, 19, 23–26, 28, 29, 32

Day 2 30, 31, 33, 36, 37, 39, 41, 43, 45–47

Problem-Set Notes and Strategies

- In problem **6b**, as either x or y increases, the other variable decreases, but the product of x and y is not a constant indicating a constant rate of decrease. Therefore, this is not an inverse variation.
- Problem **15** is a good example of inverse variation in practice in a mechanical problem. If possible look at the gears inside a clock to get some idea of the types of gearing necessary to keep correct time. Another good example is the gearing on a bicycle. Ask students which gear the rear wheel of a 10 or 12 speed the chain must be on to go the fastest. They will answer the smallest one.

Additional Answers

Answers for problems **1**, **9**, **10**, **11**, and **12** can be found in the answer pages beginning on page **A** 1.

Part Three: Exercises and Problems

Warm-up Exercises

1 Graph $xy = -4$.

P **In problems 2–5, *I* varies inversely as *M*. Find the constant of variation.**

P **2** $I = 12$ when $M = 18$. 216 **3** $I = 18$ when $M = 12$. 216

4 $I = 9$ when $M = 24$. 216 **5** $I = 3$ when $M = 8$. 24

P **6** Determine whether the tables represent inverse relationships.

a

x	3	2	1	4	6	8	10	12
y	8	12	24	6	4	3	2.4	2

Yes

b

x	1	2	3	4	5	6
y	10	8	6	4	2	0

No

c

x	$-\frac{5}{4}$	$-\frac{3}{2}$	$-\frac{1}{3}$	5	4	$\frac{8}{5}$	10
y	$\frac{4}{5}$	$\frac{2}{3}$	3	$-\frac{1}{5}$	$-\frac{1}{4}$	$-\frac{5}{8}$	$-\frac{1}{10}$

Yes

P **7** If b varies inversely as P and the constant of variation is 0.4, what is P when $b = 8$? 0.05

P **8** Find the vertices of the graph of $xy = 36$. (6, 6) and (−6, −6)

Problem Set

A

In problems 9–12, graph each equation.

P **9** $xy = 1$ **10** $xy = 3$ **11** $xy = 9$ **12** $yx = -1$

P **13** Copy and complete the table to show that y varies inversely as x.

x	−2	−1	10	−5	$-\frac{1}{2}$
y	15	30	−3	6	60

P **14** Suppose that Y varies inversely as W and that $Y = 18$ when $W = 16$. Find the value of W when $Y = 24$. 12

P **15** The number of teeth in each gear is inversely proportional to its turning speed in revolutions per minute (rpm). Gear 1 has 40 teeth and turns at 440 rpm. Gear 2 has 24 teeth. Find the turning speed of gear 2. $733.\overline{3}$ rpm

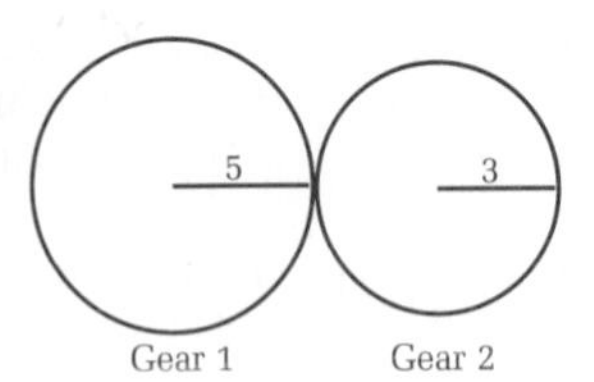

P | **16** Consider the function f such that $f(x) = \frac{16}{x}$.

a Find the equations of all asymptotes of the graph. $x = 0$ and $y = 0$

b Graph f and its asymptotes.

P | **17** Explain why the bases and the heights of all triangles with area 24 are inversely related. Are the bases and heights of triangles of any given area inversely proportional? Because $b \cdot h$ always equals 48; yes

I,R | **18** Suppose $e(x, y) = x^y$. Compute $e(1, 1)$, $e(1, 2)$, $e(1, 4)$, $e(1, -2)$, and $e(1, n)$. 1; 1; 1; 1; 1

B

In problems 19 and 20, graph each equation.

R | **19** $\frac{x^2}{9} - \frac{y^2}{36} = 1$ **20** $\frac{y^2}{9} - \frac{x^2}{36} = 1$

R,I | **21** Solve the following system for (x, y). $(x, y) = (7, -5)$

$$\begin{cases} 9x + 4y = 43 \\ 5x - 8y = 75 \end{cases}$$

P,R | **22** It costs $40 for a family to obtain a swimming pass to Lake Minian each summer. Each time the family uses the card, a punch mark is made for each person who goes swimming.

a Find the cost per punch mark if 20 punches are made and if 80 punches are made. $2.00; $0.50

b How many punches will need to be made for the cost to be less than $0.30 per punch? At least 134

P | **23** If F varies inversely as r^2, and if F is 25 when r is 3, what is the value of F when r is 6? 6.25

P | **24** Graph $xy - 2y = 4$. (Hint: Factor the left side.)

R | **In problems 25 and 26, graph each equation.**

25 $\frac{x}{4} + \frac{y}{9} = 1$ **26** $\frac{x^2}{4} + \frac{y^2}{9} = 1$

R,I | **In problems 27–29, factor each expression over the integers.**

27 $x^2 - 9x - 10$ — $(x - 10)(x + 1)$

28 $2x^3 - x^2 - 3x$ — $x(2x - 3)(x + 1)$

29 $x^2 + 9x + 7$ — Cannot be factored

P | **In problems 30 and 31, graph each function.**

30 $f(x) = \frac{12}{x}$ **31** $f(x) = \frac{12x}{x^2}$

P | **32** If y varies inversely as x and x varies inversely as z, does y vary inversely as z? Explain your answer.
No; $\frac{y}{z}$, not $y \cdot z$, equals a constant.

Problem-Set Notes and Strategies, continued

- Problem **18** introduces exponential functions with the elementary function x^y where $x = 1$. In this case, any exponent will yield an answer of 1. In Chapter **11**, exponential functions will be explored in depth.
- Problem **24** indicates a translation of the graph once it has been factored. The variable x has been replaced with $x - 2$ indicating a shift to the right of 2 units.
- In problem **32**, if you set up the variations as $y = \frac{k}{x}$ and $x = \frac{k}{z}$ and make the substitutions, the complex fraction will yield $y = kz$, a direct variation not an inverse variation (k is used generically for arbitrary constants).

Additional Answers

16b

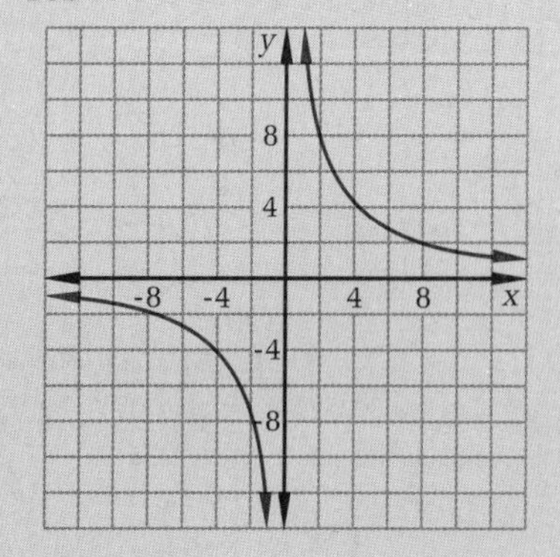

19

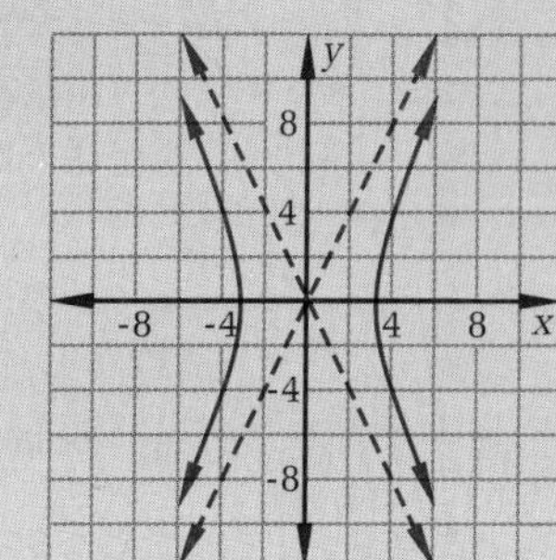

Answers for problems **20**, **24**, **25**, **26**, **30**, and **31** can be found in the answer pages beginning on page A 1.

Problem Set, *continued*

R **33** Let $y = x^2 + 1$.

a Graph the equation.
b Interchange x and y and graph the new equation.
c In the new equation, is y a function of x? No

P **34** Write an equation to represent the inverse variation represented by the graph shown. $xy = 12$

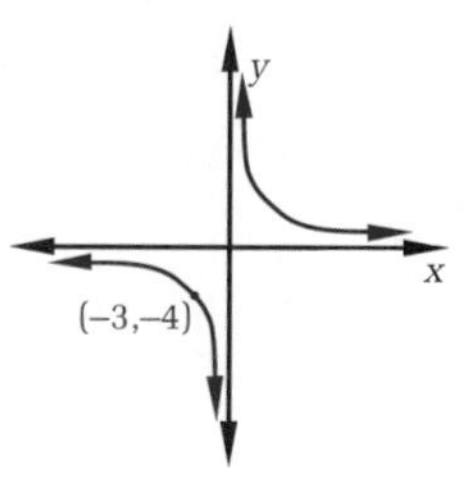

P **35** Suppose x varies inversely as y.

a Graph $xy = k$, where $k = 1$.
b Graph $xy = k$, where $k = 2$.
c Find the points of intersection of each of the curves in parts **a** and **b** with the line $y = x$.
$(1, 1), (-1, -1); (\sqrt{2}, \sqrt{2}), (-\sqrt{2}, -\sqrt{2})$

P **36** Graph $(x - 3)(y + 1) = 12$.

P,R,I **37** Solve the following system for (x, y). $\left(-3, -\frac{1}{3}\right)$

$$\begin{cases} y = \dfrac{1}{x} \\ y = \dfrac{1}{2x + 3} \end{cases}$$

P **38** Brute Force is a member of the Muscle City Health Club. His membership fee is \$160 per year.

a Find his cost per workout if he works out three days per week. \$1.03
b What would his cost per workout be if the membership fee were decreased by 20%? \$0.82

P,R **39** Consider functions f and g such that $f(x) = \frac{4}{x}$ and $g(x) = f(x + 2) + 3$.

a Graph f.
b Write the equations of the asymptotes of $y = f(x)$. $x = 0$ and $y = 0$
c Graph g.
d Write the equations of the asymptotes of $y = g(x)$.
$x = -2$ and $y = 3$

R **40** Bia Thlon rode her bike for 3 hours at a steady rate, then walked for 2 hours at a different steady rate. The next day, she rode her bike for 4 hours and walked for 1 hour at the same rates as on the previous day. She went 42 miles the first day and 51 miles the second. How much faster is Bia's cycling rate than her walking rate? 9 mph

Additional Answers

33a

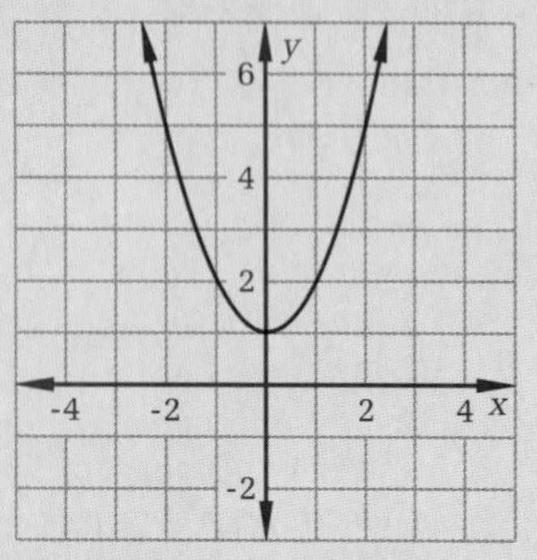

33b

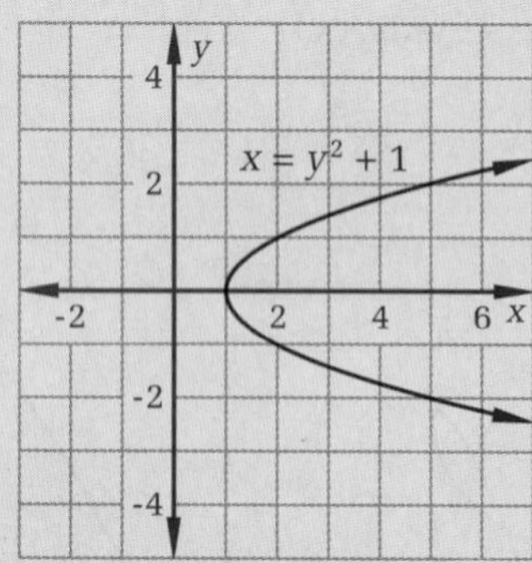

Answers for problems **35a**, **35b**, **36**, **39a**, and **39c** can be found in the answer pages beginning on page **A** 1.

- Problem **37** is most easily solved by equating the two values of *y* to one another and solving the fractional equation that results.

- Problem **40** does not ask for the two rates, just the difference between the two. Make sure the students answer the question that was asked.

P **41** The graph of $y = \frac{40}{x}$ is shown. Point P is on the graph. Find the area of the rectangle. 40

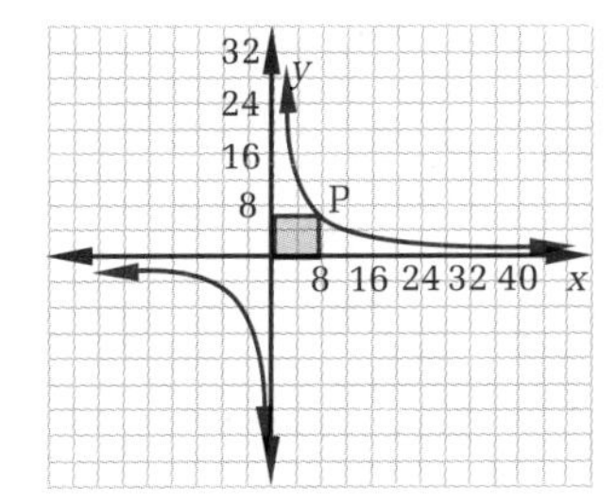

R **42** Let f be a function whose domain is the positive integers. The following are some of the values of $f(n)$.

$f(1) = 1$
$f(2) = 3 + 5 = 8$
$f(3) = 7 + 9 + 11 = 27$
$f(4) = 13 + 15 + 17 + 19 = 64$

Make a reasonable guess for each value.

a $f(5)$ 125 **b** $f(6)$ 216 **c** $f(10)$ 1000
d $f(20)$ 8000 **e** $f(100)$ 1,000,000

C P **43** The intensity with which an object is illuminated is inversely proportional to d^2, the square of its distance from the source of light. If the intensity I of light on a movie screen is 50 candles when the projector is 15 feet from the screen, what is the intensity of light on the screen when the projector is moved 5 feet farther from it? ≈28 candles

R **In problems 44 and 45, graph each equation.**

44 $x^2 + 6x + 4y^2 + 24y = 19$ **45** $x^2 - 4x - y^2 - 8y = 13$

P **46** Graph the relationship of x to y if the area of the shaded region is 576, $x > 0$, and $y > 0$.

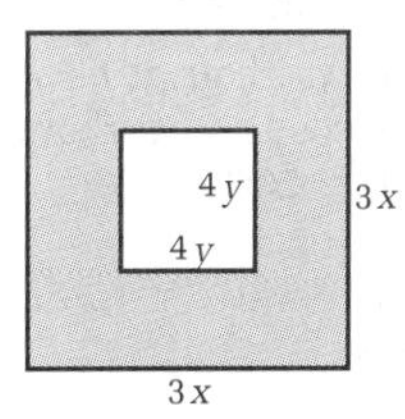

R **47** Write an equation to represent the ellipse in the diagram.
$\frac{(x-8)^2}{25} + \frac{(y-10)^2}{4} = 1$

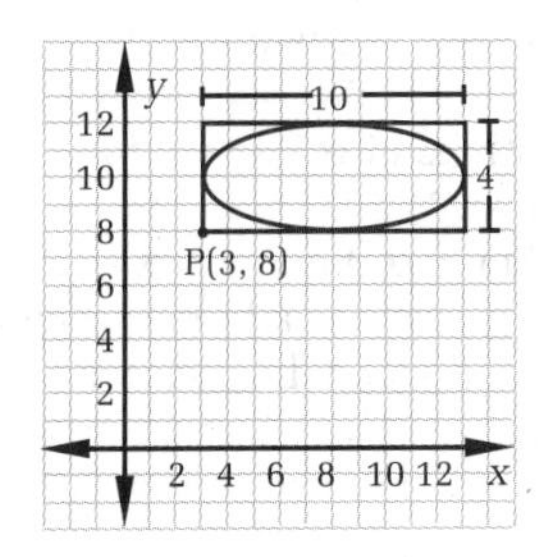

Problem-Set Notes and Strategies, continued

- In problem **41**, no coordinates were given for point P. The area of the rectangle is equal to the constant of variation, in this case $xy = 40$.
- Problem **42** is an interesting sequence that has been generated. See if students can generate a closed-form solution for this problem that is different from x^3. Then see if they can prove its equivalence to x^3.

Additional Answers

44

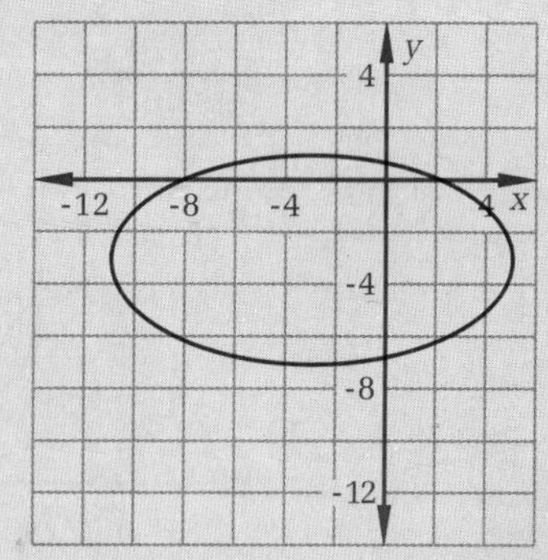

45

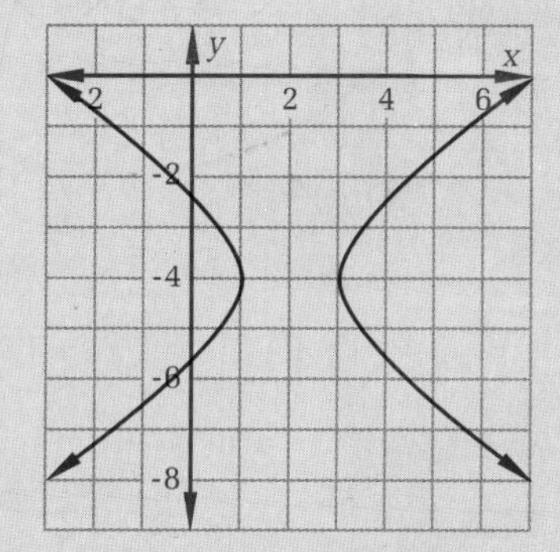

Answer for problem **46** can be found in the answer pages beginning on page **A** 1.

7.5 NONLINEAR SYSTEMS

Objective

After studying this section, you will be able to

- Solve systems containing nonlinear equations

Part One: Introduction

See p. 801

Just as you have used graphing to help solve systems of linear equations, you can use it to help solve systems that contain nonlinear equations.

Example *Solve the following system.* $\begin{cases} y^2 - x^2 = 16 \\ y^2 + x^2 = 34 \end{cases}$

First, let's sketch the graphs of these equations on the same set of axes. The graph of $y^2 - x^2 = 16$, or $\frac{y^2}{16} - \frac{x^2}{16} = 1$, is a hyperbola that is symmetrical about the y-axis, with vertices (0, 4) and (0, −4). The graph of $x^2 + y^2 = 34$ is a circle with its center at (0, 0) and a radius of $\sqrt{34}$, or ≈5.8. The sketch indicates that the system has four solutions. We can find the solutions algebraically.

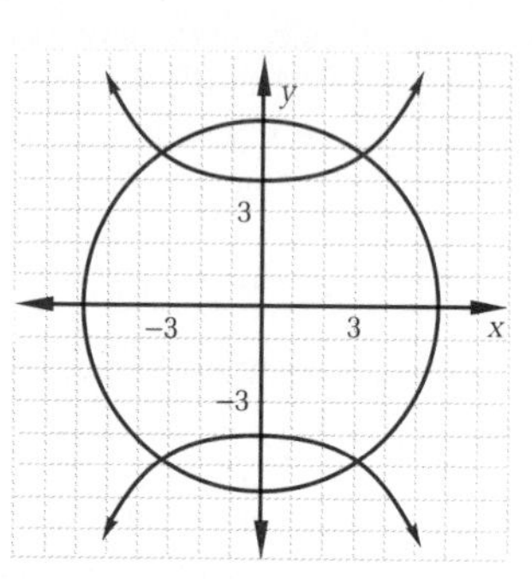

Add the equations to eliminate the x^2 term

$$\begin{aligned} \mathbf{y^2 + x^2} &= \mathbf{34} \\ \mathbf{y^2 - x^2} &= \mathbf{16} \\ \hline \mathbf{2y^2} \quad &= \mathbf{50} \\ \mathbf{y^2} &= \mathbf{25} \\ \mathbf{y = 5 \text{ or } y} &= \mathbf{-5} \end{aligned}$$

Substitute 5 and −5 for y in either of the original equations and solve for x.

$$\begin{aligned} y^2 + x^2 &= 34 \\ 5^2 + x^2 &= 34 \\ x^2 &= 9 \\ x = 3 \text{ or } x &= -3 \end{aligned} \qquad \begin{aligned} y^2 + x^2 &= 34 \\ (-5)^2 + x^2 &= 34 \\ x^2 &= 9 \\ x = 3 \text{ or } x &= -3 \end{aligned}$$

The four solutions of the system are (3, 5), (−3, 5), (3, −5), and (−3, −5). These solutions agree with our sketch.

7.5

Class Planning

Time Schedule

Average: Optional
Advanced: Optional
Honors: 1 day

Resource References

Teacher's Resource Book
- Class Opener 45
- Cooperative Learning
- Practice 47
- Enrichment 14
- Transparency 14

Class Opener

Sketch $x^2 + y^2$ 25 and $3x + 4y$ 12 on the same coordinate system.

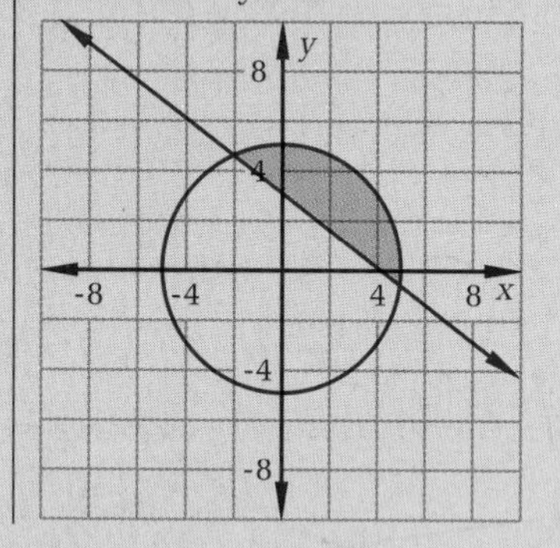

Lesson Notes

- Nonlinear systems may be solved using the same methods used to solve linear systems. The only difference is the number of solutions generated. With the conic sections, there are at most four possible solutions to a system of two distinct conics. It might be interesting for your students to verify that fact.

Part Two: Sample Problems

See p. 801

Problem 1 *Solve the following system.* $\begin{cases} 2x - 3y = 6 \\ xy = 12 \end{cases}$

Solution When we sketch the graphs of the equations, the diagram shows two solutions. Let's use substitution to solve the system algebraically. Solving the first equation for x, we have $x = \frac{3}{2}y + 3$. Now we substitute for x in $xy = 12$ and solve for y.

$$\left(\tfrac{3}{2}y + 3\right)y = 12$$
$$\tfrac{3}{2}y^2 + 3y = 12$$
$$y^2 + 2y = 8$$
$$y^2 + 2y - 8 = 0$$
$$(y - 2)(y + 4) = 0$$
$$y = 2 \text{ or } y = -4$$

If $y = 2$	If $y = -4$
$2x = 12$	$-4x = 12$
$x = 6$	$x = -3$

The two solutions of the system are (6, 2) and (−3, −4).

Problem 2 *Solve the following system.* $\begin{cases} y = 1 - x^2 \\ 4x^2 + y^2 = 16 \end{cases}$

Solution The graph of $y = -x^2 + 1$ is a parabola opening downward with a vertex at (0, 1). The graph of $4x^2 + y^2 = 16$, or $\frac{x^2}{4} + \frac{y^2}{16} = 1$, is an ellipse centered at (0, 0). The diagram shows two solutions.

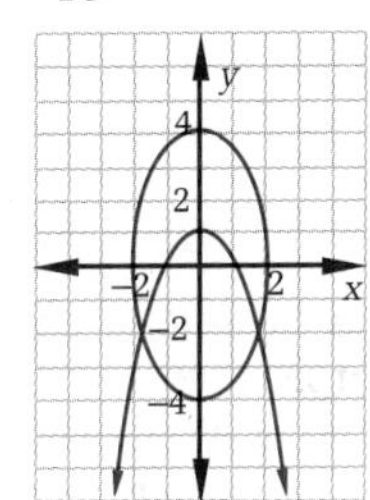

Let's solve the system by substitution. Either we can substitute $1 - x^2$ for y in $4x^2 + y^2 = 16$ to get a fourth-degree equation (try it and see), or we can substitute $1 - y$ for x^2 in $4x^2 + y^2 = 16$ to get the quadratic equation $4(1 - y) + y^2 = 16$. The latter method is easier.

$$4(1 - y) + y^2 = 16$$
$$4 - 4y + y^2 = 16$$
$$y^2 - 4y - 12 = 0$$
$$(y - 6)(y + 2) = 0$$
$$y = 6 \text{ or } y = -2$$

Our graph shows that y must be negative, so 6 is not a possible value of y. (Try substituting 6 for y in the first equation to see why.) We substitute −2 for y in the first equation and solve for x.

$$1 - x^2 = y$$
$$1 - x^2 = -2$$
$$x^2 = 3$$
$$x = \sqrt{3} \text{ or } x = -\sqrt{3}$$

The two points of intersection are $(\sqrt{3}, -2)$ and $(-\sqrt{3}, -2)$.

Lesson Notes, continued

- The substitution process is used most frequently when solving nonlinear equations and can generate higher degree equations if you are not careful. In general, try to make a substitution that will maintain the quadratic nature of the equations so that the quadratic formula can be used.

Using Computers

The *Omnifarious Plotter* and some graphing calculators allow you to graph conics. Sometimes you may have to graph the top and bottom of these non-functions of *x* separately. The ability to view the symmetries in graphs of different conics is very helpful for students. It is nice to do both the algebraic and graphical at the same time, and technology facilitates this process immeasurably.

Checkpoint

1 Solve the following system. $\begin{cases} x^2 + y^2 = 1 \\ y = 2x \end{cases}$
$\left(\frac{\sqrt{5}}{5}, \frac{2\sqrt{5}}{5}\right)$ or $\left(\frac{-\sqrt{5}}{5}, \frac{-2\sqrt{5}}{5}\right)$

2 In how many points do the graphs of $y = x^2$ and $3x^2 + 4y^2 = 5$ intersect? 2

3 Solve the following system. $\begin{cases} x^2 + y^2 = 16 \\ 4x^2 + 9y^2 = 36 \end{cases}$
No real solutions; the graphs do not intersect.

Assignment Guide

Average
Optional
Advanced
Optional
Honors
10, 12, 13, 15–17, 19, 20, 24–26, 37–39, 43, 44

Problem-Set Notes and Strategies

- Note in problems **6** and **7** that the only points of intersection of the two graphs occur at the vertices of the hyperbola and ellipse.

Stumbling Block

When solving for a system of nonlinear equations, frequently the student must solve an equation of the form x^2 = some constant. When taking the square root of both sides, care should be taken to include both of the roots and not just the positive root. If the system were graphed first to determine the number of roots, this problem would not arise.

- Problems **8**, **9**, **11** and **12** require solutions with radicals. A rough sketch will indicate how many solutions can be expected.

Part Three: Exercises and Problems

Warm-up Exercises

P **1** Estimate the solutions of the system represented in each diagram.

a

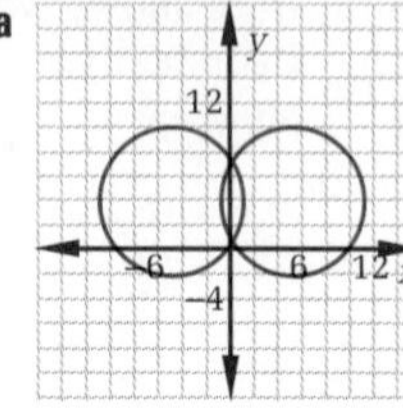

(0, 7.3), (0, 0.7)

b

No solution

c

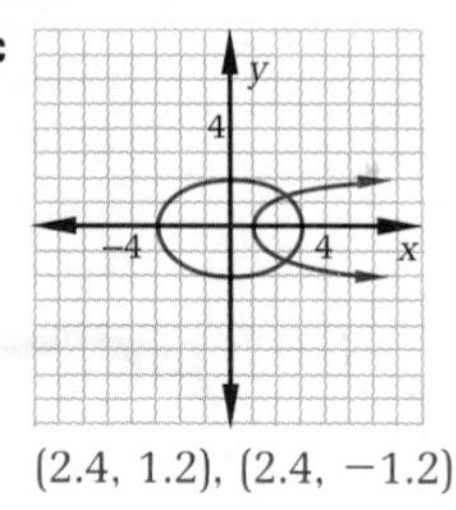

(2.4, 1.2), (2.4, −1.2)

P **2** Consider the following system.

$$\begin{cases} x^2 - y^2 = 36 \\ x^2 + y^2 = 4 \end{cases}$$

a Solve the system for (x^2, y^2). (20, −16)
b Solve the system for (x, y). No real solution

P **3** Is (x, y) = (−3, 4) a solution of the following system? Yes

$$\begin{cases} x^2 + y^2 = 25 \\ y = -2x - 2 \end{cases}$$

P **In problems 4 and 5, solve for (x, y).**

4 $\begin{cases} y = 2 \\ x^2 + y^2 = 25 \end{cases}$ $(x, y) = (\sqrt{21}, 2)$ or $(x, y) = (-\sqrt{21}, 2)$

5 $\begin{cases} y^2 - x^2 = 5 \\ x^2 + y^2 = 13 \end{cases}$

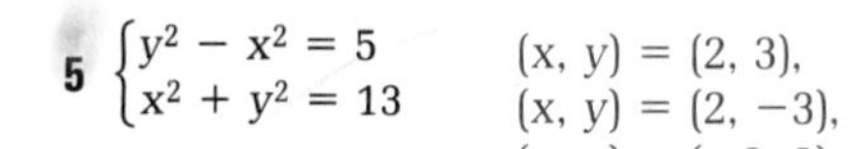

(x, y) = (2, 3), (x, y) = (2, −3), (x, y) = (−2, 3), or (x, y) = (−2, −3)

P **6** Find the points of intersection of the graphs of $\frac{x^2}{9} + \frac{y^2}{16} = 1$ and $\frac{x^2}{9} - \frac{y^2}{16} = 1$. (3, 0), (−3, 0)

P **7** Solve the following system for (x, y). $(x, y) = \left(\frac{1}{3}, 0\right)$ or $(x, y) = \left(-\frac{1}{3}, 0\right)$

$$\begin{cases} 9x^2 + 16y^2 = 1 \\ 9x^2 - 16y^2 = 1 \end{cases}$$

Problem Set

A **In problems 8–11, solve each system for (x, y).**

P **8** $\begin{cases} 2x^2 + y^2 = 36 \\ y = 2x \end{cases}$ $(x, y) = (\sqrt{6}, 2\sqrt{6})$ or $(x, y) = (-\sqrt{6}, -2\sqrt{6})$

9 $\begin{cases} x^2 + y^2 = 36 \\ 2x^2 - y^2 = 12 \end{cases}$ $(x, y) = (4, 2\sqrt{5})$, $(x, y) = (4, -2\sqrt{5})$, $(x, y) = (-4, 2\sqrt{5})$, or $(x, y) = (-4, -2\sqrt{5})$

10 $\begin{cases} x^2 + 8y = 33 \\ y = x - 9 \end{cases}$ (x, y) = (7, −2) or (x, y) = (−15, −24)

11 $\begin{cases} y = x^2 \\ x^2 + y^2 = 6 \end{cases}$ $(x, y) = (\sqrt{2}, 2)$ or $(x, y) = (-\sqrt{2}, 2)$

P **12** Find the points of intersection of the graphs of $y = x^2$ and $x^2 - y^2 = -2$. $(\sqrt{2}, 2)$, $(-\sqrt{2}, 2)$

R **13** The wavelength w, in meters, of a radio wave varies inversely as its frequency f, in kilocycles per second, according to the equation $wf = 300{,}000$.

a Find the wavelength of the transmissions of radio station WINK, which broadcasts at a frequency of 800 kilocycles per second. 375 m

b Find the wavelength of the transmissions of radio station WONK, which broadcasts at a frequency of 1120 kilocycles per second. ≈ 267.9 m

c Find the frequency, in kilocycles per second, at which radio station WAIT broadcasts, given that the wavelength of the transmissions of the station is 400 meters. 750

R **14** Graph, on the same coordinate system, the two lines described by the matrix equation.

$$\begin{bmatrix} 2 & 3 \\ -4 & -6 \end{bmatrix}\begin{bmatrix} x \\ y \end{bmatrix} = \begin{bmatrix} 12 \\ -12 \end{bmatrix}$$

R **In problems 15–17, factor each expression.**

15 $4 - x^2$ $(2 + x)(2 - x)$

16 $4 - (5x)^2$ $(2 + 5x)(2 - 5x)$

17 $4 - (5x - 3)^2$ $5(5x - 1)(1 - x)$

P,R **18** Suppose that x and y vary inversely. For positive values of x, as x increases, y also increases. For negative values of x, as x increases, y also increases. Give an example of such a relationship and sketch its graph. Possible answer: $xy = -1$

B

In problems 19–22, solve each system for (x, y).

P **19** $\begin{cases} y = x + 10 \\ x^2 + y^2 = 4 \end{cases}$ No real solution

20 $\begin{cases} x^2 - y^2 = 4 \\ y^2 = 3x \end{cases}$ $(x, y) = (4, 2\sqrt{3})$ or $(x, y) = (4, -2\sqrt{3})$

21 $\begin{cases} x^2 + y^2 = 1 \\ x^2 + y^2 = 3 \end{cases}$ No real solution

22 $\begin{cases} y^2 - 9x^2 = 16 \\ y = -5x \end{cases}$ $(x, y) = (1, -5)$ or $(x, y) = (-1, 5)$

R **23** Seven points lie on a circle. How many chords are determined by the points? 21

R,I **24** Consider the following multiplication table.

a Find the sum of the elements in each region. 1; 8; 27; 64; 125

b Find the sum of the elements in the nth such region. n^3

	1	2	3	4	5
1	1	2	3	4	5
2	2	4	6	8	10
3	3	6	9	12	15
4	4	8	12	16	20
5	5	10	15	20	25

Problem-Set Notes and Strategies, continued

- In problem **14**, ask students if they could have predicted the relationship between the equations without graphing and doing a great deal of algebra.

Additional Answer

14

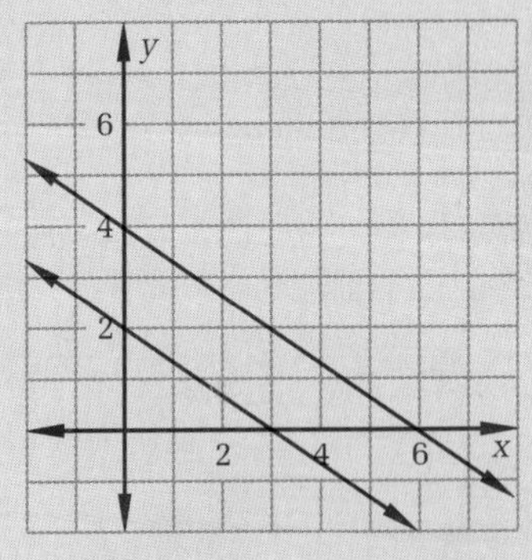

- Problems **15 – 17** illustrate several different applications of the difference of squares formula. Some students may have simplified problem **17** first and then factored.

- Graphing the equations in problems **19** and **21** may help some students realize that "no real solution" is a perfectly acceptable solution to a system.

- The formula for problem **23** is $\frac{n(n-1)}{2}$. This is the same formula for the sum of the first n integers and was used in the handshake problem at the beginning of the text.

- Pattern recognition plays a big role in mathematics and engineering applications. There is a branch of mathematics that examines patterns in problem **24**. This branch of mathematics is called Number Theory.

Problem Set, *continued*

P,R **25** The product of the x- and the y-coordinate of each point on a certain graph is 24.

a Write an equation that corresponds to the graph. $xy = 24$
b Draw the graph.

R,I **26** Find each product.

a $(a - b)(a^2 + ab + b^2)$ $a^3 - b^3$
b $(a - b)(a^2 + 2ab + b^2)$ $a^3 + a^2b - ab^2 - b^3$

R,I **27** If f is a discrete function such that $f(1) = 0$ and $f(n + 1) = 100 \cdot f(n)$, what is $f(78)$? 0

P,R **28** Two pounds of candy valued at \$3.60 per pound are mixed with eight pounds of candy valued at \$4.40 per pound.

a Find the value of the mixture. \$42.40
b Find the value per pound of the mixture. \$4.24

R **29** If $f(x) = 8x + 13$ and $g(x) = 2x - 7$, for what value of x is $f(x) = g(x)$? $-\frac{10}{3}$

P,R **In problems 30–32, factor each expression over the integers.**

30 $10x^2 - 7x + 1$ $(5x - 1)(2x - 1)$

31 $x^3 - 25x$ $x(x - 5)(x + 5)$

32 $81 - 4x^2y^4$ $(9 + 2xy^2)(9 - 2xy^2)$

R **33** A linear function f is of the form $f(x) = mx + b$, with $f(3) = 8$ and $f(-6) = 3$.

a Find m and b. $m = \frac{5}{9}$; $b = \frac{19}{3}$
b Draw the graph of f.
c Draw the graph of function n such that $n(x) = f(x - 2)$.

R **34** Suppose that x and y vary inversely and that $y = 3.7$ when $x = 5$. What is the value of x when $y = 7.4$? 2.5

R **35** Maggie Zeen sold 150 magazine subscriptions, some at \$12 and the rest at \$19, during the month of June.

a How many subscriptions of each type did she sell if the total amount of her sales was \$2059? \$12 sub.: 113; \$19 sub.: 37
b Find Maggie's commission rate if the amount of her commission was \$617.70. 30%

P,R **36** Use the process of completing the square to find the vertex of the parabola described by each function.

a $f(x) = x^2 - 12x - 16$ $(6, -52)$
b $f(x) = -x^2 - 12x - 16$ $(-6, 20)$
c $f(x) = -2x^2 - 12x - 16$ $(-3, 2)$
d $f(x) = 2x^2 - 12x - 16$ $(3, -34)$

P **37** The sum of two numbers x and y is 23, and the sum of the squares of the numbers is 289. Find (x, y). $(15, 8)$ or $(8, 15)$

Problem-Set Notes and Strategies, continued

- Problem **26b** demonstrates the most common mistake people make when they factor the difference of cubes.
- Problem **27** requires students to use some insight. The most trivial problems are sometimes the most difficult to solve. Consider the product of the following binomials: $(x-a)(x-b)(x-c) \dots (x-y)(x-z)$. This also produces zero since $(x-x)$ is one of the factors.
- It is interesting to note that the equation solving process gives both solutions of problem **37**. The geometric interpretation of the problem is the intersection of a line with a circle.

Additional Answers

25b

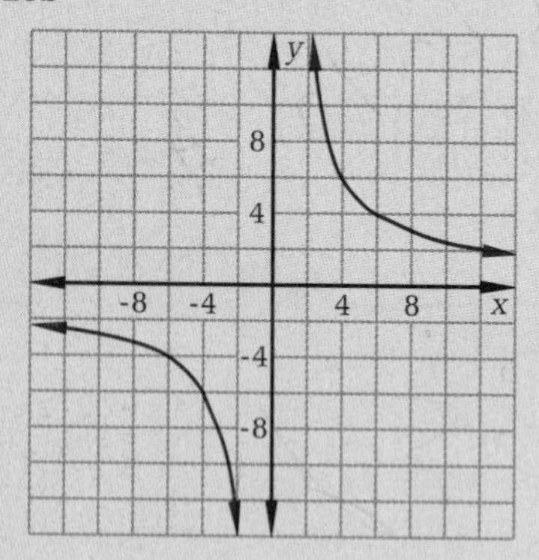

33b

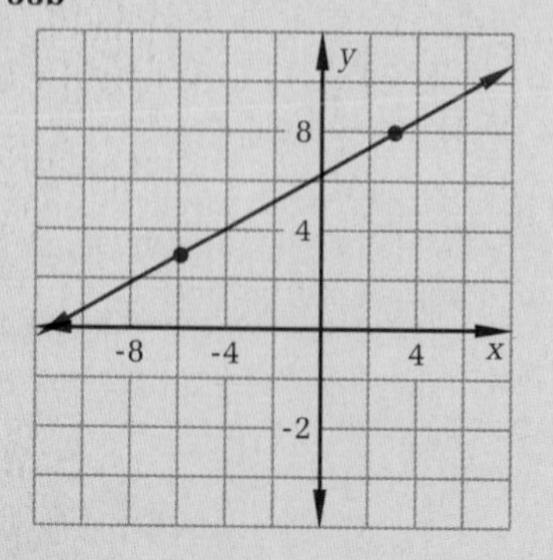

Answer for probles **33c** can be found in the answer pages beginning on page **A** 1.

P 38 Use the data displayed at the right.

a Find the mode of the data shown. 10

b Find the mean. $12\frac{1}{11}$

c Find the median. 12

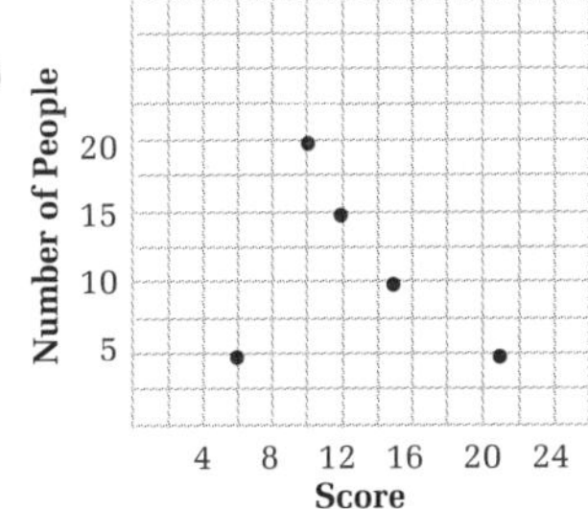

R 39 Consider the following table.

x	16			
y	8	4	2	1

a Copy and complete the table to show that x varies directly as y.

x	16	8	4	2
y	8	4	2	1

b Copy and complete the table to show that x varies inversely as y.

x	16	32	64	128
y	8	4	2	1

P 40 Find the points of intersection of the graphs of $\frac{x^2}{9} + \frac{y^2}{16} = 1$ and $\frac{x^2}{16} + \frac{y^2}{9} = 1$. $\left(\frac{12}{5}, \frac{12}{5}\right), \left(\frac{12}{5}, -\frac{12}{5}\right), \left(-\frac{12}{5}, \frac{12}{5}\right), \left(-\frac{12}{5}, -\frac{12}{5}\right)$

P,R 41 Determine the value of a, given each graph of $f(x) = ax^2$.

a $-\frac{1}{16}$

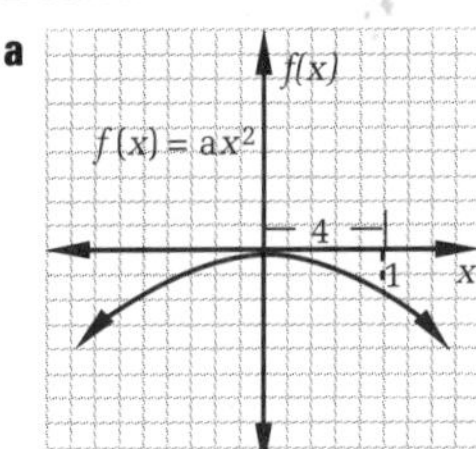

b $\frac{1}{5}$

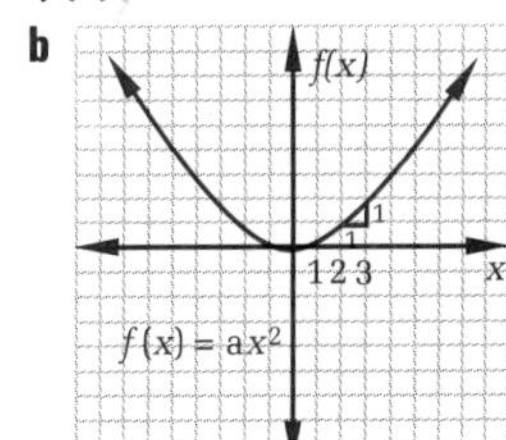

C

P 42 Solve the following system for (x, y). No real solution

$$\begin{cases} \frac{(x-2)^2}{4} + \frac{(y-3)^2}{9} = 1 \\ \frac{(x+2)^2}{4} + \frac{(y+3)^2}{9} = 1 \end{cases}$$

P 43 Solve the following system for (x, y). $(x, y) = (2, -1)$ or $(x, y) = (-1, 2)$

$$\begin{cases} x^2 + y^2 = 5 \\ x^2 + y^2 - x - y = 4 \end{cases}$$

P,R 44 Graph the following system.

$$\begin{cases} \frac{(x-4)^2}{36} + \frac{(y-3)^2}{16} \geq 1 \\ 2y + x \geq 1 \end{cases}$$

Problem-Set Notes and Strategies, continued

- Problem **38** is a good review of the statistical measures of central tendency.

- In problem **41a**, **b** remind students how a parabola's rate changes as the x value is increased.

Communicating Mathematics

Have students write a statement that identifies the most difficult concept about today's lesson. Choose a few students to read their statement. Then discuss these problem areas with the class.

- Problem **44** is a system of nonlinear inequalities. After graphing a boundary line, choose a region of the graph and test a point. If the point makes the inequality true, shade the region. If it doesn't, shade the other region.

Additional Answer

44

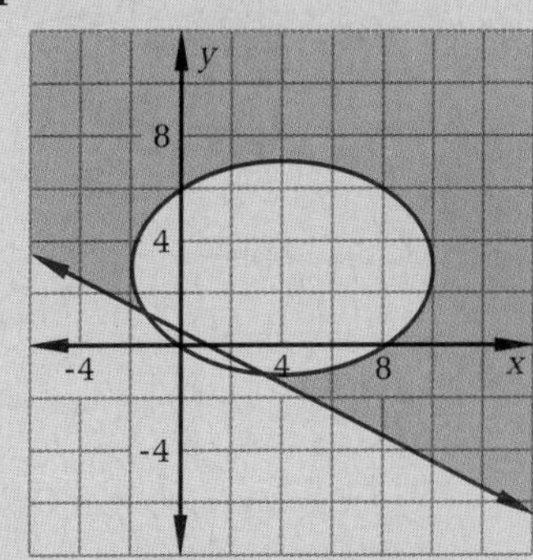

7.6

Class Planning

Time Schedule

Average: Optional
Advanced: Optional
Honors: 1 day

Resource References

Teacher's Resource Book
Class Opener 46
Practice 48

Class Opener

Solve the system: $\begin{cases} \frac{1}{x} + \frac{1}{y} = 10 \\ \frac{1}{x} - \frac{1}{y} = 4 \end{cases}$

$x = \frac{1}{7}, y = \frac{1}{3}$

Lesson Notes

- Rational expressions as part of systems of equations are examined in this section. To solve these types of systems of equations, a different kind of substitution is used which changes the problem into a simpler one that may be worked by more conventional techniques. The student must be careful to return to the original substitution and answer the question posed in the actual problem.
- Problems involving work and length of time it takes to complete a job are some of the most difficult word problems for students to master. These problems also involve rational expressions and need to be solved by our conventional methods.

7.6 OTHER SYSTEMS OF EQUATIONS

Objective

After studying this section, you will be able to

- Solve systems of equations involving rational expressions

Part One: Introduction

Systems in which one or more variables are in the denominator of a fraction can be solved by using methods similar to those you have used to solve other nonlinear systems.

Example *Solve the following system.*

$$\begin{cases} \frac{20}{x} - \frac{1}{y} = 3 \\ \frac{8}{x} + \frac{5}{y} = 12 \end{cases}$$

If we let $P = \frac{1}{x}$ and $Q = \frac{1}{y}$, we can rewrite the system as follows.

$$\begin{cases} 20P - Q = 3 \\ 8P + 5Q = 12 \end{cases}$$

To solve this system for P and Q, we multiply the first equation by 5 and add it to the second.

$$\begin{aligned} 100P - 5Q &= 15 \\ 8P + 5Q &= 12 \\ \hline 108P \quad &= 27 \\ P = \frac{27}{108} &= \frac{1}{4} \end{aligned}$$

We now substitute $\frac{1}{4}$ for P in the first equation and solve for Q.

$$\begin{aligned} 20P - Q &= 3 \\ 20\left(\tfrac{1}{4}\right) - Q &= 3 \\ 5 - 3 &= Q \\ 2 &= Q \end{aligned}$$

Since $P = \frac{1}{x}$ and $Q = \frac{1}{y}$, $x = 4$ and $y = \frac{1}{2}$.

Part Two: Sample Problems

Problem 1 *John, George, and Paul always enter the annual pie-eating contest in their hometown, Ringo. One year, John and Paul together ate 18 pies in 16 minutes. The next year, George and Paul together ate 20 pies in 16 minutes. Last year, the 3 together ate 30 pies in 16 minutes. How long does it take each to eat 1 pie?*

Solution If John eats a pie in J minutes, then his rate is $\frac{1}{J}$ pie per minute. In the same way, George's rate is $\frac{1}{G}$, and Paul's rate is $\frac{1}{P}$. The individual rates for each year add up to the combined rate for that year.

$$\begin{cases} \frac{1}{J} + \frac{1}{P} = \frac{18}{16} \\ \frac{1}{G} + \frac{1}{P} = \frac{20}{16} \\ \frac{1}{J} + \frac{1}{G} + \frac{1}{P} = \frac{30}{16} \end{cases}$$

We can clear some fractions by multiplying each equation by 16.

$$\begin{cases} \frac{16}{J} + \frac{16}{P} = 18 \\ \frac{16}{G} + \frac{16}{P} = 20 \\ \frac{16}{J} + \frac{16}{G} + \frac{16}{P} = 30 \end{cases}$$

If we now let $A = \frac{16}{J}$, $B = \frac{16}{G}$, and $C = \frac{16}{P}$, we can write the following system.

$$\begin{cases} A + C = 18 \\ B + C = 20 \\ A + B + C = 30 \end{cases}$$

By subtracting the second equation from the third equation, we find that $A = 10$. By subtracting the first equation from the third equation, we find that $B = 12$. If we substitute these values for A and B in the third equation, we find that $C = 8$. Now that we know the values of A, B, and C, we can find the values of J, G, and P.

We know that $\frac{16}{J} = A$ and $A = 10$, so $J = \frac{16}{10}$. It takes John $1\frac{3}{5}$ minutes to eat a pie.

We know that $\frac{16}{G} = B$ and $B = 12$, so $G = \frac{16}{12}$. It takes George $1\frac{1}{3}$ minutes to eat a pie.

We know that $\frac{16}{P} = C$ and $C = 8$, so $P = \frac{16}{8}$. It takes Paul 2 minutes to eat a pie.

Checkpoint

1 Solve the following system.
$$\begin{cases} \frac{1}{x} + \frac{1}{y} = 8 \\ \frac{1}{x} - \frac{1}{y} = 2 \end{cases}$$
$x = \frac{1}{5}, y = \frac{1}{3}$

2 Is the point $(\frac{1}{4}, \frac{1}{9})$ a solution of the following system?
$$\begin{cases} \frac{1}{x} - \frac{1}{y} = -5 \\ x - y = \frac{1}{5} \end{cases}$$
No

3 Jack takes 3 hours to do the yard work and his younger sister Jill takes 4 hours. If they work together, how long will it take them to do the yard work? $\frac{12}{7}$ or $1\frac{5}{7}$ hours

Problem 2 *Solve the following system.*

$$\begin{cases} x = y^2 \\ \dfrac{14}{x} = \dfrac{5}{y} + 1 \end{cases}$$

Solution We can simplify this system by multiplying the corresponding sides of the equations.

$$x\left(\frac{14}{x}\right) = y^2\left(\frac{5}{y} + 1\right)$$
$$14 = 5y + y^2$$
$$0 = y^2 + 5y - 14$$
$$0 = (y + 7)(y - 2)$$
$$y = -7 \text{ or } y = 2$$

If $y = -7$, then $x = 49$. If $y = 2$, then $x = 4$. The solutions are $(49, -7)$ and $(4, 2)$. Check these solutions in each equation of the system.

Assignment Guide

Average

Optional

Advanced

Optional

Honors

6, 8, 9, 15, 17–20, 23, 25, 31, 32, 34, 35, 37

Problem-Set Notes and Strategies

- Suggest that students begin to solve problem **1** by adding the equations of the system.
- Problem **6** can also be solved by substitution. If the first equation is factored, then the value of $x + y$ could be replaced with 4 and the remaining equations solved.

Part Three: Exercises and Problems

Warm-up Exercises

In problems 1 and 2, solve for (x, y).

P **1** $\begin{cases} \dfrac{1}{x} + \dfrac{1}{y} = 6 \\ \dfrac{1}{x} - \dfrac{1}{y} = 12 \end{cases}$ $(x, y) = \left(\frac{1}{9}, -\frac{1}{3}\right)$

2 $\begin{cases} \dfrac{1}{x} - \dfrac{2}{y} = 1 \\ \dfrac{5}{x} + \dfrac{6}{y} = -1 \end{cases}$ $(x, y) = \left(4, -\frac{8}{3}\right)$

P **3** The sum of the reciprocals of two numbers is $\frac{1}{2}$. The difference of their reciprocals is $\frac{1}{3}$. Find the two numbers. 12 and $\frac{12}{5}$

P **4** Solve the following system for (x, y). $(x, y) = (3, 3)$

$$\begin{cases} \dfrac{1}{x} + \dfrac{1}{3} = \dfrac{2}{y} \\ x = y \end{cases}$$

Problem Set

A

P **5** Solve the following system for (x, y). $(x, y) = \left(\frac{1}{3}, \frac{1}{2}\right)$

$$\begin{cases} \dfrac{3}{x} - \dfrac{2}{y} = 5 \\ \dfrac{4}{x} + \dfrac{5}{y} = 22 \end{cases}$$

R **6** Solve the following system. (Hint: Factor and divide.)

$$\begin{cases} x^2 - y^2 = 20 \\ x + y = 4 \end{cases}$$ $(x, y) = \left(\frac{9}{2}, -\frac{1}{2}\right)$

R **7** Solve the following system for (x, y). (Hint: Start by dividing the sides of one equation by the corresponding sides of the other.)

$$\begin{cases}(x + y)(x - y) = y^3 \\ x + y = y^2\end{cases}$$

$(x, y) = (0, 0)$ or $(x, y) = (6, 3)$

R **8** Teri Yaki can make a dozen egg rolls in three hours. Her sister Suki can make a dozen rolls in five hours. How long would it take them to make a dozen rolls if they worked together? $\frac{15}{8}$ hr

R,P **9** Suppose that f and g are functions such that $f(x) + g(x) = 3x^2 + 2x + 7$ and $f(x) - g(x) = 2x^2 - 3x + 1$. Write formulas to represent f and g.
$f(x) = \frac{5}{2}x^2 - \frac{1}{2}x + 4$; $g(x) = \frac{1}{2}x^2 + \frac{5}{2}x + 3$

R **10** In the figure, A is at (2, y) and B is at (3, 4). Find the value of y for which $d_1 = d_2$. ≈ 4.6

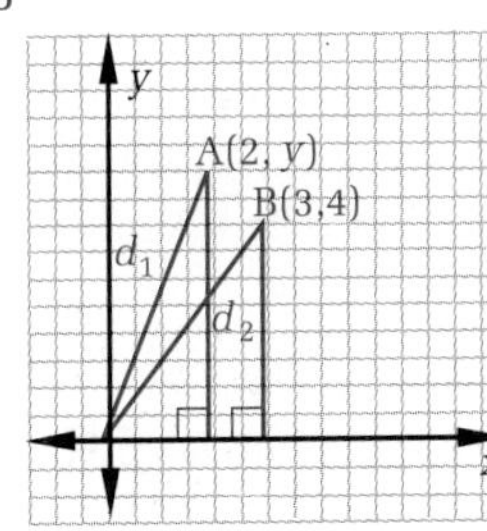

P,R **11** If it takes five workers six hours to paint an office, how long will it take eight workers to paint the office, assuming all the workers work at the same rate? 3.75 hr

R **12** Solve $q^2 + 7731q = 0$ for q. $q = 0$ or $q = -7731$

R **13** The area of a rectangle is 32 square units. Its sides have integral lengths. Find the probability that its perimeter is greater than 30 units. $\frac{2}{3}$

R,I **14** Solve $w(w - 1)(w + 2)(w - 3) = 0$ for w.
$w = 0$, $w = 1$, $w = -2$, or $w = 3$

P **15** A military squad leader wants to select three of ten soldiers for a special mission. In how many ways can the three be chosen? 120

P,R **16** If pump A can fill a pool in six hours and pump B can fill the pool in four hours, how long will filling the pool take if the two pumps work together?
2 hr 24 min

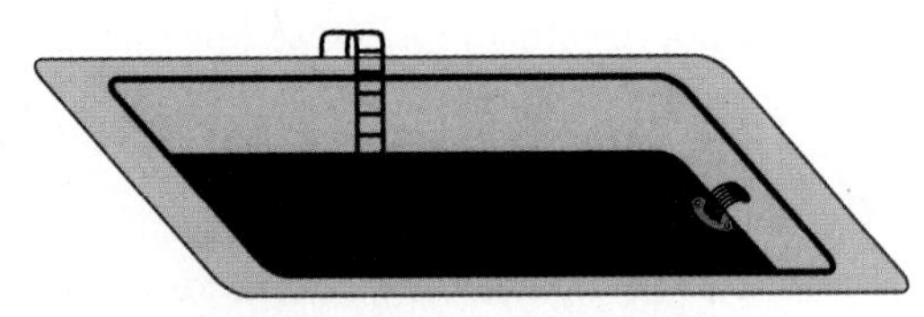

B

R **17** Consider the discrete function F for which $F(1) = 1$ and $F(n) = \frac{1}{10}[F(n - 1)]$. Find the values of the following.

a F(1), F(2), F(3), F(4), and F(5)
$1; \frac{1}{10}; \frac{1}{100}; \frac{1}{1000}; \frac{1}{10,000}$

b $\sum_{n=1}^{5} F(n)$ 1.1111

Problem-Set Notes and Strategies, continued

- Remember to compare rates and not job times in problem **8**.
- Problem **10** is a good exercise in the review of the distance formula.

Stumbling Block

Work problems are some of the most difficult to solve for students. Get students into the habit of estimating the answer. Have them think in terms of the rate at which each person is working instead of the time each takes. Rates may be added and compared. It is difficult to equate total times for a job.

- Order is not important in problem **15**, so a combination will be used.
- In problem **16**, emphasize to the students that the answer certainly must be less than 4 hours since pump *B* can do the entire job in 4 hours with no help at all from pump *A*.

Problem-Set Notes and Strategies, continued

- Problems **19** and **20** may require a substitution to simplify the equation solving process. Make sure the final answer is in terms of x and y and not the intermediate variables that were substituted for.

- Problem **21** is best solved as a two part problem. Figure out how much of the pool was filled in the first 40 minutes by using rates and then how long it would take to fill the remainder with the one pump. Remember that 40 minutes is $\frac{2}{3}$ hours and every time must be consistent.

Problem Set, *continued*

P **18** Solve the following system for (x, y). $(x, y) = \left(5, -\frac{1}{2}\right)$

$$\begin{cases} 2x + \dfrac{3}{y} = 4 \\ 5x - \dfrac{2}{y} = 29 \end{cases}$$

P **In problems 19 and 20, solve for (x, y).**

19 $\begin{cases} 3(x + 4) + 8(y - 2) = -48 \\ 2(x + 4) - 4(y - 2) = -4 \end{cases}$

$(x, y) = (-12, -1)$

20 $\begin{cases} \dfrac{2}{x + 1} + \dfrac{5}{y - 1} = 12 \\ \dfrac{5}{x + 1} + \dfrac{2}{y - 1} = 9 \end{cases}$

$(x, y) = \left(0, \frac{3}{2}\right)$

P,R **21** Pump A can fill a swimming pool in 6 hours, and pump B can fill the pool in 4 hours. The two pumps began working together to fill an empty pool, but after 40 minutes pump B broke down, and pump A had to complete the job alone. The pool was scheduled to open 4 hours after the pumping started. Was it full by then, or did the opening have to be delayed?
The opening had to be delayed. (It took 5 hr to fill the pool.)

P,R **22** The voltage across each coil of a transformer varies directly as the number of turns in the coil.

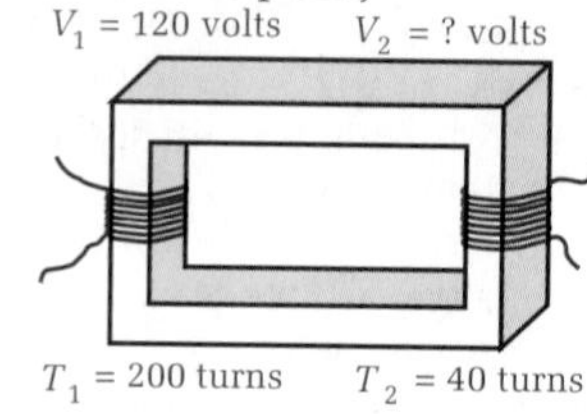

a Write a proportion to describe the relationship involving V_1, V_2, T_1, and T_2.

b V_1, the voltage across the primary coil, is 120 volts. If the primary coil consists of 200 turns and the secondary coil consists of 40 turns, what is V_2?

a $\frac{V_1}{T_1} = \frac{V_2}{T_2}$ **b** 24 v

P **23** Solve the following system for (x, y). $(x, y) = (4, 2)$

$$\begin{cases} x^2 = 8y \\ \dfrac{16}{x} = \dfrac{8}{y} \end{cases}$$

R **24** Suppose that f and g are functions such that $3[f(x)] + 4[g(x)] = 2x^2 + 3x - 5$ and $2[f(x)] + 3[g(x)] = x^2 - 2x + 7$. Write formulas to represent f and g.
$f(x) = 2x^2 + 17x - 43$; $g(x) = -x^2 - 12x + 31$

P,R **25** Solve the following system for (A, B, C, D). $(A, B, C, D) = (-4, 1, 12, 6)$

$$\begin{cases} A + B + C + D = 15 \\ A + B + C = 9 \\ A + B + D = 3 \\ A + C + D = 14 \end{cases}$$

P,R **26** Tom, Dick, and Harry paint houses. If all three work together, they finish in five days. If Tom and Dick work together, they finish in seven days. If Tom and Harry work together, they finish in ten days. How long would it take each one alone to paint a house?
Tom: $23\frac{1}{3}$ days; Dick: 10 days; Harry: $17\frac{1}{2}$ days

R **27** Suppose that $f(x) = 3x + 12$ and $g(x) = 4x - 15$. Find all values of x for which $f(x) \cdot g(x) = f(x) + g(x)$. $x \approx 4.01$ or $x \approx -3.68$

R **28** A go-cart race goes up Grand Mesa for four kilometers, across the level top for eight kilometers, and down the other side for three kilometers. Poncho Carter completed the entire race in two hours. His time going up Grand Mesa was the same as his time for the rest of the race. The sum of the time it took him to go up the mesa and the time he spent on the mesa's top was ten times the time it took him to go down the other side. How fast did Poncho go up the mesa? How fast did he go on the level top? How fast did he go down the mesa? 4 km/hr; ≈9.8 km/hr; 16.5 km/hr

P,R **29** Let $f(x) = \frac{1}{x - 3}$.

a Find INVf. INV$f(x) = \frac{1}{x} + 3$

b Show that $f \circ$ INV$f(x) = x$ and that INV$f(x) \circ f(x) = x$.

c Show graphically that the function you found in part **a** is the inverse of f.

b $f \circ \text{INV}f(x) = \frac{1}{\frac{1}{x} + 3 - 3} = x$; $\text{INV}f(x) \circ fx = \frac{1}{\frac{1}{x - 3}} + 3 = x$

R **30** For each of the following graphs, tell whether y varies inversely as x. If it does, write an equation for the inverse variation represented by the graph.

a

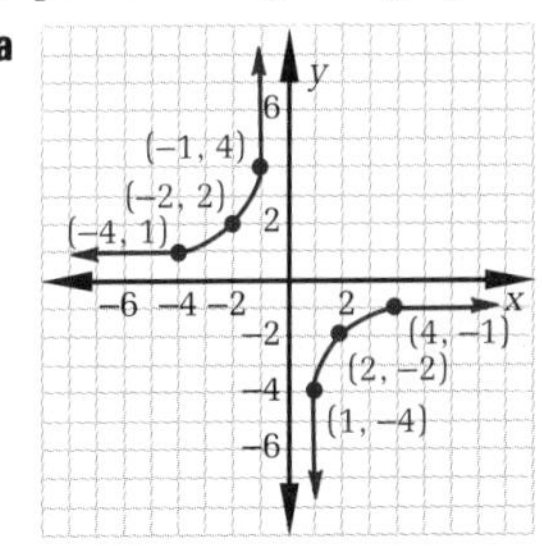

Yes; $xy = -4$

b

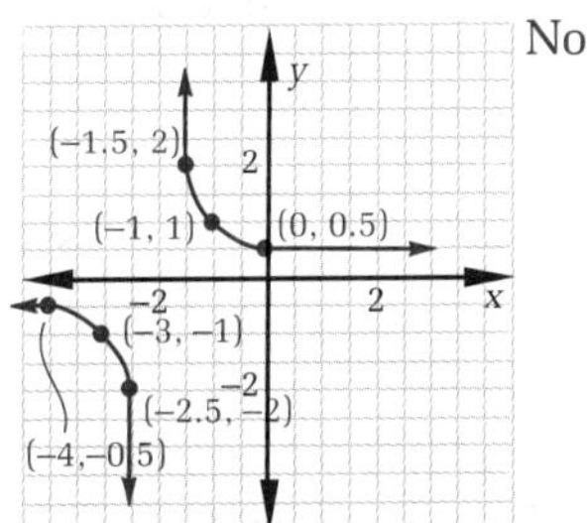

No

c

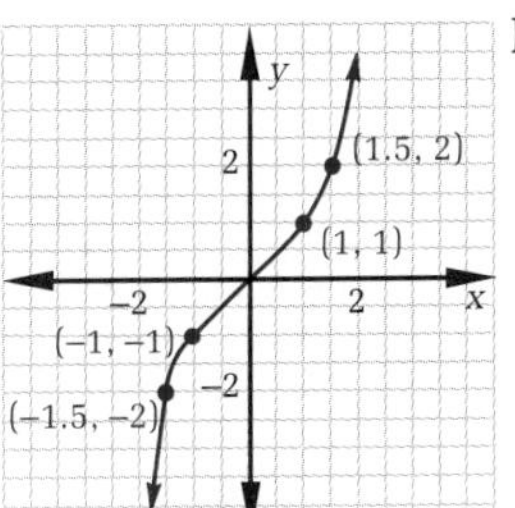

No

d

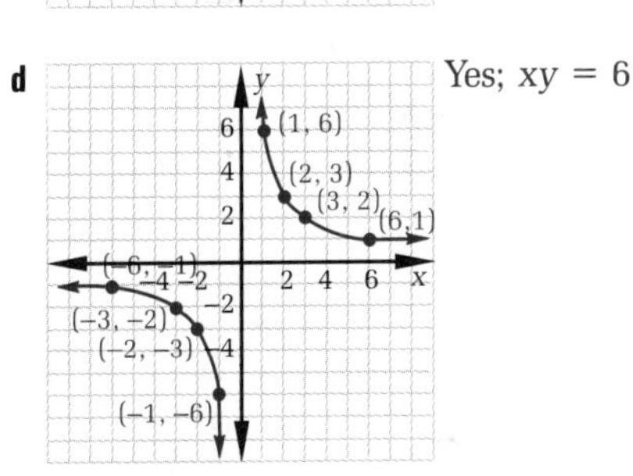

Yes; $xy = 6$

Problem-Set Notes and Strategies, continued

- Problem **26** combines an application of work and systems of equations.
- Suggest that students draw a picture for problem **28**.

Additional Answer

29c

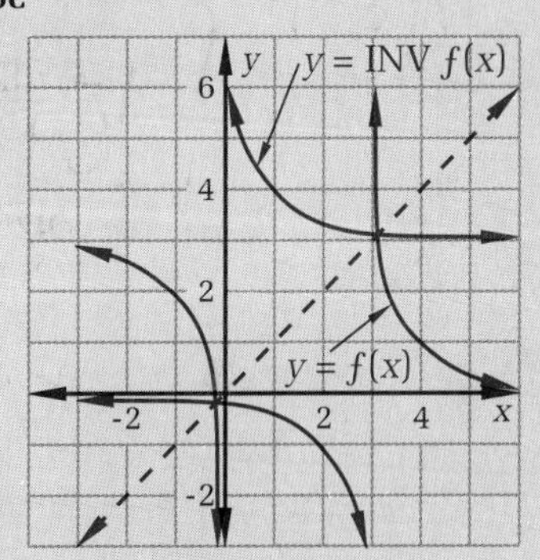

- In problem **30**, remind students that inverse variations must be symmetric about the origin. Part **b** fails that respect. Inverse variations must also be symmetric about the line $y = x$ and part **c** fails in that regard.

Problem-Set Notes and Strategies, continued

- Problem **33** will be difficult for students without a substitution of some sort.
- Notice that in problem **35** both the maximum and minimum occurred at corner points. This will be pointed out again in the short subject on linear programming and is a consequence of the fact that the region is a convex region.

Communicating Mathematics

Have students prepare a lesson plan on how to solve a system of equations with rational expressions. The lesson should include examples. Choose two or three students to present their lessons to the class.

Problem Set, *continued*

C

In problems 31–33, solve each system.

P

31 $\begin{cases} \dfrac{3}{|x|} + \dfrac{5}{|y|} = 18 \\ \dfrac{2}{|x|} - \dfrac{3}{|y|} = 50 \end{cases}$ No solution

32 $\begin{cases} x = y^2 \\ x^2 + 2x = y^3 \end{cases}$ $(x, y) = (0, 0)$

33 $\begin{cases} \dfrac{4}{x} + \dfrac{2}{y} + \dfrac{10}{z} = 2 \\ \dfrac{6}{x} + \dfrac{1}{y} + \dfrac{5}{z} = 2 \\ \dfrac{-4}{x} + \dfrac{10}{y} - \dfrac{5}{z} = -7 \end{cases}$ $(x, y, z) = (4, -2, 5)$

R **34** Consider the following system.

$\begin{cases} \sqrt{x} + 2\sqrt{y} = 7 \\ 2\sqrt{x} + \sqrt{y} = 8 \end{cases}$

a Solve the system for $(\sqrt{x}, \sqrt{y})$. $(\sqrt{x}, \sqrt{y}) = (3, 2)$
b Solve the system for (x, y). $(x, y) = (9, 4)$

I **35** Suppose that the function $f(x, y) = 5x + 3y$ has as its domain the set of values (x, y) represented by the lattice points in the shaded region shown.

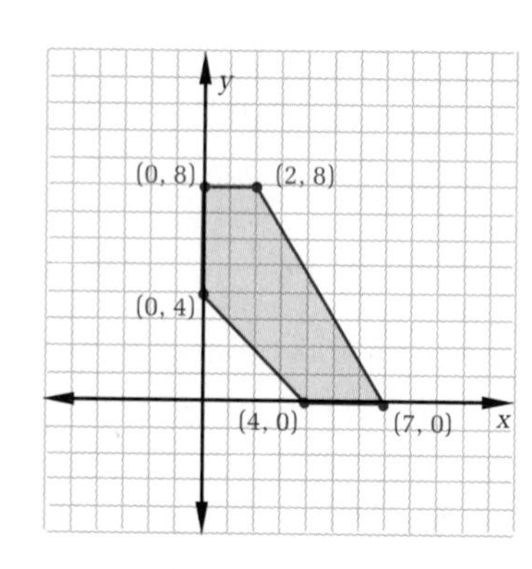

a What is the greatest possible output of function f? 35
b What is the least possible output of function f? 12

P **36** Solve the following system for (x, y). $(x, y) = \left(-\frac{3}{4}, \frac{3}{20}\right)$

$\begin{cases} \dfrac{10}{x} + \dfrac{5}{y} = 20 \\ \dfrac{1}{4x} - \dfrac{2}{5y} = -3 \end{cases}$

R **37** The following data represent the mean daily high temperatures (in degrees Fahrenheit) in Edinburgh, Scotland, for 12 months.

43, 43, 47, 50, 55, 62, 65, 64, 60, 53, 47, 44

a Find the mean, the mode, and the median of the data. 52.75; 43 and 47; 51.5
b Find the range of the data. 22
c Find the mean deviation of the data. ≈ 7.08

7

Chapter Summary

Concepts and Procedures

After studying this chapter, you should be able to

- Use systems of equations to solve problems (7.1)
- Solve systems of equations in three unknowns (7.2)
- Graph equations that contain three variables (7.2)
- Graph systems of inequalities (7.3)
- Solve inverse-variation problems (7.4)
- Graph equations of the form $xy = k$ (7.4)
- Solve systems containing nonlinear equations (7.5)
- Solve systems of equations involving rational expressions (7.6)

Vocabulary

inverse variation (7.4)
trace (7.2)

7

Class Planning

Time Schedule

All levels: 1 day

Assignment Guide

Average
2–4, 7–9, 14, 16, 18, 20, 21, 23
Advanced
2–4, 7, 9, 12–14, 16, 18, 20, 23, 24
Honors
1–3, 5–7, 9, 12, 13, 15, 16, 18, 20, 23, 25

Problem-Set Notes and Strategies

- Problem **3** is an excellent problem in mechanical applications of mathematics. In later chapters of trig, linear and angular velocity will also be discussed.
- Remember in problem **7b** that completely means to examine each step and make sure that no factor may still be factored further.

Additional Answer

1b

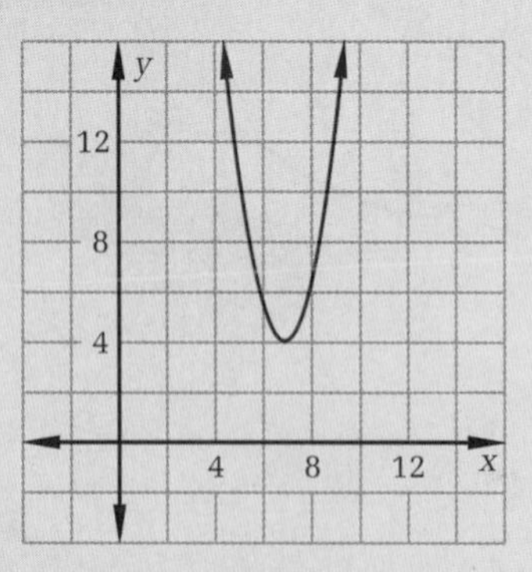

7 Review Problems

A

P,R **1** Consider the parabolic function $f(x) = 2(x - h)^2 + k$, with $f(1) = 76$, $f(9) = 12$, and $h > 0$.

a Find the values of h and k. $h = 7$; $k = 4$

b Draw the graph of $f(x)$.

P,R **2** Solve the following matrix equation for (x, y) and check your answer by matrix multiplication. $(x, y) = (4, -1)$

$$\begin{bmatrix} 3 & -4 \\ 2 & 5 \end{bmatrix}\begin{bmatrix} x \\ y \end{bmatrix} = \begin{bmatrix} 16 \\ 3 \end{bmatrix}$$

P **3** The speeds (in revolutions per minute) of two pulleys are inversely proportional to the pulleys' diameters. One pulley's diameter (D_1) is 12, and that pulley's speed (S_1) is 160 rpm. If the other pulley's diameter (D_2) is 4, what is S_2? 480 rpm

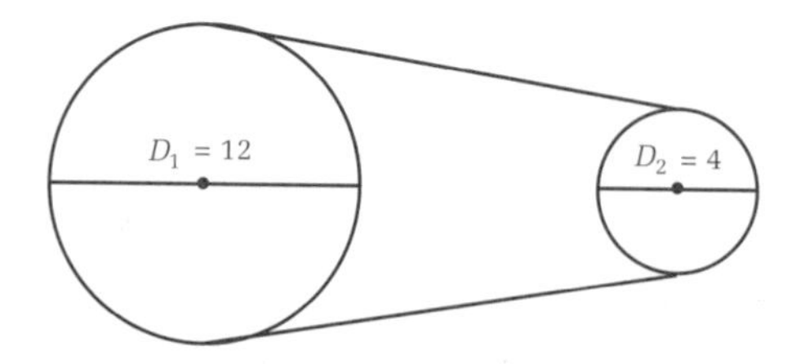

P **4** An airplane flies 480 miles in 3.2 hours with a tail wind. The return trip, against the same wind, takes $5\frac{1}{3}$ hours.

a What is the speed of the plane in still air? 120 mph

b What is the wind speed? 30 mph

P **5** The sum of the squares of two numbers is 88. The difference of the squares of the two numbers is 10. What might the two numbers be? 7 and $\sqrt{39}$, 7 and $-\sqrt{39}$, -7 and $\sqrt{39}$, or -7 and $-\sqrt{39}$

P **6** Solve the following system for (x, y). $(x, y) = (5, 35)$ or $(x, y) = (-2, 0)$

$$\begin{cases} y = x^2 + 2x \\ y = 5x + 10 \end{cases}$$

R **7** Factor each expression completely.

a $x^2 - y^2$ $(x + y)(x - y)$

b $x^4 - y^4$ $(x^2 + y^2)(x + y)(x - y)$

P **8** When Jim works in the ticket booth, he can collect $50 in 30 minutes. When Jim and Tony work together, they can collect $50 in 10 minutes. How long would it take Tony alone to collect $50? 15 min

P **9** Solve the following system for (x, y, z). $(x, y, z) = \left(-5, \frac{7}{2}, 4\right)$

$$\begin{cases} x - 4y + 5z = 1 \\ -2x + 6y - 3z = 19 \\ 7x + 10y + z = 4 \end{cases}$$

B

P **10** In a swimming-and-running biathlon, participants swim for 2 kilometers and then run for 10 kilometers. Suzy Swift completed the biathlon in 2.9 hours. If she had swum at her running rate and run at her swimming rate, she would have finished in 8.7 hours. How fast can Suzy run and swim?
Run: ≈8.33 km/hr; swim: ≈1.18 km/hr

R **11** Find the equation of the axis of symmetry of the graph of each equation.

a $y = 2(x - 4)^2 + 5$ $x = 4$ **b** $y = \frac{1}{2}(x + 7)^2 - 3$ $x = -7$
c $y = x^2 - 6x + 5$ $x = 3$

R **12** The data 2x, 2x + 7, 2x + 13, 2x + 15, and 3x − 10 have a range of 24. Find the mean of the data. 79.8 or 7.2

P **13** Consider the following system.

$$\begin{cases} x \cdot y = 36 \\ y = x \end{cases}$$

a Solve the system for (x, y). (x, y) = (6, 6) or (x, y) = (−6, −6)
b What do you notice about your answer to part **a?** The solutions represent the vertices of the hyperbolic graph of the first equation of the system.

R **14** Write a system of inequalities to describe the shaded region in the diagram.

$$\begin{cases} y \ge \frac{2}{5}(x - 8) \\ y \le -\frac{2}{5}(x - 8) \\ x \ge -2 \end{cases}$$

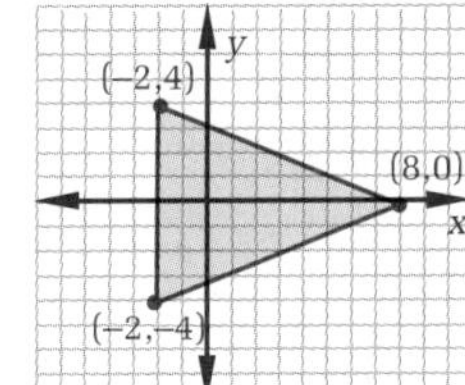

P **15** Pop and Zeb have gone fishing in their small motorboat. They traveled 19.5 miles downstream to a fishing hole in 3 hours. After 3 hours on the way back upstream, they have traveled only 8.7 miles.

a How much longer will it take them to return to their starting point? ≈3.7 hr
b How fast is the current? What is the speed of the boat in still water? 1.8 mph; 4.7 mph

P,R **16** Solve the matrix equation AX = B for X.

$$A = \begin{bmatrix} 6 & 9 & -1 \\ 3 & 4 & 2 \\ 0 & 5 & 8 \end{bmatrix} \quad X = \begin{bmatrix} p \\ q \\ r \end{bmatrix} \quad B = \begin{bmatrix} 16 \\ 19 \\ 55 \end{bmatrix} \quad X = \begin{bmatrix} -1 \\ 3 \\ 5 \end{bmatrix}$$

Problem-Set Notes and Strategies, continued

- In problem **11** students must remember that the axis of symmetry of a parabola is vertical and passes through the vertex.
- Set up a system of equations for problem **16** and solve using any appropriate method.

Additional Answers

19a

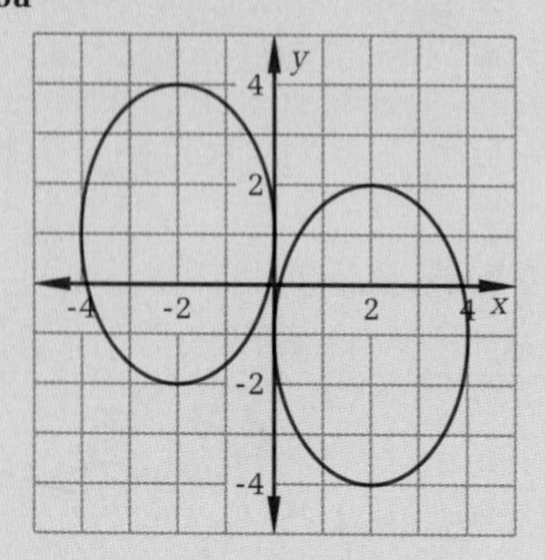

22

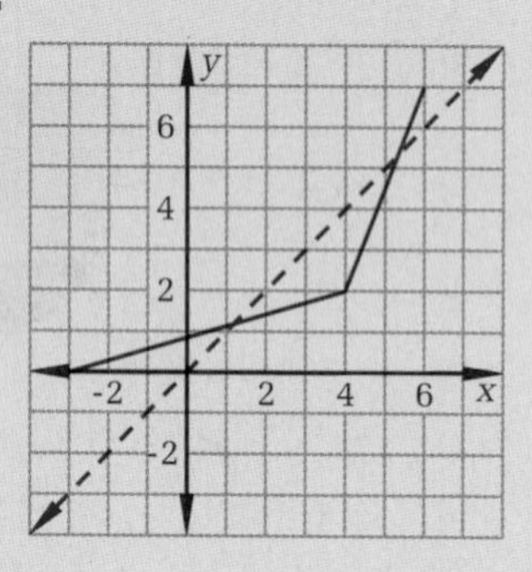

■ Since none of the points of problem **25** are identified as the vertex, a system of equations in a, b and c must be set up and solved.

Review Problems, *continued*

P,R **17** The product of two numbers is 12. If one number is increased by 20%, the second number must be decreased by n% for the product to remain 12. What is the value of n? $16.\overline{6}$

R **18** Factor each expression.

a $x^2 - 5x - 14$ $(x - 7)(x + 2)$
b $(3x)^2 - 5(3x) - 14$ $(3x - 7)(3x + 2)$
c $(x + 4)^2 - 5(x + 4) - 14$ $(x - 3)(x + 6)$

P **19** Graph $\frac{(x-2)^2}{4} + \frac{(y+1)^2}{9} = 1$ and $\frac{(x+2)^2}{4} + \frac{(y-1)^2}{9} = 1$ on the same coordinate system.

R **20** The measure of an exterior angle of a regular polygon varies inversely as the number of the polygon's sides. We know that each exterior angle of an equilateral triangle is 120°.

a Find the constant of variation for this inverse relationship. 360
b Write a formula that can be used to find the measure E of an exterior angle of a polygon with n sides. $E = \frac{360}{n}$
c What is the measure of each exterior angle of a dodecagon (12-sided polygon)? 30

P,R **21** Solve the following matrix equation for (x, y). $(x, y) = (-8, -4)$

$$\begin{bmatrix} 2 & 9 \\ -3 & 1 \end{bmatrix}\begin{bmatrix} x \\ y \end{bmatrix} = \begin{bmatrix} -52 \\ 20 \end{bmatrix}$$

R **22** The diagram shows the graph of $y = f(x)$. Sketch the reflection of this graph over the line $y = x$.

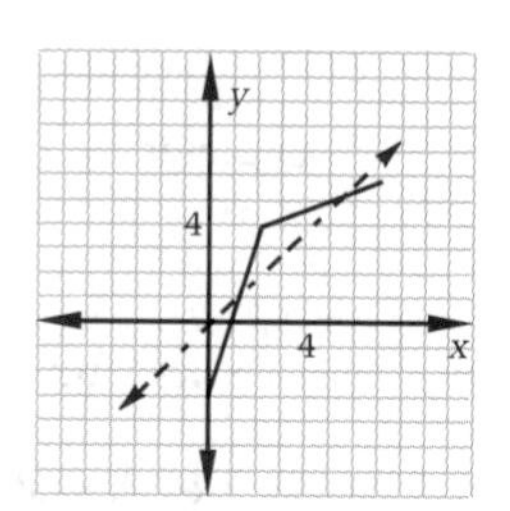

P **23** Buns Might Candy Company sells a four-pound mixture of peanuts and cashews that is 25% cashews for \$14.20. It also sells a four-pound mixture of peanuts and cashews that is 40% cashews and costs \$17.50. What would you expect to pay for one pound of cashews at Buns Might? \$7.68

C P **24** The numerator of a fraction is five more than the denominator. The sum of the numerator and denominator is 73. Find the fraction. $\frac{39}{34}$

P,R **25** The points $(-4, 6)$, $(-1, -3)$, and $(1, 21)$ are on a parabola. Find an equation of the parabola. $y = 3x^2 + 12x + 6$

7 CHAPTER TEST

1 Cool Cash has 125 coins, all dimes and quarters, in a milk bottle. If he added 10 quarters and 12 dimes, the total value of the coins would be \$24.45. How many dimes are in the bottle? 70

2 Solve the following system for (x, y, z). $(x, y, z) = (3, -1, 2)$

$$\begin{cases} 2x - y + 3z = 13 \\ 4x + 3y - 2z = 5 \\ x - y - 4z = -4 \end{cases}$$

3 Graph the equation $3x + 2y + 4z = 12$.

In problems 4 and 5, graph the solution set of each system.

4 $\begin{cases} 2x + 3y \le 6 \\ 2x - y \ge 4 \end{cases}$

5 $\begin{cases} y \le x + 3 \\ y > (x - 1)^2 - 5 \end{cases}$

In problems 6–8, graph each equation.

6 $xy = -12$

7 $xy = 9$

8 $|xy| = 4$

9 A *power function* is a function of the form $f(x) = ax^n$, where $a \neq 0$. Newton's law of gravitation says that two objects of unit mass attract each other with a gravitational force F, which varies inversely as the square of the distance d between the objects. The constant of variation is called Newton's gravitational constant, g. Express F as a power function of d. $F = gd^{-2}$

In problems 10 and 11, solve each system of equations for (x, y).

10 $\begin{cases} x^2 + y^2 = 32 \\ x^2 + y^2 = 4x + 4y \end{cases}$ $(x, y) = (4, 4)$

11 $\begin{cases} xy = 2 \\ x^2 + y^2 = 5 \end{cases}$ $(x, y) = (-2, -1)$, $(x, y) = (-1, -2)$, $(x, y) = (1, 2)$, or $(x, y) = (2, 1)$

In problems 12 and 13, solve each system for (x, y).

12 $\begin{cases} \dfrac{3}{x} - \dfrac{1}{y} = 2 \\ \dfrac{1}{x} - \dfrac{2}{y} = -10 \end{cases}$ $(x, y) = \left(\frac{5}{14}, \frac{5}{32}\right)$

13 $\begin{cases} \dfrac{2}{x - 5} + \dfrac{7}{y + 3} = 9 \\ \dfrac{3}{x - 5} - \dfrac{4}{2y + 6} = 1 \end{cases}$ $(x, y) = (6, -2)$

14 Find the vertices of the region defined by the inequalities $x \ge 0$, $y \ge 2x - 2.5$, and $y \le \frac{1}{2}x + 2$. $(0, -2.5)$, $(3, 3.5)$, $(0, 2)$

Resource References
Evaluation
Test Form 1, 2, 3

Additional Answers

3

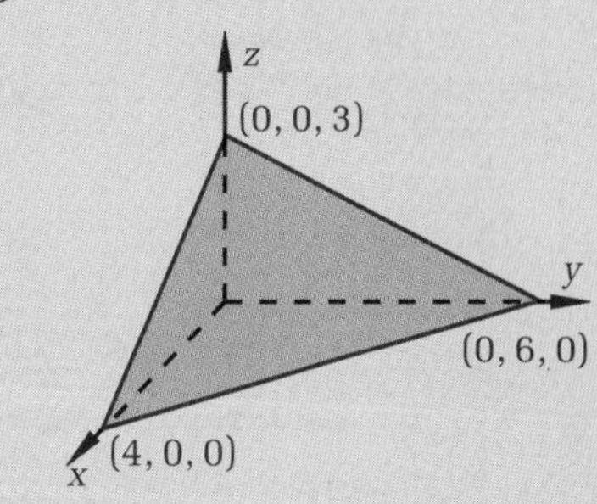

4

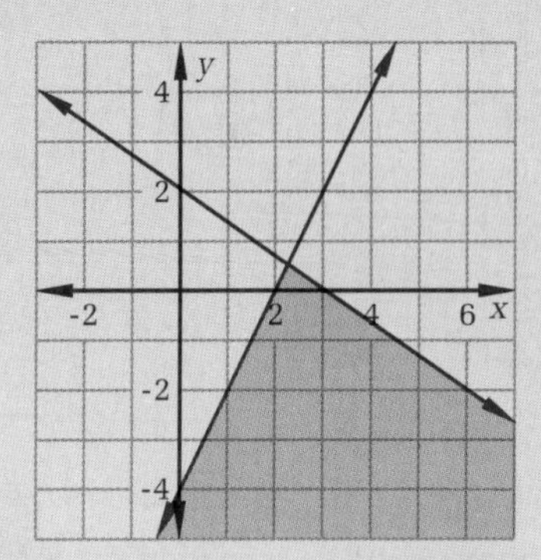

5

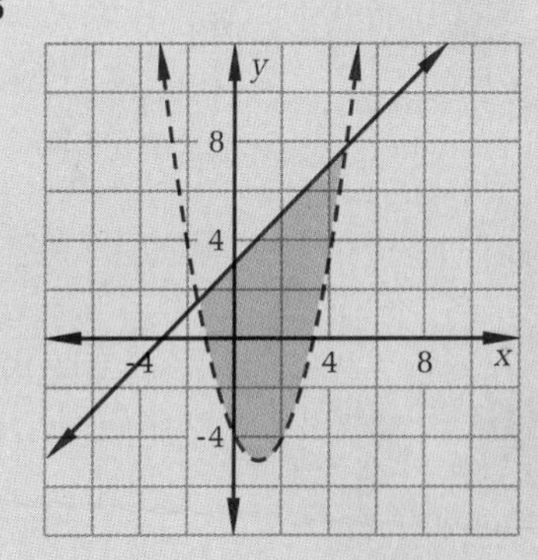

Answers for problems **6**, **7**,and **8** can be found in the answer pages beginning on page **A** 1.

Class Planning

Time Schedule

Average: 1 day
Advanced: 2 days
Honors: 3 days

Resource References

Teacher's Resource Book
Class Opener 47
Practice 49, 50

Class Opener

1 Solve for x
$x^4 - 8x^2 + 16 = 0$
$x = 2$ or $x = -2$

2 $x^4 - 9x^2 + 16 = 0$
$x^4 - 9x^2 + 16$
$= x^4 - 8x^2 + 16 - x^2$
$= (x^2 - 4)^2 - x^2$
$= (x^2 - 4 - x)(x^2 - 4 + x)$
$x = \frac{1 \pm \sqrt{17}}{2}$ or $x = \frac{-1 \pm \sqrt{17}}{2}$

Lesson Notes

- This short subject introduces several different methods of factoring. Factoring many higher degree polynomials can be identical to factoring quadratics if they fit a form where the variable has powers in the ratio 2 : 1 and there is a constant term. Factoring using the difference of squares can be extended to the difference of any two even powers by a grouping technique. Other methods of grouping are also available to enable factoring by methods already learned.

SHORT SUBJECT

Factoring to Solve Equations and Inequalities

Solving Equations by Factoring

In Section 4.4, you found that you could solve a number of quadratic equations by factoring and applying the Zero Product Property. Some cubic and quartic equations can be solved in a similar fashion.

Example *Solve* $5x^3 - 10x^2 + 9x - 18 = 0$ *for* x.

To factor the polynomial on the left side, we group the terms in twos, remove a common factor from each group, and use the Distributive Property of Multiplication over Addition or Subtraction in reverse.

	$5x^3 - 10x^2 + 9x - 18 = 0$
Group terms	$(5x^3 - 10x^2) + (9x - 18) = 0$
Remove common factors	$5x^2(x - 2) + 9(x - 2) = 0$
Distributive Property of Mult. over Add.	$(5x^2 + 9)(x - 2) = 0$
Zero Product Property	$5x^2 + 9 = 0$ or $x - 2 = 0$
	$x^2 = -\frac{9}{5}$ $\quad$ $x = 2$

Since there is no real value of x for which x^2 is negative, the only real solution of the equation is $x = 2$.

Example *Solve* $y^4 - 5y^2 + 4 = 0$ *for* y.

We can factor the degree-four polynomial in this equation by thinking of it as a quadratic with a variable of y^2.

$$y^4 - 5y^2 + 4 = 0$$
$$(y^2)^2 - 5(y^2) + 4 = 0$$
$$(y^2 - 4)(y^2 - 1) = 0$$

Difference of squares $$(y - 2)(y + 2)(y - 1)(y + 1) = 0$$

$$y - 2 = 0 \text{ or } y + 2 = 0 \text{ or } y - 1 = 0 \text{ or } y + 1 = 0$$
$$y = 2 \qquad y = -2 \qquad y = 1 \qquad y = -1$$

The solution set is $\{-2, -1, 1, 2\}$.

Example *Solve* $y^4 + 3y^2 + 4 = 0$ *for* y.

This equation appears similar to the equation in the preceding example, but a little experimentation will show that the polynomial cannot be factored in the same way. To factor this polynomial, we will complete the square by adding and subtracting y^2.

$$y^4 + 3y^2 + 4 = 0$$

Add and subtract y^2 $$y^4 + 3y^2 + y^2 + 4 - y^2 = 0$$

$$(y^4 + 4y^2 + 4) - y^2 = 0$$
$$(y^2 + 2)^2 - y^2 = 0$$

Difference of squares $$[(y^2 + 2) - y][(y^2 + 2) + y] = 0$$

$$(y^2 - y + 2)(y^2 + y + 2) = 0$$

Since neither $y^2 - y + 2 = 0$ nor $y^2 + y + 2 = 0$ has real solutions (the discriminants of both equations are negative), the original equation has no real solutions.

In Chapter 4, you were introduced to two useful factoring patterns—the perfect-square-trinomial pattern and the difference-of-squares pattern. The factoring patterns for sums and differences of perfect cubes are also useful in the solution of certain equations.

Factors of a Sum or a Difference of Two Cubes

$$x^3 + a^3 = (x + a)(x^2 - ax + a^2)$$
$$x^3 - a^3 = (x - a)(x^2 + ax + a^2)$$

Example *Solve* $z^3 = -64$ *for* z.

$$z^3 = -64$$
$$z^3 + 64 = 0$$
$$z^3 + 4^3 = 0$$

Sum of cubes $$(z + 4)(z^2 - 4z + 16) = 0$$

$$z + 4 = 0 \text{ or } z^2 - 4z + 16 = 0$$
$$z = -4$$

The equation $z^2 - 4z + 16 = 0$ has no real solutions, so the only solution is $z = -4$.

Lesson Notes, continued

- Formulas for the sum and difference of cubes are introduced and should be memorized by your students. Suggest that they write the binomial as a sum or difference of cubes before using the formula. For example, if $8x^3 - 27y^3$ is written as $(2x)^3 - (3y)^3$, using the formula will be a lot less complicated.

Lesson Notes, continued

- Solving inequalities by factoring is the last topic included. First the polynomial is solved for zero since that is the boundary point for positive and negative numbers. Then each region of the remaining number line is tested to determine whether the numbers in the region make the inequality true or false.

Solving Inequalities by Factoring

As the accompanying graph shows, the value of a polynomial $P(x)$ changes sign only at certain values of x—the values for which $P(x) = 0$. For inputs in each of the intervals bounded by these values, the value of $P(x)$ is always positive or always negative. This characteristic suggests that higher-degree polynomial inequalities can be solved by a boundary-value method similar to the methods used to solve linear inequalities.

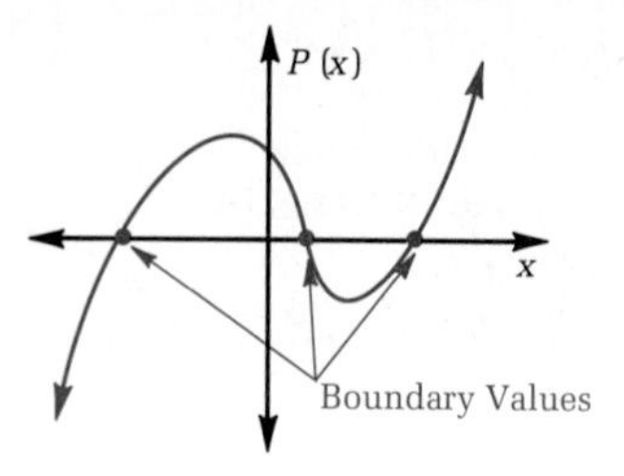

Example *Find the solution set of* $x^2 + 3x > 10$.

We first rewrite the inequality in the form $P(x) > 0$ and factor $P(x)$.

$$x^2 + 3x > 10$$
$$x^2 + 3x - 10 > 0$$
$$(x + 5)(x - 2) > 0$$

The boundary values are the values of x for which $x + 5 = 0$ or $x - 2 = 0$—that is, -5 and 2. We can plot these values on a number line, using open dots because neither value satisfies the inequality.

We choose a value from each of the three regions determined by the boundary points and test it to see whether it satisfies the original inequality. We will use -7, 0, and 5.

$\mathbf{x^2 + 3x}$	**> 10?**
$(-7)^2 + 3(-7) = 49 - 21 = 28$	Yes
$0^2 + 3(0) = 0$	No
$5^2 + 3(5) = 25 + 15 = 40$	Yes

We shade the regions containing -7 and 5. The solution set of the inequality is $\{x{:}x < -5 \text{ or } x > 2\}$, as shown in the resulting graph.

The method used in the preceding example can be summarized as follows.

Boundary Algorithm for Solving Polynomial Inequalities

1. **Rewrite the inequality so that one side is zero, then factor the polynomial on the other side.**
2. **Identify each value of the variable for which a factor equals zero. Plot these boundary values on a number line, using solid dots if the values satisfy the inequality or open dots if they do not.**
3. **Test a value from each region determined by the boundary points to see whether it satisfies the original inequality.**
4. **Shade each region in which a test value satisfies the inequality.**

Sample Problems

Problem 1 *Solve* $3x^3 + 5x^2 - 27x - 45 = 0$ *for* x.

Solution

$$3x^3 + 5x^2 - 27x - 45 = 0$$

Factor by grouping $\quad x^2(3x + 5) - 9(3x + 5) = 0$

$$(x^2 - 9)(3x + 5) = 0$$

Difference of squares $\quad (x - 3)(x + 3)(3x + 5) = 0$

$$x - 3 = 0 \text{ or } x + 3 = 0 \text{ or } 3x + 5 = 0$$

$$x = 3 \qquad x = -3 \qquad x = -\frac{5}{3}$$

Problem 2 *Solve* $81b^3 = 24$ *for* b.

Solution

$$81b^3 = 24$$

$$81b^3 - 24 = 0$$

Remove common factor $\quad 3(27b^3 - 8) = 0$

$$3[(3b)^3 - 2^3] = 0$$

Difference of cubes $\quad 3(3b - 2)(9b^2 + 6b + 4) = 0$

Since $9b^2 + 6b + 4$ has no real zeros, the only solution is given by $3b - 2 = 0$. The solution is $b = \frac{2}{3}$.

Problem 3 *Solve* $x^4 - 12x^2 + 16 \le 0$ *for* x.

Solution

$$x^4 - 12x^2 + 16 \le 0$$

Add and subtract $4x^2$ $\quad x^4 - 12x^2 + 4x^2 + 16 - 4x^2 \le 0$

$$(x^4 - 8x^2 + 16) - 4x^2 \le 0$$

$$(x^2 - 4)^2 - 4x^2 \le 0$$

Difference of squares $\quad [(x^2 - 4) - 2x][(x^2 - 4) + 2x] \le 0$

$$(x^2 - 2x - 4)(x^2 + 2x - 4) \le 0$$

By applying the quadratic formula to each trinomial factor, we identify the boundary values as $-1 \pm \sqrt{5}$ and $1 \pm \sqrt{5}$, or ≈ -3.24, ≈ -1.24, ≈ 1.24, and ≈ 3.24. Each of these values satisfies the original inequality, but when we test the values -4, -2, 0, 2, and 4 (from the regions determined by the boundary values), we find that only -2 and 2 satisfy the inequality. The solution set can therefore be represented as follows.

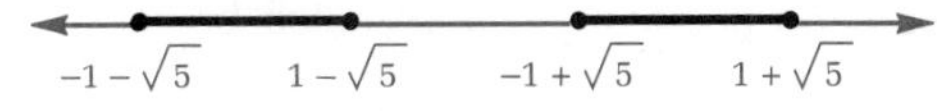

Problem Set

A
P

In problems 1–6, find the real zeros of each function.

1 $P(x) = (x + 6)(x - 2)(x + 2)$ $-6, -2, 2$

2 $Q(x) = (x - 1)(x^2 - 5)(x + 3)$ $1, \pm\sqrt{5}, -3$

3 $R(x) = (x + 4)(x + 2)(x^2 - 7)$ $-4, -2, \pm\sqrt{7}$

4 $S(x) = (x - 3)(x - 5)(x^2 + 4)$ $3, 5$

5 $T(x) = (x - 7)(x^2 - 12x + 33)$ $7, 6\pm\sqrt{3}$

6 $U(x) = (x - 2)^2(x^2 - 17)$ $-\sqrt{17}, 2, \sqrt{17}$

Assignment Guide

Average	
1, 3, 4, 7–13, 18, 27, 28	
Advanced	
Day 1	1, 2, 4, 7, 8, 11, 13, 17, 21, 22, 24, 28
Day 2	3, 5, 6, 10, 12, 15, 16, 18, 25, 27, 32, 34
Honors	
Day 1	1, 2, 4, 7, 10, 14, 18, 19, 24, 38
Day 2	5, 10, 13, 16, 17, 20, 23, 25, 26, 30
Day 3	8, 11, 15, 22, 27, 29, 34–36, 39

Problem-Set Notes and Strategies

- Problems **1 – 6** are good review of factoring skills previously learned.
- Formulas for the sum and difference of cubes are introduced and should be memorized by your students. Suggest that they write the binomial as a sum or difference of cubes before using the formula. For example, if $8x^3 - 27y^3$ is written as $(2x)^3 - (3y)^3$, using the formula will be a lot less complicated.

Problem-Set Notes and Strategies, continued

- Problem **7** illustrates the concept of the boundary solving algorithm well.
- Problem **18** provides a nice illustration of how a two-dimensional graph can help solve inequalities and explain the significance of the inequality. Make sure that students see these as two separate problems so they are not confused about two different graphs for the same problem.
- Factoring by grouping and the difference of squares are both needed to solve problem **20**.
- Problems **22** and **23** both provide a nice use of sigma notation to display polynomials. This notation will be convenient to work with when the Binomial Theorem is introduced in Chapter **9.5**.

Additional Answers

7a 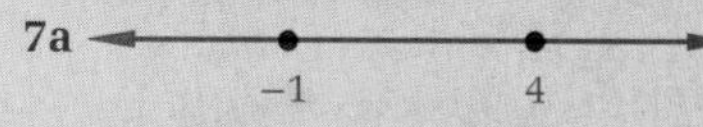

7b 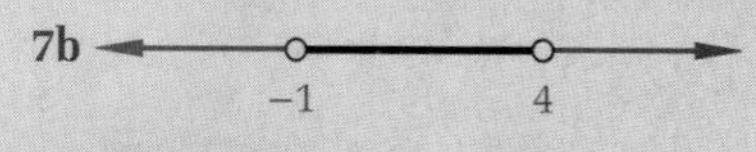

7c −1 4

18a $(x^2 - 5)(x - 1)$

18b $\{x: x < -\sqrt{5} \text{ or } 1 < x < \sqrt{5}\}$

19 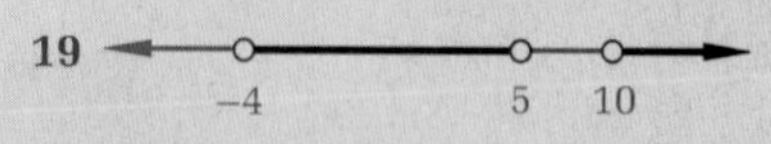

20 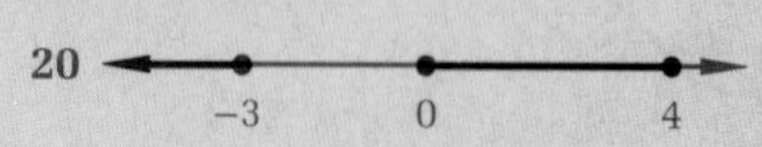

Problem Set, *continued*

P **7** Graph the solution set of each equation or inequality on a number line.

a $x^2 - 3x - 4 = 0$ **b** $x^2 - 3x - 4 < 0$ **c** $x^2 - 3x - 4 > 0$

P **8** Factor $y^4 - 7y^3 + 10y^2$ completely. $y^2(y - 2)(y - 5)$

P **9** Solve $(x^2 - 4)(x^2 + 6x + 8) = 0$ for x. $x = \pm 2$ or $x = -4$

P **In problems 10–17, factor each expression over the integers.**

10 $3x^3 + 6x^2 - x - 2$ $(3x^2 - 1)(x + 2)$

11 $x^3 + 1$ $(x + 1)(x^2 - x + 1)$

12 $x^4 - 81$ $(x^2 + 9)(x + 3)(x - 3)$

13 $8x^3 - 125$ $(2x - 5)(4x^2 + 10x + 25)$

14 $x^4 - 4x^2 - 5$ $(x^2 - 5)(x^2 + 1)$

15 $x^3 - 12x^2 + 36x$ $x(x - 6)^2$

16 $(x - 3)^2 - 64$ $(x - 11)(x + 5)$

17 $x^4 + x^2 + 25$ $(x^2 - 3x + 5)(x^2 + 3x + 5)$

P **18** The diagram shows the graph of the function $P(x) = x^3 - x^2 - 5x + 5$.

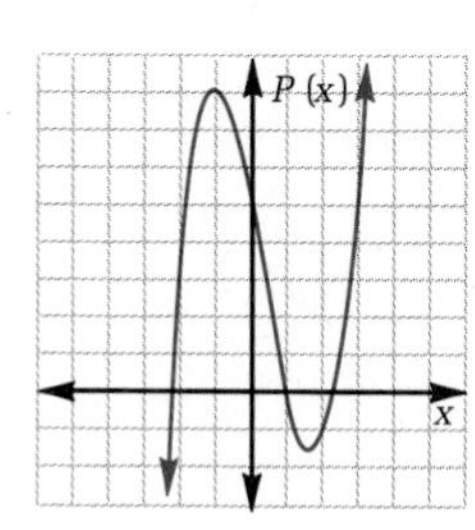

a Factor $P(x)$ over the integers.
b Find the solution set of $P(x) < 0$.
c What is the relationship between your answer to part **b** and the graph of P? The solutions of the inequality are the values of x for which the graph of P lies below the x-axis.

P **In problems 19 and 20, graph the solution set of each inequality on a number line.**

19 $(x - 5)(x - 10)(x + 4) > 0$ **20** $x^3 - 4x^2 + 3x^2 - 12x \le 0$

B P,R **21** **a** Write an expression for the sum of the volumes of the two cubes shown.
b Factor the expression you wrote in part **a**.
a $a^3 + (a + 2)^3$
b $2(a + 1)(a^2 + 2a + 4)$

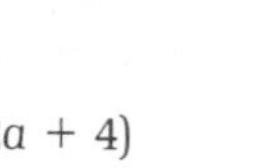

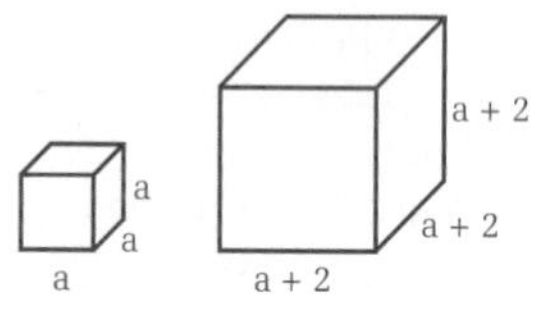

P,R **In problems 22 and 23, factor the polynomial represented by each expression.**

22 $\sum_{n=0}^{3} 2^n x^n$ $(4x^2 + 1)(2x + 1)$

23 $\sum_{n=1}^{4} (-1)^n x^{n-1}$ $(x^2 + 1)(x - 1)$

P **24** What is the solution set of $a^4 + 32 \ge 12a^2$? $\{a: a \le -2\sqrt{2} \text{ or } -2 \le a \le 2 \text{ or } a \ge 2\sqrt{2}\}$

P | **In problems 25–30, solve each equation for z.**

25 $z^4 + 10z^2 + 25 = 0$ No real solutions

26 $4z^3 + 6z^2 - 12z - 18 = 0$ $z = \pm\sqrt{3}$ or $z = -\frac{3}{2}$

27 $z^4 - 13z^2 + 36 = 0$ $z = \pm 2$ or $z = \pm 3$

28 $6z^4 = 150$ $z = \pm\sqrt{5}$

29 $z^4 - 45z^2 = -4$ $z \approx \pm 6.7$ or $z \approx \pm 0.3$

30 $15z^3 - 1 = 5z^2 - 3z$ $z = \frac{1}{3}$

P | **31** Factor $b^4 + 3b^3 - 27b - 81$ over the integers. $(b + 3)(b - 3)(b^2 + 3b + 9)$

P,R | **32** **a** Write an expression for the area of the shaded region of the circle.
b Write an expression for the area of the rectangle.
c Show that the ratio of the area of the shaded region to the area of the rectangle is π:1.

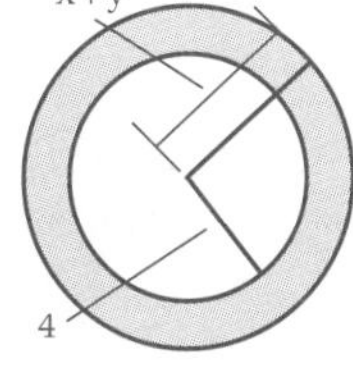

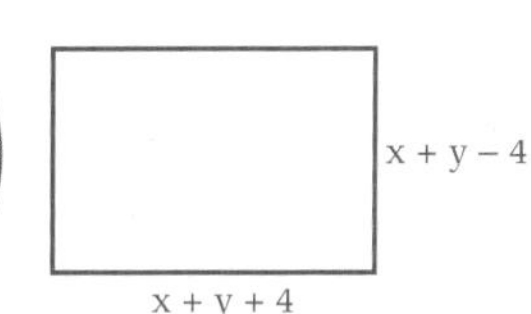

a $\pi(x + y)^2 - 16\pi$ **b** $(x^2 + 2xy + y^2) - 16$ or $(x + y)^2 - 16$ **c** $\frac{\pi[(x + y)^2 - 16]}{(x + y)^2 - 16} = \frac{\pi}{1}$

P,R | **33** Let $f(x) = x^2 + ax + p$, where p is prime and a is an integer.
a For what values of a is $f(x)$ factorable? $\{a{:}a = \pm(p + 1)\}$
b If $f(x)$ is factorable, can a be prime? Explain your answer. Yes; a can be 3 if $p = 2$.

P,I | **34** George made graphs of the following equations. Unfortunately, he forgot to write the equations alongside their corresponding graphs. Which equation corresponds to each graph?

$y = 3x^4 - 2x^2 - 1$ $\quad$ $y = x^3 - 2x^2 - 8x$

$y = x^3 - 1$ $\quad$ $y = x^3 - 3x^2 - x + 3$

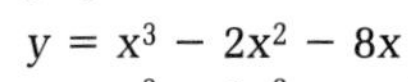

a

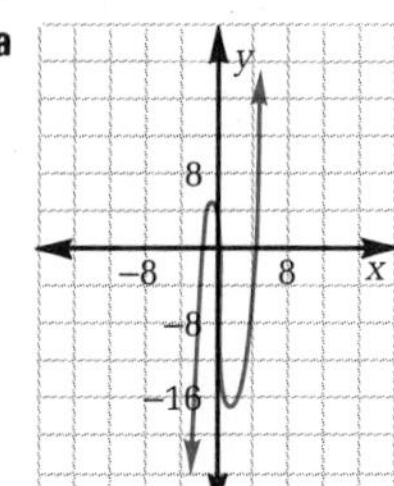

$y = x^3 - 2x^2 - 8x$

b

$y = 3x^4 - 2x^2 - 1$

c

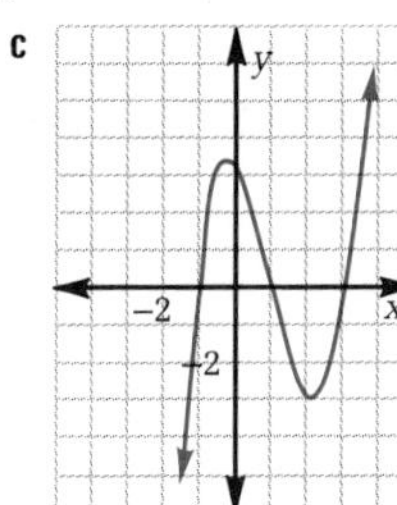

$y = x^3 - 3x^2 - x + 3$

d

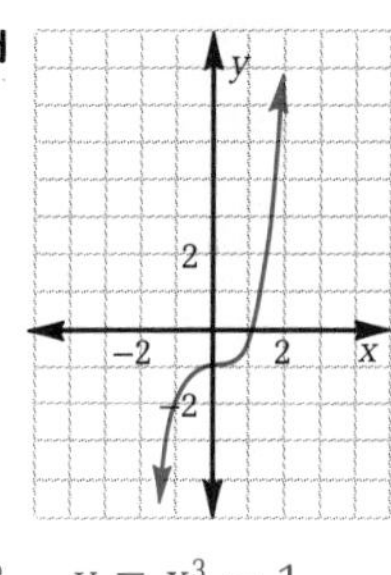

$y = x^3 - 1$

C P,I | **In problems 35–38, solve each inequality for x. Express each solution in set notation and as a number-line graph.**

35 $125x^3 - 1000 > 0$ $\{x{:}x > 2\}$

36 $x^4 + 9x^2 + 18 \geq 0$ $\{x{:}x \in \mathscr{R}\}$

37 $2x^3 + 5x^2 - 8x \leq 20$ $\left\{x{:}x \leq -\frac{5}{2} \text{ or } -2 \leq x \leq 2\right\}$

38 $x^3 - \frac{1}{8} + x^2 - \frac{1}{4} < 0$ $\left\{x{:}x < \frac{1}{2}\right\}$

P,R | **39** Solve each of the following for y and graph each solution on a number line.
a $|y|^2 - |y| = 0$ $y = 0$ or $y = \pm 1$
b $|y|^2 - 5|y| + 4 = 0$ $y = \pm 1$ or $y = \pm 4$
c $|y|^2 - 5|y| + 4 \geq 0$ $\{y{:}y \leq -4 \text{ or } -1 \leq y \leq 1 \text{ or } y \geq 4\}$

Problem-Set Notes and Strategies, continued

- Problem **33** answers the question of how does one tell if a trinomial is factorable. This is one method that is an alternative to using the quadratic formula. It is an intuitive method but not usually remembered.

Additional Answers

35

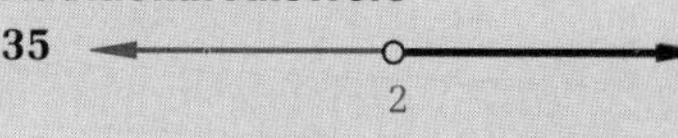

36

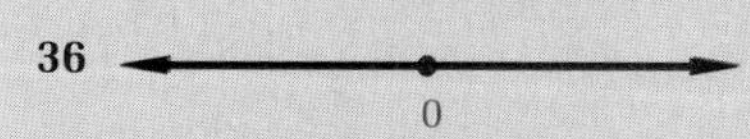

37

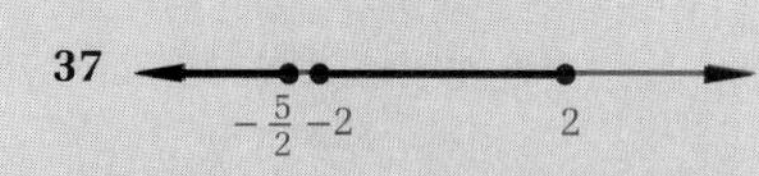

38

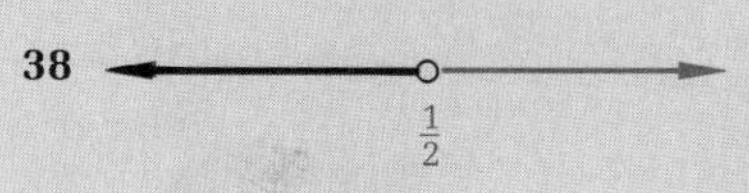

39a −1 0 1

39b −4 −1 1 4

39c −4 −1 1 4

- Students will need to remember that each problem with an absolute value such as in problem **39** will have to be solved with two different problems.

8.1

CHAPTER

8 EXTENDING THE REAL-NUMBER SYSTEM

IN CHAPTER **8**, students extend the properties of exponents to expressions with fractional exponents. Initially, this concept is introduced using expressions with nonnegative bases, then expressions with negative bases are included. Calculator usage is encouraged to help familiarize students with rational exponents.

Students will see that the strategy used to solve equations containing square roots or rational exponents is similar to the strategy used to solve linear equations — isolate the radical or the exponential term on one side of the equation. Encourage students to check their solutions in order to identify extraneous roots.

Imaginary numbers are defined and used to introduce complex numbers. Now students can better understand the difference between the phrases "no solution" and "no real solution".

The algebra of complex numbers is patterned after the algebra of binomials. The simplification of i^n where n is a whole number, is presented through patterning. You may want students to extend the pattern to include negative integral values for n.

At this time, pictures and graphics are used to introduce standard deviation and the related percentages of area under the normal curve.

This print by Paul Klee presents a combination of realistic and imaginary forms that is symbolic of the extended number system, which includes imaginary numbers.

8.1

Rational Exponents

Objectives

After studying this section, you will be able to

- Apply the properties of exponents to expressions with fractional exponents
- Understand the restrictions that apply to exponential expressions with negative bases

Part One: Introduction

An Investigation of Fractional Exponents

If we look at the first-quadrant sections of the graphs of the power functions $y = x^2$, $y = x^3$, $y = x^4$, . . . , we see that as the exponent gets larger, the graph becomes steeper and lies closer to the y-axis. The spaces between these graphs, moreover, suggest that graphs of fractional-power functions, such as $y = x^{2.5}$ and $y = x^{3.5}$, can be inserted between the graphs of the integral-power functions.

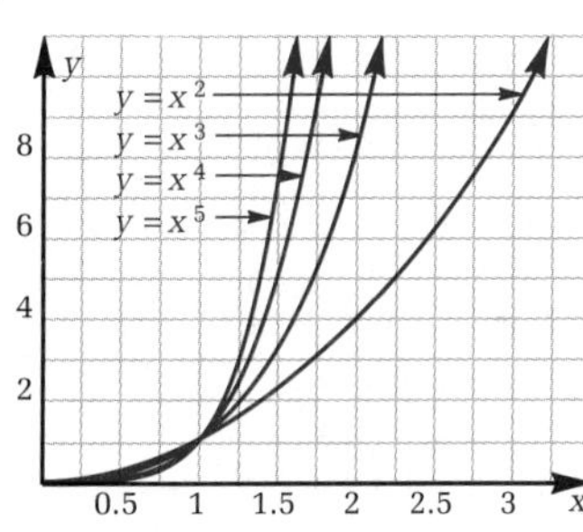

To learn more about fractional exponents, let's assume that the rules of exponents still apply and look at a particular case. For now, we will focus on nonnegative values of x.

Example *Evaluate* $9^{\frac{1}{2}}$.

According to the Power Rule, $\left(9^{\frac{1}{2}}\right)^2 = 9^{\frac{1}{2} \cdot 2} = 9^1 = 9$. Since we know that $\left(\sqrt{9}\right)^2$ is also equal to 9, we can conclude that $9^{\frac{1}{2}} = \sqrt{9}$. Therefore, the value of $9^{\frac{1}{2}}$ is 3.

This example indicates that $\left(x^{\frac{1}{2}}\right)^2 = x$. Similarly, the Power Rule tells us that $\left(x^{\frac{1}{3}}\right)^3 = x$, so $x^{\frac{1}{3}} = \sqrt[3]{x}$, the cube root of x, and $\left(x^{\frac{1}{4}}\right)^4 = x$, so $x^{\frac{1}{4}} = \sqrt[4]{x}$, and so on. In general, $x^{\frac{1}{n}} = \sqrt[n]{x}$.

We can apply this reasoning to other fractional exponents. The Product Rule implies that $\left(x^{\frac{1}{n}}\right)^m = x^{\frac{1}{n} \cdot m} = x^{\frac{m}{n}}$. This observation leads us to a general definition of rational exponentiation.

If $x \geq 0$ and m and n are positive integers, then $x^{\frac{m}{n}} = \left(\sqrt[n]{x}\right)^m$.

8.1

Class Planning

Time Schedule

Average: 2 days
Advanced: $1\frac{1}{2}$ days
Honors: $1\frac{1}{2}$ days

Resource References

Teacher's Resource Book
Class Opener 48
Practice 51

Class Opener

If students have access to a graphics calculator, have them graph the following.

1 $y = x$ 2 $y = x^2$
3 $y = x^3$ 4 $y = x^4$
5 $y = x^{\frac{1}{2}}$ 6 $y = x^{\frac{3}{2}}$
7 $y = x^{\frac{4}{3}}$

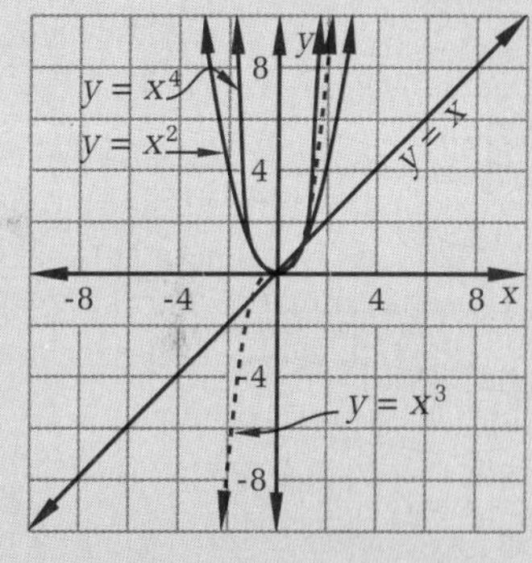

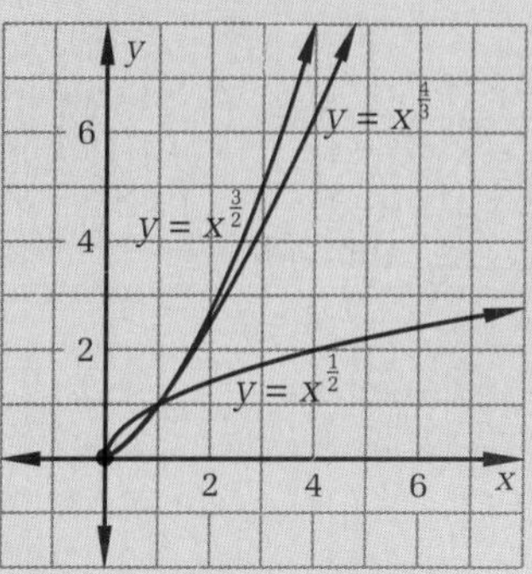

Vocabulary
root index
radicand

8.1

Class Opener, continued

If students have a calculator, have them evaluate the following.

1 $4^{\frac{1}{2}}$ 2 $9^{\frac{1}{2}}$ 3 $16^{\frac{1}{2}}$
4 $27^{\frac{1}{3}}$ 5 $125^{\frac{1}{3}}$ 6 $216^{\frac{1}{3}}$
7 $625^{\frac{1}{4}}$ 8 $1331^{\frac{1}{3}}$

1 2 2 3 3 4
4 3 5 5 6 6
7 5 8 11

Additional problems are to evaluate the following.

1 x^2x^4 2 $\frac{x^6}{x^3}$ 3 x^{3^4}

1 x^6 2 x^3 3 x^{12}

Lesson Notes

- A rational exponent is evaluated in two parts. The numerator of a rational exponent is evaluated as any other exponent.
- A denominator of a rational exponent is the index of a radical expression where the base of the exponential expression is the radicand.
- As in the past, even roots of negative numbers cannot be evaluated in the real number system.
- Rational exponents can be written as a product of the numerator and $\frac{1}{\text{denominator}}$. Since multiplication is commutative, the order of evaluation is left to the discretion of the user and can be changed to facilitate the evaluation.

The number n in an expression of the form $\sqrt[n]{x}$ is called the ***root index***, and x is called the ***radicand***.

Example *Evaluate* $16^{\frac{5}{4}}$.

$$16^{\frac{5}{4}}$$
$$= (\sqrt[4]{16})^5$$
$$= 2^5$$
$$= 32$$

Exponential Expressions with Negative Bases

So far, we have looked at expressions with nonnegative bases. What happens when a negative base is raised to a fractional power?

According to the definition of rational exponentiation, $(-4)^{\frac{1}{2}} = \sqrt{-4}$, but there is no real number equal to $\sqrt{-4}$. Indeed, for any negative value of x, $x^{\frac{1}{2}}$ is not defined over the real numbers. However, $(-8)^{\frac{1}{3}} = -2$, since $(-2)^3 = -8$. In fact, whenever a negative base is raised to the third power, the result is a negative number. It follows that negative numbers will always have real cube roots.

By extension, a negative number has a real root when the root index is odd but does not have a real root when the index is even. In general, if the value of y in an expression of the form $x^{\frac{z}{y}}$ is odd, we include negative values of x, but if the value of y is even or is unspecified, we must restrict the value of x to $\{x{:}x \geq 0\}$.

Trying to evaluate an expression such as $(-3)^{\frac{2}{5}}$ by entering 3 [±] [xʸ] .4 [=] on a calculator will often produce an error message. A better way to proceed is to determine whether the result should be positive or negative, and then use a calculator to evaluate the corresponding expression with a positive base.

Example *Evaluate each of the following expressions.*

a $(-3)^{\frac{2}{5}}$

$(-3)^{\frac{2}{5}} = (\sqrt[5]{-3})^2$

An even power of a negative real number is a positive number.

When we enter 3 [xʸ] .4 [=], the output is ≈ 1.55, so $(-3)^{\frac{2}{5}} \approx 1.55$.

b $(-3)^{\frac{3}{5}}$

$(-3)^{\frac{3}{5}} = (\sqrt[5]{-3})^3$

An odd power of a negative real number is a negative number.

When we enter 3 [xʸ] .6 [=], the output is ≈ 1.93, so $(-3)^{\frac{3}{5}} \approx -1.93$.

c $(-3)^{\frac{3}{4}}$

$(-3)^{\frac{3}{4}} = (\sqrt[4]{-3})^3$

An even root of a negative real number is not a real number. This expression has no real equivalent.

Part Two: Sample Problems

Problem 1 *Evaluate* $\left(\frac{5}{3}\right)^{-3} \cdot \left(\frac{3}{5}\right)^{\frac{3}{2}}$.

Solution We can evaluate this expression in two ways.

Method 1

$$\left(\frac{5}{3}\right)^{-3} \cdot \left(\frac{3}{5}\right)^{\frac{3}{2}}$$
$\approx 0.216(0.464758)$
≈ 0.1003877

Method 2

$$\left(\frac{5}{3}\right)^{-3} \cdot \left(\frac{3}{5}\right)^{\frac{3}{2}}$$
$$= \left(\frac{3}{5}\right)^{3} \cdot \left(\frac{3}{5}\right)^{\frac{3}{2}}$$
$$= \left(\frac{3}{5}\right)^{\frac{9}{2}}$$
$= 0.6^{4.5}$
≈ 0.1003877

Simplifying the expression first (Method 2) often produces an expression that is easier to evaluate with a calculator.

Problem 2 *The temperature of an egg that has just been hard-boiled is 98°C. If this egg is put into water with a constant temperature of 18°C, we can use the formula* $T(x) = 18 + 80(2.73)^{-0.28x}$ *to find the temperature of the egg after x minutes in the water. What is the temperature of the egg*

a *After 5 minutes?* **b** *After 10 minutes?* **c** *After 15 minutes?*

Solution We can use a calculator to find $T(5)$, $T(10)$, and $T(15)$.

a Enter 18 [+] 80 [×] 2.73 [x^y] [(] .28 [±] [×] 5 [)] [=].
The output is ≈37.609, so the temperature after 5 minutes is about 38°C.

b Enter 18 [+] 80 [×] 2.73 [x^y] [(] .28 [±] [×] 10 [)] [=].
The output is ≈22.807, so the temperature after 10 minutes is about 23°C.

c Enter 18 [+] 80 [×] 2.73 [x^y] [(] .28 [±] [×] 15 [)] [=].
The output is ≈19.178, so the temperature after 15 minutes is about 19°C.

Notice that in the first 5 minutes the temperature decreases by 60°C, in the next 5 minutes it decreases by 15°C, and in the last 5 minutes it decreases by 4°C. This is an example of *exponential change.*

Lesson Notes, continued

- Point out that the ability of calculators is limited to approximations of answers. Students may have a difficult time evaluating negative roots of any radical. They need to understand the significance of the sign when using calculators and make sure they know when the root is defined.

Stumbling Block

Students frequently don't stop to think about the problem before trying to solve it. For example, to evaluate $[(3^{\frac{2}{3}})^6 (4^{\frac{1}{3}})^{12}]^{\frac{1}{4}}$ it is much easier to first simplify the exponents and get $3 \cdot 4 = 12$ instead of reaching for your calculator. All the properties of exponents still work with rational exponents.

Communicating Mathematics

Have students write a paragraph that explains how to calculate the 0.4 power of 32 using a calculator and without a calculator.

Part Three: Exercises and Problems

Warm-up Exercises

1 Evaluate each expression.

a $6^{\frac{1}{2}} \cdot 6^{\frac{1}{2}}$ 6 **b** $4^{\frac{1}{3}} \cdot 4^{\frac{1}{3}} \cdot 4^{\frac{1}{3}}$ 4 **c** $(2^8)^{\frac{1}{2}}$ 16
d $(2^{\frac{1}{2}})^6$ 8 **e** $36^{-\frac{1}{2}}$ $\frac{1}{6}$ **f** $(\sqrt[3]{-8})^2$ 4

P

Checkpoint

1 Evaluate $16^{\frac{3}{4}}$. 8
2 Use a calculator and find $3^{2.5}$. ≈15.6
3 Evaluate $3^{\frac{2}{3}} \cdot 3^{\frac{4}{3}}$. 9
4 Write the fifth root of 8 to the fourth power using rational exponents. $8^{\frac{4}{5}}$

Assignment Guide

Average

Day 1 6–10, 14, 15, 19, 21, 23–28

Day 2 11, 12, 16–18, 20, 22, 29, 30, 37

Advanced

Day 1 7, 8, 10, 13, 14, 18, 20–26, 29, 30, 32, 34

Day 2 ($\frac{1}{2}$ day) 9, 11, 12, 15, 19, 27, 28, 31

($\frac{1}{2}$ day) Section 8.2 7, 14–16, 22, 24, 25, 30, 37, 39, 45

Honors

Day 1 10, 13, 15, 16, 21, 25–28, 30–32, 35, 38

Day 2 ($\frac{1}{2}$ day) 9, 14, 17, 20, 22, 29, 33, 34, 36, 37

($\frac{1}{2}$ day) Section 8.2 6, 10, 13, 14, 25, 30, 35, 45, 47

Problem-Set Notes and Strategies

- Problem **5** is good practice on estimating graphs. Make sure that students can relate to the fact that since $\frac{5}{4}$ is between 1 and 2, the graph of $x^{\frac{5}{4}}$ is between x and x^2. This is a property of exponential functions that will be used in Chapter **11**.
- Problem **9** should help students recall the different subsets of the real numbers. If they have forgotten, now might be a good time to review before going into the field of complex numbers later in the chapter.

P **2** Rewrite each expression as a fractional power of x.

a $\sqrt{x} \cdot x$ $x^{\frac{3}{2}}$ **b** $\sqrt[3]{x^2} \cdot \sqrt[4]{x}$ $x^{\frac{11}{12}}$

P **3** Evaluate each expression without using a calculator.

a $\sqrt{(-3)^2}$ 3 **b** $\sqrt[3]{8^2}$ 4 **c** $\sqrt[3]{-1000}$ −10

P **4** Use a calculator to evaluate each expression. Round your answers to the nearest hundredth.

a -1.35^6 ≈ -6.05 **b** $5200\,(1.03)^{2.5}$ ≈ 5598.82 **c** $2^{-5.3}$ ≈ 0.03

P,I **5** Which of the lettered curves could be the graph of $y = x^{\frac{5}{4}}$? c

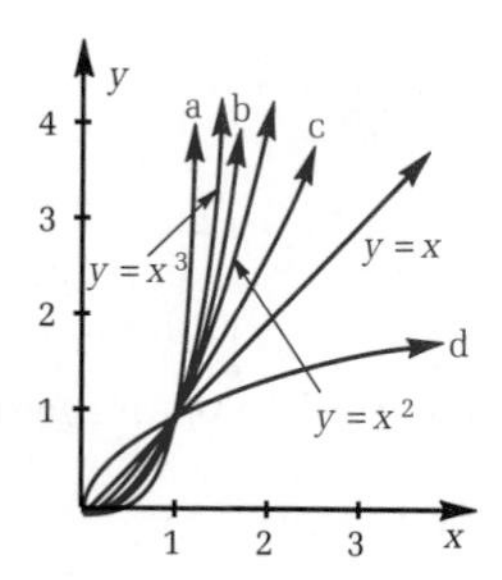

Problem Set

P **A** **6** Explain how you would determine the value of $27^{\frac{5}{3}}$.

$27^{\frac{5}{3}} = (3^3)^{\frac{5}{3}} = 3^5 = 243$

P **In problems 7 and 8, simplify each expression.**

7 $\sqrt[3]{x^6}$ x^2 **8** $7^{1.6} \cdot 7^{0.4}$ 49

R **9** Match each number with the set or sets of which it is a member.

a −6 **ii, iii,** and **iv**
b $\frac{22}{7}$ **ii** and **iv**
c $\sqrt{2}$ **iv** and **v**
d π **iv** and **v**
e $0.\overline{45}$ **ii** and **iv**

i Natural numbers
ii Rational numbers
iii Integers
iv Real numbers
v Irrational numbers
vi Whole numbers

P,R **10** Evaluate each expression.

a $\sqrt[4]{81}$ 3 **b** $81^{\frac{3}{4}}$ 27 **c** $81^{-\frac{3}{4}}$ $\frac{1}{27}$

P,R **11** Find the length of $\overline{AB}$.

$2\sqrt{37}$, or ≈ 12.17

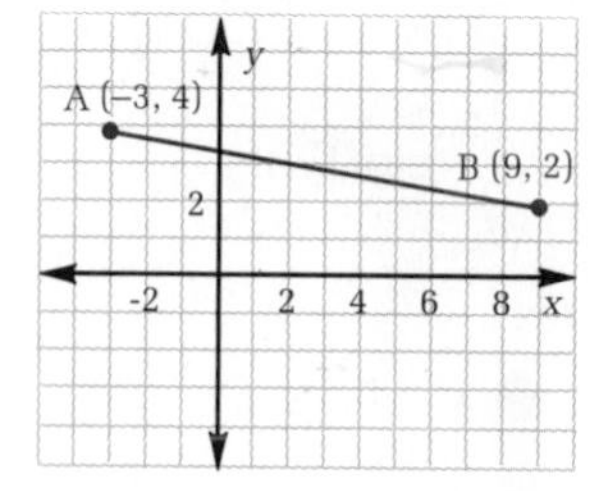

P,R | **12** If one of the expressions below is chosen at random, what is the probability that it is negative? That it is undefined? $0; \frac{3}{5}$

$(-5)^{\frac{2}{3}}$ $\quad (-3)^{\frac{3}{4}}$ $\quad \sqrt[6]{(-7)^2}$ $\quad \left(\sqrt[6]{-7}\right)^8$ $\quad (-22.2)^{\frac{5}{2}}$

P,R | **13** Consider the cube in the figure.

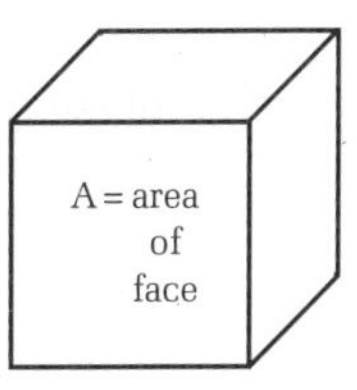

a Write an expression for the volume of the cube in terms of A. $V = A^{\frac{3}{2}}$

b Calculate the cube's volume to the nearest tenth if the area of each face is 15 square centimeters. $\approx 58.1\ \text{cm}^3$

B

In problems 14–17, solve each equation for n.

P | **14** $\left(3^{\frac{5}{7}}\right)^n = 3 \quad n = \frac{7}{5}$

15 $\left(\frac{2}{3}\right)^n = \left(\frac{3}{2}\right)^2 \quad n = -2$

16 $(-27)^n = 9 \quad n = \frac{2}{3}$

17 $4\left(16^{\frac{1}{3}}\right) = 2^n \quad n = \frac{10}{3}$

P | **In problems 18 and 19, evaluate each expression.**

18 $16^{0.25}$ $\quad 2$

19 $25^{1.5}$ $\quad 125$

P,I | **20** Suppose that $\ell(a, b)$ means "the power of a that produces an expression equal to b." For example, $\ell(3, 9) = 2$, since $3^2 = 9$; and $\ell(2, 16) = 4$, since $2^4 = 16$. Evaluate each of the following.

a $\ell(2, 8)$ $\quad 3$

b $\ell(2, 64)$ $\quad 6$

c $\ell(4, 64)$ $\quad 3$

d $\ell(4, 2)$ $\quad \frac{1}{2}$

e $\ell(5, 1)$ $\quad 0$

P | **In problems 21 and 22, evaluate each expression.**

21 $\left(7^{\frac{1}{2}} + 5^{\frac{1}{2}}\right)\left(7^{\frac{1}{2}} - 5^{\frac{1}{2}}\right) \quad 2$

22 $4^{-\frac{2}{3}}\left(4^{\frac{2}{3}} + 4^{\frac{5}{3}}\right) \quad 5$

R | **23** Factor each expression over the integers.

a $x^2 - 13x - 30 \quad (x - 15)(x + 2)$

b $y^2 - 289 \quad (y + 17)(y - 17)$

c $3a^2 + 33a + 84 \quad 3(a + 4)(a + 7)$

d $5n^2 - 16n - 16 \quad (5n + 4)(n - 4)$

P | **24** Rewrite each expression as a rational power of x.

a $\sqrt[3]{x^5} \quad x^{\frac{5}{3}}$

b $\sqrt[5]{x^{20}} \quad x^4$

P | **In problems 25–28, simplify each expression.**

25 $\left(8x^6y^9\right)^{\frac{2}{3}} \quad 4x^4y^6$

26 $\left[\left(x^{\frac{2}{3}}\right)^6\left(y^{\frac{1}{3}}\right)^{12}\right]^{\frac{1}{2}} \quad x^2y^2$

27 $\dfrac{3x^{-1}y}{2xy^{-1}} \quad \frac{3y^2}{2x^2}$

28 $\dfrac{5x^2y^7}{10x^{-2}y^{10}} \quad \frac{x^4}{2y^3}$

P,R | **29** Which of the following expressions are equivalent to $x^{\frac{4}{7}}$? **a, b, c,** and **f**

a $\left(x^4\right)^{\frac{1}{7}}$

b $\left(x^{\frac{1}{7}}\right)^4$

c $\left(\sqrt[7]{x}\right)^4$

d $\left(\sqrt[4]{x}\right)^7$

e $\sqrt[4]{x^7}$

f $\sqrt[7]{x^4}$

Problem-Set Notes and Strategies, continued

- In problems **14 – 17**, point out that if the bases are equal, the exponents must be equal.
- If problems **18** and **19** are changed into fractional exponents instead of decimals, they are easier to solve.
- Problem **20** introduces the concept of logarithms. These are explored in detail in Chapter **11**.

Problem-Set Notes and Strategies, continued

- If any of your students are studying physics, point out that problem **31** is an example of Newton's Law of Cooling.
- Problem **34** will confirm the idea presented in **30e**.
- Problems **35** and **36** illustrate formulas for the sum of cubes and the difference of fourths.
- Limits are previewed in problem **37**.
- For problem **38c** students may need an enlarged illustration to emphasize that the number of sodium atoms is the same as in part **b**.

Problem Set, *continued*

R,I **30** Indicate whether each statement is True or False.

a Every rational number is a real number. True
b Every irrational number is a real number. True
c The number 0 is both rational and irrational. False
d The sum of two rational numbers is always a rational number. True
e The sum of two irrational numbers is always an irrational number. False

P **31** A pot of 90°C coffee is placed in a room with an air temperature of 30°C. The formula $T(x) = 30 + 60(2.72)^{-0.07x}$ can be used to find the temperature of the coffee x minutes after it is placed in the room.

a Find the temperature of the coffee after 5 minutes. ≈72°C
b Find the temperature of the coffee after 20 minutes. ≈45°C
c To the nearest minute, how long will it take the coffee to cool to 33°C? ≈43 min

P **In problems 32 and 33, copy each equation, supplying the missing terms.**

32 $x^{-12} + x^{-5} = x^{-12}(? + ?)$
1; x^7

33 $x^{\frac{1}{2}} + x^{\frac{5}{2}} = x^{\frac{1}{2}}(? + ?)$
1; x^2

R **34** When the number 0.01001000100001 . . . is added to an irrational number x, the sum is rational. Find at least two possible values of x. Possible answers: 0.10110111011110 . . . and 0.21221222122221 . . .

C **In problems 35 and 36, expand and simplify.**

P **35** $\left(x^{\frac{1}{3}} + y^{\frac{1}{3}}\right)\left(x^{\frac{2}{3}} - x^{\frac{1}{3}}y^{\frac{1}{3}} + y^{\frac{2}{3}}\right)$ $x + y$

36 $\left(x^{\frac{1}{4}} + y^{\frac{1}{4}}\right)\left(x^{\frac{3}{4}} - x^{\frac{2}{4}}y^{\frac{1}{4}} + x^{\frac{1}{4}}y^{\frac{2}{4}} - y^{\frac{3}{4}}\right)$ $x - y$

P,I **37** Consider $50^{\frac{n}{n+1}}$, where n is a positive integer.

a Compute decimal approximations of $50^{\frac{1}{2}}$, $50^{\frac{2}{3}}$, $50^{\frac{3}{4}}$, $50^{\frac{4}{5}}$, $50^{\frac{100}{101}}$, and $50^{\frac{10,000}{10,001}}$. ≈7.0711; ≈13.5721; ≈18.8030; ≈22.8653; ≈48.1004; ≈49.9804
b What seems to happen to the value of $50^{\frac{n}{n+1}}$ as the value of n increases? It approaches, but does not reach, 50.

P,R **38** Table salt (NaCl) is a compound consisting of sodium ions (Na^+) and chloride ions (Cl^-) bound in a cubic arrangement.

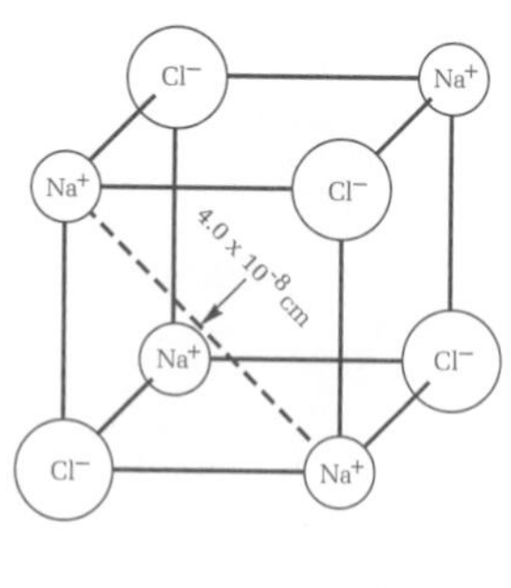

a Find the volume of the cube pictured.
b How many such cubes would be needed to make a grain of table salt (0.1 cubic millimeter)?
c About how many sodium atoms does a grain of salt contain? (Hint: The answer is not four times your answer in part **b**.)

a $\approx 2.26 \times 10^{-23}$ cm³ **b** $\approx 4.42 \times 10^{18}$ **c** $\approx 2.21 \times 10^{18}$

8.2

8.2 SOLVING RADICAL EQUATIONS

Objectives

After studying this section, you will be able to
- Solve equations that contain square roots
- Solve equations that contain rational exponents

Part One: Introduction

Equations That Contain a Square Root

The strategy we use to solve equations that contain radical expressions is similar to the strategy you use to solve linear equations.

Strategy for Solving Radical Equations

1. **Isolate the radical expression on one side of the equation.**
2. **Square both sides to eliminate the radical.**
3. **Solve the resulting equation for the variable.**
4. **Check the solutions.**

As you will see, checking is an essential step in solving equations that contain radicals or fractional exponents.

Example *Solve* $9 + 2\sqrt{4 + 3q} = 29$ *for* q.

	$9 + 2\sqrt{4 + 3q} = 29$
Add −9 to each side	$2\sqrt{4 + 3q} = 20$
Divide by 2 to isolate the radical	$\sqrt{4 + 3q} = 10$
Square both sides	$4 + 3q = 100$
Solve for q	$3q = 96$
	$q = 32$

Check

Left Side	**Right Side**
$9 + 2\sqrt{4 + 3q}$	$= 29$
$= 9 + 2\sqrt{4 + 3(32)}$	
$= 9 + 2\sqrt{100}$	
$= 9 + 20$	
$= 29$	

If $q = 32$, the sides are equal, so the answer checks.

Vocabulary
extraneous root

Class Planning

Time Schedule
Average: 2 days
Advanced: $1\frac{1}{2}$ days
Honors: $1\frac{1}{2}$ days

Resource References
Teacher's Resource Book
Class Opener 49
Cooperative Learning 23
Practice 52

Class Opener

Solve the equation.
$\sqrt{x + 7} = 25$
$x = 618$

Lesson Notes

- To solve equations that contain radicals, first isolate the radical. Then, use the inverse operation of squaring to eliminate the radical. Note that if the radical is not isolated before squaring, the squaring process will yield more radicals.
- The squaring process may yield extra or extraneous roots. All potential roots should be checked before the final solution is obtained. For exam-ple, the equation $\sqrt{x} = -1$ should illustrate this. Square both sides to obtain $x = 1$, which is not a root of the original equation.
- Reiterate the fact that by squaring both sides you are producing an equivalent equation. As long as you do the same thing on both sides, the equations are equivalent. This is one of the several cases where students must check for extraneous roots.

Using Computers

One interesting way of solving rational equations is to graph the left- and right-hand sides of the equation separately, each as function in two dimensions. Then determine where the two graphs intersect and focus on the x-coordinates of the points of intersection. Students can see visually why extraneous solutions appear.

■

Stumbling Block

In checking solutions to an equation, students sometimes substitute their answers in an intermediate step because the arithmetic is easier to perform. Make sure students check their solutions in the original equation to verify them.

■

Communicating Mathematics

Have students write a paragraph that explains why it is necessary to check the solutions of radical equations.

Example *Solve $m + \sqrt{m} = 90$ for m.*

Method 1 $\quad m + \sqrt{m} = 90$

Isolate the radical $\quad \sqrt{m} = 90 - m$

Square both sides $\quad m = (90 - m)^2$

$m = 8100 - 180m + m^2$

$0 = 8100 - 181m + m^2$

Factor $\quad 0 = (81 - m)(100 - m)$

Zero Product Property $\quad m = 81$ or $m = 100$

Check

$m = 81$		$m = 100$	
Left Side	**Right Side**	**Left Side**	**Right Side**
$m + \sqrt{m}$	$= 90$	$m + \sqrt{m}$	$= 90$
$= 81 + \sqrt{81}$		$= 100 + \sqrt{100}$	
$= 81 + 9$		$= 100 + 10$	
$= 90$		$= 110$	

If $m = 81$, the sides are equal. If $m = 100$, the sides are not equal. Therefore, $m = 81$ is the only correct answer. One reason why it is essential to check answers is that squaring both sides of an equation may introduce an ***extraneous root***—an apparent solution that does not in fact satisfy the original equation.

Method 2 $\quad m + \sqrt{m} = 90$

Let $p = \sqrt{m}$ $\quad p^2 + p = 90$

$p^2 + p - 90 = 0$

Factor $\quad (p - 9)(p + 10) = 0$

Zero Product Property $\quad p = 9$ or $p = -10$

Substitute $\sqrt{m}$ for p $\quad \sqrt{m} = 9$ or $\sqrt{m} = -10$

Since $\sqrt{m}$ cannot be negative, $\sqrt{m} = 9$—that is, $m = 81$—is the only solution. Check as before to ensure that 81 is the correct answer.

Note If the rewritten equation cannot be solved by factoring, use the quadratic formula.

Equations That Contain Rational Exponents

We solve equations that contain rational exponents in much the same way as those containing radicals: isolating the exponential term on one side of the equation and raising both sides to the power of the reciprocal of the exponent.

Example *Solve $(2c + 1)^{\frac{3}{4}} = 8$ for c.*

$(2c + 1)^{\frac{3}{4}} = 8$

Raise both sides to the $\frac{4}{3}$ power $\quad \left[(2c + 1)^{\frac{3}{4}}\right]^{\frac{4}{3}} = 8^{\frac{4}{3}}$

Use the Product Rule $\quad (2c + 1)^{\frac{3}{4} \cdot \frac{4}{3}} = 8^{\frac{4}{3}}$

$2c + 1 = 16$

$c = 7.5$

Check

Left Side	**Right Side**
$(2c + 1)^{\frac{3}{4}}$	$= 8$
$= [2(7.5) + 1]^{\frac{3}{4}}$	
$= 16^{\frac{3}{4}}$	
$= 2^3$	
$= 8$	

If $c = 7.5$, the sides are equal, so the answer checks.

Example *Solve* $5 + (j^2 - 1)^{\frac{-2}{3}} = 21$ *for* j.

$$5 + (j^2 - 1)^{-\frac{2}{3}} = 21$$

Isolate the exponential term

$$(j^2 - 1)^{-\frac{2}{3}} = 16$$

Raise both sides to the inverse power

$$(j^2 - 1)^{\left(-\frac{2}{3}\right)\left(-\frac{3}{2}\right)} = 16^{-\frac{3}{2}}$$

Use absolute-value symbols, since the inverse power has an even denominator

$$|j^2 - 1| = \frac{1}{64}$$

$$j^2 - 1 = \frac{1}{64} \text{ or } j^2 - 1 = -\frac{1}{64}$$

$$j^2 = \frac{65}{64} \text{ or } j^2 = \frac{63}{64}$$

$$j = \frac{\sqrt{65}}{8} \text{ or } j = -\frac{\sqrt{65}}{8} \text{ or } j = \frac{\sqrt{63}}{8} \text{ or } j = -\frac{\sqrt{63}}{8}$$

$$j \approx 1.0078 \text{ or } j \approx -1.0078 \text{ or } j \approx 0.9922 \text{ or } j \approx -0.9922$$

Let's check an approximate answer, 1.0078, and the corresponding exact value, $\frac{\sqrt{65}}{8}$.

Check

Left Side	**Right Side**	**Left Side**	**Right Side**
$5 + \left[\left(\frac{\sqrt{65}}{8}\right)^2 - 1\right]^{-\frac{2}{3}}$	$= 21$	$5 + (1.0078^2 - 1)^{-\frac{2}{3}}$	$= 21$
$= 5 + \left(\frac{65}{64} - 1\right)^{-\frac{2}{3}}$		$= 5 + (1.01566084 - 1)^{-\frac{2}{3}}$	
$= 5 + \left(\frac{1}{64}\right)^{-\frac{2}{3}}$		$= 5 + 0.01566084^{-\frac{2}{3}}$	
$= 5 + 16$		$= 5 + 15.9755799$	
$= 21$		$= 20.9755799$	

When we check the approximation 1.0078, the result is close to, but not equal to, 21. In many cases, it is difficult to determine whether such a discrepancy is due to rounding or indicates an incorrect solution. It is therefore better to check exact solutions whenever possible. Checking will verify that the other three results are also solutions. The solution set is $\left\{\frac{\sqrt{65}}{8}, -\frac{\sqrt{65}}{8}, \frac{\sqrt{63}}{8}, -\frac{\sqrt{63}}{8}\right\}$.

Checkpoint

1. Solve the equation $\sqrt{x} = x - 2$. 4
2. Solve $(4x + 3)^{\frac{2}{3}} = 9$ for x. 6
3. Solve for y: $\sqrt{(y - 13)} = 5$ 38

Assignment Guide

Average

Day 1 6, 8, 10, 12–16, 22–25, 30, 32, 38, 39, 45

Day 2 7, 9, 11, 17–21, 26, 27, 31, 33, 35, 40, 43, 46

Advanced

Day 1 ($\frac{1}{2}$ day) Section 8.1 9, 11, 12, 15, 19, 27, 28, 31 ($\frac{1}{2}$ day) 7, 14–16, 22, 24, 25, 30, 37, 39, 45

Day 2 6, 9, 13, 17–19, 21, 26, 27, 31–33, 35, 43, 46–48, 51

Honors

Day 1 ($\frac{1}{2}$ day) Section 8.1 9, 14, 17, 20, 22, 29, 33, 34, 36, 37 ($\frac{1}{2}$ day) 6, 10, 13, 14, 25, 30, 35, 45, 47

Day 2 7, 9, 11, 17, 19, 27, 31–34, 36, 37, 40, 43, 44, 46, 48, 50–52, 54, 55

Part Two: Sample Problems

Problem 1 *Solve* $3\sqrt{v^3 - v + 1} + v = 4\sqrt{v^3 - v + 1}$ *for* v.

Solution

$$3\sqrt{v^3 - v + 1} + v = 4\sqrt{v^3 - v + 1}$$

Isolate the radical $\quad v = \sqrt{v^3 - v + 1}$

Square both sides $\quad v^2 = v^3 - v + 1$

$$0 = v^3 - v^2 - v + 1$$

Factor by grouping $\quad 0 = v^2(v - 1) - 1(v - 1)$

$$0 = (v^2 - 1)(v - 1)$$

$$0 = (v + 1)(v - 1)^2$$

$$v = -1 \text{ or } v = 1$$

Check

If $v = -1$, the left side is equal to 2 and the right side is equal to 4, so $v = -1$ is not a solution. If $v = 1$, both sides are equal to 4, so $v = 1$ is the only solution.

Problem 2 *Solve* $\sqrt{z + 9} + \sqrt{z} = 9$ *for* z.

Solution

$$\sqrt{z + 9} + \sqrt{z} = 9$$

Isolate one radical term $\quad \sqrt{z + 9} = 9 - \sqrt{z}$

Square both sides $\quad z + 9 = 81 - 18\sqrt{z} + z$

Isolate the remaining radical term $\quad 18\sqrt{z} = 72$

$$\sqrt{z} = 4$$

Square both sides $\quad z = 16$

Check

Left Side

$\sqrt{z + 9} + \sqrt{z}$
$= \sqrt{16 + 9} + \sqrt{16}$
$= 5 + 4$
$= 9$

Right Side

$= 9$

If $z = 16$, the sides are equal, so the answer checks.

Part Three: Exercises and Problems

Warm-up Exercises

1 Solve for y.

a $\sqrt{y} = 9$ $\quad y = 81$

b $\sqrt{y + 1} = 3$ $\quad y = 8$

c $2\sqrt{y + 3} = 5$ $\quad y = \frac{13}{4}$

2 Solve for k.

a $\sqrt{k + 1} = 10$ $\quad k = 99$

b $\sqrt{k} = -15$ No real solutions

P
P

Problem-Set Notes and Strategies

- Problem **2b** is an example that shows why the roots of a rational must be checked.

P **3** Solve for p.

a $\sqrt{p} = -5$ No real solutions
b $\sqrt[3]{p} = 5$ $p = 125$
c $\sqrt[3]{p} = -5$ $p = -125$

P **4** Solve for x.

a $x^{\frac{1}{3}} = 2$ $x = 8$
b $x^{\frac{1}{3}} = \sqrt[3]{5}$ $x = 5$
c $2\sqrt{x} + 3 = 5$ $x = 1$
d $\sqrt[3]{x + 2} - 3 = 4$ $x = 341$

P,R **5** If the area of the larger rectangle is 91 square feet, what is the area of the smaller rectangle?
$10\sqrt{5}$ ft²

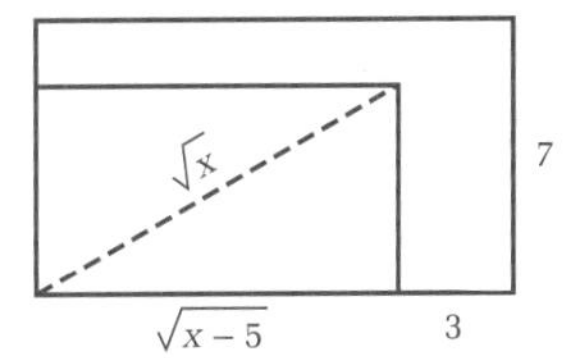

Problem Set

P **A**

6 Find a if $(x^{\frac{3}{7}})^a = x$. $a = \frac{7}{3}$

P **7** Find b if $(y^{-\frac{3}{4}})^b = y$. $b = -\frac{4}{3}$

R **8** Refer to the diagram. Find the value of x for which $AB = \sqrt{58}$.
$x = 10$

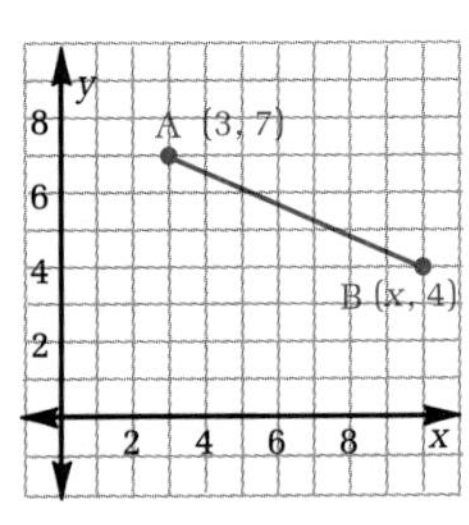

P,R **9** Suppose that $\ell(p, q)$ means "the power of p that produces an expression equal to q." For example, $\ell(16, 2) = \frac{1}{4}$, since $16^{\frac{1}{4}} = 2$. Evaluate each of the following.

a $\ell\left(2, \frac{1}{16}\right)$ -4
b $\ell(x, x)$ 1
c $\ell(10, 1000)$ 3
d $\ell(4, 8)$ $\frac{3}{2}$
e $\ell(2, 8)$ 3
f $\ell(x, x^5)$ 5

Problem-Set Notes and Strategies, continued

- Problem **9** previews the concept of logarithms that will be discussed in Chapter **11**.

P **In problems 10 and 11, solve for x.**

10 $(2x + 1)^{\frac{2}{3}} = 4$ $x = \frac{7}{2}$

11 $\sqrt[3]{1 - 2x} = -3$ $x = 14$

R **12** Refer to the circle.

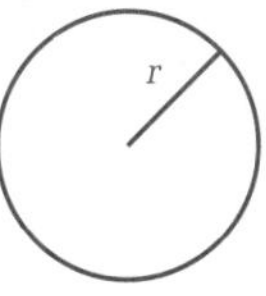

a Find the radius r if the area of the circle is 36π. $r = 6$
b Find the radius r if the area of the circle is 36.
$r \approx 3.39$

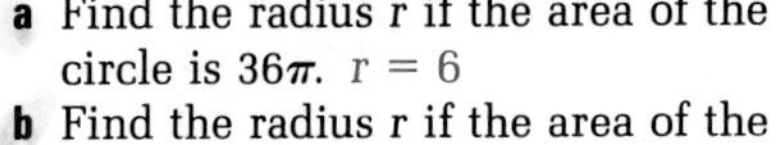

Problem-Set Notes and Strategies, continued

- Problems **12 – 19** are excellent calculator problems. Students may have difficulty with the necessary keystrokes and the proper order. Encourage them to read the manual that came with the calculator.

- Problems **29 – 31** and **34** use the 30°-60°-90° and the 45°-45°-90° right triangle theorems that students learned in geometry. These will be necessary for trigonometry later in the text.

- In problem **37**, point out that not only is the value close to 7, but it is larger than 7.

Problem Set, *continued*

P,R **13** If the volume of the sphere shown is 1773, what is the value of m? ≈ 36.37

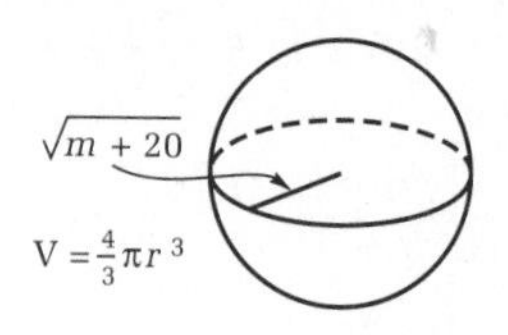

P **In problems 14–19, solve for x.**

14 $x^{\frac{1}{5}} = 2$ $x = 32$ **15** $x^{\frac{1}{2}} = 16$ $x = 256$ **16** $x^{\frac{1}{7}} = 1.64$ $x \approx 31.91$

17 $x^{0.15} = 2$ $x \approx 101.59$ **18** $x^{\frac{5}{6}} = 32$ $x = 64$ **19** $x^{-\frac{3}{4}} = 8$ $x = \frac{1}{16}$

R **In problems 20–23, write each expression in standard radical form.**

20 $\sqrt{54}$ $3\sqrt{6}$ **21** $\frac{\sqrt{3}}{\sqrt{5}}$ $\frac{1}{5}\sqrt{15}$ **22** $\sqrt{288}$ $12\sqrt{2}$ **23** $\frac{10}{\sqrt{2}}$ $5\sqrt{2}$

P,R,I **In problems 24–27, solve for y.**

24 $2^y = 32$ $y = 5$ **25** $8^y = 2$ $y = \frac{1}{3}$ **26** $27^y = 9$ $y = \frac{2}{3}$ **27** $x^y = \sqrt[5]{x^3}$ $y = \frac{3}{5}$

P,R **28** Find the value of n for which the distance from the point A(n, 5) to the point B(1, −3) is 17. 16 or −14

P,R **In problems 29–31, determine side length x for each triangle.**

29 22

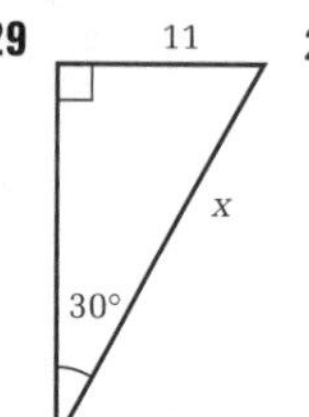

30 $\frac{10\sqrt{3}}{3}$

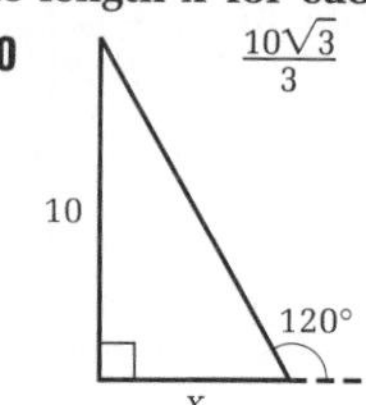

31 $6\sqrt{2}$

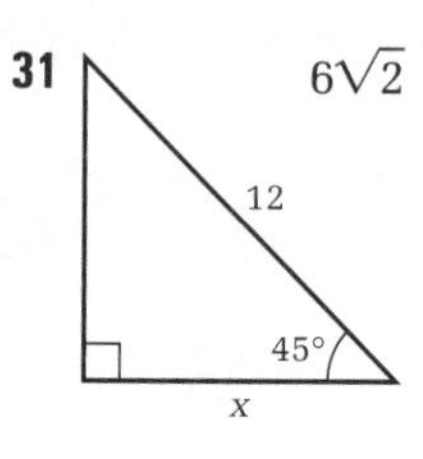

P **32** Solve $\sqrt{x^2 + 1} = x - 3$ for x. No solution

P,R **33** Determine the height of an equilateral triangle that has a perimeter of 18 centimeters. $3\sqrt{3}$ cm

R **34** Determine the perimeter of the square shown. $30\sqrt{2}$ in.

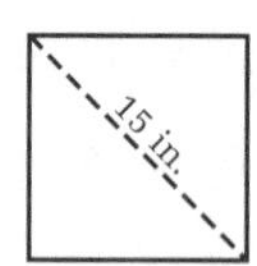

P **35** Solve $(\sqrt{n+1} - 5)(\sqrt{n+1} - 7) = 0$ for n. $n = 24$ or $n = 48$

P,R  **36** If the volume of the box shown is 66 cubic centimeters, what is the area of the right face to the nearest square centimeter? ≈ 10 cm^2

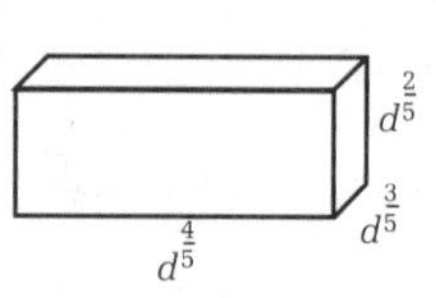

R **37** Estimate the value of $\frac{50}{\sqrt{48}}$ without using a calculator. ≈ 7

R **38** Write each expression in standard radical form.

a $\sqrt{16x^{16}}$ $4x^8$ **b** $\sqrt{25x^{25}}$ $5x^{12}\sqrt{x}$ **c** $\sqrt{32x^{12}}$ $4x^6\sqrt{2}$ **d** $\sqrt{24x^{10}y^7}$ $2|x^5y^3|\sqrt{6y}$

P,R **39** Determine, to the nearest hundredth, the length of the hypotenuse of the triangle. ≈ 6.18

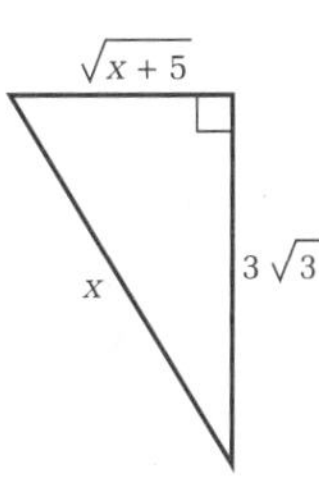

B

P,R **40** Find the geometric mean between $5\sqrt{2}+1$ and $5\sqrt{2}-1$. (Recall that the geometric mean between two positive numbers a and b is equal to $\pm\sqrt{ab}$.) ± 7

P,R **41** Explain why $(\sqrt[4]{x})^2 = \sqrt{x}$. $(\sqrt[4]{x})^2 = x^{\frac{2}{4}} = x^{\frac{1}{2}} = \sqrt{x}$

P,R **42** Use the distance formula to prove that the points $(-2, 3)$, $(0, 6)$, and $(3, 4)$ are vertices of a right triangle.

P,R,I **43** On your calculator is a key labeled log. To apply the log function to a numerical input, enter the number and then press the log key.

a Evaluate log 100. 2
b Evaluate log 10^2. 2
c Evaluate log 10^5. 5
d Evaluate log $\sqrt{10}$. 0.5
e What do you suppose *log* means?
The power of 10 that is equivalent to the input value

R **44** If the area of the rectangle shown is 15 square microns, what are the dimensions of the rectangle?
$\approx 3.19\ \mu$ by $4.70\ \mu$

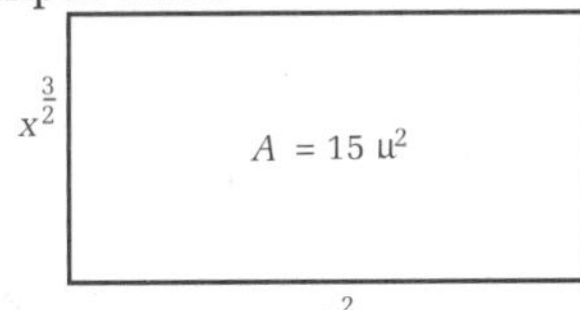

P **In problems 45–48, solve for x.**

45 $x + \sqrt{x} = 30$ $x = 25$

46 $x - 8\sqrt{x} + 15 = 0$ $x = 25$ or $x = 9$

47 $x + 4\sqrt{x} + 3 = 0$
No real solution

48 $x = 5\sqrt{x} + 24$
$x = 64$

P,R **49** Evaluate $\left(12^{\frac{1}{4}} + 6^{\frac{1}{4}}\right)\left(12^{\frac{1}{4}} - 6^{\frac{1}{4}}\right)\left(12^{\frac{1}{2}} + 6^{\frac{1}{2}}\right)$. 6

P,R **50** Consider the equation $S = 1 + 3 + 3^2 + 3^3 + \ldots + 3^{10}$.

a What equation do you obtain if you multiply each side of this equation by 3? $3S = 3 + 3^2 + 3^3 + 3^4 + \ldots + 3^{11}$
b What is the result of subtracting the given equation from the equation you obtained in part **a**? $2S = -1 + 3^{11}$
c Use your answer to part **b** to solve for S. $S = 88{,}573$

Problem-Set Notes and Strategies, continued

- One way to solve problem **39** is to use the Pythagorean Theorem and the method of squaring equations twice.
- The properties of exponents are used to solve problem **41.** All properties of exponents hold for rational exponents.
- Logarithms are previewed in problem **43.** Students should understand that this process is the opposite of finding powers.

Additional Answer

42

$$\sqrt{(-2-0)^2 + (3-6)^2} = \sqrt{13}$$
$$\sqrt{(0-3)^2 + (6-4)^2} = \sqrt{13}$$
$$\sqrt{[3-(-2)]^2 + (4-3)^2} = \sqrt{26}$$

Since $(\sqrt{13})^2 + (\sqrt{13})^2 = (\sqrt{26})^2$, the triangle determined by the three points is a right triangle.

- Geometric series are previewed in problem **50.** Chapter **13** will examine geometric series in greater detail and utilize this type of argument to prove formulas.

Problem-Set Notes and Strategies, continued

- In problem **51**, factor GCF first and then factor the remaining expression.
- In problem **53**, make sure students recognize that $(x-4)$ can never equal zero because it is in the denominator of a fraction.
- The properties of exponents are used in problem **55**. Since $15 = \frac{(6)(5)}{2}$, the exponent must be $a + b - 1$.

Cooperative Learning

Problem **56** involves formulas that are significant to triangles. Consider the right triangle with sides of length $a = 3$, $b = 4$ and $c = 5$. Have students work in small groups to answer the following questions.

1. Calculate the area of the triangle.
2. Calculate the perimeter of the triangle.
3. Calculate the semiperimeter of the triangle.
4. Use the value of the semiperimeter of the triangle for s in the formula for A given in problem **56**. What did you discover?
5. Generate another triangle that you know the area of and check the formula for A.
6. What do you think of the formula for A? See if you can generalize the formula for R as well.

Problem Set, *continued*

P **51** Solve for x.

a $x^{\frac{2}{3}} + x^{\frac{1}{3}} - 12 = 0$
$x = -64$ or $x = 27$

b $-\sqrt{x^2 + 9x + 3} = x + 5$
$x = -22$

P,R **52** Factor $x^{2n+1} + 5x^{n+1} + 4x$. $x(x^n + 4)(x^n + 1)$

C

In problems 53 and 54, solve for x.

P **53** $(x - 4)^{-2}(x + 2)^3(x - 3)^2 = 0$
$x = -2$ or $x = 3$

54 $15x^{-2} - 16x^{-1} = 15$
$x = \frac{3}{5}$ or $x = -\frac{5}{3}$

P,R **55** Suppose that $2^a = 5$ and $2^b = 6$. Determine, in terms of a and b, the value of x for which $2^x = 15$. $x = a + b - 1$

P,R,I **56** Suppose that $R = \sqrt{\frac{(s-a)(s-b)(s-c)}{s}}$ and $A = \sqrt{s(s-a)(s-b)(s-c)}$.

a What is the value of A if $R = 3$ and $s = 18$? 54

b What is the value of R if $A = 24$ and $s^2 = 144$? 2

HISTORICAL SNAPSHOT

When Is a Number Not Real?

The complex number system

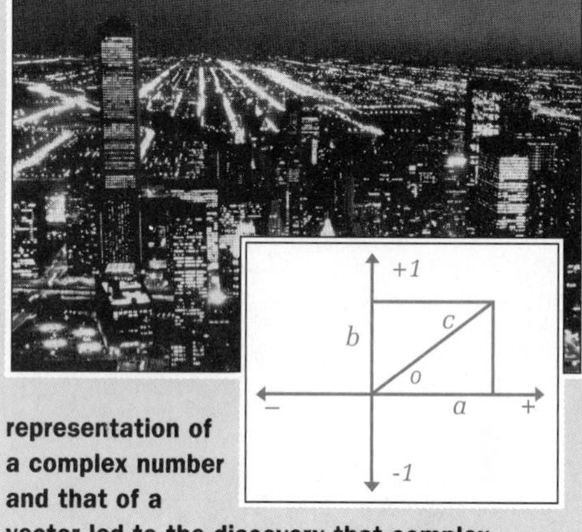

Why do we bother with complex numbers, since they are based on an imaginary quantity, the square root of −1? If mathematics arose out of real needs—to count herds, measure fields, and keep accounts—then what factors created a need for complex numbers?

While the earliest need for these numbers is unclear, we do know that around A.D. 100, Hero of Alexandria identified numbers that were the even roots of negative numbers. Girolamo Cardano (1501–1576) included calculations using imaginary numbers in his *Ars Magna* (Great Art). Leonhard Euler (1707–1783) gave us the notation *i* for the square root of −1 used in representing complex numbers.

In 1806, the Swiss mathematician Jean Robert Argand developed a means of graphically representing complex numbers. In the Argand diagram, one axis represents the imaginary component of a number, and the other axis, the real component. The similarity between the representation of a complex number and that of a vector led to the discovery that complex numbers could be used to solve problems related to alternating electrical current. The discovery and use of alternating current has made possible most of our present-day uses of electricity, in powering everything from household appliances to sound systems, from lighting to food processors and computers.

8.3

8.3 INTRODUCTION TO COMPLEX NUMBERS

Objectives

After studying this section, you will be able to

- Identify pure imaginary numbers
- Perform operations on pure imaginary numbers
- Identify complex numbers
- Determine the conjugate of a complex number

Part One: Introduction

Pure Imaginary Numbers

Quadratic equations such as $x^2 - 6x - 16 = 0$ have two distinct real roots—in this case, $x = 8$ and $x = -2$. Other quadratics, such as $x^2 - 4x + 4 = 0$, have two equal real roots—in this case, $x = 2$. Some quadratic equations, however, have no real roots.

Example *Solve the following equations for x.*

a $x^2 - 6x + 16 = 0$ **b** $x^2 + 4x + 5 = 0$

Let's use two different techniques to solve these equations.

The Quadratic Formula

$$x^2 - 6x + 16 = 0$$
$$x = \frac{6 \pm \sqrt{36 - 64}}{2}$$
$$x = \frac{6 \pm \sqrt{-28}}{2}$$

Complete the Square

$$x^2 + 4x + 5 = 0$$
$$x^2 + 4x + 4 = -5 + 4$$
$$(x + 2)^2 = -1$$
$$x + 2 = \pm \sqrt{-1}$$
$$x = -2 \pm \sqrt{-1}$$

No real number has a negative square, so the square roots of numbers such as -28 and -1 are not real numbers. These equations have no real roots.

In order to work with the solutions of equations such as those in the preceding example, mathematicians use the symbol i to represent the nonreal quantity $\sqrt{-1}$. Since $i = \sqrt{-1}$, we can say that $i^2 = -1$ and define the nonreal number i such that $i = \sqrt{-1}$ and $i^2 = -1$. If a is a positive real number, then $\sqrt{-a} = i\sqrt{a}$.

Class Planning

Time Schedule

Average: 2 days
Advanced: 2 days
Honors: 1 day

Resource References

Teacher's Resource Book
Class Opener 50
Cooperative Learning 24
Practice 53
Enrichment 15
Evaluation
Quiz Forms 1, 2, 3

Class Opener

Solve the equation.
$x^2 + 6x + 27 = 0$

$x = \frac{-6 \pm \sqrt{-72}}{2}$ or $\frac{-6 \pm 6\sqrt{-2}}{2}$ or $-3 \pm 3\sqrt{-2}$

The answer is impossible (until the introduction of i).

Lesson Notes

- Complex and imaginary numbers are introduced in this section. This is a good opportunity to continue the discussion of René DesCartes, one of the predominant forces in the development of complex numbers.
- Imaginary numbers are defined to handle the problem of taking the even roots of negative numbers.

Vocabulary

pure imaginary number
complex number
real part
imaginary part
conjugate

8.3

Lesson Notes, continued

- It was only after careful study that mathematicians discovered these numbers held many of the same properties of the real numbers.
- Pure imaginary and real numbers are both subsets of a larger set, the complex numbers. Emphasize that every real number is complex and every imaginary number is also complex. Polynomials are now solvable in the field of complex numbers because we can take square roots of negative numbers.
- Some of the properties of the complex numbers are presented in this section and the next section. Simple powers of i are used in this section and defined for all powers in the next section.

Stumbling Block

Order of operations must be revised a little when dealing with complex numbers. Numbers such as $\sqrt{-2}$ must first be simplified to $\sqrt{2}i$ before any computations are performed; otherwise discrepancies may occur in the final answer. Students often forget this order and do the algebra first before changing to complex form.

Communicating Mathematics

Have students write a paragraph to explain the significance of the complex numbers in solving equations.

Using these definitions, we can evaluate expressions that contain square roots of negative numbers. For example,

$$\sqrt{-4} = i\sqrt{4} = 2i \qquad \sqrt{-100} = i\sqrt{100} = 10i$$

$$(2i)^2 = -4 \qquad (10i)^2 = -100$$

The Arithmetic of Pure Imaginary Numbers

Numbers of the form ki, where k is real and nonzero—such as $6i$, $\sqrt{5}i$, and $-12.5i$—form a set of nonreal numbers known as the ***pure imaginary numbers.*** Do the properties of real numbers also apply to pure imaginary numbers? Let's look at the property stating that $\sqrt{x}\sqrt{y} = \sqrt{xy}$, where $x \geq 0$ and $y \geq 0$, and see whether it applies to roots of negative as well as positive real numbers.

If this property holds for negative radicands, then

$$\sqrt{-9}\sqrt{-4} = \sqrt{-9(-4)} = \sqrt{36} = 6$$

but by definition,

$$\sqrt{-9}\sqrt{-4} = 3i \cdot 2i = 6i^2 = -6$$

This is one property of radicals that does not extend to radical expressions that represent pure imaginary numbers. In fact, neither do the other properties of radicals. It is therefore important to rewrite such expressions in the form ki before performing operations.

Example *Simplify each expression.*

a $\sqrt{-2}\sqrt{-8}$

$$\begin{aligned}&\sqrt{-2}\sqrt{-8}\\ &= \sqrt{2}\sqrt{-1}\sqrt{8}\sqrt{-1}\\ &= \sqrt{2}i \cdot \sqrt{8}i\\ &= \sqrt{16}i^2\\ &= 4(-1)\\ &= -4\end{aligned}$$

b $\sqrt{\dfrac{-4}{9}}$

$$\begin{aligned}&\sqrt{\frac{-4}{9}}\\ &= \sqrt{\frac{4}{9}}\sqrt{-1}\\ &= \frac{2}{3}i\end{aligned}$$

c $\sqrt{\dfrac{24}{-25}}$

$$\begin{aligned}&\sqrt{\frac{24}{-25}}\\ &= \sqrt{\frac{24}{25}}\sqrt{-1}\\ &= \frac{2\sqrt{6}}{5}i\end{aligned}$$

Example *Simplify each expression.*

a $2i + 5i - 9i$

$$\begin{aligned}&2i + 5i - 9i\\ &= -2i\end{aligned}$$

b $(5i)(-5i)$

$$\begin{aligned}&(5i)(-5i)\\ &= -25i^2\\ &= -25(-1)\\ &= 25\end{aligned}$$

c $\sqrt{-16} - \sqrt{-9}$

$$\begin{aligned}&\sqrt{-16} - \sqrt{-9}\\ &= i\sqrt{16} - i\sqrt{9}\\ &= 4i - 3i\\ &= i\end{aligned}$$

Complex Numbers

Let's look again at the two quadratic equations with nonreal roots we saw at the beginning of this section. We can now simplify the roots of these equations.

$$x^2 - 6x + 16 = 0$$

$$x = \frac{6 \pm \sqrt{-28}}{2} = \frac{6 \pm \sqrt{28}i}{2} = \frac{6 \pm 2\sqrt{7}i}{2} = 3 \pm \sqrt{7}i$$

$$x^2 + 4x + 5 = 0$$

$$x = -2 \pm \sqrt{-1} = -2 \pm i$$

These roots are examples of ***complex numbers.*** A number that can be expressed in the form $a + bi$, where $i = \sqrt{-1}$ and a and b are real numbers, is called a ***complex number.***

The term a in a complex number $a + bi$ is known as the number's ***real part,*** and the term bi is known as the number's ***imaginary part.*** Both the real numbers and the pure imaginary numbers are subsets of the complex-number system. For example, we can think of 7 as $7 + 0i$, and we can think of $3i$ as $0 + 3i$.

Complex Conjugates

For every complex number there exists another complex number called its ***conjugate.*** The ***conjugate*** of a complex number $a + bi$ is $a - bi$.

Complex Number	Real Part	Imaginary Part	Conjugate
$-6 + 2i$	-6	$2i$	$-6 - 2i$
$-4i$	0	$-4i$	$4i$
10	10	$0i$	10
$5 - 2i$	5	$-2i$	$5 + 2i$
$3i - 7$	-7	$3i$	$-7 - 3i$

Conjugates play an important role in the arithmetic of complex numbers, as you will see in the next section.

Cooperative Learning

Have students work in small groups to explore the number i. Ask students to determine whether or not the trichotomy principle of real numbers also holds for the complex numbers. Then ask them to investigate whether the imaginary numbers have order. Which is bigger $3i$ or $7i$?

Checkpoint

1. Simplify i^3. $-i$
2. Factor $4x^2 + 9y^2$ $(2x + 3yi)(2x - 3yi)$
3. Simplify the expression $-\sqrt{-100}$. $-10i$
4. Multiply the numbers $3i, 2i, -4i$ and i. -24

Part Two: Sample Problems

Problem 1 Simplify $\frac{2}{3i}$.

Solution Although this expression is in a simple form, pure imaginary numbers are usually written in the form ki, where k is a nonzero real number. We can rewrite $\frac{2}{3i}$ in this form by multiplying both the numerator and the denominator by i.

$$\frac{2}{3i} = \frac{2}{3i} \cdot \frac{i}{i} = \frac{2i}{3i^2} = \frac{2i}{3(-1)} = -\frac{2}{3}i$$

Assignment Guide

Average

Day 1 6–13, 16, 18, 20, 24

Day 2 15, 17, 19, 21–23, 25, 26, 28, 29, 33, 35, 36

Advanced

Day 1 6, 10–13, 15–17, 19–21

Day 2 7, 9, 18, 22–25, 27, 28, 30, 33, 35, 37, 39

Honors

7, 9–11, 13, 15–17, 19, 20, 23, 24, 27, 29, 31, 32, 34, 36, 38, 39, 43

Problem-Set Notes and Strategies

- Problem 2 is good practice in changing the square root of a negative number to a complex number.

Problem 2 *Factor $x^2 + 36$ over the complex numbers.*

Solution Since we can now interpret square roots of negative numbers, we can rewrite this sum of two squares as a difference of two squares and factor it accordingly.

$$\begin{aligned} & x^2 + 36 \\ &= x^2 - (-36) \\ &= x^2 - (6i)^2 \\ &= (x + 6i)(x - 6i) \end{aligned}$$

Problem 3 *Solve $x^2 + 25 = 0$ for x.*

Solution

$$x^2 + 25 = 0$$
$$x^2 = -25$$

Take the square root of each side $\quad x = \pm\sqrt{-25}$

$$x = \pm 5i$$

The solutions are $x = 5i$ and $x = -5i$. We also could have solved this equation by factoring $x^2 + 25$ into the expression $(x + 5i)(x - 5i)$ and using the Zero Product Property to determine that $x = \pm 5i$.

Part Three: Exercises and Problems

Warm-up Exercises

P **1** Copy and complete the following table.

Complex Number	Real Part	Imaginary Part	Conjugate
2 + 5i	2	5i	2 − 5i
3 − 4i	3	−4i	3 + 4i
−2 + 7i	−2	7i	−2 − 7i
5 − 4i	5	−4i	5 + 4i
9i + 3	3	9i	3 − 9i
−6 − 5i	−6	−5i	−6 + 5i
−4i	0	−4i	4i
16	16	0i	16

P **2** Simplify each expression.

a $\sqrt{-64}$ $8i$ **b** $-\sqrt{-25}$ $-5i$ **c** $-\sqrt{-4} + \sqrt{-9}$ i

d $\sqrt{\frac{-9}{25}}$ $\frac{3}{5}i$ **e** $\sqrt{-12}$ $2\sqrt{3}i$ **f** $\sqrt{-3}\sqrt{-3}$ -3

g $\sqrt{\frac{-1}{4}}$ $\frac{1}{2}i$ **h** $3i \cdot 5i$ -15 **i** $(7i)^2$ -49

P,I **3** Does calling the number i imaginary mean that such a number does not exist? No

P | **4** Simplify each expression.

a $\frac{1}{i}$ $-i$ **b** $\frac{-1}{i}$ i **c** $\frac{2}{i}$ $-2i$ **d** $\frac{-2}{i}$ $2i$

P,R | **5** Multiply $\begin{bmatrix} i & 0 \\ 0 & i \end{bmatrix}\begin{bmatrix} i & 0 \\ 0 & i \end{bmatrix}$. $\begin{bmatrix} -1 & 0 \\ 0 & -1 \end{bmatrix}$

Problem Set

P **A** | **6** Simplify each expression.

a $-\sqrt{25}$ -5 **b** $\sqrt{-20}$ $2\sqrt{5}i$ **c** $-\sqrt{-40}$ $-2\sqrt{10}i$ **d** $2\sqrt{-36}$ $12i$

P | **7** Rewrite each expression in the form ki.

a $\frac{4}{2i}$ $-2i$ **b** $\frac{-9}{3i}$ $3i$ **c** $\frac{-5}{-4i}$ $-\frac{5}{4}i$

R | **8** Find the distance between O and A. $4\sqrt{2}$

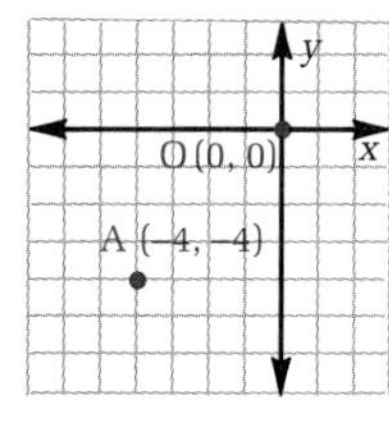

P | **9** Explain why $i^4 = 1$. $i^4 = (i^2)^2 = (-1)^2 = 1$

P | **10** Simplify each expression.

a $\sqrt{\frac{-25}{16}}$ $\frac{5}{4}i$ **b** $-\sqrt{\frac{-25}{16}}$ $-\frac{5}{4}i$ **c** $(\sqrt{-4})^2$ -4

P | **11** Find the conjugate of each number.

a $i(2i - 7)$ $-2 + 7i$ **b** $4i + 3$ $-4i + 3$ **c** $7 - i^2$ 8

P,R | **12** If a number is selected at random from the set $\{-6i^2, 8i, -4, -\sqrt{-4}, 8\sqrt{-1}i, -\sqrt{9}\}$, what is the probability that the number is a pure imaginary number? $\frac{1}{3}$

P | **13** Find each product.

a $(x + 3i)(x - 3i)$ $x^2 + 9$ **b** $(x + i)(x - i)$ $x^2 + 1$ **c** $(x + i)(x + i)$ $x^2 + 2ix - 1$

R | **14** Use the graph of $y = f(x) = x^2 - 4$ to help you sketch the graph of each of the following equations.

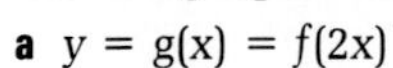

a $y = g(x) = f(2x)$
b $y = h(x) = f(x - 3)$
c $y = s(x) = f(x) + 5$
d $y = t(x) = \frac{1}{2}[f(x)] + 1$

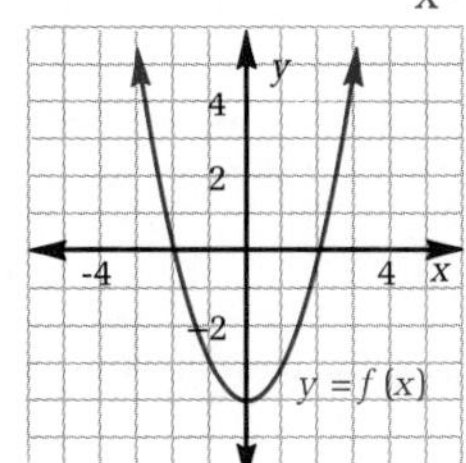

Problem-Set Notes and Strategies, continued

- Refer to problem **5**. Matrix operations with complex numbers are the same as matrix operations with real numbers.
- Problem **7** involves rationalizing the denominator as if the denominator contained a radical. Remember that *i* is the square root of –1 and is to be treated as a square root.
- Problem **9** has a significant result and will be explored in the next section on the algebra of complex numbers.
- Problem **11c** illustrates the importance of simplifying first.

Additional Answer

14a

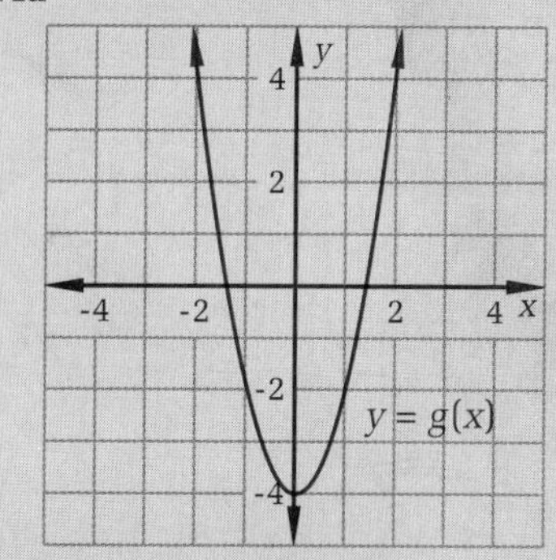

14b

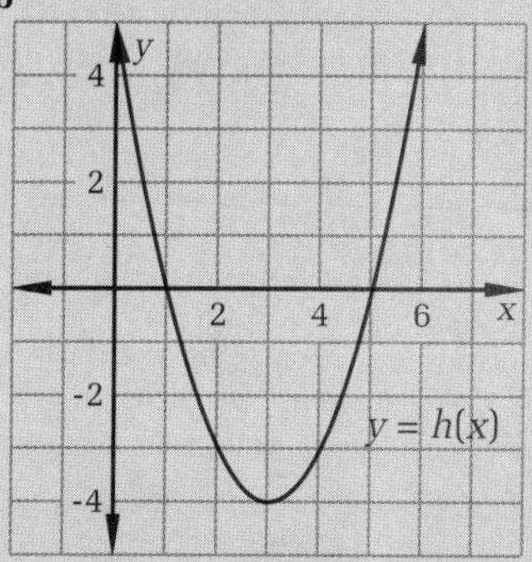

Answers for problems **14c** and **d** can be found in the answer pages beginning on **A1**.

Problem-Set Notes and Strategies, continued

- In problem **18**, both had incorrect methods, but Marianne had the correct answer. Always change the number to a complex form first and then do the algebra.
- Even when factoring in the complex numbers, the key word is *completely* and the GCF is factored out first. Be careful of this when factoring problem **22**.

Additional Answer

18 Ezra was right, since the property stating that

$$\frac{\sqrt{x}}{\sqrt{y}} = \sqrt{\frac{x}{y}}$$

does not apply to roots with negative radicands. The work should read

$$\sqrt{-\tfrac{1}{9}} = \sqrt{\tfrac{1}{9}}\,\sqrt{-1} = \tfrac{1}{3}i.$$

Problem Set, *continued*

P,R **15** Factor each expression over the complex numbers.

a $y^2 + 4$ $(y + 2i)(y - 2i)$
b $9 + 4x^2$ $(3i + 2x)(-3i + 2x)$

P **16** Solve for x.

a $x^2 = 25$ $x = \pm 5$
b $x^2 = -36$ $x = \pm 6i$
c $x^2 - 49 = 0$ $x = \pm 7$
d $x^2 + 49 = 0$ $x = \pm 7i$

P **17** Simplify each expression.

a $4\sqrt{-3}\sqrt{-12}$ -24
b $-2i + 6i - 7i$ $-3i$
c $-(\sqrt{-3})^2$ 3
d $(\sqrt{-8})^2$ -8
e $2i(4i + 3\sqrt{2}i)$ $-8 - 6\sqrt{2}$
f $\sqrt{2}(\sqrt{-8} - \sqrt{-27})$ $(4 - 3\sqrt{6})i$

P **18** Tom said that $\sqrt{-\frac{1}{9}} = \frac{\sqrt{1}}{\sqrt{-9}} = \frac{1}{3i} = \frac{1}{3i} \cdot \frac{i}{i} = \frac{i}{-3}$.

Marianne said that $\sqrt{-\frac{1}{9}} = \frac{\sqrt{-1}}{\sqrt{9}} = \frac{i}{3}$. Ezra said that both Tom and Marianne had done the work wrong. Who was right, and why?

P,R **19** Multiply $\begin{bmatrix} 2i & 3i \\ -i & 4i \end{bmatrix}\begin{bmatrix} i & 0 \\ 0 & i \end{bmatrix} \cdot \begin{bmatrix} -2 & -3 \\ 1 & -4 \end{bmatrix}$

R **20** If one of the following equations is selected at random, what is the probability that its graph is a hyperbola? $\frac{5}{6}$

a $9x^2 - 4y^2 = 36$
b $\left(\frac{x}{3}\right)^2 = \left(\frac{y}{2}\right)^2 + 1$
c $y^2 - \frac{x^2}{9} = 1$
d $x^2 - y^2 = 16$
e $(x - y)(x + y) = 25$
f $\frac{x^2}{4} + \frac{y^2}{9} = 1$

P **21** Match each number with an equivalent expression.

a 4 ii
b $4i$ iv
c -4 i
d $-4i$ iii

i $(-2i)^2$
ii $-(2i)^2$
iii $\frac{12}{3i}$
iv $\sqrt{-16}$

P,R **22** Factor each expression over the complex numbers.

a $x^2 - 25$ $(x + 5)(x - 5)$
b $x^2 + 25$ $(x + 5i)(x - 5i)$
c $4x^2 - 49$ $(2x + 7)(2x - 7)$
d $4x^2 + 49$ $(2x + 7i)(2x - 7i)$
e $4x^2 - 16$ $4(x + 2)(x - 2)$
f $4x^2 + 16$ $4(x + 2i)(x - 2i)$

B R **23** What, in terms of a, is the volume of the cone produced by rotating the shaded triangle about the x-axis?
$\frac{1}{3}\pi a^7$

y
$(0, a^2)$
$(a^3, 0)$
x

P **24** Explain why $i^{-1} = -i$. $i^{-1} = \frac{1}{i} = \frac{1}{i} \cdot \frac{i}{i} = \frac{i}{i^2} = \frac{i}{-1} = -i$

P **25** Simplify each expression.
a $\sqrt{-0.04} + \sqrt{-0.09}$ $0.5i$
b $\sqrt{-6.25}$ $2.5i$
c $6i(-6i)$ 36
d $2i(3i - \sqrt{2}i)$ $-6 + 2\sqrt{2}$

P **In problems 26–30, solve for x.**
26 $x^2 + 7 = 0$ $x = \pm\sqrt{7}i$
27 $x^2 + 12 = 0$ $x = \pm 2\sqrt{3}i$
28 $4x^2 + 4x = 0$ $x = 0$ or $x = -1$
29 $x^2 + 8x + 17 = 0$ $x = -4 \pm i$
30 $x^2 - 10x + 29 = 0$ $x = 5 \pm 2i$

P **31** Find a complex number $x + yi$ such that the value of x is twice the value of y and $2x - 3y = 5$. $10 + 5i$

R **32** Let $f(x) = \frac{x(x - 5)(x + 6)}{x(x - 4)(x - 5)(x + 1)}$.
a What are the equations of the vertical asymptotes of the graph of f? $x = -1$ and $x = 4$
b What are the x-coordinates of the holes in the graph? 0 and 5

P **In problems 33–37, solve for x.**
33 $-2x = 16i$ $x = -8i$
34 $-2ix = 16i$ $x = -8$
35 $ix = 12$ $x = -12i$
36 $3i + 12x = 4x - 15i$ $x = -\frac{9}{4}i$
37 $12 - 3ix = 2ix - 8$ $x = -4i$

P,R **38** Solve each system for (x, y).
a $\begin{cases} x + 2y = 3i \\ x - 2y = i \end{cases}$ $(x, y) = \left(2i, \frac{1}{2}i\right)$
b $\begin{cases} 2x + yi = 4 \\ 3x - yi = 1 \end{cases}$ $(x, y) = (1, -2i)$

P **39** **a** Write an equation with roots $10i$ and $-10i$. $x^2 + 100 = 0$
b Write an equation with roots 0, $5i$, and $-5i$. $x^3 + 25x = 0$

I **40** Find the sum and the product of the roots of $x^2 - 8x + 9 = 0$. 8; 9

R **41** The area of an ellipse is equal to πab, where a is half the length of the major axis and b is half the length of the minor axis. Find the areas of the base and the elliptical cross section of the cylinder shown. 16π; 20π

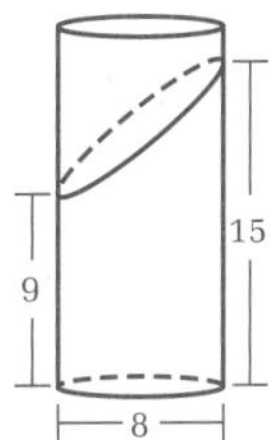

P **42** Solve $x^3 + 100x = 0$ for x. $x = 0$, $x = 10i$, or $x = -10i$

P **43** If 3 is a root of $x^3 - 3x^2 + 4x - 12 = 0$, what are the other two roots? $x = 2i$ and $x = -2i$

C

P **In problems 44 and 45, solve for x.**
44 $x^4 + 13x^2 + 36 = 0$ $x = \pm 3i$ or $x = \pm 2i$
45 $(3^x)^2 - 12(3^x) + 27 = 0$ $x = 2$ or $x = 1$

Problem-Set Notes and Strategies, continued

- Notice how complex roots occur in conjugate pairs and yield real polynomials in problem **39**.
- You might want to refer to the cooperative learning exercise in Section **7.1** to help answer problem **40**. This concept was originally explored in problem **36** of Section **7.1**.
- In problem **41**, one approach is to use the Pythagorean Theorem and draw the appropriate picture to find the major and minor axes.
- Problem **43** can be factored over the complex numbers after factoring out $x - 3$.
- Problem **45** is a quadratic-like equation with exponential terms as the quadratic variable.

8.4

Class Planning

Time Schedule
All levels: 2 days

Resource References
Teacher's Resource Book
Class Opener 51
Cooperative Learning 25
Practice 54
Enrichment 16
Transparency 15

Class Opener

Find the product. $(3 + 2i)(4 - 3i)$
$18 - i$

Lesson Notes

- A complex number is a number composed of two parts, a real part and an imaginary part. The operations on complex numbers are similar to operations on polynomials, except for division. To add or subtract polynomials we added or subtracted like terms. To add or subtract complex numbers, add or subtract the like parts.

8.4 Algebra of Complex Numbers

$(3 + 2i)(2 - i) = 8 + i$

Objectives

After studying this section, you will be able to
- Identify equal complex numbers
- Perform operations on complex numbers

Part One: Introduction

Equal Complex Numbers

What does it mean to say that two complex numbers are equal? The complex numbers $a + bi$ and $c + di$ (where a, b, c, and d are real) are equal if and only if their real parts are equal ($a = c$) and their imaginary parts are equal ($bi = di$).

One reason for expressing complex numbers in the form $a + bi$, known as ***standard complex form***, is to make comparison easier. Thus, $2 + 7i = \frac{6}{3} + \sqrt{49}i$, since $2 = \frac{6}{3}$ and $7i = \sqrt{49}i$. However, $2 + 7i \neq 7 + 2i$ because $2 \neq 7$ and $7i \neq 2i$.

Example *Find a pair of real numbers* (x, y) *for which* $6y + 2xi = -30 + 18i$.

Each side of the equation represents a complex number. We can use the definition of equal complex numbers to solve for (x, y).

Equal Real Parts	**Equal Imaginary Parts**
$6y = -30$	$2x = 18$
$y = -5$	$x = 9$

Thus, $(x, y) = (9, -5)$.

The Arithmetic of Complex Numbers

Arithmetic operations can be performed on complex numbers in much the same way that they are performed on real monomials and binomials.

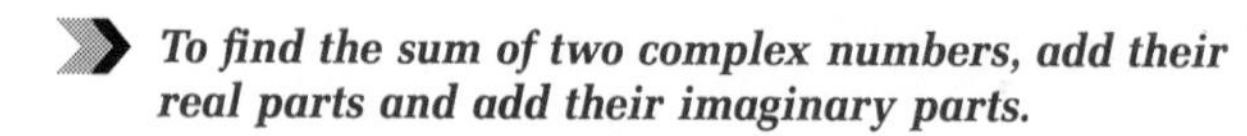

To find the sum of two complex numbers, add their real parts and add their imaginary parts.

For example, $(3 + 7i) + (-8 + 4i) = -5 + 11i$

Vocabulary
standard complex form

To find the difference of two complex numbers, subtract their real parts and subtract their imaginary parts.

For example, $(2 - 6i) - (4 - 9i) = 2 - 6i - 4 + 9i$
$= -2 + 3i$

To find the product of two complex numbers, multiply them as you would monomials or binomials.

For example,
$$\begin{aligned}(2 + 3i)(4 - 5i) &= 8 - 10i + 12i - 15i^2\\ &= 8 - 10i + 12i - 15(-1)\\ &= 8 - 10i + 12i + 15\\ &= 23 + 2i\end{aligned}$$

To find the quotient of two complex numbers, multiply the numerator and the denominator by the conjugate of the denominator.

For example,
$$\begin{aligned}\frac{2 - 3i}{3 - 4i} &= \frac{2 - 3i}{3 - 4i} \cdot \frac{3 + 4i}{3 + 4i}\\ &= \frac{6 + 8i - 9i - 12i^2}{9 + 12i - 12i - 16i^2}\\ &= \frac{18 - i}{25}\\ &= \frac{18}{25} - \frac{1}{25}i\end{aligned}$$

This method of finding a quotient is based on the fact that the product of a complex number and its conjugate is always a real number. Can you prove that this is so?

Example *Simplify each expression.*

a $2(3 - 7i) - 6(-2 + 5i)$
$$\begin{aligned}2(3 - 7i) - 6(-2 + 5i) &= 6 - 14i + 12 - 30i\\ &= 18 - 44i\end{aligned}$$

b $(-5 + 4i)(-5 - 4i)$
$$\begin{aligned}(-5 + 4i)(-5 - 4i) &= 25 + 20i - 20i - 16i^2\\ &= 25 + 20i - 20i - 16(-1)\\ &= 41\end{aligned}$$

Notice once again that the product of a complex number and its conjugate is a real number.

c i^{59}

To simplify i^{59}, let's try some simpler cases and look for a pattern.

$i^0 = 1$
$i^1 = i$
$i^2 = -1$
$i^3 = i^2i = -i$
$i^4 = i^2i^2 = -1(-1) = 1$
$i^5 = i^4i = 1i = i$
$i^6 = i^4i^2 = 1(-1) = -1$
$i^7 = i^4i^3 = 1(-i) = -i$

Lesson Notes, continued

- Multiplying complex numbers is similar to multiplying binomials. Stress that each term of the second complex number is multiplied by each term of the first.

- To divide complex numbers, rationalize the denominator.

- Powers of i follow an interesting pattern which allows us to divide the positive exponent by 4, discard the quotient, and consider only the remainder. Since the quotient will always be a factor of i^4, it will equal 1 and therefore can be discarded. Negative exponents will be considered in a cooperative learning exercise.

Communicating Mathematics

Discuss the following statement with your class: The complex numbers are closed under the operations of addition and multiplication.

Cooperative Learning

Have students work in small groups and list the powers of i beginning at zero and proceeding through ten.

1 Encourage students to draw a conclusion. Students should recognize that a pattern has evolved that repeats every four terms: $i^0 = 1$, $i^1 = i$, $i^2 = -1$, $i^3 = -i$.

2 Now have students calculate i^{-1}, i^{-2}, i^{-3}, i^{-4}, i^{-5} and i^{-6}. Ask them to generate a pattern of negative powers of i.
$i^{-1} = -i$, $i^{-2} = -1$, $i^{-3} = i$, $i^{-4} = 1$, $i^{-5} = -i$, $i^{-6} = -1$

3 Now ask them to simplify i^{-163}. After a minute or two ask them to multiply this number by i^{164} and draw a conclusion.
$i^{-163} = i$
$i^{-163} \cdot i^{164} = i \cdot 1 = i$

The powers of i generate a pattern of four repeating numbers $(1, i, -1, -i, 1, i, -1, -i, 1, i, \ldots)$ with i raised to the fourth power being equal to 1. To simplify i^{59}, we can divide 59 by 4 and use the result and the remainder to rewrite i^{59} in an easily evaluated form.

$$i^{59} = (i^4)^{14} \cdot i^3 = 1^{14}(-i) = -i$$

Part Two: Sample Problems

Problem 1 *Find the complex zeros of the function* $f(x) = x^3 + 8$.

Solution Remember, to find the zeros of a function, we set the function equal to zero and solve the resulting equation.

$$x^3 + 8 = 0$$
$$(x + 2)(x^2 - 2x + 4) = 0$$
$$x + 2 = 0 \text{ or } x^2 - 2x + 4 = 0$$
$$x = -2 \text{ or } x = \frac{2 \pm \sqrt{4 - 16}}{2}$$
$$= \frac{2 \pm \sqrt{-12}}{2}$$
$$= \frac{2 \pm \sqrt{12}\sqrt{-1}}{2}$$
$$= \frac{2 \pm 2\sqrt{3}i}{2}$$
$$= 1 \pm \sqrt{3}i$$

The zeros are -2, $1 + \sqrt{3}i$, and $1 - \sqrt{3}i$.

Problem 2 *Find a complex number x such that* $3x = 4ix - 10$.

Solution We rearrange the x terms so that they are on one side of the equation, then solve for x.

$$3x = 4ix - 10$$
$$3x - 4ix = -10$$
$$(3 - 4i)x = -10$$
$$x = \frac{-10}{3 - 4i}$$
$$= \frac{-10}{3 - 4i} \cdot \frac{3 + 4i}{3 + 4i}$$
$$= \frac{-10(3 + 4i)}{9 - 16i^2}$$
$$= \frac{-30 - 40i}{25}$$
$$= -\frac{6}{5} - \frac{8}{5}i$$

Problem 3 *Solve $x^2 - 3ix - 2 = 0$ for x.*

Solution We can use the quadratic formula, with $a = 1$, $b = -3i$, and $c = -2$.

$$x = \frac{3i \pm \sqrt{(-3i)^2 - 4(1)(-2)}}{2(1)}$$

$$= \frac{3i \pm \sqrt{9i^2 + 8}}{2}$$

$$= \frac{3i \pm \sqrt{-9 + 8}}{2}$$

$$= \frac{3i \pm \sqrt{-1}}{2}$$

$$= \frac{3i \pm i}{2}$$

Thus, $x = 2i$ or $x = i$. We could also have solved this problem by using the Zero Product Property, rewriting $x^2 - 3ix - 2$ as $x^2 - 3ix + 2i^2$, which can be factored into $(x - 2i)(x - i)$.

Problem 4 *Write an equation whose roots are $2 + i$ and $2 - i$.*

Solution For an equation of the form $ax^2 + bx + c = 0$ to have roots of $2 + i$ and $2 - i$, the polynomial on the left side must have factors of $x - (2 + i)$ and $x - (2 - i)$. We can therefore set the product of these two factors equal to zero and then multiply and simplify to find such an equation.

$$[x - (2 + i)][x - (2 - i)] = 0$$
$$[(x - 2) - i][(x - 2) + i] = 0$$
$$(x - 2)^2 - i^2 = 0$$
$$x^2 - 4x + 4 - i^2 = 0$$
$$x^2 - 4x + 4 - (-1) = 0$$
$$x^2 - 4x + 5 = 0$$

Checkpoint

1. Add the two numbers $3 + 4i$ and $2 - 3i$. $5 + i$
2. Write the expression $(3 + i)i$ in standard form. $-1 + 3i$
3. Simplify the expression i^{35}. $-i$

Assignment Guide

Average	
Day 1	7–9, 11, 12, 15, 16, 20, 22
Day 2	6, 10, 13, 14, 17–19, 21, 23, 24, 26, 28, 29, 32, 33
Advanced	
Day 1	6c,e, 7–10, 12, 13, 15, 17–19, 21–23
Day 2	20, 24–26, 27a,b, 28, 29, 32, 33, 35, 36a,b, 42
Honors	
Day 1	6, 9, 13, 14, 17, 21, 24, 26, 27b, 28, 34
Day 2	30, 32, 33, 35–39, 41, 43–45, 47

Part Three: Exercises and Problems

Warm-up Exercises

P,R

1 Copy and complete the following table.

Complex Number	a + bi Form	Real Part	Imaginary Part	Conjugate
$-5i - 7$	$-7 - 5i$	-7	$-5i$	$-7 + 5i$
$19i$	$0 + 19i$	0	$19i$	$-19i$
$2(5 + 3i)$	$10 + 6i$	10	$6i$	$10 - 6i$
27	$27 + 0i$	27	$0i$	27
$\sqrt{3} + \sqrt{-3}$	$\sqrt{3} + \sqrt{3}i$	$\sqrt{3}$	$\sqrt{3}i$	$\sqrt{3} - \sqrt{3}i$

P,R **2** Factor each expression over the complex numbers.

a $x^2 + 121$ $(x + 11i)(x - 11i)$ **b** $x^2 - 121$ $(x + 11)(x - 11)$
c $9x^2 + 4$ $(3x + 2i)(3x - 2i)$ **d** $9x^2 + 36$ $9(x + 2i)(x - 2i)$

P,R **3** Solve each equation for x.

a $ix = 3i$ $x = 3$ **b** $ix = 3$ $x = -3i$

P **4** Solve each equation for (x, y).

a $x + iy = 5 + 7i$ $(x, y) = (5, 7)$ **b** $x + iy = (6 + 7i)(9 - 4i)$ $(x, y) = (82, 39)$
c $x + yi = -2(3 - 4i)$ $(x, y) = (-6, 8)$ **d** $x + yi = (2 - i)(2 + i)$ $(x, y) = (5, 0)$

P,R **5** Write a quadratic equation with the given roots.

a 7 and −5 $x^2 - 2x - 35 = 0$ **b** $5i$ and $-5i$ $x^2 + 25 = 0$

Problem Set

Problem-Set Notes and Strategies

- Problem **6** is a good exercise in the operations and simplification of complex numbers.
- In problem **10** coin tossing is an example of the binomial probability distribution. This is an important discrete distribution and is discussed in detail in Section **11.6**.

P **A** **6** Rewrite each expression as a number in standard complex form.

a $\frac{5}{2i}$ $0 - \frac{5}{2}i$ **b** $\frac{2}{3 - i}$ $\frac{3}{5} + \frac{1}{5}i$ **c** $i^{10} + i^6 + i^5 - i^3$ $-2 + 2i$
d $(4 - 5i)(2 - i)$ $3 - 14i$ **e** $3i(2 - 4i)$ $12 + 6i$

P,R **7** Evaluate each expression.

a i^4 1 **b** i^{24} 1 **c** i^{37} i **d** i^{803} $-i$

P **8** Perform the indicated operations.

a $(8 + 7i) + (4 - 2i)$ $12 + 5i$ **b** $(3 - 6i) - (8 - 2i)$ $-5 - 4i$
c $(3 + 4i)(3 - 4i)$ 25 **d** $\frac{2 + 3i}{3 + 2i}$ $\frac{12}{13} + \frac{5}{13}i$

P **9** Simplify each expression.

a $2i^3 - 3i^9 + 4i^{15}$ $-9i$ **b** $(2i)^3 - (5i)$ $-13i$ **c** $\frac{i^5 - i}{-i^3}$ 0

R,I **10** The graph shows the results of an experiment in which five coins were tossed a number of times.

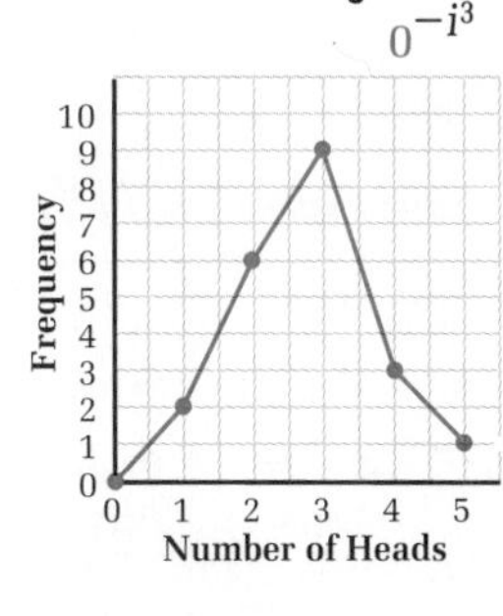

a How many heads were recorded all together? 58
b In what percentage of the trials were more than two heads recorded? ≈62%
c In what percentage of the trials was at least one tail recorded? ≈95%

P,R **11** Simplify each expression.

a $\sqrt{\frac{-16}{9}}$ $\frac{4}{3}i$ **b** $\sqrt{\frac{16}{-9}}$ $\frac{4}{3}i$ **c** $\sqrt{-\frac{16}{9}}$ $\frac{4}{3}i$

R | **12** Solve each equation for x by factoring.

a $4x^2 + 9 = 0$ $x = \pm\frac{3}{2}i$ **b** $x^2 + 49 = 0$ $x = \pm 7i$

P,R | **13** Solve the system for (x, y). $(x, y) = \left(2 + 3i, -\frac{2}{3} + i\right)$

$$\begin{cases} x + 3y = 6i \\ x - 3y = 4 \end{cases}$$

P | **In problems 14–16, simplify each expression.**

14 $(3i)^2 - (4i)^2 - 6i^2$ 13

15 $3^2i^2 + (2i)^2$ -13

16 $\sqrt{-8}\sqrt{-3}$ $-2\sqrt{6}$

P,R | **17 a** What is the discriminant of $x^2 - 4x + 8 = 0$? -16

b What are the roots of $x^2 - 4x + 8 = 0$? $x = 2 \pm 2i$

P,R | **18** Solve each equation for x.

a $x^2 + 49 = 0$ $x = \pm 7i$ **b** $x^2 - 2x + 2 = 0$ $x = 1 \pm i$ **c** $x^2 - 6x + 9 = -25$ $x = 3 \pm 5i$

P | **19** Perform the indicated operations. Write your answers in standard complex form.

a $(3 - 4i)(2 + i)$ $10 - 5i$ **b** $(6 - 4i) - (-3 - 9i)$ $9 + 5i$

c $(2 + 6i) \div (3 - i)$ $0 + 2i$

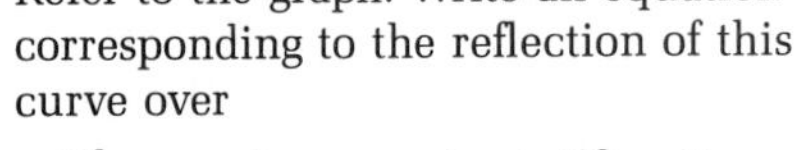

R,I | **20** Refer to the graph. Write an equation corresponding to the reflection of this curve over

a The x-axis $y = -(x + 5)^2 - 2$

b The y-axis $y = (x - 5)^2 + 2$

c The origin $y = -(x - 5)^2 - 2$

d The line x = y $x = (y + 5)^2 + 2$

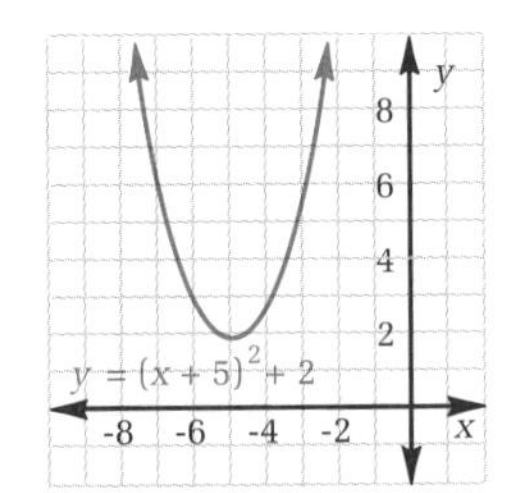

P | **In problems 21–26, simplify each expression.**

21 $i^{17} + i^{15} + i^{13} + i^{11}$ 0

22 $2i^7 - 3i^{13}$ $-5i$

23 $(2 + i) - 3(5 - 7i)$ $-13 + 22i$

24 $(1 + i)^2 - (1 - i)^2$ $4i$

25 $4i - 8i^3 + 16i^5 - 10i^7$ $38i$

26 $(2i)^2 - 4i^2 - (-3i)^2$ 9

P | **27** Find the real numbers x and y that satisfy each equation.

a $(x - 7) + 4yi = 9 + (y - 21)i$ $x = 16; y = -7$

b $(x + 2y) + (x - y)i = 11 + i$ $x = \frac{13}{3}; y = \frac{10}{3}$

c $(x - 3y)i = 4x - 12y$ $\{(x, y): x = 3y\}$

P,R | **28** Let z and $\bar{z}$ be conjugates.

a Find $z + \bar{z}$ if $z = 2 - 3i$. 4

b Find $z - \bar{z}$ if $z = -3 - 4i$. $-8i$

c Find $\bar{z} - z$ if $z = 6 - 3i$. $6i$

d Find $z \cdot \bar{z}$ if $z = 3 - 4i$. 25

e Find $z \cdot \bar{z}$ if $z = \sqrt{2} - \sqrt{3}i$. 5

Problem-Set Notes and Strategies, continued

- Problem **13** is solved as any other system of linear equations. The answers may be complex, but the method is identical.
- In problem **17**, note that if the discriminant is negative, the roots exist and are complex. In past chapters it was stated that no real solutions exist. Roots do exist, but they are complex and not real.
- For problem **20**, students may have forgotten the reflections that cause symmetries. Have them review Section **6.3** because reflections will be needed in graphing trigonometric functions.
- Equating real and imaginary parts in problems **27b** and **c** yields a system of equations that must be solved. Part **c** yields two dependent equations and therefore the answers are infinite along a line $x = 3y$.

Problem-Set Notes and Strategies, continued

- Negative exponents of i, as in problem **29**, will be explored further in a cooperative learning exercise.
- Students should be able to generate a pattern for the sum of complex numbers as they did for the powers. This pattern will be useful in solving problem **36**.
- The recursive function g in problem **43** also generates a repetitive pattern.

Problem Set, *continued*

P **29** Simplify each expression.

a $4i^{-3} + i^{-1}$ $3i$ **b** $5i^{-5} - 7i^{-1}$ $2i$

B

In problems 30–33, find all the complex solutions of each equation.

P,I **30** $x^2 - 6x + 25 = 0$ $x = 3 \pm 4i$

31 $2x^2 + 3x + 4 = 0$ $x = -\frac{3}{4} \pm \frac{\sqrt{23}}{4}i$

32 $5x^2 + 45 = 0$ $x = \pm 3i$

33 $2x^2 = 6x - 5$ $x = \frac{3}{2} \pm \frac{1}{2}i$

34 Solve $\frac{5x + 3}{2 - 5i} = \frac{2 + 5i}{2}$ for x. $x = \frac{23}{10}$

P,I **35** Find the complex zeros of $f(x) = x^2 + 20$. $\pm 2\sqrt{5}i$

P,R **36** Evaluate each expression.

a $\sum_{k=1}^{8} i^k$ 0 **b** $\sum_{k=1}^{10} i^k$ $-1 + i$ **c** $\sum_{k=1}^{100} i^k$ 0 **d** $\sum_{k=1}^{102} i^k$ $-1 + i$

P,I **37** Find the sum and the product of the roots of the equation $x^2 + 10x + 26 = 0$. -10; 26

P **38** For what values of (x, y) is $x + yi = 3(2 - 4i) - 6(x - yi)$? $\left(\frac{6}{7}, \frac{12}{5}\right)$

R **39** Write an equation that describes the conic section shown.

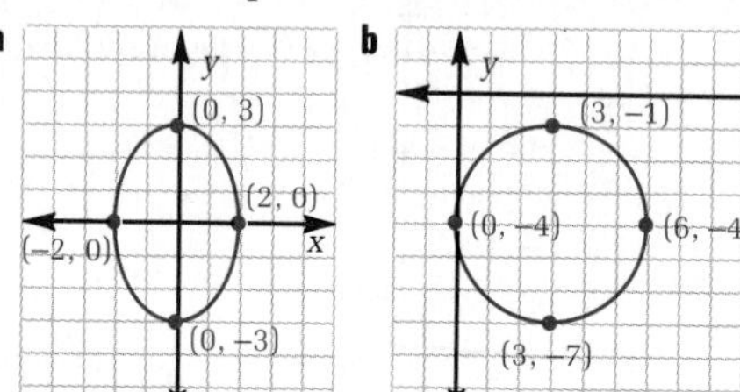

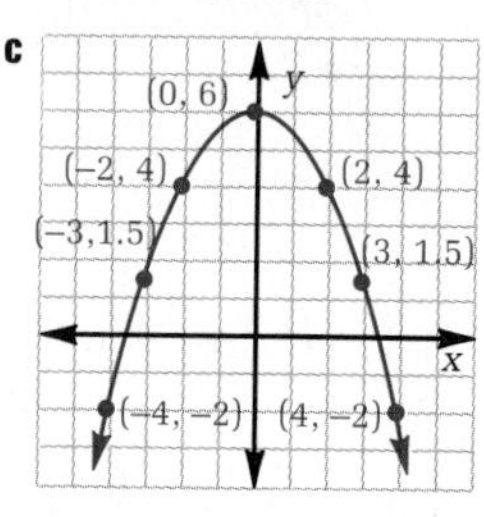

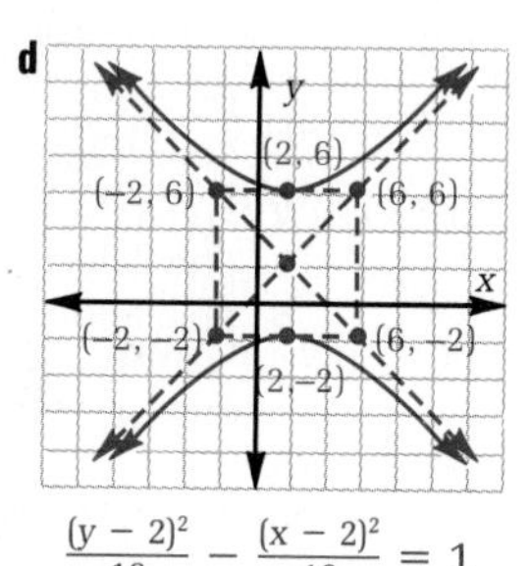

a $\frac{x^2}{4} + \frac{y^2}{9} = 1$ **b** $(x - 3)^2 + (y + 4)^2 = 9$ **c** $y = -\frac{1}{2}x^2 + 6$ **d** $\frac{(y - 2)^2}{16} - \frac{(x - 2)^2}{16} = 1$

P **40** Find the real values of x and y that solve each equation.

a $12x + 5y + xi - yi = -9 + 12i$ $x = 3$; $y = -9$

b $4x - 3y + 2 = (x - 2y + 13)i$ $x = 7$; $y = 10$

P,R **41** Solve the system for (x, y). $(x, y) = \left(3 - 2i, -1 + \frac{3}{2}i\right)$

$$\begin{cases} ix + 2y = 6i \\ ix - 2y = 4 \end{cases}$$

P **42** The Riddler said, "I'm thinking of two complex numbers that have a sum of $1 + 3i$ and a difference of $7 - 5i$." What is the product of the Riddler's two numbers? $-8 + 19i$

P **43** If $g(1) = i$ and $g(n + 1) = i^{n+1} + g(n)$ for all positive integers n, what are the values of $g(2)$, $g(3)$, $g(4)$, $g(5)$, $g(6)$, $g(7)$, $g(8)$, and $g(57)$? $-1 + i$; -1; 0; i; $-1 + i$; -1; 0; i

C

P,R — **44** Solve each equation for x.

a $x - 8 = 0$ $x = 8$
b $x^2 - 8 = 0$ $x = \pm 2\sqrt{2}$
c $x^3 - 8 = 0$ $x = 2$ or $x = -1 \pm \sqrt{3}i$
d $x^2 - 10ix + 9i^2 = 0$ $x = 9i$ or $x = i$
e $x^2 - 6ix - 5 = 0$ $x = 5i$ or $x = i$

P,R — **45** Find the complex zeros of $f(x) = x^4 + 8x^2 + 16$. $\pm 2i$

P — **46** Find each product.

a $[6 + (3 + i)][6 + (3 - i)]$ 82
b $[5 + (4 - 2i)][5 + (4 + 2i)]$ 85

P — **47** Find all values of a and b for which $(a + bi)(3 - 2i)$ is a real number. $\{(a, b): 3b = 2a\}$

Problem-Set Notes and Strategies, continued

- In problem **44**, note that a polynomial has the same number of roots as its degree. In the field of complex numbers, this is always true, as long as multiplicity is considered.

CAREER PROFILE

USING A CURVE TO TRACE DISEASE

Dr. Richard Dicker applies statistics to epidemics

"I'm the director of the Epidemic Intelligence Service, a two-year program of service and training for people who want to explore a career in epidemiology," says Richard Dicker, an epidemiologist with the Center for Disease Control.

Epidemiologists try to find the origins of epidemics (such as of flu and measles), to identify who is at risk, and to take action to protect susceptible people from the disease. One model of an epidemic's progression is the *point source* model, based on the hypothesis that the epidemic originated at a specific time and location.

For example, explains Dr. Dicker, "there was a measles outbreak in Chicago that was traced to an emergency room. A child with measles had been brought to a hospital emergency room, and other children who were there at the same time caught the measles."

A normal curve can be used in identifying the time and location of the origin of the epidemic. Dr. Dicker explains, "The epidemic curve from a point source is a skewed normal curve where, if you plot time on the x-axis and onset (number of cases) on the y-axis, there is a sharp rise in the number of cases followed by a trailing off. If you know the incubation period of the disease, you can figure out the time at which a great number of people became exposed to the disease. By finding out where they were at that time, it is possible to identify the epidemic's origin."

Dr. Dicker is a physician with training in preventive medicine. He has a bachelor's degree in chemistry from Tufts University in Medford, Massachusetts, a medical degree from the University of Massachusetts at Amherst, and a master's degree in epidemiology from Harvard.

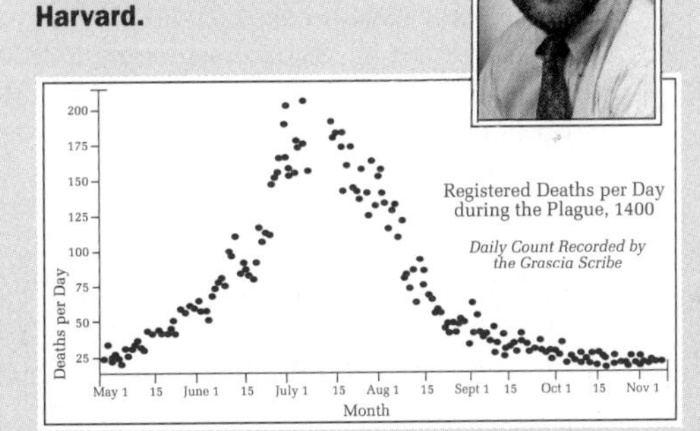

From Ann G. Carmichael, *Plague and the Poor in Renaissance Florence*. Copyright © 1986 by Cambridge University Press.

8.5

Class Planning

Time Schedule
All levels: 1 day

Resource References
Teacher's Resource Book
Class Opener 52
Practice 55
Manipulatives and Applications 13
Transparency 16

Class Opener

1 Find the sum of the roots of $x^2 + 3x + 5 = 0$

The roots are $\frac{-3 \pm \sqrt{-11}}{2}$.

The sum of the roots is -3.

2 Find the product of the roots of $x^2 + 3x + 5 = 0$ 5

Lesson Notes

- The standard deviation is used to divide the normal curve into segments that are 1 standard deviation wide. The areas of these segments is given. Notice that all areas are between 0 and 1 and the sum of the areas is also 1. This satisfies the properties of a probability distribution and can aid in discussing probabilities and percentiles.

Communicating Mathematics

Have students find examples of a distribution of values that follow a normal distribution curve. Why do you suppose that the curve is called normal?

8.5 The Normal Curve

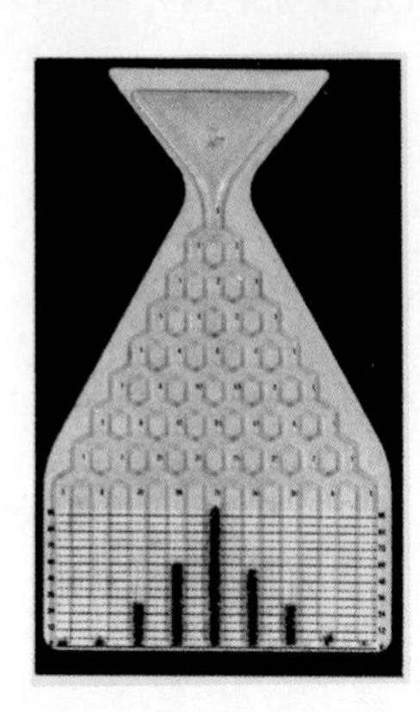

Objective

After studying this section, you will be able to

- Recognize some characteristics of a normal distribution of data

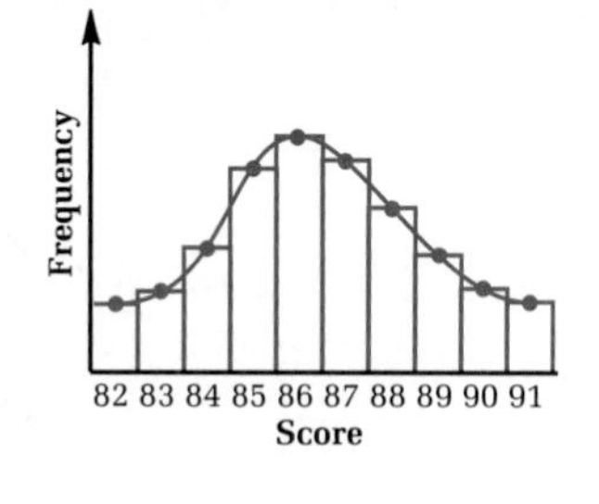

The diagram at the right represents a group of test scores. The smooth curve drawn over the data has a shape that statisticians call a *bell-shaped curve*. This curve serves as a basis for much of the formal work in statistics.

In many large collections of data, the frequency of individual data values decreases symmetrically about a single value that corresponds to the mean, the mode, and the median of the data. This is known as a ***normal distribution*** of data, and the graph is a symmetrical bell-shaped curve called a ***normal curve.***

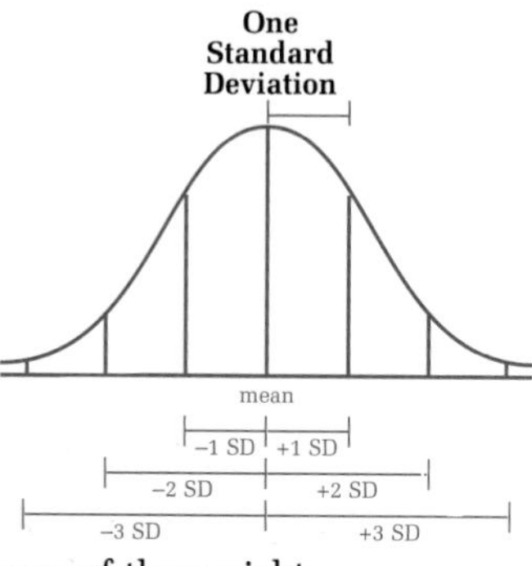

By using the mean and a nonnegative measure of dispersion known as the ***standard deviation*** (SD), we can identify some of the characteristics of data in a normal distribution. (In the next chapter, you will learn how to calculate the standard deviation of a set of data.)

Notice that the area under the curve has been divided by vertical segments at points that correspond to the mean and to values that are one, two, and three standard deviations on either side of the mean.

Mathematicians have calculated the relative areas of these eight regions, and this information is used to analyze the distribution of values in a set of data. The following diagram shows what percentage of the entire area under the curve lies within each of the regions. For example, the area of region III is approximately 13.6% of the total area under the curve, so about 13.6% of the data fall between 1 and 2 standard deviations below the mean. The combined areas of regions IV and V make up about 68.2% of the total area, so about 68.2% of the data lie within 1 standard deviation of the mean.

Vocabulary
normal distribution
normal curve
standard deviation

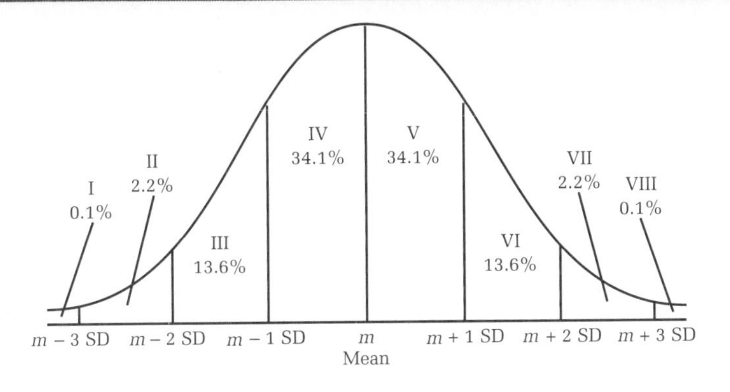

Using Computers

A statistical software package can be used effectively here. Such software enables you to obtain means, standard deviations, and other statistics from normal curves. Many statistical packages are available commercially. Among them are *TruStat* from True Basic, Inc. and *MYSTAT* from SYSTAT, Inc.

Sample Problem

Problem *A set of normally distributed data has a mean of 26 and a standard deviation of 4.*

a *What values are 1, 2, and 3 standard deviations from the mean?*
b *What percentage of the total area under the curve lies between 18 and 30?*
c *What percentage of the values are between 18 and 30?*
d *What percentile rank does a value of 30 have?*

Solution

a We sketch a representative normal curve and label the horizontal axis. The values 22 and 30 are 1 SD from the mean of 26. The values 18 and 34 are 2 SD from the mean, and the values 14 and 38 are 3 SD from the mean.

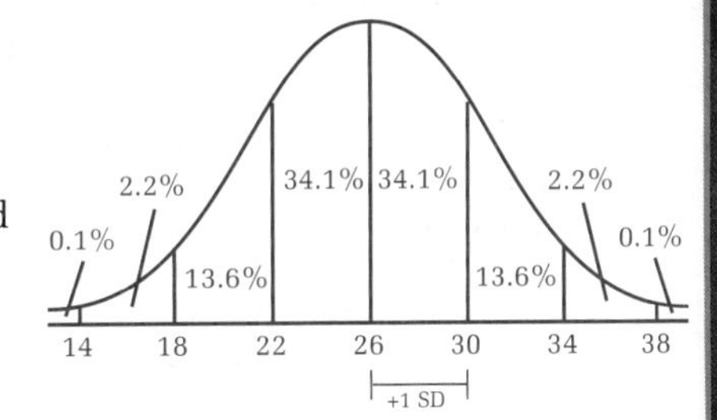

b The region between 18 and 22 occupies 13.6% of the total area under the curve, and the regions between 22 and 26 and between 26 and 30 occupy 34.1% of the total area each. The region between 18 and 30 constitutes approximately 13.6 + 34.1 + 34.1, or 81.8, percent of the total area.

c The percentage of values between 18 and 30 is the same as the corresponding percentage of area under the curve. Thus, about 81.8% of the data are between 18 and 30.

d The percentile rank of a value of 30 is the percentage of the values that are less than or equal to 30. Since 30 is 1 SD to the right of the mean, approximately 50 + 34.1, or 84.1, percent of the area under the curve is to the left of 30. Thus a value of 30 has a percentile rank of about 84.

Checkpoint

1. What percentile ranking would be given to a score that is one standard deviation below the mean?
   The 16th percentile
2. What is the area that is between 2 standard deviations below the mean and 1 standard deviation above the mean? 81.8%

Assignment Guide

Average	1–4, 6, 9, 10
Advanced	1–6, 8, 10
Honors	2–8, 10, 11

Problem-Set Notes and Strategies

- Problem **2** requires students to use the percentages listed on the previous page. These percentages are constant for any normal distribution that is listed in terms of standard deviation. Other values for intermediate standard deviations are listed in tables found in any statistics textbook.
- Problems **3** and **4** are good problems for reading and interpreting statistical graphs.

Problem Set

A

P **1** In the normal distribution shown, region 1 has an area of 50 square units; region 2, an area of 200 square units; region 3, an area of 250 square units; region 4, an area of 350 square units; and region 5, an area of 150 square units.

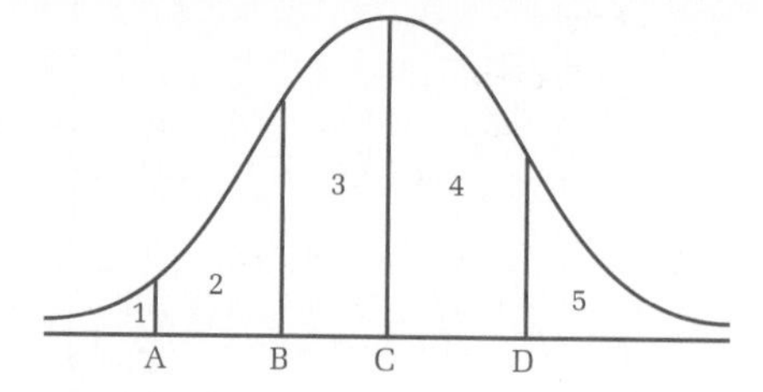

a Find the total area under the curve. 1000 square units
b What percentage of the total area is to the left of B? 25%
c What is the percentile rank of the value at B? 25
d What percentage of the total area is to the left of D? 85%
e What is the percentile rank of the value at D? 85

P **2** The area under the normal curve shown is 400. Find the areas of regions I, II, III, and IV.
0.4; 8.8; 54.4; 136.4

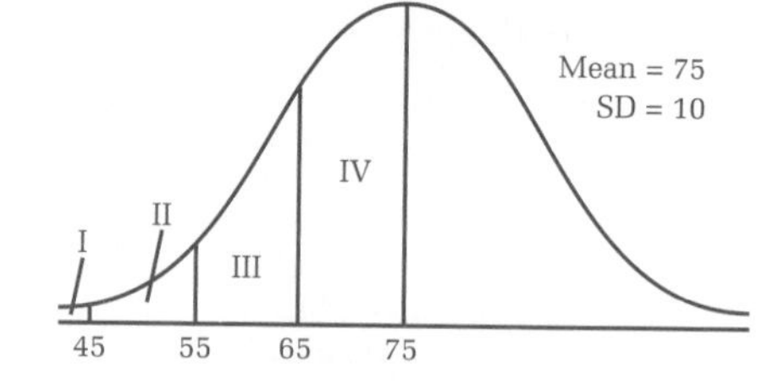

P **3** The data represented by the normal curve have a mean of 50 and a standard deviation of 10.

a What percentage of the total area under the curve lies in the shaded region? 84.1%
b What is the percentile rank of a value of 60? 84

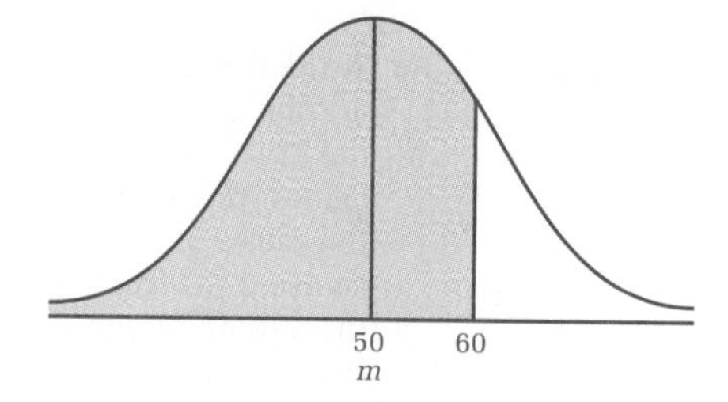

P **4** Copy and complete the following table for a normal distribution of data having a mean of 500 and a standard deviation of 100.

Data value	200	300	400	500	600	700	800
Percentile rank	1	2	15	50	84	97	99

B

P **5** What percentage of the area under the normal curve lies in the shaded region?
65.9%

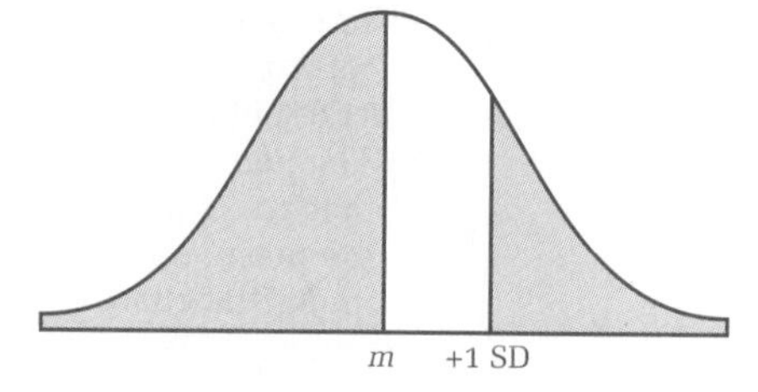

P **6** A normal distribution has a mean of 60 and a standard deviation of 8.

a What percentage of the area under the distribution's graph is between the values of 44 and 68? 81.8%

b What percentage of the data have a value greater than 52? 84.1%

P **7** A set of normally distributed data has a mean of 220 and a standard deviation of 60. Find the percentile rank of each of the following values.

a 100 2 **b** 160 15 **c** 280 84 **d** 340 97 **e** 400 99

C **In problems 8–11, find the percentage of the total area under the curve contained by the shaded region.**

P **8** 84.1%

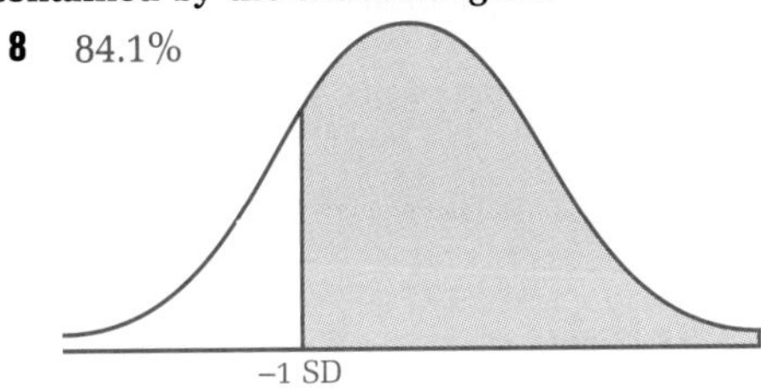

P **9** 99.9%

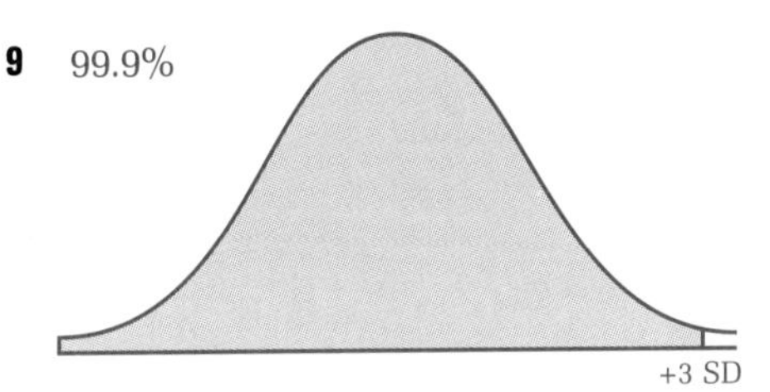

P **10** 97.7%

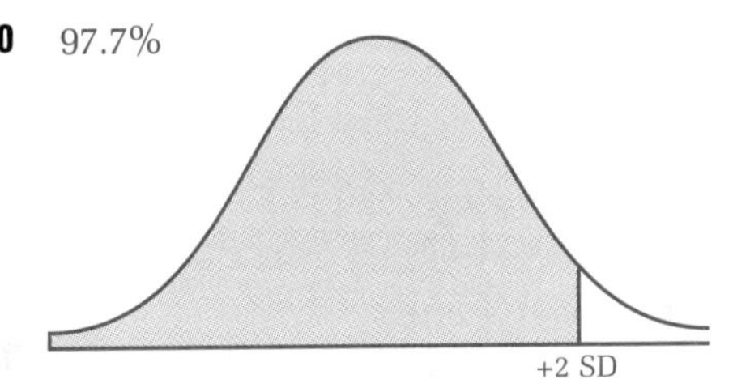

P **11** 95.4%

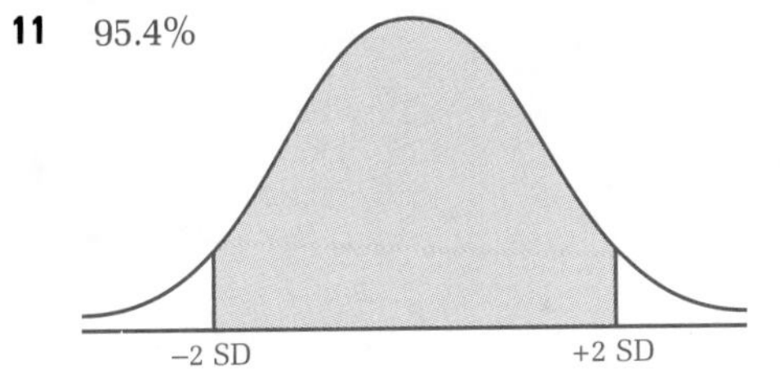

Problem-Set Notes and Strategies, continued

- Problems **8 – 11** are exercises in finding areas. These values correspond to the probability of these events occurring as well. Students may question why the area is so important. You can reassure them that the probabilities are the same as the area and that by finding probabilities, they can make predictions of data values with some confidence.

8

Chapter Summary

Concepts and Procedures

After studying this chapter, you should be able to

- Apply the properties of exponents to expressions with fractional exponents (8.1)
- Understand the restrictions that apply to exponential expressions with negative bases (8.1)
- Solve equations that contain square roots (8.2)
- Solve equations that contain rational exponents (8.2)
- Identify pure imaginary numbers (8.3)
- Perform operations on pure imaginary numbers (8.3)
- Identify complex numbers (8.3)
- Determine the conjugate of a complex number (8.3)
- Identify equal complex numbers (8.4)
- Perform operations on complex numbers (8.4)
- Recognize some characteristics of a normal distribution of data (8.5)

Vocabulary

complex number (8.3)
conjugate (8.3)
extraneous root (8.2)
imaginary part (8.3)
normal curve (8.5)
normal distribution (8.5)
pure imaginary number (8.3)
radicand (8.1)
real part (8.3)
root index (8.1)
standard complex form (8.4)
standard deviation (8.5)

REVIEW PROBLEMS

A

P **In problems 1–6, evaluate each expression.**

1 $5^{\frac{1}{3}} \cdot 5^{\frac{2}{3}}$ 5 **2** $(2^6)^{\frac{1}{3}}$ 4 **3** $\left(2^{\frac{1}{4}}\right)^8$ 4

4 $49^{-\frac{1}{2}}$ $\frac{1}{7}$ **5** $81^{-\frac{3}{4}}$ $\frac{1}{27}$ **6** $4^{-\frac{5}{2}}$ $\frac{1}{32}$

P **7** A circle with a radius of $3^{\frac{1}{4}}$ is inscribed in a square as shown. What is the area of the shaded portion of the figure?
$(4 - \pi)\sqrt{3}$, or ≈ 1.49

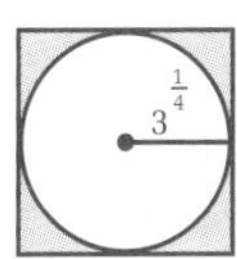

P **8** Solve $\sqrt[3]{5x + 7} = -2$ for x. $x = -3$

P **9** Rewrite each expression as a number in standard complex form.

a $\frac{4}{-3i}$ $0 + \frac{4}{3}i$ **b** $2i(3 + i)$ $-2 + 6i$ **c** $\frac{5}{6 + i}$ $\frac{30}{37} - \frac{5}{37}i$ **d** $(12i^2)^2$ $144 + 0i$

P **In problems 10–12, evaluate each expression.**

10 $2 \cdot 3^{\frac{1}{3}} \cdot 5 \cdot 3^{-\frac{4}{3}}$ $\frac{10}{3}$ **11** $\left(2^{\frac{2}{5}} \cdot 2^{\frac{1}{5}}\right)^{10}$ 64 **12** $\left(4^{\frac{3}{2}} \cdot 4^{\frac{5}{2}}\right)^{\frac{1}{2}}$ 16

P **13** A normally distributed set of data has a mean of 480 and a standard deviation of 50. Which of the data values are

a 1.5 SD from the mean? 405 and 555 **b** 2.4 SD from the mean? 360 and 600

P **In problems 14–19, simplify each expression.**

14 $\sqrt{-2}\sqrt{-32}$ -8 **15** $\frac{10}{7i}$ $-\frac{10}{7}i$ **16** $-7i + 6i - 3i$ $-4i$

17 $4\left(\sqrt{-49} - \sqrt{-16}\right)$ $12i$ **18** $-\left(\sqrt{-5}\right)^2$ 5 **19** $3i\left(i + 2\sqrt{7}i\right)$ $-3 - 6\sqrt{7}$

P **20** Find the conjugate of each complex number.

a $2 - 9i$ $2 + 9i$ **b** $\frac{1}{2} + \sqrt{3}i$ $\frac{1}{2} - \sqrt{3}i$ **c** $15i$ $-15i$

d $-3i - 5$ $-5 + 3i$ **e** $\sqrt{23}$ $\sqrt{23}$ **f** $\frac{5}{7} - \frac{5}{7}i$ $\frac{5}{7} + \frac{5}{7}i$

P **21** Simplify each expression.

a $x^{\frac{1}{3}} \cdot x^{\frac{4}{5}}$ $x^{\frac{17}{15}}$ **b** $y^{3.7}\left(y^{-5.2}\right)$ $y^{-1.5}$ **c** $\left(a^{9.5}\right)^2$ a^{19}

Class Planning

Time Schedule

All levels: 1 day

Assignment Guide

Average

1–6, 8, 9a,b, 11, 13, 14–16, 18, 20a,b,c, 21–24, 27, 28, 31, 33, 34, 41

Advanced

2–4, 8–11, 13–15, 17, 20b,c,d,e, 21, 22, 25, 26, 29, 30, 31, 34–36, 39a, 41–43

Honors

7, 9, 12, 13, 17–22, 25, 26, 29, 30–32, 34–36, 39, 40, 42, 43

Problem-Set Notes and Strategies

- Problem **7** involves several geometric and algebraic concepts.
- Problems **10 – 12** are easier if they are simplified first before using a calculator.
- Problem **21** has a review of the properties of exponents extended to rational exponents.

Review Problems, *continued*

P **22** What percentage of the total area under the normal curve lies in the shaded portion of the diagram?
18.2%

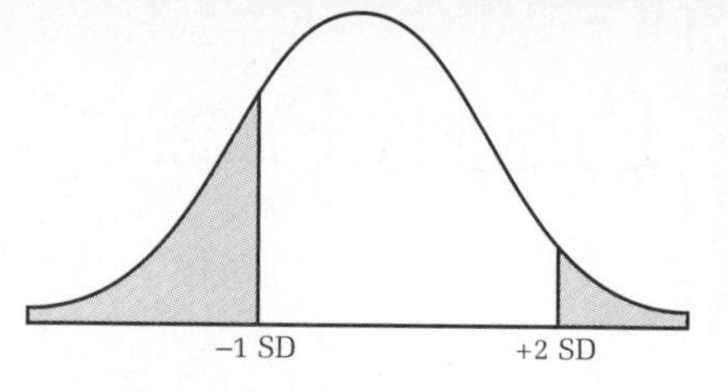

B P **In problems 23–26, perform the indicated operations. Write your answers in standard complex form.**

23 $(6 + 3i) - (-1 + 5i)$ $7 - 2i$

24 $(3 - 3i)(5 - i)$ $12 - 18i$

25 $\frac{2 + 5i}{5 + 2i}$ $\frac{20}{29} + \frac{21}{29}i$

26 $(2 - 3i)^2 + (4 + 12i)$
$-1 + 0i$

P **In problems 27–30, solve each equation for (x, y).**

27 $2x + 7iy = 14 - 28i$ $(x, y) = (7, -4)$

28 $(2y + 3) - 5i = 3 - 10xi$ $(x, y) = \left(\frac{1}{2}, 0\right)$

29 $\frac{x}{5} - 3i = (y + 5)i - 14$
$(x, y) = (-70, -8)$

30 $(x + y) + (2x - 3y)i = 15 + 15i$
$(x, y) = (12, 3)$

P **31** What are the complex zeros of the function $g(x) = x^4 - 16$?
2, −2, 2i, and −2i

P **32** Simplify $\begin{bmatrix} 3i & 8i \\ -8i & i \end{bmatrix} + \begin{bmatrix} 7i & i \\ -3i & -4i \end{bmatrix}\begin{bmatrix} 0 & 2i \\ i & 0 \end{bmatrix}$. $\begin{bmatrix} -1 + 3i & -14 + 8i \\ 4 - 8i & 6 + i \end{bmatrix}$

P **In problems 33–38, solve each equation for *a*.**

33 $\sqrt{3a - 7} + 6 = 10$ $a = \frac{23}{3}$

34 $5(a + 1)^{\frac{4}{3}} = 80$ $a = 7$

35 $15 = 12 + (a^2 - 7)^{\frac{1}{5}}$ $a = \pm 5\sqrt{10}$

36 $\sqrt{a - 3} = \sqrt{a + 3} - 3$ No real solutions

37 $(a^2 - 100)^{\frac{2}{3}} = 16$
$a = \pm 6$ or $a = \pm 2\sqrt{41}$

38 $3a + 2\sqrt{a} = 10$
$a = \frac{32 - 2\sqrt{31}}{9}$

P **39** Write an equation that has the given roots.

a 11i and −11i
$x^2 + 121 = 0$

b 3, 2 + 2i, and 2 − 2i
$x^3 - 7x^2 + 20x - 24 = 0$

C P **In problems 40–43, find the complex solutions of each equation.**

40 $5x^2 + 3x + 1 = 0$ $x = -\frac{3}{10} \pm \frac{\sqrt{11}}{10}i$

41 $x^2 - 6x + 18 = 0$ $x = 3 \pm 3i$

42 $4x^3 = 25x$
$x = 0$ or $x = \pm\frac{5}{2}$

43 $x^4 + 14x^2 = -49$
$x = \pm\sqrt{7}i$

P,I **44** If a right triangle has the side measures shown, what are the possible values of x?
5

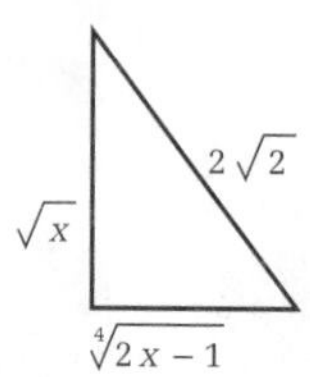

Problem-Set Notes and Strategies, continued

- In problem **31**, remind students that all real and imaginary numbers are subsets of the complex numbers. Students have a tendency to give the values 2*i* and −2*i* thinking that these are the only complex zeros.
- Problem **44** involves both the Pythagorean Theorem and solving equations with radicals. The index of one of the radicals is 4. Students may overlook this fact or not know how to square this radical. Encourage them to change all the radicals to fractional exponents.

CHAPTER TEST

Resource References
Evaluation
Test Forms 1, 2, 3

1 When is $(\sqrt[n]{x})^n = \sqrt[n]{x^n}$? When n is even and nonzero and x is nonnegative or when n is odd and x is any real number

In problems 2 and 3, simplify each expression.

2 $\dfrac{10x^{\frac{8}{5}} + 30x^{\frac{4}{5}}}{5x^{\frac{6}{5}} + 15x^{\frac{2}{5}}}$ $2x^{\frac{2}{5}}$

3 $\dfrac{10x^{\frac{9}{4}} + 25x^{\frac{5}{4}}}{-5x^{\frac{1}{4}}}$ $-2x^2 - 5x$

In problems 4 and 5, find the real roots of each equation.

4 $\sqrt{x^2 + 9} - 1 = x$
$x = 4$

5 $9x^{\frac{2}{3}} + 3x^{\frac{1}{3}} = 6$
$x = \frac{8}{27}$ or $x = -1$

In problems 6–8, match each number with all the phrases that correctly describe it.

6 $3 - 2i$ **b, c**
7 $2i - 3$ **b, d**
8 $\dfrac{\sqrt{3}}{5}i$ **a, b**

a Pure imaginary number
b Complex number
c Conjugate of $2i + 3$
d Additive inverse of $3 - 2i$

In problems 9–11, simplify each expression.

9 $3i - 2i + \dfrac{7i}{-7}$ 0

10 $(5i)(-3i)$ 15

11 $i + i^2 + i^3$ -1

In problems 12–14, simplify each expression. Write your answer in the form *a* + *bi*.

12 $(6 - i)(3 - 2i)$
$16 - 15i$

13 $(8 - 7i) - (3 + 6i)$
$5 - 13i$

14 $\dfrac{8 + 3i}{1 - 2i}$
$\frac{2}{5} + \frac{19}{5}i$

15 Find the values of x_1, x_2, x_3, and x_4.
7.75; 11.5; 14; 32

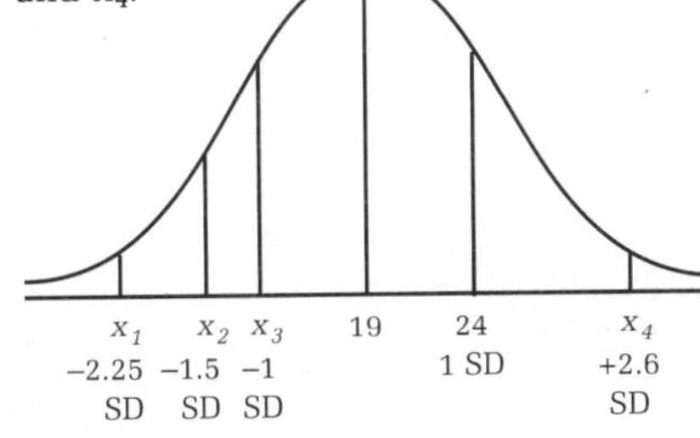

16 Is there a complex number that is equal to the square of its conjugate? If so, how many such complex numbers are there, and what numbers are they?
Yes; two; $-\frac{1}{2} + \frac{\sqrt{3}}{2}i$ and $-\frac{1}{2} - \frac{\sqrt{3}}{2}i$

9.1

THIS CHAPTER COVERS much of the traditional work on the theory of polynomial functions. It begins by introducing the technologically efficient method of writing a polynomial in nested form. This form is used to motivate several of the theorems of this chapter. The Factor and the Remainder Theorems are also introduced.

Then, synthetic division and the Rational-Zeros Theorem are presented. At this time, you may want to remind students that if there are no rational zeros it does not mean that there are no real zeros for the polynomial function.

Section **9.3** presents a method for finding accurate approximations to all real zeros of any polynomial function. First, the upper and lower bounds of the function are established. Then the bisection process is used to increase the accuracy of approximated zeros.

The Fundamental Theorem of Algebra, the Conjugate-Zeros Theorem and the Binomial Theorem are also presented in this chapter.

In the data analysis section, students will calculate the standard deviation of a set of data, work with Chebyshev's Theorem, then establish minimum percentages of data in a distribution within k standard deviations of the mean.

CHAPTER 9

POLYNOMIALS AND POLYNOMIAL FUNCTIONS

The curves in this photograph of a roller coaster resemble the graphs of polynomials.

The Remainder and Factor Theorems

9.1

Objectives

After studying this section, you will be able to

- Write a polynomial in nested form
- Divide a polynomial by a binomial
- Apply the remainder and factor theorems

Part One: Introduction

The Nested Form of a Polynomial

We usually write a polynomial such as $6x - 5x^2 - 4 + 2x^3$ in *descending order of powers*—$2x^3 - 5x^2 + 6x - 4$. A polynomial can be written in another useful form, called ***nested form.*** The following example shows how x was repeatedly factored out of all the x terms.

Example *Rewrite $7x^4 - 3x^3 + 2x^2 + 9x - 5$ in nested form.*

$$\begin{aligned} & 7x^4 - 3x^3 + 2x^2 + 9x - 5 \\ &= (7x^3 - 3x^2 + 2x + 9)x - 5 \\ &= ((7x^2 - 3x + 2)x + 9)x - 5 \\ &= (((7x - 3)x + 2)x + 9)x - 5 \\ &= ((((7)x - 3)x + 2)x + 9)x - 5 \end{aligned}$$

The following procedure can be used to write a polynomial in nested form.

Converting a Polynomial to Nested Form

1. **Write the polynomial in descending order of powers.**
2. **Factor *x* out of all terms in which it appears.**
3. **Repeat step 2 for the expression in parentheses. Continue the process until no power of *x* greater than the first appears.**

Class Planning

Time Schedule

All levels: 2 days

Resource References

Teacher's Resource Book

- Class Opener 53
- Cooperative Learning 26
- Practice 56
- Enrichment 17

Transparency 17

Class Opener

1. Evaluate.
   $(((x + 3)x - 4)x + 2)x + 7$
   if $x = -3.26$
   ≈ -33.02
2. Evaluate.
   $x^4 + 3x^3 - 4x^2 + 2x + 7$
   if $x = -3.26$
   ≈ -33.02

Lesson Notes

- Rearranging polynomials into nested form will allow for easier computation of polynomials at various roots.
- The remainder and factor theorems allow you to determine if a binomial is a factor of a given polynomial, and if not, what the remainder is when the polynomial is divided by that binomial.
- Most of the polynomials students have been exposed to have factored nicely with little effort. The Factor and Remainder Theorems will make an entire new class of polynomials accessible to them.

Vocabulary

nested form
Remainder Theorem
Factor Theorem

9.1

Checkpoint

1. Write the polynomial $x^3 - 3x^2 + 4x - 5$ in nested form. $(((x)x - 3)x + 4)x - 5$
2. Evaluate the polynomial $x^3 - 3x^2 + 4x - 5$ when $x = 2$. -1
3. What is the remainder when $x^3 - 3x^2 + 4x - 5$ is divided by $x + 1$. -13
4. Is $x + 1$ a factor of $x^3 - 3x^2 + 4x - 5$? Why or why not? No, since the remainder is -13 when the division is carried out.

Assignment Guide

Average	
Day 1	6, 7, 9, 14–16, 22, 24a, c, 25, 27a, 31
Day 2	8, 10, 12, 13, 20, 26, 32–35, 40, 43, 46
Advanced	
Day 1	6, 7, 9, 12, 16, 18, 20, 23, 24a, c, 26, 29, 31, 33, 35
Day 2	14, 15, 19, 25, 27b, 32, 36, 38, 40, 41, 44, 45, 47
Honors	
Day 1	6, 8, 12, 13, 16, 18, 19, 24a, c, 26, 27b, 31, 33, 34
Day 2	17, 22, 23, 25, 30, 36, 40, 41, 47, 49–51

Nested form allows us to write an efficient computer program to evaluate polynomials for values of x. One such program is given at the beginning of this section's problem set.

Example *If* $P(x) = (((5)x - 3)x - 4)x + 9$, *what is* $P(2)$?

Method 1

$$\begin{aligned} P(x) &= (((5)x - 3)x - 4)x + 9 \\ P(2) &= (((5)2 - 3)2 - 4)2 + 9 \\ &= ((7)2 - 4)2 + 9 \\ &= (10)2 + 9 \\ &= 29 \end{aligned}$$

Method 2

We can also use a calculator to evaluate this function. First we store the value 2 in the calculator's memory by entering 2 [Min]. With the value of x stored in the memory, we can press [MR] any time we need this value. To evaluate $P(x)$ for $x = 2$, we enter

5 [×] [MR] [−] 3 [=] [×] [MR] [−] 4 [=] [×] [MR] [+] 9 [=]

Once again, the output is 29. The [Min] and [MR] keys may have different names on your calculator, such as [M+] and [RM] or [STO] and [RCL].

Long Division of a Polynomial by a Binomial

Long division of polynomials is like long division of real numbers. Study and compare the following two procedures.

Division of 298 by 14

```
     2 1
14 | 2 9 8
     2 8
       1 8
       1 4
         4
```

Division of $2x^2 + 9x + 8$ by $x + 4$

```
          2x + 1
x + 4 | 2x² + 9x + 8
        2x² + 8x
               x + 8
               x + 4
                   4
```

Recall that dividend = divisor · quotient + remainder. For these two examples,

$298 = 14 \cdot 21 + 4$ $\qquad$ $2x^2 + 9x + 8 = (x + 4)(2x + 1) + 4$

Example *Divide* $2x^3 - 19x^2 + 29x - 40$ *by* $x - 8$.

```
             2x² −  3x +  5
x − 8 | 2x³ − 19x² + 29x − 40
        2x³ − 16x²
              −3x² + 29x
              −3x² + 24x
                      5x − 40
                      5x − 40
                            0
```

Thus, $2x^3 - 19x^2 + 29x - 40$ can be written as $(x - 8)(2x^2 - 3x + 5) + 0$. Since the remainder is zero, $x - 8$ is a factor of $2x^3 - 19x^2 + 29x - 40$.

Example *Divide* $x^3 - 12$ *by* $x - 2$.

Because the dividend has no terms of degree one or two, we save places for them by rewriting the polynomial as $x^3 + 0x^2 + 0x - 12$.

$$
\begin{array}{r|l}
 & x^2 + 2x + 4 \\
x - 2 & x^3 + 0x^2 + 0x - 12 \\
 & \underline{x^3 - 2x^2} \\
 & \quad 2x^2 + 0x \\
 & \quad \underline{2x^2 - 4x} \\
 & \qquad 4x - 12 \\
 & \qquad \underline{4x - 8} \\
 & \qquad\quad -4
\end{array}
$$

Thus, $x^3 - 12 = (x - 2)(x^2 + 2x + 4) - 4$.

The Remainder Theorem and the Factor Theorem

There is an interesting relationship between long division and evaluating a polynomial in nested form. See if you can recognize it in the following example.

Example *Let* $P(x) = 5x^3 - x^2 + 3x - 4$.

a *Find* $P(2)$ *by using the nested form of the polynomial.*

$$
\begin{aligned}
P(x) &= (((5)x - 1)x + 3)x - 4 \\
P(2) &= (((\mathbf{5})2 - 1)2 + 3)2 - 4 \\
&= ((\mathbf{9})2 + 3)2 - 4 \\
&= (\mathbf{21})2 - 4 \\
&= \mathbf{38}
\end{aligned}
$$

b *Divide* $P(x)$ *by* $x - 2$.

$$
\begin{array}{r|l}
 & \mathbf{5}x^2 + \mathbf{9}x + \mathbf{21} \\
x - 2 & 5x^3 - x^2 + 3x - 4 \\
 & \underline{5x^3 - 10x^2} \\
 & \qquad 9x^2 + 3x \\
 & \qquad \underline{9x^2 - 18x} \\
 & \qquad\quad 21x - 4 \\
 & \qquad\quad \underline{21x - 42} \\
 & \qquad\qquad \mathbf{38}
\end{array}
$$

The numbers in red in part **a** match the numerical values in $5x^2 + 9x + 21$—the quotient when $P(x)$ is divided by $x - 2$. Moreover, the remainder when $P(x)$ is divided by $x - 2$ is equal to the numerical value of $P(2)$. This suggests the following theorem, called the ***Remainder Theorem.***

 ***When a polynomial*** $P(x)$ ***is divided by*** $x - a$, ***the remainder is equal to*** $P(a)$.

Communicating Mathematics

Have students write a lesson plan on the concept of nested polynomials. The lesson plan should include examples. Choose two or three students to present their lessons to the class.

Using Computers

At the beginning of the problem set for this section is a BASIC program that evaluates a polynomial for any given value and gives the coefficients and remainder in its nested form. (In fact, these are simply the values obtained when synthetic division is used.) You might type in this program and use it for many of the problems in the problem set. The calculator also is ideal for polynomial evaluation. Many commercially-available programs such as *Derive* and the *Mathematics Exploration Toolkit* also evaluate polynomials. You also can construct a spreadsheet to evaluate polynomials.

One conclusion we draw from the Remainder Theorem is that if $P(a) = 0$, $x - a$ is a factor of $P(x)$. This conclusion is known as the ***Factor Theorem.***

If $P(x)$ is a polynomial and $P(a) = 0$, then $x - a$ is a factor of $P(x)$.

Example *Is $x - 3$ a factor of $2x^2 + x - 21$?*

$$\begin{aligned} P(x) &= ((2)x - 1)x - 21 \\ P(3) &= ((2)3 + 1)3 - 21 \\ &= (7)3 - 21 \\ &= 0 \end{aligned}$$

Since $P(3) = 0$, $x - 3$ is a factor of $2x^2 + x - 21$. By referring to the red numbers in the evaluation of $P(3)$, we see that the other factor is $2x + 7$.

Part Two: Sample Problems

Problem 1 *Rewrite $x^2 - 4$ in nested form.*

Solution This polynomial has no first-degree term, so we rewrite it as $x^2 + 0x - 4$.

$$\begin{aligned} &x^2 + 0x - 4 \\ = &(x + 0)x - 4 \\ = &((1)x + 0)x - 4 \end{aligned}$$

Problem 2 *Let $P(x) = 2x^3 + 5x^2 - 2x - 7$. Find the quotient and the remainder when $P(x)$ is divided by $x + 3$.*

Solution We rewrite $P(x)$ in nested form, as $P(x) = (((2)x + 5)x - 2)x - 7$, and evaluate $P(-3)$.

$$\begin{aligned} P(x) &= (((2)x + 5)x - 2)x - 7 \\ P(-3) &= (((2)(-3) + 5)(-3) - 2)(-3) - 7 \\ &= ((-1)(-3) - 2)(-3) - 7 \\ &= (1)(-3) - 7 \\ &= -10 \end{aligned}$$

The quotient is $2x^2 - x + 1$, and the remainder is -10. Since the remainder is not zero, we know that $x + 3$ is not a factor of $P(x)$.

If we use a calculator to evaluate $P(-3)$, we need to record the output each time we press the [=] key. First we store -3 in the memory by entering 3 [±] [Min]. Then we enter

2 [×] [MR] [+] 5 [=]	[×] [MR] [−] 2 [=]	[×] [MR] [−] 7 [=]
2	−1	1
		−10

Again, the quotient is $2x^2 - x + 1$, and the remainder is -10.

Part Three: Exercises and Problems

Warm-up Exercises

In problems 1 and 2, write each polynomial in nested form.

P **1** $x^2 - 3x + 5$ $((1)x - 3)x + 5$

2 $3x^4 - 2x^2 + 5x - 9$ $((((3)x - 0)x - 2)x + 5)x - 9$

P **3** Write the polynomial $(((4)x - 3)x + 2)x - 7$ in descending order of powers. $4x^3 - 3x^2 + 2x - 7$

P **4** Consider the polynomial $2x^3 - 13x^2 - 7x + 4$.

a Use long division to find the quotient and the remainder when $2x^3 - 13x^2 - 7x + 4$ is divided by $x - 5$. Quotient: $2x^2 - 3x - 22$; remainder: -106

b Evaluate $P(x) = 2x^3 - 13x^2 - 7x + 4$ when $x = 5$ using the nested form of $P(x)$. How does your answer compare with your answer to part **a**? -106; the value of $P(x)$ when $x = 5$ is equal to the remainder for $[P(x)] \div (x - 5)$.

P **5** Use long division to divide $x^5 - y^5$ by $x - y$. $x^4 + x^3y + x^2y^2 + xy^3 + y^4$

Problem Set

The following BASIC program evaluates a polynomial $P(x)$ for a given value of x. If you have access to a computer, you may use it to solve some of the problems in this chapter.

```
10 REM This program evaluates polynomials for various values
20 REM of x. It also gives quotient coefficients and remainder
30 REM during the evaluation.
40 GOSUB 190
50 REM ---X Value Data Entry---
60 INPUT "Enter a value of x. ==>";X
70 REM ---Nested-Form Polynomial Evaluation---
80 LET Y = A(N)
90 PRINT: PRINT Y,
100 FOR I = N - 1 TO 0 STEP -1
110 LET Y = Y * X + A(I)
120 PRINT Y,
130 NEXT I
140 PRINT
150 PRINT "When x is ";X;", the polynomial's value is ";Y;"."
160 INPUT "Do you want to try another x value? ==>";A$
170 IF A$ = "Y" OR A$ = "y" THEN 50
180 END
190 REM ---Polynomial Input Section---
200 INPUT "Enter the degree of the polynomial. ==>";N: DIM A(N)
210 FOR I = N TO 0 STEP -1
220 PRINT "Enter the coefficient of the term"
230 PRINT "of degree ";I;". ==>";: INPUT A(I)
240 NEXT I
250 RETURN
```

Problem-Set Notes and Strategies

- Students have been exposed to the difference of squares and cubes. Problem **5** leads to the generalization that there may also be formulas for the difference of fifths, sevenths, etc.

Cooperative Learning

Have students work in small groups to solve problem **5**. Use this problem as an introduction to formulas for the differences of powers. Have students explore the sum and differences of both even and odd powers and decide which are factorable and which are not. This is also a good exercise on the laws of exponents as something like $x^6 + y^6$ may easily be written as $(x^2)^3 + (y^2)^3$ and is now the sum of cubes.

Stumbling Block

Students should be careful when nesting polynomials that have terms missing. When a polynomial is missing a term, such as the cubic, it is helpful to add a $0x^3$ to the polynomial before nesting.

Problem-Set Notes and Strategies, continued

- Problem **6** illustrates the conclusion of the Factor Theorem.
- Problems **13** and **14** are good practice in evaluating nested polynomials. For additional practice you might have the student divide both polynomials by $x - 2$.
- Problems **16** and **17** lead to factoring larger degree polynomials. These are quadratic like, but some generalizations may be drawn from these two problems.

Additional Answer

18

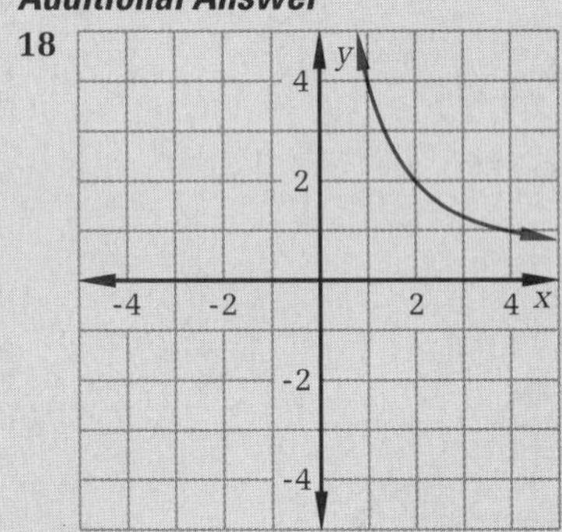

- Problems **20 – 23** lead to a formula for difference of cubes. Note that problem **22** is different from the others.

Problem Set, *continued*

P **A** **6** If $x + 7$ is a factor of the polynomial $Q(x)$, what is the value of $Q(-7)$? 0

P **In problems 7 and 8, rewrite each polynomial in nested form.**

7 $5x^3 + 2x^2 + 6x + 1$ — $(((5)x + 2)x + 6)x + 1$

8 $7x^3 - 3x^2 - 6x + 5$ — $(((7)x - 3)x - 6)x + 5$

P **In problems 9–12, rewrite each expression as a polynomial in descending order of powers.**

9 $(((3)x + 6)x - 7)x + 9$ — $3x^3 + 6x^2 - 7x + 9$

10 $((((3)x + 5)x + 0)x + 7)x - 9$ — $3x^4 + 5x^3 + 7x - 9$

11 $(((2)x + 0)x - 5)x - 7$ — $2x^3 - 5x - 7$

12 $((((4)y - 9)y + 2)y + 5)y - 2$ — $4y^4 - 9y^3 + 2y^2 + 5y - 2$

P **In problems 13 and 14, evaluate each expression for $x = 2$.**

13 $((7)x + 5)x + 1$ 39

14 $(((2)x + 5)x + 3)x + 9$ 51

P,R **15** Consider the polynomial $((((3)x_1 + 2)x_1 - 4)x_1 + 5)x_1 - 6$.

a What is the degree of the polynomial? 4
b What is the coefficient of its linear term? 5
c What is its constant term? -6
d What is its quadratic term? $-4x_1^2$
e What is its cubic term? $2x_1^3$
f What is its leading coefficient? 3

R,I **In problems 16 and 17, factor each expression.**

16 $x^4 - 8x^2 - 9$ — $(x^2 + 1)(x + 3)(x - 3)$

17 $x^4 - 6x^2 + 9$ — $(x^2 - 3)^2$

R **18** The area of the shaded region of the diagram is 12. Draw a graph that shows the relationship between x and y.

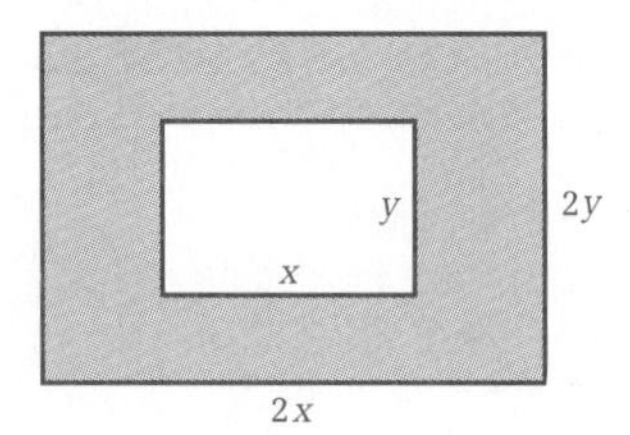

P **19** Divide $2x^3 - 5x^2 + 2x + 20$ by $2x + 3$. $x^2 - 4x + 7$, remainder -1

P,R,I **In problems 20–23, factor each expression.**

20 $x^3 - 64$ — $(x - 4)(x^2 + 4x + 16)$

21 $2x^3 - 128$ — $2(x - 4)(x^2 + 4x + 16)$

22 $4x^3 - 64$ — $4(x^3 - 16)$

23 $8x^3 - 27$ — $(2x - 3)(4x^2 + 6x + 9)$

P **24** What is the remainder when $4x^3 - 3x^2 - x - 7$ is divided by each binomial?

a $x - 1$ -7　**b** $x + 1$ -13　**c** $x - 3$ 71

P,R **B**

25 Consider the rectangles shown in the diagram.

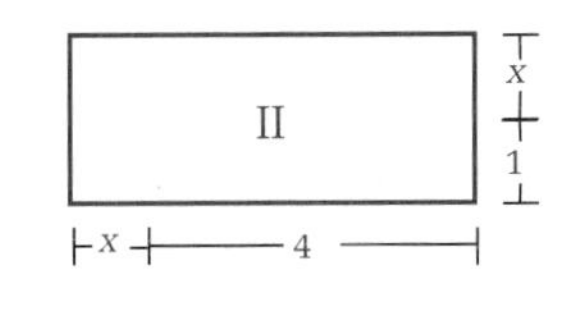

a Write an expression for the area of each rectangle. $x^2 + 5x + 6$; $x^2 + 5x + 4$
Which has the greater area? I
How much greater? 2 square units

b Divide your expression for the area of the larger rectangle by $x + 1$. $x + 4$, remainder 2

c How does the remainder in your answer to part **b** relate to your answer to part **a**? Since $x^2 + 5x + 6 = (x + 1)(x + 4) + 2$, the area of rectangle I is 2 square units greater.

P **26** If the polynomial $P(x)$ is divided by $x - 4$, the degree of the quotient is 5 and the remainder is 2.

a What is the degree of $P(x)$? 6 **b** What is $P(4)$? 2

R **27** Find the cubic term and the constant term of each product.

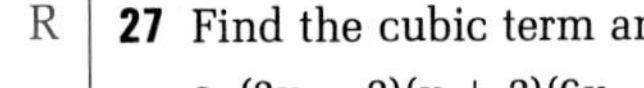

a $(3x - 2)(x + 2)(6x - 5)$ $18x^3$; 20 **b** $(3x + 5)(2x + 1)(3x + 4)$ $18x^3$; 20
c $(x - 2)(x - 10)(x + 1)$ x^3; 20

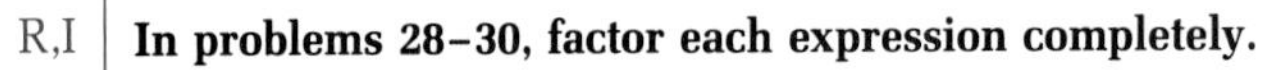

R,I **In problems 28–30, factor each expression completely.**

28 $10x^2 + x - 2$
$(5x - 2)(2x + 1)$

29 $27x^2 - 36x - 15$
$3(3x - 5)(3x + 1)$

30 $(2x - 1)^2 - 9y^2$
$(2x + 3y - 1)(2x - 3y - 1)$

P **31** Divide $x^3 - 2x^2 - 3x - 5$ by $x^2 - 2$. $x - 2$, remainder $-x - 9$

R,I **32** What are the roots of $(x - 4)(2x + 3)(3x - 5) = 0$? 4, $-\frac{3}{2}$, $\frac{5}{3}$

P,I **33** Use the following table of values to approximate the zeros of f.

x	−4	−3	−2	−1	0	1	2	3	4
$f(x)$	1.5	1	−1	−3	−2	4	8	1	−5

$-3 < z_1 < -2$; $0 < z_2 < 1$; $3 < z_3 < 4$

P **34** Let $P(x) = 2x^4 - 9x^2 + 5x + 4$ and $Q(x) = x + 2$. Find the quotient and the remainder when $P(x)$ is divided by $Q(x)$.
Quotient: $2x^3 - 4x^2 - x + 7$; remainder: -10

R **35** Write an expression for the total area of the two rectangles shown. $x^2 + 8x + 6$

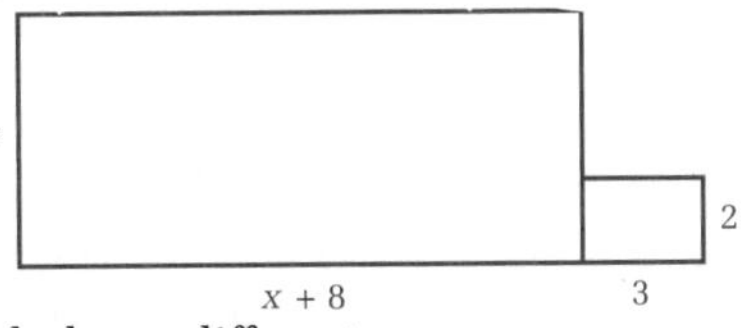

R,I **36** There are 28 students in a geometry class. Each day, a different group of 4 students goes to the board to demonstrate problems. How many classes must there be before one of these groups is forced to repeat? 20,475 classes

P,R **In problems 37–39, factor each expression over the integers.**

37 $x^4 - 25$
$(x^2 + 5)(x^2 - 5)$

38 $x^4 - 10x^2 + 9$
$(x + 3)(x - 3)(x + 1)(x - 1)$

39 $x^4 - 16$
$(x^2 + 4)(x + 2)(x - 2)$

Problem-Set Notes and Strategies, continued

- Problem **25** is a good visual geometric interpretation of the remainder and difference of polynomials.
- Problem **26** previews the Fundamental Theorem of Algebra.
- Problem **33** utilizes the Intermediate Value Theorem from calculus. If a function is continuous, and every polynomial function is, then it must assume values between the points evaluated. Since the points (−3, 1) and (−2, −1) are on the graph, it must have crossed zero somewhere in between. This is an important consequence and may require drawing pictures to convince the student.
- Problem **36** requires the use of combinations.

Problem Set, *continued*

P **40 a** Write a nested polynomial expression for the sum of the volumes of the two solids shown. $(((4)x - 3)x + 8)x + 42$

b Rewrite your answer to part **a** as a polynomial in descending order of powers.
$4x^3 - 3x^2 + 8x + 42$

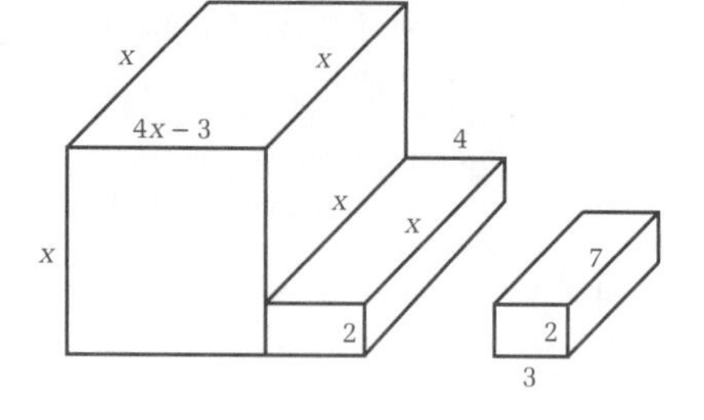

P,I **41** Let $P(x) = x^{15} + 1$.

a Find $P(-1)$. 0

b Is $P(x)$ factorable over the integers? Explain your answer.
Yes; there is an integer n such that $P(n) = 0$.

R **42** Let $P(x) = (3x - 2)(4x + 5)(x - 6)$.

a Find the constant term of $P(x)$. 60

b Find the leading coefficient of $P(x)$. 12

P **43** Divide $2x^3 - 8x^2 + 6x + 5$ by $x + 3$.
$2x^2 - 14x + 48$, remainder -139

P,R **44** Solve $\frac{3}{40}x^2 + \frac{1}{4}x - \frac{1}{5} = 0$ for x. (Hint: Multiply both sides by 40.)
$x = -4$ or $x = \frac{2}{3}$

P,I **45** The zeros of a cubic function Q are 3, −2, and 2. What is a possible factored form of the polynomial $Q(x)$?
$(x - 3)(x + 2)(x - 2)$

P,I **46** Find the quotient and the remainder for $(x^3 - 8) \div (x - 2)$.
Quotient: $x^2 + 2x + 4$; remainder: 0

P,R **47** Suppose that $P(x) = ax^2 + bx + c$ and that $P(0) = 4$, $P(1) = 8$, and $P(-1) = 12$. Find (a, b, c). $(6, -2, 4)$

P **48** Find the quotient and the remainder when the polynomial $ax^3 + bx^2 + bx + a$ is divided by $x - 1$.
Quotient: $ax^2 + (a + b)x + (a + 2b)$; remainder: $2a + 2b$

C P **49** Let $P(x) = ax^5 + bx^3 + cx$, where a, b, and c are negative. Explain why P cannot have any negative zeros.

P,I **50** The diagram shows the graphs of $y = P(x)$ and $y = g(x)$, where P is a cubic function and g is a quadratic function.

a Write a formula for function P.

b Write a formula for function g.

c Write a function H such that $H(x) = P(x) - g(x)$.

d Show that −2 is a zero of H.

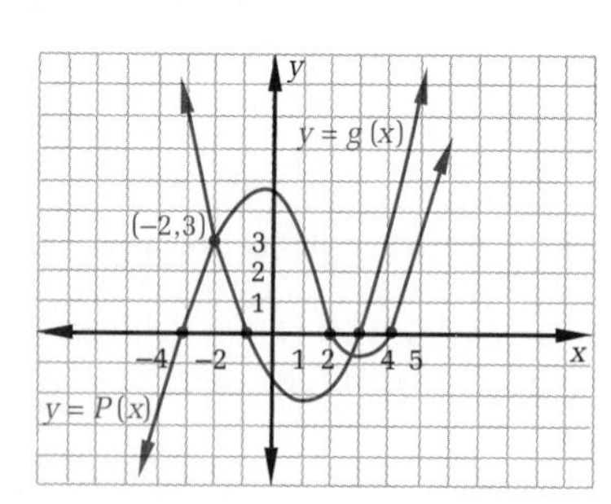

P **51** When polynomial $P(x)$ is divided by $x - 1$, the remainder is 3. When $P(x)$ is divided by $x - 2$, the remainder is 5. Find the remainder when $P(x)$ is divided by $x^2 - 3x + 2$. $2x + 1$

Problem-Set Notes and Strategies, continued

- Problem **45** is a good indicator of whether or not students have a grasp of the relationship between factors and zeros of a polynomial.
- The difference of cubes formula is being used in problem **46**.
- The relationship of functions and their graphs is explored in problem **50**. Capitalize on this opportunity and have students try to graph the function H by subtracting ordinates. It is important that they see the difference of two functions as a graph. This will be useful in graphing the trig functions in the last three chapters.

Additional Answers

49 For any negative value of x, the powers x^5, x^3, and x^1 will be negative. Since a, b, and c are also negative, each term of $P(x)$ will represent a positive number when x is negative. The sum of three positive numbers cannot be 0, so $P(x) \neq 0$ for any negative x.

50a $P(x) = \frac{1}{8}x^3 - \frac{3}{8}x^2 - \frac{5}{4}x + 3$

b $g(x) = \frac{3}{5}x^2 - \frac{6}{5}x - \frac{9}{5}$

c $H(x) = \frac{1}{8}x^3 - \frac{39}{40}x^2 - \frac{1}{20}x + \frac{24}{5}$

d $40[H(x)] = 5x^3 - 39x^2 - 2x + 192$; $40[H(-2)] = 0$, so $H(-2) = 0$.

9.2

THE RATIONAL-ZEROS THEOREM

Objectives

After studying this section, you will be able to

- Perform synthetic division
- Find zeros of functions by factoring and using the quadratic formula
- Apply the Rational-Zeros Theorem

Part One: Introduction

Synthetic Division

Before computers and calculators were developed, mathematicians frequently evaluated polynomials by using ***synthetic division.*** Synthetic division is closely related to nested-form evaluation. Synthetic division also provides the quotient and the remainder when a polynomial is divided by a binomial.

Example *Let* $P(x) = 3x^3 - 4x^2 + 5x - 7$. *Use nested-form evaluation and synthetic division to divide* $P(x)$ *by* $x - 2$.

Nested Form

$P(x) = (((3)x - 4)x + 5)x - 7$

$P(2) = (((3)2 - 4)2 + 5)2 - 7$

$= ((2)2 + 5)2 - 7$

$= (9)2 - 7$

$= \mathbf{11}$

Synthetic Division

$P(x) = 3x^3 - 4x^2 + 5x - 7$

2	3	−4	5	−7
		6	4	18
	3	**2**	**9**	**11**

The results of both methods are the same. When $P(x)$ is divided by $x - 2$, the quotient is $3x^2 + 2x + 9$ and the remainder is 11.

The following steps make up the synthetic-division algorithm for dividing a polynomial $P(x)$ by a binomial of the form $x - a$.

Synthetic-Division Algorithm

1. **Write the value of *a*, then write the coefficients and constant of *P*(*x*) in a row to the right of that value.**
2. **Bring down the first coefficient below the addition line.**
3. **Multiply this number by *a*, and write the product under the next coefficient.**
4. **Add, writing the sum below the addition line.**
5. **Repeat steps 3 and 4 until all the coefficients and the constant have been dealt with. The numbers below the addition line are the quotient coefficients and the remainder.**

Class Planning

Time Schedule

All levels: 2 days

Resource References

Teacher's Resource Book

Class Opener 54

Practice 57

Transparency 18

Class Opener

1. Rewrite $P(x)$ in nested form. $P(x) = 4x^3 - 6x^2 + 2x + 7$

   $(((4)x - 6)x + 2)x + 7$
2. Use the nested form of $P(x)$ and divide $P(x)$ by $x - 3$.

   $4x^2 + 6x + 20$, R 67

Lesson Notes

- Division by binomials can be shortened to a three-line process.
- The Rational-Zeros Theorem generates a long list, but at least it is finite and exhaustive. If there are any rational roots, they *must* be a member of this list.
- Techniques will be presented in the next sections that enable the student to pare the list of possible roots even further. Finding rational roots is a very mechanical process and can be easily programmed on a computer. You may want to have students work in small groups to write an algorithm that finds all the rational roots of a function and checks each to see if it is a root.

Vocabulary

synthetic division

Rational-Zeros Theorem

Using Computers

A student might write a program to test all possible rational zeros of a polynomial.

Using Factoring and the Quadratic Formula to Find Zeros

In previous chapters, we solved linear and quadratic functions by setting the function equal to 0 and solving the resulting equation. Remember that the zeros of a function are the input values for which the output of the function is zero.

When the degree of a function is greater than two, finding the function's zeros is usually more difficult. In Section 6.4, we approximated the zeros of such functions by graphing them and estimating the values of the graphs' x-intercepts. In this section and the next, we introduce algebraic methods that are more accurate than graphing.

Example *One of the zeros of* $P(x) = x^3 - 2x^2 - 4x + 5$ *is* 1. *Find two other zeros of P.*

Since $P(1) = 0$, we know that $x - 1$ is a factor of $P(x)$. To find the other factor, we use synthetic division.

$$\begin{array}{r|rrrr} 1 & 1 & -2 & -4 & 5 \\ & & 1 & -1 & -5 \\ \hline & 1 & -1 & -5 & 0 \end{array}$$

← The zero remainder shows that 1 really is a zero of P.

Thus, $P(x) = (x - 1)(x^2 - x - 5)$. To find other zeros of P, we use the quadratic formula to solve $x^2 - x - 5 = 0$. We find $x = \frac{1 + \sqrt{21}}{2}$ and $x = \frac{1 - \sqrt{21}}{2}$.

If you can rewrite a polynomial function as a product of linear and quadratic factors, you can find its zeros. Set each of the factors equal to 0 and solve the resulting equations.

The Rational-Zeros Theorem

In the preceding example, we could find two irrational zeros of function P because we already knew a rational zero of the function. The ***Rational-Zeros Theorem*** will help us find all the rational zeros of any polynomial function with integral coefficients.

 ***If F is a polynomial function with integral coefficients and if $\frac{p}{q}$ (a rational number in lowest terms) is a zero of F, then p is a factor of the constant term of F(x) and q is a factor of the leading coefficient of F(x).***

Proof of the Rational-Zeros Theorem:
Let $F(x) = a_nx^n + a_{n-1}x^{n-1} + \ldots + a_1x + a_0$, where each value of a_i is an integer. If $\frac{p}{q}$ is a zero of F,

$$a_n\left(\frac{p}{q}\right)^n + a_{n-1}\left(\frac{p}{q}\right)^{n-1} + \ldots + a_1\left(\frac{p}{q}\right) + a_0 = 0$$

We multiply both sides of this equation by q^n.

$$a_np^n + a_{n-1}p^{n-1}q + \ldots + a_1pq^{n-1} + a_0q^n = 0$$
$$a_{n-1}p^{n-1}q + \ldots + a_1pq^{n-1} + a_0q^n = -a_np^n$$
$$q(a_{n-1}p^{n-1} + \ldots + a_1pq^{n-2} + a_0q^{n-1}) = -a_np^n$$

Because q is a factor of the left side, it must also be a factor of the right side. Since $\frac{p}{q}$ is in lowest terms, q is not a factor of p^n, so q must be a factor of a_n, the leading coefficient of $F(x)$. Now let's reconsider the equation in which both sides were multiplied by q^n.

$$a_np^n + a_{n-1}p^{n-1}q + \ldots + a_1pq^{n-1} + a_0q^n = 0$$
$$a_np^n + a_{n-1}p^{n-1}q + \ldots + a_1pq^{n-1} = -a_0q^n$$
$$p(a_np^{n-1} + a_{n-1}p^{n-2}q + \ldots + a_1q^{n-1}) = -a_0q^n$$

Applying the same argument for p that we used for q, we can show that p is a factor of a_0, the constant term of $F(x)$.

All this may seem a little abstract to you. The important thing, however, is to be able to apply the Rational-Zeros Theorem to identify the possible zeros of a function.

Communicating Mathematics

Have students write a paragraph that explains the reasons why the only possible rational zeros of a polynomial are the ones guaranteed by the Rational-Zeros Theorem. Also, have students explain the significance of first factoring out the greatest common factor.

Example *Let $F(x) = 3x^4 - x^3 - 15x^2 - x + 2$. Find all the rational zeros of F and, if possible, all the other zeros of F.*

According to the Rational-Zeros Theorem, if there is a rational zero $\frac{p}{q}$ of F, then p must be a divisor of 2 and q must be a divisor of 3. The only possible values of p are ±1 and ±2, and the only possible values of q are ±1 and ±3. The possibilities for $\frac{p}{q}$ are therefore ±1, ±2, $\pm\frac{1}{3}$, and $\pm\frac{2}{3}$.

Let's evaluate $F(1)$ and $F(-1)$.

$$F(1) = 3(1^4) - 1^3 - 15(1^2) - 1 + 2 = -12$$
$$F(-1) = 3(-1)^4 - (-1)^3 - 15(-1)^2 - (-1) + 2 = -8$$

Neither 1 nor -1 is a zero. Let's evaluate $F(-2)$ next, using synthetic division.

$$\begin{array}{r|rrrrr} -2 & 3 & -1 & -15 & -1 & 2 \\ & & -6 & 14 & 2 & -2 \\ \hline & 3 & -7 & -1 & 1 & 0 \end{array} \leftarrow \text{A zero!}$$

So $F(x)$ can be factored as $(x + 2)(3x^3 - 7x^2 - x + 1)$. Applying the Rational-Zeros Theorem again, we find that the only possible rational zeros of $3x^3 - 7x^2 - x + 1$ are $\pm\frac{1}{3}$ and ±1. We eliminate 1 and -1, since these are not zeros of F.

Let's try $\frac{1}{3}$.

$$\begin{array}{r|rrrr} \frac{1}{3} & 3 & -7 & -1 & 1 \\ & & 1 & -2 & -1 \\ \hline & 3 & -6 & -3 & 0 \end{array} \leftarrow \text{Another zero!}$$

Now we know that $F(x) = (x + 2)\left(x - \frac{1}{3}\right)(3x^2 - 6x - 3)$, or $F(x) = 3(x + 2)\left(x - \frac{1}{3}\right)(x^2 - 2x - 1)$. We can use the quadratic formula to solve $x^2 - 2x - 1 = 0$ and find $x = 1 \pm \sqrt{2}$. The four zeros of F are -2, $1 - \sqrt{2}$, $\frac{1}{3}$, and $1 + \sqrt{2}$.

Checkpoint

1. Use synthetic division to divide $x^3 + x^2 + x - 1$ by $x + 1$.
   $x^2 + 1$, remainder -2
2. Find all possible rational roots of the polynomial $4x^3 - 3x + 9$.
   $\pm1, \pm3, \pm9, \pm\frac{1}{2}, \pm\frac{3}{2}, \pm\frac{9}{2}, \pm\frac{1}{4}, \pm\frac{3}{4}, \pm\frac{9}{4}$
3. In the synthetic division problem below, what is the dividend and what is the remainder.

2	3	−1	−14	7
		6	10	−8
	3	5	−4	−1

$3x^3 - x^2 - 14x + 7$; -1

Assignment Guide

Average	
Day 1	6–10, 12–14, 20, 21, 23
Day 2	11, 15–18, 22, 24–26, 29, 30a, 38
Advanced	
Day 1	6–14, 17, 19, 21
Day 2	15, 16, 18, 22, 23, 25, 27, 29, 30, 36–38
Honors	
Day 1	6, 7, 9–12, 15, 17, 19, 24, 27
Day 2	13, 14, 16, 18, 23, 25, 28, 30, 32, 33–35, 38

Part Two: Sample Problems

Problem 1 *Use synthetic division to find the quotient and the remainder when $2x^3 - 3x^2 + 100$ is divided by $x + 4$.*

Solution

−4	2	−3	0	100
		−8	44	−176
	2	−11	44	−76

The quotient is $2x^2 - 11x + 44$. The remainder is -76.

Problem 2 *Let $P(x) = \frac{1}{15}x^4 - \frac{1}{5}x^3 - x^2 - x + \frac{10}{3}$. Write a polynomial function Q that has integral coefficients and the same zeros as P.*

Solution The zeros of P are the solutions of the equation $P(x) = 0$. When we multiply both sides of this equation by a number, the right side remains zero, so we can multiply $P(x)$ by any number without changing the zeros. The least common denominator of the terms of $P(x)$ is 15, so if we let $Q(x) = 15[P(x)]$, function Q will have integral coefficients and the same zeros as P. Thus, $Q(x) = x^4 - 3x^3 - 15x^2 - 15x + 50$.

Problem 3 *Write a cubic polynomial function whose zeros are -2, 3, and $\frac{2}{3}$.*

Solution According to the Factor Theorem, a polynomial function with the factors $x + 2$, $x - 3$, and $x - \frac{2}{3}$ will have the required zeros.

$$P(x) = (x + 2)(x - 3)\left(x - \frac{2}{3}\right)$$
$$= x^3 - \frac{5}{3}x^2 - \frac{16}{3}x + 4$$

Another answer is $P(x) = 3x^3 - 5x^2 - 16x + 12$. There are an infinite number of polynomial functions with these three zeros.

Part Three: Exercises and Problems

Warm-up Exercises

1 Consider the following example of synthetic division.

−4	2	3	−18	5
		−8	20	−8
	2	−5	2	−3

a What polynomial is the dividend?
b What is the binomial divisor?
c What is the remainder?
d What is the quotient?

2 What are the quotient and the remainder when $3x^2 - 6x + 7$ is divided by $x - 4$? 3x+6 R = 31

P **3** Use synthetic division to divide $2x^2 - 3x + 5$ by $x - 1$.
$2x - 1$, remainder 4

P **4** Find all the rational zeros of the function
$f(x) = 3x^4 - 7x^3 - x^2 + 7x - 2$. $-1, \frac{1}{3}, 1, 2$

R **5** The zeros of a function P are 4, -3, $\frac{2}{3}$, and 7. Write an expression with integral coefficients that might represent $P(x)$.
$3x^4 - 26x^3 + x^2 + 262x - 168$

Problem Set

P **A** **6** Use synthetic division to divide $3x^3 + x^2 - 2x + 6$ by $x + 1$.
$3x^2 - 2x$, remainder 6

P **7** List the possible rational zeros of $P(x) = 2x^3 + ax + 5$, where a is an integer. $\pm1, \pm5, \pm\frac{1}{2}, \pm\frac{5}{2}$

P,R **8** Find the product $(x - 1)(x - 2)(x - 3)$, then use the product to write a polynomial function having the zeros 1, 2, and 3.
$x^3 - 6x^2 + 11x - 6$; $P(x) = x^3 - 6x^2 + 11x - 6$

P **9** Solve $4x^3 + 12x^2 - 9x - 27 = 0$ for x.
$x = -3$, $x = -\frac{3}{2}$, or $x = \frac{3}{2}$

P,R **10** Write an expression to represent the base of the rectangle shown.
$x^2 - 3x + 2$

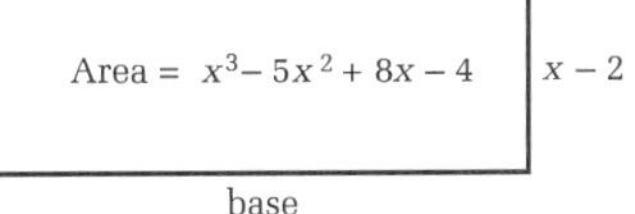

I **11** Use the graph to approximate the zeros of f.
$\approx -3.7, \approx 0.5, \approx 2.6$

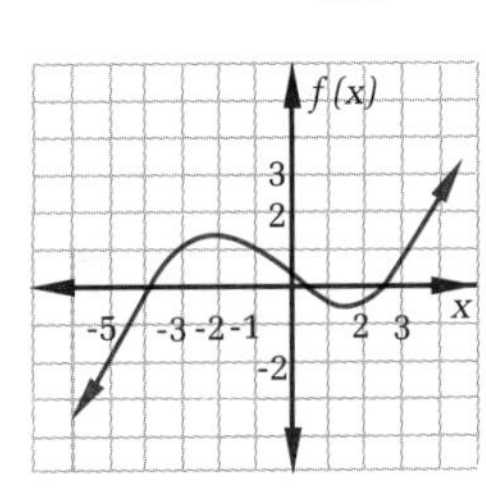

B P,R **12** The difference between the numerical value of the volume of the cylinder and the numerical value of the area of one of its bases is $\frac{4}{27}\pi$. Find the radius of the cylinder's base. $\frac{2}{3}$

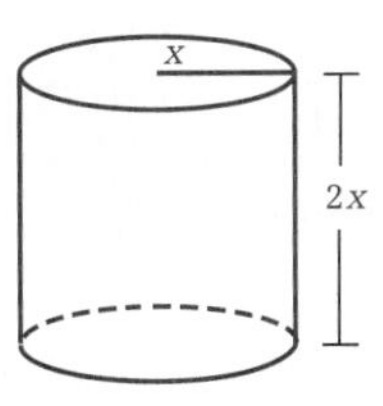

R **13** Write an equation whose roots are -2, 4, and 9.
$x^3 - 11x^2 + 10x + 72 = 0$

P **14** Use synthetic division to find the remainder for $(3x^2 - 5x + 7) \div (x - 2)$. 9

P **15** List the possible rational zeros of the function $P(x) = 4x^3 + ax^2 + bx + 6$, where a and b are integers.
$\pm1, \pm2, \pm3, \pm6, \pm\frac{1}{2}, \pm\frac{3}{2}, \pm\frac{1}{4}, \pm\frac{3}{4}$

Problem-Set Notes and Strategies

- Problem **4** will be longer than most students are used to since they must check 8 roots. Remind them that as soon as they get a quadratic, the quadratic formula or factoring can be used to finish the problem.
- Problem **5** is a good exercise in working backwards. Note also that the answer is not unique. Any integral multiple of this answer will work. You may want to point out that it is possible to multiply by another polynomial and still retain the roots. More real roots may be added but the current roots do not change.

Stumbling Block

When setting up a synthetic division problem, students frequently forget to leave a place for any missing powers of a variable in the polynomial. A zero must be inserted as a place holder to avoid error.

Problem-Set Notes and Strategies, continued

- Problem **17** previews the Fundamental Theorem of Algebra. If students relate this concept to the division of real numbers, it will make more sense.
- Problems **24** and **25** illustrate the importance of factoring out the greatest common factor first and then proceeding with factoring.
- Problem **31** may be done by using Pascal's Triangle or by using combinations. These numbers will be needed in Section **9.5** on the Binomial Theorem.
- Problem **32** may be solved by factoring or by using the quadratic formula.

Problem Set, *continued*

P **16** Explain why the following statement is false: According to the Rational-Zeros Theorem, in any rational root $\frac{p}{q}$ of the function $P(x) = 9x^3 - \frac{2}{3}x^2 + 4x - 5$ the numerator p will be a factor of 9 and the denominator q will be a factor of -5. The Rational-Zeros Theorem does not apply to function P, since P(x) contains a nonintegral coefficient.

P,R **17** When polynomial s is divided by $x + 4$, the quotient is $4x^2 - 7x + 5$ and the remainder is -4. What is the polynomial?
$s = 4x^3 + 9x^2 - 23x + 16$

R,I **18** The graph of $y = P(x)$ is shown. Between what pairs of consecutive integers can the zeros of P be found?
-2 and -1, 1 and 2, 2 and 3, and two between 5 and 6

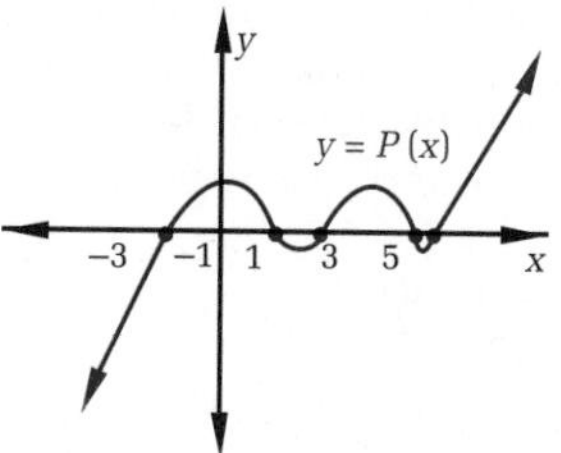

R **19** If $P(x) = x^3 - 4x^2 + cx + 6$, for what value of c is $P(3) = 0$? 1

R **In problems 20–22, solve for x in terms of the other variables.**

20 $ax = c - b$ $x = \frac{c-b}{a}$ **21** $(a + b)x = 7$ $x = \frac{7}{a+b}$ **22** $ax - bx = 9$ $x = \frac{9}{a-b}$

P,R **23** Show that $x - 2$ is a factor of $x^3 - 5x^2 + 7x - 2$. When $x = 2$, the value of the cubic polynomial is 0, so by the Factor Theorem, $x - 2$ is a factor of the polynomial.

P,R **In problems 24 and 25, factor each expression completely.**

24 $12x^3 - 21x^2 - 6x$
$3x(4x + 1)(x - 2)$

25 $18x^4 - 50x^2$
$2x^2(3x - 5)(3x + 5)$

P,R **26** Write a function whose zeros are 2, 3, -2, and -3.
$f(x) = x^4 - 13x^2 + 36$

P **27** Let $P(x) = x^3 - 4x^2 - 6x + 9$. Given that $x - 1$ is a factor of P(x), find all the zeros of P. $1, \frac{3+3\sqrt{5}}{2}, \frac{3-3\sqrt{5}}{2}$

P **28** Solve $6y^4 + 7y^3 - 37y^2 - 8y + 12 = 0$ for y.
$y = -3, y = -\frac{2}{3}, y = \frac{1}{2}$, or $y = 2$

P,R **29** A certain polynomial is evenly divisible by $x - 3$, with the quotient being $2x^2 - 5x + 7$. What is the polynomial?
$2x^3 - 11x^2 + 22x - 21$

R **30** Write a cubic polynomial function that has integral coefficients and the given zeros.

a 2, -5, and -1
$P(x) = x^3 + 4x^2 - 7x - 10$

b $\frac{2}{3}, \frac{1}{3}$, and 1
$P(x) = 9x^3 - 18x^2 + 11x - 2$

R **31** The numbers in one of the rows of Pascal's triangle are 1, 7, 21, 35, 35, 21, 7, 1. Determine the numbers in the next row.
1, 8, 28, 56, 70, 56, 28, 8, 1

P **32** What are the zeros of $P(x) = x^4 - 12x^2 + 35$?
$-\sqrt{7}, -\sqrt{5}, \sqrt{5}, \sqrt{7}$

P,I **33** A quadratic equation has two roots, one of which is twice the other. If the sum of the roots is 30, what is the equation? $x^2 - 30x + 200 = 0$

P **34** Let $P(x) = (x - 4)(x^2 - 7)$.

a What are the zeros of P? $-\sqrt{7}, \sqrt{7}, 4$

b If $Q(x) = P(x - 6)$, what are the zeros of Q? $6 - \sqrt{7}, 6 + \sqrt{7}, 10$

c If $R(x) = P(2x)$, what are the zeros of R? $-\frac{\sqrt{7}}{2}, \frac{\sqrt{7}}{2}, 2$

P **C**

In problems 35 and 36, solve for x.

35 $x^2 + ax + bx + ab = 0$ $x = -a$ or $x = -b$

36 $x^3 - 6x^2 + 12x - 9 = 0$ $x = 3$

P,R **37** What is the domain of the function $f(x) = \frac{2x + 3}{x^3 - 3x^2 - 4x + 12}$? $\{x{:}x \neq -2, x \neq 2, x \neq 3\}$

R **38** Let $Q(x) = 6x^3 - 5x^2 + kx - 2$.

a List all the possible rational zeros of Q. $\pm 1, \pm 2, \pm\frac{1}{2}, \pm\frac{1}{3}, \pm\frac{2}{3}, \pm\frac{1}{6}$

b If 2 is a zero of Q, what is the value of k? -13

Problem-Set Notes and Strategies, continued

- Problem **33** is a system of equations, but the answer to that system must be used as roots and then changed into factors of a polynomial to find the equation.

CAREER PROFILE

Statistical Analysis of Success

Leonard Ramist assesses students' prospects

Your score on a college entrance examination is a function of the mean and the standard deviation for the scores of all the students taking the test. The purpose of such examinations is to help predict how you will do in college.

One career in statistics involves determining how the test results themselves will be evaluated and how the scores will be used. This is Leonard Ramist's career. He is a Program Director in the Admission Testing Program of the Educational Testing Service in Princeton, New Jersey.

"One of the major mathematical aspects of my job is to find an equation based on SAT scores and grades in high school that will best predict a student's success as a freshman at a particular college," says Ramist. "We provide that information to colleges, as well as a statistical program that helps them determine how well grades and SAT's work in predicting success at that college."

The statistical tools Ramist uses most often are correlation and regression analysis. Regression analysis involves finding a function that best fits a set of data.

Mr. Ramist has a bachelor's degree in economics and accounting and a master's degree in operations research, both from the University of Pennsylvania. "Operations research," he says, "is a combination of mathematics and statistics and involves modeling real-life situations."

9.3

Class Planning

Time Schedule
All levels: 2 days

Resource References
Teacher's Resource Book
Class Opener 55
Cooperative Learning 27
Practice 58
Evaluation
Quiz Forms 1, 2, 3

Class Opener

1 Mario made a list of all possible rational zeros of $P(x) = x^4 - 2x^3 - 7x^2 + 12$. Mario then determined which of the possibilities were actually zeros. If one of the possible zeros were selected at random, what is the probability that it would, in fact, be a zero of the polynomial. 0

2 What would it mean if the answer to problem **1** were zero?
There are no rational roots.

Lesson Notes

- In this section, students will study how to find the upper and lower bounds for the zeros of polynomials. This process will eliminate work in solving equations. Emphasize that synthetic division is the key.
- Roots other than rational roots also exist. This section explores the bisection method to find irrational roots to an accuracy determined by a computer or calculator.

9.3 Real Zeros of Polynomial Functions

Objectives

After studying this section, you will be able to

- Identify upper and lower bounds of the zeros of a polynomial function
- Determine the approximate values of real zeros

Part One: Introduction

Upper and Lower Bounds of Zeros

In the preceding section, you learned how to find rational zeros of a polynomial function. What happens if real zeros are not rational numbers?

A useful way to begin to approximate any real zero of a function is to identify an ***upper bound***—a value of x above which there are definitely no zeros—and a ***lower bound***—a value of x below which there are definitely no zeros. We can then search for zeros between these two bounds.

Consider the function $P(x) = 2x^3 - 11x^2 - 10x + 55$. Let's try to find a positive value of x that is an upper bound. Choosing $x = 7$, we evaluate $P(7)$ using synthetic division.

7	2	−11	−10	55
		14	21	77
	2	3	11	132

When $P(x)$ is divided by $x - 7$, the quotient is $2x^2 + 3x + 11$ and the remainder is 132. In other words,

$$P(x) = (x - 7)(2x^2 + 3x + 11) + 132.$$

Since all the terms of $2x^2 + 3x + 11$ are positive, its value will be positive for any positive value of x. In addition, $x - 7$ will have a positive value for any x greater than 7. We have identified an upper bound of the function's zeros.

When a polynomial $P(x)$ is divided by $x - a$, where $a > 0$, if all the terms of the quotient are nonnegative and the remainder is nonnegative, then a is an upper bound of the zeros of P.

Vocabulary
upper bound
lower bound
bisection

Now let's find a lower bound of the zeros of P by looking at negative values of x. We will try x = −3.

−3	2	−11	−10	55
		−6	51	−123
	2	−17	41	−68

Thus, $P(x) = (x + 3)(2x^2 - 17x + 41) - 68$. In this case, the value of $2x^2 - 17x + 41$ will be positive for any negative value of x, but x + 3 will be negative for any value of x less than −3. Therefore, P(x) will be negative for any x less than −3. We have identified a lower bound of the function's zeros.

 ***When a polynomial P(x) is divided by x − a, where a < 0, if the terms of the quotient and the remainder alternate in sign, then a is a lower bound of the zeros of P.***

To simplify our calculations, in the examples in this section we will work with a function that has a positive leading coefficient. If a polynomial function P has a negative leading coefficient, we can always write a function Q, Q(x) = −[P(x)]. This function Q will have a positive leading coefficient and, as the graph shows, the same zeros as P.

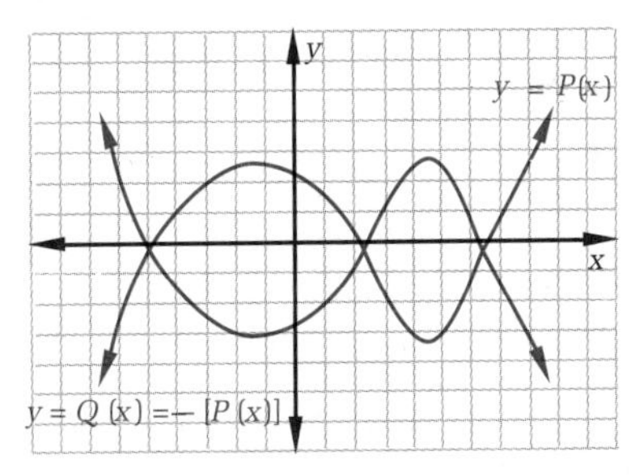

Example *Let* $P(x) = 2x^4 + 5x^3 - 13x^2 - 25x + 15$. *Find an upper bound and a lower bound of the zeros of P.*

When we evaluate P(1) or P(2), some of the terms in the quotient are negative or the remainder is negative. Let's try evaluating the function for x = 3.

3	2	5	−13	−25	15
		6	33	60	105
	2	11	20	35	120

Since all the terms of the quotient and the remainder are positive, 3 is an upper bound of the zeros.

When we test negative values of x to find a lower bound, we find that for x = −1 through x = −4 the quotient terms and remainder do not alternate in sign. Let's try x = −5.

−5	2	5	−13	−25	15
		−10	25	−60	425
	2	−5	12	−85	440

Since the terms of the quotient and the remainder alternate in sign, −5 is a lower bound of the zeros.

Lesson Notes, continued

- Have a polynomial with an irrational root ready and show students the graph. It is easy to see that one exists and the method is intuitive. Explain that bisection is a method commonly used to find roots of any continuous functions, but it is slow. There are other methods. If you have a computer and programs available, you might want to compare the bisection method with Newton's method from calculus and show the relative speed. Most commercial software packages have the two methods available.

Checkpoint

1. Find a lower bound for the polynomial $P(x) = x^3 - 8x^2 - 44x - 48$.
   Root is −2; lower bound is anything less that is a factor of 48.
2. For the polynomial $P(x) = x^4 + 5x^3 + 12x^2 + 2x - 10$, find an interval where a possible root exists.
   Between 0 and 1, since $P(0) = -10$ and $P(1) = 10$, there must be a zero between them.
3. Why must 1 be an upper bound of $x^3 + 6x^2 + 11x + 6$?
   When the polynomial is divided, the terms of quotient and the remainder are all positive.

9.3

Assignment Guide

Average	
Day 1	5, 6, 7, 10–12, 16, 27, 32
Day 2	4, 9, 15, 19a, c, 20, 22–24, 30, 31
Advanced	
Day 1	4–7, 10–12, 16, 17, 19a, c, 22–24
Day 2	13, 15, 20, 26a, b, 27, 28, 31, 32, 34
Honors	
Day 1	5–7, 12, 13, 16, 18, 22–24, 26, 27
Day 2	15, 19b, d, 20, 28, 29, 32–34

Communicating Mathematics

Your best friend was absent from class today. Write a paragraph that explains the bisection method to him/her. Include ideas about where to start and how to end the process.

Approximating the Real Zeros of a Function

Once we find upper and lower bounds of the zeros of *P*, we can make a table of values and graph *P* over the domain defined by these bounds. (A BASIC program that generates sets of ordered pairs for polynomial functions is included in the problem set.)

See p. 789

Example *Roughly estimate the values of the zeros of the function* $P(x) = 2x^4 + 5x^3 - 13x^2 - 25x + 15$.

From the previous example, we know that a lower bound of the zeros of this function is −5 and an upper bound is 3. We make a table for inputs between these bounds and sketch the graph.

x	$P(x)$
−5.0000000	440.00000
−4.6000000	263.73120
−4.2000000	142.57920
−3.8000000	64.947200
−3.4000000	20.467200
−3.0000000	0.00000000
−2.6000000	−4.3648000
−2.2000000	0.69120000
−1.8000000	9.7152000
−1.4000000	18.483200
−1.0000000	24.000000
−0.60000000	24.499200
−0.20000000	19.443200
0.20000000	9.5232000
0.60000000	−3.3408000
1.0000000	−16.000000
1.4000000	−24.076800
1.8000000	−21.964800
2.2000000	−2.8288000
2.6000000	41.395200
3.0000000	120.00000

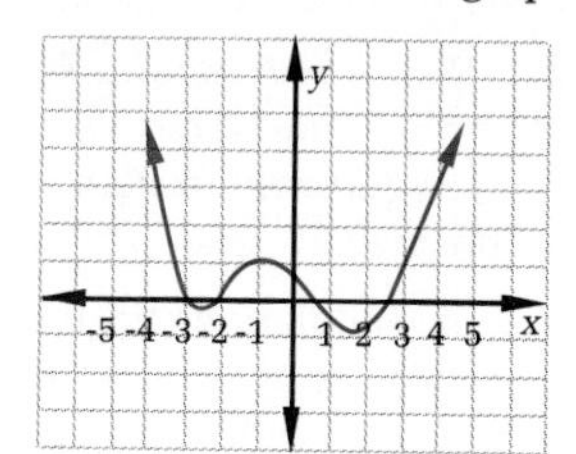

The table and the graph show that −3 is one of the zeros of *P*. There is another zero between −2.6 and −2.2, a third between 0.2 and 0.6, and a fourth between 2.2 and 2.6.

Now we can increase the accuracy of the approximations by using a process known as ***bisection.*** As you have seen, if the values of $P(x)$ have different signs for two inputs x_1 and x_2, a zero of *P* must lie somewhere between x_1 and x_2. Bisection is a technique that uses this fact repeatedly to approach closer approximations of a zero.

The Bisection Method of Locating a Zero

1. **Evaluate the function for two *x* values, one slightly to the left of the zero on the x-axis and one slightly to the right of the zero.**
2. **Find the average of the two *x* values and evaluate the function for this third value.**
3. **Of the three *x* values, choose the two on either side of the zero.**
4. **Repeat steps 1-3 until the desired accuracy is reached.**

A BASIC program that approximates zeros by bisection appears in the problem set.

Example *Let* $P(x) = 2x^4 + 5x^3 - 13x^2 - 25x + 15$. *Use the bisection method to find, to the nearest thousandth, the value of the zero of P that lies between 2.2 and 2.6.*

We first define the following variables.

x_L = the x value to the left of the zero
x_R = the x value to the right of the zero
x_M = the average of x_L and x_R
P_L = the value of P(x) at x_L
P_R = the value of P(x) at x_R
P_M = the value of P(x) at x_M

The following table summarizes our calculations.

x_L	x_M	x_R	P_L	P_M	P_R	Bounds of Zero
2.2	2.4	2.6	−2.828	15.595	41.395	x_L, x_M
2.2	2.3	2.4	−2.828	5.533	15.595	x_L, x_M
2.2	2.25	2.3	−2.828	1.148	5.533	x_L, x_M
2.2	2.225	2.25	−2.828	−0.890	1.148	x_M, x_R
2.225	2.2375	2.25	−0.890	0.117	1.148	x_L, x_M
2.225	2.3125	2.2375	−0.890	−0.390	0.117	x_M, x_R
2.23125	2.23438	2.2375	−0.390	−0.137	0.117	x_M, x_R
2.23438	2.23594	2.2375	−0.137	−0.011	0.117	x_M, x_R
2.23594	2.23672	2.2375	−0.011	0.053	0.117	x_L, x_M
2.23594	2.23633	2.23672	−0.011	0.021	0.053	x_L, x_M
2.23594	2.23614	2.23633	−0.011	0.005	0.021	x_L, x_M

The value of the zero to three decimal places is 2.236.

Part Two: Sample Problem

Problem *If the box shown has a volume of 50 cubic units, what is the value of x to three decimal places?*

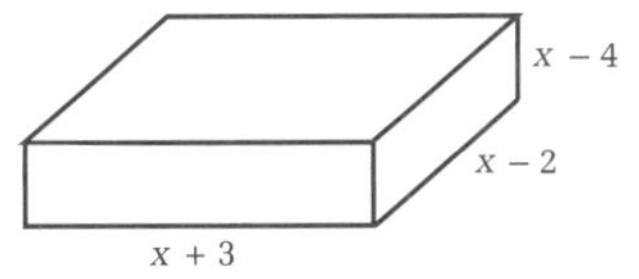

Solution We write the equation $(x + 3)(x - 2)(x - 4) = 50$ and simplify to obtain $x^3 - 3x^2 - 10x - 26 = 0$. The real roots of this equation are the same as the real zeros of $P(x) = x^3 - 3x^2 - 10x - 26$. We use the procedures shown in the examples.

When we evaluate $P(x)$ for various values of x we find that −3 is a lower bound and 6 is an upper bound. Making a table of values for inputs between these bounds shows us that P has only one real zero and that this zero is between 5.4 and 5.6. We then use the bisection method to approximate this zero to three decimal places and find that $x \approx 5.589$.

Stumbling Block

The bisection method is a great formula for approximating roots of polynomials. Its power can be extended beyond polynomials, to any continuous function. As yet, the students have not been exposed to the definition of continuity. Caution them on the use of the bisection process with functions that may not be continuous, such as rational functions.

Cooperative Learning

Have students work in small groups to solve the following problem. If $500 is invested in a bank account paying 6% compounded quarterly, the amount of money at any given time is given by the equation $A(t) = 500(1 + \frac{.06}{4})^{4t}$ where *t* is in years. Let students use the bisection method and calculate to two decimal places the number of years it will take for the money to grow to $1000. If they have trouble, have them graph the equation $A = 1000 - 500(1 + \frac{.06}{4})^{4t}$ and try to find values where it is positive and negative. The zero must lie between the two. The solution using logarithms will be presented in Chapter **11**.

Problem-Set Notes and Strategies

- In problem **2**, as well as finding the upper bound and lower bound, you might have the student find the least upper bound and greatest lower bound. These values should be unique.
- Students may know that the answer to problem **6** is the square root of 2 and may not want to use the bisection method.

Part Three: Exercises and Problems

Warm-up Exercises

P **1** Let $P(x) = 3x^4 + 5x^3 - x^2 - 4x - 10$. Explain why -2 is a lower bound of the zeros of P. When $P(x)$ is divided by $x + 2$, the terms of the quotient and the remainder alternate in sign.

P **2** Find an upper bound and a lower bound of the zeros of $P(x) = x^3 - 6x^2 - 10x + 60$.
Answers will vary. Possible answer: 8; -4

P **3** Let $P(x) = 2x^4 + 5x^3 - 13x^2 - 25x + 15$. (This is the function used in the examples in this section.) Use the bisection method to calculate the zero located between -2.6 and -2.2 to two decimal places. ≈ -2.24

Problem Set

If you have access to a computer, you may use the following two programs to solve some of the problems in this set.

```
 10 REM This program prints lists of ordered pairs for
 20 REM polynomial functions
 30 GOSUB 170
 40 GOSUB 240
 50 REM --- Ordered Pair Loop ---
 60 FOR X = LB TO UB STEP INC
 70 GOSUB 110
 80 PRINT "(";X;", ";Y;")"
 90 NEXT X
100 END
110 REM --- Nested-Form Polynomial Evaluation ---
120 LET Y = A(N)
130 FOR I = N - 1 TO 0 STEP -1
140 LET Y = Y * X + A(I)
150 NEXT I
160 RETURN
170 REM --- Polynomial Input Section ---
180 INPUT "Enter the degree of the polynomial. ==>";N: DIM A(N)
190 FOR I = N TO 0 STEP -1
200 PRINT "Enter the coefficient of"
210 PRINT "the term of degree ";I;". ==>";: INPUT A(I)
220 NEXT I
230 RETURN
240 REM --- Loop Data Input Section ---
250 INPUT "Enter the lower-bound value of x. ==>";LB
260 INPUT "Enter the upper-bound value of x. ==>";UB
270 INPUT "Enter the increment. ==>";INC
280 RETURN
```

```
 10 REM This program finds values of zeros by bisection
 20 GOSUB 140
```

```
 30 GOSUB 210
 40 REM --- Bisection Calculations ---
 50 PRINT "X-MID", "Y-MID"
 60 LET X = XL: GOSUB 270: LET YL = Y
 70 LET X = XR: GOSUB 270: LET YR = Y
 80 IF (XR - XL) < A THEN PRINT "The zero is ";XL".": END
 90 LET X = (XR + XL)/2: GOSUB 270
100 PRINT X, Y
110 IF Y * YL < 0 THEN XR = X
120 IF Y * YL >= 0 THEN XL = X
130 GOTO 60
140 REM --- Polynomial Input Section ---
150 INPUT "Enter the degree of the polynomial. ==>";N: DIM A(N)
160 FOR I = N TO 0 STEP -1
170 PRINT "Enter the coefficient of"
180 PRINT "the term of degree ";I;". ==>";: INPUT A(I)
190 NEXT I
200 RETURN
210 REM --- Initial Approximations Data Entry ---
220 INPUT "Enter an x value to the left of the zero. ==>";XL
230 INPUT "Enter an x value to the right of the zero. ==>";XR
240 INPUT "To how many places do you want the zero (1-5)? ==>";A
250 LET A = 1/10 ^ (A + 1)
260 RETURN
270 REM --- Nested-Form Polynomial Evaluation ---
280 LET Y = A(N)
290 FOR I = N - 1 TO 0 STEP -1
300 LET Y = Y * X + A(I)
310 NEXT I
320 RETURN
```

A

P **4** Let $P(x) = 2x^4 + 5x^3 - 16x^2 - 3x - 1$. Are there any real roots of P greater than

a 1? Yes **b** 2? No **c** 3? No **d** 4? No

P **5** Find an upper bound and a lower bound of the zeros of $f(x) = 2x^3 - 9x^2 - 16x + 72$. Answers will vary. Possible answer: 6; -3

P **6** Determine the zeros of $f(x) = x^2 - 2$ to the nearest thousandth. ≈ -1.414, ≈ 1.414

P **7** The function $f(x) = x^3 - x - 2$ has a zero between 1 and 2. Approximate the value of this zero to three decimal places. ≈ 1.522

P **8** Given that 5 is an upper bound of the zeros of a polynomial function P, identify 12 other upper bounds for P. Answers will vary. Any 12 numbers greater than 5 are possible answers.

R **In problems 9–11, factor each expression over the integers.**

9 $12y^2 - 3y^4$
$3y^2(2 + y)(2 - y)$

10 $a^4 - 1$
$(a^2 + 1)(a + 1)(a - 1)$

11 $8 - b^3$
$(2 - b)(4 + 2b + b^2)$

Problem Set, *continued*

R **12** Find the quotient and the remainder when $x^3 - 7x^2 + 9x + 8$ is divided by $x - 5$. Quotient: $x^2 - 2x - 1$; remainder: 3

B **In problems 13 and 14, approximate the real zeros of each function to the nearest thousandth.**

P **13** $f(x) = 3x^4 - x^3 + 5x - 1$
≈ -1.149, ≈ 0.201

14 $g(x) = x^4 - 2x^3 - 7x^2 + 6x + 12$
≈ -1.732, ≈ -1.236, ≈ 1.732, ≈ 3.236

P,R **15** Solve $x^3 + 5x^2 - 13x - 65 = 0$ for x.
$x = -5$, $x \approx -3.606$, or $x \approx 3.606$

P **16** The volume of the box shown is 140 cubic units. Find integral upper and lower bounds of the possible values of x.
Answers will vary. Possible answer: 6; 2 (since $x \not\leq 2$)

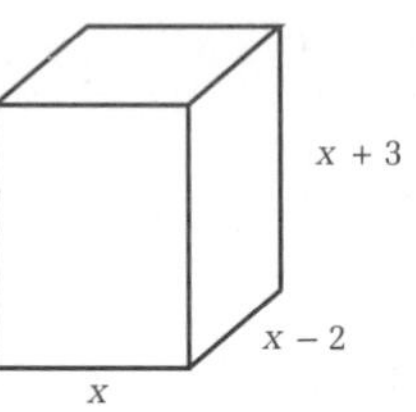

R **17** Solve the following system for (x, y). $(x, y) = (-2, 0)$ or $(x, y) = (2, 0)$
$$\begin{cases} x^2 - y^2 = 4 \\ 3x^2 + y^2 = 12 \end{cases}$$

R **18** Suppose that y varies inversely as the reciprocal of x and that $y = 36$ when $x = 12$. What is the value of y when $x = 6$? 18

P **19** Find the greatest nonpositive integral lower bound of the zeros of each function.
a $f(x) = x^4 - 2x^3 + x^2 - 3x + 4$ 0
b $g(x) = x^4 - 2x^3 + x^2 - 3x - 5$ -1
c $h(x) = x^4 - 2x^3 + x^2 - 3x - 35$ -2
d $k(x) = x^4 - 2x^3 + x^2 - 3x - 120$ -3

R,I **20** Expand each exponential expression.
a $(x + y)^0$ 1
b $(x + y)^1$ $x + y$
c $(x + y)^2$ $x^2 + 2xy + y^2$
d $(x + y)^3$ $x^3 + 3x^2y + 3xy^2 + y^3$

R **21** Let $f(z) = \frac{z^2 + z}{z}$ and $g(z) = z + 1$. Find each of the following.
a $f(3)$ 4
b $f(2)$ 3
c $f(-1)$ 0
d $g(3)$ 4
e $g(2)$ 3
f $g(-1)$ 0

R **22** Let $P(x) = 5x^2 - 7x + 11$.
a Rewrite $P(x)$ in nested form.
$((5)x - 7)x + 11$
b Find $P(2x)$.
$20x^2 - 14x + 11$

P **23** If the volume of the cylinder shown is 100 cubic units, what is the value of x to two decimal places? ≈ 1.72

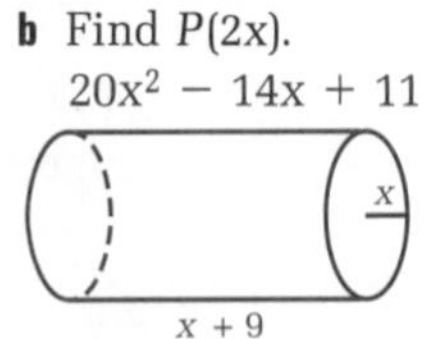

Problem-Set Notes and Strategies, continued

- A programmable calculator or a computer will help students solve problems **13 – 15**.

Using Computers

Graphing software such as the *Omnifarious Plotter* or a graphing calculator can be used to find real zeros of polynomials by zooming in around the intersection of the graphs with the x-axis. Software such as *Derive* or the *Mathematics Exploration Toolkit*, or calculators such as the *HP28s*, have built-in root-finding routines. These can be used effectively with this section.

- Problem **19** extends the concept of lower bound to a greatest lower bound. Must the greatest lower bound always exist? Good question for your students to ponder.
- Problem **20** is leading up to the discussion of the Binomial Theorem in Section **9.5**.

R | **In problems 24 and 25, solve each inequality for x.**

24 $x(x^2 - 6x + 8) < 0$ $\{x{:}x < 0 \text{ or } 2 < x < 4\}$

25 $x^2 \geq 6x$ $\{x{:}x \leq 0 \text{ or } x \geq 6\}$

R | **26** Sketch the graphs of the following equations on the same coordinate system.

a $\frac{x^2}{4} + \frac{y^2}{9} = 1$ **b** $\frac{x^2}{4} - \frac{y^2}{9} = 1$ **c** $\frac{y^2}{9} - \frac{x^2}{4} = 1$

P,R | **27** Let P be a polynomial function such that $P(2) = -8$ and $P(3) = 12$.

a Explain why P must have at least one zero between 2 and 3.
b Draw a diagram to show that P could have more than one zero between 2 and 3.
c Can P have exactly two zeros between 2 and 3? No

C P,R | **28** Let $P(x) = 5x^3 - 31x^2 + 49x - 11$.

a Between what pairs of consecutive integers can the real zeros of P be found? 0 and 1, 2 and 3, 3 and 4
b Use the Rational-Zeros Theorem and the quadratic formula to find the exact values of the zeros of P. $2 - \sqrt{3}, \frac{11}{5}, 2 + \sqrt{3}$

P | **29** If $P(x) = 2x^4 - 4x^3 - 5x^2 + ax - 6$ and if 3 is an upper bound of the zeros of P, what are the possible values of a? $\{a{:}a \geq -1\}$

P | **30** Clem divided $x^4 - 3x^3 + 4x^2 - 7x + 12$ by $x - 5$ and obtained a quotient of $x^3 + 2x + 14x + 63$ with no remainder. Did Clem divide correctly? How do you know? No; answers will vary—possible answer is that $63 \cdot (-5) \neq 12$.

P | **31** Let $P(x) = x^4 + 2x^3 - 12x^2 - 16x + 32$.

a Find $P(2)$. -16 **b** Find $P(3)$. 11
c Why must there be a zero of P between $x = 2$ and $x = 3$? Since $P(2)$ is negative and $P(3)$ is positive, $P(x)$ must equal 0 for some value of x between 2 and 3.

P | **32** Let $P(x) = ((((2)x + 4)x - 13)x - 5)x - 2$ and $Q(x) = 2x^4 + 4x^3 - 13x^2 - 5x - 3$.

a Find $P(2)$. 0 **b** Find $Q(2)$. -1
c Is $x - 2$ a factor of $P(x)$? Yes **d** Is $x - 2$ a factor of $Q(x)$? No

R | **33** Suppose that $f(x) = ax^3 + bx^2 + cx + d$ and that $f(0) = 0$.

a Find the value of d. 0
b If $f(1) = f(-1) = 0$ and $a = 1$, what is (b, c)? $(0, -1)$
c If $f(1) = f(-1) = 0$, what is b? 0
d If $f(w) = f(-w) = 0$, where $w \neq 0$, what is b? 0

P | **34** Let $P(x) = 2x^4 - 3x^2 + 2x - 8$.

a Demonstrate that there is a zero of P between 1 and 2.
b Find this zero to two decimal places. ≈ 1.57

a $P(1) = 2 - 3 + 2 - 8 = -7$ and $P(2) = 32 - 12 + 4 - 8 = 16$; since the value of $P(x)$ changes sign between $x = 1$ and $x = 2$, there must be at least one zero between these values.

Problem-Set Notes and Strategies, continued

- Problem **27** explores some more results of continuous functions. In part **c**, could a function have exactly two zeros if it were not continuous?
- Discuss problem **30** in class. Each student may see this in a different way. Make sure that they recognize that 5 cannot be a rational zero, since it is not a factor of 12.
- Problem **34** is a good example of polynomial equation solving.

Additional Answers

26

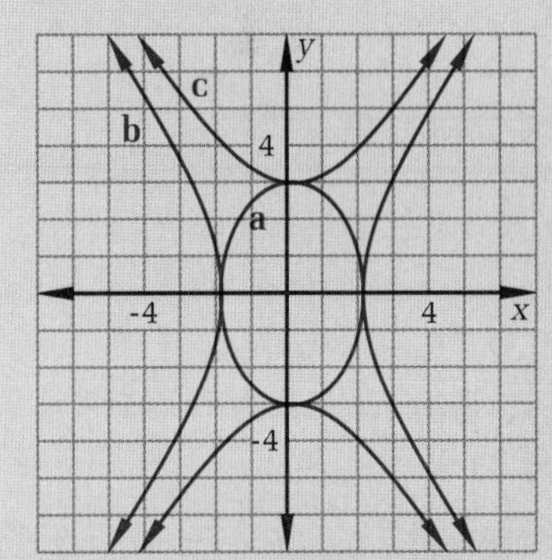

27a

$P(2)$ is negative, and $P(3)$ is positive, so the graph of P must cross the x-axis (that is, $P(x)$ must equal 0) somewhere between $x = 2$ and $x = 3$.

27b

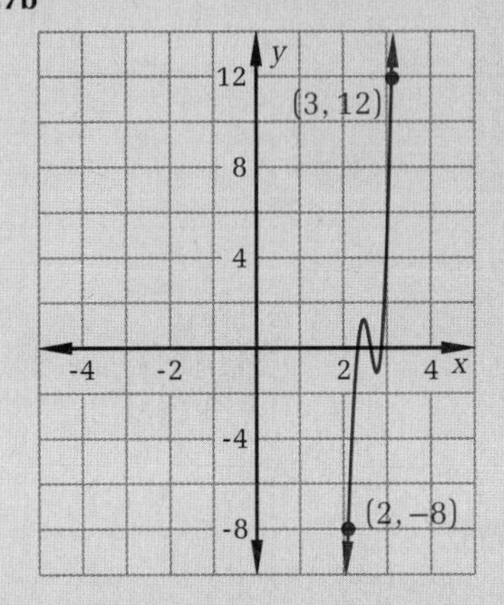

9.4 Four Useful Theorems

Class Planning

Time Schedule
All levels: 2 days

Resource References
Teacher's Resource Book
Class Opener 56
Cooperative Learning 28
Practice 59

Class Opener

Find an upper bound and lower bound for the zeros of $P(x) = (((2)x - 3)x + 4)x - 5$.
There are many answers. The best integer answers are 2 and 1.

Lesson Notes

- Four important theorems are explored in this section. The first guarantees the existence of a root to a polynomial of positive degree. As students progress in mathematics, they will encounter more existence theorems and should recognize them as such. This theorem does not find the root, just states that it exists.
- The second theorem goes one step further and states that the existence of a certain number of roots is dependent on the degree of the polynomial. This does not say anything about uniqueness, which is another important aspect in mathematics. All the roots could be the same, but there are n roots to a nth degree polynomial.
- The third theorem deals with the sum and product of the roots.

Objectives

After studying this section, you will be able to

- Determine the number of complex zeros of a polynomial function
- Determine the sum and the product of the roots of a quadratic equation
- Apply the Conjugate-Zeros Theorem

Part One: Introduction

The Fundamental Theorem of Algebra

By factoring over the integers, we find that the function $f(x) = x^2 - 3x - 4$ has two distinct real zeros, 4 and -1. By factoring over the complex numbers, we find that the function $g(x) = x^2 + 4$ has two distinct complex zeros, $-2i$ and $2i$.

Example *How many zeros does $h(x) = x^2 - 6x + 9$ have?*

We set $h(x)$ equal to zero and solve for x.

$$x^2 - 6x + 9 = 0$$
$$(x - 3)(x - 3) = 0$$
$$x = 3 \text{ or } x = 3$$

The function has two identical zeros, 3 and 3. We say that 3 is a zero of ***multiplicity*** two.

One of the greatest mathematicians of all time, Carl Friedrich Gauss (1777–1855), proved the following theorem, known as the ***Fundamental Theorem of Algebra.***

> ***Every polynomial function of positive degree with complex coefficients has at least one complex zero.***

Based on this theorem is the following, more useful theorem.

> ***Every polynomial function of positive degree n with complex coefficients has exactly n complex zeros (counting multiplicities).***

Since $f(x) = x^4 - 16$ is a fourth-degree function, we know that it has exactly four complex zeros. We also know that $g(x) = ix^3 + 7x - 2 + 3i$, a cubic function, has exactly three complex zeros. Although this theorem tells us the *number* of zeros a function has, it may be difficult or impossible to determine the exact *values* of all the zeros. Generalized techniques have been developed

Vocabulary
multiplicity
Fundamental Theorem of Algebra

for finding the complex zeros of linear, quadratic, cubic, and quartic functions, but the Norwegian mathematician Niels Henrik Abel proved that there is no generalized technique for finding the complex zeros of a function of degree greater than four. We can, however, approximate the zeros of such functions.

The Sum and the Product of the Roots of a Quadratic Equation

The following theorem states an important characteristic of the roots of a quadratic equation.

 ***The sum of the roots of a quadratic equation $ax^2 + bx + c = 0$ is equal to $-\frac{b}{a}$, and the product of the equation's roots is equal to $\frac{c}{a}$.***

Example *Find the sum and the product of the roots of $x^2 - 4x + 5 = 0$.*

In this equation, $a = 1$, $b = -4$, and $c = 5$, so the sum of the roots $\left(-\frac{b}{a}\right)$ is $-\frac{-4}{1}$, or 4, and the product of the roots $\left(\frac{c}{a}\right)$ is $\frac{5}{1}$, or 5.

We can check these answers by using the quadratic formula to find that the equation's roots are $2 + i$ and $2 - i$. Since $(2 + i) + (2 - i) = 4$ and $(2 + i)(2 - i) = 4 - i^2 = 5$, our results are correct.

We can demonstrate the preceding theorem by dividing both sides of $ax^2 + bx + c = 0$ by a, producing $x^2 + \frac{b}{a}x + \frac{c}{a} = 0$, or

$$x^2 - \left(-\frac{b}{a}\right)x + \frac{c}{a} = 0 \qquad (1)$$

Now we write a quadratic equation with the two roots r_1 and r_2.

$$(x - r_1)(x - r_2) = 0$$
$$x^2 - (r_1 + r_2)x + r_1r_2 = 0 \qquad (2)$$

Comparing equations (1) and (2), $r_1 + r_2 = -\frac{b}{a}$ and $r_1r_2 = \frac{c}{a}$.

The Conjugate-Zeros Theorem

Let's look at some functions and their zeros.

Function	**Zeros**
$p(x) = x^2 - 4x + 20$	$2 + 4i, 2 - 4i$
$q(x) = x^3 - 2x^2 + 9x - 18$	$2, 3i, -3i$
$r(x) = x^4 + 5x^2 + 4$	$i, -i, 2i, -2i$
$s(x) = x^4 + x^3 - 11x^2 - 31x - 20$	$4, -1, -2 + i, -2 - i$

Notice that whenever a nonreal complex number is a zero, its conjugate is also a zero. This is not always the case—the zeros of $f(x) = x^2 - 3ix - 2$, for example, are $2i$ and i, which are not conjugates. But it *is* always the case for polynomial functions with real coefficients.

 ***If $P(x)$ is a polynomial with real coefficients and if $a + bi$ (where a and b are real and $b \neq 0$) is a zero of P, then $a - bi$ is also a zero of P.***

Lesson Notes, continued

- The final theorem states that complex roots of polynomials with real coefficients always occur in conjugate pairs. This guarantees that if you found a complex root, there exists a second root that is its conjugate. This does not apply if the polynomial has complex coefficients. Most of those encountered in this class will have real coefficients.

Using Computers

Any or all of the methods described for Section **9.3** above can greatly enhance work in this section.

Cooperative Learning

Have students work in small groups to examine the polynomial $x^2 - 6x + 9$.

1. How many times did the sign change between terms when the polynomial was written in descending order?
2. Calculate the roots of the polynomial. They are 3 and 3 or 3 with multiplicity 2. Both of them are positive. This is the same number as the number of sign changes.

9.4

Checkpoint

1. If $P(x)$ is a polynomial with real coefficients and $3 + 2i$ is one root of the polynomial, state another root. $3 - 2i$
2. How many zeros, either real or complex, does the polynomial $x^3 + x^2 + x - 1$ have? Three
3. Find the sum and product of the roots of the quadratic equation $x^2 - 5x + 15$. Sum is 5; product is 15
4. Write a cubic polynomial with real coefficients when two of the roots are 0 and i. $x^3 + x$

Assignment Guide

Average	
Day 1	5, 8, 9, 11, 12, 14, 17, 18, 21, 24a, b
Day 2	6, 7, 10, 13, 19, 23, 25, 28, 33, 34a, 36
Advanced	
Day 1	5, 6, 8, 10 – 14, 16, 20, 23, 24a, c
Day 2	15, 18, 19, 21, 22, 25, 27, 29, 33, 34a, 36, 38
Honors	
Day 1	5, 6, 8, 10 – 13, 15, 19 – 23, 36
Day 2	24 – 27, 29, 31 – 35, 37, 38

Problem-Set Notes and Strategies

- Problems 7 and 8 could be solved by using a system of equations rather than the quadratic formula. This suggests an alternate method of solving quadratics if the equation does not factor.

Example *If 3 and $-2 + 5i$ are zeros of a cubic function with real coefficients, what is the third zero?*

Since the coefficients in the cubic function are real, the conjugate of $-2 + 5i$ must also be a zero of the function. The third zero is $-2 - 5i$.

Part Two: Sample Problems

Problem 1 *Write a quadratic equation whose roots are $3 + 2i$ and $3 - 2i$.*

Solution We could use the Zero Product Property in reverse, but it is simpler to use the sum-and-product theorem. The given roots have a sum of 6 and a product of 13, so the equation $x^2 - 6x + 13 = 0$ will have these two roots.

Problem 2 *One of the zeros of $P(x) = x^4 - 3x^3 + 5x^2 - 27x - 36$ is $-3i$. What are the other zeros of P?*

Solution Since the polynomial $P(x)$ has real coefficients, $3i$ (the conjugate of $-3i$) must also be a zero of P. Since the numbers $-3i$ and $3i$ are the zeros of the quadratic function $f(x) = x^2 + 9$, we divide $P(x)$ by $x^2 + 9$, finding the quotient $x^2 - 3x - 4$. The value of this polynomial is 0 when $x = 4$ or $x = -1$, so the four zeros of $P(x)$ are $-3i$, $3i$, 4, and -1.

Part Three: Exercises and Problems

Warm-up Exercises

P **1** How many complex zeros does the function $f(x) = 6x^2 - 3x + 5$ have? 2

P **2** Find the sum and the product of the roots of the equation $x^2 + 15x + 50 = 0$. -15; 50

P **3** Find all the zeros of $q(x) = 25x^2 - 20x + 4$. $\frac{2}{5}$

P **4** If $a + bi$ is a root of $c_4x^4 + c_3ix^3 + c_2x^2 + c_1ix + c_0 = 0$, is $a - bi$ necessarily a root of the equation?
No, since the equation contains nonreal coefficients

Problem Set

A

In problems 5 and 6, determine how many complex zeros each function has.

P **5** $f(x) = x^7 - 6x^5 + 3$ 7

6 $g(x) = x^2(x^2 + 4x + 1)(x - 3)^4$ 8

P **In problems 7 and 8, find the sum and the product of the roots of each equation.**

7 $2x^2 - 3x - 6 = 0$ $\frac{3}{2}$; -3

8 $5x^2 + 6x - 7 = 0$ $-\frac{6}{5}$; $-\frac{7}{5}$

R **In problems 9–12, write a quadratic equation with the given roots.**

9 5 and −5 $x^2 - 25 = 0$

10 $2 + \sqrt{5}$ and $2 - \sqrt{5}$ $x^2 - 4x - 1 = 0$

11 $5i$ and $-5i$
$x^2 + 25 = 0$

12 $2 + i$ and $2 - i$
$x^2 - 4x + 5 = 0$

P **13** Suppose that the equation $x^2 + 12x + c = 0$ has a root of multiplicity two.

a Find this root. −6

b What is the value of c? 36

R **14** Write each expression as a number in standard complex form.

a $i^7 \cdot i^5$ $1 + 0i$

b $i^{11} \cdot i^9$ $1 + 0i$

c $i^8 \cdot i^{12}$ $1 + 0i$

R **15** Solve $z = 3iz + 50$ for z. $z = 5 + 15i$

P **16** Write a quadratic equation that has complex coefficients and the roots $4 + i$ and $-i$. $x^2 - 4x + (1 - 4i) = 0$

B

R **17** Consider the following two matrices.

$$A = \begin{bmatrix} 3 + 2i & 4 - i \\ 6 + 5i & 6 - 6i \end{bmatrix} \qquad B = \begin{bmatrix} 6i & -4 \\ 9 - 3i & -5 - i \end{bmatrix}$$

a Find A + B. $\begin{bmatrix} 3 + 8i & -i \\ 15 + 2i & 1 - 7i \end{bmatrix}$

b Find A − B. $\begin{bmatrix} 3 - 4i & 8 - i \\ -3 + 8i & 11 - 5i \end{bmatrix}$

I **18** There are 20 elements in set A and 16 elements in set B. If there are 30 elements in A ∪ B, how many elements are in A ∩ B? 6

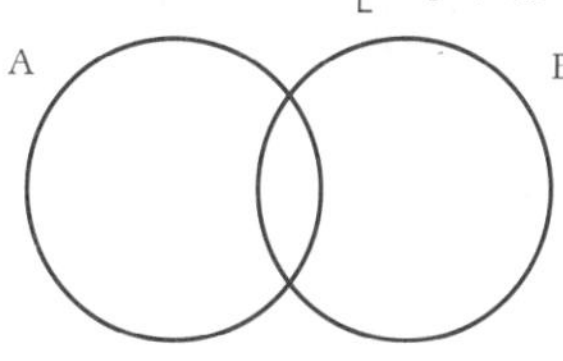

P **19** Write a quartic (degree-four) equation with integral coefficients such that two of the equation's roots are $3 - i$ and $2i$.
$x^4 - 6x^3 + 14x^2 - 24x + 40 = 0$

P **In problems 20 and 21, determine the degree of the polynomial in each equation and the number of complex roots the equation has. Then list each root with its multiplicity.**

20 $x^2 (x + 5)(x - 3) = 0$
4; 4; −5, 0 (multiplicity 2), 3

21 $(x - 2)(x^2 + 4)(x^2 - 4) = 0$
5; 5; −2, 2 (multiplicity 2), $-2i$, $2i$

R **22** Solve the matrix equation for (x, y). $\{(x, y): x - iy = -i\}$

$$\begin{bmatrix} i & 1 \\ 1 & -i \end{bmatrix} \begin{bmatrix} x \\ y \end{bmatrix} = \begin{bmatrix} 1 \\ -i \end{bmatrix}$$

R **23** Claire has a penny, a nickel, a dime, a quarter, a half dollar, and a silver dollar. If she chooses three of the coins at random, how many different amounts of money might she choose? 20

R **24** Identify the real part and the imaginary part of each complex number.

a $2(8 - 3i)$ 16; $-6i$

b $\frac{16 - 5i}{2}$ 8; $-\frac{5}{2}i$

c $\frac{5 + 3i}{i}$ 3; $-5i$

Problem-Set Notes and Strategies, continued

- Problem **14** provides a good review of the powers of *i*. Remind the student that it is a cyclic process.
- In problem **15**, you may have to remind students that *z* is a variable that typically represents a complex number.

Communicating Mathematics

Have students write a paragraph that explains the meaning of the term *multiplicity* and its significance in the process of solving equations.

- In problem **19**, the word integral implies integers and real numbers so that roots must occur in conjugate pairs. This is a good exercise in complex multiplication.
- Notice in problem **22** that since the system of equations involves complex numbers, the answers are not unique. It is not a linear system because of the inclusion of *i*.

9.4

Problem-Set Notes and Strategies, continued

- Problem **26** has an interesting result. The product of two complex matrices whose entries are conjugates of each other can be the identity matrix.

Stumbling Block

In solving equations such as $x^4 + 4x^3 + 6x^2 + 4x + 1 = 0$, students frequently find that -1 is a root of the equation by using synthetic division and the Rational-Zeros Theorem. At this point, students may still be looking for other roots because they are sure that there must be 4 roots, since the degree of the polynomial is 4. Remind them of the multiplicity of roots and the fact that they need not be unique. In this case, -1 is a root of the equation 4 times.

- In problem **32** point out that the quadratic formula will work with complex numbers as well as with real numbers.

Problem Set, *continued*

P,R **25** If one of the following pairs of numbers is selected at random, what is the probability that the numbers are the zeros of a quadratic function with at least one nonreal coefficient? $\frac{1}{5}$

a -3 and -6
b $2 + i$ and $2 - i$
c $1 + \sqrt{3}$ and $1 - \sqrt{5}$
d $2i$ and $-2i$
e $2 + 3i$ and $3 - 2i$

R **26** Multiply $\begin{bmatrix} i & 0 \\ 0 & i \end{bmatrix}\begin{bmatrix} -i & 0 \\ 0 & -i \end{bmatrix}$. $\begin{bmatrix} 1 & 0 \\ 0 & 1 \end{bmatrix}$

R **27** The six points are equally spaced on the circle shown.

a How many triangles can be formed by choosing three of these points as vertices? 20

b If one of these triangles is chosen at random, what is the probability that it is equilateral? $\frac{1}{10}$

c If one of the triangles is chosen at random, what is the probability that it is isosceles? $\frac{2}{5}$

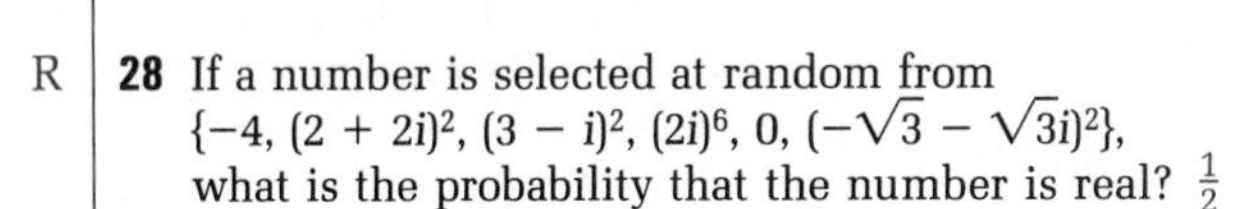

R **28** If a number is selected at random from $\{-4, (2 + 2i)^2, (3 - i)^2, (2i)^6, 0, (-\sqrt{3} - \sqrt{3}i)^2\}$, what is the probability that the number is real? $\frac{1}{2}$

P **29** Consider the equation $(x^2 + 9)^2(x^2 - 9)(x^2 + 6x + 9) = 0$.

a How many roots does the equation have? 8

b Find each root and its multiplicity. -3 (multiplicity 3), 3, $-3i$ (multiplicity 2), $3i$ (multiplicity 2)

R **30** Multiply $\begin{bmatrix} 3 & 2i \\ 4 & 5i \\ -2 & i \\ 8 & -3i \end{bmatrix}\begin{bmatrix} 2 \\ 3i \end{bmatrix}$. $\begin{bmatrix} 0 \\ -7 \\ -7 \\ 25 \end{bmatrix}$

R **31** Solve each equation for z.

a $iz + iz = 6$ $z = -3i$
b $2iz - 6i = iz + 10i$ $z = 16$

P,R **32** Solve $ix^2 + 5x + 24i = 0$ for x. $x = -3i$ or $x = 8i$

P **33** Find the sum and the product of the zeros of the function $P(x) = \sqrt{2}x^2 - \sqrt{6}x + \sqrt{10}$. $\sqrt{3}$; $\sqrt{5}$

P,R **34** Solve each equation for w.

a $\sqrt{w^2 + 5} = 2$
$w = -i$ or $w = i$

b $\sqrt{w^2 + 5}\sqrt{w^2 + 8} = 2$
$w = -2i$ or $w = 2i$

R | **35** Find the complex number $x + yi$ such that $(x + yi)^2 = i$. $x + yi = -\frac{\sqrt{2}}{2} - \frac{\sqrt{2}}{2}i$ or $x + yi = \frac{\sqrt{2}}{2} + \frac{\sqrt{2}}{2}i$

C P | **36** The graph of the function $P(x) = x^3 - 3x^2 + x - 3$ intersects the x-axis at only one point.

a How many real zeros does P have? 1
b How many complex zeros does P have? 3
c What are the zeros of P? $3, -i, i$

P | **37** Write a degree-four equation that has real coefficients and the roots $-3i$ and $6i$. $x^4 + 45x^2 + 324 = 0$

P | **38** Find the sum and the product of the roots of the equation $ix^2 - 6ix + 8 = 0$. $6; -8i$

Problem-Set Notes and Strategies, continued

- In problem **35**, it is necessary to set up a system of equations. In order to be equal, the real parts must be equal as are the imaginary parts.

HISTORICAL SNAPSHOT

HOW MUCH PROOF IS ENOUGH?

A life of work dismissed

Niels Henrik Abel (1802–1829) spent much of his short life in obscurity and poverty. But although his accomplishments went largely unnoticed during his lifetime, he made many important contributions to mathematics. One of Abel's most significant contributions was his demonstration that Leonhard Euler's proof of the binomial theorem applied to rational powers only. Euler (1707–1783) was one of the greatest mathematicians of all time and made many lasting contributions to mathematics. Abel proved the binomial theorem for all cases, including for irrational powers (such as π and the square root of 2).

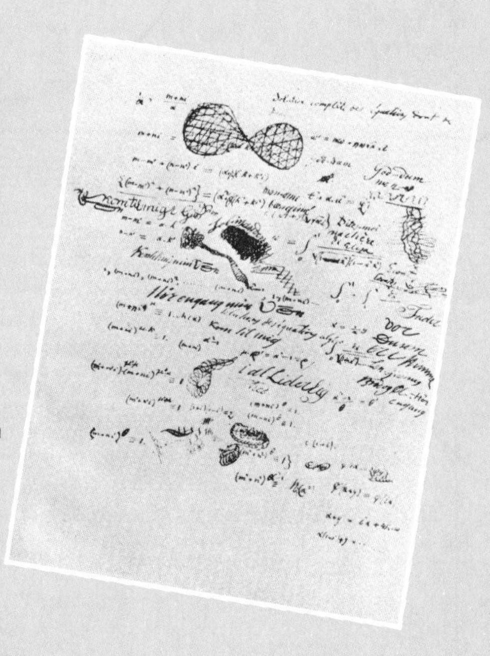

Abel went on to prove that no solution is possible for the quintic equation. He published a series of papers on equation theory, function theory, and power series. In Paris, he presented a theorem on integrals of algebraic functions, but the paper he presented was mislaid, and not until after his death was what is now called Abel's Theorem acknowledged.

While ill with tuberculosis, Abel wrote papers on elliptic functions, a generalization of trigonometric functions.

9.5

Class Planning

Time Schedule
All levels: 2 days

Resource References
Teacher's Resource Book
Class Opener 57
Practice 60
Manipulatives and Applications 14
Enrichment 18

Class Opener

1 Write a polynomial with integer coefficients if $3 + 2i$ and $-3i$ are both zeros of the polynomial.
There are many answers. One such polynomial is
$x^4 - 6x^3 + 22x^2 - 54x + 117$.

2 Expand $(x + 1)^3$.
$x^3 + 3x^2 + 3x + 1$

3 Use your answer to problem **2** to find $(x + 1)^4$
$x^4 + 4x^3 + 6x^2 + 4x + 1$

4 Use your answer to problem **3** to find $(x + 1)^5$
$x^5 + 5x^4 + 10x^3 + 10x^2 + 5x + 1$

5 If Pascal's Triangle is

1
1 2 1
1 3 3 1
1 4 6 4 1
1 5 10 10 5 1
1 6 15 20 15 6 1

predict what the expanded form of $(x + 1)^6$ would be without performing the multiplication.
$x^6 + 6x^5 + 15x^4 + 20x^3 + 15x^2 + 6x + 1$

9.5 The Binomial Theorem

Objective

After studying this section, you will be able to
- Use the binomial theorem to expand powers of binomials

Part One: Introduction

Remember that $(x + y)^n = (x + y)(x + y) \cdot \ldots \cdot (x + y)$, with the $x + y$ a factor n times in the product. For example,

$$\begin{aligned}&(x + y)^4\\ &= (x + y)(x + y)(x + y)(x + y)\\ &= (x + y)(x + y)(x^2 + 2xy + y^2)\\ &= (x + y)(x^3 + 3x^2y + 3xy^2 + y^3)\\ &= x^4 + 4x^3y + 6x^2y^2 + 4xy^3 + y^4\end{aligned}$$

Let's look for a pattern in the expansions of powers of a binomial.

$$\begin{aligned}(x + y)^0 &= 1\\ (x + y)^1 &= 1x + 1y\\ (x + y)^2 &= 1x^2 + 2xy + 1y^2\\ (x + y)^3 &= 1x^3 + 3x^2y + 3xy^2 + 1y^3\\ (x + y)^4 &= 1x^4 + 4x^3y + 6x^2y^2 + 4xy^3 + 1y^4\\ (x + y)^5 &= 1x^5 + 5x^4y + 10x^3y^2 + 10x^2y^3 + 5xy^4 + 1y^5\\ (x + y)^6 &= 1x^6 + 6x^5y + 15x^4y^2 + 20x^3y^3 + 15x^2y^4 + 6xy^5 + 1y^6\end{aligned}$$

Some of the following patterns can be observed.

1 There are $n + 1$ terms in the expansion of $(x + y)^n$. For example, the expansion of $(x + y)^4$ consists of five terms.
2 The degree of each term in the expansion of $(x + y)^n$ is n. For example, in the expansion of $(x + y)^4$, the sum of the exponents in each term is 4.
3 In each successive term in the expansion of $(x + y)^n$, the exponent of x decreases by 1 and the exponent of y increases by 1. The first term is always x^n and the last is always y^n.

The pattern of coefficients is known as ***Pascal's triangle.***

Row 0 1
Row 1 1 1
Row 2 1 2 1
Row 3 1 3 3 1
Row 4 1 4 6 4 1
Row 5 1 5 10 10 5 1
Row 6 1 6 15 20 15 6 1

Vocabulary
Pascal's triangle
binomial theorem

Each row of Pascal's triangle begins and ends with the number 1. Each interior element of the row is the sum of the two adjacent elements in the preceding row.

Example **a** *Write row 7 of Pascal's triangle.*

The first and last elements of the row are 1's, and each of the other elements is the sum of two elements of row 6. Therefore, row 7 is 1, 7, 21, 35, 35, 21, 7, 1.

b *Expand* $(x + y)^7$.

$$(x + y)^7$$
$$= x^7 + 7x^6y + 21x^5y^2 + 35x^4y^3 + 35x^3y^4 + 21x^2y^5 + 7xy^6 + y^7$$

Of course, we can find the terms of an expansion by actually multiplying all the factors.

$$(x + y)^4 = \underset{1}{(x + y)}\underset{2}{(x + y)}\underset{3}{(x + y)}\underset{4}{(x + y)}$$

The terms in the product of the four factors of $(x + y)^4$ consist of all the ways that a group of four x's and y's can be formed, taking one variable from each factor. The x^3y terms, for example, are found as shown.

Factor Number

1	2	3	4		
x	x	x	y	=	x^3y
x	x	y	x	=	x^3y
x	y	x	x	=	x^3y
y	x	x	x	=	x^3y
		Total			$4x^3y$

This process is the same as selecting one y from the four factors in every possible way. In Chapter 5, you saw that the number of groups of r items that can be selected from a set of n items is given by the combination function C.

$$C(n, r) = \frac{n!}{(n - r)!r!}, \text{ so } C(4, 1) = \frac{4!}{(4 - 1)!1!} = 4.$$

Thus, the coefficient of x^3y in the expansion is 4.

Let's try using the combination function for the other terms. (Remember that $0! = 1$.)

$$C(4, 0) = \frac{4!}{(4 - 0)!0!} = 1 \text{ (the coefficient of } x^4)$$

$$C(4, 2) = \frac{4!}{(4 - 2)!2!} = 6 \text{ (the coefficient of } x^2y^2)$$

$$C(4, 3) = \frac{4!}{(4 - 3)!3!} = 4 \text{ (the coefficient of } xy^3)$$

$$C(4, 4) = \frac{4!}{(4 - 4)!4!} = 1 \text{ (the coefficient of } y^3)$$

Lesson Notes

- This section involves the method of expanding a binomial to powers and being able to determine specific terms of the expansion without finding all of them. Methods use both Pascal's Triangle and combinations to find the coefficients. It is helpful to write out the first five or six expansions on the board and have the students make generalizations about the results. Patterns may be observed that can be generalized to obtain almost the complete theorem.
- You may also want to consider using sigma notation to write the theorem in the form

$$(x + y)^n = \sum_{i=0}^{n} [C(n, i)]x^{n-i}y^i$$

Using Computers

Symbol-manipulator software is nice to use in conjunction with this section. With this software, you can discover the patterns involved in expanding binomials to higher and higher powers. Either *Derive* or *The Mathematics Exploration Toolkit* can be used effectively. Scientific calculators capable of doing combinations and factorials also can be used here. The ideal is to let students see patterns and discover relationships. For example, have students calculate $(x + y)^n$ for increasing values of n and then substitute $x = 1$ and $y = 1$ and see what happens.

Communicating Mathematics

Have students write a paragraph that explains the first pattern of the binomial expansions. In other words, why does a binomial raised to the nth power have $n + 1$ terms in it?

Checkpoint

1. Find the third number in row 10 of Pascal's Triangle. 45
2. Expand $(2a - b)^4$. $16a^4 - 32a^3b + 24a^2b^2 - 8ab^3 + b^4$
3. What is the sixth term of the expansion of $(4x + 3y)^8$? $870{,}912x^3y^5$

Assignment Guide

Average	
Day 1	5, 6, 8, 11, 12, 14, 16–18
Day 2	7, 9, 10, 13, 15, 19, 20, 22, 25, 26, 47
Advanced	
Day 1	5, 6, 8, 9, 11–13, 15–18
Day 2	7, 10, 14, 19, 21, 23, 25–29, 30, 31, 47
Honors	
Day 1	6–9, 11–13, 15, 17–19, 30, 32
Day 2	20–23, 25b, 26, 27, 29, 31, 33, 40, 42, 43, 45

Stumbling Block

There is a formula for calculating combinations to find the coefficient of a particular term in the binomial expansion. However, students frequently forget to raise the coefficient of the variable to the same power as the variable in the expansion. For example, $(2x)^5$ is written as $2x^5$ instead of $32x^5$. Have students write out each step of the process, including the distribution of the exponent over a product such as $2x$.

A general rule is: The coefficient of the $(r + 1)$th term of the expansion of $(x + y)^n$ is equal to $C(n, r)$.

Example *Find the fourth term of the expansion of* $(x + y)^{12}$.

We are looking for the fourth term, $r + 1 = 4$, so $r = 3$. The coefficient of this term is equal to $C(12, 3)$.

$$C(12, 3) = \frac{12!}{(12 - 3)!3!} = 220$$

In the fourth term, the exponent of y is 3—the value of r—and the exponent of x is 9. (To check this, study the expansions of consecutive binomial powers listed at the beginning of this section and review patterns 2 and 3.) The fourth term of the expansion of $(x + y)^{12}$ is $220x^9y^3$.

All of our observations about expansions of powers of binomials can be summarized in the ***binomial theorem.***

For any positive integer n, the expansion of $(x + y)^n$ ***is*** $[C(n, 0)]x^n + [C(n, 1)]x^{n-1}y + [C(n, 2)]x^{n-2}y^2 + \dots + [C(n, n)]y^n$.

Part Two: Sample Problems

Problem 1 *Expand* $(3x - 2y)^4$.

Solution If we think of $(3x)$ as the x term in the binomial and $(-2y)$ as the y term, we can use the binomial theorem.

$$\begin{aligned}
&(3x - 2y)^4 \\
&= [(3x) + (-2y)]^4 \\
&= [C(4, 0)](3x)^4 + [C(4, 1)](3x)^3(-2y) + [C(4, 2)](3x)^2(-2y)^2 \\
&\quad + [C(4, 3)](3x)(-2y)^3 + [C(4, 4)](-2y)^4 \\
&= 1(81x^4) + 4(27x^3)(-2y) + 6(9x^2)(4y^2) + 4(3x)(-8y^3) + 1(16y^4) \\
&= 81x^4 - 216x^3y + 216x^2y^2 - 96xy^3 + 16y^4
\end{aligned}$$

Problem 2 *Show that for any nonnegative integral value of n,* $C(n, 0) + C(n, 1) + C(n, 2) + \dots + C(n, n) = 2^n$.

Solution According to the binomial theorem,

$$(x + y)^n = [C(n, 0)]x^n + [C(n, 1)]x^{n-1}y + [C(n, 2)]x^{n-2}y^2 + \dots + [C(n, n)]y^n$$

If we let $x = 1$ and $y = 1$, we have

$$\begin{aligned}
(1 + 1)^n &= [C(n, 0)](1^n) + [C(n, 1)](1^{n-1})(1) + [C(n, 2)](1^{n-2})(1^2) + \dots + [C(n, n)](1^n) \\
2^n &= C(n, 0) + C(n, 1) + C(n, 2) + \dots + C(n, n)
\end{aligned}$$

Part Three: Exercises and Problems

Warm-up Exercises

1 Write rows 8–10 of Pascal's triangle.

P **2** Expand $(2x + 5y)^5$.
$32x^5 + 400x^4y + 2000x^3y^2 + 5000x^2y^3 + 6250xy^4 + 3125y^5$

P **3** What is the tenth term of the expansion of $(x - 2y)^{13}$?
$-366{,}080x^4y^9$

P **4** Show that $C(n, r + 1) = C(n, r) \cdot \frac{n-r}{r+1}$. How can this relationship be used in writing the expansion of $(x + y)^n$?

Problem Set

A

In problems 5 and 6, expand each expression.

P,R **5** $(p + q)^3$ $p^3 + 3p^2q + 3pq^2 + q^3$ **6** $(2x + y)^3$ $8x^3 + 12x^2y + 6xy^2 + y^3$

P **7** Consider the expression $(x + y)^{12}$. Write the first four terms and the last four terms of the expansion of the expression.
First four: $x^{12} + 12x^{11}y + 66x^{10}y^2 + 220x^9y^3$; last four: $220x^3y^9 + 66x^2y^{10} + 12xy^{11} + y^{12}$

P **8** Find the sixth term of the expansion of $(y - 3)^{11}$. $-112{,}266y^6$

P **9** What term of the expansion of $(x^2 + 2)^5$ contains x^6?
Third term, $40x^6$

R **10** Match each factored expression with the correct expansion.

a $(a + b)^3$ iii **i** $a^3 + b^3$
b $(a - b)^3$ ii **ii** $a^3 - 3a^2b + 3ab^2 - b^3$
c $(a + b)(a^2 - ab + b^2)$ i **iii** $a^3 + 3a^2b + 3ab^2 + b^3$
d $(a - b)(a^2 + ab + b^2)$ iv **iv** $a^3 - b^3$

R **11** Indicate whether each statement is True or False.

a $(a + b)^4 = a^4 + b^4$ False
b $a^3 + b^3 = (a + b)(a^2 + b^2)$ False
c $(a + b)^3 = a^3 + 3a^2b + 3ab^2 + b^3$ True
d $a^3 + b^3 = (a + b)(a^2 + 2ab + b^2)$ False

B

R **12** Find, to the nearest thousandth, the value of x for which the volume of the box shown is 50.
≈ 3.519

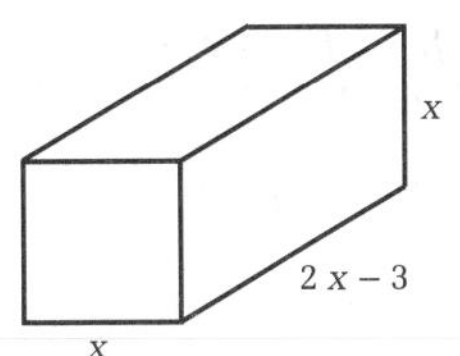

P,R **13** If $f(x) = (x + 5)^3$ and $g(x) = x^3 + 5^3$, for what values of x is $f(x) = g(x)$? 0, −5

Problem-Set Notes and Strategies

- In problem **1**, make sure that students recognize the pattern and are not computing each of these values using the combination formula.
- In problem **4**, the powers of the preceding term and the number of the term are used to determine the coefficient without using the combination formula. If possible take some time to explain this in class.
- Problem **12** requires some approximation method, such as the bisection method, to find the answer.

Additional Answers

4
$$C(n, r+1) = \frac{n!}{[n-(r+1)]!(r+1)!} = \frac{n!}{\frac{(n-r)!}{n-r} r!(r+1)} = \frac{n!}{(n-r)!r!} \cdot \frac{n-r}{r+1} = C(n, r) \cdot \frac{n-r}{r+1}$$

The relationship tells us that we can find the coefficient of each successive term by multiplying the coefficient of the preceding term by $\frac{n-r}{r+1}$, where n is the power of the binomial and the preceding term is the $(r + 1)$th term of the expansion.

1 Row 8: 1, 8, 28, 56, 70, 56, 28, 8, 1
Row 9: 1, 9, 36, 84, 126, 126, 84, 36, 9, 1
Row 10: 1, 10, 45, 120, 210, 252, 210, 120, 45, 10, 1

Problem-Set Notes and Strategies, continued

- Problem **16** shows only one of the many patterns available for Pascal's Triangle. If you have time it is worthwhile to look at the patterns and see what conjectures can be made.
- You may prefer to discuss problem **17** in class.
- Problem **26** requires approximation methods. If the computer is available, you might encourage the students to type in the algorithm for the bisection method and use it.
- Problem **27** brings up another use of the binomial theorem. See how students react to the expansion of 1.05^5 using it in the form $(1 + .05)^5$. See if they can figure out how many terms they need to expand to get four digit accuracy.

Additional Answers

29

a RRR, RRL, RLR, LRR, LLR, LRL, RLL, LLL

b RRRR, RRRL, RRLR, RLRR, LRRR, RRLL, RLRL, RLLR, LLRR, LRLR, LRRL, LLLR, LLRL, LRLL, RLLL, LLLL

c The numbers of ways in which he can reach the various junctions are the same as the numbers in Pascal's triangle, and the number of ways in which he can reach the nth level is the sum of the numbers in the $(n-1)$th row of Pascal's triangle -- that is, 2^{n-1}.

Problem Set, *continued*

R **In problems 14 and 15, factor each expression over the integers.**

14 $x^3 + 3x^2y + 3xy^2 + y^3$ $(x + y)^3$

15 $x^4 + 4x^3y + 6x^2y^2 + 4xy^3 + y^4$ $(x + y)^4$

P **16** Consider the first six rows of Pascal's triangle.

a Find the sum of the numbers in each row. 1, 2, 4, 8, 16, 32

b Predict the sum of the numbers in the twelfth row. 4096

P **17** Let C be the combination function and n be a nonnegative integer. Explain why each of the following statements is true.

a $C(n, 0) = 1$ $C(n, 0) = \frac{n!}{n!0!} = 1$

b $C(n, n) = 1$ $C(n, n) = \frac{n!}{0!n!} = 1$

P **18** Consider the expansion of $(a + b)^{22}$.

a What is the eighteenth term? $26{,}334a^5b^{17}$

b What is the sixth term? $26{,}334a^{17}b^5$

P **19** Write the first three terms of $(3x - 5y)^{10}$. $59{,}049x^{10} - 984{,}150x^9y + 7{,}381{,}125x^8y^2$

P **In problems 20–23, expand each expression.**

20 $(x + 2y)^5$ **21** $(x - 2y)^5$ **22** $(2x + y)^5$ **23** $(2x - y)^5$

R **24** Factor $x^2 + 24x + 144$ completely. $(x + 12)^2$

R **25** Divide each of the following expressions by $x + 2$.

a $x^2 + 4$ $x - 2$, remainder 8

b $x^4 + 16$ $x^3 - 2x^2 + 4x - 8$, remainder 32

R **26** The polynomial function $P(x) = x^3 - x^2 - 14x + 18$ has three real zeros. Find these zeros to two decimal places. ≈ -3.85, ≈ 1.33, ≈ 3.52

P,R **27** Find the sum of all the terms in the expansion of $\left(\frac{1}{3} + \frac{2}{3}\right)^5$. 1

P **28** Write the first four terms of $(2w + 3)^9$. $512w^9 + 6912w^8 + 41{,}472w^7 + 145{,}152w^6$

P,R **29** Adam Ant starts at level 1 and travels downward through the maze by moving to the left (L) or to the right (R). To reach level 2, for example, he can move R or move L. To reach level 3, he can move RR, RL, LR, or LL.

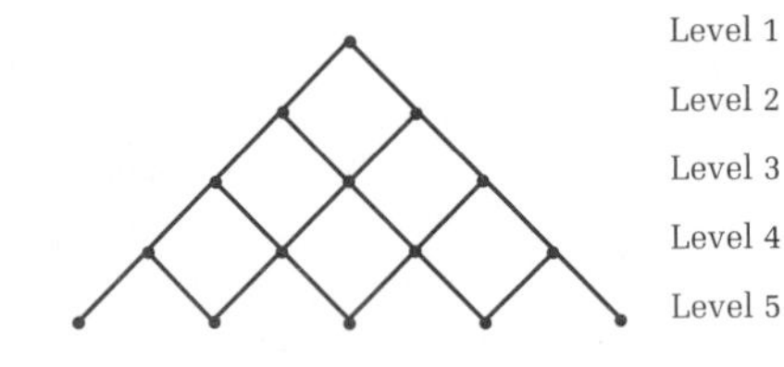

a List all the ways in which he can reach level 4.

b List all the ways in which he can reach level 5.

c What is the relationship between the number of ways in which Adam can reach the various levels and the numbers in Pascal's triangle?

20 $x^5 + 10x^4y + 40x^3y^2 + 80x^2y^3 + 80xy^4 + 32y^5$

21 $x^5 - 10x^4y + 40x^3y^2 - 80x^2y^3 + 80xy^4 - 32y^5$

22 $32x^5 + 80x^4y + 80x^3y^2 + 40x^2y^3 + 10xy^4 + y^5$

23 $32x^5 - 80x^4y + 80x^3y^2 - 40x^2y^3 + 10xy^4 - y^5$

P,R **30** Solve $(x + 2)^3 = (x + 2)^4$ for x. $x = -2$ or $x = -1$

R **31** **a** Write a formula for the volume V of the cube shown. $V = x^3 + 9x^2 + 27x + 27$

b If the volume is 100, what is the value of x (to the nearest hundredth)? ≈ 1.64

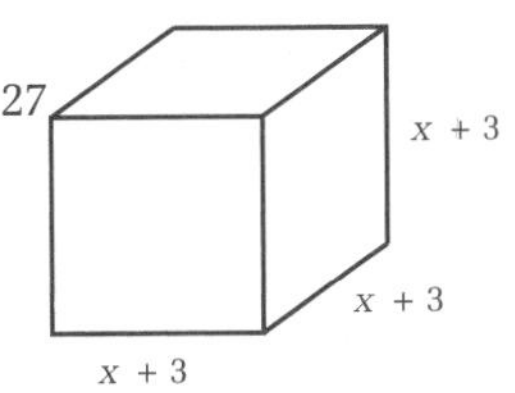

R **32** In how many different orders can seven golfers tee off at the first hole on a course? 5040

P **33** Expand $\left(2x + \frac{1}{2}y\right)^5$. $32x^5 + 40x^4y + 20x^3y^2 + 5x^2y^3 + \frac{5}{8}xy^4 + \frac{1}{32}y^5$

R **In problems 34–39, factor each expression over the integers.**

34 $x^3 + 5x^2 + 4x + 20$ $(x^2 + 4)(x + 5)$

35 $x^3 + 2x^2y - 16x - 32y$

36 $x^3 - x^2 - 25x + 25$ $(x + 5)(x - 5)(x - 1)$

37 $x^2 - 4y^2 + 3x - 6y$

38 $x^2 - y^2 - x - y$ $(x + y)(x - y - 1)$

39 $x^3 + 2x^2 - 6x - 12$ $(x^2 - 6)(x + 2)$

R **40** Consider the following system.

$$\begin{cases} x + y = 5 \\ xy = 2 \end{cases}$$

a Find the value of $x^2 + y^2$. 21

b Find the value of $x^3 + y^3$. 95

R **41** Multiply $(x - a)(x - b)(x - c)(x - d) \ldots (x - z)$. 0

P **In problems 42 and 43, find the sum of the coefficients in the expansion of each binomial power.**

42 $(2x + y)^5$ 3^5

43 $(3x + 2y)^{20}$ 5^{20}

P **C** **44** Show that for any positive integral value of n, $C(n, 0) - C(n, 1) + C(n, 2) + \ldots + (-1)^n[C(n, n)] = 0$.

P **45** Show that $C(n, r) = C(n, n - r)$. $C(n, r) = \frac{n!}{(n-r)!r!} = \frac{n!}{r!(n-r)!} = \frac{n!}{[n-(n-r)]!(n-r)!} = C(n, n - r)$

R **46** Consider the number pattern shown.

a Describe the procedure by which the successive rows of the pattern are generated, and find the next three rows.

b What is the largest number that will occur in the pattern? 3

1
1 1
2 1
1 2 1 1
1 1 1 2 2 1
3 1 2 2 1 1
1 3 1 1 2 2 2 1

R **47** If the product of $(x - 4)(x + 2)^5$ is written in descending order of powers, what is its second term? $6x^5$

Problem-Set Notes and Strategies, continued

- Problem **29** yields the same values as problem **16**.
- Problem **41** will seem insurmountable for some students until they realize that one factor is $(x - x)$.

Additional Answers

35 $(x + 2y)(x + 4)(x - 4)$

37 $(x - 2y)(x + 2y + 3)$

46a The numbers in each row (after the first) describe the arrangement of numbers in the preceding row—for example, the third row consists of one 2 and one 1, so the fourth row is 1, 2, 1, 1, which consists of one 1, one 2, and two 1's, so the fifth row is 1, 1, 1, 2, 2, 1, and so forth. The next three rows are:
1113213211
31131211131221
13211311123113112211

44 $0 = (1 - 1)^n$
$= [C(n, 0)][1^n] + [C(n, 1)][1^{n-1}](-1) + [C(n, 2)][1^{n-2}](-1)^2 + \ldots + [C(n, n)](-1)^n$
$= C(n, 0) - C(n, 1) + C(n, 2) + \ldots + (-1)^n[C(n, n)]$

9.6

Class Planning

Time Schedule
All levels: 1 day

Resource References
Teacher's Resource Book
Class Opener 58
Practice 61

Class Opener

Write $(2x + y)^5$ in expanded form.
$32x^5 + 80x^4y + 80x^3y^2 + 40x^2y^3 + 10xy^4 + y^5$

Lesson Notes

- The concept of standard deviation and one of its uses are explored in this section. The table on page **732** shows cumulative areas and percentile ranks corresponding to various numbers of standard deviations from the mean. Remind the student that this table is used only when the distribution of the data is known to be normal, either by graphing the data or by some other analytic means. Time should be spent on how to read the table given the standard deviation score or the percentage.

- Chebyshev's Rule is discussed for those cases where the data are not distributed normally, or when the distribution of the data is unknown. Inform the student that this is certainly a worst case estimate as it must work for the most deformed distribution as well as for the normal. Typically the percentages will be larger than predicted, but these predictions are worst case estimates.

9.6 Standard Deviation and Chebyshev's Theorem

Objectives

After studying this section, you will be able to
- Calculate the standard deviation of a set of data
- Apply Chebyshev's theorem

Calculating Standard Deviation

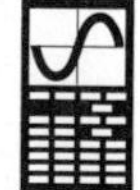

See p. 796

In Section 8.5, you were introduced to an extremely useful measure of dispersion—standard deviation. In this section, you will find out how to calculate the standard deviation of a set of data and will explore the relationship between the standard deviation and the data.

Suppose that Wanda the waitress collected the following amounts in tips (to the nearest dollar) during ten days of work at the Longbranch Steak House.

19, 23, 29, 27, 35, 16, 22, 28, 29, 42

The mean of these data is $\frac{270}{10}$, or 27. The following table will help us calculate the standard deviation of the data.

Value	Value − Mean	(Value − Mean)²
19	$19 - 27 = -8$	$(-8)^2 = 64$
23	$23 - 27 = -4$	$(-4)^2 = 16$
29	$29 - 27 = 2$	$2^2 = 4$
27	$27 - 27 = 0$	$0^2 = 0$
35	$35 - 27 = 8$	$8^2 = 64$
16	$16 - 27 = -11$	$(-11)^2 = 121$
22	$22 - 27 = -5$	$(-5)^2 = 25$
28	$28 - 27 = 1$	$1^2 = 1$
29	$29 - 27 = 2$	$2^2 = 4$
42	$42 - 27 = 15$	$15^2 = 225$
		Total 524

Vocabulary
variance
Chebyshev's theorems

The average of the (value − mean)^2 quantities is called the ***variance*** of the data. The variance of these data is $\frac{524}{10}$, or 52.4. We can now use the following rule to calculate the standard deviation of the data.

Standard deviation = $\sqrt{\textit{variance}}$.

In this case,

$$SD = \sqrt{52.4} \approx 7.239$$

Many calculators have built-in functions for mean and standard deviation. (Some use $\bar{x}$ to symbolize mean and σ_n to stand for standard deviation.) Almost all mathematicians use calculators or computers for statistical work. They are great time savers!

What does the standard deviation mean in terms of Wanda's data? Let's draw and label a number line.

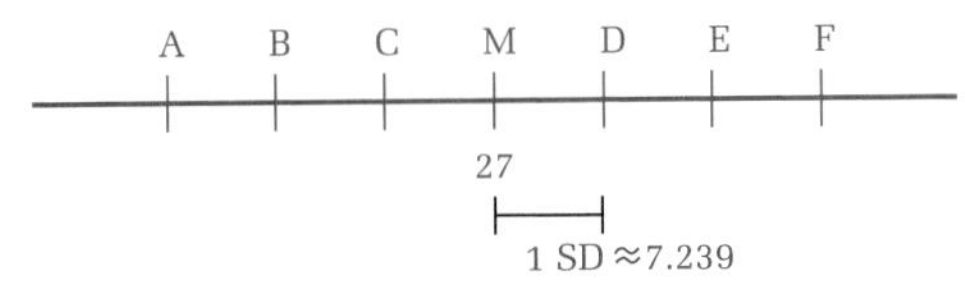

Point	Coordinate
A	Mean − 3 SD ≈ 27 − 3(7.239) = 5.283
B	Mean − 2 SD ≈ 27 − 2(7.239) = 12.522
C	Mean − 1 SD ≈ 27 − 1(7.239) = 19.761
M	Mean = 27
D	Mean + 1 SD ≈ 27 + 1(7.239) = 34.239
E	Mean + 2 SD ≈ 27 + 2(7.239) = 41.478
F	Mean + 3 SD ≈ 27 + 3(7.239) = 48.717

If we associate the number line with a bell-shaped curve representing a normal distribution, we can use the area under the curve to make predictions about Wanda's income from tips.

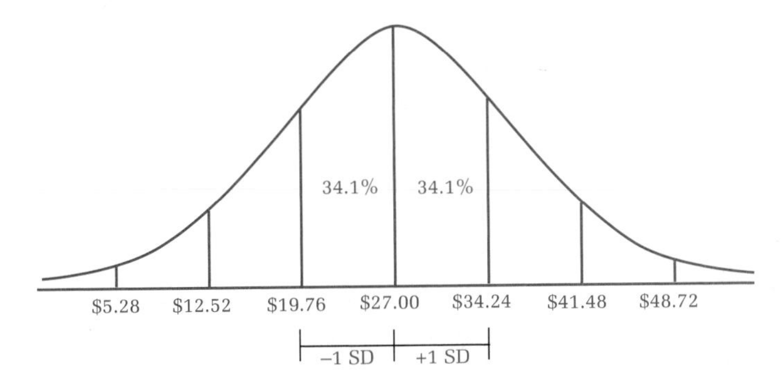

Wanda can expect to earn between $19.75 and $34.25 per day in tips approximately 68% of the time.

Using Computers

A statistics package would be good to use here. Consider the two referenced in the notes for Section **8.5**. Both can calculate the mean and standard deviations.

Checkpoint

1 Given a normal distribution with mean of 100 and standard deviation of 12, find the percentile rank of a score of 116. 90

2 Refer to the information given in problem **1**. Suppose you weren't sure of the distribution and had to use Chebyshev's Rule. What is the percentile rank now? ≥43

3 Refer to problems **1** and **2**. Find the 75th percentile for the same scores if it is known to be a normal distribution. 108

Assignment Guide

Average
2–4, 7–9, 11
Advanced
1–4, 6–10
Honors
1–3, 5–8, 10, 12

Problem

Suppose that scores on the mathematics section of the Test of Basic Understanding (TBU) are normally distributed, with a mean of 500 and a standard deviation of 100.

a *Draw a bell-shaped curve that represents the data, indicating scores that are 1 SD, 2 SD, and 3 SD from the mean.*
b *Find the percentile rank of a score that is 1 SD below the mean.*
c *Find the percentile ranks of scores of 450 and 650.*
d *If a score of 660 is k SD from the mean, what is the value of k?*
e *Which scores have percentile ranks of less than 78?*

Solution

a

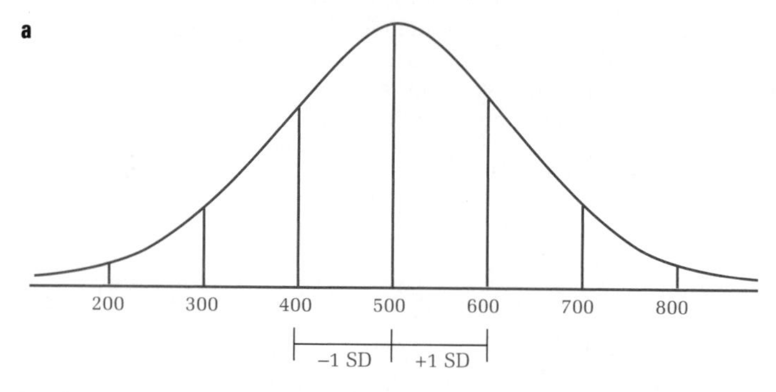

b The Normal Distribution Table on page 732 shows the area under the curve to the left of a data point and the associated percentile rank for values at various distances from the mean. According to the table, a value −1 SD from the mean (in this case, a score of 400) has a percentile rank of 15.
c A score of 450 is −0.5 SD from the mean, so according to the table, its percentile rank is 30. A score of 650 is +1.5 SD from the mean and has a percentile rank of 93.
d Since a score of 660 is 160 points, or 1.6 SD, above the mean, $k = 1.6$.
e According to the Normal Distribution Table, a score that is +0.8 SD from the mean has a percentile rank of 78. Thus, any score less than 0.8 SD above the mean will have a percentile rank less than 78. Since in this case m + 0.8 SD is 500 + 0.8(100), or 580, scores less than 580 have percentile ranks less than 78.

Chebyshev's Theorem

If the value of the standard deviation of a distribution is small, we can conclude that many of the data values are clustered close to the mean. A large standard deviation, on the other hand, indicates the data values are more widely dispersed from the mean. In any case, as you saw in Section 8.5, almost all the values in a normal distribution lie within 3 SD of the mean.

To get a sense of the dispersion of values in *any* distribution of data, normal or not, mathematicians use the following rule, known as ***Chebyshev's theorem.***

In any distribution, the percentage of the data values that lie within k standard deviations of the mean (where k > 1) is at least

$$\frac{k^2 - 1}{k^2} \cdot 100$$

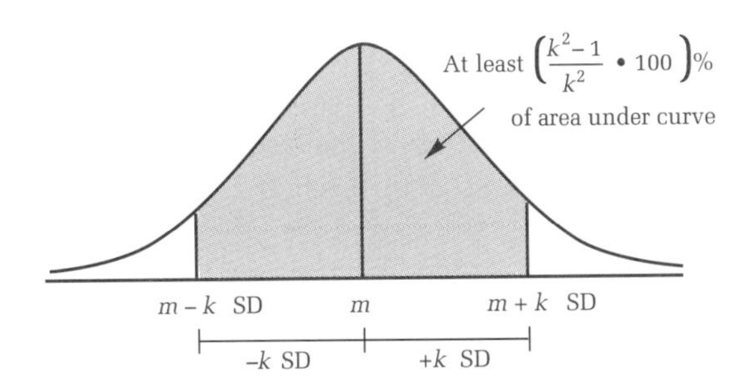

Problem *A nonnormal distribution of data has a mean of 60 and a standard deviation of 15.*

a *What data values are within 2 SD of the mean?*
b *What percentage of the values must be between 37.5 and 82.5?*
c *If at least 40% of the data lie within k SD of the mean, what are the possible values of k?*

Solution **a** The standard deviation is 15, so 2 SD = 30. Thus, all values from 30 to 90 are within 2 SD of the mean.

b Since 37.5 is $\frac{37.5 - 60}{15}$, or −1.5, SD from the mean and 82.5 is $\frac{82.5 - 60}{15}$, or +1.5, SD from the mean, according to Chebyshev's theorem at least

$$\frac{1.5^2 - 1}{1.5^2} \cdot 100, \text{ or } \approx 55.56,$$

percent of the distribution lies between these values.

c The following diagram will help us visualize the problem.

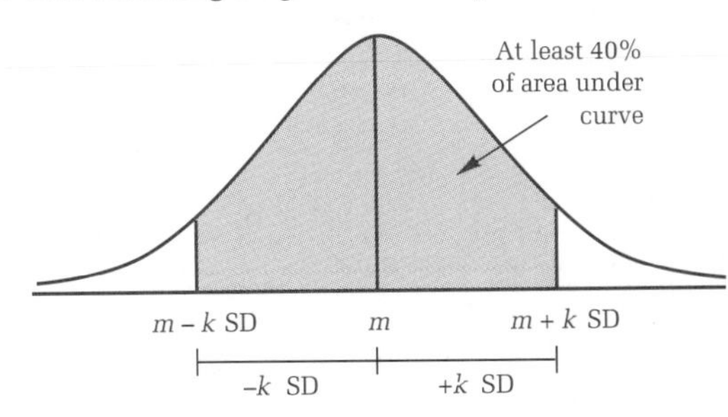

Stumbling Block

Remind students that the standard deviation is the measure of the average distance that a randomly chosen data point is from the mean. Its units are in terms of the units of the data. A point that is one standard deviation from the mean is not one unit from the mean. Likewise, a point's being −1 standard deviation from the mean does not indicate that its value is negative, only that its value is below the mean.

Communicating Mathematics

Have students write a paragraph that explains the following statement: You are a person who is consistently two standard deviations above the mean. What difference will it make to you if you also know that the distribution of values is normal?

We use Chebyshev's theorem.

$$\frac{k^2 - 1}{k^2} \cdot 100 \geq 40$$

Solve for k

$$100k^2 - 100 \geq 40k^2$$
$$60k^2 \geq 100$$
$$k^2 \geq \frac{5}{3}$$
$$k \leq -\frac{\sqrt{15}}{3} \text{ or } k \geq \frac{\sqrt{15}}{3}$$

Since Chebyshev's theorem specifies that $k > 1$, we are interested only in the positive solutions of the inequality. At least 40% of the data will be within k SD of the mean if $k \geq \frac{\sqrt{15}}{3}$, or ≈ 1.291.

Problem-Set Notes and Strategies

- Problem **1** is a good review of the formula for the mean.
- In problem **3**, make sure that the students are computing the proper area. To find the area to the right, the student must subtract the area to the left of the point from 1. The total area under the curve is equal to 1.
- Students may use their calculators for problems **5** and **6**. If they have keys for statistical entries, you might explain that the key marked σ_n is the population standard deviation and the key labeled σ_{n-1} is the sample standard deviation.

Problem Set

A

R **1** For what value of x is the mean of the following data 30? 41

10, 25, x, 17, 90, 21, 35, 6, 42, 13

P,R **2** Refer to the distribution curve.

a How many SD from the mean is a data value of 95? 0.75 SD

b How many SD from the mean is a data value of 30? −2.5 SD

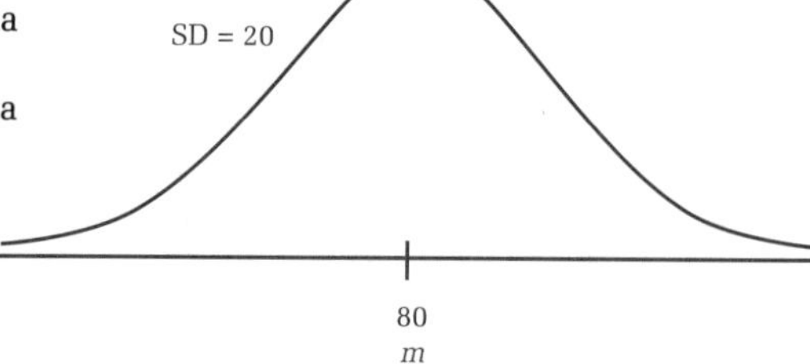

P **3** The diagram represents a normal distribution.

a Use the Normal Distribution Table to find the percentage of the area under the curve that is to the left of point A. 11.51%

b What percentage of the area under the curve is to the *right* of point B? 0.82%

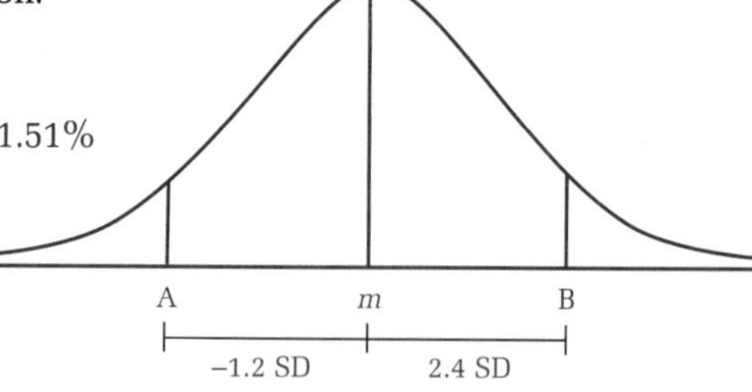

P **4** Consider the data values 6, 8, 12, 14, 18, and 20.

a What is the variance of the data? 25

b What is the standard deviation of the data? 5

P **5** Find the mean and the standard deviation of the following data. ≈43.29; ≈16.92

17, 23, 39, 47, 51, 59, 67

P **6** Determine the data values that correspond to points m, A, B, and C in the diagram.
43; 50; 57; 32.5

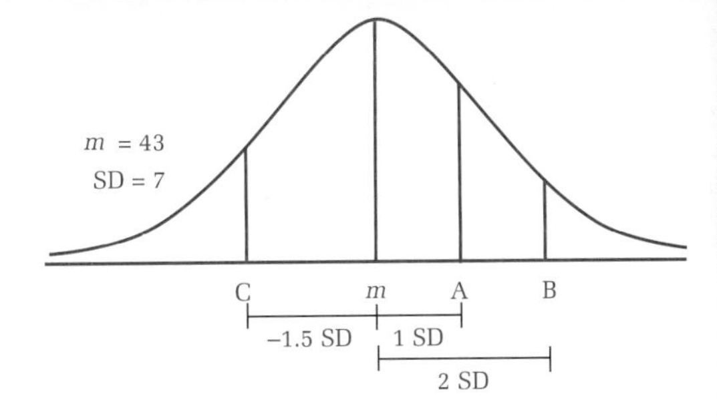

B P,R **7** If the scores on an examination are normally distributed and Amy's score is located at $m + 1.7$ SD, what is the percentile rank of her score? 95

P **8** In a nonnormal distribution of data, the mean of the data is 140 and the standard deviation is 30. What percentage of the data must lie between the values in each of the following pairs?

a 50 and 230 $\approx$88.89%

b 150 and 170 Cannot be determined (To use Chebyshev's theorem, k must be greater than 1.)

P **9** What percentage of the area under the normal curve shown is in the shaded region?
$\approx$42%

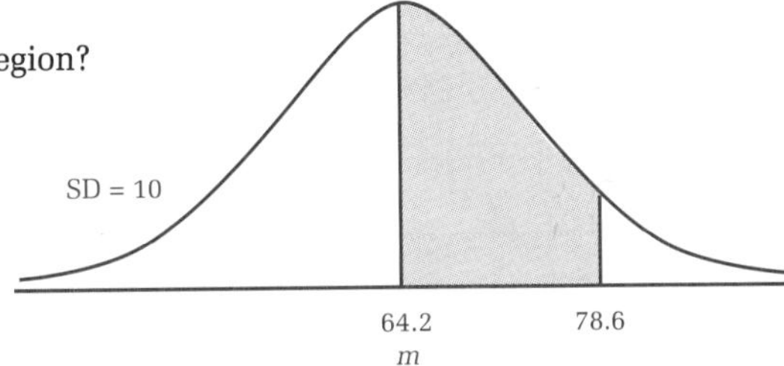

P **10** If the shaded region contains at least 60% of the area under the curve shown, what is the value of k?
$\frac{\sqrt{10}}{2}$, or $\approx$1.581

−k SD 0 +k SD

P **11** The diagram represents a set of normally distributed test scores. What percentage of the people taking the test scored between 400 and 650?
$\approx$77.45%

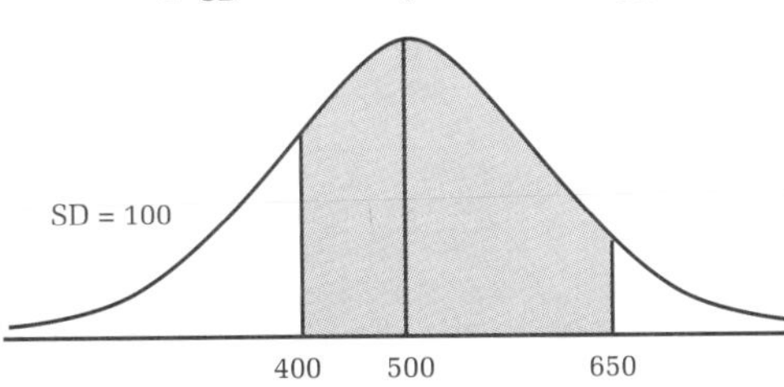

C P,I **12** A study showed that automobiles arrive at a certain intersection at an average rate of 10 per minute. If the data collected in the study form a normal distribution with a standard deviation of 2.64, what is the probability that 15 or more cars will arrive at the intersection in any given minute? $\approx$0.03

Problem-Set Notes and Strategies, continued

- In problem **8**, reiterate that the percentages may certainly be greater than these values but this is a worst possible estimate so it must be low.
- Problems such as problem **11** can also be related to probability. If a person is chosen at random, there is a 77.45% chance that he/she scored between 400 and 650.

9 Chapter Summary

Concepts and Procedures

After studying this chapter, you should be able to

- Write a polynomial in nested form (9.1)
- Divide a polynomial by a binomial (9.1)
- Apply the remainder and factor theorems (9.1)
- Perform synthetic division (9.2)
- Find zeros of functions by factoring and using the quadratic formula (9.2)
- Apply the Rational-Zeros Theorem (9.2)
- Identify upper and lower bounds of the zeros of a polynomial function (9.3)
- Determine the approximate values of real zeros (9.3)
- Determine how many complex zeros a polynomial function has (9.4)
- Determine the sum and the product of the roots of a quadratic equation (9.4)
- Apply the Conjugate-Zeros Theorem (9.4)
- Use the binomial theorem to expand powers of binomials (9.5)
- Calculate the standard deviation of a set of data (9.6)
- Apply Chebyshev's theorem (9.6)

Vocabulary

binomial theorem (9.5)
bisection (9.3)
Chebyshev's theorem (9.6)
Factor Theorem (9.1)
Fundamental Theorem of Algebra (9.4)
lower bound (9.3)
multiplicity (9.4)
nested form (9.1)
Pascal's triangle (9.5)
Rational-Zeros Theorem (9.2)
Remainder Theorem (9.1)
synthetic division (9.2)
upper bound (9.3)
variance (9.6)

9 Review Problems

P **A** **1** Use the graph of $y = f(x)$ to estimate the zeros of f.
Answers will vary. Possible answer: ≈ -4.5, ≈ 2.3, ≈ 6.5

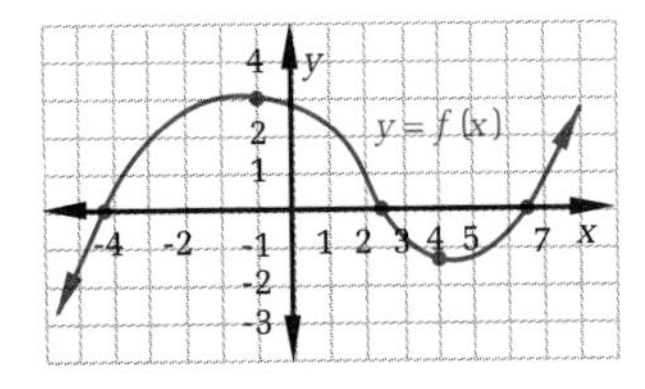

P **In problems 2 and 3, rewrite each polynomial in nested form.**

2 $2x^5 - x^3 + 4x + 15$
$(((((2)x + 0)x - 1)x + 0)x + 4)x + 15$

3 $x - 3x^3 - 4x^2 + 7$
$(((-3)x - 4)x + 1)x + 7$

P **4** What is the remainder when $x^4 - 2x^3 + 4x^2 - 8x + 16$ is divided by $x - 2$? 16

P **In problems 5 and 6, use synthetic division to find each quotient and remainder.**

5 $(4x^4 - 6x^3 + x^2 - 8x - 15) \div (x - 3)$ Quotient: $4x^3 + 6x^2 + 19x + 49$; remainder: 132

6 $(x^5 + 2x^3 - 5x^2 + 3x + 4) \div (x + 2)$
Quotient: $x^4 - 2x^3 + 6x^2 - 17x + 37$; remainder: -70

P **7** If the total area under the normal curve shown is 86 square units, what is the area of the shaded region? ≈ 37.26

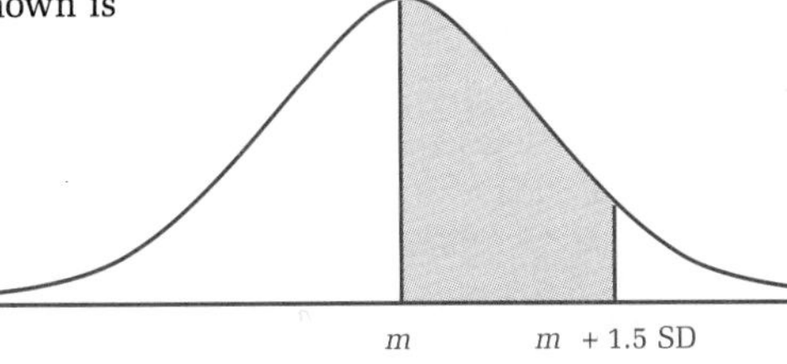

P **8** Is $x + 2$ a factor of $2x^4 + x^3 - 16x^2 - 27x - 14$? Yes

P **In problems 9–14, write a polynomial function with integral coefficients and the given zeros.**

9 -3, 1, and 4
$f(x) = x^3 - 2x^2 - 11x + 12$

10 -2, -1, 1, and 2
$f(x) = x^4 - 5x^2 + 4$

11 $3 - \sqrt{7}$ and $3 + \sqrt{7}$
$f(x) = x^2 - 6x + 2$

12 $5 - 3i$ and $5 + 3i$
$f(x) = x^2 - 10x + 34$

13 -6, $-\frac{1}{4}$, and $\frac{2}{3}$
$f(x) = 12x^3 + 67x^2 - 32x - 12$

14 $\frac{1}{2}$, 3, $-i$, and i

P **15** Let $f(x) = x^3 - 6x^2 - 13x + 42$. Use synthetic division to find an integral lower bound and an integral upper bound of the zeros of f.
Answers will vary. Possible answer: -3, 8

P **16** List all the rational numbers that might be zeros of the function $g(x) = 2x^4 + 5x^3 - 25x^2 + x + 5$. Then find all the zeros of g.
$\pm 1, \pm 5, \pm\frac{1}{2}, \pm\frac{5}{2}$; $-5, 1 - \sqrt{2}, \frac{1}{2}, 1 + \sqrt{2}$

Class Planning

Time Schedule

All levels: 1 day

Assignment Guide

Average

1, 2, 4, 5, 7–9, 14–18, 20, 23–25, 27, 28

Advanced

1, 2, 4, 6, 8, 9, 11, 13–17, 19, 20, 22–25, 27–29

Honors

2, 4, 6, 8, 11, 12, 14–17, 20–25, 27–29

Problem-Set Notes and Strategies

- In problem **1** have students estimate the answers and indicate that many times an estimate is the only necessary answer until more detailed measurements are made.
- Problem **4** may be done by synthetic division, long division, substitution, or writing in nested form and substituting.
- Problem **7** is an excellent example of using the normal table and setting up a proportion.
- The answers may vary for problems **9 – 14**. Any integral multiple of these will yield the same zeros.
- Problem **16** reminds students that answers may also be irrational or complex. Once the quadratic form is reached, the quadratic formula may be used.

Additional Answers

Answer for problem **14** can be found in the answer pages beginning on **A** 1.

Problem-Set Notes and Strategies, continued

- Problem **21** may be done with upper bounds or with Descartes's rule of signs.
- In problem **26b**, some students may try to substitute −2 into the polynomial and evaluate. Remind them that if $x + 2$ is a factor, then −2 must be a root.

Additional Answer

21 Since the coefficients and constant are positive, each term will be positive for any positive value of x. No set of positive numbers has a sum of 0, so there is no positive value of x for which $P(x) = 0$.

Review Problems, *continued*

P **17** How many complex roots does the equation $2x^4 - ix^2 - 1 + 3i = 0$ have? 4

P,R **In problems 18 and 19, expand each expression.**

18 $(x + y)^4$
$x^4 + 4x^3y + 6x^2y^2 + 4xy^3 + y^4$

19 $(x - y)^4$
$x^4 - 4x^3y + 6x^2y^2 - 4xy^3 + y^4$

P **B** **20** One of the zeros of $P(x) = 2x^3 - 8x^2 - 7x + 25$ is between 4 and 4.4. Use the bisection method to approximate the value of this zero to two decimal places. ≈ 4.11

P **21** Let $P(x) = ax^4 + bx^3 + cx^2 + dx + e$, where a, b, c, d, and e are positive. Explain why P cannot have any positive zeros.

P **22** The scores on a standardized test are normally distributed, with a mean of 500 and a standard deviation of 100. What is the percentile rank of a score of 670? 95

P,R **23** Consider the equation $x^3 - 3x^2 + 3x - 1 = 0$.

a What is the degree of the polynomial in the equation? 3

b List the roots of the equation with their multiplicities.
1 (multiplicity 3)

P **24** Find the greatest nonpositive integral lower bound of the zeros of $f(x) = x^4 - 2x^3 + x^2 - 3x - 160$. −4

P **25** Calculate the mean, the variance, and the standard deviation of the following set of data. 20; 43.25; ≈ 6.58

17, 25, 9, 31, 25, 22, 15, 16

P,R **26** Consider the polynomial $x^5 + 32$.

a Divide $x^5 + 32$ by $x + 2$. $x^4 - 2x^3 + 4x^2 - 8x + 16$

b Is −2 a root of the equation $x^5 + 32 = 0$? Yes

P **27** Write the first four terms of the expansion of $(3x + 5y)^9$.
$19{,}683x^9 + 295{,}245x^8y + 1{,}968{,}300x^7y^2 + 7{,}654{,}500x^6y^3$

P,R **28** Find all the complex roots of $x^4 - 1 = 0$.
$-1, 1, -i, i$

P **C** **29** The diagram shown represents a nonnormal distribution. What minimum percentage of the total area under the curve is in the shaded region? 84%

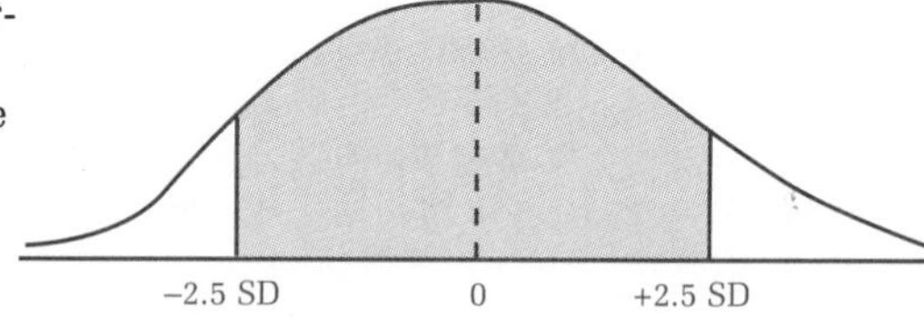

P **30** Find the least nonnegative integral upper bound of the zeros of $f(x) = x^4 + 2x^3 + x^2 + 3x - 160$. 4

Chapter Test

1 Rewrite $-10x^3 + 7x^2 + \frac{3}{2}x - 8$ in nested form.
$(((-10)x + 7)x + \frac{3}{2})x - 8$

2 Find the quotient and remainder when $x^3 - 3x^2 - 5x + 15$ is divided by $x - 3$. Quotient: $x^2 - 5$; remainder: 0

3 Show that $x - 1$ is a factor of $x^n - 1$ for every positive integer n.

4 Use synthetic division to show that $x - 2$ is *not* a factor of $x^4 + 6x^2 - 7x - 6$.

5 Find all of the rational roots of $3x^3 + x^2 + x - 2 = 0$. $\frac{2}{3}$

6 Find all of the real roots of $2x^3 + 5x^2 - 7x + 2 = 0$. $\frac{-3 - \sqrt{17}}{2}, \frac{1}{2}, \frac{-3 + \sqrt{17}}{2}$

7 Find the least nonnegative integral upper bound of the zeros of the function $q(x) = x^4 + 2x^3 + x^2 + 3x + 1$. 0

8 Use a calculator to help you find the real zero of $f(x) = 4x^3 - 3x^2 + 2$ to the nearest tenth. -0.6

9 Counting multiplicities, how many complex zeros are there for the polynomial function $P(x) = 3x^5 - x^3 + 2x^2 + 6$? 5

10 Find the sum and product of the roots of $-2x^2 + 3x + 1 = 0$. $\frac{3}{2}; -\frac{1}{2}$

11 A polynomial function $P(x)$ of degree three has 7 and $3i$ as two of its zeros. What is the other zero, and what polynomial represents the function? $-3i$; $x^3 - 7x^2 + 9x - 63$

12 What is the fourth term in the expansion of $(2x + 3)^5$? $1080x^2$

13 In a normal distribution of data, the mean of the data is 140 and the standard deviation is 30. What percentage of the data must lie between the values 35 and 245? $\approx 91.8\%$

In problems 14–17, write a quadratic equation with the given roots.

14 12 and -3 $x^2 - 9x - 36 = 0$

15 $3 + 2i$ and $3 - 2i$ $x^2 - 6x + 13 = 0$

16 $3 + \sqrt{2}$ and $3 - \sqrt{2}$
$x^2 - 6x + 7 = 0$

17 $\sqrt{2} + \sqrt{5}i$ and $\sqrt{2} - \sqrt{5}i$
$x^2 - 2\sqrt{2}x + 7 = 0$

18 Let $P(x) = 2x^5 - 7x^4 + 9x^3 + x^2 + 2x - 4$.

a Find the quotient and the remainder when $P(x)$ is divided by $x - 1$. Quotient: $2x^4 - 5x^3 + 4x^2 + 5x + 7$; remainder: 3

b Find $P(1)$. 3

Resource References

Evaluation

Test Form 1, 2, 3

Additional Answers

3 Answers will vary. Possible answer

1	1	0	0	0	...	0	−1
		1	1	1	...	1	1
	1	1	1	1	...	1	0

4

2	1	0	6	−7	−6
		2	4	20	26
	1	2	10	13	20

Since the remainder (20) is nonzero, $x - 2$ is not a factor of the polynomial.

10.1

MANY OF THE concepts that apply to rational expressions can be related to numeric fractions. Throughout the chapter, you may want to make connections between previously learned mathematics concepts and rational expressions.

Although there are software programs and claculators capable of simplifying rational expressions, students should be able to recognize a rational expression and identify restrictions on the values of the variable.

Students will multiply and divide (Section **10.1**) and add and subtract (Sections **10.3** and **10.4**) rational expressions.

Finding the real zeros of a rational function can be related to finding the real zeros of polynomial functions. Students can apply previously learned strategies to solving rational equations and inequalities.

In Section **10.5**, two methods for simplifying complex fractions are given. Some students may prefer one method over the other.

CHAPTER 10

RATIONAL EXPRESSIONS, EQUATIONS, AND FRACTIONS

Like the forms used in this construction, the elements of rational expressions are simple, but the expressions can be complex and varied.

10.1 Multiplying and Dividing Rational Expressions

Objectives

After studying this section, you will be able to

- Recognize rational expressions
- Simplify rational expressions
- Multiply and divide rational expressions

Part One: Introduction

Recognizing Rational Expressions

A ***rational expression*** is any expression that can be written as a ratio of polynomials. Here are some examples of rational expressions.

$$\frac{x+2}{x-3} \qquad 1+\frac{t}{h} \qquad \frac{x^2-10x-75}{x+5} \qquad \frac{y_2-y_1}{x_2-x_1} \qquad x^2+3$$

When working with rational expressions, we need to check for restrictions. ***Restricted values*** are values that make the denominator of a rational expression equal to zero. For example, if $f(x) = \frac{x+2}{x-3}$, the domain of f must exclude $x = 3$, because $f(3)$ is undefined. Any other real number can be used. The restrictions on the variables of the first four examples above are $x \neq 3$, $h \neq 0$, $x \neq -5$, and $x_1 \neq x_2$.

Simplifying Rational Expressions

When the numerator and denominator of a rational expression have no factor in common, the expression is ***simplified.*** To simplify a rational expression, we factor the numerator and the denominator and look for any common factors.

Example *Give all restrictions and simplify* $\frac{4x^2+6x}{2x^2+x-3}$.

Factor both the numerator and the denominator to find common factors.

$$\frac{4x^2+6x}{2x^2+x-3} = \frac{2x(2x+3)}{(x-1)(2x+3)};\ x \neq 1,\ x \neq -1.5$$

$$= \frac{2x}{x-1} \cdot 1, \text{ since } \frac{2x+3}{2x+3} = 1$$

$$= \frac{2x}{x-1};\ x \neq 1,\ x \neq -1.5$$

Class Planning

Time Schedule

Average: 2 days
Advanced: 2 days
Honors: 1 day

Resource References

Teacher's Resource Book
Class Opener 59
Practice 62

Class Opener

Find the sum.

$\frac{3}{5} \cdot \frac{2}{3} + \frac{5}{8} \div \frac{3}{4}$

$\frac{37}{30}$

Lesson Notes

- For many students, rational expressions are fractions and are not a favorite topic in algebra. Remind them of the definitions of rational numbers and suggest that since they have mastered fractions, this topic is familiar.

Stumbling Block

Students will frequently forget to restrict the domain of a problem until the end and then remember as an afterthought. This leads to many errors. Have the students identify the restricted values as the first step of the process. Emphasize the importance of this step and suggest students make it an integral part of the problem solving process.

Vocabulary

rational expression
restricted value
simplify

10.1

Lesson Notes, continued

- Multiplying rational numbers was easiest when factors could be cancelled and made simpler. This holds true for rational expressions as well. Caution students to be sure to state the domain before doing any simplifying as they may include a solution that is not in the domain of the original problem.

- Division of fractions and complex fractions is an area where students often stumble. When they ask how to divide, tell them that we don't. Remind them to change the problem into a multiplication problem. Again, caution them to first state the domain before doing any simplifying.

Communicating Mathematics

Write a paragraph explaining the reasoning why division is equivalent to multiplying by the reciprocal. You may use examples of real numbers or other operations to justify your reasoning.

It is very important to remember that although the simplified expression has only one value that makes it undefined, *all* restrictions found in the problem are included in the answer. Only under these conditions are the original expression and the simplified expression equal.

Multiplying and Dividing Rational Expressions

Multiplying and dividing rational expressions is similar to multiplying and dividing numerical fractions. However, we must always note restrictions when using variables.

Example *Multiply the following rationals.*

a $\frac{256}{81} \cdot \frac{27}{16}$

$$\frac{256}{81} \cdot \frac{27}{16}$$

Factor $= \frac{16 \cdot 16}{3 \cdot 27} \cdot \frac{27}{16}$

Find forms of 1 $= \frac{16}{16} \cdot \frac{27}{27} \cdot \frac{16}{3}$

$= (1)(1)\frac{16}{3}$

$= \frac{16}{3}$

b $\frac{x^2 - 4}{x^2y} \cdot \frac{xy^2}{x + 2}$

The restrictions on the rational expression are $x \neq 0$, $y \neq 0$, and $x \neq -2$.

$$\frac{x^2 - 4}{x^2y} \cdot \frac{xy^2}{x + 2}$$

$= \frac{(x + 2)(x - 2)}{x^2y} \cdot \frac{xy^2}{x + 2}$

$= \frac{x + 2}{x + 2} \cdot \frac{xy}{xy} \cdot \frac{y(x - 2)}{x}$

$= (1)(1)\frac{y(x - 2)}{x}$

$= \frac{y(x - 2)}{x}$; $x \neq 0, x \neq -2, y \neq 0$

To divide rational expressions, we multiply the dividend by the reciprocal of the divisor.

Example *Divide the following.*

a $\frac{46}{9}$ by $\frac{23}{3}$

$$\frac{46}{9} \div \frac{23}{3}$$

Multiply by the reciprocal of the divisor $= \frac{46}{9} \cdot \frac{3}{23}$

$= \frac{2 \cdot 23 \cdot 3}{3^2 \cdot 23}$

$= \frac{2}{3}$

b $\frac{2x + 12}{x^2}$ by $\frac{x + 6}{x}$

$\frac{2x + 12}{x^2} \div \frac{x + 6}{x}$; $x \neq 0$

$= \frac{2x + 12}{x^2} \cdot \frac{x}{x + 6}$; $x \neq 0, x \neq -6$

$= \frac{2(x + 6)x}{x^2(x + 6)}$

$= \frac{2}{x}$; $x \neq 0, x \neq -6$

In this case we have two restrictions, $x \neq 0$ and $x \neq -6$. The second restriction, $x \neq -6$, is necessary when we multiply by $\frac{x}{x+6}$, the reciprocal of the divisor.

Part Two: Sample Problems

Problem 1 *Divide $\frac{x^2 - 49}{x^2}$ into $3x + 21$.*

Solution Since we are to divide $\frac{x^2 - 49}{x^2}$ into $3x + 21$, $\frac{x^2 - 49}{x^2}$ is the divisor, and the problem is rewritten as $(3x + 21) \div \left(\frac{x^2 - 49}{x^2}\right)$.

Write $3x + 21$ as a fraction $\quad \frac{3x + 21}{1} \div \frac{x^2 - 49}{x^2}; x \neq 0$

Multiply by the reciprocal of the divisor $\quad = \frac{3x + 21}{1} \cdot \frac{x^2}{x^2 - 49}$

Factor $\quad = \frac{3(x + 7)}{1} \cdot \frac{x^2}{(x + 7)(x - 7)}; x \neq 7, x \neq -7$

Multiply $\quad = \frac{3(x + 7)x^2}{1(x - 7)(x + 7)}$

Simplify $\quad = \frac{3x^2}{x - 7}; x \neq 0, x \neq 7, x \neq -7$

There are three restrictions because the solution uses the divisor and its reciprocal. Neither the numerator nor the denominator of the divisor can be zero.

Problem 2 *Multiply and give all restrictions.*

$$\frac{x^2 - 2x - 15}{x^2y - 25y} \cdot \frac{x^2 + 5x}{x^3 + 27}$$

Solution Remember: Before simplifying, check for restrictions by finding the zeros of each polynomial in a denominator.

$$\frac{x^2 - 2x - 15}{x^2y - 25y} \cdot \frac{x^2 + 5x}{x^3 + 27}$$

$$= \frac{(x - 5)(x + 3)}{y(x + 5)(x - 5)} \cdot \frac{x(x + 5)}{(x + 3)(x^2 - 3x + 9)}; y \neq 0, x \neq -5, x \neq 5, x \neq -3$$

We use the quadratic formula to determine further restrictions on x.

Let $x^2 - 3x + 9 = 0$.

$$x = \frac{3 + \sqrt{-27}}{2} \text{ and } x = \frac{3 - \sqrt{-27}}{2}$$

$$x = \frac{3 + i \cdot 3\sqrt{3}}{2} \text{ and } x = \frac{3 - i \cdot 3\sqrt{3}}{2}$$

Checkpoint

1 Find any restricted values of the rational expression $\frac{x + 1}{x^2 + 3x + 2}$. $\quad$ −1 and −2

2 Multiply and give any restrictions for the following: $\frac{x + 1}{x - 3} \cdot \frac{x^2 - 4x + 3}{x^2 - 1}$.

1, restrictions $x \neq -1$, 1 or 3

3 Divide $\frac{x + 5}{3x}$ into $3x + 15$.

$9x$, restrictions $x \neq 0$ or −5

10.1

Assignment Guide

Average	
Day 1	11–15, 17, 22, 24, 28, 30, 32, 33, 38
Day 2	16, 18, 20, 23, 25–27, 29, 31, 40, 43
Advanced	
Day 1	11, 13, 14, 16, 18, 20, 22, 24, 26, 28, 30, 32, 33
Day 2	21, 25, 27, 29, 31, 34, 36, 39, 41–44
Honors	
	11, 16, 21, 24, 25, 27, 31, 33, 35–37, 39, 41, 42, 44, 45, 47

Problem-Set Notes and Strategies

- Problems **1 – 8** are opportunities to review the definition of a polynomial.
- In problem **9**, students may wonder why a can not be zero at first. Have them replace both variables with zero and evaluate the result.

Now we can simplify.

$$\frac{(x-5)(x+3)}{y(x+5)(x-5)} \cdot \frac{x(x+5)}{(x+3)(x^2-3x+9)}$$

$$= \frac{x}{y(x^2-3x+9)};\ y \neq 0,\ x \neq -5,\ x \neq 5,\ x \neq -3,$$

$$x \neq \frac{3+3\sqrt{3}i}{2},\ x \neq \frac{3-3\sqrt{3}i}{2}$$

Part Three: Exercises and Problems

Warm-up Exercises

In problems 1–8, tell whether each expression is rational.

P

1 $n+5$ Yes

2 $\frac{\sqrt{x}}{x+3}$ No

3 $\frac{1}{\frac{a}{b}}$ Yes

4 $\frac{\sqrt{(x-y)^2}}{x}$ No

5 $\frac{n+4}{n+2}$ Yes

6 $\frac{x^2-9}{x+3}$ Yes

7 $\frac{5x^2-10x}{x^2+x-6}$ Yes

8 $\frac{a^2+b^2}{a+b}$ Yes

P **9** Divide 1 by $\frac{a}{b}$ and give restrictions. $\frac{b}{a}$; $a \neq 0$, $b \neq 0$

P **10** Multiply $\frac{x-5}{x+2}$ by $x^2-5x-14$ and give restrictions.
$x^2-12x+35$; $x \neq -2$

Problem Set

A R,I

11 If one of the following functions is selected at random, what is the probability that it is a rational expression? $\frac{4}{5}$

a $f(x) = \frac{x+5}{2\sqrt{x}}$

b $g(x) = \frac{x^2+2x-\sqrt{3}}{x}$

c $h(x) = 15 + \frac{3}{x}$

d $j(x) = \left[\frac{2}{3}\left(x+\frac{3}{5}\right)x+\frac{5}{8}\right]x$

e $k(x) = \frac{x^2-1}{1-x^2}$

P **In problems 12–15, simplify and state restrictions.**

12 $\frac{3-5}{5-3}$
-1

13 $\frac{x-y}{y-x}$
-1; $y \neq x$

14 $\frac{x-5}{5-x}$
-1; $x \neq 5$

15 $\frac{(x+5)^4}{(x+5)^2}$
$(x+5)^2$; $x \neq -5$

P **16** Find the domain of f. $\{x{:}x \neq 0, x \neq -2, x \neq -4\}$

$$f(x) = \frac{3x+4}{3x^3+18x^2+24x}$$

P | **In problems 17–20, simplify and state restrictions.**

17 $\frac{xy^2z}{xy}$ $yz;\ x \neq 0,\ y \neq 0$

18 $\frac{x}{y} \cdot \frac{x^2}{y}$ $\frac{x^3}{y^2};\ y \neq 0$

19 $\frac{3xy}{6y^2}$ $\frac{x}{2y};\ y \neq 0$

20 $\frac{x}{y} \div \frac{x^2}{y}$ $\frac{1}{x};\ x \neq 0,\ y \neq 0$

R | **21** Study the example, then copy and complete the chart.

Expressions	Greatest Common Factor	Least Common Multiple
4, 6, 8	**2**	**24**
9, 12, 6	3	36
$4x^2$, $6xy$, $2x^2y$	2x	$12x^2y$
$x(x-2)$, $4(x-2)$	$x - 2$	$4x(x-2)$

B

In problems 22–25, simplify and state restrictions.

P | **22** $\frac{(x+2)(x+3)}{(x+1)(x+2)}$ $\frac{x+3}{x+1};\ x \neq -1,\ x \neq -2$

23 $\frac{x^3 - 3x^2}{x}$ $x^2 - 3x;\ x \neq 0$

24 $\frac{x^2 - 9}{x^2 + 2x - 3}$ $\frac{x-3}{x-1};\ x \neq -3,\ x \neq 1$

25 $\frac{2x^3 + 12x^2 + 16x}{x^3 + 4x}$ $\frac{2(x+4)(x+2)}{x^2+4};\ x \neq 0,\ x \neq 2i,\ x \neq -2i$

P | **26** Multiply $\frac{x+5}{x-2} \cdot \frac{x+3}{x+5} \cdot \frac{x-2}{x+3}$ and write your answer in simplest form. $1;\ x \neq 2,\ x \neq -5,\ x \neq -3$

P | **27** Divide $\frac{x^2 + 9x + 20}{x^2 - 9x + 20}$ into $\frac{x^2 + 10x + 25}{x^2 - 16}$ and write your answer in simplest form. $\frac{x^2 - 25}{(x+4)^2};\ x \neq 4,\ x \neq -4,\ x \neq 5,\ x \neq -5$

R | **28** The length of a rectangle is $x + 2$, and its area is $x^2 + 7x + 10$. Find its width. $x + 5$

P | **29** Let $f(x) = \frac{x^2 - 25}{x + 2}$ and $g(x) = \frac{x^2 + 4}{x + 5}$. Simplify $f(x) \cdot g(x)$. $\frac{(x^2+4)(x-5)}{x+2};\ x \neq -5,\ x \neq -2$

R | **30** Let $f(x) = \frac{6x^2(x+2)(x-3)}{(x^2-1)(x+2)(x+3)}$. Evaluate each of the following.

a $f(0)$ 0
b $f(3)$ 0
c $f(-2)$ Undefined
d $f(-3)$ Undefined
e $f(2)$ $\frac{-8}{5}$

P | **31** Given $f(x) = \frac{x+1}{x+2}$ and $g(x) = \frac{x+2}{x+1}$, find the domains of functions h and j.

a $h(x) = f(x) \cdot g(x)$ $\{x{:}x \neq -1,\ x \neq -2\}$

b $j(x) = \frac{g(x)}{f(x)}$ $\{x{:}x \neq -1,\ x \neq -2\}$

R | **32** Find all values of y that make each statement true.

a $\frac{1}{y} = \frac{1}{5}$ 5
b $\frac{2}{y} = \frac{8}{12}$ 3
c $\frac{y}{2} = \frac{2}{y}$ ±2

R,P | **33** Determine whether each statement is True or False when $a \neq 0$ and $b \neq 0$.

a $a\left(\frac{1}{a}\right) = 1$ True

b $a\left(\frac{1}{a} + \frac{1}{b}\right) = 1 + \frac{1}{b}$ False

c $ab\left(\frac{1}{a} + \frac{1}{b}\right) = b + a$ True

Problem-Set Notes and Strategies, continued

- Problem **21** will aid the students in the next several sections by helping them see how to find common denominators for polynomials. Emphasize that the key is to factor the expressions into their prime components.

- Problem **26** works out to such a nice answer that it is easy to forget the restrictions. Again, reiterate the importance of finding the restrictions before simplifying.

Problem-Set Notes and Strategies, continued

- In problem **34**, have students work through the translations of each part slowly and identify a variable for the denominator first.
- There are several ways to solve problem **35**. One way is to multiply by the LCD and simplify. Remind students to first find the restrictions, or an answer may not be valid. This is explored further in the next section.
- There is a slight difference in problem **37**. Set up a variable for the denominator and take it slowly through the translation from English into math. More of these problems will be done in the next section.
- You may want to remind students that the area of a trapezoid in problem **40** is the average length of the bases times the height.
- Problem **43** introduces the next section. Some students have had difficulty with these problems in Algebra 1. Have students first estimate an answer. They will soon realize that the answer is not the average of the two times which is a common mistake.

Problem Set, *continued*

P,I **34** The numerator of a fraction is three more than the denominator. If both the numerator and the denominator are increased by three and the resulting fraction is multiplied by the original fraction, the product is equal to $\frac{7}{5}$. Find the original fraction. $\frac{18}{15}$

I **35** Solve $\frac{x^2 + 11x + 28}{x - 5} \cdot \frac{x^2 - 8x + 15}{x + 7} = 2x$ for x. $x = 4$ or $x = -3$

R,P **36** Find all vertical asymptotes of the graph of $f(x) = \frac{x + 1}{x^2 - 6x + 8}$. $x = 2, x = 4$

I **37** The numerator of a fraction is five more than the denominator. Both the numerator and the denominator are increased by five, and the resulting fraction is multiplied by the original one. What is the relationship between the numerator and the denominator of the product? Numerator is ten more than denominator.

R **38** Expand $(2x + 3y)^3$. $8x^3 + 36x^2y + 54xy^2 + 27y^3$

I **39** Solve $\frac{2x^3 - x^2 - 6x}{x^2 - 2x} = x - 1$ for x. $x = -4$

P,R **40** In the trapezoid shown, $b_1 = 2x^2 + 2$, $b_2 = 4x + 6$, and the area is $x^3 - 8$. Find the height h. $h = x - 2$

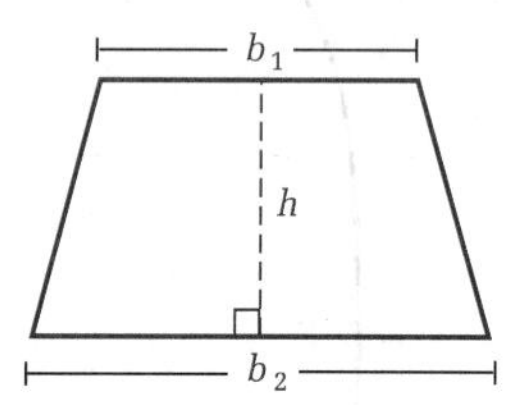

P **41** By what rational expression should we multiply $\frac{x^2 - 3x - 10}{x^2 - 16}$ to get an answer of $\frac{x + 2}{x - 4}$? $\frac{x + 4}{x - 5}$

P **42** Simplify $\frac{x^2 - 4}{x^2 + 2x + 3}$ and state restrictions. $\frac{x^2 - 4}{x^2 + 2x + 3}$; $x \neq -1 + \sqrt{2}i, x \neq -1 - \sqrt{2}i$

I **43** Working alone, Private I can dig a hole in 20 minutes. Working alone, Corporal Punishment can dig a hole in 15 minutes. How long will it take them if Sergeant Pepper has them work together to dig the hole? $8\frac{4}{7}$ min

C, P **44** Explain why f(x) and g(x) are not the same if $f(x) = \frac{x(x + 2)}{(x - 1)(x + 2)}$ and $g(x) = \frac{x}{x - 1}$. g is defined at $x = -2$, but f is not.

R **45** Write $\left(x + \frac{1}{x}\right)^6$ in expanded form. $x^6 + 6x^4 + 15x^2 + 20 + \frac{15}{x^2} + \frac{6}{x^4} + \frac{1}{x^6}$

P **46** Simplify $\frac{x^2 + 3xy + 2x + 6y}{x + 3y}$. $x + 2$; $x \neq -3y$

R **In problems 47 and 48, use the computer programs in Section 9.3 to estimate the roots of each equation to the nearest hundredth.**

47 $x^5 - 2x^4 + 2x^3 + 30x^2 - 60x - 60 = 0$ $x \approx -2.81$, $x \approx 2.34$, $x \approx -0.75$

48 $2x^4 + 4x^3 + 3x^2 - 6x - 9 = 0$ $x \approx -1.22$, $x \approx 1.22$

R **49** If $x + \frac{1}{x} = 5$, what is the value of $x^2 + \frac{1}{x^2}$? (Hint: Square both sides of the original equation.) 23

CAREER PROFILE

TEACHING MATHEMATICS

Joan Hairston communicates the beauty and joy of mathematics

"I emphasize that mathematics is a language," says Joan Hairston, a mathematics teacher at Dorsey Senior High School in Los Angeles, California. "I not only give my students problems to solve but also ask the students to write about how they have solved a given problem. They must describe both the problem and the solution. This enables me to see the depth of their understanding and their thinking process."

One of Ms. Hairston's primary goals is to nurture her students' analytical abilities and to help them see that patience and persistence can lead them to a solution to almost any problem.

As a teacher coordinator of the Urban Math Collaborative, a national organization, Ms. Hairston is working for positive change in the mathematics curriculum. She is trying to promote more critical thinking and greater use of language activities—discussions and writing—in the study of mathematics. She also gives her students assignments to research the contributions made in science and mathematics by members of underrepresented minorities.

As a teacher, she keeps herself alert to mathematical relationships that exist in the world and shares them with her students. She teaches all levels of high school mathematics, from general mathematics through algebra, geometry, and advanced algebra. "I try to make my geometry class a lab, with opportunities for exploration that allow students to see the beauty of the structure of geometry," she says. "I also like giving the students graphs and asking why they think certain things occur." In short, she wants her students to want to learn and to share her excitement about mathematics.

Ms. Hairston has a bachelor's degree from Smith College in Northampton, Massachusetts, with a double major in mathematics and psychology. She received her teaching credentials through the California state university system.

Class Planning

Time Schedule

Average: 3 days
Advanced: 2 days
Honors: 2 days

Resource References

Teacher's Resource Book
Class Opener 60
Practice 63

Class Opener

Solve the equation.

$$\frac{9+6x+x^2}{x^2+3x} \cdot \frac{x^2}{x^2-9} = 4$$

$x = 4$

Lesson Notes

- The only way that a fraction may be equal to zero is if the numerator is equal to zero. If more than one rational expression is involved in an equation, the expressions may be simplified or the equation may be multiplied by the least common denominator to eliminate the rational expressions. The resulting equation may then be solved by conventional methods.

10.2 Rational Equations and Inequalities

Objectives

After studying this section, you will be able to

- Find zeros of rational functions
- Solve rational equations and inequalities

Part One: Introduction

Zeros of Rational Functions

Finding the real zeros of rational functions is similar to finding the real zeros of polynomial functions.

Example *Find the real zeros of* $f(x) = \frac{x^2 - 5x - 36}{x - 3}$.

There is a restriction on the domain of f—$x \neq 3$.

Recall that $f(x) = 0$ only when the *numerator* of the rational expression is zero. If we find the zeros of the polynomial in the numerator and if function f is defined at these values, then these values are the zeros of f.

$$x^2 - 5x - 36 = 0$$
$$(x - 9)(x + 4) = 0$$
$$x = 9 \text{ or } x = -4$$

The rational function f is defined at these two values of x, so the zeros of f are 9 and -4.

What happens if an expression cannot be factored?

Example *Find all complex zeros of* $f(x) = \frac{x^3 - 5x^2 + 9x - 6}{x - 3}$.

The domain of f has one restriction—$x \neq 3$. We need to find the values of x that make the numerator 0. If f is defined at these values, the values are zeros of f.

According to the Rational-Zeros Theorem, ±1, ±2, ±3, or ±6 could be zeros. Using synthetic division, we check 2.

$$\begin{array}{r|rrrr} 2 & 1 & -5 & 9 & -6 \\ & & 2 & -6 & 6 \\ \hline & 1 & -3 & 3 & 0 \end{array}$$

Vocabulary

rational equation
rational inequality

The remainder is 0, so 2 is a zero. The quotient, $x^2 - 3x + 3$, cannot be factored. We use the quadratic formula to find its roots:

$$x = \frac{3 + \sqrt{3}i}{2} \text{ and } x = \frac{3 - \sqrt{3}i}{2}$$

The zeros of f are 2, $\frac{3}{2} + \frac{\sqrt{3}}{2}i$, and $\frac{3}{2} - \frac{\sqrt{3}}{2}i$.

Solving Rational Equations and Inequalities

In our work with rational functions, we often need to solve ***rational equations.*** Rational equations are equations that contain at least one rational expression.

Example *If $f(x) = \frac{1}{x}$ and $g(x) = \frac{2}{x^2}$, for what values of x does the sum of f(x) and g(x) equal 3?*

We are asked to find the values of x for which $f(x) + g(x) = 3$.

Substituting $\quad \frac{1}{x} + \frac{2}{x^2} = 3$

Multiply both sides by x^2 $\quad \frac{x^2}{x} + \frac{2x^2}{x^2} = 3x^2$

Simplify terms $\quad x + 2 = 3x^2$

$$0 = 3x^2 - x - 2$$

$$0 = (3x + 2)(x - 1)$$

$$x = -\frac{2}{3} \text{ or } x = 1$$

The functions $f(x)$ and $g(x)$ are both defined at these values of x. The sum of $f(x)$ and $g(x)$ is 3 at $x = -\frac{2}{3}$ and $x = 1$.

Example *Given $f(x) = \frac{x^2 - 5x + 6}{x^2 - 4}$, find all values of x such that $f(x) \le 0$.*

We are asked to solve the inequality $\frac{x^2 - 5x + 6}{x^2 - 4} \le 0$. To do this, we factor the rational expression and graph the solutions of the resulting ***rational inequality.***

$$\frac{x^2 - 5x + 6}{x^2 - 4} \le 0$$

Factor $\quad \frac{(x - 2)(x - 3)}{(x - 2)(x + 2)} \le 0$

Each factor gives a boundary value. The boundary values are $x = 2$, $x = 3$, and $x = -2$. At $x = -2$ and $x = 2$, the left side of the inequality is undefined. Only $x = 3$ makes the inequality a true statement. So only this point is marked with a solid dot on the number line below.

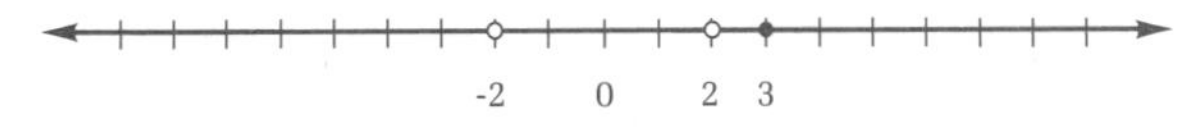

Lesson Notes, continued

- It is imperative that the restrictions on the variable be identified before the equation is multiplied by the LCD or the answers may include a result that is not valid in the original equation.

- Inequalities are solved by methods similar to equalities. Remind students that values that make the inequality equal to zero or undefined are called boundary values. These values divide the real numbers into regions, each of which must be tested. If any element in a particular region satisfies the inequality, all values in the region will satisfy the inequality. Emphasize that we include undefined values as these are also boundary values.

10.2

Communicating Mathematics

To solve rational inequalities only boundary values need be found and the regions defined by those values tested. Explain in your own words why these boundary values are so important and what is significant about them.

Stumbling Block

In solving rational inequalities, the first step is to find the boundary values, those values where the sign of the rational expression changes. This is another implication of the Zero Product Property.

Using Computers

A symbolic manipulator can solve virtually any of the rational equations in this section. If you do not have a symbolic manipulator, you can use graphing software to graph each side of the rational equation and then look at the x-coordinates of the points of intersection for your solutions. Restrictions on variables often show up clearly in graphs. On the other hand, holes in graphs are not displayed by graphing software.

We choose a test point from each region determined by the boundary values. Let's test -3, 0, 2.5, and 5.

Test Point	Left Side	Is $f(x) \le 0$?
$x = -3$	6	No, $6 > 0$
$x = 0$	-1.5	Yes, $-1.5 \le 0$
$x = 2.5$	$-\frac{1}{9}$	Yes, $-\frac{1}{9} \le 0$
$x = 5$	$\frac{2}{7}$	No, $\frac{2}{7} > 0$

The solutions of the inequality are shown on the following graph.

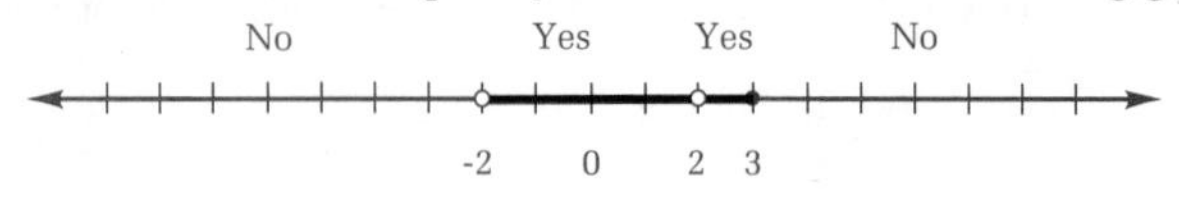

Let's look at the graph of f on the coordinate plane. Notice that $f(x) \le 0$ for the same values of x that were graphed on the number line.

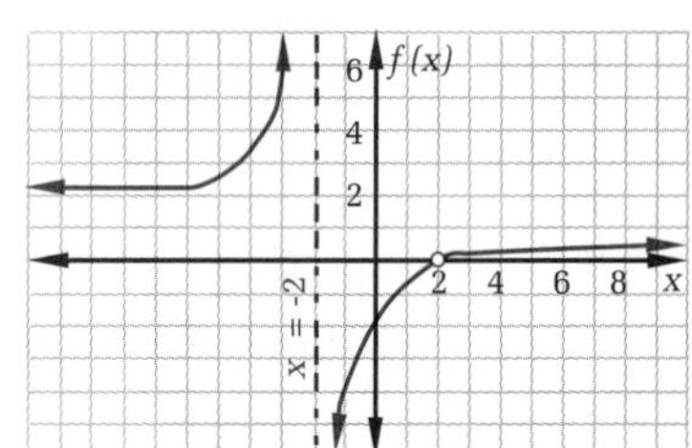

Part Two: Sample Problems

Problem 1 *If* $f(x) = \frac{3x - 1}{x} + \frac{3}{x - 3}$ *and* $g(x) = \frac{9}{x^2 - 3x}$, *what are the values of x for which* $f(x) = g(x)$?

Solution The restrictions are $x \neq 0$ and $x \neq 3$.
We are asked to find the values of x that make $f(x) = g(x)$.

$$\frac{3x - 1}{x} + \frac{3}{x - 3} = \frac{9}{x^2 - 3x}$$

Factor the denominator

$$\frac{3x - 1}{x} + \frac{3}{x - 3} - \frac{9}{x(x - 3)} = 0$$

Multiply both sides by the least common multiple (LCM) of the denominators, $x(x - 3)$.

$$\left(\frac{3x - 1}{x} + \frac{3}{x - 3} - \frac{9}{x(x - 3)}\right) \cdot x(x - 3) = 0 \cdot x(x - 3)$$

$$(3x - 1)(x - 3) + 3x - 9 = 0$$

$$3x^2 - 10x + 3 + 3x - 9 = 0$$

$$3x^2 - 7x - 6 = 0$$

$$(3x + 2)(x - 3) = 0$$

$$x = -\frac{2}{3} \text{ or } x = 3$$

Since $f(x)$ and $g(x)$ are undefined at $x = 3$, there is only one solution, $x = -\frac{2}{3}$.

Problem 2 *Bill and Tom washed cars to raise money for a school fund drive. It took Tom three minutes longer to wash a car than it took Bill. Together they washed nine cars per hour. How long did it take Bill to wash a car by himself?*

Solution If m = the number of minutes it takes Bill to wash a car, then $m + 3$ = the number of minutes it takes Tom to wash a car. The number of cars Bill and Tom washed in 60 minutes is $\frac{60}{m}$ and $\frac{60}{m+3}$ respectively.

$$\frac{60}{m} + \frac{60}{m+3} = 9$$

Multiply both sides by $m(m + 3)$

$$60(m + 3) + 60m = 9m(m + 3)$$

$$60m + 180 + 60m = 9m^2 + 27m$$

$$0 = 9m^2 - 93m - 180$$

$$0 = 3m^2 - 31m - 60$$

By the quadratic formula, $m = \frac{-5}{3}$ or $m = 12$. Since m must be positive, it took Bill 12 minutes to wash a car.

Problem 3 *A baseball team has a record of 47 wins and 39 losses. How many games must the team win in a row in order for its winning percentage to be above .650?*

Sports Page

Central Division	W	L
Centerville Cardinals	47	39
Beantown Bombers	45	41

Solution Let w = the number of games the team must win in a row. Then $\frac{47 + w}{47 + 39 + w}$ is the ratio of $\frac{\text{total wins}}{\text{total games}}$.
Find the value of w when this ratio is greater than 0.650.

$$\frac{47 + w}{86 + w} > 0.650$$

Multiply both sides by $(86 + w)$

$$\frac{(47 + w)(86 + w)}{(86 + w)} > 0.650(86 + w)$$

$$47 + w > 55.9 + 0.65w$$

$$0.35w > 8.9$$

$$w > \frac{8.9}{0.35} \approx 25.43$$

The team must win at least 26 games in a row for its winning percentage to be above .650.

Checkpoint

1. Find those values for which $\frac{x^2 + 5x + 6}{x + 2} = 0$.
   $x = -3$. The value of -2 is not included as it is not in the domain.
2. Solve the equation $\frac{3}{x} - \frac{28}{x^2} = -1$. $x = -7$ and 4
3. Find the LCD between the fractions $\frac{1}{x}, \frac{3x+4}{x-2}, \frac{5x-1}{x^2-4}$.
   $x(x^2 - 4)$
4. Solve the inequality $x(x - 5) > 0$.
   $x < 0$ and $x > 5$

Assignment Guide

Average	
Day 1	11, 14, 19, 24–26, 31, 36, 46, 53
Day 2	12, 15, 18, 23, 27, 29, 32, 40, 43, 51, 54
Day 3	13, 17, 20, 21, 28, 33, 37, 42, 52, 55, 58
Advanced	
Day 1	12, 14, 18, 23, 27, 33, 36, 46, 51, 56, 58
Day 2	11, 13, 15, 19, 21, 25, 28, 35, 38, 43, 52, 59a, b, 61
Honors	
Day 1	11–13, 18, 25, 28, 30, 36, 39, 43, 47, 51, 57
Day 2	23, 27, 32–34, 41, 42, 48, 50, 52, 56, 60, 61

Problem-Set Notes and Strategies

- Students may need some guidance to set up problem **8**.
- In problems **9** and **10**, students may be tempted to multiply both sides by the denominator. Caution them that they may be multiplying by a negative number.
- Remind students that –3 is not a solution of problem **21** since it is a restricted value.

Part Three: Exercises and Problems

Warm-up Exercises

R **In problems 1–4, find the LCM of each pair of numbers.**

1 6, 7 — 42

2 x, y — xy

3 x, xy — xy

4 $x + 1, x^2 - 1$ — $x^2 - 1$

P **In problems 5 and 6, solve each equation.**

5 $\frac{4}{x} + \frac{3}{x^2} = 1$ — $x = 2 \pm \sqrt{7}$

6 $\frac{27}{(x+1)^2} - \frac{3}{x+1} = 4$ — $x = \frac{5}{4}$ or $x = -4$

P **7** Find the zeros of $f(x) = \frac{(x+2)(2x-3)(3x+1)}{(x-2)(3x-1)}$. $-2, \frac{3}{2}, -\frac{1}{3}$

P **8** Bill can word-process text in three-fourths the time it takes Marsha. Bill worked on a project for two hours, and then Marsha joined him. Working together, they finished in one hour.

a How long did Bill work? 3 hr
b How long did Marsha work? 1 hr
c How long would it take Marsha to do the project alone? 5 hr
d How long would it take Bill to do the project alone? $3\frac{3}{4}$ hr
e How long would it take if they worked together the entire time? $2\frac{1}{7}$ hr

P **9** Solve $\frac{x}{x+1} > 5$ for x. $-\frac{5}{4} < x < -1$

P,R **10** Solve $\frac{x-3}{x-5} \geq 0$ for x and graph the solution on a number line.

Problem Set

P **A** **11** Solve $\frac{x^2 + x - 12}{x + 5} = 0$ for x. $x = -4$ or $x = 3$

R **In problems 12–15, solve for *x*.**

12 $(x + 3)(x - 5) = 0$ $x = -3$ or $x = 5$

13 $(x + 3) \div (x - 5) = 0$ $x = -3$

14 $(x + 3) + (x - 5) = 0$ $x = 1$

15 $(x + 3) \cdot (x - 5) = 9$ $x = 6$ or $x = -4$

P **In problems 16–19, solve for *x*.**

16 $\frac{1}{x} + \frac{2}{x} = 1$ $x = 3$

17 $\frac{2}{x} = \frac{3}{x+4}$ $x = 8$

18 $\frac{2}{3x+5} + \frac{6}{3x+5} = 1$ $x = 1$

19 $\frac{1}{2} + \frac{3}{x+2} = 2$ $x = 0$

R **20** Solve $\frac{a}{x} = \frac{3}{4}$ for a in terms of x. $a = \frac{3x}{4}$

P **21** Find the zeros of the function $g(x) = \frac{(x+2)(x+3)^2}{(x+3)(x+4)}$. -2

R **In problems 22–24, simplify and state restrictions.**

22 $\frac{x^2 + x}{x}$ — $x + 1; x \neq 0$

23 $\frac{x^3 - 5x^2 - 4x + 20}{(9 - x^2)(4 - x^2)}$

24 $\frac{x^2 - 5x - 6}{x + 1}$ — $x - 6; x \neq -1$

Additional Answers

10
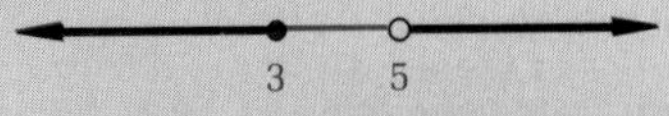

23 $\frac{x-5}{x^2-9}$; $x \neq 3, x \neq -3, x \neq 2, x \neq -2$

P,R **25** **a** If Bob can do a job in 10 minutes, what portion of the job can he do in 1 minute? $\frac{1}{10}$

b If Bill can do a job in 20 minutes, what portion of the job can he do in 1 minute? $\frac{1}{20}$

c If Jill can do a job in x minutes, what portion of the job can she do in 1 minute? $\frac{1}{x}$

B

In problems 26–29, solve for x.

P **26** $\frac{x}{6} - \frac{3x}{4} = \frac{5}{2}$ $x = -\frac{30}{7}$

27 $\frac{5}{x} + \frac{7}{x^2} = 2$ $x = \frac{7}{2}$ or $x = -1$

28 $\frac{6}{x+2} + \frac{8}{x+2} = 1$ $x = 12$

29 $\frac{6}{x-5} = \frac{8}{x-5} - \frac{1}{2}$ $x = 9$

P **30** Find the zeros of the function $h(x) = \frac{x^2 + 9x - 36}{x^2 - 36}$. $-12, 3$

R **In problems 31 and 32, divide.**

31 $\frac{x^2 - 5x}{x} \div (3x - 15)$ $\frac{1}{3}$; $x \neq 0, x \neq 5$

32 $\frac{x^2 + x + 1}{x^2 - 1} \div \frac{x^3 - 1}{(x-1)^2}$

P **In problems 33–35, solve for x.**

33 $\frac{5}{x} + \frac{7}{x} < 1$ $x > 12$ or $x < 0$

34 $\frac{2x^2 - x + 5}{3x^2 + x + 7} < \frac{2}{3}$ $x > \frac{1}{5}$

35 $\frac{5}{x} \leq \frac{7}{x} + \frac{2}{3}$ $x \leq -3$ or $x > 0$

R **36** Alex can finish a job in six hours, and Biff can do the same job in eight hours. What fraction of the job is completed by both working together for

a One hour? $\frac{7}{24}$ **b** Two hours? $\frac{7}{12}$ **c** Three hours? $\frac{7}{8}$

R **In problems 37 and 38, simplify.**

37 $\frac{42x^6}{21x^4 + 63x^2}$ $\frac{2x^4}{x^2+3}$; $x \neq 0, x \neq \sqrt{3}i, x \neq -\sqrt{3}i$

38 $\frac{16x^4 + 32x^3}{10x^2 + 20x}$ $\frac{8x^2}{5}$; $x \neq 0, x \neq -2$

P **In problems 39–41, solve and graph the solution on a number line.**

39 $\frac{(x-3)(x+4)}{(x+1)} \geq 0$

40 $\frac{5-x}{x^2} \leq 0$

41 $\frac{x}{x+5} < 0$

P **42** Solve $\frac{9}{x-3} - \frac{6}{x} = \frac{57}{x(x-3)}$ for x. $x = 13$

P,R **43** Machine 1 can do a job in three hours, machine 2 can do the same job in four hours, and machine 3 can do the job in five hours. How many hours will the job take if all three machines work together? $1\frac{13}{47}$ hr

R **44** Solve $A = p + prt$ for p. $p = \frac{A}{1 + rt}$

R **45** Find the value(s) of c so that $x^3 + 2x^2 + cx + 4$ is divisible by $x - 1$. -7

Problem-Set Notes and Strategies, continued

- Problems **33 – 35** remind students of the potential error in multiplying inequalities by a variable since the sign of the variable is unknown.
- In problem **44**, the equation is the one used to calculate the amount of money present in an account paying simple interest of r percent.

Additional Answers

32 $\frac{1}{x+1}$; $x \neq 1, x \neq -1, x \neq -\frac{1}{2} + \frac{\sqrt{3}}{2}i, x \neq -\frac{1}{2} - \frac{\sqrt{3}}{2}i$

$-4 \leq x < -1$ or $x \geq 3$

39 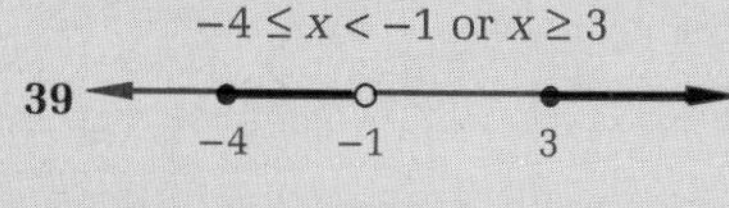

$x \geq 5$

40 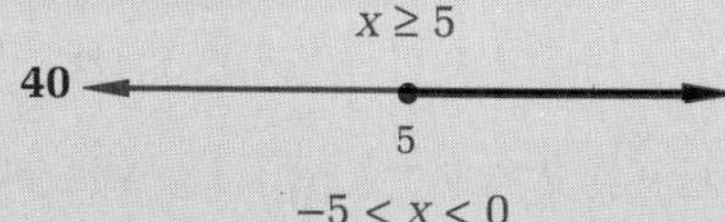

$-5 < x < 0$

41 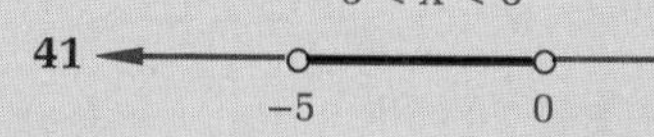

Problem Set, *continued*

P,R **46** Find all complex zeros of $f(x) = x^4 + x^3 + x^2 + 3x - 6$.
$1, -2, \pm i\sqrt{3}$

P **47** A number less the product of its reciprocal and 12 is 4. Find the number. 6 or −2

P **In problems 48 and 49, solve for x.**

48 $\frac{5}{x-7} + \frac{3}{x+4} = \frac{8x-1}{(x-7)(x+4)}$ $\{x: x \neq -4, x \neq 7\}$

49 $\frac{3}{2x+1} + \frac{5}{x-3} = \frac{13x-5}{(2x+1)(x-3)}$ No real solution

P,R **50** Find all complex zeros of $p(x) = x^4 + 5x^3 + 9x^2 - x - 14$. $-2, 1, -2 \pm i\sqrt{3}$

P **51** The Cubics won x games during the month of April and lost three fewer than they won. The ratio of wins to losses was 5:4.

a How many wins did they have in April? 15
b How many games did they play in April? 27

R **52** Refer to the diagram.

a Find the coordinates of A.
b Find the coordinates of B.
c Find the slope m of $\overline{AB}$.
a $(1, 2)$
b $(a, a^2 + 1)$
c $m = a + 1$

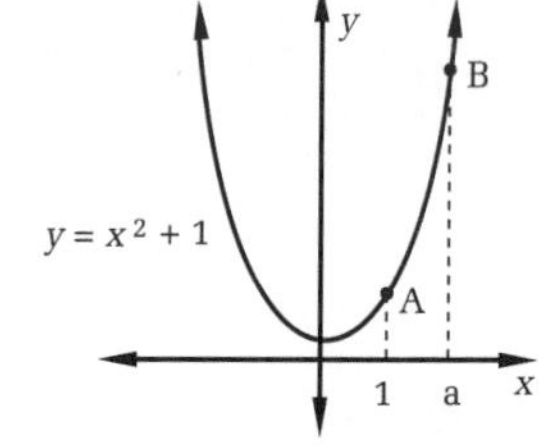

C

R **In problems 53–55, solve for x. (Hint: Factor each side.)**

53 $ax + 2x = a^2 - 4a - 12$ $x = a - 6$

54 $ax + a^2 = 16 - 4x$ $x = 4 - a$

55 $ax - 3x = a^3 - 27$ $x = a^2 + 3a + 9$

Problem-Set Notes and Strategies, continued

- Problems **53 – 55** are practice in factoring. Remind students that each side of the equation is a separate factoring problem.

P,R **56** For what values of a is $x = 2$ a solution of the inequality $\frac{3a + x}{x - 3a} < 1$? $\left\{a: a < 0, a > \frac{2}{3}\right\}$

P,R **57** One personal computer has a printer that can print a page in 30 seconds. A second computer has a printer that can print a page in 20 seconds.

a How long will it take to print a 500-page document if both printers work together? 100 min
b If the first printer starts on page 1, on what page should the second printer start in order for them to finish at approximately the same time? Page 201

P **58** The sum of the reciprocals of two consecutive integers is $-\frac{9}{20}$. Find the integers. −4 and −5

P,R **59** Let $f(x) = x^2 + x$ and $g(x) = x^2 - x$. Simplify each expression.

a $\dfrac{f(x)}{g(x)}$ $\frac{x+1}{x-1}$; $x \neq 0, x \neq 1$ **b** $\dfrac{f(x-2)}{g(x-2)}$ $\frac{x-1}{x-3}$; $x \neq 2, x \neq 3$ **c** $\dfrac{f(3x)}{g(3x)}$ $\frac{3x+1}{3x-1}$; $x \neq 0, x \neq \frac{1}{3}$

R **60** Let $(x + 3) \cdot f(x) = x^2 - 9$.

a Solve for $f(x)$.
$f(x) = x - 3$

b Graph $y = f(x)$.

R **61** Find the values of k so that $f(x) = \frac{x^2 + kx - 32}{x - 2}$ is a polynomial function. 14 (Technically this has no solution, since by definition f is not a polynomial function but a rational function.)

Additional Answer

60b

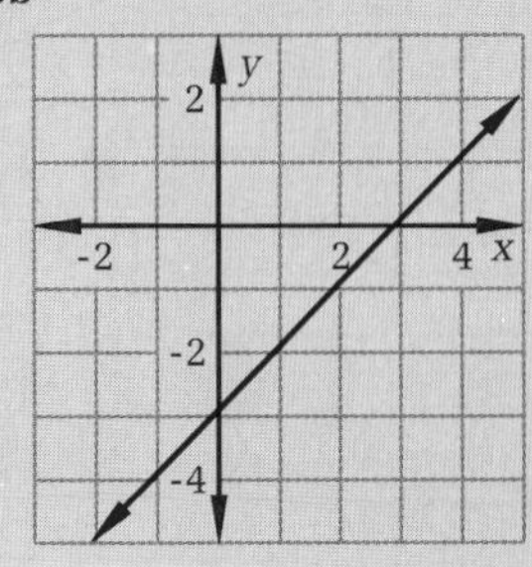

HISTORICAL SNAPSHOT

Ancient Problems

They don't get easier

The Rhind Papyrus is an Egyptian text on mathematics that was recorded by a scribe named Ahmes around 1650 B.C. It was named after Henry Rhind, a Scottish antiquary who sold the document to the British Museum, where it is now displayed. The text, written on papyrus, is believed to be a revised copy of an earlier treatise.

The first part of the work involves unit fractions—fractions with a numerator of 1. In this treatise, any fraction other than a unit fraction or $\frac{2}{3}$ is expressed as the sum of unit fractions with unlike denominators. For example, the fraction $\frac{2}{29}$ would be written as $\frac{1}{24} + \frac{1}{58} + \frac{1}{174} + \frac{1}{232}$. The fraction $\frac{2}{3}$ was expressed directly as "$\frac{2}{3}$".

The second part of the work consists of eighty-five problems. Many of the problems had practical applications, but others apparently were written to be solved for fun.

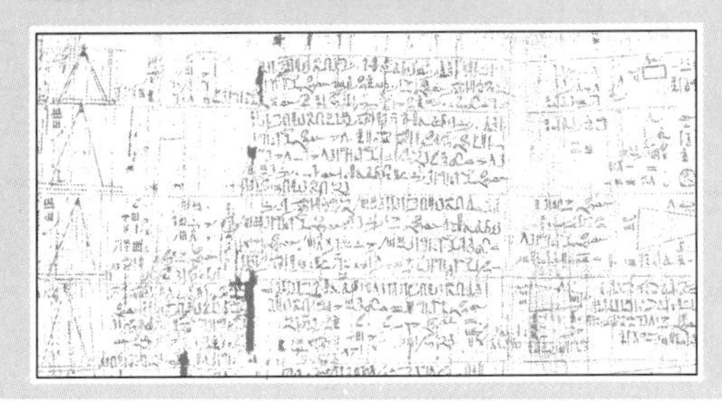

Egyptian fractions were always one part of something. The sign for 'one part of' was ⬭. Fractions shown below are $\frac{1}{10}$, $\frac{1}{9}$, $\frac{1}{8}$ and $\frac{1}{7}$.

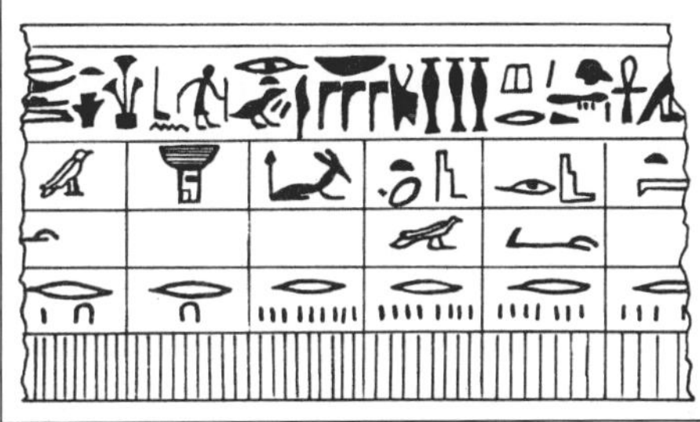

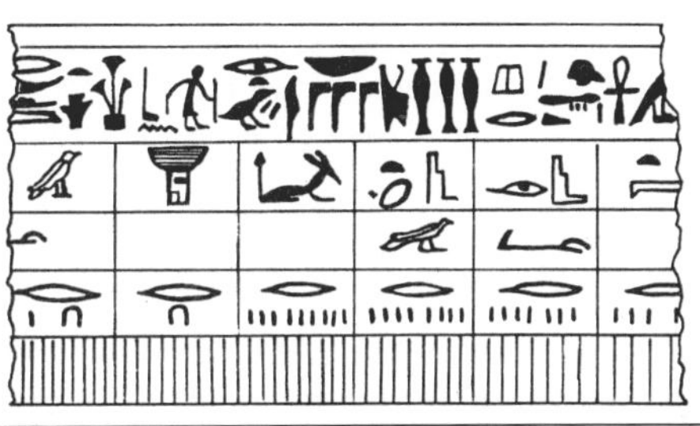

Here are three examples of problems from the papyrus. The examples show how algebra problems were written before Diophantus developed his system of algebraic notation. (See the Historical Snapshot on page 134.) For each problem, write an equation that you think represents what is described in the problem. Then solve the equation.

1 **"A quantity and its seventh added together give 19. What is the quantity?"**

2 **"A quantity and its two thirds are added together, one third of this is added, then one third of this sum is taken, and the result is 10. What is the quantity?"**

3 **"I have gone three times into the hekat measure, my one third has been added to me, [and I return], having filled the hekat measure. What is it, that this says?"**

Class Planning

Time Schedule

Average: 2 days
Advanced: $1\frac{1}{2}$ days
Honors: 1 day

Resource References

Teacher's Resource Book
Class Opener 61
Cooperative Learning 29
Practice 64
Enrichment 19
Evaluation
Quiz Forms 1, 2, 3

Class Opener

Evaluate $\frac{x+3}{x} - \frac{x-3}{x}$ for $x \in \{-3, -2, -1, 0, 1, 2, 3\}$.
$\{-2, -3, -6, \text{undef.}, 6, 3, 2\}$

Lesson Notes

- Adding and subtracting rational expressions is the identical process that is used when adding or subtracting rational numbers. As long as the denominator is the same, the numerators can be added.

10.3 Adding and Subtracting Rational Expressions with Like Denominators

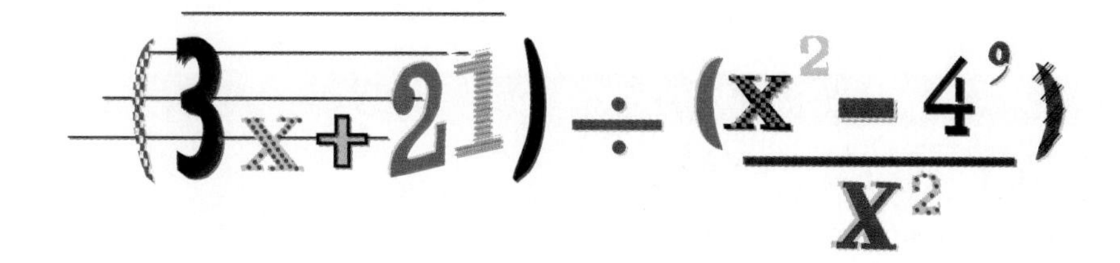

Objectives

After studying this section, you will be able to
- Add rational expressions with like denominators
- Subtract rational expressions with like denominators

Part One: Introduction

Addition of Rational Expressions

To add two or more rational expressions with like denominators, we add the numerators, just as we do with numeric fractions. We still must note restrictions, however.

Example *Add the following.*

a $\frac{2}{9} + \frac{5}{9}$

$$\frac{2}{9} + \frac{5}{9}$$
$$= \frac{2+5}{9}$$
$$= \frac{7}{9}$$

b $\frac{2x}{x+3} + \frac{3}{x+3}$

$$\frac{2x}{x+3} + \frac{3}{x+3}$$
$$= \frac{2x+3}{x+3};\ x \neq -3$$

The denominators remain unchanged.

When we multiplied rational expressions, we factored first, which guaranteed that the answers were in simplified form. When we add rational expressions, the sum may need to be simplified further. Consider the following example.

Example *Add the following.*

a $\frac{2}{9} + \frac{4}{9}$

$\frac{2}{9} + \frac{4}{9}$

$= \frac{2 + 4}{9}$

$= \frac{6}{9}$

$= \frac{2}{3}$

b $\frac{2x}{x + 3} + \frac{6}{x + 3}$

$\frac{2x}{x + 3} + \frac{6}{x + 3}$

$= \frac{2x + 6}{x + 3}$

$= \frac{2(x + 3)}{x + 3}$

$= 2; x \neq -3$

Again, we note restrictions.

Subtraction of Rational Expressions

To subtract rational expressions with like denominators, subtract their numerators. The answer will often need to be simplified further.

Example *Subtract* $\frac{x + 2}{x + 1}$ *from* $\frac{x^2}{x + 1}$.

Rewrite the problem $\frac{x^2}{x + 1} - \frac{x + 2}{x + 1}$

Subtract numerators $= \frac{x^2 - (x + 2)}{x + 1}$

Distribute $= \frac{x^2 - x - 2}{x + 1}$

Factor $= \frac{(x + 1)(x - 2)}{x + 1}$

Simplify $= x - 2; x \neq -1$

Example *Subtract* $\frac{x^3 - x + 5}{x + 5}$ *from* $\frac{x^3 + x^2 - 15}{x + 5}$.

Note that $x \neq -5$ is a restriction.

$\frac{x^3 + x^2 - 15}{x + 5} - \frac{x^3 - x + 5}{x + 5}$

$= \frac{(x^3 + x^2 - 15) - (x^3 - x + 5)}{x + 5}$

Remember to subtract every term of the polynomial being subtracted.

$= \frac{x^3 + x^2 - 15 - x^3 + x - 5}{x + 5}$

$= \frac{x^2 + x - 20}{x + 5}$

$= \frac{(x + 5)(x - 4)}{x + 5}$

$= x - 4; x \neq -5$

Lesson Notes, continued

- Subtraction may be changed into an addition problem by adding the opposite just as with numerical expressions.
- As with fractions, rational expressions may need to be simplified after finding the sum or difference. When simplifying, make sure that any restrictions that were placed on the variables in the original problem are carried through to the answer. All solutions must be valid in the original problem.

Communicating Mathematics

Explain how you would explain how to subtract the two fractions $\frac{3x + 4}{x - 7}$ and $\frac{2x - 5}{x - 7}$ to a new student in your school who has never seen subtraction of rationals before but is familiar and comfortable with addition of fractions.

Checkpoint

1. Add the fractions $\frac{1}{x}$ and $\frac{5}{x}$ and state any restrictions that must be placed on the solution.
   $\frac{6}{x}, x \neq 0$
2. Add the expressions $\frac{12}{x-4}$ and $\frac{-3x}{x-4}$.
   $-3, x \neq 4$
3. Simplify the expression $\frac{5x}{x-5} - \frac{25}{x-5}$.
   $5, x \neq 5$

Part Two: Sample Problems

Problem 1 Let $f(x) = \frac{(x + 4)^2}{x - 3}$ and $g(x) = \frac{2x^2 - 12x - 31}{x - 3}$.

a *Express the function h, the sum of f and g, in simplified form.*
b *Graph the function h.*

Solution **a** $h(x) = f(x) + g(x)$

$$= \frac{(x + 4)^2}{x - 3} + \frac{2x^2 - 12x - 31}{x - 3}$$

$$= \frac{x^2 + 8x + 16 + 2x^2 - 12x - 31}{x - 3}$$

$$= \frac{3x^2 - 4x - 15}{x - 3}$$

$$= \frac{(3x + 5)(x - 3)}{x - 3}$$

$$= 3x + 5;\ x \neq 3$$

b

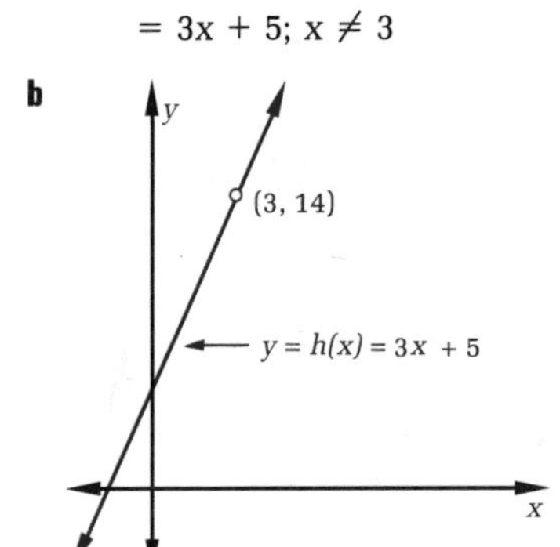

Problem 2 *Given the functions* $f(x) = \frac{x^2 - 4x + 3}{(x - 3)(x^2 - x - 1)}$ *and* $g(x) = \frac{-3x + 9}{(x - 3)(x^2 - x - 1)}$, *find the zeros of h, where* $h(x) = f(x) - g(x)$.

Solution $f(x) - g(x) = h(x)$ and $h(x) = 0$, so

$$\frac{x^2 - 4x + 3}{(x - 3)(x^2 - x - 1)} - \frac{-3x + 9}{(x - 3)(x^2 - x - 1)} = 0$$

$$\frac{x^2 - 4x + 3 - (-3x + 9)}{(x - 3)(x^2 - x - 1)} = 0$$

$$\frac{x^2 - 4x + 3 + 3x - 9}{(x - 3)(x^2 - x - 1)} = 0$$

$$\frac{x^2 - x - 6}{(x - 3)(x^2 - x - 1)} = 0$$

$$\frac{(x - 3)(x + 2)}{(x - 3)(x^2 - x - 1)} = 0$$

The numerator is zero when $x = 3$ or $x = -2$. Our work seems to be finished. But be careful! What are the restrictions on h's domain?

We can use the quadratic formula to find the restrictions.

$$x \neq 3, x \neq \frac{1 + \sqrt{5}}{2}, x \neq \frac{1 - \sqrt{5}}{2}$$

Since $h(3)$ is undefined, $x = 3$ must be rejected. So $x = -2$ is the only zero of h.

Part Three: Exercises and Problems

Warm-up Exercises

P

1 Match each expression on the left with an equivalent expression on the right.

a $\frac{x + 4}{2}$ ii

b $\frac{9}{2} + \frac{\sqrt{3}}{2}$ iv

c $\frac{7}{x}$ i

d $\frac{5}{3}$ iii

i $\frac{3}{x} + \frac{4}{x}$

ii $\frac{x}{2} + \frac{4}{2}$

iii $\frac{x + 5}{3} - \frac{x}{3}$

iv $\frac{9 + \sqrt{3}}{2}$

P,R

In problems 2–6, simplify.

2 $\frac{9}{5} - \frac{3}{5}$ $\frac{6}{5}$

3 $\frac{7}{3} + \frac{8}{3}$ 5

4 $\frac{x}{y} - \frac{1}{y}$ $\frac{x-1}{y}$; $y \neq 0$

5 $\frac{7x}{3} + \frac{8x}{3}$ 5x

6 $\frac{7}{x} + \frac{8}{x}$ $\frac{15}{x}$; $x \neq 0$

P

In problems 7–9, solve each equation and simplify each expression.

7 $\frac{2x}{3} + \frac{x}{3}$
x

8 $\frac{5}{x} - \frac{x - 7}{x} = \frac{2x - 12}{x}$
$x = 8$

9 $\frac{21}{x - 5} - \frac{5x - 4}{x - 5}$
-5; $x \neq 5$

Problem Set

A

R,I

10 If two of the following expressions are chosen at random, what is the probability that they are equal? $\frac{3}{5}$

$\frac{7}{3} + \frac{4}{3}$ $\frac{11}{6}$ $\frac{11}{3}$ $\frac{10}{3} + \frac{1}{3}$ $3 + \frac{2}{3}$

P,R

In problems 11–16, solve each equation and simplify each expression.

11 $\frac{4}{3x} = \frac{x}{3x}$ $x = 4$

12 $\frac{5x}{2y} + \frac{3x}{2y} - \frac{14x}{2y}$ $\frac{-3x}{y}$; $y \neq 0$

13 $\frac{4}{3x} + \frac{x}{3x}$ $\frac{4 + x}{3x}$; $x \neq 0$

14 $\frac{15x}{3x + 5} + \frac{25}{3x + 5}$
5; $x \neq -\frac{5}{3}$

15 $\frac{6}{5y} = \frac{3y}{5y}$
$y = 2$

16 $\frac{4x + 2}{6x - 3} - \frac{6 - 4x}{6x - 3}$
$\frac{4}{3}$; $x \neq \frac{1}{2}$

Assignment Guide

Average	
Day 1	10–12, 16, 19, 23, 24, 29, 37, 39
Day 2	15, 18, 25a, 27, 28, 30, 38, 45, 48
Advanced	
Day 1	10–12, 16, 19, 24, 26, 27, 29, 33, 39
Day 2	($\frac{1}{2}$ day) 15, 18, 25, 28, 30, 43, 48
	($\frac{1}{2}$ day) Section 10.4 11, 13, 18, 24, 27
Honors	
	11, 15, 16, 19, 24, 28–30, 38, 39, 43, 49, 51, 53

Problem-Set Notes and Strategies

- Remind students in problems **2 – 9** that any restrictions on the variable must be identified. Restrictions are often forgotten.

- In problem **16**, remind students that the subtraction sign in front of the second fraction applies to the $-4x$ as well. Students can forget to distribute it with the entire numerator.

Problem Set, *continued*

P,R **In problems 17 and 18, solve for x.**

17 $\frac{7}{a} + x = \frac{4}{a}$ $x = -\frac{3}{a}$

18 $x + \frac{a}{2b} = \frac{3a}{2b}$ $x = \frac{a}{b}$

Stumbling Block

Remind students that the restricted values must be established as the problem is worked and not when it is finished, since in the process of adding and subtracting fractions, factors may cancel.

R **19** **a** Find the ratio of the perimeter of rectangle A to that of rectangle B. $\frac{x+2}{x-2}$

b For what value of x will the ratio be 2:1? 6

R **In problems 20–22, simplify and give the restrictions.**

20 $\frac{-5+14}{5-14}$ -1

21 $\frac{-2x+3}{2x-3}$ $-1;\ x \neq \frac{3}{2}$

22 $\frac{x+3}{2x-8} - \frac{3x-5}{2x-8}$ $-1;\ x \neq 4$

P,R **23** Find the value of three fourths divided by one half. $\frac{3}{2}$

R **24** Simplify and give any restrictions.

a $x^2 \cdot x^3$ x^5

b $\frac{x^2}{\left(\frac{1}{x^3}\right)}$ $x^5;\ x \neq 0$

c $x^2 \div \frac{1}{x^3}$ $x^5;\ x \neq 0$

P **25** Simplify and give the restrictions.

a $\frac{x^2}{x+3} + \frac{2x-3}{x+3}$ $x - 1;\ x \neq -3$

b $\frac{x^2}{x+3} - \frac{2x-3}{x+3}$ $\frac{x^2-2x+3}{x+3};\ x \neq -3$

R,I **26** Find the least common multiple of each group.

a 4, 7, 14 — 28

b $x,\ x-1,\ x+1$ — $x(x^2-1)$

c $x-1,\ x+1,\ x^2-1$ — x^2-1

P **27** Let $f(x) = \frac{x}{x^2+1}$ and $g(x) = \frac{-1}{x^2+1}$.

a Find $f(x) + g(x)$. $\frac{x-1}{x^2+1};\ x \neq i,\ x \neq -i$

b Find x such that $f(x) - g(x) = 0$. $x = -1$

- Problems **20 – 22** practice the case in which the numerator and the denominator differ by a product of negative one.
- In problem **30**, remind students that the boundary points are those that make the denominator zero as well as the numerator.
- Problems **35** and **36** preview what will be done in the next section by finding the LCD before adding. Again, remind students of restrictions on the variables.

B

P,R **In problems 28–30, solve each equation or inequality and simplify each expression.**

28 $\frac{x^2}{x-7} - \frac{49}{x-7}$ $x + 7;\ x \neq 7$

29 $\frac{x^2}{x-7} = \frac{49}{x-7}$ $x = -7$

30 $\frac{x^2}{x-7} \le \frac{49}{x-7}$ $x \le -7$

P,R **In problems 31–34, solve each equation and simplify each expression.**

31 $\frac{4y}{3x} - \frac{2y}{3x} + \frac{10y}{3x}$ $\frac{4y}{x};\ x \neq 0$

32 $\frac{6}{x-y} + \frac{5}{x-y} + \frac{3}{y-x}$ $\frac{8}{x-y};\ x \neq y$

33 $\frac{4}{x-4} = \frac{6}{x-4} + 1$ $x = 2$

34 $\frac{x^2}{x^2-8x} - \frac{4x}{x^2-8x}$ $\frac{x-4}{x-8};\ x \neq 0,\ x \neq 8$

I **35** Simplify $\frac{1}{x^2} + \frac{9}{x}$. $\frac{9x+1}{x^2};\ x \neq 0$

R,I **36** Solve $\frac{1}{x^2} + \frac{9}{x} + 8 = 0$. $x = -\frac{1}{8}$ or $x = -1$

R **In problems 37 and 38, find the zeros of each function.**

37 $f(x) = \frac{x^2 - 49}{x} \cdot \frac{x^2 + x}{x + 7}$ 7, −1

38 $g(x) = \frac{x^2 - 4x - 21}{x^2 - 9}$ 7

R **39** The distance between a point (x_1, y_1) and a line in the form $ax + by + c = 0$ is given by the rational expression

$$\frac{|ax_1 + by_1 + c|}{\sqrt{a^2 + b^2}}$$

Find the distance between the line $3x - 4y + 6 = 0$ and the point $(-6, 2)$. 4

P **In problems 40 and 41, simplify.**

40 $\frac{2x}{x^2 - 3x - 18} - \frac{12}{x^2 - 3x - 18}$ $\frac{2}{x + 3}$; $x \neq 6, x \neq -3$

41 $\frac{x^2}{x^3 - 27} + \frac{9}{27 - x^3}$ $\frac{x + 3}{x^2 + 3x + 9}$; $x \neq 3$, $x \neq \frac{-3 + 3\sqrt{3}i}{2}$, $x \neq \frac{-3 - 3\sqrt{3}i}{2}$

R,I **42** Solve $\frac{9}{2x} + \frac{4}{x} = 34$ for x. $x = \frac{1}{4}$

P **43** Use the function $h(x) = f(x) - g(x)$, where $f(x) = \frac{x^3 - 2x^2}{x + 3}$ and $g(x) = \frac{9x - 18}{x + 3}$.

a Find the zeros of h. 3, 2

b Graph $y = h(x)$.

P **In problems 44–46, add or subtract and simplify.**

44 $\frac{3x - 1}{8x^3 + 1} - \frac{x - 2}{8x^3 + 1}$

45 $\frac{x^3 - x^2}{x + 2} - \frac{4x - 4}{x + 2}$

46 $\frac{9}{x + 3} - \frac{x^2}{x + 3}$ $3 - x$; $x \neq -3$

44 $\frac{1}{4x^2 - 2x + 1}$; $x \neq \frac{-1}{2}$, $x \neq \frac{1 + \sqrt{3}i}{4}$, $x \neq \frac{1 - \sqrt{3}i}{4}$ 45 $(x - 1)(x - 2)$; $x \neq -2$

P **47** Simplify $p(x) - q(x)$, where $p(x) = \frac{3}{x + 7}$ and $q(x) = \frac{3 - x}{x + 7}$. $\frac{x}{x + 7}$; $x \neq -7$

R **48** Hose A fills a pool in 1 hour 25 minutes. A second hose, B, fills the pool in 2 hours. How long will it take to fill the pool if both hoses are used? $\approx 49\frac{3}{4}$ min

R **49** Solve the inequality $\frac{x^2 + 2x + 1}{x^2 - 1} > 0$. $x > 1$ or $x < -1$

P **C** **50** Simplify $\frac{x^3 - 5x^2}{x^3 - 5x^2 - 14x} + \frac{4x^2}{x^3 - 5x^2 - 14x} + \frac{x^2 + 8}{x^3 - 5x^2 - 14x}$. $\frac{x^2 - 2x + 4}{x(x - 7)}$; $x \neq -2, x \neq 0, x \neq 7$

R,I **51** Solve $\frac{x + 4}{x + 1} + \frac{2x}{x - 1} = 6$ for x. $x = -\frac{1}{3}$ or $x = 2$

P **52** Let $f(x) = \frac{x^2 + 9}{x^3 + 6x^2 + 18x + 27}$ and $g(x) = \frac{6x}{x^3 + 6x^2 + 18x + 27}$.

a Find $f(x) + g(x)$. $\frac{x + 3}{x^2 + 3x + 9}$; $x \neq -3$, $x \neq \frac{-3 + 3\sqrt{3}i}{2}$, $x \neq \frac{-3 - 3\sqrt{3}i}{2}$

b Find x such that $f(x) - g(x) = 0$. $x = 3$

R **In problems 53 and 54, simplify.**

53 $\frac{(x + 1)^2 - 2(x + 1)}{(x + 1)^2 + (x + 1)}$ $\frac{x - 1}{x + 2}$; $x \neq -1, x \neq -2$

54 $\frac{(x + 2)(x^2 - 3x - 7) + 3(x + 2)}{(x - 4)(x^2 + 3x + 7) - 5(x - 4)}$ 1; $x \neq 4, x \neq -2, x \neq -1$

Problem-Set Notes and Strategies, continued

- Problem **41** may puzzle some students until they realize that the denominators are opposites and factoring $x - 1$ can help. They may also need to remember or look up factoring the difference of cubes.
- Encourage students to spend the time to solve problem **52.** It may take some time. The answer involves complex numbers.

Cooperative Learning

Problem **52** in the problem set involves several different techniques in order to solve and find all the restrictions. If students work in groups of three or four, assign each a different job, some to add the fractions, some to factor and some to find the restrictions. It might help to have students compare answers.

Additional Answer

43b

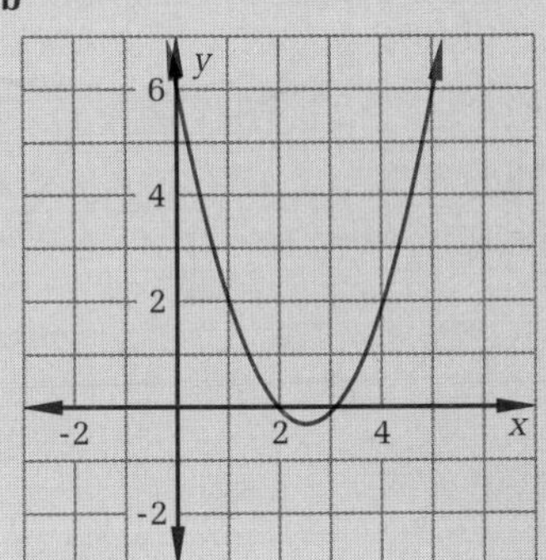

10.4

Class Planning

Time Schedule
Average: 2 days
Advanced: $2\frac{1}{2}$ days
Honors: 2 days

Resource References
Teacher's Resource Book
Class Opener 62
Cooperative Learning 30
Practice 65
Enrichment 20
Transparency 19

Class Opener

Evaluate $\frac{x+3}{x} - \frac{x+4}{x+1}$ for $x \in \{-3, -2, -1, 0, 1, 2, 3\}$.

$\{\frac{1}{2}, \frac{3}{2}, \text{undef.}, \text{undef.}, \frac{3}{2}, \frac{1}{2}, \frac{1}{4}\}$

Lesson Notes

- The words Least Common Denominator take on a whole new meaning in this section. Students are familiar with adding fractions and finding LCD's for numerical denominators. Now they are required to use the same techniques and find a LCD for a group of polynomials. Some students need to be lead through each step. Group work can help since some students will ask questions in a small group more comfortably than in a large group.

10.4 Adding and Subtracting Rational Expressions with Unlike Denominators

Objectives

After studying this section, you will be able to
- Add rational expressions with unlike denominators
- Subtract rational expressions with unlike denominators

Part One: Introduction

Adding and Subtracting Rational Expressions with Unlike Denominators

To add or subtract rational expressions with different denominators, we first write them as equivalent fractions that have the same, or common, denominators. Then we can add or subtract numerators as we did in the last section.

To find a common denominator, we factor the denominators and look for a common multiple. In some cases, our work may go more quickly if we use the least common denominator, the smallest multiple of the denominators. But often we just find a common denominator. Then we multiply each term by a form of 1. Let's look at an example.

Example *Add the numeric fractions $\frac{5}{7}$ and $\frac{1}{21}$.*

$$\frac{5}{7} + \frac{1}{21}$$

Factor denominators
$$= \frac{5}{7} + \frac{1}{3 \cdot 7}$$

A common denominator is 21.
Multiply the first term by $\frac{3}{3}$.
$$= \frac{5}{7} \cdot \frac{3}{3} + \frac{1}{21}$$

$$= \frac{15}{21} + \frac{1}{21}$$

$$= \frac{16}{21}$$

Example *Add the rational expressions $\frac{3}{2x}$ and $\frac{1}{xy}$.*

$$\frac{3}{2x} + \frac{1}{xy}$$

A common denominator is 2xy $= \frac{3}{2x} \cdot \frac{y}{y} + \frac{1}{xy} \cdot \frac{2}{2}$

$$= \frac{3y + 2}{2xy};\ x \neq 0,\ y \neq 0$$

Example *Add* $\frac{5}{x + 2} + \frac{3}{x - 2}$.

A common denominator is $(x + 2)(x - 2)$.

$$\frac{5}{x + 2} + \frac{3}{x - 2} = \frac{5}{x + 2} \cdot \frac{x - 2}{x - 2} + \frac{3}{x - 2} \cdot \frac{x + 2}{x + 2}$$

$$= \frac{5(x - 2)}{(x + 2)(x - 2)} + \frac{3(x + 2)}{(x + 2)(x - 2)}$$

$$= \frac{5x - 10}{(x + 2)(x - 2)} + \frac{3x + 6}{(x + 2)(x - 2)}$$

Add numerators $= \frac{8x - 4}{(x + 2)(x - 2)}$

$$= \frac{4(2x - 1)}{(x + 2)(x - 2)};\ x \neq 2,\ x \neq -2$$

Part Two: Sample Problems

Problem 1 *Simplify* $\frac{16}{x^2 - 2x} - \frac{1}{2 - x}$.

Solution

$$\frac{16}{x^2 - 2x} - \frac{1}{2 - x}$$

Factor denominators $= \frac{16}{x(x - 2)} - \frac{1}{2 - x}$

Notice opposite factors $(x - 2)$ and $(2 - x)$ $= \frac{16}{x(x - 2)} - \frac{1}{-(x - 2)}$

Multiply by a form of 1 $= \frac{16}{x(x - 2)} + \frac{1}{x - 2} \cdot \frac{x}{x}$

$$= \frac{16}{x(x - 2)} + \frac{x}{x(x - 2)}$$

$$= \frac{16 + x}{x(x - 2)};\ x \neq 0,\ x \neq 2$$

Problem 2 *Simplify* $\frac{4}{x^2} - \frac{16}{x^2 - 2x} - \frac{1}{x - 2}$.

Solution

$$\frac{4}{x^2} - \frac{16}{x^2 - 2x} - \frac{1}{x - 2}$$

Factor denominators $= \frac{4}{x(x)} - \frac{16}{x(x - 2)} - \frac{1}{x - 2}$

Checkpoint

1. Add the fractions $\frac{1}{x} + \frac{2}{x^2}$. $\frac{(x + 2)}{x^2}$
2. Find the LCD between the expressions x, $x^2 - 16$ and $x^2 + 8x + 4$. $x(x - 4)(x + 4)^2$
3. Simplify $\frac{3}{x - 2} + \frac{5x}{x^2 - 3x + 2}$. $\frac{8x - 3}{x^2 - 3x + 2}$, $x \neq 1$ and $x \neq 2$

Stumbling Block

If students find the LCD, suggest that they list the LCD in factored form and make sure that each denominator is included by circling it or identifying it in some other manner. Factors may be used for more than one denominator and there should be none left over that is uncircled. If there is, that is a superfluous factor and can be eliminated from the LCD.

Communicating Mathematics

Explain the process of finding the *least* common denominator and when it is helpful to have the LCD rather than *any* common denominator.

Assignment Guide

Average	
Day 1	11, 13, 18, 21, 24, 27, 28, 30, 33, 36, 42
Day 2	12, 14–17, 19, 29, 31, 34, 35, 41, 48, 52
Advanced	
Day 1	($\frac{1}{2}$ day) Section 10.3 15, 18, 25, 28, 30, 43, 48 ($\frac{1}{2}$ day) 11, 13, 18, 24, 27
Day 2	12, 14–17, 19, 21, 23, 28, 32, 35, 37, 47
Day 3	25, 29, 31, 34, 38, 42, 44, 48, 50, 52, 53, 60
Honors	
Day 1	15–17, 19, 21, 23, 27, 31, 35, 37, 42, 47, 50, 51, 60
Day 2	18, 20, 22, 29, 36, 38, 45, 49, 52, 55, 58, 61, 62

Problem-Set Notes and Strategies

- Remind students of the differences between LCM and GCF in problems **1 – 4**.
- Problem **14** will yield many different answers. This is good practice for students in communicating ideas and can lead to interesting discussion.

In this case, it is most convenient to use the least common denominator (LCD), the smallest multiple of the denominators $x(x)$, $x(x - 2)$, and $x - 2$. The smallest multiple of the denominators is $x(x)(x - 2)$.

$$= \frac{4}{x^2} \cdot \frac{(x-2)}{(x-2)} - \frac{16}{x(x-2)} \cdot \frac{x}{x} - \frac{1}{x-2} \cdot \frac{x^2}{x^2}$$

$$= \frac{4(x-2)}{x^2(x-2)} - \frac{16x}{x^2(x-2)} - \frac{x^2}{x^2(x-2)}$$

$$= \frac{4x - 8 - 16x - x^2}{x^2(x-2)}$$

$$= \frac{-8 - 12x - x^2}{x^2(x-2)}; x \neq 0, x \neq 2$$

Part Three: Exercises and Problems

Warm-up Exercises

P **In problems 1 and 2, find the least common multiple.**

1 12, 16, 24 48

2 $x^2 - x$, $x^2 + x$ $x(x - 1)(x + 1)$

P **In problems 3 and 4, find the greatest common factor.**

3 12, 16, 24 4

4 $x^2 - x$, $x^2 + x$ x

P **In problems 5–8, find a common denominator.**

5 $\frac{x^2}{x(x-3)} + \frac{9}{x-3}$ $x(x - 3)$

6 $\frac{x-3}{27} - \frac{x}{3}$ 27

7 $\frac{3x-4}{x^2-5x+6} - \frac{2x}{x^2-2x}$ $x(x - 2)(x - 3)$

8 $\frac{x}{4x+8} + \frac{16}{8x^2-32}$ $4(x - 2)(x + 2)$

P **In problems 9 and 10, add or subtract.**

9 $\frac{x^2+3x-6}{x^2-2x-3} - \frac{3}{x-3}$ $\frac{x+3}{x+1}$; $x \neq -1$, $x \neq 3$

10 $\frac{x^2+25}{x^2-25} + \frac{-25}{x^2+5x}$ $\frac{x^2-5x+25}{x(x-5)}$; $x \neq -5$, $x \neq 0$, $x \neq 5$

Problem Set

A **In problems 11–13, add or subtract.**

P **11** $\frac{3}{x} + \frac{4}{y}$ $\frac{3y+4x}{xy}$; $x \neq 0$, $y \neq 0$

12 $\frac{x}{4} + \frac{2}{x}$ $\frac{x^2+8}{4x}$; $x \neq 0$

13 $\frac{3x}{2y} - \frac{4y}{x}$ $\frac{3x^2-8y^2}{2xy}$; $x \neq 0$; $y \neq$

P **14** Explain why it is necessary to have a common denominator for fractions before they can be added or subtracted. A common denominator establishes a needed basis for comparison.

R **In problems 15–17, simplify.**

15 $\frac{x \cdot x - 3 \cdot 3}{x - 3}$ $x + 3;\ x \neq 3$

16 $\frac{x \cdot x \cdot x - 3 \cdot 3 \cdot 3}{x - 3}$ $x^2 + 3x + 9;\ x \neq 3$

17 $\frac{x \cdot x \cdot x - 3 \cdot 3 \cdot 3}{x \cdot x - 3 \cdot 3}$ $\frac{x^2 + 3x + 9}{x + 3};\ x \neq 3,\ x \neq -3$

Problem-Set Notes and Strategies, continued

- To avoid errors in problems **15–17**, suggest students write the parts of the fraction in their prime factors before simplifying.

R **18** Find the total surface area of the rectangular box.
$77\frac{1}{3}$

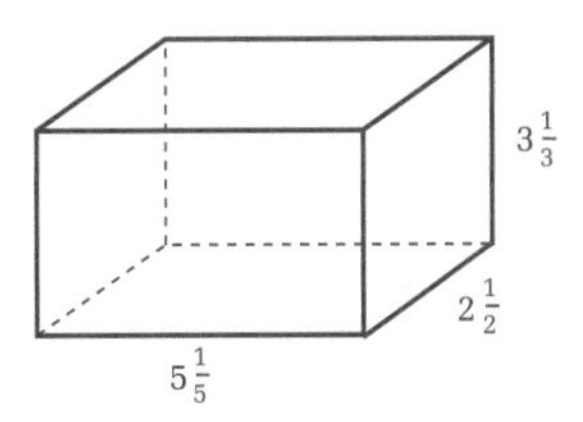

P **In problems 19–22, find a common denominator.**

19 $\frac{2}{x(x + 1)} - \frac{3x}{x + 1}$ $x(x + 1)$

20 $\frac{1}{x^2 + 2x} + \frac{3}{x^2 - 4}$ $x(x + 2)(x - 2)$

21 $\frac{2}{xy^3} + \frac{5}{x^2y}$
x^2y^3

22 $\frac{6}{x^3 - x^2} + \frac{1}{x^2 + x}$
$x^2(x + 1)(x - 1)$

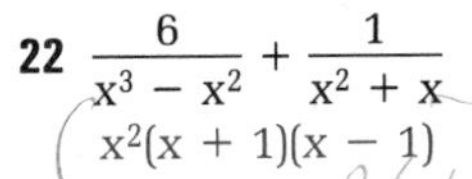

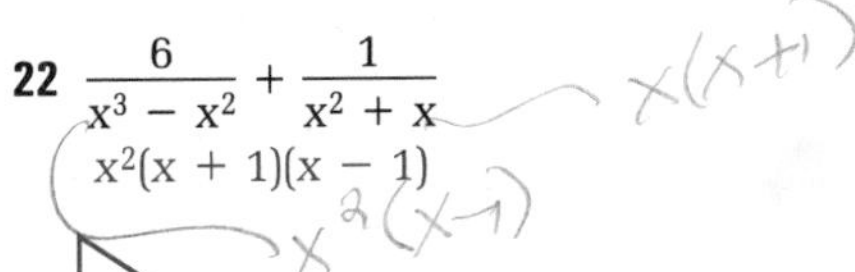

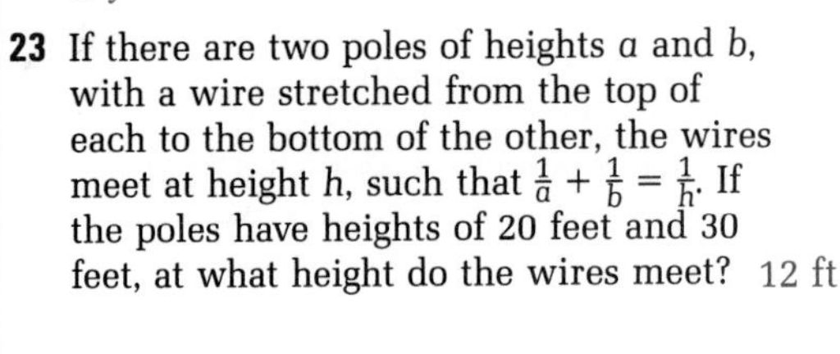

P,R **23** If there are two poles of heights a and b, with a wire stretched from the top of each to the bottom of the other, the wires meet at height h, such that $\frac{1}{a} + \frac{1}{b} = \frac{1}{h}$. If the poles have heights of 20 feet and 30 feet, at what height do the wires meet? 12 ft

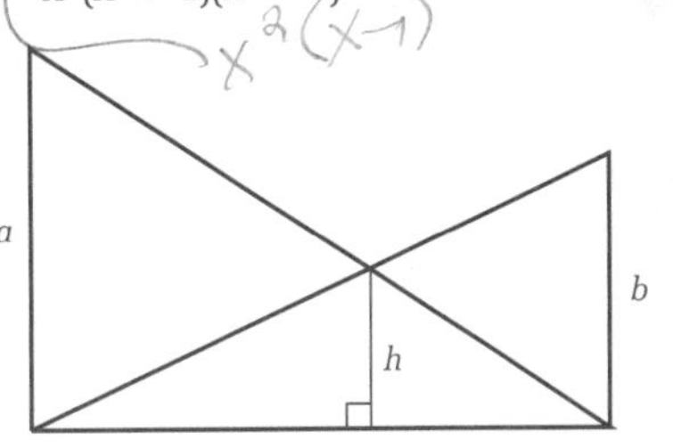

- An extension of problem **23** is for students to derive the formula from similar triangles. This is explored again in problem **62** of this section.

P **In problems 24 and 25, add or subtract.**

24 $\frac{4}{x - 3} + \frac{5}{3 - x}$ $\frac{-1}{x - 3};\ x \neq 3$

25 $\frac{2}{x^2 - 3x} - \frac{5}{3 - x}$ $\frac{5x + 2}{x(x - 3)};\ x \neq 0,\ x \neq 3$

R **26** Multiply the matrix by the scalar.

$$\frac{x}{x - 1} \cdot \begin{bmatrix} x^2 - 1 & x^2 - 2x + 1 \\ \frac{x - 1}{x} & \frac{x + 1}{x} \end{bmatrix} \quad \begin{bmatrix} x(x + 1) & x(x - 1) \\ 1 & \frac{x + 1}{x - 1} \end{bmatrix};\ x \neq 0,\ x \neq 1$$

- Problem **26** is scalar multiplication. Remind students that the fractions need to be simplified.

R **27** Solve $\frac{1}{x} = \frac{1}{x + 1}$ for x. No solution

- Some students will solve problem **27** by inspection. This can lead to an interesting discussion.

R **28** Find the product $(x + y)^3$. $x^3 + 3x^2y + 3xy^2 + y^3$

B

In problems 29–31, simplify.

P **29** $\frac{4}{x + 3} + \frac{2}{3x - 2}$
$\frac{14x - 2}{(x + 3)(3x - 2)};$
$x \neq -3,\ x \neq \frac{2}{3}$

30 $\frac{1}{y} - \frac{1}{x} + \frac{2}{xy}$
$\frac{x - y + 2}{xy};$
$x \neq 0,\ y \neq 0$

31 $\frac{4}{x + 3} - \frac{2}{3x - 2}$
$\frac{10x - 14}{(x + 3)(3x - 2)};$
$x \neq -3,\ x \neq \frac{2}{3}$

Problem-Set Notes and Strategies, continued

- Problems **33** and **35** introduce complex fractions, the subject of the next section.
- In problem **38** what may seem an obvious solution leads to an incorrect answer. Urge students to go through all steps in the simplifying process.

Additional Answer

39
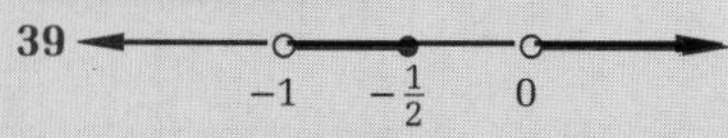

Cooperative Learning

Problem **39** combines methods introduced in several sections. It is a good problem to initiate discussion when worked in groups. You might remind students to be aware of the restrictions on the variables since they count as boundary values as well. After groups solve this problem, change the zero on the right side to any other constant such as 5 and ask how this affects the solution.

- Problem **47** results in a cubic equation. Remember the restrictions on x, $x \neq \pm 1$.

Problem Set, *continued*

R **32** Fill in the missing factor. ________ $\frac{(x-3)(x+2)}{x^2(x-1)} = \frac{x-1}{x+2}$ $\frac{x^2(x-1)^2}{(x-3)(x+2)^2}$

I **33** Find the reciprocal of the sum of the reciprocals of 2 and 3. $\frac{6}{5}$

R **34** The ratio of the area of square A to the area of square B is 3:2. Find x. $x \approx 8.90$

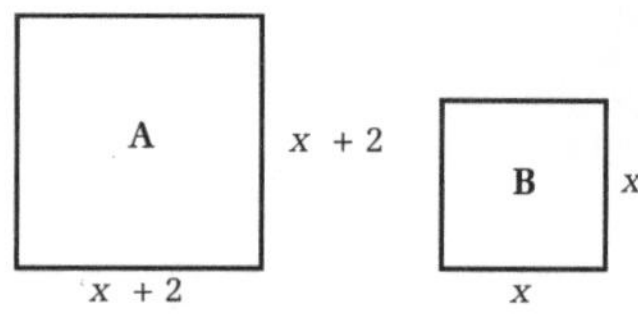

I **35** Find the reciprocal of the sum of the reciprocals of x and y. $\frac{xy}{x+y}$

P,R **In problems 36 and 37, solve for x.**

36 $\frac{5}{x+2} = \frac{7}{x-4}$ $x = -17$

37 $\frac{1}{2x} + \frac{5}{6x} = \frac{3}{4x^3}$ $x = \pm\frac{3}{4}$

P,R,I **38** Solve $x\left(\frac{1}{x_1} + \frac{1}{x_2}\right) = 1$ for x. $x = \frac{x_1x_2}{x_1 + x_2}$

P,R **39** Solve $\frac{1}{x} + \frac{1}{x+1} \geq 0$ for x and graph the solution on a number line. $-1 < x \leq -\frac{1}{2}$ or $x > 0$

P,R **In problems 40 and 41, solve for x.**

40 $\frac{3}{x} + \frac{5}{2x} = \frac{7}{4}$ $x = \frac{22}{7}$

41 $2ax + 3x = 4a^2 - 9$ $x = 2a - 3$

R **42** $\triangle ABC \sim \triangle DEF$.

a Set up a correct proportion.
b Solve your proportion for x.
a $\frac{x}{x+7} = \frac{x+2}{5x-1}$ **b** $x = \frac{7}{2}$

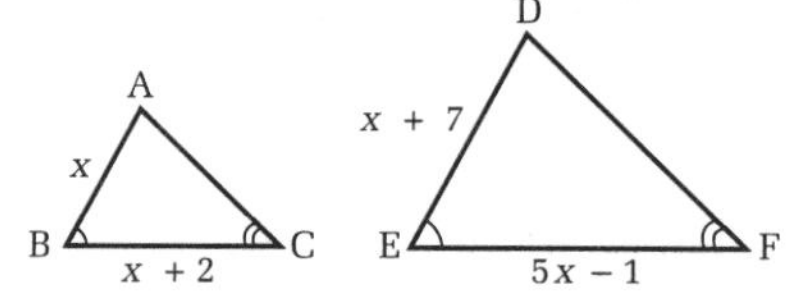

P **In problems 43–46, simplify.**

43 $\frac{9}{x^2 - 2x} - \frac{3}{2-x}$ $\frac{3(x+3)}{x(x-2)}$; $x \neq 0, x \neq 2$

44 $\frac{3}{x} + \frac{5}{x-1}$ $\frac{8x-3}{x(x-1)}$; $x \neq 0, x \neq 1$

45 $\frac{2}{x} + \frac{5}{x} + \frac{7}{x+1}$ $\frac{7(2x+1)}{x(x+1)}$; $x \neq 0, x \neq -1$

46 $\frac{x^2 - xy - 6y^2}{x^2 - 9y^2} \cdot \frac{4x + 12y}{x^2 + 4xy + 4y^2}$ $\frac{4}{x+2y}$; $x \neq -2y, x \neq 3y, x \neq -3y, x \neq 2y$

P,R **47** Solve $\frac{x}{x-1} - \frac{2}{x+1} + x = 4$ for x. $x = 3$, $x = \sqrt{2}$, or $x = -\sqrt{2}$

R **48** Find a linear function in the form $f(x) = ax + b$ such that $f(2) = 9$ and $f(4) = 17$. $f(x) = 4x + 1$

R **49** Solve the inequality $\frac{x^2 - 25}{x^2 - 9} \leq 0$. $3 < x \leq 5$ or $-5 \leq x < -3$

P,R **In problems 50 and 51, let $f(x) = x^2$ and $g(x) = 3x + 10$. Find the following.**

50 $\frac{f(x) - g(x)}{x+2}$ $x - 5$; $x \neq -2$

51 $\frac{f(x+3) - g(x+3)}{(x+3)+2}$ $x - 2$; $x \neq -5$

R **52** Refer to the diagram.

a Find the coordinates of A and B.

b Find the slope of $\overleftrightarrow{AB}$.

c For what values of a is the slope greater than 2?

a $A(a, a^3)$, $B(-a, -a^3)$
b a^2
c $\{a: a > \sqrt{2}\}$

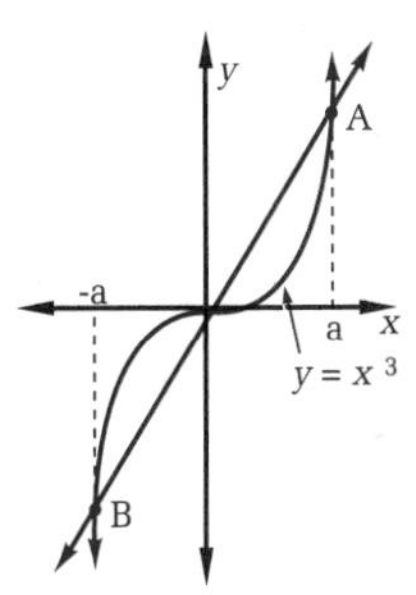

P,I **In problems 53–55, simplify.**

53 $\dfrac{x-1}{x+1} + \dfrac{x+1}{x-1}$

54 $\dfrac{x-1}{x+1} - \dfrac{x+1}{x-1}$

55 $\dfrac{\dfrac{x-1}{x+1} + \dfrac{x+1}{x-1}}{\dfrac{x-1}{x+1} - \dfrac{x+1}{x-1}}$

R **56** Graph $f(x) = \frac{x-1}{x(x+1)}$.

P **In problems 57 and 58, let $f(x) = \frac{1}{x}$. Find each value.**

57 $f(x + 3) - f(x)$ $\frac{-3}{x(x+3)}$; $x \neq 0$, $x \neq -3$

58 $f(x + h) - f(x)$ $\frac{-h}{x(x+h)}$; $x \neq 0$; $x = -h$

P,R **59** Solve $\frac{1}{x} + \frac{1}{x+1} = \frac{2x+1}{x(x+1)}$ for x. $\{x: x \neq 0, x \neq -1\}$

R **60** It takes machine A twice as long to produce 10,000 bolts as it takes machine B. Together they do the job in 8 hours. How long will it take each machine to produce 5000 bolts if each works separately? machine A: 12 hr; machine B: 6 hr

C
P,R **61** The expression $\frac{x}{x+5} + \frac{2}{x+2} - \frac{15}{(x+5)(x+2)}$ can be simplified to $\frac{B(x)}{x+2}$; $x \neq -2$, $x \neq -5$.

a Find the binomial $B(x)$. $B(x) = x - 1$

b The result $\frac{B(x)}{x+2}$ has the obvious restriction $x \neq -2$, but why must we also restrict the answer to $x \neq -5$?
Restrictions are determined by the original expression.

P,R **62** Refer to the diagram.

a Use congruent triangles to find a and b if $\frac{h}{h_1} = \frac{a}{x+y}$ and $\frac{h}{h_2} = \frac{b}{x+y}$. $a = y$; $b = x$

b Add the equations in part **a** and solve for h. $h = \frac{h_1h_2}{h_1 + h_2}$

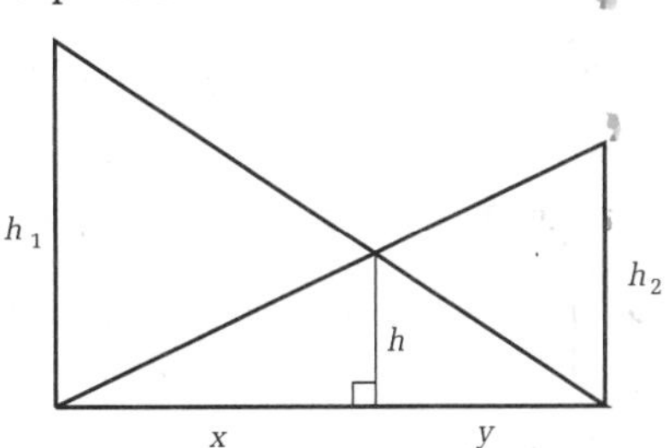

P,R **63** Find all ordered pairs of positive integers (a, b), with $a \geq 3$ and $b \geq 3$, so that $\frac{1}{a} + \frac{1}{b} > \frac{1}{2}$.
(3, 3), (3, 4), (3, 5), (4, 3), (5, 3)

Problem-Set Notes and Strategies, continued

- Problem **55** is a good example of planning a strategy on what looks difficult. One way to work is to simplify the numerator and denominator and then divide. This problem previews the next section.

- Problem **62** is the identical picture used in problem **23**. The similarities and formulas are worked out in this problem. You might go back and redo problem **23** with this new rule.

Additional Answers

53 $\frac{2(x^2+1)}{x^2-1}$; $x \neq 1$, $x \neq -1$

54 $\frac{-4x}{x^2-1}$; $x \neq 1$, $x \neq -1$

55 $\frac{x^2+1}{-2x}$; $x \neq 0$, $x \neq 1$, $x \neq -1$

56

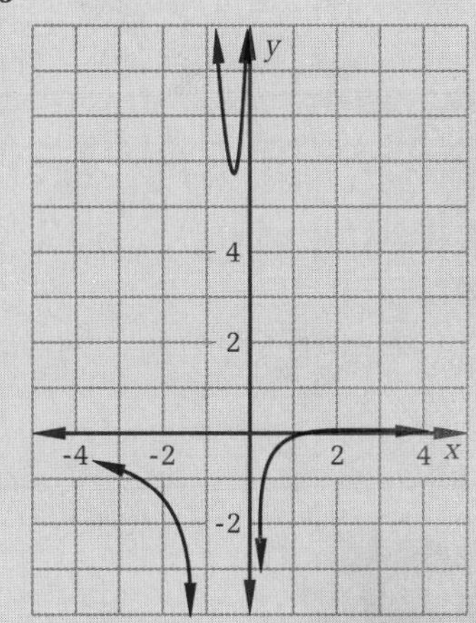

10.5

Class Planning

Time Schedule
Average: Optional
Advanced: Optional
Honors: 2 days

Resource References
Teacher's Resource Book
Class Opener 63
Practice 66
Manipulatives and Applications 15
Transparency 20

Class Opener

Evaluate $\dfrac{1+\frac{1}{x}}{1-\frac{1}{x}}$ for

$x \in \{-3, -2, -1, 0, 1, 2, 3\}$.

$\{\frac{1}{2}, \frac{1}{3}, 0, \text{undef.}, \text{undef.}, 3, 2\}$

Lesson Notes

- Two different methods of simplifying complex fractions are identified in this section. Point out to the students that even though they may find a favorite method for themselves, it is helpful to have both methods at their command as each has its place in solving these problems.
- For many students, Method 1 seems to be the easiest to handle. This simplifies each part of the complex fraction before dividing.

10.5 Complex Fractions

Objective

After studying this section, you will be able to

- Simplify complex fractions

Part One: Introduction

Simplify Complex Fractions

In later work with mathematics, you will often see ***complex fractions.*** ***Complex fractions*** are rational expressions that have fractions in the numerator, the denominator, or both. Here are three examples.

$$\frac{1+\frac{5}{n}}{n-\frac{1}{3}} \qquad \frac{\frac{a}{b}-\frac{b}{a}}{\frac{a}{b^2}-\frac{b}{a^2}} \qquad \frac{1}{1-\frac{1}{x}}$$

In this section, you will learn two methods for simplifying complex fractions into simpler rational expressions.

Example *Simplify* $\dfrac{1+\frac{5}{n}}{n-\frac{1}{3}}$.

Method 1

The numerator is a sum of rational expressions, and the denominator is a difference of rational expressions. We can add and subtract respectively. The result will be a division problem.

$$\frac{1+\frac{5}{n}}{n-\frac{1}{3}} = \frac{1\cdot\frac{n}{n}+\frac{5}{n}}{n\cdot\frac{3}{3}-\frac{1}{3}}$$

Simplify numerator
Simplify denominator
$$= \frac{\frac{n+5}{n}}{\frac{3n-1}{3}}$$

$$= \frac{n+5}{n} \div \frac{3n-1}{3}$$

Multiply by reciprocal of divisor
$$= \frac{n+5}{n} \cdot \frac{3}{3n-1}$$

$$= \frac{3(n+5)}{n(3n-1)} \text{ or } \frac{3n+15}{3n^2-n};\ n \neq 0,\ n \neq \frac{1}{3}$$

Vocabulary
complex fraction

Method 2

In this method we find a common denominator of the fractions in the numerator and denominator of the complex fraction and multiply the complex fraction by an appropriate form of 1.

We multiply by this form of 1 because 3n is a common multiple of n and 3

$$\frac{1 + \frac{5}{n}}{n - \frac{1}{3}} = \frac{1 + \frac{5}{n}}{n - \frac{1}{3}} \cdot \frac{3n}{3n}$$

Distribute

$$= \frac{3n + \frac{15n}{n}}{3n^2 - \frac{3n}{3}}$$

Simplify

$$= \frac{3n + 15}{3n^2 - n} \text{ or } \frac{3(n + 5)}{n(3n - 1)};$$

$$n \neq 0, n \neq \frac{1}{3}$$

In summary, in Method 1 we simplify the numerator and denominator and then divide. In Method 2 we multiply by a form of 1 and simplify. The form of 1 is determined by finding a common multiple of all denominators within the complex fraction.

Part Two: Sample Problems

Problem 1 *Simplify* $\frac{\frac{1}{b} - \frac{1}{a}}{\frac{1}{b^2} - \frac{1}{a^2}}$.

Solution We will use Method 2.

$$\frac{\frac{1}{b} - \frac{1}{a}}{\frac{1}{b^2} - \frac{1}{a^2}} \cdot \frac{a^2b^2}{a^2b^2}$$

$$= \frac{\frac{a^2b^2}{b} - \frac{a^2b^2}{a}}{\frac{a^2b^2}{b^2} - \frac{a^2b^2}{a^2}}$$

$$= \frac{a^2b - b^2a}{a^2 - b^2}$$

$$= \frac{ab(a - b)}{(a + b)(a - b)}$$

$$= \frac{ab}{a + b};\ a \neq 0, b \neq 0, a \neq -b, a \neq b$$

Lesson Notes, continued

- When the LCD is relatively straightforward, both parts of the complex fraction are multiplied by this LCD and the complex fraction is reduced to a fraction which can be simplified with rules from the past sections. In either case, it is necessary to carry through any restrictions from the original problem.

Stumbling Block

Complex fractions can look very complicated. Remind students that complex fractions are easiest solved one part at a time. Method 1 looks at each part separately and divides only when the numerator and denominator are fully simplified.

Communicating Mathematics

Explain in your own words why one of the methods of solving complex fractions might seem more appropriate than the other. What criteria would you use to decide which method to employ on a given problem?

Checkpoint

1 Simplify the fraction $\frac{1+\frac{1}{2}}{1-\frac{1}{2}}$.

3

2 Simplify the fraction $\frac{\frac{1}{x}+\frac{1}{x^2}}{\frac{1}{x^2}-1}$.

$\frac{x+1}{1-x^2}=\frac{1}{1-x}$; $x \neq 0, x \neq \pm 1$

3 Solve the equation $\frac{\frac{1}{x}+\frac{1}{x^2}}{\frac{1}{x^2}-1}=0$.

No solution since −1 is the only value that makes the numerator zero but is a restricted value.

Problem 2 Solve $\frac{1}{1-\frac{1}{x}} = x$ *for* x.

Solution

$$\frac{1}{1-\frac{1}{x}} = x;\ x \neq 0;\ x \neq 1$$

$$\frac{1}{1-\frac{1}{x}} \cdot \frac{x}{x} = x$$

$$\frac{x}{x-1} = x;\ x \neq 1$$

$$\frac{x}{x-1} \cdot (x-1) = x \cdot (x-1)$$

$$x = x^2 - x$$
$$0 = x^2 - 2x$$
$$0 = x(x-2)$$

This implies that $x = 0$ or $x = 2$. Since $x \neq 0$ is a restriction, $x = 0$ is not a solution. The only solution is $x = 2$.

Problem 3 *Given* $f(x) = \frac{1}{x}$; $x > 0$. *Write a general formula for the slope of* $\overline{AB}$.

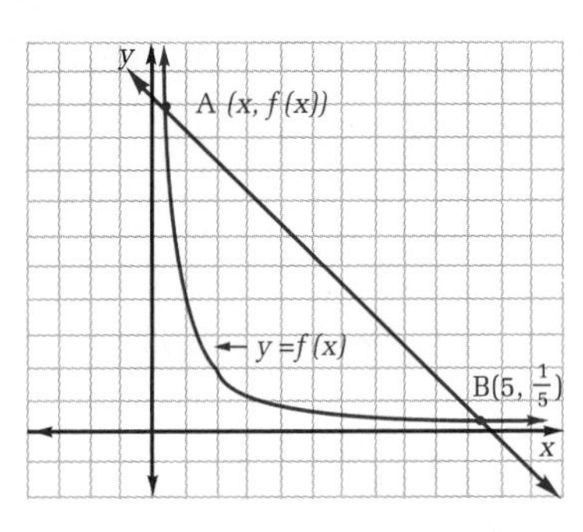

Solution

$$\textbf{Slope of } \overline{AB} = \frac{\frac{1}{x}-\frac{1}{5}}{x-5};\ x \neq 5$$

Multiply by $\frac{5x}{5x}$

$$= \frac{\frac{1}{x}-\frac{1}{5}}{x-5} \cdot \frac{5x}{5x}$$

$$= \frac{\frac{5x}{x}-\frac{5x}{5}}{(x-5)5x}$$

$$= \frac{5-x}{(x-5)5x}$$

$$= \frac{-(x-5)}{(x-5)5x}$$

$$= \frac{-1}{5x};\ x \neq 5,\ x > 0$$

Part Three: Exercises and Problems

Warm-up Exercises

In problems 1–7, simplify.

P

1 $\dfrac{\frac{1}{2}}{3}$ $\frac{3}{2}$ **2** $\dfrac{\frac{1}{a}}{b}$ $\frac{b}{a}$ **3** $\dfrac{\frac{2}{3}}{\frac{5}{2}}$ $\frac{4}{15}$ **4** $\dfrac{\frac{6}{5}}{\frac{2}{5}}$ 3

5 $\dfrac{1+\frac{2}{5}}{1-\frac{3}{10}}$ 2   **6** $\dfrac{\frac{1}{2}-\frac{1}{5}}{\frac{1}{10}}$ 3   **7** $\dfrac{2-\frac{3}{5}}{1-\frac{1}{15}}$ $\frac{3}{2}$

P

In problems 8–10, write a common denominator in order to use Method 2.

8 $\dfrac{\frac{1}{x}+\frac{1}{5}}{1+\frac{5}{x}}$ $5x$ **9** $\dfrac{1-\frac{6}{x-2}}{1+\frac{5}{x+2}}$ $(x-2)(x+2)$ **10** $\dfrac{\frac{x}{x-3}+\frac{2}{x}}{1-\frac{4}{x-3}}$ $x(x-3)$

Problem Set

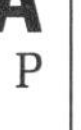

In problems 11–14, simplify.

P

11 $\dfrac{1+\frac{1}{x}}{1-\frac{1}{x}}$ $\frac{x+1}{x-1}$; $x \neq 0, x \neq 1$ **12** $\dfrac{1+\frac{1}{x^2}}{1-\frac{1}{x^2}}$ $\frac{x^2+1}{x^2-1}$; $x \neq -1, x \neq 0, x \neq 1$

13 $\dfrac{1+\frac{x}{x+1}}{1-\frac{x}{x+1}}$ $2x+1$; $x \neq -1$ **14** $\dfrac{1+\frac{2}{3}}{1-\frac{2}{3}}$ 5

P,R

15 State all restrictions for the expression $\dfrac{\frac{3}{x}+\frac{1}{x+2}}{\frac{1}{2x}-\frac{1}{x+1}}$.
$x \neq 0, x \neq -2, x \neq -1$

R

16 a Find the ratio of the area of the triangle to its hypotenuse. $\frac{6}{5}x$
b What are the restrictions on x? $x > 0$

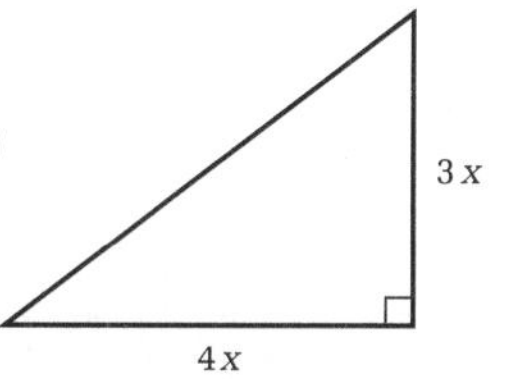

P

In problems 17 and 18, simplify, using either Method 1 or 2.

17 $\dfrac{x-3}{\frac{1}{x}-\frac{1}{3}}$ $-3x$; $x \neq 0, x \neq 3$ **18** $\dfrac{3+\frac{1}{x-2}}{3-\frac{1}{x-2}}$ $\frac{3x-5}{3x-7}$; $x \neq 2, x \neq \frac{7}{3}$

Assignment Guide

Average	
Optional	
Advanced	
Optional	
Honors	
Day 1	11, 14, 17, 21, 23, 24, 26, 30, 33, 35, 40, 45
Day 2	15, 16, 18, 25, 28, 31, 38, 42, 46, 48, 51–53

Problem-Set Notes and Strategies

- Problems **1 – 7** are straightforward. The process can be generalized to more difficult problems.
- Problems **11 – 14** do not specify a method.
- For many students, method 2 will be the choice for problem **18**. Problem **17** may be more easily solved by method 1. Encourage students to first decide on the most appropriate method rather than to depend on one of the two.

Problem Set, *continued*

Problem-Set Notes and Strategies, continued

- In problems **25 – 27**, either method may be used, but remind students to state the restrictions on the variables.
- Problem **28** is a nice problem for students taking physics or those interested in photography. It lends itself to practical examples.

Additional Answers

29

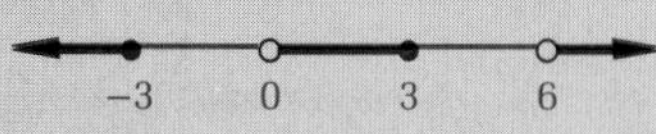

30

−3 0 3

R **In problems 19 and 20, simplify or solve.**

19 $\frac{2x - 3}{4x} - \frac{2x - 1}{2x}$ $\frac{-2x - 1}{4x}$; $x \neq 0$

20 $\frac{2x - 3}{4x} - \frac{2x - 1}{2x} = 0$ $x = -\frac{1}{2}$

B P **21** The harmonic mean of two numbers a and b is defined as $\frac{2}{\frac{1}{a} + \frac{1}{b}}$.

The harmonic mean of three numbers a, b, and c is $\frac{3}{\frac{1}{a} + \frac{1}{b} + \frac{1}{c}}$.

a Find the harmonic mean of 3 and 4. $\frac{24}{7}$

b Find the harmonic mean of 2, 3, and 4. $\frac{36}{13}$

c Find the harmonic mean of 2, 3, 4, and 5. $\frac{240}{77}$

R **In problems 22 and 23, simplify.**

22 $\frac{4}{x - 2} + \frac{6}{2 - x}$ $\frac{-2}{x - 2}$; $x \neq 2$

23 $\frac{5}{2(x + 1)} + \frac{3}{x + 1}$ $\frac{11}{2(x + 1)}$; $x \neq -1$

P,R **24** Given $f(x) = \frac{x + 3}{2x + 5}$. Find the slope of the line joining $(2, f(2))$ to $(4, f(4))$. $-\frac{1}{117}$

P **In problems 25–27, simplify.**

25 $\frac{\frac{1}{x} + \frac{1}{y}}{\frac{1}{x} - \frac{1}{y}}$ $\frac{y + x}{y - x}$; $x \neq 0$, $y \neq 0$, $x \neq y$

26 $\frac{\frac{1}{x^3} + 8}{\frac{1}{x} + 2}$ $\frac{1 - 2x + 4x^2}{x^2}$; $x \neq 0$, $x \neq -\frac{1}{2}$

27 $\frac{\frac{5}{x} - \frac{1}{2}}{\frac{25}{x^2} - \frac{1}{4}}$ $\frac{2x}{10 + x}$; $x \neq -10$, $x \neq 0$, $x \neq 10$

P,R **28** The focal length f of a thin lens is given by

$$\frac{1}{f} = \frac{1}{d_o} + \frac{1}{d_i}$$

where d_o is the distance between the object and the lens and d_i is the distance between the image and the lens.

a Solve for f. $f = \frac{d_o d_i}{d_o + d_i}$

b Solve for d_i. $d_i = \frac{f d_o}{d_o - f}$; $d_o \neq f$

R **In problems 29 and 30, solve for x and graph the solution on a number line.**

29 $\frac{x^2 - 9}{x^2 - 6x} \geq 0$ $x \leq -3$, $0 < x \leq 3$, or $x > 6$

30 $\frac{x^2 - 9}{x^2 + 3x} \geq 0$ $x < -3$, $-3 < x < 0$, or $x \geq 3$

P **31** Let $f(x) = \frac{1}{x - 3}$. Find the general formula for the slope of the line joining $A(x, f(x))$ to $B(x + 5, f(x + 5))$ when $x > 3$. $m = \frac{-1}{(x + 2)(x - 3)}$

P **32** Simplify $\frac{x - \frac{3}{x + 2}}{x - \frac{4}{x + 3}}$. $\frac{(x + 3)^2}{(x + 2)(x + 4)}$; $x \neq -4$, $x \neq -3$, $x = -2$, $x \neq 1$

P,R **33** Solve $\frac{1}{1 + \frac{2}{x}} = x$. $x = -1$

R | **34** Given $f(x) = 3 + \frac{2}{x}$, write a simplified expression for $f(f(x))$. $\frac{11x+6}{3x+2}$; $x \neq 0$, $x \neq -\frac{2}{3}$

R | **35** The sum of the slopes of lines l_1 and l_2 is $\frac{3}{4}$. Find the slope of the steeper line. $\frac{3}{7}$

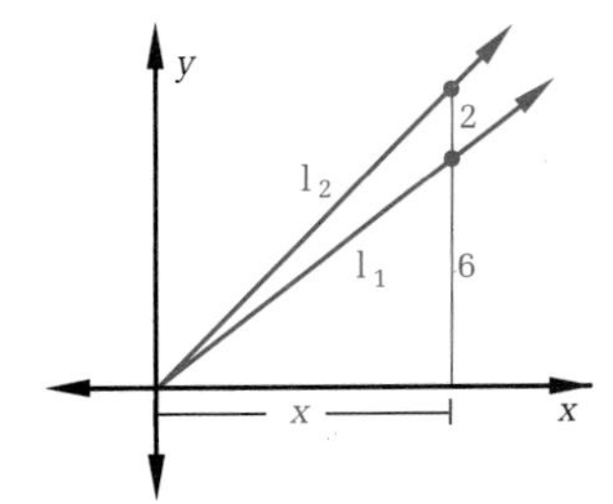

P | **36** Let $g(x) = \frac{2-3x}{9x-6}$. Write a simplified expression for $g\left(\frac{1}{x+3}\right)$. $-\frac{1}{3}$; $x \neq -3$, $x \neq \frac{-3}{2}$

P,R | **In problems 37–39, simplify.**

37 $\frac{2}{x} + \frac{2 + \frac{4}{x}}{3 + \frac{6}{x}}$ $\frac{2x+6}{3x}$; $x \neq -2$, $x \neq 0$

38 $\frac{\frac{6}{x^2} + \frac{1}{x} - 1}{x - \frac{9}{x}}$ $-\frac{x+2}{x^2+3x}$; $x \neq -3$, $x \neq 0$, $x \neq 3$

39 $\left(1 - \frac{1}{x}\right)\left(\frac{4x^2}{x^2-1}\right)$ $\frac{4x}{x+1}$; $x \neq -1$, $x \neq 0$, $x \neq 1$

P,R | **In problems 40 and 41, find the values of x that make each statement true.**

40 $x + 4 + \frac{6}{x-1} = 0$ $-1, -2$

41 $\frac{\frac{1}{x} + \frac{1}{2}}{\frac{1}{x^2} - \frac{1}{4}} = 1$ $\frac{2}{3}$

R | **42** After a rainstorm, pipeline A fills a reservoir in two hours. Pipeline B can fill the same reservoir in five hours, and pipeline C drains the reservoir in three hours.

a If the reservoir is empty and all three pipelines are operating, how full will the reservoir be after one hour? After two hours? $\frac{11}{30}$; $\frac{11}{15}$

b Under the same conditions, after how long will the reservoir be filled? $2\frac{8}{11}$ hr

P,R | **43** Given $c^2 - s^2 = 2$, simplify $\frac{\frac{2s}{c}}{1 - \frac{s^2}{c^2}}$. sc; $c \neq 0$

R | **44** Given $f(x) = \frac{x}{x+1}$, simplify.

a $f(f(x))$ $\frac{x}{2x+1}$; $x \neq -1$, $x \neq -\frac{1}{2}$

b $\begin{bmatrix}1&0\\1&1\end{bmatrix}\begin{bmatrix}1&0\\1&1\end{bmatrix}$ $\begin{bmatrix}1&0\\2&1\end{bmatrix}$

c $f(f(f(x)))$ $\frac{x}{3x+1}$; $x \neq -1$, $x \neq -\frac{1}{2}$, $x \neq -\frac{1}{3}$

d $\begin{bmatrix}1&0\\1&1\end{bmatrix}\begin{bmatrix}1&0\\1&1\end{bmatrix}\begin{bmatrix}1&0\\1&1\end{bmatrix}$ $\begin{bmatrix}1&0\\3&1\end{bmatrix}$

e $f(f(f(f(x))))$ $\frac{x}{4x+1}$; $x \neq -1$, $x \neq -\frac{1}{2}$, $x \neq -\frac{1}{3}$, $x \neq -\frac{1}{4}$

f $\begin{bmatrix}1&0\\1&1\end{bmatrix}\begin{bmatrix}1&0\\1&1\end{bmatrix}\begin{bmatrix}1&0\\1&1\end{bmatrix}\begin{bmatrix}1&0\\1&1\end{bmatrix}$ $\begin{bmatrix}1&0\\4&1\end{bmatrix}$

Problem-Set Notes and Strategies, continued

- Problems **37 – 39** are nice problems to assign to some of the students to put their solutions on transparencies.

- Problem **42** is a variation on the now familiar work problem with one pipeline working in the opposite direction of the other two. The set up is the same except for the sign of the opposite one. From here the solution is found by the same methods.

- An extension for problem **44** is to ask students: "Could you predict the value of the function f composed with itself 10 times?" Notice the pattern.

Problem-Set Notes and Strategies, continued

- You might want to point out that problem **46**, part **c** involves the use of the method called partial fractions, used extensively in a calculus class. The method breaks a fraction down into the sum or difference of smaller less complex fractions.
- Depending on your class, you might want to point out that problem **47** is part of the definition of the derivative from calculus. Missing is the limit as h approaches zero. These last two problems indicate the type of algebra necessary to solve elementary calculus problems.
- Problem **52** suggests the concept of limits in calculus. These concepts will aid in graphing expressions.
- Problems involving infinite values as in problem **53** will be explored in Section **13.4** on geometric series.

Additional Answers

Answers for problems **50** and **51** can be found in the answer pages beginning on page **A 1**.

Problem Set, *continued*

P,R **45** Find the slope of $\overline{AB}$. $\frac{-10}{x(x+h)}$; $0 < x < h, h > 0$

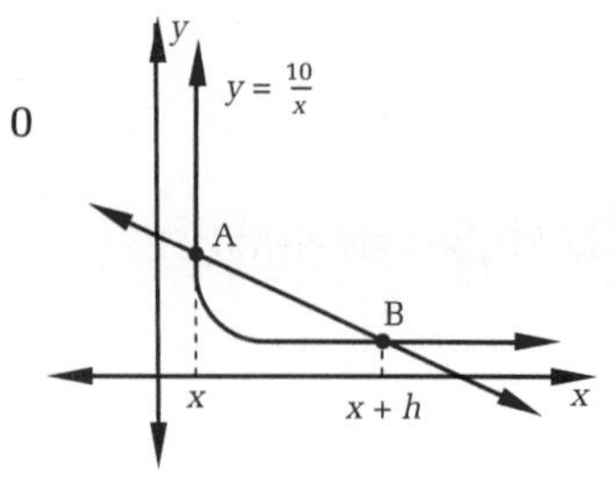

R **46** Find real numbers A, B, and C if

a $x^2 - 2x + 5 = Ax^2 + Bx + C$ $A = 1; B = -2; C = 5$

b $6x + 9 = (A + B)x + (2A - B)$ $A = 5; B = 1$

c $\frac{A}{x-4} + \frac{B}{x+3} = \frac{3x+2}{(x-4)(x+3)}$ $A = 2; B = 1$

P,R **47** Simplify $\frac{\frac{4}{x+h} - \frac{4}{x}}{h}$. $\frac{-4}{x(x+h)}$; $x \neq 0, h \neq 0, x \neq -h$

R **48** In the expansion of $\left(x + \frac{1}{x}\right)^6$, what is the coefficient of the term that does not contain x? 20

P,R **49** Let $f(x) = \frac{x}{x+3}$. Find the value of a so that the slope of the line through $(a, f(a))$ and $(-2, f(-2))$ is 1. 0

R **In problems 50 and 51, graph each equation.**

50 $x^2 - y^2 - 4(x - y) = 0$

51 $\frac{x^2 - y^2}{x - y} = 4$

C

P **52** Consider $f(x) = \frac{\frac{5}{x} + 3}{6 + \frac{2}{x}}$.

a As positive values of x get extremely large, what happens to the value of $f(x)$? $f(x)$ approaches $\frac{1}{2}$.

b As positive values of x get closer to zero, what happens to the value of $f(x)$? (Hint: Simplify the expression first.) $f(x)$ approaches $\frac{5}{2}$.

P,I **53** The expression $n = 2 + \cfrac{3}{2 + \cfrac{3}{2 + \cfrac{3}{2 + \cdots}}}$ is called an infinite continued fraction. The value of n can be determined by observing that $n = 2 + \frac{3}{n}$.

a Solve for n and check your result. Why does only one answer work? $n = 3$; n must be positive.

b Find the value of $n = 1 + \cfrac{6}{1 + \cfrac{6}{1 + \cfrac{6}{1 + \cfrac{6}{1 + \cdots}}}}$. 3

10

Chapter Summary

Concepts and Procedures

After studying this chapter, you should be able to

- Recognize rational expressions (10.1)
- Simplify rational expressions (10.1)
- Multiply and divide rational expressions (10.1)
- Find zeros of rational functions (10.2)
- Solve rational equations and inequalities (10.2)
- Add and subtract rational expressions with like denominators (10.3)
- Add and subtract rational expressions with unlike denominators (10.4)
- Simplify complex fractions (10.5)

Vocabulary

complex fraction (10.5)
rational equation (10.2)
rational expression (10.1)
rational inequality (10.2)
restricted value (10.1)
simplify (10.1)

Class Planning

Time Schedule

All levels: 1 day

Assignment Guide

Average

1, 3, 10, 12, 13, 15, 18, 20, 21, 23, 26, 27, 30, 33, 52

Advanced

1, 2, 5, 10, 12, 14, 18, 23, 26, 30, 31, 33–35, 38, 40, 52

Honors

1, 2, 5, 10, 12, 14, 18, 26, 29, 31, 33–37, 40, 45, 47, 51

Problem-Set Notes and Strategies

- In problems **2–5**, remind students not to forget the restrictions on the variables in the original problem.

10 Review Problems

A

P **1** Which of the following are rational expressions? **b, c, d, e, f**

a $\sqrt{x} + \dfrac{1}{x + 2}$ **b** $\dfrac{x_1 + x_2 + x_3 + x_4}{4}$ **c** $\dfrac{x + 2}{1 + x^2}$

d $\dfrac{\sqrt{2} + x}{\sqrt{3} + x}$ **e** $\dfrac{5x^2 - 7x}{x + 4}$ **f** $x^3 - x + 1$

P **In problems 2–5, add or subtract.**

2 $\dfrac{5x(x + 1)}{3x - 5} + \dfrac{(x - 5)(x + 5)}{3x - 5}$ $2x + 5;\ x \neq \frac{5}{3}$

3 $\dfrac{25x^2 + 6x}{5x + 7} - \dfrac{6x + 49}{5x + 7}$ $5x - 7;\ x \neq -\frac{7}{5}$

4 $\dfrac{x^2}{x^2 - 6x + 5} - \dfrac{4x + 5}{x^2 - 6x + 5}$ $\frac{x + 1}{x - 1};\ x \neq 5,\ x \neq 1$

5 $\dfrac{2x + 1}{(x + 3)(x - 3)} - \dfrac{x - 2}{(x + 3)(x - 3)}$ $\frac{1}{x - 3};\ x \neq -3,\ x \neq 3$

P **In problems 6–9, simplify.**

6 $\dfrac{\frac{1}{3} + \frac{1}{4}}{\frac{1}{6} - \frac{1}{2}}$ $-\frac{7}{4}$

7 $\dfrac{2 + \frac{3}{5}}{4 - \frac{1}{5}}$ $\frac{13}{19}$

8 $\dfrac{\frac{4}{5} - \frac{2}{3}}{\frac{1}{7} + \frac{1}{2}}$ $\frac{28}{135}$

9 $\dfrac{2 \cdot 5 + 3}{4 \cdot 5 - 1}$ $\frac{13}{19}$

P,R **10** Refer to the diagram.

a Find the ratio of the area of rectangle A to the area of rectangle B. $\frac{x + 2}{x}$

b For what values of x will the ratio be equal to 2? 2

R **11** Study the example, then copy and complete the chart.

Expressions	Greatest Common Factor	Least Common Multiple
$2(x - 1),\ (x + 1)(x - 1)$	$x - 1$	$2(x + 1)(x - 1)$
$3(x + 4),\ 6(x + 4)(x - 2)$	$3(x + 4)$	$6(x + 4)(x - 2)$
$3x - 21,\ x^2 - 49$	$x - 7$	$3(x + 7)(x - 7)$
$x^2 + 3x,\ x^2 + 6x + 9$	$x + 3$	$x(x + 3)^2$

P **In problems 12–15, solve for *y*.**

12 $\dfrac{5}{y} - \dfrac{9}{y} = 2$ $y = -2$

13 $\dfrac{y(y - 2)}{y - 4} = 0$ $y = 0$ or $y = 2$

14 $\dfrac{2}{y} + \dfrac{3}{y + 1} = \dfrac{-8}{y(y + 1)}$ $y = -2$

15 $\dfrac{y(y - 2)}{y - 2} = 0$ $y = 0$

B **In problems 16 and 17, solve for x.**

P **16** $\frac{12}{x + 2} = \frac{3}{x - 2} + \frac{6}{x^2 - 4}$ $x = 4$

17 $\frac{8}{x - 8} + \frac{2}{5} = \frac{x}{x - 8}$ No solution

P **18** Simplify and state restrictions.

$\frac{x^2 - 4}{x + 3} \cdot \frac{x^2 + 5x + 6}{x + 1} \cdot \frac{x^2 + 3x + 2}{x - 1} \cdot \frac{x^2 + x - 2}{x - 2}$

$(x + 2)^4$; $x \neq -3$, $x \neq -1$, $x \neq 1$, $x \neq 2$

P **19** Find the zeros of $f(x) = \frac{x^2 + x - 20}{x - 3}$. 4 and −5

P **In problems 20–22, simplify.**

20 $3 + \frac{4}{7}$ $\frac{25}{7}$

21 $x + \frac{3}{x}$ $\frac{x^2 + 3}{x}$; $x \neq 0$

22 $x - 5 + \frac{2}{x + 5}$ $\frac{x^2 - 23}{x + 5}$; $x \neq -5$

P **23** Find the zeros of the function $f(x) = \frac{3x^2 - 2x - 8}{x - 2}$. $-\frac{4}{3}$

P **In problems 24–26, simplify.**

24 $\frac{2x - 3}{3 - 2x}$ -1; $x \neq \frac{3}{2}$

25 $\frac{(2 - 3x)^2}{(3x - 2)^2}$ 1; $x \neq \frac{2}{3}$

26 $\frac{(2 - 3x)^3}{(3x - 2)^2}$ $2 - 3x$; $x \neq \frac{2}{3}$

P **In problems 27 and 28, perform the indicated operations and state restrictions.**

27 $\frac{x(x + 1)}{x + 2} \div \frac{x(x - 1)}{x + 2}$

$\frac{x + 1}{x - 1}$; $x \neq 0$, $x \neq 1$, $x \neq -2$

28 $\frac{3x^2 + 15x}{x^2 + 2x - 15} \cdot \frac{x^2 - 9}{x + 3}$

$3x$; $x \neq -5$, $x \neq -3$, $x \neq 3$

R **29** Graph $\frac{2}{x} + \frac{3}{y} = \frac{12}{xy}$.

R **30** The volume of the box shown is $x^3 + 2x^2 - 8x$, its length is $x + 4$, and its width is x. Find its height. $h = x - 2$

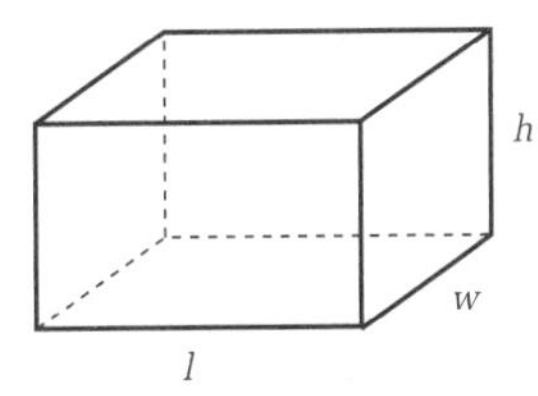

P **In problems 31–33, simplify.**

31 $\frac{1}{x + y} + \frac{1}{x - y}$ $\frac{2x}{x^2 - y^2}$; $x \neq y$, $x \neq -y$

32 $\frac{x}{x + y} + \frac{y}{x - y}$ $\frac{x^2 + y^2}{x^2 - y^2}$; $x \neq y$, $x \neq -y$

33 $\frac{x}{x + y} - \frac{y}{x - y}$ $\frac{x^2 - 2xy - y^2}{x^2 - y^2}$; $x \neq y$, $x \neq -y$

R **34** Find all vertical asymptotes of the graph of $f(x) = \frac{x^2 + 2x}{x^2 - 3x - 10}$.

$x = 5$

R **35** Solve for (x, y, z). $(x, y, z) = (6, -4, 3)$

$$\begin{cases} 3x + y - 2z = 8 \\ 2x - y + 3z = 25 \\ x - 3y - 4z = 6 \end{cases}$$

Problem-Set Notes and Strategies, continued

- In problem **23**, some students may name 2 as a zero. Remind them that 2 is not in the domain of the original problem.

Additional Answer

29

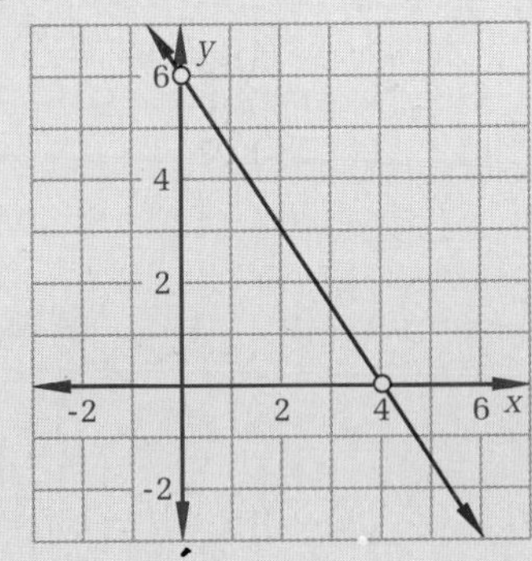

Problem-Set Notes and Strategies, continued

- Problem **36** involves the quadratic formula and complex roots.
- In problem **40**, remind students to include values that make the denominator zero as boundary values.
- Problem **49** can be solved by setting up a system of equations.

Review Problems, *continued*

P **36** Find all complex zeros of $f(x) = \frac{x^3 - 7x^2 + 16x - 16}{x - 5}$. $4, \frac{3}{2} + \frac{\sqrt{7}}{2}i, \frac{3}{2} - \frac{\sqrt{7}}{2}i$

P **37** Simplify $\dfrac{x + 1 - \frac{9}{x+1}}{x + 2 - \frac{6}{x+1}}$. $\frac{x-2}{x-1}$; $x \neq -4, x \neq -1, x \neq 1$

P **38** Simplify $\frac{3 + x}{2 - 3x} \cdot \frac{3x^2 - 2x}{x^2 + 6x + 9}$ and note restrictions. $\frac{-x}{x+3}$; $x \neq \frac{2}{3}, x \neq -3$

P **39** The denominator of a fraction is two more than its numerator. If the fraction is added to its reciprocal, the sum will be twice the original numerator. Find the original numerator.
$-1, \sqrt{2}, -\sqrt{2}$

P **40** Let $f(x) = \frac{x^2 - 7x + 12}{x^2 - 9}$. Find all values of x such that $f(x) \leq 0$.
$\{x: -3 < x < 3 \text{ or } 3 < x \leq 4\}$

P,R **In problems 41–44, solve for x.**

41 $\frac{15}{x^2} - \frac{2}{x} - 1 = 0$ $x = -5$ or $x = 3$

42 $\frac{5}{x - 3} - \frac{4}{x + 4} = 0$ $x = -32$

43 $\frac{4}{(x - 3)^2} - \frac{5}{x - 3} = -1$
$x = 4$ or $x = 7$

44 $6ax = 5ax + 16a^2$
$x = 16a$

P,R **C** **45** Let $x + y = 1$. Write $\dfrac{\frac{1}{x} + 1}{\frac{1}{x} - x}$ in terms of y. $\frac{1}{y}$

P **In problems 46 and 47, let $f(x) = \frac{2x + 3}{3x - 1}$ and $g(x) = \frac{2x - 1}{3x + 4}$.**

46 Simplify $f(g(x))$. $\frac{13x + 10}{3x - 7}$; $x \neq -\frac{4}{3}, x \neq \frac{7}{3}$

47 Simplify $g(f(x))$. $\frac{x + 7}{18x + 5}$; $x \neq -\frac{5}{18}, x \neq \frac{1}{3}$

P,R **48** Let $s^2 + c^2 = 1$. Simplify $\frac{1}{s^2} + \frac{1}{c^2}$. $\frac{1}{s^2c^2}$; $s \neq 0, c \neq 0$

P,R **49** Let $f(x) = \frac{x^2 + cx + b}{x + 4}$. Find b and c if $f(-3) = 31$ and $f(-5) = -59$. $b = 4; c = -6$

P **50** Divide $f(x + 3)$ by $f(x - 3)$ if $f(x) = x^2 - 9$. $\frac{x + 6}{x - 6}$; $x \neq 0, x \neq 6$

R **51** N. Vestor purchased $20,000 worth of bonds, some at 10% and the rest at 9%. His investment yields $1930 a year.

a Determine how much he invested at each rate. $13,000 at 10%; $7,000 at 9%

b Explain why he might not have wanted to invest all $20,000 at the higher rate of return. Answers will vary. Such considerations as maturity time, bond security, and bond rating could guide Mr. Vestor's decision.

R **52** Of the 57 test questions Irma Wisdom has answered, she has 48 right.

a If she makes no more mistakes, how many more questions must she answer so that she will have answered 90% or more correctly? 33 or more

b If the exam has 75 questions, can she get a score higher than 90%? No

CHAPTER TEST

In problems 1–4, simplify and state restrictions.

1 $\frac{(x + 3)(x + 4)}{(x + 2)(x + 3)}$ $\frac{x+4}{x+2}$; $x \neq -2, x \neq -3$

2 $\frac{x^3 + 2x^2}{x}$ $x^2 + 2x$; $x \neq 0$

3 $\frac{x^2 - 4}{x^2 + 3x + 2}$ $\frac{x-2}{x+1}$; $x \neq -2, x \neq -1$

4 $\frac{3x^3 + 15x^2 + 18x}{x^3 - 9x}$ $\frac{3x+6}{x-3}$; $x \neq -3, x \neq 0, x \neq 3$

5 Multiply $\frac{x+7}{3-x} \cdot \frac{x+4}{x+7} \cdot \frac{x-3}{x-4}$ and write your answer in simplest form.
$-\frac{x+4}{x-4}$; $x \neq 3, x \neq -7, x \neq 4$

In problems 6 and 7, solve for x.

6 $\frac{6}{x-2} = \frac{5}{x-3}$ $x = 8$

7 $\frac{3}{x^2 - 4} = \frac{-2}{5x + 10}$ $x = -\frac{11}{2}$

8 A baseball team has a record of 35 wins and 20 losses. How many games must the team win in a row in order for its winning percentage to be above .750?
At least 26 games in a row

9 Let $f(x) = \frac{x^2 + 16}{x^2 + 9x + 20}$ and $g(x) = \frac{8x}{x^2 + 9x + 20}$.

a Find $f(x) + g(x)$. $\frac{x+4}{x+5}$; $x \neq -4, x \neq -5$

b Find x such that $f(x) - g(x) = 0$. $x = 4$

In problems 10–12, perform the indicated operations.

10 $\frac{2x}{x-7} + 3$ $\frac{5x-21}{x-7}$; $x \neq 7$

11 $x + \frac{3}{x}$ $\frac{x^2+3}{x}$; $x \neq 0$

12 $x - 5 + \frac{2}{x+5}$ $\frac{x^2-23}{x+5}$; $x \neq -5$

In problems 13 and 14, add or subtract.

13 $\frac{x}{x-y} - \frac{2x}{x+y} - \frac{2xy}{x^2 - y^2}$
$\frac{-x}{x+y}$; $x \neq y, x \neq -y$

14 $\frac{2}{x^2 - y^2} + \frac{2}{(x+y)^2}$
$\frac{4x}{(x+y)^2(x-y)}$; $x \neq y, x \neq -y$

In problems 15 and 16, simplify.

15 $\frac{1 + \frac{x}{y}}{1 - \frac{x}{y}}$ $\frac{y+x}{y-x}$; $y \neq 0, y \neq x$

16 $\frac{\frac{1}{x-y} - 1}{2 - \frac{x}{x-y}}$ $\frac{1-x+y}{x-2y}$; $x \neq y, x \neq 2y$

17 Simplify $\frac{\frac{x+3}{x-3} + \frac{x-3}{x+3}}{\frac{x+3}{x-3} - \frac{x-3}{x+3}}$. $\frac{x^2+9}{6x}$; $x \neq -3, x \neq 0, x \neq 3$

18 Let $f(x) = \frac{2}{x}$. Simplify $\frac{f(x+h) - f(x)}{h}$. $\frac{-2}{x(x+h)}$; $x \neq 0, h \neq 0, x \neq -h$

Resource References
Evaluation
Test Form 1, 2, 3

Class Planning

Time Schedule

Average: Optional
Advanced: 2 days
Honors: 2 days

Resource References

Teacher's Resource Book
Class Opener 64
Practice 67, 68

Class Opener

Refer to the diagram.

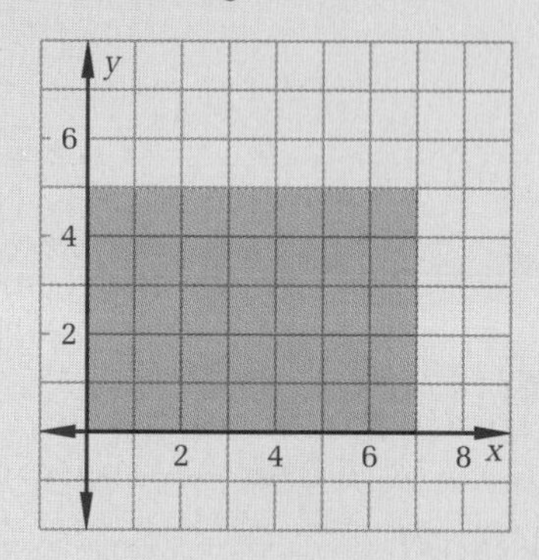

1 Write a system of inequalities that represents the shaded region.

$$\begin{cases} 0 \le x \le 7 \\ 0 \le y \le 5 \end{cases}$$

2 Now consider the set of lines that intersect this region and have a slope of –1. Which of these line's intercepts have the greatest value?
The line passing through (7, 5)

SHORT SUBJECT

Linear Programming

In Section 7.3, you saw how the solutions of systems of inequalities can be graphically represented on the coordinate plane. Now consider the following system of inequalities and the graph of its solution set (the shaded region in the diagram).

$$\begin{cases} x \ge 3 \\ y \ge 2 \\ x \le 3y \\ x + y \le 12 \end{cases}$$

Suppose we define the function $p(x, y) = 6x + 3y$ in such a way that the function's domain is restricted to values of x and y that are in the solution set of this system. Is there a particular input (x, y) that yields the maximum possible output of function p? Is there another input for which the value of p is at a minimum? Let's test the coordinates of some of the points in the region that represents the domain.

Point	(x, y)	6x + 3y
A	(3, 9)	6(3) + 3(9) = 45
B	(4, 6)	6(4) + 3(6) = 42
C	(7, 5)	6(7) + 3(5) = 57
D	(5, 3)	6(5) + 3(3) = 39
E	(9, 3)	6(9) + 3(3) = 63
F	(3, 2)	6(3) + 3(2) = 24
G	(6, 2)	6(6) + 3(2) = 42

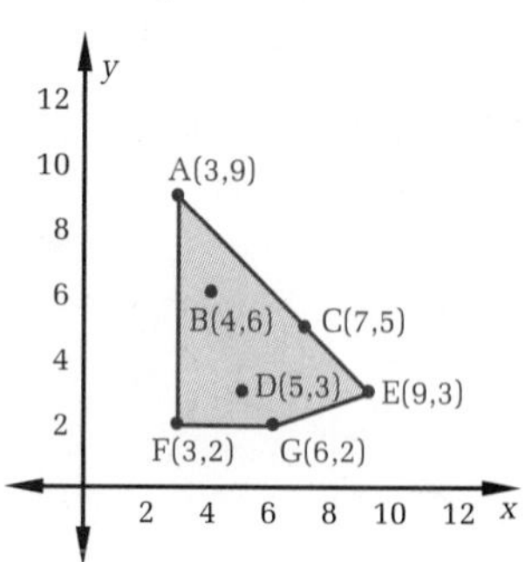

Vocabulary

linear programming
constraint

This listing, of course, is not exhaustive—there are infinitely many points in the solution set of the system, and we cannot test them all. We do notice, however, that both the greatest output we found (63) and the least output we found (24) were produced by inputs—(9, 3) and (3, 2)—that correspond to vertices of the shaded region. In fact, it has been proved that in cases such as this, the maximum and minimum values of a function will always correspond to vertices of the region representing the function's domain.

> ***If a linear function has a domain of ordered pairs (x, y) that corresponds to a polygonal region of the coordinate plane, the maximum and minimum values of the function will occur at vertices of the region.***

The preceding theorem is the basis of a technique, called ***linear programming,*** that is frequently used by businesses to determine the best way of allocating their resources to minimize costs or to maximize profits.

Example *Charlotte Russe earns money by crafting and selling wooden toys. She makes two kinds of toys—trains, on which she makes a profit of $6 each, and trucks, on which she makes a profit of $3 each. To satisfy the demand, she must produce at least 3 trains and 2 trucks each day, and at least one fourth of the toys she produces must be trucks. If she can produce no more than 12 toys in a day and if she can sell all that she produces, how many of each kind of toy should Charlotte make each day to maximize her profits?*

This problem can be modeled by the graph that we have already considered. We let x = the number of trains Charlotte makes and y = the number of trucks she makes. She must make at least 3 trains and 2 trucks, so $x \geq 3$ and $y \geq 2$. At least one fourth of the toys must be trucks, so $x \leq 3y$; and because her maximum daily output is 12 toys, $x + y \leq 12$. Thus, the restrictions on Charlotte's production, known as ***constraints,*** correspond to the system of inequalities presented at the beginning of this section.

Charlotte's profits can be represented by the function $p(x, y) = 6x + 3y$. As we have seen, under the given constraints the maximum value of this function is yielded by the input (9, 3), so Charlotte should produce 9 trains and 3 trucks each day to maximize her profits.

Linear-programming problems that involve only a few constraints in two variables can be solved by simply graphing the corresponding system of inequalities and testing the coordinates of the vertices in the function to be maximized or minimized. Most real-world applications of linear programming, however, involve so many variables that they must be dealt with by computer programs utilizing matrix operations.

Lesson Notes

- Linear programming is one of many topics in a field of Operations Research. This, as well as many other optimization algorithms, are the tools of the operations research team. It has been stated that 10% of all computer time is spent on linear programming and optimization problems. They constitute a great variety of applied mathematics techniques and offer many job opportunities for the mathematically inclined student.
- Linear constraints produce regions that have their maximum and minimum values at the vertices. These regions are called convex polygons in two dimensions as the segment between any two points in the polygon is also contained in the polygon. This may have been studied in earlier classes.
- Problems in this section are done by graphing since they contain only two variables. In higher dimensions a matrix equation is set up with n variables and solved by an iterative process. The process was discovered in 1947 and is a recent discovery in the history of mathematics. The discoverer, George Dantzig, is still an active mathematician teaching at Stanford University.

Sample Problems

Problem 1 *The Kepler Model Company is planning to market a variety of electric racing-car sets. Each set will contain at least 8 sections of curved track and 4 sections of straight track. No set will contain more than 36 sections in all or more than 20 sections of either type. If the company makes a profit of $0.40 on each straight section and $0.65 on each curved section, what combination of track sections will be most profitable for the company?*

Solution We let x = the number of curved sections in a set and y = the number of straight sections. Since there will be at least 8 and no more than 20 curved sections, $8 \le x \le 20$. Since there will be at least 4 and no more than 20 straight sections, $4 \le y \le 20$. The greatest number of sections will be 36, so $x + y \le 36$. The following system and graph, therefore, represent the constraints.

$$\begin{cases} 8 \le x \le 20 \\ 4 \le y \le 20 \\ x + y \le 36 \end{cases}$$

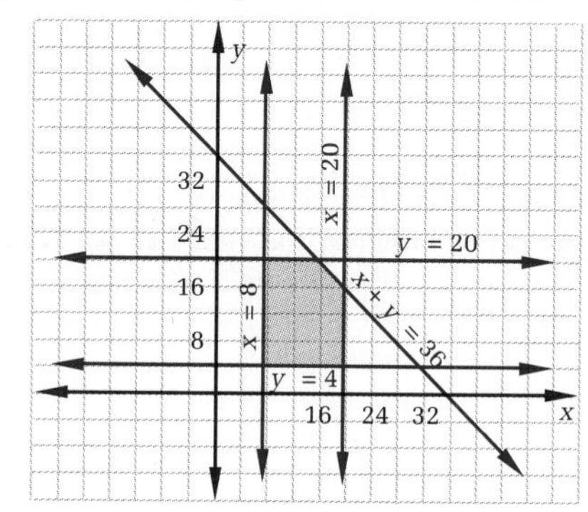

The profit function is $p(x, y) = 0.65x + 0.4y$, which we evaluate for the coordinates of the five vertices of the region—(8, 4), (8, 20), (16, 20), (20, 16), and (20, 4).

Vertex	**0.65x + 0.40y**
(8, 4)	$0.65(8) + 0.4(4) = 6.8$
(8, 20)	$0.65(8) + 0.4(20) = 13.2$
(16, 20)	$0.65(16) + 0.4(20) = 18.4$
(20, 16)	$0.65(20) + 0.4(16) = 19.4$
(20, 4)	$0.65(20) + 0.4(4) = 14.6$

The profit will be greatest ($19.40) for a combination of 20 curved sections and 16 straight sections.

Problem 2 *A certain publisher ships 300–450 books each week to a national chain of bookstores. Some of the books are shipped from the publisher's eastern warehouse, and some are shipped from the publisher's western warehouse, but at least one third of them must be shipped from each warehouse. The shipping cost per book is $0.37 from the eastern warehouse and $0.55 from the western warehouse. Find the minimum weekly shipping cost for these orders.*

Solution If we let x = the number of books shipped from the eastern warehouse and y = the number of books shipped from the western warehouse, we can represent the constraints as follows.

$$\begin{cases} x + y \geq 300 \\ x + y \leq 450 \\ x \leq 2y \\ y \leq 2x \end{cases}$$

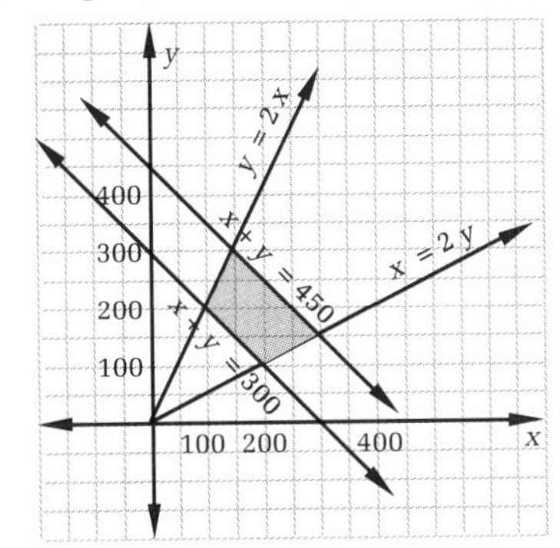

The vertices of the shaded region are at (100, 200), (150, 300), (300, 150), and (200, 100). We test these coordinates in the cost function, $c(x, y) = 0.37x + 0.55y$.

Vertex	$0.37x + 0.55y$
(100, 200)	0.37(100) + 0.55(200) = 147
(150, 300)	0.37(150) + 0.55(300) = 220.5
(300, 150)	0.37(300) + 0.55(150) = 193.5
(200, 100)	0.37(200) + 0.55(100) = 129

The minimum cost is $129, incurred when 200 books are shipped from the eastern warehouse and 100 are shipped from the western warehouse.

Problem Set

P A **1** Graph the following system of inequalities, indicating the coordinates of the vertices of the region that represents the solution set.

$$\begin{cases} 0 \leq y \leq 8 \\ x \geq 1 \\ 2x + y \leq 16 \end{cases}$$

P **2** Refer to the graph shown.

a What are the coordinates of points A, B, C, D, and E?

b Use your answers to part **a** to find the maximum value of $f(x, y) = 3x + 2y$ for the domain represented by the shaded region. 345

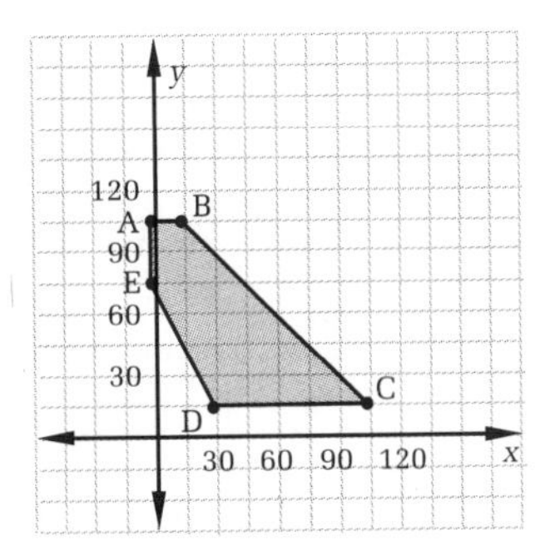

Assignment Guide

Average	
Optional	
Advanced	
Day 1	1–5
Day 2	6–10
Honors	
Day 1	1–6
Day 2	7–12

Problem-Set Notes and Strategies

- As in problem **1**, most regions of linear programming problems lie in the first quadrant. Make sure that the appropriate regions are shaded to produce a convex polygon.
- In problem **2**, the function $3x + 2y$ is called the objective function as the object of the problem is to find values to maximize that function.

Additional Answers

1

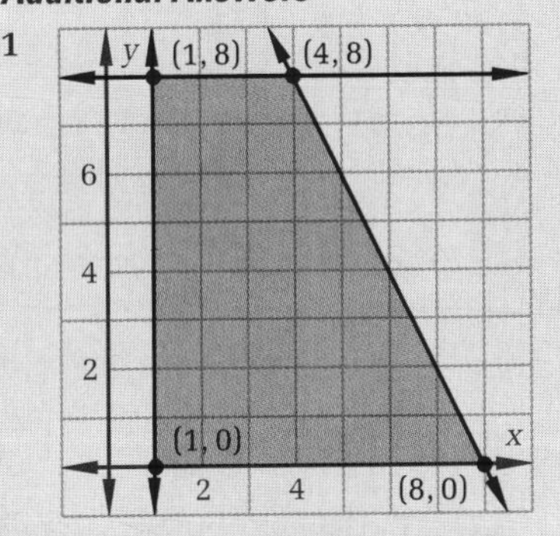

2a (0, 105); (15, 105); (105, 15); (30, 15); (0, 75)

Problem-Set Notes and Strategies, continued

- Problem **4** is asking the student to write one of many constraints that a problem may have. These constraints define the convex polygon called the feasible region.
- Problems **7**, **9**, **11**, and **12** are classic L.P. problems with objective functions and constraints all stated in words and not equations. Be prepared to spend some time on each of these problems.

Additional Answers

7a $\begin{cases} x \ge 0 \\ y \ge 0 \\ 12x + 6y \le 120 \\ 9x + 18y \le 144 \end{cases}$

8

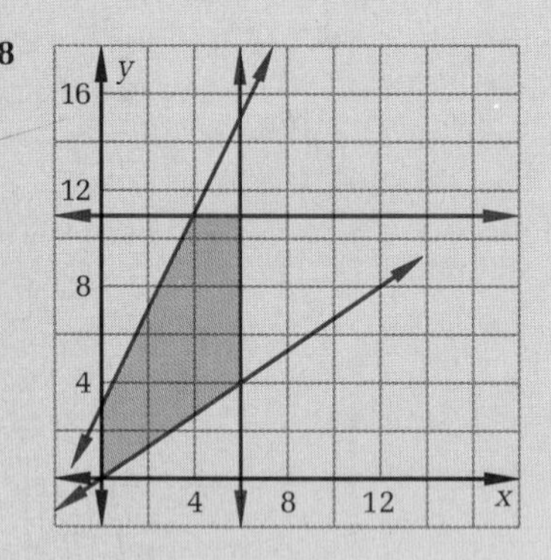

Problem Set, *continued*

R **3** If line ℓ has a slope of $-\frac{3}{2}$ and a y-intercept of 2 and line m has a slope of -2 and a y-intercept of 6, what are the coordinates of the point where ℓ and m intersect? $(8, -10)$

R **4** A milling machine can be used to produce either widgets or gimcracks. It takes 5 minutes to produce a widget and 6 minutes to produce a gimcrack, and the machine is in operation no more than 55 hours a week. Write a linear inequality to describe this situation, identifying the meaning of each variable used. $\frac{x}{12} + \frac{y}{10} \le 55$, where x = the number of widgets produced per week and y = the number of gimcracks produced per week

P **5** Let $c(x, y) = 15x - 5y$, with the following restrictions.

$x \ge 1 \qquad y \ge 0 \qquad x \le 4 \qquad y \le 2x + 5$

a Find the maximum output of function c. 60
b Find the minimum output of function c. -20

P **6** Maximize and minimize the function $g(x, y) = 3x - 4y$ for the values of (x, y) in the shaded region.
Max.: 250; min.: -625

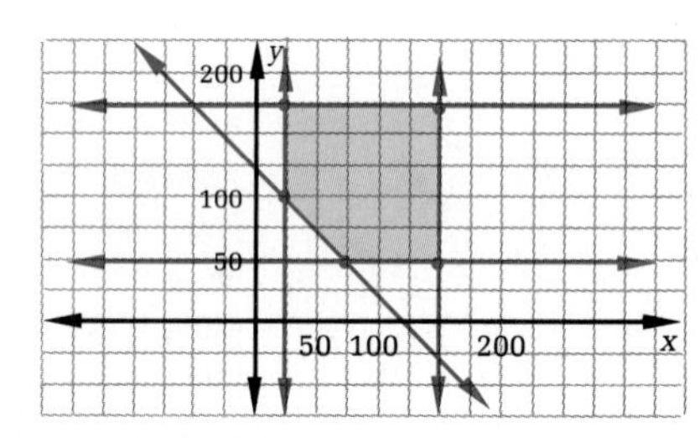

P **7** A jewelry craftsman produces bracelets and necklaces that contain both gold and platinum. The following table shows the amounts of precious metals used in the production of these items.

	Metal Content (Grams) Bracelet	Metal Content (Grams) Necklace	Grams Available per Week
Gold	12	6	120
Platinum	9	18	144

a Write a system of inequalities that represents the given constraints, using x to represent the number of bracelets produced and y to represent the number of necklaces produced.
b Suppose that the craftsman makes a profit of \$130 on each bracelet and \$90 on each necklace. Write an expression that represents his total profit on the two items. $130x + 90x$
c Use your answers to parts **a** and **b** to calculate the craftsman's maximum weekly profit. \$1400
d How many of each item should he produce per week to obtain the maximum profit? 8 bracelets, 4 necklaces

B P **8** Graph the region represented by the following system of inequalities. What are the coordinates of the region's vertices? (0, 0), (0, 3), (4, 11), (6, 11), (6, 4)

$$\begin{cases} 0 \le x \le 6 \\ 0 \le y \le 11 \\ y \le 2x + 3 \\ 3y \ge 2x \end{cases}$$

P **9** An electronics company manufactures personal tape players, on which it makes a profit of $28 each, and personal CD players, on which it makes a profit of $33 each. The company intends to produce at least 60 tape players and 100 CD players per day, but its factory is not equipped to assemble more than a total of 200 of these items each day. What should the company's daily production of these items be if profits are to be as great as possible? 60 tape players, 140 CD players

P **10** If $z = 0.60x + 0.95y$ for the set of values of (x, y) represented by the shaded polygonal region, what are the maximum and minimum values of z? 153; 52.5

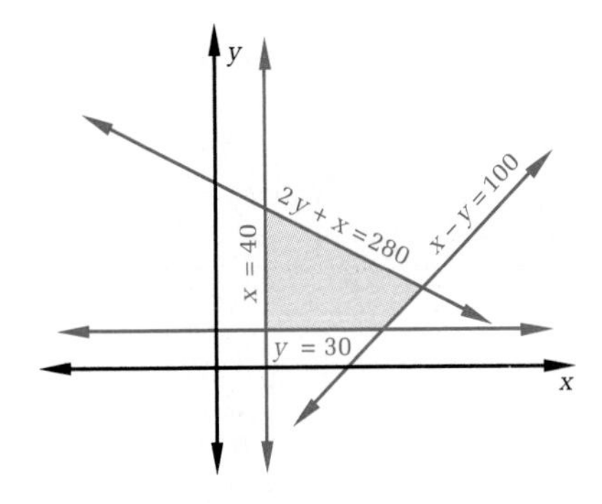

P **11** Waste Schleppers, Inc., transports sealed containers of aluminum and glass refuse to a recycling plant. Each full container of aluminum weighs 800 pounds and occupies 60 cubic feet of space, and each full container of glass weighs 900 pounds and takes up 75 cubic feet. The firm charges $100 per container for hauling aluminum and $80 per container for hauling glass. If the Waste Schleppers truck has a load limit of 22,000 pounds and only 1740 cubic feet of cargo space, what is the firm's maximum revenue for a single haul? $2700

C P **12** Penny Black, a stamp collector, purchases mixed lots of stamps from two suppliers, Commemorative Corner and House of Postage. She has found that each 100-stamp lot is likely to be made up of the following types of stamps.

	United States Stamps	Foreign Stamps
Commemorative Corner	75%	25%
House of Postage	40%	60%

If Commemorative Corner charges $5.50 per lot and House of Postage charges $4.75 per lot, and if Penny wants to add at least 400 United States stamps and 300 foreign stamps to her collection, what is the minimum amount she should expect to pay? $41

11.1

THE EMPHASIS IN this chapter is on the conceptual understanding of the logarithmic function and its inverse, the exponential function. In the first section, students are introduced to each of these functions, their properties and graphs.

The two bases used most often with the exponential and logarithmic functions, *e* and 10 are introduced.

In Section **11.3**, four properties of logarithms are derived from the definition of logarithm and the proofs of three other properties are given.

The Base-Change Formula is presented. Students will also solve exponential and logarithmic equations.

Since many real-world situations are modeled by the exponential and logarithmic functions, we encourage you to spend several days on the application problems in Section **11.5**. Data is the starting point of most real-world problems. Students will transform data into a linear form by using logarithms. In addition, curve fitting an exponential function via linear regression on the transformed data points is explored.

The last section of the chapter provides students with realistic types of applications involving and relating binomial expansions and probability.

CHAPTER 11

EXPONENTIAL AND LOGARITHMIC FUNCTIONS

A spiral, such as this winding staircase, is one representation of the Golden Ratio.

11.1

DEFINITIONS AND GRAPHS OF THE EXPONENTIAL AND LOGARITHMIC FUNCTIONS

Objectives

After studying this section, you will be able to

- Graph exponential functions
- Graph logarithmic functions
- Use the definition of the logarithmic function

Part One: Introduction

The Graph of the Exponential Function

Recall the definition of an exponential function.

> ***An exponential function is a function that can be written in the form $f(x) = a^x$, where a is a positive constant and $a \neq 1$. The positive constant a is called the base.***

Let's graph several exponential functions on the same coordinate system and look for patterns.

Shown are the graphs of $f(x) = 2^x$, $g(x) = 3^x$, and $h(x) = 4^x$.

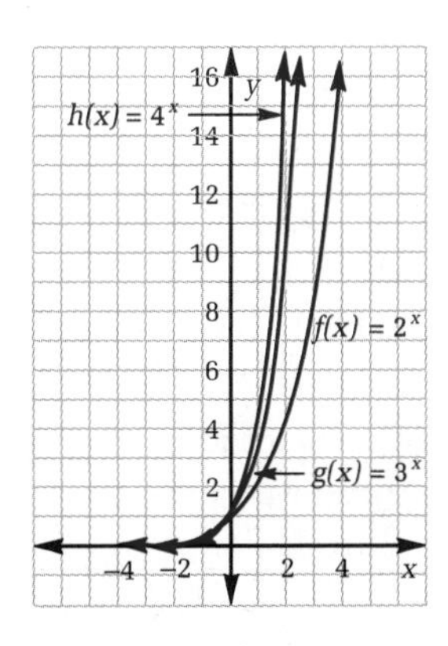

Notice the following similarities among these exponential functions.

All have domain $\{x : x \in \mathcal{R}\}$.
All have range $\{y : y > 0\}$.
All have y-intercepts of 1.
They all have $y = 0$ as a horizontal asymptote.
None of the graphs has an x-intercept.

So far the base a has always been chosen greater than 1. What happens if we use a base a such that $0 < a < 1$? Again look for patterns as we graph $f(x) = \left(\frac{1}{2}\right)^x$, $g(x) = \left(\frac{1}{3}\right)^x$, and $h(x) = \left(\frac{1}{4}\right)^x$ on the same coordinate system. Notice that all five patterns seen when $a > 1$ are also true when $0 < a < 1$.

11.1

Class Planning

Time Schedule
All levels: 1 day

Resource References
Teacher's Resource Book
Class Opener 65
Cooperative Learning 31
Practice 69
Transparency 21

Class Opener

1 Graph $y = 2^x$.

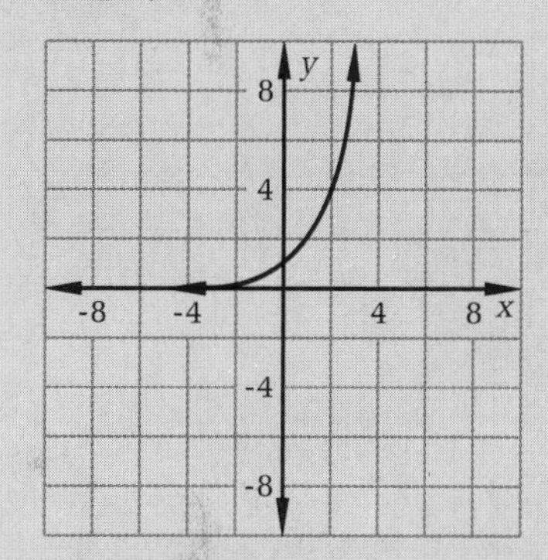

2 Make a table of values for the inverse of $y = 2^x$.

x	$\frac{1}{4}$	$\frac{1}{2}$	1	2	4	8
y	−2	−1	0	1	2	3

3 Sketch the graph of the inverse of $y = 2^x$.

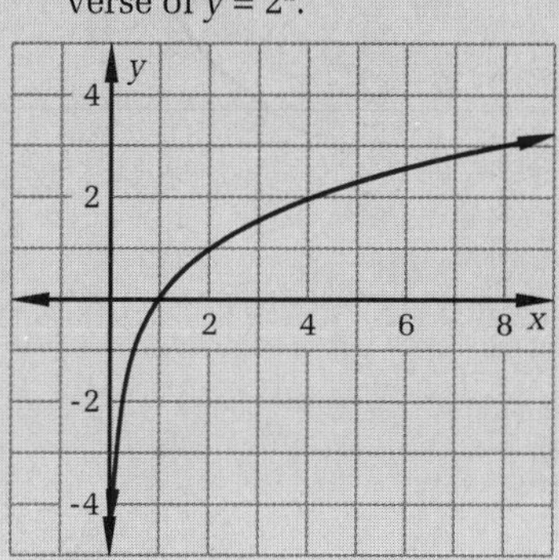

Vocabulary
exponential function
base
logarithmic function

Lesson Notes

- This chapter may be the first exposure to exponential and logarithmic functions for many students. Many of the exercises in exponential and logarithmic functions require algebraic manipulation of symbols. Encourage students to try many exercises since the ability to manipulate algebraic symbols is directly proportional to the number of exercises worked.
- The definition of a logarithm is introduced in this first section. Students need to be able to change from an exponential form to the logarithmic form of an equation and have the understanding to know when to change.
- Emphasize the relationship between logs and exponential equations. Properties will be introduced in Section **11.3** and if the students fully comprehend the connection, the properties make better sense.

Communicating Mathematics

Write a paragraph to explain the relationship between the exponential equation $a^x = y$ and the logarithmic equation $\log_a y = x$. Be sure to include the domain and range of each and the relationships of their graphs.

When $a > 1$, the function $f(x) = a^x$ increases as x increases. The larger the value of a the more rapidly the function increases.

When $0 < a < 1$, the function $f(x) = a^x$ decreases when x increases. The smaller the value of a the more rapidly the function decreases.

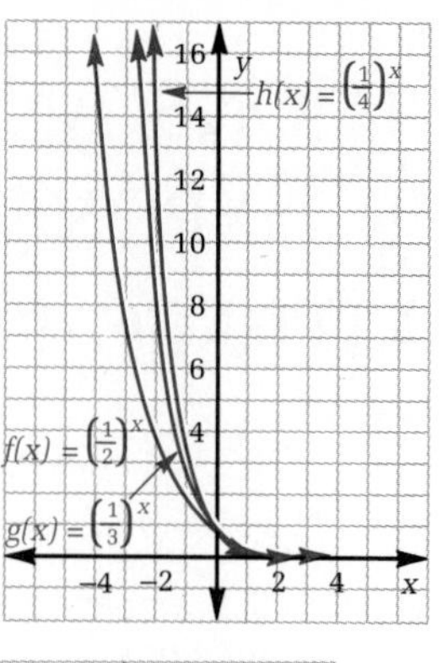

Notice that $f(x) = 2^x$ and $g(x) = \left(\frac{1}{2}\right)^x$ are reflections over the y-axis.

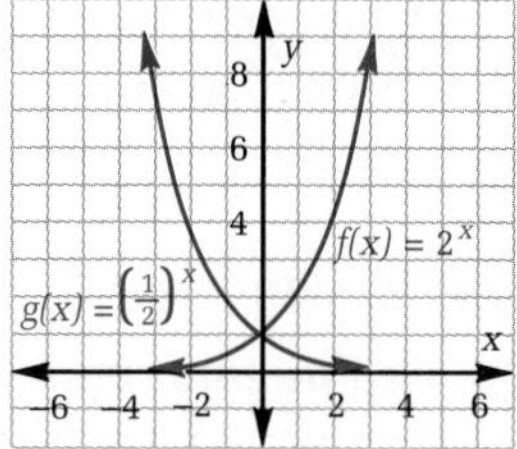

The Graph of the Logarithmic Function

The inverse of the exponential function is called the ***logarithmic function.*** To graph the inverse of a function of x we reflect the graph of the function about the line $y = x$. Thus, to graph the logarithmic function we reflect the exponential function about the line $y = x$.

Algebraically, we reverse the coordinates of the ordered pairs (x, y) of the original function to obtain the inverse function. So if $y = a^x$ is our original function, the equation with x and y reversed is $x = a^y$. Let's graph the exponential function $y = 2^x$ and its inverse, the logarithmic function $x = 2^y$.

Notice the following about the logarithmic function.

The domain is $\{x : x > 0\}$.

The range is $\{y : y \in \mathcal{R}\}$.

The x-intercept is 1.

There is no y-intercept.

$x = 0$ is a vertical asymptote.

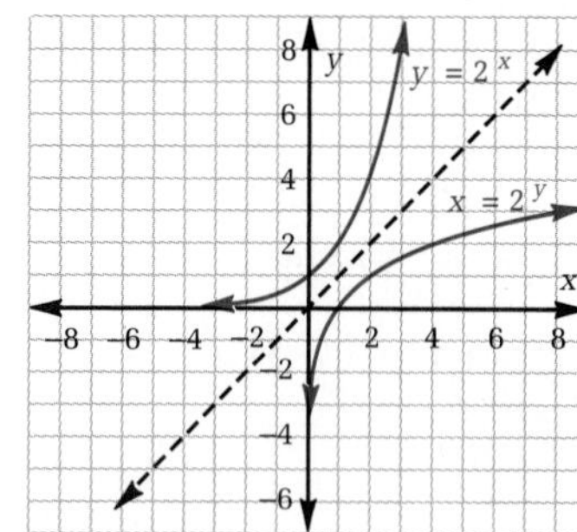

The Definition of the Logarithmic Function

Normally, when we write the equation of the inverse of $y = a^x$ we would reverse the roles of x and y to get $x = a^y$ and then solve for y. Unfortunately, we do not yet know how to solve this equation for y. Instead we state the following very important definition.

The equation $y = \log_a(x)$ is equivalent to the exponential equation $x = a^y$. $y = \log_a(x)$ is read "y equals the logarithm to the base a of the number x." The base a is always positive, and $a \neq 1$.

Note Sometimes we write $\log_a(x)$ as $\log_a x$, without parentheses. We will use parentheses when they clarify the expression.

An important observation: Notice that since $y = \log_a x$ means $x = a^y$, a logarithm is an exponent. It is the exponent that the base a requires in order to equal the number x.

Example *Evaluate y in each equation.*

a $y = \log_2 8$

$y = \log_2 8$ is equivalent to $8 = 2^y$, so $y = 3$.

b $y = \log_4 \left(\frac{1}{16}\right)$

$y = \log_4 \left(\frac{1}{16}\right)$ means $\frac{1}{16} = 4^y$, so $y = -2$.

c $y = \log_{10} -10$

$y = \log_{10} -10$ is equivalent to $-10 = 10^y$. No matter what value is chosen for y, 10^y is greater than zero. *There is no real-number logarithm for a nonpositive number.*

Stumbling Block

Students are not yet familiar with the use of the logarithmic functions and the new symbols involved. Many forget that the logarithmic function is an exponential function in disguise, with the variables x and y reversed. Emphasize that the definition of a logarithm is two-way. Have them practice with many different functions, converting from logarithms to exponents and from exponents to logarithms.

Part Two: Sample Problems

Problem 1 *Solve for the indicated variable.*

a $\log_b 16 = 2$ **b** $\log_{\frac{1}{2}} c = -3$ **c** $\log_x 9 = \frac{2}{3}$

Solution

a $\log_b 16 = 2$ means $b^2 = 16$. Therefore, $b = 4$ or $b = -4$. Since the base must be positive, the only correct answer is $b = 4$.

b $\log_{\frac{1}{2}} c = -3$ means $c = \left(\frac{1}{2}\right)^{-3}$, or $c = \frac{1^{-3}}{2^{-3}} = 2^3 = 8$. Therefore, $c = 8$.

c $\log_x 9 = \frac{2}{3}$ is equivalent to $x^{\frac{2}{3}} = 9$. Raise each side to the $\frac{3}{2}$ power.

$$\left(x^{\frac{2}{3}}\right)^{\frac{3}{2}} = 9^{\frac{3}{2}}$$
$$x = 27$$

Problem 2 *Evaluate the following for $a > 0$ and $a \neq 1$.*

a $\log_a a$ **b** $\log_a 1$

Solution

a Let $y = \log_a a$. This is equivalent to $a = a^y$, so $y = 1$. Therefore, $\log_a a = 1$ for $a > 0$, $a \neq 1$.

b Let $y = \log_a 1$. This is equivalent to $1 = a^y$. Since $a^0 = 1$, we have $y = 0$. Therefore, $\log_a 1 = 0$ for $a > 0$, $a \neq 1$.

Using Computers

Using graphing software or a graphing calculator, many graphs can be drawn quickly; the effects of changing parameters are easily explored. The inverse relationship between the logarithmic and exponential functions is apparent in the functions' graphs. The *Omnifarious Plotter*, which can graph all of $y = a^x$, $x = a^y$ and $y = \log_a x$ on the same set of axes, is especially effective. Graphing software or graphing calculators can help students in solving many of the problems in the problem sets.

Checkpoint

1 Solve the following equation for x: $3^x = 3^4$. $x = 4$
2 Solve the following equation for x: $2^{2x} = 8^2$. $x = 3$
3 Change the exponential equation $2^3 = 8$ into a logarithmic equation. $\log_2 8 = 3$
4 Express the following logarithmic equation as an exponential equation: $\log_3 243 = 5$. $3^5 = 243$

Assignment Guide

Average
7–11, 14–17, 23, 28–31
Advanced
9–12, 14, 16–21, 28–31, 38–40
Honors
9–12, 14, 16–21, 26, 28, 29, 31, 33, 34, 35a, b, 37, 40, 41, 45

Problem-Set Notes and Strategies

- Problem **1** is deceptively simple, yet so important for students to understand.
- Problems **4**, **7**, and **8** will help students to see the relationship between logarithms and exponents.

Problem 3 *Write each of the following exponential equations in logarithmic form.*

a $3^y = 12$ **b** $a^5 = 100$ **c** $4^{\log_4 16} = x$

Solution

a $3^y = 12$ is equivalent to $\log_3 12 = y$.
b $a^5 = 100$ is equivalent to $\log_a 100 = 5$.
c $4^{\log_4 16} = x$ is equivalent to $\log_4 x = \log_4 16$. In this case, we see that x must equal 16.

Problem 4 *Evaluate each of the following, where $a > 0$ and $a \neq 1$.*

a $a^{\log_a x}$ **b** $\log_a (a^x)$

Solution

a We convert to logarithmic form. We let $y = a^{\log_a x}$, which is equivalent to $\log_a x = \log_a y$. Therefore, y must equal x. We have shown that $a^{\log_a x} = x$ for all $a > 0$, $a \neq 1$.
b We convert to exponential form. We let $y = \log_a (a^x)$, which means the same as $a^y = a^x$. Therefore, y must equal x. We have shown that $\log_a (a^x) = x$ for all $a > 0$, $a \neq 1$.

Part Three: Exercises and Problems

Warm-up Exercises

P **1** Complete the following statement with one word. A logarithm is an ______. Exponent

P **2** Rewrite $x^y = z$ in logarithmic form. $\log_x z = y$

P **3** Rewrite $\log_m n = w$ in exponential form. $m^w = n$

P **4** Solve for w.
a $\log_6 w = 3$ $w = 216$ **b** $\log_5 w = 3$ $w = 125$ **c** $\log_2 w = -3$ $w = \frac{1}{8}$

P,R **5** If a function is randomly selected from the list below, what is the probability that it is an exponential function? $\frac{2}{5}$
$g(x) = 3x$ $h(x) = x^4$ $k(x) = 4^x$ $j(x) = 3^2$ $m(x) = 10^x$

P **6** Copy and complete the following tables.

a

x	0	1	2	3	4	−1	−2	−3	−4
10^x	1	10	100	1000	10,000	0.1	0.01	0.001	0.0001

b

x	1	10	100	1000	10,000	0.1	0.01	0.001	0.0001
$\log_{10} x$	0	1	2	3	4	−1	−2	−3	−4

Problem Set

P **A**

7 Solve for x.

a $\log_2 x = \log_2 16$ 16 **b** $\log_2 x = \log_4 16$ 4

P **8** Solve for *M*.

a $M = \log_7 7$ 1 **b** $M = \log_7 1$ 0

P **9** Evaluate.

a $2^{\log_2 8}$ 8 **b** $10^{\log_{10} 1000}$ 1000 **c** $3^{\log_3 81}$ 81

P **10** Evaluate.

a $\log_2 2^8$ 8 **b** $\log_5 5^3$ 3 **c** $\log_{10} 10^{-4}$ −4

P **11** Solve for y.

a $y = \log_5 125$ 3 **b** $5 = \log_{125} y$ $125^5 = 30{,}517{,}578{,}125$ **c** $3 = \log_y 125$ 5

P 

12 Solve for x.

a $\log_x 125 = 3$ 5 **b** $\log_x 0.25 = -2$ 2

c $\log_4 x = 12$ 16,777,216 **d** $x = \log_8 -8$ No real solution

P,R **13** Complete each table.

a

x	0	1	2	3	−1	−2	−3	5	−4
2^x	1	2	4	8	$\frac{1}{2}$	$\frac{1}{4}$	$\frac{1}{8}$	**32**	$\frac{1}{16}$

b

x	1	2	4	8	$\frac{1}{2}$	$\frac{1}{4}$	$\frac{1}{8}$	32	$\frac{1}{16}$
$\log_2 x$	0	1	2	3	−1	−2	**−3**	**5**	**−4**

R **14** Write 3^{12} as a power of each of the following.

a 9 9^6 **b** 27 27^4 **c** 81 81^3

P **15** Solve each equation for y.

a $y = \log_8 2$ $\frac{1}{3}$ **b** $y = \log_2 \frac{1}{8}$ −3 **c** $y = \log_8 \frac{1}{2}$ $-\frac{1}{3}$

R 

16 Simplify $8^{n-2} \cdot 2^{n+5} \cdot 4^{3n}$. 2^{10n-1}

P,R **17 a** Graph $f(x) = 10^x$ and $g(x) = \log_{10} x$ on the same coordinate plane.

b The graphs are symmetrical with respect to what line? $y = x$

P **In problems 18–21, evaluate each expression.**

18 $\log_6 216$ 3 **19** $\log_{0.01} 100$ −1 **20** $\log_7 (7^{-9})$ −9 **21** $\log_2 -4$ No real value

P **In problems 22–25, write the equivalent logarithmic form.**

22 $4^3 = 64$ $\log_4 64 = 3$

23 $3^{-4} = \frac{1}{81}$ $\log_3 \left(\frac{1}{81}\right) = -4$

24 $x = 2^{4.5}$ $\log_2 x = 4.5$

25 $2^n = 363$ $\log_2 363 = n$

Problem-Set Notes and Strategies, continued

- After solving problems similar to problem **9**, explain again that logarithms and exponents of the same base are inverse operations. One function undoes the other and so yields the original number.
- If students have difficulty with problem **10**, suggest that they set the expression equal to *y* and use the definition of logarithms to change the problem into an exponential equation.
- In problem **12d**, emphasize that the domain of the logarithm function is the set of positive real numbers.
- Problem **16** is a review of the properties of exponents which will be needed to develop the properties of logarithms in Section **11.3**.

Additional Answer

17a

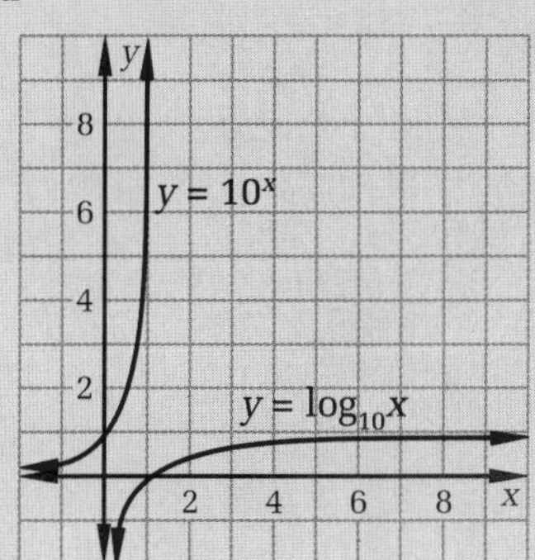

Problem-Set Notes and Strategies, continued

- Some of the properties of logarithms are hinted at in problem **29**. Help students find some relationships in these expressions. These properties are formally introduced in Section **11.3**.
- The graphs of problems **34** and **35** will provide some insight into the properties of logarithms and the translations studied earlier.
- In problem **36**, students may guess the answer. Exploring these ideas introduce a need for the properties.

Cooperative Learning

Problem **29** in this section leads to the properies of logarithms introduced in Section **11.3**. If time permits, you might have students work this problem and examine similar problems and try to make conjectures about the upcoming properties. Use problem **35a** and **b** as additional motivation. These graphs will provide additional insight. If some of the properties are discovered, ask the students to try to prove them using the exponential properties from the past chapters.

Additional Answers

34a

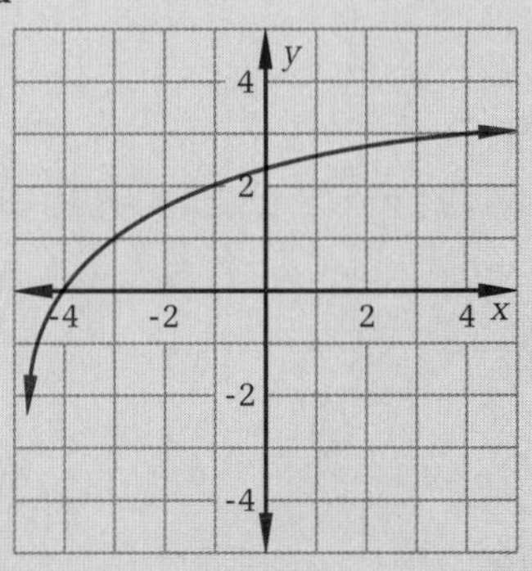

Problem Set, *continued*

P **26** Solve $\log_4 (4^{10}) = x$ for x. 10

R **27** Rationalize the denominator. $\frac{5 + \sqrt{3}}{5 - \sqrt{3}}$ $\frac{14 + 5\sqrt{3}}{11}$

P **28** Explain what is wrong with the following solution.

Given: $\log_x 16 = 2$ Solution: $x^2 = 16$, so $x = 4$ or $x = -4$

-4 is not a solution, since a base cannot be negative.

P,I **29** Evaluate each expression.

a $\log_{10} 10 + \log_{10} 100 + \log_{10} 1000$ 6

b $\log_{10} 1{,}000{,}000$ 6

c $\log_3 81 - \log_3 9$ 2

d $\log_3 \left(\frac{81}{9}\right)$ 2

P **30** Match each graph with its corresponding function.

a $f(x) = 2^x$ iv

b $g(x) = \log_2 x$ i

c $h(x) = 2^{-x}$ iii

d $k(x) = -\log_2 x$ ii

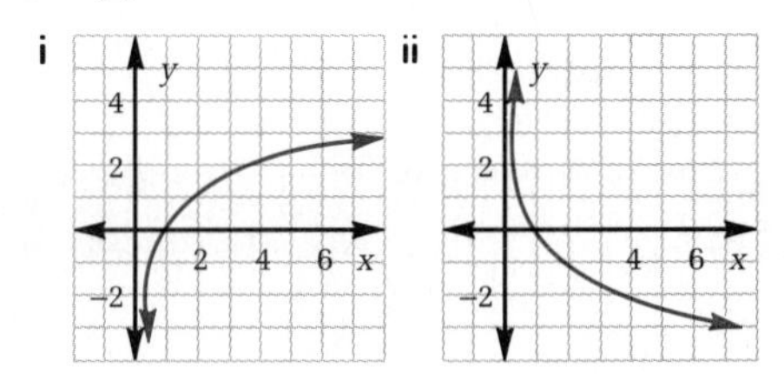

iii

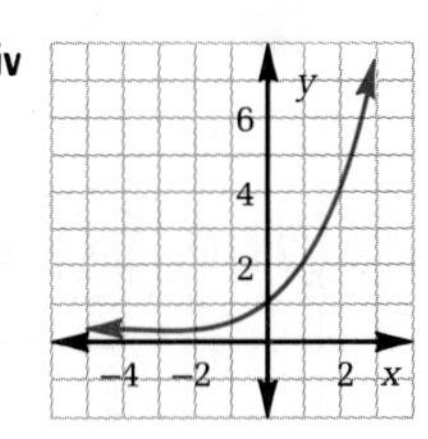

P,R,I **31** Evaluate.

a $\log_2 \sqrt{64}$ 3

b $\frac{1}{2}(\log_2 64)$ 3

c $\sqrt{\log_2 64}$ $\sqrt{6} \approx 2.4495$

P **32** Rewrite $5^7 = 78{,}125$ in logarithmic form. $\log_5 78{,}125 = 7$

R **33** The volume of the larger cube exceeds the volume of the smaller cube by 10. Find x.

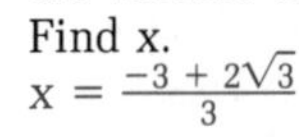
$x = \frac{-3 + 2\sqrt{3}}{3}$

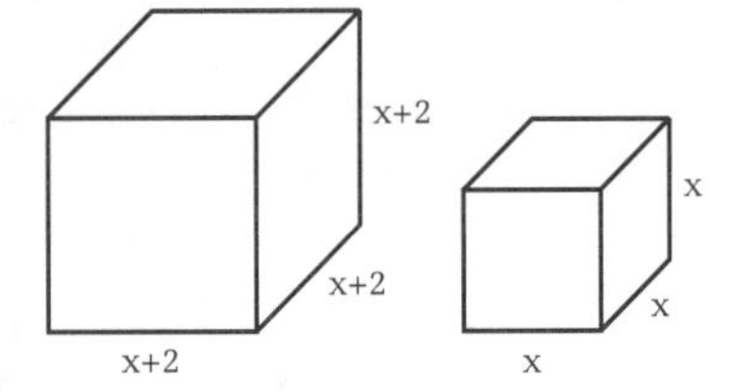

P,R **34** Sketch the graph of each logarithmic function.

a $y = \log_2 (x + 5)$

b $y = (\log_2 x) + 5$

P,R,I **35** Sketch the following functions on the same coordinate system.

a $f(x) = \log_2 x + \log_2 8$

b $h(x) = \log_2 (8x)$

c $g(x) = \log_2 (x + 8)$

I **36** Suppose $\log_a 2 \approx 0.333$ and $\log_a 3 \approx 0.528$. Approximate $\log_a 6$. ≈ 0.861

Answers for problems **34b** and **35** can be found in the answer pages beginning on page **A** 1.

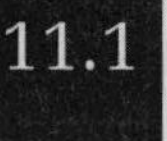

P,R **37** Specify the domain and range of each function.

a $f(x) = \log_3 x$ **b** $g(x) = 3^x$ **c** $h(x) = \log_5 |x|$

P,R **38** Graph $f(x) = \log_2 x$ and the inverse of $f(x)$.

P,R **39** Which of the following is the graph of the inverse of $f(x) = 2^x$? c

a

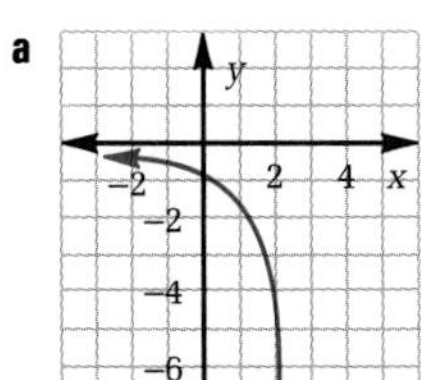

b

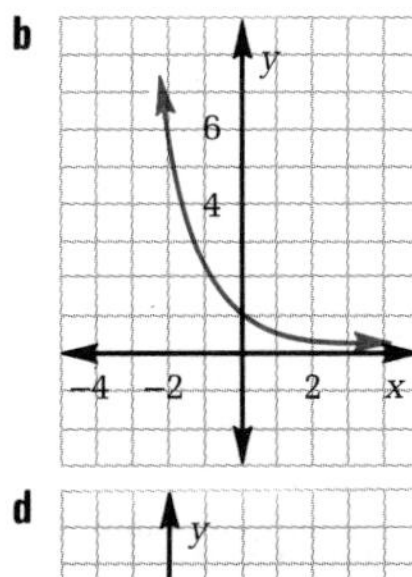

c

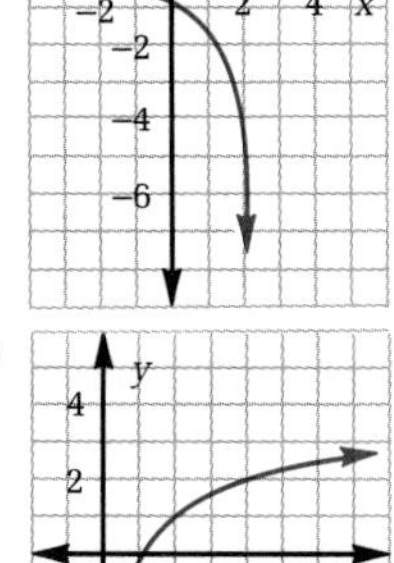

d

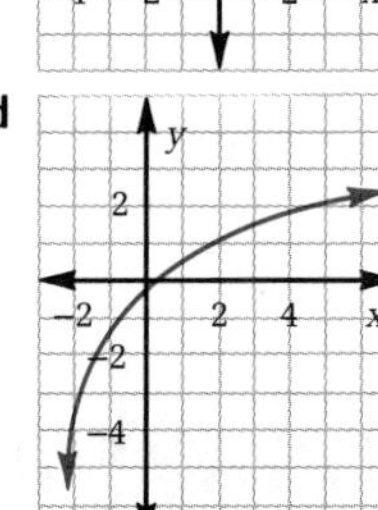

R **C** **40** Simplify $\left(\frac{3^2 - 3^0}{9^{\frac{3}{2}}}\right)^{\frac{-1}{3}}$. $\frac{3}{2}$

P **41** Graph $y = \log_{\frac{1}{4}} x$.

P,R **42** **a** Graph $f(x) = 5^{-x}$ and $g(x) = \log_{\frac{1}{5}} x$ on the same plane.

b Explain why the two graphs are symmetrical with respect to the line $y = x$.

P,R **43** Find the x- and y-intercepts and the asymptotes of the graphs of the following functions.

a $y = 7^x$ No x-intercept; y-intercept: (0, 1); horizontal asymptote: $y = 0$

b $y = \log_7 x$ No y-intercept; x-intercept: (1, 0); vertical asymptote: $x = 0$

R **44** The graph of $y = f(x)$ is shown. Sketch a graph of each of the following equations.

a $y = -f(x)$

b $y = f(-x)$

c $y = |f(x)|$

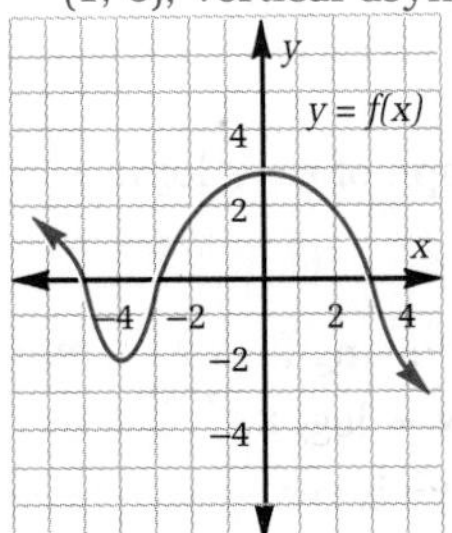

P **45** Solve $\log_x 16 = \frac{3}{4}$ for x. $x \approx 40.32$

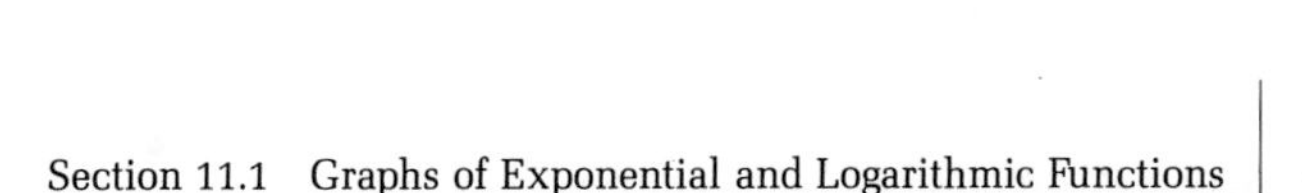

Problem-Set Notes and Strategies, continued

- Problem **40** utilizes many of the exponential properties discussed in past chapters. None are difficult in themselves, but the combination of the properties may confuse some students. Urge them to go step-by-step.
- Problem **45** is a nice exercise in the use of a calculator. First however, students need to transform the equation into its exponential form and solve for x. Then the calculator may be used to find the answer.

Additional Answers

37a Domain: $\{x: x > 0\}$; range: $\{f(x): f(x) \in \mathfrak{R}\}$

37b Domain: $\{x: x \in \mathfrak{R}\}$; range: $\{g(x): g(x) > 0\}$

37c Domain: $\{x: x \neq 0\}$; range: $\{h(x): h(x) \in \mathfrak{R}\}$

38

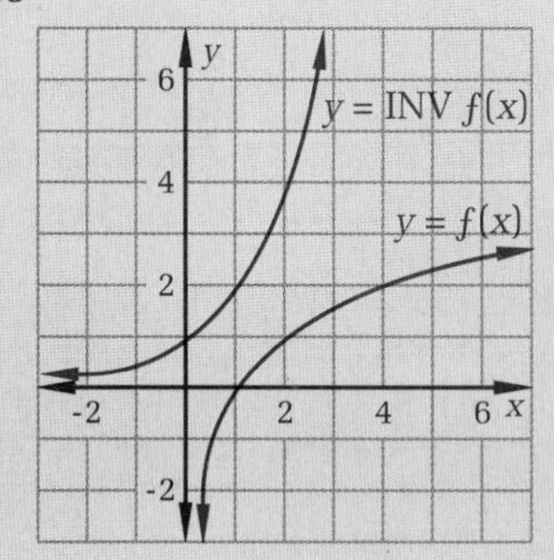

41

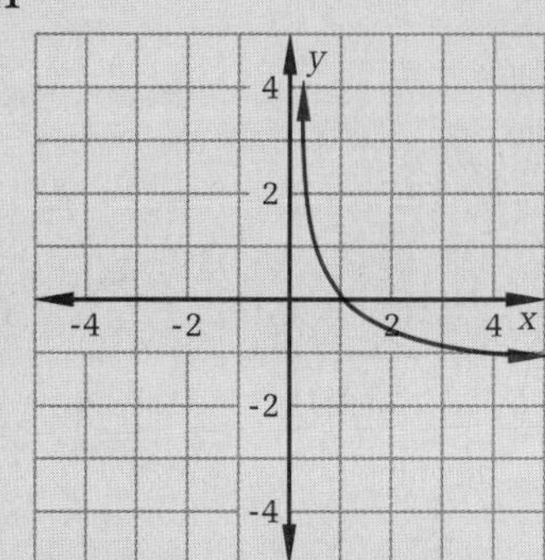

Answers for problems **42a–b**, and **44a–c** can be found in the answer pages beginning on page **A** 1.

11.2 TWO IMPORTANT EXPONENTIAL AND LOGARITHMIC FUNCTIONS

Objectives

After studying this section, you will be able to

- Solve problems involving exponential and logarithmic functions with base 10
- Solve problems involving exponential and logarithmic functions with base *e*

Part One: Introduction

Exponential and Logarithmic Functions with Base 10

In this section, we study the most often used logarithmic functions, those with bases of 10 and e.

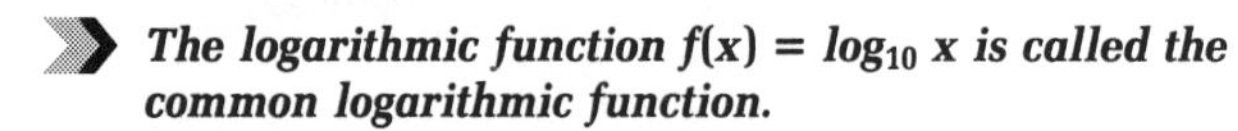

The logarithmic function $f(x) = \log_{10} x$ is called the common logarithmic function.

In fact, this logarithmic function is so common that the base 10 is often omitted. The function is written as $f(x) = \log x$.

Example *Evaluate each of the following expressions.*

a *log 100*

$y = \log 100$ is equivalent to $10^y = 100 = 10^2$, so $y = 2$. Therefore, $\log 100 = 2$.

b *log 0.001*

$y = \log 0.001$ is equivalent to $10^y = 0.001 = 10^{-3}$, so $y = -3$. Therefore, $\log 0.001 = -3$.

c *log 27*

$y = \log 27$ means $10^y = 27$. The solution for y is somewhere between 1 and 2, since $10^1 = 10$ and $10^2 = 100$. Fortunately, we can obtain a very close approximation to log 27 using a scientific calculator's [LOG] key, which evaluates base 10 logarithms. To find log 27 we enter 27 [LOG]. The display shows [1.431363764]. The exact value of log 27 is an irrational number, so we say, $\log 27 \approx 1.431363764$.

11.2

Class Planning

Time Schedule

All levels: 1 day

Resource References

Teacher's Resource Book
Class Opener 66
Cooperative Learning 32
Practice 70

Class Opener

If you have access to a scientific calculator, find each value.

1 log 10,000 4
2 log 1000 3
3 log 100 2
4 What will be the input if the output is 1? log 10
5 What do you think "log" means?
Answers will vary. A possible answer will convey that "log" is a power of 10.

In problems **6–9**, find each value.

6 ln 10,000 ≈9.21034
7 ln 1000 ≈6.907755
8 ln 100 ≈4.605170
9 ln 10 ≈2.302585
10 What seems to be the relationship between problems **6–9**?
Answers will vary. A possible answer describes the answers to **6–8** as multiples of the answer to **9**.

Lesson Notes

- This section introduces the common and natural logarithms. Natural logarithms are used in applications of all types of natural phenomenon.

Vocabulary

common logarithmic function
antilogarithm
natural logarithm
natural logarithmic function

Example *Find log 324.78.*

$y = \log 324.78$ is equivalent to $10^y = 324.78$. The answer must be between 2 and 3. Why? Using a calculator to find a closer approximation, we enter 324.78 [LOG]. The calculator displays [2.511589277]. So $\log 324.78 \approx 2.511589277$. This answer means that $324.78 \approx 10^{2.511589277}$. Let's check using either the calculator's [x^y] key or its [10^x] key. We enter 2.511589277 [10^x] and get [324.7799996]. This is very close to 324.78.

Example *If* $\log x = 2.4516$, *what is* x?

$\log x = 2.4516$ is equivalent to $10^{2.4516} = x$. We know that x must be between 100 and 1000. Why? To find a closer approximation, we use our calculator and enter 2.4516 [10^x]. The display shows [282.8785388]. Therefore, if $\log x = 2.4516$, then $x \approx 282.88$.

Note In the last example, we knew the logarithm of a number and were asked to find the number. This is sometimes called finding an ***antilogarithm.***

Exponential and Logarithmic Functions with Base e

Logarithms and powers using a base of 10 are important because of our base-10 numeration system. Logarithms and powers in formulas that model real-world situations often involve another important base. This base, denoted by the letter e, is an irrational number whose value is approximately 2.718281828. Logarithms that use base e are called ***natural logarithms.*** Natural logarithms are used extensively in economics, engineering, physics, astronomy, anthropology, and biology.

The logarithmic function $f(x) = \log_e (x)$, usually written $f(x) = \ln x$, is called the natural logarithmic function. The irrational number e has an approximate value of **2.718281828.**

Remember, logarithms can have any positive number that is not 1 as a base. The definition of a logarithm works just the same for the irrational number e as it does for 10 or any other base. When the letter e is used in the rest of this chapter, it will always represent the number e.

Example *Evaluate the following expressions.*

a $\ln e^4$
Remember, $\ln x$ really means $\log_e x$. Let $y = \ln e^4$, which means $e^y = e^4$, so $y = 4$. Therefore, $\ln e^4 = 4$.

b $e^{\ln 2}$
Let $y = e^{\ln 2}$. Now we convert this to logarithmic form using the definition of a logarithm in reverse.
$\ln y = \ln 2$, so $y = 2$.

c $\ln 1$
$y = \ln 1$ is equivalent to $e^y = 1$. Since any number to the zero power is 1, we have $y = 0$. Therefore, $\ln 1 = 0$.

Lesson Notes, continued

- Students often confuse *e* with a variable since it is a letter. Explain that *e* is a special number like π that has special applications, but is a unique number nonetheless. The number *e* was studied extensively by a mathematician named Leonard Euler (pronounced Oiler) and it is thought to be the reason we use the letter *e*.
- Several applications of the natural logarithm and natural exponential function will appear in this section, but will be discussed further in Section **11.5.**

Stumbling Block

The number *e* is frequently confused for a variable. Have students identify the button on their calculator for e^x and evaluate several powers of *e* including 0 and 1. If they do this several times for the same number, the answer will always be the same, indicating that *e* must be a constant.

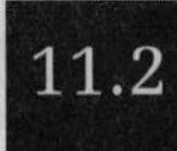

Communicating Mathematics

In your own words, explain the relationship between a common logarithm and a natural logarithm. Identify the similarities and differences between the two.

Scientific calculators have the keys ln and e^x, often as inverses on the same key, to calculate base e logarithms and antilogarithms respectively. Enter 1 e^x to see the value of e in the display of your calculator.

Example *Solve for x.*

a $x = \ln 100$

Remember, ln x means $\log_e x$. $x = \ln 100$ is equivalent to $e^x = 100$. Since e is less than 3 and $3^4 = 81$, we expect the value of x to be somewhat more than 4. We can use a calculator for a closer approximation by entering 100 ln. The display is 4.605170186. Therefore, $x \approx 4.6052$.

b $\ln x = 4$

Since $\ln x = 4$ is equivalent to $e^4 = x$, to find x we use the e^x key and enter 4 e^x. The display shows 54.59815003 so $x \approx 54.6$.

Part Two: Sample Problems

Problem 1 *Solve each of the following for x.*

a $400 = 50 \log (10x)$

b $900 = 15e^{12x}$

Solution

a $400 = 50 \log (10x)$

$8 = \log (10x)$

Convert to exponential form

$10^8 = 10x$

$10^7 = x$

b $900 = 15e^{12x}$

$60 = e^{12x}$

Convert to logarithmic form

$\ln 60 = 12x$

$\frac{\ln 60}{12} = x$

$0.3412 \approx x$

Problem 2 *The loudness L, measured in decibels (so called after Alexander Graham Bell), of a sound of intensity I is given by the logarithmic function $L(I) = 10 \log \left(\frac{I}{I_0}\right)$, where I_0, the threshold of hearing, is the faintest audible sound.*

a *If the faintest audible sound is made, how many decibels is that sound?*

b *Noise on a city street measures approximately 75 decibels. How many times as intense is this sound than the faintest audible sound?*

c *The noise produced in a thunderstorm is approximately 10^{11} times as intense as the faintest audible sound. What is the loudness of the sound in decibels?*

Solution

a Since the faintest sound has intensity $I = I_0$, we have $L(I_0) = 10 \log\left(\frac{I_0}{I_0}\right) = 10 \log 1 = 10 \cdot 0 = 0$. The faintest possible sound has a loudness of 0 decibels.

b $75 = 10 \log\left(\frac{I}{I_0}\right)$, or $7.5 = \log\left(\frac{I}{I_0}\right)$. This is equivalent to $10^{7.5} = \frac{I}{I_0}$, or $10^{7.5} I_0 = I$.
Therefore, the intensity is $10^{7.5}$, or $\approx 31{,}622{,}777$, times as great as that of the faintest audible sound.

c The intensity is $I = 10^{11} I_0$ for a thunderstorm.
Therefore, $L = 10 \log\left(\frac{10^{11} I_0}{I_0}\right)$, or $L = 10 \log(10^{11})$
$= 10 \cdot 11$
$= 110$ decibels

Problem 3 *The Euler Bank compounds interest continuously.*

a *If \$5000 is invested in the Euler Bank at 7.85%, how much money will be in the account after 12 years?*

b *Susan invests \$10,000 at the Euler Bank at 8.65%. In how many years will she have \$250,000 in the bank?*

Solution

The formula for computing the amount in an account when interest is compounded continuously is

$A = Pe^{rt}$, where P = the amount of the original deposit,
r = the yearly interest rate, and
t = the number of years the money has been in the account.

This is an example of the base e occurring in a business formula.

a
$$\begin{aligned} A &= Pe^{rt} \\ &= 5000e^{0.0785(12)} \\ &= 5000e^{0.942} \\ &\approx 5000(2.5651) \\ &\approx \$12{,}825.53 \end{aligned}$$

b
$$\begin{aligned} A &= Pe^{rt} \\ 250{,}000 &= 10{,}000e^{0.0865t} \\ 25 &= e^{0.0865t} \\ \ln 25 &= 0.0865t \\ 3.2189 &\approx 0.0865t \\ 37.2 &\approx t \end{aligned}$$

Susan will have \$250,000 in the bank after approximately 37.2 years.

Part Three: Exercises and Problems

Warm-up Exercises

P

1 Arrange in order from smallest to largest. **c, a, d, b**

a $\log 100$ **b** $10 \cdot \log 10$ **c** $(\log 10) \cdot (\log 10)$ **d** $\log(10^5)$

P,I

2 Evaluate.

a $\ln\left(\frac{e^5}{e^2}\right)$ 3 **b** $\ln e^5 - \ln e^2$ 3 **c** $\ln e^3$ 3

Checkpoint

1. Solve the following for x: $\log 100x = 5$. 1000
2. Which is the larger number $\log 10$ or $\ln 10$ and why? $\ln 10$ since the base of the logarithm is smaller.
3. Use your calculator to put the following numbers in ascending order: log 52, ln 98, log 3170 and ln 25. log 52, ln 25, log 3170, ln 98
4. Evaluate the following: $\ln x = 82.3$. $x = 5.526 \times 10^{35}$

Assignment Guide

Average
13, 14, 17, 22, 27–33, 35, 37
Advanced
13, 14, 17, 22, 24, 27, 28, 31–35, 38
Honors
13, 14, 17, 22, 26, 27, 30–33, 36, 37, 40, 45, 46

Problem-Set Notes and Strategies

- Problems **2** and **3** lead to the properties of logarithms discussed in the next section. Some students will intuitively understand why the answers are identical.

Problem-Set Notes and Strategies, continued

- Problem **4** will check student understanding of logarithms and the use of the definition to evaluate expressions.
- Some of the results in problem **13** are negative. Point out to students that while the range of a log function includes negative numbers, the domain is the set of positive real numbers.
- Problems **20** and **26** again suggest the properties of logarithms in the next section.

P,R **3** Evaluate the following expressions.

a log 1 0 **b** ln 1 0 **c** $\log_2 1$ 0 **d** $\log_8 1$ 0

P **In problems 4–7, evaluate each expression without using a calculator.**

4 log 10,000 4 **5** log 0.0001 −4 **6** $\ln\left(e^{\frac{1}{2}}\right)$ $\frac{1}{2}$ **7** $e^{\ln 5}$ 5

P  **In problems 8–11, use a calculator to evaluate each expression.**

8 log 0.000157 ≈−3.8041 **9** log 353 ≈2.5478 **10** ln 0.0015 ≈−6.5023 **11** ln 53.4 ≈3.9778

P **12** The noise produced by normal conversation is about 10^6 times as loud as the faintest audible sound. What is the loudness of this sound in decibels? (Hint: Use the formula in Sample Problem 2.) 60 db

Problem Set

A

13 Approximate each of the following to four decimal places.

P 

a log 27 ≈1.4314 **b** log 3 ≈0.4771 **c** $\log\frac{1}{3}$ ≈−0.4771 **d** $\log\frac{1}{27}$ ≈−1.4314

e log 9 ≈0.9542 **f** log 1 0 **g** $\log\frac{1}{9}$ ≈−0.9542

P  **14** Solve for x.

a x = ln 50, x ≈ 3.9120 **b** x = log 50, x ≈ 1.6990 **c** ln x = 6, x ≈ 403.4288 **d** log x = 6, x = 1,000,000

P  **15** Solve for y.

a y = ln −6, No solution **b** −6 = ln y, y ≈ 0.0024788

P,R **16** Rectangle ABCD ~ rectangle PQRS
AB = 40, AD = 35, PQ = 8

a Find the ratio AD:PS. 5:1
b Find the ratio of the area of ABCD to the area of PQRS. 25:1
c Find the ratio of the log of the area of ABCD to the log of the area of PQRS. ≈1.8:1

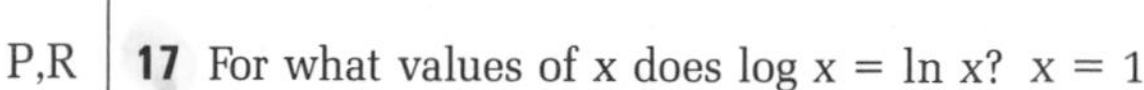

A B D C P Q S R

P,R **17** For what values of x does log x = ln x? x = 1

P,R **18** Solve $b = \log_4 \sqrt{2}$ for b. $b = \frac{1}{4}$

B

P **19** If log x ≈ 72.514, how many digits does x have to the left of the decimal place? 73

P,I **20** Evaluate each expression to five decimal places.

a 4(log 3) ≈1.90849 **b** 2(log 9) ≈1.90849 **c** log 81 ≈1.90849 **d** $8(\log\sqrt{3})$ ≈1.90849

P  **21** Find the base e antilogarithm of 2.17 to the nearest thousandth. ≈8.758

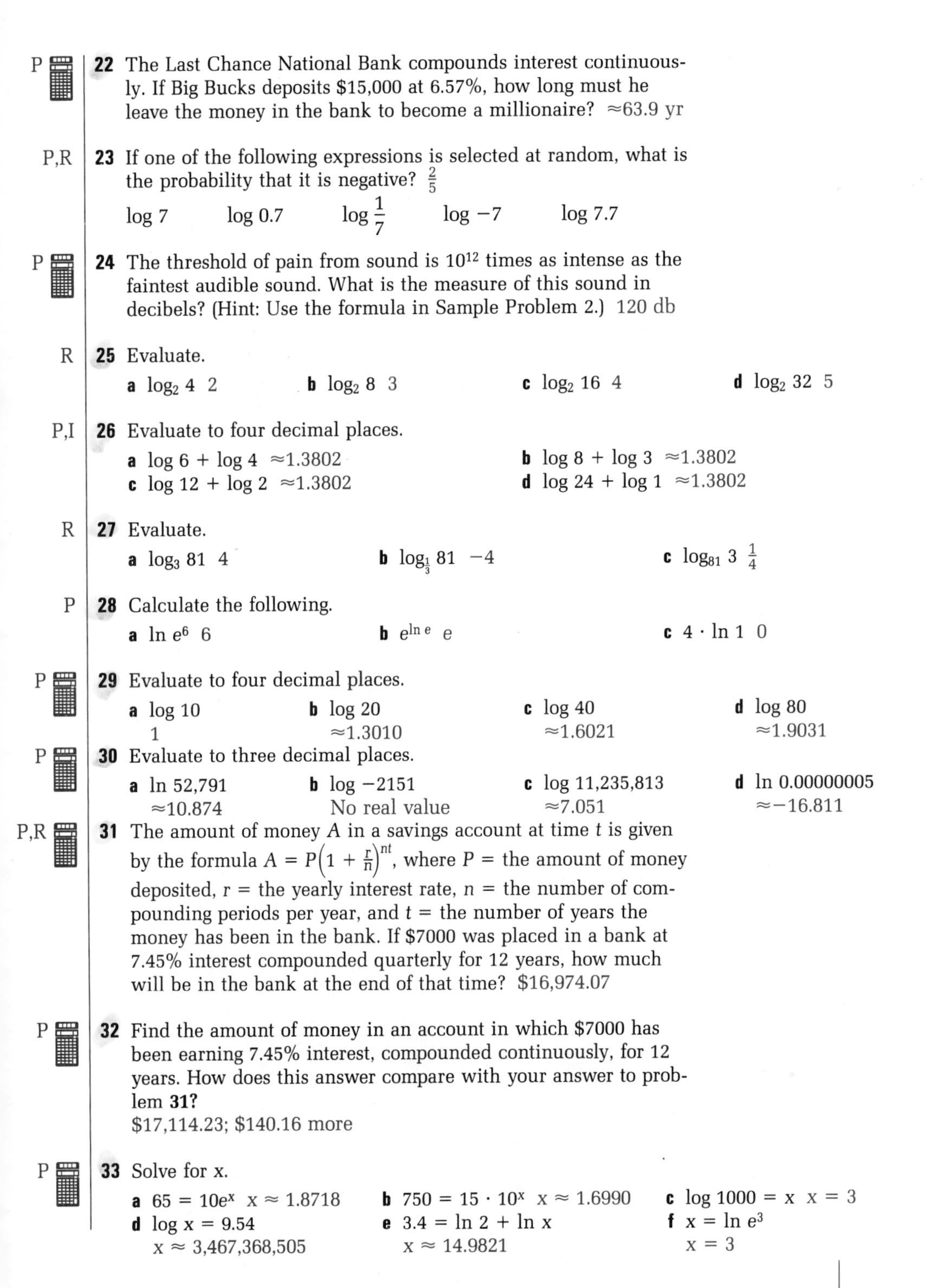

P **22** The Last Chance National Bank compounds interest continuously. If Big Bucks deposits \$15,000 at 6.57%, how long must he leave the money in the bank to become a millionaire? $\approx$63.9 yr

P,R **23** If one of the following expressions is selected at random, what is the probability that it is negative? $\frac{2}{5}$

$\log 7$ $\quad$ $\log 0.7$ $\quad$ $\log \frac{1}{7}$ $\quad$ $\log -7$ $\quad$ $\log 7.7$

P **24** The threshold of pain from sound is 10^{12} times as intense as the faintest audible sound. What is the measure of this sound in decibels? (Hint: Use the formula in Sample Problem 2.) 120 db

R **25** Evaluate.

a $\log_2 4$ 2 $\quad$ **b** $\log_2 8$ 3 $\quad$ **c** $\log_2 16$ 4 $\quad$ **d** $\log_2 32$ 5

P,I **26** Evaluate to four decimal places.

a $\log 6 + \log 4$ $\approx$1.3802 $\quad$ **b** $\log 8 + \log 3$ $\approx$1.3802
c $\log 12 + \log 2$ $\approx$1.3802 $\quad$ **d** $\log 24 + \log 1$ $\approx$1.3802

R **27** Evaluate.

a $\log_3 81$ 4 $\quad$ **b** $\log_{\frac{1}{3}} 81$ -4 $\quad$ **c** $\log_{81} 3$ $\frac{1}{4}$

P **28** Calculate the following.

a $\ln e^6$ 6 $\quad$ **b** $e^{\ln e}$ e $\quad$ **c** $4 \cdot \ln 1$ 0

P **29** Evaluate to four decimal places.

a $\log 10$ 1 $\quad$ **b** $\log 20$ $\approx$1.3010 $\quad$ **c** $\log 40$ $\approx$1.6021 $\quad$ **d** $\log 80$ $\approx$1.9031

P **30** Evaluate to three decimal places.

a $\ln 52{,}791$ $\approx$10.874 $\quad$ **b** $\log -2151$ No real value $\quad$ **c** $\log 11{,}235{,}813$ $\approx$7.051 $\quad$ **d** $\ln 0.00000005$ ≈-16.811

P,R **31** The amount of money A in a savings account at time t is given by the formula $A = P\left(1 + \frac{r}{n}\right)^{nt}$, where P = the amount of money deposited, r = the yearly interest rate, n = the number of compounding periods per year, and t = the number of years the money has been in the bank. If \$7000 was placed in a bank at 7.45% interest compounded quarterly for 12 years, how much will be in the bank at the end of that time? \$16,974.07

P **32** Find the amount of money in an account in which \$7000 has been earning 7.45% interest, compounded continuously, for 12 years. How does this answer compare with your answer to problem **31?**
\$17,114.23; \$140.16 more

P **33** Solve for x.

a $65 = 10e^x$ $x \approx 1.8718$ $\quad$ **b** $750 = 15 \cdot 10^x$ $x \approx 1.6990$ $\quad$ **c** $\log 1000 = x$ $x = 3$
d $\log x = 9.54$ $x \approx 3{,}467{,}368{,}505$ $\quad$ **e** $3.4 = \ln 2 + \ln x$ $x \approx 14.9821$ $\quad$ **f** $x = \ln e^3$ $x = 3$

Problem-Set Notes and Strategies, continued

- Problems **31**, **32**, and **35** identify the significance of the number of times per year that interest is compounded. It is interesting to have students find the interest rates and compounding times for several banks in the area and compute which is the best investment in the long run. It isn't always the higher rate or the bank that compounded more frequently. The combination of the two provide the best investment.

Cooperative Learning

Students have the two formulas for interest $A = P(1 + \frac{r}{n})^{nt}$ and for compounding continuously the formula $A = Pe^{rt}$. Have students work in groups of 2 or 3 and evaluate the expression $(1 + \frac{1}{n})^n$ for values of n = 1, 10, 100, 1000, 10000, 100000 and 1000000. Make a conjecture about the number that they get for each answer. Next, examine the formula $(1 + \frac{r}{n})^n$ and choose an arbitrary value for r that is not too large (say r = 3) and recompute the expression with the same 7 numbers. What can be said about the expression $(1 + \frac{r}{n})^n$ as n gets larger?

Problem Set, *continued*

P **34** **a** Find x if ln (3x + 4) = 2. $x \approx 1.1297$
b Find x if log (3x − 700) = 3. $x = 566\frac{2}{3}$

P,R **35** Suppose $10,000 has been earning interest at 6% for eight years. Use the compound interest formula (see problem **31**) and the continuous compound interest formula $A = Pe^{rt}$ to calculate the amount in the bank

a If interest is compounded quarterly $16,103.24
b If interest is compounded monthly $16,141.43
c If interest is compounded daily $16,160.11
d If interest is compounded continuously $16,160.74

P,R,I C **36** Solve $(\log_{10} x)^2 = 5 \log_{10} x + 6$ for x. $x = 10^6$ or $x = 10^{-1}$

Problem-Set Notes and Strategies, continued

- One way for students to approach problem **36** is as a quadratic type of equation and solve by factoring. Then each factor can be solved using the logarithmic formulas.

P **37** The noise in an office is 3162 times as intense as the faintest audible sound. The noise on a city street is 10 million times as intense as the faintest audible sound.

a Find the number of decibels for each sound. Office: 35 db; street: 70 db
b How many times as intense is the noise on a city street than the noise in an office? ≈3162.6 times

P **38** Find the base 5 logarithm of 0.008. −3

- In problem **38**, if students realize that 0.008 is equal to $\frac{1}{125}$, the problem is more accessible.

P **39** Find y by finding the antilogarithm of the following.

a log y = 3.2465 $y \approx 1764.006$
b log y = −3.9764 $y \approx 0.0001056$
c log y = 1.5796 $y \approx 37.7839$
d log y = −2.75 $y \approx 0.001778$

P **40** Solve $(\log_9 x)^2 = 81$ for x. $x = 387{,}420{,}489$ or $x \approx 0.00000000258117$

- In problem **40**, students may forget that when they take the square root of an equation there are two answers.

P **41** Find ln x if $\log_{10} x = 3$. $x \approx 6.9078$

R **42** Solve $\sqrt{w} + \sqrt{w + 3} = 3$ for w. $w = 1$

P **In problems 43–46, use a calculator to solve for x.**

43 $40 = 10e^x$ $x \approx 1.386$
44 $0.815 = \log x$ $x \approx 6.531$
45 $9000 = 5 \cdot 10^x$ $x \approx 3.255$
46 $3.4 = \ln (2x)$ $x \approx 14.98$

P,R **47** Using the data in Sample Problem 3, part **a**, calculate the amount that would be in the account if interest were compounded monthly using the formula $A = P\left(1 + \frac{r}{n}\right)^{nt}$, where
P = amount originally deposited,
r = the yearly interest rate,
n = the number of compoundings per year, and
t = the number of years. $12,786.25

11.3

Properties of Logarithms

Objective

After studying this section, you will be able to

- Use seven properties of logarithms

Part One: Introduction

In this section we develop some properties of logarithms. We assume the base $a > 0$ and $a \neq 1$. We also assume we are finding logarithms of only positive numbers.

Since $a^0 = 1$ and $a^1 = a$, we have the following.

Property 1: $\log_a 1 = 0$ **Property 2:** $\log_a a = 1$

Recall that logarithmic and exponential functions are inverses and that if f and g are inverse functions, then $g[f(x)] = x$ and $f[g(x)] = x$. So if $f(x) = a^x$ and $g(x) = \log_a x$, then $\log_a (a^x) = a^{\log_a x} = x$. This gives us the following two properties.

Property 3: $\log_a (a^x) = x$ **Property 4:** $a^{\log_a x} = x$

We now develop three additional theorems that give us very important properties of logarithms.

Example *Let $M = 10^4$ and $N = 10^3$. Evaluate $\log (M \cdot N)$ and $\log M + \log N$.*

$$\begin{aligned} &\log (M \cdot N) \\ &= \log (10^4 \cdot 10^3) \\ &= \log (10^7) \\ &= 7 \end{aligned} \qquad \begin{aligned} &\log M + \log N \\ &= \log 10^4 + \log 10^3 \\ &= 4 + 3 \\ &= 7 \end{aligned}$$

This example suggests that the logarithm of the product of two numbers is equal to the sum of the logarithms of the two numbers.

Property 5: $\log_a (M \cdot N) = \log_a M + \log_a N$

Proof of Property 5: Let $x = \log_a M$ and let $y = \log_a N$.
Then $a^x = M$ and $a^y = N$.
Therefore, $M \cdot N = a^x \cdot a^y$
$M \cdot N = a^{x+y}$
Using the definition of a logarithm, we have

$$\begin{aligned} \log_a (M \cdot N) &= x + y \\ &= \log_a M + \log_a N \end{aligned}$$

11.3

Class Planning

Time Schedule

All levels: 1 day

Resource References

Teacher's Resource Book
Class Opener 67
Cooperative Learning 33
Practice 71
Evaluation
Quiz Forms 1, 2, 3

Class Opener

Tell which statements are True or False. Try some examples. a is any positive real number ≠ 0.

1. $\log_a(a) = 1$ True
2. $\log_a(1) = a$ False
3. $\log_a(0) = 1$ False
4. $\log_a(1) = 0$ True
5. $\log_a(xy) = \log_a(x) + \log_a(y)$ True
6. $\log_a(x + y) = \log_a(x) + \log_a(y)$ False
7. $\log_a(x + y) = \log_a(x) \cdot \log_a(y)$ False
8. $\log_a(xy) = \log_a(x) \cdot \log_a(y)$ False
9. $\log_a[x^y] = [\log_a(x)]^y$ False
10. $\log_a[x^y] = y[\log_a(x)]$ True

Lesson Notes

- Seven properties of logarithms are introduced in this section and should be understood and memorized by the student. The first two properties are clear when transformed into exponents by the definition of the logarithm and should be remembered as they are encountered frequently in applications.

Lesson Notes, continued

- Properties 3 and 4 explain the inverse relationship of logarithms and exponents. If f and g are inverse functions, then $f(g(x)) = x = g(f(x))$. This is identical to the statement of properties 3 and 4. Students may need to be reminded of these facts until they are comfortable with these ideas.
- Properties 5 and 6 have counterparts in the exponential properties. The log of a product is the sum of the logs and the log of a quotient is the difference of the logs.
- The seventh property introduces a method for students to solve for the exponent in an equation. This property opens up an entire new class of problems that were inaccessible to the student previously.

■

Stumbling Block

The fifth and sixth properties of logarithms are sometimes confused. Students may say that $(\log x)(\log y) = \log x + \log y$. Have students state and write the property in words, "the log of a product is the sum of the logs." The expression $(\log x)(\log y)$ is a product of logs. This is a good time to go over the order of operations with logarithms. Remind students that logs are exponents and should be evaluated in the same order as exponents.

■

Example *Find* $\log(4.75 \times 10^5)$.

$$\log(4.75 \times 10^5) = \log 4.75 + \log(10^5) \approx 0.6767 + 5 \approx 5.6767$$

Thus, $\log(4.75 \times 10^5) \approx 5.6767$.

Now consider the following example, which will lead to our next property of logarithms.

Example *Let* $M = 10^9$ *and* $N = 10^3$. *Evaluate* $\log\left(\frac{M}{N}\right)$ *and* $\log M - \log N$.

$$\log\left(\frac{M}{N}\right) = \log\left(\frac{10^9}{10^3}\right) = \log(10^6) = 6$$

$$\log M - \log N = \log 10^9 - \log 10^3 = 9 - 3 = 6$$

This illustrates that the logarithm of the quotient of two numbers is equivalent to the difference of the logarithms of the two numbers.

Property 6: $\log_a\left(\frac{M}{N}\right) = \log_a M - \log_a N$

Proof of Property 6: Let $x = \log_a M$ and let $y = \log_a N$.
Then $a^x = M$ and $a^y = N$.
Therefore, $\frac{M}{N} = \frac{a^x}{a^y}$, or $\frac{M}{N} = a^{x-y}$.
Using the definition of a logarithm, we write

$$\log_a\left(\frac{M}{N}\right) = x - y, \text{ or } \log_a\left(\frac{M}{N}\right) = \log_a M - \log_a N$$

Example *Evaluate.*

a $\ln\left(\frac{2}{e^5}\right)$

$$\ln\left(\frac{2}{e^5}\right) = \ln 2 - \ln(e^5) \approx 0.6931 - 5 \approx -4.3069$$

b $\log 500 - \log 5$

$$\log 500 - \log 5 = \log\left(\frac{500}{5}\right) = \log 100 = 2$$

The final theorem presented in this section can be seen in the following example.

Example *Let* $M = 10^6$ *and* $p = 2$. *Evaluate* $\log(M^p)$ *and* $p \cdot \log M$.

$$\log(M^p) = \log(10^6)^2 = \log(10^{12}) = 12$$

$$p \cdot \log M = 2 \cdot \log(10^6) = 2 \cdot 6 = 12$$

This suggests that the logarithm of a number raised to a power is equal to the power times the logarithm of the number.

Property 7: $\log_a (M^p) = p \cdot \log_a M$

Proof of Property 7: Let $x = \log_a M$. Then $M = a^x$.
Therefore, $M^p = (a^x)^p$, or $M^p = a^{px}$.
Using the definition of a logarithm, we write

$\log_a (M^p) = px$, or $\log_a (M^p) = p \cdot \log_a M$

Example *Evaluate* $\log 4^5$.

$$\log 4^5 = 5 \cdot \log 4$$
$$\approx 5 \cdot (0.60206)$$
$$\approx 3.0103$$

Notice that since logarithms are exponents, each property of logarithms has a corresponding property of exponents. For example, if $x = \log_a M$ and $y = \log_a N$, the following are true.

Logarithmic Property	Exponent Property
$\log_a (M \cdot N) = \log_a M + \log_a N$	$a^x \cdot a^y = a^{x+y}$
$\log_a \left(\frac{M}{N}\right) = \log_a M - \log_a N$	$\frac{a^x}{a^y} = a^{x-y}$
$\log_a (M^p) = p \cdot \log_a M$	$(a^x)^y = a^{xy}$

Part Two: Sample Problems

Problem 1 *Let* $\log 2 \approx 0.3010$ *and* $\log 3 \approx 0.4771$. *Evaluate the following by using the properties instead of a calculator.*

a $\log 2^3$ **b** $\log 12$ **c** $\log \sqrt[3]{4}$ **d** $\log 50$

Solution

a $\log 2^3$
$= 3 \log 2$
$\approx 3(0.3010)$
≈ 0.9030

b $\log 12$
$= \log (2^2 \cdot 3)$
$= 2 \log 2 + \log 3$
$\approx 2(0.3010) + 0.4771$
≈ 1.0791

c $\log \sqrt[3]{4}$
$= \log (4)^{\frac{1}{3}}$
$= \log (2^2)^{\frac{1}{3}}$
$= \log (2^{\frac{2}{3}})$
$= \frac{2}{3} \log 2$
$\approx \frac{2}{3}(0.3010)$
≈ 0.2007

d $\log 50$
$= \log \frac{100}{2}$
$= \log 100 - \log 2$
$\approx 2 - 0.3010$
≈ 1.6990

11.3

Checkpoint

1 If $\log 5 \approx 0.6990$, find the following without using a calculator: log 25, log 50, log 125 and log 500.
1.3980, 1.6990, 2.0970 and 2.6990

2 Simplify the following:
$\log x + \log y - 2 \log z$
$\log \frac{xy}{z^2}$

3 Evaluate $2 \log_a a^3$. 6

Communicating Mathematics

Your best friend was home with a bad cold today. What notes would you have taken for her/him and how would you explain the importance of the seventh property to your friend?

Assignment Guide

Average

8, 9, 10, 12, 14, 15, 17–22, 24, 28, 30, 33, 38

Advanced

8–12, 14, 15, 17–23, 28–30, 33, 34, 36–39

Honors

8–12, 14, 15, 17–22, 28–30, 32, 34, 36, 37, 39–42

Problem-Set Notes and Strategies

- In problems **1 – 6** students may take a few minutes to figure out that if each number is transformed into an equivalent number using 2, 3 or 10, the logarithm may be found. This is good practice in the factoring of numbers.

Problem 2 *Write $2 \ln x + \ln (x + 2) - 3 \ln \pi$ as a single logarithm.*

Solution

$$\begin{aligned} &2 \ln x + \ln (x + 2) - 3 \ln \pi \\ &= \ln (x^2) + \ln (x + 2) - \ln (\pi^3) \\ &= \ln [x^2(x + 2)] - \ln (\pi^3) \\ &= \ln \left[\frac{x^2(x + 2)}{\pi^3}\right] \end{aligned}$$

Problem 3 *The formula for the depreciated value V of a car is $V = C(0.87)^t$, where C is the original price and t is the age of the car.*

a *What is the value of a five-year-old car that originally cost \$11,451?*
b *A car has a current blue-book value of \$4700, and the original price of the car was \$13,450. How old is the car?*

Solution

a $V = C(0.87)^t$; therefore
$$\begin{aligned} V &= 11{,}451(0.87)^5 \\ &\approx 11{,}451(0.4984) \\ &\approx \$5707 \end{aligned}$$
The approximate value of the car is \$5707.

b $V = C(0.87)^t$, therefore
$$\begin{aligned} 4700 &= 13{,}450(0.87)^t \\ 0.3494 &\approx (0.87)^t \\ \log 0.3494 &\approx \log (0.87)^t \\ \log 0.3494 &\approx t \cdot \log 0.87 \\ -0.4566 &\approx t(-0.0605) \\ 7.55 &\approx t \end{aligned}$$
The car is about $7\frac{1}{2}$ years old.

Note In part **b** above, we used the fact that if two quantities are equal, then the logarithms of these two quantities must also be equal.

Part Three: Exercises and Problems

Warm-up Exercises

In problems 1–6, use the values $\log 2 \approx 0.3010$ and $\log 3 \approx 0.4771$ to evaluate each expression without using a calculator.

P

1 $\log 6$ ≈0.7781
2 $\log 18$ ≈1.2552
3 $\log \frac{100}{6}$ ≈1.2219
4 $\log 4^3$ ≈1.8060
5 $\log \sqrt[4]{2}$ ≈0.07525
6 $\frac{\log 3}{\log 10}$ ≈0.4771

P

7 Match each lettered expression with an equivalent expression.

a $\log_3 (5 \cdot 7)$ iii
b $\log_3 5^7$ i
c $\log_3 \left(\frac{5}{7}\right)$ v
d $\log_3 3^{57}$ ii
e $\log_{357} 357$ iv

i $7 \cdot \log_3 5$
ii 57
iii $\log_3 5 + \log_3 7$
iv 1
v $\log_3 5 - \log_3 7$

Problem Set

P A **8** Find the exact value of each expression.

a $\log_2 96 - \log_2 3$ 5 **b** $\log_6 2 + \log_6 18$ 2 **c** $\log_6 6^5$ 5

P,R **9** The coordinates of A are (log 4, log 5). AB = log 20 and BC = log 250. Find the coordinates of C. (3, 2)

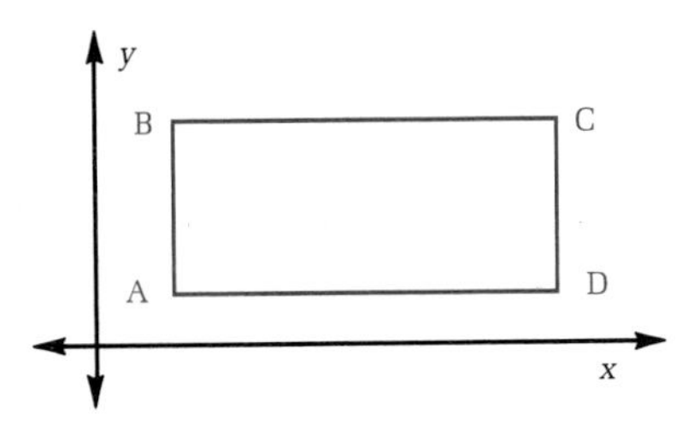

P **10** Write as a single logarithm.

a $\log_2 7 + \log_2 6 - \log_2 4$ $\log_2\left(\frac{21}{2}\right)$ **b** $3\log_4 5 - \frac{1}{2}\log_4 12$ $\log_4\left(\frac{125}{2\sqrt{3}}\right)$

c $2\log_4 (9) - 1$ $\log_4\left(\frac{81}{4}\right)$ **d** $4\log_2 3 + \log_2 9 - \log_2 12$ $\log_2\left(\frac{243}{4}\right)$

P,R **11** Arrange the following expressions in order from smallest to largest. **d, b, a = c**

a $\log (4 \cdot 3)$ **b** $\log 4 \cdot \log 3$ **c** $\log 4 + \log 3$ **d** $\log 4 - \log 3$

P **12** Evaluate $(\log 2 + \log 5)^6$. 1

P,R **In problems 13 and 14, write each expression as a sum and difference of logarithms.**

13 $\log_2 \frac{x^3\sqrt{x-3}}{y}$ $3\log_2 x + \frac{1}{2}\log_2 (x-3) - \log_2 y$

14 $\ln \frac{x}{x^5(x-7)}$ $-4\ln x - \ln(x-7)$

P **15** Write $\frac{1}{2}(3\ln x - \ln z)$ as a single logarithm. $\ln\left(\sqrt{\frac{x^3}{z}}\right)$

R,I **16** If $A = P\left(1 + \frac{r}{n}\right)^{nt}$ and $A = \$1{,}000{,}000$, $P = \$35{,}000$, $n = 4$, and $r = 8.5\%$, what is t? $t \approx 39.9$ yr

P **In problems 17–22, evaluate each expression.**

17 $\log_2 (2^5)$ 5 **18** $(\log_2 2)^5$ 1 **19** $\log_3\left(\frac{243}{27}\right)$ 2

20 $\frac{\log_3 243}{\log_3 27}$ $\frac{5}{3}$ **21** $\log_2 (8^5)$ 15 **22** $(\log_2 8)^5$ 243

P,R **23** Find the average of $\log_4 8$, $\log_4 2$, and $\log_4 4$. 1

P,R,I **24** If two of the following expressions are selected at random, what is the probability that they will be equal? 1

log 4 + log 16 6 log 2 log 128 − log 2
3 log 4 2 log 8 log 64

Problem-Set Notes and Strategies, continued

- Problems **10**, **12**, **13**, **14**, and **15** indicate whether a student understands the properties of logarithms.
- In problem **16**, the variable to solve for is in the exponent. Other methods such as guess and check will eventually yield the answer, but property 7 will simplify the work.
- Problem **24** illustrates that there are many ways to identify a specific number when using logarithms. One of these may be more helpful than the others when solving a problem. It is useful to know how to transform one into another.

11.3

Problem-Set Notes and Strategies, continued

- Remind students that in problem **26**, $(-2)^4 \neq -2^4$.

- Problem **30** is an example of the application problems that will be encountered in Section **11.5**. This one is a population growth example. The value of P_0 is the initial population.

- Problem **31** identifies a method for changing the base of a logarithm into natural logarithms. This method is explored in more detail in the next section and will allow logarithms of any base to be solved.

- Problems **32** and **34** are a nice application of the properties of logarithms in the general case.

Problem Set, *continued*

P,R **B**

25 Multiply.

$$\begin{bmatrix} \log 2 & \log 3 \\ \log 5 & \log 2 \end{bmatrix} \cdot \begin{bmatrix} 2 \\ 3 \end{bmatrix} \quad \begin{bmatrix} \log 108 \\ \log 200 \end{bmatrix}$$

R **26** Simplify the following expressions.

a $\dfrac{(-2)^4}{-2^3}$ -2 **b** $\dfrac{-x^{60}}{x^2 \cdot x^{15}}$ $-x^{43}$ **c** $\dfrac{(x^2)^3 (x^2 \cdot x^4)}{x^{20}}$ x^{-8}

P,R **27** Evaluate.

a $\log_2 64$ 6 **b** $\log_2 8$ 3 **c** $\log_8 64$ 2

d Is $(\log_2 8)(\log_8 64) = \log_2 64$ a true statement? Yes

P,R **28** Find the area of the shaded region.
16

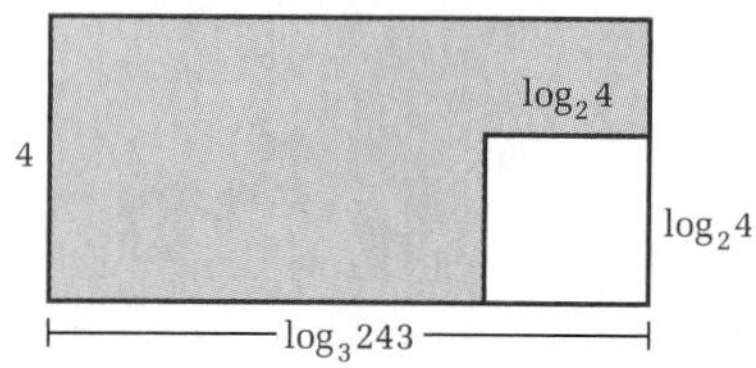

R **29** Let $f(x) = \log x$ and $g(x) = 10^x$.

a Find $f[g(x)]$. x

b Find $g[f(x)]$. x

c Is $f[g(x)] = g[f(x)]$? (Hint: Think about domains.)
No. $g[f(x)]$ has as its domain positive reals. $f[g(x)]$ has all reals as its domain.

R,I **30** The population P of a town is growing exponentially according to the equation $P = P_0 e^{\frac{t}{30}}$.

a If at $t = 0$ there are 10,000 people in the town, what is P_0? $P_0 = 10{,}000$

b How many people are in the town when $t = 15$? $P \approx 16{,}487$ people

c When will there be 35,000 people in the town? When $t \approx 37.6$

I **31** $y = \log_a b$ is equivalent to $a^y = b$. Use the equation $a^y = b$ and natural logarithms to express y in terms of the quotient of two natural logarithms. $y = \frac{\ln b}{\ln a}$

P **32** Let $\log 2 = c$ and $\log 3 = d$. Find the following values in terms of c and d.

a $\log 12$ $2c + d$ **b** $\log \frac{1}{8}$ $-3c$ **c** $\log 81$ $4d$ **d** $\log \frac{4}{27}$ $2c - 3d$

P **33** Write the following expression as a single logarithm.
$\ln \sqrt{x^2 - 1} - \ln \sqrt{x + 1}$ $\frac{1}{2} \ln (x - 1)$

P **34** Suppose $\log_2 3 = a$ and $\log_2 5 = b$. Find each of the following values in terms of a and b.

a $\log_2 15$ $a + b$ **b** $\log_2 6$ $a + 1$ **c** $\log_2 2.5$ $b - 1$

P **35** Evaluate each expression to three decimal places.

a $\log(2 \times 10^5)$ ≈ 5.301 **b** $\log(2 \times 10^{38})$ ≈ 38.301
c $\log(2 \times 10^7)$ ≈ 7.301 **d** $\log(2 \times 10^{472})$ ≈ 472.301

36 The graph of the function $f(x) = \log_6 x$ is shown.

Find the area of the trapezoid ABCD. ≈ 132.44

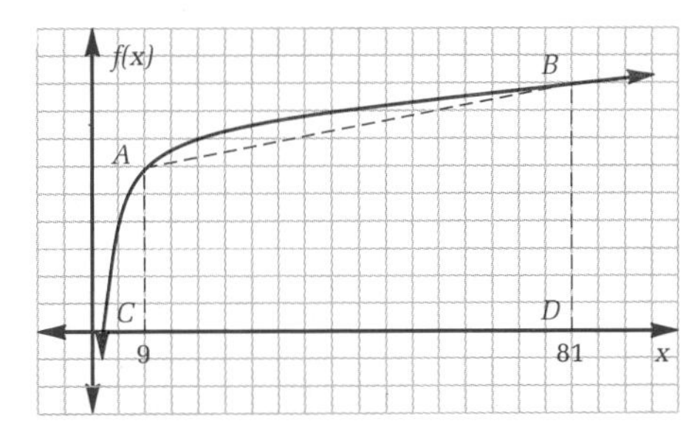

P,R **37** Write the following expression as a single logarithm.

$\ln\left(\frac{x}{x^2 - 1}\right) + \ln\left(\frac{x^2 + 2x + 1}{3x}\right)$ $\ln\left[\frac{x + 1}{3(x - 1)}\right]$

I **38** Solve $12 = 5^{(x+3)}$ for x. $x \approx -1.456$

R,I **39** Boris and Emil each inherited \$8000 from Uncle Horace. Boris deposited his inheritance in a bank that paid 6.85% interest, compounded continuously, whereas Emil deposited his in a bank that paid 7.6% interest, compounded continuously. How long will it be until Emil has twice as much money as Boris? How long would it be if Uncle Horace had given them \$4000 instead?
$t \approx 92.4$ yr; $t \approx 92.4$ yr

R **40** Sketch each function on the same coordinate system.

a $f(x) = \log_2 x + \log_2 8$ **b** $f(x) = \log_2(x + 8)$ **c** $f(x) = \log_2 8x$

P **C** **41** Evaluate $\log\frac{1}{2} + \log\frac{2}{3} + \log\frac{3}{4} + \log\frac{4}{5} + \ldots + \log\frac{9}{10}$. -1

P,R **42** Simplify. 8
$(\log 4)^3 + 3(\log 4)^2(\log 25) + 3(\log 4)(\log 25)^2 + (\log 25)^3$

I **43** Determine whether each statement is True or False.

a $\log_5 x + \log_8 x = \log_{13} x$ False **b** $\log_5 x + \log_8 x = \log_{40} x$ False
c $\log_5 x + \log_8 x = \frac{\ln 5 + \ln 8}{(\ln 5)(\ln 8)} \cdot \ln x$ True

R,I **44** One measure of the quality of an egg is the Haugh unit. An egg's Haugh rating is given by

$$100 \log\left[H - \frac{1}{100}\sqrt{32.2}\,(30w^{0.37} - 100) + 1.9\right]$$

where w is the weight of the egg in grams and H is the height of the albumen in millimeters when the egg is broken on a flat surface. Find the Haugh rating of an egg that weighs 48.7 grams if $H = 5$ millimeters. ≈ 73.3

Problem-Set Notes and Strategies, continued

- Students may likely notice some similarity in the answers to problem **35**. Have them then use the properties of logarithms to confirm their results.
- Problems **38**, **39**, and **44** are applications of the properties of logarithms in solving problems. Many more applications will be explored in Section **11.5**.
- Problem **41** may be solved with a calculator and confirmed using the properties of logs. Ask students which was easier and quicker to evaluate.
- Problem **43c** requires a change of base using the result from problem **31** and several properties including the arithmetic of fractions. Students may try to confirm this on the calculator. Remind them that the calculator gives only approximate answers for logarithms and this is insufficient to prove correctness.

Cooperative Learning

Problem **43c** is difficult enough for a student to solve that it might be a good problem to work on in groups. Remind students that they will need the result of problem **31** to be able to change the base into natural logarithms. They will also have to manipulate the fractions that result in order to produce the final result. After they verify this identity, you might have them try to find the equivalent form of $\log_5 x - \log_8 x$.

Additional Answer

Answer for problem **40** can be found in the answer pages beginning on page **A** 1.

11.4

Class Planning

Time Schedule
All levels: 1 day

Resource References
Teacher's Resource Book
Class Opener 68
Practice 72
Enrichment 21

Class Opener

Find the smallest x so that
$3^x \geq 1{,}000{,}000$
$x \approx 12.57541965$

Lesson Notes

- This section identifies the techniques used to solve both exponential and logarithmic equations. The change of base formula is presented to allow any logarithm to be transformed into a common or natural logarithm and then be evaluated on a calculator or with a log table.

11.4 Solving Exponential and Logarithmic Equations

Objectives

After studying this section, you will be able to

- Use the base-change formula
- Solve exponential equations
- Solve logarithmic equations

Part One: Introduction

The Base-Change Formula

As seen earlier, a calculator can be used to compute logarithms and powers when the base is either 10 or *e*. We can also calculate powers of any positive base by using the [xʸ] key.

Example *Use a calculator to find* $2^{4.53}$.

Enter 2 [xʸ] 4.53 [=]. The display is [23.10286713]
Therefore, $2^{4.53} \approx 23.10$.

How can we use the calculator to find logarithms with bases other than 10 or *e*? Consider the following example.

Example *Evaluate.*

a $\log_2 23.10$
From the example above, we know that $\log_2 23.10 \approx 4.53$.

b $\dfrac{\log 23.10}{\log 2}$

$$\frac{\log 23.10}{\log 2} \approx \frac{1.36361198}{0.301029995} \approx 4.529820957$$

c $\dfrac{\ln 23.10}{\ln 2}$

$$\frac{\ln 23.10}{\ln 2} \approx \frac{3.139832617}{0.69314718} \approx 4.529820949$$

All three of these answers are the same. This is not just a coincidence. This example illustrates another property, known as the base-change formula.

Vocabulary
base-change formula
exponential equation

Property 8:
The Base-Change Formula

$$\log_a M = \frac{\log_b M}{\log_b a}, \text{ where } b \text{ can be any legitimate base.}$$

Proof of Property 8: Let $x = \log_a M$. This is equivalent to

$$a^x = M$$

Take $\log_b$ of both sides $\quad \log_b a^x = \log_b M$

Use Property 7 $\quad x \cdot \log_b a = \log_b M$

$$x = \frac{\log_b M}{\log_b a}$$

Now substitute for x $\quad \log_a M = \dfrac{\log_b M}{\log_b a}$

Example *Evaluate.*

a $\log_5 36$

$$\log_5 36 = \frac{\log 36}{\log 5} \approx 2.2266$$

b $\log_{\frac{1}{2}} 0.007$

$$\log_{\frac{1}{2}} 0.007 = \frac{\ln 0.007}{\ln 0.5} \approx 7.1584$$

Use $\boxed{x^y}$ to check:

$5^{2.2266} \approx 36.002$ $\qquad$ $0.5^{7.1584} \approx 0.007$

Note We used base 10 and base *e* because we could then use a calculator to calculate the logarithms. It doesn't matter which of these two bases you use. Try each problem using alternate bases.

Solving Exponential Equations

An ***exponential equation*** is any equation with a variable in an exponent. To solve an exponential equation we usually take the logarithm, to an appropriate base, of both sides of the equation.

Example *Solve* $5^x = 12$ *for* x.

$$5^x = 12$$

Take the base-5 log of both sides $\quad \log_5 (5^x) = \log_5 12$

$$x \log_5 5 = \log_5 12$$

Base-change formula $\quad x = \dfrac{\log 12}{\log 5} \approx 1.54396$

Check
$5^x \approx 5^{1.54396} \approx 12.00001331 \approx 12$. For a closer approximation, use the value of x that is stored in the calculator after dividing log 12 by log 5.

Lesson Notes, continued

- Exponential equations are solved by simplifying all the exponents until there is a single exponential expression and finally taking logarithms of both sides to solve for the exponent.

Communicating Mathematics

Explain in your own words why the base-change formula works. An example might be: It makes no difference what base logarithm is used. The value of this fraction, $\dfrac{\log_a 100}{\log_a 25}$, will always be the same.

Example *Solve $4e^{3x} = 100$ for x.*

$$4e^{3x} = 100$$
$$e^{3x} = 25$$
Take ln of both sides
$$\ln (e^{3x}) = \ln 25$$
$$3x = \ln 25$$
$$3x \approx 3.2189$$
$$x \approx 1.0730$$

Note Here it makes sense to use base e logarithms, since the exponential equation involves the number e.

Check
$4e^{3x} \approx 4e^{3 \cdot 1.0730} \approx 4e^{3.219} \approx 100.0124183 \approx 100$

Solving Logarithmic Equations

We solve a logarithmic equation by using the properties of logarithms to obtain a form where we can eliminate the logarithms. Check your answers. Remember that we can find logarithms only of positive numbers and that the base must be positive and not 1.

Example *Solve $\log_4 x + \log_4 (x + 6) = 2$ for x.*

$$\log_4 x + \log_4 (x + 6) = 2$$
$$\log_4 [x(x + 6)] = 2$$
Change to exponential form
$$x(x + 6) = 4^2$$
$$x^2 + 6x = 16$$
$$x^2 + 6x - 16 = 0$$
$$(x + 8)(x - 2) = 0$$
$$x + 8 = 0 \text{ or } x - 2 = 0$$
$$x = -8 \text{ or } x = 2$$

$x = -8$ is not a solution of the original logarithmic equation, since we cannot find logarithms of negative numbers. The only solution is $x = 2$. You should check it yourself.

Example *Solve $\ln P - \ln P_0 = kt$ for P.*

$$\ln P - \ln P_0 = kt$$
$$\ln \frac{P}{P_0} = kt$$
Write in equivalent exponential form
$$\frac{P}{P_0} = e^{kt}$$
$$P = P_0e^{kt}$$

Note This is an important formula in science. It describes such things as population growth and radioactive decay.

Lesson Notes, continued

- Logarithmic equations are solved by first simplifying all logarithms into single logarithmic expressions and then employing the definition of the logarithm to eliminate the log and find the answer. All of these methods require first simplifying to obtain a single exponential or logarithmic expression which may then be solved.

Part Two: Sample Problems

Problem 1 *Solve* $4^{x^2} \cdot 4^{-2x} = 7$ *for* x.

Solution We first use the properties of exponents.

$$4^{x^2} \cdot 4^{-2x} = 7$$

$$4^{x^2 - 2x} = 7$$

Take $\log_4$ of both sides $\quad \log_4 (4^{x^2 - 2x}) = \log_4 7$

Base-change formula $\quad x^2 - 2x = \frac{\log 7}{\log 4}$

$$x^2 - 2x \approx 1.404$$

$$x^2 - 2x - 1.404 \approx 0$$

Quadratic formula $\quad x \approx \frac{2 \pm \sqrt{4 + 5.615}}{2}$

$$x \approx 2.550 \text{ or } x \approx -0.550$$

Checking each solution gives approximately $4^{1.4025} \approx 6.989$.

Problem 2 *Solve* $(\log x)^2 + \log (x^3) - 12 = 0$ *for* x.

Solution $(\log x)^2 + \log (x^3) - 12 = 0$
$(\log x)^2 + 3 \log x - 12 = 0$

This is a quadratic equation in log x. Using the quadratic formula, we solve for log x and obtain $\log x \approx 2.28$ or $\log x \approx -5.28$.

$x \approx 10^{2.28}$ or $x \approx 10^{-5.28}$
$x \approx 190.55$ or $x \approx 5.25 \times 10^{-6}$

Problem 3 *Solve* $\log_5 x = \log_5 (x + 5) - \log_5 2$ *for* x.

Solution We first use properties of logarithms.

$$\log_5 x = \log_5 (x + 5) - \log_5 2$$

$$\log_5 x = \log_5 \left(\frac{x + 5}{2}\right)$$

Since the logarithms are equal, the quantities must be equal.

$$x = \frac{x + 5}{2}$$

$$2x = x + 5$$

$$x = 5$$

You should check to see whether the answer is correct.

Using Computers

Exponential and logarithmic equations often can be solved by using graphing software or a graphing calculator. The base change formula becomes important because most computer programs and graphing software are only capable of performing operations on base-10 and base-*e* logarithms. Learning to change the base in order to use the software is a good experience in itself.

Checkpoint

1 Solve the equation $3^x = 15$ for x. ≈ 2.4650
2 Solve the equation $3 \log x = 10$ for x. ≈ 2154.43
3 Solve the equation $7^{3x-2} = 100$ for x. ≈ 1.4555
4 Solve the equation $\log (x - 3) + \log x = 1$ for x. 5

Assignment Guide

Average
9, 10, 12, 15, 19, 21–23, 24, 27, 29, 31, 32, 35
Advanced
9, 10, 12, 15, 19, 21–23, 24, 27, 28a, 31, 32, 33c–f, 37
Honors
11, 12, 19, 21, 23, 24, 26, 27, 29, 31, 33, 35–38

Problem-Set Notes and Strategies

- Problems **1 – 5** are applications of the properties of exponents and logarithms.
- In problem **8**, a picture may help students to find the solution.
- Problems **11**, **12**, and **17** require factoring and properties of logarithms. Remind students of the domain of the logarithm function so that they do not get too many answers.
- The solution in **13** may not be clear until students discover that the second logarithm should be changed to base 2.

Part Three: Exercises and Problems

Warm-up Exercises

In problems 1–5, solve for x. Do not use a calculator.

P
1 $3^{2x-1} = 9$ $x = \frac{3}{2}$
2 $4^x \cdot 2^{-3} = 32$ $x = 4$
3 $\log_3 (x + 2) = 2$ $x = 7$
4 $(\ln x)^2 - \ln x = 0$ $x = 1$ or $x = e$
5 $4^{x^3} = 2^{16}$ $x = 2$

P,R
6 Show that the two lines are parallel. $\log_2 13 = \frac{\log_5 13}{\log_5 2}$

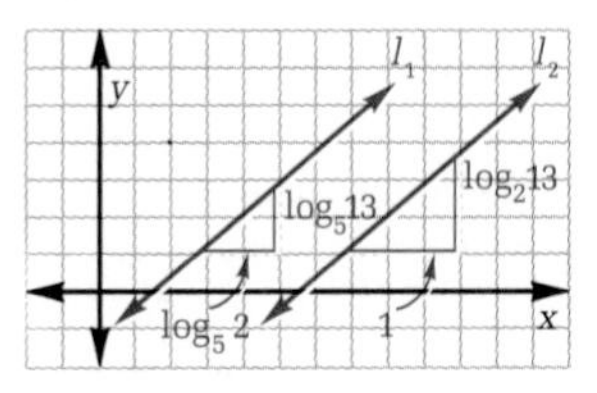

P,I 
7 The number of bacteria b on a petri dish is given by $b = 500e^{0.04t}$, where t is time measured in minutes.
a How many bacteria are there at $t = 0$? 500
b How many bacteria will there be when $t = 30$ minutes? ≈ 1660

P
8 A square with side 2^x has area 512. What is the area of a square with side x? 20.25

Problem Set

A

9 Use a calculator to find the values of the following expressions.
a $\log_2 28$ ≈ 4.8074
b $\log_{\frac{1}{2}} 36$ ≈ -5.1699
c $\log_5 130$ ≈ 3.0244

P,R
10 Solve for b.
a $3^b = 19$ $b \approx 2.6801$
b $7^b = \frac{3}{5}$ $b \approx -0.2625$
c $\log b = \frac{2}{5}$ $b \approx 2.5119$

P
In problems 11 and 12, solve for x. Do not use a calculator.
11 $\ln (x + 5) - \ln x = \ln \left(\frac{4}{3}\right)$ $x = 15$
12 $\log_3 x + \log_3 (x - 2) = 1$ $x = 3$

P
13 Evaluate $\log_2 3 \cdot \log_3 64$. 6

P
14 Prove that $\log_{a^n} x = \frac{\log_a x}{n}$.

P,R
15 Find the area of the rectangle. $\frac{3}{4}$

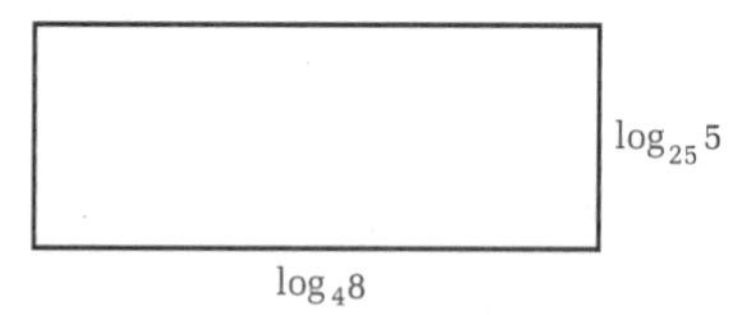

P
16 Solve for x.
a $2^x = \log 12$ $x \approx 0.1099$
b $\log_7 x = \log 125$ $x \approx 59.1690$

Additional Answer

Answer for problem **14** can be found in the answer pages beginning on page A 1.

P | **17** Solve $\log_5 x + \log_5 (x + 20) = 3$ for x. $x = 5$

R | **18** Evaluate each of the following expressions.

a $\log_2 4^5$ 10 **b** $\log_3 (81 \cdot 27)$ 7 **c** $\log_5 \left(\frac{625}{25}\right)$ 2

B

P | **19** **a** Solve $\log_7 x = \log_7 (x + 9) - \log_7 3$ for x. $x = 4.5$

b Solve $\log_7 x = \log_7 (x + a) - \log_7 3$ for x in terms of a. $x = \frac{a}{2}$, $a > 0$

R | **20** Simplify $(\log 5 + \log 2)(\log 20 - \log 2)$. 1

P | **21** Write in simple form.

a $\log_2 3 \cdot \log_3 4$ 2

b $\log_2 3 \cdot \log_3 4 \cdot \log_4 5 \cdot \log_5 6 \cdot \log_6 7 \cdot \log_7 8$ 3

P | **22** Place in order from smallest to largest.

$\log_6 21$ $\log_{10} 55$ $\log_4 10$ $\log_5 15$

$\log_4 10$, $\log_5 15$, $\log_6 21$, $\log_{10} 55$

P,R | **23** Find the point at which the graphs intersect.

$(\approx 1.54, 12)$

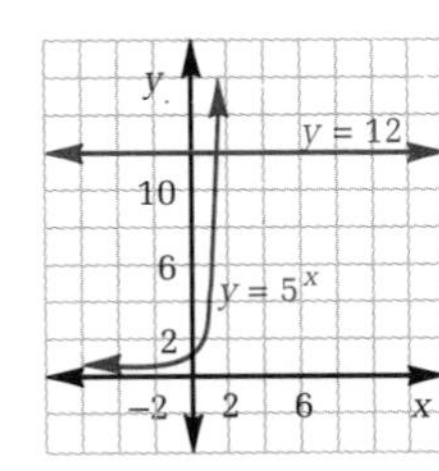

P,I | **24** The base sticker price of an automobile rises at about 6% per year. The formula $B = B_0(1.06)^t$ relates the base sticker price B to time t in years.

a At $t = 0$ (1950) the base sticker price was \$1500. Find B_0. $B_0 = \$1500$

b What would the base sticker price of the same 1950 car have been in 1970? $B \approx \$4810.70$

P,R | **25** **a** Solve $\log (\log x) = 3$ for x. $x = 10^{1000}$

b How many digits does x have? 1001

P,R | **26** Solve $\log_{x^2+6} 2 + \log_{x^2+6} 5 = 1$ for x. $x = 2$ or $x = -2$

P,R | **27** Solve $5^{x^2} \cdot 5^{-9x} = 5^{-18}$ for x. $x = 6$ or $x = 3$

R | **28** Write as a sum or difference of logarithms.

a $\ln \frac{x^2y}{x + 5}$ $2 \ln x + \ln y - \ln (x + 5)$ **b** $\ln \sqrt{\frac{x^2}{y}}$ $\ln |x| - \frac{1}{2} \ln y$

P | **29** Solve $x = \log_4 9 \cdot \log_9 12 \cdot \log_{12} 16$ for x. $x = 2$

P | **30** Solve each of the following equations for x.

a $6^x = 12$ ≈ 1.387 **b** $5e^{3x} = 10$ ≈ 0.231 **c** $16 \cdot 10^{x^2} = 144$ $\approx \pm 0.977$

Problem-Set Notes and Strategies, continued

- Problem **18** can be factored and expressed in terms of the logarithms base.
- In problem **21**, suggest students look for a pattern. The solution may appear without performing all the computations.
- Exponential growth is used in solving problem **24**. These will be explored in the next section.
- If problem **21** is understood, problem **29** should be clear. Otherwise, the computations must be carried out.

Problem-Set Notes and Strategies, continued

- Remind students to consider the domain when working problems like **32**.
- Problem **35** involves quadratic equations containing logarithms. It is to be hoped that students consider factoring and remember the domain of logarithmic functions when finding the solutions.
- In problem **37**, students can use the techniques of solving radical equations and properties of logarithms.

Problem Set, *continued*

P,I **31** The formula $P = P_0e^{kt}$ can be used to predict the number of grams P of a radioactive element remaining after t hours. P_0 is a constant and is the number of grams of the element at $t = 0$. k is a constant specific to a given radioactive element. Consider the following data.

t	0	1	2	3	4	...
P	5	≈3.533	2.5	≈1.7678	1.25	...

a What is P_0? 5
b Find k. ≈−0.3466
c How much of the element remains after nine hours? ≈0.2210 grams

P **32** Solve $\log_9 x + \log_9 (x + 6) = \frac{3}{2}$ for x. x = 3

R,I **33** The function f is defined so that $f(0) = 1$ and $f(n + 1) = f(n) \cdot 3$ for every integer n.

a Find $f(2)$. 9
b If $f(n) = 19{,}683$, what is n? 9
c Find $f(3)$. 27
d Find $f(-3)$. $\frac{1}{27}$
e Find $f(4)$. 81
f If $f(n) = 30{,}000$, what is n? Impossible. f is defined for integers only.

P,R **34** Solve $\log_2 [\log_3 (\log_4 x)] = 0$ for x. x = 64

P,R **35** Solve for x.

a $(\log_4 x)^2 = 16$
$x = 256$ or $x = 4^{-4} \approx 0.003906$
b $(\log_4 x)^2 - 15 \log_4 x = 16$
$x = 4^{16} = 4{,}294{,}967{,}296$ or $x = 0.25$

P,R **36** The area of the trapezoid is 12 square units. Find b. b = 2

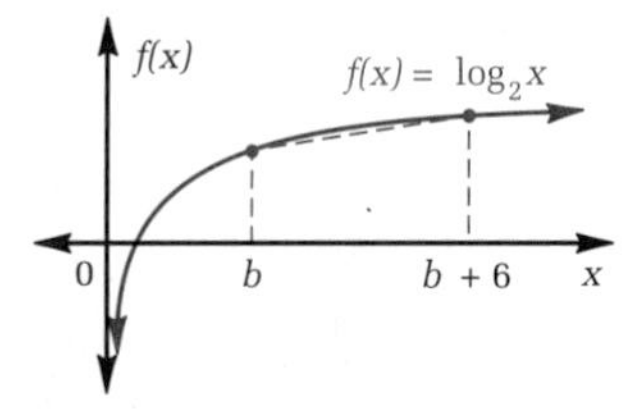

P C **37** Solve $\sqrt{\log x} = \log \sqrt{x}$ for x. $x = 10^4$ or $x = 1$

R **38** The graphs of $y = e^x$ and $y = e^{-x}$ are pictured. Note that they intersect at (0, 1).

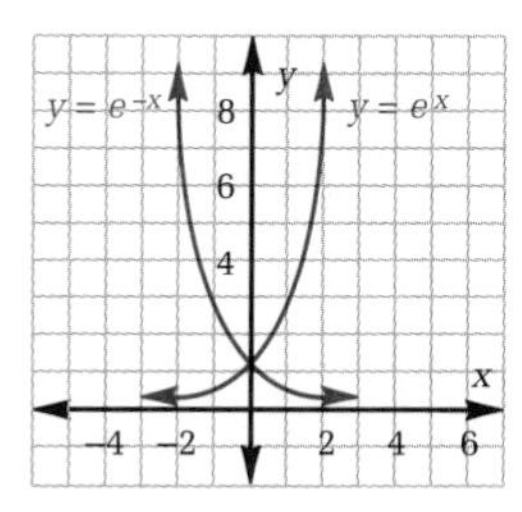

a If the curve $y = e^x$ is shifted three units to the left and the curve $y = e^{-x}$ is shifted three units to the right, where do the curves intersect? $(0, e^3)$
b If the curve $y = e^x$ is shifted three units to the left and the curve $y = e^{-x}$ is shifted five units to the right, where do the curves intersect? $(1, e^4)$

11.5

Applications of Exponential and Logarithmic Functions

Objectives

After studying this section, you will be able to

- Use exponential and logarithmic functions as mathematical models
- Analyze exponentially related data

Part One: Introduction

Applications as Mathematical Models

In many real-world situations the mathematical model that best describes the relationship between two quantities is an exponential or logarithmic function.

Example *An earthquake's magnitude, as measured by the Richter energy number R, is the base-10 logarithm of the intensity I of the quake's vibrations. The equation relating the Richter number R and the intensity of the quake I is* $R = \log\left(\frac{I}{I_0}\right)$, *where* I_0 *is the minimum detectable intensity of an earthquake.*

a *Write the Richter earthquake equation in exponential form.*

$R = \log\left(\frac{I}{I_0}\right)$ is equivalent to $10^R = \left(\frac{I}{I_0}\right)$, so $I = 10^R I_0$.

b *The atomic bomb tested on Bikini Atoll had a Richter number of 5.0, and the San Francisco earthquake of 1906 had a Richter number of 8.25. How many times more intense were the vibrations of the San Francisco quake than those at Bikini?*

$I(\text{Bikini}) = 10^5 I_0$ and $I(\text{San Fran.}) = 10^{8.25} I_0$

$$\frac{I(\text{San Fran.})}{I(\text{Bikini})} = \frac{10^{8.25} I_0}{10^5 I_0} = 10^{3.25} \approx 1778$$

Therefore, $I(\text{San Fran.}) \approx 1778 \cdot I(\text{Bikini})$. The San Francisco earthquake was about 1778 times as intense as the atomic bomb vibrations at Bikini Atoll.

11.5

Class Planning

Time Schedule

All levels: 1 day

Resource References

Teacher's Resource Book

Class Opener 69
Practice 73
Enrichment 22
Transparency 22

Class Opener

Solve for k. $100 = 18e^{4k}$
$k \approx 0.4287$

Lesson Notes

- This section elaborates on several of the applications of exponential and logarithmic equations. An equation where the rate of growth or decay is proportional to the amount present is an exponential equation of the form $A = P_0e^{kt}$. This equation holds for many significant applications.
- Another application of the natural exponential function is that all exponential expressions of the form a^{rt} can be written as e^{kt} with the appropriate choice of k. This means that all exponential equations and problems may be written as base e just as all logarithmic equations may be changed into natural logarithms using the change of base formula.

Communicating Mathematics

Have students write a paragraph that explains the concept of exponential growth.

Example *Population growth is very often exponentially related to time. That is, the population P is given by the equation $P = P_0e^{kt}$, where P_0 and k are constants. The prairie dog population of Dry Gulch was 24 in 1975 and 90 in 1980.*

a *From the given data, determine P_0 and k.*

Let $t = 0$ in 1975. Then $24 = P_0e^{k(0)}$

$$24 = P_0$$

In 1980, 5 years later, there were 90 prairie dogs, so

$$\begin{aligned} 90 &= 24e^{k(5)} \\ 3.75 &= e^{5k} \\ \ln 3.75 &= \ln \left(e^{5k}\right) \\ 1.322 &\approx 5k \\ 0.2644 &\approx k \end{aligned}$$

The prairie dog population equation is $P \approx 24e^{0.2644t}$.

b *Assuming continued exponential growth of the prairie dog population, what is the expected population of prairie dogs in Dry Gulch in the year 2000?*

Since $t = 0$ in 1975, in the year 2000, $t = 25$.

$$\begin{aligned} P &\approx 24e^{0.2644(25)} \\ P &\approx 24 \cdot 742.48 \\ P &\approx 17{,}820 \end{aligned}$$

There will be about 17,820 prairie dogs in the year 2000.

Analyzing Exponentially Related Data

We have seen several formulas of the form $y = ka^{mx}$. When such a formula occurs, we say that x and y are exponentially related. If we let $a = 10$, so that we have $y = k10^{mx}$, and take the log of both sides, an interesting thing happens.

$$\begin{aligned} y &= k10^{mx} \\ \log y &= \log (k10^{mx}) \\ \log y &= \log k + \log (10^{mx}) \\ \log y &= \log k + mx \end{aligned}$$

If we let $Y = \log y$ and $b = \log k$, we obtain

$$Y = mx + b, \text{ a linear equation!}$$

This means that the ordered pairs (x, y) are exponentially related by $y = k10^{mx}$ if and only if the ordered pairs (x, log y) are linearly related by $\log y = mx + b$. Graphically, this says that the pairs (x, y) form an exponential curve if and only if the pairs (x, log y) form a straight line.

See p. 798

Example *Are the data in the table exponentially related?*

x	10	20	30	40	50	60
y	910	390	170	72	31	14

We will make a table of x and log y and then graph the ordered pairs.

x	log y
10	≈2.96
20	≈2.59
30	≈2.23
40	≈1.86
50	≈1.49
60	≈1.15

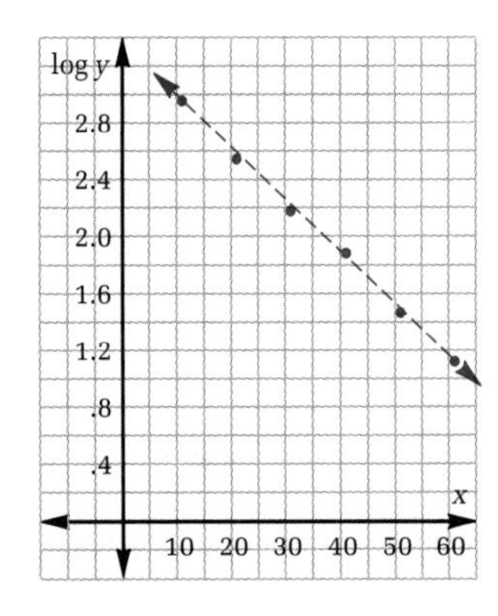

Since the graph of the ordered pairs (x, log y) is approximately a straight line, the original data are exponentially related.

Using Computers

A variety of types of software and calculator functions are appropriate for use in this section. Graphing software or graphing calculators can be used in connection with some of the problems, and symbolic manipulation software for others. A spreadsheet set up to do linear regression, or a graphing calculator capable of linear regression, can be used for problems involving exponentially-related data. *Quattro* and *Derive* both can do linear regression. In addition, *Derive* can do scatter plots. Most graphing calculators can do both scatter plots and linear regression.

Using technology in this section can contribute much to students' ability to find exponential equations that represent so many real-world phenomena.

Part Two: Sample Problems

Problem 1 *At 8:00 A.M. on a spring day, with the temperature constant at 20°C, Sherlock Holmes found a corpse lying in an alley. To determine the time of death, Holmes had Watson measure the body temperature immediately and again one hour later. The body temperature at 8:00 A.M. was 32°C, and the temperature at 9:00 A.M. was 31°C. Determine the approximate time of death.*

Solution An equation known as Newton's law of cooling enables us to determine the time of death.

The equation is $\ln \frac{T - A}{T_0 - A} = -kt$, where

T = the temperature of the body at any time after death
T_0 = the temperature of the body at time of death. In our problem this is normal body temperature, 37°C.
A = the temperature of the surrounding medium, known as ambient temperature. In our problem it is 20°C.
k = a constant that corresponds to the human body's cooling rate
t = the length of time that the body has been cooling

At the time of death, $t = 0$, $T_0 = 37°C$, and $A = 20°C$.
At 8:00 A.M. let time = t.

$$T = 32°C, \text{ so} \quad \ln \frac{32 - 20}{37 - 20} = -kt$$

$$-0.3483 \approx -kt \qquad (1)$$

Checkpoint

1. The 1906 earthquake in San Francisco had a Richter reading of 8.25. The San Francisco earthquake of 1989 had a Richter reading of 7.1 What was the difference in the magnitude of the two earthquakes?
   The 1906 quake was approximately 14 times as intense as the 1989 quake.
2. If the amount of radioactive material of a substance decreases exponentially and there was 100 grams 6 months ago and only 30 grams left now, what is the half life of the substance?
   ≈ 3.45 months
3. Rabbits have a tendency to reproduce quickly since they have a relatively short gestation period. If the state of Arizona had 10,000 rabbits in 1975 and 15,000 rabbits in 1980, how many rabbits will be in the state of Arizona in the year 2025? ≈ 576,650

At 9:00 A.M. time $= t + 1$.

$$T = 31°C, \text{ so } \quad \ln \frac{31 - 20}{17} = -k(t + 1)$$

$$-0.4353 \approx -k(t + 1) \qquad (2)$$

Dividing equation (2) by equation (1) yields

$$\frac{-0.4353}{-0.3483} \approx \frac{-k(t + 1)}{-kt}, \qquad \text{or } 1.25 \approx \frac{t + 1}{t}$$

$$1.25t \approx t + 1$$
$$0.25t \approx 1$$
$$t \approx 4$$

At 8:00 A.M. the person had been dead for about four hours. The time of death was approximately 4:00 A.M.

See p. 798

Problem 2 *The following table shows the height h, in inches, of a plant as a function of w weeks.*

w	1	2	3	4	5	6	7	8	9	10
h	0.5	0.7	1.1	1.5	2.4	3.5	5.0	7.4	10.0	17.0

a *Find an exponential function to fit the data.*
b *Predict the height of the plant by week 12.*

Solution

a Let's assume an equation of the form $h = k10^{mw}$ and take the log of both sides to obtain $\log h = \log k + mw$. Using the first and last columns of data, we have the following linear system.

Using $(w, h) = (10, 17)$ $\quad$ $\mathbf{1.2304 \approx \log k + 10m}$

Using $(w, h) = (1, 0.5)$ $\quad$ $\mathbf{-0.3010 \approx \log k + 1m}$

Subtract $\quad$ $\mathbf{1.5314 \approx 9m}$

$\mathbf{0.1702 \approx m}$

Substituting the value of m into the second equation yields

$$-0.3010 \approx \log k + 0.1702$$
$$-0.4712 \approx \log k$$
$$10^{-0.4712} \approx k$$
$$0.3379 \approx k$$

The exponential function for the data is $h \approx 0.3379 \cdot 10^{0.1702w}$. Let's see whether this equation does approximately fit the data by checking $(w, h) = (6, 3.5)$.

$$h \approx 0.3379 \cdot 10^{0.1702(6)}$$
$$\approx 0.3379(10.5)$$
$$\approx 3.55 \approx 3.5$$

b We use the equation $h \approx 0.3379 \cdot 10^{0.1702w}$, where $w = 12$.

$$h \approx 0.3379 \cdot 10^{0.1702(12)}$$
$$h \approx 0.3379(110.255)$$
$$h \approx 37.3 \text{ inches}$$

Part Three: Exercises and Problems

Warm-up Exercises

P

1 The earthquake at Hebgen Lake, Montana, in 1959 was 12,589,254 times as intense as the minimum detectable earthquake. What was its Richter energy number? 7.1

P

2 The atmospheric pressure P decreases exponentially with the height above sea level. The equation relating the pressure P and the height h is $P = 14.7e^{-kh}$, where k is a positive constant.

a Find k if the pressure is 11.9 pounds per square inch at an altitude of 5000 feet. $k \approx 0.0000423$

b Find the atmospheric pressure outside an airplane that is flying at an altitude of 22,000 feet. ≈ 5.8 lb/in.^{2}

P

3 The mass of a radioactive substance decreases exponentially with time according to the equation $M = M_0e^{-kt}$, where M_0 is the amount of the radioactive substance at $t = 0$ and k is a positive constant that depends on the type of radioactive substance.

a Find k for carbon-14 if the half-life of carbon-14 is 5600 years. (Note that half-life is the time it takes for half of the mass M_0 to decay. So $M = 0.5M_0$ when $t = 5600$ years for this problem.) $k \approx 0.0001238$

b A bone is found with 30% of its carbon-14 remaining. How old is the bone? ≈ 9727 yr old

P

4 The following table gives the temperature of a jar of water in a room after t minutes.

t	0	10	20	30	40	50	60	70	80
Temp.	125	115	106	100	95	91	87	85	83

a Are the data exponentially related? Yes

b Predict the temperature after two hours. 76°

Problem Set

A

P

5 The largest Richter number for an earthquake was 8.9. How many times more intense was this earthquake than the San Francisco earthquake of 1906? (See the example on page 501.) ≈ 4.5

P

6 Using $P = 14.7e^{-kh}$, find the atmospheric pressure at the top of Mt. Everest, which is 29,028 feet high. (See problem **2**, above.) About 4.3 lb/in.^{2}

P 

7 The number of bacteria B in a culture increases according to the equation $B = B_0e^{kt}$. There were 400 bacteria at time $t = 0$ and 700 bacteria at time $t = 3$ hours. How many bacteria will there be after 12 hours? ≈ 3752

Assignment Guide

Average
6–8, 11, 12a, b, 13, 17, 21, 23, 24
Advanced
6–8, 10–12, 14, 17, 21, 23, 24
Honors
5, 8, 10, 11, 14, 16, 17, 19, 21, 23–25

Problem-Set Notes and Strategies

- Suggest to students that in exponential problems such as **2**, a helpful first step is to find the constant k. These are proportion problems where the rate of change is proportional to the amount present. The constant k yields the constant of proportionality.
- Problem **3** relates to many archeological applications. Other dating methods besides carbon-14 are used at this time, but all depend on the same process. A question to raise is: "Will the carbon-14 ever be gone completely?"

Problem Set, *continued*

Problem-Set Notes and Strategies, continued

- The equation in problem **8** may not be unique depending on the accuracy with which the students read the graph.
- Newton's Law of Cooling is used in problem **10**. Students who are taking physics may have used this formula already or will soon. The explanation of the formula is given in the examples.
- Problem **16** introduces the counting arguments that will be studied in the next chapter on counting and probability.

P **8** The data shown represent the sales S for each week w for a recently released Rolling Rocks album.

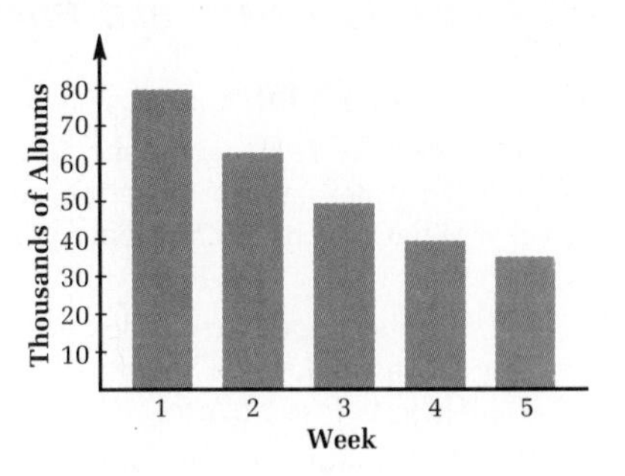

a Determine an exponential function that fits the data. $S \approx 100{,}000(10^{-0.1w})$

b What would you predict the sales of the Rolling Rocks album will be for the eighth week? ≈15,849

R **9** Rationalize the denominator.

$\frac{3}{\sqrt{2} - 1}$ $3\sqrt{2} + 3$

P **10** A tray of water is placed in a freezer in order to make ice cubes. The water is initially at 60°F. The freezer is at 0°F. If the ice cubes are formed in two hours, what is the constant k in Newton's cooling formula? (Hint: Ice freezes at 32°F.) $k \approx 0.3143$

P **11** The following data show the increasing cost of certain goods from 1980 to 1985.

Year	1980	1981	1982	1983	1984	1985
Cost	$100	$105.30	$110.88	$116.76	$122.95	$129.46

a Is the cost exponentially related to time? Yes

b If the cost continues to rise according to the same pattern, how much will the goods cost in 1999? $266.77

R **B** **12** Solve for x.

a $6^{x^2-2} = 1$ $x = \pm\sqrt{2}$

b $6^{x^2-2} = \frac{1}{36}$ $x = 0$

c $6^{x^2-2} = 36$ $x = \pm 2$

d $6^{x^2-2} = \frac{1}{216}$ No real solutions

P,R **13** Humberto deposited $785 in an account that pays $8\frac{3}{4}$% interest, compounded continuously. How long will it take until Humberto has $1000? About 2.77 yr

R **14** Multiply $(x + 3 + \sqrt{5})(x + 3 - \sqrt{5})$ and combine terms. $x^2 + 6x + 4$

R **15** Multiply $(x + y + 3)(x + y - 3)$. $x^2 + 2xy + y^2 - 9$

I **16** Four girls are trying out for the female lead in the school play, and three boys are trying out for the male lead. How many different female-male lead combinations are there? 12

P **17** The graph shows the average number of items in a supermarket for each year since 1978. Assume the data are exponentially related according to $I = I_0e^{kt}$, where I = the number of items in the store, I_0 = the number of items in 1978, and t = the number of years since 1978.

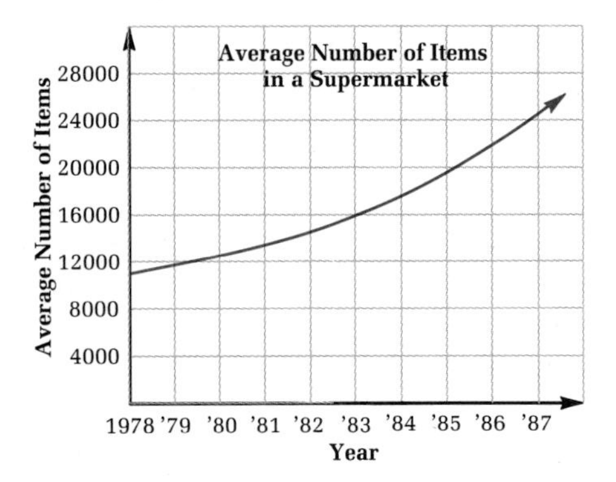

a Use the data from 1978 and 1987 to find I_0 and k. $I_0 \approx 11{,}000$; $k \approx 0.09$

b At this rate of growth, approximately how many items will be in the supermarket in 2025?
≈756,000 items

P,R **18** A hard-boiled egg has a temperature of 98°C. If it is put into a sink that maintains a temperature of 18°C, its temperature x minutes later is given by

$$T(x) = 18 + 80e^{-0.28x}$$

Hilda needs her egg to be at exactly 30°C for decorating. How long should she leave it in the water?
About 6 min 47 sec

P **19** A satellite has a radioisotope power supply. The power output in watts is given by the equation $P = 50e^{\frac{-t}{250}}$, where t is the time in days since the power supply was placed in service.

a How much power will be available at the end of one year? ≈11.61 watts

b What is the half-life of the power supply? ≈173.3 days

c The equipment aboard the satellite requires 10 watts of power to operate properly. What is the operational life of the satellite?
≈402.4 days

R **20** Consider the graph of $y = \frac{1}{x}$. $A(1, 1)$, $B\left(10, \frac{1}{10}\right)$, $C\left(100, \frac{1}{100}\right)$, and $D\left(1000, \frac{1}{1000}\right)$ are all points on the graph corresponding to the points E(1, 0), F(10, 0), G(100, 0), and H(1000, 0) on the x-axis. Find the area of each trapezoid.

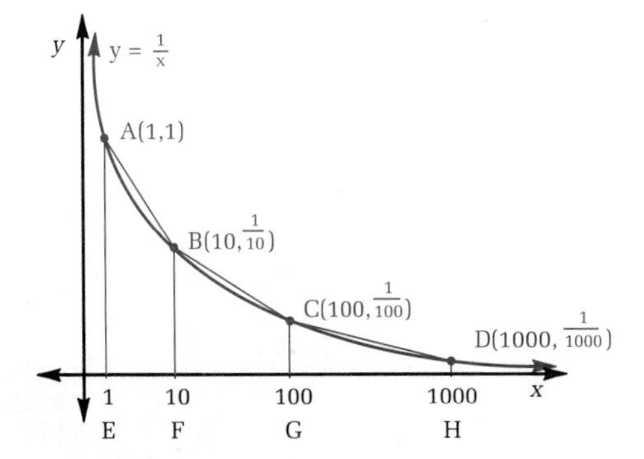

a ABFE 4.95

b BCGF 4.95

c CDHG 4.95

P **21** A 35-millimeter camera has f-stops numbered as follows: 1.4, 2, 2.8, 4, 5.6, 8, 11, 16, 22. These numbers relate to the amount of light that strikes the film; the smaller the number, the larger the lens aperture.

a Make a table of the base-10 logarithms of these numbers.

b Are these numbers exponentially related to the stops they designate? Yes

Problem-Set Notes and Strategies, continued

■ Problem **20** has some interesting results in parts **a**, **b** and **c**. Have the students see if they can explain this phenomenon using what was discussed about logarithmic and exponential graphs.

Additional Answer

21a

Stop	n	$\log n$
1	1.4	≈0.1461
2	2	≈0.3010
3	2.8	≈0.4472
4	4	≈0.6021
5	5.6	≈0.7482
6	8	≈0.9031
7	11	≈1.0414
8	16	≈1.2041
9	22	≈1.3424

Problem-Set Notes and Strategies, continued

- Problem **25** requires translations of the graph and patience.

Problem Set, *continued*

R,P **22** Multiply (1 + log 2)(1 − log 2). ≈0.9094

R C **23** Solve $x^2 + (\log 20)x + \log 2 = 0$ for x.
$x = -1$ or $x = -\log 2 \approx -0.3010$

P **24** The number of people in the Denver suburban area has grown exponentially since 1950 according to the equation $P = P_0e^{kt}$. In 1950 there were 350,000 people. In 1960 there were 600,000 people. How many people lived in the Denver suburban area in 1989? About 2,864,129 people

R **25** The graphs show $y = 3^x$ and $y = 2^{-x}$. They intersect at (0, 1). If the graph of $y = 3^x$ is shifted four units to the right and the graph of $y = 2^{-x}$ is shifted five units to the left, where will the curves meet?
(≈0.5183, ≈0.0218)

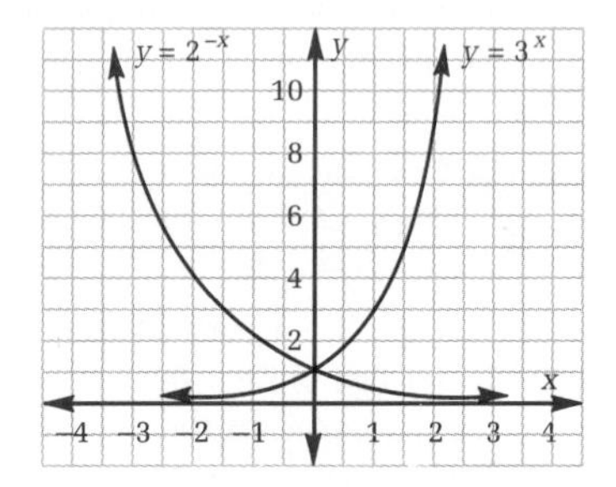

MATHEMATICAL EXCURSION

SOMETHING EARTHSHAKING

The magnitude of an earthquake

On Tuesday, October 17, 1989, at about 5:00 P.M. Pacific time, people all over the United States saw their television screens go blank as an earthquake interrupted the third game of the World Series. Measuring 7.1 on the Richter scale, the earthquake was one of the largest United States history.

The Richter scale, the system we use for reporting earthquake magnitudes, was developed by Charles Francis Richter (1900–1985). His scale describes an earthquake's magnitude according to change in ground motion.

Although no magnitude greater than 8.7 ha been recorded, the Richter scale is an open-ended logarithmic scale. (See the Caree Profile on page 518.) The Richter magnitude an earthquake is the log to the base 10 that corresponds to the amount of ground motion measured in the earthquake. This means that each unit of increase on the Richter scale represents a tenfold increase in ground motio

Compare the amounts of ground motion in Richter magnitudes 7.7 and 4.7.

11.6 Binomials and Binomial Distributions

Class Planning

Time Schedule
All levels: 1 day

Resource References
Teacher's Resource Book
Class Opener 70
Cooperative Learning 34
Practice 74
Manipulatives and Applications 16

Class Opener

1 If you tossed a coin 10 times, how many times would you expect to have the coin land heads up?
Answers will vary. Many students may answer 5.

2 Toss a coin 10 times. Record your results. How many times did the coin land heads up?
Answers will vary. It is unlikely that many students will find the coin landed heads up 5 times. Ask how many students found heads 4 times, then 6 times. Ask how many students found heads 3 times, then 7 times; 2 times, then 8 times; 1 time, then 9 times; 0 times, then 10. This should provide an insight into probability and set the stage for the binomial theorem.

Objectives

After studying this section, you will be able to
- Recognize a binomial probability experiment
- Find the mean, variance, and standard deviation of a binomial distribution

Binomials and Their Distributions

In earlier sections we have worked with binomials of the form $(a + b)^n$. We now turn our attention to binomials of the form $(pa + qb)^n$.

We begin our discussion by considering the expansion of $\left(\frac{2}{5}h + \frac{3}{5}t\right)^3$. We write

$$\left(\frac{2}{5}h + \frac{3}{5}t\right)^3 = \left(\frac{2h + 3t}{5}\right)^3 = \frac{8h^3t^0 + 36h^2t^1 + 54h^1t^2 + 27h^0t^3}{125}$$

$$= \frac{8}{125}h^3t^0 + \frac{36}{125}h^2t^1 + \frac{54}{125}h^1t^2 + \frac{27}{125}h^0t^3$$

The expansion above models the tossing of a weighted coin three times. In the binomial $\left(\frac{2}{5}h + \frac{3}{5}t\right)$, the coefficient of h, $\frac{2}{5}$, is the probability of the weighted coin's landing heads. The coefficient of t, $\frac{3}{5}$, is the probability of the weighted coin's landing tails. The exponent 3 on the binomial models the tossing of the coin three times.

Let us now look at one of the terms of the expansion, say $\frac{54}{125}h^1t^2$. The expression h^1t^2 models the outcome where the coin lands heads once and tails twice. The coefficient, $\frac{54}{125}$, is the probability of this outcome. In a like manner, the term $\frac{27}{125}h^0t^3$ tells us that the probability is $\frac{27}{125}$ that the coin will land heads zero times and tails three times when tossed three times. What is the sum of the probabilities of the four outcomes? Why?

The graphs on the next page represent the four outcomes of tossing the weighted coin three times and the corresponding probabilities.

Vocabulary
binomial probability experiment
expected value
expected nunber
mean of a binomial distribution
variance of a binomial distribution
standard deviation of a binomial distribution

Lesson Notes

- The Binomial Distribution Function is the most important of the discrete distribution functions in statistics. This section will explore what makes up a binomial experiment and show how to compute probabilities of binomial experiments.

Communicating Mathematics

Cite several examples of how different experiments can be redefined in order to make them have two outcomes (be binomial). For example, rolling a pair of dice has 11 different outcomes, but suppose we are trying to find the probability of an even outcome as opposed to an odd outcome.

Stumbling Block

Binomial experiments require two outcomes, n repeated trials, and independence of the trials. The first two are fairly simple to confirm. The independence is more difficult. Ask the question, "Does the outcome of one trial affect the outcome of any other trial?" If the answer is always no, the trials are independent. If the answer is yes, or some of the time, then the trials are dependent and this method may not be used.

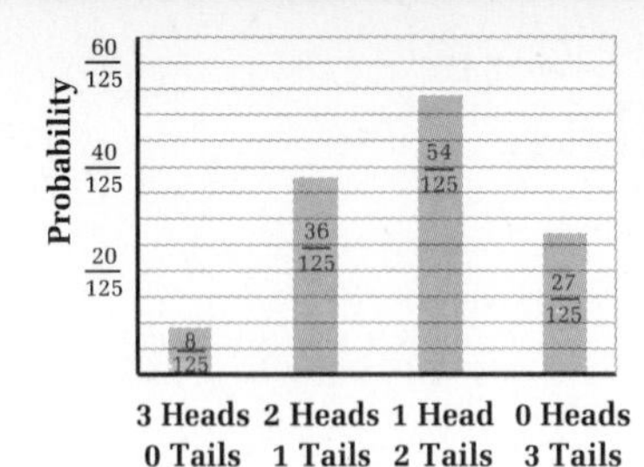

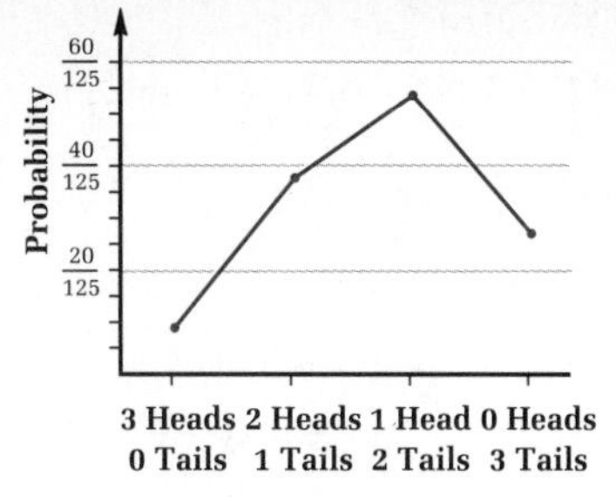

The above is an example of a ***binomial probability experiment.***

In order for an experiment to be a binomial probability experiment, the following conditions must be satisfied:

1 There must be a fixed number n of repeated trials whose outcomes are independent. (In our experiment, there were three tossings of a coin. The outcome of each coin toss was independent of the other outcomes.)

2 Each of the repeated trials must result in one of two possible outcomes. Each outcome must have the same probability as the other trials. (In our experiment, each trial had the probability of tossing a head as $\frac{2}{5}$ and the probability of tossing a tail as $\frac{3}{5}$.)

Observe that $(pa + qb)^n$ has probability p for outcome a and probability q for outcome b and that $p + q$ must equal 1.

Example *In any shipment of automobile tires from the Badmonth Tire Company, it is known that the probability of a tire's being defective is .3.*

a *Find the probability that at least two of four purchased tires are defective.*

Let $(.3d + .7n)^4$ be the binomial representation that models the purchase of four tires. The variable d models defective tires, while the variable n models nondefective tires.

The expansion of $(.3d + .7n)^4$ yields

$.0081d^4n^0 + .0756d^3n^1 + .2646d^2n^2 + .4116d^1n^3 + .2401d^0n^4$

The probability that at least two of the tires are defective is $.0081 + .0756 + .2646 = .3483$, or at least two of four tires are defective in 34.83% of cases.

b *Draw the binomial distribution graph.*

The graph for the distribution is shown below.

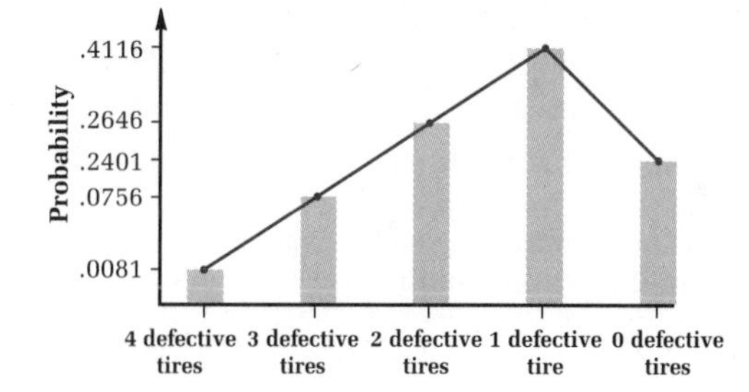

Mean and Standard Deviation of a Binomial Distribution

We have worked with the mean and standard deviation of a normal distribution. The tire example will aid us in working with the mean and standard deviation of a binomial distribution.

Example *The Keepon Trucking Company bought 300 tires from the Badmonth Tire Company. According to the preceding example, how many defective tires might they expect to receive?*

Finding the ***expected value*** or ***expected number*** of defective tires is the same as answering the question "What is the mean of the binomial distribution where the probability of getting a single defective tire is .3 and the experiment is repeated 300 times?"

The expected number of defective tires is found by multiplying .3 (the probability of getting a defective tire) by 300 (the number of tires to be purchased). The expected number (mean) of defective tires is $.3(300) = 90$.

Summarizing: The ***mean of a binomial distribution*** is found by multiplying the number of trials n by the probability p of success on each trial.

 ***Mean of a binomial distribution = np***

The ***variance of a binomial distribution*** represented by $(pa + qb)^n$ is defined to be npq.

 ***Variance = npq***

The ***standard deviation of a binomial distribution*** is defined to be $\sqrt{\text{variance}}$, or $\sqrt{npq}$.

 ***Standard deviation*** $= \sqrt{\textit{variance}} = \sqrt{npq}$

The standard deviation for the distribution of defective tires when 300 are purchased is $\sqrt{300(.3)(.7)} = \sqrt{63} \approx 7.94$.

Example *Find the probability that exactly 100 tires out of a purchase of 300 tires will be defective.*

We will use the Normal Distribution Table to aid us in solving this problem. We can associate the following bell-shaped curve with the problem.

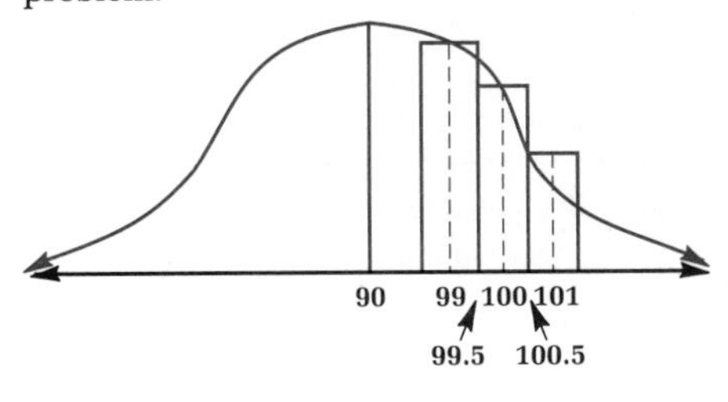

Lesson Notes, continued

- The mean (expected value) and variance of the binomial are stated in their simplest forms and are used to calculate probabilities utilizing the normal approximation. Criteria are stated as to when this is a reasonable approximation.
- The values that are used on the normal table must be changed from the discrete values given in the binomial case. A value of 0.5 must be added and subtracted from the appropriate binomial values to enable the approximation to be accurate. If a graph is drawn, these values are easier to obtain and are more intuitive.

Using Computers

A symbolic manipulator can be used effectively here for the expansion of many of the binomials in this section.

Checkpoint

1 In the coin-tossing experiment at the beginning of the section, what is the probability of obtaining at least one head? $\frac{98}{125}$

2 If a fair coin were tossed three times, what would be the probability of getting 2 heads? $\frac{3}{8}$

3 If tossing a fair coin three times, what is the expected number of heads and what is the standard deviation of that expected value?
Expected value = 1.5, standard deviation = 0.8660.

Assignment Guide

Average
1, 3, 5, 6, 8
Advanced
1–5, 7, 9
Honors
3–5, 7–10

Problem-Set Notes and Strategies

- Problem 1 will take several minutes using a calculator and will be almost impossible without one.

Observe that all the values that are greater than or equal to 99.5 and less than 100.5 would round to the value 100. We need to determine what percentage of the area under the curve is between the data values 99.5 and 100.5.

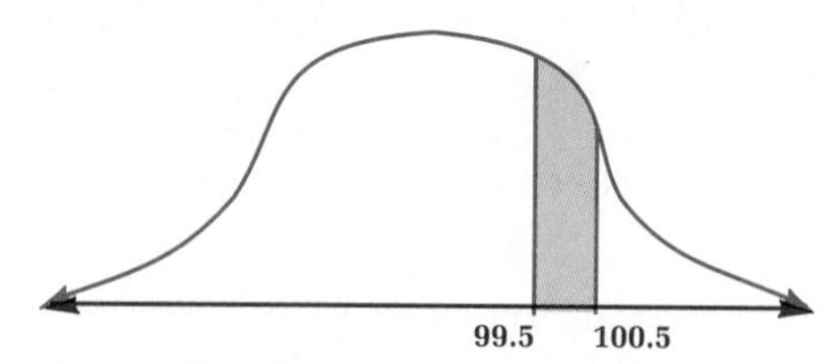

We first need to find how many standard deviations 99.5 and 100.5 are from the mean, 90, when the standard deviation is 7.94.

$$\frac{99.5 - 90}{7.94} \approx 1.2 \text{ SD} \qquad \frac{100.5 - 90}{7.94} \approx 1.3 \text{ SD}$$

The Normal Distribution Table indicates that 0.8849 and 0.9032 of the area under the curve is to the left of $m + 1.2$ SD and $m + 1.3$ SD respectively.

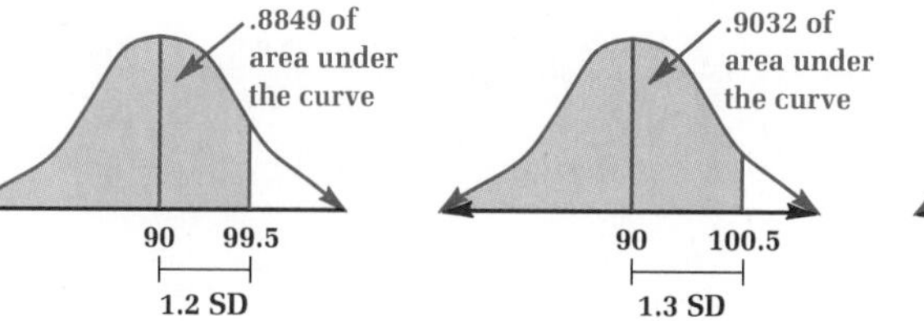

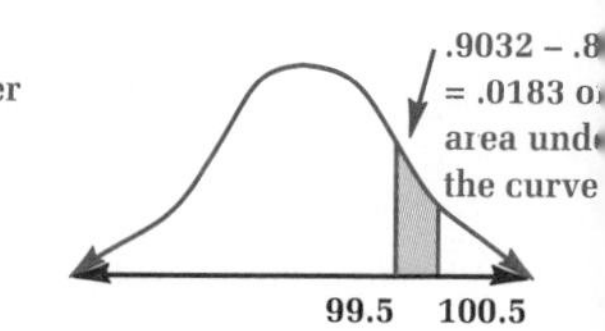

Thus, the area between the values 99.5 and 100.5 is approximately 0.9032 − 0.8849, or 0.0183, of the total area under the curve. We conclude that there is a probability of ≈.02 that exactly 100 of the tires will be defective.

Perhaps you wondered how it can be possible to approximate a binomial distribution by using a normal distribution. It has been shown mathematically that such approximations are quite accurate as long as $p \leq .5$ and $np > 5$, or $p > .5$ and $nq > 5$. In our example, $n = 300$, $p = .7$, and $q = .3$. Since

$$nq = 300(.3) > 5$$

our approximations are acceptable.

Problem Set

A

P 

1 a Write out the five terms of $(0.7W + 0.3L)^4$. $0.2401W^4 + 0.4116W^3L^2 + 0.2646W^2L^2 + 0.0756WL^3 + 0.0081L^4$

b If .7 is the probability of winning a game and .3 is the probability of losing a game, what is the probability of winning two games and losing two games in a four-game series?
.2646

P **2** A random-number generator has an equal probability of generating a 1, 2, 3, or 4. Find the probability that five 4's are generated out of ten numbers. ≈.058

P **3** On a 100-item true-false test, Randy Number knows 64 of the answers and guesses on the rest. What is her expected score? 82

B

P,R **4** Find the coefficient of the term containing a^4b^2 in the expansion of $\left(\frac{1}{4}a + \frac{3}{4}b\right)^6$. $\frac{135}{4096}$

I **5** Two marbles are to be drawn at random from a bag containing eight red marbles and five green marbles. Use the tree diagram shown to determine the probability that one of the marbles will be red and one will be green. $\frac{20}{39}$, or ≈.51

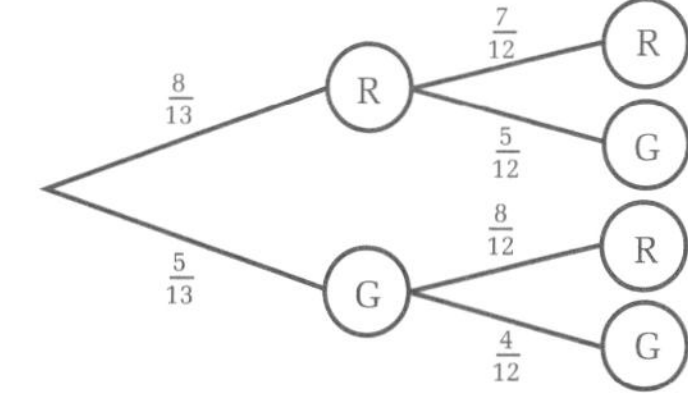

P **6** A certain coin is weighted so that when it is tossed, it will turn up heads more frequently than tails. If the coin turns up heads, on the average, two thirds of the time and if it is tossed five times, what is the probability that it will turn up heads

a Exactly four times? $\frac{80}{243}$ **b** At least four times? $\frac{112}{243}$

P  **7** Ten percent of a batch of manufactured items are defective. If five items are selected at random from this batch, what is the probability that

a None is defective? .59049
b Exactly one is defective? .32825
c At least three are defective? .00856
d Less than three are defective? .99144

P  **8** Roo Lett has an $\frac{18}{38}$ probability of winning any one game. If she plays seven games, what is the probability that Roo

a Loses all seven? ≈.0112 **b** Loses six of seven? ≈.0705

P  **9** A weighted coin comes up heads $\frac{3}{5}$ of the time. The coin is flipped 270 times.

a Find the expected number of heads (the mean). 162
b Find the standard deviation of the number of heads. ≈8.05

C

P  **10** On a true-false test, Gil guessed on every problem. The test had 80 questions.

a Find the probability that he gets each question right. $\approx 8.2718 \times 10^{-25}$
b Find the expected number of correct answers. 40
c Use a normal distribution to find the standard deviation of his number of correct responses. ≈4.47
d What percentage of the time will he get at least 35 correct? ≈86.43%

Problem-Set Notes and Strategies, continued

- The two outcomes in problem **2** must be defined as getting a four or not getting a four with probability $\frac{1}{4}$ and $\frac{3}{4}$ respectively.
- In problem **5**, the first marble was not replaced before the second was chosen. This makes the drawings dependent and not binomial. These counting arguments will be discussed more in the next chapter.
- Problems **7 – 10** are calculator intensive and take considerable time.

Cooperative Learning

An alternative method for computing the binomial probabilities is to use the formula $P(x) = C(n, x)p^xq^{n-x}$ where $C(n, x)$ is the combination of choosing x objects from n. Have students use this formula in groups of three or four and recompute the answers to problems 1, 7, 8 and 10.

11

Chapter Summary

Concepts and Procedures

After studying this chapter, you should be able to

- Graph exponential functions (11.1)
- Graph logarithmic functions (11.1)
- Use the definition of the logarithmic function (11.1)
- Solve problems involving exponential and logarithmic functions with base 10 (11.2)
- Solve problems involving exponential and logarithmic functions with base e (11.2)
- Use seven properties of logarithms (11.3)
- Use the base-change formula (11.4)
- Solve exponential equations (11.4)
- Solve logarithmic equations (11.4)
- Use exponential and logarithmic functions as mathematical models (11.5)
- Analyze exponentially related data (11.5)
- Recognize a binomial probability experiment (11.6)
- Find the mean, variance, and standard deviation of a binomial distribution (11.6)

Vocabulary

antilogarithm (11.2)
base (11.1)
base-change formula (11.4)
binomial probability experiment (11.6)
common logarithmic function (11.2)
expected number (11.6)
expected value (11.6)
exponential equation (11.4)
exponential function (11.1)
logarithmic function (11.1)
mean of a binomial distribution (11.6)
natural logarithm (11.2)
natural logarithmic function (11.2)
standard deviation of a binomial distribution (11.6)
variance of a binomial distribution (11.6)

Properties

$\log_a 1 = 0$

$\log_a a = 1$

$\log_a (a^x) = x$

$a^{\log_a x} = x$

$\log_a (M \cdot N) = \log_a M + \log_a N$

$\log_a \left(\frac{M}{N}\right) = \log_a M - \log_a N$

$\log_a (M^p) = p \cdot \log_a M$

$\log_a M = \frac{\log_b M}{\log_b a}$

11 REVIEW PROBLEMS

A **In problems 1–3, write the equivalent exponential form.**

P **1** $\log_3 81 = 4$ $3^4 = 81$ **2** $\log_5 (5^{10}) = x$ $5^x = 5^{10}$ **3** $\log_a (a^t) = t$ $a^t = a^t$

P **In problems 4–6, solve for x.**

4 $\log_x 32 = 5$ $x = 2$ **5** $\log_5 0.008 = x$ $x = -3$ **6** $9^{\log_9 x} = 35$ $x = 35$

P,R **7** Rectangle ABCD ~ rectangle PQRS. AB = 90, BC = 48, and PQ = 15

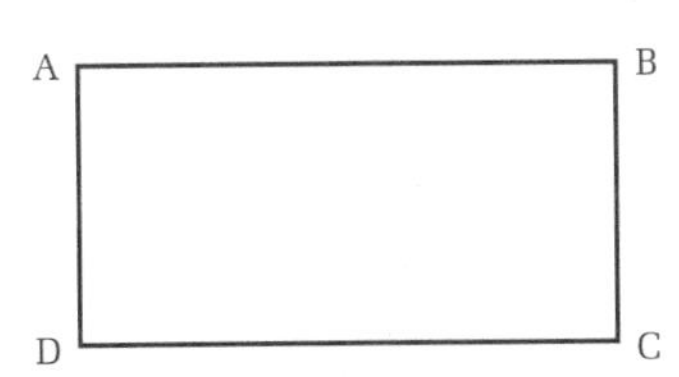

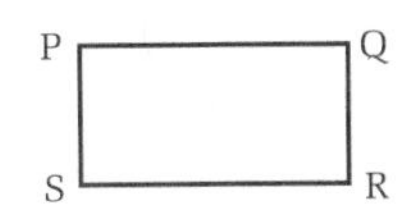

a Find QR. 8
b Find the ratio of the perimeter of ABCD to the perimeter of PQRS. 6:1
c Find the ratio of the area of ABCD to the area of PQRS. 36:1
d Find the ratio of the log of your answer to part **c** to the log of your answer to part **b.** 2:1

P **8** Approximate to three decimal places.

a $\frac{\log 30}{\log 20}$ ≈ 1.135 **b** $\frac{\ln 30}{\ln 20}$ ≈ 1.135

P **9** Find $\frac{x}{y}$ if $\log_{10} x = 2.5$ and $\ln y = 4$. ≈ 5.792

P **10** Determine whether each statement is True or False.

a $\ln A + \ln B = \ln (A + B)$ False
b $\ln A + \ln B = \ln (A \cdot B)$ True
c $(\ln A)(\ln B) = \ln (A \cdot B)$ False
d $(\ln A)(\ln B) = \ln B^{\ln A}$ True
e $(\ln A)(\ln B) = (\ln B)^{\ln A}$ False

P,R **11** Solve for x.

a $x^2 + 5x - 6 = 0$ $x = -6$ or $x = 1$
b $(9^x)^2 + 5(9^x) - 6 = 0$ $x = 0$
c $(\ln x)^2 + 5 \ln x - 6 = 0$ $x = e$ or $x \approx 0.0025$

P **12** Radium has a half-life of 1660 years. If the initial amount of radium is 200 grams, how much will remain in 500 years? About 162.3 g

Class Planning

Time Schedule

All levels: 1 day

Assignment Guide

Average
1–6, 10–12, 17–20, 23, 24, 27, 32, 34, 35a, b

Advanced
1–6, 10–14, 17–21, 23–25, 27, 29, 32–35

Honors
1–6, 10–12, 17, 20, 21, 23, 24, 27–29, 31–35

Problem-Set Notes and Strategies

- Problems **1 – 6** are practice working with the manipulation of the exponential and logarithmic expressions.
- Students may not see the difference in parts **d** and **e** in problem **10.** Have them identify the base of the exponential expression for both expressions on the right.

Problem-Set Notes and Strategies, continued

- If students do not see the pattern in problems **15** and **16**, you may want to suggest that they review the properties of logarithms once again.
- Problem **21** requires Newton's Law of Cooling. Students may need to refer back to this formula unless they keep it in their notes.

Additional Answer

19b

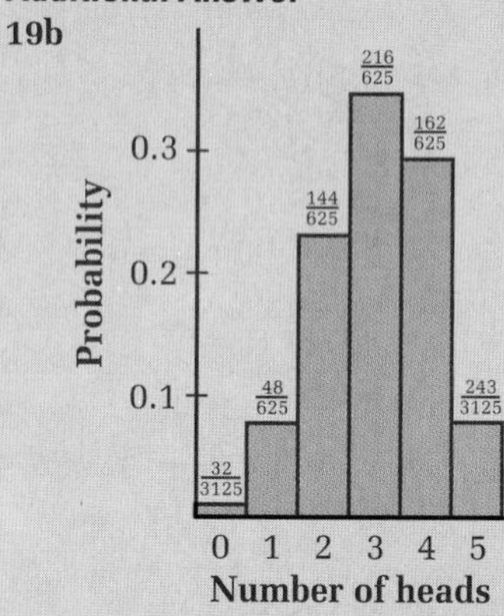

Review Problems, *continued*

P **B** **13** Find the perimeter of the rectangle. 4

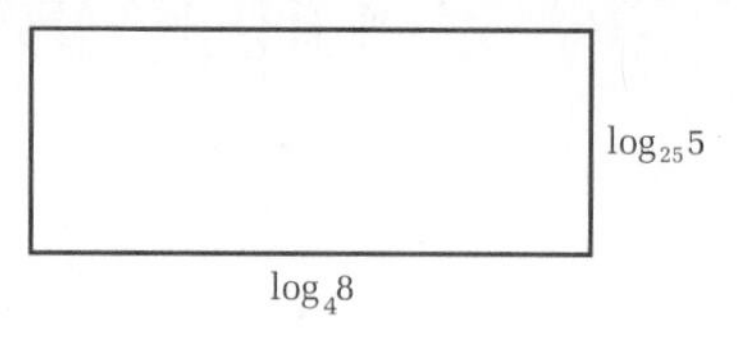

P **14** Find the coefficient of the d^2n^6 term of $\left(\frac{2}{3}d + \frac{1}{3}n\right)^8$. $\frac{112}{6561}$

P **15** Approximate each of the following expressions to six decimal places.

a log 500 ≈2.698970 **b** log 5 ≈0.698970 **c** log 0.05 ≈−1.301030
d log 50 ≈1.698970 **e** log 0.5 ≈−0.301030

P **16** Evaluate to three decimal places.

a log 60 − log 10 ≈0.778 **b** log 12 − log 2 ≈0.778 **c** log 6 − log 1 ≈0.778
d log 36 − log 6 ≈0.778 **e** log 42 − log 7 ≈0.778

P **17** The cost of 4 years of education at a private college is now approximately $64,000. How much would Chip's parents have had to invest at 6.5% 18 years ago to have that amount today, assuming the interest was compounded continuously? $19,863.49

P **18** Sales of a new breakfast cereal, Toasty Twigs, increased according to the formula $S = S_0 \log 5M$, where S_0 is the number of boxes sold when the cereal was first placed on the market, M is the number of months since the cereal was introduced, and S is the number of boxes sold during the Mth month. How many months did it take for sales of Toasty Twigs to double? 20 mo

P **19** A certain weighted coin has a $\frac{3}{5}$ probability of turning up heads each time it is tossed. The coin is tossed five times.

a Copy and complete the following table.

Number of heads	**0**	**1**	**2**	**3**	**4**	**5**
Probability	$\frac{32}{3125}$	$\frac{48}{625}$	$\frac{144}{625}$	$\frac{216}{625}$	$\frac{162}{625}$	$\frac{243}{3125}$

b Construct a bar graph showing the probabilities for this experiment.

P **20** If $\log n \approx 115.11$, how many digits are to the left of the decimal point in the number n? 116

P **21** A cup of coffee is poured when its temperature is 125°F. The room temperature is 68°F. After 5 minutes, the coffee has cooled to 105°F. What is the temperature of the coffee after 12 minutes? ≈88.2°F

P **22** Copy and complete the tables.

a

x	1	2	4	8	64	$\frac{1}{2}$	$\frac{1}{4}$	16	$\frac{1}{8}$
$\log_2 x$	0	1	2	3	6	−1	−2	**4**	**−3**

b

x	0	1	2	3	−2	−1	−3	4	−6
2^x	1	2	4	8	$\frac{1}{4}$	$\frac{1}{2}$	$\frac{1}{8}$	**16**	$\frac{1}{64}$

P **23** Solve $\log_2 x = \log_2 (x + 8) - \log_2 5$ for x. $x = 2$

P  **24** Most airplanes fly at altitudes from 30,000 to 35,000 feet above the ground. Using the equation for atmospheric pressure, $P = 14.7e^{-kh}$, and the fact that at 5000 feet the atmospheric pressure is 11.9 pounds per square inch, find the range of pressures outside a plane flying at the given altitudes. Between 4.137 lb/in.^2 and 3.349 lb/in.^2

P **25** If each question on a multiple-choice test has five choices, there is a 20% probability of answering a question correctly by guessing. Suppose that Lois Kohr guesses on all 60 questions of such a test.

a What is her expected score? 12

b What is the probability that she will answer at least 20 of the questions correctly? $\approx .005$

P **26** **a** Graph $y = 2^x$. **b** Graph $y = x^2$.

P **27** Solve $2^w = 7^{w+3}$ for w. $w \approx -4.65988$

C P  **28** Approximate log 1 + log 2 + log 3 + log 4 + . . . + log 39 to the nearest tenth. (Hint: Using the factorial key on a calculator will save you a lot of time on this one.) ≈ 46.3

P **29** Solve $(\log_4 x)^2 = 4$ for x. $x = 16$ or $x = \frac{1}{16}$

P,R  **30** The formula for compound interest is $A = P\left(1 + \frac{r}{n}\right)^{nt}$. If you deposit $25,000 at 7%, compounded quarterly, how long will it take to become a millionaire? ≈ 53.2 yr

P,I **31** A fair coin is tossed eight times. Find the probability that four of the coin tosses land heads and four of the tosses land tails. $\frac{35}{128}$

P **32** A certain virus has infected 10,000 people in a geographical area. The treatment for the virus can reduce the number of infected people by 10% each week, so the decrease is exponential. The following data have been collected for the first three weeks. How many weeks will pass until only 10% of the original 10,000 have the virus? About 22 wk

t	0	1	2	3
N	10,000	9000	8100	7300

Problem-Set Notes and Strategies, continued

- The normal approximation to the binomial should be used in problem **25**. Remind students to correct for the fact that they are using a continuous distribution. They will need to calculate the probability of at least 19.5 on the normal graph.
- If students graph correctly, problem **26** should convince students of the differences between x^2 and 2^x.
- Problem **27** may take some time until students discover that it is easier to use logs first to solve for w.
- Problem **31** is related to the cooperative learning lesson in Section **11.6** of the text.

Additional Answers

26a

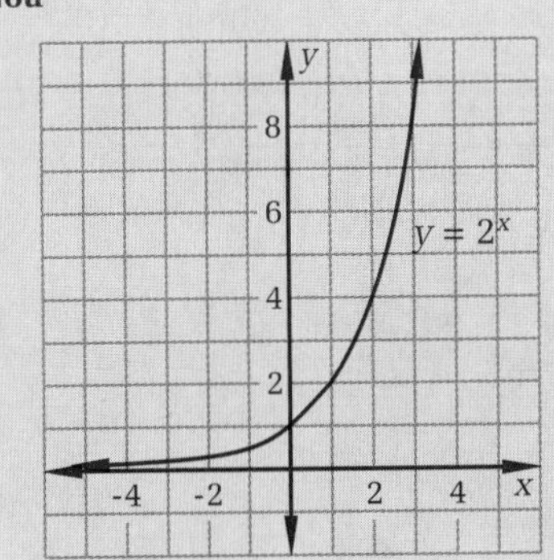

26b

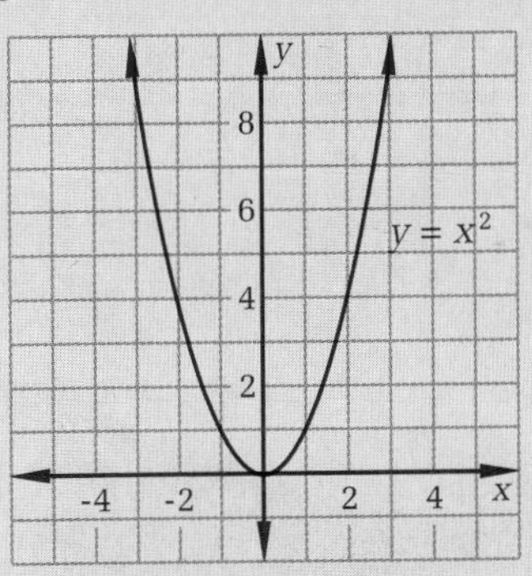

Problem-Set Notes and Strategies, continued

- Problem **34** is an exercise in working with logarithmic properties.

Review Problems, *continued*

33 The area of the rectangle is 500 square units. Find a. $a \approx 3.57794$

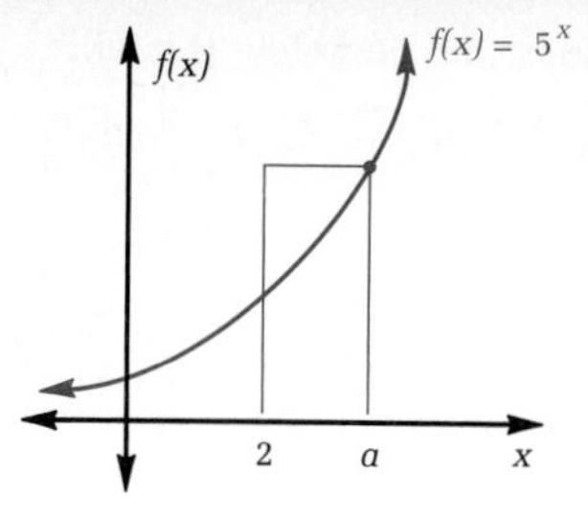

P **34** Write as a single logarithm.

a $3 \ln x + \ln \pi - 2 \ln (x - 2)$ $\ln\left[\frac{\pi x^3}{(x-2)^2}\right]$ **b** $\frac{1}{2} \ln x + \frac{1}{2} \ln y - \ln (x - 5)$ $\ln\left(\frac{\sqrt{xy}}{x-5}\right)$

P **35** Solve for x.

a $2^{x+2} = 4^x$ 2 **b** $2^{x+2} = 3^x$ ≈ 3.419 **c** $2^{x+2} = 2^x$ No solution

36 The decline in the buffalo population was exponential in the 1800's. In 1800 it was estimated that there were 6,000,000 buffalo roaming the plains. By 1830 this number had declined to 500,000. Assuming the same rate of decline, how many buffalo were there in 1930? (Use $B = B_0 e^{-kt}$.) ≈ 126

CAREER PROFILE

LOGARITHMIC INTUITION

Richard Harris reports science on the radio

"One thing people often ask me is how much more powerful the San Francisco earthquake of October 1989 was than another particular quake," says Richard Harris. Mr. Harris, as National Public Radio's science correspondent, spent ten days in San Francisco reporting on that earthquake.

Mr. Harris explains that the earthquake that occurred in southern California in February 1990, for example, measured 5.5. To compare that one with the 1989 San Francisco quake, which measured 7.1, you would divide $10^{7.1}$ by $10^{5.5}$. The quotient, $10^{1.6}$, is about 40. This means there was about forty times as much ground motion in the San Francisco quake as in the more recent California quake. (See the Mathematical Excursion on p. 508.)

"What I like best about my job is that I do different things every day," Mr. Harris says. Within one year he traveled to San Francisco, then to Brazil to do a series of stories on the Amazon rain forest, then to Florida for a space shuttle launch.

Mr. Harris graduated from the University of California at Santa Cruz. "I was a biology major, but I studied a little bit of everything. I took mathematics through calculus in order to understand my other science courses," he says. "To report a story involving science, I have to understand math."

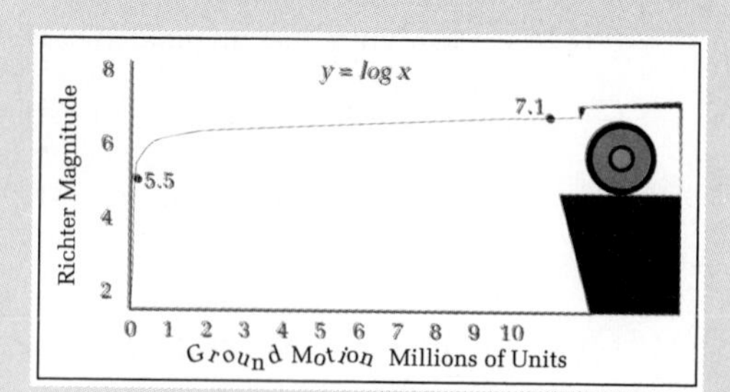

11 CHAPTER TEST

Resource References
Evaluation
Test Form 1, 2, 3

1 Solve for P.

a $P = \log_{10} 10{,}000{,}000$ 7
b $P = \log_8 64$ 2
c $P = \log_2 256$ 8
c $P = \log_{\frac{1}{2}} 8$ −3

2 Approximate to three decimal places.

a $\log 20 + \log 30$ ≈2.778
b $\log (20 + 30)$ ≈1.699
c $\log 20 + 30$ ≈31.301
d $20 + \log 30$ ≈21.477

3 Evaluate each of the following expressions.

a $\log_2 48 - \log_2 3$
4
b $\log_3 72 - 3 \log_3 2$
2
c $1 - \log_2 32$
−4

4 Write each expression as a sum and difference of logarithms.

a $\ln \frac{e^2x}{(x-2)}$ $\ln x - \ln (x - 2) + 2$
b $\log \sqrt[3]{\frac{x^5}{y^2}}$ $\frac{5}{3} \log x - \frac{2}{3} \log y$

5 If $10,000 is invested at 7.85%, compounded continuously, how much money will be in the account after six years? $16,015.95

6 Solve $6^{x^2 + 3x} = 25$ for x. $x \approx 0.5116$ or $x \approx -3.5116$

7 Solve $x = \log_3 6 \cdot \log_6 8 \cdot \log_8 27$ for x. $x = 3$

8 Noise from a Chicago elevated train measures 95 decibels. Noise in an automobile at highway speed measures 50 decibels. How many times more intense is the sound from the elevated train than the sound in the automobile? ≈31,622.78

9 The atomic bomb dropped on Hiroshima had a Richter number of 5.7. The Alaska earthquake of 1964 was 631 times as intense as the Hiroshima bomb. What was the Richter number for the Alaska earthquake? ≈8.5

10 A weighted coin comes up heads $\frac{4}{5}$ of the time. The coin is flipped 180 times.

a Find the expected number of heads (the mean). 144
b Find the standard deviation of the distribution. ≈5.37

11 Write as a single logarithm.

$\frac{1}{2} \ln y - 3 \ln x^2$ $\ln \left(\frac{\sqrt{y}}{x^6}\right)$

12.1

THIS CHAPTER BEGINS by introducing the Fundamental Counting Principle. Students studied permutations and combinations in Chapter **5**. These topics are reviewed and elaborated on in the first section of this chapter.

Sample spaces with equal elemental probabilities and probabilities of events are defined in the second section.

The relation between disjunctions and unions of sets is explained. Students will calculate probabilities of mutually exclusive events, intersections of independent events, and conditional probabilities.

The significance of probabilities and how Monte Carlo methods are used to estimate probabilities is also explained.

CHAPTER

12 COUNTING AND PROBABILITY

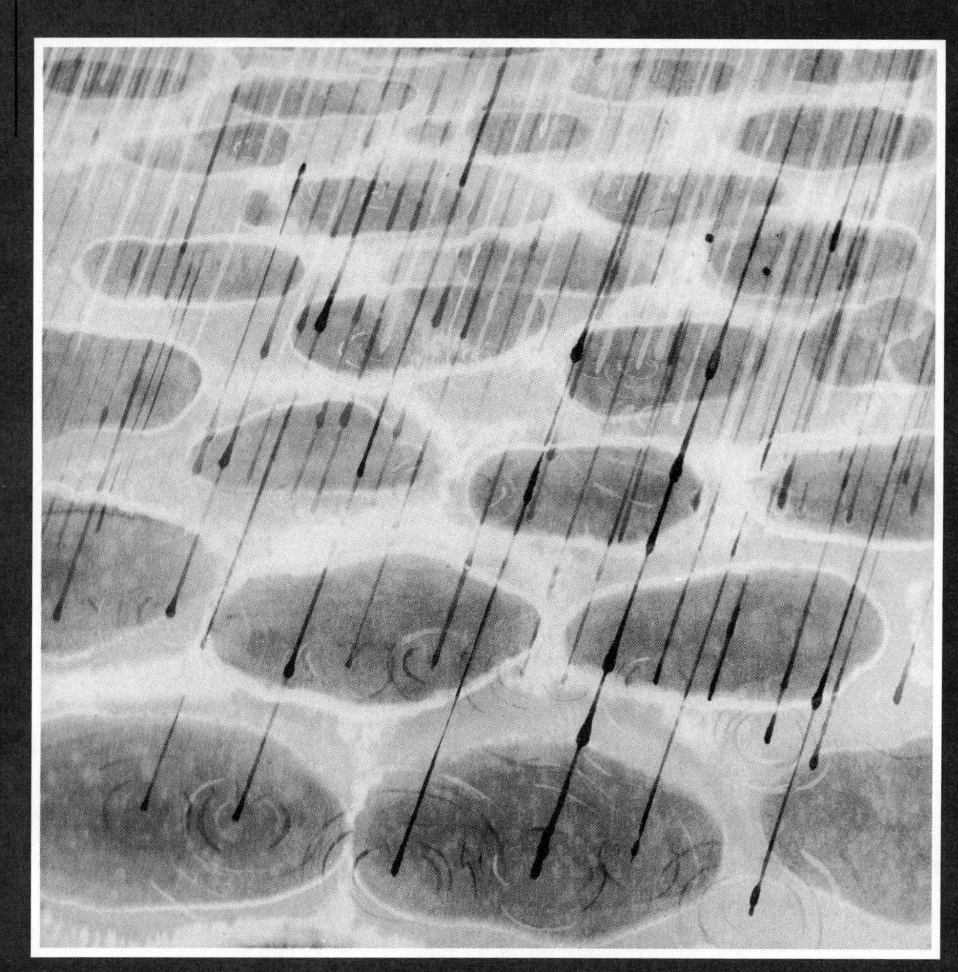

While enjoying David Hockney's beautiful *Japanese Rain on Canvas* for its own sake, we are nonetheless tempted to speculate on probabilities related to the raindrops and the spots.

12.1 Systematic Counting

Objectives

After studying this section, you will be able to

- Apply the Fundamental Counting Principle
- Relate the numbers of elements in two sets to the numbers of elements in their union and intersection
- Use the combination and permutation functions to solve counting problems

Part One: Introduction

The Fundamental Counting Principle

In Chapter 1, you explored various ways of solving the handshake problem. That problem and others like it are called counting problems or problems in ***combinatorics.*** Here is another:

Suppose Moe picks a number from 1 to 4, Larry picks a number from 5 to 7, and Curly picks the number 8 or 9. How many sets of three numbers are possible?

For each of Moe's four possible choices, Larry has three possible choices, and Curly has two. The total number of possible triples is therefore $4 \cdot 3 \cdot 2$, or 24. This solution illustrates the ***Fundamental Counting Principle.***

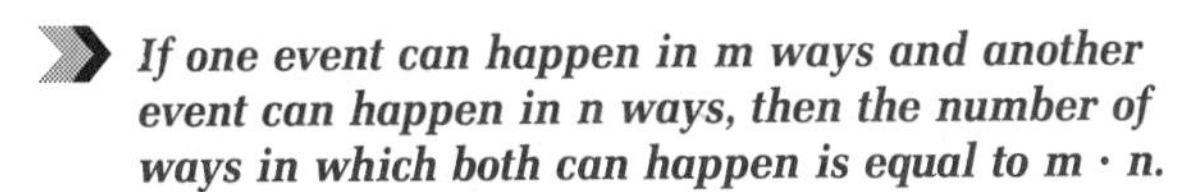

If one event can happen in m ways and another event can happen in n ways, then the number of ways in which both can happen is equal to m · n.

One way of visualizing this principle is to use a tree diagram to represent the possibilities. The following diagram shows each of Moe's four choices, Larry's three choices, and Curly's two choices in the problem described above. By counting the possible paths in the diagram, we can verify that the total number of possibilities is 24.

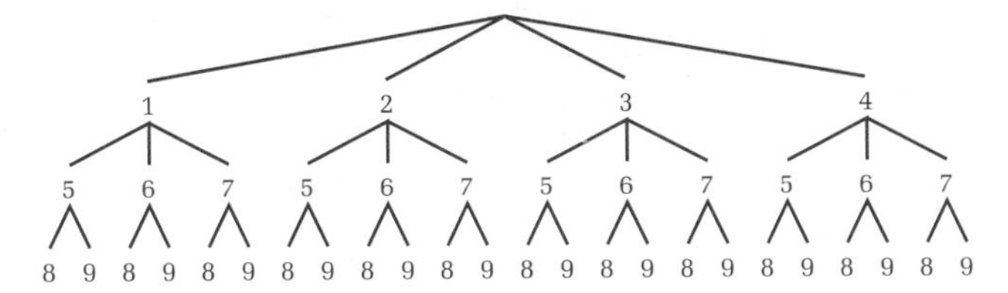

Class Planning

Time Schedule

Average: $1\frac{1}{2}$ days
Advanced: 1 day
Honors: 1 day

Resource References

Teacher's Resource Book
Class Opener 71
Cooperative Learning 35
Practice 75
Enrichment 23
Transparency 23

Class Opener

At Northside High School, 2700 students take math and science. There are 2900 students who take science. There are 3200 students who take either math or science. How many students take only math? 300

Lesson Notes

- This section presents some new and exciting techniques for counting. First, the fundamental counting principle is introduced. Some texts call this the multiplication rule for counting. Tree diagrams are an excellent illustration of this concept and should be encouraged when the size of the experiment is reasonable.

Vocabulary

combinatorics
Fundamental Counting Principle
Venn diagram

Lesson Notes, continued

- Venn diagrams are used to once again explain the difference between intersection and union. It is likely that students will have seen this several times before, but the equation relating the number of elements in *A* plus *B* to the number of elements in the intersection plus the union will be new for them. Emphasize that any of the numbers may be solved for given the other three.
- Permutations and combinations are once again reviewed (from Chapter **5**). You might want to point out that the difference here is order. If order is important, it is a permutation. If order is not important, it is a combination.

Checkpoint

1. You have three clean shrits and seven pair of socks that will complement the shirt. How many different shirt and sock combinations can you wear? 21
2. Nine different textbooks are available to put on the library bookshelf, but the space on the shelf is only big enough for four. How many different arrangements these books are possible? 3,024
3. When rolling a pair of fair dice, how many ways are there of obtaining a sum of ten or doubles on the dice? 8

Unions and Intersections

Consider the sets A = {1, 2, 3, 4} and B = {3, 4, 5, 6, 7}. The union of the sets, A ∪ B, is {1, 2, 3, 4, 5, 6, 7} and the intersection of the sets, A ∩ B, is {3, 4}. (Recall that A ∪ B consists of the elements that are in either A *or* B and that A ∩ B consists of the elements that are in both A *and* B.)

A useful way of representing this situation is a ***Venn diagram.*** The region that is included in both circles represents the intersection of the sets. The entire region enclosed by the circles represents the union of the sets.

We use the notation #(S) to refer to the number of elements in a set S. In this case, #(A) = 4, #(B) = 5, #(A ∪ B) = 7, and #(A ∩ B) = 2. Notice, moreover, that 4 + 5 = 7 + 2. This relationship suggests an important theorem that can be used to solve a variety of counting problems.

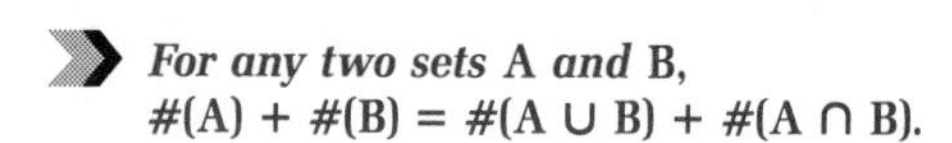

***For any two sets* A *and* B,**
#(A) + #(B) = #(A ∪ B) + #(A ∩ B).

Example *How many cards in a standard deck are either aces or hearts?*

If we let A = {aces} and B = {hearts}, we know that #(A) = 4 and #(B) = 13. Since A ∩ B = {aces and hearts} and the only element in this set is the ace of hearts, #(A ∩ B) = 1.

We want the number of elements in A ∪ B. According to the theorem presented in this section,

$$\#(A) + \#(B) = \#(A \cup B) + \#(A \cap B)$$
$$4 + 13 = \#(A \cup B) + 1$$
$$16 = \#(A \cup B)$$

Alternative solution: Let's see which cards satisfy the conditions of the problem. Certainly all 13 hearts do. There are 4 aces, but since 1 is the ace of hearts, we only count 3. Therefore, 16 cards are either aces or hearts.

Permutations and Combinations

In Chapter 5, you were introduced to two functions that simplify counting. The combination function, $C(n, r) = \frac{n!}{(n-r)!r!}$, is used to determine the number of *groups* of *r* elements that can be selected from a set of *n* elements. We can apply this function whenever we are dealing with a situation in which the order of selection does not matter.

The permutation function, $P(n, r) = \frac{n!}{(n-r)!}$, is used to determine the number of *arrangements* of *r* elements that can be selected from a set of *n* elements. We use this function when the order of selection does matter.

Example *How many different three-digit numbers can be formed from the digits 1, 2, 3, 4, 5, 6, and 7 if no digit is used twice?*

Because order does matter (324 is different from 234, for example), we use permutations. We are looking for the number of arrangements of three items chosen from a set of seven.

$$P(7, 3) = \frac{7!}{(7-3)!} = \frac{7!}{4!} = 7 \cdot 6 \cdot 5 = 210$$

There are 210 such three-digit numbers.

Part Two: Sample Problems

Problem 1 *How many different 13-card bridge hands can be dealt from a standard deck of 52 cards?*

Solution We are not concerned with the order in which the cards are dealt, so we use the combination function.

$$C(52, 13) = \frac{52!}{(52-13)!13!} = \frac{52!}{39! \cdot 13!} \approx 6.35 \times 10^{11}$$

There are over 600 billion possible bridge hands.

See p. 779

Problem 2 *A box contains 20 muffins, 5 of which are blueberry and the rest of which are raisin.*

a *How many groups of 2 blueberry muffins and 3 raisin muffins can be chosen from the box?*
b *In how many ways can a person choose 2 blueberry muffins and 3 raisin muffins if the first choice must be a raisin muffin and the two kinds must be chosen alternately?*

Solution **a** We are looking for the number of ways in which 2 muffins can be chosen from 5 and 3 muffins can be chosen from 15. The order of choosing the muffins does not matter, so we use combinations and apply the Fundamental Counting Principle.

$C(5, 2) \cdot C(15, 3) = 10 \cdot 455 = 4550$

b In this case, the order of choice does matter, so we cannot use combinations. The only order that satisfies the given conditions is (1) raisin, (2) blueberry, (3) raisin, (4) blueberry, (5) raisin.

There are 15 raisin muffins from which the first muffin can be chosen. There are 5 blueberry muffins from which the second can be chosen. There are 14 raisin muffins left for the third selection. There are 4 blueberry muffins left for the fourth selection. There are 13 raisin muffins left for the fifth choice. The number of ways in which the muffins can be chosen is therefore $15 \cdot 5 \cdot 14 \cdot 4 \cdot 13$, or 54,600.

Assignment Guide

Average

Day 1 9–11, 13, 15, 18, 19, 21, 22, 25, 28, 30

Day 2 ($\frac{1}{2}$ day) 14, 16, 17, 20, 23, 26, 29

($\frac{1}{2}$ day) Section 12.2 8, 10, 11, 13, 15, 22

Advanced

9, 10, 12, 14a–d, 16, 18–21, 23, 25–27, 29, 31, 32

Honors

12–14, 16, 18–20, 23, 25–29, 31, 32

Cooperative Learning

Have students work in groups of three or four and consider the following experiment: Choose three marbles from a bag of twenty marbles consisting of ten red, seven green and three blue.

1. If sampling is done with replacement, how many possible combinations of marbles can you find?
2. In how many ways could one marble of each color be chosen?
3. How many possible ways are there to get three red?

Now repeat the entire experiment without replacement. A tree diagram might be useful. Recognize that the diagram will have 27 branches.

Problem-Set Notes and Strategies

- Many students have keys on their calculators that will handle permutations and combinations. When discussing problems **4** and **5** you might ask your students to examine their calculators to see if these are available and help them learn the use of these keys.

Communicating Mathematics

The only difference between the formulas for combination and permutation is the $r!$ in the denominator of the combination formula. Explain why this term is present and the role it plays in finding the appropriate values.

Stumbling Block

The Fundamental Counting Principle is new to the students and seems intuitive but is often misused. When doing repeated sampling from the same population, an often overlooked criterion is whether the sampling is done with or without replacement. Encourage students to draw tree diagrams whenever the sample size is feasible. Then use the counting principle on the branches of the diagram.

Problem 3 *Bali High School has a student population of 2400. Of these students, 2135 are taking math, 1857 are taking science, and 1664 are taking both math and science. How many students are taking neither math nor science?*

Solution We can draw a chart to clarify the way in which the students are grouped. The first row represents the students taking math, and the second row represents those not taking math. The first column represents those taking science; the second column, those not taking science. We know that 1664 students take both courses.

	Science	Not Science
Math	1664	
Not Math		

Now we can begin to fill in the other boxes. The total taking math is 2135, and 2135 − 1664 = 471; so 471 students are taking math but not science. There are 1857 students taking science, so 1857 − 1664, or 193, students take science but not math.

	Science	Not Science	Total
Math	1664	471	2135
Not Math	193		
Total	1857		

Since the total number of students is 2400 and the three categories filled in so far total 2328, there must be 72 students who are taking neither math nor science.

	Science	Not Science
Math	1664	471
Not Math	193	72

Part Three: Exercises and Problems

Warm-up Exercises

In problems 1–3, identify each situation as involving combinations, permutations, or neither.

P

1 Picking a committee of seven members of the U.S. Senate Combinations

2 Lining up in a cafeteria line Permutations

3 Lining up members of a baseball team according to height Neither

P,R

In problems 4 and 5, evaluate each expression.

4 $P(10, 2)$ 90

5 $C(15, 4)$ 1365

P

6 Let A = {students in your math class who also take history} and B = {students in your math class who also take a foreign language}. Find $\#(A)$, $\#(B)$, $\#(A \cap B)$, and $\#(A \cup B)$. Do your results agree with the theorem presented in this section? Answers will vary; yes

P **7** A basketball team has three centers, four forwards, and five guards on its roster. How many different squads of one center, two forwards, and two guards can the coach put on the court? 180

P **8** Goo Lash Caterers provided a seven-course dinner for a party. Unfortunately, they forgot to label the containers of food, so the courses had to be served in a random order. In how many different orders could the courses have been served? 5040

Problem Set

A

P,R **9** Let C be the combination function and P be the permutation function.

a Evaluate C(12, 7). 792

b Evaluate P(12, 7). 3,991,680

P **10** If #(A) = 16, #(B) = 31, and #(A ∪ B) = 39, what is #(A ∩ B)? 8

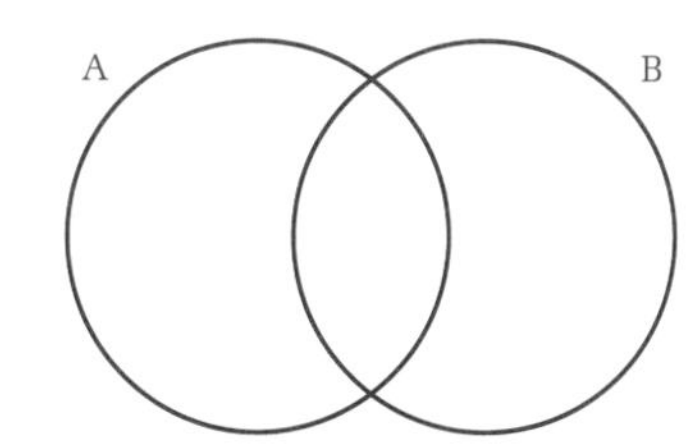

P **11** Pete's Pizza Parlor offers ten pizza toppings. Pete's Special includes a choice of any three toppings. How many three-topping combinations can be ordered? 120

P **12** In how many orders can the numbers 1, 2, 3, 5, 8, 13, and 21 be listed? 5040

P,R **13** If $m_1 \in \{3, 5, 7, 9\}$ and $m_2 \in \left\{-3, -5, -\frac{1}{7}, -9\right\}$, for how many pairs (m_1, m_2) will the graphs of $y = m_1x$ and $y = m_2x$ be perpendicular? 1

P,R **14** Evaluate each expression.

a C(10, 3) 120
b C(10, 7) 120
c C(15, 2) 105
d C(15, 13) 105
e C(18, 4) 3060
f C(18, 14) 3060

B

P **15** Superstitious Sandy, the coach of a baseball team, starts the same nine players in every game but arranges them in a different batting order each time. How many games must the team play in order for Sandy to use all possible batting orders? 362,880

P **16** There are 85 athletes at Wassamatta U. Sixteen are on the basketball team, 75 are on the football team, and 12 are on both these teams. How many are on neither the basketball team nor the football team? 6

Problem-Set Notes and Strategies, continued

- Problem **7** combines the combination formula with the fundamental counting principle. Students quickly see that counting can be an involved process.
- Problem **8** involves order so a permutation applies here.
- You might ask the students if the numbers in problem **12** have any significance. Some may remember them as the Fibonacci numbers.
- A Venn diagram might help with problem **16**.

Problem Set, *continued*

P **17** Use the Venn diagram to find each number.

a #(A) 5
b $\#(A \cap B)$ 2
c $\#(A \cap B \cap C)$ 1
d $\#(B \cap C)$ 3

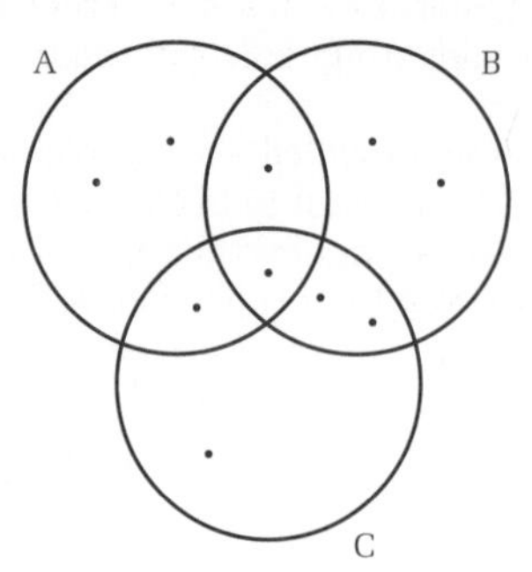

Problem-Set Notes and Strategies, continued

- Reiterate that choosing between combinations and permutations in problems **19** and **20** depends on whether order is important or not.
- Problem **22** introduces an extension of probability that will be discussed in more detail in the next section.
- In problem **23**, the customer may ask for both scoops to be the same flavor. This means that permutations may not be used. Apply the fundamental counting principle in this case.

Additional Answers

21

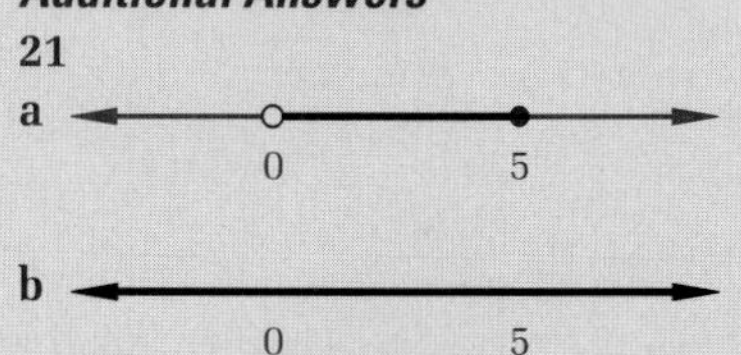

P **18** Radio station WOLD plays ten requests between 10:00 and 11:00. In how many different orders can each group of ten songs be played? 3,628,800

P **19** A sergeant needs to select 6 "volunteers" from his unit of 52 soldiers for KP duty. To find the number of ways in which the KP detail can be selected, should Sarge use combinations, permutations, neither, or both? Combinations

P **20** Eight horses are competing in a race. The first prize is $500, the second is $400, and the third is $200. To determine the number of ways in which the prizes could be awarded, would you use combinations, permutations, neither, or both? Permutations

R **21** Let $A = \{x{:}x \le 5\}$ and $B = \{x{:}x > 0\}$. Graph the following sets on a number line.

a $A \cap B$ **b** $A \cup B$

I **22** If a point within the larger square is chosen at random, what is the probability that the point is in the smaller, shaded square?
$\frac{1}{4}$

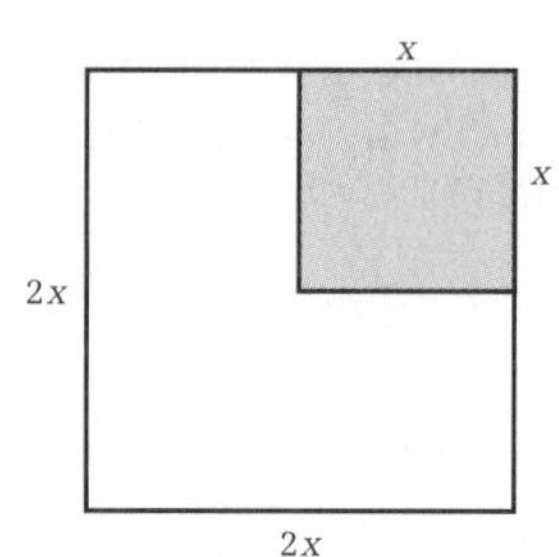

P **23** The Thirty-seven Flavors ice cream shop offers 37 different kinds of ice cream. How many different 2-scoop cones can be ordered if customers are given a choice of plain or sugar cones? (Assume that customers asking for 2 different flavors can specify the order in which the flavors are put on the cone.) 2738

P | **24** A key has five zones of indentation, each of which can be indented to any of four depths. How many different keys can be made? 1024

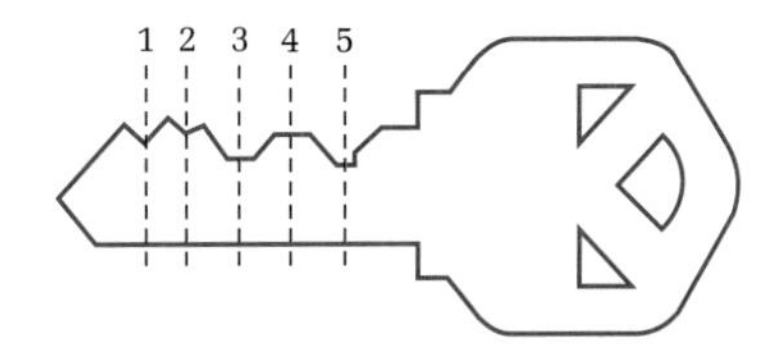

P | **25** **a** How many seven-digit numbers can be formed from the digits 0–9 if each digit can be used more than once? 10,000,000
b Are all of these possible telephone numbers in the United States? No

P,R | **26** If $m \in \{2, 4, 6, 7, 9\}$ and $n \in \{1, 3, 4, 6\}$, for how many pairs (m, n) will i^{mn} be a real number? 16

P | **27** If #(I) = 16, #(II) = 10, and #(III) = 20, what is #(A ∪ B)? 46

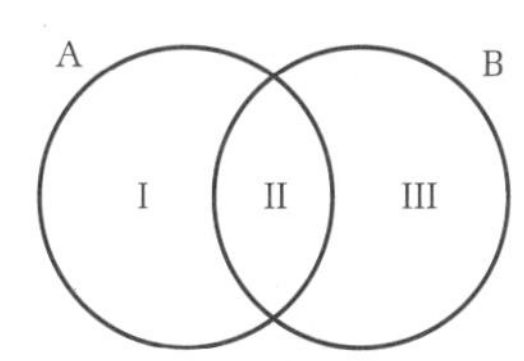

P | **28** In poker, a flush is a hand containing five cards of the same suit. How many different flushes can be dealt from a 52-card deck? 5148

P,I | **29** Consider the sets A = {1, 3, 5, 7, 9} and B = {2, 4, 6, 8, 10, 12, 14}.
a Find #(A). 5
b Find #(B). 7
c Find #(A ∩ B). 0
d Find #(A ∪ B). 12
e Is #(A) + #(B) = #(A ∪ B)? Why? Yes, because A ∩ B is empty

P,R | **30** If $a \in \{-2, 0, 2, 4\}$ and $b \in \{1, 3, 5\}$, for how many ordered pairs (a, b) is $a^b > 0$? 6

C P | **31** Ten boys and eight girls are in a class. In how many ways can four of the boys and three of the girls line up in alternate order? 1,693,440

P | **32** Explain why #(A) + #(B) = #(A ∪ B) + #(A ∩ B) is a true statement.

P,R | **33** How many triangles are in the figure shown? 51

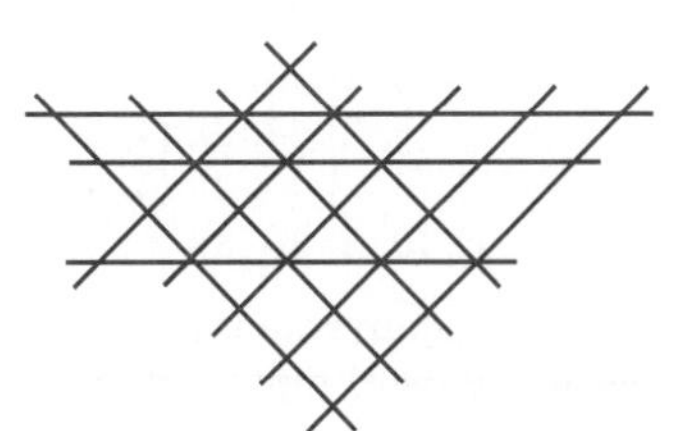

P | **34** Explain why a set consisting of n elements has 2^n subsets.

Problem-Set Notes and Strategies, continued

- Problem **26** doesn't seem to fit any of the formulas, but may easily be enumerated to find the answer. This technique is used frequently with small sample sets.
- Problem **28** requires both the combination formula and the fundamental counting principle.
- Problem **31** involves both the permutation formula and the fundamental counting principle. There are other methods of solving this, but they are more complicated.
- Problem **33** may be familiar to puzzle solvers. A systematic counting procedure will help in finding the solution.
- In problem **34** students are asked to provide a proof of a concept. Be prepared for a wide variety of answers.

Additional Answers

32 The union of A and B contains one element for every element in A and B, counting the elements in the intersection only once. By adding the number of elements in the intersection, we include the elements that need to be counted twice.

34 For every subset, each of the n elements is either in the subset or not. We therefore have two choices for each element ("in" or "not in") in forming a subset, so there are $2_1 \cdot 2_2 \cdot 2_3 \cdot \ldots \cdot 2_n$, or 2^n, possible subsets.

12.2

Class Planning

Time Schedule

Average: $1\frac{1}{2}$ days
Advanced: 1 day
Honors: 1 day

Resource References

Teacher's Resource Book
Class Opener 72
Cooperative Learning 36
Practice 76

Class Opener

If a coin is tossed three times, what is the probability that it will come up heads at least twice?
$\frac{1}{2}$

Lesson Notes

- Section **2** reviews some of the basic ideas of probability and introduces the proper vocabulary. Sample space, events, equally likely, and equal elemental probabilities are all terms which students need to learn.
- It is helpful if students get into the habit of thinking about the sample space and counting the number of elements that are contained in the sample space.
- The basic probability formula is introduced and will be intuitive to students. Thinking about events instead of particular outcomes may take a little more work. Try to help students understand that an event is a set of possible outcomes and could be anything from the empty set to the entire sample space.

12.2 Sample Spaces, Events, and Probability

Objectives

After studying this section, you will be able to
- Determine the sample space for an experiment
- Identify events and calculate their probabilities

Part One: Introduction

Sample Spaces

When you have solved probability problems in the past, you have started by determining the number of possible outcomes. The set of all possible outcomes of an experiment is called the ***sample space*** for that experiment. Although the outcomes in a sample space need not be equally likely, *in this book we will work only with sample spaces made up of equally likely outcomes.* Such a space is said to have ***equal elemental probabilities.***

Example *When two dice are rolled, the possible sums of the numbers on the upturned faces of the dice are* {2, 3, 4, 5, 6, 7, 8, 9, 10, 11, 12}. *Is this an appropriate sample space for the experiment of rolling two dice?*

No—the elements are not equally likely. (Rolling a 6, for example, is more likely than rolling a 2.) The appropriate sample space is the following set of 36 equally likely outcomes.

{(1, 1), (1, 2), (1, 3), (1, 4), (1, 5), (1, 6),
(2, 1), (2, 2), (2, 3), (2, 4), (2, 5), (2, 6),
(3, 1), (3, 2), (3, 3), (3, 4), (3, 5), (3, 6),
(4, 1), (4, 2), (4, 3), (4, 4), (4, 5), (4, 6),
(5, 1), (5, 2), (5, 3), (5, 4), (5, 5), (5, 6),
(6, 1), (6, 2), (6, 3), (6, 4), (6, 5), (6, 6)}

Sometimes it is inconvenient to list all the elements of a sample space, so other methods of representation are used. We could, for example, show the 36 elements of the sample space for the rolling of two dice in any of the following forms.

Vocabulary

sample space
equal elemental probabilities
event

A Lattice
(Each dot represents an element.)

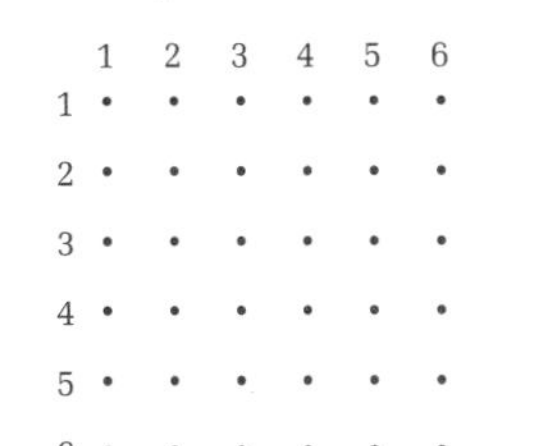

A Chart
(Each box represents an element.)

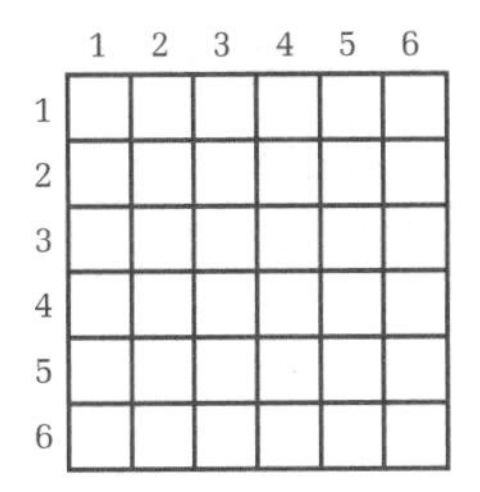

A Tree Diagram
(Each path represents an element.)

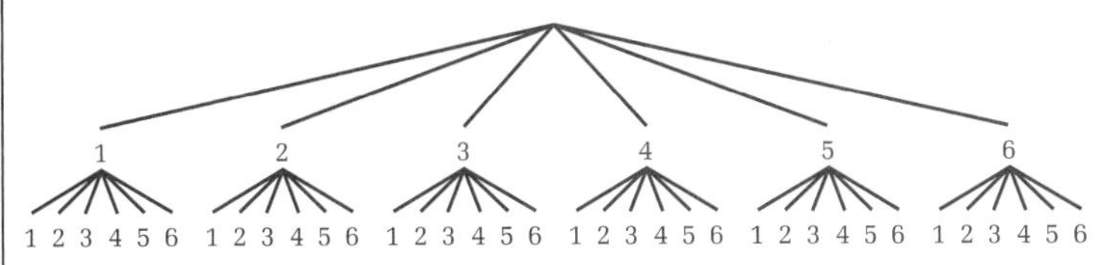

Checkpoint

1. When rolling a fair die, what is the probability of getting a number greater than 7? greater than 0?
   0 and 1
2. If a coin is flipped twice, what is the probability of getting 2 heads? A head and a tail?
   $\frac{1}{4}$ and $\frac{1}{2}$
3. What is the probability that a randomly chosen number between 0 and 1 will have its square less than $\frac{1}{4}$? $\frac{1}{2}$

Probabilities of Events

Suppose that we are going to roll two dice and want to know the probability of rolling a 6. We have already determined the sample space for the rolling of two dice, but we now need to identify the ***event***—the element or elements of the sample space that satisfy the given conditions.

In this case, the event consists of every ordered pair whose sum is 6. We can represent this event in any of the ways we used to represent the sample space. A lattice representation is shown at the right. The equivalent set notation is

{(1, 5), (2, 4), (3, 3), (4, 2), (5, 1)}

If an event A is a subset of a sample space S, then we can find the probability of event A, symbolized P(A), by means of the following formula.

$$P(A) = \frac{\#(A)}{\#(S)}$$

Example *What is the probability that when two dice are rolled, the sum of the numbers on the upturned faces will be* 6?

As we have seen, the event (a roll of 6) contains 5 elements and the sample space contains 36 elements. Therefore, the probability is $\frac{5}{36}$.

12.2

Assignment Guide

Average	
Day 1	($\frac{1}{2}$ day) Section 12.1 14, 16, 17, 20, 23, 26, 29
	($\frac{1}{2}$ day) 8, 10, 11, 13, 15, 22
Day 2	9, 12, 14, 17–21, 24, 26, 27, 31

Advanced
8–12, 15–18, 21, 22, 26–28, 31, 32

Honors
9, 11, 12, 14, 16, 18, 21, 25, 27–29, 32, 33

Example *A die is rolled and a coin is flipped. Find the probability that the number on the die is even and the coin shows heads.*

We can represent the sample space by means of a lattice and circle the elements that satisfy the given conditions (the event). The probability is $\frac{3}{12}$, or $\frac{1}{4}$.

	H	T
1	•	•
2	⊙	•
3	•	•
4	⊙	•
5	•	•
6	⊙	•

Part Two: Sample Problems

Problem 1 *A bag contains six red marbles and five blue marbles. Three marbles are drawn at random from the bag. Find the probability that all three are red.*

Solution The sample space—the set of all groups of 3 marbles that can be drawn from 11—consists of C(11, 3), or 165, elements. The event is made up of the groups of 3 marbles that can be drawn from the 6 red ones. There are C(6, 3), or 20, such groups, so the probability is $\frac{20}{165}$, or $\frac{4}{33}$.

Problem 2 *Five cards are drawn at random from a standard deck. Find the probability that two are hearts and three are spades.*

Solution The number of groups of five cards that can be chosen from a deck is C(52, 5), or 2,598,960. The number of possible groups of two hearts is C(13, 2), or 78, and the number of possible groups of three spades is C(13, 3), or 286. To find the number of elements in the event, we apply the Fundamental Counting Principle, multiplying these two results. The event therefore consists of 78 · 286, or 22,308, elements. Thus, the probability is $\frac{22{,}308}{2{,}598{,}960}$, or $\approx .0086$.

Note It is frequently most convenient to express probabilities in decimal form. This probability is a little less than .01, so there is less than one chance in a hundred of drawing two hearts and three spades.

Problem 3 *Five different cards are drawn at random, one at a time, from a standard deck. Find the probability that the selection consists of two hearts and three spades chosen alternately.*

Solution Since we are looking for the probability of choosing cards in a certain order, the sample space consists of all the possible *arrangements* of 5 cards chosen from a deck of 52. Thus, there are P(52, 5), or 311,875,200, elements in the sample space.

The event is the set of outcomes in which a spade, a heart, a spade, a heart, and a spade are chosen in sequence. There are 13 choices for the first card, 13 for the second, 12 for the third, then 12, then 11, so the event contains $13 \cdot 13 \cdot 12 \cdot 12 \cdot 11$, or 267,696, elements. The probability is $\frac{267,696}{311,875,200}$, or $\approx .00086$.

Problem 4 *Two real numbers between −1 and 1 are randomly chosen. Find the probability that the sum of their squares is less than 1.*

Solution We can represent the sample space and the event on a coordinate system. Each point in the square represents an ordered pair (x, y) of numbers between −1 and 1.

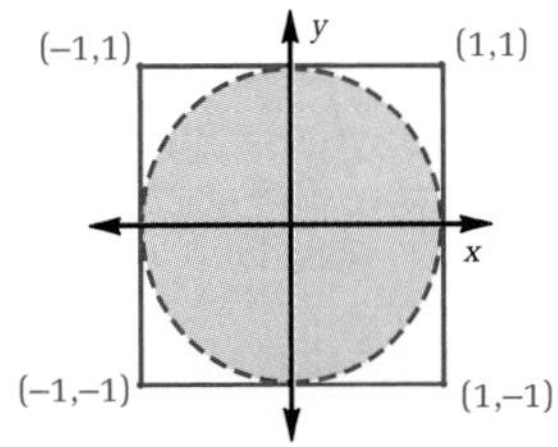

The points for which $x^2 + y^2 < 1$ are those in the interior of the dashed circle.

In this case we can't count the elements of the sample space and the event (each has infinitely many elements). The probability, however, does correspond to the ratio of the regions' areas. Since the area of the region corresponding to the event (the circle) is π and the area of the region corresponding to the sample space (the square) is $2 \cdot 2$, or 4, the probability is $\frac{\pi}{4}$, or $\approx .785$.

Part Three: Exercises and Problems

Warm-up Exercises

In problems 1–3, explain why each statement is true.

P

1 For any event A, $0 \leq P(A) \leq 1$.

2 $P(\phi) = 0$, where ϕ represents the empty set.

3 For any events A and B, $P(A) + P(B) = P(A \cup B) + P(A \cap B)$.

P  

4 In poker, a flush is a hand consisting of five cards of the same suit. If you are dealt five cards, what is the probability that you have a diamond flush? $\frac{33}{66,640}$, or $\approx .0005$

P,R

5 A dart thrown at the dart board shown lands in a random spot on the board. What is the probability that it lands in the bull's-eye? $\frac{1}{9}$

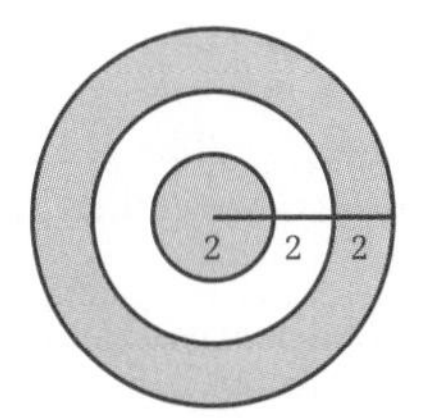

P

6 Represent, in as many ways as you can, the sample space for rolling one die and tossing two coins.

Problem-Set Notes and Strategies

- Problems **1**, **2**, and **3** identify some of the rules of basic probability.
- Problem **6** may generate a variety of answers. You might consider having several students present their answers. A lively discussion can result.

Additional Answers

1 Event A is a subset of some sample space S, so $0 \leq \#(A) \leq \#(S)$. Therefore, $\frac{0}{\#(S)} \leq \frac{\#(A)}{\#(S)} \leq \frac{\#(S)}{\#(S)}$, and hence $0 \leq P(A) \leq 1$.

2 $P(\phi) = \frac{\#(A)}{\#(S)}$. Since $\#(A) = 0$, $\frac{\#(A)}{\#(S)} = \frac{0}{\#(S)} = 0$. Therefore, $P(\phi) = 0$.

3 Since $\#(A) + \#(B) = \#(A \cup B) + \#(A \cap B)$, $\frac{\#(A)}{\#(S)} + \frac{\#(B)}{\#(S)} = \frac{\#(A \cup B)}{\#(S)} + \frac{\#(A \cap B)}{\#(S)}$. Therefore, $P(A) + P(B) = P(A \cup B) + P(A \cap B)$.

Answers for problem **6** can be found in the answer pages beginning on **A 1**.

P 7 In the game Castles and Creatures, one 12-sided die (with sides numbered 1–12) and one 4-sided die (with sides numbered 1–4) are used.

a Represent the sample space for rolling both dice.
b When a player rolls the dice, his or her character receives the powers of a wizard if the sum of the numbers on the dice is greater than 10. Find the probability of such a roll. $\frac{3}{8}$

Problem-Set Notes and Strategies, continued

- Problem **8** was enumerated in the example on page 528. The numbers 2 – 12 are the random variable. If you have time and feel it is appropriate with your class, now is a good time to introduce the definition of a random variable.
- Problem **10** involves the union of sets and will be explored in greater detail in the next section.
- Problem **11** may be intuitive to some students. How many ways are there to choose 1 element from n?
- Problem **12** can be done by counting or with the binomial probability distribution of the last chapter.
- In problem **14**, some students may need help with the counting.

Additional Answer
Answer for problem **7a** can be found in the answer pages beginning on **A** 1.

Problem Set

A

P,R **8** Two dice are rolled.

a How many possible outcomes are there? 36
b How many distinct sums of the numbers on the upturned faces are possible? 11

P **9** Three of the elements of {1, 2, 3, 5, 7, 9, 11} are chosen at random.

a What is the probability that their sum is even? $\frac{3}{7}$
b What is the probability that their product is even? $\frac{6}{35}$

I **10** In a standard deck, how many cards are aces or kings or hearts? 19

R **11** Let C be the combination function.

a Find $C(8, 1)$. 8
b Find $C(15, 1)$. 15
c Find $C(32, 1)$. 32
d Find $C(13, 1)$. 13
e What conclusion can you draw from your answers to parts **a–d?** $C(n, 1) = n$

P **12** David has twelve 25-cent stamps and eight 13-cent stamps in his drawer. He reaches in without looking and selects five stamps. What is the probability that he takes three 25-cent stamps and two 13-cent stamps? $\approx .4$

P,R **13** A square is inscribed in a circle as shown.

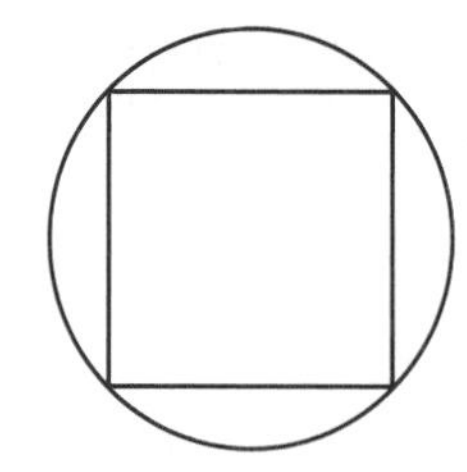

a What is the probability that a randomly chosen point in the interior of the circle also lies inside the square? $\frac{2}{\pi}$
b What is the probability that a randomly chosen point inside the square is also inside the circle? 1

P **14** You have 10 tickets for a raffle. If 120 raffle tickets have been sold and 2 tickets are drawn at random to determine the winners of 2 prizes, what is

a The probability that you will win no prize? $\approx .8396$
b The probability that you will win one prize? $\approx .1541$
c The probability that you will win both prizes? $\approx .0063$
d What is the sum of your answers to parts **a–c?** What is the meaning of this sum? 1; one of these outcomes must occur.

R | **15** Let A = {1, 2, 3, 5, 7} and B = {perfect squares less than 30}.

a Find A ∪ B. {1, 2, 3, 4, 5, 7, 9, 16, 25}
b Find A ∩ B. {1}

P | **16** A bag contains eight American League baseballs and five National League baseballs. Umpire Streichsbach reaches in and selects three balls without looking.

a What is the probability that all three are American League baseballs? ≈.196
b What is the probability that all three are National League baseballs? ≈.035

P,R | **17** Suppose that segments are drawn to points S and R from a point somewhere on $\overline{PQ}$, forming a triangle with base $\overline{RS}$. If a point inside square PQRS is selected at random, what is the probability that it is also inside the triangle? $\frac{1}{2}$

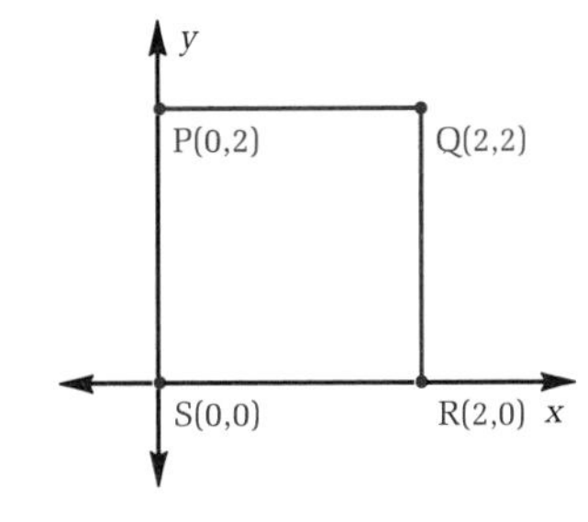

P  | **18** On the TV game show *Jewel of Fortune*, the winner gets to select, without looking, 3 gems from a bag containing 5 diamonds, 10 emeralds, and 20 zircons.

a What is the probability that a winner will select 3 zircons? ≈.17
b What is the probability that a winner will select a gem of each type? ≈.15

P | **19** James is gluing legs onto a table. Five red and eight blue legs are available. James randomly picks four of these legs. Find the probability that James's table will have four red legs. ≈.007

R | **20** Let M = {prime integers less than 50} and N = {odd positive integers less than 50}.

a Find #(N). 25
b Find #(M). 15
c Find #(M ∩ N). 14
d Find #(M ∪ N). 26
e Is #(N) + #(M) = #(M ∪ N) + #(M ∩ N)? Yes

P  | **21** A bag contains ten red marbles, four green marbles, and eight blue marbles. If four marbles are chosen at random from the bag, what is the probability that two are red and two are blue? ≈.17

B P | **22** Two congruent squares overlap as shown. What is the probability that a randomly chosen point in the interior of the figure is in the shaded region? $\frac{1}{7}$

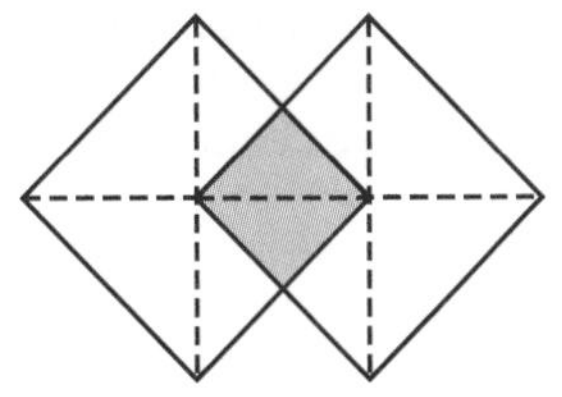

Cooperative Learning

Problem **16** involves the hypergeometric distribution which is modeled with the formula $P(x) = \frac{C(r, x)\, C(g, n - x)}{C(r + g, n)}$ where r is the number of American League baseballs, g is the number of National League baseballs, x is the number of American League baseballs chosen and n is the total number of baseballs chosen. This formula is one way students can verify the results obtained earlier. Now try this formula on problem **19** and verify its earlier result.

Problem-Set Notes and Strategies, continued

- Problem **16** is an example of the hypergeometric distribution and will be found in most statistics texts. It is within students' understanding here.
- Problems **18**, **19**, and **21** may be modeled with a tree diagram. This gives a graphic representation of the sample space and can help students find the probabilities.

Communicating Mathematics

You are a substitute teacher and must present this lesson to the next class because the teacher has become ill. How would you explain the difference between an outcome and an event?

Problem-Set Notes and Strategies, continued

- Problem **27** involves sigma notation. Students may want to write out each of the elements of the sum and calculate the answer.
- Students may guess at the pattern in problem **29** but not know it as an arithmetic sequence. These are explored in detail in the next chapter.
- Students may want to draw the picture for problem **30**. The inequality is the interior and boundary of a circle.
- Problem **31** may be a familiar logic puzzle.
- In problem **32**, students may miss the sixth square not realizing that the orientation could be changed.

Additional Answer

Answer for problem **26a** can be found in the answer pages beginning on **A** 1.

Problem Set, *continued*

P **23** Six cards are drawn at random from a standard deck. Find the probability that four are hearts and two are clubs. $\approx .003$

R **24** What is the greatest number of ancestors a person can have in the five generations preceding his or her own? 62

P **25** An aviary contained 12 birdies and 15 eagles. Bogey left the window open, and before Paar shut it, 7 of the aviary's inhabitants had flown out. Find the probability that 4 birdies and 3 eagles had escaped in alternate order.
$\approx .007$

P,R **26** Two different numbers are randomly chosen from $\{1, \sqrt{2}, \sqrt{3}, 2, \sqrt{5}, \sqrt{6}\}$.

a Represent the sample space for this experiment.
b Find the probability that at least one of the numbers selected is rational. $\frac{3}{5}$

R,I **27** Solve $\sum_{k=1}^{3}(kx + 5) = 20$ for x. $x = \frac{5}{6}$

P,R **28** A natural number x is randomly selected from set A, and a natural number y is randomly selected from set B. Find the probability that $|x - y| \geq 3$. .85

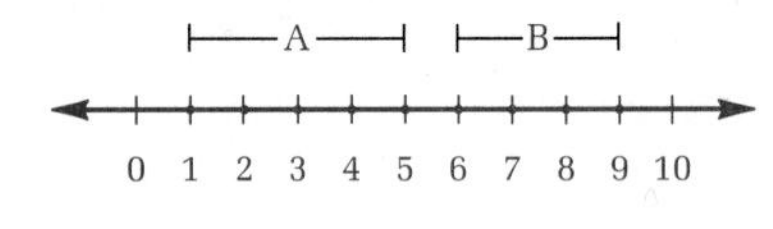

P,I **29** Consider the sequence 3, 7, 11, 15, 19, 23, ... If the pattern continues and three of the next five numbers in the sequence are chosen at random, what is the probability that none of the three is a multiple of 3? $\frac{1}{10}$

R **30** How many lattice points (x, y) satisfy the inequality $x^2 + y^2 \leq 25$? 81

C R,I **31** Bade Woggs has 25 red socks and 30 white socks in a drawer. Since it is dark, he can't tell them apart.

a How many socks must he take from the drawer to be sure that he has a matching pair of socks? 3
b How many socks must he take to be sure that he has a pair of red socks? 32

P **32** If four points are selected at random from the square lattice shown, what is the probability that the four points are vertices of a square? $\frac{1}{21}$

P **33** If a Christmas pie is 9 inches in diameter and contains 30 spherical plums, each of which is $\frac{3}{4}$ inch in diameter, what is the probability that Jack Horner will prick a plum if he sticks a pin into the pie at random? $\approx .21$

12.3 Probabilities of Unions

Objectives

After studying this section, you will be able to

- Understand disjunctions and unions of sets
- Calculate probabilities of mutually exclusive events

Part One: Introduction

Disjunctions

Suppose that one card is drawn at random from a standard deck. What is the probability that it is either an ace or a heart?

A statement in which two possible outcomes are connected by the word *or* is called a ***disjunction.*** When we are asked to find the probability of a disjunction, the event is the union of the two sets of outcomes, since the union consists of all the elements that are in one set *or* in the other set.

If we let A = {aces} and B = {hearts}, we know that

$$P(A \cup B) = \frac{\#(A \cup B)}{\#(\text{sample space})} = \frac{\#(A \cup B)}{52}$$

In Section 12.1, we found that $\#(A \cup B) + \#(A \cap B) = \#(A) + \#(B)$. In this case, $\#(A) = 4$, $\#(B) = 13$, and $\#(A \cap B) = 1$, so $\#(A \cup B) = 16$. The probability of drawing either an ace or a heart is $\frac{16}{52}$, or $\frac{4}{13}$.

When we apply the theorem presented in Section 12.1 to probabilities, we obtain the following theorem.

 ***For any events* A *and* B,**
$P(A \cup B) + P(A \cap B) = P(A) + P(B)$.

Example *If a single die is rolled, what is the probability that the roll is a multiple either of 2 or of 3?*

If we let A = {multiples of 2} and B = {multiples of 3}, we need to find the probability of $A \cup B$. The sample space is {1, 2, 3, 4, 5, 6}. Since A = {2, 4, 6} and B = {3, 6} and the intersection of A and B is {6}, $P(A) = \frac{3}{6}$, $P(B) = \frac{2}{6}$, and $P(A \cap B) = \frac{1}{6}$.

$$P(A \cup B) + P(A \cap B) = P(A) + P(B)$$

$$P(A \cup B) + \frac{1}{6} = \frac{3}{6} + \frac{2}{6}$$

$$P(A \cup B) = \frac{2}{3}$$

12.3

Class Planning

Time Schedule

Average: 2 days
Advanced: 2 days
Honors: 1 day

Resource References

Teacher's Resource Book
Class Opener 73
Practice 77
Evaluation
Quiz Forms 1, 2, 3

Class Opener

If a card is randomly drawn from a standard deck, what is the probability that it is a heart or a face card?

$\frac{11}{26}$

Lesson Notes

- One of the key ideas in the study of probability is explained in this section. Mutually exclusive events are those that have no chance of occurring at the same time. The formula $P(A \cap B) = 0$ says this statement symbolically. If two or more events are mutually exclusive, the probability of one or the other is the sum of the respective probabilities. This comes from the relationship first explored in Section **12.1**. The next section explores intersections (and) and independent events and the relationships that hold between these types of events.

Vocabulary

disjunction
mutually exclusive

12.3

Checkpoint

1. If you roll a pair of fair dice, what is the probability that you roll a sum of 7 or doubles? $\frac{1}{3}$
2. Six red and four green marbles are in a bag. If you reach inside and choose two marbles with replacement, what is the probability that they are both red or both green? $\frac{36}{100} + \frac{16}{100} = \frac{13}{25}$
3. What is the probability that you will choose both an ace and a jack when drawing one card from an ordinary 52 card poker deck and why? Probability is 0 since the events are mutually exclusive.

Assignment Guide

Average	
Day 1	5–7, 9, 11, 12, 14–16, 18, 20, 24, 30
Day 2	8, 10, 13, 17, 21, 23, 26, 28, 29, 31, 34
Advanced	
Day 1	5–7, 9, 11, 12, 14, 16, 18, 20, 24, 28, 30
Day 2	8, 10, 13, 17, 19–21, 23, 26, 28, 29, 31, 34, 36
Honors	
	6, 8, 10, 12–14, 17, 21, 23, 24–27, 31, 32, 36

Alternative Solution: Which of the elements of the sample space satisfy the given conditions?

1 is not a multiple of 2 or 3.
2 is a multiple of 2.
3 is a multiple of 3.
4 is a multiple of 2.
5 is not a multiple of 2 or 3.
6 is a multiple of both 2 and 3.

Four of the six rolls are multiples either of 2 or of 3, so the probability is $\frac{4}{6}$, or $\frac{2}{3}$.

Mutually Exclusive Events

When two events A and B cannot occur together, they are said to be ***mutually exclusive.*** In such a case, $P(A \cap B) = 0$, since the intersection of A and B is the empty set. For mutually exclusive events, therefore, the preceding theorem has the following form.

 ***If A and B are mutually exclusive events, then*** $P(A \cup B) = P(A) + P(B)$.

Example *If a card is randomly chosen from a standard deck, what is the probability that it is either an ace or a king?*

Let A = {aces} and B = {kings}. These events are mutually exclusive (the card cannot be both an ace and a king), so

$$P(A \cup B) = P(A) + P(B) = \frac{1}{13} + \frac{1}{13} = \frac{2}{13}$$

Part Two: Sample Problems

Problem 1 *A bag contains three chicken-salad sandwiches, four Limburger cheese sandwiches, and five ham sandwiches. If you reach into the bag and pick one sandwich at random, what is the probability that it is a chicken-salad sandwich or a ham sandwich?*

Solution $P(\text{chicken salad}) = \frac{3}{12}$, and $P(\text{ham}) = \frac{5}{12}$. Since choosing a chicken-salad sandwich and choosing a ham sandwich are mutually exclusive events (you can take only one sandwich), we add the two probabilities.

$$P(\text{chicken salad or ham}) = \frac{3}{12} + \frac{5}{12} = \frac{8}{12} = \frac{2}{3}$$

Problem 2 *Robin Finch has an aviary containing two kinds of birds, jays and bulbuls. Each bird is either gray or blue. She has 23 gray birds and 15 blue birds. Of the blue birds, 10 are bulbuls, and there are 18 bulbuls all together. One day, Robin's friend Martin came by and randomly selected a bird. What is the probability that he selected a gray bird or a jaybird?*

Solution If we let A = {gray birds} and B = {jays}, what we are trying to find is P(A ∪ B). Let's use a diagram to organize the information in the problem.

We start by entering the given values. Note that there are 23 + 15, or 38, birds in all.

	Gray	Blue	Total
Jays			
Bulbuls		10	18
Total	23	15	

Since there are 18 bulbuls and 10 of them are blue, 8 of them must be gray. Since 8 of the 23 gray birds are bulbuls, the other 15 must be jays. The remaining 5 jays are blue.

The first column in the diagram represents set A, and the first row represents set B. Therefore, #(A ∪ B) = 8 + 15 + 5, or 28, and $P(A \cup B) = \frac{28}{38}$, or ≈.74.

	Gray	Blue	Total
Jays	15	5	
Bulbuls	8	10	18
Total	23	15	

Communicating Mathematics

Find two events that you are sure are mutually exclusive and explain this in your own words. How would you convince others that they are mutually exclusive? Can you think of mutually exclusive relationships in terms of things other than probability events? If so, what?

Stumbling Block

Students seem to do fine with the definition of *mutually exclusive* until the definition of *independence*. Then they sometimes get confused. Sometimes it helps if they remember that "mutually exclusive" means that there is no chance of the events occurring at the same time. This means that $P(A \cap B) = 0$.

Part Three: Exercises and Problems

Warm-up Exercises

P **1** Suppose that two cards are picked from a deck. Are the following events mutually exclusive?

a {black cards} and {kings} No
b {spades} and {diamonds} Yes

P **2** Suppose that two dice are rolled. Are the following events mutually exclusive?

a {even numbers} and {odd numbers} Yes
b {multiples of 3} and {multiples of 2} No
c {1} and {prime numbers} and {composite numbers} Yes

P **3** At the annual dance of the Superheroes Club one year, each member showed up dressed either as Batman or as Superman. Of the 100 members attending, 76 were dressed as Batman, and 44 of these wore masks. All together, 51 members wore masks. What is the probability that a randomly selected member at the dance was dressed as Superman and wore no mask? $\frac{17}{100}$

Problem-Set Notes and Strategies

- Problem **3** can be solved by the process of elimination or the pigeonhole principle.

Problem-Set Notes and Strategies, continued

- In problem **5**, if time permits, have students make copies and cut up the rectangle. In this way, the problem is clearer.
- Once the ace of spades is removed in problem **6**, the events described are mutually exclusive and the probabilities may be summed.
- In problem **8**, you may want to remind students about the relationships in a 30°-60°-90° triangle.
- Problem **10** requires counting all the various possibilities and dividing that into the number of alternating possibilities.

Additional Answers

11a {(1, 1), (1, 2), (1, 3), (1, 4), (1, 5), (1, 6), (2, 1), (2, 2), (2, 3), (2, 4), (2, 5), (2, 6), (3, 1), (3, 2), (3, 3), (3, 4), (3, 5), (3, 6), (4, 1), (4, 2), (4, 3), (4, 4), (4, 5), (4, 6), (5, 1), (5, 2), (5, 3), (5, 4), (5, 5), (5, 6), (6, 1), (6, 2), (6, 3), (6, 4), (6, 5), (6, 6)}

11b {(2, 2), (2, 4), (2, 6), (4, 2), (4, 4), (4, 6), (6, 2), (6, 4), (6, 6)}

P,R **4** The figure at the right is known as Lewis Carroll's squares. If a point inside the figure is picked at random, what is the probability that it is in the shaded area? $\frac{6}{19}$

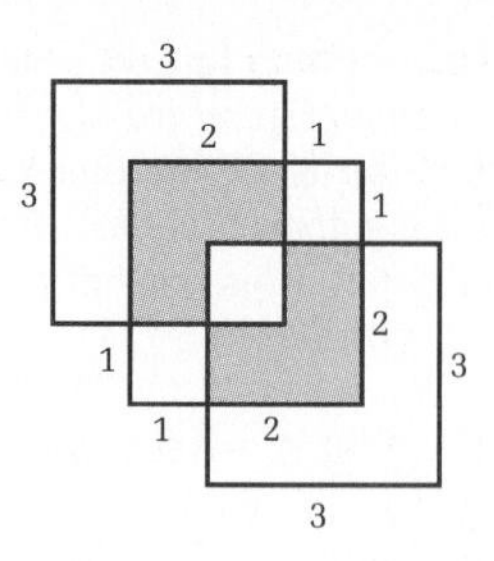

Problem Set

A

P **5** In rectangle ABCD, AB = 2(AD) and M and N are the midpoints of $\overline{AB}$ and $\overline{DC}$. If a point inside ABCD is selected at random, what is the probability it is inside △DMC or inside △ANB? $\frac{3}{4}$

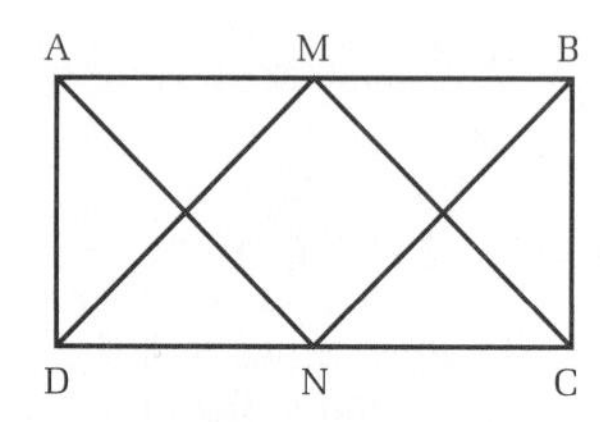

P **6** Suppose that the ace of spades is removed from a standard deck and a card is then selected at random. Find the probability that the card is either an ace or a spade. $\frac{15}{51}$, or ≈.294

P,R **7** If two of the numbers $3 + 2\sqrt{5}$, $4 - 2\sqrt{5}$, $5 + \sqrt{5}$, $4 + 2\sqrt{5}$, $3\sqrt{5}$, $3 - 2\sqrt{5}$, and $-3\sqrt{5}$ are chosen at random, what is the probability that they are

a Conjugates? $\frac{1}{7}$
b Opposites? $\frac{1}{21}$
c Opposites or conjugates? $\frac{1}{7}$
d Opposites and conjugates? $\frac{1}{21}$

P,I **8** PQRS is a 2 × 2 square. PQM and SNR are equilateral triangles. If a point inside PQRS is selected at random, what is the probability that it is inside either △PQM or △SNR? $\frac{6 - \sqrt{3}}{6}$

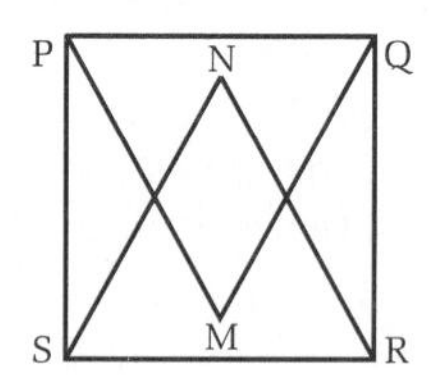

P,R **9** If four dice are rolled, what is the probability that the sum of the numbers on the dice is either odd or even? 1

R **10** A group of six adults and seven children was waiting in line for popcorn. If they had lined up randomly, what is the probability that the sequence was child, adult, child, adult, child, etc.? ≈.00058

P,R **11** **a** What is the sample space for the rolling of two dice?
b If the dice are loaded so that each always shows an even number, what is the sample space?
c If the two dice described in part **b** are rolled, what is the probability that at least one of them shows a 2? $\frac{5}{9}$

P,I | **12** If two dice—one red, the other green—are rolled, what is the probability of each event?

a {red 2 or green 5} $\frac{11}{36}$
b {red 2 and sum of 6} $\frac{1}{36}$
c {red 2 and green greater than 2} $\frac{1}{9}$
d {red 2 and sum of 2} 0

P | **13** If $P(A) = \frac{8}{15}$ and $P(B) = \frac{2}{15}$, is $P(A \cup B) = \frac{10}{15} = \frac{2}{3}$?
No, unless A and B are mutually exclusive

R,I | **14** Lectoral College has 2800 students, of whom 1910 take English, 1400 take math, and 820 take both math and English. If one of the students is chosen at random, what is the probability that he or she takes neither math nor English?
≈.111

R | **15** Seven rays have their endpoints at the origin of a coordinate system and lie in Quadrant I. How many acute angles are formed by the rays? 21

P,R,I | **16** A die is rolled. Let A = {multiples of 4} and B = {multiples of 2}. Find each of the following probabilities.

a $P(A)$ $\frac{1}{6}$ **b** $P(B)$ $\frac{1}{2}$ **c** $P(A \cup B)$ $\frac{1}{2}$ **d** $P(A \cap B)$ $\frac{1}{6}$

R  | **17** As a practical joke, Calvin took the labels off all 27 cans of food his family had in the cupboard. There were 13 cans of vegetables, 8 cans of soup, and 6 cans of baking soda. His father selected 3 cans from the shelf to use in preparing a dinner. What is the probability he got one can of soup and two of vegetables or two cans of soup and one of vegetables?
≈.338

R,I | **18** Suppose that $f(1) = 2$ and $f(n + 1) = 3[f(n)]$. Evaluate the following.

a $f(1)$ 2 **b** $f(2)$ 6 **c** $f(3)$ 18 **d** $f(4)$ 54

R,I | **19** Evaluate each of the following expressions.

a $\sum_{k=1}^{10} (\sqrt{-1})^k$ $-1 + i$ **b** $\sum_{k=1}^{11} (\sqrt{-1})^k$ -1

P,R | **20** Let A = {positive integers less than or equal to 20}. If three positive integers are randomly selected from A, what is the probability that all will be odd? ≈.1053

B
R,I | **21** Suppose that the odds of winning a game are 5 to 3. (This means that over a long series of trials, you will win 5 times for every 3 times you lose.)

a If you play the game once, what is the probability that you will win? $\frac{5}{8}$
b What is the probability that you will lose? $\frac{3}{8}$

Problem-Set Notes and Strategies, continued

- Problems **12** and **14** explore the events A *and* B instead of *or*. The sample space can be listed and counted. Further probabilities with the word *and* will be explored in the next section.

- Problems **18** and **19** introduce sequences and series, the subject of the next chapter. At this point, students can work them out and try to observe some patterns.

- Problem **21** shows the relationships between odds and probabilities. Some students may follow sports and the associated odds.

Problem Set, *continued*

Problem-Set Notes and Strategies, continued

- Problem **23** can be related to **21**.

R **22** How many triangles are in the diagram shown? 10

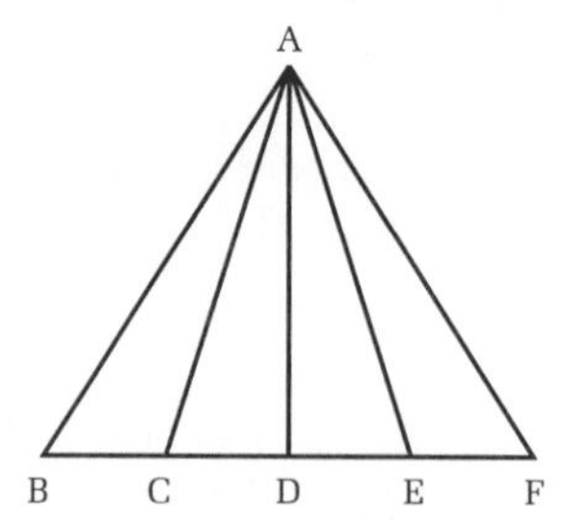

R,I **23** If the probability of a certain NFL team's winning the Super Bowl is $\frac{4}{11}$, what are the odds that the team will win the Super Bowl? 4 to 7

P  **24** The penny gum-ball machine at the Big Store contains 55 gum balls and 10 prizes. Groucho puts 3 cents into the machine and gets 3 items. Find the probability that they are either all prizes or all gum balls. ≈.6

R **25** If three of the ten points shown are selected at random, what is the probability that the points are collinear? .125

P **26** Namor's Submarine Sandwich Shop offers a "special surprise" meal for $0.50. You pick one sandwich at random from a box containing 8 liverwurst sandwiches, 2 ham sandwiches, and 4 roast beef sandwiches.

a What is the probability that you will get a ham sandwich or a roast beef sandwich for your $0.50? $\frac{3}{7}$

b How much would you have to spend to guarantee getting at least one ham or roast beef sandwich? $4.50

R  **27** At Thirty-seven Flavors, 37 kinds of ice cream, 6 kinds of syrup, and 4 kinds of nuts are available. If a person can order a sundae with or without whipped cream, how many different sundaes containing 3 scoops of ice cream and 1 kind of syrup can be ordered? (Assume that the order of stacking the scoops does not matter.) Possible answer: 303,918

P,R,I **28** A card is chosen from a standard deck. Let A = {jacks} and B = {queens}.

a Find P(A). $\frac{1}{13}$ **b** Find P(B). $\frac{1}{13}$ **c** Find P(A ∪ B). $\frac{2}{13}$ **d** Find P(A ∩ B). 0

P **29** Draw a Venn diagram to represent

a Two events, A and B, that are mutually exclusive

b Two events, A and B, that are not mutually exclusive

Additional Answers

29a

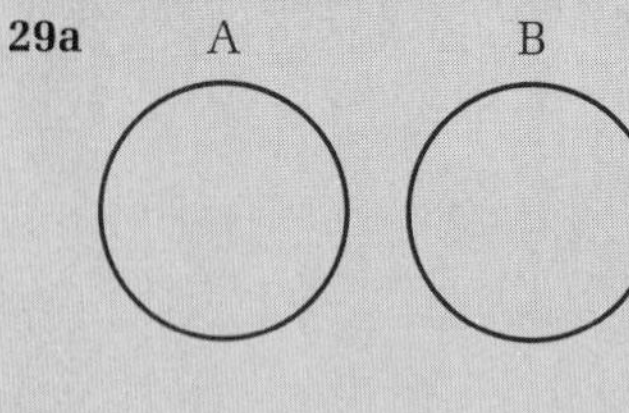

29b

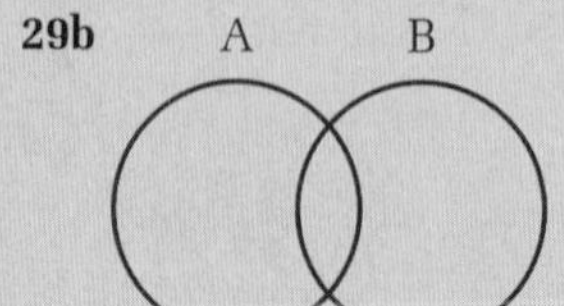

R | **30** If three of the points shown are selected at random, what is the probability that they are collinear? ≈.1545

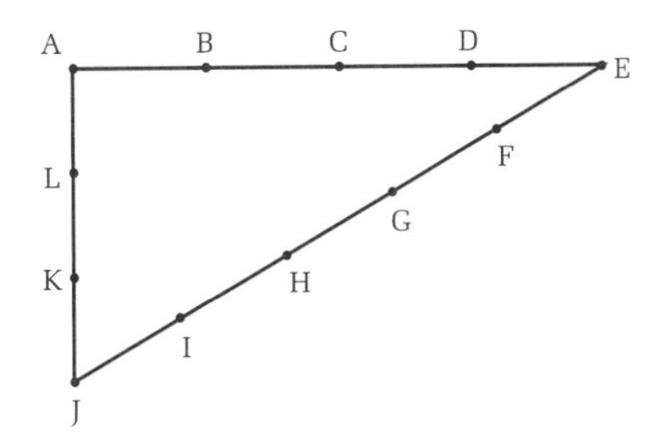

I | **31** Find the next two numbers in each sequence.

a 3, 9, 15, 21, 27, ___?___, ___?___ 33; 39
b 4, 9, 19, 39, 79, ___?___, ___?___ 159; 319
c 1, 1, 2, 3, 5, 8, 13, 21, 34, ___?___, ___?___ 55; 89

R | **32** If two of the six binomials $a + b$, $x + 2y$, $3x - y$, $x + y$, $a - b$, and $2y - x$ are chosen at random, what is the probability that their product is a difference of two squares? $\frac{2}{15}$

R | **33** A rectangle is inscribed in a square in such a way that four isosceles triangles are formed. What is the probability that

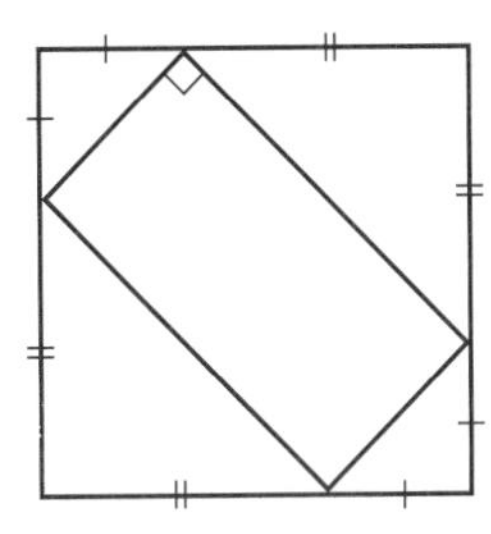

a The area of the rectangle is greater than half the area of the square? 0
b The area of the rectangle is less than half the area of the square? 1
c The area of the rectangle is equal to half the area of the square? 0

C R | **34** Consider the words *rhyme, slime, time, crime, lime, name, fame, game, same, tame,* and *blame.*

a If 2 of the 11 words are chosen at random, what is the probability that they rhyme? ≈.45
b If 2 of the 11 words are chosen at random, what is the probability that they have the same number of letters? ≈.49

R | **35** A number a is selected at random from {3, 7, 10, 12, 15}. Then a different number, b, is selected from the set, again at random. What is the probability that $\log_a b > 1$? $\frac{1}{2}$

R | **36** Let A = {multiples of 3 between 0 and 100}, B = {multiples of 5 between 0 and 100}, and C = {multiples of 3 or 5 between 0 and 100}. Find each number.

a #(A) 33 **b** #(B) 19 **c** #(C) 46

R | **37** Consider eight coplanar points, no three of which are collinear and no four of which lie on the same circle. How many circles pass through three of the eight points? 56

Problem-Set Notes and Strategies, continued

- Problem **31** asks students to match patterns and recognize the next few terms. This will be explored further in the next chapter. This can be an interesting discussion problem.
- Geometric concepts of a 45°-45°-90° triangle are needed to solve problem **33** and students may need a hint.
- When students figure out what problem **37** is asking, they can see a combination problem. Often, seeing the question is the most difficult part of the problem.

Class Planning

Time Schedule

All levels: 2 days

Resource References

Teacher's Resource Book
Class Opener 74
Cooperative Learning 37
Practice 78
Enrichment 24
Transparency 24

Class Opener

If two cards are drawn at random from a standard deck, one at a time and not replaced, what is the probability that both cards are spades?

$\frac{1}{17}$

Lesson Notes

- This section identifies the difference between independent and dependent events and also the probability of events *A* and *B* occurring. The use of the word *and* suggests the multiplication rule which works only if the events are independent. Students may become a little confused at the difference between independent, dependent and mutually exclusive. They may need some time and many examples to become comfortable with the definitions.
- If events are dependent, a conditional probability needs to be considered. An example may help. One example is: ask the probability of being selected for a five person team if there are seven possibilities to choose from. What is the probability of being chosen the first time? The second time? The third time? The probability changes each time.

12.4 PROBABILITIES OF INTERSECTIONS

Objectives

After studying this section, you will be able to

- Determine probabilities of intersections of independent events
- Solve problems involving conditional probabilities

Part One: Introduction

Intersections of Independent Events

Suppose that a die is rolled and a coin is tossed. What is the probability that the die shows 4 and the coin shows heads?

A statement in which two possible outcomes are connected by the word *and* is called a ***conjunction.*** The event corresponding to a conjunction is the intersection of the two sets of outcomes, since the intersection consists of all the elements that are in one set *and* in the other set.

If we let A = {4} and B = {heads}, then P(4 and heads) is P(A ∩ B). It is important to recognize that in this case A and B are ***independent events***—the rolling of the die has no effect on the tossing of the coin, and the tossing of the coin has no effect on the rolling of the die. When events are independent, we can find the probability of their intersection by simply multiplying the probabilities of the individual events.

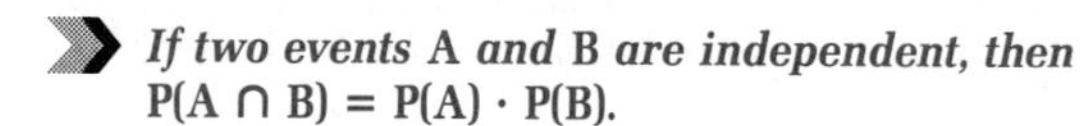

If two events A and B are independent, then **P(A ∩ B) = P(A) · P(B).**

Thus, since P(4) = $\frac{1}{6}$ and P(heads) = $\frac{1}{2}$, the probability that the die shows 4 and the coin shows heads is $\frac{1}{6} \cdot \frac{1}{2}$, or $\frac{1}{12}$. Remember, *this "multiplication rule" applies only to independent events.*

Example *Two cards are drawn at random from different standard decks. Find the probability that the first is an ace and the second is a king.*

We let A represent the event in which the first card is an ace and B represent the event in which the second is a king. The events are independent, since the choice of a card from one deck will have no effect on the choice of a card from the other deck. Thus,

$$P(A \cap B) = P(A) \cdot P(B) = \frac{4}{52} \cdot \frac{4}{52} = \frac{1}{13} \cdot \frac{1}{13} = \frac{1}{169}$$

Vocabulary

conjunction
independent events
dependent events
conditional probability

Dependent Events and Conditional Probabilities

Now suppose we draw a card from a standard deck and then, without replacing that card, we draw a second card from the deck. What is the probability that the first is an ace and the second is a king?

This situation is different from that in the preceding example. The choosing of the first card affects the probability for the choosing of the second—there are only 51 cards, rather than 52, from which the second can be chosen. If we let A represent the event in which the first card is an ace and B represent the event in which the second is a king, we say that A and B are ***dependent events.*** We want to find the probability that the first card is one of the four aces in the entire deck of 52 cards and that the second card is one of the four kings in the remaining 51 cards. Thus,

$$P(A \cap B) = \frac{4}{52} \cdot \frac{4}{51} = \frac{4}{663}$$

The factor $\frac{4}{51}$ represents the ***conditional probability*** that B will occur, *given that A has occurred,* denoted $P(B|A)$. To find the probability of the intersection of two dependent events, therefore, we use the following rule.

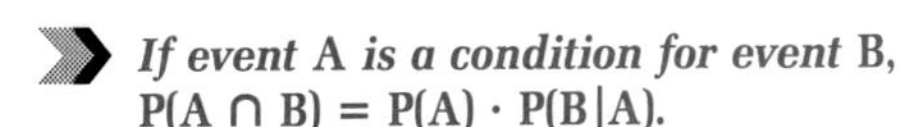

If event A is a condition for event B,
$P(A \cap B) = P(A) \cdot P(B|A).$

Note that when A and B are independent events, $P(B|A)$ is simply $P(B)$, so this rule becomes equivalent to the previously discussed multiplication rule. Notice also that when A and B are mutually exclusive events (a special case of dependent events), $P(B|A) = 0$, so $P(A \cap B) = 0$.

If we rewrite the formula $P(A \cap B) = P(A) \cdot P(B|A)$ in the form $P(B|A) = \frac{P(A \cap B)}{P(A)}$, we can use it to find conditional probabilities.

Example *Imagine two men, each of whom tells the truth only two thirds of the time. If both men say that it is raining, what is the probability that it really is raining?*

It might appear that the probability is merely the product of the probability that one man is telling the truth and the probability that the other is telling the truth, $\frac{2}{3} \cdot \frac{2}{3}$, or $\frac{4}{9}$. The problem, however, indicates that the men *agree* that it is raining, so what we really need to find is the probability that one is telling the truth (that is, the probability that it is raining), *given that* they agree.

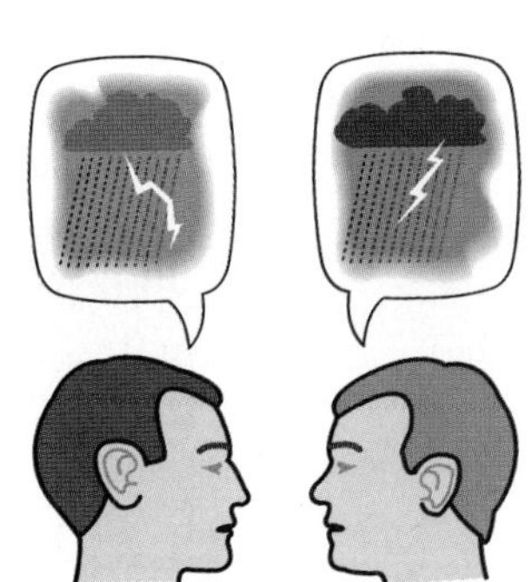

Checkpoint

1. If a black and white die are rolled, what is the probability of rolling doubles and of getting an even number on the white die? $\frac{1}{12}$
2. Are the events in the last question, i.e. doubles and even number on the white die, independent? Yes
3. When drawing cards from an ordinary deck, find the probability of getting an ace. Now find the probability of getting two aces, knowing that the first card was an ace. $\frac{1}{13}$ and $\frac{1}{221}$

Assignment Guide

Average	
Day 1	8–10, 13, 14, 20, 22, 23, 25, 26
Day 2	11, 12, 15, 17, 18, 21, 24, 27, 29, 30
Advanced	
Day 1	8–10, 13, 14, 19, 20, 22, 23, 25, 26, 29
Day 2	11, 12, 15–18, 21, 24, 27, 28, 30
Honors	
Day 1	8, 9, 11, 13, 14, 19–21, 23, 25, 26, 29, 30
Day 2	10, 12, 15–18, 22, 24, 27, 28, 31

If we let A represent the event in which the men agree and B represent the event in which one of them tells the truth, therefore, we are looking for $P(B|A)$. Now, $P(A)$ is the probability that both men are telling the truth or both are lying, so $P(A) = \frac{2}{3} \cdot \frac{2}{3} + \frac{1}{3} \cdot \frac{1}{3} = \frac{5}{9}$. $P(A \cap B)$ is $\frac{4}{9}$, the probability that both men are telling the truth. Thus,

$$P(B|A) = \frac{P(A \cap B)}{P(A)} = \frac{\frac{4}{9}}{\frac{5}{9}} = \frac{4}{5}$$

Alternative solution: We can use a chart to model the problem. On the average, each man tells the truth two times out of three. Since we know that the men agree, we do not need to consider the four cases that are crossed out. Of the remaining five cases, four involve the men telling the truth. The probability is therefore $\frac{4}{5}$.

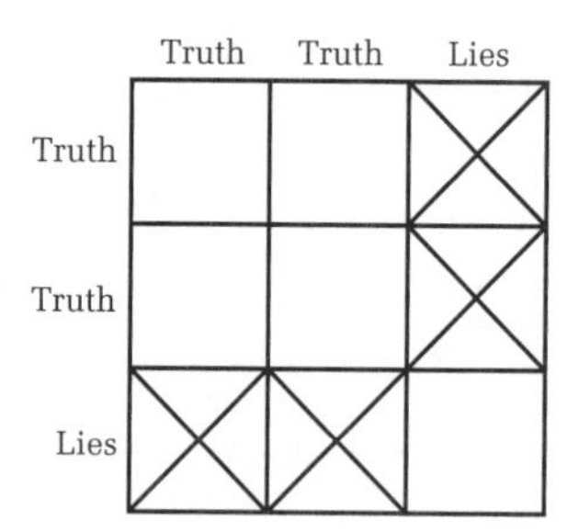

Stumbling Block

Events are mutually exclusive when $P(A \cap B) = 0$ and independent when $P(B|A) = P(B)$. Students can confuse these definitions. For many students, it helps to see the literal translations of these symbolic formulas to verify the difference. In the case of mutually exclusive events, the formula says that A and B cannot happen at the same time. In the case of independent events, the occurence of A has no effect on the probability of B's occurring.

Communicating Mathematics

Explain the relationship between independent events and mutually exclusive events. If A and B are known to be mutually exclusive, can they also be independent? If they are independent, can they be mutually exclusive? If they are dependent, what can you conclude? If they are not mutually exclusive, what can you conclude?

Part Two: Sample Problem

Problem *Tom, Dick, and Harry independently answer a question on a test. The probability that Tom answers correctly is .8, the probability that Dick answers correctly is .6, and the probability that Harry answers correctly is .3. What is the probability that*

a *All three answer the question correctly?*
b *At least two of them answer the question correctly?*
c *At least one answers the question correctly?*

Solution We will let T = {cases in which Tom answers correctly}, D = {cases in which Dick answers correctly}, and H = {cases in which Harry answers correctly}. These three events are independent, and we know that $P(T) = .8$, $P(D) = .6$, and $P(H) = .3$. Also, the probability that Tom answers incorrectly [denoted $P(T')$] is $1 - .8$, or .2. Similarly, $P(D') = .4$ and $P(H') = .7$.

a Here we want the probability of the conjunction of T and D and H—that is, $P(T \cap D \cap H)$.

$$\begin{aligned} &P(T \cap D \cap H) \\ &= P(T) \cdot P(D) \cdot P(H) \\ &= .8(.6)(.3) \\ &= .144 \end{aligned}$$

b This part is tricky. There are several ways in which at least two can answer the question correctly—the correct answerers could be Tom and Dick, Tom and Harry, Dick and Harry, or all three. We can represent these possibilities as the following four mutually exclusive events.

1 Tom and Dick answer correctly, but not Harry—$T \cap D \cap H'$.
2 Tom and Harry answer correctly, but not Dick—$T \cap H \cap D'$.
3 Dick and Harry answer correctly, but not Tom—$D \cap H \cap T'$.
4 All three answer correctly—$T \cap D \cap H$.

Therefore,

$$\begin{aligned}
&P(\text{at least two answer correctly})\\
&= P[(T \cap D \cap H') \cup (T \cap H \cap D') \cup (D \cap H \cap T') \cup (T \cap D \cap H)]\\
&= P(T \cap D \cap H') + P(T \cap H \cap D') + P(D \cap H \cap T') + P(T \cap D \cap H)\\
&= P(T) \cdot P(D) \cdot P(H') + P(T) \cdot P(H) \cdot P(D')\\
&\quad + P(D) \cdot P(H) \cdot P(T') + P(T) \cdot P(D) \cdot P(H)\\
&= .8(.6)(.7) + .8(.3)(.4) + .6(.3)(.2) + .8(.6)(.3)\\
&= .336 + .096 + .036 + .144\\
&= .612
\end{aligned}$$

c
$$\begin{aligned}
P(\text{at least one is correct}) &= 1 - P(\text{all are wrong})\\
&= 1 - P(T' \cap D' \cap H')\\
&= 1 - .2(.4)(.7)\\
&= 1 - .056\\
&= .944
\end{aligned}$$

Problem-Set Notes and Strategies

- Problems **1 – 4** can give an indication whether students grasp the definition of independence and what makes two events independent.

Part Three: Exercises and Problems

Warm-up Exercises

In problems 1–4, indicate whether events A and B are independent.

P

1 One die is rolled; A = {5} and B = {4}. No

2 Two dice are rolled; for the first die, A = {5}, and for the second die, B = {4}. Yes

3 Two cards are drawn from a standard deck; for the first card, A = {kings}, and for the second card, B = {kings}. No

4 One card is drawn from a standard deck; A = {kings} and B = {spades}. Yes

P **5** If the probability that it will rain today is .6 and the probability that it will rain tomorrow if it rains today is .35, what is the probability that it will rain both today and tomorrow? .21

P **6** Two cards are chosen at random from a standard deck. Find the probability that the first card is a king and the second is a spade. $\frac{1}{52}$, or ≈.02

Problem-Set Notes and Strategies, continued

- Problem **8** can help identify independent and mutually exclusive events. One approach is to go through the probabilities and show that the formulas produce the same results.
- In problem **9**, the events are dependent, but the independent probability is also ≈ 0.0005. You may want students to write their answer in fraction form or to five decimal places.
- Problem **10** is theoretically $\frac{1}{2}$. The actual answer may depend on some genetic traits of the parents. Students might raise this question. Explain that we are interested in theoretical results only.
- You may want to help students recognize the multiple answers for parts **c** and **d** in problem **11**. There is more than one way to have at least one musician play a C.

P,R

7 On a multiple-choice test, each item includes four choices, of which one is correct. If the test consists of six items, what is the probability that you can answer all of them correctly by guessing? $\frac{1}{4096}$

Problem Set

A

P,R

8 A die is rolled and a coin is tossed. Indicate whether each outcome is best described as consisting of independent events, dependent events, or mutually exclusive events.

a The die shows 5 and the coin shows heads. Independent

b The die shows 5 and the die shows an even number. Mutually exclusive

c The coin shows heads, given that the coin shows heads and the die shows 3. Dependent

d The die shows 2 and the die shows 3. Mutually exclusive

P

9 Three cards are chosen from a standard deck. Find the probability that the first is a king, the second is a queen, and the third is a jack. ≈.0005

P

10 All eight children in the P. R. Olific family are female. If Mr. and Mrs. Olific have a ninth child, what is the probability that it will be female? $\frac{1}{2}$

P,R

11 Each of three musicians plays a note independently. The probability that the first plays a C is $\frac{2}{3}$; the probability that the second plays a C is $\frac{1}{4}$; the probability that the third plays a C is $\frac{3}{8}$. What is the probability that

a All three play a C? $\frac{1}{16}$

b None of the three plays a C? $\frac{5}{32}$

c At least one of the three plays a C? $\frac{27}{32}$

d At least two of the three play a C? $\frac{37}{96}$

B

R

12 A gum-ball machine contains 45 gum balls and 5 prizes. If you put in 6 cents and receive 6 items, what is the probability that you will get

a All gum balls? ≈.51

b 4 gum balls and 2 prizes? ≈.009

P

13 Two points inside the figure shown are chosen at random. Find the probability that the first is in square A and the second is in rectangle B. $\frac{1}{18}$

P | **14** A die is rolled. Consider the events A = {odd numbers}, B = {even numbers}, and C = {2}.

a Find $P(A \cap B)$. 0
b Find $P(A \cap C)$. 0
c Find $P(B \cap C)$. $\frac{1}{6}$

P | **15** If $P(A) = \frac{1}{2}$ and $P(A|B) = \frac{2}{3}$, what is $P(A \cap B)$?
Cannot be determined

P | **16** Suppose that a card is drawn from a standard deck. Indicate whether each of the following outcomes consists of independent events or dependent events. Explain your answers.

a The card is a heart and the card is a 4. Independent
b The card is a heart and the card is a spade. Dependent
c The card is a queen and the card is a king. Dependent

P | **17** Suppose that two cards are drawn from a standard deck. Indicate whether each of the following outcomes consists of independent events or dependent events. Explain your answers.

a The first card is a jack and the second card is an ace. Dependent
b The first card is a club and the second card is black.
Dependent

P | **18** The Buffalo Chips Snack Food Co. produces bags containing a mixture of 35 potato chips, 24 taco chips, and 11 banana chips. Dale reaches into the bag and grabs 15 chips. Find the probability that he gets 5 of each kind. $\approx .0088$

R,I | **19** Evaluate each expression.

a $\sum_{k=1}^{n} (\sqrt{-1})^k$, where n is odd i or -1

b $\sum_{k=1}^{n} (\sqrt{-1})^k$, where n is even $i - 1$ or 0

R,I | **20** **a** Evaluate $\frac{1}{1 \cdot 2}$. $\frac{1}{2}$

b Evaluate $\frac{1}{1 \cdot 2} + \frac{1}{2 \cdot 3}$. $\frac{2}{3}$

c Evaluate $\frac{1}{1 \cdot 2} + \frac{1}{2 \cdot 3} + \frac{1}{3 \cdot 4}$. $\frac{3}{4}$

d Evaluate $\frac{1}{1 \cdot 2} + \frac{1}{2 \cdot 3} + \frac{1}{3 \cdot 4} + \frac{1}{4 \cdot 5}$. $\frac{4}{5}$

e Evaluate $\frac{1}{1 \cdot 2} + \frac{1}{2 \cdot 3} + \frac{1}{3 \cdot 4} + \frac{1}{4 \cdot 5} + \frac{1}{5 \cdot 6}$. $\frac{5}{6}$

f Use the pattern you developed in parts **a–e** to predict the value of $\frac{1}{1 \cdot 2} + \frac{1}{2 \cdot 3} + \frac{1}{3 \cdot 4} + \ldots + \frac{1}{98 \cdot 99} + \frac{1}{99 \cdot 100}$. $\frac{99}{100}$

P,R | **21** A bag contains 15 marbles, of which 6 are red and 9 are green. Suppose that 5 marbles are drawn at random from the bag.

a What is the probability that all 5 are the same color? $\approx .044$
b What is the probability that all 5 are green? $\approx .042$
c If all 5 are the same color, what is the probability that they are green? $\approx .995$

Problem-Set Notes and Strategies, continued

- Problem **19** is an example of a series. Students can list the first several terms and look for a pattern. The next chapter is devoted to analyzing series and sequences.
- Problem **20** is also a series, but it is written in a different form. After the first few, students may see a pattern and predict the results for the last few cases.
- Problem **21** is another example of the hypergeometric distribution outlined in a cooperative learning lesson in Section **12.2**. A formula was introduced there to allow easy computation of the probabilities. The problem can be solved without the formula also.

Problem Set, *continued*

Problem-Set Notes and Strategies, continued

- Problem **24** is a review of the 30°-60°-90° triangle. This result will be used extensively in the trigonometry chapters.
- In problem **25** remember that *at least one* means the first, the second or both.
- Problem **26** utilizes the formulas for conditional probabilities applied twice.
- Problems **27** and **28** may take some time.
- Problem **29** is a review of one of the rules of a probability distribution.

Cooperative Learning

Problems **26** and **27** require several concepts and problem **28** calls for geometric knowledge. Suggest that students work in groups of three or four to solve these together.

P **22** If $P(A) = \frac{2}{3}$ and $P(B) = \frac{1}{2}$, is $P(A \cap B) = \frac{2}{3} \cdot \frac{1}{2} = \frac{1}{3}$?
Only if A and B are independent

P **23** When the Wassamatta U. Mooses score a touchdown, they kick for an extra point 80% of the time. On the average, 75% of their kick attempts are successful. About what percentage of the Mooses' touchdowns are followed by a successfully kicked extra point? 60%

I **24** For each of the following equilateral triangles, find the values of x and y and the measure of $\angle\theta$.

a

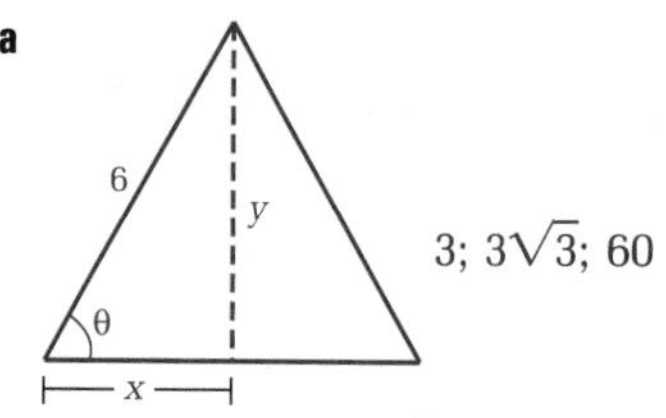

3; $3\sqrt{3}$; 60

b

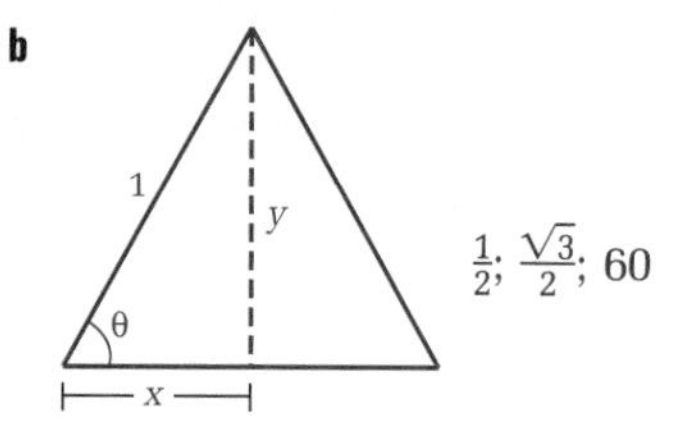

$\frac{1}{2}$; $\frac{\sqrt{3}}{2}$; 60

P **25** The probability that Mary Curd will score on a free-throw attempt is .85. If she is fouled and is awarded two free throws, what is the probability that she makes at least one free throw? .9775

P **26** If $P(A) = .8$, $P(A|B) = .64$, and $P(B|A) = .4$, what is $P(B)$?
.5

P,R **27** A piggy bank contains six pennies, five nickels, four dimes, three quarters, and two worthless slugs. If the bank is shaken until three of the items drop out, what is the probability that their total value is $0.25? ≈.03

P **28** On the dart board shown, the smallest of the concentric circles has a radius of 2 inches. Each of the other circles has a radius 2 inches greater than that of the next smaller circle. What is the probability that two darts hitting the board at random will score exactly 25 points? ≈.17

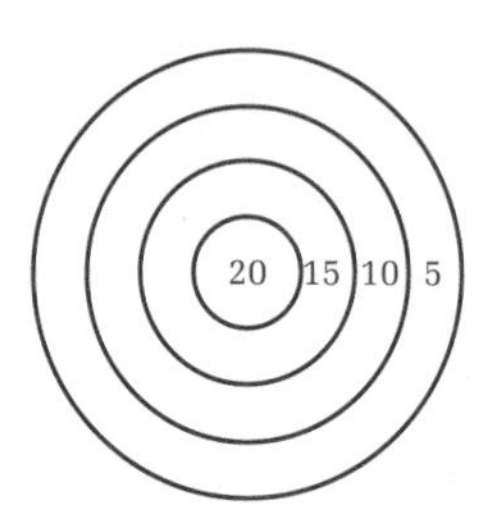

R **29** A bag contains red, green, and blue marbles. Is it possible that the probability of choosing a red marble is $\frac{1}{3}$, the probability of choosing a green marble is $\frac{1}{2}$, and the probability of choosing a blue marble is $\frac{1}{4}$? Why or why not?
No; P(red) + P(green) + P(blue) must be equal to 1.

C P,R **30** Ten decks of cards (520 cards in all) are shuffled together, and 5 cards are dealt. Find the probability that at least 2 of the cards are of the same suit. 1

P | **31** A dart board for the game of skill darts is shown. If a player's first dart lands inside the larger square, the player throws another. If the second dart lands inside the circle, the player throws a third time. If the third dart lands inside the smaller square, the player wins. If all of a player's throws land somewhere inside the larger square, what is the probability that the player wins?

$\approx .39$

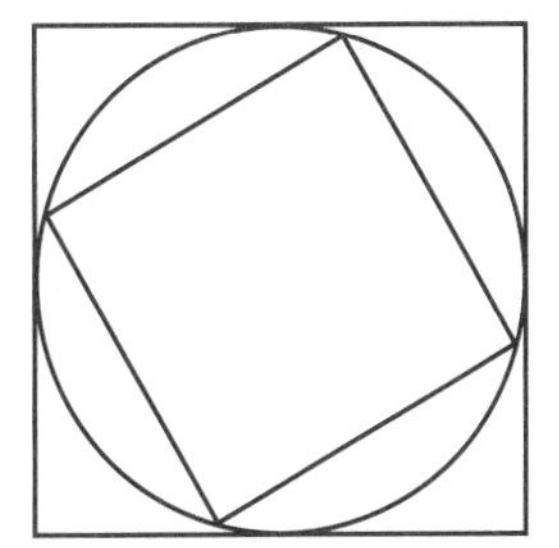

Problem-Set Notes and Strategies, continued

- Problem **31** will require recollection of geometric skills with 45°-45°-90° triangles and the area of circles. A calculator is a help.

HISTORICAL SNAPSHOT

A Lucky Collaboration

Fermat and Pascal: A co-theory of probabilities

What are the odds that two people, neither of whose interests were exclusively mathematical, would develop through correspondence the fundamentals of future probability theory? It happened in France in the 1600's.

Pierre de Fermat (1602?–1665) was a magistrate by profession and a mathematician at his leisure. Blaise Pascal (1623–1662) is well known not only as a mathematician but also as a religious philosopher and especially as a scientist. The two wrote letters back and forth in which they developed the fundamental principles of probabilities. Their correspondence provided each with a sounding board. Their exchange of ideas helped them revise hypotheses, spot logical errors, and correct them. Although their solutions to the same problems varied in details, both are credited as the originators of a single theory.

Fermat and Pascal began their work together at the request of a gambler. Interest in probability later grew as the insurance industry developed. Probability theory now has many applications, particularly in business and science. In short, a theory with seedy beginnings has become a respected area of mathematics.

Pascal himself did not develop the triangle shown below, but because of the applications he developed for it, the triangle now bears his name.

Use Pascal's triangle to find the possible outcomes of tossing a coin. Continue the triangle to show ten tosses. Do you see a pattern?

12.5

Class Planning

Time Schedule
All levels: 1 day

Resource References
Teacher's Resource Book
Class Opener 75
Cooperative Learning 38
Practice 79
Manipulatives and Applications 17

Class Opener

The weatherman predicts a 45% probability of rain tomorrow; a math teacher predicts a 25% probability that two coin tosses will result in two heads; Varhard University predicts there is a 30% probability that the average applicant will be admitted; a newspaper states that the probability of a win for the home team is 60% and the probability of a lottery ticket winning the grand prize is 1 in 1,000,000. Write three to five statements discussing the differences and similarities between the uses of probability.
Answers will vary. They should focus on the idea that some probabilities are empirical, some theoretical, some misleading(in Varhard's prediction, there is no average student, probability changes for each student).

Lesson Notes

- This section tries to give meaning to the idea of probability. Remind students that there are no guarantees in probability, just the likelihood that an event will occur.

12.5 INTERPRETING PROBABILITIES

Objectives

After studying this chapter, you will be able to

- Understand the significance of probabilities
- Understand how Monte Carlo methods are used to estimate probabilities
- Judge sampling techniques

Part One: Introduction

What Does a Probability Mean?

Imagine that there is an outbreak of influenza among the students in your school and your doctor has told you that if you continue to attend classes, there is a probability of $\frac{1}{10}$ that you will catch the flu. What does this probability mean? Does it mean that out of any group of ten students, one will certainly get sick? Does it mean that if you go to school nine days without catching the flu, you are sure to catch it on the next day? Does it mean that if two of the twenty students in your homeroom are sick, you have nothing to worry about?

Actually, it means none of these things. What your doctor is telling you is that she estimates, on the basis of her previous experience with the disease, that about one tenth of the students attending your school will catch the flu. She has no way of knowing whether you will be one of them.

It is important to realize that a probability indicates what fraction of a great number of cases are likely to involve a certain outcome; it does not predict what the outcome will be in any individual case or group of cases.

Example *A coin is tossed 13 times and turns up heads each time. What is the probability that the next toss will result in heads?*

The probability is $\frac{1}{2}$. The outcomes of the first 13 tosses do not change the probability of heads on the fourteenth toss, since each toss is independent of the preceding tosses.

Vocabulary
sample

Example *A coin is tossed 14 times. What is the most likely distribution of heads and tails?*

The most likely distribution is 7 heads and 7 tails. Although this outcome has a probability of only ≈.209, it is still more likely than any of the other possible outcomes. Since for any given toss P(heads) = $\frac{1}{2}$ and P(tails) = $\frac{1}{2}$, the most likely outcome will be half heads and half tails whenever a coin is tossed an even number of times.

Monte Carlo Methods

In the early days of gambling casinos, sophisticated techniques for calculating probabilities had not yet been developed. How did the casino owners determine the odds of winning a game?

What they did was to play the game thousands of times and record the results. Then they used the data they had collected to estimate the probability. Such methods of estimating probabilities are called Monte Carlo techniques.

Example *Suppose that we didn't know the probabilities of the possible outcomes of rolling two dice. How could we approximate them?*

We could have a computer simulate the rolling of two dice several thousand times. Suppose the computer produced the following data.

Roll	2	3	4	5	6	7	8	9	10	11	12
Frequency	453	912	1332	1815	2262	2710	2259	1822	1330	899	444

We might notice that rolls of 2 and 12 occurred about the same number of times, as did rolls of 3 and 11, 4 and 10, 5 and 9, and 6 and 8. Moreover, the number of 2's divided by the total number of trials is approximately $\frac{1}{36}$, the number of 3's divided by the total number of trials is about $\frac{2}{36}$, and so on. By repeated experimentation, we could closely approximate the probability of each outcome.

Many real-life probability problems are so complex that they cannot be solved by means of formulas and theorems. In such cases, mathematicians use Monte Carlo techniques to obtain accurate estimates of the probabilities. Of course, the mathematicians usually use computers to model the experiments instead of carrying out actual trials.

Lesson Notes, continued

- In the case of sampling, the important thing to remember is that a representative sample must be obtained. Some statisticians spend years trying to devise ways of getting a good sample. It is not an easy task in general. Look for all possibilities and be aware of unusual results, trying to determine the reasons for these results. A probability or inference based on a sample is only as good as the sample representing the population. If the sample is biased, the inference or probability will also be biased.

12.5

Checkpoint

1 If you were charged with obtaining the average age of the faculty of your campus, how would you collect a sample of data?
Answers will vary. Make sure they are representative.

2 The odds of the Oakland A's winning the world series are 7 to 5. What does this indicate.
The probability they will win is $\frac{7}{12}$ or a little better than half.

3 The probability of getting a heart when drawing a card from an ordinary poker deck is $\frac{1}{4}$. What does this mean if you have to draw one card right now?
Answers will vary.

Assignment Guide

Average
4–8, 10, 13–16, 19, 21
Advanced
4–9, 12–15, 18, 19–21
Honors
4, 6–8, 11, 13–15, 17–20, 22

Sampling

You have probably seen polls that predict the results of political elections. How are these predictions made? The pollsters clearly cannot ask every voter in the country whom he or she will vote for. Instead, they select a group of people they hope will be representative of the entire population. Such a group is called a ***sample.***

Example *Wanting to know how people in the United States feel about mass transit, we decide to stop people at a commuter train station and ask their opinion. Will our sample be a good one?*

Probably not. We would be questioning only people who actually use a mass-transit facility. The sample would not include people who prefer to ride a bike, drive a car, or walk.

Example *In 1936, the Literary Digest sent out questionnaires asking people whether they were going to vote for Franklin D. Roosevelt or Alf Landon. The magazine reported that Landon would win by a three-to-two margin. Instead, Roosevelt won by a landslide. What might have been wrong with the magazine's polling technique?*

In fact, only 23% of those to whom the questionnaires were sent mailed them back to the magazine, and most of these were wealthy people more likely to vote for the Republican Landon. To be sure of accurate results, the magazine should have tried to poll a sample that was more representative of the whole electorate.

Example *The results of political elections are frequently projected by means of exit polls—that is, by asking voters leaving polling places in selected precincts whom they voted for. Usually, such projections are fairly accurate. Why?*

There are two main reasons. First, the samples consist only of people who actually voted. Second, the pollsters select precincts in which, according to the results of previous elections, the voters are reflective of the electorate as a whole.

Part Two: Sample Problem

Problem *A cereal manufacturer puts a prize in each box of cereal. There are ten different prizes, and the prizes are randomly distributed. How might we determine the average number of boxes of cereal a person would have to buy to get all ten prizes?*

Solution One way would be to write the numbers 1–10 on slips of paper to represent the ten prizes. We could pick slips at random, recording the number on each and then replacing it with the others before drawing again, until we had recorded each number at least once. By repeating this experiment many times and calculating the mean number of draws necessary to obtain all ten numbers, we would obtain an approximate value for the number of cereal boxes one would need to buy.

Part Three: Exercises and Problems

Warm-up Exercises

P **1** Suppose that you wished to conduct a poll to determine the average number of hours students study each night. How would you conduct the poll to be sure that your sample was not biased?

P **2** A die is to be rolled repeatedly and a running total of the outcomes kept until the total exceeds 30. Design a Monte Carlo experiment to determine the probability that the total will be 32.

P **3** If the makers of Sneezarrest capsules claimed that in a poll of doctors, 90% of the physicians surveyed preferred their product, why might you be skeptical of their claim?

Problem Set

A

P **4** A high-school counselor sent a letter to each member of the preceding year's graduating class, asking how well he or she was doing in college. Of the 52% of the graduates that responded, 93% said that they were "having a great year" or "doing well." What conclusions can be drawn from this survey?

R **5** Sandy enters 6 [×] into her calculator. She then enters a random digit and hits [=]. What is the probability that the units digit of the displayed value is

a 6? $\frac{1}{5}$ **b** 8? $\frac{1}{5}$ **c** 3? 0

P **6** Flip four coins, recording the outcome, then repeat the experiment.

a Were the results of your two trials the same? Answers will vary.
b What is the most probable outcome of the flipping of four coins? 2 heads and 2 tails
c Would you expect the most probable outcome to occur each time you flipped four coins? Why or why not? No; probability indicates only the relative likelihood of an outcome.

P **7** A school's librarian was concerned because not many students were using the library. To find out why, she carefully composed a questionnaire about library use and during one week asked every student entering the library to fill out the questionnaire. Discuss the validity of her approach.

R,I **8** Evaluate $\sum_{k=1}^{7}(3k - 2)$. 70

B

P **9** If you were in charge of marketing a new hair-coloring product and needed to know how many people currently use such a product, how would you go about obtaining this information?

Problem-Set Notes and Strategies

- In problems **1 – 3**, help students recognize the importance of a representative sample. The poll or Monte Carlo experiment will only be as good as the sample.
- Problem **7** is also not a representative sample. Ask students how the librarian might have acquired a good sample.
- Problem **8** is a series. Students can calculate each term and add them together to arrive at the answer. The next chapter will discuss series in more detail.

Additional Answers

1 Answers will vary. Students may mention that subjects should be randomly selected from groups representing all grades and a variety of course loads.

Answers for problems **2**, **3**, **4**, **7**, and **9** can be found in the answer pages beginning on **A** 1.

Cooperative Learning

Use problem **6** as a group lesson to discuss the Monte Carlo method. Have each student flip a coin four times and record the results as the problem indicates and then repeat the experiment. Ask for students who got results that were significantly different from expected. Combine all the students results into a single sample and compare the empirical results to the theoretical probabilities.

Problem-Set Notes and Strategies, continued

- Problem **12** introduces a sequence. Sequences are discussed in more detail in the next chapter.
- Problem **14** asks the student to find the mean or expected value of a hypergeometric distribution. You may want to check that students understand that the mean and expected value are the same.
- Problem **15** is another example of the hypergeometric distribution. The cooperative learning lesson in Section **12.2** outlined a general formula to compute these probabilities.
- Problem **17** uses a margin of error instead of the standard deviation. Explain that since the margin was ± 50, the time and money spent to obtain the new score was not justified unless Judy learned and retained what she learned.

Additional Answers

11 Answers will vary. Students may mention the necessity of random sampling, including both those who have read and those who have not read Dickens's works.

17 Answers will vary. Students may note that, with the margin for error taken into account, Judy's two scores overlap over ≈43% of their range, so the advantage gained by retaking the test may have been minimal.

Answer for problem **18** can be found in the answer pages beginning on **A** 1.

Problem Set, *continued*

R **10** A box contains four bananas, three apples, five pears, and a pomegranate. If one fruit is chosen at random from the box, what is the probability that it is an apple? $\frac{3}{13}$

P **11** If you were a pollster who wanted to determine people's feelings about the writings of Charles Dickens, how would you go about conducting a survey?

R,I **12** Let $a(n) = n^2 - 1$.

a Find $a(1)$. 0 **b** Find $a(2)$. 3 **c** Find $a(4)$. 15
d Find $a(5)$. 24 **e** Find $a(k)$. $k^2 - 1$ **f** Find $a(k + 1)$. $k^2 + 2k$

R **13** Every box of Defeaties, the Breakfast of Also-rans, contains a prize. There are three types of prizes, and these are randomly distributed. If Mr. I.M.A. Loser buys three boxes of Defeaties, what is the probability that he gets one prize of each type? $\frac{2}{9}$

P **14** A bag contains 5 red marbles and 15 green marbles. Four of the marbles are chosen at random, and the number that are green is recorded. This process is repeated several thousand times. How many green marbles would you expect to be chosen, on the average, in each trial? 3

R **15** A bag contains five red marbles and eight green marbles. If four of the marbles are chosen at random, what is the probability that

a All are red? ≈.007 **b** All are green? ≈.098
c Two are red and two are green? ≈.39

R **16** Each of the squares shown has sides 4 units long. If a point inside the figure is chosen at random, what is the probability that it lies in a shaded region? $\frac{4}{7}$

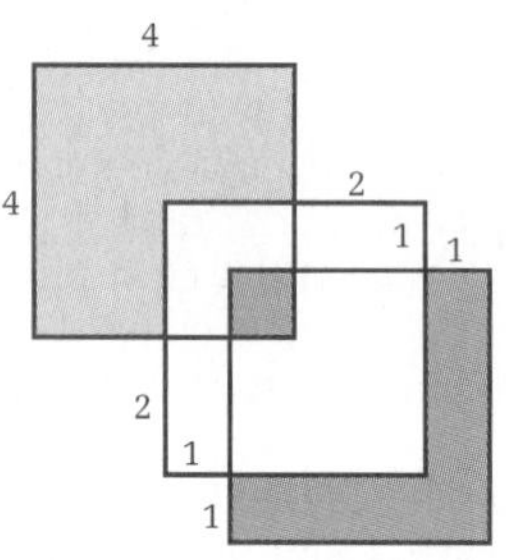

P **17** The scores for a certain college aptitude test are assigned a margin of error of ±50. Judy took the test and received a score of 1670. She decided to study for the test and retake it. On her second try, she scored 1710. Explain whether the time and money invested in the retake were worthwhile.

P **18** What does a weather forecaster mean by saying that there is a 30% chance of rain?

P | **19** Every package of cigarettes sold in the United States must bear a warning, such as "Cigarette Smoking May Be Hazardous to Your Health." The law requiring such warnings is a result of a report in which the Surgeon General concluded that there is a 99.9% certainty that a link exists between smoking and the incidence of cancer. Which of the following best describes the meaning of the Surgeon General's conclusion? d

a Out of each 1000 smokers, 999 have contracted cancer.
b Out of 1000 randomly selected smokers, 999 will get cancer.
c On the average, 999 out of every thousand smokers will get cancer.
d The probability is .999 that people who smoke are more likely to get cancer than people who do not.
e The probability is .999 that smoking causes cancer.

R | **20** In the addition table shown, a, b, c, d, e, and f represent real numbers. Each of the other numbers is the sum of the numbers heading its row and column (for example, $7 = a + e$). Find $x + y$. 5

+	a	b	c
d	1	−5	y
e	7	x	10
f	0	−6	3

P,R | **21** In Monte Cristo County, 15% of the voters are Republicans; the rest are Democrats. Of the Republicans, 80% earn over $100,000 a year. Only 5% of the Democrats earn over $100,000 a year. If one person who earns over $100,000 a year is chosen at random from the county's voters, what is the probability that the person is a Republican? ≈.74

C P,R | **22** A 3 × 3 × 3 cube is painted on all six faces, then cut into 27 identical 1 × 1 × 1 cubes. If three of the small cubes are selected at random, what is the probability that one of them has three painted faces and the other two have exactly two painted faces?
≈.18

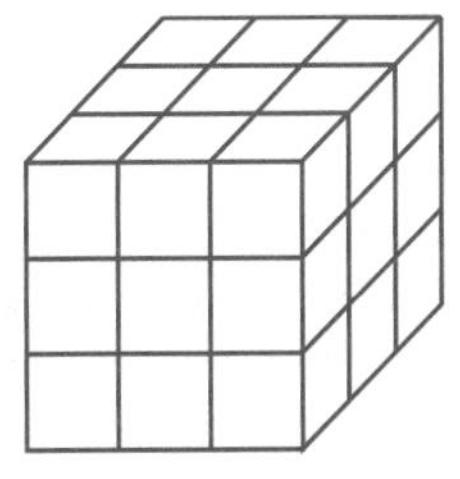

P | **23** Investigate the sampling method used to determine at least one of the following. Discuss the significance of the results obtained. Answers will vary.

a The Nielsen ratings of television programs
b *Rolling Stone*'s list of the top 100 songs
c The results of a Gallup poll
d The winners of the Academy Award
e The members of the Baseball Hall of Fame
f The *New York Times*'s list of best-selling books

Communicating Mathematics

You have been told that a representative sample is necessary for accurate predictions. Write a paragraph on the meaning of *representative*.

12

CHAPTER SUMMARY

CONCEPTS AND PROCEDURES

After studying this chapter, you should be able to

- Apply the Fundamental Counting Principle (12.1)
- Relate the numbers of elements in two sets to the numbers of elements in their union and intersection (12.1)
- Use the combination and permutation functions to solve counting problems (12.1)
- Determine the sample space for an experiment (12.2)
- Identify events and calculate their probabilities (12.2)
- Understand the relationship between disjunctions and unions of sets (12.3)
- Calculate probabilities of mutually exclusive events (12.3)
- Determine probabilities of intersections of independent events (12.4)
- Solve problems involving conditional probabilities (12.4)
- Understand the significance of probabilities (12.5)
- Understand how Monte Carlo methods are used to estimate probabilities (12.5)
- Judge sampling techniques (12.5)

VOCABULARY

combinatorics (12.1)
conditional probability (12.4)
conjunction (12.4)
dependent events (12.4)
disjunction (12.3)
equal elemental probabilities (12.2)
event (12.2)
Fundamental Counting Principle (12.1)
independent events (12.4)
mutually exclusive (12.3)
sample (12.5)
sample space (12.2)
Venn diagram (12.1)

Review Problems

A
P,R

1 Evaluate each expression.

a 5! 120 **b** 4! 24 **c** 3! 6 **d** 2! 2 **e** 1! 1 **f** 0! 1

P **2** Copy the chart shown, which represents the 36 possible outcomes when two dice are rolled.

	1	2	3	4	5	6
1						
2						
3						
4						
5						
6						

a Put an X in each box that represents a roll with a sum of 5.
b Put an O in each box that represents a roll in which at least one die shows an even number.
c How many boxes contain both an X and an O? 4

P **3** A single die is rolled.

a What is the sample space? {1, 2, 3, 4, 5, 6}
b Find the probability that the roll is a multiple of 2. $\frac{1}{2}$
c Find the probability that the roll is a multiple of 3. $\frac{1}{3}$
d Find the probability that the roll is a multiple of either 2 or 3. $\frac{2}{3}$
e Find the probability that the roll is a multiple of 6 (that is, of both 2 and 3). $\frac{1}{6}$
f What is the sum of your answers to parts **b** and **c**? $\frac{5}{6}$
g What is the sum of your answers to parts **d** and **e**? $\frac{5}{6}$

P **4** Copy the Venn diagram shown, shading the region that represents the specified set.

a $(A \cap B) \cup (B \cap C)$
b $(A \cup B) \cap (B \cup C)$

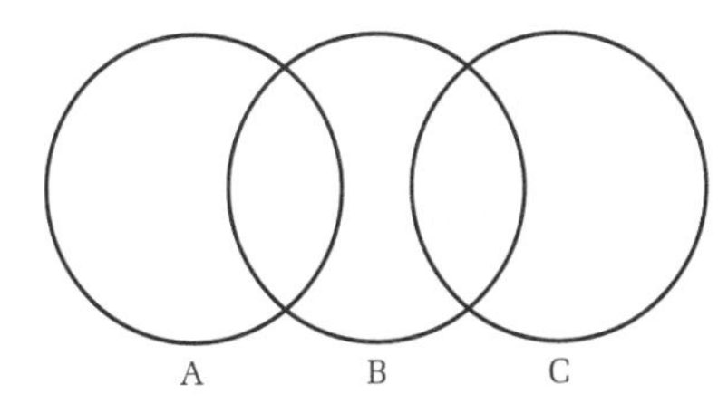

P **5** How many different radio-station call signs are possible in which the first letter is a K or a W and exactly four different letters are used? 27,600

4b

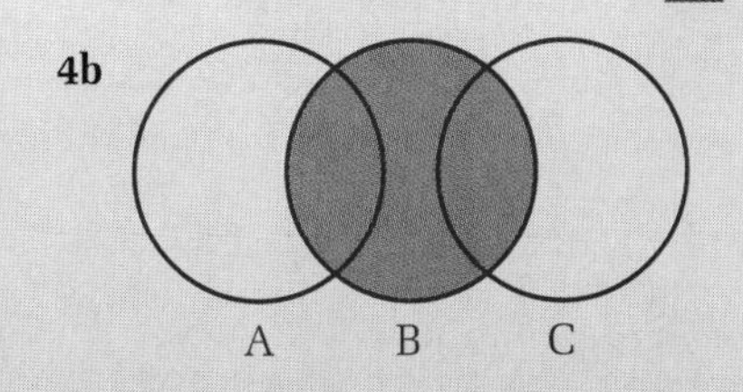

Class Planning

Time Schedule

All levels: 1 day

Assignment Guide

Average
1, 3–8, 11–16, 18, 22, 23
Advanced
1, 3–9, 11–16, 18, 19, 22–24
Honors
3–9, 11–17, 19, 21–24

Problem-Set Notes and Strategies

- Ask students if they can see any patterns in problem **3**, and have them try to explain the results using the formulas generated from mutually exclusive and independent events.

Additional Answers

2a & b

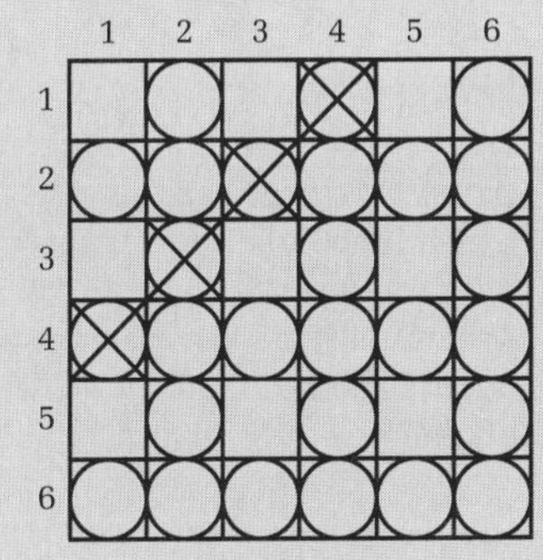

4a

A B C

Review Problems, *continued*

P **6** A red die and a green die are rolled. What is the probability that

a The red one shows 5 and the green one shows 2? $\frac{1}{36}$
b The red one shows an even number and the green one shows 3 or 4? $\frac{1}{6}$
c The red one shows a number less than 3 and the green one shows an odd number? $\frac{1}{6}$
d The red one shows 5 and the total roll is 4? 0
e The red one shows 3 and the total roll is 10? 0
f The red one shows 4 and the total roll is 6? $\frac{1}{36}$

P **7** If you randomly choose a number from {2, 3, 4, 5, 6, 7, 8, 9, 10, 11, 12, 13, 14, 15, 16}, two of the possible events are A = {multiples of 5} and B = {multiples of 2}.

a Find P(A). $\frac{1}{5}$ **b** Find P(B). $\frac{8}{15}$ **c** Find $P(A \cup B)$. $\frac{2}{3}$
d Find $P(A \cap B)$. $\frac{1}{15}$ **e** Is $P(A) + P(B) = P(A \cup B) + P(A \cap B)$? Yes

P **8** The drawing for the Lot o' Money Lottery consists of five numbers chosen from {1, 2, 3, . . . , 50}. How many different five-number choices are possible? 2,118,760

P **9** Point P is somewhere in the interior of rectangle ABCD. If another point in the interior of ABCD is selected at random, what is the probability that it is inside either △APD or △BCP? $\frac{1}{2}$

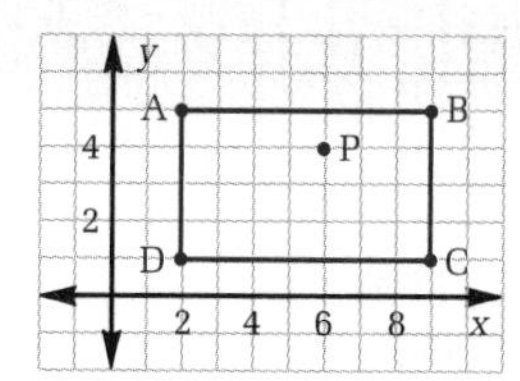

P **10** If two of the numbers in {1, 3, 5, 7, 9, 11, 13, 15} are chosen at random, what is the probability that their sum is even? 1

P **11** If a card is chosen at random from a standard deck, what is the probability that it is either a heart or the jack of diamonds? $\frac{7}{26}$

P **12** If P(A) = .5 and $P(A \cap B) = .4$, what is $P(B|A)$? .8

P **13** The service manager of an automobile dealership reported to the dealership's owner that 95% of those who bought new cars at the dealership during 1989 were "extremely satisfied" with the service department. Why might the owner have cause for concern?

P **14** A die is rolled and a coin is tossed. Find the probability that the die shows an odd number and the coin shows heads. $\frac{1}{4}$

B P **15** Consider the sets A = {3, 5, 7, 9, 11} and B = {3, 6, 9, 12}.

a Find $A \cup B$. **b** Find $A \cap B$. **c** Find $\#(A \cup B)$.
d Find $\#(A \cap B)$. **e** Find #(A) + #(B).

a {3, 5, 6, 7, 9, 11,12} **b** {3, 9} **c** 7 **d** 2 **e** 9

Problem-Set Notes and Strategies, continued

- Problems **6** and **7** are reviews of the probability formulas. It is a good place to check that students fully understand the difference between mutually exclusive and independent.
- Problem **9** is a nice geometry problem. See if students can prove the result. The proof is somewhat intuitive from the answer.

Additional Answer

13 Answers will vary. Students may note that if 95% of the customers needed to have their cars serviced within a year of buying them, the cars may have been mechanically unsound.

P  **16** A lottery pays $500 to the holders of the first number drawn, $300 to the holders of the second number, $200 to the holders of the third number, and $100 to the holders of the fourth number. If the numbers are chosen from the set {1, 2, 3, . . . , 50}, how many possible outcomes are there? 5,527,200

P,R **17** How many lines can be drawn through the point (1, 5) in such a way that one intercept is three greater than the other? 4

P,R **18** If n is a randomly chosen natural number and $i = \sqrt{-1}$, what is the probability that the following sum is a real number? $\frac{1}{2}$

$$\sum_{k=1}^{n} i^k$$

P **19** Suppose that Joe and Mary play a game in which Joe can win $1, $2, $3, or $4 and Mary can win $1, $2, or $3.

a List all of the possible outcomes.

b List all the sums of money that can be won. $2, $3, $4, $5, $6, $7

P **20** In the diagram, D, E, and F are the midpoints of $\overline{AC}$, $\overline{CB}$, and $\overline{BA}$.

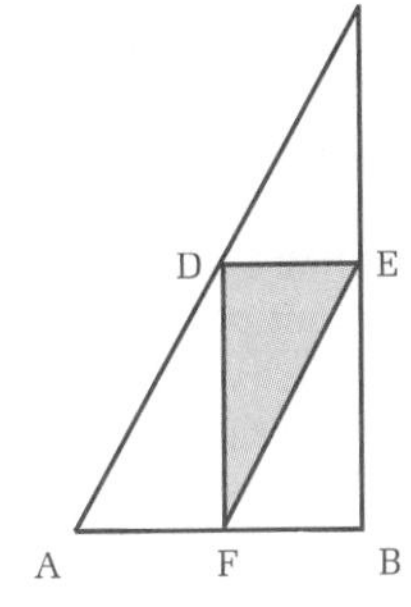

a If a point in the interior of right triangle ABC is selected at random, what is the probability that it is in the interior of △DEF? $\frac{1}{4}$

b If a point in the interior of △DEF is selected at random, what is the probability that it is in the interior of △ABC? 1

P **21** On the average, 1.5% of the widgets manufactured at a certain plant are defective. If an inspector selects two of the widgets at random for examination, what is the probability that neither is defective? ≈.97

P **22** A baseball team has five starting pitchers and four relief pitchers. In how many ways can one starter and two relievers be selected to pitch in a game? 30

P **23** A bag contains eight red marbles and three blue marbles. If two marbles are drawn at random from the bag, what is the probability that both are blue? $\frac{3}{55}$, or ≈.05

C P  **24** Find the probability that exactly six cards in a ten-card gin-rummy hand will be hearts. ≈.009

P,R **25** How many subsets does {1, 2, 3, 4, 5, 6, 7} have? 128

Problem-Set Notes and Strategies, continued

- Since the order is important in problem **16**, it is a permutation.

- Problem **24** is an example of a binomial probability. One approach is to use the formula in Section **11.6** to find the probability.

Additional Answer

19a {(Joe's winnings, Mary's winnings)} = {($1,$1), ($1,$2), ($1,$3), ($2,$1), ($2,$2), ($2,$3), ($3,$1), ($3,$2), ($3,$3), ($4,$1), ($4,$2), ($4,$3)}

Review Problems, *continued*

P **26** If a point in the interior of rectangle ABCD is selected at random, what is the probability that it is in Quadrant I or Quadrant III? $\frac{23}{45}$, or $\approx .51$

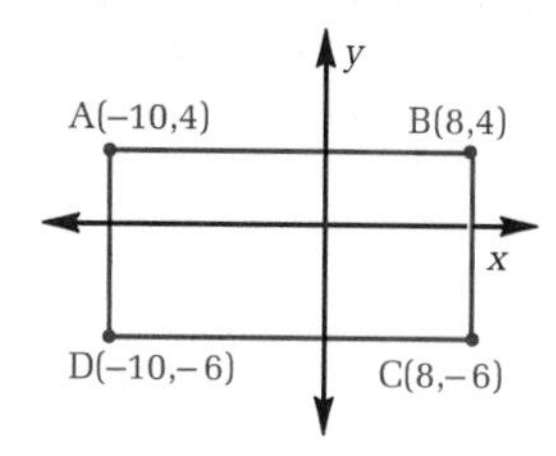

CAREER PROFILE

STATISTICS AND THE PUBLIC MIND

Dave Petts crunches numbers to find out how we feel

Television and electronic communications have generated an overabundance of numbers and statistics calling attention to what Americans think about many different subjects. Political polling is the science of representing as accurately as possible the central tendencies in the public's opinions about politics. Hence, solid statistical methods are essential to Dave Petts, a political pollster with a Washington, D.C., polling firm.

"A lot of statistics is involved because statistics and probability theory undergird political polling," says Mr. Petts. His work primarily involves recording and analyzing voters' attitudes toward the candidates represented by his firm. He employs factor analysis and regression analysis "to try to understand what's driving the voter."

The greatest challenge Mr. Petts faces is "to make decisions very quickly, based on less than perfect information. Are the voters hearing the candidate's message? What is causing misperceptions? Or, if she's doing well in polls, what do we need to do to keep it that way?"

Do you think labor unions today have too much power?

Too much power	**40%**
Not enough power	**22%**
Right amount	**33%**

Do you think American workers still need labor unions?

Still need them	**73%**
No longer necessary	**22%**

Should private-sector workers have the right to strike without losing their jobs or is this the chance they take?

Have the right to strike	**47%**
It's the chance they take	**50%**

After the strike, should Greyhound be required to rehire the striking workers?

Yes	**57%**
No	**31%**

From a telephone poll of 500 adult Americans taken for TIME/CNN on March 14 by Yankelovich Clancy Shulman. Sampling error plus or minus 4.5%.

TIME Chart by Steve Hart

12

CHAPTER TEST

1 At Slick Sal's car dealership, customers are given a choice of four different colors, three different models, six different body styles, and two different radios. How many different kinds of cars can be ordered? 144

2 There are 385 students in the junior class at Hyde Park High, of which 80% take both math and English. Fifteen juniors take math but not English, and 55 take English but not math.

a How many juniors take neither math nor English? 7
b How many juniors take math? 323
c How many juniors take English? 363

3 S. Peller is playing Skrabull and has the letters *F*, *E*, *I*, *R*, *S*, and *T* in her rack. In how many ways can she select three letters to play? 20

4 A die is rolled and a coin is tossed. Find the probability that the number on the die is odd and the coin shows heads. $\frac{1}{4}$

5 Scarlett threw 5 balloons into the air. Rhett threw 9 balloons into the air. Two of the balloons flew off, but the other 12 fell to the ground. Find the probability that both of the balloons that were gone with the wind were thrown either by Rhett or by Scarlett. $\approx .51$

6 After scoring a touchdown, the Decatur Sallies attempt to kick for an extra point 85% of the time. The other 15% of the time, they attempt a running or passing play for two extra points. They are successful in 90% of their tries for one extra point and 40% of their tries for two extra points. If the Sallies score a touchdown, what is the probability that they add at least one extra point? .825

7 Suppose two dice are rolled at the same time two coins are tossed. What is the probability that both coins show heads and the dice show a total of 10? $\frac{1}{48}$

8 A newspaper columnist wanted to find out how most people felt about gun control. So she invited her readers to write a letter, telling her whether they were for it or against it and explaining why. Why should we be skeptical of the results of her poll?

P **9** How many positive integers are divisors of 7007? (Hint: $7007 = 7^2 \cdot 11 \cdot 13$) 12

Resource References
Evaluation
Test Form 1, 2, 3

Additional Answer
8 Answers will vary. Students may mention that only people who were strongly in favor of gun control or strongly opposed to it were likely to respond.

*T*HIS CHAPTER PRESENTS an introduction to series and sequences. Emphasis is given to geometric and arithmetic sequences and series, but several other types of series are introduced for comparison and contrast. Both iterative and recursive formulations for the definitions of sequences and series are included.

The definition of a sequence is given in terms of a function whose domain is the set of natural numbers. Multiple representations of sequences are used throughout the chapter so that students are familiar with and comfortable using all notations. Students also learn how to form sequences of partial sums.

Sigma notation is important in this chapter as well as in later mathematics courses. A finite series is defined in terms of a sequence of partial sums and infinite series are evaluated.

The third section of this chapter discusses the arithmetic sequence, one of the two most common sequences. Students will find any term in an arithmetic sequence, compute arithmetic means, and evaluate arithmetic series.

The last section of this chapter focuses on geometric sequences and series. Students will find any term in a geometric sequence, compute geometric means, and evaluate finite and infinite geometric series.

CHAPTER

13 SEQUENCES AND SERIES

The sizes of the pipes in an organ form a geometric series such that the larger the pipe, the lower the tone it produces.

13.1 SEQUENCES

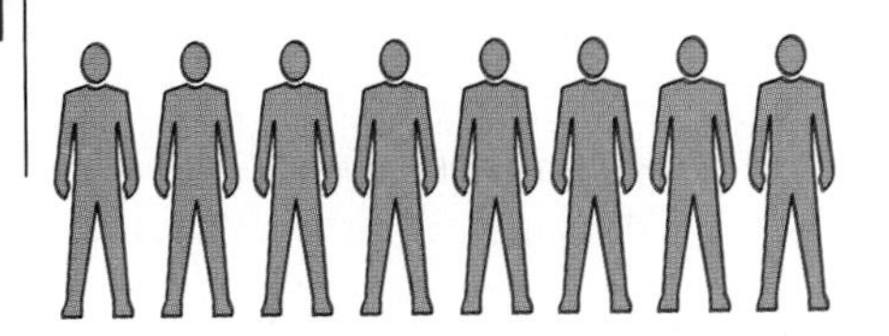

Objectives

After studying this section, you will be able to

- Apply the definition of a sequence
- Describe a sequence
- Form sequences of partial sums

Part One: Introduction

The Definition of a Sequence

Consider the following infinite list of numbers.

$$2, 5, 8, 11, 14, \ldots, 3n - 1, \ldots$$

The ellipsis (. . .) indicates that the list continues using the $3n - 1$ pattern. This list actually consists of the values of the function a,

$$a(n) = 3n - 1, \text{ where } n \text{ is chosen from the set of natural numbers, } \{1, 2, 3, \ldots\}$$

In this chapter, the letter n will always represent a positive integer.

We often use a_n instead of $a(n)$ when discussing a function whose domain is the natural numbers.

For the function $a(n) = 3n - 1$

$$a_1 = a(1) = 3(1) - 1 = 2$$
$$a_2 = a(2) = 3(2) - 1 = 5$$
$$a_3 = a(3) = 3(3) - 1 = 8$$
$$\vdots$$
$$a_n = a(n) = 3(n) - 1 = 3n - 1$$

The function a is an example of a ***sequence.*** A ***sequence*** is a function whose domain is the set of natural numbers. The values of this function are called ***terms.*** Sometimes a sequence is called a ***progression.***

The notation $\{a_n\}$ is used to denote the sequence $a_1, a_2, a_3, \ldots, a_n, \ldots$, where the nth term of the sequence is a_n.

Class Planning

Time Schedule

All levels: 2 days

Resource References

Teacher's Resource Book

Class Opener 76
Cooperative Learning 39
Practice 80
Enrichment 25

Class Opener

If $f(1) = 1$, $f(2) = 2$ and $f(n + 2) = \frac{f(n+1)}{f(n)}$, find $f(1{,}000{,}000)$.

n	1	2	3	4	5	6	7	8	9	10	11	12	13
$f(n)$	1	2	2	1	$\frac{1}{2}$	$\frac{1}{2}$	1	2	2	1	$\frac{1}{2}$	$\frac{1}{2}$	1

Answers are in cycles of 6: $6\overline{)1{,}000{,}000}$ has remainder 4, thus $f(1{,}000{,}000) = 1$.

Lesson Notes

- It may be surprising for students to realize that a sequence is a list. Nothing more, nothing less. The formal definition of a sequence can be introduced once students are convinced that it is not mysterious. Every list has a first element, second element, etc. This prepares for the rigorous definition of a function whose domain is the set of natural numbers.

Vocabulary

sequence
term
progression
array
sequence of partial sums

Lesson Notes, continued

- The definition of an array may be slightly different for some of the students, especially those with some computer background.
- The sequence of partial sums may seem a little abstract for some of the students at first. It may help if they see that it is just another list of numbers whose entries are sums of the first list. For any students going into math or engineering, these sequences of partial sums will appear again and again.

Stumbling Block

Students may have difficulty distinguishing between a sequence and a sequence of partial sums. Both are sequences and therefore abstractly the same. It may help to reiterate that a sequence is formed with a general term or expression. A sequence of partial sums is generated by adding the terms of another sequence.

Communicating Mathematics

Sequences and patterns can be recognized in everyday life. Write a paragraph describing a pattern you have observed in the last few days and see if you can generate an expression that will model that pattern.

Example *Write the first three terms and the nth term of the sequence $\{a_n\}$, where $a_n = n^2 - 1$.*

$$a_1, a_2, a_3, \ldots, a_n, \ldots = 0, 3, 8, \ldots, n^2 - 1, \ldots$$

All sequences have an infinite number of terms. In some cases, however, we want to consider only a finite number of terms. We call such a function—whose domain is a finite subset of the natural numbers—an ***array.***

Example *Write the array consisting of the first six terms of the sequence $\{a_n\} = \{5n - 3\}$.*

n	1	2	3	4	5	6
a_n	2	7	12	17	22	27

Arrays are commonly used in computer programming.

Ways to Describe a Sequence

What is the next term in the sequence 2, 3, 5, . . . ?

"Seven," said Aiko. "It is the sequence of prime numbers."

"Eight," said Bob. "Each term is the sum of the preceding two terms--$a_1 = 2$, $a_2 = 3$, $a_n = a_{n-1} + a_{n-2}$."

"Nine," said Cathy. "The sequence is $\{a_n\} = \{2^{n-1} + 1\}$."

All three answers are reasonable, which means that the sequence 2, 3, 5, . . . was not well-defined.

A sequence must be described so that it is possible to determine any term accurately. Aiko described a sequence in *words.* Bob described a different sequence *recursively.* Cathy gave a *closed-form* formula for the *n*th term. All three ways of describing a sequence are acceptable. Listing a few terms is *not* a good way to describe a sequence, because different interpretations are always possible.

Example *Find the fourth term of each of the following sequences.*

a $5, 25, 125, \ldots, 5^n, \ldots$

The formula that describes the function for this sequence is $a_n = 5^n$. Since the fourth term occurs when $n = 4$, $a_4 = 5^4 = 625$.

b $\{a_n\} = \{n^2 - 2n\}$

The formula that describes the function for the sequence is $a_n = n^2 - 2n$. The fourth term occurs when $n = 4$, so $a_4 = 4^2 - 2 \cdot 4 = 8$.

c {*nth prime number*}

There is no known formula for obtaining the prime numbers, so this sequence is described verbally. The first eight prime numbers are 2, 3, 5, 7, 11, 13, 17, and 19. The fourth prime is 7.

Sequences of Partial Sums

Associated with every sequence $\{a_n\}$ is another sequence, $\{S_n\}$, called the ***sequence of partial sums*** of the original sequence. For example, looking at the figure on the next page, find the pattern that describes the sequence $\{a_n\}$ and the sequence of partial sums, $\{S_n\}$.

If $\{a_n\} = \{2n - 1\} = 1, 3, 5, \ldots, 2n - 1, \ldots$, then

$S_1 = a_1 = \mathbf{1}$
$S_2 = a_1 + a_2 = 1 + 3 = \mathbf{4}$
$S_3 = a_1 + a_2 + a_3 = 1 + 3 + 5 = \mathbf{9}$
$S_4 = a_1 + a_2 + a_3 + a_4 = 1 + 3 + 5 + 7 = \mathbf{16}$
.
.
.

$$S_{n-1} = a_1 + a_2 + a_3 + \ldots + a_{n-1}$$
$$= 1 + 3 + 5 + \ldots + [2(n - 1) - 1]$$
$$= (n - 1)^2$$
$$S_n = a_1 + a_2 + a_3 + \ldots + a_{n-1} + a_n$$
$$= 1 + 3 + 5 + \ldots + (2n - 1)$$
$$= n^2$$

In the sequence of partial sums, $\{S_n\}$, the nth term is the sum of the first n terms of the original sequence, $\{a_n\}$.

If $\{a_n\}$ is any sequence, then the sequence $\{S_n\}$, defined by $S_n = a_1 + a_2 + a_3 + \ldots + a_n$, is called the sequence of partial sums of the sequence $\{a_n\}$. This is a closed-form definition.

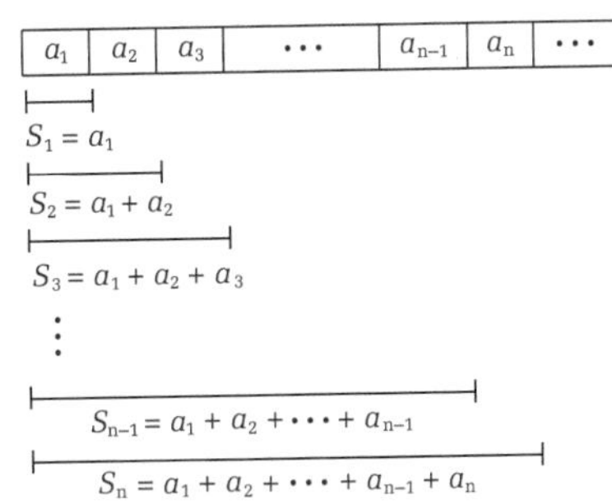

If we look at the pattern for the terms S_{n-1} and S_n, we see

$$S_{n-1} = a_1 + a_2 + a_3 + \ldots + a_{n-1}$$
$$S_n = a_1 + a_2 + a_3 + \ldots + a_{n-1} + a_n$$

Since $S_n = S_{n-1} + a_n$, we can define $\{S_n\}$ recursively. If $\{a_n\}$ is any sequence, then the sequence $\{S_n\}$, defined by $S_1 = a_1$ and $S_n = S_{n-1} + a_n$, is called the sequence of partial sums of the sequence $\{a_n\}$. This is a recursive definition.

Example *If $\{S_n\} = \{n^2\}$ is the sequence of partial sums of a sequence $\{a_n\}$, what is the formula describing a_n?*

We know that $S_n = n^2$, $S_{n-1} = (n - 1)^2$, and $S_n = S_{n-1} + a_n$

If $\quad S_n = S_{n-1} + a_n,$
then $\quad n^2 = (n - 1)^2 + a_n$
That is, $\quad a_n = n^2 - (n - 1)^2$
$$a_n = n^2 - (n^2 - 2n + 1)$$
$$a_n = n^2 - n^2 + 2n - 1$$
$$a_n = 2n - 1$$

Cooperative Learning

Some students have seen the proof that 0.9999... = 1. Have them write this number as the sequence of partial sums of the sequence whose general term is $a_n = 9(10)^{-n}$. Then look at the partial sums. They approach 1 as n gets larger.

Checkpoint

1. List the first five terms of the sequence $\{2n + 1\}$.
3, 5, 7, 9, 11
2. Find a general term for the sequence 1, 1, 2, 6, 24, ...
$\{(n - 1)!\}$
3. In the sequence of problem **2**, find the first 6 terms of the sequence of partial sums.
1, 2, 4, 10, 34, 154
4. Write the first six terms of the sequence $\{(1 + \frac{1}{n})^n\}$ to three decimal places.
2.000, 2.250, 2.370, 2.441, 2.488, 2,522

13.1

Assignment Guide

Average	
Day 1	10–12, 14, 15, 19, 21, 23, 31, 34
Day 2	13, 16, 18, 20, 24, 25 ,26, 30, 36, 37
Advanced	
Day 1	10–13, 15, 20, 24, 31, 33, 34, 37
Day 2	14, 16, 17, 21, 23, 25 ,28, 32, 35, 36
Honors	
Day 1	10–13, 15, 18, 20, 22, 26, 29, 33, 38
Day 2	17, 21, 24, 25, 31, 32, 34, 35, 37, 39

Problem-Set Notes and Strategies

- Students may recognize patterns in the sequences in problems **1 – 3**. Several of these patterns are discussed in detail in Sections **13.3** and **13.4**.
- Students may need to list and add the sequence for problem **4**.
- Problems **7**, **8**, and **13** are recursively defined. Each term requires knowledge of a previous term to find its value. If time permits, suggest that students find a closed form formula for these sequences.

Part Two: Sample Problem

Problem *A cell divides into two cells, each of which divides, to produce a total of four cells. This process continues as shown in the diagram. Each level of the diagram is called a generation. How many cells are there in the 100th generation?*

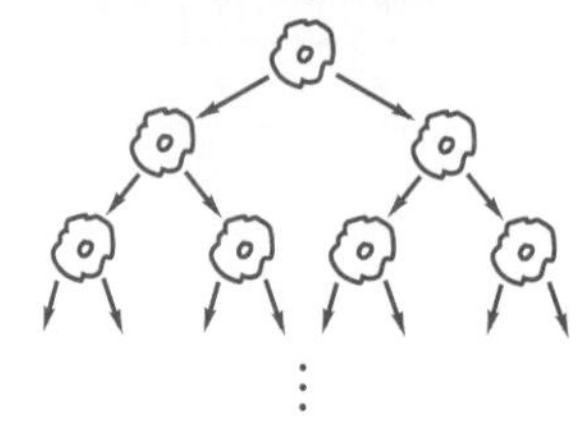

First Generation

Second Generation

Third Generation

Solution Let $\{G_n\}$ be the sequence of the number of cells in each generation n.

$G_1 = 1 = 2^0$, $G_2 = 2 = 2^1$, $G_3 = 4 = 2^2$, $G_4 = 8 = 2^3, \ldots$

The number of cells in each generation is a power of 2. The nth generation has 2^{n-1} cells. Since $G_n = 2^{n-1}$,

$G_{100} = 2^{100-1} = 2^{99} \approx 6.34 \times 10^{29}$ cells

How could you describe this sequence recursively?

Part Three: Exercises and Problems

Warm-up Exercises

In problems 1–3, list the first six terms of each sequence.

P **1** $\{2n - 7\}$
$-5, -3, -1, 1, 3, 5$

2 $\{n^2 - 1\}$
0, 3, 8, 15, 24, 35

3 {the nth prime number}
2, 3, 5, 7, 11, 13

P **4** If $\{a_n\} = \{2^n\}$ and $\{S_n\}$ is the sequence of partial sums for $\{a_n\}$, what are the first five terms of $\{S_n\}$?
2, 6, 14, 30, 62

P **5** $\{a_n\}$ is a sequence and $\{S_n\}$ is its sequence of partial sums. If $S_{10} = 35$ and $S_{11} = 31$, what is a_{11}? -4

P **In problems 6–9, give the fourth through the eighth terms of the sequence.**

6 $32, 8, 2, \ldots, 2^{7-2n}, \ldots$ $\frac{1}{2}, \frac{1}{8}, \frac{1}{32}, \frac{1}{128}, \frac{1}{512}$

7 $a_1 = 3$, $a_n = 2a_{n-1}$ 24, 48, 96, 192, 384

8 $a(1) = 1$, $a(n) = n + a(n - 1)$
10, 15, 21, 28, 36

9 $\{a_n\} = \{n^2 - 2n\}$
8, 15, 24, 35, 48

Problem Set

A P,R **10** The diagram illustrates the first four *triangular numbers*. Write the next four triangular numbers.
15, 21, 28, 36

1 3 6 10

P | **11** Find a formula that describes the terms a_n of the array. $a_n = 4n$

n	1	2	3	4
a_n	4	8	12	16

P | **In problems 12 and 13, list the fourth through eighth terms of the sequences.**

12 $\{n^2\}$ 16, 25, 36, 49, 64

13 $a_1 = 5$, $a_n = 2a_{n-1}$ 40, 80, 160, 320, 640

P  **14** Use a calculator to find a_1, a_{10}, and a_{100} to the nearest ten-thousandth if $a_n = \frac{3n + 1}{2n + 1}$. ≈ 1.3333, ≈ 1.4762, ≈ 1.4975

P,I | **In problems 15–17, compute the sum.**

15 $\sum_{n=1}^{5} n^2$ 55

16 $\sum_{n=1}^{5} n$ 15

17 $\sum_{n=1}^{5} (n^2 + n + 3)$ 85

P,I | **18** The approximate monthly cost in dollars of producing the nth bicycle is given by the formula $c = 25 - 0.01n^2$.

a Find the cost of producing the first bike, the second bike, and the third bike. \$24.99; \$24.96; \$24.91

b Find the total cost of producing the first three bikes. \$74.86

P | **In problems 19 and 20, write the first five terms of each sequence.**

19 $\left\{\frac{n}{n + 1}\right\}$ $\frac{1}{2}, \frac{2}{3}, \frac{3}{4}, \frac{4}{5}, \frac{5}{6}$

20 $\left\{\frac{n^2}{n + 1}\right\}$ $\frac{1}{2}, \frac{4}{3}, \frac{9}{4}, \frac{16}{5}, \frac{25}{6}$

P,I | **21** Life is found on the planet Trips. On Trips, each generation of cells divides into three cells instead of two.

a How many cells will be in the fifth generation? 81

b In what generation will the number of cells first exceed 3000? 9th

B

P | **22** Let $a_n = n$ and $b_n = \frac{n(n + 1)}{2}$.

a Find the first four terms of $\{a_n \cdot b_n\}$. 1, 6, 18, 40

b Find the first four terms of the sequence $\{S_n\}$ of partial sums for $\{a_n \cdot b_n\}$. 1, 7, 25, 65

P | **23** In the sample problem, after how many cell divisions will the total population first exceed 1 million? 20

P | **24** Refer to the diagram.

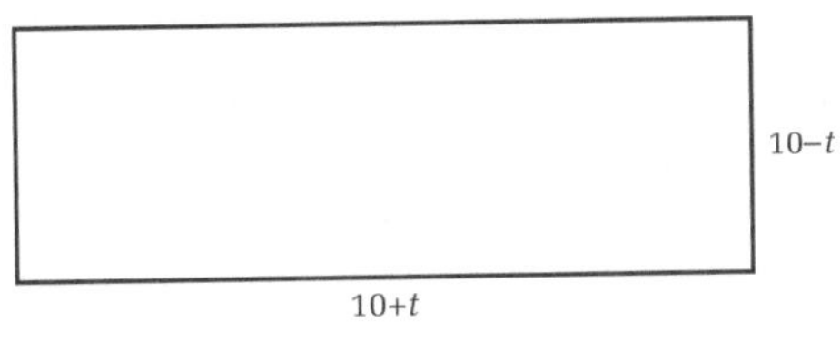

a Write the area of the rectangle as a function $A(t)$, where t is a natural number less than 10. $A(t) = 100 - t^2$

b Write the sequence of area values $A(t)$. 99, 96, 91, 84, 75, 64, 51, 36, 19

Problem-Set Notes and Strategies, continued

- Problem **14** previews limits. Ask the students what is happening to the fraction as n gets larger and larger.
- Problems **15 – 17** review summations and preview series defined in the next section.
- Problems **18** and **21** both preview series which will be done in the next section. Students can first generate the terms, then add them together for the sum of the terms of a sequence.

Problem-Set Notes and Strategies, continued

- Problem **26** is a review of sigma notation.
- Solving problem **27** may provide an indication of which students understand the sequence of partial sums.
- Problems **29** and **32** practice going from recursive forms to closed forms. Students may need help with this until they are more comfortable. Some will begin to see the benefit of a closed form definition. This is also likely in problem **32**.

Problem Set, *continued*

P,R **25** In the figure at the right, a_n is the distance from (0, 1) to (n, 0).

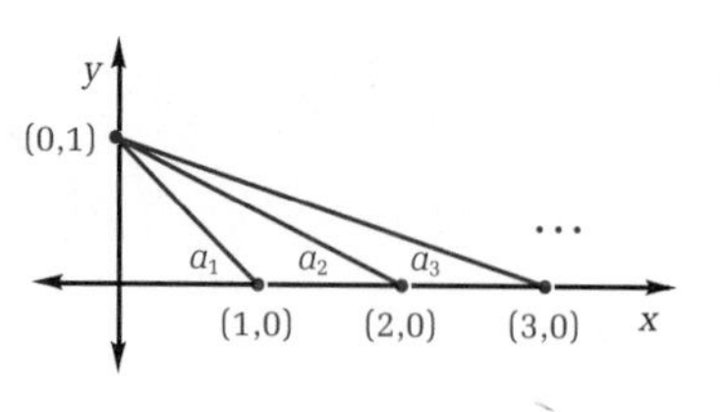

a Calculate the first four terms of the sequence $\{a_n\}$. $\sqrt{2}, \sqrt{5}, \sqrt{10}, \sqrt{17}$

b Predict the next three terms without actually calculating them. $\sqrt{26}, \sqrt{37}, \sqrt{50}$

P,R **26** Solve $\sum_{n=1}^{5}(2nj + 3) = 105$ for j. $j = 3$

P **27** If $\{S_n\} = \{3n - 2\}$ is the sequence of partial sums for $\{a_n\}$, what is a formula that describes $\{a_n\}$? $a_1 = 1, a_{n+1} = 3$

P **28** Find a_1, a_2, a_3, and a_4 so the levers are in balance.

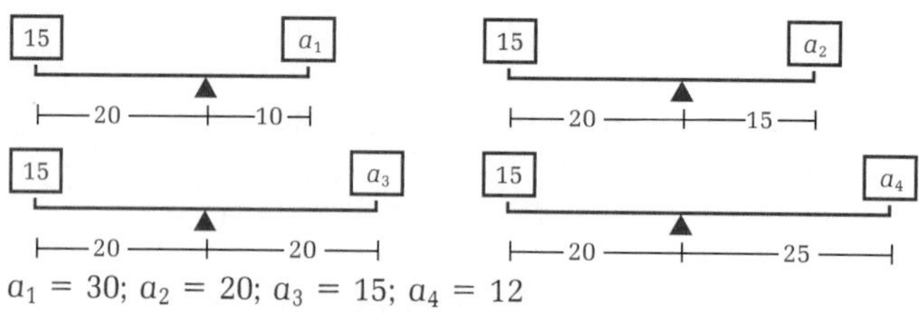

$a_1 = 30$; $a_2 = 20$; $a_3 = 15$; $a_4 = 12$

P **29** To find the next term of the sequence pictured, find the area of a square whose sides are half as long as the sides of the previous square.

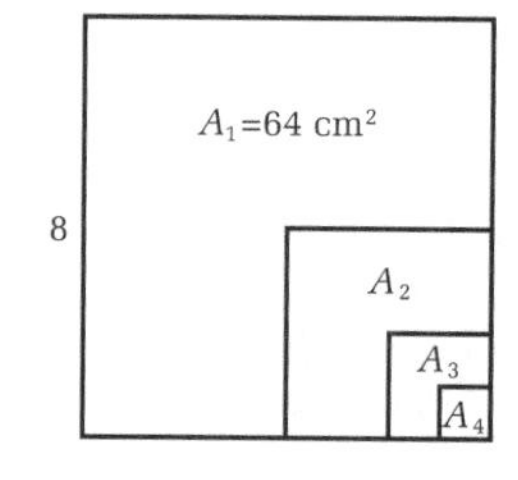

a Describe this sequence by the recursive method. $A_1 = 64, A_n = \frac{A_{n-1}}{4}$

b Describe this sequence by the closed-form method. $A_n = \frac{64}{4^{n-1}}$

c What is the 25th term? $A_{25} = \frac{64}{4^{24}} \approx 2.27 \times 10^{-13}$ cm²

P **30** The sequence $\{a_n\} = \{1000(1 + 0.06)^n\}$ represents the amounts of money in a bank after n years if \$1000 is deposited at 6% interest. Find a_1, a_2, a_3, a_4, and a_5.
\$1060; \$1123.60; \$1191.02; \$1262.48; \$1338.23

P,R **31** An equilateral triangle has sides with length 1024. The midpoints of the sides are connected to form another equilateral triangle. This second triangle's midpoints are connected to form a third equilateral triangle, and so on. Write the array consisting of the areas of the first eight triangles.

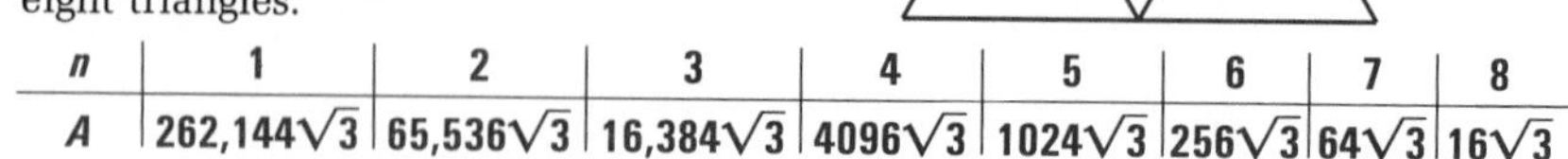

n	1	2	3	4	5	6	7	8
A	$262{,}144\sqrt{3}$	$65{,}536\sqrt{3}$	$16{,}384\sqrt{3}$	$4096\sqrt{3}$	$1024\sqrt{3}$	$256\sqrt{3}$	$64\sqrt{3}$	$16\sqrt{3}$

P | **32** 11, 101, 1001, 10,001, . . . has the pattern that each successive term places one more zero between the 1's.

a Write a recursive formula for the sequence. $a_1 = 11, a_{n+1} = 10a_n - 9$

b Write a closed-form formula for the nth sequence. $\{a_n\} = \{10^n + 1\}$

P | **33** Write the first five terms of $\{a_n\}$, where $a_1 = 1$, $a_2 = 3$, and $a_n = \frac{a_{n-1}}{a_{n-2}}$. 1, 3, 3, 1, $\frac{1}{3}$

P | **In problems 34 and 35, find a formula to describe the terms of each array.**

34

n	1	2	3	4	5
a_n	2	5	10	17	26

$a_n = n^2 + 1$

35

n	1	2	3	4	5	6
a_n	5	−10	15	−20	25	−30

$a_n = 5n(-1)^{n+1}$

C

P,R | **36** Describe as many sequences as you can whose first three terms are 4, 6, and 8.
Answers will vary. Possible answers: $\{a_n\} = \{2n + 2\}$; $a_1 = 4, a_{n+1} = a_n + 2$

P,I | **37** Consider the sequence $\{a_n\} = \left\{3\left(\frac{1}{10}\right)^n\right\}$.

a Write the first six terms of this sequence. 0.3, 0.03, 0.003, 0.0003, 0.00003, 0.000003

b Write the first six terms of $\{S_n\}$, the sequence of partial sums of $\{a_n\}$. 0.3, 0.33, 0.333, 0.3333, 0.33333, 0.333333

c What value do the values of S_n approach as n gets larger? $\frac{1}{3}$

P | **38** For the sequence pattern 1, 8, 15, . . . , find as many functions as you can that match the first three terms listed. What is the fourth term for each of your functions? Give a valid description of your sequence.
Answers will vary. Possible answer: $a_1 = 1, a_{n+1} = a_n + 7$; 22

P,R | **39** In the game Tower of Hanoi, the object is to move the rings from Peg 1 to Peg 3. The rules allow you to move one ring at a time. You may use Peg 2, and you must never place a wider ring on top of a smaller ring. If a_n is the number of moves required to move a stack of n rings from Peg 1 to Peg 3, what are a_1, a_2, a_3, a_4, and a_5? Guess a formula for a_n.
$a_1 = 1$; $a_2 = 3$; $a_3 = 7$; $a_4 = 15$; $a_5 = 31$; $a_n = 2^n - 1$

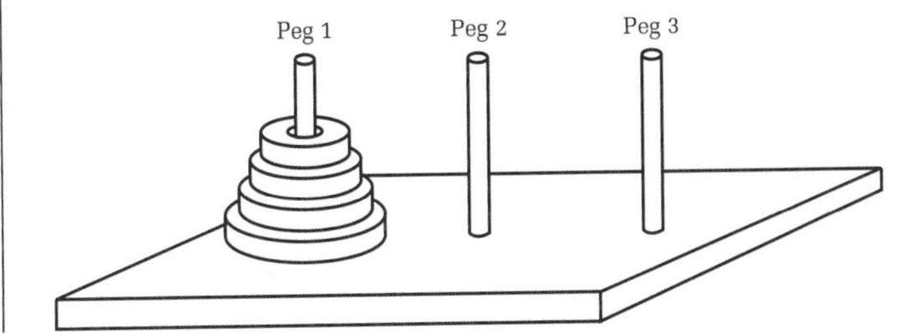

Problem-Set Notes and Strategies, continued

- The answers to problems **34** and **35** may not be unique and can solicit an interesting discussion.
- Problem **37** involves a geometric sequence and the corresponding series. The sum of the infinite series is hinted at with part **c**. You might comment on the fact that it seems that the sum is finite for an infinite summation.
- If the students can play the first few games in problem **39**, the answer becomes clearer.

13.2

Class Planning

Time Schedule
All levels: 2 days

Resource References
Teacher's Resource Book
Class Opener 77
Cooperative Learning 40
Practice 81
Transparency 25

Class Opener

Evaluate $\sum_{k=1}^{n}\left(\frac{1}{2}\right)^k$ for $n = 3$, $n = 6$, $n = 10$, $n = 15$.

$\frac{7}{8}, \frac{63}{64}, \frac{1023}{1024}, \frac{32{,}767}{32{,}768}$

Lesson Notes

- You may want to emphasize to students that a sequence is a list of terms and a series is a sum of the terms in that list. Each partial sum computed in the last section was a series, usually a finite series.
- The sigma notation introduced earlier will be utilized again in this section to denote both finite as well as infinite series. Students will have practice writing series from sigma notation and sigma notation from a series.

13.2 Series

Objectives
After studying this section, you will be able to
- Compute finite series
- Use sigma notation to express a series
- Find the values of infinite series

Part One: Introduction

Finite Series
In common usage, the words *sequence* and *series* are almost interchangeable. In mathematics, however; the words mean different things. Consider these examples.

Sequence: $2, 4, 6, \ldots, 2n, \ldots$
Related finite series: $2 + 4 + 6 + 8$
Related infinite series: $2 + 4 + 6 + \ldots + 2n + \ldots$

You may recognize that the finite series above is a partial sum of the sequence $\{a_n\} = \{2n\}$. Every ***finite series*** represents a partial sum S_n of a sequence $\{a_n\}$. We often use the term *series* when we really mean *finite series*.

Example *If $\{a_n\} = \left\{\frac{n}{2}\right\}$, what is the sum of the finite series S_5 formed by adding the first five terms of the sequence $\{a_n\}$?*

$$S_5 = a_1 + a_2 + a_3 + a_4 + a_5$$
$$= \frac{1}{2} + 1 + \frac{3}{2} + 2 + \frac{5}{2} = 7.5$$

Sigma Notation and Series
The sigma notation introduced in Chapter 1 is particularly useful for describing series. The nth partial sum, S_n, of a sequence $\{a_n\}$ can be written as

$$S_n = \sum_{k=1}^{n} a_k = a_1 + a_2 + a_3 + \ldots + a_n$$

The Σ means to sum the terms, the $k = 1$ means that the first term to add is a_1, and the n means to continue summing terms up to and including a_n.

Vocabulary
finite series
infinite series
sum of an infinite series

Example *Evaluate the finite series* $\sum_{k=1}^{6}(2k + 1)$.

$$\sum_{k=1}^{6}(2k + 1) = 3 + 5 + 7 + 9 + 11 + 13$$
$$= 48$$

Infinite Series

What would happen if we were to add all the terms of an infinite sequence? While it may be physically impossible to add an infinite number of terms, we can often determine what the sum would be.

For a sequence $\{a_n\}$, the series $\sum_{n=1}^{\infty} a_n$ is called an ***infinite series.***

The symbol ∞, representing infinity, is used to indicate that all terms of the sequence are to be added together.

Example *Evaluate* $\sum_{n=1}^{\infty} 3 \cdot \left(\frac{1}{2}\right)^{n-1}$.

This is an infinite series equivalent to

$$3 + \frac{3}{2} + \frac{3}{4} + \frac{3}{8} + \ldots + \frac{3}{2^{n-1}} + \ldots$$

Let's calculate a few partial sums, S_n, and make a graph of the ordered pairs (n, S_n) for $n = 1, 2, \ldots, 9$.

n	S_n
1	3
2	4.5
3	5.25
4	≈5.63
5	≈5.81
6	≈5.91
7	≈5.95
8	≈5.98
9	≈5.99

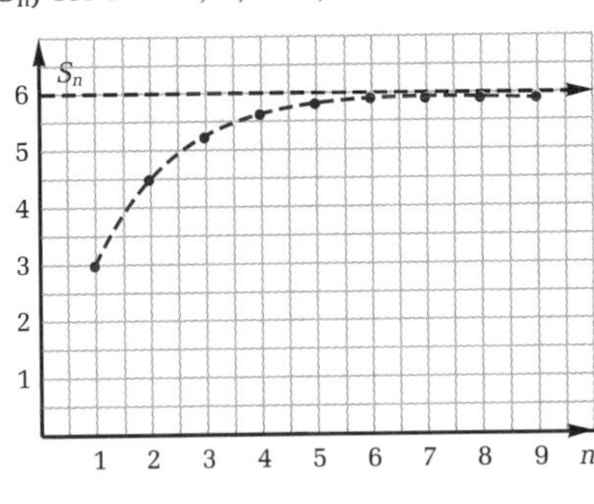

The ordered pairs described in the table above seem to approach a horizontal asymptote. The sum of the terms of $\{a_n\}$ gets closer and closer to 6.

If the graph of the sequence of partial sums, $\{S_n\}$, of a sequence $\{a_n\}$ approaches a horizontal asymptote $y = S$, we write $S = \sum_{n=1}^{\infty} a_n$ and call S the ***sum of the infinite series.*** In the preceding example, we would say that $6 = \sum_{n=1}^{\infty} 3 \cdot \left(\frac{1}{2}\right)^{n-1}$. If the graph of a sequence of partial sums does not approach a horizontal asymptote, then we say that the infinite series does not have a sum.

Lesson Notes, continued

- The difference between a sequence and a series can be confused by students. One way to differentiate is to remember that a sequence is a list and a series is a sum.

Using Computers

Symbolic manipulation software, notably *Derive*, has the capability of using sigma (Σ), notation. This type of software can help students gain a further understanding both of this notation and of the relationship between a series and its associated sequence of partial sums. Symbolic manipulation software also is capable of obtaining formulas for the sum of n terms of a given series. If symbolic manipulation software is not available, a spreadsheet package such as *Quattro* also can help students gain further experience with series and partial sums. The use of tables of data gotten from series, especially convergent geometric series, can help students see patterns and gain insights into the fact that the value of the sum of the series is all its terms added together.

Communicating Mathematics

Your best friend has missed today's lecture in math. It is up to you to help him/her understand the differences between a sequence and a series. Your notes seem clear except for those describing the sequence of partial sums. Prepare several examples to illustrate that a sequence of partial sums and a series are two different concepts.

Checkpoint

1 Write the following series in sigma notation: $1 + 5 + 9 + 13 + 17 + 21$.
$\sum_{i=1}^{6} (4i - 3)$

2 Calculate the following sum: $\sum_{i=0}^{4} (2i^3 + 5)$
$5 + 7 + 21 + 59 + 133$ or 225

3 Evaluate the sum $\sum_{i=1}^{\infty} \left(\frac{1}{2}\right)^i$ if it exists.
Sum appears to be 1.

Assignment Guide

Average	
Day 1	8–10, 13, 14a, b, 15, 18, 24, 28, 36, 38
Day 2	11, 14c, 17, 19, 22, 25, 26, 35, 37
Advanced	
Day 1	8, 10, 13, 14a, c, 16, 19, 24, 26, 29, 35
Day 2	11, 15, 17, 21, 22, 25, 30, 35, 36, 38
Honors	
Day 1	10, 13–15, 19, 22, 24, 30, 33, 42
Day 2	11, 16, 20, 25–27, 35, 39, 40, 44

Part Two: Sample Problems

Problem 1 *If $\sum_{k=1}^{n} a_k = n(n + 1)$, what is the nth term of the original sequence $\{a_n\}$?*

Solution **Method 1**

Let's look for patterns.

Number of Terms (n)	Series $\sum_{k=1}^{n} a_k$	Sum of Series $n(n + 1)$	a_n
1	a_1	$1(1 + 1) = 2$	$a_1 = 2$
2	$a_1 + a_2$	$2(2 + 1) = 6$	$a_2 = 4$
3	$a_1 + a_2 + a_3$	$3(3 + 1) = 12$	$a_3 = 6$
4	$a_1 + a_2 + a_3 + a_4$	$4(4 + 1) = 20$	$a_4 = 8$

The pattern in the last column suggests that $a_n = 2n$.

Method 2

Let $S_n = \sum_{k=1}^{n} a_k$, so $S_n = n(n + 1)$ and $S_{n-1} = (n - 1)n$.

We know that $S_n = S_{n-1} + a_n$

Substituting $n(n + 1) = (n - 1)n + a_n$

or $n(n + 1) - (n - 1)n = a_n$

$$n^2 + n - n^2 + n = a_n$$

$$2n = a_n$$

Problem 2 *Estimate the sum $\sum_{n=1}^{\infty} \left(\frac{1}{n} - \frac{1}{n + 1}\right)$.*

Solution

$$S_1 = \left(1 - \frac{1}{2}\right) = \frac{1}{2}$$

$$S_2 = \left(1 - \frac{1}{2}\right) + \left(\frac{1}{2} - \frac{1}{3}\right) = 1 - \frac{1}{3} = \frac{2}{3}$$

$$S_3 = \left(1 - \frac{1}{2}\right) + \left(\frac{1}{2} - \frac{1}{3}\right) + \left(\frac{1}{3} - \frac{1}{4}\right) = 1 - \frac{1}{4} = \frac{3}{4}$$

$$S_4 = 1 - \frac{1}{5} = \frac{4}{5}$$

$$S_5 = 1 - \frac{1}{6} = \frac{5}{6}$$

$$\vdots$$

$$S_n = 1 - \frac{1}{(n + 1)} = \frac{n}{n + 1}$$

Now graph the ordered pairs (n, S_n). The points approach a horizontal asymptote of $S_n = 1$. The sum of the infinite series appears to be 1.

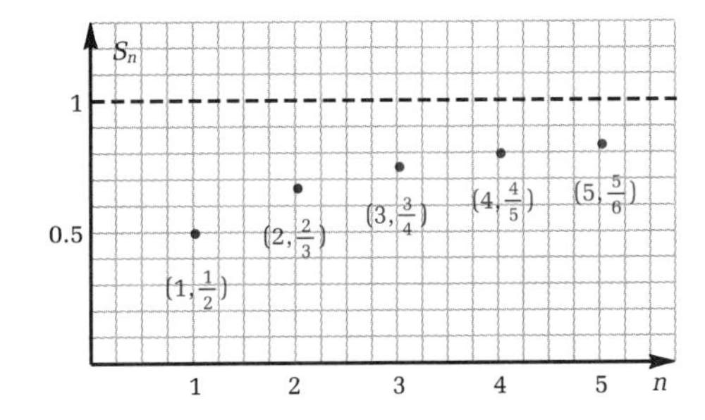

Part Three: Exercises and Problems

Warm-up Exercises

P,R **1** Let $\{a_n\} = \{n!\}$. Use sigma notation to express the fourth term of the sequence of partial sums for $\{a_n\}$. Evaluate your expression. $\sum_{n=1}^{4} n!$; 33

P,R **2** Given the sequence $\{a_n\}$, where a_n is the perimeter of a rectangle with sides n and $2n$.

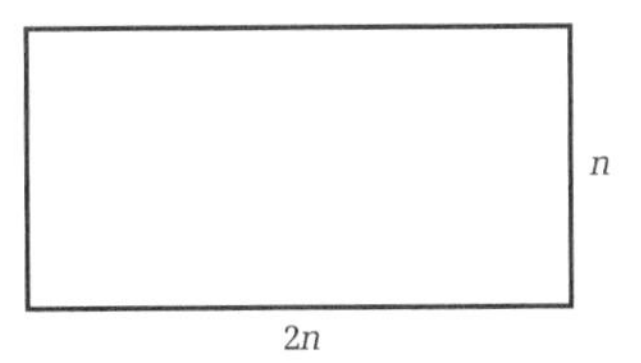

a Find a_1, a_2, a_3, a_4, and a_5. 6; 12; 18; 24; 30

b If $S_5 = \sum_{n=1}^{5} a_n$, what is S_5? 90

P,R **3** Let $\{a_n\} = \{3n - 1\}$. Write the first five terms of the sequence of partial sums of $\{a_n\}$. 2, 7, 15, 26, 40

P,I **In problems 4 and 5, evaluate each finite series.**

4 $\sum_{n=1}^{4} 6n$ 60

5 $\sum_{n=1}^{10} (2n - 1)$ 100

P,R **6** If $\{S_n\}$ is the sequence of partial sums of $\{a_n\}$ and $S_n = n^2 + 2n$, what is $\{a_n\}$? Write its first five terms. $\{2n + 1\}$; 3, 5, 7, 9, 11

R **7** If $\sum_{k=1}^{n} a_k = n^2$, what is the nth term of the sequence $\{a_n\}$? $2n - 1$

Problem Set

A P,R

In problems 8–11, evaluate each finite series.

8 $\sum_{n=1}^{7} (2n + 3)$ 77

9 $\sum_{n=1}^{8} 3n$ 108

10 $\sum_{n=1}^{10} (-1)^n$ 0

11 $\sum_{n=1}^{5} \log n$ log 120, or ≈2.08

Stumbling Block

A series is defined as the sum of a sequence. A partial sum is defined as a sum of a given number of terms of a sequence. Students may have difficulty with the difference. It may be helpful to organize the ideas as follows:

A sequence is a list.

A series is the list as a sum.

A partial sum is the total of the terms of a sequence up to a given term.

Problem-Set Notes and Strategies

- For the series in this section, the sums will be calculated by writing the individual terms and adding them together. In the next two sections, patterns will be employed and formulas produced.
- It may be difficult for students to discover the general term in problems **6** and **7**. Encourage them to look for patterns and keep trying. This is a new skill for many.

Cooperative Learning

Problem **14d** is a good example of how calculating some sums can be easier if we change the limits of the summation. Have students try to figure out **14d** working in groups. Then assign the additional problems:

$\sum_{s=5}^{10} s^2 + 2s + 1 = \sum_{s=1}^{6} ?$ $(s+5)^2$

$\sum_{r=-2}^{5} r^2 + 2r + 1 = \sum_{r=1}^{8} ?$ $(r-2)^2$

Problem-Set Notes and Strategies, continued

- Part **d** of problem **14** and problem **16** will not be obvious and may take some time for students to discover. These are good problems for a group activity.
- Students might need a little review on the area of an equilateral triangle for problem **22**.
- Problem **25** is an example of a geometric series. At this point the sum approaches 2.0. In Section **13.4**, it will be shown that the infinite sum is equal to 2 in the limiting process.
- Problem **26** just grows without bound. Help the students see that the sum is increasing at a rapid rate.

Problem Set, *continued*

P,R,I **In problems 12 and 13, write the first five terms of the sequence.**

12 $\{a_n\} = \{3n\}$ 3, 6, 9, 12, 15

13 $\{a_n\} = \{(-1)^{n+1}\}$ 1, −1, 1, −1, 1

P **14** Evaluate the following series.

a $\sum_{n=1}^{10} n^2$ 385

b $\sum_{n=2}^{11} n^2$ 505

c $\sum_{n=1}^{11} (n^2 + 2n + 1)$ 649

d What method did you use to evaluate part **c?** What is another method? Added terms; $\sum_{n=1}^{11} (n+1)^2 = \sum_{n=2}^{12} n^2 = 505 + 12^2 = 649$

P,I **15** Evaluate $\sum_{j=1}^{4} a_j$ if $a_j = 3j + 1$. 34

P,R **16** Draw the fourth and fifth terms of the sequence of points. Write a formula in closed form for the sequence. Number of dots in the nth figure $= \frac{n(3n-1)}{2}$

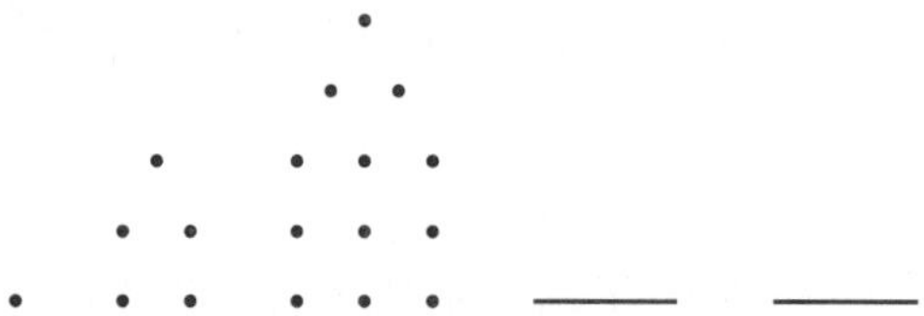

P,I **In problems 17–20, evaluate each finite series.**

17 $\sum_{n=1}^{5} [(n+2)(n+1)]$ 110

18 $\sum_{n=2}^{4} (4.1n + 2.5)$ 44.4

19 $\sum_{n=1}^{5} 6\left(\frac{1}{10}\right)^{n-1}$ 6.6666

20 $\sum_{n=3}^{5} n^n$ 3408

P,R **21** Find x if $\sum_{j=1}^{3} (5 - 2jx) = 51$. $x = -3$

P,R **22** Let a_n = area of an equilateral triangle with side n.

a Write the first four terms of the sequence $\{a_n\}$. $\frac{\sqrt{3}}{4}, \sqrt{3}, \frac{9\sqrt{3}}{4}, 4\sqrt{3}$

b Find the value of the finite series formed by adding the first four terms of the sequence. $\frac{15\sqrt{3}}{2}$

P **23** Evaluate $\sum_{n=1}^{20} 7\left(\frac{1}{10}\right)^n$. $\approx 0.\overline{7}$

R **24** Let $\{c_n\} = \{a_n + b_n\}$, where $\{a_n\} = \{n - 2\}$ and $\{b_n\} = \{n!\}$. Write the first five terms of $\{c_n\}$. 0, 2, 7, 26, 123

B

P **In problems 25 and 26, find the sum of each infinite series, if that sum exists.**

25 $\sum_{n=1}^{\infty} 18\left(\frac{1}{10}\right)^n$ 2.0

26 $\sum_{n=1}^{\infty} n$ No sum

R | **27** The eighth term of the sequence $\{a_n\}$ is 7, and $a_n = a_{n-1} + 2$ for $n \geq 2$. Determine a_7, a_6, and a_1. 5; 3; −7

P,R | **28** The sequence $\{a_n\}$ is pictured in triangle ABC.

a Find the first four terms. $1, \frac{1}{\sqrt{3}}, \frac{1}{3}, \frac{1}{3\sqrt{3}}$

b What is the geometric interpretation of $\sum_{n=1}^{\infty} a_n$? $\sum_{n=1}^{\infty} a_n = AB$

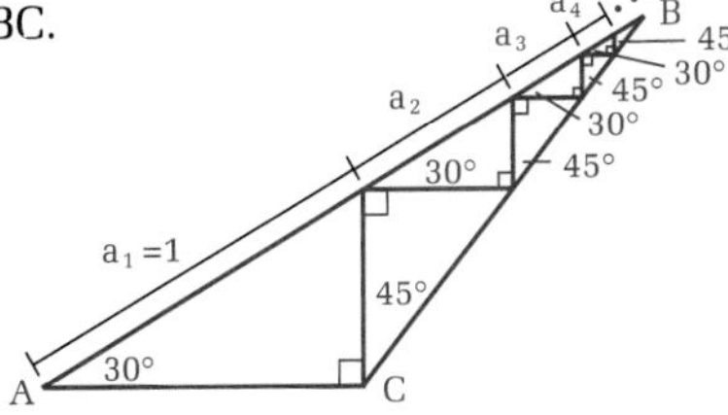

P,R | **29** Let a_1 = area of equilateral triangle with side 8, a_2 = area of equilateral triangle with side 4, and so on.

a Find $a_1, a_2, \ldots, a_5$. $a_1 \approx 27.713$; $a_2 \approx 6.928$; $a_3 \approx 1.732$; $a_4 \approx 0.433$; $a_5 \approx 0.108$

b Find S_5. $S_5 \approx 36.914$

P,R | **In problems 30 and 31, evaluate each series.**

30 $\sum_{n=1}^{5} [(n-3)(n-1)]$ 10

31 $\sum_{n=4}^{10} \log n$ log 604,800, or ≈5.78

P,R | **In problems 32–35, let $\{S_n\}$ be the sequence of partial sums of $\{a_n\}$. Graph the points (n, S_n) for $n = 1, 2, 3, \ldots, 7$. Does the associated infinite series appear to have a sum? If so, what is the sum?**

32 $\{a_n\} = \left\{\left(\frac{1}{2}\right)^{n-1}\right\}$ Yes; 2

33 $\{a_n\} = \{1^n\}$ No sum

34 $\{a_n\} = \{3(0.1)^{n-1}\}$ Yes; $3\frac{1}{3}$

35 $\{a_n\}$ = {nth digit in π} No sum

P | **In problems 36 and 37, evaluate.**

36 $\sum_{n=1}^{\infty} \left(-\frac{1}{2}\right)^{n-1}$ $\frac{2}{3}$

37 $\sum_{n=1}^{\infty} 0^n$ 0

P,R,I | **38** The array 5, 10, 15, 20, 25 has an obvious pattern—the multiples of 5. Find both a closed form and a recursive form for the general formula of the sequence.
Recursive: $a_1 = 5$, $a_n = a_{n-1} + 5$, $n = 2, 3, \ldots$; closed: $a_n = 5n$, $n = 1, 2, \ldots$

P,I | **39** If $\sum_{k=1}^{n} a_k = n(2n+1)$, what is a_n? $4n - 1$

P C | **40** Let $a_6 + a_5 = 29$, $a_6 - a_5 = 7$, and $a_n = a_{n-1} + a_{n-2}$ for $n \geq 3$. Find $\sum_{k=1}^{6} a_k$. 44

P | **In problems 41–43, find the sum of each series, if possible.**

41 $\sum_{n=1}^{\infty} \left(\frac{1}{5}\right)^{n-1}$ 1.25

42 $\sum_{n=1}^{\infty} 2^{n-1}$ No sum

43 $\sum_{n=1}^{15} i^n$, where $i = \sqrt{-1}$ −1

P | **44** Estimate $\sum_{n=1}^{\infty} 4\left(\frac{1}{2n+1} - \frac{1}{2n+3}\right)$. $\frac{4}{3}$

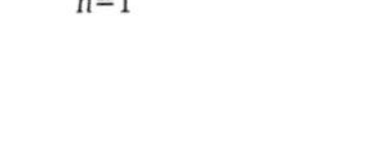

Problem-Set Notes and Strategies, continued

- Problem **28** is an interesting geometric interpretation of a geometric series. Remember that students will not have seen geometric series yet. Some of them may begin to recognize the pattern at this point. This is another infinite series that converges.
- Problem **38** also introduces an arithmetic sequence, which will be introduced in the next section. All arithmetic sequences may be defined each way.
- Plan to spend a little time working problem **40**. It isn't that difficult, just time consuming.
- Problem **44** may tax the algebra skills of some of your students, especially those that are trying to find a pattern to infinite series that converge. This convergence will be formalized in Section **13.4**.

13.3

Class Planning

Time Schedule
All levels: 2 days

Resource References
Teacher's Resource Book
Class Opener 78
Practice 82
Evaluation
Quiz Forms 1, 2, 3

Class Opener

Xavier's homework was completed before his cat scratched the paper. What was readable was the first three and last two numbers of a sequence. The first three terms were 3, 7, 11 and the last two were 355, 359. He forgot how many terms there had been. Help him. Find how many terms were in the sequence. Then find the 25th term.
90 terms; 99

Lesson Notes

- The arithmetic sequence and series are the focus of this section. Formulas are developed for both the recursive and closed form definition of the sequence. Each of these forms is useful for solving for parts of the sequence. This is a case when memorizing formulas will reduce confusion.

13.3 Arithmetic Sequences and Series

Objectives

After studying this section, you will be able to
- Compute any term in an arithmetic sequence
- Compute arithmetic means
- Evaluate arithmetic series

Part One: Introduction

Arithmetic Sequences

Consider the sequence 2, 5, 8, 11, 14, . . . , $3n - 1$, . . . The difference between any two successive terms is 3. Therefore, we obtain any term of the sequence by adding 3 to the previous term. This is an example of an ***arithmetic sequence.*** An ***arithmetic sequence*** is any sequence $\{a_n\}$ in which the difference between each two successive terms is the same. The difference, d, is called the ***common difference.***

The definition can be translated into a recursive formula for the nth term.

Definition: $a_n - a_{n-1} = d$
Recursive formula: $a_n = a_{n-1} + d$

To go from the first term to the nth term, we add d, the difference, $(n - 1)$ times to the first term. This process can be translated into a closed-form formula.

Closed-form formula: $a_n = a_1 + (n - 1)d$

We use the closed form to derive an equation relating any two elements in the series, a_n and a_{n-k}.

Formula for a_n	$a_n = a_1 + (n - 1)d$
Formula for a_{n-k}	$a_{n-k} = a_1 + (n - k - 1)d$
Subtract	$a_n - a_{n-k} = kd$
	$a_n = a_{n-k} + kd$

Example *Find the value of the 48th term of an arithmetic sequence with first term −20 and common difference 3.*

In this case, $a_1 = -20$, $d = 3$, and $n = 48$. The 48th term is a_{48}.

Closed form
$$a_n = a_1 + (n - 1)d$$
$$a_{48} = -20 + (48 - 1)3$$
$$= -20 + 141$$
$$a_{48} = 121$$

Vocabulary
arithmetic sequence
common difference
arithmetic means
arithmetic series

Example *The 18th term of an arithmetic sequence is 54 and the 24th term is 36. Find the first term and the common difference.*

Use the formula $a_n = a_{n-k} + kd$.

$$a_{24} = a_{18} + 6d$$
$$36 = 54 + 6d$$
$$-18 = 6d$$
$$-3 = d$$

Now use the closed form.

$$a_n = a_1 + (n - 1)d$$
$$a_{18} = a_1 + 17d$$
$$54 = a_1 + 17(-3)$$
$$54 = a_1 - 51$$
$$105 = a_1$$

The first term is 105 and the common difference is -3.

Arithmetic Means

You may be familiar with the arithmetic mean as an average of two numbers. The term has a broader definition based on arithmetic sequences. In an arithmetic sequence, ***arithmetic means*** are the terms between any two designated terms.

Example *A portion of an arithmetic sequence is given. Fill in the blanks.*

12, ____, ____, ____, 26

In this problem, we need to find three arithmetic means between 12 and 26. Let $a_1 = 12$; then $a_5 = 26$.

$$a_5 = a_1 + 4d$$
$$26 = 12 + 4d$$
$$14 = 4d$$
$$3.5 = d$$

The sequence is 12, 15.5, 19, 22.5, 26.

Arithmetic Series

We can derive a formula for the partial sum of an ***arithmetic series.*** Let's consider the following example.

Example *Find the partial sum S_{10} of the first ten terms of the arithmetic sequence $\{a_n\} = \{2 + (n - 2)3\}$.*

We could solve this by adding up all the terms.

$S_{10} = -1 + 2 + 5 + 8 + 11 + 14 + 17 + 20 + 23 + 26$

Let's try a more creative way. We write down the terms twice—once from first to last and once from last to first—and then add vertically.

$$S_{10} = -1 + 2 + 5 + 8 + 11 + 14 + 17 + 20 + 23 + 26$$
$$S_{10} = 26 + 23 + 20 + 17 + 14 + 11 + 8 + 5 + 2 + (-1)$$
$$2 \cdot S_{10} = 25 + 25 + 25 + 25 + 25 + 25 + 25 + 25 + 25 + 25$$
$$= 10 \cdot 25$$
$$S_{10} = \frac{10 \cdot 25}{2}$$
$$= 125$$

Lesson Notes, continued

- Finding the arithmetic means is finding equally spaced numbers between two endpoints. The terminology may be new to the student, but the concept is not. An example such as, "If a group were to travel for 500 miles and make four stops at equally spaced intervals, where would they stop? " is an illustration of the arithmetic mean that can make the idea clearer.
- The formulas for the arithmetic series are also introduced here. A common story is that Gauss discovered these formulas when he was in the second grade. It helps if students see the relevance of the division by 2 in the example and also the substitution in the second formula for a_n.

Using Computers

See Using Computers in Section 13.2.

Communicating Mathematics

Explain in your own words the significance of the terminology *common difference.*

Checkpoint

1. The first term of an arithmetic sequence is 5 and the common difference is 3, find the first six terms.
   5, 8, 11, 14, 17, 20
2. If the first term of an arithmetic sequence is 1 and the sixth term is 11, find the common difference and the sum of the first 50 terms.
   $d = 2$, $S_{50} = 2500$
3. Insert four arithmetic means between 16 and 40.
   20.8, 25.6, 30.4, 35.2
4. An arithmetic sequence has a first term of 3 and a common difference of 7. Find the sum of the first 25 terms.
   2,175

Assignment Guide

Average	
Day 1	10, 11, 14, 17, 19, 20, 25, 27, 28, 36
Day 2	12, 13, 15, 18, 22, 23, 31, 33, 38, 40
Advanced	
Day 1	11, 12, 15–17, 19, 23, 25, 27, 34, 37, 40
Day 2	13, 18, 20, 22, 24, 26, 29, 32, 33, 36, 41
Honors	
Day 1	21, 22, 25, 28, 30, 32, 37, 39, 42, 43
Day 2	13, 16, 18, 23, 33, 35, 36, 40, 41, 44

In the example, 25 was the sum of the first and tenth terms of the sequence. We can make the following generalization.

The sum of the first n terms, S_n, of the arithmetic sequence a_n with common difference d is

$$S_n = n\left(\frac{a_1 + a_n}{2}\right)$$

Proof: We proceed as in the example.

$$\begin{aligned}
S_n &= a_1 + (a_1 + d) + (a_1 + 2d) + (a_1 + 3d) + \ldots + a_n \\
S_n &= a_n + (a_n - d) + (a_n - 2d) + (a_n - 3d) + \ldots + a_1 \\
\hline
2S_n &= (a_1 + a_n) + (a_1 + a_n) + (a_1 + a_n) + (a_1 + a_n) + \ldots + (a_1 + a_n) \\
2S_n &= n \cdot (a_1 + a_n) \\
S_n &= n\left(\frac{a_1 + a_n}{2}\right)
\end{aligned}$$

Since $a_n = a_1 + (n - 1)d$, we can substitute for a_n.

$$S_n = n\left(\frac{2a_1 + (n - 1)d}{2}\right)$$

Example *An arithmetic sequence has $a_1 = -10$ and a common difference of 0.25. Find the sum of the first 75 terms of this sequence.*

$$S_n = n\left(\frac{2a_1 + (n - 1)d}{2}\right)$$

$$S_{75} = 75\left(\frac{2(-10) + 74(0.25)}{2}\right)$$

$$= -56.25$$

Part Two: Sample Problem

Problem *The sum of three consecutive terms of an arithmetic sequence is 27, and their product is 405. Find the three terms.*

Solution Let a be the second of the three terms, and let d be the common difference. Then $a - d$ is the first term and $a + d$ is the third term.

The sum is 27

$$\begin{aligned}
(a - d) + a + (a + d) &= 27 \\
3a &= 27 \\
a &= 9
\end{aligned}$$

The product is 405

$$\begin{aligned}
(9 - d)(9)(9 + d) &= 405 \\
9(81 - d^2) &= 405 \\
81 - d^2 &= 45 \\
36 &= d^2 \\
d = 6 \text{ or } d &= -6
\end{aligned}$$

When $d = 6$, the terms are 3, 9, 15.
When $d = -6$, the terms are 15, 9, 3.

Part Three: Exercises and Problems

Warm-up Exercises

In problems 1 and 2, a_n represents the nth term of an arithmetic sequence and d represents the common difference. Write the first five terms of each sequence.

P **1** $a_1 = 10, d = 3$
10, 13, 16, 19, 22

2 $a_4 = 6, d = -3$
15, 12, 9, 6, 3

P **3** An arithmetic series S is written twice below, once first to last and once last to first. Add vertically and solve for S.

$$S = 2 + 4 + 6 + \ldots + 22 + 24$$
$$S = 24 + 22 + 20 + \ldots + 4 + 2$$

$2S = 26 + 26 + 26 + \ldots + 26 + 26; S = 156$

P **4** Insert two arithmetic means between 12 and 22. $\frac{46}{3}, \frac{56}{3}$

P **5** Which of the following are arithmetic sequences? a

a $7, -4, -15, -26, \ldots$

b $2 + \sqrt{3}, 4 + \sqrt{3}, 8 + \sqrt{3}, \ldots$

P **6** Find the common difference of each arithmetic sequence.

a $7, 10, 13, 16, \ldots$ 3

b $3\sqrt{5}, -\sqrt{5}, -5\sqrt{5}, -9\sqrt{5}, \ldots$ $-4\sqrt{5}$

P **In problems 7 and 8, find the seventh term of the arithmetic sequence.**

7 $2 + \pi, 3 + 3\pi, 4 + 5\pi, \ldots$
$8 + 13\pi$

8 $-10, -8, -6, \ldots$
2

Problem Set

A

In problems 9–11, write the first four terms of each sequence.

P **9** $\{5n\}$
5, 10, 15, 20

10 $\{3n + 2\}$
5, 8, 11, 14

11 $\{-2n - 4\}$
−6, −8, −10, −12

P **In problems 12 and 13, a_n represents the nth term in an arithmetic sequence and d represents the common difference. Write the first five terms of each arithmetic sequence.**

12 $a_1 = -4, d = \frac{1}{2}$
−4, −3.5, −3, −2.5, −2

13 $a_{19} = 21, d = \frac{1}{4}$
16.5, 16.75, 17, 17.25, 17.5

P **In problems 14–16, list the next three terms of each arithmetic sequence.**

14 3, 9, 15, . . .
21, 27, 33

15 −5, −1, 3, . . .
7, 11, 15

16 $\frac{1}{2}, \frac{1}{3}, \frac{1}{6}, \ldots$
$0, -\frac{1}{6}, -\frac{1}{3}$

P **17** Consider the arithmetic series $5 + 7 + 9 + \ldots + 99 + 101$.

a What is the first term? 5

b What is the common difference? 2

c How many terms are there? 49

d Compute the sum. 2597

Problem-Set Notes and Strategies

- Problem **3** illustrates the technique reportedly discovered by Gauss in the second grade. Now might be a good time for a little history on Gauss.

- Problems **6 – 16** are straightforward applications of the formulas for arithmetic sequences and series. Many can be solved by inspection and intuition.

Problem-Set Notes and Strategies, continued

- Problem **22** requires using the formulas in the reverse to find the number of terms. It isn't difficult, but may be confusing for some students.
- Problem **25** may be done with formulas or by using intuition.
- Problem **26** is easy if students understand an arithmetic sequence.
- The series in problems **28 – 30** may take some time until students discover the common difference and use the appropriate formulas. These are more troublesome to do by adding each term.

Problem Set, *continued*

P **18** Find the first five partial sums—S_1, S_2, S_3, S_4, and S_5—of $\{a_n\} = \{3n + 1\}$. 4, 11, 21, 34, 50

B **In problems 19–21, find the fiftieth term of each arithmetic sequence.**

P **19** 3, 9, 15, . . . 297 **20** $-5, -1, 3, \ldots$ 191 **21** $\frac{1}{2}, \frac{1}{3}, \frac{1}{6}, \ldots$ $\frac{-23}{3}$

P **22** The first term of an arithmetic series is 2 and the common difference is 3. The sum is 187. How many terms are in the series? 11

P,R **23** Let A_n be the surface area of the box shown.

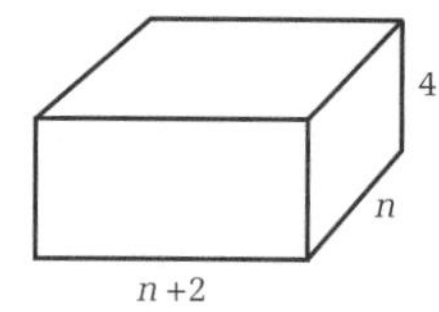

a Find A_1, A_2, A_3, A_4, and A_5.
b Write an expression for A_n.
a 38; 64; 94; 128; 166
b $2(n^2 + 10n + 8)$

R,P **24** What values are printed by the following BASIC program? 3, 9, 18, 30

```
10 LET Sum = 0
20 FOR K = 1 TO 4
30 LET Sum = Sum + 3 * K
40 PRINT Sum
50 NEXT K
60 END
```

P **25** The seventh term of an arithmetic sequence is 80, and the eleventh term is 90. Find the ninth term. 85

P **26** The average of three successive terms of an arithmetic sequence is 47.2. What is the second of the three terms? 47.2

P **27** Find the three arithmetic means between 16 and 32. 20, 24, 28

R,P **In problems 28–30, find each sum.**

28 $\sum_{k=1}^{15} 3k$ 360 **29** $\sum_{k=1}^{100} (2k - 1)$ 10,000 **30** $\sum_{k=8}^{40} (3k + 1)$ 2409

P **In problems 31 and 32, find the 25th term and the sum of the first 25 terms of the arithmetic sequence, where a_n is the nth term and d is the common difference.**

31 $a_1 = 4$, $d = 2$ 52; 700 **32** $a_{10} = 32$, $a_{15} = 44.5$ 69.5; 987.5

P **33** Find five arithmetic means between 15 and 39. 19, 23, 27, 31, 35

P **In problems 34–36, rewrite each sum, using sigma notation.**

34 $1 + 2 + 3 + 4 + \ldots + 162$ **34** $\sum_{k=1}^{162} k$

35 $1^2 + 2^2 + \ldots + 15^2$ **35** $\sum_{k=1}^{15} k^2$

36 $3^4 + 5^4 + 7^4 + \ldots + 43^4$ **36** $\sum_{k=1}^{21} (2k + 1)^4$

P **37** "Diamond" Jim McGillicuddy has \$150 in his piggy bank. He puts \$15 in the bank, and each successive week he puts \$15 more than the week before into the piggy bank. How much is in the bank at the end of two years? \$82,050

P **In problems 38 and 39, identify the first term, the common difference, and the sum of each arithmetic series.**

38 $\sum_{k=1}^{10} 8 + 5(k - 1)$ 8; 5; 305

39 $\sum_{k=1}^{8} \left(-5k\sqrt{2} + 12\sqrt{2}\right)$ $7\sqrt{2}$; $-5\sqrt{2}$; $-84\sqrt{2}$

R,P **40** Let A_n be the area of the shaded region. Find the first six terms of $\{A_n\}$.
3π, 5π, 7π, 9π, 11π, 13π

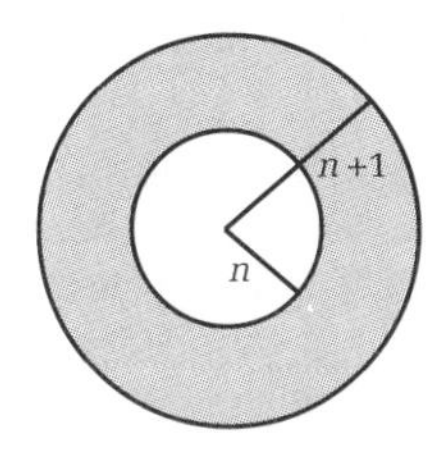

P **41** The first term of an arithmetic sequence is 325, and the 82nd term is 352. Find the common difference and the sum of the 12th through the 82nd terms. $d = \frac{1}{3}$; $\approx 24{,}163.7$

P,R **42** Let A_n be the sum of the shaded areas.

a Find A_1, A_2, A_3, and A_4.

b Write an expression for A_n.

a 10π, 14π, 18π, 22π **b** $\pi(4n + 6)$

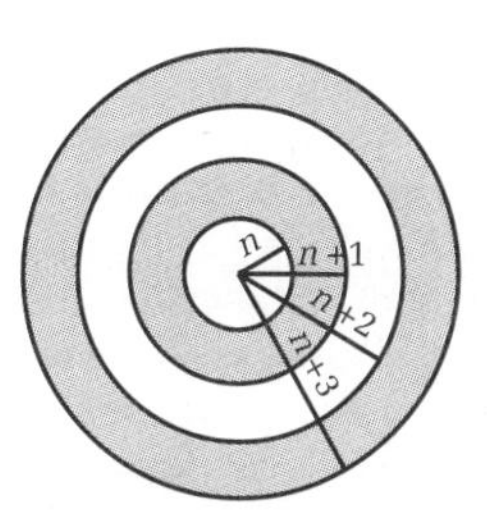

P **43** The sum of two consecutive terms of an arithmetic sequence is 2. Their product is -24. The first of the two terms is smaller than the second. Find the two terms. -4 and 6

C P,R **44** The first six terms of an arithmetic sequence are 2, 5, 8, 11, 14, and 17. Three of the six numbers are selected at random and placed in the order of selection. Find the probability that these three numbers form an arithmetic progression. $\frac{1}{10}$

Problem-Set Notes and Strategies, continued

- Problem **37** is a familiar variant of the problem: Which salary would you rather receive for 30 days work, \$450 or a dollar a day with an increase of a dollar a day for each of the 30 days?

- Because of the fractions involved, problem **41** may take some time.

- The fact that the terms in problem **43** are from an arithmetic series is not significant. In fact, any two terms can come from an arithmetic sequence.

- Some counting arguments may need to be formulated for problem **44**. It helps if students are sure all the possibilities are counted.

Problem Set, *continued*

P,R **45** The lengths of a sequence of rectangles have common difference 3, and the widths have common difference 1. The first rectangle is a 4-by-1 rectangle.

a What are the dimensions of the tenth rectangle in the sequence? 31 by 10

b What is the sum of the perimeters of the first ten rectangles? 460

c What is the sum of the areas of the first ten rectangles? 1210

CAREER PROFILE

STATISTICS IN PREDICTING LONGEVITY

C. David Williams calculates retirement needs

For most people who work, either they contribute to a pension plan that will provide money for them when they retire, or their employers contribute money on their behalf. An actuary, like C. David Williams of Detroit, applies mathematics in predicting how much money a pension plan will need to pay out over a period of time many years in the future. Into these calculations go statistical data that help determine how long people will live, what will happen to interest rates, and other factors.

An actuary advises companies in formulating retirement plans. The effects of those plans often will not be felt for as long as another forty years. Calculus, a form of mathematics that describes a curve at any given point on the curve, helps the actuary predict how much money will be available in a given year on a given investment and to how many people that money must be distributed.

"We evaluate risk with respect to mortality, economics, and demographics," says Mr. Williams. "We employ elaborate models to make sure a plan can pay according to its terms."

Furthermore, the benefits from most plans are determined by law. This places a burden on actuaries to determine the employer and employee contributions necessary to support the benefits.

Mr. Williams holds a master's degree in mathematics from Northeastern University in Boston. As an example of the importance of mathematics to the work of an actuary: to become certified, Mr. Williams had to pass five mathematics examinations.

13.4

Geometric Sequences and Series

Objectives

After studying this section, you will be able to

- Compute any term in a geometric sequence
- Compute geometric means
- Evaluate finite and infinite geometric series

Part One: Introduction

Geometric Sequences

In the preceding section, we looked at arithmetic sequences, formed by repeated addition. Now let's consider a sequence formed by repeated multiplication.

$$3, 6, 12, 24, 48, \ldots, 3 \cdot 2^{n-1}, \ldots$$

Each term is twice the preceding one, and the *ratio* of any two successive terms is always 2. This is an example of a ***geometric sequence.*** A ***geometric sequence*** is any sequence $\{a_n\}$ with the property $\frac{a_n}{a_{n-1}} = r$, where r is a constant and n is an integer larger than 1. The number r is called the ***common ratio.***

Let's write some terms of the geometric sequence with first term $a_1 = 4$ and common ratio $r = 3$ and look for patterns.

$$a_1 = 4 = 4 \cdot 3^0$$
$$a_2 = a_1 \cdot r = 4 \cdot 3 = 4 \cdot 3^1$$
$$a_3 = a_2 \cdot r = (4 \cdot 3^1) \cdot 3 = 4 \cdot 3^2$$
$$a_4 = a_3 \cdot r = (4 \cdot 3^2) \cdot 3 = 4 \cdot 3^3$$
$$a_5 = a_4 \cdot r = (4 \cdot 3^3) \cdot 3 = 4 \cdot 3^4$$

The power of the common ratio is always one less than the number of the term.

In general, the nth term of a geometric sequence $\{a_n\}$ with common ratio r is given by

$$a_n = a_1 \cdot r^{n-1} \text{ (closed form) or}$$
$$a_n = a_{n-1} \cdot r^1 \text{ (recursive form)}$$

Using the recursive form, we can develop an alternative recursive form.

13.4

Class Planning

Time Schedule

All levels: 2 days

Resource References

Teacher's Resource Book
- Class Opener 79
- Practice 83
- Manipulatives and Applications 18
- Enrichment 26

Transparency 26

Class Opener

Poor Xavier! His cat scratched his homework again. This time the first three terms were 3, 6, 12 and the last two were 6144, 12288. How many terms were in the sequence? 13

Lesson Notes

- To students, geometric sequences and series may seem a bit more difficult to identify, but as they work with these ideas, students will become more comfortable. Both recursive and closed forms of these are included.

Communicating Mathematics

A common ratio is used to identify whether a sequence is geometric or not. Explain the meaning of the term *common ratio.*

Vocabulary

geometric sequence
common ratio
geometric means
geometric series

13.4

Lesson Notes, continued

- Finding geometric means uses the same algorithm as the arithmetic case with the formulas for geometric sequences. The meaning and uses of the geometric means may not be as apparent as in the arithmetic case and may not seem intuitive.

Using Computers

See Using Computers in Section 13.2.

$$a_n = a_{n-1} \cdot r^1$$
$$a_n = (a_{n-2} \cdot r^1)r^1 = a_{n-2} \cdot r^2$$
$$a_n = (a_{n-3} \cdot r^1)r^2 = a_{n-3} \cdot r^3$$
$$\vdots$$
$$a_n = a_{n-k} \cdot r^k$$

Example *The fifth term of a geometric sequence is 48 and the ninth term is 768. Find the first term and the common ratio.*

Use the recursive form of nth term

$$\mathbf{a_n = a_{n-k} \cdot r^k}$$
$$\mathbf{a_9 = a_5 \cdot r^4}$$
$$\mathbf{768 = 48 \cdot r^4}$$
$$\mathbf{16 = r^4}$$
$$\mathbf{\pm 2 = r}$$

Now use the closed form

$$\mathbf{a_n = a_1 \cdot r^{n-1}}$$
$$\mathbf{a_5 = a_1 \cdot r^4}$$
$$\mathbf{48 = a_1 \cdot (\pm 2)^4}$$
$$\mathbf{48 = a_1 \cdot 16}$$
$$\mathbf{3 = a_1}$$

Geometric Means

Geometric means have the same relationship to geometric sequences that arithmetic means have to arithmetic sequences.

Example *Insert one positive geometric mean between* 3 *and* 12.

Think of a geometric sequence 3, ___, 12, or 3, $3r$, $3r^2$.

Solve for r

$$\mathbf{3r^2 = 12}$$
$$\mathbf{r^2 = 4}$$
$$\mathbf{r = \pm 2}$$

The positive geometric mean is $3 \cdot 2 = 6$.

A single positive geometric mean between two terms is the square root of the product of the two numbers. In the preceding example, the mean is $\sqrt{3 \cdot 12} = \sqrt{36} = 6$.

As you know from geometry, the altitude to the hypotenuse of a right triangle is the geometric mean between the two segments of the hypotenuse. Also, the length of either leg of the right triangle is the geometric mean between the entire hypotenuse and the portion of the hypotenuse adjacent to that leg.

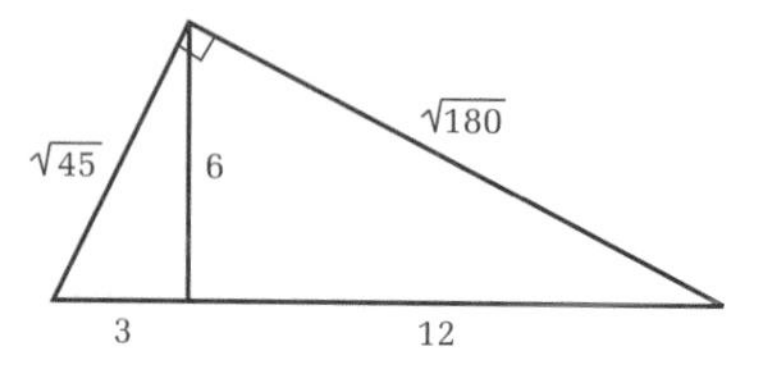

Example *Insert three real geometric means between 96 and 6.*

The geometric sequence is 96, __, __, __, 6, with $a_1 = 96$ and $a_5 = 6$.

$$a_5 = a_1 \cdot r^4$$
$$6 = 96 \cdot r^4$$
$$\frac{1}{16} = r^4$$
$$\pm\frac{1}{2} = r$$

If $r = \frac{1}{2}$, the geometric sequence is 96, 48, 24, 12, 6.

If $r = -\frac{1}{2}$, the geometric sequence is 96, −48, 24, −12, 6.

Lesson Notes, continued

- One source of motivation for geometric series is compound interest problems. The infinite case is especially interesting as it involves students in limits and provides topics for discussion.

Geometric Series

When we find the sum of the first n terms of a geometric sequence $\{a_n\}$ we are finding S_n, the value of a finite ***geometric series*** of n terms.

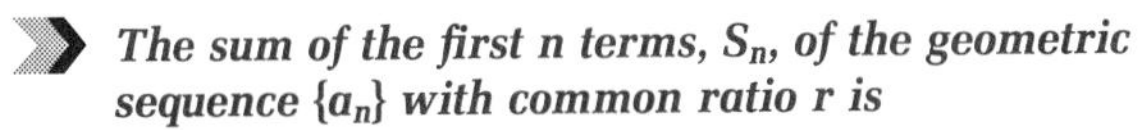

The sum of the first n terms, S_n, of the geometric sequence $\{a_n\}$ with common ratio r is

$$S_n = \frac{a_1(r^n - 1)}{r - 1}$$

Proof: $S_n = a_1 + a_1 \cdot r + a_1 \cdot r^2 + a_1 \cdot r^3 + \ldots + a_1 \cdot r^{n-1}$

$rS_n = a_1 r + a_1 \cdot r^2 + a_1 \cdot r^3 + \ldots + a_1 \cdot r^{n-1} + a_1 \cdot r^n$

Subtracting the first equation from the second, we get

$$rS_n - S_n = a_1 \cdot r^n - a_1$$
$$(r - 1)S_n = a_1(r^n - 1)$$
$$S_n = \frac{a_1(r^n - 1)}{r - 1}$$

Example *Evaluate* $\sum_{n=1}^{7} 96\left(\frac{1}{4}\right)^{n-1}$.

$$\sum_{n=1}^{7} 96\left(\frac{1}{4}\right)^{n-1} = 96 + 24 + 6 + \ldots + \frac{3}{128}$$

This is a geometric series with $a_1 = 96$ and $r = \frac{1}{4} = 0.25$.

$$S_n = \frac{a_1(r^n - 1)}{r - 1}$$
$$S_7 = \frac{96(0.25^7 - 1)}{0.25 - 1}$$
$$\approx \frac{96(0.000061035 - 1)}{-0.75}$$
$$\approx 127.992$$ The sum of the seven terms is about 127.992.

13.4

Stumbling Block

Students may use the formula for the sum of an infinite series indiscriminately. Help them understand that the absolute value of the common ratio, r, must be less than 1 in order for the formula to work. It might be informative to graph several sequences, both convergent and divergent, with positive and negative ratios.

Checkpoint

1. What is the common ratio in the sequence 128, –64, 32, –16, 8, ... $-\frac{1}{2}$
2. Find the eighth term of the sequence with first term 5 and common ratio 3. 10,935
3. Find the sum of the first five terms of the above sequence. 605
4. Find the sum of the infinite series in the first problem, if it exists. $85\frac{1}{3}$

Infinite Geometric Series

The expression $\sum_{n=1}^{\infty} (a_1 \cdot r^{n-1})$ represents an infinite geometric series.

Remember that the sum of an infinite series is defined only if the graph of the sequence of partial sums, $\{S_n\}$, approaches a horizontal asymptote.

Example *Evaluate* $\sum_{n=1}^{\infty} 96\left(\frac{1}{4}\right)^{n-1}$.

The related sequence $\{a_n\}$ is 96, 24, 6, $\frac{3}{2}, \frac{3}{8}, \frac{3}{32}, \frac{3}{128}, \ldots$ We generate a table of values of n and S_n, the sum of the terms of $\{a_n\}$, and graph the ordered pairs (n, S_n) for $n = 1, 2, \ldots, 7$.

n	S_n
1	96
2	120
3	126
4	127.5
5	127.875
6	127.96875
7	127.9921875

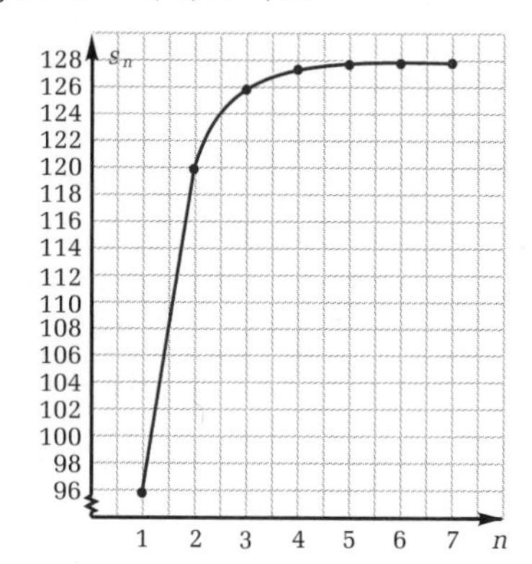

The points appear to approach an asymptote at 128. Therefore, the sum of the infinite series is 128.

Look at the formula for the sum, S_n, of n terms of a geometric sequence.

$$S_n = \frac{a_1(r^n - 1)}{r - 1}$$

What happens when n is large and $|r| < 1$? Let's examine what happens when $a_1 = 200$ and $r = 0.9$.

n	1	10	100	1000
r^n	0.9	≈0.35	≈2.66×10^{-5}	≈1.75×10^{-46}
S_n	200	≈1302.64	≈1999.94	≈2000

We notice from the table that if $|r| < 1$ and n is very large, the value of r^n is very small, so S_n approaches $\frac{a_1(0 - 1)}{r - 1}$, or $\frac{a_1}{1 - r}$. When $|r| < 1$, therefore, the sum, S, of the infinite geometric series $\sum_{n=1}^{\infty} a_1 \cdot r^{n-1}$ is given by $S = \frac{a_1}{1 - r}$.

Part Two: Sample Problems

Problem 1 *A rubber ball is shot vertically to a height of 20 feet and allowed to drop. Each bounce is 80% as high as the previous bounce. What is the total distance the ball travels?*

Solution The ball travels up and down on each bounce. On each successive bounce it travels 80% (0.8) as high as on the previous bounce.

Bounce	Distance Up (in feet)	Up-Down Total (in feet)
1	20	40
2	$20(0.8) = 16$	32
3	$16(0.8) = 12.8$	25.6
4	$12.8(0.8) = 10.24$	20.48
		⋮

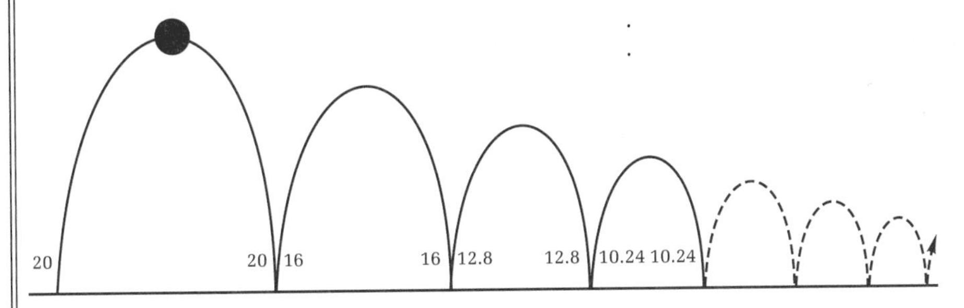

The up-down total is a geometric series with first term $a_1 = 40$ and common ratio $r = 0.8$. The total distance traveled is the sum of all these numbers.

$$S = \frac{a_1}{1 - r} = \frac{40}{1 - 0.08} = 200$$

The ball travels 200 feet.

Problem 2 *Evaluate* $\sum_{k=1}^{\infty} \frac{2}{3}\left(\frac{4}{3}\right)^k$.

Solution This is an infinite geometric series with a common ratio of $\frac{4}{3}$. Since $\frac{4}{3} > 1$, the sum does not exist.

Part Three: Exercises and Problems

Warm-up Exercises

In problems 1–3, find the seventh term and the sum of the first seven terms of the following geometric sequences, where a_n is the nth term and r is the common ratio.

P

1 $a_1 = \sqrt{2}$, $r = \sqrt{2}$
$8\sqrt{2}$; $14 + 15\sqrt{2}$

2 $a_3 = 24$, $a_7 = 1.5$
1.5; 190.5

3 $a_1 = 3i$, $r = i$
$-3i$; -3

Assignment Guide

Average	
Day 1	8, 11, 13, 14, 16, 19, 21, 23, 26
Day 2	10, 12, 15, 17, 20, 22, 25, 27, 28
Advanced	
Day 1	8, 12, 13, 15–17, 25, 28, 32
Day 2	10, 14, 18, 20, 21, 25, 26, 31, 33
Honors	
Day 1	9, 13, 14, 17, 20, 21, 25, 26, 28, 29
Day 2	8, 15, 16, 18, 19, 30–33, 35

Problem-Set Notes and Strategies

- Problems **1 – 12** are applications of the formulas. Problem **3** has a slight twist since it includes complex numbers.

P 4 Find the sum of each of the following infinite series.

a $\sum_{n=1}^{\infty} 4\left(\frac{1}{2}\right)^n$ 4

b $\sum_{n=1}^{\infty} 4(2^n)$ No sum

P 5 Find the sum of the infinite geometric series whose third term is 125 and whose sixth term is 1. 3906.25

P 6 Find the geometric mean between 12 and 8. $\approx \pm 9.80$

P 7 A ball is dropped from a building 65 feet tall. If on each bounce the ball rebounds to 75% of the height of the previous bounce, how far does it travel? 455 ft

Problem Set

A

P **In problems 8–10, write the first six terms of each geometric sequence, given the first term and common ratio.**

8 $a_1 = 6$, $r = 2$ 6, 12, 24, 48, 96, 192

9 $a_1 = 5$, $r = 0.3$
5, 1.5, 0.45, 0.135, 0.0405, 0.01215

10 $a_1 = 16$, $r = \frac{1}{2}$ 16, 8, 4, 2, 1, $\frac{1}{2}$

P **In problems 11 and 12, find the next three terms of each geometric sequence.**

11 4, 2, 1, . . . $\frac{1}{2}, \frac{1}{4}, \frac{1}{8}$

12 8, −12, 18, . . . −27, 40.5, −60.75

P,R **In problems 13–15, state whether each sum is an arithmetic or a geometric series, then give the first term and the common difference or common ratio.**

13 $\sum_{k=1}^{10} (5k + 3)$
Arith.; 8; 5

14 $\sum_{k=3}^{23} 6\left(\frac{1}{2}\right)^k$
Geom.; $\frac{3}{4}$; $\frac{1}{2}$

15 $\sum_{k=1}^{8} \left(\frac{1}{k}\right)$
Neither; first term: 1

P 16 A 12-inch-by-12-inch sheet of paper, 0.004 inch thick, is folded as shown.

a How thick is the folded paper after the eighth fold? 1.024 in.

b What is the top area after the eighth fold? 0.5625 in.^{2}

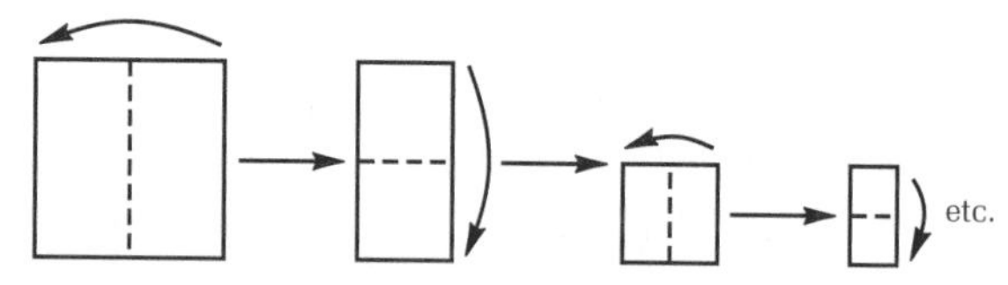

Problem-Set Notes and Strategies, continued

- Problem **17** might call for a little geometry review. These relationships will be necessary in the chapters on trigonometry.

P,R 17 Let d_n = the length of a diagonal of a square with side n.

a Find d_1, d_2, d_3, d_4, and d_5.
$\sqrt{2}$; $2\sqrt{2}$; $3\sqrt{2}$; $4\sqrt{2}$; $5\sqrt{2}$

b Find $\left\{\frac{d_{n+1}}{d_n}\right\}$ for $n = 1, 2, 3, 4$. 2, $\frac{3}{2}, \frac{4}{3}, \frac{5}{4}$

P **In problems 18–20, find the sum of each series.**

18 $\sum_{k=1}^{4} 3\left(\frac{1}{10}\right)^k$ 0.3333

19 $\sum_{k=1}^{\infty} 3\left(\frac{1}{10}\right)^k$ $\frac{1}{3}$

20 $\sum_{n=1}^{\infty} 8\left(\frac{1}{2}\right)^{n-1}$ 16

P | **In problems 21 and 22, find three geometric means between the given numbers.**

21 16 and 0.0625
$-4, 1, -0.25$; or $4, 1, 0.25$

22 8 and 16.5888 9.6, 11.52, 13.824; or -9.6, 11.52, -13.824

P | **23** 7, 7, 7, . . . is an arithmetic sequence. Could it also be a geometric sequence? Yes, with common ratio 1

P,R | **24** The lengths of the sides of a right triangle form an arithmetic sequence with common difference 3. Find the lengths of the sides. 9, 12, 15

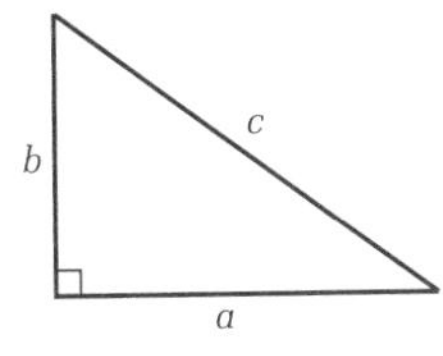

R | **25** Solve $\sum_{n=1}^{3} (nx + 30) = x + 15$ for x. $x = -15$

R  | **26** Find the average value of the first four terms of the sequence $\{a_n\} = \{\log (n + 1)\}$. $\frac{\log 120}{4}$, or ≈ 0.52

R,I | **27** Given: AC = 4,
DC = 9
Find AB. 24.25

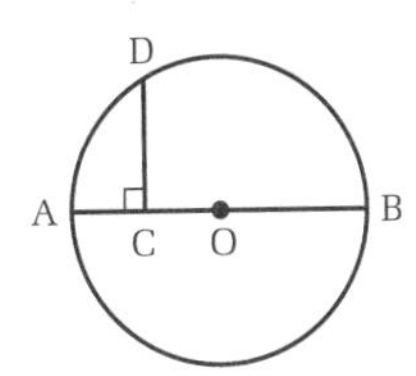

P | **28** Ann T. Gravity has developed Wacko Putty, which, if dropped, rebounds to 108% of its previous height. A ball of Wacko Putty is dropped from a height of 6 feet.

a How high does it bounce on the tenth bounce? $\approx$12.95 ft

b How far has it traveled up and down from when it was dropped to the top of the tenth bounce? $\approx$180.79 ft

In problems 29 and 30, find each sum.

P | **29** $5 - \frac{5}{2} + \frac{5}{4} - \frac{5}{8} + \frac{5}{16} - \dots$ $\frac{10}{3}$

30 $\sum_{r=1}^{\infty} \left(\frac{7}{4^r}\right)$ $\frac{7}{3}$

R,I | **31** Given: BC = 8,
CD = 10;
$\overline{AB}$ is tangent to circle O.
Find x. 12

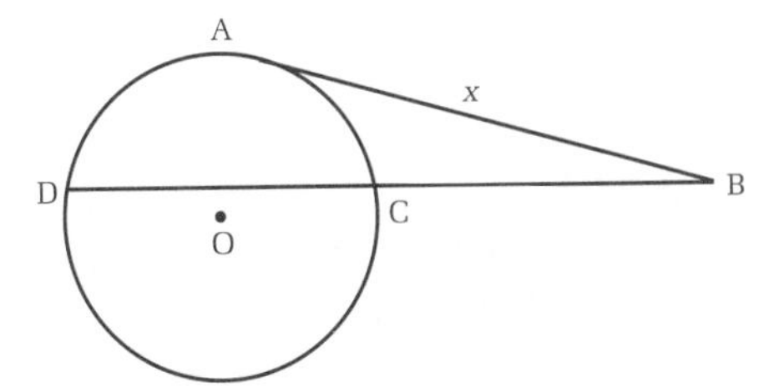

P,R | **32** The positive geometric mean of two numbers is 12, and the arithmetic mean is 12.5. Find the two numbers. 9, 16

P | **33** Insert three geometric means between $2i$ and $162i$, if $i^2 = -1$.
$6i, 18i, 54i$; or $-6i, 18i, -54i$; or $-6, -18i, 54$; or $6, -18i, -54$

Problem-Set Notes and Strategies, continued

- Problem **24** can be generalized to any arithmetic sequence using the Pythagorean numbers 3, 4 and 5. You might rephrase this question several times until students get the connection.
- A little geometry and work on geometric means will help in problem **27**. Students may have questions on this problem.
- Problem **31** will also require a short review of geometric concepts.
- A system of equations may be necessary to solve problem **32**. This is a review of some past methods.
- The answers to problem **33** are not unique. Have students compare answers with each other and see what was done differently to obtain the various results.

Problem Set, *continued*

C P **34** Explain why if one term in a geometric sequence is zero, then all terms but the first term must also be zero. (Hint: Try to find two geometric means between 4 and 0.)

R **35** The 452nd term of an arithmetic sequence is −893 and the 84th is 27. Find the sum of the first five terms of the sequence. 1147.5

R,I **36** Right triangle PLM is inscribed in circle O.

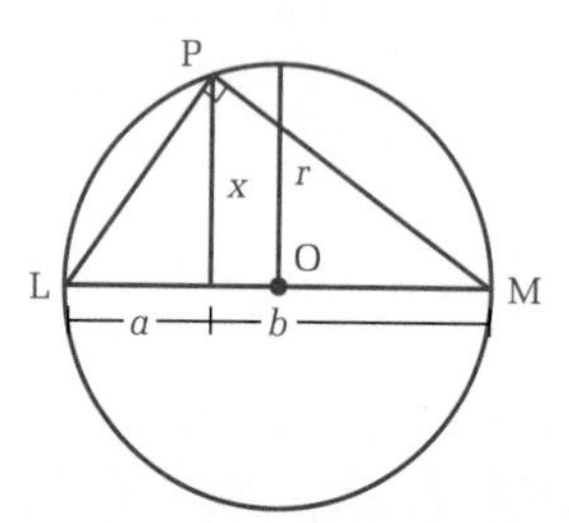

a Under what circumstances, if any, would the altitude x and the radius r be equal? If $a = b$

b Write an expression for r in terms of a and b. $r = \frac{a + b}{2}$

c Write an expression for x in terms of a and b. $x = \sqrt{ab}$

Problem-Set Notes and Strategies, continued

- Problem **34** should be straightforward, but may cause some difficulty at first. There is no formula necessary to solve this, only some understanding and logic.
- Problem **36** also requires some knowledge about circles and triangles. This problem introduces the next three chapters on trigonometry.

Additional Answer

34 Since $a_n = a_1r^{n-1}$, whenever $a_n = 0$ and $a_1 \neq 0$, $r^{n-1} = 0$. But $r^{n-1} = 0$ only if $r = 0$. Therefore, when $a_n = 0$ for one value of n, $a_n = 0$ for all $n > 1$.

HISTORICAL SNAPSHOT

CONVERGING ON π

Mathematicians rise to a challenge

Writing a series that approaches the value of π seemed to be the challenge of the day for seventeenth-century mathematicians. Two mathematicians who addressed themselves to that problem were the Scotsman James Gregory (1638–1675) and the German Gottfried Wilhelm Leibniz (1646–1716). Working independently, each came up with the series

$$\pi = 4\left(1 - \frac{1}{3} + \frac{1}{5} - \frac{1}{7} + \frac{1}{9} - \frac{1}{11} \cdots\right)$$

This series converges slowly. The value for the terms given above is about 2.98.

Two other series that converge on π are

$$\pi = 2\left(\frac{2}{1} \cdot \frac{2}{3} \cdot \frac{4}{3} \cdot \frac{4}{5} \cdot \frac{6}{5} \cdot \frac{6}{7} \cdot \frac{8}{7} \cdot \frac{8}{9} \cdots\right)$$

$$\pi = 4 \cdot \cfrac{1}{1 + \cfrac{1^2}{2 + \cfrac{3^2}{2 + \cfrac{5^2}{2 + 9^2}}}}$$

Following the discovery of these series, mathematicians applied themselves to proving that the value of pi could be expressed as a finite series. If it could, of course, π would be rational, which it is not. In 1761, a German physicist named Johann Heinrich Lambert (1728–1777) proved that π is irrational and can only be expressed as an infinite series.

Do a number of calculations in each of the series above to convince yourself that the series do indeed converge on π. Which series converges on π most quickly? Describe how each series converges on π.

"SORRY I'M LATE — I WAS WORKING OUT PI TO 5,000 PLACES."

13

Chapter Summary

Concepts and Procedures

After studying this chapter, you should be able to

- Apply the definition of a sequence (13.1)
- Describe a sequence (13.1)
- Form sequences of partial sums (13.1)
- Compute finite series (13.2)
- Use sigma notation to express a series (13.2)
- Find the values of infinite series (13.2)
- Compute any term in an arithmetic sequence (13.3)
- Compute arithmetic means (13.3)
- Evaluate arithmetic series (13.3)
- Compute any term in a geometric sequence (13.4)
- Compute geometric means (13.4)
- Evaluate finite and infinite geometric series (13.4)

Vocabulary

arithmetic means (13.3)
arithmetic sequence (13.3)
arithmetic series (13.3)
array (13.1)
common difference (13.3)
common ratio (13.4)
finite series (13.2)
geometric means (13.4)
geometric sequence (13.4)
geometric series (13.4)
infinite series (13.2)
progression (13.1)
sequence (13.1)
sequence of partial sums (13.1)
sum of an infinite series (13.2)
term (13.1)

13

Class Planning

Time Schedule
All levels: 1 day

Assignment Guide

Average
1, 2, 5, 6, 9, 10, 12, 14–16, 18, 20, 25, 30, 33, 34, 38, 47
Advanced
1, 2, 5, 6, 10, 12, 14–16, 18–20, 22, 25, 28, 30, 33, 34, 39, 42
Honors
2, 4, 5, 8, 12, 14, 15, 18–21, 25, 28–30, 33, 40, 43–45

Problem-Set Notes and Strategies

- By this time students should be comfortable with problems **1 – 13**. Number **9** may take a little more time.
- Problem **14** is a straightforward example of an arithmetic sequence with a negative common difference. You might ask students what would happen if there was interest being paid. It becomes a combination of an arithmetic and geometric sequence and can get complicated quickly.
- Problem **19** is more easily approached by arithmetic than by formulas which will work, but take more time since only four terms are asked for and they are adjacent to the given term.

13 Review Problems

A

P **In problems 1–4, list the first five terms of each sequence.**

1 $3, 6, \ldots, n^2 + 2, \ldots$ 3, 6, 11, 18, 27

2 $\{a_n\} = \{5n - 1\}$ 4, 9, 14, 19, 24

3 $\{a_n\} = \{n\text{th prime number}\}$ 2, 3, 5, 7, 11

4 $a_1 = 2$, $a_{n+1} = 9 - 2a_n$ 2, 5, −1, 11, −13

P **In problems 5–8, list the first five partial sums of each sequence.**

5 $\{a_n\} = \{n^2 + 2\}$ 3, 9, 20, 38, 65

6 $\{a_n\} = \{5n - 1\}$ 4, 13, 27, 46, 70

7 $\{a_n\} = \{n\text{th prime number}\}$ 2, 5, 10, 17, 28

8 $a_1 = 2$, $a_{n+1} = 9 - 2a_n$ 2, 7, 6, 17, 4

P **9** Let $S_7 = 15$ and $S_8 = 9$, where $\{S_n\}$ is the sequence of partial sums of $\{a_n\}$.

a Find a_8. −6

b If $\{a_n\}$ is an arithmetic sequence, what is a_1? 8.25

P,R **In problems 10–13, find the sum.**

10 $\sum_{n=1}^{10} \left(\frac{1}{n}\right)$ ≈2.93

11 $\sum_{n=1}^{10} \left(\frac{1}{n^2}\right)$ ≈1.55

12 $\sum_{k=5}^{8} (k^3 + 3k^2)$ 1718

13 $\sum_{k=1}^{10} 6k$ 330

P,R **14** Suppose that at the beginning of the year, your savings account contains $4250. If you withdraw $350 per month from the account, how much will it contain at the end of the year? $50

P,R **15** A ball is dropped from a height of 6 feet. Each time it bounces, it rises to 90% of the height from which it fell.

a How high does it rise on the fourth bounce? ≈3.9 ft

b How far has it traveled up and down by the top of the fourth bounce? ≈39.2 ft

P **16** Given $a_1 = 8$, $a_n = a_{n-1} + 3$, where $n \geq 2$.

a Write the first four terms of the sequence. 8, 11, 14, 17

b Find a_9. 32

c Find S_9. 180

P **17** Write $\sum_{k=1}^{6} (-1)^{k+1}k$ in expanded form. $1 - 2 + 3 - 4 + 5 - 6$

P **18** Insert five arithmetic means between 17 and 47. 22, 27, 32, 37, 42

P **19** The common ratio of a geometric sequence is −4 and the third term is 400. Find the first five terms of the sequence. 25, −100, 400, −1600, 6400

P **20** Insert four geometric means between −2 and 64. 4, −8, 16, −32

P **21** How many factors of 4 does the fifteenth term of the geometric series 3, 12, 48, . . . have? 14

P **22** Find the sum of the following geometric series.
$3 + \frac{3}{4} + \frac{3}{16} + \frac{3}{64} + \ldots$ 4

P **23** Write a formula that describes the terms of this array. $a_n = 4n - 1$

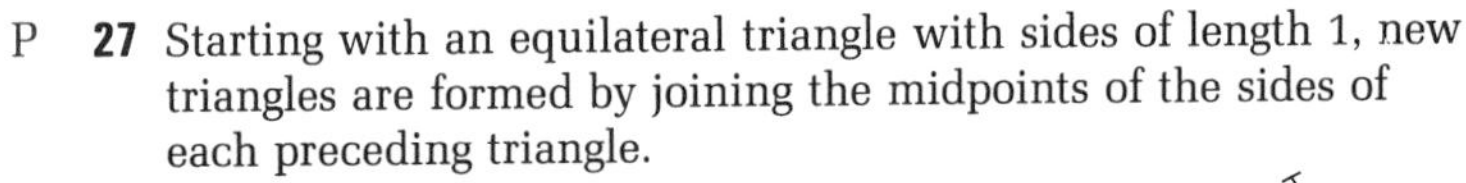

n	1	2	3	4	5
a_n	3	7	11	15	19

P **24** Find x if $10 = 1 + x + x^2 + x^3 + \ldots$ 0.9

B P,R **25** Construct a table of values for $\sum_{n=1}^{\infty}\left(\frac{1}{2^n}\right)$, listing values of S_n for $\{n : n \le 7\}$. Then graph the ordered pairs (n, S_n). What is the sum of the infinite series? 1

P **26** Let $S_n = \sum_{k=1}^{n} a_k = \frac{2n}{n + 2}$.
a Find a_4. $\frac{2}{15}$
b Find a formula for a_k in terms of k.
c Plot (n, S_n) for n = 2, 4, 6, . . . , 12. If $S = \sum_{k=1}^{\infty} a_k$ seems to exist, describe what you think it is. If it does not, explain why you think it does not. 2

b $a_k = \frac{4}{(k + 1)(k + 2)}$

P **27** Starting with an equilateral triangle with sides of length 1, new triangles are formed by joining the midpoints of the sides of each preceding triangle.
a What is the length of a side of the sixth triangle? $\frac{1}{32}$
b What is the sum of the perimeters of the first six triangles? ≈5.91
c What is the sum of the areas of all triangles if the process is repeated forever? $\frac{\sqrt{3}}{3}$, or ≈0.57735

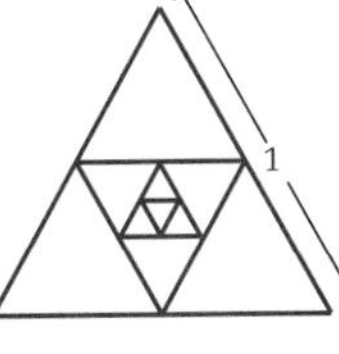

P **28** Find the sum of the infinite geometric series $1 + \frac{3}{2} + \frac{9}{4} + \ldots$ if possible. No sum

P **29** Find the sums of all multiples of 3 between 27 and 276, inclusive. 12,726

P **In problems 30–32, simplify each sum if possible.**

30 $\sum_{k=1}^{n} k$ $\frac{n(n + 1)}{2}$

31 $\sum_{k=1}^{n}\left(\frac{1}{k} - \frac{1}{k + 1}\right)$ $\frac{n}{n + 1}$

32 $\sum_{k=1}^{n} \log \frac{k}{k + 1}$ $-\log(n + 1)$

P **33** Find p if $\sum_{j=1}^{5} pj = 14$. $p = \frac{14}{15}$

Problem-Set Notes and Strategies, continued

- If students understand the concept, problem **21** is quickly answered.
- Once students recognize that the infinite polynomial of problem **24** is a geometric series, the problem is almost solved.

Additional Answer

25

n	S_n
1	0.5
2	0.75
3	0.875
4	0.9375
5	0.96875
6	0.984375
7	0.9921875

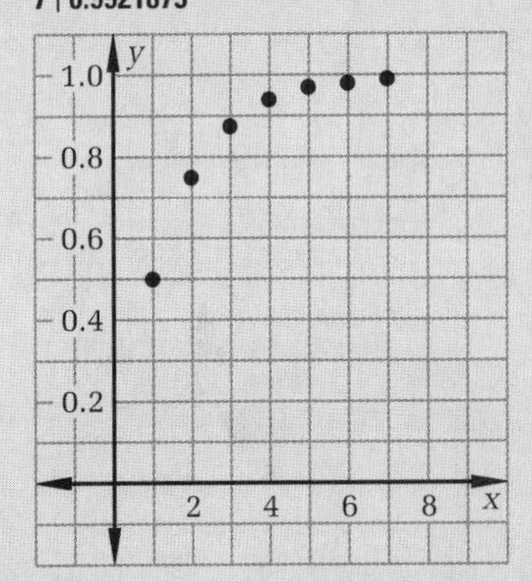

- Part **b** of **26** can present problems. Finding a closed form definition for the terms of a sequence is still a new skill. This problem is an excellent candidate for group work.
- Once students realize that an arithmetic series is needed for problem **29**, the problem goes more smoothly.
- Problems **30 – 32** may require patience and looking for patterns. This recognition process is often proportional to the number of problems students work.

Review Problems, *continued*

Problem-Set Notes and Strategies, continued

- Answers to problem **36** are not unique and can generate interesting discussions.
- Problem **44** is an example of a geometric series. Most population growth problems fall into this category.
- Problem **46** is an example of geometric sequences in the real world. You might ask students to figure out the same problem but with an initial deposit of \$2,000. It is interesting to note that the account grows at the same rate. After ten years the account is now \$5,627.89, the same percentage growth as with 2 million dollars.
- Problems **47** and **48** are examples of a snowflake called the Koch snowflake. This curve is an interesting phenomenon of a curve that is continuous everywhere and differentiable nowhere. If you are also teaching calculus, you might use this example. If not, mention that it has significance in other areas of mathematics.

P **In problems 34 and 35, tell whether each statement is True or False.**

34 $\sum_{k=2}^{10} a_k = \sum_{k=4}^{12} a_{k-2}$ True

35 $\sum_{t=1}^{4} [(t^2 - 2)(t - 1)] = \sum_{r=-1}^{4} (r^2 - 1)(r)$ False

P **36** Write $x^2 + x^3 + x^4 + \ldots$ in sigma notation. $\sum_{k=1}^{\infty} x^{k+1}$

P **In problems 37–40, find the first term and common ratio of a geometric sequence matching the given information.**

37 $a_6 = 42$, $a_7 = 24$ $a_1 \approx 689.35$; $\frac{4}{7}$

38 $a_4 = 1$, $a_8 = \frac{1}{81}$ $a_1 = 27$; $\frac{1}{3}$

39 $\frac{a_7}{a_9} = 9$, $r < 0$, $a_8 = 243$ $a_1 = -531{,}441$; $-\frac{1}{3}$

40 $a_1 + a_2 + a_3 = 18 + 12i$, $a_1 + a_2 + a_3 + a_4 = 18 + 9i$, $a_2 = 12i$ $a_1 = 24$; $\frac{i}{2}$

P **41** Find the general term of a sequence whose first four terms are $x, \frac{x^2}{2}, \frac{x^3}{3}, \frac{x^4}{4}, \ldots$ $a_n = \frac{x^n}{n}$

P **For problems 42 and 43, find the sum of each geometric series if possible.**

42 $\sum_{k=1}^{\infty} \left(\frac{2}{3}\right)^k$ 2

43 $\sum_{n=1}^{8} 12\left(-\frac{1}{2}\right)^{n-1}$ ≈ 7.97

P **44** A fungus culture growing under controlled conditions triples in size each day. How many units will the culture contain after six days if it originally contained five units? 3645 units

P **45** A town with a population of 2500 people has a predicted yearly growth of 5% over the preceding year for the next ten years. How many people are expected to live in that town at the end of ten years? $\approx$4072

C P **46** Phil T. Rich deposits \$2 million in a money market account at 10.9% interest, compounded annually. For ten years he allows the interest to accumulate.

a How much money does he have at the end of ten years? \$5,627,888.15

b Do the amounts of money in the bank each year form an arithmetic sequence, an arithmetic series, a geometric sequence, or a geometric series? Geometric sequence

P **47** The snowflake curve is a type of fractal. To make a snowflake curve, begin with an equilateral triangle. Divide each side into three equal segments and replace the middle segment of each side with an equilateral triangle. Find the area enclosed by the snowflake curve if the process is allowed to continue indefinitely.

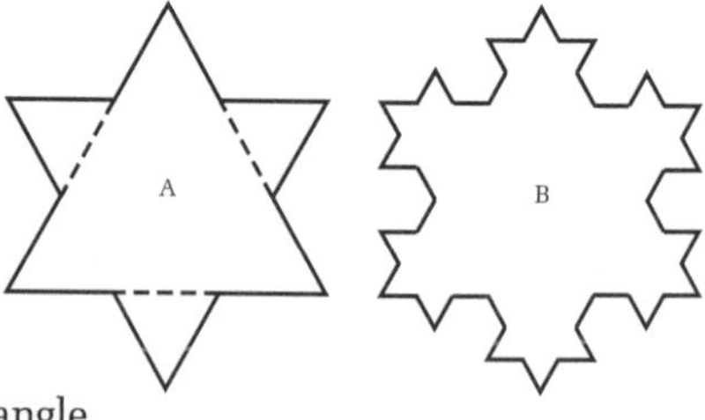

$\frac{2s^2\sqrt{3}}{5}$, where s = length of side of original triangle

P **48** Find the perimeter of the snowflake curve described in problem **47**. Infinite

13 CHAPTER TEST

Resource References
Evaluation
Test Form 1, 2, 3

In problems 1 and 2, list the first five terms of the sequence.

1 $\{a_n\} = \{n^3 - n\}$
0, 6, 24, 60, 120

2 $a_1 = 6$, $a_n = 2a_{n-1} + 3$
6, 15, 33, 69, 141

3 Find a formula to describe the terms a_n of the array below. $a_n = 3n - 1$

n	1	2	3	4
a_n	2	5	8	11

4 $\{S_n\}$ is the sequence of partial sums of $\{a_n\}$. If $S_7 = 63$ and $S_8 = 70$, what is a_8? 7

In problems 5 and 6, evaluate the finite series.

5 $\sum_{n=1}^{5} (2^n - n)$ 47

6 $\sum_{n=5}^{7} (-1)^n(n - 1)$ −5

7 Determine whether each sequence is arithmetic, geometric, or neither.

a 1, 3, 9, 27, 81, . . . Geometric

b $\{a_n\} = \{5n + 3\}$ Arithmetic

c $\{a_n\} = \{\sqrt{n}\}$ Neither

d $a_1 = 5$, $a_n = 7a_{n-1}$ Geometric

8 Give the closed-form formula and the eighth term for each sequence.

a The arithmetic sequence 5, 9, 13, . . . $a_n = 5 + 4(n - 1)$; 33

b The geometric sequence 2, 6, 18, . . . $a_n = 2 \cdot 3^{n-1}$; 4374

9 Insert five arithmetic means between 14 and 50.
20, 26, 32, 38, 44

10 Insert two positive geometric means between 40 and 320.
80; 160

11 Given: $m\angle ADC = 90°$,
AB = 10,
BC = 3
Find DB. $\sqrt{30}$

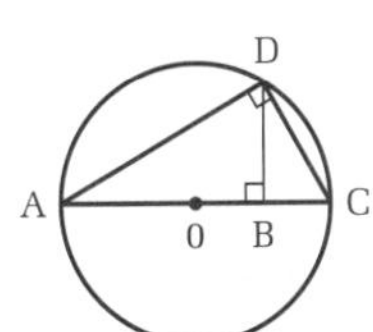

In problems 12 and 13, evaluate the series.

12 $\sum_{n=1}^{8} (3 \cdot 2^{n-1})$ 765

13 $\sum_{k=1}^{40} (5k - 8)$ 3780

In problems 14 and 15, find the sum of each series, if it exists.

14 $\sum_{n=1}^{\infty} 3\left(\frac{4}{3}\right)^{n-1}$ No sum

15 $\sum_{n=1}^{\infty} 25\left(\frac{4}{5}\right)^{n-1}$ 125

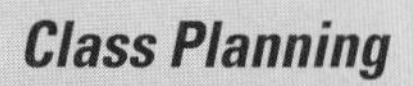

Class Planning

Time Schedule
Average: Optional
Advanced: Optional
Honors: 2 days

Resource References
Teacher's Resource Book
Class Opener 80
Practice 84, 85

Class Opener

Write a recursive definition for the arithmetic sequence formed by the set of natural numbers.
$f(1) = 1$
$f(n + 1) = n + 1$

Lesson Notes

- The method of mathematical induction may be difficult for some students. It is built upon the recursive definition of the natural numbers. The level of abstraction of the process will also be new, especially the concept of infinity.
- Both parts of the method must be verified. Part 1 is the foundation upon which everything else is built. The second part takes the foundation and applies recursive steps to generate an infinite proof which is exhaustive of all natural numbers.

SHORT SUBJECT

Mathematical Induction

Recall the handshake problem that was discussed in Chapter 1. In Section 1.4, we determined that the number of handshakes exchanged by any number n of people when each person shakes hands with each of the others is equal to

$$\sum_{k=2}^{n}(k - 1)$$

that is, to the sum $1 + 2 + 3 + \ldots + (n - 1)$, which is in turn equal to $0 + 1 + 2 + \ldots + (n - 1)$. In Section 1.3, on the other hand, we solved the handshake problem geometrically, finding that the number of handshakes exchanged by n people is equal to $\frac{n^2 - n}{2}$. Therefore, the following statement should be true.

$$0 + 1 + 2 + \ldots + (n - 1) = \frac{n^2 - n}{2}$$

Suppose, however, that you were asked to *prove* that this statement is true for every natural number n. It is easy to show that it is true when $n = 1$, when $n = 2$, when $n = 3$, and so forth, but the set of natural numbers is infinite, so it is impossible to prove the statement by testing it for each possible value of n. What is needed is a general demonstration—one that will apply to the set of natural numbers as a whole.

To see how such a demonstration might be developed, we need to look at the set of natural, or counting, numbers in a new way. As you know, these numbers can be represented by $\{1, 2, 3, \ldots\}$. This set can, however, also be defined recursively:

If $\mathcal{N}$ is a set such that

1. ***The number 1 is an element of $\mathcal{N}$***
2. ***The number $n + 1$ is an element of $\mathcal{N}$ whenever n is an element of $\mathcal{N}$***

then $\mathcal{N}$ is the set of natural numbers.

In the definition, statement 1 establishes that 1 is in $\mathcal{N}$. Statement 2 establishes that since 1 is in $\mathcal{N}$, 2 is also in $\mathcal{N}$. Similarly, since 2 is in $\mathcal{N}$, statement 2 establishes that 3 is also in $\mathcal{N}$. We can continue in this manner, using statement 2 recursively to verify that any natural number is in set $\mathcal{N}$.

Vocabulary
mathematical induction

By applying this recursive definition, we can determine whether a given statement is true of every member of the natural numbers.

Example *Prove that* $0 + 1 + 2 + \ldots + (n - 1) = \frac{n^2 - n}{2}$ *for any natural number* n.

Our recursive definition of *natural number* suggests the given statement will be true for every member of the set of natural numbers if

1 It is true when $n = 1$
2 It is true for $n + 1$ whenever it is true for n

To prove that the statement is true when $n = 1$, we simply substitute 1 for n in the statement.

$$0 + 1 + 2 + \ldots + (n - 1) = \frac{n^2 - n}{2}$$

$$1 - 1 = \frac{1^2 - 1}{2}$$

$$0 = 0$$

Since $0 = 0$ is a true statement, the given statement is true for $n = 1$.

Now, to show that the given statement is true for $n + 1$ whenever it is true for n, we *assume* that it is true for some value of n, then see whether it is also true for $n + 1$. In this case, we assume that $0 + 1 + 2 + \ldots + (n - 1) = \frac{n^2 - n}{2}$ and try to show that $0 + 1 + 2 + \ldots + [(n + 1) - 1] = \frac{(n + 1)^2 - (n + 1)}{2}$.

Given
$$\mathbf{0 + 1 + 2 + \ldots + (n - 1) = \frac{n^2 - n}{2}}$$

Add n to each side
$$\mathbf{+ n \qquad + n}$$

$$\mathbf{0 + 1 + 2 + \ldots + n = \frac{n^2 - n}{2} + n}$$

$$\mathbf{0 + 1 + 2 + \ldots + [(n + 1) - 1] = \frac{n^2 - n}{2} + \frac{2n}{2}}$$

$$\mathbf{= \frac{n^2 + 2n + 1 - n - 1}{2}}$$

$$\mathbf{= \frac{(n + 1)^2 - (n + 1)}{2}}$$

Since the given statement is true when $n = 1$ and is true for $n + 1$ whenever it is true for n, it must be true for the entire set of natural numbers.

The method of proof used in the preceding example is known as ***mathematical induction.*** It is a powerful technique of demonstration, much used in advanced mathematics. The principle on which proofs by mathematical induction are based can be stated formally as follows.

Principle of Mathematical Induction

If S is a statement that applies to the set $\mathcal{N}$ of natural numbers and

1. S is true when $n = 1$

2. When S is true for some natural number n, it is also true for $n + 1$

then S is true for every natural number n.

It is important to remember that *both* part 1 and part 2 of the principle must be verified for a proof by mathematical induction to be valid. In Sample Problems 3 and 4 you will see cases in which a statement satisfies only one of the parts and is therefore false.

Sample Problems

Problem 1 *Prove, by mathematical induction, that for every natural number n,* $\sum_{k=1}^{n} 2k = n(n + 1)$.

Solution When $n = 1$, the equation becomes $2(1) = 1(1 + 1)$, or $2 = 2$, which is a true statement, so the given statement is true for $n = 1$.

To verify part 2 of the principle of mathematical induction, we assume that the given statement is true for some n and show that $\sum_{k=1}^{n+1} 2k = (n + 1)[(n + 1) + 1]$.

Given $$\sum_{k=1}^{n} 2k = n(n + 1)$$

Add $2(n + 1)$ to each side $$\sum_{k=1}^{n} 2k + 2(n + 1) = n(n + 1) + 2(n + 1)$$

$$\sum_{k=1}^{n+1} 2k = (n + 1)(n + 2)$$

$$= (n + 1)[(n + 1) + 1]$$

By the principle of mathematical induction, the given statement is true for any natural number n.

Problem 2 *Prove that if n is a natural number, then $n^5 - n$ is divisible by 10.*

Solution The first step is to show that $n^5 - n$ is divisible by 10 when $n = 1$. Since $1^5 - 1 = 0$ and 0 is divisible by 10, the statement is true for $n = 1$.

Next, we assume that $n^5 - n$ is divisible by 10 for some value of n and show that $(n + 1)^5 - (n + 1)$ is also divisible by 10.

Expand, using the binomial theorem	$(n + 1)^5 - (n + 1)$
	$= (n^5 + 5n^4 + 10n^3 + 10n^2 + 5n + 1) - n - 1$
Regroup	$= (n^5 - n) + (5n^4 + 5n) + (10n^3 + 10n^2)$
	$= (n^5 - n) + 5n(n^3 + 1) + 10n^2(n + 1)$
	$= (n^5 - n) + 5n(n + 1)(n^2 - n + 1) + 10n^2(n + 1)$

Examining the final expression, we notice that (1) according to the given statement, $n^5 - n$ is divisible by 10, (2) since either n or $n + 1$ must be an even number, $5n(n + 1)(n^2 - n + 1)$ contains factors of 5 and 2 and therefore is divisible by 10, and (3) $10n^2(n + 1)$ is divisible by 10. According to the distributive property, since each term of the sum contains a factor of 10, the sum itself must contain a factor of 10, so $(n + 1)^5 - (n + 1)$ is divisible by 10.

By the principle of mathematical induction, then, $n^5 - n$ is divisible by 10 whenever n is a natural number.

Problem 3 *Show that the statement $n + 2 = 3n$ satisfies part 1 of the principle of mathematical induction but is not true for every natural number n.*

Solution When $n = 1$, $n + 2 = 1 + 2 = 3$ and $3n = 3(1) = 3$, so the statement is true for $n = 1$.

Given that the statement is true for some value of n, we now can try to show that $(n + 1) + 2 = 3(n + 1)$.

$$(n + 1) + 2 = 3(n + 1)$$
$$n + 3 = 3n + 3$$
$$n = 3n$$

Since this statement is false for every real value of n except 0, the given statement does not satisfy part 2 of the principle. The given statement is not true for every natural number, even though it is true when $n = 1$.

Problem 4 *Show that the statement $n = n + 1$ satisfies part 2 of the principle of mathematical induction but is not true for every natural number n.*

Solution Clearly, $1 \neq 1 + 1$, so the statement fails to satisfy part 1 of the principle.

If there were some value of n for which the statement were true, however, $n + 1 = (n + 1) + 1$ would also be a true statement.

Given	$n = n + 1$
Add 1 to each side	$n + 1 = n + 2$
	$= (n + 1) + 1$

Therefore, the given statement is false, even though it satisfies part 2 of the principle.

Assignment Guide

Average	Optional
Advanced	Optional
Honors	Day 1 1, 2, 6, 8, 12, 14, 16, 19
	Day 2 4, 7, 10, 11, 13, 15, 17, 22, 25, 26

Additional Answers

1 $(1 + h)^1 \overset{?}{\geq} 1 + h(1)$

$1 + h \geq 1 + h$

2 $\sum_{k=1}^{1} [2(1) - 1] \overset{?}{=} 1^2$

$2 - 1 \overset{?}{=} 1$

$1 = 1$

3 $1 \overset{?}{<} 2^1$

$1 < 2$

4 $1^2 - 1 + 6 = 6$, which is divisible by 2.

5*i* $(1 + h)^{n+1} \geq 1 + h(n + 1)$

5*ii* $\sum_{k=1}^{n+1} (2k - 1) = (n + 1)^2$

5*iii* $n + 1 < 2^{n+1}$

5*iv* $(n + 1)^2 - (n + 1) + 6$ is divisible by 6.

7 When $n = 1$,

$n^3 - n = 1^3 - 1$
$= 0,$

which is divisible by 3.

Assuming that $n^3 - n$ is divisible by 3 for some n,

$(n + 1)^3 - (n + 1)$
$= n^3 + 3n^2 + 3n + 1 - n - 1$
$= (n^3 - n) + (3n^2 + 3n)$
$= (n^3 - n) + 3n(n + 1)$

Since $n^3 - n$ is assumed to be divisible by 3 and since $3n(n + 1)$ also contains a factor of 3, $(n + 1)^3 - (n + 1)$ is divisible by 3.

9 When $n = 1$, $x^1 \cdot y^1 = xy = (xy)^1$

Assuming that $x^n \cdot y^n = (xy)^n$ for some n,

$x^{n+1} \cdot y^{n+1} = x \cdot x^n \cdot y \cdot y^n$
$= x^n y^n \cdot xy$
$= (xy)^n \cdot (xy)^1$
$= (xy)^{n+1}$

Answers for problems **6**, **8**, **10**, **11**, **12**, **13**, and **14a – c** can be found in the answer pages beginning on **A1**.

Problem Set

In problems 1–4, show that each statement is true when $n = 1$.

1 $(1 + h)^n \geq 1 + hn$

2 $\sum_{k=1}^{n} (2k - 1) = n^2$

P,R **A**

3 $n < 2^n$

4 $n^2 - n + 6$ is divisible by 2.

P,R **5** Rewrite the statements in problems 1–4, replacing n with $n + 1$.

P **6** Prove that $1 + 5 + 9 + \ldots + (4n - 3) = n(2n - 1)$ for every natural number n.

P **7** Prove that $n^3 - n$ is divisible by 3 for $\{n{:}n \in \mathcal{N}\}$.

P **8** Prove that the sum of the first n natural numbers is equal to $\frac{n(n + 1)}{2}$.

P **9** Use mathematical induction to show that $x^n \cdot y^n = (xy)^n$, where x and y are real numbers and n is a natural number. (Hint: $a^{n+1} = a \cdot a^n$)

P **10** Is $n + 5 > (n - 2)^2$ for every natural number n? Explain your answer.

P,R **11** Prove, by means of mathematical induction, that the sum of the first n terms of a geometric sequence $\{a_n\}$ can be found with the formula

$$S_n = \frac{a_1(r^n - 1)}{r - 1}$$

where r is the common ratio.

P **12** Prove that the number of diagonals that can be drawn in an n-sided polygon, where $n \geq 3$, is equal to $\frac{n(n - 3)}{2}$. (Hint: As your first step, prove the statement true for $n = 3$.)

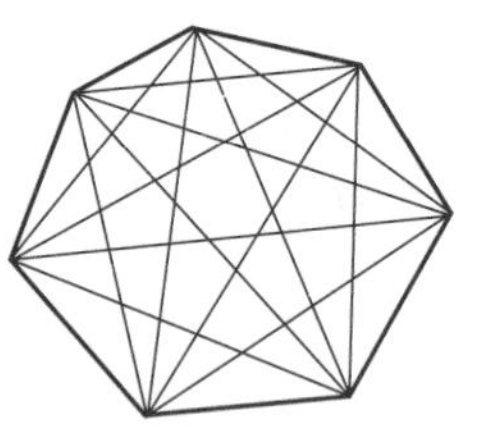

P **B**

13 Prove that $a^n - b^n$ always contains a factor of $a - b$ if a and b are real numbers and n is a natural number.

P **14** Show that each of the following properties of exponents applies to all positive integral exponents m and n.

a $a^m \cdot a^n = a^{m+n}$

b $(a^m)^n = a^{mn}$

c $\frac{a^m}{a^n} = a^{m-n}$

P | **In problems 15–18, use mathematical induction to determine whether each statement is True or False for $\{n:n \in \mathcal{N}\}$.**

15 $2^n \geq n + 1$ True

16 $n! + (n + 1)! + (n + 2)!$ is divisible by $(n + 1)^2$. False

17 $1^3 + 2^3 + 3^3 + \ldots + n^3 = n^2(n + 1)^2 - 4$ False

18 $1(1!) + 2(2!) + 3(3!) + \ldots + n(n!) = (n + 1)! - 1$ True

P | **19** Prove that $n^2 \geq \frac{n(n + 1)}{2}$ for every natural number n.

P | **20** Prove that the number represented by $n^2 + n$ contains a factor of 2 for every positive integral value of n.

P,R | **21** Prove that for every element of $\{n:n \in \mathcal{N}$ and $n \geq 2\}$,
$\log(a_1 \cdot a_2 \cdot \ldots \cdot a_n) = \log a_1 + \log a_2 + \ldots + \log a_n$
where each value of a_k is a positive real number.

P,R | **22** Let z represent a complex number and $\bar{z}$ represent its conjugate. Prove that $\overline{z^n} = \bar{z}^n$ for every natural number n. (Hint: For any two complex numbers v and w, $\bar{v} \cdot \bar{w} = \overline{vw}$.)

P | **23** Prove that for all positive integral values of n, $5^n - 1$ is divisible by 4.

P | **24** Prove that if n is a natural number greater than 2, the sum of the measures of the interior angles of an n-sided polygon is equal to $180(n - 2)$.

C P | **25** Prove that for every natural number n,

$$\sum_{k=1}^{n} \left(\frac{1}{(2k - 1)(2k + 1)}\right) = \frac{n}{2n + 1}.$$

P,R | **26** Refer to the description of the game Tower of Hanoi in Section 5.6, problem 28. Use mathematical induction to prove that the minimum number of moves needed to transfer an n-disk tower is equal to $2^n - 1$.

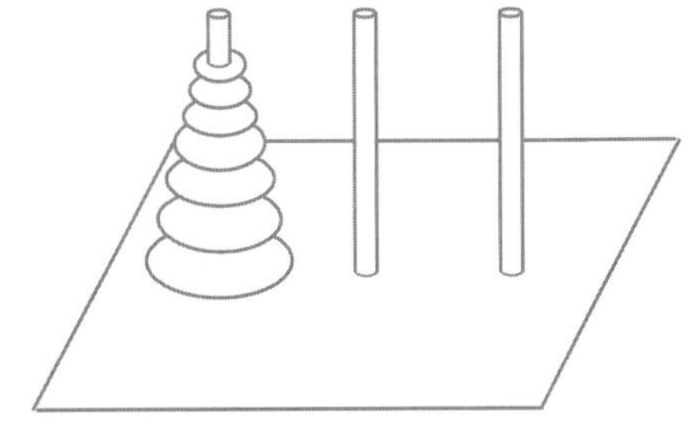

P,R | **27** Use mathematical induction to prove the binomial theorem. That is, show that for all real values of x and y and all positive integral values of n,

$$(x + y)^n = [C(n, 0)]x^n + [C(n, 1)]x^{n-1}y + [C(n, 2)]x^{n-2}y^2 + \ldots + [C(n, n)]y^n$$

Additional Answers

19

When $n = 1$, $1^2 \geq \frac{1(1 + 1)}{2}$, or $1 \geq 1$.

We assume that $n^2 \geq \frac{n(n + 1)}{2}$ for some n. Since n is positive by definition, $2n > n$, so we can add $2n + 1$ to the left side and $n + 1$ to the right side without invalidating the inequality.

$$n^2 + 2n + 1 \geq \frac{n(n + 1)}{2} + n + 1$$
$$(n + 1)^2 \geq \frac{n^2 + n + 2n + 2}{2}$$
$$\geq \frac{n^2 + 3n + 2}{2}$$
$$\geq \frac{(n + 1)(n + 2)}{2}$$
$$\geq \frac{(n + 1)[(n + 1) + 1]}{2}$$

20

When $n = 1$, $n^2 + n = 1^2 + 1 = 2$, which contains a factor of 2. Assuming that $n^2 + n$ contains a factor of 2 for some n,

$(n + 1)^2 + (n + 1)$
$= n^2 + 2n + 1 + n + 1$
$= n^2 + n + 2n + 2$
$= (n^2 + n) + 2(n + 1)$

Since $n^2 + n$ contains a factor of 2 by our assumption and since $2(n + 1)$ also contains a factor of 2, $(n + 1)^2 + (n + 1)$ must contain a factor of 2.

22

When $n = 1$, $\overline{z^1} = \bar{z}^1$, or $\bar{z} = \bar{z}$, which is a true statement. Assuming that $\overline{z^n} = \bar{z}^n$ for some n, we multiply both sides by $\bar{z}$.

$\overline{z^n} \cdot \bar{z} = \bar{z}^n \cdot \bar{z}$
$\overline{z^n \cdot z} = \bar{z}^{n+1}$
$\overline{z^{n+1}} = \bar{z}^{n+1}$

Answers for problems **21**, **23**, **24**, **25**, **26** and **27** can be found in the answer pages beginning on A1.

14.1

THE UNIT CIRCLE is introduced in the first section of this chapter. This may be the first time students encounter negative angles or angles whose measure is greater than 180.

The sine and cosine functions are defined with respect to the unit circle. The other four trig functions are defined in terms of sine and cosine. Students will also be asked to prove some of the Pythagorean Identities.

The trig functions are later redefined in terms of coordinates and in terms of the right triangle. Students can see that the three definitions are equivalent.

First-quadrant reference angles, inverse trigonometric functions and their ranges, the cofunction identities, and odd and even trigonometric functions are also defined in this chapter.

Students will express the measure of angles in terms of radians. They will also convert from degrees to radians and vice versa.

CHAPTER 14

INTRODUCTION TO TRIGONOMETRY

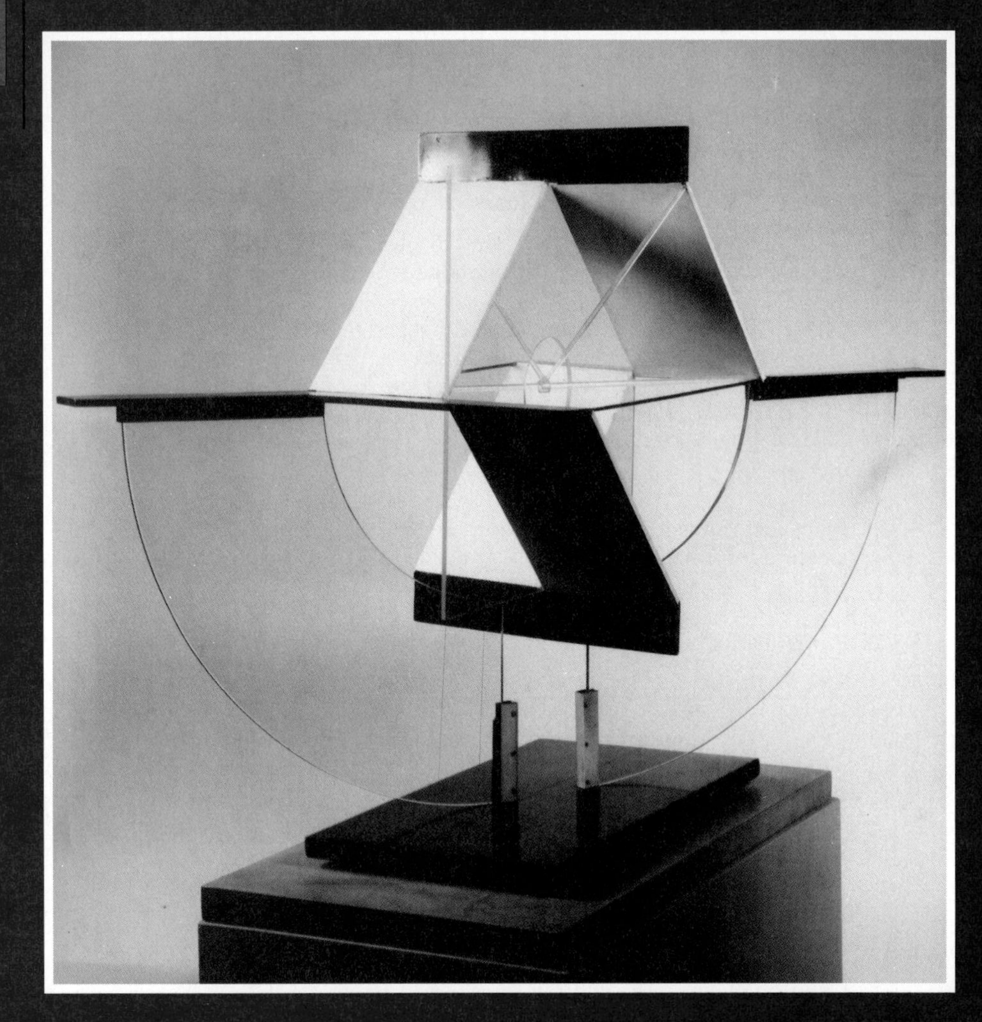

The triangles associated with circles in Naum Gabo's mystical *Construction in Space with Balance on Two Points* represent the basic principles of trigonometry.

14.1 THE UNIT CIRCLE

Objectives

After studying this section, you will be able to

- Measure angles as rotations
- Determine points of intersections of rays and the unit circle

Part One: Introduction

Recall that the distance around a circle is called the *circumference* and that a portion of the circle is an *arc*. An angle with its vertex at the center of the circle is called a ***central angle.***

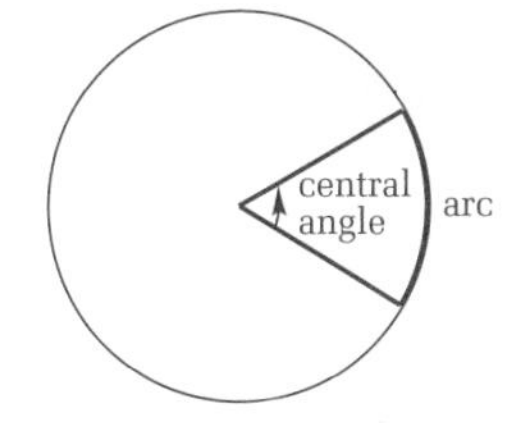

Example *The circle shown has a radius of* 12 *and central* $\angle AOB = 45°$. *Find the length of arc AB.*

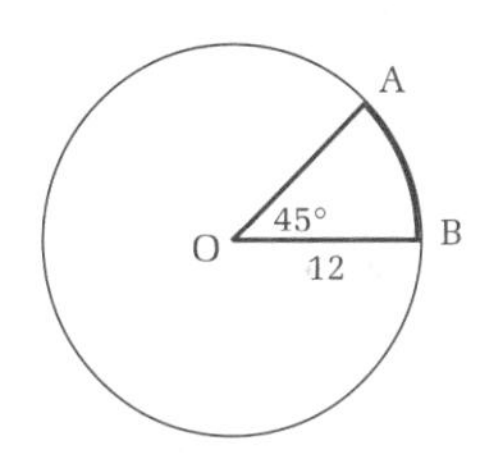

Circumference = $2\pi \cdot$ radius, so the circumference of the circle is 24π. Since the central angle is 45°, arc AB is a fractional part of the circle. It is $\frac{45}{360}$, or $\frac{1}{8}$, of the circle. The length of arc AB is $\frac{1}{8} \cdot 24\pi$, or 3π.

In geometry, an angle was defined to be two rays with a common endpoint. In trigonometry, *an angle is a rotation* from one ray to another. The rotation starts at the ***initial ray*** and stops at the ***terminal ray.***

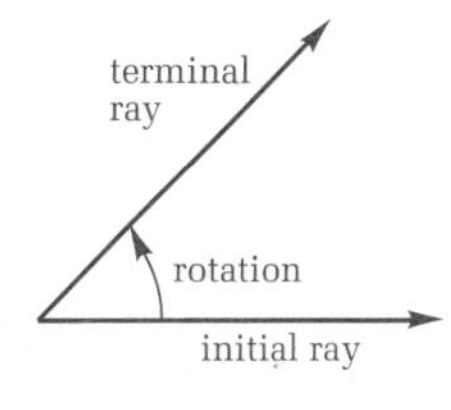

In this book, the initial ray of an angle will be the positive x-axis unless stated otherwise. Such an angle is said to be in ***standard position.*** For an angle in standard position, we only need to show the terminal ray and the rotation.

The angle shown as θ is 210° and is in standard position. (Greek letters are commonly used for angles. The letter θ is called theta.)

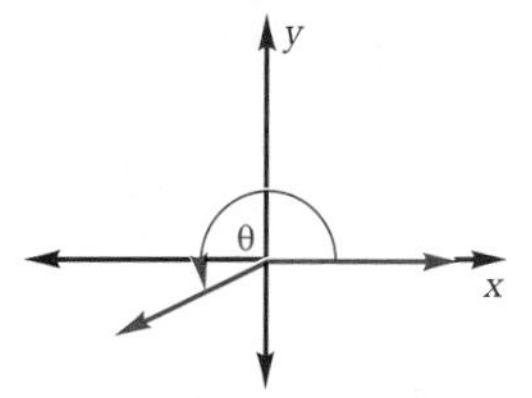

Class Planning

Time Schedule

Average: 2 days
Advanced: $1\frac{1}{2}$ days
Honors: $1\frac{1}{2}$ days

Resource References

Teacher's Resource Book
Class Opener 81
Cooperative Learning 41
Practice 86

Class Opener

1 Draw a circle with its center at the origin and radius 1.

a Find all points on the circle that are equidistant from the *x*- and *y*-axes.
$\left(\frac{1}{\sqrt{2}}, \frac{1}{\sqrt{2}}\right), \left(\frac{1}{\sqrt{2}}, -\frac{1}{\sqrt{2}}\right),$
$\left(-\frac{1}{\sqrt{2}}, \frac{1}{\sqrt{2}}\right), \left(-\frac{1}{\sqrt{2}}, -\frac{1}{\sqrt{2}}\right)$

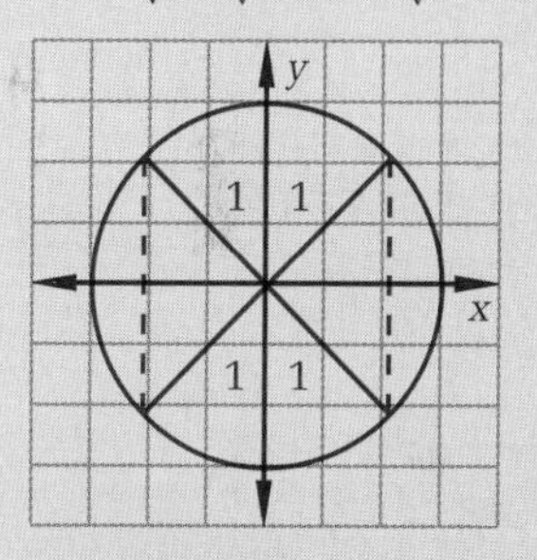

b Connect the points to the origin. See art above.

c Determine the angles formed by the lines and axes. 45°

Lesson Notes

- Chapter **14** introduces trigonometry through the unit circle. From geometry, students may be familiar with the definitions of a central angle and an arc.

Vocabulary

central angle
initial ray
terminal ray
standard position
coterminal angles
unit circle

Lesson Notes, continued

Some may also remember how to find arc lengths and the correspondence between the length of an arc and the measure of the central angle. In Section **14.6**, radian measure will be introduced as an alternative to degrees for measuring angles and arcs.

- In this section, standard angles are presented with their initial sides on the positive x-axis. This should be reasonable since the students are familiar with the rectangular coordinate system and this reinforces the relationship between algebra and trigonometry. The quadrants and the positive angles are numbered counterclockwise. Although points in the coordinate system are unique, angles are not since there may be many rotations until the terminal side is finally reached.

- Another relationship between algebra and trigonometry is introduced with the unit circle. This was studied briefly in the chapter on graphing, in the short subject on conics, and in solving systems involving conics. The points where the terminal side of the angle intersects the unit circle will become increasingly important as the sine and cosine are discussed in the next section.

Using rotations, we can have angles greater than 180°. We can also have negative angles. We define counterclockwise rotations as positive and clockwise rotations as negative.

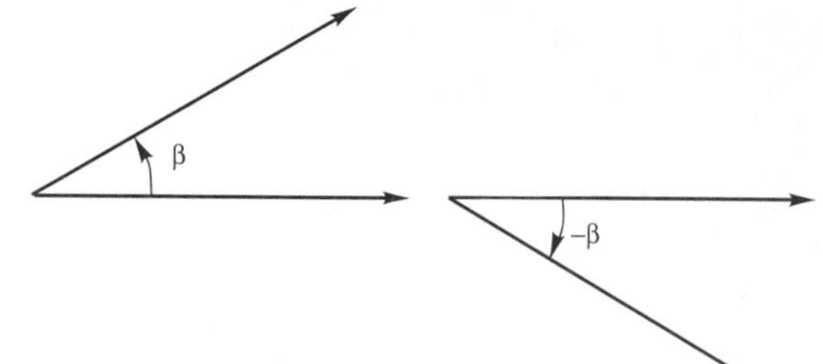

By choosing negative angles, and by adding multiples of 360° to an angle, it is possible to find more than one angle with the same terminal ray.

Example *The diagram shows a 30° angle. Name two other angles with the same terminal ray.*

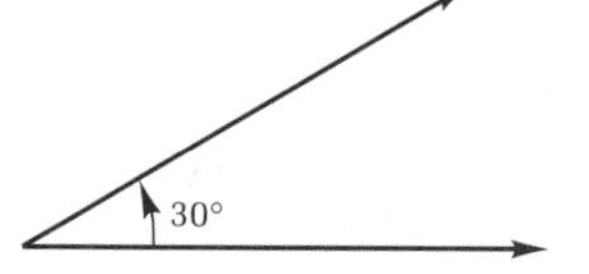

We can arrive at the same terminal ray by rotating in the opposite direction. The angle measuring −330° has the same terminal ray.

Another way is to go around more than once. One revolution is 360°, so the angle measuring (30 + 360)°, or 390°, also has the same terminal ray.

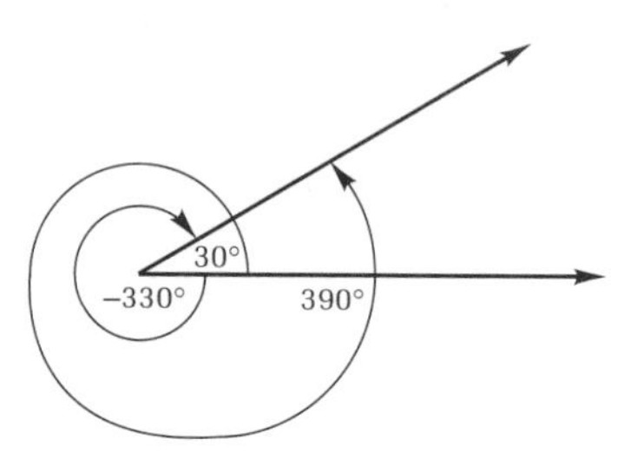

Thus the angles measuring 30°, −330°, and 390° all have the same terminal ray. Angles in standard position that have the same terminal ray are called ***coterminal angles.***

Points on the Unit Circle

The study of trigonometry is based on the circle with equation $x^2 + y^2 = 1$, called the ***unit circle.*** The unit circle has its center at the origin and a radius of 1 unit.

Example *Find the point of intersection of the unit circle and the ray* $\left\{(x, y): y = \frac{4}{3}x,\ x \geq 0\right\}$.

To find the point of intersection of the circle and ray, we solve the system of equations

$$\begin{cases} x^2 + y^2 = 1 \\ y = \dfrac{4}{3}x \end{cases}$$

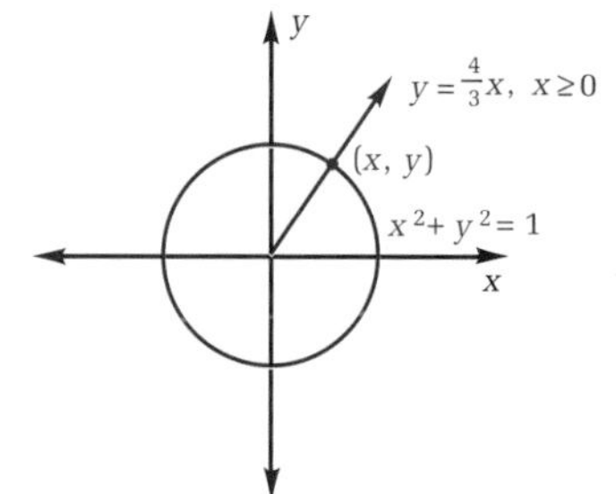

By substitution, we get $x^2 + \left(\frac{4}{3}x\right)^2 = 1$

$$x^2 + \frac{16}{9}x^2 = 1$$

$$\frac{25}{9}x^2 = 1$$

$$x^2 = \frac{9}{25}$$

$$|x| = \frac{3}{5}$$

Since the point is in the first quadrant, $x = \frac{3}{5}$. Then $y = \frac{4}{3} \cdot \frac{3}{5} = \frac{4}{5}$. The point is $\left(\frac{3}{5}, \frac{4}{5}\right)$.

Example *Find the coordinates of the point of intersection of the unit circle and the terminal ray of a 120° angle.*

Drop a perpendicular to the x-axis to form a 30°-60°-90° triangle. The hypotenuse is 1, the shorter leg is $\frac{1}{2}$, and the longer leg is $\frac{1\sqrt{3}}{2}$. Since the ray is in the second quadrant, the x-coordinate will be negative and the y-coordinate will be positive. The coordinates of the point of intersection are $\left(-\frac{1}{2}, \frac{\sqrt{3}}{2}\right)$.

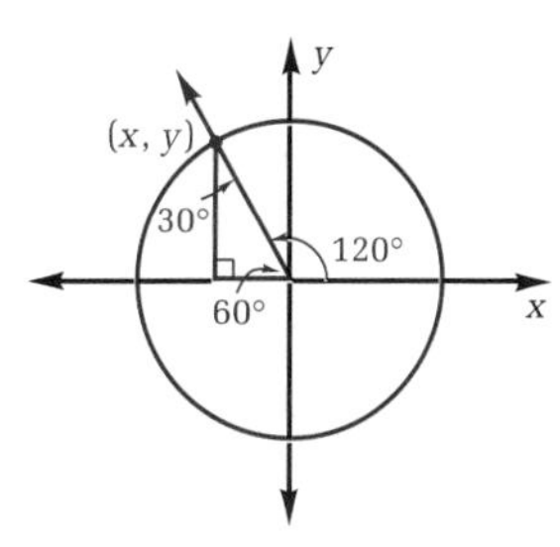

Part Two: Sample Problems

Problem 1 *Find the point of intersection of the unit circle and the angle whose terminal ray contains the point P(5, −12).*

Solution **Method 1**

By the Pythagorean Theorem, OP = 13. Drop perpendiculars to the x-axis as shown. The result is two similar right triangles and $\frac{5}{x} = \frac{12}{y} = \frac{13}{1}$. Solving, we find $x = \frac{5}{13}$ and $y = \frac{12}{13}$, but these are only distances. Since the point is in the fourth quadrant, the y-coordinate must be *negative*. Thus the point of intersection is $\left(\frac{5}{13}, \frac{-12}{13}\right)$.

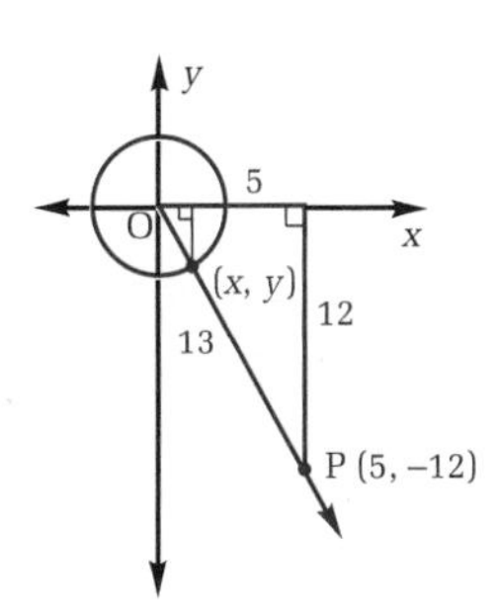

Method 2

The equation of the ray is $y = \frac{-12}{5}x$. The equation of the circle is $x^2 + y^2 = 1$. We could solve the system of equations as in the example at the bottom of the preceding page.

Lesson Notes, continued

- The Pythagorean Theorem is used to find points where the terminal side of the angle and the unit circle intersect. You may want to point out that ideas and skills presented in algebra and geometry are integrated in the study of trigonometry.

Checkpoint

1. Find the length of the arc subtended in the unit circle by a central angle of measure 135°. $\frac{3\pi}{4}$
2. Find the point of intersection of the unit circle and the angle whose terminal side contains the point (−7, −24). $\left(-\frac{7}{25}, -\frac{24}{25}\right)$
3. The terminal side of an angle contains the point (5, −5). Find the measure of the angle. 315°
4. The line $y = -\frac{9}{40}x$ intersects the unit circle in the second quadrant. Find the point of intersection. $\left(-\frac{40}{41}, \frac{9}{41}\right)$

14.1

Assignment Guide

Average	
Day 1	11–16, 18–20, 23, 25, 26
Day 2	17, 21, 22, 24, 27, 28, 30, 31
Advanced	
Day 1	11–16, 18–21, 23, 25, 26, 30
Day 2	($\frac{1}{2}$ day) 17, 22, 27, 29–31 ($\frac{1}{2}$ day) Section 14.2 9a, b, c, 11, 12, 15, 16, 20
Honors	
Day 1	17, 18, 20, 21, 23–26, 30
Day 2	($\frac{1}{2}$ day) 19, 22, 27, 29, 31, 32 ($\frac{1}{2}$ day) Section 14.2 9, 11, 12, 15, 16, 20, 21

Problem-Set Notes and Strategies

- In problem **2** and **5 – 10**, students will need the relationships of the 30°-60° -90° triangle. This might be a good time to review those as well as the 45°-45°-90° triangle. These properties will be used in the next several chapters also.
- Problems **11 – 16** involve angles with the terminal side on one of the axes. These angles are called quadrantal angles and are often easy for students to remember. Remind them of the connection between the degree measure and the quadrantal angles.

Problem 2 *The circle shown has radius 9, and arc AB has length 2π. Find the measure of central angle AOB.*

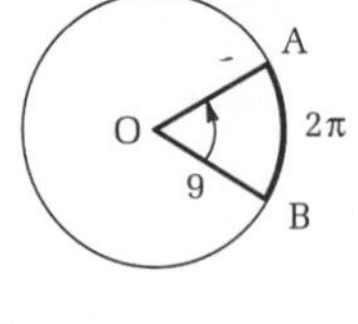

Solution The circumference of the circle is 18π. Arc AB is $\frac{2\pi}{18\pi}$, or $\frac{1}{9}$, of the circle. This means that $\angle AOB = \frac{1}{9}(360°) = 40°$.

Part Three: Exercises and Problems

Warm-up Exercises

P **1** Find the value of r. 24

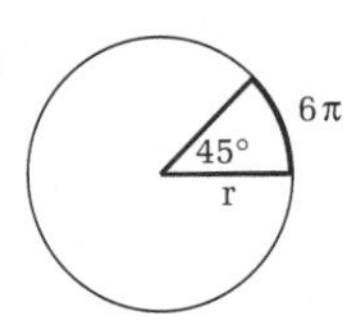

P **2** Find the coordinates of the point of intersection of the unit circle and the terminal ray of a 210° angle. $\left(-\frac{\sqrt{3}}{2}, -\frac{1}{2}\right)$

P **3** Find the value of x. $\sqrt{7}$

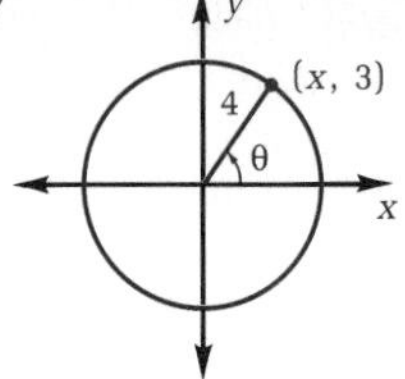

P **4** Find (x, y) in the third quadrant. $\left(-\frac{15}{17}, -\frac{8}{17}\right)$

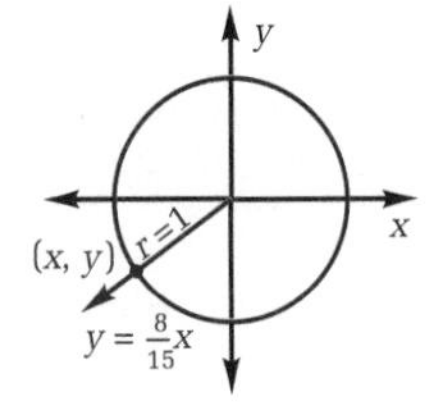

P,I **In problems 5–10, find the point of intersection of the unit circle and the terminal ray of each angle.**

5 30° $\left(\frac{\sqrt{3}}{2}, \frac{1}{2}\right)$ **6** 45° $\left(\frac{1}{\sqrt{2}}, \frac{1}{\sqrt{2}}\right)$ **7** 60° $\left(\frac{1}{2}, \frac{\sqrt{3}}{2}\right)$

8 −30° $\left(\frac{\sqrt{3}}{2}, -\frac{1}{2}\right)$ **9** −45° $\left(\frac{1}{\sqrt{2}}, -\frac{1}{\sqrt{2}}\right)$ **10** −60° $\left(\frac{1}{2}, -\frac{\sqrt{3}}{2}\right)$

Problem Set

A

In problems 11–16, find the point of intersection of the unit circle and the terminal ray of each angle.

P,I **11** 90° (0, 1) **12** 180° (−1, 0) **13** 270° (0, −1)

14 360° (1, 0) **15** 0° (1, 0) **16** −90° (0, −1)

P | **17** Refer to the diagram.

a Find the length of arc AB.
b Find the length of arc ACB.
a $\frac{25}{36}\pi$, or ≈ 2.182
b $\frac{47}{36}\pi$, or ≈ 4.102

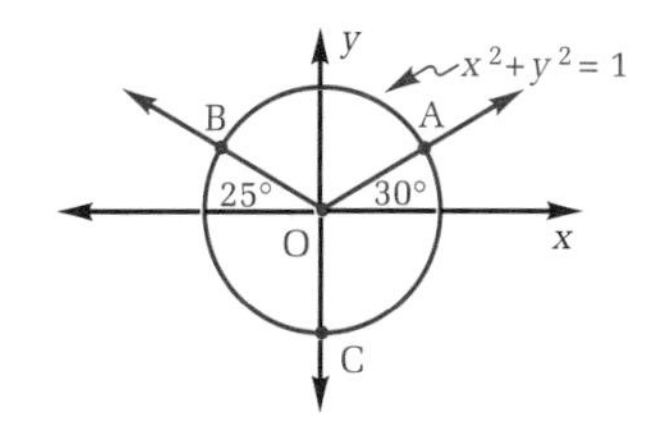

P | **18** The point $(-3, 4)$ is on the terminal ray of an angle in standard position. At what point does the terminal ray of the angle intersect the unit circle? $\left(-\frac{3}{5}, \frac{4}{5}\right)$

P | **19** Refer to the diagram.

a Find the number of degrees in angle θ.
b Find the length of arc AB if the circle is a unit circle.
a $-120°$
b $\frac{2\pi}{3}$, or ≈ 2.094

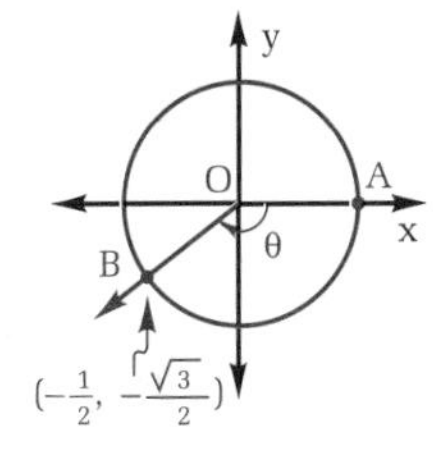

Problem-Set Notes and Strategies, continued

- Students may forget that angles measured clockwise as in problem **19a** are negative. In part **b**, the minor arc is meant unless stated otherwise.

P | **20** Find the value of β and γ if $\alpha = 75°$.
$\beta = 435°$; $\gamma = -285°$

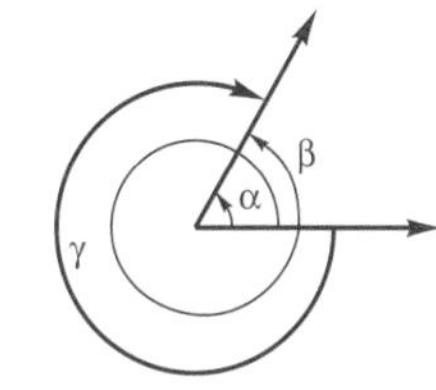

Stumbling Block

Students may have a little trouble with angles in standard position at first. The more intuitive notion of a positive angle would be in the clockwise direction. Have them relate the direction of the angle to the way that the quadrants are numbered, i.e. counterclockwise. After a little practice, this difficulty should disappear.

P **B** | **21** Find the point Q on the unit circle that lies on the terminal ray OP of angle ROP. $\left(\frac{8}{17}, \frac{-15}{17}\right)$

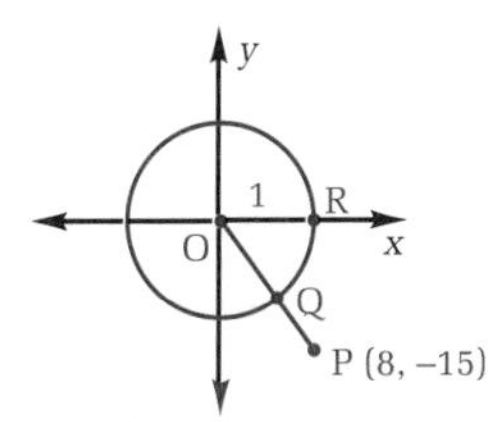

P | **22** Name the coordinates of points A, B, and C.
$A = \left(-\frac{1}{2}, \frac{\sqrt{3}}{2}\right)$; $B = \left(-\frac{1}{2}, -\frac{\sqrt{3}}{2}\right)$; $C = \left(\frac{1}{2}, -\frac{\sqrt{3}}{2}\right)$

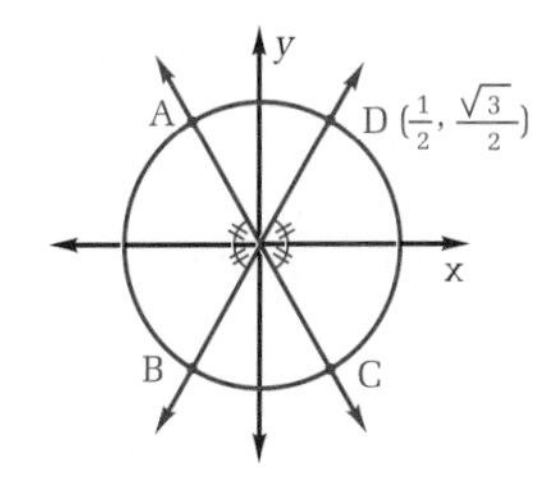

- Problem **22** illustrates the symmetry in the unit circle. Symmetry is one way students can organize trigonometric functions here and in later sections.

P | **23** Find the point of intersection of the unit circle and the ray $y = \frac{5}{12}x$, $x \geq 0$. $\left(\frac{12}{13}, \frac{5}{13}\right)$

Problem-Set Notes and Strategies, continued

- Problem **24** introduces the concept of radian measure without using the name radian. It is helpful if students discover that the ratio is the same independent of the size of the radius. It is the ratio of the length of the arc to the length of the radius.
- Problem **26** illustrates the correspondence of complementary angles. This correspondence will be helpful when the cofunctions of trigonometry are defined.
- If possible take some time to go through problem **28**. Most students can understand this problem and it combines many different concepts.
- Problem **29** is similar to **24** and again uses radian measure.

Cooperative Learning

Problems **24** and **29** both introduce radian measure. Have the students look at several circles of different sizes and generate a reason for why the ratio is always the same regardless of the size of the circle. You might have them make several circles using construction paper to illustrate this concept. Once they see that they are measuring with a different scale each time, the concept of radian measure can be introduced effectively.

Problem Set, *continued*

P,I **24** **a** If OA = 4, what is $\dfrac{\text{length of } \widehat{AB}}{OA}$?

b If OA = 20, what is $\dfrac{\text{length of } \widehat{AB}}{OA}$?

c If OA = 1, what is $\dfrac{\text{length of } \widehat{AB}}{OA}$?

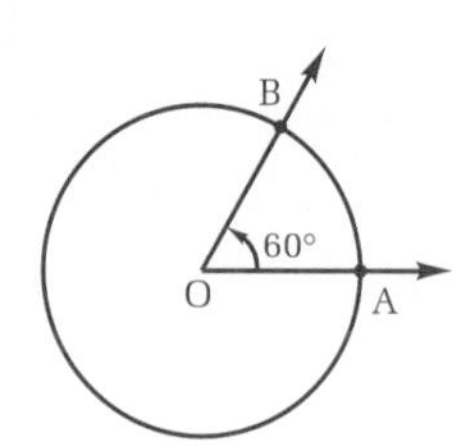

a $\frac{\pi}{3}$, or ≈1.047
b $\frac{\pi}{3}$, or ≈1.047
c $\frac{\pi}{3}$, or ≈1.047

P **25** Find all other angles θ in standard position, where $-360° \le \theta \le 360°$, such that the point of intersection of the terminal ray of the angle and the unit circle has the same x-coordinate as that of the 50° angle shown. −310°, −50°, 310°

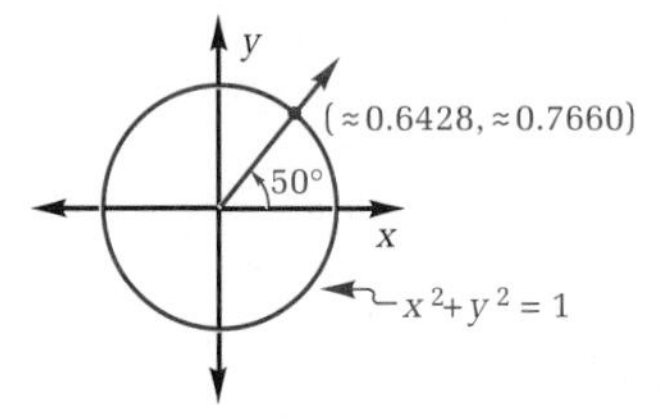

P **26** **a** Name the coordinates of point B.
b Name the coordinates of point C.
a (−b, a)
b (−a, −b)

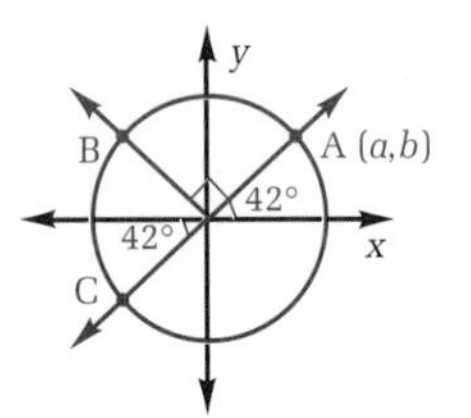

P,R **27** Solve the system. $(x, y) = \left(\frac{2}{\sqrt{5}}, \frac{1}{\sqrt{5}}\right)$ or $(x, y) = \left(\frac{-2}{\sqrt{5}}, \frac{-1}{\sqrt{5}}\right)$

$$\begin{cases} x^2 + y^2 = 1 \\ y = \frac{1}{2}x \end{cases}$$

R **28** A bicycle wheel has a diameter of 27 inches. There are 14 teeth on the gear of the rear wheel. There are 48 teeth on the pedal gear. How many revolutions (to the nearest whole revolution) does the pedal gear make when the bike travels one mile? ≈218 revolutions

P,R,I **29** In the circles shown, find each of the following.

a The length of $\widehat{AB}$. ≈1.3
b The length of $\widehat{CD}$. ≈7.9
c The ratio of the length of $\widehat{CD}$ to the length of $\widehat{AB}$. 6:1
d The ratio O′D:OB. 6:1

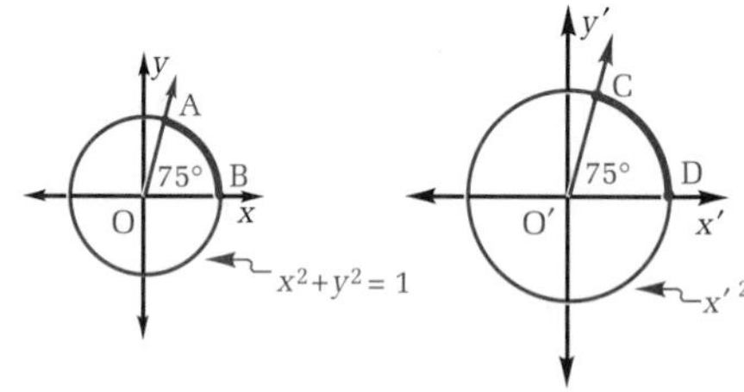

P **30** In a circle, a central angle of 60° cuts off an arc of length 10π. Find the radius of the circle. 30

C P,R **31** Find the point(s) on the unit circle $x^2 + y^2 = 1$ where the line $y = \frac{2}{5}x$ intersects the circle. $\left(\frac{5}{\sqrt{29}}, \frac{2}{\sqrt{29}}\right), \left(\frac{-5}{\sqrt{29}}, \frac{-2}{\sqrt{29}}\right)$

P,I **32** Arc AB is 14 inches long. O is the center of the circle. What is the length of $\overline{AB}$? $\approx$13.37 in.

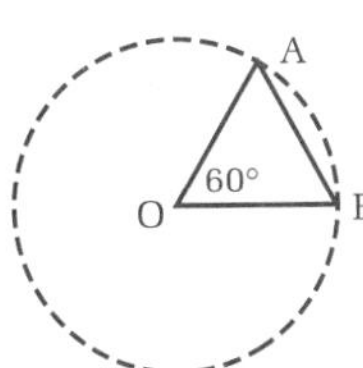

Problem-Set Notes and Strategies, continued

- Problem **32** will reward students who remember their geometry skills. Since the central angle is 60°, the triangle is equilateral and the length of *AB* is the radius of the circle.

Communicating Mathematics

There is a great deal of interplay between algebra, geometry, and trigonometry. Write a short paragraph describing what you know about the correspondences between these three seemingly different subjects.

MATHEMATICAL EXCURSION

SYSTEMS AND PROFITABILITY

The break-even point

The goal of a business is to make a profit. Even a not-for-profit venture such as a day-care center or a theater company needs to have income that at least equals expenses. Income that exceeds expenses enables a not-for-profit organization to expand its services to the community.

The Parabolic Theater Company is a not-for-profit organization that tries to provide quality productions at the lowest possible ticket prices. It wants to raise money for its productions by selling original T-shirts. The company decides to hire a designer to design and make a silk screen for the shirts. What it pays for the design and for the silk screen will be fixed costs—one-time costs that will never vary, no matter how many T-shirts are sold. The company will also rent a studio where company members will print the T-shirts each Saturday. The rent will be another fixed cost. The company decides to spend $500 on fixed costs.

There will also be costs that will vary as a function of how many T-shirts are sold. These variable costs are the costs of the shirts and the ink. The company can buy plain T-shirts for $4.00 each and has found that the cost of ink per T-shirt will be $.50. Therefore, the variable, or unit, cost of each T-shirt will be $4.50. The company decides to sell its T-shirts for $10.00 each.

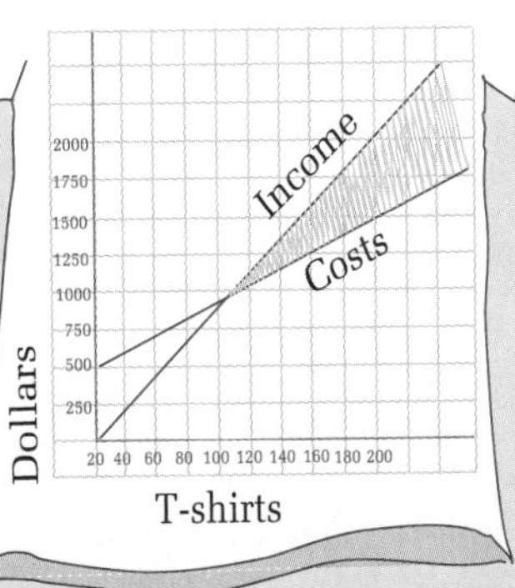

The income and expenses for this venture can be represented as a system of linear equations where number of T-shirts is graphed against dollars, as shown at the left.

How many T-shirts must the company sell simply to break even—that is, to recover its investment in the design and the silk screen? How many must it sell to make $1000 for its theater productions?

14.2

Class Planning

Time Schedule

Average: 2 days
Advanced: $1\frac{1}{2}$ days
Honors: $1\frac{1}{2}$ days

Resource References

Teacher's Resource Book
Class Opener 82
Cooperative Learning 42
Practice 87
Transparency 27

Class Opener

1 Find the coordinates of all points on the unit circle that are $\frac{1}{2}$ unit from the x-axis.

$(\frac{1}{2}, \frac{\sqrt{3}}{2}), (\frac{1}{2}, -\frac{\sqrt{3}}{2}), (-\frac{1}{2}, \frac{\sqrt{3}}{2}), (-\frac{1}{2}, -\frac{\sqrt{3}}{2})$

2 Determine the angle created when these points are connected to the origin. 30°

Lesson Notes

- The six trigonometric functions are defined according to the point of intersection, (x, y), of the terminal side with the unit circle. If students learn these relationships quickly, it is less likely that they will have difficulty as they work through this and the following chapters.

14.2 THE CIRCULAR FUNCTIONS

Objectives

After studying this section, you will be able to

- Calculate the sine, cosine, tangent, cotangent, secant, and cosecant of an angle
- Use the Pythagorean identities

Part One: Introduction

Sine and Cosine

In the previous section, you saw how the terminal ray of an angle intersects the unit circle at exactly one point. It is extremely useful to be able to compute the x- and y-coordinates of the point of intersection for any angle θ. We define the function ***cosine*** of θ—abbreviated $\cos(\theta)$—to be the x-coordinate, and ***sine*** of θ—abbreviated $\sin(\theta)$—to be the y-coordinate. Sine and cosine are two of six trigonometric functions.

 ***If the terminal ray of angle θ in standard position contains (x, y) on the unit circle, then $\cos(\theta) = x$ and $\sin(\theta) = y$, or $(\cos(\theta), \sin(\theta)) = (x, y)$.***

Example *Use the circle shown to evaluate $\cos(\theta)$ and $\sin(\theta)$.*

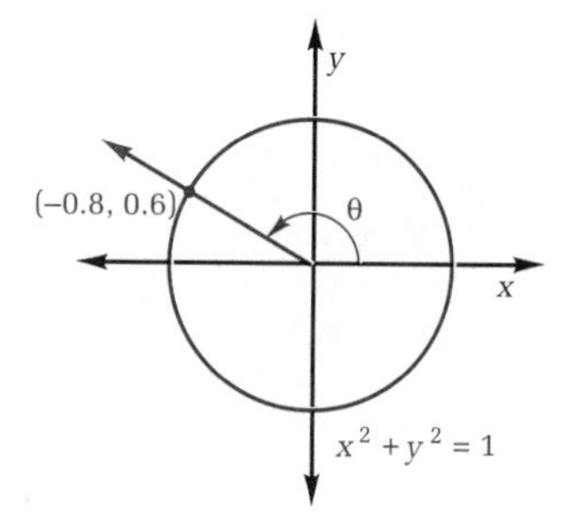

The coordinates of the points of intersection of the terminal ray of θ and the unit circle are $(-0.8, 0.6)$. This tells us that $\cos(\theta) = -0.8$ and $\sin(\theta) = 0.6$.

The sine and cosine functions take an angle as input and then output a coordinate from the unit circle. Thus, the range of these two functions is $-1 \leq \cos(\theta) \leq 1$ and $-1 \leq \sin(\theta) \leq 1$.

It is helpful to use a calculator to evaluate trigonometric functions.

Example *Approximate sin(39°) and cos(39°).*

Be sure that your calculator is in ***degree mode*** and enter 39 [sin]

Your calculator should read [0.62932039], so $\sin(39°) \approx 0.629$.

(We round off to the accuracy that is appropriate for each problem.)

Then we enter 39 [cos] and read that $\cos(39°) \approx 0.777$.

Vocabulary

cosine
sine
degree mode
tangent
cotangent
secant
cosecant
Pythagorean identities

If a calculator is not available, you can look up sines and cosines in the table on pages 738–743.

In the preceding example, we were given the angle and asked to find the sine and cosine. Now, we try the reverse—given the sine or cosine, what is the angle?

Example *If* $\sin(\theta) = 0.8296$, *with* $0° < \theta < 90°$, *what is* θ?

To solve this, we use a key labeled either [INV] or [ARC] or [2nd] or [$\sin^{-1}$] on most calculators. (We will use the [INV] notation.) When we enter .8296 [INV] [sin], the display is [56.0576702]. Thus, angle $\theta \approx 56°$. (Again, the degree of accuracy of rounding off depends on the application of the answer.)

To use the table on pages 738–743, we look for 0.8292 in the sine column. The closest value is 0.8290, which corresponds to sin (56°), so $\theta \approx 56°$.

Tangent, Cotangent, Secant, and Cosecant

Four other functions are defined in terms of cosine and sine.

Name of Function	Abbreviation and Definition
Tangent	$\tan(\theta) = \frac{\sin(\theta)}{\cos(\theta)} = \frac{y}{x}$
Cotangent	$\cot(\theta) = \frac{\cos(\theta)}{\sin(\theta)} = \frac{x}{y}$
Secant	$\sec(\theta) = \frac{1}{\cos(\theta)} = \frac{1}{x}$
Cosecant	$\csc(\theta) = \frac{1}{\sin(\theta)} = \frac{1}{y}$

The definitions of the six trigonometric functions are so important that you should memorize them.

Example *The terminal ray of* ϕ *contains* $(4, -3)$. *Evaluate the six trigonometric functions of* ϕ.

By similar triangles, the point of intersection of the terminal ray and the unit circle is $\left(\frac{4}{5}, -\frac{3}{5}\right)$. From the definitions, $\cos(\phi) = \frac{4}{5}$, $\sin(\phi) = -\frac{3}{5}$, $\tan(\phi) = -\frac{3}{4}$, $\cot(\phi) = -\frac{4}{3}$, $\sec(\phi) = \frac{5}{4}$, and $\csc(\phi) = -\frac{5}{3}$.

Example *Find* $\csc(128°)$.

Most calculators have no [csc] button, but by definition, $\csc(128°)$ is the reciprocal of $\sin(128°)$. Be sure your calculator is in degree mode. Enter 128 [sin] [1/x] to find that $\csc(128°) \approx 1.269$.

Lesson Notes, continued

- A calculator is the primary method used to evaluate the trigonometric functions although tables are supplied in the back of student texts. It might be helpful to go through several different exercises to familiarize the student with his/her own calculator in degree mode. You may want to reinforce that the trigonometric functions are functions as each angle has a unique sine, cosine, etc. The inverse trigonometric functions are also examined using a calculator, but formal definitions of the inverse functions are left to a later section. Some students may observe that the relationships of the trig functions and the inverses are not unique.

Using Computers

Using the *Omnifarious Plotter*, students can graph the unit circle and, using the zoom and trace features, explore the relationship between points and angles on the unit circle. Such an exploration can be highly effective if done in conjunction with a calculator, as instructed in the examples on pages 610 – 611.

Lesson Notes, continued

- The first identities are introduced here. You may want to restate the definition of an identity, that it is true for any value of the variable. You may need to go through the derivation of the first Pythagorean identity to convince students it is true. It will save time and may eliminate later confusion if students memorize each of the three identities in this section or at least how to derive them. They will be used often in this text.

Communicating Mathematics

The three identities introduced so far are called the Pythagorean identities, yet were introduced using the unit circle. Write a short paragraph (include pictures if necessary) to illustrate the use of the Pythagorean Theorem in these identities.

Example *If $0° < \theta < 90°$ and $\cot(\theta) = 1.92$, what is θ?*

This is a two-stage problem. There is no cot button, so we use the reciprocal of tangent. The reciprocal of 1.92 is the value of $\tan(\theta)$. Then to find the angle, we use the inverse key and the tangent key.

Entering 1.92 1/x INV tan , we find that $\theta \approx 27.5°$.

Since sin, cos, tan, cot, sec, and csc are functions, we have used parentheses around the arguments of the functions. Often, we leave out the parentheses. For example, we write sin 35° instead of sin(35°).

Pythagorean Identities

The coordinates of the points (x, y) on the unit circle satisfy the equation $x^2 + y^2 = 1$ and $(\cos\theta, \sin\theta)$ is a point on the unit circle. These two relationships lead us to the following theorem.

For any angle θ, $(\cos\theta)^2 + (\sin\theta)^2 = 1$.

To simplify the writing of exponents with trigonometric functions, we usually write $\cos^2\theta + \sin^2\theta = 1$. From this first identity it is possible to prove the following identities. (Problem **24** in the Problem Set asks you to develop these proofs.)

For any angle θ, $1 + \tan^2\theta = \sec^2\theta$.
For any angle θ, $1 + \cot^2\theta = \csc^2\theta$.

These three theorems are called the ***Pythagorean identities.***

Part Two: Sample Problems

Problem 1 *If $\cos\alpha = \frac{2}{3}$, what is the exact value of $\sin\alpha$?*

Solution There are two points on the unit circle with x-coordinate $\frac{2}{3}$. Using the equation of the unit circle, and substituting,

$$\left(\frac{2}{3}\right)^2 + y^2 = 1$$

$$y = \pm\frac{\sqrt{5}}{3}$$

$$\sin\alpha = \frac{\sqrt{5}}{3} \text{ or } \sin\alpha = -\frac{\sqrt{5}}{3}$$

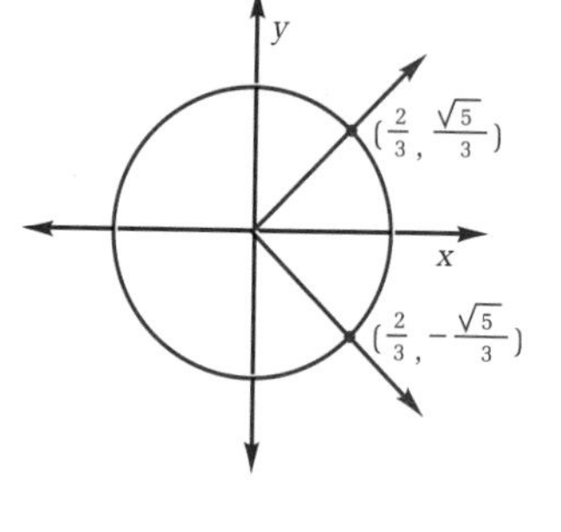

Problem 2 *Find the exact value of csc 135°.*

Solution The triangle formed in the second quadrant is a 45°-45°-90° triangle. The point on the unit circle is $\left(\frac{-1}{\sqrt{2}}, \frac{1}{\sqrt{2}}\right) = (\cos 135°, \sin 135°)$. Thus, $\csc 135° = \frac{1}{\sin 135°} = \sqrt{2}$. Since we are asked to find the exact value of csc 135°, the answer is $\sqrt{2}$. A decimal value would be an approximation.

Problem 3 *The terminal ray of β is $\{(x, y): y = 3x \text{ and } x \leq 0\}$. Evaluate $\tan \beta$.*

Solution Let's find the point of intersection of the line and the unit circle by solving a system.

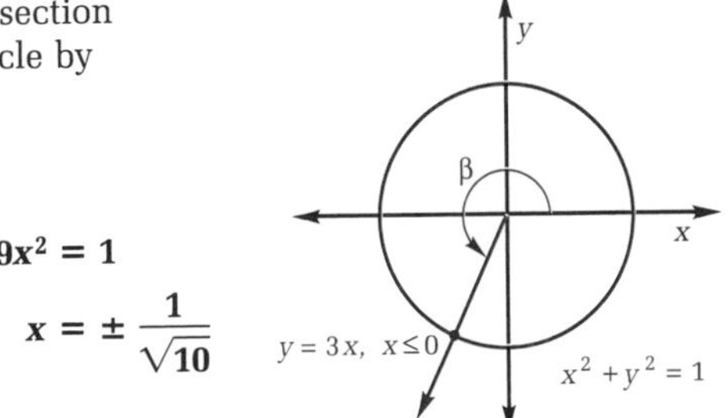

$$\begin{cases} y = 3x \\ x^2 + y^2 = 1 \end{cases}$$

Substitute for y $\quad \mathbf{x^2 + 9x^2 = 1}$

$$\mathbf{x = \pm \frac{1}{\sqrt{10}}}$$

Since $x \leq 0$, $x = -\frac{1}{\sqrt{10}}$.

The point of intersection is $\left(\frac{-1}{\sqrt{10}}, \frac{-3}{\sqrt{10}}\right)$ and $\tan \beta = \frac{\sin \beta}{\cos \beta} = \frac{y}{x} = 3$.

The tangent value is the same as the slope of the line. This is not a surprise, since the definition of tangent is $\frac{y}{x}$.

Problem 4 *The pitch of the roof of a house is $\frac{4}{12}$. What angle does the roof make with the horizontal?*

Solution The slope of the roof, $\frac{4}{12}$, is the tangent of angle θ, so angle $\theta \approx 18°$.

Problem 5 *Estimate θ to the nearest degree if $\sin \theta = -0.214$, the terminal ray of θ is in Quadrant III, and $\theta > 0°$.*

Solution When we enter 0.214 [±] [INV] [sin] into a calculator, we read an angle of about $-12°$. But the figure shows that the angle we want is in Quadrant III, so it must be greater than 180°. We know that two angles can have the same y-coordinates. The angle we want must be the same distance below the x-axis as $-12°$. So angle θ is about 180 + 12, or 192, degrees.

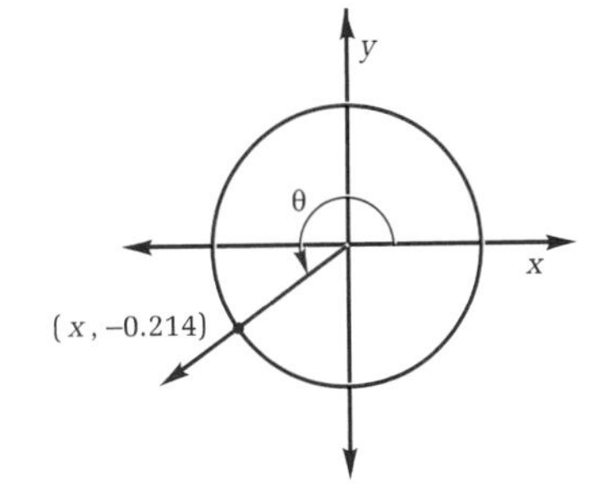

Problem 6 *Write $\cos \theta$ in terms of $\tan \theta$.*

Solution We are asked to write $\cos \theta$ as an expression that includes only "$\tan \theta$" and constants.

Remember that $\cos \theta = \frac{1}{\sec \theta}$, so $\cos^2 \theta = \frac{1}{\sec^2 \theta}$.

We use the identity $1 + \tan^2 \theta = \sec^2 \theta$ to write

$$\cos^2 \theta = \frac{1}{1 + \tan^2 \theta}$$

$$\cos \theta = \pm\sqrt{\frac{1}{1 + \tan^2 \theta}}$$

$$= \pm\frac{1}{\sqrt{1 + \tan^2 \theta}}$$

Checkpoint

1. Using a calculator or the tables in the text, find the exact value of sec 60°. 2
2. If the sine of an angle in the fourth quadrant is −0.3456, find the measure of the angle to the nearest degree. −20° or 340°
3. If the sine of an angle is $\frac{3}{4}$ and the angle is in the second quadrant, find the exact value of the cosine of the angle. $-\frac{\sqrt{7}}{4}$
4. Estimate the value of the sine of an angle if the tangent is 1.5 and the angle is in the third quadrant. Round your answer to four decimal places. −0.8321

14.2

Assignment Guide

Average

Day 1 9a–c, 11, 12, 15, 16, 18, 20, 21, 27, 28, 30

Day 2 9d–f, 10, 13, 14, 19, 22, 23, 24a, 26, 29, 31a, b

Advanced

Day 1 ($\frac{1}{2}$ day) Section 14.1 17, 22, 27, 29–31

($\frac{1}{2}$ day) 9a–c, 11, 12, 15, 16, 20

Day 2 9d, f, 10, 14, 18, 19, 21, 23, 24, 26, 28–30, 31a, b

Honors

Day 1 ($\frac{1}{2}$ day) Section 14.1 19, 22, 27, 29, 31, 32

($\frac{1}{2}$ day) 9, 11, 12, 15, 16, 20, 21

Day 2 10, 14, 17–19, 23–26, 28, 29, 31b, c, 32c, d, 33

Problem-Set Notes and Strategies

- Problem **1** can be a way to identify any student having difficulty with the definitions of the six trig functions. Verify that students have correct signs for the functions.
- In problem **3b**, students can relate slope of the line and the tangent of the angle. You may want to point out that this is a useful relationship to remember for later math classes.
- Problem **9** is a good place to verify that students are using their calculators properly and have them in degree mode.

Part Three: Exercises and Problems

Warm-up Exercises

P **1** Find all six trig function values of θ. (The coordinates shown are approximate.)
$\cos\theta \approx -0.82$; $\sin\theta \approx -0.57$; $\tan\theta \approx 0.70$; $\sec\theta \approx -1.22$; $\csc\theta \approx -1.75$; $\cot\theta \approx 1.44$

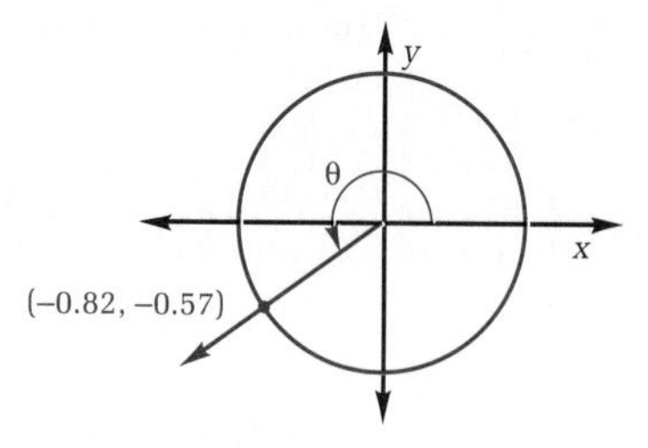

P,R **2** Find the point on the unit circle that is on the terminal ray of each angle.
a 60° $\left(\frac{1}{2}, \frac{\sqrt{3}}{2}\right)$ **b** 135° $\left(-\frac{\sqrt{2}}{2}, \frac{\sqrt{2}}{2}\right)$ **c** −30° $\left(\frac{\sqrt{3}}{2}, -\frac{1}{2}\right)$ **d** 100° (to nearest hundredth) $(\approx -0.17, \approx 0.98)$

P **3** Refer to the figure.
a Find the coordinates of point A to the nearest hundredth. $(\approx -0.77, \approx 0.64)$
b Find the slope of the ray pictured. ≈ -0.84

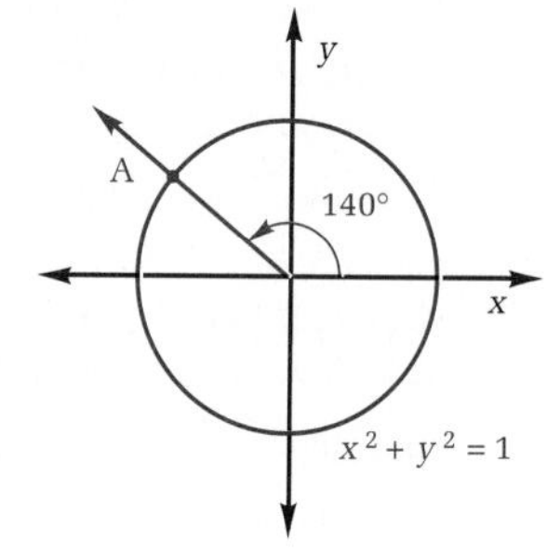

P **In problems 4–6, use the Pythagorean identities to find the value of each trig function.**

4 $\sin\alpha$, if $\cos\alpha = \frac{1}{4}$ $\pm\frac{\sqrt{15}}{4}$ **5** $\sec\theta$, if $\tan\theta = \sqrt{3}$ ± 2 **6** $\cot\beta$, if $\csc\beta = \frac{5}{3}$ $\pm\frac{4}{3}$

P **7** Match each angle with the point through which its terminal ray passes.

a −62° iii
b 118° i
c 62° ii
d 242° iv

i $(\approx -0.47, \approx 0.88)$
ii $(\approx 0.47, \approx 0.88)$
iii $(\approx 0.47, \approx -0.88)$
iv $(\approx -0.47, \approx -0.88)$

P **8** What is the value of $\sin\theta$ if $\cos\theta = \frac{3}{4}$? $\pm\frac{\sqrt{7}}{4}$

Problem Set

A

P **9** Find each first-quadrant angle described.

a $\sin\alpha = 0.3874$ $\approx 22.79°$ **b** $\cos\beta = \frac{8}{17}$ $\approx 61.93°$ **c** $\sin\theta = \frac{1}{7}$ $\approx 8.21°$

d $\tan\lambda = \frac{4}{3}$ $\approx 53.13°$ **e** $\cos\delta = \frac{1}{10}$ $\approx 84.26°$ **f** $\sin\omega = 1.342$ Impossible

P **10** **a** Find $\cos\theta$ if $\tan\theta = 2\frac{1}{8}$. $\pm\frac{8}{\sqrt{353}}$ **b** Find $\cos\theta$ if $\tan\theta = \frac{\sqrt{3}}{3}$. $\pm\frac{\sqrt{3}}{2}$

c Find $\tan\theta$ if $\sin\theta = \frac{1}{3}$. $\pm\frac{1}{2\sqrt{2}}$

P **11** Given that $\theta = 72°$, calculate each value to the nearest thousandth.

a $\sin\theta$ ≈ 0.951 **b** $\cos(90° - \theta)$ ≈ 0.951
c $\tan\theta$ ≈ 3.078 **d** $\cot(90° - \theta)$ ≈ 3.078

P,I **12** The terminal ray of θ contains $(7, -24)$. Evaluate the six trig functions of θ. $\sin\theta = -\frac{24}{25}$; $\cos\theta = \frac{7}{25}$; $\tan\theta = -\frac{24}{7}$; $\cot\theta = -\frac{7}{24}$; $\sec\theta = \frac{25}{7}$; $\csc\theta = -\frac{25}{24}$

P **13** The area of triangle OAC is 5. Find each of the following.

a The coordinates of point C $(1, 10)$
b OC $\sqrt{101}$
c The coordinates of point B $\left(\frac{1}{\sqrt{101}}, \frac{10}{\sqrt{101}}\right)$
d $m\angle AOC$ ≈ 84.29

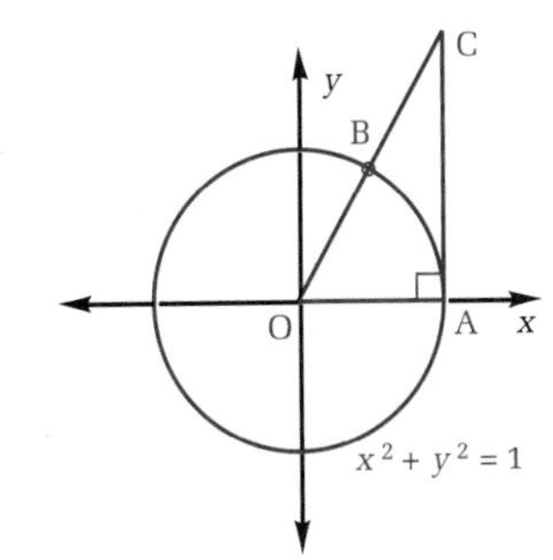

P,I **14** Find all six trig function values for θ.
$\sin\theta = -\frac{15}{17}$; $\cos\theta = -\frac{8}{17}$;
$\tan\theta = \frac{15}{8}$; $\cot\theta = \frac{8}{15}$;
$\sec\theta = -\frac{17}{8}$; $\csc\theta = -\frac{17}{15}$

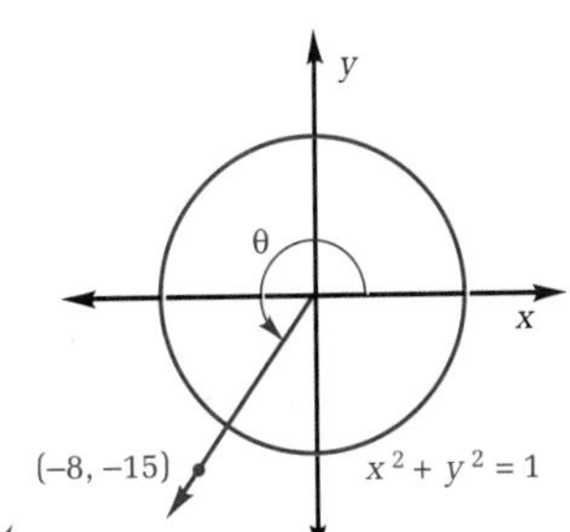

P,I **15** Refer to the figure.

a Find the value of y. $2\sqrt{21}$
b Find $\cos\theta$. $\frac{2}{5}$
c Find $\sin\theta$. $\frac{\sqrt{21}}{5}$
d Find the measure of θ. ≈ 66.42

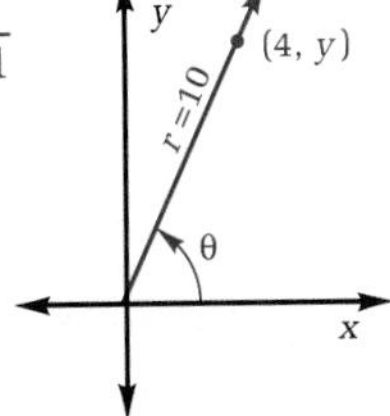

P,R **16** Find the first ten terms of the sequence $\{\cos(10k)°\}$ to the nearest hundredth. ≈ 0.98, ≈ 0.94, ≈ 0.87, ≈ 0.77, ≈ 0.64, 0.50, ≈ 0.34, ≈ 0.17, 0.00, ≈ -0.17

Problem-Set Notes and Strategies, continued

- In problem **11**, the relationship of the cofunctions as functions of complementary angles is previewed. Help students see that x and $90° - x$ are complementary angles.
- If students draw or complete the triangle and find the point of intersection with the unit circle in problems **12 – 15**, the Pythagorean Theorem can be also used to find the trigonometric functions. The triangle trig relationships will be introduced in Section **14.5**.

Problem-Set Notes and Strategies, continued

- In problem **18**, students may overlook that the angle is in the second quadrant. A trig table and a calculator will yield an answer in the first quadrant which must be changed to fit this case. You might point out that the table and calculator result is called a reference angle, the angle made with the nearest horizontal axis.

- The fact that the circle in problem **19** is not a unit circle may cause some problems at first. By drawing the appropriate triangles, students can find similar triangles and can proceed using the ratios of the sides.

Stumbling Block

Make sure that students work through problem **24** and understand the relationship between the three Pythagorean identities. Students frequently confuse one identity with another. If they remember the one with sine and cosine and the derivation of the others, it minimizes this confusion.

Cooperative Learning

Consider using problem **24** as an introduction to identities. These two identities should be reasonable once students see the relationship. Have students work in a group to generate other identities. Suggest that groups exchange and verify identities.

Problem Set, *continued*

P,R **17** Find each value to the nearest thousandth.

a The coordinates of point A
b The area of triangle AOB
a (≈0.342, ≈0.940)
b ≈0.470

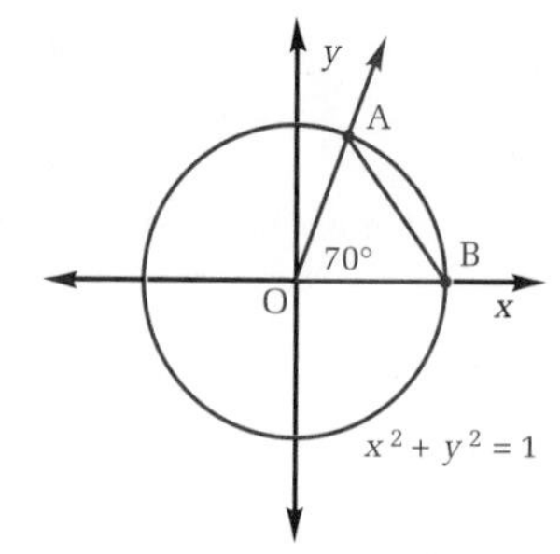

P **18** Find the value of θ if $\sin\theta \approx 0.4695$. 152°

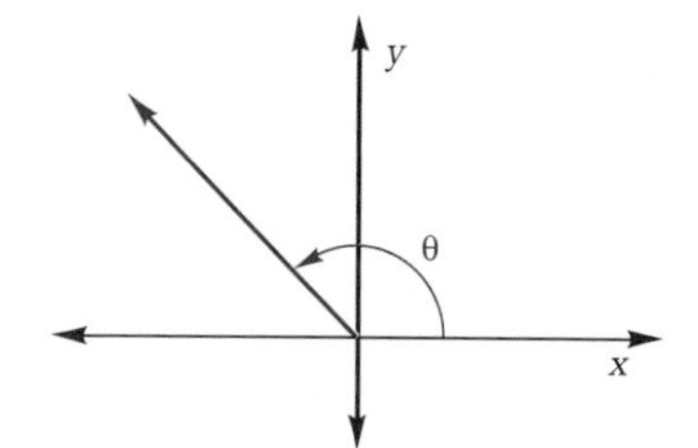

B  P **19** Find each value if $\sin\theta = -0.469$.

a θ ≈208.00
b The coordinates of point A
(≈−3.53, ≈−1.88)

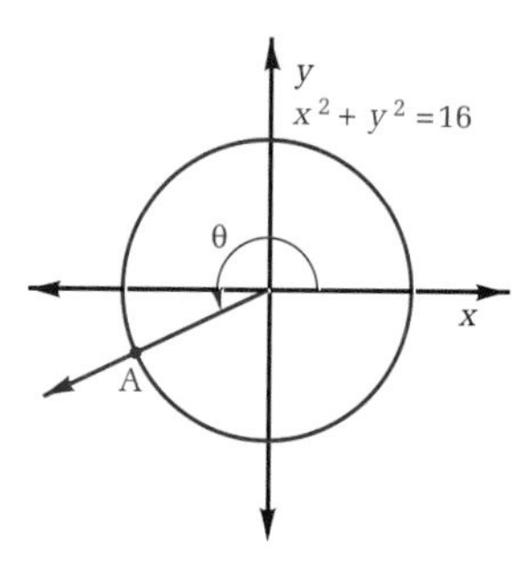

P **20** Find each of the following for $0° \le \theta \le 360°$.

a The value of θ for which $\sin\theta$ is largest 90°
b The value of θ for which $\sin\theta$ is smallest 270°
c The values of θ for which $\sin\theta = 0$ 0°, 180°, 360°

P,R **21** For each point on the unit circle, find the measure of the angle whose terminal ray passes through the point.

a (0, −1) 270
b $\left(-\frac{\sqrt{2}}{2}, -\frac{\sqrt{2}}{2}\right)$ 225
c $\left(\frac{4}{5}, -\frac{3}{5}\right)$ ≈323.13
d (≈0.819, ≈0.573) ≈35

P **22** Evaluate all six trigonometric functions for the angle whose terminal ray passes through (−0.5, ≈0.866) on the unit circle. sine: ≈0.866; cosine: −0.5; tangent: ≈1.732; cotangent: ≈0.577; secant: ≈2; cosecant: ≈1.155

P  **23** If $\cos\theta = 0.695$, what is the value of each of the following?

a $\sin\theta$ ≈±0.719
b $\tan\theta$ ≈±1.035
c $\sec\theta$ ≈1.439

P | **24** Use the statement $\cos^2\theta + \sin^2\theta = 1$ to prove each of the following.

a $1 + \tan^2\theta = \sec^2\theta$ **b** $1 + \cot^2\theta = \csc^2\theta$

P | **25** Explain why it is not possible for $\sin\theta$ to be 0.625 and $\cos\theta$ to be 0.54 if $0° \le \theta \le 360°$.
For any θ, $\sin^2\theta + \cos^2\theta = 1$, but $0.625^2 + 0.54^2 \approx 0.682$.

P,I | **26** Find the point(s) on the unit circle through which the terminal ray of each angle θ might pass.

a $\tan\theta = 1$ $\left(\frac{\sqrt{2}}{2}, \frac{\sqrt{2}}{2}\right)$ or $\left(-\frac{\sqrt{2}}{2}, -\frac{\sqrt{2}}{2}\right)$ **b** $\sin\theta = -\frac{\sqrt{2}}{2}$ $\left(-\frac{\sqrt{2}}{2}, -\frac{\sqrt{2}}{2}\right)$ or $\left(\frac{\sqrt{2}}{2}, -\frac{\sqrt{2}}{2}\right)$

c $\cos\theta = -1$ $(-1, 0)$ **d** $\sec\theta = -2$ $\left(-\frac{1}{2}, \frac{\sqrt{3}}{2}\right)$ or $\left(-\frac{1}{2}, -\frac{\sqrt{3}}{2}\right)$

P,I | **27** Let $\sec\theta = 4.25$. Find all possible values of θ such that $0° \le \theta \le 360°$. $\approx 76.39°$ or $\approx 283.61°$

I | **28** Name the quadrant that contains the terminal ray of angle θ, given the specified conditions.

a $\cos\theta < 0$ and $\sin\theta > 0$ II **b** $\tan\theta > 0$ and $\sin\theta < 0$ III

R,I | **29** Write an equation of the line that passes through (0, 0) and forms the given angle.

a 30° with the positive x-axis $y = \frac{x}{\sqrt{3}}$ **b** 150° with the positive x-axis $y = -\frac{x}{\sqrt{3}}$

P | **30** Refer to the figure and find each value.

a $\cos\theta$ a **b** $\sin\theta$ b
c $\tan\theta$ $\frac{b}{a}$ **d** $\cos(\theta + \alpha)$ c
e $\sin(\theta + \alpha)$ d **f** $\tan(\theta + \alpha)$ $\frac{d}{c}$

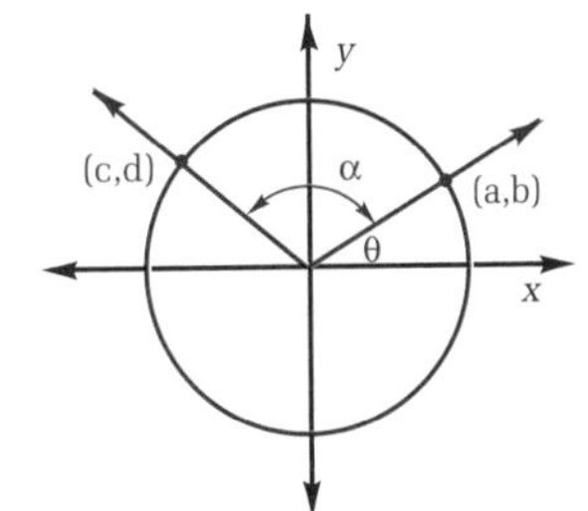

C

P,I | **31** Simplify each expression in terms of a single trig function.

a $(1 - \sin\theta)(1 + \sin\theta)$ $\cos^2\theta$ **b** $\sin^2\theta + \frac{\sin^2\theta}{\cos^2\theta} + \cos^2\theta$ $\sec^2\theta$ **c** $(\sec\theta)\left(\cos\theta - \frac{1}{\cos\theta}\right)$ $-\tan^2\theta$

P | **32** Solve for θ.

a $\tan\theta = -1$, $90° < \theta < 270°$ 135° **b** $\cos\theta = \frac{1}{2}$, $-90° < \theta < 90°$ 60° or −60°

c $\cos\theta = \frac{1}{2}$, $0° < \theta < 360°$ 60° or 300° **d** $\sec\theta = \frac{-2}{\sqrt{2}}$, $0° < \theta < 360°$ 135° or 225°

P | **33** Given that $0° \le \theta \le 360°$, find θ for each of the following.

a $\sin\theta = 1$ 90° **b** $\sin\theta = -1$ 270°
c $\sin^2\theta = 1$ 90° or 270° **d** $\cos^2\theta = 1$ 0°, 180°, or 360°

P,I | **34** Write an expression for $\tan\theta$ in terms of $\cos\theta$.
Depending on the value of θ, $\tan\theta = \frac{\sqrt{1-\cos^2\theta}}{\cos\theta}$ or $\tan\theta = -\frac{\sqrt{1-\cos^2\theta}}{\cos\theta}$.

Problem-Set Notes and Strategies, continued

- When students see the proof of problem **24** for both cases, they are more likely to remember the derivation of the identities rather than memorize them.

Additional Answers

24a $\cos^2\theta + \sin^2\theta = 1$

$$\frac{\cos^2\theta + \sin^2\theta}{\cos^2\theta} = \frac{1}{\cos^2\theta}$$

$$\frac{\cos^2\theta}{\cos^2\theta} + \frac{\sin^2\theta}{\cos^2\theta} = \frac{1}{\cos^2\theta}$$

$$1 + \left(\frac{\sin\theta}{\cos\theta}\right)^2 = \left(\frac{1}{\cos\theta}\right)^2$$

$$1 + \tan^2\theta = \sec^2\theta$$

24b $\cos^2\theta + \sin^2\theta = 1$

$$\frac{\cos^2\theta + \sin^2\theta}{\sin^2\theta} = \frac{1}{\sin^2\theta}$$

$$\frac{\cos^2\theta}{\sin^2\theta} + \frac{\sin^2\theta}{\sin^2\theta} = \frac{1}{\sin^2\theta}$$

$$\left(\frac{\cos\theta}{\sin\theta}\right)^2 + 1 = \left(\frac{1}{\sin\theta}\right)^2$$

$$1 + \cot^2\theta = \csc^2\theta$$

- For problems **26** and **27** make sure that both points or angles are found. It is easy to find one and forget that there might be another that also satisfies the conditions stated. These concepts will be explored in greater detail later in the chapter.

- Once again, the slope of the line is related to the tangent of the angle in problem **29**.

- Problem **31** is a precursor to proving and using identities, the main focus of Chapter **15**.

- Problems **32** and **33** each have multiple answers for several of the questions. Suggest that students draw some pictures and reaffirm the symmetry that leads to the multiple answers.

14.3

Class Planning

Time Schedule

Average: 2 days
Advanced: $1\frac{1}{2}$ days
Honors: $1\frac{1}{2}$ days

Resource References

Teacher's Resource Book
Class Opener 83
Practice 88
Enrichment 27
Evaluation
Quiz Form 1

Class Opener

Draw the ray from the origin to the point (12, –16). Determine the sine, cosine and tangent of the angle formed by the ray and the positive *x*-axis.

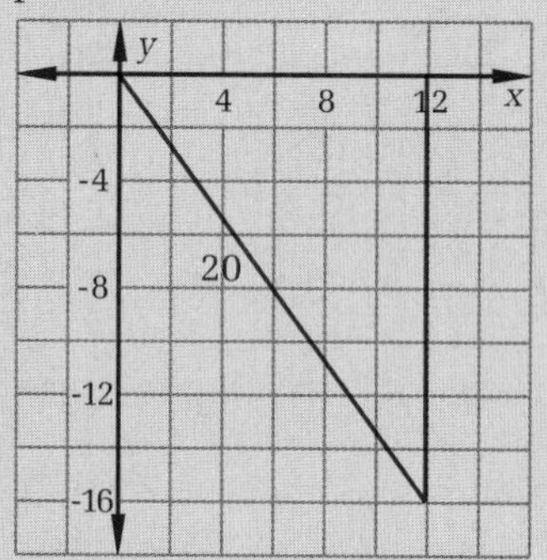

sine = –.8, cos = .6, tan = $-1.\overline{3}$

Lesson Notes

- Defining the trigonometric functions in terms of the coordinates of any point on the terminal side of the angle is the same as using a point on a circle of radius *r*. In this way, the trig functions can be generated using the same formulas, but generalized to incorporate the new radius.

14.3 Coordinate Definitions

Objectives

After studying this section, you will be able to

- Define the trigonometric functions in terms of coordinates
- Use special angles to compute trigonometric functions

Part One: Introduction

Defining Trig Functions in Terms of Coordinates

In the last section, we used points on the unit circle to define the trig functions. However, we can evaluate the trigonometric functions given *any* point on the terminal ray. The point need not be on the unit circle.

Example *The terminal ray of θ contains the point (12, 5). Evaluate the trig functions of θ.*

The point (12, 5) is 13 units from the origin. A point on this ray that is only 1 unit from the origin (on the unit circle) is $\frac{1}{13}$ of the distance to the point (12, 5), and so its coordinates will be $\left(\frac{12}{13}, \frac{5}{13}\right)$. Thus,
$\cos\theta = \frac{12}{13}$, $\sin\theta = \frac{5}{13}$, $\tan\theta = \frac{5}{12}$,
$\cot\theta = \frac{12}{5}$, $\sec\theta = \frac{13}{12}$, and $\csc\theta = \frac{13}{5}$.

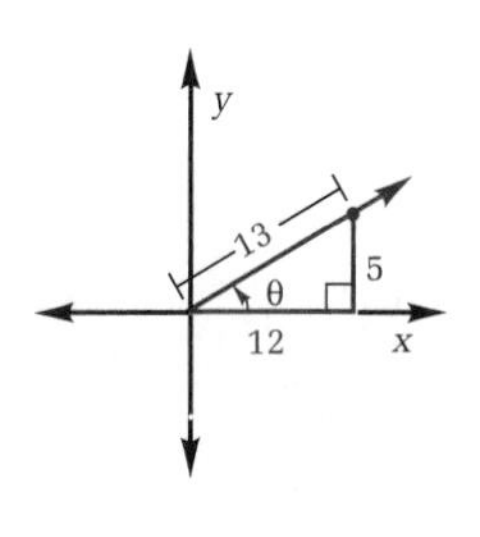

In general, if the terminal ray of an angle θ contains the point (x, y), and if the distance from the origin to (x, y) is r, then

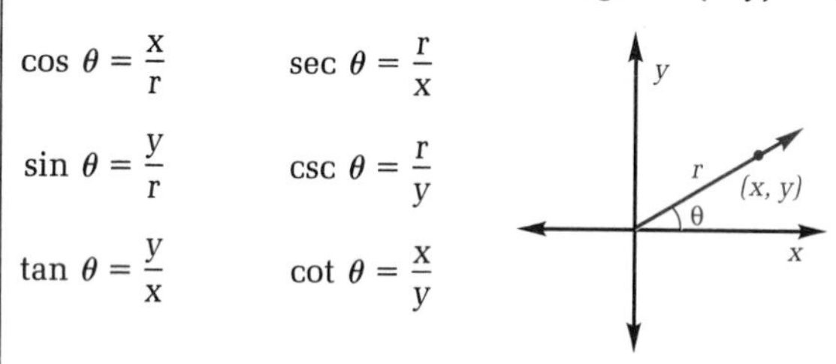

$$\cos\theta = \frac{x}{r} \qquad \sec\theta = \frac{r}{x}$$
$$\sin\theta = \frac{y}{r} \qquad \csc\theta = \frac{r}{y}$$
$$\tan\theta = \frac{y}{x} \qquad \cot\theta = \frac{x}{y}$$

By the Pythagorean Theorem, $r = \sqrt{x^2 + y^2}$.

Vocabulary
reference angle

Quadrants

The trigonometric functions are based on the position of the terminal ray. The functions will be positive or negative depending on which quadrant the terminal ray lies in.

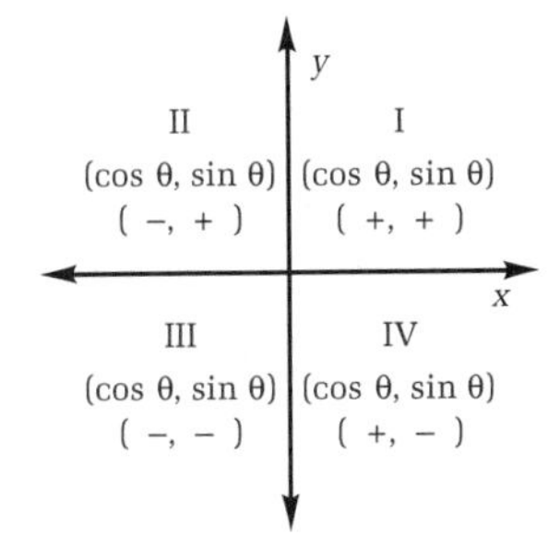

Example *If an angle α terminates in Quadrant III, will its tangent be positive or negative?*

$\tan \alpha = \frac{y}{x}$. In Quadrant III, x and y are both negative. The quotient of two negatives is positive, so $\tan \alpha$ is positive.

Special Angles and Reference Angles

For angles that are multiples of 30° or 45°, we can figure out the exact coordinates of the points on the unit circle by using the special 30°-60°-90° or 45°-45°-90° triangles as shown in the figures. Since these values are used quite often, it is helpful to memorize the values in the first quadrant. To find the values in the other quadrants, we use the symmetry of the circle as shown.

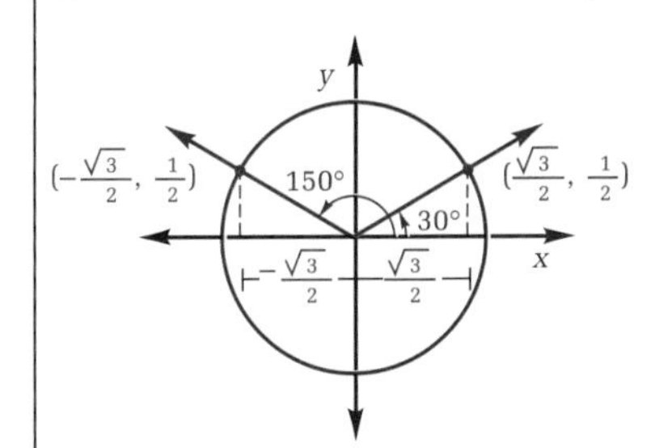

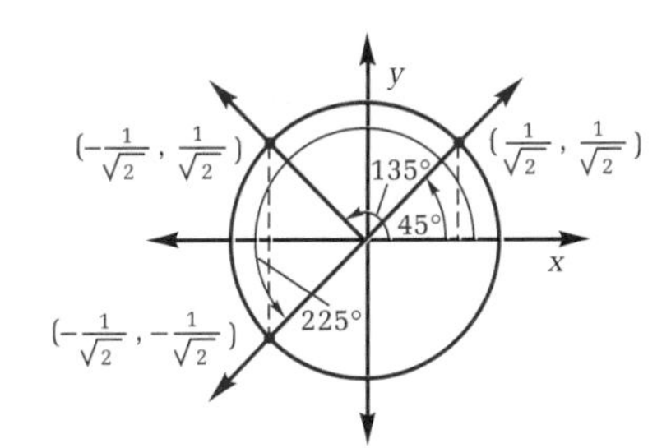

The first-quadrant angles are called the ***reference angles*** for the other angles. The ***reference angle*** for an angle θ is the first-quadrant angle (between 0° and 90°) that forms the same acute angle with the x-axis as θ.

Example *Find the value of each function.*

a cos 150°

$$\cos 150° = -\cos 30° = -\frac{\sqrt{3}}{2}$$

b *sin* 135°

$$\sin 135° = \sin 45° = \frac{1}{\sqrt{2}}$$

Lesson Notes, continued

- The relationships in a 30°-60°-90° triangle and in a 45°-45°-90° triangle are reviewed and the trig functions of these special angles are enumerated. If they have forgotten these relationships you might show students how to derive them since they will be used in almost every math class following trigonometry.

- Reference angles are also introduced as the angle formed by the terminal side and the nearest horizontal axis. These reference angles will be necessary in order to use the tables or a calculator to evaluate trigonometric functions.

Using Computers

Students can continue to use graphing software in conjunction with a calculator to explore the numerical and graphical properties of the unit circle and trigonometric functions.

Checkpoint

1 Find the tangent of an angle in Quadrant II if the sine of the angle is $\frac{3}{8}$. $-\frac{3}{\sqrt{55}}$
2 What is the reference angle for 257°? 77°
3 What is the exact value of cot 210°? $\sqrt{3}$
4 If $\cos x = 0.1234$, find the value(s) of sine x to four decimal places.
0.9924 or −0.9924

c *sin 225°*

$$\sin 225° = -\sin 45°$$
$$= -\frac{1}{\sqrt{2}}$$

For an angle that is not a multiple of 30° or 45°, we proceed as before.

Consider angles of 25°, 155°, 205°, and 335°. Each terminal ray forms an acute angle of 25° with the x-axis. By symmetry, the coordinates on the unit circle for the 25° angle are (a, b), the coordinates for 155° are $(-a, b)$, for 205° they are $(-a, -b)$, and for 335° they are $(a, -b)$. The reference angle is 25°.

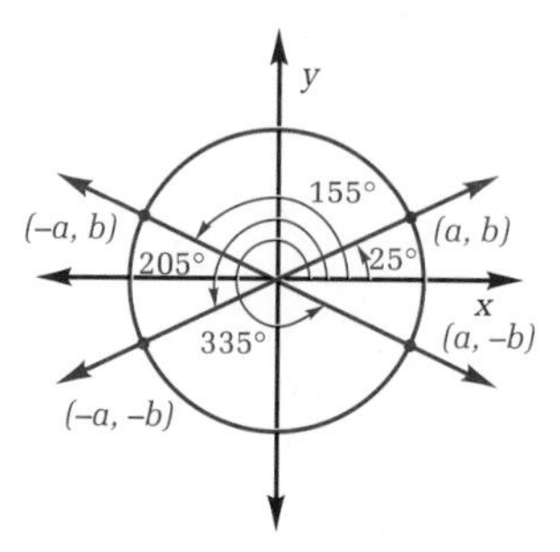

If we know the cosine and sine of the reference angle, we can adjust the sign of one or both coordinates for the cosine or sine of the symmetric angles in the other quadrants.

Example *Find the reference angle for 250°.*

The figure shows that the terminal ray for 250° makes a 70° angle with the negative x-axis. So the reference angle for 250° is 70°.

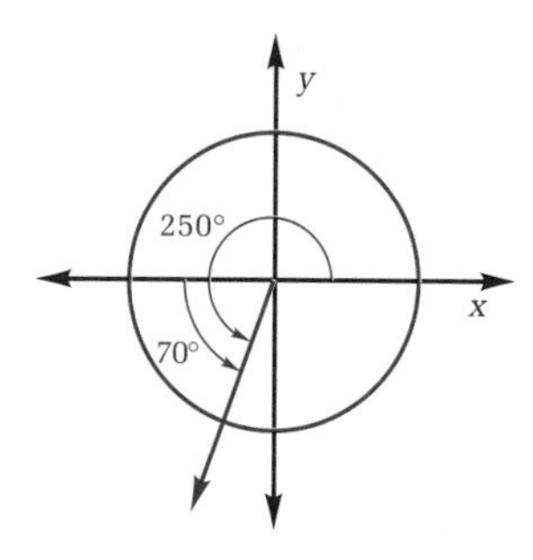

Part Two: Sample Problems

Problem 1 *Find the exact values of sin θ and cos θ if tan θ = $\frac{3}{5}$.*

Solution There are two quadrants in which tangent is positive: Quadrants I and III. Let's first consider Quadrant I.

Remember, $\tan\theta = \frac{y}{x}$. The terminal ray of angle θ passes through point (5, 3). The distance from the origin to (5, 3) is $\sqrt{34}$.

$$\cos\theta = \frac{x}{r} = \frac{5}{\sqrt{34}}$$

$$\sin\theta = \frac{y}{r} = \frac{3}{\sqrt{34}}$$

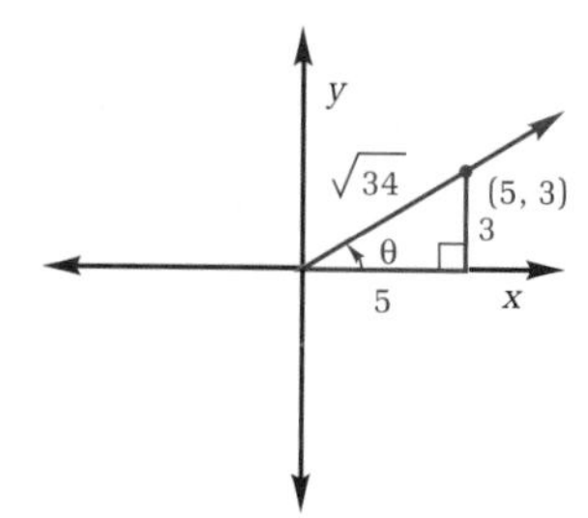

If θ is in Quadrant III, the only change is in the signs of the coordinates—
$\sin \theta = -\frac{3}{\sqrt{34}}$ and $\cos \theta = -\frac{5}{\sqrt{34}}$.

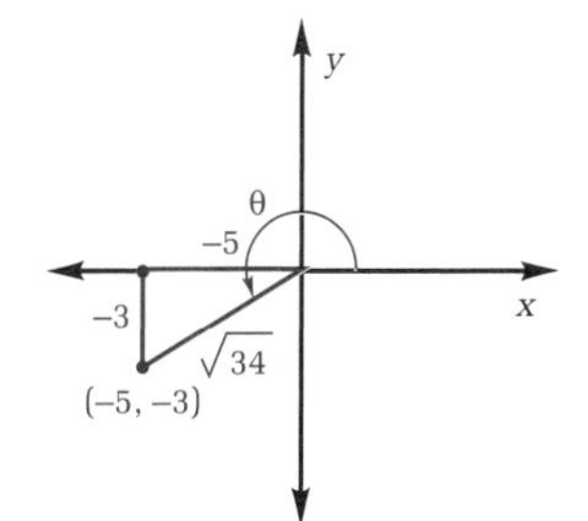

Problem 2 *Evaluate tan 90°.*

Solution The terminal ray of a 90° angle contains (0, 1). By definition, $\tan 90° = \frac{y}{x} = \frac{1}{0}$, which is undefined. So tan 90° is undefined.

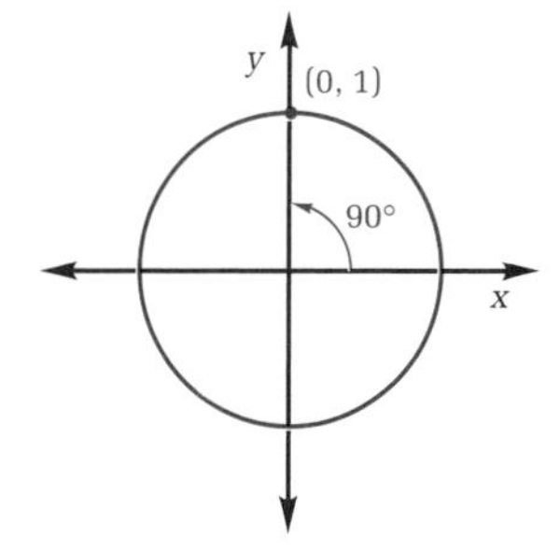

Problem 3 *Point P is ten units from the origin on the terminal ray of a 72° angle. To the nearest hundredth, find the coordinates of P.*

Solution Let (x, y) be the coordinates of P. We use the definitions of sine and cosine to solve for x and y.

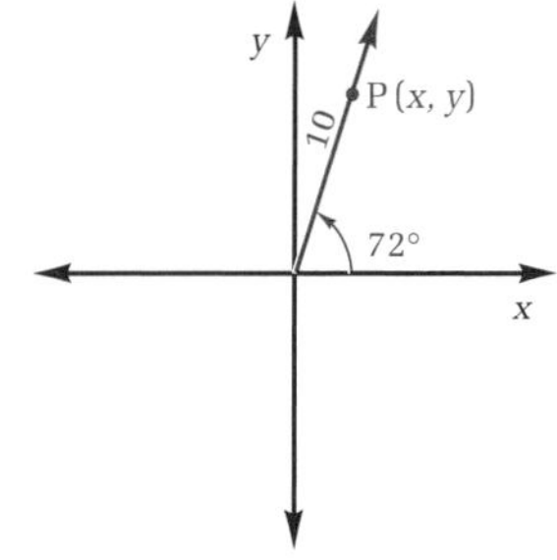

$$\cos 72° = \frac{x}{10} \qquad \sin 72° = \frac{y}{10}$$

$$0.309017 \approx \frac{x}{10} \qquad 0.9510565 \approx \frac{y}{10}$$

$$3.09 \approx x \qquad 9.51 \approx y$$

Point P is at approximately (3.09, 9.51).

Part Three: Exercises and Problems

Warm-up Exercises

P **1** State the reference angle for each angle.

a 245° 65° **b** 320° 40° **c** −160° 20°
d 142° 38° **e** 57° 57° **f** 272° 88°

P **2** Find the exact value of each of the following.

a cos 210° $-\frac{\sqrt{3}}{2}$ **b** sin 240° $-\frac{\sqrt{3}}{2}$ **c** tan 150° $-\frac{\sqrt{3}}{3}$
d sec 300° 2 **e** cot 180° Undefined **f** sin −30° $-\frac{1}{2}$

Communicating Mathematics

Every formula for the trigonometric functions introduced in this section has a counterpart in the last section, where the unit circle was used. Write a paragraph describing the similarities and differences between the corresponding formulas and identify how you might prove that the formulas yield the correct results.

Assignment Guide

Average
Day 1 12–14, 16, 18, 20, 21, 22a, b, 24, 31
Day 2 11, 15, 17, 19, 22c, d, 23, 27a, 32

Advanced
Day 1 11, 12, 15, 16, 18, 19, 21, 22a, b, 23, 27a
Day 2 ($\frac{1}{2}$ day) 13, 14, 20, 24, 25, 27b, 32, 33
($\frac{1}{2}$ day) Section 14.4 6–9, 14, 16

Honors
Day 1 11, 15–18, 20–22, 24, 32
Day 2 ($\frac{1}{2}$ day) 13, 19, 25–28, 33
($\frac{1}{2}$ day) Section 14.4 7–9, 14–16

Problem-Set Notes and Strategies

- Some students may be uncomfortable leaving the radical in the denominator as in the answer to **2c**. In past courses, students may have heard that this form is not simplified.

14.3

Problem-Set Notes and Strategies, continued

- Student answers in problem **6** may look different until they simplify the radical expression.
- Problems **9 – 11** are algebraic in nature and can be solved with straightforward application of formulas.

P **In problems 3–5, let (x, y) be a point on the terminal ray of θ, and let (x, y) be r units from the origin.**

3 Find sin θ if x = 5 and r = 13. $\pm\frac{12}{13}$

4 Find cos θ if x = 8 and y = 15. $\frac{8}{17}$

5 Find tan θ if x = 8 and r = 10. $\pm\frac{3}{4}$

R **In problems 6 and 7, refer to the figure.**

6 Find the value of sin θ, cos θ, and tan θ. $\frac{1}{\sqrt{5}}; \frac{2}{\sqrt{5}}; \frac{1}{2}$

7 Find the value of θ. ≈26.57°

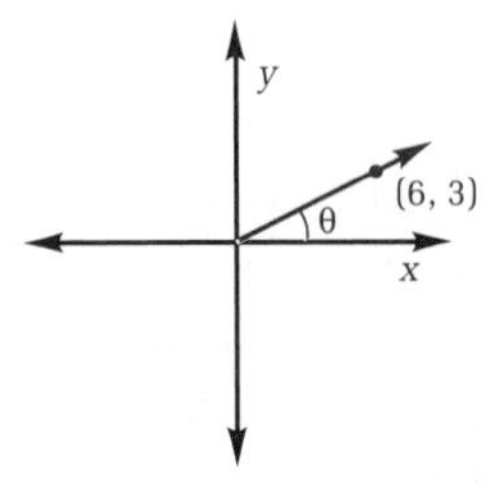

P **8** In what quadrant is the terminal ray of angle θ if θ is in standard position?

a sin θ > 0, cos θ > 0 I
b sin θ > 0, tan θ < 0 II
c sec θ < 0, csc θ < 0 III
d tan θ < 0, sec θ > 0 IV

P,R **9** Let (−7, 24) be a point on the terminal ray of an angle θ in standard position. Find the six trig functions of θ. $\sin\theta = \frac{24}{25}$; $\cos\theta = -\frac{7}{25}$; $\tan\theta = -\frac{24}{7}$; $\cot\theta = -\frac{7}{24}$; $\sec\theta = -\frac{25}{7}$; $\csc\theta = \frac{25}{24}$

P **10** Refer to the figure.

a Find cos θ, sin θ, and tan θ.
b Find the value of θ.

a $-\frac{6}{\sqrt{85}}; \frac{7}{\sqrt{85}}; -\frac{7}{6}$
b ≈130.60°

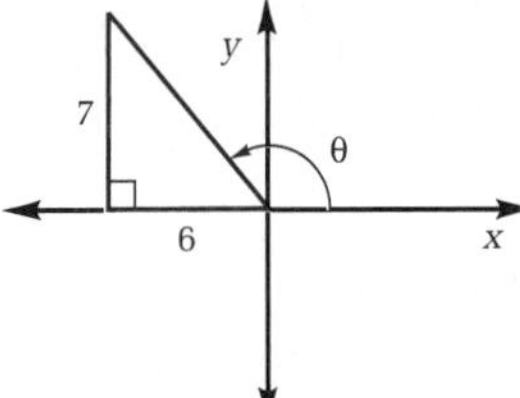

Problem Set

A

P **11** Let 0° < θ < 90°.

a If sec θ = 5, what is csc θ? $\frac{5}{2\sqrt{6}}$

b If $\tan\theta = \frac{1}{3}$ and $\sin\theta = \frac{1}{\sqrt{10}}$, what is cos θ? $\frac{3}{\sqrt{10}}$

P **12** If θ = 208°, what are the coordinates of point A on the unit circle? (≈−0.88, ≈−0.47)

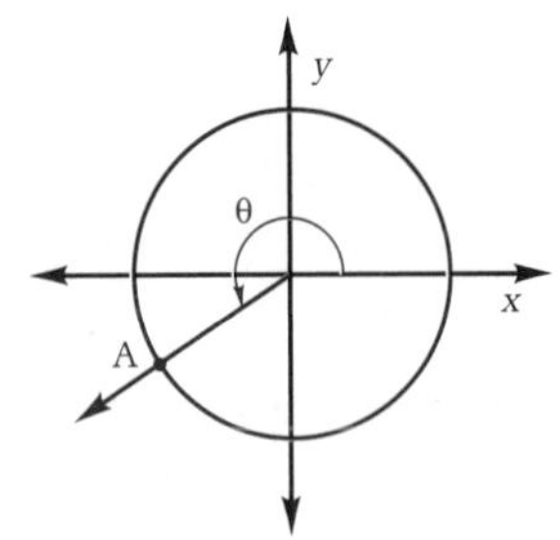

P,I **13** Which of the six trigonometric functions have undefined values? For what angles are each of these undefined? Tangent and secant for {. . . , −270°, −90°, 90°, 270°, . . . }, cotangent and cosecant for {. . . , −360°, −180°, 0°, 180°, 360°, . . . }

P,R **14** If $\tan\theta = 4$, what are the coordinates of point A?
(≈2.91, ≈11.64)

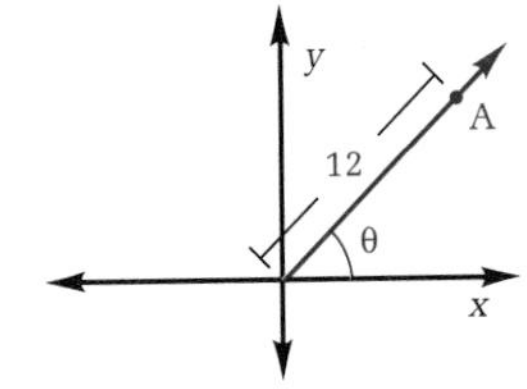

R **15** Solve for y, where $0° \le y \le 360°$.

a $\tan^2 y = 1$
45°, 135°, 225°, or 315°

b $\sec^2 y = 1$
0°, 180°, or 360°

c $\sin y = \cos y$
45° or 225°

P,R **16** Angle $\theta = 35°$.

a Find the coordinates of point A.

b Find the area of △ABC.

a (≈8.19, ≈5.74)
b ≈23.49

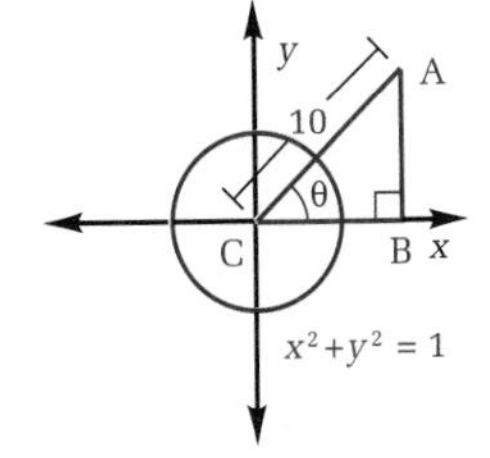

P **17** If $\cot\alpha = \frac{2}{3}$ and α is in the third quadrant, what is sin α? $-\frac{3}{\sqrt{13}}$

P,R **18** In the diagram, $\alpha = \angle POR$, $\beta = \angle QOR$, and the y-axis bisects $\overline{QP}$.

a What is the relationship between sin α and sin β? Equal

b What is the relationship between cos α and cos β?

c What is the relationship between α and β? Supplementary

d What is the measure of θ? α or 180 − β

b Equal in absolute value but different in sign

P,R,I **19** Find an expression in terms of a and b for each of the following.

a $\tan\theta$ $\frac{b}{a}$

b $\sin(180° - \theta)$ b

c $\cos(360° - \theta)$ a

d $\tan(\theta + 180°)$ $\frac{b}{a}$

e $\cos(180° + \theta)$ −a

f $\sin(360° - \theta)$ −b

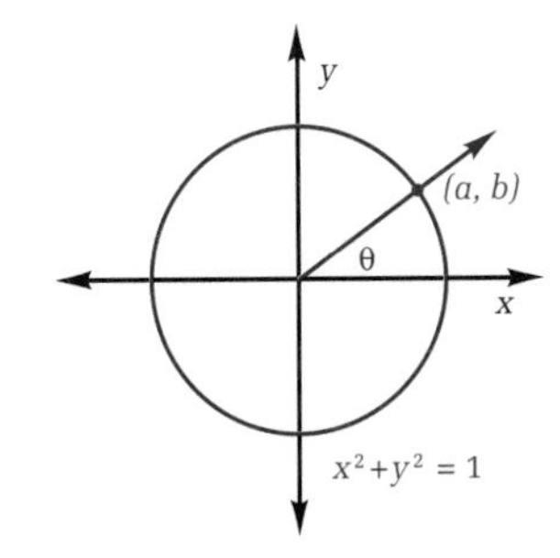

P **20** Find all values of α for which $-360° < \alpha < 360°$ and

a $\cos\alpha = \frac{-\sqrt{2}}{2}$
−225°, −135°, 135°, 225°

b $\tan\alpha = -1$
−225°, −45°, 135°, 315°

Problem-Set Notes and Strategies, continued

- Problem **14** may take some time until students make the connection between the tangent function and the x and y values. Then, armed with the Pythagorean Theorem, the solution should go smoothly.
- Answers to problem **15a** and **b** may be incomplete if students forget the ± when taking the square root.
- Some geomety may be necessary to establish the relationships between α, β, and θ in problem **18**.

Problem Set, *continued*

P,R **21** Find the exact coordinates of point Q. $(-4, 4\sqrt{3})$

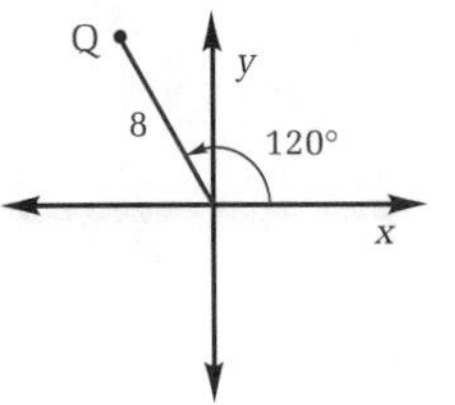

B

P,R **22** If $0° \le \theta \le 360°$, what values of θ satisfy each equation?

a $\cos\theta = \frac{1}{2}$ 60° or 300°

b $\cos^2\theta = \frac{1}{2}$ 45°, 135°, 225°, or 315°

c $\cos^2\theta - 1 = 0$ 0°, 180°, or 360°

d $\cos^2\theta - \cos\theta = 0$ 0°, 90°, 270°, or 360°

P **23** Sketch a large copy of the unit circle shown and label the coordinates of each indicated point on the circle. Explain why you need only determine the coordinates of the points in the first quadrant in order to identify the coordinates of the others.

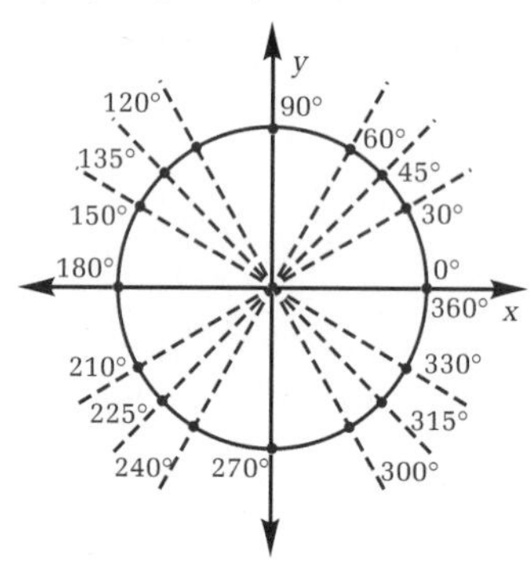

P,R **24** Solve each equation for the indicated variable.

a $2S^2 + S - 1 = 0$ for S $S = \frac{1}{2}$ or $S = -1$

b $2(\sin x)^2 + \sin x - 1 = 0$ for sin x (not for x) $\sin x = \frac{1}{2}$ or $\sin x = -1$

c $2\sin^2 x + \sin x - 1 = 0$ for x, where $0° \le x \le 360°$ $x = 30°$, $x = 150°$, or $x = 270°$

P **25** Find the exact values of sec α for which $\tan\alpha = \frac{12}{13}$. $\frac{\sqrt{313}}{13}$ or $-\frac{\sqrt{313}}{13}$

I **26** Circle O has a radius of 6, and $\sin\angle ACB = 0.87$. Find the ratio of the area of $\triangle ADC$ to the area of $\triangle ADB$. ≈ 0.32

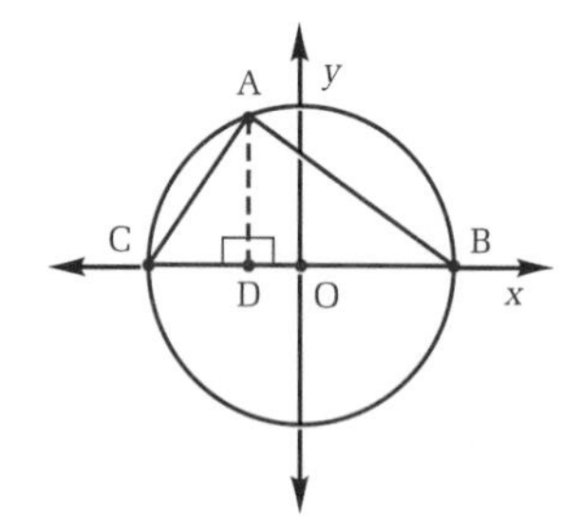

P,R **27** Find, to the nearest tenth, all angles that satisfy the given conditions.

a $-360° < \alpha < 360°$ and $\cos\alpha = 0.88$ $\approx -331.6°$, $\approx -28.4°$, $\approx 28.4°$, $\approx 331.6°$

b $-360° < \theta < 360°$ and $\csc\theta = 10$ $\approx -354.3°$, $\approx -185.7°$, $\approx 5.74°$, $\approx 174.26°$

Problem-Set Notes and Strategies, continued

- Urge students to keep the sketch of problem **23** for reference and review. You might have a student make a large poster for the classroom with this sketch on it.

Additional Answer

23

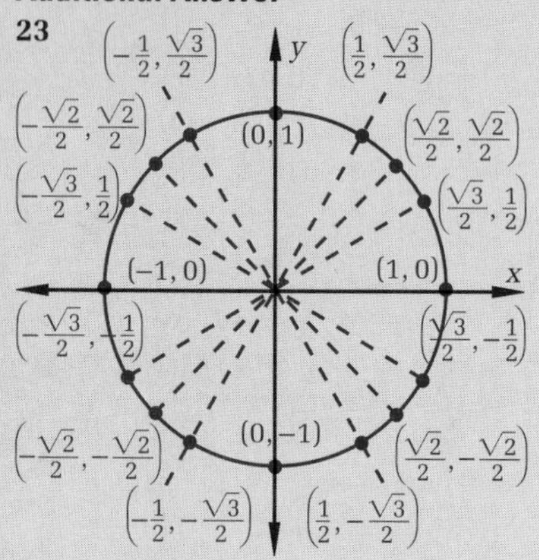

The coordinates of the points in Quadrants II, III, and IV are identical in absolute value to those of the corresponding points in Quadrant I, and the signs of the coordinates can be determined by the points' directions from the x- and y-axis.

- Going through the steps of problem **24** illustrates how algebra and trigonometry are both necessary to solve many trigonometric equations. Many more problems of this type will be looked at later in the chapter.

- Some geometric concepts are necessary to solve problem **26**. The triangle trigonometric ratios presented formally in Section **14.5** can emerge.

I | **28** Chuck owns a lot that borders Dempster Street and Benson Avenue.

a What is the frontage of his lot on Benson Avenue? ≈ 386.37 ft

b What is the frontage of his lot on Dempster Street? ≈ 373.21 ft

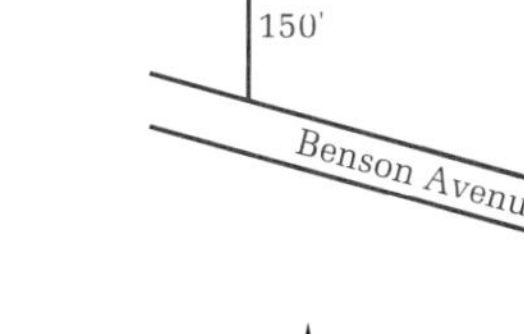

C P,I | **29** MNPQ is a square. If AO = 4 and $\theta = 145°$, what is the ratio MA:AQ? ≈ 0.18

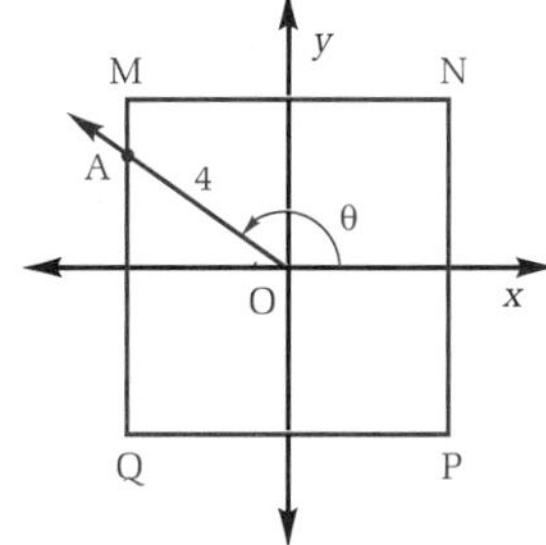

I | **30** The area of a triangle can be computed if the measures of two sides (a and b) and the included angle (θ) are known.

a Find the area, to the nearest hundredth, of a triangle with sides 8 and 18 and included angle 70°. ≈ 67.66

b At what angle, to the nearest degree, should the same two sides intersect so that the area would be 50? $\approx 44°$ or $\approx 136°$

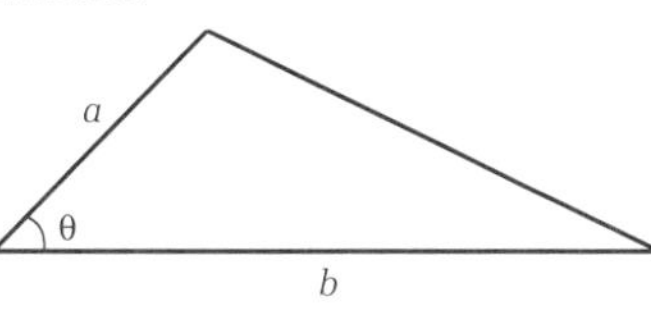

P | **31** Find the exact value of the area of $\triangle OAB$. $9\sqrt{3}$

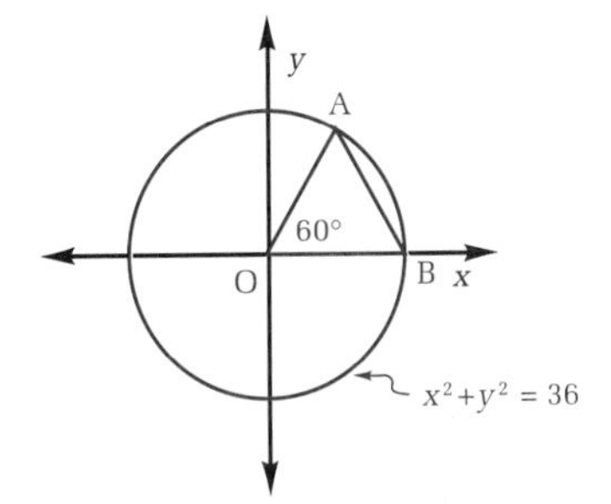

P,I | **32** After a storm, Pete noticed that a tree had been blown 20° from the vertical and that when he stood directly under the top of the tree, he was 3 meters from its base. How tall had the tree been? How far above the ground was the top of the tree after the storm? ≈ 8.77 m; ≈ 8.24 m

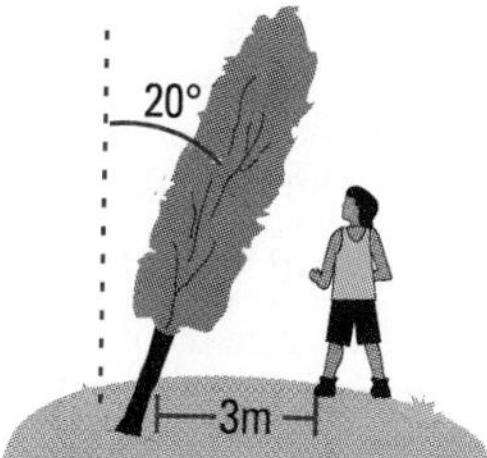

R | **33** Express each of the six trig functions of x, where $0° < x < 90°$, in terms of sin x. For example, $\cot x = \frac{\cos x}{\sin x} = \frac{\sqrt{1 - \sin^2 x}}{\sin x}$.
$\sin x = \sin x$; $\cos x = \sqrt{1 - \sin^2 x}$; $\tan x = \frac{\sin x}{\sqrt{1 - \sin^2 x}}$; $\cot x = \frac{\sqrt{1 - \sin^2 x}}{\sin x}$; $\sec x = \frac{1}{\sqrt{1 - \sin^2 x}}$; $\csc x = \frac{1}{\sin x}$

Problem-Set Notes and Strategies, continued

- Some auxiliary lines in the illustration for problem **28** will make the problem easier to solve. Students often can see how the triangle ratios develop, building a foundation for later presentation.
- Problem **30** may require subdividing the triangle and a few intermediate calculations until a formula can be developed. Once the formula has been discovered, it is a straightforward use of a calculator. This holds true for part **b** as well although an inverse trig function is necessary and yields 2 answers.
- Some students may remember the formula for the area of an equilateral triangle from geometry. If not, now is a good time to generate it again.
- A picture can be a problem-solver boost and in the case of problem **32** it may be worth drawing to help find the answer. Encourage students to draw pictures whenever possible.

14.4

Class Planning

Time Schedule

Average: 2 days

Advanced: $1\frac{1}{2}$ days

Honors: $1\frac{1}{2}$ days

Resource References

Teacher's Resource Book

Class Opener 84

Cooperative Learning 43

Practice 89

Evaluation

Quiz Form 2

Class Opener

Draw the ray from the origin to the point (–7, 24). Determine the smallest positive angle and smallest negative angle made by this ray and the positive x-axis.

106° 15′ 37″ ; –253° 44′ 23″

Lesson Notes

- You may want to point out that in order for the inverse functions to be defined as functions, the domain of the trigonometric functions must be restricted. If the domain is all real numbers, the results are not unique. Calculators and tables use the appropriate restrictions. Students may need to memorize the restrictions and realize that a problem may have an infinite number of solutions, but a table or calculator will yield a single answer. Using a function with a capital *A*, such as *Arcsin x*, implies a function and the answer must be unique and therefore between –90° and 90°.

14.4 Inverse Trigonometric Functions

Objective

After studying this section, you will be able to

- Compute the inverse trigonometric functions

Part One: Introduction

In earlier sections, you have seen that many angles can have the same terminal ray. Therefore, many angles can have the same sine and cosine.

Example *Let θ be a first-quadrant angle with $\sin \theta = \frac{1}{2}$. Determine the measure of θ.*

The obvious answer is $\theta = 30°$, but this is *not* the only answer. For example, 390°, −330°, 750°, and infinitely many others all have the same terminal ray. Because they differ by multiples of 360°, we can write

$$\theta = 30° + n \cdot 360°$$

where n is any integer.

In the previous example we restricted θ to the first quadrant. What if we remove the restriction?

Example *Find all angles α such that $\sin \alpha = \frac{1}{2}$.*

Two points on the unit circle have y-coordinate $\frac{1}{2}$. The angles corresponding to those points include 30°, 150°, and any angle of the form $30° + n \cdot 360°$ or $150° + n \cdot 360°$, where n is any integer.

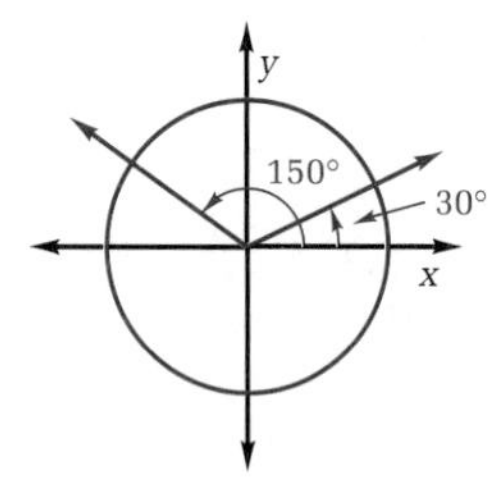

An equation like $\sin \theta = 0.6$ has infinitely many solutions for θ. In most cases, however, one solution is enough. If you use the INV sin keys, the calculator will always display one angle between −90° and 90°. We will define Arcsin, Arccos, and Arctan as the respective ***inverse functions*** of sine, cosine, and tangent. Recall that a function yields one output for each input. Therefore, the output of each function is one angle within the following ranges.

Vocabulary

inverse trigonometric function

$$-90° \le \text{Arcsin}(x) \le 90°$$
$$0° \le \text{Arccos}(x) \le 180°$$
$$-90° < \text{Arctan}(x) < 90°$$

Try some values on your calculator and see what happens.

Example *Find the exact value of* $\text{Arcsin}\left(\frac{1}{2}\right) + \text{Arccos}\left(\frac{1}{2}\right)$.

$\text{Arcsin}\left(\frac{1}{2}\right) + \text{Arccos}\left(\frac{1}{2}\right) = 30° + 60° = 90°$

Stumbling Block

It is easy to say that all problems have the restriction from −90° to 90°. This however is incorrect. Go back to the unit circle and the quadrants where the *x* and *y* coordinates are positive and negative. Both positive and negative values must be included in the inverse functions. This idea, coupled with the fact that the sine is the *y*-coordinate and the cosine is the *x*-coordinate, should make it easier to remember the quadrants where the inverse trigonometric functions are defined.

Part Two: Sample Problems

Problem 1 *Angle* θ *contains* $(-3, -3\sqrt{3})$. *Find all possible values of* θ.

Solution $\tan\theta = \frac{-3\sqrt{3}}{-3} = \sqrt{3}$, so θ has reference angle 60°. But $(-3, -3\sqrt{3})$ is in Quadrant III. The angle in Quadrant III is 240°. So $\theta = 240° + n \cdot 360°$.

Problem 2 *Romeo and Juliet met but had an argument. Romeo drove due east at 40 mph, and Juliet drove approximately northeasterly at 52 mph. After exactly 45 minutes, each stopped and looked for the other. Amazingly, Juliet was exactly due north of Romeo. To the nearest degree, at what angle from due east did Juliet travel?*

Solution In 45 minutes, Romeo traveled 30 miles, while Juliet traveled 39 miles. Let the positive x-axis represent Romeo's direction, and let P represent Juliet's stopping point. The x-coordinate of P is 30, and the distance from the origin is 39. So $\cos\theta = \frac{x}{r} = \frac{30}{39}$ and $\text{Arccos}\left(\frac{30}{39}\right) \approx 40°$. Juliet traveled about 40° north of due east.

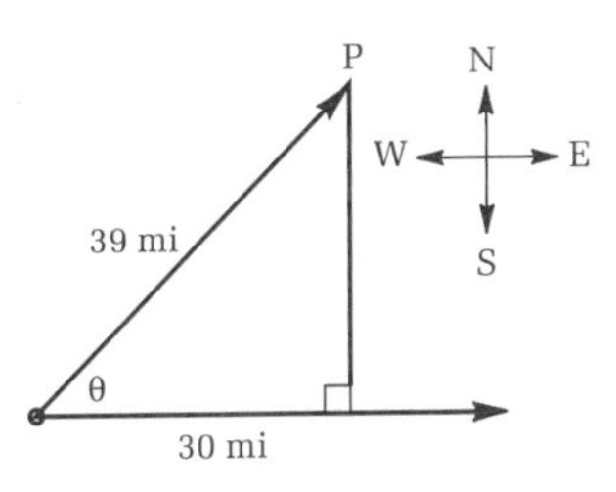

Problem 3 *Find each of the following.*

a *Arccos* $\frac{\sqrt{3}}{2}$ **b** *arccos* $\frac{\sqrt{3}}{2}$

Solution

a Recall that Arccos $\frac{\sqrt{3}}{2}$ refers to a function. We are looking for one angle in the first quadrant whose cosine is $\frac{\sqrt{3}}{2}$. Arccos $\frac{\sqrt{3}}{2} = 30°$.

b This is *not* the same as part **a.** The lowercase *a* in *arccos* indicates that we seek a *relation*, not a function. In this case, we are looking for all the first-quadrant and all the fourth-quadrant angles whose cosine is $\frac{\sqrt{3}}{2}$. Thus, arccos $\frac{\sqrt{3}}{2} = 30° + n \cdot 360°$ or $330° + n \cdot 360°$, where *n* is any integer.

Checkpoint

1. Find Arccos $\left(\frac{1}{2}\right)$. 60°.
2. True or False, Arctan (−1) = 225°. False, it is −45°
3. Find Arcsin (sin 150°). 30°
4. Find Arccos (tan 315°). 180°

14.4

Assignment Guide

Average

Day 1 6–8, 10, 12, 14, 16–18, 20, 26, 27

Day 2 5, 9, 13, 19, 21, 24a, b, c, 25, 29, 31

Advanced

Day 1 ($\frac{1}{2}$ day) Section 14.3 13, 14, 20, 24, 25, 27b, 32, 33 ($\frac{1}{2}$ day) 6–9, 14, 16

Day 2 5, 10, 13, 17, 18, 22, 24, 27, 29, 31, 35, 39

Honors

Day 1 ($\frac{1}{2}$ day) Section 14.3 13, 19, 25–28, 33 ($\frac{1}{2}$ day) 7–9, 14–16

Day 2 17, 18, 22–24, 28–30, 33, 35, 37, 38

Problem-Set Notes and Strategies

- Problem **3** is an exercise in the restrictions of the inverse functions. A discussion on part **d** may help students.
- Point out the difference between the answers to parts **a** and **b**, and **c** and **d** respectively in problem **4**. With the capital letter, uniqueness and a function exists.

Part Three: Exercises and Problems

Warm-up Exercises

P **1** Which of the following denote the set of all angles for which $\tan \theta = -1$? a and b

a $\{\theta:\theta = 135° + (360k)°$ or $\theta = 315° + (360k)°$, k is an integer$\}$

b $\{\theta:\theta = 135° + (180k)°$, k is an integer$\}$

c $\{\theta:\theta = 45° + (90k)°$, k is an integer$\}$

P **2** Write single expressions, in terms of integers k, to represent every angle coterminal with those in the diagrams.

a

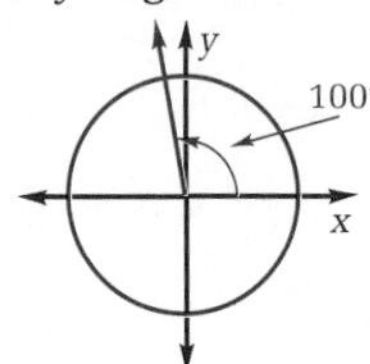

$100° + 360° \cdot k$

b

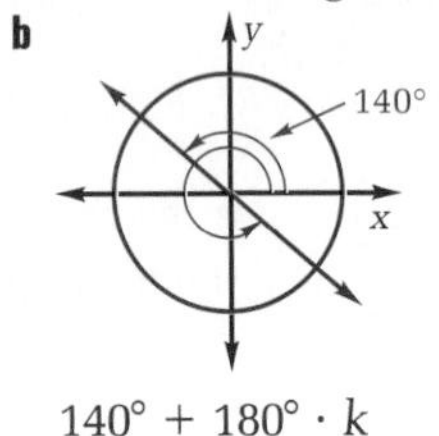

$140° + 180° \cdot k$

c

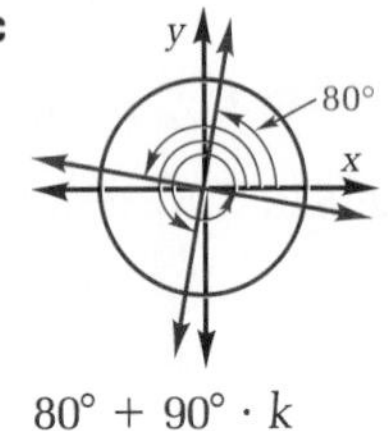

$80° + 90° \cdot k$

P **3** **a** Find Arccos (cos 60°). 60°

b Find Arccos (cos 120°). 120°

c Find Arccos (cos 225°). 135°

d Does Arccos (cos x) always equal x? Why or why not?
No; the Arccos function's range is restricted to angles from 0° to 180°.

P **4** Find each of the following.

a Arccos $\frac{\sqrt{2}}{2}$ 45°

b arccos $\frac{\sqrt{2}}{2}$ $45° + 360° \cdot n$ or $315° + 360° \cdot n$

c Arcsin $\frac{\sqrt{3}}{2}$ 60°

d arcsin $\frac{\sqrt{3}}{2}$ $60° + 360° \cdot n$ or $120° + 360° \cdot n$

Problem Set

A

P,R **5** Evaluate each of the following.

a sin (Arccos 0.67) ≈0.7424

b csc [Arcsin (−0.62)] ≈−1.6129

P **6** **a** List five positive angles coterminal with 80°. 440°, 800°, 1160°, 1520°, 1880°

b List four negative angles coterminal with 80°.
−280°, −640°, −1000°, −1360°

P **7** Find each of the following.

a Arctan −1 −45°

b arctan −1 $135° + 180° \cdot n$

P,R **8** **a** Find $\tan \theta$. $\frac{7}{2}$

b Find θ to the nearest hundredth of a degree.
≈74.05°

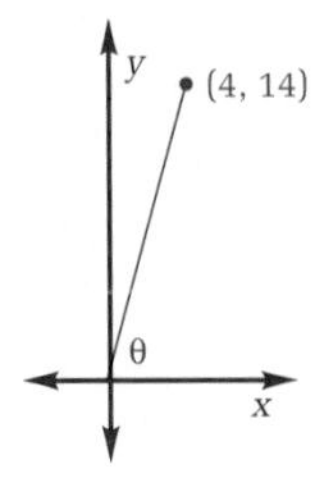

P | **9** Find all positive measures of θ less than 1000° for which $\cos\theta = \frac{1}{2}$. 60°, 300°, 420°, 660°, 780°

P,R | **10** Find all angles θ, where $0° \le \theta \le 360°$, that make each equation true.

a $\sin\theta = 0.75$ 48.59° or 131.41°

b $\sec\theta = 2.45$ 65.91° or 294.09°

c $\cos\theta = -0.43$ 115.47° or 244.53°

d $\csc\theta = \frac{2\sqrt{3}}{3}$ 60° or 120°

P | **11** Arc-function outputs are usually restricted to angles terminating in certain quadrants in order to have a connected region over which all values of the function—both positive and negative—may be obtained. Copy each diagram and write the name of an Arc function near the diagram to which it applies.

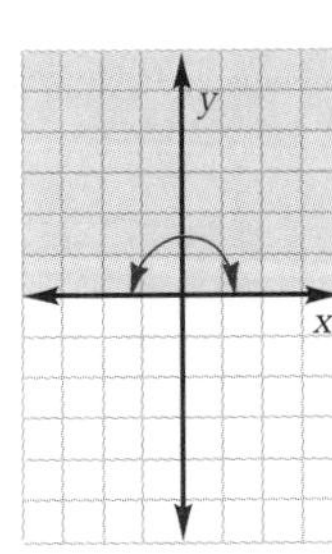

Arccos

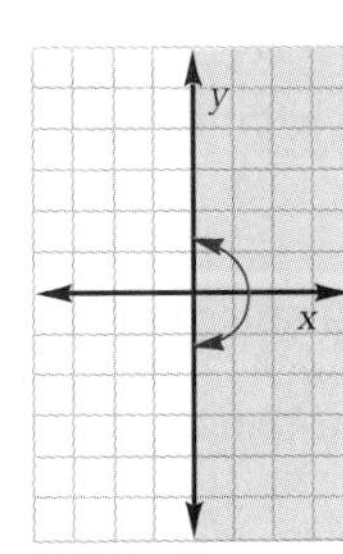

Arcsin

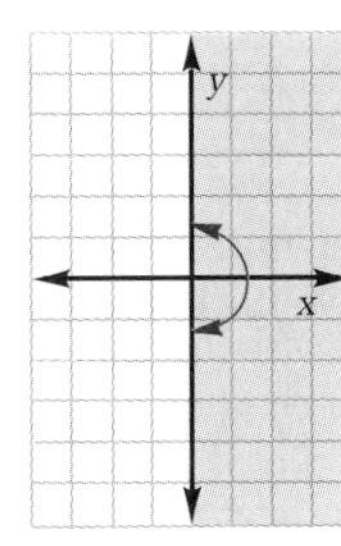

Arctan

P | **12** Angle θ contains (6, 8). Find all possible values of θ.
$\theta \approx 53.13° + n \cdot 360°$

P | **13** Find all θ, where $0° < \theta < 360°$, for which $|\sin\theta| = \frac{\sqrt{3}}{2}$. 60°, 120°, 240°, 300°

B

P | **14** Evaluate.

a 0.63 [INV] [cos] ≈50.95°

b 0.63 [INV] [cos] [sin] ≈0.7766

c Write trigonometric expressions to describe parts **a** and **b.**
Arccos 0.63; sin (Arccos 0.63)

P | **15** $\sin A = \frac{9}{13}$ and $\sin B = \frac{3}{13}$.

a Find $\frac{\sin A}{\sin B}$. 3

b Find $\frac{A}{B}$, where A and B are both acute. ≈3.2838

P | **16** Find the exact value of $\text{Arccos}\left(-\frac{1}{2}\right) + \text{Arcsin}\left(\frac{\sqrt{3}}{2}\right)$. 180°

R | **17** Find θ. ≈38°

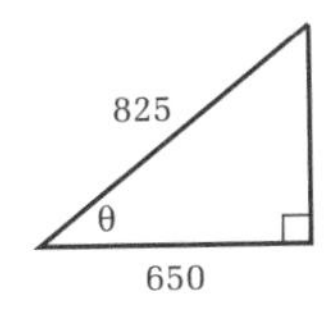

P | **18** Evaluate.

a Arccos (cos 140°) 140°

b Arctan (tan 210°) 30°

c Arcsin (cos 60°) 30°

d Arccos (sin 120°) 30°

e Arcsin (cos 120°) −30°

Problem-Set Notes and Strategies, continued

- If students will remember the pictures in problem **11**, the restrictions necessary for the inverse functions may be easier to remember.
- Problem **14**, especially part **c,** is an exercise in the order of operations necessary to communicate with your calculator.
- A drawing may be helpful for problems **18** and **19**. Any time the inverse trig function is evaluated, an angle is the result.

Problem-Set Notes and Strategies, continued

- One possible extension to generalize problem **20** is to ask what the relationship of 24π is to the problem.

- A picture may make the solution to problem **22** clearer.

- Urge students to draw some triangles for problem **24**. See the cooperative learning exercises for more work on problems of this type.

Cooperative Learning

Problem **24** provides an opportunity for students to transform trig and inverse trig functions. Have students evaluate several trig functions where the arguments are not standard angles, i.e. Find sin (Arccos $\frac{3}{5}$). Students need to realize that Arccos ($\frac{3}{5}$) is an angle and may need to draw a picture to find the sine. Next, ask students to find sin (Arccos x) where x is appropriately restricted to either positive or negative numbers. These concepts will be used in later math classes and help establish the relationship between algebra and trigonometry.

Problem Set, *continued*

P **19** Evaluate.

a $\cos\left(\text{Arccos } \frac{1}{2}\right)$ $\frac{1}{2}$ **b** Arccos (cos 300°) 60°

R **20** **a** Find the length of arc AB if $\theta = 120°$. 8π
b Find the length of arc AB if $\theta = 480°$. 8π
c By what amount does your answer to part **b** exceed your answer to part **a?** 0

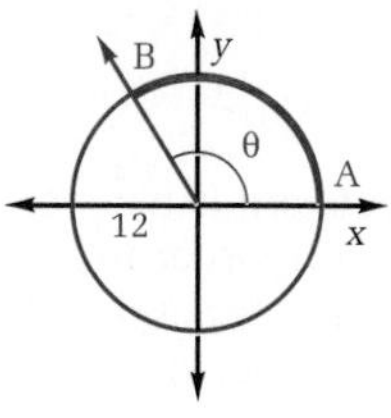

P **21** tan A = 6 and tan B = 3, where A and B are acute.

a Find $\frac{\tan A}{\tan B}$. 2 **b** Find $\frac{A}{B}$. ≈1.1254

P  **22** Bonnie and Clyde left town. Bonnie drove due west at 55 miles per hour and Clyde drove northwesterly at 64 miles per hour. After 3 hours Clyde was directly north of Bonnie. To the nearest second, what angle did Clyde's path form with Bonnie's? ≈30°45′13″

P  **23** Find each of the following.

a sin (Arccos 0.65) ≈0.7599 **b** cos [Arctan (−4.62)] ≈0.2116
c sec (2 Arcsin 0.39) ≈1.4372 **d** tan [Arccos (−0.43)] ≈−2.0996

P **24** Find the exact value of each of the following.

a $\sin\left(\text{Arccos } \frac{1}{2}\right)$ $\frac{\sqrt{3}}{2}$ **b** $\cos\left(2 \text{ Arcsin } \frac{\sqrt{2}}{2}\right)$ 0

c tan [Arctan (−1)] −1 **d** $\sec\left[\text{Arcsin}\left(-\frac{1}{2}\right)\right]$ $\frac{2}{\sqrt{3}}$

e $\sin\left[\text{Arccos}\left(-\frac{\sqrt{3}}{2}\right)\right]$ $\frac{1}{2}$

P,R  **25** Find an expression, using the variable n to stand for any integer, that gives all angles θ for which

a $\cos\theta = \frac{2}{3}$ **b** $\tan\theta = -1$ $180° \cdot n + 135°$

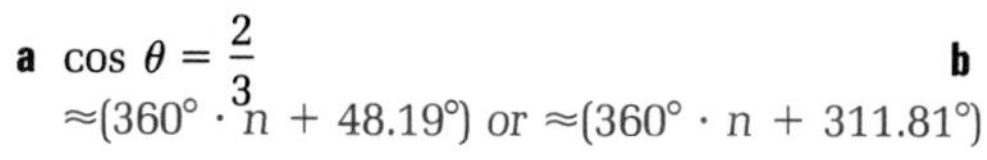
$\approx(360° \cdot n + 48.19°)$ *or* $\approx(360° \cdot n + 311.81°)$

P **26** Evaluate Arccos 0.0123 to the nearest minute. ≈89°18′

P **27** Find the measure of angle θ if $90° \le \theta \le 180°$. ≈143.14°

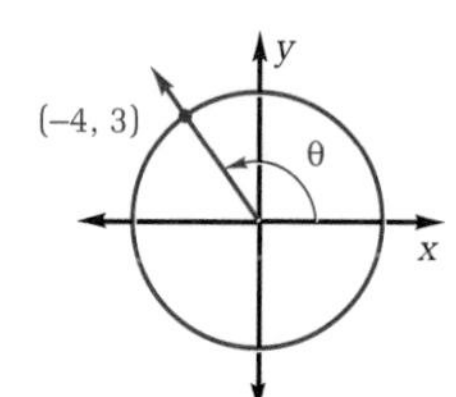

P | **28** Find all values of θ between 0° and 720° for which $\sin(\theta - 20°) = \frac{1}{2}$. 50°, 170°, 410°, 530°

P,I | **29** **a** Find the value of the expression tan (arctan 2 + arctan 3). -1

b Find the value of $\frac{2 + 3}{1 - 2 \cdot 3}$. -1

c What do you notice about these two values? The values are equal.

P | **30** **a** Solve $10(\sin x)^2 + (\sin x) - 3 = 0$ for sin x.

b Solve the equation in part **a** for x, where $0° \le x \le 360°$.

a $\sin x = -\frac{3}{5}$ or $\sin x = \frac{1}{2}$

b $x = 30°$, $x = 150°$, $x \approx 216.87°$, or $x \approx 323.13°$

P | **31** The terminal ray of angle θ contains $(1, \sqrt{3})$. Find all possible values of θ.
$60° + n \cdot 360°$

P,R | **32** Evaluate cos (arctan 2 + arctan 3). $-\frac{\sqrt{2}}{2}$, or ≈ -0.7071

P,R | **33** Find θ. $\approx 63.43°$

θ 90 150 120

P | **34** Find all values of θ for which $\tan(\theta + 10°) = 1$. $35° + 180° \cdot n$

P | **35** Without using a calculator, evaluate Arccos 0.643 + Arcsin 0.643 to the nearest minute. 90°0′

C | **36** Using a calculator, try different values of x in the function $f(x) = \arccos(\sec x)$. For what values is the function defined? Why is this so? sec $x \ge 1$ for all values of x, and the domain of the arccos function is between -1 and 1 inclusive, so arccos (sec x) exists only for values of x for which sec $x = \pm 1$.

R | **37** The length of arc BAC is 16.1 units. Find the measure of angle θ. ≈ 230.62

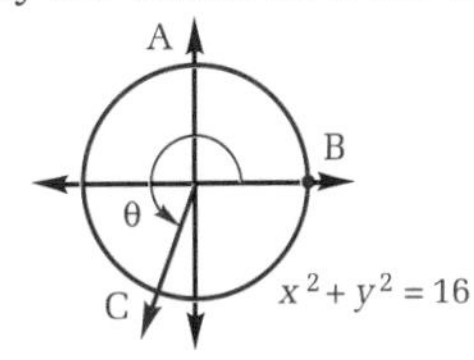

R | **38** Find an angle θ, where $0° \le \theta \le 360°$, such that cos θ is twice as large as sin θ. $\approx 26.57°$ or $\approx 206.57°$

R | **39** **a** Find θ in $\triangle ABC$.

b Find DB.

a $\approx 33.69°$

b ≈ 3.328

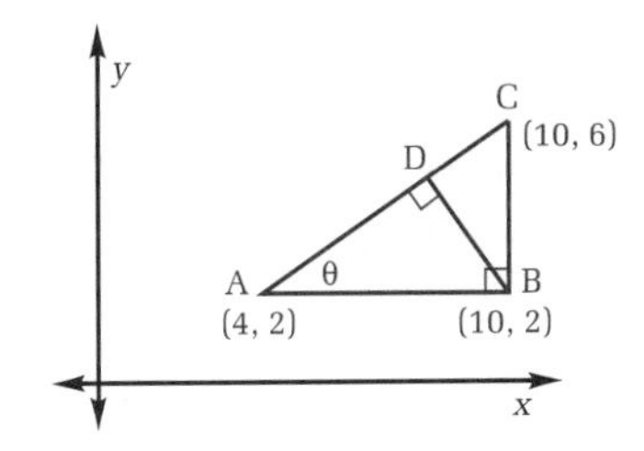

Problem-Set Notes and Strategies, continued

- Problem **29** introduces a relationship for the tangent of a sum. This identity will be explored further in Chapter **15**.
- Problem **33** may take some work. An auxiliary line can help.
- Students who understand the complementary functions should be able to proceed with problem **35**. Some students will need to evaluate in order to see the relationship.
- Problem **38** can be done several ways. One way is to evaluate the Arctan 0.5. Students will proceed differently depending on the familiarity they have with the identities and definitions.
- A little geometry will be useful in problem **39**. Similar triangles and corresponding ratios can be used.

Communicating Mathematics

You and your brother are both taking trigonometry this year. He was on a field trip today and missed his trig class describing inverse trigonometric functions. In a paragraph, describe the essence of today's lesson, with all formulas and necessary restrictions for memorizing.

14.5

Class Planning

Time Schedule
All levels: 2 days

Resource References
Teacher's Resource Book
Class Opener 85
Practice 90
Enrichment 28
Transparency 28
Evaluation
Quiz Form 3

Class Opener

Use what you know about trig to determine the measure of the largest acute angle of an 8-15-17 triangle.
61° 55′ 39″ If students work for a while and are still frustrated, suggest they use graph paper. This should give them the hint to put the angle at the origin.

Lesson Notes

- In defining ratios in terms of right triangles, they remain unchanged from preceding sections. The method of determining the lengths of the sides is the only element that changes. However, now opposite and adjacent sides must be identified as well as the hypotenuse.
- Now, the relationship between cofunctions and complementary angles is formalized and students can organize the relationships. If a student knows that the sine of 30° is $\frac{1}{2}$, they now also know that the cosine of 60° is $\frac{1}{2}$.

14.5 Right Triangles

Objectives
After studying this section, you will be able to
- Define the trigonometric functions in terms of right triangles
- Use the cofunction identities

Part One: Introduction

Trigonometric Functions and Right Triangles

The three sides of a right triangle can be classified in terms of either acute angle. Refer to △ABC in the diagram.

$\overline{AB}$ is the hypotenuse.
$\overline{BC}$ is the leg opposite ∠A.
$\overline{AC}$ is the leg adjacent to ∠A.

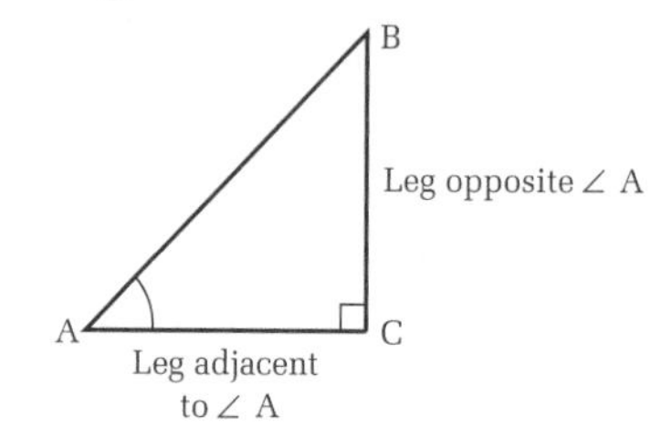

Example *Refer to the triangle in the diagram and identify each of the following.*

a *The hypotenuse*
$\overline{PQ}$ is the hypotenuse.

b *The leg adjacent to ∠P*
$\overline{PR}$ is the leg adjacent to ∠P.

c *The leg opposite ∠Q*
$\overline{PR}$ is the leg opposite ∠Q.

d *The leg opposite ∠P*
$\overline{QR}$ is the leg opposite ∠P.

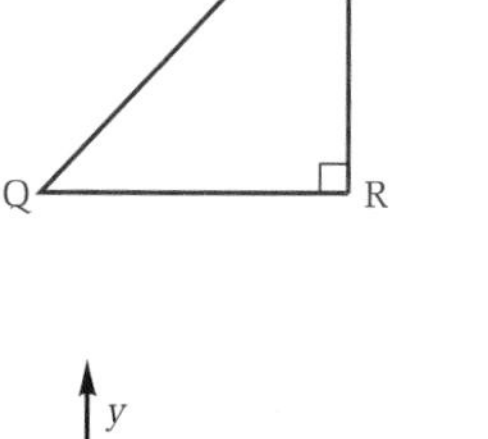

By our previous definitions, in right triangle ABC on the coordinate plane, $\sin \angle A = \frac{BC}{AC}$, $\cos \angle A = \frac{AB}{AC}$, and so on.

Now, let's remove the axes. Since the angles and sides are unchanged, the values of the sine and cosine shouldn't change. Furthermore, the position of the triangle shouldn't matter.

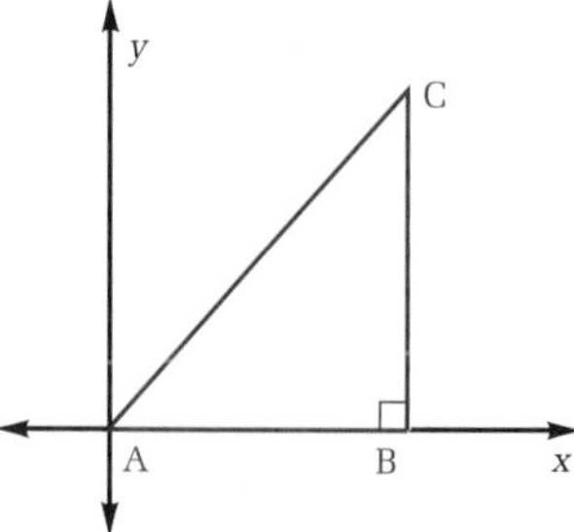

Vocabulary
cofunction

We define the following trigonometric relationships for an acute angle θ in a right triangle.

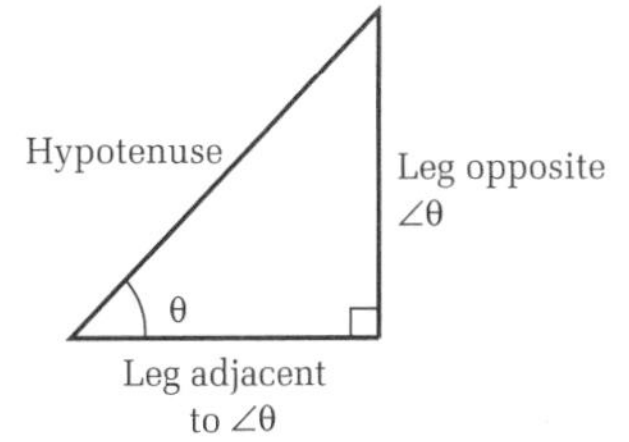

$$\sin \theta = \frac{\text{leg opposite } \theta}{\text{hypotenuse}} \qquad \cos \theta = \frac{\text{leg adjacent to } \theta}{\text{hypotenuse}} \qquad \tan \theta = \frac{\text{leg opposite } \theta}{\text{leg adjacent to } \theta}$$

From these we can write the definitions for cot θ, sec θ, and csc θ.

Example *Refer to the right triangle and evaluate each of the following.*

a *cos ∠P*

By the Pythagorean Theorem, PR = 13.

$$\cos \angle P = \frac{\text{adjacent leg}}{\text{hypotenuse}} = \frac{5}{13}$$

b *tan ∠R*

$$\tan \angle R = \frac{\text{opposite leg}}{\text{adjacent leg}} = \frac{5}{12}$$

c *sin ∠R*

$$\sin \angle R = \frac{\text{opposite leg}}{\text{hypotenuse}} = \frac{5}{13}$$

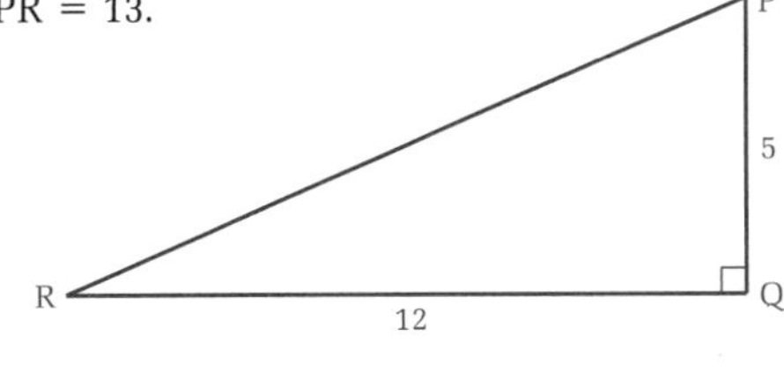

Cofunction Identities

In the preceding example cos ∠P and sin ∠R had the same value and ∠P and ∠R were complementary. Previously, we saw that $\cos 30° = \sin 60° = \frac{\sqrt{3}}{2}$ and $\cos 60° = \sin 30° = \frac{1}{2}$. In general, we have the following theorem.

 ***If two angles α and β are complementary, then $\sin \alpha = \cos \beta$.***

We can see this by looking at a right triangle. The two acute angles are complementary, and the leg *opposite* one acute angle is *adjacent* to the other acute angle.

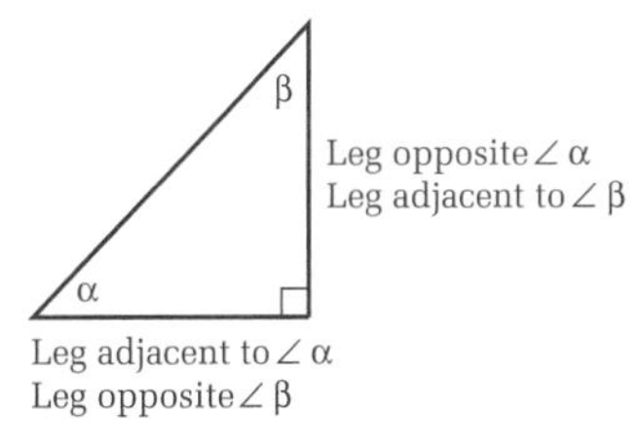

The following theorem expresses a similar relationship.

 ***If two angles α and β are complementary, then $\tan \alpha = \cot \beta$ and $\sec \alpha = \csc \beta$.***

The pairs of functions that have the same values for complementary angles are sine and **co**sine, tangent and **co**tangent, and secant and **co**secant. These pairs are called ***cofunctions.***

Here is another way of expressing these cofunction theorems.

$$\cos \theta = \sin (90° - \theta) \qquad \cot \theta = \tan (90° - \theta) \qquad \csc \theta = \sec (90° - \theta)$$

Checkpoint

1. A right triangle has sides 3, 4 and 5. Find the sine of the angle opposite the side of length 3. $\frac{3}{5}$
2. A ship leaves port and heads due north at 8 miles per hour at the same time that another ship leaves port and heads due west at 15 miles per hour. What is the angle from the second ship to the first ship measured from the line of sight to port? ≈ 28.1°
3. One angle in a right triangle is 37° and the hypotenuse is 10. Find the other angle and two sides. 53°, 6.02 and 7.99

Assignment Guide

Average	
Day 1	5, 6, 9, 11, 12, 14, 17, 19, 27
Day 2	7, 13, 16, 18, 21, 23, 25, 26, 31
Advanced	
Day 1	5, 6, 9, 11, 12, 14, 17, 19, 26, 27
Day 2	7, 10, 13, 18, 21, 23, 25, 30–32
Honors	
Day 1	6, 7, 11, 12, 14, 15, 17–19, 25, 26
Day 2	10, 13, 20–23, 29–31, 33, 34

Problem-Set Notes and Strategies

- If students understand the identities in problem **2**, they will be better prepared to proceed.

Part Two: Sample Problems

Problem 1 *A plane at a height of 5000 feet begins descending in a straight line toward a runway. The horizontal distance between the plane and the runway is 45,000 feet. What angle of depression must the plane use?*

Solution Since $\tan\theta = \frac{\text{opposite}}{\text{adjacent}} = \frac{5000}{45{,}000} \approx 0.1111$, Arctan $0.1111 \approx 6°$. The angle of depression and θ are congruent alternate interior angles. Therefore, the plane should use about a 6° angle of depression.

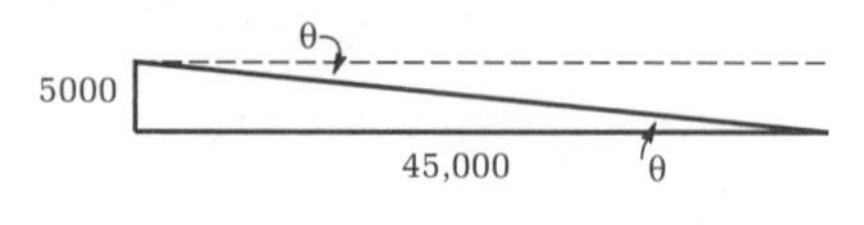

Problem 2 *Two forest-ranger stations are 20 miles apart on a straight east-west ranger road. A forest fire is spotted. Each station spots the fire and gives its angle from the other station. From station 1, the angle is 38° north of east. From station 2, the angle is 133° measured clockwise from west. How far is the fire from the ranger road?*

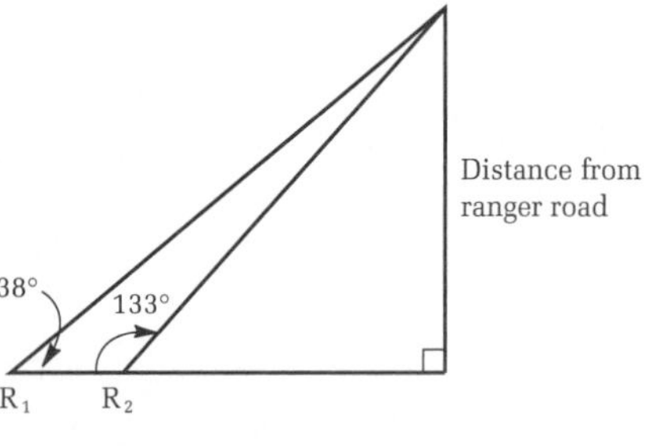

Solution There are two right triangles in the diagram, so we can get two equations. We have $\tan 38° = \frac{d}{x + 20}$ and $\tan 47° = \frac{d}{x}$. From the first equation, $x(\tan 38°) + 20(\tan 38°) = d$. From the second equation, $x = \frac{d}{\tan 47°} = d(\cot 47°)$. Substitute the second equation into the first, and solve.

$$d(\cot 47°)(\tan 38°) + 20(\tan 38°) = d$$
$$0.72856d + 15.6257 \approx d$$
$$d \approx 57.566 \text{ miles}$$

Part Three: Exercises and Problems

Warm-up Exercises

1 If $\cos \angle D = \frac{4}{5}$, what is EF? 45

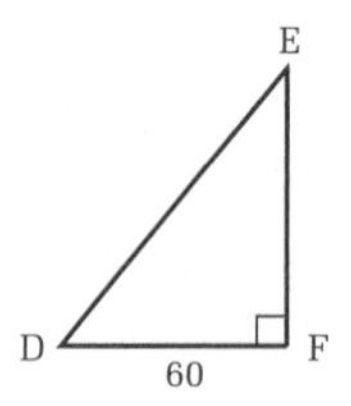

P **2** Simplify each expression to a single trigonometric function in a single variable.

a $\tan(90° - \theta)$ cot θ
b $\cos(90° - \theta)$ sin θ
c $\sec(90° - \theta)$ csc θ
d $\cot(90° - \theta)$ tan θ

P,R 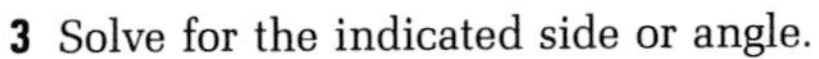

3 Solve for the indicated side or angle.

a

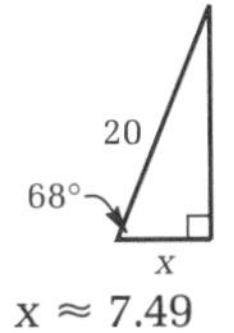

$x \approx 7.49$

b

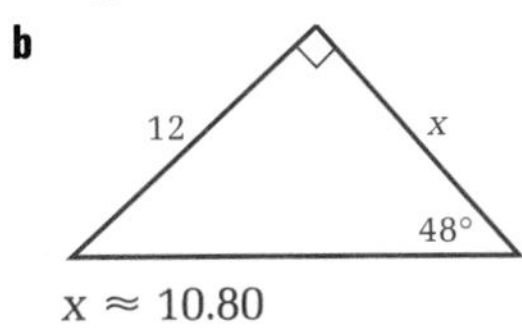

$x \approx 10.80$

c $\theta \approx 38.68°$

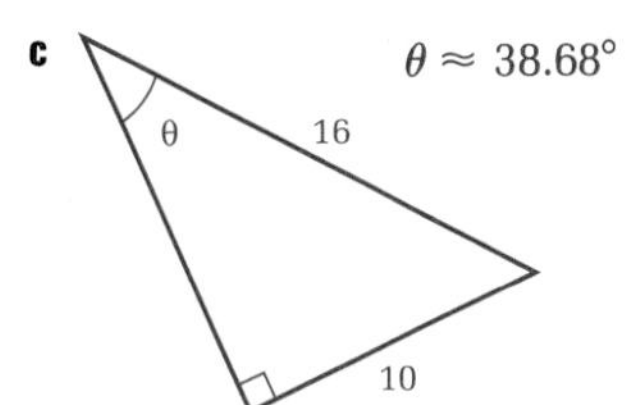

P 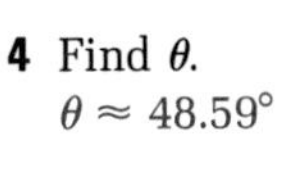

4 Find θ.
$\theta \approx 48.59°$

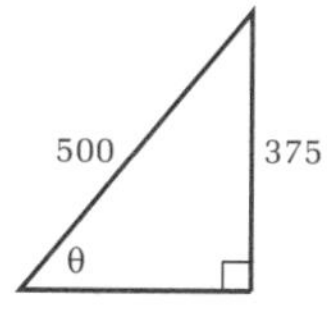

Problem-Set Notes and Strategies, continued

- Problems **3**, **4**, **6**, and **7** are fairly straighforward applications of the triangle trig formulas. A calculator will be helpful for some of the inverse measures.

Problem Set

A

P **5** Given that $\sin \theta = \frac{3}{5}$ and that $90° < \theta < 180°$, without using a calculator, find

a $\cos \theta$ $\frac{-4}{5}$ **b** $\tan \theta$ $-\frac{3}{4}$ **c** $\cos (90° - \theta)$ $\frac{3}{5}$
d $\cot (90° - \theta)$ $-\frac{3}{4}$ **e** $\sin (90° - \theta)$ $\frac{-4}{5}$

P,R **6** Find m∠ABC.
≈38.66°

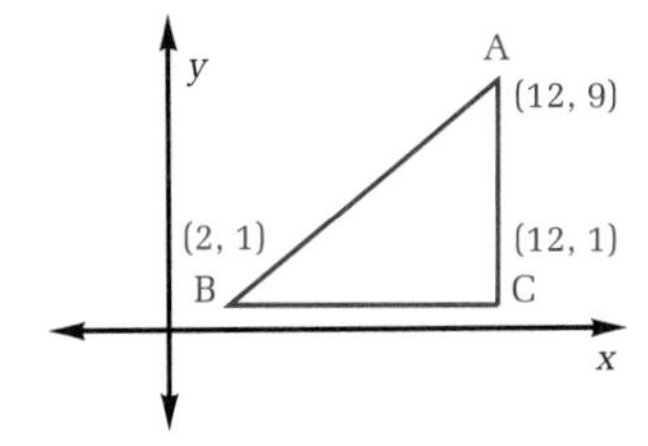

P **7** In △FGH, ∠G is a right angle and $\tan \angle F = \frac{3}{4}$.
a Find sin ∠F. $\frac{3}{5}$ **b** Find cos ∠H. $\frac{3}{5}$

I **8** Given $\tan \theta = 2$, find all values of θ to the nearest tenth of a degree. $\approx 63.4° + 180° \cdot n$

P 

9 What is the height of the flagpole?
≈85 ft

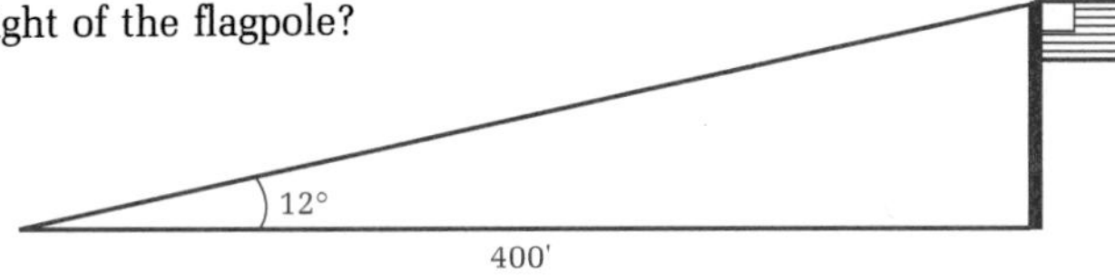

- Applications such as measuring height in problem **9** are useful illustrations of the trigonometric formulas.

P,R **10** Find θ. $\approx 53.13°$

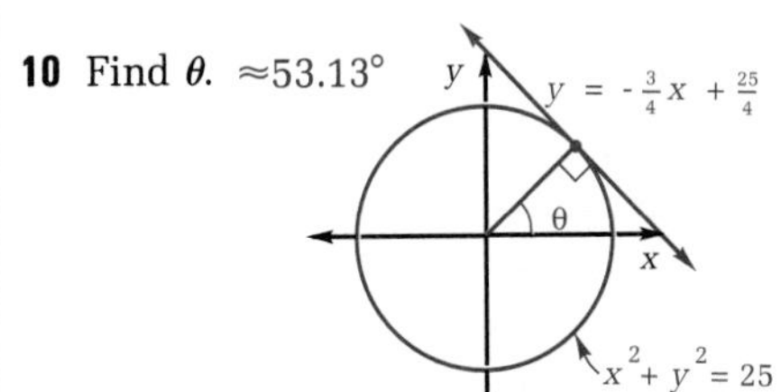

Problem-Set Notes and Strategies, continued

- In problem **14**, students may assume they are to find a ratio rather than a length.

Problem Set, *continued*

P **11** Refer to the diagram.

a If $\angle BAO = 55°$, what is $\sin \angle ABO$? ≈ 0.5736

b If $\cos \angle BAO = \frac{5}{8}$, what is $\sin \angle ABO$? $\frac{5}{8}$

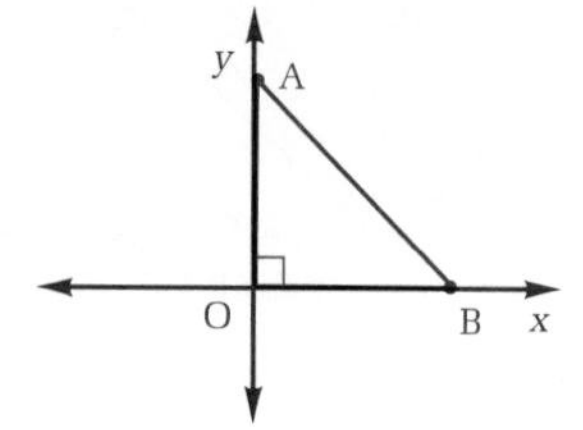

P,R **12** **a** Find BC. ≈ 16.78

b Calculate the area of $\triangle ACB$. ≈ 167.8

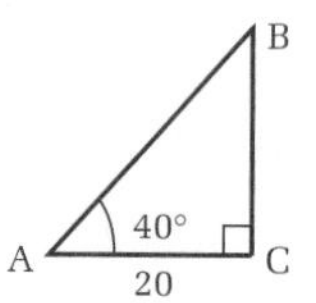

P,R **13** $\triangle ABC \sim \triangle PQR$.

a If $\sin \angle A = 0.8$, what is $\sin \angle P$? 0.8

b If BC = 8, QR = 12, and PR = 9, why is $\tan \angle A = \tan \angle P$? The ratio BC:AC is equal to the ratio QR:PR.

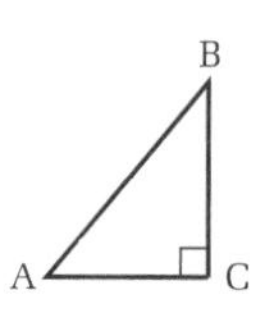

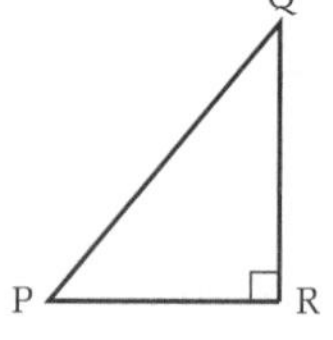

P **14** How much longer is $\overline{CD}$ than $\overline{BE}$? ≈ 5.5

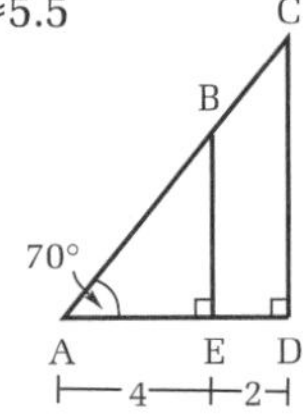

P **15** Find AC if AB = 52.1. ≈ 19.5

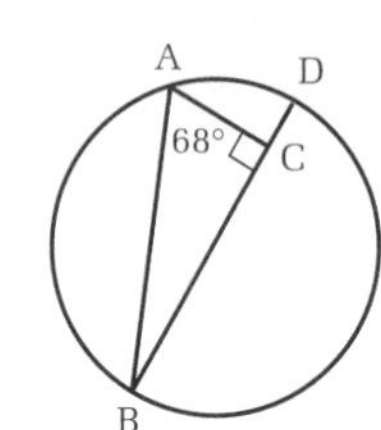

R **B** **16** What is the shortest side of this figure? $\overline{DC}$

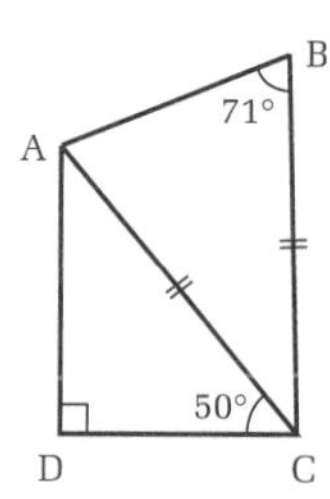

P **17** Find θ to the nearest tenth of a degree. $\approx 1.8°$

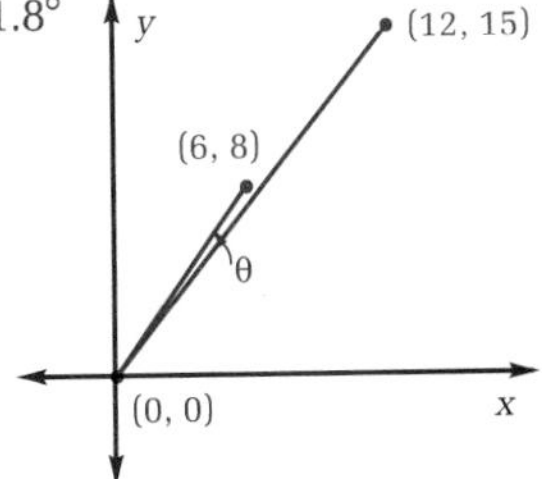

P **18** Jamie is 6 feet tall and is standing 500 feet from the base of a building. He looks up and sees Jason and Jessie looking out the windows of their apartments. Jason is 100 feet above the ground, and Jessie is 20 feet directly above him. Find θ. $\approx 2.20°$

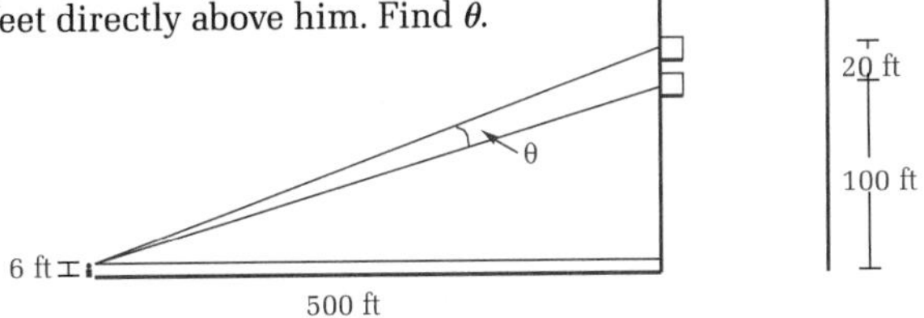

P,R **19** Find the length of arc AB. ≈ 13.9

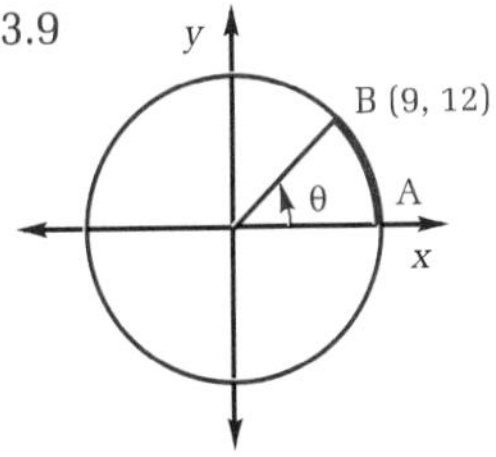

P,R **20** Segments AD and DC are tangent to circle O. $\widehat{ABC} = 200°$. The radius of circle O is 10 units. Find the area of AOCD.
≈ 567.13 sq. units

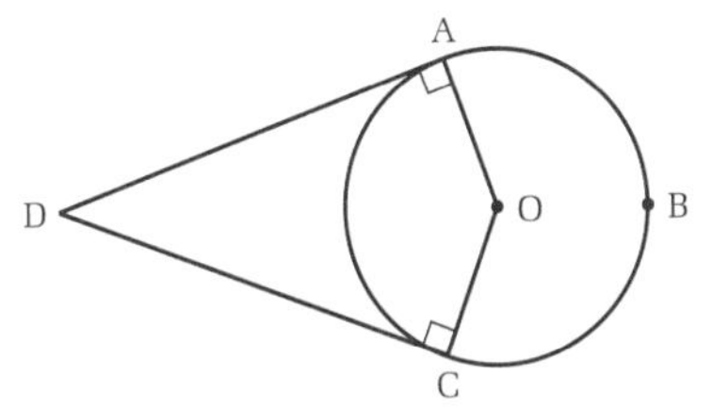

P,R **21** Consider regular pentagon PENTA. If the perimeter is 20, what is the length of apothem $\overline{GO}$? ≈ 2.75

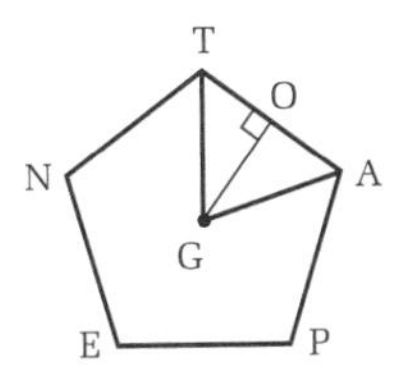

Problem-Set Notes and Strategies, continued

- Some auxiliary lines may be helpful in solving problem **17**. A calculator will be necessary.

- Problem **20** can be solved as a geometry problem if the figure is divided into two right triangles.

Problem Set, *continued*

P,R

22 If AF = 2, AE = 3, ∠B = 20°, and ∠C = 10°, what is the area of ABDC?
≈29.71

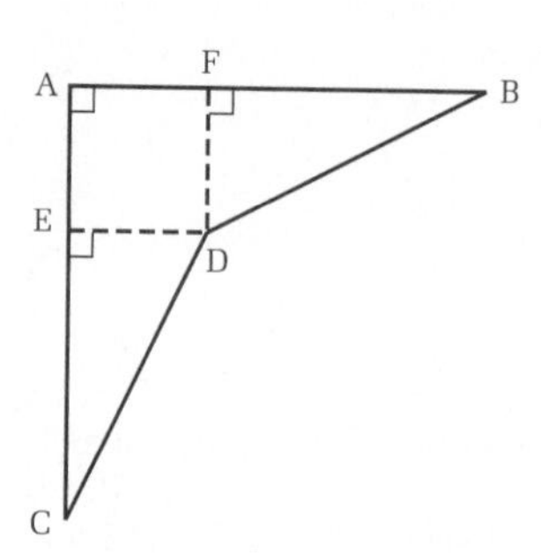

P,R

23 Find the ratio h_1:h_2 of the heights of the two buildings, given the angles of elevation shown. If the height h_2 is 250 feet, what is the height h_1?
≈1.42; ≈355 ft

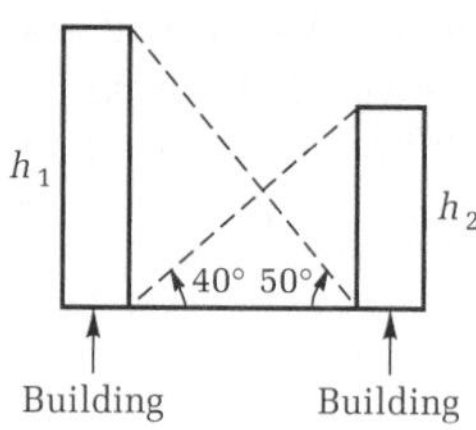

P,R

24 **a** Find sin θ_1 and sin θ_2.
b Find $\frac{\sin \theta_1}{\sin \theta_2}$.
c Find θ_1 and θ_2.
d Find $\frac{\theta_1}{\theta_2}$.
a ≈0.3511; 0.6
b ≈0.5852
c ≈20.56°; ≈36.87°
d ≈0.5575

P,R

25 Starting at point A on Skull's treasure map, face due east. Rotate 45° counterclockwise. Now take 20 paces to arrive at point B. Rotate another 45° counterclockwise and take 50 paces to point C. The buried treasure is located at the midpoint of $\overline{AC}$. If you start at point A, how many degrees should you rotate counterclockwise, and then how many paces should you take, to arrive directly at the point where the treasure is buried?
≈77.57°; ≈33 paces

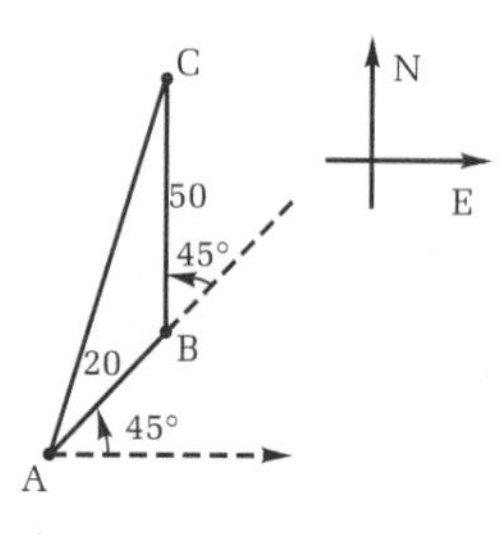

P,R

26 The equation of a line is $y = \frac{3}{7}x + 2$. What is the angle between the line and the x-axis? ≈23.2°

Problem-Set Notes and Strategies, continued

- If students remember that the slope of the line is the tangent of the angle, problem **26** will be easier.

P,R

27 ABCD is a parallelogram. What is the area of ABCD?
≈180.42

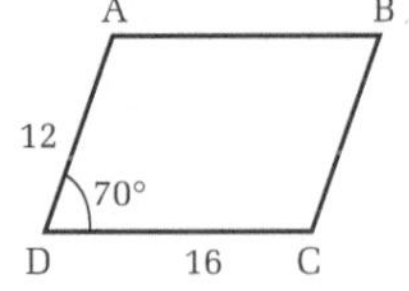

P,R  **28** Find the measure of $\angle\theta$.
≈ 46.40

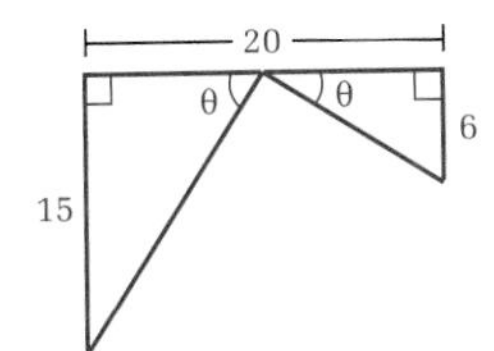

P,R **29** Find $m\angle\theta$.
≈ 78.56

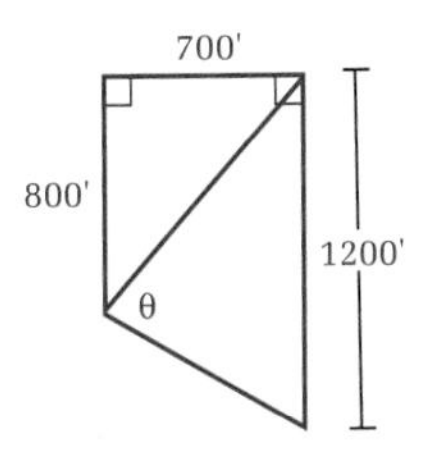

P,R **30** **a** Find $\angle DBC$. $\approx 45.10°$
b Find the area of $\triangle BCD$.
≈ 72.25

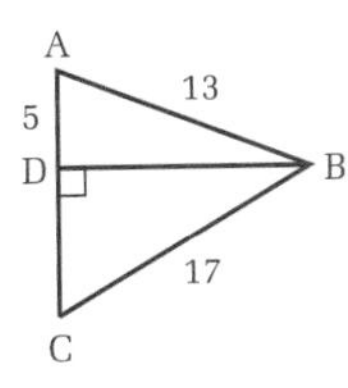

P,R **31** $\overline{PQ}$ is a tangent to circles M and N, whose radii are 6 and 3 respectively. Find $m\angle POM$.
≈ 19.47

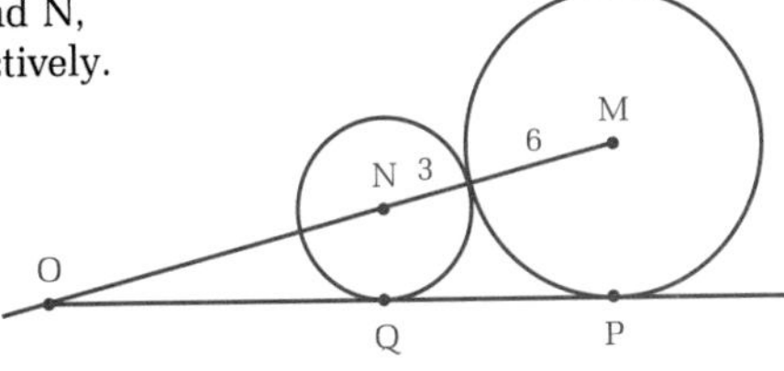

C R **32** Find $\sin\theta$ if O is the center of the circumscribed circle of regular pentagon ABCDE.
≈ 0.9511

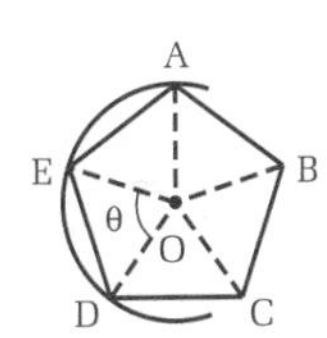

P,R **33** The graph of $\frac{x^2}{25} + \frac{y^2}{9} = 1$ is shown.
a Find $\angle ABC$. $\approx 118.07°$
b Find $\angle ABF$ if F is a focus of the ellipse. $\approx 5.91°$

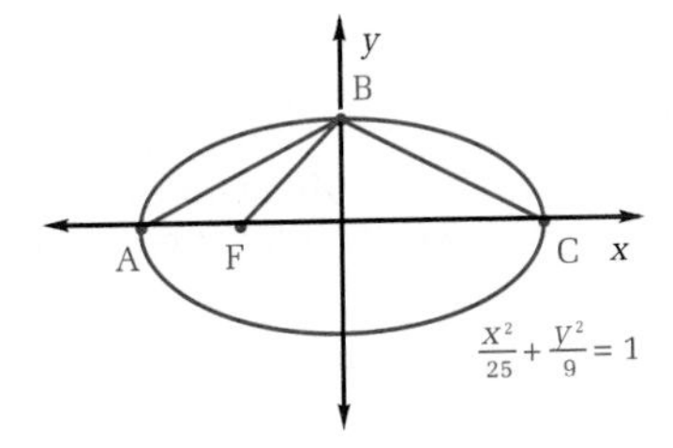

P **34** A rhombus has diagonals of lengths 16 and 22. Find the angle measures of the rhombus. $\approx 107.95°$ and $\approx 72.05°$

Problem-Set Notes and Strategies, continued

- Problem **28** requires a mix of geometry and some algebra skills. One approach is to generate and solve a system of equations.
- Additional lines may make the solution to problems **29** and **31** more obvious.

Communicating Mathematics

The relationship between cofunctions can make evaluating trigonometric functions much easier. Write a paragraph explaining this relationship and then write the same relationships using symbols as well (e.g., $\sin x = \cos(90° - x)$).

- Students may need to review some of the concepts of ellipses in order to answer problem **33**.

- For problem **34**, it helps if students recall that the diagonals of a rhombus are perpendicular bisectors of one another.

14.6

Class Planning

Time Schedule
Average: 2 days
Advanced: $1\frac{1}{2}$ days
Honors: $1\frac{1}{2}$ days

Resource References
Teacher's Resource Book
Class Opener 86
Practice 91

Class Opener

Adam Ant, crawling around a circle with a diameter of 18 inches, had gone a distance of 12 inches. Timmy Turtle, lumbering around a circle with diameter 48 inches, had gone 32 inches.

a Which had a greater percentage of their journey around the circle yet to complete?
Adam had gone $\frac{2}{3\pi}$ of the way. Timmy had also gone $\frac{2}{3\pi}$ of the way. They both had the same percentage to complete.

b If Adam crawls 6 inches per minute and Timmy crawls 20 inches per minute, who will finish first?
Adam has ≈7.4 minutes left.
Timmy has ≈5.9 minutes left.
Timmy will finish first.

14.6 Radian Measure

Objective

After studying this section, you will be able to

- Express angles in radians

Part One: Introduction

Arc Lengths and Radians

Until now, we have used degrees to measure angles. In this section, we will learn another way to measure angles.

In the diagram, we denote the radius of the circle by r, the central angle by θ, and the arc length by s.

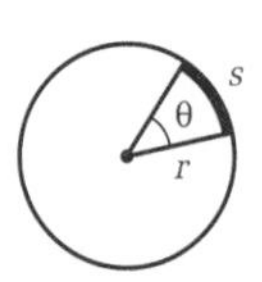

Example *If r = 1 and s = 1, what is θ?*

The circumference C is 2π. Therefore, the arc is $\frac{1}{2\pi}$ of the circle. So the angle is $\frac{1}{2\pi} \cdot 360° = \left(\frac{180}{\pi}\right)^{\circ} \approx 57.2958°$.

In this example, using degrees as units was extremely inconvenient. With the radius of 1 and an arc length of 1, it seems that the central angle should also be a simple number.

We will define a new angle-measuring unit. With this new unit, the measure of the central angle in the above example is 1 unit. We call this unit a ***radian***.

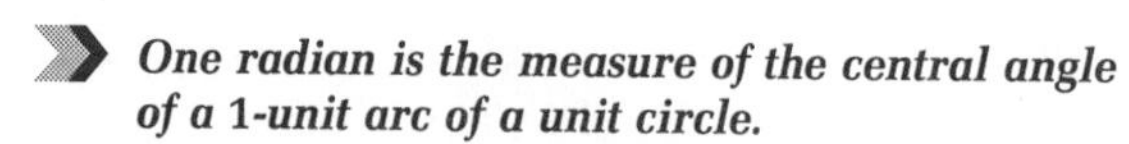

One radian is the measure of the central angle of a 1-unit arc of a unit circle.

What if we don't have a *unit* circle? Suppose the radius were 2. Then the arc length of the same central angle would double. In general, for radius r the arc length is multiplied by r.

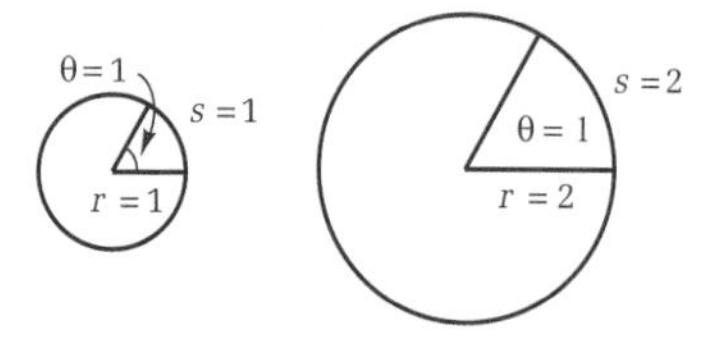

 ***In a circle with radius r, the relation between an arc s and its central angle θ is $s = r\theta$, where θ is in radians.***

Compare this formula with the formula in degrees, $s = \frac{\theta°}{360°}(2\pi r)$. You will find that using radians will make your work much easier. Also, radians are used in almost all mathematical applications and most computers.

Vocabulary
radian

Here are some basic facts about radian measure.

- A full circle is 2π radians (360°). Half a circle is π radians (180°).
- A right angle is $\frac{\pi}{2}$ radians (90°).
- We can write 2^R to denote two radians, but usually we just write 2, with no symbol. For example, we write sin 2, instead of sin 2^R. Be careful. If you write an angle measure with no degree symbol, it will be interpreted as radians.
- Your calculator has a radian mode. Be sure that you know how to put your calculator in this mode.

Example *Evaluate.*

a *sin 1*

With your calculator in radian mode, enter 1 [sin] The display on the calculator should read about 0.84147.

b *sin 1°*

With your calculator in degree mode, enter 1 [sin] The display on the calculator should read about 0.01745.

Example *What radian measure corresponds to 30°?*

Method 1 Since 30° is $\frac{1}{12}$ of a circle, the radian measure is $\frac{1}{12} \cdot 2\pi$, or $\frac{\pi}{6}$.

Method 2 The portion of the circle is the same in degrees as in radians, so $\frac{30°}{360°} = \frac{\theta}{2\pi}$. If we solve the equation, we get $\theta = \frac{\pi}{6}$.

In general, to convert between radians and degrees, we use the proportion $\frac{\theta°}{360°} = \frac{\theta^R}{2\pi^R}$.

Special Angles

The divisions between the quadrants in degrees are 0°, 90°, 180°, 270°, and 360°. In radians, these angles are 0, $\frac{\pi}{2}$, π, $\frac{3\pi}{2}$, and 2π.

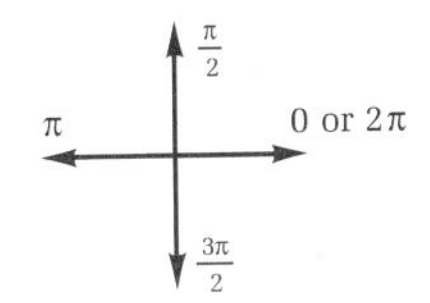

The special angles we studied before were the multiples of 30° and 45°. We now wish to express these in radians, as shown in the figure. You will need to learn the values in the first quadrant and to understand how the reference angles relate to the angles in the other quadrants.

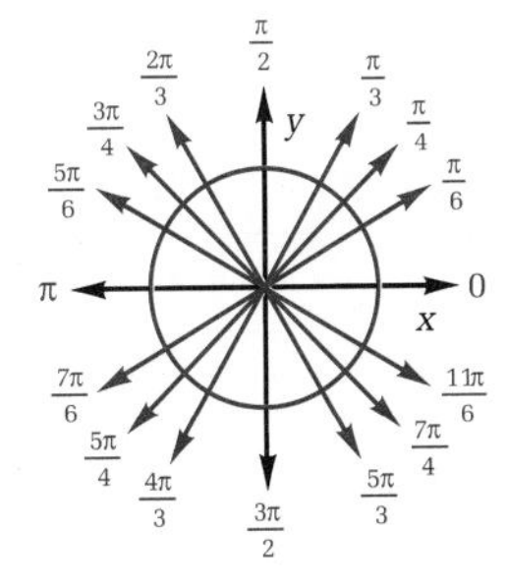

Lesson Notes

- Radians are presented as an alternative measure for angles. In trigonometry, in future math classes, and in work with computers, radians will be the preferred angle measure. Radians have the advantage of being unitless. You might want to point out that since θ in radians is defined as $\frac{s}{r}$, where s is the arclength and r is the radius, the units in the ratio cancel. A radian is in fact a ratio. It is a measure of the arclength using the length of the radius as our unit of measure.The book, *3.14 and all that* available from Dover Publications would be worthwhile for students to read.

Checkpoint

1. What is the radian measure of an angle that measures 225°? $\frac{5\pi}{4}$
2. Find the exact value of Arcsec (2) using radians. $\frac{\pi}{3}$
3. Find the length of the arc subtended in a circle of radius 6 with a central angle of 50°. 5.24
4. A CD revolves at a rate of 500 RPM. How many radians is that every minute? 1000π or ≈ 3141.59

14.6

Assignment Guide

Average	
Day 1	4, 5, 6a, b, d, 8–15, 27, 29, 31
Day 2	7, 25, 28, 30, 32, 36–39, 42, 45
Advanced	
Day 1	5, 6a, b, d, 8–15, 25, 29, 31, 39, 40
Day 2	($\frac{1}{2}$ day) 7, 24, 28, 30, 32, 41–43, 45 ($\frac{1}{2}$ day) Section 14.7 8, 10–12, 14, 17
Honors	
Day 1	6c, d, e, 8, 12–18, 25, 29, 30, 32, 39–41
Day 2	($\frac{1}{2}$ day) 24, 28, 33, 43, 44, 47 ($\frac{1}{2}$ day) Section 14.7 8, 10, 11, 14, 15, 17, 18

Problem-Set Notes and Strategies

- For problems **1** and **2**, students may have calculators which will convert between degrees to radians. Remind students that they will find decimal approximations and not exact values in terms of π.
- One way to deal with the problem of how different calculators do conversions, as in problem **6** is to have students work in groups and compare how calculators work.

Part Two: Sample Problems

Problem 1 *Evaluate cos (3.14^R).*

Solution Put your calculator in radian mode. Now enter 3.14 [cos]. The result should be approximately −0.99999872. It makes sense that cos 3.14 is close to −1, since the angle is very close to π. Therefore, the x-coordinate of the point on the unit circle is very close to −1.

Problem 2 *What is the range of values for Arcsin x in radians?*

Solution In degrees, we had $-90° \le \text{Arcsin } x \le 90°$. Changing to radians, we have $\frac{-\pi}{2} \le \text{Arcsin } x \le \frac{\pi}{2}$.

Problem 3 *Using radians, find the exact value of Arccos $\left(\frac{-1}{2}\right)$.*

Solution In radians, Arccos will be between 0 and π. So Arccos $\left(\frac{-1}{2}\right) = \frac{2\pi}{3}$.

Part Three: Exercises and Problems

Warm-up Exercises

P **1** Convert the following angles measured in degrees to radian measure. Leave your answer in terms of π.

a 60° $\frac{\pi}{3}$ **b** 210° $\frac{7\pi}{6}$ **c** −75° $-\frac{5\pi}{12}$ **d** 450° $\frac{5\pi}{2}$ **e** −30° $-\frac{\pi}{6}$

P **2** Convert the following angles given in radians to degree measure.

a $\frac{\pi}{4}$ 45° **b** $\frac{-7\pi}{6}$ −210° **c** $\frac{24\pi}{9}$ 480° **d** $\frac{-\pi}{3}$ −60° **e** $\frac{-2\pi}{3}$ −120°

P **3** Find the length of arc AB if $\theta = 110°$. ≈1.92

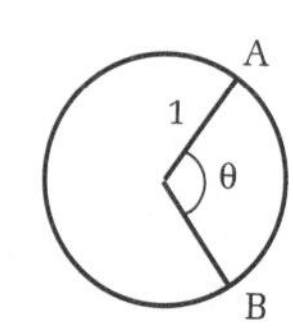

Problem Set

A R **4** A bicycle wheel is 27 inches in diameter. How far does the wheel travel in 15 revolutions? ≈1272.35 in.

R **5** Give the coordinates of points A and B.
$\left(\frac{\sqrt{3}}{2}, \frac{1}{2}\right)$; $\left(-\frac{1}{2}, \frac{\sqrt{3}}{2}\right)$

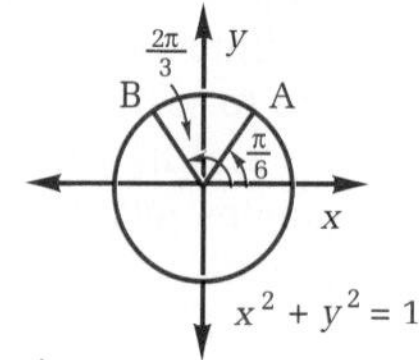

P **6** Convert each angle in degrees to its decimal approximation in radians. Give your answer to the nearest hundredth.

a 50° ≈0.87 **b** −400° ≈−6.98 **c** 25° ≈0.44 **d** 206° ≈3.60 **e** −75° ≈−1.31

R 7 Find sin ∠B.
≈0.8192

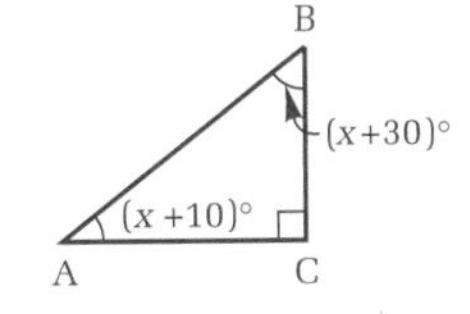

P 8 Express each angle in radians.

a Full rotation 2π

b $\frac{1}{2}$ rotation π

c $\frac{1}{3}$ rotation $\frac{2\pi}{3}$

d $\frac{2}{5}$ rotation $\frac{4\pi}{5}$

P **In problems 9–23, find the value of each.**

9 $\sin \frac{3\pi}{2}$ -1 **10** $\cos \pi$ -1 **11** $\tan \frac{\pi}{4}$ 1 **12** $\cos\left(-\frac{\pi}{2}\right)$ 0 **13** $\cot \frac{\pi}{6}$ $\sqrt{3}$

14 $\sin\left(-\frac{5\pi}{6}\right)$ $-\frac{1}{2}$ **15** $\cos \frac{3\pi}{4}$ $-\frac{\sqrt{2}}{2}$ **16** $\sin \frac{4\pi}{3}$ $-\frac{\sqrt{3}}{2}$ **17** $\tan\left(\frac{-\pi}{6}\right)$ $-\frac{\sqrt{3}}{3}$ **18** $\sec \frac{\pi}{4}$ $\sqrt{2}$

19 $\csc \frac{2\pi}{3}$ $\frac{2\sqrt{3}}{3}$ **20** $\tan\left(-\frac{7\pi}{6}\right)$ $-\frac{\sqrt{3}}{3}$ **21** $\sec \frac{5\pi}{3}$ 2 **22** $\cos\left(\frac{-5\pi}{2}\right)$ 0 **23** $\sin \frac{11\pi}{3}$ $-\frac{\sqrt{3}}{2}$

P **24** 2^R is approximately how many degrees? ≈114.5916°

R **25** Find the length of arc s.
≈4.89

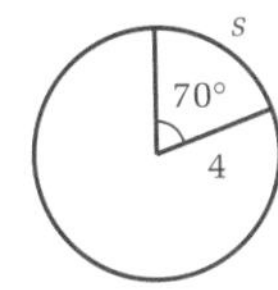

R **26** a Find tan ∠CBD and tan ∠ABD.

b Find $\frac{\tan \angle CBD}{\tan \angle ABD}$.

c Find ∠CBD and ∠ABD.

d Find $\frac{\angle CBD}{\angle ABD}$.

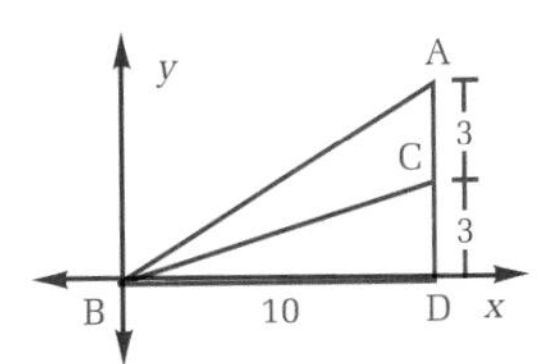

a $\frac{3}{10}$; $\frac{3}{5}$
b $\frac{1}{2}$
c ≈16.70°; ≈30.96°
d ≈0.5393

R **27** Find the length of arc AB.
6π

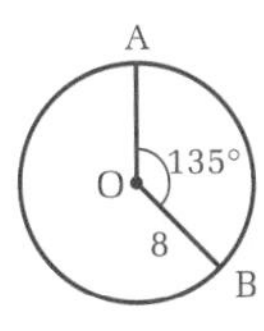

P **28** Find the exact value of each of the following.

a $\sin \frac{\pi}{6}$ $\frac{1}{2}$ b $\tan \frac{\pi}{3}$ $\sqrt{3}$ c $\cos \pi$ -1 d $\sec\left(\frac{-\pi}{6}\right)$ $\frac{2\sqrt{3}}{3}$ e $\cot \frac{\pi}{4}$ 1

P **29** Evaluate.

a $\sin 0.7^R$ ≈0.6442 b $\cos 1.2^R$ ≈0.3624 c $\tan 4.2^R$ ≈1.7778

P,R B **30** Find x in radians.

a Arcsin 0.1357 = x ≈0.1361 b Arctan −0.487 = x ≈−0.4532

Problem-Set Notes and Strategies, continued

- Problems **9 – 23** practice the use of radians. The practice will help students familiarize themselves with this new and abstract concept. Save some time at the end of the period for them to start these exercises and ask for help where necessary.

Stumbling Block

Radian measure is foreign to students when first presented. Make it a habit to use radian measure whenever possible from now on. Have a chart available or have students produce one, giving both the degree measure and the radian measure of frequently encountered angles. Use the arc-length formula frequently to show its ease of use in radian mode and how difficult it would be to produce the same answer with degrees.

- For problems **29 – 32**, remind students to use calculators in radian mode.

Problem Set, *continued*

P,R **31** If $\sin \theta^R = 0.8479$, what is the measure at θ in degrees, where $0° < \theta < 90°$? $\approx 57.98°$

P **32** Solve for x in each triangle. Angles are in radian measure.

a 12, x, 0.6 — $x \approx 21.25$

b 10, x, 1^R — $x \approx 15.57$

c 15, x, $\frac{\pi}{6}$ — $x = 7.5$

d x, 20, $\frac{\pi}{4}$ — $x = 10\sqrt{2}$

e x, 20, $\frac{\pi}{5}$ — $x \approx 11.76$

R **33** Solve $x \text{ Arcsin } 0.5 = (x - 1) \text{ Arccos } 0.5$ for x. $x = 2$

P **34** AB = AC. Find θ^R and θ in degrees.
≈ 0.2618 and 15°

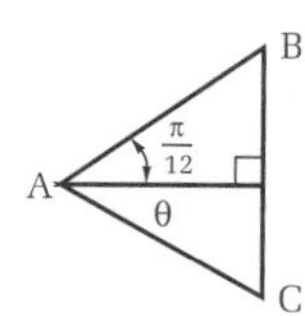

P **In problems 35–38, consider an angle θ whose terminal ray passes through $\left(\frac{7}{25}, \frac{24}{25}\right)$.**

35 Explain why the point $\left(\frac{7}{25}, \frac{24}{25}\right)$ must lie on the unit circle.

36 Explain why $\cos \theta = \frac{7}{25}$ if $\sin \theta = \frac{24}{25}$ and $0° \le \theta \le 90°$.

37 Find the value of $\tan \theta$. $\frac{24}{7}$

38 Find the value of $\sec \theta$. $\frac{25}{7}$

R **39** Which triangle has the greater area?
Neither; both areas are ≈ 38.30 sq. units.

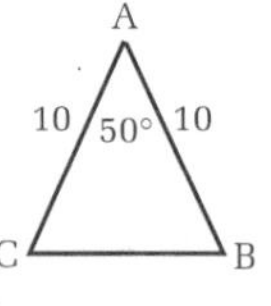

10, 10, 130°

P,R,I **40** Find the perimeter of the shaded region.
38.4

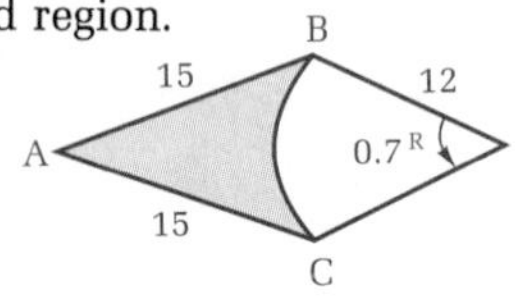

Additional Answers

35 Because $\left(\frac{7}{25}\right)^2 + \left(\frac{24}{25}\right)^2 = 1^2$ (Pythagorean Theorem)

36 If $\sin \theta = \frac{24}{25}$ and $0° \le \theta \le 90°$, the terminal ray of θ in standard position intersects the unit circle at $\left(\frac{7}{25}, \frac{24}{25}\right)$. Since $\cos \theta$ is equivalent to the x-coordinate of this point, $\cos \theta = \frac{7}{25}$.

Problem-Set Notes and Strategies, continued

- After students find the areas in problem **39**, you might ask if this is true of all isosceles triangles or if the angles made a difference. What is the relationship between the angles?
- Using the arclength formula and a little help from geometry, may make problem **40** more easily solved.

P,R **41** Solve for $0 \le x \le 2\pi$.

a $\sin^2 x = \frac{3}{4}$ $\frac{\pi}{3}, \frac{2\pi}{3}, \frac{4\pi}{3}, \frac{5\pi}{3}$ **b** $2\cos^2 x + \cos x = 0$ $\frac{\pi}{2}, \frac{2\pi}{3}, \frac{4\pi}{3}, \frac{3\pi}{2}$ **c** $\sqrt{3}\sin x = \cos x$ $\frac{\pi}{6}, \frac{7\pi}{6}$

R **42** A line has a slope of $-\frac{2}{3}$ and passes through the origin. What angle does it form with the positive x-axis, measured counterclockwise? $\approx 146.31°$

P,R **43** The graph shown represents the equation $y = \sin x$. Find the values of y indicated by the letters A, B, C, D, E, and F.
$1; \frac{\sqrt{2}}{2}; \frac{1}{2}; -\frac{\sqrt{2}}{2}; -\frac{\sqrt{3}}{2}; -1$

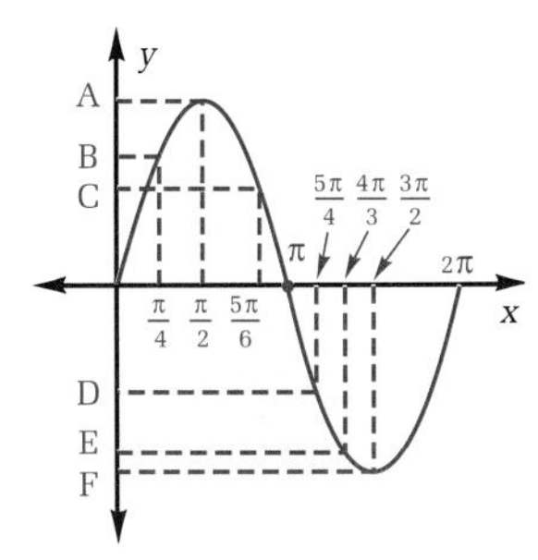

C **44** Find ∠ABC.
≈ 26.57

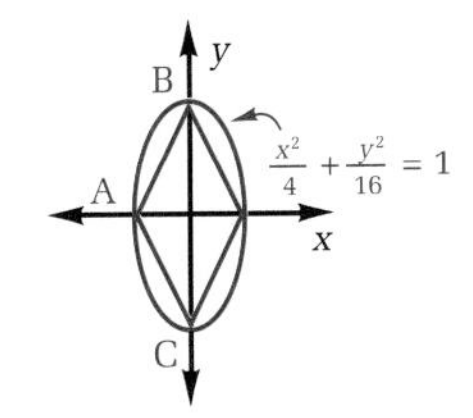

R **45** **a** Find $\cos\theta$. $\frac{\sqrt{d^2 - 1}}{d}$
b Find $(\cos\theta)^2 + (\sin\theta)^2$. 1

P,R **46** Arc AQ is an arc of circle O. Arc ABQ is an arc of circle P.
Length of arc AQ = 16″
OA = 24″
∠APQ = 2.5 ∠AOQ
Find the area of the shaded region.
≈ 23.9 in.^{2}

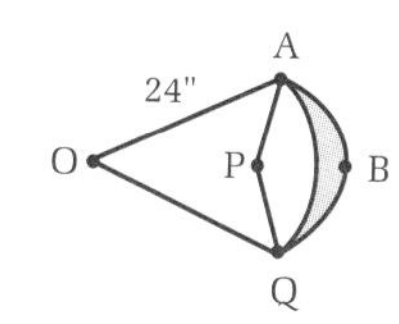

P **47** If $\angle A = 1.2^R$, AB = AC, and A is the center of arc BC, what is the length of $\overline{BC}$?
≈ 13.17 in.

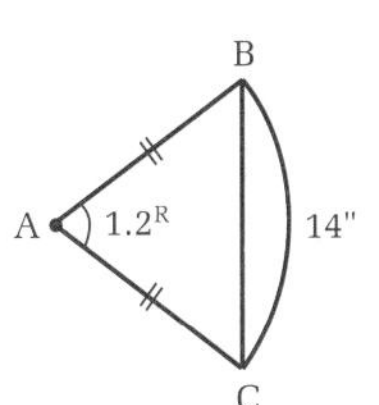

Problem-Set Notes and Strategies, continued

- Part **b** of problem **41** can be approached via factoring. Part **c** can be converted into tangents or cotangents.

Communicating Mathematics

Radian measure is important for physical applications of trigonometry. Explain in your own words how to find the radian measure of an angle and how to convert back to degrees. Next write a short statement explaining why no units of measure are used with radian measurements.

- Problem **45a** is a general exercise to work showing how the Pythagorean Theorem applies to trig problems.

- Problem **47** can be solved with some additional lines and the arclength formula to determine AB.

14.7

Class Planning

Time Schedule
All levels: $1\frac{1}{2}$ days

Resource References
Teacher's Resource Book
Class Opener 87
Cooperative Learning 44
Practice 92

Class Opener

Draw right triangle ABC, with angle A the right angle. Angle C measures $\frac{\pi}{6}$ radians.

a Determine cos B and sin C.
$\cos B = \frac{1}{2}$, $\sin C = \frac{1}{2}$

b Determine cos C and sin B.
$\cos C = \frac{\sqrt{3}}{2}$; $\sin B = \frac{\sqrt{3}}{2}$

c Find the value of cos –B and sin –B.
$\cos -B = \frac{1}{2}$; $\sin -B = -\frac{\sqrt{3}}{2}$

d Find the value of sin(π +B) and cos(π + B).
$\sin(\pi + B) = -\frac{\sqrt{3}}{2}$;
$\cos(\pi + B) = -\frac{1}{2}$

14.7 Trigonometric Relationships

Objectives
After studying this section, you will be able to
- Recognize odd and even functions
- Identify relationships between trigonometric functions of special angles

Part One: Introduction

Odd and Even Functions

In the figure, we have marked the angles θ and $-\theta$. The points of intersection of these angles and the unit circle have the same x-coordinates and opposite y-coordinates. This relationship suggests the following theorems.

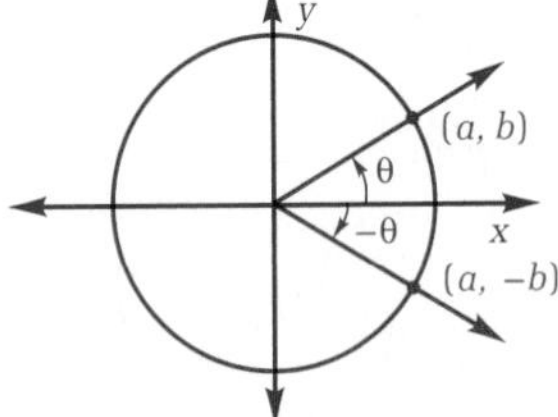

For all angles θ, $\cos(-\theta) = \cos\theta$, and $\sin(-\theta) = -\sin\theta$.

For all angles θ (where $\cos\theta \neq 0$), $\tan(-\theta) = -\tan\theta$ and $\sec(-\theta) = \sec\theta$.

For all angles θ (where $\sin\theta \neq 0$), $\cot(-\theta) = -\cot\theta$ and $\csc(-\theta) = -\csc\theta$.

Mathematicians sometimes distinguish between ***odd functions*** and ***even functions.*** A function f is said to be odd if $f(-x) = -f(x)$ for all values of x. The graph of an odd function is symmetric about the origin. In an even function, $f(-x) = f(x)$ for all values of x. The graph of an even function is symmetric about the y-axis.

The preceding theorems indicate that of the trigonometric functions, sine, tangent, cotangent, and cosecant are odd, whereas cosine and secant and cosecant are even. You will find a knowledge of these properties useful when you are introduced to the graphs of the trigonometric functions in Chapter 16.

Example *Simplify the expression $\cos(-\theta)\tan(-\theta)$.*

$$\cos(-\theta)\tan(-\theta)$$
$$= \cos(-\theta) \cdot \frac{\sin(-\theta)}{\cos(-\theta)}$$
$$= \sin(-\theta)$$
$$= -\sin\theta$$

Vocabulary
odd function
even function

Supplementary and Complementary Angles

Let's restate the cofunction identities in terms of radians.

$\sin \theta = \cos\left(\frac{\pi}{2} - \theta\right)$ $\tan \theta = \cot\left(\frac{\pi}{2} - \theta\right)$ $\sec \theta = \csc\left(\frac{\pi}{2} - \theta\right)$

The cofunctions of complementary angles are equal.

Now let's look at supplementary angles. Remember, two angles are supplementary if the sum of their measures is π. In the figure shown, angles α and β are supplementary. For these angles, the x-coordinates on the unit circle are opposites, and the y-coordinates are equal.

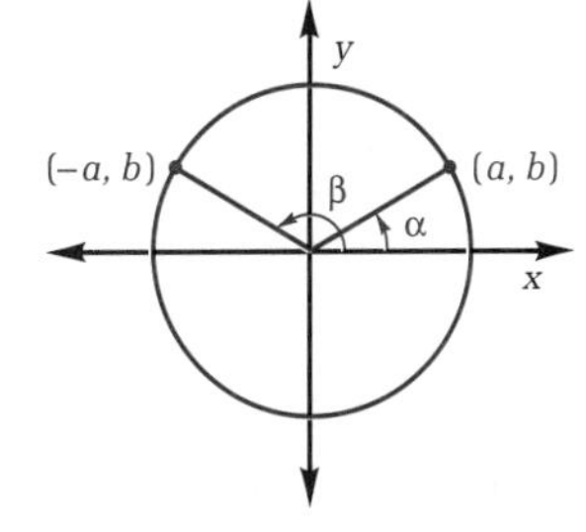

For all θ, $\sin(\pi - \theta) = \sin \theta$ and $\cos(\pi - \theta) = -\cos \theta$.

Can you state the appropriate theorems for the other four trigonometric functions?

Other Relationships

Is there a relationship between $\cos \theta$ and $\cos(\theta + \pi)$?

In the figure, we see that the angles differ by π, or half a circle. Thus, the x-coordinates are opposite and $\cos(\theta + \pi) = -\cos \theta$.

Many such relationships exist. The idea is not to memorize all of them but to understand them and be able to look at the unit circle and work them out for yourself when you need them.

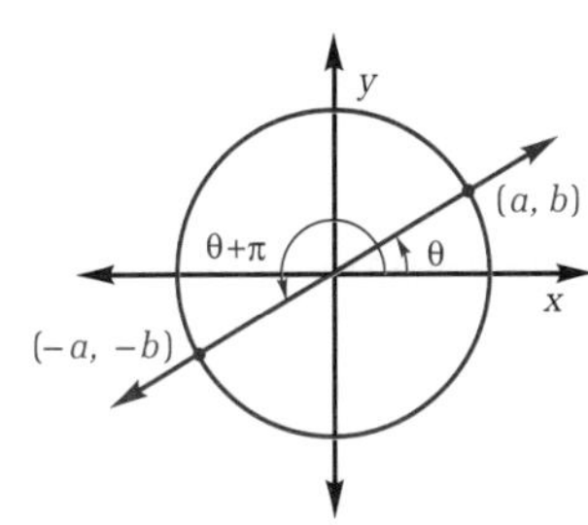

Lesson Notes

- By using the unit circle and the definition of the trigonometric functions, many different identities are available. The concept of odd and even functions are introduced as being symmetric to the origin and the y-axis. These symmetries are useful when solving equations and also to graph equations. After their earlier work with parabolas, students will appreciate that any such symmetries that can be identified before the graph is drawn will eliminate work in plotting points. Each of the trig functions is either odd or even and can be remembered by the effects on the unit circle, rather than by rote memorization.

Checkpoint

1. If $\sin x = 0.5629$, find $\sin(-x)$. -0.5629
2. True or False, $\csc x = \csc(-x)$? False.
3. Since $\cos x = \cos(-x)$, what can be said about $\sec x$? $\sec x = \sec(-x)$

Part Two: Sample Problems

Problem 1 *Evaluate $\tan\left(\frac{13\pi}{7}\right) - \tan\left(\frac{13\pi}{7} + 2\pi\right)$.*

Solution The two angles differ by 2π. Thus, they are coterminal and have the same tangent. This difference is 0.

Problem 2 *Use congruent triangles to justify that $\sin \theta = \cos\left(\frac{\pi}{2} - \theta\right)$.*

Solution If we draw a diagram as shown, we see that both triangles are right triangles having an acute angle θ and congruent hypotenuses of length 1. Thus, the two triangles are congruent. The y-coordinate of point A is therefore equal to the x-coordinate of point B, so $\sin \theta = \cos\left(\frac{\pi}{2} - \theta\right)$.

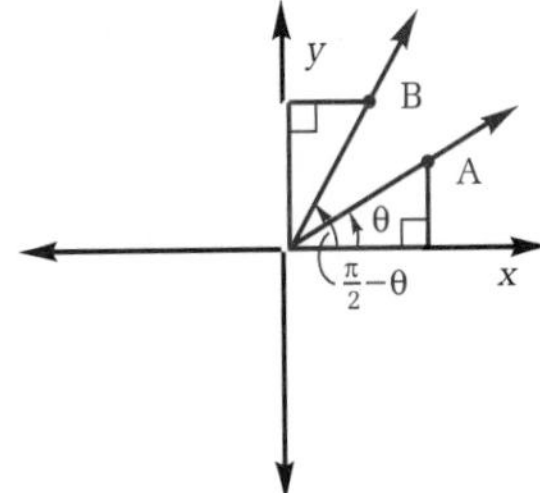

14.7

Assignment Guide

Average

Day 1	8, 10–14, 17, 18, 23, 24a–d, 27
Day 2	($\frac{1}{2}$ day) 9, 15, 16, 20, 22, 25 ($\frac{1}{2}$ day) Review Problems 1–3, 7, 8, 11, 12, 14

Advanced

Day 1	($\frac{1}{2}$ day) Section 14.6 7, 24, 28, 30, 32, 41–43, 45 ($\frac{1}{2}$ day) 8, 10–12, 14, 17
Day 2	9, 13, 15, 16, 18, 20–22, 24a–d, 25, 27

Honors

Day 1	($\frac{1}{2}$ day) Section 14.6 24, 28, 33, 43, 44, 47 ($\frac{1}{2}$ day) 8, 10, 11, 14, 15, 17, 18
Day 2	9, 13, 16, 19–24, 27–29

Problem-Set Notes and Strategies

- Problems **3** and **4** illustrate the even and odd nature of the sine and cosine functions. The others can follow from these two.
- Problem **7** may identify which students are having difficulty with the concepts of complementary angles, odd and even functions, and angles in general.

Additional Answer

Answer for problem **7** can be found in the answer pages beginning on page **A** 1.

Part Three: Exercises and Problems

Warm-up Exercises

P **1** Use a calculator to decide whether each of the following is True or False for $\theta = 125°$.

a $\cos(-\theta) = \cos\theta$ True **b** $\sin(-\theta) = -\sin\theta$ True **c** $\tan(-\theta) = -\tan\theta$ True
d $\sec(-\theta) = \sec\theta$ True **e** $\cot(-\theta) = -\cot\theta$ True

P **2** Use the unit circle shown to evaluate each of the following in terms of a and b.

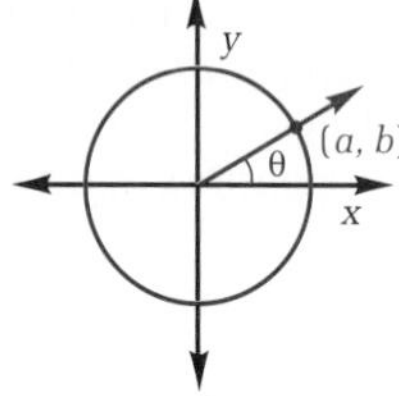

a $\cos\theta$ a **b** $\cos(\pi - \theta)$ $-a$
c $\sin\theta$ b **d** $\sin(\pi - \theta)$ b
e $\sin(\theta + \pi)$ $-b$ **f** $\cos(\theta + \pi)$ $-a$
g $\cos(-\theta)$ a **h** $\sin(-\theta)$ $-b$

P,R **3** Evaluate each of the following. What do you observe? $\cos(-\theta) = \cos\theta$

a $\cos\left(\frac{\pi}{6}\right)$ $\frac{\sqrt{3}}{2}$ **b** $\cos\left(\frac{-\pi}{6}\right)$ $\frac{\sqrt{3}}{2}$ **c** $\cos\left(\frac{\pi}{4}\right)$ $\frac{\sqrt{2}}{2}$
d $\cos\left(\frac{-\pi}{4}\right)$ $\frac{\sqrt{2}}{2}$ **e** $\cos\left(\frac{-2\pi}{3}\right)$ $-\frac{1}{2}$ **f** $\cos\left(\frac{2\pi}{3}\right)$ $-\frac{1}{2}$

P,R **4** Evaluate each of the following. What do you observe? $\sin(-\theta) = -\sin\theta$

a $\sin\left(\frac{\pi}{6}\right)$ $\frac{1}{2}$ **b** $\sin\left(\frac{-\pi}{6}\right)$ $-\frac{1}{2}$ **c** $\sin\left(\frac{\pi}{4}\right)$ $\frac{\sqrt{2}}{2}$
d $\sin\left(\frac{-\pi}{4}\right)$ $-\frac{\sqrt{2}}{2}$ **e** $\sin\left(\frac{-2\pi}{3}\right)$ $-\frac{\sqrt{3}}{2}$ **f** $\sin\left(\frac{2\pi}{3}\right)$ $\frac{\sqrt{3}}{2}$

P,R **5** Refer to the figure.

a Find $\sin(-\theta)$.
b Find $\cos(-\theta)$.
c Find $\tan(-\theta)$.

a $-\frac{11}{5\sqrt{5}}$
b $-\frac{2}{5\sqrt{5}}$
c $\frac{11}{2}$

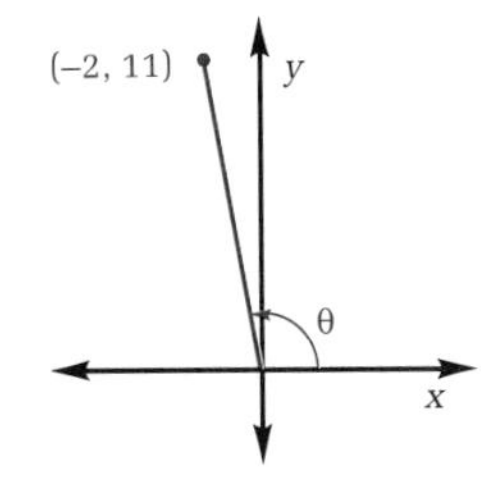

P **6** Indicate whether each statement is True or False.

a $\sin(-65°) = -\sin 65°$ True **b** $\tan 140° = -\tan(-140°)$ True
c $\cos 47° = -\cos(-47°)$ False **d** $\sin 62° + \sin(-57°) = -\sin(-62°) - \sin 57°$ True

P **7** Copy the figure and sketch $\pi - \theta$, $\pi + \theta$, and $-\theta$.

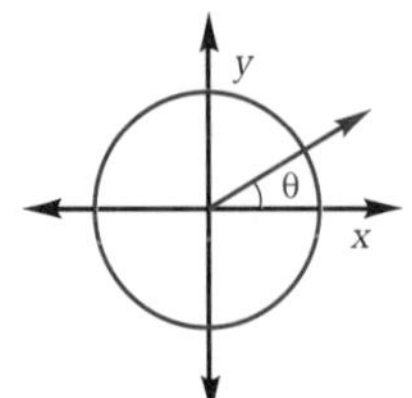

Problem Set

A

P,R

8 Rewrite each expression in terms of $\sin\theta$ or $\cos\theta$.

a $\sin(90° - \theta)$ $\cos\theta$ **b** $\cos(270° - \theta)$ $-\sin\theta$

c $\sin\left(\frac{\pi}{2} - \theta\right)$ $\cos\theta$ **d** $\cos\left(\frac{\pi}{2} + \theta\right)$ $-\sin\theta$

e $\sin(\pi + \theta)$ $-\sin\theta$

P

9 Evaluate each of the following for $\sin\alpha = 0.2$.

a $\sin(\pi - \alpha)$ 0.2 **b** $\sin(\pi + \alpha)$ -0.2 **c** $\sin(2\pi + \alpha)$ 0.2

d $\sin(4\pi + \alpha)$ 0.2 **e** $\sin(6\pi + \alpha)$ 0.2

P,I

10 For each graph, indicate whether the function represented is even or odd.

a

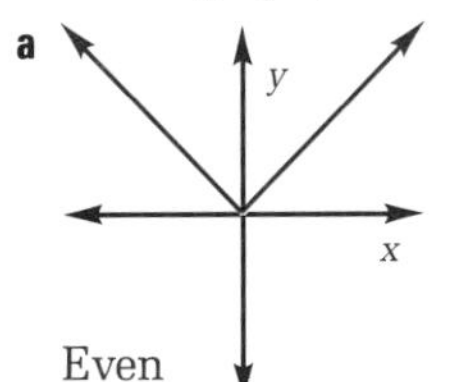

Even

b

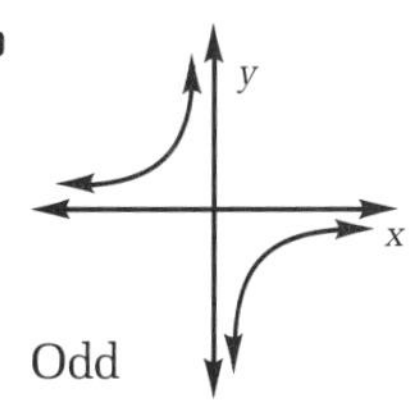

Odd

c

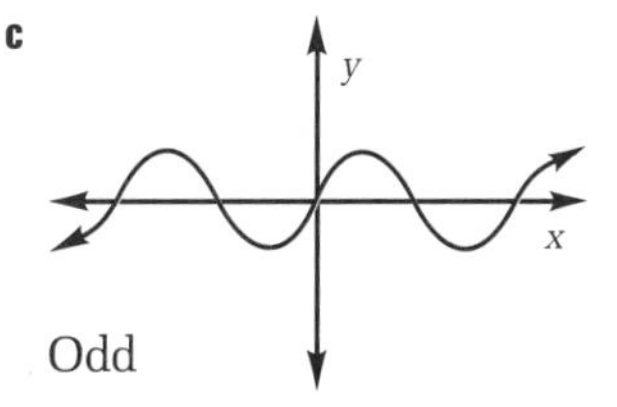

Odd

d

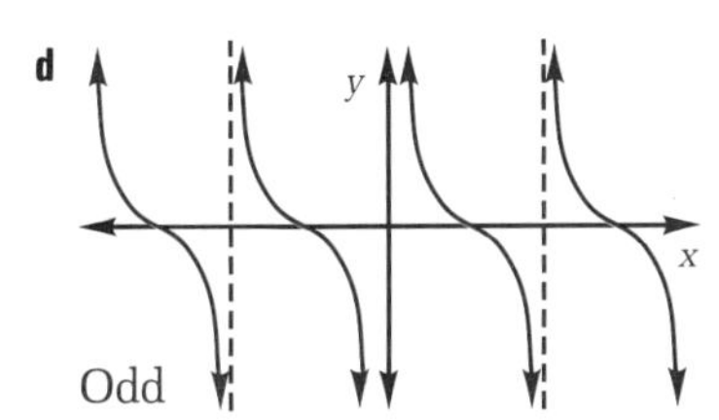

Odd

e

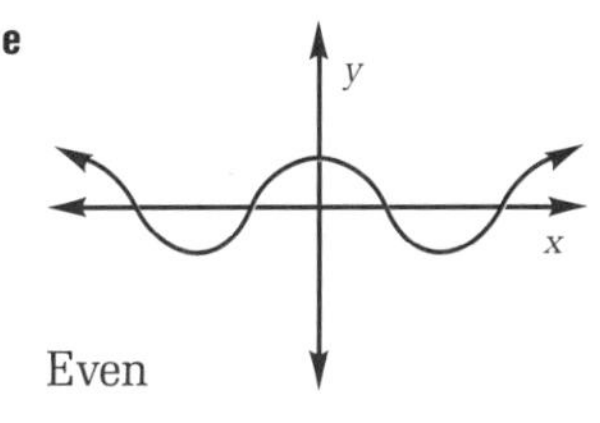

Even

f

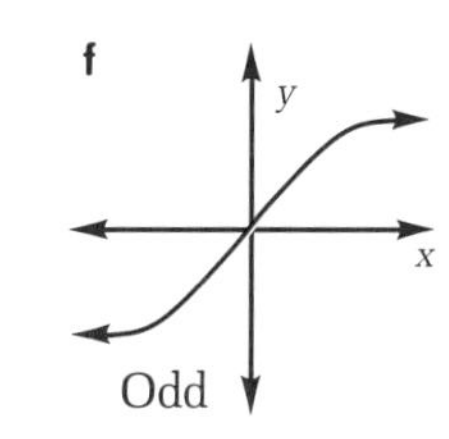

Odd

P,R

11 Write each of the following as a function of a positive acute angle.

a $\frac{\sin(-35°)}{\cos 35°}$ $-\tan 35°$ **b** $\tan(-80°)$ $-\tan 80°$

R

12 Refer to the figure to find each of the following.

a $\cos\alpha$ $\frac{b}{c}$

b $\sin\beta$ $\frac{b}{c}$

c $\alpha + \beta$ $\frac{\pi}{2}$ radians, or 90°

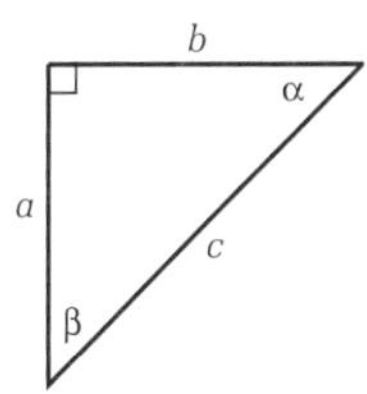

P,R

13 Copy and complete the table.

θ	2π	$\frac{13\pi}{6}$	$\frac{9\pi}{4}$	$\frac{7\pi}{3}$	$\frac{5\pi}{2}$
$\sin\theta$	0	$\frac{1}{2}$	$\frac{\sqrt{2}}{2}$	$\frac{\sqrt{3}}{2}$	1
$\cos\theta$	1	$\frac{\sqrt{3}}{2}$	$\frac{\sqrt{2}}{2}$	$\frac{1}{2}$	0
$\tan\theta$	0	$\frac{\sqrt{3}}{3}$	1	$\sqrt{3}$	Undef.
$\cot\theta$	Undef.	$\sqrt{3}$	1	$\frac{\sqrt{3}}{3}$	0

Problem-Set Notes and Strategies, continued

- Encourage students to use a picture to establish the relationships in problems **8** and **9**.

Problem Set, *continued*

P,R **14** Refer to the figure.

a Express α in terms of θ.
b Find the value of $\sin \theta$.
c Find $\angle AOB$.
d Find the value of $\cos \alpha$.

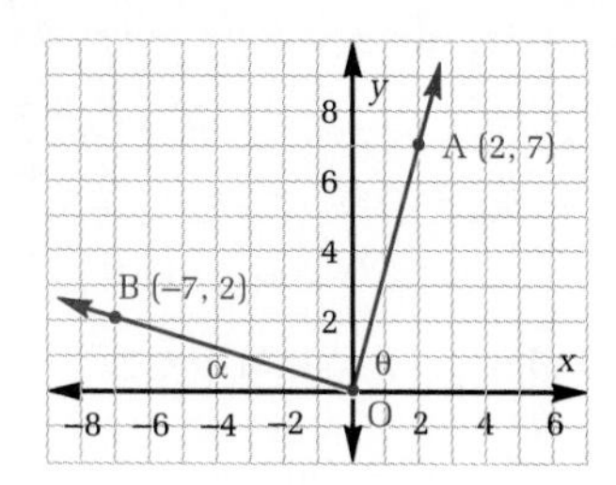

a $\alpha = \frac{\pi}{2} - \theta$, or $\alpha = 90° - \theta$
b $\frac{7}{\sqrt{53}}$
c $\frac{\pi}{2}$ radians, or 90°
d $\frac{7}{\sqrt{53}}$

P,R **15** Find the exact value of each of the following.

a CE $\sqrt{61}$
b $\cos \angle BAC$ $\frac{6}{\sqrt{61}}$
c $\tan \angle ACB$ $\frac{6}{5}$
d $\sin \angle AEC$ $\frac{9}{\sqrt{142}}$

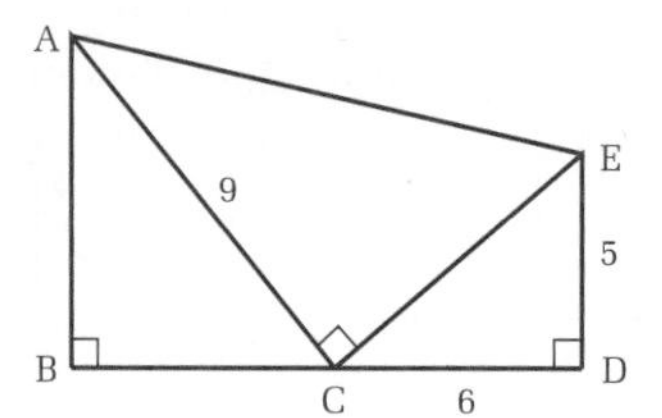

P **16** Rewrite each expression in terms of $\sin \theta$.

a $3 \sin \theta + 5 \cos \left(\frac{\pi}{2} - \theta\right)$
$8 \sin \theta$

b $4 \sin (\pi - \theta) - 2 \sin (\pi + \theta)$
$6 \sin \theta$

P,I **17** Show that each statement is true.

a $\sin (90° - \theta) = \sin 90° \cos \theta - \cos 90° \sin \theta$
b $\cos (90° - \theta) = \cos 90° \cos \theta + \sin 90° \sin \theta$

Problem-Set Notes and Strategies, continued

- The identities in problem **17** are both true and will be proven in the next chapter. These formulas will be generalized to include any angles.

Additional Answers

17a $\sin 90° \cos \theta - \cos 90° \sin \theta$
$= 1 \cos \theta - 0 \sin \theta$
$= \cos \theta$
$= \sin (90° - \theta)$

17b $\cos 90° \cos \theta + \sin 90° \sin \theta$
$= 0 \cos \theta + 1 \sin \theta$
$= \sin \theta$
$= \cos (90° - \theta)$

B

P,R **18** Use the diagram to find each of the following.

a $\tan (180° - \theta)$ $-\frac{5}{12}$
b $\tan (180° + \theta)$ $\frac{5}{12}$
c $\tan \theta$ $\frac{5}{12}$
d $\tan (-\theta)$ $-\frac{5}{12}$

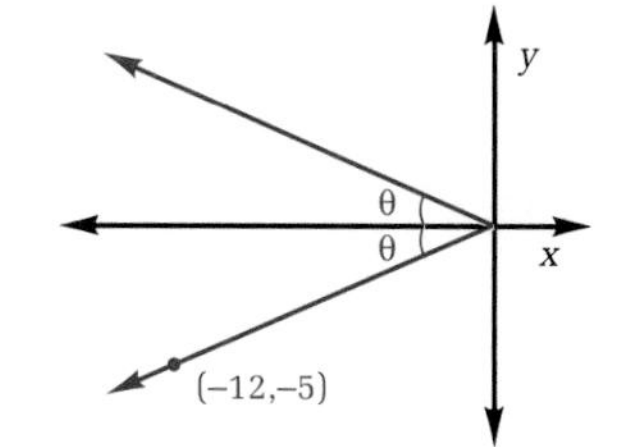

R **19** Which is longer—$\overline{BC}$ or $\overline{CD}$?
$\overline{CD}$

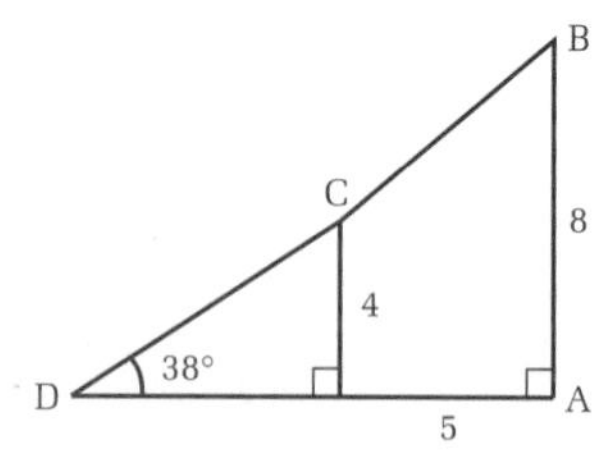

- A few algebra skills will help complete problem **19**. If students label the parts with variables, they can set up a system of equations.

P **20** Evaluate each of the following.

a $2 \sin 170° + \sin 10° + 3 \cos 80°$ ≈ 1.0419
b $\cos 200° + \cos 20° + 2 \sin 50° + 3 \cos 40°$ ≈ 3.8302

P | **21** Prove that $\sin\left(\theta + \frac{\pi}{3}\right)$ and $\cos\left(\frac{\pi}{6} - \theta\right)$ are equal.

P,R | **22** The terminal ray of θ contains the point $(-3, 4)$.

a What point on the same circle is on the terminal ray of $90° + \theta$? $(-4, -3)$

b Find $\cos(90° + \theta)$, $\sin(90° + \theta)$, and $\tan(90° + \theta)$. $-\frac{4}{5}; -\frac{3}{5}; \frac{3}{4}$

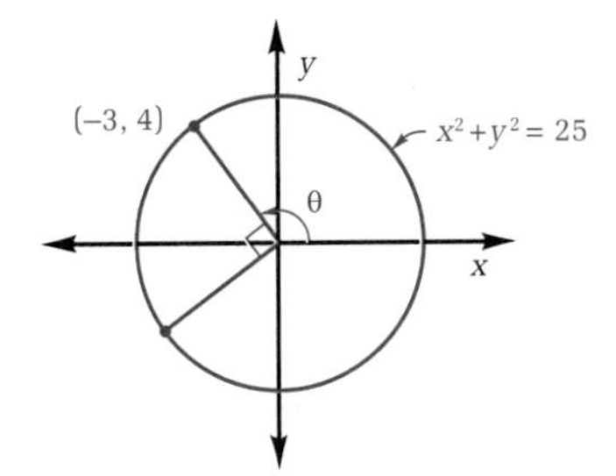

P,R | **23** Refer to the figure.

a Identify a point on the terminal side of $270° + \theta$. $(5, -12)$

b Find the exact values of $\cos(270° + \theta)$ and $\sin\theta$. $\frac{5}{13}; \frac{5}{13}$

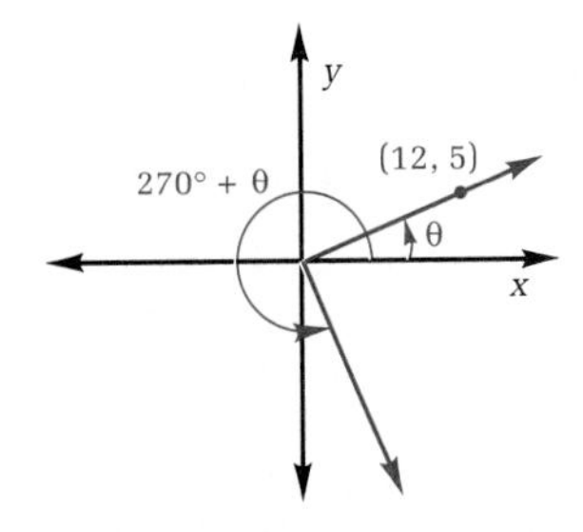

P | **24** State whether each of the following functions is even or odd.

a $f(x) = |x^2 - 4|$ Even
b $f(x) = x^3 - 2x$ Odd
c $f(x) = x^2 + \cos x$ Even
d $f(x) = \cos^2 x$ Even
e $f(x) = \sin x^2$ Even
f $f(x) = x + \sin x$ Odd

P | **25** Express each of the following as an equivalent function of an angle between 0 radians and $\frac{\pi}{2}$ radians.

a $\sin\left(\frac{1991\pi}{4}\right)$ $-\sin\left(\frac{\pi}{4}\right)$
b $\cos\left(\frac{-542\pi}{12}\right)$ $-\cos\left(\frac{\pi}{6}\right)$
c $\tan\left(\frac{-32\pi}{9}\right)$ $\tan\left(\frac{4\pi}{9}\right)$

R | **26** Simplify $\sum_{n=1}^{9} \sin[(-1)^n x]$. $-\sin x$

P | **27** Find the following for $\theta = \frac{\pi}{6}$.

a $\sin\theta$ $\frac{1}{2}$
b $\cos\theta$ $\frac{\sqrt{3}}{2}$
c $\sin\left(\frac{\pi}{2} - \theta\right)$ $\frac{\sqrt{3}}{2}$
d $\cos\left(\frac{\pi}{2} - \theta\right)$ $\frac{1}{2}$
e $\sin(\pi + \theta)$ $-\frac{1}{2}$
f $\cos(\pi + \theta)$ $-\frac{\sqrt{3}}{2}$

C R | **28** A tunnel is being dug through a mountain, a distance of two miles. The work team starting at A digs directly toward B at a rate of 4 feet per hour and an angle of depression of θ. The team at B, progressing 5 feet per hour, miscalculates and digs at an angle (α) that is $\left(\frac{1}{60}\right)°$ smaller than θ. What is the vertical error at the point where the tunnels meet?
≈ 1.71 ft

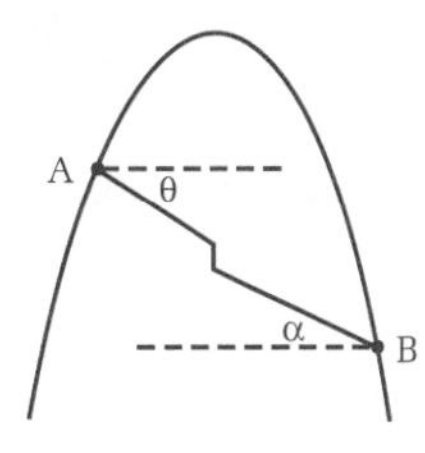

Problem-Set Notes and Strategies, continued

- Had problem **21** been stated as $\sin(x + 60°)$ and $\cos(30° - x)$, it would be easier to see that the angles are complementary. When stated in radians, it may not be as obvious. Radians will become more familiar.

Additional Answer

21 $\sin\left(\theta + \frac{\pi}{3}\right)$
$= \sin\left(\theta + \frac{\pi}{2} - \frac{\pi}{6}\right)$
$= \sin\left[\frac{\pi}{2} - \left(\frac{\pi}{6} - \theta\right)\right]$
$= \cos\left(\frac{\pi}{6} - \theta\right)$

- Many students will generalize about odd and even functions by now and be able to predict by looking at exponents of polynomials. Problem **24** contains some variants of polynomials and trigonometric equations.

- Problem **28** may need drawing some additional pictures and drawing manipulation of the trigonometric equations. To use a calculator, convert $\frac{1}{60°}$ to a decimal equivalent. It may help to spend some class time on this.

Problem-Set Notes and Strategies, continued

- Problems **29** and **30** each have a series notation for a trigonometric function that has been terminated after five or six terms. These are part of an infinite series and may be used to calculate sines and cosines of angles in radians. It also shows one is odd and the other even.

Additional Answers

Answers for problems **29** and **30** can be found in the answer pages beginning on page **A** 1.

Cooperative Learning

Use exercises **29** and **30** to evaluate the sine and cosine of various angles. Remind students to use radians. One possible question to ask is: How many terms are needed to get at least three digit accuracy (or four, five, etc)?

Communicating Mathematics

The ideas of odd and even functions will be useful when graphing any function, not just the trigonometric functions. Use the definitions of the odd and even functions to make a general statement concerning polynomials and which are odd or even.

Problem Set, *continued*

P **29** An approximation for sin x, where x is in radians, is $\sin x \approx f(x) = \frac{x}{1!} - \frac{x^3}{3!} + \frac{x^5}{5!} - \frac{x^7}{7!} + \frac{x^9}{9!}$. Prove that f is an odd function.

P **30** An approximation for cos x, where x is in radians, is $\cos x \approx f(x) = 1 - \frac{x^2}{2!} + \frac{x^4}{4!} - \frac{x^6}{6!} + \frac{x^8}{8!} - \frac{x^{10}}{10!}$. Prove that f is an even function.

CAREER PROFILE

GRAPHING THE WAVES OF THE HEART

Beba Ugarriza provides an image of electrical impulses

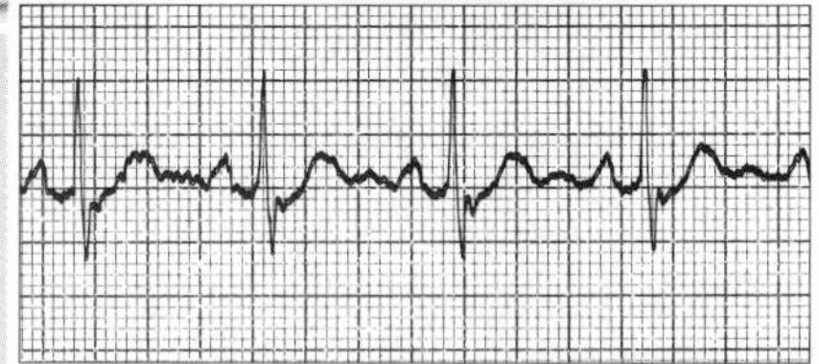

Beba Ugarriza is manager of the noninvasive cardiovascular laboratory at a Miami-area hospital. Her position involves overseeing the work of *electrocardiogram* (EKG) specialists. One of an EKG specialist's responsibilities is to operate an *electrocardiograph,* a machine that records the electrical impulses associated with the contraction and relaxation of the heart. The electrocardiogram is a graph of the amount of voltage generated by the heart against the time required for that voltage to travel through the heart. The units represented on the graph are thousandths of volts against tenths of seconds. Like the graphs of trigonometric functions, a normal EKG is periodic.

"The EKG specialist's primary responsibility, which is crucial," says Ms. Ugarriza, "is to ensure that the recordings are diagnostically pure and consistently accurate. Each EKG machine must be calibrated so that a one-millivolt standardization signal produces a deflection of exactly ten millimeters."

Some of the problems that an EKG can help identify include blocked vessels, high blood pressure, thyroid disease, and even certain kinds of malnutrition. An EKG can both identify and locate enlargement or irregularity.

To ensure accuracy, the specialist must be able to recognize variations from normal that might be caused by improper calibration of the electrocardiograph or improper administration of the EKG. Hence, the specialist must be familiar with the several distinct waves produced by each heartbeat. In a healthy heart, these are the P, Q, R, S, and T waves. The P wave represents the contraction of the atrium, which pushes blood into the ventricle. Waves Q, R, and S represent contraction of the ventricle, which pushes blood out into the body. The T wave represents the *repolarization* of the heart as the cells recharge themselves in preparation for another impulse. Sometimes, a U wave follows the T wave. A U wave can indicate low levels of potassium.

Ms. Ugarriza studied pharmaceutical science at the University of Havana. Her education and training have been rich in mathematics.

14

Chapter Summary

Concepts and Procedures

After studying this chapter, you should be able to

- Measure angles as rotations (14.1)
- Determine intersections of rays with the unit circle (14.1)
- Calculate the sine, cosine, tangent, cotangent, secant, and cosecant of an angle (14.2)
- Use the Pythagorean identities (14.2)
- Define the trigonometric functions by coordinates (14.3)
- Use special angles and reference angles to compute trigonometric functions (14.3)
- Compute the inverse trigonometric functions (14.4)
- Define the trigonometric functions in terms of right triangles (14.5)
- Use the cofunction identities (14.5)
- Describe angles in radians (14.6)
- Recognize odd and even functions (14.7)
- Identify relationships between trigonometric functions of special angles (14.7)

Vocabulary

central angle (14.1)
cofunction (14.5)
cosecant (14.2)
cosine (14.2)
cotangent (14.2)
coterminal angles (14.1)
degree mode (14.2)
even function (14.7)
initial ray (14.1)
inverse trigonometric function (14.4)
odd function (14.7)
Pythagorean identities (14.2)
radian (14.6)
reference angle (14.3)
secant (14.2)
sine (14.2)
standard position (14.1)
tangent (14.2)
terminal ray (14.1)
unit circle (14.1)

Theorems

$\cos^2 \theta + \sin^2 \theta = 1$
$\sec^2 \theta = 1 + \tan^2 \theta$
$\csc^2 \theta = 1 + \cot^2 \theta$
$\cos \theta = \sin (90° - \theta)$
$\cot \theta = \tan (90° - \theta)$
$\csc \theta = \sec (90° - \theta)$
$s = r\theta$

$\cos (-\theta) = \cos \theta$
$\sin (-\theta) = -\sin \theta$
$\tan (-\theta) = -\tan \theta$
$\sec (-\theta) = \sec \theta$
$\cot (-\theta) = -\cot \theta$
$\csc (-\theta) = -\csc \theta$
$\sin (\pi - \theta) = \sin \theta$
$\cos (\pi - \theta) = -\cos \theta$

14

Class Planning

Time Schedule
Average: $1\frac{1}{2}$ days
Advanced: 1 day
Honors: 1 day

Assignment Guide

Average	
Day 1	($\frac{1}{2}$ day) Section 14.7 9, 15, 16, 20, 22, 25
	($\frac{1}{2}$ day) 1–3, 7, 8, 11, 12, 14
Day 2	6, 9, 13, 15–17, 19–22, 23a, b, c, 24a, b, 25

Advanced
1–3, 6–8, 11, 12, 14, 15, 17, 19, 20–25

Honors
2–6, 8, 10, 12, 14, 15, 17, 18, 20–25

Problem-Set Notes and Strategies

- Problem **4** can be clearer with a picture. If students use the information to label their drawing, the balance should be reasonable.
- Students should have little difficulty with problem **7**. After they find the solution, see if they can generalize a formula for the area of a rectangle.

14 REVIEW PROBLEMS

A

P **1** The terminal ray of an angle passes through the point (−5, 12). Find the values of the six trigonometric functions for this angle.
P Sine: $\frac{12}{13}$; cosine: $-\frac{5}{13}$; tangent: $-\frac{12}{5}$; cotangent: $-\frac{5}{12}$; secant: $-\frac{13}{5}$; cosecant: $\frac{13}{12}$

P **2** Find all angles between −360° and 360° that have a reference angle of 48°. −312°, −228°, −132°, −48°, 48°, 132°, 228°, 312°

P **3** Evaluate each of the following.

a $\sin\left(\arcsin \frac{1}{2}\right)$ $\frac{1}{2}$
b $\sin\left(\text{Arcsin } \frac{1}{2}\right)$ $\frac{1}{2}$
c $\cos\left(\text{Arcsin } \frac{1}{2}\right)$ $\frac{\sqrt{3}}{2}$
d $\cos\left(\arcsin \frac{1}{2}\right)$ $\pm\frac{\sqrt{3}}{2}$

P **4** The angle of elevation from a boat to the top of a cliff is 23°. A sign is directly on top of the cliff. The angle of elevation from the boat to the top of the sign is 24°. If the sign is 40 feet high, how far is the boat from the cliff? ≈1927 ft

P,I **5** The graph represents the function y = cos x, where each x is an angle measure in degrees. What values of x are represented by points A, B, C, D, E, F, G, and H?
90; 180; 270; 360; 45; 300; 315; 150

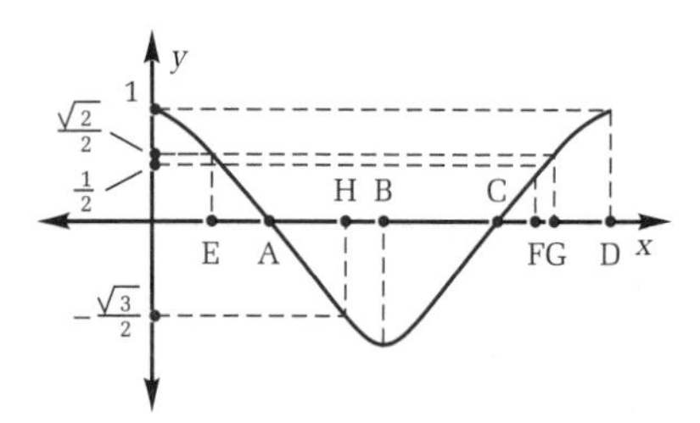

P **6** Evaluate each of the following.

a $\cos\left(\frac{35\pi}{7}\right) - \sin 12\pi$ −1
b $\tan\left(2\pi + \frac{5\pi}{4}\right)$ 1
c $\sec\left(\frac{15\pi}{2}\right)$ Undefined
d $\tan\left(\frac{13\pi}{2}\right)$ Undefined

P **7** Find the area of the rectangle shown. ≈3076.04

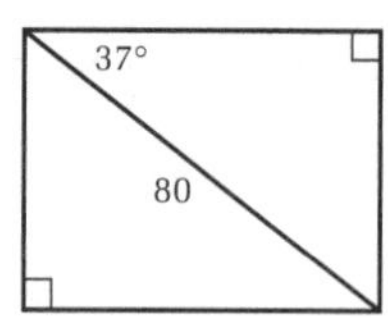

P **8** A point on the terminal side of angle θ is (−18, 15). If $0° \le \theta < 360°$, what is the measure of θ? ≈140.19

P **B** **9** Find the value of each function.

a sin ∠CAN 1
b cos ∠CAN 0
c sin ∠PAN $\frac{\sqrt{2}}{2}$
d cos ∠PAN $-\frac{\sqrt{2}}{2}$
e tan ∠PAN −1
f sin ∠CAP $\frac{\sqrt{2}}{2}$

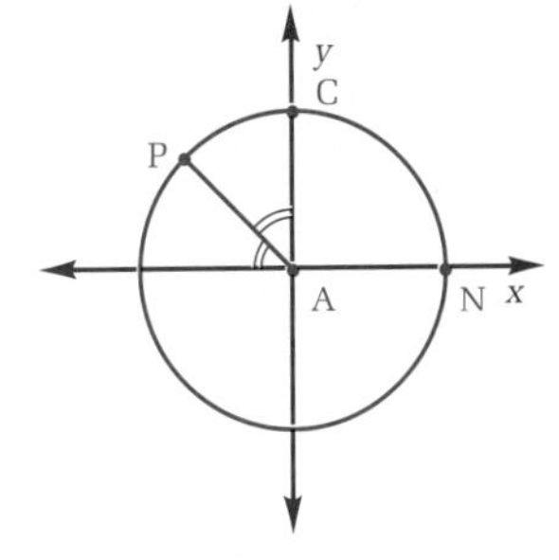

P **10** If $\ell_1 \parallel \ell_2$, what is sin θ? ≈0.9962

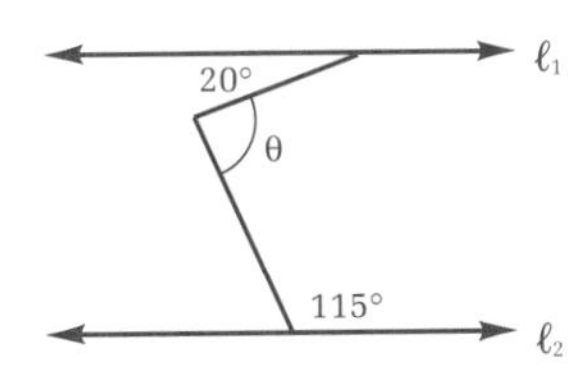

P **11** The function y = f(x), graphed at right, is odd.

a Find f(−4). −6
b Find f(1). 2

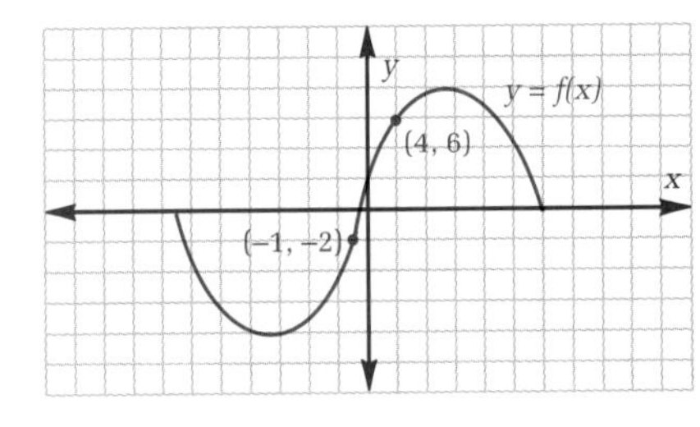

P **12** In circle O with radius 18, arc AB has a length of 3π. Find the measure of central angle AOB. 30

P **13** Suppose that sin θ = −0.3542.

a Let 270° < θ < 360°. Find cos θ without actually finding the value of θ. ≈0.9352
b If 180° < θ < 270°, what is cos θ? ≈−0.9352

P **14** Sig elevates his telescope 55° to spot the top of a building from a point on the ground 200 feet away.

a What trigonometric function of the 55° angle is equal to $\frac{h}{200}$? Tangent
b Use this function to find the height of the building. ≈286 ft

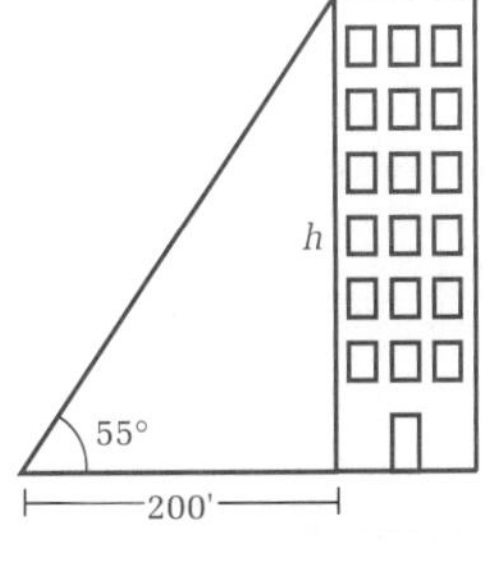

P **15** Convert each of the following radian measures to degrees. Round your answers to the nearest hundredth.

a 4.56 ≈261.27°
b −3.79 ≈−217.15°
c 15.28 ≈875.48°
d −5.25 ≈−300.80°
e −1 ≈−57.30

Problem-Set Notes and Strategies, continued

- Some geometry will help with problem **10**.

Review Problems, *continued*

P **16** If the pendulum shown swings through a $10°$ arc, $\theta = 5°$. If the pendulum is on a 33-inch long string, what is the vertical difference between its position at the bottom of a swing and at the end of a swing? ≈ 0.13 in.

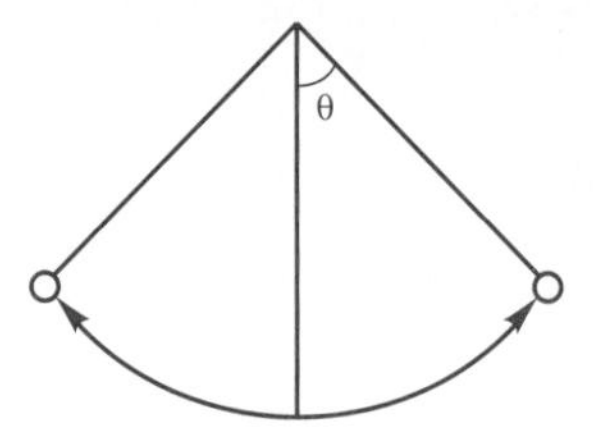

P **17** Find the points on the unit circle, $x^2 + y^2 = 1$, that are also on the graph of $y = -\frac{3}{4}x$. $\left(-\frac{4}{5}, \frac{3}{5}\right)$ and $\left(\frac{4}{5}, -\frac{3}{5}\right)$

P **18** If $\sin x = \frac{1}{3}$ and $\tan x$ is negative, what is $\cos x$? $-\frac{2\sqrt{2}}{3}$

P **19** If $\cos x = \frac{4}{5}$ and $\csc x < 0$, what is $\tan x$? $-\frac{3}{4}$

Problem-Set Notes and Strategies, continued

- Problems **18 – 20** use some of the identities from this chapter.

P **20** Simplify each of the following in terms of the given function of θ.

a $3 \tan \theta - 5 \tan(-\theta)$ $8 \tan \theta$
b $-6 \sin(-\theta) + 3 \sin \theta$ $9 \sin \theta$
c $4 \cos(-\theta) - 2 \cos \theta$ $2 \cos \theta$

P **21** For each triangle, find the radian measure of $\angle P$ in terms of π.

a $\frac{\pi}{6}$

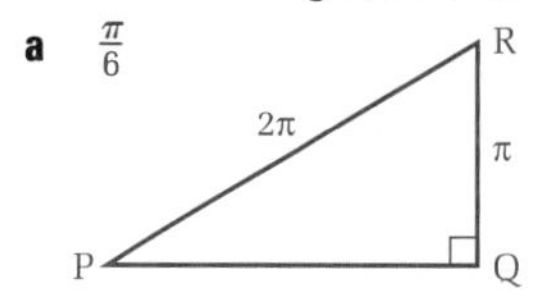

b $\frac{3\pi}{10}$

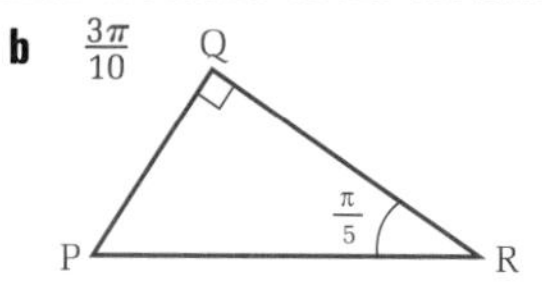

c $\frac{\pi}{4}$

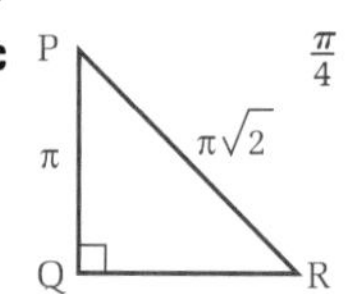

P,I **C** **22** Write an expression representing the values of θ for which $\tan 2\theta = 1$. $(22.5 + 90n)°$, where n is an integer

P **23** Find the exact value of each of the following.

a $\sin 45°$ $\frac{\sqrt{2}}{2}$
b $\tan 135°$ -1
c $\cos 225°$ $-\frac{\sqrt{2}}{2}$
d $\sin 315°$ $-\frac{\sqrt{2}}{2}$
e $\cos -135°$ $-\frac{\sqrt{2}}{2}$
f $\tan 225°$ 1

P **24** Find the missing side lengths in each triangle.

a

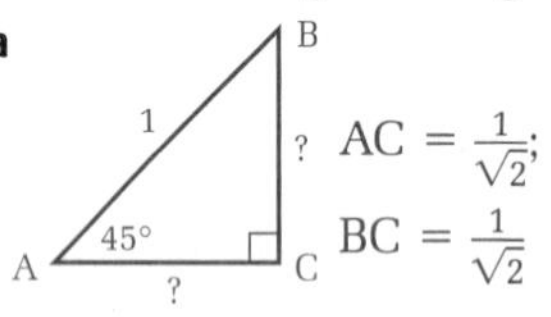

$AC = \frac{1}{\sqrt{2}}$; $BC = \frac{1}{\sqrt{2}}$

b

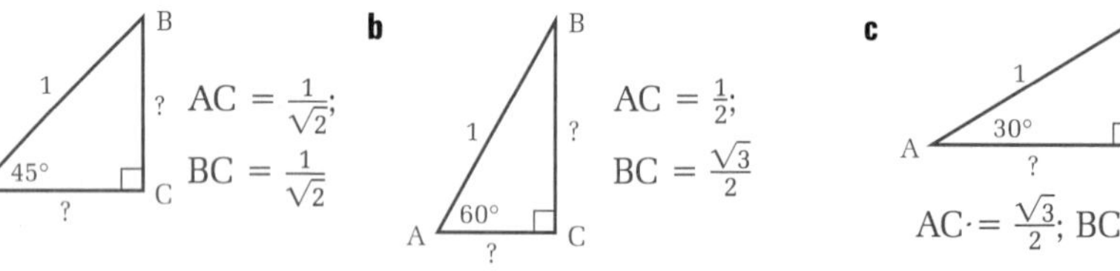

$AC = \frac{1}{2}$; $BC = \frac{\sqrt{3}}{2}$

c

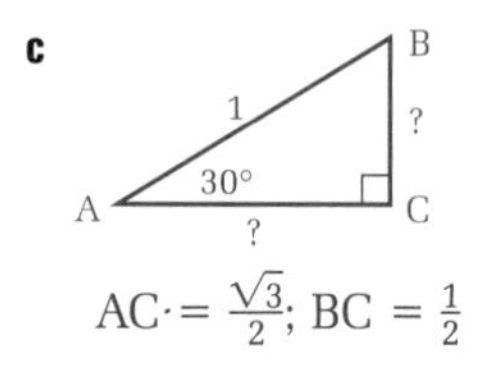

$AC = \frac{\sqrt{3}}{2}$; $BC = \frac{1}{2}$

P **25** If $\cos \theta = 0.76$ and $0° \le \theta < 90°$, what is the degree measure of θ?
≈ 40.54

14 CHAPTER TEST

Resource References
Evaluation
Test Form 1, 2, 3

1 Find the value of r. 12

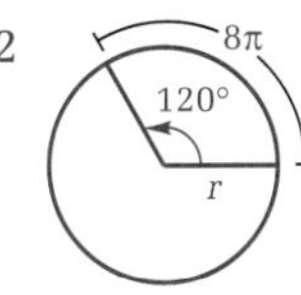

2 Find the values of the six trigonometric functions for $\theta = 200°$. $\sin\theta \approx -0.3420$; $\cos\theta \approx -0.9397$; $\tan\theta \approx 0.3640$; $\cot\theta \approx 2.747$; $\sec\theta \approx -1.064$; $\csc\theta = -2.924$

3 Which function has a greater value?

a $\sin\theta$ or $\csc\theta$ (where $0° < \theta < 90°$) $\csc\theta$

b $\sin\theta$ or $\cos\theta$ (where $90° < \theta < 180°$) $\sin\theta$

c $\tan\theta$ or $\tan(180° + \theta)$ They are equal.

4 Refer to the figure to find the following.

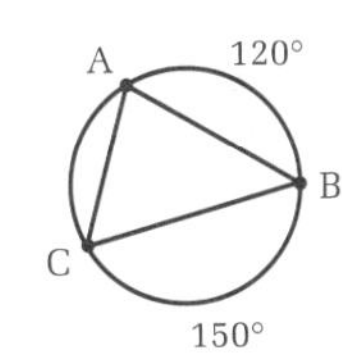

a $\sin\angle A$ and $\sin\angle B$ ≈ 0.9659; ≈ 0.7071

b $\dfrac{\sin\angle A}{\sin\angle B}$ ≈ 1.3660

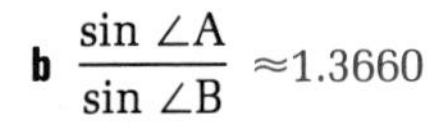

c $\dfrac{m\angle A}{m\angle B}$ $\frac{5}{3}$

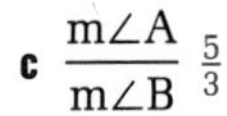

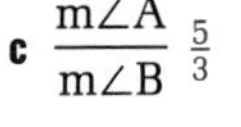

5 Find the measure of θ.
≈ 72.35

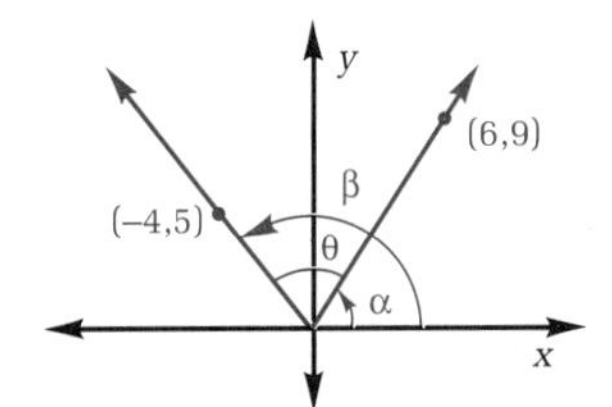

6 Explain why $\cos\theta = \sin\alpha$.
Slope of $\overline{OA} = \frac{4-0}{10-0} = \frac{2}{5}$ and slope of $\overline{OB} = \frac{5-0}{-2-0} = -\frac{5}{2}$.

Since their slopes are opposite reciprocals, $\overline{OA} \perp \overline{OB}$, so $\alpha = 90° - \theta$. Therefore, $\cos\theta = \sin\alpha$ [since $\cos\theta = \sin(90° - \theta)$].

7 Find the coordinates of point P to the nearest hundredth.
(≈ 7.66, ≈ 6.43)

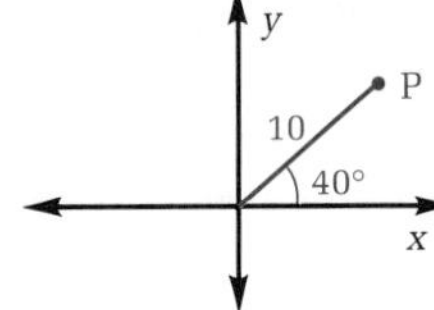

15.1

*I*N THE FIRST section of this chapter, students learn to prove trigonometric identities. A confusing concept to some students is the two uses of equality found in trigonometry when working with an identity and a trigonometric equation. In an identity, equality means identically equal — the expressions on the left and right of the equal sign are true for all possible values of the variable. On the other hand, for trigonometric equations that are not identities, the form of equality is, "For what values of the variable are the left side and right side equal?" You might want to show students the difference graphically. Graph the left and right sides of the equality separately. If the two graphs are coincident, the equation is an identity.

Students will be able to derive the double-angle and half-angle identities (Section **15.3**) from the sum and difference identities (Section **15.2**). The angle between two given lines can be found using the tangent formula for the difference of two angles.

In the last section of this chapter the product-to-sum and the sum-to-product identities are presented and students will learn to solve trigonometric equations.

CHAPTER 15

TRIGONOMETRIC IDENTITIES AND EQUATIONS

Trigonometry could be used to calculate dimensions of the houses in this peaceful scene.

15.1 BASIC TRIGONOMETRIC IDENTITIES

Objective

After studying this section, you will be able to

- Prove trigonometric identities

Part One: Introduction

By using the identities presented in Chapter 14 together with some algebra, it is possible to prove many additional identities involving the trigonometric functions. Working these proofs will show you how to transform complicated trigonometric expressions into simpler forms.

To prove that an equation is an identity, we must show that the equation holds for *every* angle θ for which both sides of the equation are defined. That is, the object is to show that the expression on the left side of the equation really does equal the expression on the right side of the equation for all possible values of θ. You may do this by performing any valid operations on these separate expressions. You may *not* use the algebraic properties of equations, however. Consider the following examples.

Example *Prove that* $\frac{1}{\sec^2\theta} + \frac{1}{\csc^2\theta} = 1$.

Often it is a good idea to change the problem into terms involving only $\sin\theta$ and $\cos\theta$.

	Left Side	**Right Side**
	$\frac{1}{\sec^2\theta} + \frac{1}{\csc^2\theta}$	$= 1$
Use the definitions of $\sec\theta$ and $\csc\theta$	$= \cos^2\theta + \sin^2\theta$	
Use a Pythagorean identity	$= 1$	

We have shown the left side to be the same as the right side, so the equation is an identity.

Class Planning

Time Schedule

All levels: 2 days

Resource References

Teacher's Resource Book

Class Opener 88

Cooperative Learning 45

Practice 93

Transparency 29

Class Opener

Find the value of θ, between 0 and $\frac{\pi}{2}$, for which $\sec^2\theta - \tan^2\theta$ is largest. Hint: try some numbers and look for a pattern.

$\sec^2\theta - \tan^2\theta = 1$. For all θ where both are defined there is no largest.

Lesson Notes

- It is important to emphasize that the algebraic properties of equations may *not* be used to verify identities. Each expression must be simplified separately and made to look like the other in order to verify.
- Changing everything to sine and cosine functions is a reasonable alternative for most of the students.

Communicating Mathematics

Have students write a paragraph explaining the importance of being able to verify identities (trigonometric or otherwise). Include examples of where this procedure might be necessary to solve a specific problem.

Using Computers

Graphing, symbolic manipulation, and equation-solving software all can be used effectively throughout the chapter. Using graphing software or a graphics calculator, students can explore and contrast trigonometric equations and trigonometric identities. One of the first things to check when asked to prove an identity is whether or not it really is an identity. By graphing the left- and right-hand sides of the identity, the student can see if the graphs coincide and therefore whether or not it really is an identity. You can also give the students a graph and ask for an equation whose graph they think will coincide, therefore having students create their own identities. One method of solving a trigonometric equation is to graph the left- and right-hand sides of the equation and determine their points of intersection. Alternatively the student can set the trigonometric equation equal to zero and then find its zeros graphically. In fact many trigonometric equations that were very hard if not impossible to solve algebraically are quite easily handled using technology. Simply consider the following:

Solve $10 \sin(x) = x + 6$

A spreadsheet program also can be used effectively throughout this chapter. Make a table of the left side and the right side of a suspected identity. Do the values in each table coincide? If they do, it is an identity. If not, it is simply a trigonometric equation. A spreadsheet program can be used effectively to locate the zeros for a trigonometric equation over a specified domain.

Example *Prove that $(\sin\theta)(\csc\theta - \sin\theta) = \cos^2\theta$. Note any restrictions.*

	Left Side	**Right Side**
	$(\sin\theta)(\csc\theta - \sin\theta)$	$= \cos^2\theta$
Distribute	$= \sin\theta \cdot \csc\theta - \sin^2\theta$	
Use the definition of $\csc\theta$	$= \sin\theta \cdot \dfrac{1}{\sin\theta} - \sin^2\theta$	
	$= 1 - \sin^2\theta$	
Use a Pythagorean identity	$= \cos^2\theta$	

Since the left side is identical to the right side, the equation is an identity. Note that because $\csc\theta = \frac{1}{\sin\theta}$, the left side of the original equation is not defined when $\sin\theta = 0$. So $\sin\theta \neq 0$ is a restriction on the identity.

Example *Prove the identity $\frac{2\cos^2\theta - \cos\theta - 1}{\cos\theta - 1} = \frac{\sec\theta + 2}{\sec\theta}$. Note any restrictions.*

Since both sides of the equation are fairly complicated, it makes sense to simplify each side individually.

Left Side	**Right Side**
$\dfrac{2\cos^2\theta - \cos\theta - 1}{\cos\theta - 1}$	$= \dfrac{\sec\theta + 2}{\sec\theta}$
$= \dfrac{(2\cos\theta + 1)(\cos\theta - 1)}{\cos\theta - 1}$	$= \dfrac{\frac{1}{\cos\theta} + 2}{\frac{1}{\cos\theta}}$
$= 2\cos\theta + 1$	$= \left(\dfrac{1}{\cos\theta} + 2\right) \cdot \cos\theta$
	$= 1 + 2\cos\theta$

Note that the left side is undefined for $\cos\theta = 1$ and that because $\sec\theta = \frac{1}{\cos\theta}$, the right side is undefined for $\cos\theta = 0$. So the restrictions are $\cos\theta \neq 0$ and $\cos\theta \neq 1$.

We summarize the process of proving identities with some helpful hints.

A Strategy for Proving Identities

1. **Separate the equation into a left-side expression and a right-side expression and work with these separately.**
2. **Simplify the more complicated side of the equation first, looking for ways to use the basic trigonometric identities.**
3. **After simplifying one side as much as you can, work on the other side until both sides are the same.**
4. **If no other approach comes to mind, try expressing both sides in terms of sines and cosines.**

Part Two: Sample Problems

Problem 1 *Prove that $(\cos\theta)(1-\cos^2\theta) = \cot\theta \cdot \sin^3\theta$. Note any restrictions.*

Solution

Left Side	Right Side
$(\cos\theta)(1-\cos^2\theta)$	$= \cot\theta \cdot \sin^3\theta$
$= \cos\theta \cdot \sin^2\theta$	$= \frac{\cos\theta}{\sin\theta} \cdot \sin^3\theta$
	$= \cos\theta \cdot \sin^2\theta$

The right side of the identity is undefined for $\sin\theta = 0$, so $\sin\theta \neq 0$ is a restriction.

Problem 2 *Prove that $\frac{\sin\theta}{1-\cos\theta} = \csc\theta + \cot\theta$. Note any restrictions.*

Solution Multiplying a fraction by a form of 1 is often a useful technique in proving identities.

Left Side

$$\frac{\sin\theta}{1-\cos\theta}$$

Multiply by a form of 1 $$= \frac{\sin\theta}{1-\cos\theta} \cdot \frac{1+\cos\theta}{1+\cos\theta}$$

$$= \frac{(\sin\theta)(1+\cos\theta)}{1-\cos^2\theta}$$

Use a Pythagorean identity $$= \frac{(\sin\theta)(1+\cos\theta)}{\sin^2\theta}$$

Simplify $$= \frac{1+\cos\theta}{\sin\theta}$$

Right Side

$$\csc\theta + \cot\theta$$

Use the definitions of $\csc\theta$ and $\cot\theta$ $$= \frac{1}{\sin\theta} + \frac{\cos\theta}{\sin\theta}$$

Add the fractions $$= \frac{1+\cos\theta}{\sin\theta}$$

The left side of the original identity is undefined for $\cos\theta = 1$. The right side is undefined for $\sin\theta = 0$. Therefore, the restrictions are $\cos\theta \neq 1$ and $\sin\theta \neq 0$.

Problem 3 *Show that the equation $\frac{\sin\theta}{1+\sin\theta} = \frac{1-\sin\theta}{\sin\theta}$ is not an identity.*

Solution To show that an equation is not an identity, we need only find one value of θ for which the left side and the right side are not equal. Let's try $\theta = 30°$.

Checkpoint

1. Determine whether $\sin\theta = \frac{1}{\sec\theta}$ False
2. If $\sin\theta = \frac{1}{3}$ and θ is in the second quadrant, use an appropriate identity to find $\cos\theta$. $-\frac{\sqrt{8}}{3}$
3. Find another expression that is equivalent to $\sin\theta\cos^2\theta$. Answers will vary, but most common will be $\sin\theta - \sin^3\theta$.

Assignment Guide

Average	
Day 1	8, 9, 11, 12, 14, 16, 17a, b, 19
Day 2	10, 18a, b, 21, 22a, c, 24a, c, 25c, 27
Advanced	
Day 1	9a, c, 10, 12–14, 16, 17, 19, 27
Day 2	11, 18, 21, 22a, b, c, 23, 25a, b, 29a, c
Honors	
Day 1	9, 10, 12–14, 16, 17, 19, 27
Day 2	18, 20, 21, 23–26, 28, 29a, c

Stumbling Block

Students have been trained in the use of the algebraic properties to solve equations and are now told not to use them to prove identities. Have them look at the equation $\sin\theta = \sqrt{1-\cos^2\theta}$. Squaring both sides yields the basic Pythagorean identity, but the left side can be both positive and negative and the right side only nonnegative. This is only an identity if the domain is restricted. Make sure that students work on each expression independently without regard to the other side of the equation.

Problem-Set Notes and Strategies

- To verify an equation for a specific value as in problems **4** and **5** means to just substitute that value and see if it is true. An equation may be true for a specific value and still not be an identity.
- Problem **8** introduces the sum identity for cosine with specific values. This will be generalized in the next section.

Additional Answers

1 $\csc^4\theta$
$= (\csc^2\theta)^2$
$= (1+\cot^2\theta)^2$; $\sin\theta \neq 0$

Answers for problems **2**, **6**, and **7** can be found in the answer pages beginning on **A** 1.

Left Side	Right Side
$\dfrac{\sin\theta}{1+\sin\theta}$	$= \dfrac{1-\sin\theta}{\sin\theta}$
$= \dfrac{\sin 30^\circ}{1+\sin 30^\circ}$	$= \dfrac{1-\sin 30^\circ}{\sin 30^\circ}$
$= \dfrac{\frac{1}{2}}{1+\frac{1}{2}}$	$= \dfrac{1-\frac{1}{2}}{\frac{1}{2}}$
$= \dfrac{1}{3}$	$= 1$

Since the expressions are not equal for $\theta = 30^\circ$, the equation is not an identity.

Part Three: Exercises and Problems

Warm-up Exercises

1 Prove that $\csc^4\theta = (1+\cot^2\theta)^2$. State restrictions.

P,R **2** Prove that $(\cos^2\theta)(\tan^2\theta + 1) = 1$. State restrictions.

P,R **3** Determine whether each statement is True or False.
a $\csc^2\theta - \cot^2\theta = 1$ True
b $\tan^2\theta = (\sec\theta + 1)(\sec\theta - 1)$ True
c $\sin^2\theta = \cos^2\theta - 1$ False

P,R,I **4** Determine whether each statement is true for $\theta = \frac{\pi}{6}$.
a $\tan\theta \cdot \cos\theta = \sin\theta$ Yes
b $\csc^2\theta - 1 = \cot^2\theta$ Yes
c $(\sin\theta + \cos\theta)^2 = 1 + \sin(2\theta)$ Yes
d $\cos(2\theta) = \cos^2\theta - \sin^2\theta$ Yes

P,R **5** Determine whether each statement is true for $\theta = \frac{\pi}{2}$.
a $\cos\theta \cdot \sec\theta = 1$ No
b $\dfrac{\sin\theta}{\cos\theta} = \tan\theta$ No

P,R **6** Prove that $\frac{1}{\cos^2\theta} - \frac{\sin^2\theta}{\cos^2\theta} = 1$. State restrictions.

P,R **7** Prove that $\frac{\cos^2\theta - \sin^2\theta}{\cos\theta - \sin\theta} = \cos\theta + \sin\theta$. State restrictions.

Problem Set

P,R **A** **8** Determine whether each statement is True or False.
a $\cos(30^\circ + 60^\circ) = \cos 30^\circ + \cos 60^\circ$ False
b $\cos(30^\circ + 60^\circ) = \cos 30^\circ \cos 60^\circ - \sin 30^\circ \sin 60^\circ$ True

P,I **9** Use identities to evaluate each of the following.

a $1 + \cot^2 30°$ 4
b $(\sec 10° - \tan 10°)(\sec 10° + \tan 10°)$ 1
c $\sin 15° \cdot \cot 15° \cdot \sec 15°$ 1

P,R **10** Prove that $(\csc \theta + \cot \theta)^2(\csc \theta - \cot \theta)^2 = 1$. State restrictions.

P,R **11** Simplify each of the following to a single trigonometric function ($\sin x$, $\cos x$, $\tan x$, $\sec x$, $\csc x$, or $\cot x$).

a $\dfrac{\cos x}{1 - \sin^2 x}$ $\sec x$ **b** $\sin x + \dfrac{\cos^2 x}{\sin x}$ $\csc x$

c $\dfrac{\cos x - \sin x}{\cot x - 1}$ $\sin x$ **d** $1 - \dfrac{\sin^2 x}{1 + \cos x}$ $\cos x$

R **12** **a** Find the coordinates of P. $\left(\frac{\sqrt{3}}{2}, \frac{1}{2}\right)$
b Find the coordinates of Q. $\left(\frac{1}{2}, \frac{\sqrt{3}}{2}\right)$
c Find distance PQ to the nearest thousandth. ≈ 0.518

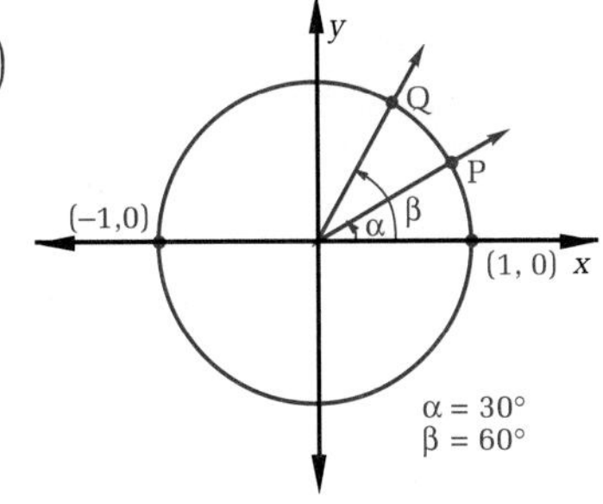

P,R **13** Prove that $\frac{\cos^3 \theta - \sin^3 \theta}{\cos \theta - \sin \theta} = 1 + \cos \theta \sin \theta$. State restrictions.

P **14** Prove that $\frac{\sin \theta}{1 + \cos \theta} = \frac{\cos \theta}{1 + \sin \theta}$ is *not* an identity.

R **15** Use a calculator to evaluate the following.

a $\sin^2 \frac{\pi}{10} + \cos^2 \frac{\pi}{10}$ 1 **b** $\sin \frac{\pi}{10} + \cos \frac{\pi}{10}$ ≈ 1.2601

P,R **16** Ida Entity supplied the following proof that $\sin \theta = \sqrt{1 - \cos^2 \theta}$: "If we square both sides, we get $\sin^2 \theta = 1 - \cos^2 \theta$, which is a valid identity. Therefore, $\sin \theta = \sqrt{1 - \cos^2 \theta}$ is an identity."

a Show that $\sin \theta = \sqrt{1 - \cos^2 \theta}$ is not an identity by showing that it is false for $\theta = 210°$.
b What did Ida do wrong?

P,R,I **17** Determine whether each statement is True or False.

a $\cos 60° = 2 \cdot \cos 30°$ False **b** $\cos 60° = \cos^2 30° - \sin^2 30°$ True
c $\sin 90° = 2 \cdot \sin 45°$ False **d** $\sin 90° = 2 \cdot \sin 45° \cos 45°$ True

B

R **18** Evaluate each of the following without using a calculator.

a $\sin\left(\text{Arcsin } \frac{1}{3}\right)$ $\frac{1}{3}$ **b** $\sin\left(\text{Arccos } \frac{1}{3}\right)$ $\frac{2\sqrt{2}}{3}$ **c** $\sin\left(\text{Arccos } \frac{3}{5}\right)$ $\frac{4}{5}$

Problem-Set Notes and Strategies, continued

- Problem **9** illustrates that it is sometimes easier to simplify an expression using identities rather than immediately reaching for a calculator. This is especially true for cases where the numerical results of the intermediate steps are irrational, but the final solution is rational.

Additional Answers

10 $(\csc \theta + \cot \theta)^2(\csc \theta - \cot \theta)^2$
$= (\csc^2 \theta - \cot^2 \theta)^2$
$= 1^2 = 1$; $\sin \theta \neq 0$

14 When $\theta = 30°$, $\frac{\sin 30°}{1 + \cos 30°} \approx 0.2679$ and $\frac{\cos 30°}{1 + \sin 30°} \approx 0.5774$. Since the expressions are not equal for $\theta = 30°$, the equation is not an identity.

Answers for problems **13**, **16a** and **16b** can be found in the answer pages beginning on **A** 1.

- To prove an equation is not an identity is to merely find one value for which the equation is not true. This may be used to solve problem **14**.

- In problem **15**, make sure that the students are using radians and not degrees. This is the most common error made when evaluating an expression.

- Problem **17b** and **d** preview the double angle identities.

- Diagrams are especially helpful in problem **18**. These will be very important especially to those students who go on into calculus classes.

Problem-Set Notes and Strategies, continued

- Problem **19** can be done with a simple plotting of points. Take ordinates from the graph of $f(x)$ and square them. With enough points, the second graph is fairly simple.
- The sum formula for the tangent function is previewed in problem **20**.
- Problem **24** is an exercise in relating algebraic functions to trigonometric functions. Once part **a** is completed, part **b** is a simple quadratic equation. Make sure they realize the answer must be negative to solve the original equation.
- Problem **26** will take some work and time.

Additional Answers

19

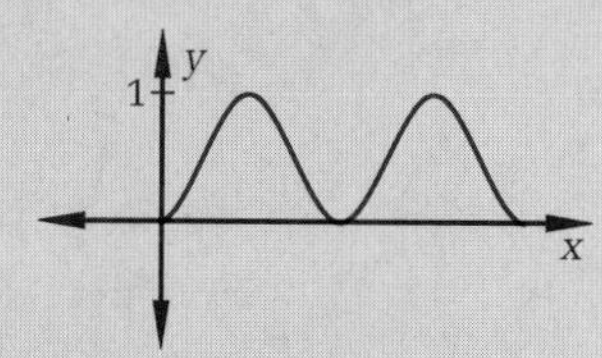

21b $\dfrac{(a+b)^2}{a^2+b^2}$

21c $\dfrac{a^2+b^2}{a^2+b^2}$, or 1

26b

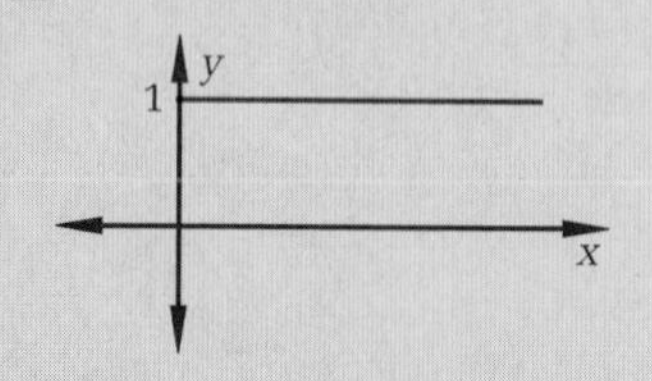

Answers for problems **21a**, **23a – c**, **26a**, and **26b** can be found in the answer pages beginning on **A** 1.

Problem Set, *continued*

R,I **19** The graph of $y = f(x)$ is shown. Draw the graph of $y = [f(x)]^2$.

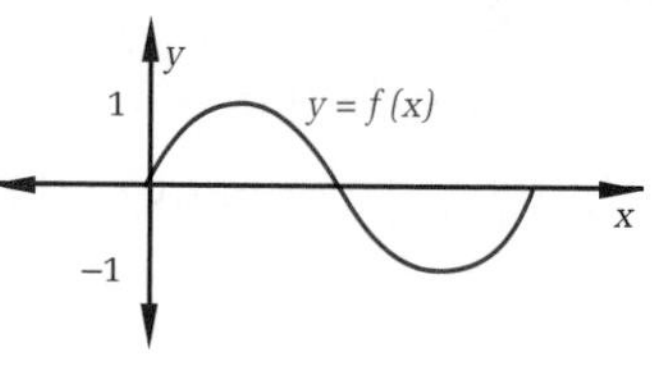

P,R,I **20** Determine whether each equation is True or False.

a $\tan(30° + 120°) = \tan 30° + \tan 120°$ False

b $\tan(30° + 120°) = \dfrac{\tan 30° + \tan 120°}{1 - \tan 30° \tan 120°}$ True

P,R **21** **a** Use the figure to show that $\dfrac{\sin\theta}{1-\cos\theta} = \csc\theta + \cot\theta$.

b Evaluate $(\sin\theta + \cos\theta)^2$.

c Evaluate $(\sin\theta)^2 + (\cos\theta)^2$.

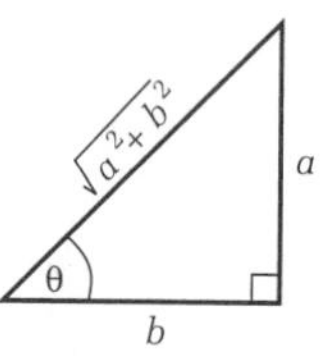

P,R **22** Simplify to a single trigonometric function.

a $\dfrac{\tan^2 x + 1}{\tan x + \cot x}$ tan x

b $\dfrac{\sin\theta + \tan\theta}{1 + \sec\theta}$ sin x

c $\dfrac{\csc x}{\cot x + \tan x}$ cos x

d $\dfrac{\sin x}{(\sec x + 1)(1 - \cos x)}$ cot x

P,R **23** Prove each identity. State restrictions.

a $\dfrac{\tan x - 1}{\tan x + 1} = \dfrac{1 - \cot x}{1 + \cot x}$

b $\cot^2 x - \cos^2 x = \cot^2 x \cdot \cos^2 x$

c $(\tan x + \sec x)^2 = \dfrac{1 + \sin x}{1 - \sin x}$

R **24** Let $\sin x = a + 1$.

a Find $\sin x + \csc x$. $\dfrac{a^2 + 2a + 2}{a + 1}$

b Find the value of a for which $\sin x + \csc x = 3$. ≈ -0.618

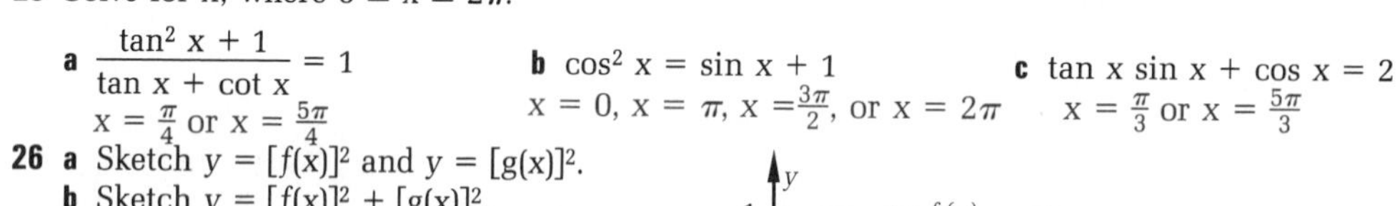

P,R,I **25** Solve for x, where $0 \le x \le 2\pi$.

a $\dfrac{\tan^2 x + 1}{\tan x + \cot x} = 1$
$x = \frac{\pi}{4}$ or $x = \frac{5\pi}{4}$

b $\cos^2 x = \sin x + 1$
$x = 0, x = \pi, x = \frac{3\pi}{2}$, or $x = 2\pi$

c $\tan x \sin x + \cos x = 2$
$x = \frac{\pi}{3}$ or $x = \frac{5\pi}{3}$

R,I **26** **a** Sketch $y = [f(x)]^2$ and $y = [g(x)]^2$.

b Sketch $y = [f(x)]^2 + [g(x)]^2$.

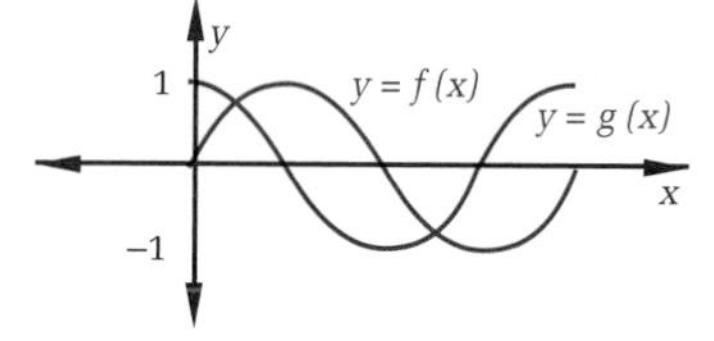

P,R **27** Prove that $\log_2 (\tan \theta) + \log_2 (\cos \theta) = \log_2 (\sin \theta)$. State restrictions.

P,R C **28** Prove that $\frac{2 \sin^2 \theta + 5 \sin \theta \cos \theta + 2 \cos^2 \theta}{\sin \theta + 2 \cos \theta} = \frac{2 \sec \theta + \csc \theta}{\sec \theta \csc \theta}$.
State restrictions.

P,R,I **29** Solve for $\{x:0 \le x \le 2\pi\}$.

a $(2 \cos x - \sqrt{2})(\tan x - \sqrt{3}) = 0$ $\left\{\frac{\pi}{4}, \frac{7\pi}{4}, \frac{\pi}{3}, \frac{4\pi}{3}\right\}$

b $\sin^2 (3x) = \frac{1}{2}$ $\left\{\frac{\pi}{12}, \frac{\pi}{4}, \frac{5\pi}{12}, \frac{7\pi}{12}, \frac{3\pi}{4}, \frac{11\pi}{12}, \frac{13\pi}{12}, \frac{5\pi}{4}, \frac{17\pi}{12}, \frac{19\pi}{12}, \frac{7\pi}{4}, \frac{23\pi}{12}\right\}$

c $2 \cos^2 x + 7 \cos x + 3 = 0$ $\left\{\frac{2\pi}{3}, \frac{4\pi}{3}\right\}$

d $\tan^2 x = 3(\sec x - 1)$ $\left\{\frac{\pi}{3}, \frac{5\pi}{3}, 0, 2\pi\right\}$

e $\sin x + \cos x = 1$ $\left\{0, \frac{\pi}{2}, 2\pi\right\}$

Problem-Set Notes and Strategies, continued

- Problem **27** is a good review of the properties of logarithms. It also illustrates that logarithms and trigonometric functions may be combined.

Cooperative Learning

Have students work in small groups to solve problems **22**, **23**, **27**, and **28**. Then choose one or two students from each group to present the group's solution to the class. It is useful to hear other students' reasoning. There is seldom a single right method. An answer usually results from the way an individual initially views the problem.

Additional Answers

Answers for problems **27** and **28** can be found in the answer pages beginning on **A 1**.

HISTORICAL SNAPSHOT

THE ORIGIN OF TRIGONOMETRY

Astronomers develop a mathematical tool

Trigonometry was invented by astronomers as a means to an end. It grew out of their attempts to describe the paths of the sun, moon, planets, and stars.

Hipparchus of Nicaea (*c.* 180 B.C.–*c.* 125 B.C.) generally is credited with compiling the first trigonometric table, which gives corresponding arc and chord values for a series of angles. Hipparchus' other contributions to astronomy include a star catalog, improvements in astronomical constants, and the discovery of the *precession of the equinoxes*, a phenomenon caused by a wobbling of the earth on its axis.

Many of Hipparchus' achievements are presented in the works of Claudius Ptolemy (*c.* A.D. 150). Ptolemy also contributed to the development of trigonometry. In his *Almagest*, he gives a formula for the sine and cosine of the sum and difference of two angles. The *Almagest* also contains a table of chords belonging to different angles, ascending by half angles. By applying his own half-angle formula, Ptolemy could approximate π as 3.1416.

CLAVDII PTOLEMAEI
PHELVDIENSIS ALEXANDRINI
ALMAGESTVM SEV MAGNAE CONSTRVCTIONIS
MATHEMATICAE OPVS PLANE DIVINVM
LATINA DONATVM LINGVA
AB GEORGIO TRAPEZVNTIO VSQVEQVAQ.
DOCTISSIMO.
PER LVCAM GAVRICVM NEAPOLIT. DIVINAE
MATHESEOS PROFESSOREM EGREGIVM
IN ALMA VRBE VENETA ORBIS REGINA
RECOGNITVM
ANNO SALVTIS M D XXVIII LABENTE

L A

Nequispiam alius Calcographus, Venetiis aut usquã locorum Venetæ ditionis impune Almagestum hunc imprimat per Decennium, Senatus Veneti Decreto cautum est.

Complete the trigonometric table below, using the sum and differences and the half-angle formulas.

Degree	Sin	Cos
0° 00'	0.0000	1.000
0° 30'		
1° 00'	0.0175	0.9998
33° 00'	0.5446	0.8387
33° 30'		

15.2

Class Planning

Time Schedule
All levels: 2 days

Resource References
Teacher's Resource Book
Class Opener 89
Practice 94
Evaluation
Quiz Forms 1, 2, 3

Class Opener

Prove or disprove the following:

$$\frac{\sec^2\theta - \tan^2\theta}{\sec\theta - \tan\theta} = \frac{\cos\theta}{1+\sin\theta}$$

$$\frac{\sec^2\theta - \tan^2\theta}{\sec\theta - \tan\theta} = \frac{(\sec\theta - \tan\theta)(\sec\theta + \tan\theta)}{\sec\theta - \tan\theta}$$

$$= \sec\theta + \tan\theta = \frac{1+\sin\theta}{\cos\theta}$$

$$= \frac{(1+\sin\theta)}{\cos\theta}\cdot\frac{(1-\sin\theta)}{(1-\sin\theta)} = \frac{\cos\theta}{1-\sin\theta}$$

Lesson Notes

- Formulas for the sums and differences of the sine, cosine and tangent functions are developed in this section.

15.2 The Sum and Difference Identities

Objectives

After studying this section, you will be able to
- Use the sum and difference identities
- Determine the angle between two graphed lines

Part One: Introduction

The Sum and Difference Identities

Does $\cos(60° + 90°) = \cos 60° + \cos 90°$? Let's check and see.

Left Side	**Right Side**
$\cos(60° + 90°)$	$= \cos 60° + \cos 90°$
$= \cos 150°$	$= \frac{1}{2} + 0$
$= \frac{-\sqrt{3}}{2}$	$= \frac{1}{2}$

The two expressions are not equal. We will develop a correct identity for $\cos(\alpha + \beta)$.

In the diagram, $\alpha > 0$ and $\beta > 0$. The coordinates of D are (1, 0). The coordinates of A and B are $(\cos\alpha, \sin\alpha)$ and $(\cos(\alpha+\beta), \sin(\alpha+\beta))$ respectively, from the definitions of the trigonometric functions. The coordinates of C are $(\cos\beta, -\sin\beta)$, since $\cos(-\beta) = \cos\beta$ and $\sin(-\beta) = -\sin\beta$. Using the distance formula, we have

$$(BD)^2 = [\cos(\alpha+\beta) - 1]^2 + \sin^2(\alpha+\beta)$$
$$= \cos^2(\alpha+\beta) - 2\cos(\alpha+\beta) + 1 + \sin^2(\alpha+\beta)$$
$$= 2 - 2\cos(\alpha+\beta)$$

$$(AC)^2 = (\cos\alpha - \cos\beta)^2 + (\sin\alpha + \sin\beta)^2$$
$$= \cos^2\alpha - 2\cos\alpha\cdot\cos\beta + \cos^2\beta + \sin^2\alpha + 2\sin\alpha\cdot\sin\beta + \sin^2\beta$$
$$= 2 - 2\cos\alpha\cdot\cos\beta + 2\sin\alpha\cdot\sin\beta$$

Since the central angles for chord BD and chord AC are equal, the chords must be the same length, so $(BD)^2 = (AC)^2$. Therefore,

$$2 - 2\cos(\alpha+\beta) = 2 - 2\cos\alpha\cdot\cos\beta + 2\sin\alpha\cdot\sin\beta$$
$$-2\cos(\alpha+\beta) = -2\cos\alpha\cdot\cos\beta + 2\sin\alpha\cdot\sin\beta$$
$$\cos(\alpha+\beta) = \cos\alpha\cdot\cos\beta - \sin\alpha\cdot\sin\beta$$

$\cos(\alpha + \beta) = \cos\alpha\cos\beta - \sin\alpha\sin\beta$

Vocabulary
sum and difference identities

Example *Verify that* $\cos(60° + 90°) = \cos 60° \cos 90° - \sin 60° \sin 90°$.

Left Side

$\cos(60° + 90°)$

$= \cos 150°$

$= \dfrac{-\sqrt{3}}{2}$

Right Side

$= \cos 60° \cos 90° - \sin 60° \sin 90°$

$= \dfrac{1}{2} \cdot 0 - \dfrac{\sqrt{3}}{2} \cdot 1$

$= \dfrac{-\sqrt{3}}{2}$

We can replace β with $-\beta$ to get the identity for the cosine of a difference.

$$\cos(\alpha - \beta) = \cos\alpha \cos\beta + \sin\alpha \sin\beta$$

To develop the identity for the sine of a sum, we use the cofunction identity $\sin x = \cos\left(\frac{\pi}{2} - x\right)$.

$$\sin(\alpha + \beta) = \cos\left[\frac{\pi}{2} - (\alpha + \beta)\right]$$

$$= \cos\left[\left(\frac{\pi}{2} - \alpha\right) - \beta\right]$$

$$= \cos\left(\frac{\pi}{2} - \alpha\right) \cdot \cos\beta + \sin\left(\frac{\pi}{2} - \alpha\right) \cdot \sin\beta$$

$$= \sin\alpha \cdot \cos\beta + \cos\alpha \cdot \sin\beta$$

Similarly, we can show that $\sin(\alpha - \beta) = \sin\alpha \cdot \cos\beta - \cos\alpha \cdot \sin\beta$.

$$\sin(\alpha + \beta) = \sin\alpha \cos\beta + \cos\alpha \sin\beta$$

$$\sin(\alpha - \beta) = \sin\alpha \cos\beta - \cos\alpha \sin\beta$$

Example *Find the exact value of* $\cos\left(\text{Arccos}\,\frac{3}{5} + \text{Arcsin}\,\frac{-5}{13}\right)$.

Let $\alpha = \text{Arccos}\,\frac{3}{5}$ and $\beta = \text{Arcsin}\,\frac{-5}{13}$.

$$\cos\left(\text{Arccos}\,\frac{3}{5} + \text{Arcsin}\,\frac{-5}{13}\right)$$

$= \cos(\alpha + \beta)$

$= \cos\alpha \cdot \cos\beta - \sin\alpha \cdot \sin\beta$

$= \dfrac{3}{5} \cdot \dfrac{12}{13} - \dfrac{4}{5} \cdot \dfrac{-5}{13}$

$= \dfrac{56}{65}$

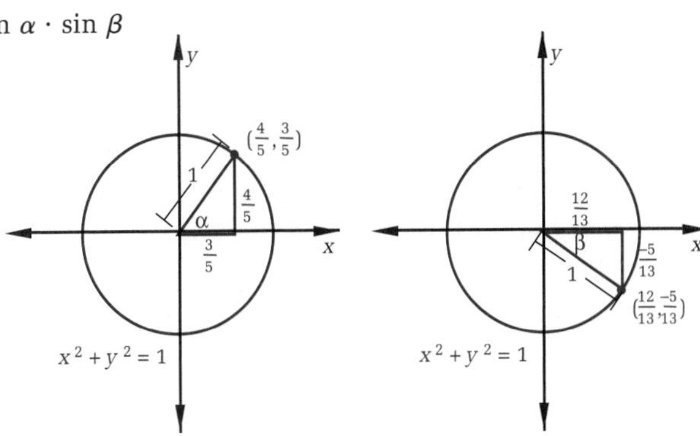

Lesson Notes, continued

- An additional application of the tangent identity allows the angle between two lines to be found using the slope of the lines. Slopes are related to tangents. You might tell students that slopes are extremely important as rates of change and they will be spending more time studying slopes later.

Stumbling Block

Each of the formulas for sines and cosines of sums and differences contains a sum or a difference . Students frequently mistake the signs in the respective formulas. In the case of the sine, the signs are the same, i.e. for the sine of a sum it is the sum of functions. For the sine of a difference, it is the difference of functions. When the cosine function is considered, the signs are reversed.

Communicating Mathematics

In this section a formula was developed to find the angle between two lines. Have students find examples of where it might be necessary to know the angle between two lines and explain its significance.

Checkpoint

1. Find the exact value of the sine of 75°. $\frac{\sqrt{2}+\sqrt{6}}{4}$
2. Find the angle between the line $y = 3x + 4$ and the line $y = x - 2$. 26.6°
3. Evaluate $\sin(\text{Arcsin } x - \text{Arccos } y)$ if Arcsin x and Arccos y are both first quadrant angles. $xy - \sqrt{(1-x^2)(1-y^2)}$
4. Use the difference identity to prove that $\sin(-x) = -\sin x$. Write $\sin(-x)$ as $\sin(0 - x)$ and simplify.

Example *Find the exact value of sin 15°.*

$$\begin{aligned} &\sin 15° \\ &= \sin(45° - 30°) \\ &= \sin 45° \cdot \cos 30° - \cos 45° \cdot \sin 30° \\ &= \frac{\sqrt{2}}{2} \cdot \frac{\sqrt{3}}{2} - \frac{\sqrt{2}}{2} \cdot \frac{1}{2} \\ &= \frac{\sqrt{6} - \sqrt{2}}{4} \end{aligned}$$

The Angle Between Two Lines

We can derive an identity for the tangent of a sum as follows.

$$\begin{aligned} \tan(\alpha + \beta) &= \frac{\sin(\alpha + \beta)}{\cos(\alpha + \beta)} \\ &= \frac{\sin\alpha\cos\beta + \cos\alpha\sin\beta}{\cos\alpha\cos\beta - \sin\alpha\sin\beta} \\ &= \frac{\sin\alpha\cos\beta + \cos\alpha\sin\beta}{\cos\alpha\cos\beta - \sin\alpha\sin\beta} \cdot \frac{\frac{1}{\cos\alpha\cos\beta}}{\frac{1}{\cos\alpha\cos\beta}} \\ &= \frac{\tan\alpha + \tan\beta}{1 - \tan\alpha\tan\beta} \end{aligned}$$

Similarly, we can obtain an identity for $\tan(\alpha - \beta)$.

$$\tan(\alpha + \beta) = \frac{\tan\alpha + \tan\beta}{1 - \tan\alpha\tan\beta}$$

$$\tan(\alpha - \beta) = \frac{\tan\alpha - \tan\beta}{1 + \tan\alpha\tan\beta}$$

A line whose equation is $y = mx + b$ has slope m. Let θ be the angle that the line forms with the x-axis. From Chapter 14 we know that $\tan\theta$ is equal to $\frac{\text{opposite side}}{\text{adjacent side}}$, which is also the slope of the line. Therefore, $\tan\theta = m$.

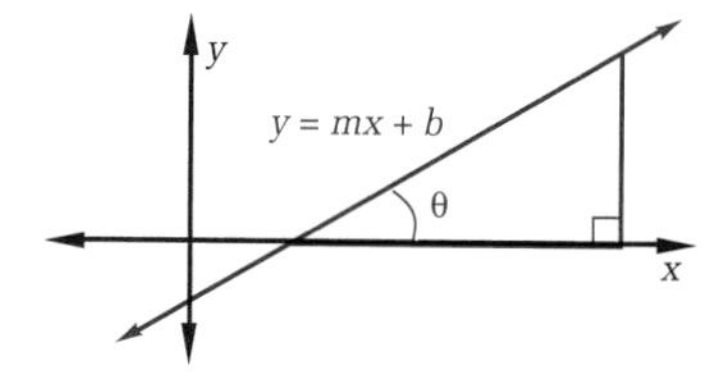

If two lines $y = m_1x + b_1$ and $y = m_2x + b_2$ form angles of θ_1 and θ_2 with the x-axis and if $\theta_2 - \theta_1 > 0$, then the angle between the lines is $\theta_2 - \theta_1$. We know that $\tan\theta_2 = m_2$ and $\tan\theta_1 = m_1$.

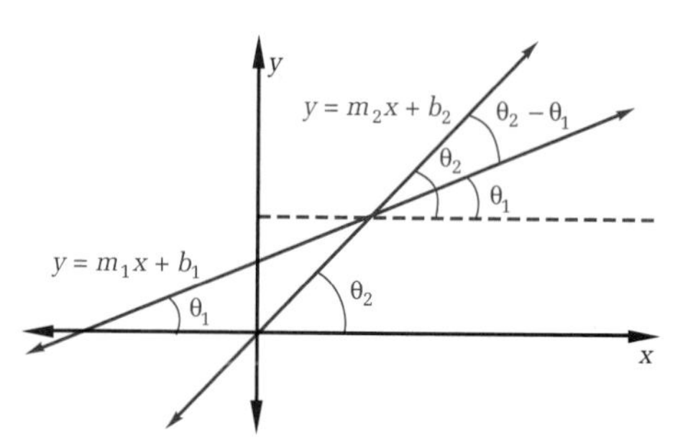

$$\text{So } \tan(\theta_2 - \theta_1) = \frac{\tan\theta_2 - \tan\theta_1}{1 + \tan\theta_2 \tan\theta_1}$$
$$= \frac{m_2 - m_1}{1 + m_2 m_1}$$

If θ is the angle measured counterclockwise from a line with slope m_1 to a line with slope m_2, then

$$\tan\theta = \frac{m_2 - m_1}{1 + m_1 m_2}$$

Example *Find the acute angle formed by the lines whose equations are $y = 5x + 1$ and $y = 0.5x + 3$.*

In this case, we let $m_2 = 5$ and $m_1 = 0.5$.

$$\tan\theta = \frac{5 - 0.5}{1 + 5(0.5)} \approx 1.2857$$

Using a trigonometric table or the INV tan keys on a calculator, we find that $\theta \approx 52.12°$.

Part Two: Sample Problems

Problem 1 *Prove the identity $\tan\left(\theta + \frac{\pi}{4}\right) = \frac{\tan\theta + 1}{\tan\theta - 1}$.*

Solution

Left Side	Right Side
$\tan\left(\theta + \frac{\pi}{4}\right)$	$= \frac{\tan\theta + 1}{1 - \tan\theta}$

$$= \frac{\tan\theta + \tan\frac{\pi}{4}}{1 - \tan\theta\tan\frac{\pi}{4}}$$

$$= \frac{\tan\theta + 1}{1 - \tan\theta}$$

Since the left side equals the right side, the equation is an identity.

Problem 2 *A man is 150 feet from the base of a 90-foot-high cliff. The man's eyes are 5 feet 6 inches above the ground. On the top of the cliff is a 25-foot-high sign. From the man's viewpoint, what is the angle of elevation from the bottom of the sign to the top of the sign?*

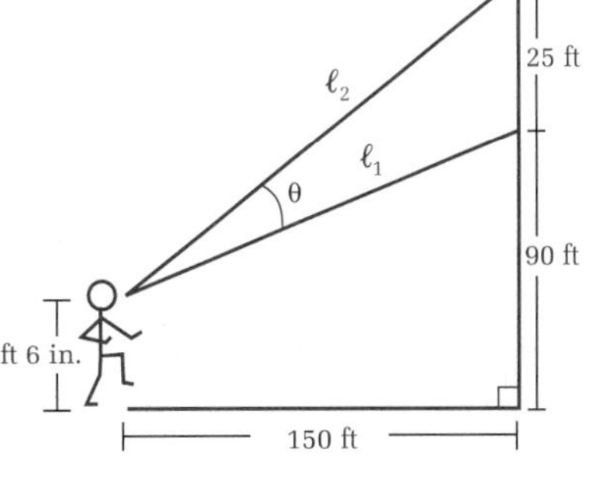

Solution The slope of line ℓ_1 is $\frac{90 - 5.5}{150}$, or ≈ 0.563.
The slope of line ℓ_2 is $\frac{115 - 5.5}{150}$, or 0.73.
Therefore, $\tan\theta \approx \frac{0.73 - 0.563}{1 + 0.563(0.73)} \approx 0.118$, and $\theta \approx 6.75°$.

Assignment Guide

Average	
Day 1	4–6, 7a, 8a, c, 11, 15, 18, 19
Day 2	7b, 8b, d, 13, 16, 20, 21a–d, 22, 23a
Advanced	
Day 1	4–8, 11–15, 20
Day 2	16, 18, 19, 21–24
Honors	
Day 1	4, 6–8, 11–15, 18–20
Day 2	16, 17, 21–27

Using Computers

See Using Computers for Section **15.1**

15.2

Problem-Set Notes and Strategies

- In problem **3**, have students evaluate these tangents several different ways to convince them of the relationship between tangents and slopes.
- Problem **4** is an example where each of the formulas is used in reverse.
- The identity in problem **5** allows a product of trig functions to be written as a sum and vice versa.
- Problem **8** is a proof of some of the intuitive ideas of the trig functions. Draw some pictures to illustrate these concepts.

Part Three: Exercises and Problems

Warm-up Exercises

P,R **1** Find each of the following.

a $\sin \alpha$ $\frac{5}{13}$
b $\sin \beta$ $\frac{12}{13}$
c $\sin (\alpha + \beta)$ 1

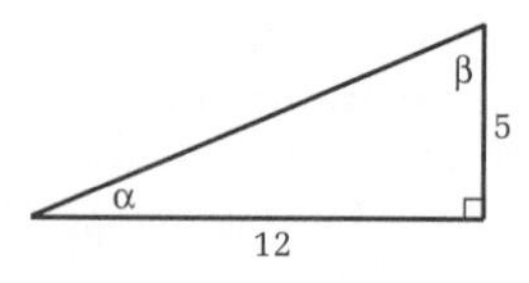

P **2** Determine whether each of the following statements is True or False.

a $\sin 70° = \sin 30° \cos 40° + \cos 30° \sin 40°$ True
b $\sin 70° = \sin 10° \cos 60° + \cos 10° \sin 60°$ True
c $\sin 70° = \sin 30° + \sin 40°$ False

P **3** Find each of the following.

a $\tan \angle POR$ -3
b $\tan \angle QOR$ 2
c $\tan \angle POQ$ 1

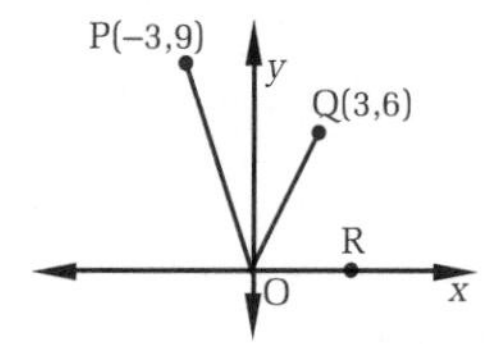

Problem Set

A

P **4** Simplify.

a $\cos 40° \cos 30° + \sin 40° \sin 30°$ $\cos 10°$
b $\sin 20° \cos 30° + \cos 20° \sin 30°$ $\sin 50°$
c $\sin 50° \cos 40° + \cos 40° \sin 50°$ $2 \sin 50° \cos 40°$

P **5** Prove that $\sin (\theta + \phi) + \sin (\theta - \phi) = 2 \sin \theta \cos \phi$.

P **6** Line L_1 has a slope of $\frac{2}{3}$, and line L_2 has a slope of $\frac{3}{2}$.

a Find $\tan \theta$. $\frac{5}{12}$
b Approximate θ to the nearest degree. $\approx 23°$

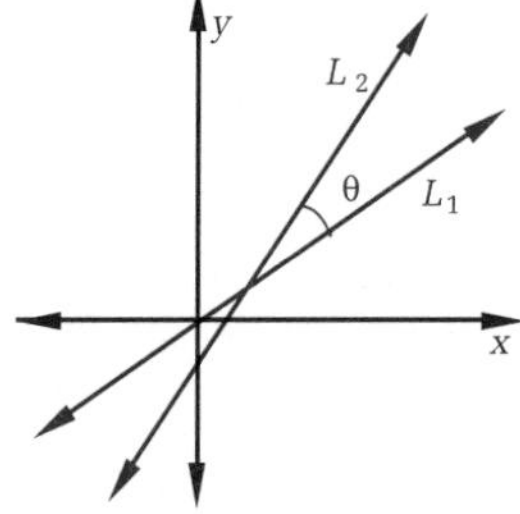

R **7** **a** Compute the exact value of $\cos \left(\text{Arcsin } \frac{4}{5}\right)$. $\frac{3}{5}$
b Use a calculator to evaluate $\cos \left(\text{Arcsin } \frac{4}{5}\right)$. 0.6
c Are your answers to parts **a** and **b** the same? Yes

P **8** Use the sum and difference identities to simplify.

a $\sin (x + \pi)$ $-\sin x$
b $\cos (x + \pi)$ $-\cos x$
c $\sin (x + 90°)$ $\cos x$
d $\sin (x + 60°)$ $\frac{1}{2} \sin x + \frac{\sqrt{3}}{2} \cos x$

Additional Answer

5 $\sin (\theta + \phi) + \sin (\theta - \phi)$
$= \sin \theta \cos \phi + \cos \theta \sin \phi + \sin \theta \cos \phi - \cos \theta \sin \phi$
$= 2 \sin \theta \cos \phi$

P **9** Determine whether each of the following statements is True or False.

a $\sin 20° = 2 \sin 10° \cos 10°$ True **b** $\sin 20° = 2 \sin 10°$ False
c $\tan 20° = \cot 70°$ True

R **10** **a** Use a calculator to compute Arcsin (sin 30°). 30°
b Use a calculator to evaluate Arcsin (sin 150°). 30°
c Explain the relationship between your answers to parts **a** and **b**. The principal value of arcsin $\frac{1}{2}$ is 30°.

B

P **11** Suppose that $\sin \alpha = \frac{3}{5}$ and $\cos \beta = \frac{1}{4}$ for acute angles α and β. Find the following.

a $\sin(\alpha + \beta)$ $\frac{3 + 4\sqrt{15}}{20}$ **b** $\cos(\alpha + \beta)$ $\frac{4 - 3\sqrt{15}}{20}$ **c** $\tan(\alpha + \beta)$ $\frac{3 + 4\sqrt{15}}{4 - 3\sqrt{15}}$

P,R **12** Find each of the following.

a $\sin \angle AOC$ $\frac{7}{25}$
b $\cos \angle AOC$ $-\frac{24}{25}$
c $\sin \angle BOC$ $\frac{3}{5}$
d $\cos \angle BOC$ $\frac{4}{5}$
e $\sin \angle AOB$ $\frac{4}{5}$
f $\cos \angle AOB$ $-\frac{3}{5}$

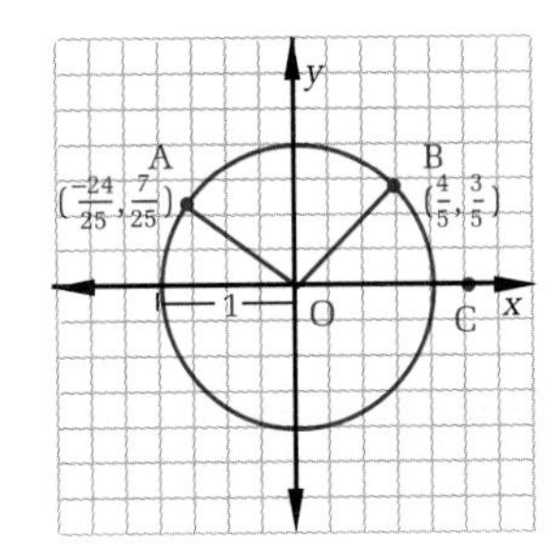

P **13** Prove that $\cot(\alpha + \beta) = \frac{\cot \alpha \cot \beta - 1}{\cot \alpha + \cot \beta}$.

R **14** Find the exact value of $\sin\left[\arcsin\left(\frac{5}{13}\right) - \arccos\left(\frac{-4}{5}\right)\right]$. $-\frac{56}{65}$ or $\frac{16}{65}$

P **15** Find the acute angle formed by the lines with equations $y = \frac{5}{8}x + 2$ and $y = -\frac{1}{4}x - 5$. $\approx 46.04°$

P **16** A lighthouse keeper is 50 meters above sea level and sights a ship that is 200 meters from the base of the lighthouse. The ship moves away from the lighthouse at 3 meters per second. After 1 minute, by how many degrees (to the nearest hundredth) has the angle of depression changed? $\approx 6.54°$

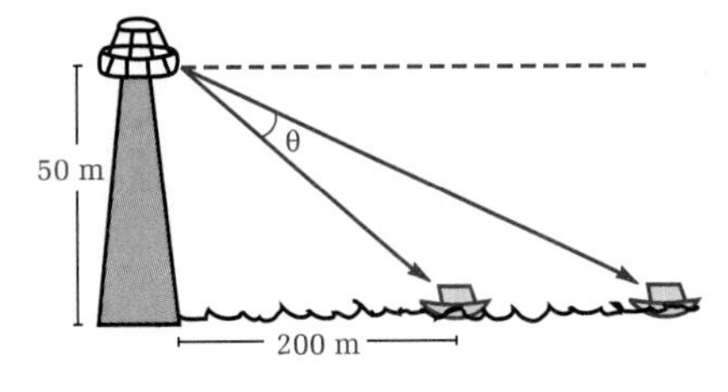

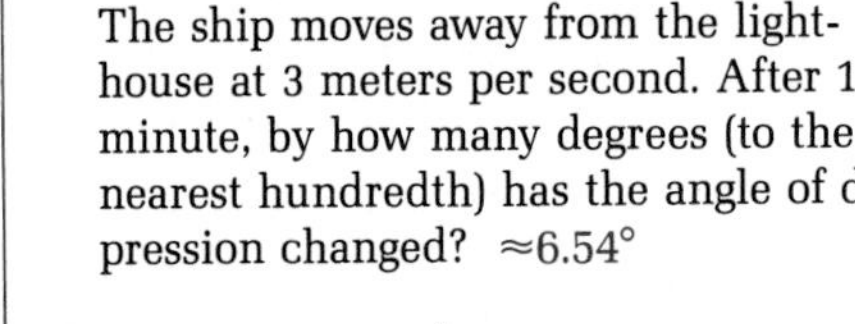

R **17** **a** Solve $\sin x = \frac{1}{2}$ for x.
$x \in \left\{\frac{\pi}{6} + n \cdot 2\pi, \frac{5\pi}{6} + n \cdot 2\pi\right\}$
b Solve $\sin y = 2$ for y.
No solution

P **18** Evaluate the following, using the sum and difference identities.

a sin 75° (Hint: 75° = 45° + 30°) $\frac{\sqrt{6} + \sqrt{2}}{4}$ **b** cos 75° $\frac{\sqrt{6} - \sqrt{2}}{4}$
c cos 15° $\frac{\sqrt{6} + \sqrt{2}}{4}$ **d** sin 105° $\frac{\sqrt{6} + \sqrt{2}}{4}$

P **19** Prove that $\frac{\sin(x + y)}{\sin x \sin y} = \cot y + \cot x$.

Problem-Set Notes and Strategies, continued

- Make sure the students understand the significance of part **c** in problem **10**. This will be needed by any student who uses a calculator.

- If students use a calculator for problem **18**, their answers will be in decimals and not exact. Inform them that these are just approximations and not what was asked for.

- Try changing everything to sine and cosine in problem **19**. Then use the sum identity.

Additional Answers

19 $\frac{\sin(x + y)}{\sin x \sin y}$

$= \frac{\sin x \cos y + \cos x \sin y}{\sin x \sin y}$

$= \frac{\sin x \cos y}{\sin x \sin y} + \frac{\cos x \sin y}{\sin x \sin y}$

$= \frac{\cos y}{\sin y} + \frac{\cos x}{\sin x}$

$= \cot y + \cot x$

Answer for problem **13** can be found in the answer pages beginning on **A 1**.

Problem-Set Notes and Strategies, continued

- Problem **22** also has some useful identities which students may want to memorize.
- Problem **25** uses the sum identity for sine twice.
- Trigonometric functions may be evaluated with a series. See if students can identify the pattern generating in problem **27**. This is a series for sine and is the method that may be used to generate the trig tables. This method was used on computers and calculators until more efficient methods were found.

Problem Set, *continued*

R **20** Evaluate each expression.

a $\log_2\left(\sin\frac{\pi}{6}\right)$ -1 **b** $\log_2\left(\sin\frac{\pi}{2}\right)$ 0

P,I **21** Find each of the following.

a $\tan\frac{\pi}{8}$ $\sqrt{2}-1$ **b** $\tan 75°$ $2+\sqrt{3}$

c $\dfrac{\tan\frac{5\pi}{8}+\tan\frac{\pi}{8}}{1-\tan\frac{5\pi}{8}\tan\frac{\pi}{8}}$ -1 **d** $\dfrac{\tan 70° - \tan 25°}{1+\tan 70° \tan 25°}$ 1

e $\dfrac{1-\tan 20° \tan 10°}{\tan 20° + \tan 10°}$ $\sqrt{3}$

P **22** Recall that $\cos(\theta+\phi) = \cos\theta\cos\phi - \sin\theta\sin\phi$ and $\cos(\theta-\phi) = \cos\theta\cos\phi + \sin\theta\sin\phi$.

a Simplify $\cos(\theta+\phi)+\cos(\theta-\phi)$ $2\cos\theta\cos\phi$
b Simplify $\cos(\theta+\phi)-\cos(\theta-\phi)$ $-2\sin\theta\sin\phi$
c Simplify $\cos(\theta+60°)+\cos(\theta-60°)$ $\cos\theta$
d If $2\cos 110°\cos 70° = \cos\alpha + \cos\beta$, what is (α, β)? $(180°, 40°)$

P **23** Find the exact value of each of the following.

a $\sin 75°$ $\frac{\sqrt{2}+\sqrt{6}}{4}$ **b** $\sin 105°$ $\frac{\sqrt{2}+\sqrt{6}}{4}$ **c** $\sin(-75°)$ $\frac{-\sqrt{2}-\sqrt{6}}{4}$

P,R **24** Evaluate each of the following.

a $\cos\left(\text{Arcsin}\frac{3}{5} - \text{Arccos}\frac{12}{13}\right)$ $\frac{63}{65}$ **b** $\tan\left(\text{Arcsin}\frac{3}{5} + \text{Arctan}\frac{1}{2}\right)$ 2

c $\sin(\text{Arcsin } 2 + \text{Arccos } 2)$ No real value

P,I C **25** Prove that $\sin 3\theta = 3\sin\theta - 4\sin^3\theta$.

P,R,I **26** Solve for y.
$y \approx 4.12$

P,R  **27** Use a calculator to compute the following for $\alpha = \frac{\pi}{6}$ (radian mode).

a $\sin\alpha$ 0.5 **b** $\alpha - \frac{\alpha^3}{3!} + \frac{\alpha^5}{5!}$ ≈ 0.500002133

c $\alpha - \frac{\alpha^3}{3!} + \frac{\alpha^5}{5!} - \frac{\alpha^7}{7!}$ ≈ 0.499999992 **d** $\alpha - \frac{\alpha^3}{3!} + \frac{\alpha^5}{5!} - \frac{\alpha^7}{7!} + \frac{\alpha^9}{9!}$ ≈ 0.5

Additional Answer

Answer for problem **25** can be found in the answer pages beginning on **A** 1.

5.3

The Double- and Half-Angle Identities

Objectives

After studying this section, you will be able to
- Apply the double-angle identities
- Apply the half-angle identities

Part One: Introduction

The Double-Angle Identities

How does sin 60° compare with sin 30°? Since $\sin 60° = \frac{\sqrt{3}}{2}$ and $\sin 30° = \frac{1}{2}$, we see that $\sin 60° \neq 2 \cdot \sin 30°$. We will now develop the formulas for $\sin 2\theta$, $\cos 2\theta$, and $\tan 2\theta$, using the sum identities we derived in the last section.

$$\begin{aligned}\sin 2\theta &= \sin(\theta + \theta)\\ &= \sin\theta \cdot \cos\theta + \cos\theta \cdot \sin\theta\\ &= 2(\sin\theta \cdot \cos\theta)\end{aligned}$$

$$\sin 2\theta = 2 \sin\theta \cos\theta$$

$$\begin{aligned}\cos 2\theta &= \cos(\theta + \theta)\\ &= \cos\theta \cdot \cos\theta - \sin\theta \cdot \sin\theta\\ &= \cos^2\theta - \sin^2\theta\end{aligned}$$

$$\cos 2\theta = \cos^2\theta - \sin^2\theta$$

We can obtain two other forms of the preceding identity by remembering that $\sin^2\theta + \cos^2\theta = 1$ and substituting for either $\cos^2\theta$ or $\sin^2\theta$ in the identity.

$$\cos 2\theta = 1 - 2\sin^2\theta$$

$$\cos 2\theta = 2\cos^2\theta - 1$$

The double-angle identity for the tangent function is derived by using $\tan 2\theta = \tan(\theta + \theta)$. You are asked to prove this identity in Warm-up Exercise 6.

$$\tan 2\theta = \frac{2\tan\theta}{1 - \tan^2\theta}$$

15.3

Class Planning

Time Schedule
All levels: 2 days

Resource References
Teacher's Resource Book
Class Opener 90
Practice 95
Enrichment 29
Transparency 30

Class Opener

Use the sum identity to expand $\cos(\theta + \theta)$.
$\cos^2\theta - \sin^2\theta$

Lesson Notes

- Now that the sum identities have been developed, the double and half-angle identities may be derived. Each of these is a variant of the sum identity and there will be one for each of the sine, cosine and tangent functions.
- All these identities should also be memorized by the students with an emphasis on the double angle identity for cosine. Note that it has a linear trig function on one side and a quadratic trig function on the other side. This is the only simple identity that allows changing of the power of the trig function.

Vocabulary
double-angle identities
half-angle identities

15.3

Using Computers

See Using Computers for Section **15.1**.

Communicating Mathematics

The double-angle formula for cosine may be written in three different ways. Have students write a paragraph explaining the significance and types of problems that may be solved using each form of the identity.

Example *Find all possible values of θ between 0 and 2π for which $\sin 2\theta + \sin \theta = 0$.*

$$\mathbf{\sin 2\theta + \sin \theta = 0}$$

Use the identity for $\sin 2\theta$

$$\mathbf{2 \sin \theta \cdot \cos \theta + \sin \theta = 0}$$

$$\mathbf{\sin \theta (2 \cos \theta + 1) = 0}$$

Use the Zero Product Property

$$\mathbf{\sin \theta = 0 \text{ or } 2 \cos \theta + 1 = 0}$$

$$\mathbf{\cos \theta = -\frac{1}{2}}$$

Since $\sin \theta = 0$ when $\theta = 0$, $\theta = \pi$, or $\theta = 2\pi$ and since $\cos \theta = -\frac{1}{2}$ when $\theta = \frac{2\pi}{3}$ or $\theta = \frac{4\pi}{3}$, the solutions are $\left\{0, \frac{2\pi}{3}, \pi, \frac{4\pi}{3}, 2\pi\right\}$.

The Half-Angle Identities

The identities for $\sin \frac{\theta}{2}$, $\cos \frac{\theta}{2}$, and $\tan \frac{\theta}{2}$ are derived by solving a $\cos 2\theta$ identity for the appropriate function and then substituting $\frac{\theta}{2}$ for θ.

We first derive the identity for $\sin \frac{\theta}{2}$.

$$\cos 2\theta = 1 - 2 \sin^2 \theta$$
$$2 \sin^2 \theta = 1 - \cos 2\theta$$
$$\sin^2 \theta = \frac{1 - \cos 2\theta}{2}$$
$$|\sin \theta| = \sqrt{\frac{1 - \cos 2\theta}{2}}$$

By substituting $\frac{\theta}{2}$ for θ in the preceding equation, we obtain the following identity.

$$\sin \frac{\theta}{2} = \pm\sqrt{\frac{1 - \cos \theta}{2}}$$

When you use this identity with a given angle you will have to determine the appropriate sign to use.

The identity for $\cos \frac{\theta}{2}$ can be derived in the same manner by beginning with $\cos 2\theta = 2 \cos^2 \theta - 1$. The derivation is asked for in Warm-up Exercise 4.

$$\cos \frac{\theta}{2} = \pm\sqrt{\frac{1 + \cos \theta}{2}}$$

Example *Find the exact value of* $\cos 195°$.

Since 195° is a third-quadrant angle, $\cos 195°$ will be negative.

$$\cos 195° = -\sqrt{\frac{1 + \cos 390°}{2}}$$
$$= -\sqrt{\frac{1 + \cos 30°}{2}}$$

$$= -\sqrt{\frac{1 + \frac{\sqrt{3}}{2}}{2}}$$

$$= -\frac{\sqrt{2 + \sqrt{3}}}{2}$$

Finally, we derive an identity for $\tan \frac{\theta}{2}$.

$$\tan \frac{\theta}{2} = \frac{\sin \frac{\theta}{2}}{\cos \frac{\theta}{2}} = \pm\frac{\sqrt{\frac{1 - \cos \theta}{2}}}{\sqrt{\frac{1 + \cos \theta}{2}}} = \pm\sqrt{\frac{1 - \cos \theta}{1 + \cos \theta}}$$

By multiplying both the numerator and the denominator by $\sqrt{1 + \cos \theta}$, we can obtain another form of this identity, $\tan \frac{\theta}{2} = \frac{\sin \theta}{1 + \cos \theta}$. (The derivation is asked for in problem **19.**) Notice that no $\pm$ appears in this identity because $1 + \cos \theta$ is never negative and $\sin \theta$ and $\tan \frac{\theta}{2}$ always have the same sign.

A third form of the $\tan \frac{\theta}{2}$ identity can be obtained by multiplying the numerator and the denominator of the preceding form by $1 - \cos \theta$.

$$\tan \frac{\theta}{2} = \frac{\sin \theta}{1 + \cos \theta} \cdot \frac{1 - \cos \theta}{1 - \cos \theta} = \frac{1 - \cos \theta}{\sin \theta}$$

$$\tan \frac{\theta}{2} = \pm\sqrt{\frac{1 - \cos \theta}{1 + \cos \theta}}$$

$$\tan \frac{\theta}{2} = \frac{\sin \theta}{1 + \cos \theta}$$

$$\tan \frac{\theta}{2} = \frac{1 - \cos \theta}{\sin \theta}$$

Part Two: Sample Problems

Problem 1 *Find the exact value of* $\tan\left(2 \text{ Arcsin } \frac{8}{17}\right)$.

Solution Let $\alpha = \text{Arcsin } \frac{8}{17}$.

$$\tan\left(2 \text{ Arcsin } \frac{8}{17}\right) = \tan 2\alpha$$

$$= \frac{2 \tan \alpha}{1 - \tan^2 \alpha}$$

$$= \frac{2\left(\frac{8}{15}\right)}{1 - \left(\frac{8}{15}\right)^2} = \frac{240}{161}$$

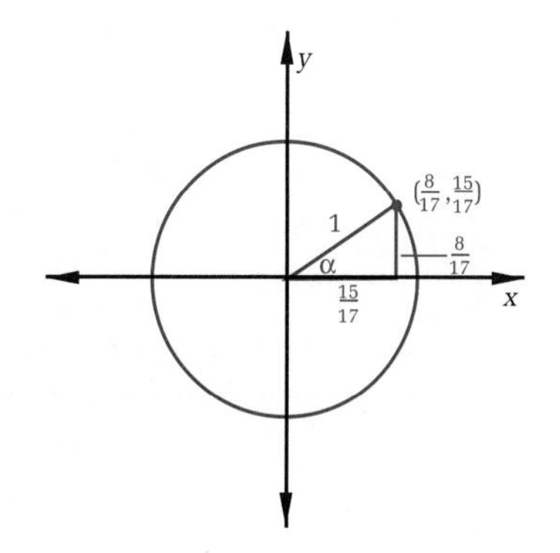

Checkpoint

1 Find the sine of 15° using a half-angle identity.

$$\sqrt{\frac{1 - \frac{\sqrt{3}}{2}}{2}}$$

$$= \sqrt{\frac{2 - \sqrt{3}}{4}} = \frac{\sqrt{2 - \sqrt{3}}}{2}$$

2 Write the $\sin^2 \theta$ in terms of $\cos \theta$.

$\frac{1 - \cos 2\theta}{2}$

3 Solve $\sin 2x + \sin x = 0$ for x between 0° and 360°.

0°, 120°, 180°, 240°, 360°

Assignment Guide

Average	
Day 1	8, 11, 13, 16, 17, 19, 20, 23
Day 2	10, 12, 14a, c, 15, 18a, 22a
Advanced	
Day 1	8, 9, 12, 13, 14a, c, 18a, b, 21, 23
Day 2	10, 11, 15–17, 18c, 19, 20, 22a
Honors	
Day 1	8–10, 13, 14b, c, 18a, b, 19, 21
Day 2	12, 16, 20, 22, 24a, b, 25, 26

Problem 2 *Find the value of x and the value of angle θ in degrees.*

18 12 2θ θ x

Solution From the diagram, $\frac{x}{18} = \cos 2\theta$ and $\frac{x}{12} = \cos \theta$. Using the identity $\cos 2\theta = 2\cos^2\theta - 1$, we have

$$\frac{x}{18} = 2\left(\frac{x}{12}\right)^2 - 1$$

$$\frac{x}{18} = \frac{x^2}{72} - 1$$

$$4x = x^2 - 72$$

$$0 = x^2 - 4x - 72$$

Using the quadratic formula, we find that $x \approx 10.72$ or $x \approx -6.72$. Since the value of x must be positive, the value of x is approximately 10.72. Since $\cos\theta = \frac{x}{12}$, we have $\cos\theta \approx \frac{10.72}{12}$. Therefore, $\theta \approx 26.7°$.

Problem 3 *Prove the identity $\tan\frac{\theta}{2} \cdot (\sec\theta + 1) = \tan\theta$. Note any restrictions.*

Solution

Left Side | **Right Side**

$$\tan\frac{\theta}{2}\cdot(\sec\theta + 1) \qquad = \tan\theta$$

$$= \frac{\sin\theta}{1+\cos\theta}\cdot\left(\frac{1}{\cos\theta} + 1\right)$$

$$= \frac{\sin\theta}{1+\cos\theta}\cdot\frac{1+\cos\theta}{\cos\theta}$$

$$= \frac{\sin\theta}{\cos\theta}$$

$$= \tan\theta$$

The left side of the identity is undefined for $\cos\theta = -1$ and $\cos\theta = 0$. The right side is also undefined for $\cos\theta = 0$. So the restrictions are $\cos\theta \neq -1$ and $\cos\theta \neq 0$.

Problem 4 *Solve $\cos 2\theta = \cos\theta$ for all values of θ such that $0 \le \theta \le 360°$.*

Solution We have three choices for an identity for $\cos 2\theta$. Let's choose $\cos 2\theta = 2\cos^2\theta - 1$ so that the resulting equation will be entirely in terms of $\cos\theta$.

$$\cos 2\theta = \cos\theta$$

$$2\cos^2\theta - 1 = \cos\theta$$

$$2\cos^2\theta - \cos\theta - 1 = 0$$

$$(2\cos\theta + 1)(\cos\theta - 1) = 0$$

$$2\cos\theta + 1 = 0 \text{ or } \cos\theta - 1 = 0$$

$$\cos\theta = \frac{-1}{2} \qquad \cos\theta = 1$$

The solutions are {0°, 120°, 240°, 360°}.

Part Three: Exercises and Problems

Warm-up Exercises

P **1** Determine whether each statement is True or False.

a $\cos^2\theta = \dfrac{1+\cos 2\theta}{2}$ True **b** $\sin^2\theta = \dfrac{1-\cos 2\theta}{2}$ True **c** $\sin^2\theta = \dfrac{1+\sin 2\theta}{2}$ False

P **2** Prove that $(\sin\theta + \cos\theta)^2 = 1 + \sin 2\theta$.

R **3** Find $\tan \angle ABC$. ≈ 8.57

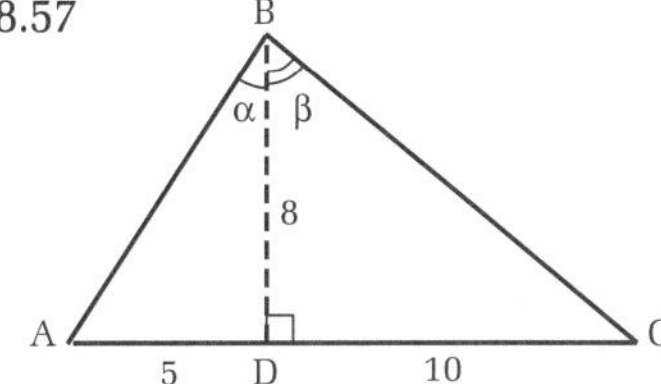

P **4** Prove that $\cos\frac{\theta}{2} = \pm\sqrt{\frac{1+\cos\theta}{2}}$.

P **5** Determine whether each statement is True or False.

a $\sin\left(\dfrac{\alpha}{8}\right) = 2\sin\left(\dfrac{\alpha}{4}\right)\cos\left(\dfrac{\alpha}{4}\right)$ False **b** $\cos\left(\dfrac{\alpha}{10}\right) = \cos^2\left(\dfrac{\alpha}{20}\right) - \sin^2\left(\dfrac{\alpha}{20}\right)$ True

P **6** Prove that $\tan 2\theta = \frac{2\tan\theta}{1-\tan^2\theta}$.

P **7** Determine whether each statement is True or False.

a $\cos 8\theta = \cos^2 4\theta - \sin^2 4\theta$ True

b $\sin\left(\dfrac{1}{2}\alpha\right) = 2\sin\left(\dfrac{\alpha}{4}\right)\cos\left(\dfrac{\alpha}{4}\right)$ True

c $\tan 3\theta = \dfrac{2\tan\left(\dfrac{3\theta}{2}\right)}{1-\tan^2\left(\dfrac{3\theta}{2}\right)}$ True

Problem Set

8 If $\cos 6\theta = 0.5400$, what is $\cos 3\theta$? ≈ 0.8775

P,R **9** Prove that $(\sin\theta + \cos\theta)^{\csc^2\theta - \cot^2\theta + 1} = 1 + \sin 2\theta$.

P,R **10** Find the exact value of each of the following.

a $\tan 22.5°$ $\sqrt{2} - 1$ **b** $\tan 75°$ $2 + \sqrt{3}$

P **11** Determine whether each statement is True or False.

a $\cos 105° = \sqrt{\dfrac{1+\cos 210°}{2}}$ False **b** $\sin 105° = \sqrt{\dfrac{1-\cos 210°}{2}}$ True

Problem-Set Notes and Strategies

- Problem **1** illustrates the importance of the cosine identity for double angles. It allows the power of the trig function to be changed.
- Problem **2** is just an algebra problem in disguise with the basic Pythagorean identity.
- Problem **7** illustrates the variants of these formulas that may be made. Different interpretations produce different formulas.
- Problem **9** is just problem **2** all over again with the exponent reworded in an equivalent form.

Additional Answers

2 $(\sin\theta + \cos\theta)^2$
$= \sin^2\theta + 2\sin\theta\cos\theta + \cos^2\theta$
$= 1 + 2\sin\theta\cos\theta$
$= 1 + \sin 2\theta$

4 $\cos 2\left(\frac{\theta}{2}\right) = 2\cos^2\frac{\theta}{2} - 1$
$2\cos^2\frac{\theta}{2} = 1 + \cos\theta$
$\cos^2\frac{\theta}{2} = \dfrac{1+\cos\theta}{2}$
$\cos\frac{\theta}{2} = \pm\sqrt{\dfrac{1+\cos\theta}{2}}$

6 $\tan 2\theta$
$= \tan(\theta + \theta)$
$= \dfrac{\tan\theta + \tan\theta}{1 - \tan\theta\tan\theta}$
$= \dfrac{2\tan\theta}{1-\tan^2\theta}$

Answer for problem **9** can be found in the answer pages beginning on **A** 1.

Problem-Set Notes and Strategies, continued

- Problem **18** involves the use of these identities to solve equations. Spend some time with this problem.

Cooperative Learning

Have students work in small groups to solve problem **20**.

Additional Answers

22a

$\sin 4\theta$
$= \sin 2\,(2\theta)$
$= 2 \sin 2\theta \cos 2\theta$
$= 2\,(2 \sin \theta \cos \theta)\,(\cos^2\theta - \sin^2\theta)$
$= 4 \sin \theta \cos^3\theta - 4 \sin^3\theta \cos \theta$

Answers for problems **19** and **22b** can be found in the answer pages beginning on **A 1**.

Problem Set, *continued*

P **12** Find the exact value of $\sin 75°$. $\frac{\sqrt{2 + \sqrt{3}}}{2}$

P,R **13** The slope of line $\mathcal{M}$ is $\frac{2}{3}$. Find the exact value of the slope of line ℓ. Check your result by estimating α and doubling it, then finding the slope of ℓ.
$m_\ell = \frac{12}{5}$; $\alpha \approx 33.69°$, so $\tan 2\alpha = \frac{12}{5}$.

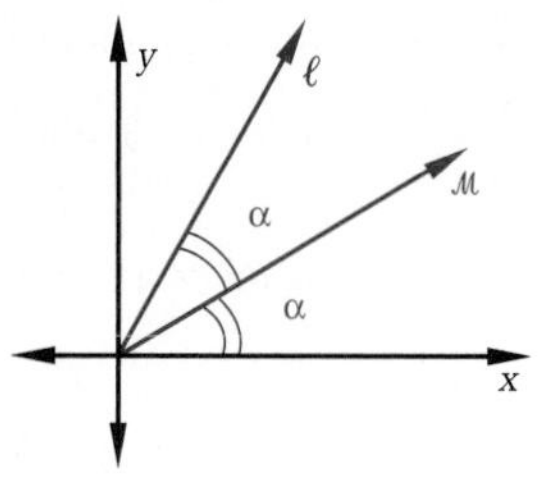

R **14** Solve for x, where $0 \le x \le 2\pi$.

a $(2 \sin x + 1)(\sin x - 1) = 0$ $\left\{\frac{\pi}{2}, \frac{7\pi}{6}, \frac{11\pi}{6}\right\}$

b $(2 \cos x + \sqrt{3})(\cos x + 1) = 0$ $\left\{\frac{5\pi}{6}, \pi, \frac{7\pi}{6}\right\}$

c $(2 \sin x - 1)(2 \cos x + 1) = 0$ $\left\{\frac{\pi}{6}, \frac{2\pi}{3}, \frac{5\pi}{6}, \frac{4\pi}{3}\right\}$

R **15** Angles α and β are as shown. Find the following.

a $\sin \alpha$ and $\cos \alpha$ $\frac{3}{5}; \frac{4}{5}$

b $\sin \beta$ and $\cos \beta$ $\frac{8}{17}; \frac{15}{17}$

c $\sin(\alpha + \beta)$ and $\cos(\alpha + \beta)$ $\frac{77}{85}; \frac{36}{85}$

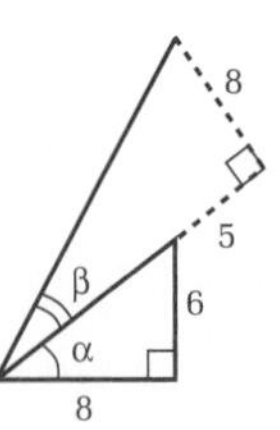

P,R **16** Let $\cos x = \frac{4}{5}$. Find each of the following.

a $\sin \frac{x}{2}$ $\approx \pm 0.3162278$ **b** $\cos \frac{x}{2}$ $\approx \pm 0.9486833$ **c** $\tan \frac{x}{2}$ $\pm\frac{1}{3}$

R **17** If $\tan \gamma = \frac{1}{2}$ and $\tan \beta = \frac{1}{3}$, what is $\tan \alpha$? 1

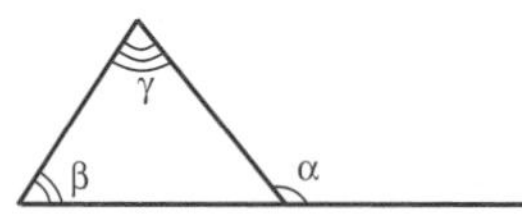

B

P **18** Solve for x, where $0 \le x \le 2\pi$.

a $\sin 2x = 2 \sin x$ $\{0, \pi, 2\pi\}$

b $\cos 2x + \cos x = 0$ $\left\{\frac{\pi}{3}, \pi, \frac{5\pi}{3}\right\}$

c $\tan 2x = \frac{3}{4}$ $\{\approx 0.32, \approx 1.89, \approx 3.46, \approx 5.03\}$

P,R **19** Prove that $\tan \frac{\theta}{2} = \frac{\sin \theta}{1 + \cos \theta}$.

P,R **20** Simplify $\cos 5\theta \cos 3\theta + \sin 5\theta \sin 3\theta$ in terms of functions of θ. $\cos^2 \theta - \sin^2 \theta$

P  **21** Find all values of x for which $\cos 2x = 2 \cos x$.
$\{\ldots, \approx -248.5°, \approx -111.5°, \approx 111.5°, \approx 248.5°, \ldots\}$

P,R **22** Prove the following identities.

a $\sin 4\theta = 4 \sin \theta \cos^3 \theta - 4 \sin^3 \theta \cos \theta$

b $\cos 4\theta = 8 \cos^4 \theta - 8 \cos^2 \theta + 1$

R | **23** Find cos $(\theta + \phi)$ and sin $(\theta + \phi)$.
$\cos(\theta + \phi) = \frac{63}{65}$; $\sin(\theta + \phi) = -\frac{16}{65}$

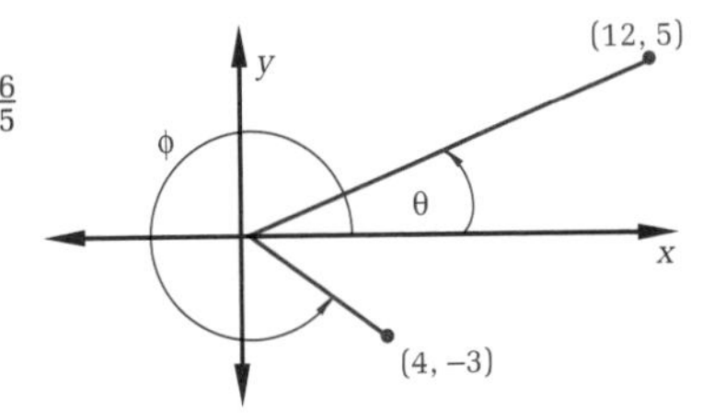

P,R | **24** Solve for θ, where $0 \le \theta < 2\pi$.

a $2\cos^2\theta - 3\cos x - 2 = 0$ $\left\{\frac{2\pi}{3}, \frac{4\pi}{3}\right\}$

b $\sin^2\theta + 2\cos\theta = 0$ $\{\approx 2.00, \approx 4.29\}$

c $\cos 2\theta = 1 - 2\sin^2\theta$ $\{\theta: 0 \le \theta < 2\pi\}$

C P,R | **25 a** Use a half-angle identity to find an expression for sin 15°. $\frac{\sqrt{2-\sqrt{3}}}{2}$

b Use a difference identity to find sin 15°. $\frac{\sqrt{6}-\sqrt{2}}{4}$

c Show that your answers to parts **a** and **b** are equal.

P,R | **26** Prove the identity $\sin 3\theta = (\sin\theta)(2\cos\theta - 1)(2\cos\theta + 1)$.

CAREER PROFILE

TRIGONOMETRY IN WINDSURFING

Pat LeMehaute designs from physics

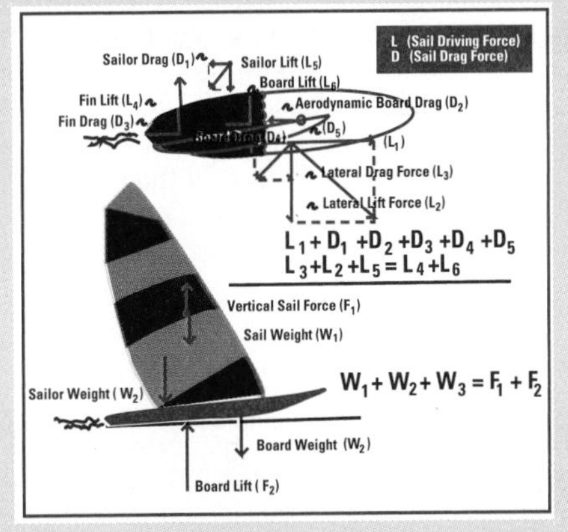

Pat LeMehaute is a writer, product tester, and windsurfing enthusiast who designs windsurfing boards. When he found little topical material on windsurfing, Mr. LeMehaute did his own experimentation and mathematical analysis to provide the basis for his designs. Windsurfing is governed by physical laws, so Mr. LeMehaute "used trigonometry and calculus and worked from physical theory," he says.

Mr. LeMehaute's first concern in constructing his board was to itemize all the forces on the board, such as the amount of board in the water, the lift and drag on the sail and fin, and the drag on the board and sailor. One method he uses is a "free-body diagram," applied so that "sailing along in a straight line and in a constant wind speed, all the forces acting on you and the sail board add up to zero," he says. "I analyze sines, tangents, and vectors. In the free-body diagram, the forces are represented by arrows and their magnitudes by the arrows' lengths. The lift on the sail is roughly one-hundred pounds, and the drag is about twenty pounds, so the lift arrow is five times as long as the drag arrow."

Applying mathematical concepts to board sailing helps Mr. LeMehaute "obtain high speeds, be more scientific, and better understand the concepts of board sailing." Mr. LeMehaute holds a bachelor's degree in engineering science from the University of Florida. His high-tech background has more than scientific or mathematical benefits, flowing into personal satisfaction and aesthetic expression.

Problem-Set Notes and Strategies, continued

- In problem **23**, two different identities are needed to change the function into one involving just θ.
- Problem **25** involves some interesting identities that are not necessary to memorize. It is sufficient to be able to do the algebra involved to prove them.

Additional Answers

25c

$$\frac{\sqrt{6}-\sqrt{2}}{4} = \frac{\sqrt{(\sqrt{6}-\sqrt{2})^2}}{4} = \frac{\sqrt{6-2\sqrt{12}+2}}{4} = \frac{\sqrt{8-4\sqrt{3}}}{4} = \frac{2\sqrt{2-\sqrt{3}}}{4} = \frac{\sqrt{2-\sqrt{3}}}{2}$$

Answer for problem **26** can be found in the answer pages beginning on **A** 1.

15.4

Class Planning

Time Schedule
All levels: 2 days

Resource References
Teacher's Resource Book
Class Opener 91
Practice 96
Enrichment 30

Class Opener

Find all values of x so that $\sin^2 x \le \sin x$.

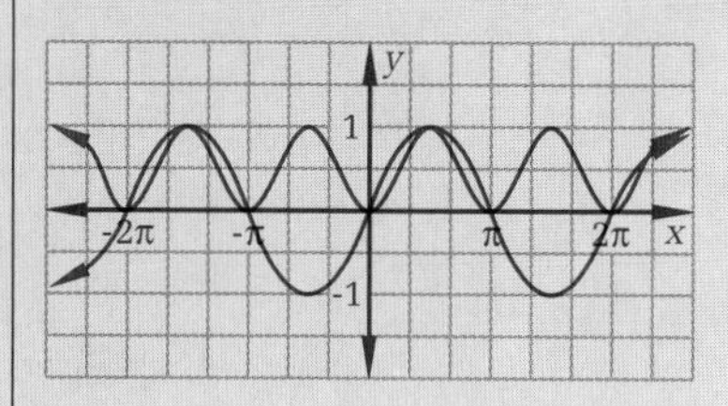

$0 \le x \le \pi$, $2\pi \le x \le 3\pi \ldots 2k\pi \le x \le (2k+1)\pi$, k is an integer.

Lesson Notes

- The main purpose of the identities in this section is to let the student know that identities such as these exist and if they are ever needed, they can be looked up in a reference text. Several examples and problems illustrate the type of equations where these are most useful.

15.4 Final Identities and Trigonometric Equations

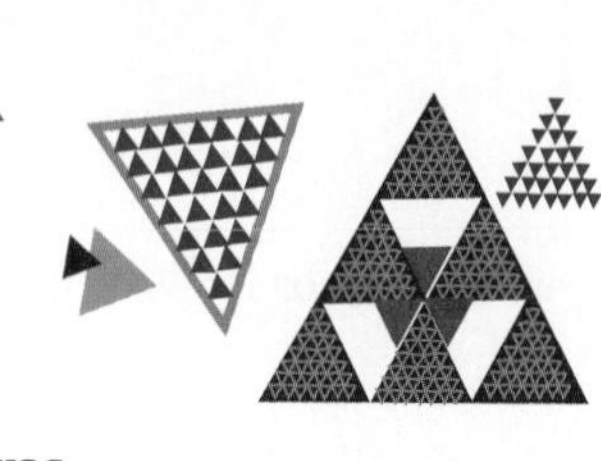

Objectives

After studying this section, you will be able to
- Use the product-to-sum and sum-to-product identities
- Solve trigonometric equations

Part One: Introduction

Product-to-Sum and Sum-to-Product Identities

There are times when it is necessary to convert a product of trigonometric functions to a sum of trigonometric functions or vice versa. The following identities, however, are not used as often as the others presented in this chapter.

To convert a product to a sum, we can use the identities expressed in the following theorems.

$$\sin \alpha \cos \beta = \frac{1}{2}[\sin(\alpha+\beta) + \sin(\alpha-\beta)]$$

We can prove this theorem by showing that the right side of the equation is equal to the left side.

Left Side	**Right Side**
$\sin \alpha \cos \beta$	$= \frac{1}{2}[\sin(\alpha+\beta) + \sin(\alpha-\beta)]$
	$= \frac{1}{2}(\sin \alpha \cos \beta + \cos \alpha \sin \beta + \sin \alpha \cos \beta - \cos \alpha \sin \beta)$
	$= \frac{1}{2}(2 \sin \alpha \cos \beta)$
	$= \sin \alpha \cos \beta$

We have proved the theorem by verifying the identity. The following theorems can be proved in a similar manner.

$$\sin \alpha \sin \beta = \frac{1}{2}[\cos(\alpha-\beta) - \cos(\alpha+\beta)]$$

Vocabulary
sum-to-product identities
product-to-sum identities

$$\cos\alpha\cos\beta = \frac{1}{2}[\cos(\alpha - \beta) + \cos(\alpha + \beta)]$$

Example *Find the exact value of* $\cos 75° \cos 15°$.

$$\cos 75° \cos 15° = \frac{1}{2}[\cos(75° - 15°) + \cos(75° + 15°)]$$

$$= \frac{1}{2}(\cos 60° + \cos 90°)$$

$$= \frac{1}{2}\left(\frac{1}{2} + 0\right)$$

$$= \frac{1}{4}$$

To convert a sum of trigonometric functions to a product, we can use the identities expressed in the following theorems.

$$\sin\alpha + \sin\beta = 2\sin\frac{\alpha + \beta}{2}\cos\frac{\alpha - \beta}{2}$$

$$\cos\alpha + \cos\beta = 2\cos\frac{\alpha + \beta}{2}\cos\frac{\alpha - \beta}{2}$$

Example *Prove the identity* $\frac{\cos 6\theta + \cos 2\theta}{\sin 6\theta + \sin 2\theta} = \cot 4\theta$.

Left Side	**Right Side**
$\frac{\cos 6\theta + \cos 2\theta}{\sin 6\theta + \sin 2\theta}$	$= \cot 4\theta$
$= \frac{2\cos\left(\frac{6\theta + 2\theta}{2}\right)\cos\left(\frac{6\theta - 2\theta}{2}\right)}{2\sin\left(\frac{6\theta + 2\theta}{2}\right)\cos\left(\frac{6\theta - 2\theta}{2}\right)}$	
$= \frac{2\cos 4\theta\cos 2\theta}{2\sin 4\theta\cos 2\theta}$	
$= \cot 4\theta$	

Trigonometric Equations

We show several examples of solving trigonometric equations that are not identities.

Example *Solve* $-4\cos^2\theta - 8\sin\theta + 7 = 0$ *for* θ.

We first convert the equation to an equation in terms of only $\sin\theta$ by substituting $1 - \sin^2\theta$ for $\cos^2\theta$.

$$-4\cos^2\theta - 8\sin\theta + 7 = 0$$
$$-4(1 - \sin^2\theta) - 8\sin\theta + 7 = 0$$
$$-4 + 4\sin^2\theta - 8\sin\theta + 7 = 0$$
$$4\sin^2\theta - 8\sin\theta + 3 = 0$$

Lesson Notes, continued

- Trigonometric equations are also solved using the formal rules of algebra combined with the necessary identities from trigonometry that have been learned in this chapter. The algebraic techniques used to solve trigonometric equations are the same as for non-trig equations. Remember the stumbling blocks that occur when using them, such as squaring both sides and introducing extraneous answers. Check solutions and remember to find all solutions, except when a specific interval is requested.

Using Computers

See Using Computers for Section **15.1**.

Communicating Mathematics

Have students write a paragraph containing at least one example of how an algebraic problem could be changed into a trigonometric problem and how a trigonometric equation could be changed into an equivalent algebraic equation.

Stumbling Block

Using algebraic techniques frequently leads to inconsistent answers such as $\sin x = 2$ or $\csc x$ used in an equation with an answer of π. Make sure that students check the answers of their equations before they are finished with the problem. Also make them aware that they must be familiar with the domains of functions in the problem before the final solutions are stated, or some solutions may be included that are not in the domain.

Now we factor the quadratic.

$$(2 \sin \theta - 3)(2 \sin \theta - 1) = 0$$
$$2 \sin \theta - 3 = 0 \text{ or } 2 \sin \theta - 1 = 0$$
$$\sin \theta = \frac{3}{2} \qquad \sin \theta = \frac{1}{2}$$

The equation $\sin \theta = \frac{3}{2}$ has no solution, since for any value of θ, $-1 \le \sin \theta \le 1$. The equation $\sin \theta = \frac{1}{2}$ has the solutions $\left\{\ldots, -\frac{11\pi}{6}, -\frac{7\pi}{6}, \frac{\pi}{6}, \frac{5\pi}{6}, \frac{13\pi}{6}, \ldots\right\}$. These solutions can be written as $\left\{2n\pi + \frac{\pi}{6}, 2n\pi + \frac{5\pi}{6}\right\}$, where n is any integer.

Example *If $\sin \theta - \cos \theta = 1$ and $0 \le \theta < 2\pi$, what is θ?*

We begin by squaring both sides. In doing this, we may introduce extraneous solutions, so we will need to check our answers.

$$\sin \theta - \cos \theta = 1$$
$$(\sin \theta - \cos \theta)^2 = 1^2$$
$$\sin^2 \theta - 2 \sin \theta \cdot \cos \theta + \cos^2 \theta = 1$$
$$1 - 2 \sin \theta \cdot \cos \theta = 1$$
$$-2 \sin \theta \cdot \cos \theta = 0$$
$$\sin \theta \cdot \cos \theta = 0$$
$$\sin \theta = 0 \text{ or } \cos \theta = 0$$

We know that $\sin \theta = 0$ when $\theta = 0$ or $\theta = \pi$ and that $\cos \theta = 0$ when $\theta = \frac{\pi}{2}$ or $\theta = \frac{3\pi}{2}$. Let's check these apparent solutions.

Check

θ	$\sin \theta - \cos \theta$	= 1?
0	$\sin 0 - \cos 0 = 0 - 1 = -1$	No
π	$\sin \pi - \cos \pi = 0 - (-1) = 1$	Yes
$\frac{\pi}{2}$	$\sin \frac{\pi}{2} - \cos \frac{\pi}{2} = 1 - 0 = 1$	Yes
$\frac{3\pi}{2}$	$\sin \frac{3\pi}{2} - \cos \frac{3\pi}{2} = -1 - 0 = -1$	No

The only solutions of the equation $\sin \theta - \cos \theta = 1$ are $\theta = \pi$ and $\theta = \frac{\pi}{2}$.

Part Two: Sample Problems

Problem 1 *Solve $\cos 6x \cos 2x - \sin 6x \sin 2x = 0.25$ for x, where $0° \le x \le 90°$.*

Solution This looks like the identity for $\cos(\alpha + \beta)$, with $\alpha = 6x$ and $\beta = 2x$.

$$\cos 6x \cos 2x - \sin 6x \sin 2x = 0.25$$
$$\cos(6x + 2x) = 0.25$$
$$\cos 8x = 0.25$$

Using the INV cos keys on a calculator, we find that the principal value of $(8x)^\circ$ is $\approx 75.52^\circ$. Since $0^\circ \le x^\circ \le 90^\circ$, we know that $0^\circ \le (8x)^\circ \le 720^\circ$. The terminal side of this angle can be in Quadrant I or IV.

Possible values of 8x	**Values of x**
$8x \approx 75.52^\circ$	$x \approx 9.44^\circ$
$8x \approx 360^\circ - 75.52^\circ$	$x \approx 35.56^\circ$
$8x \approx 360^\circ + 75.52^\circ$	$x \approx 54.44^\circ$
$8x \approx 720^\circ - 75.52^\circ$	$x \approx 80.56^\circ$

The approximate solutions are 9.44°, 35.56°, 54.44°, and 80.56°.

Problem 2 *Solve* $\sin 2x = \sin x$ *for* x.

Solution

$$\sin 2x = \sin x$$
$$2 \sin x \cos x = \sin x$$
$$2 \sin x \cos x - \sin x = 0$$
$$\sin x(2 \cos x - 1) = 0$$
$$\sin x = 0 \text{ or } \cos x = \frac{1}{2}$$

The possible values of x are $\{\ldots, -2\pi, -\pi, 0, \pi, 2\pi, \ldots\}$ or $\left\{\ldots, -\frac{7\pi}{3}, -\frac{5\pi}{3}, -\frac{\pi}{3}, \frac{\pi}{3}, \frac{5\pi}{3}, \frac{7\pi}{3}, \ldots\right\}$—that is, $\left\{n\pi, 2n\pi + \frac{\pi}{3}, 2n\pi - \frac{\pi}{3}\right\}$, where n is any integer.

It is important that in line 2 of our solution we did not divide both sides of the equation by sin x. If we had, we would have lost the solutions of the form $n\pi$. Do not divide by quantities that could equal 0.

Problem 3 *If* $AD = 25$ *and* $\sin\theta + \cos 2\theta = \frac{3}{4}$, *what is the length of altitude* $\overline{BD}$ *in triangle* ABC?

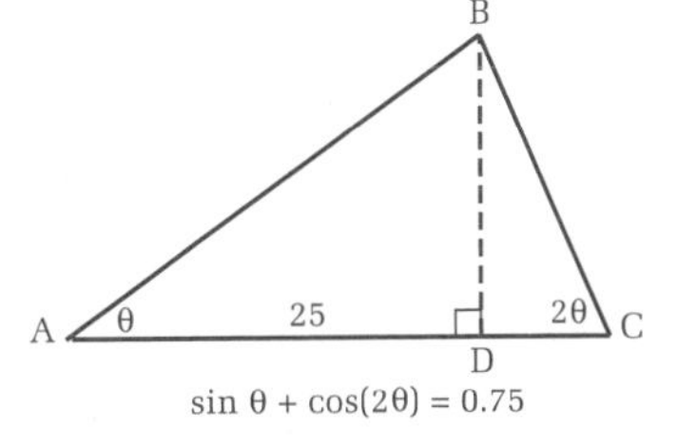

Solution

$$\sin\theta + \cos 2\theta = \frac{3}{4}$$
$$\sin\theta + 1 - 2\sin^2\theta = \frac{3}{4}$$
$$8\sin^2\theta - 4\sin\theta - 1 = 0$$

This is a quadratic equation in $\sin\theta$. We use the quadratic formula, with $a = 8$, $b = -4$, and $c = -1$, to find that $\sin\theta \approx 0.683$ or $\sin\theta \approx -0.183$. Since θ is certainly between 0° and 180°, we must use $\sin\theta \approx 0.683$. Using the INV sin keys on a calculator, we find that $\theta \approx 43.1^\circ$. Since $\frac{BD}{AD} = \tan\theta$,

$$\frac{BD}{25} \approx \tan 43.1^\circ$$
$$BD \approx 25 \cdot \tan 43.1^\circ$$
$$BD \approx 23.4$$

Checkpoint

1. Solve the equation $\sin^2 x - 1 = 0$ in the interval $0 \le x < 2\pi$.
   $\frac{\pi}{2}$ and $\frac{3\pi}{2}$
2. Write $2 \sin 6x \cos x$ as a sum of trigonometric functions.
   $\sin 7x + \sin 5x$
3. Solve the equation $\sin x + \cos 2x = 1$ in the interval $0 \le x < 2\pi$.
   $0, \frac{\pi}{6}, \frac{5\pi}{6}, \pi$

Assignment Guide

Average	
Day 1	4, 6, 12a, 13a, b, 14, 18a, b, 20a
Day 2	5, 8, 12b, 13c, d, 15, 18c, 21, 23
Advanced	
Day 1	6b, 7, 8, 10, 18a,c, 19a, 20a, 21
Day 2	4, 5a, 12, 14a, 18b, 19b, 20b, 23, 25
Honors	
Day 1	5b, 6b, 7, 8, 10, 14b, 15, 19, 20b
Day 2	9, 11a, c, 12a, 14a, 18a, c, 21, 22, 24, 25

Problem 4 *Find the values of θ for which* $\frac{1}{\sec 2\theta} + 1 = 2 - 2\sin^2\theta$.

Solution

$$\frac{1}{\sec 2\theta} + 1 = 2 - 2\sin^2\theta$$

$$\cos 2\theta + 1 = 2 - 2\sin^2\theta$$
$$2\cos^2\theta - 1 + 1 = 2 - 2\sin^2\theta$$
$$2\cos^2\theta + 2\sin^2\theta = 2$$
$$2 = 2$$

The original equation must be an identity. Therefore, the equation is true for all possible values of θ. Since there are no restrictions on the values of θ, the solution is $\{\theta: \theta \in \mathcal{R}\}$.

Part Three: Exercises and Problems

Warm-up Exercises

P

1 Determine whether each of the following is True or False.

a $\sin 50° \cos 20° = \frac{1}{2}(\sin 70° + \sin 30°)$ True

b $\sin 50° \sin 20° = \frac{1}{2}(\sin 70° + \sin 30°)$ False

c $\cos 50° \cos 20° = \frac{1}{2}(\sin 70° + \sin 30°)$ False

P

2 Write each sum as a product.

a $\sin 3\theta + \sin 5\theta$ $2 \sin 4\theta \cos \theta$

b $\cos 2\theta + \cos \theta$ $2 \cos \frac{3\theta}{2} \cos \frac{\theta}{2}$

P,R 

3 Refer to the inequality $\sin\theta + \cos\theta > 1$.

a Is $\theta = 70°$ a solution? Yes

b Is $\theta = 50°$ a solution? Yes

c Is $\theta = 90°$ a solution? No

Problem Set

R **A**

4 Find (x, y) if $\sin\theta = \frac{3}{5}$.
(6, 8)

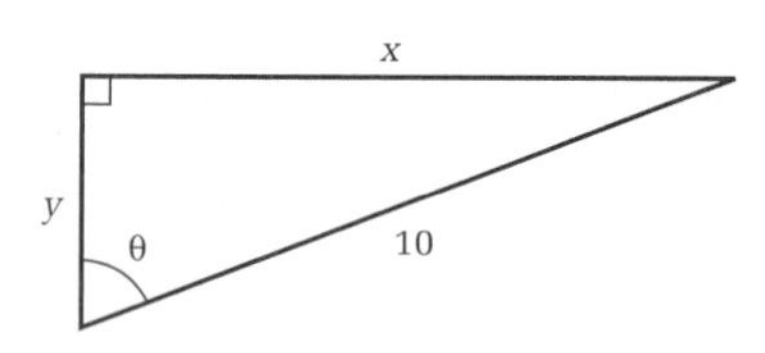

P,R

5 Simplify each expression.

a $\frac{\cos 3x \cos 2x - \sin 3x \sin 2x}{\sin 5x}$ cot 5x

b $\frac{\sin 3x \cos x + \cos 3x \sin x}{\cos^2 2x - \sin^2 2x}$ tan 4x

P,R **6** Solve for θ, where $0 \le \theta < 2\pi$.

a $\sin\theta = \cos\theta$ $\left\{\frac{\pi}{4}, \frac{5\pi}{4}\right\}$

b $\sin\theta \cdot \cos\theta = 0$ $\left\{0, \frac{\pi}{2}, \pi, \frac{3\pi}{2}\right\}$

P **7** Write each sum as a product.

a $\sin 17^\circ + \sin 3^\circ$
$2\sin 10^\circ \cos 7^\circ$

b $\cos 50^\circ + \cos 14^\circ$
$2\cos 32^\circ \cos 18^\circ$

P,R **8** Refer to the equation $2\cos^2\theta + \cos\theta - 1 = 0$.

a Solve for $\cos\theta$. $\cos\theta = \frac{1}{2}$ or $\cos\theta = -1$

b Solve for θ if $0 \le \theta < 2\pi$. $\left\{\frac{\pi}{3}, \pi, \frac{5\pi}{3}\right\}$

P **9** Prove the identity $\frac{\sin 4\theta + \sin 3\theta}{\cos 4\theta + \cos 3\theta} = \tan\frac{7\theta}{2}$.

B

P **10** Write each product as a sum.

a $2 \cdot \cos 10^\circ \cdot \cos 4^\circ$ $\cos 14^\circ + \cos 6^\circ$

b $2 \cdot \sin 70^\circ \cos 66^\circ$ $\sin 136^\circ + \sin 4^\circ$

c $2 \cdot \sin 13^\circ \cdot \sin 9^\circ$ $\cos 4^\circ - \cos 22^\circ$

P **11** Find the exact value of each of the following.

a $\sin 165^\circ \cdot \sin 15^\circ$ $\frac{2-\sqrt{3}}{4}$

b $\sin 255^\circ \cos 15^\circ$ $\frac{-2-\sqrt{3}}{4}$

c $\cos 105^\circ \cos(-15^\circ)$ $-\frac{1}{4}$

P **12** Simplify each expression.

a $\frac{\sin 6\theta + \sin 2\theta}{\cos 6\theta + \cos 2\theta}$ $\tan 4\theta$

b $\frac{\cos 10\theta + \cos 2\theta}{\sin 10\theta - \sin 2\theta}$ $\cot 4\theta$

P **13** Write each expression as a product.

a $\cos x + \cos y$ $2\cos\frac{x+y}{2}\cos\frac{x-y}{2}$

b $\cos 6\theta + \cos 2\theta$ $2\cos 4\theta \cos 2\theta$

c $\cos 12\theta + \cos 8\theta$ $2\cos 10\theta \cos 2\theta$

d $\cos x - \cos y$ $-2\sin\frac{x+y}{2}\sin\frac{x-y}{2}$

P **14** Solve each equation for θ, where $0 \le \theta < 2\pi$.

a $2\sin^2\theta + 3\sin\theta + 1 = 0$ $\left\{\frac{7\pi}{6}, \frac{3\pi}{2}, \frac{11\pi}{6}\right\}$

b $2\cos^2\theta + 3\sin\theta - 3 = 0$ $\left\{\frac{\pi}{6}, \frac{\pi}{2}, \frac{5\pi}{6}\right\}$

P  **15** Find the measure of angle θ if $0^\circ \le \theta < 90^\circ$.
≈ 38.17

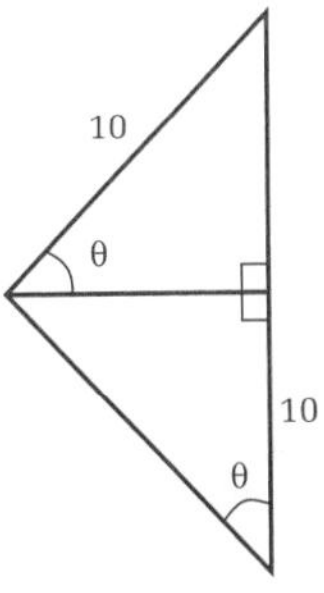

P,R **16** Solve for x.

a $\frac{\cos x - \sin x}{\cos x + \sin x} = 1$ $\{n\pi\}$, where n is an integer

b $\frac{\cos x - \sin x}{\cos x + \sin x} = \frac{1 - \tan x}{1 + \tan x}$ $\{x{:}x \in \mathcal{R}\}$

Problem-Set Notes and Strategies

- For problem **6**, have students consult the unit circle and the answers will seem obvious. Some students relate very well to this visual representation of the solution and its corresponding algebraic equivalent.
- Remind students problem **8** is quadratic and the quadratic formula can be used. The answers must be converted into angles by using trigonometry, but the formula still applies to any quadratic equation.
- Problems **9 – 13** are straightforward applications of this section's formulas.
- Problem **14** is quadratic and factors nicely. If an expression cannot be factored, the quadratic formula could be applied.
- In problem **16b**, suggest that students change everything to sine and cosine.

Additional Answer

9 $\frac{\sin 4\theta + \sin 3\theta}{\cos 4\theta + \cos 3\theta}$

$$= \frac{2\sin\frac{4\theta+3\theta}{2}\cos\frac{4\theta-3\theta}{2}}{2\cos\frac{4\theta+3\theta}{2}\cos\frac{4\theta-3\theta}{2}}$$

$$= \frac{\sin\frac{7\theta}{2}}{\cos\frac{7\theta}{2}}$$

$$= \tan\frac{7\theta}{2}$$

Problem-Set Notes and Strategies, continued

- Be careful with part **c** of problem **18**. It might be best done by factoring using the difference of squares. Otherwise it is tempting to lose a couple of solutions with the square root.
- Problem **20** involves combining several identities to solve a single equation.

Additional Answers

20a $\dfrac{\cos 5\theta + \cos \theta}{\sin 5\theta + \sin \theta}$

$= \dfrac{2 \cos \frac{5\theta+\theta}{2} \cos \frac{5\theta-\theta}{2}}{2 \sin \frac{5\theta+\theta}{2} \cos \frac{5\theta-\theta}{2}}$

$= \dfrac{\cos 3\theta}{\sin 3\theta}$

$= \cot 3\theta$

Answers for problems **19a**, **19b**, **20b**, and **24** can be found in the answer pages beginning on **A** 1.

- The quadratic formula might be necessary to solve problem **21**. In any case, a calculator is very helpful.
- Problem **22** is an identity which may be proved using the results of the last section.
- Remind students to use radian measure in problem **25**. Most graphing calculators have a trace function available. Have them use this to find the point of intersection and check it.
- Upper and lower bounds for x in problem **26** should be easy. To find the points of intersection, use a calculator with the trace function or a computer program.

Problem Set, *continued*

R **17** Find the exact value of tan 75° tan 75°. Show your method. $\frac{2+\sqrt{3}}{2-\sqrt{3}}$

P,R **18** Solve for θ, where $0 \le \theta < 2\pi$.
a $2 \sin^2 \theta + \sin \theta - 1 = 0$ **a** $\left\{\frac{\pi}{6}, \frac{5\pi}{6}, \frac{3\pi}{2}\right\}$ **b** $\left\{\frac{3\pi}{2}\right\}$ **c** $\left\{\frac{\pi}{3}, \frac{2\pi}{3}, \frac{4\pi}{3}, \frac{5\pi}{3}\right\}$
b $\sin^2 \theta - \sin \theta - 2 = 0$
c $4 \sin^2 \theta - 3 = 0$

P **19** Prove the following identities.
a $\tan \alpha = \dfrac{\sin (\alpha + \beta) + \sin (\alpha - \beta)}{\cos (\alpha - \beta) + \cos (\alpha + \beta)}$
b $\cos \theta \cos (\theta - 90°) = \dfrac{1}{2} \cos (2\theta - 90°)$

P,R **20** Prove the following identities.
a $\dfrac{\cos 5\theta + \cos \theta}{\sin 5\theta + \sin \theta} = \cot 3\theta$
b $\dfrac{\sin 10\theta + \sin 6\theta}{\cos^2 \theta - \sin^2 \theta} = 2 \sin 8\theta$

P,R **21** Solve for $\{x: 0° \le x \le 360°\}$.
a $\sin^2 x - 4 \cos^2 x = 0$ $\{\approx 63.43°, \approx 116.57°, \approx 243.43°, \approx 296.57°\}$
b $12 \cos^2 x = 15 - 20 \sin x$ $\{\approx 9.59°, \approx 170.41°\}$

P **22** Find all values of θ for which $\sin 4\theta = 2 \sin 2\theta - 8 \sin^3 \theta \cos \theta$. $\{\theta : \theta \in \mathfrak{R}\}$

P,R **23** In the diagram, $0° < \theta < 45°$.
a Find θ. $\approx 41.41°$
b Find x. $x \approx 7.937$

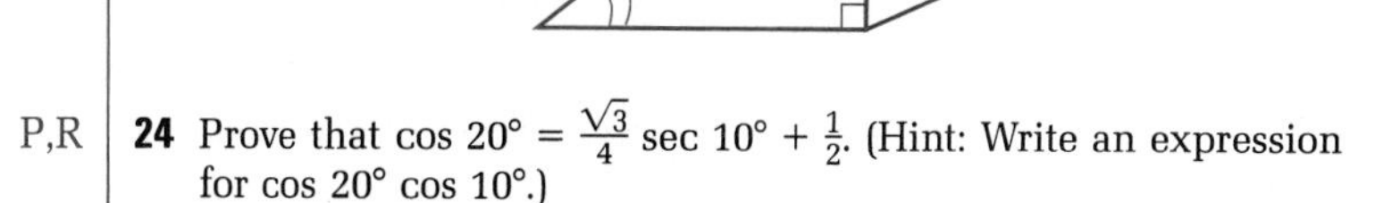

P,R **24** Prove that $\cos 20° = \frac{\sqrt{3}}{4} \sec 10° + \frac{1}{2}$. (Hint: Write an expression for cos 20° cos 10°.)

,R C **25** **a** Using a graphing program, determine the number of solutions the equation $\log x = \cos x$ has. 3
b Find the greatest value of x satisfying this equation. ≈ 6.87

P,R **26** Refer to the equation $\sin x = \frac{1}{4}x$.
a Find an upper bound and a lower bound of the solutions of this trigonometric equation. 4; −4
b Using a graphing calculator or computer program, find the approximate solutions. $x = -2.47$, $x = 0$, or $x = 2.47$

15

Chapter Summary

Concepts and Procedures

After studying this chapter, you should be able to

- Prove trigonometric identities (15.1)
- Use the sum and difference identities (15.2)
- Determine the angle between two graphed lines (15.2)
- Apply the double-angle identities (15.3)
- Apply the half-angle identities (15.3)
- Use the product-to-sum and sum-to-product identities (15.4)
- Solve trigonometric equations (15.4)

Vocabulary

double-angle identities (15.3)
half-angle identities (15.3)
product-to-sum identities (15.4)
sum and difference identities (15.2)
sum-to-product identities (15.4)

Theorems and Identities

$\sin 2\theta = 2 \sin \theta \cos \theta$

$\cos 2\theta = \cos^2 \theta - \sin^2 \theta$

$\cos 2\theta = 1 - 2 \sin^2 \theta$

$\cos 2\theta = 2 \cos^2 \theta - 1$

$$\tan 2\theta = \frac{2 \tan \theta}{1 - \tan^2 \theta}$$

$$\sin \frac{\theta}{2} = \pm\sqrt{\frac{1 - \cos \theta}{2}}$$

$$\cos \frac{\theta}{2} = \pm\sqrt{\frac{1 + \cos \theta}{2}}$$

$$\tan \frac{\theta}{2} = \pm\sqrt{\frac{1 - \cos \theta}{1 + \cos \theta}}$$

$$\tan \frac{\theta}{2} = \frac{1 - \cos \theta}{\sin \theta}$$

$$\tan \frac{\theta}{2} = \frac{\sin \theta}{1 + \cos \theta}$$

$\sin (\alpha + \beta) = \sin \alpha \cos \beta + \cos \alpha \sin \beta$

$\sin (\alpha - \beta) = \sin \alpha \cos \beta - \cos \alpha \sin \beta$

$\cos (\alpha + \beta) = \cos \alpha \cos \beta - \sin \alpha \sin \beta$

$\cos (\alpha - \beta) = \cos \alpha \cos \beta + \sin \alpha \sin \beta$

$$\tan (\alpha + \beta) = \frac{\tan \alpha + \tan \beta}{1 - \tan \alpha \tan \beta}$$

$$\tan (\alpha - \beta) = \frac{\tan \alpha - \tan \beta}{1 + \tan \alpha \tan \beta}$$

$$\sin \alpha \cos \beta = \frac{1}{2}[\sin (\alpha + \beta) + \sin (\alpha - \beta)]$$

$$\sin \alpha \sin \beta = \frac{1}{2}[\cos (\alpha - \beta) - \cos (\alpha + \beta)]$$

$$\cos \alpha \cos \beta = \frac{1}{2}[\cos (\alpha - \beta) + \cos (\alpha + \beta)]$$

$$\sin \alpha + \sin \beta = 2 \sin \frac{\alpha + \beta}{2} \cos \frac{\alpha - \beta}{2}$$

$$\cos \alpha + \cos \beta = 2 \cos \frac{\alpha + \beta}{2} \cos \frac{\alpha - \beta}{2}$$

15

Class Planning

Time Schedule

All levels: 2 days

Using Computers

See Using Computers for Section 15.1

Assignment Guide

Average	
Day 1	2, 3a, 4a, 5, 7, 8a, 9a, 10, 12, 14, 17
Day 2	3b, 4b, 6, 8b, 9b, 13a, 16, 20, 21a, c
Advanced	
Day 1	2, 3a, 4a, 5, 7, 8a, 9a, b, 11a, 13a, 15, 20
Day 2	3b, 4b, 6, 8b, 10, 11b, 13b, 16, 21a, c, 22b, 25
Honors	
Day 1	2, 3a, 4a, b, 5, 10, 11a, b 13a, 15, 19, 21a, c, 24
Day 2	3b, 4c, 7, 11c, 13b, 16, 17a, b, 20, 22a, b, 25

Problem-Set Notes and Strategies

- In problem **1**, relate cofunctions to complementary angles.
- Problem **7** reviews some geometry formulas.

Additional Answer

Answer for problem **5** can be found in the answer pages beginning on **A** 1.

15 Review Problems

P,R **1** Determine whether each statement is True or False.

a $\cos 30° = \sin 60°$ True
b $1 + (\cot 30°)^2 = (\sec 60°)^2$ True
c $\cos^2 40° - 1 = \csc^2 40°$ False

P,R **2** Find each of the following.

a $\sin \alpha$ 0.8
b $\sin \beta$ ≈ 0.857
c $\sin (\alpha + \beta)$ ≈ 0.926

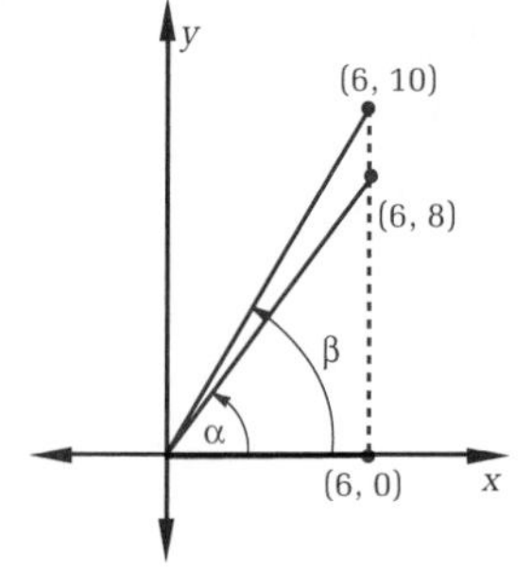

P **3** Evaluate.

a $\sin\left(2 \text{ Arcsin } \frac{4}{5}\right)$ $\frac{24}{25}$
b $\cos\left(2 \text{ Arcsin } \frac{4}{5}\right)$ $-\frac{7}{25}$

P **4** Write each product as a sum.

a $\sin 4\theta \cos 3\theta$ $\frac{1}{2}(\sin 7\theta + \sin \theta)$
b $\cos 4\theta \cos 6\theta$ $\frac{1}{2}(\cos 2\theta + \cos 10\theta)$
c $\sin 2\theta \cos 8\theta$ $\frac{1}{2}(\sin 10\theta - \sin 6\theta)$

P **5** Prove that $\frac{\cos \theta}{1 - \sin \theta} = \sec \theta + \tan \theta$. State restrictions.

P **6** Determine whether each equation is True or False.

a $\sin (40° + 50°) = \sin 90°$ True
b $\sin 40° + \sin 50° = \sin 90°$ False
c $\sin 40° \cos 50° + \cos 40° \sin 50° = \sin 90°$ True

R  **7** **a** Express the volume of each figure as a function of θ. $V_{cyl.} = 36\pi \cos \theta$; $V_{cone} = 72\pi \sin^2 \theta \cos \theta$
b If both figures have the same volume, what is θ? $\frac{\pi}{4}$

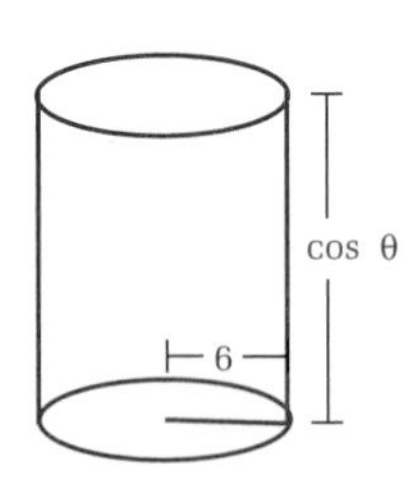

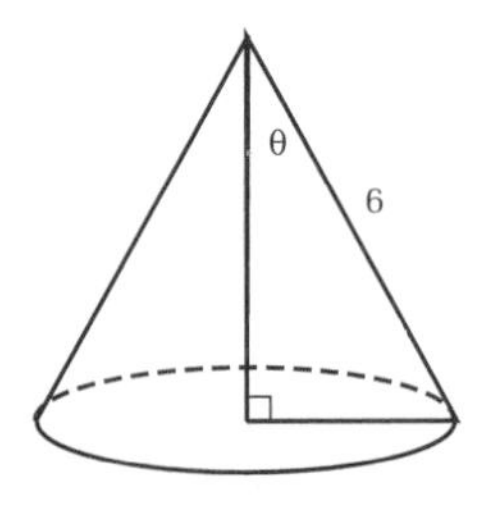

P **8** Verify each identity.

a $\tan\theta + \cot\theta = \sec\theta \cdot \csc\theta$

b $\sec^2\theta + \csc^2\theta = \sec^2\theta \cdot \csc^2\theta$

c $\dfrac{\cos\theta}{1+\sin\theta} = \dfrac{1-\sin\theta}{\cos\theta}$

P **9** Determine whether each of the following is True or False.

a $\cos 90° = 2\cos 45°$ False

b $\cos 90° = \cos 30° + \cos 60°$ False

c $\cos 90° = \cos^2 45° - \sin^2 45°$ True

P **10** Solve $\cos 2x + \sin^2 x = 1$ for x, where $0 \le x \le 2\pi$. $\{0, \pi, 2\pi\}$

P **11** Prove each identity.

a $\cos\left(\theta - \dfrac{\pi}{2}\right) = \sin\theta$

b $\sin\left(\theta - \dfrac{\pi}{2}\right) = -\cos\theta$

c $\sin\left(\theta + \dfrac{\pi}{2}\right) = \cos\theta$

d $\cos\left(\theta + \dfrac{\pi}{2}\right) = -\sin\theta$

P **12** Determine whether the following statement is True or False.

$\sin 15° = \sqrt{\dfrac{1-\cos 30°}{2}}$ True

P **13** Use identities to express each of the following in terms of functions of x.

a $\sin 3x$
$3\sin x - 4\sin^3 x$

b $\cos 3x$
$4\cos^3 x - 3\cos x$

P **14** Determine whether each of the following is True or False.

a $\tan\alpha = \dfrac{\sin\alpha}{\cos\alpha}$ True

b $\cot\alpha = \dfrac{\cos\alpha}{\sin\alpha}$ True

c $\sin\alpha = \dfrac{1}{\csc\alpha}$ True

P **15** Prove that $\cos^4\theta - \sin^4\theta = \cos^2\theta - \sin^2\theta$.

P,R **B** **16** If $\tan\beta = \frac{3}{4}$ and $\tan\alpha = -\frac{4}{3}$, what is $\tan(\alpha - \beta)$? Explain your answer.

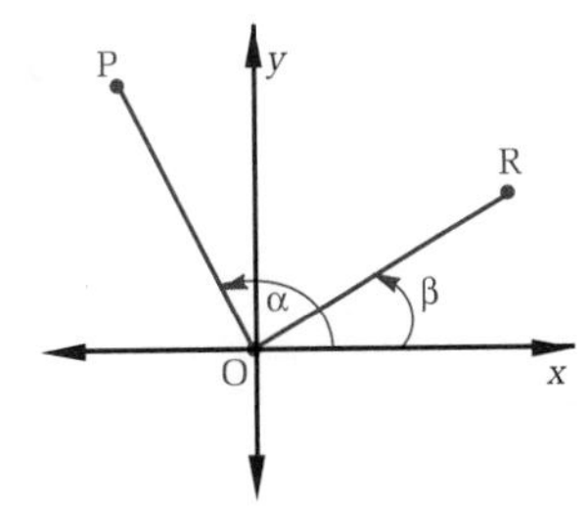

P **17** Change each sum to a product.

a $\sin x + \sin y$
$2\sin\frac{x+y}{2}\cos\frac{x-y}{2}$

b $\sin 3\theta + \sin\theta$
$2\sin 2\theta\cos\theta$

c $\sin 10\theta - \sin 4\theta$
$2\sin 3\theta\cos 7\theta$

Problem-Set Notes and Strategies, continued

- In problem **11**, have the student draw the picture of the sine and the cosine and look at these identities in relation to the graph.
- Problem **15** is a simple difference of squares.

Additional Answers

8a $\tan\theta + \cot\theta$

$= \dfrac{\sin\theta}{\cos\theta} + \dfrac{\cos\theta}{\sin\theta}$

$= \dfrac{\sin^2\theta + \cos^2\theta}{\cos\theta\sin\theta}$

$= \dfrac{1}{\cos\theta\sin\theta}$

$= \dfrac{1}{\cos\theta} \cdot \dfrac{1}{\sin\theta}$

$= \sec\theta \cdot \csc\theta$

8b $\sec^2\theta + \csc^2\theta$

$= \dfrac{1}{\cos^2\theta} + \dfrac{1}{\sin^2\theta}$

$= \dfrac{\sin^2\theta + \cos^2\theta}{\cos^2\theta\sin^2\theta}$

$= \dfrac{1}{\cos^2\theta\sin^2\theta}$

$= \dfrac{1}{\cos^2\theta} \cdot \dfrac{1}{\sin^2\theta}$

$= \sec^2\theta \cdot \csc^2\theta$

Answers for problems **8c**, **11a – d**, **15** and **16** can be found in the answer pages beginning on **A** 1.

Additional Answers

22a

$$\frac{\cos x - \sin x}{\cos x + \sin x}$$

$$= \frac{(\cos x - \sin x)(\cos x + \sin x)}{(\cos x + \sin x)(\cos x + \sin x)}$$

$$= \frac{\cos^2 x - \sin^2 x}{\cos^2 x + 2 \cos x \sin x + \sin^2 x}$$

$$= \frac{\cos 2x}{1 + 2 \sin x \cos x}$$

$$= \frac{\cos 2x}{1 + \sin 2x}$$

22b

$$1 + \tan x \tan 2x$$

$$= 1 + (\tan x)\left(\frac{2 \tan x}{1 - \tan^2 x}\right)$$

$$= 1 + \frac{2 \tan^2 x}{1 - \tan^2 x}$$

$$= \frac{1 - \tan^2 x + 2 \tan^2 x}{1 - \tan^2 x}$$

$$= \frac{1 + \tan^2 x}{1 - \tan^2 x}$$

$$= \frac{1 + \frac{\sin^2 x}{\cos^2 x}}{1 - \frac{\sin^2 x}{\cos^2 x}}$$

$$= \frac{\frac{\cos^2 x + \sin^2 x}{\cos^2 x}}{\frac{\cos^2 x - \sin^2 x}{\cos^2 x}} = \frac{1}{\cos 2x} = \sec 2x$$

23a $(-7, 24)$; $(-24, -7)$; $(7, -24)$

25

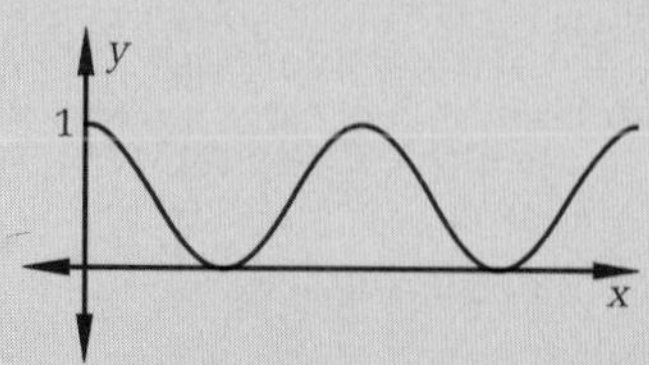

Answers for problems **22c** and **23b** can be found in the answer pages beginning on A1.

Review Problems, *continued*

P **18** Determine whether each of the following is True or False.

a $2 \sin 30° \cos 60° = \sin 90° - \sin 30°$ True

b $\sin 90° + \sin 30° = 2 \sin 60° \cos 30°$ True

P **19** Use the diagram to find each of the following.

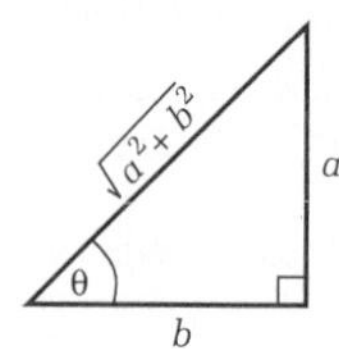

a $\dfrac{\sin \theta}{1 + \sin \theta}$ $\dfrac{a}{\sqrt{a^2 + b^2} + a}$

b $\dfrac{1 - \sin \theta}{\sin \theta}$ $\dfrac{\sqrt{a^2 + b^2} - a}{a}$

P **20** Solve $\tan 2x = \tan x$ for x. $\{n\pi\}$, where n is an integer

P,R **21** Evaluate without using a calculator.

a $\sin 40° \cos 5° + \cos 40° \sin 5°$ $\frac{\sqrt{2}}{2}$

b $\cos 18° \cos 12° - \sin 18° \sin 12°$ $\frac{\sqrt{3}}{2}$

c $\cos \frac{5\pi}{8} \cdot \cos \frac{3\pi}{8} + \sin \frac{5\pi}{8} \cdot \sin \frac{3\pi}{8}$ $\frac{\sqrt{2}}{2}$

P,R **22** Prove the following identities.

a $\dfrac{\cos x - \sin x}{\cos x + \sin x} = \dfrac{\cos 2x}{1 + \sin 2x}$

b $1 + \tan x \tan 2x = \sec 2x$

c $\sin (x + y) \sin (x - y) = \sin^2 x - \sin^2 y$

R **23** The segment OA is rotated $\frac{\pi}{2}$, π, and $\frac{3\pi}{2}$ radians to form segments OA′, OA″, and OA‴.

a Determine (x_1, y_1), (x_2, y_2), and (x_3, y_3).

b Find the sine, the cosine, and the tangent of θ, $\theta + \frac{\pi}{2}$, $\theta + \pi$, and $\theta + \frac{3\pi}{2}$.

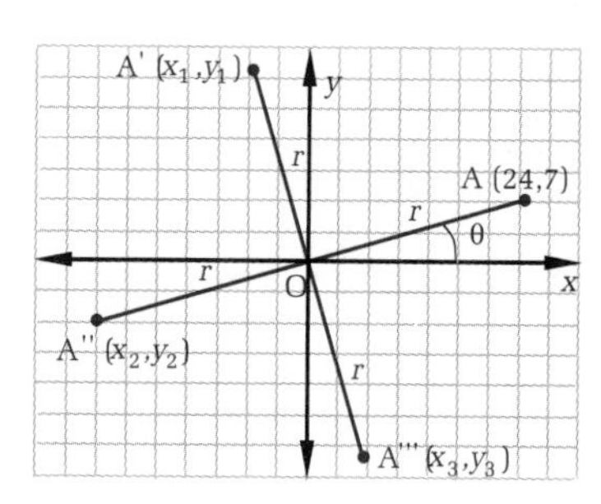

C P **24** Evaluate the following expression. 1

$$\frac{\sin 10° \cdot \sin 20° \cdot \sin 30° \cdot \sin 40° \cdot \sin 50° \cdot \sin 60° \cdot \sin 70° \cdot \sin 80°}{\cos 10° \cdot \cos 20° \cdot \cos 30° \cdot \cos 40° \cdot \cos 50° \cdot \cos 60° \cdot \cos 70° \cdot \cos 80°}$$

P,R **25** The graph of $y = f(x)$ is shown. Draw the graph of $y = [f(x)]^2$.

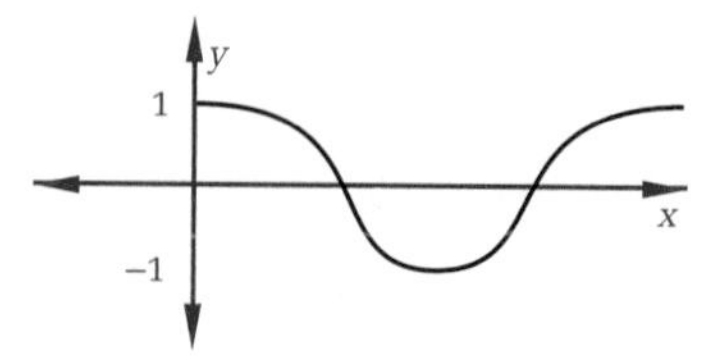

15 CHAPTER TEST

In problems 1–3, simplify each expression in terms of a single trigonometric function.

1 $\dfrac{\sin\theta + \cos\theta}{1 + \tan\theta}$ $\cos\theta$ **2** $(1 + \cos\theta)(\csc\theta - \cot\theta)$ $\sin\theta$ **3** $\dfrac{\sin\beta + \tan\beta}{\cos\beta + 1}$ $\tan\beta$

4 If $\sin\theta = \frac{-8}{17}$ and θ is in the third quadrant, what is $\cot\theta$? $\frac{15}{8}$

5 Find, to the nearest degree, the acute angle formed by the graphs of $y = \frac{-1}{2}x + 4$ and $y = \frac{4}{3}x - 2$. $\approx 80°$

In problems 6–8, solve each equation.

6 $3\tan^2\theta = 1$, where $-2\pi \le \theta \le \pi$ $\left\{-\frac{11\pi}{6}, -\frac{7\pi}{6}, -\frac{5\pi}{6}, -\frac{\pi}{6}, \frac{\pi}{6}, \frac{5\pi}{6}\right\}$

7 $4\sin\theta \cdot \cos\theta = 1$, where $0 \le \theta \le 2\pi$ $\left\{\frac{\pi}{12}, \frac{5\pi}{12}, \frac{13\pi}{12}, \frac{17\pi}{12}\right\}$

8 $\cos 2\theta = \cos^2\theta$ $\{n\pi\}$, where n is an integer

9 If $\cos\theta = \frac{3}{5}$, what is $\cos 2\theta$? $\frac{-7}{25}$

10 Find $\cos(\beta - \alpha)$ if β is in Quadrant II and $\sin\beta = \frac{2}{3}$ and if α is in Quadrant I and $\cos\alpha = \frac{3}{4}$. $\frac{2\sqrt{7} - 3\sqrt{5}}{12}$

11 If $\sec\alpha = \frac{3}{2}$ and $\frac{3\pi}{2} < \alpha < 2\pi$, what is $\sin 2\alpha$? $-\frac{4\sqrt{5}}{9}$

12 Prove that $\frac{\cos\theta}{1 + \sin\theta} + \frac{1 + \sin\theta}{\cos\theta} = 2\sec\theta$.

13 Use the diagram to find $\sin\theta$. $\frac{3}{5}$

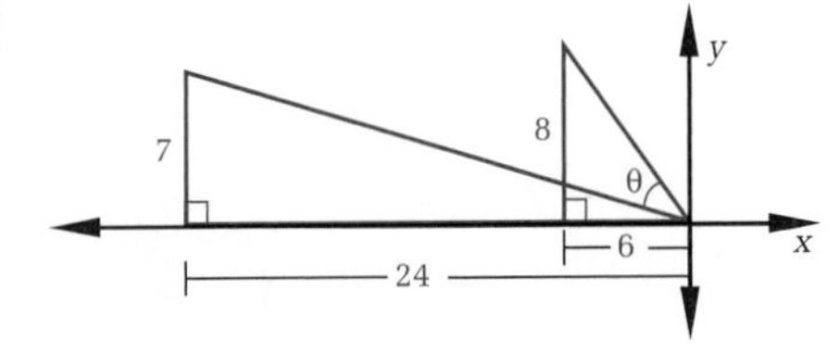

14 If $\tan 2\alpha = \frac{-24}{7}$, what is $\tan\alpha$? $-\frac{3}{4}$ or $\frac{4}{3}$

15 Prove that $\sin 2\alpha = \frac{2\tan\alpha}{1 + \tan^2\alpha}$.

P **16** Suppose that $\sin\alpha = -\frac{3}{5}$, where $\pi < \alpha < \frac{3\pi}{2}$, and that $\cos\beta = \frac{24}{25}$, where $\frac{3\pi}{2} < \beta < 2\pi$. Find the following.

a $\cos(\beta - \alpha)$ $-\frac{3}{5}$ **b** $\sin(\alpha + \beta)$ $-\frac{44}{125}$ **c** $\tan(\alpha + \beta)$ $\frac{44}{117}$

P **17** Prove that $\frac{\cos(\alpha + \beta) + \cos(\alpha - \beta)}{\sin(\alpha + \beta) + \sin(\alpha - \beta)} = \cot\alpha$.

Resource References

Evaluation

Test Form 1, 2, 3

Additional Answers

12 $\dfrac{\cos\theta}{1 + \sin\theta} + \dfrac{1 + \sin\theta}{\cos\theta}$

$= \dfrac{\cos^2\theta + (1 + \sin\theta)^2}{(1 + \sin\theta)(\cos\theta)}$

$= \dfrac{\cos^2\theta + 1 + 2\sin\theta + \sin^2\theta}{(1 + \sin\theta)(\cos\theta)}$

$= \dfrac{2 + 2\sin\theta}{(1 + \sin\theta)(\cos\theta)}$

$= \dfrac{2(1 + \sin\theta)}{(1 + \sin\theta)(\cos\theta)}$

$= \dfrac{2}{\cos\theta}$

$= 2\sec\theta$

15 $\sin 2\alpha$

$= 2\sin\alpha\cos\alpha$

$= \dfrac{2\sin\alpha\cos^2\alpha}{\cos\alpha}$

$= \dfrac{2\sin\alpha}{\cos\alpha} \div \dfrac{1}{\cos^2\alpha}$

$= 2\tan\alpha \div \dfrac{\cos^2\alpha + \sin^2\alpha}{\cos^2\alpha}$

$= 2\tan\alpha \div \left(1 + \dfrac{\sin^2\alpha}{\cos^2\alpha}\right)$

$= \dfrac{2\tan\alpha}{1 + \tan^2\alpha}$

Answer for problem **17** can be found in the answer pages beginning on **A** 1.

16.1

IN THE FIRST two sections of the chapter, the Law of Sines and the Law of Cosines are introduced. This enables students to solve application problems that require other than right-triangle trigonometry.

The remaining sections of the chapter concentrate on graphing the six trigonometric functions. Students will sketch variations of these graphs by shifting the standard graphs horizontally and vertically and by changing their amplitude and period.

CHAPTER 16

TRIGONOMETRIC GRAPHS AND APPLICATIONS

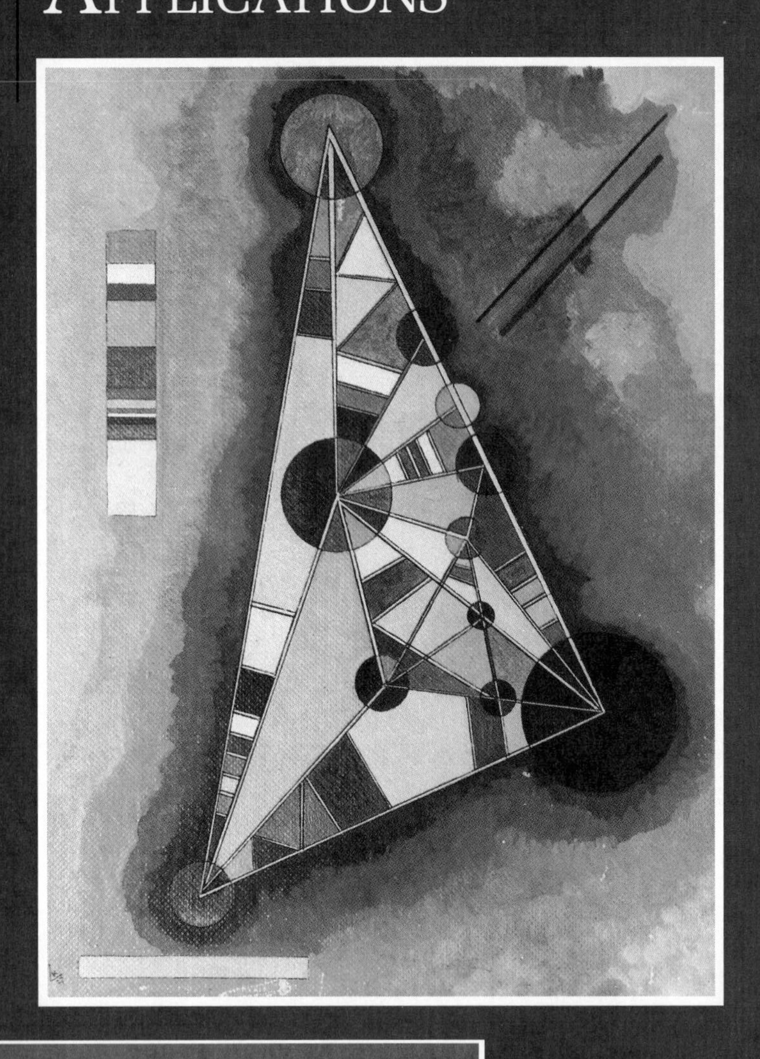

Relationships among the sides and angles of a triangle, the basis of trigonometry, are illustrated in this painting by Kandinsky.

16.1 THE LAW OF SINES

Objective

After studying this section, you will be able to

- Apply the Law of Sines

Part One: Introduction

Solving Oblique Triangles

You have seen how to solve right triangles using the trigonometric functions. Now we will discuss solving triangles that are not right triangles.

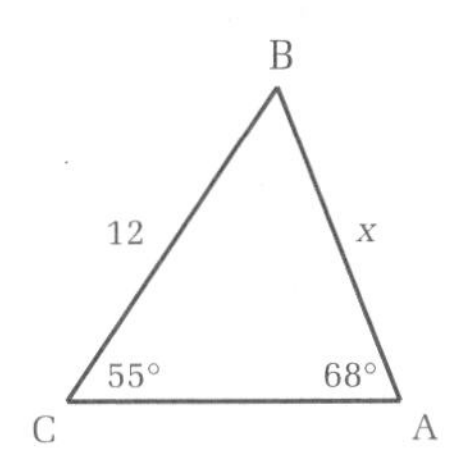

In triangle ABC, we are given two angles and a side. By the AAS theorem, which you may remember from geometry, the two angles and the side determine a unique triangle.

We drop a perpendicular (altitude) from B to side $\overline{CA}$. Now we can set up two equations involving sine.

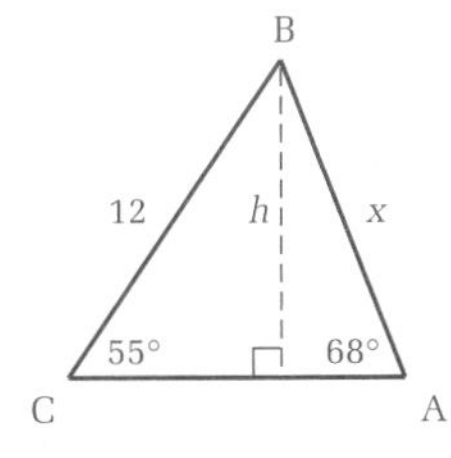

$$\sin 68° = \frac{h}{x} \qquad \sin 55° = \frac{h}{12}$$

$$x \sin 68° = h \qquad 12 \sin 55° = h$$

Thus, $x \sin 68° = 12 \sin 55°$.

$$x = \frac{12 \sin 55°}{\sin 68°} \approx \frac{12(0.8192)}{(0.9272)} \approx 10.6$$

In future problems, we can avoid the extra steps of putting in h and then removing it. Let's analyze the equation

$$x \sin 68° = 12 \sin 55°$$

This equation can be written as a proportion.

$$\frac{\sin 68°}{12} = \frac{\sin 55°}{x}$$

Notice that 12 is the measure of the side opposite the 68° angle and that x is the measure of the side opposite the 55° angle. The general form of this equation is called the ***Law of Sines.***

Class Planning

Time Schedule

All levels: 2 days

Resource References

Teacher's Resource Book

Class Opener 92
Cooperative Learning 46
Practice 97

Class Opener

Find the area of $\triangle ABC$ if $\angle A = 37°$, $AB = 13$, and $AC = 19$.

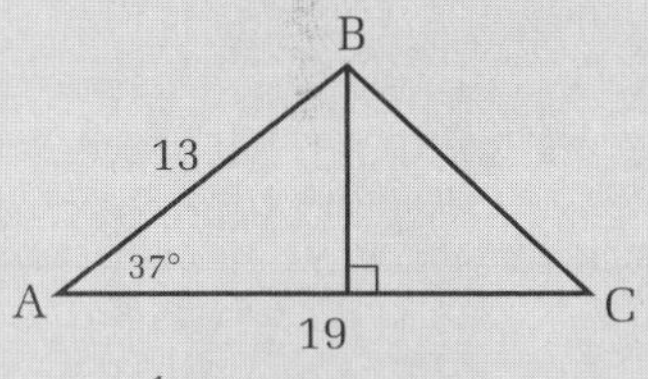

$\text{Area} = \frac{1}{2}(13)(\sin 37°)(19) \approx 74.32$

Lesson Notes

- The first section of the chapter introduces the Law of Sines. This, along with the Law of Cosines, introduced in the next section, will allow students to solve problems in triangles without right angles. Students may recognize that the Law creates right triangles by drawing auxiliary lines. Algebraic methods are used to solve for the desired relationships. Once again, we see the application of algebra to complement the trigonometry and produce the desired results.

Vocabulary

Law of Sines

Lesson Notes, continued

- The Law of Sines is in two forms with the sine of the angle either in the numerator or denominator. Examples show that either form can be used and one may be more advantageous depending on how the problem is stated.
- The ambiguous case of the Law of Sines is briefly discussed. You may want to go through examples showing cases where there are two solutions to illustrate the ambiguity.

Communicating Mathematics

The Law of Sines is a major tool of trigonometry, as it lets people solve triangles given only limited information. In addition, the triangles do not have to contain a right angle. Write the Law of Sines in your own words rather than symbols and explain the conditions necessary for it to be used.

Stumbling Block

The Law of Sines is a major result of trigonometry and should be remembered as "the ratio of a side to the sine of the opposite angle is constant in a given triangle." As with any ratio problem, the ratios can be reworked many ways to enhance the formula depending on the problem. Make sure the students understand that the law involves ratios that obey all the rules of ratios, such as cross-multiplying, inverting both sides of the ratio, etc.

The Law of Sines

If a, b, and c are the lengths of the sides opposite angles A, B, and C respectively in triangle ABC, then

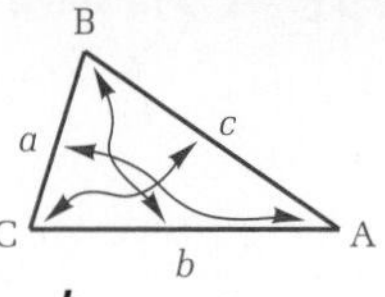

$$\frac{\sin\angle A}{a} = \frac{\sin\angle B}{b} = \frac{\sin\angle C}{c} \quad \text{or} \quad \frac{a}{\sin\angle A} = \frac{b}{\sin\angle B} = \frac{c}{\sin\angle C}$$

Example *Use the information in the figure to find lengths p and q.*

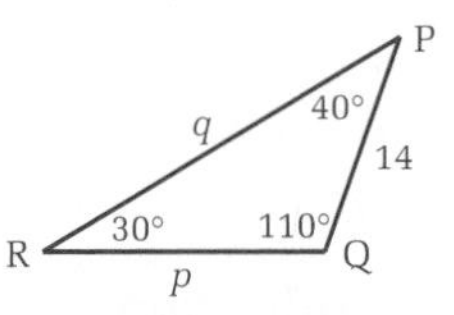

Notice that the pattern of given information is ASA (angle-side-angle), so our answers will be unique. Notice also that $\angle R$ is 30°.

By the Law of Sines, $\frac{p}{\sin\angle P} = \frac{q}{\sin\angle Q} = \frac{r}{\sin\angle R}$, so $\frac{p}{\sin 40°} = \frac{14}{\sin 30°}$ and $\frac{q}{\sin 110°} = \frac{14}{\sin 30°}$. Therefore, $p = \frac{14\sin 40°}{\sin 30°} \approx 18.0$ and $q = \frac{14\sin 110°}{\sin 30°} \approx 26.3$.

Note The Law of Sines can also be used to determine the measure of an angle.

The Law of Sines gives unique answers when the given information follows an ASA or AAS pattern. (For SAS and SSS, you need a method presented in the next section.) When the pattern of information is SSA, however, there may be one solution, two solutions, or no solution at all. This is because either one triangle, two triangles, or no triangle may be constructed using the given information.

Part Two: Sample Problems

Problem 1 *Let $a = 20$, $b = 25$, and $\angle A = 50°$. Using the Law of Sines, find $\angle B$. How many triangles can be constructed on the basis of the given information?*

Solution According to the Law of Sines, $\frac{20}{\sin 50°} = \frac{25}{\sin\angle B}$

$$\sin\angle B = \frac{25\sin 50°}{20} \approx 0.9576$$

Arcsin 0.9576 ≈ 73°, so one solution is $\angle B \approx 73°$. However, the equation $\sin\angle B \approx 0.9576$ has two solutions. The other solution is $\angle B \approx (180 - 73)°$, or $\angle B \approx 107°$. Therefore, two triangles can be constructed on the basis of the given information.

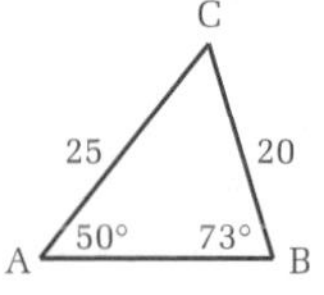

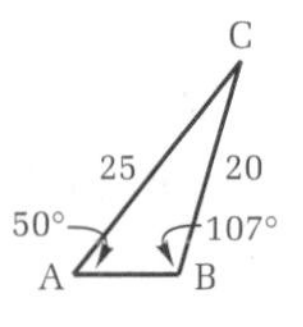

Problem 2 *Let $a = 20$, $b = 25$, and $\angle A = 72°$. Using the Law of Sines, find $\angle B$. How many triangles can be constructed on the basis of the given information?*

Solution According to the Law of Sines,

$$\frac{20}{\sin 72°} = \frac{25}{\sin \angle B}$$

$$\sin \angle B = \frac{25 \sin 72°}{20} \approx 1.1888$$

Pressing [INV] [sin] 1.1888 results in an error message on a calculator. (Why? What is the largest value $\sin \angle B$ can have?) This is therefore an impossible case. There is no solution. No triangle corresponds to the given information.

Part Three: Exercises and Problems

Warm-up Exercises

In problems 1 and 2, evaluate to the nearest thousandth.

P **1** $\dfrac{16.4 \sin 69.8°}{22.7}$ ≈0.678

2 $\dfrac{15.9 \sin 47.6°}{\sin 67.5°}$ ≈12.709

P **3** Determine $\angle A$ to the nearest degree. ≈17°

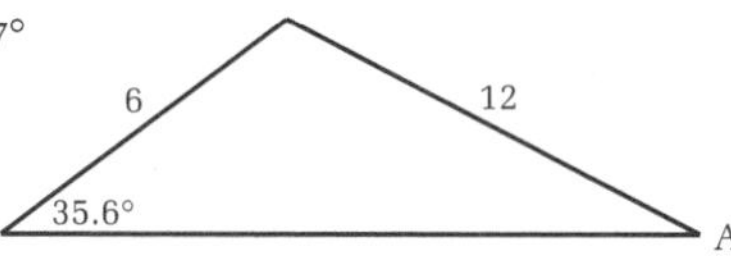

P **4** Find lengths x and y, rounded to the nearest hundredth.
x ≈ 24.25; y ≈ 27.58

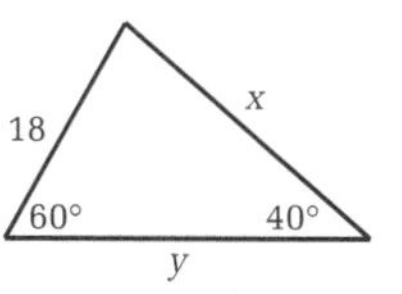

P **In problems 5–7, find each angle or side.**

5 $\angle P$ 64°

6 r ≈7.09

7 q ≈18.87

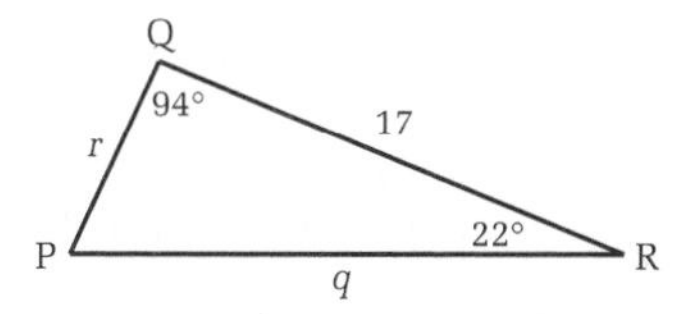

R **8** Find the area of equilateral $\triangle ABC$ shown.
≈1363.68

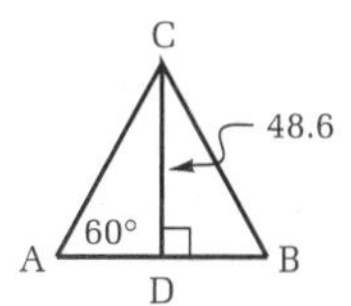

Checkpoint

1 In triangle ABC, side $b = 10$, $c = 15$ and $\angle C = 30°$. How many different triangles can be formed with this information? 1

2 Let $a = 10$, $b = 5$, and $\angle A = 55°$. Find the measure of $\angle B$. 24°

3 In the last problem, if $\angle B$ were 55°, what would be the measure of $\angle A$? No solution.

Assignment Guide

Average	
Day 1	9–12, 15, 16, 18, 20, 22–25
Day 2	13, 14, 17, 19, 21, 27, 28, 30
Advanced	
Day 1	9–12, 15, 16, 18, 20, 22–25, 27
Day 2	13, 14, 17, 19, 21, 28, 29, 33
Honors	
Day 1	9, 12–14, 16, 19–21, 22–26
Day 2	15, 17, 27, 29, 31–35

16.1

Problem-Set Notes and Strategies

- Calculators are definitely a help in this and the next section. If tables are used, a calculator is still a help with extrapolating.
- Using the relationships of 30°-60°-90° triangles wil make solving problem **8** quicker.
- In problems **10** and **11**, students evaluate expressions that will be encountered in the next section.
- Problem **12** illustrates the periodicity of the sine function. This will be useful later in the chapter when graphing is introduced.
- Points are plotted in problem **13** as an aid in plotting the sine function. The general graph will be dilated and translated in later sections.
- Problem **15** may be more difficult if students have not already solved problem **14**. If both are assigned, students will have an exact value for the sin 15°.
- Problem **17** shows how the ratios in the Law of Sines can be used to evaluate other ratios.

Additional Answer

13b

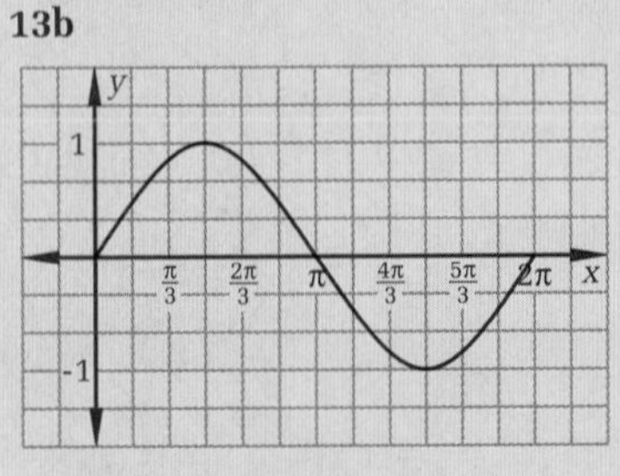

Problem Set

A

R **9 a** Find x and y to the nearest integer. (283, ≈490)

b Using (x, y) from part **a,** find the length of $\overline{PD}$ to the nearest integer. ≈868

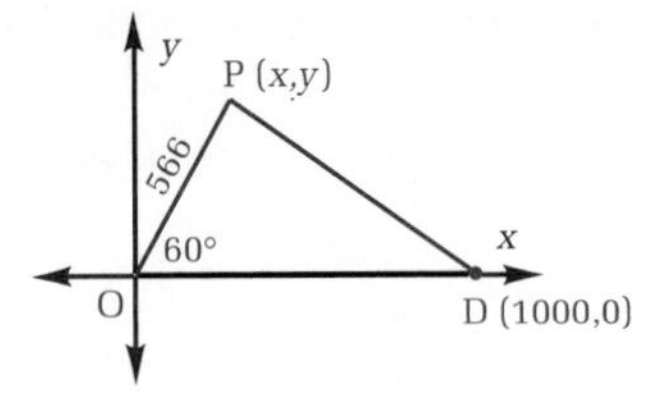

R,I **In problems 10 and 11, evaluate.**

10 $3^2 + 5^2 - 2 \cdot 3 \cdot 5 \cdot \frac{1}{2}$ 19

11 $3^2 + 5^2 - 2 \cdot 3 \cdot 5 \cdot \cos 10°$ ≈4.46

R,I **12** Evaluate.

a $\sin \frac{\pi}{2}$ 1

b $\sin\left(\frac{\pi}{2} + 2\pi\right)$ 1

c $\sin\left(\frac{\pi}{2} + 4\pi\right)$ 1

d $\sin\left(\frac{\pi}{2} + 6\pi\right)$ 1

R,I **13 a** Copy and complete the table for y = sin x, rounding to the nearest thousandth.

x	0	$\frac{\pi}{6}$	$\frac{\pi}{4}$	$\frac{\pi}{3}$	$\frac{\pi}{2}$	$\frac{2\pi}{3}$	$\frac{3\pi}{4}$	π	$\frac{3\pi}{2}$	2π
y	0	0.5	≈0.707	≈0.866	1	≈0.866	≈0.707	0	−1	0

b Sketch the graph of y = sin x.

P,R **14** Use sin (α − β) to find sin 15°, where α = 45° and β = 30°. $\frac{\sqrt{6} - \sqrt{2}}{4}$

P,R **15** Determine the exact value of x.
$5\sqrt{6} - 5\sqrt{2}$

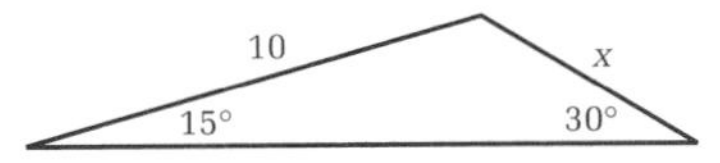

P **16** Two ranger stations are 100 miles apart. A forest fire is spotted. The angle measured at station 1 between station 2 and the fire is 61°. The angle measured at station 2 between station 1 and the fire is 48°. How far is it from station 1 to the fire?
≈78.6 mi

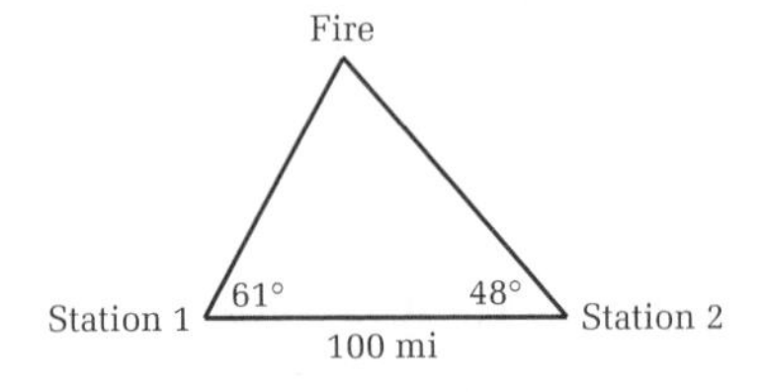

P,R **17** What is the ratio of PQ to QR in △PQR?
≈0.88

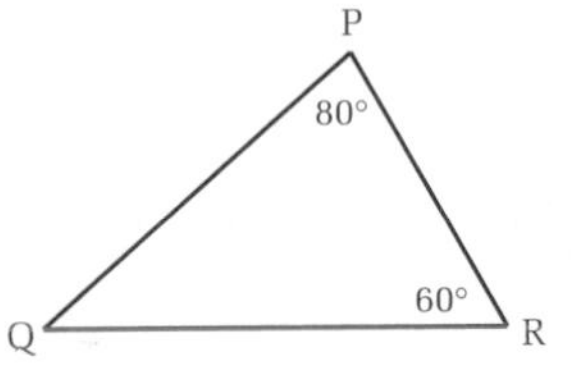

P,R

In problems 18–21, find all possible values of θ to the nearest degree.

18 $\approx 26°$

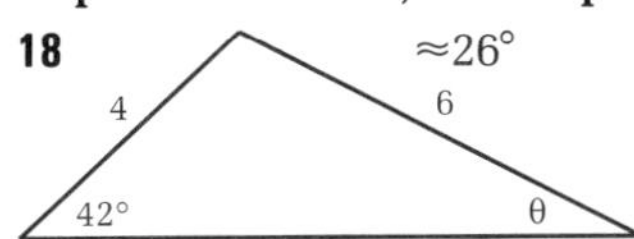

19 $\approx 63°$ or $\approx 117°$

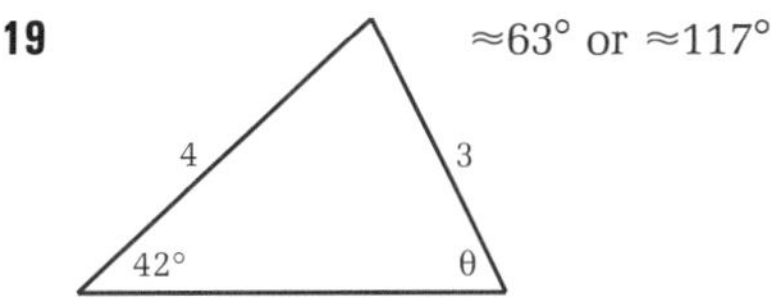

20 $90°$

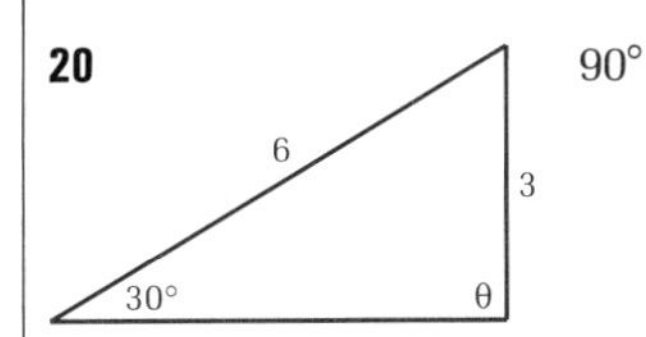

21 No solution

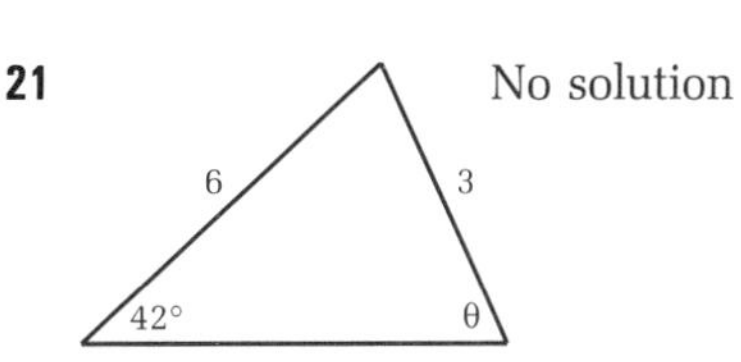

R,I

In problems 22–25, find the largest and smallest possible values of each expression.

22 $\sin \theta$ 1; -1

23 $4 \sin \theta$ 4; -4

24 $\sin 5\theta$ 1; -1

25 $4 \sin 5\theta$ 4; -4

B

P

26 Assume that $0° < \theta < 90°$. Explain each statement.

a If $b = a \sin \theta$, then exactly one triangle is determined.

b If $a \sin \theta < b < a$, then two triangles are determined.

c If $b \geq a$, then exactly one triangle is determined.

d If $b < a \sin \theta$, no triangle is determined.

R

27 From a certain point, the angle of elevation to the top of a radio tower is 48°. From a point 100 feet closer to the tower, the angle of elevation is 72°. Find the height of the tower. ≈ 173.77 ft

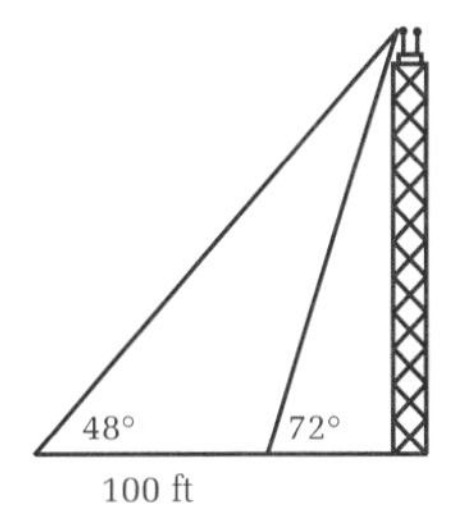

P

28 For the triangles in the diagram, $\sin \alpha = 0.5876$.

a Find all possible values of α to the nearest degree. $\approx 36°$ or $\approx 144°$

b Let $\theta = 32°$. Find all possible values of β. If $\alpha \approx 36°$, $\beta \approx 112°$; if $\alpha \approx 144°$, $\beta \approx 4°$.

c If $\theta = 60°$, what happens to the possible values of β? If $\alpha \approx 36°$, $\beta \approx 84°$; if $\alpha \approx 144°$, the sum of α and θ exceeds 180°, so no such triangle exists.

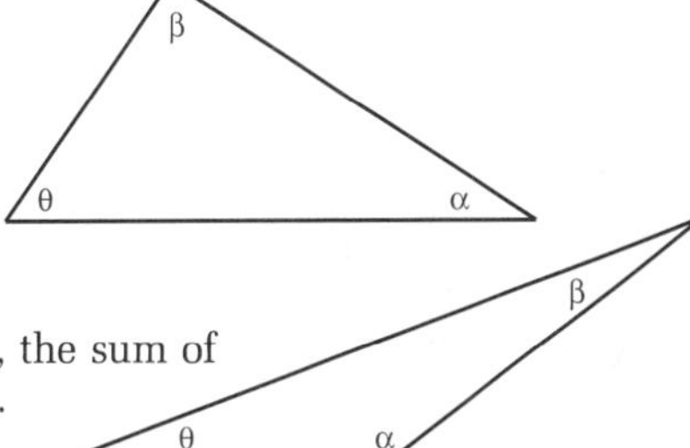

Problem-Set Notes and Strategies, continued

- Problems **22 – 25** illustrate the way amplitude is affected. These dilations will be seen in detail in Section **16.3**.
- Problem **26** provides a way to assess student understanding of the ambiguous case of the Law of Sines.

Cooperative Learning

Problem **26** is a nice problem for a group excercise to generate conditions for each case. Assign each part, **a – d** to different groups. When the groups finish, have those groups that solved part **a** or **b** etc, compare their explanation and see where they are similar and different.

Additional Answers

26a If $b = a \sin \theta$, the triangle is a right triangle (since $a \sin \theta$ is the length of the vertical altitude of the triangle shown). Therefore, since there is only one such altitude, only one triangle is determined.

26b If $a \sin \theta < b < a$, the side with length b can meet the base either to the left or to the right of the altitude, so two triangles are determined.

26c If $b \geq a$, the side with length b can only meet the base to the right of the altitude, so only one triangle is determined.

26d If $b < a \sin \theta$, the side with length b is shorter than the altitude–an impossibility, since the altitude is the shortest possible segment connecting the upper vertex and the base. Therefore, no triangle is determined.

Problem Set, *continued*

P **29** Jenny and Jill are 2000 feet apart. Jenny spots José at a 40° angle, and Jill spots him at a 65° angle. How much farther from José is Jenny than Jill? ≈546 ft

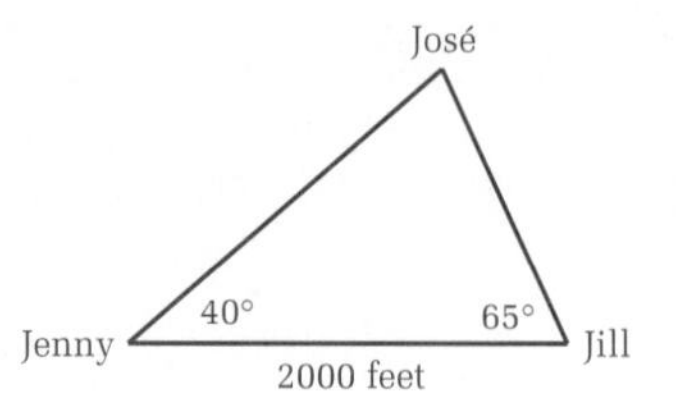

R **30** Philly can reach his second-floor window with an 18-foot ladder at a 75° angle to the ground. Willy claims that Philly could reach the window with a 16-foot ladder. Justify or disprove Willy's claim. Willy is wrong. The height of the window is 18 sin 75°, or ≈17.39, feet. A 16-foot ladder is therefore too short to reach it.

P **31** A pendulum swings between $\overline{AP}$ and $\overline{AQ}$. If AP = AQ = 18 cm, what are the measures of angles α_1 and α_2?
$\alpha_1 \approx 58.34$; $\alpha_2 \approx 121.66$

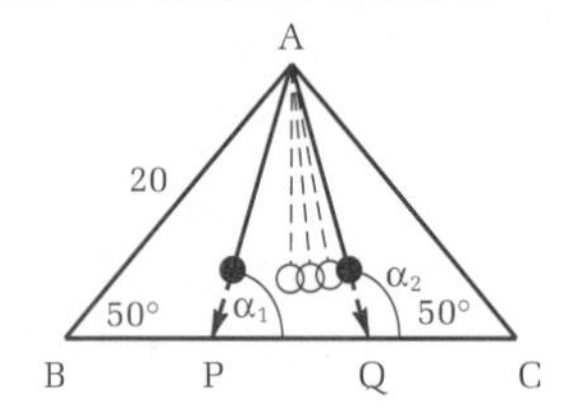

P **32** If $\overline{BD}$ bisects ∠ABC, what is the length of $\overline{BD}$?
≈6.44

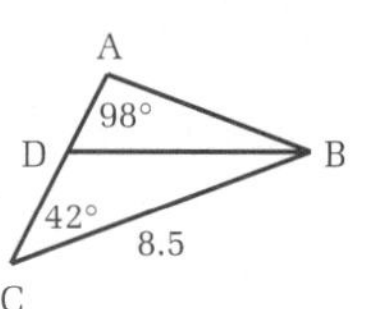

R **33** Prove that the area k of △ABC is given by $k = \frac{1}{2}ab \sin \angle C$.

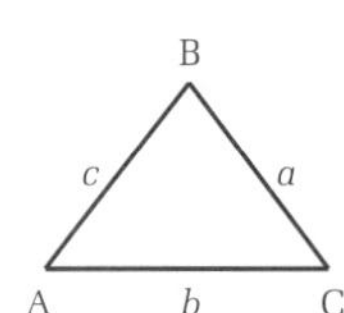

P,R **34** Find the radius R of the circle.
$R \approx 17.01$

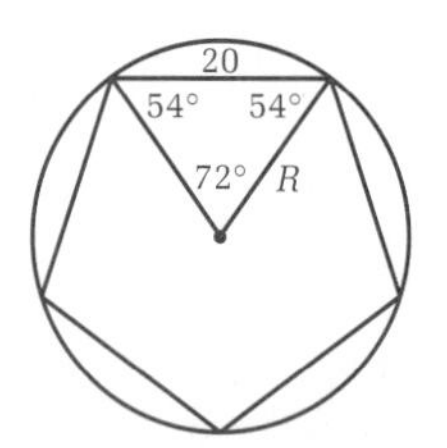

C P,R **35** Prove that for any △ABC, $\frac{a}{\sin \angle A}$ is the diameter of the circle circumscribed about the triangle.

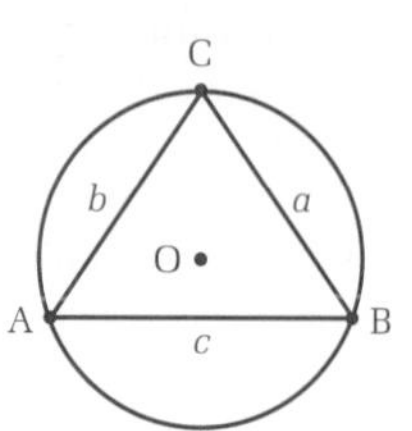

Problem-Set Notes and Strategies, continued

- Problem **35** may take some work. It is helpful to remember that any triangle inscribed in a circle with one side as the diameter is a right triangle.

Additional Answers

33 The length of the altitude from B to $\overline{AC}$ is $a \sin \angle C$. Since $k = \frac{1}{2}bh$,
$$k = \frac{1}{2} \cdot b \cdot a \sin \angle C = \frac{1}{2} ab \sin \angle C.$$

35 Draw the diameter from C through O. Label its other endpoint A′, and draw $\overline{A'B}$. ∠A′ = ∠A, since they are inscribed angles intercepting the same arc. ∠A′BC = 90°, since this angle is inscribed in a semicircle. In right triangle A′BC, $\sin \angle A' = \frac{a}{CA'}$, so sin ∠A is also equal to $\frac{a}{CA'}$. Therefore, CA′ (the diameter) is equal to $\frac{a}{\sin \angle A}$.

16.2 THE LAW OF COSINES

Objective

After studying this section, you will be able to

- Apply the Law of Cosines

Part One: Introduction

Solving SAS and SSS Triangles

When we cannot use the Law of Sines to find the dimensions of a triangle, we can draw an altitude and use right-triangle methods.

Example *Find the length of the third side in the triangle.*

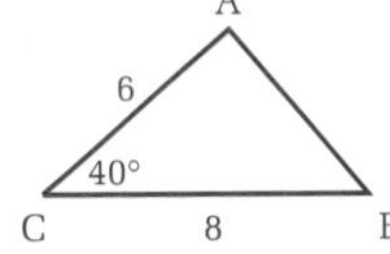

We are given SAS information, so the triangle is uniquely determined. However, the Law of Sines does not help. If we draw an altitude h, then by the Pythagorean Theorem, $6^2 = x^2 + h^2$ and $c^2 = (8 - x)^2 + h^2$.

If $c^2 = (8 - x)^2 + h^2$, then

$c^2 = 8^2 - 16x + x^2 + h^2$

We substitute 6^2 for $x^2 + h^2$.

$c^2 = 8^2 - 16x + 6^2$

$= 8^2 + 6^2 - 16x$

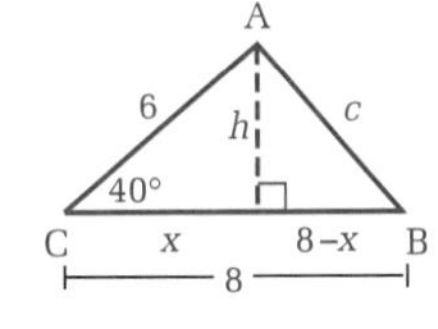

By the definition of trigonometric functions in right triangles, $\cos 40° = \frac{x}{6}$, so $x = 6 \cos 40°$. Therefore,

$c^2 = 8^2 + 6^2 - 16 \cdot 6 \cos 40°$

≈ 26.46

$c \approx 5.1$

To get a general formula, we substitute a for 8, b for 6, and $\angle C$ for the 40° angle. The result is the ***Law of Cosines.***

The Law of Cosines

If *a*, *b*, and *c* are the lengths of the sides opposite angles A, B, and C in a triangle, then

$$c^2 = a^2 + b^2 - 2ab \cos \angle C$$

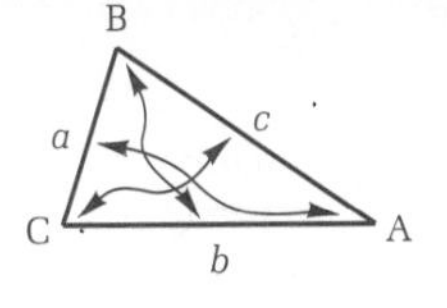

Class Planning

Time Schedule

All levels: 2 days

Resource References

Teacher's Resource Book

Class Opener 93

Practice 98

Enrichment 31

Class Opener

Find the length of BC in ΔABC if $\angle A = 72°$, $AB = 18$, $AC = 24$.

$25.16 \approx BC$

Lesson Notes

- The Law of Cosines can be applied in the case of triangles where only certain facts are given. It is a generalized form of the Pythagorean Theorem. Students can confirm that it also works for right triangles since $\cos 90° = 0$.
- Armed with both laws, students should now be able to solve almost any triangle problem given ASA, AAS, SSA, SSS or SAS.

Communicating Mathematics

Write the Law of Cosines in your own words rather than symbols. Explain the similarity between this law and the Pythagorean Theorem. What types of problems might this law be useful to solve?

Vocabulary

Law of Cosines

Stumbling Block

In using the Law of Cosines, the most common mistake is the choice of angle to substitute in the formula. The angle in the formula is always the included angle between the sides chosen for the $-2ab$ part. It is sometimes helpful to memorize the formula as $c^2 = a^2 + b^2 - 2ab$ times the cosine of the angle between a and b. The choice of a, b, and c is totally arbitrary.

Checkpoint

1. Find the angles of a triangle where the sides measure 6, 7, and 8.
   46.6°, 57.9°, and 75.5°
2. Two sides of a triangle measure 15 and 23 and the included angle measures 78°. Find the length of the third side. 24.7
3. A rhombus has sides of length 10 and one of the angles measure 50°. Find the lengths of the diagonals.
   8.45 and 18.13

The following alternative forms of the Law of Cosines may be used.

$a^2 = b^2 + c^2 - 2bc \cos \angle A$ $\qquad$ $b^2 = a^2 + c^2 - 2ac \cos \angle B$

We use the Law of Cosines when we are given the lengths of all three sides of a triangle (SSS) and must find the measure of an angle, or when we are given the lengths of two sides and the measure of the included angle of a triangle (SAS) and want to find the length of the third side.

Part Two: Sample Problems

Problem 1 *The sides of a triangle are* 8, 10, *and* 12. *Find the measure of the largest angle.*

Solution Since we are given the lengths of three sides of a triangle (SSS), we can use the Law of Cosines. The largest angle is opposite the longest side. Let's label the largest angle C and the longest side c. The other two sides are a and b.

$$c^2 = a^2 + b^2 - 2ab \cos \angle C$$

$$\cos \angle C = \frac{a^2 + b^2 - c^2}{2ab}$$

$$= \frac{64 + 100 - 144}{160}$$

$$= 0.125$$

Since $\cos \angle C = 0.125$, $\angle C \approx 82.8°$.

Problem 2 *ABCD is a parallelogram.* AD = 40, CD = 200, *and* ∠D = 115°. *Find the length of* $\overline{BD}$ *and the measure of* ∠BDC.

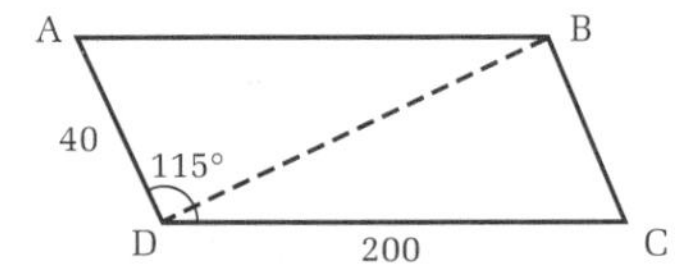

Solution Since opposite sides of a parallelogram are equal in length, BC = 40. Since consecutive angles of a parallelogram are supplementary, m∠C = 65. We can apply the Law of Cosines because △CBD has the SAS pattern.

$$(BD)^2 = 40^2 + 200^2 - 2(40)(200)(\cos 65°)$$

$$\approx 1600 + 40{,}000 - 16{,}000(0.4226)$$

$$\approx 34{,}838$$

$$BD \approx 187$$

We can use the Law of Sines to determine the measure of ∠BDC.

$$\frac{\sin \angle BDC}{40} \approx \frac{\sin 65°}{187}$$

$$\sin \angle BDC \approx \frac{40 \sin 65°}{187}$$

$$\approx 0.1939$$

Since $\sin \angle BDC \approx 0.1939$, $\angle BDC \approx 11.18°$.

16.2

Part Three: Exercises and Problems

Warm-up Exercises

P **1** Solve for a, b, and c to the nearest tenth.

$a^2 = 3^2 + 9^2 - 2 \cdot 3 \cdot 9 \cos 47°$ $a \approx 7.3$
$b^2 = 5^2 + 6^2 - 2 \cdot 5 \cdot 6 \cos 120°$ $b \approx 9.5$
$c^2 = 4^2 + 3^2 - 2 \cdot 4 \cdot 3 \cos 60°$ $c \approx 3.6$

P **In problems 2–4, find angles α, β, and γ to the nearest degree.**

2 $\approx 72°$

3 $\approx 59°$

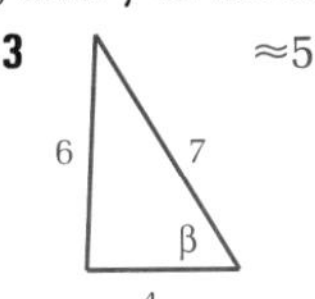

4 $\approx 18°$

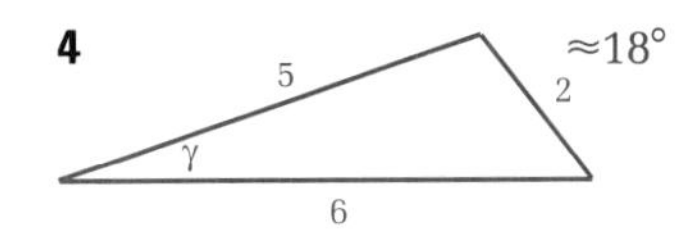

P **In problems 5–7, find lengths x, y, and z to the nearest tenth.**

5 ≈ 11.6

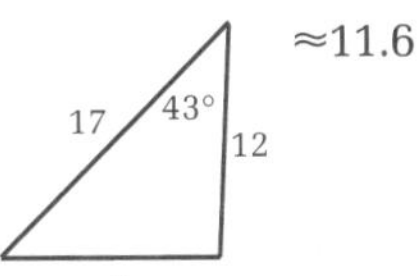

6 ≈ 8.1

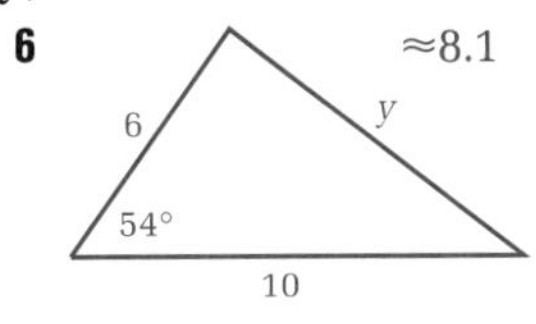

7 ≈ 47.3

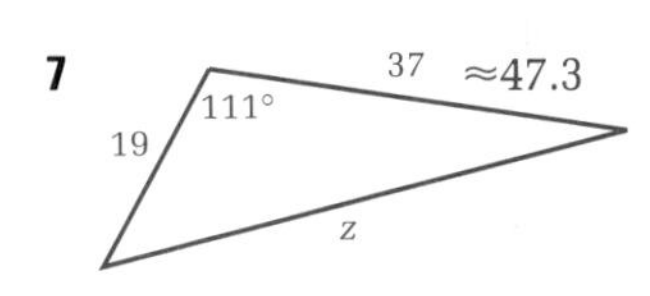

Problem Set

P,R **A** **8** Find the length of $\overline{AC}$ and the measures of $\angle A$ and $\angle C$.
$AC \approx 3.5$; $m\angle A \approx 48$; $m\angle C \approx 117$

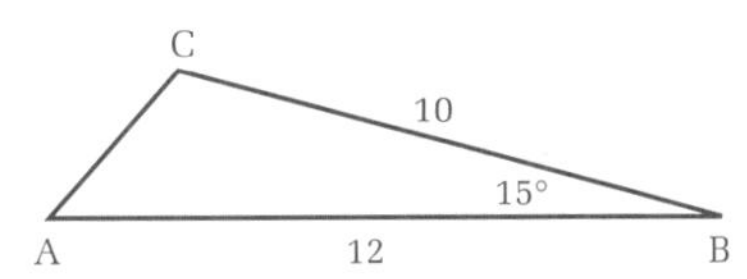

R **9** Sketch the graph of each equation.

a $y = x^2$ **b** $y = x^2 + 4$ **c** $y = (x + 4)^2$

R **In problems 10 and 11, evaluate.**

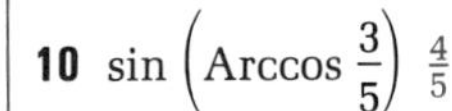

10 $\sin\left(\text{Arccos}\,\dfrac{3}{5}\right)$ $\frac{4}{5}$

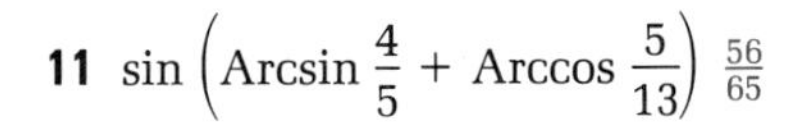

11 $\sin\left(\text{Arcsin}\,\dfrac{4}{5} + \text{Arccos}\,\dfrac{5}{13}\right)$ $\frac{56}{65}$

R **12** The lines shown are parallel. Write an equation of ℓ.
$y = 2(x - 3)$

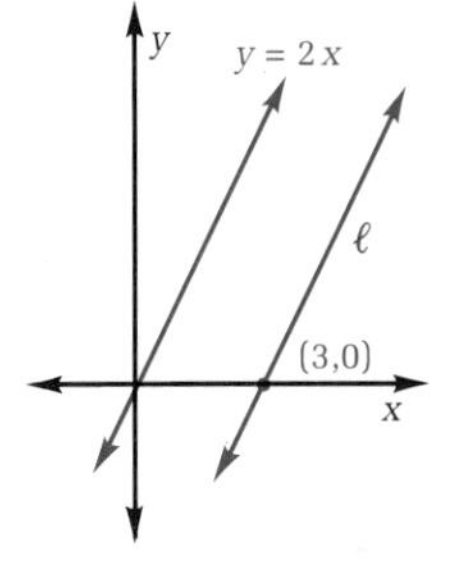

Assignment Guide

Average	
Day 1	8–10, 14, 16, 17a, b, 19, 27
Day 2	12, 13, 15, 17c, 18, 20, 21, 23, 26, 30
Advanced	
Day 1	8–10, 14, 16, 17, 19, 23, 27, 30, 32
Day 2	12, 13, 15, 18, 20, 21, 24–26, 31
Honors	
Day 1	8, 10, 13, 14, 17, 19, 23, 28, 30, 32
Day 2	12, 15, 18, 24–27, 29, 31

Problem-Set Notes and Strategies

- Once again, calculators are almost a necessity for problems **1 – 7**. If students own programmable calculators, encourage them to program in the Law of Cosines. It may be written to solve for a side or an angle. Programming the formula may reinforce understanding the way the formula works.
- Problems **9** and **12** are a review of translations. These will be necessary in the discussion of phase shifts and vertical translations of trigonometric graphs in the next several sections.

Additional Answers

Answers for problems **9a**, **9b**, and **9c** can be found in the answer pages beginning on page **A** 1.

Problem-Set Notes and Strategies, continued

- Problem **16** is an application of surveying techniques. Trigonometry has helped to measure many distances that would otherwise be inaccessible to measure.

Cooperative Learning

Suggest that students work together in small groups on problems **20 – 22**. When they have completed their sketches, suggest that groups compare their results. This provides a review of transforming functions before students apply this concept to graphs of trigonometric functions.

Additional Answers

13

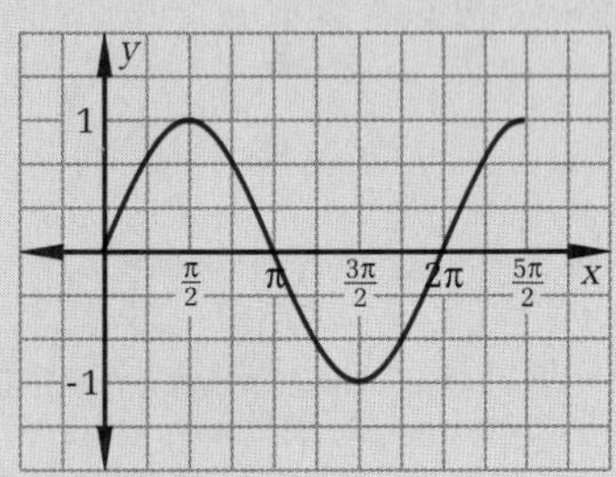

Answers for problems **20**, **21**, and **22** can be found in the answer pages beginning on page **A** 1.

- Problems **20 – 22** are reviews of translations and reflections of functions. These will be applied to trigonometric graphs in the next several sections.

Problem Set, *continued*

R,I **13** Let y = sin x. Copy and complete the table, then graph the equation.

x	0	$\frac{\pi}{4}$	$\frac{\pi}{2}$	$\frac{3\pi}{4}$	π	$\frac{5\pi}{4}$	$\frac{3\pi}{2}$	$\frac{7\pi}{4}$	2π	$\frac{9\pi}{4}$	$\frac{5\pi}{2}$
y	0	$\frac{\sqrt{2}}{2}$	1	$\frac{\sqrt{2}}{2}$	0	$\frac{-\sqrt{2}}{2}$	−1	$\frac{-\sqrt{2}}{2}$	0	$\frac{\sqrt{2}}{2}$	1

P **14** Find the measure of ∠B to the nearest degree. ≈83

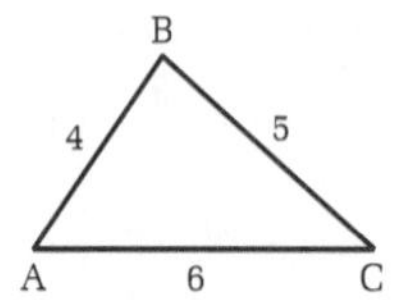

B P **15** At noon, two cars leave the junction of two straight roads. One car travels at 50 miles per hour along one road; the other car travels at 41 miles per hour along the other road. If the cars' paths form an angle of 72°, how far apart (in a straight line) are the cars at 3 P.M.? ≈162 mi

P **16** The distance from point A to point B cannot be measured directly. Use the information given in the diagram to find AB. ≈836.74 ft

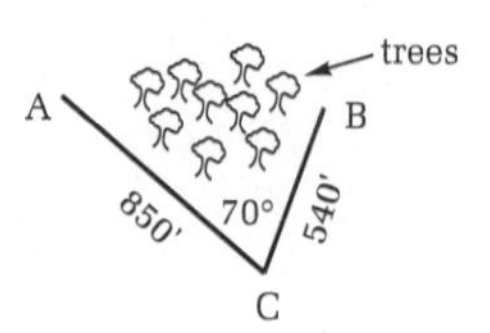

R **17** Find all possible values of x.

a sin x = 1
$x = \frac{\pi}{2} + n \cdot 2\pi$

b sin x = 0
$x = 0 + n \cdot \pi$

c sin x = −1
$x = \frac{3\pi}{2} + n \cdot 2\pi$

P,R **18** Find the perimeter of regular pentagon ABCDE. ≈94.05

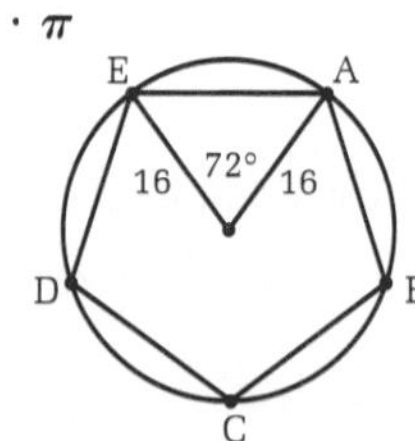

P,R **19** **a** Find the lengths of $\overline{AB}$, $\overline{BC}$, and $\overline{AC}$.
b Find θ to the nearest degree. ≈37°
a 10; 25; ≈18.03

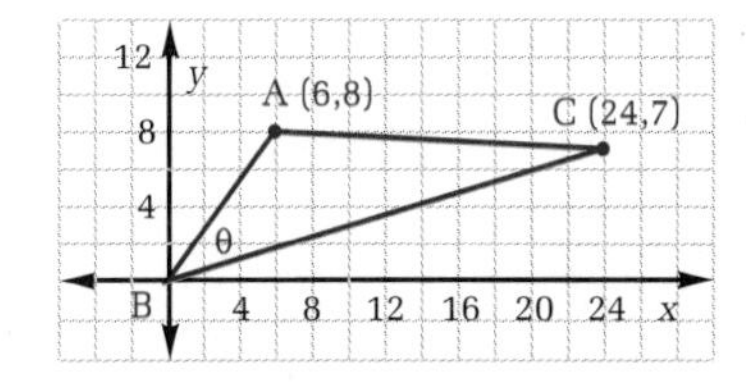

I **In problems 20–22, refer to the graph of y = f(x).**

20 Sketch y = f(x − 2).
21 Sketch y = −f(x).
22 Sketch y = f(x + 2).

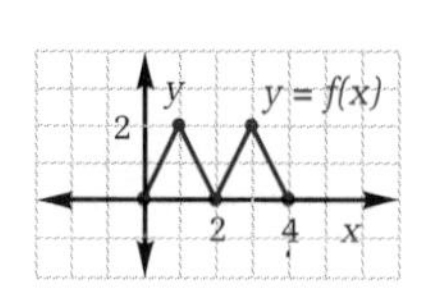

P **23** A 30-foot ladder and a 26-foot ladder both reach the top of a wall. The 30-foot ladder makes a 50° angle with the ground. Find angle θ to the nearest degree. (Remember to consider both cases.) $\approx 62°$ or $\approx 118°$

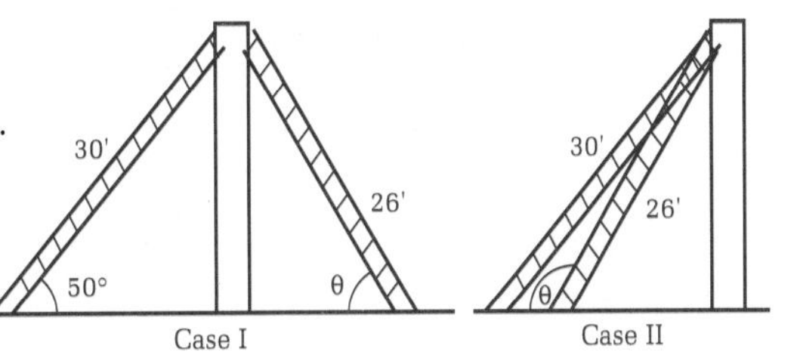

P,R **24** Find a so that $m\angle A = 60$. Impossible

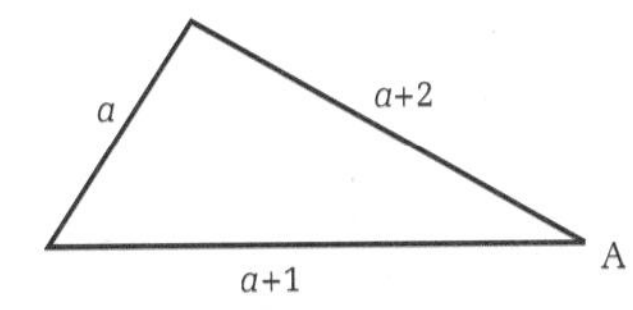

P **25** Find the length of a chord of a circle whose radius is 20 if the central angle of the chord is 18°. ≈ 6.3

P,R **26** Refer to the diagram.

a Find OB. ≈ 20.62
b Find OA. ≈ 26.93
c Find AB. ≈ 22.36
d Find $m\angle A$. ≈ 48.37

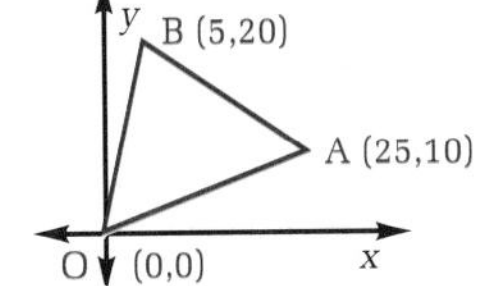

P,R **27** The lengths of the boundaries of a triangular plot of land are 80 feet, 100 feet, and 120 feet. Find the measure of the largest angle formed by the boundaries. ≈ 82.8

P,R **28** Two sides and a diagonal of a rhombus form a triangle whose perimeter is 50 centimeters. Find the measure of the angle between the sides of the rhombus if the perimeter of the rhombus is 72 centimeters. ≈ 45.77

P,R **29** The dimensions of the rectangular box shown are 10 inches by 8 inches by 5 inches.

a Find $m\angle FBH$. ≈ 76.29
b Find $m\angle CHA$. ≈ 68.67

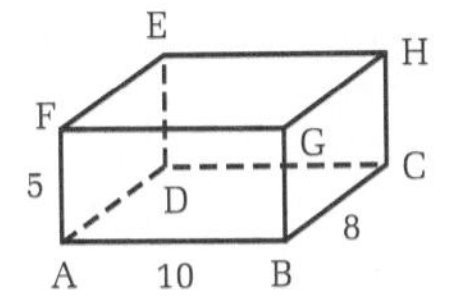

P **30** The area k of a triangle is given by the formula $k = \frac{1}{2}ab \sin \angle C$. The lengths of two sides of the triangle are 8 and 15.

a Find the angle between the two given sides. 30° or 150°
b Find the length of the third side of the triangle. ≈ 9 or ≈ 22.3

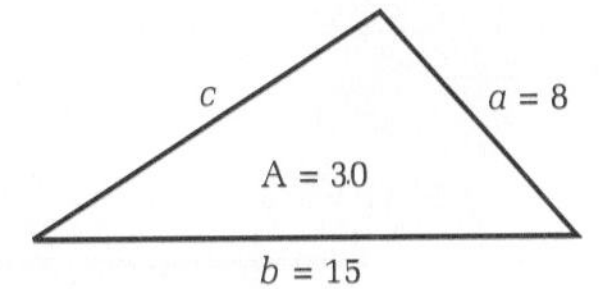

Problem-Set Notes and Strategies, continued

- Some algebraic work with the symbols will help solve problem **24**. Working through can show how values cancel and that the solution is impossible with the bounds on the cosine function.

- We hope students will remember that the largest angle is opposite the largest side. If they don't, the solution of problem **27** will take considerably longer.

- Problem **29** could take some time as a three dimensional figure is involved. If students copy the figure and label the angles to be found before proceeding they may see the problem more clearly.

Problem-Set Notes and Strategies, continued

- The expressions in problem **32** will be examined in more detail in the next sections on graphing. Both the amplitude and the phase shifts are affected here. It is important that students realize the difference depends on where the constant is added.
- The result of problem **33** is known as Heron's Formula. It allows the area of a triangle to be found given only its three sides. The quantity s is called the semi-perimeter.

Additional Answer

Answer for problem **33** can be found in the answer pages beginning on page **A** 1.

Problem Set, *continued*

P,R **31** The lengths of the sides of a triangle are 8, 10, and 14. Find the length of the median drawn to the longest side of the triangle. ≈ 5.74

C I **32** Find the largest and smallest values of each expression.

a $\sin\theta$ 1; −1
b $\sin\theta + 5$ 6; 4
c $\sin(\theta - 4)$ 1; −1
d $\sin(\theta - 4) + 5$ 6; 4

P,R **33** Suppose that k is the area of $\triangle ABC$ and $s = \frac{a+b+c}{2}$. Prove that $k = \sqrt{s(s-a)(s-b)(s-c)}$.

CAREER PROFILE

Mathematics In Medicine

Dr. Sharon Unti uses mathematics on the spot

Sharon Unti, associate director of pediatric education at Children's Memorial Hospital in Chicago, wears two buttons on her gray doctor's jacket. One reads, "I love kids." The other displays her name in a colorful simulation of a child's printing. Besides her love of children, Dr. Unti's qualifications include her medical education and three years in a pediatric residency. Her college coursework—like most premedical programs—included calculus and advanced algebra. Now Dr. Unti instructs residents at the hospital—doctors receiving advanced training in pediatric medicine.

"The way I use mathematics directly," Dr. Unti says, "is in calculating dosages of medicine and amounts of fluids to give to children, such as children who have lost fluids due to a trauma or who are dehydrated. Such calculations are based on formulas. Most have been done for all realistic values of a child's height and weight and are available in reference tables. The calculations themselves often involve exponentials, logarithms, or ratios.

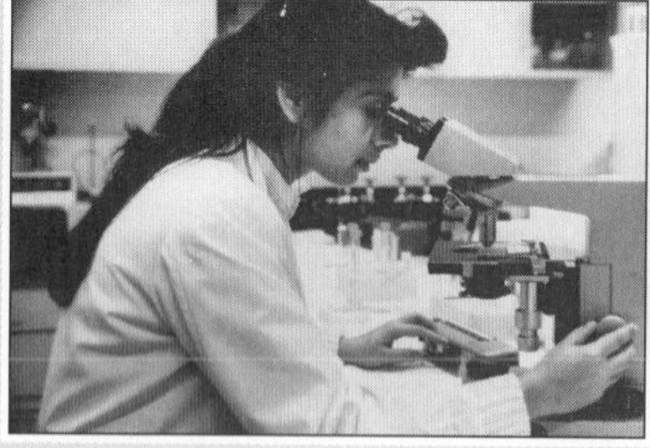

"One use of logarithms occurs when a child has an infection and we don't know exactly what bacteria may be present," Dr. Unti explains. "We coat a plate with five different antibiotics and grow five cultures from the infection site. Since the uninhibited growth of the culture is exponential, the resistance of the bacteria to each of the five antibiotics is reported as the log of the area of the plate covered by the bacteria after 48 or 72 hours. So if the culture has not grown at all at the end of the time period, its resistance to the antibiotic is 0, and we would use it on the child's infection."

Dr. Unti earned her bachelor's and medical degrees from Northwestern University in Evanston, Illinois, and Chicago.

16.3

16.3 GRAPHING SINE FUNCTIONS

Objectives

After studying this section, you will be able to

- Graph the sine function
- Translate the sine graph vertically and horizontally
- Change the amplitude of the sine graph

Part One: Introduction

The Graph of the Sine Function

The graph of $y = \sin x$ is illustrated below for $-\pi \le x \le 4\pi$. In this equation, x is the size of the angle in radians.

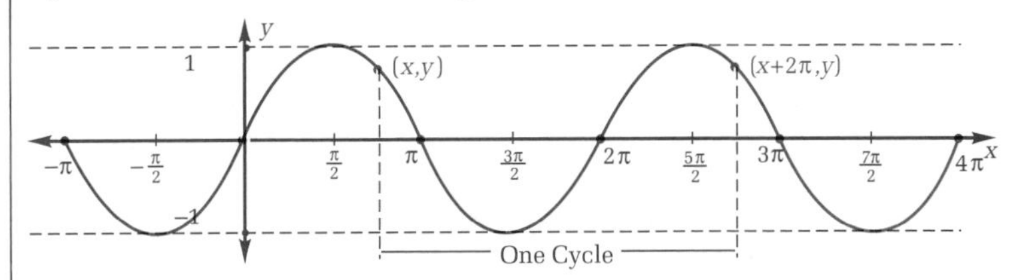

Notice some important features of the graph of the sine function.

1. The x-intercepts occur at integer multiples of π.
2. The range is $\{y: -1 \le y \le 1\}$.
3. The graph is ***periodic.*** That is, the graph repeats the same pattern over and over. The pattern is called a ***cycle.*** For the sine function, each cycle has the same length, called a ***period.***
4. The maximum (highest) points occur when x is $\frac{\pi}{2}, \frac{5\pi}{2}, \frac{9\pi}{2}, \ldots, \frac{\pi}{2} + 2k\pi$, where k is an integer.
5. The minimum (lowest) points occur when x is $\frac{-\pi}{2}, \frac{3\pi}{2}, \frac{7\pi}{2}, \ldots$ (again, in jumps of 2π).

If you start at any point (x, y) on the sine graph and move left or right a distance of 2π, you will arrive at a point that is on the graph and has the same height, y. Thus, since $y = \sin x = \sin(x \pm 2\pi) = \sin(x \pm 4\pi) = \ldots$, the period of $y = \sin x$ is 2π.

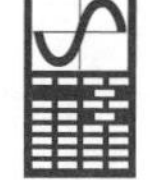
See p. 807

Translating the Sine Graph

Once you have memorized the basic graph for $y = \sin x$, you are ready to apply your earlier work on horizontal and vertical shifts.

Class Planning

Time Schedule

Average: 2 days
Advanced: 2 days
Honors: 1 day

Resource References

Teacher's Resource Book
Class Opener 94
Practice 99
Transparency 31
Evaluation
Quiz Forms 1, 2, 3

Class Opener

Find the missing side of ΔABC if $\angle A = 27°$, $AB = 16$, $BC = 12$.
$AC \approx 23.8$ or ≈ 4.7

Lesson Notes

- Concepts of translation and dilation from earlier chapters are applied to the sine function in this section. New terms such as phase shift are applied to a horizontal translation. This term comes from physics and some students will have seen it before. Some students will have observed sine waves on an oscilloscope.

Vocabulary

periodic
period
phase shift
amplitude

Just as the graph of $y = (x - 2)^2$ is nothing more than the graph of $y = x^2$ shifted two units to the right, the graph of $y = \sin(x - 2)$ is the standard sine graph shifted two units to the right. In physics, this shift is called a ***phase shift***.

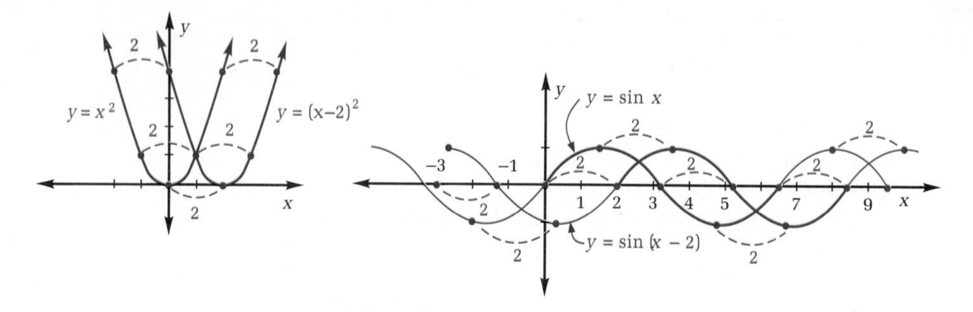

The graph of $y = \sin x - 3$ is the standard sine graph shifted down three units. *Each point* of the graph has been shifted down three units. Thus, the maximum value of this function is -2 and the minimum is -4.

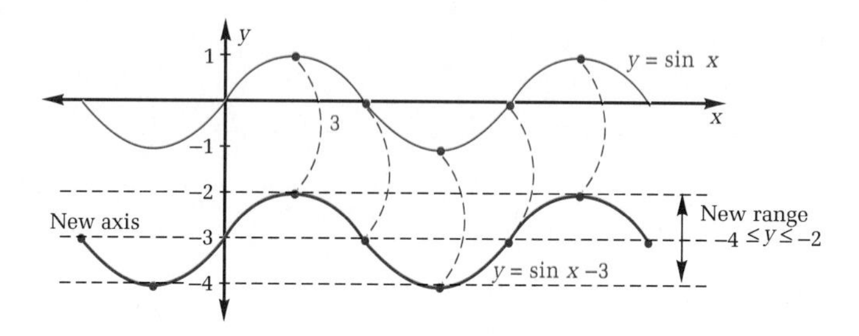

Amplitude Changes

The ***amplitude*** of a sine wave is its amount of deviation above and below the axis line.

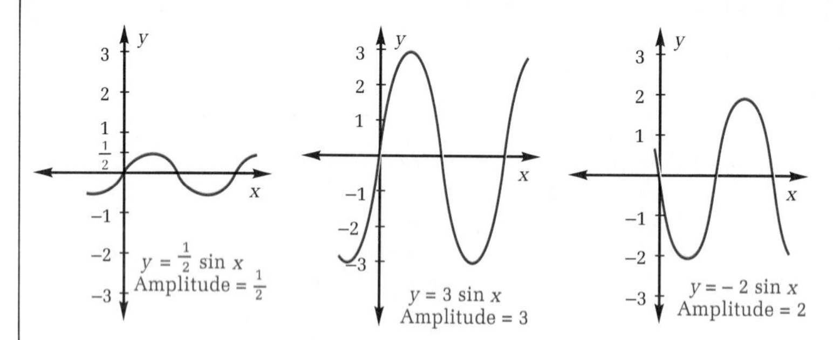

The amplitude of a sine graph is determined by the coefficient of the function. The amplitude of $y = a \sin x$ is $|a|$.

Lesson Notes, continued

- The concept of a vertical dilation is also explored. This is called the amplitude of the graph and measures the amount of deviation above and below the center of the graph. It may help if students realize the amplitude is the absolute value of the coefficient of the sine function. If the coefficient is negative, it is also a reflection through the *x*-axis. Each of these concepts has been explored with polynomials. The ideas are given different names when used with trigonometric functions.

Stumbling Block

The most common mistake in phase shifts is to go the wrong way. When shifting to produce a graph of $\sin(x - 3)$, it is too easy to say that it is a shift to the left. Have students ask themselves to find the value of x that will make the expression in parentheses equal to zero. This value will determine the phase shift, including the sign that indicates the proper direction.

Using Computers

Students can use the *Omnifarious Plotter* or other graphing software to graph variations on trigonometric functions. Many of the problems in these sections can be worked using graphing software.

Part Two: Sample Problems

Problem 1 *Refer to the graph.*

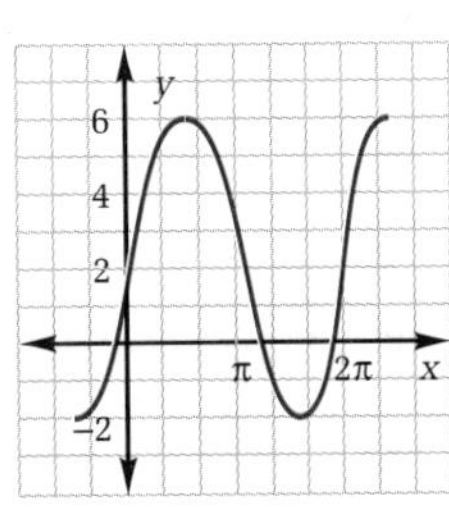

a *What are the maximum and minimum values of the function?*
b *What is the amplitude?*
c *What is the period?*
d *What is an equation of this sine graph?*

Solution
a The maximum and minimum values are 6 and −2 respectively.
b The graph oscillates between $y = -2$ and $y = 6$. The axis line must be at $y = 2$, halfway between −2 and 6. Since 6 is four units above the axis, the amplitude is 4. In general, if the maximum value of y is a and the minimum value is b, then the amplitude is $\frac{1}{2}(a - b)$.
c The graph repeats itself every 2π units, so the period is 2π.
d The graph is not shifted horizontally, but it is shifted up two units. An equation is $y = 4 \sin x + 2$.

Problem 2 *Graph* $y = \frac{3}{2} \sin\left(x - \frac{\pi}{6}\right) + 1$.

Solution The amplitude is $\frac{3}{2}$, the vertical shift is 1, and the phase shift is $\frac{\pi}{6}$ units to the right. We first locate the axis line and place the shifted zeros at $\frac{\pi}{6}$, $\frac{7\pi}{6}$, and $\frac{13\pi}{6}$. Halfway between these, we locate the maximum and minimum points 1.5 units above and 1.5 units below the axis line, $y = 1$. We connect the points with a smooth curve.

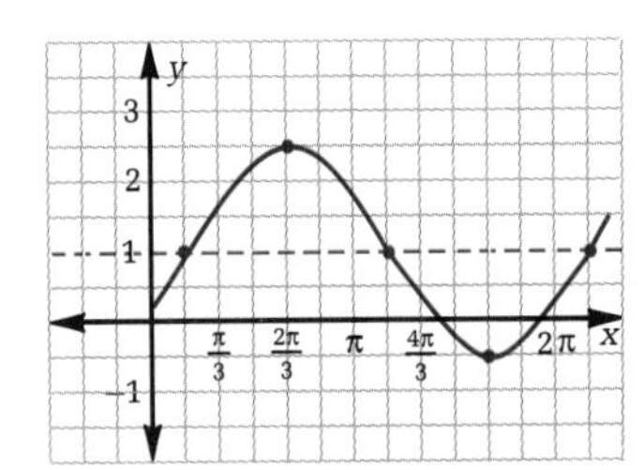

Part Three: Exercises and Problems

Warm-up Exercises

1 Refer to the graph of $y = f(x)$. What is the amplitude? 2.5

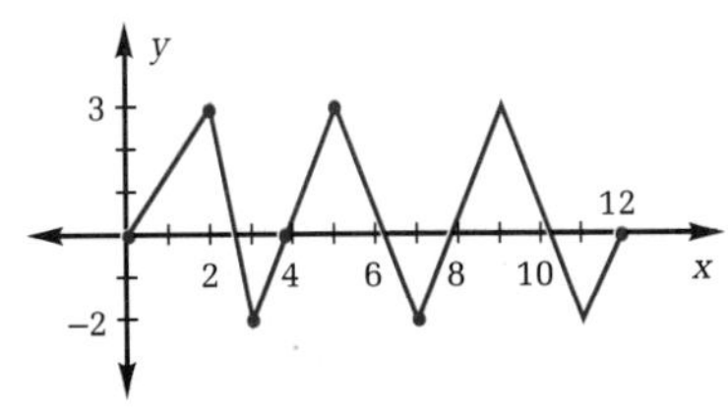

P

P **In problems 2–4, describe how each equation's graph is different from the graph of $y = \sin x$.**

2 $y = \sin\left(x - \frac{\pi}{6}\right)$
Shifted $\frac{\pi}{6}$ units to the right

3 $y = \sin x - 4$
Shifted 4 units down

4 $y = 4.5 \sin x$
Amplitude is 4.5.

Checkpoint

1 Find the amplitude of the function $y = -7 \sin x$. 7
2 The function $y = \sin\left(x + \frac{\pi}{2}\right)$ has a phase shift of how many units and in what direction? Shifted $\frac{\pi}{2}$ units to the left.
3 The sine function oscillates between the lines $y = -2$ and $y = 5$, what is the amplitude of this function? 3.5
4 Describe how the graph of $y = -2 \sin(x - 1)$ is different from $y = \sin x$.
It is reflected through the x-axis, has an amplitude of 2 and has a phase shift of 1 unit to the right.

Assignment Guide

Average	
Day 1	10, 11, 13, 15–17, 21–23
Day 2	12, 14, 18–20, 24, 26, 29
Advanced	
Day 1	10, 11, 13, 15–17, 21–23
Day 2	12, 14, 18–20, 24–26, 29, 30
Honors	
	11, 13, 14, 17–19, 23, 26, 29–31

Problem-Set Notes and Strategies

- The amplitude can be defined as one half the distance between the highest and lowest points on the graph. Using this definition, problem **1** can be more easily solved.

Problem-Set Notes and Strategies, continued

- If graphics calculators are available, encourage students to use them for problems **5 – 9** but to be careful of parentheses. The use or misuse will create different graphs. If they use the calculators, more difficult problems can be assigned and answers explained.
- In problems **11**, **12**, **14**, and **15** have students work backwards to find equations for the graphs. Some students will realize that the equations are not necessarily unique and may provide different answers. After problem **10**, they may also try to write the equation in terms of the cosine function.

Additional Answers

5

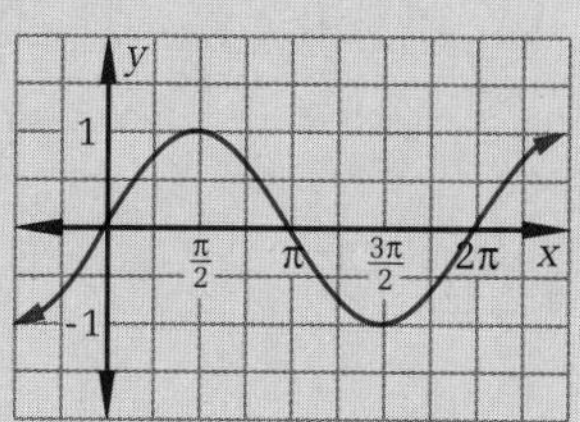

6

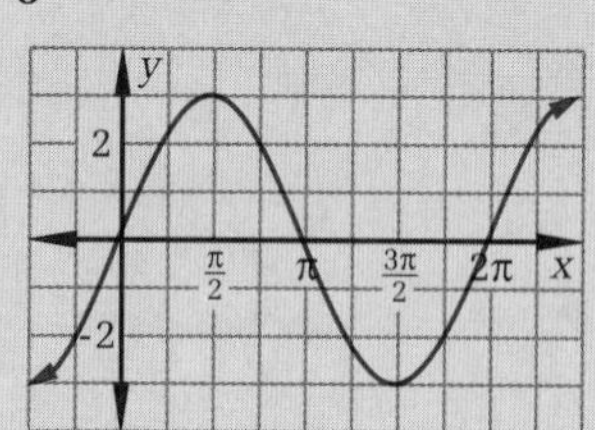

7

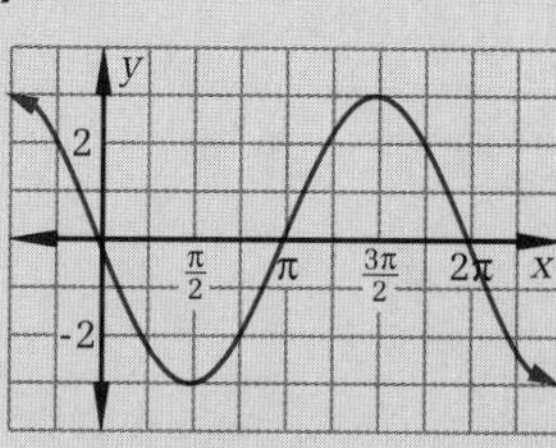

P **In problems 5–9, graph each equation.**

5 $y = \sin x$ **6** $y = 3 \sin x$ **7** $y = -3 \sin x$

8 $y = 3 \sin x + 2$ **9** $y = 3 \sin \left(x + \frac{\pi}{6}\right)$

Problem Set

A P,I **10** Copy and complete the table for $y = \cos x$, and sketch the equation's graph.

x	0	$\frac{\pi}{3}$	$\frac{\pi}{2}$	$\frac{2\pi}{3}$	π	$\frac{3\pi}{2}$	2π	$\frac{5\pi}{2}$	3π
y	1	0.5	0	−0.5	−1	0	1	0	−1

P **In problems 11 and 12, write an equation for each graph.**

11

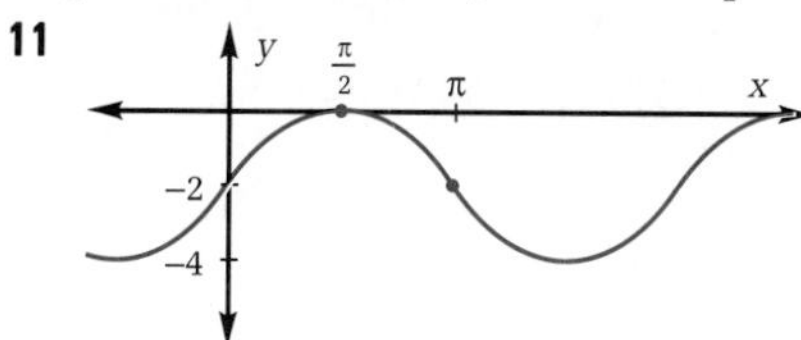

$y = 2 \sin x - 2$

12

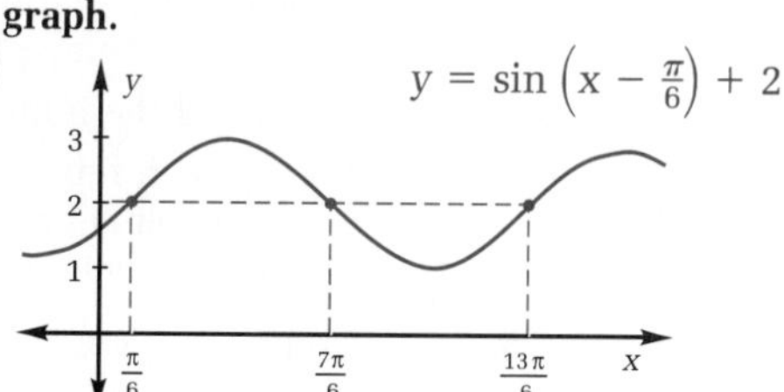

$y = \sin \left(x - \frac{\pi}{6}\right) + 2$

I **13** Find the smallest positive value of x satisfying each equation.

a $\sin x = \frac{1}{2}$ $\frac{\pi}{6}$

b $\sin (2x) = \frac{1}{2}$ $\frac{\pi}{12}$

c $\sin (3x) = \frac{1}{2}$ $\frac{\pi}{18}$

d $\sin \left(\frac{1}{2}x\right) = \frac{1}{2}$ $\frac{\pi}{3}$

P **In problems 14 and 15, write an equation for each graph.**

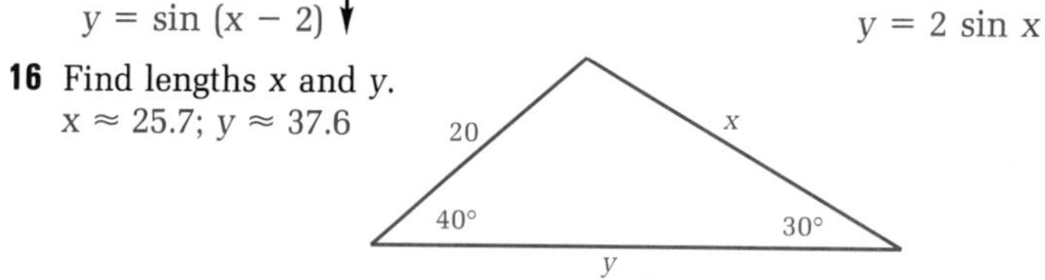

14 $y = \sin (x - 2)$

15 $y = 2 \sin x$

R **16** Find lengths x and y.
$x \approx 25.7$; $y \approx 37.6$

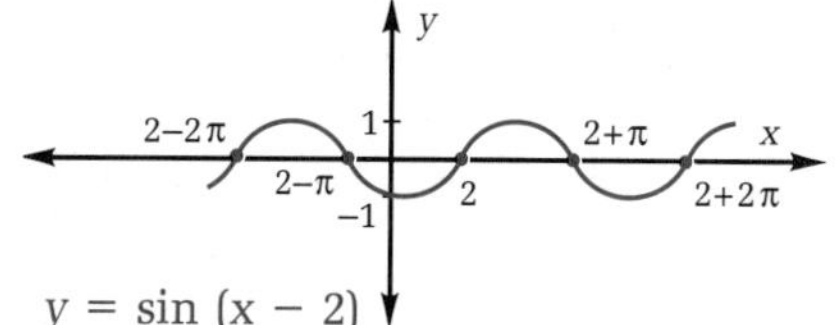

I **B** **17** **a** Find the coordinates of A, B, and C.
b What is the period of the graph?
a $A\left(\frac{2\pi}{3}, 0\right)$, $B(\pi, -3)$, $C\left(\frac{4\pi}{3}, 0\right)$ **b** $\frac{4\pi}{3}$

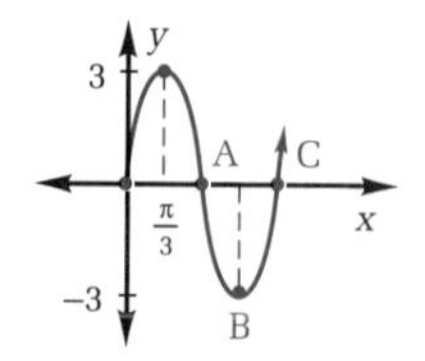

Answers for problems **8**, **9**, and **10** can be found in the answer pages beginning on page **A** 1.

P | **In problems 18–20, graph each equation.**

18 $y = 3 \sin\left(x - \frac{\pi}{4}\right) + 2$

19 $y = -3 \sin\left(x - \frac{\pi}{4}\right) + 2$

20 $y = 3 \sin\left(x - \frac{\pi}{4}\right) - 2$

P | **In problems 21 and 22, describe how the graph of each equation differs from the graph of $y = \sin x$.**

21 $y = \sin x + 1$
Shifted 1 unit up

22 $y = \sin (x + 1)$
Shifted 1 unit left

P | **23** Refer to the function $y = \pi \sin \pi x$.

a Find the period of the function. 2
b Find the amplitude of the function. π

P | **24** The amplitude of $y = A \sin x + 3$ is 4. Find the maximum value of y. 7

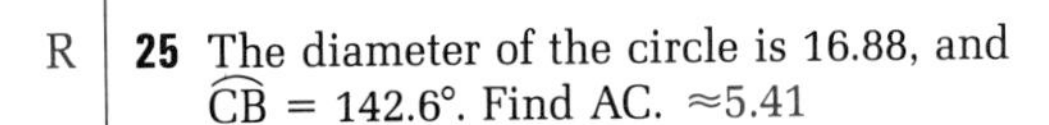

R | **25** The diameter of the circle is 16.88, and $\overset{\frown}{CB} = 142.6°$. Find AC. ≈ 5.41

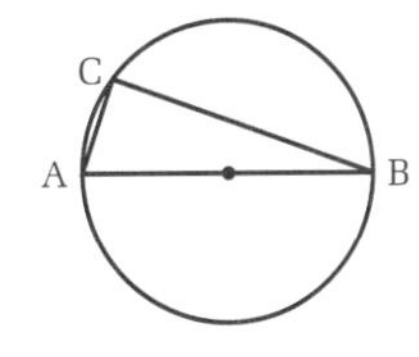

I | **26** What is the equation of a sine curve with period 3π and amplitude 5? $y = 5 \sin \frac{2}{3}x$

R | **27** Prove that $\frac{\cos 8x + \cos 2x}{\sin 8x - \sin 2x} = \cot 3x$.

R | **28** Evaluate $\sum_{k=0}^{36} \sin (10k)°$. 0

P **C** | **29** Refer to the graph to find $x_1 + x_2 + x_3 + x_4$. 4π

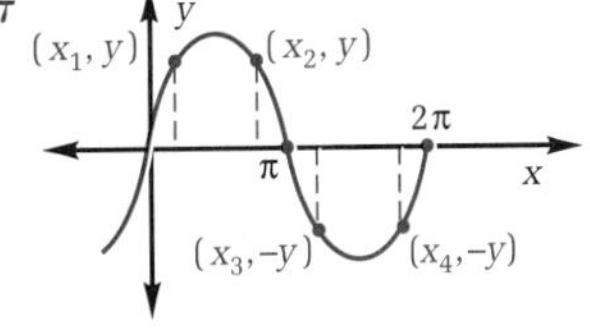

P | **30** Find the period of the graph of $y = \frac{\tan (6x) + \tan (2x)}{1 - \tan (6x) \tan (2x)}$. $\frac{\pi}{8}$

P,R | **31** Prove that the area of quadrilateral PQRS is equal to $\frac{1}{2}$(PR)(QS)(sin θ).

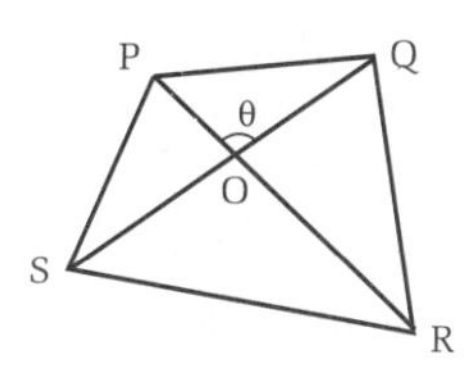

Problem-Set Notes and Strategies, continued

- Problem **26** involves a dilation in the horizontal direction. This is a period change and is discussed in detail in the next section.
- Several identities will be necessary to solve problems **27** and **30**.
- If students actually find all 37 values for problem **28**, the symmetry may then be seen and the problem seems trivial.

Communicating Mathematics

Your friend was at a cheerleading clinic today and missed the lecture on this section. Write a paragraph about how you would explain the differences and similarities between the graphing techniques discussed in this section and those in the past chapters. Establish the correspondence between terms from algebra and trigonometry.

Additional Answers

18

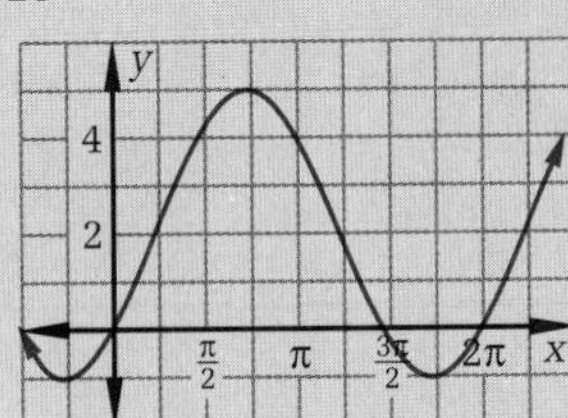

Answers for problems **19**, **20**, **27**, and **31** can be found in the answer pages beginning on page **A** 1.

16.4

Class Planning

Time Schedule
All levels: 2 days

Resource References
Teacher's Resource Book
Class Opener 95
Cooperative Learning 47
Practice 100
Transparency 32

Class Opener

Sketch the graph of $y = 2 \sin x + 3$.

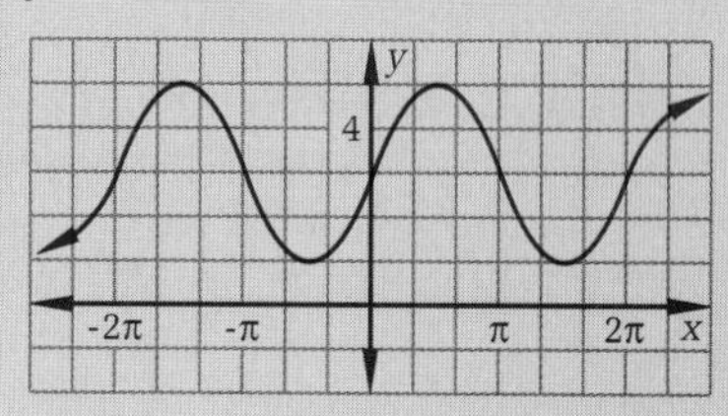

Lesson Notes

- Horizontal dilations are included in this section and identified as period changes. Both sine and cosine graphs are included with a summary of all the graphing techniques at the end of the section. Students need to know and use this information and graph any given sine or cosine curve.

16.4 CHANGING THE PERIOD

Objectives

After studying this section, you will be able to

- Graph the cosine function
- Stretch and shrink the sine and cosine graphs horizontally

Part One: Introduction

The Graph of the Cosine Function

The graph of $y = \cos x$ is shown at the right. Notice how much the cosine graph resembles the sine graph. In fact, the graph of $y = \cos x$ has exactly the same shape as the sine curve but is shifted $\frac{\pi}{2}$ units to the left. Can you prove that $\sin\left(x + \frac{\pi}{2}\right) = \cos x$?

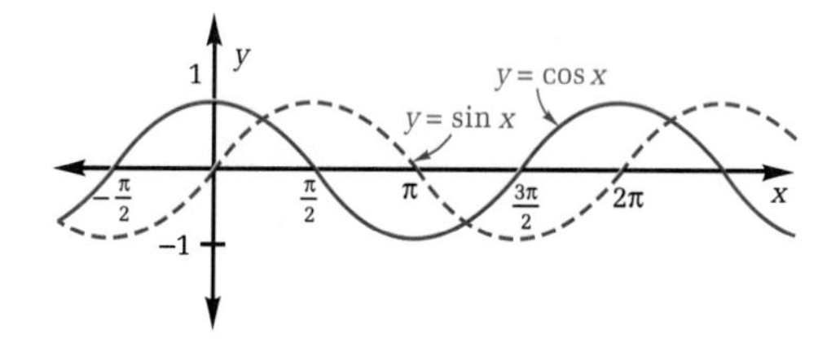

Learning to graph the cosine function is just as easy as learning to graph the sine function, but with a different set of basic points to memorize.

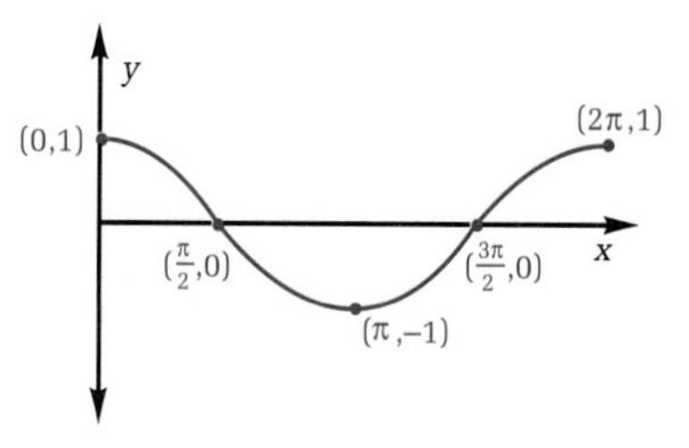

Period Changes

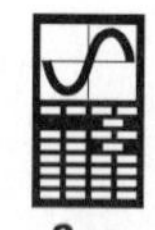

See p. 810

In the preceding section, you saw how sine graphs can be shifted upward, downward, to the left, and to the right. You also saw how the amplitude of a sine curve can be changed. We will now examine horizontal stretches and shrinks of the graphs of sine and cosine functions, which occur when the input value is multiplied by a constant.

With the help of a graphing calculator, we can graph the equation $y = \cos(2x)$.

Notice that two complete cycles fit between 0 and 2π, suggesting that the coefficient of the input variable tells us how many cycles will be compressed into the normal 2π period of the cosine curve.

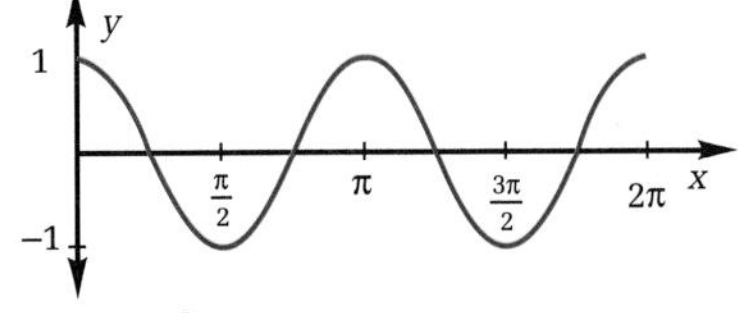

The period of the graph of $y = \cos(2x)$ is thus equivalent to $2\pi \div 2$, or π. Similarly, the graph of $y = \cos(3x)$ will have a period of $2\pi \div 3$, with three cycles in the interval from 0 to 2π, and the graph of $y = \cos(6x)$ will have a period of $2\pi \div 6$, or $\frac{\pi}{3}$, with six cycles in the interval from 0 to 2π.

What will happen in the case of $y = \cos\left(\frac{1}{2}x\right)$? Since the coefficient of x is $\frac{1}{2}$, only half a cycle will fit into the interval from 0 to 2π. The period of the graph will therefore be 4π.

Such period changes are not found only in the graphs of cosine functions; they occur in the graphs of sine functions as well. The following table summarizes the graphs of these functions.

Summary of Sine and Cosine Graphs

In the graph of $y = a\sin[b(x - c)] + d$,

1. **The amplitude is equal to $|a|$**
2. **The period is equal to $\frac{2\pi}{|b|}$**
3. **The phase shift, with respect to the graph of $y = \sin x$, is c units to the right (if $c > 0$) or c units to the left (if $c < 0$)**
4. **The vertical shift, with respect to the graph of $y = \sin x$, is d units upward (if $d > 0$) or d units downward (if $d < 0$)**

Cosine graphs have the same properties, except that shifts are with respect to the graph of $y = \cos x$.

Using Computers

Students can use the *Omnifarious Plotter* to examine changes in the periods of trigonometric graphs.

Part Two: Sample Problems

Problem 1 *Graph* $\sin\left(2x + \frac{\pi}{3}\right)$.

Solution We rewrite the equation in the form $y = a\sin[b(x - c)] + d$, as $y = \sin 2\left[x - \left(-\frac{\pi}{6}\right)\right]$, allowing us to see that the amplitude is 1; the period is $\frac{2\pi}{2}$, or π; and the phase shift is $\frac{\pi}{6}$ to the left.

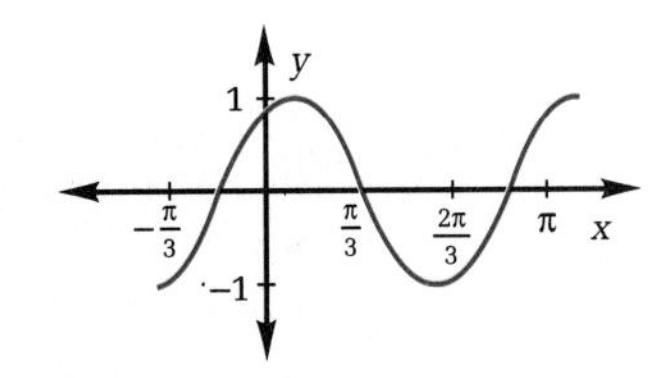

We could also have arrived at this graph by sketching the graph of $y = \sin(2x)$ and shifting it $\frac{\pi}{3} \div 2$, or $\frac{\pi}{6}$, units to the left.

Checkpoint

1 Graph the equation
$y = \cos(3x + \pi)$.

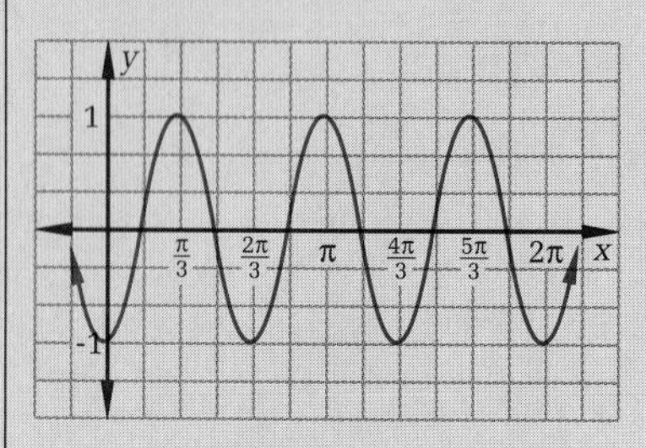

2 Graph the equation
$y = 2\sin[4(x - 1)] + 2$.

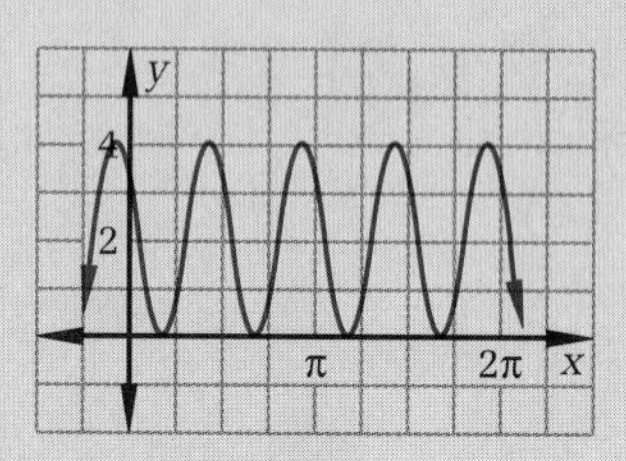

3 Find the amplitude, period, phase shift and vertical translation of the equation
$y = -3\cos[2(x + 4)] - 7$.
Amplitude = 3
Period = π
Phase shift = 4 units left.
Vertical translation = 7 units down.

Problem 2 *Graph two cycles of* $y = 3\sin\left(\frac{1}{2}x\right)$.

Solution In this case, the graph will have an amplitude of 3. The coefficient of x is $\frac{1}{2}$, so that only half a cycle will fit in the interval from 0 to 2π and the period will be $2\pi \div \frac{1}{2}$, or 4π. In order to show two cycles, therefore, we can use any interval of 8π units. We will use the interval from -4π to 4π.

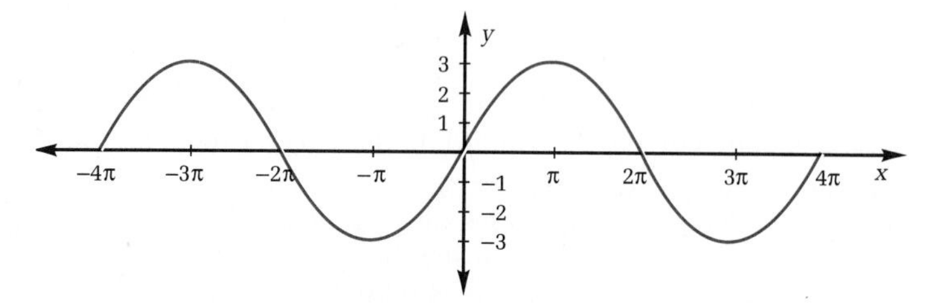

Problem 3 *Write two equations, one in terms of the sine function and one in terms of the cosine function, for each graph.*

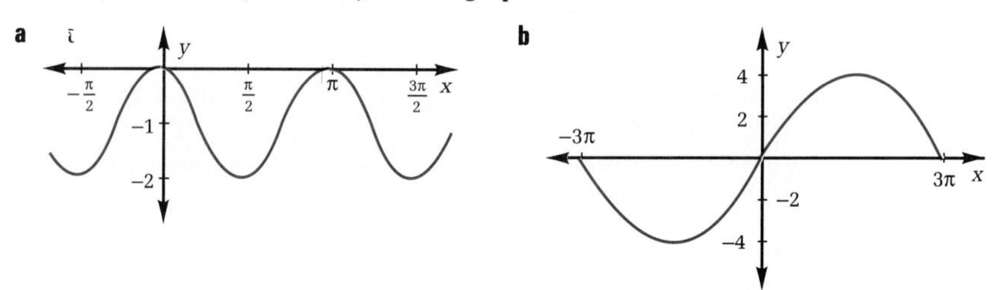

Solution

a We can begin by considering this as a cosine graph, with an equation of the form $y = a\cos[b(x - c)] + d$. Since the amplitude is 1, $a = 1$. There are two cycles between 0 and 2π, so $b = 2$. In comparison with the graph of $y = \cos x$, this graph shows no phase shift but is shifted one unit down, so $c = 0$ and $d = -1$. An equation that corresponds to the graph is therefore $y = \cos(2x) - 1$.

To represent the graph in terms of the sine function, we recall that $\cos x = \sin\left(x + \frac{\pi}{2}\right)$. Thus, the sine equation is the same as the cosine equation, but with $\frac{\pi}{2}$ added to the 2x—that is, $y = \sin\left(2x + \frac{\pi}{2}\right) - 1$, or $y = \sin\left[2\left(x + \frac{\pi}{4}\right)\right] - 1$.

b This graph differs from the graph of $y = \sin x$ in only two ways: Its amplitude is 4, and its period is 6π, so only one third of a cycle fits in the interval from 0 to 2π. It can therefore be represented by $y = 4\sin\left(\frac{1}{3}x\right)$. For a corresponding cosine equation, we subtract $\frac{\pi}{2}$ from $\frac{1}{3}x$, obtaining $y = 4\cos\left(\frac{1}{3}x - \frac{\pi}{2}\right)$, or $y = 4\cos\left[\frac{1}{3}\left(x - \frac{3\pi}{2}\right)\right]$.

Problem 4 *Graph each equation.*

a $y = \sin(-2x)$ **b** $y = \cos(-2x)$

Solution

a Since the sine function is an odd function, $\sin(-\theta) = -\sin\theta$. The graph of $y = \sin(-2x)$ is therefore the same as that of $y = -\sin(2x)$, which is the reflection of the graph of $y = \sin(2x)$ over the x-axis.

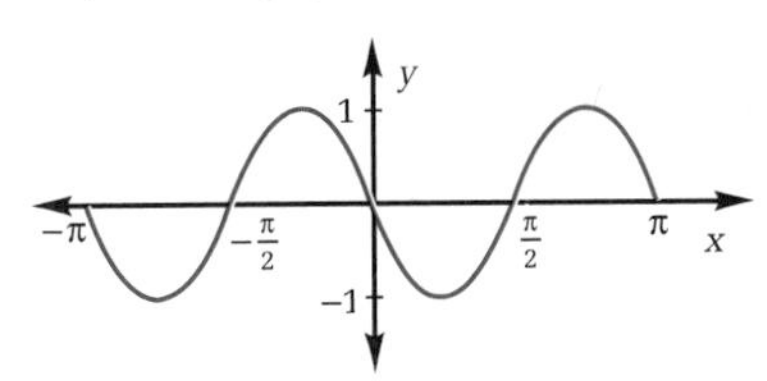

b Since the cosine function is an even function, $\cos(-\theta) = \cos\theta$. The graph of $y = \cos(-2x)$ is therefore identical to the graph of $y = \cos(2x)$.

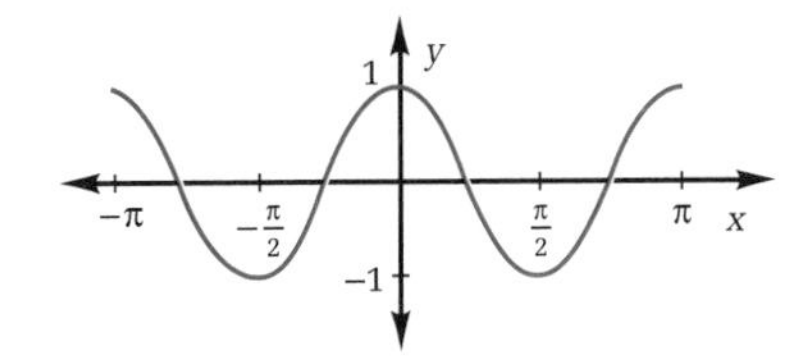

Part Three: Exercises and Problems

Warm-up Exercises

P **1** Graph each equation.

a $y = \cos x$ **b** $y = 2\cos x$ **c** $y = -2\cos x$
d $y = \cos(2x)$ **e** $y = \cos(-2x)$

P **2** One cycle of the graph of $y = 2\cos\left(x - \frac{\pi}{6}\right)$ is shown. Find the coordinates of points A, B, C, and D.
$\left(\frac{2\pi}{3}, 0\right)$; $\left(\frac{7\pi}{6}, -2\right)$; $\left(\frac{5\pi}{3}, 0\right)$; $\left(\frac{13\pi}{6}, 2\right)$

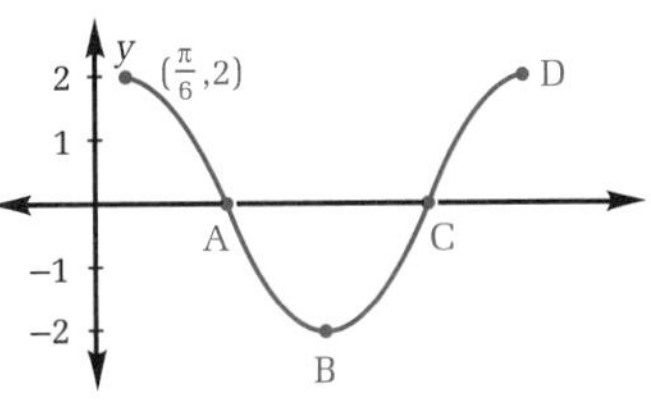

P **3** Find the period of the graph of the equation $y = \sin 8x \cos(-2x) + \sin(-2x)\cos 8x$. $\frac{\pi}{3}$

Problem Set

A

P **4** Graph each equation.

a $y = \sin x$ **b** $y = \sin(2x)$
c $y = \sin(4x)$ **d** $y = \sin\left(\frac{1}{2}x\right)$

P **5** Find the amplitude and the period of each equation's graph.

a $y = 3\sin\left(2x - \frac{\pi}{4}\right) + 4$
3; π

b $y = -5\sin(3x) - 1$
5; $\frac{2\pi}{3}$

c $y = 32\sin(\pi x)$
32; 2

P **6** Write an equation involving the cosine function to describe the curve shown.
$y = 4\cos\left(\frac{1}{4}x\right)$

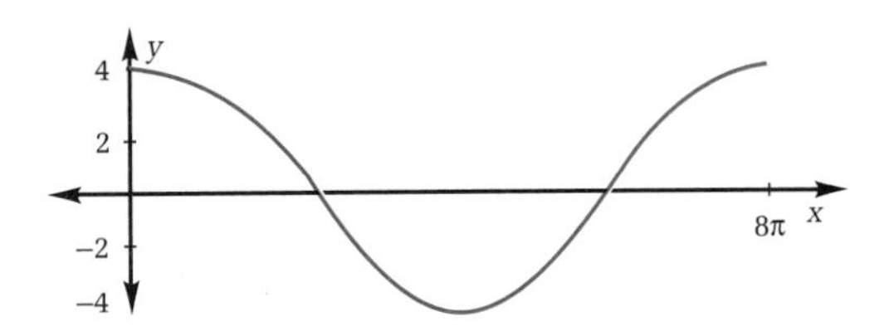

Communicating Mathematics

The graphs of the sine and the cosine function are related in several different ways. Write a short paragraph describing how the sine graph could be interpreted as a cosine graph and vice versa. Try to discover at least two different possibilities that are not just periodic changes of each other.

Stumbling Block

In the summary, everything is well laid out except what to do if the value of a is negative. The amplitude is the absolute value of a, but a negative a also reflects the graph over the x-axis. This needs to be done first before any of the translations. Reiterate this fact to the students, and they may decide to amend the summary in their notes.

Assignment Guide

Average	
Day 1	4, 5, 8–11, 13, 15
Day 2	6, 7, 12, 14, 16, 18a, b, 19–21
Advanced	
Day 1	4–6, 8–11, 13, 15, 16
Day 2	7, 12, 14, 17, 18b, c, d, 19–21
Honors	
Day 1	4, 5, 7–10, 13a, 14–16
Day 2	11, 12, 13b, 17, 18, 20–22

Additional Answers

Answers for problems **1a–e**, and **4a–d** can be found in the answer pages beginning on page A 1.

Problem-Set Notes and Strategies

- Review the graphs drawn for the tangent function in problem **8**. Help students recognizes the asymptote at $\frac{\pi}{2}$. Some graphing calculators and computer programs will try to connect the tangent function and make it continuous. This is a place where the student's knowledge is necessary to interpret a mechanically produced graph.
- Problem **10** is a review of function notation and illustrates all the graphing concepts that have been discussed in a more general form.
- Answers for problem **12** will not be unique. This can generate an interesting discussion.

Additional Answers

11a

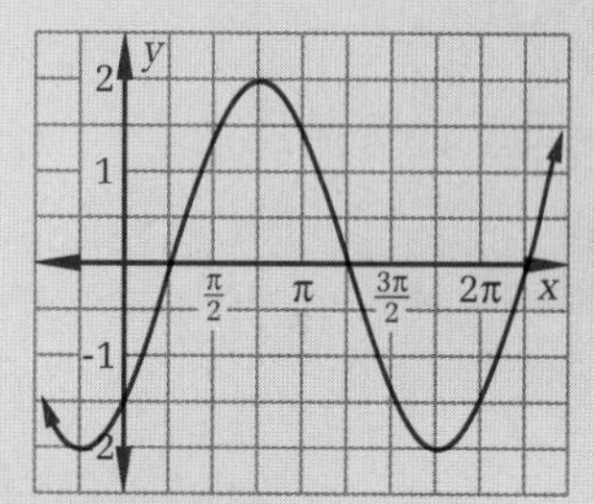

11b

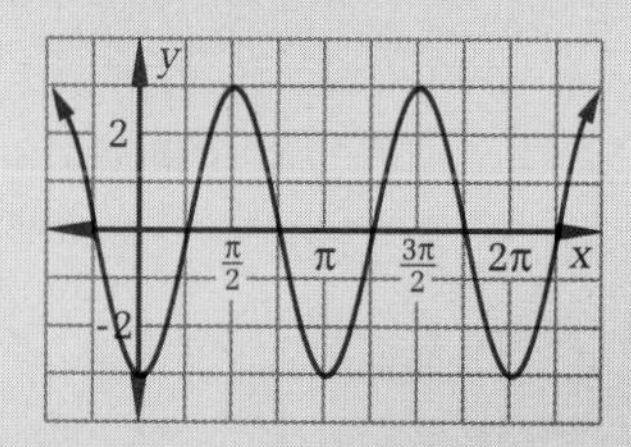

Answers for problems **8b** and **11c** can be found in the answer pages beginning on page **A** 1.

Problem Set, *continued*

P,R **7** Write an equation to represent the graph shown, using

a The sine function $y = 3 \sin\left(x + \frac{\pi}{4}\right)$

b The cosine function $y = 3 \cos\left(x - \frac{\pi}{4}\right)$

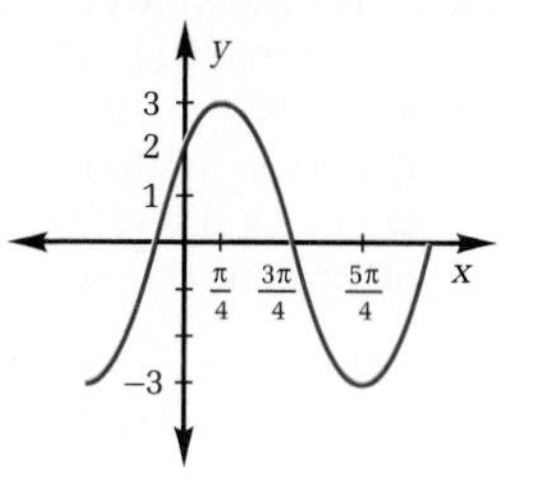

R,I **8** **a** Copy and complete the following table for $y = \tan x$.

x	0	$\frac{\pi}{6}$	$\frac{\pi}{4}$	$\frac{\pi}{3}$	$\frac{9\pi}{20}$	$\frac{\pi}{2}$	$\frac{11\pi}{20}$	$\frac{2\pi}{3}$	$\frac{3\pi}{4}$	π
y	0	$\frac{\sqrt{3}}{3}$	1	$\sqrt{3}$	≈ 6.3	Und.	≈ -6.3	$-\sqrt{3}$	-1	0

b Sketch the graph of $y = \tan x$.

R **9** Use the diagram to find

a $\cos \alpha$ $\frac{12}{13}$

b $\sin \alpha$ $\frac{5}{13}$

c $\cos \beta$ $\frac{3}{5}$

d $\sin \beta$ $\frac{4}{5}$

e $\cos(\beta - \alpha)$ $\frac{56}{65}$

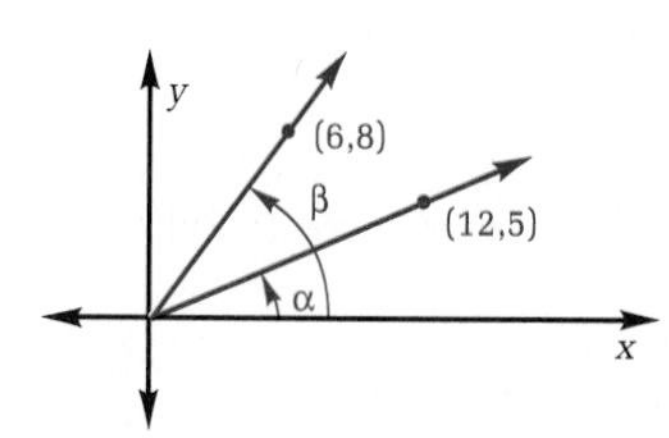

R **10** Let $f(8) = 7$.

a Find $2[f(8)]$. 14

b Find $-2[f(8)]$. -14

c Find $\frac{1}{f(8)}$. $\frac{1}{7}$

d Find the value of x for which $f(2x) = 7$. 4

e Find the value of x for which $f(x - 2) = 7$. 10

P,R **11** Graph each equation.

a $y = 2 \sin\left(x - \frac{\pi}{4}\right)$

b $y = -3 \cos(2x)$

c $y = 4 \cos(3x) + 6$

P **12** **a** Write two equations involving the sine function to describe the curve shown.

b Write two equations involving the cosine function to describe the curve shown.

a $y = 2 \sin(2x)$; $y = -2 \sin(-2x)$

b $y = 2 \cos\left[2\left(x - \frac{\pi}{4}\right)\right]$; $y = 2 \cos\left[2\left(x + \frac{3\pi}{4}\right)\right]$

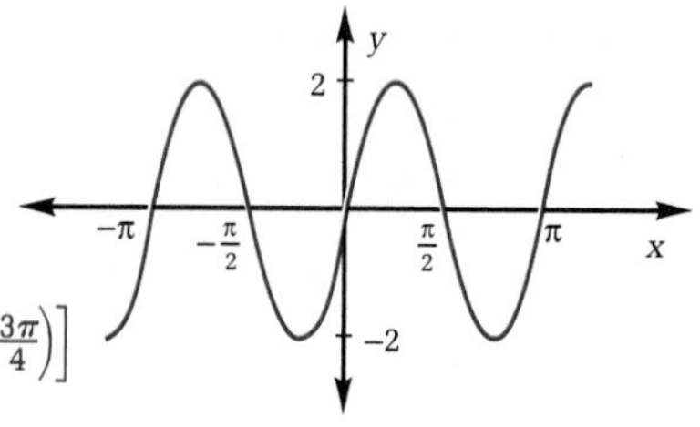

P **13** Find the period of the graph of each equation.

a $y = \sin(7x)\cos(3x) + \sin(3x)\cos(7x)$ $\frac{\pi}{5}$

b $y = \cos(9x)\cos(3x) - \sin(9x)\sin(3x)$ $\frac{\pi}{6}$

R **14** A certain triangle contains a 35° angle and a 97° angle. The length of the longest side of the triangle is 175 centimeters. Find the length of the shortest side. ≈ 101.13

B R **15** Find the area of each triangle.

a 84

b ≈13.6

65°
6
5

c ≈13.6

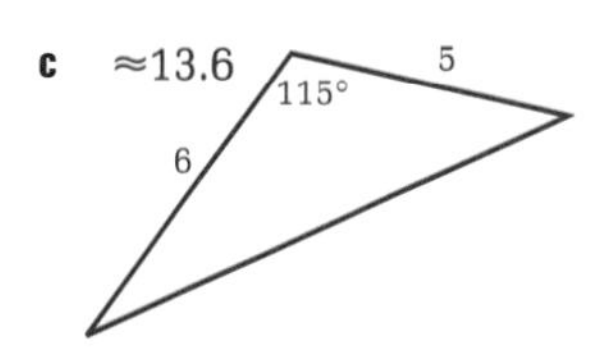

P **16** Find the domain and the range represented by four cycles (to the right of the y-axis) of each equation's graph.

a $y = 2\sin(3x) + 5$
Domain: $\left\{x{:}0 \le x \le \frac{8\pi}{3}\right\}$; range: $\{y{:}3 \le y \le 7\}$

b $y = -2\cos\left(\frac{1}{2}x\right) - 7$
Domain: $\{x{:}0 \le x \le 16\pi\}$; range: $\{y{:}-9 \le y \le -5\}$

P **17** Find the period of the graph of the equation $y = \sin(7x)\cos(-5x) + \cos(5x)\sin(-7x)$.
No period

P **18** Graph each equation.

a $y = \sin\left[2\left(x - \frac{\pi}{6}\right)\right]$

b $y = \sin\left(2x - \frac{\pi}{3}\right)$

c $y = \sin\left(2x - \frac{\pi}{3}\right) + 3$

d $y = 2\sin\left(2x - \frac{\pi}{3}\right) + 3$

I **19** Use a computer or a graphing calculator to graph each of the following equations.

a $y = \sec x$ **b** $y = \csc x$ **c** $y = \tan x$ **d** $y = \cot x$

P,R **20** The graph of $y = \cos x$ is shown for the domain $\{x{:}0 \le x \le 2\pi\}$. For what values in this domain is $\cos x \ge \frac{1}{2}$?
$\left\{x{:}0 \le x \le \frac{\pi}{3} \text{ or } \frac{5\pi}{3} \le x \le 2\pi\right\}$

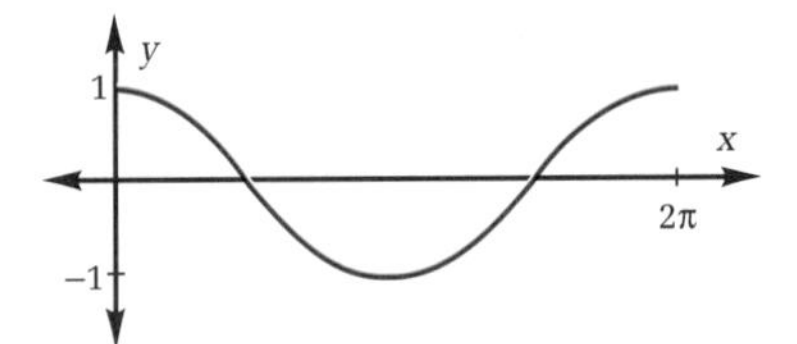

R **C** **21** If the perimeter of $\triangle PQR$ is 60, what is the length of $\overline{QR}$? ≈21.96

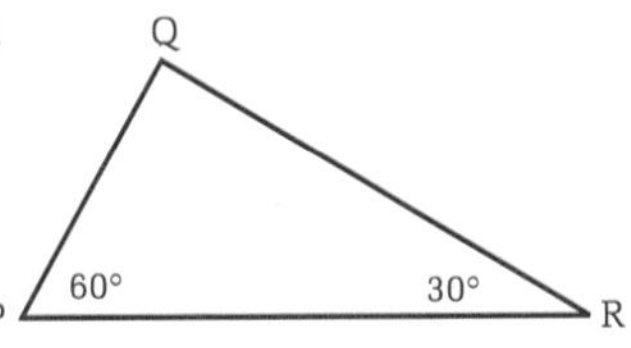

P,R **22** Evaluate $\sum_{k=0}^{360} \cos k°$. 1

R **23** Use the diagram to prove that the product of the areas of $\triangle ABE$ and $\triangle CDE$ is equal to the product of the areas of $\triangle ADE$ and $\triangle BCE$.

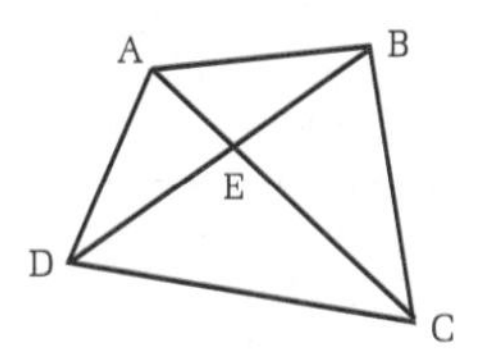

Problem-Set Notes and Strategies, continued

- Heron's formula can be used for problem **15a**. The Law of Cosines can also be used. Parts **b** and **c** are straightforward applications of the formula $k = \frac{1}{2}ab\sin C$.
- Problem **17** can be reduced to $y = 0$ by using the appropriate identities.
- Once again, students may need to check the asymptotes if a computer or graphic calculator is used in problem **19**.
- Problem **21** uses the relationships in a 30°-60°-90° triangle.
- Problem **22** is similar to problem **28** in the last section. Students may puzzle over why this answer is 1 and the answer to **28** was 0. Have them evaluate the endpoints and again check the symmetry.

Cooperative Learning

Have students work in small groups to draw a basic sine (or cosine) curve and label the axes appropriately to correspond with the amplitude and period. Once this is done, have the groups draw the final graph by performing the appropriate translations, both vertical and horizontal. This approach has the effect that the dilations and translations are separated and may be more easily understood.

Additional Answers

Answers for problems **18a–d**, **19a–d**, and **23** can be found in the answer pages beginning on page **A** 1.

16.5

Class Planning

Time Schedule
Average: 2 days
Advanced: 2 days
Honors: 1 day

Resource References
Teacher's Resource Book
Class Opener 96
Practice 101
Enrichment 32

Class Opener

Sketch the graph of
$y = 3 + \frac{1}{2}\cos(2x + 1)$.

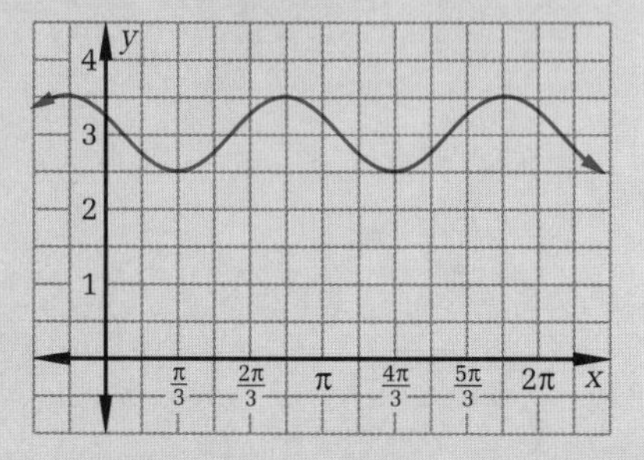

Lesson Notes

- The final four graphs are illustrated in this section. They are each related to their sine and cosine counterparts and each have asymptotes where the denominator equals zero. If students can relate these to the sine and cosine graphs, they may be more comfortable drawing a basic curve to illustrate each of them.

16.5 More Trigonometric Graphs

Objectives

After studying this section, you will be able to
- Graph the tangent function
- Graph the secant function
- Graph the cosecant and cotangent functions

Part One: Introduction

The Tangent Function: $y = \tan x$

Since $\tan x = \frac{\sin x}{\cos x}$, the tangent function is undefined whenever cos x is 0. Therefore, the graph of y = tan x has vertical asymptotes at $\ldots, -\frac{\pi}{2}, \frac{\pi}{2}, \frac{3\pi}{2}, \frac{5\pi}{2}, \ldots$

The tangent function has zeros whenever sin x is 0—that is, at $\ldots, -\pi, 0, \pi, 2\pi, \ldots$ A few convenient points occur whenever tan x is 1 or −1—that is, at $\ldots, \frac{-3\pi}{4}, -\frac{\pi}{4}, \frac{\pi}{4}, \frac{3\pi}{4}, \frac{5\pi}{4}, \frac{7\pi}{4}, \frac{9\pi}{4}, \ldots$ Notice that the tangent function is periodic and has a period of π.

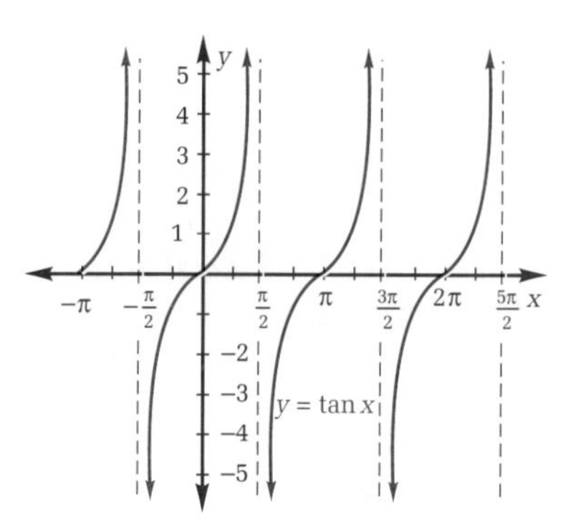

The Secant Function: $y = \sec x$

The values of the secant function are the reciprocals of the values of the cosine function.

$$y = \sec x = \frac{1}{\cos x}$$

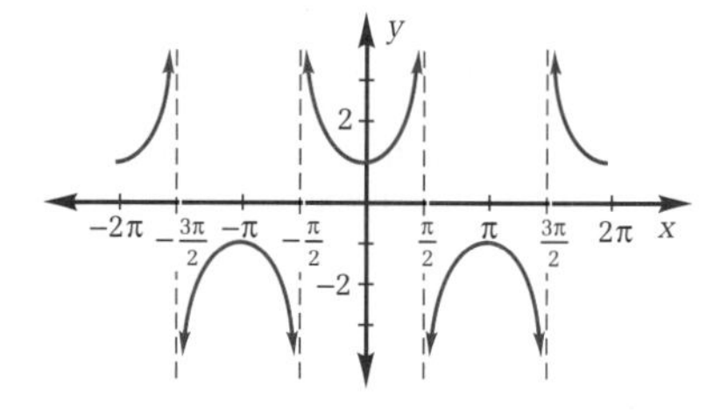

The secant function has no zeros; it does not cross the x-axis. This function is undefined whenever cos x = 0. Therefore, the graph of the secant function has vertical asymptotes at $\ldots, \frac{-3\pi}{2}, -\frac{\pi}{2}, \frac{\pi}{2}, \frac{3\pi}{2}, \ldots$ If we compare the graphs of the cosine and secant functions, we see that the cosine of an angle and the secant of the same angle are reciprocals. For example, $\cos(-\pi) = -1$ and $\sec(-\pi) = -1$, and −1 and −1 are reciprocals; cos 0 = 1 and sec 0 = 1, and 1 and 1 are reciprocals; $\cos\left(\frac{\pi}{3}\right) = \frac{1}{2}$ and $\sec\left(\frac{\pi}{3}\right) = 2$, and $\frac{1}{2}$ and 2 are reciprocals; and so forth.

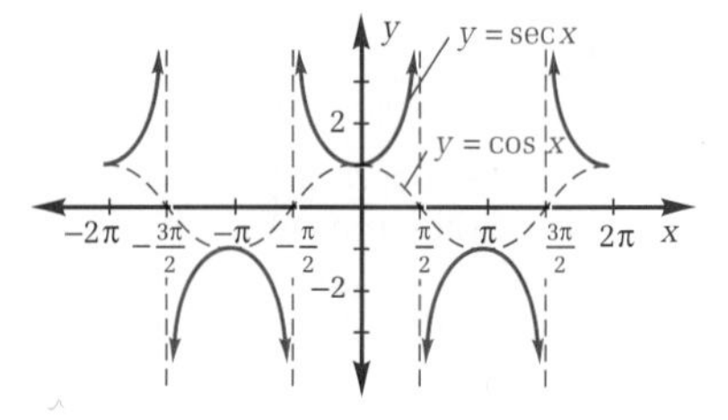

The Cosecant and Cotangent Functions: $y = \csc x$ and $y = \cot x$

Here are the graphs of the cosecant and cotangent functions.
It is a good idea to memorize the critical features of the graphs of all six trigonometric functions so that you can graph variations of them quickly.

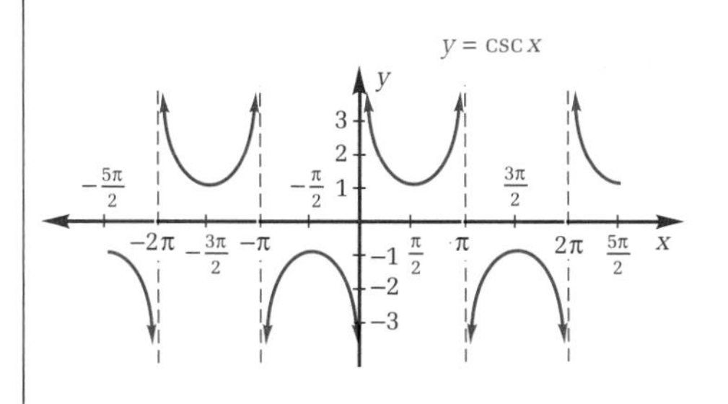

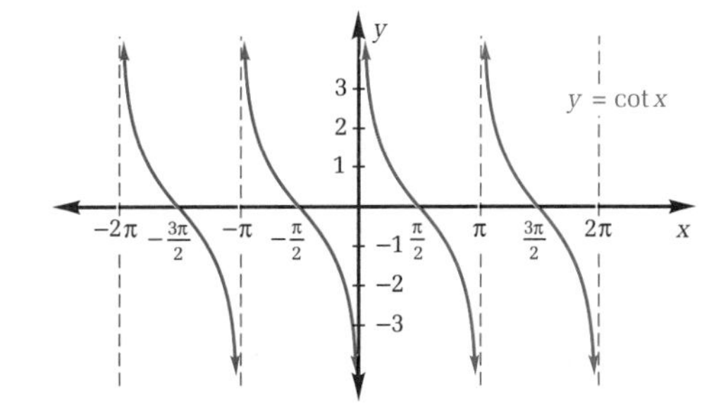

Using Computers

Students can use the *Omnifarious Plotter* to explore the graphs of the functions discussed in the section and also to solve many of the problems in the problem set.

Part Two: Sample Problems

See p. 812

Problem 1 *Graph* $y = \sec\left(x - \frac{\pi}{6}\right)$.

Solution Shift all points and the asymptotes of the graph of $y = \sec x$ to the right $\frac{\pi}{6}$ units.

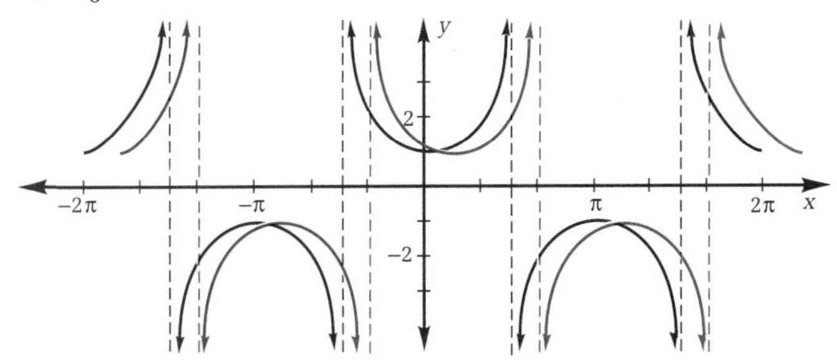

Problem 2 *Graph* $y = 0.25 \cot x$.

Solution The only difference between this graph and the graph of $y = \cot x$ is the coefficient 0.25. This is not an amplitude, since the graph does not oscillate. The coefficient changes the steepness of the graph. For example, $\cot\left(\frac{\pi}{4}\right) = 1$, so $0.25 \cot\left(\frac{\pi}{4}\right) = 0.25$; and $\cot\left(\frac{\pi}{3}\right) = \frac{1}{\sqrt{3}} \approx 0.5774$, so $0.25 \cot\left(\frac{\pi}{3}\right) \approx 0.1443$.

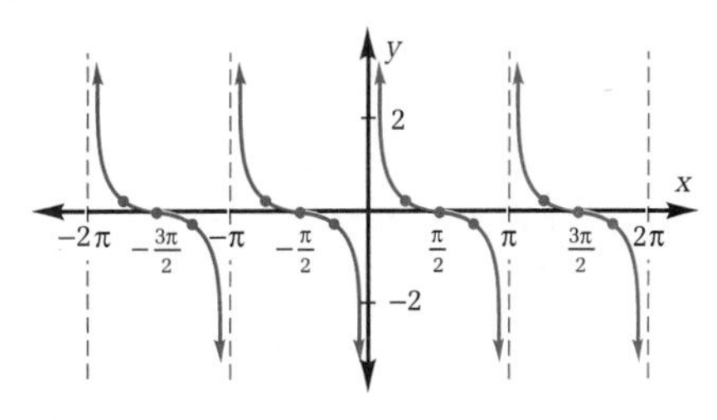

16.5

Checkpoint

1 Graph the cosecant function.

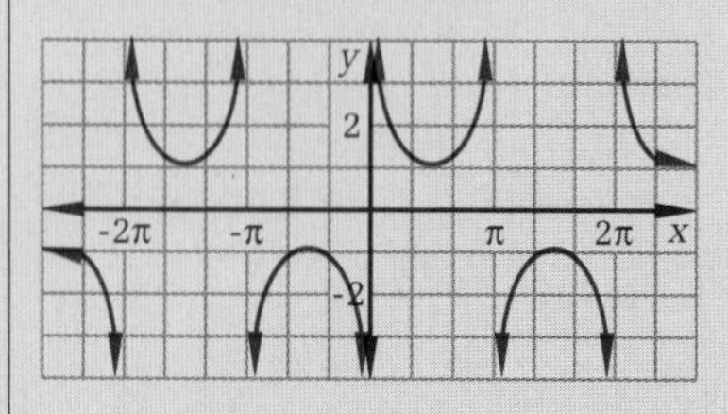

2 Graph the equation
$y = 2 \tan 2x$.

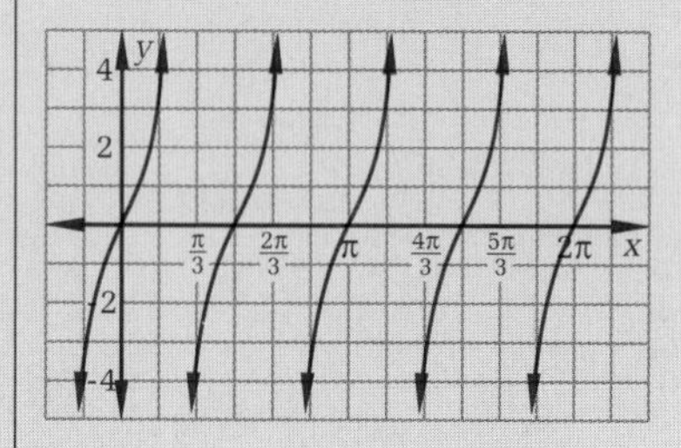

3 Graph the equation
$y = 3 \sec [2(x + 3)] - 4$.

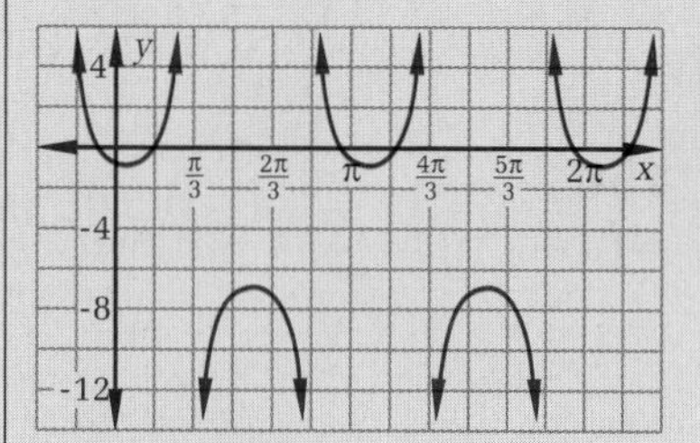

Assignment Guide

Average
Day 1 3, 6–9, 11, 12
Day 2 4a, b, d, 5, 13, 15, 16c
Advanced
Day 1 3, 6–9, 11, 12
Day 2 4a, c, e, 5, 13, 15, 16b, c
Honors
3, 4b, c, 5–7, 9c, 10, 12, 13, 15, 16b, c

Problem 3 *Graph* $y = \tan\left(\frac{1}{2}x\right)$.

Solution The normal period of $y = \tan x$ is π. In the graph of $\tan\left(\frac{1}{2}x\right)$, only half a cycle fits in the normal π interval. Therefore, the period of $y = \tan\left(\frac{1}{2}x\right)$ is stretched to $\frac{\pi}{0.5}$ or 2π. Notice that a zero remains at (0, 0) and the asymptotes are twice as far from the y-axis as in the basic tangent graph.

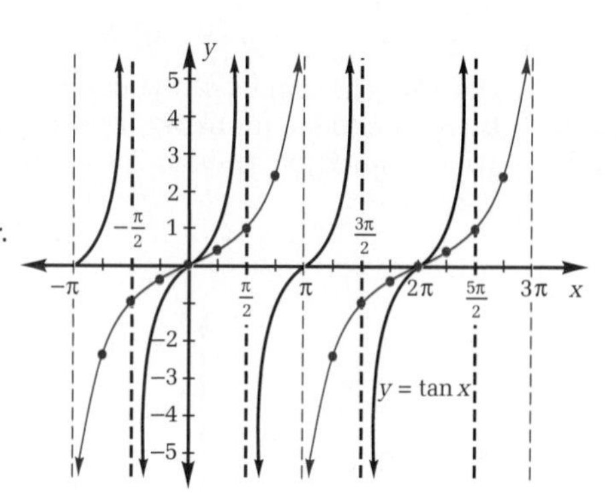

Problem 4 *The graph shown is based on the graph of* $y = \csc x$. *Find the equation of the graph.*

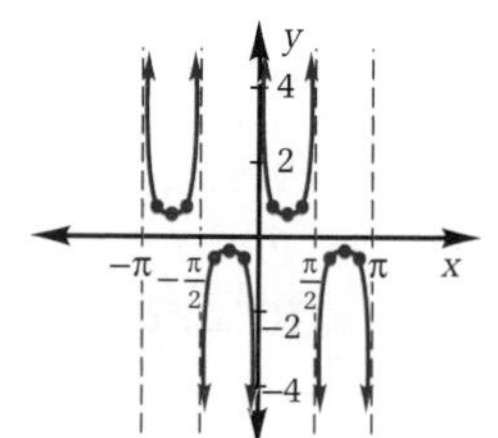

Solution The vertex of each part of the graph is 0.5 or −0.5. Since the closest this graph gets to the x-axis is 0.5 unit, the coefficient of this cosecant function is 0.5. The length of one cycle of this graph is π. Notice that the period of $y = \csc x$ has changed so that two cycles fit into the normal 2π period. Therefore, the coefficient of x is 2. The equation of the graph is $y = 0.5 \csc (2x)$.

Part Three: Exercises and Problems

Warm-up Exercises

1 Graph each equation.

a $y = \sec x$ **b** $y = 0.5 \sec x$ **c** $y = \sec (2x)$

2 Graph each equation.

a $y = \csc x$ **b** $y = 2 \csc x$ **c** $y = 2 \csc\left(x + \frac{\pi}{2}\right)$

Problem Set

3 Graph each equation.

a $y = \tan x$ **b** $y = \tan (3x)$ **c** $y = 2 \tan x$

4 Graph each equation.

a $y = \csc x$ **b** $y = 2 \csc x$ **c** $y = \csc (2x)$
d $y = -2 \csc x$ **e** $y = -2 \csc (2x)$

Additional Answers

Answers for problems **1a–c**, **2a–c**, **3a–c**, and **4a–e** can be found in the answer pages beginning on page **A** 1.

P **5** Graph each equation.

a $y = \sec\left(x - \frac{\pi}{6}\right) - 3$ **b** $y = \sec\left(x - \frac{\pi}{6}\right) + 3$

P,R **B** **6** Write a tangent equation for the graph.
$y = \tan(2x)$

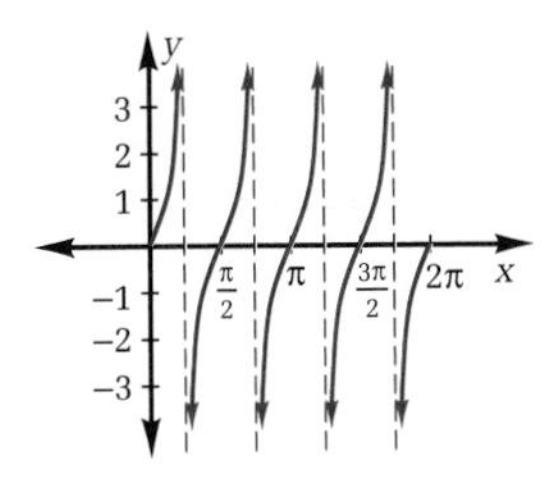

P,R **7** Find the smallest positive x value at which $y = 6 \sin(3x)$ intersects the line $y = 4$. ≈ 0.243

R **8** Patty was driving due west along a highway when she sighted a radio tower 25° to the north. Twenty minutes later, she was due south of the tower. If Patty was traveling 60 miles per hour, how far is the tower from the highway? ≈ 9.33 mi

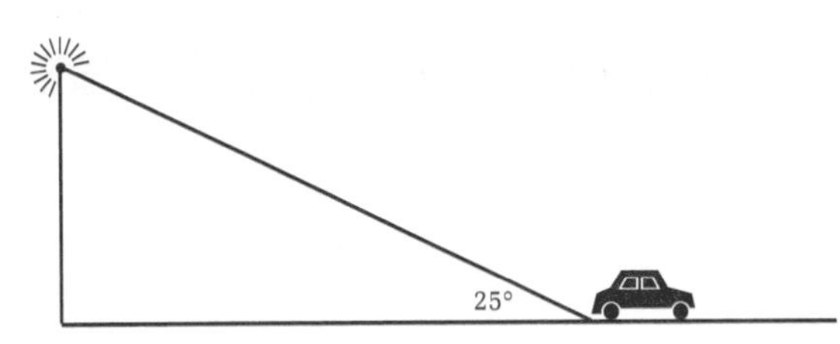

P **9** Graph each equation.

a $y = \sin x$ **b** $y = |\sin x|$ **c** $y = \sin^2 x$

R **10** An element is selected at random from
{tan 0°, tan 6°, tan 12°, . . . , tan (6*k*)°, . . . , tan 84°}.
Find the probability that the value is between $\frac{1}{2}$ and 2. $\frac{2}{5}$

R **11** The lengths of two sides of a triangle are 186 centimeters and 305 centimeters. The measure of the angle between these two sides is 42.6. Find the area of the triangle. $\approx 19{,}199.59 \text{ cm}^2$

R **12** **a** Find the length of $\overline{EF}$. $\frac{8}{3}$
b Determine all possible measures of $\angle ABC$. $0 < m\angle ABC < 90$

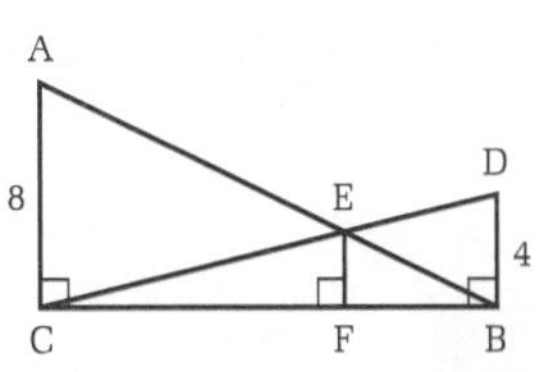

P,R **C** **13** Let $\det\begin{bmatrix} a & b \\ c & d \end{bmatrix} = ad - bc$. Graph the following equation.

$$y = \det\begin{bmatrix} \cos(3x) & \sin x \\ \sin(3x) & \cos x \end{bmatrix}$$

P,R **14** Graph $y = \cos(3x)\cos x + \sin(3x)\sin x$.

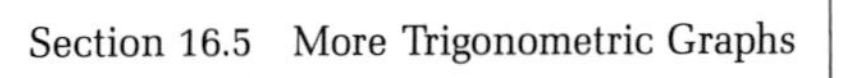

Problem-Set Notes and Strategies

- Problem **5** may be difficult. It involves many of the same concepts as the last section. With the exception of the amplitude, the translations and dilations of the sine and cosine graphs may be applied to these. Since the graphs approach infinity at the asymptotes, there is no amplitude.

- A graph of the two functions in problem **7** may aid in finding the solution.

Additional Answers

5a

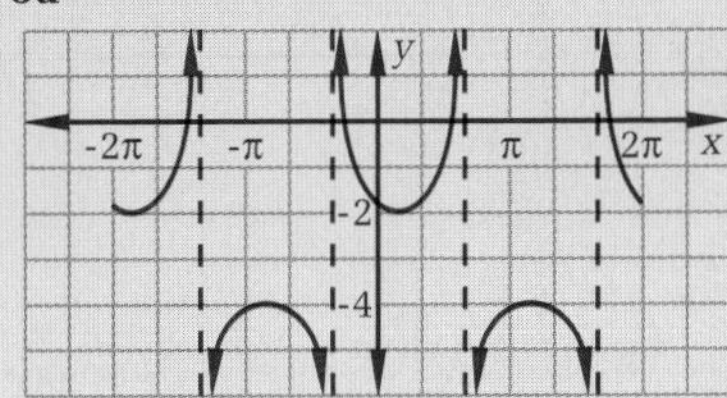

5b

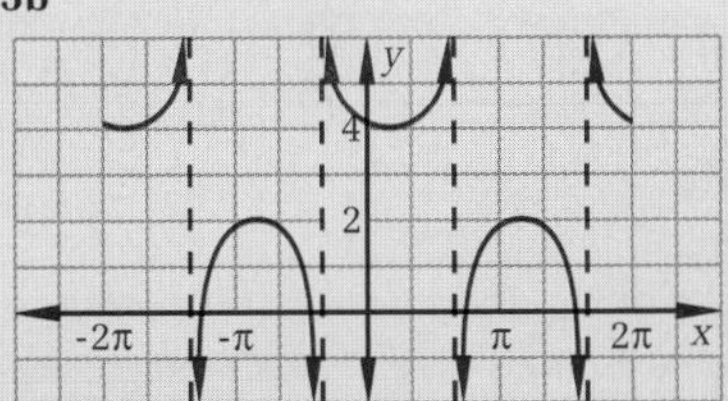

Answers for problems **9a–c**, **13**, and **14** can be found in the answer pages beginning on page **A** 1.

- Problems **13** and **14** will both require some work with identities before graphing unless a calculator or computer are used.

Problem Set, *continued*

R **15** Find y. ≈3.06

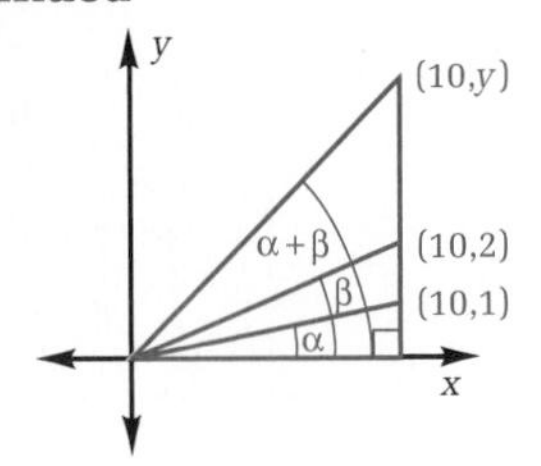

P **16** Graph each equation, using a computer or a graphing calculator.

a $y = \tan x + \sec x$ **b** $y = \sec x + \csc x$
c $y = x \sin x$ **d** $y = x^2 + \sin(4x)$

Communicating Mathematics

Explain in your own words the relationships that exist between the graphs of the six trigonometric functions. You may use co-function arguments, reciprocal arguments, etc., to help in these explanations. Use pictures if necessary.

Additional Answers

16a

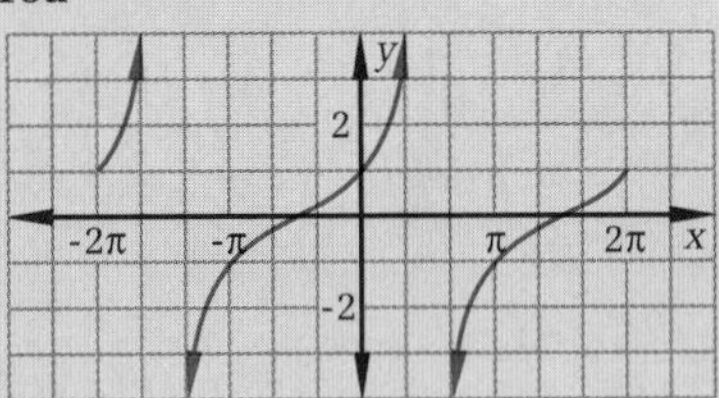

16b

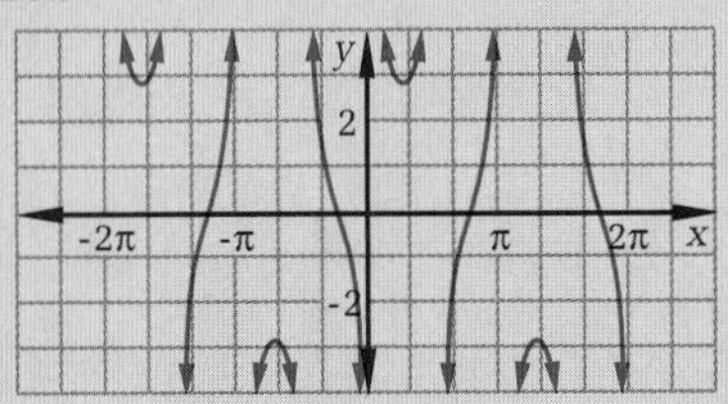

16c

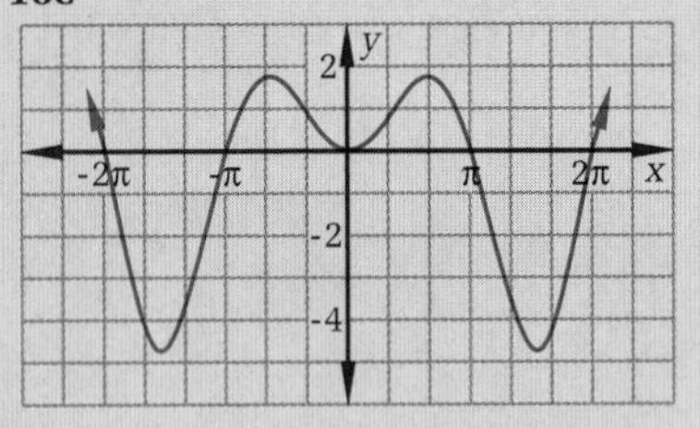

16d

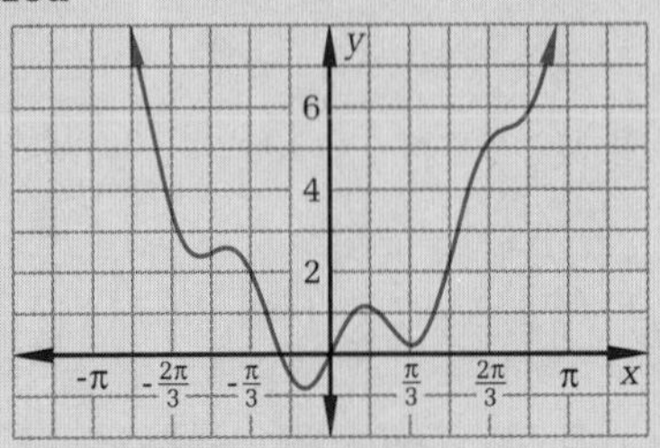

MATHEMATICAL EXCURSION

SOOTHING SOUNDS

Sines and musical harmony

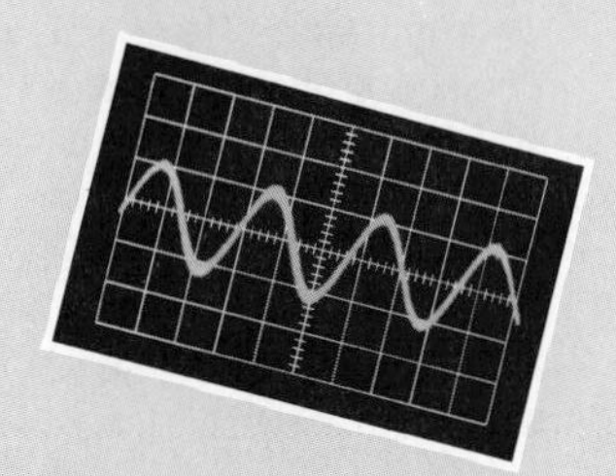

The graph of the sine function is commonly called a sine wave. One characteristic of a sine wave is that it is periodic: the peaks and valleys of the wave occur at consistent intervals. The interval between two peaks of a sine wave is called the *period.* The number of peaks that occur over a given range of *x* values is the wave's frequency.

We are inclined to think of a sound as musical when it can be represented by a wave that is periodic and when it is associated with other periodic sounds. Variations in musical tones, such as pitch and loudness, can be represented by graphing variations of the sine function. For example, the period of a sine wave can be doubled by using the equation $y = \sin\left(\frac{x}{2}\right)$.

Sound waves whose periods are not multiples of each other produce sound patterns called beats, which we hear as dissonance. Two notes sounded together whose pitches are very close produce beats of a particular frequency. The beat frequency is the number of times the peaks of the waves correspond in a given time period. The effect can be produced by sounding the same note on two instruments that are not tuned to each other or by striking adjacent black and white piano keys.

CHAPTER SUMMARY

CONCEPTS AND PROCEDURES

After studying this chapter, you should be able to

- Apply the Law of Sines (16.1)
- Apply the Law of Cosines (16.2)
- Graph the sine function (16.3)
- Translate the sine graph vertically and horizontally (16.3)
- Change the amplitude of the sine graph (16.3)
- Graph the cosine function (16.4)
- Stretch and shrink the sine and cosine graphs horizontally (16.4)
- Graph the tangent function (16.5)
- Graph the secant function (16.5)
- Graph the cosecant and cotangent functions (16.5)

VOCABULARY

amplitude (16.3)
Law of Cosines (16.2)
Law of Sines (16.1)
period (16.3)
periodic (16.3)
phase shift (16.3)

Class Planning

Time Schedule

All levels: 2 days

Assignment Guide

Average	
Day 1	2–4, 9–12, 17, 20–22
Day 2	1, 5–8, 13, 14, 16, 18, 19c, d
Advanced	
Day 1	2–4, 9–12, 17, 20–22
Day 2	1, 5–8, 13–16, 18, 19c, d, e
Honors	
Day 1	1–7, 8a, 9–12, 20, 21
Day 2	8b, 13–16, 18, 19c, d, e, 22

Problem-Set Notes and Strategies

- The graph in problem **6** may also be used to represent a radio signal. AM radio stands for amplitude modulation — changing the amplitudes of the sound wave. This can be seen in in the graph. (FM stands for frequency modulation—changing the period which is defined as $\frac{1}{\text{frequency}}$.)

Additional Answers

1a

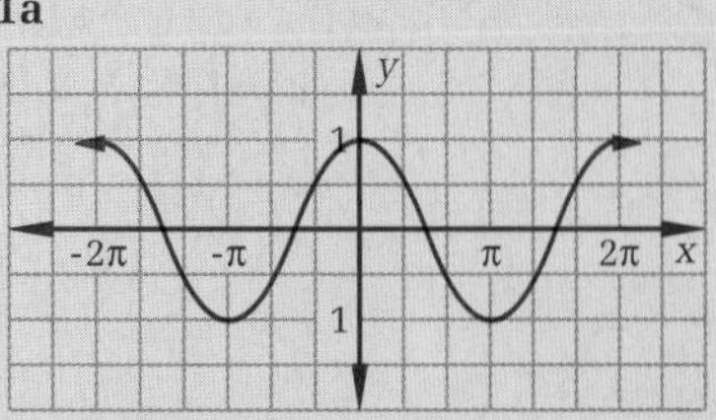

16 Review Problems

P **A** **1** Graph each of the following equations.

a $y = \cos x$ **b** $y = \cos(4x)$ **c** $y = \cos(\pi x)$

P  **2** Find the measures of angles α, β, and γ. $\alpha \approx 71.8$; $\beta \approx 58.8$; $\gamma \approx 18.2$

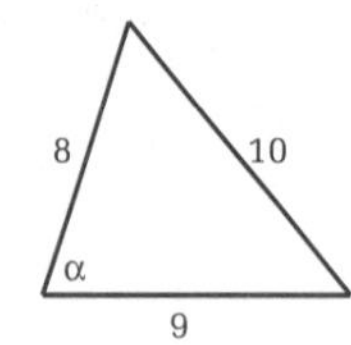

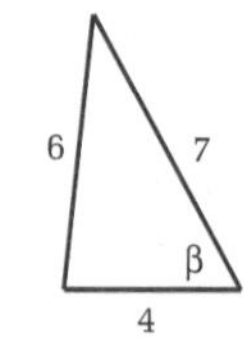

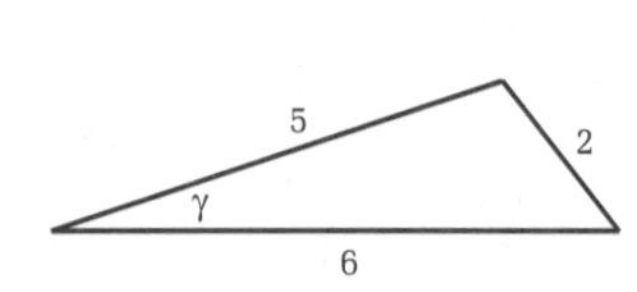

P **3** Find x and y to the nearest hundredth.
$x \approx 32.71$; $y \approx 20.10$

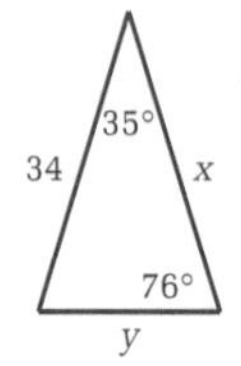

R **4** Evaluate each expression.

a $\sin \frac{\pi}{6}$ $\frac{1}{2}$ **b** $\sin\left(\frac{\pi}{6} + 2\pi\right)$ $\frac{1}{2}$ **c** $\sin\left(\frac{\pi}{6} - 2\pi\right)$ $\frac{1}{2}$

d $\sin\left(\frac{\pi}{6} + 4\pi\right)$ $\frac{1}{2}$ **e** $\sin\left(\frac{\pi}{6} - 4\pi\right)$ $\frac{1}{2}$

P  **5** Find length x.
≈ 15.665

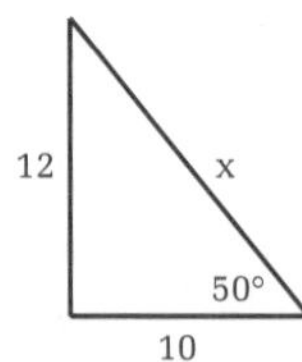

P **6** A certain sound is represented by the periodic curve shown.

a What is the curve's amplitude? 3.5

b What is its period? 8

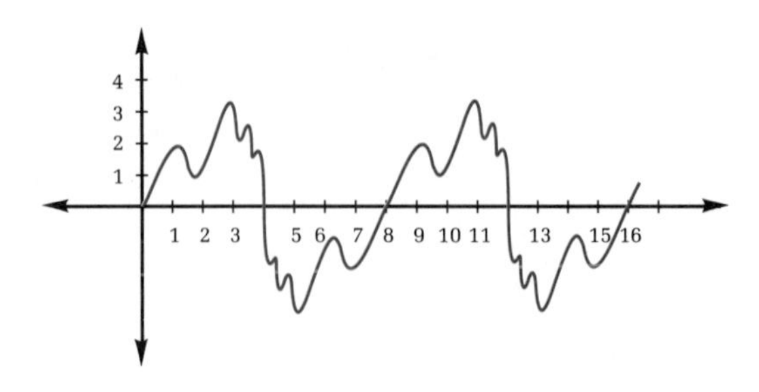

Answers for problems **1b**, and **1c** can be found in the answer pages beginning on page **A** 1.

P **7** Copy and complete the following table.

Equation	Amplitude	Period	Range
$y = -2 \sin (3x)$	2	$\frac{2\pi}{3}$	$\{y: -2 \le y \le 2\}$
$y = 4 \cos (2x)$	4	π	$\{y: -4 \le y \le 4\}$
$y = 3 \sin \left(\frac{1}{3}x\right) + 5$	3	6π	$\{y: 2 \le y \le 8\}$

P **8** Graph each equation.

a $y = 2 \sin (3x)$ **b** $y = 2 \cos (3x)$

B

P **9** How many triangles are determined by each diagram?

a 1 (6, 60°, 5) **b** 2 (5, 60°, 6) **c** 0 (3, 60°, 6)

P **10** Find the lengths of sides x, y, and z. $x \approx 11.6$; $y \approx 8.1$; $z \approx 47.3$

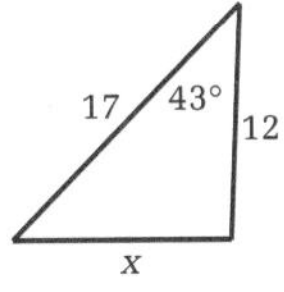

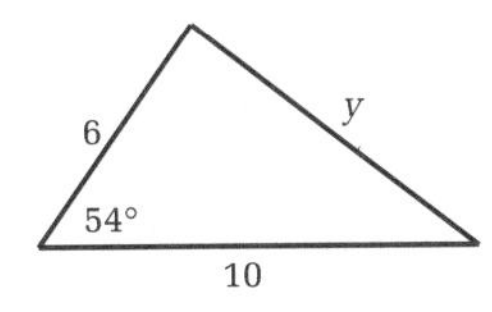

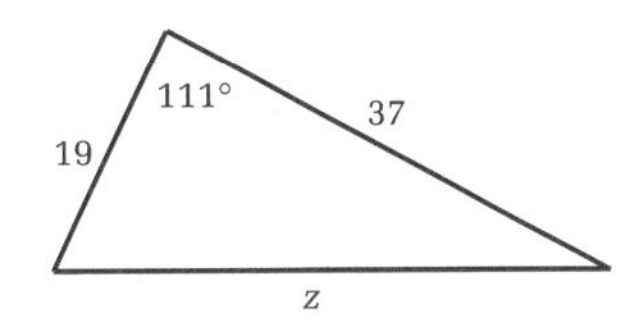

P **11** Find the measure of $\angle A$.
≈ 89

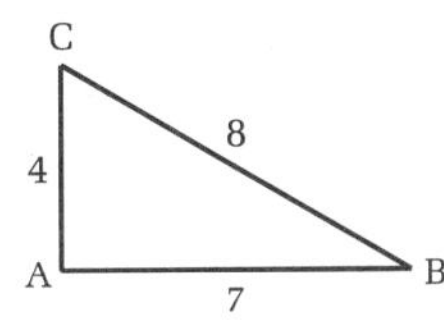

R **12** For which values of x in $\{x: 0 \le x \le 2\pi\}$ is $\sin x = \frac{1}{2}$?
$\frac{\pi}{6}, \frac{5\pi}{6}$

P **13** Write an equation for the graph.
$y = 4 \sin x$

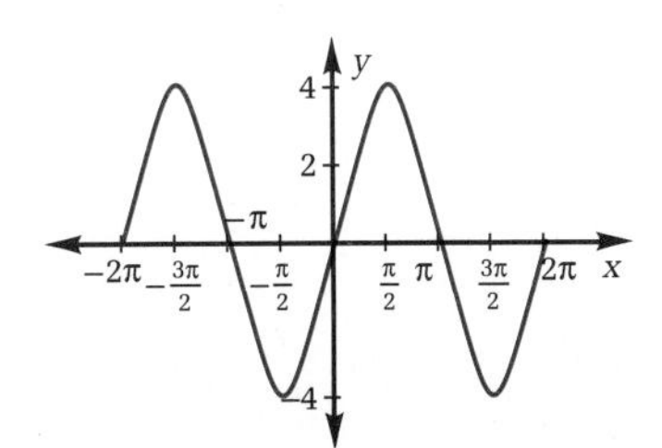

P,R **14** Using a computer or a graphing calculator, graph the following equations on the same coordinate system.

a $y = x$ **b** $y = \sin x$ **c** $y = x + \sin x$

16

Additional Answers

8a

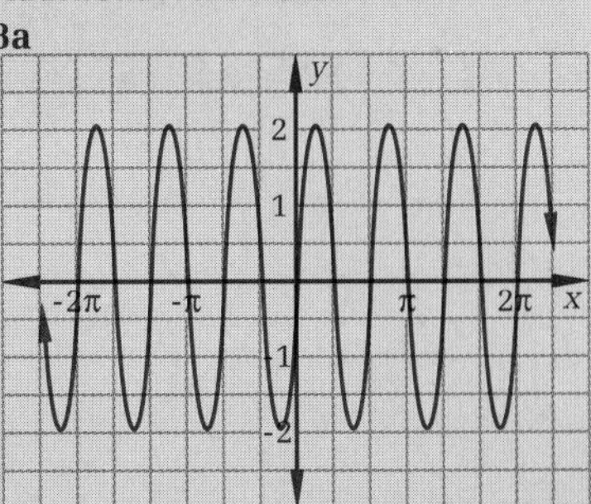

8b

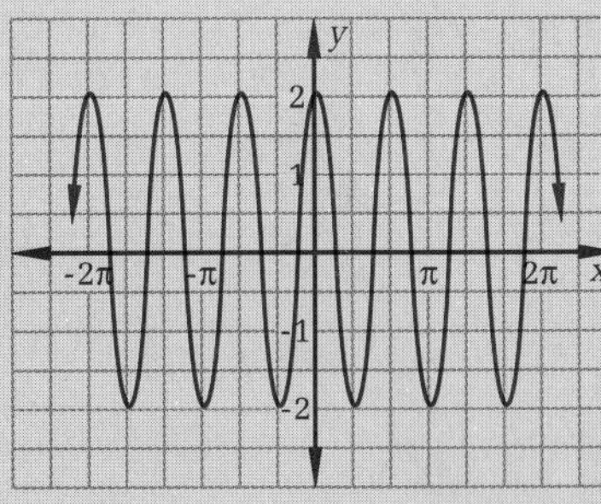

14

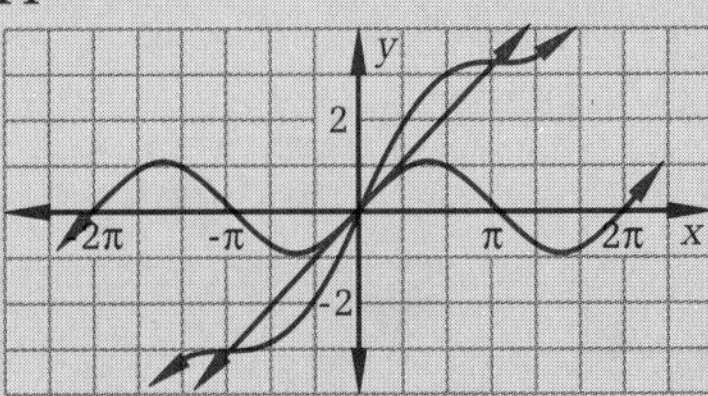

Problem-Set Notes and Strategies, continued

- The answer to problem **13** is not necessarily unique. It may be written several ways in terms of either sine or cosine.
- Once problem **14** is graphed, students may need help to see that **c** is obtained by adding the ordinates from parts **a** and **b**.

Additional Answers

19a

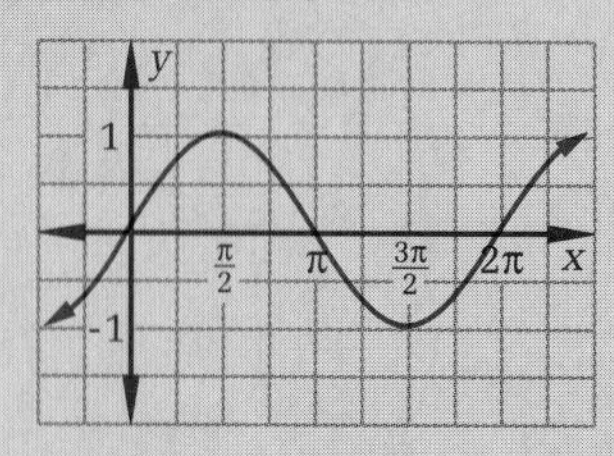

19b

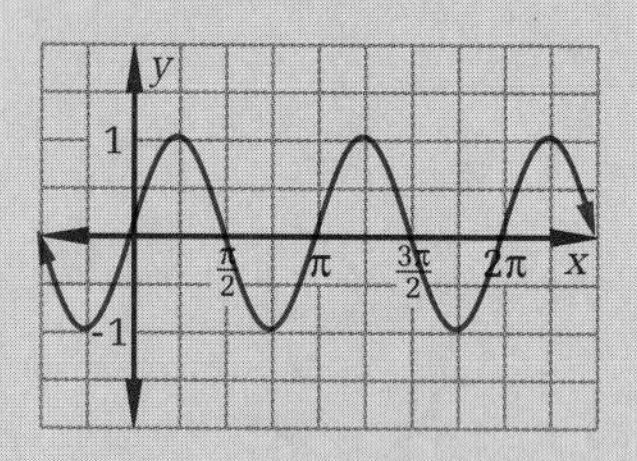

19c

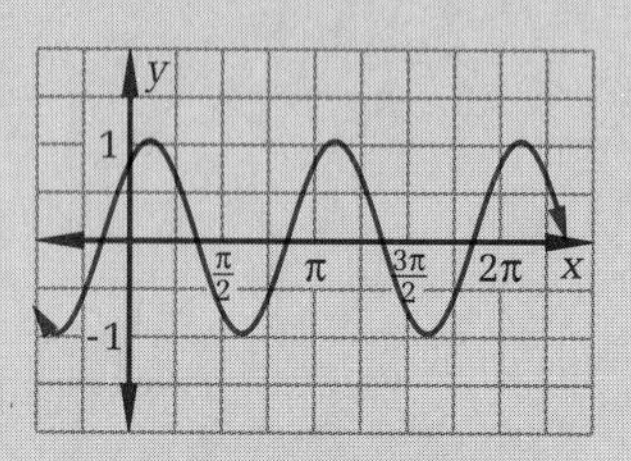

19d

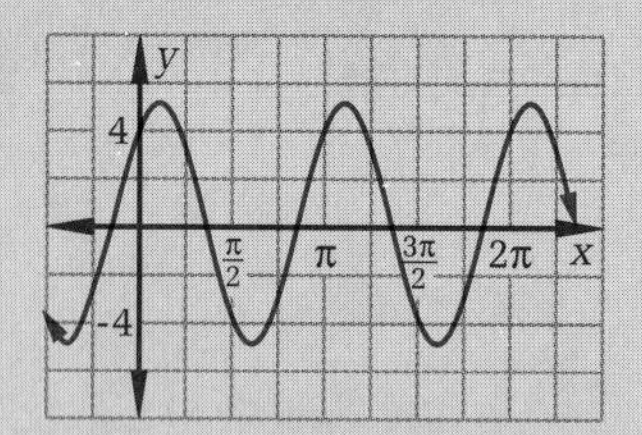

19e

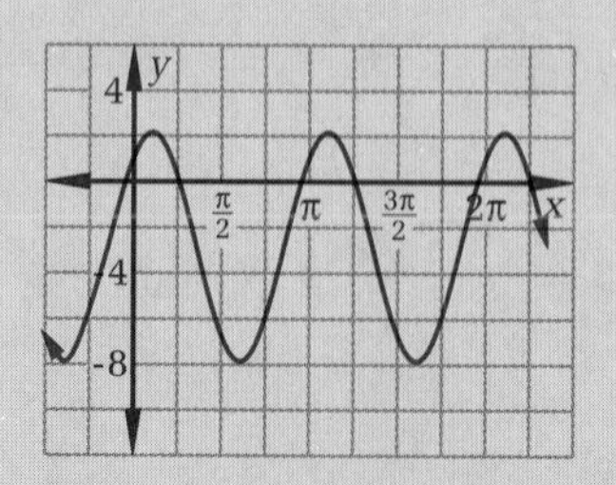

Review Problems, *continued*

P,R **15** One cycle of the graph of $y = 4 \cos (3x)$ is shown.

a Find the domain and the range represented in the graph.

b For which values of x in the domain is $y = 2$? $\frac{\pi}{9}, \frac{5\pi}{9}$

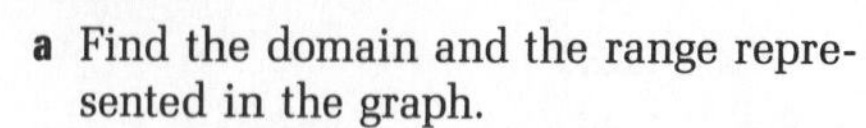

a Domain: $\left\{x{:}0 \le x \le \frac{8\pi}{3}\right\}$; range: $\{y{:}-4 \le y \le 4\}$

P **16** The period of $y = 3 \cos \beta x$ is 5. Find β. $\beta = \frac{2\pi}{5}$ or $\beta = \frac{-2\pi}{5}$

R **17** Let $A = \begin{bmatrix} 3 & 2 & 1 \\ 4 & -1 & 2 \end{bmatrix}$ and $B = \begin{bmatrix} 1 & 0 \\ 3 & 2 \\ -5 & -1 \end{bmatrix}$.

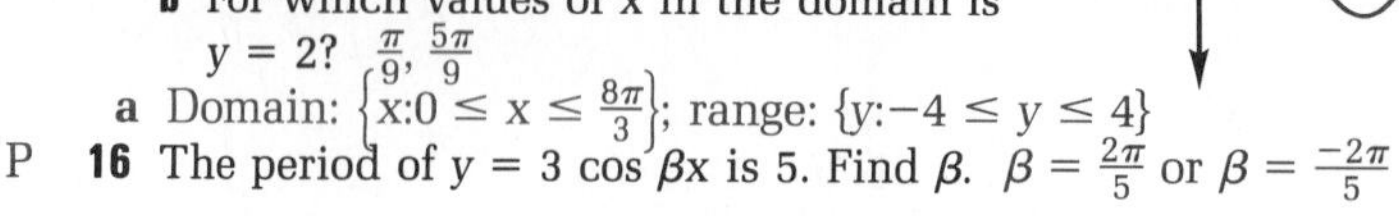

An element is selected at random from the product $A \cdot B$. Find the probability that the element is an odd number. $\frac{1}{2}$

P **18** Find the relationship between x_1 and x_2.
$x_1 + x_2 = \pi$

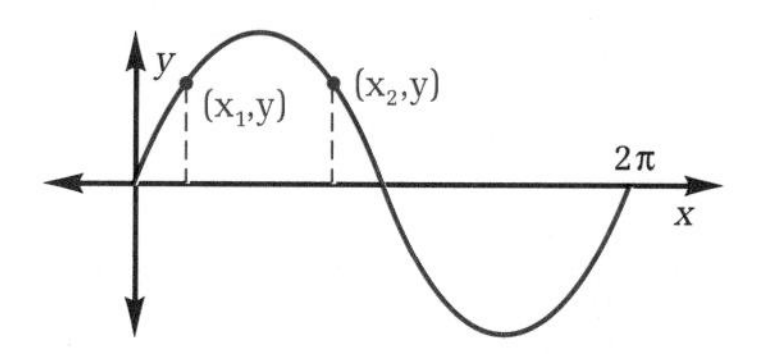

P **19** Graph each equation.

a $y = \sin x$

b $y = \sin (2x)$

c $y = \sin \left(2x + \frac{\pi}{4}\right)$

d $y = 5 \sin \left(2x + \frac{\pi}{4}\right)$

e $y = 5 \sin \left(2x + \frac{\pi}{4}\right) - 3$

R **20** What percentage of the total area under the normal curve shown is in the shaded region?
$\approx 84.1\%$

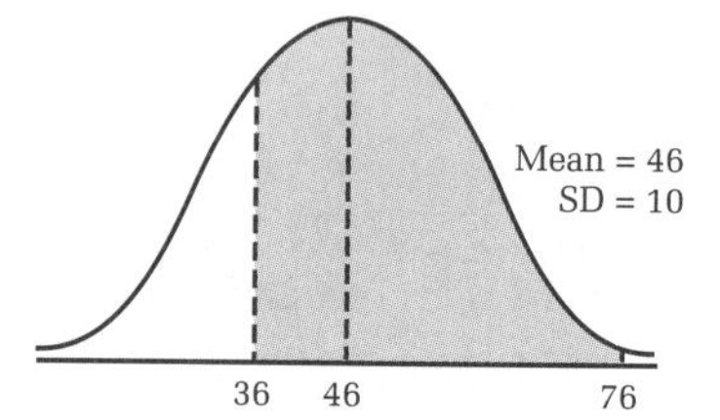

P,R **21** Given rhombus ABCD with perimeter 260 and $m\angle D = 70$, find the lengths of the rhombus's diagonals.
≈ 106.49 and ≈ 74.56

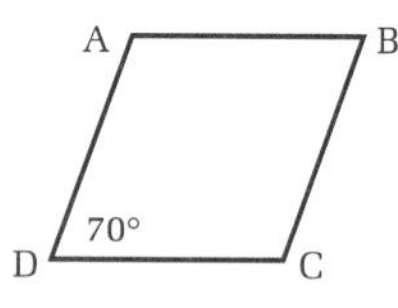

C P,R **22** A regular pentagon is inscribed in a circle with a 48-centimeter radius. Find the perimeter of the pentagon. ≈ 282.14 cm

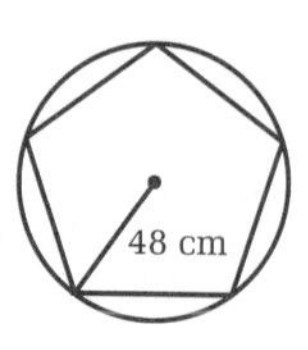

16 Chapter Test

1 If $\theta = 62°$ and $\sin \alpha = 0.9272$, what are two possible values of α and the corresponding values of β?
If $\alpha \approx 68°$, $\beta \approx 50°$; if $\alpha \approx 112°$, $\beta \approx 6°$.

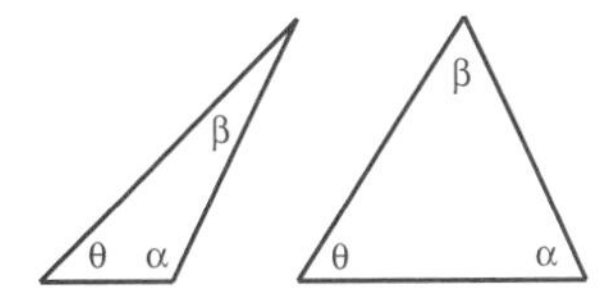

2 The measures of two angles of a triangle are 60 and 80. If the length of the longest side of the triangle is 12, what is the length of the shortest side? ≈ 7.83

3 Find the measure of acute $\angle B$.
$m\angle B \approx 8.5$

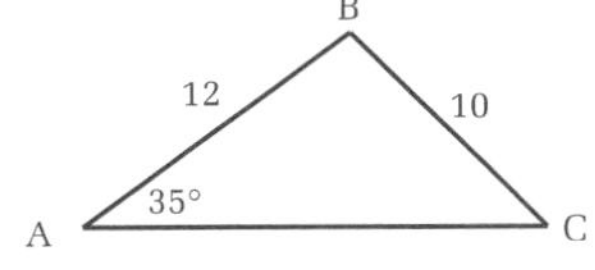

4 In $\triangle ABC$, $m\angle A = 45$ and $m\angle B = 75$. Find the ratio AB:BC.
$\sqrt{3}:\sqrt{2}$ or $\sqrt{6}:2$

5 The lengths of the sides of a parallelogram are 11 and 14. The measure of one of the parallelogram's angles is 150. Find the length of the parallelogram's shorter diagonal. ≈ 7.09

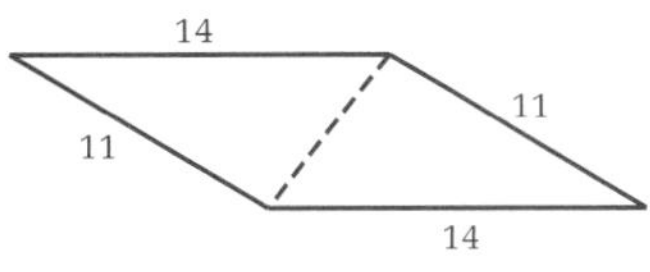

6 Find the area of the parallelogram in problem **5.** 77

7 Refer to the sine-curve graph.

a What is the amplitude? 2
b What is the period? 2π
c What is the phase shift? $\frac{2\pi}{3}$ to the right
d What is the range? $-5 \le y \le -1$
e What is an equation for the sine-curve graph? $y = 2 \sin\left(x - \frac{2\pi}{3}\right) - 3$

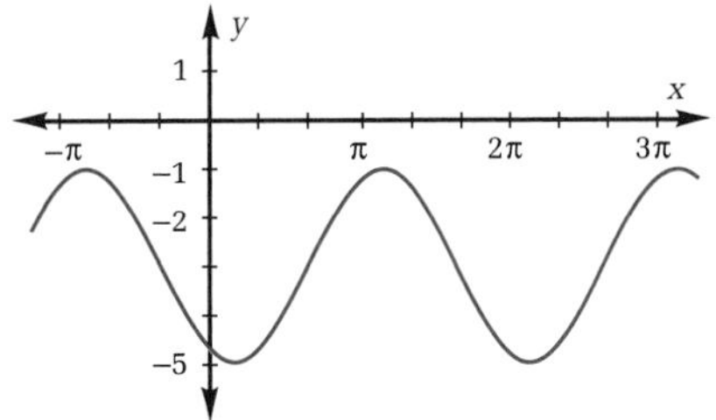

8 **a** Sketch the graph of $y = -3 \cos\left(x - \frac{\pi}{4}\right) + 2$.
b What is the amplitude? 3
c What is the phase shift? $\frac{\pi}{4}$
d What is the range? $-1 \le y \le 5$

9 Sketch the graph of $y = 2 \sin \frac{1}{3}x$.

10 At what point does the graph of $y = 2 \sin \frac{1}{3}x$ reach its peak between 0 and 4π? $\left(\frac{3\pi}{2}, 2\right)$

Resource References
Evaluation
Test Form 1, 2, 3

Additional Answers

8a

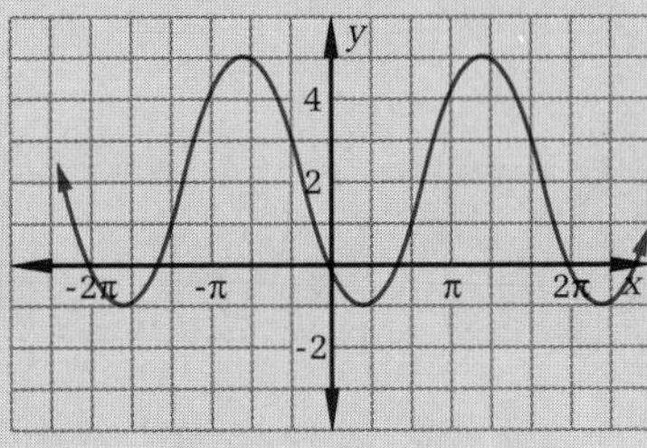

9

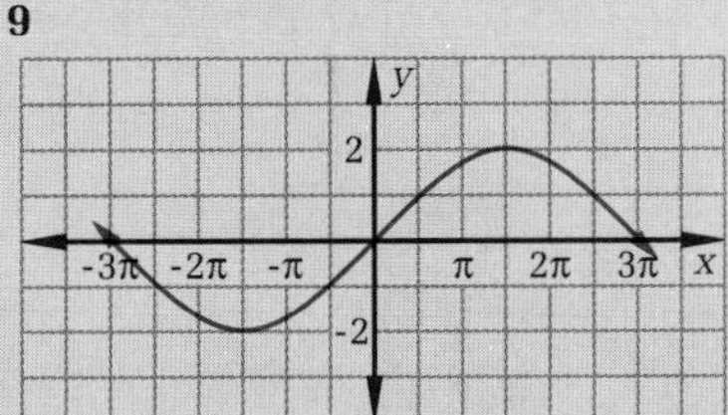

Handbook of Basic Concepts

THE REAL NUMBERS

Much of your work in algebra involves the set of ***real numbers.*** This is made up of all the numbers that can be assigned points on the real-number line.

When you first learned to count, you used the set of ***natural numbers,*** {1, 2, 3, 4, . . . }. The importance of the number zero led you to use the larger set of ***whole numbers,*** {0, 1, 2, 3, 4 . . . }. We form a still larger set by assigning ***negative numbers*** to points to the left of zero on the number line. The resulting set,

$$\{\ldots, -3, -2, -1, 0, 1, 2, 3, \ldots\}$$

is called the set of ***integers.***

The set of all integers and fractions composes the set of ***rational numbers.*** Every rational number can be written in the form $\frac{\text{an integer}}{\text{a natural number}}$. Every rational number can also be written as either a ***terminating decimal*** (for example, $\frac{3}{12} = 0.25$) or a ***repeating decimal*** (for example, $\frac{2}{3} = 0.6666666\ldots$). Repeating decimals can be indicated by placing a bar over the repeating digit or digits. For example, $\frac{2}{3} = 0.\overline{6}$.

The remaining points on the number line correspond to the ***irrational numbers.*** When expressed as a decimal, an irrational number, such as $\sqrt{5}$ or π, neither terminates nor repeats. By combining the set of rationals and the set of irrationals, we form the set of real numbers.

PROPERTIES OF REAL NUMBERS

To use the real numbers accurately, we need to know their properties with respect to the operations of addition, subtraction, multiplication, and division. The following properties apply to all real numbers a, b, and c.

Name	Property	Example
Commutative Property of Addition	$a+b=b+a$	$4+7=7+4$
Commutative Property of Multiplication	$ab=ba$	$3\cdot 6=6\cdot 3$
Associative Property of Addition	$(a+b)+c=a+(b+c)$	$(5+8)+9=5+(8+9)$
Associative Property of Multiplication	$(ab)c=a(bc)$	$(3\cdot 5)x=3(5x)$
Difference Property	$a-b=a+(-b)$	$5-9=5+(-9)$
Distributive Property of Multiplication over Addition or Subtraction	$a(b\pm c)=ab\pm ac$ $(a\pm b)c=ac\pm bc$	$2(x+7)=2x+14$ $(x-y)z=xz-yz$
Quotient Property	$a\cdot \frac{1}{b}=\frac{a}{b}$	$3\cdot \frac{1}{5}=\frac{3}{5}$
Additive Inverse Property	For every a, there is a number $-a$—called the *additive inverse,* or *opposite,* of a—such that $a+(-a)=0$.	$8+(-8)=0$
Multiplicative Inverse Property	For every $a\neq 0$, there is a number $\frac{1}{a}$, called the *multiplicative inverse* of a, such that $a\left(\frac{1}{a}\right)=1$.	$6\cdot \frac{1}{6}=\frac{6}{6}=1$
Property of the Opposite of a Quantity	$-(a+b)=-1(a+b)$ $=-a-b$ $-(a-b)=-1(a-b)$ $=-a+b$	$-(x+2)=-x-2$ $-(y-4)=-y+4$
Additive Identity Property	There is a number 0 such that $a+0=a$ and $0+a=a$.	$3+0=0+3=3$
Multiplicative Identity Property	There is a number 1 such that $1\cdot a=a\cdot 1=a$.	$1\cdot 7=7\cdot 1=7$
Multiplication Property of Zero	$0(a)=a(0)=0$	$4(0)=0(4)=0$

EQUATION SOLVING

A ***variable*** is a symbol used to represent a number. An ***expression*** is a mathematical phrase consisting of a number, a variable, or a group of numbers and variables connected by operations. The following are examples of expressions.

$$7 \qquad 5x \qquad 13y - 40 \qquad e^{-x} \qquad \frac{1}{x+1} \qquad |2z - 17|$$

An ***equation*** is a mathematical sentence stating that two expressions are equal for one or more values of the variable. Solving an equation means finding the value(s) of the variable that make the equation a true statement. The following is an example of an equation.

$$4(x + 12) = x^2 + x(7 - x)$$

An equation can be solved by applying a simple strategy. This consists of performing whatever operations are needed to isolate the variable on one side of the equation so that the result is in the form "variable = number."

Notice that we cannot carry this strategy out by means of the properties of real numbers alone. Using these properties with the sample equation yields $4x + 48 = 7x$. To proceed further, we need additional properties—the properties of equality.

Properties of Equality

Each of the following properties is true for all real numbers (represented by a, b, and c) unless otherwise noted.

Name	Property	Example
Reflexive Property	$a = a$	$21.4 = 21.4$
Symmetric Property	If $a = b$, then $b = a$.	If $y = 4$, then $4 = y$.
Transitive Property of Equality	If $a = b$ and $b = c$, then $a = c$.	If $x_1 = 9$ and $9 = x_2$, then $x_1 = x_2$.
Addition Property of Equality	If $a = b$ and $c = d$, then $a + c = b + d$.	If $x - 6 = 5$ and $6 = 6$, then $x = 11$.
Subtraction Property of Equality	If $a = b$ and $c = d$, then $a - c = b - d$.	If $x + 4 = 24$ and $4 = 4$, then $x = 20$.
Multiplication Property of Equality	If $a = b$ and $c = d$, then $ac = bd$.	If $\frac{x}{2} = 7$ and $2 = 2$, then $x = 14$.
Division Property of Equality	If $a = b$ and $c = d$, where c and d are not 0, then $\frac{a}{c} = \frac{b}{d}$.	If $3x = 15$ and $3 = 3$, then $x = 5$.
Zero Product Property	If $ab = 0$, then either $a = 0$ or $b = 0$.	If $x(x - 8) = 0$, then either $x = 0$ or $x - 8 = 0$.

We can now solve equations.

Example *Solve* $4(x + 12) = x^2 + x(7 - x)$ *for* x.

	$4(x + 12) = x^2 + x(7 - x)$
Distributive Property of Multiplication over Addition or Subtraction	$4x + 48 = x^2 + 7x - x^2$
Additive Inverse Property	$4x + 48 = 7x$
Subtraction Property of Equality	$48 = 3x$
Division Property of Equality	$16 = x$
Symmetric Property of Equality	$x = 16$

Solving Special Equations

Equations That Have No Solutions

It is possible to solve an equation accurately and yet obtain a false or meaningless result, such as $0 = 11$, $x = \sqrt{-3}$, or $|y| = -1$. Equations that lead to such results are said to have no real solutions.

Equations That Are Identities

Certain equations, such as $x + 3 = x + 3$, make true statements for all values of the variable. These equations are called ***identities.*** An identity expresses an equivalence that is always valid.

Absolute-Value Equations

An absolute-value equation is solved by isolating the absolute-value expression on one side and then applying the definition of *absolute value.*

Example *Solve* $7 + |2x| = 21$ *for* x.

	$7 + |2x| = 21$	
Subtraction Property of Equality	$|2x| = 14$	
By definition, $|a| = a$ or $|a| = -a$	$2x = 14$ or	$-2x = 14$
Division Property of Equality	$x = 7$	$x = -7$

POLYNOMIAL EXPRESSIONS

A ***monomial*** is a real number, a variable, or a product of real numbers and variables. The expressions -15, $3z$, and $-7x^3y^5z$ are all monomials. The expressions $\frac{-15}{x + 8}$, $3\sqrt{z}$, $3|z|$, and $-7x^3y^5ze^w$ are not monomials.

The ***coefficient*** of a monomial is its numerical factor. For example, the coefficient in $-2.8x^7y^3z$ is -2.8.

The ***degree of a monomial*** is the sum of the exponents of the monomial's variables. The degree of $-2.8x^7y^3z$ is 11 because $7 + 3 + 1 = 11$.

A ***polynomial*** is an expression that is either a monomial or a sum or difference of monomials. Each monomial component of a polynomial is called a ***term.*** The terms of a polynomial are usually arranged in descending order of powers. The ***leading coefficient*** of a polynomial is the coefficient of the first term when the polynomial is written in descending order of powers. For example, the leading coefficient of $-x^3 + 5x + 2$ is -1.

The ***degree of a polynomial*** is equal to the greatest of the degrees of the polynomial's terms. Polynomials are classified by their degree and by their number of terms. The polynomial $-x^3 + 5x + 2$ is classified as a cubic (degree-three) trinomial. The polynomial $y^2 - 17$ is an example of a quadratic (degree-two) binomial.

SCIENTIFIC NOTATION

Scientific notation is an efficient way of expressing very large and very small numbers. To express a number in scientific notation, we rewrite it in the form $c \times 10^n$, where $0 < |c| < 10$ and n is an integer. Numbers written in scientific notation can be multiplied or divided exactly like other exponential expressions. They may be added or subtracted, however, only after being rewritten so that they contain the same power of 10.

Example *Write* 186,000 *in scientific notation.*

$$186{,}000 = 1.86 \times 100{,}000$$
$$= 1.86 \times 10^5$$

Notice that we find the correct coefficient, 1.86, by moving the decimal point 5 places to the left and that 5 is also the correct power of 10.

Example *Write* 0.0000051 *in scientific notation.*

$$0.0000051 = \frac{5.1}{1{,}000{,}000}$$
$$= 5.1 \times 10^{-6}$$

Notice that we find the correct coefficient, 5.1, by moving the decimal point 6 places to the right and that -6 is the correct power of 10.

In general, then, the power of 10 is numerically equal to the number of places the decimal point must be moved to give the correct coefficient. The exponent is positive if the decimal point is moved to the left; the exponent is negative if the decimal point is moved to the right.

All scientific calculators allow you to enter numbers in scientific notation. The key used to enter the power of 10 may be labeled [EEX], [EE], or [EXP].

Example *Find the product* $(-3.55 \times 10^{12})(6.00 \times 10^{-23})$.

Enter the following on a calculator.

3.55 [±] [EE] 12 [×] 6 [EE] 23 [±] [=]

The display should resemble [-2.13 -10]. The product is -2.13×10^{-10}.

SPECIAL FORMULAS FROM GEOMETRY

Perimeter Formulas

To find the perimeter of a polygon, we add the lengths of all its sides. The perimeter of a circle is called the circle's ***circumference.*** We calculate a circumference by using the formula $C = \pi d$ (where d is the diameter of the circle) or $C = 2\pi r$ (where r is the radius of the circle).

Area Formulas

The formulas for area that we will use in this book are summarized in the following diagrams. Areas are always expressed in square units.

Triangle

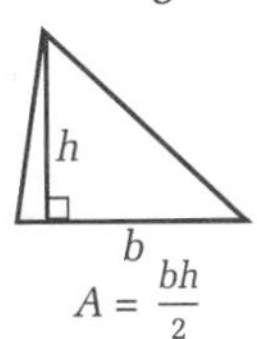

$A = \frac{bh}{2}$

Rectangle

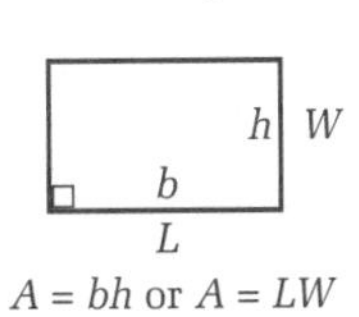

$A = bh$ or $A = LW$

Parallelogram

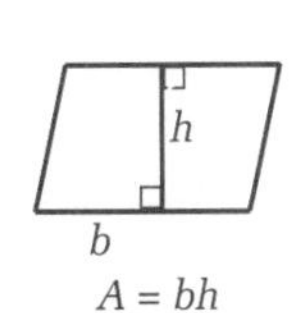

$A = bh$

Square

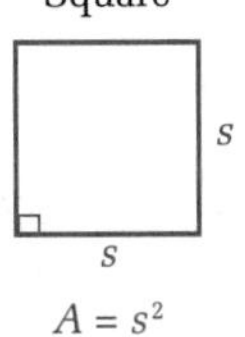

$A = s^2$

Trapezoid

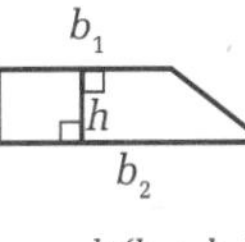

$A = \frac{h\,(b_1 + b_2)}{2}$

Circle

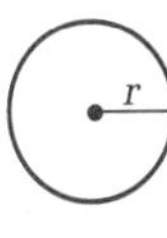

$A = \pi r^2$

Rectangular Box

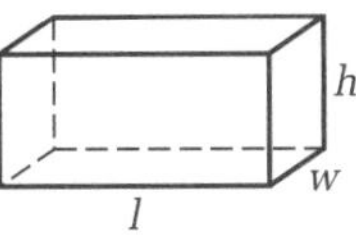

Surface Area = $2lw + 2lh + 2wh$

Sphere

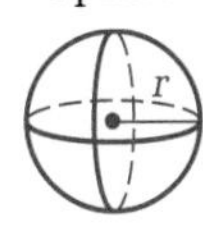

Surface Area = $4\pi r^2$

Volume Formulas

The formulas in the following diagrams are used to calculate volume, the space enclosed by a figure. Volume is always measured in cubic units.

Rectangular Prism

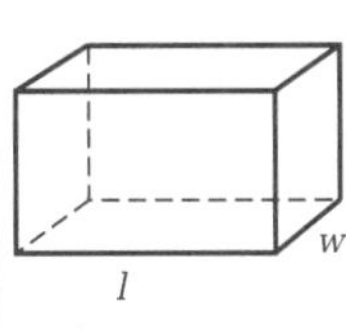

$V = lwh$

Pyramid

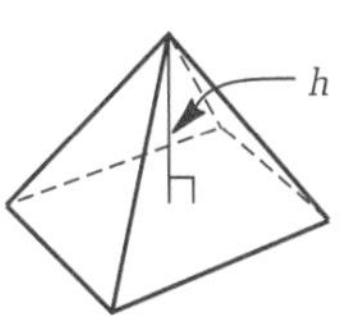

$V = \frac{Bh}{3}$, where B is the area of the base

Sphere

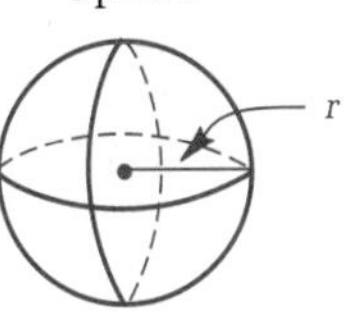

$V = \frac{4}{3}\pi r^3$

Right Circular Cylinder

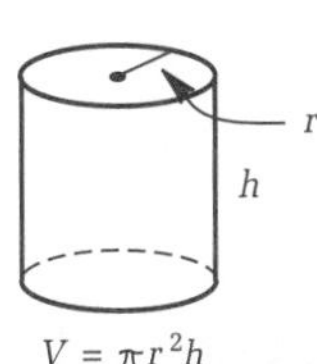

$V = \pi r^2 h$

Right Circular Cone

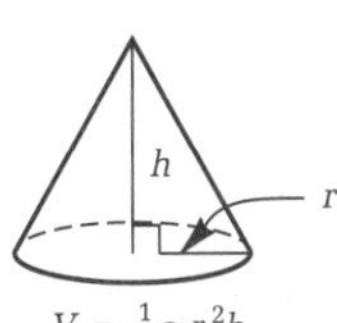

$V = \frac{1}{3}\pi r^2 h$

Normal Distribution Table

SD's from Mean	Cumulative Area	Percentile Rank	SD's from Mean	Cumulative Area	Percentile Rank
−3.4	0.0003	0	0.0	0.5000	50
−3.3	0.0005	0	0.1	0.5398	53
−3.2	0.0007	0	0.2	0.5793	57
−3.1	0.0010	0	0.3	0.6179	61
−3.0	0.0013	0	0.4	0.6554	65
−2.9	0.0019	0	0.5	0.6915	69
−2.8	0.0026	0	0.6	0.7257	72
−2.7	0.0035	0	0.7	0.7580	75
−2.6	0.0047	0	0.8	0.7881	78
−2.5	0.0062	0	0.9	0.8159	81
−2.4	0.0082	0	1.0	0.8413	84
−2.3	0.0107	1	1.1	0.8643	86
−2.2	0.0139	1	1.2	0.8849	88
−2.1	0.0179	1	1.3	0.9032	90
−2.0	0.0228	2	1.4	0.9192	91
−1.9	0.0287	2	1.5	0.9332	93
−1.8	0.0359	3	1.6	0.9452	94
−1.7	0.0446	4	1.7	0.9554	95
−1.6	0.0548	5	1.8	0.9641	96
−1.5	0.0668	6	1.9	0.9713	97
−1.4	0.0808	8	2.0	0.9772	97
−1.3	0.0968	9	2.1	0.9821	98
−1.2	0.1151	11	2.2	0.9861	98
−1.1	0.1357	13	2.3	0.9893	98
−1.0	0.1587	15	2.4	0.9918	99
−0.9	0.1841	18	2.5	0.9938	99
−0.8	0.2119	21	2.6	0.9953	99
−0.7	0.2420	24	2.7	0.9965	99
−0.6	0.2743	27	2.8	0.9974	99
−0.5	0.3085	30	2.9	0.9981	99
−0.4	0.3446	34	3.0	0.9987	99
−0.3	0.3821	38	3.1	0.9990	99
−0.2	0.4207	42	3.2	0.9993	99
−0.1	0.4602	46	3.3	0.9995	99
−0.0	0.5000	50	3.4	0.9997	99

Table of Squares and Square Roots

n	n^2	$\sqrt{n}$	n	n^2	$\sqrt{n}$	n	n^2	$\sqrt{n}$
1	1	1.000	**51**	2,601	7.141	**101**	10,201	10.050
2	4	1.414	**52**	2,704	7.211	**102**	10,404	10.100
3	9	1.732	**53**	2,809	7.280	**103**	10,609	10.149
4	16	2.000	**54**	2,916	7.348	**104**	10,816	10.198
5	25	2.236	**55**	3,025	7.416	**105**	11,025	10.247
6	36	2.449	**56**	3,136	7.483	**106**	11,236	10.296
7	49	2.646	**57**	3,249	7.550	**107**	11,449	10.344
8	64	2.828	**58**	3,364	7.616	**108**	11,664	10.392
9	81	3.000	**59**	3,481	7.681	**109**	11,881	10.440
10	100	3.162	**60**	3,600	7.746	**110**	12,100	10.488
11	121	3.317	**61**	3,721	7.810	**111**	12,321	10.536
12	144	3.464	**62**	3,844	7.874	**112**	12,544	10.583
13	169	3.606	**63**	3,969	7.937	**113**	12,769	10.630
14	196	3.742	**64**	4,096	8.000	**114**	12,996	10.677
15	225	3.873	**65**	4,225	8.062	**115**	13,225	10.724
16	256	4.000	**66**	4,356	8.124	**116**	13,456	10.770
17	289	4.123	**67**	4,489	8.185	**117**	13,689	10.817
18	324	4.243	**68**	4,624	8.246	**118**	13,924	10.863
19	361	4.359	**69**	4,761	8.307	**119**	14,161	10.909
20	400	4.472	**70**	4,900	8.367	**120**	14,400	10.954
21	441	4.583	**71**	5,041	8.426	**121**	14,641	11.000
22	484	4.690	**72**	5,184	8.485	**122**	14,884	11.045
23	529	4.796	**73**	5,329	8.544	**123**	15,129	11.091
24	576	4.899	**74**	5,476	8.602	**124**	15,376	11.136
25	625	5.000	**75**	5,625	8.660	**125**	15,625	11.180
26	676	5.099	**76**	5,776	8.718	**126**	15,876	11.225
27	729	5.196	**77**	5,929	8.775	**127**	16,129	11.269
28	784	5.292	**78**	6,084	8.832	**128**	16,384	11.314
29	841	5.385	**79**	6,241	8.888	**129**	16,641	11.358
30	900	5.477	**80**	6,400	8.944	**130**	16,900	11.402
31	961	5.568	**81**	6,561	9.000	**131**	17,161	11.446
32	1,024	5.657	**82**	6,724	9.055	**132**	17,424	11.489
33	1,089	5.745	**83**	6,889	9.110	**133**	17,689	11.533
34	1,156	5.831	**84**	7,056	9.165	**134**	17,956	11.576
35	1,225	5.916	**85**	7,225	9.220	**135**	18,225	11.619
36	1,296	6.000	**86**	7,396	9.274	**136**	18,496	11.662
37	1,369	6.083	**87**	7,569	9.327	**137**	18,769	11.705
38	1,444	6.164	**88**	7,744	9.381	**138**	19,044	11.747
39	1,521	6.245	**89**	7,921	9.434	**139**	19,321	11.790
40	1,600	6.325	**90**	8,100	9.487	**140**	19,600	11.832
41	1,681	6.403	**91**	8,281	9.539	**141**	19,881	11.874
42	1,764	6.481	**92**	8,464	9.592	**142**	20,164	11.916
43	1,849	6.557	**93**	8,649	9.644	**143**	20,449	11.958
44	1,936	6.633	**94**	8,836	9.695	**144**	20,736	12.000
45	2,025	6.708	**95**	9,025	9.747	**145**	21,025	12.042
46	2,116	6.782	**96**	9,216	9.798	**146**	21,316	12.083
47	2,209	6.856	**97**	9,409	9.849	**147**	21,609	12.124
48	2,304	6.928	**98**	9,604	9.899	**148**	21,904	12.166
49	2,401	7.000	**99**	9,801	9.950	**149**	22,201	12.207
50	2,500	7.071	**100**	10,000	10.000	**150**	22,500	12.247

COMMON LOGARITHMS OF NUMBERS

x	0	1	2	3	4	5	6	7	8	9
1.0	.0000	.0043	.0086	.0128	.0170	.0212	.0253	.0294	.0334	.0374
1.1	.0414	.0453	.0492	.0531	.0569	.0607	.0645	.0682	.0719	.0755
1.2	.0792	.0828	.0864	.0899	.0934	.0969	.1004	.1038	.1072	.1106
1.3	.1139	.1173	.1206	.1239	.1271	.1303	.1335	.1367	.1399	.1430
1.4	.1461	.1492	.1523	.1553	.1584	.1614	.1644	.1673	.1703	.1732
1.5	.1761	.1790	.1818	.1847	.1875	.1903	.1931	.1959	.1987	.2014
1.6	.2041	.2068	.2095	.2122	.2148	.2175	.2201	.2227	.2253	.2279
1.7	.2304	.2330	.2355	.2380	.2405	.2430	.2455	.2480	.2504	.2529
1.8	.2553	.2577	.2601	.2625	.2648	.2672	.2695	.2718	.2742	.2765
1.9	.2788	.2810	.2833	.2856	.2878	.2900	.2923	.2945	.2967	.2989
2.0	.3010	.3032	.3054	.3075	.3096	.3118	.3139	.3160	.3181	.3201
2.1	.3222	.3243	.3263	.3284	.3304	.3324	.3345	.3365	.3385	.3404
2.2	.3424	.3444	.3464	.3483	.3502	.3522	.3541	.3560	.3579	.3598
2.3	.3617	.3636	.3655	.3674	.3692	.3711	.3729	.3747	.3766	.3784
2.4	.3802	.3820	.3838	.3856	.3874	.3892	.3909	.3927	.3945	.3962
2.5	.3979	.3997	.4014	.4031	.4048	.4065	.4082	.4099	.4116	.4133
2.6	.4150	.4166	.4183	.4200	.4216	.4232	.4249	.4265	.4281	.4298
2.7	.4314	.4330	.4346	.4362	.4378	.4393	.4409	.4425	.4440	.4456
2.8	.4472	.4487	.4502	.4518	.4533	.4548	.4564	.4579	.4594	.4609
2.9	.4624	.4639	.4654	.4669	.4683	.4698	.4713	.4728	.4742	.4757
3.0	.4771	.4786	.4800	.4814	.4829	.4843	.4857	.4871	.4886	.4900
3.1	.4914	.4928	.4942	.4955	.4969	.4983	.4997	.5011	.5024	.5038
3.2	.5051	.5065	.5079	.5092	.5105	.5119	.5132	.5145	.5159	.5172
3.3	.5185	.5198	.5211	.5224	.5237	.5250	.5263	.5276	.5289	.5302
3.4	.5315	.5328	.5340	.5353	.5366	.5378	.5391	.5403	.5416	.5428
3.5	.5441	.5453	.5465	.5478	.5490	.5502	.5514	.5527	.5539	.5551
3.6	.5563	.5575	.5587	.5599	.5611	.5623	.5635	.5647	.5658	.5670
3.7	.5682	.5694	.5705	.5717	.5729	.5740	.5752	.5763	.5775	.5786
3.8	.5798	.5809	.5821	.5832	.5843	.5855	.5866	.5877	.5888	.5899
3.9	.5911	.5922	.5933	.5944	.5955	.5966	.5977	.5988	.5999	.6010
4.0	.6021	.6031	.6042	.6053	.6064	.6075	.6085	.6096	.6107	.6117
4.1	.6128	.6138	.6149	.6160	.6170	.6180	.6191	.6201	.6212	.6222
4.2	.6232	.6243	.6253	.6263	.6274	.6284	.6294	.6304	.6314	.6325
4.3	.6335	.6345	.6355	.6365	.6375	.6385	.6395	.6405	.6415	.6425
4.4	.6435	.6444	.6454	.6464	.6474	.6484	.6493	.6503	.6513	.6522
4.5	.6532	.6542	.6551	.6561	.6571	.6580	.6590	.6599	.6609	.6618
4.6	.6628	.6637	.6646	.6656	.6665	.6675	.6684	.6693	.6702	.6712
4.7	.6721	.6730	.6739	.6749	.6758	.6767	.6776	.6785	.6794	.6803
4.8	.6812	.6821	.6830	.6839	.6848	.6857	.6866	.6875	.6884	.6893
4.9	.6902	.6911	.6920	.6928	.6937	.6946	.6955	.6964	.6972	.6981
5.0	.6990	.6998	.7007	.7016	.7024	.7033	.7042	.7050	.7059	.7067
5.1	.7076	.7084	.7093	.7101	.7110	.7118	.7126	.7135	.7143	.7152
5.2	.7160	.7168	.7177	.7185	.7193	.7202	.7210	.7218	.7226	.7235
5.3	.7243	.7251	.7259	.7267	.7275	.7284	.7292	.7300	.7308	.7316
5.4	.7324	.7332	.7340	.7348	.7356	.7364	.7372	.7380	.7388	.7396

x	0	1	2	3	4	5	6	7	8	9
5.5	.7404	.7412	.7419	.7427	.7435	.7443	.7451	.7459	.7466	.7474
5.6	.7482	.7490	.7497	.7505	.7513	.7520	.7528	.7536	.7543	.7551
5.7	.7559	.7566	.7574	.7582	.7589	.7597	.7604	.7612	.7619	.7627
5.8	.7634	.7642	.7649	.7657	.7664	.7672	.7679	.7686	.7694	.7701
5.9	.7709	.7716	.7723	.7731	.7738	.7745	.7752	.7760	.7767	.7774
6.0	.7782	.7789	.7796	.7803	.7810	.7818	.7825	.7832	.7839	.7846
6.1	.7853	.7860	.7868	.7875	.7882	.7889	.7896	.7903	.7910	.7917
6.2	.7924	.7931	.7938	.7945	.7952	.7959	.7966	.7973	.7980	.7987
6.3	.7993	.8000	.8007	.8014	.8021	.8028	.8035	.8041	.8048	.8055
6.4	.8062	.8069	.8075	.8082	.8089	.8096	.8102	.8109	.8116	.8122
6.5	.8129	.8136	.8142	.8149	.8156	.8162	.8169	.8176	.8182	.8189
6.6	.8195	.8202	.8209	.8215	.8222	.8228	.8235	.8241	.8248	.8254
6.7	.8261	.8267	.8274	.8280	.8287	.8293	.8299	.8306	.8312	.8319
6.8	.8325	.8331	.8338	.8344	.8351	.8357	.8363	.8370	.8376	.8382
6.9	.8388	.8395	.8401	.8407	.8414	.8420	.8426	.8432	.8439	.8445
7.0	.8451	.8457	.8463	.8470	.8476	.8482	.8488	.8494	.8500	.8506
7.1	.8513	.8519	.8525	.8531	.8537	.8543	.8549	.8555	.8561	.8567
7.2	.8573	.8579	.8585	.8591	.8597	.8603	.8609	.8615	.8621	.8627
7.3	.8633	.8639	.8645	.8651	.8657	.8663	.8669	.8675	.8681	.8686
7.4	.8692	.8698	.8704	.8710	.8716	.8722	.8727	.8733	.8739	.8745
7.5	.8751	.8756	.8762	.8768	.8774	.8779	.8785	.8791	.8797	.8802
7.6	.8808	.8814	.8820	.8825	.8831	.8837	.8842	.8848	.8854	.8859
7.7	.8865	.8871	.8876	.8882	.8887	.8893	.8899	.8904	.8910	.8915
7.8	.8921	.8927	.8932	.8938	.8943	.8949	.8954	.8960	.8965	.8971
7.9	.8976	.8982	.8987	.8993	.8998	.9004	.9009	.9015	.9020	.9025
8.0	.9031	.9036	.9042	.9047	.9053	.9058	.9063	.9069	.9074	.9079
8.1	.9085	.9090	.9096	.9101	.9106	.9112	.9117	.9122	.9128	.9133
8.2	.9138	.9143	.9149	.9154	.9159	.9165	.9170	.9175	.9180	.9186
8.3	.9191	.9196	.9201	.9206	.9212	.9217	.9222	.9227	.9232	.9238
8.4	.9243	.9248	.9253	.9258	.9263	.9269	.9274	.9279	.9284	.9289
8.5	.9294	.9299	.9304	.9309	.9315	.9320	.9325	.9330	.9335	.9340
8.6	.9345	.9350	.9355	.9360	.9365	.9370	.9375	.9380	.9385	.9390
8.7	.9395	.9400	.9405	.9410	.9415	.9420	.9425	.9430	.9435	.9440
8.8	.9445	.9450	.9455	.9460	.9465	.9469	.9474	.9479	.9484	.9489
8.9	.9494	.9499	.9504	.9509	.9513	.9518	.9523	.9528	.9533	.9538
9.0	.9542	.9547	.9552	.9557	.9562	.9566	.9571	.9576	.9581	.9586
9.1	.9590	.9595	.9600	.9605	.9609	.9614	.9619	.9624	.9628	.9633
9.2	.9638	.9643	.9647	.9652	.9657	.9661	.9666	.9671	.9675	.9680
9.3	.9685	.9689	.9694	.9699	.9703	.9708	.9713	.9717	.9722	.9727
9.4	.9731	.9736	.9741	.9745	.9750	.9754	.9759	.9763	.9768	.9773
9.5	.9777	.9782	.9786	.9791	.9795	.9800	.9805	.9809	.9814	.9818
9.6	.9823	.9827	.9832	.9836	.9841	.9845	.9850	.9854	.9859	.9863
9.7	.9868	.9872	.9877	.9881	.9886	.9890	.9894	.9899	.9903	.9908
9.8	.9912	.9917	.9921	.9926	.9930	.9934	.9939	.9943	.9948	.9952
9.9	.9956	.9961	.9965	.9969	.9974	.9978	.9983	.9987	.9991	.9996

Natural Logarithms of Numbers

x	0	1	2	3	4	5	6	7	8	9
1.0	0.0000	0.0100	0.0198	0.0296	0.0392	0.0488	0.0583	0.0677	0.0770	0.0862
1.1	0.0953	0.1044	0.1133	0.1222	0.1310	0.1398	0.1484	0.1570	0.1655	0.1740
1.2	0.1823	0.1906	0.1989	0.2070	0.2151	0.2231	0.2311	0.2390	0.2469	0.2546
1.3	0.2624	0.2700	0.2776	0.2852	0.2927	0.3001	0.3075	0.3148	0.3221	0.3293
1.4	0.3365	0.3436	0.3507	0.3577	0.3646	0.3716	0.3784	0.3853	0.3920	0.3988
1.5	0.4055	0.4121	0.4187	0.4253	0.4318	0.4383	0.4447	0.4511	0.4574	0.4637
1.6	0.4700	0.4762	0.4824	0.4886	0.4947	0.5008	0.5068	0.5128	0.5188	0.5247
1.7	0.5306	0.5365	0.5423	0.5481	0.5539	0.5596	0.5653	0.5710	0.5766	0.5822
1.8	0.5878	0.5933	0.5988	0.6043	0.6098	0.6152	0.6206	0.6259	0.6313	0.6366
1.9	0.6419	0.6471	0.6523	0.6575	0.6627	0.6678	0.6729	0.6780	0.6831	0.6881
2.0	0.6931	0.6981	0.7031	0.7080	0.7130	0.7178	0.7227	0.7275	0.7324	0.7372
2.1	0.7419	0.7467	0.7514	0.7561	0.7608	0.7655	0.7701	0.7747	0.7793	0.7839
2.2	0.7885	0.7930	0.7975	0.8020	0.8065	0.8109	0.8154	0.8198	0.8242	0.8286
2.3	0.8329	0.8372	0.8416	0.8459	0.8502	0.8544	0.8587	0.8629	0.8671	0.8713
2.4	0.8755	0.8796	0.8838	0.8879	0.8920	0.8961	0.9002	0.9042	0.9083	0.9123
2.5	0.9163	0.9203	0.9243	0.9282	0.9322	0.9361	0.9400	0.9439	0.9478	0.9517
2.6	0.9555	0.9594	0.9632	0.9670	0.9708	0.9746	0.9783	0.9821	0.9858	0.9895
2.7	0.9933	0.9969	1.0006	1.0043	1.0080	1.0116	1.0152	1.0188	1.0225	1.0260
2.8	1.0296	1.0332	1.0367	1.0403	1.0438	1.0473	1.0508	1.0543	1.0578	1.0613
2.9	1.0647	1.0682	1.0716	1.0750	1.0784	1.0818	1.0852	1.0886	1.0919	1.0953
3.0	1.0986	1.1019	1.1053	1.1086	1.1119	1.1151	1.1184	1.1217	1.1249	1.1282
3.1	1.1314	1.1346	1.1378	1.1410	1.1442	1.1474	1.1506	1.1537	1.1569	1.1600
3.2	1.1632	1.1663	1.1694	1.1725	1.1756	1.1787	1.1817	1.1848	1.1878	1.1909
3.3	1.1939	1.1970	1.2000	1.2030	1.2060	1.2090	1.2119	1.2149	1.2179	1.2208
3.4	1.2238	1.2267	1.2296	1.2326	1.2355	1.2384	1.2413	1.2442	1.2470	1.2499
3.5	1.2528	1.2556	1.2585	1.2613	1.2641	1.2669	1.2698	1.2726	1.2754	1.2782
3.6	1.2809	1.2837	1.2865	1.2892	1.2920	1.2947	1.2975	1.3002	1.3029	1.3056
3.7	1.3083	1.3110	1.3137	1.3164	1.3191	1.3218	1.3244	1.3271	1.3297	1.3324
3.8	1.3350	1.3376	1.3403	1.3429	1.3455	1.3481	1.3507	1.3533	1.3558	1.3584
3.9	1.3610	1.3635	1.3661	1.3686	1.3712	1.3737	1.3762	1.3788	1.3813	1.3838
4.0	1.3863	1.3888	1.3913	1.3938	1.3962	1.3987	1.4012	1.4036	1.4061	1.4085
4.1	1.4110	1.4134	1.4159	1.4183	1.4207	1.4231	1.4255	1.4279	1.4303	1.4327
4.2	1.4351	1.4375	1.4398	1.4422	1.4446	1.4469	1.4493	1.4516	1.4540	1.4563
4.3	1.4586	1.4609	1.4633	1.4656	1.4679	1.4702	1.4725	1.4748	1.4770	1.4793
4.4	1.4816	1.4839	1.4861	1.4884	1.4907	1.4929	1.4952	1.4974	1.4996	1.5019
4.5	1.5041	1.5063	1.5085	1.5107	1.5129	1.5151	1.5173	1.5195	1.5217	1.5239
4.6	1.5261	1.5282	1.5304	1.5326	1.5347	1.5369	1.5390	1.5412	1.5433	1.5454
4.7	1.5476	1.5497	1.5518	1.5539	1.5560	1.5581	1.5602	1.5623	1.5644	1.5665
4.8	1.5686	1.5707	1.5728	1.5748	1.5769	1.5790	1.5810	1.5831	1.5851	1.5872
4.9	1.5892	1.5913	1.5933	1.5953	1.5974	1.5994	1.6014	1.6034	1.6054	1.6074
5.0	1.6094	1.6114	1.6134	1.6154	1.6174	1.6194	1.6214	1.6233	1.6253	1.6273
5.1	1.6292	1.6312	1.6332	1.6351	1.6371	1.6390	1.6409	1.6429	1.6448	1.6467
5.2	1.6487	1.6506	1.6525	1.6544	1.6563	1.6582	1.6601	1.6620	1.6639	1.6658
5.3	1.6677	1.6696	1.6715	1.6734	1.6752	1.6771	1.6790	1.6808	1.6827	1.6845
5.4	1.6864	1.6882	1.6901	1.6919	1.6938	1.6956	1.6974	1.6993	1.7001	1.7029

x	0	1	2	3	4	5	6	7	8	9
5.5	1.7047	1.7066	1.7084	1.7102	1.7120	1.7138	1.7156	1.7174	1.7192	1.7210
5.6	1.7228	1.7246	1.7263	1.7281	1.7299	1.7317	1.7334	1.7352	1.7370	1.7387
5.7	1.7405	1.7422	1.7440	1.7457	1.7475	1.7492	1.7509	1.7527	1.7544	1.7561
5.8	1.7579	1.7596	1.7613	1.7630	1.7647	1.7664	1.7682	1.7699	1.7716	1.7733
5.9	1.7750	1.7766	1.7783	1.7800	1.7817	1.7834	1.7851	1.7867	1.7884	1.7901
6.0	1.7918	1.7934	1.7951	1.7967	1.7984	1.8001	1.8017	1.8034	1.8050	1.8066
6.1	1.8083	1.8099	1.8116	1.8132	1.8148	1.8165	1.8181	1.8197	1.8213	1.8229
6.2	1.8245	1.8262	1.8278	1.8294	1.8310	1.8326	1.8342	1.8358	1.8374	1.8390
6.3	1.8405	1.8421	1.8437	1.8453	1.8469	1.8485	1.8500	1.8516	1.8532	1.8547
6.4	1.8563	1.8579	1.8594	1.8610	1.8625	1.8641	1.8656	1.8672	1.8687	1.8703
6.5	1.8718	1.8733	1.8749	1.8764	1.8779	1.8795	1.8810	1.8825	1.8840	1.8856
6.6	1.8871	1.8886	1.8901	1.8916	1.8931	1.8946	1.8961	1.8976	1.8991	1.9006
6.7	1.9021	1.9036	1.9051	1.9066	1.9081	1.9095	1.9110	1.9125	1.9140	1.9155
6.8	1.9169	1.9184	1.9199	1.9213	1.9228	1.9242	1.9257	1.9272	1.9286	1.9301
6.9	1.9315	1.9330	1.9344	1.9359	1.9373	1.9387	1.9402	1.9416	1.9430	1.9445
7.0	1.9459	1.9473	1.9488	1.9502	1.9516	1.9530	1.9544	1.9559	1.9573	1.9587
7.1	1.9601	1.9615	1.9629	1.9643	1.9657	1.9671	1.9685	1.9699	1.9713	1.9727
7.2	1.9741	1.9755	1.9769	1.9782	1.9796	1.9810	1.9824	1.9838	1.9851	1.9865
7.3	1.9879	1.9892	1.9906	1.9920	1.9933	1.9947	1.9961	1.9974	1.9988	2.0001
7.4	2.0015	2.0028	2.0042	2.0055	2.0069	2.0082	2.0096	2.0109	2.0122	2.0136
7.5	2.0149	2.0162	2.0176	2.0189	2.0202	2.0215	2.0229	2.0242	2.0255	2.0268
7.6	2.0281	2.0295	2.0308	2.0321	2.0334	2.0347	2.0360	2.0373	2.0386	2.0399
7.7	2.0412	2.0425	2.0438	2.0451	2.0464	2.0477	2.0490	2.0503	2.0516	2.0528
7.8	2.0541	2.0554	2.0567	2.0580	2.0592	2.0605	2.0618	2.0631	2.0643	2.0656
7.9	2.0669	2.0681	2.0694	2.0707	2.0719	2.0732	2.0744	2.0757	2.0769	2.0782
8.0	2.0794	2.0807	2.0819	2.0832	2.0844	2.0857	2.0869	2.0882	2.0894	2.0906
8.1	2.0919	2.0931	2.0943	2.0956	2.0968	2.0980	2.0992	2.1005	2.1017	2.1029
8.2	2.1041	2.1054	2.1066	2.1078	2.1090	2.1102	2.1114	2.1126	2.1138	2.1150
8.3	2.1163	2.1175	2.1187	2.1199	2.1211	2.1223	2.1235	2.1247	2.1258	2.1270
8.4	2.1282	2.1294	2.1306	2.1318	2.1330	2.1342	2.1353	2.1365	2.1377	2.1389
8.5	2.1401	2.1412	2.1424	2.1436	2.1448	2.1459	2.1471	2.1483	2.1494	2.1506
8.6	2.1518	2.1529	2.1541	2.1552	2.1564	2.1576	2.1587	2.1599	2.1610	2.1622
8.7	2.1633	2.1645	2.1656	2.1668	2.1679	2.1691	2.1702	2.1713	2.1725	2.1736
8.8	2.1748	2.1759	2.1770	2.1782	2.1793	2.1804	2.1815	2.1827	2.1838	2.1849
8.9	2.1861	2.1872	2.1883	2.1894	2.1905	2.1917	2.1928	2.1939	2.1950	2.1961
9.0	2.1972	2.1983	2.1994	2.2006	2.2017	2.2028	2.2039	2.2050	2.2061	2.2072
9.1	2.2083	2.2094	2.2105	2.2116	2.2127	2.2138	2.2148	2.2159	2.2170	2.2181
9.2	2.2192	2.2203	2.2214	2.2225	2.2235	2.2246	2.2257	2.2268	2.2279	2.2289
9.3	2.2300	2.2311	2.2322	2.2332	2.2343	2.2354	2.2364	2.2375	2.2386	2.2396
9.4	2.2407	2.2418	2.2428	2.2439	2.2450	2.2460	2.2471	2.2481	2.2492	2.2502
9.5	2.2513	2.2523	2.2534	2.2544	2.2555	2.2565	2.2576	2.2586	2.2597	2.2607
9.6	2.2618	2.2628	2.2638	2.2649	2.2659	2.2670	2.2680	2.2690	2.2701	2.2711
9.7	2.2721	2.2732	2.2742	2.2752	2.2762	2.2773	2.2783	2.2793	2.2803	2.2814
9.8	2.2824	2.2834	2.2844	2.2854	2.2865	2.2875	2.2885	2.2895	2.2905	2.2915
9.9	2.2925	2.2935	2.2946	2.2956	2.2966	2.2976	2.2986	2.2996	2.3006	2.3016

Trigonometric Functions

deg.	rad.	sin	cos	tan	cot	sec	csc		
0°00′	**.0000**	.0000	1.0000	.0000	—	1.000	—	**1.5708**	**90°00′**
10′	**.0029**	.0029	1.0000	.0029	343.8	1.000	343.8	**1.5679**	**50′**
20′	**.0058**	.0058	1.0000	.0058	171.9	1.000	171.9	**1.5650**	**40′**
30′	**.0087**	.0087	1.0000	.0087	114.6	1.000	114.6	**1.5621**	**30′**
40′	**.0116**	.0116	.9999	.0116	85.94	1.000	85.95	**1.5592**	**20′**
50′	**.0145**	.0145	.9999	.0145	68.75	1.000	68.76	**1.5563**	**10′**
1°00′	**.0175**	.0175	.9998	.0175	57.29	1.000	57.30	**1.5533**	**89°00′**
10′	**.0204**	.0204	.9998	.0204	49.10	1.000	49.11	**1.5504**	**50′**
20′	**.0233**	.0233	.9997	.0233	42.96	1.000	42.98	**1.5475**	**40′**
30′	**.0262**	.0262	.9997	.0262	38.19	1.000	38.20	**1.5446**	**30′**
40′	**.0291**	.0291	.9996	.0291	34.37	1.000	34.38	**1.5417**	**20′**
50′	**.0320**	.0320	.9995	.0320	31.24	1.001	31.26	**1.5388**	**10′**
2°00′	**.0349**	.0349	.9994	.0349	28.64	1.001	28.65	**1.5359**	**88°00′**
10′	**.0378**	.0378	.9993	.0378	26.43	1.001	26.45	**1.5330**	**50′**
20′	**.0407**	.0407	.9992	.0407	24.54	1.001	24.56	**1.5301**	**40′**
30′	**.0436**	.0436	.9990	.0437	22.90	1.001	22.93	**1.5272**	**30′**
40′	**.0465**	.0465	.9989	.0466	21.47	1.001	21.49	**1.5243**	**20′**
50′	**.0495**	.0494	.9988	.0495	20.21	1.001	20.23	**1.5213**	**10′**
3°00′	**.0524**	.0523	.9986	.0524	19.08	1.001	19.11	**1.5184**	**87°00′**
10′	**.0553**	.0552	.9985	.0553	18.07	1.002	18.10	**1.5155**	**50′**
20′	**.0582**	.0581	.9983	.0582	17.17	1.002	17.20	**1.5126**	**40′**
30′	**.0611**	.0610	.9981	.0612	16.35	1.002	16.38	**1.5097**	**30′**
40′	**.0640**	.0640	.9980	.0641	15.60	1.002	15.64	**1.5068**	**20′**
50′	**.0669**	.0669	.9978	.0670	14.92	1.002	14.96	**1.5039**	**10′**
4°00′	**.0698**	.0698	.9976	.0699	14.30	1.002	14.34	**1.5010**	**86°00′**
10′	**.0727**	.0727	.9974	.0729	13.73	1.003	13.76	**1.4981**	**50′**
20′	**.0756**	.0756	.9971	.0758	13.20	1.003	13.23	**1.4952**	**40′**
30′	**.0785**	.0785	.9969	.0787	12.71	1.003	12.75	**1.4923**	**30′**
40′	**.0814**	.0814	.9967	.0816	12.25	1.003	12.29	**1.4893**	**20′**
50′	**.0844**	.0843	.9964	.0846	11.83	1.004	11.87	**1.4864**	**10′**
5°00′	**.0873**	.0872	.9962	.0875	11.43	1.004	11.47	**1.4835**	**85°00′**
10′	**.0902**	.0901	.9959	.0904	11.06	1.004	11.10	**1.4806**	**50′**
20′	**.0931**	.0929	.9957	.0934	10.71	1.004	10.76	**1.4777**	**40′**
30′	**.0960**	.0958	.9954	.0963	10.39	1.005	10.43	**1.4748**	**30′**
40′	**.0989**	.0987	.9951	.0992	10.08	1.005	10.13	**1.4719**	**20′**
50′	**.1018**	.1016	.9948	.1022	9.788	1.005	9.839	**1.4690**	**10′**
6°00′	**.1047**	.1045	.9945	.1051	9.514	1.006	9.567	**1.4661**	**84°00′**
10′	**.1076**	.1074	.9942	.1080	9.255	1.006	9.309	**1.4632**	**50′**
20′	**.1105**	.1103	.9939	.1110	9.010	1.006	9.065	**1.4603**	**40′**
30′	**.1134**	.1132	.9936	.1139	8.777	1.006	8.834	**1.4573**	**30′**
40′	**.1164**	.1161	.9932	.1169	8.556	1.007	8.614	**1.4544**	**20′**
50′	**.1193**	.1190	.9929	.1198	8.345	1.007	8.405	**1.4515**	**10′**
7°00′	**.1222**	.1219	.9925	.1228	8.144	1.008	8.206	**1.4486**	**83°00′**
10′	**.1251**	.1248	.9922	.1257	7.953	1.008	8.016	**1.4457**	**50′**
20′	**.1280**	.1276	.9918	.1287	7.770	1.008	7.834	**1.4428**	**40′**
30′	**.1309**	.1305	.9914	.1317	7.596	1.009	7.661	**1.4399**	**82°30′**
		cos	**sin**	**cot**	**tan**	**csc**	**sec**	**rad.**	**deg.**

deg.	rad.	sin	cos	tan	cot	sec	csc		
7°30′	.1309	.1305	.9914	.1317	7.596	1.009	7.661	1.4399	30′
40′	.1338	.1334	.9911	.1346	7.429	1.009	7.496	1.4370	20′
50′	.1367	.1363	.9907	.1376	7.269	1.009	7.337	1.4341	10′
8°00′	.1396	.1392	.9903	.1405	7.115	1.010	7.185	1.4312	82°00′
10′	.1425	.1421	.9899	.1435	6.968	1.010	7.040	1.4283	50′
20′	.1454	.1449	.9894	.1465	6.827	1.011	6.900	1.4254	40′
30′	.1484	.1478	.9890	.1495	6.691	1.011	6.765	1.4224	30′
40′	.1513	.1507	.9886	.1524	6.561	1.012	6.636	1.4195	20′
50′	.1542	.1536	.9881	.1554	6.435	1.012	6.512	1.4166	10′
9°00′	.1571	.1564	.9877	.1584	6.314	1.012	6.392	1.4137	81°00′
10′	.1600	.1593	.9872	.1614	6.197	1.013	6.277	1.4108	50′
20′	.1629	.1622	.9868	.1644	6.084	1.013	6.166	1.4079	40′
30′	.1658	.1650	.9863	.1673	5.976	1.014	6.059	1.4050	30′
40′	.1687	.1679	.9858	.1703	5.871	1.014	5.955	1.4021	20′
50′	.1716	.1708	.9853	.1733	5.769	1.015	5.855	1.3992	10′
10°00′	.1745	.1736	.9848	.1763	5.671	1.015	5.759	1.3963	80°00′
10′	.1774	.1765	.9843	.1793	5.576	1.016	5.665	1.3934	50′
20′	.1804	.1794	.9838	.1823	5.485	1.016	5.575	1.3904	40′
30′	.1833	.1822	.9833	.1853	5.396	1.017	5.487	1.3875	30′
40′	.1862	.1851	.9827	.1883	5.309	1.018	5.403	1.3846	20′
50′	.1891	.1880	.9822	.1914	5.226	1.018	5.320	1.3817	10′
11°00′	.1920	.1908	.9816	.1944	5.145	1.019	5.241	1.3788	79°00′
10′	.1949	.1937	.9811	.1974	5.066	1.019	5.164	1.3759	50′
20′	.1978	.1965	.9805	.2004	4.989	1.020	5.089	1.3730	40′
30′	.2007	.1994	.9799	.2035	4.915	1.020	5.016	1.3701	30′
40′	.2036	.2022	.9793	.2065	4.843	1.021	4.945	1.3672	20′
50′	.2065	.2051	.9787	.2095	4.773	1.022	4.876	1.3643	10′
12°00′	.2094	.2079	.9781	.2126	4.705	1.022	4.810	1.3614	78°00′
10′	.2123	.2108	.9775	.2156	4.638	1.023	4.745	1.3584	50′
20′	.2153	.2136	.9769	.2186	4.574	1.024	4.682	1.3555	40′
30′	.2182	.2164	.9763	.2217	4.511	1.024	4.620	1.3526	30′
40′	.2211	.2193	.9757	.2247	4.449	1.025	4.560	1.3497	20′
50′	.2240	.2221	.9750	.2278	4.390	1.026	4.502	1.3468	10′
13°00′	.2269	.2250	.9744	.2309	4.331	1.026	4.445	1.3439	77°00′
10′	.2298	.2278	.9737	.2339	4.275	1.027	4.390	1.3410	50′
20′	.2327	.2306	.9730	.2370	4.219	1.028	4.336	1.3381	40′
30′	.2356	.2334	.9724	.2401	4.165	1.028	4.284	1.3352	30′
40′	.2385	.2363	.9717	.2432	4.113	1.029	4.232	1.3323	20′
50′	.2414	.2391	.9710	.2462	4.061	1.030	4.182	1.3294	10′
14°00′	.2443	.2419	.9703	.2493	4.011	1.031	4.134	1.3265	76°00′
10′	.2473	.2447	.9696	.2524	3.962	1.031	4.086	1.3235	50′
20′	.2502	.2476	.9689	.2555	3.914	1.032	4.039	1.3206	40′
30′	.2531	.2504	.9681	.2586	3.867	1.033	3.994	1.3177	30′
40′	.2560	.2532	.9674	.2617	3.821	1.034	3.950	1.3148	20′
50′	.2589	.2560	.9667	.2648	3.776	1.034	3.906	1.3119	10′
15°00′	.2618	.2588	.9659	.2679	3.732	1.035	3.864	1.3090	75°00′
		cos	sin	cot	tan	csc	sec	rad.	deg.

TRIGONOMETRIC FUNCTIONS, *continued*

deg.	rad.	sin	cos	tan	cot	sec	csc		
15°00′	**.2618**	.2588	.9659	.2679	3.732	1.035	3.864	**1.3090**	**75°00′**
10′	**.2647**	.2616	.9652	.2711	3.689	1.036	3.822	**1.3061**	**50′**
20′	**.2676**	.2644	.9644	.2742	3.647	1.037	3.782	**1.3032**	**40′**
30′	**.2705**	.2672	.9636	.2773	3.606	1.038	3.742	**1.3003**	**30′**
40′	**.2734**	.2700	.9628	.2805	3.566	1.039	3.703	**1.2974**	**20′**
50′	**.2763**	.2728	.9621	.2836	3.526	1.039	3.665	**1.2945**	**10′**
16°00′	**.2793**	.2756	.9613	.2867	3.487	1.040	3.628	**1.2915**	**74°00′**
10′	**.2822**	.2784	.9605	.2899	3.450	1.041	3.592	**1.2886**	**50′**
20′	**.2851**	.2812	.9596	.2931	3.412	1.042	3.556	**1.2857**	**40′**
30′	**.2880**	.2840	.9588	.2962	3.376	1.043	3.521	**1.2828**	**30′**
40′	**.2909**	.2868	.9580	.2994	3.340	1.044	3.487	**1.2799**	**20′**
50′	**.2938**	.2896	.9572	.3026	3.305	1.045	3.453	**1,2770**	**10′**
17°00′	**.2967**	.2924	.9563	.3057	3.271	1.046	3.420	**1.2741**	**73°00′**
10′	**.2996**	.2952	.9555	.3089	3.237	1.047	3.388	**1.2712**	**50′**
20′	**.3025**	.2979	.9546	.3121	3.204	1.048	3.356	**1.2683**	**40′**
30′	**.3054**	.3007	.9537	.3153	3.172	1.049	3.326	**1.2654**	**30′**
40′	**.3083**	.3035	.9528	.3185	3.140	1.049	3.295	**1.2625**	**20′**
50′	**.3113**	.3062	.9520	.3217	3.108	1.050	3.265	**1.2595**	**10′**
18°00′	**.3142**	.3090	.9511	.3249	3.078	1.051	3.236	**1.2566**	**72°00′**
10′	**.3171**	.3118	.9502	.3281	3.047	1.052	3.207	**1.2537**	**50′**
20′	**.3200**	.3145	.9492	.3314	3.018	1.053	3.179	**1.2508**	**40′**
30′	**.3229**	.3173	.9483	.3346	2.989	1.054	3.152	**1.2479**	**30′**
40′	**.3258**	.3201	.9474	.3378	2.960	1.056	3.124	**1.2450**	**20′**
50′	**.3287**	.3228	.9465	.3411	2.932	1.057	3.098	**1.2421**	**10′**
19°00′	**.3316**	.3256	.9455	.3443	2.904	1.058	3.072	**1.2392**	**71°00′**
10′	**.3345**	.3283	.9446	.3476	2.877	1.059	3.046	**1.2363**	**50′**
20′	**.3374**	.3311	.9436	.3508	2.850	1.060	3.021	**1.2334**	**40′**
30′	**.3403**	.3338	.9426	.3541	2.824	1.061	2.996	**1.2305**	**30′**
40′	**.3432**	.3365	.9417	.3574	2.798	1.062	2.971	**1.2275**	**20′**
50′	**.3462**	.3393	.9407	.3607	2.773	1.063	2.947	**1.2246**	**10′**
20°00′	**.3491**	.3420	.9397	.3640	2.747	1.064	2.924	**1.2217**	**70°00′**
10′	**.3520**	.3448	.9387	.3673	2.723	1.065	2.901	**1.2188**	**50′**
20′	**.3549**	.3475	.9377	.3706	2.699	1.066	2.878	**1.2159**	**40′**
30′	**.3578**	.3502	.9367	.3739	2.675	1.068	2.855	**1.2130**	**30′**
40′	**.3607**	.3529	.9356	.3772	2.651	1.069	2.833	**1.2101**	**20′**
50′	**.3636**	.3557	.9346	.3805	2.628	1.070	2.812	**1.2072**	**10′**
21°00′	**.3665**	.3584	.9336	.3839	2.605	1.071	2.790	**1.2043**	**69°00′**
10′	**.3694**	.3611	.9325	.3872	2.583	1.072	2.769	**1.2014**	**50′**
20′	**.3723**	.3638	.9315	.3906	2.560	1.074	2.749	**1.1985**	**40′**
30′	**.3752**	.3665	.9304	.3939	2.539	1.075	2.729	**1.1956**	**30′**
40′	**.3782**	.3692	.9293	.3973	2.517	1.076	2.709	**1.1926**	**20′**
50′	**.3811**	.3719	.9283	.4006	2.496	1.077	2.689	**1.1897**	**10′**
22°00′	**.3840**	.3746	.9272	.4040	2.475	1.079	2.669	**1.1868**	**68°00′**
10′	**.3869**	.3773	.9261	.4074	2.455	1.080	2.650	**1.1839**	**50′**
20′	**.3898**	.3800	.9250	.4108	2.434	1.081	2.632	**1.1810**	**40′**
30′	**.3927**	.3827	.9239	.4142	2.414	1.082	2.613	**1.1781**	**67°30′**
		cos	sin	cot	tan	csc	sec	rad.	deg.

deg.	rad.	sin	cos	tan	cot	sec	csc		
22°30′	**.3927**	.3827	.9239	.4142	2.414	1.082	2.613	**1.1781**	**30′**
40′	**.3956**	.3854	.9228	.4176	2.394	1.084	2.595	**1.1752**	**20′**
50′	**.3985**	.3881	.9216	.4210	2.375	1.085	2.577	**1.1723**	**10′**
23°00′	**.4014**	.3907	.9205	.4245	2.356	1.086	2.559	**1.1694**	**67°00′**
10′	**.4043**	.3934	.9194	.4279	2.337	1.088	2.542	**1.1665**	**50′**
20′	**.4072**	.3961	.9182	.4314	2.318	1.089	2.525	**1.1636**	**40′**
30′	**.4102**	.3987	.9171	.4348	2.300	1.090	2.508	**1.1606**	**30′**
40′	**.4131**	.4014	.9159	.4383	2.282	1.092	2.491	**1.1577**	**20′**
50′	**.4160**	.4041	.9147	.4417	2.264	1.093	2.475	**1.1548**	**10′**
24°00′	**.4189**	.4067	.9135	.4452	2.246	1.095	2.459	**1.1519**	**66°00′**
10′	**.4218**	.4094	.9124	.4487	2.229	1.096	2.443	**1.1490**	**50′**
20′	**.4247**	.4120	.9112	.4522	2.211	1.097	2.427	**1.1461**	**40′**
30′	**.4276**	.4147	.9100	.4557	2.194	1.099	2.411	**1.1432**	**30′**
40′	**.4305**	.4173	.9088	.4592	2.177	1.100	2.396	**1.1403**	**20′**
50′	**.4334**	.4200	.9075	.4628	2.161	1.102	2.381	**1.1374**	**10′**
25°00′	**.4363**	.4226	.9063	.4663	2.145	1.103	2.366	**1.1345**	**65°00′**
10′	**.4392**	.4253	.9051	.4699	2.128	1.105	2.352	**1.1316**	**50′**
20′	**.4422**	.4279	.9038	.4734	2.112	1.106	2.337	**1.1286**	**40′**
30′	**.4451**	.4305	.9026	.4770	2.097	1.108	2.323	**1.1257**	**30′**
40′	**.4480**	.4331	.9013	.4806	2.081	1.109	2.309	**1.1228**	**20′**
50′	**.4509**	.4358	.9001	.4841	2.066	1.111	2.295	**1.1199**	**10′**
26°00′	**.4538**	.4384	.8988	.4877	2.050	1.113	2.281	**1.1170**	**64°00′**
10′	**.4567**	.4410	.8975	.4913	2.035	1.114	2.268	**1.1141**	**50′**
20′	**.4596**	.4436	.8962	.4950	2.020	1.116	2.254	**1.1112**	**40′**
30′	**.4625**	.4462	.8949	.4986	2.006	1.117	2.241	**1.1083**	**30′**
40′	**.4654**	.4488	.8936	.5022	1.991	1.119	2.228	**1.1054**	**20′**
50′	**.4683**	.4514	.8923	.5059	1.977	1.121	2.215	**1.1025**	**10′**
27°00′	**.4712**	.4540	.8910	.5095	1.963	1.122	2.203	**1.0996**	**63°00′**
10′	**.4741**	.4566	.8897	.5132	1.949	1.124	2.190	**1.0966**	**50′**
20′	**.4771**	.4592	.8884	.5169	1.935	1.126	2.178	**1.0937**	**40′**
30′	**.4800**	.4617	.8870	.5206	1.921	1.127	2.166	**1.0908**	**30′**
40′	**.4829**	.4643	.8857	.5243	1.907	1.129	2.154	**1.0879**	**20′**
50′	**.4858**	.4669	.8843	.5280	1.894	1.131	2.142	**1.0850**	**10′**
28°00′	**.4887**	.4695	.8829	.5317	1.881	1.133	2.130	**1.0821**	**62°00′**
10′	**.4916**	.4720	.8816	.5354	1.868	1.134	2.118	**1.0792**	**50′**
20′	**.4945**	.4746	.8802	.5392	1.855	1.136	2.107	**1.0763**	**40′**
30′	**.4974**	.4772	.8788	.5430	1.842	1.138	2.096	**1.0734**	**30′**
40′	**.5003**	.4797	.8774	.5467	1.829	1.140	2.085	**1.0705**	**20′**
50′	**.5032**	.4823	.8760	.5505	1.816	1.142	2.074	**1.0676**	**10′**
29°00′	**.5061**	.4848	.8746	.5543	1.804	1.143	2.063	**1.0647**	**61°00′**
10′	**.5091**	.4874	.8732	.5581	1.792	1.145	2.052	**1.0617**	**50′**
20′	**.5120**	.4899	.8718	.5619	1.780	1.147	2.041	**1.0588**	**40′**
30′	**.5149**	.4924	.8704	.5658	1.767	1.149	2.031	**1.0559**	**30′**
40′	**.5178**	.4950	.8689	.5696	1.756	1.151	2.020	**1.0530**	**20′**
50′	**.5207**	.4975	.8675	.5735	1.744	1.153	2.010	**1.0501**	**10′**
30°00′	**.5236**	.5000	.8660	.5774	1.732	1.155	2.000	**1.0472**	**60°00′**
		cos	sin	cot	tan	csc	sec	rad.	deg.

TRIGONOMETRIC FUNCTIONS, *continued*

deg.	rad.	sin	cos	tan	cot	sec	csc		
30°00′	**.5236**	.5000	.8660	.5774	1.732	1.155	2.000	**1.0472**	**60°00′**
10′	**.5265**	.5025	.8646	.5812	1.720	1.157	1.990	**1.0443**	**50′**
20′	**.5294**	.5050	.8631	.5851	1.709	1.159	1.980	**1.0414**	**40′**
30′	**.5323**	.5075	.8616	.5890	1.698	1.161	1.970	**1.0385**	**30′**
40′	**.5352**	.5100	.8601	.5930	1.686	1.163	1.961	**1.0356**	**20′**
50′	**.5381**	.5125	.8587	.5969	1.675	1.165	1.951	**1.0327**	**10′**
31°00′	**.5411**	.5150	.8572	.6009	1.664	1.167	1.942	**1.0297**	**59°00′**
10′	**.5440**	.5175	.8557	.6048	1.653	1.169	1.932	**1.0268**	**50′**
20′	**.5469**	.5200	.8542	.6088	1.643	1.171	1.923	**1.0239**	**40′**
30′	**.5498**	.5225	.8526	.6128	1.632	1.173	1.914	**1.0210**	**30′**
40′	**.5527**	.5250	.8511	.6168	1.621	1.175	1.905	**1.0181**	**20′**
50′	**.5556**	.5275	.8496	.6208	1.611	1.177	1.896	**1.0152**	**10′**
32°00′	**.5585**	.5299	.8480	.6249	1.600	1.179	1.887	**1.0123**	**58°00′**
10′	**.5614**	.5324	.8465	.6289	1.590	1.181	1.878	**1.0094**	**50′**
20′	**.5643**	.5348	.8450	.6330	1.580	1.184	1.870	**1.0065**	**40′**
30′	**.5672**	.5373	.8434	.6371	1.570	1.186	1.861	**1.0036**	**30′**
40′	**.5701**	.5398	.8418	.6412	1.560	1.188	1.853	**1.0007**	**20′**
50′	**.5730**	.5422	.8403	.6453	1.550	1.190	1.844	**.9977**	**10′**
33°00′	**.5760**	.5446	.8387	.6494	1.540	1.192	1.836	**.9948**	**57°00′**
10′	**.5789**	.5471	.8371	.6536	1.530	1.195	1.828	**.9919**	**50′**
20′	**.5818**	.5495	.8355	.6577	1.520	1.197	1.820	**.9890**	**40′**
30′	**.5847**	.5519	.8339	.6619	1.511	1.199	1.812	**.9861**	**30′**
40′	**.5876**	.5544	.8323	.6661	1.501	1.202	1.804	**.9832**	**20′**
50′	**.5905**	.5568	.8307	.6703	1.492	1.204	1.796	**.9803**	**10′**
34°00′	**.5934**	.5592	.8290	.6745	1.483	1.206	1.788	**.9774**	**56°00′**
10′	**.5963**	.5616	.8274	.6787	1.473	1.209	1.781	**.9745**	**50′**
20′	**.5992**	.5640	.8258	.6830	1.464	1.211	1.773	**.9716**	**40′**
30′	**.6021**	.5664	.8241	.6873	1.455	1.213	1.766	**.9687**	**30′**
40′	**.6050**	.5688	.8225	.6916	1.446	1.216	1.758	**.9657**	**20′**
50′	**.6080**	.5712	.8208	.6959	1.437	1.218	1.751	**.9628**	**10′**
35°00′	**.6109**	.5736	.8192	.7002	1.428	1.221	1.743	**.9599**	**55°00′**
10′	**.6138**	.5760	.8175	.7046	1.419	1.223	1.736	**.9570**	**50′**
20′	**.6167**	.5783	.8158	.7089	1.411	1.226	1.729	**.9541**	**40′**
30′	**.6196**	.5807	.8141	.7133	1.402	1.228	1.722	**.9512**	**30′**
40′	**.6225**	.5831	.8124	.7177	1.393	1.231	1.715	**.9483**	**20′**
50′	**.6254**	.5854	.8107	.7221	1.385	1.233	1.708	**.9454**	**10′**
36°00′	**.6283**	.5878	.8090	.7265	1.376	1.236	1.701	**.9425**	**54°00′**
10′	**.6312**	.5901	.8073	.7310	1.368	1.239	1.695	**.9396**	**50′**
20′	**.6341**	.5925	.8056	.7355	1.360	1.241	1.688	**.9367**	**40′**
30′	**.6370**	.5948	.8039	.7400	1.351	1.244	1.681	**.9338**	**30′**
40′	**.6400**	.5972	.8021	.7445	1.343	1.247	1.675	**.9308**	**20′**
50′	**.6429**	.5995	.8004	.7490	1.335	1.249	1.668	**.9279**	**10′**
37°00′	**.6458**	.6018	.7986	.7536	1.327	1.252	1.662	**.9250**	**53°00′**
10′	**.6487**	.6041	.7969	.7581	1.319	1.255	1.655	**.9221**	**50′**
20′	**.6516**	.6065	.7951	.7627	1.311	1.258	1.649	**.9192**	**40′**
30′	**.6545**	.6088	.7934	.7673	1.303	1.260	1.643	**.9163**	**52°30′**
		cos	sin	cot	tan	csc	sec	rad.	deg.

deg.	rad.	sin	cos	tan	cot	sec	csc		
37°30′	**.6545**	.6088	.7934	.7673	1.303	1.260	1.643	**.9163**	**30′**
40′	**.6574**	.6111	.7916	.7720	1.295	1.263	1.636	**.9134**	**20′**
50′	**.6603**	.6134	.7898	.7766	1.288	1.266	1.630	**.9105**	**10′**
38°00′	**.6632**	.6157	.7880	.7813	1.280	1.269	1.624	**.9076**	**52°00′**
10′	**.6661**	.6180	.7862	.7860	1.272	1.272	1.618	**.9047**	**50′**
20′	**.6690**	.6202	.7844	.7907	1.265	1.275	1.612	**.9018**	**40′**
30′	**.6720**	.6225	.7826	.7954	1.257	1.278	1.606	**.8988**	**30′**
40′	**.6749**	.6248	.7808	.8002	1,250	1.281	1.601	**.8959**	**20′**
50′	**.6778**	.6271	.7790	.8050	1.242	1.284	1.595	**.8930**	**10′**
39°00′	**.6807**	.6293	.7771	.8098	1.235	1.287	1.589	**.8901**	**51°00′**
10′	**.6836**	.6316	.7753	.8146	1.228	1.290	1.583	**.8872**	**50′**
20′	**.6865**	.6338	.7735	.8195	1.220	1.293	1.578	**.8843**	**40′**
30′	**.6894**	.6361	.7716	.8243	1.213	1.296	1.572	**.8814**	**30′**
40′	**.6923**	.6383	.7698	.8292	1.206	1.299	1.567	**.8785**	**20′**
50′	**.6952**	.6406	.7679	.8342	1.199	1.302	1.561	**.8756**	**10′**
40°00′	**.6981**	.6428	.7660	.8391	1.192	1.305	1.556	**.8727**	**50°00′**
10′	**.7010**	.6450	.7642	.8441	1.185	1.309	1.550	**.8698**	**50′**
20′	**.7039**	.6472	.7623	.8491	1.178	1.312	1.545	**.8668**	**40′**
30′	**.7069**	.6494	.7604	.8541	1.171	1.315	1.540	**.8639**	**30′**
40′	**.7098**	.6517	.7585	.8591	1.164	1.318	1.535	**.8610**	**20′**
50′	**.7127**	.6539	.7566	.8642	1.157	1.322	1.529	**.8581**	**10′**
41°00′	**.7156**	.6561	.7547	.8693	1.150	1.325	1.524	**.8552**	**49°00′**
10′	**.7185**	.6583	.7528	.8744	1.144	1.328	1.519	**.8523**	**50′**
20′	**.7214**	.6604	.7509	.8796	1.137	1.332	1.514	**.8494**	**40′**
30′	**.7243**	.6626	.7490	.8847	1.130	1.335	1.509	**.8465**	**30′**
40′	**.7272**	.6648	.7470	.8899	1.124	1.339	1.504	**.8436**	**20′**
50′	**.7301**	.6670	.7451	.8952	1.117	1.342	1.499	**.8407**	**10′**
42°00′	**.7330**	.6691	.7431	.9004	1.111	1.346	1.494	**.8378**	**48°00′**
10′	**.7359**	.6713	.7412	.9057	1.104	1.349	1.490	**.8348**	**50′**
20′	**.7389**	.6734	.7392	.9110	1.098	1.353	1.485	**.8319**	**40′**
30′	**.7418**	.6756	.7373	.9163	1.091	1.356	1.480	**.8290**	**30′**
40′	**.7447**	.6777	.7353	.9217	1.085	1.360	1.476	**.8261**	**20′**
50′	**.7476**	.6799	.7333	.9271	1.079	1.364	1.471	**.8232**	**10′**
43°00′	**.7505**	.6820	.7314	.9325	1.072	1.367	1.466	**.8203**	**47°00′**
10′	**.7534**	.6841	.7294	.9380	1.066	1.371	1.462	**.8174**	**50′**
20′	**.7563**	.6862	.7274	.9435	1.060	1.375	1.457	**.8145**	**40′**
30′	**.7592**	.6884	.7254	.9490	1.054	1.379	1.453	**.8116**	**30′**
40′	**.7621**	.6905	.7234	.9545	1.048	1.382	1.448	**.8087**	**20′**
50′	**.7650**	.6926	.7214	.9601	1.042	1.386	1.444	**.8058**	**10′**
44°00′	**.7679**	.6947	.7193	.9657	1.036	1.390	1.440	**.8029**	**46°00′**
10′	**.7709**	.6967	.7173	.9713	1.030	1.394	1.435	**.7999**	**50′**
20′	**.7738**	.6988	.7153	.9770	1.024	1.398	1.431	**.7970**	**40′**
30′	**.7767**	.7009	.7133	.9827	1.018	1.402	1.427	**.7941**	**30′**
40′	**.7796**	.7030	.7112	.9884	1.012	1.406	1.423	**.7912**	**20′**
50′	**.7825**	.7050	.7092	.9942	1.006	1.410	1.418	**.7883**	**10′**
45°00′	**.7854**	.7071	.7071	1.000	1.000	1.414	1.414	**.7854**	**45°00′**
		cos	sin	cot	tan	csc	sec	rad.	deg.

Graphing Calculator Handbook

1 Working with Expressions

Objective To learn the basics of entering, editing, and manipulating expressions on a graphing calculator

Graphing calculators have features that make them easier to work with and better tools for learning than normal scientific calculators. The examples in this lesson highlight ways to make your calculator work for you.

Example 1

Evaluate the expression $3x^2 + 11x$ for $x = 0.5625$.

TI-81

To store $x = 0.5625$ in your calculator, press

0 . 5 6 2 5 [STO▸] [X|T] [ENTER]

Your calculator now will substitute the value 0.5625 for x every time you refer to x.

To evaluate $3x^2 + 11x$, press

3 [X|T] [x^2] [+] 1 1 [X|T] [ENTER]

Your screen should look like this:

```
0.5625→X
           .5625
3X²+11X
      7.13671875
```

TI-82

To store $x = 0.5625$ in your calculator, press

0 . 5 6 2 5 [STO▸] [X,T,θ] [ENTER]

Your calculator now will substitute the value 0.5625 for x every time you refer to x.

To evaluate $3x^2 + 11x$, press

3 [X,T,θ] [x^2] [+] 1 1 [X,T,θ] [ENTER]

Your screen should look like this:

```
0.5625→X
           .5625
3X²+11X
      7.13671875
```

Casio fx-7700GB

To store $x = 0.5625$ in your calculator, press

0 . 5 6 2 5 [→] [X,θ,T] [EXE]

Your calculator now will substitute the value 0.5625 for x every time you refer to x.

To evaluate $3x^2 + 11x$, press

3 [X,θ,T] [SHIFT] [x^2] [+] 1 1 [X,θ,T] [EXE]

Your screen should look like this:

```
0.5625→X
          0.5625
3X²+11X
      7.13671875
```

When you want to evaluate an expression for several values of the variable, it is often easier to edit the first expression you enter than to type a new expression each time.

Example 2

Evaluate the expression $\sqrt{x^2 + 9}$ for $x = 4$, $x = 7$, and $x = 10$.

TI-81

The first expression that you want to evaluate is $\sqrt{4^2 + 9}$. Press [2nd] [√] to access the square-root function. Then press [(] 4 [x^2] [+] 9 [)] [ENTER] to get the first answer.

Instead of retyping the expression to change the value of the variable to 7, just press [2nd] [ENTRY] to recall the preceding expression. Then press [◄] five times to move the cursor over the 4. Press 7 to replace the 4 with 7, and press [ENTER] to get the answer.

To change the 7 to 10, you must learn to delete and insert. After recalling the preceding expression and moving the cursor over the 7, press [DEL] to delete the 7, then press [INS] to put the calculator in insert mode. Then press 1 0 to insert 10 without replacing any other part of the expression. Press [ENTER] to get the answer.

continued on next page

TI-82

The first expression that you want to evaluate is $\sqrt{4^2 + 9}$. Press [2nd] [√] to access the square-root function. Then press [(] 4 [x^2] [+] 9 [)] [ENTER] to get the first answer.

Instead of retyping the expression to change the value of the variable to 7, just press [2nd] [ENTRY] to recall the preceding expression. Then press [◄] five times to move the cursor over the 4. Press 7 to replace the 4 with 7, and press [ENTER] to get the answer.

To change the 7 to 10, you must learn to delete and insert. After recalling the preceding expression and moving the cursor over the 7, press [DEL] to delete the 7, then press [2nd] [INS] to put the calculator in insert mode. Then press 1 0 to insert 10 without replacing any other part of the expression. Press [ENTER] to get the answer.

continued on next page

Casio fx-7700GB

The first expression that you want to evaluate is $\sqrt{4^2 + 9}$. Press [√] to access the square-root function. Then press [(] 4 [SHIFT] [x^2] [+] 9 [)] [EXE] to get the first answer.

Instead of retyping the expression to change the value of the variable to 7, just press [◄] to remove the answer and return to the preceding expression. Then press [◄] five more times to move the cursor to the 4. Press 7 to replace the 4 with 7, and press [EXE] to get the answer.

To change the 7 to 10, you must learn to delete and insert. After returning to the preceding expression and moving the cursor to the 7, press [DEL] to delete the 7, then press [SHIFT] [INS] to put the calculator in insert mode. Then press 1 0 to insert 10 without replacing any other part of the expression. Press [EXE] to get the answer.

continued on next page

TI-81

continued from previous page

Your screen should look like this:

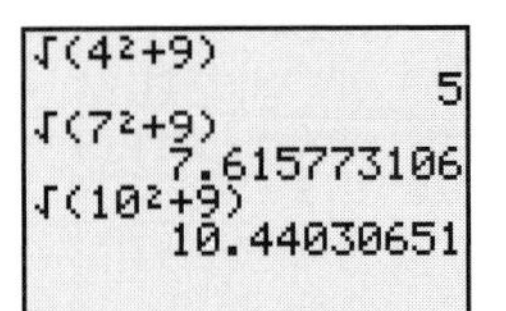

Hint

You can also recall an expression by pressing [▲] instead of [2nd] [ENTRY].

TI-82

continued from previous page

Your screen should look like this:

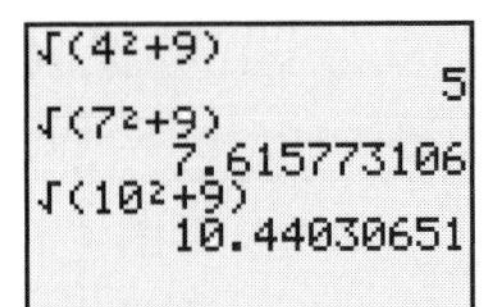

Casio fx-7700GB

continued from previous page

Your screen should look like this:

```
√(10²+9)
   10.44030651
```

Hint

When editing a lengthy expression, you can go to the beginning of the expression by pressing [▲] twice and to the end of the expression by pressing [▼] twice. This can be a lot faster than pressing the left and right arrows repeatedly.

Once you have evaluated an expression, you might need to use its value in another computation. A graphing calculator makes this easy to do without your having to reenter expressions or values.

Example 3

Find the next three terms of the sequence 1, 2, 4, 16, . . . , in which each term is the square of the term that precedes it.

TI-81

To find the fifth term, press 1 6 [x^2] [ENTER].

The answer, 256, is displayed on the screen and stored in the calculator as Ans. Notice that ANS appears in blue above the [(-)] key. To square 256, therefore, just press [2nd] [ANS] to call up the preceding answer, then press [x^2] [ENTER].

continued on next page

TI-82

To find the fifth term, press 1 6 [x^2] [ENTER].

The answer, 256, is displayed on the screen and stored in the calculator as Ans. Notice that ANS appears in blue above the [(-)] key. To square 256, therefore, just press [2nd] [ANS] to call up the preceding answer, then press [x^2] [ENTER].

continued on next page

Casio fx-7700GB

To find the fifth term, press 1 6 [SHIFT] [x^2] [EXE].

The answer, 256, is displayed on the screen and stored in the calculator as Ans. To square 256, press [ANS], then press [SHIFT] [x^2] [EXE].

Repeat this process to square the new answer.

continued on next page

TI-81	TI-82	Casio fx-7700GB
continued from previous page	*continued from previous page*	*continued from previous page*
Repeat this process to square the new answer. Your screen should look like this:	Repeat this process to square the new answer. Your screen should look like this:	Your screen should look like this:
16² 256 Ans² 65536 Ans² 4294967296	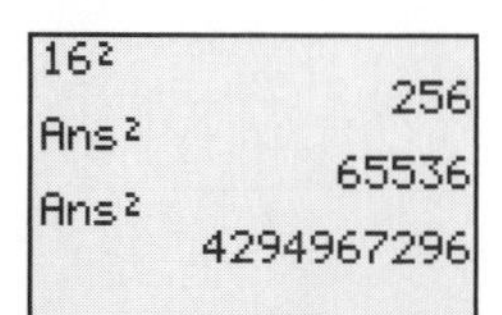	16² 256. Ans² 65536. Ans² 4294967296.

Hint

You can sometimes make Ans appear on the screen without pressing [ANS] or [2nd] [ANS]. In Example 3, for instance, just skip the steps that precede the operation of squaring. Try this yourself—it saves time!

PROBLEMS

In problems 1–3, evaluate each expression for $x = 2.053124$.

1 $x^2 + 3x$ 10.37469016

2 $x^5 + 0.25x^3 + 3$ 41.64541251

3 $\dfrac{x+3}{x^2+3}$ 0.7003328042

In problems 4–9, evaluate each expression for $x = 3$, $x = 7$, $x = 11$, and $x = \sqrt{5}$.

4 $12x^2 + 34$

5 $\sqrt{x^2 - 4}$

6 $x^3 + 11$

7 $x^2 + 7x - 13$

8 $2x^3 + 3x^2 + x$

9 $\dfrac{x^2+3}{x^2-3}$

10 Find $\sqrt{500}$. Then find the square root of the answer. Continue to find square roots of your answers until you get an answer less than 2. How many square roots did you have to find? 4 square roots

11 Repeat problem **10**, starting with $\sqrt{2{,}000{,}000}$ instead of $\sqrt{500}$. Before you begin, predict how many steps it will take you to obtain an answer less than 2. 5 square roots

12 Continue problem **11** beyond the point where you stopped. Keep track of what happens, and explain it in writing. Do the answers continue to decrease? Will they ever stop decreasing? The answers continue to decrease, becoming closer and closer to 1, although at a certain point the calculator begins to approximate each answer as 1.

4 142; 622; 1486; 94
5 2.236067977; 6.708203932; 10.81665383; 1
6 38; 354; 1342; 22.18033989
7 17; 85; 185; 7.652475842
8 84; 840; 3036; 39.59674775
9 2; 1.130434783; 1.050847458; 4

2 ADDING AND SUBTRACTING MATRICES

Objective To learn how to enter, add, and subtract matrices on a graphing calculator

You can use your graphing calculator to perform arithmetic operations on matrices. This lesson shows you how to store two matrices, then find their sum or difference.

Example 1

Store the matrix $\begin{bmatrix} 4 & 5 & 6 \\ -3 & 0 & 12 \end{bmatrix}$ as matrix A in your calculator.

TI-81

To enter the matrix, press MATRX ▶ 1. The screen will identify the matrix as [A] and prompt you to enter its dimensions. Since A has 2 rows and 3 columns, press 2 ENTER 3 ENTER.

The calculator then moves to the line showing "1 , 1 =." This prompts you to enter the first row's first element. Press 4 ENTER. The calculator then moves to the line that shows "1 , 2 =," prompting you to enter the first row's second element. Press 5 ENTER. Continue in this manner to enter all six numbers.

Press 2nd [QUIT] to clear the screen and 2nd [[A]] ENTER to view the matrix. Your screen should look like this:

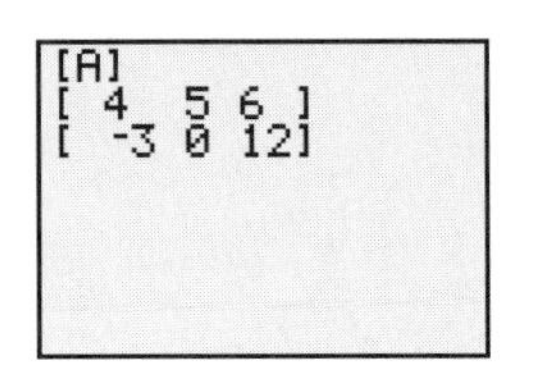

continued on next page

TI-82

To enter the matrix, press MATRX ▶ ▶ 1. The screen will identify the matrix as [A] and prompt you to enter its dimensions. Since A has 2 rows and 3 columns, press 2 ENTER 3 ENTER.

The calculator then prompts you with "1 , 1 =" at the bottom of the screen while highlighting the element in row 1, column 1. Press 4 ENTER. The calculator then prompts you with "1 , 2 =," highlighting the element in row 1, column 2. Press 5 ENTER. Continue in this manner to enter all six numbers. Your screen should look like this:

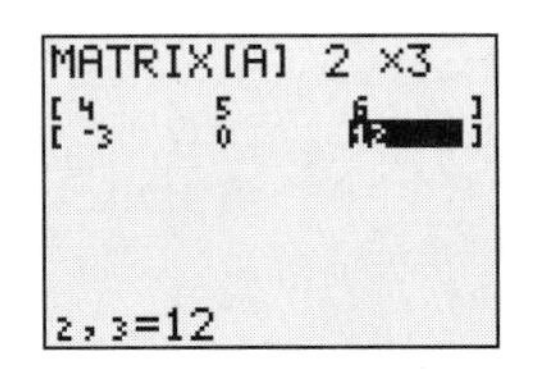

continued on next page

Casio fx-7700GB

To enter the matrix, press MODE 0 to access matrix mode, followed by F1 to access matrix A. Then press F6 F1 so that you can enter the number of rows and columns. Press 2 EXE 3 EXE.

The calculator then presents a matrix with 2 rows and 3 columns, highlighting the first element of the first row. Enter that element by pressing 4 EXE. The calculator now highlights the second element of the first row. Press 5 EXE. Continue in this manner to enter all six numbers. Your screen should look like this:

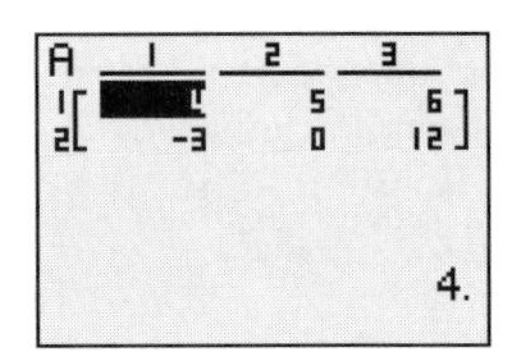

continued on next page

TI-81

continued from previous page

Hint

To erase a matrix A, press 0 [STO▶] [2nd] [[A]] [ENTER].

TI-82

continued from previous page

Press [2nd] [QUIT] to clear the screen.

Hint

To erase a matrix A, press [2nd] [MEM] 2 4 [ENTER].

Casio fx-7700GB

continued from previous page

Press [MODE] [+] [AC] to clear the screen and return to computation mode.

Hint

To erase a matrix A, press [MODE] 0 to access matrix mode. Then press [F1] [F6] [F2] [F1].

Example 2

Find the matrix sum A + B, where $A = \begin{bmatrix} 4 & 5 & 6 \\ -3 & 0 & 12 \end{bmatrix}$ and $B = \begin{bmatrix} 1 & 2 & 3 \\ 12 & 0 & -3 \end{bmatrix}$.

TI-81

You stored matrix A in your calculator in Example 1. To store matrix B, press [MATRX] [▶] 2 to access [B] and proceed as before. When you finish entering the elements of B, press [2nd] [QUIT] to clear the screen.

To find A + B, press [2nd] [[A]] [+] [2nd] [[B]] [ENTER]. The answer $\begin{bmatrix} 5 & 7 & 9 \\ 9 & 0 & 9 \end{bmatrix}$, is displayed on the screen.

Hint

When you find A + B, the sum becomes Ans. You can store the sum as a new matrix, C, by pressing [STO▶] [2nd] [[C]] [ENTER].

TI-82

You stored matrix A in your calculator in Example 1. To store matrix B, press [MATRX] [▶] [▶] 2 to access [B] and proceed as before. When you finish entering the elements of B, press [2nd] [QUIT] to clear the screen.

To find A + B, press [MATRX] 1 [+] [MATRX] 2 [ENTER]. The answer, $\begin{bmatrix} 5 & 7 & 9 \\ 9 & 0 & 9 \end{bmatrix}$, is displayed on the screen.

Hint

When you find A + B, the sum becomes Ans. You can store the sum as a new matrix, C, by pressing [STO▶] [MATRX] 3 [ENTER].

Casio fx-7700GB

You stored matrix A in your calculator in Example 1. To store matrix B, press [MODE] 0 [F2] to access matrix B, and then proceed as before.

When you finish entering the elements of B, press [PRE] to return to the main matrix menu. Then, to find A + B, press [F3]. The answer, $\begin{bmatrix} 5 & 7 & 9 \\ 9 & 0 & 9 \end{bmatrix}$, is displayed on the screen.

Hint

When you find A + B, the sum is automatically stored as matrix C.

To find A – B on the TI-81 or TI-82, simply follow the procedure for finding A + B, pressing [–] instead of [+]. On the Casio fx-7700GB, press [F4] after accessing the main matrix menu.

PROBLEMS

In problems 1–4, use your calculator to find A + B and A – B for each pair of matrices.

1 $A = \begin{bmatrix} 3 & 8 & 9 \\ -2 & 4 & -3 \end{bmatrix}$ and $B = \begin{bmatrix} 11 & 12 & 13 \\ -4 & 10 & 5 \end{bmatrix}$

2 $A = \begin{bmatrix} -11 & -7 \\ -6 & 4 \end{bmatrix}$ and $B = \begin{bmatrix} 3 & 4 \\ 5 & 6 \end{bmatrix}$

3 $A = \begin{bmatrix} 1 & 2 & 3 \\ 4 & 5 & 6 \\ 7 & 8 & 9 \end{bmatrix}$ and $B = \begin{bmatrix} 21 & 20 & 19 \\ 18 & 17 & 16 \\ 15 & 14 & 13 \end{bmatrix}$

4 $A = \begin{bmatrix} 0 & -3 \\ -4 & 0 \\ 0 & 15 \end{bmatrix}$ and $B = \begin{bmatrix} -3 & 0 \\ 0 & -4 \\ 15 & 0 \end{bmatrix}$

5 Store $\begin{bmatrix} 1 & 2 \\ 3 & 4 \end{bmatrix}$ as matrix A and $\begin{bmatrix} 5 & 6 & 7 \\ 8 & 9 & 0 \end{bmatrix}$ as matrix B. Try to find A + B. What does your calculator do? Explain. The calculator displays an error message. Because the dimensions of A and B are not the same, the matrices cannot be added.

1 $A + B = \begin{bmatrix} 14 & 20 & 22 \\ -6 & 14 & 2 \end{bmatrix}$; $A - B = \begin{bmatrix} -8 & -4 & -4 \\ 2 & -6 & -8 \end{bmatrix}$

2 $A + B = \begin{bmatrix} -8 & -3 \\ -1 & 10 \end{bmatrix}$; $A - B = \begin{bmatrix} -14 & -11 \\ -11 & -2 \end{bmatrix}$

3 $A + B = \begin{bmatrix} 22 & 22 & 22 \\ 22 & 22 & 22 \\ 22 & 22 & 22 \end{bmatrix}$; $A - B = \begin{bmatrix} -20 & -18 & -16 \\ -14 & -12 & -10 \\ -8 & -6 & -4 \end{bmatrix}$

4 $A + B = \begin{bmatrix} -3 & -3 \\ -4 & -4 \\ 15 & 15 \end{bmatrix}$; $A - B = \begin{bmatrix} 3 & -3 \\ -4 & 4 \\ -15 & 15 \end{bmatrix}$

3 Scalar Multiplication and Matrix Multiplication

Objective To learn how to use a calculator to multiply a matrix by a scalar and to multiply two matrices

Once you know how to enter matrices, you can manipulate them by using the functions built into the calculator. The time-consuming operation of matrix multiplication can be carried out quickly on any graphing calculator.

Example

Let $A = \begin{bmatrix} 5 & 3 & 9 \\ 3 & 1 & 2 \end{bmatrix}$ and $B = \begin{bmatrix} 9 & 5 \\ 0 & 8 \\ 7 & 8 \end{bmatrix}$. Find 6A and AB.

TI-81

To store the matrices, follow the procedures outlined in Lesson 2.

The expression 6A represents the multiplication of matrix A by the scalar 6. To find 6A, press

6 [2nd] [[A]] [ENTER]

The answer, $\begin{bmatrix} 30 & 18 & 54 \\ 18 & 6 & 12 \end{bmatrix}$, is displayed on the screen.

The expression AB represents the product of matrices A and B. To find AB, press

[2nd] [[A]] [×] [2nd] [[B]] [ENTER]

The answer is displayed on the screen.

continued on next page

TI-82

To store the matrices, follow the procedures outlined in Lesson 2.

The expression 6A represents the multiplication of matrix A by the scalar 6. To find 6A, press

6 [MATRX] 1 [ENTER]

The answer $\begin{bmatrix} 30 & 18 & 54 \\ 18 & 6 & 12 \end{bmatrix}$, is displayed on the screen.

The expression AB represents the product of matrices A and B. To find AB, press

[MATRX] 1 [×] [MATRX] 2 [ENTER]

The answer is displayed on the screen.

continued on next page

Casio fx-7700GB

To store the matrices, follow the procedures outlined in Lesson 2.

The expression 6A represents the multiplication of matrix A by the scalar 6. To find 6A, press [MODE] 0 to get to the main matrix menu, or press [PRE] if you are already in matrix mode but not at the main menu. Then press [F1] to access matrix A, followed by 6 [F1]. The answer, $\begin{bmatrix} 30 & 18 & 54 \\ 18 & 6 & 12 \end{bmatrix}$, is displayed on the screen as matrix C.

The expression AB represents the product of matrices A and B. To find AB, press [PRE] to return to the main matrix menu, then press [F5]. The answer is displayed on the screen as matrix C.

continued on next page

TI-81

continued from previous page

Your screen should look like this:

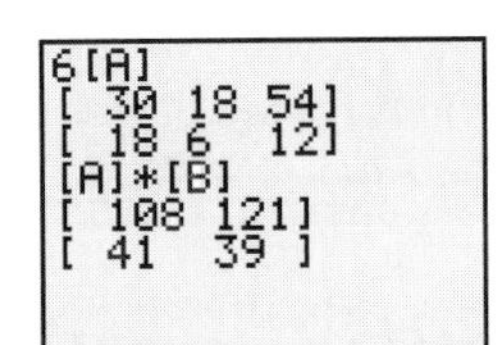

TI-82

continued from previous page

Your screen should look like this:

```
6[A]
      [[30 18 54]
       [18 6  12]]
[A]*[B]
       [[108 121]
        [41  39 ]]
```

Casio fx-7700GB

continued from previous page

Your screen should look like this:

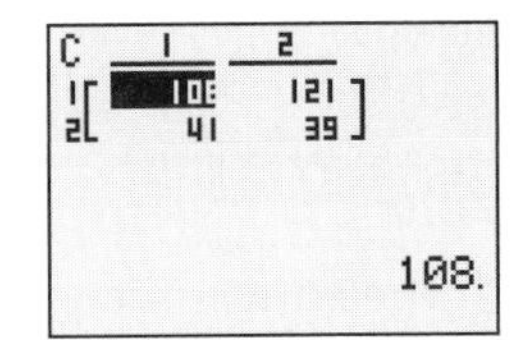

Hint

When performing matrix multiplication, you can omit pressing [×].

PROBLEMS

In problems 1–4, use your calculator to find AB, 5A, and 12B for each pair of matrices.

1 $A = \begin{bmatrix} 3 & 4 \\ 5 & 6 \end{bmatrix}$ and $B = \begin{bmatrix} 5 & 1 & 1 \\ -2 & 4 & 5 \end{bmatrix}$

2 $A = \begin{bmatrix} 1 & 2 & 3 \\ 4 & 5 & 6 \\ 7 & 8 & 9 \end{bmatrix}$ and $B = \begin{bmatrix} 1 \\ 0 \\ -1 \end{bmatrix}$

3 $A = \begin{bmatrix} 13 & 31 & 17 & 29 \end{bmatrix}$ and $B = \begin{bmatrix} -2 \\ -3 \\ 10 \\ 25 \end{bmatrix}$

4 $A = \begin{bmatrix} 1 & 0 & 0 \\ 0 & 1 & 0 \\ 0 & 0 & 1 \end{bmatrix}$ and $B = \begin{bmatrix} 3 & 1 & 5 & 4 \\ 6 & 7 & 0 & 9 \\ 8 & 9 & 5 & 7 \end{bmatrix}$

5 For any two real numbers a and b, $ab = ba$. Does this property also apply to matrices? Use several pairs, A and B, of 2 × 2 matrices to investigate AB and BA. Explain your findings.

6 Explain why AB consists of a single element in problem **3**.

7 Let $A = \begin{bmatrix} 6 & 7 \\ 8 & 9 \end{bmatrix}$ and $B = \begin{bmatrix} 7 & 0 & 8 \\ 8 & 7 & 1 \end{bmatrix}$. Use your calculator to find AB and 2A∗3B. See if you can figure out how to find 2A∗3B without entering "2A" and "3B" on your calculator. Come to class ready to report on the method you discovered.

1 $AB = \begin{bmatrix} 7 & 19 & 23 \\ 13 & 29 & 35 \end{bmatrix}$; $5A = \begin{bmatrix} 15 & 20 \\ 25 & 30 \end{bmatrix}$; $12B = \begin{bmatrix} 60 & 12 & 12 \\ -24 & 48 & 60 \end{bmatrix}$

2 $AB = \begin{bmatrix} -2 \\ -2 \\ -2 \end{bmatrix}$; $5A = \begin{bmatrix} 5 & 10 & 15 \\ 20 & 25 & 30 \\ 35 & 40 & 45 \end{bmatrix}$; $12B = \begin{bmatrix} 12 \\ 0 \\ -12 \end{bmatrix}$

3 $AB = [\,776\,]$; $5A = \begin{bmatrix} 65 & 155 & 85 & 145 \end{bmatrix}$; $12B = \begin{bmatrix} -24 \\ -36 \\ 120 \\ 300 \end{bmatrix}$

Additional Answers

4 $AB = \begin{bmatrix} 3 & 1 & 5 & 4 \\ 6 & 7 & 0 & 9 \\ 8 & 9 & 5 & 7 \end{bmatrix}$; $5A = \begin{bmatrix} 5 & 0 & 0 \\ 0 & 5 & 0 \\ 0 & 0 & 5 \end{bmatrix}$; $12B = \begin{bmatrix} 36 & 12 & 60 & 48 \\ 72 & 84 & 0 & 108 \\ 96 & 108 & 60 & 84 \end{bmatrix}$

5 In general, $AB \neq BA$. Sometimes the product AB exists, but not the product BA (as, for example, in problems **1** and **2**).

6 Since A has one row and B has one column, AB has one row and one column.

7 $AB = \begin{bmatrix} 98 & 49 & 55 \\ 128 & 63 & 73 \end{bmatrix}$; $2A*3B = \begin{bmatrix} 588 & 294 & 330 \\ 768 & 378 & 438 \end{bmatrix}$; students may discover that $2A*3B = 6AB$, so that they can obtain the answer by evaluating the latter expression.

4 GRAPHING INEQUALITIES

Objective To learn how to use a calculator to graph simple inequalities and to identify their solution sets

A graph is a handy way to record the solutions of an inequality. You can use some graphing calculators to generate graphs of linear inequalities.

Example

Graph the solution set of the inequality $16 > 3x - 5$.

By using algebraic operations, you can find that the solution set is $\{x : x < 7\}$. To graph these solutions on a number line, you shade the points to the left of 7, using an open dot at 7 to indicate that it is not a solution. By using a calculator, however, you can graph the inequality without solving it.

TI-81

First, set your calculator to graph in dot mode by pressing MODE, using the arrow keys to move the flashing cursor over "Dot" in the fifth row, then pressing ENTER. Press CLEAR to clear the screen.

To graph the inequality, press Y=, then

1 6 2nd [TEST] 3 3 X|T − 5 GRAPH

When you pressed 2nd [TEST], you accessed a menu of symbols that are used to test the truth of mathematical statements. In this case, you have tested whether 16 is greater than $3x - 5$ for any x values on the display. If the answer is yes, the calculator lets $y = 1$. If the answer is no, the calculator lets $y = 0$.

continued on next page

TI-82

First, set your calculator to graph in dot mode by pressing MODE, using the arrow keys to move the flashing cursor over "Dot" in the fifth row, then pressing ENTER. Press CLEAR to clear the screen.

To graph the inequality, press Y=, then

1 6 2nd [TEST] 3 3 X,T,θ − 5 GRAPH

When you pressed 2nd [TEST], you accessed a menu of symbols that are used to test the truth of mathematical statements. In this case, you have tested whether 16 is greater than $3x - 5$ for any x values on the display. If the answer is yes, the calculator lets $y = 1$. If the answer is no, the calculator lets $y = 0$.

continued on next page

Casio fx-7700GB

The Casio fx-7700GB does not graph inequalities.

TI-81	TI-82
continued from previous page Thus, the calculator graphs the solution set along the horizontal line $y = 1$. To read the graph, you can count the tick marks along the x-axis. Another way is to press TRACE and move the flashing cursor to the end of the ray. The x value displayed at the bottom of the screen is the approximate coordinate of the graph's endpoint.	*continued from previous page* Thus, the calculator graphs the solution set along the horizontal line $y = 1$. To read the graph, you can count the tick marks along the x-axis. Another way is to press TRACE and move the flashing cursor to the end of the ray. The x value displayed at the bottom of the screen is the approximate coordinate of the graph's endpoint.

Hint

When you have finished graphing inequalities, return your calculator to connected graphing mode by pressing MODE, moving the flashing cursor over "Connected" in the fifth row, and pressing ENTER.

PROBLEMS

In problems 1–6, use a calculator to graph each inequality, then record the inequality's solution set in set notation.

1 $2x + 5 > 11$ $\{x{:}x > 3\}$

2 $3x > x - 4$ $\{x{:}x > -2\}$

3 $-x + 3 < 5x + 15$ $\{x{:}x > -2\}$

4 $x + 2 \geq 3x - 1$ $\{x{:}x \leq 1.5\}$

5 $9 - x \leq 19 - 7x$ $\{x{:}x \leq 1.\overline{6}\}$

6 $x + 1 < x$ No solution

5 GRAPHING LINEAR EQUATIONS

Objective To learn how to use a calculator to graph linear equations

Graphing calculators can be used to graph any linear equation in the form $y =$ *expression*. You can also graph several linear equations at the same time to compare their characteristics.

Example 1

Graph $3x - y = 6$.

We start by isolating y on one side of the equation. For this equation, we get $y = 3x - 6$.

TI-81

First, press [Y=] and use the [CLEAR] key to make sure all four rows are cleared of entries. If necessary, use the [▲] key to return the cursor to the row headed "Y_1=."

To enter the equation, press 3 [X|T] [−] 6.

Then press [GRAPH]. The graph is displayed on the screen, as shown below.

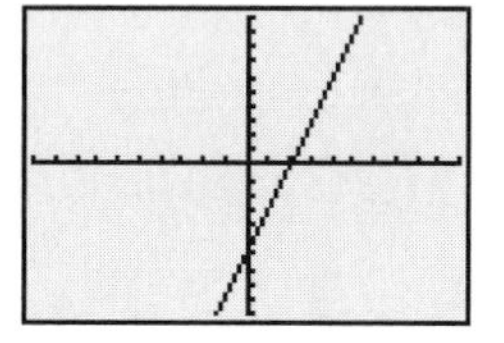

TI-82

First, press [Y=] and use the [CLEAR] key to make sure all the rows are cleared of entries. If necessary, use the [▲] key to return the cursor to the row headed "Y_1=."

To enter the equation, press 3 [X,T,θ] [−] 6.

Then press [GRAPH]. The graph is displayed on the screen, as shown below.

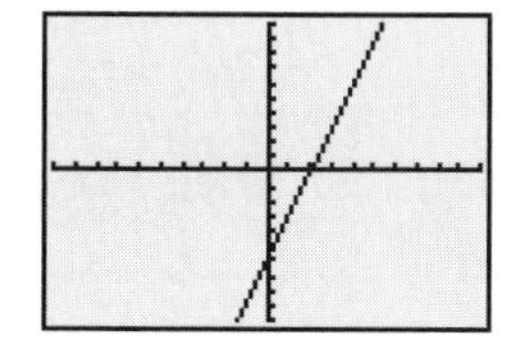

Casio fx-7700GB

Make sure you are in graphing mode by pressing [MODE] 1 [MODE] [+] [MODE] [SHIFT] [+].

To enter the equation, press [GRAPH]. "Graph Y=" will appear on the screen. Then press 3 [X,θ,T] [−] 6.

Press [EXE], and the graph is displayed on the screen, as shown below.

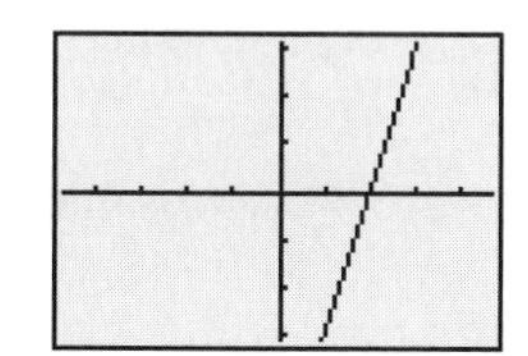

To clear the graphing screen at any time, press [SHIFT] [Cls] [EXE].

Hint

You don't have to rewrite an equation such as $3x + 4y = 8$ in slope-intercept form in order to enter it on a calculator. You may find entering this equation as $y = (8 - 3x) \div 4$ easier than taking the time to rewrite it as $y = -\frac{3}{4}x + 2$.

Example 2

Graph the family of equations $\{y = x - 3, y = x - 1, y = x + 1, y = x + 3\}$ and describe any interesting characteristics of the graphs.

TI-81

To enter the first equation, proceed as in Example 1. When you finish entering the equation, press ENTER. The cursor will move to the row headed "Y_2=." Enter the equation $y = x - 1$ there and proceed to enter the rest of the equations.

When you finish entering the equations, press GRAPH to see the four graphs. The result is a family of parallel lines, as shown below.

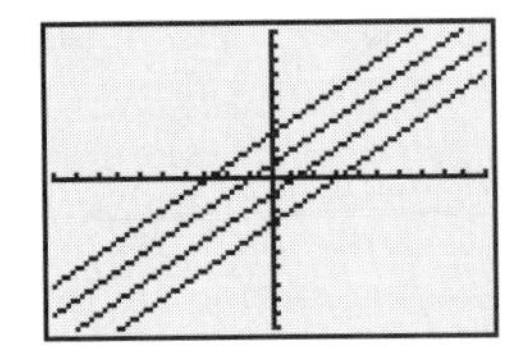

TI-82

To enter the first equation, proceed as in Example 1. When you finish entering the equation, press ENTER. The cursor will move to the row headed "Y_2=." Enter the equation $y = x - 1$ there and proceed to enter the rest of the equations.

When you finish entering the equations, press GRAPH to see the four graphs. The result is a family of parallel lines, as shown below.

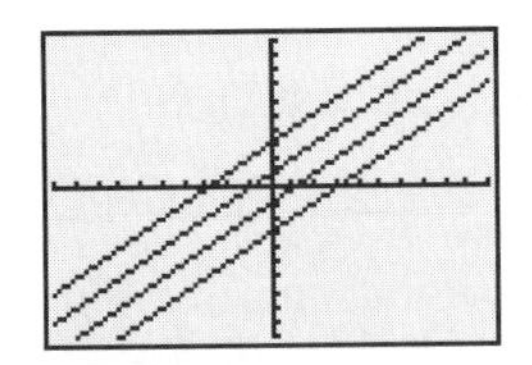

Hint

You can create tables of x and y values that are solutions of equations you are graphing. To do this for an equation you have already entered, press 2nd [TblSet]. The cursor will prompt you at "TblMin=." Type a starting x value for your table and press ENTER. Then the cursor will prompt you at "ΔTbl=." Here, enter the interval that you want between the x values in your table. If you want only integers, for example, press 1 ENTER. To see the table appear, press 2nd [TABLE].

Casio fx-7700GB

To enter the first equation, proceed as in Example 1. When you finish entering the equation, press EXE, then enter the second equation just as you did the first. Proceed to enter the rest of the equations, hitting EXE each time to see the new graph along with the ones already drawn.

The result is a family of parallel lines, as shown below.

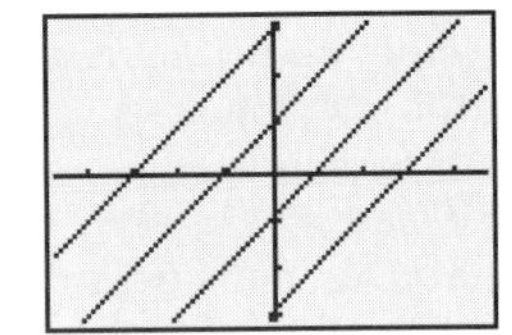

Problems

In problems 1–6, use a graphing calculator to graph each equation.

1 $y = 3x - 2$
2 $2x + 3y = 1$
3 $x - y = 4$
4 $x = 3y + 3$
5 $y = 2$
6 $x + 2y = y - x + 1$

7 Rewrite the equation $2x + 4y = 11$ in the form $y =$ *expression*. Then write down at least two sets of keystrokes you could use to enter the equation on your calculator. Which way would you choose to enter the equation? Why?

8 Rewrite the equation $x + 7y = 12 - y$ in the form $y =$ *expression*. Then write down at least two sets of keystrokes you could use to enter the equation on your calculator. Which way would you choose to enter the equation? Why?

In problems 9–11, determine whether the equation is equivalent to $2x + 3y = 6$.

9 $y = -\left(\frac{2x}{3}\right) + 2$ Yes
10 $y = -\left(\frac{2x + 2}{3}\right)$ No
11 $y = -\left(\frac{2(x - 1)}{3}\right)$ No

12 Graph the family of equations $\{y = x, y = 1.5x, y = 2x, y = 3x\}$ and describe any interesting characteristics of the graphs.

13 Graph the family of equations $\{y = x + 2, y = -x + 2, y = x - 2, y = -x - 2\}$ and describe any interesting characteristics of the graphs. (Note: On the TI-81 and TI-82, you need to press ZOOM 5 after graphing to see what is interesting about this family of equations. Do you see why this step is necessary?)

1–6 Check students' calculator displays.
7 Two possibilities are $y = (11 - 2x) \div 4$ and $y = (-1 \div 2)x + 11 \div 4$, of which most students will find the first easier to enter.
8 Two possibilities are $y = (12 - x) \div 8$ and $y = (-1 \div 8)x + 3 \div 2$, of which most students will find the first easier to enter.
12 Check students' calculator displays; students may note that all four lines intersect at the origin and that the slope increases as the coefficient of x increases.
13 Check students' calculator displays; students may note that the graphs are two pairs of parallel lines, with each pair perpendicular to the other.

6 SETTING THE VIEWING WINDOW

Objective To learn how to adjust the x-axis and the y-axis to obtain more useful graphs

In Lesson 5, you graphed equations within your calculator's standard viewing window. The dimensions of this window are illustrated below. In each case, "max" means "maximum," "min" means "minimum," and "scl" means "scale."

TI-81 and TI-82

Xmin = –10
Xmax = 10
Xscl = 1
Ymin = –10
Ymax = 10
Yscl = 1

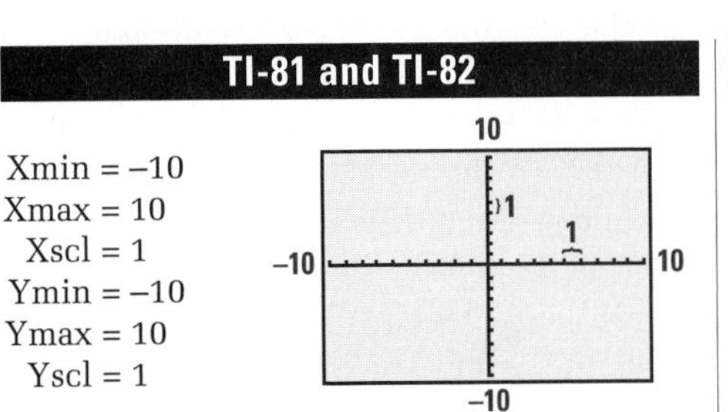

Casio fx-7700GB

Xmin = –4.7
Xmax = 4.7
Xscl = 1
Ymin = –3.1
Ymax = 3.1
Yscl = 1

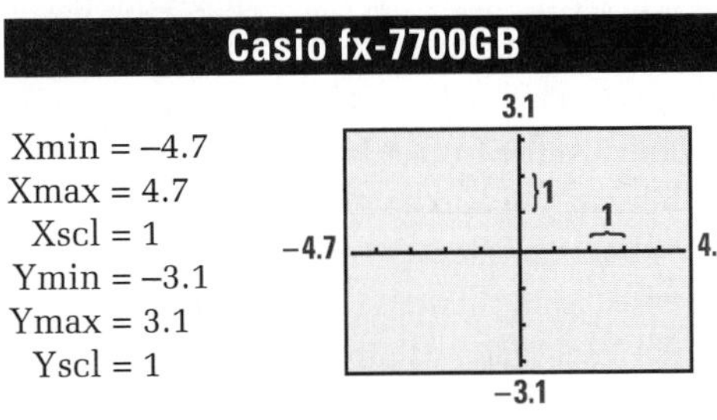

When you use your calculator to graph an equation, you can change the dimensions of the viewing window. If a graph is linear, knowing its slope and y-intercept can help you determine how to adjust the viewing window.

Example 1

Graph $y = 20x - 80$.

On no graphing calculator does the standard viewing window extend far enough to show the point, (0, –80), where this equation's graph intersects the y-axis. To make a better graph, you can adjust the window to show more of the y-axis.

TI-81

Press [ZOOM] 6 to put your calculator in standard graphing mode, then graph the equation. As you can see, the line is almost vertical, and the y-intercept is not shown.

To adjust the window, press [RANGE], then move the cursor down to the line headed "Ymin=." The value entered

continued on next page

TI-82

Press [ZOOM] 6 to put your calculator in standard graphing mode, then graph the equation. As you can see, the line is almost vertical, and the y-intercept is not shown.

To adjust the window, press [WINDOW], then move the cursor down to the line headed "Ymin=." The value entered

continued on next page

Casio fx-7700GB

Press [RANGE] [F1] [RANGE] [RANGE] to put your calculator in standard graphing mode, then graph the equation. As you can see, the line is almost vertical, and the y-intercept is not shown.

To adjust the window, press [RANGE], then move the cursor down to the line headed "Ymin:." The value entered

continued on next page

TI-81

continued from previous page

here determines the minimum y value shown on the screen. Replace –10 with –200 so that more of the graph, particularly the y-intercept, will be seen. Then press ENTER and change the "Ymax" value from 10 to 100. This sets the maximum y value shown on the screen. Press ENTER again to move down to "Yscl." The value entered here determines the interval at which tick marks will be shown on the y-axis. Change this entry to 50.

Then press GRAPH to see the graph, as shown below.

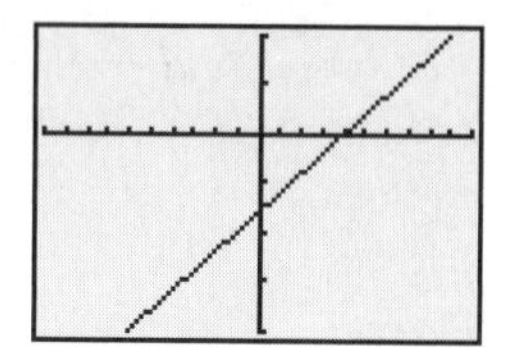

TI-82

continued from previous page

here determines the minimum y value shown on the screen. Replace –10 with –200 so that more of the graph, particularly the y-intercept, will be seen. Then press ENTER and change the "Ymax" value from 10 to 100. This sets the maximum y value shown on the screen. Press ENTER again to move down to "Yscl." The value entered here determines the interval at which tick marks will be shown on the y-axis. Change this entry to 50.

Then press GRAPH to see the graph, as shown below.

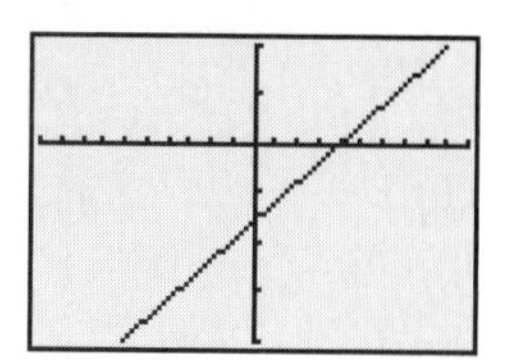

Casio fx-7700GB

continued from previous page

here determines the minimum y value shown on the screen. Replace –3.1 with –200 so that more of the graph, particularly the y-intercept, will be seen. Then press EXE and change the "max" value from 3.1 to 100. This sets the maximum y value shown on the screen. Press EXE again to move down to "scl." The value entered here determines the interval at which tick marks will be shown on the y-axis. Change this entry to 50.

Then press RANGE twice to return to the calculation screen. Press EXE to regraph the equation, as shown below.

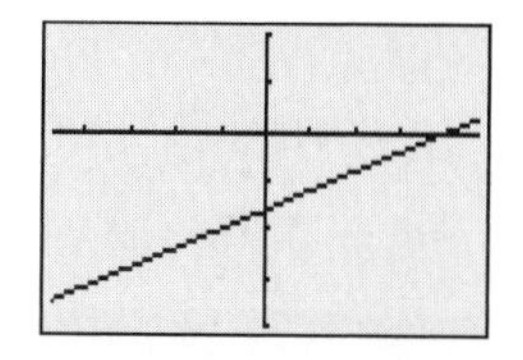

Example 2

Graph $100x + 2y = 2000$.

Solving for y, you get $y = -50x + 1000$. When you graph this equation on a calculator set for the standard viewing window, you see an empty screen! The whole graph lies outside the boundaries of the standard viewing window. To fix this problem, you can adjust the range of both axes.

TI-81

Since the y-intercept is 1000, press RANGE and change Ymin to –100, Ymax to 1200, and Yscl to 100. After you press GRAPH, your screen will look like this:

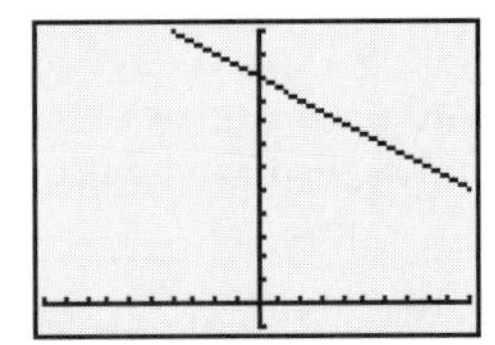

The new graph is an improvement, but it would be even better if the x-intercept were visible. Algebraically, you can find that the x-intercept is 20. Press RANGE again and change Xmax to 25 and Xscl to 5. After you press GRAPH, your screen will look like this:

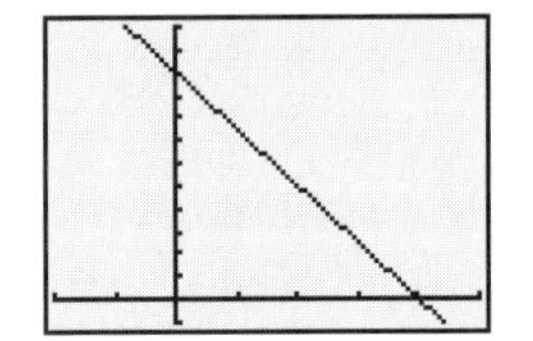

TI-82

Since the y-intercept is 1000, press WINDOW and change Ymin to –100, Ymax to 1200, and Yscl to 100. After you press GRAPH, your screen will look like this:

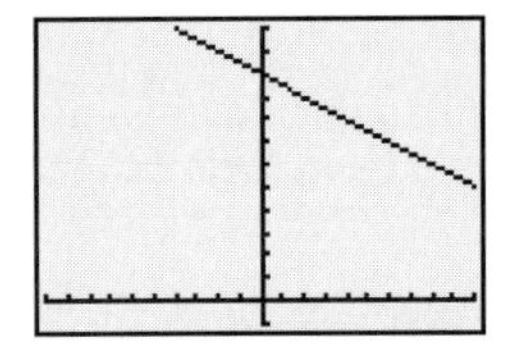

The new graph is an improvement, but it would be even better if the x-intercept were visible. Algebraically, you can find that the x-intercept is 20. Press WINDOW again and change Xmax to 25 and Xscl to 5. After you press GRAPH, your screen will look like this:

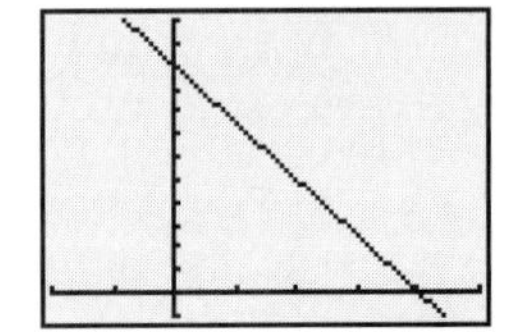

Hint

Check out the difference between dot mode and connected mode when you graph equations. Also note that you can remove all tick marks by entering 0 for Xscl and Yscl.

Casio fx-7700GB

Since the y-intercept is 1000, press RANGE and change Ymin to –100, Ymax to 1200, and Yscl to 100. Press RANGE twice, then regraph the equation. Your screen will look like this:

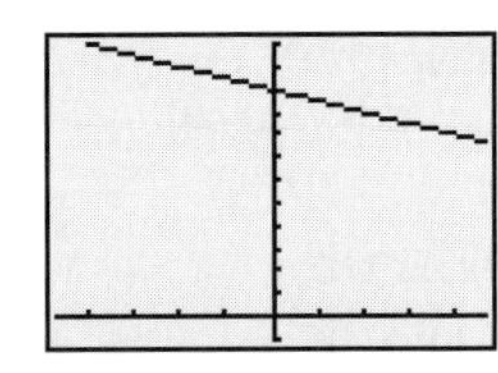

The new graph is an improvement, but it would be even better if the x-intercept were visible. Algebraically, you can find that the x-intercept is 20. Press RANGE again and change Xmax to 25 and Xscl to 5. After you regraph the equation, your screen will look like this:

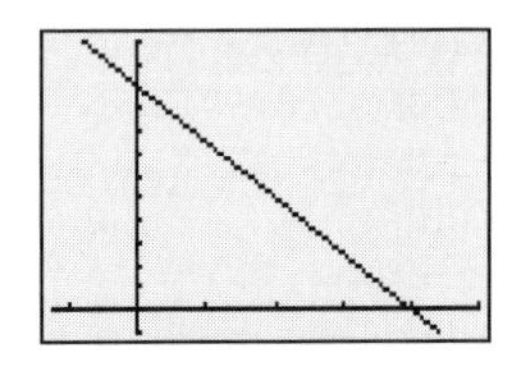

Hint

To instruct the calculator to produce connected graphs, press MODE MODE 5. For dot graphs, press MODE MODE 6.

PROBLEMS

In problems 1–6, graph each equation, using both the standard graphing window on your calculator and the window specified at right. Which window do you think is better to use? Why?

Xmin = –100
Xmax = 100
Xscl = 10
Ymin = –100
Ymax = 100
Yscl = 10

1 $y = 2x - 30$

2 $x + y = 100$

3 $x + y = 1$

4 $y = 100x$

5 $y = 5$

6 $y = x$

In problems 7–15, analyze each equation to decide upon appropriate settings for the x-axis and the y-axis. Then graph the equation on your graphing calculator and record the window settings that you used.

7 $30x + y = 60$

8 $30x - y = 60$

9 $x - 30y = 60$

10 $x + 30y = 60$

11 $y = 11x + 100$

12 $12x + 4y = 192$

13 $x = 18y + 162$

14 $6x + 7y = 420$

15 $101x + 99y = 9999$

1 Possible answer: The larger window, because the graph does not appear in the standard window
2 Possible answer: The larger window, because the graph does not appear in the standard window
3 Possible answer: The standard viewing window, because it shows the graph's y-intercept, x-intercept, and slope clearly
4 Possible answer: Neither window works well. The window needs further adjustment.
5 Possible answer: The standard window, because the position of the graph is clearer
6 Possible answer: Both work equally well.
7 Possible answer: Xmin = –10, Xmax = 10, Xscl = 1, Ymin = –100, Ymax = 100, Yscl = 10
8 Possible answer: Xmin = –10, Xmax = 10, Xscl = 1, Ymin = –100, Ymax = 100, Yscl = 10
9 Possible answer: Xmin = –100, Xmax = 100, Xscl = 10, Ymin = –3, Ymax = 3, Yscl = 1
10 Possible answer: Xmin = –100, Xmax = 100, Xscl = 10, Ymin = –3, Ymax = 3, Yscl = 1
11 Possible answer: Xmin = –20, Xmax = 10, Xscl = 2, Ymin = –100, Ymax = 200, Yscl = 20
12 Possible answer: Xmin = –100, Xmax = 100, Xscl = 10, Ymin = –100, Ymax = 100, Yscl = 10
13 Possible answer: Xmin = –100, Xmax = 300, Xscl = 20, Ymin = –20, Ymax = 10, Yscl = 2
14 Possible answer: Xmin = –100, Xmax = 100, Xscl = 10, Ymin = –100, Ymax = 100, Yscl = 10
15 Possible answer: Xmin = –150, Xmax = 150, Xscl = 10, Ymin = –150, Ymax = 150, Yscl = 10

7 Finding Statistical Means

Objective To learn how to use a calculator to find the mean of a set of data

Graphing calculators can operate in a statistical mode that allows you to enter and analyze data. Here, you will learn how to enter salaries and compute the mean salary.

Example

A law firm pays its employees the following annual salaries. Find the mean salary.

Employee	Salary	Number of Employees
Partner	$250,000	8
Associate	65,000	24
Paralegal Clerk	32,000	50
Office Clerk	25,000	40

TI-81

First, clear any stored data by pressing [2nd] [STAT] to access statistical mode, pressing [▸] twice to access the DATA menu, then pressing 2 [ENTER]. Your screen will display "Done."

Begin to enter data by pressing [2nd] [STAT] [▸] [▸] 1. The calculator will prompt you to enter $x1$. Press 2 5 0 0 0 0 [ENTER] to record $250,000 as the first salary. The cursor will move to $y1$. Press 8 [ENTER] to record 8 as the number of employees making that salary. Enter the three other pairs of numbers in the same way.

After entering all the data, press [2nd] [STAT] to access the CALC menu, then press

continued on next page

TI-82

First, clear any stored data. Press [STAT] to access statistical mode, then press 4. "ClrList" will appear on your screen. Since you will need two lists for your data, press [2nd] [L_1] [,] [L_2] [ENTER]. Your screen will display "Done."

Then press [STAT] again to access the EDIT menu. Press 1 to access the lists screen. First enter the four salaries under L_1:

2 5 0 0 0 0 [ENTER]

6 5 0 0 0 [ENTER]

3 2 0 0 0 [ENTER]

2 5 0 0 0 [ENTER]

Then press [▸] to move to L_2, and enter the corresponding numbers of employees in the same way.

continued on next page

Casio fx-7700GB

First, clear any stored data and prepare your calculator for statistical work by pressing [MODE] [×] [MODE] [SHIFT] 2, followed by [AC] [SHIFT] 3 [F2] [EXE] [PRE].

Then enter the data as follows:

2 5 0 0 0 0 [F3] 8 [F1]

6 5 0 0 0 [F3] 2 4 [F1]

3 2 0 0 0 [F3] 5 0 [F1]

2 5 0 0 0 [F3] 4 0 [F1]

By pressing [F3], you let the calculator know that the next entry represents the number of employees who had the preceding salary. By pressing [F1], you tell the calculator to input the data pair.

continued on next page

TI-81

continued from previous page
1 ENTER to compute statistical measures. On the screen, the mean is displayed as $\overline{x}$. In this case, the mean salary is ≈$50,491.80.

TI-82

continued from previous page
To find the mean, first press 2nd [QUIT] to return to a blank screen, then press 2nd [LIST] ▸ 3 2nd [L_1] , 2nd [L_2].

You have now instructed the calculator to find the mean of the values in L_1 at the frequencies listed in L_2. When you press ENTER, the mean, ≈$50,491.80, is displayed.

Casio fx-7700GB

continued from previous page
When you finish entering the data, press F4 F1 EXE.

The calculator displays the mean, ≈$50,491.80.

PROBLEMS

1 The following table shows employees' annual salaries at a department store. Find the employees' mean salary. $29,548.39

Employee	Salary	Number of Employees
Manager	$50,000	4
Sales	30,000	48
Security	21,000	4
Maintenance	18,000	6

2 The U.S. federal government paid its employees the following annual salaries in 1990. What was the mean salary of federal employees in that year? $31,239.02

Grade	Mean Salary	Number of Employees
GS-1	$10,947	921
GS-2	12,564	6,603
GS-3	14,284	44,523
GS-4	16,453	140,077
GS-5	18,699	193,018
GS-6	21,075	103,033
GS-7	23,005	146,823
GS-8	26,160	32,040
GS-9	27,793	157,679
GS-10	31,892	29,238
GS-11	33,812	196,320
GS-12	40,801	208,574
GS-13	49,003	138,266
GS-14	58,363	73,140
GS-15	70,316	34,787
GS-16	77,605	572
GS-17	78,200	88
GS-18	78,200	46

3 In problem **2,** one grade has only 46 employees, and some grades have more than 100,000. Find the mean number of employees per grade. ≈83,653

4 How could you have found the mean in the law-firm example without entering the data into lists?

5 Suppose that the law firm in the example places an advertisement in a newspaper, claiming that the average annual salary at the law firm is more than $50,000. Do you think that this claim is fair? Explain. Possible answer: The mean is indeed greater than $50,000, but since 90 out of 122 employees earn $32,000 or less and since the greatest number of employees earn $32,000, the median and the mode of the salaries are considerably less than $50,000. Therefore, the claim is somewhat unfair.

4 Possible answer: By evaluating [8(250,000) + 24(65,000) + 50(32,000) + 40(25,000)] ÷ 122

8 Solving Systems of Linear Equations

Objective To learn how to use a calculator to find the solution of a system of two linear equations

Any two linear equations can be graphed on your graphing calculator, and if the graphs intersect, their point of intersection can be found. This lesson explores ways to determine the coordinates of points of intersection.

Sometimes, a solution can be guessed from the calculator screen and checked. The first example involves such a case.

Example 1

Solve this system: $\begin{cases} y = 2x - 4 \\ y = -x + 5 \end{cases}$

TI-81

Set the standard viewing window by pressing [ZOOM] 6. Then press [Y=] and clear any equations that are stored.

Enter the two equations as explained in Lesson 5, letting $Y_1 = 2x - 4$ and $Y_2 = -x + 5$. Then press [GRAPH] to see the system. Your screen will look like this:

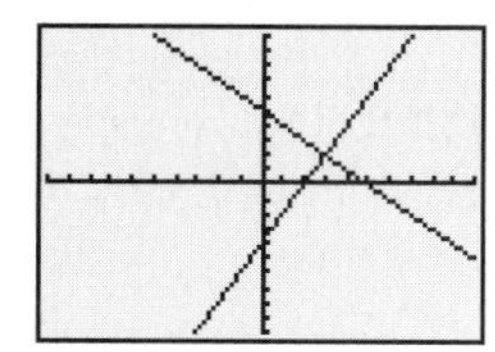

The solution appears to be (3, 2). Check it in both equations:

$2 = 2(3) - 4$ ✔

$2 = -(3) + 5$ ✔

TI-82

Set the standard viewing window by pressing [ZOOM] 6. Then press [Y=] and clear any equations that are stored.

Enter the two equations as explained in Lesson 5, letting $Y_1 = 2x - 4$ and $Y_2 = -x + 5$. Then press [GRAPH] to see the system. Your screen will look like this:

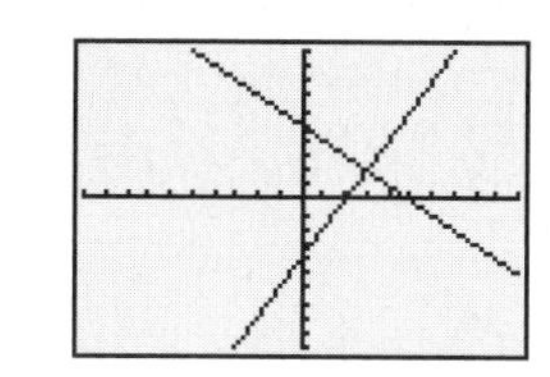

The solution appears to be (3, 2). Check it in both equations:

$2 = 2(3) - 4$ ✔

$2 = -(3) + 5$ ✔

Casio fx-7700GB

Make sure that your calculator is in graphing mode by following the steps described in Lesson 5. Then set the standard viewing window by pressing [RANGE] [F1] [RANGE] [RANGE].

Press [GRAPH] and enter the first equation as explained in Lesson 5. But instead of pressing [EXE] when you finish entering the equation, press [SHIFT] [↵]. Then enter the second equation, starting with [GRAPH] and ending with [EXE]. Your screen will look like this:

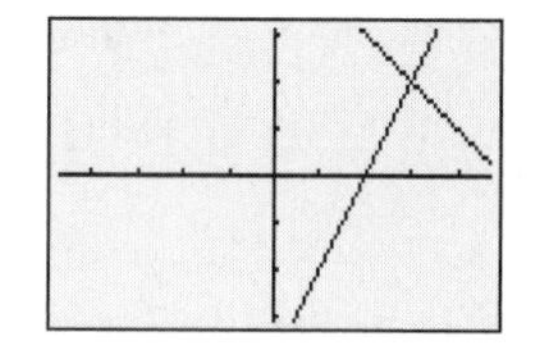

continued on next page

Casio fx-7700GB

continued from previous page

The solution appears to be (3, 2). Check it in both equations:

$$2 = 2(3) - 4 \checkmark$$

$$2 = -(3) + 5 \checkmark$$

When you have difficulty identifying the coordinates of a point of intersection, the trace and zoom features of your calculator can help you. The trace function allows you to move a cursor along a graph, with the calculator displaying the coordinates of each position. The zoom function allows you to "zoom in on" any portion of a graph so that you can more easily identify the coordinates of a point. The next example shows how these two features can help you identify solutions of systems.

Example 2

Solve this system: $\begin{cases} 2x + 3y = 4 \\ 4x + 2y = 3 \end{cases}$

When you solve each equation for y, then graph the equations as in Example 1, there appears to be a point of intersection near $x = 0$. But substituting 0 for x in the original equations yields two different values of y. You therefore have to dig deeper.

TI-81

To zoom in on the point of intersection, first press [TRACE]. A flashing cursor will appear on one of the two lines. Use the [▸] or [◂] key to move the cursor close to the point of intersection, then press [ZOOM] 2 [ENTER] to redraw the graphs automatically. The calculator will use ranges for the axes that make the point of intersection easier to identify.

continued on next page

TI-82

To determine the point of intersection, press [2nd] [CALC] 5. The calculator will prompt you to identify the first curve. With the cursor on your first graph, press [ENTER]. The calculator will then prompt you to identify the second curve. With the cursor on your second graph, press [ENTER]. The calculator then asks you for a guess.

continued on next page

Casio fx-7700GB

To zoom in on the point of intersection, first press [SHIFT] [Trace]. A flashing cursor will appear on one of the two lines. Use the [▸] or [◂] key to move the cursor close to the point of intersection, then press [SHIFT] [Zoom] [F3] to redraw the graphs automatically. The calculator will use ranges for the axes that make the point of intersection easier to identify.

continued on next page

TI-81

continued from previous page

You do not need to stop here. The trace/zoom procedure allows you to continue to zero in on the point of intersection. Try it one or two more times to see what happens.

After you have magnified the graphs several times, press TRACE once again. As before, pressing ► or ◄ moves the flashing cursor along one of your graphs. In addition, pressing ▲ or ▼ allows you to move the cursor from one graph to the other. By moving the cursor from graph to graph, you should be able to determine that both pass through the point (0.125, 1.25). You can check these coordinates in the original equations to confirm the solution algebraically.

TI-82

continued from previous page

Using ► or ◄, move the cursor close to the point of intersection and press ENTER.

The calculator will then work for a moment and display the coordinates of the point of intersection, (0.125, 1.25), at the bottom of the screen. You can check these coordinates in the original equations to confirm the solution algebraically.

If you wish to see the point of intersection more clearly, press TRACE to move the cursor onto one of your graphs. Move the cursor near the point of intersection, then press ZOOM 2 ENTER. The graph will be redrawn. You can repeat this process to zoom in further.

Hint

If you want to return to the original viewing window after zooming in, press ZOOM 3 ENTER repeatedly to zoom back out. If a point of intersection does not appear in the window when you first graph a system, reset the dimensions of the window as explained in Lesson 6.

Casio fx-7700GB

continued from previous page

You do not need to stop here. The trace/zoom procedure allows you to continue to zero in on the point of intersection. Try it one or two more times to see what happens.

After you have magnified the graphs several times, press SHIFT [Trace] once again. As before, pressing ► or ◄ moves the flashing cursor along one of your graphs. In addition, pressing ▲ or ▼ allows you to move the cursor from one graph to the other. By moving the cursor from graph to graph, you should be able to determine that both pass through the point (0.125, 1.25). You can check these coordinates in the original equations to confirm the solution algebraically.

Hint

To return to the original viewing window after zooming in, press SHIFT [Zoom] F5. If a point of intersection does not appear in the window when you first graph a system, you can use the arrow keys to reposition the window, or you can reset the window dimensions as explained in Lesson 6.

PROBLEMS

In problems 1–6, use a graphing calculator to solve each system. Check your solutions algebraically.

1 $\begin{cases} y = 4x - 3 \\ y = x + 3 \end{cases}$ (2, 5)

2 $\begin{cases} y = x + 2 \\ y = 6x - 3 \end{cases}$ (1, 3)

3 $\begin{cases} y = 8x \\ y = 6x + 1 \end{cases}$ (0.5, 4)

4 $\begin{cases} 2x + y = 12 \\ x + y = 8 \end{cases}$ (4, 4)

5 $\begin{cases} 5x + 2y = 1 \\ 20x + 50y = 27 \end{cases}$

6 $\begin{cases} 3x + 7y = 2 \\ 84x - 168y = 4 \end{cases}$

In problems 7–9, use a graphing calculator to solve each system. (Each problem requires you to adjust the viewing window's original dimensions.) Check your solutions algebraically.

7 $\begin{cases} x - y = 1 \\ x + y = 32 \end{cases}$ (16.5, 15.5)

8 $\begin{cases} 5x + 2y = 3 \\ 3x + y = -4 \end{cases}$ (−11, 29)

9 $\begin{cases} 11x + 13y = 15 \\ x + y = 15 \end{cases}$ (90, −75)

10 Find the coordinates of the vertices of the triangle formed by the graphs of $y = 8x$, $x + 2y = 20$, and $4y = x - 24$. Possible answer: ($\approx$1.176, $\approx$9.412), ($\approx$−0.774, $\approx$−6.193), ($21.\overline{3}$, $-0.\overline{6}$)

5 Possible answer: ($\approx$−0.019, $\approx$0.548)

6 Possible answer: ($\approx$0.333, $\approx$0.143)

9 FINDING A LINE OF BEST FIT

Objective To learn how to find the linear equation that best describes a set of data points on the coordinate plane

Data that involve two variables can be represented as points on a coordinate plane, and an equation obtained algebraically can be used to model the relationship between the variables. If the data points seem to cluster around a straight line, a statistical technique called regression analysis is used to find a linear regression equation for the data. This equation represents the line of best fit for the data. If you enter pairs of data values on your graphing calculator, the calculator can compute a linear regression equation for the data.

Example 1

The table shows the tuition costs at a private college in the years 1986–1994. Use your calculator to make a scattergram of the data to see if their relationship appears linear.

Year	Cost	Year	Cost	Year	Cost
1986	$7,800	1989	$ 9,000	1992	$12,500
1987	8,000	1990	10,000	1993	13,200
1988	8,500	1991	11,200	1994	13,800

TI-81

Clear your calculator of any stored data and enter the years as x values and the corresponding costs as y values, as explained in Lesson 7.

To produce an appropriate window for graphing the points, press RANGE and enter these values:

Xmin=1980
Xmax=2000
Xscl=1
Ymin=7000
Ymax=15000
Yscl=1000

continued on next page

TI-82

Clear your calculator of any stored data and enter the years in L_1 and the corresponding costs in L_2, as explained in Lesson 7.

Press 2nd [STAT PLOT] 1 to let the calculator know that you want to graph the data points. On the screen that appears, press ENTER to turn the data-display function on. Also make sure that in the line headed "Type:" the first picture is highlighted. If it is not, move the cursor over it and press ENTER.

continued on next page

Casio fx-7700GB

To set the viewing window for the scattergram, press RANGE and enter these values:

Xmin:1980
max:2000
scl:1
Ymin:7000
max:15000
scl:1000

Then press RANGE RANGE, followed by SHIFT 3 F2 EXE. Put your calculator in the appropriate mode for data entry and linear regression by pressing MODE ÷ MODE 4 MODE SHIFT 1. Then erase any stored data by pressing F2 F3 F1.

continued on next page

TI-81

continued from previous page

Press [2nd] [STAT] [▶] to access the statistical DRAW menu, then press 2 [ENTER]. The calculator displays a scattergram showing a pattern that is approximately linear.

TI-82

continued from previous page

Then press [ZOOM] 9. The calculator displays a scattergram showing a pattern that is approximately linear. Notice that the calculator automatically sets an appropriate window for the scattergram. If you want to see the dimensions it is using, press [WINDOW].

Casio fx-7700GB

continued from previous page

To let the calculator know that you want a scattergram, press [MODE] [SHIFT] 3, followed by [MODE] [SHIFT] [+].

Then enter the data pairs by pressing

1 9 8 6 [SHIFT] [,] 7 8 0 0 [F1]

1 9 8 7 [SHIFT] [,] 8 0 0 0 [F1]

and so forth.

Each time you press [F1], a point is added to the scatterplot. The final display shows a pattern that is approximately linear.

Example 2

Write a linear regression equation for the data in Example 1. Use the equation to predict the tuition in the year 2000.

TI-81

With the data from Example 1 still stored, press [2nd] [STAT] to access the statistical menu, then press 2 to select LinReg. Press [ENTER], and your screen should look like this:

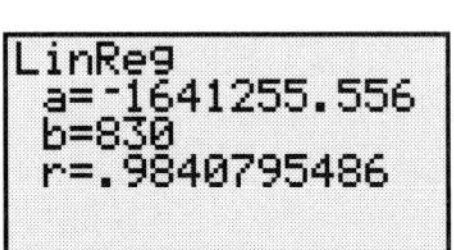

continued on next page

TI-82

With the data from Example 1 still stored, press [STAT] to access the statistical menu, then [▶] 9. "LinReg(a+bx)" will appear on the screen. Press [ENTER], and your screen should look like this:

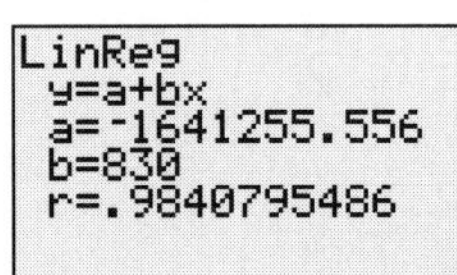

continued on next page

Casio fx-7700GB

With the data from Example 1 still stored and the calculator still in linear regression mode, press [F6] to access the regression menu. Then press [F1] [EXE] [F2] [EXE]. Your screen should look like this:

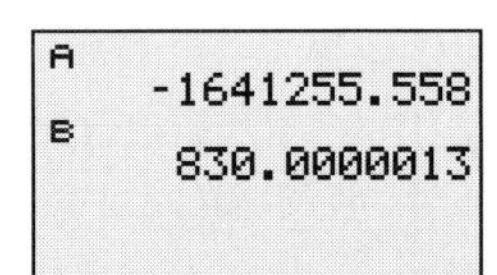

continued on next page

TI-81	TI-82	Casio fx-7700GB
continued from previous page The numbers displayed after "a=" and "b=" represent values in a regression equation of the form $y = a + bx$. Therefore, the regression equation for these data is $y = -1{,}641{,}255.556 + 830x$, where x is a year and y is the corresponding tuition.	*continued from previous page* The numbers displayed after "a=" and "b=" represent values in a regression equation of the form $y = a + bx$. Therefore, the regression equation for these data is $y = -1{,}641{,}255.556 + 830x$, where x is a year and y is the corresponding tuition.	*continued from previous page* The values displayed for A and B represent values in a regression equation of the form $y = a + bx$. Therefore, the regression equation for these data is approximately $y = -1{,}641{,}256 + 830x$, where x is a year and y is the corresponding tuition.
To predict the tuition in the year 2000, you can use the stored values of a and b, which are accessed by pressing VARS ▶ ▶, followed by 1 for a or 2 for b. To evaluate $a + b(2000)$, press VARS ▶ ▶ 1 + VARS ▶ ▶ 2 × 2 0 0 0 ENTER. The answer, ≈$18,744, is displayed on the screen.	To predict the tuition in the year 2000, you can use the stored values of a and b, which are accessed by pressing VARS 5 ▶ ▶, followed by 1 for a or 2 for b. To evaluate $a + b(2000)$, press VARS 5 ▶ ▶ 1 + VARS 5 ▶ ▶ 2 × 2 0 0 0 ENTER. The answer, ≈$18,744, is displayed on the screen.	To predict the tuition in the year 2000, you can use the stored values of a and b. To evaluate $a + b(2000)$, press F1 + F2 × 2 0 0 0 EXE. The answer, ≈$18,744, is displayed on the screen.

Hint

Once you have done the work shown in Examples 1 and 2, you can graph your regression line and your data points on the same screen to get an idea of how well the line fits the data.

Problems

In problems 1–3, use the data to draw a scattergram and write a regression equation.

1

x	1	12	4	7	5	8
y	3	30	12	18	17	20

Check students' scattergrams; $y = 2.272941176 + 2.334117647x$

2

x	12	13	14	15	16	17
y	12	13	14	15	16	30

Check students' scattergrams; $y = -24.76190476 + 2.857142857x$

3

x	−5	−3	12	23	34	34
y	44	31	−9	13	13	21

4 The parabolic graph of $y = x^2$ passes through the points (0, 0), (1, 1), (2, 4), (3, 9), and (4, 16). Write a regression equation for these five points, then graph it and the parabola on the same screen. $y = 4x - 2$; check students' graphs.

3 Check students' scattergrams; $y = 26.72648256 - 0.4985146881x$ (Students may note that the pattern of data points in this problem is clearly nonlinear, however.)

5 Repeat problem **4,** using the points (–2, 4), (–1, 1), (0, 0), (1, 1), and (2, 4) on the parabola as the basis of the regression equation. **$y = 2$; check students' graphs.**

6 Annual property taxes on a family's home are shown in the table. Write a regression equation for the data and use it to predict the property tax in the years 1998 and 2010. **Approximate equation: $y = -405{,}674 + 206.3416x$, where x is a year and y is the corresponding tax; predicted tax for 1998: \$6,596.18; predicted tax for 2010: \$9,072.28**

Year	Tax
1968	\$1000
1976	1400
1980	2300
1988	5300
1992	5400
1993	5400

7 Home-run statistics for the first five years of a baseball player's career are shown in the table. Write a regression equation for the data and use it to predict how many home runs the player will hit in the year 2000. Explain any problems you see with using the regression equation to make this prediction. **$y = -13{,}493 + 6.8x$, where x is a year and y is the corresponding number of home runs; predicted home runs in 2000: 107; possible answer: A linear model is not appropriate because a baseball player's home-run productivity does not increase every year over his entire career.**

Year	Home Runs
1985	6
1986	12
1987	15
1988	28
1989	32

10 Finding Roots of Quadratic Equations

Objective To learn how to use a calculator to find the roots of a quadratic equation

You have learned to find the roots of quadratic equations algebraically by factoring, by completing the square, and by using the quadratic formula. A fourth method, involving geometry, can be carried out on graph paper or by means of a graphing calculator.

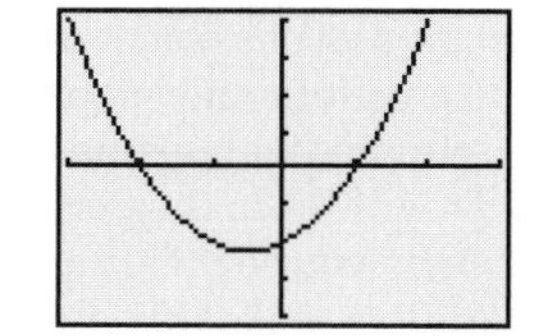

Suppose you want to find the roots of the equation $x^2 + x - 2 = 0$. The screen shown to the right displays the graph of $y = x^2 + x - 2$ (at the range settings Xmin = –3, Xmax = 3, Xscl = 1, Ymin = –4, Ymax = 4, and Yscl = 1).

The roots of $x^2 + x - 2 = 0$ correspond to the points on the graph of $y = x^2 + x - 2$ at which $y = 0$. In other words, the roots of the equation are the same as the x-intercepts of the graph. It appears that the roots of this equation are –2 and 1. Checking these values in the original equation confirms this:

$$(-2)^2 + (-2) - 2 = 0 \checkmark$$

$$(1)^2 + (1) - 2 = 0 \checkmark$$

Of course, not all quadratic equations have roots that are so easily identified. It is in more difficult cases that a graphing calculator is even more useful.

Example 1

Find the roots of $x^2 + 3x - 5 = 0$.

TI-81

Set the standard viewing window and clear any equations that are stored. Press [Y=], enter the equation $Y_1 = x^2 + 3x - 5$, then press [GRAPH]. Your screen should look like this:

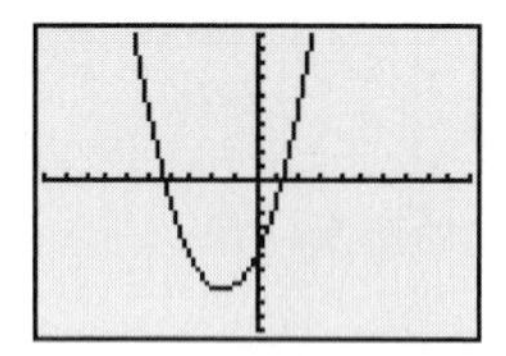

continued on next page

TI-82

Set the standard viewing window and clear any equations that are stored. Press [Y=], enter the equation $Y_1 = x^2 + 3x - 5$, and press [GRAPH]. Your screen should look like this:

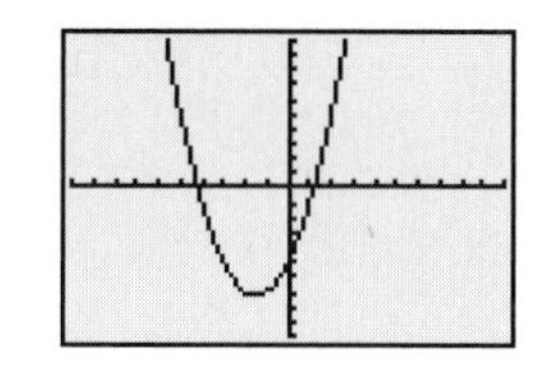

continued on next page

Casio fx-7700GB

Make sure that your calculator is in graphing mode by following the steps described in Lesson 5. Then set the standard viewing window by pressing [RANGE] [F1] [RANGE] [RANGE]. Press [GRAPH] and enter $y = x^2 + 3x - 5$.

continued on next page

TI-81

continued from previous page

It appears that the roots lie near $x = 1$ and $x = -4$. Press [TRACE] and move the cursor to a point as near $x = 1$ as you can get, then zoom in by pressing [ZOOM] 2 [ENTER]. Repeat the trace/zoom procedure a number of times; each time you will get a better estimate of the root near $x = 1$. Eventually, you will discover that this root is ≈ 1.19.

To zero in on the root near $x = -4$, first press [ZOOM] 6 to return to the standard window. Then repeat the procedure that you used to find the first root. You will find this root to be ≈ -4.19.

TI-82

continued from previous page

It appears that the roots lie near $x = 1$ and $x = -4$. To find the root near $x = 1$, press [2nd] [CALC] 2. The calculator will ask you for a lower bound—a point that lies close to the root but to the left of it. Use the arrow keys to move the cursor to a lower bound, and press [ENTER]. The calculator will then ask you for an upper bound—a point that lies close to the root but to the right of it. Use the arrow keys to move the cursor to an upper bound, and press [ENTER]. Finally, the calculator will ask you for a guess. Move the cursor as close as you can to the actual x-intercept and press [ENTER]. The calculator will display an approximation of the root, 1.1925824, as an x value.

To find the root near $x = -4$, repeat the procedure, starting again with [2nd] [CALC] 2. You should discover that the root is about -4.192582.

Casio fx-7700GB

continued from previous page

After you press [EXE], your screen should look like this:

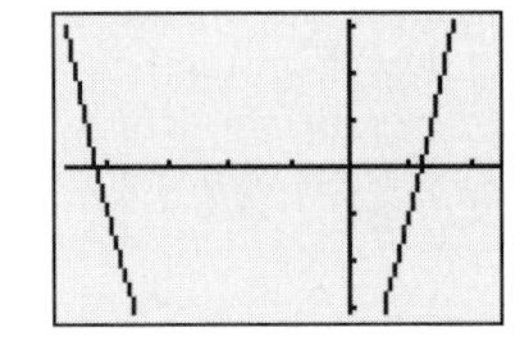

It appears that the roots lie near $x = 1$ and $x = -4$. Press [SHIFT] [Trace] and move the cursor to a point as near $x = 1$ as you can get, then zoom in by pressing [SHIFT] [Zoom] [F3]. Repeat the trace/zoom procedure a number of times; each time you will get a better estimate of the root near $x = 1$. Eventually, you will discover that this root is ≈ 1.19.

To zero in on the root near $x = -4$, first return to the original viewing window by pressing [SHIFT] [Zoom] [F5]. Then repeat the procedure that you used to find the first root. You will find this root to be ≈ -4.19.

The easiest algebraic approach to the problem in Example 1 is to use the quadratic formula. By using it, you would find that the roots are $\frac{-3 \pm \sqrt{3^2 - 4(1)(-5)}}{2}$, or ≈ 1.19258 and ≈ -4.19258. As Example 1 shows, a graphing calculator produces excellent approximations of the roots.

Example 2

Find the roots of $x^2 - 24x + 64 = 0$.

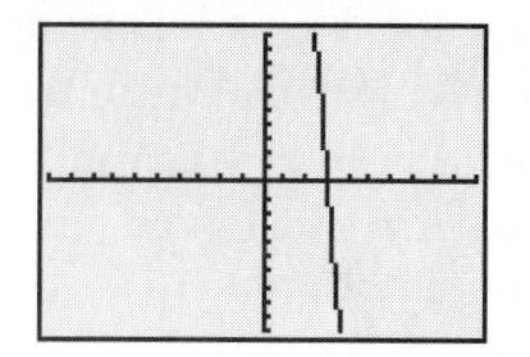

Graphing $y = x^2 - 24x + 64$ in the standard viewing window yields a graph like the one to the right. One of the roots is apparent, but the other is hidden. To locate the hidden root, you must expand the window by increasing the Xmax value—a process that involves trial and error. First try Xmax = 20. You will see that the other root is still not visible. Then try Xmax = 30. This choice makes both roots visible.

Once both roots are visible, you can proceed as in Example 1. The roots are ≈3.06 and ≈20.94.

Hint

If you cannot find any roots for a quadratic equation, it may be that the equation has no real roots.

PROBLEMS

In problems 1–12, use a graphing calculator to approximate the roots of each quadratic equation. Check your solutions algebraically.

1 $x^2 + x - 1 = 0$

2 $x^2 + 3x - 4 = 0$ 1, −4

3 $2x^2 + 3x + 1 = 0$ −0.5, −1

4 $3x^2 + 2x - 4 = 0$

5 $8x^2 - x - 1 = 0$

6 $12x^2 - 25x + 12 = 0$

7 $x^2 - 12x - 45 = 0$ 15, −3

8 $x^2 - 6x = 1$ ≈6.162, ≈−0.162

9 $x^2 = 11x + 300$

10 $100x^2 - 3 = 0$

11 $x^2 + 5x + 8 = 0$

12 $x^2 - 24x + 144 = 0$ 12

1 ≈0.618, ≈−1.618
4 ≈0.869, ≈−1.535
5 ≈0.422, ≈−0.297
6 $1.\overline{3}$, 0.75
9 ≈23.673, ≈−12.673
10 ≈0.173, ≈−0.173
11 No real roots

11 The Determinant of a Matrix

Objective To learn how to use a calculator to find the determinant of a square matrix

A determinant is a number associated with a square matrix. If $A = \begin{bmatrix} a & b \\ c & d \end{bmatrix}$, the determinant of A, written "det A," is $ad - bc$. It is easy to find the determinant of a 2×2 matrix with pencil and paper or with any scientific calculator. When a matrix is 3×3 or larger, however, calculating its determinant becomes more difficult.

Consider the following example:

$$\det \begin{bmatrix} a & b & c \\ d & e & f \\ g & h & i \end{bmatrix} = a\left(\det \begin{bmatrix} e & f \\ h & i \end{bmatrix}\right) - b\left(\det \begin{bmatrix} d & f \\ g & i \end{bmatrix}\right) + c\left(\det \begin{bmatrix} d & e \\ g & h \end{bmatrix}\right)$$

$$= aei - ahf - bdi + bgf + cdh - cge$$

For a 4×4 matrix, you would have to evaluate an expression containing 24 products. With a graphing calculator, once you have stored a matrix in the calculator, finding its determinant is simple.

Example

Use a graphing calculator to find $\det \begin{bmatrix} 1 & 6 & -4 \\ 5 & 0 & 4 \\ 7 & 2 & 11 \end{bmatrix}$.

TI-81

Follow the procedure explained in Lesson 2 to store the matrix as matrix A. Then press MATRX 5 to access the determinant function, and press 2nd [[A]] to identify the matrix. When you press ENTER, the calculator will display the answer, –210.

TI-82

Follow the procedure explained in Lesson 2 to store the matrix as matrix A. Then press MATRX ► 1 to access the determinant function, and press MATRX 1 to identify the matrix. When you press ENTER, the calculator will display the answer, –210.

Casio fx-7700GB

Follow the procedure explained in Lesson 2 to store the matrix as matrix A. Then press F3 (notice that the menu label for the determinant is |A|). The matrix will disappear from the screen, and the answer, –210, will be displayed.

Problems

In problems 1–8, use a calculator to find the determinant of each matrix.

1 $\begin{bmatrix} 1 & 2 & 3 \\ 4 & 5 & 6 \\ 7 & 8 & 9 \end{bmatrix}$ 0

2 $\begin{bmatrix} 9 & 8 & 7 \\ 6 & 5 & 4 \\ 3 & 2 & 1 \end{bmatrix}$ 0

3 $\begin{bmatrix} -12 & -3 \\ -4 & 8 \end{bmatrix}$ −108

4 $\begin{bmatrix} 0 & 2 \\ 4 & 0 \end{bmatrix}$ −8

5 $\begin{bmatrix} 3 & 5 & 8 & 9 \\ 2 & 0 & 8 & 4 \\ 6 & 3 & 8 & 1 \\ 7 & 7 & 4 & 9 \end{bmatrix}$ 1104

6 $\begin{bmatrix} -2 & -1 & 0 \\ 4 & 5 & 6 \\ 5 & 0 & 5 \end{bmatrix}$ −60

7 $\begin{bmatrix} 1 & 0 & 0 \\ 0 & 1 & 0 \\ 0 & 0 & 1 \end{bmatrix}$ 1

8 $\begin{bmatrix} 1 & 0 & 0 & 0 \\ 0 & 1 & 0 & 0 \\ 0 & 0 & 1 & 0 \\ 0 & 0 & 0 & 1 \end{bmatrix}$ 1

9 Find out how determinants are used in mathematics and science. Report to your classmates on what you find out. Answers will vary.

12 PERMUTATIONS AND COMBINATIONS

Objective To learn how to use a calculator to find numbers of permutations and combinations

In Section 5.5 of this book, you learned the permutation and combination formulas $P(n, r) = \frac{n!}{(n-r)!}$ and $C(n, r) = \frac{n!}{(n-r)!r!}$. When n is large, however, these formulas are difficult to evaluate without a special calculator or computer program. Graphing calculators have built-in functions that allow you to determine permutations and combinations easily.

Example

Use a graphing calculator to find $P(12, 6)$ and $C(12, 6)$.

TI-81

Your calculator symbolizes $P(12, 6)$ and $C(12, 6)$ as 12 nPr 6 and 12 nCr 6. To calculate $P(12, 6)$, first press 1 2. Then press MATH ► ► ► 2 to access the permutation function. Press 6 ENTER, and the answer, 665,280, will be displayed.

To calculate $C(12, 6)$, follow the same steps, but press 3 instead of 2 after pressing MATH ► ► ►. Your screen should look like this:

```
12 nPr 6
          665280
12 nCr 6
             924
```

Hint

Your calculator also has a factorial function, which can be accessed by pressing MATH 5.

TI-82

Your calculator symbolizes $P(12, 6)$ and $C(12, 6)$ as 12 nPr 6 and 12 nCr 6. To calculate $P(12, 6)$, first press 1 2. Then press MATH ► ► ► 2 to access the permutation function. Press 6 ENTER, and the answer, 665,280, will be displayed.

To calculate $C(12, 6)$, follow the same steps, but press 3 instead of 2 after pressing MATH ► ► ►. Your screen should look like this:

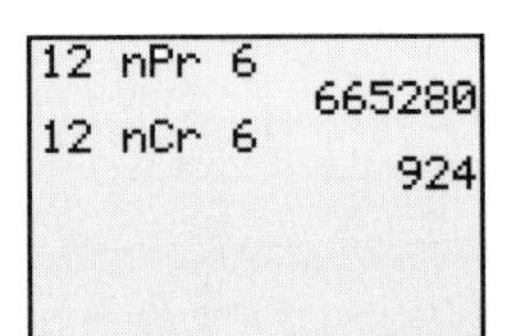

Hint

Your calculator also has a factorial function, which can be accessed by pressing MATH ► ► ► 4.

Casio fx-7700GB

Your calculator symbolizes $P(12, 6)$ and $C(12, 6)$ as 12P6 and 12C6. To calculate $P(12, 6)$, first press 1 2. Then press SHIFT [MATH] F2 F2 to access the permutation function. Press 6 EXE, and the answer, 665,280, will be displayed.

To calculate $C(12, 6)$, press 1 2 F3 6 EXE. Your screen should look like this:

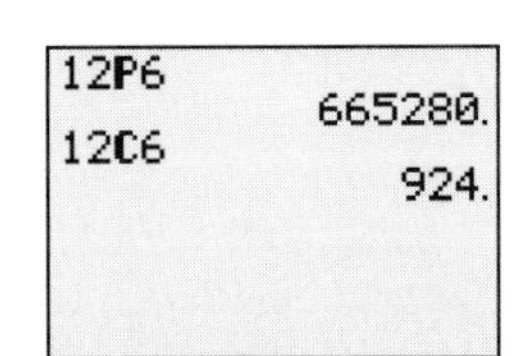

When you finish, return to computation mode by pressing PRE twice.

Hint

Your calculator also has a factorial function, which can be accessed by pressing SHIFT [MATH] F2 F1.

PROBLEMS

In problems 1–8, use a graphing calculator to evaluate each permutation or combination.

1 $P(72, 5)$ **2** $P(18, 8)$ **3** $P(10, 10)$ 3,628,800 **4** $P(33, 3)$ 32,736

5 $C(18, 12)$ 18,564 **6** $C(77, 7)$ **7** $C(10, 10)$ 1 **8** $C(24, 12)$

9 What happens if you try to use your calculator to find $P(n, r)$ or $C(n, r)$ where $n < r$? Why?

10 What happens if you try to use your calculator to find $P(n, r)$ or $C(n, r)$ for a negative value of r? Why?

11 Use your calculator to find $P(100, 10)$. Explain what the answer displayed by the calculator means.

12 In the game of hearts, a player's hand consists of 13 cards dealt from a standard deck of 52 cards. How many different hands are possible in a game of hearts? $\approx 6.35 \times 10^{11}$

1 1,678,985,280
2 1,764,322,560
6 2,404,808,340
8 2,704,156
9 The calculator displays either 0 or an error message; possible answer: A group of more than n objects cannot be selected from a group of n objects.
10 The calculator displays an error message; possible answer: A group of objects cannot contain a negative number of objects.
11 Possible answer: The calculator displays an answer in scientific notation, representing $6.281565096 \times 10^{19}$. This is an approximation, since the calculator cannot display all 20 digits of the answer.

13 GRAPHING CONIC SECTIONS

Objective To learn how to use a calculator to graph equations whose graphs are circles, ellipses, and hyperbolas

You already know how to use a calculator to produce the parabolic graphs of equations of the form $y = ax^2 + bx + c$. Because equations whose graphs are circles, ellipses, and hyperbolas are not functions, you must rewrite such an equation as two functions in order to graph it.

Example 1

Graph the equation $4x^2 + 9y^2 = 36$.

This equation has an elliptical graph. The first thing to do is to solve the equation for y.

$$4x^2 + 9y^2 = 36$$
$$9y^2 = 36 - 4x^2$$
$$y^2 = 4 - \tfrac{4}{9}x^2$$
$$y = \pm\sqrt{4 - \tfrac{4}{9}x^2}$$

The equation can now be rewritten as the two functions $y_1 = \sqrt{4 - \frac{4}{9}x^2}$ and $y_2 = -\sqrt{4 - \frac{4}{9}x^2}$.

TI-81

Clear any stored equations from your calculator and set the standard viewing window by pressing [ZOOM] 6. Press [Y=] and enter the two functions as Y_1 and Y_2. Then press [GRAPH]. Your screen should look like this:

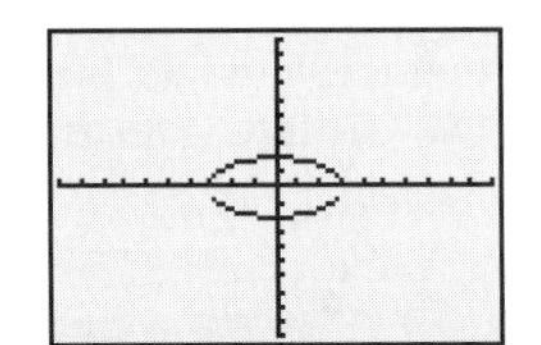

TI-82

Clear any stored equations from your calculator and set the standard viewing window by pressing [ZOOM] 6. Press [Y=] and enter the two functions as Y_1 and Y_2. Then press [GRAPH]. Your screen should look like this:

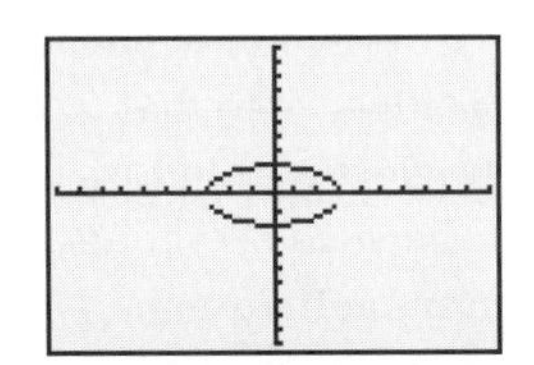

Hint

You can enter the second function in Example 1 by using the Y-VARS menu. After entering Y_1, press [(-)] [2nd] [Y-VARS], then select "Y_1" from the list of options (on the TI-82, you first must select FUNCTION). You will have entered $Y_2 = -Y_1$.

Casio fx-7700GB

Make sure your calculator is in graphing mode by pressing [MODE] 1 [MODE] [+] [MODE] [SHIFT] [+]. Press [RANGE] [F1] [RANGE] [RANGE] to set the standard viewing window and clear the graphing screen.

Then press [GRAPH], enter the first function, and press [EXE]. Graph the second function in the same way. Your screen should look like this:

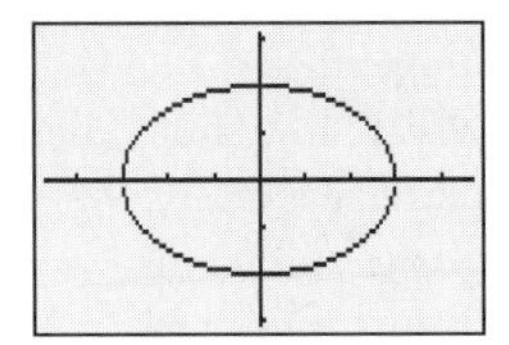

When you graph a circle in the standard viewing window of a Texas Instruments calculator, something odd happens, as the next example shows.

Example 2

Graph the equation $(x-2)^2 + (y-1)^2 = 4$.

Solving for y as in Example 1 yields

$$(x-2)^2 + (y-1)^2 = 4$$
$$(y-1)^2 = 4 - (x-2)^2$$
$$y - 1 = \pm\sqrt{4-(x-2)^2}$$
$$y = 1 \pm \sqrt{4-(x-2)^2}$$

The two functions to graph are therefore $y_1 = 1 + \sqrt{4-(x-2)^2}$ and $y_2 = 1 - \sqrt{4-(x-2)^2}$.

TI-81

When you use the procedure described in Example 1 to graph the two functions, the resulting graph looks more like an ellipse than like a circle:

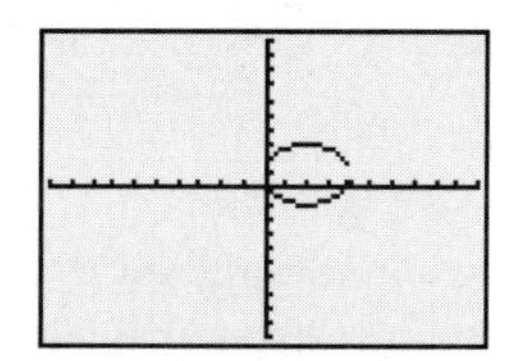

The reason is that the display screen of the TI-81 is wider than it is high, so in the standard viewing window a unit on the y-axis is shorter than a unit on the x-axis. This has the effect of stretching circles horizontally. To display the graph more accurately, first press [ZOOM] 1, move the flashing cursor near the

continued on next page

TI-82

When you use the procedure described in Example 1 to graph the two functions, the resulting graph looks more like an ellipse than like a circle:

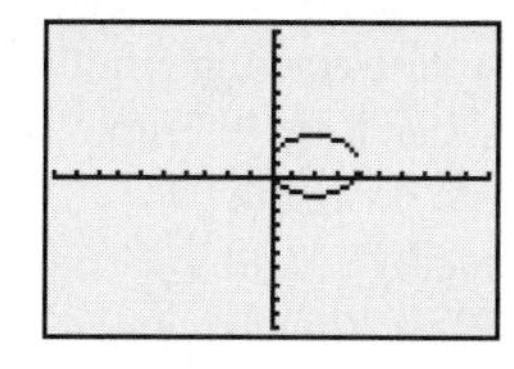

The reason is that the display screen of the TI-82 is wider than it is high, so in the standard viewing window a unit on the y-axis is shorter than a unit on the x-axis. This has the effect of stretching circles horizontally. To display the graph more accurately, first press [ZOOM] 1, move the flashing cursor near the

continued on next page

Casio fx-7700GB

When you use the procedure described in Example 1 to graph the two functions, your screen will look like this:

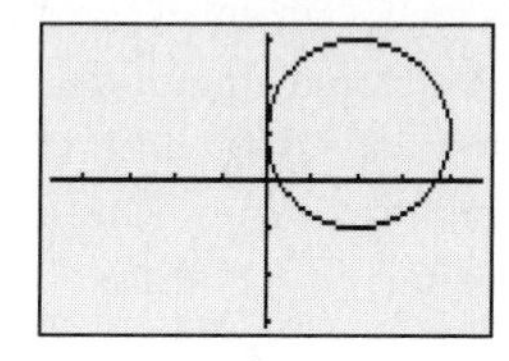

The range settings of the Casio fx-7700GB's standard viewing window were chosen to match the display screen's length-to-width ratio of about 47:31. The result is a square format, in which circles appear circular.

Whenever you choose to change the range settings on your calculator, keep the 47:31 ratio in mind if you wish to maintain a square format.

TI-81

continued from previous page
center of the "ellipse," and press ZOOM 2 ENTER to enlarge the graph. Then press ZOOM 5 to reset the window to a square format. This step changes the ranges of the axes so that the units on each are the same length. The graph should now appear circular.

TI-82

continued from previous page
center of the "ellipse," and press ZOOM 2 ENTER to enlarge the graph. Then press ZOOM 5 to reset the window to a square format. This step changes the ranges of the axes so that the units on each are the same length. The graph should now appear circular.

PROBLEMS

In problems 1–8, use a graphing calculator to graph each equation. Indicate whether the graph is a circle or an ellipse. Check students' calculator screens.

1 $9x^2 + 4y^2 = 36$ Ellipse

2 $(x-3)^2 + (y-3)^2 = 25$ Circle

3 $x^2 + y^2 = 121$ Circle

4 $\frac{(x-2)^2}{49} + \frac{(y-1)^2}{81} = 1$ Ellipse

5 $\frac{x^2}{49} + \frac{y^2}{81} = 1$ Ellipse

6 $x^2 + 25y^2 = 100$ Ellipse

7 $(x+12)^2 + (y+24)^2 = 4$ Circle

8 $4x^2 = 36 - 4y^2$ Circle

9 Figure out how to use your calculator to produce the hyperbolic graph of $y^2 - 2x^2 = 1$. Then write a step-by-step procedure for graphing the equation. Possible answer: Solve the equation for y to get $y = \pm\sqrt{1 + 2x^2}$. Then, after clearing stored equations and setting the standard viewing window on the calculator, graph $y_1 = \sqrt{1 + 2x^2}$ and $y_2 = -\sqrt{1 + 2x^2}$ on the same screen. The result is the hyperbola that represents the equation.

14 WORKING WITH ABSOLUTE VALUES

Objective To learn how to use a calculator to graph and solve equations containing absolute values

Since the ability to evaluate absolute values is programmed into your graphing calculator, you can use it to graph equations like $y = |x - 3| - 2$ and to solve equations like $|x| + |x + 1| = 2$.

Example 1

Graph $y = |x - 3| - 2$.

TI-81

Clear any stored equations from your calculator and set the standard viewing window by pressing [ZOOM] 6. Press [Y=] [2nd] [ABS] [(] [X|T] [−] 3 [)] [−] 2, then press [GRAPH].

Your screen should look like this:

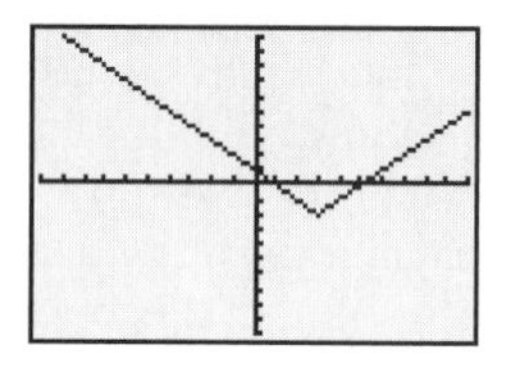

TI-82

Clear any stored equations from your calculator and set the standard viewing window by pressing [ZOOM] 6. Press [Y=] [2nd] [ABS] [(] [X,T,θ] [−] 3 [)] [−] 2, then press [GRAPH].

Your screen should look like this:

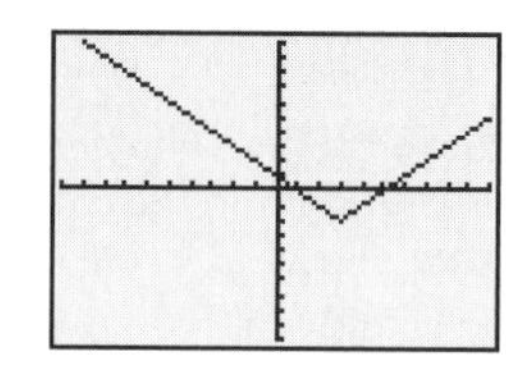

Casio fx-7700GB

Make sure your calculator is in graphing mode by pressing [MODE] 1 [MODE] [+] [MODE] [SHIFT] [+]. Press [RANGE] [F1] [RANGE] [RANGE] to set the standard viewing window and clear the graphing screen.

Press [GRAPH], then press [SHIFT] [MATH] [F3] [F1] to access the absolute-value function, and press [(] [X,θ,T] [−] 3 [)] [−] 2 [EXE].

Your screen should look like this:

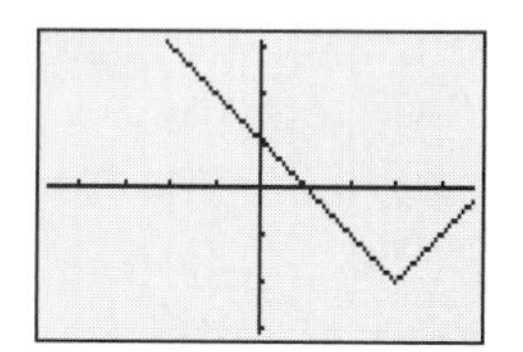

Hint

When you want to enter an absolute-value expression that involves an operation, like $|x - 3|$, it is essential to include parentheses around the expression. If you enter "abs X − 3," the calculator will interpret it as $|x| - 3$ rather than $|x - 3|$.

The next example shows how a graphing calculator can be used to find the solutions of an absolute-value equation in one variable.

Example 2

Solve the equation $|x| + |x + 1| = 2$.

TI-81

Set each side of the equation equal to y to produce the following system:

$$\begin{cases} y_1 = 2 \\ y_2 = |x| + |x+1| \end{cases}$$

After you graph these equations on your calculator, your screen should look like this:

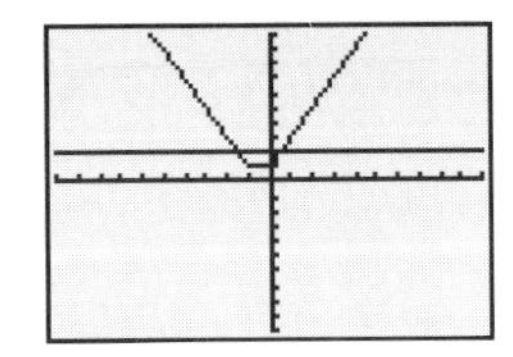

The solutions of the original equation are the x-coordinates of the points where the graphs intersect. First, zoom in by pressing [ZOOM] 2 [ENTER]. Then use the trace function to find the coordinates of these points. (Recall that [◄] and [►] move the flashing cursor along one graph, whereas [▲] and [▼] move the cursor from one graph to the other.)

The two points of intersection are (0.5, 2) and (–1.5, 2). Therefore, the original equation has the two solutions 0.5 and –1.5.

TI-82

Set each side of the equation equal to y to produce the following system:

$$\begin{cases} y_1 = 2 \\ y_2 = |x| + |x+1| \end{cases}$$

After you graph these equations on your calculator, your screen should look like this:

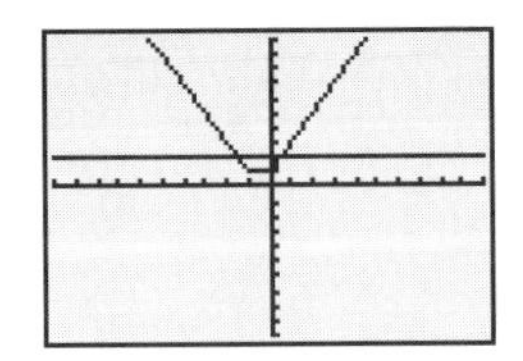

The solutions of the original equation are the x-coordinates of the points where the graphs intersect. Press [2nd] [CALC] 5 and use the intersect function, as explained in Lesson 8, to find the coordinates of one of the points of intersection. You will need to go through this process again, since there are two solutions to be found.

The two points of intersection are (0.5, 2) and (–1.5, 2). Therefore, the original equation has the two solutions 0.5 and –1.5.

Casio fx-7700GB

Set each side of the equation equal to y to produce the following system:

$$\begin{cases} y = 2 \\ y = |x| + |x+1| \end{cases}$$

After you graph these equations, your screen should look like this:

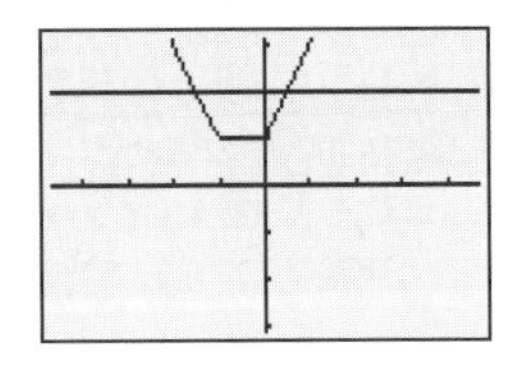

The solutions of the original equation are the x-coordinates of the points where the graphs intersect. Press [SHIFT] [Trace] and move the flashing cursor to the points of intersection to determine their coordinates.

The two points of intersection are (0.5, 2) and (–1.5, 2). Therefore, the original equation has the two solutions 0.5 and –1.5.

Hint

You must press [SHIFT] [↵] rather than [EXE] after you enter the first equation if you want to trace along both graphs. In Example 2, if you press [EXE] after entering $y = 2$, the tracing cursor will appear only on the graph of $y = |x| + |x + 1|$.

PROBLEMS

In problems 1–4, use a graphing calculator to graph each equation.

1 $y = |x - 2| + 3$

2 $y = 2|x + 1| + 1$

3 $y = |x| - |x - 1|$

4 $y = |2x| + 3$

In problems 5–8, use a graphing calculator to solve each equation for x.

5 $|x| + |x - 1| = 2$

6 $|3x - 1| = 4$

7 $|x| = |2x - 1|$

8 $|x| + |x - 1| = |x + 1| - |x - 1|$

9 Is $|x^2 + 1|$ always equal to $x^2 + 1$? Use your graphing calculator to investigate, and explain why or why not.

1–4 Check students' calculator screens.
5 $x = -0.5$ or $x = 1.5$
6 $x = -1$ or $x = 1.\overline{6}$
7 $x = 0.\overline{3}$ or $x = 1$
8 $x = 0.5$ or $x = 1.5$
9 Yes; possible answer: The graphs of $y = |x^2 + 1|$ and $y = x^2 + 1$ are identical, so $|x^2 + 1| = x^2 + 1$ for all values of x.

15 GRAPHING A FUNCTION AND ITS INVERSE

Objective To use a calculator to see the geometric relationship between a function and its inverse

You learned in Section 5.3 that the graphs of a function and its inverse are reflections of each other over the graph of $y = x$. A graphing calculator allows you to examine this relationship easily.

Example

Graph $f(x) = 2x - 3$ and its inverse.

The first step is to find the inverse by rewriting the function as $y = 2x - 3$, then interchanging x and y and solving for y:

$$x = 2y - 3$$

$$x + 3 = 2y$$

$$\frac{x+3}{2} = y$$

$$\text{INV}f(x) = \frac{x+3}{2}$$

TI-81

Clear all stored equations, and press ZOOM 5 to put the window in a square format. Then graph these equations:

$$y_1 = x$$

$$y_2 = 2x - 3$$

$$y_3 = \frac{x+3}{2}$$

continued on next page

TI-82

Clear all stored equations, and press ZOOM 5 to put the window in a square format. Then graph these equations:

$$y_1 = x$$

$$y_2 = 2x - 3$$

$$y_3 = \frac{x+3}{2}$$

continued on next page

Casio fx-7700GB

Make sure your calculator is in graphing mode by pressing MODE 1 MODE + MODE SHIFT +. Set the standard viewing window by pressing RANGE F1 RANGE RANGE.

Then graph these equations:

$$y = x$$

$$y = 2x - 3$$

$$y = \frac{x+3}{2}$$

continued on next page

TI-81	TI-82	Casio fx-7700GB
continued from previous page	*continued from previous page*	*continued from previous page*
The completed graph shows the symmetrical relationship between the function and its inverse:	The completed graph shows the symmetrical relationship between the function and its inverse:	The completed graph shows the symmetrical relationship between the function and its inverse:
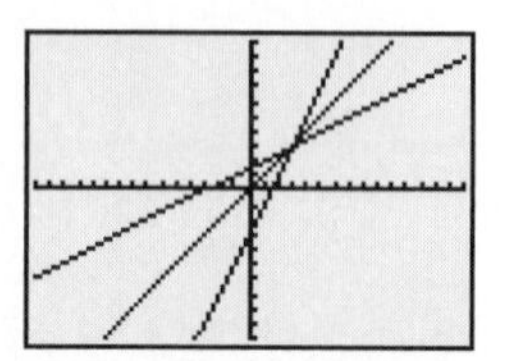	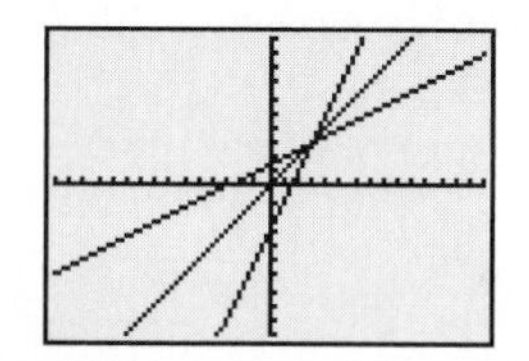	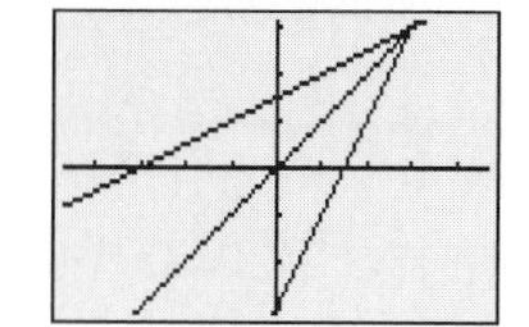

Hint

In the standard viewing window, symmetry is usually distorted because of the different scales of the axes. When investigating symmetrical graphs, therefore, you should usually set the window in a square format.

PROBLEMS

In problems 1–6, write the inverse of each function, then use a graphing calculator to graph the function and its inverse. Check students' calculator screens.

1 $f(x) = 3x - 5$ INV$f(x) = \frac{x+5}{3}$ **2** $f(x) = x + 7$ INV$f(x) = x - 7$ **3** $f(x) = \frac{x-22}{5}$

4 $f(x) = x^3 + 3$ **5** $f(x) = \sqrt{2x}$ **6** $f(x) = x^2 - 2$, where $x \geq 0$

7 Graph $f(x) = \frac{1}{x}$ and its inverse on your calculator. Comment on anything interesting you observe. Possible answer: The graphs are identical, since the function is its own inverse.

8 If the graph of a function contains a point on the line $y = x$, does the graph of its inverse necessarily contain the same point? Why or why not?

In problems 9 and 10, use a calculator to graph each pair of functions. Are the functions inverses of each other? How do you know?

9 $f(x) = 3x - 4$ and $g(x) = 4x - 3$ **10** $f(x) = x^3 + x + 1$ and $g(x) = \sqrt[3]{x-1}$

3 INV$f(x) = 5x + 22$ 4 INV$f(x) = \sqrt[3]{x-3}$

5 INV$f(x) = \frac{x^2}{2}$, where $x \geq 0$ 6 INV$f(x) = \sqrt{x+2}$

8 Yes (as long as the function has an inverse); possible answer: Since the graphs of a function and its inverse are reflections of each other over the line $y = x$, every point where one function intersects the line will also be a point where the other intersects it.

9 No, since the graphs aren't symmetric about the line $y = x$

10 No, since the graphs aren't symmetric about the line $y = x$

16 GRAPHING POLYNOMIAL FUNCTIONS

Objective To use a calculator to explore the graphs of polynomial functions of degrees greater than two

Section 6.4 of this book introduced you to the general shapes of the graphs of polynomial functions of degrees three and four. Knowing the general shape of a function's graph, however, may not give you all the information you need about the function. A graphing calculator makes it easy to produce precise graphs of polynomial functions and to locate key points on the graphs.

Example 1

Graph $f(x) = x^3 - 3x$.

TI-81

Clear all stored equations from your calculator and set the standard viewing window. Then graph the function, pressing [X|T] [^] 3 for the term x^3. Your screen should look like this:

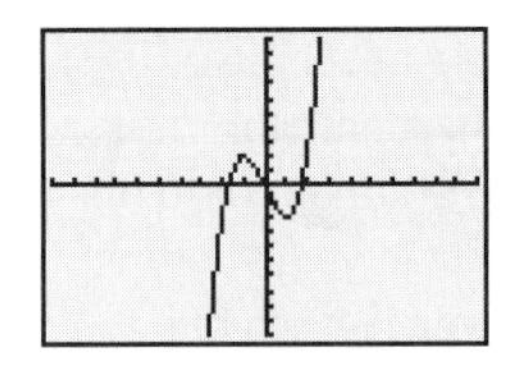

For a closer examination of the key region of the graph, press [ZOOM] 2 [ENTER] to zoom in on the origin.

TI-82

Clear all stored equations from your calculator and set the standard viewing window. Then graph the function, pressing [X,T,θ] [^] 3 for the term x^3. Your screen should look like this:

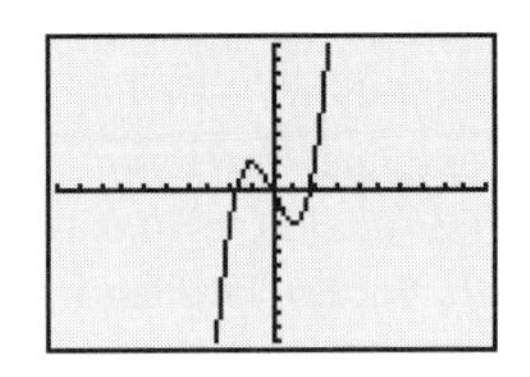

For a closer examination of the key region of the graph, press [ZOOM] 2 [ENTER] to zoom in on the origin.

Casio fx-7700GB

Make sure you are in graphing mode by pressing [MODE] 1 [MODE] [+] [MODE] [SHIFT] [+]. Press [RANGE] [F1] [RANGE] [RANGE] to set the standard viewing window.

Then graph the function, pressing [X,θ,T] [x^y] 3 for the term x^3. Your screen should look like this:

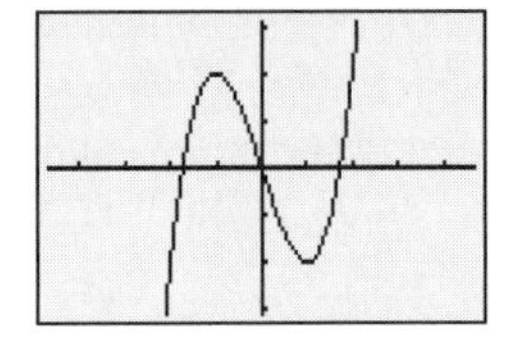

In Example 1, the graph indicates that as the value of x increases, the outputs of the function increase until reaching a "turning point" at (–1, 2), then decrease until reaching a turning point at (1, –2), then begin to increase again. The x-intercepts of the graph, however—corresponding to the real zeros of the function—do not all have integer coordinates. (In fact, two of the function's zeros are irrational.) The next example shows how to use your calculator to approximate the coordinates of such key points on the graph of a polynomial function.

Example 2

Graph $f(x) = x^3 - x^2 - x$. Identify the zeros of the function and the turning points of the graph.

When you graph this function in an appropriate viewing window, your graph will look something like the one shown here. From the graph, it appears that one of the zeros of the function is 0; and substituting 0 for x in the function shows that this is indeed the case. The two other zeros and the coordinates of the two turning points are not so easy to determine visually, but your calculator can help you.

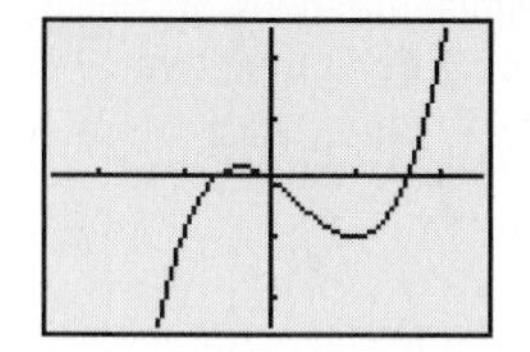

TI-81

Press TRACE and move the cursor as near as possible to the leftmost x-intercept. Then press ZOOM 2 ENTER to magnify the area around the intercept. By repeating the trace/zoom procedure several times to achieve closer approximations, you will find that the function has a zero of ≈–0.618. Press ZOOM 3 ENTER repeatedly (or press ZOOM 6 to return to the standard viewing window) so that you can see the rightmost x-intercept, then use the trace/zoom procedure again to approximate the remaining zero, ≈1.618.

Use the same procedure to approximate the coordinates of each turning point, this time zeroing in on the point where a region of the graph reaches its maximum (left-hand turning point) or minimum (right-hand turning point) height. In this case, the

continued on next page

TI-82

To determine each of the other x-intercepts, press 2nd [CALC] 2 and use the procedure explained in Lesson 10. Your calculator will display –0.618034 and 1.618034 as the approximate values of the zeros.

The method of locating the turning points depends on the fact that they are high points (maxima) and low points (minima) in regions of the graph. For the left-hand turning point, press 2nd [CALC] 4, then enter an appropriate lower bound, upper bound, and guess. Your calculator will display a set of coordinates similar to (–.3333314, .18518519). For the right-hand turning point, press 2nd [CALC] 3, then repeat the process. Your calculator will display a set of coordinates similar to (1.0000012, –1).

Casio fx-7700GB

Press F1 and move the cursor as near as possible to the leftmost x-intercept. Then press F2 F3 to magnify the region around the intercept. By repeating this trace/zoom procedure several times to achieve closer approximations, you will find that the function has a zero of ≈–0.618. Press F2 F4 repeatedly (or press F2 F5 to return to the standard viewing window) so that you can see the rightmost x-intercept, then use the trace/zoom procedure again to approximate the remaining zero, ≈1.618.

Use the same procedure to approximate the coordinates of each turning point, this time zeroing in on the point where a region of the graph reaches its maximum (left-hand turning point) or minimum (right-hand turning

continued on next page

TI-81	TI-82	Casio fx-7700GB
continued from previous page turning points are at (≈–0.333, ≈0.185) and (1, –1).		*continued from previous page* point) height. In this case, the turning points are at (≈–0.333, ≈0.185) and (1, –1).

Hint (TI-81, TI-82)

To view a "close-up" of any rectangular region of the window, press ZOOM 1, then move the cursor to a corner of the region you want to view. Press ENTER to mark the corner, and move the cursor to the opposite corner of the region (you will see the rectangle being defined as you do this). Then press ENTER, and an enlargement of the region will appear.

Hint (Casio fx-7700GB)

To view a "close-up" of any rectangular region of the window, press F2 F1, then move the cursor to a corner of the region you want to view. Press EXE to mark the corner, and move the cursor to the opposite corner of the region (you will see the rectangle being defined as you do this). Then press EXE, and an enlargement of the region will appear.

Example 3

Use a graphing calculator to produce an accurate graph of $f(x) = x^4 - 8x^3 - 8x^2 + x + 3$.

When you graph this function in a calculator's standard viewing window, your knowledge of the usual shape of a quartic function's graph should alert you that key points of the graph lie outside the window. The screen at the left below shows what a TI-81 or TI-82 displays; the screen at the right shows what a Casio fx-7700GB displays.

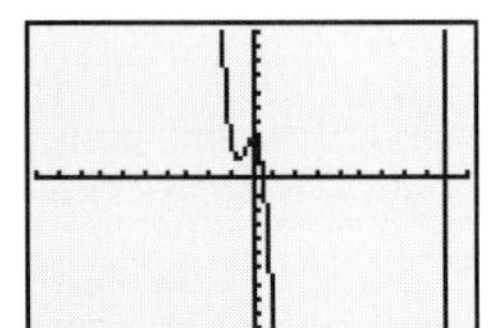

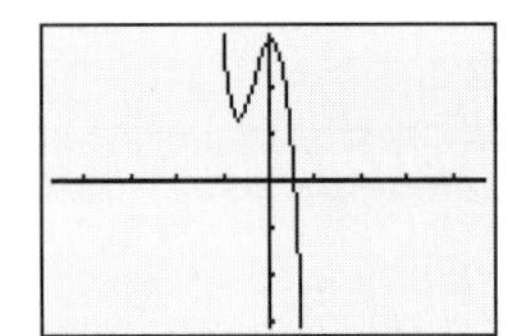

The TI version indicates that the graph has two x-intercepts and three turning points, but the third turning point is not visible. The Casio version looks more like the graph of a cubic function than the graph of a quartic function: Only one x-intercept and two turning points are visible. Here's where your sense of the general shape of the graph of a quartic function comes in handy. After a little experimentation with different dimensions of the window, you should be able to produce a graph like the one shown below. (The dimensions used are also shown.)

Xmin = −10
Xmax = 10
Xscl = 1
Ymin = −800
Ymax = 800
Yscl = 100

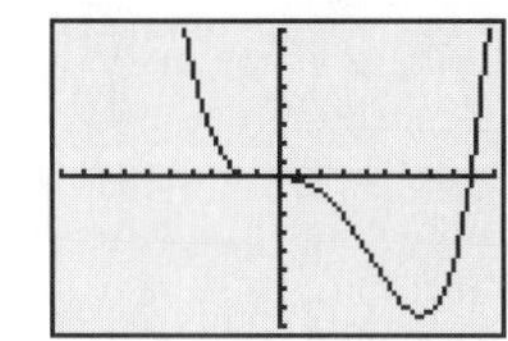

This graph allows you to investigate the third turning point. Notice, however, that the first two turning points are now obscured because of the increased vertical range. To get an accurate sense of the function's behavior, you need to consider both the big picture (where the third turning point is visible) and the small picture (where the behavior near the origin is visible). You will find that such a double view is frequently necessary when you want to locate the key points on the graph of a polynomial function.

Problems

In problems 1–6, use a graphing calculator to identify all the zeros of each function and all the turning points of its graph.

1 $f(x) = x^3 - 3x + 2$ **2** $f(x) = x^3 + x^2 + 7x - 12$ **3** $f(x) = x^4 - x^2 + 1$

4 $f(x) = x^3 + 5x^2 - 8$ **5** $f(x) = 2x^4 - x^3 - 8x^2 + 5x$ **6** $f(x) = -x^4 + x^3$

7 Does the graph of every cubic function have at least one x-intercept? If so, explain why. If not, give a counterexample.

8 Does the graph of every quartic function have at least one x-intercept? If so, explain why. If not, give a counterexample.

1 Zeros: −2, 1; turning points: (−1, 4), (1, 0)
2 Zero: ≈1.231; no turning points
3 No real zeros; turning points: (≈−0.707, 0.75), (0, 1), (≈ 0.707, 0.75)
4 Zeros: ≈−4.626, ≈−1.515, ≈1.141; turning points: (≈−3.333, ≈10.519), (0, −8)
5 Zeros: ≈−2.048, 0, ≈0.639, ≈1.909; turning points: (≈−1.389, ≈−12.255), (≈0.309, ≈0.770), (≈1.455, ≈−3.778)
6 Zeros: (0, 0), (1, 0); turning point: (0.75, ≈0.105)
7 Yes; possible answer: Since the graph of every cubic function rises without limit on one side and descends without limit on the other, the graph must cross the x-axis somewhere between.
8 No; possible counterexample: $f(x) = x^4 + 3$

17 GRAPHS WITH HOLES AND ASYMPTOTES

Objective To use a calculator to graph equations that have discontinuous graphs

Once you get used to graphing on a calculator, there is a temptation to place too much reliance on it. If you don't analyze equations before graphing them, you may be led to some strange conclusions.

Example 1

Graph $y = \frac{x^2 - 1}{x - 1}$.

Whatever calculator you use, the graph will resemble a straight line. This may surprise you, since the equation does not look like a linear equation. If you are working with a TI-81 or TI-82, the graph will appear identical to the graph of $y = x + 1$. If you are working with a Casio fx-7700GB, the graph will be like the graph of $y = x + 1$ but with a noticeable gap at (1, 2). The Casio calculator shows the gap to alert you that the graph has a hole at $x = 1$, since $\frac{x^2 - 1}{x - 1}$ is not defined for $x = 1$ (when the denominator equals zero); the TI calculators fail to alert you to the hole.

Before graphing an equation like $y = \frac{x^2 - 1}{x - 1}$, you should determine whether there are any restrictions on the value of x that will be reflected as holes or asymptotes in the graph. Analyzing the equation algebraically will prepare you to interpret what your calculator shows you:

$$y = \frac{x^2 - 1}{x - 1}$$
$$= \frac{(x + 1)(x - 1)}{(x - 1)}$$
$$= x + 1;\ x \neq 1$$

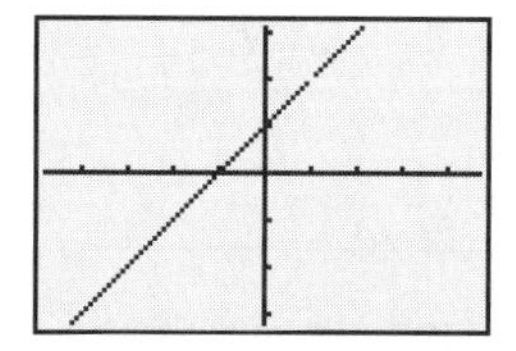

Example 2

Graph $y = \frac{x^2}{2x - 2}$. Identify all horizontal asymptotes, vertical asymptotes, and holes in the graph.

The right side of this equation is undefined when its denominator is equal to zero—that is, when $x = 1$—so there will be a break in the graph at this point. Since the numerator is not zero for this value of x, the break should take the form of a vertical asymptote.

TI-81

Graphing the equation in the standard viewing window yields an odd-looking result:

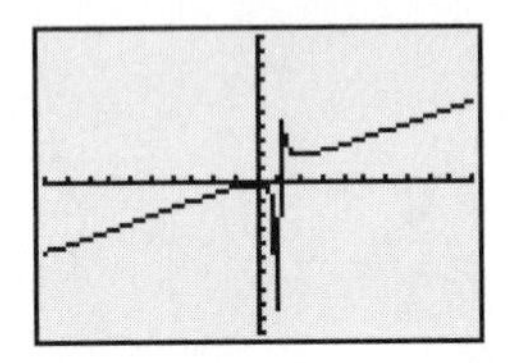

You already know, however, that there is a vertical asymptote at $x = 1$. By adjusting the dimensions of the window to focus on that region of the graph, you can get a clearer view of the graph's behavior at this point. The following display was produced with the settings Xmin = 0.5 and Xmax = 1.5.

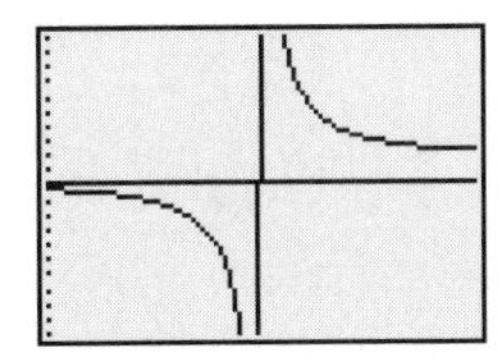

Returning to the standard window and zooming out for a larger view suggests that there is no horizontal asymptote—that as x increases and decreases without limit, so does y. Therefore, the vertical asymptote at $x = 1$ is the only discontinuity in the graph.

TI-82

Graphing the equation in the standard viewing window yields an odd-looking result:

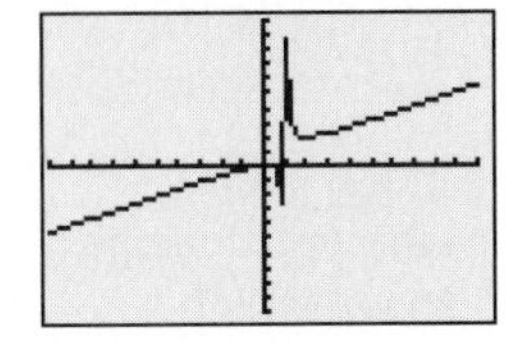

You already know, however, that there is a vertical asymptote at $x = 1$. By adjusting the dimensions of the window to focus on that region of the graph, you can get a clearer view of the graph's behavior at this point. The following display was produced with the settings Xmin = 0.5 and Xmax = 1.5.

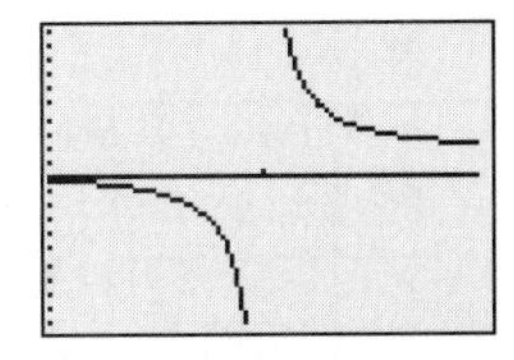

Returning to the standard window and zooming out for a larger view suggests that there is no horizontal asymptote—that as x increases and decreases without limit, so does y. Therefore, the vertical asymptote at $x = 1$ is the only discontinuity in the graph.

Casio fx-7700GB

Graphing the equation in the standard viewing window yields the following result:

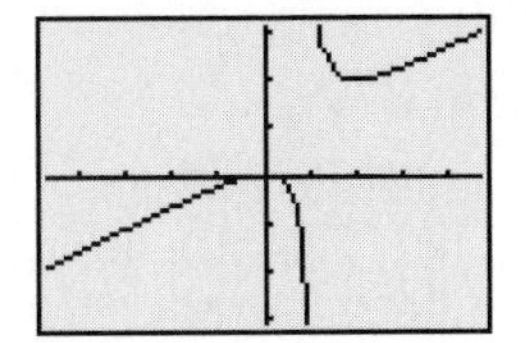

As expected, there is an asymptote at $x = 1$. To examine other areas of the graph, you can use the arrow keys to move the viewing window, and you can adjust the dimensions of the window. When you have a sense of the whole graph, you will recognize that there is no horizontal asymptote—that as x increases and decreases without limit, so does y. Therefore, the vertical asymptote at $x = 1$ is the only discontinuity in the graph.

TI-81 | TI-82

Hint

Your calculator will often display discontinuous graphs in a confusing way (as in Example 2), especially when it includes an "almost vertical" line where an asymptote should be. One way to alleviate this problem is to change the graphing mode from connected to dot. To do this, press MODE, move the flashing cursor over "Dot," and press ENTER. Then graph the equation again and see if the graph is clearer.

PROBLEMS

In problems 1–6, without using a graphing calculator, determine whether the graph of each equation has any vertical asymptotes or holes.

1 $y = \dfrac{x}{x+1}$

2 $y = \dfrac{x^2-1}{3x-3}$

3 $y = \dfrac{x^2-2}{3x-3}$

4 $y = \dfrac{1}{\sqrt{x-1}}$

5 $y = \dfrac{1}{x^2+1}$

6 $y = \dfrac{1}{x^2-1}$

In problems 7–12, use a graphing calculator to graph each equation. Identify all vertical asymptotes, horizontal asymptotes, and holes.

7 $y = \dfrac{x}{x+1}$

8 $y = \dfrac{x^2-1}{3x-3}$

9 $y = \dfrac{x^2-2}{3x-3}$

10 $y = x + \dfrac{1}{x}$

11 $y = x + \dfrac{1}{x} + \dfrac{1}{x^2}$

12 $y = \dfrac{1}{x} + \dfrac{1}{x-1}$

1 Vertical asymptote at $x = -1$
2 Hole at $x = 1$
3 Vertical asymptote at $x = 1$
4 Vertical asymptote at $x = 1$
5 No vertical asymptotes or holes
6 Vertical asymptotes at $x = -1$ and $x = 1$
7 Vertical asymptote at $x = -1$, horizontal asymptote at $y = 1$
8 Hole at $x = 1$
9 Vertical asymptote at $x = 1$
10 Vertical asymptote at $x = 0$
11 Vertical asymptote at $x = 0$
12 Vertical asymptotes at $x = 0$ and $x = 1$, horizontal asymptote at $y = 0$

18 FINDING STANDARD DEVIATIONS

Objective To learn how to use a calculator to find standard deviations

As you learned in Sections 8.5 and 9.6, statisticians use a measure of dispersion called standard deviation to express how closely the values in a set of data are clustered about the mean. Calculating standard deviations with paper and pencil is rather laborious, but graphing calculators have a function that can calculate them for you.

Example

The ten students in a mathematics class had the following test scores: 82, 98, 96, 78, 73, 68, 91, 88, 80, 85. Find the mean and the standard deviation of the scores.

TI-81

To clear any stored data, press [2nd] [STAT] [▶] [▶] 2 [ENTER]. Your screen will say "Done."

Then press [2nd] [STAT] [▶] [▶] 1. The screen will prompt you to enter $x1$. Press 8 2 [ENTER] to record 82 as the first test score. When the cursor moves to $y1$, just press [ENTER], since the value 82 appears only once in the data set. Enter the nine other test scores in the same way.

After entering the data, press [2nd] [STAT] 1 [ENTER]. The calculator displays the mean, 83.9, as $\bar{x}$ and the standard deviation, ≈9.159148432, as σx.

TI-82

To clear any stored data, press [STAT] 4, then (since you will need only one list for your data) press [2nd] [L_1] [ENTER]. Your screen will say "Done."

Press [STAT] 1 to access the data lists, and begin entering the test scores in L_1 by pressing

8 2 [ENTER]

9 8 [ENTER]

Continue in this manner until you have entered all ten scores.

Then press [STAT] [▶] 3. Under "1-Var Stats," make sure that "L_1" is highlighted in the line headed "Xlist" and "1" is highlighted in the line headed "Freq." (If necessary, use the arrow keys to move the flashing cursor to the entry you want to

Casio fx-7700GB

To clear any stored data and prepare your calculator for statistical work, press [MODE] [×] [MODE] [SHIFT] 2, then [AC] [SHIFT] 3 [F2] [EXE] [PRE].

Begin entering the test scores by pressing

8 2 [F1]

9 8 [F1]

Continue in this manner until you have entered all ten scores.

After entering the data, press [F4] [F1] [EXE]. The calculator displays 83.9 for the mean of the test scores, $\bar{x}$.

Then press [F2] [EXE]. The calculator displays ≈9.159148432 for the standard deviation of the test scores, $x\sigma n$.

continued on next page

TI-81	TI-82	Casio fx-7700GB
	continued from previous page highlight, then press ENTER.) By doing this, you are telling the calculator that L_1 is the list you want calculations performed on and that each value is to be counted once. Press STAT ▶ 1 ENTER, and the calculator displays the mean, 83.9, as $\bar{x}$ and the standard deviation, ≈9.159148432, as σx.	

Hint

When dealing with a large set of data in which values are repeated, you can save time by creating a frequency entry for each value, as shown in Lesson 7.

PROBLEMS

1 A group of six students decides that each of them will flip a coin 100 times, recording the number of times that heads comes up. They obtain the following results: 46, 42, 51, 56, 53, 47. Find the mean number of heads obtained and the standard deviation of the data.

2 When asked how long (in minutes) it took them to get to school each day, the 21 students in a class gave the following responses: 5, 5, 10, 10, 10, 10, 15, 15, 15, 25, 30, 30, 30, 30, 60, 60, 60, 60, 60, 60, 90. Find the mean and the standard deviation of the travel times.

3 The 12 students in a class received the following test scores: 78, 79, 80, 85, 85, 85, 88, 90, 90, 92, 94, 94. Find the mean and the standard deviation of the scores.

4 Predict what would happen to the mean and the standard deviation of the scores in problem **3** if each student's score were five points higher. Then use your calculator to check your prediction. Do you notice anything surprising about your results?

5 **a** A new diet program was tried by an experimental group, which achieved the following weight losses (in pounds) over the first 60 days: 3, 5, 7, 8, 8, 9, 9, 9, 9, 10, 12, 12. Compute the mean and the standard deviation of the weight reductions.

b A second diet program was tried by a different experimental group, and the following weight losses (in pounds) were achieved over the first 60 days: 0, 1, 1, 2, 2, 3, 6, 6, 20, 24, 26, 30. Compute the mean and the standard deviation for this diet program.

c Use your results from parts **a** and **b** to compare the effectiveness of the two diet programs. Use mean and standard deviation as part of your analysis.

1 Mean: $49.1\overline{6}$; standard deviation: ≈4.670
2 Mean: ≈32.857; standard deviation: ≈24.279
3 Mean: $86.\overline{6}$; standard deviation: ≈5.375

Additional Answers

4 Mean: $91.\overline{6}$; standard deviation: ≈5.375; students may note that although the mean increases by 5, the standard deviation remains the same, since the distance of each value from the mean remains the same.

5 **a** Mean: $8.41\overline{6}$; standard deviation: ≈2.465

b Mean: $10.08\overline{3}$; standard deviation: ≈10.889

c Possible answer: Although the second diet produced a greater mean weight loss, the greater standard deviation for that diet indicates that the weight reductions it produced tended to vary more from the mean. Therefore, the first diet may be more reliable in producing weight reductions.

19 CURVE FITTING

Objective To learn how to use a calculator to find the equation that best describes a set of data

Suppose you want to find an equation that describes a set of data. As you saw in Lesson 9, you can enter the data on a graphing calculator and use the calculator to determine a linear regression equation. In many cases, however, a line is not the best model for a set of data. This lesson explains how to use a graphing calculator to determine whether a line, an exponential curve, a logarithmic curve, or a power curve provides the best model for a set of data.

Your calculator can produce linear, logarithmic, exponential, and power equations to describe a set of data. Each of these types of equations is shown below, along with the characteristic shape of its graph.

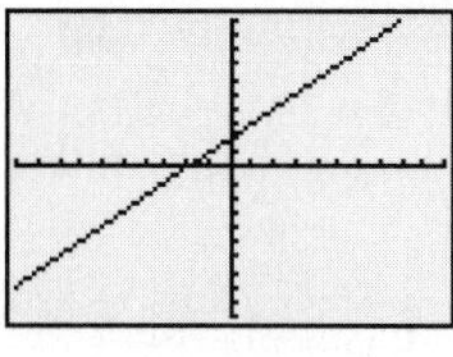

Linear Model
$y = a + bx$

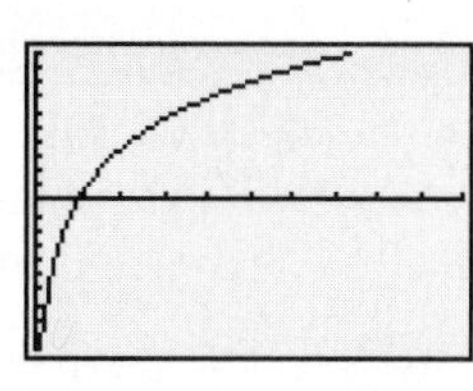

Logarithmic Model
$y = a + b \ln x$

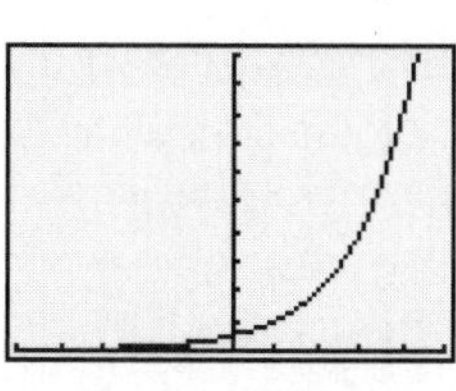

Exponential Model
$y = ab^x$

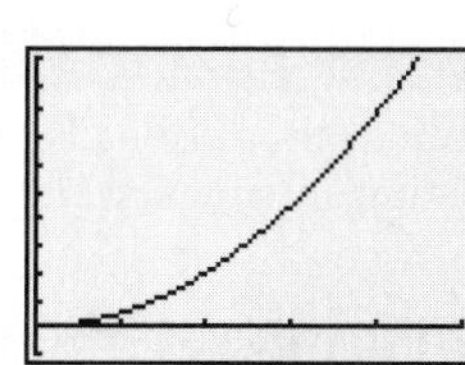

Power Model
$y = ax^b$

Example

A ball is dropped from the top of a tall building. Every half second, the total distance it has traveled is recorded, yielding the data in the table below. Use the data to write an equation expressing the relationship between the time the ball falls and the distance it travels.

Time (sec)	Distance (ft)	Time (sec)	Distance (ft)	Time (sec)	Distance (ft)
0.5	4	2.5	100	4.5	325
1.0	15	3.0	139	5.0	400
1.5	35	3.5	200	5.5	480
2.0	66	4.0	250	6.0	572

Clear your calculator of any stored data, enter the data in the table, and plot a scattergram as explained in Lesson 9. Your scattergram should be similar to the one shown (in which Xmin = 0, Xmax = 7, Ymin = 0, and Ymax = 600).

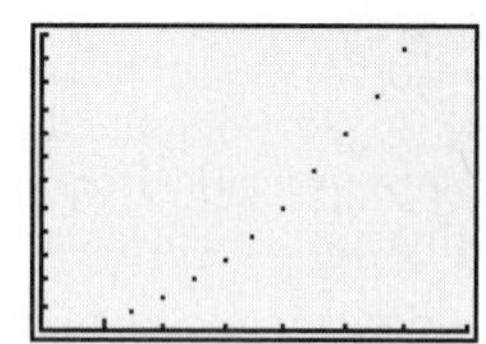

It could be argued that the pattern of points looks fairly linear, but it also shows some of the curvature of a power or exponential model. To see which model is best, you can use the following procedure.

TI-81

Press 2nd [STAT] to access the statistical CALC menu. On this menu, choices 2, 3, 4, and 5 represent the linear, logarithmic, exponential, and power regression models. When you use the calculator to determine the values of a and b in the equation for one of these models, the calculator will also display a value labeled r. This value is a measure of how closely the model fits the data—the closer r is to 1 or –1, the better the fit. By comparing the r values of the four models, you can choose the equation that best describes the data.

To test the linear model, move the cursor to choice 2, then press ENTER twice and record the value of r that is displayed. For each of the other models, reaccess the statistical CALC menu and select the appropriate choice in the same way. You should find that the best model is the power regression:

$a = 15.72454896$
$b = 2.008176094$
$r = .9998664889$

The equation that best describes the data is therefore approximately $y = 15.7x^2$, where x is time in seconds and y is distance traveled in feet.

TI-82

Press STAT ▸ to access the statistical CALC menu. On this menu, choices 9, 0, A, and B represent the linear, logarithmic, exponential, and power regression models. When you use the calculator to determine the values of a and b in the equation for one of these models, the calculator will also display a value labeled r. This value is a measure of how closely the model fits the data—the closer r is to 1 or –1, the better the fit. By comparing the r values of the four models, you can choose the equation that best describes the data.

To test the linear model, move the cursor to choice 9, then press ENTER twice and record the value of r that is displayed. For each of the other models, reaccess the statistical CALC menu and select the appropriate choice in the same way. You should find that the best model is the power regression:

$a = 15.72454896$
$b = 2.008176094$
$r = .9998664889$

The equation that best describes the data is therefore approximately $y = 15.7x^2$, where x is time in seconds and y is distance traveled in feet.

Casio fx-7700GB

After viewing your scattergram, press MODE. On the menu that appears, the four choices under "REG model," labeled 4, 5, 6, and 7, represent the linear, logarithmic, exponential, and power regression models. Associated with each of these models is a value designated r, which is a measure of how closely the model fits the data—the closer r is to 1 or –1, the better the fit. By comparing the r values of the four models, you can choose the equation that best describes the data.

Your calculator is currently set for linear regression. To calculate the corresponding r value, press SHIFT 4, then press F6 F3 EXE. Record the value the calculator displays. Press MODE 5 to switch to logarithmic regression, then press F6 F6 F3 EXE to obtain the corresponding r value. Switch to the other two regression models and find their r values in the same way. You should find that the power regression, with $r \approx 0.9998664889$, provides the best model. With the calculator in power regression mode, press F6 F6 F1 EXE F2 EXE

continued on next page

Additional Answers

2 a $y \approx 1200(1.085)^x$, where x is a number of years and y is the corresponding value in dollars; exponential
b 8.5%
c ≈$6134.46

3 a Sales: exponential; $y \approx 99{,}096(1.15)^x$, where x is a quarter and y is sales in dollars
Profit: logarithmic; $y \approx 9963 + 2016 \ln x$, where x is a quarter and y is profit in dollars
b The president's prediction of exponential growth was correct with regard to sales but not with regard to profits. Students may differ as to whether this constitutes a correct prediction.
c Possible answer: Despite the continuing increases in sales, the decreases in the quarterly growth of profits may indicate that the company is headed for a period of stagnation. If increases in overhead are retarding the growth of profits, perhaps economizing on labor costs or other production costs may produce increased growth.

Casio fx-7700GB

continued from previous page
to display the values of a and b in the regression equation.

The equation that best describes the data is approximately $y = 15.7x^2$, where x is time in seconds and y is distance traveled in feet.

PROBLEMS

1 Apply the techniques you learned in this lesson to write an equation that describes the data in the example at the top of page 503 of this book. What model does your equation represent? $y \approx 2087(0.92)^x$; exponential

2 Jack invested $1200 in a fund paying a fixed annual interest rate and left it there for twelve years. The value of his investment at the end of each year is shown in the table.

Year	Value	Year	Value	Year	Value
1	$1302.00	5	$1804.37	9	$2500.58
2	1412.67	6	1957.74	10	2713.12
3	1532.74	7	2124.14	11	2943.73
4	1663.02	8	2304.69	12	3193.94

a Write an equation that describes the data. What model does your equation represent?
b According to your equation, what was the annual interest rate of the investment?
c Use your equation to predict what the value of the investment would be at the end of twenty years.

3 Speaking to the press during the first month of a new company's operation, the president of the company declared, "I think this company is really going to go places. I see exponential growth ahead!" The sales and profit figures for each of the first twelve quarters of the company's operation are shown in the table.

Quarter	Sales	Profit	Quarter	Sales	Profit
1	$115,000	$10,000	7	$265,000	$13,900
2	130,000	11,400	8	300,000	14,200
3	150,000	12,000	9	350,000	14,400
4	175,000	12,800	10	400,000	14,600
5	200,000	13,200	11	465,000	14,800
6	230,000	13,600	12	535,000	14,950

a Identify the best models for the sales data and the profit data. Then write an equation to describe each set of data.
b Was the company president's prediction correct? Why or why not?
c On the basis of your models, what predictions can you make about the future of this company? Explain the reasoning behind your predictions.

20 SOLVING NONLINEAR SYSTEMS GRAPHICALLY

Objective To learn how to use a calculator to solve nonlinear systems

In Lesson 8, you learned how to use a graphing calculator to solve systems of linear equations. The technique for solving systems of nonlinear equations is no different, except that you might have to look for multiple solutions.

Example 1

Solve the system $\begin{cases} y = x^2 - 4 \\ y = -x^2 + 2x \end{cases}$ algebraically and graphically.

The graphs of these equations are both parabolas. To solve the system algebraically, you can set $x^2 - 4$ equal to $-x^2 + 2x$, solve for x as shown at right, then substitute these values of x in the equations to find the corresponding values of y. The solutions of this system are (2, 0) and (–1, 3).

$$x^2 - 4 = -x^2 + 2x$$
$$2x^2 - 2x - 4 = 0$$
$$(2x - 4)(x + 1) = 0$$
$$2x - 4 = 0 \text{ or } x + 1 = 0$$
$$x = 2 \text{ or } x = -1$$

To solve the system graphically, make sure your calculator is in standard graphing mode and set the standard viewing window. Then use the following procedure.

TI-81

Enter $Y_1 = x^2 - 4$ and $Y_2 = -x^2 + 2x$, and press GRAPH to see the graph of the system:

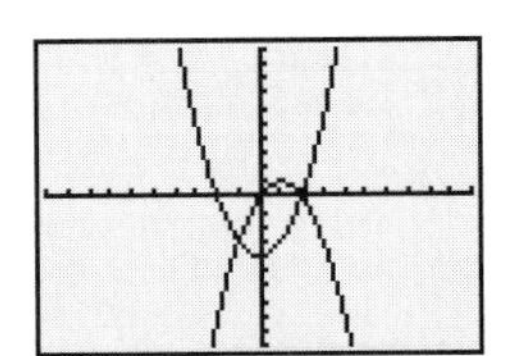

Then locate the points of intersection by using the calculator's trace function. (Remember that pressing ► or ◄ moves the flashing cursor along one of the curves and pressing ▲ or ▼ moves the cursor from one curve to the other.)

continued on next page

TI-82

Enter $Y_1 = x^2 - 4$ and $Y_2 = -x^2 + 2x$, and press GRAPH to see the graph of the system:

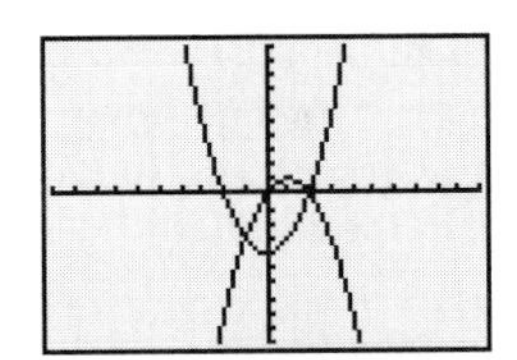

To locate a point of intersection, press 2nd [CALC] 5. The calculator will prompt you to identify the first curve. With the cursor on the first parabola, press ENTER. The calculator will then prompt you to identify the second curve.

continued on next page

Casio fx-7700GB

Press GRAPH and enter $y = x^2 - 4$, but instead of pressing EXE when you are finished, press SHIFT [↵]. Then press GRAPH again, enter $y = -x^2 + 2x$, and press EXE. Your screen should look like this:

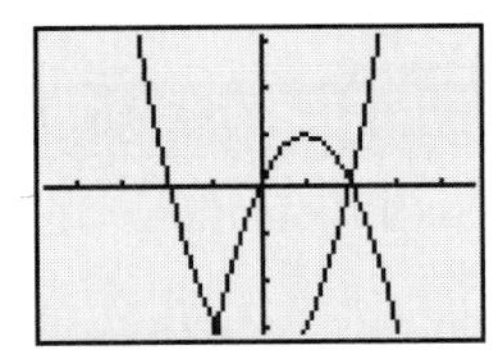

To locate the points of intersection, press F1 to access the trace function, then use ► and ◄ to move the flashing cursor along one

continued on next page

TI-81	TI-82	Casio fx-7700GB

TI-81

continued from previous page

You should be able to identify quickly that both parabolas pass through the point (2, 0). To locate the other point of intersection, you may need to zoom in on it. In the original viewing window, it appears that the parabolas intersect near (–1, 3), but the tracing cursor skips over this point on both curves. By using the trace/zoom procedure, however, along with some estimating and checking, you should be able to determine that the second solution is in fact (–1, 3).

TI-82

continued from previous page

With the cursor on the other parabola, press ENTER. The calculator then asks for a guess. Use the arrow keys to move the cursor close to one of the points of intersection, and press ENTER. The calculator will then display the coordinates of that point of intersection.

Repeat the process to locate the other point of intersection. You should find that the solutions are (–1, 3) and (2, 0).

Casio fx-7700GB

continued from previous page

of the curves and ▲ and ▼ to move the cursor from one curve to the other until you have determined that both parabolas pass through the points (2, 0) and (–1, 3).

Example 2

Solve this system: $\begin{cases} 9x^2 + 4y^2 = 36 \\ y = 2x^2 - x - 3 \end{cases}$

The first equation has an elliptical graph. To express it in a form that your calculator can graph, solve it for y, as explained in Lesson 13. You will obtain the two equations $y = 0.5\sqrt{36 - 9x^2}$ and $y = -0.5\sqrt{36 - 9x^2}$. Graph these, along with the second equation in the system, in the standard viewing window.

TI-81

Your screen should look like this:

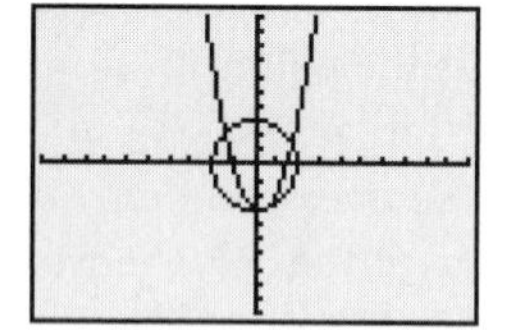

continued on next page

TI-82

Your screen should look like this:

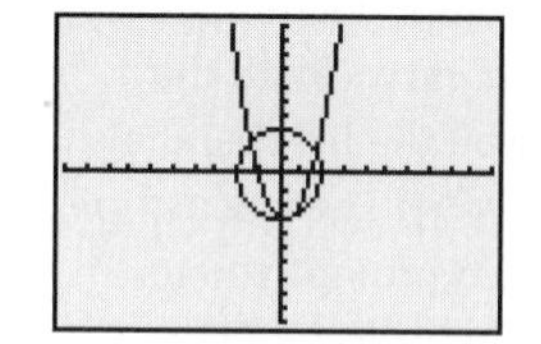

continued on next page

Casio fx-7700GB

Your screen should look like this:

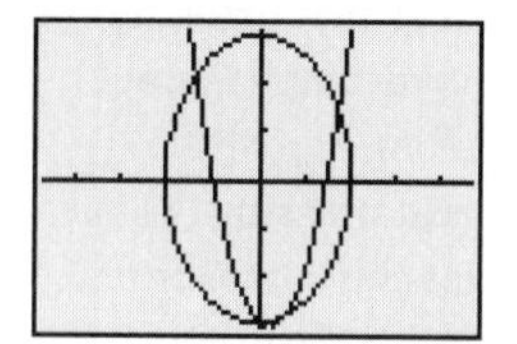

continued on next page

TI-81

continued from previous page

From this graph, it is difficult to tell whether there are two, three, or four points of intersection. The best way to magnify the region containing the intersections is to use your calculator's box feature as described in the Hint in Lesson 16. By defining an appropriate box, you should be able to obtain a graph something like this:

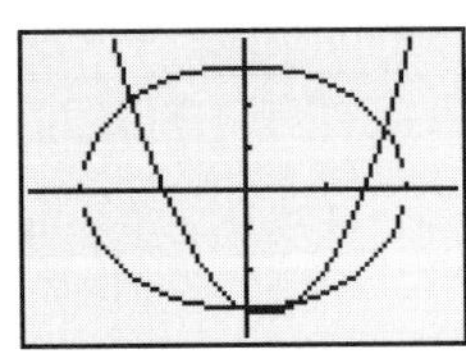

This display makes it clear that there are four points of intersection. By using the trace/zoom procedure, as in Example 1, you can determine that their coordinates—the four solutions of the system—are (≈–1.378, ≈2.175), (≈1.759, ≈1.428), (0, –3), and (≈0.619, ≈–2.853).

TI-82

continued from previous page

From the graph, it is difficult to tell whether there are two, three, or four points of intersection. The best way to magnify the region containing the intersections is to use your calculator's box feature as described in the Hint in Lesson 16. By defining an appropriate box, you should be able to obtain a graph something like this:

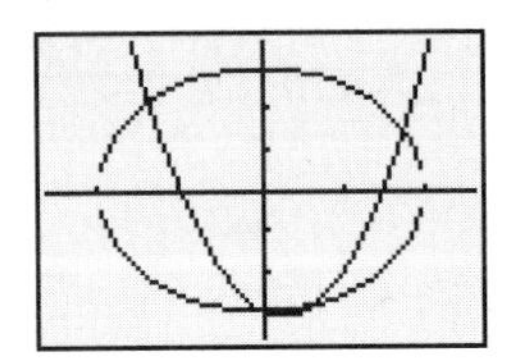

This display makes it clear that there are four points of intersection. By using your calculator's intersect function, as in Example 1, you can determine that their coordinates—the four solutions of the system—are (≈–1.378, ≈2.175), (≈1.759, ≈1.428), (0, –3), and (≈0.619, ≈–2.853).

Casio fx-7700GB

continued from previous page

The graph shows that there are four points of intersection. You can use the trace/zoom procedure described in Lesson 8 to determine the coordinates of each of these points. (To zoom in on the relevant parts of the graph, you may prefer to use your calculator's box feature as described in the Hint in Lesson 16.)

You should find that the four solutions of the system are (≈–1.378, ≈2.175), (≈1.759, ≈1.428), (0, –3), and (≈0.619, ≈–2.853).

PROBLEMS

In problems 1–12, use a graphing calculator to solve each system.

1 $\begin{cases} y = x^2 - 2 \\ y = -x^2 + 1 \end{cases}$ **2** $\begin{cases} y = 4x^2 - 3x - 1 \\ y = x^2 + 3 \end{cases}$ **3** $\begin{cases} y = 2x^2 + 5x - 7 \\ y = x - 5 \end{cases}$

4 $\begin{cases} y = x \\ x^2 + y^2 = 16 \end{cases}$ **5** $\begin{cases} y = 2x - 2 \\ 4x^2 + 16y^2 = 64 \end{cases}$ **6** $\begin{cases} y = 4x + 1 \\ y^2 = x^2 + 9 \end{cases}$

7 $\begin{cases} x^2 + y^2 = 16 \\ 4x^2 + 25y^2 = 100 \end{cases}$ **8** $\begin{cases} x^2 + y^2 = 25 \\ 4x^2 + 25y^2 = 100 \end{cases}$ **9** $\begin{cases} x^2 + y^2 = 36 \\ 4x^2 + 25y^2 = 100 \end{cases}$

10 $\begin{cases} x^2 - y^2 = 16 \\ 4x^2 + 25y^2 = 100 \end{cases}$ **11** $\begin{cases} x^2 - y^2 = 25 \\ 4x^2 + 25y^2 = 100 \end{cases}$ **12** $\begin{cases} x^2 - y^2 = 36 \\ 4x^2 + 25y^2 = 100 \end{cases}$

13 Write a system whose graph consists of a parabola and a hyperbola intersecting in four points.

1 $(\approx -1.22, -0.5), (\approx 1.22, -0.5)$
2 $(\approx -0.76, \approx 3.58), (\approx 1.76, \approx 6.09)$
3 $(\approx -2.41, \approx -7.41), (\approx 0.41, \approx -4.59)$
4 $(\approx -2.83, \approx -2.83), (\approx 2.83, \approx 2.83)$
5 $(0, -2), (\approx 1.88, \approx 1.76)$
6 $(\approx -1.04, \approx -3.18), (\approx 0.51, \approx 3.04)$
7 $(\approx -3.78, \approx 1.31), (\approx 3.78, \approx 1.31), (\approx -3.78, \approx -1.31), (\approx 3.78, \approx -1.31)$
8 $(-5, 0), (5, 0)$
9 No real solutions
10 $(\approx -4.15, \approx 1.11), (\approx 4.15, \approx 1.11), (\approx -4.15, \approx -1.11), (\approx 4.15, \approx -1.11)$
11 $(-5, 0), (5, 0)$
12 No real solutions
13 Answers will vary. Possible answer: $\begin{cases} y = x^2 - 2 \\ y^2 - x^2 = 1 \end{cases}$

21 APPLICATIONS OF QUADRATIC FUNCTIONS

Objective To use a calculator to solve real-world problems involving quadratic functions

In Section 4.6 of this book, you learned some algebraic methods of solving problems that can be modeled with quadratic functions. Often, however, it is quicker to use a calculator to graph such a function, then interpret the graph.

Example

A projectile is fired from ground level with an initial upward velocity of 1500 feet per second. What is the maximum height (in feet) that the projectile will reach? How long will it be before the projectile hits the ground?

You know from Section 4.6 that the height of a projectile at any time t is given by the function $h(t) = -\frac{1}{2}gt^2 + v_0t + h_0$, where g is the acceleration due to gravity (32 ft/sec^2), v_0 is the projectile's initial upward velocity, and h_0 is the projectile's initial height. This problem, therefore, is modeled by the function $h(t) = -16t^2 + 1500t$.

Before graphing, it is a good idea to set appropriate dimensions for the viewing window. For this problem, in which the projectile has a great initial velocity and consequently will reach a great height, Xmin = 0, Xmax = 100, Xscl = 10, Ymin = 0, Ymax = 50,000, and Yscl = 5000 are suitable settings.

When you enter the function, using x to represent t, your calculator will produce the graph shown here. You can then use the following procedure to answer the questions in the problem.

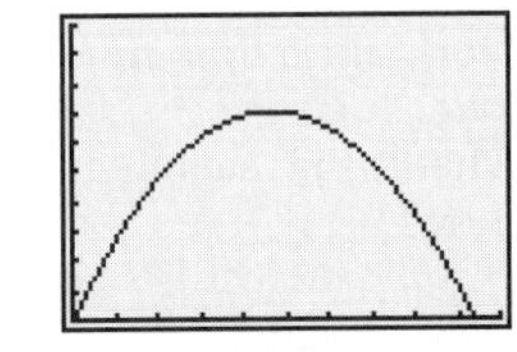

TI-81

To determine the maximum height of the projectile, press TRACE, move the flashing cursor near the parabola's vertex, and press ZOOM 2 ENTER to enlarge this region of the graph. Repeat the trace/zoom procedure several times to determine the coordinates of the vertex to a fair degree of accuracy.

continued on next page

TI-82

To determine the maximum height of the projectile, press 2nd [CALC] 4. Enter a lower bound, an upper bound, and a guess, as explained in Lesson 16. The calculator will display the coordinates of the vertex, approximately (46.88, 35,156).

continued on next page

Casio fx-7700GB

To determine the maximum height of the projectile, press F1, move the flashing cursor near the parabola's vertex, and press F2 F3 to enlarge this region of the graph. Repeat the trace/zoom procedure several times to determine the coordinates of the vertex to a fair degree of accuracy.

continued on next page

TI-81

continued from previous page

You should find that they are about (46.88, 35,156). The projectile will therefore reach a maximum height of about 35,156 feet.

To determine how long it will take the projectile to return to earth, reset the original window dimensions, then use the trace/zoom procedure to identify the right-hand x-intercept of the parabola. You should find that the parabola crosses the x-axis at (93.75, 0). The projectile will hit the ground in 93.75 seconds.

TI-82

continued from previous page

The projectile will therefore reach a maximum height of about 35,156 feet.

To determine how long it will take the projectile to return to earth, press [2nd] [CALC] 2 and enter a lower bound, an upper bound, and a guess for the right-hand x-intercept of the parabola. The calculator will display the coordinates of the intercept, (93.75, 0). The projectile will hit the ground in 93.75 seconds.

Casio fx-7700GB

continued from previous page

You should find that they are about (46.88, 35,156). The projectile will therefore reach a maximum height of about 35,156 feet.

To determine how long it will take the projectile to return to earth, reset the original window dimensions, then use the trace/zoom procedure to identify the right-hand x-intercept of the parabola. You should find that the parabola crosses the x-axis at (93.75, 0). The projectile will hit the ground in 93.75 seconds.

PROBLEMS

In problems 1–3, use a graphing calculator to answer each question.

1 A projectile is fired from ground level with an initial upward velocity of 400 feet per second.

a What is the maximum height that the projectile will reach? 2500 ft

b How long will it be before the projectile hits the ground? 25 sec

c Will the projectile's upward flight or its downward flight take the greater amount of time? Explain.

2 Mick threw a ball from the roof of a 50-foot-high building. The ball's initial upward velocity was 36 feet per second.

a What was the ball's greatest height above the ground? ≈70 ft

b How long did it take for the ball to reach the ground? ≈3.2 sec

c Did the ball's upward flight or its downward flight take the greater amount of time? Explain.

3 A gardener has 300 feet of fencing to use in enclosing a rectangular plot of ground for a garden and dividing it into three equal parts, as shown to the right. If the gardener wants the total area of the garden to be as great as possible, what dimensions should she use in laying out the rectangular plot? 75 ft × 37.5 ft

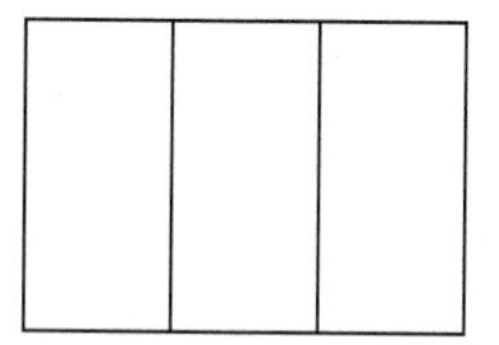

1 c They will take the same amount of time; possible answer: Because of the parabola's symmetry, the x-coordinate of its vertex is equidistant from $x = 0$ and $x = 25$.

2 c The ball's downward flight; possible answer: The vertex of the parabola is closer to $x = 0$ than to $x \approx 3.2$.

22 TRANSLATING TRIGONOMETRIC GRAPHS

Objective To learn how to use a calculator to graph a trigonometric function, along with vertical and horizontal translations of the graph

A vertical translation of a graph shifts each point of the graph the same number of units upward or downward—in other words, each y value is changed by the addition or subtraction of a constant. A graphing calculator can help you quickly graph trigonometric equations so that you can analyze how changes in an equation translate its graph.

Example 1

Graph $y = \sin x$, $y = \sin x + 3$, and $y = \sin x - 2$ on the same display.

TI-81

To verify that your calculator is ready for graphing and in radian mode, press MODE and make sure that "Rad" in the third line and "Sequence" in the sixth line are highlighted. If either is not, move the flashing cursor over it and press ENTER.

Then adjust the viewing window to appropriate dimensions. Since the period of the sine curve is 2π, make the range of the x-axis approximately -2π to 2π by setting Xmin = –6.3 and Xmax = 6.3. Set Ymin = –5 and Ymax = 5, since the amplitude of the sine curve is only 1.

To enter the first equation, press Y= SIN X|T ENTER. Enter the other two equations in similar fashion, then press GRAPH.

TI-82

To verify that your calculator is ready for graphing and in radian mode, press MODE and make sure that "Radian" in the third line and "Sequential" in the sixth line are highlighted. If either is not, move the flashing cursor over it and press ENTER.

Then adjust the viewing window to appropriate dimensions. Since the period of the sine curve is 2π, make the range of the x-axis -2π to 2π by entering (-) 2 2nd [π] for Xmin and 2 2nd [π] for Xmax. Set Ymin = –5 and Ymax = 5, since the amplitude of the sine curve is only 1.

To enter the first equation, press Y= SIN X,T,θ ENTER. Enter the other two equations in similar fashion, then press GRAPH.

Casio fx-7700GB

Prepare your calculator for graphing. To make sure your calculator is in radian mode, press SHIFT [DRG] F2 EXE.

Then adjust the viewing window to appropriate dimensions. Since the period of the sine curve is 2π, make the range of the x-axis -2π to 2π by entering SHIFT [(–)] 2 SHIFT [π] for Xmin and 2 SHIFT [π] for Xmax. Also set Ymin = –5 and Ymax = 5.

Press GRAPH SIN X,θ,T SHIFT [↵] to store the first equation. Enter the other two equations in similar fashion, but press EXE rather than SHIFT [↵] after the last one.

Your display should look like the one to the right. Notice that the graphs' shapes, periods, and amplitudes are identical; the only differences between them are a result of the vertical shifting of each point. You can investigate this further by using your calculator's trace function. Move the flashing cursor to any point on one of the graphs and press [▲] or [▼] to move it from graph to graph. The x value will remain the same, but the y value will change by the amount of vertical translation.

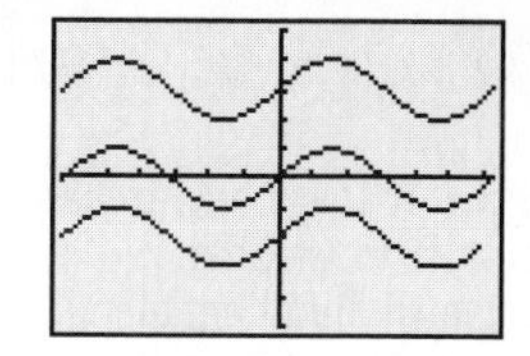

A horizontal translation—a shift to the left or right—of a trigonometric graph is produced when a value is added to or subtracted from the input of the trigonometric function.

Example 2

Graph $y = \sin x$ and $y = \sin\left(x + \frac{\pi}{2}\right)$ on the same display.

Without changing the dimensions of the viewing window, enter the equations and graph them on a clear screen. (To enter π on a TI-81 or TI-82, press [2nd] [π]; on a Casio calculator, press [SHIFT] [π].) Your display should look like the one to the right.

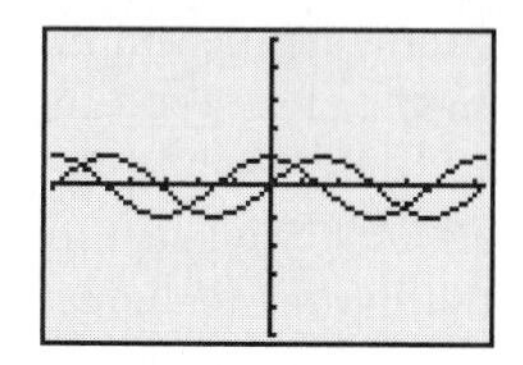

Again the graphs are identical in shape, amplitude, and period. This time, however, the second graph is a shift of the sine curve $\frac{\pi}{2}$ units to the left. To verify that the translation is $\frac{\pi}{2}$ units, you can use the trace function, or you can change the Xscl value to $\frac{\pi}{2}$ (or an approximation such as 1.57). When the calculator redraws the graphs, you will see that each x-intercept of the second is $\frac{\pi}{2}$ units to the left of the corresponding intercept of the first.

Example 3

Graph $y = \sin x$ and $y = \sin\left(x - \frac{\pi}{4}\right) + 2$ on the same display.

When you graph these equations in the same window used in Examples 1 and 2, your display should look like the one to the right.

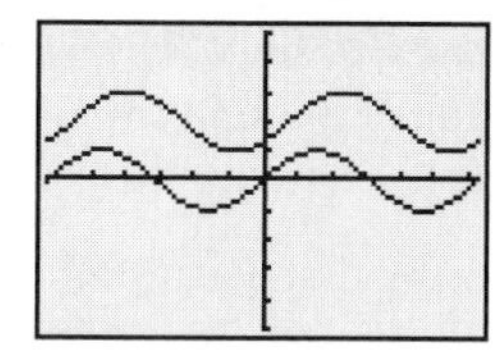

Once again the graphs are identical in shape, amplitude, and period. In this case, however, the second graph is the result of a twofold translation of the first—a shift $\frac{\pi}{4}$ unit to the right and a shift 2 units upward. The two translations can be considered as separate movements, but together they have the effect of sliding the sine curve diagonally.

PROBLEMS

In problems 1–9, use a graphing calculator to graph each equation or pair of equations along with the basic sine curve. Then describe each graph as a translation of the sine curve.

1 $y = \sin x + 1$, $y = \sin x - 1$

2 $y = \sin x + 6$, $y = \sin x + 10$

3 $y = \sin\left(x - \frac{\pi}{2}\right)$

4 $y = \sin\left(x + \frac{\pi}{4}\right)$

5 $y = \sin x + \frac{\pi}{2}$, $y = \sin\left(x + \frac{\pi}{2}\right)$

6 $y = \sin x - \pi$, $y = \sin (x - \pi)$

7 $y = \sin (x + \pi) - 5$

8 $y = \sin\left(x + \frac{\pi}{3}\right) - 10$

9 $y = \sin (x - 1) - \pi$

10 Graph $y = \sin x$ and $y = \sin (x + 2\pi)$ on the same display. Why does only one curve appear on your screen?

11 Graph $y = \sin x$, $y = -\sin x$, and $y = \sin (x + \pi)$ on the same display. Why do only two curves appear on your screen?

1 Vertical translation 1 unit upward; vertical translation 1 unit downward
2 Vertical translation 6 units upward; vertical translation 10 units upward
3 Horizontal translation $\frac{\pi}{2}$ units to the right
4 Horizontal translation $\frac{\pi}{4}$ unit to the left
5 Vertical translation $\frac{\pi}{2}$ units upward; horizontal translation $\frac{\pi}{2}$ units to the left
6 Vertical translation π units downward; horizontal translation π units to the right
7 Horizontal translation π units to the left and vertical translation 5 units downward
8 Horizontal translation $\frac{\pi}{3}$ units to the left and vertical translation 10 units downward
9 Horizontal translation 1 unit to the right and vertical translation π units downward
10 Possible answer: Since the graph of $y = \sin x$ has a period of 2π, translating it 2π units horizontally produces a graph identical to it.
11 Possible answer: Because of the symmetry and period of the sine curve, reflecting it over the x-axis and translating it π units horizontally produce identical results, so the graphs of $y = -\sin x$ and $y = \sin (x + \pi)$ are the same.

23 STRETCHING AND SHRINKING TRIGONOMETRIC GRAPHS

Objective To use a calculator to investigate vertical and horizontal stretches and shrinks of basic trigonometric graphs

A vertical stretch or shrink of the graph of a trigonometric function produces a change in the graph's amplitude. A horizontal stretch or shrink produces a change in the graph's period.

Example 1

Graph $y = \cos x$, $y = 3 \cos x$, and $y = \frac{1}{2} \cos x$ on the same display.

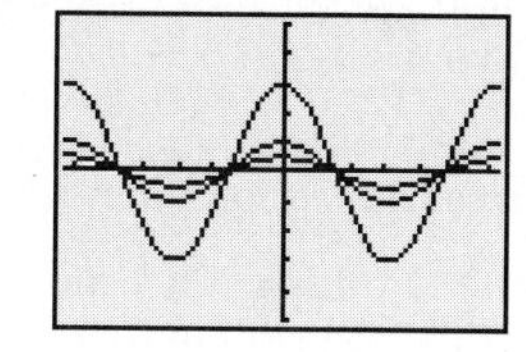

Prepare your calculator for graphing as you did in Lesson 22, using the same window dimensions. When you enter and graph the three equations, the display should appear like the one to the right.

Notice that all three curves have the characteristic cosine shape and that they have the same period. Their respective amplitudes, however, are 1, 3, and 0.5. The graphs of the second and third equations are a vertical stretch and a vertical shrink of the basic cosine curve by factors corresponding to the coefficients in the equations.

You can investigate these changes in amplitude by using your calculator's trace function. Move the flashing cursor to any point on one of the graphs, then press ▲ or ▼ to move it from graph to graph. The x value will remain the same, but the y values will reflect the factors by which the basic cosine curve is stretched and shrunk.

Example 2

Graph $y = \cos x$ and $y = \cos (2x)$ on the same display.

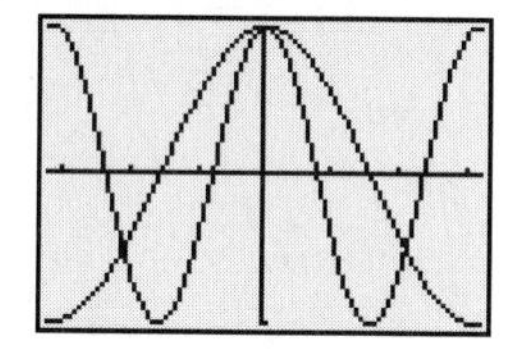

Change the window dimensions on your calculator to

Xmin = −3.2 (or $-\pi$)
Xmax = 3.2 (or π)
Ymin = −1
Ymax = 1

When you enter and graph the equations, the display should appear like the one to the right. In this case, the amplitude is unaffected, but the graphs have different periods. Whereas only one period of the first graph appears in the display, two periods of the second appear. Multiplying the input of the cosine function by 2 has halved the period of the graph, shrinking it from 2π to π.

Example 3

Graph $y = \cos x$ and $y = \cos(\frac{1}{2}x)$ on the same display.

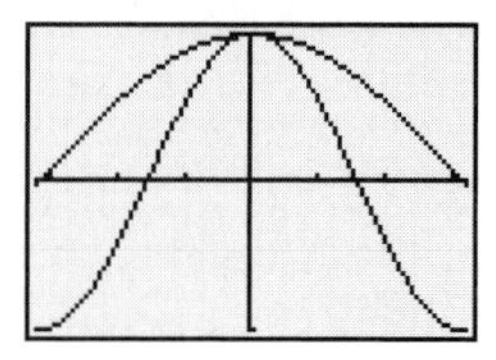

When you enter and graph these equations, using the same window dimensions as in Example 2, the display should appear like the one to the right. Again, the amplitude is unaffected. In this case, however, only half a period of the graph of $y = \cos(\frac{1}{2}x)$ appears in the display. Multiplying the input of the cosine function by $\frac{1}{2}$ has doubled the period of the graph to 4π.

Example 4

Graph $y = \cos x$ and $y = 0.5 \cos(2x)$ on the same display.

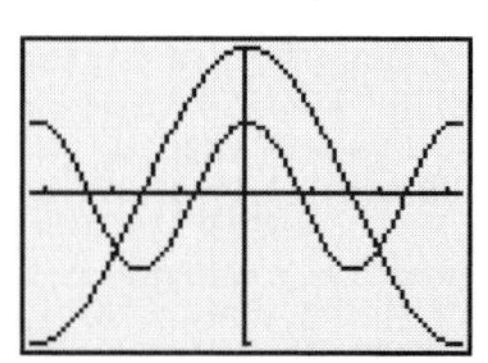

When you enter and graph these equations, again without changing the dimensions of the viewing window, the display should appear like the one to the right. This time, the second graph differs from the basic cosine curve in both period and amplitude. Changing x to $2x$ shrinks the period by a factor of $\frac{1}{2}$ (from 2π to π), so that two full periods appear in the display. Multiplying $\cos x$ by 0.5 shrinks the amplitude by a factor of $\frac{1}{2}$ (from 1 to 0.5), so that the second graph is only half as high as the first. The two factors combine to create the compound transformation shown.

Problems

In problems 1–9, use a graphing calculator to graph each equation or pair of equations along with the basic curve of the function (sine or cosine) it contains. Then describe each graph as a stretch or shrink of the basic sine or cosine curve.

1 $y = 6 \cos x$, $y = 3 \cos x$

2 $y = 2 \sin x$, $y = 0.5 \sin x$

3 $y = \pi \cos x$

4 $y = \sin(2x)$

5 $y = \cos(0.2x)$

6 $y = 2 \sin(3x)$

7 $y = 0.8 \cos(8x)$

8 $y = \pi \sin(\pi x)$

9 $y = 2 \sin\left(\frac{x}{4}\right)$

10 Graph $y = 2 \cos x$ and $y = -2 \cos x$. What effect does multiplying the output of a trigonometric function by a negative value have on its graph?

1 Amplitude stretched by a factor of 6; amplitude stretched by a factor of 3
2 Amplitude stretched by a factor of 2; amplitude shrunk by a factor of 0.5
3 Amplitude stretched by a factor of π

Additional Answers

4 Period shrunk by a factor of $\frac{1}{2}$
5 Period stretched by a factor of 5
6 Period shrunk by a factor of $\frac{1}{3}$ and amplitude stretched by a factor of 2
7 Period shrunk by a factor of $\frac{1}{8}$ and amplitude shrunk by a factor of $\frac{4}{5}$
8 Period shrunk by a factor of π and amplitude stretched by a factor of π
9 Period stretched by a factor of 4 and amplitude stretched by a factor of 2
10 The graph is reflected over the x-axis.

Answers for page 813

1 Tangent curve is translated $\frac{\pi}{2}$ units to the left, stretched vertically by a factor of 2, and translated 1 unit downward.
2 Tangent curve has its period shrunk by a factor of 3, is stretched vertically by a factor of 5, and is translated 6 units upward.
3 Tangent curve has its period shrunk by a factor of $\frac{1}{2}$, is stretched vertically by a factor of 2, and is translated 1 unit upward. (The horizontal translation of $\frac{\pi}{2}$ units does not affect the graph.)
4 Sine curve has its period shrunk by a factor of $\frac{1}{2}$, is translated $\frac{\pi}{2}$ units to the left, has its amplitude stretched by a factor of 2, and is translated 1 unit upward.
5 Cosine curve is translated π units to the left, reflected over the x-axis, and translated 1 unit downward.
6 Sine curve has its period shrunk by a factor of $\frac{1}{2}$, has its amplitude stretched by a factor of 3, is reflected over the x-axis, and is translated 4 units downward.
7 Tangent curve is translated $\frac{\pi}{4}$ unit to the right, reflected over the x-axis, and translated π units upward.
8 Sine curve has its period shrunk by a factor of π, is translated 1 unit to the left, has its amplitude stretched by a factor of π, and is translated π units upward.
9 Cosine curve has its period stretched by a factor of 2, is translated 1 unit to the right, has its amplitude shrunk by a factor of $\frac{1}{2}$, and is translated $\frac{1}{2}$ unit downward.

24 Compound Transformations of Trigonometric Graphs

Objective To use a calculator to investigate complicated transformations of basic trigonometric graphs

Frequently, the graph of even a complicated trigonometric equation—$y = 3 \tan (2x + \frac{\pi}{2}) + 1$, for example—can be interpreted as a compound transformation of the graph of one of the basic trigonometric functions—in this case, the tangent function. A graphing calculator can help you analyze such graphs.

Example

Describe the graph of $y = 3 \tan (2x + \frac{\pi}{2}) + 1$ as a compound transformation of the graph of $y = \tan x$.

First, adjust the dimensions of the viewing window. The settings Xmin = –1.57 (or $-\frac{\pi}{2}$), Xmax = 1.57 (or $\frac{\pi}{2}$), Ymin = –5, and Ymax = 5 will allow you to view one full period of the graph of $y = \tan x$.

The displays and instructions that follow show how you can develop the graph of $y = 3 \tan (2x + \frac{\pi}{2}) + 1$ step by step from the graph of $y = \tan x$. As you perform the steps on your calculator, watch carefully as the graphs appear so that you can distinguish between them.

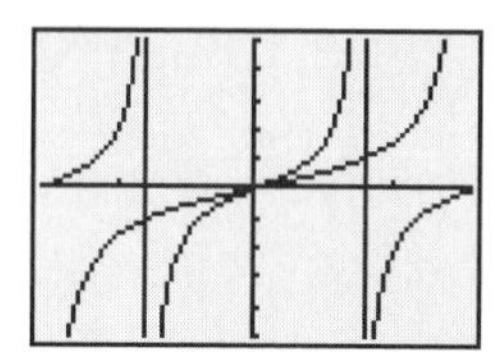

Step 1: Graph $y = \tan x$ and $y = \tan (2x)$. Multiplying the input by 2 decreases the period from π to $\frac{\pi}{2}$. This period change corresponds to a horizontal shrinking of the graph. Notice that the point (0, 0) remains fixed; it is common to both graphs.

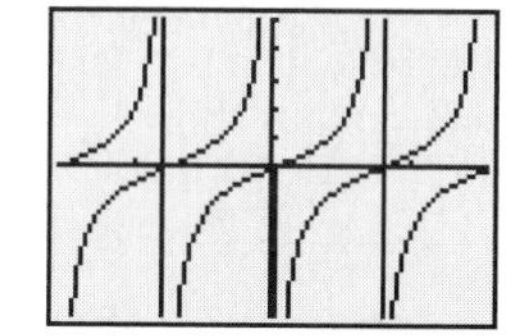

Step 2: Graph $y = \tan (2x)$ and $y = \tan (2x + \frac{\pi}{2})$. Adding $\frac{\pi}{2}$ to the input shifts the graph $\frac{\pi}{4}$ unit to the left. (A good way to determine the distance of a horizontal translation is to set the function's input equal to zero and solve for x: $2x + \frac{\pi}{2} = 0$ yields $x = -\frac{\pi}{4}$.) The point (0, 0) is moved to $(-\frac{\pi}{4}, 0)$.

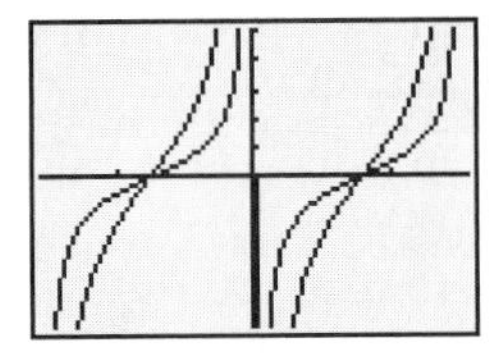

Step 3: Graph $y = \tan(2x + \frac{\pi}{2})$ and $y = 3\tan(2x + \frac{\pi}{2})$. Multiplying the output of the tangent function by 3 stretches the graph vertically by a factor of 3. (In a curve with an amplitude, this would appear as a threefold increase in the amplitude. In the tangent curve, it appears as an increase in the curve's steepness.) The point $(-\frac{\pi}{4}, 0)$ is not affected by the transformation.

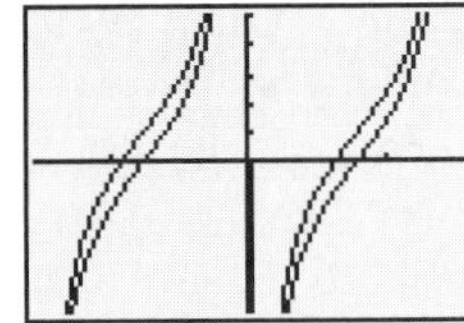

Step 4: Graph $y = 3\tan(2x + \frac{\pi}{2})$ and $y = 3\tan(2x + \frac{\pi}{2}) + 1$. Adding 1 to the output of the tangent function shifts the graph 1 unit upward. The point $(-\frac{\pi}{4}, 0)$ is moved to $(-\frac{\pi}{4}, 1)$.

The transformation of the graph of $y = \tan x$ into the graph of $y = 3\tan(2x + \frac{\pi}{2}) + 1$ can be summarized as follows: The period is shrunk by a factor of $\frac{1}{2}$, the graph is translated $\frac{\pi}{4}$ unit to the left, it is stretched vertically by a factor of 3, and it is translated 1 unit upward.

Hint

To graph the cosecant, secant, and cotangent functions on a calculator, express these functions as the reciprocals of the sine, cosine, and tangent functions:

$$\csc x = \frac{1}{\sin x} \qquad \sec x = \frac{1}{\cos x} \qquad \cot x = \frac{1}{\tan x}$$

The key labels "SIN^{-1}," "COS^{-1}," and "TAN^{-1}" on your calculator do not represent the cosecant, secant, and cotangent functions; they represent the Arcsin, Arccos, and Arctan functions.

Problems

In problems 1–12, use a calculator to develop the graph of each equation from the graph of a basic trigonometric function. Then describe the equation's graph as a compound transformation of the basic function's curve.

1 $y = 2\tan\left(x + \frac{\pi}{2}\right) - 1$

2 $y = 5\tan(3x) + 6$

3 $y = 2\tan(2x + \pi) + 1$

4 $y = 2\sin(2x + \pi) + 1$

5 $y = -\cos(x + \pi) - 1$

6 $y = -3\sin(2x) - 4$

7 $y = -\tan\left(x - \frac{\pi}{4}\right) + \pi$

8 $y = \pi\sin(\pi x + \pi) + \pi$

9 $y = \frac{1}{2}\cos\left(\frac{1}{2}x - \frac{1}{2}\right) - \frac{1}{2}$

10 $y = 2\cot\left(x + \frac{\pi}{2}\right) + 3$

11 $y = 5\csc(2x + \pi) - 2$

12 $y = -\sec(\frac{1}{2}x + \pi) + 1$

Answers (cont'd)

10 Cotangent curve is translated $\frac{\pi}{2}$ units to the left, stretched vertically by a factor of 2, and translated 3 units upward.

11 Cosecant curve has its period shrunk by a factor of $\frac{1}{2}$, is translated $\frac{\pi}{2}$ units to the left, is stretched vertically by a factor of 5, and is translated 2 units downward.

12 Secant curve has its period stretched by a factor of 2 and is translated 1 unit upward. (The combination of a translation of 2π units to the left and a reflection over the x-axis does not affect the graph.)

Selected Answers

1.1 Introduction to Problem Solving

Problem Set

1 Answers will vary. Possible answer: Draw a diagram. **2** Answers will vary. Possible answer: Build a model. **3 a** 28 **b** 56 **c** $28n$ **4 a** 24 **b** 6 **5 a** *None* **b** Possible answer:

R	W	B	G
G	B	W	R
W	R	G	B
B	G	R	W

6 a 100 **b** 90 **c** 45

1.2 Graphic Solutions

Problem Set

1 a

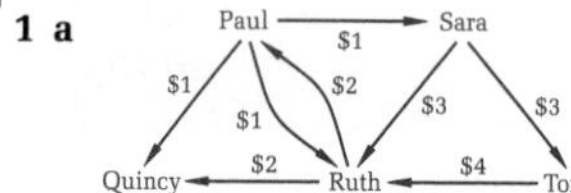

b Paul: \$9; Quincy: \$13; Ruth: \$14; Sara: \$5; Tom: \$9 **2** $\frac{n(n-1)}{2}$ **3 a** Via Lincoln, Wichita, and Albuquerque **b** $41\frac{3}{4}$ hr **c** Via Lincoln, Denver, and Albuquerque **4** Many possible answers **5 a** $p_1p_2p_5p_4$ **b** Yes: $p_1p_2p_5p_4p_3$ **c** No: a path cannot start at p_3, and once a path reaches p_3, it cannot leave. **6** Job 1: Roel; job 2: Noel; job 3: Joel; job 4: Zoel.

1.3 Geometric Modeling

Problem Set

1 275 **2** $\frac{n(n-1)}{2}$ **3 a** 15 **b** 12 **c** 9 **4** Any numbers of the form a, $6 - a$, $10 - a$, $2 + a$ **5** Only **a**, **b** and **d** can be drawn. If we call a vertex at which an even number of segments meet an even vertex and one at which an odd number of segments meet an odd vertex, then only networks with all even vertices and those with exactly two odd vertices can be traced. **6 a** 45 **b** 90 **7 a** 1 **b** 1 **c** 1 **d** 1 **e** 2 **f** 3 **g** 1 **h** 1 **i** 3 **j** 1 **k** 10 **8** Possible answers:

a

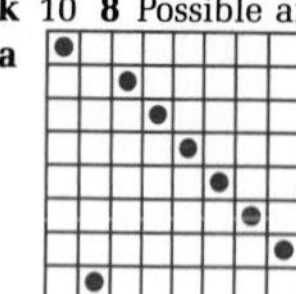

b

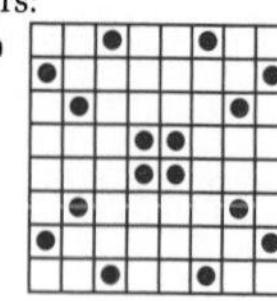

1.4 Finding a Pattern

Problem Set

1 a 13, 15; consecutive odd numbers **b** 19, 22; nth term = $3n - 2$ **c** 99, 120; nth term = $n^2 - 1$ **d** 29, 37; nth term = $\frac{1}{2}(n^2 - n) + 1$ **e** 256, 512; nth term = 2^{n-1} **f** 89, 144; each term (after first 2) is sum of 2 preceding terms. **2 a** 7, 11, 19, 35, 67 **b** 32, 64, 128, 256, 512

3 There are only 366 possible birthdays (including leap year day), so at least 2 of the 367 people must share a birthday. **4 a** 81 **b** 91 **5** 11 **6** 1, 6, 15, 20, 15, 6, 1; 1, 7, 21, 35, 35, 21, 7, 1 **7 a** 4; 9; 16; 25; the sums are consecutive perfect squares **b** 1, 4, 9, 16, 25

8 It changes the value stored in the variable NS; whereas the equation is a statement of equality, the computer line is a command. **9 a** \$3000 **b** 1200 or 1600 **c** −\$2000; it means a loss of \$2000, perhaps because the income received from the sale of so few sticks is less than production costs. **d** 700 or 2100 **e** Between 10,000 and ≈17,300 **f** Answers will vary. Possible answer: because more workers and machinery are needed to produce more sticks. **g** Answers will vary. Possible answer: 1500, to generate the greatest profit

1.5 A Recursive Solution

Problem Set

1 a $y = \sum_{x=1}^{n} (2x - 1)$ **b** $y = x^2$ **c** $S_n = S_{n-1} + 2(n - 1)$ **2** Answers will vary. Possible answer: Divide all integers from 1 to $\sqrt{n}$ into n to see which leave no remainder. **3** Answers will vary. Possible answer: Use the process devised for problem 2 to create the set of divisors, and then count the elements of the set. **4** $S_1 = 1$, $S_n = n \cdot S_{n-1}$ **5 a** 21 min **b** 22 min **c** At least 24 min 38 sec, at most 24 min 40 sec **d** 36%

1.6 Demonstrations

Problem Set

1 a The set of all numbers of the form $2n$, where n is an integer **b** Since the set of every other number is the set of even numbers only if "number" is taken to mean "integer" and if one element of the set is even **2 a** −3, 3 **b** 4.5 **c** 0, 2 **3 a** Even **b** Odd **c** Odd **d** Non-negative integer **4** "Opposite" means on the other side of zero on the number line. Only positive numbers are on the opposite side of zero from negative numbers. **5** The number of chords between 7 points on a circle is the same as the number of handshakes between 7 people, $\frac{7(7-1)}{2} = 21$. **6** 512; 336

7 **a** $x = 4$ **b** $x = 243$ **c** $x = 27$ **d** $x = -9$ or $x = 9$ **8** $(2m)(2n) = 4mn$ **9** 87.5% **10** 23; a possible explanation is "by solving $77 - x = 31 + x$."
11 Explanations will vary, e.g., adding a negative number moves left on the number line. Any number to the left of a negative number is negative. **12** 4, 0, 6; $(x, y) \to |x - y|$ **13** $6y - 18$ **14** $A = I - 1 + \frac{S}{2}$

2.1 The Language of Expressions

Warm-up Exercises

1 $2x + 8$ **2** -1 **3** $6x + 21$ **4** $a(x + y)$ **5** **a** $5(3x + 8)$; $15x + 40$ **b** $2x^2 + 4x$ **6** 31 **7** 80 **8** 5 **9** 11 **10** 8 **11** 26 **12** $32 - 12x$

Problem Set

13 $17x - 15$ **15** $(x + 3)(x + 2)$; $x^2 + 5x + 6$ **16** **a** -17 **b** $9 + 3(-6) - 8 = -17$ **17** **a** 21 **b** 17 **c** 21 **d** No **e** Yes **19** $7x + 6y$ **21** **a** $27 - 2x$ **b** $2n - 5$ **c** $4(x + 7)$ **22** **a** $5x$ **b** x^6 **23** -3 **25** 33 **27** 30 **28**

x	−4	−2	6	3	−6
y	31	21	−19	−4	41

29 **a**

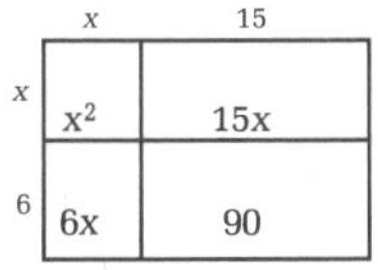

b $x^2 + 21x + 90$ **30** $22r + 12s$ **31** **a** ≈ 16.16 **b** ≈ 16.16 **33** 5 **35** $x^3 - 2x^2 + 6x - 12$ **37** $n(n + 1)$ is even since either n is even or $(n + 1)$ is even.
38 No; $n + (n + 1) = 2n + 1$, which is odd because $2n$ is always even. **39** **a** 30 **b** 33 **c** 27 **d** No **e** Yes **f** Yes **40** 86 **41** **a** 169 **b** 97 **c** 169 **d** a and c **42** 42 **43** $(x + 2)(x + 3)(x + 4)$; $x^3 + 9x^2 + 26x + 24$

2.2 Solving Equations

Warm-up Exercises

1 a or b **2** $z = 4$ or $z = 0$ **3** **a** $P = 3x + 19$ **b** $x = 15$ **4** Yes **5** Yes **6** $x = 7.5$
7

x	−4	−7	1	0	5
y	0	−3	5	4	9

8 **a** 4.9 lb **b** $0.7n$ **c** 36.2 lb **d** $W = 0.7n + 1.2$ **9** **a** No **b** Yes **c** No **d** No **e** Yes **f** No

Problem Set

11 $x = 10$ **12** **a** $x = -\frac{3}{8}$ **b** $x = -\frac{4}{3}$ **c** -4.5
13 **a** $-7, -2, 3, 4$ **b** $-4, -3, 2, 7$ **c** Reversed order **14** **a** \$2.65 **b** $V = 0.05n + 0.1d + 0.25q$ **15** **a** \$4000 **b** 250 or any number between 400 and 600 inclusive **c** Loss of \$2000 **16** **a** -2 **b** $\frac{3}{2}$ **17** $x = 4$; $y = -8$; $z = 64$ **18** **a** $\frac{-48}{5}$ **b** 16 **19** 150 cm^2 **21** 9 **22** $5x + 15$ **23** **a** 7 sec **b** 288 ft **25** $\frac{8x}{13}$ **27** Point E, since to avoid retracing, a path must begin at a point that is on the E-D loop and has 2 exit possibilities.
29 $y = -3$ or $y = 5$ **31** **a** 9, 9 **b** 4, -4 **32** **a** 13, -13 **b** 10, -10 **c** 10, -16 **33** $a = 2x - 11$ **35** 31 **37** $x = 2$ **38** **a** 18 **b** 60 **c** -312 **d** -168 **39** $x = 6$ **41** $4x^2 + 30x + 2$ **43** $x = \frac{a + 2b}{3}$
45 **a**

x	−6	−4	0	2	8	12
y	−6	−5	−3	−2	1	3

b A straight line **c** Yes **46** **a** $36x + 144$ **b** $x = 3.25$
47 Let a and b be two odd numbers, i.e. $a = 2k + 1$ and $b = 2n + 1$. The product $(2k + 1)(2n + 1)$ is equivalent to $4kn + 2k + 2n + 1$, which is odd, since $4kn + 2k + 2n$ is even and adding 1 makes it odd.
48 **a** $(0, -2)$ **b** PR = 5; RQ = 12 **c** 30 **d** 13
49 **a** 16 sec **b** 32 sec **51** $-32x^2$ **53** $\frac{23}{3}$
54

x	−5	−4	−3	−2	−1	0	1	2	3	4	5
y	−5	−3	−1	1	3	5	7	9	11	13	15

55 $3x^2 - x + 11$ **56** **a** 2 **b** -1; 5 **c** -2.5; 2; 6.5
57 $20{,}098p + 22{,}126w$

2.3 Solving Inequalities

Warm-up Exercises

1 c and **d**

2 

3 

4 $x > \frac{7}{2}$ 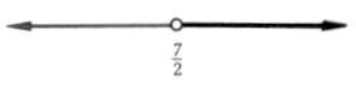

5 **a** $y + 10 < 3y$ **b** $5n^2 \neq 20$ **c** $\sqrt{x} - \frac{1}{5} \geq \sqrt{3}$ **d** $6 < 2\pi r < 8$ **6** **a** $-12 < -6$ **b** $12 > 6$ **7** **a** $-5 \leq -3$ **b** $10 \geq 4$ **8** $8(3x - 5) < 96$; $x < \frac{17}{3}$ **9** $y > \frac{3}{5}$ **10** $3k \cdot 3k = 9k^2$, which contains a factor of 9.

Problem Set

11 $x > -2$ 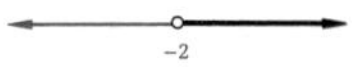

13 $z > -7$ 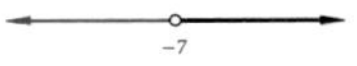

15 $x \leq 7$ 

17 $x < -1$ **19** $x \neq -5$ **21** Answers will vary. One possible answer is to think of and solve the problem as an equation with x on the left side. Then consider the inequality. **22** **a** 1.2 hr **b** 8.4 mi **c** 2 hr 6 min
23 **a** $3x + 12 < 30$ **b** $x < 6$

24 **a** Possible table:

x	−3	−2	−1	0	1	2
y	−1	1	3	5	7	9

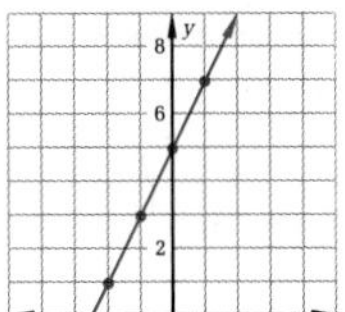

b Possible table:

x	−3	−2	−1	0	1	2
y	−2	−2	−2	−2	−2	−2

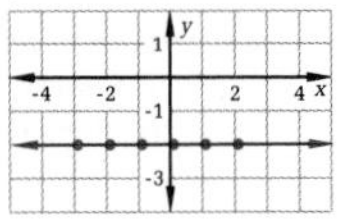

25 $14 < P < 28$ **26 a** Yes **b** No **c** Yes **27** $\left(-2, -\frac{1}{2}\right)$
29 $48x^3$ **31** $-12x^3 + 28x^2 + 20x$

33

	Small	Medium	Large
Thin crust	7.48	8.63	9.78
Thick crust	8.63	9.78	10.93

34 29 **35** $27x^2 + 6x$ **37** $3x^4y^2$ **39** 6 **40 a** $20 - x$ **b** $(0.9)x + 1.4(20 - x) = 22.00$ **c** 12 \$0.90 stamps and 8 \$1.40 stamps **41** 13 **43** $x = -3$ or $x = \frac{9}{2}$ **45** $x_3 = \frac{31}{5}$
47 $x \geq 29.6$ (29.6)

49 $n \neq 5$ (5)

51 $-1.4 < y < 1.4$ (−1.4, 1.4)

52 $A \geq 150$ cm^2 **53** $x < 2$ (2)

55 $x < 5$ (5)

56 $y = 2^x$ **57** .308, 12, 33
58 $a = 4, b = -11, c = -2$ **59** $z = 3$, $z = -\frac{1}{3}$, or $z = \frac{8}{5}$ **61** $y = \frac{1}{9}$ **63** $\frac{-x^2}{2} - 128$ **64 a** No real value **b** -3 **c** ≈ -0.293 **65 a** $y = -3x + 1$
b

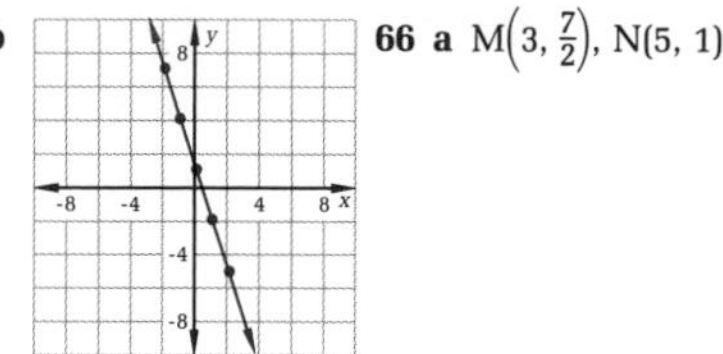

66 a $M\left(3, \frac{7}{2}\right)$, $N(5, 1)$
b $MN \approx 3.2$ **c** $AC \approx 6.4$ **67 a** $\frac{8}{27}$ **b** $\frac{2}{9}$ **c** 0 **d** $\frac{1}{27}$

2.4 Introduction to Graphing

Warm-up Exercises

1 a (11, 5) **b** 28 **2**

x	−2	−1	0	1	2
y	10	−7	−4	−1	2

3 −17 **4** 5 **5 a** D **b** E **c** A **d** C **e** B **6** 10

Problem Set

7 a −2 **b** 0 **c** −2.5 **d** 6 and 8 **8 a** 15 **b** (4, 3.5)
9 −2 **10** $\left\{x: \frac{7}{4} < x < 13\right\}$ **11** {−33, −6, 6, 39}
12 a No **b** No **c** Yes **d** Yes **13** $x = \frac{F - Q}{7}$
15 $x = \frac{c}{a + b}$ **16 a** $6x = 10x - 16$ **b** $x = 4$ **c** 4; 8 **d** 24 **17** $\frac{2}{5}$ **18 a** 60% **b** 80% **c** 79 **d** ≈58% **19 a** 2 **b** 8 **20 a** $y = -x + 10$ **b** $y = \frac{1}{x}$ **21 a** (−1, 2) **b** 6, 8 **c** 10 **22** 62 **23** (14, 19) **24** $2 \leq x \leq 7$ and $-6 \leq y \leq 4$; (2, 4), (7, 4), (7, −6), (2, −6) **25** $x = 10$ **26 a** 16 **b** 9.375 in. **c** 16:25 **27 a** 2 and 6 **b** $\{x: 2 < x < 6\}$ **c** $\{h(x): h(x) \geq -3\}$ **28** \$4; Sam and Bom each give Pim \$2 **29** 75; (12.5, 7.5) **30 a** 6 **b** 4

2.5 Matrices

Warm-up Exercises

1 $\begin{bmatrix} -2 & 6 \\ 3 & -4 \\ -8 & 3 \end{bmatrix}$ **2** $\begin{bmatrix} 20 & -16 \\ -24 & 12 \\ 24 & -8 \end{bmatrix}$ **3** $\begin{bmatrix} 8 & -2 \\ -9 & 2 \\ 4 & -1 \end{bmatrix}$ **4** 11
5 [27] **6** $\begin{bmatrix} -13 \\ 29 \end{bmatrix}$ **7 a** 2 × 3 **b** B must have three rows **c** C must have 2 columns **8** George: \$326; Gracie: \$286

Problem Set

9 $\begin{bmatrix} 20 & -1 \\ -20 & 4 \\ 7 & 0 \end{bmatrix}$ **11** $\begin{bmatrix} 23 & -11 \\ -20 & 26 \\ 19 & -17 \end{bmatrix}$ **13** No, the number of columns in P does not equal the number of rows in Q. **15** $a = -14, b = \frac{22}{3}$ **16 a** Mother: \$1500.48; father: \$2186.21; child 1: \$660.64; child 2: \$1095.65 **b** \$5442.98 **c** \$142.78 **17 a** $\begin{bmatrix} -71 & -52 \\ 36 & 62 \end{bmatrix}$
b $\begin{bmatrix} 59 & 13 \\ -4 & 42 \end{bmatrix}$ **c** $\begin{bmatrix} -52 & -71 \\ 62 & 36 \end{bmatrix}$ **19** 1.15
20 a $x^2 + 14x + 49$ **b** $(2x + 3)^2$

	2x	3
2x	$4x^2$	6x
3	6x	9

21 $16\pi < A < 100\pi$ **22** (5, 7) **23** $\frac{1}{2}$ **24 a** $x \approx 129.0$ **b** $x \approx 3.4$ or $x \approx -3.4$ **c** $x \approx 5.7$ **d** $x \approx 22.7$
25 a \$2,390,000 **b** $\begin{bmatrix} 27{,}560 & 29{,}680 & 33{,}920 & 41{,}340 \\ 31{,}270 & 37{,}100 & 39{,}220 & 43{,}460 \\ 39{,}220 & 42{,}400 & 44{,}520 & 47{,}700 \\ 44{,}520 & 47{,}700 & 50{,}880 & 51{,}940 \end{bmatrix}$
c \$143,400 **26 a** $\frac{1}{4}$ **b** $\frac{1}{4}$ **27 a** 0 **b** Undefined **c** Undefined **28** $\frac{5}{8}$ **29** $19 + \sqrt{41}$ **30 a** Yes **b** No **c** No **d** Yes **31 a** $\begin{matrix} \text{I} \\ \text{II} \\ \text{III} \end{matrix} \begin{bmatrix} 260 \\ 300 \\ 380 \end{bmatrix}$ **b** \$940 **32** The number you gave Igor **33** $\begin{bmatrix} 3 & 6 & 7 & x & x \\ 8 & 2 & 4 & 3x & x + 2 \\ 2 & 6 & 3 & 2x & 2x \\ 145 & 200 & 150 & 85x & 65x + 20 \end{bmatrix}$
34 b and e **35** (18, 14) **37** $\{x: x > 3.5\}$ **39** $\{x: x \leq 3\}$
40 $y = x - 7$
41 a (1, 5)

b (4, 6)

c (7, 13)

d (7; k, k)

42 a 5 **b** 10 **c** 10 **d** 5 **e** 1

43 $\begin{bmatrix} -13 & 15 \\ 58 & 22 \end{bmatrix}$; $\begin{bmatrix} 45 & 29 \\ -16 & -36 \end{bmatrix}$; no

2.6 Probability

Warm-up Exercises

1 0 and 1, inclusive **2** Answers will vary. One possible answer is an election. **3 a** $\frac{1}{6}$; $\frac{7}{30}$; $\frac{1}{4}$; $\frac{7}{20}$ **b** 1 **4** 1

Problem Set

5 $\frac{5}{8}$ **6 a** $\frac{7}{12}$ **b** $\frac{5}{12}$ **7** 10 **8** $n(t) = 120t$ **9 a** No **b** Yes **c** Yes **10** $\frac{5}{7}$ **11 a** -3 **b** -3 **c** -3 **12** 50 sec **13** No **14** $\left(-3, -\frac{22}{5}\right)$ **15** $(-3, 7)$

16 The lines all pass through $(0, -5)$ and the lines become more vertical as the coefficient of x increases.

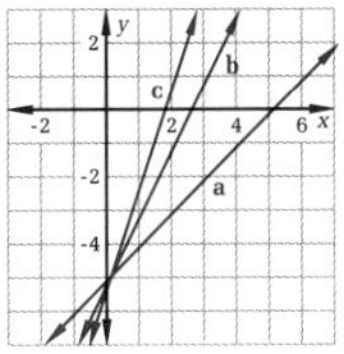

17 $\begin{bmatrix} 2 & 3 & 5 \\ 7 & -2 & -9 \end{bmatrix}$ **18 a** -30 **b** 23 **c** 1 **19** $\frac{1}{2}$ **21** $\frac{1}{2}$

23 a 2×1 **b** $\begin{bmatrix} 4 \\ -3 \end{bmatrix}$

Review Problems

1 $8x^2 - 5xy - 2y^2$ **3** $x = 2.4$ **5 a** 10 **b** .6 **c** .4
7 $x < -6$

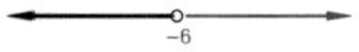

9 $x = 8, y = 12, z = -52$
11 $x^2 + 5x + 6$ **13** $x^3 - 27$ **15** $A = w(w + 7)$
17 $\approx .40$ **18 a** $\frac{7}{12}$ **b** Undefined **19 a** 230 mi **b** 3:00 P.M. **c** 3:20 P.M. **20**

x	3	5	0	−1
y	−3	−4	2	−3
z	1	1	15	−7

21 ≈ -74.52 **22** $-\frac{19}{3}$ **23** $\frac{8}{7}$ **24 a** 28 **b** 70 **c** 0
25 a -4 **b** $2x^2 + 48$
26

Hours of Overtime	6	8	4	10
Overtime Pay	90	120	60	150
Total Pay	430	460	400	490

27

29

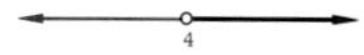

31 -152 **32** 0 **33 a** .05 **b** .2 **c** .7 **d** ≈ 45
e Answers will vary. **34 a** $B = 44 + 30h + p$ **b** $47
35 $w = 0$ **37** $y \approx 27.85$ **39** $x = 1$ **41 a** $45
b $267.50

Exponents and Radicals

Problem Set

1 $24b$ **3** True **5** False **6 a** 15,625 **b** 15,625 **c** 3125 **d** 3125 **7** $(x, y) = (5, 7)$ **9** False **11** $x^5 + 3x^2 - 2$

12 Possible table:

x	0	1	2	3	4
y	0	1	≈1.41	≈1.73	2

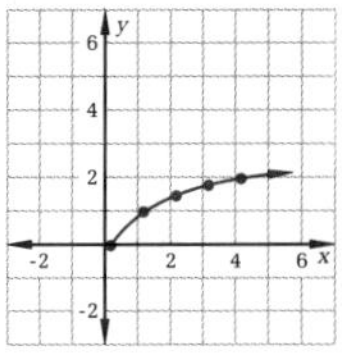

b Possible table:

x	−2	−1	0	1	2	3
y	$\frac{1}{4}$	$\frac{1}{2}$	1	2	4	8

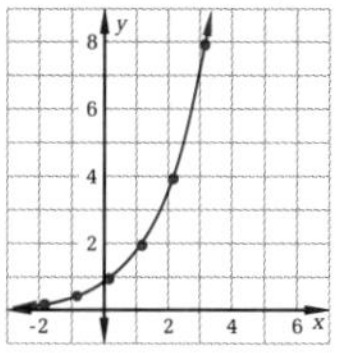

13 $2x^6$ **14 a** $\frac{1}{36}$ **b** $\frac{9}{10}$ **c** $\frac{6}{5}$ **d** 6 **15** $\frac{x^{-15}}{w^{-20}y^{-10}}$ **17** $8x^6y^{14}$
18 2197; 793; no **19** $x^3 + 8y^3$ **20** $1{:}x^9y^6z^{14}$ **21 a** 2 **b** 4 **c** 4 **d** 625 **22 a** 1 **b** -1 **c** 0
23 $s = 1 - 2x + 3x^2$ **24** $w = 125$; $x = 1$; $y = 3$; $z = 3$
25 $-\frac{1}{8}$ **26 a** i **b** iv **c** iii **d** ii **27** $\frac{y^{20}}{x^{30}}$ **29** $\frac{x^{15}}{y^{20}}$
30 $\frac{1}{\sqrt{5} + \sqrt{3}}$ **31** $-\frac{5yz}{6x^3}$ **33** x^8y^8 **34 a** $(x + 4)^2$ **b** $x^2 + 4^2$ **c** 0 **35** $\frac{1}{x + 1}$ **37** $\frac{3}{xy^{11}}$ **39** $(x, y) = (6, 9)$
41 True **43** False **44 a** $x = 49$ **b** No real solution
45 $2\sqrt{5}$ **47** $\frac{\sqrt{10}}{2}$ **49 a** $\sqrt{5}$ **b** $\sqrt{a + b}$ **51** $\frac{1}{2}$
53 $\frac{1 + \sqrt{5}}{4}$; $\frac{-1}{1 - \sqrt{5}}$ **54 a** iii **b** iv **c** ii **d** i **55** $15\sqrt[3]{2}$
56 $(x, y) = \left(\frac{9}{4}, \frac{1}{4}\right)$ **57 a** $2\sqrt{2}$ **b** $2\sqrt{3}$ **58** $n = 1$

3.1 Graphing Linear Equations

Warm-up Exercises

1 Answers will vary. Possible answers: (0, 4), (4, 6), (6, 7) **2** $\{k{:}k < 0\}$ **3** $-\frac{1}{2}$ **4** -2 **5** $a = 3$, $b = 9$, $c = 18$

7

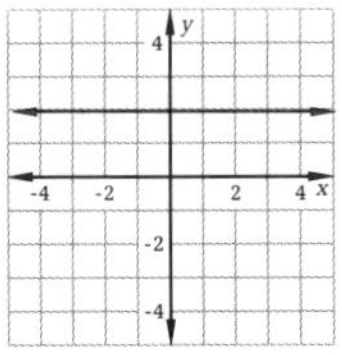

9

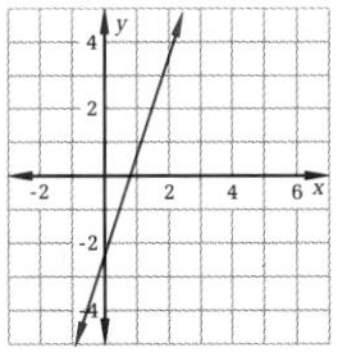

11 $\frac{1}{5}$ **12 a** -2 **b** $\frac{1}{2}$

Problem Set

13 L_4, L_2, L_3, L_1
14 $(-4, -7)$ **15** Undefined
16 a False **b** True **c** False **d** False **e** True **17** $\frac{4}{3}$
18 a $R_b = 0.6(220 - A)$; $R_a = 0.7(220 - A)$; $R_c = 0.8(220 - A)$ **b** ≈ 138 **c** $42\frac{1}{2}$ **19** $\frac{3}{5}$

20 a $y = -\frac{2}{3}x + 4$ **b** $y - 2 = \frac{3}{5}(x + 1)$

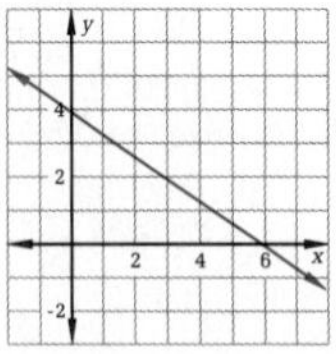

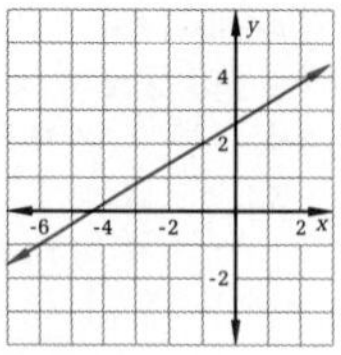

21 b **22 a** \$11.10 **b** $0.30x + 0.50(x + 3)$ **c** 63 boxes
23 a

	Sue	Sara	Kai
Le Chic	64.00	96.50	80.00
Le Pen	73.50	107.50	84.75
Bon Ton	75.25	107.25	77.75

b \$1.75 **c** Sara **24** $y - 1 = -\frac{2}{5}(x - 3)$ **25** $y = 0$

26 $k = \frac{23}{4}$ **27** 95 mi **28 a** $16B + 10G = 232{,}672$

b

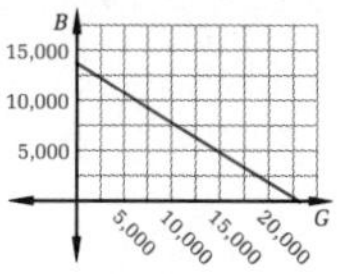

c If $B = 5015$, $G = 15243.2$, and only whole numbers of tickets can be sold; answers will vary. **29 a** B **b** A, D, E **c** C, F **30 a** The decrease in value (in dollars) per year **b** Can't tell. Table predicts −\$2400, which is impossible. \$300 might be a reasonable guess. **31 a** 4 **b** 7 **32** $\frac{1}{4}$ **33** $\frac{13}{3}$

34 $\begin{bmatrix} -1 & 3 & -16 \\ 1 & 3 & 3 \end{bmatrix}$ **35 a** 1 **b** 2

3.2 Systems of Linear Equations

Warm-up Exercises

1 It satisfies both equations, so it is a solution.
2 $(x, y) = (12, 4)$ **3 a** Graphing **b** Substitution or guess and check **c** Multiplication-addition **d** Substitution
4 (4, 2) **5** (0, 5) **7 a** (1.5, −1.3) **b** (1.5, −1.25)

Problem Set

8 $x = 8$; $y = 4$ **9 a** $(x, y) = (7, 1)$ **b** $(x, y) = (5, 2)$ **c** $(x, y) = (0.5, 4.75)$ **d** $(x, y) = (4, -5)$ **10** 85¢

11 $x \approx 2.14$; $y \approx 2.57$ **12** $\frac{3}{5}$ **13** None **14** 2.75; 3.75; −7.25; 49.75 **15 a** $W = 4.75h + 0.01s$ **b** \$2200
16 −52 **17** $(x, y) = (2, 3)$ **19** x^7 **21** $25x^{10}$
23 $3x^7 - 2x^3$ **25** $(x, y) = (5, 9)$ **26** Yes—**b** or **c**

27 \$20.08 **28**

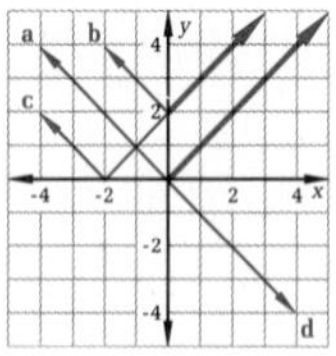

29 $\frac{5}{9}$ **30** 30

31 $w = 7$; $x = 8$; $y = 7$; $z = 0$ **32** 28 **33** 1

34 a $C_M = 12{,}895 + 1.16\left(\frac{x}{32}\right)$; $C_T = 10{,}550 + 1.16\left(\frac{x}{26}\right)$
b ≈280,322 mi **c** ≈220,552 mi and ≈340,092 mi

3.3 Graphing Inequalities

Warm-up Exercises

1

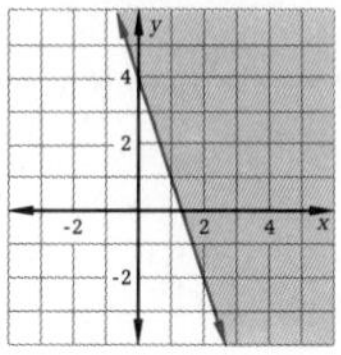

2

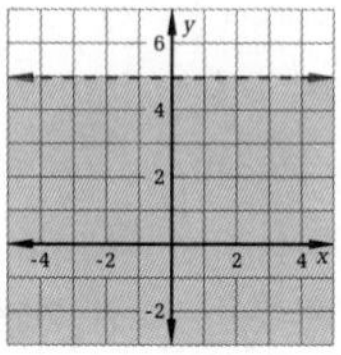

3

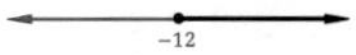

4 Yes

5 a {2, 4, 6, 7, 8, 9, 12, 13, 14, 15, 17} **b** {2, 6, 14, 15}
c {2, 4, 6, 8, 13, 14, 15} **d** {2, 4, 6, 8, 13, 14, 15}
6 a {multiples of 6} **b** {multiples of 15} **c** {multiples of 12} **7**

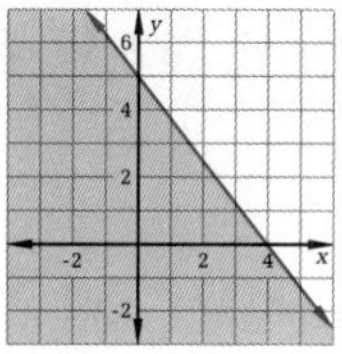

8 $6x + 5y \geq 30$

Problem Set

9

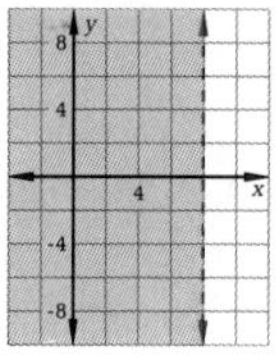

10

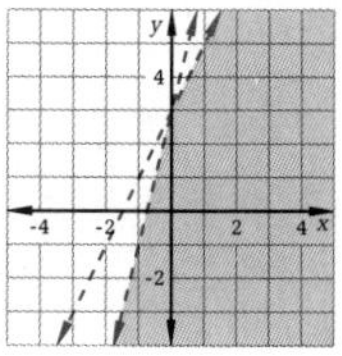

11 a

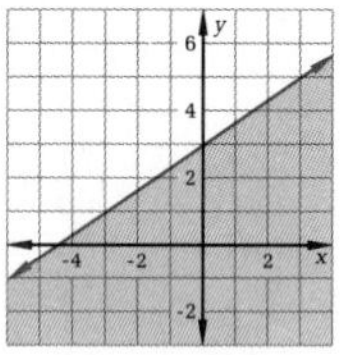

b $\left\{(x, y): y \leq \frac{2}{3}x + 3\right\}$

12 Any equations of the form $y - 7.5 = m(x - 0.5)$ and $y - 7.5 = -\frac{1}{m}(x - 0.5)$ where $m \neq 0$ **13** Answers will vary; $\left(\frac{11}{8}, \frac{51}{8}\right)$ **14 a** 176 **b** 14 **c** −6

d $\{x: x > -4\}$ **15** $\frac{1}{8}$ **16 a** 1960–1969 **b** 1940–1949
c American **d** National **17** 75 lawns, 30 fences

18 **a** 20 **b** 100 **19**

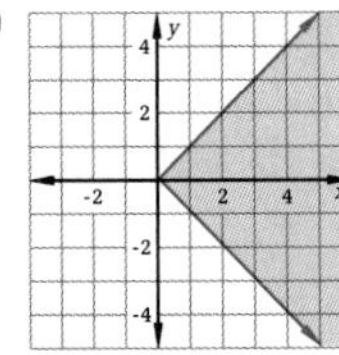

20

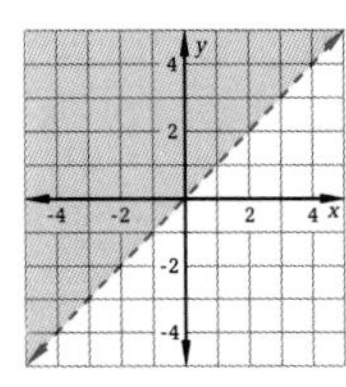

21 $\{(x, y): y \geq -2x + 11\} \cap \left\{(x, y): y \geq \frac{1}{2}x - \frac{3}{2}\right\}$

22 **a** $\frac{87}{17}$, or ≈ 5.12 **b** $-\frac{95}{2}$, or -47.5

23 **a** $12.95c + 7.95t \leq 85$; $c + t \geq 4$

b

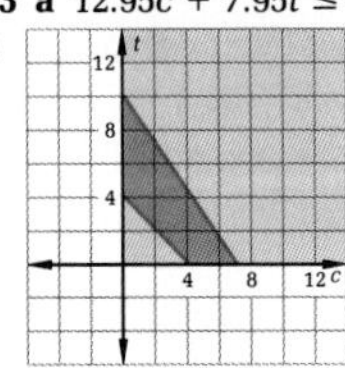

c The numbers of CD's and tapes represented by lattice points in the intersection of the inequalities' graphs **d** 1 CD, 9 tapes

25

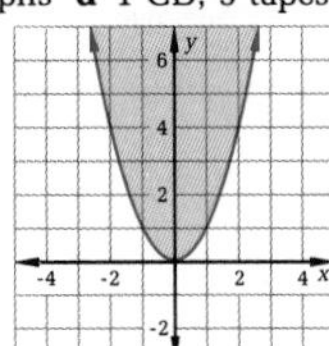

27 $m = 4$; $b = -18$

28 $k_1 = -11$; $k_2 = -31$ **29** $\frac{1}{3}$ **30**

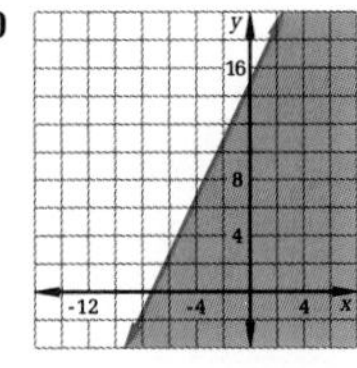

31 $x = 20$; $y = 4$ **32** $a = -\frac{1}{2}$, $b = -4$

33

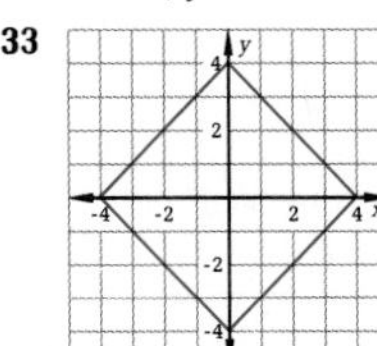

34

x	−20	−10	−6	0	1	8	10	12	20	50
$f(x)$	40	20	20	20	20	20	20	24	40	100

35 **a** 31 **b** $2(n) + 5$ **36** $\frac{2}{9}$

37 **a** $x + y > 10$, $10 + x > y$, $10 + y > x$

b

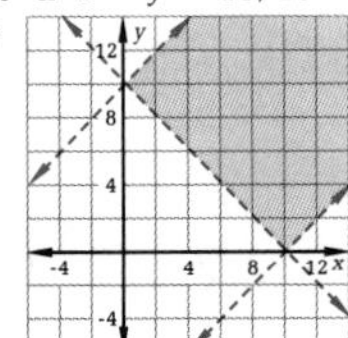

3.4 Functions and Direct Variation

Warm-up Exercises

1 \$8.75 **2** Yes **3** Yes **4** \$26 **5** Answers will vary. Possible answer: If the graph of the linear relationship passes through the origin, the data are directly proportional. **6** $x = \frac{30}{7}$ **7** $x = -\frac{16}{9}$ **8** 3

Problem Set

9 Yes **10** $17\frac{1}{2}$ **11** $(n, d) = (50, 7)$ **13** 51

15

	2^5	2^8	2^{11}
2^0	2^5	2^8	2^{11}
2^2	2^7	2^{10}	2^{13}

16 The slope of $\overleftrightarrow{PR}$ is -1, which is the opposite reciprocal of 1, the slope of $y = x$. **17** No; if x is negative, $-x$ is positive, and if $x = 0$, $-x = 0$.
19 $m = 3$, $b = 36$ **21** $-9, 4$

23 **a** $\frac{520}{18} \approx \frac{410.3}{14.2} \approx \frac{277.4}{9.6} \approx \frac{471}{16.3} \approx 28.9$ **b** ≈ 28.9 **c** Gas mileage in miles per gallon **24** 129; 53 **25** 13 quarters, 12 nickels **26** ≈ 2.569 **27** $(x, y) = (-12, 16)$

29

2

31

32 **a**

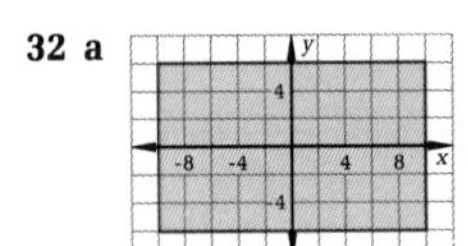

b 240 **33** **a** 52 **b** 26 **c** He begins to repeat the numbers 1, 4, and 2. **34** **a, c, e** **35** **a** 13.5 **b** 7.5 **36** 3 **37** 3.8; 4.4 **38** Colombian: ≈ 87.6 lb; Bolivian: ≈ 112.4 lb **39** 8 **40** Answers will vary. Possible answer: Animal II seems to be the predator, since its population is smaller and its numbers seem to increase only after the population of Animal I increases. **41** **a** ≈ 1.05 **b** ≈ 90.53 **c** Yes

42 $h(x) = 4x - 3$ **43** $f(x) = x + 12$ **44** $\begin{cases} y \leq 3x \\ y \geq \frac{1}{4}x \end{cases}$

45 **a** \$0.384 **b** $B = 7.5 + 0.384T$ **c** \$330.96

3.5 Fitting a Line to Data

Warm-up Exercises

1 $y - 4 = -0.6(x - 9.9)$ **2 a** Answers will vary. **b** ≈ -0.2 **c** $t = -0.2d + 16$ **3 a** 25 **b** ≈ 10.7 **c** 725 **d** 350 **4** $m \approx 1.7$; $b \approx -32.6$

Problem Set

5 a

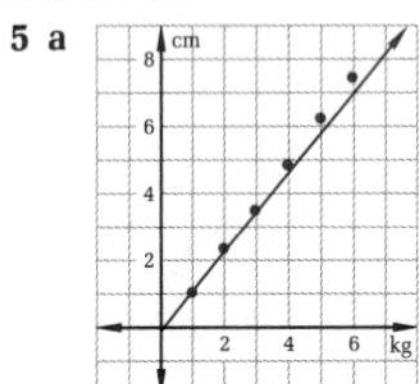

b ≈ 12.7 cm **6 a** $e = 5.50h$ **b** Yes **c** Yes **7** $y = -1527.8x + 14{,}073.6$ **9 a** ≈ 5.49 **b** ≈ 120.7 **10** (5, 0) **11** $y = 3.23x + 17.8$

12

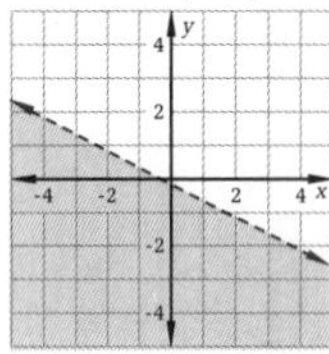

13 a Answers will vary. Possible answers: Because when the refrigerator door is opened, the temperature inside rises, and because the refrigerator's thermostat keeps the temperature within a certain range but not constant **b** ≈ 0 **c** $y = -0.05x + 37.8$ **d** The slope will decrease; answers will vary. **14** 2; 5 **15** $\{x : x \geq 7\}$ **16 a, b** **17 a** ≈ 43.56 km **b** ≈ 87.54 km **18 a** 12 **b** -1 **c** 1 **19** \$6 (by buying 2 six-packs and discarding two bottles) **20 a** 2 **b** $\frac{13}{3}$ **c** $\left(8, \frac{28}{3}\right)$ **d** $y = \frac{5}{6}x + \frac{8}{3}$

21 **23**

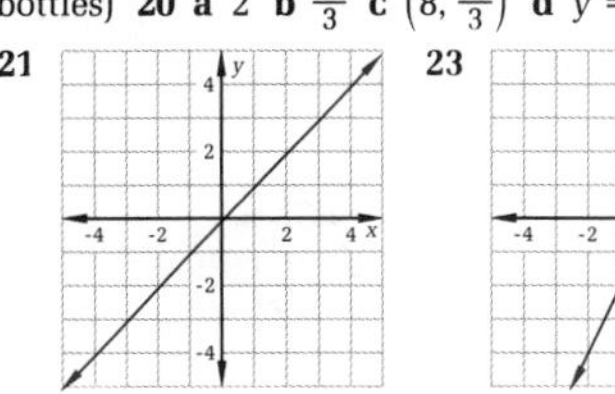

25 **27** ± 4 **29** ≈ 1.1 **31 a** $10x$

b $\frac{10x}{(x + 5)^2}$ **32 a** π **b** ≈ 150.8 in. **33** $h(x) = 4x - 3$

34 $\frac{1}{4}$ **35 a** $\begin{cases} y \leq \frac{3}{4}x + 6 \\ y > \frac{3}{4}x + 2 \end{cases}$ **b** $21\frac{1}{3}$

3.6 Compound Functions

Warm-up Exercises

1 14 **2** 3 **3** 2 **4** g(7.9) **5**

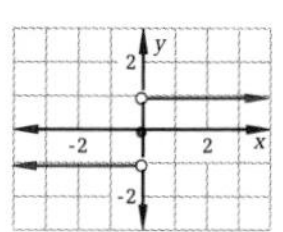

6 **7** 2 **8** 6 **9** 6

10 $w = -27$ or $w = 7$ **11** $w = -10$ or $w = 20$ **12** $w = -10$ or $w = -20$ **13 a** $c(x) = \begin{cases} 2.75x \text{ if } x < 25 \\ 2.4x \text{ if } x \geq 25 \end{cases}$

b 27.5; 66; 60 **c** It's cheaper than buying 24.

Problem Set

15 Undefined **17** 20 **19** Undefined

20 a

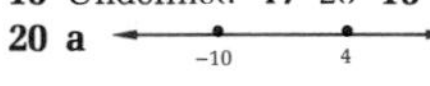

b

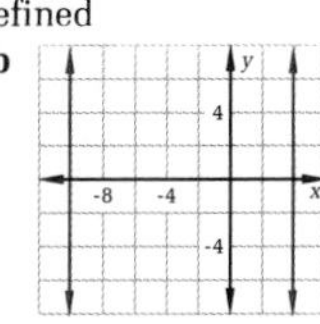

21 a **b**

c **22 a**

b

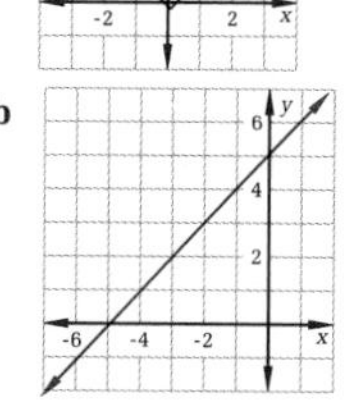

c

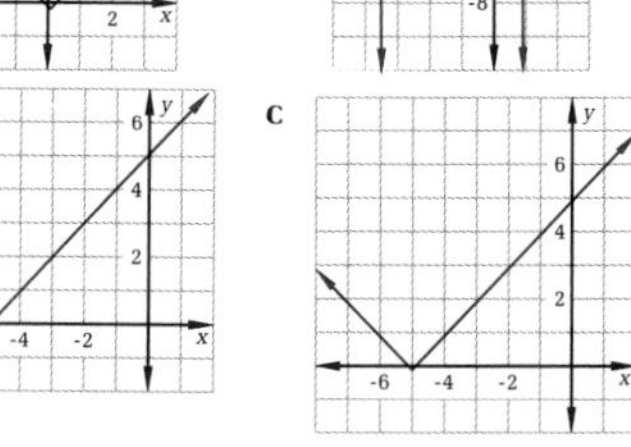

23 $x = -12$ or $x = 6$ **25** $x^2 - y^2$

26 a $c(x) = \begin{cases} 17.95x \text{ if } 1 \leq x \leq 10 \\ 15.95x \text{ if } 11 \leq x \leq 200 \\ 14.5x \text{ if } x > 200 \end{cases}$ **b** 125.65; 829.4; 5002.5 **c** 10 and 183–200 **27 a** Answers will vary. $t = 0.05d + 0.50$ (where t is toll and d is distance from eastern end) **b** $t = 0.0438d + 0.52$ **c** \$4.90 **d** ≈ 233.6 miles **28** $g(x) = 6x$ **29** $f(x) = -\frac{3}{2}x + 23$

30 $\frac{15}{2}$ **31** **a** and **c**; **b** and **d** **a** 9 **b** 21 **c** 9 **d** 21
32 **a** $f(x) = x:(12 - x)$ **b** 4:8, or 1:2 **33** **a** 3 **b** 0
c No real value **35** $p = \frac{4}{3}$ or $p = \frac{20}{3}$
37

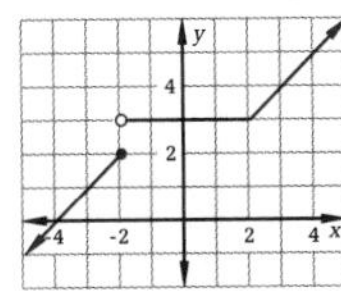

38 **a** $A(x) = 30(x + 1)$ **b** Yes **39** $\frac{5}{8}$
40 **a**

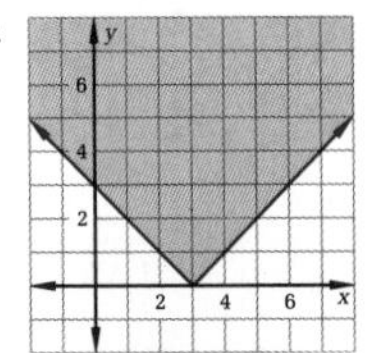

b

c

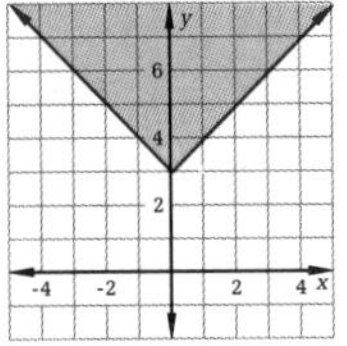

d

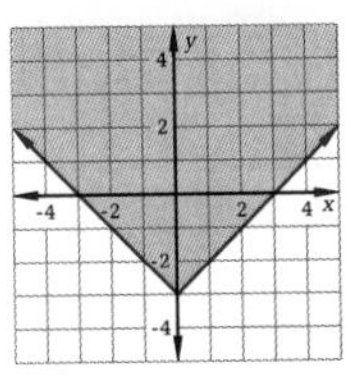

41 **a** 2 **b** 7 **42** $\begin{cases} y \le -\frac{4}{7}x + \frac{64}{7} \\ y \ge 4 \\ x \ge 2 \end{cases}$ **43** **a, c** **44** **a, b, c, d**

45 $\{x : x \in \mathcal{R}\}$ **47** $\{x : x \in \mathcal{R}\}$ **49** $\{x : x \ge 0\}$
51 $\{(x, y) : 0 \le y \le x \text{ or } x \le y \le 0\}$ **53** $\{x : x \in \mathcal{R}\}$
55 $g(x) = \begin{cases} 2x - 8 \text{ if } x \ge 4 \\ 8 - 2x \text{ if } x < 4 \end{cases}$
57 $R(x) = \begin{cases} -x - 3 \text{ if } x \ge -3 \\ x + 3 \text{ if } x < -3 \end{cases}$ **59** $g(x) = |x - 2|$
61 $R(x) = -|x - 3|$ **62** By definition, $|a| = a$ if $a \ge 0$, and $|a| = -a$ if $a < 0$. Therefore, $-|a| = a$ if $a \le 0$, $-|a| < a$ if $a > 0$, $|a| = a$ if $a \ge 0$, and $|a| > a$ if $a < 0$. Thus, $-|a| \le a \le |a|$.

Review Problems

1

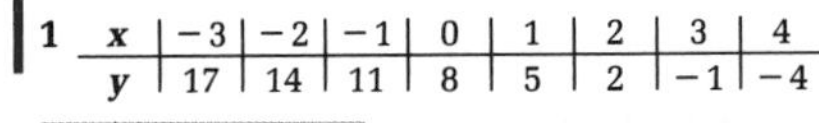

x	−3	−2	−1	0	1	2	3	4
y	17	14	11	8	5	2	−1	−4

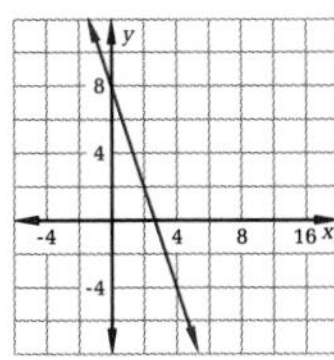

2 **a** Answers will vary. **b** Answers will vary.
c Answers will vary. **3** $x = 9.5$ or $x = -0.5$
5 $x = 11.3$ or $x = -22.7$ **7** **a** $y = -0.773x + 63.2$

b

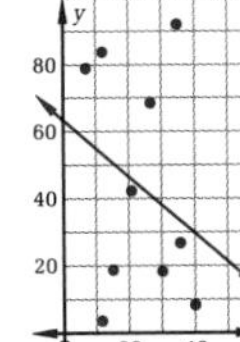

c Answers will vary. You may remark that the data do not seem to be linearly related. **8** **a** $k = 2$ **b**

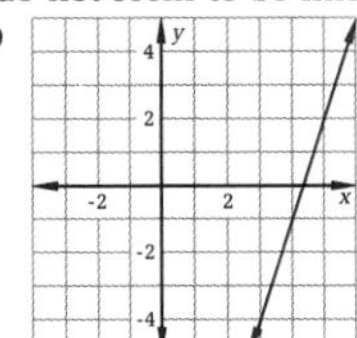

9 $(0, -3)$

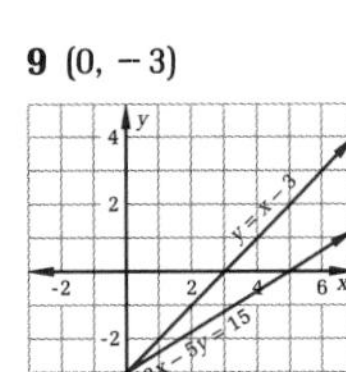

10 Bolivian: 2500 lb; Columbian: 1500 lb

11 **a**

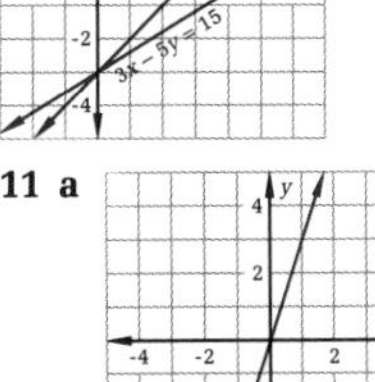

b

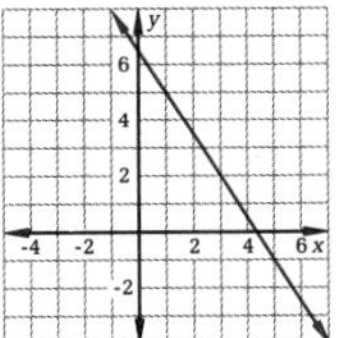

c

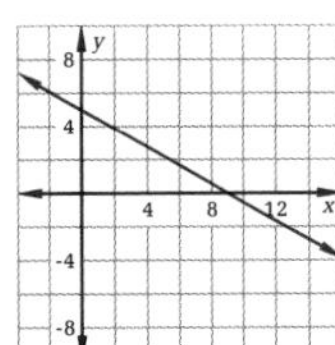

12 **a**

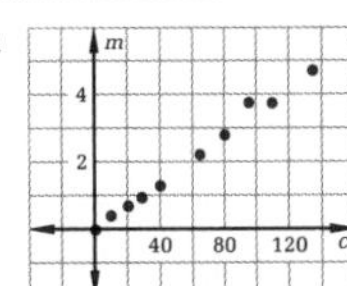

b Yes **c** $\approx \$0.035$

d $c = 0.035m$ **e** $\approx \$4.27$ **13** **a** $200 + 6t$; $100 + 9t$
b 344 ft; 316 ft **c** In ≈ 33.3 sec **14** **a** 3
b $3x + 4y = 12a$ **c** $\frac{4}{3}$ **15**

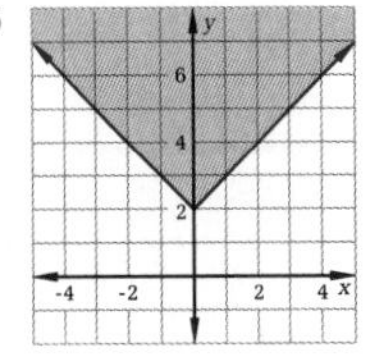

17 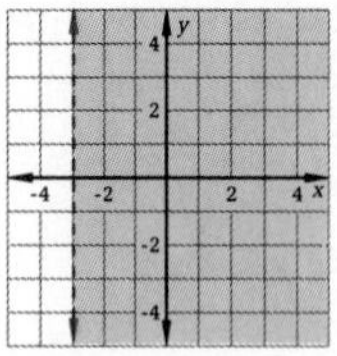

19 96 **20** x-intercept $= \frac{C}{A}$; y-intercept $= \frac{C}{B}$ **21 a** $y = 43{,}652x + 38{,}345$ **b** $\approx 43{,}829$ **c** ≈ 23 mo **d** May **22** -1 **23** $(3, -8)$ **24** No; one of the properties of absolute value is that $|a + b| \le |a| + |b|$, so in some cases $|x + y| \ne |x| + |y|$. **25** 38 **26** 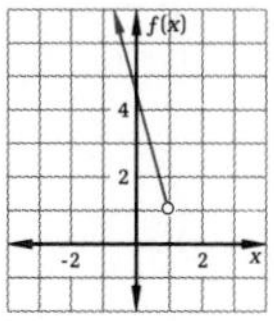

27 a $y = 3554.3x - 7{,}025{,}000$ **b** $\approx \$48{,}057$

4.1 What Are Quadratics?

Warm-up Exercises

1 a ii **b** i **c** i **d** iii **e** ii **2 a** $4x^2 - 9$ **b** $6x^2 + 5x - 6$ **c** $9x^2 - 4$ **3 a** $9y^2 + 24y + 16$ **b** $16x_1^2 - 24x_1 + 9$ **4** $18x^2 + 49x + 10$; $\left\{x : x > -\frac{2}{9}\right\}$ **5** $16x^2 + 56x + 49$ **6 a** $9x^2 + 6x + 1$ **b** 961 Because we use a base-10 number system, the digits in part **b** are the same as the coefficients and constant in part **a.**

Problem Set

7 a, b, e 8 $2x^2 + 11x + 15$ **9 a** 9 **b** 29 **c** 9 **d a** and **c 10 a** False **b** True **c** False **d** False **11** $y = -19$ **13** Fred $= \frac{20}{9}$ **15 a** $25x^2 - 9$ **b** $36x^2 + 60x + 25$ **c** $16y^6 - 36x^2$ **d** $81a^2 - 18ab + b^2$ **e** $x^4 - 1$ **16 a** $\frac{1}{x^3}$ **b** $\frac{y^7z}{x^7}$ **17** $x^2 + y^2 = 16$ **18** ≈ 29.09 **19 a** \$10.20 **b** \$4.88 **c** \$16.50 **d** 27.9 lb, 1–100 mi; 18.5 lb, 101–400 mi; 14.1 lb, 401–900 mi; 10.9 lb, 901–1500 mi; 8 lb, > 1500 mi **20 a** $(x, y) = \left(\frac{2}{11}, -\frac{16}{11}\right)$ **b** $(x, y) = \left(\frac{3}{7}, \frac{23}{7}\right)$ **21 a** $x = 3$ or $x = -4$ **b** $x = 0$ or $x = \frac{1}{3}$ **c** $x = 3$, $x = -3$, or $x = -5$ **d** $x = 3$ **22** 1; 7; 10 **23** ≈ 1.414, 1.0, error, error; the calculator works only with real numbers. **24** $d = 60t$; yes **25 a** 

b 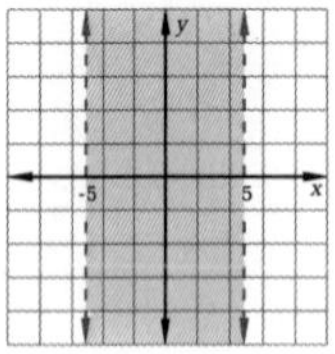

26 111 g **27** $(x + 2)^2 + (y - 5)^2 = 36$ **28** $x^2 + x$ **29** 3.5; 6.5 **30 a** $w_1 = 9$ or $w_1 = -1$ **b** $w_1 = 2$ **31 a** $\{(a, b): ab < 0\}$ **b** $\{(a, b): ab \ge 0\}$

32 a 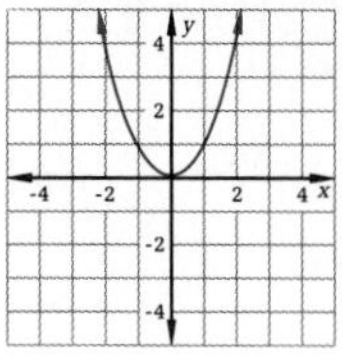

b 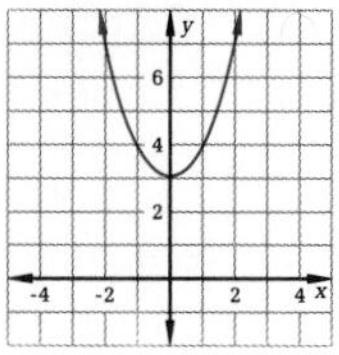

c 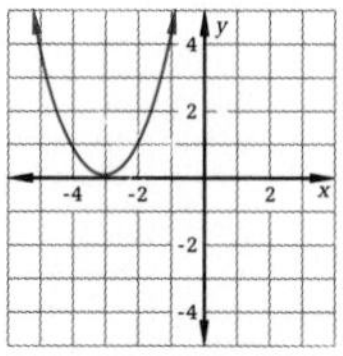

33 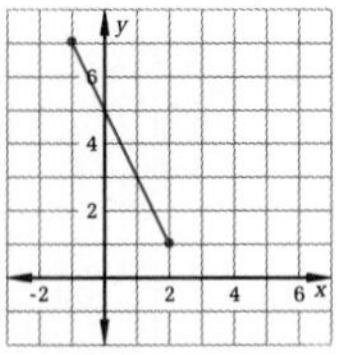

34 a $10x^2 + 2x = 16x^2$ **b** $16x = 14x + 2$ **35 a** 100 **b** 4999 **36 a** $c = 19.2x + 690$ **b** \$16,050 **c** Answers will vary. More machinery and workers might be needed, so that the relationship would no longer be linear. **37** 12 (for $y = 2$) or 204 (for $y = -14$) **38** ≈ 3.68 **39 a** $\sqrt{10x^2 + 18x + 9}$ **b** -0.9 **c** $\frac{3\sqrt{10}}{10}$

40 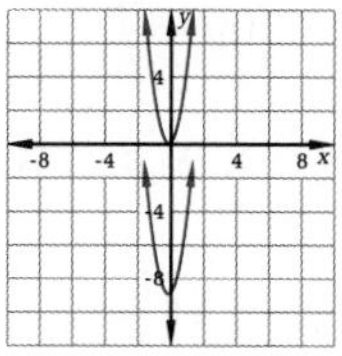

41 a $c(x) = \begin{cases} 95x \text{ if } x \le 10 \\ [95 - 0.50(x - 10)] \text{ if } x > 10 \end{cases}$

b 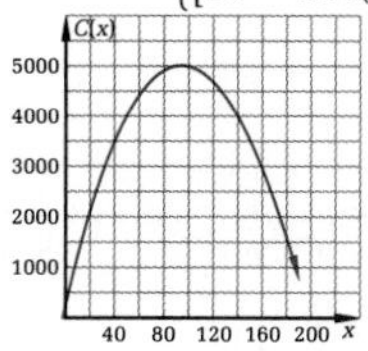

c 100 **d** 125 pieces of software

4.2 Factoring Quadratics

Warm-up Exercises

1 a i **b** i and iii **c** ii **d** iii **e** i and ii **f** iv **2** $(x + 3)(x + 4)$ **3** $(x + 7)^2$ **4** $(4x + 1)(2x - 3)$ **5** $3x + 2$ **6** $5(x - 3)^2$ **7** $2(x + 3)(x - 1)$ **8** $(x^2 + y^2)(x + y)(x - y)$ **9** $\pi(a + b)(a - b)$

Problem Set

11 $7(3y^2 + 4)$ **13 a**

	$3x$	4
$2x$	$6x^2$	$8x$
5	$15x$	20

b $(2x + 5)(3x + 4)$; $6x^2 + 23x + 20$ **14 a** False **b** False **15** \$0.78; \$12.71 **16 a** $0.\overline{1}, 0.\overline{2}, 0.\overline{3}, 0.\overline{4}$ **b** $0.\overline{7}$, $0.\overline{8}$, and 1 **17** 144 **19** $16x^6$ **20** -9; 0; $4x^2$ **21** $6x^2$ **23** $3x - 6$ **24 a** $ac + ad + bc + bd$ **b** $xy + 2y - 3x - 6$ **25 a** $P = 6x + 4y - 8$ **b** $P = 30$

26 a

x	−4	−3	−2	−1	0	1	2	3	4
y	89	48	19	2	−3	4	23	54	97

26 b

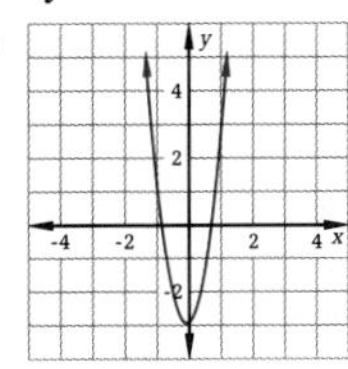

27 $w = \pm 3$; $x = 10$ or $x = -4$; $y = 2$ or $y = -12$; $z = \frac{1}{6}$ **29** $49(4x^2 + w^2)$

31 a

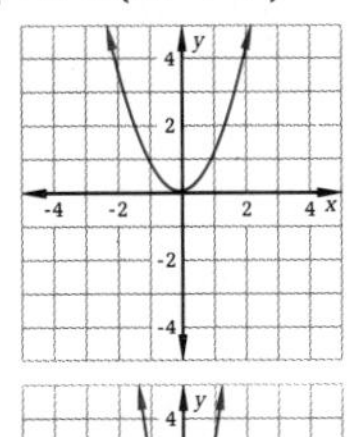

b

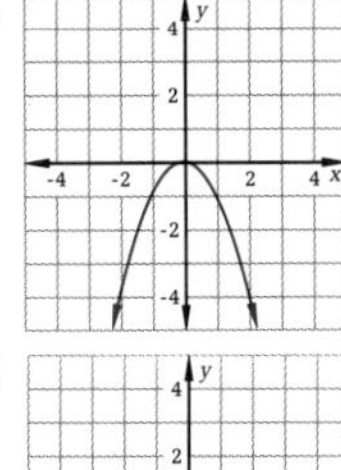

c

d

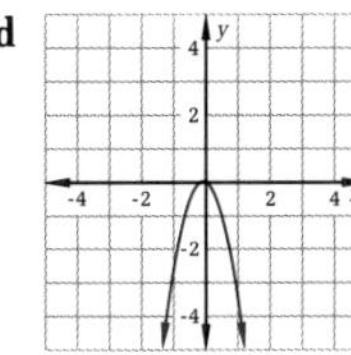

32 a $a^3 - 3a^2b + 3ab^2 - b^3$ **b** $a^3 + a^2b - ab^2 - b^3$ **c** $a^3 - 3a^2b + 3ab^2 - b^3$ **d a** and **c**; **c** is a partially expanded form of **a**.
33 a $-5, -1, 5$ **b** $x = -5$, $x = -1$, or $x = 0$
34 a $[f(x) + g(x)][f(x) - g(x)]$ **b** $(x + 7)(x - 1)$ **c** $(x + 7)(x - 1)$ **35 a** Evens are divisible by 2. **b** $2a + 2b = 2(a + b)$ **36** $\frac{8}{9}$ **37 a** $x = 3$ or $x = -3$ **b** $y = 3$ or $y = -3$ **c** $x - 3 = 3$ or $x - 3 = -3$ **d** $x = 3$ or $x = -3$ **38** $f(67)$; since (67, 0) is closer to the axis of symmetry, $f(67)$ is closer to the vertex.
39 3.5 ft by 7 ft **40 a**

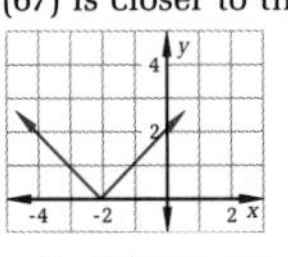

b

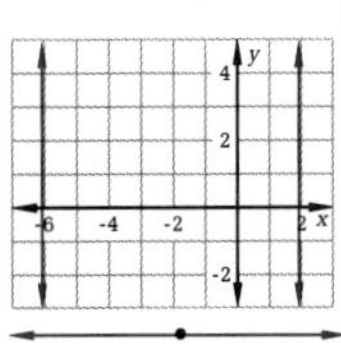

c

d

e

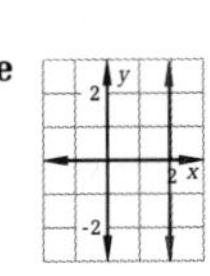

41 a $x^2(x + 1)(x - 1)$ **b** $(x^2 + y)(x^2 - y)$ **c** $y^4(x^2 + 1)(x + 1)(x - 1)$ **42** $f(x) = 2x + 1$ **43** 1040 mi **44 a** 2 **b** $\frac{13}{3}$ **c** $\left(8, \frac{28}{3}\right)$ **d** $y = \frac{5}{6}x + \frac{8}{3}$ **45** $(x^2 + 3 - y)(x^2 + 3 + y)$ **46 a** $A(x) = 100 - x^2$ **b** $\{x: 0 < x < 10\}$ **c** $5\sqrt{2}$ **47** $x = 1$, $x = -\frac{1}{2}$, or $x = \frac{3}{4}$ **48** $(x, y) = (1, 2)$ **49 a** $\frac{1}{2}$; 2.5; -2.5; undefined

b

50 $\frac{1}{2}$ **51**

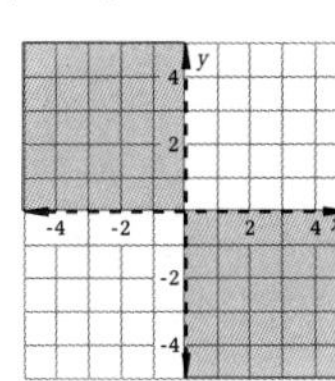

4.3 Graphing Quadratic Functions

Warm-up Exercises

1 -6 or 6 **2** $y = 2(x - 4)^2 + 1$ **3** $y = -(x + 5)^2 + 9$

Problem Set

4 a

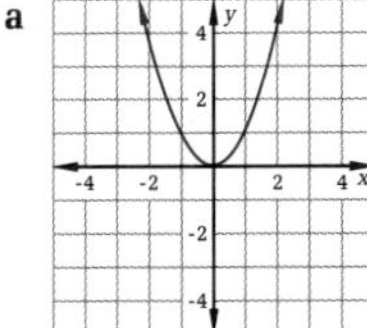

b

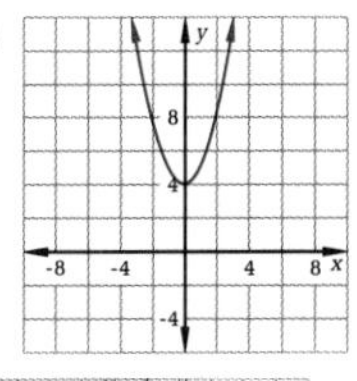

c

d

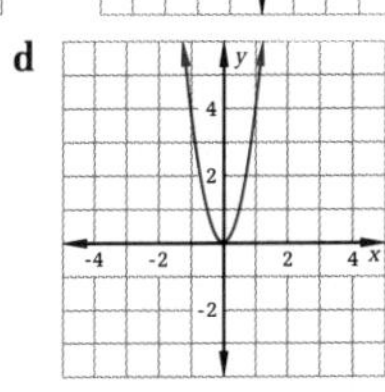

5 $a = 6$ and $b = 3$ or $a = -6$ and $b = -3$ **6 a** iv **b** iii **c** ii **d** i **7** -7 **9** $x = -1$, $x = 2$, or $x = -2$
11

13 $x = 2$ **15** $10.5w - 12.5y$ **17** $x = -23.2$
19 a -1 and 5 **b** $\{x: -1 < x < 5\}$ **c** 2
20 $n \le 435$ **21 a** $\frac{3}{2}$ **b** -5 **c** Yes **22** $(x + 2)(x + 4)$; $x^2 + 6x + 8$ **23**

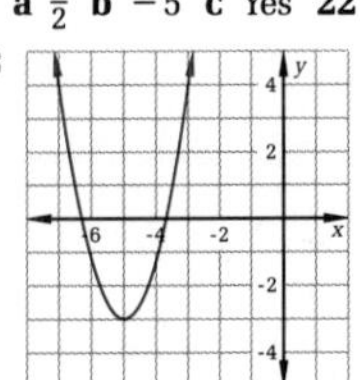

24 −3, 2 **25** True **27** False
29 a x = 4 or x = −7 **b** x = 4 or x = −7

30 $\frac{3}{11}$ **31 a** $4(t-3)^2$ **b** 36 ft **c** After 3 sec **d** After 6 sec **32 a** 2x − y **b** 2(x − y) **33 a**

b If the product of two numbers is zero, at least one of the numbers must be zero, so the graph of xy = 0 is the x- and y-axes. **34 a** $b = -\frac{8}{3}$ or b = 6 **b** {b:b ∈ ℜ}

c b = 0 **35** $\frac{1}{3}$ **36 a** $x^3 + a^3$ **b** $(x + 3)(x^2 - 3x + 9)$

37 (2x + 3)(x + 5) **39 a** $A = 6e^2$ **b** 3 **40** *a* − *b* = 0, so division by *a* − *b* is undefined.

4.4 Solving Quadratic Equations

Warm-up Exercises

1 a $\frac{1}{3}$ **b** −2 **2 a** x = 0 or x = 3 **b** y = −2 or y = 2 **c** z = 0 or z = −4 **3 a** x = 4 or x = 5 **b** y = −8 or y = 2 **c** $z = \frac{1}{2}$ or z = −3

Problem Set

5 x = 2 or $x = -\frac{3}{4}$ **7** −8 and 4 **8** −1 and 6

9

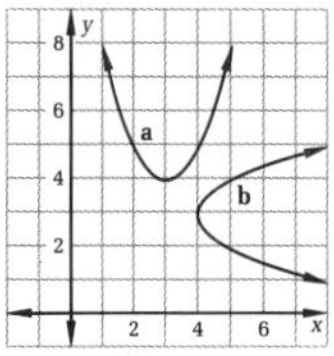

10 (3, 0) and (6, 0)

11 a $y = \frac{3}{2}$, y = −1, or y = 0 **b** y = 2, y = 1, or $y = \frac{2}{3}$

c $y = -\frac{1}{2}$, $y = \frac{1}{2}$, or y = 0 **13** $x = -\frac{1}{2}$ or x = 3

15 −1 or 4 **16** 0 or 7 **17** {x:3 < x < 18} **18 a** $\frac{1}{14}$

b $\frac{5}{28}$ **19 a** z = 7 or z = −2 **b** x = 0 or x = −9

21 $(x - 3y)^2$ **22** 15 **23** x = 0 or $x = -\frac{3}{2}$

24 4 hr 20 min **25** $2x^2 - 9x - 45 = 0$

26 $x \geq \frac{5}{2}$

27 Cashews: \$4.33; macadamias: \$9.46 **28** *f*(x) = −7 or *f*(x) = 2

29 a

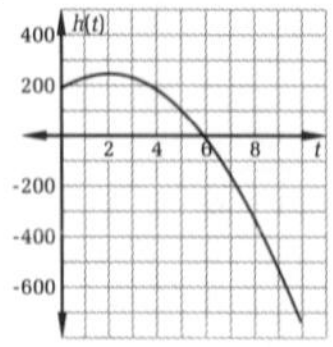

b 200 ft **c** 264 ft

d After ≈ 6.06 sec **e** {*t*:0 ≤ *t* ≤ ≈6.06} **30 a** 20 **b** 192

31

x	−1	0	1	2	3	4	5
y	2	$\sqrt{3}$	$\sqrt{2}$	1	0	Undef.	Undef.

32 27 cm by 36 cm **33** No **34 a** Answers will vary.

b y = 938x − 1,863,085 **35** x = 8 or $x = \frac{5}{2}$

37 a x = −5, x = 0, or x = 5 **b** x = −8 or x = 8

38

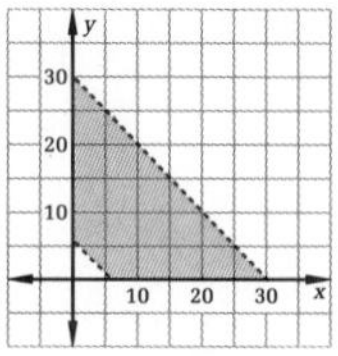

39 9x(x − 3y) **41** (*w* + *z*)(x + y) **43** In ≈22.2 min
44 a x = −*b* or x = −*a* **b** x = −2*c* or x = −3*d*
c x = −3 or x = −*a* **45 a** y = 2x

b

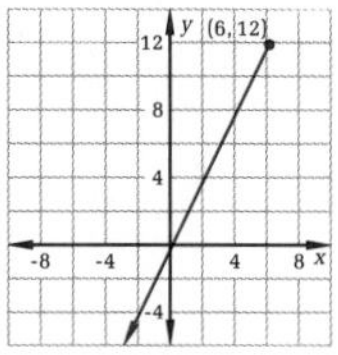

46 a $d^2 = 3250t^2 - 700t + 100$ **b** ≈62.3 **c** After ≈0.108 hr, or ≈6.5 min **d** ≈7.89 miles **47 a** It approaches zero. **b** It increases without limit. **48** 72
49 2x + *h* **50** 144.5

4.5 Completing the Square

Warm-up Exercises

1 a 16 **b** 9 **c** $\frac{81}{4}$ **d** 4 **e** $\frac{81}{16}$ **2 a** 6x **b** 16*y* **c** $8y^2x$

d 11x **3** 16 **4 a** (3, 4) **b** $\left(-\frac{1}{2}, 5\right)$ **c** (−3, 6)

d (5, −20) **5 a** x = 1 or x = −11 **b** x = 15 or x = −1 **c** $x = -3 \pm \sqrt{14}$ **d** x = 25 **6 a** 14 **b** −6

Problem Set

7 $(x - 4)^2 + 29$ **9** $2(x + 25)^2 - 1200$
11 $(x + 25)^2 - 647$

13

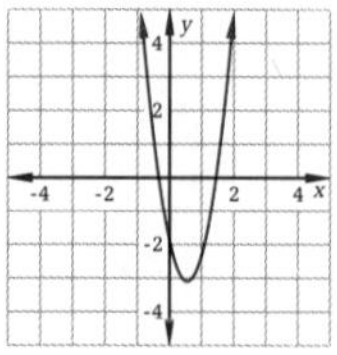

15

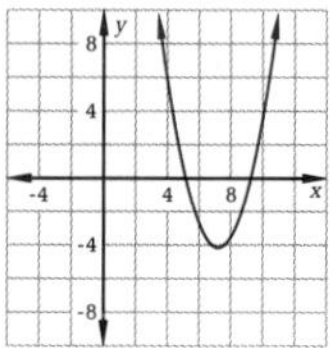

16 a y = 2 **b** y = 12 or y = −3
17 x = 19; m∠1 = 91 **18** $x^2 + 3x - 10 = 0$

19 a $A(x) = 3x^2 + 6x$ **b** $B(x) = 2x + 4$ **c** $\frac{2}{3}$ **21** ± 10 **22** $(x, y, z) = (6, 1, 161)$ **23 a** 12 **b** 7 **c** $a - 5$ **d** a **24** $x = 5, y = 5$ **25** $x = 5$ or $x = -2$ **26** $y = x^2 - 2x - 3$ **27** ± 8 **28** Any equation of the form $y = a(x - 4)^2 - 1$ **29 a** $z = 7$ or $z = -1$ **b** $x = \frac{7}{12}$ or $x = -\frac{1}{12}$ **30 a** $d(x) = 2x^2 - 4x + 4$ **b** 1 **c** 2 **31** $x = \frac{3 \pm \sqrt{5}}{2}$ **33 a** 140; 50 **b** 356.25 in. **c** 50 in. **34 a** $x^2 - 2x - 35 = 0$ **b** $2x^2 - 9x + 4 = 0$ **35** $(x, y) = (3, -1.5)$ **36 a** $y = 0, y = 3$, or $y = -3$ **b** $y = 1$ or $y = -3$ **c** $y = 4$ or $y = -4$ **37** Between 130 and 140 sheets **38** 4 **39** $x = 6$ or $x = -1$ **41** $x \neq -1 \pm \sqrt{6}$ **42** $\frac{7}{2}; \frac{11}{4}$ **43 a** $\left(-\frac{17}{8}, -\frac{481}{16}\right)$ **b** $\left(-\frac{b}{2a}, -\frac{b^2 - 4ac}{4a}\right)$ **44** ± 10 **45** $f(x) = 2^x$ **46** 244

4.6 Applications of Quadratic Functions

Warm-up Exercises

1 $v_0 = 45$ ft/sec; $h_0 = 0$ ft **2** $v_0 = 32$ m/sec; $h_0 = 2$ m **3** $v_0 = 25$ m/sec; $h_0 = 0.5$ m **4** $v_0 = 31$ ft/sec; $h_0 = 0$ ft **5 a** 164 ft **b** ≈ 23.6 m **6** \$3.00 per cassette

Problem Set

7 \$21,000 **8 a** $A(x) = x^2 + 6x - 16$ **b** $\{x : x > 2\}$ **c** 144 **d** $\{x : x > 7\}$ **9 a** In 2 sec **b** 16 ft **c** In ≈ 14.7 sec **10** {10, 11, 12, 20, 21, 30} **11** $833.\overline{3}$ ft² **12 a** 2 and 5 **b** $\frac{7}{2}$ **13** \$1.10 **14 a** $h(t) = -\frac{1}{2}(9.8)t^2 + 2.94t + 3$ **b** ≈ 3.44 m **c** ≈ 1.14 sec **15** $x \approx -46.23$ **16 a** $x = \pm 9$ **b** $x = \pm 9$ **17 a** $x = 6 + \sqrt{57}$ **b** $\sqrt{57}$ is between 7 and 8. **c** Between 13 and 14 **19** $b = -\frac{5}{2}$ or $b = 3$ **21** $x = 15$ or $x = -6$ **22 a** In ≈ 0.56 sec **b** In ≈ 1.13 sec **c** In ≈ 1.39 sec **23** $\left\{n : n > 66\frac{2}{3}\right\}$ **24** $-2, 8; -1, 4$ **25** $x = \frac{-1 + \sqrt{5}}{2}$ **26 a** ≈ 5.04 sec and ≈ 19.12 sec after launch **b** ≈ 743.16 m **c** ≈ 24.40 sec **27 a** $A = (10 + x)\sqrt{36 - x^2}$ **b** ≈ 68.19

4.7 Data Displays

Problem Set

1 a

5	3 4
4	0 1 1
3	0 0 1 2 2 3 3 4 5 5 6 7 7 9
2	0 1 1 2 3 7 9 9 9
1	8 9

2 | 1 means 21 points scored during a game

b 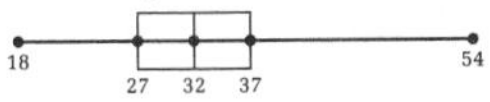

2 a 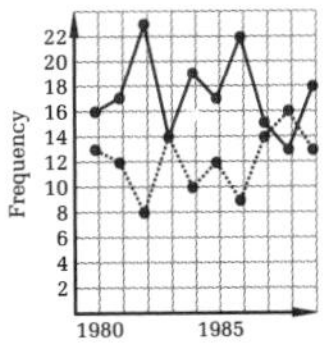

b 17.4 **c** 1986 **d** 5.4 **3 a** True **b** False **c** True **d** False **4 a** 37 **b** 8 **c** Cannot be determined **d** 8 **5** 1 **6 a** 88 and 16 **b** 72 **c** 62 and 81 **d** Those below 81 **e** Cannot be determined **7** $(x, y) = (2, 5)$ **8** Answers will vary. **9 a** $\approx 62.5\%$ **b** $\approx 18.2\%$ **c** $\approx$5,400,00 bu **10** 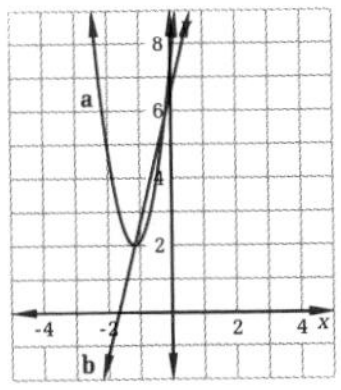

11 a $x^2 + 6x$ **b** 9 **c** $(x + 3)^2$; $x^2 + 6x + 9$ **12 a** 98 and 23 **b** 9 **c** 76 **d** 7 **e** 29 **f** Cannot be determined **13 a** $\frac{13}{90}$ **b** $\frac{1}{6}$ **14 a** In ≈ 0.66 min **b** ≈ 2752.5 ft²

Review Problems

1 $6x^3 - 12x^2$ **3** $36y^2 - z^2$ **5** $9x^2 - 30x + 25$ **7** $x = \pm 10$ **9** $z = 3$ or $z = -2$ **10** $y = \frac{1}{2}(x - 2)^2 - 5$ **11** $\frac{5}{8}$ **13** $x = \pm 4$ **15 a** $A = 2x^2 - 6x$ **b** $\{x : x > 3\}$ **c** 216 **16** $16x^3 + 4x^2 + 7x + 5$ **17** $(3y - 1)(2y - 1)$ **19** $(3x + 2)(2x + 3)$ **21** $(z - 9)(z - 8)$ **23 a** 47 **b** 56 **c** 77 **25** $x = 4 \pm \sqrt{21}$ **27** $x = \frac{-9 \pm \sqrt{87}}{3}$ **28 a** 40; 60; 84 **b** 42 **29 a** 36 ft **b** At $t = 3$ **c** At $t = -2$ and $t = 8$ **d** 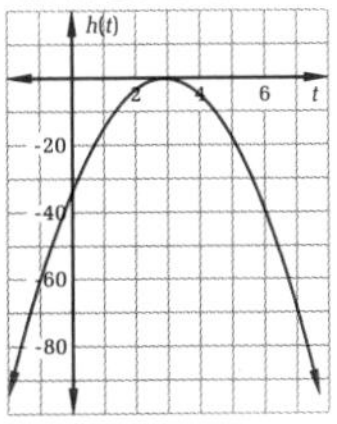

30 $\{x : x > 5\}$ 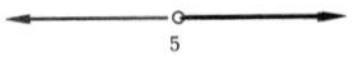

31 $x^3 + 3x + \frac{3}{x} + \frac{1}{x^3}$ **32 a** 894,000 **b** $\approx 33.2\%$; $\approx 21\%$; $\approx 8.9\%$ **33** $x = 5$ **35** $x = 3$ or $x = -1$ **36** No such points. 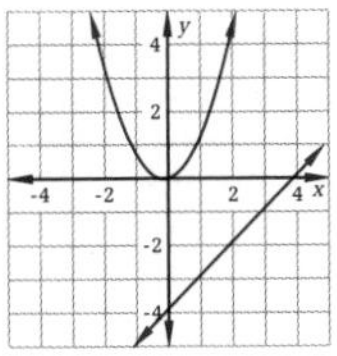

37 a $V = 6x^2 + 11x + 6$ **b** $\{x: x > 0\}$
38 a $A(x) = -2x^2 + 12x + 144$ **b** 3 **c** 162
39 $(x + 1 + y)(x + 1 - y)$ **41** 96 trees per acre
42 $4x^3 + 3x^2 + 2x + 1$

5.1 Relations and Functions

Warm-up Exercises

1 a $\{(-2, 5), (-1, 2), (0, 5), (1, 2), (2, 0)\}$

b

Input	−2	−1	0	1	2
Output	5	2	5	2	0

c

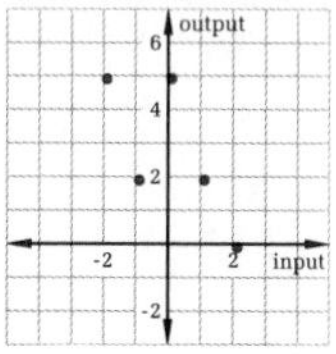

d

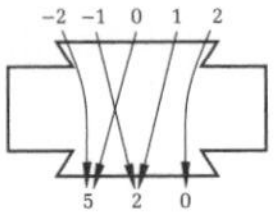

2 b, c, d **3** Yes; no **4 a** 15 **b** $f(n) = \sum_{k=1}^{n} k$
5 a $\{-4, -2, 0, 2, 5\}$ **b** $\{-4, -2, 0, 5\}$ **c** $\{(-4, -2), (-2, 0), (0, 5), (2, -4), (5, -4)\}$ **6 a** 16 **b** 4 **c** 3 **d** 9

Problem Set

7 Both **9 a** $\{(-2, 4), (0, 2), (2, 0), (4, 2), (6, 4)\}$

b

x	−2	0	2	4	6
$f(x)$	4	2	0	2	4

c

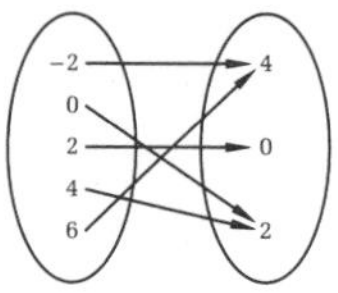

d

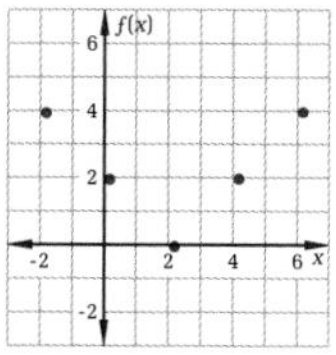

e

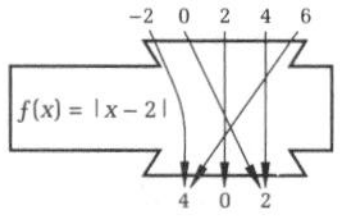

10 a 0 **b** −1 **c** $x = 3$ **d** $x = -2$ or $x = 2$ **e** $\{-2, 0, 1, 2, 3\}$ **f** $\{-3, -1, 0, 1\}$ **11** No; since each input can have only one output, no. of outputs ≤ no. of inputs. **12 a** 60 **b** Impossible figure **c** $\{x: x > 2\}$
13 $x < 2$

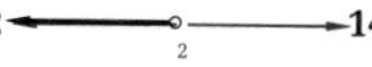

14 Answers will vary. For many scientific calculators, x must be ≤ 69. **15 a** 9 **b** 2 **c** 31 **16 a** $C(q) = 80q$ **b** 280
17 a True **b** False **18** $\{F: F < 68\}$
19

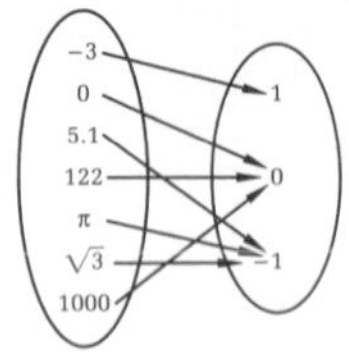

20 a $A(x) = 5x + 36$ **b** $\{x: x > -3\}$ **21 a** $p(x)$ = year in which x was first elected President of the U.S. **b** 1960; 1884 **23** No; yes
25 (6, 1) **27** $x = \frac{3 \pm \sqrt{105}}{4}$ **29** $x = 0$ or $x = 3$
30 a $M(x) = \frac{1}{5}(x + 120)$ **b**

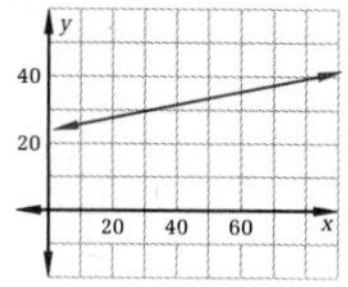

c −120 **d** 30 **31 a** $d(x) = x^2 + 5$ **b** −5 and 5
32 a 1 **b** ≈3.73 **c** ≈63.43

5.2 More About Functions

Warm-up Exercises

1 a Domain: $\{x: x \geq 0\}$; range: $\{f(x): f(x) \geq 3\}$
b Domain: $\{x: x \neq -5\}$; range is $\{g(x): g(x) \neq 1\}$
c Domain: {integers greater than 2}; range: {positive multiples of 180} **d** Domain: $\{y: y \neq -2$ and $y \neq -3\}$; range: $\{h(y): h(y) \geq 0$ or $h(y) \leq -48\}$ **2 a** 13
b $8x + 29$ **3 a** 50
b 196 **c** No **4 a** 7 **b** 7 **c** 9 **d** 9 **e** x
5 $g[f(x)] = f[g(x)] = x$ for all x **6 a** 7; 7 **b** 49; 49
c 15; 47 **d** Multiplication is commutative, but composition of functions is not.

Problem Set

7 a None **b** $x \geq 4$ **c** x; $\{x: x \geq 4\}$ **d** $|x|$; $\{x: x \in \Re\}$
e 5; 5 **f** Undefined; 4 **8 a** 11 **b** {3, 7} **c** {11, 13}
9 $8 - 15y$; $-8 - 15y$ **10** $\{x: x < -5\}$ **11 a** 4; 4 **b** 9; undefined **c** No (unless the domain of s is restricted)
12 1 or −6 **13** No; $f[g(x)] = 4x - 5$, and $g[f(x)] = 4x + 5$, so $f[g(x)] \neq g[f(x)] \neq x$. **14** −1 and 6
15 a Yes **b** No **16** $\frac{3}{2}$ **17 a** 11 **b** 0 **c** −9 **d** −6 **e** 3
f −2 **g** 25 **h** 1 **18** −2 and 2 **19 a** $m(x) = 2x + 40$
b $-2 < x < 10$ **c** It increases from 44 to 50. **d** It decreases from 24 to 15. **20** Same domain, same rule, same range **21 a** 8 **b** 8 **c** No (unless the domain of f is restricted) **22 a** 4; 4 **b** Yes **23 a** 5 **b** 5 **c** 5
24 a $\begin{bmatrix} 11 & 18 \\ 25 & 26 \end{bmatrix}$ **b** $\begin{bmatrix} 1 & 12 \\ 17 & -38 \end{bmatrix}$ **c** $\{(A, B)$: dimensions of A = dimensions of $B\}$ **d** $\{(A, B)$: no. of columns in A = no. of rows in $B\}$ **25 a** $h(x) = 0.0973x$ **b** The sales tax charged on a purchase of x hamburgers.
26 a $\frac{x^2 - 9}{x}$; $\{x: x \neq 0, x \neq -3$, and $x \neq 3\}$ **b** $\frac{x}{1 - 9x^2}$; $\left\{x: x \neq 0, x \neq \frac{1}{3}, \text{ and } x \neq -\frac{1}{3}\right\}$ **27 a** 2 **b** 3 **c** −5
d Undefined **e** $\text{INV}f(x) = f(x)$ for $\{x: x \neq 1\}$ **28 a** 16
b 20 **c** 24 **29** Domain: $\{x: 5 < x < 14\}$; range: $\{P(x): 27 < P(x) < 54\}$ **30** $\frac{4}{21}$ **31 a** $D(n) = \frac{n(n-3)}{2}$
b {Integers ≥ 3} **c** {0, 2, 5, 9, 14, . . .}
32 $T(p) = \left[\frac{p}{20}\right] + 1$ **33 a** 124,050 ft **b** 90.5 sec
c ≈178.55 sec

5.3 Inverse Functions

Warm-up Exercises

1 a (1, 0), (4, 2), (16, 4), (25, 5) **b**

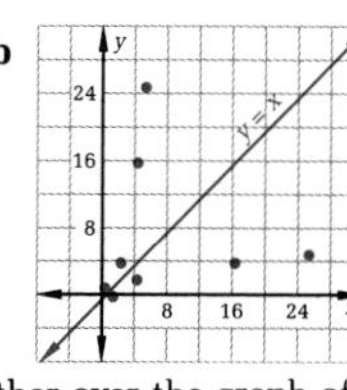

c They are reflections of each other over the graph of $y = x$. **2** $INVf(x) = \frac{x}{10}$ **3** $INVg(x) = x + 6$

4 $INVh(x) = \frac{x - 3}{6}$ **5** $INVp(x) = \sqrt{x + 1}$, where $x \geq -1$ **6** $INVg(x) = x^2 + 1$, where $x \geq 0$

7 $INVr(x) = \frac{1}{2x}$ **8 a** Domain is $\{-4, -1, 0, 6, 8\}$; range is $\{-2, -1, 1, 3, 4\}$. **b** $(-2, -4), (-1, -1), (3, 0), (1, 6), (4, 8)$ **c** Domain is $\{-2, -1, 1, 3, 4\}$; range is $\{-4, -1, 0, 6, 8\}$ **9 a**

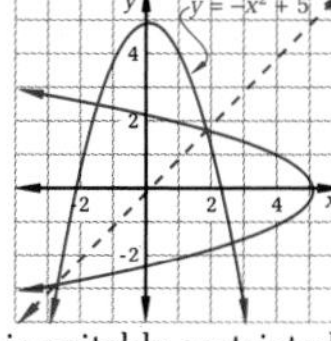

b No **c** Yes, if the domain of g is suitably restricted—for instance to $\{x: x \geq 0\}$, in which case $INVg(x) = \sqrt{-x + 5}$.

Problem Set

10 a 6 **b** 64 **c** $\frac{n - 5}{2}$ **11** $\{(-2, 6), (0, 7), (1, 5), (4, 4), (7, 3)\}$ **12 a**

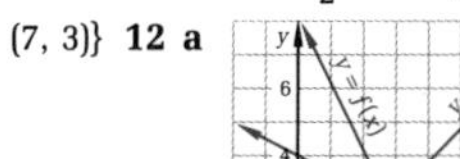

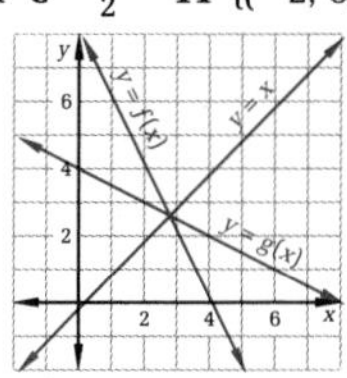

b 6; -10

c $f[g(x)] = -2\left(-\frac{1}{2}x + 4\right) + 8 = (x - 8) + 8 = x$ **d** Yes

13

x	0	1	2	3	4	5	8
f(x)	5	3	0	2	1	4	8
g(x)	2	4	3	1	5	0	8

14 $f(x) = 24 - 2x$ **15** $\frac{1}{10}$ **16** 42

17 a $\{f(x): 1 < f(x) < 100\}$ **b** $g(x) = \sqrt{x}$ **c** Domain: $\{x: 1 < x < 100\}$; range: $\{g(x): 1 < g(x) < 10\}$ **d** Yes; domain of INV*f* = range of *f*, and range of INV*f* = domain of *f*. **18 a** Output the number of sides of a polygon when the sum of its angles is input **b** $n = \frac{S}{180} + 2$ **c** {positive multiples of 180} **19** 6

20 $INVf(x) = -\sqrt{x}$ **21 a** Yes

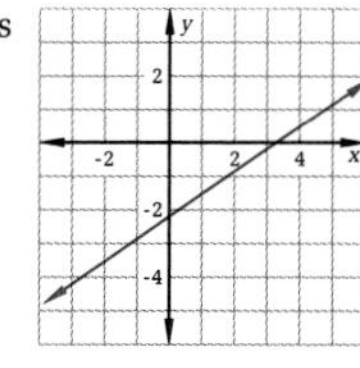

b Yes

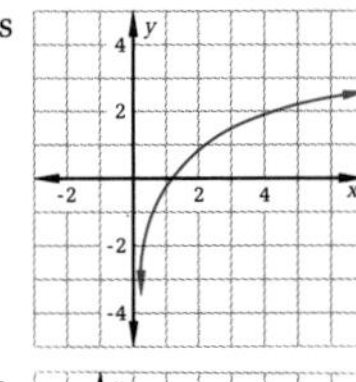

c Yes

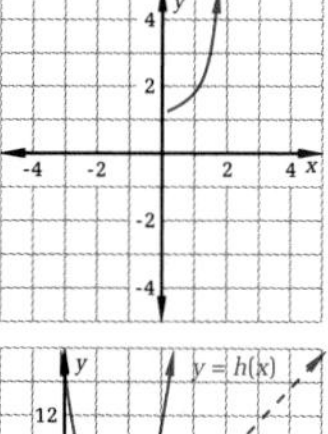

d No

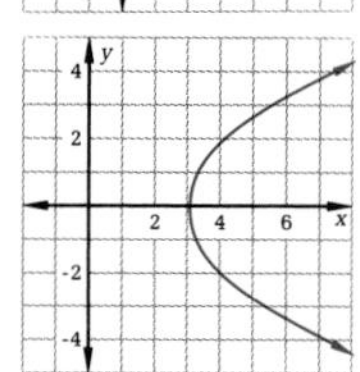

22 a

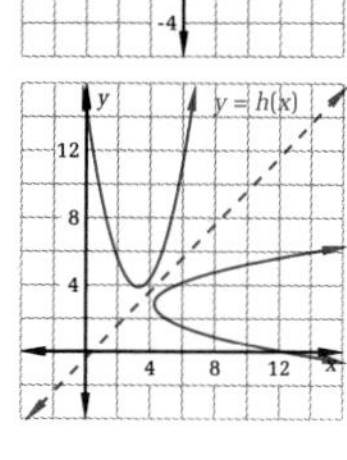

b No **c** Yes, if the domain of *h* is suitably restricted—for instance, to $\{x: x \geq 3\}$, in which case $INVh(x) = \sqrt{x - 4} + 3$

23 a $INVf(x) = \frac{x - 4}{3}$ **b** $INVf(x) = \frac{9}{x}$, where $x \neq 0$

c $INVf(x) = \frac{3}{2} - 2x$ **d** Answers will vary. Possible answer: If $x \geq -3$, $INVf(x) = \sqrt{x} - 3$. **24** ≈15.65 cm and ≈6.65 cm **25** $INVg(x) = -\frac{3}{2}x - 5$ **26 a** No

b Area of square with side length 2x **c** Total area of two squares, with side length x **27 a** Yes **b** Yes

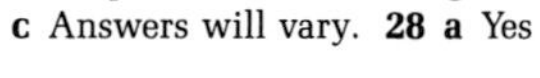

c Answers will vary. **28 a** Yes

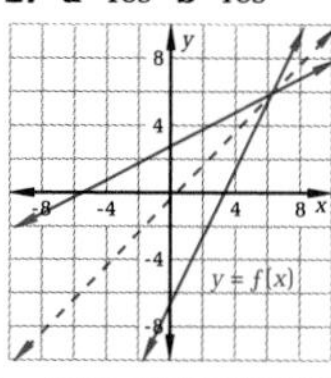

b No

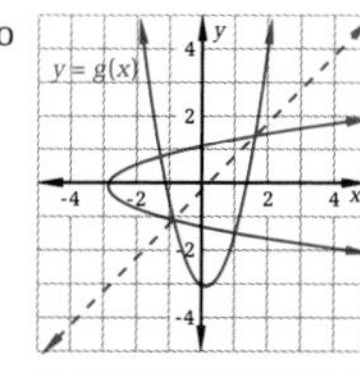

c No

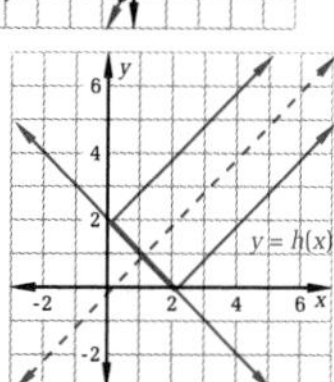

d Yes

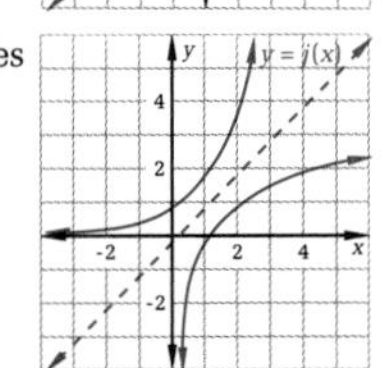

29 $x = 0$ **31** $y = 3$ **32** 10

33 They are inverses. **34 a** $INVT(x) = \frac{x}{0.07}$

b Purchase price on which a sales tax of x dollars is charged **35** $y = 1.18x + 4.69$ **36** $\frac{1}{5}$

37 $INVf(x) = \sqrt{\frac{x - k}{a}} + h$ **38 a** $\begin{bmatrix} 10 & 10 \\ 4 & 2 \end{bmatrix}$

b $INVf\left(\begin{bmatrix} a & b \\ c & d \end{bmatrix}\right) = \frac{1}{2}\begin{bmatrix} a - 4 & b \\ c & d - 4 \end{bmatrix}$

c $f \circ INVf\left(\begin{bmatrix} a & b \\ c & d \end{bmatrix}\right) = 2\left(\frac{1}{2}\begin{bmatrix} a-4 & b \\ c & d-4 \end{bmatrix}\right) + 4\begin{bmatrix} 1 & 0 \\ 0 & 1 \end{bmatrix}$

$= \begin{bmatrix} a-4 & b \\ c & d-4 \end{bmatrix} + \begin{bmatrix} 4 & 0 \\ 0 & 4 \end{bmatrix} = \begin{bmatrix} a & b \\ c & d \end{bmatrix}$

5.4 Operations on Functions

Warm-up Exercises

1 $f(x) = x - 2$ **2** $g(x) = \frac{2}{5}x$ **3 a** $-4 + \sqrt{3}$ **b** $-4\sqrt{3}$ **c** $-\frac{4}{\sqrt{3}}$ **d** $\sqrt{3} - 7$ **e** 5 **4 a** $A_r(x) = 2x^2 - 2x$ **b** $A_t(x) = \frac{x^2\sqrt{3}}{2}$ **c** It represents the total area of the figure. **d** ≈ 1106.41 **5 a** $9x - 6$ **b** $9x^2 - 24x + 9$ **c** $27x^2 - 78x + 48$ **d** $9x - 16$ **e** $36x^2 - 126x + 104$ **6** 7 **7**

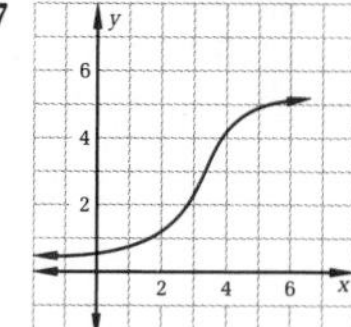

8 $h(3x)$ if $x \neq 0$

Problem Set

9 a $\sqrt{3x - 5}$ **b** $\sqrt{x}$ **c** $\sqrt{x^2 - 5}$ **d** $\sqrt{2x - 6}$ **e** $\sqrt{\sqrt{x - 5} - 5}$ **10** $f(x) = -2x$

11 a **b** **c**

12 a

x	−4	−3	−2	−1	0	1	2	3	4
f(x)	9	4	1	0	1	4	9	16	25
f(x − 1)	16	9	4	1	0	1	4	9	16
f(x + 1)	4	1	0	1	4	9	16	25	36

b

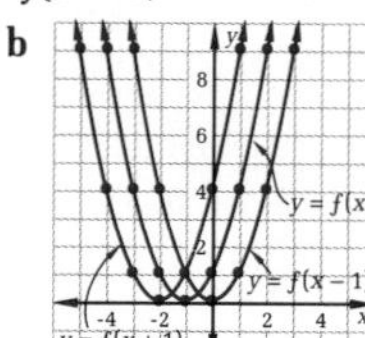

13 $g(x) = x^3$

14

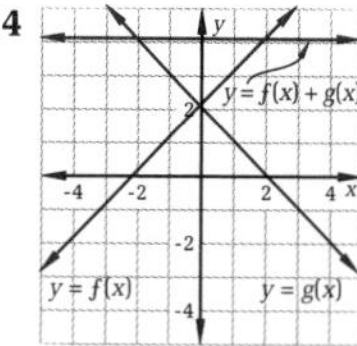

15 $\{x : x < -4 \text{ or } 4 < x < 8\}$

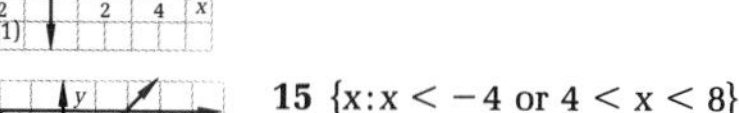

16 24 **17 a** **b**

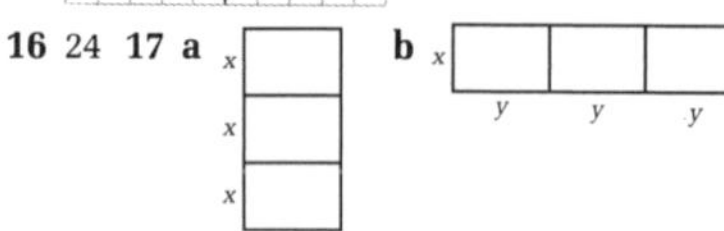

c **d**

18

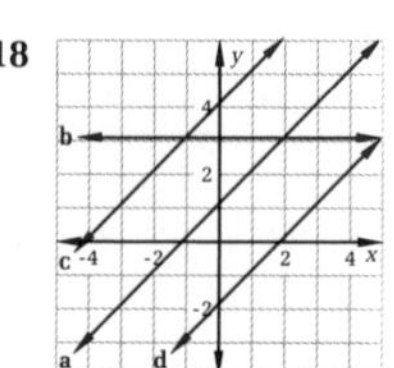

19 c **20**

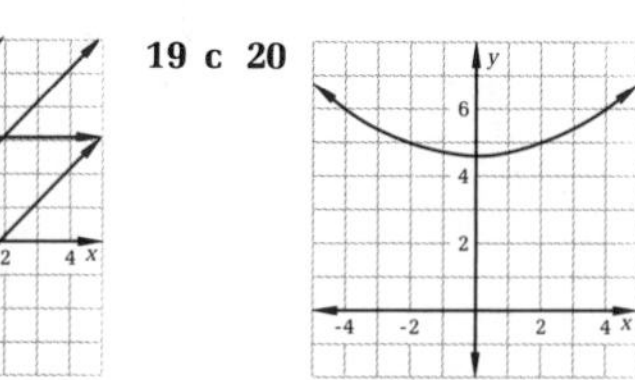

21 a $Q(x) = \frac{x^2 - x + 3}{3}$ **b** $R(x) = \frac{x^2 - 14x + 45}{x^2 - 4}$ **c** $x^2 - 14x + 45$ **d** $x^2 - 11$ **22 a** $g(x) = 2$ **b**

c $x^2 + 2$ **23 a**

b

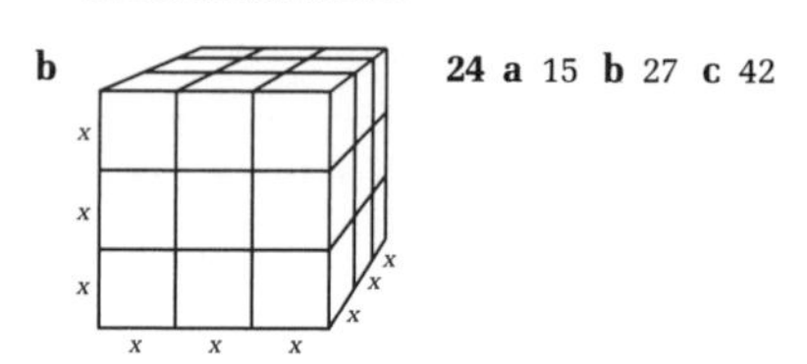

24 a 15 **b** 27 **c** 42

25 $g(x) = x^2 - 4x$ **26** -122 **27 a** 0.7781 **b** 1.0791 **c** 0.9542 **d** 1.2552 **28** $g(2x)$ if $x > 0$, $2 \cdot g(x)$ if $x < 0$, neither if $x = 0$ **30** No

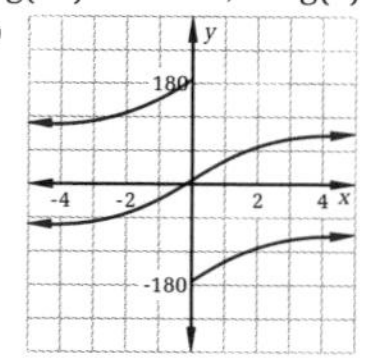

31 a $h(x) = \frac{x}{2}$ **b**

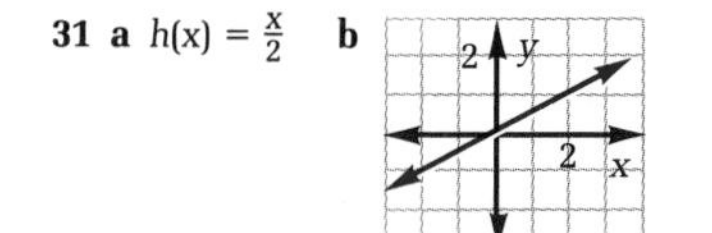

32 a $f[g(x)] = \begin{cases} \sqrt{10} \text{ if } x < 4 \\ 0 \text{ if } x = 4 \\ 9 \text{ if } x > 4 \end{cases}$ **b**

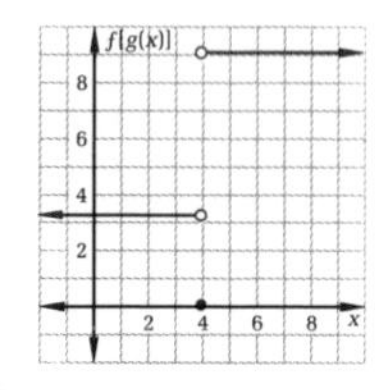

c Domain: $\{x : x \in \Re\}$; range: $\{0, \sqrt{10}, 9\}$

5.5 Discrete Functions

Warm-up Exercises

1 120 **2** 3,628,800 **3** $\approx 1.3 \times 10^{12}$ **4** $\approx 1.2 \times 10^{56}$ **5** 479,001,600 **6 a** 198,485 **b** 1,190,910 **7 a** 42 **b** 21 **c** 3,603,600 **d** 5005 **e** $P(a, b) \geq C(a, b)$ **8** 36^6, or $\approx 2.18 \times 10^9$ **9** $26^2 \cdot 10^4$, or 6.76×10^6

Problem Set

10 a 9 **b** 97 **c** n + 1 **11 a** No **b** 330 days **12 a** 35 **b** 35 **13** 5 **15** 12 **17** $C(n, n-1) = n$ **18** 4 **19** 8 **20** $\frac{1}{6}$ **21** $\frac{1}{36}$ **22** 0 **23 a** 15; 15 **b** 6; 6 **c** 28; 28 **d** 56; 56 **e** $C(n, r) = C(n, n-r)$ **24 a** $\{-2, 1, 10\}$ **b** $\{y: -2 \leq y \leq 10\}$ **25** 35 **26 a** $\text{INV}f(x) = \frac{3}{2}(x + 6)$

b

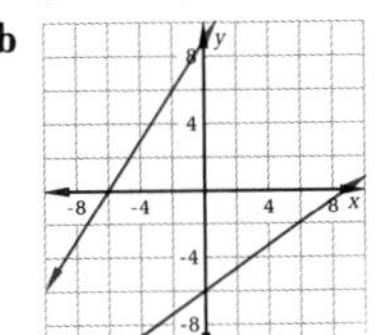

c $(-18, -18)$ **27**

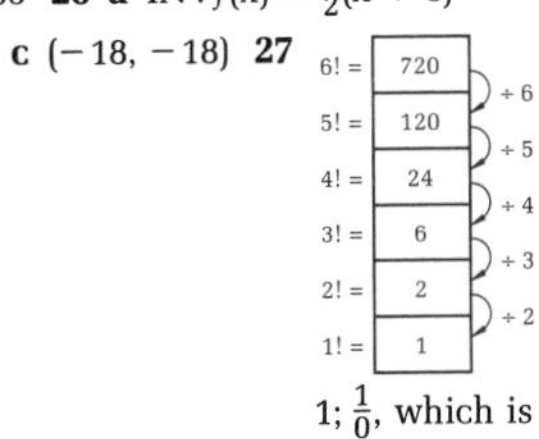

1; $\frac{1}{0}$, which is undefined

28 11 **29** Yes **30** $C(25, 2) = 300$; we need to determine the number of sets of two people that can be formed from a set of 25, and order does not matter. **31** 8 **32** 6 **33** k = 7 **34** $f(x) = 8$ or $f(x) = 4$ **35 a** 50 **b** 1175 **c** $\frac{46}{587}$ **36 a** \$0.75 **b** \$0.25 **c** $\{w: 3 < w \leq 5\}$ **d** 25, 50, 75 **37 a** 2; 6; 24; 120 **b** $(x - 1)!$ **38 a** Answers will vary. **b** y = 6.35x − 11,660 **39 a** 10 **b** 40 **40 a** 1 **b** 0 **c** $\frac{1}{\sqrt{3}}$ or ≈ 0.5774 **d** 1 **e** 1

41
$$C(n-1, k-1) + C(n-1, k)$$
$$= \frac{(n-1)!}{[n-1-(k-1)]!(k-1)!} + \frac{(n-1)!}{(n-1-k)!k!}$$
$$= \frac{(n-1)!}{(n-k)!(k-1)!} + \frac{(n-1)!}{(n-k-1)!k!}$$
$$= \frac{k \cdot (n-1)!}{(n-k)!k!} + \frac{(n-k) \cdot (n-1)!}{(n-k)!k!}$$
$$= \frac{[k + (n-k)] \cdot (n-1)!}{(n-k)!k!}$$
$$= \frac{n \cdot (n-1)!}{(n-k)!k!} = \frac{n!}{(n-k)!k!} = C(n, k)$$

5.6 Recursively Defined Functions

Warm-up Exercises

1 40; 160; 640

2

x	−3	−2	−1	0	1	2	3	4
f(x)	1	−1	−1	1	5	11	19	29

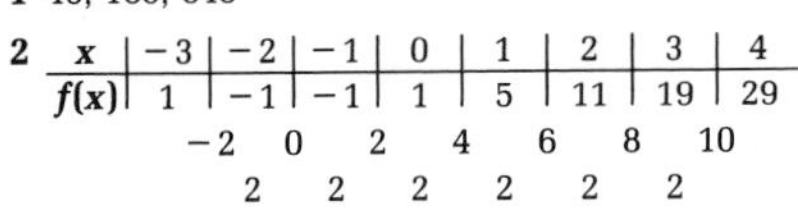

−2 0 2 4 6 8 10

2 2 2 2 2 2

3 Linear **4** 2; 6; 24; 120 **5** $f(x) = x!$, where x is a positive integer

Problem Set

6 16; 22; 30; 120

7

x	−3	−2	−1	0	1	2	3	4
f(x)	−8	−12	−14	−14	−12	−8	−2	6

−4 −2 0 2 4 6 8

2 2 2 2 2 2

8 a 15 **b** 20 **9 a** 1; −1; 1 **b** −1 **c** 1 **d** $F(x) = (-1)^x$, where x is a positive integer **10 a** $\{x: x \in \mathfrak{R}\}$ **b** $\{f(x): f(x) \geq -3\}$ **11** 35.71875 **12** 12 **13** 4x + 1

14 $\begin{bmatrix} -1 & 0 \\ 0 & -1 \end{bmatrix}$; $\begin{bmatrix} -1 & 0 \\ 0 & 1 \end{bmatrix}$; $\begin{bmatrix} -1 & 0 \\ 0 & -1 \end{bmatrix}$; $\begin{bmatrix} -1 & 0 \\ 0 & 1 \end{bmatrix}$; $\begin{bmatrix} -1 & 0 \\ 0 & -1 \end{bmatrix}$ **15 a** 41,664 **b** 249,984 **16 a** 8; 11; 15; 20 **b** Quadratic **17** $(x, y) = \left(-3, \frac{1}{2}\right)$

18 $(x, y) = \left(\frac{60}{17}, \frac{2}{17}\right)$ **19 a** $(\approx -0.5, \approx -1.5)$, $(\approx 2.5, \approx 1.5)$ **b** $x \approx -0.41$ or $x \approx 2.41$ **20 a** Slope: 4; y-intercept: −9 **b** g(0) = 10, g(n) = g(n − 1) + 3

21 a

n	0	1	2	3	4	5
F(n)	−4	2	8	14	20	26

b 6 **c** F(x) = 6x − 4, where x is a nonnegative integer **22 a** 531,441 **b** 60,480

23

n	1	2	3	4	5	6
F(n)	4	6	22	94	492	3046

24 406 **25** F(1) = 12, F(n + 1) = 3[F(n)] + 4

26

n	1	2	3	4	5	6
F(n)	4	4	4	4	4	4
g(n)	0	0	0	0	0	0

27 a

n	1	2	3	4	5	6	7	8
F(n)	1	−1	2	−2	3	−3	4	−4

b 249 **c** −2617

28 a

Number of disks	1	2	3	4
Minimum number of moves needed to finish game	1	3	7	15

b H(1) = 1, H(n) = 2[H(n − 1)] + 1 **c** H(1) = 1; H(2) = 2(1) + 1 = 3; H(3) = 2(3) + 1 = 7; H(4) = 2(7) + 1 = 15 **29 a** 1; 1; 2; 3; 5; 8 **b** F(1) = 1; F(2) = 1; F(n + 1) = F(n) + F(n − 1), where n ≥ 2

5.7 Measures of Central Tendency and Percentiles

Problem Set

1 75 **2** 260 **3**

8	0
7	0 0 1 1 2 2 2 2 3 3 3 6 9
6	1 3 3 4 5 5 5 6 6 7 8 9

7 | 2 indicates a height of 72 inches.

4 a Mean is ≈10.9;

mode is 0; median is 12. **b** Answers may vary. The median is probably best, since no student actually works the mean number of hours, and the mode doesn't indicate that a majority of students work more than 10 hours a week. **c** Mean increases to ≈ 11.3; mode is unchanged; median is unchanged. **5** A box-and-whisker plot does not show individual scores, as a stem-and-leaf plot does. **6** 6 **7** 123.2, 123.4, 123.6, 123.8, 124.0, 124.2, 124.4 **8** 9, 12, 15 **9** Answers will vary.

10 a

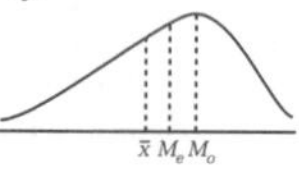

b

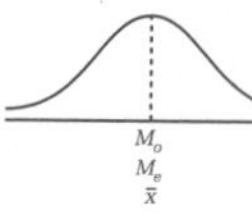

c

11 Mean is ≈ 8.1; mode is 8; median is 8. **14** (4, 0), (3, 0), (2, 0), (2, 1), (1, 1), (1, 0), (0, 2), (0, 1), (0, 0)

Review Problems

1 Yes **2 a** $18x - 5$ **b** $3x + 4$ **c** $12y + 4$ **3 a** Domain: $\{x: x \neq 0\}$; range: $\{f(x): f(x) > 0\}$ **b** Domain: $\{x: x \in \mathfrak{R}\}$; range: $\{g(x): g(x) \in \mathfrak{R}\}$ **c** Domain: $\{x: x \in \mathfrak{R}\}$; range: $\{h(x): h(x) \geq 0\}$ **4 a** -4 **b** 116 **5 a** and **c** **6** INVg(x) = $3(x + 2)$ **7** 5 **8 a** x **b** *z* **c** x, where $x \geq 0$ **d** y, where $y \neq 0$ **9 a** 3; 6; 18; 72 **b** $f(x) = 3x!$, where x is a nonnegative integer **10 a** 38; 168 **b** ≈ 4.42; 8; 4 **11** $\{x: x \neq \pm 2\}$ **13** {nonnegative integers} **15** {positive integers} **16** $\begin{bmatrix} -37 \\ -313 \end{bmatrix}$ **17 a** and **c**; **b** and **c** **18 a** 362,880 **b** 15,603,840 min, or ≈ 30 years, if played nonstop **19** 336 **21**

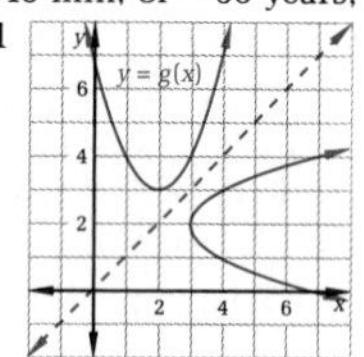

No

23

No

24 a $\left\{\frac{1}{3}, \frac{1}{4}, \frac{1}{5}, \frac{1}{6}, \frac{1}{7}, \frac{1}{8}\right\}$ **b** {0, 1, 2, 3, 4, 5} **25** $\frac{1}{32}$ **27** If $x \geq 0$, INVr(x) = $\sqrt{\frac{x + 3}{2}}$ **29** INVt(x) = $2x - \frac{5}{4}$ **30** $\frac{x - 5}{x}$; $\frac{1}{1 - 5x}$ **31 a** $P(x) = 6x + 8$ **b** $\{x: x > 0\}$; $\{P(x): P(x) > 8\}$ **c** INVP(x) = $\frac{1}{6}(x - 8)$ **d** The value of x corresponding to a given perimeter **33** $g(n) = 4)^{\frac{n+1}{2}}$, where $n \in$ {odd positive integers} **34 a** $x^2 - 2x - 2$ **b** $4 - x^2$ **c** $-x - 5$ **d** $x^2 + 12x + 33$ **e** $36x^2 - 3$ **f** $1 - 6x$ **35** h(1) = 5, h(n + 1) = h(n) + 2(n + 2) **36** $-5[p(x)]$ if $-\frac{\sqrt{3}}{3} < x < \frac{\sqrt{3}}{3}$; neither if $x = \pm\frac{\sqrt{3}}{3}$; $p(-5x)$ if $x < -\frac{\sqrt{3}}{3}$ or $x > \frac{\sqrt{3}}{3}$

6.1 Beyond Plotting Points

Warm-up Exercises

1

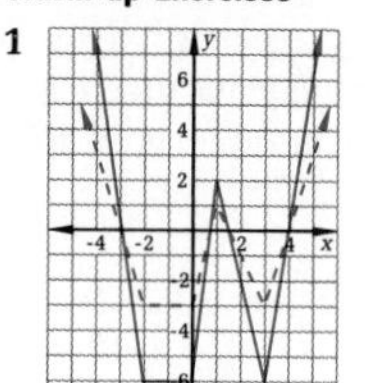

2

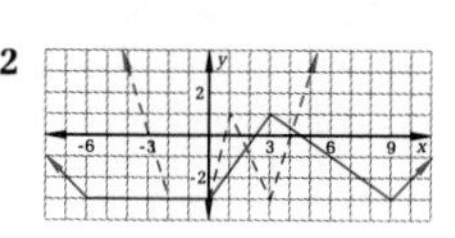

3

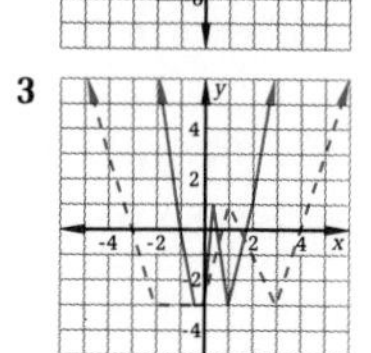

4

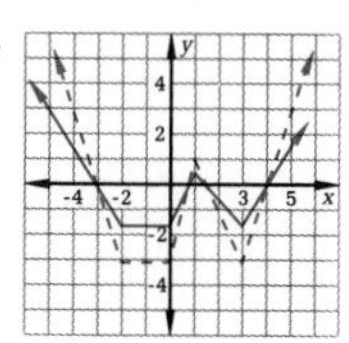

5

x	−1	0	1	1.5
$f(2x)$	7	2	−15	−40

6

x	−1	0	1	2
$\frac{1}{2}[f(x)]$	$-\frac{1}{2}$	1	$-\frac{3}{2}$	$-\frac{15}{2}$

7

x	−2	−1	0	1	2
$f\left(\frac{1}{2}x\right)$	−1	1	2	$-\frac{1}{2}$	−3

8 a 1 **b** 3 **c** $\frac{1}{3}$ **9** Stretched horizontally by a factor of 2 **10** Shrunk vertically by a factor of $\frac{1}{2}$

Problem Set

11 2 **13** 4 **15** Stretched vertically by a factor of 2.5 and reflected over the x-axis **16 a** (10, −2) **b** $\sqrt{89}$ **17** Halfway between his face and the image **18 a** {0, 2, 4, 6} **b**

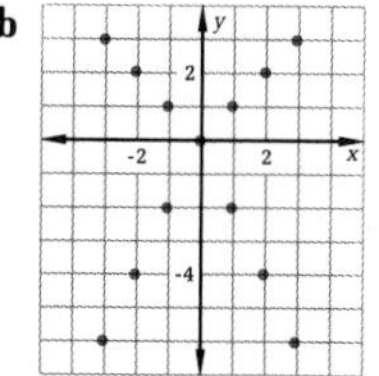

19 (2, 24) **21** $\left(2, \frac{12}{7}\right)$

22 a, b, c **23** −3 **24 a** 0 m **b** 8 sec **c** ≈ 1.6 sec and ≈ 6.4 sec **25** $\{(x, 4[f(x)]): (x, 4[f(x)]) = (-2, 0), (3, 16), (7, 48), (9, -52)\}$ **26 a** 9 **b** $\frac{4}{9}$ **27** $f[g(1)] = 33$; $g[f(1)] = 117$ **28** 12 **29** $-2[f(b)]$, $-f(b)$, $f(b)$, $2[f(b)]$, $f(2b)$, $f(-3b)$ **30 a** (−5, 25), (2, 4) **b** $\left(\frac{-5}{k}, 25\right)$, $\left(\frac{2}{k}, 4\right)$

31 a

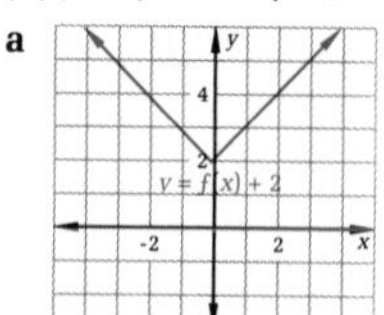

x	−3	−2	−1	0	1	2	3
f(x) + 2	5	4	3	2	3	4	5

b

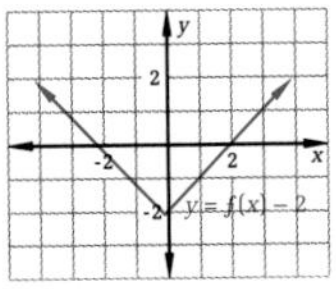

x	−3	−2	−1	0	1	2	3
f(x) − 2	1	0	−1	−2	−1	0	1

c The graph of $y = f(x) + 2$ is moved up two units, and the graph of $y = f(x) - 2$ is moved down two units. **32** $x \approx -5.3$ or $x \approx 2.3$ **33** $x = 10$ or $x = -6$ **35** Adult: 18,458; Children's: 12,325 **36 a** $4ad - 4bc$ **b** $2ad - 2bc$ **c** Part **a** is double part **b.** **37** Over the entire year, he gains the most money with option **a.** **38 a** $\{x : x \neq 2\}$ **b** $\{x : x \in \mathfrak{R}\}$ **39** $\approx 2.9 \times 10^{22}$ atoms

6.2 Translating Graphs

Warm-up Exercises

1 Right **2** Down **3** Up **4** Left **5** (20, −7) **6** (20, 6) **7** (4, 5) **8** (21, 5) **9** $y = (x - 3)^2$ **10** $y = (x + 5)^2$ **11** $y = x^2 + 4$ **12** $y = x^2 - 2$

Problem Set

13 (4, 0); lowest **15** (−2, 0); highest **17** x = 0 or x = 6 **19** x = 2 or x = 8 **21** x = 1 or x = 5 **23 a** 16; −16 **b** 4; −4 **24** $10\sqrt{2} + 5\sqrt{3}$

25 a

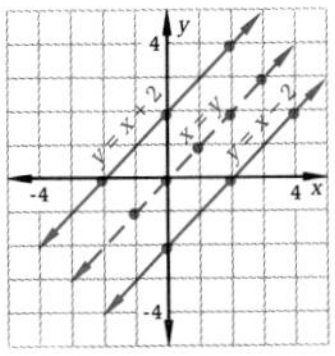

b $y = x - 2$ **c** $y = x - b$

26 −3 **27**

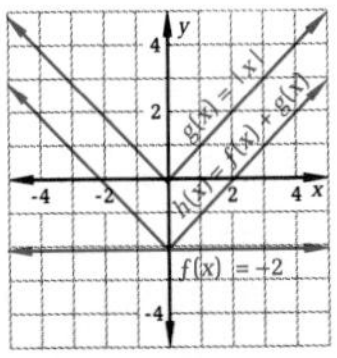

29 −11 **31** −5 **32 a** 6 **b** 24 **c** 120

33 a

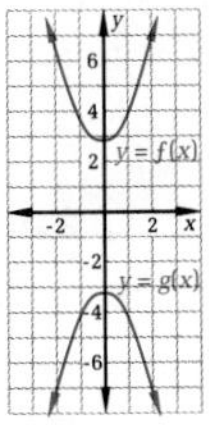

b

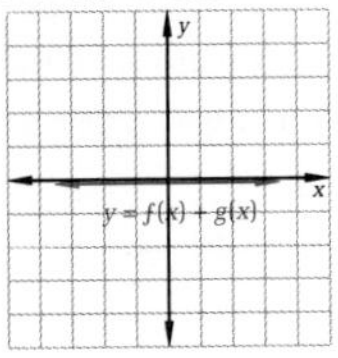

34 n = 5 **35** (−5, −3) **37** k = ±10 **38** (−3, 3)

39

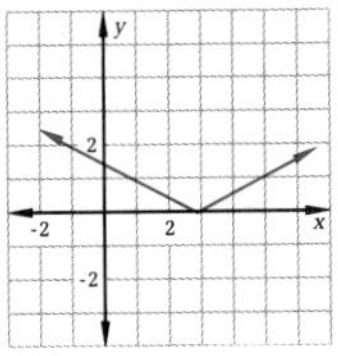

41

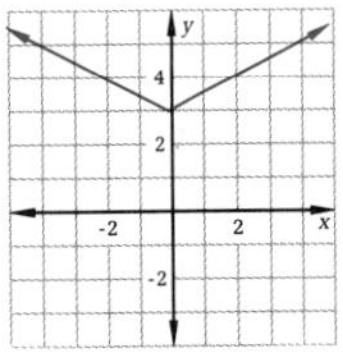

43 (x, y) = (4, −2) **44** $f(x) = \frac{1}{2}(x - 2)^2 - 5$

45

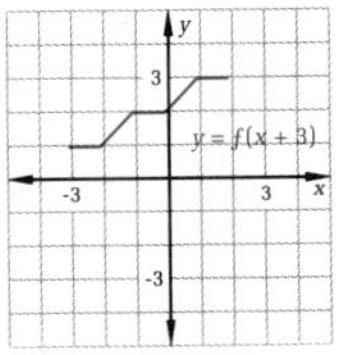

47

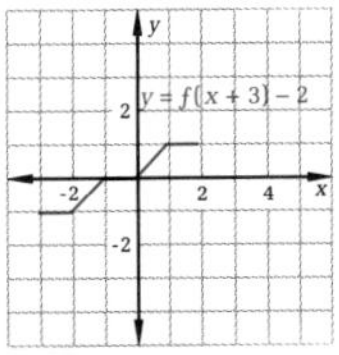

48 35 days **49** x^{6-3a} **51**

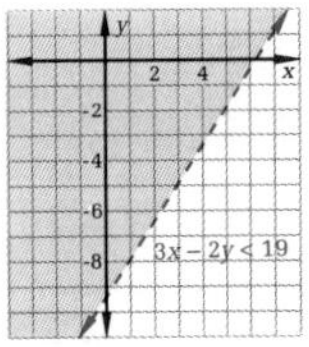

52 a (9, 15) **b** (5, 15) **53** a = −4; k = 10 **54**

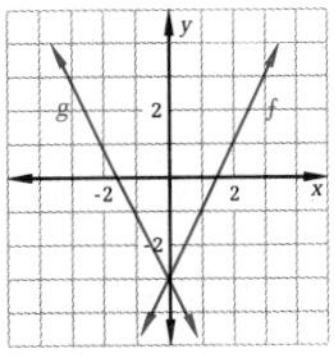

55 a

n	1	2	3	4	5	6	7	8
F(n)	1	2	4	8	16	32	64	128

b $F(n) = 2^{n-1}$

56 a $\frac{x+1}{x} = \frac{x}{1}$ **b** $x = \frac{1+\sqrt{5}}{2} \approx 1.62$ **57 a** 90 ft **b** ≈320 ft **c** When x ≈ 53.1 ft and x ≈ 266.9 ft

6.3 Reflections and Symmetry

Warm-up Exercises

1 (3, −7) **2** (2, −4) **3** (4, 2) **4** (−3, −7) **5** No; neither point has a corresponding (−x, −y) point. **6** y = 0 **7** x = 0 **8** x = 3 **9** y = x or y = −x **10** $y = -(x - 6)^2 - 4$ **11** $y = (x + 6)^2 + 4$

Problem Set

13 (−2, −5) **15** (2, 5) **17** (5, −2) **19** x = −2 **21** None **22** $f(x) = 3x + 5$ **23**

25

26 a (−2, −5), (3, −7), (6, −1) **b** (2, −5), (−3, −7), (−6, −1) **27 a** −7 **b** 7 **c** 5 **d** −7 **28** 20% **29 a**

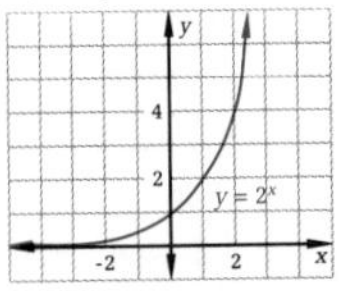

b

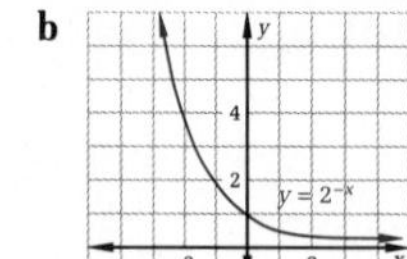

c

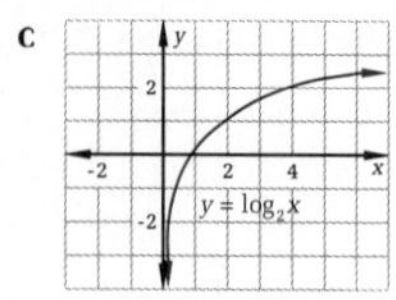

30 $(-2, 8)$
31 **a** $y = x^2$ **b** $y = (x - 2)^2$ **32** **a** $C(h) = 20h + 55$ **b** At least 4.75 hr **33** Answers will vary. Possible answer: A and A′ are reflections over $y = x$; $AB = A'B$; $\overline{AA'}$ is perpendicular to $y = x$; M is the midpoint of $\overline{AA'}$. **34**

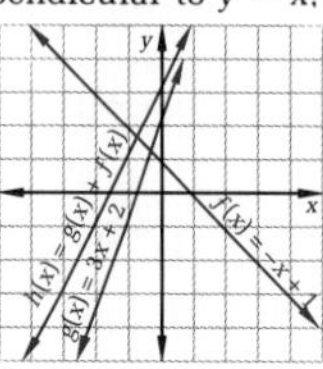

35 $(4, 3)$ **36** $f(A) = \begin{bmatrix} 4.2 & -12 \\ 18 & 2 \end{bmatrix}$ **37** $x = 4 - \sqrt{3}$

39 $y = x$ or $y = -x$ **41** $y = -\frac{1}{2}x + k, k \in \Re$ **42** **a** 8 **b** 8 **c** Yes; both are the distance between x and y on a number line. **43** **a** 20 **b** 20 **c** 25 **44** 4

6.4 Graphing Other Functions

Warm-up Exercises

1 **a**

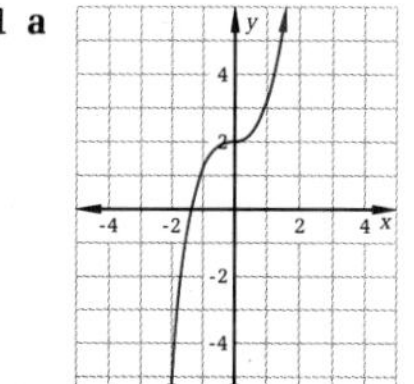

b

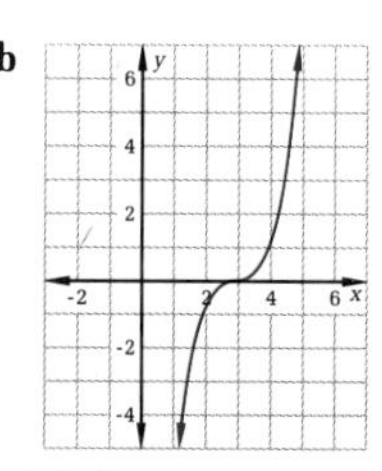

2 **a** i **b** ii **c** iii **3** **a** $-3, -1, 1$ **b** $x = -3, x = -1$, and $x = 1$ **c** $-5, -3, -1$

Problem Set

4 **a** ii **b** i **c** iii **d** iv **e** i **f** iv **g** i **h** iv **5** **a** 2 **b** 3 **c** None **6** **a** True; both equations are equivalent to $xy = 1$. **b** True; 0 is in the domain of $y = x$ but not $y = \frac{x^2}{x}$. **7** **a** $(3, 7)$ **b** $(3, 10)$ **c** $\left(3, \frac{2}{5}\right)$ **8** **a** $x = \frac{3 \pm \sqrt{41}}{4}$, or $x \approx 2.35$ or $x \approx -0.85$ **b** $\frac{5}{2}$ **9** **a** At $x = 3$ **b** At $x = \frac{3}{2}$ **10** 20 mph; 10 mph **11** **a** 4 **b** 3 **c** 4 **d** 2 **e** 0 **f** 2 **12** **a**

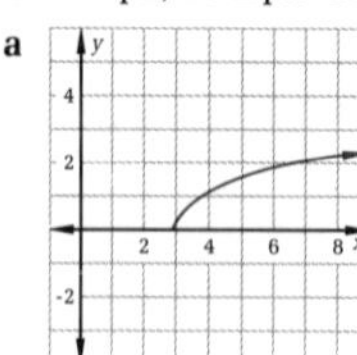

b

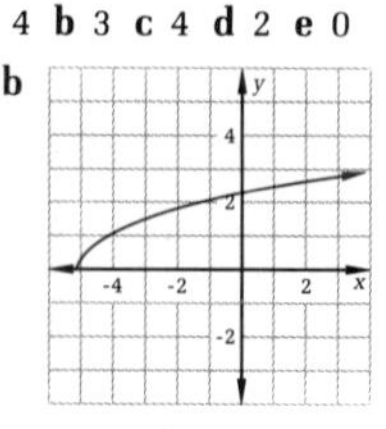

c

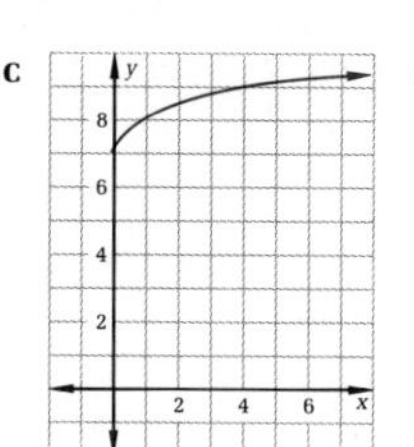

d

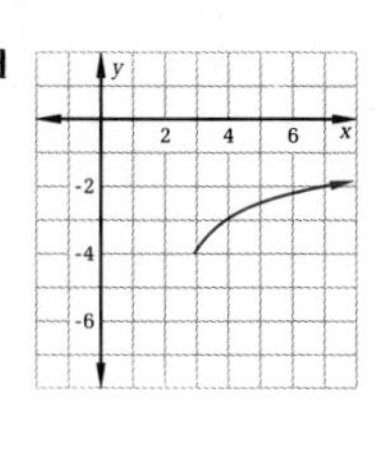

13 40 **14** a **15** $\{x : x \in \Re\}$ **17** $\{x : x \neq 4\}$
19 **a**

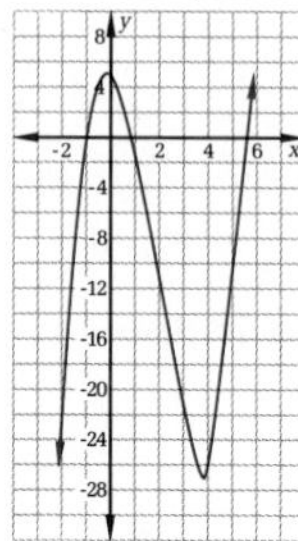

b 1, ≈5.85, and ≈−0.85

20 $x = 2$ or $x \approx 1.35$ **21** Successive differences increase by a factor of 4.

x	−4	−3	−2	−1	0	1	2	3	4
***f*(x)**	≈0.0039	≈0.0156	0.0625	0.25	1	4	16	64	256

22 **a** −2, 0, and 2 **b** 2, 4, and 6 **c** −9, −7, and −5 **d** −1, 0, and 1 **e** −8, 0, and 8 **23** $\sqrt{197}$, or ≈14.04
24 **a** $x = 3$ **b** $x = 4$ **c** $x = 1$
25 **a**

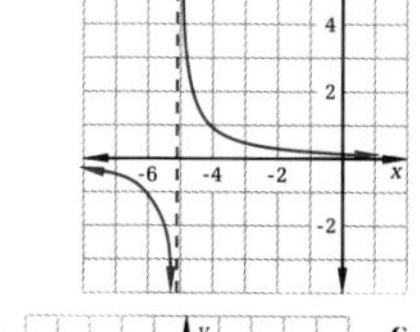

b **c**

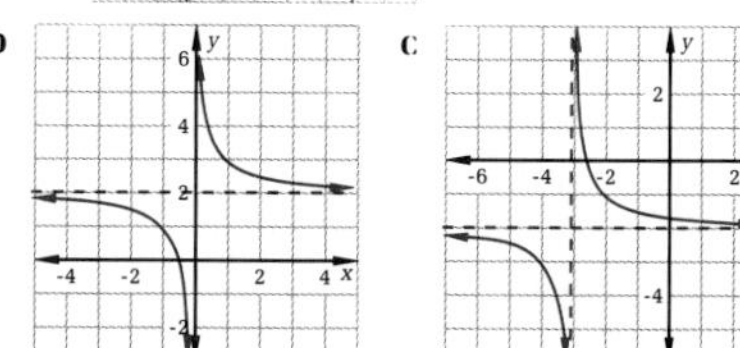

26 **a** None **b** None **c** $\frac{1}{2}$ and $-\frac{1}{2}$ **d** $\frac{1}{2}$ and $-\frac{1}{2}$ **27** **a** $\frac{1}{x}$ **b** $\frac{1}{x}$ **c** g maps x to itself, so compound f, g functions are preserved. **28** **a** 0.48 lumen **b**

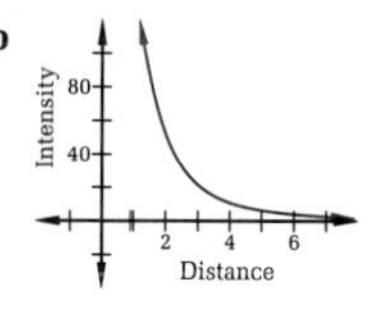

29 **a** $(0, 3)$, $(1, 4)$, and $(-1, 2)$ **b** $(4, 0)$, $(5, 1)$, and $(3, -1)$ **c** $(-3, -2)$, $(-2, -1)$, and $(-4, -3)$

6.5 Asymptotes and Holes in Graphs

Warm-up Exercises

1 $x = -2$ and $x = 3$ **2 a** $x = -5$ **b** 0 and 3 **3 a** $y = 2$ and $y = -1$ **b** $x = -4$, $x = 0$, and $y = 0$ **4** $y = 2$; $x = -1$ **5 a** ≈ 6.983 **b** ≈ 6.9983 **c** ≈ 6.99983 **d** ≈ 6.999983

Problem Set

6 $x = -7$ and $x = 3$ **7 a** $y = -1$; $x = 0$; and $x = 2$ **b** $y = 0$ and $x = 3$ **8**

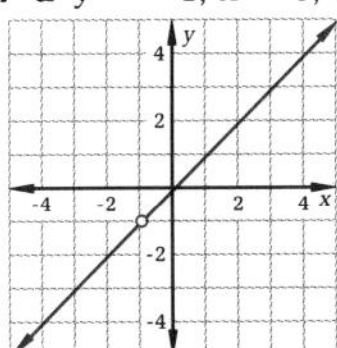

9

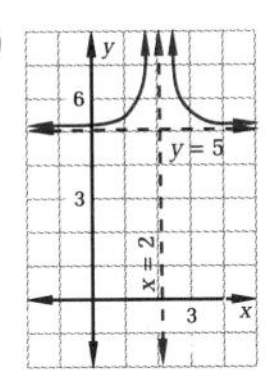

10 a

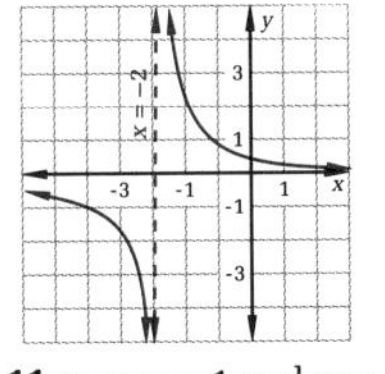

b

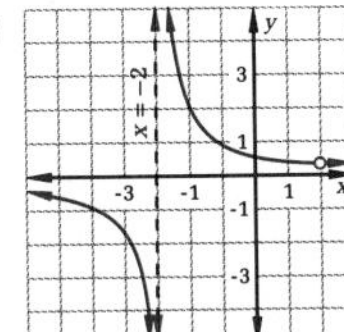

11 a $x = -1$ and $x = 2$ **b** $x = 4$ and $x = 10$ **12 a**

b $x = -4$, $x = 2$ **13** $x = 3$ **15** $x = \frac{1}{2}$ **16** Answers will vary. A possible response is shown.

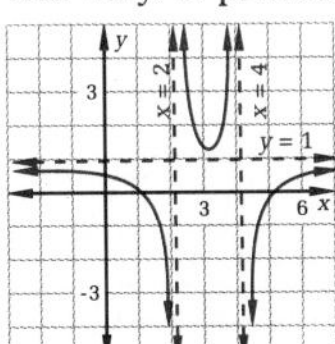

17 $(x, y) = (-2, -3)$ **18** 362,880 **19 a** $\frac{4x(x-2)}{(x+5)(x-2)}$ **b** $x = -5$ and $y = 4$ **c** 2 **20** $x = 6$ **21** $\frac{1}{\sqrt{5} + \sqrt{3}}$ **22 a** 7.5 mph **b** 2.5 mph **c** 15 mi downstream **23** $y = 2\frac{(x+2)(x-4)}{(x+2)(x-4)}$ **24** $x = 0$ and $y = 0$ **25 a** 18 **b** 9 **c** $f(x) = 3x$ **26** $(x, y) = (-2, 8)$ or $(x, y) = (2, 8)$ **27** $x = 12$; $y = 12$ **28** A(3, 4, 0), B(0, 4, 5), C(3, 0, 5), D(3, 4, 5) **29** $y = \frac{(x+1)(x-3)}{(x-3)}$ **30 a** (0, 0); (0, 0) **b** (3, 0); (0, −27) **c** (−2, 0); (0, 8) **d** (−2, 0); (0, 8) **31** $x = 6$; $x = -1$ **32** Slope of l:1; slope of m is undefined; slope of n:0

6.6 Measures of Dispersion

Problem Set

1 a 12 **b** 3 **c** Mean ≈ 1.71; mode = 1; median = 1 **2 a** 26 **b** 6 **3** Twosine

4 a

Scores	Deviation Score	Absolute Value of Deviation Score
23	−23	23
32	−14	14
50	4	4
54	8	8
71	25	25

b 14.8 **5** No **6** Mean deviation: ≈19.6; median deviation: $15.\overline{5}$ **7 a** 65 **b** Mean ≈ 70.5; median = 70; mode = 66 **c** ≈12.5

Review Problems

1

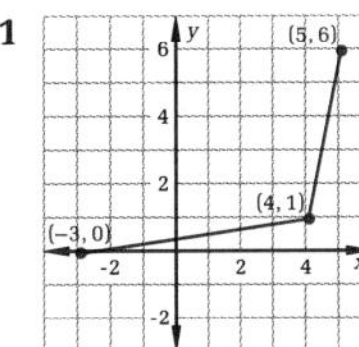

2 a 7 **b** 3.9 **3 a** 0; 80 **b** 0; 0 **c** Has hole at $x = 4$ and a vertical asymptote at $x = -12$ **4** 58 **5** 1925 mi **6 a** $\left(3, \frac{1}{2}\right)$ **b** (3, 4) **7 a** v **b** ii **c** iii **d** i **e** iv **8 a** −8; −4; −3; 4; 11 **b** 6 **9** (2, 54) **10 a** {−14, −11, −8, −5, −2, 1, 4} **b** {−56, −44, −32, −20, −8, 4, 16} **c** {−4, −1, 2, 5, 8, 11, 14} **11 a** 28.8 oz saline; 19.2 oz active ingredients **b** ≈66.9% **12** Answers will vary. Possible answer:

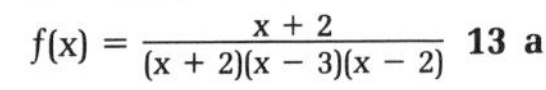

$f(x) = \frac{x+2}{(x+2)(x-3)(x-2)}$ **13 a**

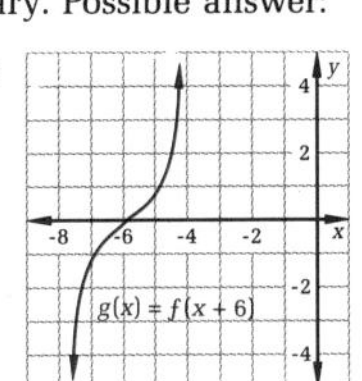

b **c**

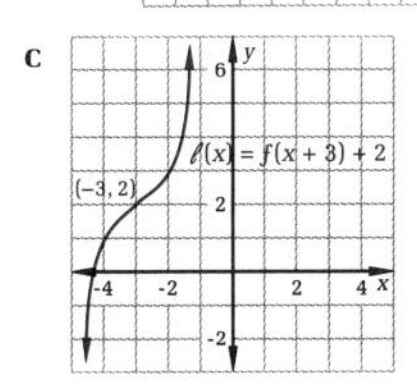

14 a (0, 9) **b** (10, 9) **c** (3, 13) **d** (3, 0) **e** (1.5, 9) **15 a** 1; 3 **b** 4; 12 **c** $\frac{1}{3}$; 1 **16** $x = -\frac{1}{2}$ or $x = 2$

17 a (3, 6) **b** $x = 3$ **c** $y = -2(x - 3)^2 - 6$
18 a (−2, 0) **b** (6, 0) **c** (0, 5) **d** (0, −7) **e** (−2, −3)
f (7, 5) **19** $y = \pm\sqrt{x + 2}$ **20** (−1, 1)

Conics

Problem Set

1 Hyperbola; center (0, 0): vertices $\left(0, \pm\frac{2\sqrt{2}}{3}\right)$, asymptotes $y = \pm\frac{2\sqrt{2}}{3}$ **3** Ellipse; center (0, 0)
5 Circle; center (0, 0); radius $7\sqrt{7}$ **7 a** $\frac{x^2}{25} + \frac{y^2}{9} = 1$
b $x = y^2$ **c** $\frac{y^2}{4} - \frac{x^2}{4} = 1$ **9** Hyperbola; center (−2, −5); opens up and down; asymptotes $y = \frac{7}{5}x - \frac{11}{5}$, $y = -\frac{7}{5}x - \frac{39}{5}$

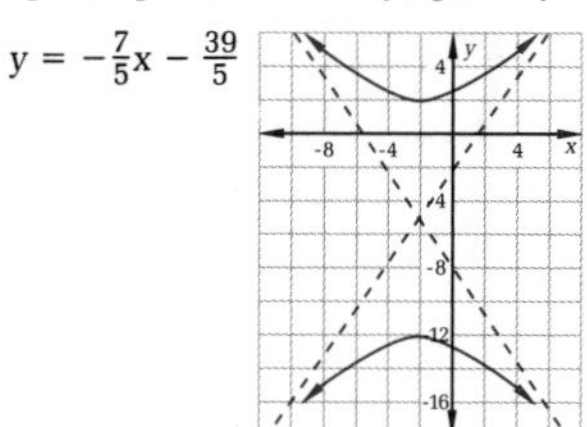

11 Circle; center (−5, 4); radius $\sqrt{41}$

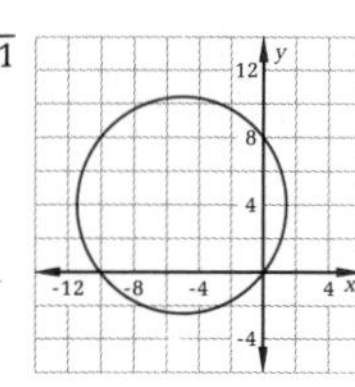

13 Parabola; vertex (4, −3); opens to right

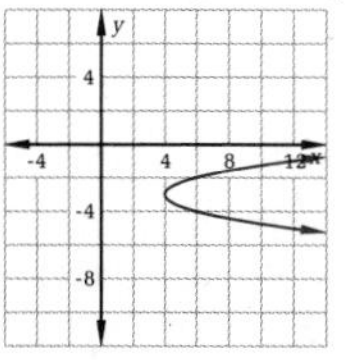

14 Circle; center (0, 0); radius 9

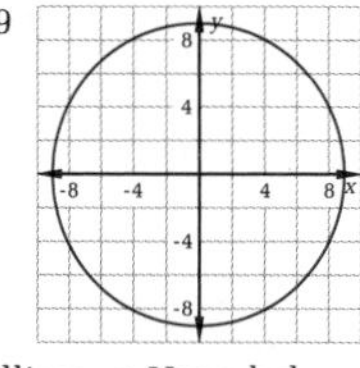

15 All except **e**. **a** Circle **b** Ellipse **c** Hyperbola **d** Parabola **f** Parabola

17

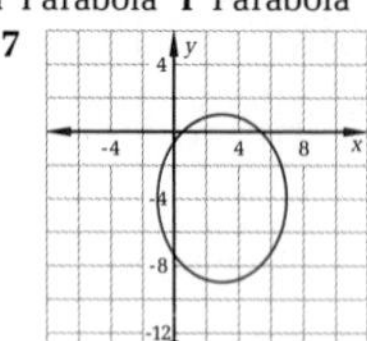

19

20 $\frac{1}{3}$ **21** $\frac{x^2}{9} - \frac{y^2}{4} = 1$ **22** $(x - 6)^2 + (y - 1)^2 = 16$

23

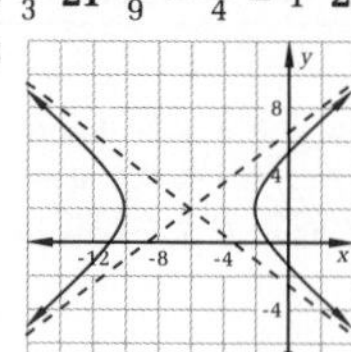

24 $(x + 3)^2 + (y - 7)^2 = 25$

25 a ±2 **b** −1, −5 **26 a–d**

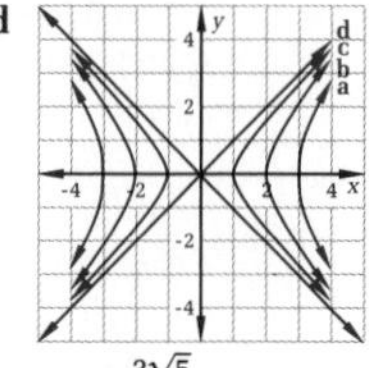

27 x-intercepts: 1, 5; y-intercepts: $\pm\frac{3\sqrt{5}}{2}$
28 a $f(x) = \left(x + \frac{3}{2}\right)^2 - \frac{57}{4}$ **b** Parabola **c** $\left(-\frac{3}{2}, -\frac{57}{4}\right)$

7.1 Solving Problems with Systems of Equations

Warm-up Exercises

1 a Yes **2** Yes **3 a** 389 mph **b** 39 mph **4** (r, c) = (10, 6) **5** One solution **6** No solution **7** Many solutions **8** Zelda is 31; Zuider is 14. **9** 13

Problem Set

11 Many solutions **13** (x, y) = (6, 4) **15** No solution
17 Mary is 6; Sam is 18. **18** $(x, y, z) = \left(\frac{13}{2}, \frac{13}{4}, \frac{91}{4}\right)$
19 a $C_d = 22.5 + 0.18m$ **b** $C_s = 28.95 + 0.15m$
c 215 mi **20** x = 4; y = 2 **21** −20 **22 a** 4.7 **b** 37
23 a ≈0.83 mph **b** ≈4.17 mph **24** (x, y) = (−2, 5)
25 a 3, −4, and 9 **b** $\frac{3}{2}$, −2, and $\frac{9}{2}$ **26** ≈±1.87
27 a 42.5% **b** ≈243.4 L **28** $(m, b) = \left(\frac{6}{5}, \frac{13}{5}\right)$ **29** No solution **31** (x, y) = (14, 6) **32** $\frac{4}{5}$ **33** x = 24; y = 6

34

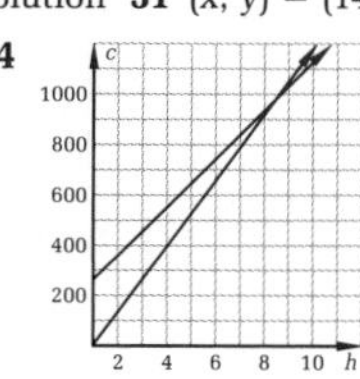

a $C_c = 255 + 85h$; $C_h = 115h$ **b** Jobs of more than 8.5 hr **35** q = 4 **36 a** $x = \frac{1}{3}$ or $x = -2$ **b** $-\frac{5}{3}$ **c** $-\frac{2}{3}$
37 (x, y) = (3, 4) **38 a** 120° **b** If the triangles are isosceles, x = 30°, y = 30°. If the triangles are not isosceles we know only that x + y = 60°. **39** ≈1.62 qt
41 No solution

7.2 Linear Systems with Three Variables

Warm-up Exercises

1 No **2** 23 **3** (4, 0, 0); (4, 5, 0); (4, 5, −6) **4** (x, y, z) = (13, −5, 11) **5** (x, y, z) = (5, 3, 7)

6 $y = -\frac{5}{8}x^2 + \frac{5}{4}x + \frac{75}{8}$ **7** $(x, y, z) = (1, 2, 4)$

Problem Set

9 $(x, y, z) = \left(\frac{16}{3}, \frac{35}{3}, 16\right)$ **10** $(x, y, z) = \left(9\frac{1}{2}, 5\frac{1}{2}, 14\frac{1}{2}\right)$

11 a 6 **b** 4 **c** 12 **d**

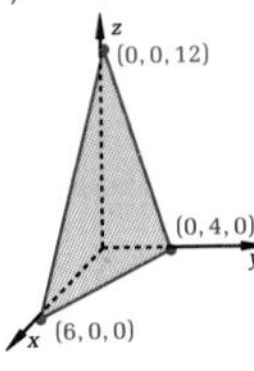

12 Bill is 16; his father is 48. **13** $y = -6x^2 + 20x - 7$
15 $(x, y, z) = (-1, 3, 5)$
16 a

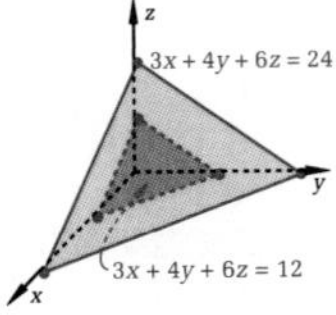

b They are parallel.

17 4 **18** $x = 3$, $y = -5$, $z = 6$ is the solution to the system represented by the matrix.

19 $\begin{bmatrix} 3 & 4 & 1 \\ 2 & -1 & 5 \\ -4 & 7 & 2 \end{bmatrix}\begin{bmatrix} x \\ y \\ z \end{bmatrix} = \begin{bmatrix} 11 \\ 9 \\ -4 \end{bmatrix}$ **20** 2,598,960

21 $\{(x, y): x + y = 5\}$ **22 a** $S(x) = (x + 2)^2$ **b** $T(x) = 3(x + 2)$ **c** $x = 1$ **23 a** $(3, 6)$ **b** $\sqrt{29}$, or ≈ 5.39 **c** $\sqrt{29}$, or ≈ 5.39 **24** \$4.50 tickets: 12,452; \$7.00 tickets: 9270; \$12.50 tickets: 2480 **25** 8 of A, 6 of B, 5 of C **26 a** Yes, $f(n + 1) - f(n)$ increases uniformly as n increases. **b** $f(n) = \frac{1}{2}n^2 - \frac{3}{2}n$ **27** 2.8 qt **28** 49

29 $(a, b, c) = \left(\frac{1}{2}, -\frac{1}{2}, 0\right)$ **30** 514 **31 a** $(0, 3, 5)$; $(4, 3, 0)$; $(4, 0, 5)$ **b** $15x + 20y + 12z = 120$
32 $(a, b, c, d, e) = (9, 2, 1, 5, 7)$

7.3 Systems of Inequalities

Warm-up Exercises

1

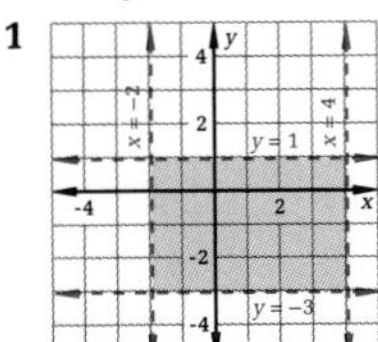

2 No solutions

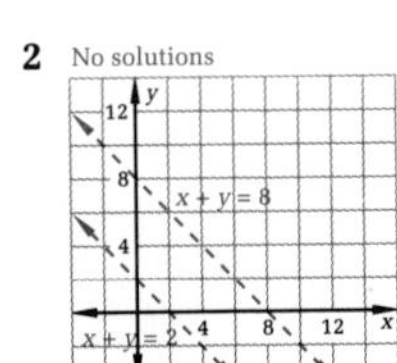

3 $y \le -\frac{3}{4}x + 6$ **4**

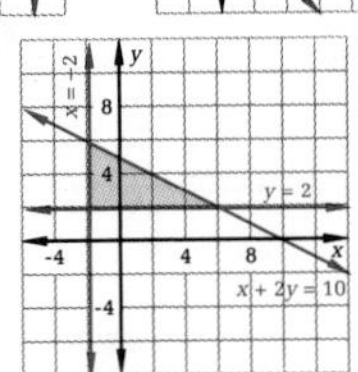

5 $(-9, -2)$, $(-9, 0)$, $(4, -2)$, $(4, 0)$

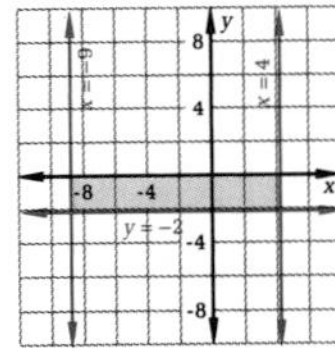

6 No **7** $55x + 87y \le 80{,}000$

Problem Set

9

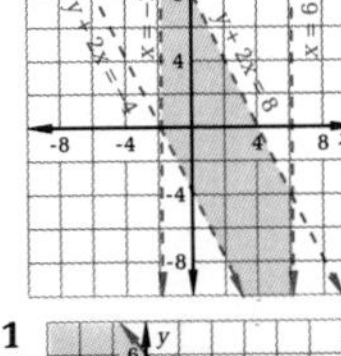

10 $(2, 1)$, $(6, 0)$, $(9, 6)$, $(2, 6)$

11

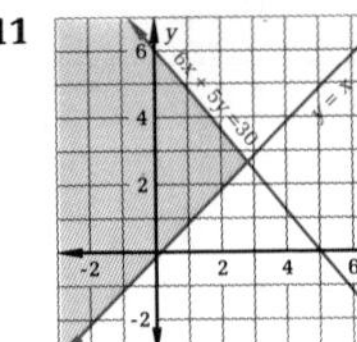

13 $(x, y, z) = (0, 7, 5)$

14 $3x + 7y \le 200$ **15**

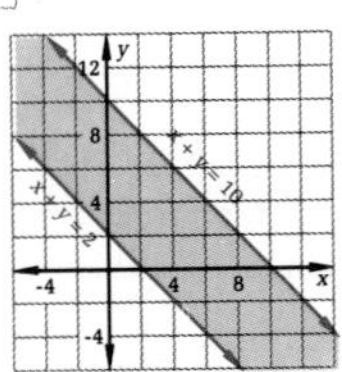

16 $y = \frac{1}{3}$ or $y = -1$ **17**

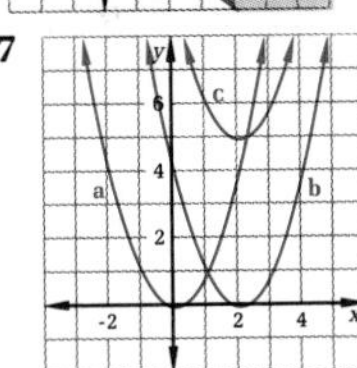

18 a $(0, 0, 3)$; $(0, 4, 3)$; $(5, 4, 3)$; $(5, 0, 3)$; $(5, 0, 0)$; $(5, 4, 0)$; $(0, 4, 0)$; $(0, 0, 0)$ **b** ABCD: $z = 3$; EFGH: $z = 0$; DCFE: $x = 5$; BCFG: $y = 4$ **19 a** $(0, 0)$, $(3, 0)$, $\left(\frac{16}{3}, \frac{7}{3}\right)$, $(0, 5)$ **b** $f(0, 0) = 0$; $f(3, 0) = 6$; $f\left(\frac{16}{3}, \frac{7}{3}\right) = \frac{53}{3}$; $f(0, 5) = 15$ **20**

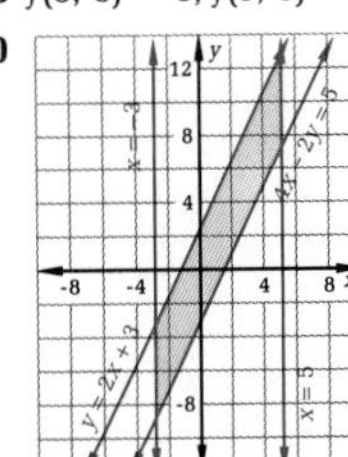

21 $\begin{cases} 2y - x \ge 1 \\ y \le -\frac{24}{25}x^2 + \frac{48}{5}x - 18 \end{cases}$ **22** 10% **23** $\frac{1}{2}$ **24** 60 mph

25 a $\frac{3}{5}$ **b** $y = \frac{3}{5}x$; $y = -\frac{3}{5}x$ **c** Since the slope of $\overleftrightarrow{BD}$ is $\frac{3}{5}$, the ratio of the change in y to the change in x between any two points on the line is $\frac{3}{5}$.

26 $\begin{cases} y \le x + 5 \\ y \le -x + 5 \\ y \ge -x - 5 \\ y \ge x - 5 \end{cases}$ **27** $x \ge 2$; $y \ge 3$; $x + y \le 10$; $200x + 375y \ge 1525$ **28** $f(-3, 1) = -9$ **29** 30 **30** $(x, y, z) = (-2, 6, -2)$ **31** A: 720 mph; B: 770 mph **32** 78 **33** $\begin{cases} y < 8 \\ x < 0 \\ y > x - 5 \\ y > -3x - 7 \end{cases}$ **34** 90

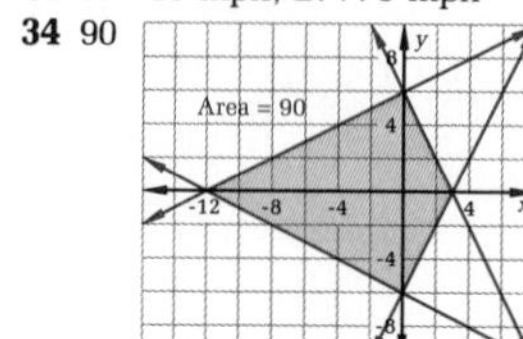

35 a $\begin{bmatrix} 1 & 0 & 1 & 1 \\ 1 & 0 & 1 & 1 \\ 1 & 1 & 0 & 0 \\ 2 & 1 & 3 & 0 \end{bmatrix}$ **b**

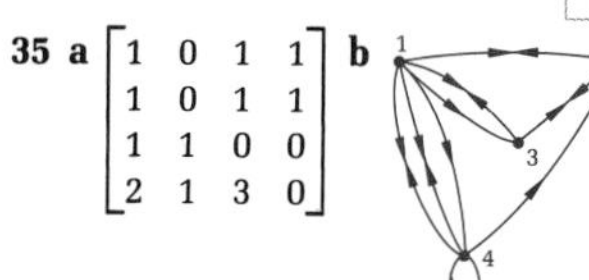

36

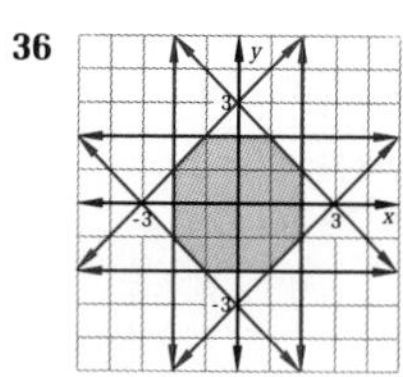

37

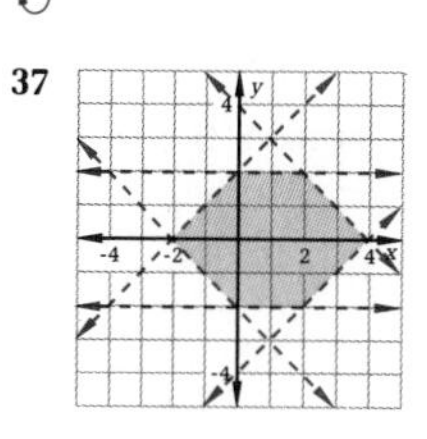

7.4 Inverse Variation

Warm-up Exercises

1

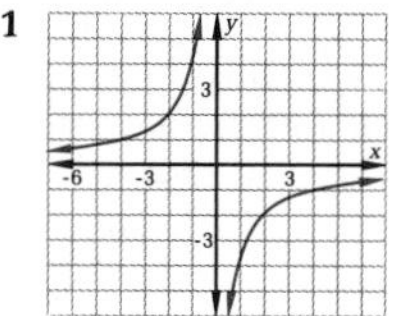

2 216 **3** 216 **4** 216 **5** 24 **6 a** Yes **b** No **c** Yes **7** 0.05 **8** (6, 6) and (−6, −6)

Problem Set

9

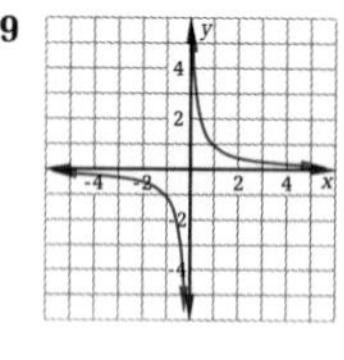

11

13

x	−2	−1	10	−5	$-\frac{1}{2}$
y	15	30	−3	6	60

14 12 **15** 733.$\overline{3}$ rpm **16 a** x = 0 and y = 0 **b**

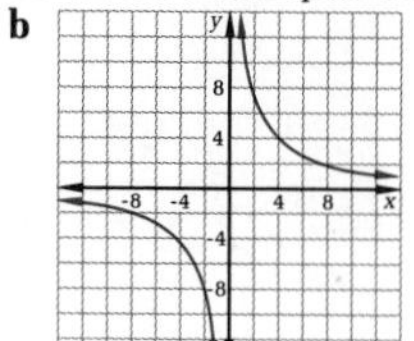

17 Because $b \cdot h$ always equals 48; yes **18** 1; 1; 1; 1; 1 **19**

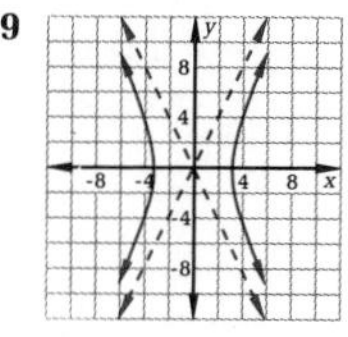

21 $(x, y) = (7, -5)$ **22 a** \$2.00; \$0.50 **b** At least 134 **23** 6.25 **24**

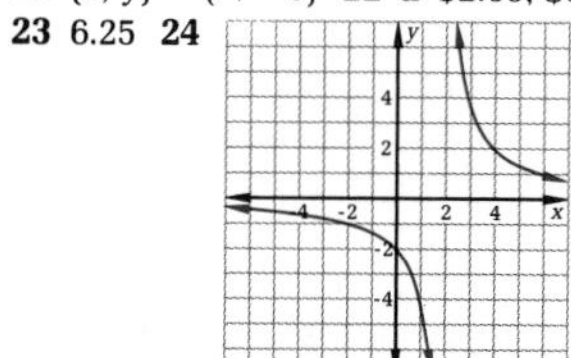

25

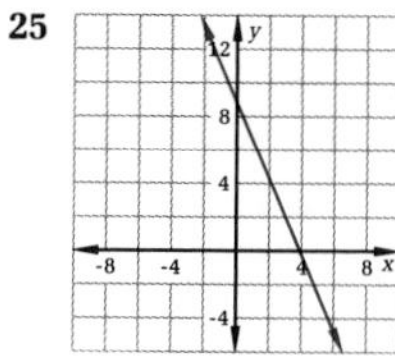

27 $(x - 10)(x + 1)$ **29** Cannot be factored **30 and 31**

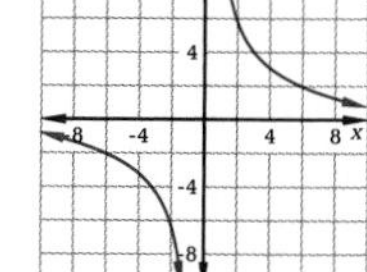

32 No; $\frac{y}{z}$,

not $y \cdot z$, equals a constant. **33 a**

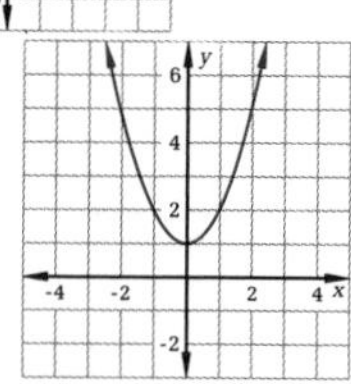

b

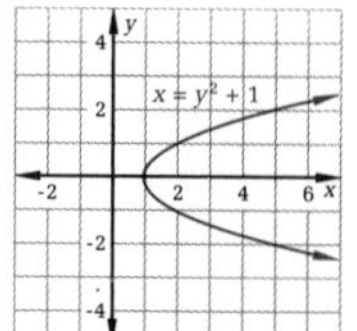

c No **34** $xy = 12$

35 a

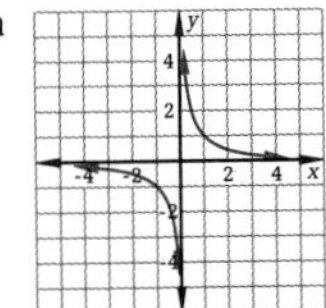

b

c (1, 1), (−1, −1); $(\sqrt{2}, \sqrt{2})$, $(-\sqrt{2}, -\sqrt{2})$

36

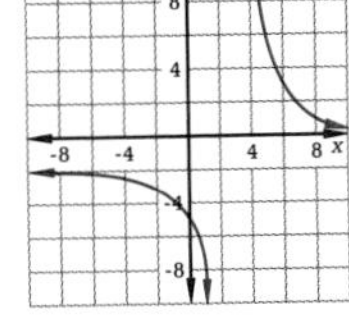

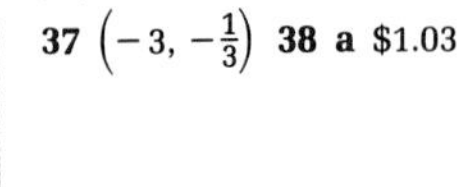

37 $\left(-3, -\frac{1}{3}\right)$ **38 a** \$1.03 **b** \$0.82 **39 a**

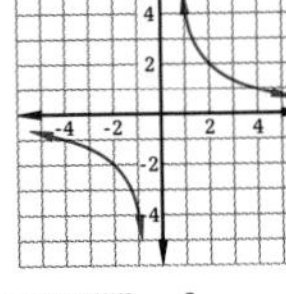

b x = 0 and y = 0

c

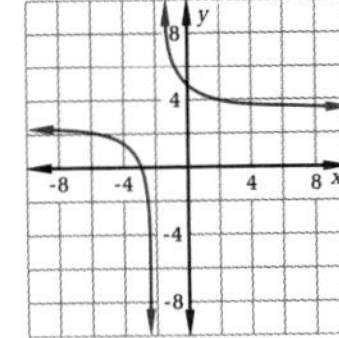

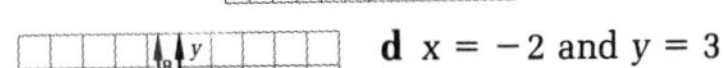

d x = −2 and y = 3

40 9 mph **41** 40 **42 a** 125 **b** 216 **c** 1000 **d** 8000 **e** 1,000,000 **43** ≈28 candles **45**

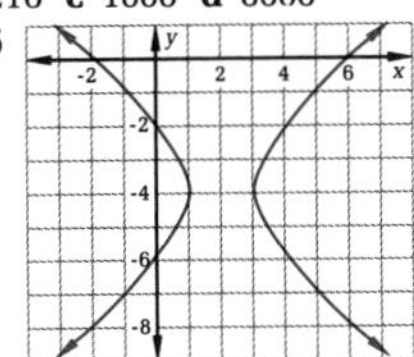

46

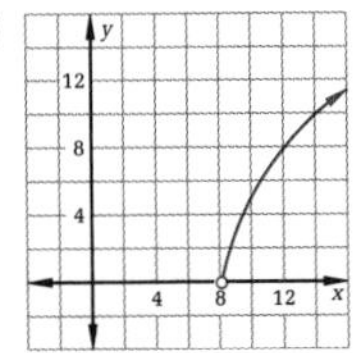

47 $\frac{(x-8)^2}{25} + \frac{(y-10)^2}{4} = 1$

7.5 Nonlinear Systems

Warm-up Exercises

1 a (0, 7.3), (0, 0.7) **b** No solution **c** (2.4, 1.2), (2.4, −1.2) **2 a** (20, −16) **b** No real solution **3** Yes **4** $(x, y) = (\sqrt{21}, 2)$ or $(x, y) = (-\sqrt{21}, 2)$ **5** (x, y) = (2, 3), (x, y) = (2, −3), (x, y) = (−2, 3), or (x, y) = (−2, −3) **6** (3, 0), (−3, 0) **7** $(x, y) = \left(\frac{1}{3}, 0\right)$ or $(x, y) = \left(-\frac{1}{3}, 0\right)$

Problem Set

9 $(x, y) = (4, 2\sqrt{5})$, $(x, y) = (4, -2\sqrt{5})$, $(x, y) = (-4, 2\sqrt{5})$ or $(x, y) = (-4, -2\sqrt{5})$ **11** $(x, y) = (\sqrt{2}, 2)$ or $(x, y) = (-\sqrt{2}, 2)$ **12** $(\sqrt{2}, 2)$, $(-\sqrt{2}, 2)$ **13 a** 375 m **b** ≈267.9 m **c** 750 **14**

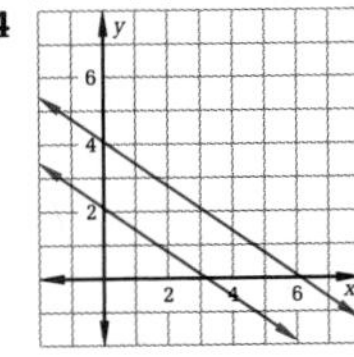

15 (2 + x)(2 − x) **17** 5(5x − 1)(1 − x) **18** Possible answer: xy = −1 **19** No real solution **21** No real solution **23** 21 **24 a** 1; 8; 27; 64; 125 **b** n^3 **25 a** xy = 24 **b**

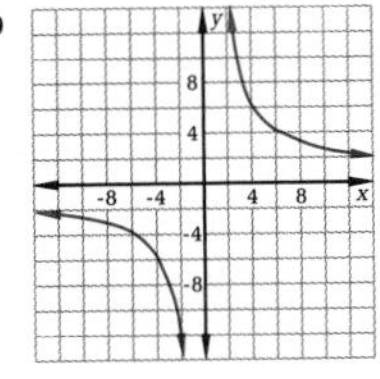

26 a $a^3 - b^3$ **b** $a^3 + a^2b - ab^2 - b^3$ **27** 0 **28 a** \$42.40 **b** \$4.24 **29** $-\frac{10}{3}$ **31** x(x − 5)(x + 5) **33 a** $m = \frac{5}{9}$; $b = \frac{19}{3}$

b

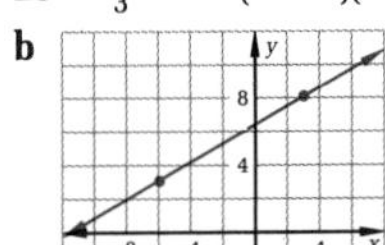

c

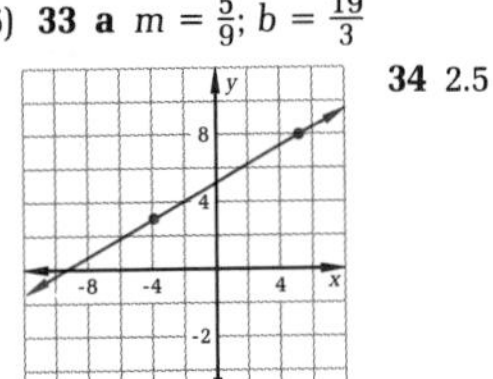

34 2.5

35 a \$12 sub.: 113; \$19 sub.: 37 **b** 30% **36 a** (6, −52) **b** (−6, 20) **c** (−3, 2) **d** (3, −34) **37** (15, 8) or (8, 15) **38 a** 10 **b** $12.\overline{09}$ **c** 12

39 a

x	16	8	4	2
y	8	4	2	1

b

x	16	32	64	128
y	8	4	2	1

40 $\left(\frac{12}{5}, \frac{12}{5}\right)$, $\left(\frac{12}{5}, -\frac{12}{5}\right)$, $\left(-\frac{12}{5}, \frac{12}{5}\right)$, $\left(-\frac{12}{5}, -\frac{12}{5}\right)$

41 a $-\frac{1}{16}$ **b** $\frac{1}{5}$ **42** No real solution **43** (x, y) = (2, −1) or (x, y) = (−1, 2)

44

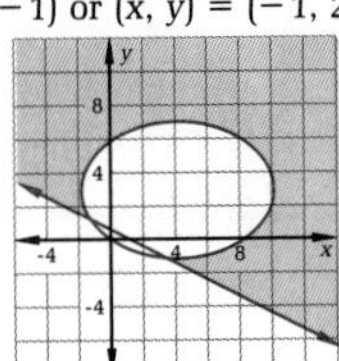

7.6 Other Systems of Equations

Warm-up Exercises

1 $(x, y) = \left(\frac{1}{9}, -\frac{1}{3}\right)$ **2** $(x, y) = \left(4, -\frac{8}{3}\right)$ **3** 12 and $\frac{12}{5}$
4 $(x, y) = (3, 3)$

Problem Set

5 $(x, y) = \left(\frac{1}{3}, \frac{1}{2}\right)$ **6** $(x, y) = \left(\frac{9}{2}, -\frac{1}{2}\right)$ **7** $(x, y) = (0, 0)$ or $(x, y) = (6, 3)$ **8** $\frac{15}{8}$ hr **9** $f(x) = \frac{5}{2}x^2 - \frac{1}{2}x + 4$; $g(x) = \frac{1}{2}x^2 + \frac{5}{2}x + 3$ **10** ≈ 4.6 **11** 3.75 hr **12** $q = 0$ or $q = -7731$ **13** $\frac{2}{3}$ **14** $w = 0, w = 1, w = -2$, or $w = 3$
15 120 **16** 2 hr 24 min **17 a** 1; $\frac{1}{10}$; $\frac{1}{100}$; $\frac{1}{1000}$; $\frac{1}{10,000}$
b 1.1111 **18** $(x, y) = \left(5, -\frac{1}{2}\right)$ **19** $(x, y) = (-12, -1)$
21 The opening had to be delayed. (It took 5 hr to fill the pool.) **22 a** $\frac{V_1}{T_1} = \frac{V_2}{T_2}$ **b** 24 v **23** $(x, y) = (4, 2)$
24 $f(x) = 2x^2 + 17x - 43$; $g(x) = -x^2 - 12x + 31$
25 $(A, B, C, D) = (-4, 1, 12, 6)$ **26** Tom: $23\frac{1}{3}$ days; Dick: 10 days; Harry: $17\frac{1}{2}$ days **27** $x \approx 4.01$ or $x \approx -3.68$ **28** 4 km/hr; ≈ 9.8 km/hr; 16.5 km/hr
29 a $\text{INV}f(x) = \frac{1}{x} + 3$ **b** $f \circ \text{INV}f(x) = \frac{1}{\frac{1}{x} + 3 - 3} = x$; $\text{INV}f \circ (fx) = \frac{1}{\frac{1}{x-3}} + 3 = x$ **c**

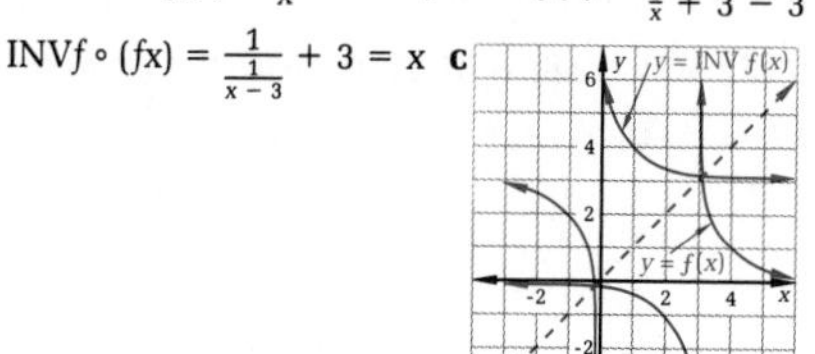

30 a Yes; $xy = -4$ **b** No **c** No **d** Yes; $xy = 6$ **31** No solution **33** $(x, y, z) = (4, -2, 5)$ **34 a** $(\sqrt{x}, \sqrt{y}) = (3, 2)$ **b** $(x, y) = (9, 4)$ **35 a** 35 **b** 12 **36** $(x, y) = \left(-\frac{3}{4}, \frac{3}{20}\right)$ **37 a** 52.75; 43 and 47; 51.5 **b** 22 **c** ≈ 7.08

Review Problems

1 a $h = 7$; $k = 4$ **b**

2 $(x, y) = (4, -1)$ **3** 480 rpm **4 a** 120 mph **b** 30 mph
5 7 and $\sqrt{39}$, 7 and $-\sqrt{39}$, -7 and $\sqrt{39}$, or -7 and $-\sqrt{39}$ **6** $(x, y) = (5, 35)$ or $(x, y) = (-2, 0)$
7 a $(x + y)(x - y)$ **b** $(x^2 + y^2)(x + y)(x - y)$ **8** 15 min
9 $(x, y, z) = \left(-5, \frac{7}{2}, 4\right)$ **10** Run: ≈ 8.33 km/hr; swim: ≈ 1.18 km/hr **11 a** $x = 4$ **b** $x = -7$ **c** $x = 3$
12 79.8 or 7.2 **13 a** $(x, y) = (6, 6)$ or $(x, y) = (-6, -6)$

b The solutions represent the vertices of the hyperbolic graph of the first equation in the system.
14 $\begin{cases} y \ge \frac{2}{5}(x - 8) \\ y \le -\frac{2}{5}(x - 8) \\ x \ge -2 \end{cases}$ **15 a** ≈ 3.7 hr **b** 1.8 mph; 4.7 mph

16 $X = \begin{bmatrix} -1 \\ 3 \\ 5 \end{bmatrix}$ **17** $16.\overline{6}$ **18 a** $(x - 7)(x + 2)$
b $(3x - 7)(3x + 2)$ **c** $(x - 3)(x + 6)$
19

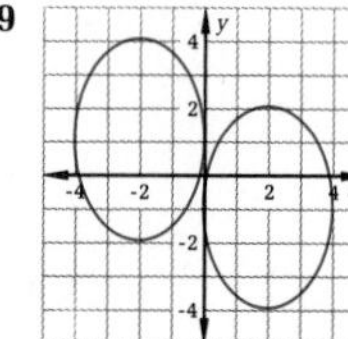

20 a 360 **b** $E = \frac{360}{n}$ **c** 30

21 $(x, y) = (-8, -4)$ **22**

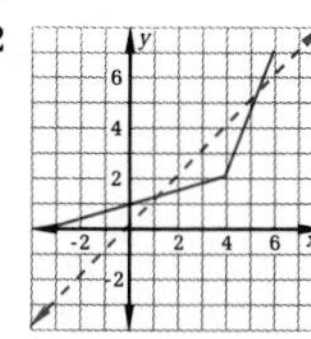

23 \$7.68 **24** $\frac{39}{34}$ **25** $y = 3x^2 + 12x + 6$

Factoring to Solve Equations and Inequalities

Problem Set

1 $-6, -2, 2$ **3** $-4, -2, \pm\sqrt{7}$ **5** $7, 6 \pm \sqrt{3}$
7 a

b

c

8 $y^2(y - 2)(y - 5)$ **9** $x = \pm 2$ or $x = -4$ **11** $(x + 1)(x^2 - x + 1)$
13 $(2x - 5)(4x^2 + 10x + 25)$ **15** $x(x - 6)^2$
17 $(x^2 - 3x + 5)(x^2 + 3x + 5)$ **18 a** $(x^2 - 5)(x - 1)$
b $\{x{:}x < -\sqrt{5}$ or $1 < x < \sqrt{5}\}$ **c** The solutions of the inequality are the values of x for which the graph of P lies below the x-axis. **19**

21 a $a^3 + (a + 2)^3$ **b** $2(a + 1)(a^2 + 2a + 4)$
23 $(x^2 + 1)(x - 1)$ **24** $\{a{:}a \le -2\sqrt{2}$ or $-2 \le a \le 2$ or $a \ge 2\sqrt{2}\}$ **25** No real solutions **27** $z = \pm 2$ or $z = \pm 3$
29 $z \approx \pm 6.7$ or $z \approx \pm 0.3$ **31** $(b + 3)(b - 3)(b^2 + 3b + 9)$
32 a $\pi(x + y)^2 - 16\pi$ **b** $(x^2 + 2xy + y^2) - 16$ or $(x + y)^2 - 16$ **c** $\frac{\pi[(x + y)^2 - 16]}{(x + y)^2 - 16} = \frac{\pi}{1}$
33 a $\{a{:}a = \pm(p + 1)\}$ **b** Yes; a can be 3 if $p = 2$.
34 a $y = x^3 - 2x^2 - 8x$ **b** $y = 3x^4 - 2x^2 - 1$
c $y = x^3 - 3x^2 - x + 3$ **d** $y = x^3 - 1$
35 $\{x{:}x > 2\}$

37 $\left\{x{:}x \le -\frac{5}{2} \text{ or } -2 \le x \le 2\right\}$

39 a $y = 0$ or $y = \pm 1$ 

b $y = \pm 1$ or $y = \pm 4$ 

c $\{y: y \le -4 \text{ or } -1 \le y \le 1 \text{ or } y \ge 4\}$

8.1 Rational Exponents

Warm-up Exercises

1 a 6 **b** 4 **c** 16 **d** 8 **e** $\frac{1}{6}$ **f** 4 **2 a** $x^{\frac{3}{2}}$ **b** $x^{\frac{11}{12}}$ **3 a** 3 **b** 4 **c** -10 **4 a** ≈ -6.05 **b** ≈ 5598.82 **c** ≈ 0.03 **5** c

Problem Set

6 $27^{\frac{5}{3}} = (3^3)^{\frac{5}{3}} = 3^5 = 243$ **7** x^2 **9 a** ii, iii, and iv **b** ii and iv **c** iv and v **d** iv and v **e** ii and iv **10 a** 3 **b** 27 **c** $\frac{1}{27}$ **11** $2\sqrt{37}$, or ≈ 12.17 **12** 0; $\frac{3}{5}$ **13 a** $V = A^{\frac{3}{2}}$ **b** $\approx 58.1 \text{ cm}^3$ **15** $n = -2$ **17** $n = \frac{10}{3}$ **19** 125 **20 a** 3 **b** 6 **c** 3 **d** $\frac{1}{2}$ **e** 0 **21** 2 **23 a** $(x - 15)(x + 2)$ **b** $(y + 17)(y - 17)$ **c** $3(a + 4)(a + 7)$ **d** $(5n + 4)(n - 4)$ **24 a** $x^{\frac{5}{3}}$ **b** x^4 **25** $4x^4y^6$ **27** $\frac{3y^2}{2x^2}$ **29** a, b, c, and f **30 a** True **b** True **c** False **d** True **e** False **31 a** $\approx 72°\text{C}$ **b** $\approx 45°\text{C}$ **c** ≈ 43 min **33** 1; x^2 **34** Possible answers: 0.10110111011110 . . . and 0.21221222122221 . . . **35** $x + y$ **37 a** ≈ 7.0711; ≈ 13.5721; ≈ 18.8030; ≈ 22.8653; ≈ 48.1004; ≈ 49.9804 **b** It approaches, but does not reach, 50. **38 a** $\approx 2.26 \times 10^{-23} \text{ cm}^3$ **b** $\approx 4.42 \times 10^{18}$ **c** $\approx 2.21 \times 10^{18}$

8.2 Solving Radical Equations

Warm-up Exercises

1 a $y = 81$ **b** $y = 8$ **c** $y = \frac{13}{4}$ **2 a** $k = 99$ **b** No real solutions **3 a** No real solutions **b** $p = 125$ **c** $p = -125$ **4 a** $x = 8$ **b** $x = 5$ **c** $x = 1$ **d** $x = 341$ **5** $10\sqrt{5} \text{ ft}^2$

Problem Set

6 $a = \frac{7}{3}$ **7** $b = -\frac{4}{3}$ **8** $x = 10$ **9 a** -4 **b** 1 **c** 3 **d** $\frac{3}{2}$ **e** 3 **f** 5 **11** $x = 14$ **12 a** $r = 6$ **b** $r \approx 3.39$ **13** ≈ 36.37 **15** $x = 256$ **17** $x \approx 101.59$ **19** $x = \frac{1}{16}$ **21** $\frac{1}{5}\sqrt{15}$ **23** $5\sqrt{2}$ **25** $y = \frac{1}{3}$ **27** $y = \frac{3}{5}$ **28** 16 or -14 **29** 22 **31** $6\sqrt{2}$ **32** No solution **33** $3\sqrt{3}$ cm **34** $30\sqrt{2}$ in. **35** $n = 24$ or $n = 48$ **36** $\approx 10 \text{ cm}^2$ **37** ≈ 7 **38 a** $4x^8$ **b** $5x^{12}\sqrt{x}$ **c** $4x^6\sqrt{2}$ **d** $2|x^5y^3|\sqrt{6y}$ **39** ≈ 6.18 **40** ± 7 **41** $(\sqrt[4]{x})^2 = x^{\frac{2}{4}} = x^{\frac{1}{2}} = \sqrt{x}$
42 $\sqrt{(-2 - 0)^2 + (3 - 6)^2} = \sqrt{13}$
$\sqrt{(0 - 3)^2 + (6 - 4)^2} = \sqrt{13}$
$\sqrt{[3 - (-2)]^2 + (4 - 3)^2} = \sqrt{26}$
Since $(\sqrt{13})^2 + (\sqrt{13})^2 = (\sqrt{26})^2$, the triangle determined by the three points is right. **43 a** 2 **b** 2

c 5 **d** 0.5 **e** The power of 10 that is equivalent to the input value **44** $\approx 3.19\ \mu$ by $\approx 4.70\ \mu$ **45** $x = 25$ **47** No real solution **49** 6 **50 a** $3S = 3 + 3^2 + 3^3 + 3^4 + \ldots + 3^{11}$ **b** $2S = -1 + 3^{11}$ **c** $S = 88{,}573$ **51 a** $x = -64$ or $x = 27$ **b** $x = -22$ **52** $x(x^n + 4)(x^n + 1)$ **53** $x = -2$ or $x = 3$ **55** $x = a + b - 1$ **56 a** 54 **b** 2

8.3 Introduction to Complex Numbers

Warm-up Exercises

1

Complex Number	Real Part	Imaginary Part	Conjugate
$2 + 5i$	2	$5i$	$2 - 5i$
$3 - 4i$	3	$-4i$	$3 + 4i$
$-2 + 7i$	-2	$7i$	$-2 - 7i$
$5 - 4i$	5	$-4i$	$5 + 4i$
$9i + 3$	3	$9i$	$3 - 9i$
$-6 - 5i$	-6	$-5i$	$-6 + 5i$
$-4i$	0	$-4i$	$4i$
16	16	$0i$	16

2 a $8i$ **b** $-5i$ **c** i **d** $\frac{3}{5}i$ **e** $2\sqrt{3}i$ **f** -3 **g** $\frac{1}{2}i$ **h** -15 **i** -49 **3** No **4 a** $-i$ **b** i **c** $-2i$ **d** $2i$ **5** $\begin{bmatrix} -1 & 0 \\ 0 & -1 \end{bmatrix}$

Problem Set

6 a -5 **b** $2\sqrt{5}i$ **c** $-2\sqrt{10}i$ **d** $12i$ **7 a** $-2i$ **b** $3i$ **c** $-\frac{5}{4}i$ **8** $4\sqrt{2}$ **9** $i^4 = (i^2)^2 = (-1)^2 = 1$ **10 a** $\frac{5}{4}i$ **b** $-\frac{5}{4}i$ **c** -4 **11 a** $-2 + 7i$ **b** $-4i + 3$ **c** 8 **12** $\frac{1}{3}$ **13 a** $x^2 + 9$ **b** $x^2 + 1$ **c** $x^2 + 2ix - 1$

14 a 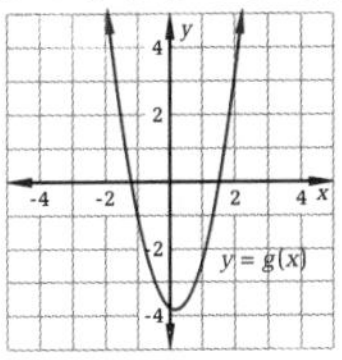

b 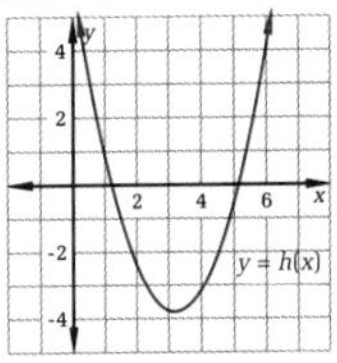

c 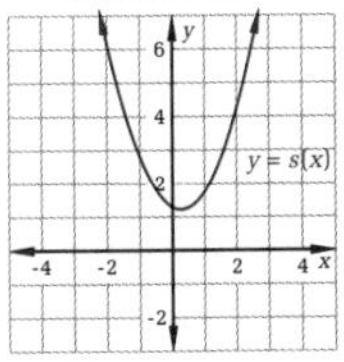

d 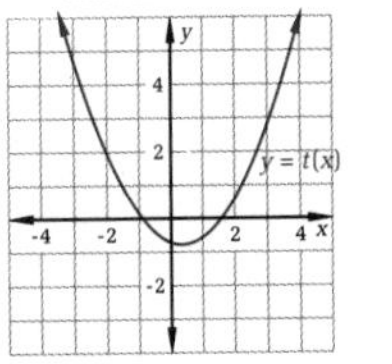

15 a $(y + 2i)(y - 2i)$ **b** $(3i + 2x)(-3i + 2x)$ **16 a** $x = \pm 5$ **b** $x = \pm 6i$ **c** $x = \pm 7$ **d** $x = \pm 7i$ **17 a** -24 **b** $-3i$ **c** 3 **d** -8 **e** $-8 - 6\sqrt{2}$ **f** $(4 - 3\sqrt{6})i$ **18** Ezra was right, since the property stating that $\frac{\sqrt{x}}{\sqrt{y}} = \sqrt{\frac{x}{y}}$ does not apply to roots with negative radicands. The work should read $\sqrt{-\frac{1}{9}} = \sqrt{\frac{1}{9}}\sqrt{-1} = \frac{1}{3}i$. **19** $\begin{bmatrix} -2 & -3 \\ 1 & -4 \end{bmatrix}$ **20** $\frac{5}{6}$ **21 a** ii **b** iv **c** i **d** iii **22 a** $(x + 5)(x - 5)$ **b** $(x + 5i)(x - 5i)$ **c** $(2x + 7)(2x - 7)$ **d** $(2x + 7i)(2x - 7i)$ **e** $4(x + 2)(x - 2)$

f $4(x + 2i)(x - 2i)$ **23** $\frac{1}{3}\pi a^7$ **24** $i^{-1} = \frac{1}{i} = \frac{1}{i} \cdot \frac{i}{i} = \frac{i}{i^2} = \frac{i}{-1} = -i$ **25 a** $0.5i$ **b** $2.5i$ **c** 36 **d** $-6 + 2\sqrt{2}$ **27** $x = \pm 2\sqrt{3}i$ **29** $x = -4 \pm i$ **31** $10 + 5i$ **32 a** $x = -1$ and $x = 4$ **b** 0 and 5 **33** $x = -8i$ **35** $x = -12i$ **37** $x = -4i$ **38 a** $(x, y) = \left(2i, \frac{1}{2}i\right)$ **b** $(x, y) = (1, -2i)$ **39 a** $x^2 + 100 = 0$ **b** $x^3 + 25x = 0$ **40** 8; 9 **41** 16π; 20π **42** $x = 0$, $x = 10i$, or $x = -10i$ **43** $x = 2i$ and $x = -2i$ **45** $x = 2$ or $x = 1$

8.4 Algebra of Complex Numbers

Warm-up Exercises

1

Complex Number	$a + bi$ Form	Real Part	Imaginary Part	Conjugate
$-5i - 7$	$-7 - 5i$	-7	$-5i$	$-7 + 5i$
$19i$	$0 + 19i$	0	$19i$	$-19i$
$2(5 + 3i)$	$10 + 6i$	10	$6i$	$10 - 6i$
27	$27 + 0i$	27	$0i$	27
$\sqrt{3} + \sqrt{-3}$	$\sqrt{3} + \sqrt{3}i$	$\sqrt{3}$	$\sqrt{3}i$	$\sqrt{3} - \sqrt{3}i$

2 a $(x + 11i)(x - 11i)$ **b** $(x + 11)(x - 11)$ **c** $(3x + 2i)(3x - 2i)$ **d** $9(x + 2i)(x - 2i)$ **3 a** $x = 3$ **b** $x = -3i$ **4 a** $(x, y) = (5, 7)$ **b** $(x, y) = (82, 39)$ **c** $(x, y) = (-6, 8)$ **d** $(x, y) = (5, 0)$ **5 a** $x^2 - 2x - 35 = 0$ **b** $x^2 + 25 = 0$

Problem Set

6 a $0 - \frac{5}{2}i$ **b** $\frac{3}{5} + \frac{1}{5}i$ **c** $-2 + 2i$ **d** $3 - 14i$ **e** $12 + 6i$ **7 a** 1 **b** 1 **c** i **d** $-i$ **8 a** $12 + 5i$ **b** $-5 - 4i$ **c** 25 **d** $\frac{12}{13} + \frac{5}{13}i$ **9 a** $-9i$ **b** $-13i$ **c** 0 **10 a** 58 **b** ≈62% **c** ≈95% **11 a** $\frac{4}{3}i$ **b** $\frac{4}{3}i$ **c** $\frac{4}{3}i$ **12 a** $x = \pm\frac{3}{2}i$ **b** $x = \pm 7i$ **13** $(x, y) = \left(2 + 3i, -\frac{2}{3} + i\right)$ **15** -13 **17 a** -16 **b** $x = 2 \pm 2i$ **18 a** $x = \pm 7i$ **b** $x = 1 \pm i$ **c** $x = 3 \pm 5i$ **19 a** $10 - 5i$ **b** $9 + 5i$ **c** $0 + 2i$ **20 a** $y = -(x + 5)^2 - 2$ **b** $y = (x - 5)^2 + 2$ **c** $y = -(x - 5)^2 - 2$ **d** $x = (y + 5)^2 + 2$ **21** 0 **23** $-13 + 22i$ **25** $38i$ **27 a** $x = 16$; $y = -7$ **b** $x = \frac{13}{3}$; $y = \frac{10}{3}$ **c** $\{(x, y): x = 3y\}$ **28 a** 4 **b** $-8i$ **c** $6i$ **d** 25 **e** 5 **29 a** $3i$ **b** $2i$ **31** $x = -\frac{3}{4} \pm \frac{\sqrt{23}}{4}i$ **33** $x = \frac{3}{2} \pm \frac{1}{2}i$ **35** $\pm 2\sqrt{5}i$ **36 a** 0 **b** $-1 + i$ **c** 0 **d** $-1 + i$ **37** -10; 26 **38** $\left(\frac{6}{7}, \frac{12}{5}\right)$ **39 a** $\frac{x^2}{4} + \frac{y^2}{9} = 1$ **b** $(x - 3)^2 + (y + 4)^2 = 9$ **c** $y = -\frac{1}{2}x^2 + 6$ **d** $\frac{(y - 2)^2}{16} - \frac{(x - 2)^2}{16} = 1$ **40 a** $x = 3$; $y = -9$ **b** $x = 7$; $y = 10$ **41** $(x, y) = \left(3 - 2i, -1 + \frac{3}{2}i\right)$ **42** $-8 + 19i$ **43** $-1 + i$; -1; 0; i; $-1 + i$; -1; 0; i **44 a** $x = 8$ **b** $x = \pm 2\sqrt{2}$ **c** $x = 2$ or $x = -1 \pm \sqrt{3}i$ **d** $x = 9i$ or $x = i$ **e** $x = 5i$ or $x = i$ **45** $\pm 2i$ **46 a** 82 **b** 85 **47** $\{(a, b): 3b = 2a\}$

8.5 The Normal Curve

Problem Set

1 a 1000 square units **b** 25% **c** 25 **d** 85% **e** 85 **2** 0.4; 8.8; 54.4; 136.4 **3 a** 84.1% **b** 84

4

Data value	200	300	400	500	600	700	800
Percentile rank	1	2	15	50	84	97	99

5 65.9% **6 a** 81.8% **b** 84.1% **7 a** 2 **b** 15 **c** 84 **d** 97 **e** 99 **8** 84.1% **9** 99.9% **10** 97.7% **11** 95.4%

Review Problems

1 5 **3** 4 **5** $\frac{1}{27}$ **7** $(4 - \pi)\sqrt{3}$, or ≈1.49 **8** $x = -3$ **9 a** $0 + \frac{4}{3}i$ **b** $-2 + 6i$ **c** $\frac{30}{37} - \frac{5}{37}i$ **d** $144 + 0i$ **11** 64 **13 a** 405 and 555 **b** 360 and 600 **15** $-\frac{10}{7}i$ **17** $12i$ **19** $-3 - 6\sqrt{7}$ **20 a** $2 + 9i$ **b** $\frac{1}{2} - \sqrt{3}i$ **c** $-15i$ **d** $-5 + 3i$ **e** $\sqrt{23}$ **f** $\frac{5}{7} + \frac{5}{7}i$ **21 a** $x^{\frac{17}{15}}$ **b** $y^{-1.5}$ **c** a^{19} **22** 18.2% **23** $7 - 2i$ **25** $\frac{20}{29} + \frac{21}{29}i$ **27** $(x, y) = (7, -4)$ **29** $(x, y) = (-70, -8)$ **31** 2, -2, $2i$, and $-2i$ **32** $\begin{bmatrix} -1 + 3i & -14 + 8i \\ 4 - 8i & 6 + i \end{bmatrix}$ **33** $a = \frac{23}{3}$ **35** $a = \pm 5\sqrt{10}$ **37** $a = \pm 6$ or $a = \pm 2\sqrt{41}$ **39 a** $x^2 + 121 = 0$ **b** $x^3 - 7x^2 + 20x - 24 = 0$ **41** $x = 3 \pm 3i$ **43** $x = \pm\sqrt{7}i$ **44** 5

9.1 The Remainder and Factor Theorems

Warm-up Exercises

1 $((1)x - 3)x + 5$ **2** $((((3)x - 0)x - 2)x + 5)x - 9$ **3** $4x^3 - 3x^2 + 2x - 7$ **4 a** Quotient: $2x^2 - 3x - 22$; remainder: -106 **b** -106; the value of $P(x)$ when $x = 5$ is equal to the remainder for $[P(x)] \div (x - 5)$. **5** $x^4 + x^3y + x^2y^2 + xy^3 + y^4$

Problem Set

6 0 **7** $(((5)x + 2)x + 6)x + 1$ **9** $3x^3 + 6x^2 - 7x + 9$ **11** $2x^3 - 5x - 7$ **13** 39 **15 a** 4 **b** 5 **c** -6 **d** $-4x_1^2$ **e** $2x_1^3$ **f** 3 **17** $(x^2 - 3)^2$ **18**

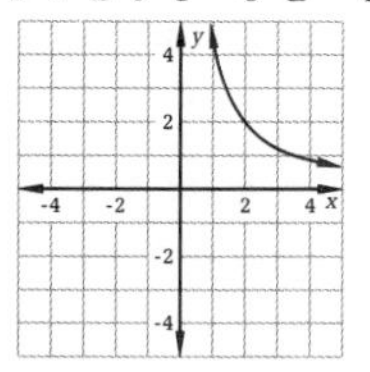

19 $x^2 - 4x + 7$, remainder -1 **21** $2(x - 4)(x^2 + 4x + 16)$ **23** $(2x - 3)(4x^2 + 6x + 9)$ **24 a** -7 **b** -13 **c** 71 **25 a** $x^2 + 5x + 6$; $x^2 + 5x + 4$; I; 2 square units **b** $x + 4$, remainder 2 **c** Since $x^2 + 5x + 6 = (x + 1)(x + 4) + 2$, the area of rectangle I is 2 square units greater. **26 a** 6 **b** 2 **27 a** $18x^3$; 20 **b** $18x^3$; 20 **c** x^3; 20 **29** $3(3x - 5)(3x + 1)$ **31** $x - 2$, remainder $-x - 9$ **32** 4, $-\frac{3}{2}$, $\frac{5}{3}$

33 $-3 < z_1 < -2$; $0 < z_2 < 1$; $3 < z_3 < 4$
34 Quotient: $2x^3 - 4x^2 - x + 7$; remainder: -10
35 $x^2 + 8x + 6$ **36** 20,475 classes **37** $(x^2 + 5)(x^2 - 5)$
39 $(x^2 + 4)(x + 2)(x - 2)$ **40 a** $(((4)x - 3)x + 8)x + 42$
b $4x^3 - 3x^2 + 8x + 42$ **41 a** 0 **b** Yes; there is an integer n such that $P(n) = 0$. **42 a** 60 **b** 12
43 $2x^2 - 14x + 48$, remainder -139 **44** $x = -4$ or $x = \frac{2}{3}$ **45** $(x - 3)(x + 2)(x - 2)$ **46** Quotient: $x^2 + 2x + 4$; remainder: 0 **47** $(6, -2, 4)$ **48** Quotient: $ax^2 + (a + b)x + (a + 2b)$; remainder: $2a + 2b$ **49** For any negative value of x, the powers x^5, x^3, and x^1 will be negative. Since a, b, and c are also negative, each term of $P(x)$ will represent a positive number when x is negative. The sum of three positive numbers cannot be 0, so $P(x) \neq 0$ for any negative x. **50 a** $P(x) = \frac{1}{8}x^3 - \frac{3}{8}x^2 - \frac{5}{4}x + 3$ **b** $g(x) = \frac{3}{5}x^2 - \frac{6}{5}x - \frac{9}{5}$ **c** $H(x) = \frac{1}{8}x^3 - \frac{39}{40}x^2 - \frac{1}{20}x + \frac{24}{5}$ **d** $40[H(x)] = 5x^3 - 39x^2 - 2x + 192$; $40[H(-2)] = 0$, so $H(-2) = 0$ **51** $2x + 1$

9.2 The Rational-Zeros Theorem

Warm-up Exercises

1 a $2x^3 + 3x^2 - 18x + 5$ **b** $x + 4$ **c** -3
d $2x^2 - 5x + 2$ **2** Quotient: $3x + 6$; remainder: 31
3 $2x - 1$, remainder 4 **4** $-1, \frac{1}{3}, 1, 2$ **5** $3x^4 - 26x^3 + 262x - 168$

Problem Set

6 $3x^2 - 2x$, remainder 6 **7** $\pm 1, \pm 5, \pm\frac{1}{2}, \pm\frac{5}{2}$
8 $x^3 - 6x^2 + 11x - 6$; $P(x) = x^3 - 6x^2 + 11x - 6$
9 $x = -3$, $x = -\frac{3}{2}$, or $x = \frac{3}{2}$ **10** $x^2 - 3x + 2$
11 $\approx -3.7, \approx 0.5, \approx 2.6$ **12** $\frac{2}{3}$ **13** $x^3 - 11x^2 + 10x + 72 = 0$ **14** 9 **15** $\pm 1, \pm 2, \pm 3, \pm 6, \pm\frac{1}{2}, \pm\frac{3}{2}, \pm\frac{1}{4}, \pm\frac{3}{4}$
16 The Rational-Zeros Theorem does not apply to function P, Since $P(x)$ contains a nonintegral coefficient. **17** $s = 4x^3 + 9x^2 - 23x + 16$ **18** -2 and -1, 1 and 2, 2 and 3, and two between 5 and 6 **19** 1
21 $x = \frac{7}{a + b}$ **23** When $x = 2$, the value of the cubic polynomial is 0, so by the Factor Theorem, $x - 2$ is a factor of the polynomial. **25** $2x^2(3x - 5)(3x + 5)$
26 $f(x) = x^4 - 13x^2 + 36$ **27** $1, \frac{3 + 3\sqrt{5}}{2}, \frac{3 - 3\sqrt{5}}{2}$
28 $y = -3$, $y = -\frac{2}{3}$, $y = \frac{1}{2}$, or $y = 2$ **29** $2x^3 - 11x^2 + 22x - 21$ **30 a** $P(x) = x^3 + 4x^2 - 7x - 10$ **b** $P(x) = 9x^3 - 18x^2 + 11x - 2$ **31** 1, 8, 28, 56, 70, 56, 28, 8, 1
32 $-\sqrt{7}, -\sqrt{5}, \sqrt{5}, \sqrt{7}$ **33** $x^2 - 30x + 200 = 0$
34 a $-\sqrt{7}, \sqrt{7}, 4$ **b** $6 - \sqrt{7}, 6 + \sqrt{7}, 10$ **c** $-\frac{\sqrt{7}}{2}, \frac{\sqrt{7}}{2}, 2$ **35** $x = -a$ or $x = -b$ **37** $\{x : x \neq -2, x \neq 2, x \neq 3\}$ **38 a** $\pm 1, \pm 2, \pm\frac{1}{2}, \pm\frac{1}{3}, \pm\frac{2}{3}, \pm\frac{1}{6}$ **b** -13

9.3 Real Zeros of Polynomial Functions

Warm-up Exercises

1 When $P(x)$ is divided by $x + 2$, the terms of the quotient and the remainder alternate in sign.
2 Answers will vary. Possible answer: 8; -4
3 ≈ -2.24

Problem Set

4 a Yes **b** No **c** No **d** No **5** Answers will vary. Possible answer: 6; -3 **6** $\approx -1.414, \approx 1.414$ **7** ≈ 1.522
8 Answers will vary. Any 12 numbers greater than 5 are possible answers. **9** $3y^2(2 + y)(2 - y)$
11 $(2 - b)(4 + 2b + b^2)$ **12** Quotient: $x^2 - 2x - 1$; remainder: 3 **13** $\approx -1.149, \approx 0.201$ **15** $x = -5$, $x \approx -3.606$, or $x \approx 3.606$ **16** Possible answer: 6; 2 (since $x \neq 2$) **17** $(x, y) = (-2, 0)$ or $(x, y) = (2, 0)$
18 18 **19 a** 0 **b** -1 **c** -2 **d** -3 **20 a** 1 **b** $x + y$
c $x^2 + 2xy + y^2$ **d** $x^3 + 3x^2y + 3xy^2 + y^3$ **21 a** 4 **b** 3
c 0 **d** 4 **e** 3 **f** 0 **22 a** $((5)x - 7)x + 11$ **b** $20x^2 - 14x + 11$ **23** ≈ 1.72 **25** $\{x : x \leq 0 \text{ or } x \geq 6\}$
26

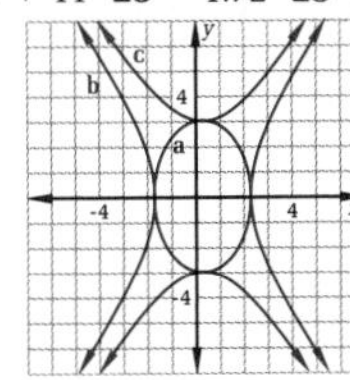

27 a $P(2)$ is negative, and $P(3)$ is positive, so the graph of P must cross the x-axis (that is, $P(x)$ must equal 0) somewhere between $x = 2$ and $x = 3$. **b**

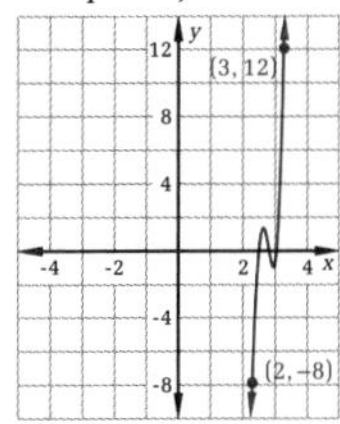

c No **28 a** 0 and 1, 2 and 3, 3 and 4 **b** $2 - \sqrt{3}, \frac{11}{5}, 2 + \sqrt{3}$ **29** $\{a : a \geq -1\}$
30 No; answers will vary—possible answer is that $63 \cdot (-5) \neq 12$. **31 a** -16 **b** 11 **c** Since $P(2)$ is negative and $P(3)$ is positive, $P(x)$ must equal 0 for some value of x between 2 and 3. **32 a** 0 **b** -1
c Yes **d** No **33 a** 0 **b** $(0, -1)$ **c** 0 **d** 0 **34 a** $P(1) = 2 - 3 + 2 - 8 = -7$ and $P(2) = 32 - 12 + 4 - 8 = 16$; since the value of $P(x)$ changes sign between $x = 1$ and $x = 2$, there must be at least one zero between these values. **b** ≈ 1.57

9.4 Four Useful Theorems

Warm-up Exercises

1 2 **2** -15; 50 **3** $\frac{2}{5}$ **4** No, since the equation contains nonreal coefficients

Problem Set

5 7 **7** $\frac{3}{2}$; -3 **9** $x^2 - 25 = 0$ **11** $x^2 + 25 = 0$
13 a -6 **b** 36 **14 a** $1 + 0i$ **b** $1 + 0i$ **c** $1 + 0i$
15 $z = 5 + 15i$ **16** $x^2 - 4x + (1 - 4i) = 0$
17 a $\begin{bmatrix} 3+8i & -i \\ 15+2i & 1-7i \end{bmatrix}$ **b** $\begin{bmatrix} 3-4i & 8-i \\ -3+8i & 11-5i \end{bmatrix}$ **18** 6
19 $x^4 - 6x^3 + 14x^2 - 24x + 40 = 0$ **21** 5; 5; -2, 2 (multiplicity 2), $-2i$, $2i$ **22** $\{(x, y): x - iy = -i\}$
23 20 **24 a** 16; $-6i$ **b** 8; $-\frac{5}{2}i$ **c** 3; $-5i$ **25** $\frac{1}{5}$
26 $\begin{bmatrix} 1 & 0 \\ 0 & 1 \end{bmatrix}$ **27 a** 20 **b** $\frac{1}{10}$ **c** $\frac{2}{5}$ **28** $\frac{1}{2}$ **29 a** 8
b -3 (multiplicity 3), 3, $-3i$ (multiplicity 2), $3i$ (multiplicity 2) **30** $\begin{bmatrix} 0 \\ -7 \\ -7 \\ 25 \end{bmatrix}$ **31 a** $z = -3i$ **b** $z = 16$
32 $x = -3i$ or $x = 8i$ **33** $\sqrt{3}$; $\sqrt{5}$ **34 a** $w = -i$ or $w = i$ **b** $w = -2i$ or $w = 2i$ **35** $x + yi = -\frac{\sqrt{2}}{2} - \frac{\sqrt{2}}{2}i$ or $x + yi = \frac{\sqrt{2}}{2} + \frac{\sqrt{2}}{2}i$ **36 a** 1 **b** 3 **c** 3, $-i$, i
37 $x^4 + 45x^2 + 324 = 0$ **38** 6; $-8i$

9.5 The Binomial Theorem

Warm-up Exercises

1 Row 8: 1, 8, 28, 56, 70, 56, 28, 8, 1; row 9: 1, 9, 36, 84, 126, 126, 84, 36, 9, 1; row 10: 1, 10, 45, 120, 210, 252, 210, 120, 45, 10, 1 **2** $32x^5 + 400x^4y + 2000x^3y^2 + 5000x^2y^3 + 6250xy^4 + 3125y^5$ **3** $-366{,}080x^4y^9$

4 $C(n, r+1) = \frac{n!}{[n - (r+1)]!(r+1)!} = \frac{n!}{\frac{(n-r)!}{n-r}r!(r+1)}$
$= \frac{n!}{(n-r)!r!} \cdot \frac{n-r}{r+1} = C(n, r) \cdot \frac{n-r}{r+1}$

The relationship tells us that we can find the coefficient of each successive term by multiplying the coefficient of the preceding term by $\frac{n-r}{r+1}$, where n is the power of the binomial and the preceding term is the $(r + 1)$th term of the expansion.

Problem Set

5 $p^3 + 3p^2q + 3pq^2 + q^3$ **7** First four: $x^{12} + 12x^{11}y + 66x^{10}y^2 + 220x^9y^3$; last four: $220x^3y^9 + 66x^2y^{10} + 12xy^{11} + y^{12}$ **8** $-112{,}266y^6$ **9** Third term, $40x^6$
10 a iii **b** ii **c** i **d** iv **11 a** False **b** False **c** True **d** False **12** ≈ 3.519 **13** 0, -5 **15** $(x + y)^4$ **16 a** 1, 2, 4, 8, 16, 32 **b** 4096 **17 a** $C(n, 0) = \frac{n!}{n!0!} = 1$
b $C(n, n) = \frac{n!}{0!n!} = 1$ **18 a** $26{,}334a^5b^{17}$ **b** $26{,}334a^{17}b^5$
19 $59{,}049x^{10} - 984{,}150x^9y + 7{,}381{,}125x^8y^2$ **21** $x^5 - 10x^4y + 40x^3y^2 - 80x^2y^3 + 80xy^4 - 32y^5$ **23** $32x^5 - 80x^4y + 80x^3y^2 - 40x^2y^3 + 10xy^4 - y^5$ **24** $(x + 12)^2$
25 a $x - 2$, remainder 8 **b** $x^3 - 2x^2 + 4x - 8$, remainder 32 **26** ≈ -3.85, ≈ 1.33, ≈ 3.52 **27** 1
28 $512w^9 + 6912w^8 + 41{,}472w^7 + 145{,}152w^6$
29 a RRR, RRL, RLR, LRR, LLR, LRL, RLL, LLL
b RRRR, RRRL, RRLR, RLRR, LRRR, RRLL, RLRL, RLLR, LLRR, LRLR, LRRL, LLLR, LLRL, LRLL, RLLL, LLLL **c** The numbers of ways in which he can reach the various junctions are the same as the numbers in Pascal's triangle, and the number of ways in which he can reach the nth level is the sum of the numbers in the $(n - 1)$th row of Pascal's triangle—that is, 2^{n-1}.
30 $x = -2$ or $x = -1$ **31 a** $V = x^3 + 9x^2 + 27x + 27$ **b** ≈ 1.64 **32** 5040 **33** $32x^5 + 40x^4y + 20x^3y^2 + 5x^2y^3 + \frac{5}{8}xy^4 + \frac{1}{32}y^5$ **35** $(x + 2y)(x + 4)(x - 4)$
37 $(x - 2y)(x + 2y + 3)$ **39** $(x^2 - 6)(x + 2)$ **40 a** 21 **b** 95 **41** 0 **43** 5^{20}

44 $0 = (1 - 1)^n$
$= [C(n, 0)](1^n) + [C(n, 1)](1^{n-1})(-1) + [C(n, 2)](1^{n-2})(-1)^2 + \ldots + [C(n, n)](-1)^n$
$= C(n, 0) - C(n, 1) + C(n, 2) + \ldots + (-1)^n[C(n, n)]$

45 $C(n, r) = \frac{n!}{(n-r)!r!} = \frac{n!}{r!(n-r)!} = \frac{n!}{[n-(n-r)]!(n-r)!}$
$= C(n, n - r)$

46 a The numbers in each row (after the first) describe the arrangement of numbers in the preceding row. For example, the third row is one 2 and one 1 so the fourth row is 1211. The next three rows are
1113213211
31131211131221
13211311123113112211 **b** 3 **47** $6x^5$

9.6 Standard Deviation and Chebyshev's Theorem

Problem Set

1 41 **2 a** 0.75 SD **b** -2.5 SD **3 a** 11.51% **b** 0.82%
4 a 25 **b** 5 **5** ≈ 43.29; ≈ 16.92 **6** 43; 50; 57; 32.5
7 95 **8 a** $\approx 88.89\%$ **b** Cannot be determined (To use Chebyshev's theorem, k must be greater than 1.)
9 $\approx 42\%$ **10** $\frac{\sqrt{10}}{2}$, or ≈ 1.581 **11** $\approx 77.45\%$ **12** ≈ 0.03

Review Problems

1 Answers will vary. Possible answer: ≈ -4.5, ≈ 2.3, ≈ 6.5 **3** $(((-3)x - 4)x + 1)x + 7$ **4** 16 **5** Quotient: $4x^3 + 6x^2 + 19x + 49$; remainder: 132 **7** ≈ 37.26 **8** Yes
9 $f(x) = x^3 - 2x^2 - 11x + 12$ **11** $f(x) = x^2 - 6x + 2$
13 $f(x) = 12x^3 + 67x^2 - 32x - 12$ **15** Answers will vary. Possible answer: -3; 8 **16** ± 1, ± 5, $\pm\frac{1}{2}$, $\pm\frac{5}{2}$; -5, $1 - \sqrt{2}$, $\frac{1}{2}$, $1 + \sqrt{2}$ **17** 4 **19** $x^4 - 4x^3y + 6x^2y^2 - 4xy^3 + y^4$ **20** ≈ 4.11 **21** Since the coefficients and constant are positive, each term will be positive for any positive value of x. No set of positive numbers has a sum of 0, so there is no positive value of x for which $P(x) = 0$. **22** 95 **23 a** 3 **b** 1 (multiplicity 3) **24** -4
25 20; 43.25; ≈ 6.58 **26 a** $x^4 - 2x^3 + 4x^2 - 8x + 16$ **b** Yes **27** $19{,}683x^9 + 295{,}245x^8y + 1{,}968{,}300x^7y^2 + 7{,}654{,}500x^6y^3$ **28** -1, 1, $-i$, i **29** 84% **30** 4

10.1 Multiplying and Dividing Rational Expressions

Warm-up Exercises

1 Yes **2** No **3** Yes **4** No **5** Yes **6** Yes **7** Yes **8** Yes **9** $\frac{b}{a}$; $a \neq 0, b \neq 0$ **10** $x^2 - 12x + 35$; $x \neq -2$

Problem Set

11 $\frac{4}{5}$ **13** -1; $y \neq x$ **15** $(x + 5)^2$; $x \neq -5$ **16** $\{x : x \neq 0, x \neq -2, x \neq -4\}$ **17** yz; $x \neq 0, y \neq 0$ **19** $\frac{x}{2y}$; $y \neq 0$

21

Expressions	GCF	LCM
4, 6, 8	2	24
9, 12, 6	3	36
$4x^2$, $6xy$, $2x^2y$	$2x$	$12x^2y$
$x(x - 2)$, $4(x - 2)$	$x - 2$	$4x(x - 2)$

23 $x^2 - 3x$; $x \neq 0$ **25** $\frac{2(x + 4)(x + 2)}{x^2 + 4}$; $x \neq 0, x \neq 2i, x \neq -2i$ **26** 1; $x \neq 2, x \neq -5, x \neq -3$ **27** $\frac{x^2 - 25}{(x + 4)^2}$; $x \neq 4, x \neq -4, x \neq 5, x \neq -5$ **28** $x + 5$ **29** $\frac{(x^2 + 4)(x - 5)}{x + 2}$; $x \neq -5, x \neq -2$ **30 a** 0 **b** 0 **c** Undefined **d** Undefined **e** $\frac{-8}{5}$ **31 a** $\{x : x \neq -1, x \neq -2\}$ **b** $\{x : x \neq -1, x \neq -2\}$ **32 a** 5 **b** 3 **c** ± 2 **33 a** True **b** False **c** True **34** $\frac{18}{15}$ **35** $x = 4$ or $x = -3$ **36** $x = 2, x = 4$ **37** Numerator is ten more than denominator. **38** $8x^3 + 36x^2y + 54xy^2 + 27y^3$ **39** $x = -4$ **40** $h = x - 2$ **41** $\frac{x + 4}{x - 5}$ **42** $\frac{x^2 - 4}{x^2 + 2x + 3}$; $x \neq -1 + \sqrt{2}i, x \neq -1 - \sqrt{2}i$ **43** $8\frac{4}{7}$ min **44** g is defined at $x = -2$, but f is not. **45** $x^6 + 6x^4 + 15x^2 + 20 + \frac{15}{x^2} + \frac{6}{x^4} + \frac{1}{x^6}$ **46** $x + 2$; $x \neq -3y$ **47** $x \approx -2.81, x \approx 2.34, x \approx -0.75$ **49** 23

10.2 Rational Equations and Inequalities

Warm-up Exercises

1 42 **2** xy **3** xy **4** $x^2 - 1$ **5** $x = 2 \pm \sqrt{7}$ **6** $x = \frac{5}{4}$, $x = -4$ **7** $-2, \frac{3}{2}, -\frac{1}{3}$ **8 a** 3 hr **b** 1 hr **c** 5 hr **d** $3\frac{3}{4}$ hr **e** $2\frac{1}{7}$ hr **9** $-\frac{5}{4} < x < -1$

10 

Problem Set

11 $x = -4$ or $x = 3$ **13** $x = -3$ **15** $x = 6$ or $x = -4$ **17** $x \neq 8$ **19** $x = 0$ **20** $a = \frac{3x}{4}$ **21** -2 **23** $\frac{-(x - 5)}{(9 - x^2)}$; $x \neq \pm 3, x \neq \pm 2$ **25 a** $\frac{1}{10}$ **b** $\frac{1}{20}$ **c** $\frac{1}{x}$ **27** $x = \frac{7}{2}$ or $x = -1$ **29** $x = 9$ **30** $-12, 3$ **31** $\frac{1}{3}$; $x \neq 0, x \neq 5$ **33** $x > 12$ or $x < 0$ **35** $x \leq -3$ or $x > 0$ **36 a** $\frac{7}{24}$ **b** $\frac{7}{12}$ **c** $\frac{7}{8}$

37 $\frac{2x^4}{x^2 + 3}$; $x \neq 0, x \neq \sqrt{3}i, x \neq -\sqrt{3}i$

39 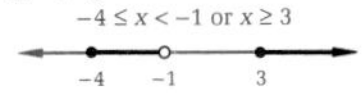

41 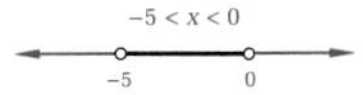

42 $x = 13$ **43** $1\frac{13}{47}$ hr **44** $p = \frac{A}{1 + rt}$ **45** -7 **46** 1, $-2, \pm i\sqrt{3}$ **47** 6 or -2 **49** No real solution **50** $-2, 1, -2 \pm i\sqrt{3}$ **51 a** 15 **b** 27 **52 a** (1, 2) **b** $(a, a^2 + 1)$ **c** $m = a + 1$ **53** $x = a - 6$ **55** $x = a^2 + 3a + 9$ **56** $\left\{a : a < 0, a > \frac{2}{3}\right\}$ **57 a** 100 min **b** Page 201 **58** -4 and -5 **59 a** $\frac{x + 1}{x - 1}$; $x \neq 0, x \neq 1$ **b** $\frac{x - 1}{x - 3}$; $x \neq 2, x \neq 3$ **c** $\frac{3x + 1}{3x - 1}$; $x \neq 0, x \neq \frac{1}{3}$ **60 a** $f(x) = x - 3$ **b**

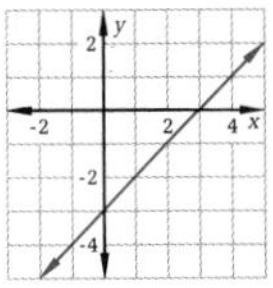

61 14 (Technically this has no solution, since by definition f is not a polynomial function but a rational function.)

10.3 Adding and Subtracting Rational Expressions with Like Denominators

Warm-up Exercises

1 a ii **b** iv **c** i **d** iii **2** $\frac{6}{5}$ **3** 5 **4** $\frac{x - 1}{y}$; $y \neq 0$ **5** 5x **6** $\frac{15}{x}$; $x \neq 0$ **7** x **8** $x = 8$ **9** -5; $x \neq 5$

Problem Set

10 $\frac{3}{5}$ **11** $x = 4$ **13** $\frac{4 + x}{3x}$; $x \neq 0$ **15** $y = 2$ **17** $x = -\frac{3}{a}$ **19 a** $\frac{x + 2}{x - 2}$ **b** 6 **21** -1; $x \neq \frac{3}{2}$ **23** $\frac{3}{2}$ **24 a** x^5 **b** x^5; $x \neq 0$ **c** x^5; $x \neq 0$ **25 a** $x - 1$; $x \neq -3$ **b** $\frac{x^2 - 2x + 3}{x + 3}$; $x \neq -3$ **26 a** 28 **b** $x(x^2 - 1)$ **c** $x^2 - 1$ **27 a** $\frac{x - 1}{x^2 + 1}$; $x \neq i, x \neq -i$ **b** $x = -1$ **29** $x = -7$ **31** $\frac{4y}{x}$; $x \neq 0$ **33** $x = 2$ **35** $\frac{9x + 1}{x^2}$; $x \neq 0$ **36** $x = -\frac{1}{8}$ or $x = -1$ **37** 7, -1 **39** 4 **41** $\frac{x + 3}{x^2 + 3x + 9}$; $x \neq 3, x \neq \frac{-3 + 3i\sqrt{3}}{2}$, $x \neq \frac{-3 - 3i\sqrt{3}}{2}$ **42** $x = \frac{1}{4}$ **43 a** 3, 2 **b**

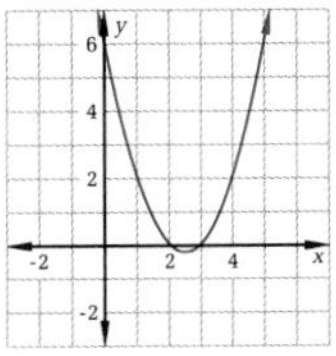

45 $(x - 1)(x - 2)$; $x \neq -2$ **47** $\frac{x}{x + 7}$; $x \neq -7$ **48** $\approx 49\frac{3}{4}$ min

49 x > 1 or x < −1
50 $\frac{x^2 - 2x + 4}{x(x - 7)}$; x ≠ −2, x ≠ 0, x ≠ 7 **51** x = $-\frac{1}{3}$ or x = 2 **52 a** $\frac{x + 3}{x^2 + 3x + 9}$; x ≠ −3, x ≠ $\frac{-3 \pm 3i\sqrt{3}}{2}$
b x = 3 **53** $\frac{x - 1}{x + 2}$; x ≠ −1, x ≠ −2

59 {x:x ≠ 0, x ≠ −1} **60** Machine A: 12 hr; machine B: 6 hr **61 a** $B(x) = x - 1$ **b** Restrictions are determined by the original expression. **62 a** $a = y$, $b = x$
b $h = \frac{h_1h_2}{h_1 + h_2}$ **63** (3, 3), (3, 4), (3, 5), (4, 3), (5, 3)

10.4 Adding and Subtracting Rational Expressions with Unlike Denominators

Warm-up Exercises

1 48 **2** x(x − 1)(x + 1) **3** 4 **4** x **5** Answers may vary. Possible answer: x(x − 3) **6** Answers may vary. Possible answer: 27 **7** Answers may vary. Possible answer: x(x − 2)(x − 3) **8** Answers may vary. Possible answer: 4(x − 2)(x + 2) **9** $\frac{x + 3}{x + 1}$; x ≠ −1, x ≠ 3
10 $\frac{x^2 - 5x + 25}{x(x - 5)}$; x ≠ −5, x ≠ 0, x ≠ 5

Problem Set

11 $\frac{3y + 4x}{xy}$; x ≠ 0, y ≠ 0 **13** $\frac{3x^2 - 8y^2}{2xy}$; x ≠ 0, y ≠ 0
14 A common denominator establishes a needed basis for comparison. **15** x + 3; x ≠ 3 **17** $\frac{x^2 + 3x + 9}{x + 3}$;
x ≠ 3, x ≠ −3 **18** $77\frac{1}{3}$ **19** x(x + 1) **21** x^2y^3 **23** 12 ft
25 $\frac{5x + 2}{x(x - 3)}$; x ≠ 0, x ≠ 3 **26** $\begin{bmatrix} x(x+1) & x(x-1) \\ 1 & \frac{x+1}{x-1} \end{bmatrix}$;
x ≠ 0, x ≠ 1
27 No solution **28** $x^3 + 3x^2y + 3xy^2 + y^3$
29 $\frac{14x - 2}{(x + 3)(3x - 2)}$; x ≠ −3, x ≠ $\frac{2}{3}$ **31** $\frac{10x - 14}{(x + 3)(3x - 2)}$;
x ≠ −3, x ≠ $\frac{2}{3}$ **32** $\frac{x^2(x - 1)^2}{(x - 3)(x + 2)^2}$ **33** $\frac{6}{5}$ **34** ≈8.90
35 $\frac{xy}{x + y}$ **37** x = $\pm\frac{3}{4}$ **38** $x = \frac{x_1x_2}{x_1 + x_2}$ **39** $-1 < x \le -\frac{1}{2}$ or x > 0 **41** x = 2a − 3
42 a $\frac{x}{x + 7} = \frac{x + 2}{5x - 1}$ **b** x = $\frac{7}{2}$ **43** $\frac{3(x + 3)}{x(x - 2)}$; x ≠ 0, x ≠ 2
45 $\frac{7(2x + 1)}{x(x + 1)}$; x ≠ 0, x ≠ −1 **47** x = 3, x = $\sqrt{2}$, x = $-\sqrt{2}$ **48** f(x) = 4x + 1 **49** 3 < x ≤ 5 or −5 ≤ x < −3 **51** x − 2; x ≠ −5 **52 a** $A(a, a^3)$, $B(-a, -a^3)$ **b** a^2 **c** {a:a > $\sqrt{2}$} **53** $\frac{2(x^2 + 1)}{x^2 - 1}$; x ≠ 1, x ≠ −1
55 $\frac{x^2 + 1}{-2x}$; x ≠ 1, x ≠ −1
56

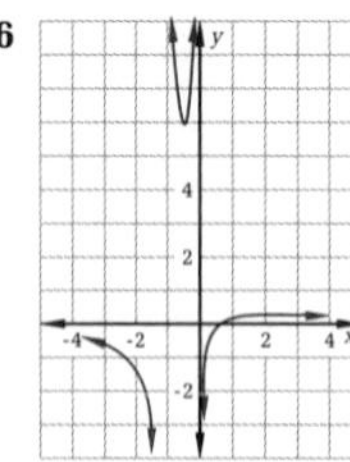

57 $\frac{-3}{x(x + 3)}$; x ≠ 0, x ≠ −3

10.5 Complex Fractions

Warm-up Exercises

1 $\frac{3}{2}$ **2** $\frac{b}{a}$ **3** $\frac{4}{15}$ **4** 3 **5** 2 **6** 3 **7** $\frac{3}{2}$ **8** 5x
9 (x − 2)(x + 2) **10** x(x − 3)

Problem Set

11 $\frac{x + 1}{x - 1}$; x ≠ 0, x ≠ 1 **13** 2x + 1; x ≠ −1 **15** x ≠ 0, x ≠ −2, x ≠ −1 **16 a** $\frac{6}{5}$x **b** x > 0 **17** −3x; x ≠ 0, x ≠ 3 **18** $\frac{3x - 5}{3x - 7}$; x ≠ 2, x ≠ $\frac{7}{3}$ **19** $\frac{-2x - 1}{4x}$; x ≠ 0
21 a $\frac{24}{7}$ **b** $\frac{36}{13}$ **c** $\frac{240}{77}$ **23** $\frac{11}{2(x + 1)}$; x ≠ −1 **24** $-\frac{1}{117}$
25 $\frac{y + x}{y - x}$; x ≠ 0, y ≠ 0, x ≠ y **27** $\frac{2x}{10 + x}$; x ≠ 0, x ≠ ±10 **28 a** $f = \frac{d_od_i}{d_o + d_i}$ **b** $d_i = \frac{fd_o}{d_o - f}$; $d_o \ne f$
29 x ≤ −3, 0 < x ≤ 3, or x > 6

31 $m = \frac{-1}{(x + 2)(x - 3)}$ **32** $\frac{(x + 3)^2}{(x + 2)(x + 4)}$; x ≠ 1, x ≠ −2, x ≠ −3, x ≠ −4 **33** x = −1 **34** $\frac{11x + 6}{3x + 2}$; x ≠ 0, x ≠ $\frac{-2}{3}$ **35** $\frac{3}{7}$ **36** $-\frac{1}{3}$; x ≠ −3, x ≠ $-\frac{3}{2}$ **37** $\frac{2x + 6}{3x}$;
x ≠ 0, x ≠ −2 **39** $\frac{4x}{x + 1}$; x ≠ ±1, x ≠ 0 **41** $\frac{2}{3}$
42 a $\frac{11}{30}$; $\frac{11}{15}$ **b** $2\frac{8}{11}$ hr **43** sc; c ≠ 0 **44 a** $\frac{x}{2x + 1}$;
x ≠ −1, x ≠ $-\frac{1}{2}$ **b** $\begin{bmatrix} 1 & 0 \\ 2 & 1 \end{bmatrix}$ **c** $\frac{x}{3x + 1}$; x ≠ $-\frac{1}{2}$, x ≠ $-\frac{1}{3}$,
x ≠ −1 **d** $\begin{bmatrix} 1 & 0 \\ 3 & 1 \end{bmatrix}$ **e** $\frac{x}{4x + 1}$; x ≠ $-\frac{1}{3}$, x ≠ $-\frac{1}{4}$, x ≠ $-\frac{1}{2}$
x ≠ −1 **f** $\begin{bmatrix} 1 & 0 \\ 4 & 1 \end{bmatrix}$ **45** $\frac{-10}{x(x + h)}$; 0 < x < h, h > 0
46 a A = 1; B = −2; C = 5 **b** A = 5; B = 1 **c** A = 2; B = 1 **47** $\frac{-4}{x(x + h)}$; x ≠ 0, h ≠ 0, x ≠ −h **48** 20 **49** 0
51

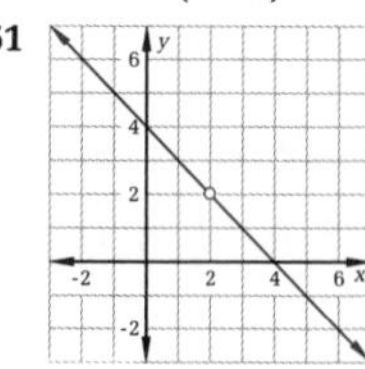

52 a f(x) approaches $\frac{1}{2}$. **b** f(x) approaches $\frac{5}{2}$. **53 a** n = 3; n must be positive. **b** 3

Review Problems

1 b, c, d, e, f **3** $5x - 7; x \neq -\frac{7}{5}$ **5** $\frac{1}{x-3}; x \neq -3, x \neq 3$ **7** $\frac{13}{19}$ **9** $\frac{13}{19}$ **10 a** $\frac{x+2}{x}$ **b** 2 **11** Greatest common factor: $3(x + 4); x - 7; x + 3$ Least common multiple: $6(x + 4)(x - 2); 3(x + 7)(x - 7); x(x + 3)^2$ **13** $y = 0$ or $y = 2$ **15** $y = 0$ **17** No solution **18** $(x + 2)^4; x \neq -3, x \neq 1, x \neq -1, x \neq 2$ **19** 4 and -5 **21** $\frac{x^2+3}{x}; x \neq 0$ **23** $-\frac{4}{3}$ **25** $1; x \neq \frac{2}{3}$ **27** $\frac{x+1}{x-1}; x \neq 0, x \neq 1, x \neq -2$

29

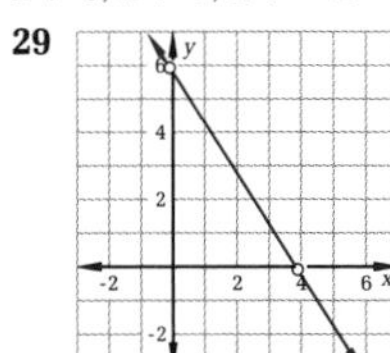

30 $h = x - 2$ **31** $\frac{2x}{(x+y)(x-y)}; x \neq y, x \neq -y$ **33** $\frac{x^2 - 2xy - y^2}{x^2 - y^2}; x \neq \pm y$ **34** $x = 5$ **35** $(x, y, z) = (6, -4, 3)$ **36** $4, \frac{3 \pm i\sqrt{7}}{2}$ **37** $\frac{x-2}{x-1}; x \neq \pm 1, x \neq -4$ **38** $\frac{-x}{x+3}; x \neq \frac{2}{3}, x \neq -3$ **39** $-1, \sqrt{2}$, or $-\sqrt{2}$ **40** $-3 < x < 3, 3 < x \leq 4$, **41** $x = -5$ or $x = 3$ **43** $x = 4$ or $x = 7$ **45** $\frac{1}{y}$ **47** $\frac{x+7}{18x+5}; x \neq \frac{1}{3}, x \neq -\frac{5}{18}$ **48** $\frac{1}{s^2c^2}; s \neq 0, c \neq 0$ **49** $b = 4; c = -6$ **50** $\frac{x+6}{x-6}; x \neq 3, x \neq 6$ **51 a** \$13,000 at 10%; \$7,000 at 9% **b** Answers will vary. Such considerations as maturity time, bond security, and bond rating could guide Mr. Vestor's decision. **52 a** 33 or more **b** No

Linear Programming

Problem Set

1

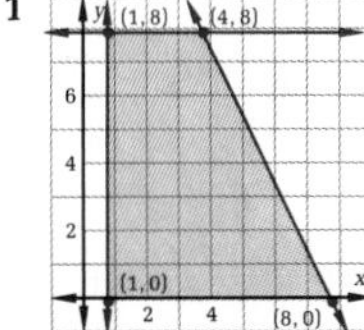

2 a (0, 105); (15, 105); (105, 15); (30, 15); (0, 75) **b** 345 **3** (8, −10) **4** $\frac{x}{12} + \frac{y}{10} \leq 55$, where x = the number of widgets produced per week and y = the number of gimcracks produced per week. **5 a** 60 **b** −20 **6** Maximum: 250; minimum: −625 **7 a** $\begin{cases} x \geq 0 \\ y \geq 0 \\ 12x + 6y \leq 120 \\ 9x + 18y \leq 144 \end{cases}$ **b** $130x + 90y$ **c** \$1400 **d** 8 bracelets, 4 necklaces

8

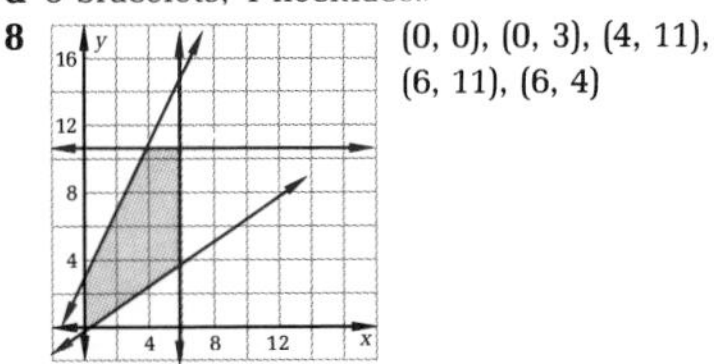

(0, 0), (0, 3), (4, 11), (6, 11), (6, 4)

9 60 tape players, 140 CD players **10** 153; 52.5 **11** \$2700 **12** \$41

11.1 Definitions and Graphs of the Exponential and Logarithmic Functions

Warm-up Exercises

1 Exponent **2** $\log_x z = y$ **3** $m^w = n$ **4 a** $w = 216$ **b** $w = 125$ **c** $w = \frac{1}{8}$ **5** $\frac{2}{5}$

6 a

x	0	1	2	3	4	−1	−2	−3	−4
10^x	1	10	100	1000	10,000	0.1	0.01	0.001	0.0001

b

x	1	10	100	1000	10,000	0.1	0.01	0.001	0.0001
$\log_{10} x$	0	1	2	3	4	−1	−2	−3	−4

Problem Set

7 a 16 **b** 4 **8 a** 1 **b** 0 **9 a** 8 **b** 1000 **c** 81 **10 a** 8 **b** 3 **c** −4 **11 a** 3 **b** $125^5 = 30{,}517{,}578{,}125$ **c** 5 **12 a** 5 **b** 2 **c** 16,777,216 **d** No real solution

13 a

x	0	1	2	3	−1	−2	−3	5	−4
2^x	1	2	4	8	$\frac{1}{2}$	$\frac{1}{4}$	$\frac{1}{8}$	32	$\frac{1}{16}$

b

x	1	2	4	8	$\frac{1}{2}$	$\frac{1}{4}$	$\frac{1}{8}$	32	$\frac{1}{16}$
$\log_2 x$	0	1	2	3	−1	−2	−3	5	−4

14 a 9^6 **b** 27^4 **c** 81^3 **15 a** $\frac{1}{3}$ **b** −3 **c** $-\frac{1}{3}$ **16** 2^{10n-1}

17 a

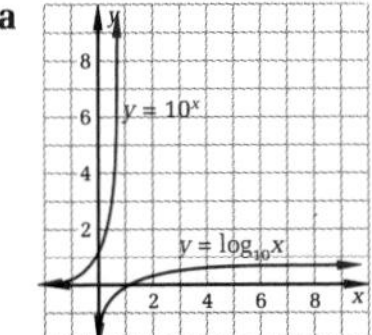

b $y = x$ **19** −1 **21** No real value **23** $\log_3\left(\frac{1}{81}\right) = -4$ **25** $\log_2 363 = n$ **26** 10 **27** $\frac{14 + 5\sqrt{3}}{11}$ **28** −4 is not a solution, since a base cannot be negative. **29 a** 6 **b** 6 **c** 2 **d** 2 **30 a** iv **b** i **c** iii **d** ii **31 a** 3 **b** 3 **c** $\sqrt{6} \approx 2.4495$ **32** $\log_5 78{,}125 = 7$ **33** $x = \frac{-3 + 2\sqrt{3}}{3}$

34 a **b**

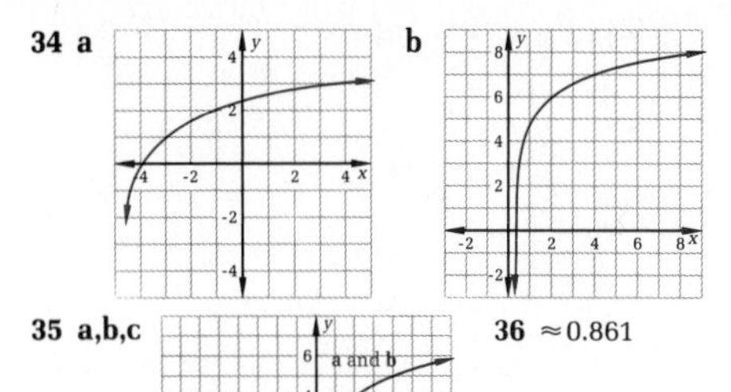

35 a,b,c **36** ≈ 0.861

37 a Domain: positive reals; Range: reals **b** Domain: reals; Range: positive reals **c** Domain: nonzero reals; Range: reals **38**

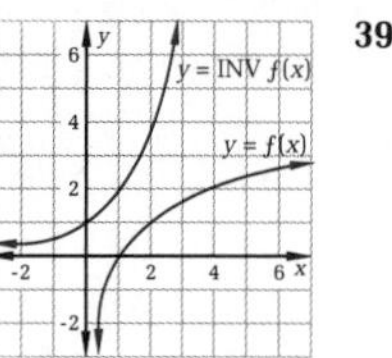

39 c **40** $\frac{3}{2}$

41 **42 a**

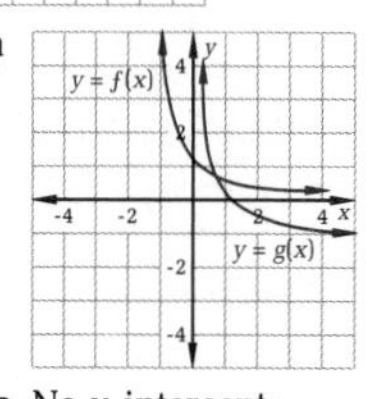

b Because $g(x) = \text{INV}f(x)$ **43 a** No x-intercept; y-intercept: (0, 1); horizontal asymptote: $y = 0$ **b** No y-intercept; x-intercept: (1, 0); vertical asymptote: $x = 0$

44 a **b**

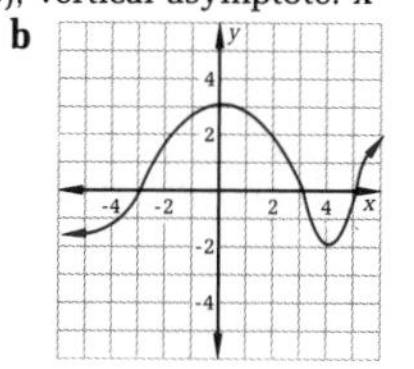

c

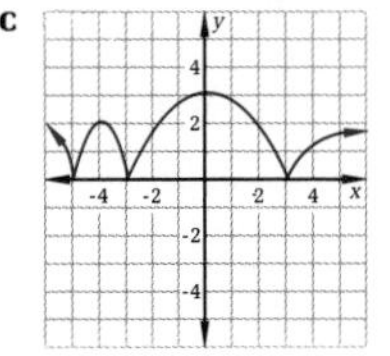

45 $x \approx 40.32$

11.2 Two Important Exponential and Logarithmic Functions

Warm-up Exercises

1 c, a, d, b **2 a** 3 **b** 3 **c** 3 **3 a** 0 **b** 0 **c** 0 **d** 0 **4** 4 **5** -4 **6** $\frac{1}{2}$ **7** 5 **8** ≈ -3.8041 **9** ≈ 2.5478 **10** ≈ -6.5023 **11** ≈ 3.9778 **12** 60 db

Problem Set

13 a ≈ 1.4314 **b** ≈ 0.4771 **c** ≈ -0.4771 **d** ≈ -1.4314 **e** ≈ 0.9542 **f** 0 **g** ≈ -0.9542 **14 a** $x \approx 3.9120$ **b** $x \approx 1.6990$ **c** $x \approx 403.4288$ **d** $x = 1{,}000{,}000$ **15 a** No solution **b** $y \approx 0.0024788$ **16 a** 5:1 **b** 25:1 **c** $\approx$1.8:1 **17** $x = 1$ **18** $b = \frac{1}{4}$ **19** 73 **20 a** ≈ 1.90849 **b** ≈ 1.90849 **c** ≈ 1.90849 **d** ≈ 1.90849 **21** ≈ 8.758 **22** $\approx$63.9 yr **23** $\frac{2}{5}$ **24** 120 db **25 a** 2 **b** 3 **c** 4 **d** 5 **26 a** ≈ 1.3802 **b** ≈ 1.3802 **c** ≈ 1.3802 **d** ≈ 1.3802 **27 a** 4 **b** -4 **c** $\frac{1}{4}$ **28 a** 6 **b** e **c** 0 **29 a** 1 **b** ≈ 1.3010 **c** ≈ 1.6021 **d** ≈ 1.9031 **30 a** ≈ 10.874 **b** No real value **c** ≈ 7.051 **d** ≈ -16.811 **31** \$16,974.07 **32** \$17,114.23; \$140.16 more **33 a** $x \approx 1.8718$ **b** $x \approx 1.6990$ **c** $x = 3$ **d** $x \approx 3{,}467{,}368{,}505$ **e** $x \approx 14.9821$ **f** $x = 3$ **34 a** $x \approx 1.1297$ **b** $x = 566\frac{2}{3}$ **35 a** \$16,103.24 **b** \$16,141.43 **c** \$16,160.11 **d** \$16,160.74 **36** $x = 10^6$ or $x = 10^{-1}$ **37 a** Office: 35 db; street: 70 db **b** $\approx$3162.6 times **38** -3 **39 a** $y \approx 1764.006$ **b** $y \approx 0.0001056$ **c** $y \approx 37.7839$ **d** $y \approx 0.001778$ **40** $x = 387{,}420{,}489$ or $x \approx 0.00000000258117$ **41** $x \approx 6.9078$ **42** $w = 1$ **43** $x \approx 1.386$ **45** $x \approx 3.255$ **47** \$12,786.25

11.3 Properties of Logarithms

Warm-up Exercises

1 ≈ 0.7781 **2** ≈ 1.2552 **3** ≈ 1.2219 **4** ≈ 1.8060 **5** ≈ 0.07525 **6** ≈ 0.4771 **7 a** iii **b** i **c** v **d** ii **e** iv

Problem Set

8 a 5 **b** 2 **c** 5 **9** (3, 2) **10 a** $\log_2\left(\frac{21}{2}\right)$ **b** $\log_4\left(\frac{125}{2\sqrt{3}}\right)$ **c** $\log_4\left(\frac{81}{4}\right)$ **d** $\log_2\left(\frac{243}{4}\right)$ **11** d, b, a = c **12** 1 **13** $3\log_2 x + \frac{1}{2}\log_2(x-3) - \log_2 y$ **15** $\ln\left(\sqrt{\frac{x^3}{z}}\right)$ **16** $t \approx 39.9$ yr **17** 5 **19** 2 **21** 15 **23** 1 **24** 1 **25** $\begin{bmatrix}\log 108\\ \log 200\end{bmatrix}$ **26 a** -2 **b** $-x^{43}$ **c** x^{-8} **27 a** 6 **b** 3 **c** 2 **d** Yes **28** 16 **29 a** x **b** x **c** No. $g[f(x)]$ has as its domain positive reals. $f[g(x)]$ has all reals as its domain. **30 a** 10,000 **b** $\approx$16,487 people **c** At $t \approx 37.6$ **31** $y = \frac{\ln b}{\ln a}$ **32 a** $2c + d$ **b** $-3c$ **c** $4d$ **d** $2c - 3d$ **33** $\frac{1}{2}\ln(x-1)$ **34 a** $a + b$ **b** $a + 1$ **c** $b - 1$ **35 a** ≈ 5.301 **b** ≈ 38.301 **c** ≈ 7.301 **d** ≈ 472.301 **36** ≈ 132.44 **37** $\ln\left(\frac{x+1}{3(x-1)}\right)$ **38** $x \approx -1.456$ **39** $\approx$ 92.4 yr; $\approx$92.4 yr **40 a, b, c** **41** -1 **42** 8

43 a False **b** False **c** True **44** ≈ 73.3

11.4 Solving Exponential and Logarithmic Equations

Warm-up Exercises

1 $x = \frac{3}{2}$ **2** $x = 4$ **3** $x = 7$ **4** $x = 1$ or $x = e$ **5** $x = 2$ **6** $\log_2 13 = \frac{\log_5 13}{\log_5 2}$ **7 a** 500 **b** ≈ 1660 **8** 20.25

Problem Set

9 a ≈ 4.8074 **b** ≈ -5.1699 **c** ≈ 3.0244 **10 a** $b \approx 2.6801$ **b** $b \approx -0.2625$ **c** $b \approx 2.5119$ **11** $x = 15$ **13** 6 **14** $\text{Log}_{a^n} x = \frac{\log_a x}{\log_a a^n} = \frac{\log_a x}{n}$ **15** $\frac{3}{4}$ **16 a** $x \approx 0.1099$ **b** $x \approx 59.1690$ **17** $x = 5$ **18 a** 10 **b** 7 **c** 2 **19 a** $x = 4.5$ **b** $x = \frac{a}{2}, a > 0$ **20** 1 **21 a** 2 **b** 3 **22** $\log_4 10 \approx 1.661$, $\log_5 15 \approx 1.682$, $\log_6 21 \approx 1.699$, $\log_{10} 55 \approx 1.740$ **23** $(\approx 1.54, 12)$ **24 a** $B_o = \$1500$ **b** $B \approx \$4810.70$ **25 a** 10^{1000} **b** 1001 **26** $x = 2$ or $x = -2$ **27** $x = 6$ or $x = 3$ **28 a** $2\ln x + \ln y - \ln(x + 5)$ **b** $\ln|x| - \frac{1}{2}\ln y$ **29** $x = 2$ **30 a** ≈ 1.387 **b** ≈ 0.231 **c** $\approx \pm 0.977g$ **31 a** 5 **b** ≈ -0.3466 **c** $\approx 0.2210g$ **32** $x = 3$ **33 a** 9 **b** 9 **c** 27 **d** $\frac{1}{27}$ **e** 81 **f** Impossible. f is defined for integers only. **34** $x = 64$ **35 a** $x = 256$ or $x = 4^{-4} \approx 0.003906$ **b** $x = 4^{16} = 4{,}294{,}967{,}296$ or $x = 0.25$ **36** $b = 2$ **37** $x = 10^4$ or $x = 1$ **38 a** $(0, e^3)$ **b** $(1, e^4)$

11.5 Applications of Exponential and Logarithmic Functions

Warm-up Exercises

1 7.1 **2 a** $k \approx 0.0000423$ **b** ≈ 5.8 lb/in.^{2} **3 a** $k \approx 0.0001238$ **b** ≈ 9727 yr old **4 a** Yes **b** $\approx 76°$

Problem Set

5 ≈ 4.7 **6** About 4.3 lb/in.^{2} **7** ≈ 3752 **8 a** $S \approx 100{,}000(10^{-0.1w})$ **b** $\approx 15{,}849$ **9** $3\sqrt{2} + 3$ **10** $k \approx 0.3143$ **11 a** Yes **b** \$266.77 **12 a** $x = \pm\sqrt{2}$ **b** $x = 0$ **c** $x = \pm 2$ **d** No real solutions **13** About 2.77 yr **14** $x^2 + 6x + 4$ **15** $x^2 + 2xy + y^2 - 9$ **16** 12 **17 a** $I_o = 11{,}000$; $k \approx 0.09$ **b** $\approx 756{,}000$ items **18** About 6 min 47 sec **19 a** ≈ 11.61 watts **b** ≈ 173.3 days **c** ≈ 402.4 days **20 a** 4.95 **b** 4.95 **c** 4.95

21 a

stop	1	2	3	4	5
n	1.4	2	2.8	4	5.6
$\log n$	≈ 0.146	≈ 0.301	≈ 0.447	≈ 0.602	≈ 0.748

6	7	8	9
8	11	16	22
≈ 0.903	≈ 1.041	≈ 1.204	≈ 1.342

b Yes **22** ≈ 0.9094 **23** $x = -1$ or $x = -\log 2 \approx -0.3010$ **24** About 2,864,129 people **25** $(\approx 0.5183, \approx 0.0218)$

11.6 Binomials and Binomial Distributions

Problem Set

1 a $0.2401W^4 + 0.4116W^3L^2 + 0.2646W^2L^2 + 0.0756WL^3 + 0.0081L^4$ **b** .2646 **2** $\approx .058$ **3** 82 **4** $\frac{135}{4096}$ **5** $\frac{20}{39}$, or $\approx .51$ **6 a** $\frac{80}{243}$ **b** $\frac{112}{243}$ **7 a** .59049 **b** .32825 **c** .00856 **d** .99144 **8 a** $\approx .0112$ **b** $\approx .0705$ **9 a** 162 **b** ≈ 8.05 **10 a** $\approx 8.2718 \times 10^{-25}$ **b** 40 **c** ≈ 4.47 **d** $\approx 86.43\%$

Review Problems

1 $3^4 = 81$ **3** $a^t = a^t$ **5** $x = -3$ **7 a** 8 **b** 6:1 **c** 36:1 **d** 2:1 **8 a** ≈ 1.135 **b** ≈ 1.135 **9** ≈ 5.792 **10 a** False **b** True **c** False **d** True **e** False **11 a** $x = -6$ or $x = 1$ **b** $x = 0$ **c** $x = e$ or $x \approx 0.0025$ **12** About 162.3 g **13** 4 **14** $\frac{112}{6561}$ **15 a** ≈ 2.698970 **b** ≈ 0.698970 **c** ≈ -1.301030 **d** ≈ 1.698970 **e** ≈ -0.301030 **16 a** ≈ 0.778 **b** ≈ 0.778 **c** ≈ 0.778 **d** ≈ 0.778 **e** ≈ 0.778 **17** \$19,863.49 **18** 20 mo

19 a

Number of heads	0	1	2	3	4	5
Probability	$\frac{32}{3125}$	$\frac{48}{625}$	$\frac{144}{625}$	$\frac{216}{625}$	$\frac{162}{625}$	$\frac{243}{3125}$

b

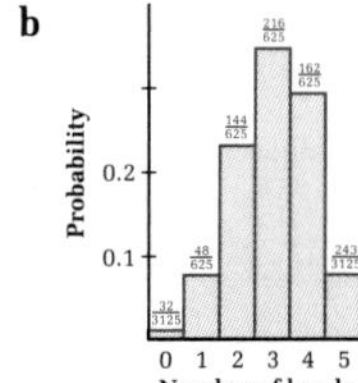

20 116 **21** $\approx 88.2°F$

22 a

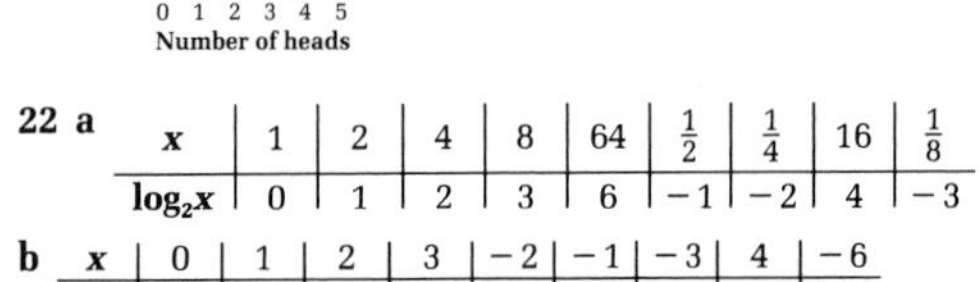

x	1	2	4	8	64	$\frac{1}{2}$	$\frac{1}{4}$	16	$\frac{1}{8}$
$\log_2 x$	0	1	2	3	6	−1	−2	4	−3

b

x	0	1	2	3	−2	−1	−3	4	−6
2^x	1	2	4	8	$\frac{1}{4}$	$\frac{1}{2}$	$\frac{1}{8}$	16	$\frac{1}{64}$

23 $x = 2$ **24** Between 4.137 lb/in.^{2} and 3.349 lb/in.^{2} **25 a** 12 **b** $\approx .005$ **26 a**

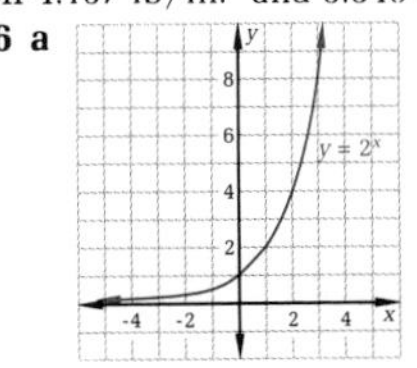

b

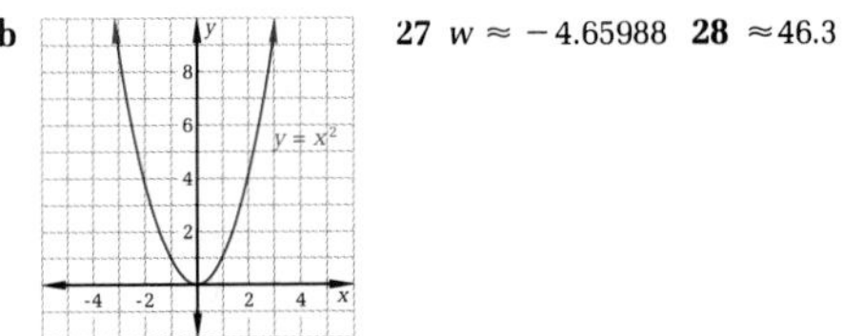

27 $w \approx -4.65988$ **28** ≈ 46.3

29 $x = 16$ or $x = \frac{1}{16}$ **30** ≈ 53.2 yr **31** $\frac{35}{128}$

32 About 22 wk **33 a** ≈ 3.57794 **34 a** $\ln\left(\frac{\pi x^3}{(x-2)^2}\right)$ **b** $\ln\left(\frac{\sqrt{xy}}{x-5}\right)$ **35 a** 2 **b** ≈ 3.419 **c** No solution **36** ≈ 126

12.1 Systematic Counting

Warm-up Exercises

1 Combinations **2** Permutations **3** Neither **4** 90 **5** 1365 **6** Answers will vary; yes **7** 180 **8** 5040

Problem Set

9 a 792 **b** 3,991,680 **10** 8 **11** 120 **12** 5040 **13** 1 **14 a** 120 **b** 120 **c** 105 **d** 105 **e** 3060 **f** 3060 **15** 362,880 **16** 6 **17 a** 5 **b** 2 **c** 1 **d** 3 **18** 3,628,800 **19** Combinations **20** Permutations

21 a,b **22** $\frac{1}{4}$

23 2738 **24** 1024 **25 a** 10,000,000 **b** No **26** 16 **27** 46 **28** 5148 **29 a** 5 **b** 7 **c** 0 **d** 12 **e** Yes, because $A \cap B$ is empty **30** 6 **31** 1,693,440 **32** The union of *A* and *B* includes one element for every element in *A* and in *B*, counting the elements in the intersection only once. If we then add the number of elements in the intersection, we include the elements that are counted twice. **33** 51 **34** Each of the *n* elements are either in the subset or not. Therefore, there are two choices for each element, or $2 \cdot 2 \cdot 2 \cdot 2 \cdot \ldots \cdot 2 = 2^n$ possible subsets.

12.2 Sample Spaces, Events, and Probability

Warm-up Exercises

1 *A* is subset of the sample space S, so $0 \le \#(A) \le (S)$. Therefore, $\frac{0}{\#(S)} \le \frac{\#(A)}{\#(S)} \le \frac{\#(S)}{\#(S)}$, and hence $0 \le P(A) \le 1$.

2 $P(\phi) = \frac{\#(A)}{\#(S)}$. Since $\#(A) = 0$, $\frac{\#(A)}{\#(S)} = \frac{0}{\#(S)} = 0$. Therefore, $P(\phi) = 0$. **3** The probability of the union of *A* and *B* includes the probability of the elements of *A* and *B*, counting the probability of the elements in the intersection only once. If we then add the probability of the elements in the intersection, we include the probability of the elements that are counted twice.

4 $\frac{33}{66{,}640}$, or $\approx .0005$ **5** $\frac{1}{9}$

6 Answers will vary. A tree diagram and a lattice are shown.

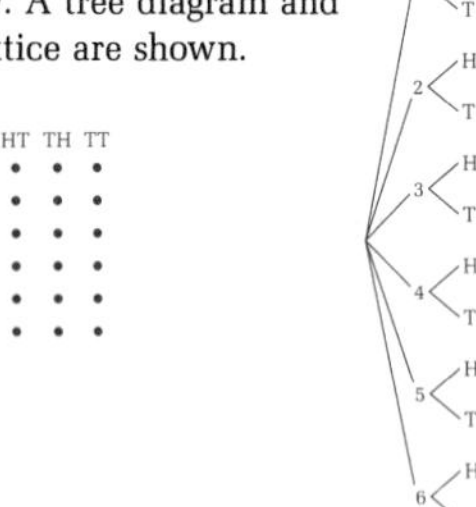

7 a

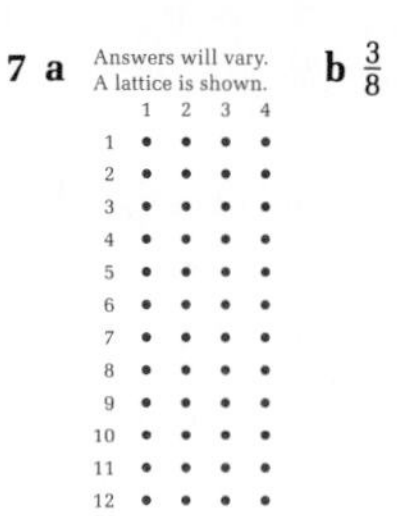

b $\frac{3}{8}$

Problem Set

8 a 36 **b** 11 **9 a** $\frac{3}{7}$ **b** $\frac{6}{35}$ **10** 19 **11 a** 8 **b** 15 **c** 32 **d** 13 **e** $C(n, 1) = n$ **12** $\approx .4$ **13 a** $\frac{2}{\pi}$ **b** 1 **14 a** $\approx .8396$ **b** $\approx .1541$ **c** $\approx .0063$ **d** 1; one of these outcomes must occur. **15 a** {1, 2, 3, 4, 5, 7, 9, 16, 25} **b** {1} **16 a** $\approx .196$ **b** $\approx .035$ **17** $\frac{1}{2}$ **18 a** $\approx .17$ **b** $\approx .15$ **19** $\approx .007$ **20 a** 25 **b** 15 **c** 14 **d** 26 **e** Yes **21** $\approx .17$ **22** $\frac{1}{7}$ **23** $\approx .003$ **24** 62 **25** $\approx .007$ **26 a** $\{(1, \sqrt{2}), (1, \sqrt{3}), (1, 2), (1, \sqrt{5}), (1, \sqrt{6}) (\sqrt{2}, \sqrt{3}), (\sqrt{2}, 2), (\sqrt{2}, \sqrt{5}), (\sqrt{2}, \sqrt{6}), (\sqrt{3}, 2), (\sqrt{3}, \sqrt{5}), (\sqrt{3}, \sqrt{6}), (2, \sqrt{5}) (2, \sqrt{6}), (\sqrt{5}, \sqrt{6})\}$ **b** $\frac{3}{5}$ **27** $x = \frac{5}{6}$ **28** .85 **29** $\frac{1}{10}$ **30** 81 **31 a** 3 **b** 32 **32** $\frac{1}{21}$ **33** $\approx .21$

12.3 Probabilities of Unions

Warm-up Exercises

1 a No **b** Yes **2 a** Yes **b** No **c** Yes **3** $\frac{17}{100}$ **4** $\frac{6}{19}$

Problem Set

5 $\frac{3}{4}$ **6** $\frac{15}{51}$, or $\approx .294$ **7 a** $\frac{1}{7}$ **b** $\frac{1}{21}$ **c** $\frac{1}{7}$ **d** $\frac{1}{21}$ **8** $\frac{6-\sqrt{3}}{6}$ **9** 1 **10** $\approx .00058$ **11 a** {(1, 1), (1, 2), (1, 3), (1, 4), (1,5), (1, 6), (2, 1), (2, 2), (2, 3), (2, 4), (2, 5), (2, 6), (3, 1), (3, 2), (3, 3), (3, 4), (3, 5), (3, 6), (4, 1), (4, 2), (4, 3), (4, 4), (4, 5), (4, 6), (5, 1), (5, 2), (5, 3), (5, 4), (5, 5), (5, 6), (6, 1), (6, 2), (6, 3), (6, 4), (6, 5), (6, 6)} **b** {(2, 2), (2, 4), (2, 6), (4, 2), (4, 4), (4, 6), (6, 2), (6, 4), (6, 6)} **c** $\frac{5}{9}$ **12 a** $\frac{11}{36}$ **b** $\frac{1}{36}$ **c** $\frac{1}{9}$ **d** 0 **13** No, unless A and B are mutually exclusive **14** $\approx .111$ **15** 21 **16 a** $\frac{1}{6}$ **b** $\frac{1}{2}$ **c** $\frac{1}{2}$ **d** $\frac{1}{6}$ **17** $\approx .358$ **18 a** 2 **b** 6 **c** 18 **d** 54 **19 a** $-1 + i$ **b** -1 **20** $\approx .1053$ **21 a** $\frac{5}{8}$ **b** $\frac{3}{8}$ **22** 10 **23** 4 to 7 **24** $\approx .6$ **25** .125 **26 a** $\frac{3}{7}$ **b** \$4.50 **27** Possible answer: 303,918 **28 a** $\frac{1}{13}$ **b** $\frac{1}{13}$ **c** $\frac{2}{13}$ **d** 0 **29 a** A B **b** A B

30 $\approx .1545$ **31 a** 33; 39 **b** 159; 319 **c** 55; 89 **32** $\frac{2}{15}$ **33 a** 0 **b** 1 **c** 0 **34 a** $\approx .45$ **b** $\approx .49$ **35** $\frac{1}{2}$ **36 a** 33 **b** 19 **c** 46 **37** 56

12.4 Probability of Intersections

Warm-up Exercises

1 No **2** Yes **3** No **4** Yes **5** .21 **6** $\frac{1}{52}$, or ≈.02 **7** $\frac{1}{4056}$

Problem Set

8 a Independent **b** Mutually exclusive **c** Dependent **d** Mutually exclusive **9** ≈.0005 **10** $\frac{1}{2}$ **11 a** $\frac{1}{16}$ **b** $\frac{5}{32}$ **c** $\frac{27}{32}$ **d** $\frac{37}{96}$ **12 a** ≈.51 **b** ≈.009 **13** $\frac{1}{18}$ **14 a** 0 **b** 0 **c** $\frac{1}{6}$ **15** Cannot be determined **16 a** Independent **b** Dependent **c** Dependent **17 a** Dependent **b** Dependent **18** ≈.0088 **19 a** i or -1 **b** $i-1$ or 0 **20 a** $\frac{1}{2}$ **b** $\frac{2}{3}$ **c** $\frac{3}{4}$ **d** $\frac{4}{5}$ **e** $\frac{5}{6}$ **f** $\frac{99}{100}$ **21 a** ≈.044 **b** ≈.042 **c** ≈.995 **22** Only if A and B are independent **23** 60% **24 a** 3; $3\sqrt{3}$; 60 **b** $\frac{1}{2}$; $\frac{\sqrt{3}}{2}$; 60 **25** .9775 **26** .5 **27** ≈.03 **28** ≈.17 **29** No; P(red) + P(green) + P (blue) must be equal to 1. **30** 1 **31** ≈.39

12.5 Interpreting Probabilities

Warm-up Exercises

1 Answers will vary. Possible answer: Conduct the poll using students from different schools. **2** Answers will vary. Possible answer: Write a computer program that randomly generates the numbers from 1 through 6, inclusive. The program should be able to keep a running total of the numbers. If the total exceeds 30, the program should acknowledge a failure and try again. **3** Answers will vary. Possible answer: The poll could have been conducted from a very small sample of doctors.

Problem Set

4 Answers will vary. Possible answer: At least approximately half of the preceding year's graduating class was doing well in college. **5 a** $\frac{1}{5}$ **b** $\frac{1}{5}$ **c** 0 **6 a** Answers will vary. **b** 2 heads and 2 tails **c** No; probability indicates only the relative likelihood of an outcome. **7** Answers will vary. Possible answer: The time span over which the survey was conducted may have been too short to include enough students. **8** 70 **9** Answers will vary. Possible answer: Gather inventory data from stores that distribute the product. **10** $\frac{3}{13}$ **11** Answers will vary. Possible answer: Distribute surveys to high school and college literature teachers who will distribute the surveys to their classes. **12 a** 0 **b** 3 **c** 15 **d** 24 **e** $k^2 - 1$ **f** $k^2 + 2k$ **13** $\frac{2}{9}$ **14** 3 **15 a** ≈.007 **b** ≈.098 **c** ≈.39 **16** $\frac{4}{7}$ **17** Possible answer: No, the margin of error was not significantly increased by retaking the test. **18** Possible answer: Based on similar previous conditions, it will probably rain over 30% of the listening area. **19** d **20** 5 **21** ≈.74 **22** ≈.18 **23** Answers will vary.

Review Problems

1 a 120 **b** 24 **c** 6 **d** 2 **e** 1 **f** 1 **2 a,b**

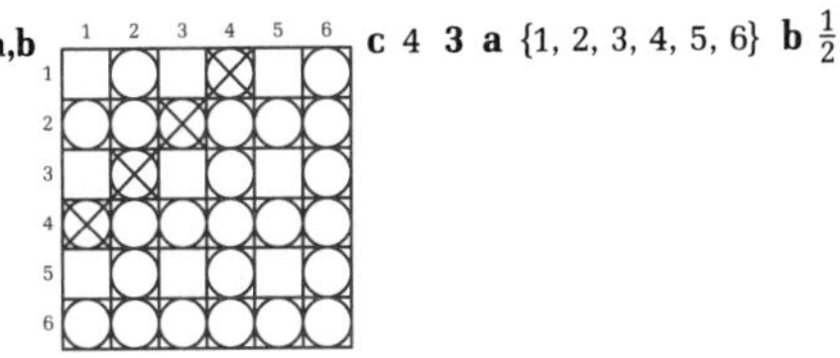

c 4 **3 a** {1, 2, 3, 4, 5, 6} **b** $\frac{1}{2}$ **c** $\frac{1}{3}$ **d** $\frac{2}{3}$ **e** $\frac{1}{6}$ **f** $\frac{5}{6}$ **g** $\frac{5}{6}$ **4 a**

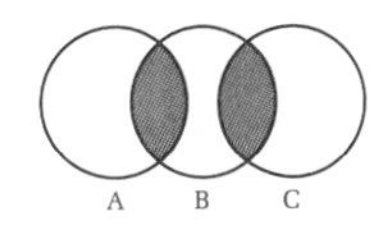

b

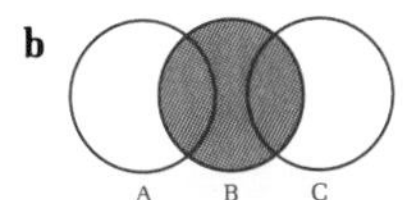

5 27,600 **6 a** $\frac{1}{36}$ **b** $\frac{1}{6}$ **c** $\frac{1}{6}$ **d** 0 **e** 0 **f** $\frac{1}{36}$ **7 a** $\frac{1}{5}$ **b** $\frac{8}{15}$ **c** $\frac{2}{3}$ **d** $\frac{1}{15}$ **e** Yes **8** 2,118,760 **9** $\frac{1}{2}$ **10** 1 **11** $\frac{7}{26}$ **12** .8 **13** Answers will vary. **14** $\frac{1}{4}$ **15 a** {3, 5, 6, 7, 9, 11, 12} **b** {3, 9} **c** 7 **d** 2 **e** 9 **16** 5,527,200 **17** 4 **18** $\frac{1}{2}$ **19 a** {($1, $1), ($1, $2), ($1, $3), ($2, $1), ($2, $2), ($2, $3), ($3, $1), ($3, $2), ($3, $3), ($4, $1), ($4, $2), ($4, $3)} **b** $2, $3, $4, $5, $6, $7 **20 a** $\frac{1}{4}$ **b** 1 **21** ≈.97 **22** 30 **23** $\frac{3}{55}$, or ≈.05 **24** ≈.009 **25** 128 **26** $\frac{23}{45}$, or ≈.51

13.1 Sequences

Warm-up Exercises

1 −5, −3, −1, 1, 3, 5 **2** 0, 3, 8, 15, 24, 35 **3** 2, 3, 5, 7, 11, 13 **4** 2, 6, 14, 30, 62 **5** −4 **6** $\frac{1}{2}, \frac{1}{8}, \frac{1}{32}, \frac{1}{128}, \frac{1}{512}$ **7** 24, 48, 96, 192, 384 **8** 10, 15, 21, 28, 36 **9** 8, 15, 24, 35, 48

Problem Set

10 15, 21, 28, 36 **11** $4n$ **13** 40, 80, 160, 320, 640 **14** ≈1.3333, ≈1.4762, ≈1.4975 **15** 55 **17** 85 **18 a** $24.99, $24.96, $24.91 **b** $74.86 **19** $\frac{1}{2}, \frac{2}{3}, \frac{3}{4}, \frac{4}{5}, \frac{5}{6}$ **21 a** 81 **b** 9th **22 a** 1, 6, 18, 40 **b** 1, 7, 25, 65 **23** 20 **24 a** $A(t) = 100 - t^2$ **b** 99, 96, 91, 84, 75, 64, 51, 36, 19 **25 a** $\sqrt{2}, \sqrt{5}, \sqrt{10}, \sqrt{17}$ **b** $\sqrt{26}, \sqrt{37}, \sqrt{50}$ **26** $j = 3$ **27** $a_1 = 1, a_{n+1} = 3$ **28** a_1 = 30kg; a_2 = 20kg; a_3 = 15kg; a_4 = 12kg; **29 a** $A_1 = 64$, $A_n = \frac{A_{n-1}}{4}$ **b** $A_n = \frac{64}{4^{n-1}}$ **c** $A_{25} = \frac{64}{4^{24}} \approx 2.27 \times 10^{-13}\text{cm}^2$ **30** $1060, $1123.60, $1191.02, $1262.48, $1338.23

31

n	1	2	3	4
A	$262{,}144\sqrt{3}$	$65{,}536\sqrt{3}$	$16{,}384\sqrt{3}$	$4096\sqrt{3}$

5	6	7	8
$1024\sqrt{3}$	$256\sqrt{3}$	$64\sqrt{3}$	$16\sqrt{3}$

32 a $a_1 = 11$, $a_{n+1} = 10a_n - 9$ **b** $\{a_n\} = \{10^n + 1\}$
33 1, 3, 3, 1, $\frac{1}{3}$ **35** $5n(-1)^{n+1}$ **36** Answers will vary. Possible answers: $\{a_n\} = \{2n + 2\}: a_1 = 4, a_{n+1} = a_n + 2$
37 a 0.3, 0.03, 0.003, 0.0003, 0.00003, 0.000003
b $s_1 = 0.3$; $s_2 = 0.33$; $s_3 = 0.333$; $s_4 = 0.3333$; $s_5 = 0.33333$; $s_6 = 0.333333$ **c** $\frac{1}{3}$ **38** Answers will vary. Possible answer: $a_1 = 1$, $a_{n+1} = a_n + 7$, 22 **39** $a_1 = 1$; $a_2 = 3$; $a_3 = 7$; $a_4 = 15$; $a_5 = 31$; $a_n = 2^n - 1$, $n = 1, 2, 3, \ldots$

13.2 Series

Warm-up Exercises

1 $\sum_{n=1}^{4} n!$; 33 **2 a** 6, 12, 18, 24, 30 **b** 90 **3** 2, 7, 15, 26, 40 **4** 60 **5** 100 **6** $\{a_n\} = \{2n + 1\}$: 3, 5, 7, 9, 11
7 $2n - 1$

Problem Set

9 108 **11** log 120, or ≈ 2.08 **13** 1, −1, 1, −1, 1
14 a 385 **b** 505 **c** 649 **d** Added terms.
$\sum_{n=1}^{11}(n + 1)^2 = \sum_{n=2}^{12} n^2 = 505 + 12^2 = 649$ **15** 34
16 Number of dots in the nth figure $= \frac{n(3n - 1)}{2}$
17 110 **19** 6.6666 **21** $x = -3$ **22 a** $\frac{\sqrt{3}}{4}, \sqrt{3}, \frac{9\sqrt{3}}{4}, 4\sqrt{3}$, **b** $\frac{15}{2}\sqrt{3}$ **23** $\approx 0.\overline{7}$ **24** 0, 2, 7, 26, 123 **25** 2.0
27 5, 3, −7 **28 a** $a_1 = 1, a_2 = \frac{1}{\sqrt{3}}, a_3 = \frac{1}{3}, a_4 = \frac{1}{3\sqrt{3}}$
b $\sum_{n=1}^{\infty} a_n = AB$ **29 a** $a_1 \approx 27.713$, $a_2 \approx 6.928$, $a_3 \approx 1.732$, $a_4 \approx 0.433$, $a_5 \approx 0.108$ **b** $S_5 \approx 36.914$
31 log 604,800, or ≈5.78 **33** No sum **35** No sum
37 0 **38** Recursive: $a_1 = 5$, $a_n = a_{n-1} + 5$, $n = 2, 3, \ldots$; Closed: $a_n = 5n$, $n = 1, 2, \ldots$ **39** $a_n = 4n - 1$
40 44 **41** 1.25 **43** −1 **44** $\frac{4}{3}$

13.3 Arithmetic Sequences and Series

Warm-up Exercises

1 10, 13, 16, 19, 22 **2** 15, 12, 9, 6, 3 **3** 156 **4** $\frac{46}{3}$ and $\frac{56}{3}$ **5 a** **6 a** 3 **b** $-4\sqrt{5}$ **7** $8 + 13\pi$ **8** 2

Problem Set

9 5, 10, 15, 20 **11** −6, −8, −10, −12 **13** 16.5, 16.75, 17.0, 17.25, 17.5 **15** 7, 11, 15 **17 a** 5 **b** 2 **c** 49
d 2,597 **18** 4, 11, 21, 34, 50 **19** 297 **21** $\frac{-23}{3}$ **22** 11
23 a 38, 64, 94, 128, 166 **b** $A_n = 2n^2 + 20n + 16$
24 3, 9, 18, 30 **25** 85 **26** 47.2 **27** 20, 24, 28
29 10,000 **31** 52; 700 **33** 19, 23, 27, 31, 35 **35** $\sum_{k=1}^{15} k^2$
37 $82,050 **39** first = $7\sqrt{2}$, difference = $-5\sqrt{2}$, sum = $-84\sqrt{2}$ **40** $3\pi, 5\pi, 7\pi, 9\pi, 11\pi, 13\pi$ **41** $d = \frac{1}{3}$ ≈24,163.7 **42 a** $10\pi, 14\pi, 18\pi, 22\pi$ **b** $\pi(4n + 6)$
44 $\frac{12}{20}$ or $\frac{1}{10}$ **45 a** 31 × 10 **b** 460 **c** 1210

13.4 Geometric Sequences and Series

Warm-up Exercises

1 $8\sqrt{2}$; $14 + 15\sqrt{2}$ **2** 1.5; 190.5 **3** $-3i$; −3 **4 a** 4 **b** No sum **5** 3906.25 **6** ≈±9.80 **7** 455 ft

Problem Set

9 5, 1.5, 0.45, 0.135, 0.0405, 0.01215 **11** $\frac{1}{2}, \frac{1}{4}, \frac{1}{8}$
13 Arith. series; first term: 8; comm. diff.: 5
15 Neither arith. nor geom.; first term: 1 **16 a** 1.024 in. **b** 0.5625 in.^{2} **17 a** $\sqrt{2}, 2\sqrt{2}, 3\sqrt{2}, 4\sqrt{2}, 5\sqrt{2}$ **b** 2, $\frac{3}{2}, \frac{4}{3}, \frac{5}{4}$ **19** $\frac{1}{3}$ **21** −4, 1, −0.25, or 4, 1, 0.25 **23** Yes, with comm. ratio 1 **24** 9, 12, 15 **25** $x = -15$
26 $\frac{\log 120}{4} \approx 0.52$ **27** 24.25 **28 a** ≈12.95 ft
b ≈180.79 ft **29** $\frac{10}{3}$ **31** $x = 12$ **32** 9, 16 **33** $6i, 18i, 54i$; or $-6i, 18i, -54i$; or $-6, -18i, 54$; or $6, -18i, -54$
34 $a_n = a_1 r^{n-1}$ If $a_n = 0$ and $a_1 \neq 0$, then $r^{n-1} = 0$. But this is true only if $r = 0$. Therefore, if $a_n = 0$ for one n value, then $a_n = 0$ for all $n > 1$. **35** 1,147.5 **36 a** If $a = b$, then $x = r$. **b** $r = \frac{a + b}{2}$ **c** $x = \sqrt{ab}$

Review Problems

1 3, 6, 11, 18, 27 **3** 2, 3, 5, 7, 11 **5** 3, 9, 20, 38, 65
7 2, 5, 10, 17, 28 **9 a** −6 **b** 8.25 **11** ≈ 1.55 **13** 330
14 $50 **15 a** ≈ 3.9 feet **b** ≈ 39.2 feet **16 a** 8, 11, 14, 17 **b** 32 **c** 180 **17** 1 − 2 + 3 − 4 + 5 − 6 **18** 22, 27, 32, 37, 42 **19** 25, −100, 400, −1600, 6400 **20** 4, −8, 16, −32 **21** 14 **22** 4 **23** $a_n = 4n - 1$ **24** 0.9
25

n	1	2	3	4	5
s_n	0.5	0.75	0.875	0.9375	0.96875

;

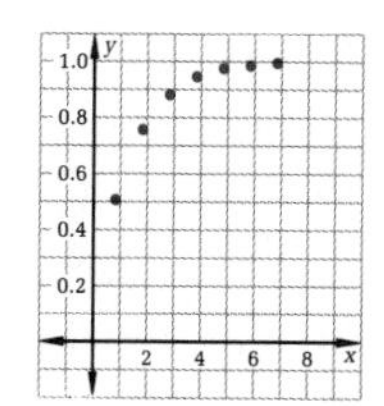

26 a $\frac{2}{15}$ **b** $a_k = \frac{4}{(k + 1)(k + 2)}$
c $a_k = \frac{4}{(k + 1)(k + 2)} = \frac{4}{k + 1} - \frac{4}{k + 2}$

$$\therefore \sum_{k=1}^{\infty} a_k = 4 \sum_{k=1}^{\infty} \frac{1}{k + 1} - \frac{1}{k + 2}$$
$$= 4\left(\frac{1}{2} - \frac{1}{3} + \frac{1}{3} - \frac{1}{4} + \frac{1}{4} - \frac{1}{5} + \ldots\right)$$
$$= 4\left(\frac{1}{2}\right)$$
$$= 2$$

27 a $\frac{1}{32}$ **b** ≈5.91 **c** $\frac{\sqrt{3}}{3}$, or ≈0.57785 **29** 12,726
31 $\frac{n}{n + 1}$ **33** $p = \frac{14}{15}$ **35** False **36** $\sum_{k=1}^{\infty} x^{k+1}$

37 $a_1 \approx 689.35; \frac{4}{7}$ **39** $a_1 = -531,441; -\frac{1}{3}$ **41** $a_n = \frac{x^n}{n}$
43 ≈ 7.97 **44** 3645 units **45** ≈ 4072 persons
46 a \$5,627,888.15 **b** Geometric sequence **47** $\frac{2s^2\sqrt{3}}{5}$ where s = length of side of original triangle **48** Since $r > 1$, in fact $r = \frac{4}{3}$, the perimeter is infinite.

Mathematical Induction

Problem Set

1 $(1 + h)^1 \overset{?}{\geq} 1 + 1h$
$1 + h \geq 1 + h$; True, since $1 + h = 1 + h$ for all h.
3 $1 \overset{?}{<} 2^1$
$1 < 2$
5 i $(1 + h)^{n+1} \geq 1 + h(n + 1)$ **iii** $(n + 1) < 2^{n+1}$
6 For n = 1, $4(1) - 3 = 1[2(1) - 1]$
$1 = 1$
For some n, $1 + 5 + 9 + \ldots + (4n - 3) = n(2n - 1)$
Add $4n + 1$ to each side.
$1 + 5 + 9 + \ldots + (4n - 3) + (4n + 1)$
$= n(2n - 1) + 4n + 1$
$1 + 5 + 9 + \ldots + [4(n + 1) - 3]$
$= 2n^2 - n + 4n + 1$
$= 2n^2 + 3n + 1$
$= (n + 1)(2n + 1)$
$= (n + 1)[2(n + 1) - 1]$

7 i) For n = 1, $1^3 - 1 = 0$; 0 *is divisible* by 3.
ii) Assume true for n:
$n^3 - n$ *is divisible* by 3.
$(n + 1)^3 - (n + 1) = n^3 + 3n^2 + 3n + 1 - n - 1$
$= (n^3 - n) + (3n^2 + 3n)$
$= (n^3 - n) + 3n(n + 1)$
According to the given statement,
$n^3 - n$ is divisible by 3.
$3n(n + 1)$ contains the factor 3, and is therefore divisible by 3.
According to the distributive property, since each term of the sum contains a factor of 3, the sum itself must contain a factor of 3, so $(n + 1)^3 - (n + 1)$ is divisible by 3.

8 i) $\frac{1(1 + 1)}{2} = 1$
ii) Assume $1 + 2 + 3 + \ldots + n = \frac{n(n + 1)}{2}$ true for n. Add n + 1 to each side.
$1 + 2 + 3 + \ldots + n + (n + 1) = \frac{n(n + 1)}{2} + n + 1$
$= \frac{n(n + 1) + 2n + 2}{2}$
$= \frac{n^2 + n + 2n + 2}{2}$
$= \frac{n^2 + 3n + 2}{2}$
$= \frac{(n + 1)(n + 2)}{2}$
$= \frac{(n + 1)[(n + 1) + 1]}{2}$

9 i) $x^1 + y^1 = (xy)^1$
ii) Assume $x^n + y^n = (xy)^n$ is true for n.
$x^{n+1} \cdot y^{n+1} = x^n \cdot x^1 \cdot y^n \cdot y^1$
$= x^n y^n \cdot x^1 y^1$
$= (xy)^n \cdot (xy)^1$
$= (xy)^{n+1}$

10 No; not true for $n \geq 6$.

11 i) $S_1 = \frac{a_1(r^1 - 1)}{r - 1} = a_1$
ii) Assume $S_n = \frac{a_1(r^n - 1)}{r - 1}$ is true for n.
Add a_1r^n to each side.
$S_n + a_1r^n = \frac{a_1(r^n - 1)}{r - 1} + a_1r^n$
$S_{n+1} = \frac{a_1(r^n - 1)}{r - 1} + \frac{a_1r^n(r - 1)}{r - 1}$
$= \frac{a_1r^n - a_1 + a_1r^{n+1} - a_1r^n}{r - 1}$
$= \frac{a_1(r^{n+1} - 1)}{r - 1}$

12 i) $\frac{3(3 - 3)}{2} = 0$. True; there are no diagonals for a triangle.
ii) Assume the number of diagonals $\frac{n(n - 3)}{2}$ is true for n. For 1 more side, there are $n + 1 - 2$ more diagonals.
$\frac{n(n - 3)}{2} + n - 1 = \frac{n(n - 3)}{2} + \frac{2n - 2}{2} =$
$\frac{n^2 - 3n + 2n - 2}{2} = \frac{n^2 - n - 2}{2} =$
$\frac{(n - 2)(n + 1)}{2} = \frac{(n + 1)[(n + 1) - 3]}{2}$

13 i) $a^1 - b^1 = a - b$; $a - b$ is a factor of $a - b$.
ii) Assume $a^n - b^n$ contains a factor of $a - b$ for some n
$a^{n+1} - b^{n+1} = a \cdot a^n - b \cdot b^n$
$= a \cdot a^n - ba^n + ba^n - b \cdot b^n$
$= a^n(a - b) + b(a^n - b^n)$
Since $a^n(a - b)$ contains a factor of $a - b$ and since $b(a^n - b^n)$ contains a factor of $a - b$ (by assumption), $a^{n+1} - b^{n+1}$ contains a factor of $a - b$.

14 a Let m be any positive integer. When n = 1, $a^m \cdot a^n = a^m \cdot a^1 = a^m \cdot a = a^{m+1}$.
Assuming that $a^m \cdot a^n = a^{m+n}$ for some n, we multiply both sides by a.
$a^m \cdot a^n \cdot a = a^{m+n} \cdot a$
$a^m \cdot a^{n+1} = a^{(m+n)+1}$
$= a^{m+(n+1)}$

b Let m be any positive integer. When $n = 1$,
$(a^m)^n = (a^m)^1 = a^m = a^{m \cdot 1}$.
Assuming that $(a^m)^n = a^{mn}$ for some n, we multiply both sides by a^m.

$$(a^m)^n \cdot a^m = a^{mn} \cdot a^m$$
$$(a^m)^n \cdot (a^m)^1 = a^{mn+m}$$
$$(a^m)^{n+1} = a^{m(n+1)}$$

c Let m be any positive integer. When $n = 1$,
$\frac{a^m}{a^n} = \frac{a^m}{a^1} = a^m \cdot a^{-1} = a^{m+(-1)} = a^{m-1}$.

Assuming that $\frac{a^m}{a^n} = a^{m-n}$ for some n, we multiply both sides by a^{-1}.

$$\frac{a^m}{a^n} \cdot a^{-1} = a^{m-n} \cdot a^{-1}$$
$$\frac{a^m}{a^n} \cdot \frac{1}{a} = a^{(m-n)+(-1)}$$
$$\frac{a^m}{a^n \cdot a} = a^{m-n-1}$$
$$\frac{a^m}{a^{n+1}} = a^{m-(n+1)}$$

15 True; i) $2^1 \geq 1 + 1$
ii) Assume $2^n \geq n + 1$ true for n.
Then $2 \cdot 2^n \geq 2(n + 1)$
$2^{n+1} \geq 2n + 2$
$\therefore 2^{n+1} \geq n + 2$, since n is a natural number.
$2^{n+1} \geq (n + 1) + 1$

17 $1^3 = 1^2(1 + 1)^2 - 4$
$= 4 - 4$
$= 0$
False; $1 \neq 0$

19 When $n = 1$, $1^2 \geq \frac{1(1 + 1)}{2}$, or $1 \geq 1$.

We assume that $n^2 \geq \frac{n(n + 1)}{2}$ for some n. Since n is positive by definition, $2n > n$, so we can add $2n + 1$ to the left side and $n + 1$ to the right side without invalidating the inequality.

$$n^2 + 2n + 1 \geq \frac{n(n + 1)}{2} + n + 1$$
$$(n + 1)^2 \geq \frac{n^2 + n + 2n + 2}{2}$$
$$\geq \frac{n^2 + 3n + 2}{2}$$
$$\geq \frac{(n + 1)(n + 2)}{2}$$
$$\geq \frac{(n + 1)[(n + 1) + 1]}{2}$$

20 When $n = 1$, $n^2 + n = 1^2 + 1 = 2$, which contains a factor of 2.
Assuming that $n^2 + n$ contains a factor of 2 for some n,
$(n + 1)^2 + (n + 1) = n^2 + 2n + 1 + n + 1$
$= n^2 + n + 2n + 2$
$= (n^2 + n) + 2(n + 1)$
Since $n^2 + n$ contains a factor of 2 by our assumption and since $2(n + 1)$ also contains a factor of 2, $(n + 1)^2 + (n + 1)$ must contain a factor of 2.

21 When $n = 2$, $\log (a_1 \cdot a_2) = \log a_1 + \log a_2$ by Property 5 of Logarithms.
Assuming that $\log (a_1 \cdot a_2 \cdot \ldots \cdot a_n) = \log a_1 + \log a_2 + \ldots + \log a_n$ for some n, we add $\log a_{n+1}$ to each side.

$$\log (a_1 \cdot a_2 \cdot \ldots \cdot a_n) + \log a_{n+1}$$
$$= \log a_1 + \log a_2 + \ldots + \log a_n + \log a_{n+1}$$
$$\log [(a_1 \cdot a_2 \cdot \ldots \cdot a_n) \cdot a_{n+1}]$$
$$= \log a_1 + \log a_2 + \ldots + \log a_{n+1}$$
$$\log (a_1 \cdot a_2 \cdot \ldots \cdot a_{n+1})$$
$$= \log a_1 + \log a_2 + \ldots + \log a_{n+1}$$

22 When $n = 1$, $\overline{z^1} = \overline{z}^1$, or $\overline{z} = \overline{z}$, which is a true statement.
Assuming that $\overline{z^n} = \overline{z}^n$ for some n, we multiply both sides by $\overline{z}$.

$$\overline{z^n} \cdot \overline{z} = \overline{z}^n \cdot \overline{z}$$
$$\overline{z^n \cdot z} = \overline{z}^{n+1}$$
$$\overline{z^{n+1}} = \overline{z}^{n+1}$$

23 When $n = 1$, $5^n - 1 = 5^1 - 1 = 4$, which is divisible by 4.
Assuming that $5^n - 1$ is divisible by 4 for some n,
$5^{n+1} - 1 = 5 \cdot 5^n - 1$
$= 5 \cdot 5^n - 5^n + 5^n - 1$
$= 5^n(5 - 1) + (5^n - 1)$
$= 4(5^n) + (5^n - 1)$
Since $4(5^n)$ is divisible by 4 and since $5^n - 1$ is also divisible by 4 (according to our assumption), $5^{n+1} - 1$ must be divisible by 4.

24 When $n = 1$, $180(n - 2) = 180 \cdot 1 = 180$, and we know from a theorem of geometry that the sum of the measures of a triangle's angles is 180.
Assuming that the sum of the interior angles of a polygon is $180(n - 2)$ for some n, we note that n is increased by 1, the sum of the angles is increased by 180. Therefore, the sum of the measures of the interior angles of a polygon with $n + 1$ sides is
$180(n - 2) + 180$
$= 180n - 360 + 180$
$= 180n - 180$
$= 180(n - 1)$
$= 180[(n + 1) - 2]$

25 When $n = 1$,

$$\sum_{k=1}^{n} \left(\frac{1}{(2k - 1)(2k + 1)}\right) = \frac{1}{(2 \cdot 1 - 1)(2 \cdot 1 + 1)} = \frac{1}{3} = \frac{1}{2(1) + 1}.$$

Assuming that $\sum_{k=1}^{n} \left(\frac{1}{(2k - 1)(2k + 1)}\right) = \frac{n}{2n + 1}$ for some n,

we add $\frac{1}{[2(n + 1) - 1][2(n + 1) + 1]}$ to each side.

$$\sum_{k=1}^{n} \left(\frac{1}{(2k - 1)(2k + 1)}\right) + \frac{1}{[2(n + 1) - 1][2(n + 1) + 1]}$$
$$= \frac{n}{2n + 1} + \frac{1}{[2(n + 1) - 1][2(n + 1) + 1]}$$
$$\sum_{k=1}^{n+1} \left(\frac{1}{(2k - 1)(2k + 1)}\right) = \frac{n}{2n + 1} + \frac{1}{(2n + 1)(2n + 3)}$$
$$= \frac{n(2n + 3) + 1}{(2n + 1)(2n + 3)}$$
$$= \frac{2n^2 + 3n + 1}{(2n + 1)(2n + 3)}$$
$$= \frac{(2n + 1)(n + 1)}{(2n + 1)(2n + 3)}$$

$= \frac{n+1}{2n+3}$

$= \frac{n+1}{2(n+1)+1}$

26 When $n = 1$, $2^1 - 1 = 1$, and it obviously only takes 1 move to transfer 1 disk.
Let $H(n)$ represent the number of moves needed to transfer an n-disk tower. In Section 5.6, problem 28, part b, it was established that $H(n) = 2[H(n-1)] + 1$. Assuming that $H(n) = 2^n - 1$ for some n,

$H(n+1) = 2(H[(n+1) - 1]) + 1$
$= 2[H(n)] + 1$
$= 2(2^n - 1) + 1$
$= 2 \cdot 2^n - 2 + 1$
$= 2^{n+1} - 1$

27 When $n = 1$, $(x + y)^n = (x + y)^1 = x + y = 1x + 1y = [C(1, 0)]x^1 + [C(1, 1)]y^1$.
Assuming that $(x + y)^n = [C(n, 0)]x^n + [C(n, 1)]x^{n-1}y + [C(n, 2)]x^{n-2}y^2 + \ldots + [C(n, n)]y^n$ for some value of n, we multiply both sides by $(x + y)$.

$(x + y)(x + y)^n = (x + y)([C(n, 0)]x^n + [C(n, 1)]x^{n-1}y + [C(n, 2)]x^{n-2}y^2 + \ldots + [C(n, n)]y^n)$
$(x + y)^{n+1}$
$= (x[C(n, 0)]x^n + x[C(n, 1)]x^{n-1}y + x[C(n, 2)]x^{n-2}y^2 + \ldots) + (y[C(n, 0)]x^n + y[C(n, 1)]x^{n-1}y + \ldots + y[C(n, n)]y^n$
$= [C(n, 0)]x^{n+1} + [C(n, 1)]x^n y + [C(n, 2)]x^{n-1}y^2 + \ldots + [C(n, 0)]x^n y + [C(n, 1)]x^{n-1}y^2 + \ldots + [C(n, n)]y^{n+1}$
$= 1x^{n+1} + (n + 1)x^n y + [C(n, 2) + C(n, 1)]x^{n-1}y^2 + \ldots + 1y^{n+1}$
$= [C(n + 1, 0)]x^{n+1} + [C(n + 1, 1)]x^{(n+1)-1}y + [C(n + 1, 2)]x^{(n+1)-2}y^2 + \ldots + [C(n + 1, n + 1)]y^{n+1}$

14.1 The Unit Circle

Warm-up Exercises

1 24 **2** $\left(-\frac{\sqrt{3}}{2}, -\frac{1}{2}\right)$ **3** $\sqrt{7}$ **4** $\left(-\frac{15}{17}, -\frac{8}{17}\right)$ **5** $\left(\frac{\sqrt{3}}{2}, \frac{1}{2}\right)$ **6** $\left(\frac{1}{\sqrt{2}}, \frac{1}{\sqrt{2}}\right)$ **7** $\left(\frac{1}{2}, \frac{\sqrt{3}}{2}\right)$ **8** $\left(\frac{\sqrt{3}}{2}, -\frac{1}{2}\right)$ **9** $\left(\frac{1}{\sqrt{2}}, -\frac{1}{\sqrt{2}}\right)$ **10** $\left(\frac{1}{2}, -\frac{\sqrt{3}}{2}\right)$

Problem Set

11 (0, 1) **13** (0, −1) **15** (1, 0) **17 a** $\frac{25}{36}\pi$, or ≈ 2.182 **b** $\frac{47}{36}\pi$, or ≈ 4.102 **18** $\left(-\frac{3}{5}, \frac{4}{5}\right)$ **19 a** −120° **b** $\frac{2\pi}{3}$, or ≈ 2.094 **20** $\beta = 435°$; $\gamma = -2.85°$ **21** $\left(\frac{8}{17}, \frac{-15}{17}\right)$ **22** $A = \left(-\frac{1}{2}, \frac{\sqrt{3}}{2}\right)$; $B = \left(-\frac{1}{2}, -\frac{\sqrt{3}}{2}\right)$; $C = \left(\frac{1}{2}, -\frac{\sqrt{3}}{2}\right)$ **23** $\left(\frac{12}{13}, \frac{5}{13}\right)$ **24 a** $\frac{\pi}{3}$, or ≈ 1.047 **b** $\frac{\pi}{3}$, or ≈ 1.047 **c** $\frac{\pi}{3}$, or ≈ 1.047 **25** −310°, −50°, 310° **26 a** $(-b, a)$ **b** $(-a, -b)$ **27** $\left(\frac{2}{\sqrt{5}}, \frac{1}{\sqrt{5}}\right)$ or $\left(\frac{-2}{\sqrt{5}}, \frac{-1}{\sqrt{5}}\right)$ **28** ≈ 218 revolutions **29 a** ≈ 1.3 **b** ≈ 7.9 **c** 6:1 **d** 6:1 **30** 30 **31** $\left(\frac{5}{\sqrt{29}}, \frac{2}{\sqrt{29}}\right)$, $\left(\frac{-5}{\sqrt{29}}, \frac{-2}{\sqrt{29}}\right)$ **32** ≈ 13.37 in.

14.2 The Circular Functions

Warm-up Exercises

1 $\cos\theta \approx -0.82$; $\sin\theta \approx -0.57$; $\tan\theta \approx 0.70$; $\sec\theta \approx -1.22$; $\csc\theta \approx -1.75$; $\cot\theta \approx 1.44$ **2 a** $\left(\frac{1}{2}, \frac{\sqrt{3}}{2}\right)$ **b** $\left(-\frac{\sqrt{2}}{2}, \frac{\sqrt{2}}{2}\right)$ **c** $\left(\frac{\sqrt{3}}{2}, -\frac{1}{2}\right)$ **d** $(\approx -0.17, \approx 0.98)$ **3 a** $(\approx -0.77, \approx 0.64)$ **b** ≈ -0.84 **4** $\pm\frac{\sqrt{15}}{4}$ **5** ± 2 **6** $\pm\frac{4}{3}$ **7 a** iii **b** i **c** ii **d** iv **8** $\pm\frac{\sqrt{7}}{4}$

Problem Set

9 a $\approx 22.79°$ **b** $\approx 61.93°$ **c** $\approx 8.21°$ **d** $\approx 53.13°$ **e** $\approx 84.26°$ **f** Impossible **10 a** $\pm\frac{8}{\sqrt{353}}$ **b** $\pm\frac{\sqrt{3}}{2}$ **c** $\pm\frac{1}{2\sqrt{2}}$ **11 a** ≈ 0.951 **b** ≈ 0.951 **c** ≈ 3.078 **d** ≈ 3.078 **12** $\sin\theta = -\frac{24}{25}$, $\cos\theta = \frac{7}{25}$; $\tan\theta = -\frac{24}{7}$; $\cot\theta = -\frac{7}{24}$; $\sec\theta = \frac{25}{7}$; $\csc\theta = -\frac{25}{24}$ **13 a** (1, 10) **b** $\sqrt{101}$ **c** $\left(\frac{1}{\sqrt{101}}, \frac{10}{\sqrt{101}}\right)$ **d** ≈ 84.29 **14** $\sin\theta = -\frac{15}{17}$, $\cos\theta = -\frac{8}{17}$; $\tan\theta = \frac{15}{8}$; $\cot\theta = \frac{8}{15}$; $\sec\theta = -\frac{17}{8}$; $\csc\theta = -\frac{17}{15}$ **15 a** $2\sqrt{21}$ **b** $\frac{2}{5}$ **c** $\frac{\sqrt{21}}{5}$ **d** ≈ 66.42 **16** ≈ 0.98, ≈ 0.94, ≈ 0.87, ≈ 0.77, ≈ 0.64, 0.50, ≈ 0.34, ≈ 0.17, 0.00, ≈ -0.17 **17 a** $(\approx 0.342, \approx 0.940)$ **b** ≈ 0.470 **18** 152° **19 a** ≈ 208.00 **b** $(\approx -3.53, \approx -1.88)$ **20 a** 90° **b** 270° **c** 0°, 180°, 360° **21 a** 270 **b** 225 **c** ≈ 323.13 **d** ≈ 35 **22** sine: ≈ 0.866; cosine: −0.5; tangent: ≈ -1.732; cotangent: ≈ -0.577; secant: ≈ -2; cosecant: ≈ 1.155 **23 a** $\approx \pm 0.719$ **b** $\approx \pm 1.035$ **c** ≈ 1.439

24 a
$1 + \tan^2\theta = \sec^2\theta$
$\frac{\cos^2\theta + \sin^2\theta}{\cos^2\theta} = \sec^2\theta$
$\frac{1}{\cos^2\theta} = \sec^2\theta$
$\sec^2\theta = \sec^2\theta$

b
$1 + \cot^2\theta = \csc^2\theta$
$\frac{\sin^2\theta + \cos^2\theta}{\sin^2\theta} = \csc^2\theta$
$\frac{1}{\sin^2\theta} = \csc^2\theta$
$\csc^2\theta = \csc^2\theta$

25 For any θ, $\sin^2\theta + \cos^2\theta = 1$, but $0.625^2 + 0.54^2 \approx 0.682$. **26 a** $\left(\frac{\sqrt{2}}{2}, \frac{\sqrt{2}}{2}\right)$ or $\left(-\frac{\sqrt{2}}{2}, -\frac{\sqrt{2}}{2}\right)$ **b** $\left(-\frac{\sqrt{2}}{2}, -\frac{\sqrt{2}}{2}\right)$ or $\left(\frac{\sqrt{2}}{2}, -\frac{\sqrt{2}}{2}\right)$ **c** (−1, 0) **d** $\left(-\frac{1}{2}, \frac{\sqrt{3}}{2}\right)$ or $\left(-\frac{1}{2}, -\frac{\sqrt{3}}{2}\right)$ **27** $\approx 76.39°$ or $\approx 283.61°$ **28 a** II **b** III **29 a** $y = \frac{x}{\sqrt{3}}$ **b** $y = -\frac{x}{\sqrt{3}}$ **30 a** a **b** b **c** $\frac{b}{a}$ **d** c **e** d **f** $\frac{d}{c}$ **31 a** $\cos^2\theta$ **b** $\sec^2\theta$ **c** $-\tan^2\theta$ **32 a** 135° **b** 60° or −60° **c** 60° or 300° **d** 135° or 225° **34** Depending on the value of θ, $\tan\theta = \frac{\sqrt{1 - \cos^2\theta}}{\cos\theta}$ or $\tan\theta = -\frac{\sqrt{1 - \cos^2\theta}}{\cos\theta}$

14.3 Coordinate Defintions

Warm-up Exercises

1 a 65° **b** 40° **c** 20° **d** 38° **e** 57° **f** 88° **2 a** $-\frac{\sqrt{3}}{2}$ **b** $-\frac{\sqrt{3}}{2}$ **c** $-\frac{\sqrt{3}}{3}$ **d** 2 **e** Undefined **f** $-\frac{1}{2}$ **3** $\pm\frac{12}{13}$ **4** $\frac{8}{17}$

5 $\pm\frac{3}{4}$ **6** $\frac{1}{\sqrt{5}}; \frac{2}{\sqrt{5}}; \frac{1}{2}$ **7** $\approx 26.57°$ **8** **a** I **b** II **c** III **d** IV
9 $\sin\theta = \frac{24}{25}$; $\cos\theta = -\frac{7}{25}$; $\tan\theta = -\frac{24}{7}$; $\cot\theta = -\frac{7}{24}$; $\sec\theta = -\frac{25}{7}$; $\csc\theta = \frac{25}{24}$ **10** **a** $-\frac{6}{\sqrt{85}}; \frac{7}{\sqrt{85}}; -\frac{7}{6}$
b ≈ 130.60

Problem Set

11 **a** $\frac{5}{2\sqrt{6}}$ **b** $\frac{3}{\sqrt{10}}$ **12** $(\approx -0.88, \approx -0.47)$ **13** Tangent and secant for $\{\ldots, -270°, -90°, 90°, 270°, \ldots\}$, cotangent and cosecant for $\{\ldots, -360°, -180°, 0°, 180°, 360°, \ldots\}$ **14** $(\approx 2.91, \approx 11.64)$ **15** **a** 45°, 135°, 225°, or 315° **b** 0°, 180°, or 360° **c** 45° or 225°
16 **a** $(\approx 8.19, \approx 5.74)$ **b** ≈ 23.49 **17** $-\frac{3}{\sqrt{13}}$ **18** **a** Equal **b** Equal in absolute value but different in sign
c Supplementary **d** α or $180 - \beta$ **19** **a** $\frac{b}{a}$ **b** b **c** a
d $\frac{b}{a}$ **e** $-a$ **f** $-b$ **20** **a** −225°, −135, 135°, 225°
b −225°, −45°, 135°, 315° **21** $(-4, 4\sqrt{3})$ **22** **a** 60° or 300°
b 45°, 135°, 225°, or 315° **c** 0°, 180°, or 360° **d** 0°, 90°, 270°, or 360° **23** The coordinates of the points in Quadrants II, III, and IV are identical in absolute value to those of the corresponding points in Quadrant I, and the signs of the coordinates can be determined by the point's directions from the x- and y-axis.

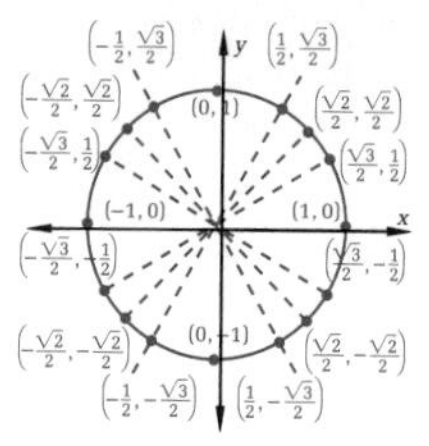

24 **a** $S = \frac{1}{2}$ or $S = -1$
b $\sin x = \frac{1}{2}$ or $\sin x = -1$ **c** $x = 30°$, $x = 150°$, or $x = 270°$ **25** $\frac{\sqrt{313}}{13}$ or $-\frac{\sqrt{313}}{13}$ **26** ≈ 0.32
27 **a** $\approx -331.6°$, $\approx -28.4°$, $\approx 28.4°$, $\approx 331.6°$
b $\approx -354.3°$, $\approx -185.7°$, $\approx 5.74°$, $\approx 174.26°$
28 **a** ≈ 386.37 ft **b** ≈ 373.21 ft **29** ≈ 0.18
30 **a** ≈ 67.66 **b** $\approx 44°$ or $\approx 136°$ **31** $9\sqrt{3}$ **32** ≈ 8.77 m; ≈ 8.24 m **33** $\sin x = \sin x$; $\cos x = \sqrt{1 - \sin^2 x}$;
$\tan x = \frac{\sin x}{\sqrt{1 - \sin^2 x}}$; $\cot x = \frac{\sqrt{1 - \sin^2 x}}{\sin x}$;
$\sec x = \frac{1}{\sqrt{1 - \sin^2 x}}$; $\csc x = \frac{1}{\sin x}$

14.4 Inverse Trigonometric Functions

Warm-up Exercises

1 a and b **2** **a** $100° + 360° \cdot k$ **b** $140° + 180° \cdot k$
c $80° + 90° \cdot k$ **3** **a** 60° **b** 120° **c** 135° **d** No; the Arccos function's range is restricted to angles from 0° to 180°. **4** **a** 45° **b** $45° + 360° \cdot n$ or $315° + 360° \cdot n$
c 60° **d** $60° + 360° \cdot n$ or $300° + 360° \cdot n$

Problem Set

5 **a** ≈ 0.7424 **b** ≈ -1.6129 **6** **a** 440°, 800°, 1160°, 1520°, 1880° **b** −280°, −640°, −1000°, −1360°
7 **a** −45° **b** $135° + 180° \cdot n$ **8** **a** $\frac{7}{2}$ **b** $\approx 74.05°$ **9** 60°, 300°, 420°, 660°, 780° **10** **a** 48.59° or 131.41° **b** 65.91° or 294.09° **c** 115.47° or 244.53° **d** 60° or 120°
11 Arccos; Arcsin; Arctan **12** $\theta \approx 53.13° + n \cdot 360°$
13 60°, 120°, 240°, 300° **14** **a** $\approx 50.95°$ **b** ≈ 0.7766
c Arccos 0.63; sin (Arccos 0.63) **15** **a** 3 **b** ≈ 3.2838
16 180° **17** $\approx 38°$ **18** **a** 140° **b** 30° **c** 30° **d** 30°
e −30° **19** **a** $\frac{1}{2}$ **b** 60° **20** **a** 8π **b** 8π **c** 0 **21** **a** 2
b ≈ 1.1254 **22** $\approx 30°45'13''$ **23** **a** ≈ 0.7599 **b** ≈ 0.2116
c ≈ 1.4372 **d** ≈ -2.0996 **24** **a** $\frac{\sqrt{3}}{2}$ **b** 0 **c** −1 **d** $\frac{2}{\sqrt{3}}$
e $\frac{1}{2}$ **25** **a** $\approx(360° \cdot n + 48.19°)$ or $\approx(360° \cdot n + 311.81°)$
b $180° \cdot n + 135°$ **26** $\approx 89°18'$ **27** $\approx 143.14°$ **28** 50°, 170°, 410°, 530° **29** **a** −1 **b** −1 **c** The values are equal. **30** **a** $\sin x = -\frac{3}{5}$ or $\sin x = \frac{1}{2}$ **b** $x = 30°$, $x = 150°$, $x \approx 216.87°$ or $x \approx 323.13°$ **31** $60° + n \cdot 360°$
32 $-\frac{\sqrt{2}}{2}$, or ≈ -0.7071 **33** $\approx 63.43°$ **34** $35° + 180° \cdot n$
35 90°0′ **36** $\sec x \geq 1$ for all values of x, and the domain of the arccos function is between −1 and 1 inclusive, so arccos (sec x) exists only for values of x for which $\sec x = \pm 1$. **37** ≈ 230.62 **38** $\approx 26.57°$ or $\approx 206.57°$ **39** **a** $\approx 33.69°$ **b** ≈ 3.328

14.5 Right Triangles

Warm-up Exercises

1 45 **2** **a** $\cot\theta$ **b** $\sin\theta$ **c** $\csc\theta$ **d** $\tan\theta$
3 **a** $x \approx 7.49$ **b** $x \approx 10.80$ **c** $\theta \approx 38.68°$ **4** $\theta \approx 48.59°$

Problem Set

5 **a** $-\frac{4}{5}$ **b** $-\frac{3}{4}$ **c** $\frac{3}{5}$ **d** $-\frac{3}{4}$ **e** $\frac{-4}{5}$ **6** $\approx 38.66°$ **7** **a** $\frac{3}{5}$
b $\frac{3}{5}$ **8** $\approx 63.4° + 180° \cdot n$ **9** ≈ 85 ft **10** $\approx 53.13°$
11 **a** ≈ 0.5736 **b** $\frac{5}{8}$ **12** **a** ≈ 16.78 **b** ≈ 167.8 **13** **a** 0.8
b The ratio BC:AC is equal to the ratio QR:PR.
14 ≈ 5.5 **15** ≈ 19.5 **16** $\overline{DC}$ **17** $\approx 1.8°$ **18** $\approx 2.20°$
19 ≈ 13.9 **20** ≈ 567.13 sq. units **21** ≈ 2.75 **22** ≈ 29.71
23 ≈ 1.42; ≈ 355 ft **24** **a** ≈ 0.3511; 0.6 **b** ≈ 0.5852
c $\approx 20.56°$; $\approx 36.87°$ **d** ≈ 0.5575 **25** $\approx 77.57°$; ≈ 33 paces **26** $\approx 23.2°$ **27** ≈ 180.42 **28** ≈ 46.40 **29** ≈ 78.56
30 **a** $\approx 45.10°$ **b** ≈ 72.25 **31** ≈ 19.47 **32** ≈ 0.9511
33 **a** $\approx 118.07°$ **b** $\approx 5.91°$ **34** $\approx 107.95°$ and $\approx 72.05°$

14.6 Radian Measure

Warm-up Exercises

1 **a** $\frac{\pi}{3}$ **b** $\frac{7\pi}{6}$ **c** $-\frac{5\pi}{12}$ **d** $\frac{5\pi}{2}$ **e** $-\frac{\pi}{6}$ **2** **a** 45° **b** −210°
c 480° **d** −60° **e** −120° **3** ≈ 1.92

Problem Set

4 ≈ 1272.35 in. **5** $\left(\frac{\sqrt{3}}{2}, \frac{1}{2}\right)$; $\left(-\frac{1}{2}, \frac{\sqrt{3}}{2}\right)$ **6** **a** ≈ 0.87

b ≈ -6.98 **c** ≈ 0.44 **d** ≈ 3.60 **e** ≈ -1.31 **7** ≈ 0.8192 **8 a** 2π **b** π **c** $\frac{2\pi}{3}$ **d** $\frac{4\pi}{5}$ **9** -1 **11** 1 **13** $\sqrt{3}$ **15** $-\frac{\sqrt{2}}{2}$ **17** $-\frac{\sqrt{3}}{3}$ **19** $\frac{2\sqrt{3}}{3}$ **21** 2 **23** $-\frac{\sqrt{3}}{2}$ **24** $\approx 114.5916°$ **25** ≈ 4.89 **26 a** $\frac{3}{10}$; $\frac{3}{5}$ **b** $\frac{1}{2}$ **c** $\approx 16.70°$; $\approx 30.96°$ **d** ≈ 0.5393 **27** 6π **28 a** $\frac{1}{2}$ **b** $\sqrt{3}$ **c** -1 **d** $\frac{2\sqrt{3}}{3}$ **e** 1 **29 a** ≈ 0.6442 **b** ≈ 0.3624 **c** ≈ 1.7778 **30 a** ≈ 0.1361 **b** ≈ -0.4532 **31** ≈ 57.98 **32 a** $x \approx 21.25$ **b** $x \approx 15.57$ **c** $x = 7.5$ **d** $x = 10\sqrt{2}$ **e** $x \approx 11.76$ **33** $x = 2$ **34** ≈ 0.2618 and 15° **35** The sum of the squares of the coordinates is 1. **37** $\frac{24}{7}$ **39** Neither; both areas are ≈ 38.30 sq. units. **40** 38.4 **41 a** $\frac{\pi}{3}, \frac{2\pi}{3}, \frac{4\pi}{3}, \frac{5\pi}{3}$ **b** $\frac{\pi}{2}, \frac{2\pi}{3}, \frac{4\pi}{3}, \frac{3\pi}{2}$ **c** $\frac{\pi}{6}, \frac{7\pi}{6}$ **42** $\approx 146.31°$ **43** 1; $\frac{\sqrt{2}}{2}$; $\frac{1}{2}$; $-\frac{\sqrt{2}}{2}$; $-\frac{\sqrt{3}}{2}$; -1 **44** ≈ 26.57 **45 a** $\frac{\sqrt{d^2-1}}{d}$ **b** 1 **46** ≈ 23.9 in.^{2} **47** ≈ 13.17 in.

14.7 Trigonometric Relationships

Warm-up Exercises

1 a True **b** True **c** True **d** True **e** True **2 a** a **b** $-a$ **c** b **d** b **e** $-b$ **f** $-a$ **g** a **h** $-b$ **3** $\cos(-\theta) = \cos\theta$ **a** $\frac{\sqrt{3}}{2}$ **b** $\frac{\sqrt{3}}{2}$ **c** $\frac{\sqrt{2}}{2}$ **d** $\frac{\sqrt{2}}{2}$ **e** $-\frac{1}{2}$ **f** $-\frac{1}{2}$ **4** $\sin(-\theta) = -\sin\theta$ **a** $\frac{1}{2}$ **b** $-\frac{1}{2}$ **c** $\frac{\sqrt{2}}{2}$ **d** $-\frac{\sqrt{2}}{2}$ **e** $-\frac{\sqrt{3}}{2}$ **f** $\frac{\sqrt{3}}{2}$ **5 a** $-\frac{11}{5\sqrt{5}}$ **b** $-\frac{2}{5\sqrt{5}}$ **c** $\frac{11}{2}$ **6 a** True **b** True **c** False **d** True

7

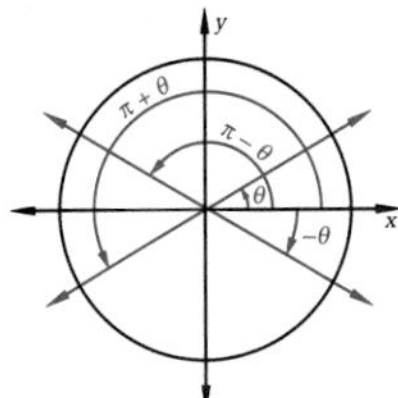

Problem Set

8 a $\cos\theta$ **b** $-\sin\theta$ **c** $\cos\theta$ **d** $-\sin\theta$ **e** $-\sin\theta$ **9 a** 0.2 **b** -0.2 **c** 0.2 **d** 0.2 **e** 0.2 **10 a** Even **b** Odd **c** Odd **d** Odd **e** Even **f** Odd **11 a** $-\tan 35°$ **b** $-\tan 80°$ **12 a** $\frac{b}{c}$ **b** $\frac{b}{c}$ **c** $\frac{\pi}{2}$ radians, or 90°

13

θ	2π	$\frac{13\pi}{6}$	$\frac{9\pi}{4}$	$\frac{7\pi}{3}$	$\frac{5\pi}{2}$
$\sin\theta$	0	$\frac{1}{2}$	$\frac{\sqrt{2}}{2}$	$\frac{\sqrt{3}}{2}$	1
$\cos\theta$	1	$\frac{\sqrt{3}}{2}$	$\frac{\sqrt{2}}{2}$	$\frac{1}{2}$	0
$\tan\theta$	0	$\frac{\sqrt{3}}{3}$	1	$\sqrt{3}$	Undef.
$\cot\theta$	Undef.	$\sqrt{3}$	1	$\frac{\sqrt{3}}{3}$	0

14 a $\alpha = \frac{\pi}{2} - \theta$, or $\alpha = 90° - \theta$ **b** $\frac{7}{\sqrt{53}}$ **c** $\frac{\pi}{2}$ radians, or 90° **d** $\frac{7}{\sqrt{53}}$ **15 a** $\sqrt{61}$ **b** $\frac{6}{\sqrt{61}}$ **c** $\frac{6}{5}$ **d** $\frac{9}{\sqrt{142}}$

16 a $8\sin\theta$ **b** $6\sin\theta$

17 a
$$\begin{aligned}\sin(90° - \theta) &= \cos\theta \\ &= 1\cos\theta - 0\sin\theta \\ &= \sin 90°\cos\theta - \cos 90°\sin\theta\end{aligned}$$

b
$$\begin{aligned}\cos(90° - \theta) &= \sin\theta \\ &= 0\cos\theta + 1\sin\theta \\ &= \cos 90°\cos\theta + \sin 90°\sin\theta\end{aligned}$$

18 a $-\frac{5}{12}$ **b** $\frac{5}{12}$ **c** $\frac{5}{12}$ **d** $-\frac{5}{12}$ **19** $\overline{CD}$ **20 a** ≈ 1.0419 **b** ≈ 3.8302 **21** $\left(\theta + \frac{\pi}{3}\right) + \left(\frac{\pi}{6} - \theta\right) = \frac{\pi}{2}$, so $\theta + \frac{\pi}{3}$ and $\frac{\pi}{6} - \theta$ are complementary angles. Therefore, since $\sin\alpha = \cos\beta$ when α and β are complementary, $\sin\left(\theta + \frac{\pi}{3}\right) = \cos\left(\frac{\pi}{6} - \theta\right)$. **22 a** $(-4, -3)$ **b** $-\frac{4}{5}$; $-\frac{3}{5}$; $\frac{3}{4}$ **23 a** $(5, -12)$ **b** $\frac{5}{13}$; $\frac{5}{13}$ **24 a** Even **b** Odd **c** Even **d** Even **e** Even **f** Odd **25 a** $-\sin\left(\frac{\pi}{4}\right)$ **b** $-\cos\left(\frac{\pi}{6}\right)$ **c** $\tan\left(\frac{4\pi}{9}\right)$ **26** $-\sin x$ **27 a** $\frac{1}{2}$ **b** $\frac{\sqrt{3}}{2}$ **c** $\frac{\sqrt{3}}{2}$ **d** $\frac{1}{2}$ **e** $-\frac{1}{2}$ **f** $-\frac{\sqrt{3}}{2}$ **28** ≈ 1.71 ft

29
$$\begin{aligned}f(-x) &= \frac{-x}{1!} - \frac{(-x)^3}{3!} + \frac{(-x)^5}{5!} - \frac{(-x)^7}{7!} + \frac{(-x)^9}{9!} \\ &= -\frac{x}{1!} + \frac{x^3}{3!} - \frac{x^5}{5!} + \frac{x^7}{7!} - \frac{x^9}{9!} \\ &= -\left(\frac{x}{1!} - \frac{x^3}{3!} + \frac{x^5}{5!} - \frac{x^7}{7!} + \frac{x^9}{9!}\right) \\ &= -f(x)\end{aligned}$$

30
$$\begin{aligned}f(-x) &= 1 - \frac{(-x)^2}{2!} + \frac{(-x)^4}{4!} - \frac{(-x)^6}{6!} + \frac{(-x)^8}{8!} - \frac{(-x)^{10}}{10!} \\ &= 1 - \frac{x^2}{2!} + \frac{x^4}{4!} - \frac{x^6}{6!} + \frac{x^8}{8!} - \frac{x^{10}}{10!} \\ &= f(x)\end{aligned}$$

Review Problems

1 Sine: $\frac{12}{13}$; cosine: $-\frac{5}{13}$; tangent: $-\frac{12}{5}$; cotangent: $-\frac{5}{12}$; secant: $-\frac{13}{5}$; cosecant: $\frac{13}{12}$ **2** $-312°$, $-228°$, $-132°$, $-48°$, 48°, 132°, 228°, 312° **3 a** $\frac{1}{2}$ **b** $\frac{1}{2}$ **c** $\frac{\sqrt{3}}{2}$ **d** $\pm\frac{\sqrt{3}}{2}$ **4** ≈ 1927 ft **5** 90; 180; 270; 360; 45; 300; 315; 150 **6 a** -1 **b** 1 **c** Undefined **d** Undefined **7** ≈ 3076.04 **8** ≈ 140.19 **9 a** 1 **b** 0 **c** $\frac{\sqrt{2}}{2}$ **d** $-\frac{\sqrt{2}}{2}$ **e** -1 **f** $\frac{\sqrt{2}}{2}$ **10** ≈ 0.9962 **11 a** -6 **b** 2 **12** 30 **13 a** ≈ 0.9352 **b** ≈ -0.9352 **14 a** Tangent **b** ≈ 286 ft **15 a** $\approx 261.27°$ **b** $\approx -217.15°$ **c** $\approx 875.48°$ **d** $\approx -300.80°$ **e** $\approx -57.30°$ **16** ≈ 0.13 in. **17** $\left(-\frac{4}{5}, \frac{3}{5}\right)$ and $\left(\frac{4}{5}, -\frac{3}{5}\right)$ **18** $-\frac{2\sqrt{2}}{3}$ **19** $-\frac{3}{4}$ **20 a** $8\tan\theta$ **b** $9\sin\theta$ **c** $2\cos\theta$ **21 a** $\frac{\pi}{6}$ **b** $\frac{3\pi}{10}$ **c** $\frac{\pi}{4}$ **22** $(22.5 + 90n)°$, where n is an integer **23 a** $\frac{\sqrt{2}}{2}$ **b** -1 **c** $-\frac{\sqrt{2}}{2}$ **d** $-\frac{\sqrt{2}}{2}$ **e** $-\frac{\sqrt{2}}{2}$ **f** 1 **24 a** $AC = \frac{1}{\sqrt{2}}$; $BC = \frac{1}{\sqrt{2}}$ **b** $AC = \frac{1}{2}$; $BC = \frac{\sqrt{3}}{2}$ **c** $AC = \frac{\sqrt{3}}{2}$; $BC = \frac{1}{2}$ **25** ≈ 40.54

15.1 Basic Trigonometric Identities

Warm-up Exercises

1 $\csc^4\theta = (\csc^2\theta)^2 = (1+\cot^2\theta)^2$; $\sin\theta \neq 0$

2 $(\cos^2\theta)(\tan^2\theta + 1) = \cos^2\theta\sec^2\theta = \frac{\cos^2\theta}{\cos^2\theta} = 1$; $\cos\theta \neq 0$ **3 a** True **b** True **c** False **4 a** Yes **b** Yes **c** Yes **d** Yes **5 a** No **b** No

6 $\frac{1}{\cos^2\theta} - \frac{\sin^2\theta}{\cos^2\theta} = \frac{1-\sin^2\theta}{\cos^2\theta} = \frac{\cos^2\theta}{\cos^2\theta} = 1$; $\cos\theta \neq 0$

7 $\frac{\cos^2\theta - \sin^2\theta}{\cos\theta - \sin\theta} = \frac{(\cos\theta+\sin\theta)(\cos\theta-\sin\theta)}{\cos\theta-\sin\theta}$
$= \cos\theta + \sin\theta$; $\cos\theta \neq \sin\theta$

Problem Set

8 a False **b** True **9 a** 4 **b** 1 **c** 1

10 $(\csc\theta + \cot\theta)^2(\csc\theta - \cot\theta)^2 = (\csc^2\theta - \cot^2\theta)^2$
$= 1^2 = 1$; $\sin\theta \neq 0$

11 a $\sec x$ **b** $\csc x$ **c** $\sin x$ **d** $\cos x$ **12 a** $\left(\frac{\sqrt{3}}{2}, \frac{1}{2}\right)$ **b** $\left(\frac{1}{2}, \frac{\sqrt{3}}{2}\right)$ **c** ≈ 0.518

13 $\frac{\cos^3\theta - \sin^3\theta}{\cos\theta - \sin\theta}$
$= \frac{(\cos\theta - \sin\theta)(\cos^2\theta + \cos\theta\sin\theta + \sin^2\theta)}{\cos\theta - \sin\theta}$
$= \cos^2\theta + \cos\theta\sin\theta + \sin^2\theta$
$= (\sin^2\theta + \cos^2\theta) + \cos\theta\sin\theta$
$= 1 + \cos\theta\sin\theta$; $\cos\theta \neq \sin\theta$

14 Let $\theta = 0°$. $\frac{\sin 0}{1+\cos 0} = 0$, but $\frac{\cos 0}{1+\sin 0} = 1$
Therefore, the equation is not an identity. **15 a** 1 **b** ≈ 1.2601 **16 a** $\sin 210° = -0.5$ but $\sqrt{1-\cos^2 210°} = 0.5$
Therefore, equation is false for $\theta = 210°$. **b** Ida used an algebraic property of equations instead of simplifying each side separately. $\sqrt{1-\cos^2\theta} = \sqrt{\sin^2\theta} = |\sin\theta|$, and $\sin\theta = |\sin\theta|$ is not an identity, since it is a true statement only for values of θ with a positive sine. **17 a** False **b** True **c** False **d** True **18 a** $\frac{1}{3}$ **b** $\frac{2\sqrt{2}}{3}$ **c** $\frac{4}{5}$ **19**

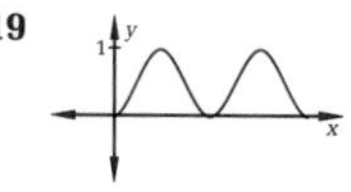

20 a False **b** True

21 a $\frac{\sin\theta}{1-\cos\theta} = \frac{\sqrt{a^2+b^2}+b}{a} = \csc\theta + \cot\theta$
b $\frac{(a+b)^2}{a^2+b^2}$ **c** $\frac{a^2}{a^2+b^2} + \frac{b^2}{a^2+b^2} = \frac{a^2+b^2}{a^2+b^2} = 1$ **22 a** $\tan x$ **b** $\sin x$ **c** $\cos x$ **d** $\cot x$

23 a $\frac{\tan x - 1}{\tan x + 1} = \frac{\frac{\sin x}{\cos x} - 1}{\frac{\sin x}{\cos x} + 1} = \frac{\frac{\sin x - \cos x}{\cos x}}{\frac{\sin x + \cos x}{\cos x}} = \frac{\sin x - \cos x}{\sin x + \cos x}$
$= \frac{\frac{\sin x - \cos x}{\sin x}}{\frac{\sin x + \cos x}{\sin x}}$
$= \frac{1 - \frac{\cos x}{\sin x}}{1 + \frac{\cos x}{\sin x}} = \frac{1-\cot x}{1+\cot x}$; $\cos x \neq 0$, $\sin x \neq 0$, $\sin x \neq -\cos x$

b $\cot^2 x - \cos^2 x = \frac{\cos^2 x}{\sin^2 x} - \cos^2 x = \left(\frac{1}{\sin^2 x} - 1\right)\cdot\cos^2 x$
$= (\csc^2 x - 1)\cdot\cos^2 x = \cot^2 x\cdot\cos^2 x$; $\sin x \neq 0$

c $(\tan x + \sec x)^2 = \left(\frac{\sin x}{\cos x} + \frac{1}{\cos x}\right)^2 = \left(\frac{\sin x + 1}{\cos x}\right)^2$
$= \frac{(\sin x + 1)^2}{1 - \sin^2 x} = \frac{(\sin x + 1)^2}{(1+\sin x)(1-\sin x)} = \frac{1+\sin x}{1-\sin x}$;
$\cos x \neq 0$, $\sin x \neq -1$, $\sin x \neq 1$

24 a $\frac{a^2+2a+2}{a+1}$ **b** ≈ -0.618 **25 a** $x = \frac{\pi}{4}$ or $x = \frac{5\pi}{4}$ **b** $x = 0$, $x = \pi$, $x = \frac{3\pi}{2}$, or $x = 2\pi$ **c** $x = \frac{\pi}{3}$ or $x = \frac{5\pi}{3}$

26 a $y = [g(x)]^2$, $y = [f(x)]^2$ **b**

27 $\log_2(\tan\theta) + \log_2(\cos\theta)$
$= \log_2(\sin\theta) - \log_2(\cos\theta) + \log_2(\cos\theta)$
$= \log_2(\sin\theta)$; $\sin\theta > 0$, $\cos\theta > 0$

28 $\frac{2\sin^2\theta + 5\sin\theta\cos\theta + 2\cos^2\theta}{\sin\theta + 2\cos\theta}$
$= \frac{2(\sin^2\theta + \cos^2\theta) + 5\sin\theta\cos\theta}{\sin\theta + 2\cos\theta}$
$= \frac{2 + 5\sin\theta\cos\theta}{\sin\theta + 2\cos\theta}$
$= \frac{2 + 5\sin\theta\cos\theta}{\sin\theta + 2\cos\theta}\cdot\frac{\sin\theta - 2\cos\theta}{\sin\theta - 2\cos\theta}$
$= \frac{2\sin\theta - 4\cos\theta + 5\sin^2\theta\cos\theta - 10\sin\theta\cos^2\theta}{\sin^2\theta - 4\cos^2\theta}$
$= \frac{2\sin\theta(1 - 5\cos^2\theta) + \cos\theta(-4 + 5\sin^2\theta)}{\sin^2\theta - 4\cos^2\theta}$
$= \frac{2\sin\theta(1 - 5\cos^2\theta) + \cos\theta(-4 + 5 - 5\cos^2\theta)}{1 - \cos^2\theta - 4\cos^2\theta}$
$= \frac{2\sin\theta(1 - 5\cos^2\theta) + \cos\theta(1 - 5\cos^2\theta)}{1 - 5\cos^2\theta}$
$= 2\sin\theta + \cos\theta$
$= \frac{2}{\csc\theta} + \frac{1}{\sec\theta}$
$= \frac{2\sec\theta + \csc\theta}{\sec\theta\csc\theta}$; $\sin\theta \neq 0$, $\cos\theta \neq 0$,
$\sin\theta \neq \pm 2\cos\theta$, $\cos\theta \neq \pm\frac{\sqrt{5}}{5}$

29 a $\left\{\frac{\pi}{4}, \frac{7\pi}{4}, \frac{\pi}{3}, \frac{4\pi}{3}\right\}$ **b** $\left\{\frac{\pi}{12}, \frac{\pi}{4}, \frac{5\pi}{12}, \frac{7\pi}{12}, \frac{3\pi}{4}, \frac{11\pi}{12}, \frac{13\pi}{12}, \frac{5\pi}{4}, \frac{17\pi}{12}, \frac{19\pi}{12}, \frac{7\pi}{4}, \frac{23\pi}{12}\right\}$ **c** $\left\{\frac{2\pi}{3}, \frac{4\pi}{3}\right\}$ **d** $\left\{\frac{\pi}{3}, \frac{5\pi}{3}, 0, 2\pi\right\}$ **e** $\left\{0, \frac{\pi}{2}, 2\pi\right\}$

15.2 The Sum and Difference Identities

Warm-up Exercises

1 a $\frac{5}{13}$ **b** $\frac{12}{13}$ **c** 1 **2 a** True **b** True **c** False **3 a** -3 **b** 2 **c** 1

Problem Set

4 a $\cos 10°$ **b** $\sin 50°$ **c** $2\sin 50°\cos 40°$

5 $\sin(\theta + \phi) + \sin(\theta - \phi)$
$= \sin\theta\ \cos\phi + \cos\theta\ \sin\phi + \sin\theta\ \cos\phi - \cos\theta\ \sin\phi$
$= \sin\theta\ \cos\phi + \sin\theta\ \cos\phi + \cos\theta\ \sin\phi - \cos\theta\sin\phi$
$= 2\sin\theta\cos\phi$

6 a $\frac{5}{12}$ **b** $\approx 23°$ **7 a** $\frac{3}{5}$ **b** 0.6 **c** Yes **8 a** $-\sin x$ **b** $-\cos x$ **c** $\cos x$ **d** $\frac{1}{2}\sin x + \frac{\sqrt{3}}{2}\cos x$ **9 a** True **b** False **c** True **10 a** 30° **b** 30° **c** The principal value of $\arcsin\frac{1}{2}$ is 30°. **11 a** $\frac{3+4\sqrt{15}}{20}$ **b** $\frac{4-3\sqrt{15}}{20}$ **c** $\frac{3+4\sqrt{15}}{4-3\sqrt{15}}$ **12 a** $\frac{7}{25}$ **b** $-\frac{24}{25}$ **c** $\frac{3}{5}$ **d** $\frac{4}{5}$ **e** $\frac{4}{5}$ **f** $-\frac{3}{5}$

13 $\cot(\alpha + \beta) = \frac{\cos(\alpha + \beta)}{\sin(\alpha + \beta)}$

$= \frac{\cos\alpha \cdot \cos\beta - \sin\alpha \cdot \sin\beta}{\sin\alpha \cdot \cos\beta + \cos\alpha \sin\beta} \cdot \frac{\frac{1}{\sin\alpha \sin\beta}}{\frac{1}{\sin\alpha \sin\beta}}$

$= \frac{\frac{\cos\alpha \cos\beta}{\sin\alpha \sin\beta} - 1}{\frac{\sin\alpha \cos\beta}{\sin\alpha \sin\beta} + \frac{\cos\alpha \sin\beta}{\sin\alpha \sin\beta}}$

$= \frac{\cot\alpha \cot\beta - 1}{\cot\beta + \cot\alpha}$

14 $-\frac{56}{65}$ or $\frac{16}{65}$ **15** $\approx 46.04°$ **16** $\approx 6.54°$

17 a $x \in \left\{\frac{\pi}{6} + n \cdot 2\pi, \frac{5\pi}{6} + n \cdot 2\pi\right\}$ **b** No solution

18 a $\frac{\sqrt{6} + \sqrt{2}}{4}$ **b** $\frac{\sqrt{6} - \sqrt{2}}{4}$ **c** $\frac{\sqrt{6} + \sqrt{2}}{4}$ **d** $\frac{\sqrt{6} + \sqrt{2}}{4}$

19 $\frac{\sin(x + y)}{\sin x \sin y} = \frac{\sin x \cos y + \cos x \sin y}{\sin x \sin y} \cdot \frac{\frac{1}{\sin x \sin y}}{\frac{1}{\sin x \cdot \sin y}}$

$= \frac{\cos y}{\sin y} + \frac{\cos x}{\sin x}$

$= \cot y + \cot x$

20 a -1 **b** 0 **21 a** $\sqrt{2} - 1$ **b** $2 + \sqrt{3}$ **c** -1 **d** 1 **e** $\sqrt{3}$ **22 a** $2\cos\theta\cos\phi$ **b** $-2\sin\theta\sin\phi$ **c** $\cos\theta$ **d** $(180°, 40°)$ **23 a** $\frac{\sqrt{2} + \sqrt{6}}{4}$ **b** $\frac{\sqrt{2} + \sqrt{6}}{4}$ **c** $\frac{-\sqrt{2} - \sqrt{6}}{4}$

24 a $\frac{63}{65}$ **b** 2 **c** No real value

25 $\sin 3\theta = \sin(\theta + 2\theta)$

$= \sin\theta \cdot \cos 2\theta + \cos\theta \sin 2\theta$

$= \sin\theta \cdot \cos(\theta + \theta) + \cos\theta \sin(\theta + \theta)$

$= \sin\theta \cdot (\cos\theta\cos\theta - \sin\theta\sin\theta) + \cos\theta\,(\sin\theta\cos\theta + \cos\theta\sin\theta)$

$= \sin\theta\,(\cos^2\theta - \sin^2\theta) + \cos\theta\,(2\cos\theta\sin\theta)$

$= \cos^2\theta\sin\theta - \sin^3\theta + 2\cos^2\theta\sin\theta$

$= 3\cos^2\theta\sin\theta - \sin^3\theta$

$= 3\sin\theta \cdot (1 - \sin^2\theta) - \sin^3\theta$

$= 3\sin\theta - 3\sin^3\theta - \sin^3\theta$

$= 3\sin\theta - 4\sin^3\theta$

26 $y \approx 4.12$ **27 a** 0.5 **b** ≈ 0.500002133 **c** ≈ 0.499999992 **d** ≈ 0.5

15.3 The Double- and Half-Angle Identities

Warm-up Exercises

1 a True **b** True **c** False

2 $(\sin\theta + \cos\theta)^2 = \sin^2\theta + 2\sin\theta\cos\theta + \cos^2\theta$

$= 1 + 2\sin\theta\cos\theta$

$= 1 + \sin 2\theta$

3 ≈ 8.57

4 $\cos 2\left(\frac{\theta}{2}\right) = 2\cos^2\frac{\theta}{2} - 1$

$2\cos^2\frac{\theta}{2} = 1 + \cos\theta$

$\cos^2\frac{\theta}{2} = \frac{1 + \cos\theta}{2}$

$\cos\frac{\theta}{2} = \pm\sqrt{\frac{1 + \cos\theta}{2}}$

5 a False **b** True

6 $\tan 2\theta = \tan(\theta + \theta) = \frac{\tan\theta + \tan\theta}{1 - \tan\theta\tan\theta}$

$= \frac{2\tan\theta}{1 - \tan^2\theta}$

7 a True **b** True **c** True

Problem Set

8 ≈ 0.8775 **9** $(\sin\theta + \cos\theta)^{\csc^2\theta - \cot^2\theta + 1}$ $= (\sin\theta + \cos\theta)^2 = \sin^2\theta + \cos^2\theta + 2\sin\theta\cos\theta = 1 + \sin 2\theta$ **10 a** $\sqrt{2} - 1$ **b** $2 + \sqrt{3}$ **11 a** False **b** True

12 $\sin 75° = \frac{\sqrt{2 + \sqrt{3}}}{2}$ **13** $m\ell = \frac{12}{5}$; $\alpha \approx 33.69°$; so $\tan 2\alpha = \frac{12}{5}$ **14 a** $\left\{\frac{\pi}{2}, \frac{7\pi}{6}, \frac{11\pi}{6}\right\}$ **b** $\left\{\frac{5\pi}{6}, \pi, \frac{7\pi}{6}\right\}$ **c** $\left\{\frac{\pi}{6}, \frac{2\pi}{3}, \frac{5\pi}{6}, \frac{4\pi}{3}\right\}$ **15 a** $\frac{3}{5}$; $\frac{4}{5}$ **b** $\frac{8}{17}$; $\frac{15}{17}$ **c** $\frac{77}{85}$; $\frac{36}{85}$ **16 a** $\approx \pm 0.3162278$ **b** $\approx \pm 0.9486833$ **c** $\pm\frac{1}{3}$

17 1 **18 a** $\{0, \pi, 2\pi\}$ **b** $\left\{\frac{\pi}{3}, \pi, \frac{5\pi}{3}\right\}$ **c** $\{\approx 0.32, \approx 1.89, \approx 3.46, \approx 5.03\}$

19 $\tan\frac{\theta}{2} = \pm\sqrt{\frac{1 - \cos\theta}{1 + \cos\theta}} \cdot \frac{\sqrt{1 + \cos\theta}}{\sqrt{1 + \cos\theta}}$

$= \pm\sqrt{\frac{1 - \cos^2\theta}{(1 + \cos\theta)^2}} = \pm\sqrt{\left(\frac{\sin\theta}{1 + \cos\theta}\right)^2}$

$= \frac{\sin\theta}{1 + \cos\theta}$

20 $\cos^2\theta - \sin^2\theta$

21 $\{\ldots, \approx -248.5°, \approx -111.5°, \approx 111.5°, \approx 248.5°, \ldots\}$

22 a $\sin 4\theta = \sin 2(2\theta)$

$= 2\sin 2\theta\cos 2\theta$

$= 2(2\sin\theta\cos\theta)(\cos^2\theta - \sin^2\theta)$

$= 4\sin\theta\cos^3\theta - 4\sin^3\theta\cos\theta$

b $\cos 4\theta$

$= \cos 2(2\theta)$

$= \cos^2 2\theta - \sin^2 2\theta$

$= (\cos^2\theta - \sin^2\theta)^2 - 4\sin^2\theta\cos^2\theta$

$= \cos^4\theta - 2\sin^2\theta\cos^2\theta + \sin^4\theta - 4\sin^2\theta\cos^2\theta$

$= \cos^4\theta + \sin^4\theta - 6\sin^2\theta\cos^2\theta$

$= \cos^4\theta + (1 - \cos^2\theta)^2 - 6\cos^2\theta \cdot (1 - \cos^2\theta)$

$= \cos^4\theta + 1 - 2\cos^2\theta + \cos^4\theta - 6\cos^2\theta + 6\cos^4\theta$

$= 8\cos^4\theta - 8\cos^2\theta + 1$

23 $\cos(\theta + \phi) = \frac{63}{65}$; $\sin(\theta + \phi) = -\frac{16}{65}$ **24 a** $\left\{\frac{2\pi}{3}, \frac{4\pi}{3}\right\}$ **b** $\{\approx 2.00, \approx 4.29\}$ **c** $\{\theta : 0 \le \theta < 2\pi\}$ **25 a** $\frac{\sqrt{2 - \sqrt{3}}}{2}$

b $\frac{\sqrt{6} - \sqrt{2}}{4}$

c $\frac{\sqrt{6} - \sqrt{2}}{4} = \frac{\sqrt{(\sqrt{6} - \sqrt{2})^2}}{4}$

$= \frac{\sqrt{6 - 2\sqrt{12} + 2}}{4}$

$= \frac{2\sqrt{2 - \sqrt{3}}}{4} = \frac{\sqrt{2 - \sqrt{3}}}{2}$

26 $\sin 3\theta = \sin(2\theta + \theta)$

$= \sin 2\theta\cos\theta + \sin\theta\cos 2\theta$

$= 2\sin\theta\cos^2\theta + \sin\theta\,(2\cos^2\theta - 1)$

$= 2\sin\theta\cos^2\theta + 2\sin\theta\cos^2\theta - \sin\theta$

$= 4\sin\theta\cos^2\theta - \sin\theta$

$= \sin\theta\,(4\cos^2\theta - 1)$

$= \sin\theta\,(2\cos\theta + 1)(2\cos\theta - 1)$

15.4 Final Identities and Trigonometric Equations

Warm-up Exercises

1 a True **b** False **c** False **2 a** $2\sin 4\theta\cos\theta$ **b** $2\cos\frac{3\theta}{2}\cos\frac{\theta}{2}$ **3 a** Yes **b** Yes **c** No

Problem Set

4 (6, 8) **5 a** $\cot 5x$ **b** $\tan 4x$ **6 a** $\left\{\frac{\pi}{4}, \frac{5\pi}{4}\right\}$

b $\left\{0, \frac{\pi}{2}, \pi, \frac{3\pi}{2}\right\}$ **7 a** $2 \sin 10° \cos 7°$ **b** $2 \cos 32° \cos 18°$

8 a $\cos \theta = \frac{1}{2}$ or $\cos \theta = -1$ **b** $\left\{\frac{\pi}{3}, \pi, \frac{5\pi}{3}\right\}$

9 $\frac{\sin 4\theta + \sin 3\theta}{\cos 4\theta + \cos 3\theta} = \tan \frac{7\theta}{2}$

$$\frac{2 \sin \frac{4\theta + 3\theta}{2} \cdot \cos \frac{4\theta - 3\theta}{2}}{2 \cos \frac{4\theta + 3\theta}{2} \cdot \cos \frac{4\theta - 3\theta}{2}} = \tan \frac{7\theta}{2}$$
$$\frac{2 \sin \frac{7\theta}{2} \cdot \cos \frac{\theta}{2}}{2 \cos \frac{7\theta}{2} \cdot \cos \frac{\theta}{2}} = \tan \frac{7\theta}{2}$$
$$\frac{\sin \frac{7\theta}{2}}{\cos \frac{7\theta}{2}} = \tan \frac{7\theta}{2}$$
$$\tan \frac{7\theta}{2} = \tan \frac{7\theta}{2}$$

10 a $\cos 14° + \cos 6°$ **b** $\sin 136° + \sin 4°$ **c** $\cos 4° - \cos 22°$ **11 a** $\frac{2 - \sqrt{3}}{4}$ **b** $\frac{-2 - \sqrt{3}}{4}$ **c** $-\frac{1}{4}$

12 a $\tan 4\theta$ **b** $\cot 4\theta$ **13 a** $2 \cos \frac{x + y}{2} \cos \frac{x - y}{2}$ **b** $2 \cos 4\theta \cos 2\theta$ **c** $2 \cos 10\theta \cos 2\theta$ **d** $-2 \sin \frac{x + y}{2} \sin \frac{x - y}{2}$ **14 a** $\left\{\frac{7\pi}{6}, \frac{3\pi}{2}, \frac{11\pi}{6}\right\}$ **b** $\left\{\frac{\pi}{6}, \frac{\pi}{2}, \frac{5\pi}{6}\right\}$ **15** ≈ 38.17

16 $\{n\pi\}$, where n is an integer **b** $\{x: x \in \mathcal{R}\}$

17 $\frac{2 + \sqrt{3}}{2 - \sqrt{3}}$; $\tan 75 \cdot \tan 75$

$$= \tan (30 + 45) \cdot \tan (30 + 45)$$
$$= \frac{\tan 30 + \tan 45}{1 - \tan 30 \tan 45} \cdot \frac{\tan 30 + \tan 45}{1 - \tan 30 \tan 45}$$
$$= \frac{\frac{\sqrt{3}}{3} + 1}{1 - \frac{\sqrt{3}}{3}(1)} \cdot \frac{\frac{\sqrt{3}}{3} + 1}{1 - \frac{\sqrt{3}}{3}(1)}$$
$$= \frac{\frac{\sqrt{3} + 3}{3}}{\frac{3 - \sqrt{3}}{3}} \cdot \frac{\frac{\sqrt{3} + 3}{3}}{\frac{3 - \sqrt{3}}{3}}$$
$$= \frac{\sqrt{3} + 3}{3} \cdot \frac{3}{3 - \sqrt{3}} \cdot \frac{\sqrt{3} + 3}{3} \cdot \frac{3}{3 - \sqrt{3}}$$
$$\frac{(\sqrt{3} + 3)^2}{(3 - \sqrt{3})^2} = \frac{(\sqrt{3} + 3)(\sqrt{3} + 3)}{(3 - \sqrt{3})(3 - \sqrt{3})} = \frac{3 + 6\sqrt{3} + 9}{9 - 6\sqrt{3} + 3}$$
$$\frac{12 + 6\sqrt{3}}{12 - 6\sqrt{3}} = \frac{2 + \sqrt{3}}{2 - \sqrt{3}}$$

18 a $\left\{\frac{\pi}{6}, \frac{5\pi}{6}, \frac{3\pi}{2}\right\}$ **b** $\frac{3\pi}{2}$ **c** $\left\{\frac{\pi}{3}, \frac{2\pi}{3}, \frac{4\pi}{3}, \frac{5\pi}{3}\right\}$

19 a $\tan \alpha = \frac{\sin (\alpha + \beta) + \sin (\alpha - \beta)}{\cos (\alpha - \beta) + \cos (\alpha + \beta)}$

$$= \frac{\sin \alpha \cos \beta + \cos \alpha \sin \beta + \sin \alpha \cos \beta - \cos \alpha \sin \beta}{\cos \alpha \cos \beta + \sin \alpha \sin \beta + \cos \alpha \cos \beta - \sin \alpha \sin \beta}$$
$$= \frac{2 \sin \alpha \cos \beta}{2 \cos \alpha \cos \beta}$$
$$= \tan \alpha$$

b $\cos \theta \cos (\theta - 90°)$

$$= \tfrac{1}{2}(\cos [\theta - (\theta - 90°)] + \cos [\theta + (\theta - 90°)])$$
$$= \tfrac{1}{2}[\cos 90° + \cos (2\theta - 90°)]$$
$$= \tfrac{1}{2}[0 + \cos (2\theta - 90°)]$$
$$= \tfrac{1}{2} \cos (2\theta - 90°)$$

20 a $\cot 3\theta = \frac{\cos 5\theta + \cos \theta}{\sin 5\theta + \sin \theta}$

$$= \frac{2 \cos \frac{5\theta + \theta}{2} \cdot \cos \frac{5\theta - \theta}{2}}{2 \sin \frac{5\theta + \theta}{2} \cdot \cos \frac{5\theta - \theta}{2}}$$
$$= \frac{2 \cos 3\theta \cdot \cos 2\theta}{2 \sin 3\theta \cdot \cos 2\theta}$$
$$= \frac{\cos 3\theta}{\sin 3\theta}$$
$$= \cot 3\theta$$

b $2 \sin 8\theta = \frac{\sin 10\theta + \sin 6\theta}{\cos^2 \theta - \sin^2 \theta}$

$$= \frac{2 \sin \frac{10\theta + 6\theta}{2} \cdot \cos \frac{10\theta - 6\theta}{2}}{\cos^2 \theta - \sin^2 \theta}$$
$$= \frac{2 \sin 8\theta \cos 2\theta}{\cos 2\theta}$$
$$= 2 \sin 8\theta$$

21 a $\{\approx 63.43°, \approx 116.57°, \approx 243.43°, \approx 296.57°\}$ **b** $\{\approx 9.59°, \approx 170.41°\}$ **22** $\{\theta : \theta \in \mathcal{R}\}$ **23 a** $\approx 41.41°$ **b** $x \approx 7.937$

24 $\cos 10° \cos 20° = \frac{1}{2}[\cos (20° - 10°) + \cos (10° + 20°)]$

$$= \tfrac{1}{2}(\cos 10° + \cos 30°)$$
$$= \tfrac{1}{2}\left(\cos 10° + \frac{\sqrt{3}}{2}\right)$$
$$\cos 20° = \frac{\frac{1}{2} \cos 10° + \frac{\sqrt{3}}{4}}{\cos 10°}$$
$$= \frac{1}{2} + \frac{\sqrt{3}}{4} \cdot \frac{1}{\cos 10°}$$
$$= \frac{1}{2} + \frac{\sqrt{3}}{4} \cdot \sec 10°$$
$$= \frac{\sqrt{3}}{4} \sec 10° + \frac{1}{2}$$

25 a 3 **b** ≈ 6.87 **26 a** 4; -4 **b** $x = -2.47$, $x = 0$, or $x = 2.47$

Review Problems

1 a True **b** True **c** False **2 a** 0.8 **b** ≈ 0.857 **c** ≈ 0.926 **3 a** $\frac{24}{25}$ **b** $-\frac{7}{25}$ **4 a** $\frac{1}{2}(\sin 7\theta + \sin \theta)$

b $\frac{1}{2}(\cos 2\theta + \cos 10\theta)$ **c** $\frac{1}{2}(\sin 10\theta - \sin 6\theta)$

5 $\sec \theta + \tan \theta$

$$= \frac{1}{\cos \theta} + \frac{\sin \theta}{\cos \theta}$$
$$= \frac{1 + \sin \theta}{\cos \theta} \cdot \frac{1 - \sin \theta}{1 - \sin \theta}$$
$$= \frac{1 - \sin^2 \theta}{\cos \theta(1 - \sin \theta)} = \frac{\cos^2 \theta}{\cos \theta(1 - \sin \theta)} = \frac{\cos \theta}{1 - \sin \theta};$$

$\cos \theta \neq 0$, $\sin \theta \neq 1$

6 a True **b** False **c** True **7 a** $V_{cyl} = 36\pi \cos \theta$; $V_{cone} = 72\pi \sin^2\theta \cos \theta$ **b** $\frac{\pi}{4}$

8 a $\tan \theta + \cot \theta = \frac{\sin \theta}{\cos \theta} + \frac{\cos \theta}{\sin \theta}$

$$= \frac{\sin^2\theta + \cos^2\theta}{\cos \theta \sin \theta}$$
$$= \frac{1}{\cos \theta \sin \theta}$$
$$= \sec \theta \csc \theta$$

b $\sec^2\theta + \csc^2\theta = \frac{1}{\cos^2\theta} + \frac{1}{\sin^2\theta}$

$$= \frac{1}{\cos^2\theta \sin^2\theta}$$
$$= \sec^2\theta \csc^2\theta; \ \sin \theta \neq 0, \cos \theta \neq 0$$

c $\frac{\cos \theta}{1 + \sin \theta} = \frac{\cos \theta}{1 + \sin \theta} \cdot \frac{1 - \sin \theta}{1 - \sin \theta}$

$$= \frac{1 - \sin \theta}{\cos \theta}; \ \sin \theta \neq 1, \sin \theta \neq -1$$

9 a False **b** False **c** True **10** $\{0, \pi, 2\pi\}$

11 a $\cos \left(\theta - \frac{\pi}{2}\right) = \cos \theta \cos \frac{\pi}{2} + \sin \theta \sin \frac{\pi}{2}$

$$= 0 \cos \theta + 1 \sin \theta$$
$$= \sin \theta$$

b $\sin\left(\theta - \frac{\pi}{2}\right) = \sin\theta\cos\frac{\pi}{2} - \cos\theta\sin\frac{\pi}{2}$
$= 0\sin\theta - 1\cos\theta$
$= -\cos\theta$

c $\sin\left(\theta + \frac{\pi}{2}\right) = \sin\theta\cos\frac{\pi}{2} + \cos\theta\sin\frac{\pi}{2}$
$= 0\sin\theta + 1\cos\theta$
$= \cos\theta$

d $\cos\left(\theta + \frac{\pi}{2}\right) = \cos\theta\cos\frac{\pi}{2} - \sin\theta\sin\frac{\pi}{2}$
$= 0\cos\theta - 1\sin\theta$
$= -\sin\theta$

12 True **13 a** $3\sin x - 4\sin^3 x$ **b** $4\cos^3 x - 3\cos x$

14 a True **b** True **c** True

15 $\cos^4\theta - \sin^4\theta = (\cos^2\theta + \sin^2\theta)(\cos^2\theta - \sin^2\theta)$
$= \cos^2\theta - \sin^2\theta$

16 $\tan(\alpha - \beta)$ is undefined. $\tan\alpha$ and $\tan\beta$ are the slopes of the lines QP and QR. Since these are negative reciprocals, the lines are perpendicular. So $\alpha - \beta = 90°$ **17 a** $2\sin\frac{x+y}{2}\cos\frac{x-y}{2}$ **b** $2\sin 2\theta\cos\theta$ **c** $2\sin 3\theta\cos 7\theta$ **18 a** True **b** True

19 a $\frac{a}{\sqrt{a^2+b^2}+a}$ **b** $\frac{\sqrt{a^2+b^2}-a}{a}$ **20** $\{n\pi\}$, where n is an integer **21 a** $\frac{\sqrt{2}}{2}$ **b** $\frac{\sqrt{3}}{2}$ **c** $\frac{\sqrt{2}}{2}$

22 a $\frac{\cos x - \sin x}{\cos x + \sin x} = \frac{(\cos x - \sin x)(\cos x + \sin x)}{(\cos x + \sin x)(\cos x + \sin x)}$
$= \frac{\cos^2 x - \sin^2 x}{\cos^2 x + 2\cos x\sin x + \sin^2 x}$
$= \frac{\cos 2x}{1 + 2\sin x\cos x}$
$= \frac{\cos 2x}{1 + \sin 2x}$

b $1 + \tan x\tan 2x = 1 + (\tan x)\left(\frac{2\tan x}{1 - \tan^2 x}\right)$
$= 1 + \frac{2\tan^2 x}{1 - \tan^2 x}$
$= \frac{1 - \tan^2 x + 2\tan^2 x}{1 - \tan^2 x}$
$= \frac{1 + \tan^2 x}{1 - \tan^2 x}$
$= \frac{1 + \frac{\sin^2 x}{\cos^2 x}}{1 - \frac{\sin^2 x}{\cos^2 x}}$
$= \frac{\frac{\cos^2 x + \sin^2 x}{\cos^2 x}}{\frac{\cos^2 x - \sin^2 x}{\cos^2 x}} = \frac{1}{\cos 2x} = \sec 2x$

c $\sin(x + y)\sin(x - y)$
$= \frac{1}{2}(\cos[(x + y) - (x - y)] - \cos[(x + y) + (x - y)])$
$= \frac{1}{2}(\cos 2y - \cos 2x)$
$= \frac{1}{2}[1 - 2\sin^2 y - (1 - 2\sin^2 x)]$
$= \frac{1}{2}(2\sin^2 x - 2\sin^2 y)$
$= \sin^2 x - \sin^2 y$

23 a $(-7, 24)$; $(-24, -7)$; $(7, -24)$ **b** $\sin\theta = \frac{7}{25}$; $\cos\theta = \frac{24}{25}$; $\tan\theta = \frac{7}{24}$; $\sin\left(\theta + \frac{\pi}{2}\right) = \frac{24}{25}$; $\cos\left(\theta + \frac{\pi}{2}\right) = -\frac{7}{25}$; $\tan\left(\theta + \frac{\pi}{2}\right) = -\frac{24}{7}$; $\sin(\theta + \pi) = -\frac{7}{25}$; $\cos(\theta + \pi) = -\frac{24}{25}$; $\tan(\theta + \pi) = \frac{7}{24}$; $\sin\left(\theta + \frac{3\pi}{2}\right) = -\frac{24}{25}$; $\cos\left(\theta + \frac{3\pi}{2}\right) = \frac{7}{25}$; $\tan\left(\theta + \frac{3\pi}{2}\right) = -\frac{24}{7}$ **24** 1

25

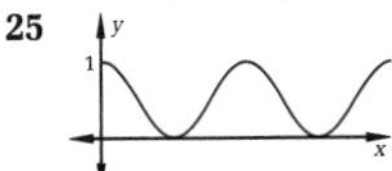

16.1 The Law of Sines

Warm-up Exercises

1 ≈ 0.678 **2** ≈ 12.709 **3** 17° **4** $x \approx 24.25$; $y \approx 27.58$ **5** 64° **6** ≈ 7.09 **7** 18.87 **8** ≈ 1363.68

Problem Set

9 a $(283, \approx 490)$ **b** ≈ 868 **11** ≈ 4.46 **12 a** 1 **b** 1 **c** 1 **d** 1 **13 a** 0; 0.5; ≈ 0.707; ≈ 0.866; 1; ≈ 0.866; ≈ 0.707; 0; -1; 0 **b** **14** $\frac{\sqrt{6} - \sqrt{2}}{4}$

15 $5\sqrt{6} - 5\sqrt{2}$ **16** ≈ 78.6 mi **17** ≈ 0.88 **19** $\approx 63°$ or $\approx 117°$ **21** No solution **23** 4; -4 **25** 4; -4 **26 a** If $b = a\sin\theta$, the triangle is a right triangle (since $a\sin\theta$ is the length of the vertical altitude of the triangle shown). Therefore, since there is only one such altitude, only one triangle is determined. **b** If $a\sin\theta < b < a$, the side with length b can meet the base either to the left or to the right of the altitude, so two triangles are determined. **c** If $b \geq a$, the side with length b can only meet the base to the right of the altitude, so only one triangle is determined. **d** If $b < a\sin\theta$, the side with length b is shorter than the altitude—an impossibility, since the altitude is the shortest possible segment connecting the upper vertex and the base. Therefore, no triangle is determined. **27** ≈ 173.77 ft **28 a** $\approx 36°$ or $\approx 144°$ **b** If $\alpha \approx 36°$, $\beta \approx 112°$; if $\alpha \approx 144°$, $\beta \approx 4°$. **c** If $\alpha \approx 36°$, $\beta \approx 84°$; if $\alpha \approx 144°$, the sum of α and θ exceeds 180°, so no such triangle exists. **29** ≈ 546 ft **30** Willy is wrong. The height of the window is 18 sin 75°, or ≈ 17.39, feet. A 16-foot ladder is therefore too short to reach it.

31 $\alpha_1 \approx 58.34$; $\alpha_2 \approx 121.66$ **32** ≈ 6.44 **33** The length of the altitude from B to $\overline{AC}$ is $a\sin\angle C$. Since $k = \frac{1}{2}bh$, $k = \frac{1}{2}\cdot b\cdot a\sin\angle C = \frac{1}{2}ab\sin\angle C$.

34 $R \approx 17.01$ **35** Draw the diameter from C through O. Label its other endpoint A′, and draw $\overline{A'B}$. $\angle A' = \angle A$, since they are inscribed angles intercepting the same arc. $\angle A'BC = 90°$, since this is an angle inscribed in a semicircle. In right triangle A′BC, $\sin\angle A' = \frac{a}{CA'}$, so $\sin\angle A$ is also equal to $\frac{a}{CA'}$. Therefore, CA′ (the diameter) is equal to $\frac{a}{\sin\angle A}$.

16.2 The Law of Cosines

Warm-up Exercises

1 $a \approx 7.3$; $b \approx 9.5$; $c \approx 3.6$ **2** $\approx 72°$ **3** $\approx 59°$ **4** $\approx 18°$ **5** ≈ 11.6 **6** ≈ 8.1 **7** ≈ 47.3

Problem Set

8 AC ≈ 3.5; m∠A ≈ 48; m∠C ≈ 117

9 a **b**

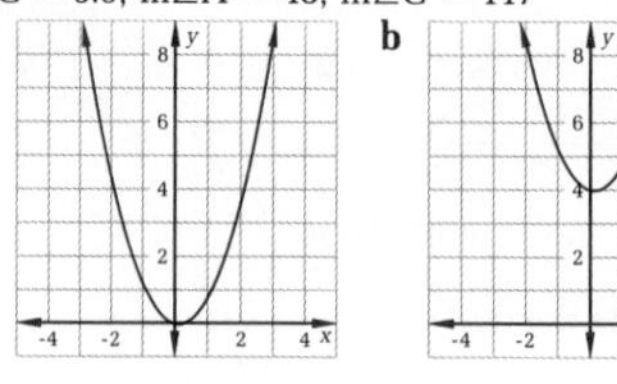

c

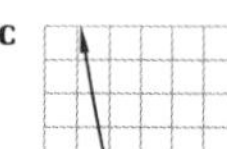

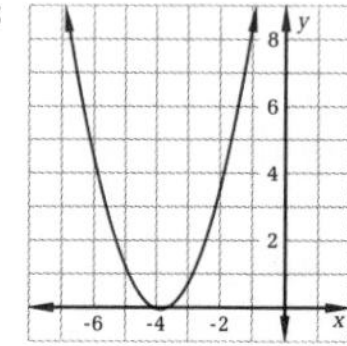

11 $\frac{56}{65}$ **12** y = 2(x − 3)

13

x	0	$\frac{\pi}{4}$	$\frac{\pi}{2}$	$\frac{3\pi}{4}$	π	$\frac{5\pi}{4}$	$\frac{3\pi}{2}$	$\frac{7\pi}{4}$	2π	$\frac{9\pi}{4}$	$\frac{5\pi}{2}$
y	0	$\frac{\sqrt{2}}{2}$	1	$\frac{\sqrt{2}}{2}$	0	$\frac{-\sqrt{2}}{2}$	−1	$\frac{-\sqrt{2}}{2}$	0	$\frac{\sqrt{2}}{2}$	1

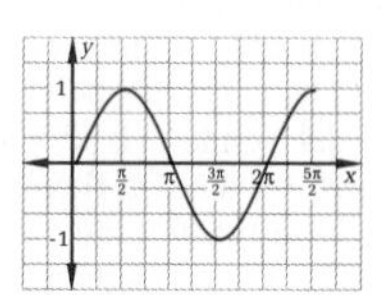

14 ≈83 **15** ≈162 mi **16** ≈836.74 ft **17 a** $x = \frac{\pi}{2} + n \cdot 2\pi$ **b** $x = 0 + n \cdot \pi$ **c** $x = \frac{3\pi}{2} + n \cdot 2\pi$ **18** ≈94.05 **19 a** 10; 25; ≈18.03 **b** ≈37° **21**

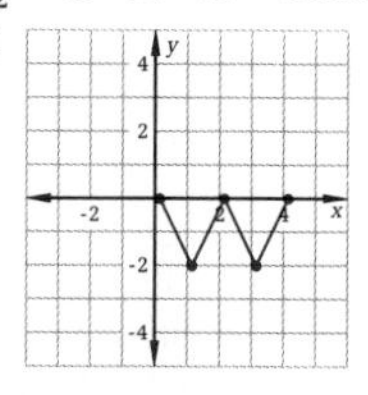

23 ≈62° or ≈118° **24** Impossible **25** ≈6.3
26 a ≈20.62 **b** ≈26.93 **c** ≈22.36 **d** ≈48.37
27 ≈82.8 **28** ≈45.77 **29 a** ≈76.29 **b** ≈68.67
30 a 30° or 150° **b** ≈9 or ≈22.3 **31** ≈5.74 **32 a** 1; −1 **b** 6; 4 **c** 1; −1 **d** 6; 4 **33** By the Pythagorean Theorem,

$h^2 = c^2 - x^2$
$h^2 = b^2 - (a - x)^2$

Therefore,

$c^2 - x^2 = b^2 - (a - x)^2$
$c^2 - x^2 = b^2 - a^2 + 2ax - x^2$
$2ax = a^2 - b^2 + c^2$
$x = \frac{a^2 + c^2 - b^2}{2a}$

But $h^2 = c^2 - x^2 = (c + x)(c - x)$

Therefore, $h^2 = \left(c + \frac{a^2 + c^2 - b^2}{2a}\right)\left(c - \frac{a^2 + c^2 - b^2}{2a}\right)$

$h^2 = \left(\frac{2ac + a^2 + c^2 - b^2}{2a}\right)\left(\frac{2ac - a^2 - c^2 + b^2}{2a}\right)$

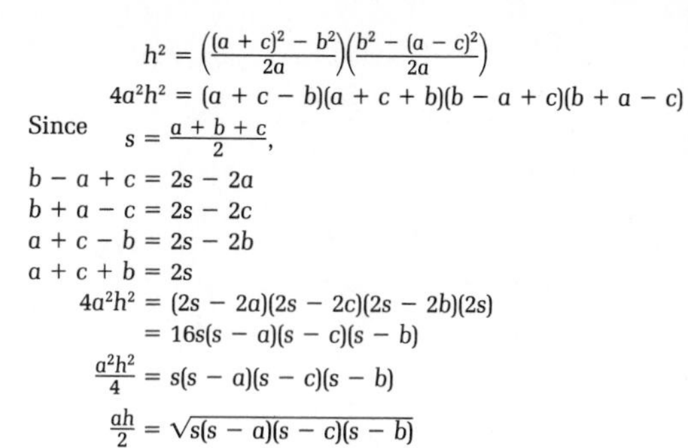

$h^2 = \left(\frac{(a + c)^2 - b^2}{2a}\right)\left(\frac{b^2 - (a - c)^2}{2a}\right)$

$4a^2h^2 = (a + c - b)(a + c + b)(b - a + c)(b + a - c)$

Since $s = \frac{a + b + c}{2}$,

$b - a + c = 2s - 2a$
$b + a - c = 2s - 2c$
$a + c - b = 2s - 2b$
$a + c + b = 2s$

$4a^2h^2 = (2s - 2a)(2s - 2c)(2s - 2b)(2s)$
$= 16s(s - a)(s - c)(s - b)$

$\frac{a^2h^2}{4} = s(s - a)(s - c)(s - b)$

$\frac{ah}{2} = \sqrt{s(s - a)(s - c)(s - b)}$

16.3 Graphing Sine Functions

Warm-up Exercises

1 2.5 **2** Shifted $\frac{\pi}{6}$ units to the right **3** Shifted 4 units down **4** Amplitude is 4.5. **5**

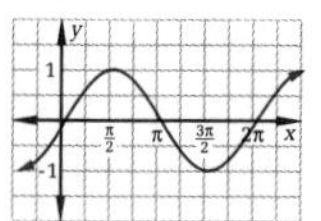

6

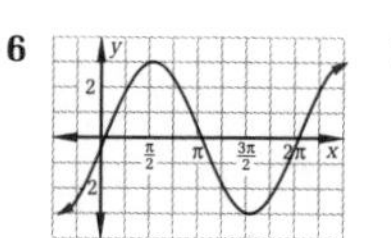

7

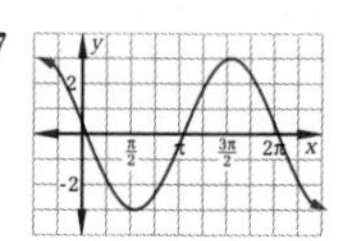

8

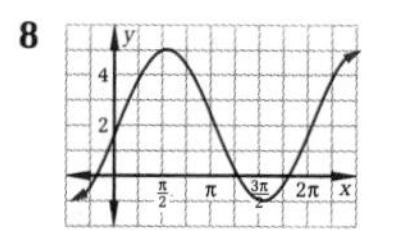

9

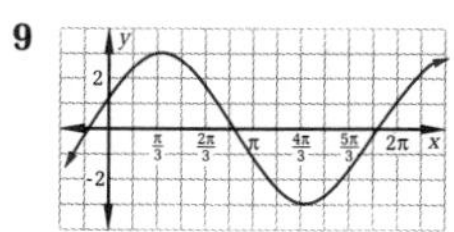

Problem Set

10

x	0	$\frac{\pi}{3}$	$\frac{\pi}{2}$	$\frac{2\pi}{3}$	π	$\frac{3\pi}{2}$	2π	$\frac{5\pi}{2}$	3π
y	1	0.5	0	−0.5	−1	0	1	0	−1

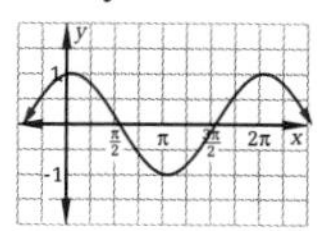

11 y = 2 sin x − 2 **13 a** $\frac{\pi}{6}$ **b** $\frac{\pi}{12}$ **c** $\frac{\pi}{18}$ **d** $\frac{\pi}{3}$ **15** y = 2 sin x **16** x ≈ 25.7; y ≈ 37.6 **17 a** $A\left(\frac{2\pi}{3}, 0\right)$, $B(\pi, -3)$, $C\left(\frac{4\pi}{3}, 0\right)$ **b** $\frac{4\pi}{3}$ **19**

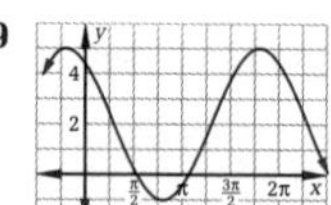

21 Shifted 1 unit up **23 a** 2 **b** π **25** ≈5.41

26 $y = 5 \sin \frac{2}{3}x$

27 $\frac{\cos 8x + \cos 2x}{\sin 8x - \sin 2x} = \frac{2 \cos 5x \cos 3x}{2 \sin 3x \cos 5x}$
$= \frac{\cos 3x}{\sin 3x} = \cot 3x$

28 0 **29** 4π **30** $\frac{\pi}{8}$

16.4 Changing the Period

Warm-up Exercises

1 a

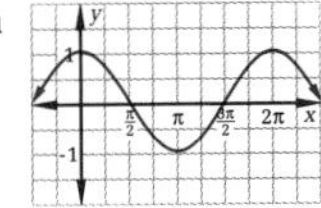

b

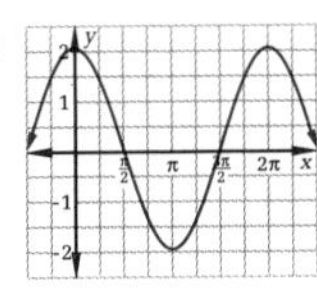

c

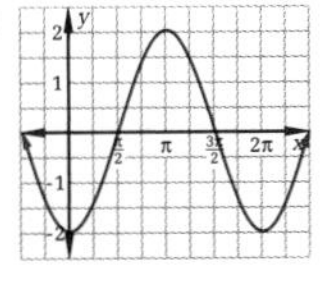

d

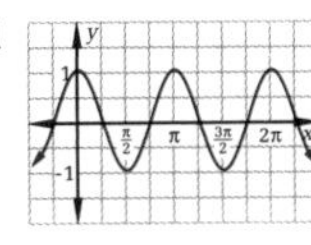

e

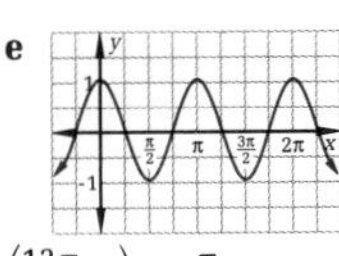

2 $\left(\frac{2\pi}{3}, 0\right)$; $\left(\frac{7\pi}{6}, -2\right)$; $\left(\frac{5\pi}{3}, 0\right)$; $\left(\frac{13\pi}{6}, 2\right)$ **3** $\frac{\pi}{3}$

Problem Set

4 a

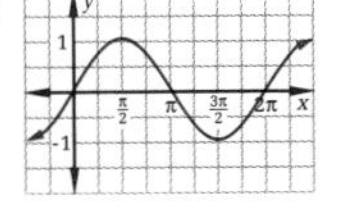

b

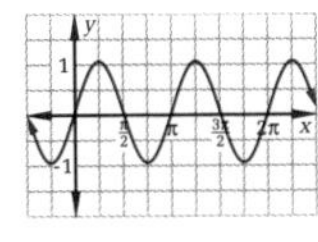

c

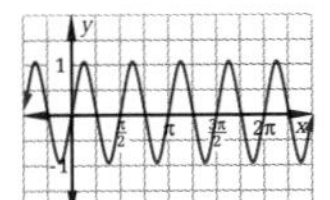

d

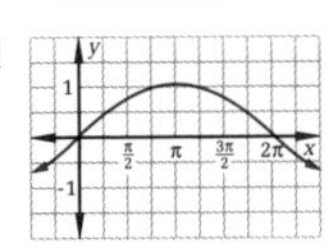

5 a 3; π **b** 5; $\frac{2\pi}{3}$ **c** 32; 2 **6** $y = 4 \cos\left(\frac{1}{4}x\right)$

7 a $y = 3 \sin\left(x + \frac{\pi}{4}\right)$ **b** $y = 3 \cos\left(x - \frac{\pi}{4}\right)$

8

x	0	$\frac{\pi}{6}$	$\frac{\pi}{4}$	$\frac{\pi}{3}$	$\frac{9\pi}{20}$	$\frac{\pi}{2}$	$\frac{11\pi}{20}$	$\frac{2\pi}{3}$	$\frac{3\pi}{4}$	π
y	0	$\frac{\sqrt{3}}{3}$	1	$\sqrt{3}$	≈ 6.3	Und.	≈ -6.3	$-\sqrt{3}$	-1	0

b

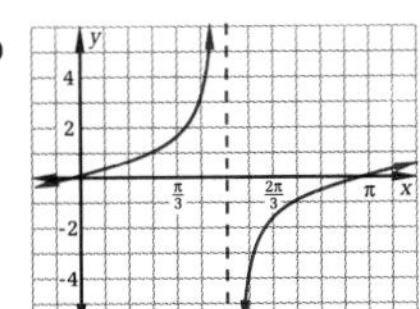

9 a $\frac{12}{13}$ **b** $\frac{5}{13}$ **c** $\frac{3}{5}$ **d** $\frac{4}{5}$ **e** $\frac{56}{65}$

10 a 14 **b** -14 **c** $\frac{1}{7}$ **d** 4 **e** 10 **11 a**

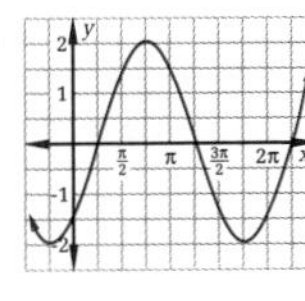

b **c**

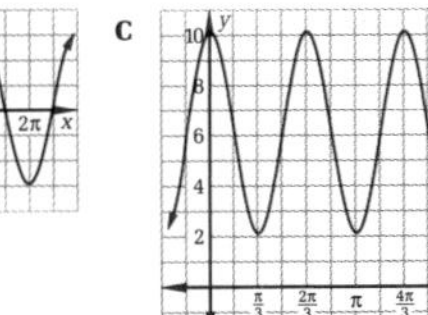

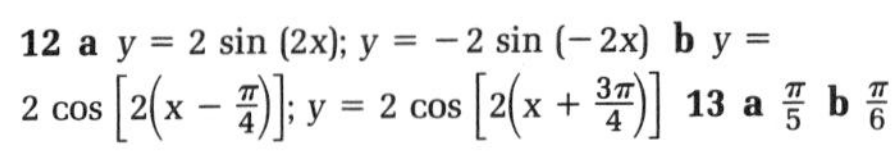

12 a $y = 2 \sin (2x)$; $y = -2 \sin (-2x)$ **b** $y = 2 \cos \left[2\left(x - \frac{\pi}{4}\right)\right]$; $y = 2 \cos \left[2\left(x + \frac{3\pi}{4}\right)\right]$ **13 a** $\frac{\pi}{5}$ **b** $\frac{\pi}{6}$

14 ≈ 101.13 cm **15 a** 84 **b** ≈ 13.6 **c** ≈ 13.6

16 a Domain: $\left\{x: 0 \le x \le \frac{8\pi}{3}\right\}$; range: $\{y: 3 \le y \le 7\}$
b Domain: $\{x: 0 \le x \le 16\pi\}$; range: $\{y: -9 \le y \le -5\}$

17 No period **18 a**

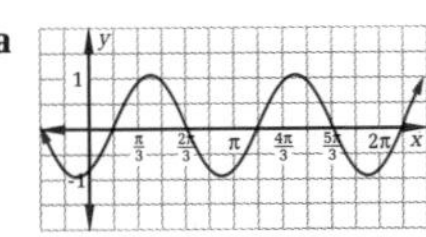

b

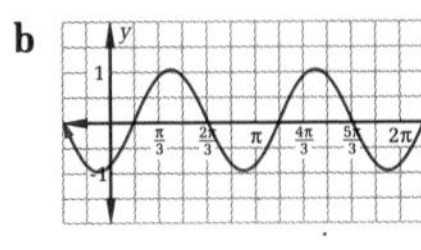

c

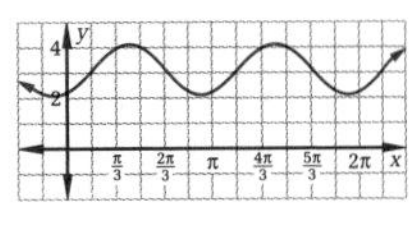

d

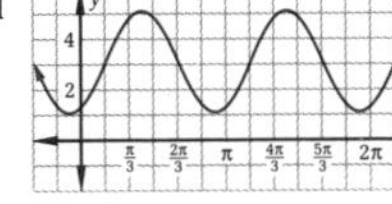

19 a

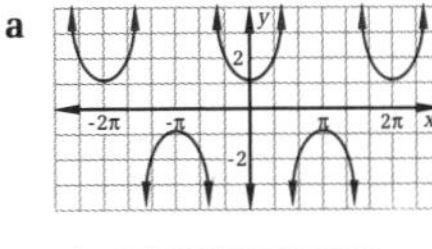

b

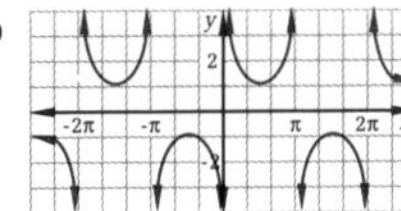

c

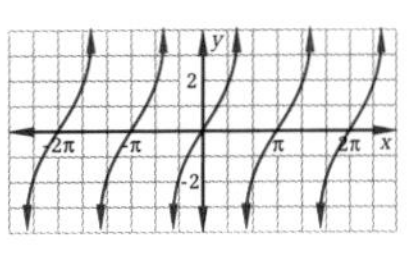

d

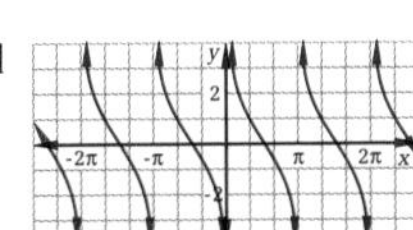

20 $\left\{x: 0 \le x \le \frac{\pi}{3} \text{ or } \frac{5\pi}{3} \le x \le 2\pi\right\}$

21 ≈ 21.96 **22** 1

23

Area of ΔABE
$= \frac{1}{2}(AE)(BE)(\sin \angle AEB)$

Area of ΔADE
$= \frac{1}{2}(AE)(DE)(\sin \angle AED)$
$= \frac{1}{2}(AE)(DE)(\sin \angle AEB)$

Area of ΔBCE
$= \frac{1}{2}(CE)(BE)(\sin \angle BEC)$
$= \frac{1}{2}(CE)(BE)(\sin \angle AEB)$

Area of ΔCDE
$= \frac{1}{2}(CE)(DE)(\sin \angle DEC)$
$= \frac{1}{2}(CE)(DE)(\sin \angle AEB)$

$A_{\Delta ABE} \cdot A_{\Delta CDE}$
$= \frac{1}{2}(AE)(BE)(\sin \angle AEB) \cdot \frac{1}{2}(CE)(DE)(\sin \angle AEB)$
$= \frac{1}{2}(AE)(DE)(\sin \angle AEB) \cdot \frac{1}{2}(CE)(BE)(\sin \angle AEB)$
$= A_{\Delta ADE} \cdot A_{\Delta BCE}$

16.5 More Trigonometric Graphs

Warm-up Exercises

1 a 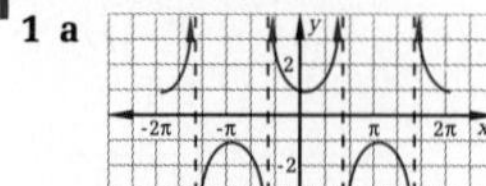 **b** 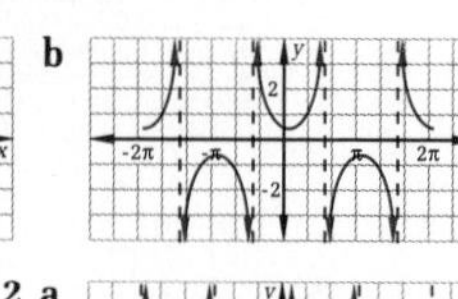

c 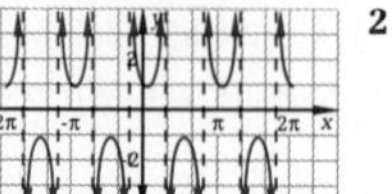 **2 a** 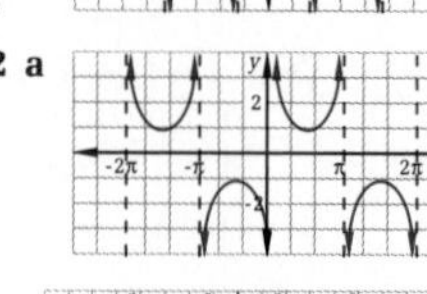

b 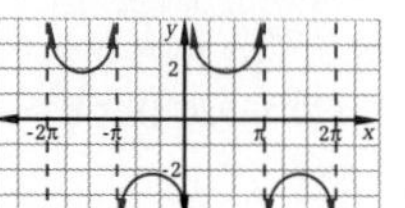 **c** 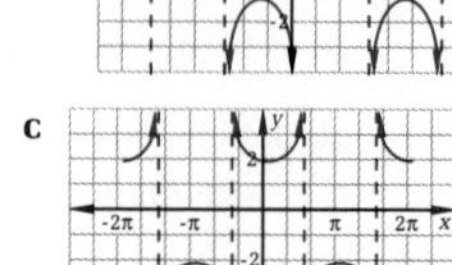

Problem Set

3 a 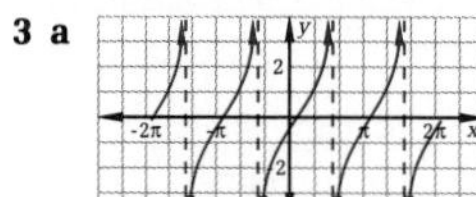 **b** 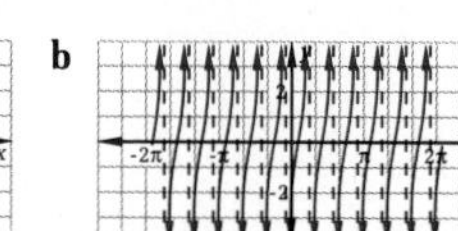 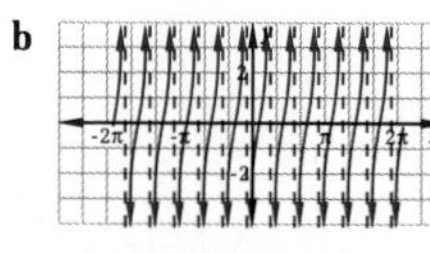

c 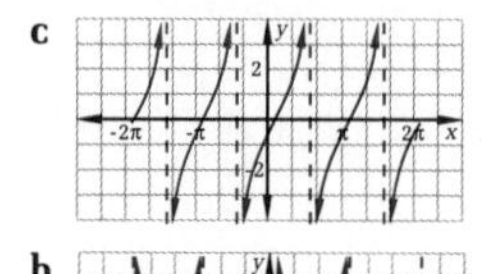 **4 a** 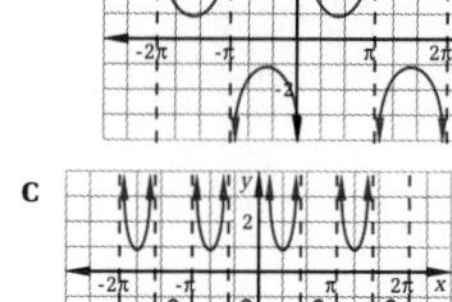

b **c**

d 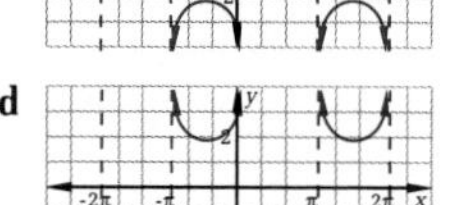 **e** 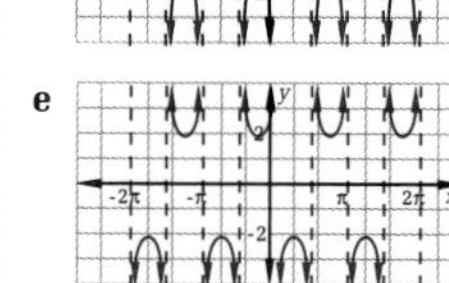

5 a 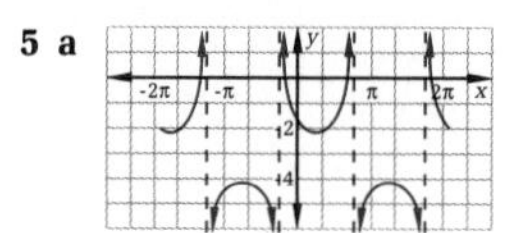 **b** 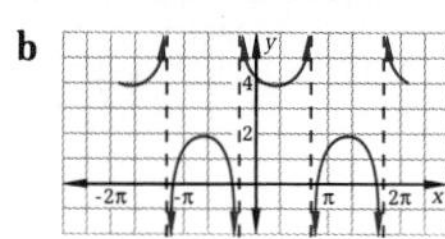

6 $y = \tan(2x)$ **7** ≈ 0.243 **8** ≈ 9.33 *mi*

9 a 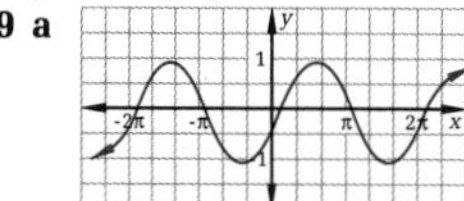 **b** 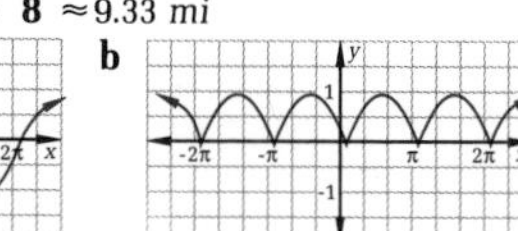

c 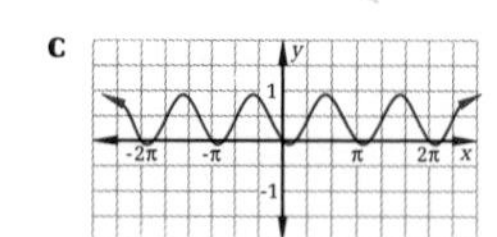

10 $\frac{2}{5}$ **11** $\approx 19{,}199.59\ cm^2$ **12 a** $\frac{8}{3}$ **b** $0 < m\angle ABC < 90$ **13** 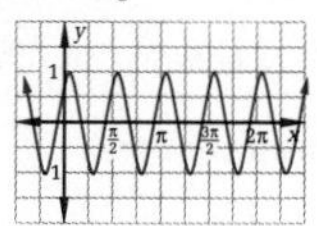

14 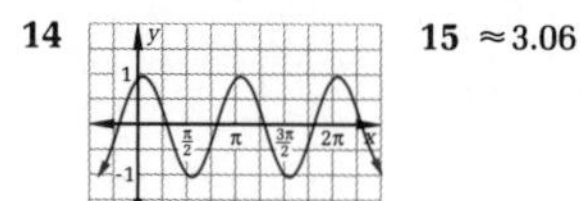 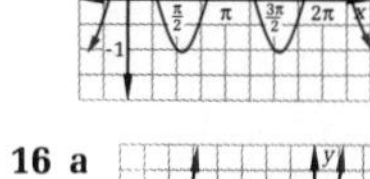 **15** ≈ 3.06

16 a 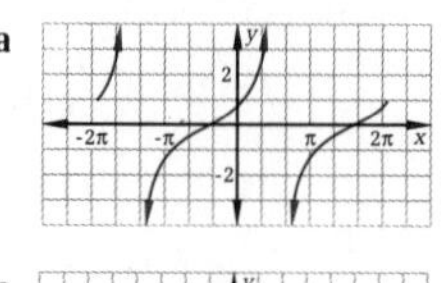 **b** 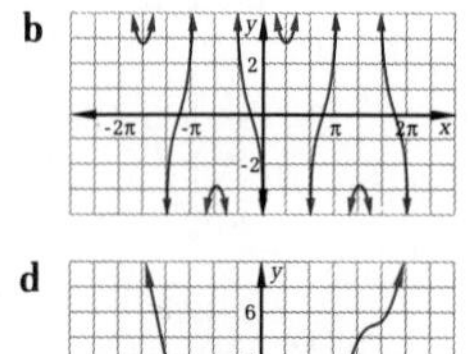

c 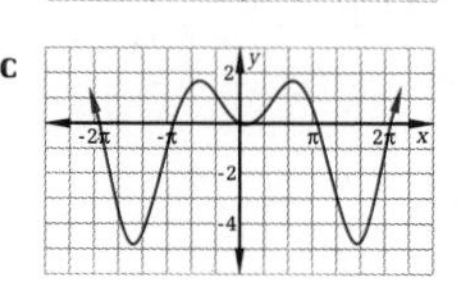 **d** 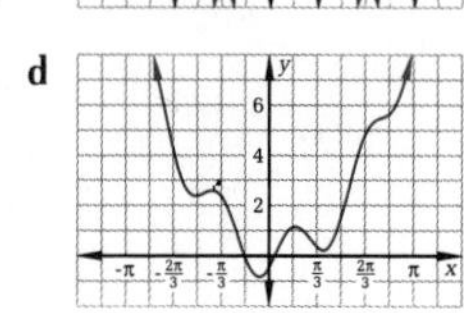

Review Problems

1 a 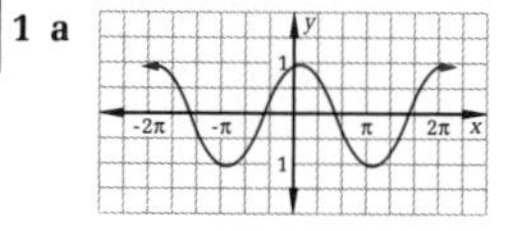 **b** 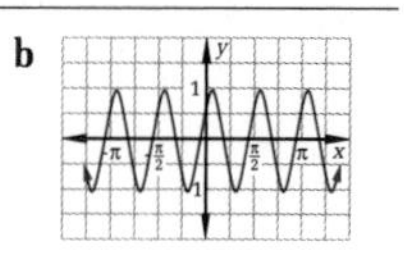

c 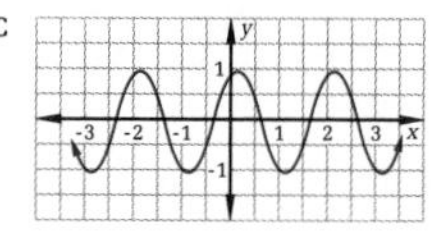

2 $\alpha \approx 71.8$; $\beta \approx 58.8$; $\gamma \approx 18.2$

3 $x \approx 32.71$; $y \approx 20.10$ **4 a** $\frac{1}{2}$ **b** $\frac{1}{2}$ **c** $\frac{1}{2}$ **d** $\frac{1}{2}$ **e** $\frac{1}{2}$

5 ≈ 15.665 **6 a** 3.5 **b** 8

7

Equation	Amplitude	Period	Range
$y = -2\sin(3x)$	2	$\frac{2\pi}{3}$	$\{y: -2 \le y \le 2\}$
$y = 4\cos(2x)$	4	π	$\{y: -4 \le y \le 4\}$
$y = 3\sin\left(\frac{1}{3}x\right) + 5$	3	6π	$\{y: 2 \le y \le 8\}$

8 a 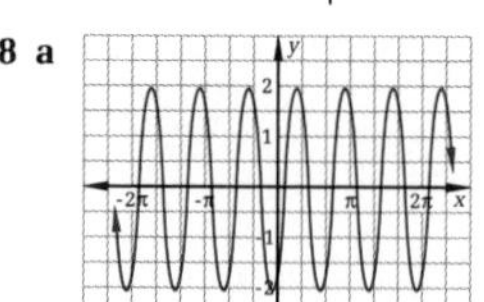 **b** 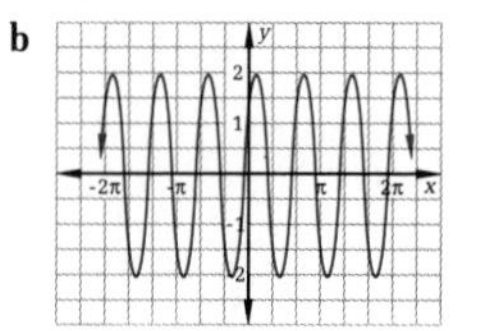

9 a 1 **b** 2 **c** 0 **10** $x \approx 11.6$; $y \approx 8.1$; $z \approx 47.3$

11 ≈ 89 **12** $\frac{\pi}{6}, \frac{5\pi}{6}$ **13** $y = 4\sin x$

14 a,b,c 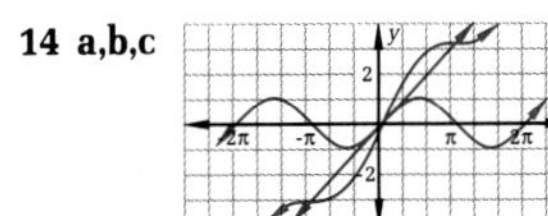

15 a Domain: $\left\{x: 0 \le x \le \frac{2\pi}{3}\right\}$; range: $\{y: -4 \le y \le 4\}$

b $\frac{\pi}{9}, \frac{5\pi}{9}$ **16** $\frac{2\pi}{5}$ or $\frac{-2\pi}{5}$ **17** $\frac{1}{2}$

18 $x_1 + x_2 = \pi$ **19 a** 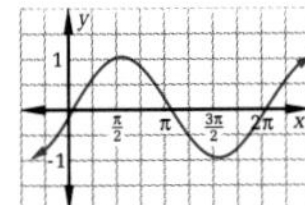

b 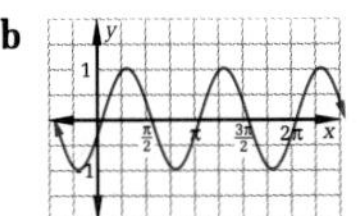 **c** 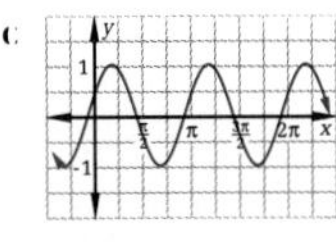

d 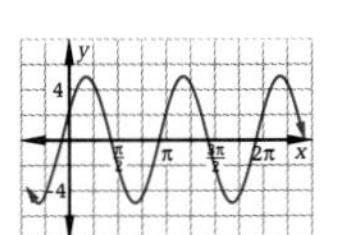 **e** 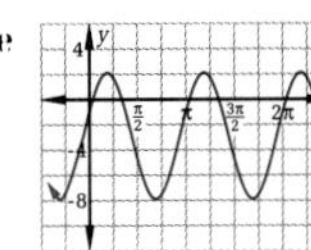 **20** $\approx 84.1\%$

21 ≈ 106.49; ≈ 74.56 **22** ≈ 282.14 cm

Glossary

abscissa The first of an ordered pair of coordinates; an x-coordinate. (p. 45)
absolute value The numerical value of an expression considered without regard to its sign. The absolute value of a number a, symbolized $|a|$, is equal to a if a is positive or zero or to $-a$ if a is negative. (p. 24)
algorithm A step-by-step procedure for solving a problem in a finite number of steps. (p. 13)
amplitude The maximum distance of a sine or cosine graph from its axis. (p. 706)
antilogarithm The number corresponding to a base raised to a logarithmic power. For example, since $\log 100 = 2$, the base-10 antilogarithm of 2 is 100. (p. 481)
arithmetic mean **1** One of the terms between two designated terms of an arithmetic sequence. (p. 577) **2** See ***mean.***
arithmetic sequence A sequence in which the difference of any two successive terms is constant. (p. 576)
arithmetic series A representation of the sum of a specified number of terms of an arithmetic sequence. (p. 577)
array A finite number of successive terms of a sequence. (p. 564)
asymptote A line that is approached but never reached by a curved graph. (p. 269)
axis of symmetry A line that divides a graph into two parts that are mirror images of each other. (p. 142)

base In an exponential expression, the quantity that is raised to a power. (p. 70)
bimodal Having two modes. (p. 229)
binomial probability experiment Any probability experiment that can be modeled by a binomial expansion. Such an experiment consists of a finite number of independent trials, with each trial resulting in one of two possible outcomes and with the probabilities of these outcomes being the same for each trial. (p. 570)
binomial theorem A formula for the expansion of expressions of the form $(x + y)^n$, where n is a positive integer. (p. 412)
bisection A method of approximating a real zero of a function by repeated averaging of values on either side of the zero. (p. 398)
boundary-value algorithm A procedure for determining the solutions of inequalities by locating the specific regions of the real-number line or the coordinate plane that represent the solutions. (p. 94)
box-and-whisker plot A statistical diagram showing the outliers and the quartiles of a set of data values. (p. 173)

central angle An angle whose vertex is the center of a circle. (p. 603)
circumference The perimeter of a circle. (p. 731)
closed-form formula A formula that shows the relationship between any input of a function and its corresponding output. (p. 221)
coefficient The numerical factor in a monomial. (p. 729)

cofunctions A pair of trigonometric functions that output equal values when their inputs are angles complementary to each other. Sine and cosine, tangent and cotangent, and secant and cosecant are cofunctions. (p. 633)

combination function A function that for a pair of inputs (n, r) outputs the number of groups of r elements that can be selected from a set of n elements. (p. 214)

common difference The difference between any two successive terms of an arithmetic sequence. (p. 576)

common factor A factor shared by all the terms of a polynomial. (p. 135)

common logarithm A base-10 logarithm. (p. 480)

common ratio The ratio between any two successive terms of a geometric sequence. (p. 583)

completing the square Rewriting a quadratic polynomial so that it contains a perfect-square trinomial. (p. 157)

complex fraction A rational expression containing a fraction in the numerator, the denominator, or both. (p. 454)

complex number Any number that can be expressed in the form $a + bi$, where $i = \sqrt{-1}$ and a and b are real numbers. The term a is known as the number's real part, and the term bi is known as its imaginary part. (p. 359)

composite function A function made up of two other functions, with the outputs of one being used as the inputs of the other. (p. 190)

compound function A function in which different rules are applied to different inputs. (p. 114)

conditional probability The probability that a certain event will occur, given that another event has previously occurred. (p. 543)

conjugates **1** A pair of expressions of the form $a + b\sqrt{c}$ and $a - b\sqrt{c}$, whose product will always be rational if a, b, and c are rational. (p. 73) **2** A pair of complex numbers of the form $a + bi$ and $a - bi$, whose product will always be a real number. (p. 359)

conjunction A statement in which two possible outcomes are connected by *and*, representing the intersection of two events. (p. 542)

constant of variation The value of the quotient of two quantities that vary directly or of the product of two quantities that vary inversely. Also called *constant of proportionality*. (p. 101)

constraint One of the restrictions that defines the domain in a linear-programming problem, represented by a linear inequality. (p. 467)

cosecant A trigonometric function equivalent to the reciprocal of an angle's sine. (p. 611)

cosine A trigonometric function equivalent to the x-coordinate of the point where the terminal ray of an angle in standard position intersects a unit circle. (p. 610)

cotangent A trigonometric function equivalent to the reciprocal of an angle's tangent. (p. 611)

coterminal angles Any angles that have the same terminal ray when in standard position. (p. 604)

degree (of a monomial) The sum of the powers to which the variables in a monomial are raised. (p. 729)

degree (of a polynomial) The greatest of the degrees of a polynomial's terms. (p. 730)

dependent events Two events such that the occurrence of one affects the probability of the occurrence of the other. (p. 543)

deviation score The difference between a particular value in a set of data and the set's mean. (p. 276)

directed graph A network whose segments indicate direction. Also called *digraph*. (p. 7)

direct variation A relationship between two variable quantities in which the quantities' quotient is constant. (p. 101)

discrete function A function whose domain is a set of numbers that can be listed. (p. 213)

discriminant The expression $b^2 - 4ac$ associated with a quadratic equation in the form $ax^2 + bx + c = 0$. (p. 150)

disjunction A statement in which two possible outcomes are connected by *or*, representing the union of two events. (p. 535)

domain The set of all inputs for which a function produces meaningful outputs. (p. 184)

ellipse A conic section that is the graph of an equation of the form

$$\frac{(x-h)^2}{a^2} + \frac{(y-k)^2}{b^2} = 1$$

where a, b, h, and k are real. (p. 284)

equation A mathematical sentence stating that two expressions have the same value. (p. 30)

even function Any function f for which $f(-x) = f(x)$ for all values of x. (p. 646)

event The element or elements in a sample space that satisfy specified conditions. (p. 529)

expand To rewrite a factored expression as a single polynomial. (p. 127)

exponent In an exponential expression, the number indicating how many times the base is used as a factor. (p. 70)

exponential equation An equation containing a variable in an exponent. (p. 495)

expression A mathematical phrase consisting of a number, a variable, or a group of numbers and variables connected by operations. (p. 23)

extraneous root A value of the variable that is obtained by solving an equation but which is not in fact a solution of the equation. (p. 350)

factor To rewrite an expression as the product of its factors. (p. 127)

factorial function A function that for a nonnegative integral input n outputs the number of ways in which n items can be arranged. (p. 213)

Fibonacci numbers The elements of the infinite sequence 1, 1, 2, 3, 5, 8, 13, . . . , in which the first two elements are 1's and each succeeding element is the sum of its two predecessors. (p. 223)

first differences The differences between successive outputs of a recursively defined function. (p. 223)

fitted line A line whose equation approximately represents a linear relationship between data values. (p. 107)

function A relation in which each input is paired with exactly one output. (p. 184)

geometric mean One of the terms between two specified terms of a geometric sequence. (p. 584)

geometric sequence A sequence in which the quotient of any two successive terms is constant. (p. 583)

geometric series A representation of the sum of a specified number of terms of a geometric sequence. (p. 585)

hyperbola A conic section that is the graph of an equation of the form

$$\frac{(x-h)^2}{a^2} - \frac{(y-k)^2}{b^2} = 1 \text{ or}$$

$$\frac{(y-k)^2}{b^2} - \frac{(x-h)^2}{a^2} = 1$$

where a, b, h, and k are real. (p. 284)

identity An equation that makes a true statement for all possible values of the variable(s). (p. 729)

imaginary number See ***pure imaginary number.***

independent events Two events such that the occurrence of one does not affect the probability of the occurrence of the other. (p. 542)

inequality A mathematical sentence stating that two expressions are not equal. (p. 38)

initial ray The ray at which the rotation of an angle begins. (p. 603)

intercept A point at which a graph crosses one of the coordinate axes. (p. 46)

intersection The set of all elements common to two sets. (p. 95)

inverse functions Any pair of functions f and g such that $f[g(x)] = g[f(x)] = x$ for all values of x in the domains of f and g. (p. 191)

inverse variation A relationship between two variable quantities in which the quantities' product is constant. (p. 313)

lattice point A point having integral coordinates. (p. 21)

leading coefficient The coefficient of the first term of a polynomial written in descending order of powers. (p. 730)

like terms Terms containing the same variables raised to the same powers. (p. 23)

linear Having a degree of one. (p. 30)

linear programming A method of finding the maximum or minimum output of a function whose domain can be represented by a polygonal region of the coordinate plane. (p. 467)

line of best fit A line whose equation, determined by the method of least squares, represents a linear relationship between data values. Also called *regression line*. (p. 108)

logarithm The power to which a given base must be raised to equal a given number. (p. 475)

lower bound of zeros A value that is less than any of the real zeros of a given polynomial function. (p. 396)

mathematical induction A method of proving a statement true for all natural numbers by showing that it is true for the first natural number and that if it is true for a number n, it is also true for $n + 1$. (p. 597)

matrix A rectangular arrangement of data. (p. 52)

mean A measure of central tendency equal to the sum of a set of data values divided by the number of values. (p. 228)

mean deviation A measure of variation equal to the sum of the absolute values of the deviation scores of a set of data values divided by the number of values. (p. 276)

median A measure of central tendency equal to the middle value (or the mean of the two middle values) in an ordered set of data values. (p. 173)

mode A measure of central tendency equal to the value that occurs the greatest number of times in a set of data values. (p. 228)

model A mathematical or diagrammatic representation of a situation. (p. 4)

monomial A real number, a variable, or a product of real numbers and variables. (p. 729)

multiplication-addition method An algebraic method of solving a system of equations. (p. 88)

multiplicity The number of times a particular value is a zero of a polynomial function. (p. 404)

mutually exclusive Having no outcome in common. (p. 536)

natural logarithm A base-e logarithm. (p. 481)

nested form A form of polynomial expression used to simplify calculator and computer calculations. (p. 381)

network A geometric model consisting of points joined by segments. (p. 10)

normal curve The graph of a normal distribution. (p. 372)

normal distribution A set of data values whose frequencies decrease symmetrically about a single value that corresponds to the data's mean, mode, and median. (p. 372)

odd function Any function f for which $f(-x) = -f(x)$ for all values of x. (p. 646)

opposite A number equal in absolute value but opposite in sign to a given number. (p. 24)

ordered pair A pair of numbers in the form (x, y), used to represent a solution of an equation in two variables or an input and corresponding output of a relation. (p. 31)
ordinate The second of an ordered pair of coordinates; a y-coordinate. (p. 45)
origin The intersection of the axes of a coordinate plane. Its coordinates are (0, 0). (p. 45)
outlier scores The greatest and the least value in a set of data values. (p. 173)

parabola A conic section that is the graph of an equation of the form
$y = a(x - h)^2 + k$
where a, h, and k are real. (p. 142)
partial sum of a sequence The sum of a specified number of terms (beginning with the first) of a sequence. (p. 564)
Pascal's triangle A triangular arrangement of numbers in which each row begins and ends with a 1 and each interior element is the sum of two adjacent elements in the preceding row. (p. 410)
percentile rank A number representing the percentage of the data values in a set that are less than a given value. (p. 229)
periodic Repeating the same pattern over and over. The repeated pattern in a periodic graph is called a cycle, and its length is called the period of the graph. (p. 705)
permutation function A function that for a pair of inputs (n, r) outputs the number of arrangements of r elements that can be selected from a set of n elements. (p. 214)
phase shift A leftward or rightward shift of the graph of a periodic function. (p. 706)
polynomial An expression consisting of a monomial or a sum or difference of monomials. (p. 729)
probability The ratio of the number of elements in a specified event to the number of elements in the sample space of which it is a part. (pp. 60, 529)
proportion An equation of the form $\frac{a}{b} = \frac{c}{d}$, where a, b, c, and d are real numbers, $b \neq 0$, and $d \neq 0$. (p. 102).
pure imaginary number A number of the form ki, where k is real and nonzero and $i = \sqrt{-1}$. (p. 358)

quadrant One of the four regions into which the axes divide the coordinate plane. (p. 45)
quadratic Having a degree of two (p. 127)
quartile A value corresponding to the 25th, the 50th, or the 75th percentile of a set of data. (p. 173)

radian A unit of angular measurement equivalent to the measure of a central angle that intercepts a 1-unit-long arc of a unit circle. (p. 640)
radicand The expression under a radical sign. (p. 344)
range **1** The set of all meaningful outputs of a function. (p. 184) **2** A measure of dispersion equal to the difference between the greatest value and the least value in a set of data. (p. 275)
rational equation An equation that contains at least one rational expression. (p. 435)
rational expression Any expression that can be written as a ratio of polynomials. (p. 427)
recursive Involving a specified beginning value and a formula for obtaining each succeeding value from its predecessor(s). (p. 16)
reference angle For an angle in standard position, the first-quadrant angle whose terminal ray forms the same acute angle with the x-axis. (p. 619)
reflection The mirror image of a graph with respect to a specified line. (p. 200)
relation A set of ordered pairs. (p. 183)
restricted value Any value of a variable for which the denominator of a rational expression is equal to zero. (p. 427)
root A solution of an equation. (p. 144)

root index In a radical expression, the number that indicates the number of times the expression must be used as a factor to produce the radicand. (p. 344)

scalar A quantity by which a matrix is multiplied. (p. 52)

sample A group selected to be representative of an entire population. (p. 552)

sample space The set of all possible outcomes of an experiment. (p. 528)

scientific notation The expression of a number in the form

$c \times 10^n$

where $0 < |c| < 10$ and n is an integer. (p. 730)

secant A trigonometric function equivalent to the reciprocal of an angle's cosine. (p. 611)

second differences The differences between the first differences of the outputs of a recursively defined function. (p. 223)

sequence The ordered outputs of a function whose domain is the set of natural numbers. Also called *progression.* (p. 563)

series A representation of a partial or total sum of the terms of a sequence. (p. 570)

sigma notation A method of expressing the sum of a specified number of terms of a sequence. (p. 12)

sine A trigonometric function equivalent to the y-coordinate of the point where the terminal ray of an angle in standard position intersects a unit circle. (p. 610)

slope The ratio of the change in y values to the change in x values between any two points on a line in the coordinate plane. (p. 79)

standard deviation A measure of dispersion equal to the square root of the variance of a set of data. (p. 417)

standard position The position of an angle whose initial ray coincides with the positive x-axis. (p. 603)

standard radical form The form $a\sqrt{b}$, where a is a rational number and b is an integer having no perfect-square factors. (p. 73)

stem-and-leaf plot A form of organized representation of a set of data. (p. 172)

substitution An algebraic method of solving a system of equations. (p. 88)

synthetic division An algorithm for determining the quotient and the remainder when a polynomial $P(x)$ is divided by a binomial of the form $(x - a)$. (p. 389)

system of equations A group of two or more equations, the solutions of which consist of the values that simultaneously make all of the equations true statements. Also called *simultaneous equations.* (p. 87)

tangent A trigonometric ratio equivalent to the quotient of an angle's sine and cosine. (p. 611)

term **1** One of the elements of a sequence. (p. 563) **2** One of the monomials of which a polynomial consists. (p. 729)

terminal ray The ray at which the rotation of an angle ends. (p. 603)

tree diagram A diagram consisting of branches, each of which represents a possible outcome or choice. (p. 7)

union The set of elements in one or both of two sets. (p. 95)

unit circle A circle with its center at the origin of a coordinate system and a radius of one unit; the graph of $x^2 + y^2 = 1$. (p. 604)

upper bound of zeros A value that is greater than any of the real zeros of a given polynomial function. (p. 396)

variable A symbol used to represent a number. (p. 728)

variance A measure of dispersion equal to the average of the squares of the deviation scores of a set of data values. (p. 417)

Venn diagram A diagram used to represent the relationships among sets. (p. 522)

x-axis The horizontal axis of a coordinate system. (p. 45)

x-coordinate A value representing the distance of a point from the y-axis on a coordinate plane. (p. 45)

y-axis The vertical axis of a coordinate system. (p. 45)

y-coordinate A value representing the distance of a point from the x-axis on a coordinate plane. (p. 45)

zero (*of a polynomial function*) An input value for which a polynomial function outputs a value of zero. The real zeros of a polynomial function correspond to the x-intercepts of the function's graph. (p. 144)

INDEX

A

AAS theorem, 693, 694
Abel, Niels Henrik, 405
Abscissa, 45. *See also* x-coordinate.
Absolute value, 24, 115, 116, 313, 351
  functions, 239
    graphs of , 115, 239–40
  properties of, 115
Acute angle, 632, 633
Addition property
  of equality, 30, 728
  of inequality, 38
Additive inverse, 24
Algorithm, 13
Algorithm, boundary-value. *See* Boundary-value algorithm.
Altitude, 605, 693, 699
Amplitude, 706, 710
Angle(s)
  acute, 620, 632, 633
  central, 603, 640, 666
  complementary, 633, 647
  coterminal, 604
  of depression, 634, 651, 671
  of elevation, 654, 669, 697
  measurement of
    degrees, 603–5, 610–12, 619–20, 626–27, 633, 640–41, 693, 694, 699
    radians, 640–41. *See also* Radian.
  rays of
    initial, 603
    terminal, 603–5, 610, 611, 618–20, 626
  reference, 619, 620, 641
  rotation, 603, 604
  solving triangles, 693–94, 699–700
  standard position of, 603, 604, 610
  supplementary, 647
  trigonometric functions and, 610, 611, 618–20, 626, 632–33, 646–47
  trigonometric identities and, 659
Antilogarithm, 481, 482
Arc, 603
  length, 640
Arccos, 626, 627, 667
Arcsin, 626, 627, 667
Arctan, 626, 627
Area, 127, 128
  of circle, 731
  of ellipse, 363
  of parallelogram, 731
  of rectangle, 731
  of square, 731
  surface
    of rectangular box, 731
    of sphere, 731
  of trapezoid, 731
  of triangle, 731
Area diagram, 136
Arithmetic mean, 577, 584. *See also* Mean.
Arithmetic sequence, 576–77, 583, 584
Arithmetic series, 577–78
Array, 564. *See also* Sequence(s).
Artistic graph, 172
ASA theorem, 694
Associative property
  of addition, 727
  of multiplication, 727
Asymptote, 286, 287, 312, 318
  horizontal, 269, 473, 571, 586
  vertical, 269, 474, 716
Axis of symmetry, 142, 143, 254. *See also* Symmetry.

B

Base, 70, 473, 475
Base-change formula, 494–95, 497
Base
  e, 481–82, 494
  10, 480, 494
Bell-shaped curve, 417, 418, 511
Bimodal, 229. *See also* Mode.
Binomial(s), 128, 143, 509, 730. *See also* Factoring.
  expansion of, 410, 509
  multiplying, 135–36
  Pascal's triangle and, 410–11
  probability and, 509–10
  theorem, 412
Binomial distribution, 509–12
Binomial theorem, 412
Bisection method for approximating real zeros, 398
  computer program for, 400–401
Boundary-value algorithm, 94
  for polynomial inequalities, 338
  for rational inequalities, 435–36
  for systems of inequalities, 306
Box-and-whisker plot, 173, 229
  median and, 173–74
  outlier scores and, 173
  quartiles and, 173, 174, 229

C

Career Profiles, 29, 93, 156, 232, 259, 305, 371, 395, 433, 518, 560, 582, 652, 679, 704
Cartesian coordinate system, 45, 531. *See also* Graph(s); Graphing.
Center (of circle), 128, 318, 603, 604
Central angle, 603, 640, 666
Chart, 172, 529
Chebyshev's Theorem, 419–20. *See also* Standard deviation.
Circle, 603, 604, 640. *See also* Unit circle.
  area of, 731
  center, 128, 318, 603, 604
  circumference of, 603, 640, 731
  as conic section, 284, 287
  origin, 128, 284, 604, 618
  radius, 128, 284, 287, 318, 603, 604, 640
Circle graph, 171
Circumference, 603, 640, 731

Closed-form formula in sequences, 564, 565, 576, 583
Closed-form function, 221
Coefficient. *See also* Factoring; Binomial(s), expansion of.
of binomials, 509
integral, 390
leading, 391, 397, 730
of linear terms, 135, 157
of monomial, 729
in trigonometric functions, 706, 711
Cofunction, 633
identities, 633, 647, 667
Combination function, 214–15, 411–12, 522
Combinatorics, 521. *See also* Counting problems.
Common difference, 576, 577, 578
Common factor, 135, 427
Common multiple, 448, 455. *See also* Multiples.
Common ratio, 583, 584, 585, 586
Commutative property
of addition, 727
of multiplication, 727
Complementary angles, 633, 647
Completing the square, 157, 158, 164, 166, 247, 337, 357
Complex fraction(s)
simplifying, 454–55
Complex number(s). *See also* Pure imaginary number.
arithmetic of, 364–65
conjugate of, 359, 365, 405
imaginary part, 359
real part, 359
standard form, 364
as zeros of polynomials, 366, 404, 405
Composite function, 190, 191
Compound function, 114, 115
Computer graphics, 172
Computer graphing program, 262, 269, 270
Computer programs (BASIC)
approximating real zeros by bisection, 400–401
handshake problem, 13
line of best fit by least-squares method, 110
ordered pairs for polynomial functions, 400
polynomial evaluation, 385
Computer programs (Pascal)
handshake problem, 17
Conditional probability, 543. *See also* Probability.
Conic sections, 284–87
Conjugate
of complex number, 359, 365, 405
of radical expression, 73
Conjugate-Zeros Theorem, 405
Conjuction, 542. *See also* Intersection; Probability.
Constant of proportionality. *See* Constant of variation.
Constant of variation, 101, 313. *See also* Direct variation.
Constraint, 467
Cosecant
complementary/supplementary angles and, 647
coordinates and, 618
graph of function, 717
identities
cofunction, 633, 647
Pythagorean, 612
odd/even functions, 646
relationship with sine, 611–12
use in proving identities, 659–60
Cosine(s)
complementary/supplementary angles and, 647
coordinates and, 618–20
graph of function, 710, 711, 716
identities
cofunction, 633, 647, 667
difference, 667
double-angle, 673
half-angle, 674
product-to-sum, 680, 681
Pythagorean, 612
sum, 666
sum-to-product, 681
with inverse trigonometric functions, 626–27
Law of, 699–700
odd/even functions, 646
solving right triangles and, 632–33
unit circle and, 610–12
use in proving identities, 659–60
Cotangent
complementary/suplementary angles and, 647
coordinates and, 618
graph of function, 717
identities
cofunction, 633, 647
Pythagorean, 612
odd/even functions, 646
relationship with sine/cosine, 611–12
use in proving identities, 661
Coterminal angles, 604
Counting numbers. *See* Natural numbers.
Counting problems, 521, 522
theorem for solving, 522
Cube root, 343, 344
Cubic
equation, 336
expression, 127
function, 260, 263, 404, 405

D

Data analysis, 502
Data displays
bar graph, 171, 173
box-and-whisker plot, 173, 229
chart, 172
frequency polygon, 171
pie graph, 171
stem-and-leaf plot, 172–73
Decimals
repeating, 726
terminating, 726
Degree (angle measurement), 603–5, 610–12, 619–20, 626–27, 633, 640–41, 693–94, 699
Demonstrations. *See* Mathematical demonstrations.
Dependent events, 543. *See also* Probability.
Descending order of powers, 381, 730
Deviation score, 275, 276
Diagonals
number in polygons with *n* sides, 9
Difference identities
cosine, 667
sine, 667, 668
tangent, 668
Difference of two squares, 136, 337. *See also* Factoring.
Difference property, 727
Digraph. *See* Graph(s).
Direct variation, 101, 102, 312
proportion, 102
Directed graph. *See* Graph(s), digraph.
Directly proportional. *See* Direct variation.
Discrete
domain, 221
function, 213
Discriminant. 150, 151, 337. *See also* Quadratics, formula.
Disjunction, 535. *See also* Probability; Union.
Distance
between a point and a line, 447

from origin to point, 618
with sine and cosine, 666
in three-dimensional space, 266
between two points, 47, 128
Distributive property
of exponentiation
over division, 71
over multiplication, 71
of multiplication
over addition, 24, 33, 135, 336, 727, 729
over subtraction, 336, 727, 729
Division property
of equality, 31, 728, 729
of inequality, 38
Domain
and asymptotes, 269
discrete, 221
in functions
exponential, 262, 473
factorial, 213
linear, 101, 184, 190
in logarithmic, 474
in permutation, 214
in reciprocal, 261
in sequences, 563, 564
restrictions on, 190, 199
in trigonometric identities, 660, 676
split, 114
Double-angle identities
cosine, 673
sine, 673
tangent, 673

E

Ellipse
area of, 363
as conic section, 284, 285, 287
major axis of, 285
minor axis of, 285
origin, 285
Ellipsis, 563
use of in sequences, 213, 220, 564, 565, 570, 571, 576, 583, 585, 586
Empty set, 300
Equal elemental probability(ies), 528. *See also* Probability.
Equality, properties of.
Equations, 30, 728. *See also* Systems of equations.
absolute-value, 729
cubic, 336
equivalent, 30
exponential, 495–96
identities, 729
linear
analyzing data, 502
general form, 81
graphing, 79–83, 84. *See also* Graph(s).
identity, 292
slope. *See* Slope.
solving, 31–32, 87–89, 728
systems of, 87–89
logarithmic, 496
and properties of equality
addition, 30, 728
division, 31, 728, 729
multiplication, 31, 728
reflexive, 728
subtraction, 30, 728, 729
transitive, 728
zero product, 31, 33, 145, 149, 336, 350, 728
quadratic, 336, 676, 682. *See also* Quadratics.
quartic, 336–37
radical, 349–51
rational, 435–36
solving, 30–32, 390, 435, 495–96, 728. *See also* Factoring, completing the square; Quadratics, formula.
trigonometric, 681–82
identities. *See* Trigonometry(ic) identities.
with no solutions, 729
Event(s), 529. *See also* Probability.
dependent, 543
independent, 542
mutually exclusive, 536, 543
Expected value (number), 511
Exponential change, 345
Exponential curve, 502
Exponential equation, 495–96
Exponential function, 262, 473–74, 501
applications of, 483, 485, 486, 502
Exponent(s), 70, 730. *See also* Functions, exponential.
negative, 70
power rule, 70, 71, 343, 489
product rule, 70, 71, 343, 350, 489
quotient rule, 70, 71, 489
rational, 343, 350
scientific notation, 730
Expressions. *See also* Equations.
cubic, 127
equivalent, 23, 25
expanded form, 127, 128. *See also* Quadratics.
language of, 23–25, 728
like terms, 23
linear, 127, 128, 135
order of operations, 25
polynomial, 127, 729–30
quadratic, 127–28
radical, 71–72
rational. *See* Rational expressions.
Extraneous root, 350
Extraneous solutions, 682. *See also* Root(s).

F

Factor(s). *See also* Factoring.
common, 135, 427
difference of two cubes, 337
sum of two cubes, 337
Theorem, 384
Factoring, 336–38, 350, 404
completing the square, 157, 158, 164, 166, 247, 337, 357
difference of two squares, 136, 337
discriminant, 150, 151, 337. *See also* Quadratics, formula.
perfect-square trinomials, 128, 136, 143, 151, 157, 337
quadratics, 135–36, 149–50. *See also* Quadratics, formula.
rational
equations/inequalities, 435–36
expressions, 427–28
functions, 434–35
Factor Theorem, 384. *See also* Factoring; Polynomials.
Fibonacci sequence, 17, 222–23
Finite, 564, 570. *See also* Finite geometric series.
Finite geometric series, 585
Finite series, 570
First differences, 223
45°-45°-90° triangle, 612, 619
Functions. *See also* Relation(s).
absolute value, 239
closed form, 221
relationship with recursive, 221
cofunctions, 633
combination, 214–15, 411–12, 522
composite, 190, 191
compound, 114, 115
cubic, 260, 262, 404, 405
degree of, 223
directly proportional, 101
discrete, 213
domain. *See* Domain.
exponential, 262, 473–74, 501
applications, 483, 485, 486, 502
factorial, 213

graphs of. *See* Graph(s), functions.
input/output, 47, 101, 184, 198, 220, 221, 610, 626
diagram, 184, 185, 198
operations on, 199, 206–7
inverse, 191, 198, 199, 474, 487
inverse trigonometric, 626. *See also* Arccos; Arcsin; Arctan.
linear, 81, 100–101, 142–43
logarithmic, 474–75, 480–82, 501. *See also* Logarithms.
maximum value, 142, 157, 164–65, 705
minimum value, 142, 157, 164–65, 705
notation of, 47, 81
odd/even, 646
permutation, 214, 522
polynomial
Conjugate-Zeros Theorem and, 405
evaluation of, 382
computer program for, 385
Factor Theorem and, 384
Fundamental Theorem of Algebra and, 404
quartic, 261
Rational-Zeros Theorem and, 390–91, 434
Remainder Theorem and, 383
synthetic division and, 389–90
zeros and, 390, 396–99, 404, 405, 434
power, 335
quadratic, 128, 165–66, 223, 390, 405. *See also* Functions, polynomial.
quartic, 261, 404, 405
range. *See* Range (functions).
rational, 434, 435
reciprocal, 261
recursive, 221
relationship with closed-form, 221–22
sequences, 563–65, 570, 571, 576, 577, 578, 583, 584
trigonometric. *See also* Cosecant, Cosine, Cotangent, Secant, Sine, Tangent.
zeros of, 390. *See also* Zero(s).
Fundamental Counting Principle, 521
Fundamental Theorem of Algebra, 404

G

Gauss, Carl Friedrich, 404
Geometric mean, 170, 584–85
Geometric model, 9, 20
Geometric sequence, 583–86
Geometric series. *See* Finite geometric series; Infinite geometric series.
Graph(s)
asymptote. *See* Asymptote.
conic sections, 284–87
digraph, 7, 311
functions
absolute value, 115, 239, 240
compound, 114–15
cubic, 260
exponential, 262, 473–74
inverse, 199–200
linear, 81, 142, 143
logarithmic, 474
polynomial, 398
quadratic, 142–44, 157, 239, 286
quartic, 261
reciprocal, 261
trigonometric, 646, 705–706, 710–11, 716–17
holes, 270
intersection, 95
linear
equations, 46, 79–83
inequalities, 39, 94–95
parallel, 82, 292
perpendicular, 82
periodic nature of, 705, 711
rational inequalities, 436
relations, 183
symmetry. *See* Symmetry.
systems, 87, 300, 306, 466, 467
transformations of
reflection, 200, 240, 252–53, 260, 474
shift, 246, 260, 705–706, 710
shrink, 240, 241, 260, 710
stretch, 240, 241, 260, 710
tree diagram, 7, 529
union, 95
Graphing
calculator/computer, 262, 269, 270, 710
fitting a line to data, 107, 108
method for organizing information, 6
number line, 39, 95
three-dimensional, 274, 300
two-dimensional. *See* Cartesian coordinate system; Graph(s).

H

Half-angle identities
cosine, 674
sine, 674
tangent, 675
Half-turn, 255
Handshake problem, 3–4, 521, 596
computer program for
BASIC, 13
Pascal, 17
Harmonic mean, 478
Historical Snapshots, 134, 251, 297, 356, 409, 441, 549, 590, 665
Hole, 270
Horizontal asymptote, 269, 473, 571, 586
Hyperbola
aysmptotes. *See* Asymptote.
as conic section, 284, 285–86, 287
origin, 285–86
symmetry. *See* Symmetry.
Hypotenuse, 632, 633

I

Identity, 292. *See also* Trigonometry(ic) identities.
properties of
additive, 727
multiplicative, 727
Independent events, 542. *See also* Probability.
Inequality(ies)
solving, 38–39, 435–36
systems of, 306, 466, 467
properties of, 38
Infinite, 223, 563, 564, 571, 596. *See also* Infinite geometric series; Infinite series.
Infinite continued fraction, 460
Infinite geometric series, 586
Infinite series, 571
sum of, 571, 586
Input. *See* Functions.
Integers, 70, 71, 72, 73, 213, 233, 726, 730
Intersection, 95
in counting problems, 522
in probability, 535, 542, 543
and unit circle, 604, 611, 646
Inverse(s)
additive, 24
finding, 198–99
functions, 191, 198, 199–200, 474, 487
trigonometric, 626. *See also* Arccos, Arcsin, Arctan.

notation, 191
properties of
additive, 727, 729
multiplicative, 727
variation, 313. *See also* Direct variation.
Irrational numbers, 726
Irrational zero, 390

L

Lattice points, 21, 529
Law of Cosines, 699–700
Law of Sines, 693–94, 699
LCD, 448, 450. *See also* Multiples.
LCM, 436. *See also* Multiples.
Least common denominator. *See* LCD.
Least common multiple. *See* LCM.
Least-squares method. *See* Method of least squares.
Leg
adjacent, 632, 633
opposite, 632, 633
Legend, 172. *See also* Stem-and-leaf plot.
Like denominators, 442–43, 448, 455
Like terms, 23
Linear
equations, 30, 46, 79–83. *See also* Equation(s).
function, 81, 100–101, 142–43
programming, 466–67
theorem of, 467
Line of best fit, 108, 109, 133, 154
Line of symmetry, 239. *See also* Symmetry.
Locus, 128
Logarithmic equation, 496
Logarithmic function, 474–75, 480–82, 501
Logarithms, 475, 480–82
antilogarithm, 481, 482
applications of, 482, 496, 501, 502, 503–504
common, 480–81
natural, 481–82
properties of, 487–89, 495, 496
Long division, 382, 383. *See also* Polynomials.
Lower quartile, 229
Lucas sequence, 17

M

Major axis, 285
Mathematical demonstrations
formal proofs, 19
informal proofs, 19
Mathematical Excursions, 120, 219, 508, 609, 720
Mathematical induction, 597, 598
Matrix(ces)
addition of, 52
with linear programming, 467
multiplication of, 52, 53–54
scalar multiplication, 52
Maximum value
in functions, 142, 157, 164–65, 705
in linear programming, 466, 467
Mean, 174, 278
arithmetic, 577, 584
in binomial distribution, 511
geometric, 170, 584–85
harmonic, 478
Mean deviation, 275, 276
Measures of central tendency, 228, 275
mean, 228
median, 228, 229. *See also* Box-and whisker plot.
mode, 176, 229
Measures of dispersion, 275
deviation score, 275, 276
mean, 416, 417, 418
mean deviation, 275, 276
median deviation, 277
range, 275
standard deviation, 275, 372, 416–20, 511–12
Median deviation, 277
Method of group averages, 112, 141
Method of least squares, 108, 109, 133, 154
line of best fit
computer program for, 100
Midpoint formula, 46
Minimum value
in functions, 142, 157, 164–65, 705
in linear programming, 466, 467
Minor axis, 285
Mode, 176, 229
Model. *See also* Graph(s).
computer, 551
geometric, 9, 20
network, 10
use in applications, 164, 501–504
Monomial, 729
degree of, 729
Monte Carlo techniques, 551
Multiple(s)
common, 448, 455
least common (LCM), 436
least common denominator (LCD), 448, 450
Multiplication-addition method of solving systems, 88, 89, 298
Multiplication property
of equality, 31, 728
of inequality, 38
of zero, 727
Multiplicity, 404
Mutually exclusive events, 536, 543. *See also* Probability.

N

Natural logarithm, 481–82. *See also* Logarithms.
Natural numbers, 563, 564, 596, 597, 726
Negative
exponent, 70
slope, 80
Nested form, 381, 382, 383, 389
Network, 10. *See also* Model, geometric.
Nonnormal distribution, 419
Normal distribution. *See* Bell-shaped curve.
Number line, 39, 95

O

Opposite. *See* Additive inverse.
Opposite of a quantity property, 727
Opposite reciprocals, 82
Ordered pair(s), 31, 45
computer program for polynomial functions, 400
exponentially related data and, 502, 503
in finding inverses, 198, 199
in finding real zeros, 398
functions and, 185, 618, 666, 705
in graphing
conic sections, 284–87
linear equations, 32, 45, 81
logarithmic functions, 474
nonlinear systems, 318
quadratic equations, 128, 142, 239, 144
reciprocal functions, 261
inverse variation and, 313
linear programming and, 466, 467
probability and, 529, 531
relations and, 183
sequences and, 571, 586
slope and, 80, 81
in solving systems, 87, 89, 299, 306

unit circle and, 604, 610, 611, 612
Order of operations, 25
Ordinate, 45. *See also* y-coordinate.
Origin
circle, 128, 284, 604, 618. *See also* Unit circle.
ellipse, 285
hyperbola, 285–86
parabola, 142, 239
reflection over, 253
symmetry with respect to, 254, 261, 284, 285, 646
Outcome(s), 528, 535, 550. *See also* Probability.
Outlier score, 173. *See also* Box-and-whisker plot.
Output. *See* Functions.

P

Parabola. *See also* Quadratics; Polynomials.
general equation of, 239, 299
graphs. *See* Graph(s), functions, quadratic; Graph(s), transformation; Graph(s), translation.
maximum value, 142, 157, 164–65
minimum value, 142, 157, 164–65
mirror images, 142
origin, 142, 239
symmetry of. *See* Symmetry.
Pascal's triangle, 410–11. *See also* Binomial(s).
Percentile rank, 229, 418
quartiles, 229. *See also* Box-and-whisker plot.
Perfect-square trinomial, 128, 136, 143, 151, 157, 337
Perimeter, 731
Permutation function, 214, 522
Phase shift, 706
Point-slope equation of a line, 80–81
Polynomial(s), 127, 729. *See also* Quadratics; Parabola.
degree of, 127, 729
descending order of powers, 381, 730
evaluation of, 382, 383, 384, 389
computer program for, 385
functions, 261, 382–84, 389, 390, 396, 398, 404, 405, 434
graph of, 398
long division, 382, 383
nested form, 381, 382, 383, 389
ordered pairs. *See* Ordered pairs, computer program for polynomial functions.
rational expression, 427–29. *See also* Rational expressions.
synthetic division, 389, 390, 434
zeros of. *See also* Zero(s).
approximating real, 398–99, 400–401
complex, 366, 404, 405
rational, 391, 396, 397, 398
Positive slope, 80
Power function, 335
Power rule, 70, 71, 343, 489
Probability, 60, 509, 510, 528, 550
basic steps, 60
binomial experiment, 509–10
conditional, 543
conjunction, 542. *See also* Intersection, in probability.
dependent events, 543
disjunction, 535. *See also* Union.
equal elemental, 528
event, 529
formula for, 60
independent events, 542
methods of representation
chart, 529
lattice, 529. *See also* Lattice points.
tree diagram, 529. *See also* Tree diagram.
Monte Carlo techniques, 551
mutually exclusive events, 536, 543
outcomes, 509, 510, 528, 535, 550
sample, 552
sample space, 528, 529, 535, 536
Problem solving
algorithm, 13
find a pattern, 12
handshake problem, 3–4. *See also* Handshake problem.
model, 4, 9, 10. *See also* Model.
organized procedure, 4
process and answer, 3
recursive technique, 16
solve simpler case, 12
systems of equations, 291–92. *See also* Systems of equations.
Products of the roots of a quadratic equation, 405. *See also* Quadratics.
Product rule, 70, 71, 343, 350, 489
Product-to-sum identities, 680–81
Progression, 563. *See also* Sequence(s).
Projectile motion, 128, 165–66
Property(ies)
difference, 727
distributive
of exponentiation, 71
of multiplication
over addition, 24, 33, 135, 336, 727, 729
over subtraction, 336, 727, 729
of equality
addition, 30, 728
division, 31, 728, 729
multiplication, 31, 728
reflexive, 728
subtraction, 30, 728, 729
transitive, 728
zero product, 31, 33, 145, 149, 336, 350, 728
of inequality
addition, 38
division, 38
multiplication, 38
subtraction, 38
of logarithms, 487–89, 495, 496
opposite of a quantity, 727
quotient, 727
triangle-inequality, 99
Proofs
formal, 19
informal, 19
Proportion, 102. *See also* Direct variation.
Pure imaginary number, 357–58. *See also* Complex number(s).
Pythagorean identities, 612, 659, 660, 673
Pythagorean Theorem, 47, 605, 618, 699

Q

Quadrant(s), 45, 46, 312, 604, 605, 619, 626, 641
Quadratics. *See also* Parabola; Polynomial(s).
applications
maximum/minimum, 164–65

projectile motion, 128, 165–66
completing the square, 157, 158, 164, 166, 247, 337, 357
equation, 127, 128, 149–51, 357, 358
expression, 127–28
factoring. See Factoring.
formula, 150, 339, 350, 357, 390, 391, 405, 435, 445
derivation of, 158
descriminant, 150, 151, 337
functions, 128, 165–66, 223, 390, 405
graphs of, 142–44, 157, 239, 286
sum and product of roots, 405
vetex form, 144, 157
Quartic function, 261, 404, 405
Quartile, 173, 174, 418. *See also* Box-and-whisker plot.
Quotient property, 727
Quotient rule, 70, 71, 489

R

Radian (angle measurement), 640, 641, 705, 710, 711
basic facts, 641
Radical(s)
conjugate of, 73
equations, 349–51
expressions, 71–72
interpreting, 71–72
powers and roots, 71. *See also* Root(s).
radicand, 344
root index, 344
sign of, 71–73, 150. *See also* Quadratics, formula.
square roots, 72. *See also* Root(s).
standard form, 73
Radicand, 344
Radius, 128, 284, 287, 318, 603, 604, 640
Range
in dispersion, 275
in functions
exponential, 262, 473
linear, 101, 184, 190
logarithmic, 474
reciprocal, 261
sine, 705
Rational exponents, 343, 350
Rational expression(s)
adding, 442–43, 448–49
complex fractions, 454–55
dividing, 428–29
equations and inequalities, 435–36
multiplying, 428
restrictions, 427, 428, 434, 442, 443
simplifying, 427–28, 442, 443, 454–55
subtracting, 443, 448
Rational function, 434, 435
Rational numbers, 726
Rational zero, 391
Rational-Zeros Theorem, 390, 391, 434
proof of, 390–91
Ray(s)
initial, 603
terminal, 603–5, 610, 611, 618–20, 626
Real numbers, 24, 70, 72, 102, 142, 358, 726
Real solutions. *See* Root(s).
Real zero
approximating, 398–99
computer program for, 400–401
in functions
cubic, 263
polynomial, 398, 404
quadratic, 144, 149
rational, 434
tangent, 716
secant, 716
Fundamental Theorem of Algebra and, 404
Reciprocal, 428, 429
Reciprocal function, 261
Rectangular coordinate system. *See* Cartesian coordinate system.
Recursive formula (sequences), 564, 565, 576, 583, 584
Recursive function, 221
Recursive technique, 16
Reference angle, 619, 620, 641
Reflection
and functions
absolute-value, 240
cubic, 260
exponential, 474
inverse, 200
half-turn, 255
and parabolas, 240, 253
and triangles, 252
Reflexive property, 728
Regression line, 108, 109, 133, 154
computer program for, 110
Relation(s), 183. *See also* Functions.
representation of, 183–84
Remainder Theorem, 383, 384
Right triangle, 632, 633, 693, 699
Root (of equation)
cubic, 260
quadratic, 144, 149, 151, 357
quartic, 337
Root(s). *See also* Radicals(s).
cube, 343, 344
extraneous, 350
fractional exponents and, 344
square, 72
Root index, 344
Rotation, 603, 604

S

SAS theorem, 694, 699, 700
Sample, 552. *See also* Probability.
Sample space, 528, 529, 535, 536. *See also* Probability.
Scalar multiplication, 52
Scientific calculator
use with
inverse trigonometric functions, 626, 627
logarithms, 480, 481, 482, 494, 495
scientific notation, 730
trigonometric functions, 610, 611, 612, 641
Scientific notation, 730
Secant
complementary/supplementary angles and, 647
coordinates and, 618
graph of function, 716
identities
cofunction, 633, 647
Pythagorean, 612
odd/even functions, 646
relationship with cosine, 611–12
use in proving identities, 659–60
Second differences, 223
Sequence(s), 223
arithmetic, 576–77, 583, 584
geometric, 583–84, 585, 586
notation of, 563–65, 570, 571, 576, 577, 583–86
of partial sums, 564–65, 570, 571, 586
terms of, 563–65, 570, 571, 576–78, 583–85
ways to describe, 564
Series, 570–71. *See also* Arithmetic series; Finite geometric series; Infinite geometric series.
Set notation, 529
intersection, 95
union, 95

Shifting (graphs), 246, 260, 705–6, 710
horizontal, 246
vertical, 246
Shrinking (graphs), 240, 260, 710
Sigma notation, 12, 13, 570–71, 585, 586, 596
Simultaneous equations, 87. *See also* Systems of equations.
Sine(s)
complementary/supplementary angles and, 647
coordinates and, 618–20
graph of function, 705–706
identities
cofunction, 633, 647, 667
difference, 667
double-angle, 673
half-angle, 674
product-to-sum, 680
Pythagorean, 612
sum, 667
sum-to-product, 681
with inverse trigonometric functions, 626–27
Law of, 693–94
odd/even functions, 646
solving right triangles and, 632–33
unit circle and, 610–12
use in proving identities, 659–60
Skewed distribution, 278
Slope
constant of variation, 101
equations for lines
point-slope, 80–81
slope-intercept, 81, 668
formula for, 80
functions, 81, 100–101, 142–43
line classification
negative, 80
parallel, 82, 292
perpendicular, 82
positive, 80
undefined, 80
zero, 80
tangent and, 668, 669
Slope-intercept equation of a line, 81, 668
Solving triangles
AAS and ASA, 693–94
Law of Cosines, 699–700
Law of Sines, 693–94
oblique, 693–94
right, 632–33
SAS and SSS, 699–700
SSA, 694
Split domain, 114
Square of a binomial, 129, 157. *See also* Factoring.
Square root, 72
SSA theorem, 694
SSS theorem, 694, 700
Standard complex form, 364
Standard deviation, 275, 372, 416–20, 511–12
Chebyshev's Theorem and, 419–20
Standard position, 603, 604, 610
Standard radical form, 73
Statistics, 417. *See also* Measures of central tendency; Measures of dispersion.
Stem-and-leaf plot, 172–73
Stretching (graphs), 240, 260, 710
Substitution method for solving systems, 88, 89, 605
Subtraction property
of equality, 30, 728, 729
of inequality, 38
Sum and difference of the same terms, 129. *See also* Factoring.
Sum identities
cosine, 666
sine, 667
tangent, 668
Sum of the infinite series. *See* Infinite series.
Sum of the roots of a quadratic equation, 405. *See also* Quadratics.
Sum-to-product identities, 681
Supplementary angles, 647
Surface area
of rectangular box, 731
of sphere, 731
Symmetric property, 728, 729
Symmetry
of angles (trigonometric functions), 620, 646
axis of, 142, 143, 254
of cubic functions, 260
of hyperbolas, 269, 312
line of, 239
of parabolas, 166, 253
of reciprocal functions, 261
Synthetic division, 389, 390, 434
Systems of equations
inequalities, 306, 466, 467
linear, 87–89, 291–92, 298–300, 604
nonlinear, 318, 604
problem solving, 291–92
solving
algebraically, 88–89, 289–300, 291–92, 318, 324, 605
graphically, 87, 291–92, 300, 306, 466, 467
types of solutions, 292, 300
with rational expressions, 324

T

Tangent
complementary/supplementary angles and, 647
coordinates and, 618–20
graph of function, 716
identities
cofunction, 633, 647
difference, 668
double-angle, 673
half-angle, 675
Pythagorean, 612
sum, 668
with inverse trigonometric functions, 626–27
odd/even functions, 646
relationship with sine/cosine, 611
slope and, 668–69
solving right triangles and, 633
unit circle and, 611–12
Term(s). *See* Sequence(s), terms of.
Terminal ray. *See* Ray(s).
30°-60°-90° triangle, 605, 619
Trace(s), 300. *See also* Graphing, three-dimensional.
Transitive property, 728
Tree diagram, 7, 529
Triangle-inequality property, 99
Trigonometry(ic)
equations, 681–82
functions
cosecant. *See* Cosecant.
cosine. *See* Cosine.
cotangent. *See* Cotangent.
graphs of, 705–6, 710–11, 716–17
inverse, 626–27, 667
secant. *See* Secant.
sine. *See* Sine.
tangent. *See* Tangent.
identities
cofunction, 622, 647, 667
difference, 667–68
double-angle, 673
half-angle, 674–75
product-to-sum, 680–81
Pythagorean, 659, 660, 673
strategy for proving, 660
sum, 666–68
sum-to-product, 681
solving triangles
Law of Cosines, 699–700
Law of Sines, 693–94
right, 632–33
Trinomial, 135–36, 150, 157, 730. *See also* Factoring.

U

Undefined slope, 80
Union, 95, 522, 535
Unit circle, 604–5
  Pythagorean identities and, 612
  radian angle measurement and, 640
  reference angle and, 620
  special triangles and
    30°-60°-90°, 619
    45°-45°-90°, 619
  trigonometric functions and, 610, 611, 618, 626, 646, 647
Unlike denominators, 448. *See also* Rational expressions.
Upper quartile, 229

V

Variable
  in expressions and equations, 23, 31, 38, 728
  input, 47
  solving for. *See* Equations, solving; Systems of equations, solving, algebraically.
Variance, 417, 511
Variation. *See* Direct variation; Inverse(s), variation.
Venn diagram, 522
Vertex. *See also* Quadratics.
  of absolute-value functions, 239
  central angle and, 603
  of ellipses, 287
  form, 157
  of hyperbolas, 287, 318
  of parabolas, 142–44, 157, 239, 286
  projectile motion and, 165–66
Vertical asymptote, 269, 474, 716
Volume, 263
  of pyramid, 731
  of rectangular prism, 731
  of right circular cone, 731
  of right circular cylinder, 731
  of sphere, 731

W–X

Whole numbers, 726
x-axis, 45, 603, 605, 619, 632, 668, 716. *See also* x-intercept.
  reflection over, 240, 252
  symmetry about, 254
x-coordinate, 605, 610, 646, 647
x-intercept, 81, 144, 149, 260, 261, 284, 285, 300, 390, 473, 474, 705

Y

y-axis, 45, 632. *See also* y-intercept.
  reflection over, 252
  shrinking along, 240
  stretching along, 240
  symmetry about, 254, 646
y-coordinate, 605, 610, 626, 646, 647
y-intercept, 46, 81, 82, 143, 284, 285, 292, 300, 306, 473, 474

Z

Zero(s)
  complex, 366, 404, 405
  irrational, 390. *See also* Zero(s), complex.
  rational, 391. *See also* Zero(s), real.
    lower bound, 396, 397
    Theorem, 390, 391, 434
    upper bound, 396
  real
    approximating, 398–99, 400–401
    of functions, 144, 149, 263, 398, 404, 434, 716
    Fundamental Theorem of Algebra and, 404
Zero product property, 31, 33, 145, 149, 336, 350, 728
Zero slope, 80
z-intercept, 300. *See also* Graphing, three-dimensional.

Credits

Cover

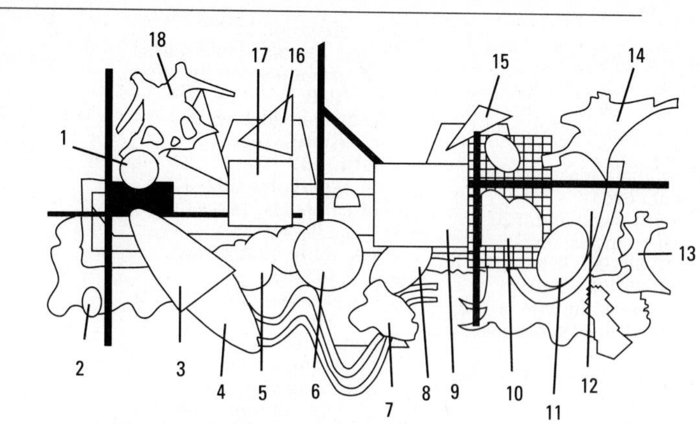

1 Titan, Saturn's largest satellite. JPL, NASA's Regional Planetary Image Facilities; **2** Warbler bird egg. © 1988 Chip Clark, Arlington, Virginia; **3** Saturn's ring system. JPL, NASA's Regional Planetary Image Facilities; **4** Egyptian papyrus *Book of the Dead* (detail). Courtesy of the Trustees of the British Museum, London; **5** Fossil ammonite. © 1988 Chip Clark, Arlington, Virginia; **6** Mesa Verde, Colorado pueblo. Galen Rowell/Mountain Light, Albany, California; **7** Galena crystals. © 1986 Chip Clark, Arlington, Virginia; **8, 12, 15, 16** *Engram* (detail) from a painting by Cathy Halstead. Acrylic on canvas 24″ × 36″, Goldman-Kraft Gallery, Ltd., Chicago; **9** *Dancers*, 1990, Carol Tornatore; **10** Test tube re-creation of Earth's primeval chemistry. B. B. Armstrong and D. W. Deamer, Department of Zoology, University of California, Davis; **11** Chicken egg. © 1988 Chip Clark, Arlington, Virginia; **13, 18** Based on medieval drawings by Villard de Honnecourt: **14** Voyager spacecraft. JPL, NASA's Regional Planetary Image Facilities; **17** Ocean tide whirlpool. Ardea, London.

Fine Art and Photographs

v Milton and Joan Mann, Cameramann International, Park Ridge, Illinois; **vi** (above) Historical Pictures Service, Chicago; (below) JPL, NASA's Regional Planetary Image Facilities; **vii** Based on medieval drawings by Villard de Honnecourt; **viii** (above) United States Postal Service, Washington, D.C.; (below, left) © 1985 Southern Illinois University/Peter Arnold, Inc., New York; (below, right) © Manfred Kage, Peter Arnold, Inc., New York; **x, xiii** © 1988 Chip Clark, Arlington, Virginia; **xi** The Bettmann Archive, New York; **2** The Metropolitan Museum of Art, Michael C. Rockefeller Memorial Collection, Bequest of Nelson A. Rockefeller, 1979 (1979.206.394); **22** (detail) Courtesy the Trustees of the British Museum, London; **29** (left) Carmine Fantasia; (above, right) The Bettmann Archive, New York; (below, right) The Granger Collection, New York; **78** The British Library, London; **93** Amy Iezzoni; **120** Illinois Bell Telephone Company/Milton and Joan Mann/Cameramann International, Park Ridge, Illinois; **126** *Bush Tucker Dreaming at Wingellina*, 1988, William Sandy (Pitjantjatjarra tribe), Courtesy John Weber Gallery, New York; **156** (left) © 1977 Sidney Harris, New Haven, Connecticut: (above, right) Michael Marsland/Yale University, New Haven, Connecticut; (below, right) Figure from *The Physics of Baseball* by Robert K. Adair, © 1990 by Robert K. Adair, Reprinted by permission of Harper & Row, Publishers, Inc., New York; **182** Algimantas Kezys, Stickney, Illinois; **232** Corporate Photographers, Inc., New York; **238** Paul A. Otto, Lake St. Louis, Missouri; **239** (detail) Paul A. Otto, Lake St. Louis, Missouri; **259** Carmine Fantasia; **290** © Georg Gerster/Comstock, New York; **297** (above) U.S. Postal Service, Washington, D.C.; (below) Library of Congress, Washington, D.C.; **305** (left) Ken Kohl; (right) © Camerique/H. Armstrong Roberts, Chicago; **342** *Ghost Chamber with the Tall Door*, 1925, Paul Klée. The Metropolitan Museum of Art, Berggruen

Klée Collection, 1987 (1987.455.16); **356** © Camerique/H. Armstrong Roberts, Chicago; **357** *Prayer Wheel*, 1954, Jean Tinguely. The Museum of Fine Arts, Houston. Museum purchase with funds provided by D. and J. de Menil; **371** Richard Dicker; **380** © Georg Gerster/Comstock, New York; **395** (left) Leonard Ramist; (right) © Henley and Savage/TSW/Click Chicago; **409** Reprinted by permission of the University of Minnesota Press, ©1957, 1974 The University of Minnesota, Renewed 1985 by The Estate of Oystein Ore; **426** *Untitled*, 1960, Lee Bontecou, The National Trust for Historic Preservation, Nelson A. Rockefeller Collection; **433** Rob Mustard Photography, Redondo Beach, California; **441** (detail) (below) Courtesy the Trustees of the British Museum, London; **472** Dennis Murphy, Dallas; **508** (left) UPI/Bettmann Newsphotos, New York; (right) © Keven Schafer/Peter Arnold, Inc., New York; **518** Stan Barough; **520** *Japanese Rain on Canvas*, 1972, David Hockney, acrylic on canvas, 48″ × 48″, © 1972 David Hockney; **549** (both) The Bettmann Archive, New York; **560** (left) Pat Linnell, Washington, D.C.; (right) © 1990 The Time, Inc. Magazine Company, New York, Reprinted by permission; **562** Weese Hickey Weese Architects, Ltd., Chicago; **582** (above) David Williams; (below) The Stock Shop, New York; **590** © 1977 Sidney Harris, New Haven, Connecticut: **602** *Construction in Space with Balance on Two Points*, 1925, Naum Gabo, Yale University Art Gallery, Gift of H. Wade White; **618** Milton and Joan Mann/Cameramann International, Park Ridge, Illinois; **652** (above) Instock/Don Rogers, Fort Lauderdale, Florida; **658** © Richard Hamilton Smith, St. Paul; **659** Hugh Sitton/TSW/Click Chicago; **665** Historical Pictures Service, Chicago; **692** *Variegation in the Triangle*, 1927, Wassily Kandinsky, Galerie Adrien Maeght, Paris; **704** (above) Carmine Fantasia; (below, left) © 1985 SIU/Peter Arnold, Inc., New York; (below, right) © Manfred Kage/Peter Arnold, Inc., New York; **720** (above) © Richard Megna/Fundamental Photographs, New York; (below) © Cecile Brunswick/Peter Arnold, Inc., New York.

Illustrations

Pawel Bodytko v, 30, 35, 45, 52, 60.

Kathleen Mattingly 295, 303, 315, 327.

Carol Tornatore xii, 12, 19, 364, 442, 680.

Robert Voigts 102, 110, 112, 113, 120, 135, 154, 162, 164, 165, 197, 198, 214, 228, 243, 250, 293, 301, 308, 336, 381, 416, 418, 432, 437 (both), 471, 516, 538, 545, 552, 553, 563, 581, 625, 710.

McDougal, Littell and Company has made every effort to locate the copyright holders for the images used in this book and to make full acknowledgment for their use.

Additional Answers

CHAPTER 1

Section 1.2

4 Possible answer:

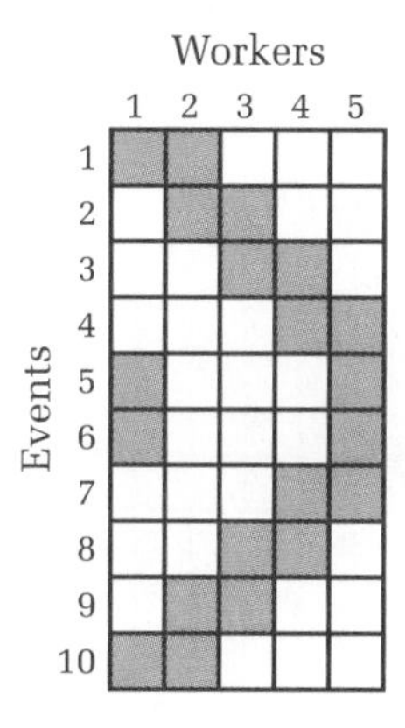

CHAPTER 2

Section 2.3

24b Possible table:

x	−3	−2	−1	0	1	2
y	−2	−2	−2	−2	−2	−2

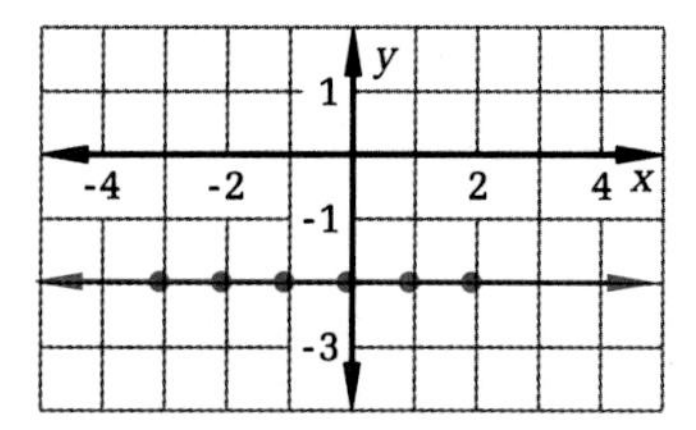

46

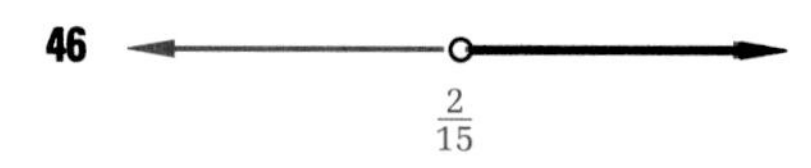

47

29.6

48

26 31

49

5

50

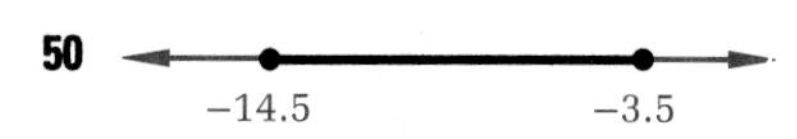

51

−1.4 1.4

53

2

54

4

55

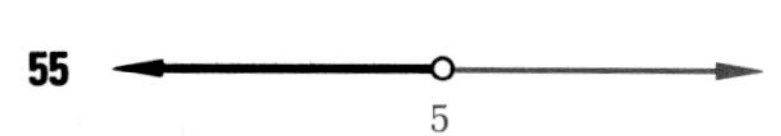

Section 2.5

17a $\begin{bmatrix} -71 & -52 \\ 36 & 62 \end{bmatrix}$ **17b** $\begin{bmatrix} 59 & 13 \\ -4 & 42 \end{bmatrix}$

17c $\begin{bmatrix} -52 & -71 \\ 62 & 36 \end{bmatrix}$

18a

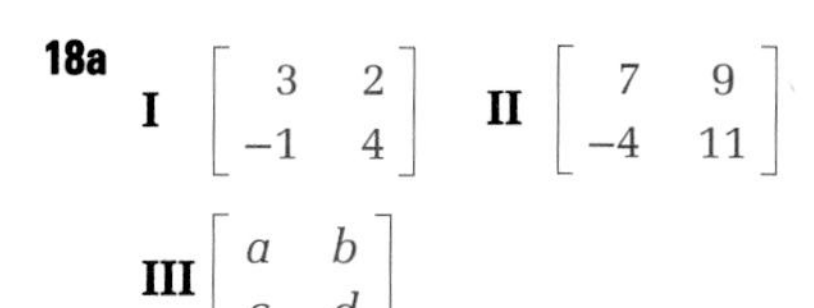

I $\begin{bmatrix} 3 & 2 \\ -1 & 4 \end{bmatrix}$ **II** $\begin{bmatrix} 7 & 9 \\ -4 & 11 \end{bmatrix}$

III $\begin{bmatrix} a & b \\ c & d \end{bmatrix}$

43a

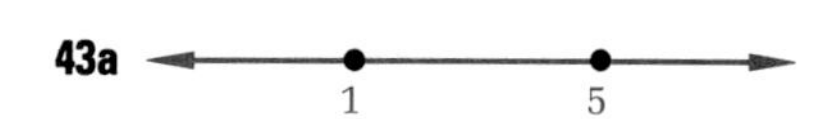

43b

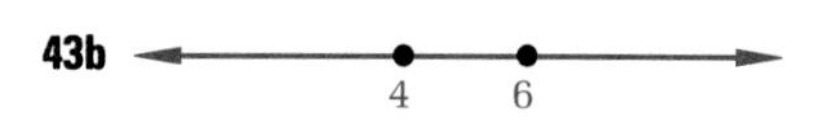

43c

43d

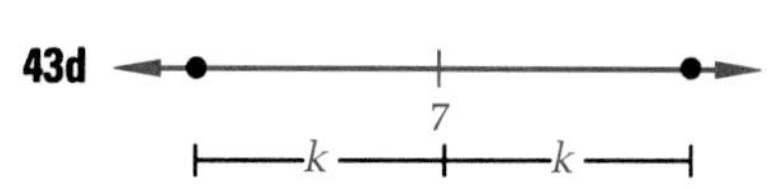

Section 2.6

16

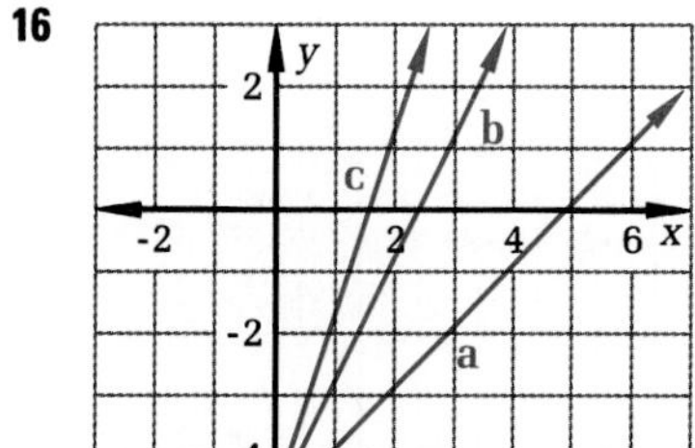

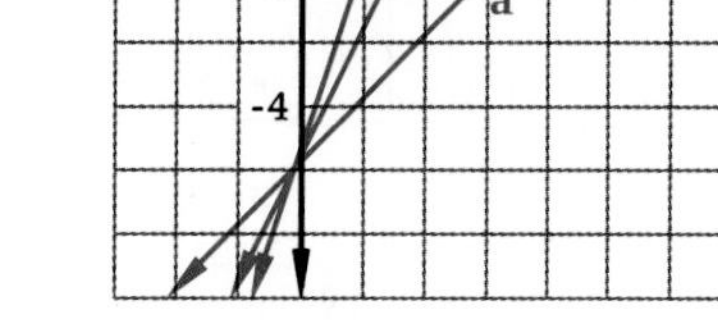

Review for Chapter 2

6

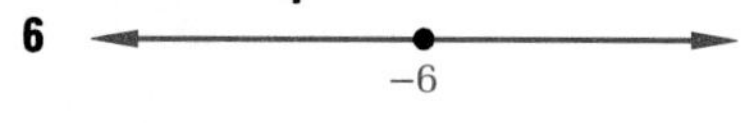

7

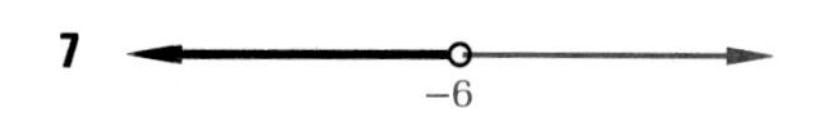

8

Short Subject: Exponents and Radicals

12a Possible table:

x	0	1	2	3	4
y	0	1	≈1.41	≈1.73	2

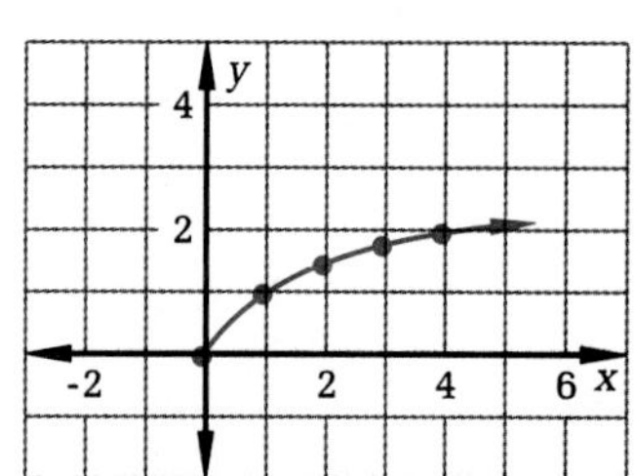

12b Possible table:

x	−2	−1	0	1	2	3
y	$\frac{1}{4}$	$\frac{1}{2}$	1	2	4	8

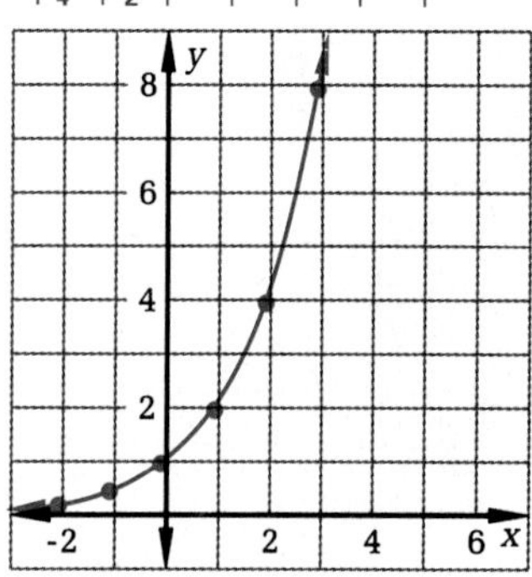

CHAPTER 3

Section 3.1

7

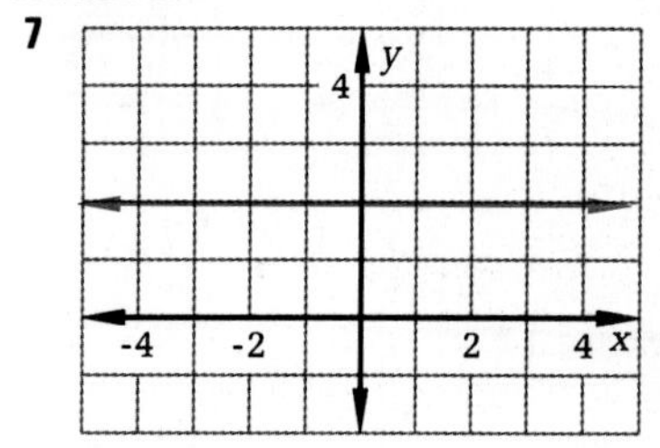

8

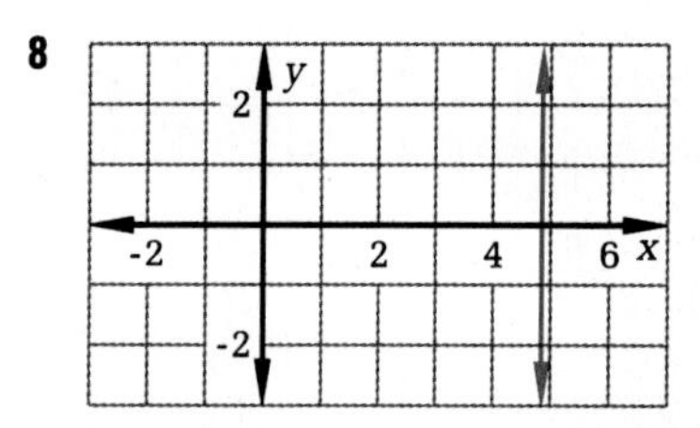

9

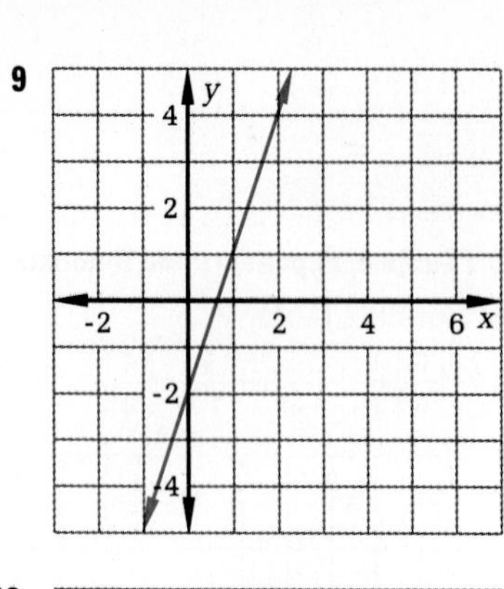

10

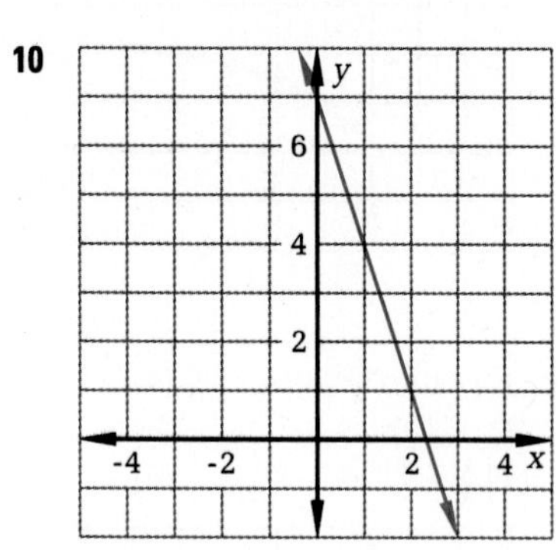

20a

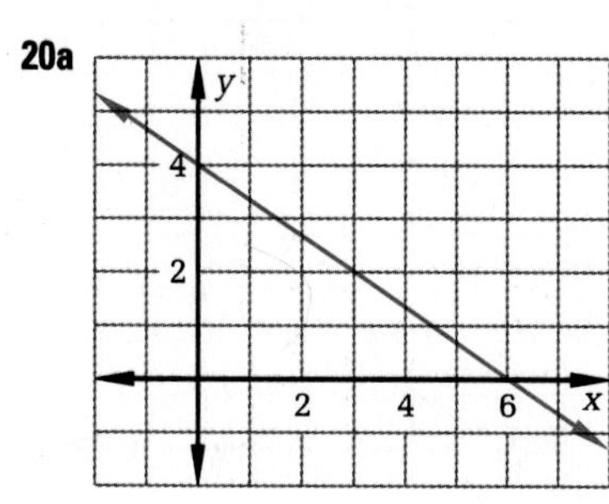

20b

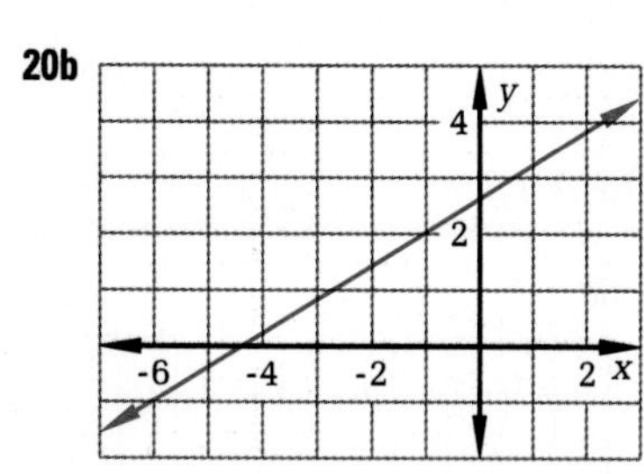

Section 3.3

1

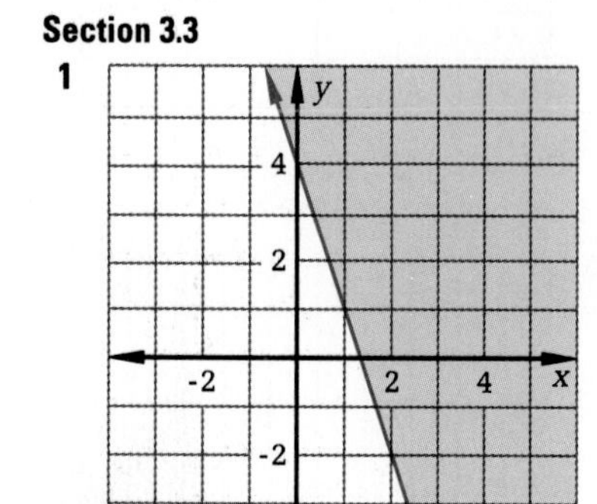

2

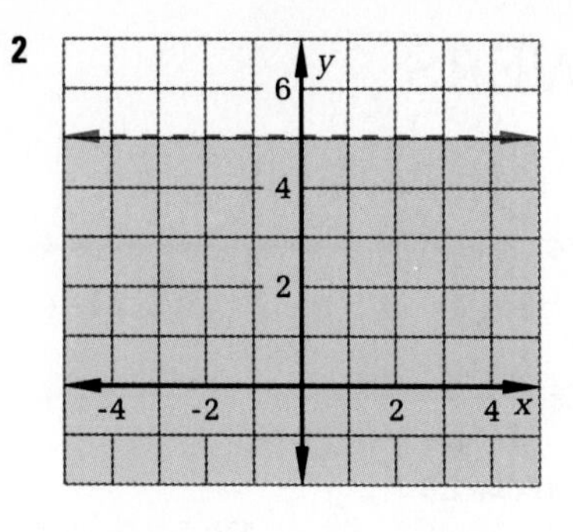

7

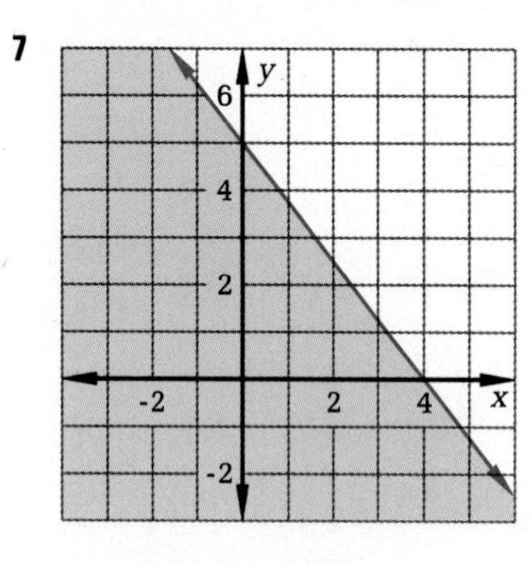

9

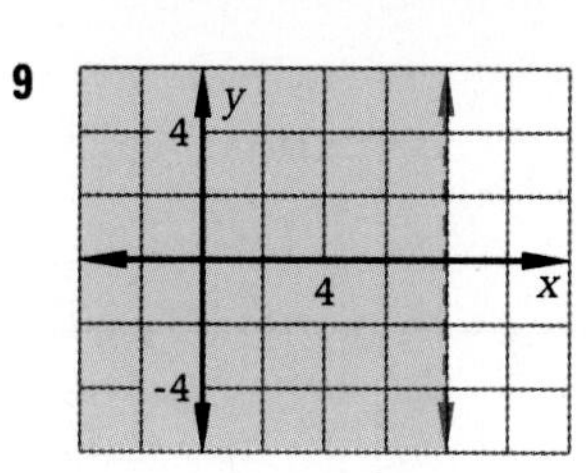

10

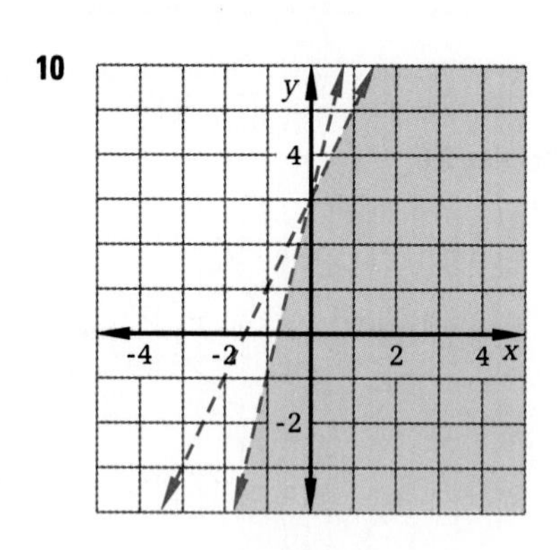

11

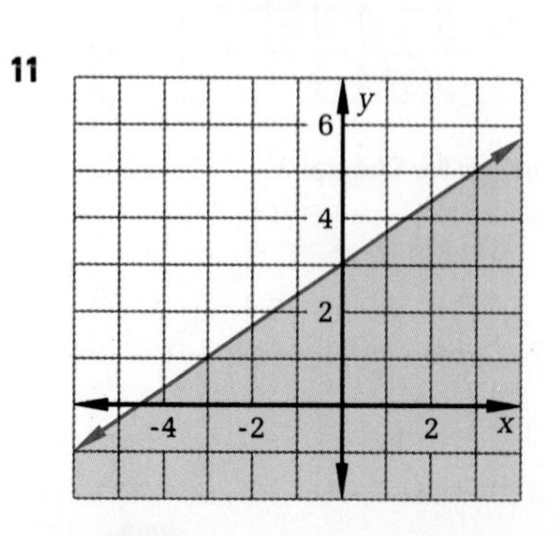

20

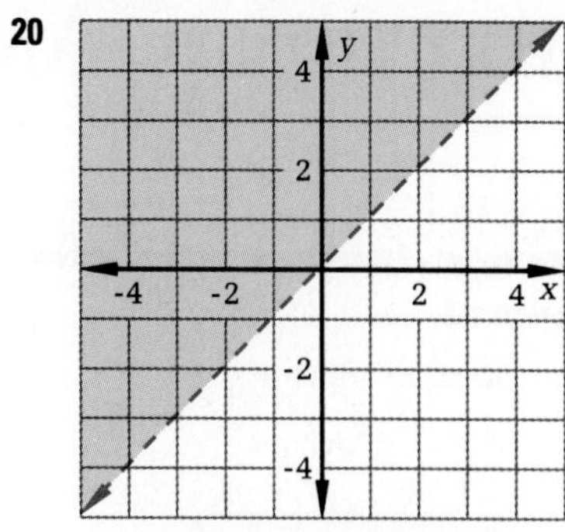

23b

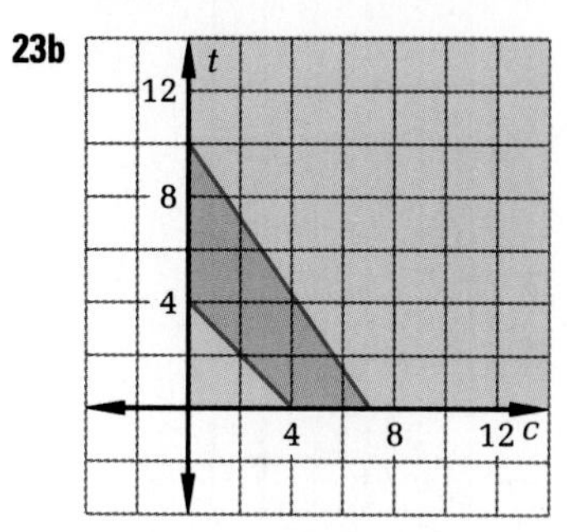

24

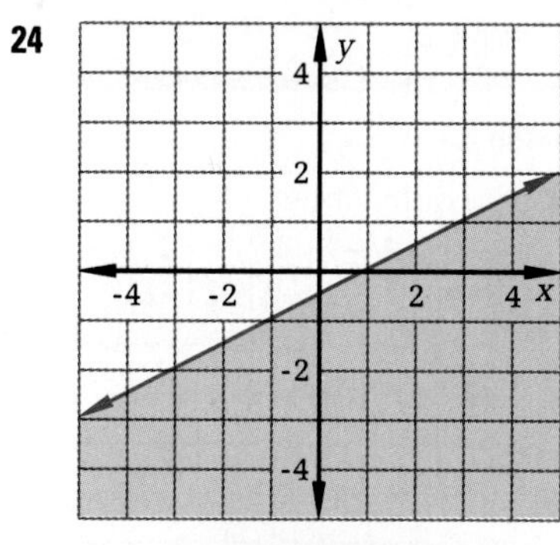

25

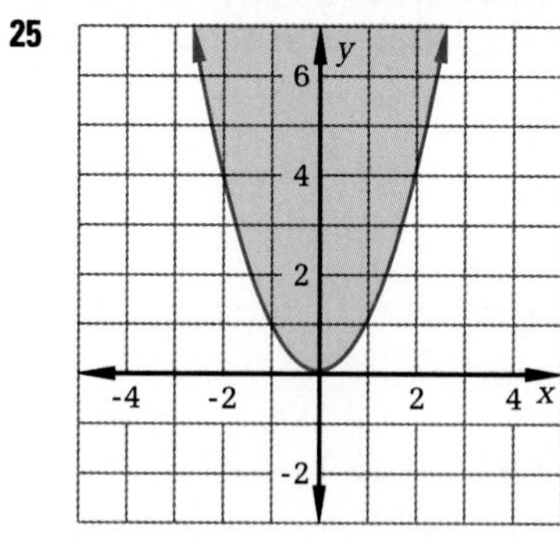

26

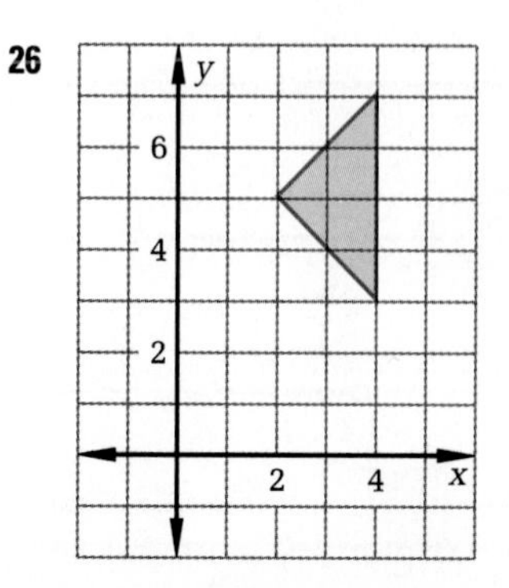

30

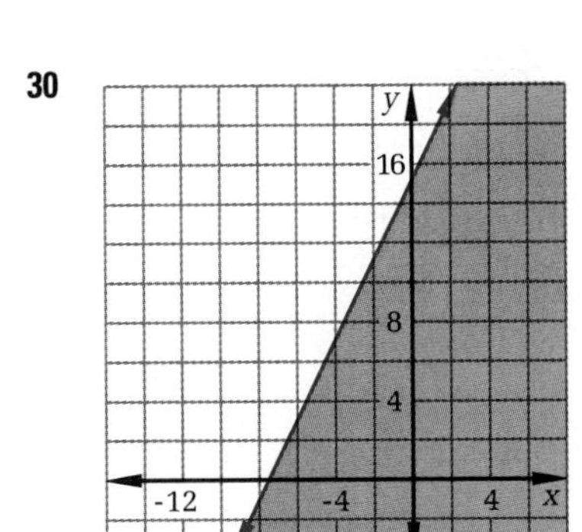

37b

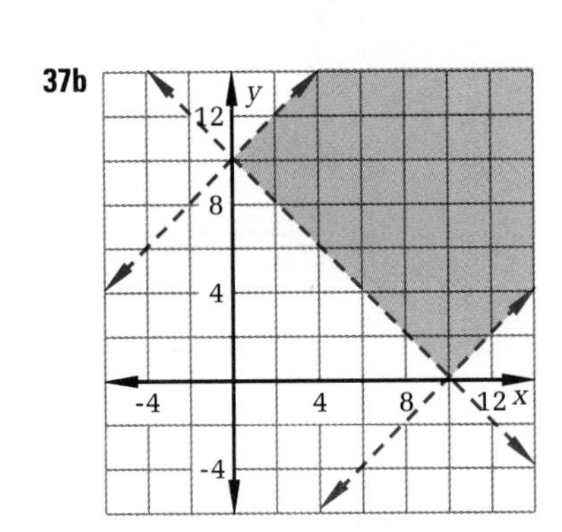

Section 3.4

28

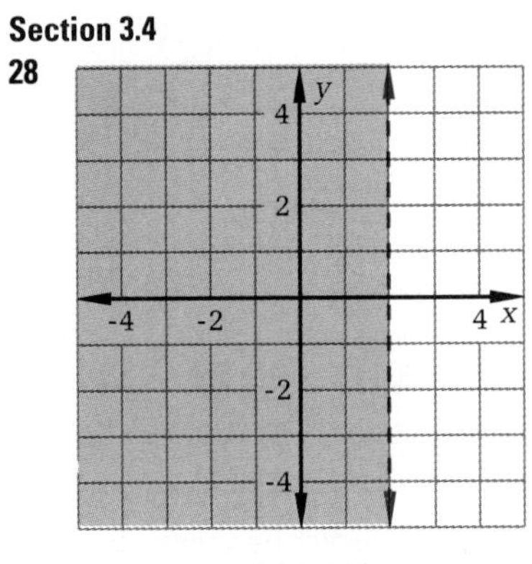

29

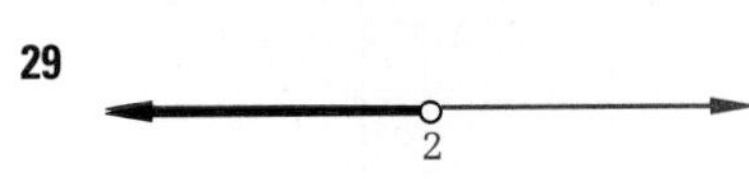

30

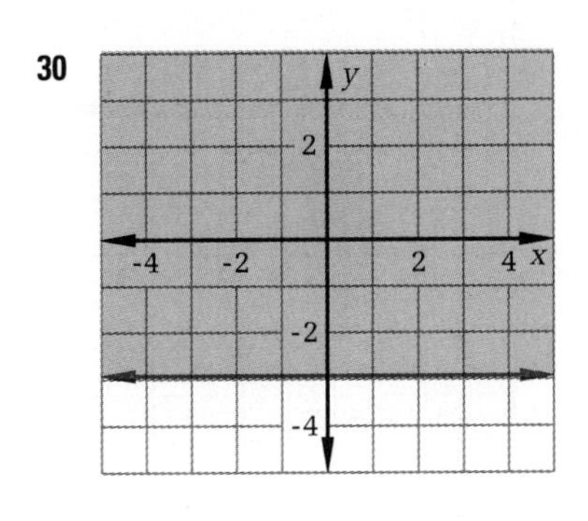

31

0

32a

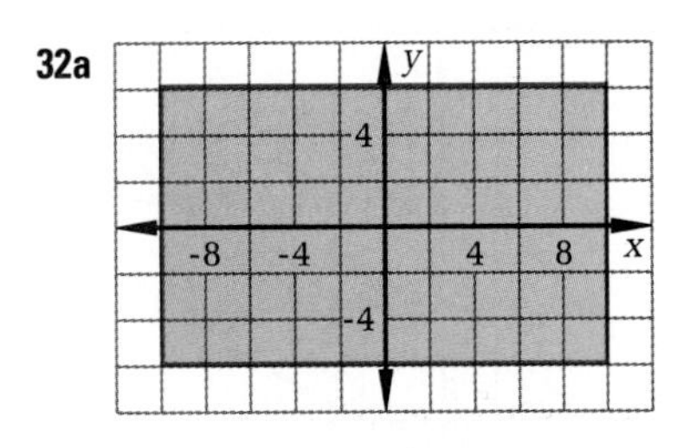

Section 3.5

23

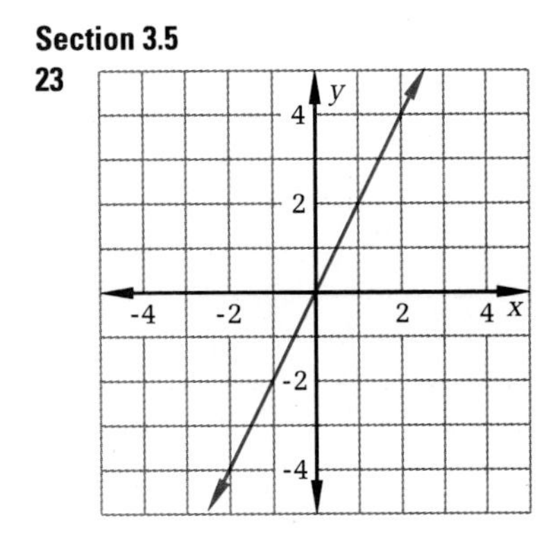

24

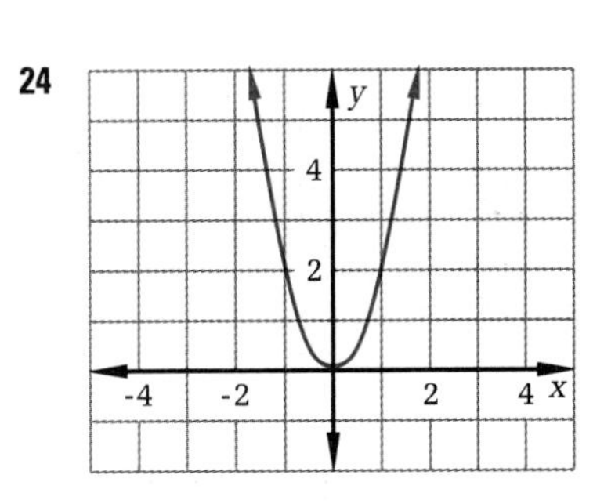

25

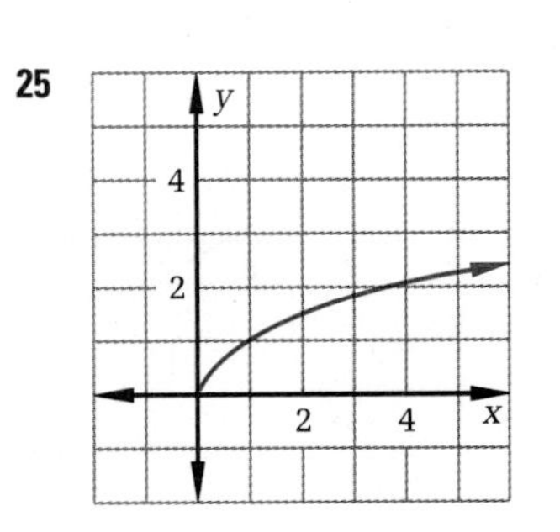

26

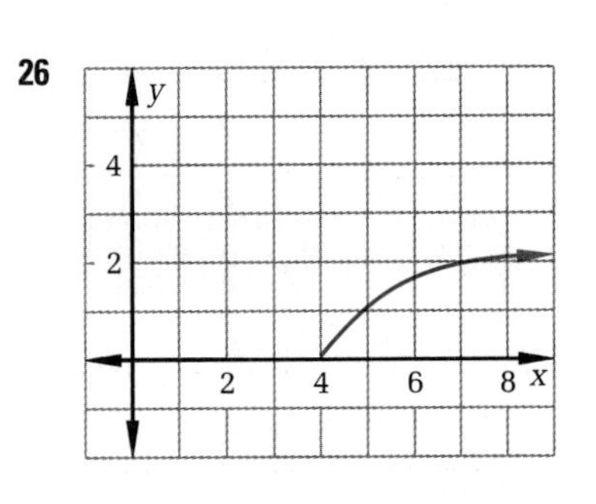

Section 3.6

6a

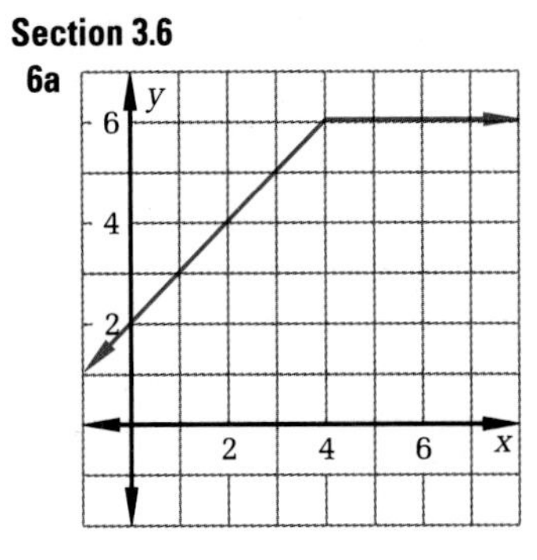

21c

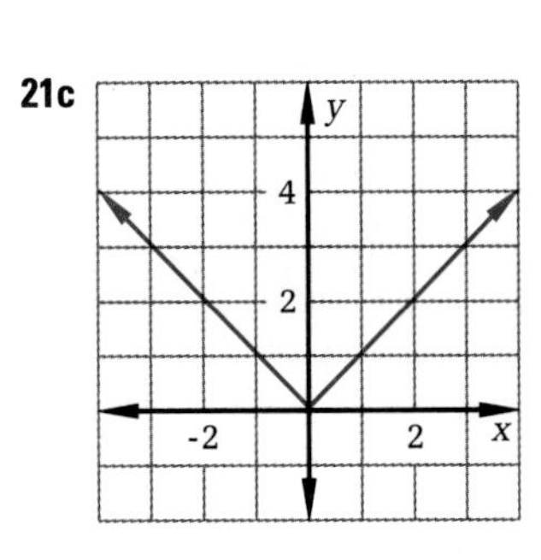

22a

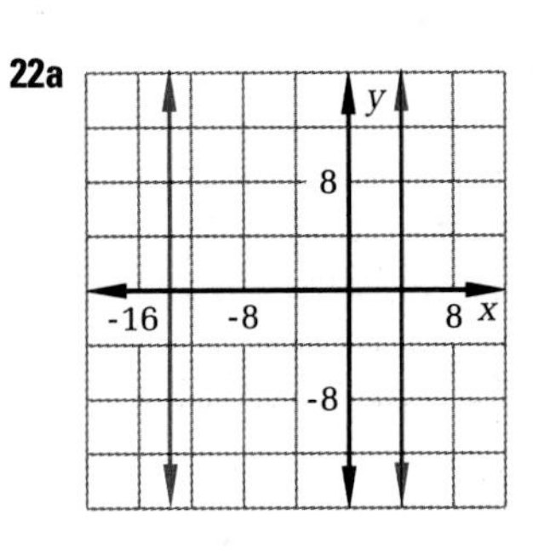

22b

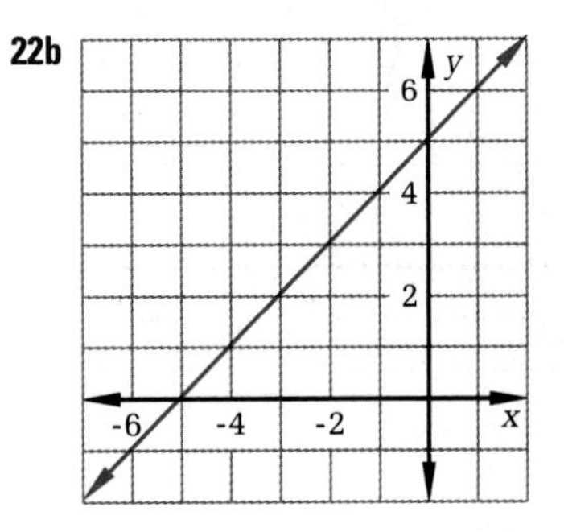

22c

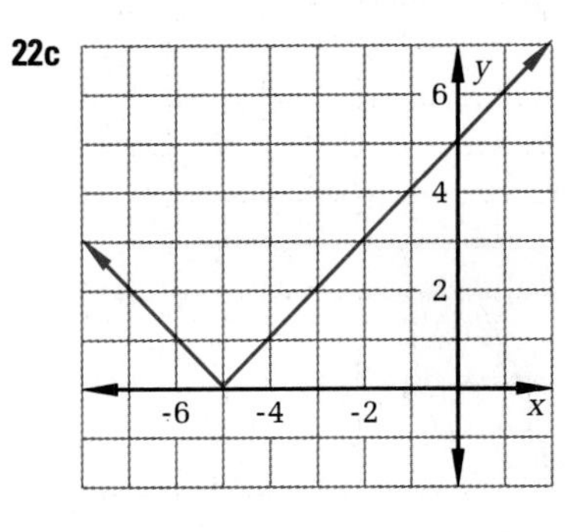

40c

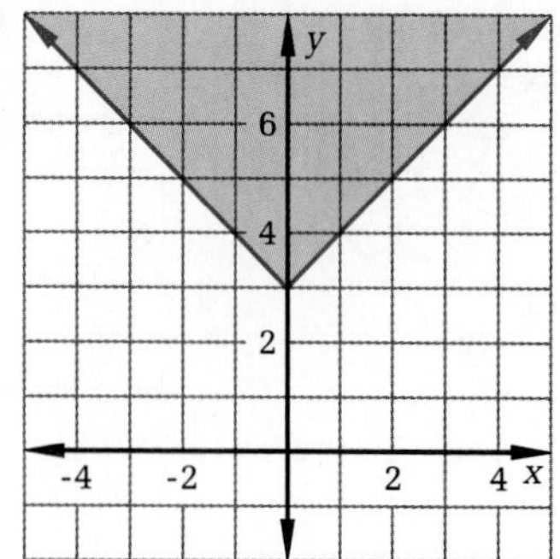

40d

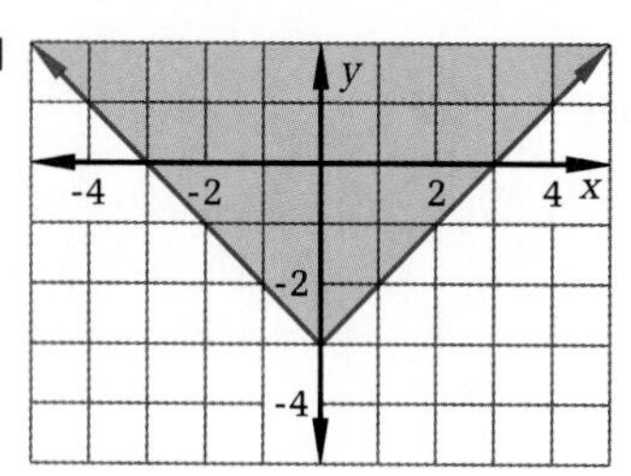

Review for Chapter 3

1

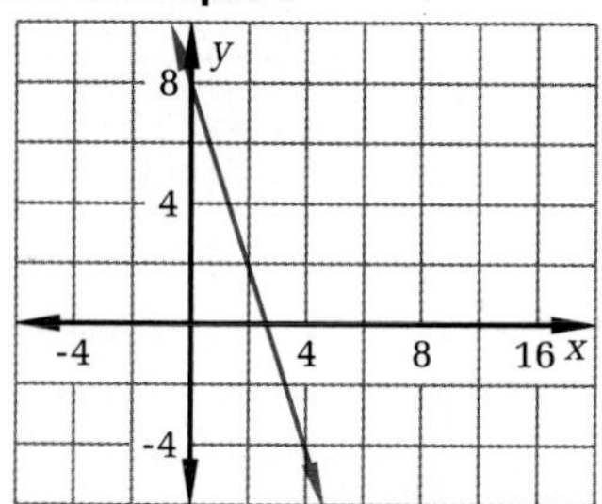

7b

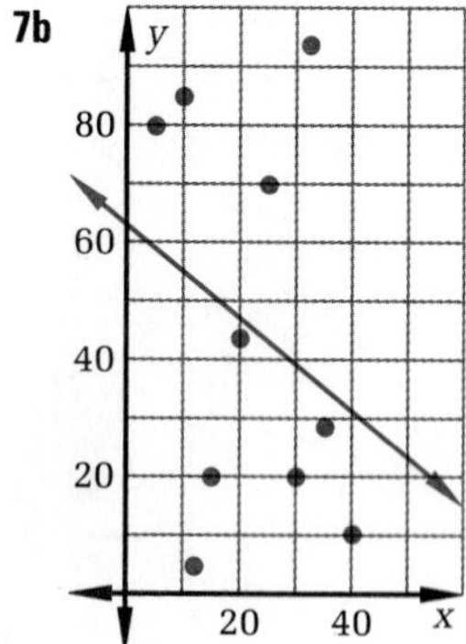

8b

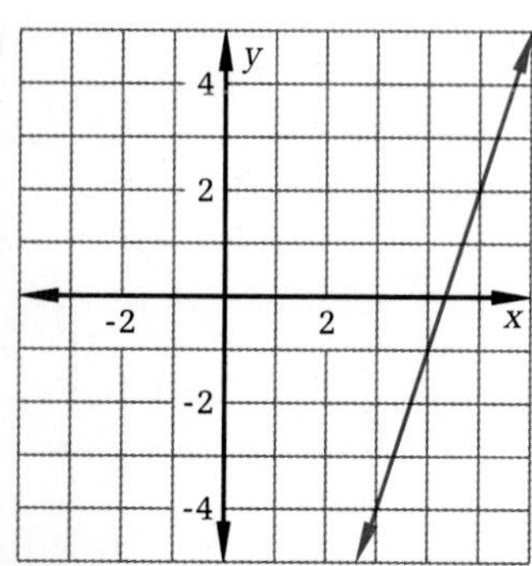

9

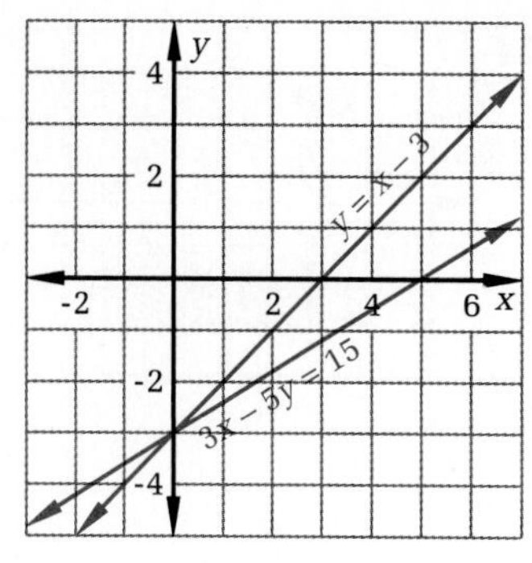

11a

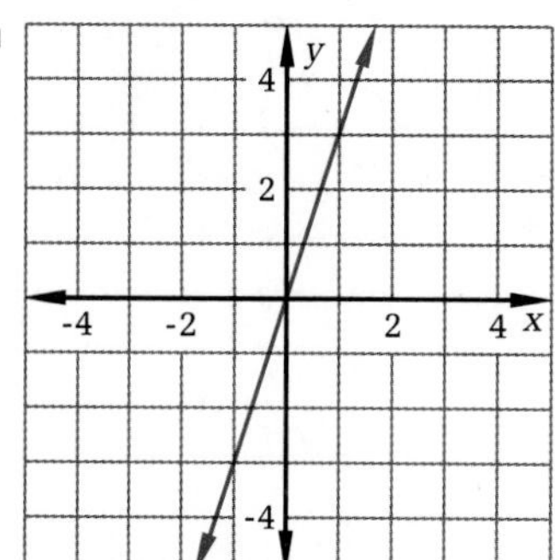

11b

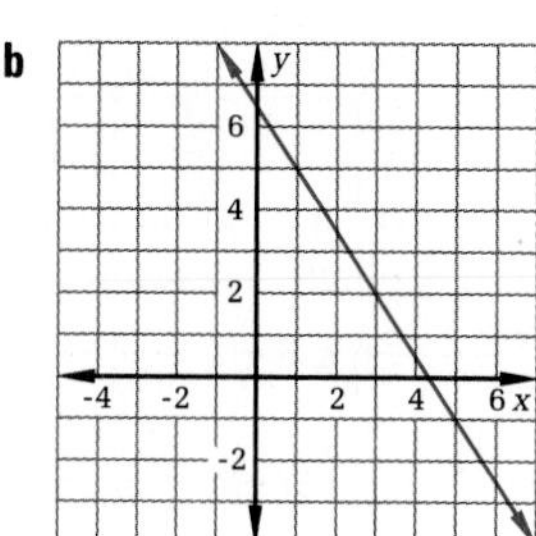

11c

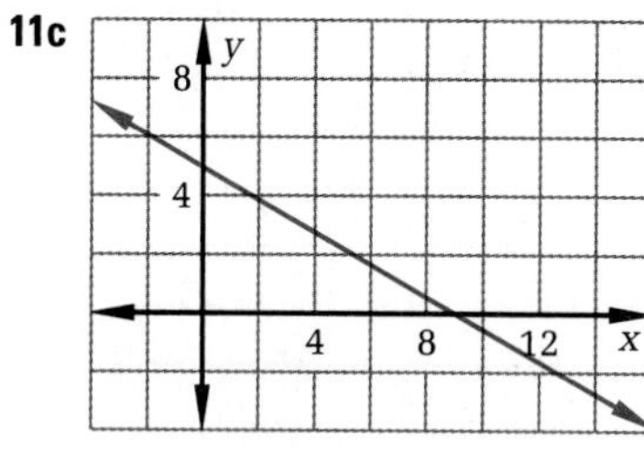

16

−5

17

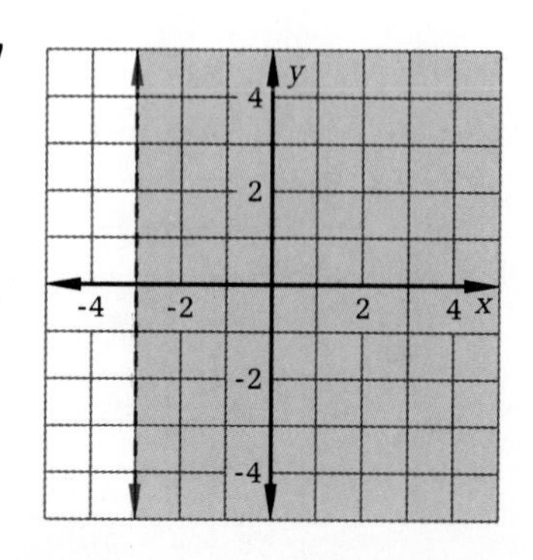

18

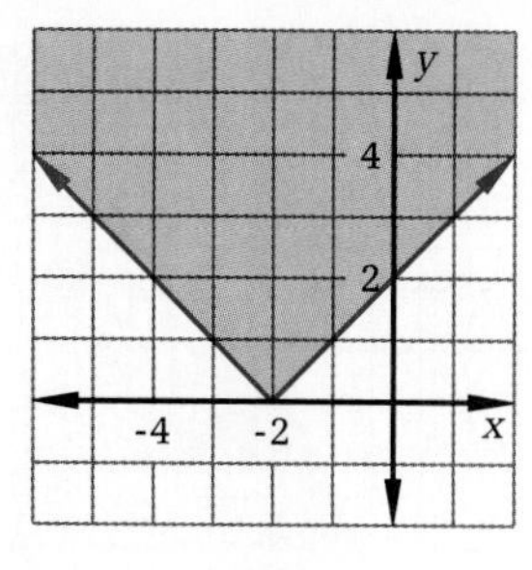

26

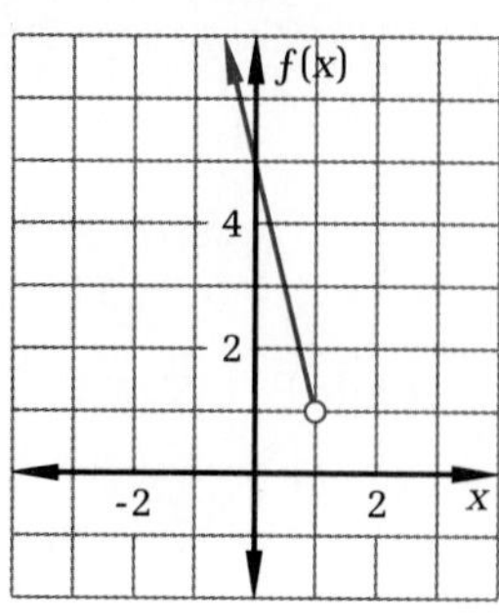

CHAPTER 4

Section 4.1

32a

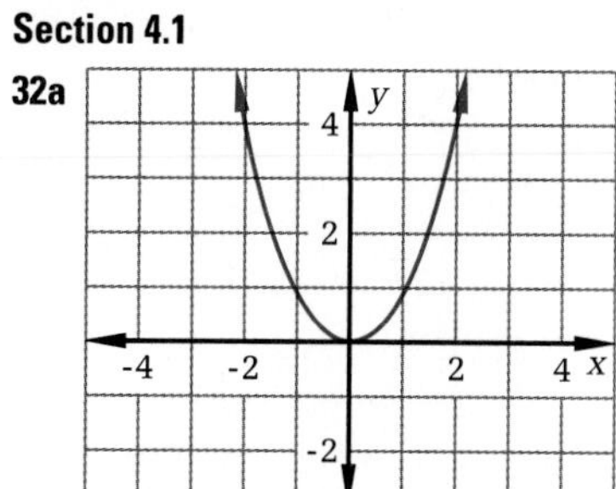

32b

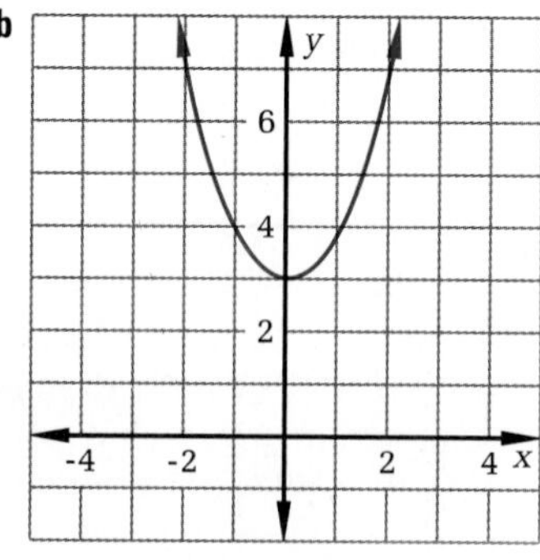

32c

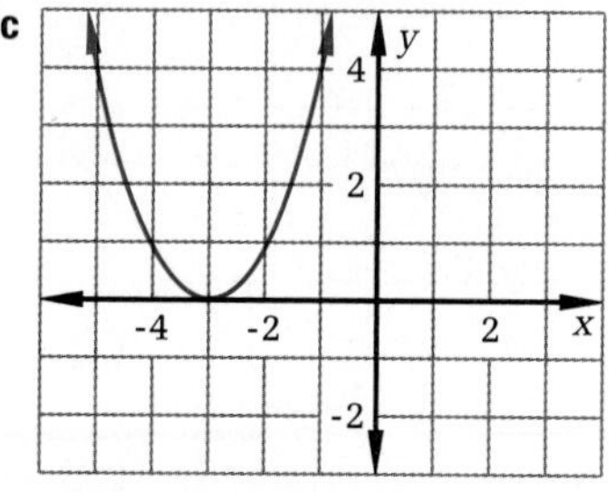

40

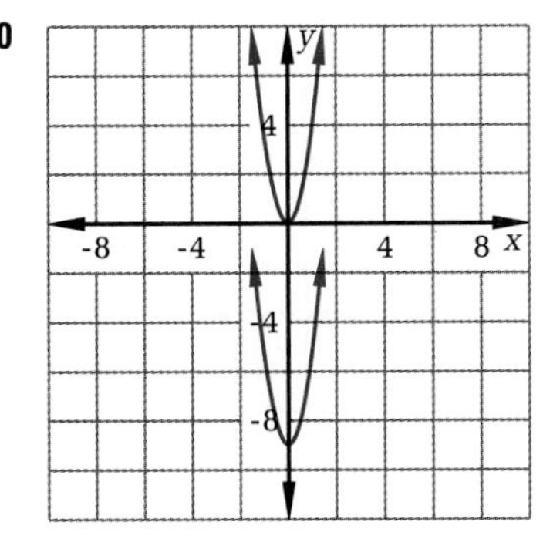
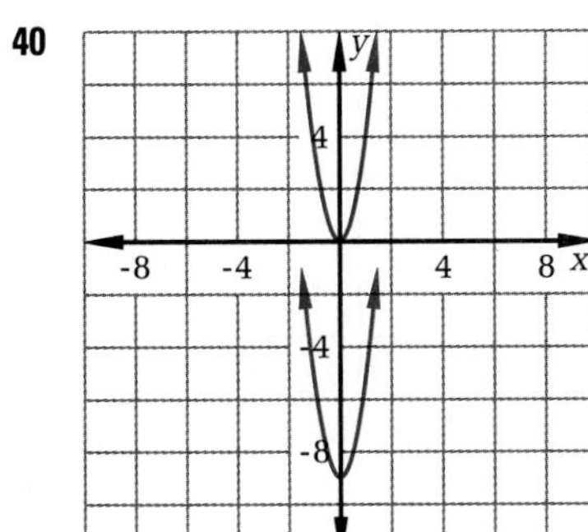
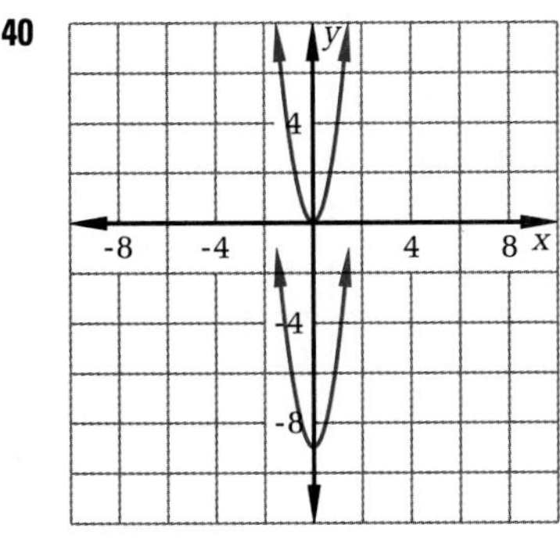

Section 4.2

31a

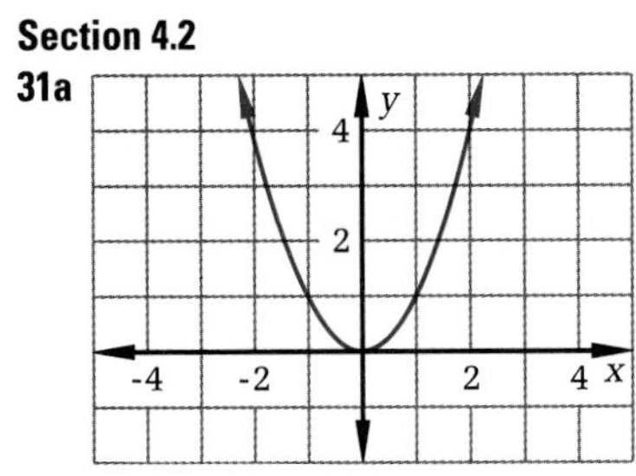

31b

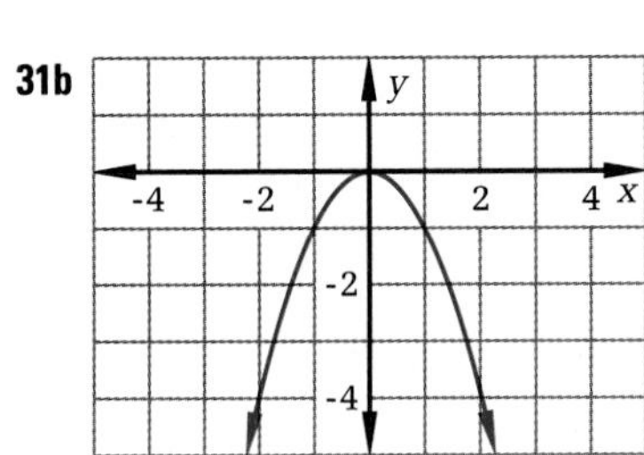

31c

31d

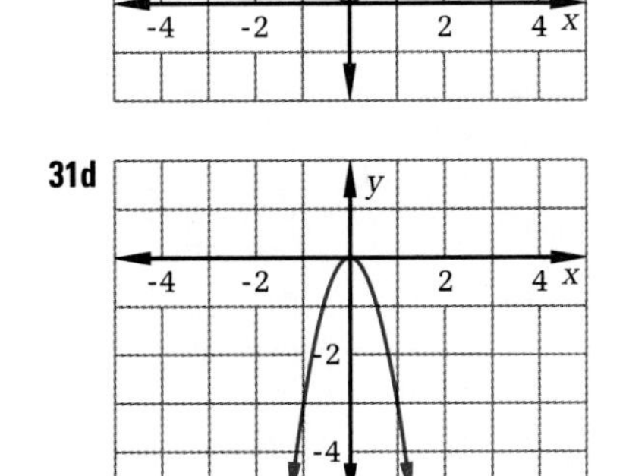

51

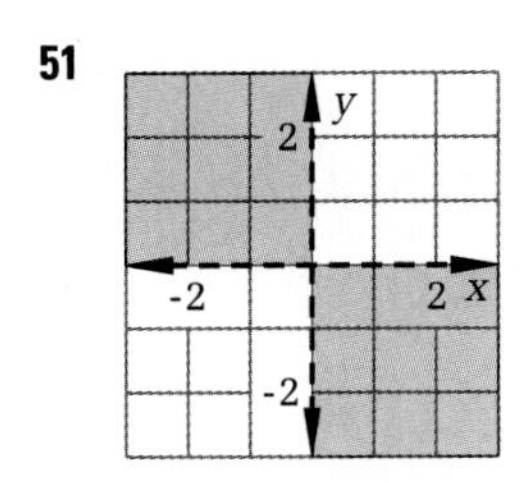

Section 4.3

4a

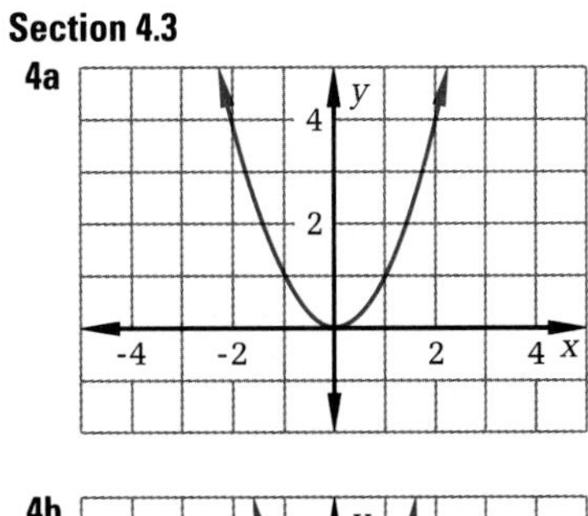

4b

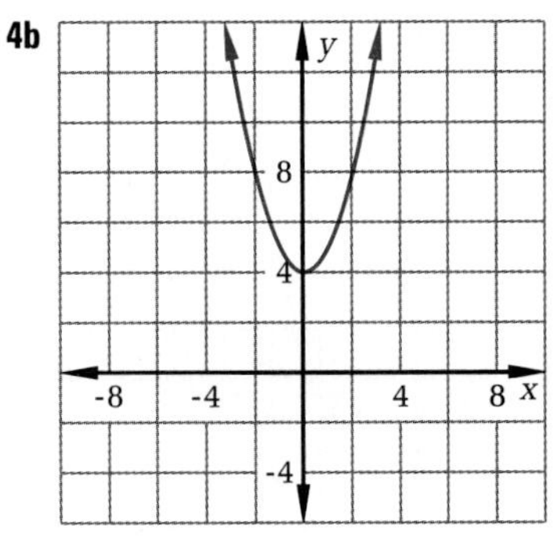

4c

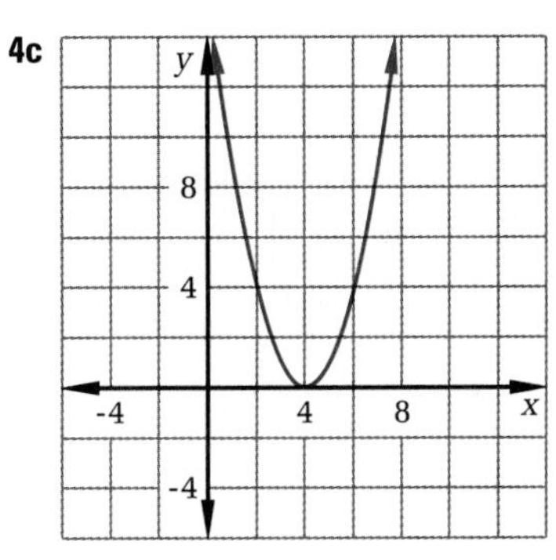

4d

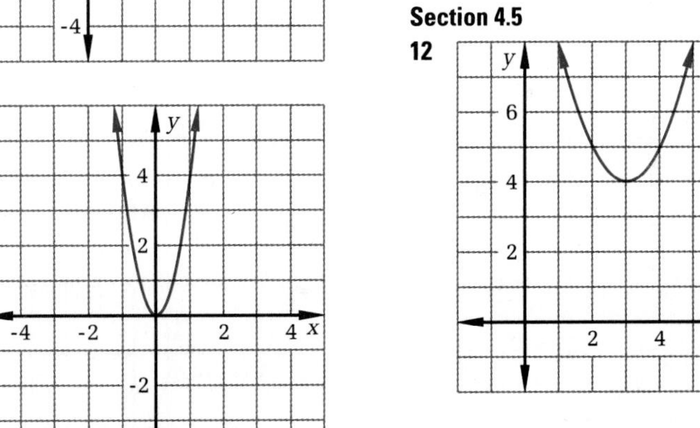

33a

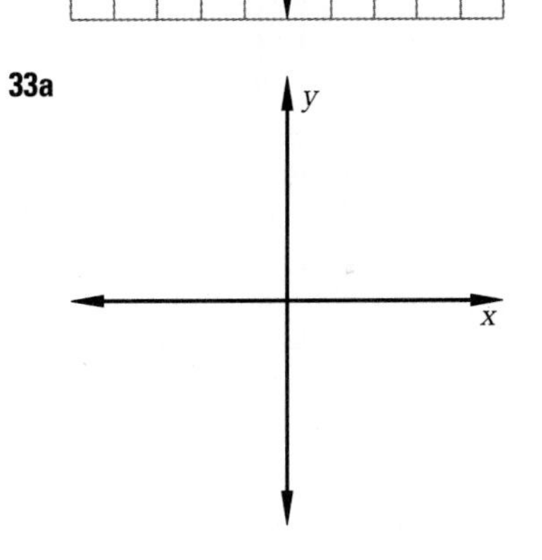

Section 4.4

9

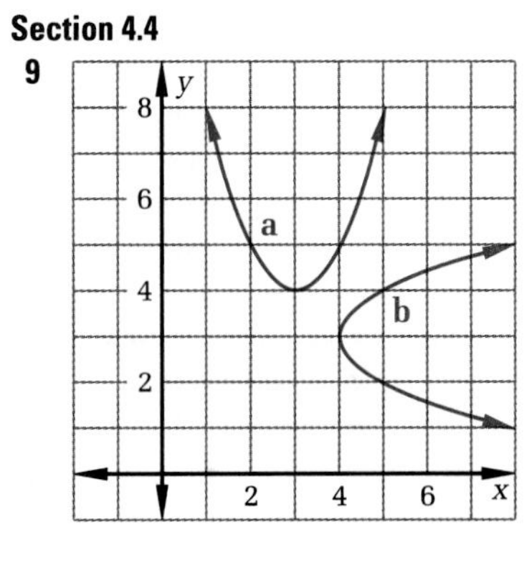

45b

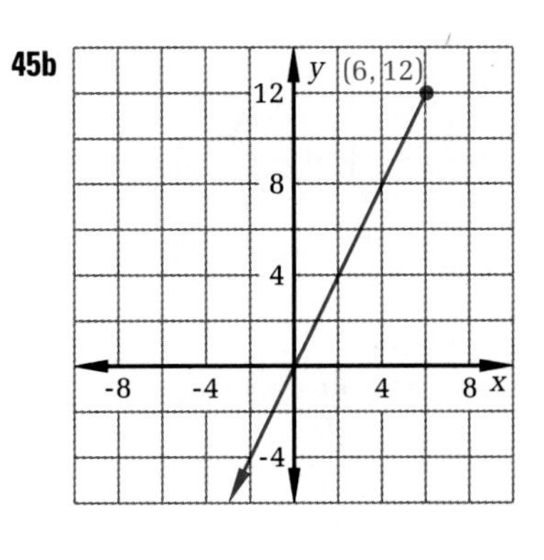

Section 4.5

12

13

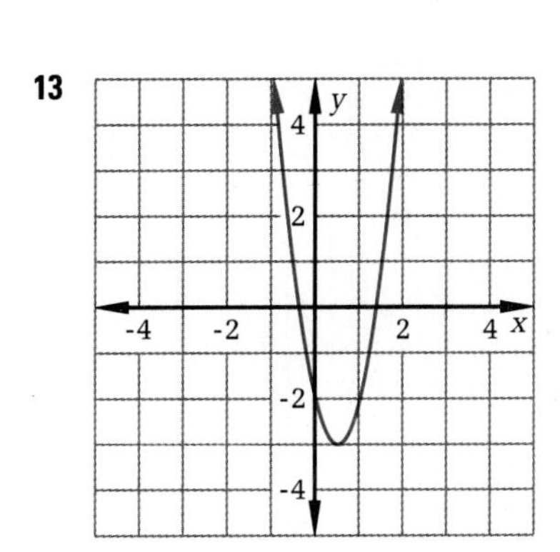

14

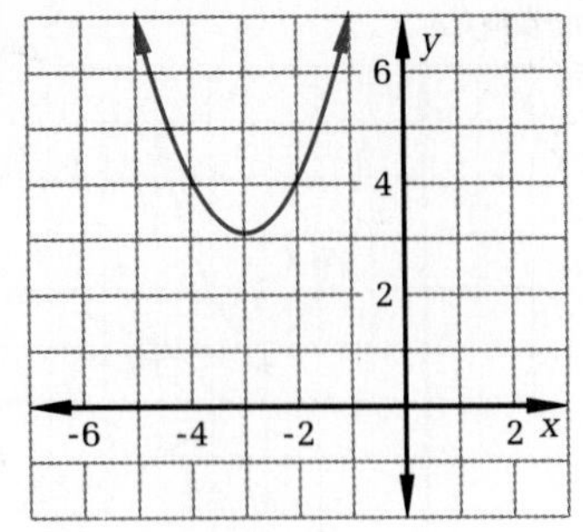

15

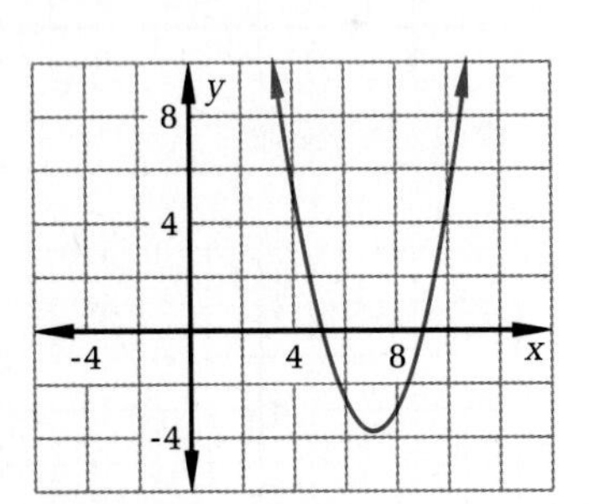

CHAPTER 5

Section 5.3

1b

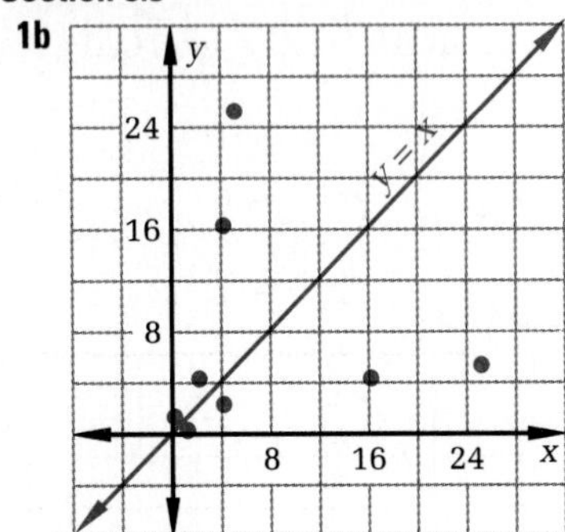

12a

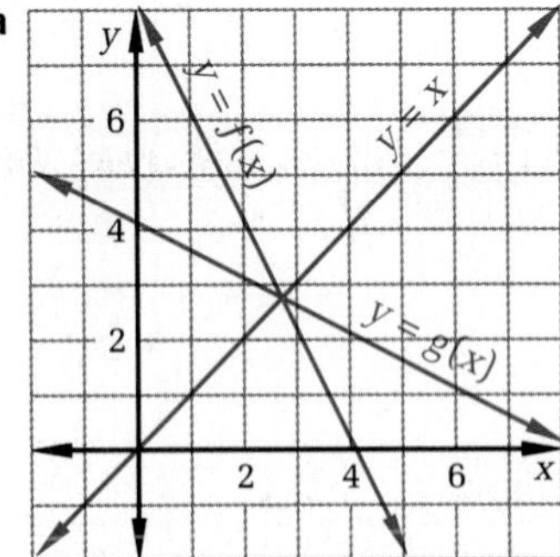

21c

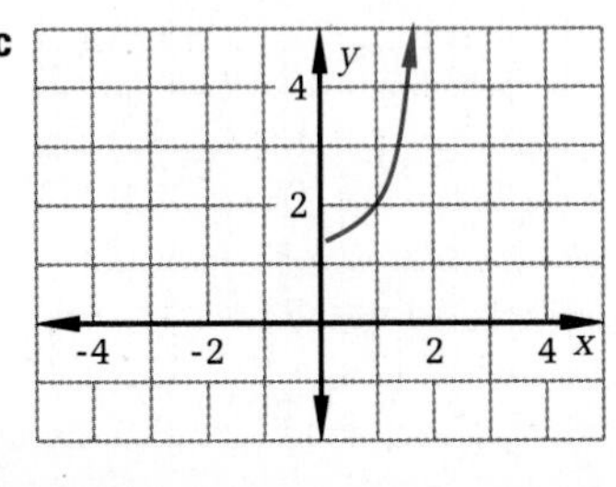

21d

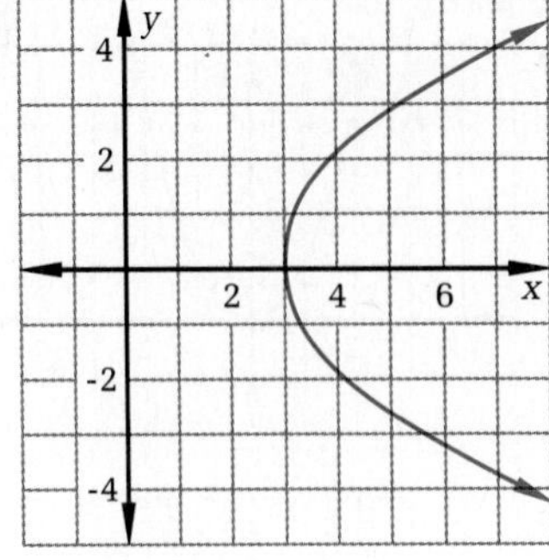

22a

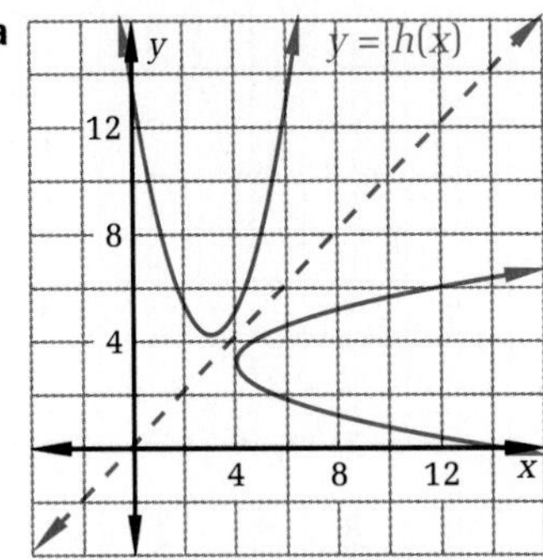

28a

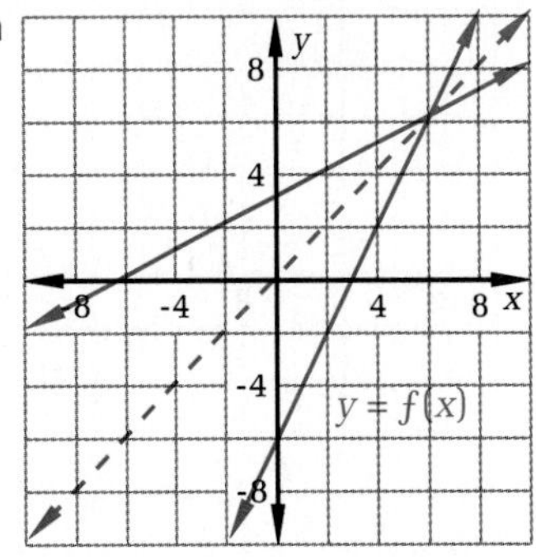

28b

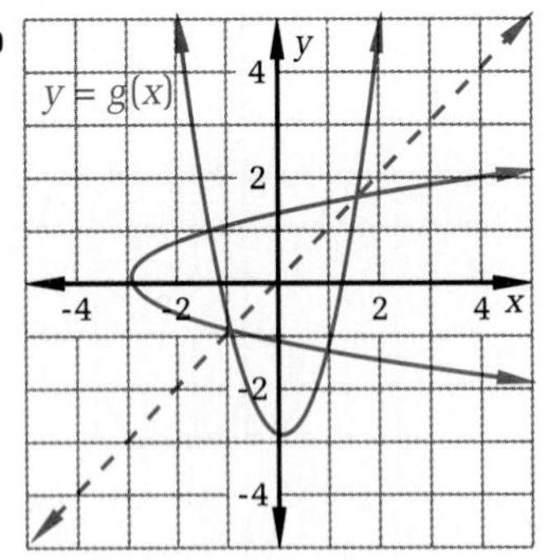

28c

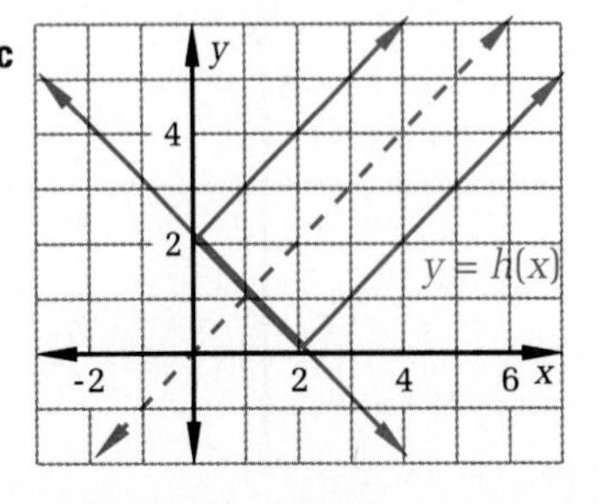

28d

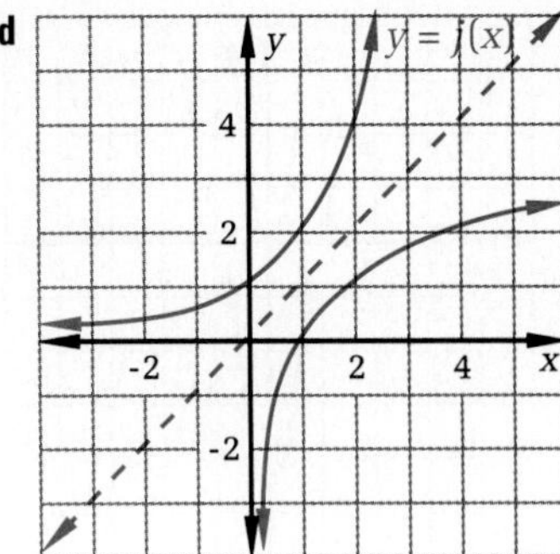

37 INV $f(x) = \sqrt{\frac{x-k}{a}} + h$

38c

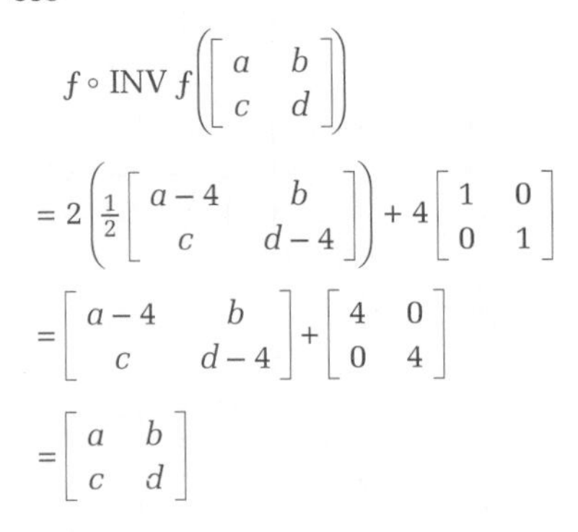

$$f \circ \text{INV } f\left(\begin{bmatrix} a & b \\ c & d \end{bmatrix}\right)$$

$$= 2\left(\frac{1}{2}\begin{bmatrix} a-4 & b \\ c & d-4 \end{bmatrix}\right) + 4\begin{bmatrix} 1 & 0 \\ 0 & 1 \end{bmatrix}$$

$$= \begin{bmatrix} a-4 & b \\ c & d-4 \end{bmatrix} + \begin{bmatrix} 4 & 0 \\ 0 & 4 \end{bmatrix}$$

$$= \begin{bmatrix} a & b \\ c & d \end{bmatrix}$$

Section 5.4

7

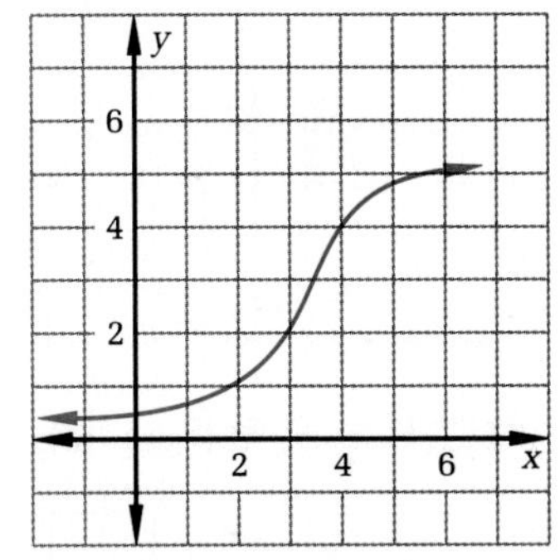

11a

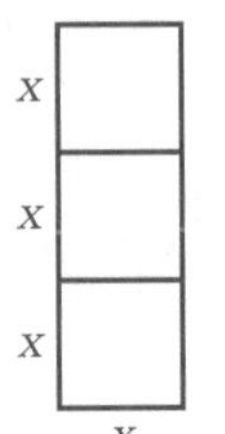

11b

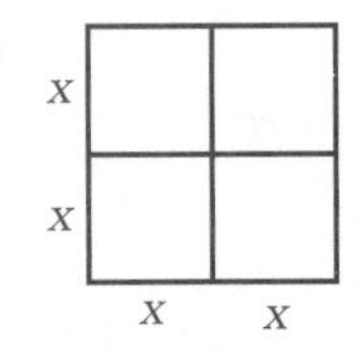

11c

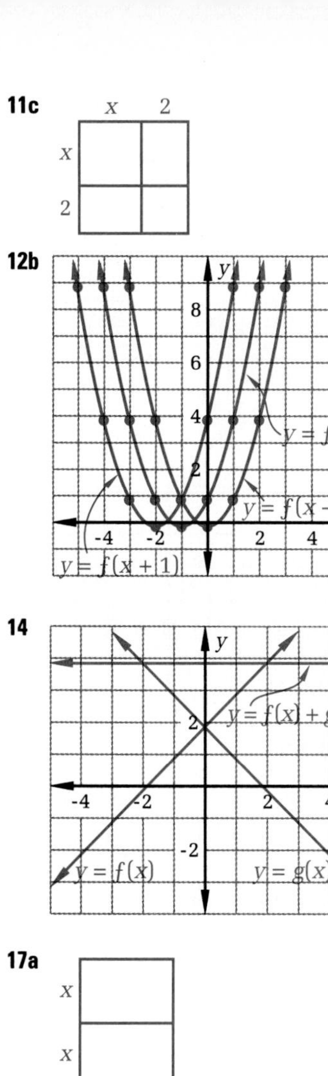

12b

14

17a

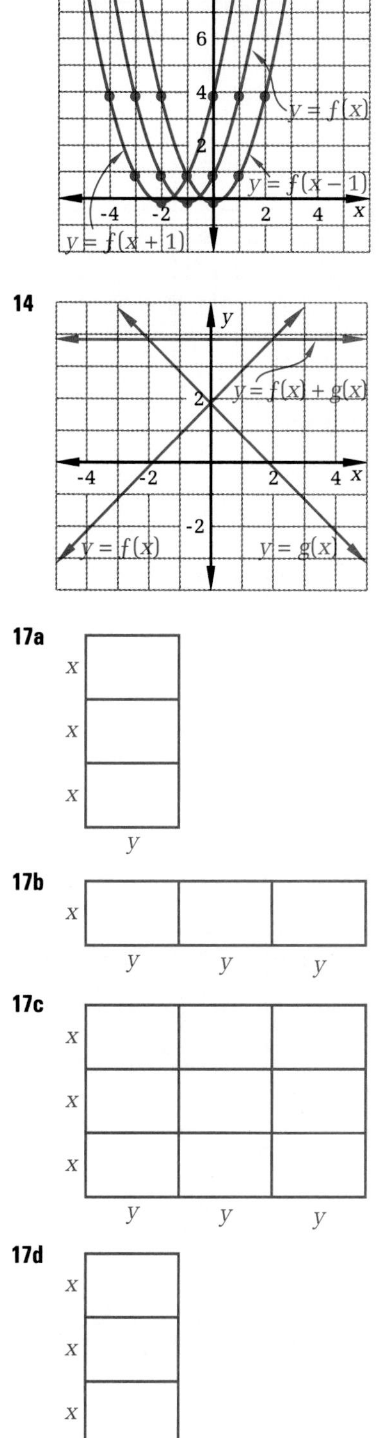

17b

17c

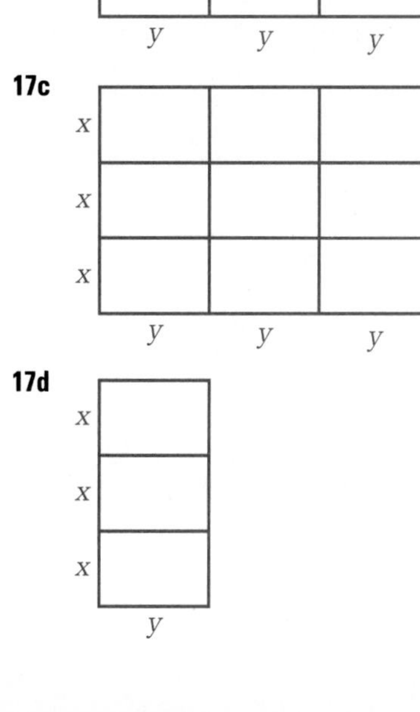

17d

18

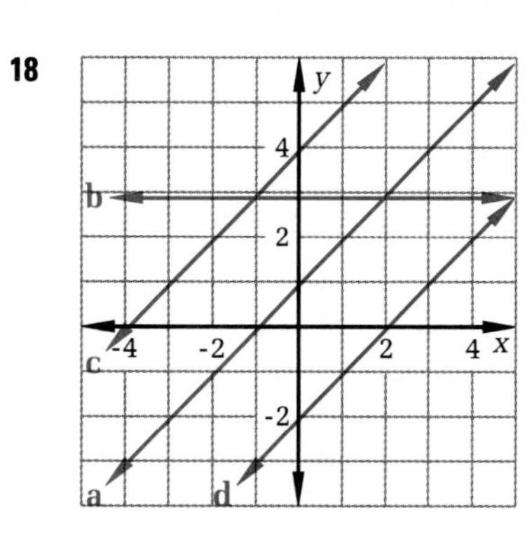

22b

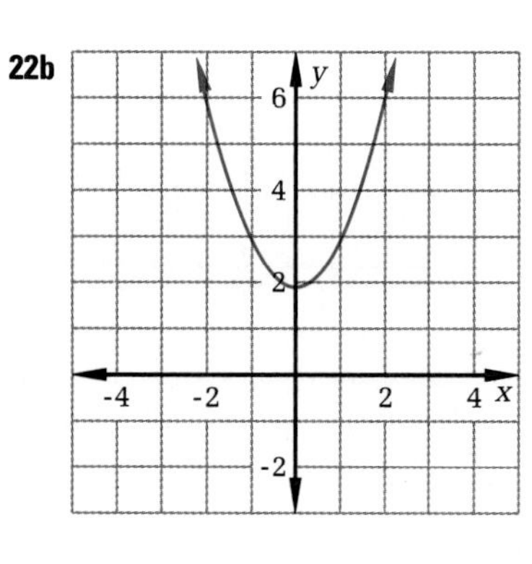

23a

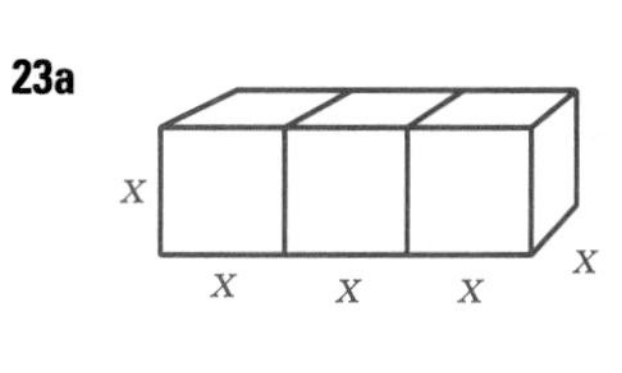

23b

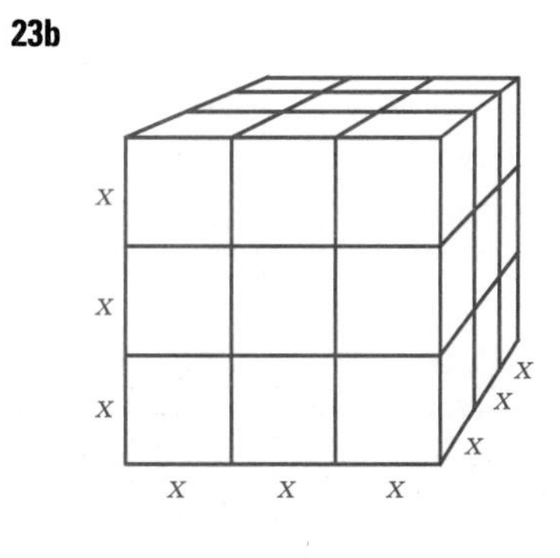

32b

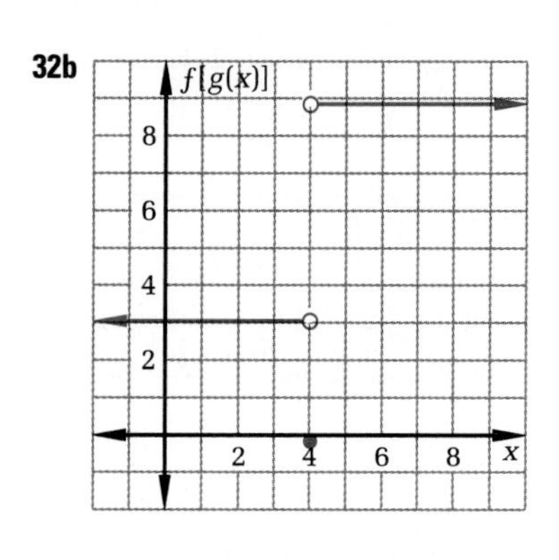

Sec tion 5.5

27

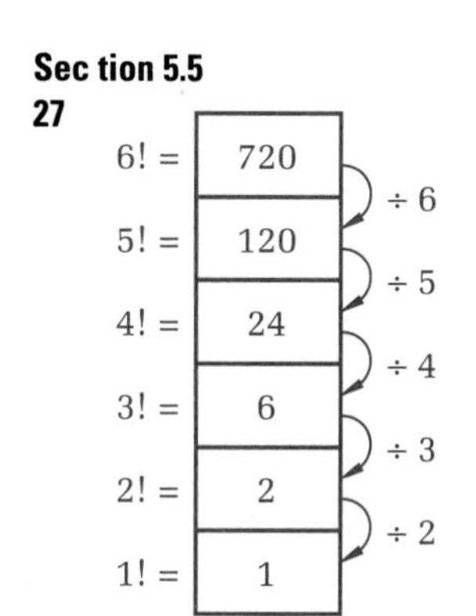

Review for Chapter 5

22

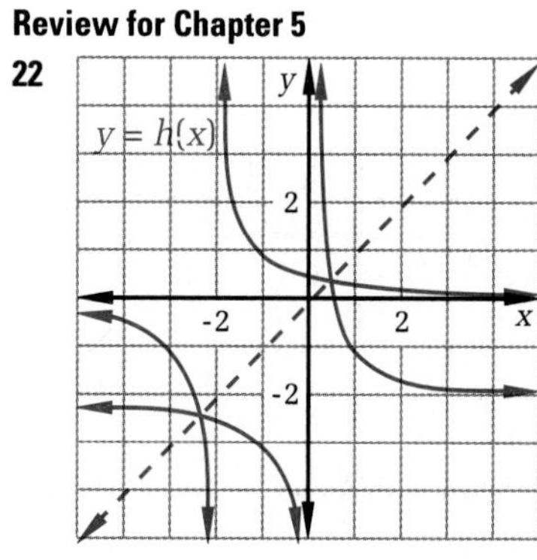

23

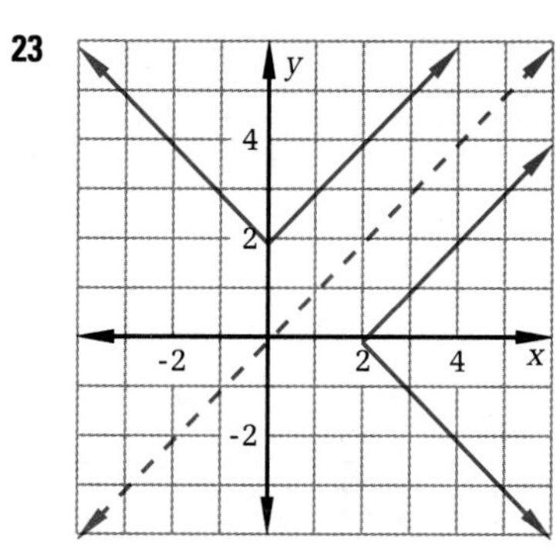

CHAPTER 6

Section 6.1

1

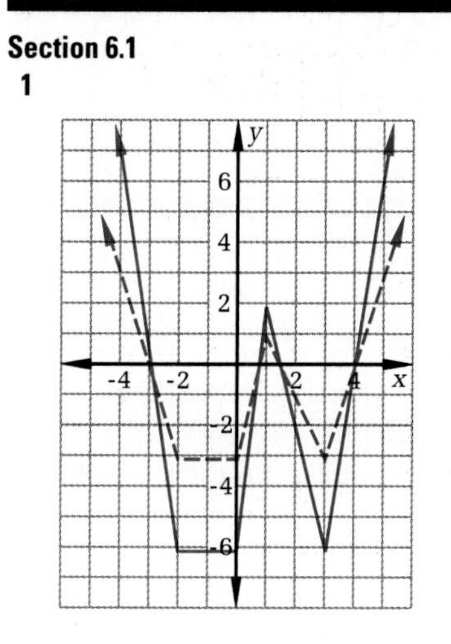

2

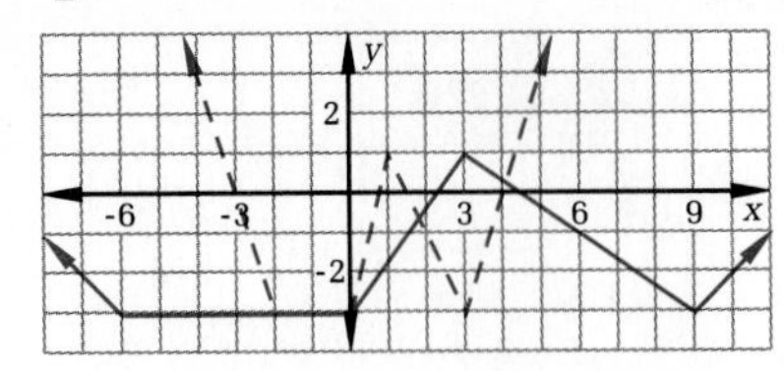

3

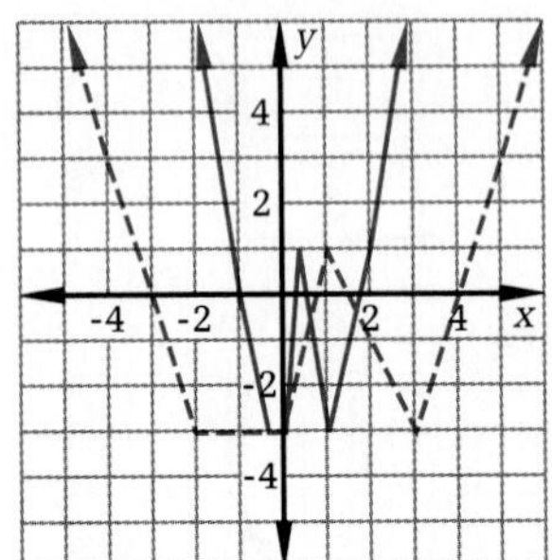

4

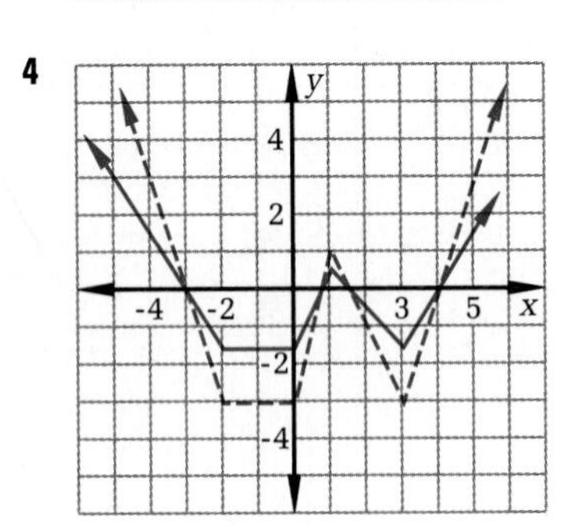

Section 6.2

25a

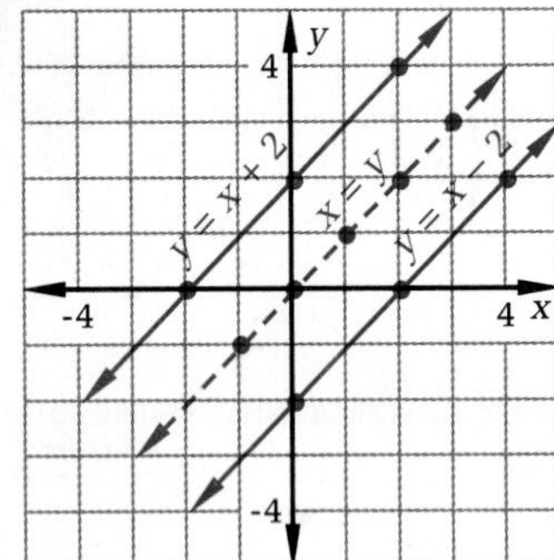

27a, b, c

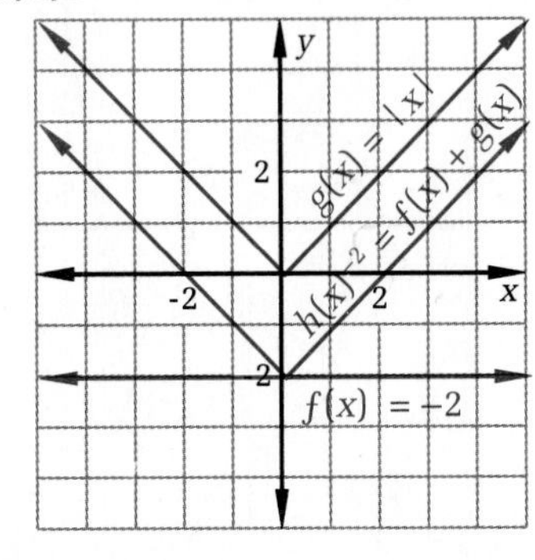

33a

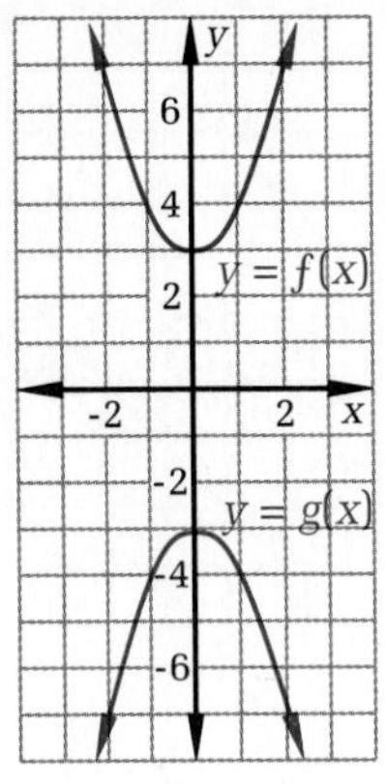

41

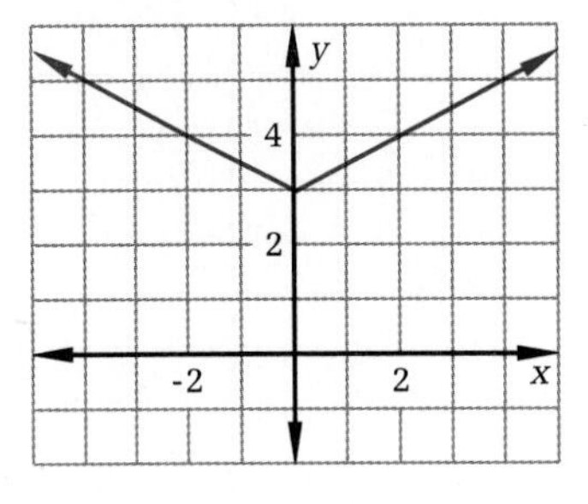

42

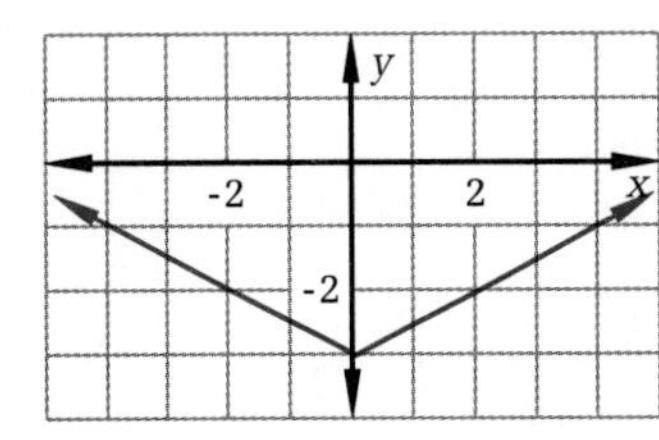

45

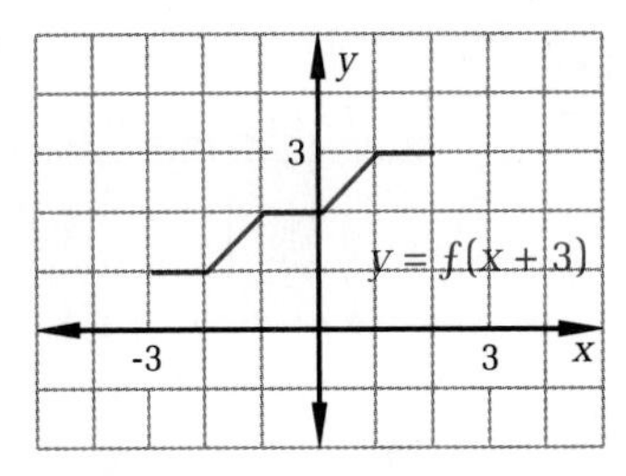

46

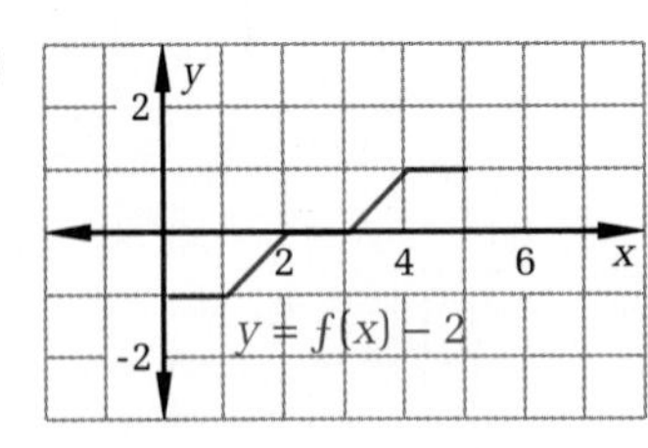

47

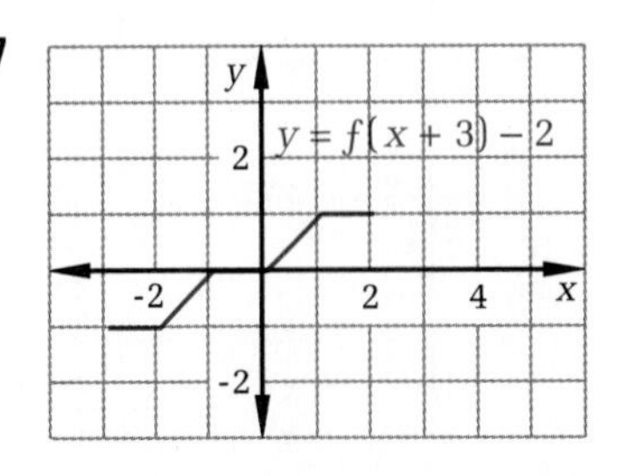

51

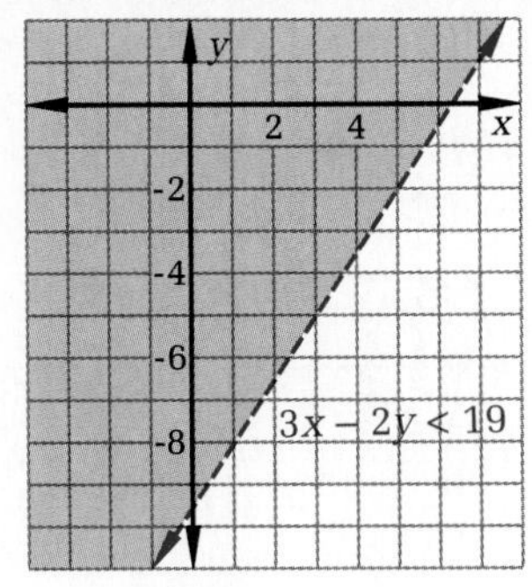

54

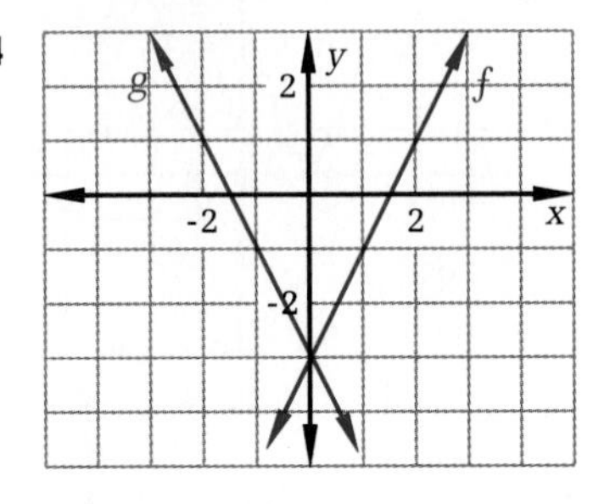

Section 6.4

12c

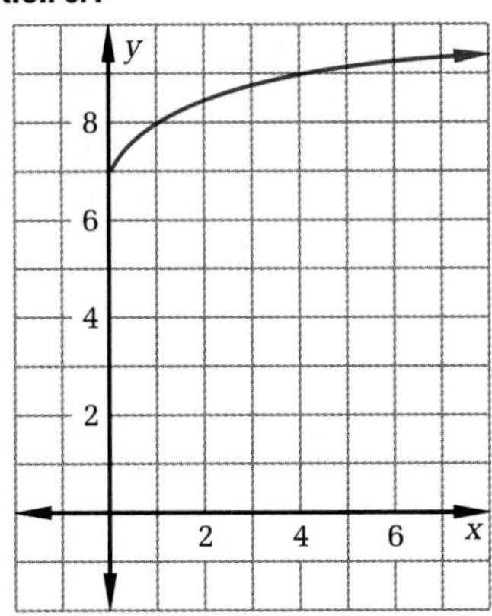

12d

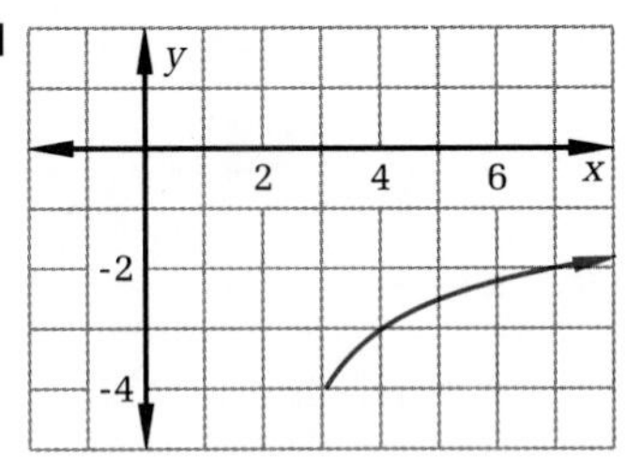

25c

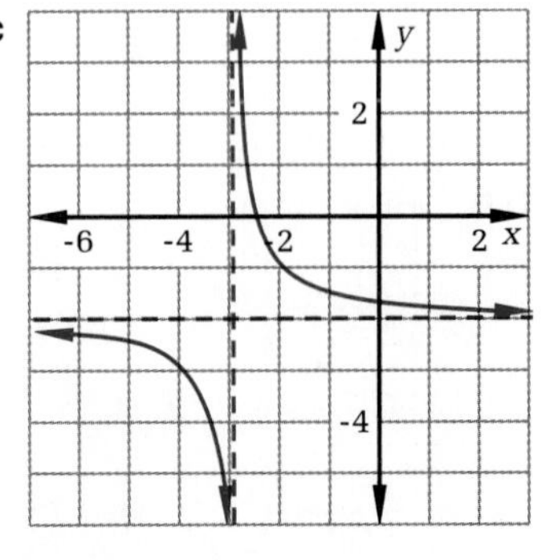

28b

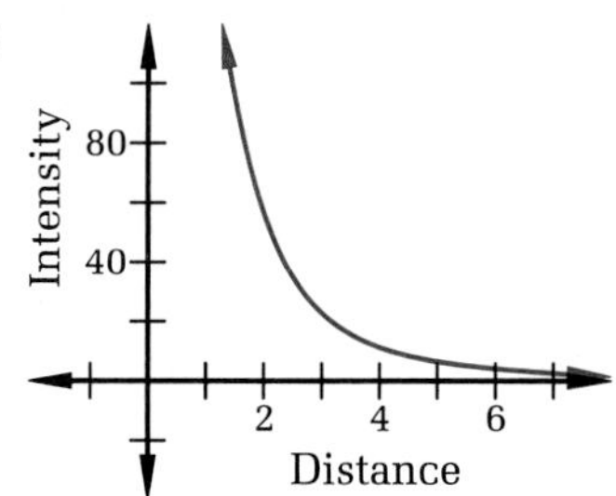

12a

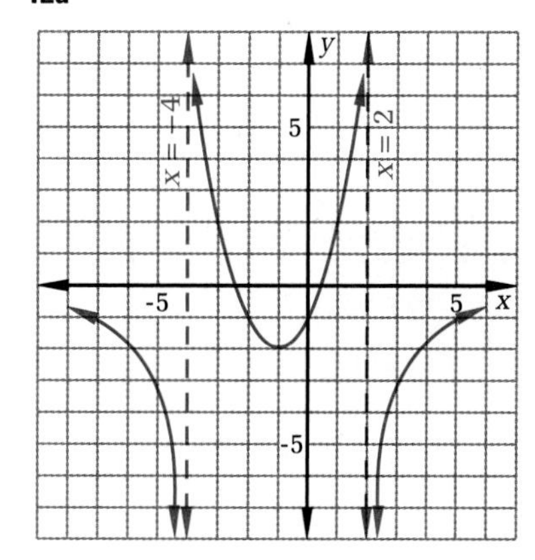

Section 6.5

9

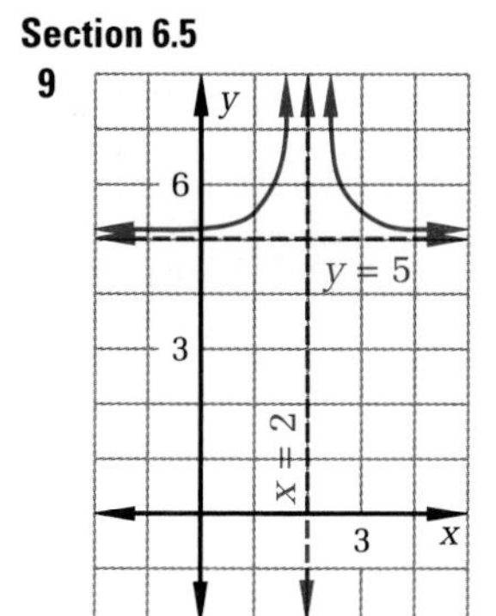

10a

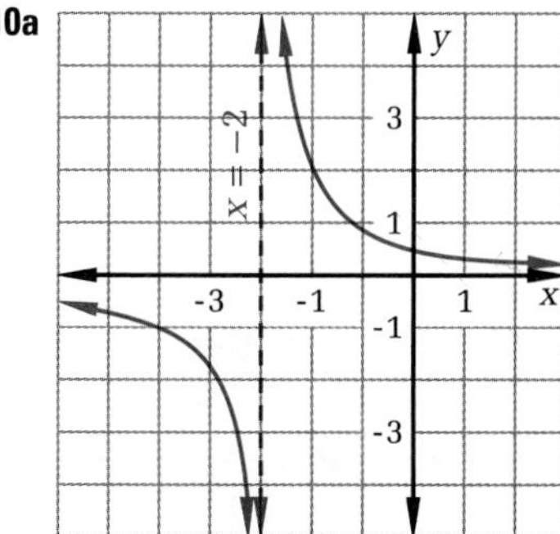

10b

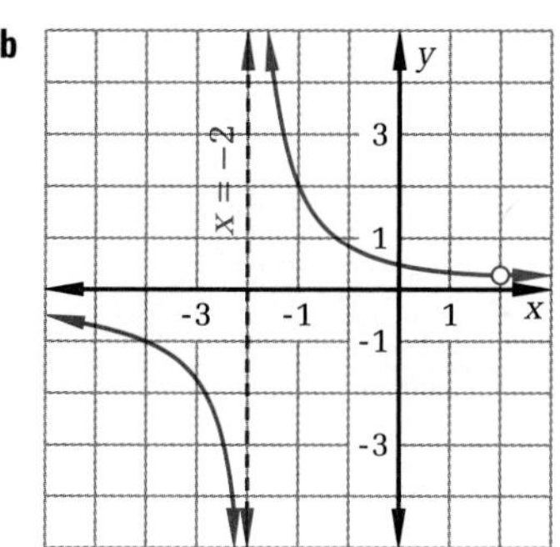

16 Graph with vertical asymptotes at $x = 2$ and $x = 4$, and horizontal asymptote at $y = 1$. (This is just an example of one possible graph.)

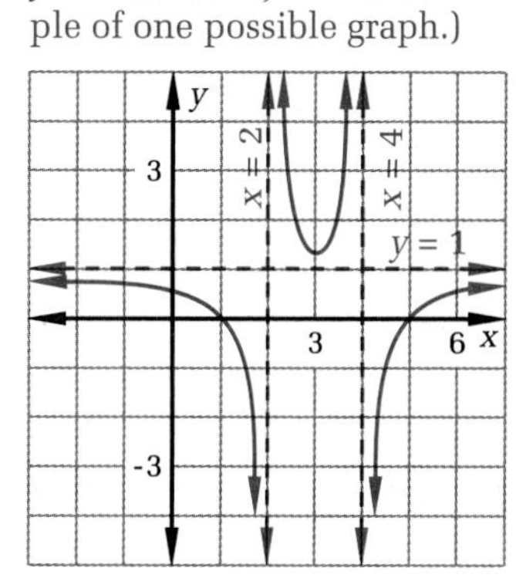

Review for Chapter 6

19

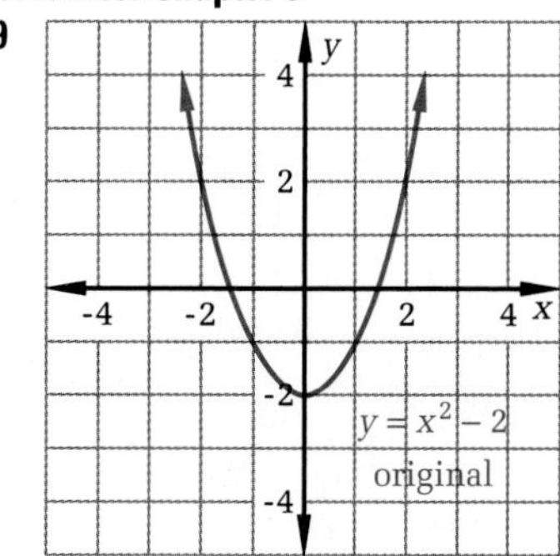

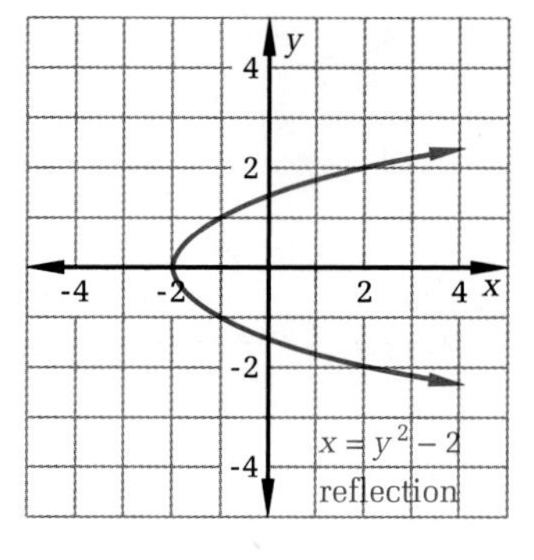

Short Subject: Conics

8 Ellipse: center (6,–2)

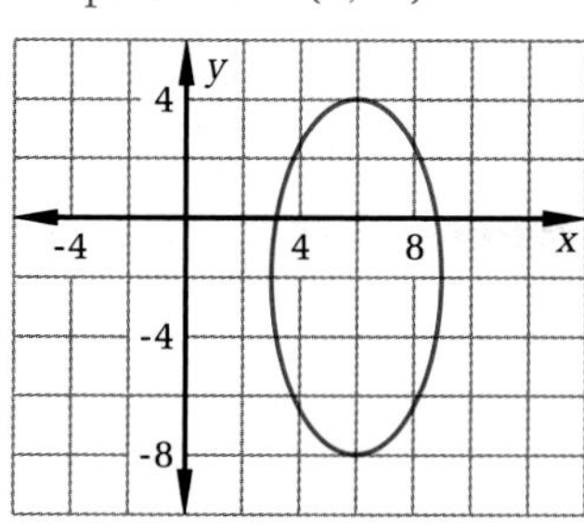

9 Hyperbola: center (–2, –5); opens up and down; asymptotes $y = \frac{7}{5}x - \frac{11}{5}$ and $y = -\frac{7}{5}x - \frac{39}{5}$

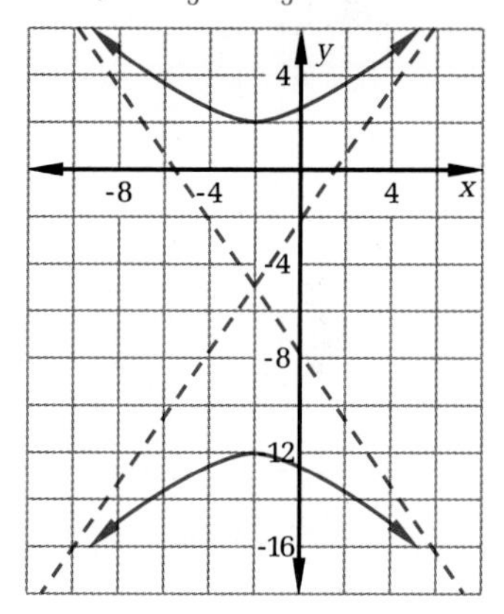

10 Hyperbola: center (0,–3); opens up and down; asymptotes $y = x - 3$ and $y = -x - 3$

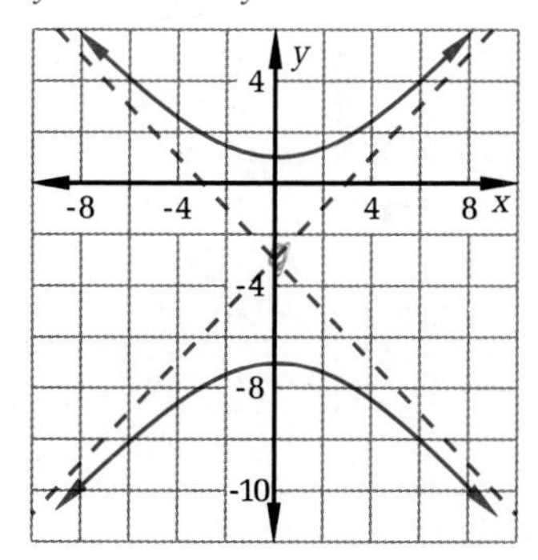

11 Circle: center (–5,4); radius $\sqrt{41}$

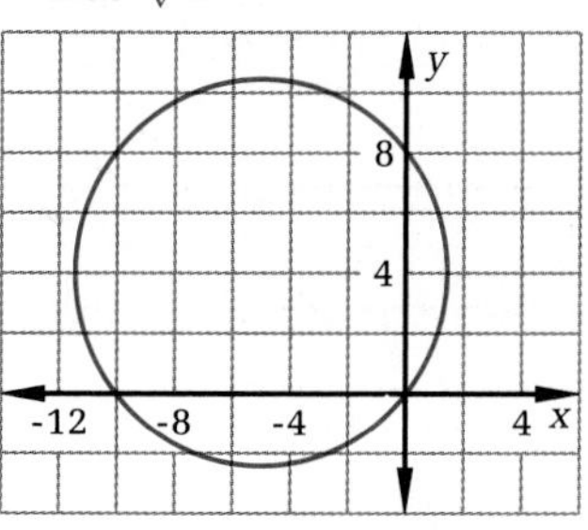

12

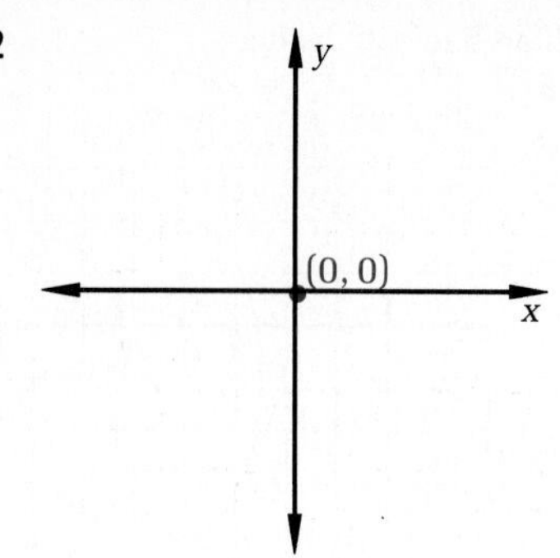

13 Parabola: vertex (4,–3); opens to right

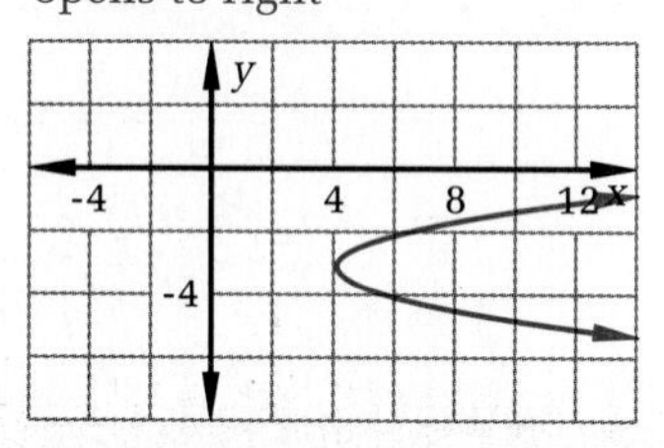

14 Circle: center (0,0); radius 9

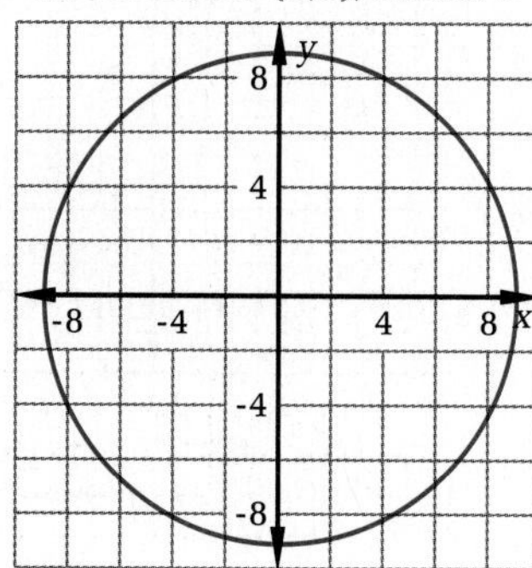

18

19 Ellipse: center (6,–2)

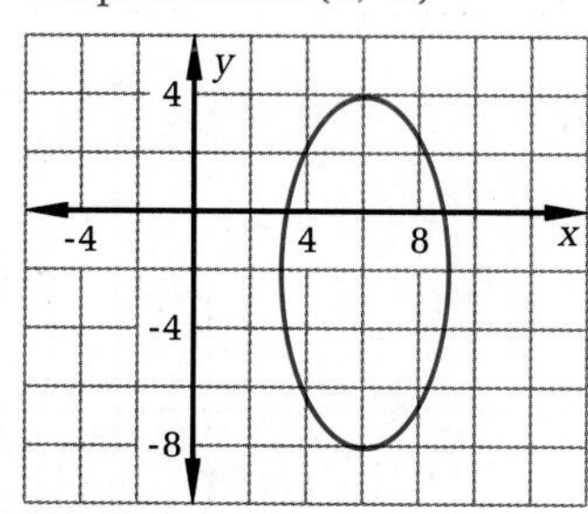

23

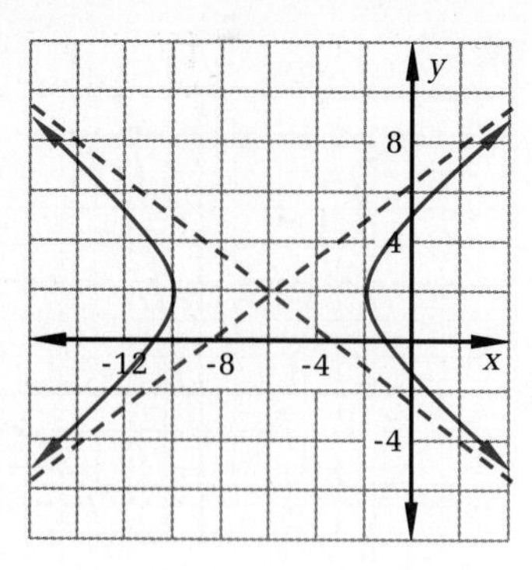

26

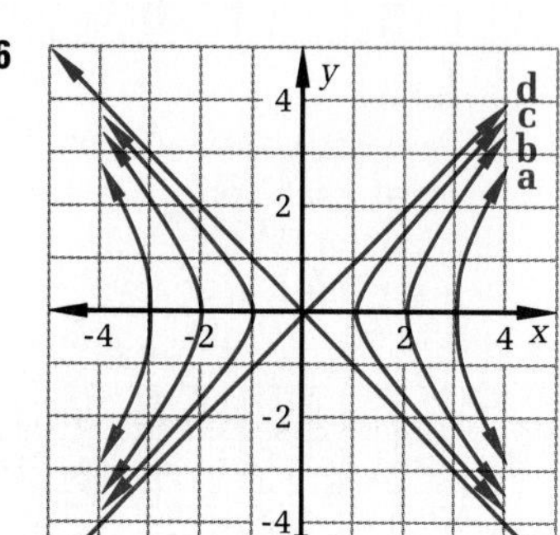

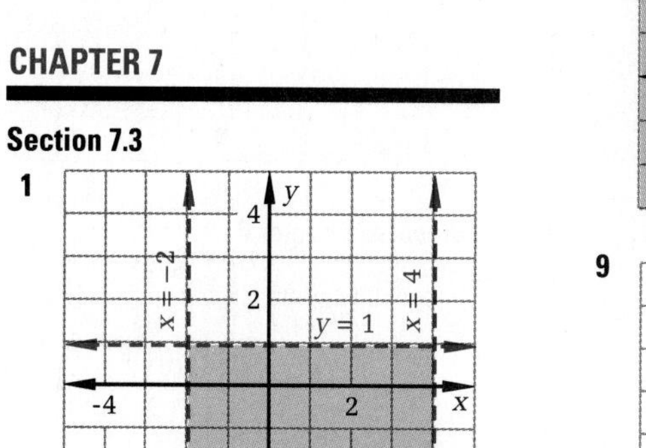

CHAPTER 7

Section 7.3

1

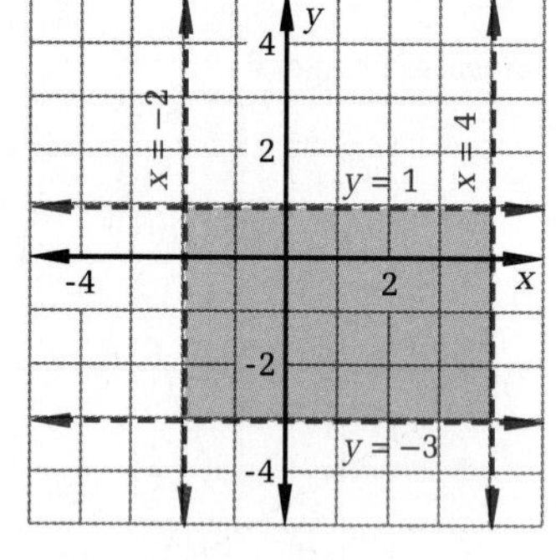

2 No solutions

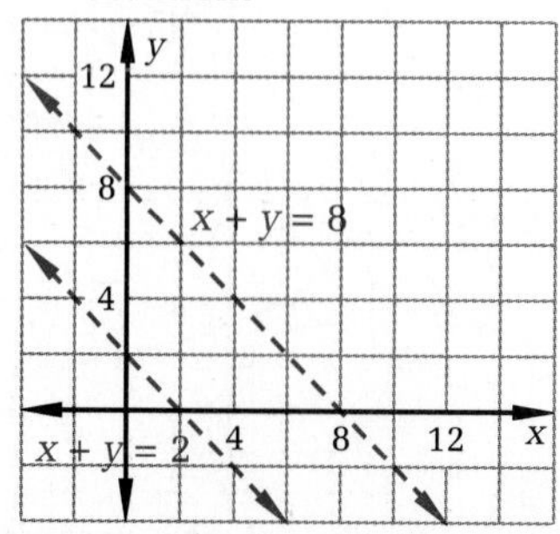

4

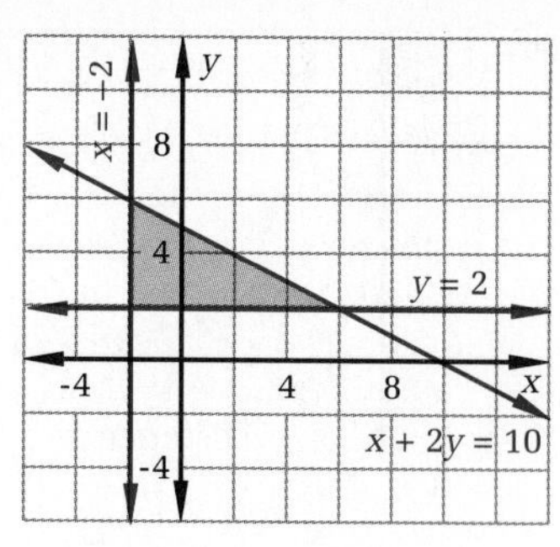

5

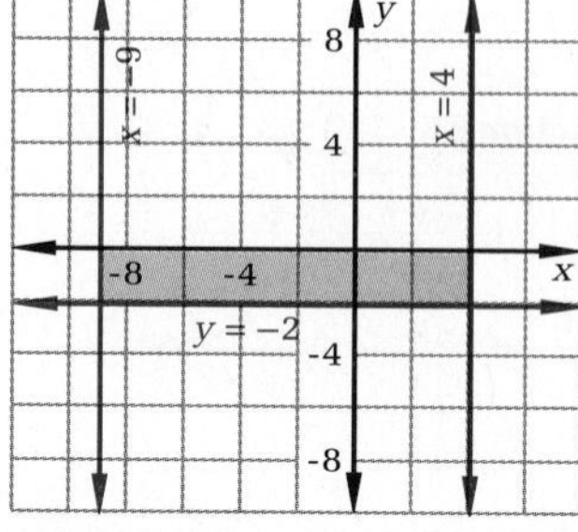

8

9

11

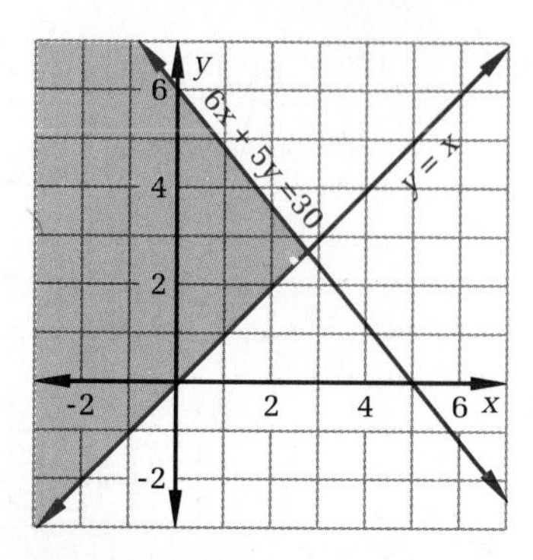

15

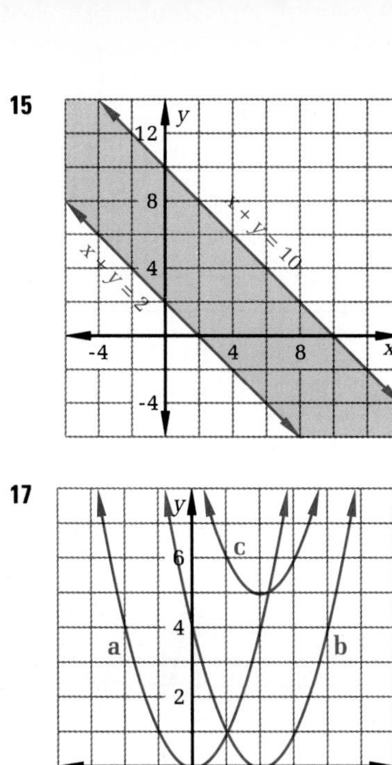

17

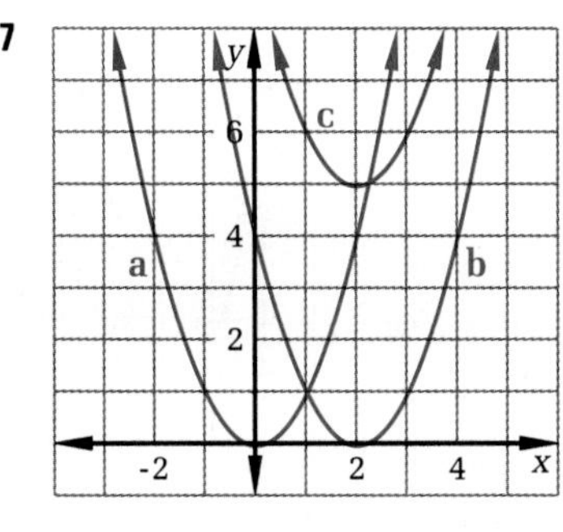

34

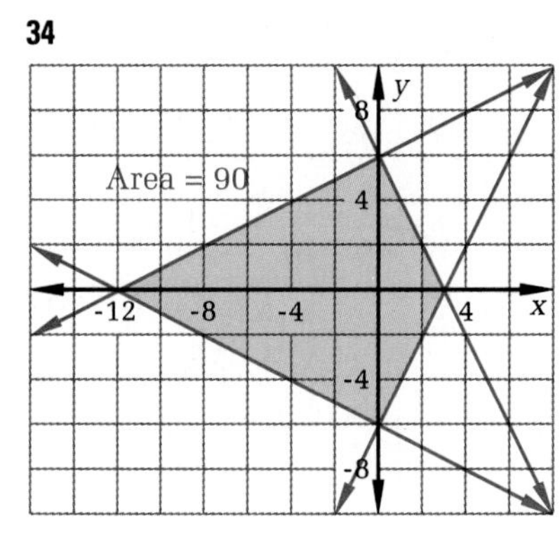

35b

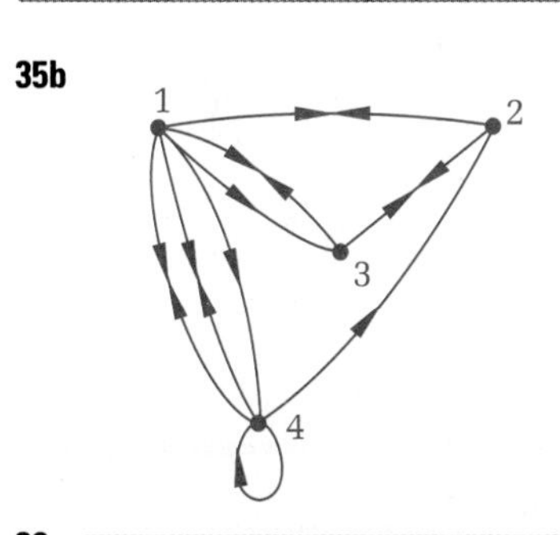

36

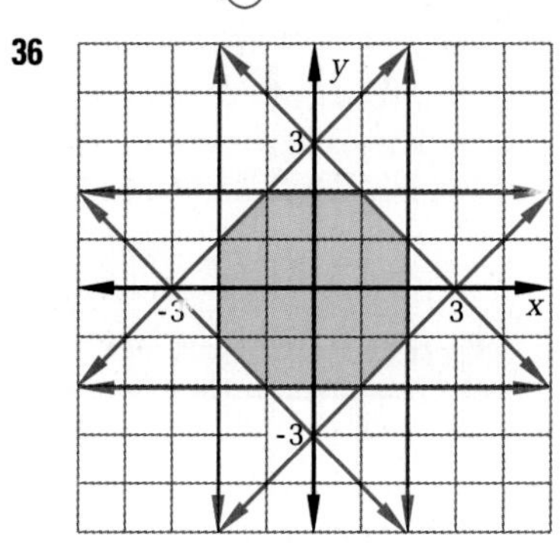

37

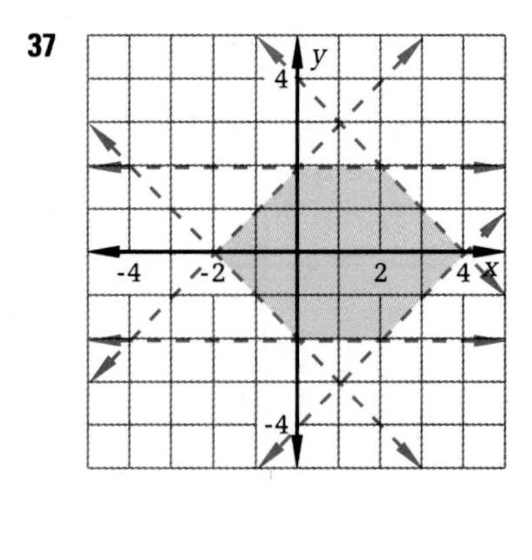

Section 7.4

1

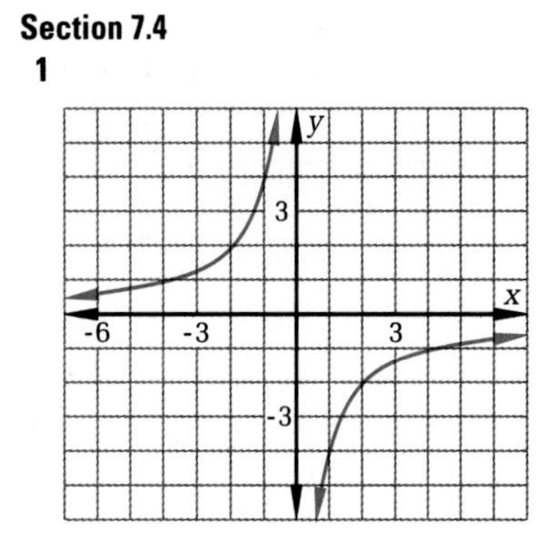

9

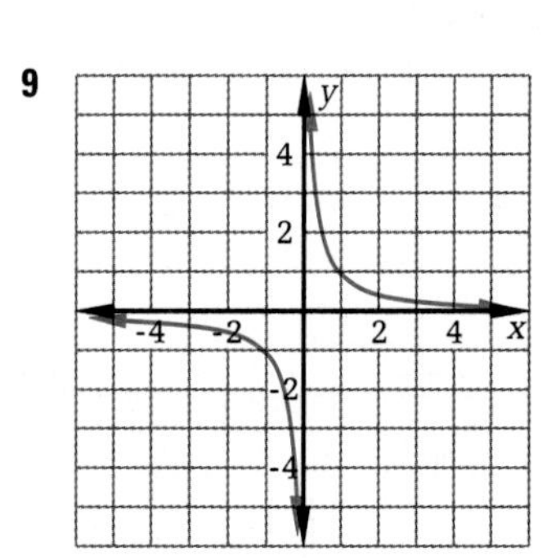

10

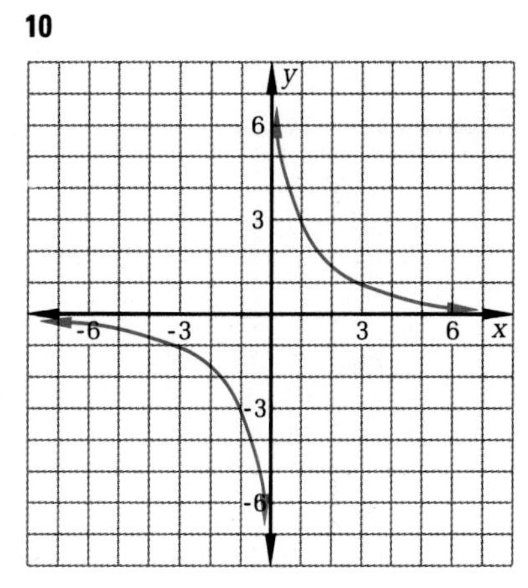

11

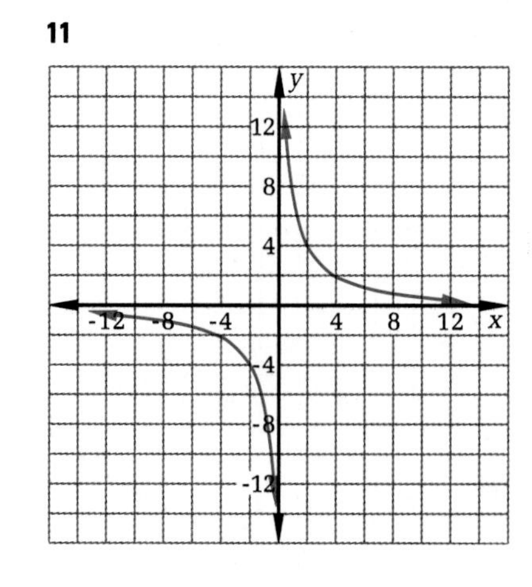

12

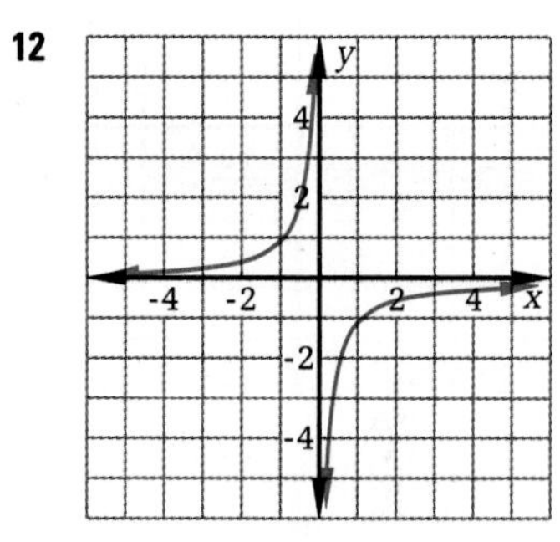

20

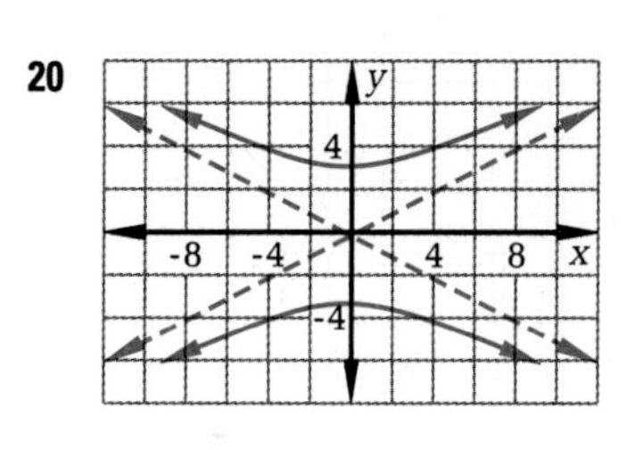

24

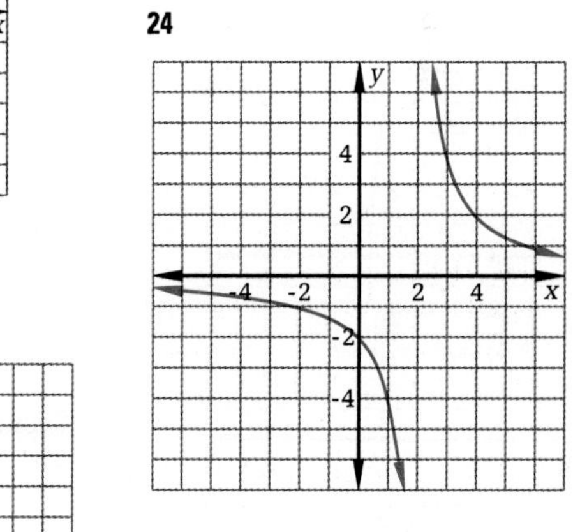

25

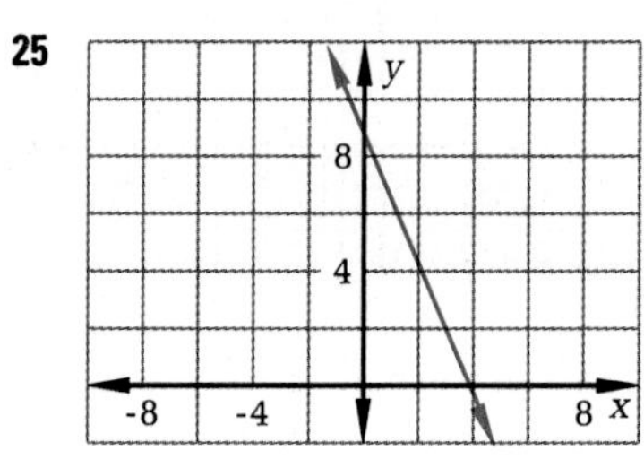

26

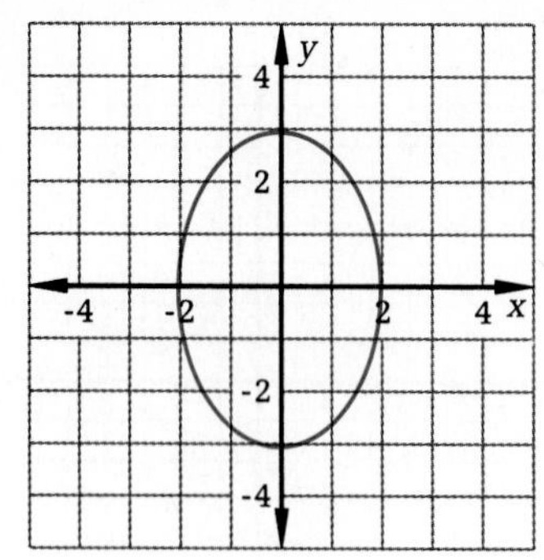

36

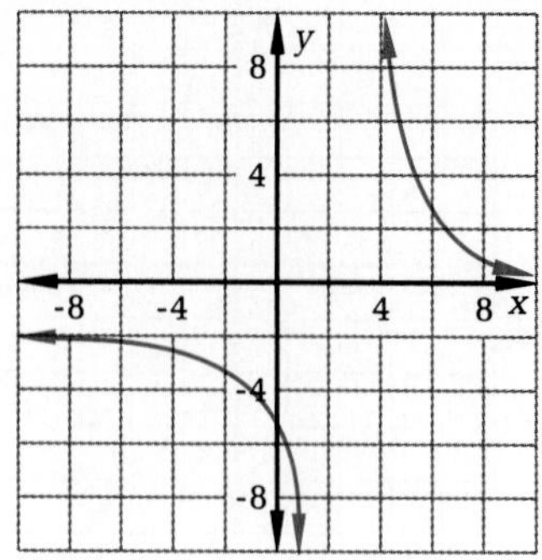

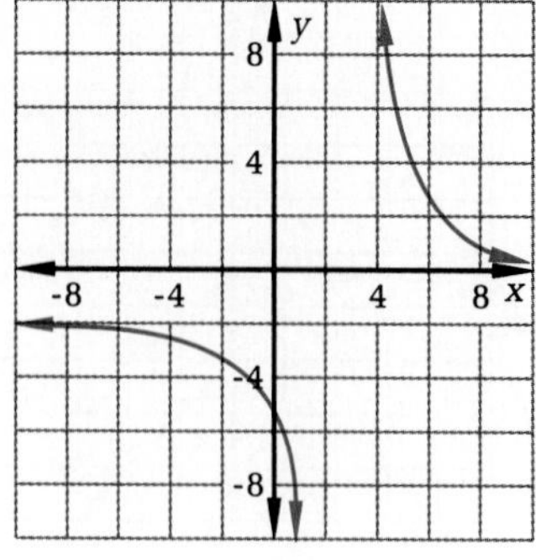

30

39a

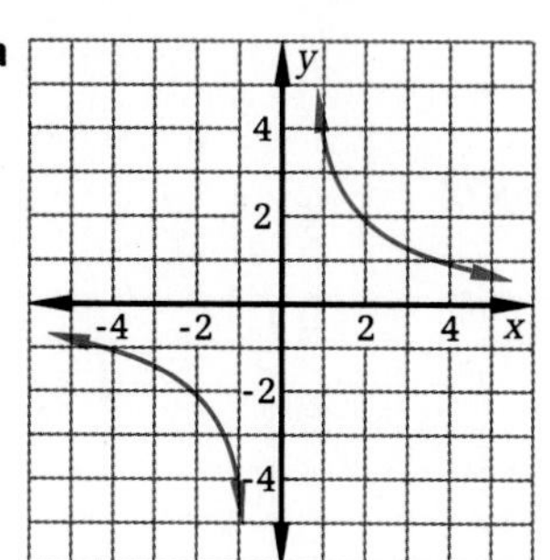

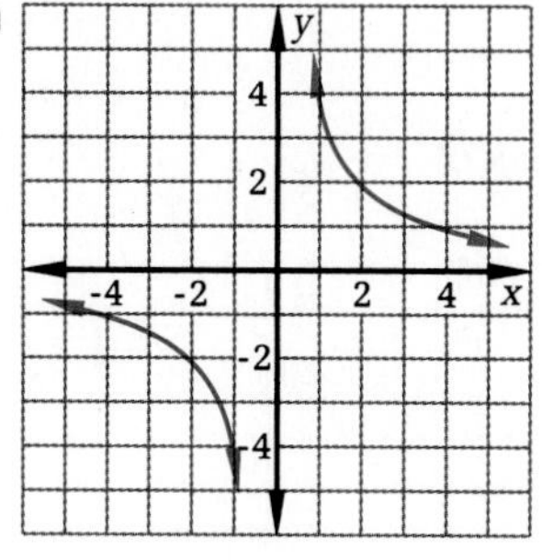

31

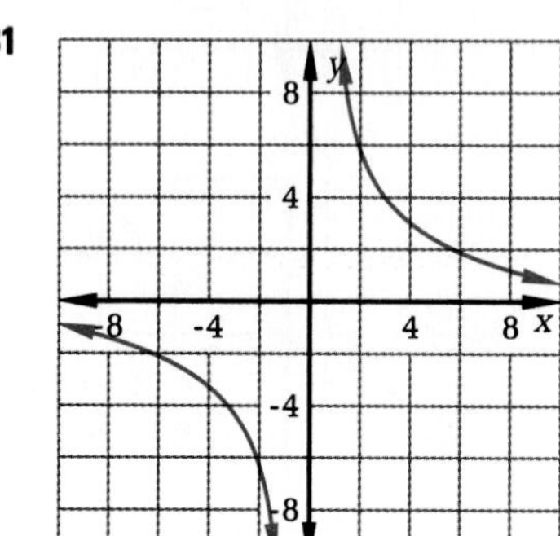

39c

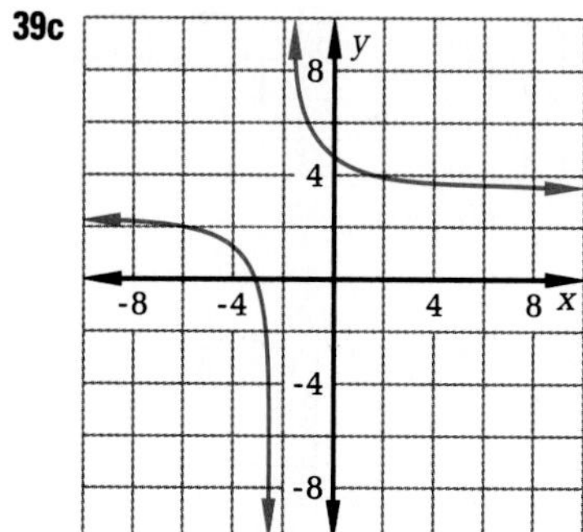

35a

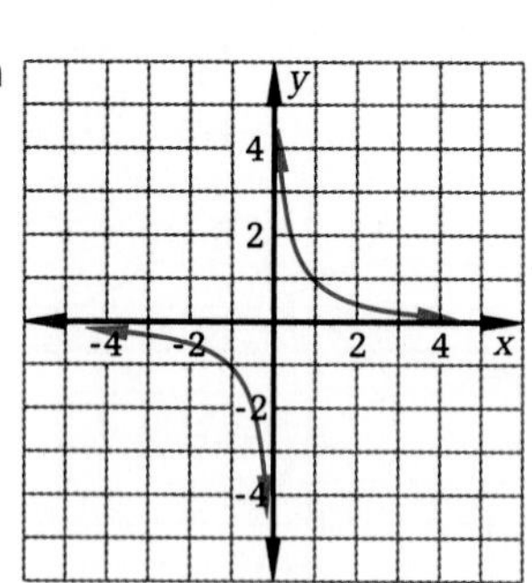

46

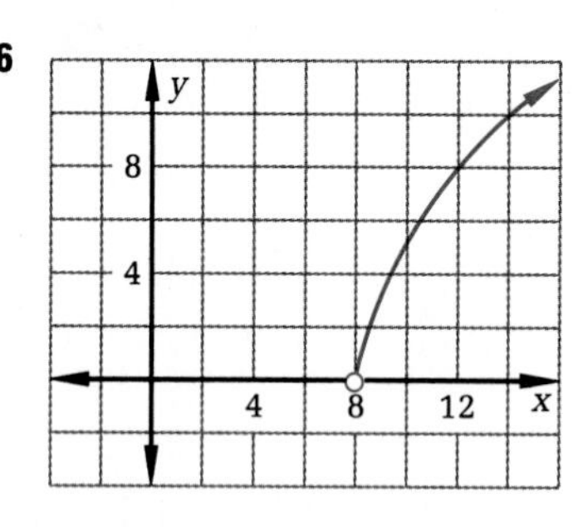

35b

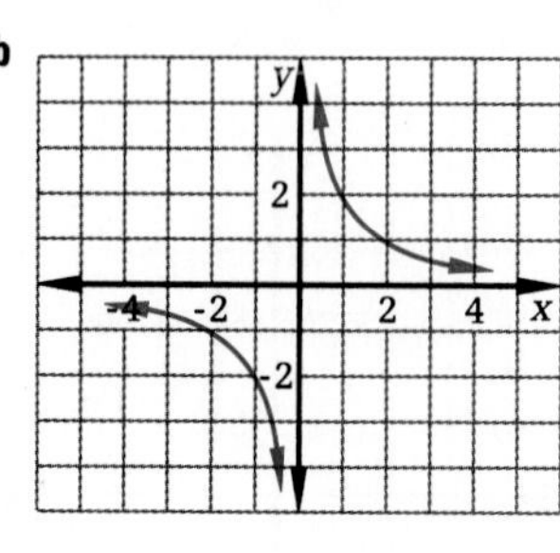

Section 7.5

33c

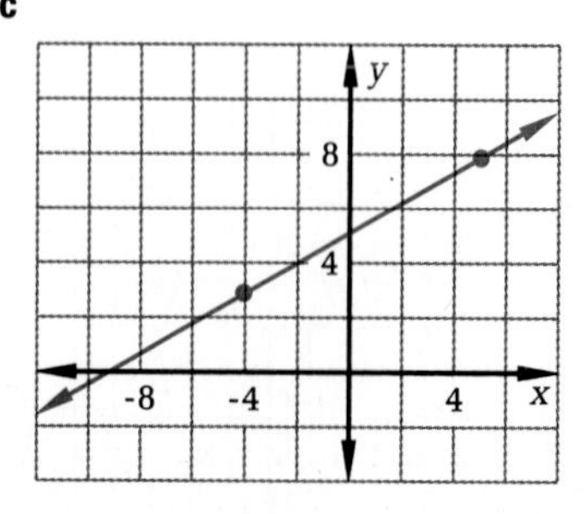

Test for Chapter 7

6

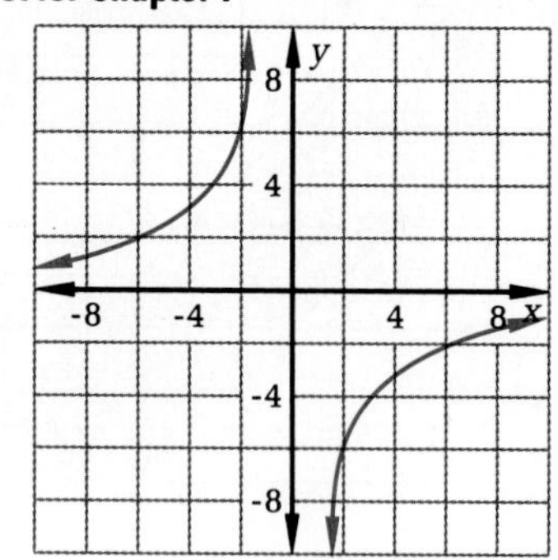

7

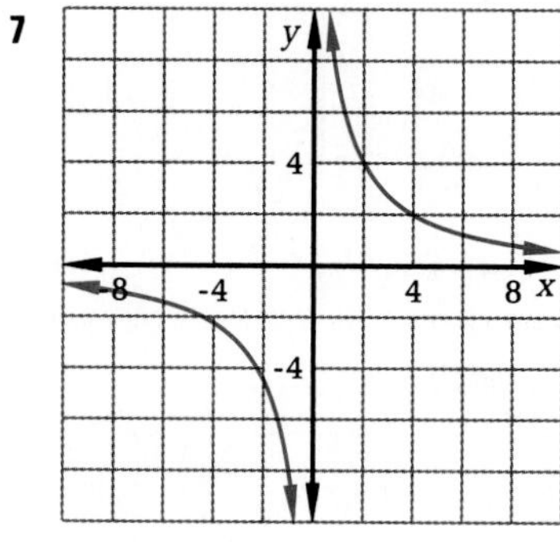

8

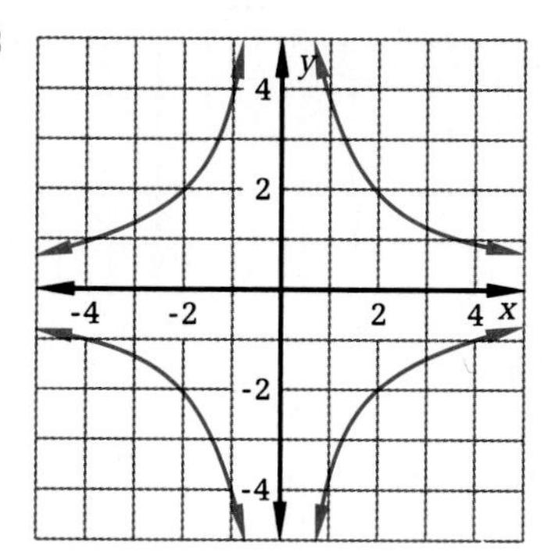

CHAPTER 8

Section 8.3

14c

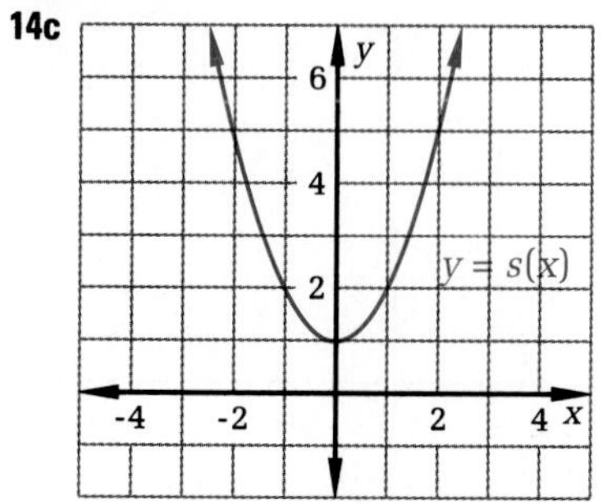

14d

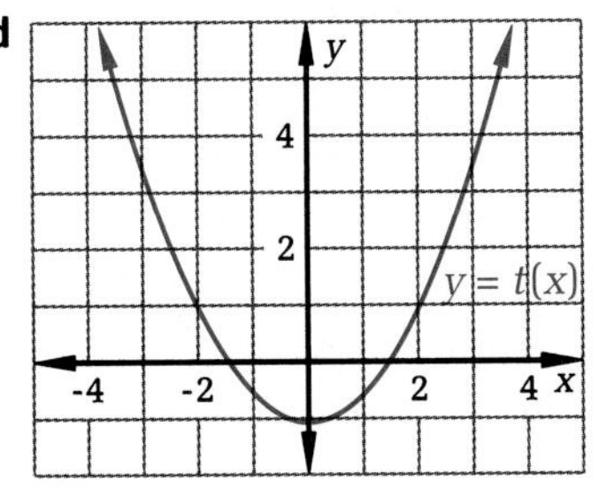

35

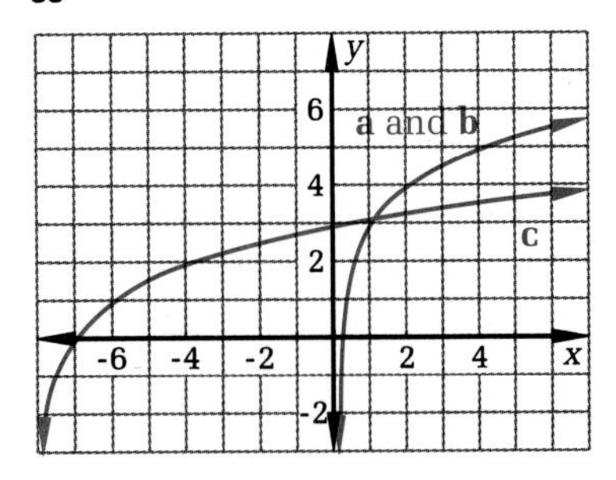

CHAPTER 9

Review for Chapter 9

9 $f(x) = x^3 - 2x^2 - 11x + 12$

14

$f(x) = 2x^4 - 7x^3 + 5x^2 - 7x + 3$

CHAPTER 10

Section 10.5

50

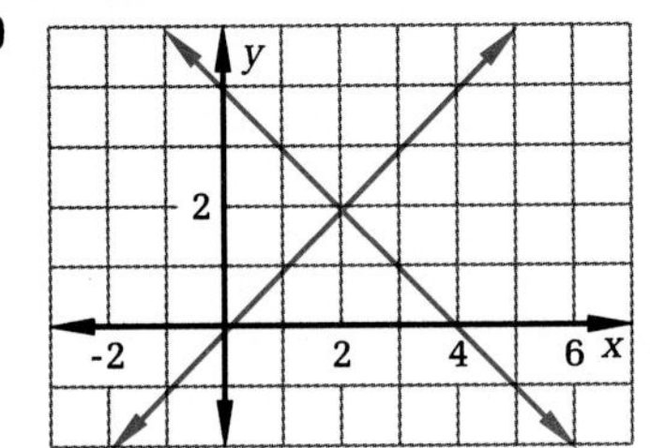

51

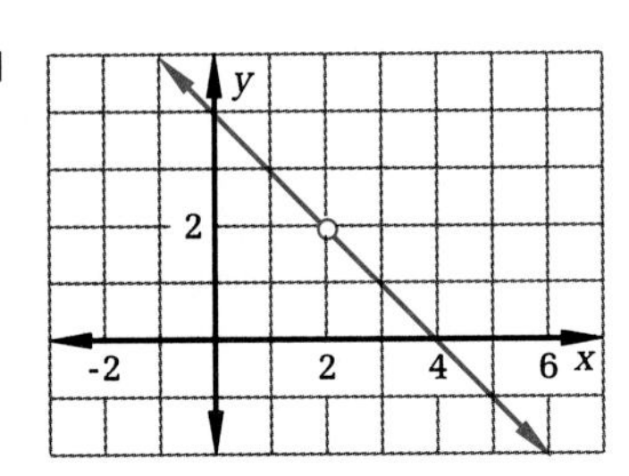

CHAPTER 11

Section 11.1

34b

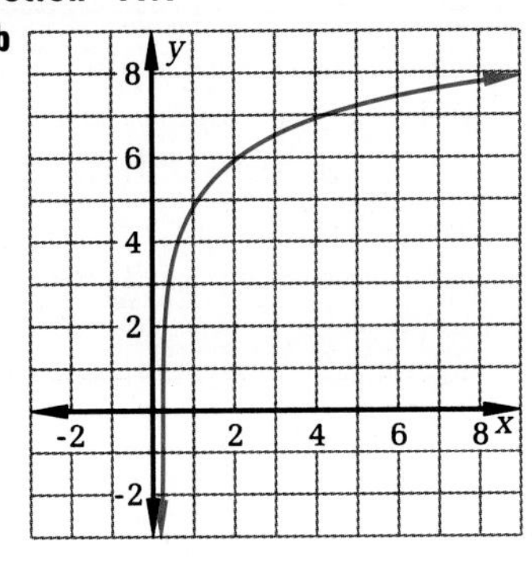

42a

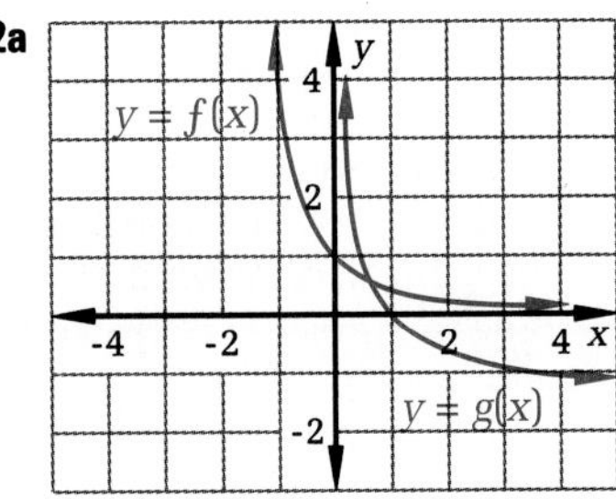

42b $\log_{\frac{1}{5}} x = y$ is equivalent to $\left[\frac{1}{5}\right]^y = x$ thus it is an inverse function and symmetric to the line $y = x$.

44a

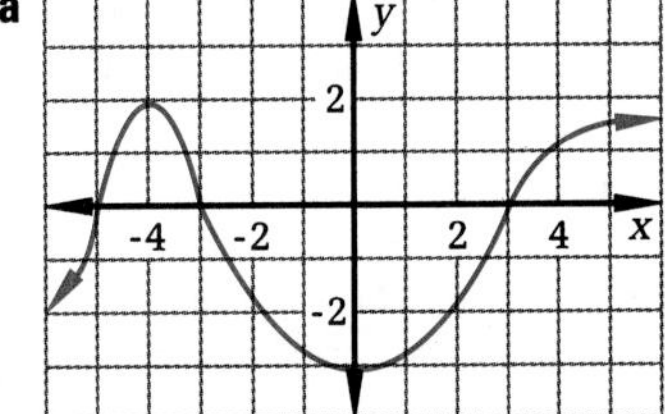

44b

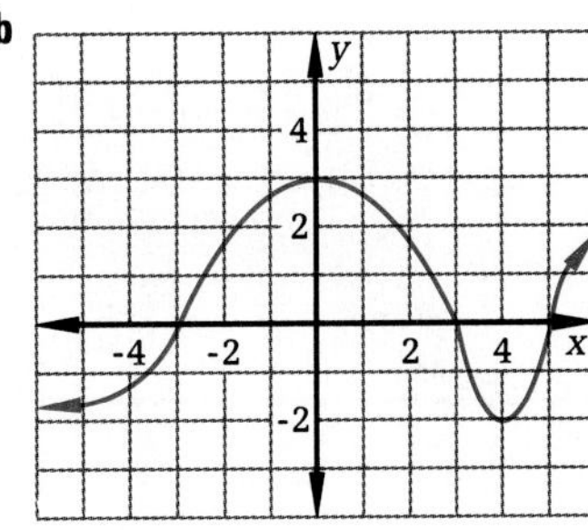

44c

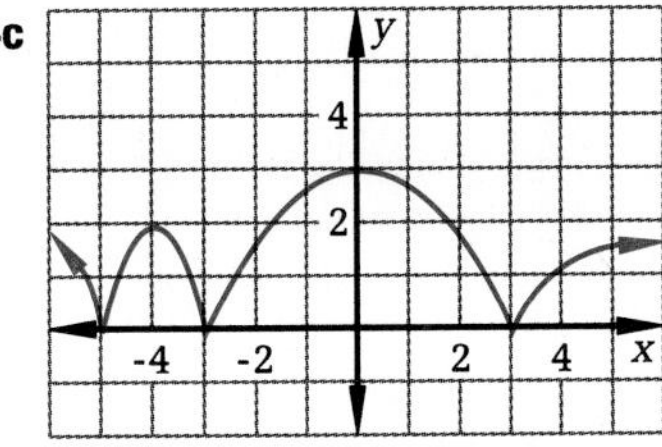

Section 11.3

40

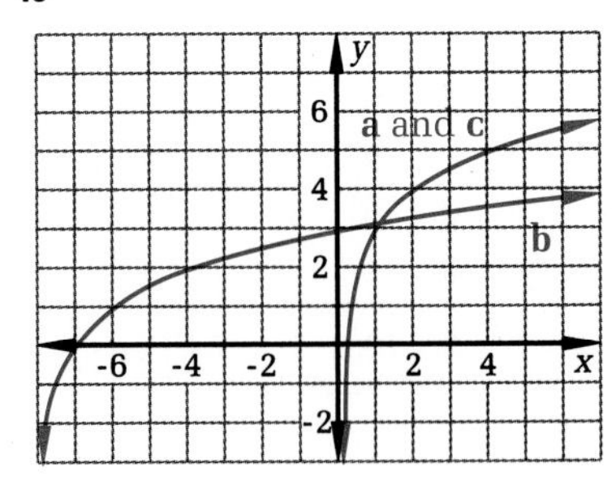

Section 11.4

14

$$\begin{aligned}\log_{a^n} x &= \frac{\log x}{\log a^n} \\ &= \frac{\log x}{n \log a} \\ &= \frac{1}{n} \cdot \frac{\log x}{\log a} \\ &= \frac{1}{n} \cdot \log_a x \\ &= \frac{\log_a x}{n}\end{aligned}$$

CHAPTER 12

Section 12.2

6 Answers will vary. A tree diagram and a lattice are shown.

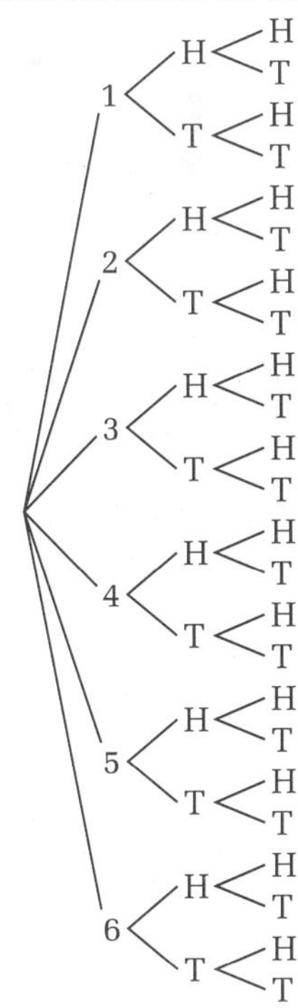

continued on next page

6 *continued*

	HH	HT	TH	TT
1	•	•	•	•
2	•	•	•	•
3	•	•	•	•
4	•	•	•	•
5	•	•	•	•
6	•	•	•	•

7a Answers will vary. A lattice is shown.

	1	2	3	4
1	•	•	•	•
2	•	•	•	•
3	•	•	•	•
4	•	•	•	•
5	•	•	•	•
6	•	•	•	•
7	•	•	•	•
8	•	•	•	•
9	•	•	•	•
10	•	•	•	•
11	•	•	•	•
12	•	•	•	•

26a Answers will vary. A set of ordered pairs representing the sample space is $\{(1, \sqrt{2}), (1, \sqrt{3}), (1, 2), (1, \sqrt{5}), (1, \sqrt{6}), (\sqrt{2}, \sqrt{3}), (\sqrt{2}, 2), (\sqrt{2}, \sqrt{5}), (\sqrt{2}, \sqrt{6}), (\sqrt{3}, 2), (\sqrt{3}, \sqrt{5}), (\sqrt{3}, \sqrt{6}), (2, \sqrt{5}), (2, \sqrt{6}), (\sqrt{5}, \sqrt{6})\}$.

Section 12.5

2 Answers will vary. Possible answer: Write a computer program that randomly generates the numbers 1 – 6, keeps a running total of their sum, and terminates when the sum exceeds 30. By running the program repeatedly and dividing the number of cases in which the sum is 32 by the total number of runs, one can approximate the probability.

3 Answers will vary. Students may mention that the claim does not specify what Sneezarest was preferred to and that the poll might have been based on a very small sample (perhaps only 10 doctors).

4 Answers will vary. Students may note that none of the graduates who did not go on to college may have responded to the letter and that the vagueness of the question and the responses makes it difficult to draw specific conclusions.

7 Anwers will vary. Students may mention that little useful information about why people were not using the library could be obtained by questioning people who were using the library.

9 Answers will vary. Possible answer: By polling randomly selected consumers and statistically analyzing the results

18 Answers will vary. Possible answer: On the basis of the expected weather conditions, the forecaster estimates that the probability is 0.3 that rain will fall at any particular spot in the region for which the forecast is made.

CHAPTER 13

Short Subject: Mathematical Induction

6

For $n = 1$, $4(1) - 3 = 1\,[2(1) - 1]$

$1 = 1$

Assuming that $1 + 5 + 9 + \ldots + (4n - 3) = n(2n - 1)$ for some n, we add $4n + 1$ to each side.

$$1 + 5 + 9 + \ldots + (4n - 3) + (4n + 1) = n(2n - 1) + 4n + 1$$
$$1 + 5 + 9 + \ldots + [4(n + 1) - 3] \quad 2n^2 - n + 4n + 1$$
$$= 2n^2 + 3n + 1$$
$$= (n + 1)(2n + 1)$$
$$= (n + 1)[2(n + 1) - 1]$$

8 When $n = 1$, $1 = \frac{1(1 + 1)}{2} = 1$.

Assuming that $1 + 2 + 3 + \ldots + n = \frac{n(n + 1)}{2}$ for some n, we add $n + 1$ to each side.

$$1 + 2 + 3 + \ldots + n + (n + 1) = \frac{n(n + 1)}{2} + n + 1$$
$$1 + 2 + 3 + \ldots + (n + 1) = \frac{n(n + 1) + 2n + 2}{2}$$
$$= \frac{n^2 + 3n + 2}{2}$$
$$= \frac{(n + 1)(n + 2)}{2}$$
$$= \frac{(n + 1)[(n + 1) + 1]}{2}$$

10 No. Although the statement is true for $n = 1$, if we add 1 to each side of $n + 5 > (n - 2)^2$ we get

$$n + 6 > (n - 2)^2 + 1$$
$$(n + 1) + 5 > n^2 - 4n + 5$$
$$> n^2 - 2n + 2 - 2n + 3$$
$$> (n - 1)^2 - 2n + 3$$
$$> [(n + 1) - 2]^2 - 2n + 3$$

so $(n + 1) + 5 \not> [(n + 1) - 2]^2$ for all values of n.

11 When $n = 1$, $S_1 = \frac{a_1(r^1 - 1)}{r - 1} = a_1$, which is true.

Assuming that $S_n = \frac{a_1(r^n - 1)}{r - 1}$ for some n, we add $a_1 r^n$ to each side.

$$S_n + a_1 r^n = \frac{a_1(r^n - 1)}{r - 1} + a_1 r^n$$
$$S_{n+1} = \frac{a_1(r^n - 1)}{r - 1} + \frac{a_1 r^n(r - 1)}{r - 1}$$
$$= \frac{a_1 r^n - a_1 + a_1 r^{n+1} - a_1 r^n}{r - 1}$$
$$= \frac{a_1(r^{n+1} - 1)}{r - 1}$$

12

When $n = 3$, $\frac{n(n - 3)}{2} = \frac{3(3 - 3)}{2} = 0$, which is true, since a triangle has no diagonals.

Assuming that the number of diagonals is $\frac{n(n - 3)}{2}$ for some n, we note that whenever n is increased by 1 the number of diagonals increases by $n - 1$. Therefore, the number of diagonals in a polygon of $n + 1$ sides is

$$\frac{n(n - 3)}{2} + (n - 1)$$
$$= \frac{n^2 - 3n + 2n - 2}{2}$$
$$= \frac{n^2 - n - 2}{2}$$
$$= \frac{(n + 1)(n - 2)}{2}$$
$$= \frac{(n + 1)[(n + 1) - 3]}{2}$$

13 When $n = 1$, $a^n - b^n = a^1 - b^1 = a - b$, which contains a factor of $a - b$.

Assuming that $a^n - b^n$ contains a factor of $a - b$ for some n,

$$a^{n+1} - b^{n+1} = a \cdot a^n - b \cdot b^n$$
$$= a \cdot a^n - ba^n + ba^n - b \cdot b^n$$
$$= a^n(a - b) + b(a^n - b^n)$$

Since $a^n(a - b)$ contains a factor of $a - b$ and since $b(a^n - b^n)$ also contains a factor of $a - b$ (according to our assumption), $a^{n+1} - b^{n+1}$ contains a factor of $a - b$.

14a Let m be any positive integer. When $n = 1$, $a^m \cdot a^n = a^m \cdot a^1 = a^m \cdot a = a^{m+1}$.

Assuming that $a^m \cdot a^n = a^{m+n}$ for some n, we multiply both sides by a.

$$a^m \cdot a^n \cdot a = a^{m+n} \cdot a$$
$$a^m \cdot a^{n+1} = a^{(m+n)+1}$$
$$= a^{m+(n+1)}$$

14b Let m be any positive integer. When $n = 1$, $(a^m)^n = (a^m)^1 = a^m = a^{m \cdot 1}$.

Assuming that $(a^m)^n = a^{mn}$ for some n, we multiply both sides by a^m.

$$(a^m)^n \cdot a^m = a^{mn} \cdot a^m$$
$$(a^m)^n \cdot (a^m)^1 = a^{mn+m}$$
$$(a^m)^{n+1} = a^{m(n+1)}$$

14c Let m be any positive integer.

When $n = 1$, $\frac{a^m}{a^n} = \frac{a^m}{a^1} = a^m \cdot a^{-1} = a^{m+(-1)} = a^{m-1}$.

Assuming that $\frac{a^m}{a^n} = a^{m-n}$ for some n, we multiply both sides by a^{-1}.

$$\frac{a^m}{a^n} \cdot a^{-1} = a^{m-n} \cdot a^{-1}$$
$$\frac{a^m}{a^n} \cdot \frac{1}{a} = a^{(m-n)+(-1)}$$
$$\frac{a^m}{a^n \cdot a} = a^{m-n-1}$$
$$\frac{a^m}{a^{n+1}} = a^{m-(n+1)}$$

23 When $n = 1$, $5^n - 1 = 5^1 - 1$
$= 4$,
which is divisible by 4.
Assuming that $5^n - 1$ is divisible by 4 for some n,
$5^{n+1} - 1$
$= 5 \cdot 5^n - 1$
$= 5 \cdot 5^n - 5^n + 5^n - 1$
$= 5^n(5 - 1) + (5^n - 1)$
$= 4(5^n) + (5^n - 1)$
Since $4(5^n)$ is divisible by 4 and since $5^n - 1$ is also divisible by 4 (according to our assumption), $5^{n+1} - 1$ must be divisible by 4.

24 When $n = 3$, $180(n - 2) = 180 \cdot 1$
$= 180$,
and we know from a theorem of geometry that the sum of the measures of a triangle's angles is 180.
Assuming that the sum of the interior angles of a polygon is $180(n - 2)$ for some n, we note that n is increased by 1, the sum of the angles is increased by 180. Therefore, the sum of the measures of the interior angles of a polygon with $n + 1$ sides is
$180(n - 2) + 180$
$= 180n - 360 + 180$
$= 180n - 180$
$= 180(n - 1)$
$= 180[(n + 1) - 2]$

26 When $n = 1$, $2^1 - 1 = 1$, and it obviously only takes 1 move to transfer 1 disk.
Let $H(n)$ represent the number of moves needed to transfer an n–disk tower. In Section 5.6, problem 28, part **b**, it was established that $H(n) = 2[H(n - 1)] + 1$.
Assuming that $H(n) = 2^n - 1$ for some n,
$H(n + 1)$
$= 2(H[(n + 1) - 1]) + 1$
$= 2[H(n)] + 1$
$= 2(2^n - 1) + 1$
$= 2 \cdot 2^n - 2 + 1$
$= 2^{n+1} - 1$

25

When $n = 1$, $\sum_{k=1}^{n} \left(\frac{1}{(2k-1)(2k+1)}\right) = \frac{1}{(2 \cdot 1 - 1)(2 \cdot 1 + 1)}$
$= \frac{1}{3}$
$= \frac{1}{2(1) + 1}$

Assuming that $\sum_{k=1}^{n} \left(\frac{1}{(2k-1)(2k+1)}\right) = \frac{n}{2n+1}$ for some n,

we add $\frac{1}{[2(n+1) - 1][2(n+1) + 1]}$ to each side.

$\sum_{k=1}^{n} \left(\frac{1}{(2k-1)(2k+1)}\right) + \frac{1}{[2(n+1) - 1][2(n+1) + 1]}$
$= \frac{n}{2n+1} + \frac{1}{[2(n+1) - 1][2(n+1) + 1]}$

$\sum_{k=1}^{n+1} \left(\frac{1}{(2k-1)(2k+1)}\right)$
$= \frac{n}{2n+1} + \frac{1}{(2n+1)(2n+3)}$
$= \frac{n(2n+3) + 1}{(2n+1)(2n+3)}$
$= \frac{2n^2 + 3n + 1}{(2n+1)(2n+3)}$
$= \frac{(2n+1)(n+1)}{(2n+1)(2n+3)}$
$= \frac{n+1}{2n+3}$
$= \frac{n+1}{2(n+1) + 1}$

27

When $n = 1$,
$(x + y)^n = (x + y)^1$
$= x + y$
$= 1x + 1y$
$= [C(1, 0)]x^1 + [C(1, 1)]y^1$.
Assuming that $(x + y)^n = [C(n, 0)]x^n + [C(n, 1)]x^{n-1}y + [C(n, 2)]x^{n-2}y^2 + \ldots + [C(n, n)]y^n$ for some value of n, we multiply both sides by $(x + y)$.
$(x + y)(x + y)^n$
$= (x + y)([C(n, 0)]x^n + [C(n, 1)]x^{n-1}y + [C(n, 2)]x^{n-2}y^2 + \ldots + [C(n, n)]y^n)$
$(x + y)^{n+1}$
$= (x[C(n, 0)]x^n + x[C(n, 1)]x^{n-1}y + x[C(n, 2)]x^{n-2}y^2 + \ldots) + (y[C(n, 0)]x^n + y[C(n, 1)]x^{n-1}y + \ldots + y[C(n, n)]y^n)$
$= [C(n, 0)]x^{n+1} + [C(n, 1)]x^n y + [C(n, 2)]x^{n-1}y^2 + \ldots + [C(n, 0)]x^n y + [C(n, 1)]x^{n-1}y^2 + \ldots + [C(n, n)]y^{n+1}$
$= 1x^{n+1} + (n + 1)x^n y + [C(n, 2) + C(n, 1)]x^{n-1}y^2 + \ldots + 1y^{n+1}$
$= [C(n + 1, 0)]x^{n+1} + [C(n + 1, 1)]x^{(n+1)-1}y + [C(n + 1, 2)]x^{(n+1)-2}y^2 + \ldots + [C(n + 1, n + 1)]y^{n+1}$

CHAPTER 14

Section 14.7

7

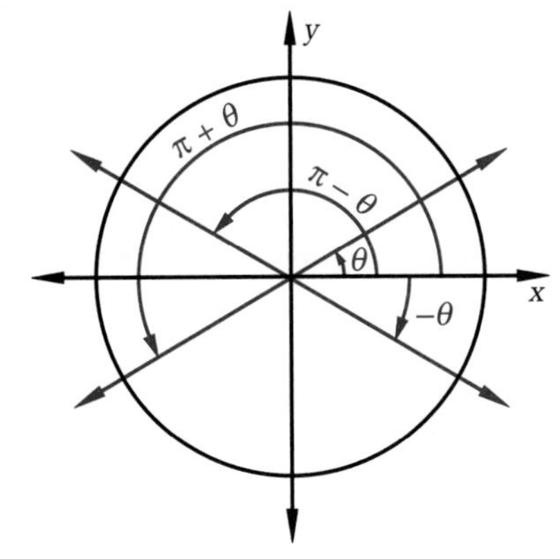

29 $f(-x) = \frac{-x}{1!} - \frac{(-x)^3}{3!} + \frac{(-x)^5}{5!} - \frac{(-x)^7}{7!} + \frac{(-x)^9}{9!}$

$= \frac{-x}{1!} - \frac{-x^3}{3!} + \frac{-x^5}{5!} - \frac{-x^7}{7!} + \frac{-x^9}{9!}$

$= -\frac{x}{1!} + \frac{x^3}{3!} - \frac{x^5}{5!} + \frac{x^7}{7!} - \frac{x^9}{9!}$

$= -\left(\frac{x}{1!} - \frac{x^3}{3!} + \frac{x^5}{5!} - \frac{x^7}{7!} + \frac{x^9}{9!}\right)$

$= -f(x)$

30 $f(-x) = 1 - \frac{(-x)^2}{2!} + \frac{(-x)^4}{4!} - \frac{(-x)^6}{6!} + \frac{(-x)^8}{8!} - \frac{(-x)^{10}}{10!}$

$= 1 - \frac{x^2}{2!} + \frac{x^4}{4!} - \frac{x^6}{6!} + \frac{x^8}{8!} - \frac{x^{10}}{10!}$

$= f(x)$

CHAPTER 15

Section 15.1

2 $(\cos^2 \theta)(\tan^2 \theta + 1)$

$= (\cos^2 \theta)(\sec^2 \theta) = \frac{\cos^2 \theta}{\cos^2 \theta} = 1; \cos \theta \neq 0$

6 $\frac{1}{\cos^2 \theta} - \frac{\sin^2 \theta}{\cos^2 \theta}$

$= \frac{1 - \sin^2 \theta}{\cos^2 \theta} = \frac{\cos^2 \theta}{\cos^2 \theta} = 1; \cos \theta \neq 0$

7 $\frac{\cos^2 \theta - \sin^2 \theta}{\cos \theta - \sin \theta}$

$= \frac{(\cos \theta + \sin \theta)(\cos \theta - \sin \theta)}{\cos \theta - \sin \theta}$

$= \cos \theta + \sin \theta; \cos \theta \neq \sin \theta$

13 $\frac{\cos^3 \theta - \sin^3 \theta}{\cos \theta - \sin \theta}$

$= \frac{(\cos \theta - \sin \theta)(\cos^2 \theta + \cos \theta \sin \theta + \sin^2 \theta)}{\cos \theta - \sin \theta}$

$= \cos^2 \theta + \cos \theta \sin \theta + \sin^2 \theta$

$= (\sin^2 \theta + \cos^2 \theta) + \cos \theta \sin \theta$

$= 1 + \cos \theta \sin \theta; \cos \theta \neq \sin \theta$

16a When $\theta = 210°$, $\sin 210° = -\frac{1}{2}$ and $\sqrt{1 - \cos^2 210°} = \frac{1}{2}$. Since the expressions are not equal for $\theta = 210°$, the expression is not an identity.

16b Ida used an algebraic property of equations instead of simplifying each side separately. $\sqrt{1 - \cos^2 \theta} = \sqrt{\sin^2 \theta} = |\sin \theta|$, and $\sin \theta = |\sin \theta|$ is not an identity, since it is true only for values of θ with a positive sine.

21a $\frac{\sin \theta}{1 - \cos \theta}$

$= \frac{\frac{a}{\sqrt{a^2 + b^2}}}{1 - \frac{b}{\sqrt{a^2 + b^2}}} = \frac{\frac{a}{\sqrt{a^2 + b^2}}}{\frac{\sqrt{a^2 + b^2} - b}{\sqrt{a^2 + b^2}}} = \frac{a}{\sqrt{a^2 + b^2} - b}$

$= \frac{a\sqrt{a^2 + b^2} + ab}{a^2 + b^2 - b^2} = \frac{\sqrt{a^2 + b^2} + b}{a}$

$= \frac{\sqrt{a^2 + b^2}}{a} + \frac{b}{a} = \csc \theta + \cot \theta$

23a $\frac{\tan x - 1}{\tan x + 1}$

$= \frac{\frac{\sin x}{\cos x} - 1}{\frac{\sin x}{\cos x} + 1} = \frac{\frac{\sin x - \cos x}{\cos x}}{\frac{\sin x + \cos x}{\cos x}}$

$= \frac{\sin x - \cos x}{\sin x + \cos x} = \frac{\frac{\sin x - \cos x}{\sin x}}{\frac{\sin x + \cos x}{\sin x}}$

$= \frac{1 - \frac{\cos x}{\sin x}}{1 + \frac{\cos x}{\sin x}} = \frac{1 - \cot x}{1 + \cot x}; \cos x \neq 0, \sin x \neq 0, \sin x \neq -\cos x$

23b $\cot^2 x - \cos^2 x$

$= \frac{\cos^2 x}{\sin^2 x} - \cos^2 x = \left(\frac{1}{\sin^2 x} - 1\right) \cdot \cos^2 x$

$= (\csc^2 x - 1) \cdot \cos^2 x = \cot^2 x \cdot \cos^2 x; \sin x \neq 0$

23c $(\tan x + \sec x)^2$

$$= \left(\frac{\sin x}{\cos x} + \frac{1}{\cos x}\right)^2 = \left(\frac{\sin x + 1}{\cos x}\right)^2$$
$$= \frac{(\sin x + 1)^2}{1 - \sin^2 x} = \frac{(\sin x + 1)^2}{(1 + \sin x)(1 - \sin x)}$$
$$= \frac{1 + \sin x}{1 - \sin x};\ \cos x \neq 0,\ \sin x \neq -1,\ \sin x \neq 1$$

26a

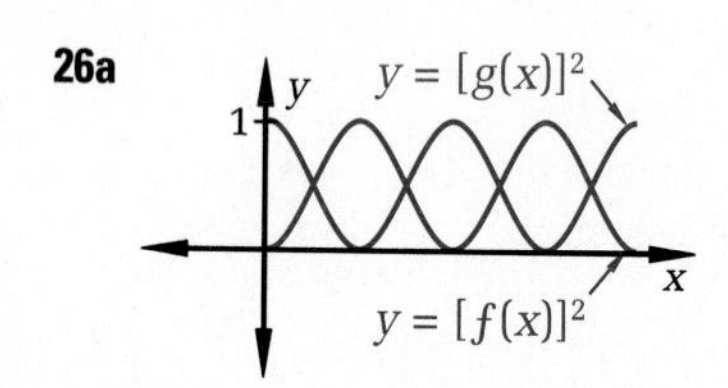

27 $\log_2 (\tan \theta) + \log_2 (\cos \theta)$

$$= \log_2 \left(\frac{\sin \theta}{\cos \theta}\right) + \log_2 (\cos \theta)$$
$$= \log_2 (\sin \theta) - \log_2 (\cos \theta) + \log_2 (\cos \theta)$$
$$= \log_2 (\sin \theta);\ \sin \theta > 0,\ \cos \theta > 0$$

28 $\dfrac{2 \sin^2 \theta + 5 \sin \theta \cos \theta + 2 \cos^2 \theta}{\sin \theta + 2 \cos \theta}$

$$= \frac{(2 \sin \theta + \cos \theta)(\sin \theta + 2 \cos \theta)}{\sin \theta + 2 \cos \theta}$$
$$= 2 \sin \theta + \cos \theta$$
$$= \frac{2}{\csc \theta} + \frac{1}{\sec \theta}$$
$$= \frac{2 \sec \theta + \csc \theta}{\sec \theta \csc \theta};\ \sin \theta \neq 0,\ \cos \theta \neq 0,\ \sin \theta \neq -2 \cos \theta$$

Section 15.2

13 $\cot (\alpha + \beta)$

$$= \frac{\cos (\alpha + \beta)}{\sin (\alpha + \beta)}$$
$$= \frac{\cos \alpha \cdot \cos \beta - \sin \alpha \cdot \sin \beta}{\sin \alpha \cdot \cos \beta + \cos \cdot \alpha \sin \beta} \cdot \frac{\frac{1}{\sin \alpha \sin \beta}}{\frac{1}{\sin \alpha \sin \beta}}$$
$$= \frac{\frac{\cos \alpha \cos \beta}{\sin \alpha \sin \beta} - 1}{\frac{\sin \alpha \cos \beta}{\sin \alpha \sin \beta} + \frac{\cos \alpha \sin \beta}{\sin \alpha \sin \beta}}$$
$$= \frac{\cot \alpha \cot \beta - 1}{\cot \beta + \cot \alpha}$$

25 $\sin 3\theta$

$$= \sin (\theta + 2\theta)$$
$$= \sin \theta \cos 2\theta + \cos \theta \sin 2\theta$$
$$= \sin \theta \cos (\theta + \theta) + \cos \theta \sin(\theta + \theta)$$
$$= \sin \theta \cdot (\cos \theta \cos \theta - \sin \theta \sin \theta) + \cos \theta (\sin \theta \cos \theta + \cos \theta \sin \theta)$$
$$= \sin \theta \cdot (\cos^2\theta - \sin^2\theta) + \cos \theta \cdot (2 \cos \theta \sin \theta)$$
$$= \sin \theta \cos^2\theta - \sin^3\theta + 2 \cos^2\theta \sin \theta$$
$$= 3 \cos^2 \theta \sin \theta - \sin^3\theta$$
$$= 3 \sin \theta \cdot (1 - \sin^2\theta) - \sin^3\theta$$
$$= 3 \sin \theta - 3 \sin^3\theta - \sin^3\theta$$
$$= 3 \sin \theta - 4 \sin^3\theta$$

Section 15.3

9 $(\sin \theta + \cos \theta)^{\csc^2 \theta - \cot^2 \theta + 1}$

$$= (\sin \theta + \cos \theta)^{\frac{1}{\sin^2 \theta} - \frac{\cos^2 \theta}{\sin^2 \theta} + 1}$$
$$= (\sin \theta + \cos \theta)^{\frac{1 - \cos^2 \theta}{\sin^2 \theta} + 1}$$
$$= (\sin \theta + \cos \theta)^2$$
$$= \sin^2 \theta + 2 \sin \theta \cos \theta + \cos^2 \theta$$
$$= 1 + \sin 2\theta$$

19 $\tan \frac{\theta}{2} = \pm\sqrt{\dfrac{1 - \cos \theta}{1 + \cos \theta}}$

$$= \pm\sqrt{\frac{1 - \cos \theta}{1 + \cos \theta}} \cdot \frac{\sqrt{1 + \cos \theta}}{\sqrt{1 + \cos \theta}}$$
$$= \pm\sqrt{\frac{1 - \cos^2 \theta}{(1 + \cos \theta)^2}}$$
$$= \pm\sqrt{\left(\frac{\sin \theta}{1 + \cos \theta}\right)^2}$$
$$= \frac{\sin \theta}{1 + \cos \theta}$$

22b $\cos 4\theta$

$$= \cos 2 (2\theta)$$
$$= \cos^2 2\theta - \sin^2 2\theta$$
$$= (\cos^2 \theta - \sin^2 \theta)^2 - 4 \sin^2 \theta \cos^2 \theta$$
$$= \cos^4 \theta - 2 \sin^2 \theta \cos^2 \theta + \sin^4 \theta - 4 \sin^2 \theta \cos^2 \theta$$
$$= \cos^4 \theta + \sin^4 \theta - 6 \sin^2 \theta \cos^2 \theta$$
$$= \cos^4 \theta + (1 - \cos^2 \theta)^2 - 6 \cos^2 \theta \cdot (1 - \cos^2 \theta)$$
$$= \cos^4 \theta + 1 - 2 \cos^2 \theta + \cos^4 \theta - 6 \cos^2 \theta + 6 \cos^4 \theta$$
$$= 8 \cos^4 \theta - 8 \cos^2 \theta + 1$$

26 $\sin 3\theta$

$$= \sin (2\theta + \theta)$$
$$= \sin 2\theta \cos \theta + \sin \theta \cos 2\theta$$
$$= 2 \sin \theta \cos \theta \cdot \cos \theta + \sin \theta \cdot (2 \cos^2 \theta - 1)$$
$$= 2 \sin \theta \cos^2 \theta + 2 \sin \theta \cos^2 \theta - \sin \theta$$
$$= 4 \sin \theta \cos^2 \theta - \sin \theta$$
$$= (\sin \theta) (4 \cos^2 \theta - 1)$$
$$= (\sin \theta) (2 \cos \theta - 1) (2 \cos \theta + 1)$$

Section 15.4

19a $\frac{\sin(\alpha+\beta)+\sin(\alpha-\beta)}{\cos(\alpha-\beta)+\cos(\alpha+\beta)}$

$= \frac{\sin\alpha\cos\beta+\cos\alpha\sin\beta+\sin\alpha\cos\beta-\cos\alpha\sin\beta}{\cos\alpha\cos\beta+\sin\alpha\sin\beta+\cos\alpha\cos\beta-\sin\alpha\sin\beta}$

$= \frac{2\sin\alpha\cos\beta}{2\cos\alpha\cos\beta}$

$= \tan\alpha$

19b $\cos\theta\cos(\theta-90°)$

$= \frac{1}{2}(\cos[\theta-(\theta-90°)]+\cos[\theta+(\theta-90°)])$

$= \frac{1}{2}[\cos 90° + \cos(2\theta-90°)]$

$= \frac{1}{2}[0+\cos(2\theta-90°)]$

$= \frac{1}{2}\cos(2\theta-90°)$

20b $\frac{\sin 10\theta+\sin 6\theta}{\cos^2\theta-\sin^2\theta}$

$= \frac{2\sin\frac{10\theta+6\theta}{2}\cos\frac{10\theta-6\theta}{2}}{\cos 2\theta}$

$= \frac{2\sin 8\theta\cos 2\theta}{\cos 2\theta}$

$= 2\sin 8\theta$

24 $\cos 20°\cos 10°$

$= \frac{1}{2}[\cos(20°-10°)+\cos(20°+10°)]$

$= \frac{1}{2}\cos 10° + \frac{1}{2}\cos 30°$

$= \frac{1}{2}\cos 10° + \frac{\sqrt{3}}{4}$

$\cos 20°$

$= \frac{\frac{1}{2}\cos 10° + \frac{\sqrt{3}}{4}}{\cos 10°}$

$= \frac{1}{2} + \frac{\sqrt{3}}{4} \cdot \frac{1}{\cos 10°}$

$= \frac{\sqrt{3}}{4}\sec 10° + \frac{1}{2}$

Review for Chapter 15

5 $\frac{\cos\theta}{1-\sin\theta}$

$= \frac{\cos^2\theta}{(\cos\theta)(1-\sin\theta)}$

$= \frac{1-\sin^2\theta}{(\cos\theta)(1-\sin\theta)}$

$= \frac{(1+\sin\theta)(1-\sin\theta)}{(\cos\theta)(1-\sin\theta)}$

$= \frac{(1+\sin\theta)}{\cos\theta}$

$= \frac{1}{\cos\theta} + \frac{\sin\theta}{\cos\theta}$

$= \sec\theta + \tan\theta;\ \cos\theta \neq 0,\ \sin\theta \neq 1$

8c $\frac{\cos\theta}{1+\sin\theta}$

$= \frac{(\cos\theta)(1-\sin\theta)}{(1+\sin\theta)(1-\sin\theta)}$

$= \frac{(\cos\theta)(1-\sin\theta)}{1-\sin^2\theta}$

$= \frac{(\cos\theta)(1-\sin\theta)}{\cos^2\theta}$

$= \frac{1-\sin\theta}{\cos\theta}$

11a $\cos(\theta-\frac{\pi}{2})$

$= \cos\theta\cos\frac{\pi}{2} + \sin\theta\sin\frac{\pi}{2}$

$= 0\cos\theta + 1\sin\theta$

$= \sin\theta$

11b $\sin(\theta-\frac{\pi}{2})$

$= \sin\theta\cos\frac{\pi}{2} - \cos\theta\sin\frac{\pi}{2}$

$= 0\sin\theta - 1\cos\theta$

$= -\cos\theta$

11c $\sin(\theta+\frac{\pi}{2})$

$= \sin\theta\cos\frac{\pi}{2} + \cos\theta\sin\frac{\pi}{2}$

$= 0\sin\theta + 1\cos\theta$

$= \cos\theta$

11d $\cos\left(\theta + \frac{\pi}{2}\right)$

$= \cos\theta \cos\frac{\pi}{2} - \sin\theta \sin\frac{\pi}{2}$

$= 0 \cos\theta - 1 \sin\theta$

$= -\sin\theta$

15 $\cos^4\theta - \sin^4\theta$

$= (\cos^2\theta + \sin^2\theta)(\cos^2\theta - \sin^2\theta)$

$= 1(\cos^2\theta - \sin^2\theta)$

$= \cos^2\theta - \sin^2\theta$

16 $\tan(\alpha - \beta)$ is undefined. Since $\tan\alpha$ and $\tan\beta$ (the slopes of $\overleftrightarrow{OP}$ and $\overleftrightarrow{OR}$) are opposite reciprocals, $\overleftrightarrow{OP}$ and $\overleftrightarrow{OR}$ are perpendicular. Therefore, $\angle POR = 90°$, and $\tan 90°$ is undefined.

22c $\sin(x + y)\sin(x - y)$

$= \frac{1}{2}\left(\cos[(x + y) - (x - y)] - \cos[(x + y) + (x - y)]\right)$

$= \frac{1}{2}(\cos 2y - \cos 2x)$

$= \frac{1}{2}[1 - 2\sin^2 y - (1 - 2\sin^2 x)]$

$= \frac{1}{2}(2\sin^2 x - 2\sin^2 y)$

$= \sin^2 x - \sin^2 y$

23b

$\sin\theta = \frac{7}{25}$; $\cos\theta = \frac{24}{25}$; $\tan\theta = \frac{7}{24}$;

$\sin\left(\theta + \frac{\pi}{2}\right) = \frac{24}{25}$; $\cos\left(\theta + \frac{\pi}{2}\right) = -\frac{7}{25}$; $\tan\left(\theta + \frac{\pi}{2}\right) = -\frac{24}{7}$;

$\sin(\theta + \pi) = -\frac{7}{25}$; $\cos(\theta + \pi) = -\frac{24}{25}$; $\tan(\theta + \pi) = \frac{7}{24}$;

$\sin\left(\theta + \frac{3\pi}{2}\right) = -\frac{24}{25}$; $\cos\left(\theta + \frac{3\pi}{2}\right) = \frac{7}{25}$; $\tan\left(\theta + \frac{3\pi}{2}\right) = -\frac{24}{7}$

Test for Chapter 15

17

$$\frac{\cos(\alpha + \beta) + \cos(\alpha - \beta)}{\sin(\alpha + \beta) + \sin(\alpha - \beta)}$$

$$= \frac{\cos\alpha\cos\beta - \sin\alpha\sin\beta + \cos\alpha\cos\beta + \sin\alpha + \sin\beta}{\sin\alpha\cos\beta + \cos\alpha\sin\beta + \sin\alpha\cos\beta - \cos\alpha\sin\beta}$$

$$= \frac{2\cos\alpha\cos\beta}{2\sin\alpha\sin\beta}$$

$$= \frac{\cos\alpha}{\sin\alpha}$$

$= \cot\alpha$

CHAPTER 16

Section 16.2

9a

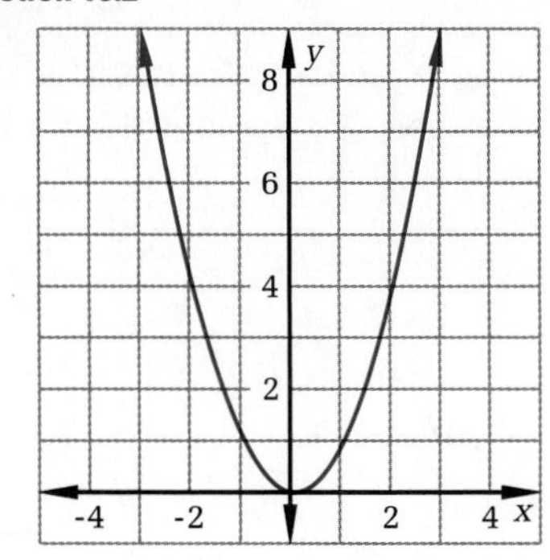

9b

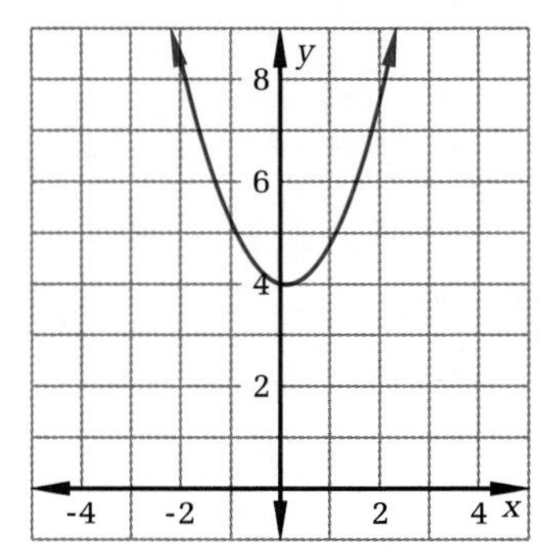

9c

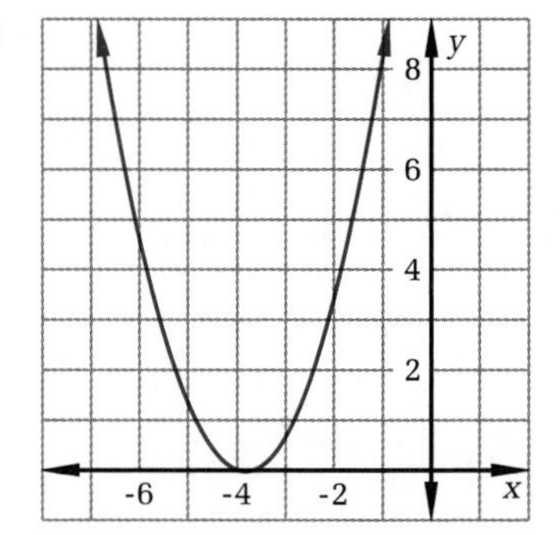

20

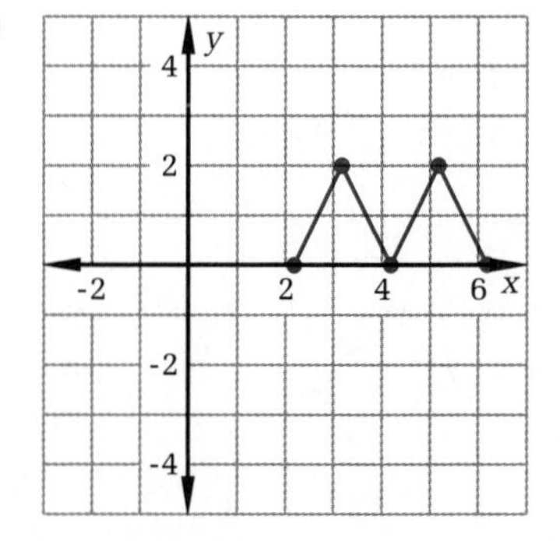

21

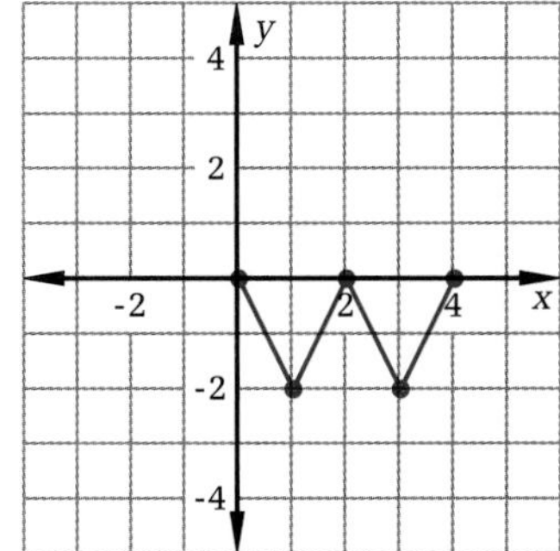

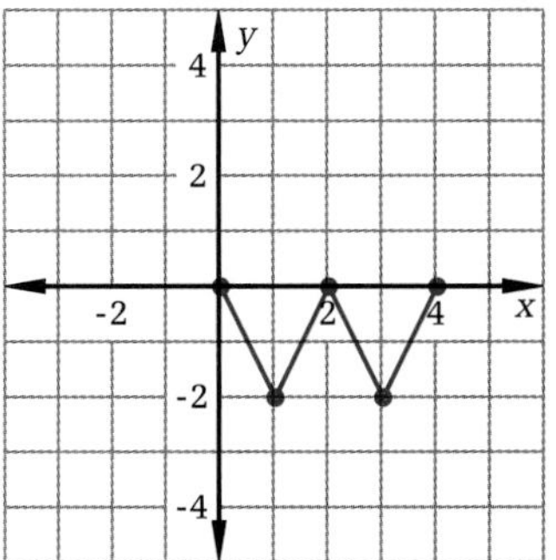

22

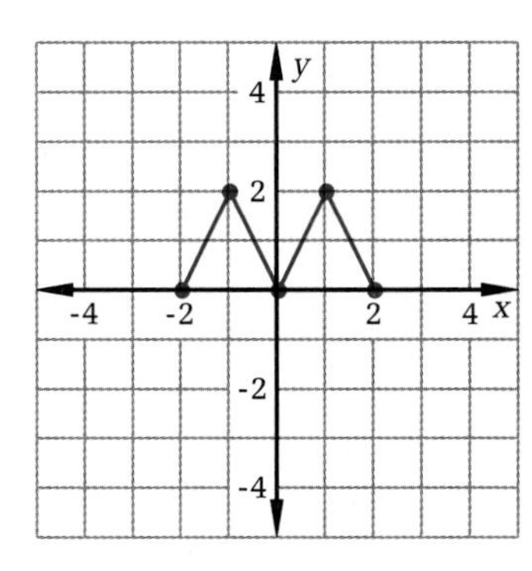

Section 16.3

8

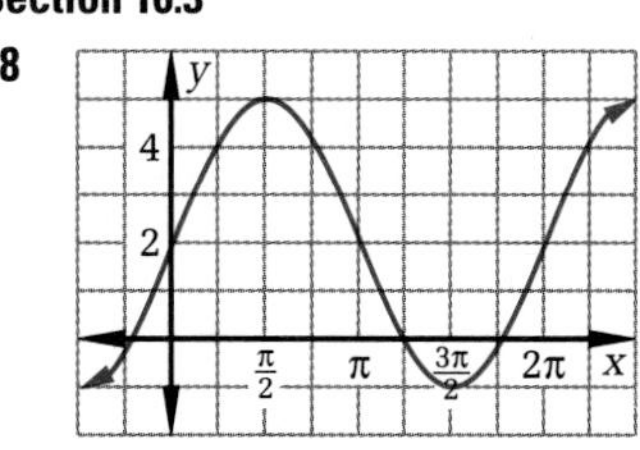

9

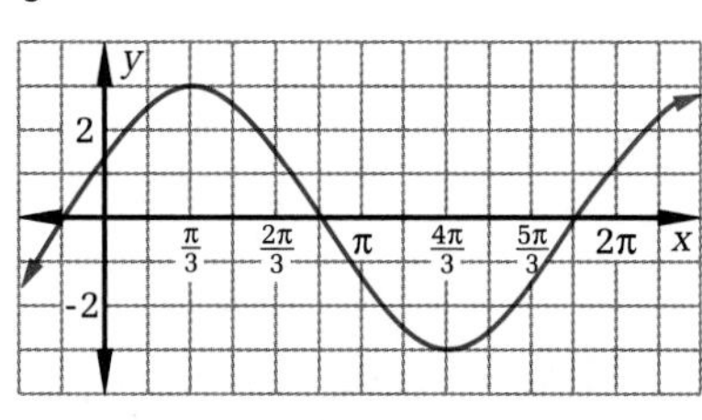

10

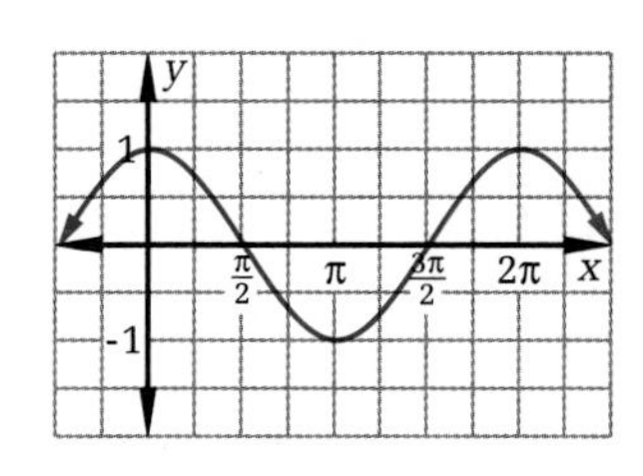

19

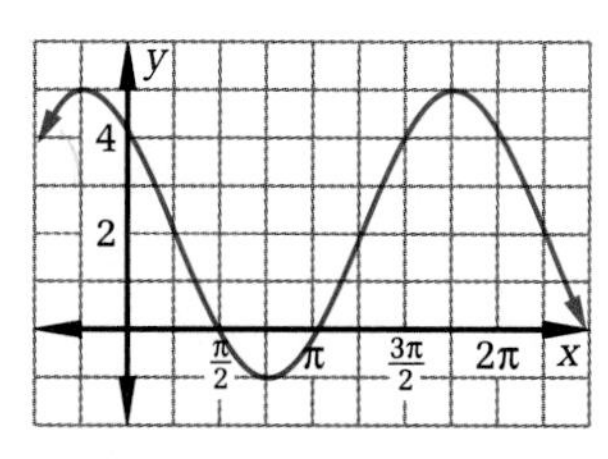

20

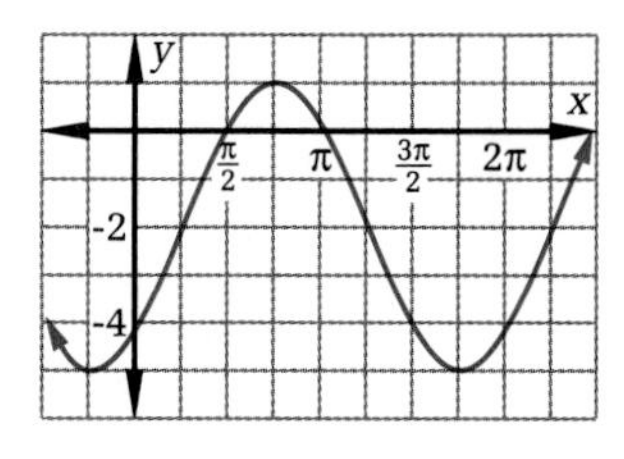

27

$$\frac{\cos 8x + \cos 2x}{\sin 8x - \sin 2x}$$

$$= \frac{2\cos\frac{8x+2x}{2}\cos\frac{8x-2x}{2}}{2\cos\frac{8x+2x}{2}\sin\frac{8x-2x}{2}}$$

$$= \frac{\cos 3x}{\sin 3x}$$

$$= \cot 3x$$

33

We know that $k = \frac{1}{2}ab\sin\angle C$ (see Section 16.1, problem 33).

$$k = \frac{1}{2}ab\sin\angle C$$

$$= \sqrt{\frac{1}{4}a^2b^2\sin^2\angle C}$$

$$= \sqrt{\frac{1}{4}a^2b^2(1-\cos^2\angle C)}$$

$$= \sqrt{\frac{1}{4}a^2b^2(1+\cos\angle C)(1-\cos\angle C)}$$

$$= \sqrt{\frac{1}{2}ab(1+\cos\angle C)\cdot\frac{1}{2}ab(1-\cos\angle C)}$$

By the Law of Cosines, $\cos\angle C = \frac{a^2+b^2-c^2}{2ab}$, so

$$k = \sqrt{\frac{1}{2}ab\left(1+\frac{a^2+b^2-c^2}{2ab}\right)\cdot\frac{1}{2}ab\left(1-\frac{a^2+b^2-c^2}{2ab}\right)}$$

$$= \sqrt{\frac{1}{2}ab\left(\frac{2ab+a^2+b^2-c^2}{2ab}\right)\cdot\frac{1}{2}ab\left(\frac{2ab-a^2-b^2+c^2}{2ab}\right)}$$

$$= \sqrt{\left(\frac{a^2+2ab+b^2-c^2}{4}\right)\left(\frac{c^2-a^2+2ab-b^2}{4}\right)}$$

$$= \sqrt{\left[\frac{(a+b)^2-c^2}{4}\right]\left[\frac{c^2-(a-b)^2}{4}\right]}$$

$$= \sqrt{\left(\frac{a+b+c}{2}\right)\left(\frac{a+b-c}{2}\right)\left(\frac{c+a-b}{2}\right)\left(\frac{c-a+b}{2}\right)}$$

Since $s = \frac{a+b+c}{2}$, it can be shown that $s-a = \frac{b+c-a}{2}$, that $s-b = \frac{a-b+c}{2}$, and that $s-c = \frac{a+b-c}{2}$. Therefore, $k = \sqrt{s(s-a)(s-b)(s-c)}$.

31

Area of ΔSPQ

$= \frac{1}{2}(PO)(QO)(\sin\theta) + \frac{1}{2}(PO)(SO)[\sin(180° - \theta)]$

$= \frac{1}{2}(PO)(QO + SO)(\sin\theta)$

$= \frac{1}{2}(PO)(QS)(\sin\theta)$

Area of ΔSRQ

$= \frac{1}{2}(RO)(SO)(\sin\theta) + \frac{1}{2}(RO)(QO)[\sin(180° - \theta)]$

$= \frac{1}{2}(RO)(SO + QO)(\sin\theta)$

$= \frac{1}{2}(RO)(QS)(\sin\theta)$

Total Area

$= \frac{1}{2}(PO)(QS)(\sin\theta) + \frac{1}{2}(RO)(QS)(\sin\theta)$

$= \frac{1}{2}(PO + RO)(QS)(\sin\theta)$

$= \frac{1}{2}(PR)(QS)(\sin\theta)$

Section 16.4

1a

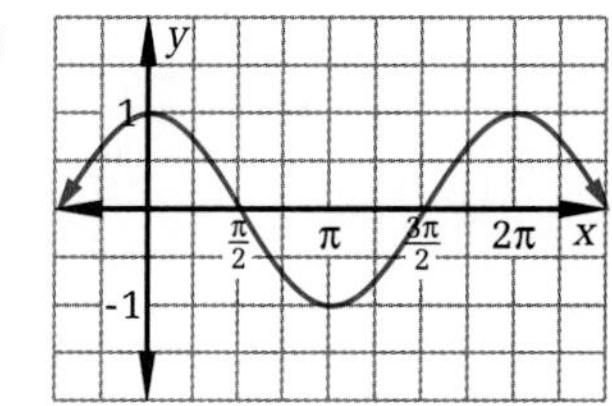

1b

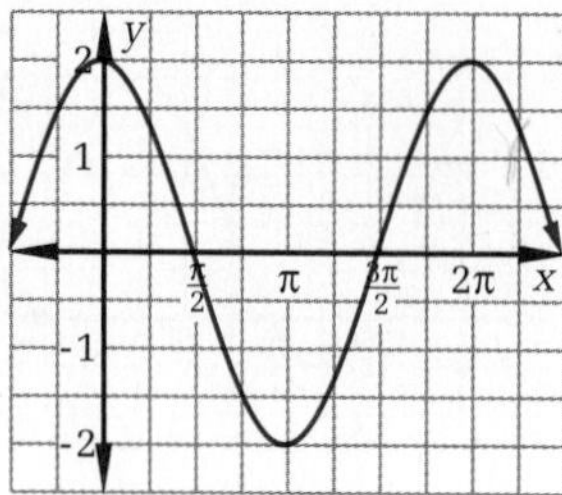

1c

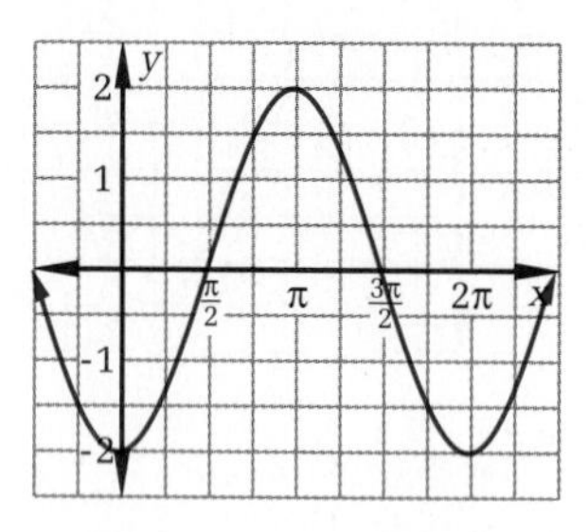

1d

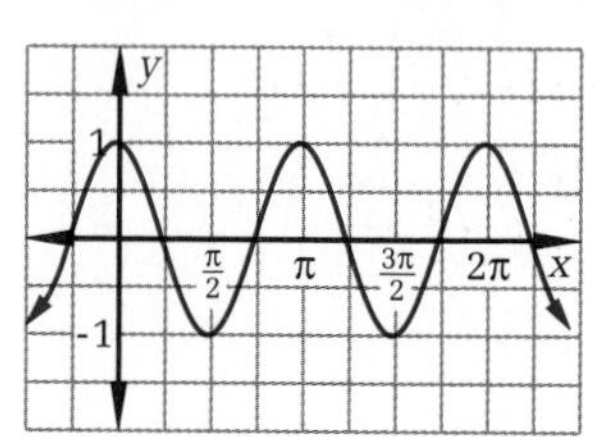

1e

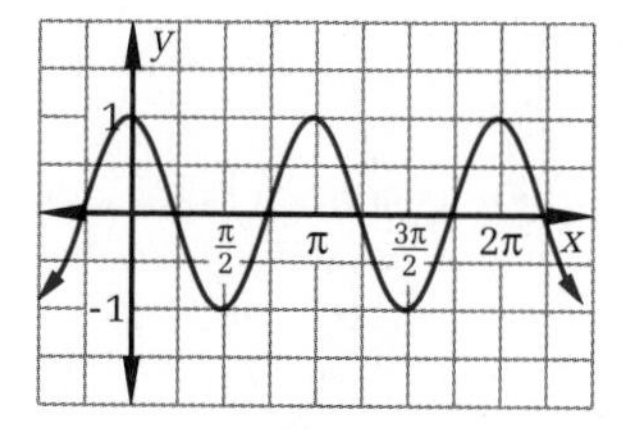

4a

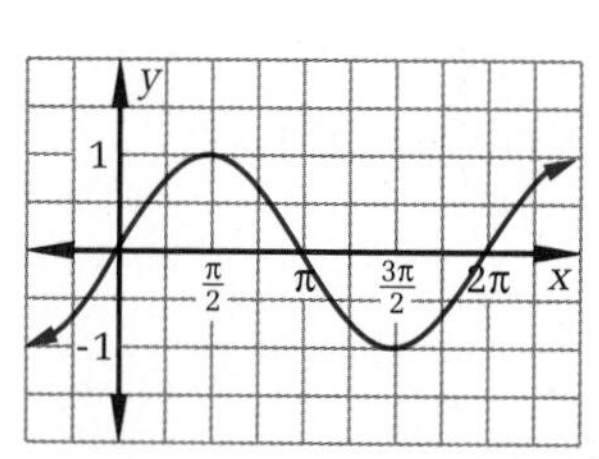

4b

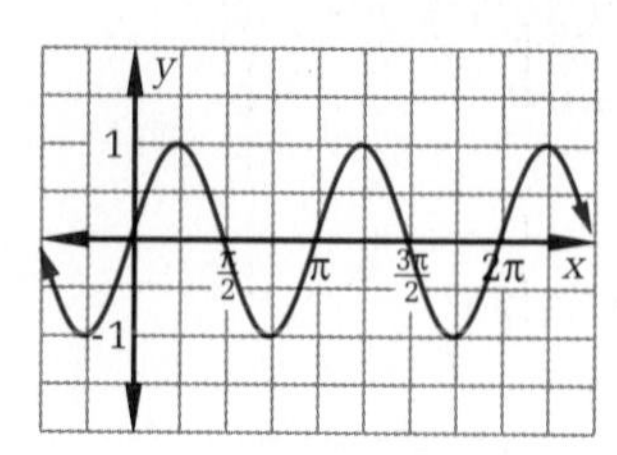

4c

4d

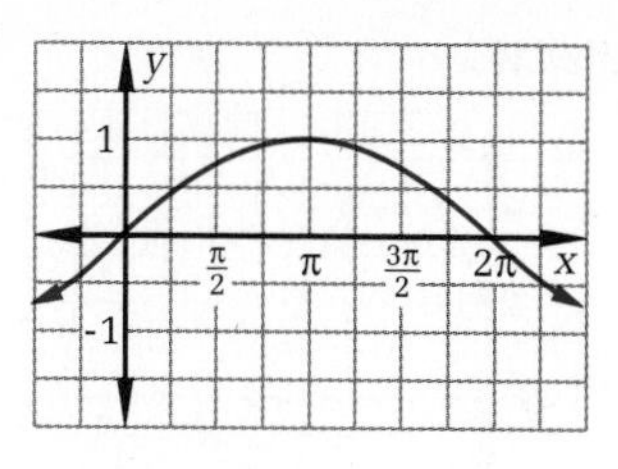

8b

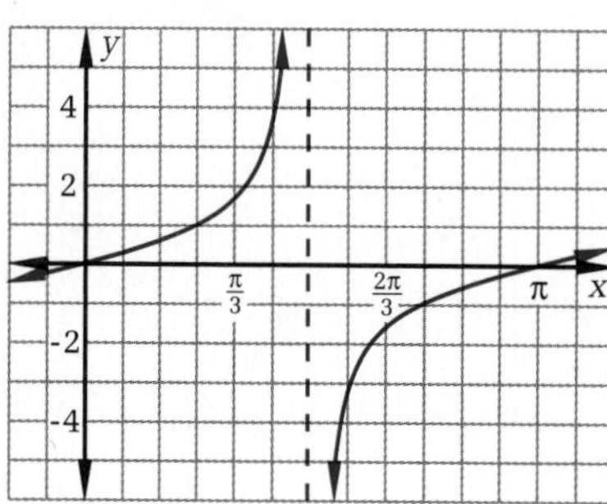

11c

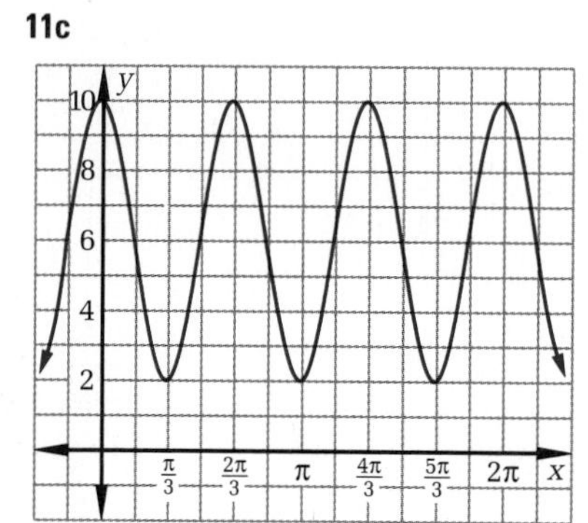

18a

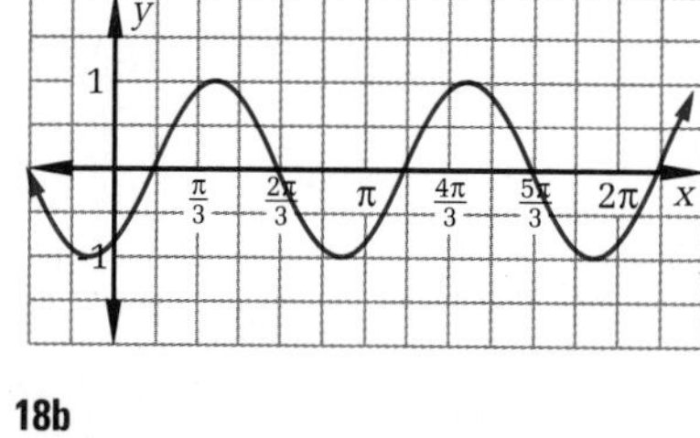

18b

18c

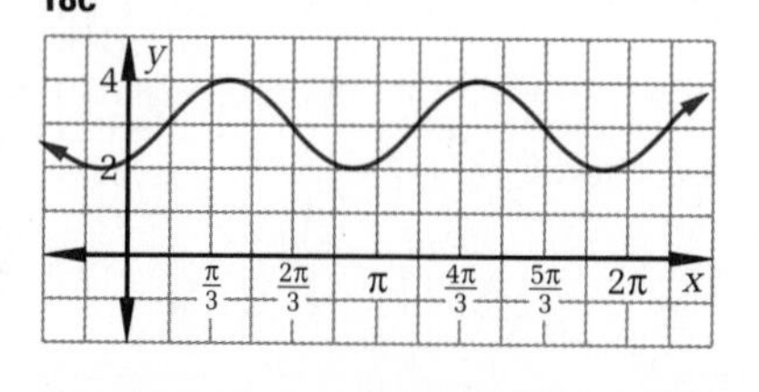

18d

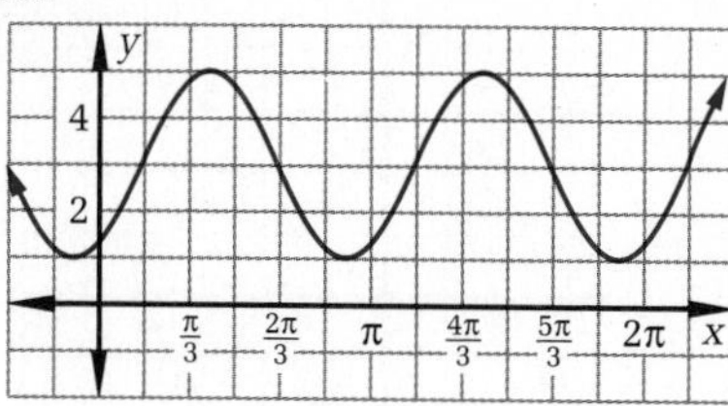

19a

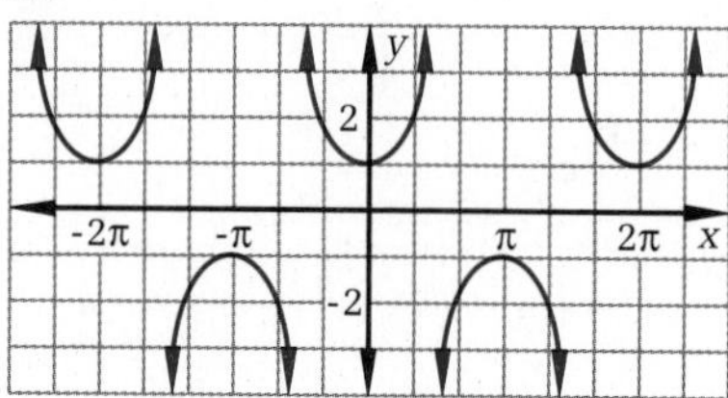

19b

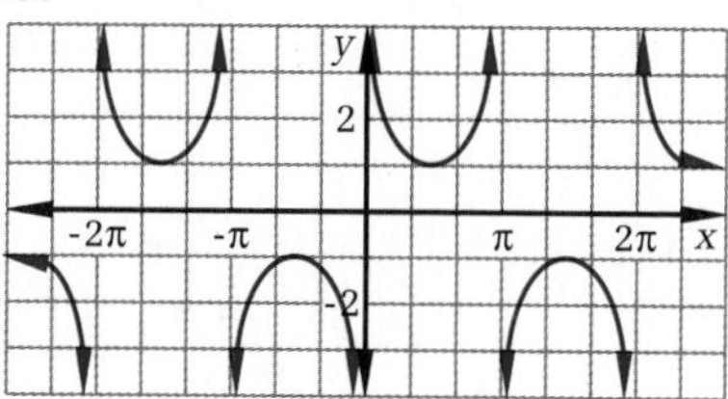

19c

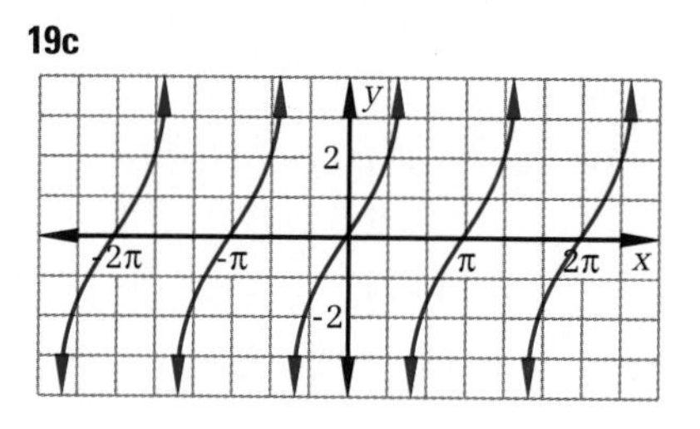

19d

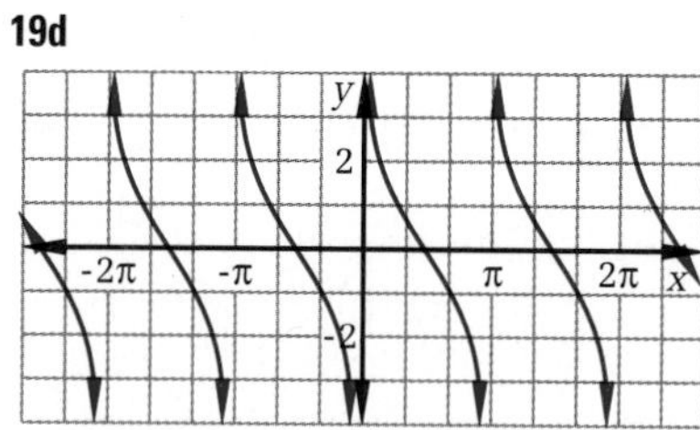

23

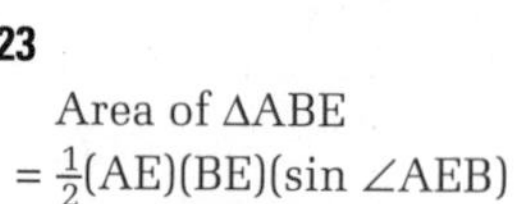

Area of ΔABE
$= \frac{1}{2}$(AE)(BE)(sin $\angle$AEB)

Area of ΔADE
$= \frac{1}{2}$(AE)(DE)(sin $\angle$AED)
$= \frac{1}{2}$(AE)(DE)(sin $\angle$AEB)

Area of ΔBCE
$= \frac{1}{2}$(CE)(BE)(sin $\angle$BEC)
$= \frac{1}{2}$(CE)(BE)(sin $\angle$AEB)

continued on next page

23 *continued*

Area of ΔCDE

$= \frac{1}{2}$(CE)(DE)(sin $\angle$DEC)

$= \frac{1}{2}$(CE)(DE)(sin $\angle$AEB)

$A_{\Delta ABE} \cdot A_{\Delta CDE}$

$= \frac{1}{2}$(AE)(BE)(sin $\angle$AEB) $\cdot \frac{1}{2}$(CE)(DE)(sin $\angle$AEB)

$= \frac{1}{2}$(AE)(DE)(sin $\angle$AEB) $\cdot \frac{1}{2}$(CE)(BE)(sin $\angle$AEB)

$= A_{\Delta ADE} \cdot A_{\Delta BCE}$

Section 16.5

1a

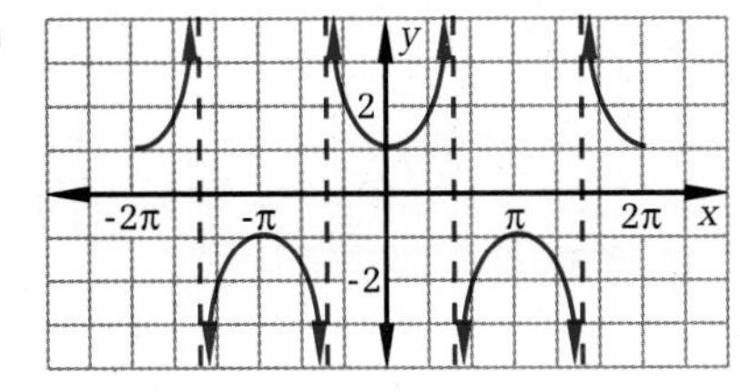

1b

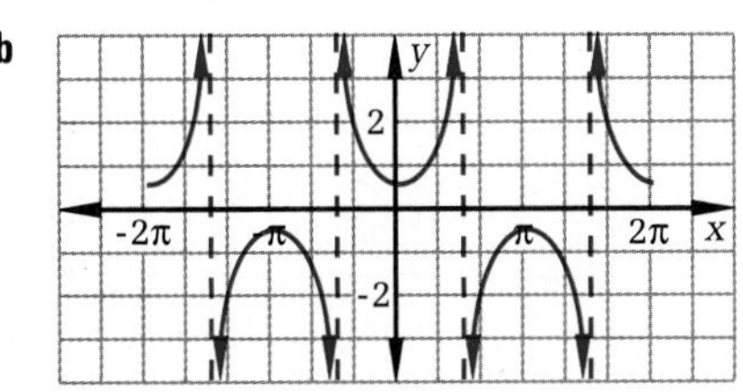

1c

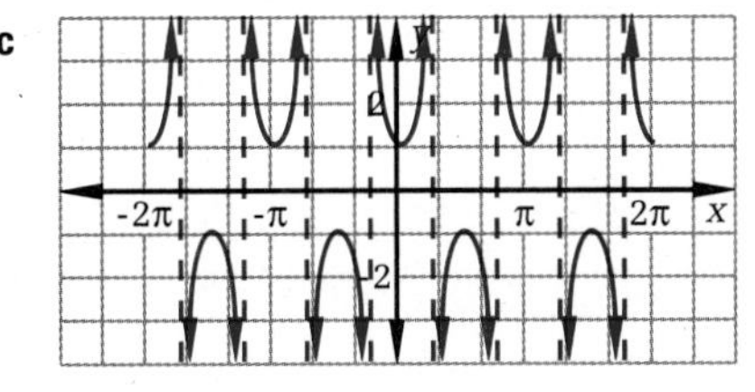

2a

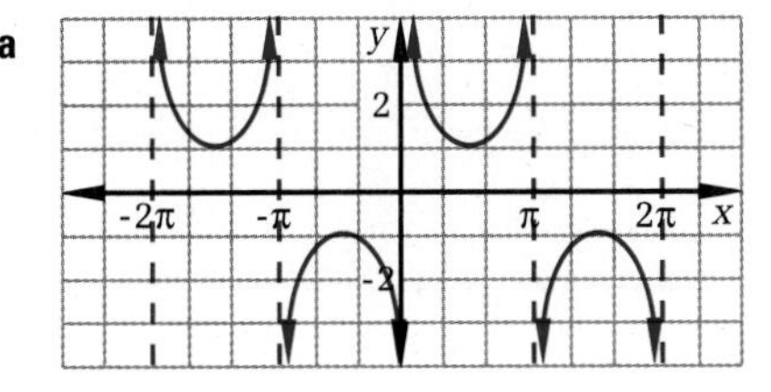

2b

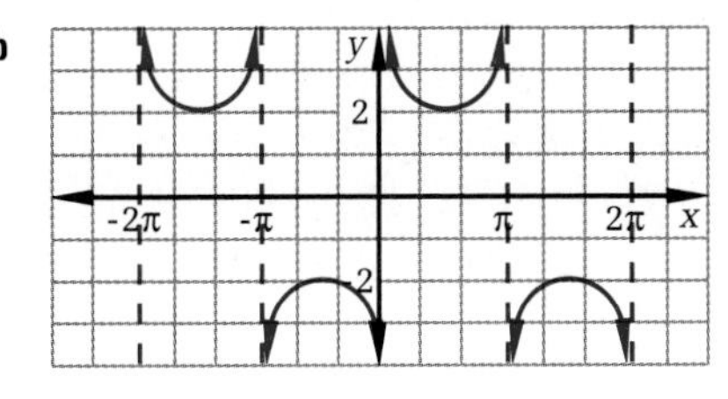

2c

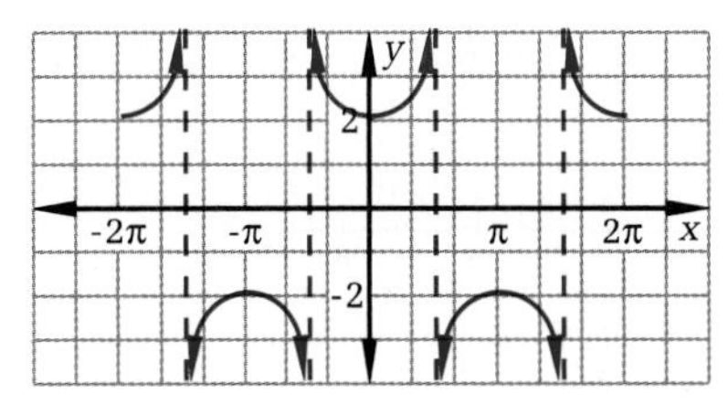

3a

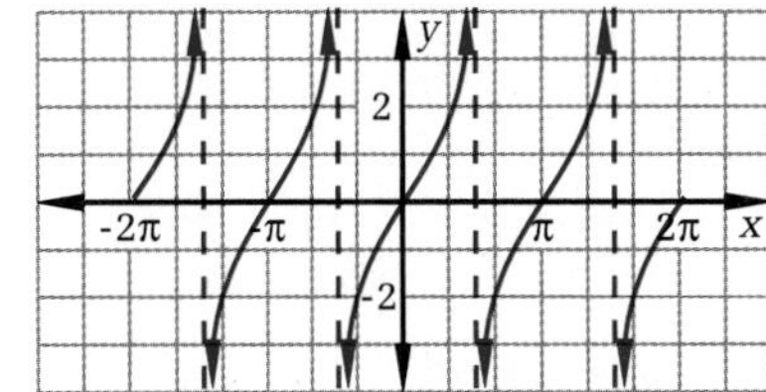

3b

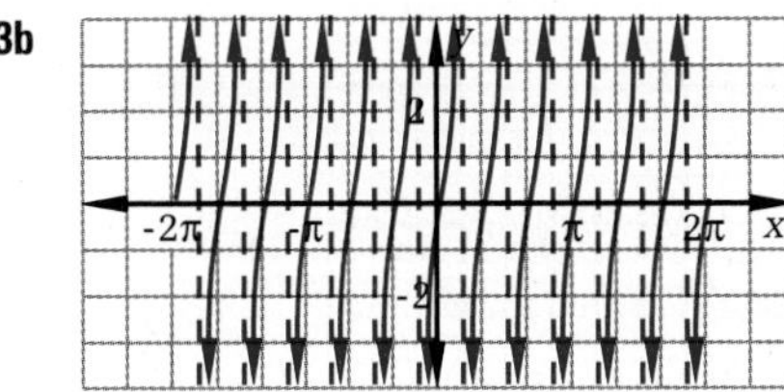

3c

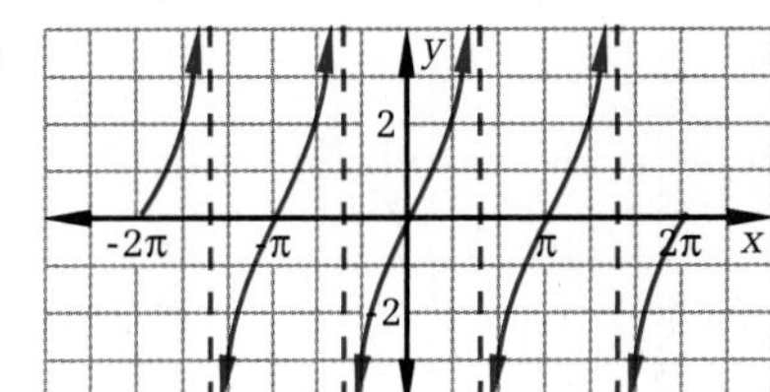

4a

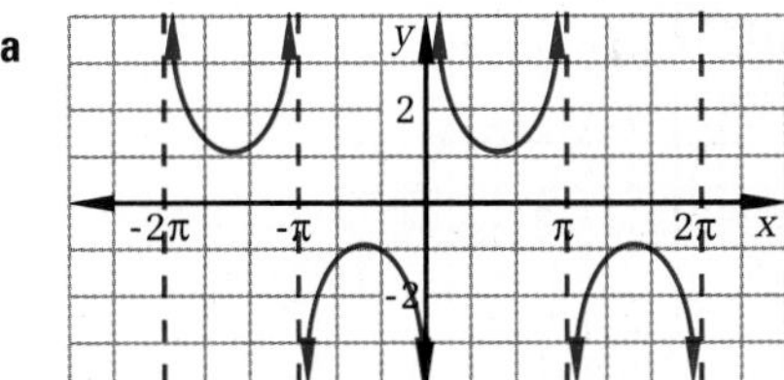

4b

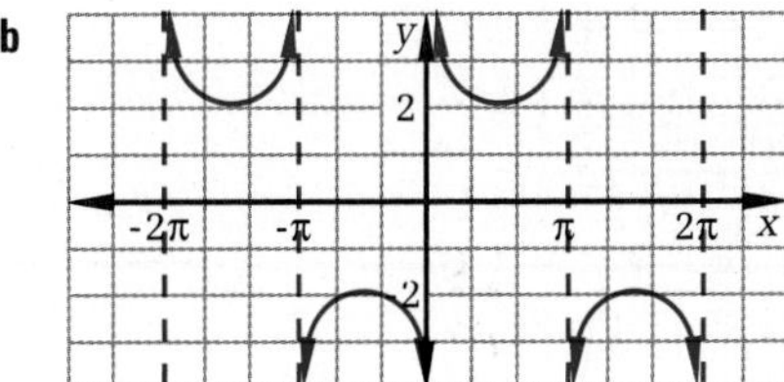

4c

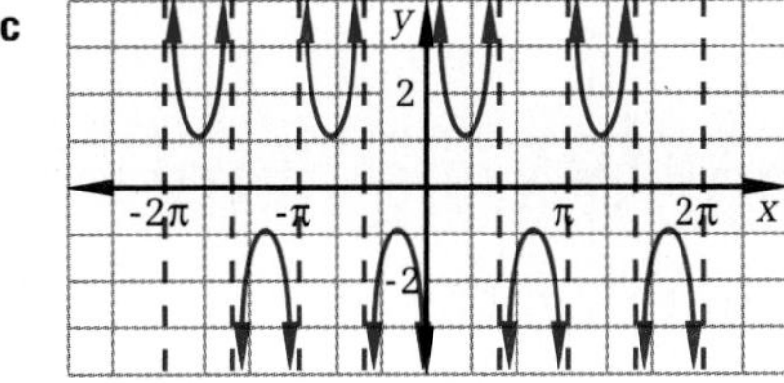

4d

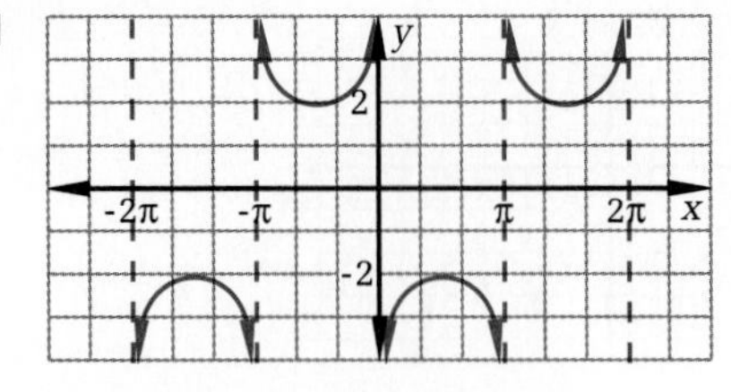

4e

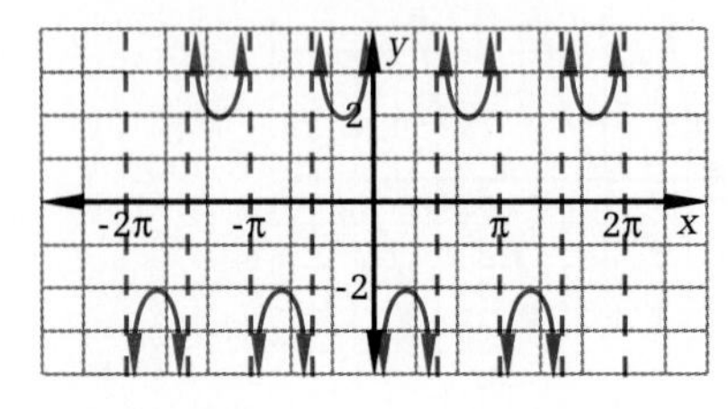

9a

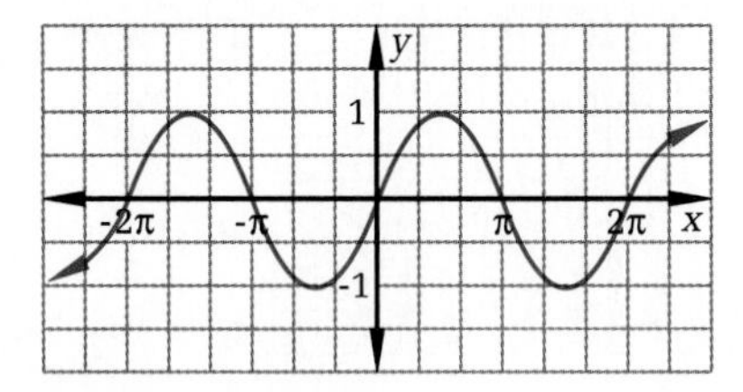

9b

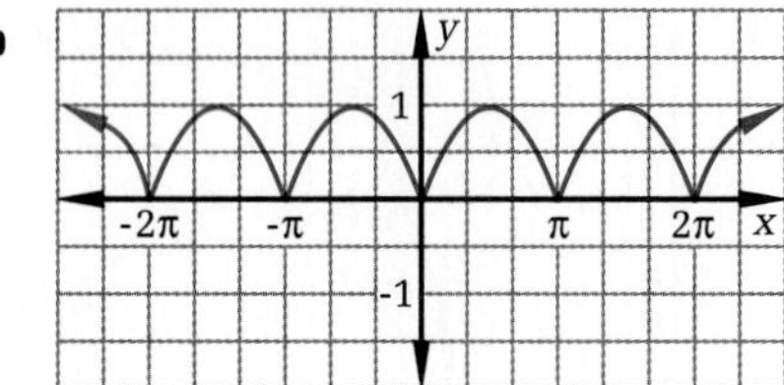

9c

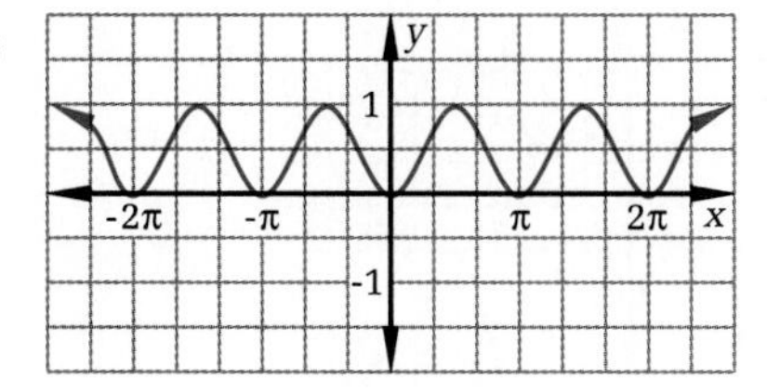

13

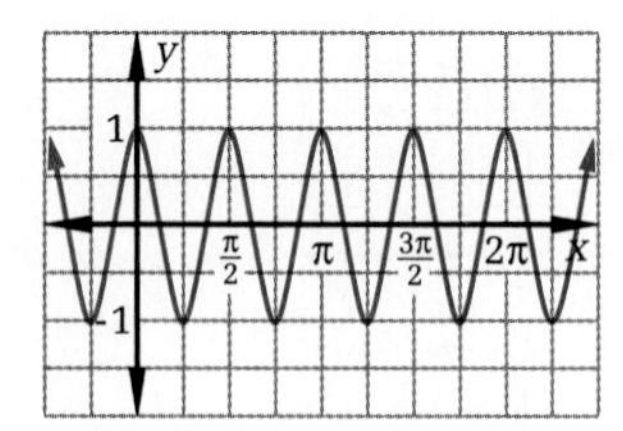

14

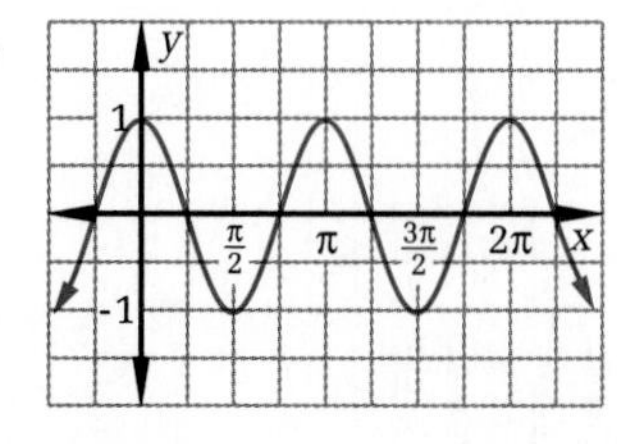

Review for Chapter 16

1b

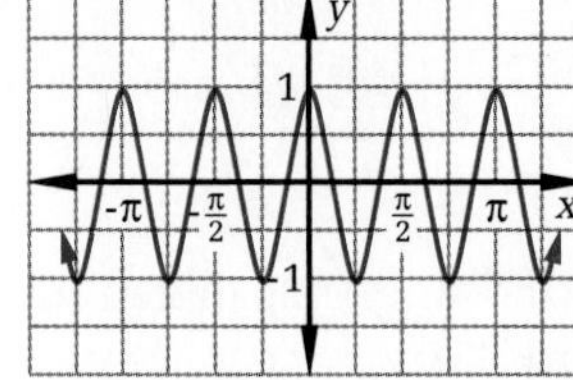

1c

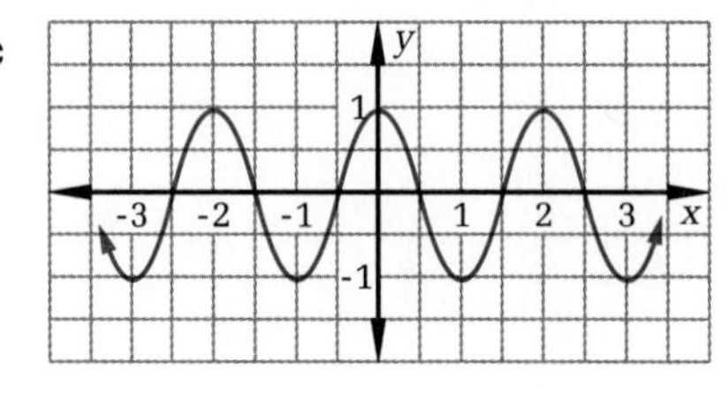